KIRK-OTHMER

ENCYCLOPEDIA OF CHEMICAL TECHNOLOGY

FOURTH EDITION

VOLUME **3**

ANTIBIOTICS
TO
BATTERIES

KIRK-OTHMER

ENCYCLOPEDIA OF CHEMICAL TECHNOLOGY

FOURTH EDITION

VOLUME **3**

ANTIBIOTICS
TO
BATTERIES

A Wiley-Interscience Publication
JOHN WILEY & SONS
New York • Chichester • Brisbane • Toronto • Singapore

Library of Congress Cataloging-in-Publication Data

Encyclopedia of chemical technology / executive editor, Jacqueline I. Kroschwitz; editor, Mary Howe-Grant.—4th ed.

p. cm.

At head of title: Kirk-Othmer.

"A Wiley-Interscience publication."

Includes index.

Contents: v. 3. Antibiotics to benzene.

ISBN 0-471-52671-1 (v. 3)

1. Chemistry, Technical—Encyclopedias. I. Kirk, Raymond E. (Raymond Eller), 1890–1957. II. Othmer, Donald F. (Donald Frederick), 1904– . III. Kroschwitz, Jacqueline I., 1942– . IV. Howe-Grant, Mary, 1943– . V. Title: Kirk-Othmer encyclopedia of chemical technology.

TP9.E685 1992 91-16789
660′.03—dc20 CIP

Printed in the United States of America

CONTENTS

EDITORIAL STAFF FOR VOLUME 3

Executive Editor: **Jacqueline I. Kroschwitz**
Editor: **Mary Howe-Grant**
Assistant Editor: **Cathleen A. Treacy**
Assistant Managing Editor: **Lindy J. Humphreys**
Copy Editor: **Linda M. Kim**

CONTRIBUTORS TO VOLUME 3

Elizabeth R. Aguinaldo, *FMC Corporation, Princeton, New Jersey,* Barium compounds

Jean-François Alary, *Université de Sherbrooke, Sherbrooke, Quebec, Canada,* Asbestos

G. F. Archer, *ASARCO, Inc., Denver, Colorado,* Antimony and antimony alloys

Brian Bannister, *The Upjohn Company, Kalamazoo, Michigan,* Lincosaminides (under Antibiotics)

John Becher, *CDC Drug Service, Atlanta, Georgia,* Anthelmintics (under Antiparasitic agents)

George Blomgren, *Eveready Battery Co., Inc., West Lake, Ohio,* Batteries, primary cells

Claudio Boffito, *Saes Getters, SpA, Milan, Italy,* Barium

Daniel P. Bonner, *Bristol-Myers Squibb Pharmaceutical Research Institute, Princeton, New Jersey,* Monobactams (under Antibiotics, β-lactams)

Ralph Brodd, *Gould, Inc., East Lake, Ohio,* Batteries, introduction

Kathryn R. Bullock, *AT&T Bell Labs, Mesquite, Texas,* Lead–acid (under Batteries, secondary cells)

S. C. Carapella, *Consultant, Tuckahoe, New York,* Antimony and antimony alloys; Arsenic and arsenic alloys

Geert Cauwenbergh, *Janssen Research Foundation, Beerse, Belgium,* Antimycotics (under Antiparasitic agents)

Ian Chopra, *American Cyanamid Company, Pearl River, New York,* Tetracyclines (under Antibiotics)

Rasik J. Chudgar, *BASF Corporation, Rensselaer, New York,* Azo dyes

Frank H. Y. Chung, *Rhône-Poulenc Inc., Cranbury, New Jersey,* Chemical leavening agents (under Bakery processes and leavening agents)

Louise W. Crandall, *Lilly Research Laboratories, Indianapolis, Indiana,* Polyethers (under Antibiotics)

Phillip DeLassus, *The Dow Chemical Company, Midland, Michigan,* Barrier polymers

Martin Dexter, *CIBA-GEIGY Corporation, Ardsley, New York,* Antioxidants

Patrick M. DiBello, *FMC Corporation, Princeton, New Jersey,* Barium compounds

G. O. Doak, *North Carolina State University, Raleigh, North Carolina,* Antimony compounds; Arsenic compounds

William G. Dukek, *Consultant, Summit, New Jersey,* Aviation and other gas turbine fuels

George A. Ellestad, *American Cyanamid Company, Pearl River, New York,* Tetracyclines (under Antibiotics)

Michael H. Fisher, *Merck, Sharp, and Dohme Research Laboratories, Rahway, New Jersey,* Avermectins (under Antiparasitic agents)

Leon D. Freedman, *North Carolina State University, Raleigh, North Carolina,* Antimony compounds; Arsenic compounds

Ashit K. Ganguly, *Schering-Plough Corporation, Bloomfield, New Jersey,* Oligosaccharides (under Antibiotics)

Roque Edward Garcia, *Technicon Instruments Corporation, Tarrytown, New York,* Hematology (under Automated instrumentation)

Kathleen F. George, *Union Carbide Chemicals and Plastics Company, Inc., South Charleston, West Virginia,* Antifreezes and deicing fluids

Paul Gherson, *Miles, Inc., Tarrytown, New York,* Clinical chemistry (under Automated instrumentation)

Paul Gifford, *Ovonic Battery Company, Troy, Michigan,* Other (under Batteries, secondary cells)

V. M. Girijavallabhan, *Schering-Plough Corporation, Bloomfield, New Jersey,* Oligosaccharides (under Antibiotics)

Robert L. Hamill, *Lilly Research Laboratories, Indianapolis, Indiana,* Polyethers (under Antibiotics)

Joseph J. Hlavka, *American Cyanamid Company, Pearl River, New York,* Tetracyclines (under Antibiotics)

James Hunter, *Eveready Battery Co., Inc., West Lake, Ohio,* Batteries, primary cells

Carmel R. Jolicoeur, *Université de Sherbrooke, Sherbrooke, Quebec, Canada,* Asbestos

Herbert Kirst, *Eli Lilly and Company, Indianapolis, Indiana,* Macrolides (under Antibiotics)

Daniel L. Klayman, *Walter Reed Army Institute of Research, Washington, D.C.,* Antiprotozoals (under Antiparasitic agents)

Martin Klein, *Rutgers University, Piscataway, New Jersey,* Alkaline (under Batteries, secondary cells)

William H. Koster, *Bristol-Myers Squibb Pharmaceutical Research Institute, Princeton, New Jersey,* Monobactams (under Antibiotics)

Howard Lanza, *Miles, Inc., Tarrytown, New York,* Clinical chemistry (under Automated instrumentation)

Robert P. Lattimer, *The BF Goodrich Co., Brecksville, Ohio,* Antiozonants

Robert W. Layer, *The BF Goodrich Co., Brecksville, Ohio,* Antiozonants

T. Li, *ASARCO, Inc., Denver, Colorado,* Antimony and antimony alloys

Klaus R. Linder, *Bristol-Myers Squibb Pharmaceutical Research Institute, Princeton, New Jersey,* Monobactams (under Antibiotics)

G. Gilbert Long, *North Carolina State University, Raleigh, North Carolina,* Antimony compounds; Arsenic compounds

James L. Manganaro, *FMC Corporation, Princeton, New Jersey,* Barium compounds

Jana B. Milford, *University of Connecticut, Storrs, Connecticut,* Atmospheric modeling

Helmut Mrozik, *Merck, Sharp, and Dohme Research Laboratories, Rahway, New Jersey,* Avermectins (under Antiparasitic agents)

Neal F. Osborne, *SmithKline Beecham Pharmaceuticals, Surrey, United Kingdom,* Carbapenems and penems (under Antibiotics, β-lactams)

Joel D. Payne, *Kansas State University, Manhattan, Kansas,* Yeast-raised products (under Bakery processes and leavening agents)

Milton Pelavin, *Miles, Inc., Tarrytown, New York,* Clinical chemistry (under Automated instrumentation)

John R. Pierson, *Johnson Controls, Inc., Milwaukee, Wisconsin,* Lead–acid (under Batteries, secondary cells)

R. J. Ponsford, *SmithKline Beecham Pharmaceuticals, Surrey, United Kingdom,* Penicillins and others (under Antibiotics, β-lactams)

Joseph G. Ponte, Jr., *Kansas State University, Manhattan, Kansas,* Yeast-raised products (under Bakery processes and leavening agents)

Nancy L. Reichenback, *Temple University School of Medicine, Philadelphia, Pennsylvania,* Nucleosides (under Antibiotics)

Ganapathi R. Revankar, *Triplex Pharmaceutical Corporation, The Woodlands, Texas,* Antiviral agents

C. K. Rhee, *The BF Goodrich Co., Brecksville, Ohio,* Antiozonants

John Roberts, *Hoffmann-La Roche, Inc., Nutley, New Jersey,* Cephalosporins (under Antibiotics, β-lactams)

Roland K. Robins, *ICN Nucleic Acid Research Institute, Costa Mesa, California,* Antiviral agents

Armistead G. Russell, *Carnegie Mellon University, Pittsburgh, Pennsylvania,* Atmospheric modeling

Alvin J. Salkind, *Rutgers University, Piscataway, New Jersey,* Alkaline (under Batteries, secondary cells)

Rosalie A. Schnick, *U.S. Fish and Wildlife Service, La Crosse, Wisconsin,* Aquaculture chemicals

Roy E. Smith, *CIBA-GEIGY Corporation, Greensboro, North Carolina,* Azine dyes

Asta Sokov, *Université de Sherbrooke, Sherbrooke, Quebec, Canada,* Asbestos

Robert Southgate, *SmithKline Beecham Pharmaceuticals, Surrey, United Kingdom,* Carbapenems and penems (under Antibiotics, β-lactams)

James G. Speight, *Western Research Institute, Laramie, Wyoming,* Asphalt

J. G. Stam, *Pfizer Central Research, Groton, Connecticut,* β-lactamase inhibitors (under Antibiotics, β-lactams)

Barbara A. Stefl, *Union Carbide Chemicals and Plastics Company, Inc., South Charleston, West Virginia,* Antifreezes and deicing fluids

Robert J. Suhadolnik, *Temple University School of Medicine, Philadelphia, Pennsylvania,* Nucleosides (under Antibiotics)

Jefferson W. Tilley, *Hoffmann-LaRoche Inc., Nutley, New Jersey,* Antiobesity drugs

John Turrell, *Technicon Instruments Corporation, Tarrytown, New York,* Hematology (under Automated instrumentation)

Dan Vlastelica, *Miles Inc., Tarrytown, New York,* Clinical chemistry (under Automated instrumentation)

Leonard E. Walp, *Humko Chemical, Division of the Witco Corporation, Memphis, Tennessee,* Antistatic agents

Edmund M. Wise Jr., *The Wellcome Research Laboratories, Research Triangle Park, North Carolina,* Peptides (under Antibiotics)

David Zelmanovic, *Technicon Instruments Corporation, Tarrytown, New York,* Hematology (under Automated instrumentation)

NOTE ON CHEMICAL ABSTRACTS SERVICE REGISTRY NUMBERS AND NOMENCLATURE

Chemical Abstracts Service (CAS) Registry Numbers are unique numerical identifiers assigned to substances recorded in the CAS Registry System. They appear in brackets in the *Chemical Abstracts* (CA) substance and formula indexes following the names of compounds. A single compound may have synonyms in the chemical literature. A simple compound like phenethylamine can be named β-phenylethylamine or, as in *Chemical Abstracts,* benzeneethanamine. The usefulness of the *Encyclopedia* depends on accessibility through the most common correct name of a substance. Because of this diversity in nomenclature careful attention has been given to the problem in order to assist the reader as much as possible, especially in locating the systematic CA index name by means of the Registry Number. For this purpose, the reader may refer to the CAS Registry Handbook Number Section which lists in numerical order the Registry Number with the *Chemical Abstracts* index name and the molecular formula; eg, **458-88-8,** Piperidine, 2-propyl-, (*S*)-, $C_8H_{17}N$; in the *Encyclopedia* this compound would be found under its common name, coniine [*458-88-8*]. Alternatively, this information can be retrieved electronically from CAS Online. In many cases molecular formulas have also been provided in the *Encyclopedia* text to facilitate electronic searching. The Registry Number is a valuable link for the reader in retrieving additional published information on substances and also as a point of access for on-line data bases.

In all cases, the CAS Registry Numbers have been given for title compounds in articles and for all compounds in the index. All specific substances indexed in *Chemical Abstracts* since 1965 are included in the CAS Registry System as are a large number of substances derived from a variety of reference works. The CAS Registry System identifies a substance on the basis of an unambiguous computer-language description of its molecular structure including stereochemical detail. The Registry Number is a machine-checkable number (like a Social Security number) assigned in sequential order to each substance as it enters the registry system. The value of the number lies in the fact that it is a concise and unique means of substance identification, which is independent of, and therefore bridges, many systems of chemical nomenclature. For polymers, one Registry Number may be used for the entire family; eg, polyoxyethylene (20) sorbitan monolaurate has the same number as all of its polyoxyethylene homologues.

Cross-references are inserted in the index for many common names and for some systematic names. Trademark names appear in the index. Names that are incorrect, misleading, or ambiguous are avoided. Formulas are given very frequently in the text to help in identifying compounds. The spelling and form used, even for industrial names, follow American chemical usage, but not always the usage of *Chemical Abstracts* (eg, *coniine* is used instead of *(S)-2-propylpiperidine, aniline* instead of *benzenamine,* and *acrylic acid* instead of *2-propenoic acid*).

There are variations in representation of rings in different disciplines. The dye industry does not designate aromaticity or double bonds in rings. All double bonds and aromaticity are shown in the *Encyclopedia* as a matter of course. For example, tetralin has an aromatic ring and a saturated ring and its structure

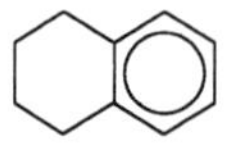

appears in the *Encyclopedia* with its common name, Registry Number enclosed in brackets, and parenthetical CA index name, ie, tetralin [*119-64-2*] (1,2,3,4-tetrahydronaphthalene). With names and structural formulas, and especially with CAS Registry Numbers, the aim is to help the reader have a concise means of substance identification.

CONVERSION FACTORS, ABBREVIATIONS, AND UNIT SYMBOLS

SI Units (Adopted 1960)

The International System of Units (abbreviated SI), is being implemented throughout the world. This measurement system is a modernized version of the MKSA (meter, kilogram, second, ampere) system, and its details are published and controlled by an international treaty organization (The International Bureau of Weights and Measures) (1).

SI units are divided into three classes:

BASE UNITS

length	meter[†] (m)
mass	kilogram (kg)
time	second (s)
electric current	ampere (A)
thermodynamic temperature[‡]	kelvin (K)
amount of substance	mole (mol)
luminous intensity	candela (cd)

SUPPLEMENTARY UNITS

plane angle	radian (rad)
solid angle	steradian (sr)

[†]The spellings "metre" and "litre" are preferred by ASTM; however "-er" is used in the *Encyclopedia*.

[‡]Wide use is made of Celsius temperature (t) defined by

$$t = T - T_0$$

where T is the thermodynamic temperature, expressed in kelvin, and $T_0 = 273.15$ K by definition. A temperature interval may be expressed in degrees Celsius as well as in kelvin.

DERIVED UNITS AND OTHER ACCEPTABLE UNITS

These units are formed by combining base units, supplementary units, and other derived units (2–4). Those derived units having special names and symbols are marked with an asterisk in the list below.

Quantity	Unit	Symbol	Acceptable equivalent
*absorbed dose	gray	Gy	J/kg
acceleration	meter per second squared	m/s^2	
*activity (of a radionuclide)	becquerel	Bq	1/s
area	square kilometer	km^2	
	square hectometer	hm^2	ha (hectare)
	square meter	m^2	
concentration (of amount of substance)	mole per cubic meter	mol/m^3	
current density	ampere per square meter	A/m^2	
density, mass density	kilogram per cubic meter	kg/m^3	g/L; mg/cm^3
dipole moment (quantity)	coulomb meter	C·m	
*dose equivalent	sievert	Sv	J/kg
*electric capacitance	farad	F	C/V
*electric charge, quantity of electricity	coulomb	C	A·s
electric charge density	coulomb per cubic meter	C/m^3	
*electric conductance	siemens	S	A/V
electric field strength	volt per meter	V/m	
electric flux density	coulomb per square meter	C/m^2	
*electric potential, potential difference, electromotive force	volt	V	W/A
*electric resistance	ohm	Ω	V/A
*energy, work, quantity of heat	megajoule	MJ	
	kilojoule	kJ	
	joule	J	N·m
	electronvolt†	eV†	
	kilowatt-hour†	kW·h†	
energy density	joule per cubic meter	J/m^3	
*force	kilonewton	kN	
	newton	N	$kg·m/s^2$

†This non-SI unit is recognized by the CIPM as having to be retained because of practical importance or use in specialized fields (1).

Quantity	Unit	Symbol	Acceptable equivalent
*frequency	megahertz	MHz	
	hertz	Hz	1/s
heat capacity, entropy	joule per kelvin	J/K	
heat capacity (specific), specific entropy	joule per kilogram kelvin	J/(kg·K)	
heat transfer coefficient	watt per square meter kelvin	$W/(m^2·K)$	
*illuminance	lux	lx	lm/m^2
*inductance	henry	H	Wb/A
linear density	kilogram per meter	kg/m	
luminance	candela per square meter	cd/m^2	
*luminous flux	lumen	lm	cd·sr
magnetic field strength	ampere per meter	A/m	
*magnetic flux	weber	Wb	V·s
*magnetic flux density	tesla	T	Wb/m^2
molar energy	joule per mole	J/mol	
molar entropy, molar heat capacity	joule per mole kelvin	J/(mol·K)	
moment of force, torque	newton meter	N·m	
momentum	kilogram meter per second	kg·m/s	
permeability	henry per meter	H/m	
permittivity	farad per meter	F/m	
*power, heat flow rate,	kilowatt	kW	
radiant flux	watt	W	J/s
power density, heat flux density, irradiance	watt per square meter	W/m^2	
*pressure, stress	megapascal	MPa	
	kilopascal	kPa	
	pascal	Pa	N/m^2
sound level	decibel	dB	
specific energy	joule per kilogram	J/kg	
specific volume	cubic meter per kilogram	m^3/kg	
surface tension	newton per meter	N/m	
thermal conductivity	watt per meter kelvin	W/(m·K)	
velocity	meter per second	m/s	
	kilometer per hour	km/h	
viscosity, dynamic	pascal second	Pa·s	
	millipascal second	mPa·s	
viscosity, kinematic	square meter per second	m^2/s	
	square millimeter per second	mm^2/s	

Quantity	Unit	Symbol	Acceptable equivalent
volume	cubic meter	m^3	
	cubic decimeter	dm^3	L (liter) (5)
	cubic centimeter	cm^3	mL
wave number	1 per meter	m^{-1}	
	1 per centimeter	cm^{-1}	

In addition, there are 16 prefixes used to indicate order of magnitude, as follows:

Multiplication factor	Prefix	Symbol	Note
10^{18}	exa	E	
10^{15}	peta	P	
10^{12}	tera	T	
10^{9}	giga	G	
10^{6}	mega	M	
10^{3}	kilo	k	
10^{2}	hecto	h[a]	[a]Although hecto, deka, deci, and centi are SI prefixes, their use should be avoided except for SI unit-multiples for area and volume and nontechnical use of centimeter, as for body and clothing measurement.
10	deka	da[a]	
10^{-1}	deci	d[a]	
10^{-2}	centi	c[a]	
10^{-3}	milli	m	
10^{-6}	micro	μ	
10^{-9}	nano	n	
10^{-12}	pico	p	
10^{-15}	femto	f	
10^{-18}	atto	a	

For a complete description of SI and its use the reader is referred to ASTM E 380 (4) and the article UNITS AND CONVERSION FACTORS which appears in Vol. 24.

A representative list of conversion factors from non-SI to SI units is presented herewith. Factors are given to four significant figures. Exact relationships are followed by a dagger. A more complete list is given in the latest editions of ASTM E 380 (4) and ANSI Z210.1 (6).

Conversion Factors to SI Units

To convert from	To	Multiply by
acre	square meter (m^2)	4.047×10^3
angstrom	meter (m)	1.0×10^{-10}†
are	square meter (m^2)	1.0×10^{2}†
astronomical unit	meter (m)	1.496×10^{11}

†Exact.

To convert from	To	Multiply by
atmosphere, standard	pascal (Pa)	1.013×10^{5}
bar	pascal (Pa)	$1.0 \times 10^{5\dagger}$
barn	square meter (m^2)	$1.0 \times 10^{-28\dagger}$
barrel (42 U.S. liquid gallons)	cubic meter (m^3)	0.1590
Bohr magneton (μ_B)	J/T	9.274×10^{-24}
Btu (International Table)	joule (J)	1.055×10^{3}
Btu (mean)	joule (J)	1.056×10^{3}
Btu (thermochemical)	joule (J)	1.054×10^{3}
bushel	cubic meter (m^3)	3.524×10^{-2}
calorie (International Table)	joule (J)	4.187
calorie (mean)	joule (J)	4.190
calorie (thermochemical)	joule (J)	$4.184^{\dagger}$
centipoise	pascal second (Pa·s)	$1.0 \times 10^{-3\dagger}$
centistokes	square millimeter per second (mm^2/s)	$1.0^{\dagger}$
cfm (cubic foot per minute)	cubic meter per second (m^3/s)	4.72×10^{-4}
cubic inch	cubic meter (m^3)	1.639×10^{-5}
cubic foot	cubic meter (m^3)	2.832×10^{-2}
cubic yard	cubic meter (m^3)	0.7646
curie	becquerel (Bq)	$3.70 \times 10^{10\dagger}$
debye	coulomb meter (C·m)	3.336×10^{-30}
degree (angle)	radian (rad)	1.745×10^{-2}
denier (international)	kilogram per meter (kg/m)	1.111×10^{-7}
	tex‡	0.1111
dram (apothecaries')	kilogram (kg)	3.888×10^{-3}
dram (avoirdupois)	kilogram (kg)	1.772×10^{-3}
dram (U.S. fluid)	cubic meter (m^3)	3.697×10^{-6}
dyne	newton (N)	$1.0 \times 10^{-5\dagger}$
dyne/cm	newton per meter (N/m)	$1.0 \times 10^{-3\dagger}$
electronvolt	joule (J)	1.602×10^{-19}
erg	joule (J)	$1.0 \times 10^{-7\dagger}$
fathom	meter (m)	1.829
fluid ounce (U.S.)	cubic meter (m^3)	2.957×10^{-5}
foot	meter (m)	$0.3048^{\dagger}$
footcandle	lux (lx)	10.76
furlong	meter (m)	2.012×10^{-2}
gal	meter per second squared (m/s^2)	$1.0 \times 10^{-2\dagger}$
gallon (U.S. dry)	cubic meter (m^3)	4.405×10^{-3}
gallon (U.S. liquid)	cubic meter (m^3)	3.785×10^{-3}
gallon per minute (gpm)	cubic meter per second (m^3/s)	6.309×10^{-5}
	cubic meter per hour (m^3/h)	0.2271

†Exact.

‡See footnote on p. xiv.

To convert from	To	Multiply by
gauss	tesla (T)	1.0×10^{-4}
gilbert	ampere (A)	0.7958
gill (U.S.)	cubic meter (m^3)	1.183×10^{-4}
grade	radian	1.571×10^{-2}
grain	kilogram (kg)	6.480×10^{-5}
gram force per denier	newton per tex (N/tex)	8.826×10^{-2}
hectare	square meter (m^2)	$1.0 \times 10^{4\dagger}$
horsepower (550 ft·lbf/s)	watt (W)	7.457×10^{2}
horsepower (boiler)	watt (W)	9.810×10^{3}
horsepower (electric)	watt (W)	$7.46 \times 10^{2\dagger}$
hundredweight (long)	kilogram (kg)	50.80
hundredweight (short)	kilogram (kg)	45.36
inch	meter (m)	$2.54 \times 10^{-2\dagger}$
inch of mercury (32°F)	pascal (Pa)	3.386×10^{3}
inch of water (39.2°F)	pascal (Pa)	2.491×10^{2}
kilogram-force	newton (N)	9.807
kilowatt hour	megajoule (MJ)	$3.6^{\dagger}$
kip	newton(N)	4.448×10^{3}
knot (international)	meter per second (m/s)	0.5144
lambert	candela per square meter (cd/m^3)	3.183×10^{3}
league (British nautical)	meter (m)	5.559×10^{3}
league (statute)	meter (m)	4.828×10^{3}
light year	meter (m)	9.461×10^{15}
liter (for fluids only)	cubic meter (m^3)	$1.0 \times 10^{-3\dagger}$
maxwell	weber (Wb)	$1.0 \times 10^{-8\dagger}$
micron	meter (m)	$1.0 \times 10^{-6\dagger}$
mil	meter (m)	$2.54 \times 10^{-5\dagger}$
mile (statute)	meter (m)	1.609×10^{3}
mile (U.S. nautical)	meter (m)	$1.852 \times 10^{3\dagger}$
mile per hour	meter per second (m/s)	0.4470
millibar	pascal (Pa)	1.0×10^{2}
millimeter of mercury (0°C)	pascal (Pa)	$1.333 \times 10^{2\dagger}$
minute (angular)	radian	2.909×10^{-4}
myriagram	kilogram (kg)	10
myriameter	kilometer (km)	10
oersted	ampere per meter (A/m)	79.58
ounce (avoirdupois)	kilogram (kg)	2.835×10^{-2}
ounce (troy)	kilogram (kg)	3.110×10^{-2}
ounce (U.S. fluid)	cubic meter (m^3)	2.957×10^{-5}
ounce-force	newton (N)	0.2780
peck (U.S.)	cubic meter (m^3)	8.810×10^{-3}
pennyweight	kilogram (kg)	1.555×10^{-3}
pint (U.S. dry)	cubic meter (m^3)	5.506×10^{-4}

†Exact.

To convert from	To	Multiply by
pint (U.S. liquid)	cubic meter (m^3)	4.732×10^{-4}
poise (absolute viscosity)	pascal second (Pa·s)	0.10†
pound (avoirdupois)	kilogram (kg)	0.4536
pound (troy)	kilogram (kg)	0.3732
poundal	newton (N)	0.1383
pound-force	newton (N)	4.448
pound force per square inch (psi)	pascal (Pa)	6.895×10^3
quart (U.S. dry)	cubic meter (m^3)	1.101×10^{-3}
quart (U.S. liquid)	cubic meter (m^3)	9.464×10^{-4}
quintal	kilogram (kg)	1.0×10^{2}†
rad	gray (Gy)	1.0×10^{-2}†
rod	meter (m)	5.029
roentgen	coulomb per kilogram (C/kg)	2.58×10^{-4}
second (angle)	radian (rad)	4.848×10^{-6}†
section	square meter (m^2)	2.590×10^6
slug	kilogram (kg)	14.59
spherical candle power	lumen (lm)	12.57
square inch	square meter (m^2)	6.452×10^{-4}
square foot	square meter (m^2)	9.290×10^{-2}
square mile	square meter (m^2)	2.590×10^6
square yard	square meter (m^2)	0.8361
stere	cubic meter (m^3)	1.0†
stokes (kinematic viscosity)	square meter per second (m^2/s)	1.0×10^{-4}†
tex	kilogram per meter (kg/m)	1.0×10^{-6}†
ton (long, 2240 pounds)	kilogram (kg)	1.016×10^3
ton (metric) (tonne)	kilogram (kg)	1.0×10^3†
ton (short, 2000 pounds)	kilogram (kg)	9.072×10^2
torr	pascal (Pa)	1.333×10^2
unit pole	weber (Wb)	1.257×10^{-7}
yard	meter (m)	0.9144†

Abbreviations and Unit Symbols

Following is a list of common abbreviations and unit symbols used in the *Encyclopedia*. In general they agree with those listed in *American National Standard Abbreviations for Use on Drawings and in Text (ANSI Y1.1)* (6) and *American National Standard Letter Symbols for Units in Science and Technology (ANSI Y10)* (6). Also included is a list of acronyms for a number of private and

†Exact.

government organizations as well as common industrial solvents, polymers, and other chemicals.

Rules for Writing Unit Symbols (4):

1. Unit symbols are printed in upright letters (roman) regardless of the type style used in the surrounding text.
2. Unit symbols are unaltered in the plural.
3. Unit symbols are not followed by a period except when used at the end of a sentence.
4. Letter unit symbols are generally printed lower-case (for example, cd for candela) unless the unit name has been derived from a proper name, in which case the first letter of the symbol is capitalized (W, Pa). Prefixes and unit symbols retain their prescribed form regardless of the surrounding typography.
5. In the complete expression for a quantity, a space should be left between the numerical value and the unit symbol. For example, write 2.37 lm, *not* 2.37lm, and 35 mm, *not* 35mm. When the quantity is used in an adjectival sense, a hyphen is often used, for example, 35-mm film. *Exception:* No space is left between the numerical value and the symbols for degree, minute, and second of plane angle, degree Celsius, and the percent sign.
6. No space is used between the prefix and unit symbol (for example, kg).
7. Symbols, not abbreviations, should be used for units. For example, use "A," not "amp," for ampere.
8. When multiplying unit symbols, use a raised dot:

   $$\mathrm{N{\cdot}m} \quad \text{for} \quad \text{newton meter}$$

   In the case of W·h, the dot may be omitted, thus:

   $$\mathrm{Wh}$$

   An exception to this practice is made for computer printouts, automatic typewriter work, etc, where the raised dot is not possible, and a dot on the line may be used.
9. When dividing unit symbols, use one of the following forms:

   $$\mathrm{m/s} \quad \textit{or} \quad \mathrm{m{\cdot}s^{-1}} \quad \textit{or} \quad \frac{\mathrm{m}}{\mathrm{s}}$$

   In no case should more than one slash be used in the same expression unless parentheses are inserted to avoid ambiguity. For example, write:

   $$\mathrm{J/(mol{\cdot}K)} \quad \textit{or} \quad \mathrm{J{\cdot}mol^{-1}\cdot K^{-1}} \quad \textit{or} \quad \mathrm{(J/mol)/K}$$

   but *not*

   $$\mathrm{J/mol/K}$$

10. Do not mix symbols and unit names in the same expression. Write:

joules per kilogram *or* J/kg *or* $J{\cdot}kg^{-1}$

but *not*

joules/kilogram *nor* joules/kg *nor* $joules{\cdot}kg^{-1}$

ABBREVIATIONS AND UNITS

A	ampere
A	anion (eg, HA)
A	mass number
a	atto (prefix for 10^{-18})
AATCC	American Association of Textile Chemists and Colorists
ABS	acrylonitrile–butadiene–styrene
abs	absolute
ac	alternating current, *n.*
a-c	alternating current, *adj.*
ac-	alicyclic
acac	acetylacetonate
ACGIH	American Conference of Governmental Industrial Hygienists
ACS	American Chemical Society
AGA	American Gas Association
Ah	ampere hour
AIChE	American Institute of Chemical Engineers
AIME	American Institute of Mining, Metallurgical, and Petroleum Engineers
AIP	American Institute of Physics
AISI	American Iron and Steel Institute
alc	alcohol(ic)
Alk	alkyl
alk	alkaline (not alkali)
amt	amount
amu	atomic mass unit
ANSI	American National Standards Institute
AO	atomic orbital
AOAC	Association of Official Analytical Chemists
AOCS	American Oil Chemists' Society
APHA	American Public Health Association
API	American Petroleum Institute
aq	aqueous
Ar	aryl
ar-	aromatic
as-	asymmetric(al)
ASHRAE	American Society of Heating, Refrigerating, and Air Conditioning Engineers
ASM	American Society for Metals
ASME	American Society of Mechanical Engineers
ASTM	American Society for Testing and Materials
at no.	atomic number
at wt	atomic weight
av(g)	average
AWS	American Welding Society
b	bonding orbital
bbl	barrel
bcc	body-centered cubic
BCT	body-centered tetragonal
Bé	Baumé
BET	Brunauer-Emmett-Teller (adsorption equation)
bid	twice daily
Boc	*t*-butyloxycarbonyl
BOD	biochemical (biological) oxygen demand
bp	boiling point
Bq	becquerel

C	coulomb
°C	degree Celsius
C-	denoting attachment to carbon
c	centi (prefix for 10^{-2})
c	critical
ca	circa (approximately)
cd	candela; current density; circular dichroism
CFR	Code of Federal Regulations
cgs	centimeter-gram-second
CI	Color Index
cis-	isomer in which substituted groups are on same side of double bond between C atoms
cl	carload
cm	centimeter
cmil	circular mil
cmpd	compound
CNS	central nervous system
CoA	coenzyme A
COD	chemical oxygen demand
coml	commercial(ly)
cp	chemically pure
cph	close-packed hexagonal
CPSC	Consumer Product Safety Commission
cryst	crystalline
cub	cubic
D	Debye
D-	denoting configurational relationship
d	differential operator
d	day; deci (prefix for 10^{-1})
d-	*dextro*-, dextrorotatory
da	deka (prefix for 10^{1})
dB	decibel
dc	direct current, *n.*
d-c	direct current, *adj.*
dec	decompose
detd	determined
detn	determination
Di	didymium, a mixture of all lanthanons
dia	diameter
dil	dilute
DIN	Deutsche Industrie Normen
dl-; DL-	racemic
DMA	dimethylacetamide
DMF	dimethylformamide
DMG	dimethyl glyoxime
DMSO	dimethyl sulfoxide
DOD	Department of Defense
DOE	Department of Energy
DOT	Department of Transportation
DP	degree of polymerization
dp	dew point
DPH	diamond pyramid hardness
dstl(d)	distill(ed)
dta	differential thermal analysis
(*E*)-	entgegen; opposed
ϵ	dielectric constant (unitless number)
e	electron
ECU	electrochemical unit
ed.	edited, edition, editor
ED	effective dose
EDTA	ethylenediaminetetraacetic acid
emf	electromotive force
emu	electromagnetic unit
en	ethylene diamine
eng	engineering
EPA	Environmental Protection Agency
epr	electron paramagnetic resonance
eq.	equation
esca	electron spectroscopy for chemical analysis
esp	especially
esr	electron-spin resonance
est(d)	estimate(d)
estn	estimation
esu	electrostatic unit
exp	experiment, experimental
ext(d)	extract(ed)
F	farad (capacitance)
F	faraday (96,487 C)
f	femto (prefix for 10^{-15})

FAO	Food and Agriculture Organization (United Nations)
fcc	face-centered cubic
FDA	Food and Drug Administration
FEA	Federal Energy Administration
FHSA	Federal Hazardous Substances Act
fob	free on board
fp	freezing point
FPC	Federal Power Commission
FRB	Federal Reserve Board
frz	freezing
G	giga (prefix for 10^9)
G	gravitational constant = 6.67×10^{11} $N{\cdot}m^2/kg^2$
g	gram
(g)	gas, only as in $H_2O(g)$
g	gravitational acceleration
gc	gas chromatography
gem-	geminal
glc	gas–liquid chromatography
g-mol wt; gmw	gram-molecular weight
GNP	gross national product
gpc	gel-permeation chromatography
GRAS	Generally Recognized as Safe
grd	ground
Gy	gray
H	henry
h	hour; hecto (prefix for 10^2)
ha	hectare
HB	Brinell hardness number
Hb	hemoglobin
hcp	hexagonal close-packed
hex	hexagonal
HK	Knoop hardness number
hplc	high performance liquid chromatography
HRC	Rockwell hardness (C scale)
HV	Vickers hardness number
hyd	hydrated, hydrous
hyg	hygroscopic
Hz	hertz
^{i} (eg, Pr^i)	iso (eg, isopropyl)
i-	inactive (eg, *i*-methionine)
IACS	International Annealed Copper Standard
ibp	initial boiling point
IC	integrated circuit
ICC	Interstate Commerce Commission
ICT	International Critical Table
ID	inside diameter; infective dose
ip	intraperitoneal
IPS	iron pipe size
ir	infrared
IRLG	Interagency Regulatory Liaison Group
ISO	International Organization Standardization
ITS-90	International Temperature Scale (NIST)
IU	International Unit
IUPAC	International Union of Pure and Applied Chemistry
IV	iodine value
iv	intravenous
J	joule
K	kelvin
k	kilo (prefix for 10^3)
kg	kilogram
L	denoting configurational relationship
L	liter (for fluids only) (5)
l-	*levo-*, levorotatory
(l)	liquid, only as in $NH_3(l)$
LC_{50}	conc lethal to 50% of the animals tests
LCAO	linear combination of atomic orbitals
lc	liquid chromatography
LCD	liquid crystal display
lcl	less than carload lots

LD_{50}	dose lethal to 50% of the animals tested
LED	light-emitting diode
liq	liquid
lm	lumen
ln	logarithm (natural)
LNG	liquefied natural gas
log	logarithm (common)
LPG	liquefied petroleum gas
ltl	less than truckoad lots
lx	lux
M	mega (prefix for 10^{6}); metal (as in MA)
M	molar; actual mass
$\overline{M}_w$	weight-average mol wt
$\overline{M}_n$	number-average mol wt
m	meter; milli (prefix for 10^{-3})
m	molal
m-	meta
max	maximum
MCA	Chemical Manufacturers' Association (was Manufacturing Chemists Association)
MEK	methyl ethyl ketone
meq	milliequivalent
mfd	manufactured
mfg	manufacturing
mfr	manufacturer
MIBC	methyl isobutyl carbinol
MIBK	methyl isobutyl ketone
MIC	minimum inhibiting concentration
min	minute; minimum
mL	milliliter
MLD	minimum lethal dose
MO	molecular orbital
mo	month
mol	mole
mol wt	molecular weight
mp	melting point
MR	molar refraction
ms	mass spectrometry
mxt	mixture
μ	micro (prefix for 10^{-6})
N	newton (force)
N	normal (concentration); neutron number

N-	denoting attachment to nitrogen
n (as n_D^{20})	index of refraction (for 20°C and sodium light)
^{n} (as Bu^{n}), *n*-	normal (straight-chain structure)
n	neutron
n	nano (prefix for 10^{9})
na	not available
NAS	National Academy of Sciences
NASA	National Aeronautics and Space Administration
nat	natural
ndt	nondestructive testing
neg	negative
NF	*National Formulary*
NIH	National Institutes of Health
NIOSH	National Institute of Occupational Safety and Health
NIST	National Institute of Standards and Technology (formerly National Bureau of Standards)
nmr	nuclear magnetic resonance
NND	New and Nonofficial Drugs (AMA)
no.	number
NOI-(BN)	not otherwise indexed (by name)
NOS	not otherwise specified
nqr	nuclear quadruple resonance
NRC	Nuclear Regulatory Commission; National Research Council
NRI	New Ring Index
NSF	National Science Foundation
NTA	nitrilotriacetic acid
NTP	normal temperature and pressure (25°C and 101.3 kPa or 1 atm)
NTSB	National Transportation Safety Board

O-	denoting attachment to oxygen
o-	ortho
OD	outside diameter
OPEC	Organization of Petroleum Exporting Countries
o-phen	*o*-phenanthridine
OSHA	Occupational Safety and Health Administration
owf	on weight of fiber
Ω	ohm
P	peta (prefix for 10^{15})
p	pico (prefix for 10^{-12})
p-	para
p	proton
p.	page
Pa	pascal (pressure)
PEL	personal exposure limit based on an 8-h exposure
pd	potential difference
pH	negative logarithm of the effective hydrogen ion concentration
phr	parts per hundred of resin (rubber)
p-i-n	positive-intrinsic-negative
pmr	proton magnetic resonance
p-n	positive-negative
po	per os (oral)
POP	polyoxypropylene
pos	positive
pp.	pages
ppb	parts per billion (10^9)
ppm	parts per million (10^6)
ppmv	parts per million by volume
ppmwt	parts per million by weight
PPO	poly(phenyl oxide)
ppt(d)	precipitate(d)
pptn	precipitation
Pr (no.)	foreign prototype (number)
pt	point; part
PVC	poly(vinyl chloride)
pwd	powder
py	pyridine
qv	quod vide (which see)
R	univalent hydrocarbon radical
(*R*)-	rectus (clockwise configuration)
r	precision of data
rad	radian; radius
RCRA	Resource Conservation and Recovery Act
rds	rate-determining step
ref.	reference
rf	radio frequency, *n.*
r-f	radio frequency, *adj.*
rh	relative humidity
RI	Ring Index
rms	root-mean square
rpm	rotations per minute
rps	revolutions per second
RT	room temperature
RTECS	Registry of Toxic Effects of Chemical Substances
^{s} (eg, Bu^s); *sec*-	secondary (eg, secondary butyl)
S	siemens
(*S*)-	sinister (counterclockwise configuration)
S-	denoting attachment to sulfur
s-	symmetric(al)
s	second
(s)	solid, only as in $H_2O(s)$
SAE	Society of Automotive Engineers
SAN	styrene–acrylonitrile
sat(d)	saturate(d)
satn	saturation
SBS	styrene–butadiene–styrene
sc	subcutaneous
SCF	self-consistent field; standard cubic feet
Sch	Schultz number
SFs	Saybolt Furol seconds
SI	Le Système International d'Unités (International System of Units)
sl sol	slightly soluble
sol	soluble
soln	solution
soly	solubility

sp	specific; species
sp gr	specific gravity
sr	steradian
std	standard
STP	standard temperature and pressure (0°C and 101.3 kPa)
sub	sublime(s)
SUs	Saybolt Universal seconds
syn	synthetic
^{t} (eg, Bu^t), *t-*, *tert-*	tertiary (eg, tertiary butyl)
T	tera (prefix for 10^{12}); tesla (magnetic flux density)
t	metric ton (tonne)
t	temperature
TAPPI	Technical Association of the Pulp and Paper Industry
TCC	Tagliabue closed up
tex	tex (linear density)
T_g	glass-transition temperature
tga	thermogravimetric analysis
THF	tetrahydrofuran
tlc	thin layer chromatography
TLV	threshold limit value
trans-	isomer in which substituted groups are on opposite sides of double bond between C atoms
TSCA	Toxic Substances Control Act
TWA	time-weighted average
Twad	Twaddell
UL	Underwriters' Laboratory
USDA	United States Department of Agriculture
USP	*United States Pharmacopeia*
uv	ultraviolet
V	volt (emf)
var	variable
vic-	vicinal
vol	volume (not volatile)
vs	versus
v sol	very soluble
W	watt
Wb	weber
Wh	watt hour
WHO	World Health Organization (United Nations)
wk	week
yr	year
(*Z*)-	zusammen; together; atomic number

Non-SI (Unacceptable and Obsolete) Units		Use
Å	angstrom	nm
at	atmosphere, technical	Pa
atm	atmosphere, standard	Pa
b	barn	cm^2
bar[†]	bar	Pa
bbl	barrel	m^3
bhp	brake horsepower	W
Btu	British thermal unit	J
bu	bushel	m^3; L
cal	calorie	J
cfm	cubic foot per minute	m^3/s
Ci	curie	Bq
cSt	centistokes	mm^2/s
c/s	cycle per second	Hz
cu	cubic	exponential form

[†]Do not use bar (10^5Pa) or millibar (10^2Pa) because they are not SI units, and are accepted internationally only for a limited time in special fields because of existing usage.

Non-SI (Unacceptable and Obsolete) Units		Use
D	debye	C•m
den	denier	tex
dr	dram	kg
dyn	dyne	N
dyn/cm	dyne per centimeter	mN/m
erg	erg	J
eu	entropy unit	J/K
°F	degree Fahrenheit	°C; K
fc	footcandle	lx
fl	footlambert	lx
fl oz	fluid ounce	m^3; L
ft	foot	m
ft•lbf	foot pound-force	J
gf den	gram-force per denier	N/tex
G	gauss	T
Gal	gal	m/s^2
gal	gallon	m^3; L
Gb	gilbert	A
gpm	gallon per minute	(m^3/s); (m^3/h)
gr	grain	kg
hp	horsepower	W
ihp	indicated horsepower	W
in.	inch	m
in. Hg	inch of mercury	Pa
in. H_2O	inch of water	Pa
in.-lbf	inch pound-force	J
kcal	kilo-calorie	J
kgf	kilogram-force	N
kilo	for kilogram	kg
L	lambert	lx
lb	pound	kg
lbf	pound-force	N
mho	mho	S
mi	mile	m
MM	million	M
mm Hg	millimeter of mercury	Pa
mμ	millimicron	nm
mph	miles per hour	km/h
μ	micron	μm
Oe	oersted	A/m
oz	ounce	kg
ozf	ounce-force	N
η	poise	Pa•s
P	poise	Pa•s
ph	phot	lx
psi	pounds-force per square inch	Pa
psia	pounds-force per square inch absolute	Pa
psig	pounds-force per square inch gage	Pa
qt	quart	m^3; L
°R	degree Rankine	K
rd	rad	Gy
sb	stilb	lx
SCF	standard cubic foot	m^3
sq	square	exponential form
thm	therm	J
yd	yard	m

BIBLIOGRAPHY

1. The International Bureau of Weights and Measures, BIPM (Parc de Saint-Cloud, France) is described in Appendix X2 of Ref. 4. This bureau operates under the exclusive supervision of the International Committee for Weights and Measures (CIPM).
2. *Metric Editorial Guide (ANMC-78-1),* latest ed., American National Metric Council, 5410 Grosvenor Lane, Bethesda, Md. 20814, 1981.
3. *SI Units and Recommendations for the Use of Their Multiples and of Certain Other Units (ISO 1000-1981),* American National Standards Institute, 1430 Broadway, New York, N.Y. 10018, 1981.
4. Based on *ASTM E 380-89a (Standard Practice for Use of the International System of Units (SI),* American Society for Testing and Materials, 1916 Race Street, Philadelphia, Pa. 19103, 1989.
5. *Fed. Regist.,* Dec. 10, 1976 (41 FR 36414).
6. For ANSI address, see Ref. 3.

R. P. LUKENS
ASTM Committee E-43 on SI Practice

A

Continued

ANTIBIOTICS

β-LACTAMS

CARBAPENEMS AND PENEMS

In the period up to 1970 most β-lactam research was concerned with the penicillin and cephalosporin group of antibiotics (1). Since that time, however, a wide variety of new mono- and bicyclic β-lactam structures have been described. The carbapenems, characterized by the presence of the bicyclic ring system (**1**, X = CH_2) originated from natural sources; the penem ring (**1**, X = S) and its derivatives are the products of the chemical synthetic approach to new antibiotics. The chemical names are: 7-oxo-(*R*)-1-azabicyclo[3.2.0]hept-2-ene-2-carboxylic acid [*78854-41-8*], $C_7H_7NO_3$, and 7-oxo-(*R*)-4-thia-1-azabicyclo[3.2.0]hept-2-ene-2-carboxylic acid [*69126-94-9*], $C_6H_5NO_3S$, respectively.

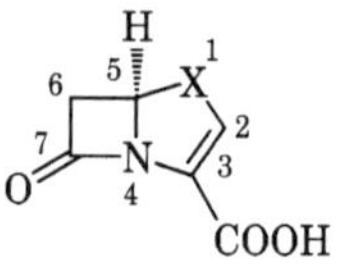

(**1**)

Carbapenems

During the screening of soil samples for inhibitors of peptidoglycan biosynthesis, thienamycin (**2**), shown in Table 1, and other related derivatives were isolated from the microorganism *Streptomyces cattleya* (2). At the same time, while screening for natural inhibitors of β-lactamases, a group of interrelated metabolites known as the olivanic acids, obtained from *Streptomyces olivaceus,* was found (3). The first three compounds isolated were the sulfated derivatives having structures (**3**) and (**4**); subsequently, several nonsulfated analogues were also isolated (4). Thienamycin has the (*R*)-configuration at the C-8 hydroxy group with a trans-arrangement of protons on the β-lactam ring. In the olivanic acids the stereochemistry is 8(*S*)- with a cis-substituted β-lactam in the sulfated series, and both *cis-* and *trans-β*-lactams in the case of the C-8 hydroxy compounds. These compounds represented a completely new family of β-lactam structures based on the 7-oxo-1-azabicyclo[3.2.0]hept-2-ene-2-carboxylic acid ring system, and which are now generally referred to as the 1-carbadethiapen-2-em or simply 1-carbapenem family of antibiotics.

Subsequently, other structural variations were reported encompassing compounds such as PS-5 (**5**) (5), carpetimycin A (**6**) (6), asparenomycin A (**7**) (7), and pluracidomycin A (**8**) (8), from a wide variety of streptomycete strains. Following these structures the simplest member of the series, having the completely unsubstituted nucleus (**1**, X = CH_2), was isolated from bacterial strains of *Serratia* and *Erwinia* (9). All other natural products reported have substituents at both the C-6 and C-2 positions of the bicyclic ring system. Differences in the nature and stereochemistry of these substituents has provided a wide variety of structures, and over forty variations have been reported and comprehensively listed (10).

Occurrence, Fermentation, and Biosynthesis. Although a large number of *Streptomyces* species have been shown to produce carbapenems, only *S. cattleya* (2) and *S. penemfaciens* (11) have been reported to give thienamycin (**2**). Generally the antibiotics occur as a mixture of analogues or isomers and are often co-produced with penicillin N and cephamycin C. Yields are low compared to other β-lactams produced by streptomycetes, and titres are of the order of 1–20 μg/mL despite, in many cases, a great deal of effort on the optimization of the media and fermentation conditions. The rather poor stability of the compounds also contributes to a low recovery in the isolation procedures. The fermentation and isolation processes for thienamycin and the olivanic acids has been reviewed in some detail (12).

Early studies on the biosynthesis of thienamycin (**2**) indicated that the pyrroline ring was derived from glutamic acid (13,14). Further reports demonstrated that the C-6 and C-7 atoms of the β-lactam ring are acetate derived, that the C-2 cysteaminyl side chain is from cysteine, and the two carbon atoms of the hydroxyethyl side chain originate from methionine (15). The incorporation of glutamate into the parent ring system (**1**, X = CH_2) has also been reported (16); in this study bacterial production of various isomers of the saturated carbapenam ring structure was observed including one isomer having stereochemistry opposite to that of thienamycin at the C-5 position. Some streptomycete carbapenem producers have also been found to produce structures in which the pyrroline ring is reduced as in (**9**) (17). The microbiological interconversion of a number of the olivanic acids has been demonstrated (4), as well as the role of carbapenems, such as (**10**, R = H, OH, or OSO_3H) possessing a pantothenyl side chain, as early precursors of other carbapenem antibiotics (18). The possible biosynthetic pathways leading to both carbapenem and carbapenam ring structures have been discussed in some detail (14,16,18).

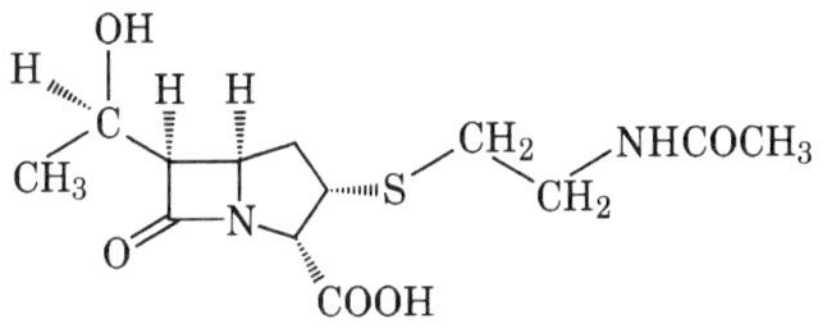

(**9**)

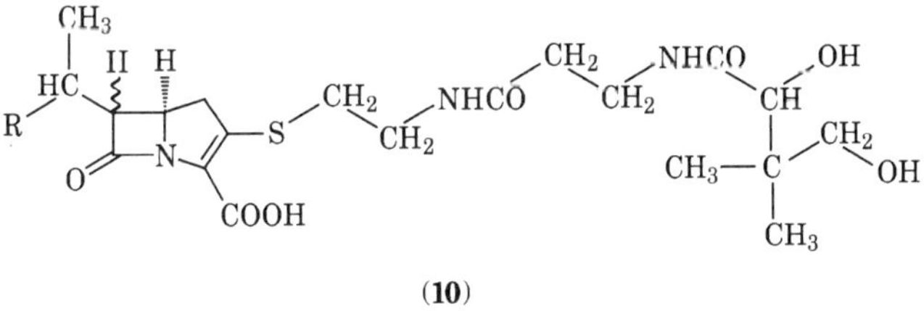

(**10**)

Properties. Thienamycin is isolated as a colorless, hygroscopic, zwitterionic solid, although the majority of carbapenems have been obtained as sodium salts and, in the case of the sulfated olivanic acids, as disodium salts (12). Concentrated aqueous solutions of the carbapenems are generally unstable, particularly at low pH. All the substituted natural products have characteristic uv absorption properties that are often used in assay procedures. The ir frequency of the β-lactam carbonyl is in the range 1760–1790 cm^{-1}.

Structure Determinations. The structural elucidation of the early carbapenems, thienamycin, and the olivanic acids, followed a fairly similar sequence making use of both spectroscopic and degradation studies (19–22). Infrared absorption spectra suggested the presence of a β-lactam ring (ν_{max} 1765 cm^{-1}) and in the case of thienamycin (**2**) a trans-arrangement of β-lactam protons was indicated by the small coupling constant ($J_{5,6}$ <3 Hz) for the β-lactam

Table 1. Carbapenems

Name	CAS Registry Number	Molecular formula	Structure number	Structure
thienamycin	[*59995-64-1*]	$C_{11}H_{16}N_2O_4S$	(**2**)	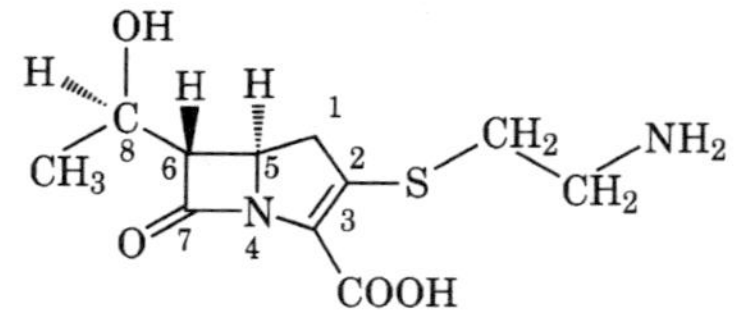
MM 4550 MM 13902	[*12795-21-0*] [*57459-82-2*]	$C_{13}H_{16}N_2O_9S_2$ $C_{13}H_{16}N_2O_8S_2$	(**3**, $n = 1$) (**3**, $n = 0$)	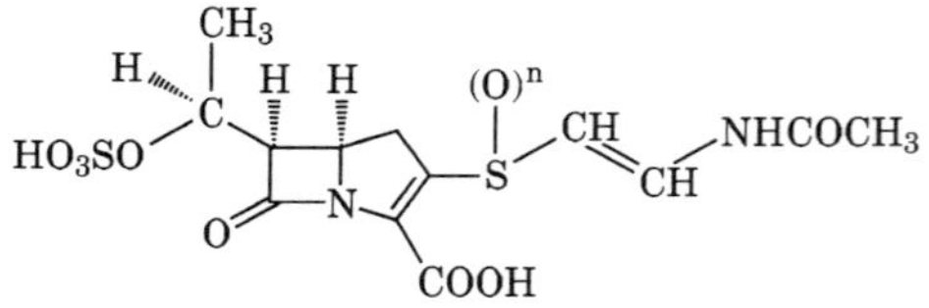
MM 17880	[*61036-81-5*]	$C_{13}H_{18}N_2O_8S_2$	(**4**)	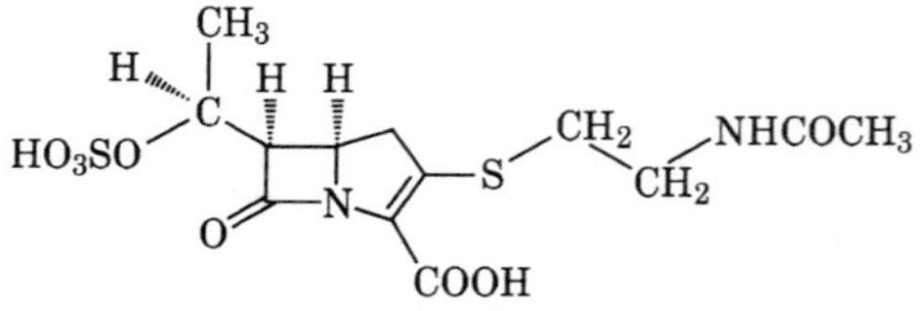

PS–5	[*67007-29-8*]	$C_{13}H_{18}N_2O_4S$	(**5**)	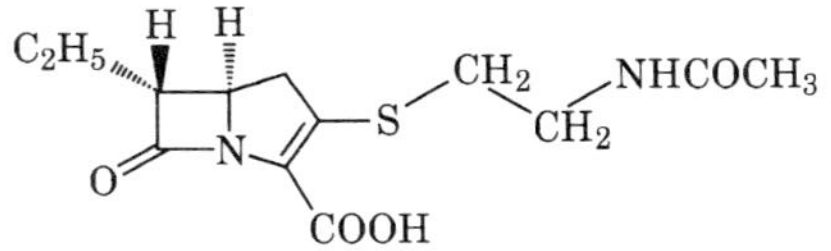
carpetimycin A	[*77209-15-5*]	$C_{14}H_{18}N_2O_6S$	(**6**)	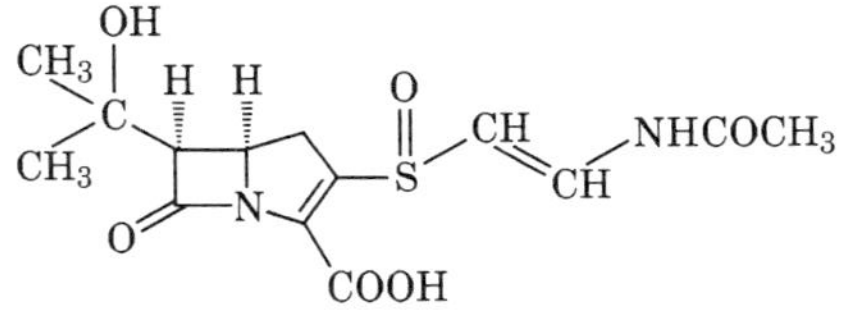
asparenomycin A	[*76466-24-5*]	$C_{14}H_{16}N_2O_6S$	(**7**)	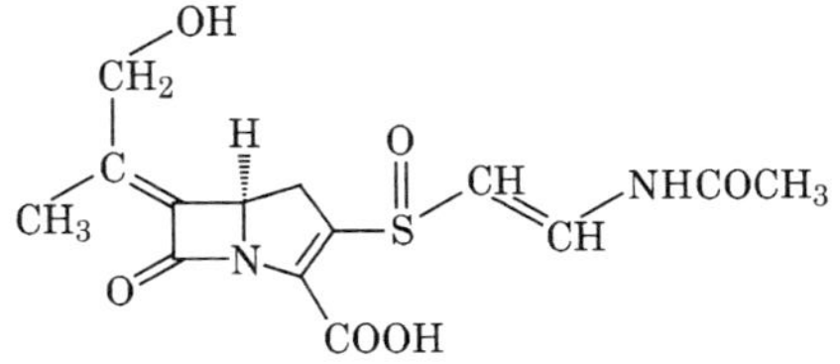
pluracidomycin A	[*82138-64-5*]	$C_9H_{11}NO_{10}S_2$	(**8**)	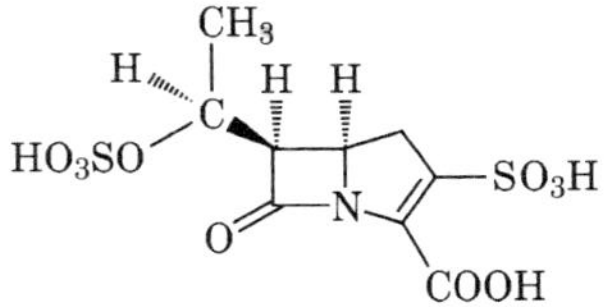

hydrogens in the nmr. For the sulfated olivanic acids, (**3**) and (**4**), the coupling constant ($J_{5,6} \approx 6$ Hz) indicated the more familiar *cis-β*-lactam stereochemistry found in the penicillins and cephalosporins. One key reaction in confirming the structure of MM 13902 (**3**, $n = 0$) is shown in Figure 1**a**. The rearrangement of the monomethyl ester (**11**) produced the pyrrole (**12**); subsequent esterification followed by elimination of the sulfate residue gave a mixture of isomers (**13**) (20). Elimination of the sulfate moiety of the diester (**14**) gave solely the (*E*)-isomer of the ethylidene derivative (**15**) indicating a stereospecific (E)-$_2$ elimination from which it was possible to assign the configuration at C-8 as (*S*). The (*E*)-geometry of the C-2 acetamidoethenyl substituent was clearly apparent from the coupling constant ($J = 14$ Hz) of the double bond protons.

In the case of thienamycin (Fig. 1**b**) the absolute stereochemistry at C-5 was unambiguously determined from the ene-lactam (**16**). The resultant (*R*)-aspartic acid (**17**) demonstrated that the absolute stereochemistry at C-5 of thienamycin is (*R*), corresponding to that found in the C-5 position of both penicillins and cephalosporins. Confirmation of the stereochemical assignments in both thienamycin (**2**) and the olivanic acid MM 13902 (**3**, $n = 0$) has been confirmed by x-ray crystallography (19,21,22). The structural determination of the nonsulfated derivatives from *S. olivaceus* (23), PS-5 (**5**) (5), the carpetimycins (6), and the asparenomycins (7) followed a similar pattern.

Reactions. Although carbapenems are extremely sensitive to many reaction conditions, a wide variety of chemical modifications have been carried out. Many derivatives of the amino, hydroxy, and carboxy group of thienamycin (**2**) have been prepared primarily to study structure–activity relationships (24). The most interesting class of *N*-derivatives are the amidines which are usually obtained in good yield by reaction of thienamycin with an imidate ester at pH 8.3. Introduction of this basic but less nucleophilic moiety maintains or improves the potency of the natural material while greatly increasing the chemical stability. Thus *N*-formimidoyl thienamycin [*64221-86-9*] (MK 0787) (**18**), $C_{12}H_{17}N_3O_4S$, (25) was chosen for clinical evaluation and development. Another reaction of thienamycin involves isomerization to the chemically stable, but biologically inactive, Δ-1 isomer (26). Reductive removal of the cysteaminyl side chain has also been reported (27).

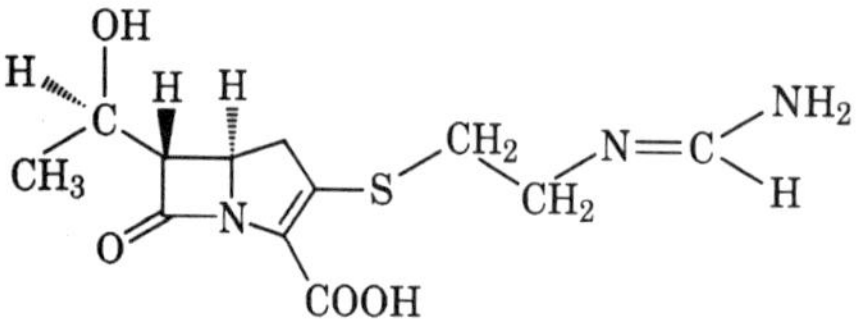

(**18**)

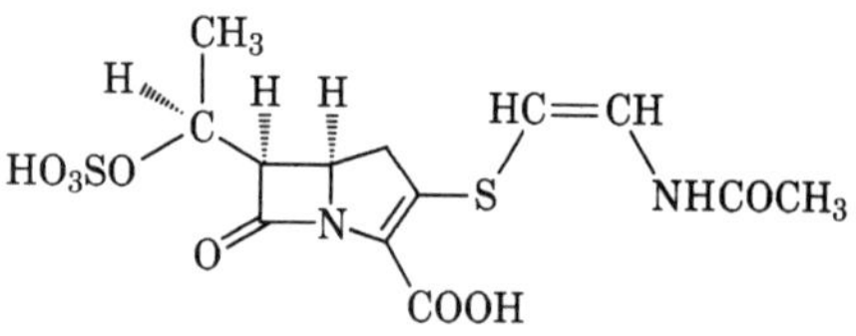

(**19**)

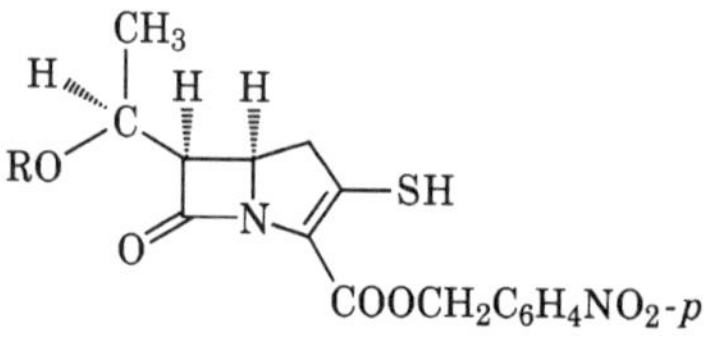

(**20**)

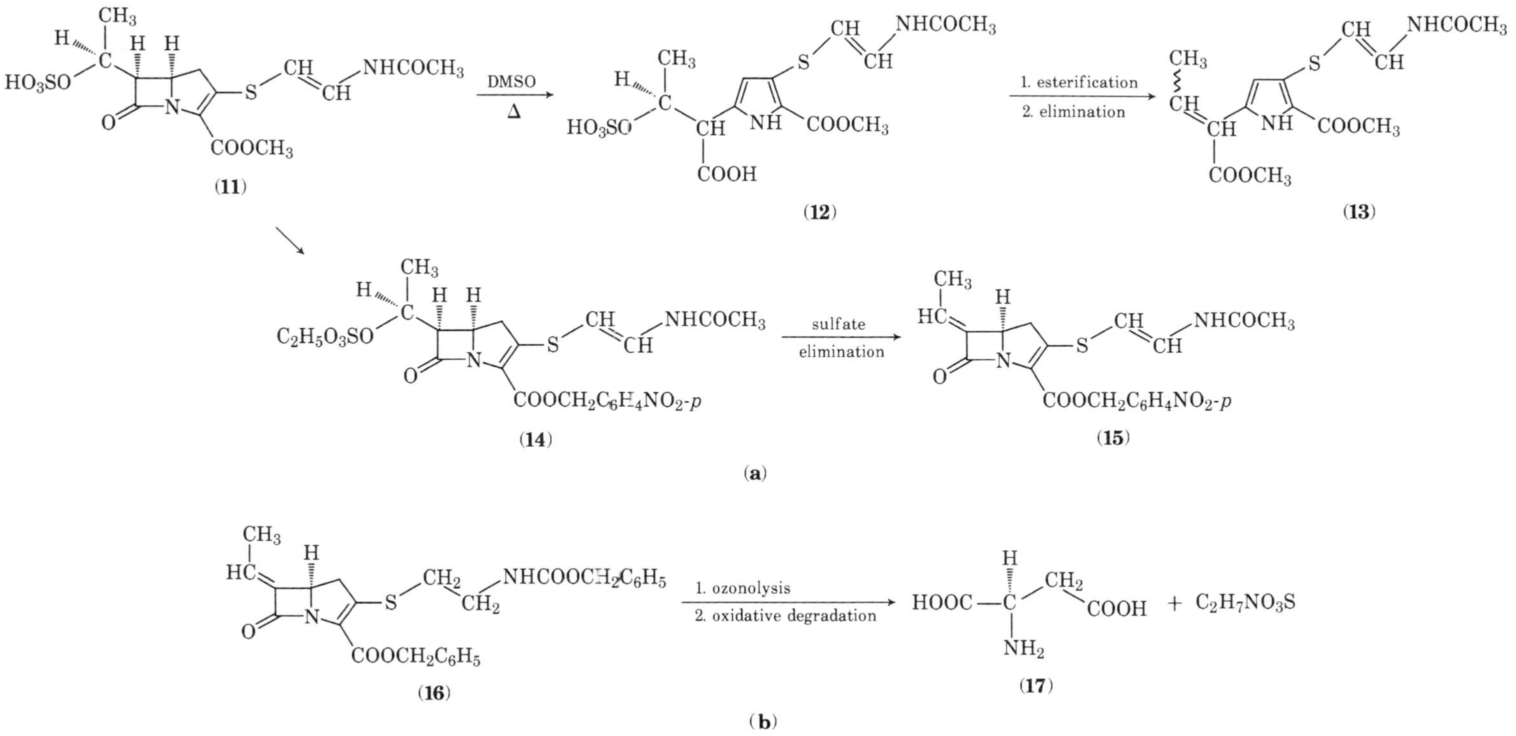

Fig. 1. Reaction sequences for steps in the structure determination of **(a)** MM 13902 (**3**, $n = 0$) and **(b)** thienamycin (**2**).

In the olivanic acid series of carbapenems the (*E*)-acetamidoethenyl grouping can be isomerized to the (*Z*)-isomer (**19**) (22) and reaction with hypobromous acid provides a bromohydrin that fragments to give a thiol of type (**20**) when R = H, SO_3H, or $COCH_3$. The thiol is not isolated but can react to provide new alkyl or alkenyl C-2 substituents (28). In the case of the nonsulfated olivanic acids, inversion of the stereochemistry at the 8(*S*)-hydroxyl group by way of a Mitsunobu reaction affords an entry to the 8(*R*)-thienamycin series (29). An alternative method for introducing new sulfur substituents makes use of a displacement reaction of a carbapenem (*S*)-oxide with a thiol (30). Microbial deacylation of the acylamino group in PS-5 (**5**) has been used to obtain the aminoethyl analogue (31).

Synthesis. One consequence of the discovery of the carbapenem natural products has been the development of new synthetic methods, the impetus for which was provided by the exceptional antibacterial potential of the compounds coupled with the extremely poor fermentation yields. Only chemical synthesis could provide the quantities of MK 0787 (**18**) necessary for clinical trials and commercial production.

The racemic form of the unsubstituted nucleus (**1**, X = CH_2) was synthesized by several groups (32–34) prior to the disclosure of the natural material. One reaction path involved an azetidinone (**21**) where R = CH_2OH or $CH{=}CH_2$ converted to the corresponding phosphorane (**22**) where R′ = *o*-NO_2-C_6H_4 when R = CH_2OH and R′ = $COCH_3$ when R=$CH{=}CH_2$.

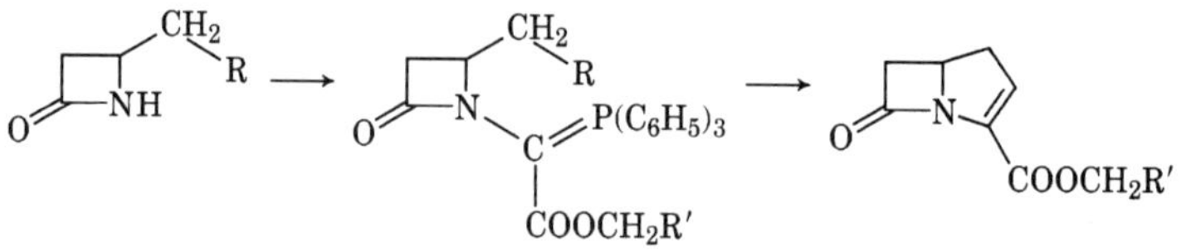

(**21**) (**22**) (**23**)

Cyclization by way of the aldehyde (**22**, R = CHO) provides the appropriate ester of the bicyclic nucleus (**23**), which on deprotection gives an unstable sodium salt. Formation of the [2,3] double bond by this intramolecular Wittig procedure has been widely used for both analogue and natural product synthesis (35). Cyclization is achievable even using a thiol ester such as (**24**) which leads to the protected olivanic acid MM 22383 [*74819-56-0*] (**25**), $C_{28}H_{26}N_4O_{11}S$ (36). Addition of thiols to the double bond of phosphorane-derived derivatives lacking a C-2 substituent (**26**) has been used for the synthesis of PS-5 (**5**) (37).

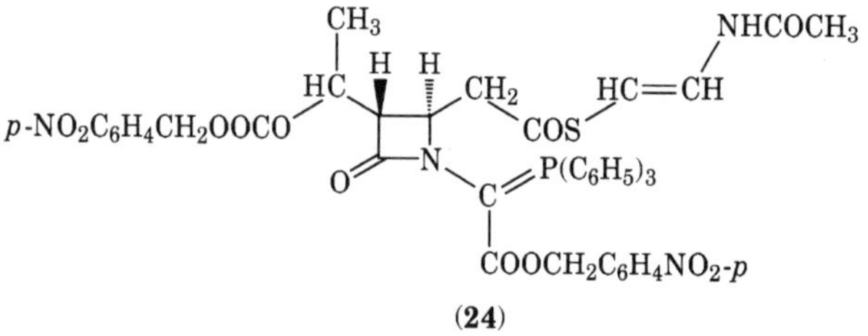

(**24**)

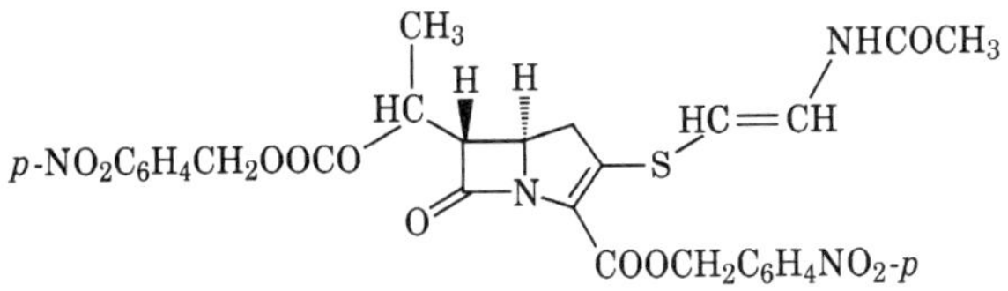

(25)

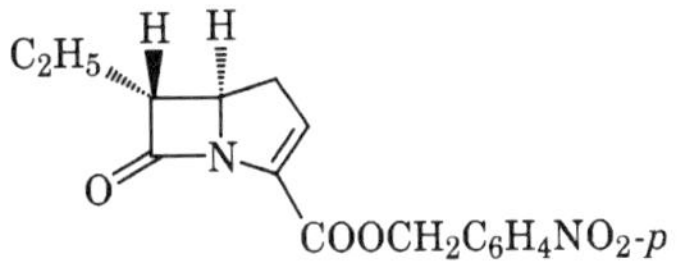

(26)

A synthetic approach that involves the [3,4] bond formation using a carbene insertion reaction has been highly successful and is illustrated by the enantioselective synthesis of (+)-thienamycin starting from L-aspartic acid [*56-84-8*], $C_4H_7NO_4$ (38). Over several steps the acid is converted to a protected form of the azetidinone (**27**) having all three contiguous chiral centers with the correct stereochemistry. Progression to the diazoketone (**28**) and cyclization using a catalytic amount of rhodium acetate provides the bicyclic ester (**29**, R = H) in high yield. Introduction of the cysteaminyl side via an activated enol derivative of (**29**) and deprotection gives (+)-thienamycin;

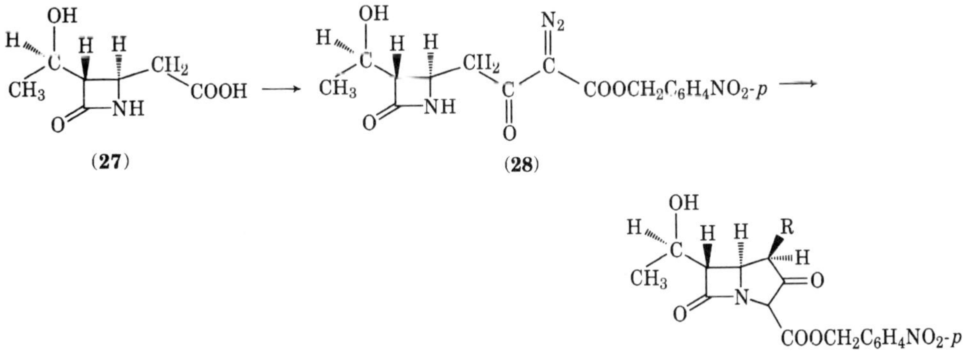

(27) **(28)** **(29)**

using the amidine side chain (**30**) MK 0787 (**18**) is directly accessible (39).

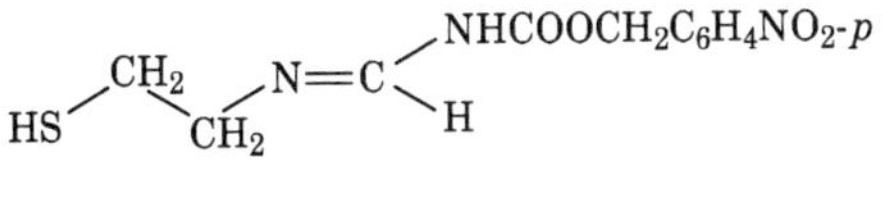

(30)

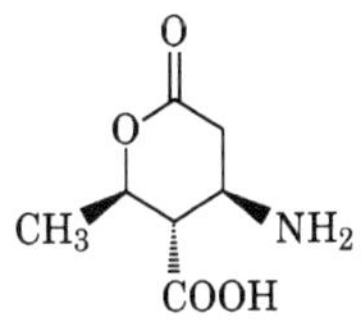

(31)

A second method makes use of the lactone (**31**) from acetone dicarboxylate (40) and for which a synthesis from (−)-carvone has been reported (41). Displacement of chlorine from the 6-aminopenicillanic acid (6-APA) derived β-lactam (**32**) by (**33**) illustrates yet another approach to the diazoketone (**28**) (42).

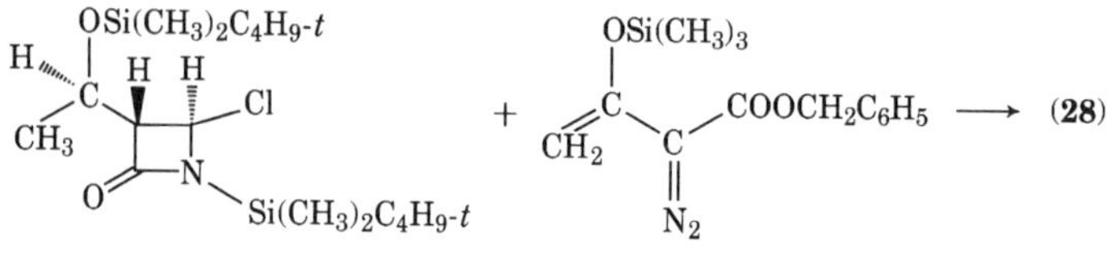

(**32**) (**33**)

Formal syntheses of thienamycin (**2**) from precursors such as carbohydrates (43–45), amino acids (46,47), isoxazolidines (48), and tricarbonyliron lactam complexes (49) have also been reported. Many other methods for carbapenem synthesis have been widely reviewed (10,50–52).

In addition to variable chemical stability the carbapenems are susceptible to β-lactam cleavage by a dehydropeptidase enzyme (DHP-I) located on the brush borders of the kidney (53). Clinically, MK 0787 (**18**) is used with an inhibitor of this enzyme, cilastatin [*78852-98-9*] (MK 0791) (**34**), $C_{16}H_{26}N_2O_5S$, which has a dramatic effect not only on the urinary recovery of the drug, but also reduces any nephrotoxic potential (52) (see ENZYME INHIBITION).

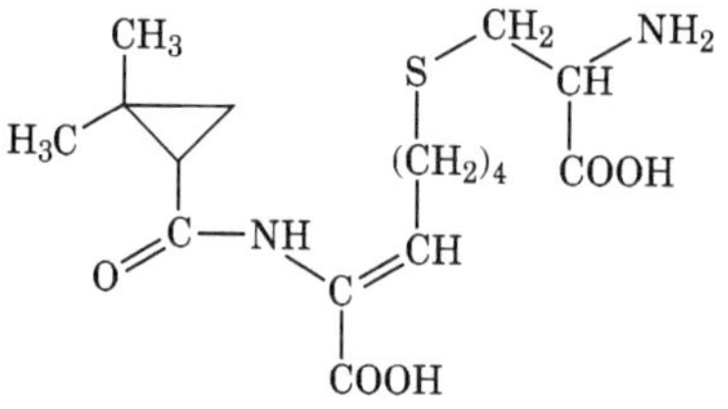

(**34**)

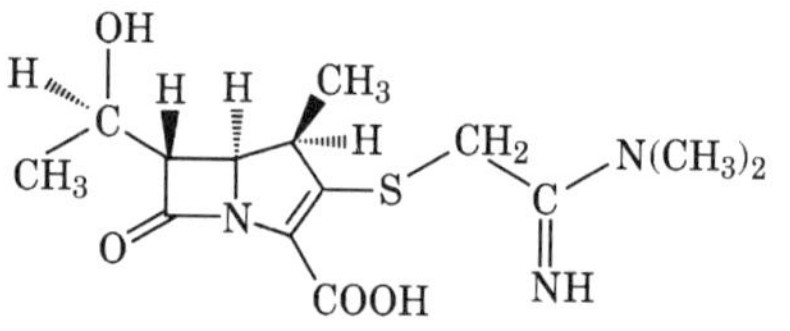

(**35**)

The disclosure that the 1β-methylcarbapenem [*90822-26-7*] (**35**), $C_{14}H_{21}N_3O_4S$, exhibits greatly increased chemical and enzymatic stability over MK 0787 (**18**) (54), while retaining the antibacterial potency of thienamycin (**2**), has caused attention to be focused on methods leading to structural analogues. The synthesis of (**35**) was by alkylation of (**27**) and elaboration via the ketoester (**29**, R = CH_3), but little stereocontrol was attainable by this procedure. Many newer approaches use the acid (**36**) derived by diastereoselective displacement at C-4 using the chiral 4-acyloxy β-lactams (**37**, R = CH_3 or C_6H_5) (55,56). Examples make use of the tin or the boron enolates producing (**36**) in yields of 72–79% with a ratio of β:α isomers ranging from 24:1 to 60:1 (57–59).

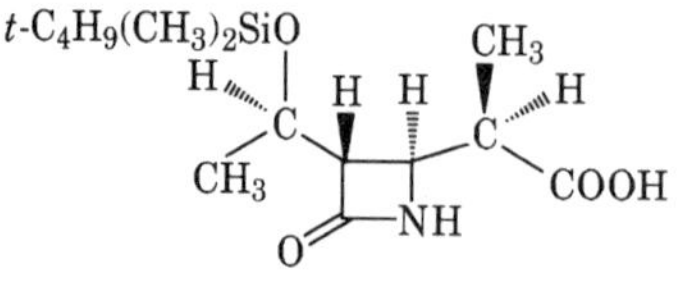

(**36**)

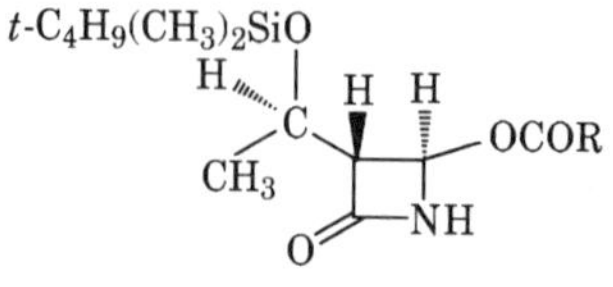

(**37**)

Other approaches to (**36**) make use of (**37**, R = CH_3) and reaction with a tributylstannyl allene (60) or 3-siloxypentadiene (61). A chemicoenzymatic synthesis for both thienamycin (**2**) and 1β-methyl analogues starts from the chiral monoester (**38**), derived by enzymatic hydrolysis of the dimethyl ester, and proceeding by way of the β-lactam (**39**, R = H or CH_3) (62,63). (*S*)-Methyl-3-hydroxy-2-methylpropanoate [*80657-57-4*] (**40**), $C_5H_{10}O_3$, has also been used as starting material for (**36**) (64), whereas 1,3-dipolar cycloaddition of a chiral nitrone with a crotonate ester affords the oxazolidine (**41**) which again can be converted to a suitable β-lactam precursor (65).

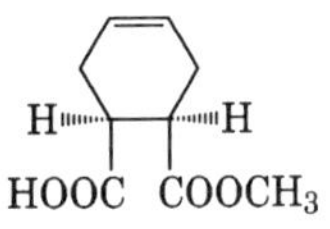

(**38**)

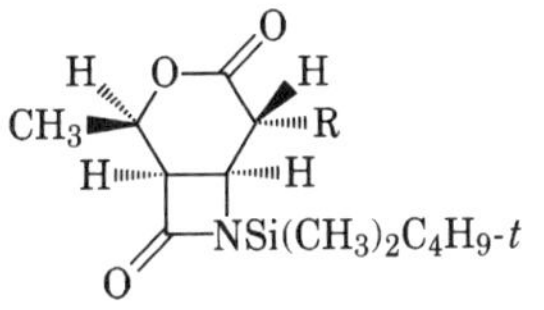

(**39**)

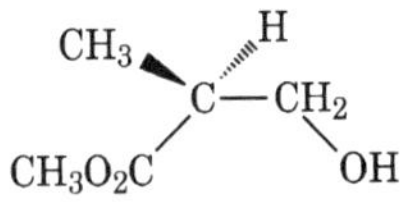

(**40**)

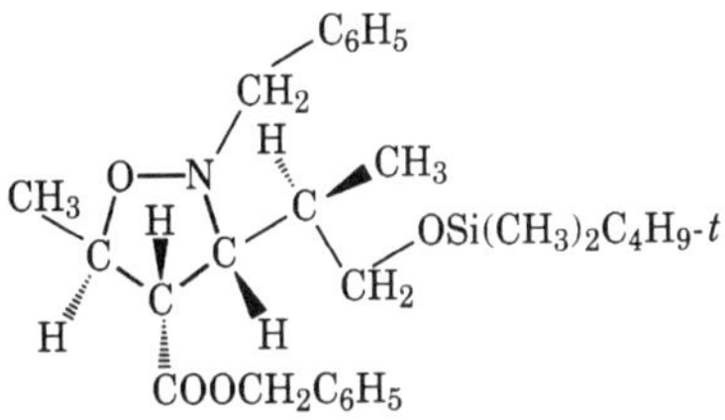

(**41**)

A number of highly potent DHP-I stable 1β-methylcarbapenems having a variety of C-2 substituents have now been described (60,66–69) including SM 7338 [*96036-03-2*] (**42**), $C_{17}H_{25}N_3O_5S$. An acylamino compound (66) and a 1β-methoxy analogue (70) provide other variations. The pyrrolidine substituted 1β-methylcarbapenem SM 7338 (**42**) is being developed as a broad-spectrum parenteral antibiotic under the name meropenem; the synthesis of (**42**) is by way of the lactone (**43**) derived by a novel Diels-Alder approach to dihydropyran precursors of (**43**) (71).

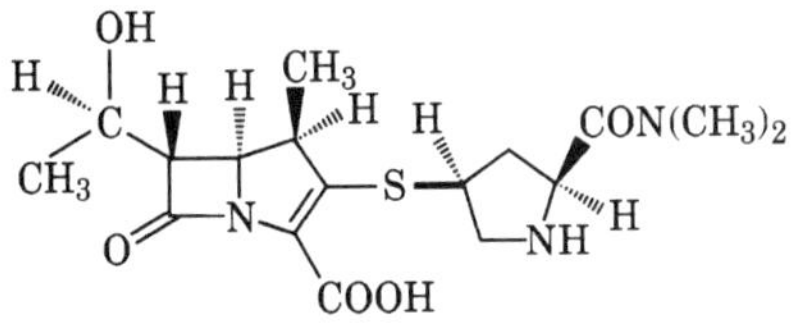

(**42**)

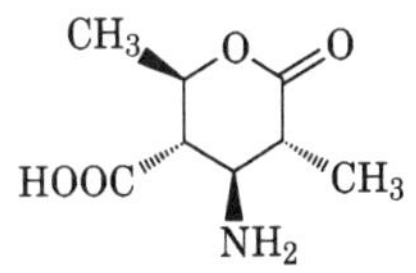

(**43**)

Biological Properties. Thienamycin, the olivanic acids, and the majority of carbapenems are highly active broad-spectrum antibiotics having good stability to β-lactamases. Of the natural products thienamycin (**2**) is the most potent having a spectrum of activity encompassing both aerobic and anaerobic gram-positive and gram-negative bacteria including *Pseudomonas* species (72). The latter activity is attributed to the presence of the basic cysteaminyl substituent at C-2. All of the *N*-acylated derivatives have much reduced antipseudomonal activity (24).

The sulfated compounds MM 13902 (**3**, $n = 0$) and MM 17880 (**4**) are also broad-spectrum agents, but not as potent as thienamycin and all lack any significant activity against *Pseudomonas* (73). Many carbapenems are excellent inhibitors of isolated β-lactamases, particularly the olivanic acid sulfoxide MM 4550 (**3**, $n = 1$) (3). The possible mechanism of action of the carbapenems as inhibitors of β-lactamases has been discussed in some detail (74). Other carbapenems such as PS-5 (**5**) (75), the carpetimycins (76), asparenomycins (77), and pluracidomycins (8) are all highly active as antibiotics or β-lactamase inhibitors. The parent nucleus itself (**1**, X = CH_2) is intrinsically active, but chemically unstable (9).

Unlike the classical penicillins and cephalosporins, a rigid adherence to the cis-stereochemistry around the β-lactam ring is not necessary for biological activity as both trans-isomers such as (**2**), and cis-isomers (**3**, $n = 1$) are active. The hydroxy olivanic acids are not as potent as the sulfates, although in this case the cis-β-lactams are more active than the trans-isomers (73). Thienamycin (**2**) is the most stable to β-lactamases, a property attributed to the presence of the 8(*R*)-hydroxy group. Several synthetic examples lacking the C-6 hydroxyethyl side chain are much more susceptible to hydrolysis by β-lactamases (78). The order of antibiotic activity in the nonsulfated hydroxyethyl series appears to be trans-substituted β-lactam with 8(*R*) stereochemistry ie, thienamycin > cis 8(*S*) > trans 8(*R*) (25). As in the case of other β-lactam antibiotics, the carbapenems inhibit bacterial growth by interfering with peptidoglycan synthesis in the cell wall. In *E. coli* thienamycin has the greatest affinity for the penicillin binding protein PBP-2 (79); most of the newer cephalosporins bind principally to PBP-3.

The amidine derivative MK 0787 (**18**), named as imipenem, shows greatly improved solution stability as compared to thienamycin (**2**). Against some 800 isolates of gram-positive and gram-negative bacteria, imipenem (**18**) inhibited the majority of organisms at concentrations below 1 μg/mL and was not hydrolyzed by plasmid mediated or chromosomal β-lactamases (80). The plasma half-life of imipenem in humans was considered satisfactory (1 h), but urinary recovery of antibiotic was extremely variable ranging from 6–40% (81) resulting from degradation by the DHP-I enzyme. This led to the development of the imipenem-cilastatin combination as the clinical product (82). The urinary recovery of the combination was established consistently at some 70% (83). The imipenem-cilastatin combination is used as a parenteral broad-spectrum antibiotic against a wide variety of bacterial infections. No other carbapenem has yet been developed as a commercial product although effort is being directed towards the DHP-I stable 1β-methyl analogues of which meropenem (**42**) would appear to be at the most advanced stage of development. This is a broad-spectrum agent similar to imipenem but with greatly enhanced stability to dehydropeptidase and should therefore not require concomitant administration of an enzyme inhibitor (69).

Penems

Historically, the development of penems is contemporary with that of the naturally occurring carbapenems and the direction of penem research has clearly been influenced by the structures of the closely related natural products. The origins of the two groups of compounds is, however, quite different. Unlike carbapenems, no penems have been found in nature. When first described (84,85) they were viewed as hybrid molecules combining structural features of penicillins and cephalosporins.

Synthesis. *Woodward's Phosphorane Route.* The first penem synthesis, shown in Figure 2, utilized an intramolecular Wittig reaction to form the [2,3] double bond of the thiazoline ring (84). Reductive acylation of the penicillin derived disulfide (**44**) gave the thioester (**45**). Ozonolysis of the latter provided the oxalimide (**46**) which on mild methanolysis gave the azetidinone (**47**). Well established methods were applied to convert (**47**) to the phosphorane (**48**) which underwent thermal cyclization to the penem ester (**49**). Catalytic hydrogenation gave the penem acid [*64370-39-4*] (**50**) which was shown to possess antibacterial activity in spite of its rather limited stability.

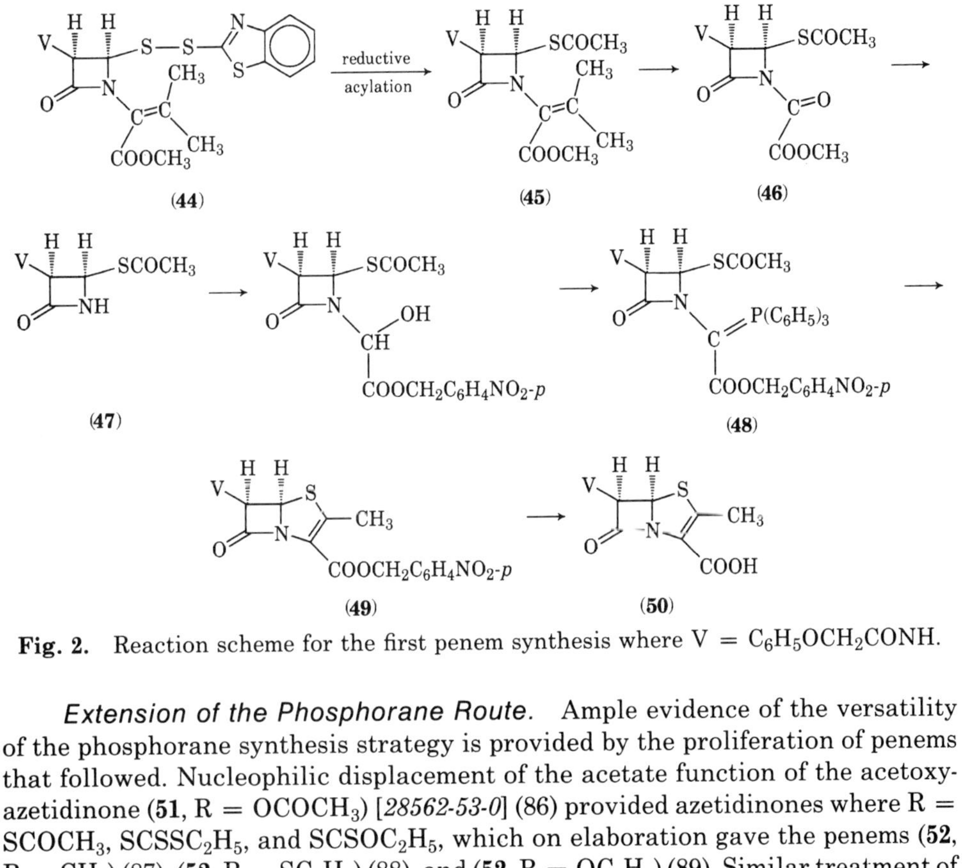

Fig. 2. Reaction scheme for the first penem synthesis where V = $C_6H_5OCH_2CONH$.

Extension of the Phosphorane Route. Ample evidence of the versatility of the phosphorane synthesis strategy is provided by the proliferation of penems that followed. Nucleophilic displacement of the acetate function of the acetoxyazetidinone (**51**, R = $OCOCH_3$) [*28562-53-0*] (86) provided azetidinones where R = $SCOCH_3$, $SCSSC_2H_5$, and $SCSOC_2H_5$, which on elaboration gave the penems (**52**, R = CH_3) (87), (**52**, R = SC_2H_5) (88), and (**52**, R = OC_2H_5) (89). Similar treatment of 3-substituted (or disubstituted) acetoxyazetidinones allowed the synthesis of a number of 2-substituted-6-alkyl- and 6,6-dialkylpenems (90).

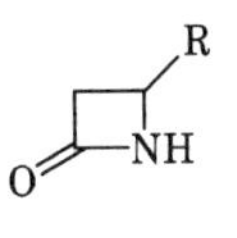

(**51**)

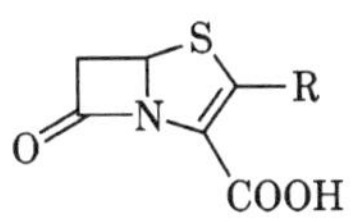

(**52**)

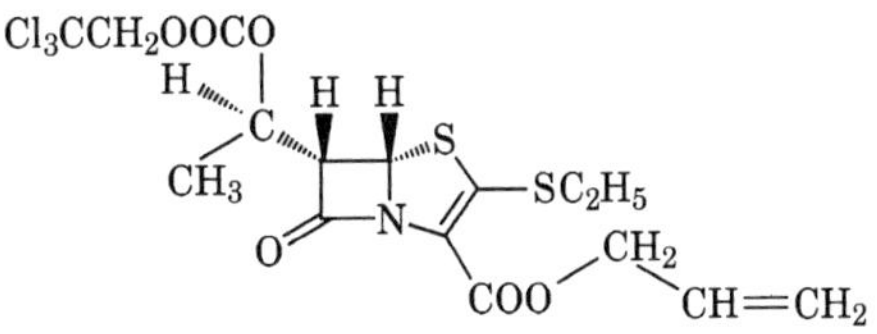

(**55**)

The naturally occurring carbapenems provided the impetus for the synthesis of the four racemates of 6-(1-hydroxyethyl)-2-ethylthiopenem-3-carboxylate (91). As seen in Figure 3, conversion of (**51**, R = $OCOCH_3$) to the *N*-protected azetidinone followed by aldol type addition gave a mixture of alcohols. Further manipulation gave the four isomers of the trithiocarbonate (**53**) each of which was progressed separately to the corresponding penems (**54**) by means of the phosphorane route. During the thermal cyclization each product gives rise to a second isomer by virtue of a novel epimerization at C-5; eg, thermolysis of the rel-5(*R*)-, 6(*S*)-, 8(*R*)-isomer [*76431-42-0*] gives (**55**) which could be shown to be the rel-5(*R*)-, 6(*R*)-, 8(*S*)-isomer [*80575-20-8*]. Double deprotection of (**54**) gave the corresponding penem salts (**56**).

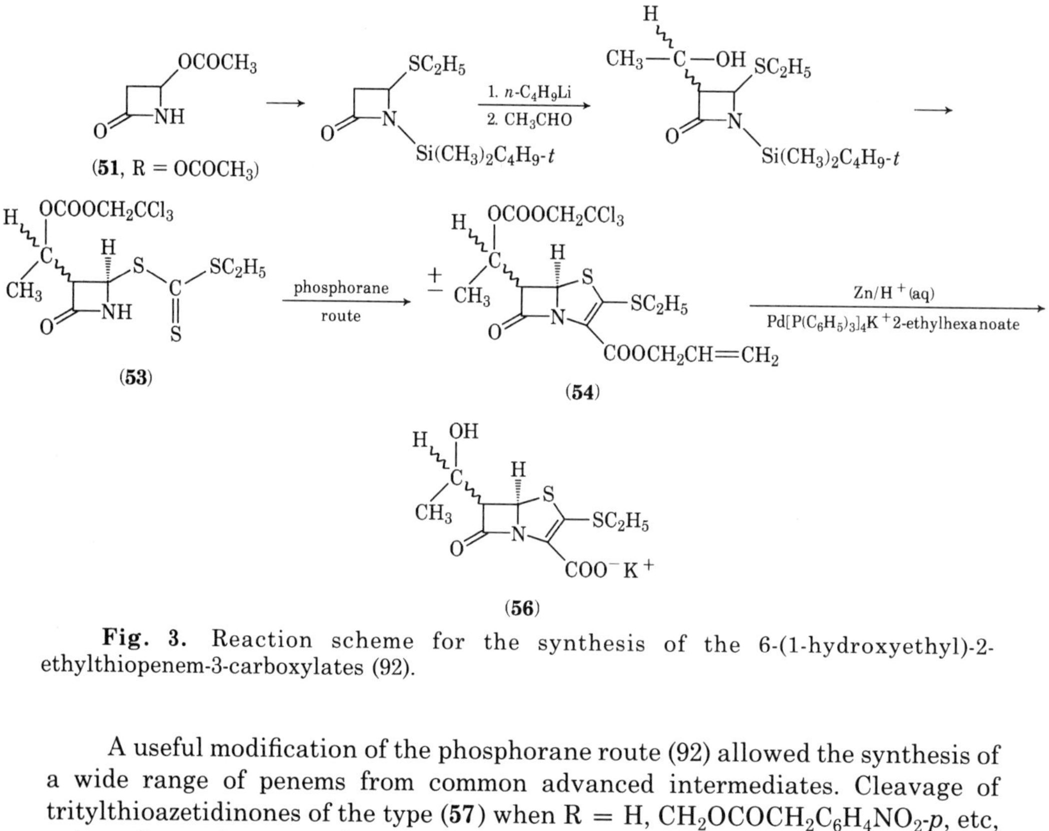

(**56**)

Fig. 3. Reaction scheme for the synthesis of the 6-(1-hydroxyethyl)-2-ethylthiopenem-3-carboxylates (92).

A useful modification of the phosphorane route (92) allowed the synthesis of a wide range of penems from common advanced intermediates. Cleavage of tritylthioazetidinones of the type (**57**) when R = H, $CH_2OCOCH_2C_6H_4NO_2$-*p*, etc, using silver nitrate–methanol afforded the silver mercaptides (**58**) which on acylation–cyclization gave a wide range of penems in which the 2-substituent was linked through carbon or sulfur.

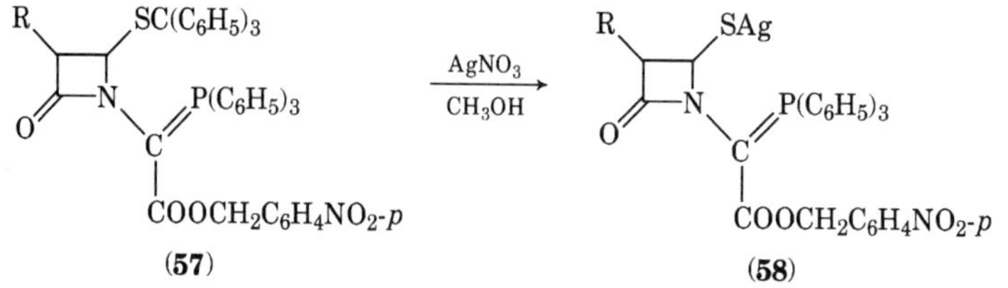

(**57**) (**58**)

In common with the naturally occurring carbapenem thienamycin (**2**), the introduction of the *trans*-6-[1-(*R*)-hydroxyethyl] group had a profound effect on the biological properties of the penems. This, together with an indication from an early study (93) that, as with other β-lactams, the 5(*R*)-enantiomer was solely responsible for antibacterial activity, provided impetus for the development of methods for the synthesis of chiral penems.

The stereochemical outcome of aldol reactions of 6,6-dibromopenicillanates was described in 1977 (94) making 6-aminopenicillanic acid [*551-16-6*] (6-APA), $C_8H_{12}N_2O_3S$, an attractive starting material for the synthesis of chiral 6-(1-hydroxyethyl)penems (Fig. 4). Metal-halogen exchange on (**59**) followed by treat-

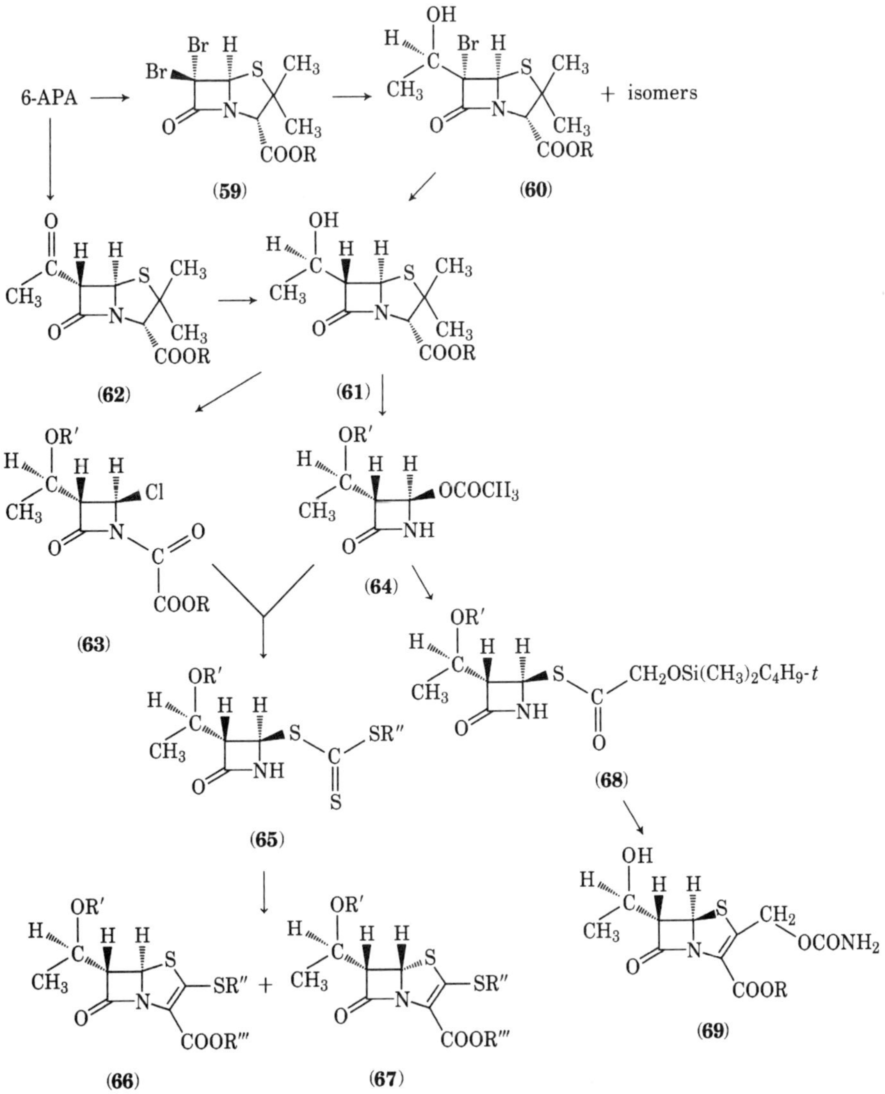

Fig. 4. Reaction scheme for the synthesis of chiral 6-(1-hydroxyethyl) penems where R′ = H, $OCOCH_2CCl_3$, $Si(CH_3)_2C_4H_9$-*t*; R″ = C_2H_5; and R‴ = $CH_2CH{=}CH_2$, $CH_2C_6H_5$.

ment with acetaldehyde gave a mixture of isomers from which the bromohydrin (**60**) having a desired 8(*R*) stereochemistry could be isolated as major product. Reductive removal of the halogen atom using zinc then gave the (5(*R*)-, 6(*S*)-, 8(*R*))-penem (**61**). Alternatively, diazotization followed by Lewis acid catalyzed reaction using acetaldehyde gave the unstable *trans*-6-acetylpenicillanate (**62**) which could be reduced stereoselectively to the alcohol (**61**) (95).

Several groups have built on this work and prepared intermediates capable of conversion to penems using the phosphorane route. In one method (96), chlorinolysis was used to prepare the chloroazetidinone (**63**), in another (97) the thiazolidine ring was cleaved using mercuric acetate, obtaining the acetoxyazetidinone (**64**) after oxidative removal of the nitrogen substituent. Reaction of either (**63**) (96) or (**64**) (98) with a sodium trithiocarbonate proceeded with retention of configuration to give the azetidinones (**65**) from which the (5(*R*)-, 6(*S*)-, 8(*R*))-penems (**66**) were obtained. Some of the unwanted *cis*-5(*S*)-penems (**67**) were also formed during cyclization. Moreover, when either pure (**66**) or (**67**) was heated the same equilibrium mixture was obtained (96,98), possibly involving the intermediacy of a betaine (98). The acetoxyazetidinone (**64**) has also been used to prepare intermediates of the type (**68**) for the synthesis of the Farmitalia clinical candidate FCE 22101 [*84845-58-9*] (**69**, R = Na), $C_{10}H_{12}N_2O_6SNa$ (99).

Other methods for cleavage of the thiazolidine ring of penicillanates with retention of the sulfur atom have been described. Silver assisted cleavage afforded mercaptides (100) whereas the sulfenic acids generated by the thermolysis of penicillanate sulfoxides have been intercepted by acetylenes (101–103) and mercaptans (104,105). All these methods have provided useful intermediates for the synthesis of penems. Total synthesis starting from L-threonine [*72-19-5*], $C_4H_9NO_3$, (106,107) and ethyl-3-(*S*)-hydroxybutanoate [*56816-01-4*], $C_6H_{12}O_3$, (108) has also been used to prepare intermediates for the preparation of penems having defined stereochemistry.

Alternative Methods for Formation of the Penem Ring. The reactivity of the carbonyl group in readily available oxalimides of the type (**70**) (Fig. 5) provides a useful alternative to the original phosphorane route. Treatment of oxalimides (**70**, X = O, R = CH_2OCONH_2) and (**70**, X = S, R = SC_2H_5) with a trialkyl phosphite at elevated temperatures gives the corresponding penems (**72**) (100,109) and (**74**) (110,111). In a detailed study (112) it was shown that for thiolesters (**70**, X = 0) carbene generation/trapping gives intermediate trialkyoxyphosphoranes (**71**) which then cyclize thermally. In contrast, cyclization of trithiocarbonates (**70**, X = S, R = SC_2H_5) does not proceed via phosphorane intermediates. Instead it has been proposed (110,111) that the carbene generated added intramolecularly to the more reactive thiocarbonyl group to form an episulfide (**73**) which was then desulfurized to give the penem. The lower temperature and shorter reaction times required for cyclization of trithiocarbonates avoided the troublesome C-5 epimerization experienced with the conventional phosphorane route. The high reactivity of the oxalimide carbonyl group has also been demonstrated by the more recently reported (113) cyclization of phosphoranes of the type (**70**, X = $P(C_6H_5)_3$).

Although the phosphorus mediated ring closures are by far the most widely used, other methods for [2,3] bond formation have been described. Treatment of the dithiocarbamates (**75**, R = H or $Si(CH_3)_3$; R′ = 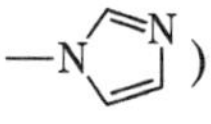) with a strong base

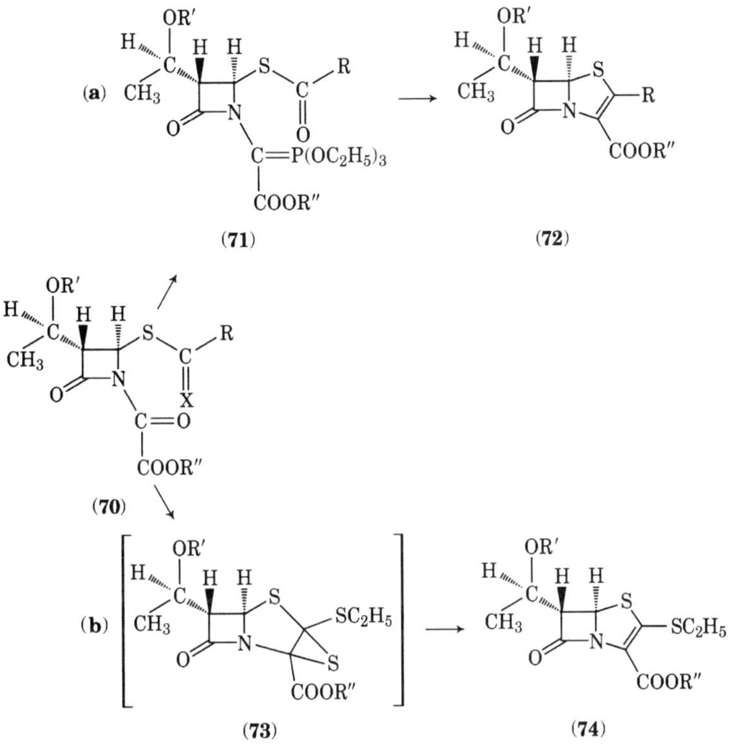

Fig. 5. An alternative method for the formation of the penem ring where R′ = $OCOCH_2CCl_3$, $Si(CH_3)_2$, or C_4H_9-*t* and R″ = CH_2CCl_3, or $CH_2C_6H_4NO_2$-*p*, CH_2CHCH_2; (**a**) X = O, R = CH_2OCONH_2, (**b**) X = S, R = SC_2H_5.

at low temperature gives thioxopenams (**76**) which are readily alkylated to give 2-alkylthiopenems (114).

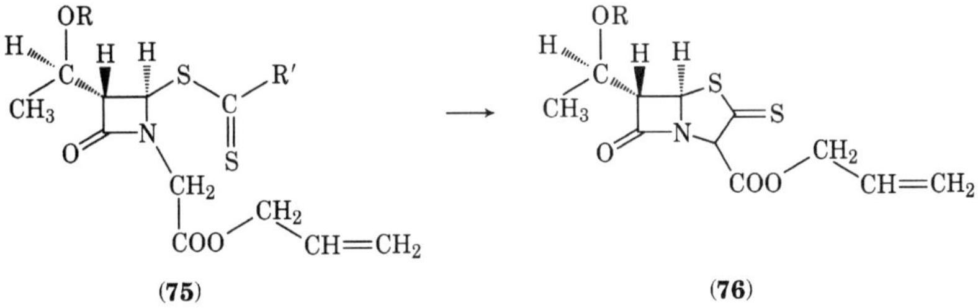

An analogous preparation of thioxopenams from dithiocarbonates (**75**, R = t-$C_4H_9(CH_3)_2Si$, R′ = OC_6H_5) has also been described (115). Additionally, an intramolecular Michael addition reaction to form the [2,3] bond has been exploited in penem synthesis to prepare FCE 22101 (**69**) (116).

Synthesis of 2-thioxopenams (**76**) has also been realized using a [sulfur, C-5] ring closure (117). Cyclization of 4-(*S*)-chloroazetidinone [*88816-44-8*] (**77**, R = R′ = $SCOCH_3$), $C_{19}H_{19}ClN_2O_8S_2$, using imidazole in aqueous dioxane proceeded stereospecifically to give the 5-(*R*)-thioxopenam [*83362-57-6*] (**78**). Using the same approach, the azetidinone (**77**, R = 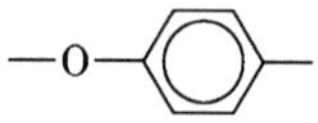; R′ = $SCOC_4H_9$-*t*), was cyclized to the corresponding 2-aryloxypenem [*85768-34-9*] (118).

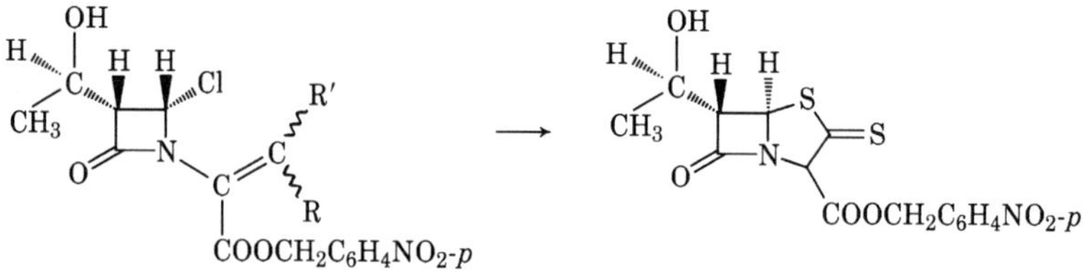

(**77**) (**78**)

Only one successful penem synthesis involving a [nitrogen, C-3] ring closure has been described (119). Cyclization of the ketomalonate (**79**) using hydrofluoric acid–pyridine afforded the penams (**80**) which were converted to the penem (**81**). Attempts to adapt the versatile diazoketoester–carbapenem cyclization for penem synthesis failed (120).

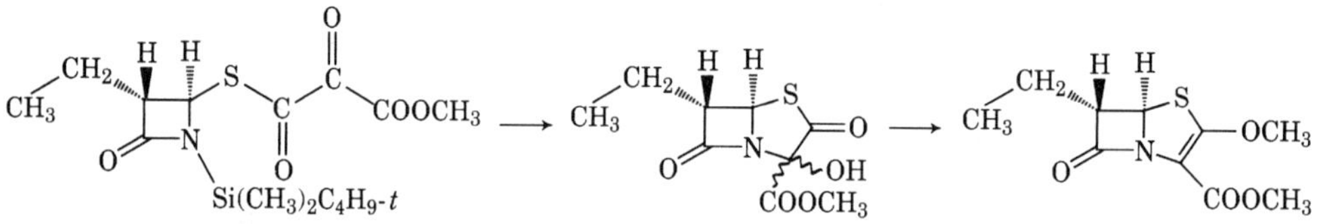

(**79**) (**80**) (**81**)

Ring contraction of 2-thiocephems has also been examined as a route to penems. Desulfurization of (**82**, $n = 0$) using triphenylphosphine gave mixtures of 5(*R*)- and 5(*S*)-penems (121). The stereochemical problem was neatly overcome by regioselective oxidation to the thiosulfonate (**82**, $n = 2$) which underwent stereospecific thermal extrusion of sulfur dioxide (122) to give the 5(*R*)-penem (**83**).

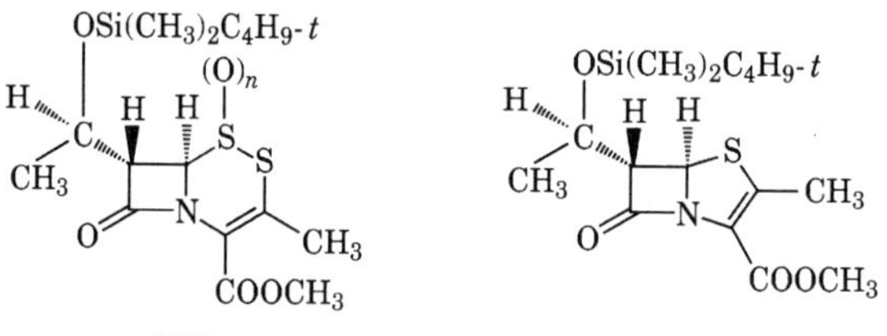

(**82**) (**83**)

Modification of Intact Penems. Functional group modification has been used by a number of researchers (123–125) to synthesize a wide range of 2-substituted penems. For example, activation of the hydroxyl group of (**84**, R = OH) followed by displacement reactions provided 2-heterocyclylthiomethylpenems (**84**, R = (*S*)-heterocyclyl) (125) and 2-(quaternary ammonio)methylpenems (**84**, R = NR'_3X) (125).

Displacement of the sulfinyl group of penems (**85**), obtained by regioselective oxidation of (**74**, R′ = $Si(CH_3)_2C_4H_7$-*t*, R″ = CH_2CHCH_2) (Fig. 5) using thiolates, allowed the synthesis of a range of thio-substituted penems from a common intermediate (126). A conceptually similar approach (127) enables the displacement of the phenolic leaving group in penems (**86**, R = CN) [*85768-38-3*] or (**86**, R = NO_2) [*114707-30-1*] using amines and thus provides a general route to 2-aminopenems (**87**) where R can be CH_3 or n-C_3H_7; R′ can be H, CH_3, $CH_2COOC_2H_5$, or 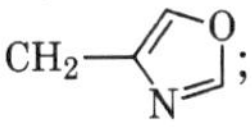; or both R and R′ can be 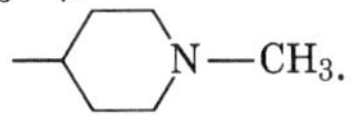

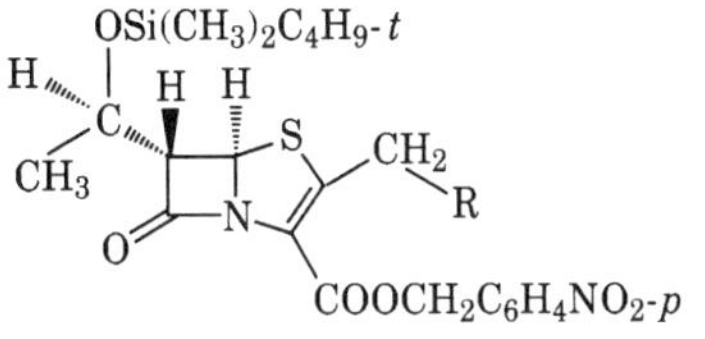

(**84**)

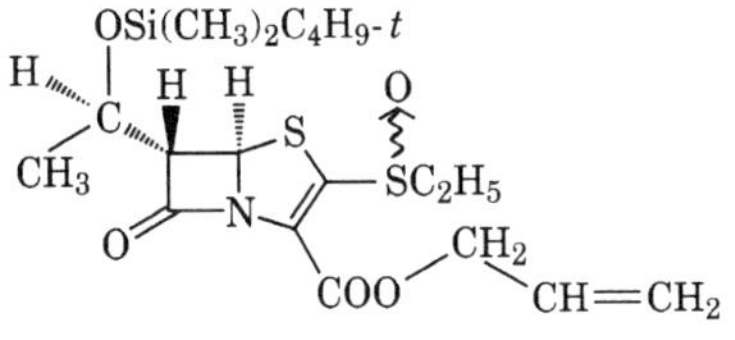

(**85**)

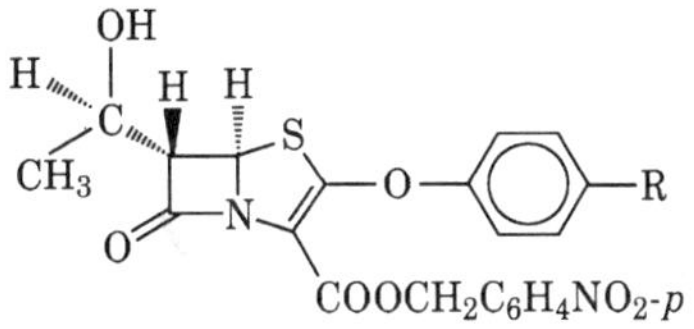

(**86**)

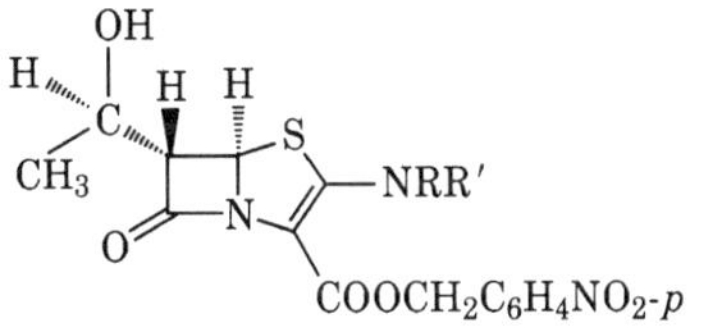

(**87**)

Although less researched than the 2-position, modifications at the 6-position of intact penems have been reported. Generation of the dianion of the penem (**52**, R = CH_3) using a strong base such as n-butyllithium or lithium diisopropylamide, followed by reaction with electrophiles yields 6-substituted 2-methylpenems in moderate yield (128). The enhanced acidity of the 6-proton in the bromopenem (**88**) [*114409-16-4*] has been exploited to prepare the Beecham β-lactamase inhibitor BRL 42715 [*102209-75-6*] (**89**, R = Na), $C_{10}H_8N_4O_3SNa$ (105). Lithium diphenylamide, a weaker base, was used to generate the anion of (88) which on sequential treatment with 1-methyl-1,2,3-triazole-4-carbaldehyde and acetic anhydride gives a mixture of diastereomers of the bromoacetate (**90**). Reductive elimination then provided the (Z)-penem (**89**, R = $CH_2C_6H_4OCH_3$-p) as major product which on Lewis acid mediated deprotection gave BRL 42715 (**89**, R = Na).

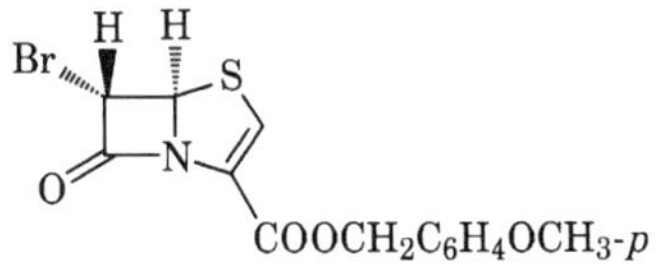

(**88**)

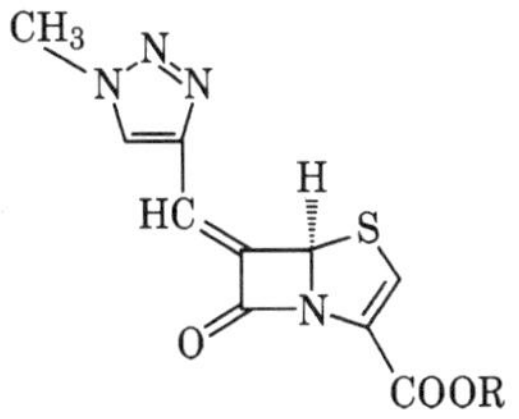

(**89**)

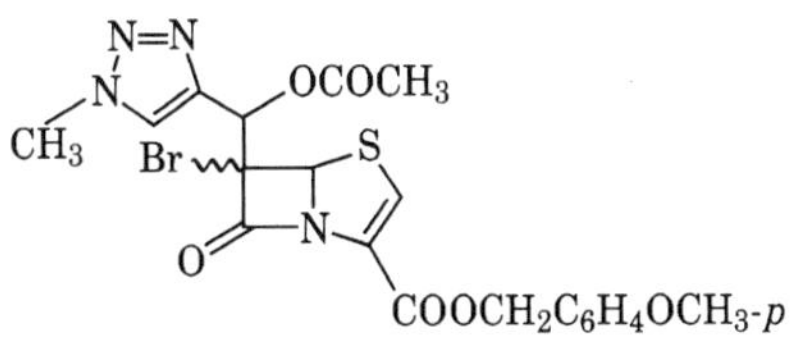

(**90**)

Table 2. Penems

Name	CAS Registry Number	Molecular formula	Structure number	Structure	Reference
SCH 29482 SCH 34343	[*77646-83-4*] [*94392-37-7*]	$C_{10}H_{13}NO_4S_2$ $C_{11}H_{14}N_2O_6S_2$	(**94**, R = C_2H_5) (**94**, R = $(CH_2)_2OCONH_2$)	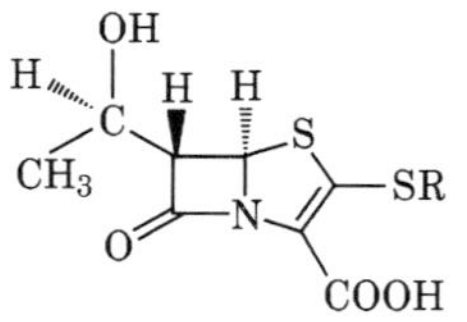	145 146
FCE 221201 FCE 22891	[*84845-58-9*] [*87238-52-6*]	$C_{10}H_{12}N_2O_6S$ $C_{13}H_{16}N_2O_8S$	(**69**, R = Na) (**69**, R = CH_2OCOCH_3)	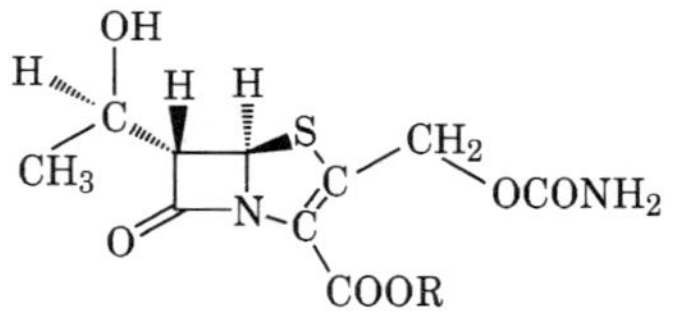	146 147
HRE 664	[*109888-39-3*]	$C_{15}H_{14}N_2O_6S$	(**95**)	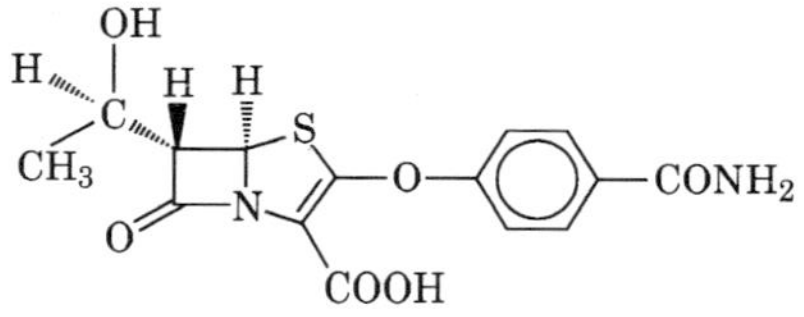	148

SUN 5555	[*106560-14-9*]	$C_{12}H_{15}NO_5S$	(**96**, R = 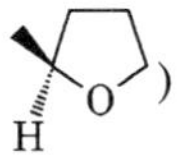)	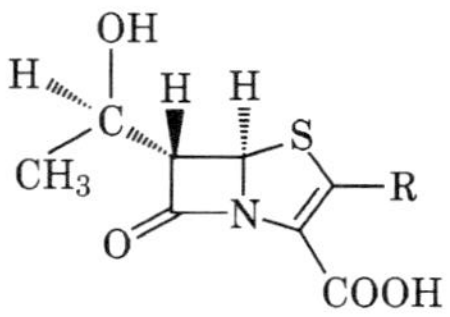	149
CGP 31608	[*97644-04-7*]	$C_9H_{12}N_2O_4S$	(**96**, R = CH_2NH_2)		150
CP 65207	[*120788-07-0*]	$C_{12}H_{15}NO_5S_3$	(**96**, R = 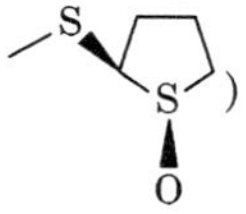)		151
FCE 25199	[*120796-34-1*]	$C_{15}H_{17}NO_7S$	(**97**)	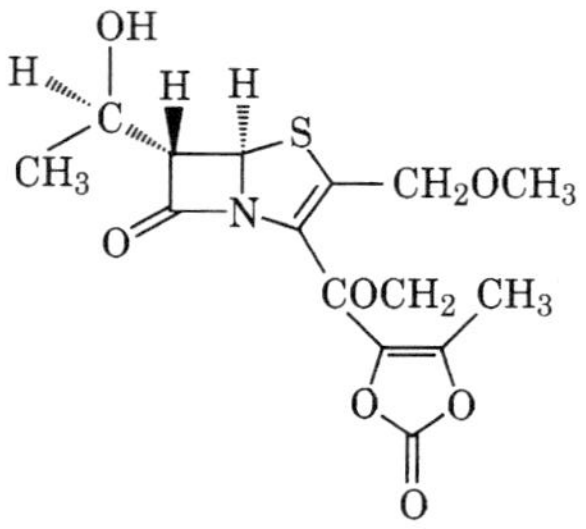	152

Biological Properties. In marked contrast to the antibacterially inactive penicillanic and cephalosporanic acids, 6-unsubstituted penems such as (**52**, R = CH_3) [*69077-00-5*] exhibit good activity against both gram-positive and gram-negative bacteria (87). The introduction of a 6β-acylamino substituent had an adverse effect on the stability of the penem nucleus, and compounds such as (**50**) were shown to possess only weak antibacterial properties (84,85). Attempts to stabilize 6β-acylaminopenems by the introduction of a 6α-methoxy or 6α-methyl group was only marginally successful in the latter case and the compounds still showed only weak activity (129). A similar reduction in stability and activity was also noted for the 6α-methoxypenem (**91**, R = OCH_3) [*78839-71-1*], whereas the 6α-methylpenem (**91**, R = CH_3) [*73660-04-5*], although considerably more stable, still only exhibited modest antibacterial activity (130). 6,6-Dialkylpenems have been shown to be extremely stable but are devoid of antibacterial reactivity (130).

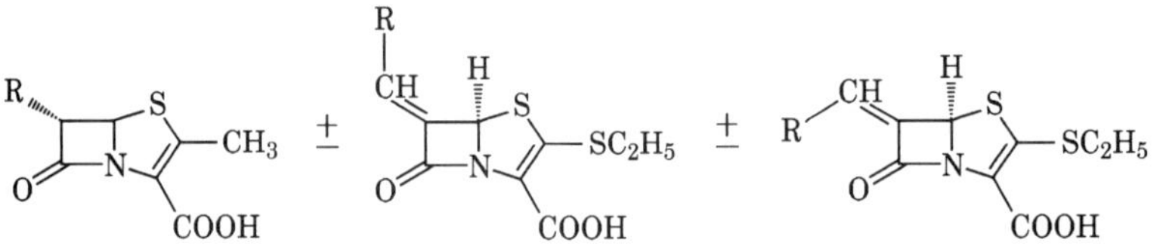

(**91**) (**92**) (**93**)

The introduction of the 1-hydroxyethyl group found in the naturally occurring carbapenems proved to be crucial for providing penems with broad-spectrum antibacterial activity and stability to β-lactamases. As with their carbapenem counterparts, biological activity showed a marked dependence on the relative stereochemistry at the three chiral centers. Evaluation of the four racemic penems (**56**) revealed that the *trans*-8(*R*)-isomer [*76431-47-5*], corresponding to thienamycin (**2**), was 20- to 30-fold more potent than the *trans*-8(*S*)-isomer [*76431-46-4*] whereas the two cis-penems [*80629-72-7*] and [*80629-98-7*] were of intermediate potency (131). Further structural modification at C-8 resulted in adverse effects on antibacterial activity. Removal of the C-8 methyl group resulted in loss of activity especially against β-lactamase producing bacteria (131) whereas an additional methyl group as in the 6-hydroxyisopropylpenem [*79583-53-2*], (**91**, R = $COH(CH_3)_2$), $C_{10}H_{13}NO_4S$, resulted in almost complete loss of activity (132). The increased bulk produced by substitution of the C-8 methyl group was also detrimental, 1-hydroxy-1-propylpenem [*75940-79-3*] (**91**, R = $CHOHC_2H_5$) and 1-hydroxy-2-phenylethylpenem [*75940-91-9*] (**91**, R = $CHOHCH_2C_6H_5$) being devoid of useful activity (132).

The introduction of unsaturation at [6,8] resulted in loss of all antibacterial activity except that against *staphylococcal* species. Both the (*Z*)- (**92**, R = CH_3) [*81463-33-4*] and the (*E*)- (**93**, R = CH_3) [*81463-37-8*] isomers of the 6-ethylidenepenem sodium salt showed potent broad spectrum β-lactamase inhibitory properties and were capable of reducing the minimum inhibitory concentration (MIC) values of sensitive β-lactams, eg, Amoxycillin, when used in combination (133). Unlike the saturated counterparts, the 6-methylenepenems were tolerant to profound changes at C-8. Improved inhibitory properties and synergism resulted from the introduction of a furan ring. However, in the case of this larger C-8 substituent, a preferred geometry for the double bond was identified and the (*Z*)-isomer (**92**, R = 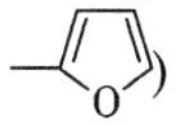) [*93853-68-0*] was more active than the (*E*)-isomer (**93**, R = 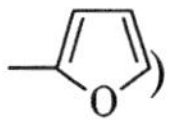) (134). Structure–activity relationships for a wide range of (*Z*)-6-heterocyclylmethylenepenems have been reported (135).

With a few exceptions the 2-substituent can be seen as a means of fine-tuning the activity, pharmacokinetics, and stability properties of the penems. The great number of structural variations (98,124,125,136–144) that can be accommodated without loss of the potent activity conveyed by the 6-[1-(*R*)-hydroxyethyl] group bears witness to the undemanding nature of this region of the molecule. These penems typically exhibit excellent activity against gram-positive bacteria including β-lactamase producing strains. The activity against gram-negative organisms, including those producing β-lactamase, is generally good although of a lower order. Except for compounds possessing a basic amino function in the 2-substituent (98,144) penems are devoid of useful activity against *pseudomonas* species.

Research carried out in a number of laboratories has revealed several members of the penem family having biological properties worthy of further investigation. Some of these compounds are given in Table 2. Extensive studies (135) have also identified BRL 42715 [*102209-75-6*] (**89**, R = Na), $C_{10}H_7N_4O_3SNa$, as a potentially useful β-lactamase inhibitor capable of potentiating the activity of a number of clinically important β-lactam antibiotics against resistant strains (153).

In terms of activity there seems little to prevent some of these compounds finding a place in therapy, especially those such as SCH 29482, SUN 5555, and FCE 25199 which have oral absorption properties. However, as is the case for the carbapenems, some penems are extensively metabolized by human renal dehydropeptidase-I enzyme (144). Although no penem has received approval for clinical use as of this writing, expectations are high that future research and development will change that.

Economic Aspects

Extensive carbapenem and penem antibiotic research has been ongoing since thienamycin was discovered in 1978. However, only the imipenem-cilastatin combination has become a commercial product. Launched in 1985 in the United States as a broad-spectrum hospital product under the name Primaxin, this product had worldwide sales of some $300 million in 1988. Sales were predicted to rise to $345 million for the year ending 1989 (154).

BIBLIOGRAPHY

"Penicillin" in *ECT* 1st ed., Vol. 9, pp. 922–943, by W. E. Brown, Squibb Institute for Medical Research; "Penicillins" in *ECT* 2nd ed., Vol. 14, pp. 652–707, by Kenneth Butler, Chas. Pfizer & Co., Inc.; "Antibiotics, β-Lactams" in *ECT* 3rd ed., Vol. 2, pp. 871–919, by J. R. E. Hoover, and C. H. Nash, Smith Kline & French Laboratories.

1. "Cephalosporins and Penicillins" in E. H. Flynn, ed., *Chemistry and Biology,* Academic Press, New York and London, 1972.
2. J. S. Kahan and co-workers, *J. Antibiot.* **32,** 1 (1979).
3. D. Butterworth, M. Cole, G. Hanscomb, and G. N. Rolinson, *J. Antibiot.* **32,** 287 (1979); J. D. Hood, S. J. Box, and M. S. Verall, *J. Antibiot.* **32,** 295 (1979).
4. S. J. Box, J. D. Hood, and S. R. Spear, *J. Antibiot.* **32,** 1239 (1979).
5. K. Yamamoto and co-workers, *J. Antibiot.* **33,** 796 (1980).
6. M. Nakayama and co-workers, *J. Antibiot.* **34,** 818 (1981).
7. N. Tsuji and co-workers, *J. Antibiot.* **35,** 24 (1982).
8. N. Tsuji and co-workers, *J. Antibiot.* **35,** 536 (1982).
9. W. L. Parker and co-workers, *J. Antibiot.* **35,** 653 (1982).
10. R. Southgate and S. Elson, *Fortschr. Chem. Org. Naturst.* (Progress in the *Chemistry of Organic Natural Products*), Springer-Verlag, New York, **47,** 1 (1985).
11. Eur. Pat. Appl. 38,534 (Apr. 15, 1981), K. Tanaka, N. Tsuji, E. Kondo, and Y. Kawamura (to Shionogi and Co., Ltd.).
12. D. Butterworth, J. D. Hood, and M. S. Verall in A. Mizrahi, ed., *Advances in Biotechnological Processes,* Vol. 1, A. R. Liss, New York, 1982, pp. 251–282.
13. G. Albers-Schönberg and co-workers, "Abstract 228" in *Abstracts of the 16th Interscience Conference on Antimicrobial Agents and Chemotherapy, 1976,* American Society for Microbiology, Washington, D.C., 1976.
14. J. M. Williamson, *CRC Crit. Rev. Biotechnol.* **4,** 111 (1986).
15. J. M. Williamson and co-workers, *J. Biol. Chem.* **260,** 4637 (1985); D. R. Houk, K. Kobayashi, J. M. Williamson, and H. G. Loss, *J. Am. Chem. Soc.* **108,** 5365 (1986).
16. B. W. Bycroft in P. H. Bentley and R. Southgate, eds., *Recent Advances in the Chemistry of β-Lactam Antibiotics,* RSC Special Publication No. 70, London, 1989, pp. 23–32.
17. T. Haneishi and co-workers, *J. Antibiot.* **36,** 1581 (1983).
18. Y. Fukagawa, M. Okabe, T. Yoshioka, and T. Ishikura in A. G. Brown and S. M. Roberts, ed., *Recent Advances in the Chemistry of β-Lactam Antibiotics,* Special Publication No. 52, London, 1985, pp. 163–182.
19. G. Albers-Schönberg and co-workers, *J. Am. Chem. Soc.* **100,** 6491 (1978).
20. A. G. Brown, D. F. Corbett, A. J. Eglington, and T. T. Howarth, *J. Chem. Soc., Chem. Commun.,* 523 (1977); 953 (1977).
21. A. G. Brown, D. F. Corbett, A. J. Eglington, and T. T. Howarth in G. I. Gregory, ed., *Recent Advances in Chemistry of β-Lactam Antibiotics,* RSC Special Publication No. 38, London, 1981, pp. 255–268.
22. A. G. Brown, D. F. Corbett, A. J. Eglington, and T. T. Howarth, *J. Antibiot.* **39,** 2551 (1983).
23. A. G. Brown, D. F. Corbett, A. J. Eglington, and T. T. Howarth, *J. Antibiot.* **32,** 961 (1979).
24. W. J. Leanza and co-workers in ref. 21, pp. 240–254.
25. W. J. Leanza, K. J. Wildonger, T. W. Miller, and B. G. Christensen, *J. Med. Chem.* **22,** 1435 (1979).
26. D. H. Shih and R. W. Ratcliffe, *J. Med. Chem.* **24,** 639 (1981).
27. D. H. Shih, J. Hannah, and B. G. Christensen, *J. Am. Chem. Soc.* **100,** 8004 (1978).
28. D. F. Corbett, *J. Chem. Soc., Chem. Commun.,* 803 (1981).
29. D. F. Corbett, S. Coulton, and R. Southgate, *J. Chem. Soc. Perkin I,* 3011 (1982).
30. K. Yamamoto and co-workers, *Tetrahedron Lett.* **23,** 897 (1982).
31. Y. Fukagawa, K. Kubo, T. Ishikura, and K. Kounu, *J. Antibiot.* **33,** 543 (1980).
32. L. D. Cama and B. G. Christensen, *J. Am. Chem. Soc.* **100,** 8006 (1978).
33. J. H. Bateson and co-workers, *J. Chem. Soc. Perkin I,* 3242 (1981).
34. H. R. Pfaendler, J. R. Gosteli, R. B. Woodward, and G. Rihs, *J. Am. Chem. Soc.* **103,** 4526 (1981).
35. J. H. Bateson and co-workers in ref. 21, pp. 291–313.

36. R. J. Ponsford and R. Southgate, *J. Chem. Soc., Chem. Commun.,* 1085 (1980).
37. J. H. Bateson and co-workers, *J. Chem. Soc., Chem. Commun.,* 1084 (1980).
38. T. N. Salzman, R. W. Ratcliffe, B. G. Christensen, and F. A. Boufford, *J. Am. Chem. Soc.* **102,** 6161 (1980).
39. I. Shinkai and co-workers, *Tetrahedron Lett.* **23,** 4903 (1982).
40. D. G. Melillo and co-workers, *Tetrahedron Lett.* **22,** 913 (1981).
41. T. Kametani and co-workers, *J. Chem. Soc., Chem. Commun.,* 646 (1989).
42. S. Karady, J. S. Amato, R. A. Reamer, and L. M. Weinstock, *J. Am. Chem. Soc.* **103,** 6765 (1981).
43. M. Miyashita, M. N. Chida, and A. Yoshikoshi, *J. Chem. Soc., Chem. Commun.,* 1354 (1982).
44. N. Ikota, O. Yoshino, and K. Koga, *Chem. Pharm. Bull.* **30,** 1929 (1982).
45. A. Knierzinger and A. Vasella, *J. Chem. Soc., Chem. Commun.,* 9 (1985).
46. H. Maruyama, M. Shiozaki, and T. Hiraoka, *Bull. Chem. Soc. Jap.* **58,** 3264 (1985).
47. M. Shiozaki, N. Ishida, H. Maruyama, and T. Hiraoka, *Tetrahedron* **39,** 2399 (1983).
48. T. Kametani, T. Nagahara, and T. Honda, *J. Org. Chem.* **50,** 2327 (1985).
49. S. T. Hodgson, D. M. Hollinshead, and S. V. Ley, *J. Chem. Soc., Chem. Commun.,* 494 (1984).
50. T. Kametani, *Heterocycles* **17,** 463 (1982).
51. T. Nagahara and T. Kametani, *Heterocycles* **25,** 729 (1987).
52. F. M. Kahan, H. Kropp, J. G. Sundelof, and J. Birnbaum, *J. Antimicrob. Chem.,* Supplement D, **12,** 1 (1983).
53. H. Kropp, J. G. Sundelof, R. H. Hajdu, and F. M. Kahan, *Antimicrob. Agents Chemother.* **22,** 62 (1982).
54. D. H. Shih, F. Baker, L. D. Cama, and B. G. Christensen, *Heterocycles* **21,** 29 (1984).
55. A. Afonso and co-workers in ref. 16, pp. 295–302.
56. D. M. Tschaen and co-workers, *Tetrahedron Lett.* **29,** 2779 (1988).
57. R. Deziel and D. Faureau, *Tetrahedron Lett.* **47,** 5687 (1986).
58. F. Shirai and T. Nakai, *J. Org. Chem.* **52,** 5491 (1987).
59. T. Shibata and Y. Sugimura, *J. Antibiot.* **42,** 374 (1989).
60. J. Haruta and co-workers, *Chem. Pharm. Bull.* **37,** 2338 (1989).
61. T. J. Sowin and A. I. Meyers, *J. Org. Chem.* **53,** 4154 (1988).
62. S. Kobayashi, K. Kamijama, T. Iimori, and M. Ohno, *Tetrahedron Lett.* **25,** 2557 (1984); **29,** 1057 (1988).
63. H. Kaga, S. Kobayashi, and M. Ohno, *Tetrahedron Lett.* **30,** 113 (1989).
64. F. Shirai and T. Nakai, *Chem. Lett.,* 445 (1989).
65. M. Ihara, M. Takahashi, M. Fukumoto, and T. Kametani, *Heterocycles* **27,** 327 (1988).
66. T. N. Salzmann and co-workers in ref. 16, pp. 171–189.
67. S. Schmitt, T. N. Salzmann, D. M. Shih, and B. G. Christensen, *J. Antibiot.* **41,** 780 (1988).
68. C. Kim, B. Y. Luh, P. F. Misco, and M. J. Hitchcock, *J. Med. Chem.* **32,** 601 (1989).
69. P. Davey and co-workers, eds., *J. Antimicrob. Chemother.* **24,** Supplement A (1989).
70. Y. Nagao and co-workers, *J. Chem. Soc., Chem. Commun.* 821 (1989).
71. R. Bayles and co-workers in ref. 16, pp. 190–195.
72. H. C. Neu, *Am. J. Med.* **78** (Supplement 6A), 33 (1985).
73. M. J. Basker, R. J. Boon, and P. A. Hunter, *J. Antibiot.* **33,** 878 (1980).
74. J. R. Knowles, *Acc. Chem. Res.* **18,** 97 (1985).
75. M. Sakamoto and co-workers, *J. Antibiot.* **32,** 272 (1979).
76. F. Kobayashi and co-workers, *Antimicrob. Agents Chemother.* **21,** 536 (1982).
77. Y. Kimura and co-workers, *J. Antibiot.* **35,** 32 (1982).
78. M. J. Basker and co-workers, *J. Antibiot.* **34,** 1224 (1981).
79. B. G. Spratt, V. Jobanputra, and W. Zimmerman, *Antimicrob. Agents Chemother.* **12,** 406 (1977).

80. H. C. Neu and P. Labthavikul, *Antimicrob. Agents Chemother.* **21,** 180 (1982).
81. S. R. Norrby, B. Björnegard, F. Ferber, and K. H. Jones, *J. Antimicrob. Chemother.* (Suppl. D), **12,** 109 (1983).
82. J. Birnbaum, F. M. Kahan, H. Kropp, and J. S. MacDonald in H. C. Neu, *Am. J. Med.* **78** (Suppl. 6A), 33 (1985).
83. S. R. Norrby and co-workers, *Antimicrob. Agents Chemother.* **23,** 300 (1983).
84. R. B. Woodward in J. Elks, ed., *Recent Advances in the Chemistry of β-Lactam Antibiotics,* The Chemical Society, 1977, pp. 167–180.
85. I. Ernest and co-workers, *J. Am. Chem. Soc.* **100,** 8214 (1978).
86. K. Claues, D. Grimm, and G. Prossel, *Justus Liebigs Ann. Chem.,* 539 (1974).
87. M. Lang and co-workers, *Am. Chem. Soc.* **101,** 6296 (1979).
88. M. Lang, K. Prasad, J. Gosteli, and R. B. Woodward, *Helv. Chim. Acta* **63,** 1093 (1980).
89. Brit. Pat. 2,042,508A (1980), E. G. Brain (to Beecham Group, Ltd.).
90. Eur. Pat. 0,000,258 (1978), J. Gosteli, I. Ernest, M. Lang, and R. B. Woodward.
91. S. W. McCombie, *Tetrahedron Lett.* **22,** 3489 (1981).
92. U.S. Pat. 4,272,437 (1981), M. Ménard and A. Martel (to Bristol-Myers).
93. H. R. Pfaendler, J. Gosteli, and R. B. Woodward, *J. Am. Chem. Soc.* **101,** 6306 (1979).
94. F. DiNinno, T. R. Beattie, and B. G. Christensen, *J. Org. Chem.* **42,** 2960 (1977).
95. S. Karady, J. S. Amato, R. A. Reamer, and L. M. Weinstock, *J. Am. Chem. Soc.* **103,** 6765 (1981).
96. V. M. Girijavallabhan and co-workers, *Tetrahedron Lett.* **22,** 3485 (1981).
97. A. Yoshida and co-workers, *Chem. Pharm. Bull.* **29,** 2899 (1981).
98. T. Hayashi and co-workers, *Chem. Pharm. Bull.* **29,** 3158 (1981).
99. G. Franceschi and co-workers, *J. Antibiot.* **36,** 938 (1983).
100. M. Alpegiani and co-workers, *J. Am. Chem. Soc.* **107,** 6398 (1985).
101. M. Foglio and co-workers, *Heterocycles* **20,** 1491 (1983).
102. M. Foglio, G. Franceschi, C. Scarafile, and P. Zini, *Heterocycles* **16,** 1919 (1981).
103. M. Foglio, C. Battistini, F. Zarini, and G. Franceschi, *Heterocycles* **19,** 485 (1982).
104. M. Alpegiani and co-workers, *Tetrahedron Lett.* **24,** 1627 (1983).
105. N. F. Osborne and co-workers, *J. Chem. Soc., Chem. Commun.,* 371 (1989).
106. H. Yanagisawa, A. Ando, M. Shiozaki, and T. Hiraoki, *Tetrahedron Lett.* **24,** 1037 (1983).
107. S. Chackalamannil and co-workers, *J. Org. Chem.* **53,** 450 (1988).
108. G. I. Georg, J. Kant, and H. S. Gill, *J. Am. Chem. Soc.* **109,** 1129 (1987).
109. C. Battistini, C. Scarafile, M. Foglio, and G. Franceschi, *Tetrahedron Lett.* **25,** 2395 (1984).
110. A. Afonso and co-workers, *J. Am. Chem. Soc.* **104,** 6138 (1982).
111. A. Yoshida and co-workers, *Chem. Pharm. Bull.* **31,** 768 (1983).
112. E. Perrone and co-workers, *Tetrahedron Lett.* **25,** 2399 (1984).
113. A. J. Baker and M. J. Jenkins in ref. 16, pp. 259–272.
114. V. M. Girijavallabhan, A. K. Ganguly, P. Pinto, and R. Versace, *J. Chem. Soc., Chem. Commun.,* 908 (1983).
115. W. J. Leanza and co-workers, *Tetrahedron* **39,** 2505 (1983).
116. S. Hanessian, A. Bedeschi, C. Battistini, and N. Mongilli, *J. Am. Chem. Soc.* **107,** 1438 (1985).
117. N. Daniels, G. Johnson, and B. C. Ross, *J. Chem. Soc., Chem. Commun.,* 1006 (1983).
118. M. D. Cooke, K. W. Moore, B. C. Ross, and S. E. Turner, *J. Chem. Soc., Chem. Commun.,* 1005 (1983).
119. H. H. Wasserman and W. T. Han, *J. Am. Chem. Soc.* **107,** 1444 (1985).
120. J. Marchand-Brynaert, L. Ghosez, and E. Cossement, *Tetrahedron Lett.* **21,** 3085 (1980).
121. E. Perrone and co-workers, *Tetrahedron Lett.* **24,** 1631 (1983).
122. E. Perrone and co-workers, *J. Org. Chem.* **51,** 3413 (1986).
123. S. Oida and co-workers, *Chem. Pharm. Bull.* **28,** 3258 (1980).

124. G. Franceschi and co-workers, *J. Antibiot.* **37,** 685 (1984).
125. E. Perrone and co-workers, *J. Antibiot.* **39,** 1351 (1986).
126. F. DiNinno, D. A. Muthard, R. W. Ratcliffe, and B. G. Christensen, *Tetrahedron Lett.* **23,** 3535 (1982).
127. A. J. Baker, M. R. Teall, and G. Johnson, *Tetrahedron Lett.* **28,** 2283 (1987).
128. Y. Ueda and co-workers, *Can. J. Chem.* **60,** 904 (1982).
129. J. Banville, P. La Pointe, B. Belleau, and M. Ménard, *Can. J. Chem.* **66,** 1390 (1988).
130. J. Gosteli, W. Holick, M. Land, and R. B. Woodward in ref. 21, pp. 359–367.
131. S. W. McCombie and co-workers, *21st Interscience Conference on Antimicrobial Agents and Chemotherapy,* Abstract 830, Chicago, 1981, American Society for Microbiology, Washington, D.C., 1981.
132. Brit. Pat. Appl. 2,042,515 A (Sept. 24, 1980), M. Ménard and A. Martel (to Bristol-Myers).
133. M. J. Basker and N. F. Osborne, *J. Antibiot.* **43,** 70 (1990).
134. N. J. P. Broom, K. Coleman, P. A. Hunter, and N. F. Osborne, *J. Antibiot.* **43,** 76 (1990).
135. N. J. P. Broom and co-workers in ref. 16, pp. 247–258.
136. V. M. Girijavallabhan and co-workers, *J. Antibiot.* **39,** 1182 (1986).
137. V. M. Girijavallabhan and co-workers, *J. Antibiot.* **39,** 1187 (1985).
138. G. Emmer and co-workers, *J. Antibiot.* **38,** 1371 (1985).
139. M. Lang and co-workers, *J. Antibiot.* **40,** 217 (1987).
140. H. G. Capraro and co-workers, *J. Antibiot.* **41,** 759 (1988).
141. M. Ishiguro and co-workers, *J. Antibiot.* **41,** 1685 (1988).
142. M. Alpegiani, E. Perrone, and G. Franceschi, *Heterocycles* **27,** 49 (1988).
143. M. Alpegiani and co-workers, *Heterocycles* **27,** 1329 (1988).
144. O. Zak and co-workers, *J. Clin. Pharmacol.* **28,** 128 (1988).
145. I. Phillips, R. Wise, and H. C. Neu, eds., *J. Antimicrob. Chemother.* **9** (Suppl. C) (1982).
146. R. Wise and I. Phillips, eds., *J. Antimicrob. Chemother.* **15** (Suppl. C) (1985).
147. D. Reeves, D. Speller, R. Spencer, and P. J. Daly, eds., *J. Antimicrob. Chemother.* **23** (Suppl. C) (1989).
148. G. Seibert and co-workers, *J. Antibiot.* **40,** 660 (1987); N. Kesel and co-workers, *J. Antibiot.* **40,** 1184 (1987).
149. T. Nishino and co-workers, *J. Antibiot.* **42,** 977 (1989).
150. *26th Interscience Conference on Antimicrobial Agents and Chemotherapy,* Abstracts 630–645, New Orleans, 1986, American Society for Microbiology, Washington, D.C., 1986.
151. T. Gootz and co-workers, *Antimicrob. Agents Chemother.* **33,** 1160 (1989).
152. *29th Interscience Conference on Antimicrobial Agents and Chemotherapy,* Abstracts 85–87, Houston, 1989, American Society for Microbiology, Washington, D.C., 1989.
153. K. Coleman, D. R. J. Griffin, J. W. J. Page, and P. A. Upshon, *Antimicrob. Agents Chemother.* **33,** 1580 (1989).
154. *Scrip Review Issue,* 1989, p. 12.

General Reference

R. W. Ratcliffe and G. Albers-Schönberg, "The Chemistry of Thienamycin and Other Carbapenem Antibitotics" and I. Ernest, "The Penems" in R. B. Morin and M. Gorman, eds., *Chemistry and Biology of β-Lactam Antibiotics,* Academic Press, New York, 1982, Chapts. 4 and 5.

ROBERT SOUTHGATE
NEAL F. OSBORNE
SmithKline Beecham Pharmaceuticals (UK)

CEPHALOSPORINS

The cephalosporins, a subgroup of β-lactam antibiotics, consist of a 4-membered lactam ring fused through the nitrogen and the adjacent tetrahedral carbon atom to a second heterocycle forming a 6-membered dihydrothiazine ring. Other structural features common to all the cephalosporins are a carboxyl group on the dihydrothiazine ring on the carbon next to the ring nitrogen and a functionalized amino group on C-7, the carbon of the β-lactam ring opposite the nitrogen. These features are evidenced in 7-aminocephalosporanic acid [*957-68-6*] (7-ACA), $C_{10}H_{12}N_2O_5S$ (**1**). Cephalosporins, like all β-lactam antibiotics, exert their antibacterial effect by interfering with the synthesis of the bacterial cell wall. These antibiotics tend to be "irreversible" inhibitors of cell wall biosynthesis and they are usually bactericidal at concentrations close to their bacteriostatic levels. Cephalosporins are widely used for treating bacterial infections. They are highly effective antibiotics and have low toxicity.

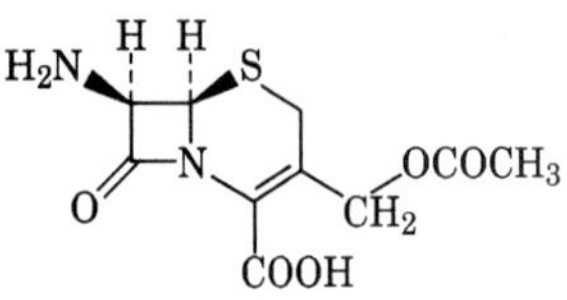

(**1**)

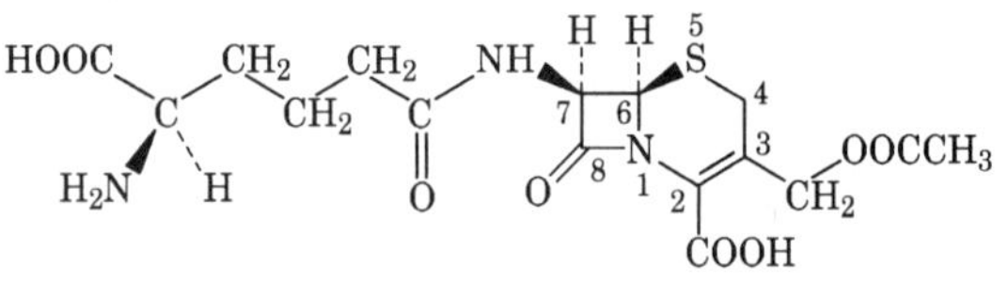

(**2**)

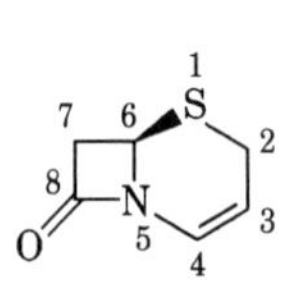

(**3**)

Nomenclature and Stereochemistry

Chemical Abstracts indexes cephalosporins as 5-thia-1-azabicyclo[4.2.0]oct-2-enes. Using this system then, cephalosporin C [*61-24-5*], $C_{16}H_{21}N_3O_8S$, (**2**) is 3-[(acetyloxy)methyl]-7-[(5-amino-5-carboxy-1-oxopentyl)amino]-8-oxo-5-thia-1-azabicyclo[4.2.0]oct-2-ene-2-carboxylic acid. Such names are too cumbersome for general use. One simplification defines the ring system of the cephalosporins as cepham (**3**) (1). Hence cephalosporins become 3-acetoxymethyl-7-acylamino-3-cephem-4-carboxylic acids. In this widely used nomenclature system, the numbering, as shown in structure (**3**), differs from that used by *Chemical Abstracts* shown in structure (**2**). In both systems, however, the important C-7, C-6, and C-3 positions remain unchanged. The cephem numbering system (**3**) is used herein.

The largest group of cephalosporins are derived from the cephem nucleus containing a 3-(substituted)methyl group. It is sometimes useful to designate the methyl group as C-3′. Thus, for example, a 3-acetoxymethyl, present in structures (**1**) and (**2**), becomes the 3′-acetoxy functionality. Another simplification in nomenclature is the use of the term cephalosporanic acid (2) as the parent name. Although convenient, this system is somewhat restrictive for cephalosporins in which the 3-acetoxymethyl grouping has been replaced. The clinically important cephalosporins all have a generic name which, by convention, starts with ceph- or cef-.

Structures of the naturally occurring cephalosporins, cephamycins, and the 7-formamido cephalosporins are as indicated in Tables 1 and 2. The amide bearing carbon, C-7, has the *R*-configuration; the nonplanar fused rings are folded along the C-6 to ring nitrogen bond; and the configuration of the fused rings is such that

Table 1. Naturally Occurring Cephalosporins

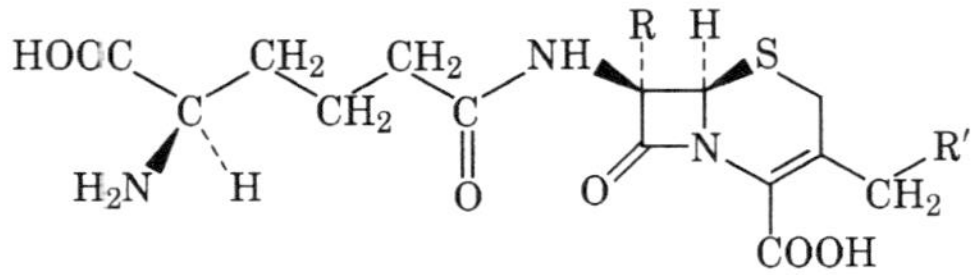

Name	CAS Registry Number	Molecular formula	R	R′	Ref.
deacetoxycephalosporin C	[*26924-74-3*]	$C_{14}H_{19}N_3O_8S$	—H	—H	3
deacetylcephalosporin C	[*1476-46-6*]	$C_{14}H_{19}N_3O_7S$	—H	—OH	3
cephalosporin C	[*61-24-5*]	$C_{16}H_{12}N_2O_5S$	—H	$—OCOCH_3$	4–6
O-carbamoyldeacetylcephalosporin C	[*32178-83-9*]	$C_{15}H_{19}N_4O_8S$	—H	$—OCONH_2$	7
3′-methylthiodeacetoxycephalosporin C (F_1)	[*60890-86-0*]	$C_{15}H_{21}N_3O_6S_2$	—H	$—SCH_3$	8
3′-sulfothiodeacetoxycephalosporin C (F_2)	[*65580-83-8*]	$C_{14}H_{19}N_3O_9S_3$	—H	$—SSO_2OH$	9
C43-219	[*60857-71-8*]	$C_{19}H_{28}N_4O_8S_2$	—H	$—SC(CH_3)_2CH(NH_2)CO_2H$	10
7α-methoxycephalosporin C	[*32178-82-8*]	$C_{17}H_{23}N_3O_9S$	$—OCH_3$	$—OCOCH_3$	7
cephamycin C	[*34279-55-1*]	$C_{16}H_{22}N_4O_9S$	$—OCH_3$	$—OCONH_2$	7,11
cephamycin A	[*34279-78-2*]	$C_{25}H_{29}N_3O_{14}S_2$	$—OCH_3$	$—OCOC(OCH_3){=}CHC_6H_4OSO_2OH$-*p*	11
cephamycin B	[*34729-77-1*]	$C_{25}H_{29}N_3O_{11}S$	$—OCH_3$	$—OCOC(OCH_3){=}CHC_6H_4OH$-*p*	11
Takeda C2801X	[*62851-50-7*]	$C_{25}H_{29}N_3O_{12}S$	$—OCH_3$	$—OCOC(OCH_3){=}CHC_6H_3(OH)_2$-*m,p*	12
SF-1623	[*56487-86-6*]	$C_{15}H_{21}N_3O_{10}S_3$	$—OCH_3$	$—SSO_2OH$	13

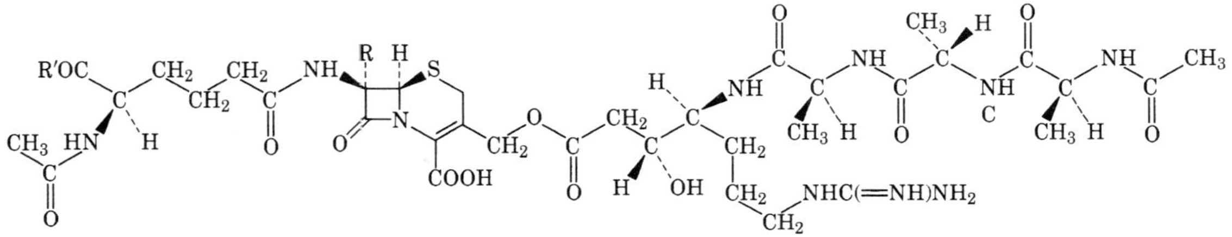

Name	CAS Registry Number	Molecular formula	R	R′	Ref.
SQ 28,516	[*92121-46-5*]	$C_{36}H_{55}N_{11}O_{15}S_1$	—NHCHO	—OH	14
SQ 28,517	[*92131-67-4*]	$C_{36}H_{56}N_{12}O_{14}S_1$	—NHCHO	$—NH_2$	14

Table 2. Naturally Occurring Cephabacin Cephalosporins[a]

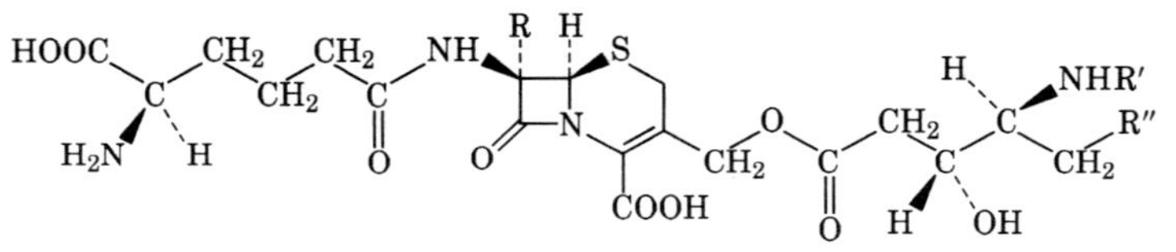

Name	CAS Registry Number	Molecular formula	R	R′	R″	Refs.
cephabacin F_1	*[95041-98-8]*	$C_{26}H_{41}N_9O_{11}S$	—NHCHO	-L-Ala	—$CH_2CH_2NHC(=NH)NH_2$	15
or chitinovorin A						16
cephabacin F_2	*[95722-76-2]*	$C_{29}H_{46}N_{10}O_{12}S$	—NHCHO	-L-Ala-L-Ala	—$CH_2CH_2NHC(=NH)NH_2$	15
or chitinovorin B						16
cephabacin F_3	*[95042-00-5]*	$C_{32}H_{51}N_{11}O_{13}S$	—NHCHO	-L-Ala-L-Ala-L-Ala	—$CH_2CH_2NHC(=NH)NH_2$	15
cephabacin F_4	*[95042-01-6]*	$C_{26}H_{41}N_9O_{12}S$	—NHCHO	-L-Ser	—$CH_2CH_2NHC(=NH)NH_2$	15
cephabacin F_5	*[95042-02-7]*	$C_{29}H_{46}N_{10}O_{14}S$	—NHCHO	-L-Ser-L-Ser	—$CH_2CH_2NHC(=NH)NH_2$	15
cephabacin F_6	*[95042-03-8]*	$C_{32}H_{51}N_{11}O_{15}S$	—NHCHO	-L-Ser-L-Ser-L-Ala	—$CH_2CH_2NHC(=NH)NH_2$	15
cephabacin F_7	*[95230-87-8]*	$C_{26}H_{41}N_7O_{12}S$	—NHCHO	-L-Ser	—$CH_2CH_2CH_2NH_2$	15
cephabacin F_8	*[95230-88-9]*	$C_{29}H_{46}N_8O_{14}S$	—NHCHO	-L-Ser-L-Ser	—$CH_2CH_2CH_2NH_2$	15
cephabacin F_9	*[95230-89-0]*	$C_{32}H_{51}N_9O_{15}S$	—NHCHO	-L-Ser-L-Ser-L-Ala	—$CH_2CH_2CH_2NH_2$	15
cephabacin H_1	*[95042-07-2]*	$C_{25}H_{40}N_8O_{10}S$	—H	-L-Ala	—$CH_2CH_2NHC(=NH)NH_2$	15
cephabacin H_2	*[95042-08-3]*	$C_{28}H_{45}N_9O_{11}S$	—H	-L-Ala-L-Ala	—$CH_2CH_2NHC(=NH)NH_2$	15
cephabacin H_3	*[95042-09-4]*	$C_{31}H_{50}N_{10}O_{12}S$	—H	-L-Ala-L-Ala-L-Ala	—$CH_2CH_2NHC(=NH)NH_2$	15
cephabacin H_4	*[95042-10-7]*	$C_{25}H_{40}N_8O_{11}S$	—H	-L-Ser	—$CH_2CH_2NHC(=NH)NH_2$	15
cephabacin H_5	*[95042-11-8]*	$C_{28}H_{45}N_9O_{13}S$	—H	-L-Ser-L-Ser	—$CH_2CH_2NHC(=NH)NH_2$	15
cephabacin H_6	*[95042-12-9]*	$C_{31}H_{50}N_{10}O_{14}S$	—H	-L-Ser-L-Ser-L-Ala	—$CH_2CH_2NHC(=NH)NH_2$	15
cephabacin M_1	*[99332-96-4]*	$C_{31}H_{50}N_8O_{13}S$	—OCH_3	-L-Val-L-Orn	—$CONH_2$	17
cephabacin M_2	*[99313-76-5]*	$C_{34}H_{55}N_9O_{15}S$	—OCH_3	-L-Val-L-Orn-L-Ser	—$CONH_2$	17
cephabacin M_3	*[99313-75-4]*	$C_{37}H_{60}N_{10}O_{16}S$	—OCH_3	-L-Val-L-Orn-L-Ser-L-Ala	—$CONH_2$	17
cephabacin M_4	*[99313-74-3]*	$C_{41}H_{69}N_{11}O_{15}S$	—OCH_3	-L-Val-L-Orn-L-Val-L-Orn	—$CONH_2$	17
cephabacin M_5	*[99332-95-3]*	$C_{44}H_{74}N_{12}O_{17}S$	—OCH_3	-L-Val-L-Orn-L-Val-L-Orn-L-Ser	—$CONH_2$	17
cephabacin M_6	*[99313-73-2]*	$C_{47}H_{79}N_{13}O_{18}S$	—OCH_3	-L-Val-L-Orn-L-Val-L-Orn-L-Ser-L-Ala	—$CONH_2$	17

[a]Structure as shown where R, R′, and R″ are as indicated.

the hydrogen on C-6 is cis to the hydrogen (or the methoxyl or formamido moiety) on C-7. Thus the absolute stereochemistry is *6R*:*7R*. The stereochemistry of substituents attached to the rings is designated using α and β. For cephalosporins, the hydrogens at C-7 and C-6 are α, the acylamino group at C-7 is β; in the cephamycins, the C-7 methoxyl is α. These stereochemical relationships and the absolute stereochemistry are also essential for antibacterial activity in the synthetic and semisynthetic cephalosporins. Antibacterial activity is lost whenever either or both of the asymmetric centers are inverted.

Discovery and Historical Development

In 1945 it was noted that a strain of *Cephalosporium acremonium* produced antibiotic material that was active against gram-negative as well as gram-positive organisms (18). The main hydrophilic component, originally called cephalosporin N, and active against both gram-negative and gram-positive organisms was found, in fact, to be a penicillin and was renamed penicillin N. A second hydrophilic antibiotic, present only in minute amounts in the original fermentation mixture and stable to acidic hydrolysis, was named cephalosporin C (**2**) (19). Chemical studies demonstrated that, like penicillin N, cephalosporin C contained a fused β-lactam ring system and a D-α-aminoadipamido substituent. Additional experiments led to the correct structure which was confirmed by x-ray crystallographic studies (4–6,20).

All cephalosporins found in nature (Tables 1 and 2) have the D-α-aminoadipic acid 7-acyl side chain (21). All of these compounds can be classified as having rather low specific activity. A substantial amount of the early work in the cephalosporin area was unsuccessfully directed toward replacing the aminoadipic acid side chain or modifying it appropriately by fermentation or enzymatic processes (6,22). A milestone in the development of cephalosporins occurred in 1960 with the discovery of a practical chemical process to remove the side chain to afford 7-ACA (**1**) (1). Several related processes were subsequently developed (22,23). The ready availability of 7-ACA opened the way to thousands of new semisynthetic cephalosporins. The cephalosporin structure offers more opportunities for chemical modification than does that of penicillins. There are two side chains that especially lend themselves to chemical manipulation: the 7-acylamino and 3-acetoxymethyl substituents.

The cephamycins, described almost simultaneously by two different groups (7,11) in 1971, are produced by the *Streptomyces* species and have the same basic cephalosporin ring system (Table 1) (24). The natural products have the 7β-D-α-aminoadipamido substituent and a methoxy substituent instead of a hydrogen in the 7α-position. These cephamycins have only weak antibacterial activity, but the methoxy group imparts good β-lactamase stability to these compounds. Semisynthetic analogues (25) have been obtained either by chemically introducing a methoxy group at the 7α-position of the cephalosporin ring system or by manipulating the 3- and 7β-substituents of the natural cephamycins (26,27). Some 7α-methoxylated cephalosporins have also been reported which have amino acid or oligopeptides at the C-3 position (Table 2) (17). The compounds are designated herein as cephabacin M (from *ceph*em *a*ntibiotics of *bac*terial origi*n* with the M designating a 7-methoxyl substituent).

In 1984 a new class of naturally occurring cephalosporins containing a 7α-formamido group, was isolated (14–16). These compounds (Table 2) called chitinovorins (16) or cephabacin F (15) where F designates a formamido group in the 7-position, also have amino acids or oligopeptides at the C-3 position. An analogous 7α-hydrogen series, named cephabacin H, containing the same oligopeptides at C-3 has also been isolated (15). The derivatives all have weak antibacterial activity. The 7α-formamido group imparts good β-lactamase stability (15) as is the case for the cephamycins. Chemical manipulation afforded a series of semisynthetic analogues having excellent broad-spectrum activity (28–30).

Biogenesis

The biosynthesis of cephalosporins and penicillins (Fig. 1) both start from the amino acids and proceed via δ-(L-α-aminoadipyl)-L-cysteinyl-D-valine [*32467-88-2*] (LLD-ACV) (**9**), $C_{14}H_{25}N_3O_6S$, (31–34), often referred to as the Arnstein tripeptide (35). Because LLD-ACV is not transported into intact cells, a cell-free system was required to determine that this intermediate is the precursor of the penicillins (36). The cell-free system, obtained fron *C. acremonium,* converts LLD-ACV into isopenicillin N [*525-94-0*] $C_{14}H_{21}N_3O_6S$ (**10**). Isopenicillin N (IPN) synthetase, the enzyme which catalyses this conversion, requires oxygen, Fe^{2+}, a reducing agent such as ascorbate, and a thiol group such as that of dithiothreitol for high activity (37,38). Isopenicillin N synthetase is present in *P. chrysogenum* and in species of *Streptomyces* as well as in *C. acremonium* (39,40). In the *Cephalosporium* species and the *Streptomyces* species, an epimerase is present that converts (**10**) into penicillin N [*58678-43-6*] (**11**), $C_{14}H_{21}N_3O_6S$, (41,42) which then undergoes a ring expansion to deacetoxycephalosporin C (**4**) (43–46). The ring expansion enzyme (REX) from *C. acremonium* appears to be bifunctional (47); it also catalyzes (48,49) the subsequent hydroxylation of (**4**) to deacetylcephalosporin C (**5**), itself a precursor of cephalosporin C (**2**) (50). The introduction of the methoxyl group is also a two-step process involving molecular oxygen (51,52). Using cephalosporin C (**2**) or the corresponding carbamoyl derivative (**6**) as a substrate, another dioxygenase catalyzes the incorporation of a 7α-hydroxy function (53). The resulting 7α-hydroxycephalosporin is then methylated using *S*-adenosylmethionine to form the corresponding 7α-methoxycephalosporin (53). The details of these steps are discussed in depth in the literature (21,39,42,48,54,55). Rapid advances in this area have been made possible by the successful cloning and expression of isopenicillin N synthetase (IPNS) and ring expansion-hydroxylase (REX) (39,48,56,57).

Physical Properties

Most cephalosporin antibiotics are white, off-white, tan, or pale yellow solids that are usually amorphous, but can sometimes be obtained crystalline. The cephalosporins do not usually have sharp melting points, but rather decompose upon

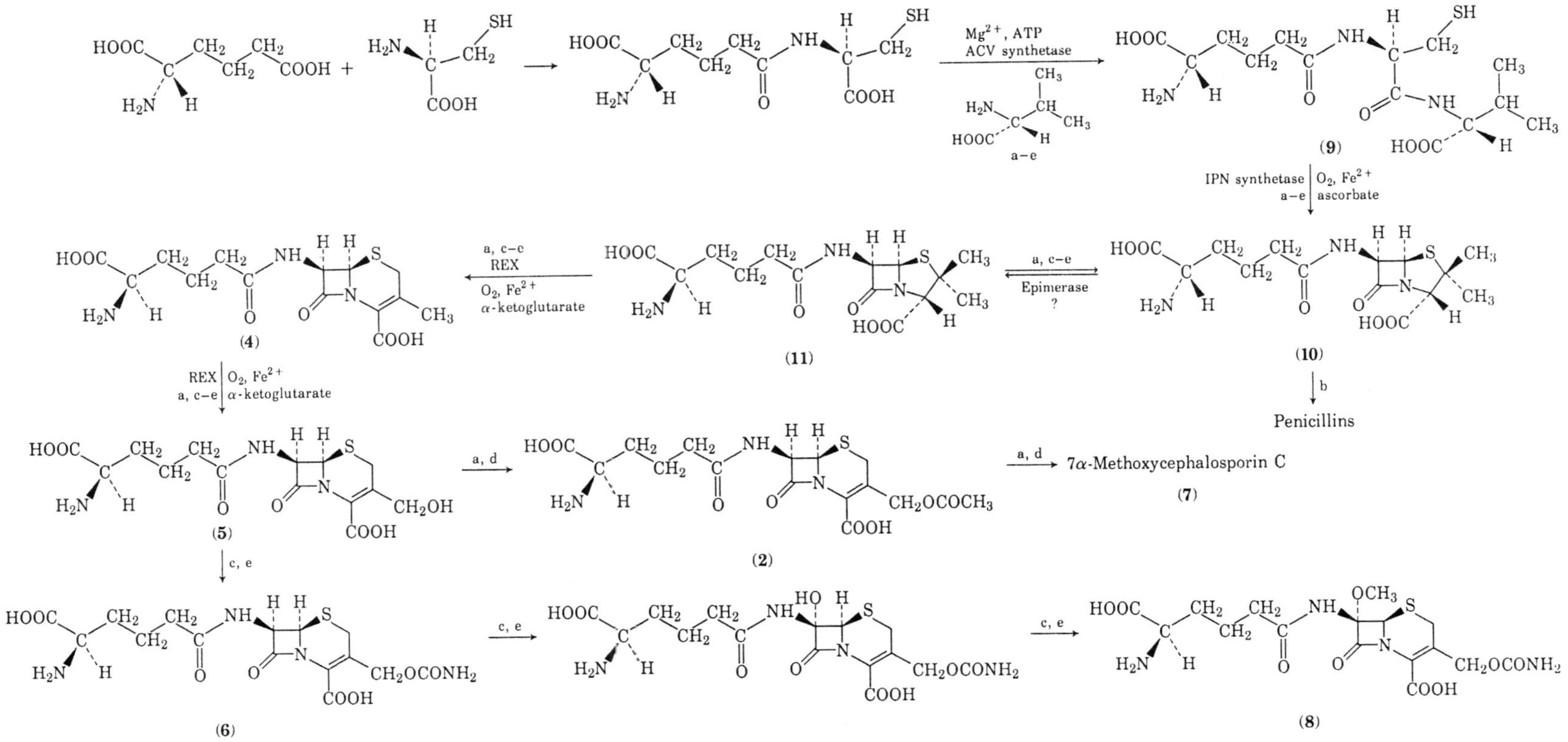

Fig. 1. Biosynthesis of cephalosporins and cephamycins. a, *Cephalosporium acremonium;* b, *Penicillium chrysogenum;* c, *Streptomyces clavuligerus;* d, *Streptomyces lipmanii;* e, *Streptomyces wadayamensis;* REX is a ring expansion enzyme (deacetoxycephalosporin C synthethase).

heating at elevated temperatures. The acid strength, pK_a, of the carboxyl group on the dihydrothiazine ring depends on environment (58). Representative pK_a values are given in Table 3, as are other physical properties.

Table 3. Properties of Cephalosporins[a]

Name	pK_a[b]	β-Lactam ir stretching frequency, cm^{-1}[c]	Uv absorption, λ max, nm (ε, $cm^{-1}M^{-1}$)[d]	Nmr absorption, ppm[e] H-7[f]	H-6[g]	$J_{6,7}$, Hz[h]
7-ACA	1.75, 4.63	1806	261(8500)	5.53d 4.83d	5.13d	4.5
cephalosporin C	<2.6, 3.1, 9.8	1780	260(8900)	5.66d	5.15d	4.7
cephalothin	5.0	1760	265(9000)	5.70d	5.14d	4.5
cephalexin	5.2, 7.3	1775	260(7750)	6.10d	5.45d	4.2
cephamycin C	4.2, 5.6, 10.4[i]	1770	264(6900) 242(5700)		5.19s	

[a]Refs. 7, 11, 58–62, 64, 65.
[b]Values for aqueous solutions unless otherwise noted, either by direct determination or by extrapolation from mixed solvents.
[c]Nujol mull.
[d]In aqueous solution.
[e]In D_2O relative to external standard tetramethyl silane (TMS); s = singlet, d = doublet.
[f]Range 5.23–6.21.
[g]Range 4.24–5.46.
[h]$J_{6,7}(cis)$ = 4–5 Hz, $J_{6,7}(trans)$ = 1.5–2 Hz, $J_{7\text{-}NH}$ = 8–11 Hz.
[i]In 66% DMF.

One of the distinguishing physical characteristics of the cephalosporins is the infrared stretching frequency of the β-lactam carbonyl. This absorption occurs at higher frequencies (1770–1815 cm^{-1}) than those of either normal secondary amides (1504–1695 cm^{-1}) or ester carbonyl groups (1720–1780 cm^{-1}). The nuclear magnetic resonance signals of the protons attached to the β-lactam ring provide information about ring integrity, the attachments to the ring, and the relative stereochemistry of substituents in these positions. These protons form an ABX system with the vicinal proton on the side-chain amide nitrogen (Table 3) (58–62). The cephalosporins display uv absorption maxima at about 260 nm, which is at a longer wavelength than expected for a normal α,β-unsaturated amide, usually about 220–230 nm. This shift may result from the presence of low lying sulfur atom d-orbitals (63).

Chemical Properties

Much of the chemical reactivity of the β-lactam antibiotics is associated with the β-lactam moiety. The geometry and the accompanying increased ring strain results in very little, if any, amide-resonance stabilization leading to a marked increase in chemical reactivity when compared to a normal amide. In fact, in many instances the reactivity of the lactam carbonyl is analogous to that of a carboxylic acid anhydride. Fused β-lactam antibiotics are readily attacked by nucleophiles with resultant ring opening and loss of biological activity. The cephalosporins are more resistant to ring opening than the penicillins. For example, although alcohols readily attack the penicillin β-lactam, cephalosporins are sufficiently stable to permit the use of methanol as a recrystallization solvent.

Cephalosporins are attacked by nucleophiles such as alkoxide or hydroxylamine, but because of the nature of the dihydrothiazine ring, products equivalent to the penicilloic or penilloic acids derived from the penicillins under similar conditions, are not obtained. Rather, studies indicate involvement of the 3′-substituent as a leaving group (66). Thus 3-deacetoxycephalosporins are found to be more stable than cephalosporins having 3′-substituents that can be readily lost. The 3′-substituents exert an electronic influence on the chemical reactivity of the β-lactam carbonyl group (67,68) which can be correlated with antibiotic activity within a series (69,70) and it has been claimed that cephalosporin antibacterial activity is related to the ease of C-3′ group expulsion (71). The deacetoxy analogues generally show lower levels of activity *in vitro* than their 3′-functionalized analogues.

It has also been proposed that reactivity is enhanced because nucleophilic attack on the β-lactam carbonyl is concerted with expulsion of the leaving group (69). However, evidence is now available which strongly suggests that in alkaline hydrolysis and alcoholysis fission of the cephalosporin β-lactam C–N bond occurs before that of the C-3′ leaving group (71–73). This would seem to be the case for interaction of cephalosporins with β-lactamase enzymes (72,74). Thus cephalosporin reactivity is probably best represented by the sequence outlined in Figure 2. In the case where the attacking nucleophile is part of the active site of an enzyme (serine hydroxyl), the acyl–enzyme complex may break down before expulsion of the leaving group or may expel the leaving group to afford a new acyl–enzyme complex (72). The chemistry of cephalosporins, as utilized in synthetic processes and manufacturing, such as deacylation and acylation of the C-7 amino, esterification of the carboxyl, modification at C-3, and introduction of functionality into

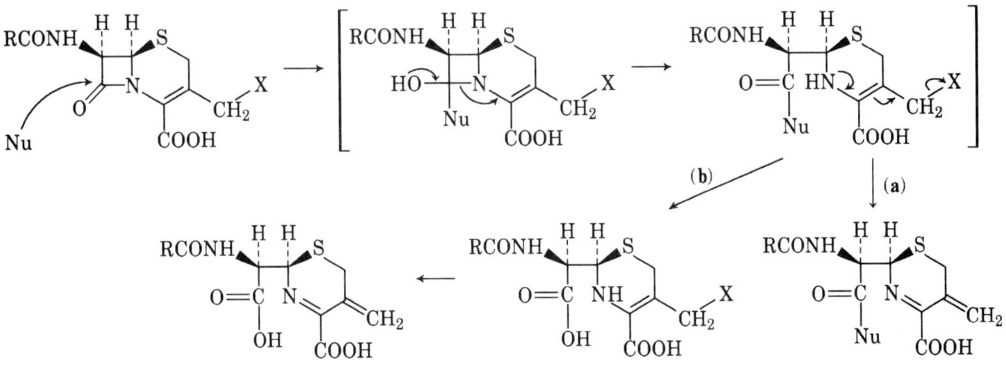

Fig. 2. Cephalosporin β-lactam reactivity where Nu is a nucleophile and X is a leaving group; (**a**) path followed for nucleophiles; (**b**) when Nu is the serine OH of an enzyme (ie, Nu = Enz–Ser–OH) deacylation may precede expulsion of the leaving group and both pathways (**a**) and (**b**) may be observed.

the C-7α position are discussed later. A discussion of other aspects of cephalosporin chemistry, such as deamination or epimerization at C-7, reactions at sulfur, isomerization of and addition to the double bond, and modification on the allylic methylene group at C-2, can be found in leading reviews (75,76).

Biological Properties

The clinical effectiveness of the cephalosporins depends on a number of properties. The antibiotic must inhibit, or preferably kill, bacteria at acceptable concentrations of the drug (*in vitro* activity); it must be capable of achieving host serum and tissue levels greater than those required to inhibit the pathogenic organism; and the selective toxicity profile must allow for safe administration to the host. Much of the chemical work that has been done on the cephalosporins has been directed toward modifying and improving the antibacterial and pharmacokinetic properties of these substances. Pharmacokinetic properties include the efficiency and rate of adsorption, the rate of metabolism and excretion, tissue distribution, serum binding, and serum and tissue inactivation. Chemical studies are thus closely coupled to extensive biological evaluation which examines *in vitro* and *in vivo* antibacterial activity expressed as minimum inhibitory concentration (MIC), minimum bactericidal concentration (MBC), and protective effectiveness in laboratory animals (PD_{50} or ED_{50}) as well as pharmacokinetic characteristics. New compounds are also tested for the ability to resist inactivation by β-lactamases (penicillinases, cephalosporinases) produced by bacteria.

Classification. As of 1991, there were approximately fifty different cephalosporins in clinical use or at an advanced stage of evaluation and development (77–79). Cephalosporins may be classified for convenience by their clinical pharmacology, β-lactamase resistance, chemical structure, or their antibacterial spectrum. The most common classification, which is somewhat arbitrary, divides the cephalosporins into three groups or "generations" (Tables 4–8), based primarily on their antibacterial spectrum (80,81). First-generation cephalosporins are characterized by good gram-positive activity and modest to weak gram-negative activity. They are usually active against aerobic gram-positive cocci, with the exception of enterococci, methicillin-resistant *Staphyloccus aureus,* and *Staphlylococcus epidermidis,* and are also active against some gram-negative bacilli such as *Escherichia coli, Klebsiella pneumoniae,* and *Proteus mirabilis,* and most anaerobic cocci and bacilli other than *Bacteroides fragilis* (80–83).

The second-generation cephalosporins have a broader spectrum and generally are more active against aerobic gram-negative organisms than the first-generation cephalosporins. Most second-generation cephalosporins have some activity against indole-positive *Proteus, Enterobacter* (except *E. cloacae*), and *Haemophilus influenzae.* Two agents included in this group, cefotetan (**26**) and cefoxitin (**23**), which are both cephamycin analogues, have significant activity against *Bacteroides fragilis* (80–83).

Third-generation cephalosporins have an expanded gram-negative spectrum and are the most active against enteric gram-negative bacilli, including penicillinase-producing strains, as well as *Serratia* and *Citrobacter.* They are also highly active against *Haemophilus influenzae* and *Neisseria gonorrhoeae, Streptococcus pneumoniae* and *pyogenes,* moderately active against *Pseudomonas aeruginosa* and anaerobes, but somewhat less active against staphylococci than the first-generation cephalosporins (80–84). A subset of the third-generation cephalosporins has somewhat improved, albeit still not ideal, activity against *Pseudomonas aeruginosa.* This group includes cefoperazone (**34**), cefpiramide (**42**), ceftazidime (**40**), and cefsulodin (**41**) (80), although cefsulodin is relatively inactive against most other organisms. Cefixime (**16**), a new oral cephalosporin touted as the first oral third-generation cephalosporin, has a spectrum between those of the second- and third-generation compounds. It has enhanced activity against *Enterobacteriaceae,* and against most common respiratory and urinary tract pathogens, but lacks activity against *Staphylococcus aureus,* enterococci, and *Pseudomonas aeruginosa.* Cefixime (**16**) would be better compared to its direct competition in the oral market, cefaclor (**13**), and the oral prodrug cefuroxime axetil (**20**).

Some other compounds tentatively labeled as examples of the as yet undefined "fourth-generation" have appeared in the literature, but these are probably best thought of as third-generation cephalosporins, having slight advantages over earlier examples of this group. This group includes cefpirome (**43**), cefepime (**45**), and others undergoing clinical trials.

It should be noted that the classification into generations is not chronological, and so some second-generation compounds came to the market relatively recently, after the third-generation was firmly established. Within each class there is also a further classification as to whether the compounds are administered orally or parenterally. A contemporary variation of the generation classification has been proposed, but has not found widespread use (85). Another proposal groups cephalosporins according to clinical indication or application (86).

The structures of selected cephalosporins on the U.S. market, or in the final stages of development, are shown in Tables 4–8 (see also 78, 87). For every cephalosporin which has made it to the marketplace, literally thousands of analogues were synthesized in order to establish the structure-activity profile and allow selection of a clinical candidate. In addition to these compounds, there is a tremendous number of cephalosporin compounds currently at various stages of development. A more extensive listing of the newer cephalosporins under preclinical or clinical evaluation may be found in a number of reviews (79,88).

***In Vitro* Antibacterial Activity and Structure-Activity Relationships.** The *in vitro* antibacterial activity of any particular cephalosporin is a combination of the degree and type of activity at the target site, the ease with which it can penetrate to the target, and the ability to resist the attack of destructive enzymes. The nature and complexity of the biochemical target(s) for β-lactam antibiotics is fairly well-established and the mechanisms of penetration are also understood to some extent. However, many factors are involved, including pharmacokinetic and pharmacodynamic properties, and the antibiotic may not perform as predicted. The *in vitro* antibacterial activity of a representative selection of the cephalosporins given in Tables 4–8 against a variety of bacterial organisms is depicted in Table 9. More extensive comparative listings of the antibacterial activity of cephalosporins may be found in the literature (77,80,81,130–132).

Structure-activity relationships can be inferred by comparison of the antibacterial properties of the clinical agents and related compounds. Different acyl side chains can result in significant changes in the antibacterial activity, both

Table 4. Oral Cephalosporins

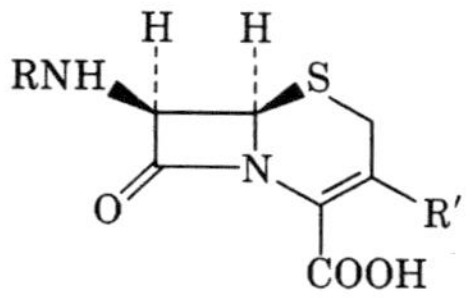

Generic name	CAS Registry Number	Molecular formula	Generation	Trade name	Company	Structure number	R	R′	Year[a]	Refs.
cephalexin	*[15686-71-2]*	$C_{16}H_{17}N_3O_4S$	[1]	Keflex	Lilly	**(12)**	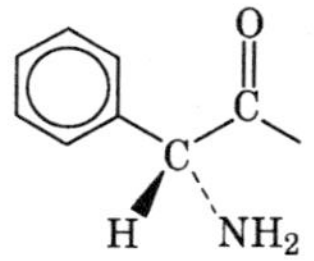	—CH_3	1970	77,89
cefaclor	*[53994-73-3]*	$C_{15}H_{14}N_3O_4S_1Cl$	[2]	Ceclor	Lilly	**(13)**	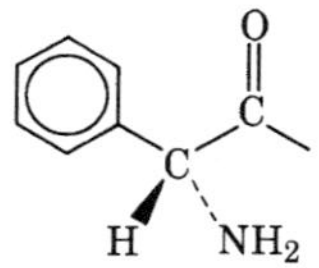	—Cl	1979	77,90
cephradine	*[3882-53-3]*	$C_{16}H_{19}N_3O_4S$	[1]	Velosef Anspor	Squibb SKF-Beecham	**(14)**	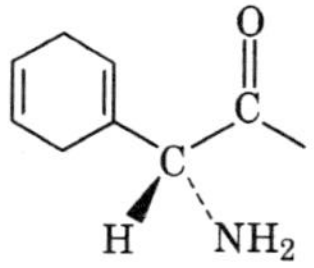	—CH_3	1972	77,91–93
cefadroxil	*[50370-12-2]*	$C_{16}H_{17}N_3O_5S$	[1]	Duricef Ultracef	Mead-Johnson Bristol-Myers Squibb	**(15)**	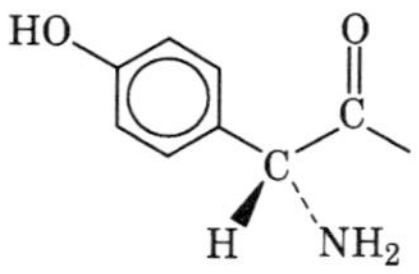	—CH_3	1980	77,94,95

cefixime	[*79350-37-1*]	$C_{16}H_{15}N_5O_7S_2$	[2–3]	Suprax Cefspan[b]	Lederle Fujisawa	**(16)**	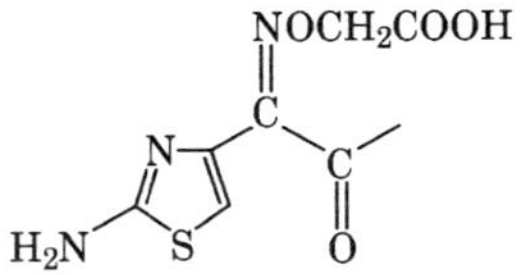	$-CH{=}CH_2$	1989 1987[c]	96
ceftibuten	[*97519-39-6*]	$C_{15}H_{14}N_4O_6S_2$	[2–3]	Sch 39720 7432-S	Schering Shionogi	**(17)**	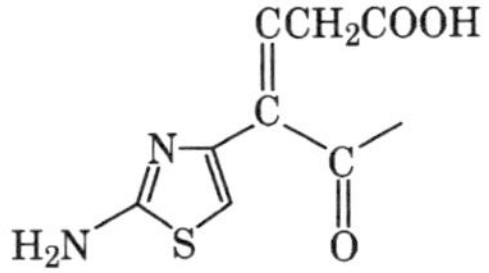	$-H$	Ph-III	97
cefprozil	[*92665-29-7*]	$C_{18}H_{19}N_3O_5S$	[2]	BMY28100	Bristol-Myers Squibb	**(18)**	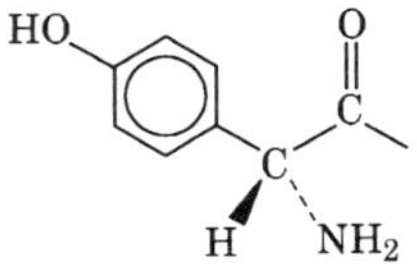	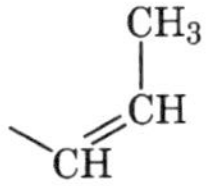	Ph-III	98
—	[*107888-46-0*]	$C_{15}H_{15}N_5O_6S_2$	[2–3]	BMY28232	Bristol-Myers Squibb	**(19)**	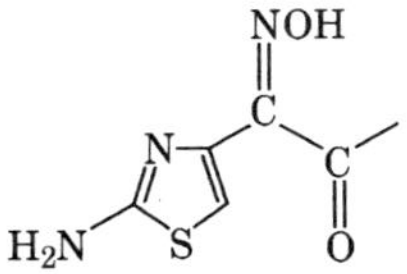	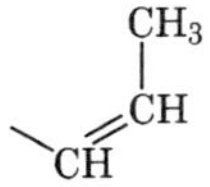	Pre-Cl	99

[a]Year is year of launch in the U.S. market unless indicated. Pre-Cl = preclinical, Ph III = phase III clinical trials (status as of Feb. 1990).
[b]Japanese trade name.
[c]Launch date in Japan.

Table 5. Oral Cephalosporins-Prodrugs

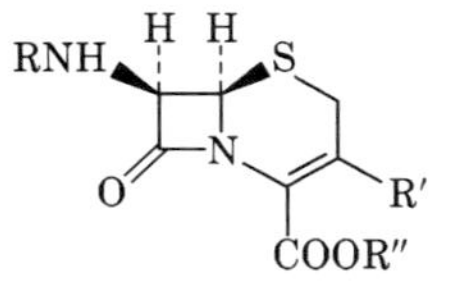

Generic name[a]	CAS Registry Number	Molecular formula	Trade name	Company	Structure number	R	R′	R″	Year[b]	Refs.
cefuroxime-axetil [2]	[*64544-07-6*]	$C_{20}H_{22}N_4O_{10}S_1$	Ceftin	Glaxo Roche	(**20**)	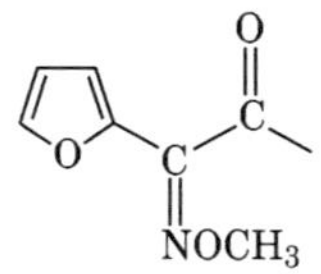	$-CH_2OCONH_2$	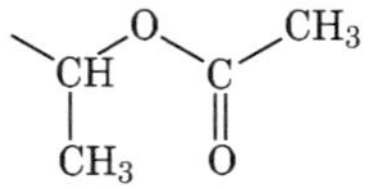	1988	100
cefpodoxime-proxetil [2–3]	[*87239-81-4*]	$C_{21}H_{27}N_5O_9S_2$	Banan	Upjohn Sankyo	(**21**)	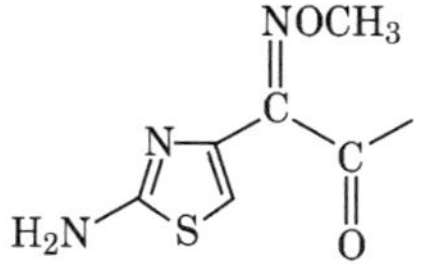	$-CH_2OCH_3$	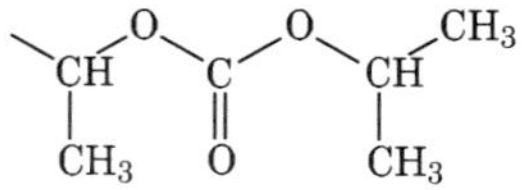	Ph-III	101
[2–3]	[*107888-50-6*]	$C_{18}H_{19}N_5O_7S_2$	BMY28271	Bristol-Myers Squibb	(**22**)	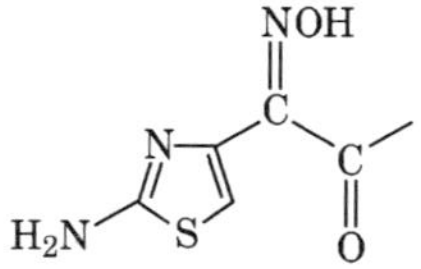	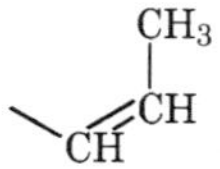	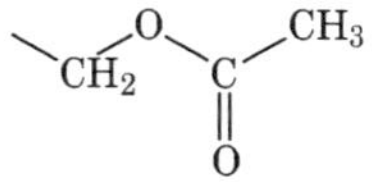	Pre-Cl	99,102

[a]Drug generation is given in brackets.
[b]Year is year of launch in the U.S. market unless indicated. Pre-Cl = preclinical and Ph III = phase III clinical trials (status as of Feb. 1990).

Table 6. Cephamycins

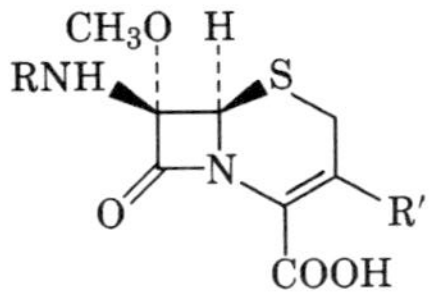

Generic name[a]	CAS Registry Number	Molecular formula	Trade name	Company	Structure number	R	R′	Year[b]	Refs.
cefoxitin [2]	*[35607-66-0]*	$C_{15}H_{15}N_3O_6$	Mefoxin	Merck	**(23)**	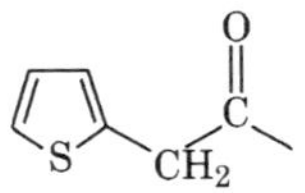	$-CH_2OCONH_2$	1978	103
cefmetazole [2]	*[56796-20-4]*	$C_{14}H_{15}N_7O_4S_3$	Zefazone Cefmetazon[c]	Upjohn Sankyo	**(24)**	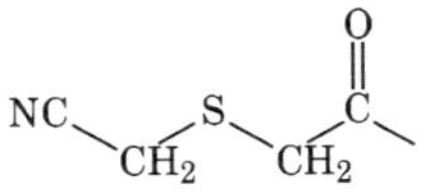	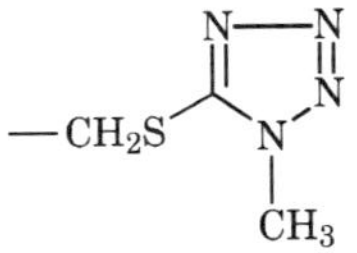	1989 1980[d]	104
cefminox [2]	*[75221-13-0]*	$C_{15}H_{19}N_7O_6S_3$	Meicelin[c]	Meiji Seika Bristol-Myers Squibb	**(25)**	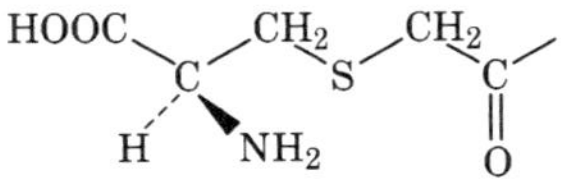	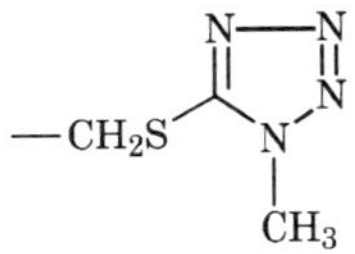	1987[d]	105
cefotetan [2]	*[69712-56-7]*	$C_{16}H_{15}N_7O_7S_4$	Ceftan Apacef[c] Yamatetan	Stuart Yamanouchi	**(26)**	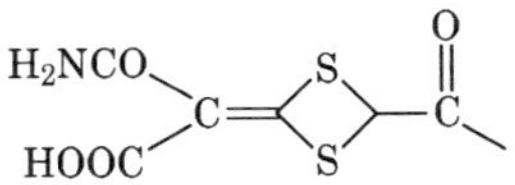	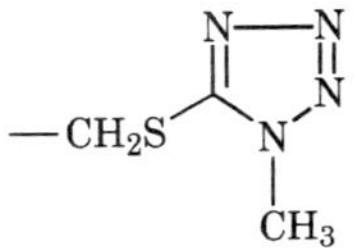	1986 1984[d]	106

[a]Drug generation is given in brackets.
[b]Year is year of launch in the U.S. market unless indicated.
[c]Japanese trade name.
[d]Launch date in Japan.

Table 7. Parenteral Cephalosporins

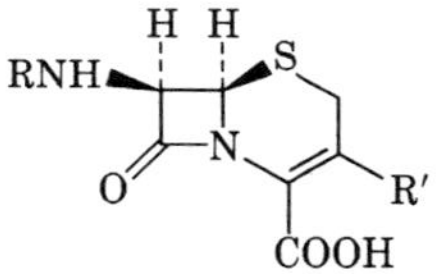

Generic name[a]	CAS Registry Number	Molecular formula	Trade name	Company	Structure number	R	R′	Year[b]	Refs.
cephalothin [1]	*[153-61-7]*	$C_{16}H_{16}N_2O_6S_2$	Keflin	Lilly	**(27)**	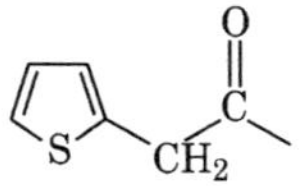	$—CH_2OCOCH_3$	1965	87,107
cephacetrile [1]	*[23239-41-0]*	$C_{13}H_{13}N_3O_6S$	Celospor	Ciba	**(28)**	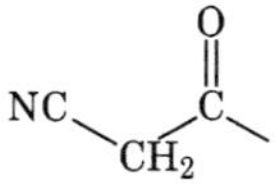	$—CH_2OCOCH_3$	1979[c]	87,108
cephapirin [1]	*[21593-23-7]*	$C_{17}H_{17}N_3O_6S_2$	Cefadyl	Bristol-Myers Squibb	**(29)**	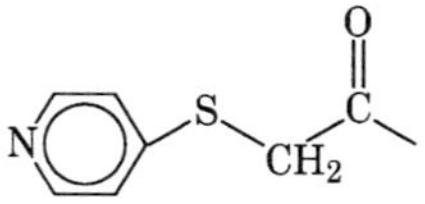	$—CH_2OCOCH_3$	1974	87,91,109
cefamandole [2]	*[34444-01-4]*	$C_{11}H_{18}N_6O_5S_2$	Mandol	Lilly	**(30)**	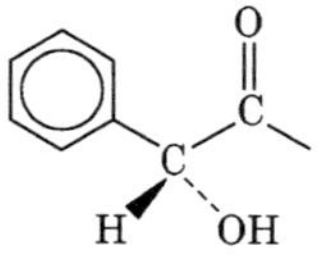	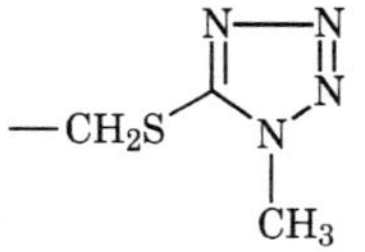	1978	87,110
cefonicid [2]	*[61270-58-4]*	$C_{18}H_{17}N_6O_8S_3Na$	Monicid	SKF-Beecham	**(31)**	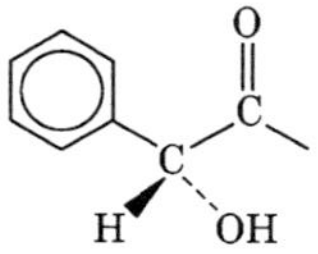	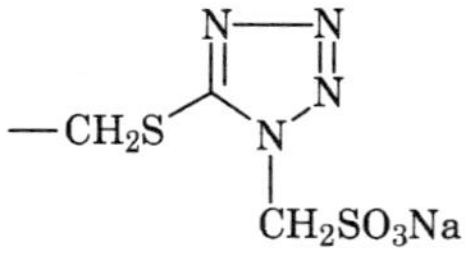	1984	87,111
cefazolin [1]	*[25953-19-9]*	$C_{14}H_{14}N_8O_4S_3$	Ancef Kefzol	SKF-Beecham Lilly	**(32)**	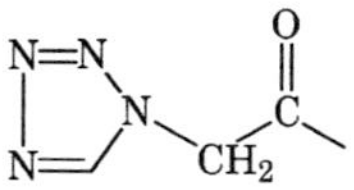	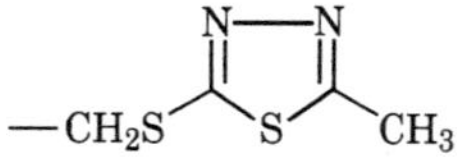	1973	87,112

ceforanide [2]	[*40851-79-4*]	$C_{20}H_{19}N_7O_6S_2$	Precef	Bristol-Myers Squibb	(**33**)	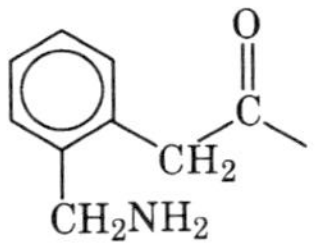	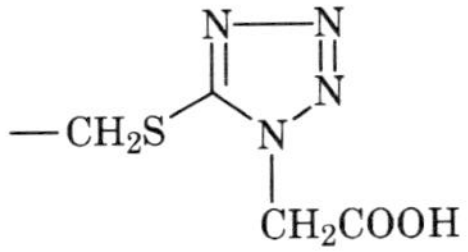	1984	87,113
cefoperazone [3]	[*62893-19-0*]	$C_{25}H_{27}N_9O_8S_2$	Cefobid	Pfizer Roerig	(**34**)	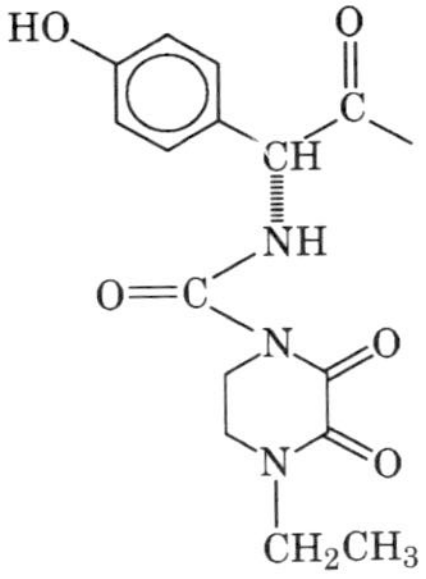	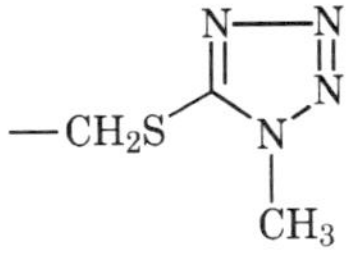	1983	87,114
cefuroxime [2]	[*55268-75-2*]	$C_{16}H_{16}N_4O_8S$	Zinacef Kefurox	Glaxo Lilly	(**35**)	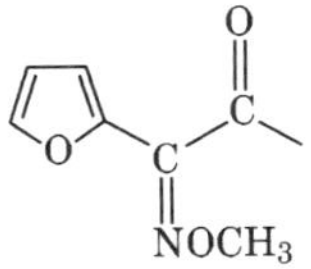	$—CH_2OCONH_2$	1983	87,115
cefotaxime [3]	[*63527-52-6*]	$C_{16}H_{17}N_5O_7S_2$	Claforan	Hoechst-Roussel	(**36**)	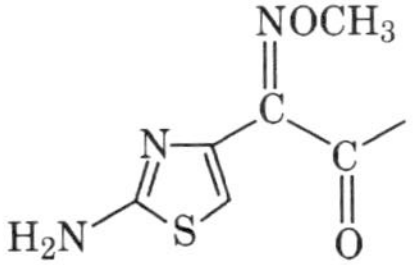	$—CH_2OCOCH_3$	1981	87,116
ceftizoxime [3]	[*68401-81-0*]	$C_{13}H_{13}N_5O_5S_2$	Cefizox	SKF-Beecham	(**37**)	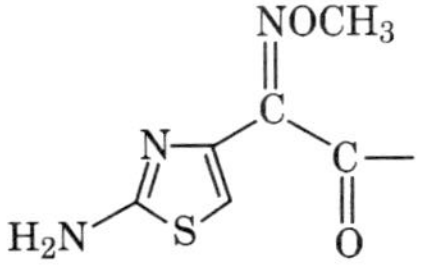	—H	1983	87,117

Table 7. *(Continued)*

Generic name[a]	CAS Registry Number	Molecular formula	Trade name	Company	Structure number	R	R′	Year[b]	Refs.
cefmenoxime [3]	*[65085-01-0]*	$C_{16}H_{17}N_9O_5S_3$	Cefmax Bestcall[e]	Abbott Takeda-Roche	**(38)**	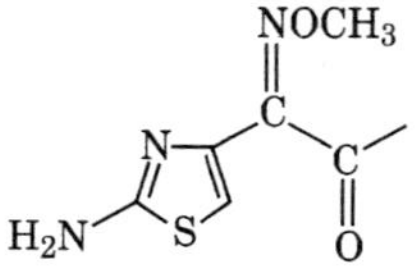	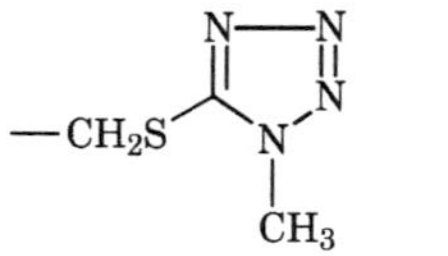	1987[d] 1983[f]	87,118
ceftriaxone [3]	*[73384-59-5]*	$C_{18}H_{17}N_8O_7S_3Na$	Rocephin	Roche	**(39)**	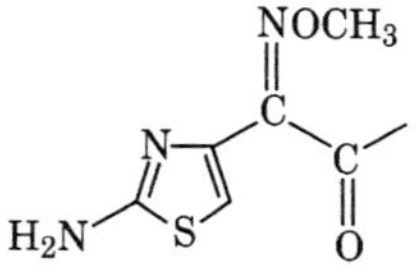	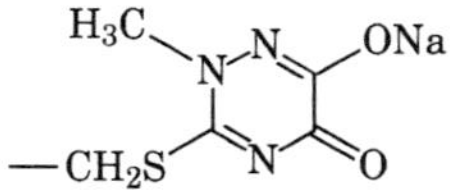	1985	87,119
ceftazidime [3]	*[72558-82-8]*	$C_{22}H_{22}N_6O_7S_2$	Fortaz Tazicef Tazidime	Glaxo SKF-Beecham Lilly	**(40)**	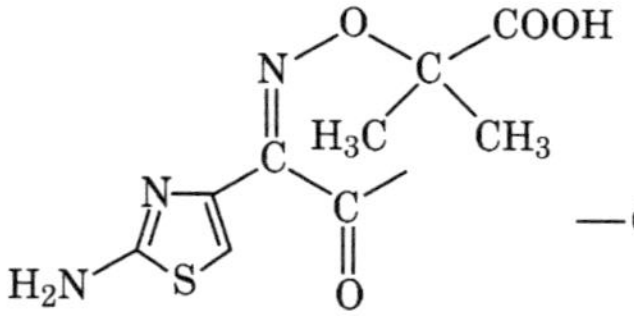	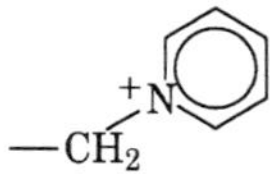	1985	87,120
cefsulodin	*[52152-93-9]*	$C_{22}H_{20}N_4O_8S_2$	Cefomonil	Takeda-Abbott	**(41)**	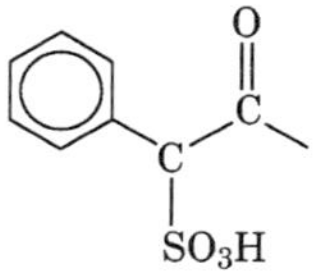	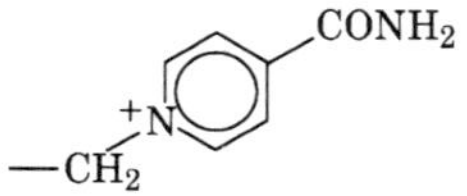	1986[d]	121

cefpiramide [3]	[*70797-11-4*]	$C_{25}H_{24}N_8O_7S_2$	Cefpiran Sepatren[e] Suncefal[e]	Wyeth-Ayerst Sumitomo Yamanouchi Banyu	(**42**)	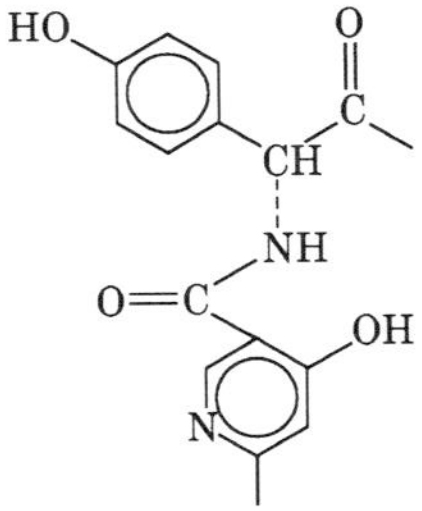	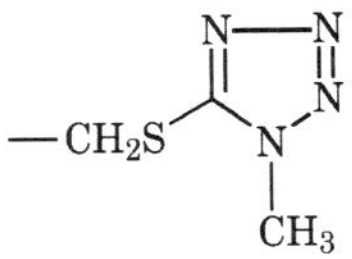	1989[d] 1985[f] 1985[f]	122
cefpirome [3–"4"]	[*84957-29-9*]	$C_{22}H_{22}N_6O_5S_2$	HR810	Hoechst-Roussel	(**43**)	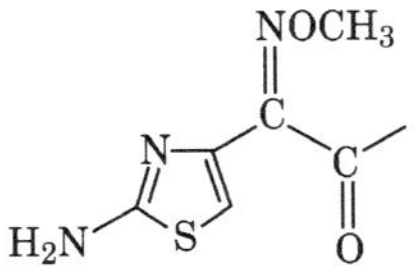	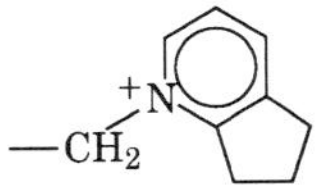	Ph-III	123
cefpimizole [3]	[*84572-33-8*]	$C_{28}H_{26}N_6O_{10}S_2$	U-63,196 Ajicef[e] Renilan[e]	UpJohn Ajinomoto Mochida	(**44**)	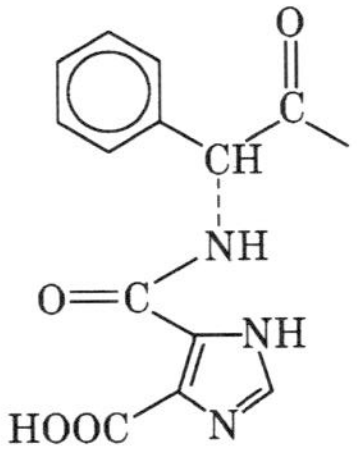	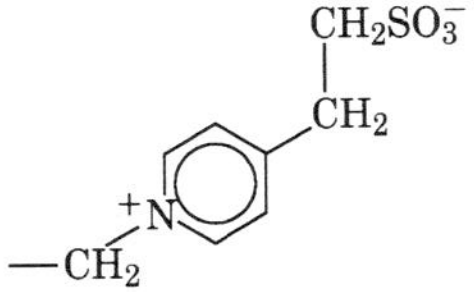	Ph-III 1987[f] 1987[f]	124
cefepime [3–"4"]	[*88040-23-7*]	$C_{19}H_{24}N_6O_5S_2$	BMY28142	Bristol-Myers Squibb	(**45**)	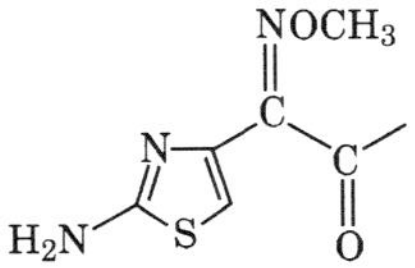	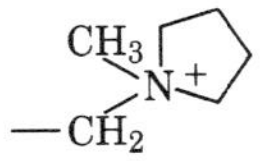	Ph-II	125

Table 7. (*Continued*)

Generic name[a]	CAS Registry Number	Molecular formula	Trade name	Company	Structure number	R	R′	Year[b]	Refs.
[3–"4"]	[*106772-82-1*]	$C_{28}H_{21}N_9O_{11}S_3Na_2$	M14659	Mochida	**(46)**	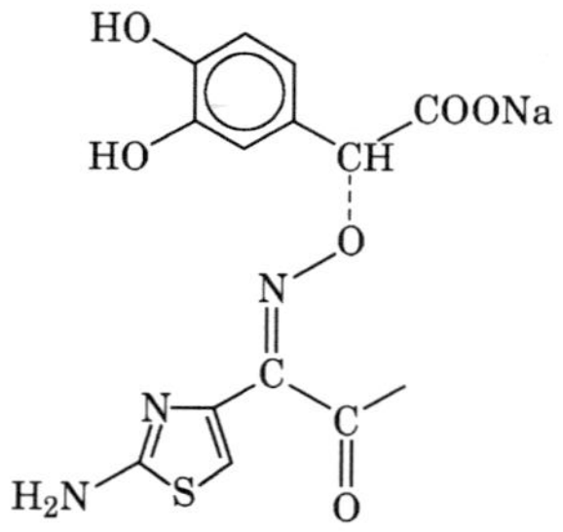	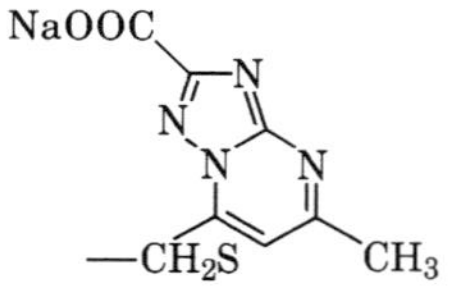	Pre-Cl	126
["4"]	[*115622-58-7*]	$C_{31}H_{31}N_8O_8S_2F_3$	Ro 23-9424	Roche	**(47)**	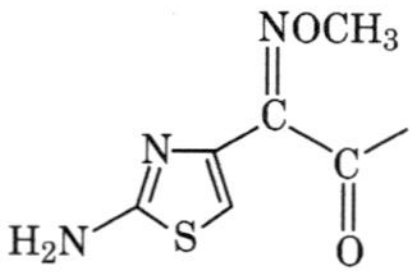	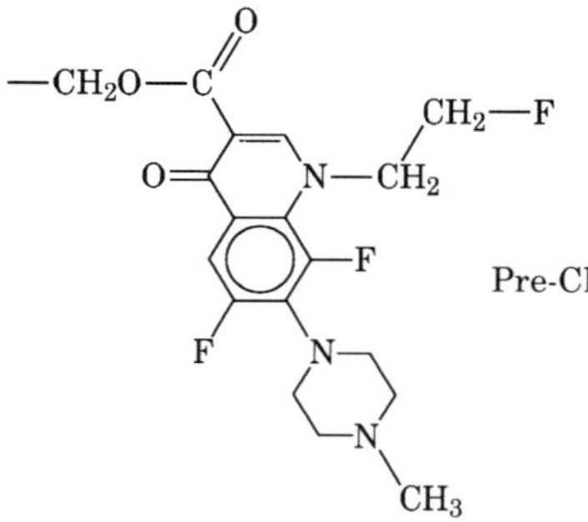	Pre-Cl	127

[a]Drug generation is given in brackets.
[b]Year is year of launch in the U.S. market unless indicated. Pre-Cl = preclinical, Ph III = phase III clinical trials, and Ph II = phase II clinical trials (status as of Feb. 1990).
[c]Launch date in Italy.
[d]New drug application (NDA) filing or approval.
[e]Japanese trade name.
[f]Launch date in Japan.

Table 8. Oxadethiacephalosporins

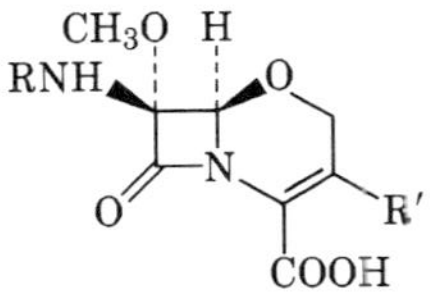

Generic name	CAS Registry Number	Molecular formula	Generation	Trade name	Company	Structure number	R	R′	Year[b]	Refs.
moxalactam latamoxef	[*64952-97-2*]	$C_{20}H_{20}N_6O_8S$	[3]	Moxam	Lilly	**(48)**	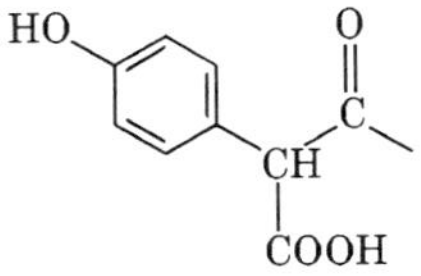	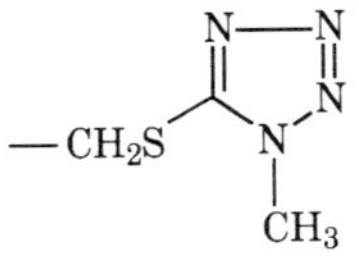	1981	128
flomoxef	[*99665-00-6*]	$C_{15}H_{18}N_6O_6S_2F_2$	[3]	Flumarin[b]	Shionogi	**(49)**	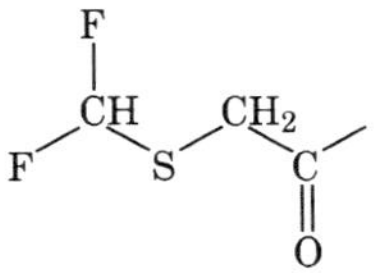	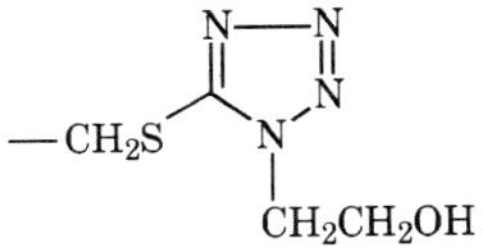	1988[c]	129

[a]Year is year of launch in the U.S. market unless indicated.
[b]Japanese trade name.
[c]Launch date in Japan.

Table 9. Cephalosporin *In Vitro* Antibacterial Activity, μg/mL, and Pharmacokinetic Data[a,b]

	(12)	(27)	(13)	(31)	(35)	(16)	(36)	(39)	(40)	(23)	(48)	(47)
Antibacterial activity												
Staph. aureus[c]	2	0.5	1	4	1	32	2	4	4	2	8	1
Strep. pyogenes	0.5	0.12	0.25	0.5	0.03	0.25	0.03	0.03	0.12	1	1	0.06
Strep. pneumoniae	2	0.12	1	1	0.12	0.5	0.12	0.25	0.25	2	1	<0.06
E. coli[d]	4	4	1	1	1	0.25	0.03	0.12	0.12	4	0.12	0.06
Kleb. pneumoniae[e]	4	4	1	4	2	0.12	0.03	0.06	0.12	2	0.12	0.06
Pr. mirabilis[d]	8	2	1	0.5	1	0.03	0.03	0.03	0.06	2	0.12	0.25
Proteus sp (ind+)	>128	>128	>128	16	8		0.12	0.12	0.12	4	0.25	0.12
Enterobacter sp[f]	>128	>128	>128	32	16	64	0.12	0.25	0.25	>128	0.12	0.06
Citrobacter sp	>128	64	>128		8		0.25	0.5	0.5		0.25	0.12
Serratia marcescens	>128	>128	>128	64	64		0.12	0.25	0.12	16	0.25	<0.12
Ps. aeruginosa	>128	>128	>128	>128	>128	>128	32	32	2	>128	16	4
H. influenzae	8	4	1	1	0.5	0.03	0.03	0.03	0.12	2	0.12	0.03
B. fragilis[g]	64	64	>128	128	32		32	64	64	4	4	128
N. gonorrhoeae[c]	2	0.5	0.5	1	0.06	0.06	<0.01	<0.01	0.06	0.5	0.06	<0.01
Pharmacokinetic data												
plasma half life, h	0.8	0.6	0.6	4.0	1.3	0.7	1.2	8.0	1.8	0.8	2.5	
protein bound, %	10	65	25	98	35	25	35	95	17	70	50	
urinary excretion, %	90	70[h]	60[i]	96	95	65	70[h]	65	85	80	80	
biliary excretion, %	0.3	0.03	0.05	1	0.5		1	35	<3	<2	20	

[a]Refs. 77, 80, 81, 130–132. Cefixime Ref. 96; Ro 24-9424 Ref. 127.
[b]Note that values are approximated from references because specific strains are not specified.
[c]Penicillin resistant.
[d]Ampicillin sensitive.
[e]Cefazolin sensitive.
[f]Cefotaxime sensitive.
[g]Penicillin sensitive.
[h]Plus 30% deacetylcephalosporin.
[i]±30% inactive.

with respect to potency and to breadth of spectrum. The highest activities are observed when the acylamino side chain at C-7 is a substituted acetic acid. Homologation of the acetic acid moiety lowers activity dramatically as exemplified by the naturally occurring cephalosporins which all have weak activity. Compared to the penicillins, there is somewhat more latitude for the type of substituent(s) attached to the α-carbon of the side chain. The phenylacetyl and phenoxyacetyl side chains are less effective in conferring activity on the cephalosporins than the pencillins. More effective are acetic acid side chains attached to heterocyclic rings, such as thiophene, pyridine, tetrazole, furan, and aminothiazole. Much smaller and nonaromatic substituents can also confer high levels of *in vitro* activity. Such substituents include the cyano, methylthio, methylsulfonyl, and trifluoromethylthio moieties. The aromatic ring seems to only be associated with effective activity when there is a second substituent on the α-carbon of the acyl group. The effect of this second substituent on the cephalosporin activity is different to what is observed for the penicillins. The carboxyl and sulfonic acid groups do not particularly improve activity overall for the cephalosporins, whereas the amino and hydroxy group seem effective in this regard. A second substituent on the α-carbon introduces a new chiral center. The D-configuration (*R*) is preferred for the phenylglycine and mandelic acid derived side chains as in both cases the L-configuration (*S*)-analogue is considerably less active.

That the aminothiazole group imparts high potency to the cephalosporins against gram-negative bacteria was first demonstrated in 1977 with the synthesis of cefotiam [*61625-34-2*] (SCE-963), $C_{18}H_{22}N_9O_4Na$ (**50**) (133). The oxime group in cefuroxime (**35**) and most of the aminothiazoyl containing cephalosporins is especially effective in further increasing potency and conferring resistance to certain β-lactamases. A marked stereochemical preference is observed: the syn-isomer where the alkoxy and acyl groups are cis to each other, is considerably more active than the corresponding antiisomer.

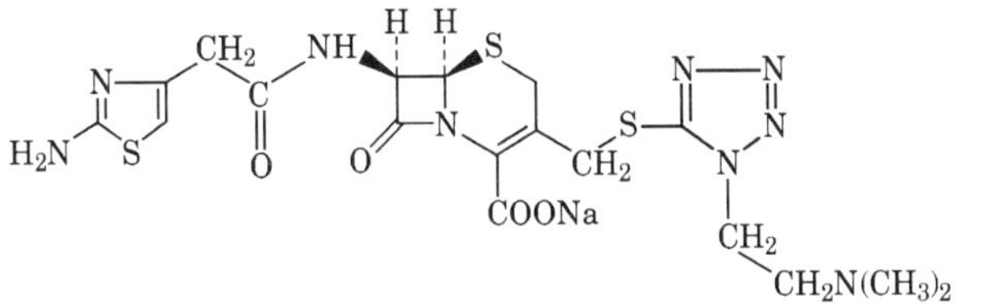

(**50**)

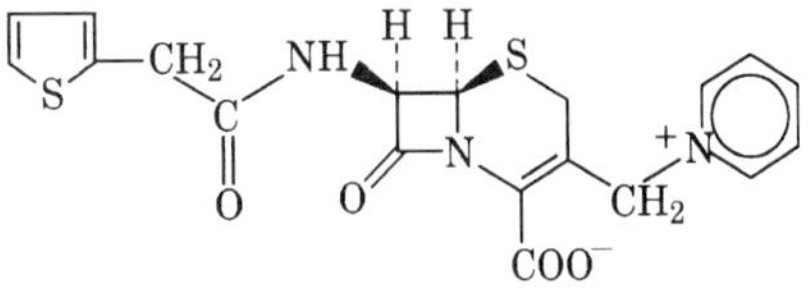

(**51**)

One of the principal deficiencies of the older cephalosporins was the lack of resistance to β-lactamases. Compounds with improved β-lactamase resistance have one or more of the following characteristics: a second, monovalent substituent on the α-carbon of the C-7 acyl group such as found in cefamandole (**30**) and cefoperazone (**34**); a syn-oxime substituent, eg, cefuroxime (**35**) and ceftriaxone (**39**); a methoxy, or formamido, substituent on the β-lactam ring at the 7α-position such as in cefoxitin (**23**). However, these various substituents have different effects, and increased resistance to one enzyme does not indicate resistance to all β-lactamases. The 7α-methoxy, or formamido, group is the most effective in imparting a broad spectrum of resistance to β-lactamases, followed by acyl side-chain oxime ether substituents. Thus cefuroxime (**35**), the first oximino–cephalosporin, has a β-lactamase resistance spectrum similar to that of cefotaxime (**36**) whereas the less resistant cefamandole (**30**) is similar to cefoperazone (**34**).

The nature of the C-3 substituent influences the pharmacokinetic properties, as well as the intrinsic antibacterial activity. Analogues having a 3-methyl substituent generally have a low level of antibacterial activity except when the 7-acyl group is D-mandelic acid, D-phenylglycine, or a related moiety. Thus an α-hydroxyl or amino group on the side chain compensates, at least in part, for the lack of a nucleophilic leaving group on the C-3 methyl. The 7-phenylglycine analogues were the only known cephalosporins that combined *in vitro* activity and efficient oral adsorption. However, more recent work has shown that a C-3 vinyl substituent permits reasonable oral adsorption even for the oximino–aminothiazole side chain, eg, cefixime (**16**), and BMY28232 (**19**), and it appears likely that other C-3 substituents will be found to facilitate oral adsorption. Changes in the 3′-position are thought to primarily effect pharmacologic activity. However some improvement in antibacterial activity can also result from such substitution. Of particular note is the improvement seen using pyridinium and other quaternary-ammonium moieties, especially against the gram-positive organisms, and certain heterocyclic thio groups. A number of the newer cephalosporins, such as M14659 (**46**), have been rationally designed to take advantage of the so called *ton B* mediated iron transport system (*vide infra*) and show considerably improved activity against *Enterobacteriaceae* and *Pseudomonas aeruginosa* in particular (126).

Another interesting approach to obtaining potent, broad-spectrum activity has been reported (127). The "dual-action" antibacterial concept involves incorporation of two moieties having complimentary antibacterial modes of action into the same molecule, and uses the mode of action of one part to release the second antibacterial at the site of action. This approach is exemplified in Ro 23–9424 (**47**) (127), which uses the mode of action and reactivity of the cephalosporin moiety (Fig. 2) to release the quinolone portion.

Intrinsic Activity and Biochemical Targets. β-Lactam antibiotics affect sensitive bacteria by inhibiting late stages in the biosynthesis of their cell wall peptidoglycan (134–136). The cell wall contains a matrix of complex macromolecules that provide rigidity and mechanical stability by way of cross-linked latticeworklike structures. The peptidoglycan is one of these polymers, and is made up of glycan strands (consisting of alternating *N*-acetylglucosamine and *N*-acetylmuramic acid units) which are interconnected by peptide cross-links in a three-dimensional array. The last step in the construction of the peptidoglycan cross-link involves a transpeptidase-catalyzed reaction in which an amide bond is formed. In the case of the peptidoglycan from *Escherichia coli* (type A1γ-peptidoglycan) the cross-linking-peptide chain is a pentapeptide attached to the *N*-acetylmuramic acid residue of the glycan repeating subunit via the lactic acid side chain. The third amino acid in the pentapeptide is *meso*-diaminopimelic acid (DAP) which is anchored into the peptide via its L-center. The cross-linking step, referred to as transpeptidation, which the β-lactam antibiotics inhibit, is the process whereby the amino group of the D-center of *meso*-DAP forms an amide link to the penultimate D-alanine of another peptide with concomitant expulsion of the final D-alanine residue. Not all peptide chains are involved in the formation of cross-links, in *E. coli* the extent of cross-linking is about 33%, and the noncross-linked chains may be modified by the removal of the terminal D-alanine unit without transpeptidation. The enzyme responsible for this "pruning" is a D,D-carboxypeptidase (135,136).

β-Lactam antibiotics inhibit the cross-linking process by forming a covalent acyl-enzyme complex with transpeptidase or carboxypeptidase (Fig. 2). A serine hydroxyl function in the active site of these enzymes is implicated in the formation of the acyl-enzyme bond. The stability of the acyl-enzyme bond in the case of the transpeptidases is such that for all intents and purposes the enzyme is irreversibly inhibited. The carboxypeptidases, although intrinsically more sensitive to the β-lactam antibiotics in general, are reversibly inhibited because the acyl-enzyme bond is readily hydrolyzed to regenerate free enzyme (135,136).

Cell wall biosynthesis is a dynamic process and during cell division new cell wall must be synthesized (135–137). Thus there must also be some cleavage of certain peptide cross-links and/or glycan chains to allow for new cell growth and division. Hence the picture is rather more complicated than the basic mechanism just described and there is a group of enzymes which has an affinity for β-lactam antibiotics. These enzymes, the so-called penicillin-binding proteins (PBPs), are embedded in the inner, or plasma, membrane. These PBPs are designated according to decreasing molecular weight (Table 10) and are species specific. Hence PBP1 of *Escherichia coli* may not have anything in common with PBP 1 of any other organism. There are invariably multiple PBPs varying in number (from 3 in gonococci to 10 in *Escherichia coli*), molecular weight, and number of copies per cell. In the case of *Escherichia coli* there are seven important PBPs, of which four, namely 1A, 1B, 2, and 3 are considered essential for cell viability (134–136,138).

In a general sense all cephalosporins are thought to form a covalent acyl-enzyme complex involving the β-lactam carbonyl and a serine hydroxyl in the enzyme active site. However, as shown by competitive binding experiments, cephalosporins differ in their affinities for the various PBPs and it is well-established that preferential binding to the various PBPs produces different morphological effects (135–138). These effects are summarized in Table 10. Preferential binding to PBP 1B of *Escherichia coli,* as exhibited by cephaloridine [*50-59-9*] (**51**), $C_{19}H_{17}N_3O_4S_2$, produces rapid lysis with little malformation of the cells. Septum cross-wall formation is controlled by PBP 3, and cephalosporins, such as cefuroxime (**35**), that primarily inhibit this protein leave growing cells incapable of cell division. Each cephalosporin inhibits the various proteins to varying extents, and the intrinsic activity depends largely on the concentration of the compound required to saturate the most susceptible PBP. The 7α-methoxycephalosporins have considerable affinity for the nonessential PBPs 4, 5, and 6, whereas cephalosporins with no 7α-substituent bind very little if at all to these proteins. The binding of selected cephalosporins to the PBPs of *Escherichia coli* is detailed in Table 11.

Resistance. Resistance to the cephalosporins may result from the alteration of target pencillin-binding sites (PBPs), decreased permeability of the bacterial cell wall and outer membrane, or by inactivation via enzyme mediated hydrolysis of the lactam ring (80,81,138–140). This resistance can be either natural or acquired. Although resistance is often attributed specifically to one of these

Table 10. Penicillin Binding Proteins (PBPs) of *E. coli*[a]

PBP	Mol wt	Number per cell[b] (%)	Proposed *in vitro* functions	Effects of inactivation
1A	92,000	200 (6.5)	transglycosylase, minor transpeptidase, can partially compensate for loss of 1B activity; essential for cylindrical cell wall synthesis	rapid lysis if both 1A and 1B are inactivated
1B	90,000	250 (8)	transglycosylase, major transpeptidase of cell elongation; essential for cylindrical cell wall synthesis	rapid lysis
2	66,000	20 (0.7)	maintenance of cylindrical rod shape; requires transpeptidase?	spherical cell formation
3	60,000	50 (1.6)	transglycosylase–transpeptidase required septum cross-wall synthesis	nonseptate cells; filament formation
4	49,000	110 (3.6)	secondary transpeptidase, D,D-carboxypeptidase, or D,D-endopeptidase acting on maturing peptidoglycan	controls transpeptidation delayed transpeptidation is absent
5	42,000	1800 (59)	D,D-carboxypeptidase	none apparent
6	40,000	600 (19.6)	D,D-carboxypeptidase	none apparent

[a]Refs. 134–136, and 138.
[b]Number in parenthesis is the % of the total PBPs per cell.

Table 11. Selective Binding of Cephalosporins to the PBPs of *E. coli*[a,b]

Antibiotic	PBP-1A	PBP-1B	PBP-2	PPB-3	PBP-4	PBP-5	PBP-6
cephalexin (**12**)	4	240	250	8	30	250	250
cephalothin (**27**)	<0.25	16	37	1	60	125	130
cephaloridine (**51**)	0.25	2.5	50	8	17	250	250
cefotaxime (**36**)	0.05	0.7	5	0.05	30	50	50
cefoxitin (**23**)	0.1	3.9	>250	5.8	7.0	0.6	0.9

[a]Refs. 134, 136, and 138.
[b]Concentration required for 50% inhibition of binding of [^{14}C]pencillin G, μg/mL.

factors, in reality it reflects the interplay of several factors. In most instances, however, resistance results from the production of a β-lactamase enzyme, which opens the β-lactam ring as depicted in Figure 2.

Bacteria produce chromosomally and R-plasmid (resistance factor) mediated β-lactamases. The plasmid-mediated enzymes can cross interspecific and intergeneric boundaries. This transfer of resistance via plasmid transfer between strains and even species has enhanced the problems of β-lactam antibiotic resistance. Many species previously controlled by β-lactam antibiotics are now resistant. The chromosomal β-lactamases are species specific, but can be broadly classified by substrate profile, sensitivity to inhibitors, analytical isoelectric focusing, immunological studies, and molecular weight determination. Individual enzymes may inactivate primarily penicillins, cephalosporins, or both, and the substrate specificity predetermines the antibiotic resistance of the producing strain. Some β-lactamases are produced only in the presence of the β-lactam antibiotic (inducible) and others are produced continuously (constitutive).

Because of the highly permeable nature of the cell wall of gram-positive organisms, they produce β-lactamases which are not only found throughout the cell wall, but also in the extracellular environment. Hence the extracellular β-lactamases can act on the antibiotic before the cell is entered. Gram-negative organisms produce cell-bound β-lactamases which reside in the periplasmic space. Thus, for gram-negative bacteria, the antibiotic must penetrate the outer cell membrane/wall before coming in contact with a β-lactamase (80,139,140).

In general, the first-generation cephalosporins are readily hydrolyzed by the inducible, chromosomally mediated β-lactamases, and to a lesser extent, by the plasmid mediated enzymes such as the common TEM enzymes. Second- and third-generation cephalosporins have greater stability to hydrolysis by these better enzymes, although the emergence of resistant organisms during therapy using second- and third-generation cephalosporins is an increasingly serious medical problem (141). Resistance is most frequently reported in cases involving species which characteristically possess inducible β-lactamase enzymes (such as the so-called "overproducers" *Enterobacter* and *Serratia*) (142). Although it has also been proposed that some β-lactamases may also protect the enzyme-producing organisms from the effects of certain, mostly third-generation, cephalosporins by "trapping" or binding the drugs without actually hydrolyzing them (141,143), experiments have cast doubts as to the validity of this proposal (144).

Resistance can also arise when target enzymes, ie, the PBPs, and in particular the transpeptidases, are modified. Target-mediated cephalosporin resistance can involve either a reduced affinity for an existing PBP, or the acquisition of a supplementary, β-lactam insensitive PBP (139).

The third general mechanism for resistance is permeability barriers, or inaccessibility of the target because of transport problems. In gram-positive organisms, the PBPs are located outside of the cytoplasmic membrane and problems from permeability barriers is minimal. However the situation with gram-negative bacteria is considerably more complex, and penetration to the target can be a problem.

Transport and Cell Penetration. One of the causes of bacterial resistance to the cephalosporins is poor transport of the antibiotic through the outer membrane of gram-negative bacteria. This lipid-bilayer membrane carries receptor proteins for the recognition and transport of essential nutrients, but provides an effective barrier to large molecules. In the case of the cephalosporins there can be a considerable difference between the concentration required to inhibit intact cells and the concentrations required to saturate the target enzymes in broken cell preparations. For example, the concentration of cefoxitin (**23**) required to inhibit the growth of whole organisms is ca 100-fold that needed to saturate the essential PBPs, whereas using cefuroxime (**35**) the minimum inhibitory concentration (MIC) is only four times the concentration required to saturate the PBPs (145).

Many small molecules can penetrate the outer cell membrane by diffusion through channels created by outer-membrane proteins called porins. Porin channels are implicated in the transport of cephalosporins because cells deficient in porins are much more impermeable than are cells that are rich in porins. The porins appear to function as a molecular sieve, allowing molecules of relatively low molecular weight to gain access to the periplasmic space by passive diffusion. In enterobacteria, a clear correlation exists between porin quantity and cephalosporin resistance, suggesting that the outer membrane is the sole barrier to permeability. However, such a relationship is not clearly defined for *Pseudomonas aeruginosa,* where additional barriers may be involved (139,144,146).

Bacteria have evolved efficient acquisition systems to meet specific needs. In the case of the transport of iron, an essential trace element, an iron-chelating ligand (siderophore) is used to sequester the iron, and this siderophore-iron complex is then recognized by specific receptors on the outer membrane. This is known as the *ton B*-dependent iron transport pathway because it appears to be under the general control of the *ton B* gene in *Escherichia coli*. The most common siderophores are catechols, α-hydroxy carboxylic acids or *N*-alkylated hydroxamic acids. In an attempt to exploit the iron-transport pathway a number of cephalosporins have been made containing catechol or the isosteric 3-hydroxypyrid-4-one group (147–150). These compounds, such as E-0702 [*77768-57-1*] (**52**), $C_{29}H_{21}N_7O_{12}S_2Na_2$, (148) and M14659 (**46**) (150), have good activity against *Pseudomonas aeruginosa* and *Enterobacteriaceae* and experiments do indeed implicate the iron-transport mechanism in the ability of these catechol cephalosporins to penetrate to their targets.

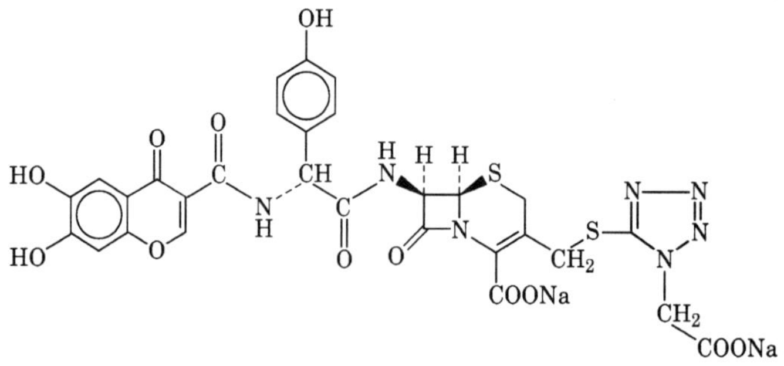

(**52**)

Pharmacokinetics. The pharmacokinetic properties of the cephalosporins depend to a large extent on the substituent at C-3. The 3′-acetoxy group is metabolized in the body to the less active 3′-alcohol. Most other substituents, including the 3′-carbamate, are metabolically stable. Most cephalosporins are eliminated rapidly, having serum half-lives of 1–2 hours. A number of the clinically useful cephalosporins contain the 1-*N*-substituted tetrazol-5-ylmercaptomethyl moiety in the C-3 position, eg, cefoperazone (**34**), moxalactam (**48**), cefmenoxime (**38**), cefamandole (**30**), and cefotetan (**26**). However, these compounds do not all have identical pharmacokinetic properties because there is always some interaction with, or contribution from, the 7-acyl group in determining the overall pharmacokinetic profile. When the tetrazoylmercaptomethyl 3′-substituent bears an acidic group such as found in cefonicid (**31**) or ceforanide (**33**), the rate of urinary excretion is reduced and higher, longer-lasting serum levels are obtained. Ceftriaxone (**39**), having an acidic group in its C-3 substituent, has extremely long-lasting serum levels. It appears that acidic C-3 substituents are associated with a very high level of serum binding which may contribute to the long-lasting serum levels. Most cephalosporins, both parenteral and oral, are renally excreted by glomerular filtration (free unbound drug) or tubular secretion (total drug). Some cephalosporins are readily secreted and have renal clearances via tubular secretion essentially independent of protein binding; others are eliminated predominantly by filtration, eg, ceftazidime (**40**), and exhibit renal clearance that is highly dependent on free drug concentration. A few of the cephalosporins, such as cefoperazone (**34**) and ceftriaxone (**39**) are also eliminated by biliary excretion. In such cases of dual excretion modes, the biliary excretion increases when renal function is impaired and, conversely, renal excretion increases when biliary excretion is impaired (80,151,152).

Manufacture and Chemical Synthesis

At present all of the cephalosporins are manufactured from one of four β-lactams, cephalosporin C (**2**), penicillin V [*87-08-1*], penicillin G [*113-98-4*], and cephamycin C (**8**), which are all produced in commercial quantities by fermentation (87). The manufacturing process consists of three steps: fermentation, isolation, and chemical modification.

Fermentation. The commercial β-lactam antibiotics which act as starting material for all of the cephalosporins are produced by submerged fermentation. The organisms used for the commercial production of the penicillins and cephalosporins are mutants of *Penicillin chrysogenum* and *Cephalosporium acremonium,* respectively (3,153,154). Both are true fungi (eucaryotes). In contrast, the cephamycins are produced by certain species of procaryotic *Streptomyces,* including *Streptomyces clavuligerus* and *Streptomyces lipmanii* (21,103).

Superior penicillin producing cultures are capable of producing in excess of 30 mg/mL of penicillin G (154). Cephalosporin producing strains, however, generally grow poorly and cephalosporin C production is not as efficient as is that of penicillin. Factors such as strain maintenance, strain improvement, fermentation development, inoculum preparation, and fermentation equipment requirements are discussed in the literature (3,154).

The β-lactam antibiotics are produced by secondary metabolic reactions that differ from those responsible for the growth and reproduction of the microorganism. In order to enhance antibiotic synthesis, nutrients must be diverted from the primary pathways to the antibiotic biosynthetic sequences. Although most media for the production of penicillins and cephalosporins are similar, they are individually designed for the specific requirements of the high yielding strains and the fermentation equipment used.

A typical fermentation medium for penicillin production contains lactose, corn steep liquor, and calcium carbonate (3,153,154). In most industrial processes the carbohydrate source, glucose, beet molasses, or lactose, is continuously added to the fermentation. The rate of glucose addition must be carefully monitored, by pH or rate of oxygen depletion, because the synthesis of penicillin is markedly reduced in the presence of excess glucose. Alternative sources of nitrogen, such as cottonseed, peanut, linseed, or soybean meal, have been substituted for corn steep liquor. Other ingredients that help to optimize penicillin production are monopotassium phosphate, for additional phosphorus and pH control, and sodium, ammonium or magnesium sulfate, for additional sulfur. Media for the large scale fermentative production of cephalosporin C generally contains fish meal, peanut meal or soybean meal, corn steep liquor, beet molasses, lard oil or methyl oleate, glucose, and methionine (155). In the cephalosporin medium, lipids, rather than glucose, provide the main source of carbon and energy. The sulfur for cephalosporin C is derived more efficiently from methionine than from sulfate. The yields of cephalosporin C are potentiated by both methionine and methyl oleate. The media and fermentation conditions for cephamycin C production are briefly discussed (3,103).

Isolation. Isolation procedures rely primarily on solubility, adsorption, and ionic characteristics of the β-lactam antibiotic to separate it from the large number of other components present in the fermentation mixture. The penicillins are monobasic carboxylic acids which lend themselves to solvent extraction techniques (154). Pencillin V, because of its improved acid stability over other penicillins, can be precipitated directly from broth filtrates by addition of dilute sulfuric acid (154,156). The separation process for cephalosporin C is more complex because the amphoteric nature of cephalosporin C precludes direct extraction into organic solvents. This antibiotic is isolated through the use of a combination of ion-exchange and precipitation procedures (157). The use of neutral, macroporous resins such as XAD-2 or XAD-4, allows for a more rapid elimination of impurities in the initial steps of the isolation (158). The isolation procedure for cephamycin C also involves a series of ion exchange treatments (103).

Chemical Modification. The chemistry and synthetic strategies used in the commercial synthesis of cephalosporins have been reviewed (87) and can be broadly divided into: (*1*) Selection of starting material; penicillin precursors must be rearranged to the cephalosporin nucleus; (*2*) cleavage of the acyl side chain of the precursor; (*3*) synthesis of the C-7 and C-3 side-chain precursors; (*4*) acylation of the C-7 amino function to introduce the desired acylamino side chain; (*5*) introduction of the C-3 substituent; and (*6*) protection and/or activation of functional groups that may be required.

For a viable commercial process, the selection of materials and the choice of synthetic route is governed primarily by cost, not by overall yield. The selection of starting material is dictated usually by the desired C-3 substituent. For cephalosporins containing 3-acetoxymethyl or 3-(substituted)methyl such as 3-thiomethyl and 3-aminomethyl derived moieties, the most direct synthetic route is from cephalosporin C, whereas pencillin V or G is the preferred starting material for the synthesis of the C-3 methyl cephalosporins. The three chemical transformations (*3*), (*5*), and (*6*) can potentially be carried out in a variety of ways, the precise sequence being determined by a balance of competing factors such as cost and optimization of yield (87).

Substituents at C-3. *3-Methyl Cephalosporins.* Direct reduction of the 3′-acetoxy group of compounds derived from cephalosporin C, either by catalytic hydrogenation (159) or by acid catalyzed reduction using trialkyl-silanes (160), leads to 3-methyl cephalosporins (87). However, the most economical route to this class of compounds involves the acid-catalyzed rearrangement of penicillin sulfoxides (161,162). Since the disclosure of this rearrangement in 1963, a tremendous number of papers dealing with the nature of the catalyst, solvent, optimal reaction parameters, and the reaction mechanism have appeared and been thoroughly reviewed (87,163–165). In addition, over twenty U.S. patents dealing with this process have been issued (87). The process requires that the carboxylic acid be protected, and high yields can be obtained by the correct choice of protecting group, solvent, and acid catalyst. The relatively stable *p*-nitrobenzyl, trichloroethyl, and benzhydryl esters of pencillin V or G sulfoxide undergo ring expansion best in either 1,4 dioxane or 1,1,2-trichloroethane with pyridine salts of phosphoric acid as the catalyst (166). This affords the cephalosporin with a carboxylic acid protecting group suitable for subsequent C-7 side-chain manipulation. Alternatively the carboxylic acid protecting group can be chosen such that deprotection occurs during isolation of the ring-expanded product. The best example is the use of trimethylsilyl protection. Thus reaction of penicillin G sulfoxide [*34104-15-9*] (**53**), $C_{16}H_{18}N_2O_5S$, with pyridinium hydrobromide (PyHBr) and bistrimethylsilylurea [*18297-63-7*] (BSU) in toluene followed by hydrolysis of the silyl ester affords the cephalosporin [*27255-72-7*] (**54**) in 97% yield (87,167).

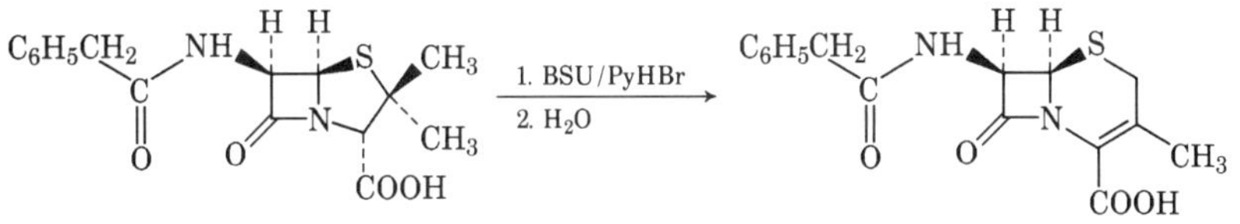

(**53**) (**54**)

3′-Heteroatom Cephalosporins. The most direct synthetic route to this class of compounds utilizes cephalosporin C (**2**) as the starting material, and thus requires the introduction of a new 3′-substituent. The 3′-acetoxy group can be readily displaced using a variety of nitrogen, sulfur, and carbon nucleophiles under both aqueous and nonaqueous conditions (75,87,168). However, oxygen nucleophiles do not readily displace the acetoxy group. These displacement reactions have been extensively studied, and tend to follow S_N kinetics (75,168). In fact under aqueous conditions, mechanistic studies show that displacement occurs via an S_N process with carboxylate stabilization of the intermediate carbonium ion (75,87,169,170). These displacement reactions generally cannot be carried out on cephalosporin esters, lactones, or sulfoxides.

Nitrogen nucleophiles used to diplace the 3′-acetoxy group include substituted pyridines, quinolines, pyrimidines, triazoles, pyrazoles, azide, and even aniline and methylaniline if the pH is controlled at 7.5. Sulfur nucleophiles include alkylthiols, thiosulfate, thio and dithio acids, carbamates and carbonates, thioureas, thioamides, and most importantly, from a biological viewpoint, heterocyclic thiols. The yields of the displacement reactions vary widely. Two general approaches for improving 3′-acetoxy displacement have been reported. One approach involves initial, or *in situ,* conversion of the acetoxy moiety to a more facile leaving group. The other approach utilizes Lewis or Brønsted acid activation (87).

Conversion to a more facile, sulfur-derived, leaving group can be achieved by treatment with sodium thiosulfate or salts of thio and dithio acids (75,87).

Under anhydrous conditions, boron tribromide converts the 3′-acetoxy group to a bromide whereas trimethylsilyl iodide gives good yields of the 3′-iodide (87,171,172). These 3′-halides are much more reactive, even when the carboxyl group is esterified, and can be displaced readily by cyano and by oxygen nucleophiles (127).

Methane sulfonic acid, trifluoroacetic acid, hydrogen iodide, and other Brønsted acids can facilitate 3′-acetoxy displacement (87,173). Displacement yields can also be enhanced by the addition of inorganic salts such as potassium thiocyanate and potassium iodide (174). Because initial displacement of the acetoxy by the added salt does not appear to occur, the role of these added salts is not clear. Under nonaqueous conditions, boron trifluoride complexes of ethers, alcohols, and acids also facilitate displacement (87,175).

An alternative to the displacement of a 3′-acetoxy group for entry into the 3′-heteroatom cephalosporins is functionalization of a 3-methyl cephalosporin. The low cost of the 3-methyl cephalosporins, derived from penicillin via the Morin rearrangement, makes this approach potentially competitive and attractive. Allylic bromination, using *N*-bromosuccinimide [*128-08-5*] (NBS) and photochemical initiation, of the esters of 3-methyl-Δ-3-cephem sulfoxides affords the desired 3′-bromo compound (87,176,177), although this reaction does not appear to work on the corresponding sulfide (75). Bromination can be carried out with a variety of substituents on the C-7 amino, but can suffer from significant bromination at C-2. However, the competing bromination at C-2 appears to be related to the nature of the substituent on the C-7 amino. Of particular note, the trichloroacetylamino group at C-7 completely suppresses the undesired C-2 bromination (176). Figure 3 outlines a particularly interesting application of this chemistry. Overbrominating, to afford a mixture of the 3′-monobromo compound [*71543-00-5*] (**55**) and corresponding C-2,3′-dibromo compound [*71542-99-9*] (**56**) is deliberate. This mixture is then treated with trimethylphosphite to effect selective reduction at C-2, and the resulting 3′-monobromo cephalosporin (**55**) can then be subjected to nucleophilic displacement (87,178).

3-Hydroxymethyl Cephalosporins. The 3-hydroxymethyls are a special case of the 3′-heteroatom cephalosporins. These deacetylcephalosporins serve as precursors for the 3′-carbamoyl cephalosporins, such as cefuroxime (**35**) and cefoxitin (**23**). Treatment of a cephalosporin acid or ester to hydrolytic deacylation under basic conditions often suffers from the problems of double bond migration and/or β-lactam cleavage. However, deacylation of 7-aminocephalosporanic acid (7-ACA) (**1**), derived from cephalosporin C (**2**) by treatment with sodium hydroxide at 0°C affords the corresponding 3′-hydroxy compound in good yield (64%) (87,179,180). This deacylation can also be achieved by enzymatic hydrolysis, using a variety of esterases and substrates (75,179–182). The mild selective conditions of enzyme catalyzed cleavage make this the process of choice for industrial production and such a process using a yeast, *Rhodotorula rubra* as the source of esterase activity is employed in the commercial synthesis of cefuroxime (**35**) (87,182).

3-Methylene Cephalosporins and 3-Heteroatom Cephalosporins. A number of clinically important cephalosporins have a hydrogen or a heteroatom at C-3 instead of a methyl or substituted methyl, requiring removal of the C-3 methyl from either of the commercially available starting materials. Removal is conven-

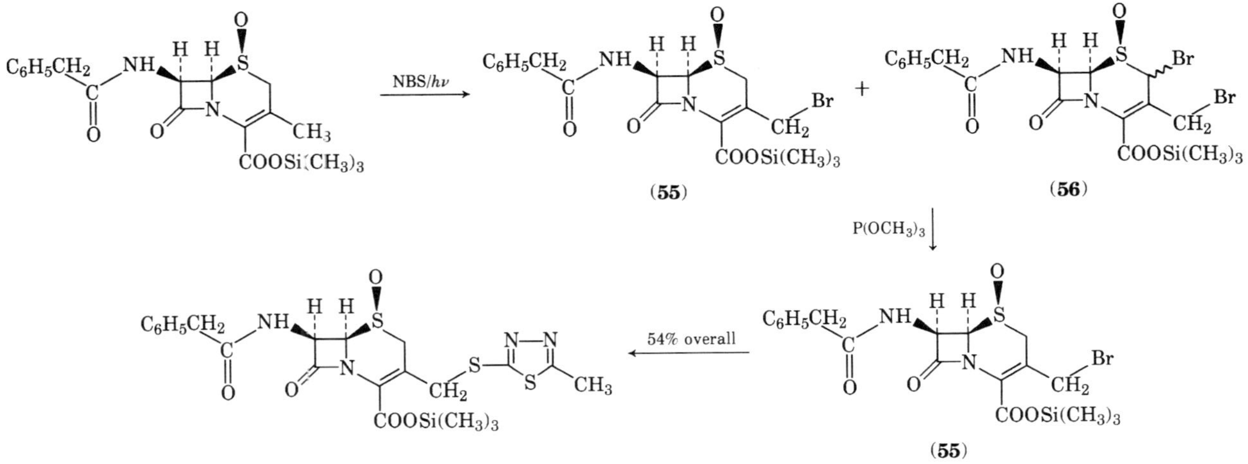

Fig. 3. Allylic bromination at C-3′; NBS = *N*-bromosuccinimide.

iently achieved via ozonolysis of an exocyclic methylene group, and a number of routes have been developed to produce such 3-methylene cephalosporins from both penicillins and cephalosporins derived from cephalosporin C. Industrially, the most viable routes to 3-methylenecephams are via penicillins, and thus the penicillins serve as the precursors for the C-3 heteroatom cephalosporins (87). This chemistry has been reviewed rather extensively, along with the subsequent conversion into heteroatom substituted cephalosporins (164,183–185). One of these routes takes advantage of advances in penicillin sulfoxide chemistry and involves the rearrangement of the sulfoxide to the 3-exomethylenecepham. This is exemplified in the synthesis of cefaclor (**13**) shown in Figure 4 (185,186).

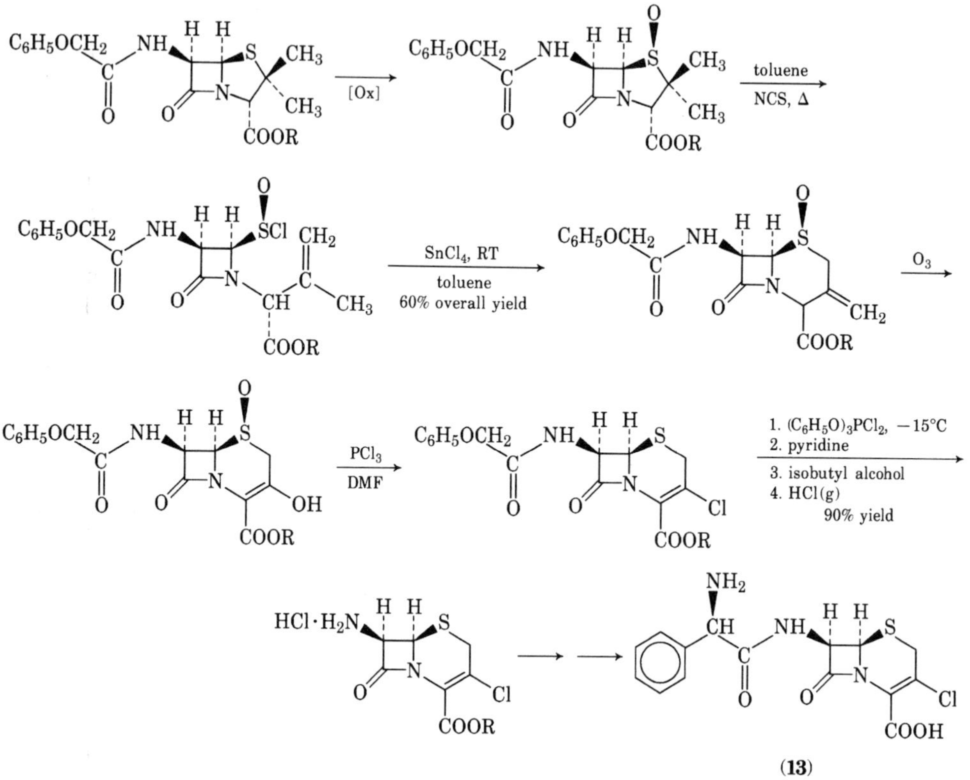

(**13**)

Fig. 4. Synthesis of cefaclor (**13**) where DMF is dimethylformamide.

7β-Substituents. *C-7 Side-Chain Cleavage.* 7-ACA (**1**), the key intermediate for the synthesis of a large number of the clinical cephalosporins, cannot be made by fermentation, nor can it be readily obtained by enzymatic removal (187) of the aminoadipic acid side chain from cephalosporin C. Because substantial quantities of 7-ACA are used, an efficient method for the chemical cleavage of the acyl side chain of cephalosporin C was sought. The first efficient method involved diazotization using nitrosyl chloride, followed by hydrolysis or alcoholysis, and gave 7-ACA in 50% yield (22,23,87,188). The reaction proceeds via the formation of the cyclic imino ether (**57**) which is easily hydrolyzed under very mild conditions.

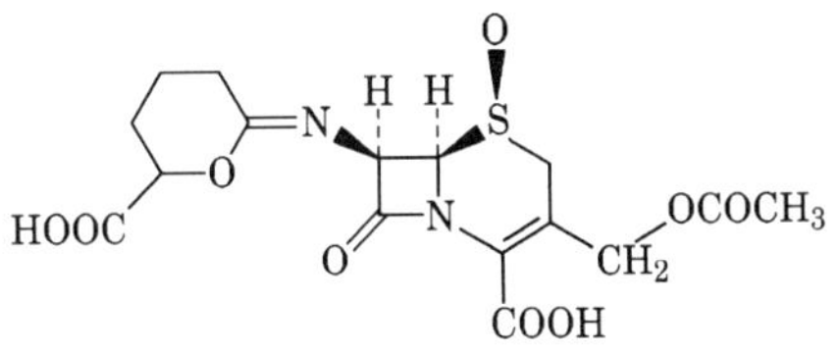

(57)

Subsequently, a general procedure for amide side-chain cleavage was developed that involves initial conversion of the amide linkage into an imino chloride, achieved under anhydrous conditions using phosphorus trichloride, or more preferably phosphorus pentachloride. Alcoholysis to the imino ether, followed by hydrolysis affords 7-ACA (22,23,87,189). The process requires protection of both the free carboxyl and the amino group. This can be achieved either by reaction with acetyl chloride before cleavage or by silylation of the anhydrous sodium salt as shown in Figure 5. In the latter case, the overall yield of 7-ACA is 92.5%. A procedure for the removal of the acyl-amino side chain of cephalosporin C, using an immobilized enzyme has been reported (187). However, this process requires either the prior removal, or protection, of the adipoyl side-chain amino group, before enzymatic cleavage of the amide bond can be effected.

C-7 Amino Acylation. Essentially all the known methods for amine acylation, used in the peptide field, have been applied to cephalosporins. These procedures generally involve acyl group activation prior to the acylation (76). Commercially, the most common activation probably involves either mixed anhydride or acid chloride formation (87). Acylations via acid chloride activation are usually carried out under Schotten-Bauman conditions eg, aqueous acetone, 0°C, sodium bicarbonate, pH 7–8. A wide variety of active esters have also been used in cephalosporin chemistry, especially heterocyclic-thio esters and *o*-acyl substituted hydroxylamines, either preformed or generated *in situ,* as well as coupling methods employing dicyclohexylcarbodiimide (DCC) and similar condensing agents. The use of active esters in the acylation step is exemplified by the synthesis of ceftriaxone (**39**). In order to circumvent the poor solubility of 7-ACA in organic solvents, prior formation of a soluble derivative, usually an ester, is required for most activation methods using nonaqueous conditions. Silylation of 7-ACA meets this requirement and the silylated moiety has the advantage of easy removed during the workup (87). Other commonly used esters include benzhydryl, *t*-butyl, and less frequently *p*-nitrobenzyl, *p*-methoxybenzyl, trichloroethyl, and allyl. However, the use of esters in the acylation step, under conditions requiring an organic base, often leads to the problematic isomerization of the ring double bond (75).

7α-Substituents. A large number of cephalosporins with a variety of substituents in the 7α-position have been prepared and their biological properties and antibacterial activities evaluated. The substituents introduced into the 7α-position include groups such as CH_3–, CH_3CH_2–, –CH_2OH, –CH_2F, –CH_2NH_2, –COOH, –CHO, –$COCH_3$, –OCH_3, –OCH_2CH_3, –OCH_2CN, –$OCH_2CH_2OCH_3$, –SCH_3, –CN, –Br, –SCH_3, –$CH(COOCH_2CH_3)_2$, –N_3, –NHCHO, –$PO(OCH_3)_2$, and –SCH_3 (25–30). Of these 7α-substituted cephalosporins, only those with the methoxyl and formamido group have interesting antibacterial activity. A number of

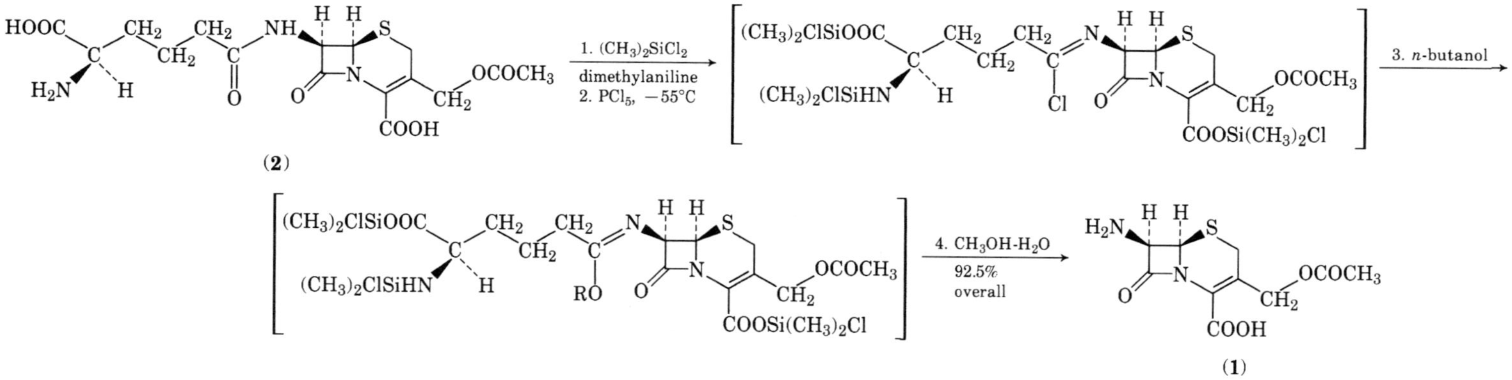

Fig. 5. C-7 side-chain cleavage.

synthetic routes to such compounds have been described and reviewed (27). The key steps for the introduction of the 7α-substituent in three of these approaches to the 7α-methoxy substituted cephalosporins are depicted in Figure 6 (27). Figure 6e outlines the introduction of the 7α-formamido substituent (28–30).

Synthetic Strategy. For a commercial process the deciding factor in a synthetic strategy is overall cost, although other factors, such as patent rights and material and facility availability also play a role. Generally the goal is to introduce the more expensive materials as late in the total synthesis as possible (87), and to use a convergent rather than a long linear approach. These factors generally favor complete construction of side chains before attachment to the cephalosporin nucleus and most routes follow this approach. The variety of routes to some of the more common, newer C-3 and C-7 side chains have been compared in a review (87). The side chains may also be built up on the cephalosporin nucleus, generally 7-ACA, and furthermore, the sequence by which side chains are either built up or introduced intact, can also be varied. The syntheses of cefotiam (**50**), ceftazidime (**40**), and ceftriaxone (**39**) are described by the reaction sequences given in Figures 7–9, respectively. These are not necessarily the commercial, or the most efficient, routes.

The schemes depicted in Figure 7 contrast two complimentary approaches to cefotiam (**50**) in which the timing of the introduction of the C-3 substituent differs. In Route A the heterocyclic thiol C-3 substituent is introduced even before the removal of the aminoadipoyl acyl side chain. The acetonylacetyl C-3 substituent was introduced because it gave considerably higher yields and cleaner product in the nucleophilic displacement step than the corresponding acetoxy, and the starting material, deacetylcephalosporin C (**5**) was readily available from a fermentation process (190,191).

In the synthesis of ceftazidime (**40**) (Fig. 8), the protected, preassembled aminothiazole side chain [*68672-66-2*] (**60**) is coupled to a protected 7-ACA first and the C-3 displacement step carried out last. By way of contrast, in the synthesis of ceftriaxone (**39**) (Fig. 9), the preformed C-3 substituent is introduced onto the cephalosporin nucleus in the first step and then the acyl-amino side chain is introduced. This last step is noteworthy for two reasons in that it demonstrates the use of an activated thio ester in the coupling step and that no protecting group chemistry is required (192,193).

The substitution of oxygen for sulfur, as found in the 1-oxadethiacephems such as moxalactam (**48**), introduces a new synthetic problem in that the oxadethiacephem nucleus is not naturally occurring. Synthetic routes to the oxadethiacephems have been investigated fairly extensively (194,195) and in all the routes published the key step is the stereocontrolled introduction of the ring oxygen function. The synthesis of moxalactam (**48**), depicted in Figure 10, efficiently retains all the carbons of its starting material, 6-aminopenicillanic acid [*551-16-6*] (6-APA) (**61**), $C_8H_{12}N_2O_2S$ (194–196).

Nuclear Analogues of Cephalosporins

In the search for improved antibacterials not only has the effect produced by the variation of the C-7 amido side chain and the 3′ substituent been studied, but so also has the more synthetically challenging question of the effect of changes in

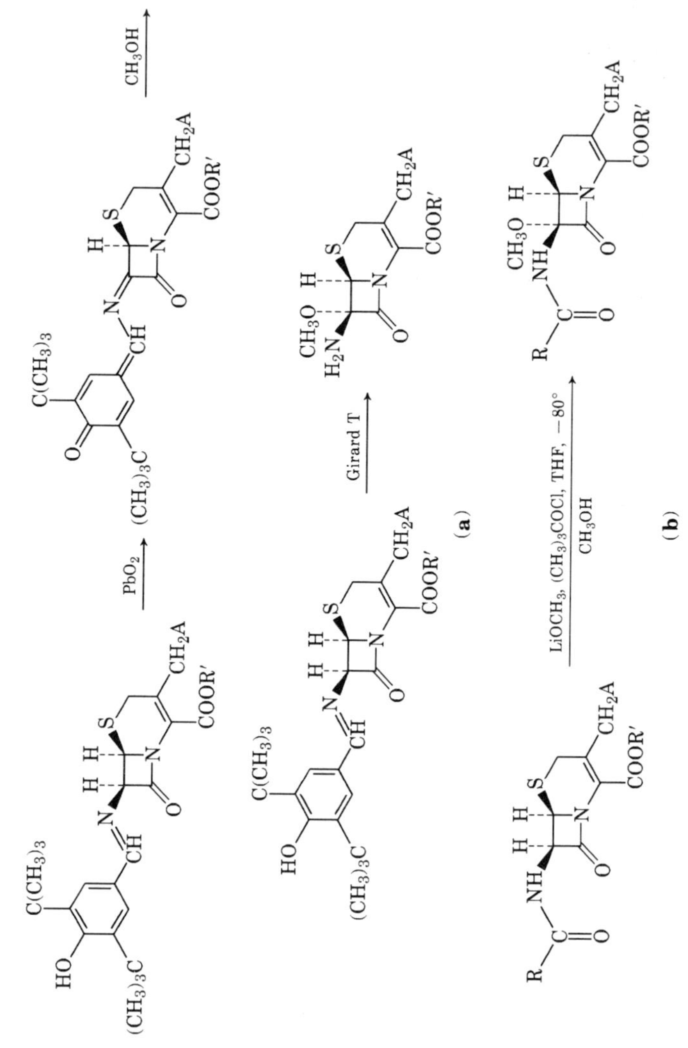
PbO2
CH3OH
Girard T
(a)
LiOCH3, (CH3)3COCl, THF, −80°
CH3OH
(b)

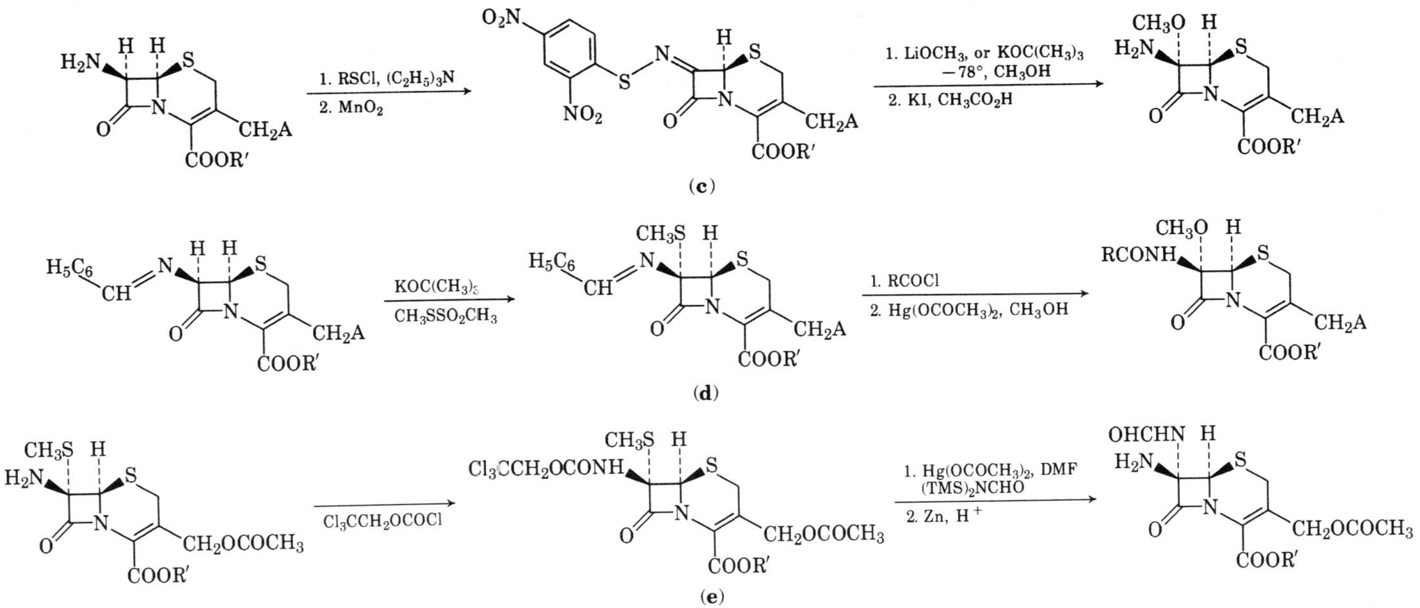

Fig. 6. Synthetic routes to (**a**)–(**d**) 7α-methoxycephalosporins and (**e**) 7α-formamidocephalosporins where THF = tetrahydrofuran, and DMF = dimethylformamide, and TMS = trimethylsilane.

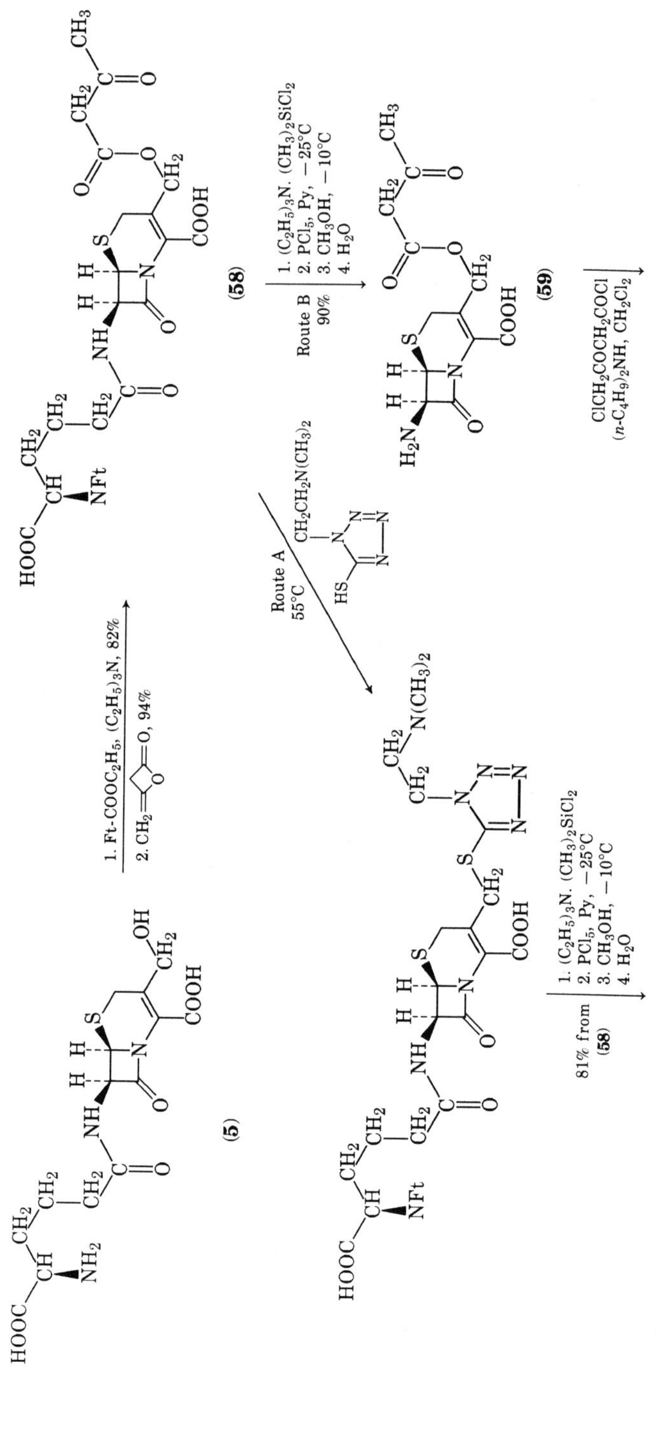
(5)
1. Ft-COOC2H5, (C2H5)3N, 82%
2. CH2=C—O, 94%
(58)
Route A
55°C
HS
CH2CH2N(CH3)2
Route B
90%
1. (C2H5)3N. (CH3)2SiCl2
2. PCl5, Py, −25°C
3. CH3OH, −10°C
4. H2O
(59)
ClCH2COCH2COCl
(n-C4H9)2NH, CH2Cl2
81% from
(58)
1. (C2H5)3N. (CH3)2SiCl2
2. PCl5, Py, −25°C
3. CH3OH, − 10°C
4. H2O

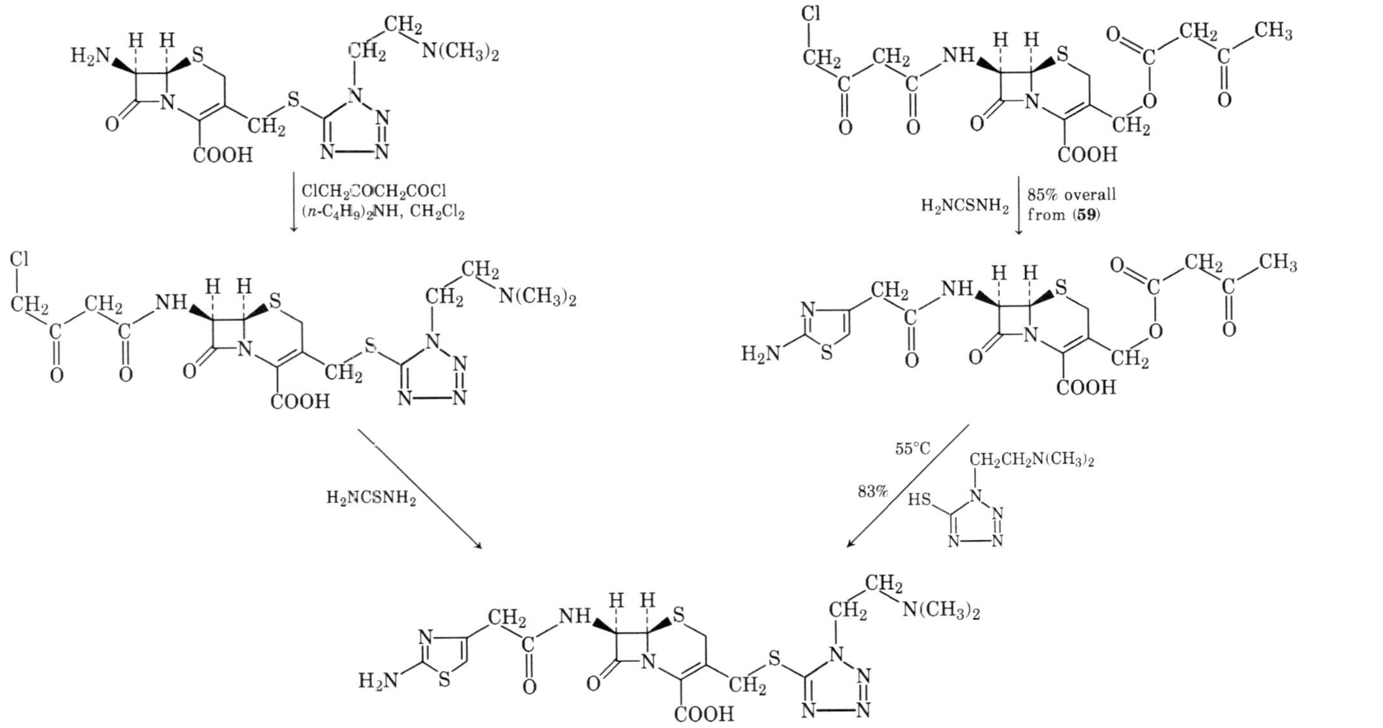

Cefotiam (**50**)

Fig. 7. Synthesis of cefotiam (**50**) where Ft is phthalimido and Py is pyridine, C_5H_5N.

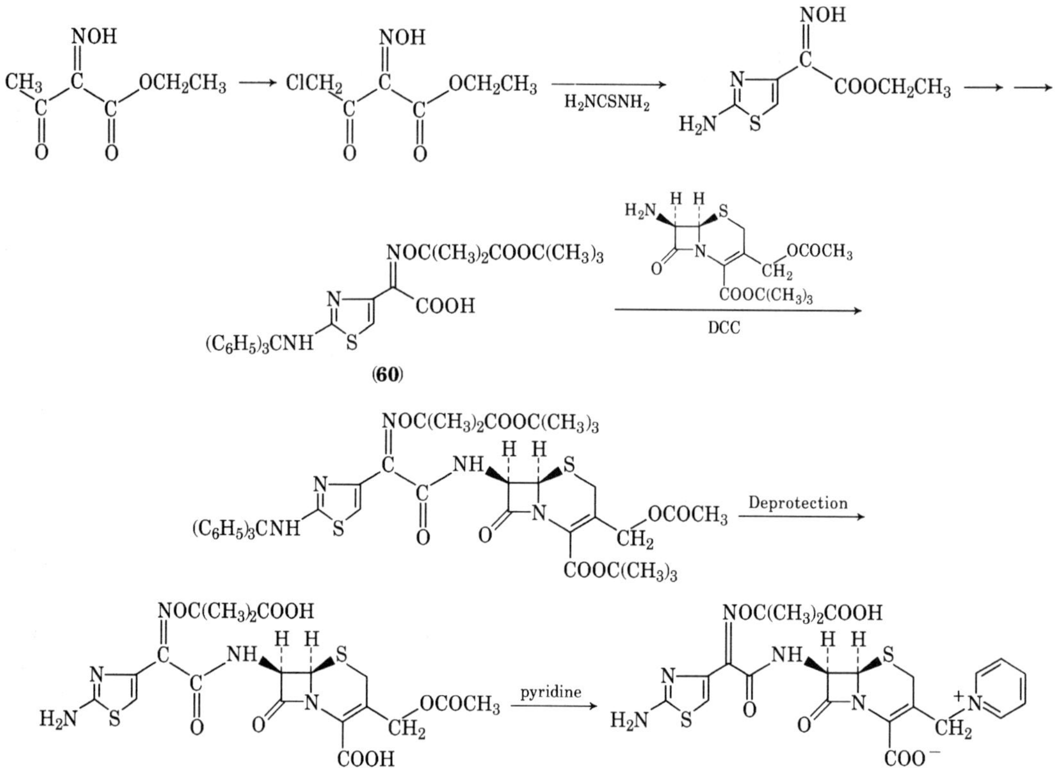

Ceftazidime (**40**)

Fig. 8. Synthesis of ceftazidime (**40**); DCC = *N,N*-dicyclohexyl-carbodiimide.

the cephem nucleus (194,197,198). Nuclear analogues have been studied since the early 1970s but only the oxacephem class has reached the marketplace.

Oxadethiacephalosporins. The first compound in which the sulfur of a cephalosporin was replaced by oxygen was 1-oxa-dethiacephalothin [*54214-83-4*] (**64**), $C_{16}H_{16}N_2O_7S$, (199) which is approximately twice as active as cephalothin (**27**) taking into account that the synthetic material was racemic. The oxygen analogue [*62504-53-4*], (**65**, X = OH, R = 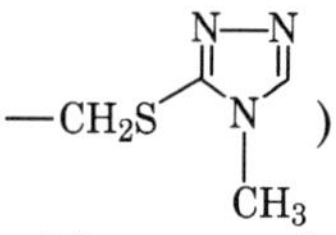) of cefamandole (**30**) was as good as, or better than, cefamandole itself, except for activity against *Proteus mirabilis* (200). Other studies have demonstrated that the 1-oxa-dethiacephalosporins are for the most part comparable to, or slightly better than, the cephalosporin counterpart in antibacterial activity, although there are some exceptions, such as 1-oxa-dethiacephalexin [*66204-07-7*] (**65**, X = NH_2, R = H), $C_{15}H_{15}N_3O_5$, which was inactive (201). Both the 1-oxa-dethiacephalosporins having a mandelamido C-7 side chain and either a methyl (**65**, X = OH, R = CH_3) or a hydrogen (**65**, X = OH, R = H) at C-3 had properties comparable to cephalexin (**12**) (201). The effect of various substituents at C-3, C-7β, acylamino side chains, and C-7α, methoxy and hydrogen, in the 1-oxa-dethiacephalosporin system have been exten-

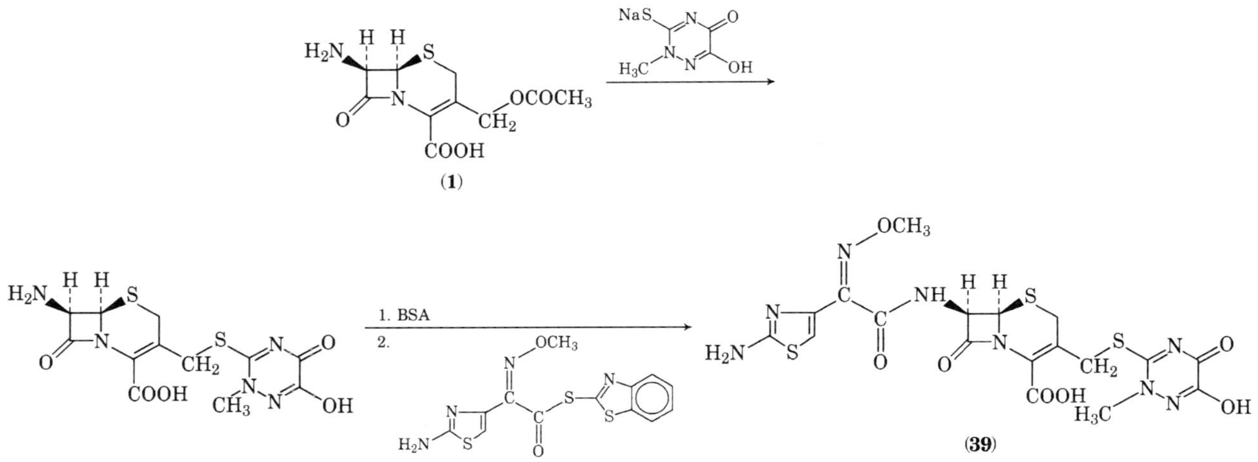

Fig. 9. Synthesis of ceftriaxone (39) where BSA is N,O-bis (trimethylsilyl) acetamide.

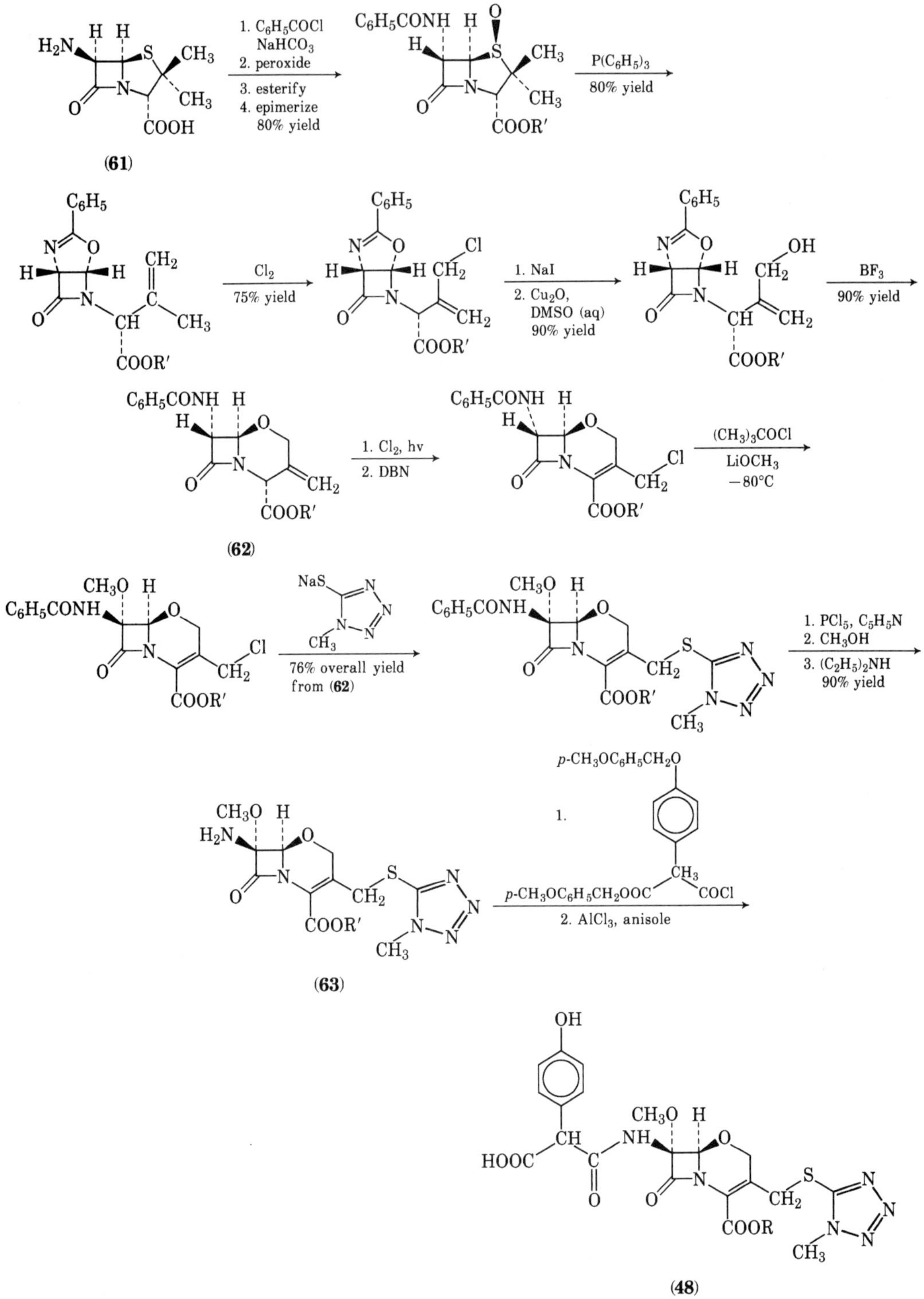

Fig. 10. Synthesis of moxalactam (48) from 6-aminopenicillanic acid (6-APA) where R′ is $CH(C_6H_5)_2$, DMSO is dimethyl sulfoxide and DBN is 1,5-diazabicyclo[4.3.0]non-5-ene. The overall yield of (63) is 26% from 6-APA in the 13 steps shown.

sively studied (194). These structure-activity investigations led to the development of moxalactam (**48**) (194,195). The substitution of oxygen for sulfur in moxalactam (**48**) affords a several-fold increase in intrinsic activity, but introduces the synthetic challenge of the oxacephem nucleus.

The antibacterial spectrum of moxalactam (Table 9) is similar in breadth and potency to that of cefotaxime (**36**). Hence, moxalactam (**48**) is classified with the third-generation cephalosporins. In general 1-oxacephalosporins are considerably more susceptible to β-lactamases than their sulfur counterparts (194). However, moxalactam is highly resistant to a broad spectrum of β-lactamases owing to the effect of the 7α-methoxy group (194,195).

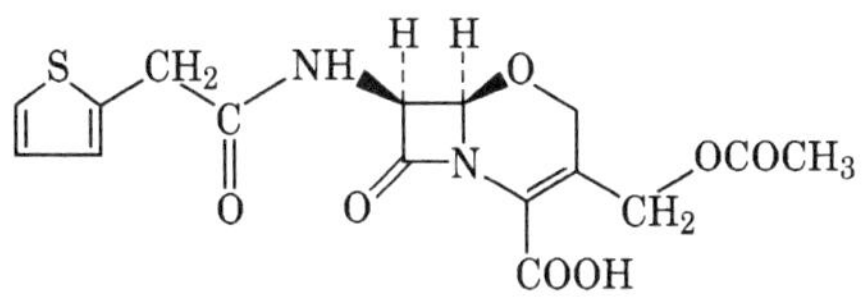

(**64**)

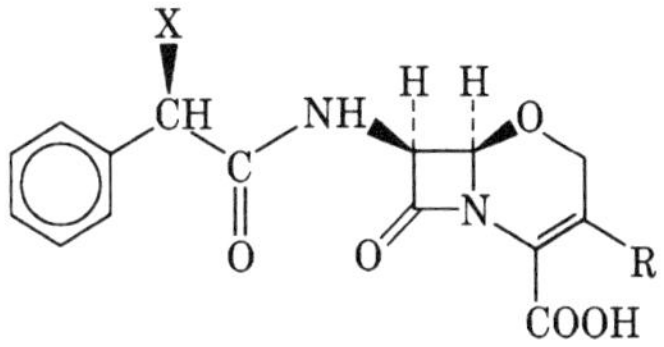

(**65**)

Isocephems. 1-Dethia-2-thia-cephalosporins, more conveniently known as isocephems, have attracted interest mostly from a synthetic perspective (197), but analogues have been found, eg, (**66**, X = S, Y = 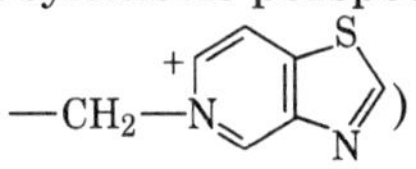) which exhibit good antibacterial activity *in vitro* (198). Structure-activity studies in the series clearly demonstrate the need for a polar, preferably positively charged potential

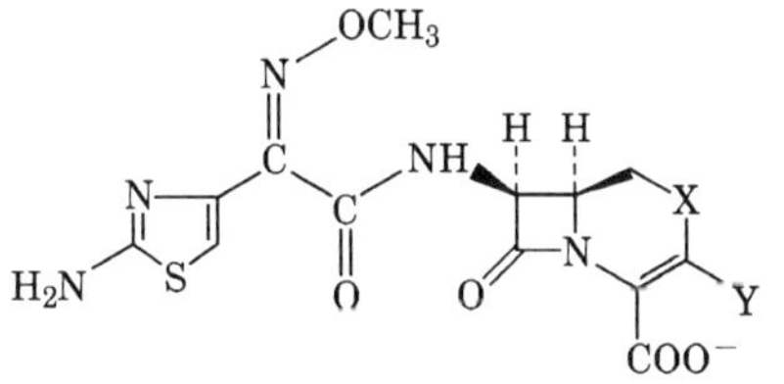

(**66**)

leaving group in the 3′-position to obtain good antibiotic activity. The insertion of a *z*-double bond between the 3- and 3′-position of the isocephems, eg, [*121037-46-5*] (**66**, X = S, Y = 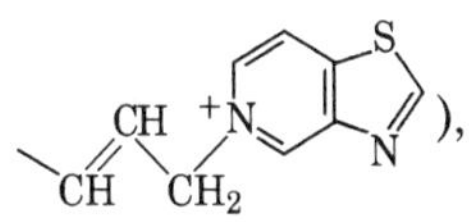), profoundly effects the antibacterial potency (198,202) especially for the gram-positive bacteria where the improvement in potency can be 10-fold. The corresponding 2-oxa-isocephems have also been investigated, eg, [*123581-13-5*] (**66**, X = O, Y = 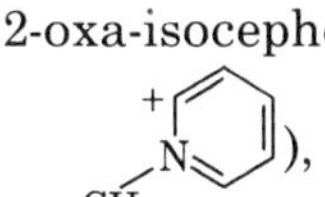), but were found to not be quite as potent as the corresponding isocephem and more susceptible to β-lactamase inactivation (198).

Carbacephems. The carbacephems in which the sulfur group is replaced by a methylene, were introduced in the 1970s (197,200,203,204). Generally these 1-carba analogues possess antibacterial activity that is similar to the cephalosporin counterpart (197,204), and have the advantage of possessing greater chemical stability (204). Thus, for example, locarbacef [*76470-66-1*] (LY163892 or KT3777) (**67**), $C_{11}H_{16}N_2O_4Cl$, has antibacterial activity similar to that of cefaclor (**13**) (205), but is 140 times more stable than cefaclor in pH 7.4 buffer (206). Furthermore, human pharmacokinetic studies demonstrate that locarbacef has a longer half life and better oral bioavailability than cefaclor (207). A new series of 3-acyl-1-carbacephems with promising activity and pharmacokinetic properties has been reported (208). Carbacephems having an iminoaminothiazole acyl side chain and a quaternary nitrogen group directly attached to the ring in the C-3 position, eg, LY258360 [*125615-06-7*] (**68**), $C_{18}H_{19}N_7O_5S$, are reported to exhibit improved potency compared to the third-generation cephalosporins, with the exception of their poor activity against *Pseudomonas* (209).

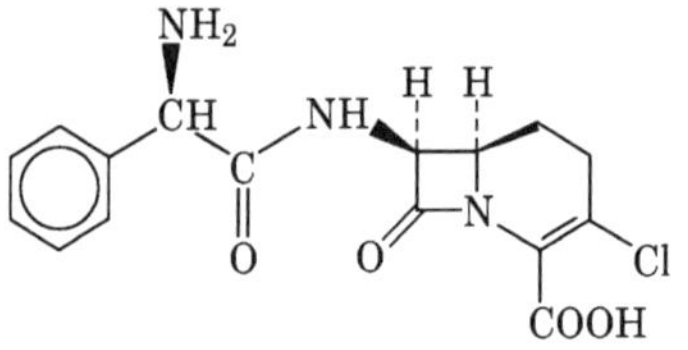

(**67**)

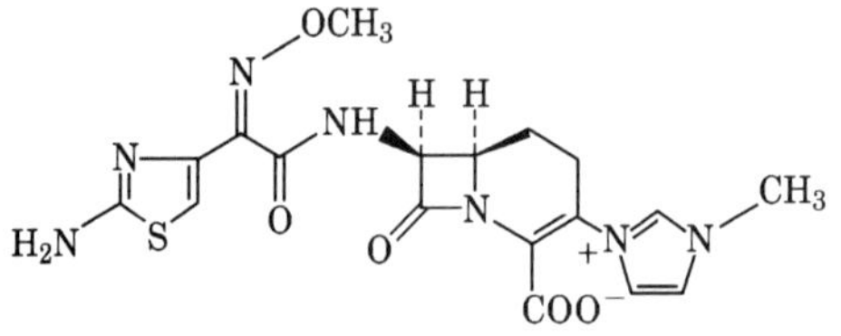

(**68**)

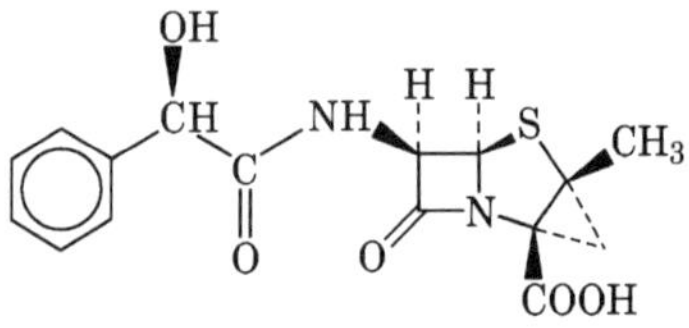

(**69**)

α-Cyclopropylpenams. Although at first glance (2,3)-α-methylenepenams such as [*109279-36-9*] (**69**), look like penicillins, these compounds can be considered as nuclear analogues of the cephalosporins. The profile of their antibacterial activity and their mode of action support this view. The penicillin nucleus in these compounds is locked in the open conformation, which is the proposed biologically active form for pencillins. However, the absolute configuration at the carboxylic acid bearing carbon (C-2) is opposite to the naturally occurring penicillins, and places the carboxyl group in a position similar to that found in cephalosporins. Thus the compounds exhibit an antibacterial profile closer to those of their cephalosporin analogues than to that of the penicillin counterpart. Consistent with this, the compounds in this series show a marked selectivity for PBP-3 of *Escherichia coli,* and are more susceptible to cephalosporinases than penicillinases (210).

Cephalosporins with Special Properties

Chromogenic Cephalosporins. A 3-substituted pyridinium cephalosporin, known as PADAC [*77449-91-3*] (**70**), $C_{27}H_{26}N_6O_4S_2$, is purple in color but on hydrolysis or treatment with β-lactamase releases the 3′-pyridinium group with concomitant loss of the purple color (211). PADAC is an example of the chromogenic cephalosporins which are useful in studying interactions with β-lactamases. Another example is nitrocefin [*41906-86-9*] (**71**), $C_{21}H_{16}N_4O_8S_2$, which is light yellow in solution, $\lambda_{max} = 386$ nm. On hydrolysis, or treatment with β-lactamase, the color changes to a deep red, $\lambda_{max} = 482$ nm, but unlike the situation with PADAC the nitrocefin 3′ group is not expelled (212) (see CHROMOGENIC MATERIALS).

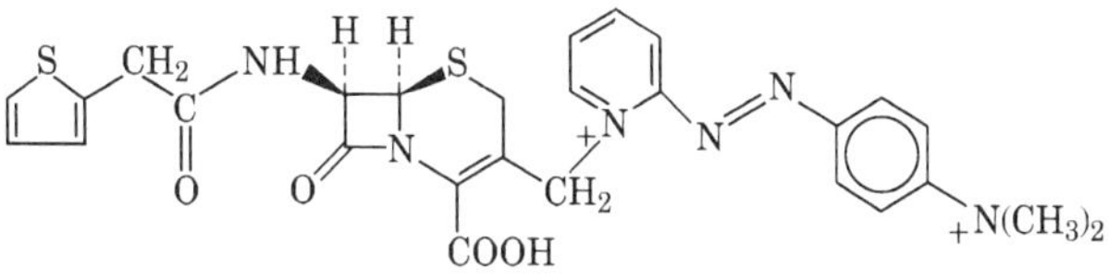

(**70**)

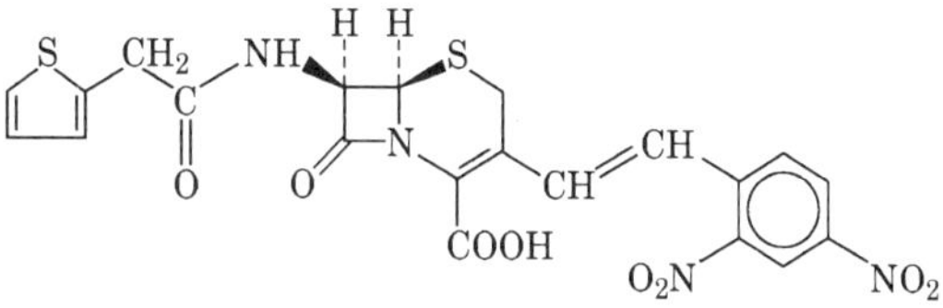

(**71**)

Uses

The cephalosporins are used for treating infectious diseases of bacterial origin in both humans and animals (80,132,213–218). First-generation cephalosporins such as cephalothin (**27**) and cephalexin (**12**) are the most active against staphylococci and nonenterococcal streptococci and are effective alternatives to the penicillins in patients with endocarditis, osteomyelitis, septic arthritis, and cellulitis (213). They are especially useful for treating patients who are allergic to the pencillins or who have mixed infections from gram-positive and gram-negative bacteria. Although these drugs have proved useful in treating infections such as bacteremias, urinary tract infections, and pneumonias, caused by gram-negative bacilli, their use as single agents in this regard is not recommended, because activity against gram-negative organisms is somewhat weak and unpredictable. The first-generation cephalosporins have been widely used for prophylaxis in cardiovascular, orthopedic, biliary, pelvic, and intra-abdominal surgery (132,214). Cefazolin (**32**), which has a longer half-life than the other first-generation compounds, is the first-generation agent of choice for surgical prophylaxis (132,213,214).

Cefuroxime (**35**) is effective against community-acquired pneumonia in which ampicillin-resistant *Haemophilus influenzae* is the probable etiologic agent. Cefoxitin (**23**) is used to treat mixed aerobic–anaerobic infections including pelvic infections, intra-abdominal infections, and nosocomial aspiration pneumonia. Cefonicid (**31**), because of its long half-life has been used in a once-a-day regimen to treat a variety of mild to moderate infections including community-acquired pneumonias, urinary tract infections, and infections of the skin and soft tissue (132,215).

Whereas third-generation cephalosporins do have some coverage against gram-positive infections, they are not the agents of choice. Similarly, most community-acquired infections are better treated with drugs other than the third-generation cephalosporins. The treatment of meningitis is an important exception (132,216). Cefotaxime (**36**), ceftriaxone (**39**), and ceftazidime (**40**) have proven effective in treating meningitis, especially in children where *Haemophilus influenzae, Streptococcus pneumoniae,* and *Neisseria meningitidis* are the leading etiologic agents. In meningitis caused by enteric gram-negative bacilli (community-acquired or nosocomial) the third-generation cephalosporins are clearly the drugs of choice. Ceftazidime (**40**) has even been used to treat patients with *Pseudomonas aeruginosa* meningitis with an 80% response rate.

The third-generation cephalosporins are effective in the treatment of bacteremias, pneumonias, urinary tract infections, intra-abdominal infections, and skin and soft tissue infections. Because of its long half-life, cetriaxone (**39**) has been used for outpatient treatment of skin and soft tissue infections. Single dose ceftriaxone (**39**) has also been effectively used in the treatment of urinary tract infections (132,215). Ceftriaxone is now the agent of choice for treating uncomplicated urethral, anorectal, or pharyngeal gonorrhea, and gonococcal opthalmia, including infections caused by penicillin-resistant strains (132,217). There is no clearly defined role that cephalosporins play in the treatment of syphilis, mycoplasma or clamydial infections, or bacterial vaginosis, but ceftriaxone may be effective in the treatment of chancroid.

Toxicity

The cephalosporins generally cause few side effects (80,132,219–221). Thrombophlebitis occurs as a result of intravenous administration of all cephalosporins. Hypersensitivity reactions related to the cephalosporins are the most common side effects observed, but these are less common than found with the penicillins. Clinically only about 5–10% of patients with allergic reactions to the penicillins manifest the same reactions to the cephalosporins, and data would contradict any true cross-reactivity to cephalosporins in patients who have previously reacted to penicillin (80,132,219).

Although immediate reactions of anaphylaxis, bronchospasm, and urticaria have been reported, most commonly patients exhibiting an adverse reaction develop a maculopapular rash, usually after several days of therapy. They may also develop fever and eosinophilia (80,219). Cefoperazone (**34**) and ceftriaxone (**39**), having greater biliary excretion than other cephalosporins, are associated with an increased risk of diarrhea, which may be caused by selection of cytotoxin producing stains of *Clostridium difficile* (219).

Generally, nephrotoxicity is not a problem. Some cephalosporins, especially those with the 3-methylthiotetrazole side chain, such as moxalactam (**48**), show a tendency to promote bleeding. This appears to be due to a reduction in the synthesis of prothrombin and can be a problem especially in elderly patients, patients with renal insufficiency, or patients suffering from malnutrition (219). The same side chain seems to promote a disulfiramlike reaction in patients consuming alcohol following a cephalosporin dose (80,219).

Economic Aspects

Cephalosporins first entered the marketplace in 1964, when cephalothin (**27**) and cephaloridine (**51**), which are both injectable, were launched. By the late 1970s, the injectable cephalosporins had become important therapeutic agents in the hospitals. Also in 1964 the first oral cephalosporin, cephaloglycin [*3577-01-3*], $C_{18}H_{19}N_3O_6S$, was launched only to be displaced by the end of the year by cephalexin (**12**). For years cephalexin was the leading oral cephalosporin on the market. It has since been displaced by cefaclor (**13**). With the advent of the more β-lactamase stable cephalosporins such as cefoxitin (**23**) and cefuroxime (**35**), and the more potent agents such as cefotaxime (**36**) and other third-generation compounds, cephalosporins now dominate the antibiotic market worldwide.

On a worldwide basis, total pharmaceutical sales for 1983 were $71 billion, antibiotic sales totaled $9.2 billion, and cephalosporins accounted for $1.2 billion, or 13% of the antibacterial market (87). By 1986 the worldwide antibiotic market was valued at around $11 billion and cephalosporins represented approximately 41% of the total. Marketing projections for antibiotics from 1986 to 1991 show an average annual growth rate of 3.7% to $13,144 million. Cephalosporin sales are projected to reach 40% or $5,234 million (222). Worldwide, in 1988, the top selling cephalosporin was the oral cephalosporin Ceclor (**13**) which had 1988 sales of $605 million and projected 1989 sales of $696 million. The second best selling, and top parenteral, cephalosporin was Rocephin (**39**) which had 1988 sales of $475.8 million and projected sales of $597 million for 1989. The third cephalosporin on the list was Claforan (**36**) which had 1988 sales of $376.2 million and a projected decline in sales for 1989 to $353 million as it is displaced by the newer cephalosporins (223).

BIBLIOGRAPHY

"Antibiotics (β-Lactams)" in *ECT* 3rd ed., Vol. 2, pp. 871–919, by J. R. E. Hoover and C. H. Nash, Smith Kline and French Laboratories; "Antibiotics, β-Lactams (Cephalosporins)" in *ECT* 3rd ed., Supplement, pp. 64–83, by C. O'Callaghan and J. Elks, Glaxo Pharmaceuticals Ltd.

1. R. B. Morin, B. G. Jackson, E. H. Flynn, and R. W. Roeske, *J. Am. Chem. Soc.* **84,** 3400 (1962).
2. B. Loder, G. G. F. Newton, and E. P. Abraham, *Biochem. J.* **79,** 408 (1961).
3. R. P. Elander and H. Aoki in R. B. Morin and M. Gorman, eds., *Chemistry and Biology of β-Lactam Antibiotics,* Vol. 3, Academic Press, New York, 1982, p. 84.
4. E. P. Abraham and G. G. F. Newton, *Biochem. J.* **79,** 377 (1961).
5. E. P. Abraham, *Pharm. Rev.* **14,** 473 (1962); E. P. Abraham, *Quart. Rev. Chem. Soc.* **21,** 231 (1967); R. B. Morin and B. E. Jackson, *Fortschr. Chem. Org. Naturst.* **28,** 343 (1970).
6. E. P. Abraham and P. B. Loder in E. H. Flynn, ed., *Cephalosporins and Penicillins: Chemistry and Biology,* Academic Press, New York, 1972, p. 1.
7. R. Nagarajan and co-workers, *J. Am. Chem. Soc.* **93,** 2308 (1971).
8. T. Kanzaki and co-workers, *J. Antibiot.* **27,** 361 (1974).
9. T. Kanzaki and Y. Fujisawa in D. Perlman, ed., *Advances in Applied Microbiology,* Vol. 20, Academic Press, New York, 1976, p. 159; Ger. Offen. 2,723,463 (1977), T. Niwa and co-workers (to Meiji Seika Kaisha Ltd.).

10. K. Kitano and co-workers, *Ann. Meet. Soc. Ferment. Tech. Jpn.*, 1975, Abstract p. 244; T. Kanzaki and co-workers, *Hakko Kogaku Zasshi* **54,** 720 (1976); Jpn. Kokai 76,488 (1976), T. Kanzaki and co-workers (to Takeda Chemical Industries).
11. E. O. Stapley and co-workers, *Eleventh Interscience Conference on Antimicrobial Agents and Chemotherapy,* 1971, Atlantic City, N.J., American Society of Microbiology, Washington, 1971, Abstract 15; E. O. Stapley and co-workers, *Antimicrob. Agents Chemother.* **2,** 122 (1972); G. Albers-Schonberg, B. H. Arison, and J. L. Smith, *Tetrahedron Lett.* **29,** 2911 (1972).
12. H. Fukase and co-workers, *J. Antibiot.* **29,** 113 (1976).
13. Jpn. Kokai 76 76,487 (1976), S. Inoue and co-workers (to Meiji Confectionary Co., Ltd.); Ger. Offen. 2,455,992 (1975), S. Inoue and co-workers (to Meiji Confectionary Readings); see also U.S. Pat. 4,010,155 (1977).
14. P. D. Singh and co-workers, *J. Antibiot.* **37,** 773 (1984).
15. H. Ono, Y. Nozaki, N. Katayama, and H. Okazaki, *J. Antibiot.* **37,** 1528 (1984); S. Harada, S. Tsubotani, H. Ono, and H. Okazaki, *J. Antibiot.* **37,** 1536 (1984); S. Tsubotani and co-workers, *J. Antibiot.* **37,** 1546 (1984); Y. Nozaki and co-workers, *J. Antibiot.* **37,** 1555 (1984).
16. J. Shoji and co-workers, *J. Antibiot.* **37,** 1486 (1984).
17. Y. Nozaki and co-workers, *J. Antibiot.* **38,** 1141 (1985); S. Tsubotani, T. Hida, H. Ono, and S. Harada, *J. Antibiot.* **38,** 1152 (1985).
18. C. G. Brotzu, *Richerche su di Nuovo Antibiotico,* Lav. Ist. Igiene Cagliari, 1948.
19. G. G. F. Newton and E. P. Abraham, *Nature (London)* **175,** 548 (1955); G. G. F. Newton and E. P. Abraham, *Biochem. J.* **62,** 651 (1956).
20. D. C. Hodgkin and E. N. Maslen, *Biochem. J.* **79,** 393 (1961).
21. S. W. Queener and N. Neuss in Ref. 3, p. 2. The only naturally occurring cephalosporins isolated not containing an aminoadipoyl side chain would appear to be metabolites arising from enzymatic oxidative deamination followed by oxidative decarboxylation. See p. 62 of this reference.
22. F. M. Huber, R. R. Chauvette, and W. G. Jackson in Ref. 6, p. 27.
23. L. D. Hatfield and co-workers in G. I. Gregory, ed., *Recent Advances in the Chemistry of β-Lactam Antibiotics: Proceedings of the Second International Symposium, 1980,* The Royal Society of Chemistry, London, 1981, p. 109.
24. A. K. Miller and co-workers, *Antimicrob. Agents Chemother.* **2,** 287 (1972) and prior articles in this series; H. Fukase and co-workers, *J. Antibiot.* **29,** 113 (1976) and related papers.
25. H. R. Onishi and co-workers, *Antimicrob. Agents Chemother.* **5,** 38 (1974) and prior articles in this series; H. Nakao and co-workers, *J. Antibiot.* **29,** 554 (1976) and references cited therein.
26. T. Hiraoka and co-workers, in J. Elks, ed., *Recent Advances in the Chemistry of β-Lactam Antibiotics: Proceedings of the First International Symposium, 1976,* The Royal Society of Chemistry, London, 1977, p. 123.
27. E. M. Gordon and R. B. Sykes in R. B. Morin and M. Gorman, eds., *Chemistry and Biology of β-Lactam Antibiotics,* Vol. 1, Academic Press, New York, 1982, p. 199.
28. A. W. Guest and co-workers, *J. Chem. Soc., Perkin Trans. 1,* 45 (1987); P. H. Milner and co-workers, *J. Chem. Soc., Chem. Commun.,* 1335 (1984).
29. C. L. Branch and co-workers, *J. Antibiot.* **40,** 646 (1987).
30. M. J. Basker and co-workers, *Twenty Seventh Interscience Conference on Antimicrobial Agents and Chemotherapy 1987,* New York, American Society for Microbiology, Washington, 1987, Abstract 810.
31. P. B. Loder and E. P. Abraham, *Biochem. J.* **123,** 471 (1971).
32. P. Adriaens and co-workers, *Antimicrob. Agents Chemother.* **8,** 638 (1975).
33. P. A. Fawcett and co-workers, *Biochem. J.* **157,** 651 (1976).

34. P. B. Loder and E. P. Abraham, *Biochem. J.* **123,** 477 (1971).
35. H. R. V. Arnstein and D. Morris, *Biochem. J.* **76,** 357 (1960).
36. P. A. Fawcett and co-workers, *J. Gen. Microbiol.* **79,** 293 (1973).
37. C.-P. Pang and co-workers, *Biochem. J.* **222,** 789 (1984).
38. J. E. Baldwin, R. L. White, E.-M. M. John, and E. P. Abraham, *Biochem. J.* **203,** 791 (1982).
39. J. E. Baldwin and E. P. Abraham, *Nat. Prod. Rep.,* 129 (1988).
40. S. A. Goulden and F. W. Chattaway, *Biochem. J.* **110,** 55 (1968).
41. T. Konomi and co-workers, *Biochem. J.* **184,** 427 (1979).
42. E. P. Abraham and co-workers in G. I. Gregory, ed., *Recent Advances in the Chemistry of β-Lactam Antibiotics: Proceedings of the Second International Symposium,* The Royal Society of Chemistry, London, 1981, p. 125.
43. M. Yoshida and co-workers, *Pro. Natl. Acad. Sci. USA* **75,** 6253 (1978).
44. J. E. Baldwin, J. W. Keeping, P. D. Singh, and C. A. Vallejo, *Biochem. J.* **194,** 649 (1981).
45. D. J. Hook, L. T. Chang, R. P. Elander, and R. B. Morin, *Biochem. Biophys. Res. Commun.* **87,** 258 (1979).
46. J. E. Baldwin and co-workers, *Biochem. J.* **245,** 831 (1987).
47. A. Scheidegger, M. T. Kuenzi, and J. Nuesch, *J. Antibiot.* **37,** 522 (1984).
48. J. E. Baldwin in P. H. Bentley and R. Southgate, eds., *Recent Advances in the Chemistry of β-Lactam Antibiotics: Proceedings of the Fourth International Symposium,* The Royal Society of Chemistry, London, 1989, p. 1.
49. M. K. Turner, J. E. Farthing, and S. J. Brewer, *Biochem. J.* **173,** 839 (1978).
50. M. Liersch, J. Nuesch, and H. J. Treichler in *Proceedings of the Second International Symposium on Genetics of Industrial Micro-organisms,* Academic Press, London, 1976, p. 179.
51. J. D. Hood and co-workers in A. G. Brown and S. M. Roberts, eds., *Recent Advances in the Chemistry of β-Lactam Antibiotics: Proceedings of the Third International Symposium,* The Royal Society of Chemistry, London, 1985, p. 52.
52. J. O'Sullivan, R. T. Aplin, C. M. Stevens, and E. P. Abraham, *Biochem. J.* **179,** 47 (1979).
53. J. O'Sullivan and E. P. Abraham, *Biochem. J.* **186,** 613 (1980).
54. J. E. Baldwin in A. G. Brown and S. M. Roberts, eds., *Recent Advances in the Chemistry of β-Lactam Antibiotics: Proceedings of the Third International Symposium,* The Royal Society of Chemistry, London, 1985, p. 62.
55. E. P. Abraham in S. F. Queener, J. A. Webber, and S. W. Queener, eds., *Beta-Lactam Antibiotics for Clinical Use,* Marcel Dekker, New York, 1986, p. 103.
56. R. D. G. Cooper in P. H. Bentley and R. Southgate, eds., *Recent Advances in the Chemistry of β-Lactam Antibiotics: Proceedings of the Fourth International Symposium,* The Royal Society of Chemistry, London, 1989, p. 196.
57. S. M. Samson and co-workers, *Nature (London)* **318,** 191 (1985).
58. P. V. Demarco and R. Nagarajan in Ref. 6, p. 311.
59. K. D. Barrow and T. M. Spotswood, *Tetrahedron Lett.,* 3325 (1965).
60. J. L. Luche and co-workers, *Tetrahedron* **24,** 1275 (1968).
61. D. A. Johnson, D. Mania, C. A. Panetta, and H. H. Silvestri, *Tetrahedron Lett.,* 1903 (1968).
62. G. E. Gutowski, *Tetrahedron Lett.,* 1779 (1970).
63. D. M. Green, A. G. Long, P. J. May, and A. F. Turner, *J. Chem. Soc.,* 766 (1964).
64. M. Windholz, ed., *The Merck Index,* 9th ed., Merck Corporation, Rahway, N.J., 1976.
65. J. P. Hou and J. W. Poole, *J. Pharm. Sci.* **60,** 503 (1971); G. F. H. Green, J. E. Page, and S. E. Staniforth, *J. Chem. Soc.,* 1595 (1965); B. Fechtig, H. Peter, H. Bickel, and E. Vischer, *Helv. Chim. Acta* **51,** 1108 (1968).

66. J. M. T. Hamilton-Miller, G. G. F. Newton, and E. P. Abraham, *Biochem. J.* **116,** 371 (1970); J. M. T. Hamilton-Miller, E. Richards, and E. P. Abraham, *Biochem. J.* **116,** 385 (1970); H. Bundgaard, *Arch. Pharm. Chem. Sci. Ed.* **3,** 94 (1975); C. H. O'Callaghan and co-workers, *J. Bacteriol.* **110,** 988 (1970).
67. J. Nishikawa and K. Tori, *J. Med. Chem.* **27,** 1657 (1984).
68. B. Coene, A. Schanck, J.-M. Dereppe, and M. Van Meerssche, *J. Med. Chem.* **27,** 694 (1984).
69. D. B. Boyd, D. K. Herron, W. H. W. Lunn, and W. A. Spitzer, *J. Am. Chem. Soc.* **102,** 1812 (1980).
70. D. B. Boyd in Ref. 27, p. 437.
71. D. B. Boyd, R. B. Hermann, E. E. Presti, and M. M. Marsh, *J. Med. Chem.* **18,** 408 (1975); D. B. Boyd and W. H. W. Lunn, *J. Med. Chem.* **22,** 778 (1979).
72. W. S. Faraci and R. F. Pratt, *J. Am. Chem. Soc.* **106,** 1489 (1984); W. S. Faraci and R. F. Pratt, *Biochemistry* **24,** 903 (1985); W. S. Faraci and R. F. Pratt, *Biochemistry* **25,** 2934 (1986); W. S. Faraci and R. F. Pratt, *J. Am. Chem. Soc.* **108,** 5328 (1986).
73. E. J. J. Grabowski, A. W. Douglas, and G. B. Smith, *J. Am. Chem. Soc.* **107,** 267 (1985).
74. D. Agathocleous, S. Buckwell, P. Proctor, and M. I. Page, in A. G. Brown and S. M. Roberts, eds., *Recent Advances in the Chemistry of β-Lactam Antibiotics: Proceedings of the Third International Symposium,* The Royal Society of Chemistry, London, 1985, p. 18; M. I. Page and P. J. Proctor, *J. Am. Chem. Soc.* **106,** 3829 (1984).
75. C. F. Murphy and J. A. Webber in Ref. 6, p. 134.
76. G. V. Kaiser and S. Kukolja in Ref. 6, p. 74.
77. G. N. Rolinson, *J. Antimicrob. Chemother.* **17,** 5 (1986).
78. J. Elks, *Drugs* **34** (Suppl. 2), 240 (1987).
79. *Ibid.,* 247 (1987).
80. G. L. Mandell in G. L. Mandell, R. G. Douglas, Jr., and J. E. Bennett, *Anti-infective Therapy,* John Wiley & Sons, Inc., New York, 1985, p. 76.
81. E. Squires and R. Cleeland, *Microbiology and Pharmacokinetics of Parenteral Cephalosporins,* Roche Laboratories, Nutley, N.J., 1985.
82. *Med. Lett.* **25,** 57 (June 10, 1983).
83. A. J. Weinstein, *Drugs* **19,** 137 (1980).
84. P. Garzone, J. Lyon, and V. L. Yu, *Drug Intel. Clin. Pharm.* **17,** 615 (1983).
85. J. D. Williams, *Drugs* **34**(Suppl. 2), 15 (1987).
86. R. Quintiliani, *Scand. J. Infect. Dis.* Suppl. **42,** 99 (1984).
87. C. A. Bunnell, W. D. Luke, and F. M. Perry, Jr., in Ref. 55, p. 255.
88. J. V. Heck, *Ann. Rep. Med. Chem.* **24,** 101 (1989); T. F. Walsh, *Ann. Rep. Med. Chem.* **23,** 121 (1988); G. L. Dunn, *Ann. Rep. Med. Chem.* **21,** 131 (1986); G. L. Dunn, *Ann. Rep. Med. Chem.* **20,** 127 (1985).
89. C. W. Ryan, R. L. Simon, and E. M. VanHeyningen, *J. Med. Chem.* **12,** 310 (1969); U.S. Pat. 3,946,002 (1976), M. Davison, D. B. Frankham, and T. W. M. Spence (to Lilly Industries Ltd., London); U.S. Pat. 3,957,773 (1976), B. Burton and W. Graham (to Glaxo Laboratories, Ltd.).
90. U.S. Pat. 3,925,372 (1975), R. Chauvette (to Eli Lilly and Company).
91. U.S. Pat. 3,503,967 (1970), H. H. Sivestri and D. Aaron (to Bristol Myers Co.).
92. J. E. Dolfini and co-workers, *J. Med. Chem.* **14,** 117 (1971).
93. U.S. Pat. 3,912,726 (1975), P. Diassi and M. S. Atwal (to E. R. Squibb & Sons).
94. U.S. Pat. 3,489,752 (1970), L. B. Crast, Jr. (to Bristol Myers Co.).
95. U.S. Pat. 3,985,741 (1976), L. B. Crast, Jr., and W. Gottstein (to Bristol Myers Co.).
96. *Drugs Future* **8,** 682 (1983); **13,** 790 (1988); *Drugs* **38,** 524 (1989).
97. *Drugs Future* **13,** 1096 (1988).
98. *Drugs Future* **14,** 507 (1989).
99. K. Tomatsu and co-workers, *Twenty-Seventh Interscience Conference on Antimicrobial*

Agents and Chemotherapy, New York, American Society for Microbiology, Washington, 1987, Abstract 806.
100. *Drugs Future* **10,** 112 (1985); **13,** 171 (1988).
101. *Drugs Future* **13,** 16 (1988); **14,** 73 (1989).
102. S. Masuyoshi and co-workers, *Twenty-Eighth Interscience Conference on Antimicrobial Agents and Chemotherapy 1988,* Los Angeles, American Society for Microbiology, Washington, 1988, Abstract 527.
103. E. O. Stapley, E. S. Inamine, and C. M. Martin in Ref. 55, p. 411.
104. *Drugs Future* **2,** 517 (1977); **6,** 505 (1981); *Drugs Today* **17,** 11 (1981).
105. *Drugs Future* **7,** 395 (1982); **11,** 502 (1986).
106. *Drugs Future* **6,** 609 (1981); **9,** 781 (1984); *Drugs Today* **19,** 335 (1983).
107. R. R. Chauvette and co-workers, *Antimicrob. Agents Chemother.* **7,** 687 (1962).
108. Review Issue, *Arneimittel Forsch.* **24,** 1445–1534 (1974); *Drugs Today* **10,** 242 (1974); **9,** 51 (1973).
109. U.S. Pat. 3,422,100 (1969), L. B. Crast, Jr. (to Bristol Myers Co.); U.S. Pat. 3,578,661 (1971), R. E. Havranek and L. B. Crast, Jr. (to Bristol Myers Co.); *Drugs Today* **10,** 299 (1974); **9,** 56 (1973).
110. U.S. Pat. 4,110,338 (1978), T. Kamiya, K. Tanaka, and Y. Shiokawa (to Fujisawa Pharmaceutical Co.); U.S. Pat. 3,641,021 (1972), C. W. Ryan (to Eli Lilly and Co.); *Drugs Today* **10,** 50 (1974); *Drugs Future* **2,** 646 (1977).
111. U.S. Pat. 4,159,373 (1979), D. A. Berges (to SmithKline Corp.); *Drugs Today* **21,** 12 (1985); *Drugs Future* **4,** 634 (1979); **10,** 777 (1985).
112. U.S. Pat. 4,338,461 (1982), K. Unuma, H. Saito, J. Inomata, and S. Takizawa (to Nippon Kasei Chemical Co., Ltd.); U.S. Pat. 3,767,667 (1973), T. Kamiya and Y. Saito (to Fujisawa Pharmaceutical Co.); U.S. Pat. 3,962,272 (1976), A. S. Katner (to Eli Lilly and Company); U.S. Pat. 3,798,229 (1974), K. Kariyone and co-workers (to Fujisawa Pharmaceutical Co., Ltd.); U.S. Pat. 3,824,246 (1974), G. W. Huffman (to Eli Lilly and Company).
113. *Drugs Future* **2,** 643 (1977); **10,** 864 (1985); *Drugs Today* **20,** 615 (1984).
114. *Drugs Future* **4,** 675 (1979); **6,** 569 (1981); *Drugs Today* **17,** 535 (1981); U.S. Pat. 4,304,909 (1981), I. Saikawa and co-workers (to Toyama Chemical Co., Ltd.).
115. *Drugs Future* **3,** 266 (1978); **4,** 302 (1979); Eur. Pat. Appl. 88,183 (1983) (to Glaxo Laboratories Ltd.); U.S. Pat. 4,093,803 (1978), M. C. Cook, G. I. Gregory, and J. Bradshaw (to Glaxo Laboratories, Ltd.); U.S. Pat. 4,013,680 (1977), M. G. Johnson, J. P. Turnbull, and H. A. Crisp (to Glaxo Laboratories, Ltd.).
116. *Drugs Future* **3,** 905 (1978); **5,** 519 (1980); *Drugs Today* **17,** 16 (1981); U.S. Pat. 4,098,888 (1978), M. Ochiai and co-workers, (to Takeda Chemical Industries, Ltd.).
117. *Drugs Future* **2,** 291 (1977); **8,** 445 (1983); *Drugs Today* **19,** 133 (1983); U.S. Pat. 4,166,115 (1979), T. Takaya and co-workers, (to Fujisawa Pharmaceutical Co., Ltd.); H. Tankasugi and co-workers, *J. Antibiot.* **37,** 846 (1983).
118. *Drugs Today* **19,** 335 (1983); *Drugs Future* **5,** 146 (1980).
119. *Drugs Future* **6,** 132 (1981); **8,** 254 (1983); *Drugs Today* **19,** 653 (1983); U.S. Pat. 4,349,672 (1982), M. Montavon and R. Reiner (to Hoffmann-La Roche, Inc.).
120. *Drugs Future* **6,** 612 (1981); **9,** 782 (1984); *Drugs Today* **20,** 555 (1984); U.S. Pat. 4,331,661 (1982), W. E. Marki, M. R. Brown, and J. E. F. Rivier (to The Salk Institute for Biological Studies).
121. *Drugs Future* **5,** 67 (1980); **6,** 110 (1981); *Drugs Today* **17,** 542 (1981).
122. *Drugs Future* **5,** 254 (1980); **11,** 417 (1986).
123. *Drugs Future* **9,** 252 (1984); **14,** 371 (1989).
124. *Drugs Future* **7,** 373 (1982); **12,** 581 (1987).
125. *Drugs Future* **10,** 805 (1985); **14,** 994 (1989).
126. *Drugs Future* **13,** 1047 (1988).

127. H. A. Albrecht and co-workers, *J. Med. Chem.* **33,** 77 (1990); R. N. Jones, A. L. Barry, and C. Thornsberry, *Antimicrob. Agents Chemother.* **33,** 944 (1989); G. Beskid and co-workers, *Antimicrob. Agents Chemother.* **33,** 1072 (1989).
128. *Drugs Future* **4,** 667 (1979); **6,** 578 (1981); *Drugs Today* **18,** 134 (1982).
129. *Drugs Future* **11,** 452 (1986); **13,** 572 (1988).
130. J. D. Williams and F. Moosdeen, *Drugs* **34**(Suppl. 2), 44 (1987).
131. R. R. Kohler, L. J. Wheat, and A. C. White in Ref. 55, p. 351.
132. G. R. Donowitz and G. L. Mandell, *N. Engl. J. Med.* **318,** 490 (1988).
133. H. Numata and co-workers, *J. Antibiotics* **31,** 1262 (1978).
134. A. Tomaz in R. K. Root and M. A. Sande, eds., *New Dimensions in Antimicrobial Therapy,* Churchill Livingstone, New York, 1984, p. 1.
135. D. J. Tipper in Ref. 55, p. 17.
136. D. J. Waxman and J. L. Strominger in Ref. 3, p. 210.
137. G. D. Shockman, L. Daneo-Moore, T. D. McDowell, and W. Wong in Ref. 3, p. 304.
138. N. H. Georgopapdakou in P. K. Peterson and J. Verhoef, eds., *Antimicrobial Agents Annual 3,* Elsevier Science Publishers, New York, 1988, p. 409.
139. D. M. Livermore, *Drugs* **34**(Suppl. 2), 64 (1987); T. D. Gootz in D. M. Bailey, ed., *Annual Reports in Medicinal Chemistry,* Vol. 20, Academic Press, New York, 1985, p. 137.
140. A. A. Medeiros and G. A. Jacoby in Ref. 55, p. 49.
141. C. C. Sanders, *J. Infect. Dis.* **147,** 585 (1983).
142. C. C. Sanders and W. E. Sanders, Jr., *Rev. Infect. Dis.* **5,** 639 (1983).
143. R. L. Then and P. Angehrn, *Antimicrob. Agents Chemother.* **21,** 711 (1982).
144. H. Nikaido, *J. Antimicrob. Chemother.* **22**(Suppl. A), 17 (1988).
145. M. H. Richmond and W. Wotten, *Antimicrob. Agents Chemother.* **10,** 219 (1976).
146. H. Nikaido in M. Stalton and G. A. Shockman, eds., *β-Lactam Antibiotics: Mode of Action, Recent Developments and Future Prospects,* Academic Press, New York, 1981, p. 249; H. Nikaido, E. Y. Rosenberg, and J. Foulds, *J. Bacteriology* **153,** 232 (1983).
147. I. Chopra in R. C. Allen, ed., *Annual Reports in Medicinal Chemistry,* Vol. 24, Academic Press, New York, 1989, p. 139.
148. N. Watanabe, T. Nagasu, K. Katsu, and K. Kitoh, *Antimicrob. Agents Chemother.* **31,** 497 (1987).
149. N. A. C. Curtis and co-workers, *Antimicrob. Agents Chemother.* **32,** 1879 (1988).
150. H. Mochizuki and co-workers, *Twenty-Eighth Interscience Conference on Antimicrobial Agents and Chemotherapy 1988,* Los Angeles, American Society for Microbiology, Washington, 1988, Abstract 428; H. Mochizuki and co-workers, *J. Antibiot.* **41,** 377 (1988).
151. K. I. Plaisance, C. H. Nightingale, and R. Quintiliani in Ref. 55, p. 285.
152. T. Bergan, *Drugs* **34**(Suppl. 2), 89 (1987).
153. P. A. Lemke and D. R. Brannon in E. H. Flynn, eds., *Cephalosporins and Penicillins: Chemistry and Biology,* Academic Press, New York, 1972, p. 370; A. L. Demain, *Lloydia* **37,** 147 (1974).
154. D. A. Lowe in Ref. 55, p. 117.
155. J. L. Ott and co-workers, *Appl. Microbiol.* **10,** 515 (1962).
156. U.S. Pat. 4,354,971 (1982), J. M. Edmundowicz and E. E. Allanson (to Eli Lilly and Co.).
157. U.S. Pat. 3,184,454 (1965), E. P. Abraham, G. G. F. Newton, and C. W. Hale (to National Research Development Corp.).
158. U.S. Pat. 3,725,400 (1973), W. Vosser (to CIBA, Ltd.).
159. R. J. Stedman, K. Swered, and J. R. E. Hoover, *J. Med. Chem.* **7,** 117 (1964).
160. U.S. Pat. 4,346,219 (1982), W. C. Vladuchick (to Eli Lilly and Co.).
161. R. B. Morin and co-workers, *J. Am. Chem. Soc.* **85,** 1896 (1963).

162. R. B. Morin and co-workers, *J. Am. Chem. Soc.* **91,** 1401 (1969).
163. R. D. G. Cooper and D. O. Spry in E. H. Flynn, ed., *Cephalosporins and Penicillins: Chemistry and Biology,* Academic Press, New York, 1972, p. 181.
164. R. D. G. Cooper and G. A. Koppel in R. B. Morin and M. Gorman, eds., *Chemistry and Biology of β-Lactam Antibiotics,* Vol. 1, Academic Press, New York, 1982, p. 1.
165. R. D. G. Cooper, L. D. Hatfield, and D. O. Spry, *Accounts Chem. Res.* **6,** 32 (1973); P. G. Sammes, *Chem. Rev.* **76,** 113 (1976); R. J. Stoodley, *Tetrahedron* **31,** 2321 (1975).
166. U.S. Pat. 3,725,399 (1973), N. V. Ellerton, W. F. Paradise, and P. E. Sanford (to Glaxo Laboratories, Ltd.); U.S. Pat. 3,953,440 (1976), L. D. Hatfield (to Eli Lilly and Co.); U.S. Pat. 3,879,379 (1975), U. Stache and co-workers (to Farbwerke Hoechst Aktiengesellschaft Vormals Meister Lucius and Bruning).
167. U.S. Pat. 4,003,894 (1977), J. Verweij, H. S. Tan, and H. J. Kooreman (to Gist-Brocades).
168. L. D. Hatfield and co-workers, *Philos. Trans. R. Soc. London* **B289,** 173 (1980).
169. J. D. Cocker and co-workers, *J. Chem. Soc.* 5015 (1965).
170. A. B. Taylor, *J. Chem. Soc.* 7020 (1965).
171. H. Yazawa, H. Nakamura, K. Tanaka, and K. Kariyone, *Tetrahedron Lett.* 3991 (1974).
172. R. Bonjouklian and M. L. Phillips, *Tetrahedron Lett.* **22,** 3915 (1981); U.S. Pat. 4,266,049 (1981), R. Bonjouklian (to Eli Lilly and Co.).
173. U.S. Pat. 4,014,874 (1977), H. Peter, B. Muller, W. Sibral, and H. Bickel (to CIBA-GEIGY Corp.).
174. U.S. Pat. 3,270,012 (1966), H. M. Higgins, Jr. (to Eli Lilly and Co.); J. L. Spencer and co-workers, *J. Org. Chem.* **32,** 500 (1967).
175. U.S. Pat. 4,317,907 (1982), I. Saikawa and co-workers, (to Toyama Chemical Co., Ltd.).
176. I. Miskolczi, F. Sztariczai, and R. Bognar, *Org. Prep. Proc. Int.* **13,** 315 (1981).
177. U.S. Pat. 4,182,870 (1979), C. A. Bruynes and J. J. DeKoning (to Gist-Brocades); U.S. Pat. 4,107,431 (1978), J. C. Clark, J. Kennedy, A. G. Long, and N. G. Weir (to Glaxo Labs. Ltd.); U.S. Pat. 4,113,591 (1978), B. Laudon. B. R. Cowley, and D. C. Humber (to Glaxo Lab. Ltd.).
178. Eur. Pat. Appl. 34,394 (1981), C. A. Bruynes and T. K. Jurriens (to Gist-Brocades).
179. N. Nomura and co-workers, *J. Med. Chem.* **17,** 1312 (1974).
180. T. Takaya, H. Takasugi, T. Murakawa, and H. Nakano, *J. Antibiot.* **34,** 1300 (1981).
181. J. D. Jeffrey, E. P. Abraham, and G. G. F. Newton, *Biochem. J.* **81**, 591 (1961).
182. E. M. Wilson, *Chem. Ind. London* 217 (1984).
183. F. A. Jung, W. R. Pilgrim, J. P. Poysen, and P. J. Siret, in P. G. Sammes, ed., *Topics in Antibiotic Chemistry,* Vol. 4, John Wiley & Sons, Inc., New York, 1980.
184. M. Gorman, R. R. Chauvette, and S. Kukolja, in J. S. Bindra and D. Lednicer, eds., *Chronicles of Drug Discovery,* Vol. 2, John Wiley & Sons, Inc., New York, 1983, p. 49.
185. S. Kukolja and R. R. Chauvette in R. B. Morin and M. Gorman, eds., *Chemistry and Biology of β-Lactam Antibiotics,* Vol. 1, Academic Press, New York, 1982, p. 93.
186. S. Kukolja and co-workers, *J. Org. Chem.* **41,** 2276 (1976).
187. Y. Shibuya, K. Matsumoto, and T. Fujii, *Agric. Biol. Chem.* **45,** 1561 (1981).
188. R. B. Morin and co-workers, *J. Am. Chem. Soc.* **91,** 1396 (1969).
189. B. Fechtig, H. Peter, and E. Vischer, *Helv. Chim. Acta.* **51,** 1108 (1968).
190. S. Tsushima and co-workers, *Chem. Pharm. Bull. Jpn.* **27,** 696 (1979).
191. K. Morita and co-workers, *Phil. Trans. R. Soc. London* **B289,** 181 (1980).
192. R. Reiner and co-workers, *J. Antibiot.* **33,** 783 (1980).
193. Eur. Pat. 037,380 (1981), G. Ascher (to Biochemie G.m.b.H.).
194. W. Ngata, M. Narisada, and Y. Yoshida in R. B. Morin and M. Gorman, eds., *Chemistry and Biology of β-Lactam Antibiotics,* Vol. 2, Academic Press, New York, 1982, p. 1.
195. B. G. Jackson, R. B. Kammer, M. Yoshioka, and H. Yamada in Ref. 55, p. 371.

196. M. Yoshioka and co-workers, *Tetrahedron Lett.* **21,** 351 (1980).
197. K. G. Holden in R. B. Morin and M. Gorman, eds., *Chemistry and Biology of β-Lactam Antibiotics,* Vol. 2, Academic Press, New York, 1982, p. 101.
198. J. Aszodi and co-workers in P. H. Bentley and R. Southgate, eds., *Recent Advances in the Chemistry of β-Lactam Antibiotics: Proceedings of the Fourth International Symposium,* The Royal Society of Chemistry, London, 1989, p. 350.
199. L. D. Cama and B. G. Christensen, *J. Am. Chem. Soc.* **96,** 7582 (1974).
200. R. A. Firestone and co-workers, *J. Med. Chem.* **20,** 551 (1977).
201. C. L. Branch and M. J. Pearson, *J. Chem. Soc. Perkins Trans. 1,* 2268 (1979).
202. J. Aszodi and co-workers, Abstract 123, and J. F. Chantot, Abstract 436, *Twenty-Eighth Interscience Conference on Antimicrobial Agents and Chemotherapy, 1988,* Los Angeles, American Society for Microbiology, Washington, 1988.
203. R. N. Guthikonda, L. D. Cama, and B. G. Christensen, *J. Am. Chem. Soc.* **96,** 7584 (1974).
204. J. A. Eudaly and co-workers in P. H. Bentley and R. Southgate, eds., *Recent Advances in the Chemistry of β-Lactam Antibiotics: Proceedings of the Fourth International Symposium,* The Royal Society of Chemistry, London, 1989, p. 333.
205. F. T. Counter and co-workers, *Twenty-Seventh Interscience Conference on Antimicrobial Agents and Chemotherapy,* New York, American Society for Microbiology, Washington, 1987, Abstract 1188.
206. J. F. Quay and co-workers in Ref. 205, Abstract 1205.
207. H. R. Black and co-workers in Ref. 202, Abstract 123.
208. J. A. Eudaly and co-workers in P. H. Bentley and R. Southgate, eds., *Recent Advances in the Chemistry of β-Lactam Antibiotics: Proceedings of the Fourth International Symposium,* The Royal Society of Chemistry, London, 1989, p. 333.
209. G. K. Cook and co-workers in Ref. 202, Abstract 532.
210. J. Christenson and co-workers in P. H. Bentley and R. Southgate, eds., *Recent Advances in the Chemistry of β-Lactam Antibiotics: Proceedings of the Fourth International Symposium,* The Royal Society of Chemistry, London, 1989, p. 33; J. G. Christenson, D. L. Pruess, M. K. Talbot, and D. D. Keith, *Antimicrob. Agents Chemother.* **32,** 1005 (1988); N. H. Georgopapadakou and co-workers, *Antimicrob. Agents Chemother.* **31,** 1069 (1987).
211. P. Schindler and P. Huber in U. Brodbeck, ed., *Enzyme Inhibitors,* Verlag Chemie, Weinheim, FRG, 1980, p. 169.
212. R. B. Sykes and M. Matthew in J. M. T. Hamilton-Miller and J. T. Smith, eds., *Beta-Lactamases,* Academic Press, New York, 1980, p. 17.
213. J. Symonds and A. M. Geddes, *Drugs* **34**(Suppl. 2), 121 (1987).
214. D. W. McEniry and S. L. Gorbach, *Drugs* **34**(Suppl. 2), 216 (1987).
215. L. O. Gentry, *Drugs* **34**(Suppl. 2), 154 (1987).
216. H. C. Neu, *Drugs* **34**(Suppl. 2), 135 (1987).
217. I. Phillips, *Drugs* **34**(Suppl. 2), 164 (1987).
218. R. Finch, *Drugs* **34**(Suppl. 2), 180 (1987).
219. S. R. Norrby, *Drugs* **34**(Suppl. 2), 105 (1987).
220. B. R. Meyers, *Am. J. Med.* **79**(Suppl. 2A), 96 (1985).
221. R. Platt, *J. Antimicrob. Chemother.* **10**(Suppl. C), 135 (1982).
222. *Chemistry and Industry* **22,** 769 (Nov. 16, 1987). Note that the figure given in this article of $1,100 million is in error. The correct figure is $11,000 million.
223. *Scrip,* Review issue, 12 (1989).

JOHN ROBERTS
Hoffmann-La Roche, Inc.

β-LACTAMASE INHIBITORS

The antibacterial effectiveness of penicillins, cephalosporins, and other β-lactam antibiotics depends upon selective acylation and consequently, inactivation, of transpeptidases involved in bacterial cell wall synthesis. This acylating ability is a result of the reactivity of the β-lactam ring (**1**). Bacteria that are resistant to β-lactam antibiotics often produce enzymes called β-lactamases that inactivate the antibiotics by catalyzing the hydrolytic opening of the β-lactam ring to give products (**2**) devoid of antibacterial activity.

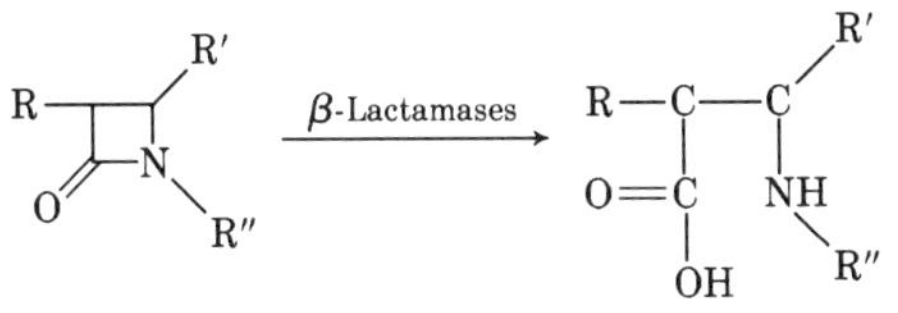

(**1**) (**2**)

active inactive

A book (1) and several general reviews (2–4) on β-lactamases have been published. Based on sequence data, it has been suggested that β-lactamases evolved from the enzymes involved in bacterial cell wall synthesis (5–7).

One approach to combating antibiotic resistance caused by β-lactamase is to inhibit the enzyme (see ENZYME INHIBITION). Effective combinations of enzyme inhibitors with β-lactam antibiotics such as penicillins or cephalosporins, result in a synergistic response, lowering the minimal inhibitory concentration (MIC) by a factor of four or more for each component. However, inhibition of β-lactamases alone is not sufficient. Pharmacokinetics, stability, ability to penetrate bacteria, cost, and other factors are also important in determining whether an inhibitor is suitable for therapeutic use. Almost any class of β-lactam is capable of producing β-lactamase inhibitors. Several reviews have been published on β-lactamase inhibitors, detection, and properties (8–15).

Classification and Occurrence

Several classification schemes for β-lactamases have been used to describe the activity of β-lactamase inhibitors (3,16,17). The β-lactamases from gram-positive *Staphylococcal* organisms have been divided into four types A, B, C, D on the basis of serological data and are referred to in general as penicillinases or penases. Gram-negative bacteria produce a much greater diversity of β-lactamases and therefore require a more complex classification scheme. The Richmond-Sykes system (18) is most frequently used to describe the activity of β-lactamase inhibitors. This system defines five classes (I–V) of β-lactamases; a sixth (VI) has been added for the β-lactamases produced by *Bacteroides* species (19). The many plasmid-derived β-lactamases of gram-negative bacteria have been separated by isoelectric focusing. They have been given names depending on plasmid origin. The clinically most important plasmid-derived β-lactamases and their prevalence, indicated in parenthesis, are (19): TEM-1 (75%), TEM-2 (3%), OXA-1 (5%), OXA-2

(3%), OXA-3 (2%), SHV-1 (2%), and others (10%). The TEM-1 enzyme stands out as an important target for a β-lactamase inhibitor.

Table 1 shows the clinically most important bacteria that produce β-lactamases, separated into gram-positive and gram-negative organisms. The prevalence of β-lactamase is indicated as is the origin and Richmond-Sykes classification. Based on these data the most important β-lactamases to inhibit clinically are the gram-positive penases, the gram-negative TEM, which are Richmond-Sykes type III, and the gram-negative chromosomal cephalosporinases–cephases which are Richmond-Sykes type I. These enzymes are subsequently referred to as penase, TEM(III), and cephase(I).

Table 1. β-Lactamase Producing Bacteria[a]

Organism	β-Lactamase producers, %[b]	Enzyme origin, %		Richmond-Sykes classification
		Chromosomal	Plasmid	
Gram-positive				
Staphylococcus aureus	80		penase	
Staphylococcus epidermidis	80		penase	
Gram-negative				
Escherichia coli	25 (16–76)	15	TEM, 80; OXA-1, 7.5	I,III,V
Haemophilus influenzae	25–60		TEM, 92; ROB-1, 8	III
Neisseria gonorrhea	1–10		TEM, 100	III
Salmonella	6		TEM, 100	III
Shigella			TEM	III
Klebsiella pneumoniae	60–90		TEM, 24; SHV-1, 76	III,IV
Enterobacter	25–30	73	TEM, 27	I,III
Citrobacter	20–50	77	TEM, 23	I,III
Pseudomones aeruginosa	23	44	TEM, 9; PSE, 10; others	I,III
Bacillus catarrhalis	87	90^+	BRO-1	
Bacillus fragilis	87	~100		VI

[a]Refs. 16, 19–21.
[b]Clinically resistant bacteria, virtually 100% of the *Enterobacteriaceae,* produce a low level of chromosomal enzyme that can clinically be selected for higher levels.

Mechanistic Aspects of β-Lactamase Inhibition

The clinically important β-lactamases, eg, the penases, TEM(III), and cephases(I), are serine proteases that form an acyl enzyme intermediate with β-lactam substrates and β-lactam-derived β-lactamase inhibitors (22–25). Mechanistic studies using several β-lactamase inhibitors have been extensively reviewed (6,26,27) and a general inhibition scheme is illustrated in Figure 1. Following Michaelis complex formation (**3**), the enzyme's serine hydroxyl reacts with the β-lactam carbonyl, forming an acyl enzyme intermediate (**4**), the structure of which depends on the inhibitor. This acyl enzyme can then follow Route B and hydrolyze to products and active enzyme, react further to give an irreversibly inactivated enzyme (Route A) or a transiently inactivated enzyme species (Route C). The transient inhibition can result from a conformational change in the enzyme or formation of a species resistant to hydrolysis. In the latter case, amino acrylatelike esters are commonly encountered. A suitably long-lived transient species can occupy the β-lactamase and protect a β-lactam from hydrolysis resulting in a synergistic effect. The transiently inhibited species can then hydrolyze to products, revert to acyl enzyme, or react further to give irreversibly inactivated enzyme. β-Lactamase inhibitors that branch away from acyl enzyme (**4**) formation have been termed branched pathway inhibitors. The exact mechanistic pathway is dependent on both the enzyme and the inhibitor.

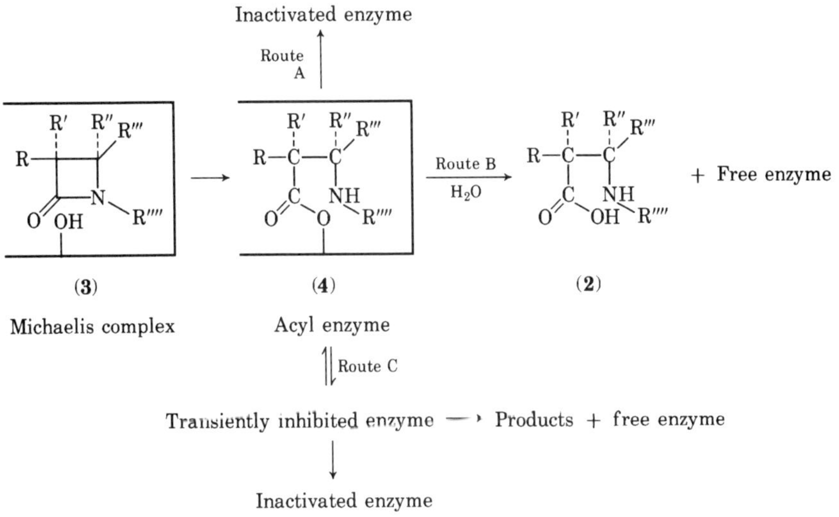

Fig. 1. Scheme for the interaction of β-lactamase inhibitors and β-lactamases where the enzyme is represented by

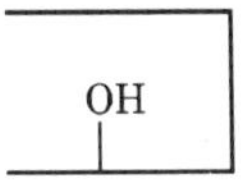

Active site directed β-lactam-derived inhibitors have a competitive component of inhibition, but once in the active site they form an acyl enzyme species which follows one or more of the pathways outlined in Figure 1. Compounds that follow Route C and form a transiently inhibited enzyme species and are subsequently hydrolyzed to products have been termed inhibitory substrates or competitive substrates. Inhibitors that give irreversibly inactivated β-lactamase (Route A) are called suicide inactivators or irreversible inhibitors. The term progressive inhibitor has also been used. An excellent review has appeared on inhibitor interactions with β-lactamases (28).

The activity of β-lactamase inhibitors is often expressed as an IC_{50} value, which is defined as the concentration of inhibitor that causes 50% inhibition of enzyme activity for a given set of conditions. IC_{50} values, which vary widely according to substrate, time of incubation, and other factors, are presented herein solely to give an indication of potency and enzyme inhibitor specificity. Values that decrease with preincubation are indicative of irreversible inhibitors.

β-Lactam-Based Inhibitors

Penicillins, Cephalosporins, and Monobactams. Early attempts at inhibiting β-lactamases using inorganics or penicillin fragments were not successful (29,30). The use of reactive penicillin species, such as diazo species from 6-aminopenicillanic acid (6-APA) or ampicillin (31) and penicillin isocyanates (32), was successful, but not practical because of the reactivity of the molecules. The discovery that cephalosporin C (33) and methicillin (34) inhibited β-lactamases resulted in the screening of numerous antibiotics and a number of β-lactamase-resistant penicillins and cephalosporins were found to be β-lactamase inhibitors. These compounds act by forming a transiently inhibited acyl enzyme species as a result of conformational change (35–37) and are inhibitory substrates (Fig. 1, Route C). No clinically useful inhibitors have been identified from this class. These efforts have been extensively reviewed (8,10,13,14,37).

Modern β-lactamase-resistant cephalosporins, primarily of the aminothiazole oxime structural type shown in Figure 2, have been reported to be inhibitors of type I cephases (38–42). β-Lactamase inhibition occurs through the transiently inhibited enzyme species (**7**), which requires a good C-3 leaving group. Cephalosporins without a leaving group at C-3 are poor inhibitors (43,44). None of the aminothiazole oxime structural types is an effective synergist.

A series of activated cephalosporins having electron withdrawing groups at C-3, such as [*128474-86-2*] (**8**, R = CHO), $C_{10}H_{11}NO_5S$, and [*123036-47-5*] (**8**, R = CN), $C_{10}H_{10}N_2O_4S$, were prepared in an attempt to increase β-lactamase inhibition (45,46). These compounds inhibited type I cephases and synergy was demon-

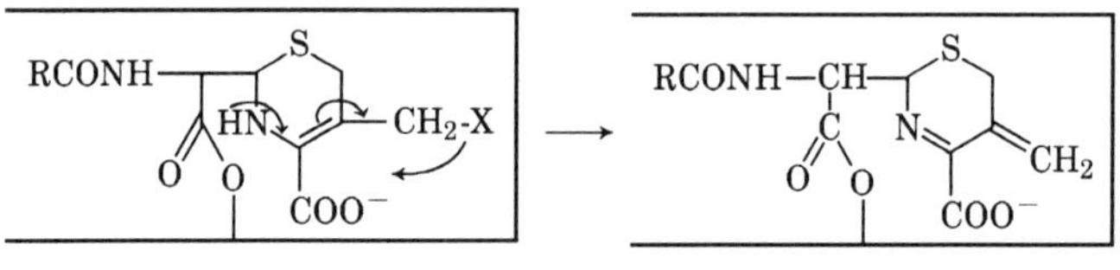

(**6**) (**7**)

Acyl enzyme Transient inhibition

Fig. 2. Transiently inhibited species for cephalosporins bound to β-lactamase. R, an aminothiazole oxime, =

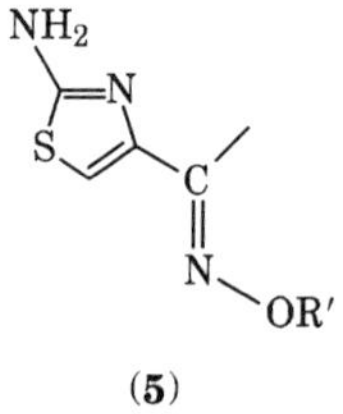

(**5**)

strated for (**8**, R = CN) with ceftizoxime. A series of 7-β-[2-(1,3-dithiolan-2-ylidene)acetamido] cephalosporins represented by [*71908-34-4*] (**9**), $C_{18}H_{20}N_5O_6S_4$, are potent inhibitors of type I cephase and give synergy with cephaloridine (47). Several monobactams have been reported to be inhibitory substrates for type I cephases (48–51). Interesting exceptions are the monophosphams such as represented by [*120020-34-0*] (**10**), $C_{12}H_{15}N_5O_8PS$, and some oligopeptide monobactams which are reported to be irreversible inhibitors (Fig. 1, Route A) (52,53). The monocyclic β-lactam [*123039-89-4*] (**11**), $C_9H_8N_2O_2$, inhibits a broad range of β-lactamases, but is too unstable for practical use (54).

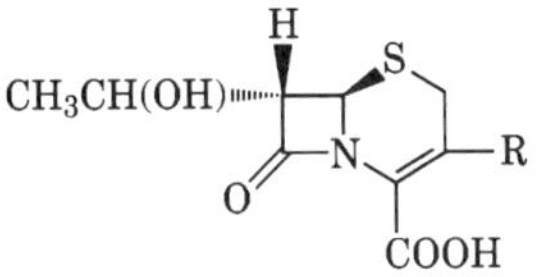

(**8**)

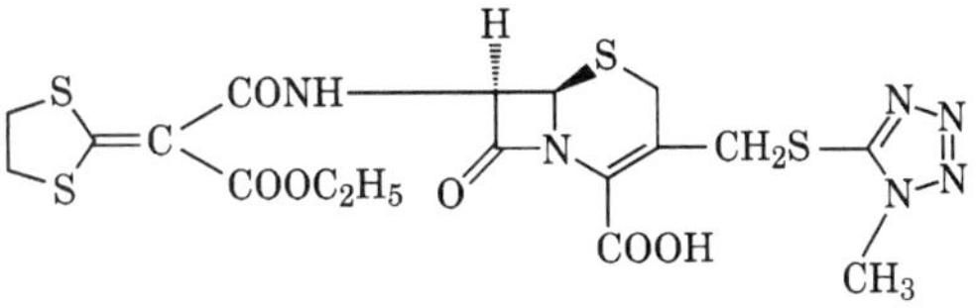

(**9**)

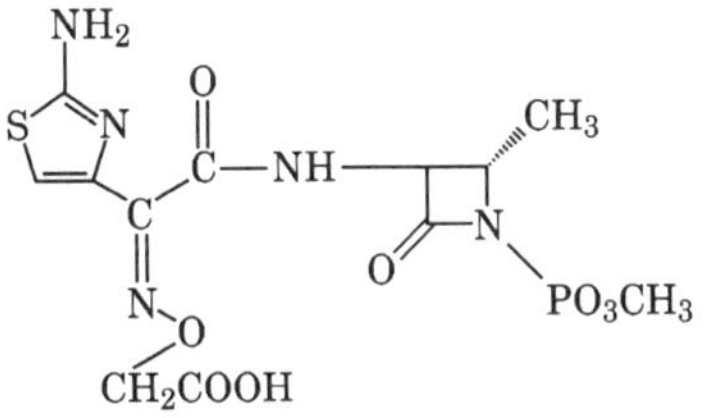

(**10**)

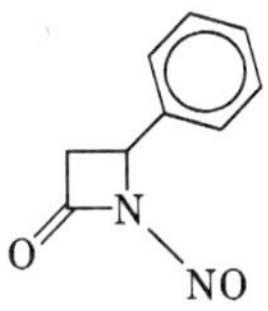

(**11**)

Clavulanic Acid Class of β-Lactamase Inhibitors. The discovery of clavulanic acid [*58001-44-8*] (**12**, R = CH_2OH, R′ = H), $C_8H_9NO_5$, in 1976 marked a turning point in β-lactamase inhibitor research (55). This natural product isolated from *Streptomyces clavuligerus,* ATCC 27064, using a cleverly designed screen, had an unusual β-lactam structure compared to classical penicillins and cephalosporins. The nucleus has the trivial name clavam: an oxygen replaces the ring sulfur and there is no C-6 substituent. The x-ray structure of clavulanic acid, the chemical name of which is [2(*R*)-(2α,3(*Z*),5α)]-3-(2-hydroxy-ethylidine)-7-oxo-4-oxa-1-azabicyclo[3.2.0.]heptane-2-carboxylic acid, has been published (56).

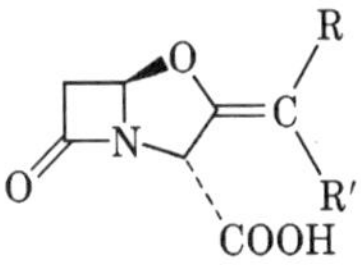

(**12**)

Clavulanic acid has only weak antibacterial activity, but is a potent irreversible inhibitor for many clinically important β-lactamases (10–14,57,58) including penases, and Richmond-Sykes types II, III, IV, V, VI (*Bacteroides*). Type I cephases are poorly inhibited. Clavulanic acid synergizes the activity of many penicillins and cephalosporins against resistant strains. The chemistry (59–63), microbiology (64,65), structure activity relationships (10,13,60–62,66), biosynthesis (67–69), and mechanism of action (6,26,27,67) have been reviewed.

Mechanistic studies (6,26,27,67) have shown that the acyl enzyme species is the ring opened compound (**13**), which can tautomerize to the transiently inhibited amino acrylate (**14**), and both of these species can react further to give irreversibly inactivated enzyme. Three inactivated forms of the enzyme have been detected. Two, according to labeling studies, retain the complete clavulanate skeleton and the other retains only the carbon chain of the β-lactam ring. Structure (**15**) has been suggested as one possible inactivated form.

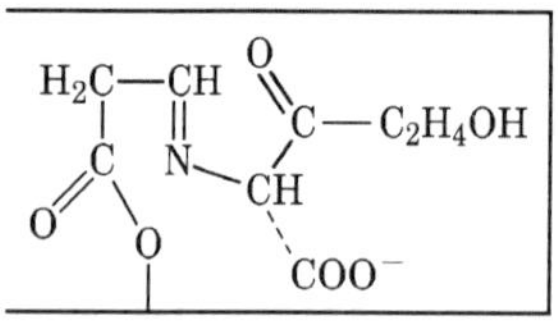
(**13**)

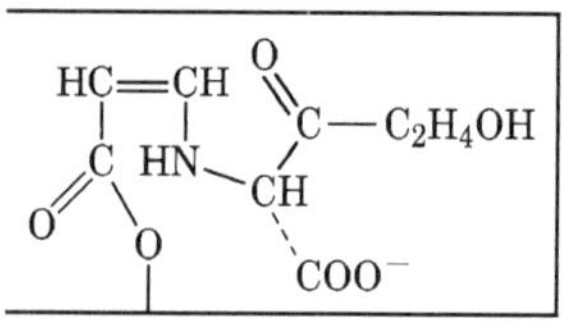
(**14**)

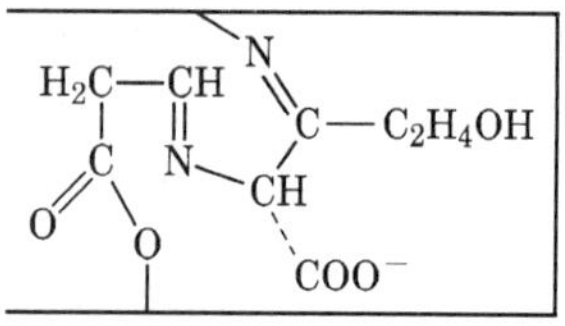
(**15**)

Consistent with the proposed mechanism is the finding that deoxyclavulanate [*69779-62-0*] (**12**, R = CH_3, R′ = H), $C_8H_9NO_4$, and the oxapenem [*78808-19-2*] (**16**), $C_8H_9NO_4$, are potent inhibitors whereas the dihydroclavulanic acid [*59897-36-2*] (**17**), $C_8H_{11}NO_5$, is devoid of useful activity. The oxapenem (**16**) is a more potent inhibitor than clavulanic acid (**12**, R = CH_2OH, R′ = H) but (**16**) is unstable.

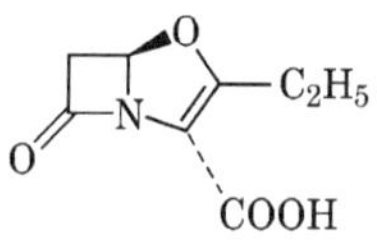
(**16**)

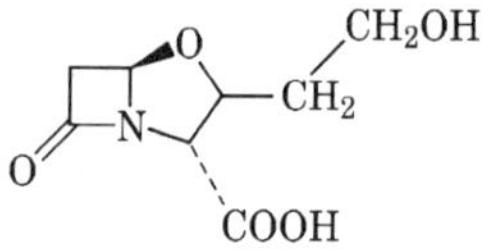
(**17**)

Analogues of clavulanic acid have been reviewed (10,11,13,60–62,66). Inhibitory activity for selected compounds is given in Table 2. A number of clavulanic acid analogues are more potent than the parent compound. Additionally, the natural (*Z*)-isomer (**12**, R = CH_2OH, R′ = H) is more potent than the (*E*)-isomer (**12**, R = H, R′ = CH_2OH) which is a general trend. One surprising feature is that the carboxyl group of clavulanic acid is not required for potent inhibition of β-lactamase (10,13,62,70,71). For antibacterial activity β-lactam antibiotics normally require an analogously positioned carboxylic acid or other acidic group. Descarboxy compounds [*86408-58-4*] (**18**, R = H, R′ = SC_6H_5), $C_{12}H_{11}NO_2S$, and [*79908-54-6*] (**18**, R = $COOCH_3$, R′ = H), $C_8H_8NO_4$, demonstrate β-lactamase inhibition. Few derivatives have been reported that have substitution at C-6. The hydroxyalkyl compounds [*66825-93-2*] (**19**) and (**20**), $C_{10}H_{13}NO_6$, have been disclosed as β-lactamase inhibitors (13) where the cis epimer (**19**) is more active than the trans (**20**).

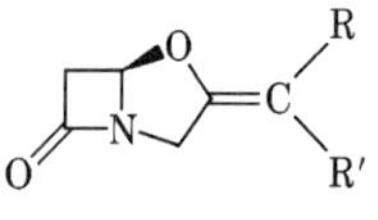
(**18**)

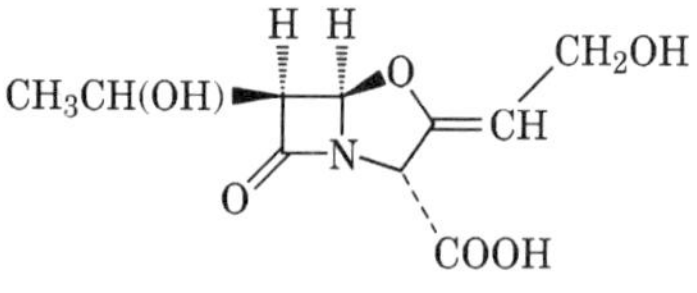
(**19**)

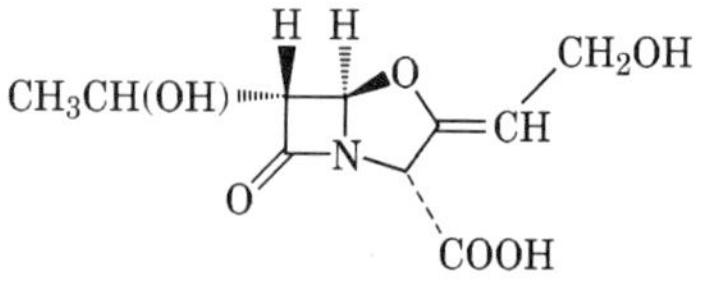
(**20**)

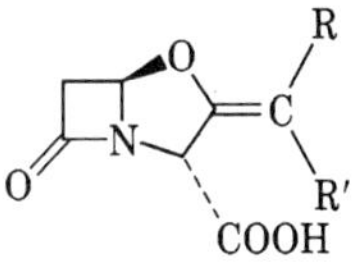

(12)

Table 2. β-Lactamase Inhibitory Activity for Clavulanic Acid and Analogues[a]

CAS Registry Number	Molecular formula	R	IC_{50}, μg/mL[b]			Refs.
			Penase	TEM(III)	Cephase	
[*58001-44-8*]	$C_8H_6NO_5$	CH_2OH	8.2[c]	1.35[c]	10[c]	60,66
			0.03	0.06		13
[*69779-62-0*]	$C_8H_9NO_4$	H	0.12	0.09	5	66
[*78739-05-6*]	$C_{10}H_{11}NO_6$	CH_2OOCCH_3	0.04		0.4	10,66
[*63617-41-4*]	$C_{10}H_{12}N_2O_6$	$CH_2OOCNHCH_3$	1.5	2.5	0.45	10,13,66
[*67529-21-9*]	$C_{10}H_{13}NO_5$	$CH_2OC_2H_5$	0.03	0.05	60	13
[*130097-27-7*]	$C_8H_9NO_8S$	CH_2OSO_3H		3.4		10
[*78739-06-7*]	$C_9H_{10}NO_4S$	CH_2SCH_3	1.6[c]	1.1[c]		10,60,62,66
[*64832-72-0*]	$C_{10}H_{14}N_2O_4S$	$CH_2SCH_2CH_2NH_2$	0.23[c]	0.05[c]		10,60,62
[*78739-07-8*]	$C_9H_9N_2O_4S_2$	$CH_2S_2CNH_2$	3.8[c]	0.08[c]		10,60,62
[*65788-55-8*]	$C_8H_{10}N_2O_4$	CH_2NH_2	0.8[c]	0.08[c]		10,60,62,72
[*75167-01-0*]	$C_9H_{12}N_2O_4$	CH_2NHCH_3	0.03	0.007	>50	13
[*67065-13-8*]	$C_9H_{10}N_2O_5$	CH_2NHCHO	0.04[c]	0.01[c]		10,60,62
[*78739-12-5*]	$C_{15}H_{15}N_3O_5$	$CH_2NHCONHC_6H_5$	0.12[c]	0.06[c]		10,60
[*78739-15-8*]	$C_9H_{12}N_2O_6S$	$CH_2NHSO_2CH_3$	1.15[c]	0.33[c]		10,60,62
[*87131-73-5*]	$C_{10}H_{13}NO_5$	$CH_2CH_2CH_2OH$	0.02	0.1		13
[*64475-40-7*]	$C_{13}H_{11}NO_4$	C_6H_5	<0.08[c]	0.02[c]		10,73
[*62319-53-3*]	$C_8H_9NO_5$	H[e]	0.6[d]	1[d]		10,13,62

[a]Structure (**12**) where R is defined and R′ is H unless otherwise indicated.
[b]The substrate is nitrocefin unless otherwise indicated.
[c]The substrate is ampicillin.
[d]The substrate is penicillin G.
[e]R′ = CH_2OH.

Carbapenem β-Lactamase Inhibitors. Carbapenems are another class of natural product β-lactamase inhibitors discovered about the same time as clavulanic acid. Over forty naturally occurring carbapenems have been identified; many are potent β-lactamase inhibitors. Carbapenem is the trivial name for the 1-azabicyclo[3.2.0]hept-2-ene ring system (**21**) shown in Table 3. The synthesis (74), biosynthesis (75), and β-lactamase inhibitory properties (13,14,66) of carbapenems have been reviewed. Carbapenems are often more potent than clavulanic acid and include type I cephases in the spectrum of inhibition. Table 3 lists the available β-lactamase inhibition data. Synergy is frequently difficult to demonstrate because the compounds are often potent antibacterials.

The mechanism of β-lactamase inhibition has been studied (6,26,27,83). Reversible inhibition is believed to result from the Δ^1-pyrroline (**22**). Certain sulfate ester containing compounds, such as SF-2103A (**21**, R = R″ = SO_3H, R′ = H), give long-lived transiently inhibited species suggesting possible conformational changes or additional ionic bonds to the enzyme (6,83,86,89). MM-22381 (**21**, R = H, R′ = CH_3, R″ = $SCH_2CH_2NHCOCH_3$) and MM-22383 (**21**, R = R′ = H, R″ =

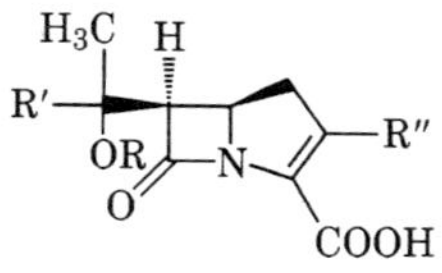

Table 3. Activity of Carbapenem β-Lactamase Inhibitors[a] (21)

Name	CAS Registry Number	Molecular formula	R	R′	R″	IC$_{50}$, μg/mL[b] Penase	TEM(III)	Cephase(I)	Refs.
trans stereochemistry in side chain R″									
C-19393S$_2$	*[76025-74-6]*	$C_{14}H_{18}N_2O_9S_2$	SO_3H	CH_3	$OSCHCHNHCOCH_3$	0.21	0.00027	0.043	76–80
C-19393S$_2$M$_1$	*[82795-81-1]*	$C_{14}H_{18}N_2O_8S_2$	SO_3H	CH_3	$SCHCHNHCOCH_3$	2.4	0.025	0.0024	76–78
MM-4550	*[76985-32-5]*	$C_{13}H_{16}N_2O_9S_2$	SO_3H	H	$OSCHCHNHCOCH_3$	0.003	0.0004	0.001	76,81,82
MM-13902	*[79057-46-8]*	$C_{13}H_{16}N_2O_8S_2$	SO_3H	H	$SCHCHNHCOCH_3$	0.008	0.003	0.0001	14,76,81–83
C-19393H$_2$	*[83916-44-3]*	$C_{14}H_{18}N_2O_6S$	H	CH_3	$OSCHCHNHCOCH_3$	0.004	0.003	0.034	76,78–80
C-19393H$_2$M$_1$	*[82502-20-3]*	$C_{14}H_{18}N_2O_6S$	H	CH_3	$SCHCHNHCOCH_3$	0.6	0.01	0.005	76
C-19393E$_5$	*[83310-72-9]*	$C_{13}H_{16}N_2O_6S$	H	H	$OSCHCHNHCOCH_3$	0.0003	0.027	0.015	76,84
epithienamycin B (MM-22382)	*[65376-20-7]*	$C_{13}H_{16}N_2O_5S$	H	H	$SCHCHNHCOCH_3$	0.012	0.32	0.001	14,76,83
cis stereochemistry in side chain R″									
	[83916-46-5]	$C_{14}H_{18}N_2O_6S$	H	CH_3	$OSCHCHNHCOCH_3$	0.59		0.014	79
	[83916-36-3]	$C_{14}H_{18}N_2O_9S_2$	SO_3H	CH_3	$OSCHCHNHCOCH_3$	0.24		0.0005	79

	[*83916-40-9*]	$C_{13}H_{16}$-$N_2O_9S_2$	SO_3H	H	$OSCHCHNHCOCH_3$	0.003		0.0002	79
	[*83916-39-6*]	$C_{14}H_{18}N_2O_8S_2$	SO_3H	CH_3	$SCHCHNHCOCH_3$	1		0.004	79
	[*83916-42-1*]	$C_{13}H_{16}N_2O_8S_2$	SO_3H	H	$SCHCHNHCOCH_3$	0.014		0.0004	79,85
	[*83916-38-5*]	$C_{14}H_{18}N_2O_6S$	H	CH_3	$OSCHCHNHCOCH_3$	0.055		0.001	79
	[*83916-41-0*]	$C_{13}H_{16}N_2O_6S$	H	H	$OSCHCHNHCOCH_3$	0.0006		0.019	79
	[*83916-37-4*]	$C_{14}H_{18}N_2O_5S$	H	CH_3	$SCHCHNHCOCH_3$	0.36		0.017	79
	[*75443-31-1*]	$C_{13}H_{16}N_2O_5S$	H	H	$SCHCHNHCOCH_3$	0.065		1.5	79
no stereochemistry in side chain R″									
C-$19393S_2M_2$	[*82795-82-2*]	$C_{14}H_{20}N_2O_8S_2$	SO_3H	CH_3	$SCH_2CH_2NHCOCH_3$	2.3	0.042	0.006	76
MM-17880	[*79057-45-7*]	$C_{13}H_{18}N_2O_8S_2$	SO_3H	H	$SCH_2CH_2NHCOCH_3$	0.021	0.005	0.0003	76,81,82
C-$19393H_2M_3$	[*82889-91-6*]	$C_{14}H_{20}N_2O_5S$	H	CH_3	$SCH_2CH_2NHCOCH_3$	0.56	0.0045	0.0054	76
epithienamycin A (MM-22380)	[*63582-78-5*]	$C_{13}H_{18}N_2O_5S$	H	H	$SCH_2CH_2NHCOCH_3$	0.037	0.69	0.003	76
pluracidomycin A (SF-2103A)	[*82138-64-5*]	$C_9H_{11}NO_1OS_2$	SO_3H	H	SO_3H	4	0.05	0.0015	14,86–88
trans β-Lactam stereochemistry									
MM-22381	[*96193-21-4*]	$C_{14}H_{20}N_2O_5S$	H	CH_3	$SCH_2CH_2NHCOCH_3$	5^c	0.03^c	0.2	14
MM-22383	[*65322-98-7*]	$C_{13}H_{16}N_2O_5S$	H	H	$SCHCHNHCOCH_3$	0.04^c	0.04^c	10	14

[a]Structure (**21**) where R, R′, and R″ are as defined.
[b]The substrate is amicillin for penicillinases and cephalothin for cephases unless otherwise indicated.
[c]The substrate is nitrocefin.

$SCHCHNHCOCH_3$), are the only compounds having trans β-lactam stereochemistry for which IC_{50} data are available. Other trans compounds, such as thienamycin, imipenem, and the PS-5 series, have been reported to be β-lactamase inhibitors (90–97), but only limited data have been published. Several totally synthetic carbapenems have also been reported to be potent β-lactamase inhibitors (98,99). Asparenomycin [*76466-24-5*] (**23**), $C_{14}H_{16}N_2O_6S$, and its analogues are reported to be irreversible inhibitors (14,100). These compounds have functionality at C-6 that could react further in a Michael fashion. However, no commercial β-lactamase inhibitor product has been developed from this class of compounds, in part because of low fermentation titers (high synthetic cost), poor stability (82,84,87), short half-life in humans (101), and degradation by renal dipeptidase (102–106).

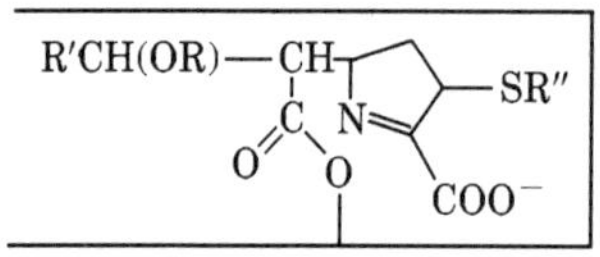

(**22**)

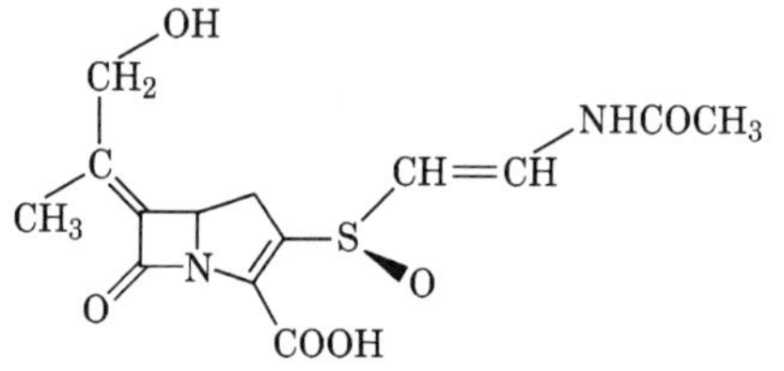

(**23**)

Penem β-Lactamase Inhibitors. The synthesis and antibacterial properties of penems, the trivial name for the 4-thia-1-azabicyclo[3.2.0]hept-2-ene ring system (**24**), have been reviewed (107,108). Like the closely related carbapenems, many of the penems are potent antibacterials. Additionally, penems are also susceptible to degradation by renal dipeptidase, but to a lesser extent. The limited β-lactamase inhibitory data available for penems are presented in Table 4. SCH-29,482 [*77646-83-4*] (**24**, R = H, R′ = $CH(OH)CH_3$, R″ = SC_2H_5), $C_{10}H_{13}NO_4S_2$, is reported to be an inhibitor of type I cephases and the OXA-2 enzyme (109). Compounds [*101803-54-7*] and [*101914-68-5*] (**24**, R = H, R′ = $CH_3CH(OH)$, R″ = 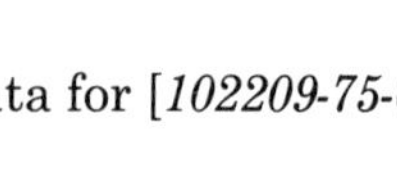), and [*130097-28-8*] (**24**, R = R′ = R″ = CH_3) are also reported to inhibit cephase(I) and some TEM inhibition was reported for the first two. 6-Methylene-substituted penems (R = R_1CH=) are potent, better β-lactamase inhibitors than clavulanic acid, and many of the compounds are potent synergists. No data have been published on the mechanism of β-lactamase inhibition. Comparative data for [*102209-75-6*] the (*Z*)-isomer of (**24**, R = 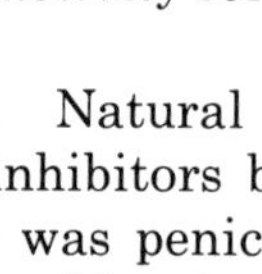, R″ = H) with other other β-lactamase inhibitors (113), and structure activity relationships have been reported (115).

Penam Sulfone β-Lactamase Inhibitors. Natural product discoveries stimulated the rational design of β-lactamase inhibitors based on the readily accessible penicillin nucleus. An early success was penicillanic acid sulfone, (2(*S*)-cis)-3,3-dimethyl-7-oxo-4,4-dioxide-4-thia-1-azabicyclo [3.2.0]heptane-2-car-

Table 4. Activity of Penem β-Lactamase Inhibitors[a]

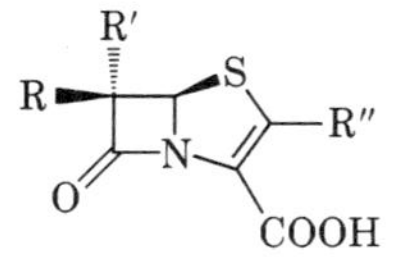

CAS Registry Number	Molecular formula	R	R′	R″	IC$_{50}$, μg/mL[b] Penase	TEM(III)	Cephase(I)	Refs.
[*77646-83-4*][c]	$C_{10}H_{13}NO_4S_2$	H	$CH_3CH(OH)$	SC_2H_5			inhib.	109
[*101803-54-7*]	$C_{13}H_{16}N_2O_6S_2$	H	$CH_3CH(OH)$	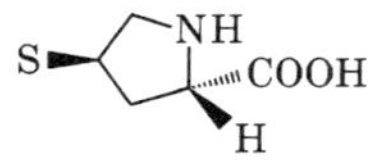	no	0.2–1.8[d]	0.07–0.25[d]	110
[*101914-68-5*]	$C_{13}H_{18}N_2O_5S_2$	H	$CH_3CH(OH)$	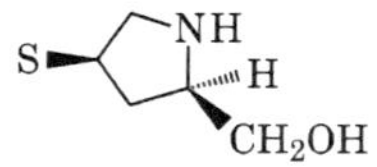				
[*130097-28-8*]	$C_9H_{11}NO_3S$	CH_3	CH_3	CH_3	no	no	inhib.	111
[*81519-84-8*]	$C_{10}H_9NO_4S$	$CH_3COCH=$	Z isomer	CH_3	claim broad spectrum			112
[*102209-75-6*]	$C_{10}H_8N_4O_3S$	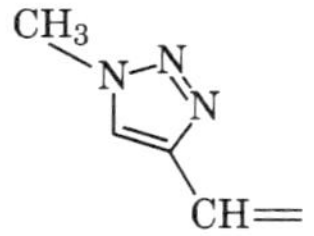	Z isomer	H	0.016	0.002	0.002	113–115
[*81519-81-5*]	$C_8H_7NO_3S$	$CH_3CH=$[e]	Z isomer	H	0.6	0.04	1	115–117
[*128658-07-1*]	$C_8H_7NO_3S$	$CH_3CH=$	E isomer	H	3.2	0.2	0.4	115,117
[*93853-93-0*]	$C_{11}H_7NO_3S_2$	2-thienyl-CH=[e]	Z isomer	H	0.4	0.008	0.02	116
[*93853-93-1*]	$C_{11}H_7NO_3S_2$	2-thienyl-CH=	E isomer	H	>10	3.5	3	116
[*93854-02-5*]	$C_{11}H_7NO_3S_2$	3-thienyl-CH=	Z isomer	H	0.012	0.13	0.003	115,116
[*93854-42-3*]	$C_{11}H_7NO_4S$	2-furyl-CH=	Z isomer	H	0.013	0.003	0.005	115,116
[*93853-72-6*]	$C_{11}H_7NO_4S$	3-furyl-CH=	Z isomer	H	0.4	0.1	0.025	116

[a]Structure (**24**) where R, R′, and R″ are as defined.
[b]The substrate is nitrocefin.
[c]Compound SCH-29482.
[d]Data is from a series of compounds where substituents on R″ vary. Best compounds are the one shown and where R″ = 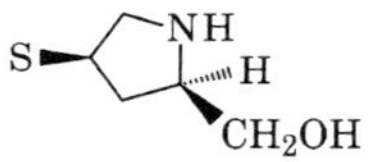
[e]Represents a series of compounds.

boxylic acid [*68373-14-8*] (sulbactam) (**25**, R = R′ = H, R″ = R‴ = CH_3), $C_8H_{11}NO_5S$. The synthesis (118), microbiology (119–121), and clinical use (122–124) of sulbactam have been reviewed. Sulbactam, with minor exceptions, is a weak antibacterial, but is a potent irreversible inactivator of many β-lactamases including penases, and Richmond-Sykes type II,III,IV,V; VI(*Bacteroides*) β-lactamases. Sulbactam is better than clavulanic acid against type I cephases and synergy is observed for combinations of many penicillins and cephalosporins. Because sulbactam is not well absorbed orally, prodrug forms have been developed (125–129). Mechanism of action studies (26,27), including labeling experiments, are consistent with formation of a thiazolidine opened acyl enzyme species (**26**) that tautomerizes to the transiently inhibited form (**27**). These intermediates, shown in Figure 3, can react further to give inactivated enzyme species such as (**28**).

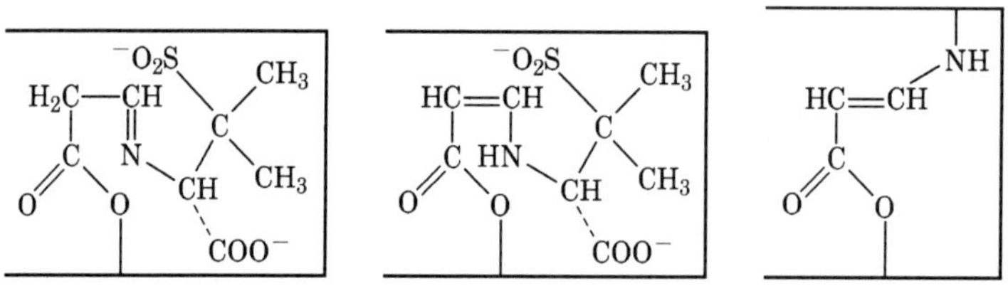

(**26**) (**27**) (**28**)

Fig. 3. Species in the inhibition pathway for sulbactam with β-lactamase.

Numerous other penicillin sulfones have been reported to be β-lactamase inhibitors, as illustrated in Table 5. The effect of C-6 substituents has been extensively explored starting with 6-APA sulfone (**25**, R = NH_2, R′ = H, R″ = R‴ = CH_3), which has modest activity. Mechanistic considerations led to preparation of sulfones of poor substrates, compounds such as methicillin, cloxacillin, nafacillin, and quinacillin sulfone (**25**, R = an aromatic moiety, R′ = H, and R″ = R‴ = CH_3), which were expected to have a long-lived acyl enzyme species (**26**) that would react further to give inactivated enzyme. The compounds all act as β-lactamase inhibitors; but they are poor synergists. Aged solutions of cloxacillin sulfone, where R = 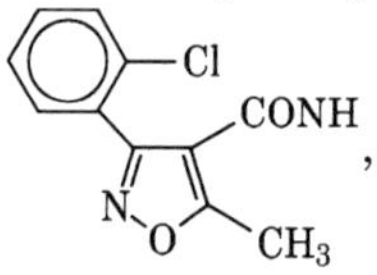, in phosphate buffer gave a product that inhibited *C. freundi* β-lactamase (153,154). Both the C-6 phthalamido [*82087-00-1*] and sulfonamido [*83053-95-6*] compounds, for example, are reported to inhibit *B. cereus* β-lactamase through a conformational change in the enzyme (138).

C-6 halo-substituted sulfones (**25**, R = H, Cl or Br, R′ = Cl or H, R″ = R‴ = CH_3) have been shown to be weak β-lactamase inhibitors and a series of C-6-heteroarylmethylene derivatives, where R = 2-pyridyl-CH or 2-thiazoyl-CH, are potent suicide β-lactamase inhibitors and synergists that are superior to clavulanic acid. Structure activity relationships are discussed in the literature (141) and based on reactions with methanol, used to mimic the serine hydroxyl of β-lactamases, the acyl enzyme species (**29**) has been suggested as being responsi-

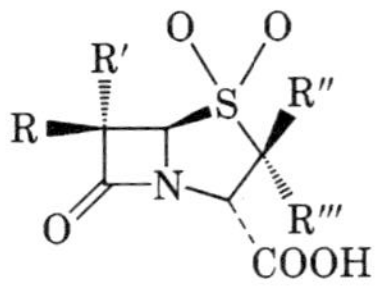

Table 5. Activity of Penam Sulfone β-Lactamase Inhibitors[a] **(25)**

Compound	CAS Registry Number	Molecular formula	R	R″	IC$_{50}$, μg/mL[b,c] Penase	TEM(III)	Cephase(I)	Refs.
sulbactam	[*68373-14-8*]	$C_8H_{11}NO_5S$	H	CH_3	1.4	1.7	5	118–120, 130–132
6-APA sulfone	[*36091-15-3*]	$C_8H_{12}N_2O_5S$	NH_2	CH_3	0.4	40	50	13
methicillin sulfone	[*76350-41-9*]	$C_{17}H_{20}N_2O_8S$	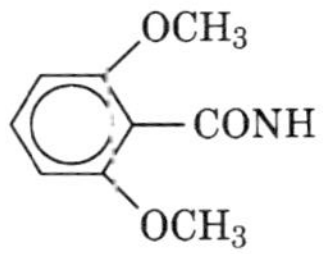	CH_3	>50	16	20	13,33,40
cloxacillin sulfone	[*76788-83-5*]	$C_{19}H_{18}ClN_3O_7S$	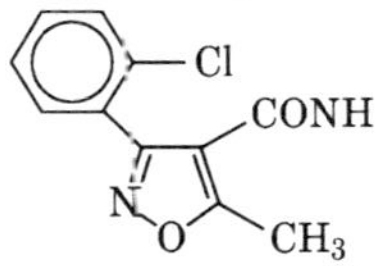	CH_3	+	+	+	133–135
nafacillin sulfone	[*77724-76-6*]	$C_{25}H_{21}N_2O_7S$	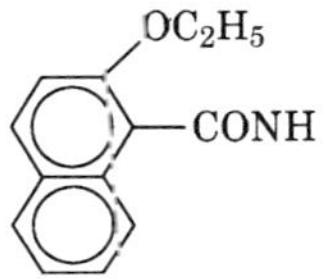	CH_3		+		133

Table 5. (Continued)

Compound	CAS Registry Number	Molecular formula	R	R″	IC$_{50}$, μg/mL[b,c] Penase	TEM(III)	Cephase(I)	Refs.
quinacillin sulfone	[*76788-82-4*]	$C_{19}H_{17}N_3O_8S$	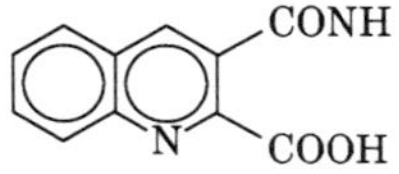	CH_3	+	+		136,137
	[*82087-00-1*]	$C_{16}H_{14}N_2O_7S$	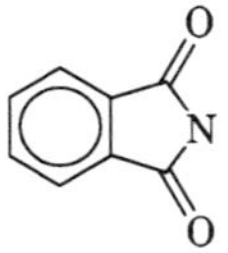	CH_3	+			138
d	[*83053-95-6*]	$C_9H_{11}F_3N_2O_7S_2$	CF_3SO_2NH	CH_3	+	+		137,138
	[*76350-36-2*]	$C_8H_{10}ClNO_5S$	H	CH_3	50	30	30	13,139
e	[*76613-60-8*]	$C_8H_{10}ClNO_5S$	Cl	CH_3	50	13	>50	13
	[*75527-87-6*]	$C_8H_{10}BrNO_5S$	Br	CH_3	4	1.5	>50	13
d,f	[*108636-76-6*]	$C_{14}H_{14}N_2O_5S$	2-pyridyl-CH=	CH_3	+	+	+	140,141
f	[*110088-75-7*]	$C_{12}H_{12}N_2O_5S_2$	2-thiazolyl-CH=	CH_3	+	0.0009[g]	+[g]	141
f	[*123361-44-4*]	$C_{11}H_{15}NO_5S$	CH_2CHCH_2	CH_3	claim broad spectrum[c]			142
	[*78494-75-4*]	$C_{15}H_{17}NO_6S$	$C_6H_5CH(OH)$	CH_3	no[g]		+[g]	142
	[*76517-56-1*]	$C_9H_{13}NO_6S$	$HOCH_2$	CH_3	+[g]	+[g]	+[g]	143
	[*87579-77-9*]	$C_8H_9NO_5S$	H	CH_2[h]	1.5	0.25	15.5	144
BLP-2013	[*79886-07-0*]	$C_8H_{10}ClNO_5S$	H	CH_2Cl	9.6	+		145–147
i	[*115095-95-9*]	$C_8H_{10}ClNO_5S$	H	CH_3	15.3[j]			146
	[*95123-11-8*]	$C_8H_{10}BrNO_5S$	H	CH_2Br	+[j]	+[j]		147
	[*95123-10-7*]	$C_8H_{11}NO_6S$	H	CH_2OH	no[j]	+[j]		147
d	[*95123-06-1*]	$C_{15}H_{15}NO_7S$	H	$CH_2OOCC_6H_5$	no[j]	+[j]		147
	[*112110-80-2*]	$C_9H_{10}N_2O_5S_2$	H	CH_2SCN	2.8[j]	+[j]	+[j]	148
	[*89051-51-4*]	$C_{10}H_{16}N_5O_5S_2$	H	CH_2N_3	1.1[j]			131
	[*112110-81-3*]	$C_{10}H_{16}N_5O_5S_2$	H	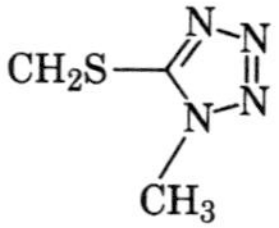	0.335[j]			148

	[112110-83-5]	$C_{15}H_{15}N_5O_5S_2$	H	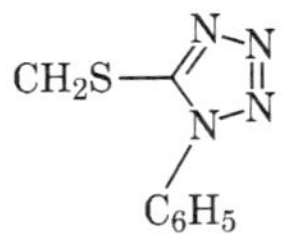	0.354[j]			148
d	*[98382-76-4]*	$C_{12}H_{16}N_4O_5S$	H	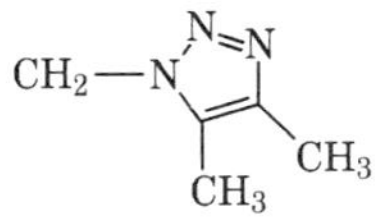	0.026[j]			149
tazobactam	*[89786-04-9]*	$C_{10}H_{12}N_4O_5S$	H	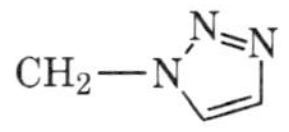	0.27	0.04[j]	0.93	131,148,150, 151,152
d	*[89785-79-5]*	$C_{11}H_{11}N_4O_7SK$	H	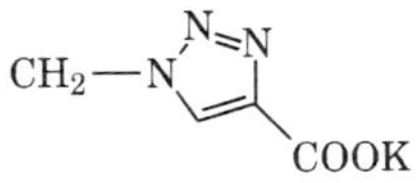	0.23[j]			131
	[89786-01-6]	$C_{12}H_{14}N_4O_7S$	H	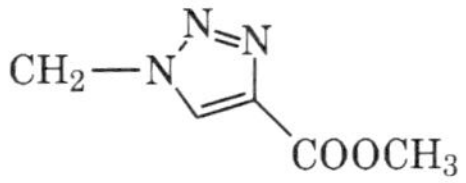	0.018[f]			131
	[98382-75-3]	$C_{10}H_{13}N_5O_5S$	H	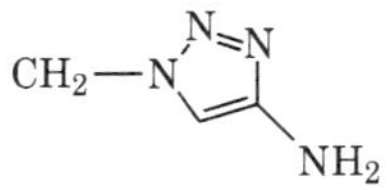	0.22[f]			131

[a]Structure (**25**) where R, R″ are as defined, and R′ = H and R‴ = CH_3 unless otherwise indicated.
[b]Substrate is nitrocefin unless otherwise indicated.
[c]+ Indicates enzyme inhibition, but no available IC_{50} data.
[d]Represents series of compounds.
[e]R′ = Cl.
[f]R′ = (*Z*)-isomer.
[g]Substrate is ampicillin.
[h]Cyclopropane ring bonded also to C-2.
[i]R‴ = CH_2Cl.
[j]Substrate is penicillin G.

ble for inactivation of the enzyme, at least for the R = 2-pyridyl-CH series. Compounds having a leaving group such as acetoxy or fluoro, which can undergo elimination to form the 6-thiazolyl methylene penam sulfone have been reported to be β-lactamase inhibitors and synergists (155). The allyl sulfone (**25**, R = CH_2CHCH_2, R′ = H, R″ = R‴ = CH_3) was reported to be a potent β-lactamase inhibitor, although no IC_{50} was provided (142). The C-6 hydroxyalkyl sulfones (**25**, R = C_6H_5CHOH or $HOCH_2$, R′ = H, R″ = R‴ = CH_3) have been reported as β-lactamase inhibitors. The compound where R = C_6H_5CHOH was primarily active against cephase(I), but the simpler species, R = $HOCH_2$, is reported to be a broad spectrum inhibitor and synergist.

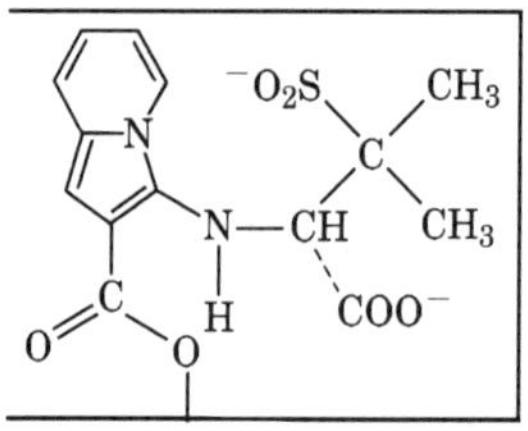

(29)

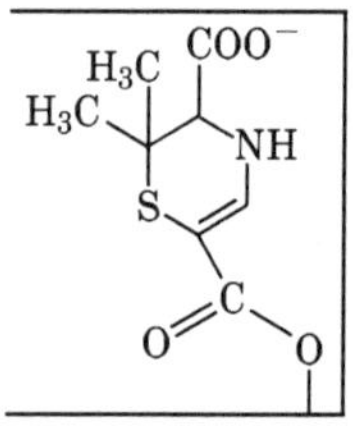

(30)

Substitution of the β-methyl group of penicillanic acid sulfone has also been extensively investigated. The cyclopropyl sulfone (**25**, R = R′ = H, R″ = CH_2, R‴ = CH_3) resembles sulbactam, but is overall a weaker inhibitor. BLP-2013 (**25**, R = R′ = H, R″ = CH_2Cl, R‴ = CH_3) has activity similar to that of sulbactam (145,147). The α-chloromethyl analogue (**25**, R = R′ = H, R″ = CH_3, R‴ = CH_2Cl) is less active. Potent activity has been demonstrated for a series of heteroaryl-substituted methyl sulfones. The compound most studied is tazobactam (YTR-830) (**25**, R = R′ = H, R″ = 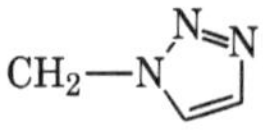, R‴ = CH_3) which has activity similar to that of clavulanic acid and inhibits many of the type I cephases. The synthesis (156), microbiological activity (152,156–162), and stability (152) of tazobactam have been reported.

Penam β-Lactamase Inhibitors. Penam is the trivial name given derivatives of the penicillin nucleus (**31**) the chemical name of which is 4-thia-1-azabicyclo[3.2.0]heptane. Table 6 gives activity data for a diverse group of penams. The report that 6-β-bromopenicillanic acid [*26631-90-3*], [2(*S*)-(2α,5α,6β)]-6-bromo-3,3-dimethyl-7-oxo-4-thia-1-azabicyclo[3.2.0]heptane-2-carboxylic acid, (**31**, R = Br, R′ = H, R″ = R‴ = CH_3) is a potent inhibitor led to intense study both of this compound and analogues. The microbiology profile of 6-β-bromopenicillanic acid has been reported (165) and the compound has progressed to clinical trials (6). Mechanistic studies have demonstrated that the dihydrothiazine derivative (**30**) is responsible for inactivation of β-lactamases (6,26,27). The iodo analogue of the bromopenicillinic acid (**31**, R = I, R′ = H, R″ = R‴ = CH_3) has comparable activity, but the chloro compound (**31**, R = Cl, R′ = H, R″ = R‴ = CH_3) is substantially less active. Mutual prodrugs of the 6-β-bromo and 6-β-iodo compounds with ampicillin have been reported (189). Analogues of 6-β-bromopenicillanic acid having a substituent in the β-methyl position have been explored, (**31**, R = Br, R′ = H, R″ = CH_2F, R‴ = CH_3) is an example, and in general offer no advantages.

The (*Z*)-isomer of 6-acetylmethylene penicillanic acid, RO 15-1903 [*83116-77-2*] (**31**, R = CH_3COCH, (*Z*)-isomer, R″ = R‴ = CH_3), [2(*S*)-(2α,5α,6(*Z*))]-3,3-dimethyl-7-oxo-6-(2-oxo-propylidene)-4-thia-1-azabicyclo[3.2.0]-heptane-2-carboxylic acid, has been reported to be a more potent β-lactamase inhibitor than clavulanic acid. Despite excellent β-lactamase inhibition, the compound gave poorer than expected *in vitro* and *in vivo* synergy results (172,173). This was subsequently attributed to poor uptake of drug by bacteria (172) and instability in serum (177). Mechanistic studies with the TEM enzyme indicate that one molecule of compound inactivates one molecule of enzyme. There is essentially no turnover as substrate. Four inactivated forms of the β-lactamase enzyme have been detected in inhibition studies using RO 15-1903. Structures (**32**) and (**33**) have been implicated in this inactivation. The corresponding (*E*)-isomer (**31**, R = CH_3COCH = (*E*)-isomer, R″ = R‴ = CH_3) and the related structures, (*Z*)-isomers where R = CH_3SO_2CH or $C_6H_5SO_2CH$, are less potent.

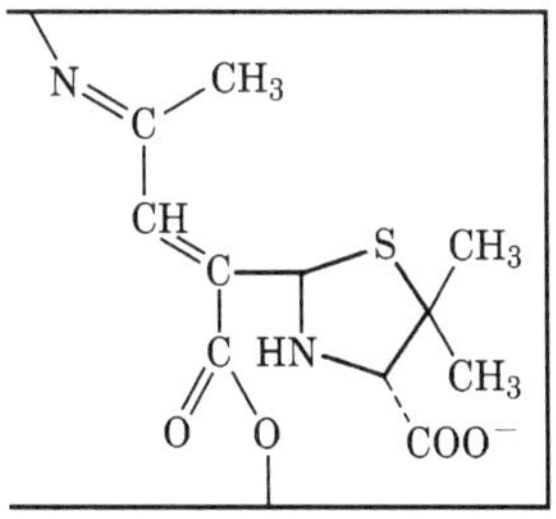

(**32**)

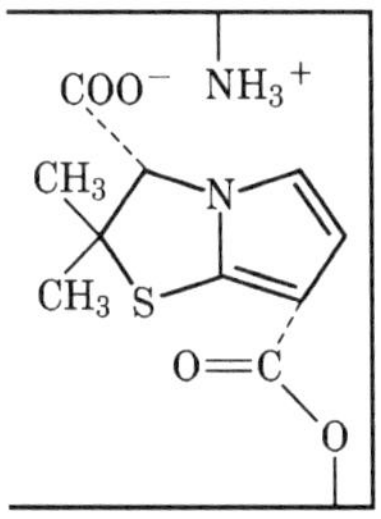

(**33**)

The 6-methoxymethylene penicillanic acid [*93040-42-7*] (**31**, R = CH_3OCH_1 (*Z*)-isomer, R″ = R‴ = CH_3) was designed to mimic the aminoacrylate species found using clavulanic acid and sulbactam. Upon the reaction of this compound with the enzyme, the potential exists for further Michael addition to inactivate the enzyme. The compound is indeed a β-lactamase inhibitor but no synergy data have been reported. The related imine structure [*73707-58-1*], R = *p*-$CH_3C_6H_5N$, (*Z*)-isomer, has also been reported to inhibit type I cephases. The 6-β-[bistrifluoromethane sulfonyl]amidopenicillanic acid [*82954-44-7*] (**31**, R = $(CF_3SO_2)_2N$), R′ = H, R″ = R‴ = CH_3) was speculated to inhibit the penase from *B. cereus* by triflation or generation of an imine that reacts further with the enzyme. Several C-6 substituted amino penams have been reported to inhibit type I cephases; the complex hydroxyalkyl penicillanic acid derivatives are reported to be potent β-lactamase inhibitors and synergists. In combination with amoxicillin, compound (**31**, R = 1-naphthylCOCH(OH), R′ = H, R″ = R‴ = CH_3) protected animals from infection. Synergy has been reported for the usual penam structures where R involves a 2-naphthylsulfoxide or a 2-naphthylphosphate against a β-lactamase producing strain of *Klebsiella*. An unusual series of ferrocenyl penicillins (190) were reported to inhibit β-lactamases.

Other Unusual β-Lactam Based Inhibitors. There are a number of other unusual β-lactams reported to have β-lactamase inhibition activity (191–194). In

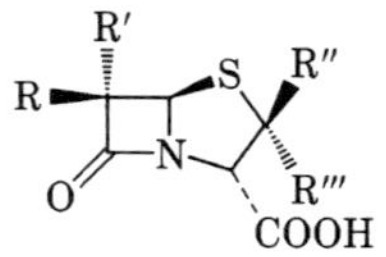

Table 6. Activity of Penam-Based β-Lactamase Inhibitors[a] (**31**)

CAS Registry Number	Molecular formula	R	R′	R″	R‴	IC$_{50}$, μg/mL[b] Penase	TEM(III)	Cephase(I)	Refs.
[*26631-90-3*]	$C_8H_{10}BrNO_3S$	Br	H	CH_3	CH_3	1.5/0.005[c]	0.15/0.005[c]	1.5/0.32[c]	13,163–167,168
[*74772-32-0*]	$C_8H_{10}INO_3S$	I	H	CH_3	CH_3	0.7	0.06	5.5	13,164,165
[*74326-78-6*]	$C_8H_{10}ClNO_3S$	Cl	H	CH_3	CH_3	0.3	50	50	13,163,164,167, 169,170
[*108934-65-2*]	$C_8H_9BrFNO_3S$	Br	H	CH_2F	CH_3		0.13	0.2	168
[*108934-66-3*]	$C_8H_9BrN_4O_3S$	Br	H	CH_2N_3	CH_3	0.2	0.04	8.1	168
[*105433-74-7*]	$C_{10}H_{12}BrNO_3S$	Br	H	$CH_2O_2CCH_3$	CH_3	3.2	0.4	50	168
[*105433-78-1*]	$C_9H_{12}BrNO_4S$	Br	H	CH_2OCH_3	CH_3	1.3	0.2	6.3	168
[*108934-67-4*]	$C_{10}H_{14}BrNO_4S$	Br	H	$CH_2OC_2H_5$	CH_3	3.2	0.4	50	168
[*105458-19-3*]	$C_9H_9BrN_2O_3S_2$	Br	H	CH_2SCN	CH_3	0.4	0.03	50	168
[*83116-77-2*][g]	$C_{11}H_{13}NO_4S$	$CH_3COCH=$	Z isomer	CH_3	CH_3	0.00015	0.0002	1.2	171–177
[*83116-56-7*]	$C_{11}H_{13}NO_4S$	$CH_3COCH=$	E isomer	CH_3	CH_3		0.002	0.012	171
[*104127-98-2*]	$C_{10}H_{13}NO_5S_2$	$CH_3SO_2CH=$	Z isomer	CH_3	CH_3	+[d]		+[d]	178
[*104127-97-1*]	$C_{15}H_{15}NO_5S_2$	$C_6H_5SO_2CH=$	Z isomer	CH_3	CH_3	+[d]		+[d]	178
[*86492-69-5*]	$C_{11}H_{14}N_2O_4S$	$CH_3CN(OH)CH=$	Z isomer	CH_3	CH_3		0.2[e]		176
[*93040-42-7*]	$C_{10}H_{13}NO_4S$	$CH_3OCH=$	Z isomer	CH_3	CH_3		+[d]		179,180
[*73707-58-1*]	$C_{15}H_{16}N_2O_3S_2$	p-CH_3-C_6H_5S-N=	Z isomer	CH_3	CH_3			0.5	181

[*82954-44-7*]	$C_{10}H_{10}F_6N_2O_7S_3$	$(CF_3SO_2)_2N$	H	CH_3	CH_3	$+^d$			182
[*83670-99-9*]	$C_8H_{12}N_2O_6S_2$	HO_3SNH	H	CH_3	CH_3			$+^d$	183
[*89012-63-5*]	$C_{12}H_{18}N_2O_5S$	$C_2H_5O_2CCH_2NH$	H	CH_3	CH_3	no[f]	no[f]	0.25[f]	184
[*89012-69-1*]	$C_{12}H_{20}N_2O_4S$	$CH_3CH(OH)(CH_2)_2NH$	H	CH_3	CH_3	no	no	0.32	184
[*89012-67-9*]	$C_{12}H_{18}N_2O_4S$	$CH_3CO(CH_2)_2NH$	H	CH_3	CH_3	no	no	0.40	184
[*118499-99-3*]	$C_{16}H_{17}N_3O_4S$	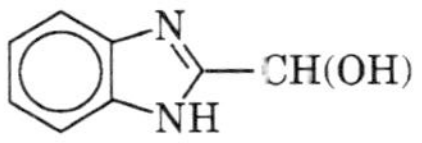	H	CH_3	CH_3				185
[*118498-71-8*]	$C_{18}H_{19}N_3O_4S$	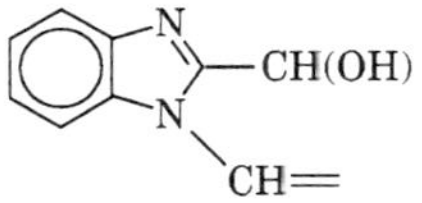	H	CH_3	CH_3				185
[*113656-11-4*]	$C_{17}H_{19}NO_5S$	$C_6H_5CH_2COCH(OH)$	H	CH_3	CH_3	$+^f$	$+^f$		186
[*113656-13-6*]	$C_{16}H_{17}NO_5S$	$C_6H_5COCH(OH)$	H	CH_3	CH_3	+	+		186
[*109990-44-5*]	$C_{20}H_{19}NO_5S$	1-naphthylCOCH(OH)	H	CH_3	CH_3	+	+	+	187
[*94723-14-5*]	$C_{18}H_{17}NO_6S_2$	2-naphthylSO_2O^f	H	CH_3	CH_3				188
[*94723-16-7*]	$C_{24}H_{22}NO_6PS$	2-naphthyl$OP(C_6H_5)O_2$	H	CH_3	CH_3				188
[*94723-39-4*]	$C_9H_{11}NO_4S$	epoxide		CH_3	CH_3				188
[*49628-20-8*]	$C_{17}H_{19}NO_5S$	$C_6H_5OCH_2COCH_2$	H	CH_3	CH_3				188

[a]Structure (**30**) where R, R′, R″, and R‴ are defined as indicated.
[b]The substrate is nitrocefin unless otherwise indicated.
[c]These data correspond to set of β-methyl analogues.
[d]Compound is active but no IC_{50} data are available.
[e]Substrate is penicillin G.
[f]Represents a series of compounds.
[g]Compound Ro 15-1903.

general these compounds are not very potent and are not irreversible inhibitors. Data are also very limited.

Economic Aspects

Although a broad range of β-lactamase inhibitors has been discovered, only clavulanic acid and sulbactam have been commercialized. Clavulanic acid (**12**, R = CH_2OH, R′ = H), manufactured by SmithKline Beecham, is sold as an oral and parenteral product in combination with amoxicillin under the trade name Augmentin. A parenteral product in combination with ticarcillin [*34787-01-4*], $C_{15}H_{16}N_2O_6S$, has the trade name, Timentin. In 1990 worldwide sales of clavulanic acid containing products were about $725 million.

Sulbactam (**25**, R = R′ = H, R″ = R‴ = CH_3) is produced by Pfizer. The oral version of sulbactam in combination with ampicillin is called Unasyn Oral which is the mutual prodrug sultamicillin. Two sulbactam parenteral products are sold, a combination product with ampicillin called Unasyn and a combination with cefoperazone [*62893-19-0*] called Sulperazon. In addition, sulbactam is sold alone for parenteral use with any β-lactam antibiotic as Betamaze. In 1990 worldwide sales of sulbactam containing products were over $280 million.

BIBLIOGRAPHY

"Antibiotics, β-Lactams, Clavulanic Acid, Thienamycin, and Others," in *ECT* 3rd ed., Supplement, pp. 83–131, by Allan Brown, Beecham Research Laboratories.

1. J. M. T. Hamilton Miller and J. T. Smith, eds., *Beta-Lactamases,* Academic Press, New York, 1979.
2. M. Richmond in M. Salton and G. D. Shockman, eds., *β-Lactam Antibiotics (Mode of Action, New Developments, and Future Prospects),* Academic Press, New York, 1981, pp. 261–274.
3. A. A. Medeiros and G. A. Jacoby in S. F. Queener, J. A. Webber, and S. W. Queener, eds., *Clinical Pharmacology 4, β-Lactam Antibiotics for Clinical Use,* Marcel Dekker Inc., New York, 1986, pp. 49–84.
4. N. H. Georgopapadakou and R. B. Sykes in A. L. Demain and N. A. Solomon, eds., *Antibiotics Containing the β-Lactam Structure,* Part II, Springer-Verlag, New York, 1983, pp. 1–78.
5. J.-M. Ghuysen, *Rev. Inf. Dis.* **10,** 726 (1988).
6. R. F. Pratt in M. Sandler and H. J. Smith, eds., *Design of Enzyme Inhibitors as Drugs,* Oxford University Press, New York, 1989, pp. 178–205.
7. R. A. Nicholas and J. L. Strominger, *Rev. Inf. Dis.* **10,** 733 (1988).
8. M. Cole in Ref. 1, pp. 205–289.
9. R. B. Sykes and K. Bush in R. B. Morin and M. Gorin, eds., *Chemistry and Biology of β-Lactam Antibiotics,* Vol. 3, Academic Press, New York, 1982, pp. 155–207.
10. J. R. E. Hoover in Ref. 4, pp. 212–226.
11. M. Cole, *Phil. Trans. R. Soc. London* **B289,** 207 (1980).
12. M. Cole, *Drugs of the Future* **6,** 697 (1980).
13. C. Reading and M. Cole, *J. Enzyme Inhibition* **1,** 83 (1986).
14. S. J. Cartwright and S. G. Waley, *Med. Res. Rev.* **3,** 341 (1983).
15. C. E. Newall in Ref. 2, pp. 287–300.

16. A. Bauernfeind, *Rev. Inf. Dis.* **8,** 5470 (1986).
17. K. Bush, *Rev. Inf. Dis.* **10,** 681 (1988).
18. R. H. Richmond and R. B. Sykes, *Adv. Microb. Physiol.* **9,** 31 (1973).
19. H. C. Neu, *Am. J. Med.* **79,** (Suppl. B.), 2 (1985).
20. B. Wiedemann, C. Kliebe, and M. Kresken, *J. Antimicrob. Chemother.* **24,** (Suppl. B), 1 (1989).
21. J. D. Williams, *Drugs* **35,** (Suppl. 7), 3 (1988).
22. A. L. Fink, *Pharm. Res.* **2,** 55 (1985).
23. S. J. Cartwright, A. K. Tan, and A. L. Fink, *Biochem. J.* **263,** 905 (1989).
24. R. Virden, A. K. Tan, and A. L. Fink, *Biochemistry* **29,** 145 (1990).
25. V. Knott-Hunziker, S. Petursson, S. G. Waley, B. Jaurin, and T. Grundstrom, *Biochem. J.* **207,** 315 (1982).
26. J. R. Knowles, *Acc. Chem. Res.* **18,** 97 (1985).
27. J. R. Knowles in F. E. Hahn, ed., *Antibiotics VI (Modes and Mechanisms of Microbial Growth Inhibitors),* Springer-Verlag, New York, 1983, pp. 90–107.
28. K. Bush and R. B. Sykes, *J. Antimicrob. Chemother.* **11,** 97 (1983).
29. R. Reid, L. C. Felton, and M. A. Pittroff, *Proc. Soc. Exp. Biol. Med.* **63,** 438 (1946).
30. O. K. Behrens and L. Garrison, *Arch. Biochem. Biophys.* **27,** 94 (1950).
31. G. V. Patil and R. A. Day, *Biochem. Biophys. Acta* **293,** 490 (1973).
32. H. Ogawara and H. Umezawa, *Biochimica Biophys. Acta.* **327,** 481 (1973).
33. E. P. Abraham and G. G. F. Newton, *Biochem. J.* **63,** 628 (1956).
34. G. N. Rolinson, S. Stevens, F. R. Batchelor, J. Cameron-Wood, and E. B. Chain, *Lancet* (2), 564 (1960).
35. N. Citri, A. Samuni, and N. Zyk, *Proc. Nat. Acad. Sci. USA* **73,** 1048 (1976).
36. A. L. Fink, K. M. Behner, and A. K. Tan, *Biochem.* **26,** 4248 (1987).
37. M. Cole in J. D. Williams, ed., *Antibotic Interactions,* Academic Press, New York, 1979, pp. 99–135.
38. M. Arisawa, J. Ohshima, and H. B. Maruyama, *Chem. Pharm. Bull.* **30,** 3333 (1982).
39. K. P. Fu and H. C. Neu, *J. Antibiot.* **32,** 909 (1979).
40. M. H. Richmond, *J. Antimicrob. Chemother.* **6,** 445 (1980).
41. K. P. Fu and H. C. Neu, *Antimicrob. Agents Chemother.* **14,** 322 (1978).
42. K. Murakami and T. Yoshida, *Antimicrob. Agents Chemother.* **27,** 727 (1985).
43. R. L. Charnas and R. L. Then, *Rev. Inf. Dis.* **10,** 752 (1988).
44. W. S. Faraci and R. F. Pratt, *Biochemistry* **24,** 903 (1985).
45. S. Nishimura and co-workers, *J. Antibiot.* **42,** 159 (1989).
46. S. Nishimura, N. Yasuda, H. Sasaki, Y. Matsmoto, and T. Kamimura, *J. Antibiot.* **43,** 114 (1990).
47. S. Ohya, T. Miyadera, and M. Yamazaki, *Antimicrob. Agents. Chemother.* **21,** 613 (1982).
48. H. C. Neu and N.-X. Chin, *Antimicrob. Agents Chemother.* **31,** 505 (1987).
49. R. L. Then, *Chemotherapy* **30,** 398 (1984).
50. K. Bush, J. S. Freudenberger, and R. B. Sykes, *Antimicrob. Agents Chemother.* **22,** 414 (1982).
51. R. B. Sykes and co-workers, *Nature* **291,** 489 (1981).
52. K. Bush, S. A. Smith, S. K. Tanaka, and D. B. Bonner, *J. Antimicrob. Chemother.* **22,** 801 (1988).
53. R. Cooper, K. Bush, P. A. Principe, W. H. Trejo, J. S. Wells, and R. B. Sykes, *J. Antibiot.* **36,** 1252 (1983).
54. K. Tanizawa, K. Santoh, and Y. Kanaoka, *FEBS Lett.* **250,** 218 (1989).
55. A. G. Brown, D. Butterworth, M. Cole, G. Hanscomb, J. D. Hood, and C. Reading, *J. Antibiot.* **29,** 668 (1976).
56. T. T. Howarth, A. G. Brown, and T. J. King, *J. Chem. Soc., Chem. Commun.,* 266 (1976).

57. C. Reading and M. Cole, *Antimicrob. Agents Chemother.* **11,** 852 (1977).
58. H. C. Neu and K. P. Fu, *Antimicrob. Agents Chemother.* **14,** 650 (1978).
59. P. C. Cherry and C. E. Newall in R. B. Morin and M. Gorman, eds., *Chemistry and Biology of β-Lactam Antibiotics,* Academic Press, New York, 1982, pp. 362–402.
60. C. E. Newall in G. I. Gregory, ed., *Recent Advances in the Chemistry of β-Lactam Antibiotics,* Royal Society of Chemistry, London, 1980, pp. 151–167.
61. P. H. Bently, P. D. Berry, G. Brooks, M. L. Gilpin, E. Hunt, and I. I. Zomaya in Ref. 60, pp. 175–183.
62. R. D. G. Cooper in P. G. Sammes, ed., *Topics in Antibiotic Chemistry,* Vol. 3, John Wiley & Sons, Inc., New York, 1980, pp. 43–73.
63. G. Brooks and co-workers in A. G. Brown and S. M. Roberts, eds., *Recent Advances in the Chemistry of β-Lactam Antibiotics,* Royal Society of Chemistry, London, 1984, pp. 222–241.
64. C. Reading and B. Slocombe in Ref. 3, pp. 527–592.
65. *J. Antimicrob. Chemother.* **24,** (Suppl. B), (1989).
66. A. G. Brown, *J. Antimicrob. Chemother.* **7,** 15 (1981).
67. C. Reading and P. Hepburn, *Biochem. J.* **179,** 67 (1979).
68. A. Basak, S. P. Salowe, and C. A. Townsend, *J. Am. Chem. Soc.* **112,** 1654 (1990).
69. S. W. Elson in P. H. Bently and R. Southgate, eds., *Recent Advances in the Chemistry of β-Lactam Antibiotics,* Royal Society of Chemistry, London, 1989, pp. 303–320.
70. C.-P. Mak, K. Prasad, and F. Turnowsky, *J. Antibiot.* **36,** 398 (1983).
71. E. Hunt, P. H. Bently, G. Brooks, and M. L. Gilpin, *J. Chem. Soc., Chem. Commun.,* 906 (1977).
72. S. W. Elson, J. Gillett, N. H. Nicholoson, and J. W. Tyler, *J. Chem. Soc., Chem. Commun.,* 979 (1988).
73. P. H. Bently, G. Brooks, M. L. Gilpin, and E. Hunt, *J. Chem. Soc., Chem. Commun.,* 905 (1977).
74. R. W. Ratcliffe and G. Alber-Schonberg in R. B. Morin and M. Gorman, eds., *Chemistry and Biology of β-Lactam Antibiotics,* Academic Press, New York, 1982, pp. 227–313.
75. Y. Fukagawa, M. Okabe, T. Yoshioka, and T. Ishikura in Ref. 63, pp. 163–182.
76. K. Okonogi, S. Harada, S. Shinagawa, A. Imada, and M. Kuno, *J. Antibiot.* **35,** 963 (1982).
77. F. Kobayashi and co-workers, *Antimicrob. Agents Chemother.* **21,** 536 (1982).
78. K. Okonogi, Y. Nozaki, A. Imada, and M. Kuno, *J. Antibiot.* **34,** 212 (1981).
79. S. Harada, S. Tsubotani, M. Asai, K. Okonogi, and M. Kondo, *J. Med. Chem.* **26,** 753 (1983).
80. M. Nakayama and co-workers, *J. Antibiot.* **33,** 1388 (1980).
81. D. Butterworth, M. Cole, G. Hanscomb, and G. N. Rolinson, *J. Antibiot.* **32,** 287 (1979).
82. J. D. Hood, S. J. Box, and M. S. Verrall, *J. Antibiot.* **32,** 295 (1979).
83. R. L. Charnas and J. R. Knowles, *Biochemistry* **20,** 2732 (1981).
84. S. Harada, Y. Nozaki, and S. Shinagawa, *J. Antibiot.* **35,** 957 (1982).
85. Y. Nakamura and co-workers, *J. Antibiot.* **39,** 956 (1986).
86. A. Yamaguchi, T. Hirata, and T. Sawai, *Antimicrob. Agents Chemother.* **25,** 348 (1984).
87. T. Yoshida, Y. Kazuno, T. Shomura, S. Murata, S. Inouye, and T. Ito, *J. Antibiot.* **39,** 956 (1986).
88. T. Ito and co-workers, *J. Antibiot.* **35,** 533 (1982).
89. T. Niwa, T. Yoshida, A. Tamura, Y. Kazuno, S. Inouye, and T. Ito, *J. Antibiot.* **39,** 943 (1986).
90. R. H. Richmond, *J. Antimicrob. Chemother.* **7,** 279 (1981).
91. M. Toda, K. Sato, H. Nakazawa, M. Inoue, and S. Mitsuhashi, *Antimicrob. Agents Chemother.* **18,** 837 (1980).

92. K. Okamura, M. Sakamoto, Y. Fukagawa, T. Isahikura, and J. Lein, *J. Antibiot.* **32,** 280 (1979).
93. K. Okamura, M. Sakamoto, and T. Isahikura, *J. Antibiot.* **33,** 293 (1980).
94. K. Okamura, M. Sakamoto, and T. Isahikura, *Biochem. J.* **185,** 177 (1980).
95. N. Shibamoto, M. Nishino, K. Okamura, Y. Fukagawa, and T. Ishikura, *J. Antibiot.* **35,** 763 (1982).
96. M. Sakamoto, T. Ishikura, and Y. Fukagawa, *J. Antibiot.* **37,** 1414 (1984).
97. H. Hirai, K. Sawada, M. Aratani, and M. Hashimoto, *Tetrahedron. Lett.* **25,** 5075 (1984).
98. H. Natsugari and co-workers, *J. Antibiot.* **36,** 855 (1983).
99. N. Tamura and co-workers, *Chem. Pharm. Bull.* **35,** 996 (1987).
100. K. Murakami, M. Doi, and T. Sawai, *Antimicrob. Agents Chemother.* **25,** 348 (1984).
101. M. J. Basker, R. J. Boon, S. J. Box, E. A. Prestige, G. M. Smith, and S. R. Specer, *J. Antibiot.* **36,** 416 (1983).
102. J. Birnbaum, F. M. Kahan, H. Kropp, and J. S. MacDonald, *Am. J. Med.* **78,** (Suppl. 6A), 3 (1985).
103. H. Mikami, M. Ogashiwa, Y. Saini, M. Inoue, and S. Mitsuhashi, *Antimicrob. Agents Chemother.* **22,** 693 (1982).
104. S. R. Norrby and co-workers, *Antimicrob. Agents Chemother.* **23,** 300 (1983).
105. N. Shibamoto, M. Sakamoto, Y. Fukagawa, and T. Ishikura, *J. Antibiot.* **35,** 729 (1982).
106. P. Schindler, G. Huber, and W. Konig, *J. Antibiot.* **36,** 876 (1983).
107. I. Ernest in R. B. Morin and M. Gorman, eds., *Chemistry and Biology of β-Lactam Antibiotics,* Academic Press, New York, 1982, pp. 315–360.
108. S. W. McCombie and A. K. Ganguly, *Med. Res. Rev.* **8,** 393 (1988).
109. J. C. Pechere, R. Letarte, R. Guay, C. Asselin, and C. Morin, *J. Antimicrob. Chemother.* **9,** Suppl. C, 123–132 (1982).
110. G. Emmer and co-workers, *J. Antibiot.* **38,** 1371 (1985).
111. R. B. Woodward, *Phil. Trans. R. Soc. London* **B289,** 239 (1980).
112. S. Adams, *Heterocycles* **22,** 1509 (1984).
113. K. Coleman, D. R. J. Griffin, J. W. J. Page, and P. A. Upshon, *Antimicrob. Agents Chemother.* **33,** 1580 (1989).
114. N. F. Osborne, N. J. P. Broom, S. Coulton, J. B. Harbridge, M. A. Harris, I. Stirling-Francois, and G. Walker, *J. Chem. Soc., Chem. Commun.,* 371 (1989).
115. N. J. P. Broom, S. Coulton, I. Francois, J. B. Harbridge, J. H. C. Nayler, and N. F. Osborne in Ref. 69, pp. 247–258.
116. N. J. P. Broom, K. Coleman, P. A. Hunter, and N. F. Osborne, *J. Antibiot.* **43,** 76 (1990).
117. M. J. Basker and N. F. Osborne, *J. Antibiot.* **43,** 70 (1990).
118. A. J. Sweetman, *Drugs of the Future* **5,** 23 (1980).
119. A. R. English, J. A. Retsema, A. E. Girard, J. G. Lynch, and W. E. Barth, *Antimicrob. Agents Chemother.* **14,** 414 (1978).
120. J. A. Retsema, A. R. English, and A. E. Girard, *Antimicrob. Agents Chemother.* **17,** 615 (1980).
121. J. Klastersky and P. Van der Auwera, *Diagn. Microbial. Inf. Dis.* **12,** 165S (1989).
122. D. M. Campoli-Richards and R. N. Brogden, *Drugs* **33,** 577 (1987).
123. D. L. Hemsell and H. H. Werner, eds., *Drugs* **31,** (Suppl. 2), 1–28 (1986).
124. *Diagn. Microbiol. Inf. Dis.* **12,** (Suppl.) 107S–215S (1989).
125. G. Foulds and co-workers, *Proceedings of the 11th International Congress and the 19th Interscience Conference on Antimicrobial Agents and Chemotherapy,* Boston, Mass., 1980, p. 353.
126. J. G. Stam and co-workers, *22nd Interscience Conference on Antimicrobial Agents and Chemotherapy,* Miami Beach, Fla., 1982, p. 510.

127. *Chemotherapy* **33,** (Suppl. 3), (1984).
128. B. Baltzen and co-workers, *J. Antibiot.* **33,** 1183 (1980).
129. A. R. English, D. Girard, V. J. Jasys, R. J. Martingano, and M. S. Kellogg, *J. Med. Chem.* **33,** 344 (1990).
130. K. P. Fu and H. C. Neu, *Antimicrob. Agents Chemother.* **15,** 171 (1979).
131. R. G. Micetich and co-workers, *J. Med. Chem.* **30,** 1469 (1987).
132. J. A. Retsema, W. U. Schelky, A. E. Girard, and A. R. English, *Drugd. Expt. Clin. Res.* **7,** 255 (1981).
133. A. L. Fink, L. M. Ellerby, and P. M. Bassett, *J. Am. Chem. Soc.* **111,** 6871 (1989).
134. T. Swai, A. Yamaguchi, and K. Tsukamato, *Rev. Inf. Dis.* **10,** 721 (1988).
135. A. Yamaguchi, A. Adachi, T. Hirata, H. Adachi, and T. Sawai, *J. Antibiot.* **38,** 83 (1985).
136. J. Fisher, R. L. Charnas, S. M. Bradley, and J. R. Knowles, *Biochemistry* **20,** 2726 (1981).
137. P. S. F. Mezes, A. J. Clark, G. I. Dmitrienko, and T. Viswanatha, *FEBS Lett.* **143,** 265 (1982).
138. G. I. Dmitrienko, C. R. Copeland, L. Arnold, M. E. Savard, A. J. Clarke, and T. Viswanatha, *Bioorg. Chem.* **13,** 34 (1985).
139. S. J. Cartwright and A. F. W. Coulson, *Nature* **278,** 360 (1979).
140. Y. L. Chen, C.-W. Chang, and K. Hedberg, *Tetrahedron Lett.* **27,** 3449 (1986).
141. Y. L. Chen and co-workers, *J. Antibiot.* **40,** 803 (1987).
142. S. Hanessian and M. Alepegiani, *Tetrahedron* **45,** 941 (1989).
143. M. G. Waters, J. A. Retsema, and M. S. Kellogg, *ASM Meeting,* Lake George, New York, 1983.
144. D. D. Kieth, J. Tengi, P. Rossman, L. Todaro, and M. Weigele, *Tetrahedron* **39,** 2445 (1983).
145. W. J. Gottstein, L. B. Crest, R. G. Graham, U. J. Haynes, and D. N. McGregor, *J. Med. Chem.* **24,** 1531 (1981).
146. S. N. Maiti, P. Spevak, K. Ogawa, and R. G. Micetich, *J. Org. Chem.* **53,** 3803 (1988).
147. W. J. Gottstein, U. J. Hayes, and D. N. McGreggor, *J. Med. Chem.* **28,** 518 (1985).
148. H. Tanaka, M. Tanaka, A. Nakai, S. Yamada, N. Ishida, T. Otani, and S. Torii, *J. Antibiot.* **41,** 579 (1988).
149. R. G. Micetich and co-workers, *Proceedings of the 14th International Congress on Chemotherapy,* Kyoto, Japan, 1985, pp. 249–250.
150. L. Gutmann, M.-D. Kitzis, S. Yamabe, and J. F. Acar, *Antimicrob. Agents Chemother.* **29,** 955 (1986).
151. F. Moosdeen, J. D. Williams, and S. Yamabe in Ref. 149, pp. 1272–1273.
152. N. Ishida, A. Hyodo, C. Hanehara, Y. Miyake, Y. Kawaguchi, and S. Yamabe in Ref. 149, pp. 1274–1275.
153. A. Yamaguchi, A. Adachi, T. Hirata, H. Adachi, and T. Sawai, *J. Antibiot.* **38,** 83 (1985).
154. A. Yamaguchi, T. Inaba, and T. Sawai, *J. Antibiot.* **41,** 690 (1988).
155. Y. L. Chen, K. Hedberg, J. F. Barrett, and J. A. Retsema, *J. Antibiot.* **41,** 134 (1988).
156. M. N. Serradell and J. Castaner, *Drugs of the Future* **11,** 588 (1986).
157. S. C. Aronoff, M. R. Jacobs, S. Johenning, and S. Yamabe, *Antimicrob. Agents Chemother.* **26,** 580 (1984).
158. M. R. Jacobs, S. C. Arnoff, S. Johenning, and S. Yamabe, *J. Antimicrob. Chemother.* **18,** 177 (1986).
159. M. R. Jacobs, S. C. Arnoff, S. Johenning, D. M. Shales, and S. Yamabe, *Antimicrob. Agents Chemother.* **29,** 980 (1986).
160. P. C. Appelbaum, M. R. Jacobs, S. K. Spangler, and S. Yamabe, *Antimicrob. Agents Chemother.* **30,** 789 (1986).

161. S. C. Arnoff, M. R. Jacobs, P. H. Labrozzi, and S. Yamabe, *J. Antimicrob. Chemother.* **18,** 271 (1986).
162. *Diagn. Microb. Inf. Dis.,* 477 (1990).
163. P. J. Claes, H. Vanderhaeghie, E. Roets, and J. Thomis, *J. Antibiot.* **37,** 75 (1985).
164. W. von Daehne, *J. Antibiot.* **33,** 451 (1980).
165. H. C. Neu, *Antimicrob. Agents Chemother.* **23,** 63 (1983).
166. R. F. Pratt and M. J. Loosemore, *Proc. Natl. Acad. Sci. USA* **75,** 4145 (1978).
167. R. Wise, J. M. Andrews, and N. Patel, *J. Antimicrob. Chemother.* **7,** 531 (1981).
168. W. von Daehne, E. T. Hansen, and N. Rastrup-Andersen, in Ref. 63, pp. 375–380.
169. J. E. G. Kemp, M. D. Closier, S. Narayanaswami, and M. H. Stefaniak, *Tetrahedron Lett.* **21,** 2991 (1980).
170. B. A. Moore and K. W. Brammer, *Antimicrob. Agents Chemother.* **20,** 327 (1981).
171. S. Adam, R. L. Then, and P. Angehrn, *J. Antibiot.* **40,** 108 (1987).
172. N.-X. Chin, M. J. McElrath, and H. C. Neu, *Chemotherapy* **34,** 318 (1988).
173. P. Angehrn and M. Arisawa, *J. Antibiot.* **35,** 1584 (1982).
174. M. Arisawa and R. L. Then, *J. Antibiot.* **35,** 1578 (1982).
175. M. Arisawa and R. L. Then, *Biochem. J.* **209,** 609 (1983).
176. M. Arisawa and S. Adam, *Biochem. J.* **211,** 447 (1983).
177. M. Arisawa and R. L. Then, *J. Antibiot.* **36,** 1372 (1983).
178. D. Habich and K. Metzger, *Heterocycles* **24,** 289 (1986).
179. D. G. Brenner, *J. Org. Chem.* **50,** 18 (1985).
180. D. G. Brenner and J. R. Knowles, *Biochemistry* **23,** 5839 (1984).
181. E. M. Gordon, H. W. Chang, C. M. Cimarusti, B. Toeplitz, and J. Z. Gougoutas, *J. Am. Chem. Soc.* **102,** 1690 (1980).
182. P. S. F. Mezes, R. W. Friesen, T. Viswanatha, and G. I. Dmitrienko, *Heterocycles* **19,** 1207 (1982).
183. M. Yamashita, S. Hashimoto, and M. Ezaki, *J. Antibiot.* **36,** 1774 (1983).
184. M. J. Calverley and M. Begtrup, *J. Antibiot.* **36,** 1507 (1983).
185. Y. L. Chen, K. G. Hedberg, and K. J. Guarino, *Tetrahedron Lett.* **30,** 1067 (1989).
186. L. A. Reed, D. A. Charleson, and R. A. Volkmann, *Tetrahedron Lett.* **28,** 3431 (1987).
187. J. A. Retsema and co-workers, *28th Interscience Conference on Antimicrobial Agents and Chemotherapy,* 122 (1988).
188. J. C. Sheehan, E. Chacko, T. J. Commons, Y. S. Lo, D. R. Ponzi, and A. Schwabacher, *J. Antibiot.* **37,** 1441 (1984).
189. W. von Daehne, B. H. Nielsen, and I. Roholt, in Ref. 149, pp. 1280–1281.
190. E. I. Edwards, R. Epton, and G. Marr, *J. Organomet. Chem.* **85,** C23 (1975).
191. R. Joyeau, H. Molines, R. Labia, and M. Wakselman, *J. Med. Chem.* **31,** 370 (1988).
192. D. Davies and M. J. Pearson, *J. Chem. Soc., Perkin Trans. 1,* 2539 (1981).
193. D. Habich, P. Naab, and K. Metzger, *Tetrahedron Lett.* **24,** 2559 (1983).
194. R. Joyeau, L. D. S. Yadav, and M. Wakselman, *J. Chem. Soc., Perkin Trans. 1,* 1899 (1987).

J. G. Stam
Pfizer Central Research

MONOBACTAMS

β-Lactam antibiotics are one of the best established classes of antimicrobial agents for the treatment of infectious diseases. Since the discovery of penicillin in 1929 and cephalosporin C in 1955, a large number of semisynthetic analogues has

Table 1. Naturally Occurring Monobactams

Organism	Monobactam	CAS Registry Number	Molecular formula	Structure number	Structure
Chromobacterium violaceum	SQ 26,180	*[79720-08-4]*	$C_6H_{10}N_2O_6S{\cdot}K$	(**3**, R = OCH_3)	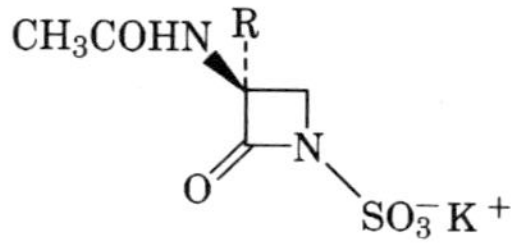
Gluconobacter oxydans	SQ 26,445	*[77912-79-9]*	$C_{12}H_{20}N_4O_9S$	(**4**, R″ = H)	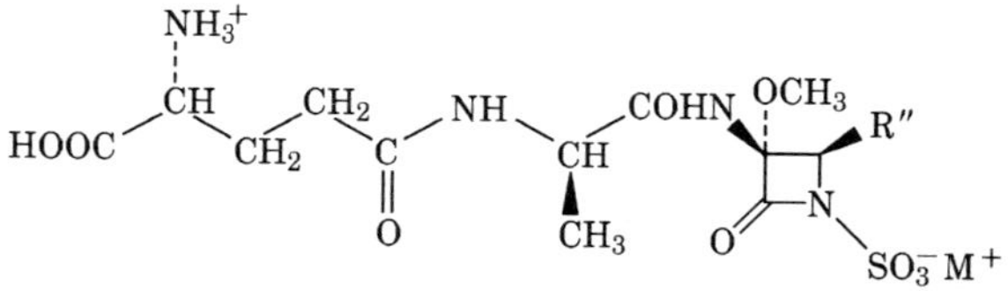
Pseudomonas acidophila	sulfazecin	*[77912-79-9]*	$C_{12}H_{20}N_4O_9S$	(**4**, R″ = H)	
Pseudomonas mesoacidophila	isosulfazecin	*[77900-75-5]*	$C_{12}H_{20}N_4O_9S$	(**5**)	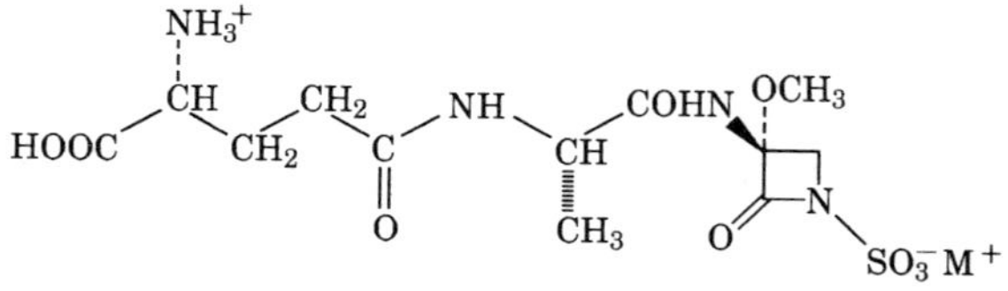

Agrobacterium radiobacter			(**6**) 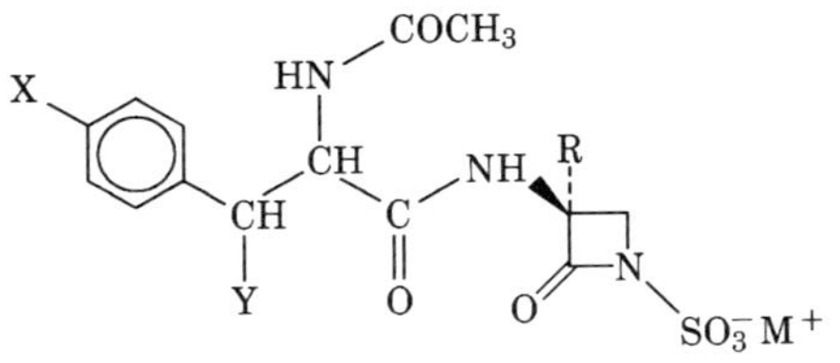
	[79720-13-1]	$C_{15}H_{19}N_3O_7S \cdot Na$	(**6**, X = H, Y = H, R = OCH_3, M = Na)
	[79720-14-2]	$C_{15}H_{19}N_3O_8S \cdot K$	(**6**, X = OH, Y = H, R = OCH_3, M = K)
	[81919-28-0]	$C_{14}H_{17}N_3O_7S \cdot K$	(**6**, X = OH, Y = H, R = H, M = K)
	[79720-16-4]	$C_{15}H_{19}N_3O_{12}S \cdot 2Na$	(**6**, X = $OSO_3^- Na^+$, Y = OH, R = OCH_3, M = Na)
	[79720-17-5]	$C_{15}H_{19}N_3O_{15}S_3 \cdot 3Na$	(**6**, X = $OSO_3^- Na^+$, Y = $OSO_3^- Na^+$, R = OCH_3, M = Na)

become available. In addition, more sophisticated screening and isolation techniques have afforded a variety of new naturally-occurring β-lactams such as the cephamycins, clavulanic acid, nocardicin, and the carbapenems. In late 1979 another entirely new class of β-lactams, characterized by the 2-oxoazetidine-1-sulfonic acid moiety (**1**, R = H or OCH_3, R′ = acyl), was discovered (1,2). To distinguish this class of monocyclic, bacterially-produced β-lactam antibiotics from the bicyclic cephalosporins and cephamycins (**2**, R = H or OCH_3, R′ = γ-linked α-aminoadipoyl), the term monobactam was introduced (1).

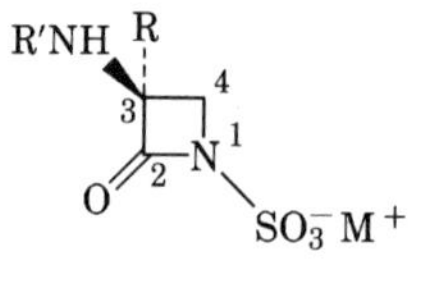

(**1**)

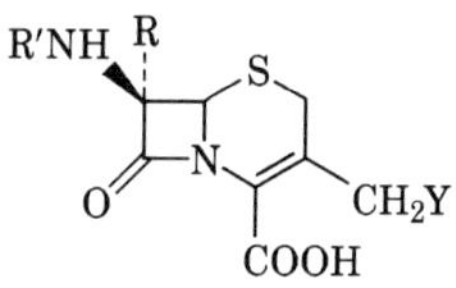

(**2**)

By screening strains of bacteria that were specifically responsive to β-lactam antibiotics, monobactams, varying in substitution at the C-3 position, were identified. The structures of the compounds along with their producing organisms are shown in Table 1. More recently, the 4β-methyl analogue (**4**, R″ = CH_3) of SQ 26,445/sulfazecin (**4**, R″ = H) was isolated using a differential antibacterial assay (3).

Structure Determination and Biological Properties

The structure and absolute stereochemistry of SQ 26,445 (sulfazecin) (**4**, R″ = H) was unequivocally established by x-ray crystallography (4) and that of SQ 26,180 (**3**, R = OCH_3) was determined by synthesis from a homochiral precursor of defined structure. Starting with 7-aminocephalosporanic acid (7-ACA), stereospecific introduction of the methoxyl group and degradation of the thiazine ring ultimately afforded the corresponding N-1 unsubstituted azetidinone. Subsequent sulfonation confirmed the structure of SQ 26,180 (**8**, R = CH_3) (Fig. 1) (5). The structure and absolute configuration of the nonmethoxylated naturally-occurring monobactam (**6**, X = OH, Y = R = H, M = K) was also proven by total synthesis (6). The monobactams (**1**) correspond in configuration to the naturally-occurring cephalosporins and cephamycins (**2**). And, as is the case with regard to bicyclic β-lactams, monobactams of opposite absolute configuration to those encountered in nature are inactive. The enantiomeric pair of monobactams shown in Table 2 provide a striking illustration of this specificity.

In penicillins, incorporation of a 6α-methoxyl substituent increases chemical stability three- to fivefold as measured by basic hydrolysis of the β-lactam moiety (7). Similar substitution of a 7α-methoxyl substituent in cephalosporins also results in a modest increase in chemical stability. In contrast, methoxylated monobactams having simple side chains are much less chemically stable than the corresponding nonmethoxylated compounds. However, the methoxylated monobactams show an increased stability to β-lactamases when compared to

Table 2. Biological Activity of (*S*)- and (*R*)-3-{[(2-amino-4-thiazolyl)(methoxyimino)acetyl]amino}-2-oxo-1-azetidinesulfonic acid

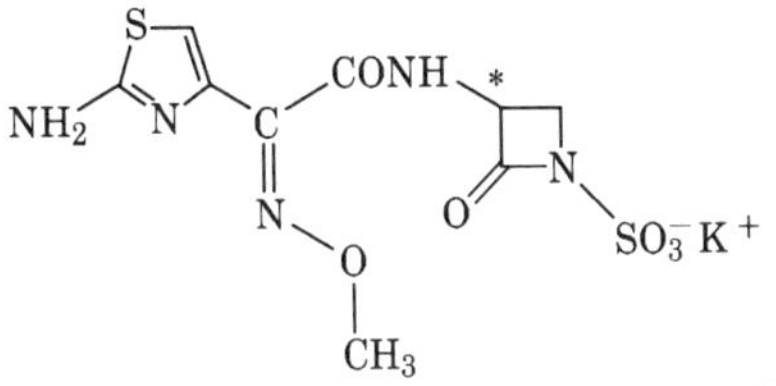

Absolute configuration[a]	CAS Registry Number	Molecular formula	MIC, μg/mL[b]
(*S*)-	[*78626-19-4*]	$C_9H_{11}N_5O_6S_2$	<0.05
(*R*)-	[*78612-61-0*]	$C_9H_{11}N_5O_6S_2$	>100

[a]At C-3.
[b]For *P. rettgeri*.

their unsubstituted analogues. This relationship is demonstrated by comparing the V_{max}, the rate at saturating substrate concentration for SQ 26,180 (**3**, R = OCH_3) with that of the nonmethoxylated counterpart SQ 26,396 [*79720-10-8*] (**3**, R = H), $C_5H_8N_2O_5SK$, against the clinically important RTEM lactamase (8). The relative V_{max} (RTEM) are <0.01 and 12, respectively compared to cephaloridine at 100. Similarly, increased β-lactamase stability was also observed for the 7-methoxy-cephalosporins (9).

The monobactams, like pencillins and cephalosporins, interfere with the synthesis of bacterial cell walls. β-Lactam antibiotics bind to a series of penicillin-binding proteins (PBPs) on the cytoplasmic membrane and their antibacterial effect is believed to result from inhibition of a subset of these PBPs known as peptidoglycan transpeptidases. These enzymes are responsible for cross-linking adjacent peptide strands within the peptidoglycan structure of the cell wall, thus improving the integrity of the bacterial protective coat. Aztreonam [*78110-38-0*], $C_{13}H_{17}N_5O_8S_2$, a clinically efficacious monobactam, specifically binds to PBP-3 thereby preventing septation and subsequently leading to cell death (10).

The biosynthetic origin of monobactams has been elucidated by fermentation experiments using radioactively labeled amino acids (qv). The monocyclic ring of the naturally-occurring C-4 unsubstituted monobactams is derived from serine. Similar techniques have shown that the methyl moiety of the methoxyl group in 3α-methoxylated monobactams is derived from methionine (11).

All of the naturally-occurring monobactams discovered as of this writing have exhibited poor antibacterial activity. However, as in the case of the penicillins and cephalosporins, alteration of the C-3 amide side chain led to many potent new compounds (12). Furthermore, the monobactam nucleus provides a unique opportunity to study the effect of structural modifications at the N-1 and C-4 positions of the azetidinone ring on biological activity. In contrast to the bicyclic β-lactams, these positions on the monocyclic ring system are readily accessible by synthesis.

Synthesis

Initial syntheses employed the sulfonation of an N-1 unsubstituted azetidinone as the key step. The natural product SQ 26,180 (**3**, R = OCH_3) (**8**, R = CH_3) as well as other methoxylated monobactams were synthesized, starting from either 7-aminocephalosporanic acid [*957-68-6*] (7-ACA) or 6-aminopenicillanic acid [*551-16-6*] (6-APA), via sulfonation of the N-1 unsubstituted intermediates (**7** or **10**) as shown in Figure 1 (5,13). Subsequently, many more C-3 side-chain analogues were prepared by this method. Examples include the sulfonation of the N-1 unsubstituted azetidinones (**10**) that carry protecting groups on the nitrogen atom attached to C-3. 3-Aminomonobactamic acid [*79720-18-6*] (3-AMA) (**12**), $C_3H_6N_2O_4S$, was readily formed by removal of the protecting group and could subsequently be acylated to afford a wide variety of side-chain monobactams (**13**). 6-APA proved to be a readily available source for 3-AMA (14,15) which alternatively could be prepared from the *O*-methylhydroxamate (**9**) of L-serine (16).

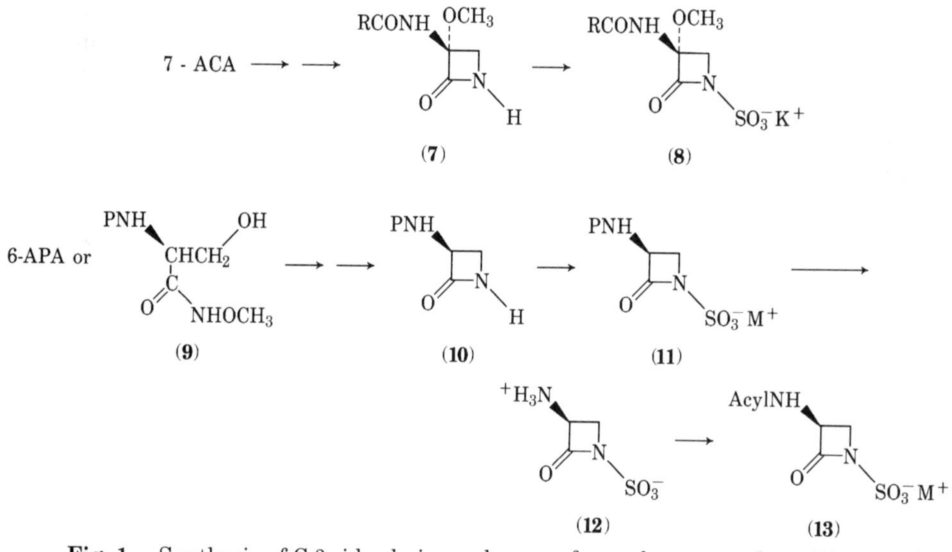

Fig. 1. Synthesis of C-3 side chain analogues of monobactams, where P is an amine protecting group and other terms are defined in text.

A second, conceptually distinct chiral synthesis of monobactams was developed from β-hydroxy amino acids. As shown in Figure 2, cyclization of the acylsulfamate of an amino-protected *O*-mesylserine derivative (**14**, R = H) leads directly to the monobactam (**15**). This methodology was also applied to the synthesis of 4α- (**15**, R = CH_3) and 4β-methyl monobactams from L-threonine and allothreonine, respectively (17). The 3(*S*)-(*trans*)-3-amino-4(*S*)-methylmonobactamic acid [*80082-65-1*] (**16**, R = CH_3), $C_4H_8N_2O_4S$, is the key intermediate to aztreonam [*78110-38-0*] (**17**), $C_{13}H_{17}N_5O_8S_2$, the first monobactam to achieve the status of a clinically useful antibiotic. A similar synthesis was utilized for the preparation of carumonam [*89638-04-8*] (**19**), $C_{12}H_{14}N_6O_{10}S_2$, the second mono-

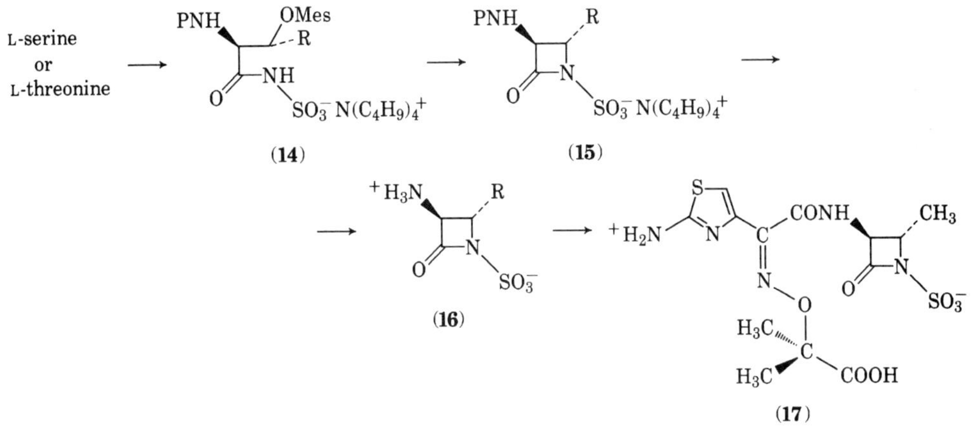

Fig. 2. Synthesis of clinically useful monobactams where R = H, CH_3; P is an amino protecting group, and Mes = mesyl is methanesulfonyl.

bactam to undergo clinical trials. The absolute stereochemistry at the C-3 and C-4 positions, as well as the functionality on the C-4 position, were controlled by starting with either L-threonate or L-(+)-tartaric acid. Thus the appropriately substitued acyl sulfamate (**18**), analogous to (**14**), was obtained for conversion to carumonam (**19**) (18).

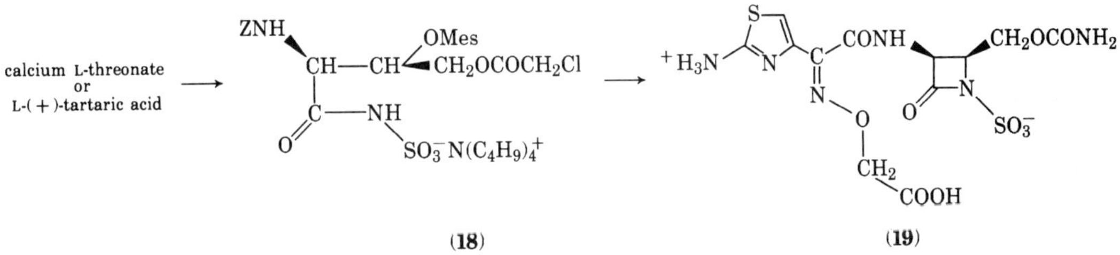

Although the activity of methoxylated monobactams could be improved by appropriate side-chain modifications, difficulty of synthesis and poor chemical stability focused attention on the nonmethoxylated analogues. Both high intrinsic activity and excellent β-lactamase stability are exhibited by monobactams that combine C-3 aminothiazole oxime side chains and 4-alkyl, 4-alkenyl, and 4-alkynyl groups (19).

Aztreonam. Aztreonam (**17**) is a totally synthetic compound having an antibacterial spectrum that is unique among β-lactam antibiotics. It exhibits potent and specific activity against a wide range of both β-lactamase-producing and nonproducing aerobic gram-negative bacteria, including *Pseudomonas aeruginosa,* but displays minimal inhibition against anaerobic and gram-positive aerobic, bacteria, eg, staphylococci and streptococci. When tested against strains of *Enterobacteriaceae,* aztreonam inhibited 50% of these isolates at concentrations <0.3 μg/mL (20). Particularly sensitive were members of *Escherichia coli, Klebsiella pneumoniae, Proteus mirabilis, Morganella morganii, Serratia marcescens,* and many *Providencia sp.*. The activity of aztreonam is comparable to that of the aminoglycosides and third generation cephalosporins against these *Enterobacteriaceae*. Similarly, 50% of strains of ampicillin-sensitive and resistant *Haemophilis influenzae* and *Neisseria gonorrhoeae* were inhibited at <0.1 μg/mL.

Most dramatic were results with strains of *Pseudomonas aeruginosa* where aztreonam inhibited 50% of the strains at 4 μg/mL, and 90% at 16 μg/mL. Cessation of growth is not the sole consequence of exposure of these organisms to aztreonam as the inhibitory concentration is also generally the lethal concentration. Aztreonam does not inhibit the majority of strains of *Acinetobacter sp.* and many strains of *Pseudomonas sp.*, eg, *P. maltophilia* and *P. cepacia*. Also little or no activity was seen against *Bacteroides fragilis*. Minimal inhibitory concentrations for the anaerobic and gram-positive aerobic bacteria were usually >100 μg/mL.

Many of the gram-negative isolates tested were multiresistant β-lactamase-producing strains that are becoming more common and are gaining increasing clinical importance, particularly in nosocomial infections. Generally resistant to first- and second-generation cephalosporins and most of the penicillins, these organisms elaborate a diverse array of β-lactamases targeted at destroying β-lactam antibiotics. Aztreonam exhibits a high degree of resistance to enzymatic hydrolysis by most of the common β-lactamases (21). Particularly insidious are the plasmid-mediated TEM β-lactamases carried by a wide range of gram-negative bacteria. The plasmid-mediated genes responsible for enzyme production pass freely among the *Enterobacteriaceae*, pseudomonads, *Neisseria gonorrhoeae*, and *Haemophilus influenzae*. Other β-lactamases capable of this same promiscuous behavior are the OXA, SHV, and PSE enzymes. Aztreonam, similar to many of the third-generation aminothiazoleoxime cephalosporins, such as cefotaxime and ceftizoxime, is stable to these β-lactamases. In contrast, compounds like cephaloridine and cefoperazone are readily hydrolyzed. Aztreonam, cefotaxime, and ceftizoxime are also generally stable to the chromosomally mediated β-lactamases, which are widely distributed among specific gram-negative bacteria, but are not transferable. These enzymes, which may coexist in the same organisms with a plasmid-mediated β-lactamase, are found in organisms such as *Klebsiella*, *Providencia*, *Enterobacter*, *Serratia*, *Proteus*, and *Bacteriodes*. The only enzyme in this group showing any appreciable destruction of aztreonam is the relatively uncommon K-1 β-lactamase produced by some strains of *Klebsiella oxytoca*.

In vivo, aztreonam shows excellent pharmacokinetic properties. One hour after parenteral administration of a 1 g dose, serum levels are 45–50 μg/mL. After a 2 g dose, levels are 90 μg/mL at 1 h and there is no apparent accumulation of the drug after multiple dosing. Aztreonam is excreted primarily by the kidneys with two-thirds of an administered dose eliminated unchanged in the urine. Renal secretion, glomerular filtration, and nonrenal mechanisms are involved in the elimination of aztreonam. Serum half-life for elimination is 1.7 h. Aztreonam penetrates well into peritoneal fluid, pleural fluid, and blister fluid. Furthermore, aztreonam undergoes biliary secretion with peak biliary concentrations greater than 40 μg/mL at 2.5 h.

In the clinic, aztreonam has demonstrated efficacy in urinary tract infections (UTI), lower respiratory tract infections, bacteremias, skin and soft tissue infections, intraabdominal and pelvic infections, leukopenic patients, and bone and joint infections resulting from susceptible gram-negative pathogens, allowing for selective therapy or combination therapy if mixed pathogens are involved. Aztreonam markedly decreases the fecal aerobic gram-negative bacteria, selectively eradicating pathogenic bacteria from the intestinal tract without significantly altering the gut anaerobes or impairing their colonization-resistance function.

Overall, aztreonam appears to be a safe agent having toxicity side effects similar to those of other β-lactams. The safety profile suggests that aztreonam may be useful as a replacement for aminoglycoside therapy (22). The biological properties of aztreonam have been extensively reviewed (23).

Alternative N-1 Activating Groups

β-Lactam antibiotics exert their antibacterial effects via acylation of a serine residue at the active site of the bacterial transpeptidases. Critical to this mechanism of action is a reactive β-lactam ring having a proximate anionic charge that is necessary for positioning the ring within the substrate binding cleft (24).

All of the naturally-occurring monobactams and aztreonam are characterized by the presence of the N-1 sulfonate group, which serves a dual function. The electron withdrawing sulfonate moiety renders the β-lactam ring more reactive toward nucleophilic attack and at the same time provides the anionic charge necessary for binding. A variety of monobactam subclasses bearing N-1 activating groups having the necessary physical properties for antibacterial activity are listed in Table 3 (25).

Monophosphams. The similar tetrahedral geometry and bond lengths of tetracoordinate phosphorus(V) compared to those of sulfur(VI) suggested that phosphonic and phosphinic acid groups might act as bioisosteres for the sulfonic acid moiety in the parent monobactams. The 2-oxoazetidine-1-phosphonates and phosphinates (monophosphams) were prepared, as shown in Figure 3**a**, by lithiation of the 3-acylamino-2-azetidinones (**20**) followed by treatment with a phosphoryl or phosphonyl halide to yield the phosphonylated, or phosphinylated, azetidinones (**21**) or (**23**). Cleavage of these esters using trimethylsilyl bromide [*2857-97-8*] in the presence of bis-trimethylsilylacetamide [*10416-59-8*] (BSA), which serves as an acid scavenger, affords the diacidic and monoacidic monophosphams (**22**) and (**24**), respectively (26,27). Alternatively, Figure 3**b** reaction of lithiated azetidinones with phosphoryl dichlorides yields azetidinone-1-phosphonyl chlorides (**26**), which are solvolyzed to give the monoacidic ester (**27**) or aminolized to ultimately afford the phosphonamide (**29**). Cleavage of the urethane protecting groups by catalytic hydrogenation, when R = $C_6H_5CH_2O$, or by trifluororacetic acid, when R = *t*-C_4H_9O, gives 3-aminomonophosphamic acids. Coupling using *N,N'*-dicyclohexylcarbodiimide (DCC)-*N*-hydroxybenzotriazole or mixed anhydride with carboxylic acids formed 3-acylamino derivatives for screening (27).

The monophosphams are less intrinsically active, but more stable to β-lactamases, than their monobactam counterparts. SQ 27,327 [*84486-64-6*] (**30**), $C_{14}H_{20}N_5O_8PS$, the most active monophospham, is generally two- to eightfold less active against aerobic gram-negative organisms than is its sulfonic acid analogue, aztreonam (**17**). SQ 27,327, however, is substantially more stable to the K-1

Table 3. Biologically Active Monobactam Subclasses

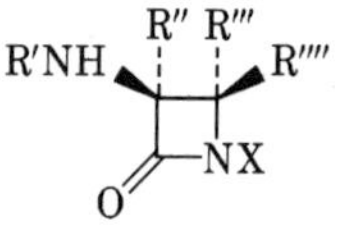

Subclass	Substituent, X
monobactams	$-SO_3H$
monophosphams	$-P(=O)(OH)R$
monosulfactams	$-OSO_3H$
monophosphatams	$-OP(=O)(OH)R$
monoxacetams (oxamazins)	$-OCH_2COOH$
monocarbams	$-C(=O)NHSO_2R$
tetrazole-activated monobactams	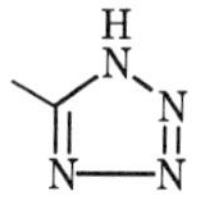

β-lactamase than aztreonam and this is reflected in the >250-fold difference in antimicrobial activity against the enzyme-producing *Klebsiella* strain (Table 4) (27).

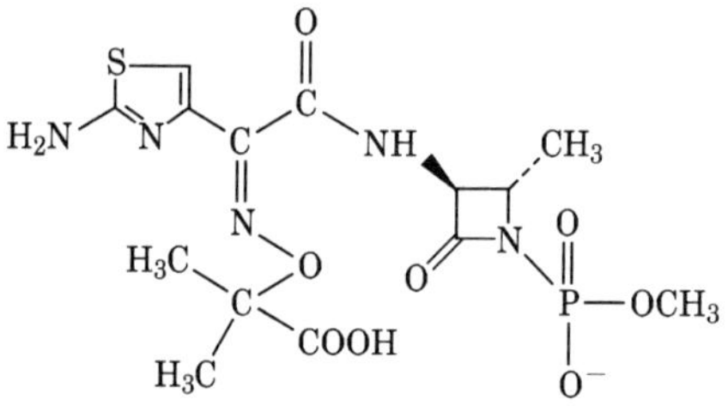

(30)

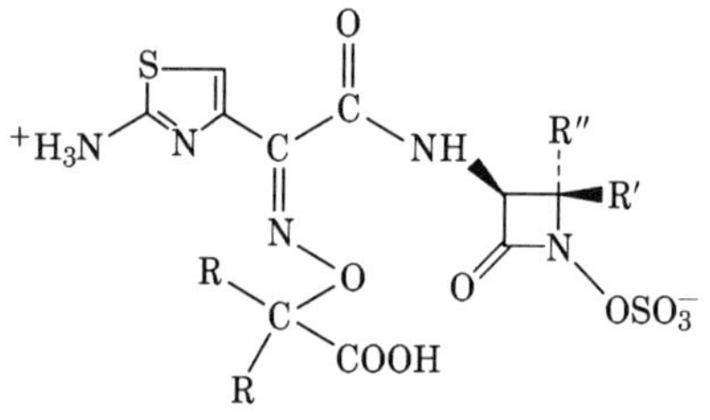

(31)

Monosulfactams. The monosulfactams are a class of monobactams activated by an *N*-1-OSO_3^- group (28) resulting in good intrinsic activity against gram-negative bacteria. However, a C-4 unsubstituted (**31**, R = CH_3, R′ = R″ = H) [*109895-38-7*] and the monomethylated [*87577-16-0*] (**31**, R = CH_3, R′ = H, R″ = CH_3), $C_{13}H_{17}N_5O_9S_2$ and [*102572-29-2*] (**31**, R = CH_3, R′ = CH_3, R″ = H),

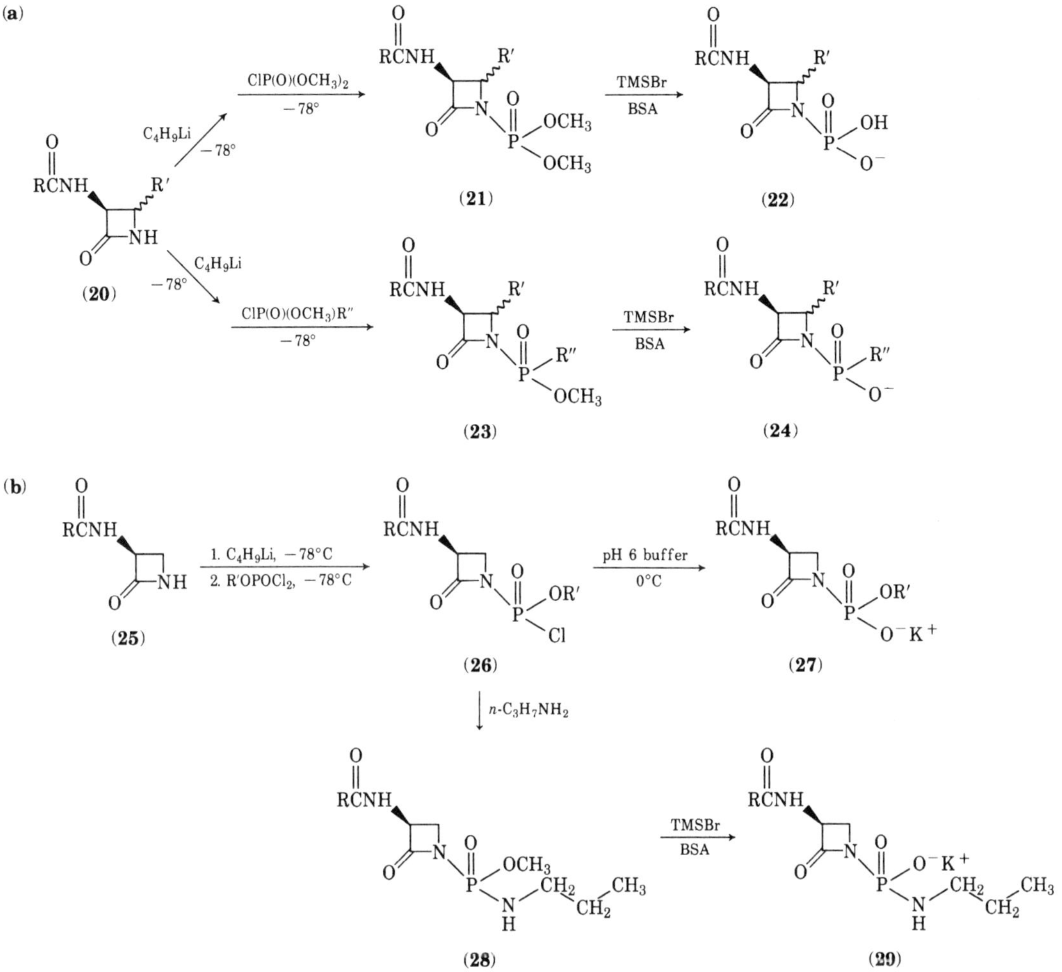

Fig. 3. Synthesis of diacidic and monoacidic N-1 azetidinone phosphonates and phosphinates where TMSBr is trimethylsilyl bromide; BSA is bis-trimethylsilylacetamide; **(a)** R = $C_6H_5CH_2O$, t-C_4H_9O, or 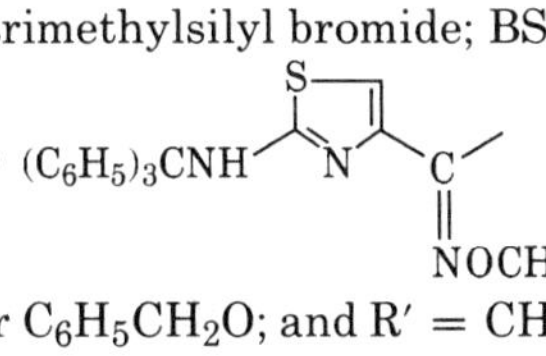

; R′ = H or CH_3; and R″ = CH_3 or C_6H_5; and **(b)** R = t-C_4H_9O or $C_6H_5CH_2O$; and R′ = CH_3, CH_2CH_2F, CH_2CF_3, n-C_4H_9, or C_6H_5.

$C_{13}H_{47}N_5O_9S_2$, monosulfactams have poor chemical and β-lactamase stability. Dimethylation at the C-4 position (**31**, R = R′ = R″ = CH_3) [*109885-32-7*] improves the chemical and β-lactamase stability remarkably, while maintaining a high degree of antibacterial activity. Furthermore, removal of the *gem*-dimethyl groups from the C-3 side chain of this C-4 dimethyl compound minimally affects β-lactamase stability (29), but affords an orally-absorbed monosulfactam SQ 30,213 [*102507-71-1*] (tigemonam) (**31**, R = H, R′ = R″ = CH_3), $C_{12}H_{15}N_5O_9S_2$ (30). The high degree of oral absorption (80% in mice) is atypical for monobactams. The

Table 4. Biological Activities of Aztreonam and N-1 Substituted Analogues

Gram-negative organisms	SC#[b]	MIC, μg/mL[a]							
		Aztreonam		SQ 27,327		Tigemonam		SQ 28,112	
		Low[c]	High[c]	Low[c]	High[c]	Low[c]	High[c]	Low[c]	High[c]
E. coli TEM+	10,404	0.1	0.2	0.4	0.8	0.4	0.8	0.4	0.8
E. coli TEM−	10,439	0.1	0.2	0.8	0.8	0.8	0.8	0.8	0.8
Ent. cloacae P99+	10,435	12.5	50	6.3	50	6.3	50	6.3	50
Ent. cloacae P99−	10,441	0.1	0.2	0.8	1.6	0.8	1.6	0.8	1.6
K. aerogenes K1+	10,436	100	>100	0.4	0.4	0.4	0.4	>100	0.4
K. aerogenes K1−	10,440	<0.05	0.1	0.4	0.8	0.4	0.8	0.8	0.8
K. pneumo.	11,066	0.4	0.8	1.6	1.6	0.8	1.6	3.1	25
C. freundii	10,204	0.1	25	0.4	3.1	0.2	0.4	0.8	50
Prot. rettgeri	8,217	<0.05	<0.05	0.2	0.2	<0.05	<0.05	<0.5	0.4
Prot. vulgaris	10,951B	<0.05	0.2	0.4	0.4	<0.05	<0.05	0.4	1.6
Ser. marcescens	9,782	0.2	0.2	1.6	3.1	0.8	1.6	0.8	3.1
Ps. aeruginosa	8,329	3.1	12.5	>100	>100	50	>100	3.1	12.5
Ps. aeruginosa	9,545	0.4	0.8	6.3	6.3	1.6	1.6	0.8	1.6

[a]Minimum Inhibitory Concentration measured by dilution of test compound in agar inoculated with microorganism.
[b]Squibb culture number.
[c]Low and high refer to the number of colony forming units (CFU) used to inoculate the media. Comparing MICs at low (10^4) and high (10^6) CFU, a large increase in MIC at the high inoculum level vs a β-lactamase producer is an indication of instability via enzymatic hydrolysis.

oral absorption and efficacy of tigemonam in animals have been confirmed in humans in Phase I and Phase II clinical studies (31).

Syntheses of the C-4 unsubstituted and monomethylated monosulfactams utilized hydroxamates derived from β-hydroxy amino acids such as serine or threonine, as precursors to the β-lactam ring (28). By analogy, synthesis of C-4 dimethylated monobactams (Fig. 4) required a protected form of (*S*)-β-hydroxyvaline [*102509-19-7*] (**33**). This unusual amino acid was prepared by reaction of the dianion of Boc-glycine benzyl ester [*54244-89-8*] (**32**) with acetone. Cleavage of the benzyl ester afforded the Boc-protected racemic valine [*102507-13-1*], $C_{10}H_{19}NO_5$, which was resolved as its phenethylamine salt to give Boc-(*S*)-β-hydroxyvaline. Coupling *O*-benzylhydroxylamine with amino acid (**33**) afforded the hydroxamate (**34**). Selective sulfonation of the hydroxyl group of (**34**) and subsequent heating of the resulting sulfate (**35**) [*107982-89-8*] in the presence of base effected cyclization at the sterically hindered carbon atom and gave the β-lactam intermediate (**36**) [*10250-25-5*]. The final steps, benzyl cleavage, sulfonation, and coupling of the side chain, yielded tigemonam (**38**) (**31**, R = H, R′ = R″ = CH_3), and related analogues (32).

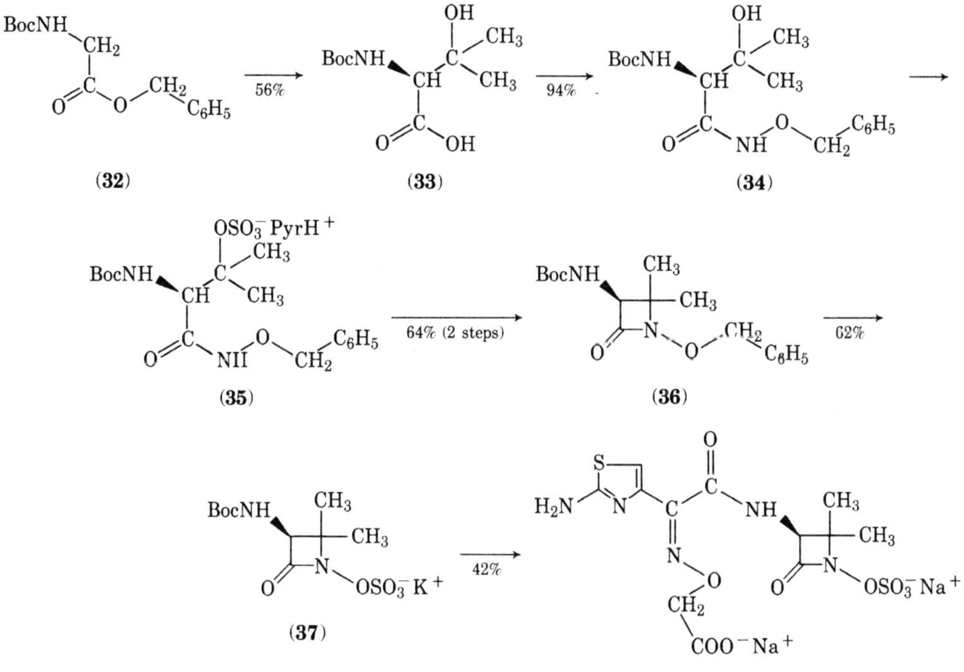

Fig. 4. Synthesis of tigemonam where Boc is the *t*-butoxycarbonyl group; $PyrH^+$ is the pyridinium ion; and the numbers given are percent yield of product.

As shown in Table 4, tigemonam is also relatively inactive vs gram-positive bacteria and anaerobes, yet exhibits a high degree of activity against gram-negative organisms as well as resistance to hydrolysis by β-lactamases. The activity of tigemonam is cidal against the *Enterobacteriaceae, Hemophilus influ-*

enzae, and *N. gonorrhoeae* but unlike aztreonam, the activity is weak against the pseudomonads (33,34).

Monophosphatams. The preparation of monophosphatams, the phosphorus analogues of the monosulfactams, is exemplified by the reaction shown in Figure 5 (35). Phosphorylation of the *N*-hydroxyazetidinone [*80542-48-9*] (**39**) using methylphosphonic or methylphosphoric dichlorides in the presence of 2,6-lutidine or triethylamine at low temperatures gave the chloro intermediates (**40**), which were hydrolyzed *in situ* at pH 3–4 to give the stable potassium salts (**41**) in from 12 to 40% yield. Removal of the *t*-butoxycarbonyl group under acidic conditions at −10°C for 1 h gave the amine salts (**42**) in good yield. The desired acyl side chains were then coupled to the amino group using an activated ester of the side chain to afford various acyl derivatives such as SQ 28,112 [*91603-73-5*] (**43**), $C_{14}H_{20}N_5O_8PS$.

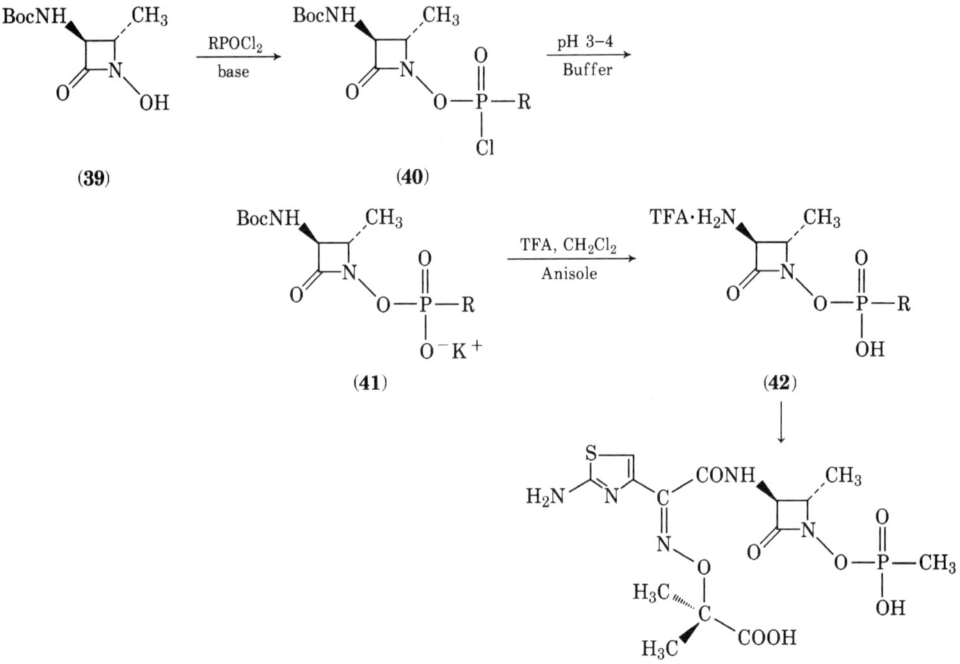

(**39**) (**40**) (**41**) (**42**) (**43**)

Fig. 5. Synthesis of monophosphatams where Boc is *t*-butoxycarbonyl; R is either CH_3 or OCH_3; and TFA is trifluoroacetic acid.

The biological activity of SQ 28,112 (**43**), shown in Table 4, is typical of the monophosphatams which possess the same spectrum of activity as aztreonam, but have comparatively lower intrinsic activity and β-lactamase stability.

Monoxacetams. Once the structural criteria for ring activation and binding of monobactams to the transpeptidases were established, researchers set out to determine if an oxygen atom could provide the requisite β-lactam ring reactivity. It was first necessary to insulate the anionic binding moiety from the oxygen atom by a "spacer" group, and using glycolate structures appended to the *N*-1 position of the β-lactam ring, structures (**44**) were synthesized (36). These 3-acylamino-1-azetidinyloxyacetic acids are known variously as monoxacetams and as oxamazins.

The oxyacetic acid residue of the monoxacetams is bioisosteric with the sulfate moiety of the monosulfactams. In addition, the presence of the carboxylate moiety provided opportunity for the preparation of esters that could act as orally absorbed prodrugs. Based on structure-activity relationships, SQ 82,291 (**45**) [*90898-90-1*], $C_{12}H_{15}N_5O_6S$, the nonacidic methoxime side chain of which was necessary to maintain oral absorption of the prodrugs, was prepared. SQ 82,291 has a high, specific affinity for the PBP-3 transpeptidase of gram-negative bacteria. However, it lacks the isobutyric acid moiety of aztreonam (**17**) on the oxime residue and whereas the activity of SQ 82,291 vs the *Enterobacteriaceae* was maintained, antipseudomonal activity was significantly diminished (37).

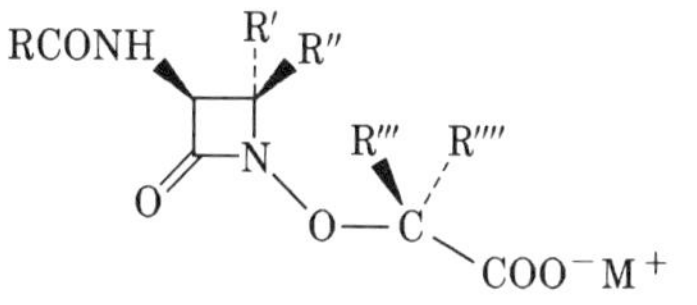

(**44**)

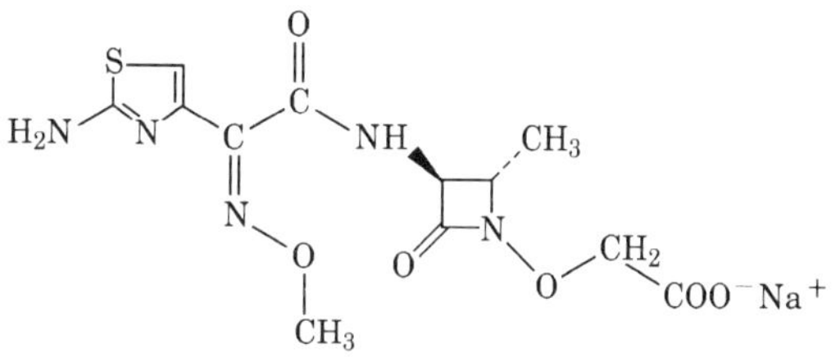

(**45**)

To synthesize the monoxacetam structures (Fig. 6), alkylation of *N*-protected 1-hydroxyazetidinones (**46**) with the appropriate haloacetic acid derivatives provided (**47**). Alternatively, (**47**) could be prepared from the acyclic hydroxamate ester (**48**). Deprotection of (**47**) furnished the zwitterionic intermediate (**49**) [*90849-16-4*], $C_6H_{10}N_2O_4$, which subsequently underwent acylation using the C-3 aminothiazole oxime side chain to afford SQ 82,291 (**45**) also known as oximonam (37).

The first oral monobactam to reach clinical trials was SQ 82,531 [*90850-05-8*] (gloximonam), (**50**), $C_{18}H_{25}N_5O_8S$, the *t*-butyl glycolate ester of oximonam. This prodrug ester could be formed in high yield by direct alkylation of SQ 82,291 or by carrying the ester from the earlier synthetic stages (**47** or **48**, R′ = —CH_2COOt-C_4H_9). On a multikilogram scale, the overall yield via the route from *N*-hydroxyazetidinone (**46**) was ca 25% (38).

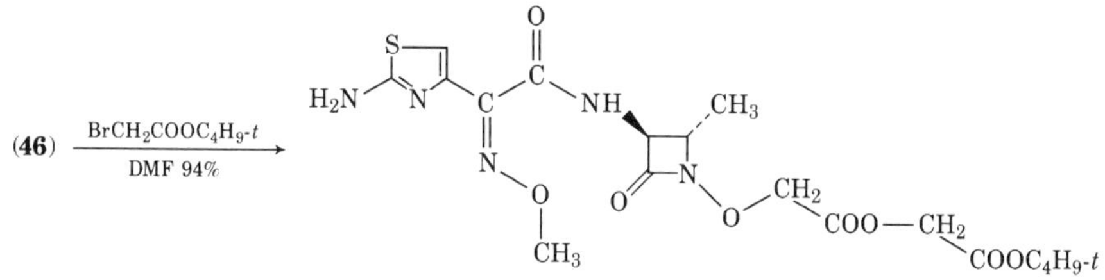

(**50**)

Gloximonam [*90850-05-8*] (**50**) was selected for clinical studies because of its desirable physical properties such as crystallinity and solubility and because it was orally absorbed in mice, rats, and monkeys, thus resulting in therapeutic serum levels of the active oximonam. Gloximonam was orally efficacious in a variety of murine gram-negative systemic infections (39) and in Phase I clinical studies gloximonam was well-absorbed. However, further clinical development

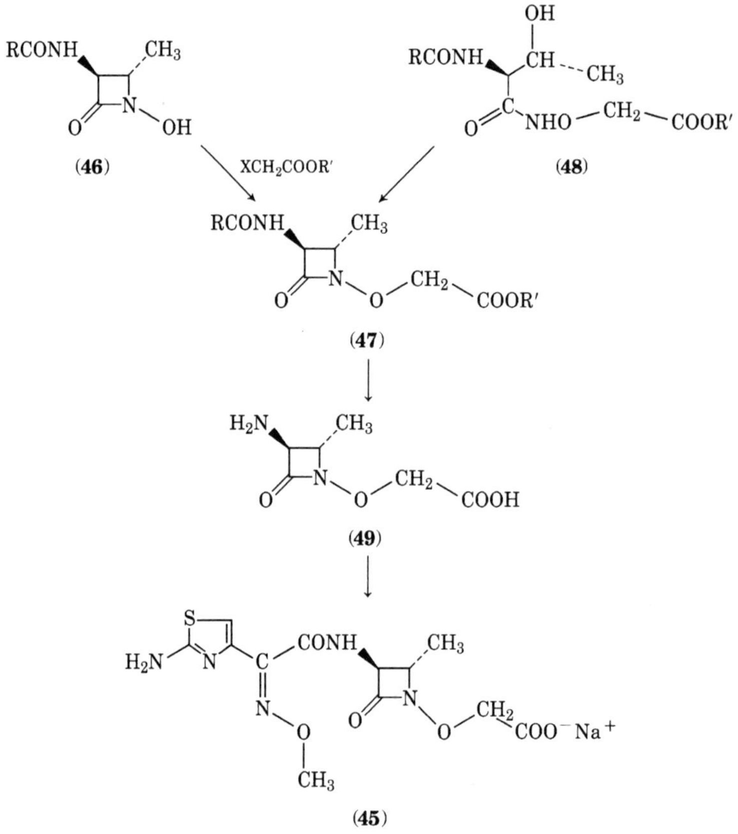

Fig. 6. Synthesis of monoxacetams.

was discontinued once it was determined that in humans the β-lactam ring underwent significant hydrolysis.

The *N*-aza (40) and *N*-thia (41) analogues of the monoxacetams have also been prepared. Within the nitrogen series, the imino acids [*121142-86-7*] (**51**, X = N=CHCOOH, R = H, CH_3) the imine stereochemistry of which is unknown, have moderate, notably less than oximonam anti-gram-negative activity. Both the *N*-glycyl [*121142-76-5*] (**51**, X = $NHCH_2COOH$, R = H) and *N*-thioacetic acid [*102652-83-5*] (**52**) derivatives were devoid of any significant biological activity.

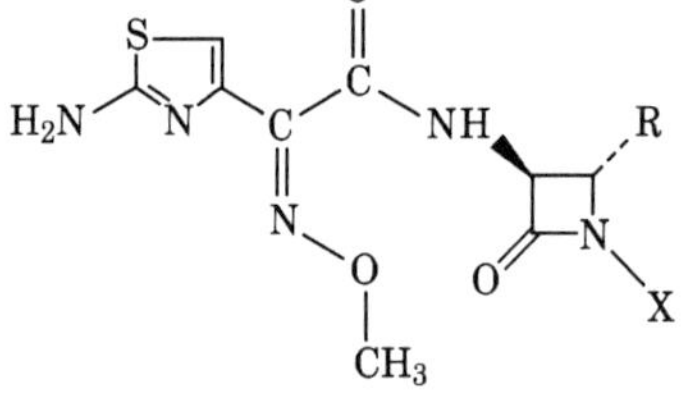

(**51**)

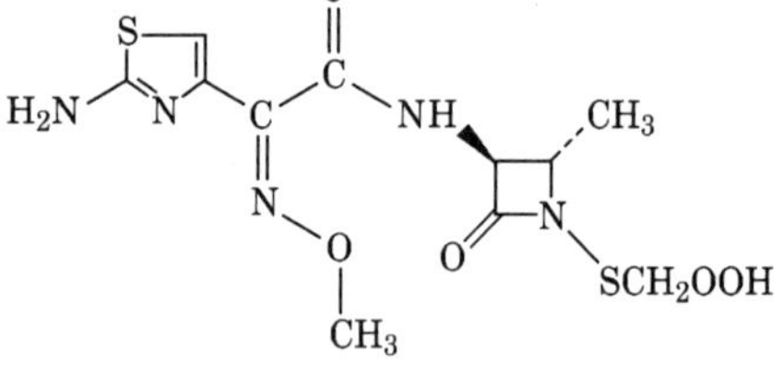

(**52**)

Monocarbams. The reaction sequence for monocarbam preparation is shown in Figure 7. Simple carboxyl *N*-1 substitution, representing one of the simplest carbonyl activating groups, led, as expected, to a carbamic acid that was hydrolytically unstable. Several monocarbams (**55**) and (**56**), having acidic sulfonylaminocarbonyl activating groups, were produced (42). The monocarbams (**55**) were prepared by sequential acylation of 3-aminoprotected 2-azetidinones (**53**) using alkyl or aryl isocyanates, removal of the protecting group, and acylation with activated side-chain acids. Reaction of (**53**) and chlorosulfonyl isocyanate (CSI) afforded the key intermediate (**54**) from which the aminosulfonyl derivatives (**56**) were prepared.

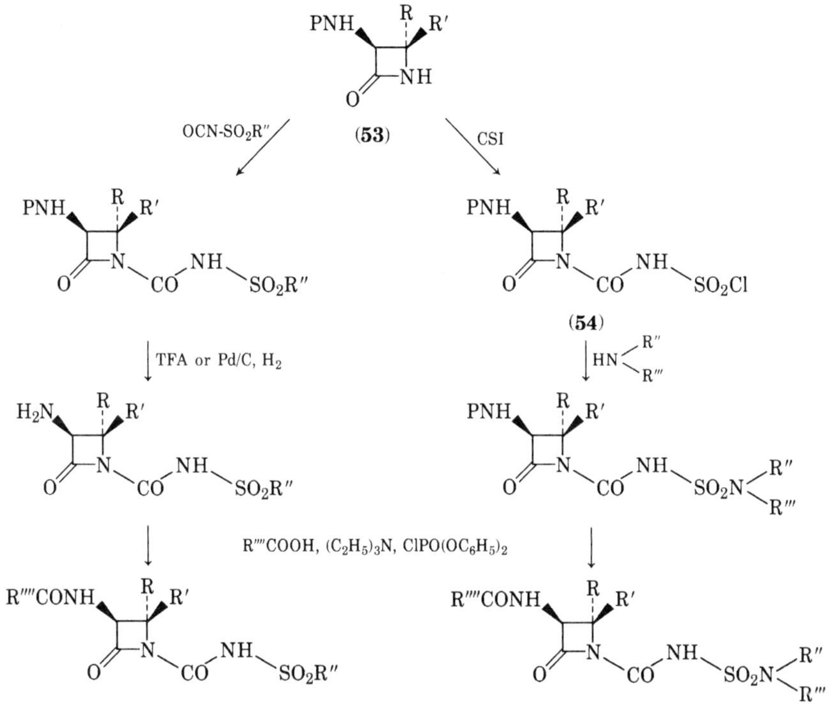

(**55**) (**56**)

Fig. 7. Synthesis of monocarbams where P is an amino protecting group and CSI is chlorosulfonyl isocyanate. TFA is trifluoroacetic acid.

Monocarbams of the general formula (**56**), in which the terminal nitrogen is incorporated in a heterocyclic group, represent a class of potent β-lactamase stable antibacterials which are specifically active against gram-negative microorganisms. SQ 82,228 [*109895-39-8*] (**57**), featuring an *N*-ethylpiperazinedione terminus, is one of the most active (42), displaying favorable pharmacokinetic properties in experimental animals and good efficacy when tested parenterally in model infections.

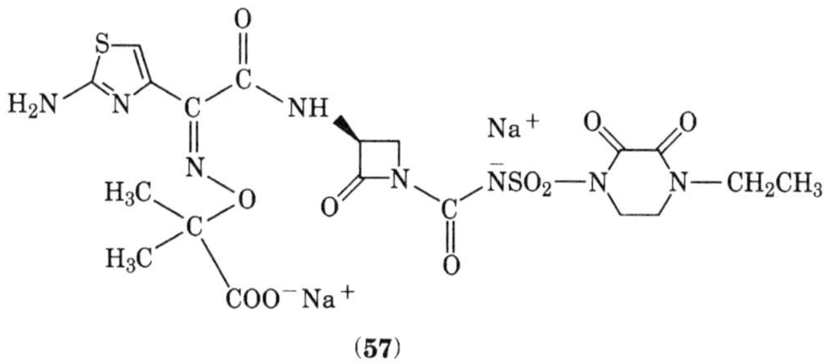

(**57**)

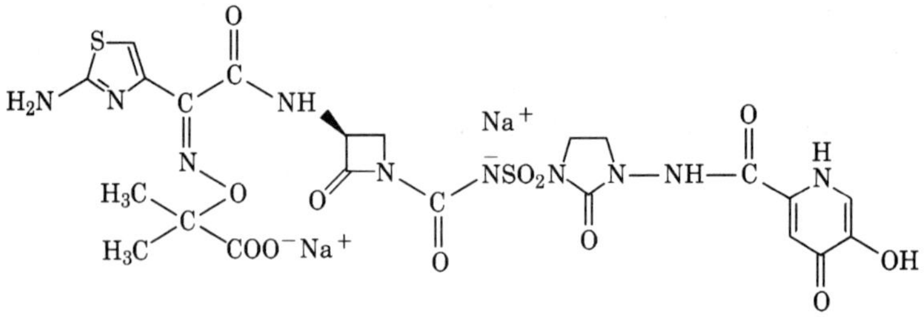

(**58**)

Continued efforts to improve the activity of the monobactams against nonfermenting gram-negative rods such as *Pseudomonas aeruginosa,* led to the discovery of SQ 83,360 [*104393-00-2*] (**58**), $C_{22}H_{24}N_{10}O_{12}S_2{\cdot}2Na$, a 3-hydroxy-4-pyridone containing monocarbam. The enhanced activity of SQ 83,360 is a direct consequence of the hydroxypyridone moiety, a catechol surrogate recognized by the bacterial *ton* B-dependent ion transport pathway. Uptake of SQ 83,360 by this mechanism increases the concentration of drug in the periplasmic space where the transpeptidases are located (43).

The synthesis of SQ 83,360 is shown in Figure 8. The heterocyclic terminus (**60**) [*112333-97-8*] of the *N*-1 substituent was constructed in five steps from kojic acid [*501-30-4*] (**59**). After persilylation, the reaction of intermediate (**60**) and the chlorosulfonyl isocyanate adduct of Z-protected 3-amino-2-azetidinone to afford monocarbam [*112333-98-9*] (**61**) occurred. Hydrogenolysis of the Z-group formed (**62**) [*12335-07-6*], followed by coupling and deprotection of the side chain in (**63**) [*112334-01-7*] to yield SQ 83,360 [*108319-07-9*] (**58**) in the acid form. A ready supply of Z-azetidinone [*80082-81-1*], $C_{11}H_{12}N_2O_3$, used in the synthesis of SQ 83,360, was obtained starting with the mesylate of Z-protected serine amide (17). The azetidinone was formed in 53–57% overall yield (44). SQ 83,360 (**58**) demonstrated exceptional activity *in vitro* as can be seen in Table 5. Against numerous clinical *Enterobacteriaceae* strains, SQ 83,360 was similar to aztreonam, yet against nonfermenting gram-negative bacilli, particularly *Ps. aeruginosa,* SQ 83,360 was significantly superior.

Table 5. Comparative *In Vitro* Activity of SQ 83,360 and Aztreonam

		$MIC_{90}{}^{b}$, μg/mL	
Organism[a]	Number of strains	SQ 83,360	Azetreonam
Enterobacteriaceae	236	<0.03–1.0	<0.03–12.5
Ps. aeruginosa	35	0.05	15.3
Pseudomonas sp.	58	6.3	90.4
Acinetobacter sp.	45	14.7	46.0

[a]All are nonfermentors except the *Enterobacteriaceae*.
[b]Minimum inhibitory concentration for 90% of the strains used.

Since the discovery of SQ 83,360, compounds such as U-78,608 [*123444-35-9*] (**64**) having different linker groups between the hydroxypyridone group and the sulfonyl residue have been reported. U-78,608 and SQ 83,360 have similar *in vitro* and *in vivo* activity (45).

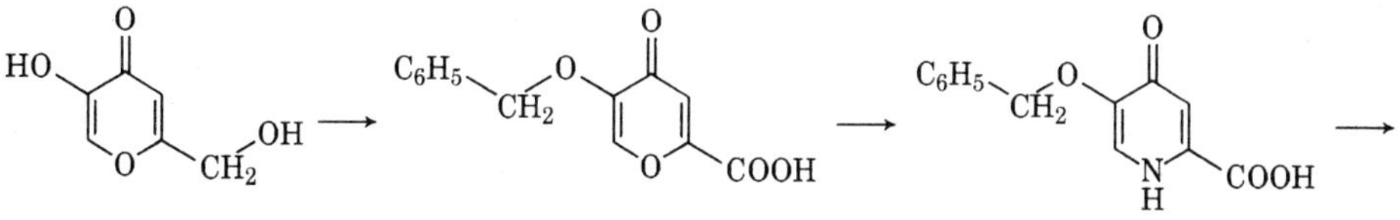

(59)

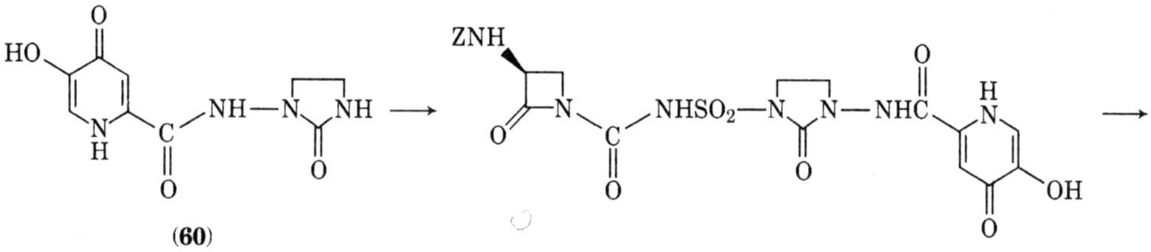

(60)

(61)

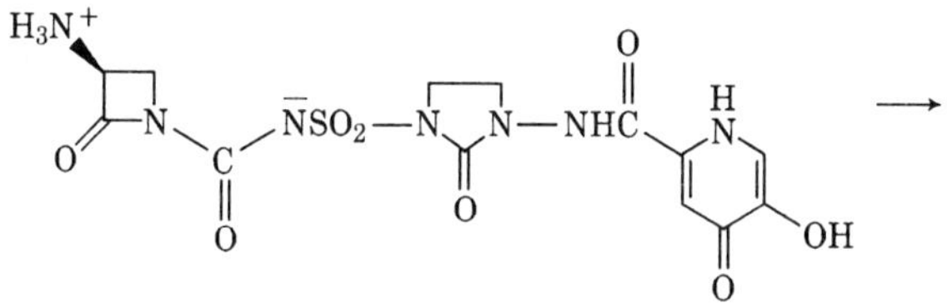

(62)

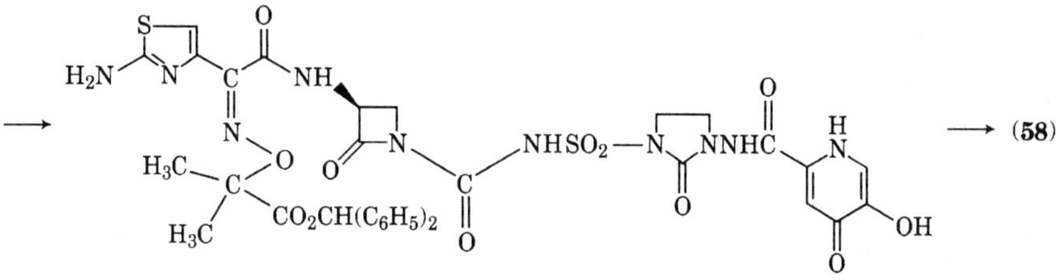

(63)

Fig. 8. Synthesis of SQ 83,360.

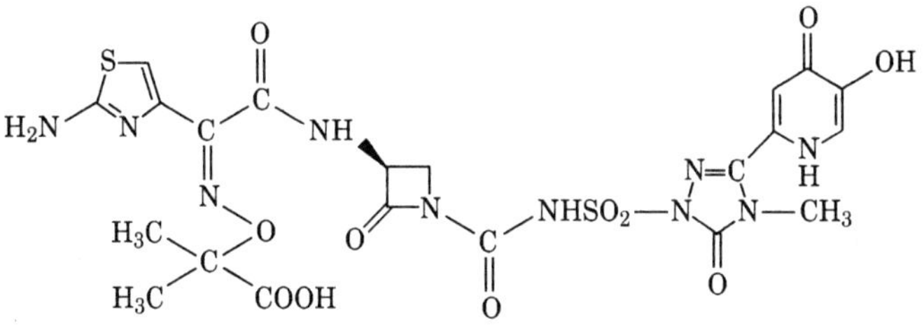

(64)

Tetrazole-Activated Monobactams. The tetrazole moiety is one of the closest isosteres to a carboxylic acid, having similar electron-withdrawing capability and acidity. The syntheses of homochiral tetrazole derivatives (**65**) from serine and threonine have been reported (46). As shown in Figure 9, cyclization of amides (**66**, R = H) [*90208-25-6*] and (**66**, R = CH_3) [*90181-42-3*] was effected using an intramolecular Mitsunobu reaction. Deprotection afforded the amine salts (**67**, R = H) [*90181-47-8*] and (**67**, R = CH_3) [*90181-49-0*], which were then coupled to afford the side-chain analogues (**65**, R′ = CH_3 or $C(CH_3)_2COOH$). Using similar methodology, other C-4 substituted monobactams were prepared whereby the configuration of R in structure (**65**) was either cis or trans and R = CH_2F, $COOCH_3$, CH_2OH, or CH_2OCONH_2 (47).

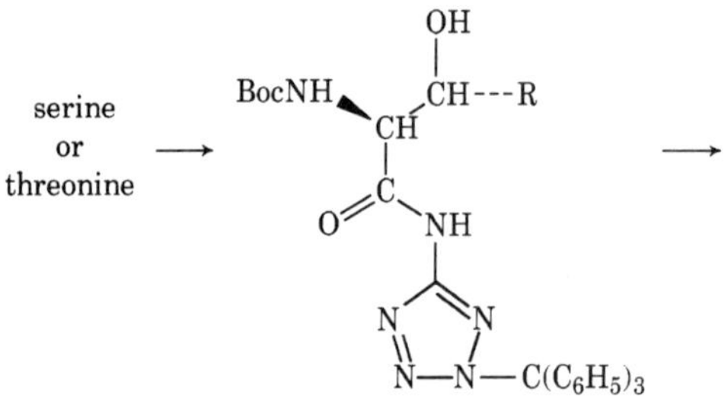

(**66**)

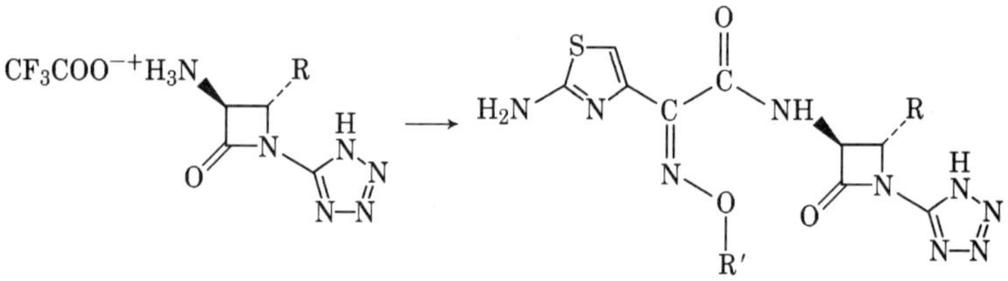

(**67**) (**65**)

Fig. 9. Synthesis of *N*-(tetrazolyl-5-yl) azetidinones where Boc is *t*-butoxycarbonyl; R = H or CH_3; and R′ = CH_3 or $C(CH_3)_2COOH$.

Alternatively, various 4-substituted derivatives have been prepared via synthesis of amino acid (**68**) by reaction of the anion formed from protected glycine and an appropriately substituted Schiff base.

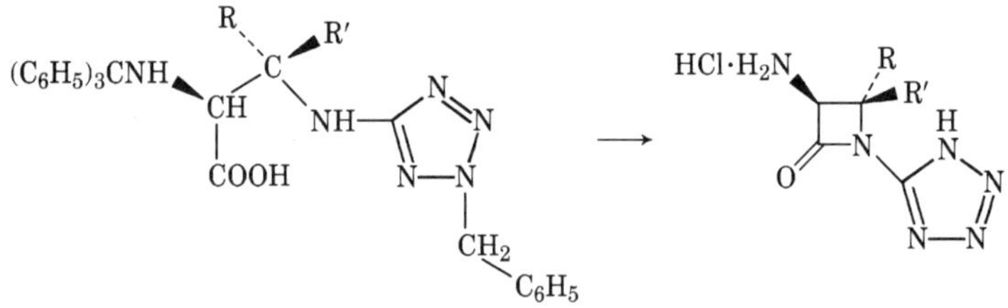

(**68**) (**69**)

Cyclization of (**68**) and deprotection then yield the monobactam nucleus (**69**) which may be coupled with various C-3 side chains (48). Direct alkylation of the *N*-unsubstituted azetidinone (**70**) using fluorotetrazole [*93607-94-4*], $C_8H_7FN_4$, also produced (**69**) after deprotection of (**71**) (49).

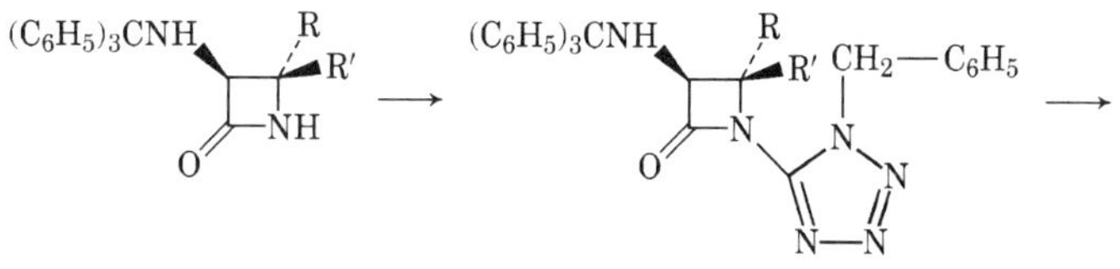

(70) (71)

(69) 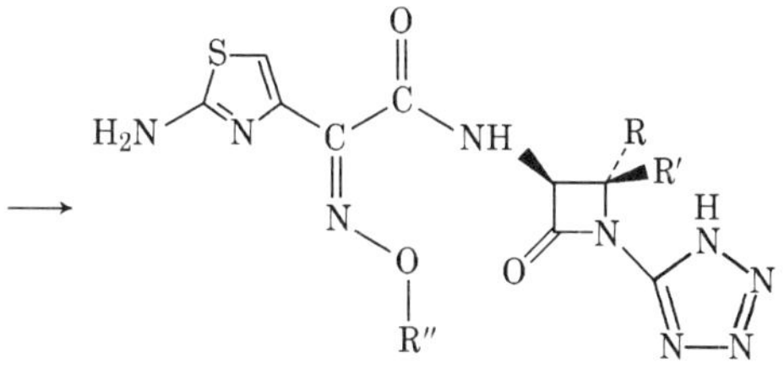

(72)

Structure (**72**) exemplifies the C-4 derivatives prepared by these routes. From this class of monobactams, RU-44790 [*110012-78-7*] (**72**, R = H, R′ = CH_2F, R″ = ) was identified as having improved β-lactamase stability and activity vs gram-negative bacilli similar to that of aztreonam (50).

Economic Aspects

Two monobactams were in clinical use as of 1990. Aztreonam (**17**) manufactured by Bristol-Myers Squibb, has the worldwide trademark of Azactam. Aztreonam has received regulatory approval for use in humans in 67 countries and has pending filings in seven additional countries. The global experience with aztreonam and its clinical acceptance signifies that the monobactams are recognized as an important new class of antibacterial agents. Carumonam, manufactured by Takeda, received approval for human usage in 1988 in Japan. Carumonam (**19**) has the trademark, Amasulin.

BIBLIOGRAPHY

"Antibiotics, β-Lactams-Monobactams" in *ECT* 3rd ed., Supplement Vol., pp. 131–144, by C. M. Cimarusti and R. B. Sykes, Squibb Institute for Medical Research.

1. R. B. Sykes and co-workers, *Nature (London)* **291,** 489 (1981).
2. A. Imada and co-workers, *Nature (London)* **289,** 590 (1981).
3. W. L. Parker, J. O'Sullivan, and R. B. Sykes, *Advances in Applied Microbiology,* Vol. 31, Academic Press, New York, 1986, pp. 181–205; A. G. Brown, *Pure Appl. Chem.* **59,** 475 (1987).
4. K. Kamiya, M. Takamoto, Y. Wada, and M. Asai, *Acta Cryst.* **B37,** 1626 (1981).
5. W. L. Parker and co-workers, *J. Antibiot.* **35,** 189 (1982).
6. W. L. Parker and M. L. Rathnum, *J. Antibiot.* **35,** 300 (1982).

7. J. M. Indelicato and W. L. Wilham, *J. Med. Chem.* **17,** 528 (1974).
8. R. B. Sykes, D. P. Bonner, N. H. Georgopapadakou, and J. S. Wells, *J. Antimicrob. Chemother.* (Suppl. E) **8,** 1 (1981).
9. C. M. Cimarusti, *J. Med. Chem.* **27,** 247 (1984).
10. N. H. Georgopapadakou, S. A. Smith, and R. B. Sykes, *Antimicrob. Agents Chemother.* **21,** 950 (1982).
11. J. O'Sullivan and co-workers, *J. Antimicrob. Agents Chemother.* **21,** 558 (1982).
12. D. P. Bonner and R. B. Sykes in C. S. F. Easmon and J. Jeljaszewicz, eds., *Medical Microbiology,* Vol. 4, Academic Press, New York, 1984, pp. 171–197; D. P. Bonner and R. B. Sykes, *J. Antimicrob. Chemother.* **14,** 313 (1984).
13. T. Matsuo and co-workers, *Chem. Pharm. Bull.* **31,** 2200 (1983).
14. W. H. Koster, C. M. Cimarusti, and R. B. Sykes in R. B. Morin and M. Gorman, eds., *Chemistry and Biology of β-Lactam Antibiotics,* Vol. 3, Academic Press, New York, 1982, pp. 339–375.
15. T. Matsuo and co-workers, *Chem. Pharm. Bull.* **31,** 1874 (1983).
16. D. M. Floyd and co-workers, *J. Org. Chem.* **47,** 5160 (1982).
17. D. M. Floyd, A. W. Fritz, and C. M. Cimarusti, *J. Org. Chem.* **47,** 176 (1982).
18. P. S. Manchand and co-workers, *J. Org. Chem.* **53,** 5507 (1988).
19. C. M. Cimarusti and co-workers, *Tetrahedron* **39,** 2577 (1983); H. Breuer, *J. Antimicrob. Chemother* (Suppl. E) **8,** 21 (1981).
20. R. B. Sykes, D. P. Bonner, K. Bush, and N. H. Georgopapadakou, *Antimicrob. Agents Chemother* **21,** 85 (1982).
21. K. Bush, J. Freudenberger, and R. S. Sykes, *Antimicrob. Agents Chemother.* **22,** 414 (1982).
22. R. E. Reese, D. E. Sentochnik, R. G. Douglas, and R. F. Betts, *Handbook of Antibiotics,* Little, Brown and Company, Boston, 1988, pp. 134–141.
23. *Reviews of Infectious Disease,* **1** (Suppl. 4) (1985); D. P. Bonner and R. B. Sykes, *J. Antimicrob. Chemother.* **14,** 313 (1984); R. B. Sykes and D. P. Bonner, *Amer. J. Med.* **78**(2A), 2 (1985); R. B. Sykes, D. P. Bonner, and E. A. Swabb, *Pharmac. Ther.* **29,** 321 (1985); *Chemotherapy* **35,** (Suppl. 1) (1989).
24. N. C. Cohen, *J. Med. Chem.* **26,** 259 (1983); J. B. Bartolone, G. J. Hite, J. A. Kelly, and J. R. Knox, in A. G. Brown and S. M. Roberts, eds., *Recent Advances in the Chemistry of β-Lactam Antibiotics,* The Royal Society of Chemistry, 1985, pp. 318–327; J. A. Kelley, J. C. Boyington, P. C. Moews, and J. R. Knox, in H. Umezawa, ed., *Frontiers of Antibiotic Research,* Academic Press, New York, 1987, pp. 327–337.
25. C. M. Cimarusti and R. B. Sykes, *Med. Res. Rev.* **4,** 1 (1984).
26. W. H. Koster and co-workers, *J. Am. Chem. Soc.* **105,** 3743 (1983).
27. W. H. Koster and co-workers, *Abstracts, 22nd Interscience Conference on Antimicrobial Agents and Chemotherapy* No. 674, American Society for Microbiology, Miami, Fla., 1982; Ref. 25, pp. 15–17.
28. E. M. Gordon and co-workers, *J. Am. Chem. Soc.* **104,** 6053 (1982).
29. S. K. Tanaka and co-workers, *Antimicrob. Agents. Chemother.* **31,** 219 (1987).
30. J. M. Clark and co-workers, *Antimicrob. Agents Chemother.* **31,** 226 (1987).
31. D. A. Conrad, A. L. Lentnek, and the Tigemonan Study Group, *29th Interscience Conference on Antimicrobial Agents and Chemotherapy,* Abstract No. 921, Houston, Tex., 1989.
32. W. A. Slusarchyk and co-workers, *Tetrahedron Lett.* **27,** 2789 (1986); J. D. Godfrey, Jr., R. H. Mueller, and D. J. Von Langen, 2793 (1986).
33. W. H. Koster and D. P. Bonner in H. Umezawa, ed., *Frontiers of Antibiotic Research,* Academic Press, New York, 1987, pp. 211–226.
34. R. B. Sykes, W. H. Koster, and D. P. Bonner, *J. Clin. Pharmacol.* **28,** 113 (1988).
35. W. A. Slusarchyk and co-workers, *Heterocycles* **21,** 191 (1984).

36. F. R. Atherton and R. W. Lambert, *Tetrahedron* **40,** 1039 (1984); H. Breuer and co-workers, *Abstract No. 135, 24th Interscience Conference on Antimicrobial Agents and Chemotherapy,* Washington, D.C., 1984; S. R. Woulfe and M. J. Miller, *Tetrahedron Lett.* **25,** 3293 (1984); S. R. Woulfe and M. J. Miller, *J. Med. Chem.* **28,** 1447 (1985).
37. H. Breuer and co-workers, *J. Antibiot.* **38,** 813 (1985).
38. C. M. Cimarusti, *Gazz. Chim. Ital.* **116,** 169 (1986).
39. W. H. Koster and co-workers, Abstract No. 136, and J. M. Clark and co-workers, Abstract No. 139, *24th Interscience Conference on Antimicrobial Agents and Chemotherapy,* Washington, D.C., 1984.
40. W. V. Curran and R. H. Lenhard, *J. Med. Chem.* **32,** 1749 (1989).
41. S. R. Woulfe and M. J. Miller, *J. Org. Chem.* **51,** 3133 (1986).
42. Ref. 25, pp. 17–20.
43. Ref. 33, pp. 211–218; H. Nikaido and E. Y. Rosenberg, *J. Bacteriol.* **172,** 1361 (1990).
44. U.S. Pat. 4,587,051 (May 6, 1986), W. H. Koster (to E. R. Squibb and Sons, Inc.); ref. 33, pp. 214–216.
45. M. R. Barbachyn and T. C. Tuominen, Abstract No. 231, and G. E. Zurenko and B. Hannon, Abstract No. 233, *Interscience Conference on Antimicrobial Agents and Chemotherapy,* Los Angeles, Cal., 1988.
46. A. Andrus, B. Partridge, J. V. Heck, and B. G. Christensen, *Tetrahedron Lett.* **25,** 911 (1984).
47. C. Yoshida and co-workers, *J. Antibiot.* **39,** 215 (1986).
48. M. Klich and G. Teutsch, *Tetrahedron Lett.* **25,** 3849 (1984).
49. M. Klich and G. Teutsch, *Tetrahedron* **42,** 2677 (1986).
50. G. Teutsch, M. Klich, and J. F. Chantot, *Abstracts, Interscience Conference on Antimicrobial Agents and Chemotherapy,* No. 843, New Orleans, La., 1986; Ref. 47.

KLAUS R. LINDNER
DANIEL P. BONNER
WILLIAM H. KOSTER
Bristol-Myers Squibb Pharmaceutical Research Institute

PENICILLINS AND OTHERS

The basic structural features of the penicillin nucleus (**1**) include a β-lactam ring fused through nitrogen and the adjacent tetrahedral carbon to a second heterocycle which, in natural penicillin is a 5-membered thiazolidine ring. Biologically active penicillins are generally characterized by a functionalized amino group in the 6β-position of the β-lactam ring and a carboxyl group in the 3-position in the thiazolidine ring.

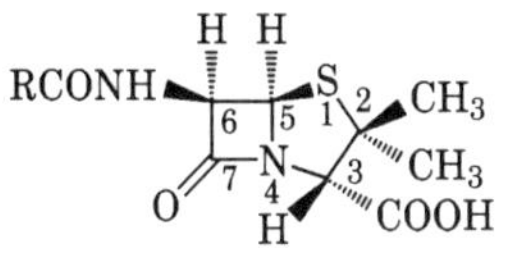

(**1**)

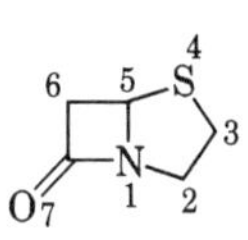

(**2**)

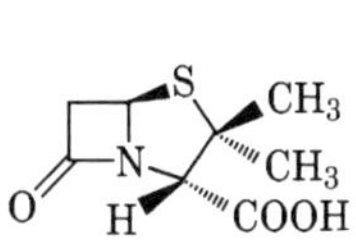

(**3**)

In general, penicillins exert their biological effect, as do the other β-lactams, by inhibiting the synthesis of essential structural components of the bacterial cell wall. These components are absent in mammalian cells so that inhibition of the synthesis of the bacterial cell wall structure occurs with little or no effect on mammalian cell metabolism. Additionally, penicillins tend to be irreversible inhibitors of bacterial cell-wall synthesis and are generally bactericidal at concentrations close to their bacteriostatic levels. Consequently penicillins have become widely used for the treatment of bacterial infections and are regarded as one of the safest and most efficacious classes of antibiotics.

Nomenclature

Chemical Abstracts indexes most penicillins as 4-thia-1-azabicyclo[3.2.0]heptane-7-ones (**2**). Using this system, penicillin G [*161-33-6*] (**1**, R = $C_6H_5CH_2$), $C_{16}H_{18}N_2O_4S$, is 6-(2-phenyl-acetamido)-3,3-dimethyl-7-oxo-4-thia-1-azabicyclo[3.2.0]heptane-2-carboxylic acid. The unsubstituted bicyclic ring system of the penicillins is designated as penam [*53908-04-6*], C_5H_7NOS, (**2**) (1) and the penicillins (**1**) are generally 6-acylamino-2,2-dimethylpenam-3-carboxylic acids. A further simplification is the use of the term penicillanic acid [*87-53-6*], $C_8H_{11}NO_3S$, to designate the penicillin ring system having the substituents indicated in (**3**). Thus the penicillins are named as the appropriate acylaminopenicillanic acid. Because the great majority of the variations of the penicillin structure are in the acyl side chain, the carbonyl of the acyl group is included in the basic moiety name penicillin.

Landmarks in the development of penicillin are its discovery in 1929 (2) and the subsequent recognition in 1940 of the potential utility of penicillin for controlling antibacterial infections in animals (3) and shortly afterward in humans (4,5). The realization that penicillin was a useful drug occurred during World War II (6).

Penicillin was first obtained as a product of the fungus *Penicillium notatum*, but the early manufacture utilizing this organism was not amenable to large-scale handling. Modifications were soon introduced which improved yields and quality of the product (7). Although there was an intense effort from 1940 on, it took until 1945 before the combined results of chemical degradation and x-ray crystallography made it possible to be certain that the penicillin structure contained the β-lactam thiazolidine ring system (8).

In 1959 isolation of the penicillin nucleus, 6-aminopenicillanic acid [*551-16-6*] (6-APA) (**1**, RCO = H), $C_8H_{12}N_2O_3S$, from fermentations deficient in side-chain precursors was reported (9). It was soon found that 6-APA could be produced more efficiently by enzymatically removing the acyl group from various penicillins using penicillin acylases. Subsequently this intermediate has also been made by chemically removing the 6-acyl group after first protecting the C-3 carboxyl group. However, for manufacturing purposes, the majority of 6-APA is currently produced by enzymatic deacylation of penicillin G or penicillin V [*87-08-1*]. The isolation of 6-aminopenicillanic acid has resulted in the synthesis and biological evaluation of many thousands of penicillin analogues derived primarily by acylation of the 6-amino group. From this vast chemical investment an important range of semisynthetic penicillins have found wide clinical use (Table 1).

Physical Properties

Penicillins have several properties that are characteristic of β-lactam antibiotics (10). They are obtained in relatively pure form as off-white, tan, or yellow freeze-dried or spray-dried solids that are usually amorphous. Alternatively they are sometimes obtained as crystalline solids, often as hydrates. Penicillins do not usually have sharp melting points, but decompose upon heating to elevated temperatures. Most natural members have a free carboxyl group and commercial preparations are generally either supplied as salts, most frequently as sodium salts, or in zwitterionic form as hydrates, eg, amoxicillin trihydrate [*61336-70-7*]. The acid strength of the carboxyl group in aqueous solution varies from pK_{a1} = 2.73 for oxacillin to pK_{a1} = 3.06 for carbenicillin. For zwitterionic penicillins amoxicillin, having pK_{a1} = 2.67 and pK_{a2} = 7.11, is typical (11).

Spectral Characteristics. The infrared stretching frequency of the penicillin β-lactam carbonyl group normally occurs at relatively high frequencies (1770–1815 cm^{-1}) as compared to the absorptions for the secondary amide (1504–1695 cm^{-1}) and ester (1720–1780 cm^{-1}) carbonyl groups. There is little difference between solution and KBr spectra. The nuclear magnetic resonance spectrum of penicillins invariably provides information about the integrity of the ring, attachments to the ring, and the stereochemistry of those attachments. The disposition of the C-5 and C-6 resonances are also characteristic. Thus the proton on the amide-bearing carbon appears as a quartet that collapses to a doublet on D_2O exchange, where δ = 5.42–6.15 ppm and J_{AB}(cis) = 4–5 Hz; J_{AB}(trans) = 1.5–2 Hz. The proton on the bridgehead carbon appears as a doublet, δ = 4.66–5.60; the C-3 proton absorbs in the range δ = 3.88–4.58 (10,12). More recently nmr spectroscopy has also proved to be a valuable tool for detecting and characterizing penicillin metabolites in biofluids (13).

The carbon-13 nmr features of penicillins have been studied and documented (10,14). The four sp^3 ring carbons give rise to resonances in order of decreasing chemical shift, C-3, C-5, C-2, and C-6. For C-3 the chemical shift ranges are 73.5–74.4 ppm for alkali metal salts in D_2O and 70.9–71.6 ppm for HCl salts having nonionizable carboxylate functions. The C-5 resonance ranges from 66.0 to 68.7 ppm, and the C-6 resonance from 57.4 to 59.1 ppm, or 60.0 to 64.6 ppm for examples with protonated side chains. The C-2 ranges from 63.9 to 66.0 ppm. The range for the 2α,2β-methyl signals are both narrow: for the high field 2-methyl the range is 26.5–28.0 ppm; and for the low field 2-methyl the range is 29.6–32.5 ppm. The carbonyl carbons give rise to resolved signals in most spectra in the range 167–176 ppm.

The analysis of penicillins by mass spectrometry (qv) has developed with the advent of novel techniques such as fast atom bombardment. The use of soft ionization techniques has enabled the analysis of thermally labile nonvolatile compounds. These techniques have proven extremely valuable in providing abundant molecular weight information from underivatized penicillins, both as free acids and as metal salts (15).

Stereochemistry. The absolute stereochemistry of the penicillins is 3(*S*):5(*R*):6(*R*) and the stereochemistry of the substituents attached to the ring is designated by the α and β-notations. Thus the β-lactam hydrogens are α, the

Table 1. Penicillins[a] in Clinical Use

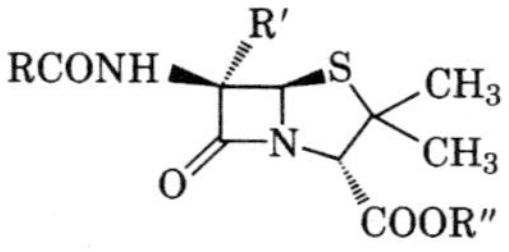

(**4**)

Name	CAS Registry Number	Molecular formula	R	R′	R″[b]	Administration: Route[c]	Administration: Av dose, g/d
Limited spectrum							
benzyl penicillin (penicillin G)[b]	[*61-33-6*]	$C_{16}H_{18}N_2O_4S$	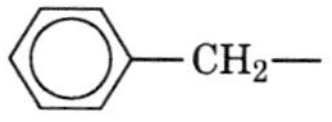	H	H	iv im po	3–12[d] 1–7[e] 0.6–6[f]
phenoxymethyl penicillin (penicillin V)	[*87-08-1*]	$C_{16}H_{18}N_2O_5S$	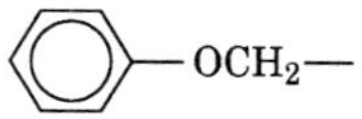	H	H	po	0.3–6
phenethicillin	[*132-93-4*]	$C_{17}H_{20}N_2O_5S{\cdot}K$	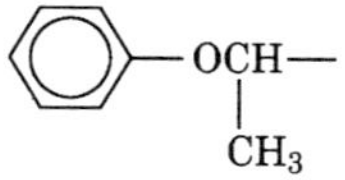	H	K	po	0.5–1
β-Lactamase stable							
methicillin	[*132-92-3*]	$C_{17}H_{20}N_2O_6S{\cdot}Na$	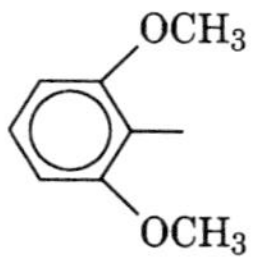	H	Na	im iv	6–12 4–6

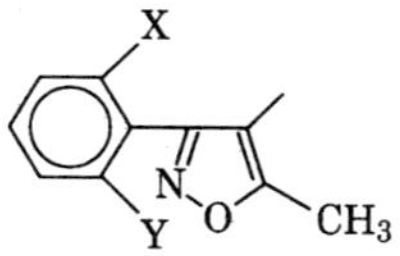

oxacillin	*[66-79-5]*	$C_{19}H_{19}N_3O_5S$	X = H, Y = H	H	H	po, im, iv	1–6
cloxacillin	*[61-72-3]*	$C_{19}H_{18}ClN_3O_5S$	X = H, Y = Cl	H	H	po	1–6
dicloxacillin	*[3116-76-5]*	$C_{19}H_{17}Cl_2N_3O_5S$	X = Cl, Y = Cl	H	H	po	1–6
flucloxacillin	*[5250-39-5]*	$C_{19}H_{17}ClFN_3O_5S$	X = Cl, Y = F	H	H	po	1–6
nafcillin	*[985-16-0]*	$C_{21}H_{22}N_2O_5SNa$	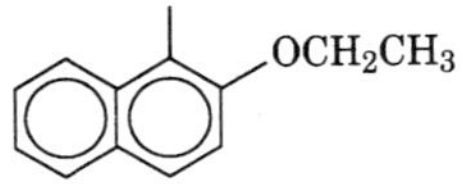	H	Na	po	1–2
Broad spectrum							
ampicillin	*[69-53-4]*	$C_{16}H_{19}N_3O_4S$	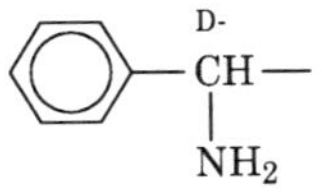	H	H	po, iv, im	1–4
hetacillin[g]	*[3511-16-8]*	$C_{19}H_{23}N_3O_4S$	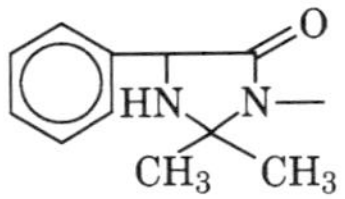	H	H		1–4
pivampicillin	*[33817-20-8]*	$C_{22}H_{29}N_3O_6S$	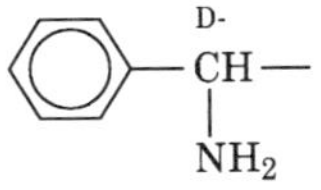	H	$CH_2OCOC(CH_3)_3$		1–4
talampicillin	*[47747-56-8]*	$C_{24}H_{23}N_3O_6S$	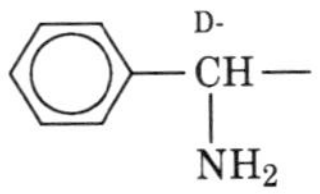	H	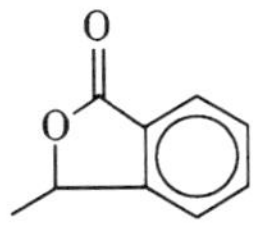		1–4
bacampicillin	*[50972-17-3]*	$C_{21}H_{27}N_3O_7S$	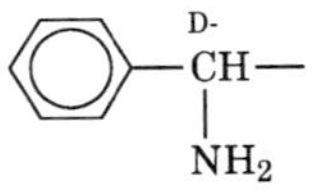	H	$-CH(CH_3)OCOOCH_2CH_3$	po, iv, im	1–4
ciclacillin	*[3485-14-1]*	$C_{15}H_{23}N_3O_4S$	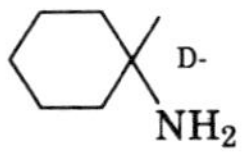	H	H	po, iv	1–4

Table 1. (Continued)

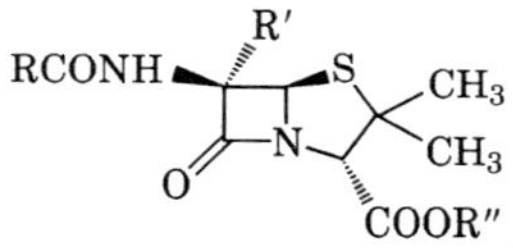

(**4**)

Name	CAS Registry Number	Molecular formula	R	R′	R''^{b}	Administration	
						Route[c]	Av dose, g/d
amoxicillin	[*26787-78-0*]	$C_{16}H_{19}N_3O_5S$	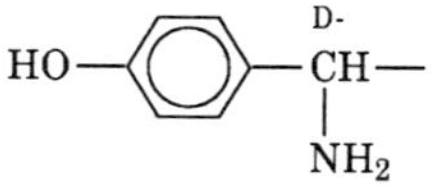	H	H	po	0.75–3
carbenicillin[h]	[*4697-36-3*]	$C_{17}H_{18}N_2O_6S$	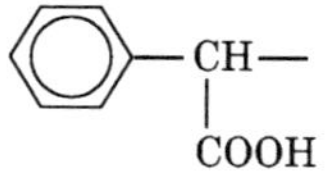	H	H	iv	to 30
ticarcillin	[*34787-01-4*]	$C_{15}H_{16}N_2O_6S_2$		H	H	iv	to 20
sulbenicillin	[*41744-40-5*]	$C_{16}H_{18}N_2O_7S_2$	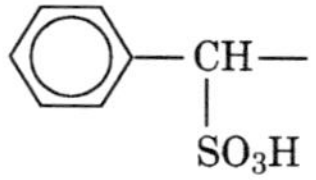	H	H	iv	to 20
azlocillin	[*37091-66-0*]	$C_{20}H_{23}N_5O_6S$	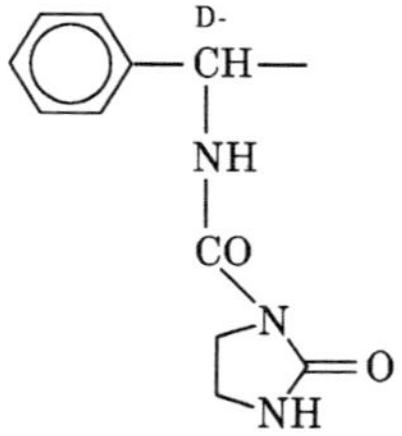	H	H	iv	to 20

mezlocillin	[*51481-65-3*]	$C_{21}H_{25}N_5O_8S_2$	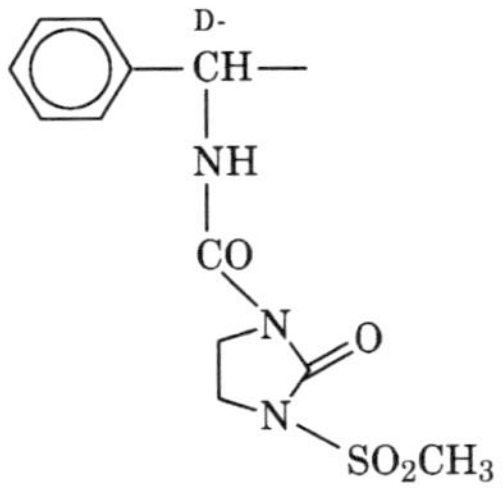	H	H	iv	to 20
piperacillin	[*61477-96-1*]	$C_{23}H_{27}N_5O_7S$	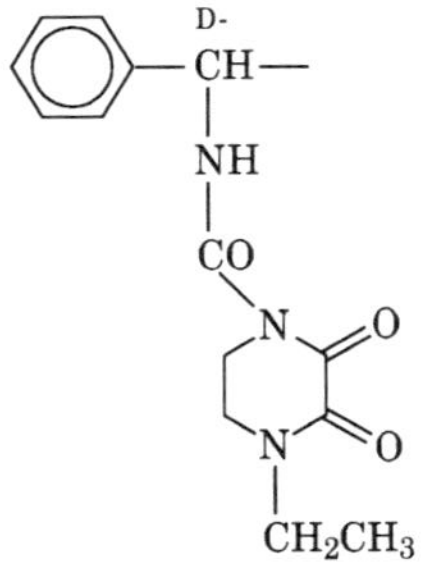	H	H	iv	to 20
Directed spectrum β-lactamase stable							
temocillin	[*66148-78-5*]	$C_{16}H_{18}N_2O_7S$	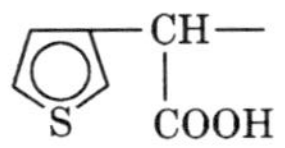	OCH_3	H	iv	to 20

[a]Structure (**4**) where R, R′, and R″ are as indicated.
[b]Whereas the carboxylic acid is shown for many of these compounds, the majority of the commercial penicillins are sodium salts. Penicillin G is available as the calcium salt [*973-53-5*], $C_{16}H_{17}N_2O_4S \cdot \frac{1}{2}Ca$, and as the potassium salt [*113-98-4*], $C_{16}H_{17}N_2O_4S \cdot K$.
[c]Where iv = intravenously, im = intramuscularly, and po = orally.
[d]Corresponds to 5–20 million units.
[e]Corresponds to 2–12 million units.
[f]Corresponds to 1–10 million units.
[g]The R group given is equivalent to the RCONH of (**4**) and attaches directly to the C-6 of the penam ring (**2**).
[h]Also available as the indanyl ester [*26605-69-6*], $C_{26}H_{26}N_2O_6SNa$.

acylamino group is β, and the penicillin carboxyl is α as in (**1**). A study of derivatives for which crystal structures have been determined has shown that the derivatives belong to two classes, either open or closed, depending on the conformation of the thiazolidine ring as shown in Figure 1 (16). In both cases puckering of the ring system is relatively pronounced and it is suggested that changes in the nature of the conformation influence biological activity.

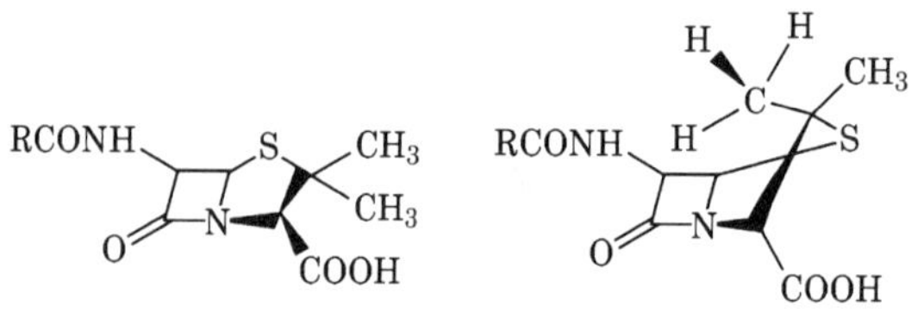

(**a**) (**b**)

Fig. 1. The penicillin thiazolidine ring: (**a**) open conformation; (**b**) closed conformation.

Chemical Properties

Penicillin to Other β-Lactam Conversions. Penicillin G (**1**, R = $C_6H_5CH_2$), penicillin V (**1**, R = $C_6H_5OCH_2$), and 6-APA (**1**, RCO = H) have proven to be cheap and versatile starting materials for a number of conversions to novel β-lactam containing systems. The commercial process for the preparation of 7-aminodeacetoxycephalosporanic acid (7-ADCA) and the deacetoxycephalosporins, cephalexin and cephradine, utilizes penicillin V (17) and the processes for conversion of penicillins into cephalosporins are well documented (18). Moxalactam [*64952-97-2*] (**5**) and related oxacephems are similarly derived from 6-APA (19).

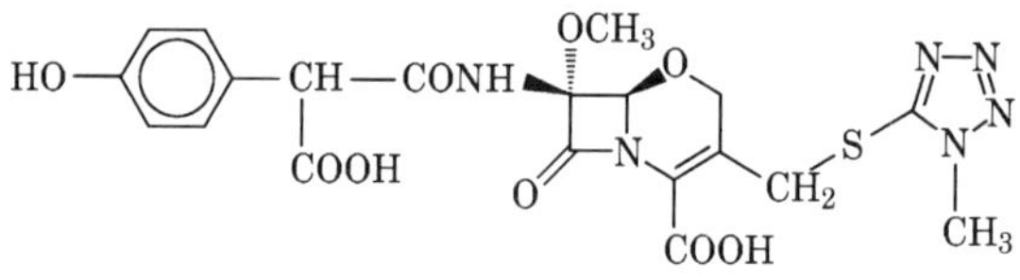

(**5**)

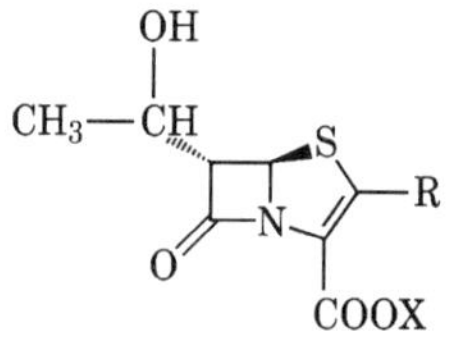

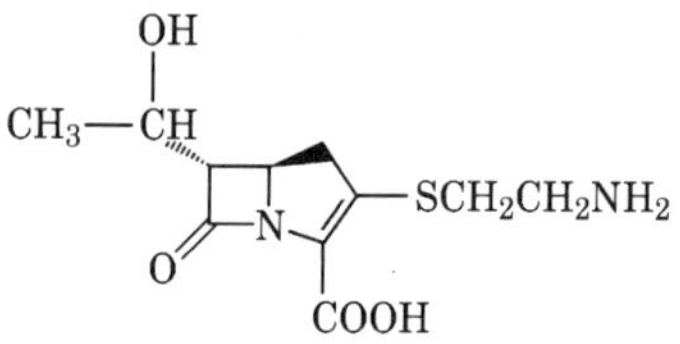

(**6**) (**7**)

Synthesis of the penem ring system (**6**) from 6-APA has been described (20) and subsequently syntheses of novel penems having potential biological utility such as FCE 22101 (**6**, R = CH_2OCONH_2, X = Na) [*84845-58-9*] (21), SCH 29482 (**6**, R = SCH_2CH_3, X = Na) [*77646-84-5*], and SCH 34343 (**6**, R = $SCH_2CH_2OCONH_2$, X = Na) [*94392-35-5*] (22) have been documented. Similarly, synthesis of the carbapenem antibiotic thienamycin [*59995-64-1*] (**7**), $C_{11}H_{16}N_2O_4S$, has been achieved from 6-APA (23).

Synthesis

The only penicillins used in their natural form are benzylpenicillin (penicillin G) and phenoxymethylpenicillin (penicillin V). The remainder of penicillins in clinical use are derived from 6-APA and most penicillins having useful biological properties have resulted from acylation of 6-APA using standard procedures.

A variety of coupling methods have been employed including: acid chorides, mixed anhydrides, mixed sulfonic acid anhydrides, *N*,*N*-dicyclohexylcarbodiimide and similar condensing agents, activated esters with *N*-hydroxysuccinimide, and *N*-hydroxybenzotriazole together with other acylating agents commonly used in peptide synthesis. The choice of coupling reaction is invariably dependent on the side chain. The efficiency of the reaction varies considerably between side chains so that optimization of the coupling procedure is often a matter of trial and error.

The use of protecting groups is common in penicillin chemistry: the amino function is normally protected by a trityl, benzyloxycarbonyl, *p*-nitrobenzyloxycarbonyl, trichloroethyloxycarbonyl, or trimethylsilyl group; and the carboxylic acid is usually protected as a benzyl, *p*-nitrobenzyl, *p*-methoxybenzyl, or trichloroethyl ester. Acylations may thus be carried out in aqueous or nonaqueous media with subsequent removal of the protecting group as required. Use of a tertiary base such as triethylamine is common in anhydrous reactions. Aqueous systems frequently utilize acetone–water mixtures in the presence of sodium bicarbonate. The resulting penicillins are usually isolated by the partitioning of the product as its free acid into a water-immiscible organic solvent and the salts or other impurities into water. The penicillins can be precipitated from organic solution as amine salts, or as alkali metal salts. Potassium or sodium ethylhexanoate dissolved in lower aliphatic alcohols may be used to effect purification. Alkali metal salts may be further purified by use of ion-exchange resins, eg, Diaion HP-20SS, and the final product isolated by freeze-drying (lyophilization), spray-drying, precipitation, or occasionally by crystallization.

After the influx of several α-amino penicillins such as ampicillin (Table 1) and related analogues, and the β-lactamase stable isoxazolyl penicillins in the 1950s and early 1960s, attention turned to identifying derivatives having a broader spectrum of activity. Carbenicillin and later, ticarcillin, reached the marketplace by the early 1970s. Effort then focused on the acyl and ureido ampicillin derivatives and culminated in the commercialization first of azlocillin and mezlocillin, then of piperacillin. Many thousands of derivatives were synthesized and some of the more active examples that have made little or no commercial impact are shown in Tables 2 and 3.

Examples of the acyl ampicillin series found in Table 2 are apalcillin (24), piridacillin (25), PL-385 (26), BLP-1908 (27), the Teijin examples TEI 1194 and TEI 2012 (28), and aspoxicillin (29). Other members of the ureido series of penicillins, of which azlocillin, mezlocillin, and piperacillin (Table 1) are commercial examples, are furazlocillin, EMD 39734 (30), and VX-VC-43 (31). During the 1970s the

Table 2. α-Amino, Acyl, or Ureido Penicillins[a] Under Investigation or Exhibiting Limited Clinical Utility

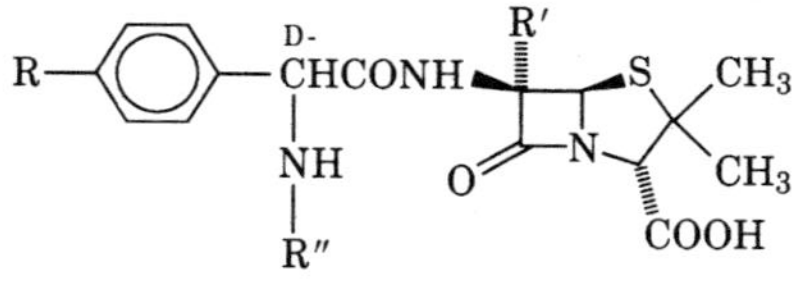

(**8**)

Name	CAS Registry Number	Molecular formula	R	R′	R″	Originating company	Status[b]
apalcillin	[*63469-19-2*]	$C_{25}H_{23}N_5O_6S$	H	H	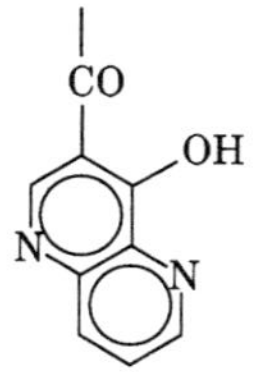	Sumitomo	C
piridacillin (Cl-867)	[*69414-41-1*]	$C_{32}H_{35}N_5O_{11}S_2$	OH	H	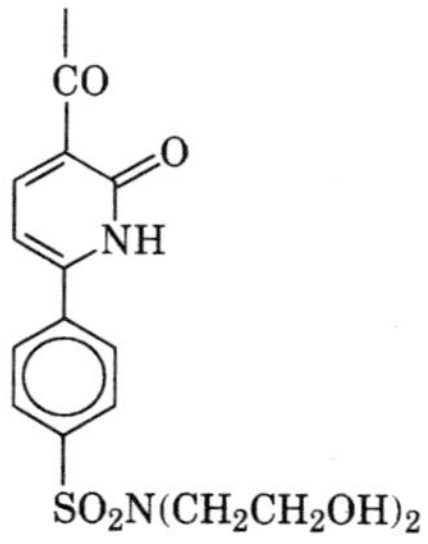	Warner-Lambert	N

PL-385	[*85550-66-9*]	$C_{29}H_{30}N_8O_8S$	OH	H	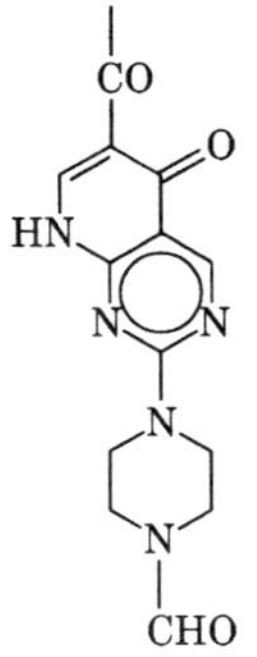	Dainippon	N
BLP-1908	[*65742-38-3*]	$C_{20}H_{20}N_6O_8S$	OH	H	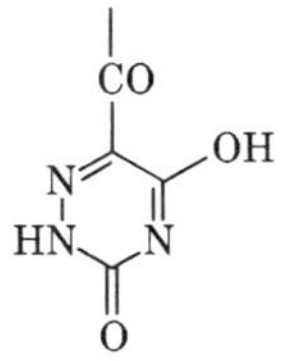	Bristol	N
					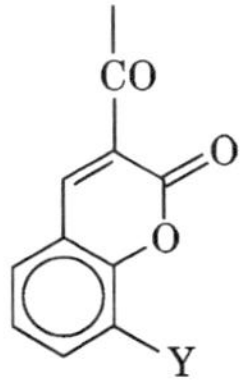		
TEI 1194	[*67609-13-6*]	$C_{26}H_{23}N_3O_7S{\cdot}Na$	H	H	Y = H	Teijin	N
TEI 2012	[*71344-34-8*]	$C_{26}H_{23}N_3O_8S{\cdot}Na$	H	H	Y = OH	Teijin	N
aspoxicillin (TA058)	[*63358-49-6*]	$C_{21}H_{27}N_5O_7S$	OH	H	—CO—CH(NH_2)—$CH_2CONHCH_3$	Tanabe	I

Table 2. ***(Continued)***

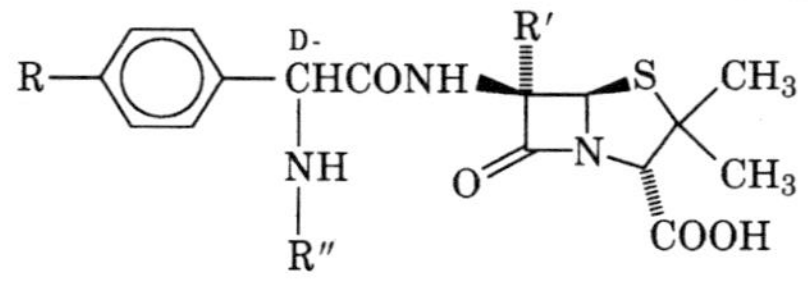

(**8**)

Name	CAS Registry Number	Molecular formula	R	R′	R″	Originating company	Status[b]
furazlocillin	[*66327-51-3*]	$C_{25}H_{26}N_6O_8S$	H	H	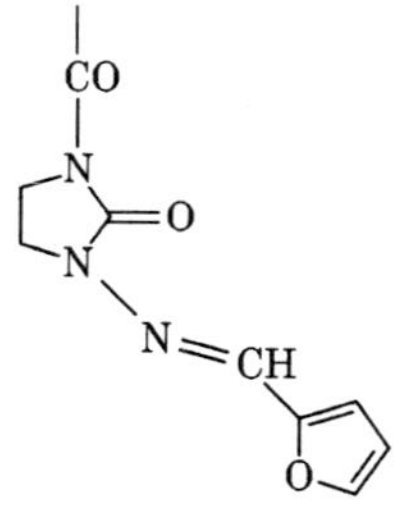	Bayer	C
EMD 39734		$C_{29}H_{29}N_7O_9S$	OH	H	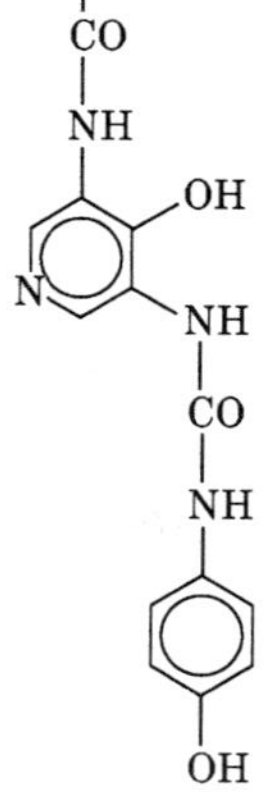	E. Merck	N

VX-VC-43	[*82509-56-6*]	$C_{27}H_{28}N_8O_9S_2$	H	H	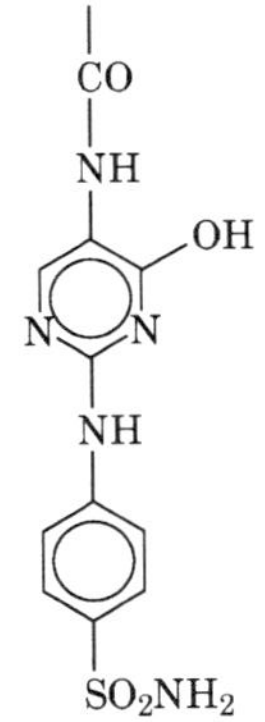	Thomae	N
lenampicillin[c]	[*86273-18-9*][c]	$C_{21}H_{23}N_3O_7S$[c]	H	H	H	Kanebo	C
formidacillin (BRL 36650)	[*94425-13-5*]	$C_{24}H_{28}N_6D_{10}S{\cdot}Na$	3,4 di-OH	NHCHO	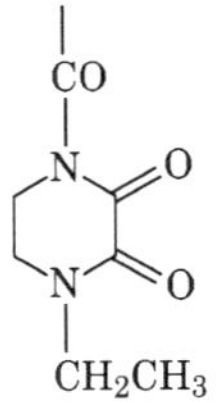	Beecham	N

[a]Structure (**8**) where R, R′, and R″ are as indicated.
[b]Drug status where C = in clinical use, I = investigational, and N = not progressed.
[c]Represents the 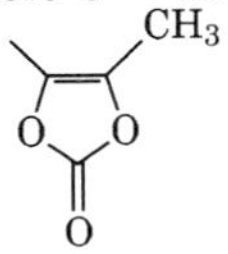 ester.

Table 3. Other Penicillins[a] Under Investigation or Exhibiting Limited Clinical Utility

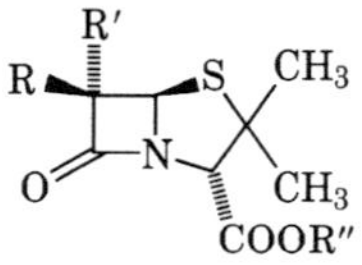

(**9**)

Name	CAS Registry Number	Molecular formula	R	R′	R″	Originating company	Status[b]
mecillinam	[*32887-01-7*]	$C_{15}H_{23}N_3O_2S$	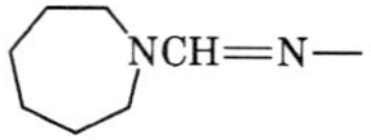	H	H	Leo	C
bacmecillinam	[*50846-45-2*]	$C_{20}H_{31}N_3O_6S$	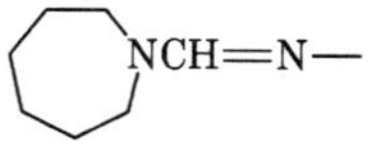	H	$—CH(CH_3)OC(=O)OC_2H_5$	Kyowa	I
BRL 28917	[*85621-68-7*]	$C_{17}H_{20}N_2O_{10}S_2$	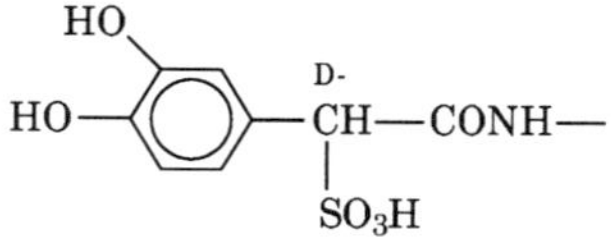	OCH_3	H	Beecham	N
BRL 20330	[*78968-21-5*]	$C_{23}H_{24}N_2O_7S_2$	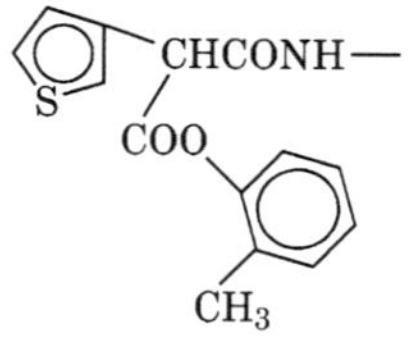	OCH_3	H	Beecham	N
BRL 44154	[*110717-44-7*]	$C_{19}H_{24}N_5O_6S_2$	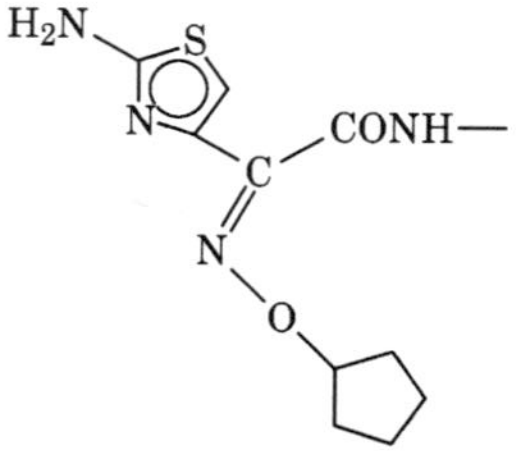	H	H	Beecham	I

[a]Structures (**9**), where R, R′, and R″ are as indicated.
[b]Drug status where C = in clinical use, I = investigational, and N = not progressed.

amidino penicillin mecillinam (Table 3), which possesses an unusual nonacyl side-chain substituent and a spectrum of activity limited to gram-negative organisms, was prepared. More recently a further example of a different type of acylated β-lactamase stable 6-APA derivative, BRL 44154, has been documented (32). BRL 44154 exhibits high activity against gram-positive organisms and methicillin-resistant staphylococci (MRSA).

Chemical Modification. Chemical modification of most positions in the penicillin nucleus have been carried out and these are summarized in Table 4. Apart from acylation of 6-APA, few of these modifications have proven profitable in terms of improving the biological properties of the derived penicillins. However, one of the modifications that has led to beneficial properties is substitution at the 6α-position.

Table 4. Chemical Modification of Penicillins

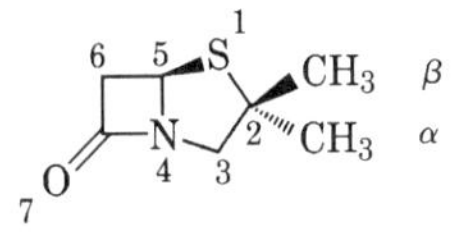

Position	Modification: From	Modification: To
1	S	SO, SO_2, $\overset{+}{S}CH_3$
2	$(CH_3)_2$	$(H)_2$, cycloalkyl, 4-piperidinyl, $=CH_2$ α-CH_2 halogen, β-CH_2 halogen, α-CHF_2, β-CHF_2 β-$COCH_3$, β-CH_2OR, β-CH_2OH, β-CHO, β-vinyl β-$CON(CH_3)_2$, β- $\overbrace{—CH—CH}^{O}$, β-$CH{=}CHCOOR$ β-$CH{=}CHCN$, β-$CH{=}CH$-tetrazole
3	COOH	H, COOR, $COOCH_2OCOR$, COR, COOOCOR, CO_2SiR_3, CON_3, CH_2OH, CH_2OR, CH_2COOH, $COCHN_2$, CN, $PO(OCH_3)OH$, $PO(OCH_3)_2$, CHO, OH, NCO, CH_2N_3, CH_2NR_2, COSH, CH_2COOH, CH_2CH_2COOH, tetrazolyl, diketopiperazinyl
2,3		cyclopropyl
6β	RCONH	H, NH_2, R′CONH, R″N=C=N, ArC=N, R_2N, RNH, $CH(OH)CH_3$, CH_2OH, CH_2NH_2, OH, halogen, 6-oxopenam
6α	H	CH_3, OCH_3, NH_2, RCONH, OH, CH_2OH, CH_2Cl, CH_2F, D, CH(OH)F, CH_2NH_2, SCH_3, $S(O)CH_3$, $CH_2CO_2CH_3$, CH_2Ar, COOH, COOR, NC, $NHCH_3$, $CH{=}CH_2$, CN, OCH_2CH_3, CH=CH-CN, CH=CHCHO, NHC_6H_5, $NHNHCOOCH_3$, $NHOCH_3$, N(CHO)NHCHO, $NHCOOCH_3$, $NHCOCF_3$, $NHSO_2CH_3$, $N(OCH_3)CHO$, $NHCONHCH_3$, $N(CH_3)_2$, $N(CH_3)CHO$, NHNHCHO, $N(OH)CH_3$, $COCH_3$, CH_2OCH_3, $CH_2OSO_2CH_3$, CHO, ONHCHO, ONH_2, C_6H_5, tetrazolyl, succinimido-oxy, $CH(OH)OCH_3$, $CONH_2$, CH(OH)Ar, CH_2CH_2CN
7	O	S

6α-Substituents. In the early 1970s the discovery of the cephamycins, stabilized to attack by β-lactamase through the presence of 7α-methoxy substituents, stimulated a parallel investigation of penicillins substituted by methoxy at the 6α-position (33). This study led to the identification of the directed spectrum β-lactamase stable temocillin, (Table 1) (34), the stability of which results from the combination of the 6α-methoxy group and the dicarboxylic acid functionality. Temocillin possessed high and prolonged serum levels in human volunteers but was directed in its spectrum of activity to gram-negative organisms. Whereas temocillin was active only by the parenteral route, the *o*-methylphenyl ester prodrug BRL 20330 (Table 3) was found to be absorbed orally (35). Replacement of the side-chain carboxylic acid by sulfonic acid led to BRL 28917 (Table 3), a β-lactamase stable penicillin with high activity against *Pseudomonas sp.* (36).

Substitution of penicillins by 6α-methoxy was found to be compatible with an α-acidic side chain in terms of antibacterial activity, but less beneficial when the side chain contained an α-acyl or α-ureido substituent. However, analogues of the ureido penicillin VX-VC-43 (Table 2) containing a 6α-methoxy substituent (**10**) were found to combine good stability to β-lactamase and relatively high antibacterial activity (37). Following an extensive program to identify other 6α-substituents that would stabilize the acyl and ureido series of penicillins, the 6α-formamido series (**11**) represented by formidacillin (BRL 36650) (Table 2) was developed (38).

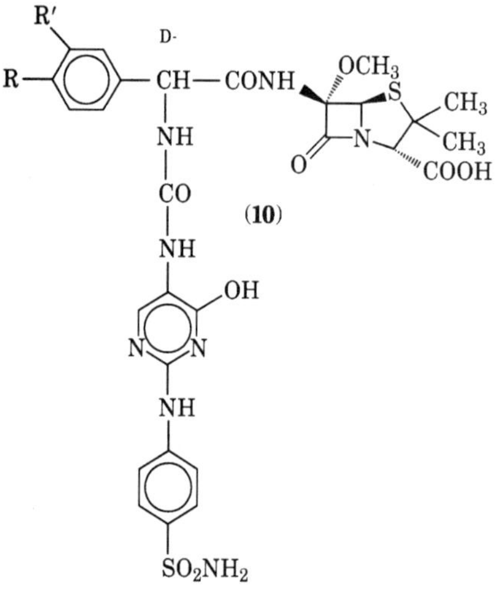

(**10**)

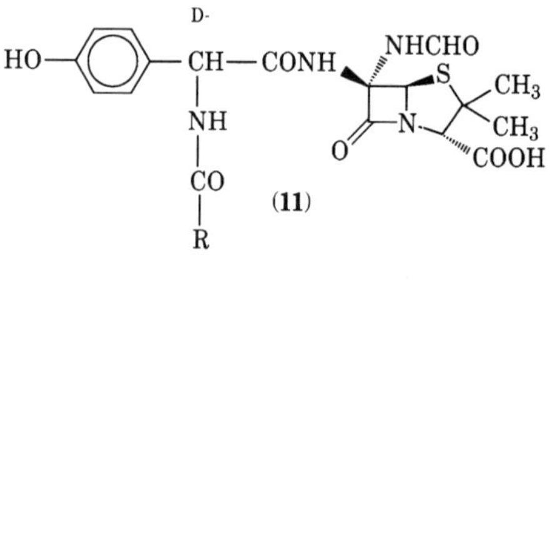

(**11**)

The early chemistry leading to these derivatives was originally carried out via the 6α-(methylthio) derivative (**17**) which was prepared by way of a Schiff's base (39). The 6α-thiomethyl group could then be displaced by various nucleophiles giving rise to 6α-methoxy or other 6α-substituted penicillins. A stereospecific one-step introduction of a methoxy group at C-6 in penicillins provided a simple entry to 6α-methoxy penicillins (40) in yields ranging from 50–62%.

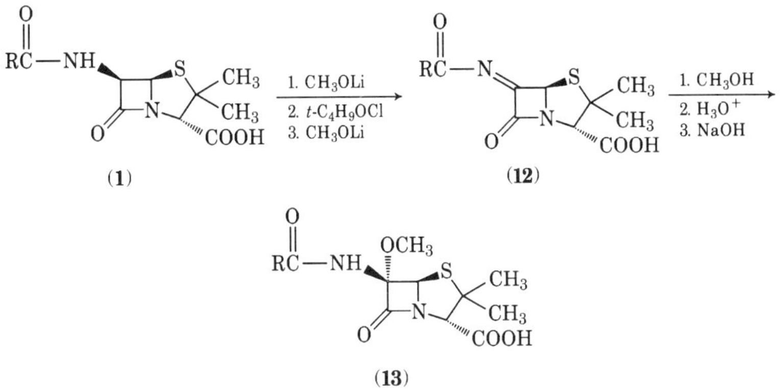

After formation of the acylimine (**12**), methanol adds to the less sterically hindered α-face of the molecule with high selectivity to provide (**13**). A further direct incorporation of a 6α-methoxy group (41) and subsequently a 6α-formamido group into penicillin has been achieved using trifluoromethanesulfonamides of type (**14**) (42).

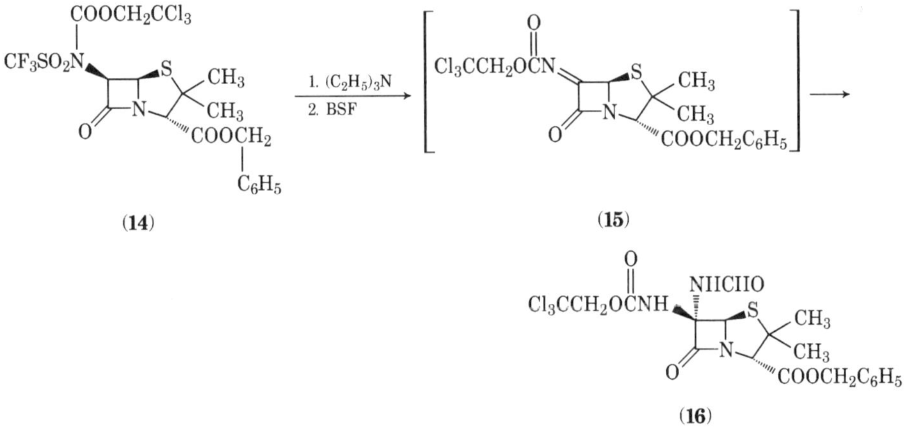

Preparation of (**14**) and treatment with *N,N*-bis(trimethylsilyl)formamide (BSF) and triethylamine provided (**16**) in 84% yield via the putative imine intermediate (**15**). The trifluoromethyl group could be replaced by other moieties such as 2,4,5 trichlorophenyl, pentafluorophenyl, or nonafluorobutyl with increasing effectiveness.

Further versatility was added to the range of substituents available for introduction into the 6α-position by use of the 6α(succinimido-oxy) derivative (**18**) prepared by treatment of the 6α-(methylthio) derivative (**17**) with *N*-hydroxysuccinimide and silver(I) acetate in dimethylformamide in virtually quantitative yield. In this way the 6α-cyanopenicillin (**19**, X = CN), 6α-vinylpenicillin (**19**, X = CH=CH_2) and 6α-phenylpenicillin (**19**, X = C_6H_5) could be prepared in high yield (43).

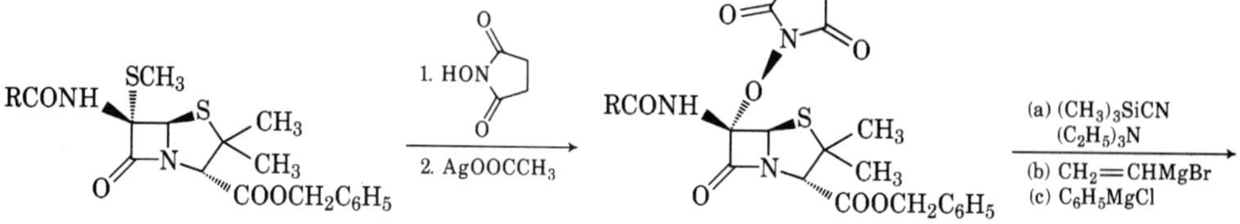

(17) (18)

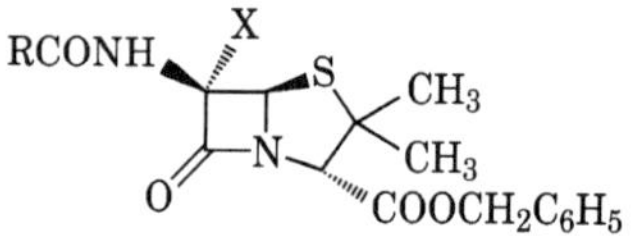

(19)

2α-2β-Substituents. Modifications at the 2-position of the penicillin nucleus has led to benefits in the biological properties of the modified penicillins. The series that has generated the most interest is the bisnorpenicillins depicted by **(22)**. Synthesis of these derivatives employs as a key intermediate the 4-mercaptoazetidin-2-one **(20)** which in turn is generated from penicillin V (**1**, R = $C_6H_5OCH_2$) (44).

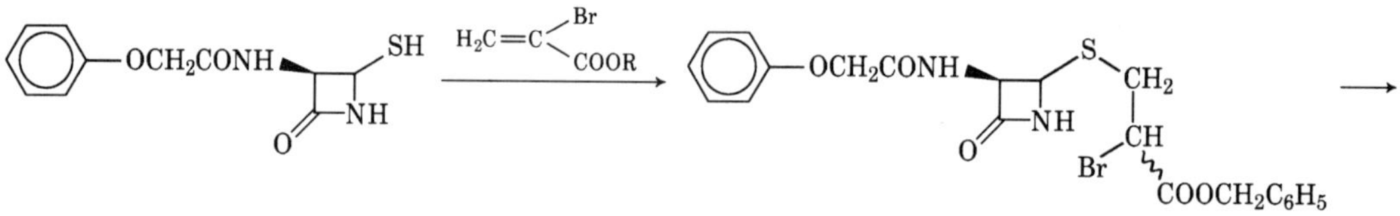

(20) (21)

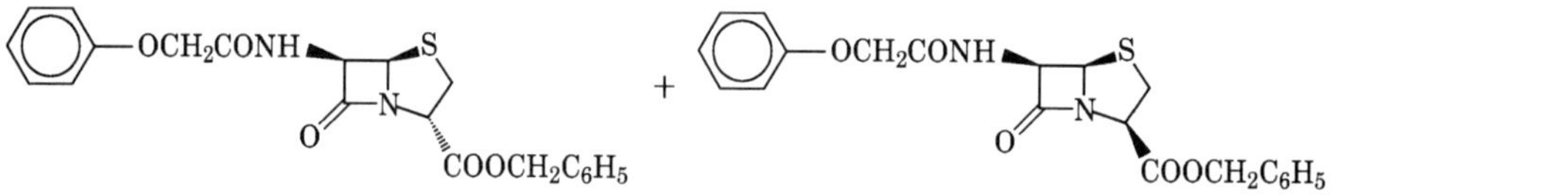

(22) (23)

Reaction of **(20)** with ethyl 2-bromoacrylate in hexamethylphosphorous triamide (HMPT) in the presence of 1,5-diazabicyclo[4.3.0]non-5-ene (DBN) yielded **(22)** in moderate yield. Use of potassium carbonate in DMF afforded a mixture of C-3 epimeric bisnorpenicillanates **(22)** and **(23)** (45). Removal of the phenoxyacetyl side chain was effected by Delft cleavage namely, treatment with phosphorus pentachloride in methylene chloride containing *N*-methylmorpholine followed by methanol and water. Acylation was then carried out using standard procedures. Bisnorampicillin was found to be less active than ampicillin, but bisnorpiperacillin proved to be similar in activity to piperacillin. Bisnorpiperacillin possesses a slightly broader spectrum of activity than the unmodified compound because of increased stability towards bacterial β-lactamases (46). Studies have shown that enhanced β-lactamase stability is a general property of acyl and ureido bisnorpenicillin derivatives (47). Synthesis of cycloalkanespiro-2-bisnorpenicillins led to a further series of nuclear modified penicillins:

spirocyclohexane, spirocyclopentane, spirocyclobutane, and spirocyclopropane analogues have been prepared. Only those containing the spirocyclopropane ring system provided analogues that had antibacterial properties comparable to those of the corresponding penicillins (48). Other modifications of the C-2 position are shown in Table 4.

2,3-Methylenepenams. The penam nucleus is also able to assume both of the ring conformations depicted in Figure 1 and the (2,3)-α-and (2,3)-β-methylenepenams shown in Figure 2 may be regarded as penicillin analogues (49). Studies involving the (2,3)-methylenepenams, when R = $CH_2C_6H_5$, and analogues revealed that the (2,3)α-methylenepenam penetrated the outer membrane of *E. coli* more readily than did the penicillin G analogue (50). Intrinsically, the alpha compound was found to have similar activity to that of penicillin G. By comparison the β-methylene analogue had poor activity.

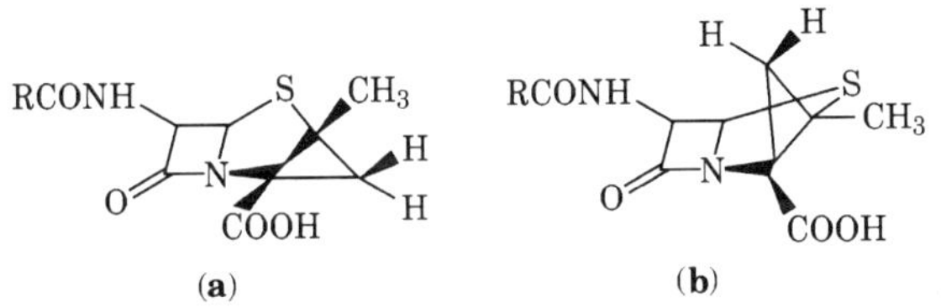

Fig. 2. Structures of (**a**) (2,3)-α-methylenepenams, and (**b**) (2,3)-β-methylenepenams.

Later investigations led to improved synthetic routes and analogues such as those shown in Figure 3 (51). Conversion of the mixed imide (**24**) to the bromide (**25**) provided a mixture of α- and β-bromomethylpenicillins in which the α-isomer predominated. 1,8-Diazabicyclo[5,4,0]undec-7-ene(DBU)-mediated cyclization led to the (2,3)-methylenepenicillanic acid (**26**) which was converted by catalytic hydrogenation followed by reverse phase chromatography to the potassium salt (**27**). The preparation of analogues (**29**) was carried out by reaction of (**26**) with 1.1 equivalents of PCl_5 and 1.9 equivalents of dry pyridine in chloroform followed by treatment with *n*-propanol and saturated aqueous sodium chloride solution. The resulting amine (**28**) was used without further purification and then transformed into derivatives (**29**) using suitably activated acids followed by hydrogenolysis using 10% palladium on carbon.

In general the (2,3)-α-methylenepenams were found to be chemically reactive and biologically active in contrast to the β-methylene counterparts supporting the hypothesis that the open conformation of penicillins is the biologically active form. In (2,3)-α-methylenepenams the absolute configuration at C-2 is opposite that of the naturally occurring penicillins placing the carboxyl group in a position more akin to that found in the cephalosporins. The compounds were found to be more susceptible to cephalosporinase than penicillinase. The (2,3)-α-methylene penicillin G analogue was found to be more selective in the inhibition of PBP 3 than was the β-analogue. It also bound more effectively to PBP 2 of *S. aureus* than the β-isomer (52).

Biochemical studies showed that the α-aminoadipoyl analogue derived from (2,3)-β-methylenepenam was not a substrate for expandase activity but rather, it was a potent reversible inhibitor of the ring expansion of α-aminoadipoylpenicillin into deacetoxycephalosporanic acid by the expandase enzyme (53).

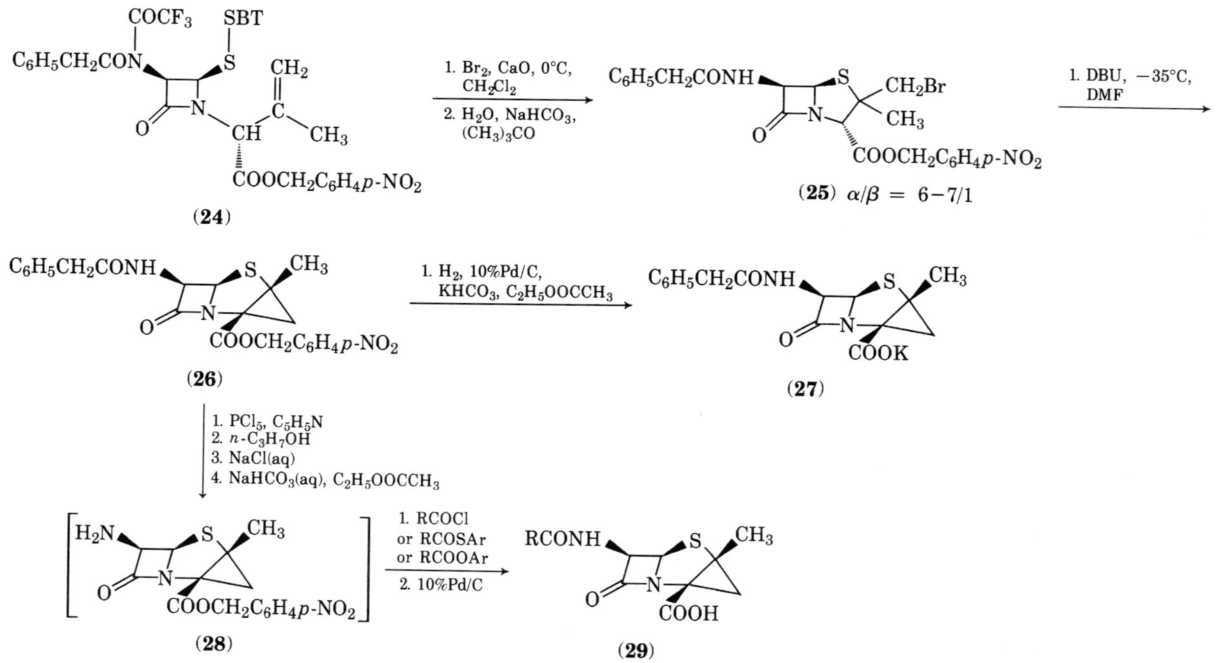

Fig. 3. Reaction scheme for the synthesis of (2,3)-α-methylenepenams where BT is 2-benzthiazolyl, DBU is 1,8-diazabicyclo[5.4.0]undec-7-ene.

Degradation. Penicillins are rapidly hydrolyzed by aqueous alkali to the corresponding penicilloic acids (**30**) which are stable as salts, but which decarboxylate on acidification to yield penilloic acids (**31**).

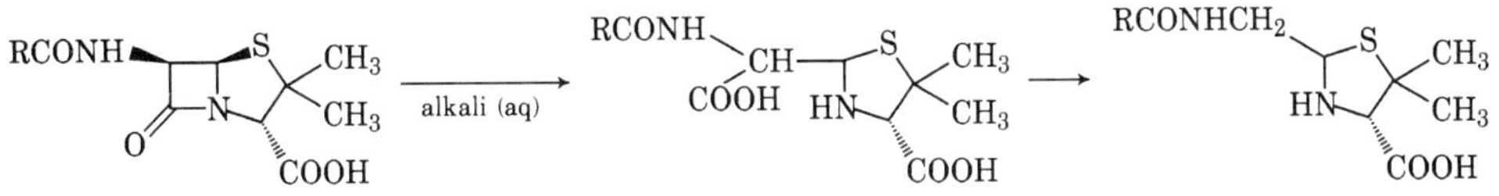

(**1**) (**30**) (**31**)

Penicillins are also degraded by aqueous acids via initial reaction of the side-chain carbonyl group with the β-lactam. Penicillenic acids (**33**) are obtained when hydrolysis is carried out at pH 4, penillic acids (**34**) at pH 2.

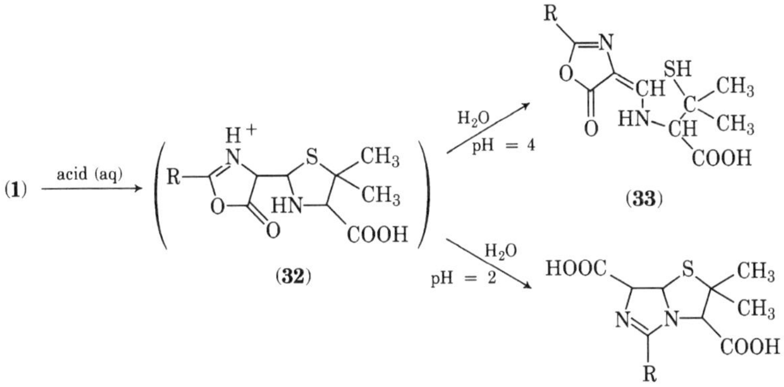

(**32**) (**33**) (**34**)

Electron-withdrawing groups at the α-position of the 6-acyl substituent stabilize the compounds to attack by acids. Thus phenoxymethylpenicillin and ampicillin are more stable to acid and more efficiently absorbed by the oral route than is benzyl penicillin. Degradation of temocillin (Table 1), a 6α-methoxypenicillin, occurs in a way similar to that shown for (**31**), (**33**), or (**34**). Although temocillin exhibits good stability in mild aqueous acid or base, in stronger acid the corresponding methoxypenillic acid is formed and under alkaline conditions the methoxypenicilloic acid is generated together with the C-5 epimer (54). Degradation of BRL 36650 (**35**) shown in Figure 4, containing a 6α-formamido substituent, occurs quite readily under mildly acidic aqueous conditions. Cleavage of the C-5–C-6 bond leads to two fragments, the formamido acid (**39**) and *N*-formyl penicillamine (**40**) presumably generated via the putative penicillenic acid intermediate (**36**) (55). The presence of the electron-withdrawing 6α-formamido group serves to accelerate degradation and leads to products that have been detected for penicillins in aqueous solution left over a longer period of time.

Biological Properties

Structure-Activity Relationships. Biological evaluation of penicillins yields information such as *in vitro* and *in vivo* antibacterial activities, minimum inhibitory concentration (MIC), minimum bactericidal concentration (MBC),

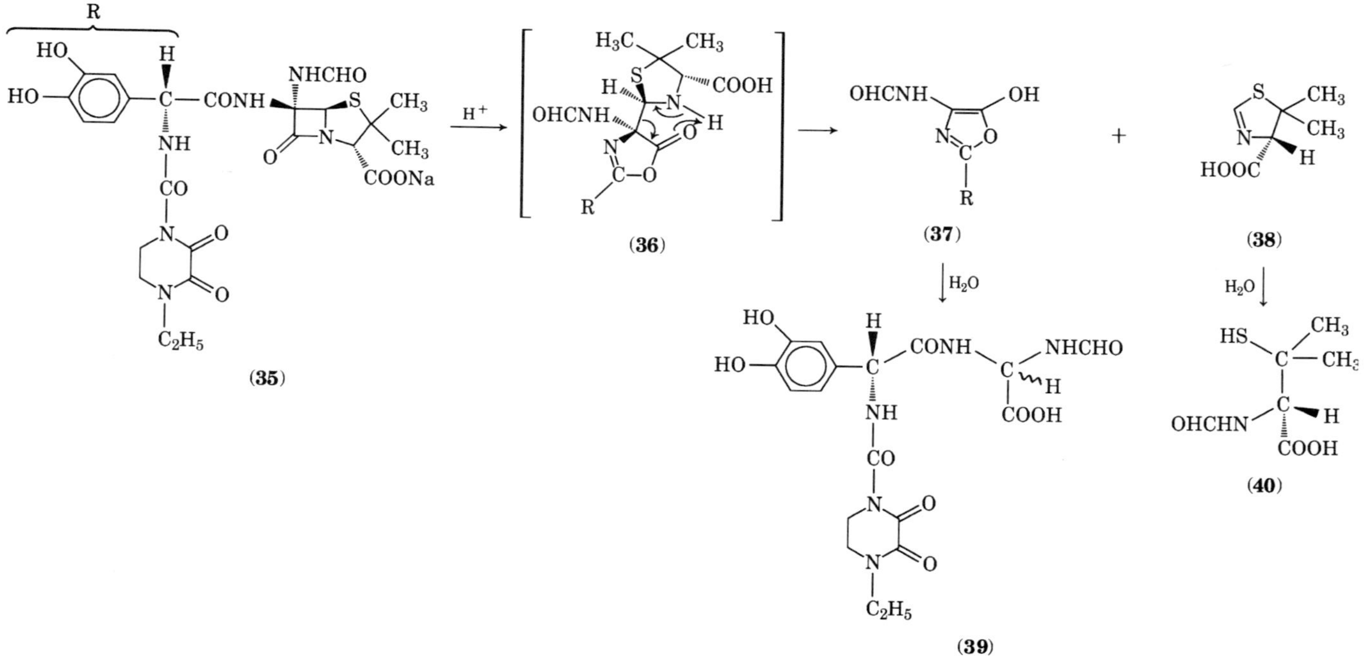

Fig. 4. Acid degradation of BRL 36650.

protective effectiveness in laboratory animals (PD_{50}), and pharmacokinetic characteristics including efficiency of absorption, serum levels, tissue distribution, urinary excretion, metabolism, serum binding, serum and tissue activation, biliary excretion, recycling, etc. Penicillins are also tested for ability to resist inactivation by β-lactamase produced by both gram-positive and gram-negative bacteria.

Table 5 gives a range of activities for some modified penicillins. Penicillin G remains probably the most active penicillin against gram-positive organisms. However, the majority of *Staphylococcus aureus* strains are resistant to penicillin by virtue of β-lactamase production. The β-lactamase resistant penicillins such as oxacillin, cloxacillin, flucloxacillin, and nafcillin are active against most penicillin-resistant *Staphylococci* but lack activity against methicillin-resistant *Staphylococci* (MRSA) and the majority of gram-negative organisms.

Ampicillin and its congeners amoxicillin, bacampicillin, and ciclacillin, have largely similar antibacterial spectra, exhibiting activity against both penicillin sensitive gram-positive and gram-negative microorganisms. The susceptibility of ampicillin or amoxicillin to β-lactamase may be overcome by combination with a β-lactamase inhibitor. Clavulanic acid and sulbactam are two such β-lactamase inhibitors in clinical use. The products Augmentin, a combination of clavulanic acid and amoxycillin for oral use primarily, and Unasyn, a combination of sulbactam and ampicillin for use by injection, are highly active against both gram-positive aerobic and anaerobic organisms in addition to many important gram-negative pathogens. Carbenicillin and ticarcillin possess moderate broad-spectrum activity and were the first penicillins to show activity against *Pseudomonas* species. Azlocillin, mezlocillin, and piperacillin possess largely the same characteristics but have slightly greater potency against many gram-negative species. Temocillin is the first penicillin to possess a high level of activity against gram-negative organisms, notably the *Enterobacteriaceae,* as well as excellent β-lactamase stability. The introduction of the 6α-methoxy group and the concomitant β lactamase stability compromises activity against both gram-positive organisms and *Pseudomonas* species. Similarly BRL 36650, having the 6α-formamido substituent, exhibits high bacterial β-lactamase stability. Some activity is retained against most species of *Streptococci* but the compound is inactive against *Staphylococci.* BRL 36650 is highly potent against most gram-negative organisms including refractory gram-negative species such as *Acinetobacter* and *Pseudomonas.* The level of potency is much greater than either that of ticarcillin or piperacillin (56).

The pharmacology of penicillins differs markedly from compound to compound but has been well reviewed (57). The majority of derivatives, including penicillin G and the antipseudomonal penicillins, are unstable in gastric acid and are not available orally. The isoxazolyl penicillins are relatively acid stable but not consistently well absorbed by the oral route. Nafcillin and oxacillin are poorly absorbed orally; cloxacillin, dicloxacillin, and flucloxacillin are more reliable. Penicillin V, ampicillin, and particularly amoxicillin are relatively well absorbed orally. Esters of ampicillin such as bacampicillin, pivampicillin, and talampicillin improve the level of oral absorption of ampicillin to that achieved by amoxicillin. Absorption can be diminished by food after oral administration, however, and peak blood levels, usually achieved after 1 to 2 h, are somewhat delayed after ingestion of food.

Table 5. Antibacterial Activity[a] of Modified Penicillins

Organism	Penicillin G	Ampicillin	Amox+clav[b]	Ox/Naf[c]	BRL 44154	Carb/Ticar[d]	Azlo/Mezlo/Piper[e]	Temocillin	BRL 36650
Gram-positive									
Staphylococcus (−)[f]	++	++	++	++	++	+	++	−	−
Staphylococcus (+)[g]	−	−	++	++	++	−	−	−	−
Staphylococcus (MRSA)[h]	−	−	−	−	+	−	−	−	−
Streptococcus pyogenes	++	++	++	+	++	+	++	−	+
Streptococcus pneumoniae	++	++	++	+	++	+	++	−	+
Enterococcus	++	++	++	−	−	+	++	−	−
Bacterioides fragilis	−	+	++	−	−	+	+	−	−
other anaerobes	−	+	++	−	−	+	+	−	−
Gram-negative									
E. coli	−	+	++	−	+	+	++	++	++
Klebsiella	−	−	++	−	−	−	+	++	++
Proteus mirabilis	−	++	++	−	++	++	++	++	++
other *Proteus*	−	−	−	−	−	+	++	++	++
Enterobacter	−	−	−	−	−	+	++	++	++
Pseudomonas aeruginosa	−	−	−	−	±	+	+±	−	++
other *Pseudomonas* species	−	−	−	−	−	−	+	−	++

[a]++ = very active at low concentrations; + = moderately active against most strains or active only at high concentrations; and − = inactive against the majority of strains.
[b]A mixture of amoxicillin and clavulanic acid.
[c]Activity represents that of both oxacillin and nafcillin.
[d]Activity represents that of both carbenicillin and ticarcillin.
[e]Activity represents that of azlocillin, mezlocillin, and piperacillin.
[f]*Staphylococcus* (−) = β-lactamase negative.
[g]*Staphylococcus* (+) = β-lactamase positive.
[h]*Staphylococcus* (MRSA) = methicillin-resistant *Staphylococcus aureus*.

Parenteral penicillins generally are virtually 100% available and half-lives vary from one-half hour for nafcillin and oxacillin to over an hour for amoxicillin and ticarcillin. The exception to this rule is temocillin which has an unusually long half-life of 4.5–5 h. Long-acting preparations of penicillin G are available as the procaine and benzathine derivatives. When these derivatives are administered intramuscularly, they provide effective serum levels for 12–24 and 3–4 weeks, respectively. Penicillins are eliminated by active secretion by renal tubular epithelial cells. This secretion can be blocked by probenicid [*57-66-9*], $C_{13}H_{19}NO_4S$, resulting in higher blood levels and prolonged serum half-lives. Most penicillins distribute well to body cavities and kinetics are usually linear. Urinary concentrations of penicillins are normally high, even in the presence of renal failure. Intracellular penetration of penicillins because of their lipid insolubility is invariably low. They also penetrate poorly into the central nervous system and the eye. Biliary concentrations of penicillins generally exceed serum concentrations in the presence of adequate bile flow.

The penicillins in general, are renowned for their lack of toxicity. The most common adverse effect of the use of penicillins is an allergic reaction which can change from a mild rash to fatal anaphylactic shock in rare cases. All penicillins cross the placenta and are excreted in maternal milk. However, the relative freedom from toxicity renders these compounds valuable agents during pregnancy and lactation.

Biosynthesis. The microbial synthesis of penicillins from eukaryotes such as *C. acremonium* and *P. chrysogenum* has been comprehensively reviewed (58). In essence the biosynthesis of penicillins is described in Figure 5 although certain stages have yet to be fully characterized. Products are isopenicillin N, penicillin G, and penicillin N [*525-94-0*], $C_{14}H_{21}N_3O_6S$.

The key enzyme in this sequence, isopenicillin N synthase (IPNS), has been purified from *E. coli* (59) and the recombinant enzyme shown to be a single polypeptide of 336 amino acids containing two cysteines, numbers 106 and 255 from the *N*-terminus, and probably a ferrous ion in a nonheme environment. The enzyme has been crystallized and studies undertaken to obtain suitably sized crystals for diffraction studies.

The IPNS enzyme has also been shown to recognize modified tripeptides. The synthesis of a range of tripeptides, other than aminoadipoyl cysteinyl valine (ACV) (Table 6), has given rise to a selection of modified penicillins using IPNS as a means of cyclizing the tripeptide (58).

Mode of Action. Penicillins exert their antibacterial effect by inhibiting the high molecular weight penicillin binding proteins (PBPs) that are implicated in the final stages of peptidoglycan synthesis (60). These PBPs have been extensively studied in *E. coli* in terms of the relative affinity of the drug for the three physiologically important proteins. Inhibition of PBP-1 causes rapid cell lysis whereas inhibition of PBP-2 and PBP-3 results in the formation of spherical and filamentous cells, respectively. PBP-1 may be further subdivided into PBPs 1*a* and 1*b*. PBP-1*a* is usually highly sensitive to penicillins and it is the effective inhibition of PBP-1*b* that is regarded as the more crucial stage leading to cell lysis. The low molecular weight PBPs 4, 5, and 6 of *E. coli* are believed to be nonessential for cell survival and therefore not considered of major importance in the killing mechanism of penicillins (61).

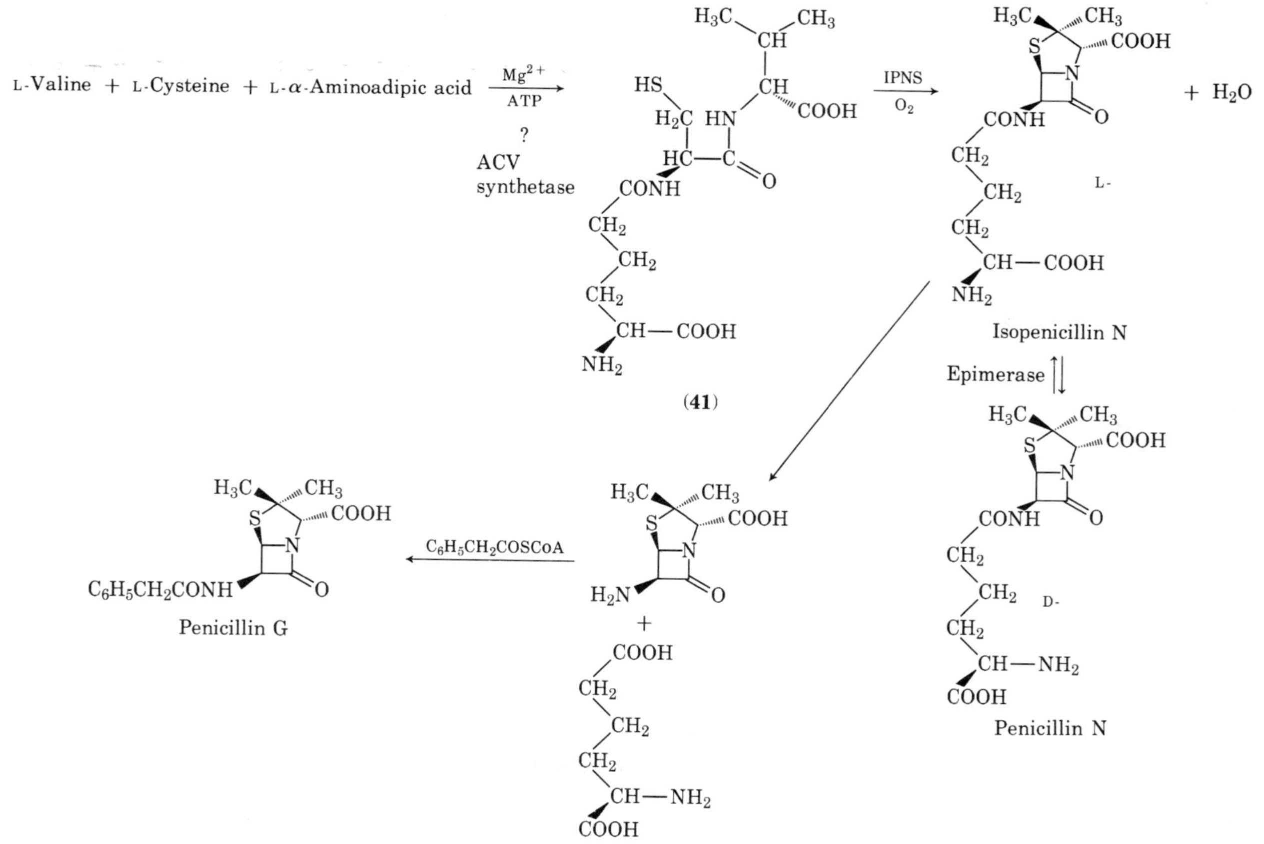

Fig. 5. Biosynthesis of penicillins when ACV is aminoadipoly cysteinyl valine and IPNS is isopenicillin N synthase and $C_6H_5CH_2COSCoA$ represents benzyl coenzyme A. ACV synthetase is thought to catalyze the first step of this reaction sequence. Courtesy of J. E. Baldwin, Royal Society of Chemistry.

Table 6. Modified Penicillins[a] Prepared Using IPNS[b]

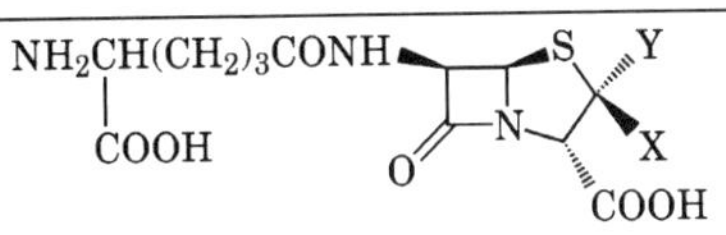

(**42**)

X	Y	X	Y
CH_3	H	CH_3	CH_2OH
H	CH_3	H	CH_2OH
CH_3	C_2H_5	CH_3	OCH_3
C_2H_5	CH_3		
$-CH{=}CH_2$	H	H	C_2H_5
H	$-CH{=}CH_2$	H	—◁
CH_3	$-CH{=}CH_2$	H	$-C{=}C{=}CH_2$

[a]Structure (**42**) where X and Y are as indicated.
[b]Ref. 58.

The changes that occur when an organism is exposed to different concentrations of penicillin provides a characteristic morphological response that can be particularly well demonstrated in the scanning electron microscope. Penicillin G binds preferentially to PBP 3 causing filamentation. Most penicillins bind principally to PBPs 1 and 3 leading to filamentous forms. Exceptions are amoxicillin and mecillinam. Amoxicillin binds to PBPs 1 and 3 but at low concentrations forms spheroplasts and at higher concentrations, filaments. Mecillinam binds only to PBP 2 producing ovoid and later spherical cells. The novel penicillin BRL 44154 has been shown to bind to PBPs 1 and 3 and also to PBP 2′, the protein responsible for the resistance in methicillin-resistant *Staphylococcus aureus* (MRSA) (62).

Rapid progress is being made in the protein crystallography of low molecular weight PBPs and demonstration of the feasibility of obtaining water-soluble forms of high molecular weight PBPs suggests that crystallization of the latter proteins can be achieved. These enzymes are inserted into the cytoplasmic membrane only at their amino termini and water-soluble forms are expected to be suitable for crystallization and x-ray analysis (57). Preliminary investigations involving a β-lactam-sensitive, bifunctional D-alanyl-carboxypeptidase–transpeptidase (C Pase–T Pase) from *Streptomyces* R61 have identified the three-dimensional structure and catalytic site of interaction with penicillins (63).

Production, Manufacture, and Processing

Most methods used for the production of the commercially important α-amino penicillins, such as ampicillin and amoxicillin, are based on modifications of an enamine process employing the appropriate phenylglycine and methylacetoacetate followed by coupling with 6-APA (64). Other aspects of the fermentation, strain maintenance, equipment, inoculum development, media, and procedures used in the production of penicillin are well covered in previous editions of the *Encyclopedia*. Developments in these areas have been reviewed (65).

Economic Aspects

Worldwide retail antibiotic sales in 1986 were approximately $11 billion of which penicillins comprised approximately $2.5 billion (66). Preparations containing ampicillin are estimated at $800 million whereas those containing amoxicillin are estimated at $1,020 million. Total sales include sales of ampicillin esters and the ureidopenicillins derived from ampicillin. Total sales of penicillins, including semisynthetic penicillins in the United States in 1987, were $25 million (67). Sales of amoxicillin, the largest single selling penicillin, were approximately $350 million.

The volume of bulk ampicillin produced was 3,700 t, valued at $260 million, and of bulk amoxicillin, 2,200 t valued at $190 million, both as the trihydrate. It was predicted that the market for bulk ampicillin was growing at 3% annually and would reach 4,160 t by 1990. Sales of amoxicillin, growing at 10% annually were predicted to reach a consumption of 2,800 t in 1990 with an average growth rate of 8%. The demand for both products is expected to reach 4,500 t by 2000 (68).

The cost of 6-APA, ampicillin, and amoxicillin is invariably linked to that of penicillin G. 6-APA remains one of the cheapest raw starting materials available for the preparation of semisynthetic penicillins and other related β-lactam antibiotics.

BIBLIOGRAPHY

"Penicillin," in *ECT* 1st ed., Vol. 9, pp. 922–943, by W. E. Brown, Squibb Institute for Medical Research; in *ECT* 2nd ed., Vol. 14, pp. 652–707, K. Butler, Chas. Pfizer and Co.; "Antibiotics (β-Lactams)," in *ECT* 3rd ed., Vol. 2, pp. 871–919, by J. R. E. Hoover and C. H. Nash, SmithKline and French Laboratories.

1. J. C. Sheehan, K. R. Henery-Logan, and D. A. Johnson, *J. Am. Chem. Soc.* **75,** 3292 (1953).
2. A. Fleming, *Br. J. Exp. Pathol.* **10,** 226 (1929).
3. E. Chain and co-workers, *Lancet* (II), 177 (1941).
4. E. P. Abraham, and co-workers, *Lancet* **1977**(II) (1941).
5. H. W. Florey and co-workers, *Antibiotics,* Vol. 11, Oxford University Press, London, 1949.
6. H. T. Clarke, J. R. Johnson, and R. Robinson, eds., *The Chemistry of Penicillin,* Princeton University Press, N.J., 1949.
7. A. J. Moyer and R. D. Coghill, *J. Bacteriol.* **53,** 329 (1947).
8. D. Crowfoot, C. W. Bunn, B. W. Rogers-Low, and A. Turner-Jones in Ref. 6, p. 310; H. T. Clarke, J. R. Johnson, and R. Robinson in Ref. 6, p. 5; Ref. 5, p. 667.
9. F. R. Batchelor, F. P. Doyle, J. H. C. Nayler, and G. N. Rolinson *Nature (London)* **183,** 257 (1959), F. R. Batchelor, E. B. Chain, T. L. Hardy, K. R. L. Mansford, and G. N. Rolinson, *Proc. R. Soc. London* Ser.B. **154,** 498 (1961).

10. E. H. Flynn, ed., *Cephalosporins and Penicillins, Chemistry and Biology,* Academic Press, Inc., New York, 1972.
11. M. Tsuji and Y. Yamada in S. Mitsuhashi, ed., *β-Lactam Antibiotics,* Japan Scientific Society's Press, Tokyo, 1981, p. 235.
12. S. K. Branch, A. F. Casy, and E. M. A. Ominde, *J. Pharm. Biomed. Anal.* **5,** 73 (1987).
13. J. R. Everett, J. W. Tyler, and G. Woodnutt, *J. Pharm. Biomed. Anal.* **7,** 397 (1989).
14. S. K. Branch, A. F. Casy, A. Lipczynski, and E. M. A. Ominde, *Magn. Reson. Chem.* **24,** 465 (1986).
15. M. Barber and co-workers, *Biomed. Mass Spectrom.* **9,** 11 (1982).
16. A. Balsamo, P. Domiano, B. Macchia, F. Macchia, and N. Nardelli, *Eur. J. Med. Chem.* **15,** 559 (1980); N. C. Cohen, *J. Med. Chem.* **26,** 259 (1983).
17. R. D. G. Cooper and D. O. Spry in Ref. 10, p. 181.
18. F. A. Jung, W. R. Pilgrim, J. P. Poyser, and P. Sirel in P. G. Sammes, ed., *Topics in Antibiotic Chemistry,* Vol. 4, John Wiley & Sons, Inc., New York, 1980, Chapt. 4.
19. Y. Hamashima and co-workers, in G. I. Gregory, ed., *Recent Advances in the Chemistry of β-Lactam Antibiotics,* Special Publication No. 38, The Royal Society of Chemistry, 1980, p. 57.
20. I. Ernest and co-workers, *J. Am. Chem. Soc.* **100,** 8214 (1978).
21. M. Foglio, C. Battistini, F. Zarini, C. Scarafile, and G. Francheschi, *Heterocycles* **20,** 1491 (1983).
22. A. Alfonso, A. K. Ganguly, V. Girijavallabhan, and S. McCombie in A. G. Brown and S. M. Roberts, eds., *Recent Advances in the Chemistry of β-Lactam Antibiotics,* The Royal Society of Chemistry, Special Publication No. 52, 1984, p. 266.
23. M. Aratani, K. Sawada, and M. Hashimoto, *Tetrahedron Lett.,* 3921 (1982); H. Maruyama and T. Hiraoka, *J. Org. Chem.* **51,** 399 (1986).
24. H. Tobiki and co-workers, *Yakugaku Zasshi* **100,** 49 (1980). *Chem. Abstr.* **93,** 95179x (1980).
25. J. S. Kaltenbronn and co-workers, *J. Antibiot.* **32,** 621 (1979).
26. S. Nakamura and co-workers, *18th Interscience Conference on Antimicrobial Agents and Chemotherapy,* Atlanta, Ga., 1978, Abstract No. 30.
27. P. C. Fuchs and co-workers, *J. Antibiot.* **30,** 1098 (1977).
28. Y. Suzuki and co-workers, *J. Antibiot.* **32,** 711 (1979).
29. T. Nishino, N. Ishii, T. Tanino, S. Ohshima, and T. Yamaguchi, *19th Interscience Conference on Antimicrobial Agents and Chemotherapy,* Boston, Mass., 1979, Abstr. No. 241.
30. R. Gottschlich, V. Rudolph, H. Wahling, and E. Poetsh, *20th Interscience Conference on Antimicrobial Agents and Chemotherapy,* New Orleans, La., 1980, Abstr. No. 146.
31. B. Wetzel and co-workers in Ref. 19, p. 26.
32. P. C. A. Chapman and co-workers, *29th Interscience Conference on Antimicrobial Agents and Chemotheraphy,* Abstracts 89–91, Houston, Tex., 1989.
33. J. E. Baldwin, F. J. Urban, R. D. G. Cooper, and F. L. Jose. *J. Amer. Chem. Soc.* **95,** 2401 (1973); G. A. Koppel and R. E. Koehler, *J. Amer. Chem. Soc.,* **95,** 2403 (1973); R. A. Firestone and B. G. Christensen, *J. Org. Chem.* **38,** 1436 (1973); W. A. Slusarchyk and co-workers, *J. Org. Chem.* **38,** 943 (1973).
34. B. Slocombe and co-workers, *Antimicrob. Agents Chemother.* **20,** 38 (1981).
35. R. J. Ponsford, D. J. Merrikin, B. Slocombe, M. J. Basker, and T. C. G. Tasker, *23rd Interscience Conference on Antimicrobial Agents and Chemotherapy,* Las Vegas, Nev., 1983, Abstr. No. 415; M. J. Basker, D. J. Merrikin, R. J. Ponsford, B. Slocombe, and T. C. G. Tasker, *J. Antimicrob. Chemother.* **18,** 399 (1986).
36. G. Burton, D. J. Best, R. A. Dixon, R. F. Kenyon, and A. G. Lashford, *J. Antibiot.* **39,** 1419 (1986).
37. D. J. Best, G. Burton, F. P. Harrington, and J. Lowther, *J. Antibiot.* **39,** 483 (1986).

38. R. J. Ponsford and co-workers, in Ref. 22, p. 32; M. J. Basker and co-workers, *Antimicrob. Agents Chemother.* **26,** 734 (1984).
39. T. Jen, J. Frazee, and J. R. E. Hoover, *J. Org. Chem.* **38,** 2857 (1973).
40. P. Feyen and W. Schröck, *Angew. Chem. Int. Ed.* **20,** 808 (1981).
41. P. S. F. Mézes, R. W. Friesen, T. Viswanatha, and G. I. Dmitrienko, *Heterocycles* **19,** 1207 (1982).
42. C. L. Branch, M. J. Pearson, and T. C. Smale, *J. Chem. Soc. Perkin 1,* 401 (1988).
43. A. V. Stachulski, *J. Chem. Soc. Chem. Commun.* 401 (1986).
44. N. F. Osborne, *J. Chem. Soc., Perkin 1,* 146 (1980).
45. Ref. 44, p. 150.
46. M. J. Driver and co-workers, *J. Antibiot.* **37,** 297 (1984).
47. M. J. Driver, J. Lowther, and A. W. Taylor, *J. Antibiot.* **38,** 550 (1985).
48. N. F. Osborne, *J. Chem. Soc., Perkin 1,* 1435 (1982).
49. D. D. Keith, J. Tengi, P. Rossman, L. Todaro, and M. Weigele, *Tetrahedron* **39,** 2445 (1983).
50. J. S. Chapman and N. H. Georgopapadakou, *Antimicrob. Agents Chemother.* **31,** 1994 (1987).
51. J. Christenson and co-workers in P. H. Bentley and R. Southgate, eds., *Recent Advances in the Chemistry of β-Lactam Antibiotics,* The Royal Society of Chemistry, Special Publication No. 70, 1988 p. 33; J. G. Christenson, D. L. Pruess, M. K. Talbot, and D. D. Keith, *Antimicrob. Agents Chemother.* **32,** 1005 (1988).
52. N. F. Georgopapadakou ánd co-workers, *Antimicrob. Agents Chemother.* **31,** 1069 (1987).
53. J. E. Baldwin and co-workers, *Biochem. J.* **245,** 831 (1987).
54. G. Burton, C. R. Dobson, J. R. Everett, *J. Pharm. Pharmacol.* **38,** 758 (1986).
55. E. A. Cutmore and co-workers, *J. Chem. Soc. Chem. Commun.* 21 (1987).
56. M. J. Basker and co-workers, *Antimicrob. Agents Chemother.* **26,** 734 (1984).
57. M. F. Parry, *Med. Clin. North Am., Update on Antibiotics 1, The Penicillins* **71,** 1093 (1987).
58. J. E. Baldwin in Ref. 51, p. 1; R. D. G. Cooper in Ref. 51, p. 196; J. E. Baldwin and E. P. Abraham, *Natural Product Rpts.* **5,** 129 (1988).
59. S. M. Samson and co-workers, *Nature* **318,** 191 (1985).
60. W. B. Pratt and R. Fekety, eds., *The Antimicrobial Drugs,* Oxford University Press, London, 1986.
61. B. W. Bycroft and R. E. Shute, *Pharm. Res.* **1,** 3 (1985).
62. B. G. Spratt and K. D. Cromie, *Revs. Inf. Dis.* **10,** 699 (1988).
63. J. B. Bartolone, G. J. Hite, J. A. Kelly, and J. R. Knox in Ref. 22, p. 318; J. R. Knox and J. A. Kelly, in S. M. Roberts, ed., *Chemical and Biochemical Problems in Molecular Recognition,* The Royal Society of Chemistry, London, 1989; J. A. Kelly, J. R. Knox, H. Zhao, J.-M. Frère, and J-M. Ghuysen, *J. Mol. Biol.* **209,** 281 (1989).
64. A. A. W. Long, J. H. C. Nayler, H. Smith, T. Taylor, and N. Ward, *J. Chem. Soc.* (**C**), 1920 (1971).
65. J. O. Neway, ed., *Fermentation Process Development of Industrial Organisms,* Marcel Dekker Inc., New York, 1989.
66. Scrip No. **1238,** 24 (1987).
67. *Synthetic Organic Chemicals, United States Production and Sales 1987,* U.S. International Trade Commission Publication 2118, U.S. Government Printing Office, Washington, D.C., 1987.
68. Scrip No. **1206,** 24 (1987).

R. J. PONSFORD
SmithKline Beecham Pharmaceuticals

LINCOSAMINIDES

Lincomycin [*154-21-2*] (**1**, R = OH, R′ = H), $C_{18}H_{34}N_2O_6S$, the first lincosaminide antibiotic to which a structure was assigned, is defined chemically as methyl 6,8-dideoxy-6-(1-methyl-*trans*-4-propyl-L-pyrrolidin-2-ylcarbonylamino)-1-thio-D-erythro-D-galacto-octopyranoside. Both lincomycin and the semisynthetic clindamycin [*18323-44-9*] (**1**, R = H, R′ = Cl), $C_{18}H_{33}ClN_2O_5S$, are widely used in clinical practice. The trivial name of the sugar fragment of this antibiotic, methyl α-thiolincosaminide, has lent itself to the other members of this family, whether produced as secondary metabolites of soil microorganisms or derived semisynthetically by chemical modification. Thus celesticetin [*2520-21-0*] (**2**, R = 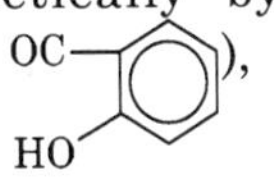), $C_{24}H_{36}N_2O_9S$, and desalicetin [*19246-70-9*] (**2**, R = H), $C_{17}H_{32}N_2O_7S$, are also lincosaminides.

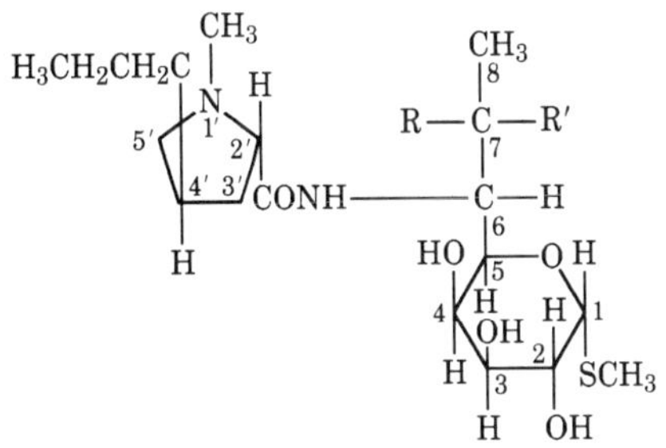

(**1**)

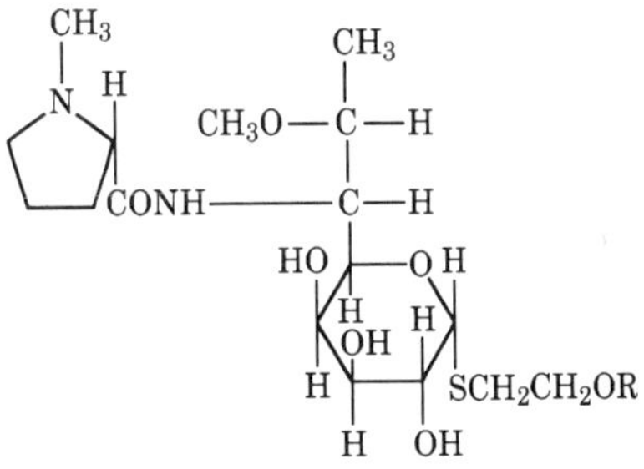

(**2**)

Lincomycin

The discovery and biological properties of lincomycin (**1**, R = OH, R′ = H) were described in 1962 (1). This antibiotic is active *in vitro* and *in vivo* against most of the common gram-positive pathogens. Resistance by Staphylococci is developed slowly in a stepwise manner, based on *in vitro* serial subculture experiments, and its activity is not influenced by body fluids up to concentrations of 50% in the assay medium (2).

Lincomycin was produced originally by the actinomycete *Streptomyces lincolnensis* var. *lincolnensis,* NRRL 2936. Subsequently, it has been produced by a variety of *Streptomyces* strains (3–8) and by strain 1146 of *Actinomyces roseolus* (9). Extraction by standard procedures using *Sarcina lutea* as the assay organism on agar trays (10) leads to a crystalline hydrochloride having molecular formula $C_{18}H_{34}N_2O_6S{\cdot}HCl{\cdot}\frac{1}{2}\,H_2O$ (11). A second, more dense form exists as the monohydrate (12), and this latter form of lincomycin hydrochloride is the Upjohn trademarked Lincocin available commercially. This salt is highly soluble in water, and moderately soluble in methanol and ethanol. It has one basic function, $pK_a = 7.5$, and the specific rotation $[\alpha]_D^{25} = +137°$ (1.0 mg/mL in H_2O). The structure of lincomycin was determined by classical chemical degradation studies, together with the interpretation of proton magnetic resonance and mass spectra (13–18).

Biosynthesis. The terminal *C*-methyl of the propyl side chain, the *S*-methyl, and the *N*-methyl groups are derived from methionine (19). *trans*-4-Propyl-L-proline [*31101-27-6*] was shown to accumulate when *Streptomyces lincolnensis* is grown in media deficient in sulfur, and the addition of L-tyrosine or L-dihydroxyphenylalanine (DOPA) was shown to stimulate this production. From a comparison of the incorporation of label from L-[1-^{14}C]tyrosine and L[U-^{14}C]tyrosine, it was demonstrated that all of the carbon atoms of the propylproline other than the *C*-terminal methyl group were derived from tyrosine and, by the use of L-[^{15}N]tyrosine, it was determined that the proline nitrogen atom was derived from the amino–nitrogen atom of tyrosine (20).

Later work (21), using ^{2}H- and ^{13}C-labeled precursors, in combination with ^{13}C nmr and mass spectral analyses, showed that glucose is converted into tyrosine and thence into DOPA which, via an initial 2,3-extradiol cleavage, cyclization, loss of two carbon atoms, and introduction of a *C*-methyl group from *S*-adenosylmethionine, gives propylproline. The biosynthesis of methyl α-thiolincosaminide [*14810-93-6*], $C_9H_{19}NO_5S$, has also been demonstrated to proceed from glucose via a C-5 and a C-3 unit, using both specifically ^{13}C-labeled substrates and uniformily labeled D-(^{13}C) glucose followed by the analysis of the ^{13}C spin coupling patterns (22).

Mechanism of Action. The earliest studies on the mechanism of action of lincomycin showed that lincomycin had the immediate effect on *Staphylococcus aureus* of complete inhibition of protein synthesis (23). This inhibition results from the blocking of the peptidyltransferase site of the 50S subunit of the bacterial ribosome (24). Little effect on DNA and RNA synthesis was observed.

Resistance to Lincomycin. Resistance to lincomycin is developed slowly, and is usually caused by modification of 23S ribosomal RNA, which leads to coresistance to macrolide, lincosaminide, and streptogramin B antibiotics (25). Inactivation of lincomycin by clinical isolates of strains of *Staphylococcus aureus* and *Staphylococcus haemolyticus,* though retention of sensitivity to macrolides (see ANTIBIOTICS, MACROLIDES) and streptogramins (see ANTIBIOTICS, PEPTIDES), has been found to be the consequence of the conversion of the antibiotic into its 3-(5′-adenylate) (26).

Pharmacology and Uses. Lincomycin hydrochloride (Lincocin) is available in oral dosage forms and as a sterile solution for injection. After oral administration to adults, mean peak serum levels were achieved between 2 and 6 h. Single 500 and 1000 mg doses gave peak levels of 1.8–5.3 and 2.5–6.7 μg/mL, respectively (27–31). The half-life is 4.2–5.4 h (32). Intramuscular administration gave peak levels at 0.5–2 h. After 300 and 600 mg doses, mean peak levels were 11.7–15 and 9.3–18.5 μg/mL, respectively (27,28,31,33). Lincomycin activity is widely distributed in human body tissues and fluids. Significant concentrations were found in bile, peritoneal fluid, pleural fluid, the eye, bone, and brain (31,34–37). It is excreted both in the urine and in the feces (27–31).

Lincomycin has found use in the treatment of diseases of the ear, throat, nose, respiratory tissue, skin and soft tissue, bone, joint, dental, and septicemic infections caused by staphylococci, pneumonococci, and streptococci (other than enterococci). It has also been used in the treatment of diphtheria and a variety of anaerobic infections, including actinomycosis (38).

Celesticetin and Other Lincosaminide Metabolites

The production (39) and isolation (40) of the antibiotic celesticetin (**2**, R = 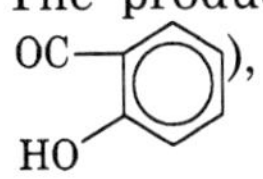), a salicylate ester of the β-hydroxyethylthio-substituent was reported as early as the 1950s, although its structure was not determined until 1968 (41). Reexamination of the fermentation showed the presence of desalicetin (**2**, R = H) and a variety of other β-hydroxyethylthio esters, together with similar esters of 7-de-*O*-methyldesalicetin [*39032-07-0*], (**3**) $C_{16}H_{30}N_2O_7S$ (42,43). One auxotrophic mutant of *S. caelestis* produced 7-de-*O*-methylcelesticetin [*39032-05-8*] (**4**, R = CH_3, R′ = H), $C_{23}H_{34}N_2O_9S$ (44), whereas a second produced both de-*N*-methylcelesticetin [*40736-31-0*] (**4**, R = H, R′ = CH_3), $C_{23}H_{34}N_2O_9S$, and de-*N*-methyl-7-de-*O*-methylcelesticetin, (**4**, R = R′ = H), $C_{22}H_{32}N_2O_9S$ (45).

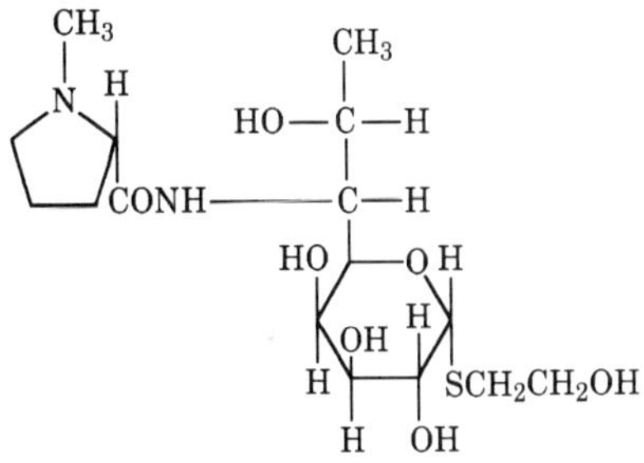

(**3**)

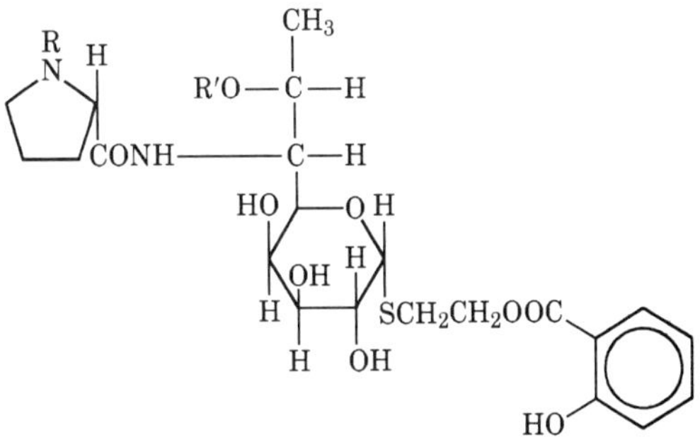

(**4**)

Antibiotic Bu-2545, 7-*O*-methyl-4′-depropyllincomycin [*75007-09-9*] (**1**, R = OCH_3, R′ = H but lacking the 4′-propyl group), $C_{16}H_{30}N_2O_6S$, produced by *Streptomyces* strain No. H 230-5, possesses structural features in common with both celesticetin and lincomycin (46,47).

Accompanying lincomycin in *S. lincolnensis* fermentations is a small amount of the analogue 4′-depropyl-4′-ethyllincomycin [*2520-24-3*], $C_{17}H_{32}N_2O_6S$ (48), which has considerably lower antibacterial activity than the parent compound. Extension of the normal six-day fermentation to twelve days resulted in the formation of lincomycin sulfoxide [*23444-78-2*], $C_{18}H_{34}N_2O_7S$, and 1-demethylthio-1-hydroxylincomycin [*22099-01-0*], (lincomycose) $C_{17}H_{32}N_2O_7$, (**5**, R = CH_3, R′ = H, R″ = OH) (49), both of which have greatly reduced antibacterial activity. The addition of D,L-ethionine and of methyl α-thiolincosaminide, the sugar fragment of lincomycin, produces the *S*-ethyl analogue of lincomycin, [*14042-43-4*], $C_{19}H_{36}N_2O_6S$ (50,51) and de-*N*-methyllincomycin [*16843-70-2*] (**5**, R = R′ = H, R″ = SCH_3), $C_{17}H_{32}N_2O_6S$, (52) respectively, whereas the addition of ethyl α-thiolincosaminide [*6734-39-0*], $C_{10}H_{21}NO_5S$, the sugar fragment of the *S*-ethyl analogue, produces the de-*N*-methyl-*S*-ethyl analogue (**5**, R = R′ = H, R″ = SC_2H_5), $C_{18}H_{34}N_2O_6S$ (50).

Rapid inactivation of added lincomycin was found to result from the growth of *Streptomyces rochei* in a synthetic medium; the antibiotic was converted into lincomycin 3-phosphate [*23670-99-7*] (**5**, R = CH_3, R′ = PO_3H_2, R″ = SCH_3), $C_{18}H_{35}N_2O_9PS$, readily cleaved back to the antibiotic upon treating with alkaline phosphatase (53).

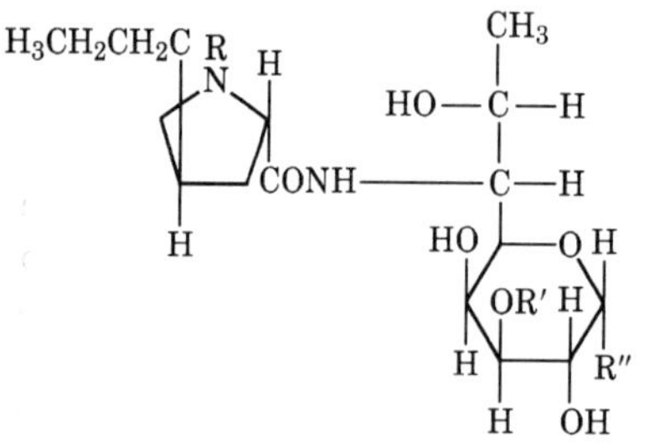

(**5**)

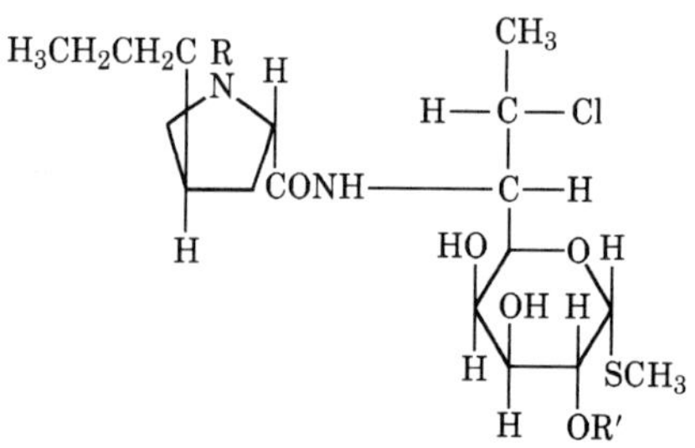

(**6**)

Clindamycin

Clindamycin, 7(*S*)-7-chloro-7-deoxylincomycin [*18323-44-9*], (**1**, R = H, R′ = Cl), also known as Cleocin, first resulted from the reaction of lincomycin and thionyl chloride (54); improved synthetic methods involve the reaction of lincomycin and triphenylphosphine dichloride or triphenylphosphine in carbon tetrachloride (55). Clindamycin is significantly more active than lincomycin against gram-positive bacteria *in vitro,* and is absorbed rapidly following oral administration. Clindamycin 2-palmitate [*36688-78-5*], (**6**, R = CH_3, R′ = $OC(CH_2)_{14}CH_3$), $C_{34}H_{63}ClN_2O_8S$, the 2-palmitate ester of clindamycin, is tasteless and antibacterially inactive. However, following oral administration, esterase cleavage occurs to give good blood levels of clindamycin (56), and this ester has been developed as a pediatric formulation of the antibiotic (Cleocin Pediatric). Given intramuscularly, clindamycin hydrochloride causes pain at the site of injection; the 2-phosphate ester (**6**, R = CH_3, R′ = PO_3H_2) (57) (Cleocin Phosphate), antibacterially inactive, is much better tolerated on injection, and is hydrolyzed by phosphatase, giving good levels of clindamycin.

Pharmacology and Uses. Oral administration of clindamycin hydrochloride gives peak serum levels of 1.9–2.7, 3.6, and 5.6 μg/mL one hour after single doses of 150, 300, and 450 mg, respectively. The half-life is 2.0–3.8 h (58–61). Clindamycin palmitate was also absorbed rapidly, following oral administration, but serum concentrations were lower than after clindamycin hydrochloride (58). Intramuscular injection of single 300, 450, and 600 mg doses of clindamycin phosphate gave mean peak serum levels of 3.8–4.9, 5.3, and 6.2–6.3 μg/mL respectively, 2–4 h after administration (59,62,63). Applied topically, clindamycin 2-phosphate [*24729-96-2*] (Cleocin T), $C_{18}H_{34}ClN_2O_8PS$, is effective in the treatment of acne vulgaris (64). Clindamycin, like lincomycin, is distributed widely in human body tissues, except that no significant levels are found in the brain or eye (38).

Antibacterial activity of clindamycin is found both in urine and feces after administration of clindamycin. This activity is a consequence of the presence of both clindamycin and its metabolite, de-*N*-methylclindamycin [*22431-45-4*] (**6**, R = R′ = H). Unlike de-*N*-methyllincomycin, the de-*N*-methyl analogue is as active *in vitro* as clindamycin. The analogue has been isolated from the urine of humans who had received clindamycin, and its presence in serum has been detected (65).

Clindamycin has found use in the treatment of common infections caused by gram-positive cocci. It is also efficacious in the treatment of anaerobic infections, including actinomycosis (38). Clindamycin has been shown to be active against strains of *Plasmodium* in animals (66–68).

Resistance to Clindamycin. Cross-resistance between lincomycin and clindamycin is complete (64), and co-resistances of lincomycin also apply to clindamycin. However, the inactivation of clindamycin by clinical isolates of *Staphylococcus haemolyticus* and *Staphylococcus aureus* is caused by adenylylation at the 4-position to form clindamycin 4-(5′-adenylate) [*29752-38-3*] (**7**) in contrast to the lincomycin 3-(5′-adenylate) [*117785-83-8*] (**8**) that forms (26).

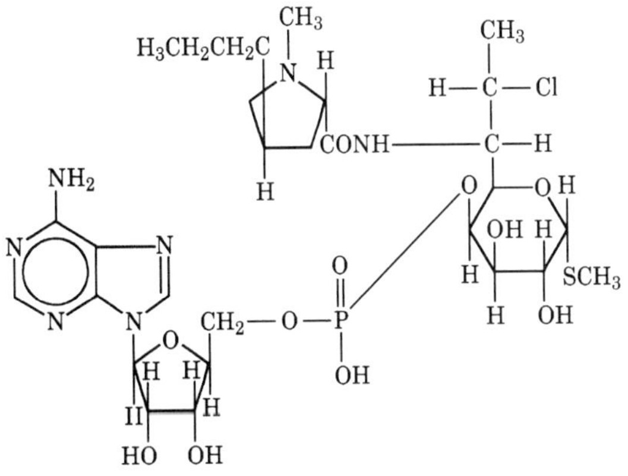

(**7**)

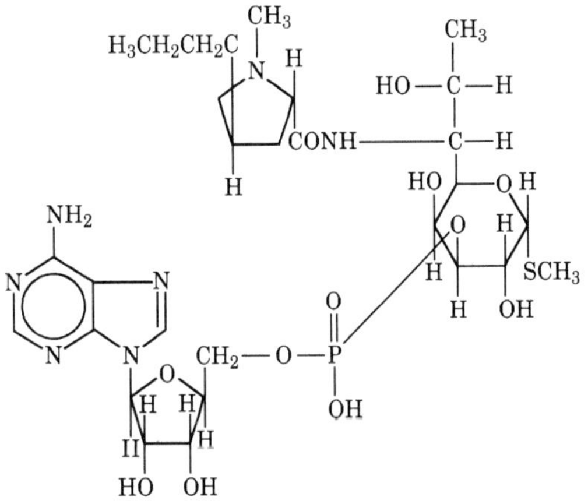

(**8**)

Biomodification of Clindamycin. When added to fermentations of *Streptomyces punipalus,* clindamycin is converted into de-*N*-methylclindamycin (**6**, R = R′ = H). However, when clindamycin is incubated with *Streptomyces armentosus,* clindamycin sulfoxide [*68366-52-9*], $C_{18}H_{33}ClN_2O_6S$, which has low antibacterial activity, is formed (69). Clindamycin 3-phosphate [*28708-34-1*], antibacterially inactive *in vitro,* and the ribonucleotides clindamycin 3-(5′-cytidylate) [*31186-90-0*], clindamycin 3-(5′-adenylate) [*31186-91-1*], clindamycin 3-(5′-uridylate) [*36010-69-2*], and clindamycin 3-(5′-guanylate) [*36010-70-5*], all inactive *in vitro,* can be generated (70). All of these derivatives protect mice infected with *Staphylococcus aureus,* however, presumably because of biotransformation into clindamycin (71–73).

Other Chemically Modified Lincosaminides

Modification of the Thioglycoside. Reaction of lincomycin (**1**, R = OH, R′ = H) and bromine in aqueous solution results in the replacement of the methylthio moiety with a hydroxy group producing lincomycose (**5**, R = CH_3, R′ = H, R″ = OH). Treatment of lincomycose with alkanethiols in the presence of strong acids affords a mixture of dialkyl dithioacetals (**9**) which are inactive and de-*S*-methyl-*S*-alkyllincomycins (**10**), which are equivalent to lincomycin in activity if the alkyl groups are small, but are less active when larger alkyl substituents are present (54). Inversion of the configuration of the methylthio-substituent from α- to β- results in the total loss of activity (74).

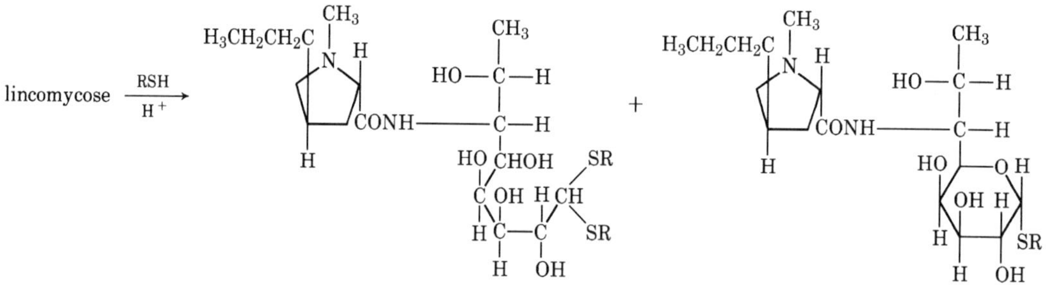

(**9**) (**10**)

Modification of the Carbohydrate Ring Substituents. Replacement of lincomycin's 2-hydroxy group by methoxy or by hydrogen (74), or inversion of the configuration of the hydroxy group at C-4 or at C-2 destroys activity (75).

Modification of the Carbohydrate Side Chain. The importance to the antibacterial activity of the C-7 substituents in lincosaminides has been shown not only by the increased activity of the 7(*S*)-chloro-analogue, clindamycin, (**1**, R = H, R′ = Cl) and of the corresponding 7(*S*)-bromo-7-deoxylincomycin [*18464-22-7*] (**1**, R = H, R′ = Br), $C_{18}H_{33}BrN_2O_5S$, and iodo, $C_{18}H_{33}IN_2O_5S$ [*26390-05-6*] (**1**, R = H, R′ = I) analogues, but also by the minimal activity of 7-deoxylincomycin [*72496-56-1*] (**1**, R = R′ = H), $C_{18}H_{34}N_2O_5S$, and 7-deoxy-7-ketolincomycin [*17057-59-9*] (**1**, R = R′ = O), $C_{18}H_{32}N_2O_6S$, and the diminished activity of 7-epilincomycin [*17017-22-0*] (**1**, R = H, R′ = OH), $C_{18}H_{34}N_2O_6S$, and of the 7(*R*)-7-chloro-deoxylincomycin [*78788-75-7*] (**1**, R = Cl, R′ = H), $C_{18}H_{33}ClN_2O_5S$, derived from it (76). 7-*O*-Methyllincomycin [*39679-82-8*] (**1**, R = OCH_3, R′ = H), $C_{19}H_{36}N_2O_6S$, possesses somewhat enhanced activity (77) and the 7(*S*)-methoxy analogue (**1**,R = H, R′ = OCH_3) is significantly more active than lincomycin, albeit with the same spectrum of activity, but activity decreases rapidly as the size of the alkoxy-substitutent increases (78).

Whereas introducing a thiol moiety at C-7 markedly reduced the antibacterial activity relative to lincomycin (79), the 7(*S*)-7-deoxy-7-alkylthiolincomycins exhibited considerably enhanced antibacterial activity without apparent regard for the size of the alkyl group (80–82). A marked increase in gram-negative activity was shown when the 7(*S*)-substitutent contained a 2- or 3-hydroxy or amino group, but this activity was insufficient to be effective in infected mice (83–85).

Modification of the Proline Fragment. Besides the 4′-ethyl analogue of lincomycin, co-produced in fermentations of *S. lincolnensis* (48), other 4′-alkyl analogues were chemically synthesized. A peak of activity is reached with the *trans*-4′-pentyl and hexyl analogues (**11**), the cis-epimers (**12**) possessing about one-half of the activity of the trans-. Equivalent activities are shown by *N*-ethyl lincosaminides, but larger *N*-alkyl substituents show decreased activity (86).

Similar structure activity relationships were found in the 4′-alkyl analogues of clindamycin (87). The de-*N*-methylclindamycin intermediates including de-*N*-methylclindamycin itself, but unlike de-*N*-methyllincomycin, are also highly active antibacterially. In addition, they are active *in vivo* as antimalarial agents (66–68).

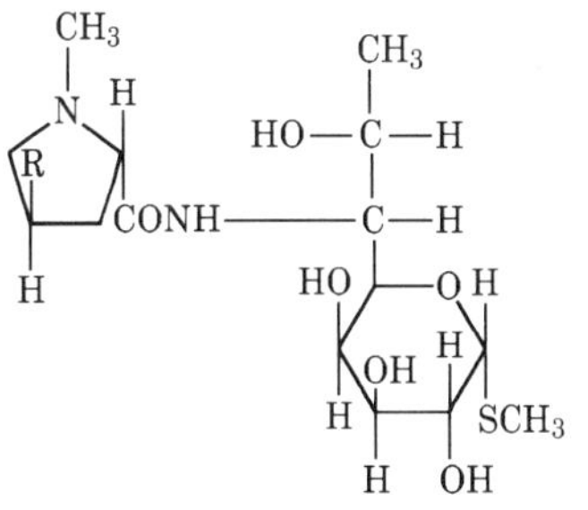

(**11**)

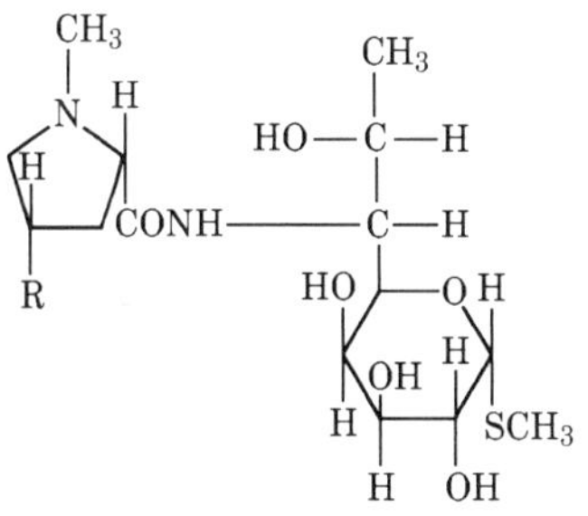

(**12**)

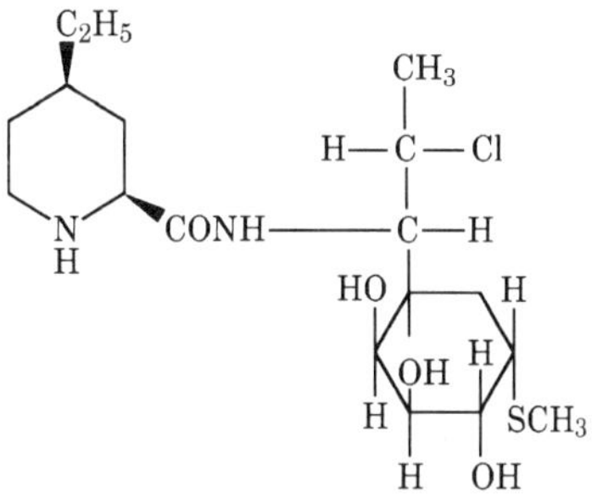

(**13**)

Other Changes in the Amino Acid Fragment

The influence of the size of the cyclic amino acid in analogues of clindamycin has been examined. The (2-(*S*)-*cis*)-4-ethylpipecolic acid analogue (6-ring), $C_{17}H_{31}ClN_2O_5S$, [*78822-40-9*] (**13**) was highly active both *in vitro* and *in vivo,* but the *cis*-(*R*)-isomer [*80081-63-6*] had minimal activity. Significant activity was shown for the azepine (7-ring) analogue, $C_{16}H_{29}ClN_2O_5S$, [*88015-22-9*], but the azetidine (4-ring) analogue, $C_{13}H_{23}ClN_2O_5S$ [*88000-11-7*], showed little activity (88).

Synthesis of the Carbohydrate Fragment of Lincomycin. The first formal synthesis of methyl α-thiolincosaminide started with D-galactose and was reported in 1970 (89). Other syntheses beginning with carbohydrates may also be found in the literature (90–96). Additionally, the total synthesis of methyl β-lincosaminide, albeit racemic, has been reported, utilizing the cyclocondensation of activated dienes and aldehydes, catalyzed by Lewis acids, to yield functionalized pyran rings (97). The synthesis of methyl 1-thio-α-lincosaminide from methyl α-D-galactopyranoside involving an elegant stereocontrolled introduction of the amino–alcohol substituents at C-6 and C-7 has also been reported (98).

Economic Aspects

It is estimated (99) that U.S. sales of clindamycin in 1989 were $45 million. Clindamycin was the seventh most widely used of all prescription products in U.S. hospitals (100). Figures are not available for worldwide sales of clindamycin, or for sales of lincomycin, which are virtually all outside the United States.

The composition of matter patents in the United States issued to The Upjohn Company on clindamycin phosphate and hydrochloride expired at the end of 1986 and in early 1987, respectively. Since then, these compounds have been available generically from more than two dozen companies in the United States alone (101,102).

BIBLIOGRAPHY

"Antibiotics (Lincosaminides)" in *ECT* 3rd ed., Vol. 2, pp. 930–936, by Thomas E. Eble, The Upjohn Co.

1. D. J. Mason, A. Dietz, and C. DeBoer, *Antimicrob. Agents Chemother.* **1962,** 544 (1963).
2. C. Lewis, H. W. Clapp, and J. E. Grady, *Antimicrob. Agents Chemother.* **1962,** 570 (1963).
3. E. L. Patterson, J. H. Hash, M. Lincks, P. A. Miller, and N. Bohonos, *Science* **146,** 1691 (1964).
4. U.S. Pat. 3,697,380 (Oct. 10, 1972), A. D. Argoudelis, J. H. Coats, and T. R. Pyke (to The Upjohn Co.).
5. U.S. Pat. 3,833,475 (Sept. 3, 1974), F. Reusser and A. D. Argoudelis (to The Upjohn Co.).
6. U.S. Pat. 3,726,766 (Apr. 10, 1973), A. D. Argoudelis and J. H. Coats (to The Upjohn Co.).
7. U.S. Pat. 3,812,014 (May 21, 1974), A. D. Argoudelis and J. H. Coats (to The Upjohn Co.).
8. Belg. Pat. 830,068 (Dec. 10, 1975), M. E. Bergy, J. H. Coats, and V. S. Malik (to The Upjohn Co.).
9. USSR Pat. 287,743 (Feb. 18, 1975), G. F. Gause and co-workers (to The University of Moscow).
10. L. J. Hanka, D. J. Mason, M. R. Burch, and R. W. Treick, *Antimicrob. Agents Chemother.* **1962,** 565 (1963).
11. R. R. Herr and M. E. Bergy, *Antimicrob. Agents Chemother.* **1962,** 560 (1963).
12. U.S. Pat. 3,313,695 (Apr. 11, 1967), D. VanOverloop (to The Upjohn Co.).
13. H. Hoeksema, *Abstracts, Papers, 149th Meeting of the American Chemical Society,* Detroit, Mich., 1965, p. 9-C.
14. H. Hoeksema and co-workers, *J. Am. Chem. Soc.* **86,** 4223 (1964).
15. R. R. Herr and G. Slomp, *J. Am. Chem. Soc.* **89,** 2444 (1967).
16. W. Schroeder, B. Bannister, and H. Hoeksema, *J. Am. Chem. Soc.* **89,** 2448 (1967).
17. G. Slomp and F. A. MacKellar, *J. Am. Chem. Soc.* **89,** 2454 (1967).
18. B. J. Magerlein, R. D. Birkenmeyer, R. R. Herr, and F. Kagan, *J. Am. Chem. Soc.* **89,** 2459 (1967).
19. A. D. Argoudelis, T. E. Eble, J. A. Fox, and D. J. Mason, *Biochemistry* **8,** 3408 (1969).
20. D. F. Witz, E. J. Hessler, and T. L. Miller, *Biochemistry* **10,** 1128 (1971).
21. N. M. Brahme and co-workers, *J. Am. Chem. Soc.* **106,** 7873 (1984).
22. N. M. Brahme and co-workers, *J. Am. Chem. Soc.* **106,** 7878 (1984).

23. J. J. Josten and P. M. Allen, *Biochem. Biophys. Res. Commun.* **14,** 241 (1964).
24. E. F. Gale, ed., *The Molecular Basis of Antibiotic Action,* John Wiley & Sons, Inc., 2nd ed., New York, 1981, pp. 478–480.
25. B. Weisblum, *J. Antimicrob. Chemother. Suppl. A,* **16,** 63 (1985).
26. A. Brisson-Noël, P. Delrieu, D. Samain, and P. Courvalin, *J. Biol. Chem.* **263,** 15,880 (1988).
27. J. J. Vavra, W. T. Sokolski, and J. B. Lawson, *Antimicrob. Agents Chemother.* **1963,** 176 (1964).
28. K. Kaplan, W. H. Chew, and L. Weinstein, *Am. J. Med. Sci.* **250,** 137 (1965).
29. C. E. McCall, N. H. Steigbigel, and M. Finland, *Am. J. Med. Sci.* **254,** 144 (1967).
30. R. F. McGehee, Jr., C. B. Smith, C. Wilcox, and M. Finland, *Am. J. Med. Sci.* **256,** 279 (1968).
31. A. Medina, N. Fiske, I. Hjelt-Harvey, C. D. Brown, and A. Prigot, *Antimicrob. Agents Chemother.* **1963,** 189 (1964).
32. J. G. Wagner, J. I. Northam, and W. T. Sokolski, *Nature (London)* **207,** 201 (1965).
33. P. Ma, M. Lim, and J. H. Nodine, *Antimicrob. Agents Chemother.* **1963,** 183 (1964).
34. P. A. Thomas and P. C. Jolly, *Am. Rev. Respir. Dis.* **96,** 1044 (1967).
35. E. F. Becker, *Am. J. Ophthalmol.* **67,** 963 (1969).
36. W. McL. Davis, Jr., *Oral Surg.* **27,** 688 (1969).
37. V. J. Nemecek, M. Hejzlar, and V. Vacek, *Arzneim. Forsch.* **19,** 1760 (1969).
38. R. J. Fass in E. M. Kagan, ed., *Antimicrobial Therapy,* 2nd ed., W. B. Saunders Co., Philadelphia, Pa., 1974, p. 83.
39. C. DeBoer, A. Dietz, J. R. Wilkins, C. Lewis, and G. M. Savage, *Antibiot. Ann.* 831 (1954–1955).
40. H. Hoeksema, G. F. Crum, and W. H. DeVries, *Antibiot. Ann.* 837 (1954–1955).
41. H. Hoeksema, *J. Am. Chem. Soc.* **90,** 755 (1968).
42. A. D. Argoudelis and T. F. Brodasky, *J. Antibiot.* **25,** 194 (1972).
43. A. D. Argoudelis and T. F. Brodasky, *J. Antibiot.* **27,** 642 (1974).
44. A. D. Argoudelis, J. H. Coats, P. G. Lemaux, and O. K. Sebek, *J. Antibiot.* **25,** 445 (1972).
45. A. D. Argoudelis, J. H. Coats, P. G. Lemaux, and O. K. Sebek, *J. Antibiot.* **26,** 7 (1973).
46. M. Hanada, M. Tsunakawa, K. Tomita, H. Tsukiwa, and H. Kawaguchi, *J. Antibiot.* **33,** 751 (1980).
47. S. Toda, S. Nakagawa, T. Naito, and H. Kawaguchi, *J. Antibiot.* **34,** 596 (1981).
48. A. D. Argoudelis, J. A. Fox, and T. E. Eble, *Biochemistry* **4,** 698 (1965).
49. A. D. Argoudelis and D. J. Mason, *J. Antibiot.* **22,** 289 (1969).
50. A. D. Argoudelis, J. A. Fox, D. J. Mason, and T. E. Eble, *J. Am. Chem. Soc.* **86,** 5044 (1964).
51. A. D. Argoudelis and D. J. Mason, *Biochemistry* **4,** 704 (1965).
52. A. D. Argoudelis, J. A. Fox, and D. J. Mason, *Biochemistry* **4,** 710 (1965).
53. A. D. Argoudelis and J. H. Coats, *J. Antibiot.* **22,** 341 (1969).
54. R. D. Birkenmeyer, *Abstract, 5th Interscience Conference on Antimicrobial Agents and Chemotherapy,* 1965, p. 18.
55. R. D. Birkenmeyer and F. Kagan, *J. Med. Chem.* **13,** 616 (1970).
56. A. A. Sinkula, W. Morozowich, and E. L. Rowe, *J. Pharm. Sci.* **62,** 1106 (1973).
57. W. Morozowich, D. J. Lamb, R. M. DeHaan, and J. E. Gray, *Abstracts, Papers, American Pharmaceutical Association Meeting,* Washington, D.C., 1970, p. 63.
58. R. M. DeHaan, C. M. Metzler, D. Schellenberg, W. D. VandenBosch, and E. L. Masson, *Int. J. Clin. Pharmacol.* **62,** 105 (1972).
59. C. M. Metzler, R. M. DeHaan, D. Schellenberg, and W. D. VandenBosch, *J. Pharm. Sci.* **62,** 591 (1973).
60. B. R. Meyers, K. Kaplan, and L. Weinstein, *Appl. Microbiol.* **17,** 653 (1969).

61. J. G. Wagner, E. Novak, N. C. Patel, C. G. Chidester, and W. L. Lummis, *Am. J. Med. Sci.* **256,** 25 (1968).
62. R. M. DeHaan, C. M. Metzler, D. Schellenberg, and W. D. VandenBosch, *J. Clin. Pharmacol.* **13,** 190 (1973).
63. R. J. Fass and S. Saslaw, *Am. J. Med. Sci.* **263,** 369 (1972).
64. J. L. LeFrock, A. Molavi, and R. A. Prince, *Med. Clin. N. Am.* **66,** 103 (1982).
65. T. F. Brodasky, A. D. Argoudelis, and T. E. Eble, *J. Antibiot.* **21,** 327 (1968).
66. C. Lewis, *J. Parasitol.* **54,** 169 (1968).
67. K. G. Powers, *Amer. J. Trop. Med. Hyg.* **18,** 485 (1969).
68. L. H. Schmidt, J. Harrison, R. Ellison, and P. Worcester, *Amer. J. Trop. Med. Hyg.* **19,** 1 (1970).
69. A. D. Argoudelis, J. H. Coats, D. J. Mason, and O. K. Sebek, *J. Antibiot.* **22,** 309 (1969).
70. J. H. Coats and A. D. Argoudelis, *J. Bacteriol.* **108,** 459 (1971).
71. A. D. Argoudelis and J. H. Coats, *J. Am. Chem. Soc.* **93,** 534 (1971).
72. A. D. Argoudelis, J. H. Coats, and S. A. Mizsak, *J. Antibiot.* **30,** 474 (1977).
73. V. P. Marshall, T. E. Patt, and A. D. Argoudelis, *J. Indust. Microbiol.* **1,** 17 (1986).
74. B. Bannister, *J. Chem. Soc. Perkin Trans.* **1,** 3025 (1972).
75. B. Bannister, *J. Chem. Soc. Perkin Trans.* **1,** 3031 (1972).
76. B. J. Magerlein, *Adv. Appl. Microbiol.* **14,** 203 (1971).
77. B. Bannister, *J. Chem. Soc. Perkin Trans.* **1,** 1676 (1973).
78. B. Bannister, *J. Chem. Soc. Perkin Trans.* **1,** 360 (1974).
79. B. J. Magerlein and F. Kagan, *J. Med. Chem.* **12,** 974 (1969).
80. B. Bannister, *J. Chem. Soc. Perkin Trans.* **1,** 1057 (1977).
81. B. Bannister, *J. Chem. Soc. Perkin Trans.* **1,** 274 (1978).
82. B. Bannister, *J. Chem. Soc. Perkin Trans.* **1,** 967 (1978).
83. B. Bannister, *J. Chem. Soc. Perkin Trans.* **1,** 540 (1980).
84. B. Bannister, *Tetrahedron* **40,** 1633 (1984).
85. B. Bannister, *J. Chem. Res.* (*S*) 90 (1989); *J. Chem. Res.* (*M*) 701 (1989).
86. B. J. Magerlein, R. D. Birkenmeyer, and F. Kagan, *J. Med. Chem.* **10,** 355 (1967).
87. B. J. Magerlein and F. Kagan, *J. Med. Chem.* **12,** 780 (1969).
88. R. D. Birkenmeyer, S. J. Kroll, C. Lewis, K. F. Stern, and G. E. Zurenko, *J. Med. Chem.* **27,** 216 (1984).
89. B. J. Magerlein, *Tetrahedron Lett.* **33** (1970).
90. G. B. Howarth, W. A. Szarek, and J. K. N. Jones, *J. Chem. Soc.* (*C*) 2218 (1970).
91. H. Saeki and E. Ohki, *Chem. and Pharm. Bull. (Japan)* **18,** 789 (1970).
92. T. Alsumi, T. Fukumaru, T. Ogawa, and M. Matsui, *Agr. Biol. Chem.* **37,** 2621 (1973).
93. G. R. Woolard, E. B. Rathbone, W. A. Szarek, and J. K. N. Jones, *J. Chem. Soc. Perkin Trans.* **1,** 950 (1976).
94. A. Gateau-Olesker, A. M. Sepulchre, G. Vass, and S. D. Gero, *Tetrahedron* **33,** 393 (1977).
95. I. Hoffe and V. Schöllkopf, *Liebigs Ann. Chem.* 1474 (1980).
96. S. M. David and J.-C. Fischer, *Carbohydrate Res.* **38,** 147 (1974).
97. S. J. Danishefsky, E. Larson, and J. P. Springer, *J. Am. Chem. Soc.* **107,** 1274 (1985).
98. S. Knapp and P. J. Kukkola, *J. Org. Chem.* **55,** 1632 (1990).
99. M. I. Smith, Pharmaceutical Data Services of Scottsdale, Ariz., personal communication, 1990.
100. *Scrip,* No. 1500, March 28th, 1990, p. 18.
101. *American Druggist Blue Book, 1989–1990,* Hearst Corp., New York, 1990, pp. 128, 129.
102. *Drug Topics Red Book, 1989,* Medical Economics Co., Oradell, N.J., 1989, p. 238.

BRIAN BANNISTER
The Upjohn Company

MACROLIDES

Macrolide antibiotics are well established antimicrobial agents in both clinical and veterinary medicine. These agents can be administered orally and are generally used to treat infections in the respiratory tract, skin and soft tissues, and genital tract caused by gram-positive organisms, *Mycoplasma* species, and certain susceptible gram-negative and anaerobic bacteria.

The macrolide class is large and structurally diverse. Macrolides are produced by fermentation of soil microorganisms. Additionally, structural modifications using both chemical and microbiological means have yielded biologically active semisynthetic derivatives. A book devoted to macrolides includes work to 1984 (1). Monographs are available for older macrolides such as erythromycin (2–5) and spiramycin (6) and newer ones such as roxithromycin (7–9), azithromycin (10), and erythromycin acistrate (11). Reviews have covered the field generally (12–15) and specifically with regard to chemical modification (15,16), antimicrobial properties (17,18), and pharmacokinetics and pharmacology (19–23). These complex molecules are also targets for total synthesis (24–26).

The term macrolide was introduced to denote the class of substances produced by *Streptomyces* species containing a macrocyclic lactone ring (27). The generalized structure is a highly substituted monocyclic lactone (aglycone) to which is attached one or more saccharides glycosidically linked to hydroxyl groups on either the aglycone or another saccharide. The aglycones are derived via similar polyketide biosynthetic pathways and thus share many structural features in terms of pattern and stereochemistry of substituents (28). Traditional macrolide antibiotics are divided into three families according to the size of the aglycone which can be 12-, 14-, or 16-membered.

Because one or more aminosugars are usually present, these compounds are basic and can form acid addition salts. In addition, one or more neutral sugars are often present. A few macrolides possess no aminosugar. The saccharides share some common features: they tend to be highly deoxygenated and *N*- and/or *O*-methylated; and the amino groups are located at either positions 3 or 4. The most common sugars found in macrolides are given in Table 1.

12-Membered Ring Macrolides

The few macrolides having 12-membered rings are listed in Table 2. Methymycin (**12**, R = OH, R′ = H), isolated from culture broths of a *Streptomyces* species (29), was the first macrolide structure elucidated (30). It is comprised of the aglycone methynolide (**13**, R = OH, R′ = H) and the aminosugar desosamine (**1**, R = OH, R′ = H) (31,32). Methymycin was also the first conventional macrolide made by total synthesis (33).

Neomethymycin (**12**, R = H, R′ = OH), an isomer co-produced with methymycin, is the product of hydroxylation at C-12 rather than C-10 of the lactone (34,35). The corresponding aglycone, neomethynolide (**13**, R = H, R′ = OH), was isolated with methynolide from broths of *S. venezuelae* (36). The stereochemistry of 12(*R*)- for neomethynolide was established by total synthesis (37). YC-17 (**12**, R = R′ = H), also found in broths of *S. venezuelae*, is a possible

Table 1. Saccharides Found in Macrolide Antibiotics

Saccharide	CAS Registry Number	Molecular formula	Structure number	Structure
		Aminosaccharides		
D-angolosamine	[*14702-57-9*]	$C_8H_{17}NO_3$	(**1**, R = H, R′ = OH)	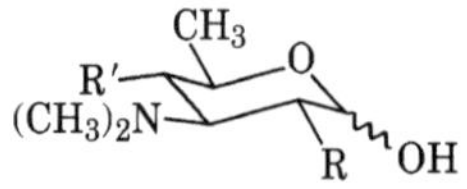
D-desosamine	[*5779-39-5*]	$C_8H_{17}NO_3$	(**1**, R = OH, R′ = H)	
D-mycaminose	[*519-21-1*]	$C_8H_{17}NO_4$	(**1**, R = R′ = OH)	
D-forosamine	[*18423-27-3*]	$C_8H_{17}NO_2$	(**2**)	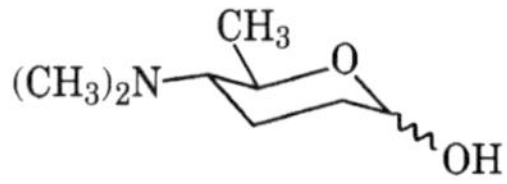
L-megosamine	[*65832-97-5*]	$C_8H_{17}NO_3$	(**3**)	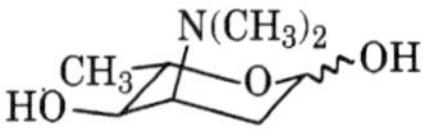
		Neutral saccharides		
D-aldgarose	[*26428-87-5*]	$C_9H_{14}O_6$	(**4**)	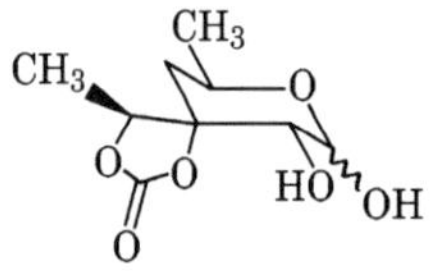
L-amicetose	[*67528-19-2*]	$C_6H_{12}O_3$	(**5**, R = H, R′ = OH)	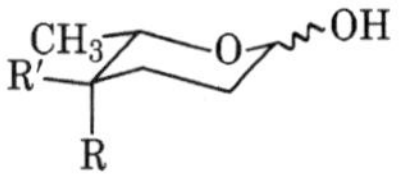
L-cinerulose	[*33985-39-6*]	$C_6H_{10}O_3$	(**5**, R + R′ = O)	
L-rhodinose	[*35903-48-1*]	$C_6H_{12}O_3$	(**5**, R = OH, R′ = H)	
L-arcanose	[*26548-40-3*]	$C_8H_{16}O_4$	(**6**)	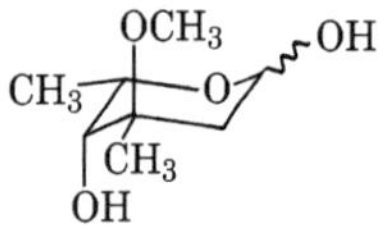
D-chalcose	[*3150-28-5*]	$C_7H_{14}O_4$	(**7**)	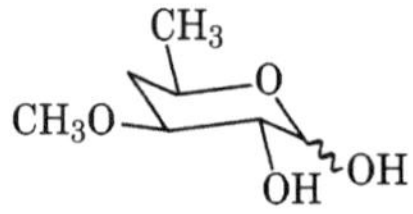
L-cladinose	[*470-12-2*]	$C_8H_{16}O_4$	(**8**, R = CH_3)	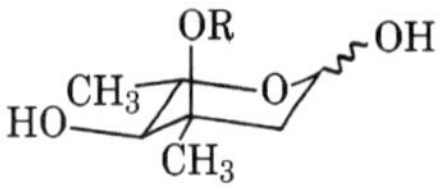
L-mycarose	[*6032-92-4*]	$C_7H_{14}O_4$	(**8**, R = H)	
D-6-deoxyallose	[*4348-84-9*]	$C_6H_{12}O_5$	(**9**, R = R′ = H)	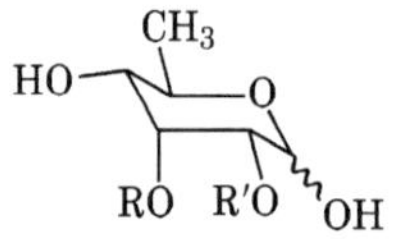
D-javose	[*921-90-4*]	$C_7H_{14}O_5$	(**9**, R = H, R′ = CH_3)	
D-mycinose	[*21967-31-7*]	$C_8H_{16}O_5$	(**9**, R = R′ = CH_3)	
D-4,6-dideoxy-3-dehydroallose		$C_6H_{10}O_4$	(**10**)	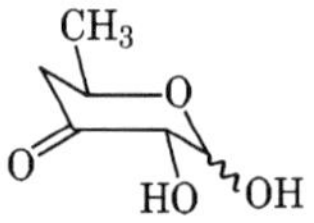
L-oleandrose	[*6786-76-1*]	$C_7H_{14}O_4$	(**11**, R = CH_3)	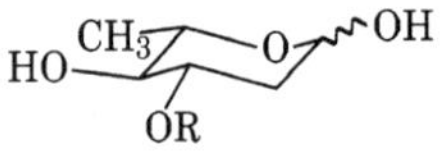
L-olivose	[*25029-50-9*]	$C_6H_{12}O_4$	(**11**, R = H)	

Table 2. Macrolides Having a 12-Membered Ring

Macrolide	CAS Registry Number	Molecular formula	Structure number	Structure
methymycin	*[497-72-3]*	$C_{25}H_{43}NO_7$	(**12**, R = OH, R′ = H)	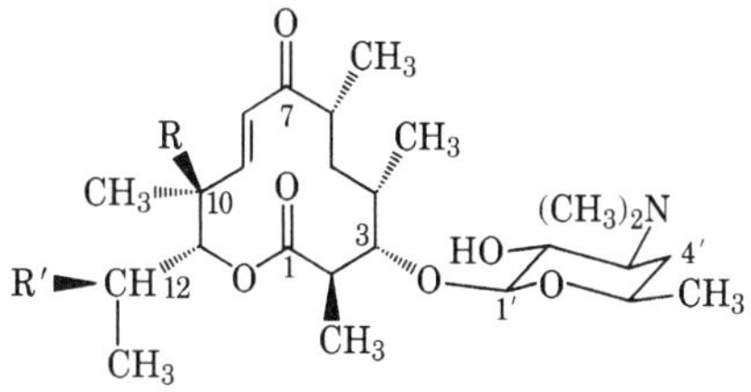
neomethymycin	*[497-73-4]*	$C_{25}H_{43}NO_7$	(**12**, R = H, R′ = OH)	
YC-17	*[11091-33-1]*	$C_{25}H_{43}NO_6$	(**12**, R = R′ = H)	
methynolide	*[534-32-7]*	$C_{17}H_{28}O_5$	(**13**, R = OH, R′ = H)	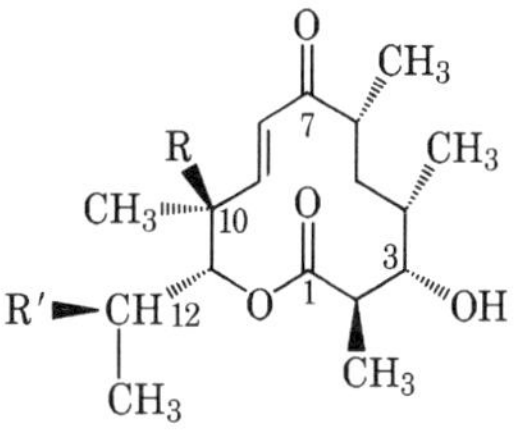
neomethynolide	*[497-51-8]*	$C_{17}H_{28}O_5$	(**13**, R = H, R′ = OH)	

precursor of methymycin and neomethymycin. The hydroxyl groups at C-12 and C-10 are probably added as late steps in the biosynthesis (38).

14-Membered Ring Macrolides

Natural Products. Naturally occurring 14-membered macrolides are given in Table 3.

Erythromycins. Erythromycin A (**14**, R = OH, R′ = CH_3, R″ = H), the most widely used macrolide antibiotic, was the principal product found in culture broths of *Streptomyces erythreus* (39), now reclassified as *Saccharopolyspora erythraea* (40). It contains a highly substituted aglycone, erythronolide A, (**16**, R = R′ = OH) to which desosamine (**1**, R = OH, R′ = H) and cladinose (**8**, R = CH_3) are attached (41). The complete stereochemistry of erythromycin A was established by x-ray analysis of its hydroiodide dihydrate (42); total synthesis of erythromycin A was a landmark achievement (43), a task previously considered hopeless (44).

Several minor factors are co-produced with erythromycin A. Erythromycin B (**14**, R = R″ = H, R′ = CH_3) lacks a hydroxyl group at C-12 (45). Erythromycin C (**14**, R = OH, R′ = R″ = H) contains mycarose (**8**, R = H) as the neutral sugar (46). Erythromycin D (**14**, R = R′ = R″ = H), which simultaneously lacks the C-12 hydroxyl group and contains mycarose (47), plus three of its 3″- and 4″-esters were isolated from culture broths of a *Nocardia* species (48). Erythromycin F (**14**, R = R″ = OH, R′ = CH_3), which contains a 2-hydroxymethyl group (49,50), may be the biosynthetic precursor of erythromycin E (**15**), which possesses an ortho ester linkage (51,52). Several possible biosynthetic intermediates or shunt metabolites

Table 3. Naturally Occurring 14-Membered Macrolides

Macrolide	CAS Registry Number	Molecular formula	Structure number	Structure
			Erythromycins and erythronolides	
erythromycin A	[*114-07-8*]	$C_{37}H_{67}NO_{13}$	(**14**, R = OH, R′ = CH_3, R″ = H)	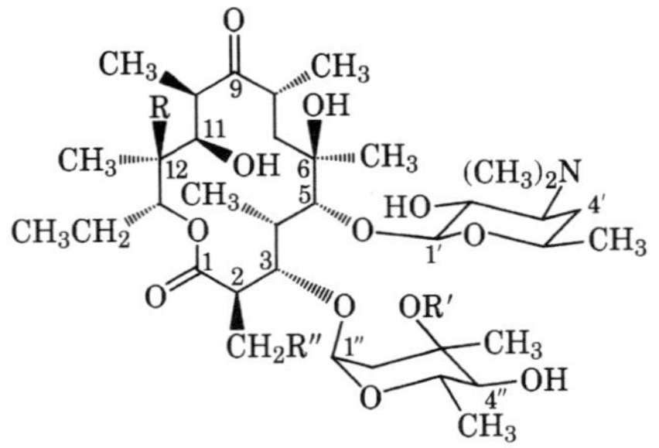
erythromycin B	[*527-75-3*]	$C_{37}H_{67}NO_{12}$	(**14**, R = R″ = H, R′ = CH_3)	
erythromycin C	[*1675-02-1*]	$C_{36}H_{65}NO_{13}$	(**14**, R = OH, R′ = R″ = H)	
erythromycin D	[*33442-56-7*]	$C_{36}H_{65}NO_{12}$	(**14**, R = R′ = R″ = H)	
erythromycin F	[*82230-93-1*]	$C_{37}H_{67}NO_{14}$	(**14**, R = R″ = OH, R′ = CH_3)	
erythromycin E	[*41451-91-6*]	$C_{37}H_{65}NO_{14}$	(**15**)	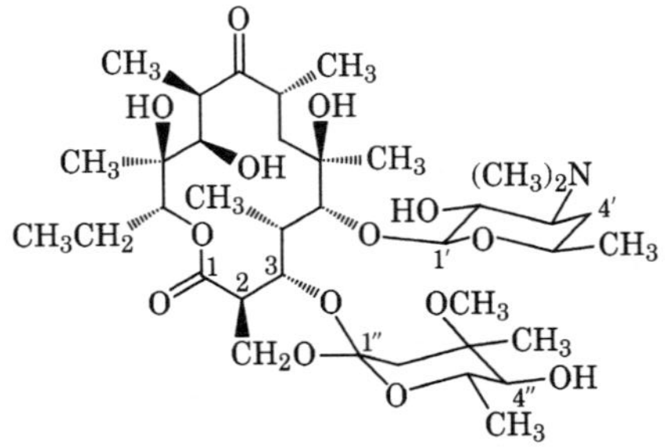
erythronolide A	[*26754-37-0*]	$C_{21}H_{38}O_8$	(**16**, R = R′ = OH)	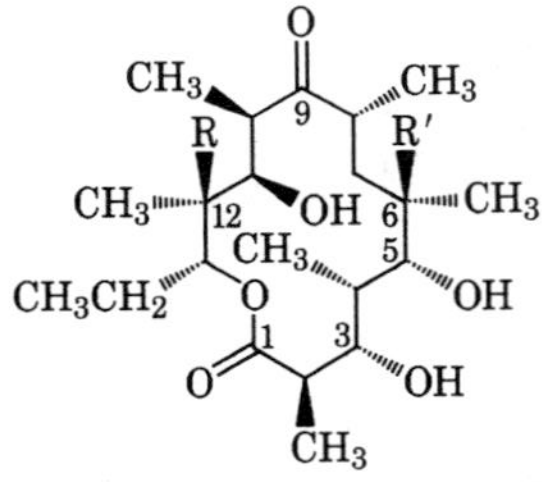
erythronolide B	[*3225-82-9*]	$C_{21}H_{38}O_7$	(**16**, R = H, R′ = OH)	
6-deoxyerythronolide B	[*15797-36-1*]	$C_{21}H_{38}O_6$	(**16**, R = R′ = H)	

Oleandomycin and related compounds

oleandomycin	*[3922-90-5]*	$C_{35}H_{61}NO_{12}$	(**17**, R = CH_3)
oleandomycin Y	*[64743-16-4]*	$C_{34}H_{59}NO_{12}$	(**17**, R = H)

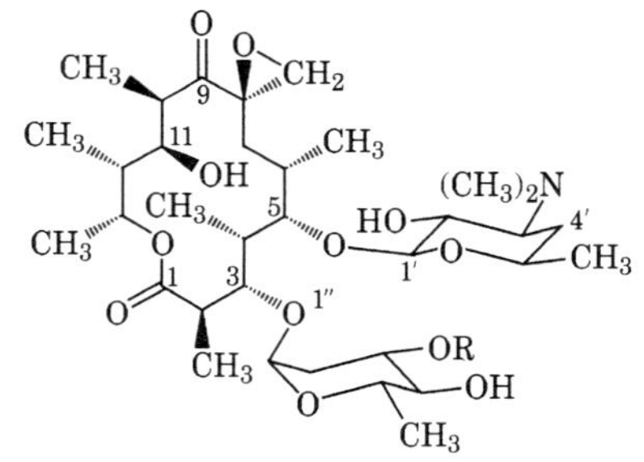

oleandolide	*[68540-16-9]*	$C_{20}H_{34}O_7$	(**18**, X = 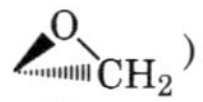)
8,8a-deoxyoleandolide	*[53428-54-9]*	$C_{20}H_{36}O_6$	(**18**, X = )

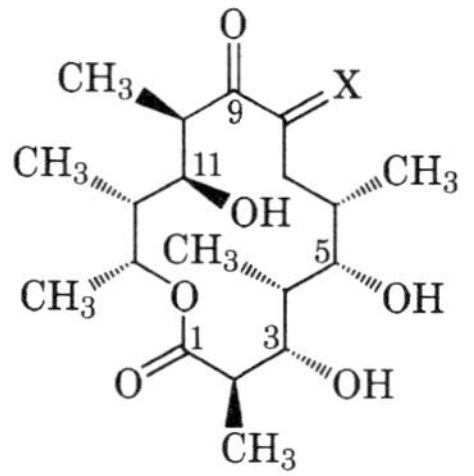

Pikromycin and related compounds

pikromycin	*[19721-56-3]*	$C_{28}H_{47}NO_8$	(**19**, R = OH, R′ = H)
narbomycin	*[6036-25-5]*	$C_{28}H_{47}NO_7$	(**19**, R = R′ = H)
5-*O*-mycaminosylnarbonolide	*[61435-20-9]*	$C_{28}H_{47}NO_8$	(**19**, R = H, R′ = OH)

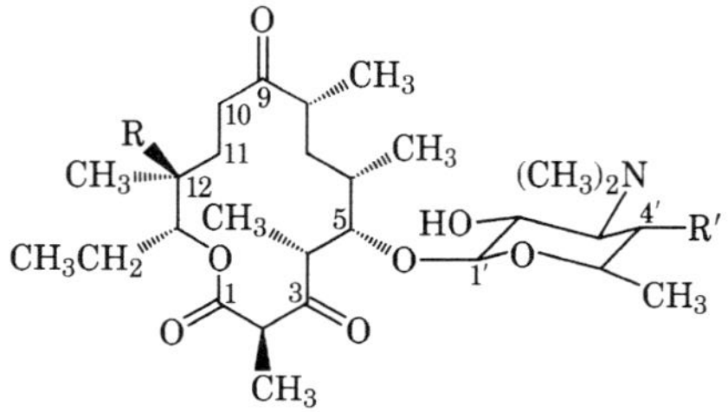

Table 3. (*Continued*)

Macrolide	CAS Registry Number	Molecular formula	Structure number	Structure
pikronolide	[*51724-55-1*]	$C_{20}H_{32}O_6$	(**20**, R = OH)	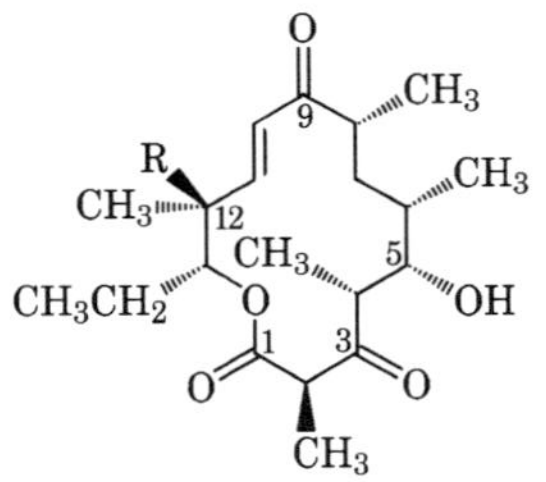
narbonolide	[*32885-75-9*]	$C_{20}H_{32}O_5$	(**20**, R = H)	
10,11-dihydropikromycin	[*27656-56-0*]	$C_{28}H_{49}NO_8$	(**21**, R = OH, R′ = H)	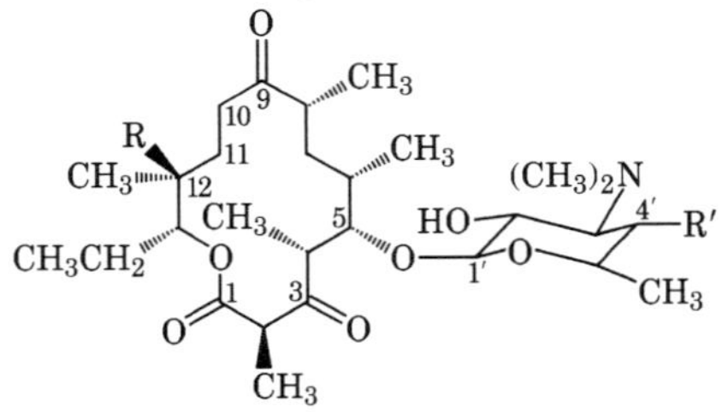
kayamycin[a]	[*102907-96-0*]	$C_{28}H_{49}NO_8$	(**21**, R = H, R′ = OH)	
12-deoxykromycin	[*55527-99-6*]	$C_{20}H_{30}O_4$	(**22**, R = H)	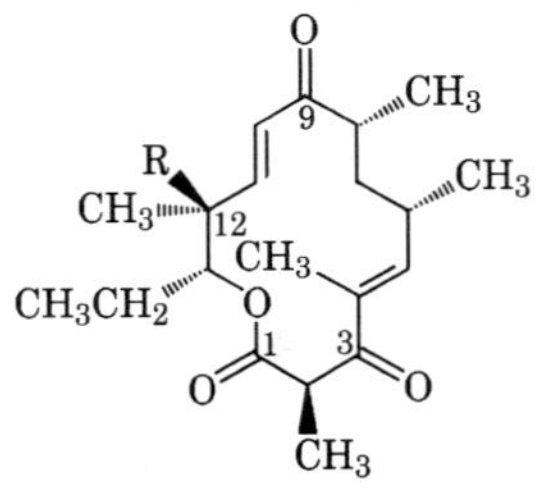
kromycin	[*20509-23-3*]	$C_{20}H_{30}O_5$	(**22**, R = OH)	

Megalomicins

megalomicin A[b]	[*28022-11-9*]	$C_{44}H_{80}N_2O_{15}$	(**23**, R = R′ = H)
megalomicin B[c]	[*49669-75-2*]	$C_{46}H_{82}N_2O_{16}$	(**23**, R = H, R′ = $OCCH_3$)
megalomicin C_1[d]	[*49669-76-3*]	$C_{48}H_{84}N_2O_{17}$	(**23**, R = R′ = $OCCH_3$)
megalomicin C_2[e]	[*49669-77-4*]	$C_{49}H_{86}N_2O_{17}$	(**23**, R = $OCCH_3$, R′ = $OCCH_2CH_3$)
XK-41-B_2	[*50692-34-7*]	$C_{47}H_{84}N_2O_{16}$	(**23**, R = H, R′ = $OCCH_2CH_3$)

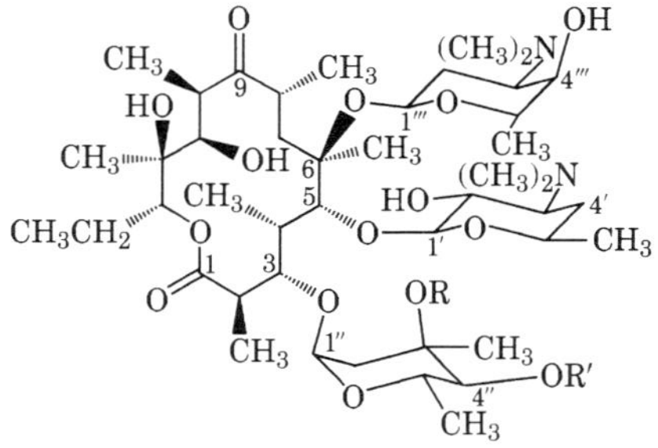

Lankamycin and related compounds

lankamycin (kujimycin B)	[*30042-37-6*]	$C_{42}H_{72}O_{16}$	(**24**, R = $OCCH_3$, R′ = OH, R″ = H)
kujimycin A	[*33955-27-0*]	$C_{40}H_{70}O_{15}$	(**24**, R = H, R′ = OH, R″ = H)
15-dehydrolankamycin	[*65423-08-7*]	$C_{42}H_{70}O_{16}$	(**24**, R = $OCCH_3$, R′ + R″ = O)
15-*O*-(4-*O*-acetylarcanosyl)-lankamycin	[*65423-07-6*]	$C_{52}H_{88}O_{20}$	(**24**, R = $OCCH_3$, R′ = *O*-4-*O*-acetylarcanosyl, R″ = H)

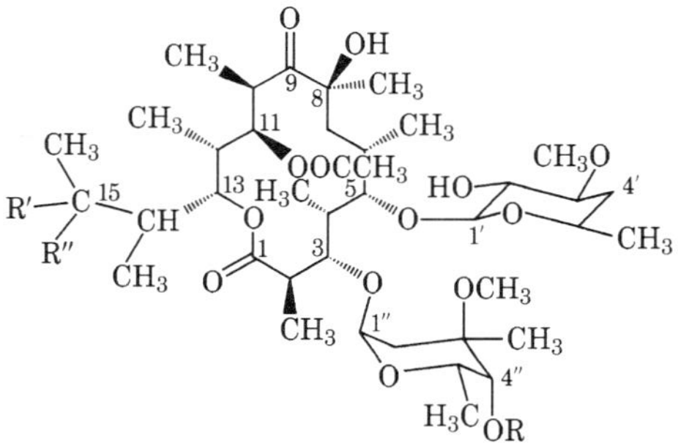

3″-*O*-demethyl-2″,3″-anhydrolankamycin	[*60462-94-4*]	$C_{41}H_{68}O_{15}$	(**25**)

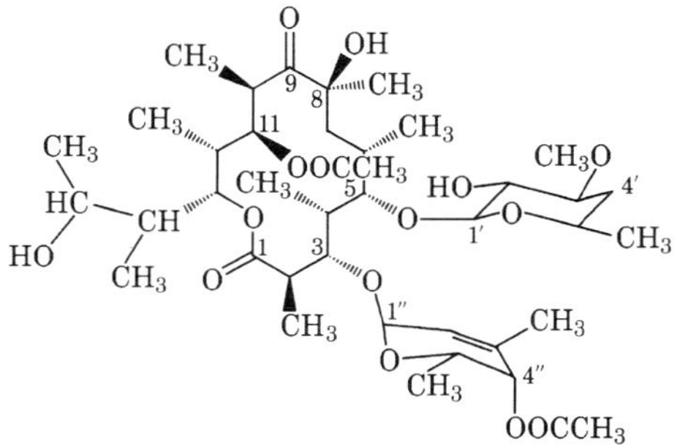

Table 3. (*Continued*)

Macrolide	CAS Registry Number	Molecular formula	Structure number	Structure
23672 RP	[*37359-09-4*]	$C_{48}H_{82}O_{20}$	(**26**, R = α-hydroxyisovaleryl)	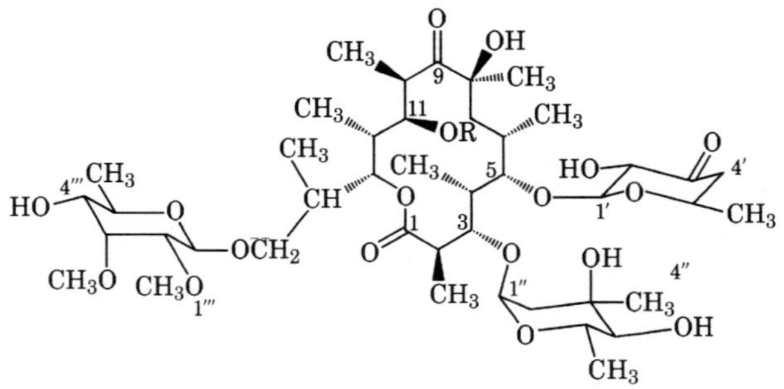

[a]Also known as A1-R6-4.
[b]Also known as XK-41-C.
[c]Also known as XK-41-B_1.
[d]Also known as XK-41-A_2.
[e]Also known as XK-41-A_1.

of erythromycin have been obtained after single-step fermentations or two-step, precursor-feeding experiments utilizing biosynthetically blocked mutants of the erythromycin-producing organism (53–59). Additional compounds related to erythromycin have been discovered (50,52,60–62).

Chemical degradation of erythromycin A yielded its aglycone, erythronolide A (**16**, R = R′ = OH), whereas erythronolide B (**16**, R = H, R′ = OH) was obtained from fermentation (63,64). Biosynthesis of erythromycin proceeds via 6-deoxyerythronolide B (**16**, R = R′ = H) and then erythronolide B (64,65). The first total synthesis of erythromycin-related compounds was erythronolide B (66); syntheses of erythronolide A and 6-deoxyerythronolide B soon followed (67,68).

Oleandomycin. Oleandomycin (**17**, R = CH_3) was the primary factor in culture broths of *S. antibioticus* (69,70). Its aglycone, oleandolide (**18**, X = $\overset{O}{\triangle}CH_2$), differs from the erythronolides at several centers, most notably by the epoxide at C-8. In addition, the neutral sugar in oleandomycin is oleandrose (**11**, R = CH_3) (71–73). A by-product, oleandomycin Y (**17**, R = H), contains the neutral sugar L-olivose (**11**, R = H) (74). Oleandolide has been obtained by total synthesis and chemical degradation (75,76). The aglycone lacking the epoxide function, 8,8a-deoxyoleandolide (**18**, X = $\overset{H}{\blacktriangle}CH_3$), was produced by a mutant of the erythromycin-producing organism (59).

Pikromycin. Pikromycin (**19**, R = OH, R′ = H), the first macrolide discovered (77,78), is produced by *S. felleus.* Pikromycin is identical to amaromycin and albomycetin (79–81) and may be identical to proactinomycin (82–84). The structure of pikromycin was determined from chemical degradation, mass spectrometry, nmr, and x-ray crystallography (85–90). Its aglycone, pikronolide (**20**, R = OH), was produced by *S. venezuelae* (36). A derivative, kromycin (**22**, R = OH), was formed from pikromycin under acidic conditions (87,88,91) and more drastic conditions produced an intramolecular spiroketal of pikronolide named kromin (89,91). 10,11-Dihydropikromycin (**21**, R = OH, R′ = H), was also produced by *S. venezuelae* (92).

Narbomycin (**19**, R = R′ = H), produced by *S. narbonensis* (93), is 12-deoxypikromycin (94). Its aglycone, narbonolide, (**20**, R = H) is produced by *S. venezuelae* and bioconverted into narbomycin and then pikromycin (95–97). The stereochemistry of narbonolide was established by total synthesis (98). 12-Deoxykromycin (**22**, R = H) was formed from narbonolide in a manner analogous to that yielding kromycin (99). Two derivatives of narbonolide containing mycaminose (**1**, R = R′ = OH) rather than desosamine (**1**, R = OH, R′ = H) were produced by species of *Nocardiopsis:* kayamycin (**21**, R = H, R′ = OH) (100) and 5-*O*-mycaminosylnarbonolide (**19**, R = H, R′ = OH). This latter macrolide and its 9-dihydro derivative had previously been obtained by feeding narbonolide to strains of *S. platensis* (101,102).

Megalomicins. The megalomicins, the first macrolides obtained from *Micromonospora* species (*M. megalomicea*) (103), were later obtained from culture broths of *M. inositola* (104,105). Structures were determined by nmr and x-ray crystallographic studies (106,107). Megalomicins are the only 14-membered macrolides containing two aminosugars: desosamine and a novel aminosugar, megosamine (**3**) glycosidically linked to the 6-hydroxyl group. The megalomicins also differ from erythromycin in that the neutral sugar is either mycarose (**8**, R =

H) or its acyl derivative instead of cladinose (**8**, R = CH_3). The aglycones of megalomicin and erythromycin A are identical (104).

Lankamycins/Kujimycins. The lankamycin/kujimycin group lack amino-sugars but contain the neutral sugars D-chalcose (**7**) and L-arcanose (**6**) or their derivatives. Lankamycin (**24**, R = $OCCH_3$, R′ = OH, R″ = H) was found in culture broths of *S. violaceoniger* (108) and its structure revised after detailed nmr studies (109,110). Kujimycin A (**24**, R = H, R′ = OH, R″ = H) and kujimycin B, identical to lankamycin, were isolated from broths of *S. spinichromogenes* (111,112). Related factors have been found which possess an unsaturated sugar (**25**), or a C-15 ketone or a 4-*O*-acetylarcanosyl moiety on the C-15 hydroxyl group (113,114). In 23672 RP (**26**) the C-11 hydroxyl group is acylated by an α-hydroxyvaleryl group (115).

Semisynthetic Derivatives. Erythromycin has been the principal subject of modification of 14-membered macrolides; some of the derivatives are being commercially launched (116).

Decomposition of erythromycin under acidic conditions generates erythromycin 8,9-anhydro-6,9-hemiketal [*33396-29-1*] (**27**) and erythromycin 6,9;9,12-spiroketal [*23893-13-2*] (**28**) (117–119) as shown in Figure 1. The ring-contracted product [*105882-69-7*] (**29**), derived from trans-acylation in (**27**), was a minor component in mother liquors from which erythromycin was isolated (60), and was also found in pharmaceutical formulations of erythromycin (50). Syntheses of (**29**) have been published (120,121). A related pair of ring contractions was observed in oleandomycin, yielding a 10-membered bicyclic product (122). In all of these examples, the bicyclic product apparently stabilizes the 10- and 12-membered lactones. In contrast, translactonization of 14-hydroxy-6-*O*-meth-

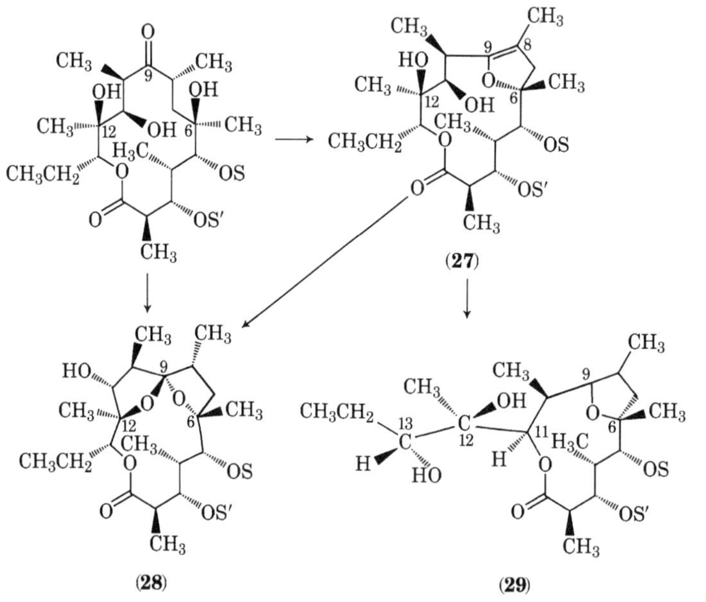

Fig. 1. Intramolecular cyclization reactions of erythromycin A where S = desosamine (**1**, R = OH, R′ = H) and S′ = cladinose (**8**, R = CH_3).

ylerythromycin A yielded a ring-expanded 15-membered macrolide which existed in two tautomeric forms possessing either a 9-ketone or a 9,12-hemiketal (123).

The acid-instability of erythromycin makes it susceptible to degradation in the stomach to intramolecular cyclization products lacking antimicrobial activity. Relatively water-insoluble, acid-stable salts, esters, and/or formulations have therefore been employed to protect erythromycin during passage through the stomach, to increase oral bioavailability, and to decrease the variability of oral absorption. These various derivatives and formulations also mask the very bitter taste of macrolides.

Ester derivatives of erythromycin, shown in Table 4, were synthesized soon after its discovery (124–126). They are readily prepared by acylation of the 2′-hydroxyl group; the neighboring 3′-dimethylamino group directs acylation to this site. Commonly used esters of erythromycin are propionate, acetate, ethyl succinate, and ethyl carbonate (**30**). 2′-Esters do not bind bacterial ribosomes, however, and consequently must be hydrolyzed back to the parent to exert antibacterial activity (127). Ester derivatives are still being prepared (128,129). 2′-*O*-Acetylerythromycin stearate, also known as erythromycin acistrate, is an example undergoing clinical trial (11,130,131). Two salts of 2′-*O*-propionylerythromycin, *N*-acetylcysteinate (erythromycin stinoprate) and mercaptosuccinate, were prepared to combine antibiotic and mucolytic properties in a single agent (132–134).

Ester derivatives of oleandomycin (**17**, R = CH_3) were also pursued leading to commercial development of (**31**) also known as troleandomycin (TOA), which improved oral bioavailability and taste as compared to the parent (135).

Another successful strategy for derivatization of erythromycin employed modification of functional groups involved in intramolecular cyclizations. The C-9 ketone, C-6 hydroxyl group, C-8 proton, and/or C-11,12-diol of erythromycin were converted into functional groups which participate poorly, if at all, in intramolecular cyclizations. Some derivatives which have been extensively evaluated in preclinical and clinical trials exhibit such desirable properties as better stability under acidic conditions, greater oral bioavailability, and higher and more prolonged concentrations of antibiotic in serum and tissues.

The 11,12-carbonate of erythromycin (**32**) is an older cyclic ester which had greater stability and antibiotic activity by diminishing formation of intramolecular enol ether (**27**) (136,137). A later analogue, the 11-*N*-12-*O*-cyclic carbamate of 11-deoxy-11-(2-dimethylamino)ethylamino-6-*O*-methylerythromycin (A-62514), was selected for further study (138,139).

Several structural modifications of the C-9 ketone of erythromycin have been explored; oximes and hydrazones are less prone to intramolecular cyclization, but they often have less antibiotic activity than erythromycin (140). Synthesis of more complex oxime derivatives resulted in the development of roxithromycin, the 9-[*O*-(2-methoxyethoxy)methyl]oxime (**33**) (141). Reduction of the oximes and hydrazones produced 9(*S*)-erythromycylamine (**34**) as the principal product, with minor amounts of the 9(*R*)-isomer (140); however, clinical studies showed that 9(*S*)-erythromycylamine and its *N*-benzylidene derivative were poorly absorbed in humans (142). Evaluation of more complex oxazine derivatives of erythromycylamine led to dirithromycin, the 2-(2-methoxyethoxy)ethylidene oxazine derivative (**35**) (143). A third route to modification of the ketone utilized a

Table 4. Derivatives of Erythromycin and Oleandonycin

Macrolide	CAS Registry Number	Molecular formula	Structure number	Structure
		Erythromycin esters		
2′-*O*-acetylerythromycin	[*992-69-8*]	$C_{39}H_{69}NO_{14}$	(**30**, R = $COCH_3$)	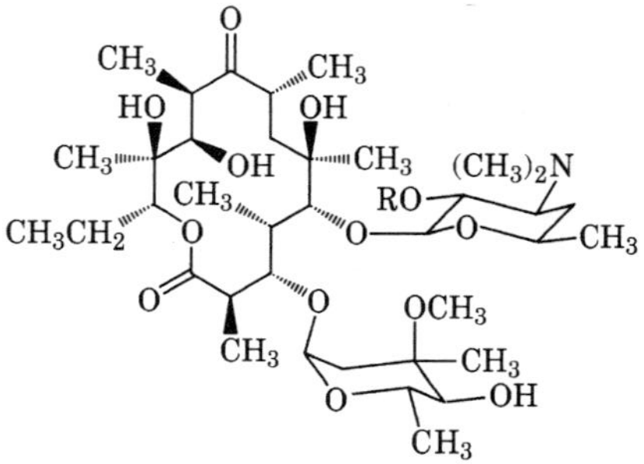
2′-*O*-propionylerythromycin	[*134-36-1*]	$C_{40}H_{71}NO_{14}$	(**30**, R = $COCH_2CH_3$)	
erythromycin ethyl carbonate	[*7218-80-6*]	$C_{40}H_{71}NO_{15}$	(**30**, R = $COOCH_2CH_3$)	
erythromycin ethyl succinate	[*1264-62-6*]	$C_{43}H_{75}NO_{16}$	(**30**, R = $CO(CH_2)_2COOCH_2CH_3$)	
		Other derivatives[a]		
tri-*O*-acetyloleandomycin	[*2751-09-9*]	$C_{41}H_{67}NO_{15}$	(**31**)	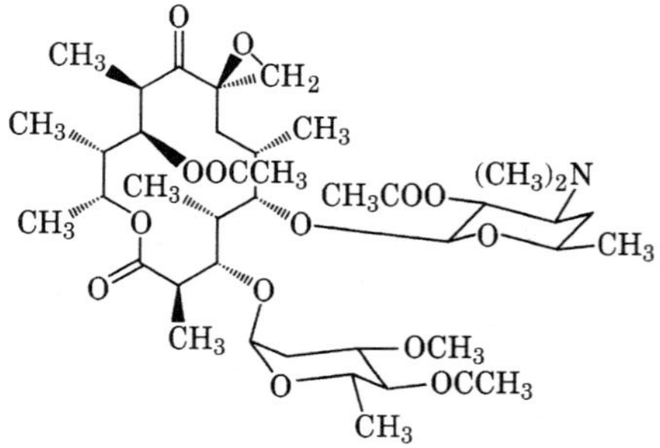

erythromycin-11,12-carbonate	*[55224-05-0]*	$C_{38}H_{65}NO_{14}$	**(32)**	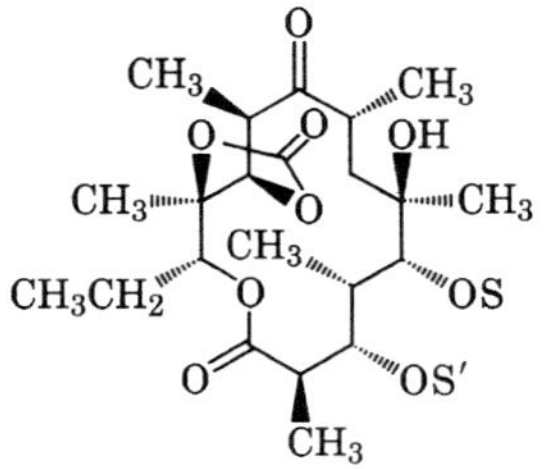
roxithromycin	*[80214-83-1]*	$C_{41}H_{76}N_2O_{15}$	**(33)**	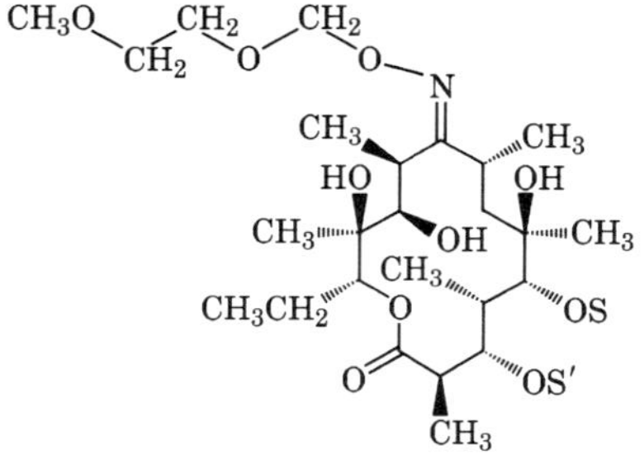
9(*S*)-erythromycylamine	*[26116-56-3]*	$C_{37}H_{70}N_2O_{12}$	**(34)**	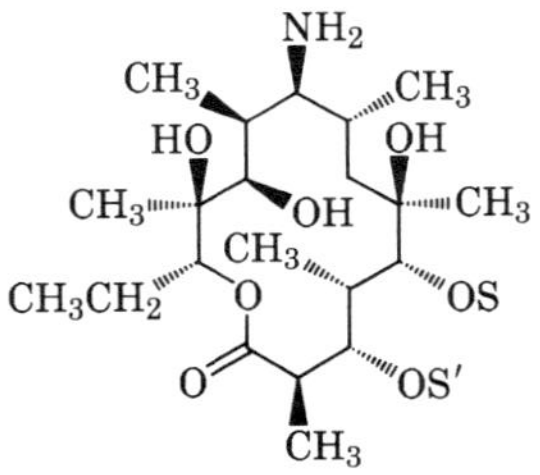
dirithromycin	*[62013-04-1]*	$C_{42}H_{78}N_2O_{14}$	**(35)**	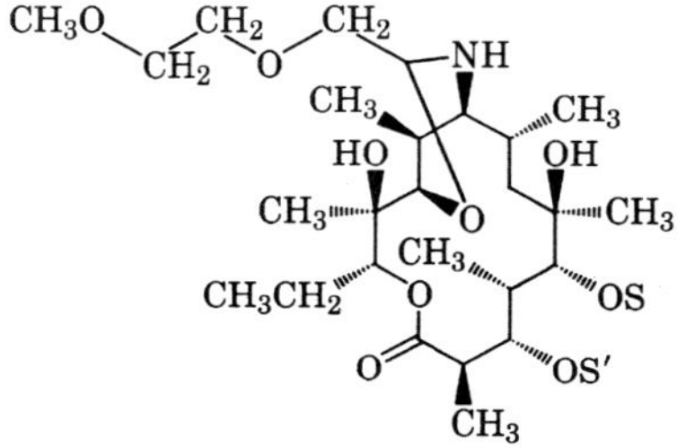

Table 4. (*Continued*)

Macrolide	CAS Registry Number	Molecular formula	Structure number	Structure
azithromycin	[*83905-01-5*]	$C_{38}H_{72}N_2O_{12}$	(**36**)	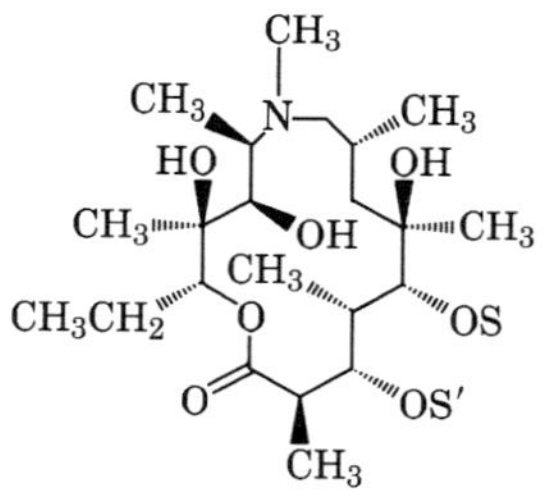
clarithromycin	[*81103-11-9*]	$C_{38}H_{69}NO_{13}$	(**37**)	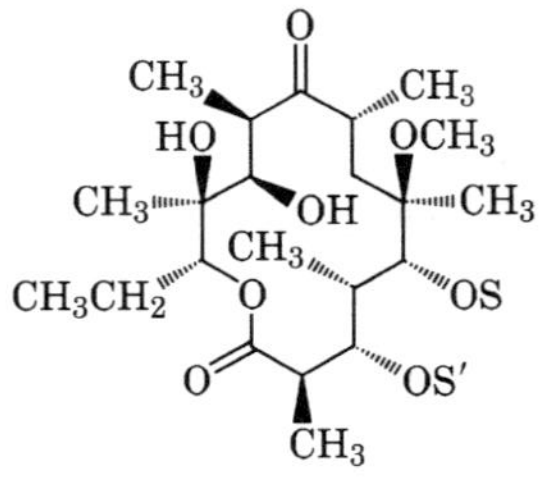 (**37**)
flurithromycin	[*82664-20-8*]	$C_{37}H_{66}FNO_{13}$	(**38**)	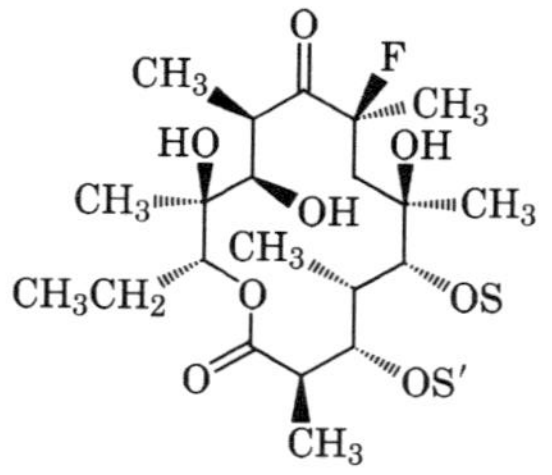 (**38**)

[a]S = desosamine (**1**, R = OH, R′ = H); S′ = cladinose (**8**, R = CH_3).

Beckmann rearrangement of the 9-oxime to expand the 14-membered ring to a 15-membered intermediate, which was subsequently reduced and *N*-methylated to yield azithromycin (**36**) (144,145). The term azalide was proposed to denote these 15-membered azalactones (10,145).

Other approaches to inhibiting intramolecular cyclizations of erythromycin have also proven successful. From a series of *O*-alkyl derivatives of erythromycin, clarithromycin (6-*O*-methylerythromycin) (**37**) was selected for clinical development (146,147). Another approach replaced the C-8 proton of erythromycin with fluorine, which was accomplished by both chemical and bioconversion methods to yield flurithromycin (**38**) (148).

A few bioconversions of erythromycin have been reported (149). Certain modifications of erythromycin at both its 9-ketone and 11-hydroxyl groups inhibited hydrolysis of the lactone ring by an esterase isolated from *Escherichia coli* which might be involved in bacterial resistance to macrolides (150). The 9-methoxime-11-[(2-dimethylaminoethyl)oxymethyl] derivative of erythromycin, ER 42859, showed good antibacterial activity and pharmacokinetics in animals, but it gave lower blood levels than erythromycin when administered to humans (151). Other 9,11-cyclic derivatives of erythromycin have been synthesized (152,153). Series of 9-*N*-alkyl derivatives of both 9(*S*)- and 9(*R*)-erythromycylamine have been prepared which improved activity *in vivo* over erythromycin (154,155).

A novel series of 9,12-epoxy derivatives synthesized from (**27**) showed unexpectedly good antibiotic activity; A-69334, which possesses an 11-amino-9,12-epoxy moiety, exhibited pharmacokinetic advantages in animals (156). Some 4″-*O*-acyl derivatives of erythromycin did not induce resistance to other macrolide, lincosamide (see ANTIBIOTICS, LINCOSAMINIDES), and streptogramin B (see ANTIBIOTICS, PEPTIDES) antibiotics (MLS-resistance) in bacteria which were inducibly-resistant to erythromycin (157,158); furthermore, certain derivatives of erythromycin modified at the 4″-hydroxyl and/or 11,12-diol groups demonstrated activity against constitutively MLS-resistant bacteria (159,160). Some 4″-*O*-acyl derivatives promoted less gastrointestinal motility in dogs compared to erythromycin, and 4″-*O*-carbamoyl-6-*O*-methylerythromycin (A-63075) was selected for further study (161).

16-Membered Ring Macrolides

Natural Products. 16-Membered macrolides are divided into leucomycin- and tylosin-related groups, which differ in the substitution pattern of their aglycones. Multi-factor complexes are usually produced and some compounds have been isolated from culture broths of different organisms and then been given different names.

The leucomycin complex of ten factors, Type I in Table 5, occasionally called kitasamycin, was produced by *S. kitasatoensis* (162), later reclassified as *Streptoverticillium* (163). After extensive studies, a 16-membered lactone substituted by an aminosugar and a neutral sugar was proposed, with individual factors differing in acylation of the 3- and 4″-hydroxyl groups (164–172). X-ray crystallography of demycarosylisoleucomycin A_3 hydrobromide confirmed these structures

Table 5. Leucomycins and Related Compounds

Macrolide	CAS Registry Number	Molecular formula	R	R′

Type I: C-9 to C-13 is dienol

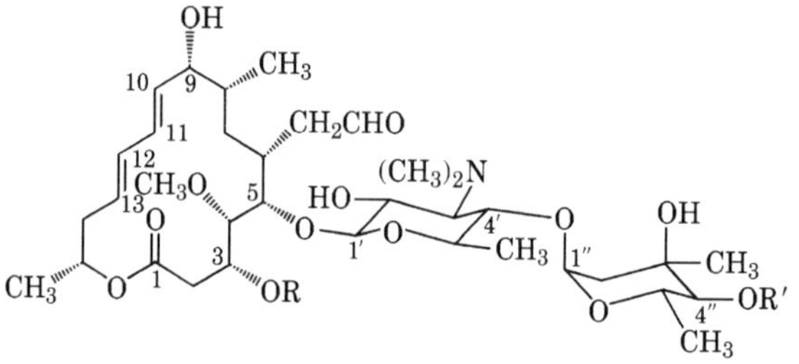

(**39**)

Macrolide	CAS Registry Number	Molecular formula	R	R′
leucomycin A_1	*[16846-34-7]*	$C_{40}H_{67}NO_{14}$	H	isovaleryl
leucomycin A_5	*[18361-45-0]*	$C_{39}H_{65}NO_{14}$	H	butyryl
leucomycin A_7	*[18361-47-2]*	$C_{38}H_{63}NO_{14}$	H	propionyl
leucomycin A_9	*[18361-49-4]*	$C_{37}H_{61}NO_{14}$	H	acetyl
leucomycin V	*[22875-15-6]*	$C_{35}H_{59}NO_{13}$	H	H
leucomycin A_3 (josamycin)	*[16846-24-5]*	$C_{42}H_{69}NO_{15}$	acetyl	isovaleryl
leucomycin A_4	*[18361-46-1]*	$C_{41}H_{67}NO_{15}$	acetyl	butyryl
leucomycin A_6	*[18361-48-3]*	$C_{40}H_{65}NO_{15}$	acetyl	propionyl
leucomycin A_8	*[18361-50-7]*	$C_{39}H_{63}NO_{15}$	acetyl	acetyl
leucomycin U	*[31642-61-2]*	$C_{37}H_{61}NO_{14}$	acetyl	H
platenomycin A_1	*[40615-47-2]*	$C_{43}H_{71}NO_{15}$	propionyl	isovaleryl
midecamycin A_2	*[35457-81-9]*	$C_{42}H_{69}NO_{15}$	propionyl	butyryl
midecamycin A_1 (espinomycin A_1), (platenomycin B_1)	*[35457-80-8]*	$C_{41}H_{67}NO_{15}$	propionyl	propionyl
platenomycin C_2 (espinomycin A_3)	*[35867-32-4]*	$C_{40}H_{65}NO_{15}$	propionyl	acetyl
DHP	*[35906-56-0]*	$C_{38}H_{63}NO_{14}$	propionyl	H
espinomycin A_2	*[40867-12-7]*	$C_{42}H_{69}NO_{15}$	propionyl	isobutyryl
platenomycin A_0	*[52310-62-0]*	$C_{44}H_{73}NO_{15}$	butyryl	isovaleryl

Type II: C-9 to C-13 is epoxyenol

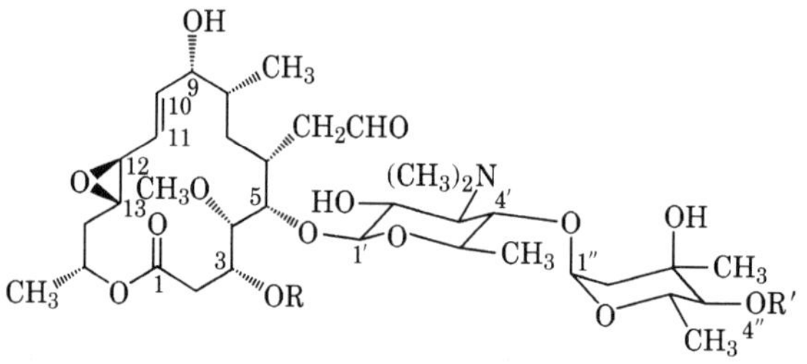

(**40**)

Macrolide	CAS Registry Number	Molecular formula	R	R′
maridomycin I (platenomycin C_3)	*[35908-44-2]*	$C_{43}H_{71}NO_{16}$	propionyl	isovaleryl
maridomycin VII	*[56078-81-0]*	$C_{42}H_{69}NO_{16}$	propionyl	butyryl

Table 5. (*Continued*)

Macrolide	CAS Registry Number	Molecular formula	R	R′
maridomycin III (platenomycin C_1)	[*35775-82-7*]	$C_{41}H_{67}NO_{16}$	propionyl	propionyl
maridomycin V	[*35942-57-5*]	$C_{40}H_{65}NO_{16}$	propionyl	acetyl
maridomycin II (platenomycin C_4)	[*35908-45-3*]	$C_{42}H_{69}NO_{16}$	acetyl	isovaleryl
maridomycin IV	[*35942-56-4*]	$C_{40}H_{65}NO_{16}$	acetyl	propionyl
maridomycin VI	[*35775-66-7*]	$C_{39}H_{63}NO_{16}$	acetyl	acetyl

Type III: C-9 to C-13 is epoxyenone

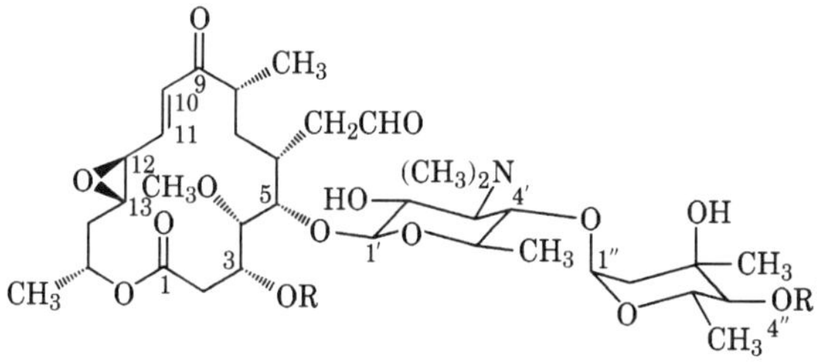

(41)

Macrolide	CAS Registry Number	Molecular formula	R	R′
carbomycin A (deltamycin A_4)	[*4564-87-8*]	$C_{42}H_{67}NO_{16}$	acetyl	isovaleryl
deltamycin A_3	[*58880-24-3*]	$C_{41}H_{65}NO_{16}$	acetyl	butyryl
deltamycin A_2	[*58880-23-2*]	$C_{40}H_{63}NO_{16}$	acetyl	propionyl
deltamycin A_1	[*58880-22-1*]	$C_{39}H_{61}NO_{16}$	acetyl	acetyl
deltamycin X, (EOA)	[*40625-70-5*]	$C_{37}H_{59}NO_{15}$	acetyl	H
EOP	[*41688-43-1*]	$C_{38}H_{61}NO_{15}$	propionyl	H

Type IV: C-9 to C-13 is dienone

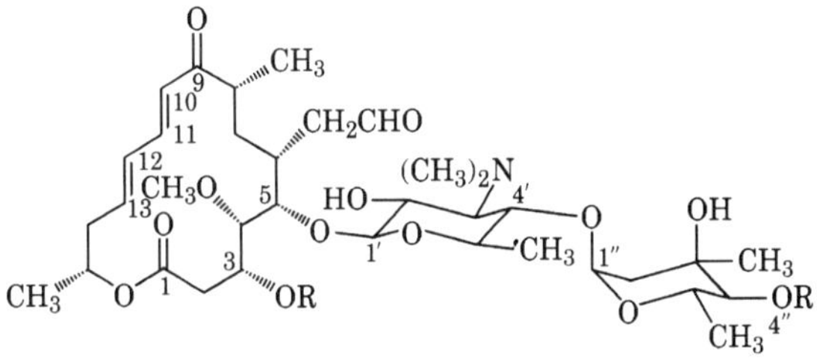

(42)

Macrolide	CAS Registry Number	Molecular formula	R	R′
carbomycin B	[*21238-30-2*]	$C_{42}H_{67}NO_{15}$	acetyl	isovaleryl
platenomycin W_1	[*35867-31-3*]	$C_{43}H_{69}NO_{15}$	propionyl	isovaleryl
platenomycin W_2	[*52310-61-9*]	$C_{44}H_{71}NO_{15}$	butyryl	isovaleryl
niddamycin	[*20283-69-6*]	$C_{39}H_{63}NO_{14}$	H	butyryl
midecamycin A_4	[*36083-82-6*]	$C_{42}H_{67}NO_{15}$	propionyl	butyryl
midecamycin A_3	[*36025-69-1*]	$C_{41}H_{65}NO_{15}$	propionyl	propionyl
DOA	[*52278-66-7*]	$C_{37}H_{59}NO_{14}$	acetyl	H
DOP	[*40580-80-1*]	$C_{38}H_{61}NO_{14}$	propionyl	H

except in stereochemistry at C-9 (173,174). Interconversion between leucomycin and isoleucomycin had suggested a β configuration for the C-9 hydroxyl group (175,176), but this was later revised to α (177) and confirmed by additional x-ray studies (178). Conformational studies of the leucomycins have been published (179). Josamycin (**39**, R = acetyl, R′ = isovaleryl) was obtained from culture broths of *S. narbonensis* var. *josamyceticus* (180). Although initially reported as a new macrolide, it was later shown to be identical to leucomycin A_3 (181). Josamycin was commercially launched in Japan in 1970 and is now one of the most widely used 16-membered macrolides (116).

The maridomycin complex of seven factors (**40**), Type II in Table 5, was obtained from culture broths of *S. hygroscopicus* (182–188). The principal difference from the leucomycins is a 12,13-epoxide. Confirmation of structure was obtained from x-ray crystallography and spectroscopy of 9-*O*-acyl derivatives of maridomycin III (188,189).

Carbomycin A (**41**, R = acetyl, R′ = isovaleryl), Table 5, Type III, the first 16-membered macrolide discovered, was isolated from culture broths of *S. halstedii* (also called magnamycin or M-4209) (190–192). The 17-membered lactone originally proposed was later revised to a 16-membered lactone (27,193,194). Its absolute stereochemistry was established by correlations with the maridomycins (187,188). Carbomycin is also produced by *S. thermotolerans* (195,196). The deltamycins, produced by *S. halstedii* subsp. *deltae,* are five related factors which differ in their 4″-*O*-acyl group (197,198). Deltamycin A_4 is the same as carbomycin A. Deltamycins differ from the leucomycins and maridomycins by having a 12,13-epoxy-9-ketone. De-epoxidation and isomerization products have been prepared by chemical and microbiological methods (199). Carbomycin B (**42**, R = acetyl, R′ = isovaleryl), Type IV, Table 5, a minor component in the fermentation broths of *S. halstedii,* contains a C-9 to C-13 dienone (200).

Many related complexes, such as the platenomycin complex (initially YL-704) from *S. platensis* subsp. *malvinus,* have been obtained (201–206). A series designated by abbreviations (DHP, DOA) was obtained from another strain of *S. platensis* (207); the midecamycin complex (initially SF-837) was produced by *S. mycarofaciens* (208–212); the espinomycin factors were produced by *S. fungicidicus* var. *espinomyceticus* (213,214); the turimycin series of fifteen components, isolated from culture broths of *S. hygroscopicus,* are identical to other members of the leucomycin group (215). Niddamycin (**42**, R = H, R′ = butyryl) was isolated from cultures of *S. djakartensis* (216).

Macrolides containing only neutral sugars were obtained from a platenomycin-producing organism (217,218). Four demycarosyl derivatives of platenomycins were isolated from biosynthetically blocked mutants of *S. platensis;* two of these possessed a methyl group rather than an aldehyde (219). A pair of novel compounds related to carbomycin were isolated in which one contained an unusual 10,11-dihydro-12,13-diol moiety, the other a 14-hydroxy-epoxyenone moiety (220).

The spiramycin complex, also discovered as foromacidine, was isolated from culture broths of *S. ambofaciens* (221–224). The spiramycins, shown in Table 6, are distinguished by a second aminosugar, forosamine (**2**), attached to the 9-hydroxyl group by a β-glycosidic linkage (225). The three primary factors, spiramycin I, II, and III, differ in their 3-*O*-acyl substituent. Two minor factors containing a

Table 6. Spiramycins and 16-Membered Aglycones

Macrolide	CAS Registry Number	Molecular formula	Structure number	Structure
spiramycin I (foromacidine A)	[*24916-50-5*]	$C_{43}H_{74}N_2O_{14}$	(**43**, R = H, R′ = mycarosyl)	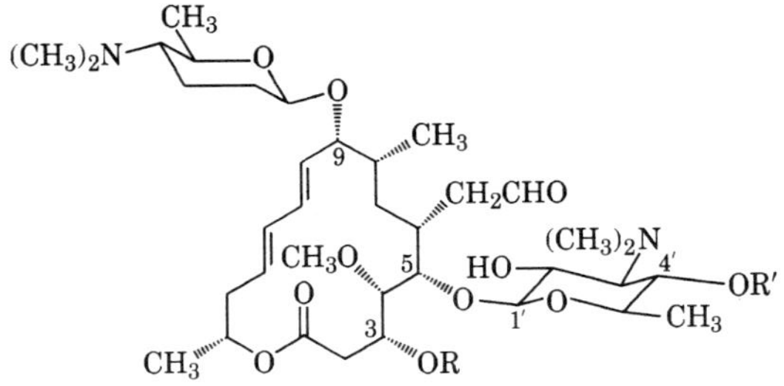
spiramycin II (foromacidine B)	[*24916-51-6*]	$C_{45}H_{76}N_2O_{15}$	(**43**, R = acetyl, R′ = mycarosyl)	
spiramycin III (foromacidine C)	[*24916-52-7*]	$C_{46}H_{78}N_2O_{15}$	(**43**, R = propionyl, R′ = mycarosyl)	
neospiramycin I	[*70253-62-2*]	$C_{36}H_{62}N_2O_{11}$	(**43**, R = R′ = H)	
leuconolide A_1	[*56201-85-5*]	$C_{20}H_{32}O_7$	(**44**)	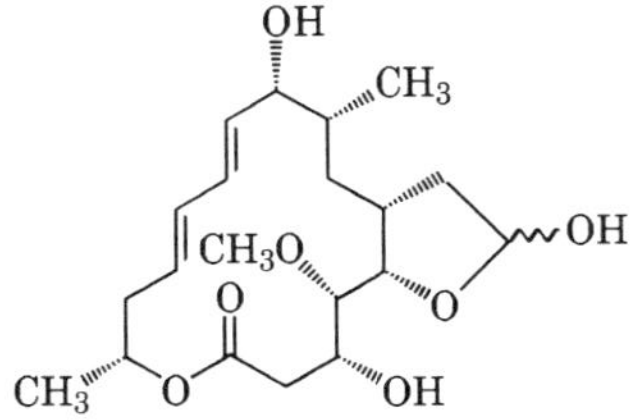
spiramycin U	[*87708-39-2*]	$C_{42}H_{71}NO_{16}$	(**45**, R = CHO, R′ = mycarosyl)	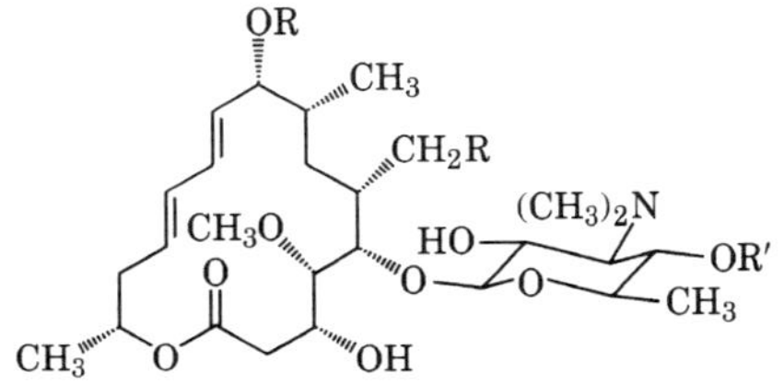
spiramycin S	[*87718-67-0*]	$C_{42}H_{73}NO_{16}$	(**45**, R = CH_2OH, R′ = mycarosyl)	
forocidin I	[*70282-40-5*]	$C_{28}H_{47}NO_{10}$	(**45**, R = CHO, R′ = H)	
platenolide I	[*52212-94-9*]	$C_{20}H_{32}O_6$	(**46**, R, R′ = O)	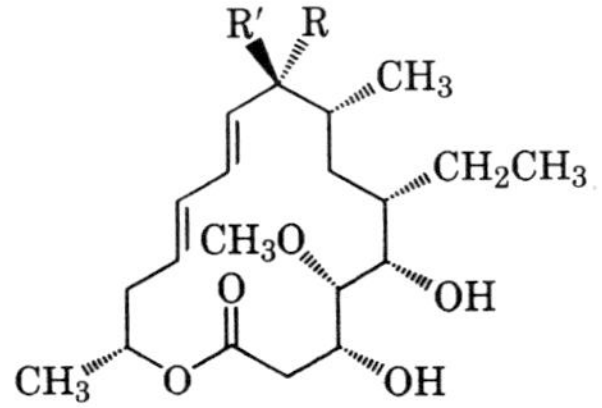
platenolide II	[*52212-95-0*]	$C_{20}H_{34}O_6$	(**46**, R = OH, R′ = H)	

neutral sugar instead of forosamine, spiramycin U and S, have been found (226). Selective hydrolysis of the terminal sugar yields the respective neospiramycin factors (227); subsequent hydrolysis of forosamine produces the forocidin factors (228).

Aglycones of 16-membered macrolides exist as intramolecularly cyclized hemiketals as a result of the close proximity of the aldehyde and 5-hydroxyl group. Aglycones can differ in their oxidation level at C-9 and C-12,13, types I–IV, Table 5, and their 3-*O*-substituent. The names parallel those of the parent macrolides; for example, leuconolide A_1 (**44**) is the aglycone from any of five leucomycin factors. The Polonovski reaction is generally used to cleave the aminosugar; acid hydrolysis destroys the aglycone before the sugar is hydrolyzed (229–231). Two reduced aglycones (**46**) were isolated from either a mutant strain or producing strain of *S. platensis* (232,233); these compounds are biosynthetic precursors of platenomycin; oxidation to the aldehyde occurs as a late step in the fermentation (234).

A second large group of 16-membered macrolides differs from the leucomycins in the substitution pattern of the aglycone. One difference is a methyl or hydroxymethyl group at C-14. If hydroxymethyl is present, it may be glycosidically substituted by a neutral sugar such as mycinose (**9**, R = R′ = CH_3), or an analogue. The most prominent member of this group is tylosin shown in Table 7, an important veterinary antibiotic produced by *S. fradiae* (235,236). The structure was deduced by chemical methods (237,238) and nmr (225). Absolute configuration was established by spectroscopic comparison with leucomycins (239) and later confirmed by correlations based upon x-ray crystallography of biosynthetic precursors (240,241).

Relomycin (Table 7) was found in culture broths of *S. hygroscopicus* (242) and shown to be 20-dihydrotylosin by chemical and microbial reductions (242,243). Macrocin (3‴-*O*-demethyltylosin), was found as a minor factor in tylosin (244). It is the immediate biosynthetic precursor of tylosin and can be produced by a biosynthetically blocked mutant of *S. fradiae* (245–248). 6-Deoxy-2-*O*-methyl-D-allose [*921-90-4*], a neutral sugar in macrocin, is a constituent of cardiac glycosides from *Antiaris toxicaria* and named javose (**9**, R = H, R′ = CH_3) (249). 2‴-*O*-Demethylmacrocin (DOMM) obtained from another mutant of *S. fradiae* (246,250–252), is the penultimate precursor of tylosin (245–248). Other mutant strains of *S. fradiae* yield shunt metabolites of tylosin in which mycinose is altered (251,252). CP-56,064 [*88378-53-4*] is an analogue of tylosin in which mycarose is replaced by amicetose (**5**, R = H, R′ = OH) (253).

The 2-deoxy sugar, mycarose, is selectively hydrolyzed from tylosin under acidic conditions to produce demycarosyltylosin known as desmycosin (236). Although analogous hydrolyses have been performed in the leucomycin–spiramycin group, desmycosin possesses greater antibiotic activity than demycarosyl compounds from the other group. Hydrolyses of mycarose from macrocin and 3‴-*O*-demethylmacrocin produce lactenocin and 2″-*O*-demethyllactenocin (DOML), respectively (246,250). Desmycosin was also produced by mutant strains of *S. fradiae* possessing a biosynthetic block in synthesis and/or attachment of mycarose (246). Other mutants of *S. fradiae* which are blocked in oxidation of the C-23 methyl group, attachment of a neutral sugar onto the C-23 hydroxyl group, or oxidation of the C-20 methyl group provide a convenient source of tylosin-related

Table 7. Tylosin and Related Compounds

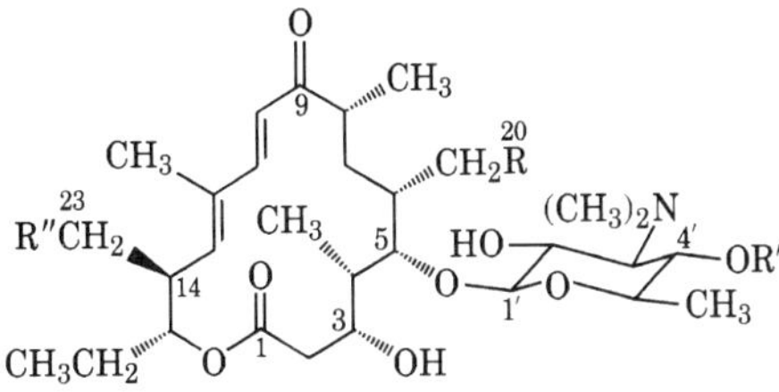

(**47**)

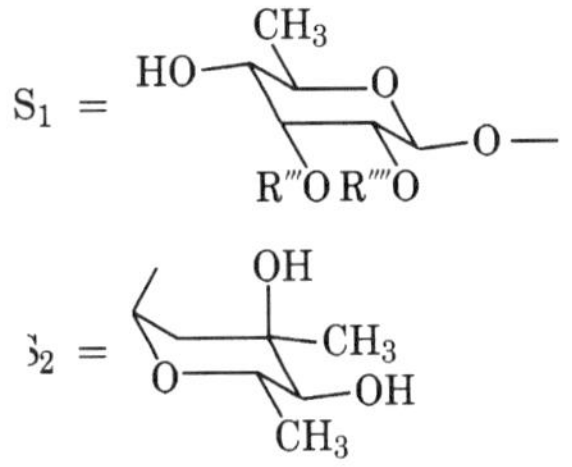

Macrolide	CAS Registry Number	Molecular formula	R	R′	R″	R‴	R⁗
tylosin	[*1401-69-0*]	$C_{46}H_{77}NO_{17}$	CHO	S_2	S_1	CH_3	CH_3
relomycin (20-dihydrotylosin)	[*62902-06-1*]	$C_{46}H_{79}NO_{17}$	CH_2OH	S_2	S_1	CH_3	CH_3
macrocin	[*11049-15-3*]	$C_{45}H_{75}NO_{17}$	CHO	S_2	S_1	H	CH_3
O-demethylmacrocin	[*79404-98-1*]	$C_{44}H_{73}NO_{17}$	CHO	S_2	S_1	H	H
desmycosin	[*11032-98-7*]	$C_{39}H_{65}NO_{14}$	CHO	H	S_1	CH_3	CH_3
lactenocin	[*110469-05-1*]	$C_{38}H_{63}NO_{14}$	CHO	H	S_1	H	CH_3
O-demethyllactenocin (DOML)	[*81557-35-9*]	$C_{37}H_{61}NO_{14}$	CHO	H	S_1	H	H
23-*O*-demycinosyltylosin	[*79592-92-0*]	$C_{38}H_{63}NO_{13}$	CHO	S_2	OH		
23-(demycinosyloxy)tylosin	[*79404-97-0*]	$C_{38}H_{63}NO_{12}$	CHO	S_2	H		
GS-77-1	[*86293-46-1*]	$C_{38}H_{65}NO_{11}$	CH_3	S_2	H		
GS-77-3	[*81661-90-7*]	$C_{38}H_{65}NO_{12}$	CH_3	S_2	OH		

macrolides in which mycarose is present and mycinose is absent (246,250,251). Some of these are shown in Table 7.

Other macrolides in this group have a 12,13-epoxide. Angolamycin (**48**, S = α-L-mycarosyl), produced by *S. eurythermus* (254) and identical to shincomycin A from *S. flavochromogenes* (255), contains the 2-deoxy aminosugar, angolosamine (**1**, R = H, R′ = OH) (256,257). Its structure is given in Table 8. Demycarosylangolamycin was isolated from cultures of *Streptomyces* sp. AS-NG-16 and named staphcoccomycin (**48**, S = H) (258). CP-56,063 [*88378-52-3*] and its 20-dihydro derivative CP-56,678, contain amicetose (**5**, R = H, R′ = OH) as the terminal neutral sugar (253). Acumycin (**49**, S = α-L-cinerulosyl) was isolated from culture broths of *S. griseoflavus* (259) and shown by x-ray crystallography to contain cinerulose (**5**, R = O) as the terminal sugar (260). This same compound was isolated from culture broths of *S. fradiae* var. *acinicolor* and designated cirramycin B or B-58941 (261–263). A6888C, produced by *S. flocculus,* contains dihydrocinerulose; its 20-dihydro derivative, A6888X, was also isolated (264). Cirramycin F-1, identical to A6888C, established that the neutral sugar was L-rhodinose (**5**, R = OH, R′ = H) (265). Cirramycin F-2 possesses the 4-epimeric sugar, L-amicetose (**5**, R = H, R′ = OH) (265). M119-a (**49**, S = L-cladinosyl) produced by an alkalophilic actinomycete, differs from the others in the terminal neutral sugar (266).

Table 8. Angolamycin and Related Compounds

Macrolide	CAS Registry Number	Molecular formula	Structure number	Structure
angolamycin (shincomycin A)	*[1402-83-1]*	$C_{46}H_{77}NO_{17}$	(**48**, S = α-L-mycarosyl)	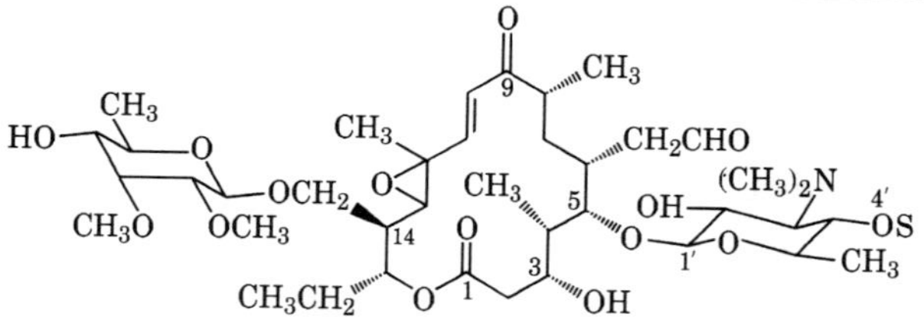
staphcoccomycin	*[73047-31-1]*	$C_{39}H_{65}NO_{14}$	(**48**, S = H)	
acumycin (cirramycin B)	*[25999-30-8]*	$C_{37}H_{59}NO_{12}$	(**49**, S = α-L-cinerulosyl)	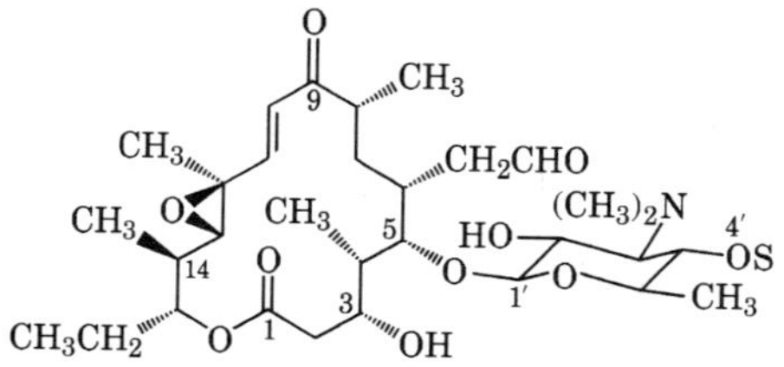
cirramycin F-1 (A6888C)	*[74918-31-3]*	$C_{37}H_{61}NO_{12}$	(**49**, S = α-L-rhodinosyl)	
cirramycin F-2	*[120851-46-9]*	$C_{37}H_{61}NO_{12}$	(**49**, S = α-L-amicetosyl)	
M119-a	*[111205-12-0]*	$C_{39}H_{65}NO_{13}$	(**49**, S = α-L-cladinosyl)	

Many 16-membered macrolides possess a tylosin-type aglycone with only one saccharide, an aminosugar that is either mycaminose or desosamine. These compounds, given in Table 9, also differ in their degree of oxidation at C-20, C-23, and C-9 to C-13 (dienone or epoxyenone). The first macrolide found in this group was cirramycin A_1 (**50**, R = CHO, R′ = OH, R″ = H), obtained from a complex of

Table 9. Cirramycin A_1, Rosaramicin, and Related Compounds

Macrolide	CAS Registry Number	Molecular formula	R	R′	R″
		C-9 to C-13, epoxyenone			
		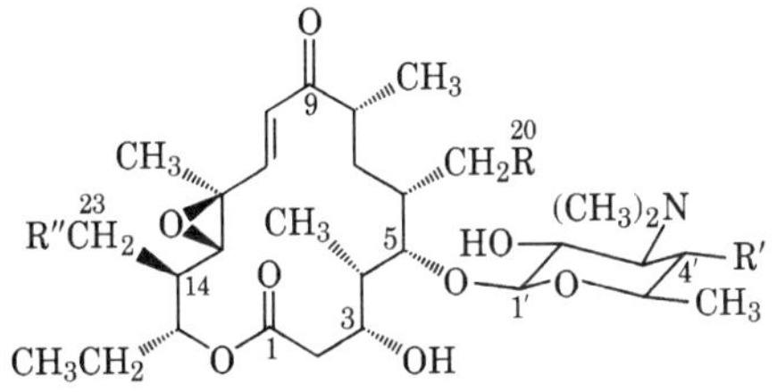 (**50**)			
cirramycin A_1	*[25339-90-6]*	$C_{31}H_{51}NO_{10}$	CHO	OH	H
rosaramicin	*[35834-26-5]*	$C_{31}H_{51}NO_9$	CHO	H	H
izenamicin A_3	*[86132-03-8]*	$C_{31}H_{51}NO_{10}$	CHO	H	OH
juvenimicin A_4	*[61475-37-4]*	$C_{31}H_{53}NO_9$	CH_2OH	H	H
M-4365 A_1	*[59227-84-8]*	$C_{31}H_{53}NO_8$	CH_3	H	H
juvenimicin A_2	*[61417-47-8]*	$C_{30}H_{51}NO_8$	H	H	H
		C-9 to C-13, dienone			
		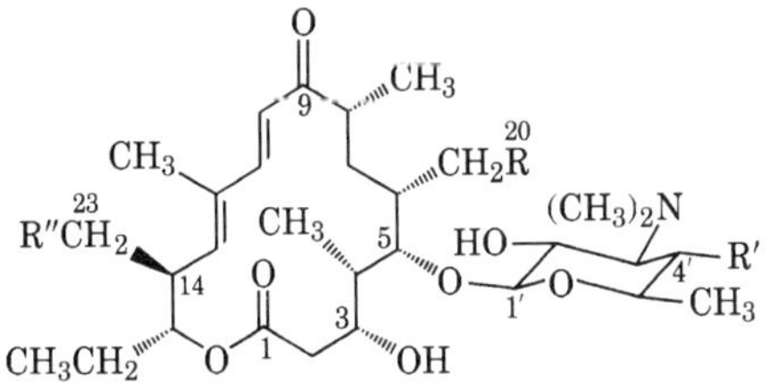 (**51**)			
5-*O*-mycaminosyl-tylonolide (OMT)	*[61257-02-1]*	$C_{31}H_{51}NO_{10}$	CHO	OH	OH
23-deoxy-5-*O*-mycaminosyltylonolide (DOMT)	*[115033-64-2]*	$C_{31}H_{51}NO_9$	CHO	OH	H
izenamicin B_3	*[80240-61-5]*	$C_{31}H_{51}NO_9$	CHO	H	OH
M-4365 G_2	*[56689-42-0]*	$C_{31}H_{51}NO_8$	CHO	H	H
juvenimicin B_3	*[61417-48-9]*	$C_{31}H_{53}NO_9$	CH_2OH	H	OH
juvenimicin B_1	*[58947-83-4]*	$C_{31}H_{53}NO_8$	CH_2OH	H	H
GS-77-4	*[80830-18-8]*	$C_{31}H_{53}NO_9$	CH_3	OH	OH
GS-77-2	*[80830-17-7]*	$C_{31}H_{53}NO_8$	CH_3	OH	H
M-4365 G_1	*[59227-83-7]*	$C_{31}H_{53}NO_7$	CH_3	H	H
izenamicin B_2	*[86131-94-4]*	$C_{30}H_{51}NO_8$	H	H	OH

several factors produced by *S. cirratus* (267–269). It was later prepared by hydrolysis of cinerulose from cirramycin B (**49**, S = α-L-cinerulosyl) (261). Rosaramicin, formerly rosamicin, was produced by *M. rosaria* and shown to be 4′-deoxycirramycin A_1 (270,271). Its x-ray crystal structure has been published (272). Structures of four related compounds from the juvenimicin complex of eight factors, A_1–A_4 and B_1–B_4, isolated from culture broths of *M. chalcea* var. *izumensis* (273,274), have been determined. Shortly thereafter, the M-4365 complex of six factors (A_1–A_3 and G_1–G_3), produced by *M. capillata* (275,276), was reported. From the izenamicin complex of seven factors produced by a *Micromonospora* species, three were new fermentation products (277). Many compounds isolated are identical: juvenimicin A_3, M-4365 A_2, and izenamicin A_1 are the same as rosaramicin. Structures have been proven by chemical interconversions (261,277,278) and microbial transformations (279).

Hydrolysis of both neutral sugars from tylosin yields 5-*O*-mycaminosyltylonolide (OMT) (**51**, R = CHO, R′ = R″ = OH) (238). However, this compound is more conveniently obtained by hydrolysis of mycarose from the fermentation-derived demycinosyltylosin. Analogous hydrolyses of mycarose from DMOT, GS-77-3, and GS-77-1 yield, respectively, 23-deoxy-5-*O*-mycaminosyltylonolide (DOMT), 20-deoxo-5-*O*-mycaminosyltylonolide (GS-77-4), and 5-*O*-mycaminosyltylactone (GS-77-2), all given in Table 9 (246,250,280).

The mycinamicin complex, also found as AR-5 complex, was isolated from culture broths of *M. griseorubida* (281–284). These compounds are listed in Table 10. They contain a 2,3-double bond and a methyl group rather than a two-carbon substituent at C-6 of the aglycone. A hydroxyl group is present at C-14 in mycinamicins II and V. X-ray crystallography established the stereochemistry and absolute configuration of the aglycone as identical to tylosin (**47**) (284–287). Minor factors were isolated where the C-14 hydroxymethyl group is either unsubstituted or glycosidated (288), and where the aminosugar was lacking (289). These products were also found after either acid hydrolysis or Polonovski reaction of the parent mycinamicins (290,291). Mycinamicin II (**52**, R = R′ = CH_3, R″ = OH), now named miporamicin, is being investigated in pre-clinical studies (292).

A few neutral 16-membered macrolides have been isolated. The structure of chalcomycin (**54**, R = CH_3, R′ = OH), produced by *S. bikiniensis,* was established as a 16-membered lactone having two neutral sugars, D-chalcose (**7**) and D-mycinose (**9**, R = R′ = CH_3) (293–295). Its aglycone resembles that of the mycinamicins, but differs by a hydroxyl group, the stereochemistry of which is uncertain, at C-8 and a methyl group at C-15 (296–298). X-ray crystallographic data has been reported (299). The closely-related neutramycin (**54**, R = H, R′ = OH), produced by *S. rimosus* (300–302), lacks a substituent at C-6 of the aglycone. These compounds are given in Table 11. CP-70,662, produced by *S. hirsutus,* may be 8-deoxychalcomycin (303,304). CP-70,663, its *de*-desosaminyl derivative, was also obtained from this culture (303).

Also shown in Table 11 are the aldgamycins, produced by *S. lavendulae,* containing a novel bicyclic sugar, aldgarose (**4**) (305). Structures of aldgamycin F (**55**, R = OH) and E (**56**) were established by chemical degradation and mass spectrometry and differ by the presence or absence, respectively, of a hydroxyl group at C-8 and a double bond at C-10,11 of the aglycone (306–308). 8-Deoxyaldgamycin F (**55**, R = H), was named aldgamycin G, from *S. avidinii,* or

Table 10. The Mycinamicins

Macrolide	CAS Registry Number	Molecular formula	R	R′	R″
		C-9 to C-13, epoxyenone			
		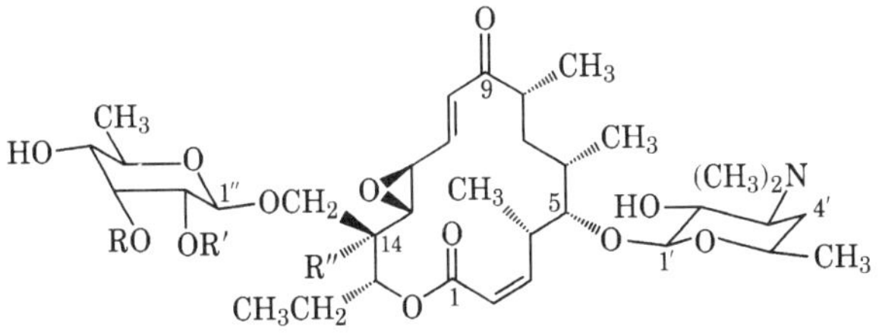 (**52**)			
mycinamicin I	[*73665-15-3*]	$C_{37}H_{61}NO_{12}$	CH_3	CH_3	H
mycinamicin II (miporamicin)	[*73684-69-2*]	$C_{37}H_{61}NO_{13}$	CH_3	CH_3	OH
		C-9 to C-13, dienone			
		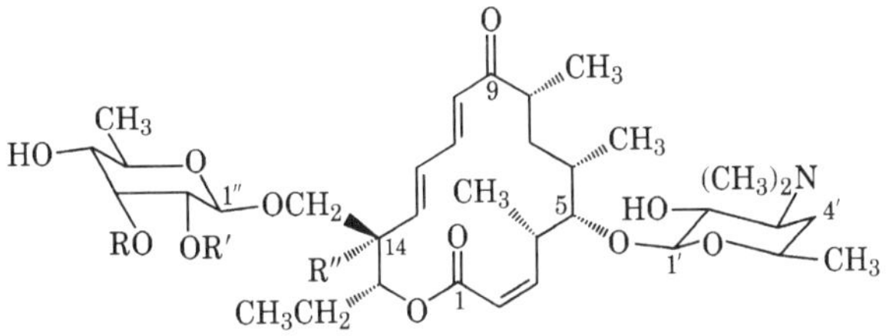 (**53**)			
mycinamicin III	[*73684-70-5*]	$C_{36}H_{59}NO_{11}$	H	CH_3	H
mycinamicin IV	[*73684-71-6*]	$C_{37}H_{61}NO_{11}$	CH_3	CH_3	H
mycinamicin V	[*73684-72-7*]	$C_{37}H_{61}NO_{12}$	CH_3	CH_3	OH
mycinamicin VI	[*85687-36-1*]	$C_{35}H_{57}NO_{11}$	H	H	H
mycinamicin VII[a]	[*77704-58-6*]	$C_{29}H_{47}NO_7$			H

[a]Mycinamicin VII has a hydrogen where the sugar is at the C-14 position.

CP-70,661, from *S. hirsutus* (303,309). Its *des*-epoxy analogue was isolated from culture broths of *S. anandii* subsp. *swalpus* and named swalpamycin (**57**) (310,311).

As is the case for the leucomycin group, several aglycones exist in the tylosin group which differ in the pattern of oxidation and substitution of the lactone. Some of these are listed in Table 12. If the aglycone contains an aldehyde, cleavage of the aminosugar yields the hemiketal, such as tylonolide (**58**) and rosaranolide (**60**) (231). The aldehyde-reduced aglycone (**59**), is a biosynthetic intermediate to these macrolides (240,246); it was synthesized or produced from biosynthetically blocked mutants of *S. fradiae* (246,312). Four possible mycinolides and corresponding protomycinolides (deoxymycinolides) are exemplified by mycinolide IV (**61**, R = OH) and protomycinolide IV (**61**, R = H) (313).

Table 11. Chalcomycin, Aldgamycin, and Related Compounds

Macrolide	CAS Registry Number	Molecular formula	Structure number	Structure
chalcomycin neutramycin CP-70,662	[*20283-48-1*] [*1404-08-6*] [*125225-83-4*]	$C_{35}H_{56}O_{14}$ $C_{34}H_{54}O_{14}$ $C_{35}H_{56}O_{13}$	(**54**, R = CH_3, R′ = OH) (**54**, R = H, R′ = OH) (**54**, R = CH_3, R′ = H)	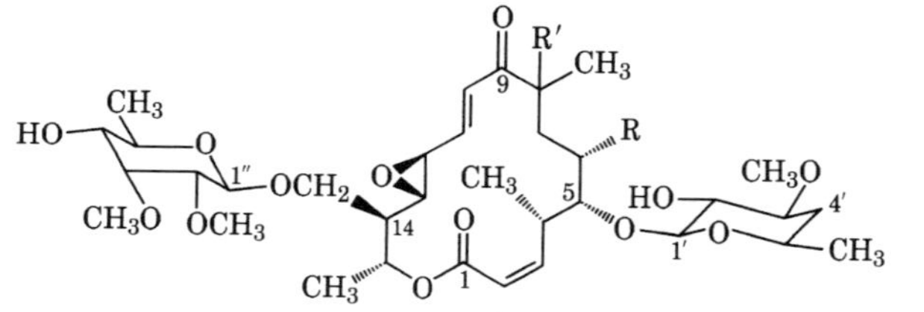
aldgamycin F aldgamycin G (8-deoxyaldgamycin F)	[*55141-41-8*] [*107745-56-2*]	$C_{37}H_{56}O_{16}$ $C_{37}H_{56}O_{15}$	(**55**, R = OH) (**55**, R = H)	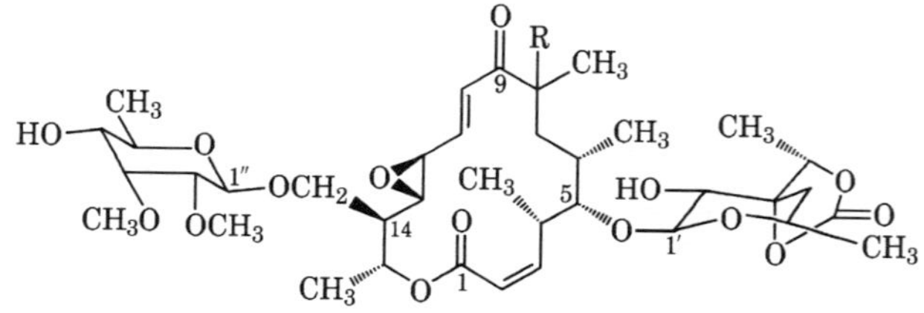
aldgamycin E	[*11011-06-6*]	$C_{37}H_{58}O_{15}$	(**56**)	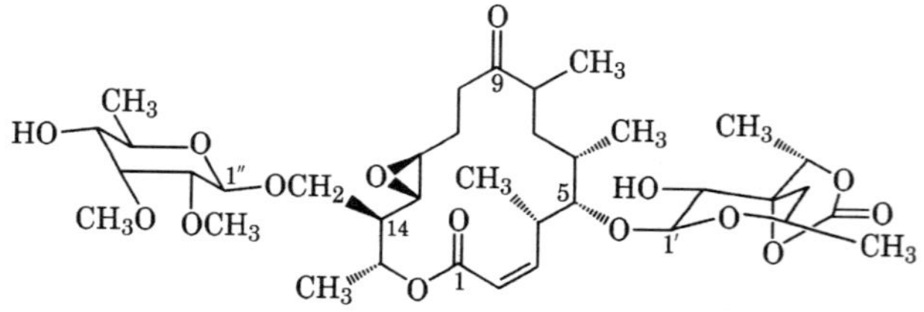
swalpamycin	[*112008-27-2*]	$C_{37}H_{56}O_{14}$	(**57**)	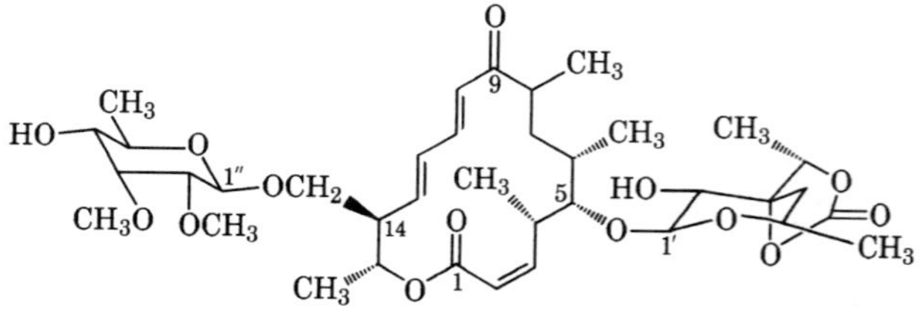

Table 12. Tylosin-Type Aglycones

Macrolide	CAS Registry Number	Molecular formula	Structure number	Structure
tylonolide	[*61219-81-6*]	$C_{23}H_{36}O_7$	(**58**)	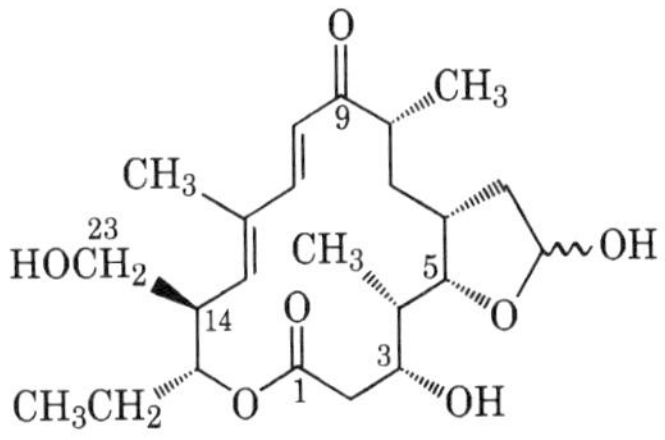
tylactone (protylonolide)	[*74758-60-4*]	$C_{23}H_{38}O_5$	(**59**)	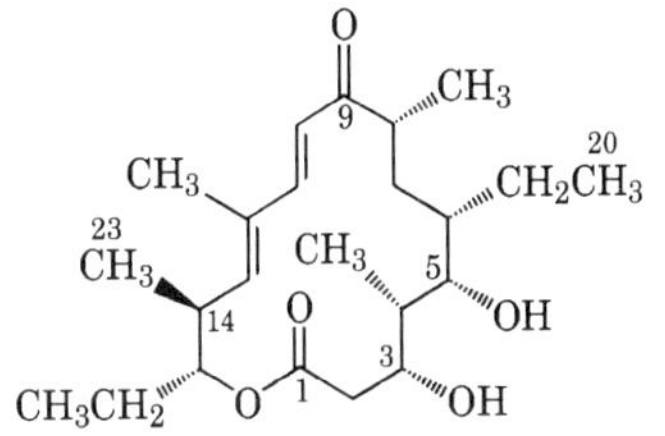
rosaranolide	[*81560-69-2*]	$C_{23}H_{36}O_7$	(**60**)	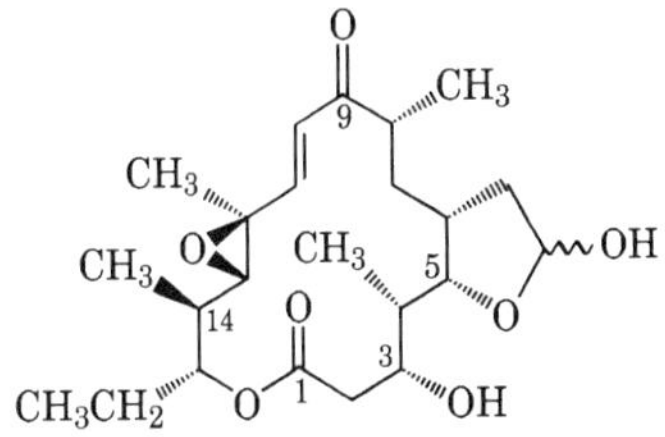
mycinolide IV protomycinolide IV	[*77704-61-1*] [*79495-87-7*]	$C_{21}H_{32}O_5$ $C_{21}H_{32}O_4$	(**61**, R = OH) (**61**, R = H)	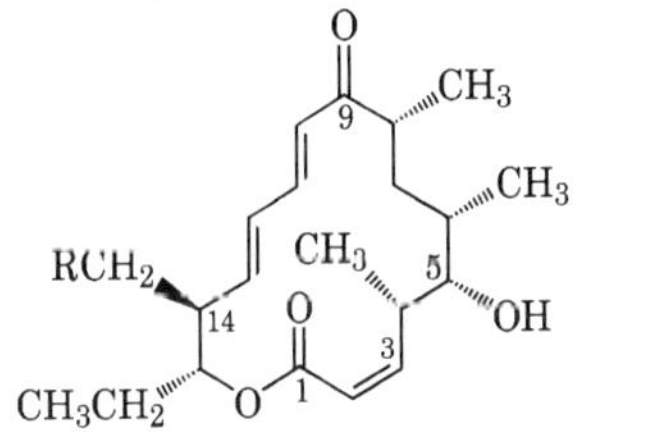

Semisynthetic Derivatives. No significant improvements in activity have resulted from modifications of the 3-, 9-, 2′- and/or 4″-hydroxyl groups (314). 3″-*O*-Acyl derivatives have not been found via fermentation, but chemical acylation of the 3″-hydroxyl group yields products having good antibiotic activity and better pharmacokinetics than the parent macrolides. Two such compounds have been developed (315,316): 3″-*O*-propionyl-leucomycin A_5 (rokitamycin) [*74014-51-0*], $C_{42}H_{69}NO_{15}$, formerly TMS-19-Q, (317,318); and 9,3″-di-*O*-acetylmidecamycin (miokamycin) [*55881-07-7*], $C_{45}H_{71}NO_{17}$, (319–321). At least part of the *in vivo* improvement was attributed to slower elimination of active metabolites from serum (322–324). These compounds are shown in Figure 2. More recently, 3″- and 4″-*O*-substituted derivatives of spiramycin were synthesized which showed similar *in vivo* improvements in activity (325–327). Rokitamycin and miokamycin have been commercially launched in Japan (116).

The enhanced activity from acylation of the 3- and 4″-hydroxyl groups of

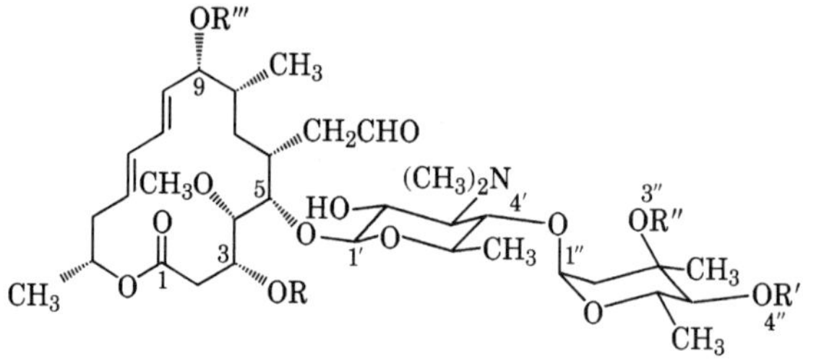

(**62**)

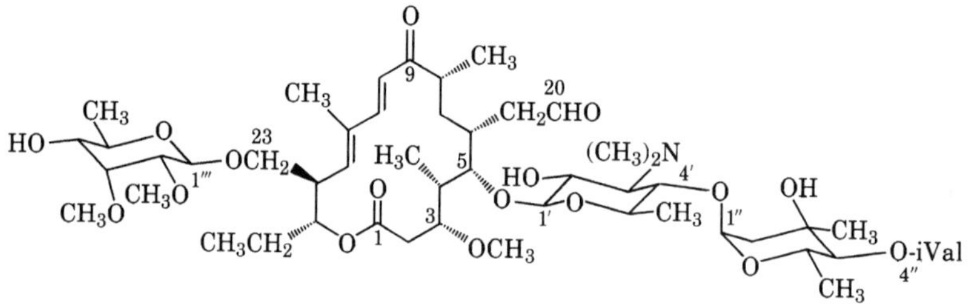

(**63**)

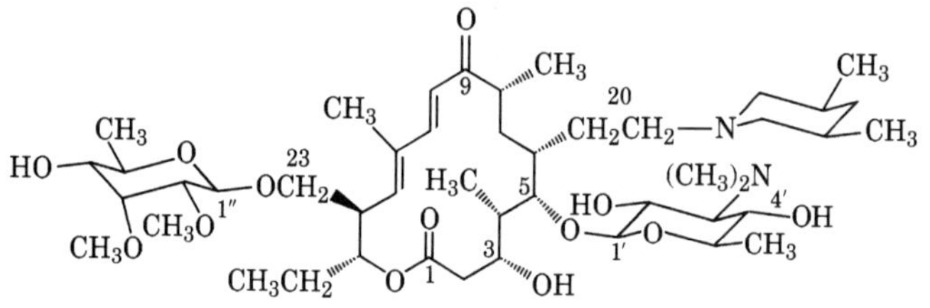

(**64**)

Fig. 2. Semisynthetic derivatives of 16-membered macrolides rokitamycin (**62**, R = R‴ = H, R′ = butyryl, R″ = propionyl); miokamycin (**62**, R = R′ = propionyl, R″ = R‴ = acetyl); AIV-tylosin (**63**); and tilmicosin (**64**).

leucomycin prompted analogous studies of tylosin. Bioconversion of tylosin by *S. thermotolerans* yielded 3- and/or 4″-*O*-acyl derivatives possessing increased activity against certain resistant microorganisms and higher concentrations of antibiotic in serum after oral administration (328–331). From this study, 3-*O*-acetyl-4″-*O*-isovaleryltylosin (AIV-tylosin) [*63409-12-1*], $C_{53}H_{87}NO_{19}$, (**63**) was developed as a new veterinary antibiotic (332,333).

Analogous bioconversions of tylosin-related macrolides have been reported (334). This work was extended by cloning the 4″-*O*-acylase gene and expressing it in *S. lividans* by bioconversion of spiramycin to 4″-*O*-isovalerylspiramycin [*129204-11-1*], $C_{48}H_{82}N_2O_{15}$, (335,336). Chemical methods for selective 4″-*O*-substitution of tylosin were also developed (337–339). Derivatives more stable to liver esterases have been prepared, such as 4″-*O*-(4-methoxyphenylacetyl)tylosin, [*103360-00-5*], $C_{55}H_{85}NO_{19}$, which is undergoing preclinical studies (338–340). Other studies have explored synthesis and structure–activity relations of 4″-*O*-acyl derivatives of tylosin-related macrolides (334,341,342). Modification of the

4″-*O*-acyl group in the deltamycins also produces derivatives having good antibiotic activity (343,344).

Another successful approach utilized the modification of the tylosin aldehyde moiety. Although reduction of tylosin reduced activity, other modifications yielded compounds having good antibiotic potency, better oral bioavailability, and higher and more prolonged concentrations of antibiotic activity in serum (345). These modifications were especially promising when applied to desmycosin (Table 7) (345–348). Similar pharmacokinetic improvements were observed for the structurally related mycinamicins (282). As a result of these advantages, mycinamicin II (miporamicin) (**52**, R = R′ = CH_3, R″ = H) and 19-deformyl-4′-deoxydesmycosin (TMC-016) [*85382-79-2*], $C_{38}H_{65}NO_{12}$, are being studied for clinical development (292,347,349). Reductive amination of the aldehyde in desmycosin yielded another useful series (350–354), from which 20-deoxo-20-(3,5-dimethylpiperidinyl)desmycosin (tilmicosin) [*108050-54-0*] (**64**), $C_{46}H_{80}N_2O_{13}$, was selected as a therapeutic agent to treat pneumonia in cattle and pigs because of its activity against *Pasteurella* species, oral bioavailability, and prolonged concentrations *in vivo* (355–357).

Derivatives of 5-*O*-mycaminosyltylonolide (**51**, R = CHO, R′ = R″ = OH) have been extensively investigated. Modifications of its 23-hydroxyl group, especially replacements by secondary amino groups, yielded derivatives having excellent activity *in vitro* but lacking good oral efficacy or bioavailability in animals (346,358–361). Clinical trials of the structurally related macrolide, rosaramicin (**50**, R = CHO, R′ = R″ = H), were unsuccessful (362). 4′-Deoxy-5-*O*-mycaminosyltylonolide and 4′-deoxydesmycosin were twice as potent as the corresponding compounds containing a 4′-hydroxyl group (363,364). Tylosin was modified by substituting or replacing mycinose (342,349,365,366) and some unusual 14,15-cyclopropyl derivatives were prepared (361,365). Modifications of the dienone of tylosin and desmycosin also yielded active derivatives (348,367–369).

Hybrid Macrolides

Other macrolides have been prepared which represent hybrids of structures within the 14-membered family, within the 16-membered family, or between the two families. These hybrids have been made by chemical, bioconversion, or genetic manipulations. The 9-*O*-[(2-methoxyethoxy)methyl]oxime of tylosin (Table 7) was synthesized, using the oxime found in roxithromycin (**33**) (369). 3-*O*-Cladinosyl derivatives of 16-membered macrolides were synthesized in which the neutral sugar of erythromycin (Table 3) was attached to tylosin derivatives at their 3-hydroxyl group, analogous to its position in erythromycin (370).

An aglycone obtained from one macrolide has been bioconverted by a microorganism producing either the same or a different macrolide; such studies have been important for production of new compounds and elucidation of biosynthetic pathways (102,279,289,291,371–378). These bioconversions were pioneered by use of cerulenin [*17397-89-6*], $C_{12}H_{17}NO_3$, an inhibitor of fatty acid synthesis, to block biosynthesis of the aglycone normally produced by the bioconverting organism (379). For example, feeding tylactone (protylonolide) (**59**) to the spiramycin-producer, *S. ambofaciens,* yielded four 9-dihydro-9-*O*-forosaminyl derivatives named chimeramycins (371). A summary of microbial conversions of macrolides has been published (149).

The advent of molecular biology has opened new possibilities for producing hybrid macrolides. Insertion of DNA from the oleandomycin-producer into a mutant strain of the erythromycin-producer produced derivatives of erythromycin lacking the 2-methyl group (380). Genetic manipulations of biosynthetic pathways in macrolide-producing microorganisms complement traditional chemical and microbiological approaches (381–386). For example, inactivation of the *Ery F* gene coding for a 6-hydroxylase in *Sa. erythraea* produced 6-deoxyerythromycin (385).

Biological Properties

Antimicrobial Properties. Macrolides inhibit growth of gram-positive bacteria, *Mycoplasma* species, and certain gram-negative and anaerobic bacteria. Susceptible gram-positive bacteria include many species of *Staphylococcus* and *Streptococcus;* susceptible gram-negative bacteria include *Bordetella pertussis, Legionella pneumophila, Moraxella catarrhalis* (formerly *Branhamella*), and *Haemophilus ducreyi.* Although erythromycin has some *in vitro* activity against *Haemophilus influenzae,* it does not exhibit a high level of *in vivo* efficacy. An important susceptible anaerobe is *Propionibacterium acnes.* Comparative evaluations have been published (387–392), and monographs devoted to particular macrolides and reviews are available (2–14).

Azithromycin (**36**) expanded the traditional *in vitro* spectrum because of increased activity against gram-negative bacteria, including *H. influenzae* (393). *In vivo* efficacy against this organism was demonstrated in a middle ear infection in gerbils (394); however, azithromycin failed to eradicate *H. influenzae* in an initial study of chronic bronchitis (395). Miokamycin (**62**), rokitamycin (**62**), clarithromycin (**37**), and roxithromycin (**33**) showed increased activity *in vitro* against *L. pneumophila* (387,388); the latter two were more efficacious than erythromycin against *Legionella* in gerbils (396–398). *Helicobacter pylori* (formerly *Campylobacter*) is very susceptible to macrolides *in vitro,* but treatment of gastric disorders with macrolides has not been established (399,400). Spiramycin (Table 6) is used to treat *Toxoplasma gondii,* but more effective agents are being sought (401). Some macrolides were active at high doses against acute toxoplasmosis or toxoplasmic encephalitis in mice (402–404), and new methods for *in vitro* testing have been reported (405). Azithromycin (**36**) was active *in vivo* against *Borrelia burgdorferi,* causative agent of Lyme disease (406,407). *In vitro* inhibition of *Chlamydia pneumoniae* (formerly *Chlamydia* TWAR) (408) and several species of mycobacteria, including *M. avium,* have been reported (409–414). These examples represent efforts to expand the spectrum of macrolides and establish clinical efficacy against organisms susceptible *in vitro.*

Macrolides inhibit growth of bacteria by inhibiting protein synthesis on ribosomes (17,415,416). Bacterial resistance to macrolides is often accompanied by cross-resistance to lincosamide and streptogramin B antibiotics (MLS-resistance), which can be either inducible or constitutive (417). 14-Membered macrolides generally induce resistance to themselves, whereas 16-membered macrolides do not; consequently, one advantage of the latter is their activity against bacteria which are inducibly resistant to erythromycin. Both 14- and 16-membered macrolides lack activity against constitutively resistant strains (387,388).

Bacterial resistance to antibiotics usually results from modification of a target site, enzymatic inactivation, or reduced uptake into bacterial cells. Modification of the macrolide's ribosomal binding site involves methylation of the 23S fraction of ribosomal RNA (418). Modifications of 23S rRNA are under investigation (419–421). Derivatives of erythromycin have been found which do not induce resistance and also inhibit strains of both inducibly and constitutively resistant bacteria (157–160). Enzymatic inactivations of macrolides include esterases (422), phosphorylases (423,424), and glycosidases (425), and reduced uptake into cells has been reported (426,427). The possible origin of bacterial inactivating enzymes from macrolide-producing organisms has been proposed (422).

Pharmacokinetics and Pharmacology. Older macrolides such as erythromycin (Table 7) exhibit relatively low serum concentrations, short *in vivo* half-lives, highly variable oral absorption, and low oral bioavailability. Improvements in these pharmacokinetic parameters have been accomplished for newer derivatives. Pharmacokinetic parameters for individual compounds are available (2–14,19,20,428–430). A beneficial feature of macrolides is the high concentrations in certain tissues such as pulmonary and prostate that far exceed serum concentrations (430). Another capability of these agents is intracellular penetration, a necessary condition for activity against intracellular pathogens. Cellular uptake, intracellular antibacterial activity, and interactions of macrolides with components of the host defense system are being investigated (431–443).

The principal route of macrolide excretion is by way of the liver. Effects of macrolides on hepatic metabolic enzymes, particularly cytochrome P-450, have been studied in order to identify and reduce potential interference with metabolism of other drugs (21–23,444–447). Several macrolides are initially metabolized to products which retain antibiotic activity, thereby contributing to more persistent *in vivo* concentrations of antibiotic. Clarithromycin (**37**) is oxidized to its 14-hydroxy derivative; the 14(*R*) isomer has activity comparable to clarithromycin (448) and is readily prepared by biotransformation using *Mucor circinelloides* (449). Dirithromycin (**35**) is a pro-drug converted to erythromycylamine (**34**) *in vivo* (450). Miokamycin and rokitamycin (**62**) undergo deacylations to other compounds in the leucomycin group (322–324,451,452).

The principal side effects of macrolides are gastrointestinal problems, such as pain, indigestion, diarrhea, nausea, and vomiting (453). Erythromycin (**14**) and oleandomycin (**17**, R = CH_3) stimulate motility of the gastrointestinal tract in dogs, whereas 16-membered macrolides do not (454,455). This model has been used to select derivatives of erythromycin having reduced potential for causing gastrointestinal upset. More recently, erythromycin was shown to exert a variety of effects on the human gastrointestinal tract, including induction of migrating motor complex and increased lower esophageal sphincter pressure and gastric emptying (456,457). At least part of its mechanism of action was proposed as an agonist of motilin, an endogenous peptide in the gastrointestinal tract responsible for interdigestive contractility (458–460). The dog contractility model was also used to select and evaluate macrolide derivatives for the treatment of gastrointestinal motility disorders (461–463).

Biosynthetic Patterns and Conformational Analysis

Macrolides are obtained by controlled submerged aerobic fermentations of soil microorganisms. Although species of *Streptomyces* have dominated, species of *Saccharopolyspora, Micromonospora,* and *Streptoverticillium* are also well represented. New techniques such as enzyme-linked immunosorbent assay (ELISA) based assays may prove beneficial for discovering new structures (464).

The biosynthetic steps which convert the initially produced aglycone to final product have been elucidated by bioconversions, blocked mutants, and radiolabeled precursors (465,466). The regular pattern of methyl substituents on alternating carbon atoms (propionate or Gerzon's rule) results from methyl groups of propionate (27,467). Many studies have established that aglycones are assembled from acetate and propionate units, with an occasional butyrate, and sugars are derived from glucose (28,465,466). Enzymes involved in these processes have been reviewed (383,468), and factors regulating production have been summarized (469). Biosynthesis of aglycones and fatty acids have some common features (383,470); this analogy was exploited by blocking aglycone biosynthesis with cerulenin, an inhibitor of fatty acid biosynthesis (379). A common pattern found in the stereochemistry of substituents on aglycones (Celmer's model or rules) has been used to predict stereochemistry and identify incorrect assignments (471,472). The sequence of biosynthetic intermediates leading to aglycones (383,473–477) and deoxysugars (478,479) is under investigation.

Conformational analysis has been used to find and predict conformations which maximize antibiotic activity, using x-ray crystal structures coupled with nmr and cd spectra. An early approach utilized the Dale diamond lattice conformational model (480), which was extended to other diamond lattice models (472,481–483). Other studies have relied on nmr data (225,484–491). However, extensive correlations between conformation and biological activity have not been successful (486,492).

Economic Aspects

Macrolide antibiotics are used clinically to treat infections resulting from susceptible organisms in the upper and lower respiratory tract, skin and soft tissues, and genital tract. They are generally used orally, although they can be given intravenously. For the latter purpose, a water-soluble salt is used, because macrolides are poorly soluble in water as free bases. Commonly employed acid-addition salts include the lactobionate, gluceptate, tartrate, and phosphate. For oral administration, acid-stable esters, salts, formulations, and coatings are necessary for erythromycin, which is unstable as its unprotected free base under acidic conditions such as those in the stomach. They are not administered by intramuscular injection because of severe pain.

Reported worldwide sales of macrolides are estimated as $800,000,000 annually; approximately 25% of this figure are U.S. sales. Table 13 provides a partial list of products commercially available for clinical use (493–495).

Macrolides are regarded as among the safest of antibiotics (496). The principal side effects, which in some cases are sufficiently severe to require cessation of the drug, are gastrointestinal.

Table 13. Commercial Macrolide Products[a]

Macrolide	Trade name[b]	Clinical route	Manufacturer
erythromycin[c]	Ery-tab	oral	Abbott
erythromycin[d]	Erythromycin base filmtab	oral	Abbott
erythromycin[e]	Erythromycin delayed-release capsules	oral	Abbott
erythromycin[f]	E-Mycin	oral	Boots
erythromycin[c]	Ilotycin	oral	Dista
erythromycin[e]	Eryc	oral	Parke-Davis
erythromycin topical solution	Eryderm 2%	topical	Abbott
erythromycin topical solution	ETS-2%	topical	Paddock
erythromycin ophthalmic ointment	Ilotycin ointment	topical	Dista
erythromycin stearate	Erythrocin stearate	oral	Abbott
erythromycin lactobionate	Erythrocin lactobionate-iv	iv	Abbott
	Erythrocin piggyback	iv	Abbott
erythromycin gluceptate	Ilotycin gluceptate	iv	Dista
erythromycin ethyl succinate	E.E.S.	oral	Abbott
	Eryped	oral	Abbott
erythromycin estolate	Ilosone	oral	Dista
erythromycin acistrate	Erasis (Finland)	oral	Orion
erythromycin stinoprate	Erythrocist (Italy)	oral	Refarmed
propionylerythromycin mercaptosuccinate	Zalig (Italy)	oral	Pierrel
erythromycin-11,12-carbonate	Davercin (Poland)	oral	Tarchomin
roxithromycin	Rulid (France)	oral	Roussel-Uclaf
clarithromycin	Klaricid (Ireland)	oral	Abbott
azithromycin	Sunamed (Yugoslavia)	oral	Pliva
triacetyloleandomycin	TAO	oral	Roerig
josamycin	Josamycin (Japan)	oral	Yamanouchi
miokamycin	Miocamycin (Japan)	oral	Meiji Seika
rokitamycin	Ricamycin (Japan)	oral	Toyo Jozo

[a]Abstracted from refs. 116, 495; updates of 116 and 495, and 494.
[b]Country in parenthesis represents launch outside the United States.
[c]Enteric-coated tablets.
[d]Film-coated tablets.
[e]Enteric-coated pellets.
[f]Coated tablets.

Relatively few macrolides are used in veterinary medicine (497). The most important is tylosin (Tylan, Elanco Products), which is used to control chronic respiratory disease caused by *Mycoplasma gallisepticum* in poultry. It is also used to treat and control infections in pigs resulting from gram-positive bacteria and *Mycoplasma,* and as a growth promotant for pigs and poultry. Other macrolides in veterinary use include erythromycin, oleandomycin, and spiramycin (497). Newer macrolides being developed for veterinary applications are 3-*O*-acetyl-4″-*O*-isovaleryltylosin and tilmicosin. The pharmacokinetics of macrolides in animals have been reviewed (428,429), and a book devoted to pharmacokinetics of veterinary antibiotics is scheduled (498).

BIBLIOGRAPHY

"Antibiotics" in *ECT* 1st ed., Vol. 2, pp. 7–37, by J. S. Kiser and H. B. Woodruff; in *ECT* 2nd ed., Vol. 2, pp. 533–540, by W. E. Brown, The Squibb Institute for Medical Research; "Macrolide Antibiotics" in *ECT* 2nd ed., Vol. 12, pp. 632–660, by R. Morin and M. Gorman, Eli Lilly and Company; "Antibiotics, Macrolides" in *ECT* 3rd ed., Vol. 2, pp. 937–957, by A. K. Mallams, Schering-Plough Corporation.

1. S. Omura, ed., *Macrolide Antibiotics: Chemistry, Biology, and Practice,* Academic Press, Orlando, Fla., 1984.
2. D. C. Brittain, *Med. Clin. North Am.* **71,** 1147 (1987).
3. J. D. Nelson, ed., *Pediatr. Infect. Dis.* **5,** 118 (1986).
4. J. A. Washington and W. R. Wilson, *Mayo Clin. Proc.* **60,** 189 (1985).
5. *Ibid.,* p. 271.
6. P. Davey, J.-C. Pechere, and D. Speller, eds., *J. Antimicrob. Chemother.* **22**(Suppl. B) (1988).
7. H. C. Neu and J. F. Acar, eds., *Brit. J. Clin. Pract.* **42**(Suppl. 55) (1988).
8. I. Phillips, J.-C. Pechere, A. Davies, and D. Speller, eds., *J. Antimicrob. Chemother.* **20**(Suppl. B) (1987).
9. J. P. Butzler, ed., *Macrolides: A Review with an Outlook on Future Developments,* Excerpta Medica, Amsterdam, The Netherlands, 1986.
10. D. A. Leigh and G. L. Ridgway, eds., *J. Antimicrob. Chemother.* **25**(Suppl. A) (1990).
11. P. Davey and R. Williams, eds., *J. Antimicrob. Chemother.* **21**(Suppl. D) (1988).
12. H. A. Kirst, *Ann. Rep. Med. Chem.* **25,** Chapt. 13 (1990).
13. H. A. Kirst and G. D. Sides, *Antimicrob. Agents Chemother.* **33,** 1413 (1989).
14. *Ibid.,* p. 1419.
15. Y. Fukagawa, K. Kiyoshima, and T. Yoshioka, *Life Chemistry Reports* **6,** 267 (1988).
16. H. A. Kirst, in G. Lukacs and M. Ohno, eds., *Recent Progress in the Chemical Synthesis of Antibiotics,* Springer-Verlag, Heidelberg, Germany, 1990, p. 39.
17. I. Phillips and J. D. Williams, eds., *J. Antimicrob. Chemother.* **16**(Suppl. A) (1985).
18. R. W. Auckenthaler, A. Zwahlen, and F. A. Waldvogel, in P. K. Peterson and J. Verhoef, eds., *Antimicrobial Agents Annual 3,* Elsevier Science Publishers, Amsterdam, The Netherlands, 1988, p. 122.
19. P. Periti, T. Mazzei, E. Mini, and A. Novelli, *Clin. Pharmacokinetics* **16,** 193 (1989).
20. *Ibid.,* p. 261.
21. T. M. Ludden, *Clin. Pharmacokinetics* **10,** 63 (1985).
22. J. Descotes, P. Andre, and J. C. Evreux, *J. Antimicrob. Chemother.* **15,** 659 (1985).
23. H. C. Neu, in J. C. Petrie, ed., *Gastrointestinal, Haematological and Infectious Disease Therapy,* Elsevier Science Publishers, Amsterdam, The Netherlands, 1985, p. 215.
24. K. Tatsuta in Ref. 16, p. 1.
25. R. K. Boeckman and S. W. Goldstein, in J. ApSimon, ed., *The Total Synthesis of Natural Products,* Vol. 7, John Wiley & Sons, Inc., New York, 1988, p. 1.
26. I. Paterson and M. M. Mansuri, *Tetrahedron* **41,** 3569 (1985).
27. R. B. Woodward, *Angew. Chem.* **69,** 50 (1957).
28. D. O'Hagan, *Natl. Prod. Reports* **6,** 205 (1989).
29. M. N. Donin, J. Pagano, J. D. Dutcher, and C. M. McKee, *Antibiot. Annu. 1953–1954,* 179 (1953).
30. C. Djerassi and J. A. Zderic, *J. Am. Chem. Soc.* **78,** 2907 (1956).
31. *Ibid.,* p. 6390.
32. D. G. Manwaring, R. W. Rickards, and R. M. Smith, *Tetrahedron Lett.,* 1029 (1970).
33. S. Masamune, H. Yamamoto, S. Kamata, and A. Fukuzawa, *J. Am. Chem. Soc.* **97,** 3513 (1975).
34. C. Djerassi and O. Halpern, *J. Am. Chem. Soc.* **79,** 2022 (1957).

35. C. Djerassi and O. Halpern, *Tetrahedron* **3,** 255 (1958).
36. I. Maezawa, A. Kinumaki, and M. Suzuki, *J. Antibiot.* **27,** 84 (1974).
37. J. Inanaga, Y. Kawanami, and M. Yamaguchi, *Chem. Lett.,* 1415 (1981).
38. A. Kinumaki and M. Suzuki, *J. Chem. Soc., Chem. Commun.,* 744 (1972).
39. J. M. McGuire and co-workers, *Antibiot. Chemother.* **2,** 281 (1952).
40. D. P. Labeda, *Int. J. Syst. Bacteriol.* **37,** 19 (1987).
41. P. F. Wiley and co-workers, *J. Am. Chem. Soc.* **79,** 6062 (1957).
42. D. R. Harris, S. G. McGeachin, and H. H. Mills, *Tetrahedron Lett.,* 679 (1965).
43. R. B. Woodward and co-workers, *J. Am. Chem. Soc.* **103,** 3215 (1981).
44. R. B. Woodward, in A. Todd, ed., *Perspectives in Organic Chemistry,* Interscience Publishers, New York, 1956, p. 160.
45. P. F. Wiley and co-workers, *J. Am. Chem. Soc.* **79,** 6070 (1957).
46. P. F. Wiley, R. Gale, C. W. Pettinga, and K. Gerzon, *J. Am. Chem. Soc.* **79,** 6074 (1957).
47. J. Majer, J. R. Martin, R. S. Egan, and J. W. Corcoran, *J. Am. Chem. Soc.* **99,** 1620 (1977).
48. J. Tone and co-workers, *Proc. 24th Intersci. Conf. Antimicrob. Agents Chemother.,* Oct. 8–10, 1984, Washington, D.C., Abstr. No. 796.
49. J. R. Martin and co-workers, *J. Antibiot.* **35,** 426 (1982).
50. Th. Cachet, E. Roets, J. Hoogmartens, and H. Vanderhaeghe, *J. Chromatogr.* **403,** 343 (1987).
51. J. R. Martin, R. S. Egan, A. W. Goldstein, and P. Collum, *Tetrahedron* **31,** 1985 (1975).
52. Th. Cachet and co-workers, *J. Chromatogr.* **445,** 290 (1988).
53. J. R. Martin, T. J. Perun, and R. L. Girolami, *Biochemistry* **5,** 2852 (1966).
54. J. R. Martin and W. Rosenbrook, *Biochemistry* **6,** 435 (1967).
55. J. R. Martin and T. J. Perun, *Biochemistry* **7,** 1728 (1968).
56. J. R. Martin and R. S. Egan, *Biochemistry* **9,** 3439 (1970).
57. J. R. Martin, T. J. Perun, and R. S. Egan, *Tetrahedron* **28,** 2937 (1972).
58. J. R. Martin, R. S. Egan, T. J. Perun, and A. W. Goldstein, *Tetrahedron* **29,** 935 (1973).
59. J. R. Martin and co-workers, *J. Antibiot.* **27,** 570 (1974).
60. I. O. Kibwage and co-workers, *J. Antibiot.* **40,** 1 (1987).
61. I. O. Kibwage and co-workers, *J. Chromatogr.* **346,** 309 (1985).
62. Th. Cachet and co-workers, *J. Chromatogr.* **409,** 91 (1987).
63. R. A. LeMahieu and co-workers, *J. Med. Chem.* **17,** 953 (1974).
64. P. P. Hung, C. L. Marks, and P. L. Tardrew, *J. Biol. Chem.* **240,** 1322 (1965).
65. S. Masamune, G. S. Bates, and J. W. Corcoran, *Angew. Chem. Int. Eng. Ed.* **16,** 585 (1977).
66. E. J. Corey and co-workers, *J. Am. Chem. Soc.* **100,** 4620 (1978).
67. E. J. Corey and co-workers, *J. Am. Chem. Soc.* **101,** 7131 (1979).
68. S. Masamune and co-workers, *J. Am. Chem. Soc.* **103,** 1568 (1981).
69. B. A. Sobin, A. R. English, and W. D. Celmer, *Antibiot. Annu. 1954–1955,* 827 (1955).
70. W. D. Celmer, H. Els, and K. Murai, *Antibiot. Annu. 1957–1958,* 476 (1958).
71. F. A. Hochstein and co-workers, *J. Am. Chem. Soc.* **82,** 3225 (1960).
72. W. D. Celmer, *J. Am. Chem. Soc.* **87,** 1797 (1965).
73. H. Ogura, K. Furuhata, Y. Harada, and Y. Iitaka, *J. Am. Chem. Soc.* **100,** 6733 (1978).
74. W. D. Celmer, *Pure Appl. Chem.* **28,** 413 (1971).
75. K. Tatsuta, Y. Kobayashi, and H. Gunji, *J. Antibiot.* **41,** 1520 (1988).
76. K. Tatsuta, Y. Kobayashi, H. Gunji, and H. Masuda, *Tetrahedron Lett.* **29,** 3975 (1988).
77. H. Brockmann and W. Henkel, *Naturwissenschaften* **37,** 138 (1950).
78. H. Brockmann and W. Henkel, *Chem. Ber.* **84,** 284 (1951).
79. T. Hata and co-workers, *J. Antibiot. Ser. A* **8,** 9 (1955).
80. B. Takahashi, *J. Antibiot. Ser. A* **7,** 149 (1954).
81. H. Ogura, A. Otagoshi, Y. Sano, and T. Hata, *Chem. Pharm. Bull.* **15,** 682 (1967).

82. A. D. Gardner and E. Chain, *Brit. J. Exp. Pathol.* **23,** 123 (1942).
83. H. Brockmann, H. Genth, and R. Strufe, *Chem. Ber.* **85,** 426 (1952).
84. S. A. Waksman, *The Actinomycetes,* Vol. 2, Williams & Wilkins, Baltimore, Md., 1961, p. 215.
85. H. Muxfeldt, S. Shrader, P. Hansen, and H. Brockmann, *J. Am. Chem. Soc.* **90,** 4748 (1968).
86. R. W. Rickards, R. M. Smith, and J. Majer, *J. Chem. Soc., Chem. Commun.,* 1049 (1968).
87. R. E. Hughes, H. Muxfeldt, C. Tsai, and J. J. Stezowski, *J. Am. Chem. Soc.* **92,** 5267 (1970).
88. C. Tsai, J. J. Stezowski, and R. E. Hughes, *J. Am. Chem. Soc.* **93,** 7286 (1971).
89. H. Ogura, K. Furuhata, and H. Kuwano, *Tetrahedron Lett.,* 4715 (1971).
90. A. Furusaki, T. Matsumoto, K. Furuhata, and H. Ogura, *Bull. Chem. Soc. Japan* **55,** 59 (1982).
91. H. Brockmann and R. Oster, *Chem. Ber.* **90,** 605 (1957).
92. J. Majer, J. B. McAlpine, R. S. Egan, and J. W. Corcoran, *J. Antibiot.* **29,** 769 (1976).
93. R. Corbaz and co-workers, *Helv. Chim. Acta* **38,** 935 (1955).
94. V. Prelog, A. M. Gold, G. Talbot, and A. Zamojski, *Helv. Chim. Acta* **45,** 4 (1962).
95. T. Hori, I. Maezawa, N. Nagahama, and M. Suzuki, *J. Chem. Soc., Chem. Commun.,* 304 (1971).
96. I. Maezawa, T. Hori, and M. Suzuki, *Agric. Biol. Chem.* **38,** 91 (1974).
97. *Ibid.,* p. 539.
98. T. Kaiho, S. Masamune, and T. Toyoda, *J. Org. Chem.* **47,** 1612 (1982).
99. H. Ogura, K. Furuhata, H. Kuwano, and N. Harada, *J. Am. Chem. Soc.* **97,** 1930 (1975).
100. Jpn. Pat. Appl. Kokai 86-234,791 (Oct. 20, 1986).
101. *Meiji Seika Kenkyu Nenpo* (24), 52 (1985); *Chem. Abstr.* **106,** 152654 (1987).
102. I. Maezawa, A. Kinumaki, and M. Suzuki, *J. Antibiot.* **29,** 1203 (1976).
103. J. Marquez and co-workers, *J. Antibiot.* **22,** 259 (1969).
104. I. Kawamoto and co-workers, *J. Antibiot.* **27,** 493 (1974).
105. R. S. Egan and co-workers, *J. Antibiot.* **27,** 544 (1974).
106. T. T. Thang and co-workers, *J. Am. Chem. Soc.* **100,** 663 (1978).
107. P. Bartner and co-workers, *J. Chem. Soc., Perkin Trans. 1,* 1600 (1979).
108. E. Gaumann and co-workers, *Helv. Chim. Acta* **43,** 601 (1960).
109. G. Roncari and W. Keller-Schierlein, *Helv. Chim. Acta* **49,** 705 (1966).
110. R. S. Egan and J. R. Martin, *J. Am. Chem. Soc.* **92,** 4129 (1970).
111. S. Omura and co-workers, *J. Antibiot.* **22,** 500 (1969).
112. S. Omura and co-workers, *J. Antibiot.* **22,** 629 (1969).
113. J. R. Martin and co-workers, *Helv. Chim. Acta* **59,** 1886 (1976).
114. J. R. Martin and co-workers, *Helv. Chim. Acta* **60,** 2559 (1977).
115. B. Arnoux, C. Pascard, L. Raynaud, and J. Lunel, *J. Am. Chem. Soc.* **102,** 3605 (1980).
116. J. Hardcastle, ed., *Pharmaprojects (Therapeutic Categories, Main Volume),* Vol. 10, PJB Publications, Richmond, UK, 1989, p. a616, and monthly updates.
117. P. Kurath, P. H. Jones, R. S. Egan, and T. J. Perun, *Experientia* **27,** 362 (1971).
118. Th. Cachet and co-workers, *Int. J. Pharm.* **55,** 59 (1989).
119. C. Vinckier and co-workers, *Int. J. Pharm.* **55,** 67 (1989).
120. I. O. Kibwage and co-workers, *J. Org. Chem.* **52,** 990 (1987).
121. H. A. Kirst, J. A. Wind, and J. W. Paschal, *J. Org. Chem.* **52,** 4359 (1987).
122. A. A. Nagel and co-workers, *J. Org. Chem.* **51,** 5397 (1986).
123. T. Adachi, *J. Org. Chem.* **54,** 3507 (1989).
124. H. W. Murphy, *Antibiot. Annu. 1953–1954,* 500 (1953).
125. V. C. Stephens, *Antibiot. Annu. 1953–1954,* 514 (1953).
126. R. K. Clark and E. L. Varner, *Antibiot. Chemother.* **7,** 487 (1957).

127. P. L. Tardrew, J. C. H. Mao, and D. Kenney, *Appl. Microbiol.* **18,** 159 (1969).
128. L. Dall'Asta and co-workers, *J. Antibiot.* **41,** 139 (1988).
129. B. Steffansen and H. Bundgaard, *Int. J. Pharm.* **56,** 159 (1989).
130. A. Gordin and co-workers, *Antimicrob. Agents Chemother.* **32,** 1019 (1988).
131. P. Saloranta, L. Roos, E. Elonen, and H. Allonen, *J. Antimicrob. Chemother.* **24,** 455 (1989).
132. M. DeBernardi, F. Feletti, G. Gazzani, and G. B. Fregnan, *Int. J. Clin. Pharm. Ther. Tox.* **26,** 444 (1988).
133. G. Ricevuti and co-workers, *Chemotherapy (Basel)* **34,** 374 (1988).
134. E. Concia and co-workers, *J. Int. Med. Res.* **14,** 137 (1986).
135. W. D. Celmer, *Antibiot. Annu. 1958–1959,* 277 (1959).
136. A. Neszmelyi and H. Bojarska-Dahlig, *J. Antibiot.* **31,** 487 (1978).
137. P. T. Mannisto and co-workers, *Proc. 13th Int. Cong. Chemother.,* Vienna, Austria, Aug. 28–Sept. 2, 1983, Vol. 5, Part 107, p. 5.
138. W. R. Baker and co-workers, *Proc. 27th Intersci. Conf. Antimicrob. Agents Chemother.,* New York, Oct. 4–7, 1987, Abstr. No. 221.
139. W. R. Baker, J. D. Clark, R. L. Stephens, and K. H. Kim, *J. Org. Chem.* **53,** 2340 (1988).
140. H. Sakakibara and S. Omura, in Ref. 1, p. 89.
141. J.-F. Chantot, A. Bryskier, and J.-C. Gasc, *J. Antibiot.* **39,** 660 (1986).
142. E. H. Massey, B. S. Kitchell, L. D. Martin, and K. Gerzon, *J. Med. Chem.* **17,** 105 (1974).
143. P. Luger and R. Maier, *J. Cryst. Mol. Struct.* **9,** 329 (1979).
144. S. Djokic and co-workers, *J. Chem. Res. (S),* 152 (1988).
145. G. M. Bright and co-workers, *J. Antibiot.* **41,** 1029 (1988).
146. S. Morimoto and co-workers, *J. Antibiot.* **43,** 286 (1990).
147. S. Morimoto and co-workers, *J. Antibiot.* **43,** 295 (1990).
148. L. Toscano, G. Fioriello, S. Silingardi, and M. Inglesi, *Tetrahedron* **40,** 2177 (1984).
149. P. F. Wiley, *Dev. Indust. Microbiol.* **26,** 97 (1985).
150. J. Wilson, J. Durodie, and M. Foulstone, *J. Antimicrob. Chemother.* **22,** 84 (1988).
151. J. M. Wilson, P. C. T. Hannan, C. Shillingford, and D. J. C. Knowles, *J. Antibiot.* **42,** 454 (1989).
152. J. S. Davies, E. Hunt, and I. I. Zomaya, *J. Chem. Soc., Perkin Trans. 1,* 1409 (1990).
153. E. Hunt and co-workers, *J. Antibiot.* **42,** 293 (1989).
154. H. A. Kirst and co-workers, *Proc. 29th Intersci. Conf. Antimicrob. Agents Chemother.,* Sept. 17–20, 1989, Houston, Tex., Abstr. No. 1021.
155. R. N. Swanson and co-workers, in Ref. 154, Abstr. No. 1025.
156. L. A. Freiberg and co-workers, *Proc. 9th Internat. Symp. Future Trends Chemother.,* March 26–28, 1990, Palexpo (Geneva), Switzerland, p. 30.
157. N. E. Allen, *Antimicrob. Agents Chemother.* **11,** 669 (1977).
158. S. Kadam, C. C. Doran, and R. C. Goldman, *Proc. 88th Mtg. Amer. Soc. Microbiol.,* May 8–13, 1988, Miami, Fla., Abstr. No. A-11.
159. P. B. Fernandes, W. R. Baker, L. A. Freiberg, D. J. Hardy, and E. J. McDonald, *Antimicrob. Agents Chemother.* **33,** 78 (1989).
160. R. C. Goldman and S. K. Kadam, *Antimicrob. Agents Chemother.* **33,** 1058 (1989).
161. L. A. Freiberg and co-workers, in Ref. 138, Abstr. No. 224.
162. T. Hata and co-workers, *J. Antibiot.* **6,** 87 (1953).
163. T. Arai, S. Kuroda, and Y. Mikami, in T. Arai, ed., *Actinomycetes: The Boundary Microorganisms,* Toppan Co. Ltd., Tokyo, Japan, 1976, p. 632.
164. S. Omura, H. Ogura, and T. Hata, *Tetrahedron Lett.,* 609 (1967).
165. *Ibid.,* p. 1267.
166. T. Hata and co-workers, *Chem. Pharm. Bull.* **15,** 358 (1967).
167. S. Omura, M. Katagiri, and T. Hata, *J. Antibiot.* **20,** 234 (1967).

168. S. Omura, M. Katagiri, H. Ogura, and T. Hata, *Chem. Pharm. Bull.* **15,** 1529 (1967).
169. S. Omura, M. Katagiri, and T. Hata, *J. Antibiot.* **21,** 199 (1968).
170. *Ibid.,* p. 272.
171. S. Omura, M. Katagiri, H. Ogura, and T. Hata, *Chem. Pharm. Bull.* **16,** 1167 (1968).
172. *Ibid.,* p. 1181.
173. M. Hiramatsu and co-workers, *Bull. Chem. Soc. Japan* **40,** 2982 (1967).
174. M. Hiramatsu and co-workers, *Bull. Chem. Soc. Japan* **43,** 1966 (1970).
175. S. Omura and co-workers, *Chem. Pharm. Bull.* **16,** 1402 (1968).
176. S. Omura and co-workers, *Chem. Pharm. Bull.* **18,** 1501 (1970).
177. L. A. Freiberg, R. S. Egan, and W. H. Washburn, *J. Org. Chem.* **39,** 2474 (1974).
178. A. Ducruix, C. Pascard, A. Nakagawa, and S. Omura, *J. Chem. Soc., Chem. Commun.,* 947 (1976).
179. S. Omura and co-workers, *Tetrahedron* **28,** 2839 (1972).
180. T. Osono and co-workers, *J. Antibiot. Ser. A* **20,** 174 (1967).
181. S. Omura, Y. Hironaka, and T. Hata, *J. Antibiot.* **23,** 511 (1970).
182. M. Muroi, M. Izawa, and T. Kishi, *Experientia* **28,** 129 (1972).
183. M. Muroi and co-workers, *Experientia* **28,** 501 (1972).
184. M. Muroi and co-workers, *Experientia* **28,** 878 (1972).
185. H. Ono, T. Hasegawa, E. Higashide, and M. Shibata, *J. Antibiot.* **26,** 191 (1973).
186. M. Muroi and co-workers, *J. Antibiot.* **26,** 199 (1973).
187. M. Muroi, M. Izawa, and T. Kishi, *Chem. Pharm. Bull.* **24,** 450 (1976).
188. *Ibid.,* p. 463.
189. K. Kamiya, Y. Wada, and K. Kondo, *Chem. Pharm. Bull.* **26,** 1753 (1978).
190. F. W. Tanner, A. R. English, T. M. Lees, and J. B. Routien, *Antibiot. Chemother.* **2,** 441 (1952).
191. J. F. Pagano, M. J. Weinstein, and C. M. McKee, *Antibiot. Chemother.* **3,** 899 (1953).
192. R. L. Wagner and co-workers, *J. Am. Chem. Soc.* **75,** 4684 (1953).
193. M. E. Kuehne and B. W. Benson, *J. Am. Chem. Soc.* **87,** 4660 (1965).
194. R. B. Woodward, L. S. Weiler, and P. C. Dutta, *J. Am. Chem. Soc.* **87,** 4662 (1965).
195. R. Okamoto and co-workers, *J. Ferment. Technol.* **57,** 519 (1979).
196. J. K. Epp, S. G. Burgett, and B. E. Schoner, *Gene* **53,** 73 (1987).
197. Y. Shimauchi and co-workers, *J. Antibiot.* **31,** 261 (1978).
198. Y. Shimauchi and co-workers, *J. Antibiot.* **32,** 878 (1979).
199. M. Sakamoto and co-workers, *J. Antibiot.* **37,** 130 (1984).
200. F. A. Hochstein and K. Murai, *J. Am. Chem. Soc.* **76,** 5080 (1954).
201. M. Suzuki and co-workers, *Tetrahedron Lett.,* 435 (1971).
202. M. Suzuki and co-workers, *J. Antibiot.* **24,** 904 (1971).
203. T. Furumai and co-workers, *J. Antibiot.* **27,** 95 (1974).
204. A. Kinumaki and co-workers, *J. Antibiot.* **27,** 102 (1974).
205. A. Kinumaki and co-workers, *J. Antibiot.* **27,** 107 (1974).
206. A. Kinumaki and co-workers, *J. Antibiot.* **27,** 117 (1974).
207. Jpn. Pat. Appl. Kokai 73-10,288 (Feb. 2, 1973).
208. T. Niida and co-workers, *J. Antibiot.* **24,** 319 (1971).
209. T. Tsuruoka and co-workers, *J. Antibiot.* **24,** 452 (1971).
210. S. Inouye and co-workers, *J. Antibiot.* **24,** 460 (1971).
211. T. Tsuruoka and co-workers, *J. Antibiot.* **24,** 476 (1971).
212. T. Tsuruoka and co-workers, *J. Antibiot.* **24,** 526 (1971).
213. U.S. Pat. 3,819,834 (June 25, 1974).
214. Jpn. Pat. Appl. Kokai 72-25,384 (Oct. 20, 1972).
215. I. Haupt, H. Fricke, J. Cerna, and I. Rychlik, *J. Antibiot.* **29,** 1314 (1976).
216. G. Huber and co-workers, *Arzneim.-Forsch.* **12,** 1191 (1962).
217. U. Grafe and co-workers, *J. Antibiot.* **33,** 566 (1980).

218. U. Grafe and co-workers, *J. Antibiot.* **33,** 574 (1980).
219. T. Furumai and M. Suzuki, *J. Antibiot.* **28,** 775 (1975).
220. Jpn. Pat. Appl. Kokai 86-257,198 (Nov. 14, 1986).
221. S. Pinnert-Sindico, L. Ninet, J. Preud'Homme, and C. Cosar, *Antibiot. Annu. 1954–1955,* 724 (1955).
222. R. Corbaz and co-workers, *Helv. Chim. Acta* **39,** 304 (1956).
223. R. Paul and S. Tchelitcheff, *Bull. Soc. Chim. France,* 650 (1965).
224. S. Omura and co-workers, *J. Am. Chem. Soc.* **91,** 3401 (1969).
225. S. Omura and co-workers, *J. Am. Chem. Soc.* **97,** 4001 (1975).
226. Jpn. Pat. Appl. Kokai 83-74,695 (May 6, 1983).
227. R. Paul and S. Tchelitcheff, *Bull. Soc. Chim. France,* 443 (1957).
228. R. J. Adamski and co-workers, *J. Med. Chem.* **9,** 932 (1966).
229. S. Omura and co-workers, *J. Antibiot.* **27,** 147 (1974).
230. N. N. Girotra and N. L. Wendler, *Tetrahedron Lett.,* 227 (1975).
231. S. Masamune, Y. Hayase, W. K. Chan, and R. L. Sobczak, *J. Am. Chem. Soc.* **98,** 7874 (1976).
232. T. Furumai and M. Suzuki, *J. Antibiot.* **28,** 783 (1975).
233. U. Grafe and co-workers, *J. Antibiot.* **33,** 663 (1980).
234. T. Furumai, K. Takeda, and M. Suzuki, *J. Antibiot.* **28,** 789 (1975).
235. J. M. McGuire and co-workers, *Antibiot. Chemother.* **11,** 320 (1961).
236. R. L. Hamill, M. E. Haney, M. Stamper, and P. F. Wiley, *Antibiot. Chemother.* **11,** 328 (1961).
237. R. B. Morin and M. Gorman, *Tetrahedron Lett.,* 2339 (1964).
238. R. B. Morin, M. Gorman, R. L. Hamill, and P. V. Demarco, *Tetrahedron Lett.,* 4737 (1970).
239. S. Omura, A. Nakagawa, M. Machida, and H. Imai, *Tetrahedron Lett.,* 1045 (1977).
240. S. Omura and co-workers, *J. Antibiot.* **33,** 915 (1980).
241. N. D. Jones and co-workers, *J. Antibiot.* **35,** 420 (1982).
242. H. A. Whaley and co-workers, *Antimicrob. Agents Chemother.,* 45 (1963).
243. L. I. Feldman and co-workers, *Antimicrob. Agents Chemother.,* 54 (1963).
244. R. L. Hamill and W. M. Stark, *J. Antibiot., Ser. A* **17,** 133 (1964).
245. A. J. Kreuzman, J. R. Turner, and W.-K. Yeh, *J. Biol. Chem.* **263,** 15,626 (1988).
246. R. H. Baltz and E. T. Seno, *Antimicrob. Agents Chemother.* **20,** 214 (1981).
247. E. T. Seno and R. H. Baltz, *Antimicrob. Agents Chemother.* **20,** 370 (1981).
248. R. H. Baltz, E. T. Seno, J. Stonesifer, and G. M. Wild, *J. Antibiot.* **36,** 131 (1983).
249. P. Muhlradt, E. Weiss, and T. Reichstein, *Helv. Chim. Acta* **47,** 2164 (1964).
250. H. A. Kirst and co-workers, *J. Antibiot.* **36,** 376 (1983).
251. R. Okamoto and co-workers, *J. Antibiot.* **35,** 921 (1982).
252. K. Kiyoshima and co-workers, *J. Antibiot.* **40,** 1123 (1987).
253. Y. Yamauchi and co-workers, *Proc. 7th Int. Symp. Biol. Actinomycetes,* May 22–26, 1988, Tokyo, Japan, Abstr. No. P4-14.
254. R. Corbaz and co-workers, *Helv. Chim. Acta* **38,** 1202 (1955).
255. N. Nishimura and co-workers, *J. Antibiot.* **18,** 251 (1965).
256. M. Brufani and W. Keller-Schierlein, *Helv. Chim. Acta* **49,** 1962 (1966).
257. A. Kinumaki and M. Suzuki, *J. Antibiot.* **25,** 480 (1972).
258. I. R. Shimi, S. Shoukry, and F. T. Ali, *J. Antibiot.* **32,** 1248 (1979).
259. H. Bickel and co-workers, *Helv. Chim. Acta* **45,** 1396 (1962).
260. J. Clardy, J. Finer-Moore, L. Weiler, and D. C. Wiley, *Tetrahedron* **37**(Suppl. 1), 91 (1981).
261. T. Suzuki, *Bull. Chem. Soc. Jpn.* **43,** 292 (1970).
262. T. Suzuki, N. Sugita, and M. Asai, *Chem. Lett.,* 789 (1973).
263. T. Suzuki, E. Mizuta, and N. Sugita, *Chem. Lett.,* 793 (1973).

264. S. M. Nash and co-workers, in J. D. Nelson and C. Grassi, eds., *Current Chemotherapy and Infectious Disease,* Vol. 1, American Society for Microbiology, Washington, D.C., 1980, p. 462.
265. Y. Sawada and co-workers, *J. Antibiot.* **42,** 242 (1989).
266. T. Kawasaki and co-workers, in Ref. 138, Abstr. No. 227.
267. H. Koshiyama and co-workers, *J. Antibiot. Ser. A* **16,** 59 (1963).
268. H. Koshiyama and co-workers, *J. Antibiot.* **22,** 61 (1969).
269. H. Tsukiura, M. Konishi, M. Saka, T. Naito, and H. Kawaguchi, *J. Antibiot.* **22,** 89 (1969).
270. G. H. Wagman and co-workers, *J. Antibiot.* **25,** 641 (1972).
271. H. Reimann and R. S. Jaret, *J. Chem. Soc., Chem. Commun.,* 1270 (1972).
272. A. K. Ganguly and co-workers, *Tetrahedron Lett.,* 4699 (1980).
273. K. Hatano, E. Higashide, and M. Shibata, *J. Antibiot.* **29,** 1163 (1976).
274. T. Kishi, S. Harada, H. Yamana, and A. Miyake, *J. Antibiot.* **29,** 1171 (1976).
275. T. Furumai and co-workers, *J. Antibiot.* **30,** 443 (1977).
276. A. Kinumaki, K. Harada, T. Suzuki, M. Suzuki, and T. Okuda, *J. Antibiot.* **30,** 450 (1977).
277. H. Imai and co-workers, *J. Antibiot.* **42,** 1000 (1989).
278. H. Tsukiura and co-workers, *J. Antibiot.* **22,** 100 (1969).
279. S. Omura, H. Ikeda, H. Matsubara, and N. Sadakane, *J. Antibiot.* **33,** 1570 (1980).
280. H. A. Kirst and co-workers, *J. Antibiot.* **35,** 1675 (1982).
281. M. Hayashi, M. Ohno, and S. Satoi, *J. Chem. Soc., Chem. Commun.,* 119 (1980).
282. S. Satoi, N. Muto, M. Hayashi, T. Fujii, and M. Otani, *J. Antibiot.* **33,** 364 (1980).
283. M. Hayashi and co-workers, *J. Antibiot.* **34,** 276 (1981).
284. A. K. Ganguly, Y-T. Liu, A. T. McPhail, and O. Sarre, *J. Chem. Soc., Chem. Commun.,* 853 (1982).
285. M. Hayashi, K. Kinoshita, S. Satoi, and K. Nakatsu, *J. Antibiot.* **35,** 1243 (1982).
286. K. Kinoshita and co-workers, *J. Antibiot.* **38,** 522 (1985).
287. K. Kinoshita, S. Satoi, M. Hayashi, and K. Nakatsu, *J. Antibiot.* **42,** 1003 (1989).
288. M. Hayashi and co-workers, *J. Antibiot.* **36,** 175 (1983).
289. J. Lotvin and co-workers, *J. Antibiot.* **35,** 1407 (1982).
290. M. Hayashi and co-workers, *J. Antibiot.* **34,** 346 (1981).
291. A. K. Ganguly and co-workers, *J. Chem. Soc., Chem. Commun.,* 855 (1982).
292. M. Sasaki and co-workers, *Jpn. J. Antibiot.* **42,** 2412 (1989).
293. P. W. K. Woo, H. W. Dion, and Q. R. Bartz, *J. Am. Chem. Soc.* **83,** 3352 (1961).
294. H. W. Dion, P. W. K. Woo, and Q. R. Bartz, *J. Am. Chem. Soc.* **84,** 880 (1962).
295. P. W. K. Woo, H. W. Dion, and L. F. Johnson, *J. Am. Chem. Soc.* **84,** 1066 (1962).
296. P. W. K. Woo, H. W. Dion, and Q. R. Bartz, *J. Am. Chem. Soc.* **84,** 1512 (1962).
297. P. W. K. Woo, H. W. Dion, and Q. R. Bartz, *J. Am. Chem. Soc.* **86,** 2724 (1964).
298. *Ibid.,* p. 2726.
299. J. Krc and R. B. Scott, *Microscope* **23,** 15 (1975).
300. D. V. Lefemine and co-workers, *Antimicrob. Agents Chemother.,* 41 (1963).
301. M. P. Kunstmann and L. A. Mitscher, *Experientia* **21,** 372 (1965).
302. L. A. Mitscher and M. P. Kunstmann, *Experientia* **25,** 12 (1969).
303. W. Cullen and co-workers, in Ref. 253, Abstr. No. P4-13.
304. J. R. Hauske, J. Dibrino, M. Guadliana, and G. Kostek, *J. Org. Chem.* **51,** 2808 (1986).
305. M. P. Kunstmann, L. A. Mitscher, and E. L. Patterson, *Antimicrob. Agents Chemother.,* 87 (1964).
306. H. Achenbach and W. Karl, *Chem. Ber.* **108,** 759 (1975).
307. *Ibid.,* p. 772.
308. *Ibid.,* p. 780.

309. S. Mizobuchi, J. Mochizuki, H. Soga, H. Tanba, and H. Inoue, *J. Antibiot.* **39,** 1776 (1986).
310. C. M. M. Franco, J. N. Gandhi, S. Chatterjee, and B. N. Ganguli, *J. Antibiot.* **40,** 1361 (1987).
311. S. Chatterjee and co-workers, *J. Antibiot.* **40,** 1368 (1987).
312. S. Omura, C. Kitao, and H. Matsubara, *Chem. Pharm. Bull.* **28,** 1963 (1980).
313. M. Hayashi and co-workers, *J. Antibiot.* **34,** 1075 (1981).
314. S. Omura and co-workers, *J. Antibiot.* **21,** 532 (1968).
315. J. R. Prous, ed., *Drugs of the Future* **10,** 486 (1985).
316. J. R. Prous, ed., *Drugs of the Future* **8,** 790 (1983).
317. H. Sakakibara, O. Okekawa, T. Fujiwara, M. Otani, and S. Omura, *J. Antibiot.* **34,** 1001 (1981).
318. H. Sakakibara, O. Okekawa, T. Fujiwara, M. Aizawa, and S. Omura, *J. Antibiot.* **34,** 1011 (1981).
319. S. Omoto, K. Iwamatsu, S. Inouye, and T. Niida, *J. Antibiot.* **29,** 536 (1976).
320. S. Inouye, S. Omoto, K. Iwamatsu, and T. Niida, *J. Antibiot.* **33,** 61 (1980).
321. K. Kawaharajo, Y. Sekizawa, and M. Inoue, *J. Antibiot.* **34,** 436 (1981).
322. K. Fukaya and co-workers, *Chemotherapy (Japan)* **29,** 448 (1981).
323. T. Shomura, S. Someya, S. Murata, K. Umemura, and M. Nishio, *Chem. Pharm. Bull.* **29,** 2413 (1981).
324. T. Yoshida and co-workers, *Jpn. J. Antibiot.* **35,** 1462 (1982).
325. H. Sano and co-workers, *J. Antibiot.* **37,** 750 (1984).
326. *Ibid.,* p. 760.
327. H. Sano and co-workers, *J. Antibiot.* **38,** 1350 (1985).
328. R. Okamoto and co-workers, *J. Antibiot.* **33,** 1300 (1980).
329. R. Okamoto and co-workers, *J. Antibiot.* **33,** 1309 (1980).
330. M. Tsuchiya and co-workers, *J. Antibiot.* **34,** 305 (1981).
331. M. Tsuchiya, T. Sawa, T. Takeuchi, H. Umezawa, and R. Okamoto, *J. Antibiot.* **35,** 673 (1982).
332. B. J. Skelly, D. Andersen, M. Pruss, and R. Pellegrino, *Avian Diseases* **30,** 505 (1986).
333. T. M. Jacks, F. R. Judith, S. D. Feighner, and R. O. Likoff, *Am. J. Vet. Res.* **47,** 2325 (1986).
334. H. A. Kirst and co-workers, *J. Antibiot.* **39,** 1724 (1986).
335. J. K. Epp, M. L. Huber, J. R. Turner, and B. E. Schoner, in C. L. Hershberger, S. W. Queener, and G. Hegeman, eds., *Genetics and Molecular Biology of Industrial Microorganisms,* American Society for Microbiology, Washington, D.C., 1989, p. 35.
336. J. K. Epp, M. L. B. Huber, J. R. Turner, T. Goodson, and B. E. Schoner, *Gene* **85,** 293 (1989).
337. M. Tsuchiya and co-workers, *J. Antibiot.* **35,** 661 (1982).
338. T. Yoshioka and co-workers, *J. Antibiot.* **41,** 1617 (1988).
339. K. Kiyoshima and co-workers, *Chem. Pharm. Bull.* **37,** 861 (1989).
340. T. Takeuchi and co-workers, *J. Antibiot.* **40,** 1358 (1987).
341. K. Kiyoshima and co-workers, *J. Antibiot.* **42,** 1661 (1989).
342. A. G. Fishman, A. K. Mallams, and R. R. Rossman, *J. Chem. Soc., Perkin Trans. 1,* 787 (1989).
343. Y. Shimauchi and co-workers, *J. Antibiot.* **33,** 284 (1980).
344. Y. Shimauchi, M. Sakamoto, K. Hori, T. Ishikura, and J. Lein, *J. Antibiot.* **34,** 245 (1981).
345. H. A. Kirst and co-workers, *J. Med. Chem.* **31,** 1631 (1988).
346. H. Matsubara, K. Miyano, A. Nakagawa, and S. Omura, *Chem. Pharm. Bull.* **30,** 97 (1982).

347. T. Fujiwara, H. Watanabe, Y. Kogami, Y. Shiritani, and H. Sakakibara, *J. Antibiot.* **42,** 903 (1989).
348. A. G. Fishman, A. K. Mallams, M. S. Puar, R. R. Rossman, and R. L. Stephens, *J. Chem. Soc., Perkin Trans. 1,* 1189 (1987).
349. T. Fujiwara, A. Sakai, and H. Sakakibara, *J. Antibiot.* **43,** 327 (1990).
350. H. A. Kirst and co-workers, *J. Antibiot.* **42,** 1673 (1989).
351. M. Debono and co-workers, *J. Antibiot.* **42,** 1253 (1989).
352. H. A. Kirst and co-workers, *J. Antibiot.* **41,** 938 (1988).
353. H. Matsubara, J. Inokoshi, A. Nakagawa, H. Tanaka, and S. Omura, *J. Antibiot.* **36,** 1713 (1983).
354. S. Omura, K. Miyano, H. Matsubara, and A. Nakagawa, *J. Med. Chem.* **25,** 271 (1982).
355. E. E. Ose, *J. Antibiot.* **40,** 190 (1987).
356. E. E. Ose and L. V. Tonkinson, *Vet. Rec.* **123,** 367 (1988).
357. R. N. Gourlay, L. H. Thomas, S. G. Wyld, and C. J. Smith, *Res. Vet. Sci.* **47,** 84 (1989).
358. A. Tanaka, T. Tsuchiya, S. Umezawa, M. Hamada, and H. Umezawa, *J. Antibiot.* **34,** 1377 (1981).
359. A. Tanaka and co-workers, *J. Antibiot.* **35,** 113 (1982).
360. H. A. Kirst and co-workers, *J. Antibiot.* **39,** 1108 (1986).
361. H. A. Kirst and co-workers, *J. Antibiot.* **40,** 823 (1987).
362. R. C. Brunham, C-C. Kuo, C. E. Stevens, and K. K. Holmes, *Ann. Int. Med.* **97,** 216 (1982).
363. A. Tanaka, T. Tsuchiya, S. Umezawa, and H. Umezawa, *J. Antibiot.* **34,** 1374 (1981).
364. A. Tanaka, A. Watanabe, T. Tsuchiya, S. Umezawa, and H. Umezawa, *J. Antibiot.* **34,** 1381 (1981).
365. A. K. Mallams and R. R. Rossman, *J. Chem. Soc., Perkin Trans. 1,* 775 (1989).
366. J. P. Leeds and H. A. Kirst, *Synth. Commun.* **18,** 777 (1988).
367. S. Omura, H. Matsubara, K. Tsuzuki, and A. Nakagawa, *J. Antibiot.* **37,** 1007 (1984).
368. K. Tsuzuki, H. Matsubara, A. Nakagawa, and S. Omura, *J. Antibiot.* **39,** 1784 (1986).
369. C. Ruggeri and co-workers, *J. Antibiot.* **42,** 1443 (1989).
370. A. K. Mallams and R. R. Rossman, *J. Chem. Soc., Perkin Trans. 1,* 799 (1989).
371. S. Omura, N. Sadakane, Y. Tanaka, and H. Matsubara, *J. Antibiot.* **36,** 927 (1983).
372. S. Omura, N. Sadakane, and H. Matsubara, *Chem. Pharm. Bull.* **30,** 223 (1982).
373. N. Sadakane, Y. Tanaka, and S. Omura, *J. Antibiot.* **35,** 680 (1982).
374. N. Sadakane, Y. Tanaka, and S. Omura, *J. Antibiot.* **36,** 921 (1983).
375. S. Omura, C. Kitao, and N. Sadakane, *J. Antibiot.* **33,** 911 (1980).
376. S. Omura, N. Sadakane, C. Kitao, H. Matsubara, and A. Nakagawa, *J. Antibiot.* **33,** 913 (1980).
377. B. K. Lee and co-workers, *J. Antibiot.* **36,** 742 (1983).
378. L. Toscano, G. Fioriello, R. Spagnoli, and L. Cappelletti, *J. Antibiot.* **36,** 1585 (1983).
379. S. Omura, *Bacteriol. Rev.* **40,** 681 (1976).
380. J. B. McAlpine and co-workers, *J. Antibiot.* **40,** 1115 (1987).
381. D. A. Hopwood, *Phil. Trans. Royal Soc. London, B* **324,** 549 (1989).
382. C. R. Hutchinson, C. W. Borell, S. L. Otten, K. J. Stutzman-Engwall, and Y. Wang, *J. Med. Chem.* **32,** 929 (1989).
383. J. A. Robinson, *Chem. Soc. Rev.* **17,** 383 (1988).
384. E. T. Seno and R. H. Baltz, in S. Shapiro, ed., *Regulation of Secondary Metabolism in Actinomycetes,* CRC Press, Boca Raton, Fla., 1989, p. 1.
385. S. Donadio and co-workers, in Ref. 335, p. 53.
386. J. M. Weber, B. Schoner, and R. Losick, *Gene* **75,** 235 (1989).
387. D. J. Hardy and co-workers, *Antimicrob. Agents Chemother.* **32,** 1710 (1988).
388. P. B. Fernandes and D. J. Hardy, *Drugs Exptl. Clin. Res.* **14,** 445 (1988).

389. A. L. Barry, R. N. Jones, and C. Thornsberry, *Antimicrob. Agents Chemother.* **32,** 752 (1988).
390. J. E. Hoppe and A. Eichhorn, *Eur. J. Clin. Microbiol. Infect. Dis.* **8,** 653 (1989).
391. P. Le Noc, J. Croize, A. Bryskier, D. Le Noc, and J. Robert, *Pathol. Biol.* **37,** 553 (1989).
392. N. Bornstein, C. Roudier, and J. Fleurette, *J. Antimicrob. Chemother.* **15,** 17 (1985).
393. J. Retsema and co-workers, *Antimicrob. Agents Chemother.* **31,** 1939 (1987).
394. A. E. Girard and co-workers, *Antimicrob. Agents Chemother.* **31,** 1948 (1987).
395. B. I. Davies, F. P. V. Maesen, and R. Gubbelmans, *J. Antimicrob. Chemother.* **23,** 743 (1989).
396. P. B. Fernandes, R. N. Swanson, D. J. Hardy, E. J. McDonald, and N. Ramer, *Drugs. Exptl. Clin. Res.* **14,** 441 (1988).
397. S. Kohno and co-workers, *J. Antimicrob. Chemother.* **24,** 397 (1989).
398. R. B. Fitzgeorge and A. S. R. Featherstone, *J. Antimicrob. Chemother.* **23,** 462 (1989).
399. D. J. Hardy, C. W. Hanson, D. M. Hensey, J. M. Beyer, and P. B. Fernandes, *J. Antimicrob. Chemother.* **22,** 631 (1988).
400. J. A. Garcia-Rodriguez, J. E. Garcia Sanchez, M. I. Garcia Garcia, E. Garcia Sanchez, and J. L. Munoz Bellido, *Antimicrob. Agents Chemother.* **33,** 1650 (1989).
401. R. McCabe and J. S. Remington, *N. Eng. J. Med.* **318,** 313 (1988).
402. B. J. Luft, *Eur. J. Clin. Microbiol. Infect. Dis.* **6,** 479 (1987).
403. H. R. Chang, F. C. Rudareanu, and J.-C. Pechere, *J. Antimicrob. Chemother.* **22,** 359 (1988).
404. F. G. Araujo, D. R. Guptill, and J. S. Remington, *Antimicrob. Agents Chemother.* **32,** 755 (1988).
405. S. Chamberland, H. A. Kirst, and W. L. Current, in Ref. 154, Abstr. No. 1036.
406. V. Preac-Mursic, B. Wilske, G. Schierz, E. Suss, and B. Gross, *Eur. J. Clin. Microbiol. Infect. Dis.* **8,** 651 (1989).
407. R. C. Johnson, C. Kodner, and M. Russell, in Ref. 138, Abstr. No. 235.
408. K. Chirgwin, P. M. Roblin, and M. R. Hammerschlag, *Antimicrob. Agents Chemother.* **33,** 1634 (1989).
409. N. Khardori, K. Rolston, B. Rosenbaum, S. Hayat, and G. P. Bodey, *J. Antimicrob. Chemother.* **24,** 667 (1989).
410. S. Naik and R. Ruck, *Antimicrob. Agents Chemother.* **33,** 1614 (1989).
411. C. B. Inderlied, P. T. Kolonoski, M. Wu, and L. S. Young, *J. Infect. Dis.* **159,** 994 and 1095 (1989).
412. P. B. Fernandes, D. J. Hardy, D. McDaniel, C. W. Hanson, and R. N. Swanson, *Antimicrob. Agents Chemother.* **33,** 1531 (1989).
413. N. Ramasesh, J. L. Krahenbuhl, and R. C. Hastings, *Antimicrob. Agents Chemother.* **33,** 657 (1989).
414. E. A. Gorzynski, S. I. Gutman, and W. Allen, *Antimicrob. Agents Chemother.* **33,** 591 (1989).
415. M. DiGiambattista, G. Chinali, and C. Cocito, *J. Antimicrob. Chemother.* **24,** 485 (1989).
416. M. A. Arevalo, F. Tejedor, F. Polo, and J. P. G. Ballesta, *J. Med. Chem.* **32,** 2200 (1989).
417. S. Mitsuhashi and M. Inoue, in L. E. Bryan, ed., *Antimicrobial Drug Resistance,* Academic Press, Orlando, Fla., 1984, p. 279.
418. B. Weisblum, *Brit. Med. Bull.* **40,** 47 (1984).
419. S. Douthwaite, T. Powers, J. Y. Lee, and H. F. Noller, *J. Mol. Biol.* **209,** 655 (1989).
420. M. Zalacain and E. Cundliffe, *J. Bacteriol.* **171,** 4254 (1989).
421. S. K. Kadam, *J. Bacteriol.* **171,** 4518 (1989).
422. M. Arthur, A. Brisson-Noel, and P. Courvalin, *J. Antimicrob. Chemother.* **20,** 783 (1987).

423. V. P. Marshall, W. F. Liggett, and J. I. Cialdella, *J. Antibiot.* **42,** 826 (1989).
424. K. O'Hara, T. Kanda, K. Ohmiya, T. Ebisu, and M. Kono, *Antimicrob. Agents Chemother.* **33,** 1354 (1989).
425. M.-S. Kuo and co-workers, *Antimicrob. Agents Chemother.* **33,** 2089 (1989).
426. B. C. Lampson, W. von David, and J. T. Parisi, *Antimicrob. Agents Chemother.* **30,** 653 (1986).
427. G. Palu, S. Valisena, M. F. Barile, and G. A. Meloni, *Eur. J. Epidemiology* **5,** 146 (1989).
428. G. E. Burrows, *J. Am. Vet. Med. Assoc.* **176,** 1072 (1980).
429. G. S. Duthu, *J. Pharm. Sci.* **74,** 943 (1985).
430. F. Fraschini, M. Falchi, and V. Copponi, *New Perspectives Clin. Microbiol.* **4,** 79 (1981).
431. F. O'Grady, *Symp. Soc. Gen. Microbiol.* **38,** 341 (1985).
432. L. Madgwick, S. Mayer, and P. Keen, *J. Antimicrob. Chemother.* **24,** 709 (1989).
433. M. Raghoebar, E. Lindeyer, W. B. Van den Berg, and C. A. M. Van Ginneken, *Biochem. Pharm.* **37,** 3221 (1988).
434. M. L. Villa, F. Valenti, F. Scaglione, M. Falchi, and F. Fraschini, *J. Antimicrob. Chemother.* **24,** 765 (1989).
435. A. M. Cuffini, N. A. Carlone, V. Tullio, and M. Borsotto, *Microbios* **58,** 27 (1989).
436. R. P. Gladue, G. M. Bright, R. E. Isaacson, and M. F. Newborg, *Antimicrob. Agents Chemother.* **33,** 277 (1989).
437. R. Anderson, G. Joone, and C. E. J. van Rensburg, *J. Antimicrob. Chemother.* **22,** 923 (1988).
438. M. Ishiguro and co-workers, *J. Antimicrob. Chemother.* **24,** 719 (1989).
439. M. T. Labro and C. Babin-Chevaye, *J. Antimicrob. Chemother.* **24,** 731 (1989).
440. B. Scorneaux and P. M. Tulkens, in Ref. 154, Abstr. No. 157.
441. R. Anderson, *J. Infect. Dis.* **159,** 966 (1989).
442. A. Naess and C. O. Solberg, *Acta Pathol. Microbiol. Immunol. Scand.* **96,** 503 (1988).
443. K. Takeshita and co-workers, *Drugs Exptl. Clin. Res.* **15,** 527 (1989).
444. D. Pessayre, *Int. J. Clin. Pharm. Res.* **3,** 449 (1983).
445. E. Sartori, M. Delaforge, and D. Mansuy, *Biochem. Pharm.* **38,** 2061 (1989).
446. M. Tinel and co-workers, *J. Pharmacol. Expt. Ther.* **250,** 746 (1989).
447. T. Miura and co-workers, *J. Antimicrob. Chemother.* **24,** 551 (1989).
448. T. Adachi and co-workers, *J. Antibiot.* **41,** 966 (1988).
449. J. Sasaki, K. Mizoue, S. Morimoto, T. Adachi, and S. Omura, *J. Antibiot.* **41,** 908 (1988).
450. G. Bozler, G. Heinzel, U. Lechner, K. Schumacher, and U. Busch, *Proc. 28th Intersci. Conf. Antimicrob. Agents Chemother.,* Oct. 23–26, 1988, Los Angeles, Calif., Abstr. No. 924.
451. T. Watanabe and co-workers, *Drugs Exptl. Clin. Res.* **10,** 371 (1984).
452. A. Sakai and co-workers, in Ref. 450, Abstr. No. 913.
453. B. L. Carter, J. C. Woodhead, K. J. Cole, and G. Milavetz, *Drug Intelligence and Clin. Pharm.* **21,** 734 (1987).
454. Z. Itoh, T. Suzuki, M. Nakaya, M. Inoue, and S. Mitsuhashi, *Antimicrob. Agents Chemother.* **26,** 863 (1984).
455. X. Y. Qin, M-A. Pilot, H. H. Thompson, and J. P. Maskell, *J. Antimicrob. Chemother.* **18,** 747 (1986).
456. T. Tomomasa, T. Kuroume, H. Arai, K. Wakabayashi, and Z. Itoh, *Dig. Dis. Sci.* **31,** 157 (1986).
457. AGA Abstracts, *Gastroenterology* **98,** A64, A342, A345, A358, and A375 (1990).
458. Y. Kondo, K. Torii, S. Omura, and Z. Itoh, *Biochem. Biophys. Res. Commun.* **150,** 877 (1988).
459. T. Peeters and co-workers, *Am. J. Physiol.* **257,** G470 (1989).
460. C. H. S. McIntosh and J. C. Brown, *Adv. Metabolic Disorders* **11,** 439 (1988).

461. K. Tsuzuki and co-workers, *Chem. Pharm. Bull.* **37,** 2687 (1989).
462. T. Sunazuka and co-workers, *Chem. Pharm. Bull.* **37,** 2701 (1989).
463. N. Inatomi and co-workers, *J. Pharm. Expt. Ther.* **251,** 707 (1989).
464. R. C. Yao and D. F. Mahoney, *Appl. Environ. Microbiol.* **55,** 1507 (1989).
465. S. Omura and Y. Tanaka, in Ref. 1, p. 199.
466. S. Omura and A. Nakagawa, in J. W. Corcoran, ed., *Antibiotics,* Vol. 4, Springer-Verlag, Berlin, Germany, 1981, p. 175.
467. K. Gerzon and co-workers, *J. Am. Chem. Soc.* **78,** 6396 (1956).
468. J. Neuzil and Z. Hostalek, *Folia Microbiol.* **31,** 402 (1986).
469. S. Omura and Y. Tanaka, in H. Pape and H.-J. Rehm, volume eds., *Biotechnology,* Vol. 4, VCH Publishers, Weinheim, Germany, 1986, p. 359.
470. D. A. Hopwood and D. H. Sherman, *Ann. Rev. Genetics* **24,** 1990.
471. W. D. Celmer, *J. Am. Chem. Soc.* **87,** 1801 (1965).
472. W. D. Celmer, *Pure Appl. Chem.* **28,** 413 (1971).
473. S. Yue, J. S. Duncan, Y. Yamamoto, and C. R. Hutchinson, *J. Am. Chem. Soc.* **109,** 1253 (1987).
474. D. E. Cane and C-C. Yang, *J. Am. Chem. Soc.* **109,** 1255 (1987).
475. K. Kinoshita, S. Takenaka, and M. Hayashi, *J. Chem. Soc., Chem. Commun.,* 943 (1988).
476. S. Takano, Y. Sekiguchi, Y. Shimazaki, and K. Ogasawara, *Tetrahedron Lett.* **30,** 4001 (1989).
477. M. L. B. Huber and co-workers, *Antimicrob. Agents Chemother.* **34** (1990).
478. J. Vara, M. Lewandowska-Skarbek, Y-G. Wang, S. Donadio, and C. R. Hutchinson, *J. Bacteriol.* **171,** 5872 (1989).
479. O. Han and H-W. Liu, *J. Am. Chem. Soc.* **110,** 7893 (1988).
480. J. Dale, *Angew. Chem. Int. Eng. Ed.* **5,** 1000 (1966).
481. R. S. Egan, J. R. Martin, T. J. Perun, and L. A. Mitscher, *J. Am. Chem. Soc.* **97,** 4578 (1975).
482. H. Ogura, K. Furuhata, H. Kuwano, and M. Suzuki, *Tetrahedron* **37**(Suppl. 1), 165 (1981).
483. T. H. Keller, E. G. Neeland, S. Rettig, J. Trotter, and L. Weiler, *J. Am. Chem. Soc.* **110,** 7858 (1988).
484. J. R. Everett, I. K. Hatton, and J. W. Tyler, *Magn. Res. Chem.* **28,** 114 (1990).
485. J. R. Everett, I. K. Hatton, E. Hunt, J. W. Tyler, and D. J. Williams, *J. Chem. Soc., Perkin Trans. 2,* 1719 (1989).
486. J. R. Everett and J. W. Tyler, *J. Chem. Soc., Perkin Trans. 2,* 325 (1988).
487. *Ibid.,* 1987, p. 1659.
488. M. S. Puar, R. Brambilla, and R. Jaret, *J. Antibiot.* **35,** 381 (1982).
489. M. S. Puar, *J. Antibiot.* **34,** 602 (1981).
490. J. G. Nourse and J. D. Roberts, *J. Am. Chem. Soc.* **97,** 4584 (1975).
491. D. J. Ager and C. K. Sood, *Mag. Res. Chem.* **25,** 948 (1987).
492. K. H. Kim and Y. C. Martin, *Prog. Clin. Biol. Res.* **291,** 325 (1989).
493. I. Nakayama, in Ref. 1, p. 261.
494. *Physicians Desk Reference,* 44th ed., Medical Economics Co., Oradell, N.J., 1990.
495. *Scrip's Antibacterial Report,* PJB Publications, Richmond, UK, 1989.
496. R. E. Brummett and K. E. Fox, *Antimicrob. Agents Chemother.* **33,** 791 (1989).
497. R. C. Wilson, in Ref. 1, p. 302.
498. J. E. Riviere, A. L. Craigmill, and S. F. Sundlof, eds., *Comparative Pharmacokinetics and Residues of Veterinary Antimicrobials,* CRC Press, Boca Raton, Fla., 1990, Chapt. 5.

H. A. KIRST
Eli Lilly and Company

NUCLEOSIDES AND NUCLEOTIDES

The naturally occurring nucleoside and nucleotide antibiotics listed alphabetically in Table 1 exist as either the *C*- or *N*-glycosides (see NUCLEIC ACIDS). These antibiotics contain a variety of purine and pyrimidine rings, including the diazepin, maleimide, indole, imidazole, pyrrolopyrimidine, pyrazolpyrimidine, pyrazole, oxazine, triazene, hydantoin, and the purine ring having a 6-*N*-phosphoramidate substituent. The structures of the carbohydrate moieties also vary. Some nucleosides have additional carbon atoms attached to the ribosyl moiety and in some cases ribose has been replaced by other sugars such as allulose, olefinic sugars, 2-, 3-, 4-, or 5-amino sugars, 4-fluororibose, 4-aminohexoses, aminouronic acids, disaccharides, or a tricyclic dodecose.

The naturally occurring nucleoside/nucleotide antibiotics, which have been isolated from bacteria, fungi, blue-green algae, and marine sponges, have proven to be useful biochemical probes in eucaryotic, procaryotic, viral, fungal, and plant systems. For example, the tunicamycins [*11089-65-9*] and mureidomycins are valuable because of their importance in peptidoglycan synthesis, mucopolysaccharide synthesis, and inhibition of the synthesis of the oligosaccharide moiety of the glycoproteins. Antibiotic A201A, which is structurally similar to puromycin, is an inhibitor of protein synthesis. 2′-Deoxycoformycin has found dramatic application in the treatment of hairy cell leukemia (see CHEMOTHERAPEUTIC AGENTS, ANTICANCER) and as a means to increase the therapeutic efficacy of cordycepin, formycin, and arabinofuranosyladenine (ara-A) through inhibition of adenosine deaminase. The 4-aminohexose nucleoside antibiotics and 2′-amino-2′-deoxyguanosine have been useful in elucidating the steps in protein synthesis, whereas psicofuranine and decoyinine have found use as inducers of sporulation in bacteria. Showdomycin is an excellent probe of membrane transport. The arabinosylpyrimidine nucleosides, ara-T and ara-U, have antiviral activities (see ANTIVIRAL AGENTS). Several nucleoside analogues have herbicidal activity (see HERBICIDES). Some excellent reviews on the nucleoside/nucleotide antibiotics are available (1–4) (see also ANTIPARASITIC AGENTS).

C-Nucleosides

The naturally occurring *C*-nucleosides containing *C*-glycosyl linkages are shown in Table 1.

Ezomycins. The ezomycins [*58572-97-7*] (**1**–**6**) are naturally occurring pyrimidine nucleosides produced by the *Streptomyces* (1–4). The ezomycins, which are comprised of ezomycin A [*60182-24-3*], ezomycin B [*60182-25-4*], and ezomycin C, inhibit the phytopathogenic fungi, *Sclerotinia* and *Botrytis* sp., but not bacteria and yeast. Ezomycins B_1, B_2, C_1, and C_2 are pseudouridine-type *C*-nucleosides, whereas ezomycins A_1 and A_2 are *N*-nucleosides. Ezomycins A_1, B_1, and C_1, but not A_2, B_2, and C_2, contain the L-cystathionine moiety. Ezomycins A_1, A_2, B_1, and B_2 are β-nucleosides; C_1 and C_2 are α-nucleosides.

Showdomycin. Showdomycin (2-β-D-ribofuranosylmaleimide) (**7**) is a maleimide *C*-nucleoside antibiotic synthesized by *S. showdoensis*; isoshowdomycin (**8**) and maleimycin (**9**) have also been isolated (1–6). Showdomycin is not phosphorylated by nucleoside kinase and is not a substrate for nucleoside phosphorylase. Once (**7**) enters the cell, it blocks the uptake of glucose and other nutrients.

Oxazinomycin. Oxazinomycin (5-β-D-ribofuranosyl-1,3-oxazine-2,4-dione) (**10**), also known as minimycin, has been isolated from *S. hygroscopicus* and *S. mediocidicus* (1–4). It inhibits gram-positive and gram-negative bacteria and tumor growth, but not yeast or fungi. The biosynthesis of (**6**) proceeds from [5-^{14}C,3-^{3}H]glutamic acid with carbons 4, 5, and 6 of the oxazine ring of (**10**) arising from carbons 5, 4, and 3 of glutamic acid (4). The naturally occurring *C*-nucleoside antibiotics showdomycin (**7**), formycin (**12**), oxazinomycin (**10**), and pyrazomycin (**11**) utilize acetate and glutamate as the carbon source for their aglycons (4,7–10).

Pyrazomycin. Pyrazomycin (**11**), 3-(1-β-D-ribofuranosyl)-4-hydroxy-pyrazole 5-carboxamide, is isolated from *S. candidus* (1–4,9,10). The incorporation of [2-^{13}C]acetate and [1- and U-^{14}C]glutamate into the four contiguous carbons of pyrazomycin has been reported (11,12). Pyrazomycin 5′-phosphate inhibits orotidylic acid decarboxylase. Pyrazomycin inhibits adenosine phosphorylation and decreases the incorporation of deoxyuridine into DNA of Novikoff hepatoma cells in culture. It also inhibits the growth of tumor cells and the cytopathic effects of vaccinia, herpes simplex, vesicular stomatitis, Newcastle disease, measles, Sindbis, polio, hepatitis A, and coxsackie viruses (13,14). The inhibitory action of (**11**) on viral multiplication is reversed by uridine.

Pyrazolopyrimidine Nucleosides. Formycin (**12**), formycin B (**13**), and oxoformycin B (**14**) are naturally occurring pyrazolopyrimidine nucleosides isolated from *Nocardia interforma, S. lavendulae,* and *S. gunmaences* (1–4,15). Coformycin, which inhibits adenosine deaminase, is also isolated from *N. interforma*. Glutamate is the precursor for the biosynthesis of formycin (10). Compound (**12**), but not (**13**), is converted to its 5′-mono-, di-, and triphosphates (3). Compound (**12**) inhibits DNA and protein synthesis, but not RNA synthesis (1–4). Formycin 5′-triphosphate (FTP) is incorporated into RNA and codes as an adenosine 5′-triphosphate (ATP) with bacterial, viral, plant, RNA, and DNA polymerases (16). It is also an important probe of nucleoside transport (17–21). The tight binding of formycin 5′-phosphate to adenosine monophosphate (AMP) deaminase has been reported (22). Compound (**12**) has been used to elucidate ribosome-transfer RNA (tRNA) interactions and it has been reported that the stacking of tRNA^{PheCCF}, where the tail is phenylalanine–cytosine–cytosine–formycin, is weaker than that of tRNA^{PheCCA} (23,24). Donor activity for peptide bond formation of formycin-containing tRNA is much lower than that of normal tRNA. Compound (**12**) is rapidly deaminated by erythrocytic adenosine deaminase; the absence of this deaminase in erythrocytes of adenosine deaminase (ADA)-deficient patients results in the conversion of (**12**) to FTP in quantities three times greater than ATP (25). An analogue of inosine [*58-63-9*] (**13**), inhibits tumors and viruses, but does not appear to act via formation of the 5′-phosphates. It has been reported that (**12**) inhibits L5178Y cell growth as a competitive inhibitor of the nicotinamide adenine dinucleotide cation (NAD$^+$) and showed that (**7**) and (**13**) inhibit poly(ADP-ribose) polymerase (26,27). The formycin analogue of NAD$^+$ has been shown to inhibit DNA synthesis (28).

Most recently the structure of pyrrolosine has been shown to be an isomeric *C*-nucleoside analogue of 9-deazainosine, which is 7-(β-D-ribofuranosyl)-4-oxo-

Table 1. *C*-Nucleoside Antibiotics

Name	CAS Registry Number	Molecular formula	Structure number	R	Structure
ezomycin A_1	[*39422-19-0*]	$C_{26}H_{38}N_8O_{15}S$	(**1**)	β-1-cytosine	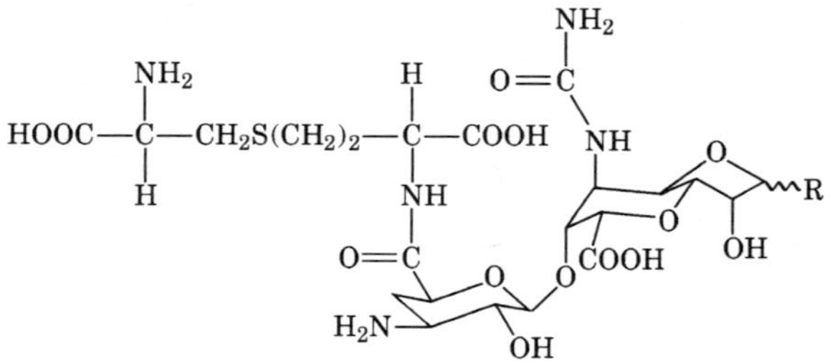
ezomycin B_1	[*39422-20-3*]	$C_{26}H_{39}N_7O_{17}S$	(**2**)	β-5-uracil	
ezomycin C_1	[*58002-06-5*]	$C_{26}H_{37}N_7O_{16}S$	(**3**)	α-5-uracil	
ezomycin A_2	[*54328-22-2*]	$C_{19}H_{26}N_6O_{12}$	(**4**)	β-1-cytosine	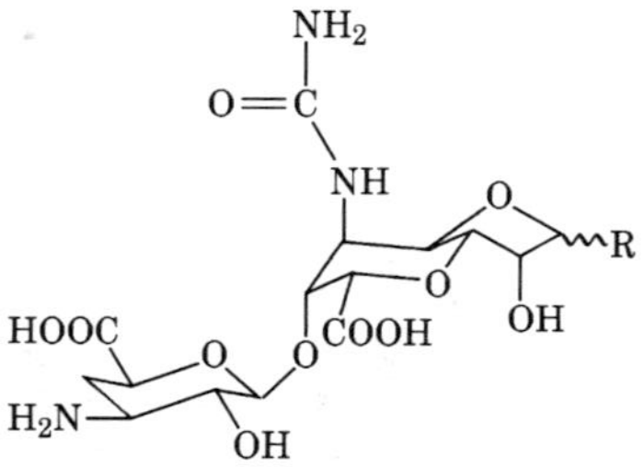
ezomycin B_2	[*57973-16-7*]	$C_{19}H_{25}N_5O_{13}$	(**5**)	β-5-uracil	
ezomycin C_2	[*57973-14-5*]	$C_{19}H_{25}N_5O_{13}$	(**6**)	α-5-uracil	
showdomycin	[*16755-07-0*]	$C_9H_{11}NO_6$	(**7**)	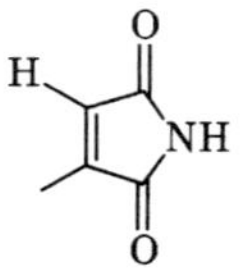	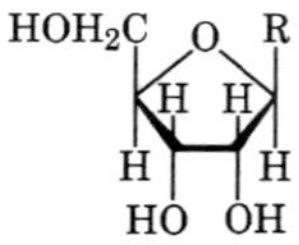

isoshowdomycin	[*37760-84-2*]	$C_9H_{11}NO_6$	**(8)**	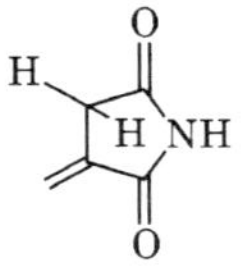	
maleimycin	[*50988-16-4*]	$C_7H_7NO_3$	**(9)**		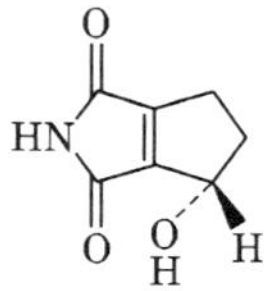
oxazinomycin	[*32388-21-9*]	$C_9H_{11}NO_7$	**(10)**	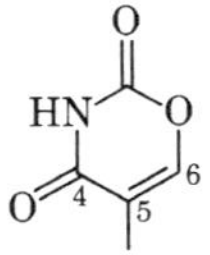	
pyrazomycin (pyrazofurin)	[*30868-30-5*]	$C_9H_{13}N_3O_6$	**(11)**	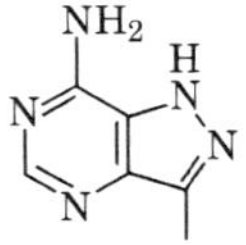	
formycin formycin B	[*6742-12-7*] [*13877-76-4*]	$C_{10}H_{13}N_5O_4$ $C_{10}H_{12}N_4O_5$	**(12)** **(13)**	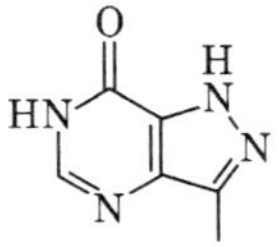	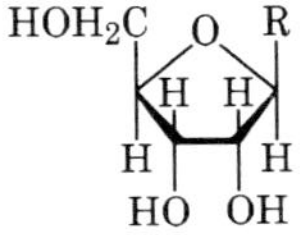
oxoformycin B	[*19246-88-9*]	$C_{10}H_{12}N_4O_6$	**(14)**	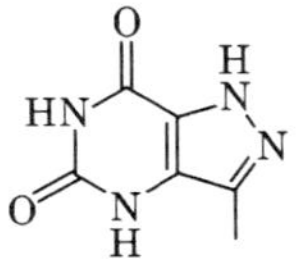	

3*H*,5*H*-pyrrolo[3,2-*d*]pyrimidine (29). Pyrrolosine inhibits development of starfish embryos.

N-Nucleosides

The naturally occurring nucleoside analogues discussed in this section contain the *N*-glycosyl linkage and either purine, pyrimidine, imidazole, diazepin, or indole rings. The purine nucleosides inhibit protein synthesis, RNA and DNA synthesis, and methyltransferases; they have antimycoplasmal, antiviral, hypotensive, antifungal, antimycobacterial, and antitumor activities and induce sporulation (1–4). The pyrimidine nucleosides inhibit protein synthesis, virus replication, RNA and DNA synthesis, and cAMP phosphodiesterase. The imidazole nucleosides inhibit nucleic acid synthesis. The diazepin nucleosides inhibit adenosine deaminase (ADA). The indole nucleosides inhibit bacteria, yeast, fungi, and viruses.

Purine *N*-Nucleosides. The purine *N*-nucleoside antibiotics are given in Tables 2 and 3.

2′-Amino-2′-deoxypurines. 2′-Amino-2′-deoxyadenosine (**15**) is a naturally occurring *N*-nucleoside isolated from *Actinomadura* that shows antimycoplasmal activity (1,4). Adenosine is the direct precursor for its biosynthesis (30). 2′-Amino-2′-deoxyguanosine (**16**), isolated from a strain of *Enterobacter cloacae* (1,4), shows the growth of HeLa S3 cells and Sarcoma 180 *in vivo* and has been tested for antibacterial activity.

3′-Amino-3′-deoxyadenosine. 3′-Amino-3′-deoxyadenosine (**17**) is elaborated by *Cordyceps militaris, Aspergillus nidulans,* and *Helminthosporium* (3,4). The biosynthesis proceeds directly from adenosine. Compound (**17**) inhibits RNA polymerase, but not DNA polymerase, and replaces the adenosyl residue at the 3′-terminus of tRNA. Phenylalanyl-(3′-amino-3′-deoxyadenosyl)-tRNA has acceptor but not donor activity (31,32). Compound (**17**) also inhibits retroviral RNA-dependent DNA polymerase (33).

Cordycepin (3′-Deoxyadenosine). Cordycepin (**18**), also known as 3′-deoxyadenosine, is isolated from *C. militaris* and *A. nidulans* (1–4). The biosynthesis utilizes adenosine. It was originally proposed to be an analogue of adenine having the branched chain sugar, cordycepose. A functional analogue of adenosine, (**18**) is a most versatile biological probe for studying many reactions in procaryotes, eucaryotes, plants, and viruses. Initial studies indicated that (**18**) was converted to its 5′-phosphates and inhibits RNA, but not DNA, synthesis. A chain terminator in RNA synthesis, (**18**) inhibits poly(A) and tRNA methylation of the 5′-terminus of mRNA and blocks the production of RNA viruses. It inhibits 2′-*O*-methylation, forms the 3′-deoxy analogues of *S*-adenosylmethionine (SAM) and *S*-adenosylhomocysteine (SAH), and inhibits initiation of HSV DNA synthesis (34). Both SAM and SAH analogues inhibit nucleic acid methylation. When (**18**) replaces adenosine at the 3′-end of the $tRNA^{Phe}$, the addition of phenylalanine at the 2′-hydroxyl position results in a charged tRNA that is not a substrate for poly(U)-directed polyphenylalanine synthesis. 3′*d*ATP inhibits RNA and DNA synthesis in permeabilized *E. coli.* 3′*d*ATP has been reported to be a substrate for 2′,5′-oligoadenylate (2-5A) synthetase. The resulting 2′,5′-cordycepin trimer

5′-triphosphate inhibits translation in lysed rabbit reticulocytes better than authentic 2-5A (35). The inhibition of HIV-1 replication in infected MT-2 cells, and of HIV-1 reverse transcriptase by the cordycepin and phosphorothioate analogues of 2-5A has also been reported (36,37). Compound (**18**) damages messenger RNA (mRNA) processing in brain tissue, affects mRNA processing in yeast and *Xenopus* oocytes, increases mRNA stability in HL60 cells, affects mobility of nucleoside transport in cells, disrupts microtubule networks, and has been used to detect mRNAs in nonheat shocked thymic lymphocytes that code for noncordycepin inhibited heat shock proteins (38–47).

Puromycin. Puromycin (**19**), elaborated by *S. alboniger* (1–4), inhibits protein synthesis by replacing aminoacyl-tRNA at the A-site of peptidyltransferase (48,49). Photosensitive analogues of (**19**) have been used to label the A-site proteins of peptidyltransferase and $tRNA^{Phe}$ (50). Compound (**19**) and its carbocyclic analogue have been used to study the accumulation of glycoprotein-derived free sialooligosaccharides, accumulation of mRNA, methylase activity, enzyme transport, rat embryo development, the acceptor site of human placental 80S ribosomes, and gene expression in mammalian cells (51–60).

Three other 3′-aminoacyl-3′-deoxyadenosine nucleosides, homocitrullylaminoadenosine (**20**), lysylaminoadenosine (**21**), and chryscandin (**22**), have been reported (61). Unlike (**19**), chryscandin has strong inhibitory activity against *C. albicans* (61). Antibiotic FR900403, which is structurally similar to (**22**), has been isolated from the fungus, *Kernia* sp. F-19849; it inhibits *C. albicans* (62).

Antibiotic A201A. Antibiotic A201A (**23**), produced by *S. capreolus,* is an N^6-dimethyladenine nucleoside structurally similar to puromycin (**19**). Compound (**23**) which contains an aromatic acid and monosaccharide residues (1,4), inhibits the incorporation of amino acids into proteins but has no effect on RNA or DNA synthesis. Compound (**23**) does not accept polypeptides as does (**19**), and does appear to block formation of the initiation complex of the 50S subunit. It may block formation of a puromycin-reactive ribosome.

Ketohexose Nucleosides. The ketohexose nucleosides, psicofuranine (**24**) and decoyinine (**25**), are isolated from *S. hygroscopicus* (1–4). Psicofuranine, 6-amino-9-(β-D-psicofuranosyl)purine, and decoyinine, 9-β-D-(5,6-psicofuranoseenyl)-6-aminopurine, angustmycin A, inhibit XMP aminase noncompetitively. Unlike many other nucleoside antibiotics, phosphorylation is not necessary for inhibitory action. Compound (**25**) effectively induces sporulation; (**24**) inhibits GMP synthesis and subsequently AMP synthesis. Therefore (**24**) is sufficient to cause sporulation of *B. subtilis*. Guanine and adenine can reverse this inhibition. Compounds (**24**) and (**25**) regulate sporulation by induction of the alarmones in bacteria, ie, lowering levels of nucleotides and/or folate, enzyme promoters, and early sporulation genes (63–70).

Arabinosyladenine. Three naturally occurring arabinosyl purine and pyrimidine nucleoside antibiotics (ara-A, ara-T, and ara-U) have been isolated from the sponge, *Cryptotethia crypta* (1–4). 9-β-D-Arabinofuranosyladenine (**26**), also known as ara-A and Vidarabine, has also been isolated from *S. antibioticus* and *S. herbaceus*. (The adenosine deaminase inhibitor, 2′-deoxycoformycin (**59**), is also isolated from *S. antibioticus*). The structural elucidation and chemical syntheses of (**26**), its phosphate triester, and 2-halo and 5′-*O*-acyl derivatives have been reported (1–4,71–73). An enzymatic synthesis of (**26**) via transarabinosylation of

Table 2. Purine *N*-Nucleoside Antibiotics[a]

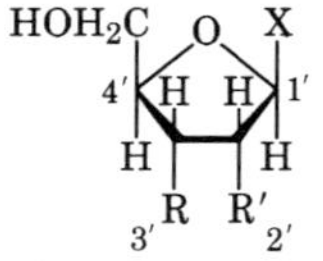

Name	CAS Registry Number	Molecular formula	Structure number	R	R′	X[b]
2′-amino-2′-deoxy-adenosine	[*10414-81-0*]	$C_{10}H_{14}N_6O_3$	(**15**)	OH	NH_2	A
2′-amino-2′-deoxy-guanosine	[*60966-26-9*]	$C_{10}H_{14}N_6O_4$	(**16**)	OH	NH_2	G
3′-amino-3′-deoxy-adenosine	[*2504-55-4*]	$C_{10}H_{14}N_6O_3$	(**17**)	NH_2	OH	A
cordycepin (3′-deoxyadenosine)	[*73-03-0*]	$C_{10}H_{13}N_5O_3$	(**18**)	H	OH	A
puromycin	[*53-79-2*]	$C_{22}H_{29}N_7O_5$	(**19**)	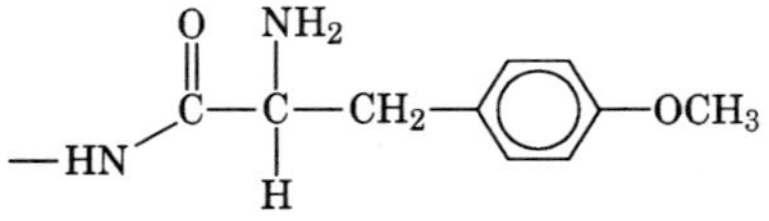	OH	*N*,*N*-dimethyl-A
homocitrullylamino-adenosine	[*59204-62-5*]	$C_{17}H_{27}N_9O_5$	(**20**)	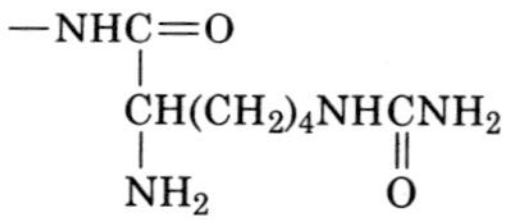	OH	A

lysylaminoadenosine	[*31518-64-6*]	$C_{16}H_{26}N_8O_4$	(**21**)	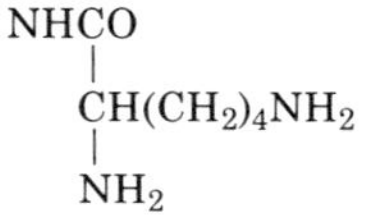		OH	A
chryscandin	[*86936-90-5*]	$C_{20}H_{23}N_7O_6$	(**22**)	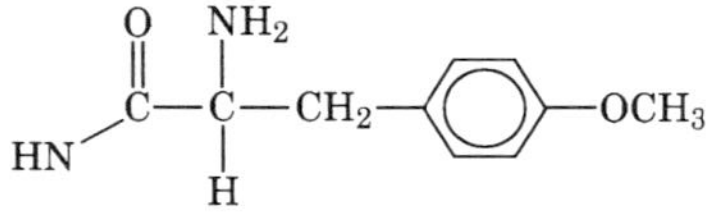		OH	A
A201A		$C_{37}H_{50}N_6O_{14}$	(**23**)	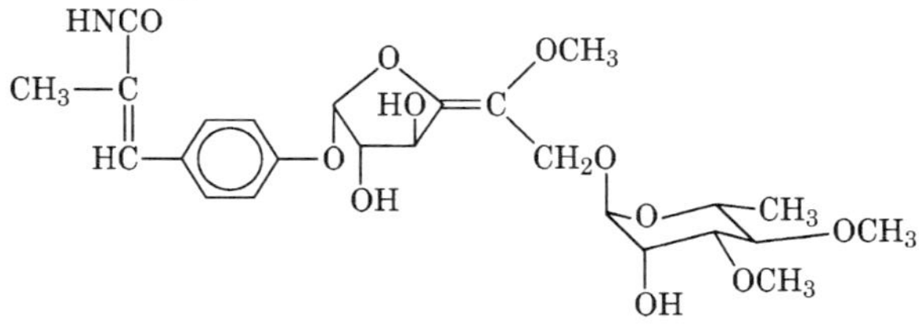		OH	*N,N*-dimethyl-A
psicofuranine[c]	[*1874-54-0*]	$C_{11}H_{15}N_5O_5$	(**24**)		OH	OH	A
decoyinine[c,d] (angustmycin A)	[*2004-04-8*]	$C_{11}H_{13}N_5O_4$	(**25**)		OH	OH	A

[a]Structure as shown unless otherwise indicated where R, R′, and X are as defined.

[b]A = adenine 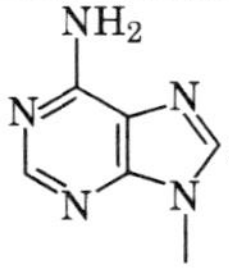; G = guanine 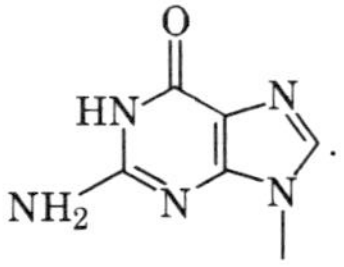.

[c]A CH_2OH group replaces the H at position 1′.

[d]The CH_2OH and H at position 4′ are replaced by a CH_2 group.

Table 3. Other Purine *N*-Nucleoside Antibiotics, Sugar-1′-X

Name	CAS Registry Number	Molecular formula	Structure number	Sugar[a]	X[b]
arabinofuranosyladenine (ara-A, vidarabine)	*[5536-17-4]*	$C_{10}H_{13}N_5O_4$	**(26)**	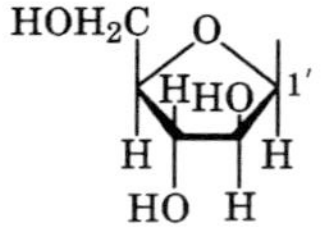	A
sinefungin	*[58944-73-3]*	$C_{15}H_{23}N_7O_5$	**(27)**	S, R = $CH_2CHNH_2(CH_2)_2CHNH_2COOH$	A
A9145A			**(28)**	S, R = $CH_2CHNH_2(CH_2)_2CHNH_2CONH_2$	A
A9145C	*[66753-47-7]*	$C_{15}H_{21}N_7O_5$	**(29)**	S, R = $CHCHNH_2(CH_2)_2CHNH_2COOH$	A
herbicidins	*[77699-46-8]*			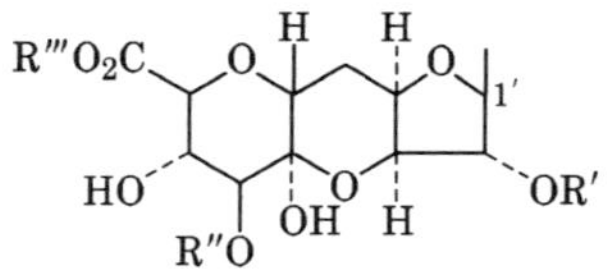	
				R′ R″ R‴	
herbicidin A	*[55353-31-6]*		**(30)**	CH_3 $CO(CH_2OH))C{=}CHCH_3$ CH_3	A
herbicidin B	*[55353-32-7]*		**(31)**	CH_3 H CH_3	A
herbidicin E	*[72067-15-3]*	$C_{23}H_{30}N_5O_{10}$	**(32)**	CH_3 $COCH(CH_3)_2$ CH_3	A
herbicidin F	*[72283-61-5]*		**(33)**	CH_3 $CO(CH_3)C{=}CHCH_3$ CH_3	A
herbicidin G	*[72283-62-6]*		**(34)**	H $CO(CH_3)C{=}CHCH_3$ H	A
doridosine	*[73027-05-1]*	$C_{11}H_{15}N_5O_5$	**(35)**	S, R = CH_2OH	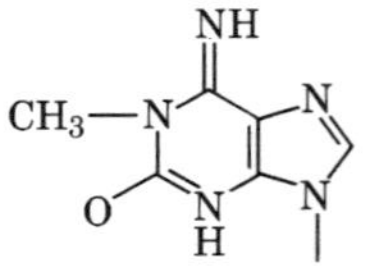
aristeromycin	*[19186-33-5]*	$C_{11}H_{15}N_5O_3$	**(36)**	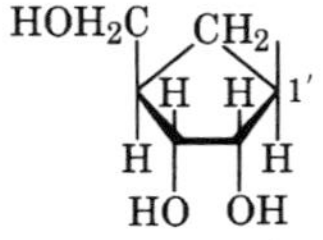	A
neplanocin A	*[72877-50-0]*	$C_{11}H_{13}N_5O_3$	**(37)**	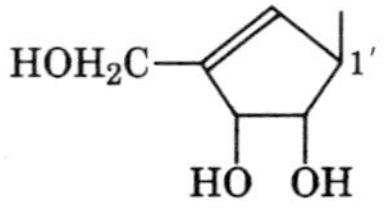	A

neplanocin B	[*72877-49-7*]	$C_{11}H_{13}N_5O_4$	(**38**)	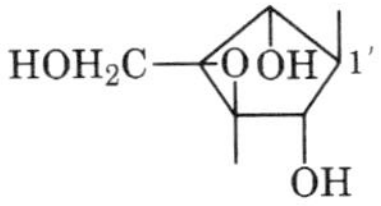	A
neplanocin C	[*72877-48-6*]	$C_{11}H_{13}N_5O_4$	(**39**)	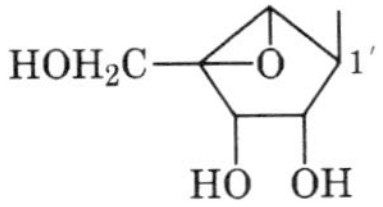	A
neplanocin D	[*72877-47-5*]	$C_{11}H_{12}N_4O_4$	(**40**)	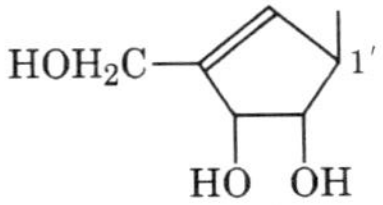	G
neplanocin F	[*72877-51-1*]	$C_{11}H_{13}N_5O_3$	(**41**)	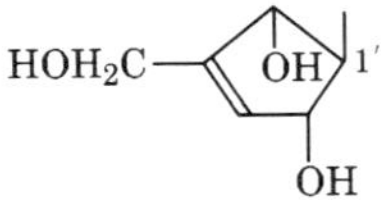	A
nucleocidin	[*24751-69-7*]	$C_{10}H_{13}FN_6O_6S$	(**42**)	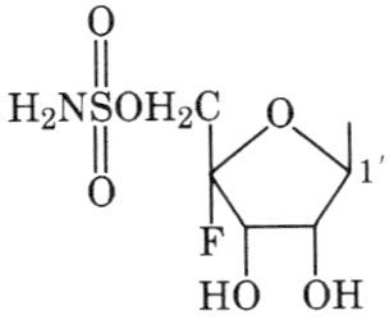	A
ascamycin	[*91432-48-3*]	$C_{13}H_{18}ClN_7O_7S$	(**43**)	S, R = 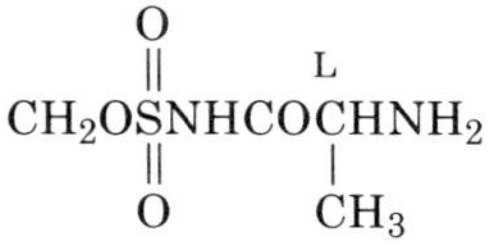	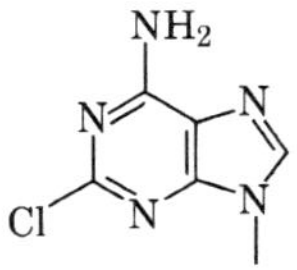
nebularine	[*550-33-4*]	$C_{10}H_{12}N_4O_4$	(**44**)	S, R = CH_2OH	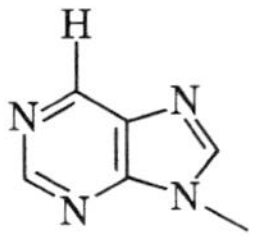
crotonoside	[*3373-53-3*]	$C_5H_5N_5O$	(**45**)	S, R = CH_2OH	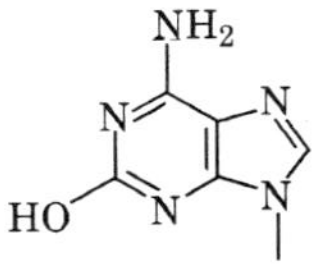

Table 3. (*Continued*)

Name	CAS Registry Number	Molecular formula	Structure number	Sugar[a]	X[b]
griseolic acid	[*79030-08-3*]	$C_{14}H_{13}N_5O_8$	(**46**)	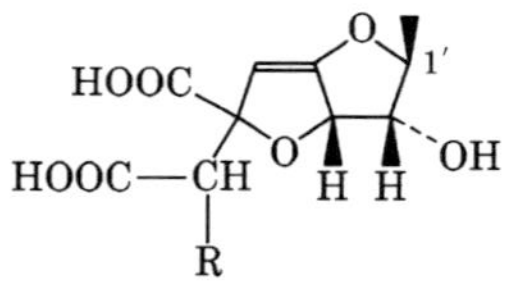 , R = H	A
griseolic acid B	[*98808-01-8*]	$C_{14}H_{13}N_5O_9$	(**47**)	, R = OH	A
griseolic acid C	[*100242-49-7*]	$C_{14}H_{15}N_5O_7$	(**48**)	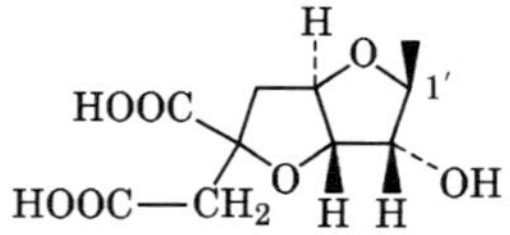	A
oxetanocin A	[*103913-16-2*]	$C_{10}H_{13}N_5O_3$	(**49**)	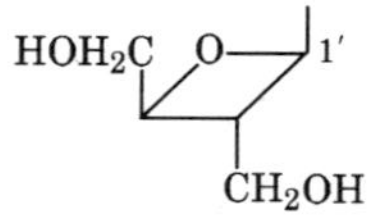	A
oxanosine	[*80394-72-5*]	$C_{10}H_{12}N_4O_6$	(**50**)	S, R = CH_2OH	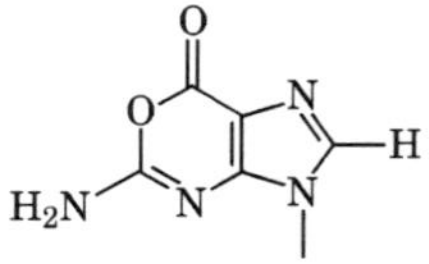

[a]S = 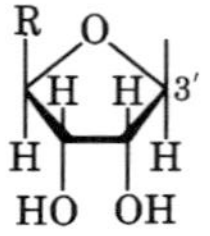 and R is as indicated.

[b]A = adenine 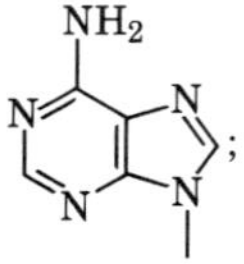; G = guanine 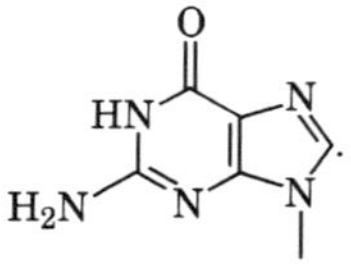.

ara-U has been reported (74) as has the biosynthesis of (**26**) from [^{14}C, ^{3}H, 2′-^{18}O and 3′-^{18}O]adenosine has been reported (75). Ara-A is a potent antiviral drug that shows great selectivity in its influence on host cell metabolism and virus growth (76–78). Ara-A and its 5′-monophosphate are effective antiviral analogues with known activity against herpes simplex keratitis, herpes simplex encephalitis, neonatal herpes, herpes zoster, and mucocutaneous herpes in humans. Ara-ATP is a competitive inhibitor of 2′*d*ATP. Compound (**26**) is incorporated into HSV DNA and it is an S-phase specific inhibitor of tumor cell growth (79). Many studies have reported the inhibition of DNA and RNA synthesis in procaryotes, eucaryotes, and viruses by ara-A and ara-ATP (1–4,80). Eucaryotic and procaryonuclear poly(A) polymerase, poly(ADP-ribose) polymerase, and nuclear poly(A) polymerase are not inhibited by ara-ATP (26,81,82). Although ara-ATP does not inhibit the DNA-dependent DNA polymerase from *E. coli,* it does inhibit the DNA-dependent DNA polymerase from mammalian cells. HSV-1 induced DNA polymerase is more sensitive to ara-ATP than is DNA polymerase-α in uninfected rabbit kidney cells (83). In L5178Y cells, (**26**) is deaminated to ara-hypoxanthine which is converted to ara-inosinetic phosphate ara-ITP; ara-ITP does not affect RNA or protein synthesis, but does inhibit DNA polymerase-α in L5178Y cells. In permeabilized mouse leukemic cells, the DNA replicating complex is competitively inhibited by ara-ATP (4). Ara-A inhibits DNA synthesis by blocking formation of new replicons and inhibits DNA repair, nucleoside uptake, and DNA strand elongation (84–87). 2′-Deoxycoformycin (**59**) potentiates the growth inhibition of (**26**) in cultured human cell lines (88). Inhibition of virus replication occurs when (**26**) is converted to ara-ATP (89).

Sinefungins. *S. griseolus* produces the nucleosides, sinefungin, A9145, A9145A, and A9145C (**27**–**29**) (1,4) which are structural analogues of SAM and SAH and inhibit methyltransferase reactions (90). The biosynthesis of (**27**) was reported to utilize L-arginine and ATP (91). It was demonstrated that adenosine and ornithine are converted to (**27**) (92) and it has been shown that adenine, ribose, and ornithine are the direct precursors of (**27**) (93). Regulation of the biosynthesis of (**27**) has been studied using auxotrophic mutants of *S. incarratus;* mutants with blocks in arginine synthesis were not able to produce (**27**) (94). Compounds (**27**–**29**) have antifungal, antiparasitic, and antiviral activity, and inhibit transformed cells in culture (95–97). They are potent inhibitors of the SAM:*O*-methyltransferases I and III (98,99). Compound (**27**) is the most potent inhibitor of protein methylase I activity and is comparable to SAH with respect to its inhibitory effect (100). It inhibits photolabeling of restriction enzyme EcoR-II methyltransferase, protein methylation, EcoR-I adenine DNA methylase, and myelin basic protein (arginine) methyltransferase (101–104). Methyltransferase inhibitors may provide a therapeutic strategy in myasthenia gravis (105).

Herbicidins. Five adenine nucleosides, herbicidins A, B, E, F, and G (**30**–**34**), have been isolated from the culture filtrates of *S. sagamonensis* No. 4075 (1,4) which also produces ara-A and 2′-deoxycoformycin. The herbicidins contain a tricyclic dodecose. They inhibit dicotyledonous plants, germination of plant seeds, growth of blue-green algae, and fungi.

Doridosine. Doridosine, N^1-methylisoguanosine, (**35**) was isolated from the dorid nudibranchs of *Anisodoris nobilis* and the sponge, *Tedania* (106,107). The injection of (**35**) into the saphenous vein of anesthetized rats produces hypotension and bradycardia almost immediately. The observed changes in the electrocardiograms are minor and indicate little interference with conduction of the impulse within the heart (see CARDIOVASCULAR AGENTS).

Thraustomycin. Thraustomycin and β-thraustomycin are isolated from *S. exfoliatus* (4). Although their structures have not been totally elucidated, hydrolysis of thraustomycin shows that it contains equimolar quantities of adenine, L-leucine, and a tetrahydroxymonocarboxylic acid. Thraustomycin is a potent inhibitor of the fungus, *M. hiemallis* (+), but does not inhibit bacteria.

Aristeromycin. Aristeromycin (**36**), the first carbocylic analogue of adenosine, was isolated from the culture filtrates of *S. citricolor* as part of a search for inhibitors of bacterial leaf blight (1–4). A herbicidally active hypoxanthine analogue of (**36**), coaristeromycin, has also been isolated (108). Several chemical syntheses of (**36**) have appeared (1–4,109). It inhibits *Xanthomonas oryzae* and *Pyricularia oryzae,* bacterial leaf blight, blast disease of rice plants, and leaf mold of tomato plants, by inhibiting both cell division and elongation. An inhibitor of *S*-adenosylmethionine (SAM) hydrolase, (**36**) also affects *de novo* synthesis of purine nucleotides and RNA. Adenosine and, to a lesser degree, adenine, inosine and 2′-deoxyadenosine reverse inhibition by (**36**). Aristeromycin diphosphate binds to polynucleotide phosphorylase. The copolymer poly(A, aristeromycin) is formed when aristeromycin 5′-diphosphate is incubated with an excess of ADP. Aristeromycin derivatives of *S*-adenosyl homocysteine (SAH) have been synthesized in an effort to characterize the SAH binding site on mRNA (guanine-7)-methyltranferases that are required for efficient translation and mRNA-ribosome binding. Whereas the aristeromycin derivatives of SAH showed only marginal inhibitory activity, *S*-tubercidinyl-L-homocysteine was a potent inhibitor. The metabolism of analogues of (**36**) and neplanocin A (**37**) have been compared (110). Both nucleosides are converted to the 5′-triphosphates; however, only (**36**) was converted to a guanosine nucleotide derivative. A carbocyclic inosine analogue of (**36**) has potent antileishmanial activity (111). The 3-deazaaristeromycin derivative is an inhibitor of SAH hydrolase and has been widely used in studies related to RNA synthesis in transformed cells and viruses and cell differentiation (112–118). 3-Deazaaristeromycin-resistant clones of human β-lymphoblasts have been reported to display elevated SAM levels (119). The biosynthesis of the cyclopentane ring of (**36**) proceeds by carbon–carbon bond formation between C-2 and C-6 of D-glucose (120).

Neplanocins. Neplanocins A–D and F (**37**–**41**) are carbocyclic nucleoside antibiotic products of *Ampullariella regularis* (1,4) that are structurally related to (**36**) in that they contain either a cyclopentene or epoxy cyclopentane ring (121,122). The chemical syntheses of (**37**–**41**) and the 3-deazaneplanocins have been reported (123–126). Compound (**37**), which is converted to its 5′-triphosphate, has potent antitumor and antiviral activities (127–129). It strongly inhibits SAM in cells and viruses (128–131) and is converted to the 3′-keto derivative by *S*-adenosylhomocysteine hydrolase (132,133).

Nucleocidin and Ascamycin. Nucleocidin, 4′-fluoro-5′-*O*-sulfamoyladenosine, (**42**) is isolated from the culture filtrates of *S. clavus* (1–4). Among the nucleoside antibiotics (**42**) is unique in that it contains a fluoride group at carbon-4′. Compound (**42**) and its chemically synthesized analogues are inhibitors of bacteria and are extremely toxic to mammals. Compound (**42**) inhibits protein synthesis by blocking peptide elongation. Ascamycin, 5′-*O*-L-alanylsulfamoyl-2-chloroadenosine, (**43**) is similar to (**42**) in that it contains a sulfamoyl group (134,135). A specific amino-peptidase that hydrolyzes the alanyl group of (**43**) has been purified (136). Two additional 5′-*O*-sulfamoylnucleosides that are isolated from the *Streptomyces* are 5′-*O*-sulfamoyladenosine and 5′-*O*-sulfamoyltubercidin (137).

Nebularine. Nebularine (**44**) is a naturally occurring purine riboside isolated from *S. yokosukanensis* (1,3,4). It is phosphorylated, and inhibits purine biosynthesis and RNA synthesis, but is not incorporated into RNA by *E. coli* RNA polymerase. It has also found application as a transition state analogue for treatment of schistosomiasis and as a substrate for the restriction endonuclease, HindII (138–141).

Crotonoside. Crotonoside, also called isoguanosine, (**45**) has been isolated from *Croton tiglium* (4). Ip administration of [2-^{14}C]crotonoside to rats results in incorporation into nucleic acids (qv). It inhibits the inducible binding sites, inhibits glutamic acid dehydrogenase, and accumulates cAMP.

Griseolic Acids. The three griseolic acids, A, B, and C (**46**–**48**), have been isolated from *S. griseoaurantiacus* (1,142). They are potent inhibitors of cAMP phosphodiesterase (143,144). Compound (**46**) stimulates glycogen degradation in mice.

Oxetanocins. Oxetanocin A (**49**), formerly oxetanocin, is the first naturally occurring oxetanose derivative and is isolated from *Bacillus megaterium* (1,145). It inhibits gram-positive bacteria, herpes viruses, and human immunodeficiency virus (HIV) (146). The chemical synthesis of (**49**) and several derivatives has been reported (147).

Oxanosine. Oxanosine (**50**), which is isolated from *S. capreolus* (1,148), is a potent inhibitor of GMP synthetase and nucleic acid synthesis (149,150).

Pyrrolopyrimidine Nucleosides. The pyrrolopyrimidine *N*-nucleoside antibiotics are listed in Table 4. Tubercidin (**51**), toyocamycin (**52**), sangivamycin (**53**), cadeguomycin (**54**), and kanagawamicin (**55**) are pyrrolopyrimidine nucleoside antibiotics isolated from strains of *Streptomyces* (1–4,151). In addition, the 5′-α-D-glucopyranosyl derivatives of tubercidin and toyocamycin have been isolated from blue-green algae (152). The structural elucidation, physical and chemical properties, chemical syntheses, and biological activities of these nucleosides and their derivatives have been reported (1–4,153–172). GTP, but not ATP, is a precursor for the biosynthesis of the pyrrolopyrimidine ring of (**51**–**53**) (161). The nitrile group of (**52**) is enzymatically converted to the carboxamide group in the formation of (**53**) (162). Tubercidin 5′-triphosphate is not a substrate for either adenosine deaminase or nucleoside phosphorylase. Compound (**51**) is incorporated into RNA, inhibits the processing of RNA, but does not inhibit the transcription of 45S precursor RNA to 28S and 18S RNA; DNA synthesis is inhibited. The importance of (**51**) as a methylase inhibitor has been studied in detail (163,164). *S*-Tubercidinyl homocysteine inhibits methylation (163,164). It has been used to elucidate the requirements for the catalytic site of RNA polymerase with respect to initiation, elongation, and termination. Rous sarcoma virus and SV40 replication are inhibited by 5-bromotubercidin. Another promising application of (**51**) is as an anthelmintic agent. It inhibits purine nucleotide synthesis in *Schistosoma mansoni* and xylotubercidin inhibits HSV-2 in mice (165,166).

Table 4. Pyrrolopyrimidine, Diazepin, and Other *N*-Nucleoside Antibiotics, Sugar-1′-X

Name	CAS Registry Number	Molecular formula	Structure number	Sugar[a]	X
Pyrrolopyrimidine N-nucleosides					
tubercidin	[*69-33-0*]	$C_{11}H_{14}N_4O_4$	**(51)**	S	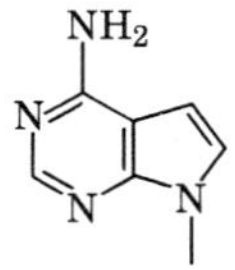
toyocamycin	[*606-58-6*]	$C_{12}H_{13}N_5O_4$	**(52)**	S	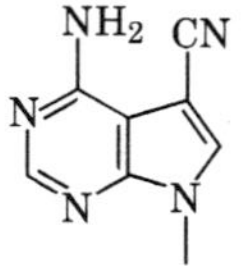
sangivamycin	[*18417-89-5*]	$C_{12}H_{15}N_5O_5$	**(53)**	S	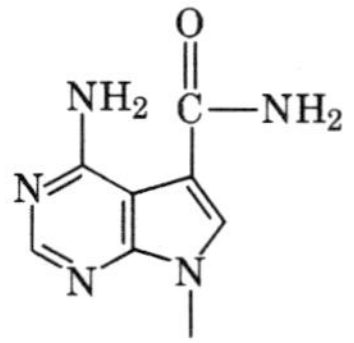
cadeguomycin	[*81645-08-1*]	$C_{12}H_{14}N_4O_7$	**(54)**	S	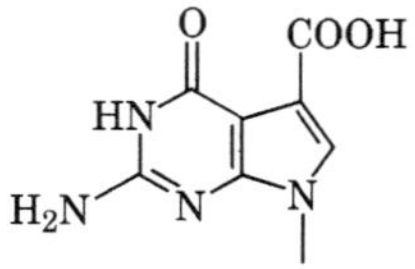
kanagawamicin	[*84873-16-5*]	$C_{13}H_{17}N_5O_6$	**(55)**	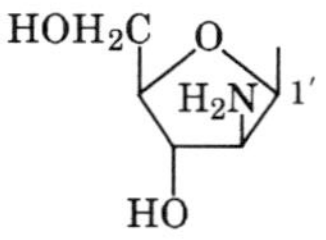	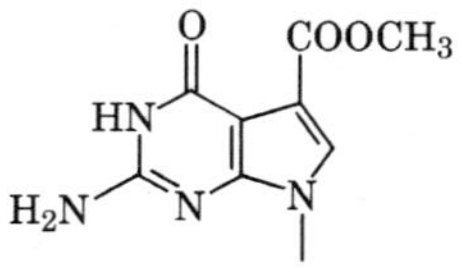
mycalisine A mycalisine B	[*98890-73-4*] [*98890-72-3*]	$C_{13}H_{13}N_5O_3$ $C_{13}H_{12}N_4O_4$	**(56)** **(57)**	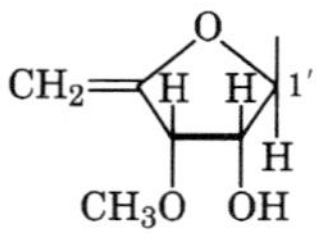	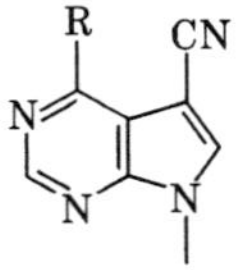 R = NH_2 R = OH

Diazepin N-nucleosides					
coformycin	[*11033-22-0*]	$C_{11}H_{16}N_4O_5$	**(58)**	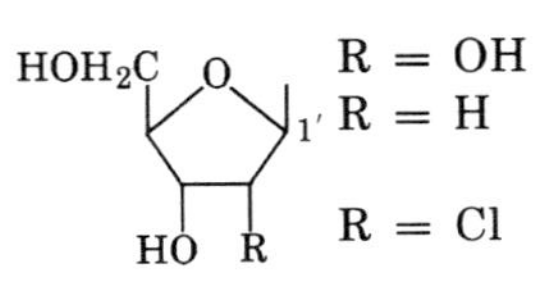	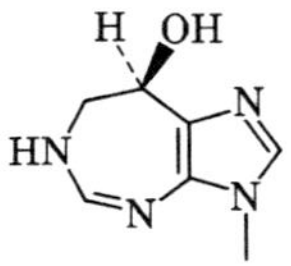
2′-deoxycoformycin (co-vidarabine, pentostatin)	[*53910-25-1*]	$C_{11}H_{16}N_4O_4$	**(59)**		
adechlorin (2′-chloro-2′-deoxycoformycin)	[*96328-17-5*]	$C_{11}H_{15}ClN_4O_4$	**(60)**		
adecypenol	[*104493-13-2*]	$C_{12}H_{16}N_4O_4$	**(61)**	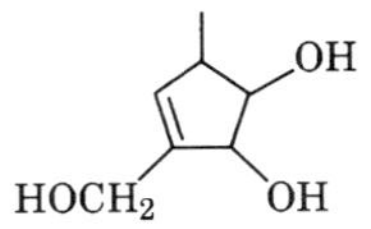	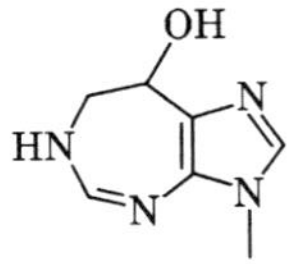
		Others			
bredinin	[*50924-49-7*]	$C_{19}H_{13}N_3O_6$	**(62)**	S	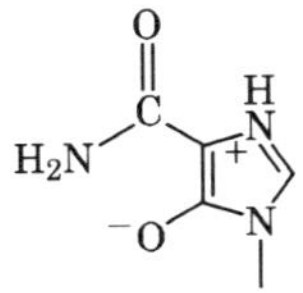
neosidomycin	[*72033-44-4*]	$C_{17}H_{20}N_2O_6$	**(63)**	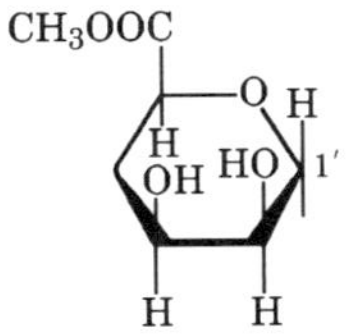	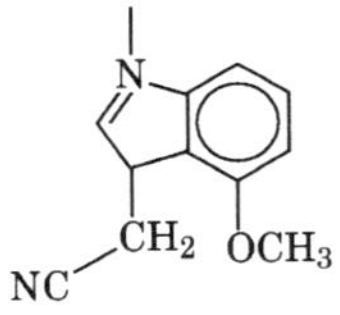
SF-2140	[*91284-30-9*]	$C_{18}H_{20}N_2O_6$	**(64)**	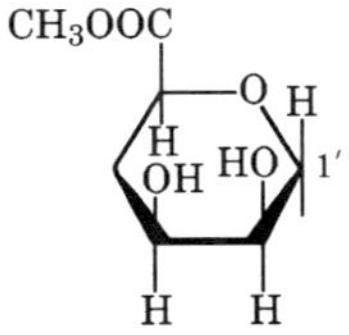	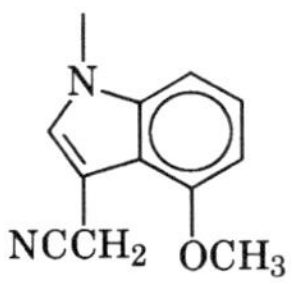

[a] S = 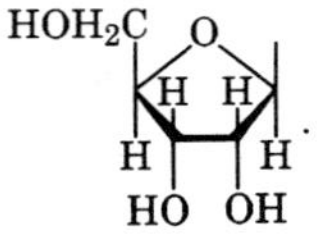

Toyocamycin (**52**), a cytotoxic analogue of adenosine, is selectively incorporated into the 45S ribosomal RNA precursor and prevents cleavage into 28S and 18S RNA (1,4). When Ehrlich ascites cells are incubated with (**52**) in a medium rich in amino acids, the high transcription rate of rRNA is lowered. Toyocamycin is incorporated into HeLa cell 45S RNA, blocks polyadenylation of heterogenous nuclear RNA (hnRNA), and inhibits Friend, Rous Sarcoma, and vesicular stomatitis virus replication. Sangivamycin (**53**), which is phosphorylated but not deaminated, is incorporated into RNA and DNA of normal tissue, inhibits *de novo* purine synthesis in L1210 cells in culture, competitively inhibits amino acid activation, charging of tRNA, and can replace ATP in the synthesis of RNA catalyzed by *E. coli* RNA polymerase. The molecular effects of (**53**) and its thio analogue on RNA and DNA synthesis in L1210 cells and as an inhibitor of rhodopsin kinase have been studied (167,168). Compound (**54**) potentiates the cytotoxicity of ara-C and has been used in studies on the T-cell immunostimulatory properties of certain peptides (169–171). Compound (**55**) has antitumor and antibacterial activity (172).

Two other pyrrolopyrimidine nucleoside antibiotics, mycalisines A and B (**56, 57**), have been isolated from the sponge *Mycale* sp. (173). These bioactive marine metabolites inhibit cell division of starfish eggs.

Diazepin Nucleosides. Four naturally occurring diazepin nucleosides, coformycin (**58**), 2′-deoxycoformycin (**59**), adechlorin or 2′-chloro-2′-deoxycoformycin (**60**), and adecypenol (**61**), have been isolated (1–4,174,175). The biosynthesis of (**59**) and (**60**) have been reported to proceed from adenosine and C-1 of D-ribose (30,176,177). They are strong inhibitors of adenosine deaminase and AMP deaminase (178). Compound (**58**) protects adenosine and formycin (**12**) from deamination by adenosine deaminase. Advanced hairy cell leukemia has shown rapid response to (**59**) with or without α- or β-interferon treatment (179–187). In addition, (**59**) affects interleukin-2 production, receptor expression on human T-cells, DNA repair synthesis, immunosuppression, natural killer cell activity, and cytokine production (188–194).

Bredinin, Neosidomycin, and SF-2140. Bredinin (**62**), isolated from the culture filtrates of *Eupenicillium brefeldianum* (1,4), inhibits the multiplication of L5178Y, HeLa S3, RK-13, mouse L-cells, and Chinese hamster cells. GMP can reverse the inhibition by (**62**), but (**62**) is not incorporated into the nucleic acids. The inhibition of nucleic acid synthesis and chromosomal damage in the S and G_2 phases that is caused by (**62**), is reversed by GMP. It blocks the conversion of IMP to XMP and XMP to GMP. In combination with GMP, (**62**) interferes with intracellular cAMP levels and thereby inhibits cell division.

Neosidomycin (**63**) and SF-2140 (**64**) are indole *N*-glycosides produced by *S. hygroscopicus* and *Actinomadura,* respectively (195,196). A revised structure for (**63**) has appeared (197). Compound (**64**) contains an α-*N*-glycoside linkage. Both (**63**) and (**64**) show activity against gram-positive bacteria, yeast, fungi, and viruses.

Pyrimidine–Nucleoside Antibiotics. *Fatty Acyl Nucleosides.* The nucleoside antibiotics with fatty acyl groups containing adenine, uracil, and dihydrouracil aglycons, the tunicamycins [*11089-65-9*] (**65**–**74**), streptovirudins (**75**–**84**), corynetoxins (**85**–**97**), are given in Table 5. The structure of liposidomycin (**98**), septacidins [*62362-59-8*] (**99**), spicamycin [*62362-59-8*] (**100**), and FR900848 are given in Figure 1. The chemistry of the tunicamycins has been extensively reviewed (198). The structural elucidation of (**75**–**78**) has been described (199). The isolation and structural elucidation of liposidomycins A, B, and C, septacidins, corynetoxins, and spicamycins have been described (200–205). Compounds (**85**–**97**), produced by *Corynebacterium rathayi,* infect rye grass which is toxic to grazing animals (205). A recently discovered fatty acyl nucleoside antibiotic with antifungal activity, antibiotic FR900848 [*120500-69-8*] (**101**), contains one isolated and four serial cyclopropane rings (206).

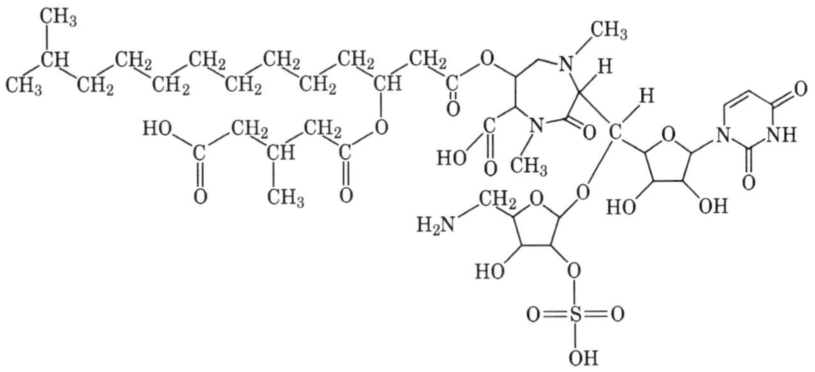

Liposidomycin B (**98**)

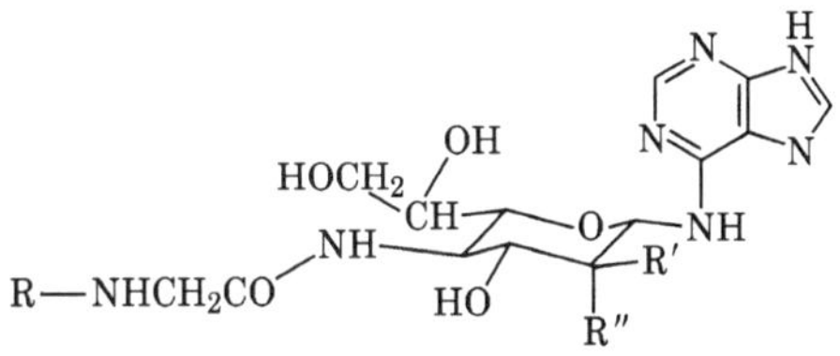

Septacidin (**99**, R′ = OH, R″ = H) R = $(CH_3)_2CH(CH_2)_nCO-$,
Spicamycin (**100**, R′ = H, R″ = OH) $CH_3CH_2CH_2(CH_2)_nCO-$
$n = 8-14$

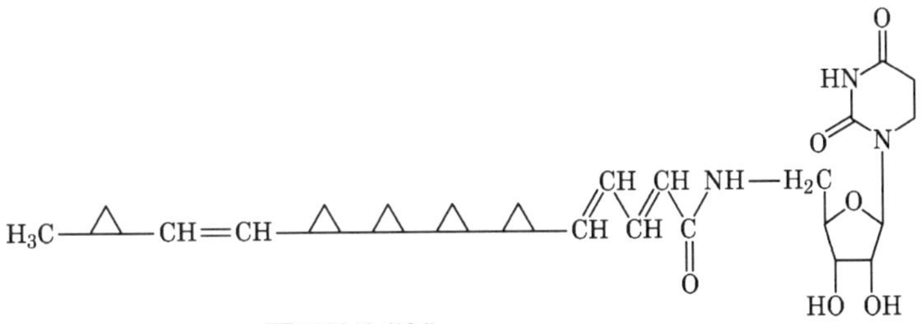

FR900848 (**101**)

Fig. 1. Structures of fatty acyl *N*-nucleosides.

The tunicamycins, streptovirudins, corynetoxins, and liposidomycins inhibit the formation of glycoproteins and peptidoglycans. Oligosaccharide synthesis for glycoproteins requires a polyisoprenoid carrier followed by transfer to the nascent polypeptide (207–209). The tunicamycins and streptovirudins are

Table 5. Pyrimidine Fatty Acyl *N*-Nucleoside

Name	CAS Registry Number	Molecular formula	Structure number	R
		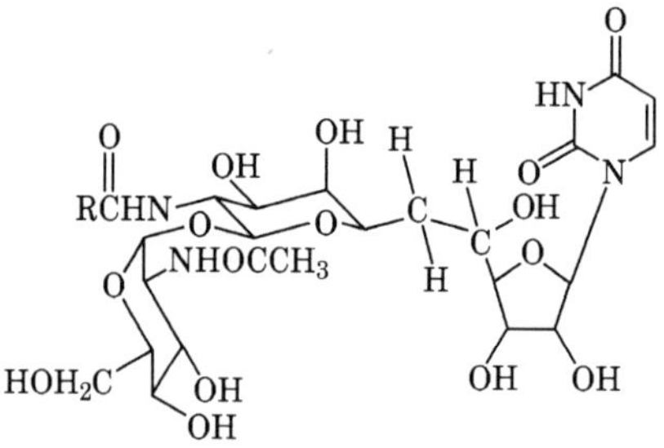		
Tunicamycins [11089-65-9]				
I(A_0)	[*73942-10-6*]	$C_{36}H_{58}N_4O_{16}$	**(65)**	$(CH_3)_2CH(CH_2)_7\overset{H}{C}=\underset{H}{C}-$
II(C)	[*66081-37-6*]	$C_{37}H_{60}N_4O_{16}$	**(66)**	$(CH_3)_2CH(CH_2)_8\overset{H}{C}=\underset{H}{C}-$
III(A_2)	[*76544-45-1*]	$C_{37}H_{60}N_4O_{16}$	**(67)**	$CH_3(CH_2)_{10}\overset{H}{C}=\underset{H}{C}-$
IV(B_2)	[*73942-09-3*]	$C_{38}H_{62}N_4O_{16}$	**(68)**	$CH_3(CH_2)_{11}\overset{H}{C}=\underset{H}{C}-$
V(A)	[*66054-36-2*]	$C_{38}H_{62}N_4O_{16}$	**(69)**	$(CH_3)_2CH(CH_2)_9\overset{H}{C}=\underset{H}{C}-$
VI	[*88263-43-8*]	$C_{38}H_{62}N_4O_{16}$	**(70)**	$(CH_3)_2CH(CH_2)_{11}-$
VII(B)	[*66081-36-5*]	$C_{39}H_{64}N_4O_{16}$	**(71)**	$(CH_3)_2CH(CH_2)_{10}\overset{H}{C}=\underset{H}{C}-$
VIII(C_2)	[*73942-07-1*]	$C_{39}H_{64}N_4O_{16}$	**(72)**	$CH_3(CH_2)_{12}\overset{H}{C}=\underset{H}{C}-$
IX(D_1)	[*73942-08-2*]	$C_{40}H_{66}N_4O_{16}$	**(73)**	$CH_3(CH_2)_{13}\overset{H}{C}=\underset{H}{C}-$
X(D)	[*66081-38-7*]	$C_{40}H_{66}N_4O_{16}$	**(74)**	$(CH_3)_2CH(CH_2)_{11}\overset{H}{C}=\underset{H}{C}-$
Streptovirudins (A_1–D_1)[a]				
A_1	[*51330-28-0*]	$C_{35}H_{58}N_4O_{16}$	**(75)**	$(CH_3)_2CH(CH_2)_6CH=CH-$

Table 5. ***(Continued)***

Name	CAS Registry Number	Molecular formula	Structure number	R
B_1	[*51330-30-4*]	$C_{36}H_{60}N_4O_{16}$	**(76)**	$CH_3CH_2\overset{CH_3}{\overset{\vert}{C}}H(CH_2)_6CH{=}CH-$
B_{1a}	[*81093-26-7*]	$C_{36}H_{60}N_4O_{16}$	**(77)**	$(CH_3)_2CH(CH_2)_7CH{=}CH-$
C_1	[*51330-32-6*]	$C_{37}H_{62}N_4O_{16}$	**(78)**	$(CH_3)_2CH(CH_2)_8CH{=}CH-$
D_1	[*51330-34-8*]	$C_{38}H_{64}N_4O_{16}$	**(79)**	$CH_3CH_2\overset{CH_3}{\overset{\vert}{C}}H(CH_2)_8CH{=}CH-$
Streptovirudins (A_2–D_2)[b]				
A_2	[*51330-29-1*]	$C_{35}H_{56}N_4O_{16}$	**(80)**	$(CH_3)_2CH(CH_2)_6CH{=}CH-$
B_2	[*51330-31-5*]	$C_{36}H_{58}N_4O_{16}$	**(81)**	$CH_3CH_2\overset{CH_3}{\overset{\vert}{C}}H(CH_2)_6CH{=}CH-$
B_{2a}	[*73942-10-6*]	$C_{36}H_{58}N_4O_{16}$	**(82)**	$(CH_3)_2CH(CH_2)_7CH{=}CH-$
C_2	[*51330-33-7*]	$C_{37}H_{60}N_4O_{16}$	**(83)**	$(CH_3)_2CH(CH_2)_9CH{=}CH-$
D_2	[*51330-35-9*]	$C_{38}H_{62}N_4O_{16}$	**(84)**	$CH_3CH_2\overset{CH_3}{\overset{\vert}{C}}H(CH_2)_5CH{=}CH-$
Corynetoxins[c]				
H16*i*	[*82138-73-6*]	$C_{39}H_{66}N_4O_{17}$	**(85)**	$(CH_3)_2CH(CH_2)_{10}\overset{OH}{\overset{\vert}{C}}HCH_2-$
H18*i*	[*82138-74-7*]	$C_{41}H_{70}N_4O_{17}$	**(86)**	$(CH_3)_2CH(CH_2)_{12}\overset{OH}{\overset{\vert}{C}}HCH_2-$
H17*a*	[*82151-55-1*]	$C_{40}H_{68}N_4O_{17}$	**(87)**	$CH_3CH_2\overset{CH_3}{\overset{\vert}{C}}H(CH_2)_{10}\overset{OH}{\overset{\vert}{C}}HCH_2-$
H19*a*	[*82151-56-2*]	$C_{42}H_{72}N_4O_{17}$	**(88)**	$CH_3CH_2\overset{CH_3}{\overset{\vert}{C}}H(CH_2)_{12}\overset{OH}{\overset{\vert}{C}}HCH_2-$
U16*i*	[*66081-36-5*]	$C_{39}H_{64}N_4O_{16}$	**(89)**	$(CH_3)_2CH(CH_2)_{10}CHCH-$
U18*i*	[*82138-75-8*]	$C_{41}H_{68}N_4O_{16}$	**(90)**	$(CH_3)_2CH(CH_2)_{12}CHCH-$
U17*a*	[*82151-57-3*]	$C_{40}H_{66}N_4O_{16}$	**(91)**	$CH_3CH_2\overset{CH_3}{\overset{\vert}{C}}H(CH_2)_{10}CHCH-$
U19*a*	[*82138-76-9*]	$C_{42}H_{70}N_4O_{16}$	**(92)**	$CH_3CH_2\overset{CH_3}{\overset{\vert}{C}}H(CH_2)_{12}CHCH-$
S16*i*	[*82138-77-0*]	$C_{39}H_{66}N_4O_{16}$	**(93)**	$(CH_3)_2CH(CH_2)_{12}-$
S18*i*	[*82138-78-1*]	$C_{41}H_{70}N_4O_{16}$	**(94)**	$(CH_3)_2CH(CH_2)_{14}-$
S15*a*	[*82138-79-2*]	$C_{38}H_{64}N_4O_{16}$	**(95)**	$CH_3CH_2\overset{CH_3}{\overset{\vert}{C}}H(CH_2)_{10}-$
S17*a*	[*82138-80-5*]	$C_{40}H_{68}N_4O_{16}$	**(96)**	$CH_3CH_2\overset{CH_3}{\overset{\vert}{C}}H(CH_2)_{12}-$
S19*a*	[*82138-87-6*]	$C_{42}H_{72}N_4O_{16}$	**(97)**	$CH_3CH_2\overset{CH_3}{\overset{\vert}{C}}H(CH_2)_{14}-$

[a]Dihydrouracil incorporated, ie, uracil C=C bond is saturated.
[b]Uracil as shown in structure.
[c]The H stands for β-hydroxy, S for saturated, and U for α–β unsaturated; i = iso; a = anteiso.

lipophilic nucleoside analogues of UDP-*N*-acetylglucosamine and inhibit the first transferase reaction in the lipid-linked saccharide pathway, ie, the formation of dolichyl pyrophosphoryl-*N*-acetylglucosamine, by inhibiting dolichyl-P:*N*-acetylglucosamine-1-*P* transferase. Compounds (**65**–**74**) are also biological probes for cell wall polymer synthesis in bacteria, fungi, immunoglobulin synthesis, and glycoprotein biosynthesis in insects and plants. By blocking synthesis of the phosphomannosyl residues on lysosomal enzymes with tunicamycin, the secretion of the lysosomal enzymes is inhibited as is the synthesis of rhodopsin and the conversion of procollagen to collagen (210–212). The effect of the tunicamycins and streptovirudins on adherence of group B streptococci to influenza virus-infected kidney cells has been investigated (213). In hen oviduct microsomes, tunicamycin abolished *N*-glycosylation, but had no effect on *O*-glycosylation (214). Tunicamycin inhibits peptidoglycan and 1,3-poly(glycerol phosphate) teichoic acid in *B. licheniformis*, the formation of undecaprenylacetyl-glucosaminyl-lipid and the formation of the peptidoglycan and teichuronic acid, *N*-acetylglucosamine-lipid formation in plants, glycosylation of human interferon, glycosaminoglycan synthesis in cultured embryonic chick fibroblasts, and polysaccharide chain formation of corneal keratan sulfate (215–218). Tunicamycin has also been used to investigate virus assembly. The reduction of glycosamine incorporation caused by tunicamycin is accompanied by a decrease in infectious virus particle production (219). In a model system for *N*-glycosylation of the intracellular transport of rabies virus, tunicamycin blocks surface expression and accumulation of G protein (220), the first step in the lipid cycle of *E. coli* and *S. cerevisiae* (221) and IgE binding protein (222). Tunicamycin inhibits the incorporation of *O*-linked sugars on the CD45 phosphotyrosine phosphatase (223), glycosylation of tumor necrosis factor (224), causes accumulation of dolicholphosphoryl-mannose, an essential anchor to CHO cell mutants (225), and inhibits lipoprotein lipase complex mannose *N*-linked oligosaccharides (226). Monoclonal antibodies to HSV glycoproteins did not react with cell extracts of tunicamycin treated cells (227). The inhibition of *N*-glycosylation processing and transport surface expression also has been reported (227–231).

Peptidyl *N*-Nucleoside Antibiotics. *Polyoxins and Neopolyoxins.* The polyoxins [*11113-80-7*] (**102**–**113**), and neopolyoxins (**114**–**116**), shown in Figure 2, are peptidylpyrimidine nucleoside antibiotics that have achieved use as agricultural fungicides (1–4,232) (see FUNGICIDES, AGRICULTURAL). Polyoxins A–M, which have the pyrimidine chromophore, have been isolated from *S. cacaoi* var *asoensis* and *S. piomogenus*. Polyoxin N [*37362-29-1*] (**113**), $C_{16}H_{23}N_5O_{12}$, neopolyoxin A [*75044-69-8*] [*72864-26-7*] (**114**), $C_{20}H_{25}N_5O_{10}$, and neopolyoxin B [*75005-71-9*] (**115**), $C_{20}H_{25}N_5O_{11}$, have the imidazole chromophore and neopolyoxin C [*75044-70-1*] (**116**), $C_{20}H_{25}N_5O_{10}$, contains uracil and a 3-hydroxypyridine rings (1–4). The thymine chromophore of the polyoxins is formed by a pathway independent of thymidylate synthetase (233). The aminouronic acid moiety is not directly synthesized from glucose, but rather via phosphoenolpyruvate and the 5′-aldehyde of uridine to give the intermediate octofuranose uronic acid nucleoside (1). Oxidative elimination of the two terminal carbons (carbon 7′ and 8′) followed by the addition of the amino group on carbon-5′ results in the formation of the polyoxin nucleoside. Azetidine carboxylic acid (polyoximic acid) is formed from the carbon–nitrogen skeleton of isoleucine and carbamoylpolyoxamic acid, is formed

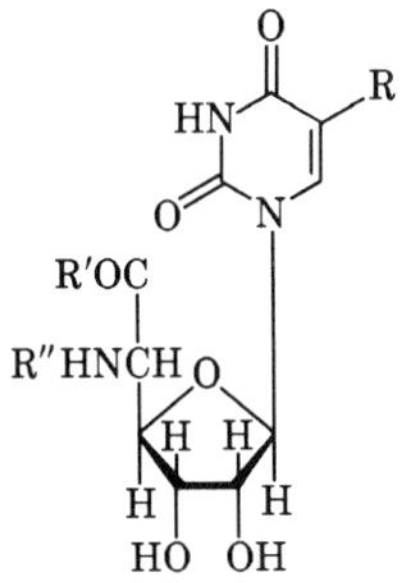

Polyoxins A–M (**102–112**)
R = -CH_2OH, -COOH, -CH_3, -H

R′ = -OH, 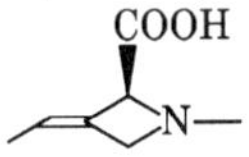

R″ = -H, 5-*O*-carbamoyl-2-amino-2-deoxy-L-xylonic acid or the 3-deoxy derivative

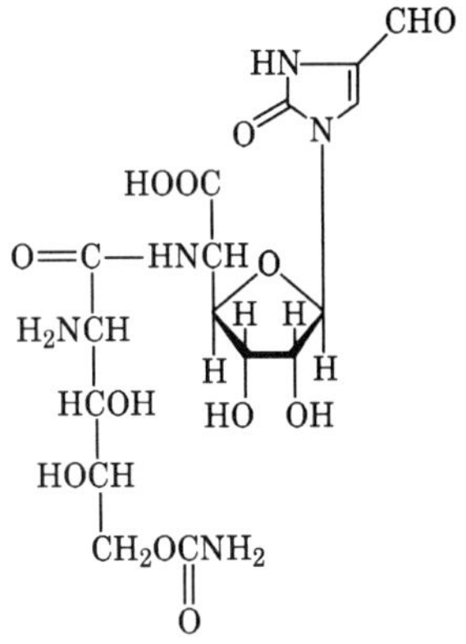

Polyoxin N (**113**)

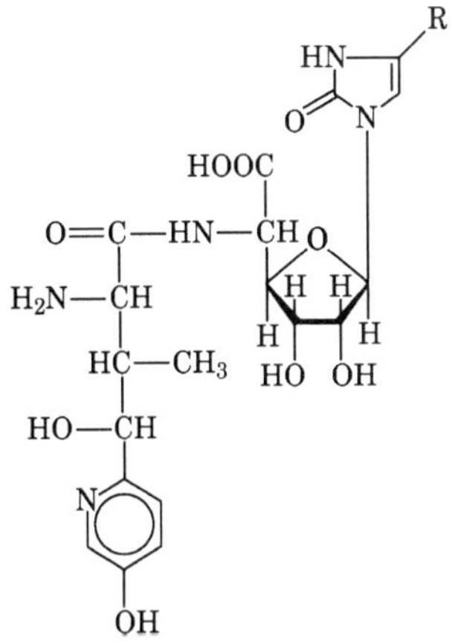

Neopolyoxin A (**114**, R = CHO)
Neopolyoxin B (**115**, R = COOH)

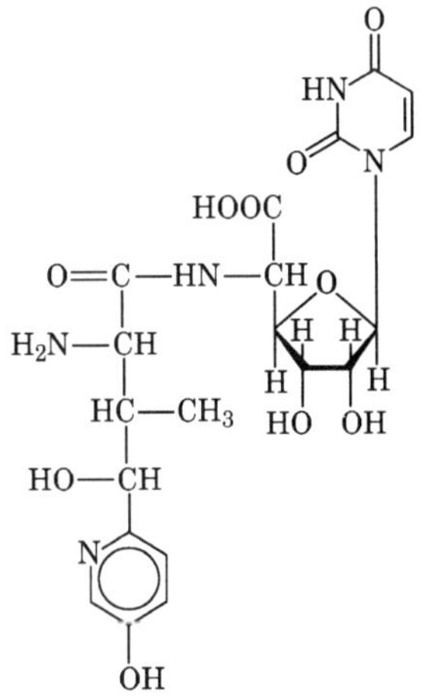

Neopolyoxin C (**116**)

Fig. 2. Structures of the polyoxins. Polyoxin A [*19396-03-3*] [*11016-31-2*], $C_{23}H_{32}N_6O_{14}$; polyoxin B [*19396-06-6*] [*11016-32-3*] [*24695-50-9*], $C_{17}H_{25}N_5O_{13}$; polyoxin C [*21027-33-8*] [*11043-74-6*], $C_{11}H_{15}N_3O_8$; polyoxin D [*22976-86-9*] [*33401-46-6*] [*11043-75-7*] [*29762-31-0*], $C_{17}H_{23}N_5O_{14}$; polyoxin E [*22976-87-0*] [*11043-76-8*], $C_{17}H_{23}N_5O_{13}$; polyoxin F [*23116-76-9*] [*11043-77-9*], $C_{23}H_{30}N_6O_{15}$; polyoxin G [*22976-88-1*] [*11043-78-0*], $C_{17}H_{25}N_5O_{12}$; polyoxin H [*24695-54-3*] [*11043-79-1*], $C_{23}H_{32}N_6O_{13}$; polyoxin I [*22886-33-5*] [*11043-80-4*], $C_{17}H_{22}N_4O_9$; polyoxin J [*22976-89-2*] [*25546-71-8*], $C_{17}H_{25}N_5O_{12}$; polyoxin K [*22886-46-0*], $C_{22}H_{30}N_6O_{13}$; polyoxin L [*22976-90-5*] [*100566-82-3*] [*25546-73-0*] [*59519-82-3*], $C_{16}H_{23}N_5O_{12}$; and polyoxin M [*34718-88-2*], $C_{16}H_{23}N_5O_{11}$.

from glutamate. Compounds (**102–116**) inhibit sheath-blight disease of rice crops, ie, the pathogenic fungus *Pellicularia philamentosa f. sasakii.* The polyoxins, which are structurally similar to UDP-*N*-acetylglucosamine, inhibit chitin synthetase (1). Polyoxin D (**104**) is a competitive inhibitor of UDP-*N*-acetylglucosamine. Polyoxin-resistant mutants of *A. kikuchiana* have a decreased uptake of the polyoxins. To overcome the resistance of these mutants, three approaches have been used: transnucleosidation, biosynthesis of a polyoxin with the 5-fluorouracil moiety, and decarboxylation of the 5-carboxyuracil polyoxins.

Mureidomycins and Pacidamycins. The mureidomycins and pacidamycins are listed in Table 6. The four peptidylnucleosides, mureidomycins A–D (**117**–**120**), are produced by *S. flavidovirens* (234). They contain two moles of *m*-tyrosine, one mole of 2-amino-3-*N*-methylaminobutyric acid, and one mole of methionine (235). The mureidomycins inhibit *Pseudomonas aeruginosa* (236). Treatment of *P. aeruginosa* with mureidomycin C results in cell lysis. The mureidomycins show low toxicity to mice and protect them from infection with *P. aeruginosa*. Mureidomycin A inhibits peptidoglycan synthesis and lipid-intermediate formation from UDP-*N*-acetylmuramyl (MurNAc)-pentapeptide and UDP-*N*-

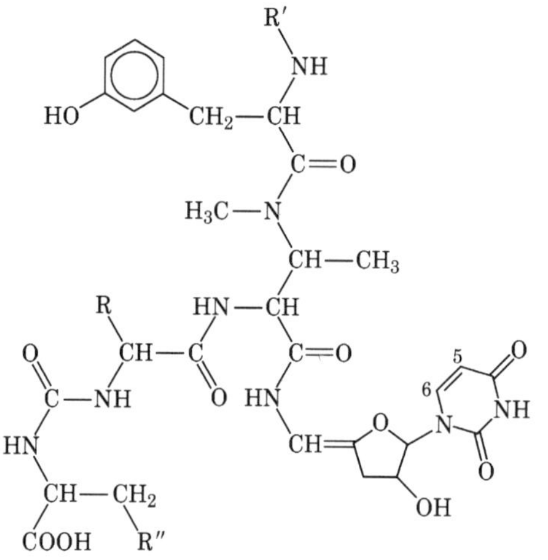

Table 6. The Mureidomycins[a] and the Pacidamycins[b]

Name	CAS Registry Number	Molecular formula	Structure number	R′	R″
Mureidomycins					
mureidomycin A	[*114797-04-5*]	$C_{38}H_{48}N_8O_{12}S$	(**117**)	H	*m*-hydroxyphenyl
mureidomycin B[c]	[*114797-05-6*]	$C_{38}H_{50}N_8O_{12}S$	(**118**)	H	*m*-hydroxyphenyl
mureidomycin C	[*114797-06-7*]	$C_{40}H_{51}N_9O_{13}S$	(**119**)	glycyl	*m*-hydroxyphenyl
mureidomycin D[c]	[*114797-07-8*]	$C_{40}H_{53}N_9O_{13}S$	(**120**)	glycyl	*m*-hydroxyphenyl
Pacidamycins					
pacidamycin 1	[*121264-05-9*]		(**121**)	alanyl	indol-3-yl
pacidamycin 2	[*121264-06-0*]	$C_{39}H_{49}N_9O_{12}$	(**122**)	alanyl	phenyl
pacidamycin 3	[*121280-49-7*]	$C_{39}H_{49}N_9O_{13}$	(**123**)	alanyl	*m*-hydroxyphenyl
pacidamycin 4			(**124**)	H	indol-3-yl
pacidamycin 5	[*121255-43-0*]	$C_{36}H_{44}N_8O_{11}$	(**125**)	H	phenyl
pacidamycin 6			(**126**)	glycyl	indol-3-yl
pacidamycin 7		$C_{38}H_{47}N_9O_{12}$	(**127**)	glycyl	phenyl

[a]R is $(CH_2)_2SCH_3$.
[b]R is CH_3.
[c]Positions 5, 6 of the uracil moiety are saturated to give dihydrouracil.

acetylglucosamine (237). Mureidomycin A inhibits translocase, thereby blocking formation of undecaprenyl-pyrophosphoryl-MurNAc-pentapeptide.

The seven pacidamycins (**121**–**127**), which are isolated from the culture filtrates of *S. coeruleorubidus,* are structurally similar to the mureidomycins. These seven peptidyl nucleoside antibiotics differ in the terminal amino acid residues (238). The biosynthesis of these nucleoside antibiotics is markedly affected by the amino acids added to the culture medium (239).

4-Aminohexose Nucleosides. The 4-aminohexose nucleosides (**128**–**140**) are listed in Table 7 (1–4,240–242). A biosynthetic relationship between the 4-aminohexose peptidyl nucleoside antibiotics and the pentopyranines has been proposed (1). The 4-aminohexose pyrimidine nucleoside antibiotics block peptidyl transferase activity and inhibit transfer of amino acids from aminoacyl-tRNA to polypeptides. Hikizimycin, gougerotin, amicetin, and blasticidin S bind to the peptidyl transferase center at overlapping sites (243).

Nikkomycins. The nikkomycins (**141**–**159**), isolated from *S. tendae,* are nucleoside-peptide antibiotics (1,4,244,245) as shown in Table 8. Nikkomycins X and Z are structurally identical to neopolyoxins A and C, respectively. Compound (**141**) is a competitive inhibitor of chitin synthetase. Two new nikkomycins, nikkomycin pseudo-Z and pseudo-J (**158**, **159**), contain a *C*-glycosidic bond between C-5 of uracil and C-1′ of 5-amino-5-deoxy-D-*allo*furanuronic acid (245). Other peptidyl nucleosides, the albomycins [*1414-39-7*] (**160**–**162**), amipurimycin (**163**), and miharamycins A and B (**164, 165**), are shown in Table 9 (246–248). These nucleosides are either iron chelators or inhibitors of plant pathogens (244,246–248).

Glycosyl Nucleosides. There are five glycosyl antibiotics with either the adenine, uracil, or 7-deazaguanine aglycon given in Table 10. The glycosylated moieties are either *N*-linked disaccharides, unsaturated uronic acid, or a sulfated sugar. The antibacterial and antifungal activities of these compounds have been described (249–253).

Octosyl Acids. Three octosyl uronic acid nucleosides, produced by *S. cacaoi* sub sp. *asoensis* are shown in Figure 3. The biosynthesis of (**172**) and (**173**) has been reported (1). The replacement of the pyrimidine chromophore of (**171**) with adenine results in a nucleoside analogue that is a competitive inhibitor of cAMP.

Arabinosylpyrimidine Nucleosides. 1-β-D-Arabinofuranosylthymine [*605-23-2*] (ara-T) (**174**), $C_{10}H_{14}N_2O_6$, and 1-β-D-arabinofuranosyluracil [*3083-77-0*] (ara-U) (**175**), $C_9H_{12}N_2O_6$, also known as spongouridine, were first isolated from

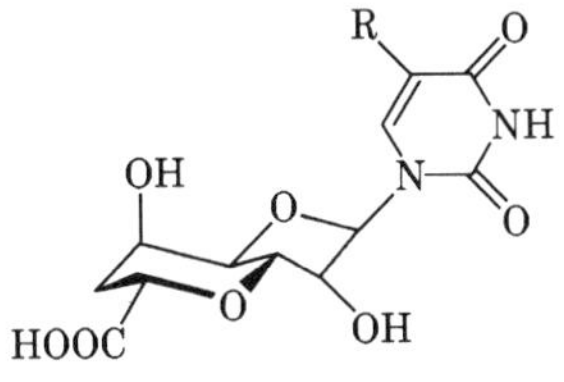

Octosyl acid A (**171**, R = COOH)
Octosyl acid B (**172**, R = CH_2OH)

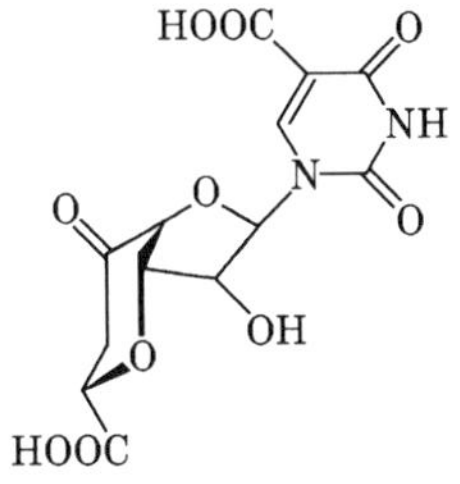

Octosyl acid C (**173**)

Fig. 3. Octosyl acids.

Table 7. 4-Aminohexose Nucleosides

Name	CAS Registry Number	Molecular formula	Structure number	Structure
gougerotin	[*2096-42-6*]	$C_{16}H_{25}N_7O_8$	**(128)**	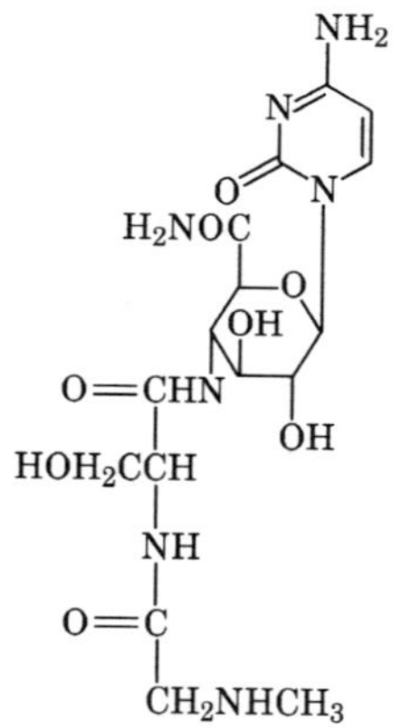
blasticidin S	[*2079-00-7*]	$C_{17}H_{26}N_8O_5$	**(129)**	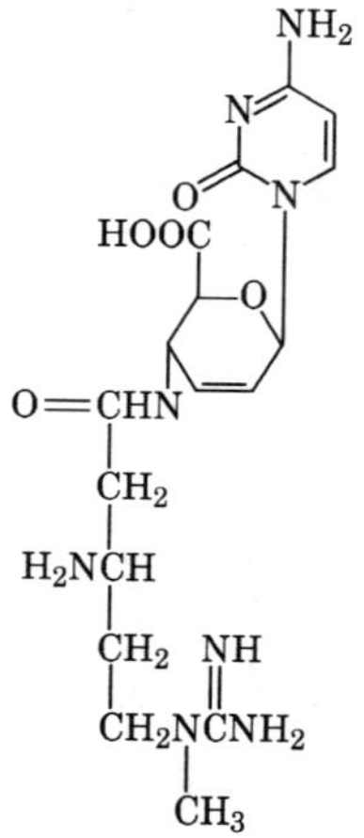

				R	R′	R″
amicetin	[*17650-86-1*]	$C_{29}H_{42}N_6O_9$	(**130**)	α-methylseryl	H	$N(CH_3)_2$
bamicetin	[*43043-14-7*]	$C_{28}H_{40}N_6O_9$	(**131**)	α-methylseryl	H	$NHCH_3$
oxamicetin	[*52665-75-5*]	$C_{29}H_{42}N_6O_{10}$	(**132**)	α-methylseryl	OH	$N(CH_3)_2$
plicacetin	[*43043-15-8*]	$C_{25}H_{35}N_5O_7$	(**133**)	H	H	$N(CH_3)_2$
norplicacetin	[*64040-47-7*]	$C_{24}H_{33}N_5O_7$	(**134**)	H	H	$NHCH_3$

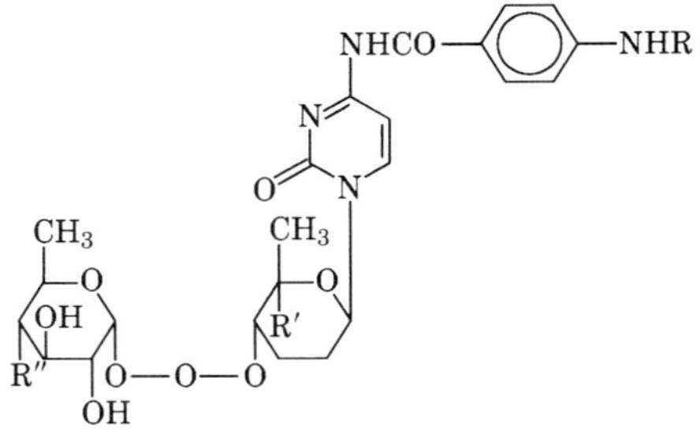

hikizimycin (anthelmycin) [*12706-94-4*] $C_{21}H_{37}N_5O_{14}$ (**135**)

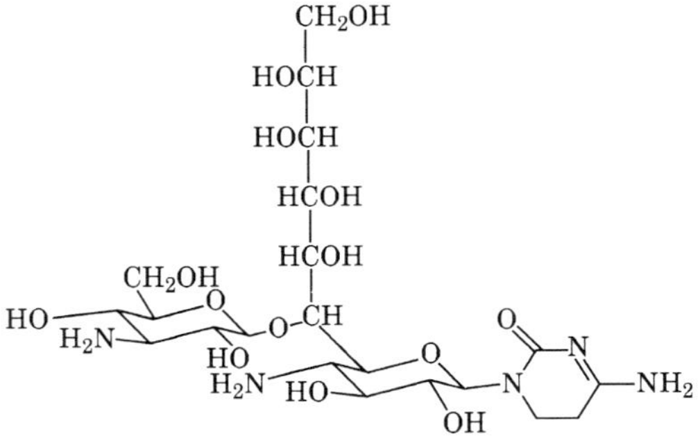

Table 7. (*Continued*)

Name	CAS Registry Number	Molecular formula	Structure number	Structure
bagougeramine A	[*104840-35-9*]	$C_{17}H_{28}N_{10}O_7$	**(136)**	R = H 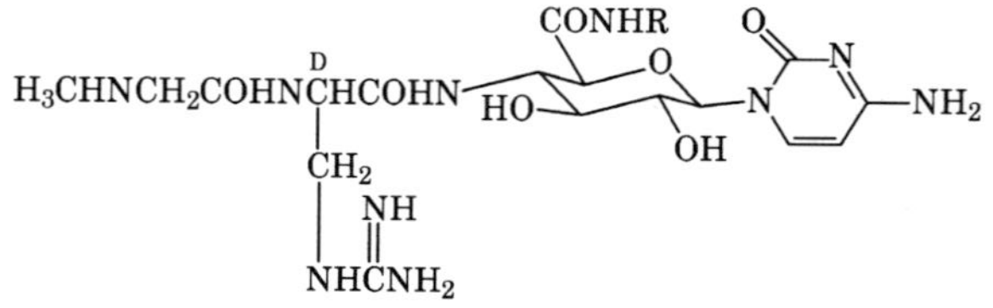
bagougeramine B	[*104840-34-8*]	$C_{24}H_{44}N_{12}O_7$	**(137)**	R = $(CH_2)_3NH(CH_2)_4NH_2$
arginomycin	[*106133-33-9*]	$C_{18}H_{28}N_8O_5$	**(138)**	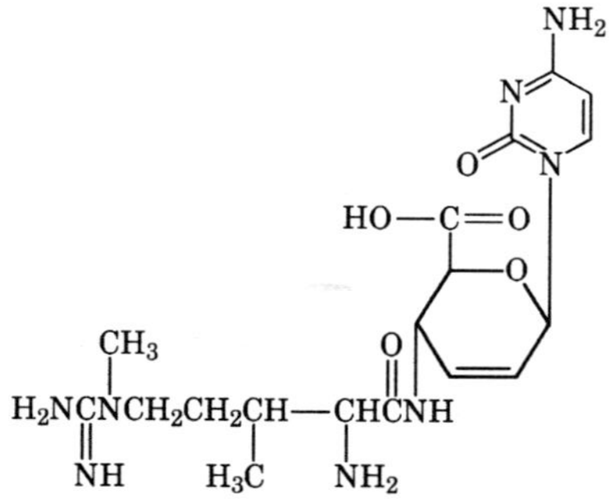

mildiomycin	[*67527-71-3*]	$C_{19}H_{30}N_8O_9$	(**139**)	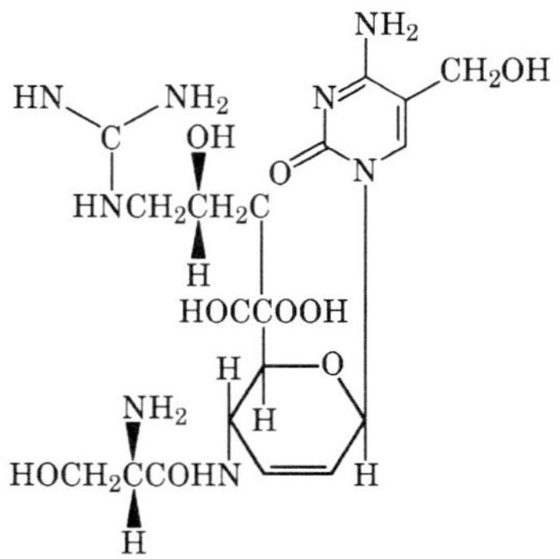
SCH 36605 (rodaplutin)	[*113667-06-4*]	$C_{24}H_{39}N_9O_7$	(**140**)	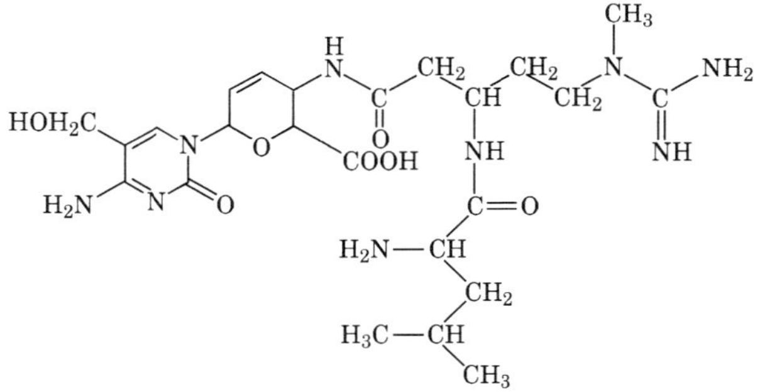

Table 8. The Nikkomycins

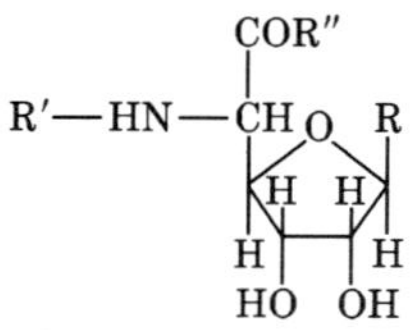

Nikkomycin	CAS Registry Number	Molecular formula	Structure number	R	R′	R″
Z	*[59456-70-1]* *[75044-70-1]*	$C_{20}H_{25}N_5O_{10}$ $C_{20}H_{25}N_5O_{10}$	**(141)**	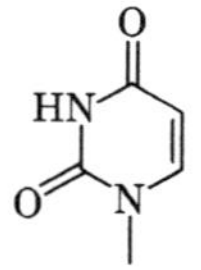	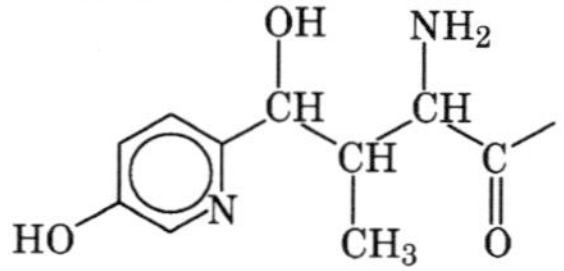	OH
X	*[72864-26-7]*	$C_{20}H_{25}N_5O_{10}$	**(142)**	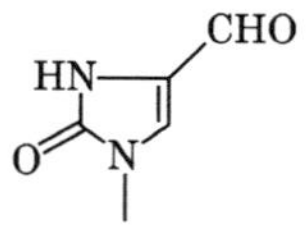	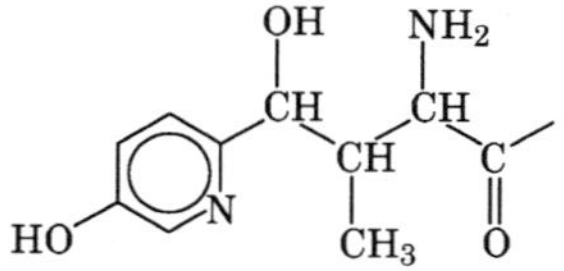	OH
J	*[77368-59-3]*	$C_{25}H_{32}N_6O_{13}$	**(143)**	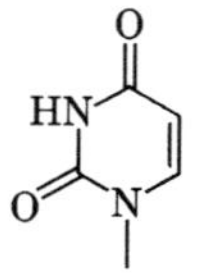	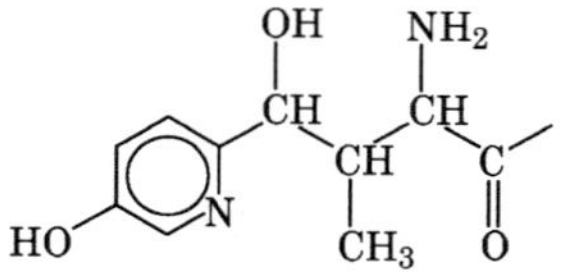	Glu
I	*[77368-60-6]*	$C_{25}H_{32}N_6O_{13}$	**(144)**	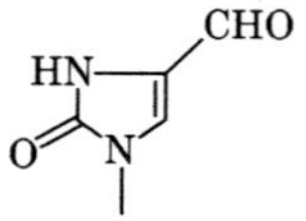	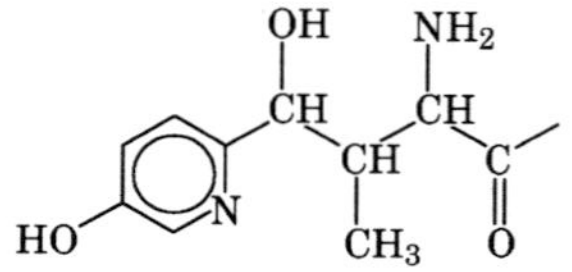	Glu

B_z			**(145)**	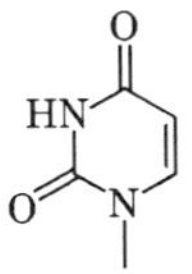	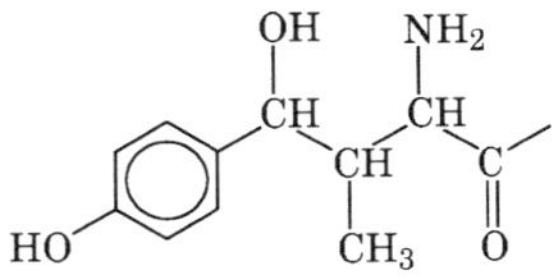	OH
B_x	*[75410-71-8]*	$C_{21}H_{26}N_4O_{10}$	**(146)**	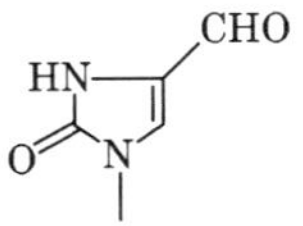	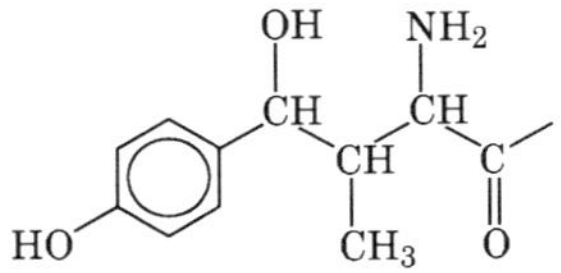	OH
K_z	*[95259-45-3]*	$C_{19}H_{23}N_5O_9$	**(147)**	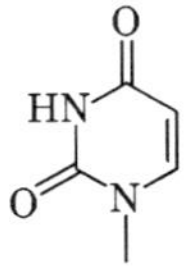	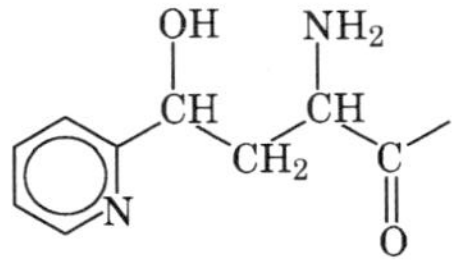	OH
K_x	*[95259-46-4]*	$C_{19}H_{23}N_5O_9$	**(148)**	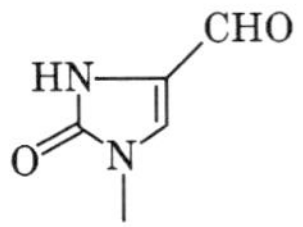	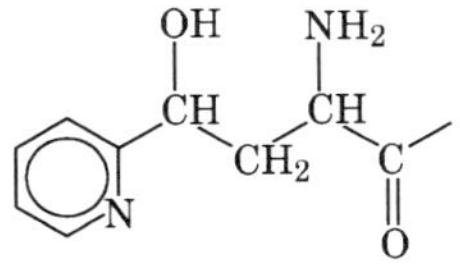	OH
O_z	*[95259-47-5]*	$C_{19}H_{23}N_5O_{10}$	**(149)**	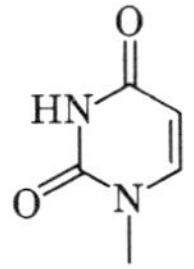	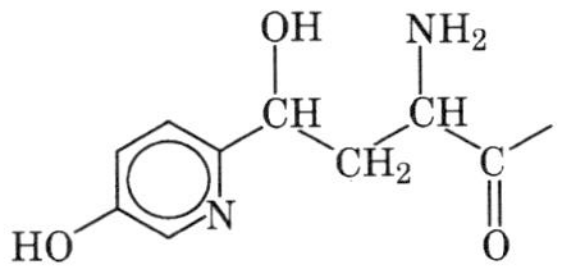	OH
O_x	*[95259-49-7]*	$C_{19}H_{23}N_5O_{10}$	**(150)**	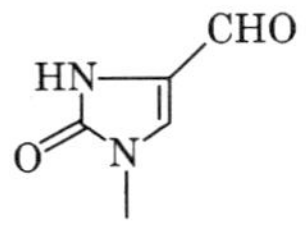	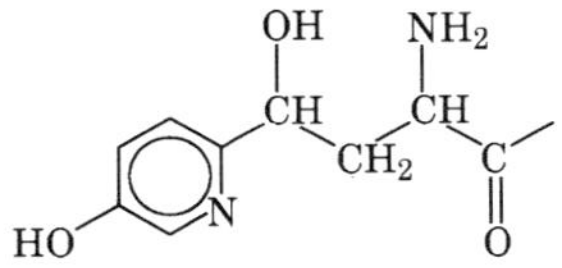	OH

Table 8. (*Continued*)

Nikkomycin	CAS Registry Number	Molecular formula	Structure number	R	R′	R″
Q_z	[*95259-50-0*]	$C_{24}H_{32}N_6O_{12}$	**(151)**	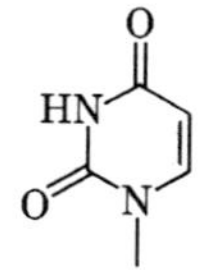	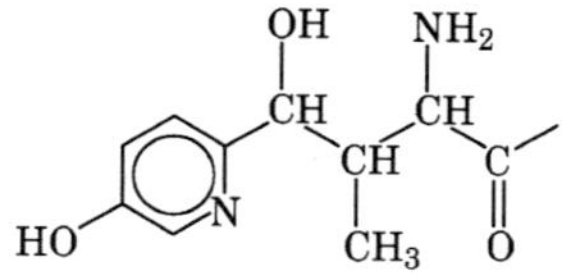	Homo-Ser
Q_x	[*95259-48-6*]	$C_{24}H_{32}N_6O_{12}$	**(152)**	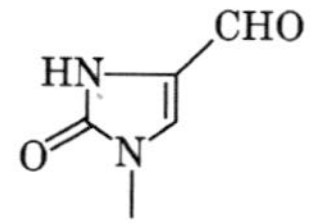	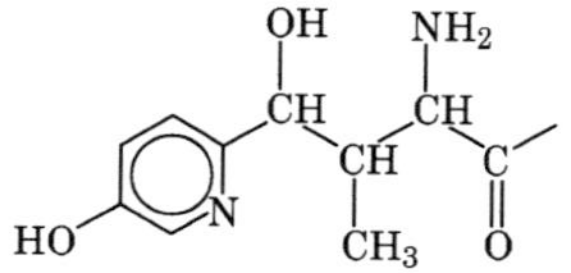	Homo-Ser
P_x		$C_{20}H_{25}N_5O_9$	**(153)**	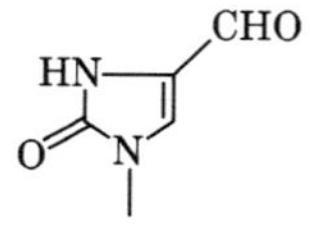	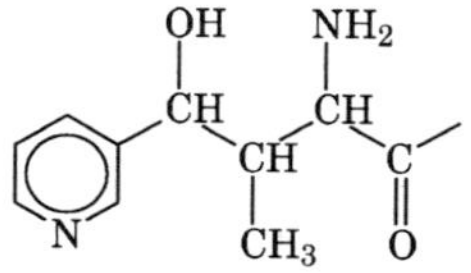	OH
R_z		$C_{25}H_{32}N_6O_{12}$	**(154)**	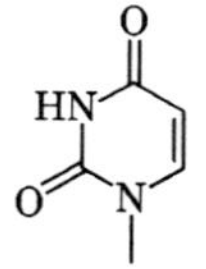	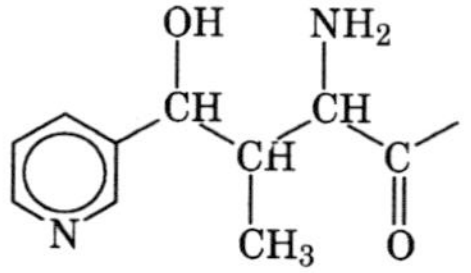	Glu

R_x		$C_{25}H_{32}N_6O_{12}$	(**155**)	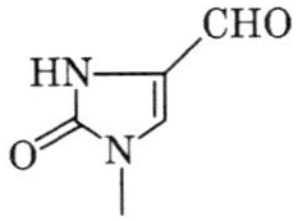	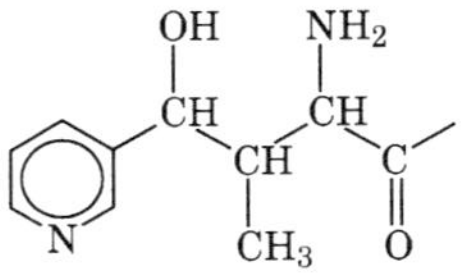	Glu
W_z		$C_{19}H_{21}N_4O_9$	(**156**)	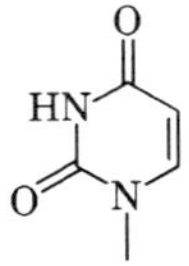	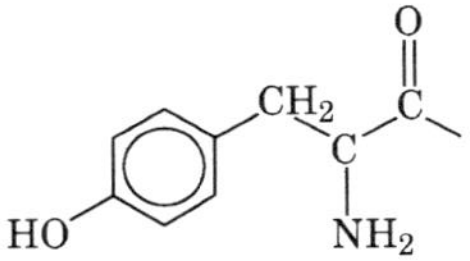	OH
W_x		$C_{19}H_{21}N_4O_9$	(**157**)	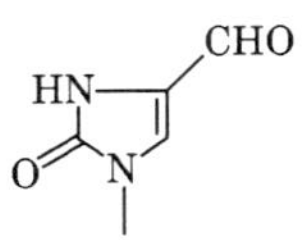	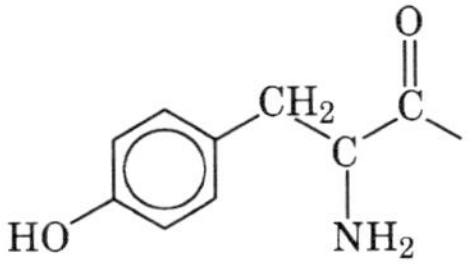	OH
pseudo-Z	[*120796-21-6*]	$C_{20}H_{25}N_5O_{10}$	(**158**)	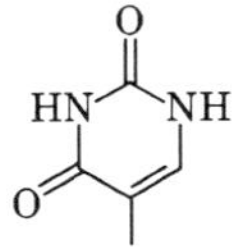	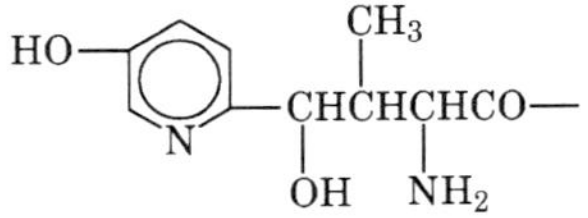	OH
pseudo-J	[*120796-22-7*]	$C_{25}H_{32}N_6O_{13}$	(**159**)	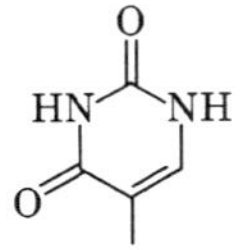	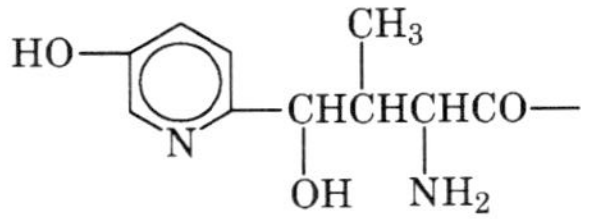	NHCHCOOH with $(CH_2)_2COOH$ on the CH

Table 9. Other Peptidyl Nucleosides

Name	CAS Registry Number	Molecular formula	Structure number	X	Structure
albomycin δ_2	[*82113-67-5*]	$C_{37}H_{60}N_{12}O_{18}S$	**(160)**	$NCONH_2$	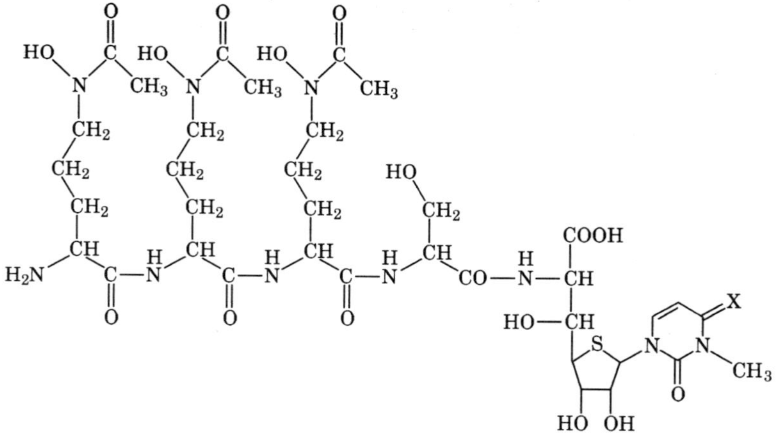
albomycin ϵ	[*82113-68-6*]	$C_{36}H_{59}N_{11}O_{17}S$	**(161)**	NH	
albomycin δ_1	[*82113-69-7*]	$C_{36}H_{58}N_{10}O_{18}S$	**(162)**	O	
amipurimycin	[*61991-08-0*]	$C_{20}H_{29}N_7O_8$	**(163)**		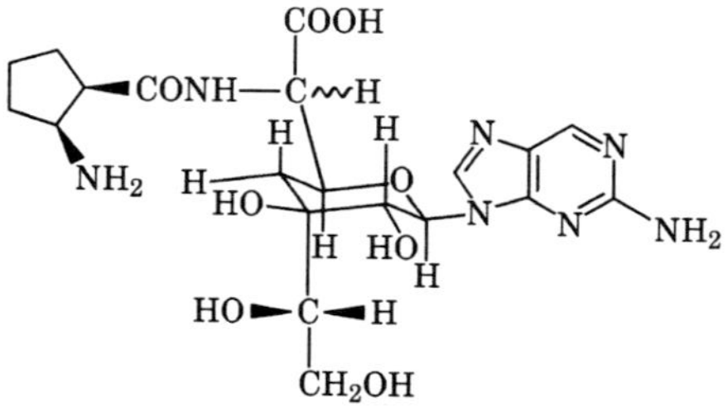
miharamycin A	[*12626-16-3*]	$C_{20}H_{30}N_{10}O_9$	**(164)**	OH	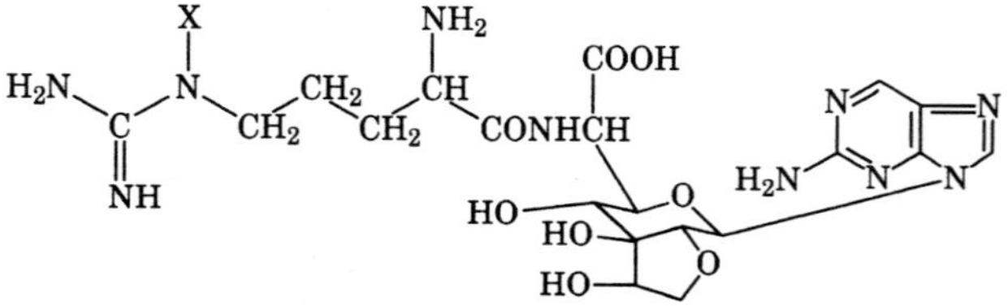
miharamycin B	[*12626-15-2*]	$C_{20}H_{30}N_{10}O_8$	**(165)**	H	

Table 10. Glycosyl Nucleosides

Name	CAS Registry Number	Molecular formula	Structure number	R	Structure
dapiramicin A	[*67298-15-1*]	$C_{21}H_{29}N_5O_{10}$	**(166)**		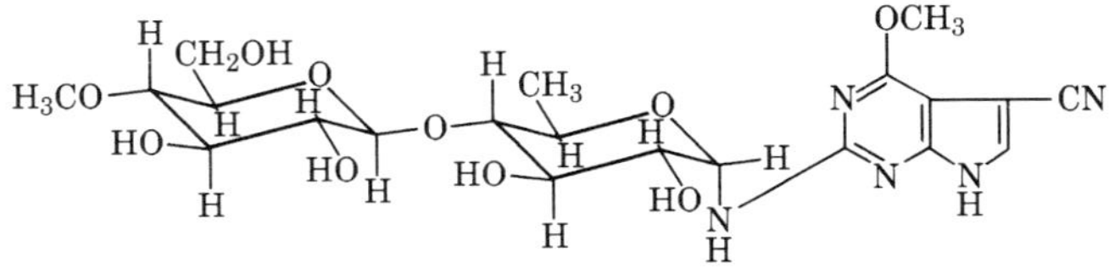
epidapiramicin A dapiramicin B	[*90044-19-2*] [*90044-18-1*]	$C_{21}H_{29}N_5O_{10}$ $C_{21}H_{29}N_5O_{11}$	**(167)** **(168)**	H OH	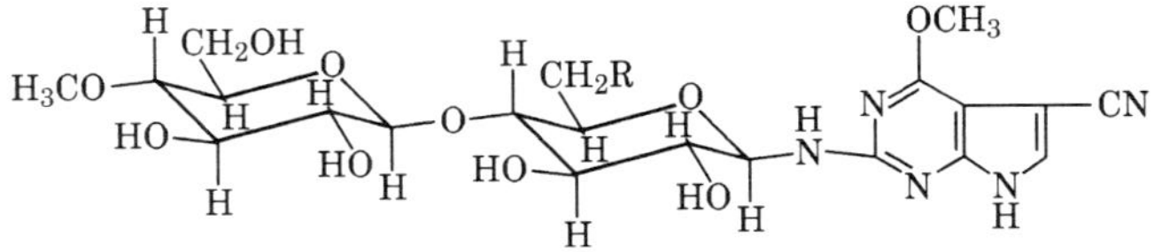
capuramycin	[*102770-00-3*]	$C_{23}H_{31}N_5O_{12}$	**(169)**		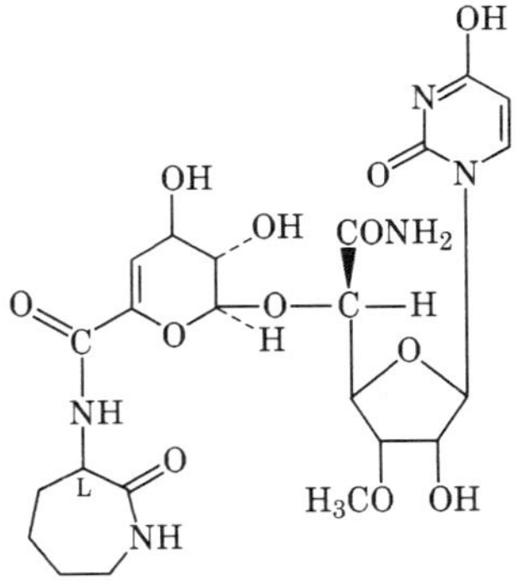
adenomycin	[*76174-56-6*]	$C_{25}H_{39}N_7O_{18}S$	**(170)**	$-\overset{O}{\overset{\parallel}{C}}\underset{NH_2}{\underset{\mid}{C}}HCH_2OH$	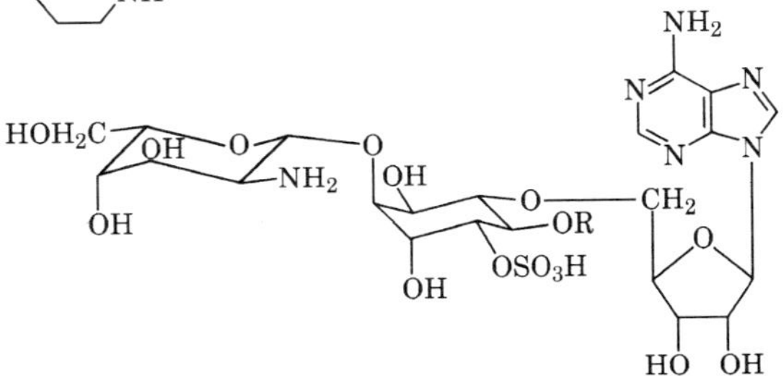

the sponge *C. crypta* (1–4). The structures of these compounds are shown in Figure 4. Compound (**174**) moderately inhibits *E. coli* and is not toxic to mammalian cells. It inhibits the replication of HSV-1 and varicella zoster virus and is converted to the 5′-triphosphate which inhibits HSV-induced deoxypyrimidine kinase. The chemical synthesis of (**174**) from thymidine has been reported (254). Ara-TTP competitively inhibits DNA polymerase-α and -β isolated from L5178Y cells with respect to dTTP; thymidine and uridine reverse the inhibition by (**174**). RNA polymerases are not inhibited by ara-TTP. It is phosphorylated to its 5′-triphosphate in noninfected and HSV-infected cells. Ara-U (**175**) can satisfy the uracil requirement of an auxotroph of *E. coli* and can be phosphorylated to its 5′-phosphate. Although (**175**) has little or no antiviral activity, the 5-alkyl derivatives of (**175**) show antiherpesviral activity and inhibition of cell growth in human

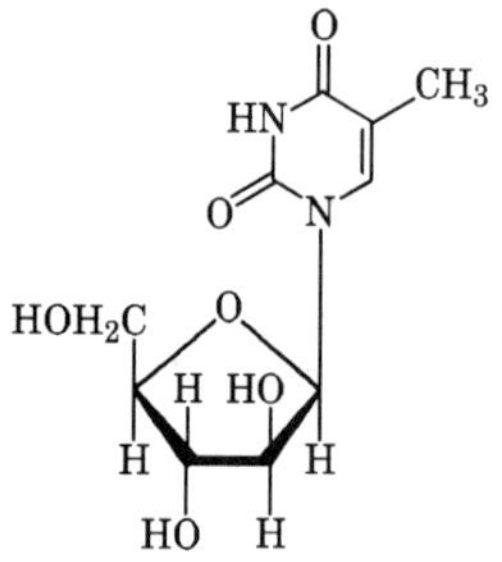

Ara-T (**174**)

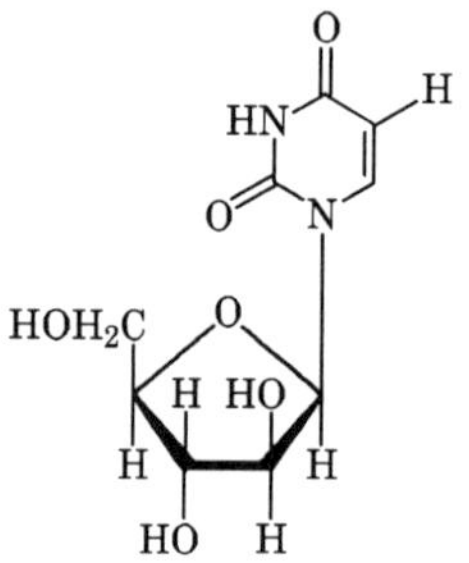

Ara-U (**175**)

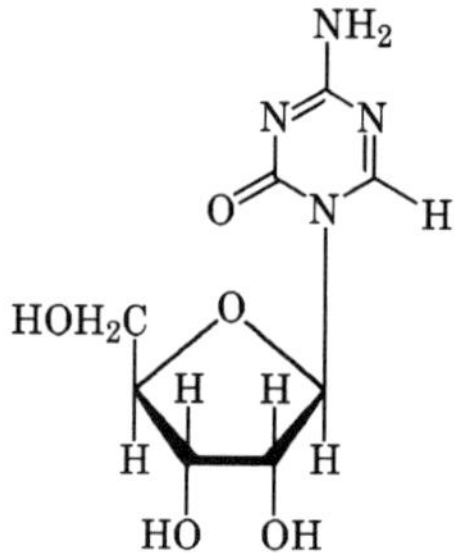

5-Azacytidine (**176**)

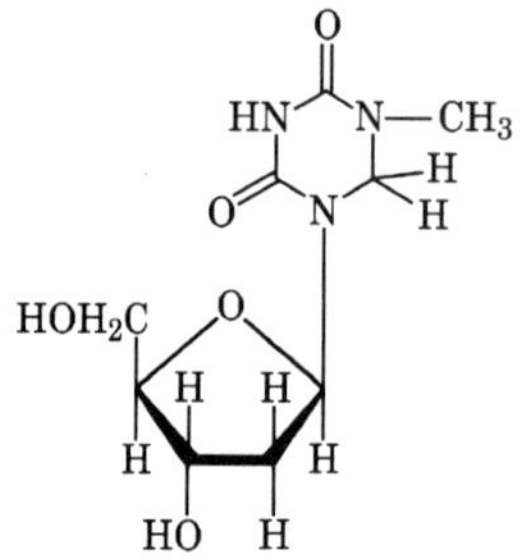

5,6-Dihydro-5-azathymidine (**177**)

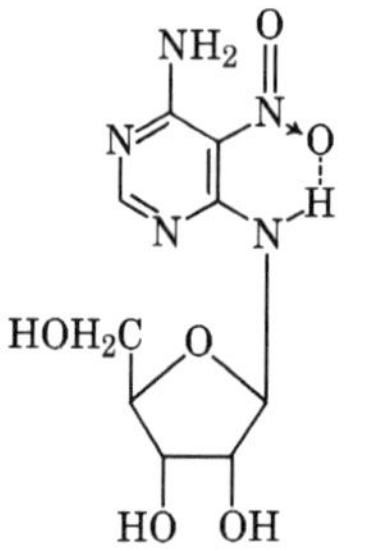

Clitocine (**178**)

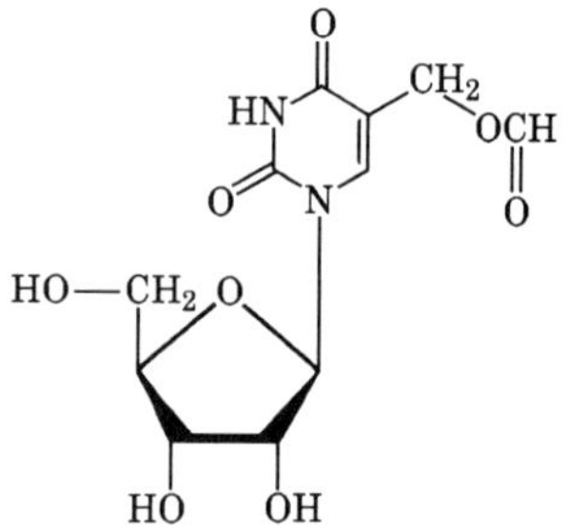

5-Formyloxymethyluridine (**179**)

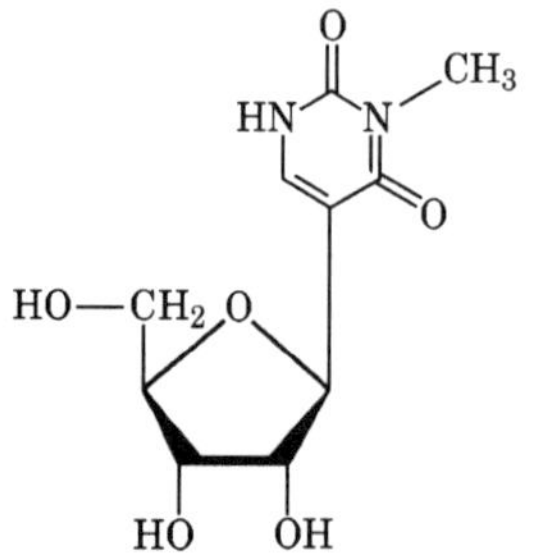

3-Methylpseudouridine (**180**)

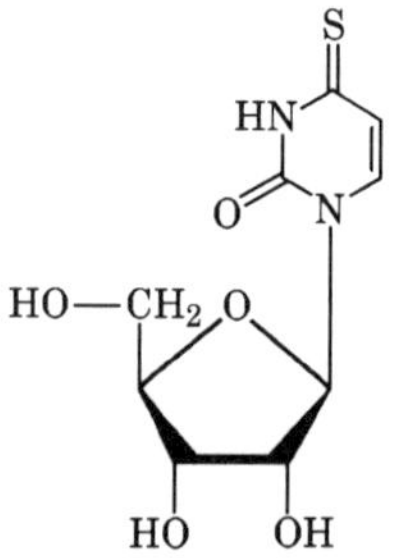

4-Thiouridine (**181**)

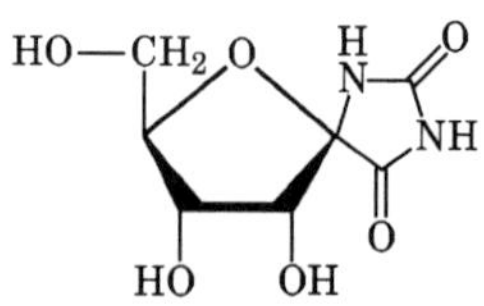

Hydantocidin (**182**)

Fig. 4. Structures of pyrimidine nucleosides.

lung embryonic fibroblasts (255). The 5-halogenated derivatives of ara-UTP inhibit DNA polymerase-α, -β, and -γ and reverse transcriptase.

5-Azacytidine. 5-Azacytidine [*320-67-2*] (**176**), $C_8H_{12}N_4O_5$, 4-amino-1-β-D-ribofuranosyl-*S*-triazine-2-(1*H*-one), was chemically synthesized in 1964 and subsequently isolated from culture filtrates of *S. ladakanus* (1–4). Compound (**176**) and 2′-deoxy-5-azacytidine, which are cytotoxic analogues of cytidine, are potent anticancer agents that inhibit DNA methylation (256). Compound (**176**) is incorporated into RNA and tRNA and subsequently inhibits protein synthesis. It inhibits the maturation of 28S and 18S RNA, but not 13S RNA. Polysome degradation increases in the presence of (**176**) with an accumulation of monosomes and an inhibition of protein synthesis. The incorporation of 2′-deoxy-5-azacytidine 5′-monophosphate into DNA causes chromosome breakage. 5-Azacytidine induces 2-5A synthetase in human lymphoid cells (257). This hypomethylating nucleoside is a useful probe to analyze specific gene products in the earliest stages of chemical carcinogenesis (258,259). It can retransform cells by a mechanism other than DNA methylation (260). In addition, (**176**) induces cytidine deaminase in HL-60 cells, is incorporated into RNA and DNA, inhibits HIV-1 replication, upregulates Epstein-Barr virus nuclear antigen-2, suppresses ADP-ribosylation, and causes chromosomal aberrations (261–268). Compound (**176**) has been used in the treatment of thalassemia (269). The biochemistry of (**176**) has been reviewed (270).

5,6-Dihydro-5-azathymidine. 5,6-Dihydro-5-azathymidine [*57350-36-4*] (**177**), $C_9H_{15}N_3O_5$, contains the *s*-triazine ring and is isolated from the culture filtrates of *S. platensis* var. *clarensis* (4). The inhibition of viral replication by (**177**) can be reversed by either thymidine, deoxyuridine, or deoxycytidine. Compound (**177**) protects mice following intracerebral inoculation with HSV-1 and appears to inhibit the phosphorylation of thymidine rather than its incorporation into DNA. Resistance to (**177**) by *E. coli* appears to result from a change in thymidine phosphorylase.

Clitocine. Clitocine [*105798-74-1*] (**178**), $C_9H_{13}N_5O_6$, isolated from the mushroom, *Clitocybe inversa*, is 5-nitro-4-(β-D-ribofuranosylamino)pyrimidine-6-amine (271). The crystal structure indicates that the nitro group is hydrogen bonded to the 4-amino hydrogen atom as shown in Figure 4 (272). The base is in the *anti* conformation and the sugar moiety is disordered. Chemical syntheses of (**178**) have been described (273,274). Clitocine shows strong insecticidal activity against *Pectinophora gossypiella* (271) and is a substrate and inhibitor of adenosine kinase (274).

Uridine Analogues. 5-Formyloxymethyluridine (**179**), $C_{11}H_{14}N_2O_8$, produced by *Serratia plymuthica* (274), inhibits bacterial growth. 3-Methylpseudouridine (**180**), $C_{10}H_{14}N_2O_6$, has been isolated and identified from the culture filtrates of *Nocardia lactamdurans* (275). 1-Methylpseudouridine was previously isolated from *S. platensis* (276). 4-Thiouridine (**181**), $C_9H_{12}N_2O_5S$, has been isolated from an actinomycete resembling *S. hygroscopicus* (277).

Hydantocidin. Hydantocidin (**182**), $C_7H_{10}N_2O_6$, is elaborated by *S. hygroscopicus* (278). It is unique in that the anomeric carbon of the ribosyl moiety forms the spiro bond of hydantoin (279). The ribofuranose moiety which has been reported to be in a C_2-*endo* conformation (279) has been synthesized (280,281). Hydantocidin is a herbicidal nucleoside with activity against monocotyledenous and dicotyledenous plants.

Table 11. ***N*****-Nucleoside Antibiotics**

Name	CAS Registry Number	Molecular formula	Structure number	Structure
agrocin 84	[*59111-78-3*]	$C_{21}H_{34}N_6O_{16}P_2$	**(183)**	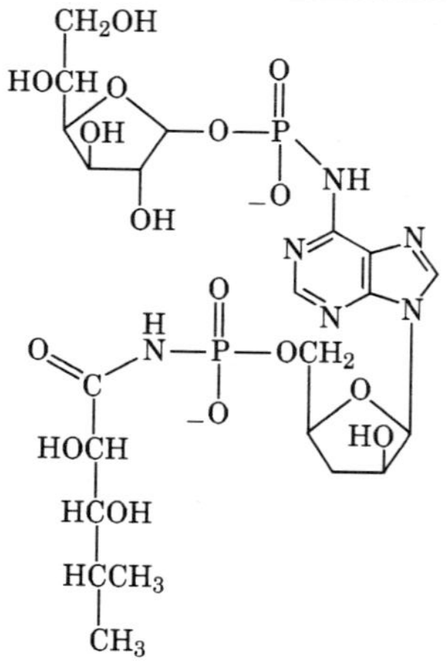
thuringiensin	[*23526-02-5*]	$C_{22}H_{30}N_5O_{19}P$	**(184)**	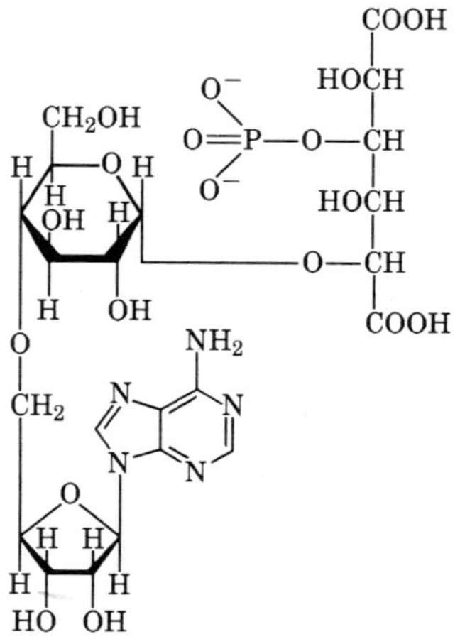

phosmidosine		$C_{16}H_{24}N_7O_8P$	**(185)**	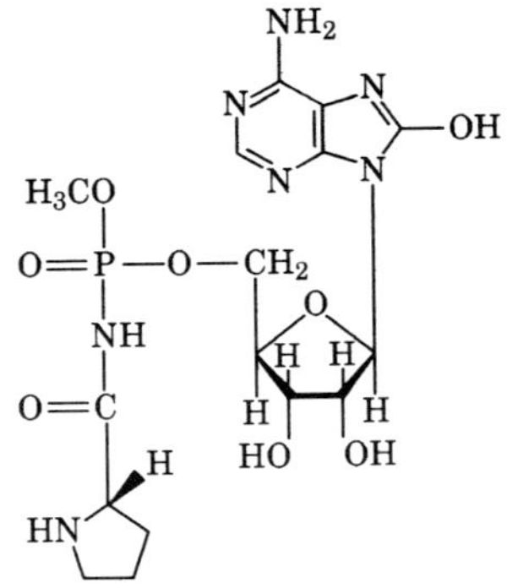
fosfadecin	[*120909-50-4*]	$C_{13}H_{19}N_5O_{10}P_2$	**(186)**	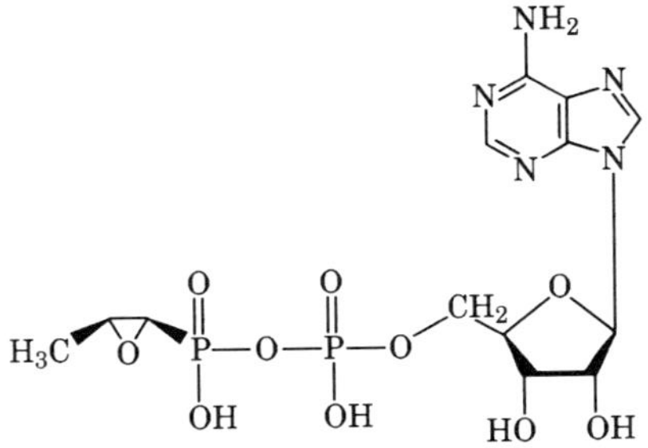
fosfocytocin	[*126986-25-2*]	$C_{12}H_{20}N_4O_{13}P_2$	**(187)**	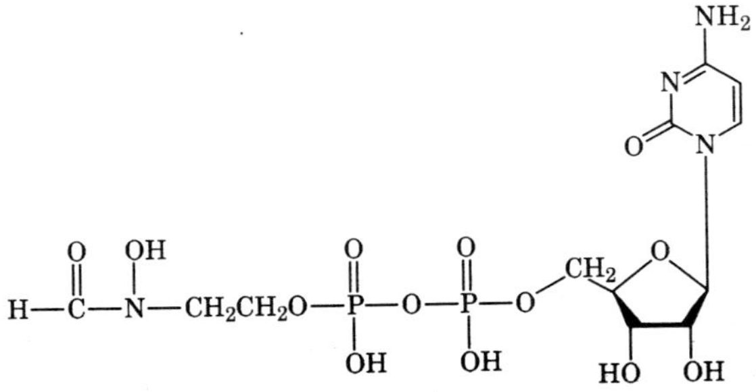

N-Nucleotides

The *N*-nucleotide antibiotics are given in Table 11. Agrocin 84 (**183**) is an adenine 6-*N*-phosphoramidate nucleotide analogue that contains adenine, phosphate, and D-glucose in a 1:2:1 ratio (1,4). It is produced by *Agrobacterium radiobacter* strain K-84. Crown gall tumors are induced in a number of dicotyledonous plants following inoculation at wound sites with virulent *A. tumefaciens*. Crown gall tumors are caused by the incorporation of part of a virulent plasmid carried by *A. tumefaciens*. Proof that a large plasmid was essential is that the virulent strain C-58 can be converted to a stable avirulent strain. Crown gall formation by *A. tumefaciens* represents the first case where there is a transfer of genetic information from *A. tumefaciens* into the genome of the plant. This type of interaction has been termed "genetic colonization."

Thuringiensin (**184**), produced by *B. thuringiensis* (1,4), is a β-exotoxin that exerts its toxic action on insects and mammals through the inhibition of RNA polymerases.

The structure of the nucleotide antibiotic, phosmidosine (**185**), isolated from the culture filtrates of *Streptomyces* sp. RK-16 has been elucidated (282). Phosmidosine inhibits pore formation of *Botrytis cinerea* and *Aspergillus niger*.

Fosfadecin (**186**) and fosfocytocin (**187**) are adenine and cytosine nucleotide antibiotics isolated from the culture filtrates of *Pseudomonas viridiflava* PK-5 and *P. fluorescens* PK-52, respectively (283). Hydrolysis produces fosfoxacin which is also isolated from the culture filtrates. Compounds (**186**) and (**187**) inhibit gram-positive and gram-negative bacteria.

BIBLIOGRAPHY

"Antibiotics (Nucleosides)," in *ECT* 3, Vol. 2, pp. 962–986, by A. Bloch, Roswell-Park Memorial Institute.

1. K. Isono, *J. Antibiot.* **41,** 1711 (1988).
2. K. Isono, *Pharmacol. Therapeut.,* in press, 1992.
3. R. J. Suhadolnik, *Nucleoside Antibiotics,* John Wiley & Sons, Inc., New York, 1970.
4. R. J. Suhadolnik, *Nucleosides as Biological Probes,* John Wiley & Sons, Inc., New York, 1979.
5. A. P. Reuter, J. A. Figueiredo, and I. M. Ismael, *Carbohydrate Res.* **188,** 19 (1989).
6. E. F. Elstner and co-workers, *Biochemistry* **12,** 4992 (1973).
7. K. Isono and R. J. Suhadolnik, *J. Antibiot.* **30,** 272 (1977).
8. E. F. Elstner, R. J. Suhadolnik, and A. Allerhand, *J. Biol. Chem.* **248,** 5385 (1973).
9. J. G. Buchanan, A. Stobie, and R. H. Wightman, *J.C.S. Chem. Commun.,* 916 (1980).
10. C. R. Petrie, III and co-workers, *J. Med. Chem.* **29,** 268 (1986).
11. R. J. Suhadolnik and N. L. Reichenbach, *Biochemistry* **20,** 7042 (1981).
12. J. G. Buchanan, M. R. Hamblin, G. R. Sood, and R. H. Wightman, *J.C.S. Chem. Commun.,* 917 (1980).
13. J. M. Crance, E. Biziagos, J. Passagot, H. van Cuyck-Gandrie, and R. Deloince, *J. Med. Virol.* **31,** 155 (1990).
14. M. E. Sant and co-workers, *Cancer Res.* **49,** 2645 (1989).
15. Y. Kang, S. B. Larson, R. K. Robins, and G. R. Revenkar, *J. Med. Chem.* **32,** 1547 (1989).

16. P. Laquel, M. Castrouviejo, and S. Litvak, *Nucleic Acids Res.* **18,** 4867 (1990).
17. C. R. Crawford, C. Y. Ng, L. D. Noel, and J. A. Belt, *J. Biol. Chem.* **265,** 9732 (1990).
18. P. G. Plagemann and C. Woffendin, *Biochim. Biophys. Acta* **1010,** 7 (1989).
19. D. Vijayalakshmi and J. A. Belt, *J. Biol. Chem.* **263,** 19,419 (1988).
20. P. G. Plagemann, J. M. Aran, and C. Woffendin, *Biochim. Biophys. Acta* **1022,** 93 (1990).
21. E. S. Jakobs, D. J. Van Os-Corby, and A. R. P. Paterson, *J. Biol. Chem.* **265,** 22,210 (1990).
22. V. L. Giranda, H. M. Berman, and V. L. Schramm, *Biochemistry* **27,** 5813 (1988).
23. A. Maelicke, M. Sprinzl, F. Von der Haar, T. A. Khwaja, and F. Cramer, *Eur. J. Biochem.* **43,** 617 (1974).
24. E. Baksht, N. de Groot, M. Sprinzl, and F. Cramer, *FEBS Letters* **55,** 105 (1975).
25. R. P. Agarwal and co-workers, *J. Clin. Invest.* **57,** 1025 (1976).
26. W. E. G. Müller and co-workers, *Cancer Res.* **35,** 3673 (1975).
27. W. E. G. Müller and R. K. Zahn, *Experientia* **31,** 1014 (1975).
28. R. J. Suhadolnik and co-workers, *J. Biol. Chem.* **252,** 4134 (1977).
29. B. A. Otter, S. A. Patil, R. S. Klein, and S. E. Ealick, *J. Am. Chem. Soc.* **114,** 668 (1992).
30. R. J. Suhadolnik, S. Pornbanlualap, D. C. Baker, K. N. Tiwari, and A. K. Hebbler, *Arch. Biochem. Biophys.* **270,** 374 (1989).
31. M. Sprinzl and F. Cramer, *Nature New Biol.* **245,** 3 (1973).
32. T. H. Frazer and A. Rich, *Proc. Natl. Acad. Sci. USA* **70,** 2671 (1973).
33. N. Van der Heyden, C. Rodi, and L. Ratner, *AIDS Res. Human Retroviruses* **5,** 647 (1989).
34. N. M. Kredich, *J. Biol. Chem.* **255,** 7380 (1980).
35. P. Doetsch, J. M. Wu, Y. Sawada, and R. J. Suhadolnik, *Nature* **291,** 355 (1981).
36. D. C. Montefiori and co-workers, *Proc. Natl. Acad. Sci. USA* **86,** 7191 (1989).
37. W. E. G. Müller and co-workers, *Biochemistry* **30,** 2027 (1991).
38. D. A. Kuznetsov and N. I. Musajev, *Int. J. Neurosci.* **51,** 53 (1990).
39. J. S. Butler, P. P. Sadhale, and T. Platt, *Mol. Cell. Biol.* **10,** 2599 (1990).
40. P. A. Mathew, L. K. Ellis, and G. P. Studzinski, *J. Cell. Physiol.* **140,** 212 (1989).
41. P. G. Plagemann, J. M. Aran, R. M. Wohlhueter, and C. Woffendin, *Biochim. Biophys. Acta* **1022,** 103 (1990).
42. L. L. McGrew, E. Dworkin-Rastl, and J. D. Richter, *Genes Dev.* **3,** 803 (1989).
43. J. McLauchlan, C. L. Moore, S. Simpson, and J. B. Clements, *Nuc. Acids Res.* **16,** 5323 (1988).
44. S. K. Pramanik, W. J. Meadus, and J. Bag, *Eur. J. Biochem.* **172,** 355 (1988).
45. G. W. Zieve, R. J. Feeney, and E. J. Roemer, *Cell Motility Cytoskeleton* **7,** 337 (1987).
46. R. A. Colbert and D. A. Young, *J. Biol. Chem.* **262,** 9939 (1987).
47. D. Zarkower and M. Wickens, *EMBO J.* **6,** 177 (1987).
48. J. Remme, T. Margus, R. Villems, and K. H. Nierhaus, *Eur. J. Biochem.* **183,** 281 (1989).
49. D. Synetos and C. Coutsogeorgopoulos, *Eur. J. Biochem.* **184,** 47 (1989).
50. C. C. Hall, D. Johnson, and B. S. Cooperman, *J. Biochem.* **27,** 3983 (1988).
51. S. Inoue, M. Iwasaki, K. Ishii, K. Kitajima, and Y. Inoue, *J. Biol. Chem.* **264,** 18,520 (1989).
52. A. Santos, A. Perez-Castillo, N. C. Wong, and J. H. Oppenheimer, *J. Biol. Chem.* **262,** 16,880 (1987).
53. K. S. Park and co-workers, *J. Biol. Chem.* **262,** 14,702 (1987).
54. T. J. Flynn, L. Friedman, and T. N. Black, *Experientia* **43,** 433 (1987).
55. U. Geigenmuller, T. P. Hausner, and K. H. Nierhaus, *Eur. J. Biochem.* **16,** 715 (1986).
56. L. Kaczmarek, E. Surmacz, and R. Baserga, *Cell Biol. Int. Rep.* **10,** 455 (1986).
57. R. Vince, S. Daluge, and J. Brownell, *J. Med. Chem.* **29,** 2400 (1986).
58. J. A. Vara, A. Portela, J. Ortin, and A. Jimenez, *J. Nuc. Acids Res.* **14,** 4617 (1986).

59. C. J. Weitzmann and B. S. Cooperman, *Biochemistry* **29,** 3450 (1990).
60. D. M. Graifer, O. S. Fedorova, and G. G. Karpova, *FEBS Lett.* **277,** 4 (1990).
61. M. Yamashita and co-workers, *J. Antibiot.* **37,** 1279 (1984).
62. T. Iwamoto and co-workers, *J. Antibiot.* **43,** 1183 (1990).
63. A. Cumaraswamy and H. Henney, Jr., *Biochem. Cell Biol.* **68,** 769 (1990).
64. A. Fouet and A. L. Sonenshein, *J. Bacteriol.* **172,** 835 (1990).
65. U. Bai, M. Lewandoski, E. Dubnau, and I. Smith, *J. Bacteriol.* **172,** 5432 (1990).
66. M. A. Glazebrook, J. L. Doull, C. Stuttard, and L. C. Vining, *J. Gen. Microbiol.* **136,** 581 (1990).
67. W. L. Nicholson and G. H. Chambliss, *J. Bacteriol.* **169,** 5867 (1987).
68. K. Ochi, *J. Bacteriol.* **169,** 3608 (1987).
69. E. J. Dubnau, K. Cabane, and I. Smith, *J. Bacteriol.* **169,** 1182 (1987).
70. C. E. Rohlman and R. G. Matthews, *J. Bacteriol.* **172,** 7200 (1990).
71. C. McGuigan, S. M. Tollerfield, and P. A. Riley, *Nucleic Acids Res.* **17,** 6065 (1989).
72. J. A. Secrist, III, A. T. Shortnacy, and J. A. Montgomery, *J. Med. Chem.* **31,** 405 (1988).
73. D. C. Baker, T. H. Haskell, S. R. Putt, and B. J. Sloan, *J. Med. Chem.* **22,** 273 (1979).
74. T. Utagawa and co-workers, *FEBS Letters* **109,** 261 (1980).
75. R. J. Suhadolnik, S. Pornbanlualap, J. M. Wu, D. C. Baker, and A. K. Hebbler, *Arch. Biochem. Biophys.* **270,** 363 (1989).
76. W. E. G. Müller, R. K. Zahn, K. Bittlingmaier, and D. Falke, *Ann. N.Y. Acad. Sci.* **284,** 34 (1977).
77. J. P. Verheyden, *Rev. Infec. Dis.* **10,** Suppl. 477 (1988).
78. D. Pavan-Langston, R. A. Buchanan, and C. A. Alford, Jr., eds., *Adenine Arabinoside: An Antiviral Agent,* Raven Press, New York, 1975.
79. T. Takimoto and co-workers, *Leukemia* **2,** 443 (1988).
80. T. Yoshida, T. Suzutani, and M. Azuma, *Tohoku J. Exp. Med.* **156,** 279 (1988).
81. T. Spector, *J. Biol. Chem.* **250,** 7372 (1975).
82. W. E. G. Müller, *Experientia* **32,** 1572 (1976).
83. W. E. G. Müller and R. K. Zahn, in P. Chandra, ed., *Antiviral Mechanisms in the Control of Neoplasia,* Plenum Publishing Corp., New York, 1979, p. 553.
84. D. E. Bell and A. Fridland, *Biochim. Biophys. Acta* **606,** 57 (1980).
85. G. Iliakis and R. Seaner, *Rad. Environ. Biophys.* **27,** 295 (1988).
86. M. Kubota and co-workers, *Anticancer Res.* **8,** 339 (1988).
87. Y. Ohno, D. Spriggs, A. Matsukage, T. Ohno, and D. Kufe, *J. Cancer Res.* **49,** 2077 (1989).
88. Y. Kuroki, M. Shimoyama, S. Inaba, and M. Hirose, *Japanese J. Cancer Res.* **80,** 482 (1989).
89. J. Balzarini and E. DeClercq, *Biochem. Biophys. Res. Commun.* **173,** 781 (1990).
90. N. M. Mozier, K. P. McConnell, and J. L. Hoffman, *J. Biol. Chem.* **263,** 4527 (1988).
91. H. Malina, M. H. Tempete, and M. Robert-Gero, *J. Antibiot.* **40,** 505 (1987).
92. D. R. Berry and B. J. Abbott, *J. Antibiot.* **31,** 185 (1978).
93. A. Parry, *J. Am. Chem. Soc.* **113,** 4704 (1991).
94. H. Malina and M. Robert-Gero, *Applied Environ. Micro.* **54,** 1109 (1988).
95. W. K. Long, G. E. Fronko, R. G. Lindemeyer, B. Wu, and E. E. Henderson, *J. Virol.* **61,** 221 (1987); L. L. Nolan, *Antimicrob. Agents Chemother.* **31,** 1542 (1987).
96. R. A. Neal, M. V. Iwabi, and M. Robert-Gero, *Comptes Rendus Acad. Sci.* **308,** 485 (1989).
97. B. L. Mirkin, R. F. O'Dea, and H. P. C. Hogenkamp, *Cancer Res.* **47,** 3650 (1987).
98. R. T. Borchardt, L. E. Eiden, B-S. Wu, and C. O. Rutledge, *Biochem. Biophys. Res. Commun.* **89,** 919 (1979).
99. J. Najbauer and D. W. Oswald, *J. Biol. Chem.* **265,** 12,717 (1990).
100. W.-K. Paik and S. Kim, in A. Meister, ed., *Protein Methylation (Biochemistry: A Series of Monographs),* Vol. I, John Wiley & Sons, Inc., New York, 1980, p. 197.

101. S. Som and S. Friedman, *J. Biol. Chem.* **265,** 4278 (1990).
102. N. O. Reich and N. Mashhoon, *J. Biol. Chem.* **265,** 8966 (1990).
103. C. Barbis, J. Sanchez, M. Robert-Gero, and C. Hardisson, *FEMS Microbiol. Lett.* **57,** 239 (1990).
104. S. G. Amur, G. Shanker, J. M. Cochran, H. S. Ved, and R. A. Pieringer, *J. Neuroscience Res.* **16,** 367 (1986).
105. D. B. Drachman and R. Adams, *Proc. Natl. Acad. Sci. USA* **85,** 4032 (1988).
106. F. A. Fuhrman, G. J. Fuhrman, Y. H. Kim, L. A. Pavelka, and H. S. Mosher, *Science* **207,** 193 (1980).
107. R. P. Gregson, R. J. Quinn, and A. F. Cook, *Chem. Abst.* **91,** 39,792 (1979).
108. B. G. Issac, S. W. Ayer, L. J. Letendre, and R. J. Stonard, *J. Antibiot.* **44,** 729 (1991).
109. A. K. Saksena, *Tetrahedron Lett.* **21,** 133 (1980).
110. L. L. Bennett, Jr., P. W. Allan, L. M. Rose, R. M. Comber, and J. A. Secrist, III, *Molec. Pharmacol.* **29,** 383 (1986).
111. O. Hiraoko and co-workers, *Biochem. Biophys. Res. Commun.* **134,** 1114 (1986).
112. A. A. Fischer, K. Muller, and C. Scholtissek, *J. Virol.* **177,** 523 (1990).
113. R. Djurhuus, A. M. Svardal, and P. M. Ueland, *J. Cancer Res.* **49,** 324 (1989).
114. K. Dalaker and H. Prydz, *J. Biochem. Pharm.* **35,** 3433 (1986).
115. J. Aarbakke, G. A. Miura, and P. S. Prytz, *J. Cancer Res.* **46,** 5469 (1986).
116. P. S. Backlund, Jr., D. Carotti, and G. L. Cantoni, *Eur. J. Biochem.* **160,** 245 (1986).
117. A. M. Svardal, R. Djurhuus, H. Refsum, and P. M. Ueland, *J. Cancer Res.* **46,** 5095 (1986).
118. R. G. Liteplo, *Cancer Lett.* **39,** 319 (1988).
119. M. L. Greenberg, S. Chaffee, and M. S. Hershfield, *J. Biol. Chem.* **264,** 795 (1989).
120. R. J. Parry, V. Bornemann, and R. Subramanian, *J. Am. Chem. Sic.* **111,** 5819 (1989).
121. S. Yaginuma, N. Muto, T. Tsujino, and M. Hayashi, *J. Antibiot.* **34,** 359 (1981).
122. M. Hayashi, S. Yaginuma, H. Yoshioka, and K. Nakatsu, *J. Antibiot.* **34,** 675 (1981).
123. M.-L. Lim and V. E. Marquez, *Tetrahedron Lett.* **24,** 5559 (1983).
124. J. Michael, G. Oflenbaecher, and J. Retey, *Helv. Chim. Acta* **66,** 1915 (1983).
125. M. Arita, K. Adachi, Y. Ito, H. Sawai, and M. Ohno, *J. Am. Chem. Soc.* **105,** 4049 (1983).
126. C. K. Tseng and co-workers, *J. Med. Chem.* **32,** 1442 (1989).
127. E. DeClerq and co-workers, *Antimicrob. Agents Chemother.* **33,** 1291 (1989).
128. E. DeClerq, *Biochem. Pharmacol.* **36,** 2567 (1987).
129. R. M. Ransohoff, P. Narayan, D. F. Ayers, F. M. Rottman, and T. W. Nilsen, *J. Antiviral Res.* **7,** 317 (1987).
130. M. Cools and E. DeClerq, *Biochem. Pharmacol.* **40,** 2259 (1990).
131. D. L. Kramer, C. W. Porter, R. T. Borchardt, and J. R. Sufrin, *J. Cancer Res.* **50,** 3838 (1990).
132. S. D. Paisley, M. S. Wolfe, and R. T. Borchardt, *J. Med. Chem.* **32,** 1415 (1989).
133. M. Inaba, K. Nagashima, S. Tsukagoshi, and Y. Sakurai, *Cancer Res.* **46,** 1063 (1986).
134. K. Isono and co-workers, *J. Antibiot.* **37,** 670 (1985).
135. E. Takahashi and T. Beppu, *J. Antibiot.* **35,** 939 (1982).
136. M. Osada and K. Isono, *Antimicrob. Agents Chemother.* **27,** 230 (1985).
137. H. Osada and K. Isono, *Biochem. J.* **233,** 459 (1986).
138. W. M. Kati and R. Wolfenden, *Science* **243,** 1591 (1989).
139. M. H. el Kouni, N. J. Messier, and S. Cha, *J. Biochem. Pharm.* **36,** 3815 (1987).
140. M. H. el Kouni and S. Cha, *J. Biochem. Pharm.* **36,** 1099 (1987).
141. J. Jiricny, S. G. Wood, D. Martin, and A. Ubasawa, *Nuc. Acids Res.* **14,** 6579 (1986).
142. F. Nakagawa, T. Okazaki, A. Naito, Y. Iijima, and M. Yamazaki, *J. Antibiot.* **38,** 823 (1985).
143. S. Takahashi and co-workers, *J. Antibiot.* **38,** 830 (1985).
144. S. Takahashi, F. Nakagawa, and S. Sato, *J. Antibiot.* **41,** 705 (1988).
145. N. Shimada and co-workers, *J. Antibiot.* **39,** 1623 (1986).

146. H. Hoshino, N. Shimizu, N. Shimada, T. Takita, and T. Takeuchi, *J. Antibiot.* **40,** 1077 (1987).
147. F. X. Wilson and co-workers, *Tetrahedron Lett.* **31,** 6931 (1990).
148. N. Shimada and co-workers, *J. Antibiot.* **34,** 1216 (1981).
149. N. Yagisawa and co-workers, *J. Antibiot.* **35,** 755 (1982).
150. Y. Uehara, M. Hasegawa, M. Hori, and H. Umezawa, *Cancer Res.* **45,** 5230 (1985).
151. S. Naruto, H. Uno, A. Tanaka, H. Kotani, and Y. Takase, *Heterocycles* **20,** 27 (1983); B. D. Yuan and co-workers, *J. Antibiot.* **38,** 1588 (1985).
152. J. B. Stewart and co-workers, *J. Antibiot.* **41,** 1048 (1988).
153. C. R. Loomis and R. M. Bell, *J. Biol. Chem.* **263,** 1682 (1988).
154. T. T. Tsunoda and K. Isono, *J. Antibiot.* **42,** 102 (1989).
155. P. K. Gupta and co-workers, *J. Med. Chem.* **32,** 402 (1989).
156. E. DeClercq, J. Balzarini, D. Madej, F. Hansske, and M. J. Robins, *J. Med. Chem.* **30,** 481 (1987).
157. S. H. Krawczyk and co-workers, *J. Med. Chem.* **33,** 3160 (1990).
158. H. Osada, T. Sonoda, K. Tsunoda, and K. Isono, *J. Antibiot.* **42,** 102 (1989).
159. N. K. Saxena, L. A. Coleman, J. C. Drach, and L. B. Townsend, *J. Med. Chem.* **33,** 1980 (1990).
160. P. K. Gupta and co-workers, *J. Med. Chem.* **32,** 402 (1989).
161. E. F. Elstner and R. J. Suhadolnik, *J. Biol. Chem.* **246,** 6973 (1971).
162. T. Uematsu and R. J. Suhadolnik, *Arch. Biochem. Biophys.* **162,** 614 (1974).
163. M. Kaehler, J. Coward, and F. Rottman, *Biochemistry* **16,** 5770 (1977).
164. G. Seibert, A. Maidhof, R. K. Zahn, and W. E. G. Müller, *Gann* **69,** 739 (1978).
165. E. DeClercq and M. J. Robins, *Antimicrob. Agents Chemother.* **30,** 719 (1986).
166. E. DeClerq and M. J. Robins, *Nuc. Acids Res.* **15,** 3059 (1987).
167. R. I. Glazer and A. L. Peale, *Biochem. Pharmacol.* **29,** 305 (1980).
168. K. Palczewski, N. Kahn, and P. A. Hargrave, *Biochemistry* **29,** 6276 (1990).
169. K. Ramasamy, B. S. Sharma, W. B. Jolley, R. K. Robins, and G. R. Revankar, *J. Med. Chem.* **32,** 1905 (1989).
170. S. H. Kim, T. Okabe, N. Tanaka, and H. Suzuki, *J. Antibiot.* **40,** 1776 (1987).
171. H. Suzuki and co-workers, *Cancer Res.* **47,** 713 (1987).
172. S. Naruto, H. Uno, A. Tanaka, H. Kotani, and Y. Takase, *Heterocycles* **20,** 27 (1983).
173. Y. Kato, N. Fusetani, S. Matsunaga, and K. Hashimoto, *Tetrahedron Lett.* **26,** 3483 (1985).
174. S. Omura and co-workers, *J. Antibiot.* **38,** 1008 (1985).
175. S. Omura, H. Tanaka, H. Kuga, and N. Imamura, *J. Antibiot.* **39,** 309 (1986).
176. J. C. Hanvey, J. K. Hardman, R. J. Suhadolnik, and D. C. Baker, *Biochemistry* **23,** 904 (1984).
177. J. C. Hanvey and co-workers, *Biochemistry* **26,** 5636 (1987).
178. D. J. Merkler, M. Brenowitz, and V. L. Schramm, *Biochemistry* **29,** 8358 (1990).
179. M. R. Grever and co-workers, *J. Clin. Oncol.* **3,** 1196 (1985).
180. H. M. Golomb and M. J. Ratain, *N. Engl. J. Med.* **316,** 870 (1987).
181. A. S. Spiers and co-workers, *N. Engl. J. Med.* **316,** 825 (1987).
182. J. B. Johnston and co-workers, *Cancer Res.* **46,** 2179 (1986).
183. B. C. Lembersky, M. J. Ratain, C. Westbrook, and H. M. Golomb, *Am. J. Hematol.* **27,** 60 (1988).
184. A. D. Ho and co-workers, *Blood* **72,** 1884 (1988).
185. S. Durrleman, J. L. Grem, and B. D. Cheson, *Eur. J. Haematol.* **43,** 297 (1989).
186. J. P. Dutcher, K. M. Salva, and P. H. Wiernik, *Am. J. Clin. Oncol.* **13,** 290 (1990).
187. R. M. Lowenthal, J. S. Wiley, K. F. Rooney, D. R. Challis, and G. M. Woods, *Brit. J. Haematol.* **76,** 555 (1990).
188. T. J. Ruers, W. A. Buurman, and C. J. van der Linden, *J. Immunol.* **138,** 116 (1987).
189. S. S. Matsumoto, J. Yu, and A. L. Yu, *J. Biol. Chem.* **263,** 7153 (1988).

190. A. Begleiter, L. Pugh, L. G. Israels, and J. B. Johnston, *J. Cancer Res.* **48,** 3981 (1988).
191. F. Lamballe, P. V. LePrise, E. LeGall, and J. C. David, *J. Leukemia* **3,** 97 (1989).
192. E. H. Kraut, J. C. Neff, B. A. Bouroncle, D. Gochnour, and M. R. Grever, *J. Clin. Oncol.* **8,** 848 (1990).
193. N. Lahat and co-workers, *Scand. J. Immunol.* **32,** 205 (1990).
194. A. D. Ho and co-workers, *Leukemia* **4,** 584 (1990).
195. R. Furuta, R. S. Naruto, A. Tamura, and K. Yokogawa, *Tetrahedron Lett.,* 1701 (1979).
196. T. Ito and co-workers, *J. Antibiot.* **37,** 931 (1984).
197. J. G. Buchanan, J. Stoddart, and R. H. Wightman, *J. Chem. Soc., Chem. Commun.,* 823 (1989).
198. G. Tamura, ed., *Tunicamycin,* Japan Scientific Society Press, Tokyo, 1982.
199. E. Eckardt, W. Ihn, D. Tresselt, and D. Krebs, *J. Antibiot.* **34,** 1631, 1632 (1981).
200. K. Isono and co-workers, *J. Antibiot.* **38,** 1617 (1985).
201. M. Ubukata, K. Isono, K. Kimura, C. C. Nelson, and J. A. McCloskey, *J. Am. Chem. Soc.* **110,** 4416 (1988).
202. Y. Hayakawa and co-workers, *Agric. Biol. Chem.* **49,** 2685 (1985).
203. J. D. Dutcher, M. H. von Saltza, and F. E. Pansy, *Antimicrob. Agents Chemother.* **1963,** 83 (1964).
204. H. L. Agahigrian and co-workers, *J. Org. Chem.* **30,** 1085 (1965).
205. J. F. Edgar and co-workers, *J. Chem. Soc., Chem. Commun.,* 222 (1982).
206. M. Yoshida and co-workers, *J. Antibiot.* **43,** 748 (1990).
207. I. Tabas and S. Kornfeld, *J. Biol. Chem.* **254,** 11,655 (1979).
208. C. J. Waechter and W. J. Lennarz, *Annu. Rev. Biochem.* **45,** 95 (1976).
209. R. T. Schwarz and R. Datema, *Trends Biochem. Sci.* **5,** 65 (1980).
210. K. VonFigura, M. Rey, R. Prinz, B. Voss, and K. Ullrich, *Eur. J. Biochem.* **101,** 103 (1979).
211. E. L. Kean, J. J. Plantner, and L. Poncz, *Fed. Proc.* **39,** 1899 (1980).
212. D. Duksin and P. Bornstein, *J. Biol. Chem.* **252,** 955 (1977).
213. Y. T. Pan, J. W. Schmitt, B. A. Sanford, and A. D. Elbein, *J. Bacteriol.* **139,** 507 (1979).
214. J. A. Hanover, W. J. Lennarz, and J. D. Young, *J. Biol. Chem.* **255,** 6713 (1980).
215. G. E. Bettinger, A. N. Chaterjee, and F. E. Young, *J. Biol. Chem.* **252,** 4118 (1977).
216. H. R. Perkins, S. J. Thorpe, and C. A. Brown, *Biochem. Soc. Trans.* **8,** 163 (1980).
217. W. T. Forsee, G. Valkovich, and A. D. Elbein, *Arch. Biochem. Biophys.* **174,** 469 (1976).
218. J. Fujisawa, Y. Iwakura, and Y. Kawade, *J. Biol. Chem.* **253,** 8677 (1978).
219. G. W. Hart and W. J. Lennarz, *J. Biol. Chem.* **253,** 5795 (1978).
220. S. R. Burger and co-workers, *J. Gen. Virol.* **72,** 359 (1991).
221. M. Ikeda, M. Wachi, H. K. Jung, F. Ishino, and M. Matsuhashi, *J. Bacteriol.* **173,** 1021 (1991).
222. P. J. Walker, K. A. Byrne, D. H. Cybinski, D. L. Doolan, and Y. H. Wang, *J. Gen. Virol.* **72,** 67 (1991).
223. D. L. Hogue, K. C. Hodgson, and C. E. Cass, *Biochem. Cell Biol.* **68,** 199 (1990).
224. V. Gross and co-workers, *J. Biol. Chem.* **371,** 231 (1990).
225. A. Giraud and J. L. Franc, *Eur. J. Cell Biol.* **48,** 128 (1989).
226. R. Konig, G. Ashwell, and J. A. Hanover, *J. Biol. Chem.* **263,** 9502 (1988).
227. U. Blank, C. S. Ra, and J. P. Kinet, *J. Biol. Chem.* **266,** 2639 (1991).
228. B. Sherry, D. M. Jue, A. Zentella, and A. Cerami, *Biochem. Biophys. Res. Commun.* **173,** 1072 (1990).
229. N. Singh and A. M. Tartakoff, *Molec. Cell. Biol.* **11,** 391 (1991).
230. A. J. Hoogewerf and A. Bensadoun, *J. Biol. Chem.* **266,** 1048 (1991).
231. J. J. York, S. Sonza, M. R. Brandon, and K. J. Fahey, *Arch. Virol.* **115,** 147 (1990).
232. M. Uramoto, C.-J. Kim, K. Shin-ya, H. Kusakabe, and K. Isono, *J. Antibiot.* **44,** 375 (1991).
233. K. Isono and R. J. Suhadolnik, *Arch. Biochem. Biophys.* **173,** 141 (1976).

234. M. Inukai and co-workers, *J. Antibiot.* **42,** 662 (1989).
235. F. Isono, M. Inukai, S. Takahashi, and T. Haneishi, *J. Antibiot.* **42,** 667 (1989).
236. F. Isono, T. Katayama, M. Inukai, and T. Haneishi, *J. Antibiot.* **42,** 674 (1989).
237. F. Isono and M. Inukai, *Antimicrob. Agents Chemother.* **35,** 234 (1991).
238. R. H. Chen, A. M. Buko, D. N. Whittern, and J. B. McAlpine, *J. Antibiot.* **42,** 512 (1989).
239. J. P. Karwowski and co-workers, *J. Antibiot.* **42,** 506 (1989).
240. V. Gullo and co-workers, *J. Antibiot.* **41,** 20 (1988).
241. R. Cooper, M. Conover, and M. Patel, *J. Antibiot.* **41,** 123 (1988).
242. H. Dellweg and co-workers, *J. Antibiot.* **41,** 1145 (1988).
243. A. Gonzalez, D. Vazquez, and J. Jimenez, *Biochim. Biophys. Acta* **561,** 403 (1979).
244. H.-P. Fiedler, ed., *Chemische Biologie der Mikroorganismen, Ergebnisse des Sonderforschungsbreichs 76 der Deutschen Forschungsmeinschaft 1972–1985.*
245. H. Heitsch and co-workers, *J. Antibiot.* **42,** 711 (1989).
246. H. Maehr and R. G. Pitcher, *J. Antibiot.* **24,** 830 (1971).
247. G. Benz and co-workers, *Angew Chem. Int. Ed. Engl.* **21,** 527 (1982).
248. G. Benz, *Liebigs Ann. Chem.,* 1399 (1984).
249. T. Ogita and co-workers, *Tetrahedron Lett.* **21,** 3203 (1980).
250. H. Yamaguchi and co-workers, *J. Antibiot.* **39,** 1047 (1986).
251. H. Seto and co-workers, *Tetrahedron Lett.* **29,** 2343 (1988).
252. T. Shimura and co-workers, *J. Antibiot.* **36,** 1300 (1983).
253. N. Nishizawa and co-workers, *J. Antibiot.* **37,** 1 (1984).
254. J. J. McGowan, C. P. Allen, J. M. Barnett, and G. A. Gentry, *Virology* **101,** 516 (1980).
255. H. Machida and co-workers, *Antimicrob. Agents Chemother.* **16,** 158 (1979).
256. J. J. Harrison, A. Anisowicz, I. K. Gade, M. Raffeld, and R. Sager, *Proc. Natl. Acad. Sci. USA* **80,** 6606 (1983).
257. N. Fujii and K. Oguma, *J. Gen. Virol.* **67,** 2521 (1986).
258. B. Richardson, L. Kahn, E. J. Lovett, and J. Hudson, *J. Immunol.* **137,** 35 (1986).
259. S. Rainier and A. P. Feinberg, *Proc. Natl. Acad. Sci. USA* **85,** 6384 (1988).
260. N. Jahroudi, R. Foster, J. Price-Haughey, G. Beitel, and L. Gedamu, *J. Biol. Chem.* **265,** 6506 (1990).
261. R. L. Momparler and J. Lalibert, *Leukemia Res.* **14,** 751 (1990).
262. J. Bouchard and co-workers, *Antimicrob. Agents Chemother.* **34,** 206 (1990).
263. R. A. Winegar, J. W. Phillips, L. H. Lutze, and W. F. Morgan, *Somatic Cell Molec. Genet.* **16,** 251 (1990).
264. C. Bonnerot, S. Amigorena, W. H. Fridman, J. Even, and M. Dameron, *J. Immunol.* **144,** 323 (1990).
265. M. C. John and R. M. Amasino, *Molec. Cell. Biol.* **9,** 4298 (1989).
266. J. Bouchard, *Leukemia Res.* **13,** 715 (1989).
267. V. I. Avramis, W. C. Powell, and R. A. Mecum, *Cancer Chemother. Pharmacol.* **24,** 155 (1989).
268. J. Hoshino, J. Frahm, and H. Kroger, *Biochem. Biophys. Res. Commun.* **142,** 468 (1987).
269. C. Dunbar, W. Travis, Y. W. Kan, and A. Nienhuis, *Brit. J. Haematol.* **72,** 467 (1989).
270. A. B. Glover and B. Leyland-Jones, *Cancer Treatment Reports* **71,** 959 (1987).
271. I. Kubo and M. Kim, *Tetrahedron Lett.* **27,** 4277 (1986).
272. S. B. Larson, R. J. Moss, G. D. Kini, and R. K. Robins, *Acta Cryst.* **C44,** 1076 (1988).
273. T. Kamikawa, S. Fujie, Y. Yamagiwa, M. Kim, and H. Kawaguchi, *J. Chem. Soc. Chem. Commun.,* 195 (1988).
274. K. Isshiki and co-workers, *J. Antibiot.* **43,** 584 (1990).
275. J. B. K. Nielsen and B. H. Arison, *J. Antibiot.* **42,** 1248 (1989).
276. A. D. Argoudelis and S. A. Mizsak, *J. Antibiot.* **29,** 818 (1976).
277. T. Kurokawa and co-workers, *Japan Kokai Tokyo Koho,* JP 90312596 (1990).
278. M. Nakajima and co-workers, *J. Antibiot.* **44,** 293 (1991).

279. H. Haruyama and co-workers, *J. Chem. Soc. Perkin Trans. 1,* 1637 (1991).
280. S. Mio, R. Ichinose, K. Goto, S. Sugai, and S. Sato, *Tetrahedron* **47,** 2111 (1991).
281. S. Mio, Y. Kumagawa, and S. Sugai, *Tetrahedron* **47,** 2133 (1991).
282. N. Katayama, S. Tsubotani, Y. Nozaki, S. Harada, and H. Ono, *J. Antibiot.* **43,** 238 (1990).
283. M. Uramoto and co-workers, *J. Antibiot.* **44,** 375 (1991).

ROBERT J. SUHADOLNIK
NANCY L. REICHENBACH
Temple University School of Medicine

OLIGOSACCHARIDES

Oligosaccharide antibiotics represented by the everninomicins (**1**) (Table 1) (1–13), flambamycins (**2**) (Table 2) (22,23), avilamycins (**2**) (Table 2) (14,15), and curamycins (**2**) (Table 2) (18,19) have complex and unique structural features. They are sometimes referred to as orthosomycins because characteristically these oligosaccharides possess two acid sensitive ortho-ester linkages, cleavage of which results in the complete loss of antibiotic activity (3,22). Another structural feature common to all the oligosaccharide antibiotics is the presence of a substituted phenol ester derived from dichloroisoeverninic acid attached to the sugar ring B at C-13. This acidic phenolic group, which could form salts with organic or inorganic bases, eg, the sodium or *N*-methylglucamine salt, is also essential for the antibiotic activity (3).

An additional structural feature found in everninomicin D is the presence of a nitro sugar residue, evernitrose (**3**) linked to ring B at C-12. Everninomicins B (1), C (2), and 13-384 Component 1 (13) also have this unique nitro sugar. Chemical modification of the nitro group has led to the preparation of a number of highly active derivatives (6,7).

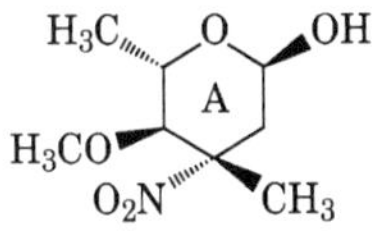

(**3**)

The first member of the oligosaccharide antibiotics to have its complete chemical structure elucidated was everninomicin D (5). The structure of everninomicin D then represented a template for the assignment of the structures of flambamycin (23), avilamycins (14), curamycins (19), and, more recently, everninomicin 13-384 Components 1 and 5 (13).

Properties

Physical Properties. Oligosaccharide antibiotics are colorless solids, which are often crystalline and have defined melting points and optical rotations, $[\alpha]_D$ (see Tables 1 and 2). They have a characteristic uv spectrum resulting from

Table 1. The Everninomicins[a]

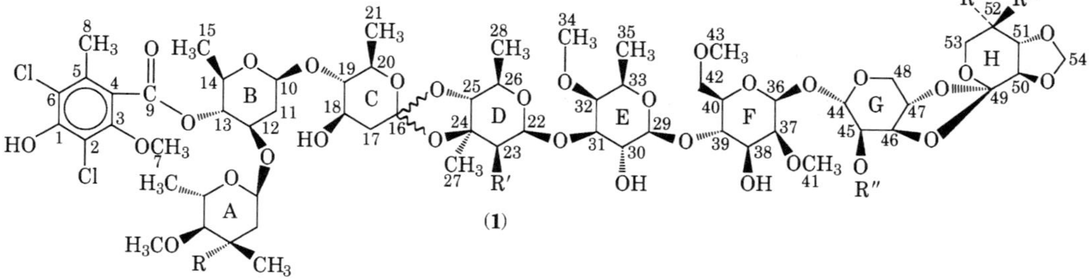

(1)

Compound	CAS Registry Number	Molecular formula	Mp, °C	$[\alpha]_D$	R	R′	R″	R‴	R⁗	Reference
everninomicin B	*[53296-30-3]*	$C_{66}H_{99}Cl_2NO_{36}$	184–185	−33.1	NO_2	OH	CH_3	(S)-$CH(CH_3)OCH_3$	OH	1
everninomicin C	*[53296-29-0]*	$C_{63}H_{93}Cl_2NO_{34}$	181–184	−33.7	NO_2	H	CH_3	OH	H	2
everninomicin D	*[39340-46-0]*	$C_{66}H_{99}Cl_2NO_{35}$	169–171	−34.2	NO_2	H	CH_3	(S)-$CH(CH_3)OCH_3$	OH	3–5,6–11

everninomicin 2	[*116217-29-9*]	$C_{58}H_{86}Cl_2O_{31}$	212–216	0.5	[b]	H	CH_3	$(S)\text{-}\underset{\displaystyle CH_3}{\underset{\vert}{C}}HOCH_3$	OH	12
everninomicin 3	[*64743-83-5*]	$C_{66}H_{98}Cl_2O_{33}$	157–158	−29.8	[c]	H	CH_3	$(S)\text{-}\underset{\displaystyle CH_3}{\underset{\vert}{C}}HOCH_3$	OH	11
everninomicin 7	[*64597-16-6*]	$C_{66}H_{100}Cl_2O_{34}$	173–175	−35.6	OH	H	CH_3	$(S)\text{-}\underset{\displaystyle CH_3}{\underset{\vert}{C}}HOCH_3$	OH	11
everninomicin-13-384-Component 1		$C_{70}H_{97}Cl_2NO_{38}$		−47.2	NO_2	OH	H	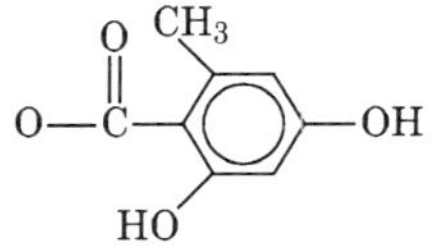	H	13
everninomicin-13-384-Component 5		$C_{70}H_{99}Cl_2NO_{36}$		−42.3	NH_2	OH	H	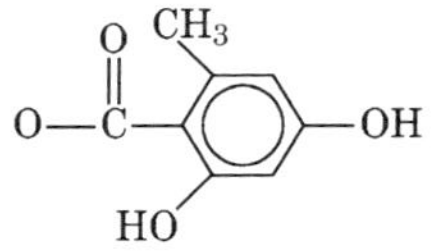	H	13

[a]The numbering is as in everninomicin (13).
[b]Ring A is absent in everninomicin 2, ie, Ring A = H.
[c]There is a double bond in everninomicin 3 at R.

Table 2. The Avilamycins (R = H), Curamycin (R = H), and Flambamycin (R = OH).

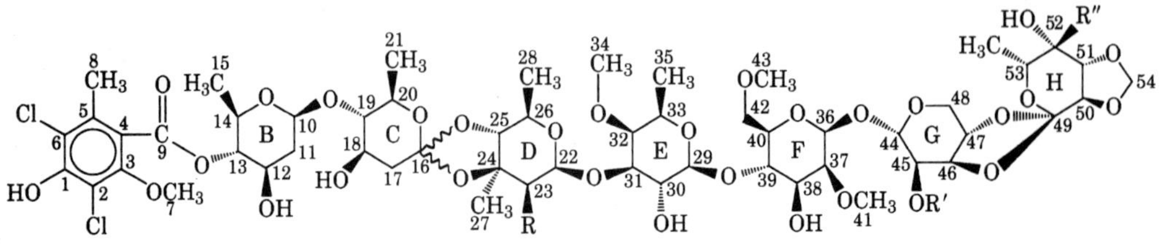

(**2**)

Compound	CAS Registry Number	Molecular formula	R′	R″	References
avilamycin A	*[69787-79-7]*	$C_{61}H_{88}Cl_2O_{32}$	$COCH(CH_3)_2$	$COCH_3$	14–17
avilamycin A_1			$COCH_2CH_3$	H	
avilamycin B[a]	*[73240-30-9]*	$C_{59}H_{84}Cl_2O_{32}$	$COCH_3$	$COCH_3$	
avilamycin C[b]	*[69787-80-0]*	$C_{61}H_{90}Cl_2O_{32}$	$COCH(CH_3)_2$	$CH(OH)CH_3$	
avilamycin D_1	*[82278-49-7]*	$C_{57}H_{82}Cl_2O_{31}$	H	$COCH_3$	
avilamycin D_2			$COCH_3$	$CH(OH)CH_3$	
avilamycin E			H	$CH(OH)CH_3$	
avilamycin F[c]	*[104748-55-2]*	$C_{60}H_{87}ClO_{32}$	$COCH(CH_3)_2$	$COCH_3$	
avilamycin G	*[104748-53-0]*	$C_{62}H_{90}Cl_2O_{32}$	COC_4H_9	$COCH_3$	
avilamycin H[d]	*[104748-54-1]*	$C_{61}H_{89}ClO_{32}$	$COCH(CH_3)_2$	$COCH_3$	
avilamycin I	*[104765-14-2]*	$C_{60}H_{86}Cl_2O_{32}$	$COCH_2CH_3$	$COCH_3$	
avilamycin J[e]	*[104748-48-3]*	$C_{60}H_{86}Cl_2O_{32}$	$COCH(CH_3)_2$	$COCH_3$	
avilamycin K[f]	*[104748-51-8]*	$C_{61}H_{88}Cl_2O_{33}$	$COCH(CH_3)_2$	$COCH_3$	
avilamycin L	*[104748-52-9]*	$C_{60}H_{86}Cl_2O_{32}$	$COCH(CH_3)_2$	CHO	
avilamycin M[g]	*[104748-50-7]*	$C_{60}H_{86}Cl_2O_{32}$	$COCH(CH_3)_2$	$COCH_3$	
avilamycin N[h]	*[104748-49-4]*	$C_{60}H_{86}Cl_2O_{32}$	$COCH(CH_3)_2$	$COCH_3$	
curamycin A[i,a]	*[73240-30-9]*	$C_{59}H_{84}Cl_2O_{32}$	$COCH_3$	$COCH_3$	18–21
flambamycin	*[42617-24-3]*	$C_{61}H_{88}Cl_2O_{33}\cdot H_2O$	$COCH(CH_3)_2$	$COCH_3$	22–26

[a]Avilamycin B and curamycin A appear to be identical.
[b]$[\alpha]_D = -4.8$.
[c]Avilamycin F has an H at C-2.
[d]Avilamycin H has an H at C-6.
[e]Avilamycin J has an H in place of the C_{43} CH_3 group.
[f]Avilamycin K has a CH_2OH in place of the C_{35} CH_3 group.
[g]Avilamycin M has an H in place of the C_{35} CH_3 group.
[h]Avilamycin N has an H in place of the C_{41} CH_3 group.
[i]$[\alpha]_D = 5.3$.

the phenolic ester residue, $\lambda_{max} \approx 285$ nm (ϵ ca 22), which shifts under basic conditions to $\lambda_{max} \approx 295$ nm (ϵ ca 80). The infrared spectrum shows absorptions for the hydroxyl and carbonyl moieties, and for other special features such as the nitro group which occurs at 1538 cm^{-1}. Molecular ions can be readily generated using modern mass spectral conditions such as fast atom bombardment (fab) (see MASS SPECTROMETRY) and the fingerprint fragmentation pattern has been employed for structure elucidation (8,13,14). X-ray analyses have been performed for the confirmation of partial structures of everninomicin (9,16) and avilamycin (27).

Modern nmr spectroscopic methods have also contributed heavily to the structure assignments (see MAGNETIC SPIN RESONANCE).

Chemical Properties and Structure. Oligosaccharide antibiotics are sensitive to acid pH because of the ortho-ester linkages. Yet all of them have an acidic phenolic group which makes the molecule relatively unstable to handling conditions. The ortho-ester connecting the C and D rings (Table 1) is comparatively more sensitive to acid pH than the one linking the G and H residues. Other chemically reactive groups in these compounds are the glycosidic linkages, the hydroxyl and carbonyl groups, and any nitro groups present.

Everninomicins

Everninomicin D is the principal component from cultures of *Micromonospora carbonacae* (10). Its structure (5) was elucidated using extensive chemical degradation coupled with spectroscopic analysis and it was the first reported instance of a natural product containing a tertiary nitrosugar. X-ray analyses of both the olgose residue (9) and the nitrosugar (16) have been reported as has a complete mass spectral analysis of everninomicin D (8).

Chemical modification (6) of the nitro group in everninomicin D has resulted in the formation of amino, mono- and dialkylamino, *N*-acylamino, and *N*-hydroxylamino (3) (and its nitrone derivatives) everninomicin D, all of which possess great antibiotic activity (6) against gram-positive bacteria. Aminoeverninomicin D can be converted to the known (11) everninomicins 3 and 7 under buffered conditions (6). Electrochemical reduction (11) of nitro and nitroso everninomicin D also produced everninomicins 3 and 7. Phosphite reduction of nitroso everninomicin causes fragmentation of sugar residue A leading to everninomicin 2 (12).

Everninomicin B (1), though a secondary component, has been extensively investigated because of its improved pharmacokinetic properties over those of everninomicin D (3,22). Everninomicin 13-384 Components 1 and 5, produced by *Micromonospora carbonacae* (24,28) are newer members of the everninomicin family. The structures (13) were assigned through extensive use of nmr and fab ms techniques. As in everninomicin B, C, and D, 13-384 Component 1 has a nitrosugar, whereas Component 5 has an amino sugar in its place, shown by conversion of Component 1 to 5 by hydrogenation (13). ^{13}C nmr confirmed the characteristic ortho-ester carbon atoms at 119.2 and 120.4 ppm and an extra aromatic residue in ring H at C-52. Nmr and high resolution fab ms confirmed this aromatic residue to be the 2,4-dihydroxy-6-methylbenzoyloxy group. The two aromatic residues on either end of this molecule appear much closer in conformational proximity, as observed by nmr (13) in $CDCl_3$ solution. Chemical modification of Components 1 and 5 has provided a number of active derivatives (7).

Flambamycin, Curamycin, and Avilamycins

Flambamycin. Flambamycin, produced by *Streptomyces hygroscopicus* DS 23230, also has undergone substantial chemical degradative experiments during structure elucidation (22,23,25). Nmr interpretation of the structure was reported later (20,26). The structure, shown in Table 2, chemical reactions, and biological properties have been thoroughly reviewed (22); it melts at 202–203°C.

Curamycin. Curamycin A, presented in Table 2, is the primary component of the culture *Streptomyces curacoi* (21) and preliminary chemical degradation studies (18) established it to be an oligosaccharide antibiotic. Later when the complete structure was assigned (17,19), the close relationship to flambamycin was evident. Curamycin A melts at 192–199°C.

Avilamycins. At least 16 avilamycins have been reported (14,15,29) and are listed in Table 2. Avilamycins A (mp 181–182) and C (mp 188–189) are the primary components (14) produced by the strain *Streptomyces viridochromogenes*. Structures have been established based on chemical degradation, nmr, and x-ray analysis. Structures for the rest of the compounds have been assigned based on nmr and fab ms (14) interpretations. Avilamycins F and H have only one chlorine atom in the phenolic ring. Relative hplc retention times for all the avilamycins have been recorded (14).

Other Oligosaccharides and Relevant Work

Sporacuracin A [*58205-91-7*], $C_{63}H_{94}Cl_2O_{35}$, mp 145–148°C, $[\alpha]_D$ −26.3, and sporacuracin B [*58205-92-8*], $C_{63}H_{94}Cl_2O_{35}$, mp 162–165°C, $[\alpha]_D$ −18.8, are other probable members of the oligosaccharide antibiotic family (22,30).

In all of the oligosaccharides discussed herein the stereochemistry of the ortho-ester in ring C of C-16 has not as of this writing been unequivocally established. Based on mechanistic studies using simple model reactions, a proposal has been put forward (31) to define the possible configuration at the ortho-ester linkage of oligosaccharide antibiotics. Experimental evidence suggests that the axially oriented oxygen atom (Fig. 1) with respect to ring C undergoes cleavage under acidic pH. The total synthesis of curacin [*20585-97-1*], $C_{15}H_{18}Cl_2O_7$, corresponding to the A B ring fragment, has been reported (32) and has been compared to its synthetic regioisomer, isocuracin.

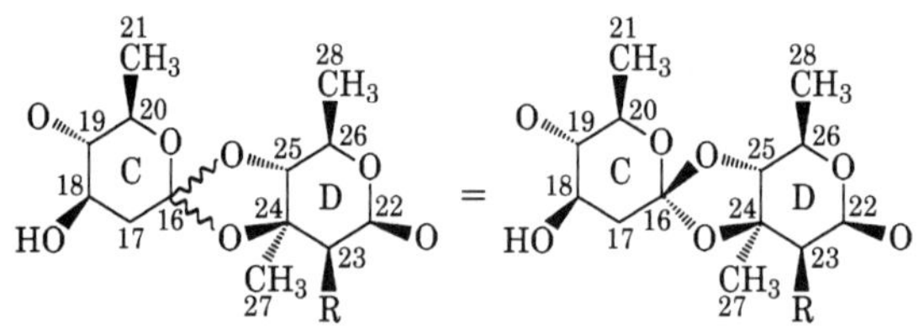

Fig. 1. Proposed ortho-ester configuration for ring C of the oligosaccharide antibiotics.

Biological Properties. The *in vitro* activity is such that oligosaccharides, in general, are highly potent, but are narrow-spectrum antibiotics (3,4,22). Everninomicins are active against a wide variety of gram-positive aerobes and anaerobes; *Neisseria, Mycoplasma,* and some *Mycobacteria*. These antibiotics are active against methicillin (see ANTIBIOTICS, PENICILLINS AND OTHERS): resistant *Staphylococci,* and compare to vancomycin (see ANTIBIOTICS, PEPTIDES). They lack cross-resistance with vancomycin because of different modes of action and their anaerobe activity is comparable to clindamycin (see ANTIBIOTICS, LINCOSAMINIDES). Everninomicin 13-384 Component 5 is said to have measurable *Pseudomonas* activity, but other members are devoid of any gram-negative activity. Comparatively, flambamycin is less potent than everninomicin D (22).

In Vivo Activity. Although highly active *in vitro,* everninomicin D posed pharmacokinetic problems when tested *in vivo* on mice and dogs. Everninomicin B, a close relative, on the other hand, gave better blood levels and *in vivo* efficacy. Similarly, Sch 23199, a semisynthetic *N*-acetyl analogue of everninomicin D also had useful activity and pharmacokinetic profile. In preliminary experiments in animals, everninomicin 13-384 Component 1 shows good *in vivo* activity. Finally, potential for use of oligosaccharide antibiotics in both human and animal health care has been claimed in various patents (7,24,28,30,33).

BIBLIOGRAPHY

"Antibiotics, Oligosaccharides" in *ECT* 3rd ed., Vol. 2, pp. 986–990, by A. K. Ganguly, Schering Corporation.

1. A. K. Ganguly and A. K. Saksena, *J. Antibiot.* **28,** 707 (1975).
2. A. K. Ganguly and S. Szmulewicz, *J. Antibiot.* **28,** 710 (1975).
3. A. K. Ganguly, in P. Sammes, ed., *Oligosaccharide Antibiotics, Topics in Antibiotic Chemistry,* Vol. 2, Ellis Horwood Publishers, Chichester, UK 1978, p. 49.
4. A. K. Ganguly, in M. Grayson, ed., *Oligosaccharides, Antibiotics, Chemotherapeutics, and Antibacterial Agents for Disease Control,* John Wiley & Sons, Inc., New York, 1982, p. 205.
5. A. K. Ganguly, O. Z. Sarre, D. Greeves, and J. Morton, *J. Am. Chem. Soc.* **97,** 1982 (1975).
6. A. K. Ganguly, V. M. Girijavallabhan, G. H. Miller, and O. Z. Sarre, *J. Antibiot.* **35,** 561 (1982).
7. U.S. Pat. 4,622,314 (Dec. 12, 1978), A. K. Ganguly, O. Z. Sarre, and V. M. Girijavallabhan (to Schering-Plough Corp.).
8. A. K. Ganguly and co-workers, *J. Antibiot.* **38,** 808 (1985).
9. A. K. Ganguly, O. Z. Sarre, A. T. McPhail, and K. D. Onan, *J. Chem. Soc. Chem. Commun.* **9,** 313 (1977).
10. M. J. Weinstein, G. M. Luedemann, E. M. Oden, and G. H. Wagman, *Antimicrob. Agents Chemother.* 24 (1964).
11. A. K. Ganguly and co-workers, *J. Chem. Soc. Chem. Commun.* **2,** 56 (1980).
12. A. K. Ganguly, S. Szmulewicz, O. Z. Sarre, and V. M. Girijavallabhan, *J. Chem. Soc. Chem. Commun.* **15,** 609 (1976).
13. A. K. Ganguly and co-workers, *Heterocycles* **28,** 83 (1989).
14. J. L. Mertz and co-workers, *J. Antibiot.* **39,** 877 (1986).
15. W. Keller-Schierlein, W. Heilman, W. D. Ollis, and C. Smith, *Helv. Chim. Acta* **62,** 7 (1979).
16. A. K. Ganguly, O. Z. Sarre, A. T. McPhail, and R. W. Miller, *J. Chem. Soc. Chem. Commun.* **1,** 22 (1979).
17. A. K. Ganguly, A. K. Bose, and N. F. Cappuccino, *J. Antibiot.* **32,** 1213 (1979).
18. E. G. Gros, V. Deulofeu, O. L. Galmarini, and B. Frydman, *Experientia* **24,** 323 (1968).
19. N. F. Capuccino, A. K. Bose, J. B. Morton, and A. K. Ganguly, *Heterocycles* **15,** 1621 (1981).

20. L. Valente and co-workers, *Tetrahedron Lett.* **13,** 1149 (1979).
21. O. L. Galmarini and V. Deulofeu, *Tetrahedron* **15,** 76 (1961).
22. D. E. Wright, *Tetrahedron,* Report No. 62, **35,** 1207 (1979).
23. W. D. Ollis, C. S. Smith, and D. E. Wright, *Tetrahedron* **35,** 105 (1979).
24. U.S. Pat. 4,735,903 (Apr. 5, 1988), J. A. Waitz and A. Horan (to Schering-Plough Corp.).
25. L. Ninet and co-workers, *Experientia* **30,** 1270 (1974).
26. W. D. Ollis and co-workers, *Tetrahedron* **35,** 993 (79).
27. E. Kupfer, K. Neupert-Laves, M. Dobler, and W. Keller-Schierlein, *Helv. Chim. Acta* **65,** 3 (1982).
28. U.S. Pat. 4,597,968 (July 1, 1986), J. A. Waitz and co-workers (to Schering-Plough Corp.).
29. F. Buzzetti and co-workers, *Experientia* **24,** 320 (1968).
30. Jpn. Pat. 75 125,094 (Oct. 1, 1975), A. Tamura and co-workers (to Dainippon); *J. Antibiot.,* Suppl. to Index 30, 77-111 and 77-112 (1977).
31. J-M Beau, G. Jaurand, J. Esnault, and P. Sinay, *Tetrahedron Lett.* **28,** 1105 (1987).
32. P. Jutten, J. Dornhagen, and H. Scharf, *Tetrahedron* **43,** 4133 (1987).
33. U.S. Pat. 4,436,734 (Mar. 13, 1984), E. E. Ose (to Eli Lilly Corp.).

V. M. Girijavallabhan
Ashit K. Ganguly
Schering-Plough Corporation

PEPTIDES

Peptide antibiotics are classified according to their overall shape, which can be linear or cyclic, and by the nature of the bonds joining the constituent and other amino acids (qv), which can be all amide bonds or amide plus ester bonds. Most peptide antibiotics are cyclic peptides that do not contain disulfide linkages. In homomeric cyclic peptides, all the atoms in the ring are contributed by amino acids; in heteromeric cyclic peptides such as enniatin [*11113-62-5*] and valinomycin [*2001-95-8*], $C_{54}H_{90}N_6O_{18}$, some atoms in the ring arise from nonamino acids. Homodetic homomeric cyclic peptides contain a ring of amino acids cyclized only through amide bonds. Amphomycin [*1402-82-0*], $C_{58}H_{91}N_{13}O_{20}$, and bacitracin [*1405-87-4*] are examples. The term heterodetic homomeric cyclic peptide applies if some amino acids in the ring are joined by ester bonds, eg, esterified serine or threonine hydroxyls as in virginiamycin S_1 [*23152-29-6*], $C_{43}H_{49}N_7O_{10}$. Heteromeric cyclic peptides, also called true depsipeptides, contain both amino acids and hydroxy acids in the ring. However, the term cyclic depsipeptide is often used in the literature to refer to heterodetic homomeric peptide lactones in which ester bonds to amino acid hydroxyls are present in the ring.

Peptide antibiotics differ in many respects from proteins (qv) and from peptides having hormonal or other functions in higher animals and plants (see Hormones). Among the significant differences are low molecular weight, where the vast majority of peptide antibiotics have molecular weights in the 500–1500 range. The average protein has mol wt of 40,000. Many peptide antibiotics have unusual fatty acids and amino acids, such as D-amino acids, *N*-methyl amino acids, or imino acids, and they usually lack methionine and histidine. Additionally, peptide antibiotics often have other nonamino acid moieties, such as the chromophore of dactinomycin [*50-76-0*]. Ring closure in cyclic peptide antibiotics is by amide or ester bonds, not by disulfide bonds as in normal proteins; or, if the peptide antibiotic has a tail, a diamino acid or a hydroxy amino acid provides the amide or ester bonds for ring closure. Peptide antibiotics are resistant to the usual proteases and peptidases. Also they are synthesized on multienzyme complexes much as are fatty acids, and not on ribosomes as are proteins (1). Nisin and related lantibiotics are exceptions. Most peptide antibiotics are synthesized, and are often marketed, as groups of closely related structures, usually reflecting a lack of specificity of the biosynthetic enzymes. Even the small fraction of peptide antibiotics that have therapeutic usefulness are quite toxic. For these useful peptide antibiotics the therapeutic index, the ratio of the minimum toxic dose to the maximum effective dose, is smaller than for most nonpeptide antibiotics.

Although historically most useful antibiotics have come from spore forming microorganisms, marine organisms have yielded the candidate antitumor peptide didemnin B [*77327-50-0*] (2) and cytostatic peptides such as the patellamides (3). Many of the marine peptides have little or no antimicrobial activity. Antibacterial peptides called magainins are found in frog skin (4) and antibacterial proteins called defensins are found in mammalian white blood cells (5). The commercially important insecticidal proteins from *Bacillus thuringensis* (6) are not discussed herein nor are the numerous peptide siderophores (7,8), which, except for the albomycins (9), are usually not antimicrobial.

Peptide antibiotics are not often the drugs of first choice for systemic therapy of important human disease. However, the World Health Organization, which chooses drugs especially for Third World use based on efficacy, safety, quality, price, and availability, includes as essential such peptide antibiotics as bleomycin, dactinomycin, and bacitracin (as an ointment containing neomycin), plus several β-lactams (see ANTIBIOTICS, β-LACTAMS) (10). Systemic use of peptide antibiotics is many times limited by nephrotoxicity and other toxicities. Semisynthesis or complete chemical synthesis of analogues of peptide antibiotics has most often not resulted in improved drugs. The complex structure of peptide antibiotics adds considerably to the problems of synthesis, but more recent efforts toward improved peptide antibiotics are encouraging (see FERMENTATION). Methods of bioassay and other laboratory use of economic antibiotics are available (11). Table 1 is a list of important peptide antibiotics. More extensive listings of minor peptide antibiotics have been published (32–36).

Peptide Antibacterials from Bacillus Species

Several economically useful peptide antibiotics are produced by members of the bacterial genus *Bacillus*.

Polymyxins. Polymyxin B [*1404-26-8*] and polymyxin E [*1066-17-7*] (colistin) are the only medically important antibacterials from a group of some thirty closely related cyclic decapeptide antibiotics in the polymyxin family, some of which are given in Table 2. Most family members contain a seven amino acid ring attached to a three amino acid tail, to which is attached a fatty acyl group. The octapeptins have a tail containing a single amino acid. Each antibiotic contains

Table 1. Peptide Antibiotics of Special Interest

Name	CAS Registry Number	Related to	Producing microbe	Antibiotic activity[a]	Mol wt (number of amino acids)	Number in family	Reference[b]
From bacteria of the genus Bacillus							
bacillomycin F	[*81689-97-6*]		*B. subtilis*	F	1080 (8)	2	12
bacilysin	[*29393-20-2*]		*B. subtilis, B. pumilis*	P, N	270 (2)		13
bacitracin (ayfivin)			*B. licheniformis*	P	1410 (11)	5	
circulins	[*9008-54-2*]	polymyxins	*B. circulans*	N	1150 (10)	2	
colistin (polymyxin E)	[*1066-17-7*]	polymyxins, circulins	*B. colistinus*	N	1150 (10)	10	
edeine A_1	[*27656-72-0*]		*B. brevis*	P, N, F	755 (5)	6	14
gramicidins, linear, Dubos			*B. brevis*	P	1900 (15)	6	
gramicidin S	[*113-73-5*]		*B. brevis*	P	1141 (10)	3	
iturins A	[*52229-90-0*]	bacillomycin B	*B. subtilis*	F	1050 (8)	9	15
micrococcin P_1	[*67401-56-3*]	thiocillins	*B. pumilis*	P	1144 (5)	4	16
mycobacillin	[*18524-67-9*]		*B. subtilis*	F	1528 (13)	1	17
polymyxins		circulins	*B. polymyxa*	N	1200 (10)	10	
subtilin	[*1393-38-0*]	nisin	*B. subtilis*	P	3321 (26)	1	
tyrocidines			*B. brevis*	P, N	1300 (10)	5	
tyrothricin			*B. brevis*	P, N	mixture		
From higher bacteria of the genus Streptomyces							
albomycin	[*1414-39-7*]		*S. subtropicus, S. griseus*	P	950 (5)	5	18
amphomycin (glumamycin)	[*1402-82-0*]		*S. canis, S. violaceus, S. albogriseolus*	P	1290 (10)	5	
bialaphos (SF 1293)	[*35597-43-4*]	phosphinothricin	*S. hydroscopicus*, other *S.* species	P, N,[c]	323 (3)	2	
bicozamycin (bicyclomycin)	[*38129-37-2*]		*S. sapporonensis, S. aizunensis*	N	302 (2)		
bleomycins		phleomycin, peplomycin, liblomycin	*S. verticillus*	P, N, M, F, T	1416 (12)		

bottromycin			*S. bottropensis*	P, N	810	(6)	5	19
capreomycins		viomycin, enviomycin, tuberoactinomycin	*S. capreolus*	P, N, M	670	(5)	4	
dactinomycin (actinomycin D)	[*50-76-0*]	cactinomycin	*S. parvullus*	P, T	1255	(10)	30	
daptomycin (deptomycin, LY 146032, semisynthetic)	[*103060-53-3*]		from A21978C of *S. roseosporus*	P	1619	(14)	6	
distamycin A (stallimycin)	[*636-47-5*]	netropsin	*S. distallicus*	F, T, V	482	(4)		20
enduracidins (enramycin)			*S. fungicidicus*	P, M	2355	(17)	4	
enviomycin (tuberactinomycin N)		capreomycin	*S. griseoverticilatus* var. *tuberacticus*	P, N, M	686	(6)	3	
ferrimycin A	[*15319-50-3*]	sideromycins	*S. griseoflavus*, other *S.* species	P, N	978	(6)	4	21
globomycin	[*67076-74-8*]		*S. halstedii*, *S. cinnamomeus*	N	657	(5)		22
ilamycin A (rufomycin A)	[*11006-41-2*]		*S. islandicus*, *S. atratus*	P, M	1026	(7)	5	23
neoviridogrisein II (etamycin B)	[*66002-40-2*]		*S. griseoviridis*	P	863	(7)	3	24
netropsin	[*1438-30-8*]	distamycin	*S. netropsis*, *S. chromogens*, others	P, N, M, T, V	430	(2)	8	20,25
quinomycin C	[*11001-74-4*]	triostin	*S. aureus*	P, N, T, V	1141	(8)	3	26
stendomycin	[*11006-78-3*]		*S. antimycoticus*	P, N, F	1850	(14)	3	27
streptogramin A (pristinamycin IIA, virginiamycin M1)	[*21411-53-0*]	virginiamycin M2	*S. virginiae*, others	P	527	(3)	2	
streptogramin B (pristinamycin IA)	[*3131-03-1*]	virginiamycin S_1	*S. virginiae*, others	P	867	(6)	12	
streptothricin A (polymycin A)	[*3484-67-1*]	noureothricins, racemomycins	*S. lavendulae*, others	P, N, M, V	1143	(6)	7	
telomycin	[*19246-24-3*]		*S. canus*, others	P	1272	(11)	1	28

Table 1. (*Continued*)

Name	CAS Registry Number	Related to	Producing microbe	Antibiotic activity[a]	Mol wt (number of amino acids)	Number in family	Reference[b]
thiopeptin			*S. tateyamensis*	P	1550 (6)	8	29
thiostrepton (bryamycin)	[*1393-48-2*]		*S. aureus*	P	1665 (9)	1	
valinomycin	[*2001-95-8*]		*S. fulvissimus, S. tsusimaensis*	P, N, F, T, V	1111 (6)		
viridogrisein (etamycin A)	[*299-20-7*]		*S. lavendulae*	P	879 (7)	2	30
zinostatin (neocarzinostatin)			*S. carcinostaticus*	P, T	11400 (113)	1	
From other bacteria							
nisin		subtilin	*Streptococcus lactis*	P	3354 (34)		
From fungi							
alamethicin	[*27061-78-5*]		*Trichoderma viridis*		1970 (19)	2	
cilofungin (semisynthetic)	[*79404-91-4*]	aculeacin B, sporiofungin	from echinocandin B of *Aspergillus* spp.	F	1029 (4)		
cyclosporine (cyclosporin A)		other cyclosporins	*Cylindrocarpon lucidium, Tolypocladium inflatum*	F, [c]	1203 (11)	25	
destruxin B			*Oospora destructor*, others	[d]	584 (5)		31
echinocandin B	[*54651-05-7*]		*Aspergillus rugulosus, A. nidulans* var. *roseus*	F	1060 (4)	8	
enniatin A	[*2503-13-1*]		*Fusarium orthoceras, F. sciroi*	P, M, F	682 (6)	3	
tentoxin	[*28540-82-1*]		*Alternaria tenuis, A. alternata*	[e]	415 (4)	2	

[a]Activity against gram-positive bacteria = P; gram-negative bacteria = N; mycobacteria = M; fungi = F; tumors = T; and viruses = V.
[b]Only peptides not discussed in the text of this article are given references.
[c]Active against the immune system.
[d]Active as an insecticide.
[e]Active as a herbicide.

Table 2. Polymyxins[a]

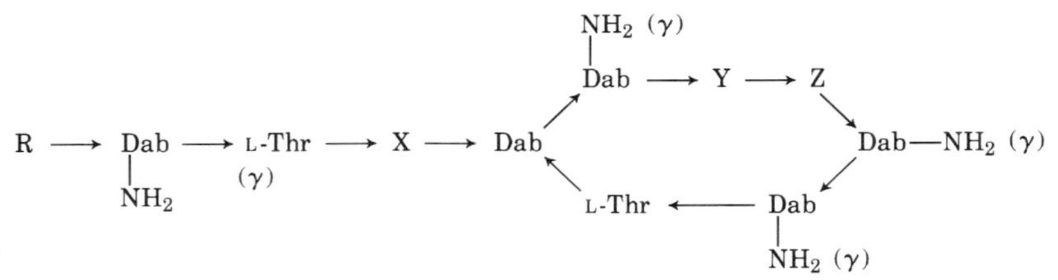

Name	CAS Registry Number	Molecular formula	R[b]	X	Y	Z
polymyxin A	[*1404-24-6*]	$C_{56}H_{98}N_{16}O_{13}$	MeOct	D-Dab	D-Leu	L-Thr
polymyxin B_1	[*4135-11-9*]	$C_{55}H_{96}N_{16}O_{13}$	MeOct	L-Dab	D-Phe	L-Leu
polymyxin B_2	[*34503-87-2*]	$C_{50}H_{93}N_{15}O_{15}$	MeHep	L-Dab	D-Phe	L-Leu
polymyxin D_1	[*10072-50-1*]	$C_{49}H_{91}N_{15}O_{15}$	MeOct	D-Ser	D-Leu	L-Thr
polymyxin D_2	[*34167-45-8*]	$C_{53}H_{100}N_{16}O_{13}$	MeHep	D-Ser	D-Leu	L-Thr
polymyxin E_1 (colistin A)	[*7722-44-3*]		MeOct	L-Dab	D-Leu	L-Leu
polymyxin E_2 (colistin B)	[*7239-48-7*]	$C_{52}H_{98}N_{16}O_{13}$	MeHep	L-Dab	D-Leu	L-Leu
circulin A	[*5854-98-8*]	$C_{53}H_{100}N_{16}O_{13}$	MeOct	L-Dab	D-Leu	L-*i*Leu
octapeptin B_1 [c,d]	[*50935-41-6*]		βOHMeDec		D-Leu	L-Phe

[a] Where the arrows represent bonds from the *C*-terminus to the *N*-terminus, and Dab = α,γ-diaminobutyric acid [*27252-32-0*], $C_4H_{10}N_2O_2$. Unless otherwise noted all Dab is L-Dab.

[b] MeHep = 6-methylheptanoic acid [*929-10-2*], $C_7H_{16}O_2$, MeOct = 6-methyloctanoic acid [*504-99-4*], $C_9H_{18}O_2$, MeDec = 8-methyldecanoic acid [*5601-60-5*], $C_{11}H_{22}O_2$, and βOHMeDec is the MeDec β-hydroxy derivative.

[c] For the octapeptins, omit the dipeptide, L-Thr-X, in the amino acid tail.

[d] The amino acid in the tail bound to R is D-Dab.

several L-α,γ-diaminobutyric acid [*1758-80-1*], $C_4H_{10}N_2O_2$, (Dab) residues, L-threonine [*72-19-5*], at least one D-amino acid, and one of a very limited list of fatty acids.

Polymyxin B sulfate [*1404-20-5*] (USP) and colistin sulfate [*1264-72-8*] (USP) are mixtures containing mostly polymyxins B_1 and B_2 and colistins A (polymyxin E_1) and B (polymyxin E_2), respectively. Polymyxin B sulfate, colistin sulfate (Coly-Mycin S), and colistin sodium methanesulfonate [*8068-28-8*], (colistimethate sodium, Coly-Mycin M), $C_{58}H_{105}N_{16}Na_5O_{28}S_5$, an injectable prodrug, at one time were important systemic agents against most gram-negative infections, especially those caused by *Pseudomonas aeruginosa*. However, in the 1990s most polymyxin is used, safely and effectively, as topical preparations. It is not absorbed through the skin. Polymyxin studies have been reviewed extensively; information regarding general basic science (37–41), chemistry (42), biosynthesis from multienzyme complexes and production (43,44), mechanism of action (43–45), and medical aspects (46,47) is available.

Polymyxin B sulfate, colistin sulfate, and colistimethate sodium are soluble in water and less soluble in organic solvents (48). Preliminary studies of polymyxin conformation in solution have been published (49). It took 15 years to elucidate the structures of the polymyxins following their discovery in 1947 (42). Before structures were elucidated, polymyxin A, later rediscovered as polymyxin M, polymyxins C and D, and the circulins were found to be more toxic than polymyxin B or E (the latter rediscovered as colistin). The structure determination of polymyxin B_1 was made possible by enzymic degradation, careful determination of D- and L-amino acids, and by final chemical synthesis of the molecule (42). There were limited but futile attempts to make medically useful improvements on polymyxins; exceptions were the sulfomethylation product of polymyxin B, no longer sold in the United States, and polymyxin E (colistimethate). Synthesis involved blocking the five amino groups of the DAB residues by treatment with formaldehyde and sodium bisulfite (50,51). The sodium salt of this *N*-blocked derivative, variously called the sodium methanesulfonate or the *N*-sulfomethyl derivative, has low antibiotic activity *in vitro* but is less nephrotoxic and less painful on injection. *In vivo,* hydrolytic removal of the methanesulfonate groups results in release of antibiotically active polymyxin B or E (colistin).

Most polymyxin B sold for human use in the United States is in dermatological, otic, and ophthalmic preparations that usually contain one or more other spectrum extending antibacterials such as bacitracin, neomycin sulfate [*1404-04-2*], $C_{23}H_{46}N_6O_{13}$, linear gramicidin, oxytetracycline [*79-57-2*], $C_{22}H_{24}N_2O$, or trimethoprim [*738-70-5*], $C_{14}H_{18}N_4O_3$. These preparations may also contain an antiinflammatory corticoid such as hydrocortisone [*50-23-7*], $C_{21}H_{30}O_5$, or hydrocortisone acetate [*50-03-3*], $C_{23}H_{32}O_6$, the analgesic diperidon [*537-12-2*], $C_{22}H_{28}ClN_3O_4$, or a detergent, eg, thonzonium bromide [*553-08-2*], $C_{32}H_{55}BrN_4O$.

Polymyxin B sulfate, whether as the bulk drug or as the final product, is sold by the unit and not by weight. One unit of polymyxin B sulfate is the weight of polymyxin B sulfate having the microbiological activity, usually against *Bordetella bronchoseptica,* of one unit of polymyxin B base. One milligram of standard polymyxin base has been arbitrarily set equal to 10,000 units. Polymyxin B sulfate USP must have the microbiological potency of at least 6000 units polymyxin B base per milligram. Most drug has a considerably higher purity.

Some polymyxins are sold for second-line systemic therapy. Polymyxin B sulfate and colistimethate sodium can be used for intravenous, intramuscular, or intrathecal administration, especially for *Pseudomonas aeruginosa* infections, but also for most other gram-negative organisms, such as those resistant to first-line antibiotics. Nephrotoxicity and various neurotoxicities are common in parenteral, but not in topical, use. Resistance to polymyxins develops slowly, involves mutation and, at least in some bacteria, adaptation, a poorly understood type of resistance that is rapidly lost on transfer to a medium free of polymyxin. Resistance can involve changes in the proteins, the lipopolysaccharides, and lipids of the outer membrane of the cell (52). Polymyxin and colistin show complete cross-resistance.

The mode of action of the polymyxins is related to the effect these molecules have on membrane integrity. Polymyxins are detergentlike, highly positively charged molecules that damage membranes of gram-negative bacteria by binding to lipopolysaccharide (endotoxin) and phospholipids of the membranes. Bactericidal action by this mechanism, which allows leakage of low molecular weight metabolites, can occur on both growing and nongrowing cells (53). Polymyxin B nonapeptide [*86408-36-8*], prepared by enzymic removal of both the fatty acid and the terminal amino acid of the tail (54–56), is a less toxic polymyxin derivative retaining considerable membrane permeabilizing activity. It does not, however, appear to be medically useful (57). Other cleavage products of polymyxins have been prepared (58).

The ability of polymyxin and its cleavage products to bind to lipopolysaccharide (endotoxin) suggests usefulness in preventing endotoxic shock caused by endotoxemia in gram-negative bacteremias. Practical use in patients has been limited, however, because of the toxicity of these agents (57). Polymyxin immobilized on a solid support is useful for removing contaminating endotoxin from genetically engineered mammalian proteins made in *E. coli* (see GENETIC ENGINEERING) (59).

Bacitracin. Bacitracin, a cyclic peptide active against gram-positive bacteria, was discovered in 1943. Bacitracin received drug certification in 1949 (60–62). Whereas human usage of bacitracin is almost exclusively topical, the vast majority of bacitracin manufactured worldwide is used as an animal feed additive. Reviews of work on bacitracin include its chemistry (63–67), comprehensive aspects (62), medical aspects (62,68), biosynthesis on large enzyme complexes and genetics (69–71), and production (71,72).

Bacitracin is produced and marketed as a mixture of at least nine water-soluble peptides. The principal (60–80%) component is bacitracin A [*22601-59-8*], $C_{66}H_{103}N_{17}O_{16}S$. The most probable structure of bacitracin A is shown in Figure 1 (67,73). Bacitracin B [*1402-99-9*] is similar to bacitracin A except that one of the isoleucines of bacitracin A is replaced by a valine. Bacitracin F [*22601-63-4*], $C_{66}H_{98}N_{16}O_{17}S$, is a relatively inactive degradation product of bacitracin A. Many other bacitracin components have been identified (64–67,74), but structure determinations are incomplete. At least some of the bacitracin F and probably some of the other bacitracins are degradation products of bacitracins A and B produced during fermentation or in manufacture and storage. Pure bacitracins are prepared by hplc (74). Bacitracin A has not been crystallized, but hydrogen–tritium exchange and nmr studies show that there is at least one internal hydrogen bond

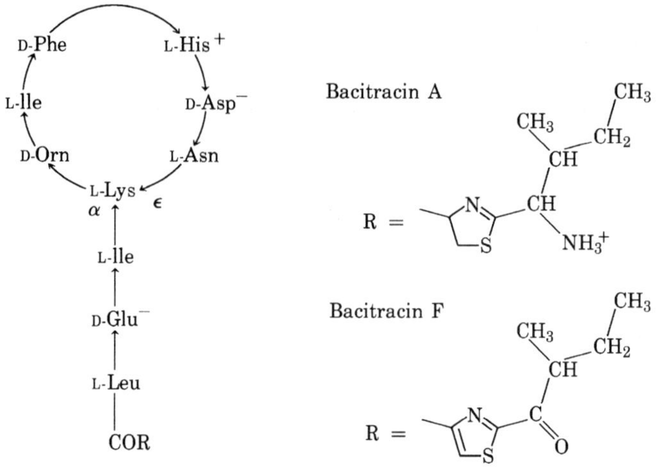

Fig. 1. Bacitracins A and F.

and that the peptide tail is folded over the ring. Only a partial chemical synthesis has been reported (75).

Bacitracin is a bitter, amorphous, white to pale buff, hygroscopic powder that is freely soluble in water and soluble in ethanol. When dry, bacitracin is stable from 25°C to 37°C. Several formulations of bacitracin have been developed. Anhydrous grease preparations of bacitracin are fairly stable, but water miscible creams are not. Zinc bacitracin, the form used in human medicine, is a white or pale yellowish gray or tan hygroscopic powder containing 2–10% zinc on a dry weight basis. It is sparingly soluble in both water and alcohol and more stable than bacitracin in troches, ointments, and tablets. Zinc may also enhance antibacterial activity. Bacitracin methylenedisalicylic acid (BMD) [*1405-88-5*], the form usually used in animal feed, is a white to grayish brown powder having a somewhat disagreeable odor. It is soluble in water, pyridine, and ethanol (67,76,77) and has increased stability over bacitracin. Both BMD and the zinc salt are used in animal feed (78,79,81,82).

One unit of bacitracin USP is defined as the activity given by 26 micrograms of the dried FDA master standard. The USP requirement for commercial bacitracin is a potency of at least 40 units per milligram in a bioassay with *Micrococcus flavus* or *Sarcina lutea*. Pure bacitracin A has a potency of about 100 units per milligram. Standards for animal feed grades of bacitracin are available (71,81,82) as are analytical methods (67).

Bacitracin given parenterally is sufficiently nephrotoxic that it is rarely used in human medicine for other than topical indications (80). Thus safe and effective use, especially as the zinc salt, is limited almost completely to ointments, sprays, and solutions for skin and ophthalmic use in concentrations of 250 to 1000 units per milliliter. Bacitracin is only rarely skin sensitizing. As in the case of polymyxin, bacitracin is usually combined with other antibiotics to enlarge its spectrum of activity, or with corticoids or analgesics to relieve pain or itching.

In the United States, 100 times more bacitracin by weight is used as a feed additive than for human medicine. Most of this is BMD but some is the zinc salt.

BMD can be used in subtherapeutic nutritional dosages for increase in feed efficiency and for growth promotion in poultry, swine, and feedlot cattle at concentrations of 2.6–33 g/t of feed (78,81,82). Prophylactic or therapeutic medicinal dosing at higher concentrations is used for necrotic enteritis in chickens, transmissable enteritis in turkeys, ulcerative enteritis in quail, dysentery in swine, and liver abcess in cattle (see FEEDS AND FEED ADDITIVES).

Bacitracin is bactericidal for many gram-positive cocci and rods, and a few gram-negative bacteria. Common gram-negative, enteric rod bacteria such as *E. coli* and *Pseudomonas,* and yeasts and other fungi are resistant. Resistant strains of bacteria do not normally arise during therapy, but an increasing number of *Staphylococcus aureus* strains are bacitracin-resistant. The mechanism of antibacterial action has been reviewed (69,70,83) and may be limited to a single step in bacterial cell wall biosynthesis. Bacitracin binds to the polyisoprenyl pyrophosphate used in bacterial cell wall biosynthesis and prevents its further cycling, thus blocking wall synthesis. There is a metal requirement for this binding and zinc is especially active. In mammalian metabolism, somewhat related polyisoprenyl pyrophosphates are found; the binding of bacitracin to these substances may be responsible for the parenteral toxicity of this antibiotic.

Tyrothricin and Gramicidin S. The linear pentadecapeptide gramicidins and the cyclic decapeptide tyrocidines, shown in Table 3, together comprise tyrothricin [*1404-88-2*], an antibiotic mixture active against many gram-positive and a few gram-negative bacteria. Tyrothricin was discovered in a strain of *Bacillus brevis* in 1939 (84). Gramicidin S [*113-73-7*], $C_{60}H_{92}N_{12}O_{10}$, which stands for gramicidin Soviet, is also called gramicidin J, for Japan. This material forms a group of cyclic decapeptides, also shown in Table 3, that is related to tyrocidines and obtained from another strain of *B. brevis* (85).

Although tyrothricin is too toxic for parenteral therapy, it was formerly sold in the United States as oral lozenges. Modern tyrothricin formulations are composed of 70–80% tyrocidines and 30–20% linear gramicidins. Tyrocidines are not as active as linear gramicidins and are too toxic for any therapeutic use by themselves. The bactericidal linear gramicidins are used in the United States solely as an ophthalmic solution in combination with polymyxin B sulfate and neomycin sulfate. The linear gramicidin is used in this aqueous product as a substitute for bacitracin, which lacks stability under such conditions.

The linear gramicidin of commerce is a mixture of at least eight closely related components, about 80% of which is valine–gramicidin A. Gramicidin USP is a white, odorless crystalline powder containing not less than 90% gramicidins on a dry weight basis. It is almost insoluble in water, but is readily soluble in methanol (86). Research on linear gramicidin's comprehensive aspects (86,87), conformations and mechanism of action (88–92), biosynthesis on multienzyme complexes (93), and chemistry (87–91) has been reviewed.

Linear gramicidins are neutral, uncharged, hydrophobic pentadecapeptides composed of alternating D- and L-amino acids. Glycine is classified here as a D-amino acid. This mixture is called gramicidin [*1405-97-6*] or gramicidin D (Dubos) [*1393-88-0*]. Gramicidin D is also used as a name for isoleucine–gramicidin A or as a name for an undefined polar fraction. Various naturally occurring and other gramicidins have been synthesized (94–96). Purified Val-gramicidin A has been shown to adopt different dimeric conformations depending on its physical

Table 3. Gramicidins and Tyrocidines

Name	CAS Registry Number	Molecular formula	R	R′
Linear gramicidins				
HCO—R—Gly—L-Ala—D-Leu—L-Ala—D-Val—L-Val—D-Val—L-Trp—D-Leu—R′—D-Leu—L-Trp—D-Leu—L-Trp—$NH_2CH_2CH_2OH$				
gramicidin A	*[11029-61-1]*			
Val-gramicidin A	*[4419-81-2]*	$C_{99}H_{140}N_{20}O_{17}$	L-Val	L-Trp
Ile-gramicidin A	*[5536-03-8]*	$C_{100}H_{142}N_{20}O_{17}$	L-Ile	L-Trp
gramicidin B	*[11041-38-6]*			
Val-gramicidin B	*[4422-52-0]*	$C_{97}H_{139}N_{19}O_{17}$	L-Val	L-Phe
Ile-gramicidin B			L-Ile	L-Phe
gramicidin C	*[9062-61-7]*			L-Tyr
Val-gramicidin C	*[58442-65-2]*	$C_{97}H_{139}N_{19}O_{18}$	L-Val	L-Tyr
Ile-gramicidin C			L-Ile	L-Tyr
Tyrocidines				
cyclo(L-Val—L-Orn—L-Leu—D-Phe—L-Pro—R—R′—L-Asn—L-Gln—L-Tyr)				
tyrocidine A	*[1481-70-5]*	$C_{66}H_{87}N_{13}O_{13}$	L-Phe	D-Phe
tyrocidine B	*[865-28-1]*	$C_{68}H_{88}N_{14}O_{13}$	L-Tyr	D-Phe
tyrocidine C	*[3252-29-7]*	$C_{70}H_{89}N_{15}O_{13}$	L-Tyr	D-Tyr
Gramicidin S				
cyclo(R—L-Orn—L-Leu—D-Phe—L-Pro—R′—L-Orn—L-Leu—D-Phe—L-Pro)				
gramicidin S_1	*[113-73-5]*	$C_{60}H_{92}N_{12}O_{14}$	L-Val	L-Val
gramicidin S_2	*[92278-12-1]*	$C_{59}H_{90}N_{12}O_{10}$	2-Aba[a]	L-Val
gramicidin S_3	*[92278-13-2]*	$C_{58}H_{88}N_{12}O_{10}$	2-Aba[a]	2-Aba[a]

[a]2-Aba is 2-aminobutanoic acid *[80-60-4]*, $C_4H_9NO_2$.

state. The pore forms seen in solvents appear to be parallel and antiparallel double helical forms. The pore forms viewed in the solid state by x-ray crystallography are antiparallel double helices (88–91,97). Dimeric gramicidin is also believed to produce channel forms that conduct monovalent cations in membrane models (88–91). The handedness of many of these helical forms is still under debate.

Tyrocidine *[8011-61-8]* is a mixture of three closely related components. Tyrocidine studies on mechanism of action (98), biosynthesis on multienzyme complexes (93,99,100), and chemistry (101) are available, and tyrothricin production is discussed (102). Although the mechanism of action of linear gramicidins has been well researched, such work on tyrocidine is more limited; it appears that tyrocidine damages membranes (103,104).

Gramicidin S is a fairly toxic mixture of at least three components, of which gramicidin S_1 is the primary component. Although it is not used in human medicine in the United States, there is some topical use of gramicidin S elsewhere. The biosynthesis of gramicidin S_1 on two multifunctional enzymes has been the subject of much work (99,100,105) and the mode of action of S_1 and analogues as membrane damaging agents also has elicited much attention (104). Gramicidin S studies are reviewed: mechanism of action (101), biosynthesis on multienzyme complexes (93,99,100), and chemistry (101).

Peptide Antibacterials and Antifungals from Other Microorganisms

Most peptide antibiotics have been isolated from *Actinomycetes* and especially from *Streptomyces*. Many β-lactams, however, and a few other useful peptide antibiotics have come from fungi. These streptomycete and fungal peptide antibiotics include many structural types having varied therapeutic indications and/or scientific applications.

Capreomycin, Viomycin, and Enviomycin. Capreomycin (Capastat, Lilly), a bacteriostatic, antimycobacterial peptide mixture isolated from *Streptomyces capreolus* was first reported in 1961 (106–108). This tuberactinomycin family member, shown in Table 4, was introduced into the U.S. market in 1971 where it has remained a useful but nephrotoxic and ototoxic second-line alternative to first-line tuberculosis therapies. Because capreomycin is somewhat less toxic than viomycin (tuberoactinomycin B [*32988-50-4*]), $C_{25}H_{43}N_{13}O_{10}$ (109,110), capreomycin has now displaced viomycin in the United States and most other markets. The structure of viomycin is shown in Figure 2. The related enviomycin (tuberactinomycin N [*33103-22-9*]), $C_{25}H_{43}N_{13}O_{10}$, is marketed in Japan (111–114). Available tuberactinomycin literature includes: capreomycin reviews (87,115, 116); capreomycin production (117); and viomycin synthesis on multienzyme complexes (118).

Capreomycin disulfate USP is a white solid, soluble in water and relatively

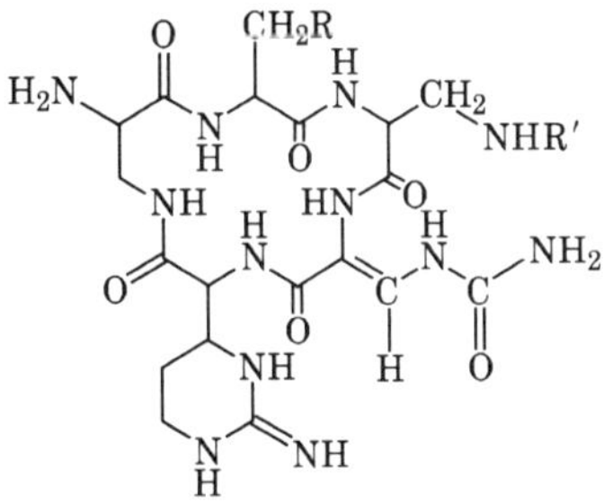

Table 4. Capreomycins

Name	CAS Registry Number	Molecular formula	R	R′
capreomycin IA	[*37280-35-6*]	$C_{25}H_{44}N_{14}O_8$	OH	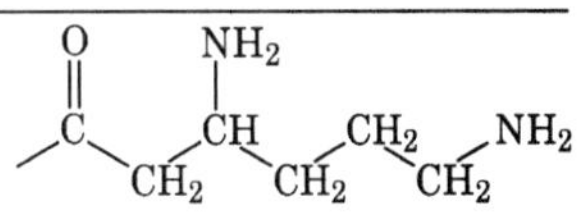
capreomycin IB	[*33490-33-4*]	$C_{25}H_{44}N_{14}O_7$	H	
capreomycin IIA	[*62639-89-8*]	$C_{19}H_{32}N_{12}O_7$	OH	H
capreomycin IIB	[*62639-90-1*]	$C_{19}H_{32}N_{12}O_6$	H	

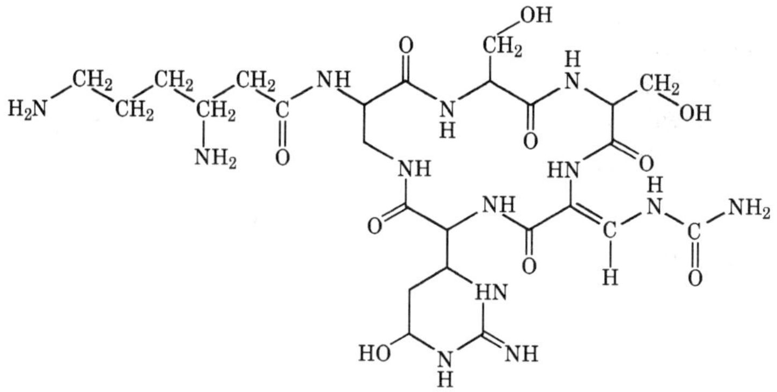

Fig. 2. Structure of viomycin (tuberoactinomycin B).

insoluble in most organic solvents. It must contain at least 90% capreomycins I. A representative lot contained 25% capreomycin IA, 67% IB, 3% IIA, and 6% IIB (119). The structures of capreomycin IA and IB have been determined and the compounds synthesized (120). The structures of capreomycin IIA and IIB are probably those shown in Table 4.

Capreomycin given intramuscularly is used to treat tuberculosis patients who do not respond to first-line drugs. Auditory, vestibular, and kidney function must be monitored during long-term therapy. Capreomycin shows cross-resistance to viomycin and some strains resistant to the more toxic second-line drug kanamycin [*8063-07-8*] (see ANTIBIOTICS, AMINOGLYCOSIDES) are also resistant to capreomycin. But capreomycin is not cross-resistant to streptomycin and other preferred therapies (121,122). As for many primary tuberculosis therapies, mutational resistance to capreomycin can develop during therapy. To slow the development of resistance, capreomycin is normally given with other drugs such as ethambutol [*74-55-5*], $C_{10}H_{24}N_2O_2$, or isoniazid [*54-85-3*], $C_6H_7N_3O$. Capreomycin is not effective with the majority of *Mycobacterium avium/intracellulare* strains that are often seen in autoimmune deficiency syndrome (AIDS) patients (123,124) (see ANTIVIRAL AGENTS; IMMUNOTHERAPEUTIC AGENTS). This latter resistance is apparently caused by poor uptake of drug into the bacterial cell. Capreomycin and other tuberactinomycins inhibit protein synthesis in sensitive bacteria at the translocation step and resistant strains have altered ribosomes (125).

Streptogramins. Virginiamycin (antibiotic 899, staphylomycin) was isolated in 1955 from a strain of *Streptomyces virginiae* and was shown to have several components inhibitory to gram-positive bacteria (126). Two years earlier a peptide antibiotic called streptogramin had been isolated (127). Streptogramin is a peptide antibiotic mixture similar to virginiamycin. Many virginiamycinlike antibiotics have since been discovered including the ostreogrycins, vernamycins, synergistins, PA 114, mikamycins, and pristinamycins. Streptogramin family antibiotics find some use in human medicines, eg, pristinamycin is used in France (128), but most streptogramin production is for use as animal feed additives. Virginiamycin is used in feeds in the United States and elsewhere and mikamycin is used in Japan (129). The literature on streptogramins, especially virginiamycins, has been reviewed (130–135).

As shown in Figure 3 and Table 5, these antibiotics are mixtures of two quite

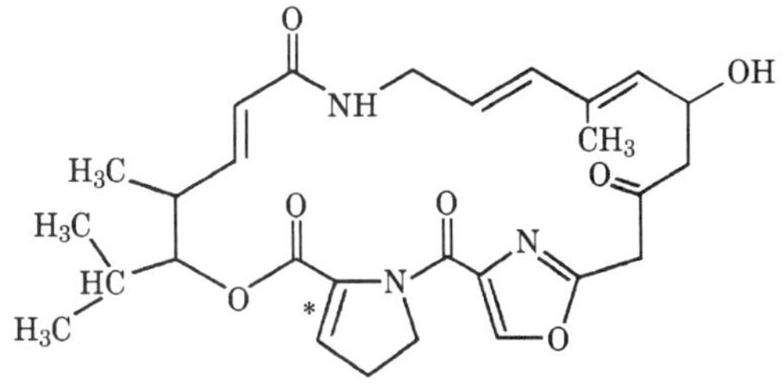

Fig. 3. Structure of streptogramin A, and when the bond denoted * is saturated, of ostreogrycin G. Synonyms for streptogramin A are mikamycin A, PA114A, pristinamycin IIA, vernamycin A, ostreogrycin A, virginiamycin M1, and staphylomycin M. Synonyms for ostreogrycin G are virginiamycin M2 and pristinamycin IIB (132).

different groups of chemical structures usually denoted A (or M) and B (or S) (136). The A components may be considered highly modified depsipeptides. Although they constitute a family of only two members, differing by one double bond, they have been given many different synonyms. The B component is a group of closely related cyclic hexadepsipeptides containing nonprotein aminoacids.

Members of the A group (Fig. 3) are less powerful antibiotics than members of the B group (Table 5), but they are powerfully synergistic with members of the B group. Singly, A- and B-group members act bacteriostatically; acting together, they are bactericidal. Group A members make up about 70% by weight, group B members about 30%, of the antibiotic yielded by a producer strain. This ratio also gives the maximum synergy and represents the approximate concentration ratio found in the marketed antibiotic. The biosynthesis of these A- and B-group substances is not well understood (137,138).

Virginiamycin is an amorphous white powder slightly soluble in water and very soluble in chloroform or methanol. Solubility and other physical data, including x-ray crystallographic, have been reviewed (180). Data for the separated components of virginiamycin and for pristinamycin are similar. The compounds that comprise virginiamycin were crystallized following separation by silica gel chromatography, and the structures were then determined using a variety of procedures (139). Virginiamycin is relatively stable at both acid and neutral pH, allowing oral dosage in animals and adequate shelf life in feed mixtures. Semisynthetic studies have produced more water-soluble derivatives of the A and B components of pristinamycin and these may allow development of an injectable formulation (140).

The production of virginiamycin by fermentation of *S. virginiae* has been detailed (133). Although crude, dry virginiamycin is relatively stable, there is a need to improve virginiamycin's stability when it is used as a feed additive, because of incompatibility with components in some feed mixtures. Virginiamycin is mixed with carboxymethylcellulose and chalk and the mixture is then granulated to a particle size of less than 1 mm. This preparation contains 50% virginiamycin and is known as Stafac 500 (SmithKline Beecham) (133).

Pristinamycin is used in France as an oral drug for human use, especially for staphylococcal infections (141); it has at least *in vitro* activity against a number of anaerobes and genital pathogens as well (142,143). U.S. rules for animal feed use permit virginiamycin in subtherapeutic nutritional dosages for increase in feed

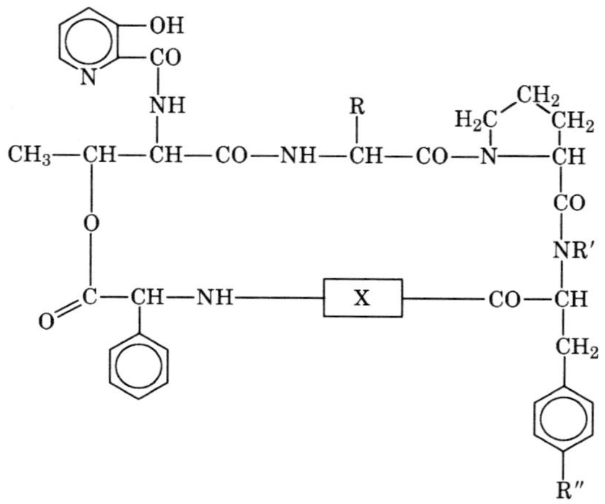

Table 5. Streptogramin B Antibiotics

Name	CAS Registry Number	Molecular formula	R	R′	R″	X
streptogramin B[a]	[*3131-03-1*]	$C_{45}H_{54}N_8O_{10}$	CH_2CH_3	CH_3	$N(CH_3)_2$	4-oxopipecolic acid
vernamycin Bβ[b]	[*57206-54-9*]	$C_{44}H_{52}N_8O_{10}$	CH_2CH_3	CH_3	$NHCH_3$	4-oxopipecolic acid
vernamycin Bγ[c]	[*28979-74-0*]	$C_{44}H_{52}N_8O_{10}$	CH_3	CH_3	$N(CH_3)_2$	4-oxopipecolic acid
vernamycin Bδ	[*29139-17-1*]	$C_{43}H_{50}N_8O_{10}$	CH_3	CH_3	$NHCH_3$	4-oxopipecolic acid
vernamycin C[d]	[*14014-70-1*]	$C_{43}H_{52}N_8O_{11}$	CH_2CH_3	CH_3	$N(CH_3)_2$	aspartic acid
ostreogrycin B_3	[*31508-69-7*]	$C_{45}H_{54}N_8O_{11}$	CH_2CH_3	CH_3	$N(CH_3)_2$	3-hydroxy-4-oxopipecolic acid
patricin A	[*2105-09-1*]	$C_{42}H_{49}N_7O_9$	CH_2CH_3	CH_3	H	proline
patricin B	[*38188-91-9*]	$C_{43}H_{51}N_7O_9$	CH_2CH_3	CH_3	H	pipecolic acid
virginiamycin S1[e]	[*23152-29-6*]	$C_{43}H_{49}N_7O_{10}$	CH_2CH_3	CH_3	H	4-oxopipecolic acid
virginiamycin S2[e]	[*33477-38-2*]	$C_{42}H_{49}N_7O_{10}$	CH_2CH_3	H	H	4-hydroxypipecolic acid
virginiamycin S3[e]	[*33477-39-3*]	$C_{43}H_{49}N_7O_{11}$	CH_2CH_3	CH_3	H	3-hydroxy-4-oxopipecolic acid
virginiamycin S4[e]	[*33477-37-1*]	$C_{42}H_{47}N_7O_{10}$	CH_3	CH_3	H	4-oxopipecolic acid

[a]Also known as mikamycin B, PA114B, pristinamycin 1A, vernamycin Bα, ostreogrycin B, and synergistin B.
[b]Also known as pristinamycin 1B and ostreogrycin B2.
[c]Also known as pristinamycin 1C and ostreogrycin B1.
[d]Also known as doricin.
[e]The virginiamycins are also known as staphylomycins.

efficiency and for growth promotion in poultry (5.5–16.5 g/t of feed) and swine (5.5–11 g/t) (81,82,144). Higher concentrations are used for prophylactic or therapeutic medicinal dosing against necrotic enteritis in turkeys and dysentery in swine (81,82).

Virginiamycinlike antibiotics inhibit protein synthesis by binding to ribosomes. The molecular mechanism of the powerful synergistic action of these drugs is complicated and is the subject of continuing research (145). Resistance to these antibiotics can occur as the result of permeability decrease, ribosomal mutation, *O*-acetylation or hydrolysis of antibiotic, or by methylation of adenyl groups in ribosomal RNA (146–149). The mechanism of resistance varies. Macrolide–lincosamide–streptogramin B (MLS) resistance determines methylation of ribosomal RNA and resistance to the streptogramin B group member (see ANTIBIOTICS, LINCOSAMINIDES; ANTIBIOTICS, MACROLIDES). MLS resistance is often plasmid-borne and transferable, and sometimes induceable. Such MLS resistance in a strain may not cause loss of synergy between A- and B-group members or loss of bacteriostatic sensitivity to the mixture of the A and B members, but it does lead to a loss of cidal activity. One concern is that the subtherapeutic use of streptograminlike antibiotics in animal husbandry could generate and cause the spread of resistance to human pathogens, resulting in the development of resistance to erythromycin and clindamycin in human pathogens.

Thiostrepton and Related Antibiotics. Thiostrepton [*1393-48-2*] (thiactin, bryamycin), $C_{72}H_{85}N_{19}O_{18}S_5$, shown in Figure 4, is a large (mol wt 1660) antibiotic discovered in 1954 and produced by *Streptomyces azureus* (150,151). It is a member of a family of large, poorly water-soluble, bacteriostatic antibiotics that are active at very low concentrations against gram-positive bacteria, inactive against most

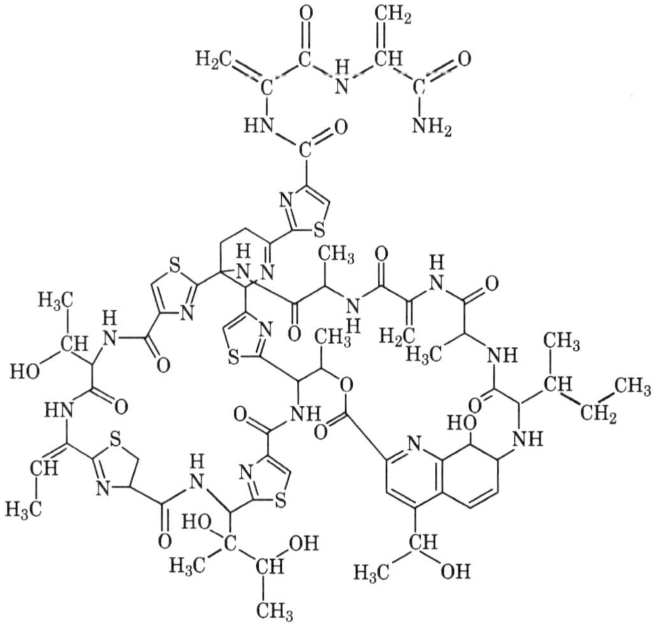

Fig. 4. Structure of thiostrepton.

gram-negative bacteria, and inactive against fungi. Thiostrepton inhibits sensitive bacteria by binding tightly to ribosomes and inhibiting protein synthesis (152,153). Other antibiotics in this family that contain multiple thiazole rings include nosiheptide [*56377-79-8*], $C_{51}H_{43}N_{13}O_{12}S_6$, (multhiomycin), actinotiocin [*39454-49-4*], micrococcin P, saramycetin [*11130-70-4*], siomycin [*11017-43-9*], $C_{64}H_{88}N_{16}O_{22}S_4$, (sporangiomycin) and the thiopeptins (154). The structure of thiostrepton was almost completely determined by a combination of chemical fragmentation and x-ray crystallography (155). The two dehydroalanines in the side chain were finally detected by ^{13}C nmr (156). The solution conformation of thiostrepton is similar to its crystalline conformation (157). Commercial thiostrepton is a white to light yellow crystalline powder, insoluble in water or diethyl ether, but very soluble in chloroform or dimethylformamide (158).

Thiostrepton family members are biosynthesized by extensive modification of simple peptides. Thus, from amino acid incorporation studies, the somewhat smaller (mol wt 1200) nosiheptide, which contains five thiazole rings, a trisubstituted indole, and a trisubstituted pyridine, is speculated to arise from a simple dodecapeptide. This work shows that the thiazole moieties arise from the condensation of serine with cysteine (159,160). Only a few reports on the biosynthesis of the thiostrepton family are available (159,160). Thiostrepton is presently used in the United States only as a polyantimicrobial veterinary ointment (Panalog, Squibb), but thiazole antibiotics have, in the past, been used as feed additives in various parts of the world. General (158) and mechanism of action (152) reviews on thiostrepton are available.

Enduracidin. Enduracidin (enramycin) is a mixture of some six cyclic lipopeptides isolated from *Streptomyces fungicidicus* in 1967 (161). As shown in Figure 5, enduracidins are composed of a ring of 16 D- and L-amino acids and an amino acid tail that is acylated with a branched-chain fatty acid. The enduracidin of commerce is a mixture of the hydrochlorides of enduracidin A [*34438-27-2*], $C_{107}H_{138}Cl_2N_{26}O_{31}$, enduracidin B [*34304-21-7*], $C_{108}H_{140}Cl_2N_{26}O_{31}$, enduracidin S_A [*40958-06-3*], $C_{107}H_{139}ClN_{26}O_{31}$, and enduracidin S_B [*40958-07-4*], $C_{108}H_{141}ClN_{26}O_{31}$ (161,162). This poorly soluble but bactericidal antibiotic is active *in vitro* on a variety of gram-positive bacteria such as *Staphylococcus aureus* and *Streptococcus pyogenes*. Mycobacteria are sensitive, but gram-negative bacteria and fungi are insensitive to enduracidin (161). Enduracidin does not show cross-resistance with other marketed antibacterials (161,163). It inhibits the transglycosylation step in bacterial cell wall synthesis (164). Enduracidin is not active when given orally, but does have a long half-life when given parenterally. It is painful at the injection site and is fairly toxic when given intravenously to animals (165). It is used only in veterinary medicine in Japan.

Bicozamycin. Bicozamycin (bicyclomycin) [*38129-37-2*], $C_{12}H_{18}N_2O_7$, shown in Figure 6, is a colorless, water-soluble diketopiperazine or cyclic dipeptide derived from leucine and isoleucine. It was simultaneously isolated in 1972 in Japan from cultures of two different *Streptomyces* strains (166). Bicozamycin is a relatively nontoxic but weak antibiotic having a minimum inhibitory concentration of 25–100 μg/mL for *Salmonella* and 5–50 μg/mL for *E. coli*. It inhibits bacterial cell wall synthesis and is bactericidial to gram-negative pathogens such as *E. coli, Salmonella,* and *Shigella. Proteus, Pseudomonas,* gram-positive bacteria, mycobacteria, and fungi are insensitive. This antibiotic is poorly absorbed when orally administered. Initial clinical trials for therapeutic

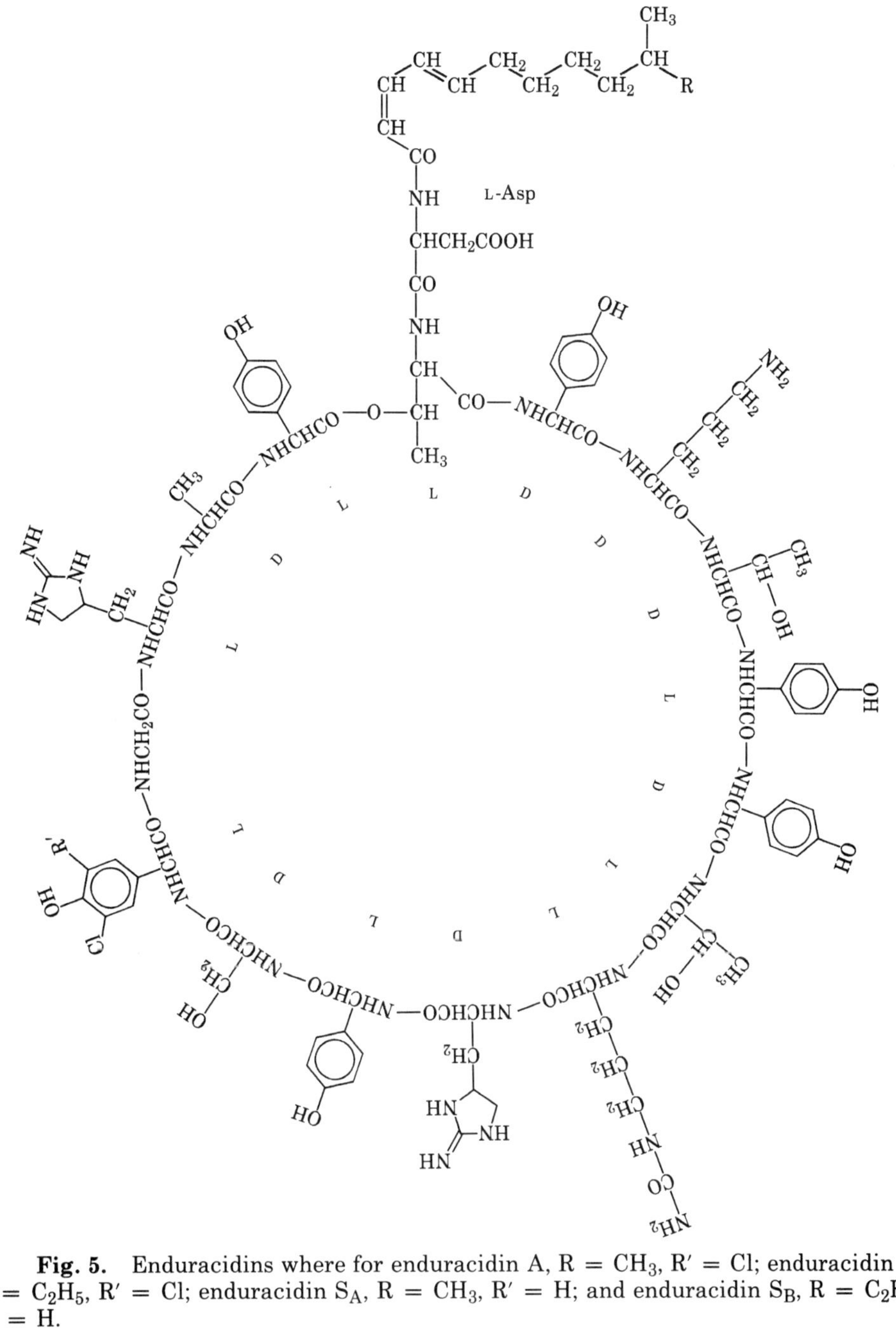

Fig. 5. Enduracidins where for enduracidin A, R = CH_3, R′ = Cl; enduracidin B, R = C_2H_5, R′ = Cl; enduracidin S_A, R = CH_3, R′ = H; and enduracidin S_B, R = C_2H_5, R′ = H.

and prophylactic use in travelers' diarrhea were promising (167), but human use was not pursued in the United States. This fermentation product is now sold in the Orient for bacterial diarrhea in calves and pigs. Chemical synthetic and analogue work, and other aspects of the bicozamycin literature, have been reviewed (168).

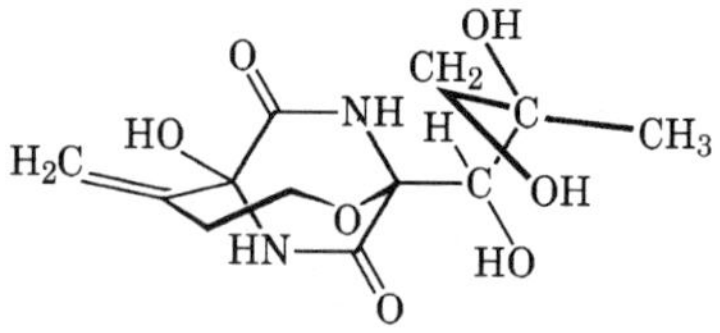

Fig. 6. Structure of bicozamycin (bicyclomycin).

Amphomycin (glumamycin). In 1953 a highly surface-active lipopeptide mixture of antibiotics, named amphomycin because of its amphoteric properties (169–172), was isolated from a strain of *Streptomyces canus*. Amphomycin [*1402-82-0*], $C_{58}H_{91}N_{13}O_{20}$, shown in Figure 7, is composed of two compounds, identical except for the different fatty acids. Aspartocin [*4117-65-1*], $C_{42}H_{64}N_{12}O_{12}S_2$, and tsushimycin [*11054-63-0*], $C_{59}H_{93}N_{13}O_{20}$, are similar to amphomycin (172). Amphomycin (glumamycin) has moderate activity toward gram-positive pathogens, but is inactive against gram-negative bacteria and fungi. Although amphomycin is toxic when given systemically, it was introduced in the United States in 1963 as the water-insoluble, crystalline calcium salt in ointments for topical use. However, it has been withdrawn, probably for economic reasons.

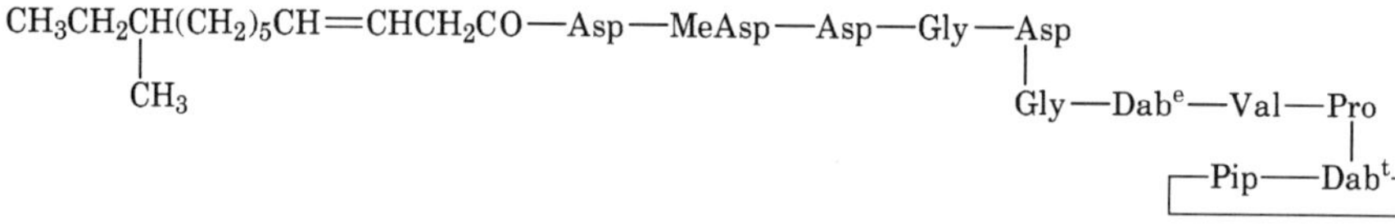

Fig. 7. Structure of one member of the amphomycin family where Dab^e = D-*erythro*-α,β-diaminobutyric acid; Dab^t = L-*threo*-α,β-diaminobutyric acid; Pip = D-pipecolic acid; MeAsp = *N*-methylaspartic acid. The amphomycin family member having (+)-3-anteisotridecenoic acid is shown; not shown is the second member containing (+)-3-isododecenoic acid.

Amphomycin inhibits bacterial cell wall synthesis at the translocase step by binding to the large isoprenoid lipid called undecaprenylphosphate (173,174). In eukaryotes, in a similar manner, amphomycin binds to the large isoprenoid lipid called dolicholphosphate thus blocking the transfer of mannose from its uridinediphosphate derivative to this lipid (173–177). Amphomycin has been patented for use as a feed additive (178).

Valinomycin and the Enniatins. Neutral ionophores such as the cyclic dodecadepsipeptide valinomycin [*2001-95-8*], $C_{54}H_{90}N_6O_{18}$, (Fig. 8) from *Streptomyces fulvissimus* (179), and the cyclic hexadepsipeptides enniatin [*11113-62-5*] and beauvericin [*26048-05-5*], $C_{45}H_{57}N_3O_9$, from fungi (180–182), are toxic to many forms of life. Valinomycin is active against gram-positive bacteria such as *Staphylococcus aureus,* whereas gram-negative bacteria are some 25 times less sensitive. Mammals are about as sensitive to valinomycin as gram-positive bacteria (179). In sharp contrast to the less toxic, charged polyether ionophores, neutral ionophores have little economic value, but they have been important in elucidating cell membrane function. Carrier ionophores, such as valinomycin and lasalocid, should be sharply distinguished from channel-forming antibiotics, such

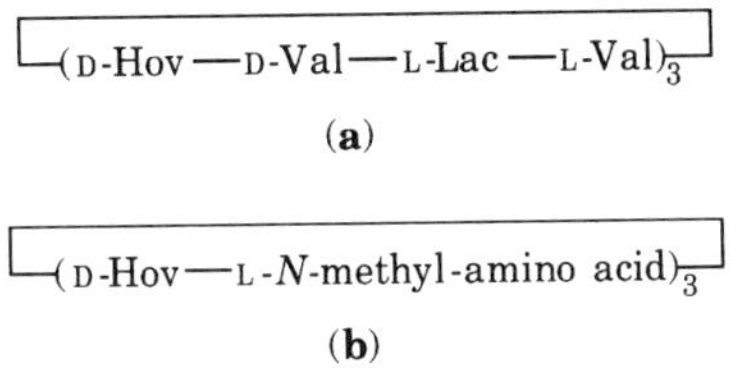

Fig. 8. Structure of (**a**) valinomycin and (**b**) the enniatins and beauvericin. Hov = α-hydroxy-isovaleric acid and Lac = lactic acid. The *N*-methylamino acid for enniatin A is isoleucine; enniatin B, valine; enniatin C, leucine; and beauvericin, phenylalanine.

as the linear gramicidins or alamethicin [*27061-78-5*], $C_{77}H_{128}N_{20}O_{22}$, that allow ion passage through membranes by forming channels through the whole width of the membrane (183). Neutral peptide ionophores have been reviewed (184–187).

Valinomycin is a cyclic dodecadepsipeptide (mol wt 1110) that forms a shell or bracelet of beta sheet that is held rigidly together by hydrogen bonds between carbonyl oxygens and amino hydrogens (188). Because these hydrogen bonds define a relatively inflexible cavity the size of a potassium ion, valinomycin is quite specific for potassium ion. Sodium binds with 3,000- to 10,000-fold lower avidity than potassium. Many valinomycin analogues have been chemically synthesized (189). The biosynthesis of valinomycin on a multienzyme complex and some genetic work have been described (190).

The enniatins enniatin A [*2503-13-1*], $C_{36}H_{63}N_3O_9$, enniatin B [*917-13-5*], $C_{33}H_{57}N_3O_9$, enniatin C [*19893-23-3*], $C_{36}H_{63}N_3O_9$, and beauvericin (Fig. 8) are 700–800 molecular weight cyclic hexadepsipeptides. They form valinomycinlike hydrophilic cavities surrounded by outer lipophilic regions, but they have more flexible structures than those seen with valinomycin and therefore have less specificity for potassium over sodium ion than valinomycin (186,187).

Nisin and Other Lantibiotics. Nisin [*1414-45-5*], $C_{143}H_{230}N_{42}O_{37}S_7$, shown in Figure 9, is the only economically useful member of a group of large (mol wt ca 3500) peptide antibiotics synthesized on ribosomes, as are normal proteins, rather than on large multienzyme complexes. These antibiotics are highly modified following translation; in particular some amino acid residues are converted

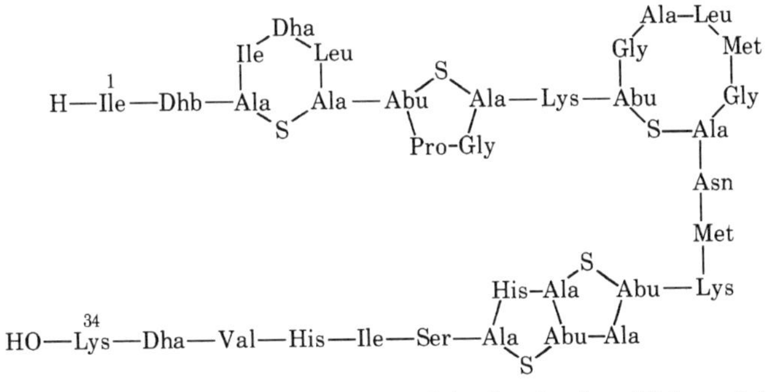

Fig. 9. Structure of nisin where Dha = dehydroalanine; Dhb = dehydrobutyrine; Abu = α-aminobutyric acid; Ala-S-Ala = *meso*-lanthionine; and Abu-S-Ala = *threo*-methyllanthionine. The first amino acid residue (Ala or Abu) of the lanthionine or methyllanthionine is a D residue; all other amino acids are L.

to lanthionine [*922-55-4*], $C_6H_{12}N_2O_4S$, and related amino acids, hence the group name, lantibiotics. The lantibiotics include nisin from *Streptococcus lactis* group N, subtilin from *Bacillus subtilis,* epidermin [*8022-54-4*], from *Staphylococcus epidermidis,* duramycin [*1391-36-2*], from *Streptoverticillium* sp., gallidermin [*117978-77-5*], ancovenin [*88201-41-6*], Ro 09-0198 [*110655-58-8*], $C_{89}H_{125}N_{25}O_{25}S_3$, pep 5 [*84931-86-2*], and cinnamycin [*1405-39-6*] (191,192).

Lantibiotics may also be classified as bacteriocins, a group of antibacterial proteins that are often bactericidal to bacteria closely related to the producing species (193). Nisin or nisin-producing bacteria have been used in Europe and elsewhere since 1960 as preservatives in canned and other foods, and as a protective agent during the manufacture of Swiss cheese. Nisin is surprisingly stable to the heat of canning processes, especially at acid pH. In 1988 the U.S. Food and Drug Administration (FDA) allowed nisin at 250 ppm to be a Generally Recognized As Safe (GRAS) additive for certain pasteurized cheese spreads to inhibit outgrowth of *Clostridium botulinum* spores, especially heat-treated ones, and toxin production (194). This may have been stimulated by the growing awareness of the undesirability of potentially carcinogenic nitrites that are added to foods for the same purpose (see FOOD ADDITIVES). Consumers normally eat small amounts of nisin because nisin-producing strains are ubiquitous in milk products. Nisin is about as toxic as sodium chloride with an oral LD_{50} in rats of 7 g/kg. Nisin literature has been reviewed (87,195).

Nisin was first isolated in 1947 (196,197). The structure was proposed in 1971 (198) and confirmed in 1988 by mass spectrometry studies (199). The first total chemical synthesis was reported in 1988 (200). Nisin is synthesized on ribosomes as a 57-amino acid prepeptide. It is then modified by enzymic elimination of water from the serines and the threonines of the prepeptide. Some of the resulting dehydroalanines and dehydrobutyrines combine with nearby cysteines in the molecule to form the thioether amino acids *meso*lanthionine and *threo*methyllanthionine. The *N*-terminal 23-amino acid leader peptide is probably cleaved off last, leaving nisin. The nisin produced by fermentation of *S. lactis* on skim milk is a semipurified mixture containing true nisin plus degradation products of nisin, some of which have antibacterial activity. The UK producer Aplin and Barrett sells a product, Nisaplin, that is 2.5% pure nisin (195,201). One gram of pure nisin contains 40,000 International or Reading units.

Nisin acts bactericidally primarily against gram-positive bacteria. It acts best at acid pH and is almost insoluble at physiological pH. Nisin and probably all lantibiotics appear to permeabilize the bacterial cell membrane to release small molecules, resulting in an immediate collapse of the membrane potential (202).

Semisynthetic Peptides in Development. Although the efforts to develop synthetic or semisynthetic improvements on naturally occurring peptide antibiotics have met with few successes, two lipopeptide antimicrobials are in development: the semisynthetic antibacterial daptomycin [*103060-53-3*], $C_{72}H_{101}N_{17}O_{26}$, and the semisynthetic antifungal cilofungin [*79404-91-4*], $C_{49}H_{71}N_7O_{17}$. As of 1991, cilofungin was less promising than originally expected.

Daptomycin (deptomycin, LY 146032), shown in Figure 10, is a cyclic peptide lactone containing a 10-amino acid ring and a 3-amino acid tail to which is attached a fatty acid. Daptomycin is derived from A21978C [*82800-76-8*], a complex of similar antibiotics produced by *Streptomyces roseosporus* (203,204). The A21978C complex is isolated and its ornithine amino groups are protected by

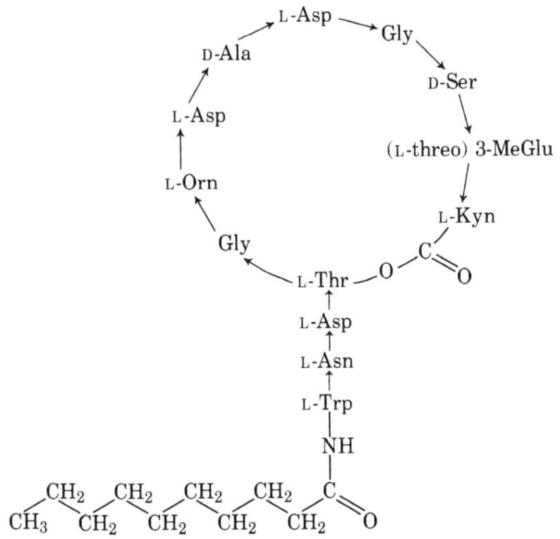

Fig. 10. Structure of daptomycin where L-Kyn is L-kynurenine and (L-*threo*) 3-MeGlu is L-*threo*-3-methylglutamic acid.

making the *tert*-butyloxycarbonyl (*t*-BOC) derivative. These blocked derivatives are then incubated with a culture of *Actinoplanes utahensis* to biologically remove the various fatty acids from the common polypeptide nucleus. The blocked polypeptide nucleus is then isolated and reacylated on the tryptophan amino group by an activated *n*-decanoyl group. The *t*-BOC group is then removed with trifluoroacetic acid, resulting in daptomycin (205,206).

Daptomycin is bactericidal to many gram-positive bacteria including *Streptococcus faecalis,* methicillin-resistant staphylococci, and certain vancomycin-resistant strains; most gram-negative bacteria are resistant (207). This antibiotic inhibits bacterial cell wall formation and the primary target, though not completely known, may be lipoteichoic acid formation (208). Daptomycin showed less toxicity in animals than the precursor A21978C (205,206) and was active in several animal models of infection (209). When it is administered once a day by injection, daptomycin may provide an alternative to vancomycin and other agents. Daptomycin was in clinical trials in the United States in 1991.

Cilofungin (LY121019), a semisynthetic cyclic lipohexapeptide shown in Figure 11, is being developed as an injectable anti-*Candida* drug. It is an analogue of the antifungal echinocandin B [*54651-05-7*], $C_{52}H_{81}N_7O_{18}$, (ECB, Lilly A30912A, Sandoz SF7810F) (see ANTIPARASITIC AGENTS, ANTIMYCOTICS). ECB, isolated from species of the fungus *Aspergillus* (210), is one of a closely related family of antifungals isolated from various *Aspergillus* species. This family includes echinocandin C [*71018-12-7*], $C_{52}H_{81}N_7O_{15}$, and echinocandin D [*71018-13-8*], $C_{52}H_{81}N_7O_{13}$, A30912H [*77368-57-1*], $C_{53}H_{83}N_7O_{16}$, aculeacin A [*58814-86-1*], mulundocandin [*108351-20-8*], $C_{48}H_{77}N_7O_{16}$, and sporiofungin [*82029-17-2*], $C_{50}H_{80}N_8O_6$ (210). ECB is produced by fermentation and is then deacylated biologically using *Actinoplanes utahensis.* The deacylated ECB is then isolated and reacylated chemically using an activated 4-*n*-octyl-oxo-benzoyl moiety (210). The resulting cilofungin is much less toxic than ECB.

Candida albicans is considerably more sensitive to cilofungin than to ECB

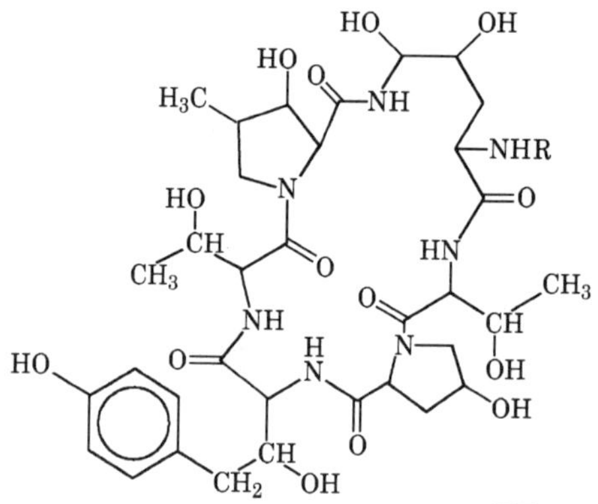

Fig. 11. Structure of cilofungin, R = —CO—C_6H_4—$O(CH_2)_7CH_3$, and echinocandin B, R = linoleoyl.

(211). Cilofungin has an *in vitro* spectrum of activity limited to *Candida albicans, C. tropicalis, Neurospora crassa,* and *Malassezia pachydermatis,* but most other fungi, including other *Candida* species are resistant (212). *Pneumocystis carinii,* an important cause of pneumonia in AIDS and other immunosuppressed patients, is sensitive to cilofungin (213). The mechanism of action of cilofungin in susceptible species appears to be by inhibition of the synthesis of 1,3,β-D-glucan, a wall component (214). Cilofungin is some twenty times less toxic than amphotericin B. In some animal infection models cilofungin is more effective than amphotericin B, but the fungicidal activity of cilofungin is not as reliable as desired (211,215). Cidal agents are especially valuable in therapy of immunosuppressed patients in whom systemic *Candida* infection is a significant problem. Cilofungin was in clinical trials in 1991. Unfortunately, initial trials showed excessive toxicity of the cilofungin preparation, much of it attributable to the vehicle required for this insoluble compound (216).

Antitumor Agents and Immunomodulators

Bleomycin and cyclosporine are the two economically most important streptomycete peptide antibiotics used as antitumor agents and immunomodulators, although dactinomycin is important medically for several tumors (see CHEMOTHERAPEUTICS, ANTICANCER; IMMUNOTHERAPEUTIC AGENTS).

Dactinomycin. Dactinomycin [*50-76-0*] (actinomycin D, actinomycin C1, Cosmegen), $C_{62}H_{86}N_{12}O_{16}$, is the only actinomycin in clinical use. Dactinomycin, an antineoplastic drug, was discovered in 1943 and is made in rather pure form by *Streptomyces parvullus.* Dactinomycin has some bacteriostatic antibacterial and antifungal activity, but high toxicity limits its use to antineoplastic therapy. It may be used alone or with other antineoplastics, with or without surgery and synergistic x-ray therapy. Dose limiting bone marrow toxicity may result in low white cell and platelet count. Intestinal mucosal damage also occurs. Reviews of more detailed chemotherapeutic information are available (217–222).

Dactinomycin forms bright red, rhomboid prisms from absolute ethanol.

Dactinomycin USP is a yellow lyophylized powder. It is soluble in water at 10°C, slightly soluble in water at 37°C, and soluble in ethanol. Dactinomycin USP must be protected from light and should be preserved in air-tight containers and stored below 40°C. Great care must be taken in handling this extremely toxic drug to avoid breathing the powder or getting dry powder or solutions on the skin. Dactinomycin is mutagenic, carcinogenic, and teratogenic. The LD_{50} in mice, when given intravenously, is about 0.7 mg/kg body weight. Other properties, fermentation, and purification methods (223–226), and studies on biosynthesis by multienzyme complexes can be found in the literature (227).

Almost all actinomycins have the same chromophore, a planar phenoxazinone dicarboxylic acid called actinocin. In dactinomycin, the structure of which is shown in Figure 12, the two pendent pentapeptide lactones are identical, but in other actinomycins these lactones may be different. In other actinomycins the first amino acid, amide linked with actinocin, is usually L-threonine, as in dactinomycin; the second position is sometimes D-allo-isoleucine instead of D-valine; the third position may be sarcosine or oxoproline; the fourth position is sarcosine; and the fifth position is sometimes *N*-methyl isoleucine instead of *N*-methylvaline. The lactone ring is always present.

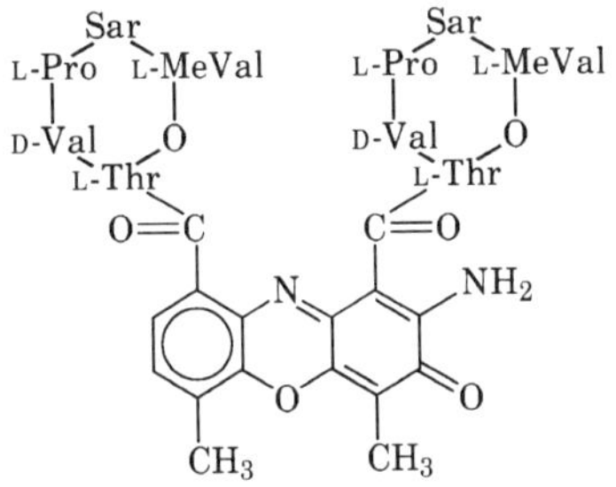

Fig. 12. Structure of dactinomycin (actinomycin D) where MeVal = *N*-methylvaline and Sar = sarcosine.

Total synthesis of dactinomycin has been accomplished, and at least thirty natural actinomycins and many synthetic and semisynthetic actinomycins have been tested (220,221,228). At one time cactinomycin [*8052-16-2*] (actinomycin C), a mixture of actinomycins D, C2, and C3, was sold as an antineoplastic.

Dactinomycin is useful against several tumors especially Wilms' tumor of the kidney and rhabdomyosarcoma. Wilms' tumor normally requires a combination of surgery, x-ray irradiation, and dactinomycin plus vincristine [*57-22-7*], $C_{46}H_{56}N_4O_{16}$, or other antitumor agents. Resistance of the multidrug type to dactinomycin can develop during therapy (229).

Dactinomycin intercalates into double stranded DNA inhibiting DNA-dependent RNA polymerase. Dactinomycin also can cause single strand breaks in DNA. Inhibition of DNA synthesis only takes place at higher concentrations of drug. A molecular model for dactinomycin-DNA interaction has been proposed based in part on x-ray crystallography of a complex containing two molecules of deoxyguanosine and one of dactinomycin (230). Conformations of dactinomycin–oligonucleotide complexes are being studied by nmr (231,232).

Bleomycin and Related Glycopeptides. Bleomycins (Fig. 13) comprise a group of related antineoplastic glycopeptide antibiotics isolated from a strain of *Streptomyces verticillus* in 1965 (233). Bleomycin sulfate [*9041-93-4*], (Blenoxane, Bristol-Myers Squibb) is mostly a mixture of bleomycins A_2 and B_2. Bleomycins differ from one another in having different R groups on the glycopeptide bleomycinic acid moiety as shown in Table 6. Hundreds of bleomycin analogues have been made either by adding different precursor amines during fermentation or by adding various amines to purified bleomycinic acid (234,235). Peplomycin (pepleomycin) (236–241) and liblomycin (238,239) are newer semisynthetic analogues having potentially useful properties. Peplomycin is marketed in Japan and was in clinical trials in the United States in 1990; liblomycin is at an early stage of development. Others include tallysomycin A [*65057-90-1*], reported in 1977, which was in clinical development at one time (240), and phleomycin [*11006-33-0*], reported in 1956 (241), which was too toxic for drug use.

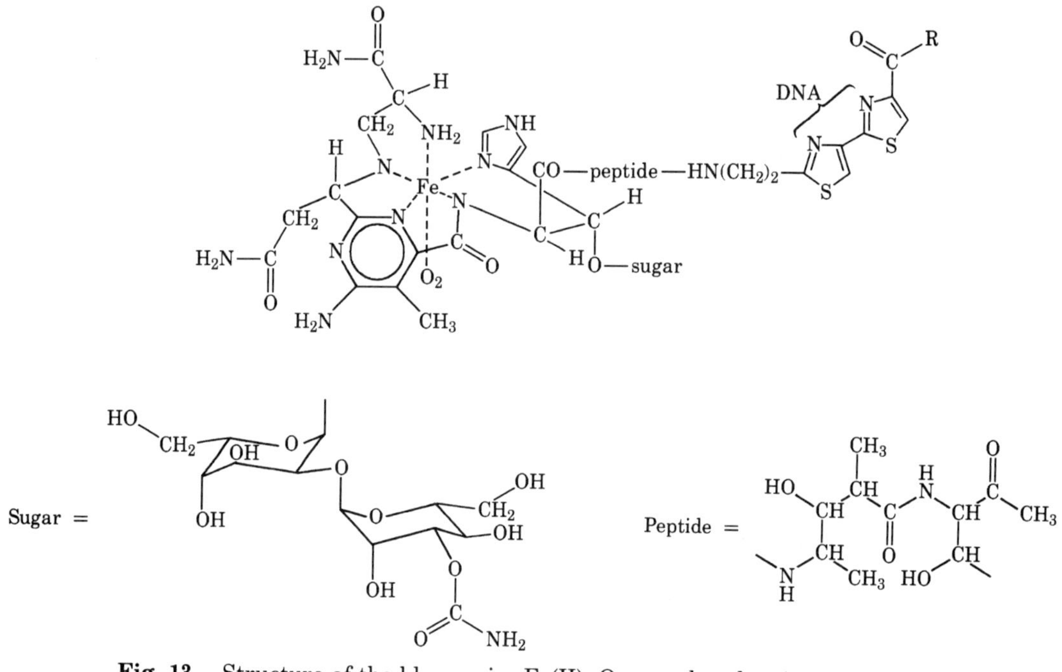

Fig. 13. Structure of the bleomycin–Fe(II)–O_2 complex showing the DNA binding region.

Bleomycin sulfate is used alone or in combination with other antineoplastic drugs. Using this DNA-cleaving drug, disease-free (possibly cured) survivors of some cancers have been seen, especially with the use of bleomycin in multidrug regimens. General reviews of bleomycin (234,242–245) and its mechanism of action (246–249) are available.

The structure of bleomycin shown in Figure 13 was reported in 1978 (250); total synthesis was reported in 1982 (251–253). The commercial fermentation and purification of bleomycin has been described (244) as has its nonribosomal synthesis using a large multienzyme complex (254).

Table 6. Bleomycins and Derivatives[a]

Name	CAS Registry Number	Molecular formula	R
bleomycinic acid	[*37364-66-2*]	$C_{50}H_{69}N_{15}O_{22}S_2$	—OH
bleomycin A_2	[*11116-31-7*]	$C_{55}H_{81}N_{16}O_{21}S_3$	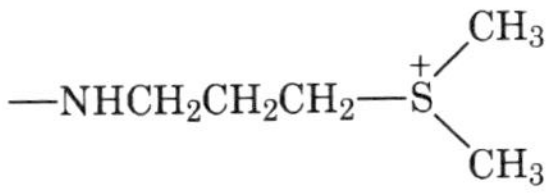
bleomycin B_2	[*9060-10-0*]	$C_{55}H_{81}N_{19}O_{21}S_2$	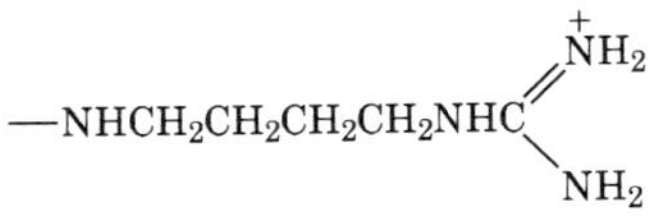
peplomycin peplomycin sulfate	[*68247-85-8*] [*70384-29-1*]	$C_{61}H_{88}N_{18}O_{21}S_2$ $C_{61}H_{90}N_{18}O_{25}S_3$	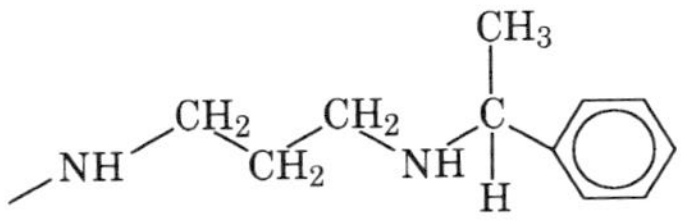
liblomycin	[*88266-67-5*]	$C_{99}H_{125}N_{19}O_{25}S_2$	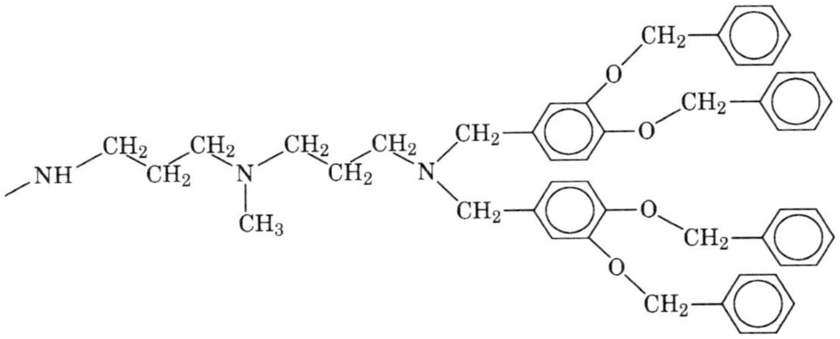

[a]See Figure 13.

Bleomycin sulfate USP, a white or yellowish powder, is very soluble in water and sparingly soluble in alcohol. It is 55–70% bleomycin A_2 and 25–32% bleomycin B_2 based on total bleomycin base content as determined by hplc. This combination was found to be more effective than use of either bleomycin alone (234).

Bleomycin has been used for the treatment of testicular cancer and malignant lymphomas and it has some activity against squamous cell carcinomas of various organs. Because of low bone marrow and immune system toxicities, bleomycin is often combined with bone marrow suppressive drugs such as vinblastine [*865-21-4*], $C_{46}H_{58}N_4O_9$, vincristine [*23214-92-8*], $C_{27}H_{29}NO_{11}$, cisplatin [*15663-27-1*], $N_2H_6Cl_2Pt$, doxorubicin [*23214-92-8*], $C_{27}H_{29}NO_{11}$, dacarbazine [*4342-03-4*], $C_6H_{10}N_{60}$, and mitomycin C [*50-07-7*], $C_{15}H_{18}N_4O_5$. Tumors often acquire resistance to bleomycin during therapy because of increased levels of bleomycin hydrolase or of DNA repair activities, or possibly by decreased transport (255). The hydrophilic bleomycin is not affected by the multidrug resistance system involving transport. The toxicity of bleomycin, which is high in lung and skin and low in bone marrow, correlates well with the level of a bleomycin hydrolase, a cysteine proteinase with cathepsin-*H*-like activity (256). This enzyme inactivates bleomycin by cleaving the alanine carboxamide moiety on the pyrimidoblamic unit.

Newer bleomycins such as peplomycin and especially liblomycin, are more resistant to bleomycin hydrolase. This results in less lung toxicity but more bone marrow toxicity, and allows for a different spectrum of antitumor action. Bleomycin is inactive orally; it is given intravenously, intramuscularly, subcutaneously, or directly into a cavity such as the pleural cavity. The majority of drug is excreted unchanged in the urine.

Much work has been done on the mechanism of action of bleomycin as an agent that causes single and double strand breaks in DNA. Bleomycin binds to cupric ions strongly, but the less strongly bound ferrous ions are probably the important ions *in vivo*. Ferrous bleomycin binds with considerable base specificity to DNA (257) probably in part by the intercalation of the bithiazole moiety between the DNA base pairs. Reductive activation of dioxygen by this antibiotic damages the strand of DNA in at least two separate ways (246–249). This mechanism of action of bleomycin is still speculative and has stimulated the design of other structures with similar action. An antitumor antibiotic called zinostatin [*9014-02-2*] (neocarzinostatin) acts somewhat similarly to bleomycin. It is composed of a protein (mol wt 11,000) having a tightly bound nonpeptide prosthetic group [*81604-85-5*] (258).

Cyclosporine. Cyclosporin A [*59865-13-3*], $C_{62}H_{111}N_{11}O_{12}$, called cyclosporine in the United States and called Sandimmun(e) by Sandoz, is a lipophilic cyclic undecapeptide, isolated from fungi, that has important immunosuppressant action. This action is directed mainly against T-lymphocytes resulting in inhibition of cell-mediated immunity and, to a lesser extent, antibody production. Cyclosporine has little effect on blood-forming tissues and phagocytic cells, thus preserving much antibacterial defense; it is not mutagenic. In 1970 culture extracts of the fungi *Cylindrocarpon lucidum* and *Tolypocladium inflatum* showed a narrow spectrum of antifungal activity *in vitro* but poor antifungal activity in whole animal models (259). Purified extracts, composed mainly of cyclosporin A, shown in Figure 14, and cyclosporin B, yielded high immunosuppressive and antiinflammatory activities, surprisingly not linked to general cytostatic effects (260). This initial work was followed by the separation and purification of some 25 naturally occurring cyclosporins, denoted A through Z, as well as the synthesis of analogues.

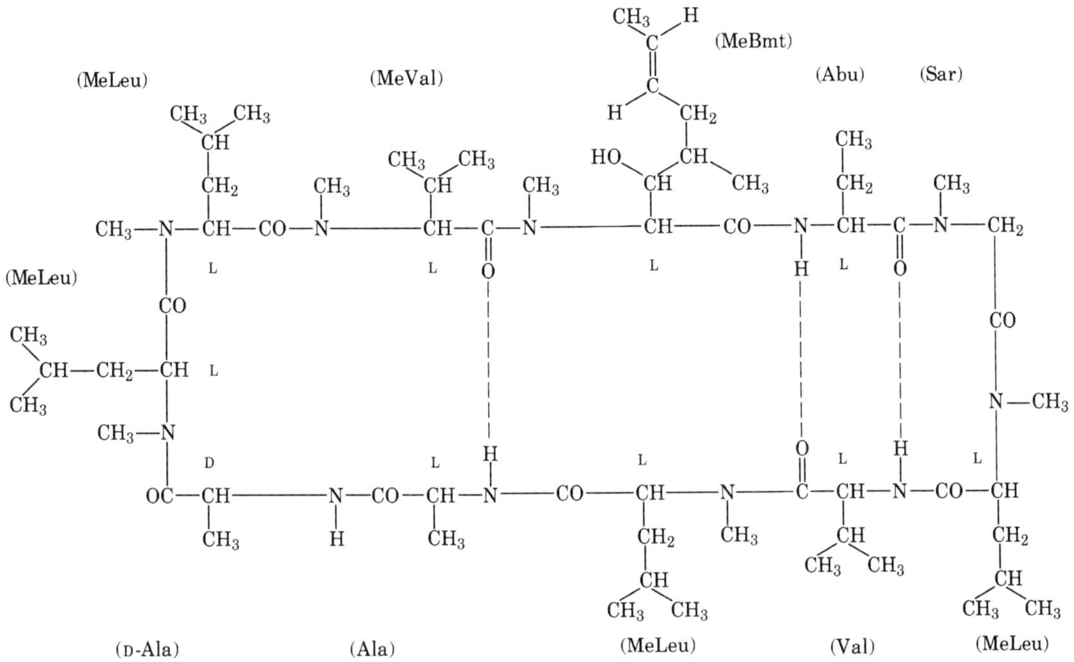

Fig. 14. Structure of cyclosporin A where MeBmt = 4-(2-butenyl)-4,*N*-dimethylthreonine; Sar = sarcosine; MeLeu = *N*-methylleucine; MeVal = *N*-methylvaline; and Abu = α-aminobutyric acid. The dotted lines indicate hydrogen bonds.

The first human kidney and bone marrow transplants using cyclosporine were reported in 1978. Oral or intravenous cyclosporine is an immunosuppressant for transplantation of these and other organs and investigations are underway for its possible use in a variety of autoimmune diseases including rheumatoid arthritis, severe psoriasis, and Crohn's disease. Dose-dependent nephrotoxicity (261–264) remains the primary limitation of the drug and necessitates close monitoring of patients, including measurement of drug levels in blood. Cyclosporine research has been reviewed (265–274).

Cyclosporin A forms white prismatic crystals from acetone and is only slightly soluble in water and saturated hydrocarbons, but is very soluble in methanol, ethanol, acetone, and diethyl ether. Optical and nmr data on cyclosporins and x-ray crystallographic data on cyclosporin A and an iododerivative have been reviewed (273,275).

Cyclosporin A contains no charged groups. Acid hydrolysis yields a mixture of amino acids. Sequence analysis and x-ray crystallography yielded the cyclic structure containing the several hydrogen bonds shown in Figure 14 (259,275).

The structures of some 25 naturally occurring cyclosporins have been determined (269). The probable conformation of cyclosporin A in aqueous media has been reviewed (273). The cyclosporins have been synthesized chemically and many analogues have been made, but no cyclosporin has been shown to be more medically useful than cyclosporin A (cyclosporine) (273,276). The production and purification of cyclosporine have been described (277). The nonribosomal biosynthesis of cyclosporine A in cell-free systems involves a single, very large multienzyme polypeptide that appears to catalyze at least 40 enzymic reactions (278).

The mechanism of action of cyclosporine is not fully known, but the main effects appear to act against T-helper cells and against the production and release of lymphokines including interleukin-2 and γ-interferon. Cyclosporine inhibits at least one form of the enzymic activity responsible for the cis-trans isomerization of the peptide bond involving the imino group of prolyl residues in proteins (274,279–282). This class of ubiquitous small enzymes, denoted peptidyl-prolyl-*cis-trans*-isomerases (rotamases or immunophilins), is composed of at least cyclophilin, to which cyclosporine binds (274,283), and FK-binding protein, to which two other immunosuppressive macrolides under development, FK506 and rapamycin, bind (see ANTIBIOTICS, MACROLIDES) (284–289). It has not yet been proven how inhibition of rotamase by cyclosporine explains the immunosuppressive action of the drug. It is hypothesized that rotamase acts as a foldase on one or more control proteins that modulate the rate of synthesis of messenger RNA for interleukins and other proteins, but this may not fully explain cyclosporine action.

Ubenimex. Ubenimex [*58970-76-6*] (bestatin), $C_{16}H_{24}N_2O_4$, is a dipeptide isolated from *Streptomyces olivoreticuli* that stimulates parts of the immune system, especially T-cell function including delayed-type hypersensitivity, and causes tumor regression in animals (290–292). It has a low toxicity but lacks antibiotic activity. Ubenimex, in the class of nonantiinfective fermentation products, has been marketed in Japan since 1987 as an oral anticancer agent to be used with other agents to maintain patients with adult acute nonlymphocytic leukemia who are in remission.

Ubenimex, [(2(*S*),3(*R*))-3-amino-2-hydroxy-4-phenylbutanoyl]-L-leucine, was isolated as an inhibitor of aminopeptidases, on which it acts as a strong, reversible transition-state analogue inhibitor (293). Analogues of ubenimex have been made and some other aminopeptidase inhibitors, not all of them peptides, have been isolated from streptomycetes (294–296).

Antimetabolite Peptides

An antimetabolite is defined as a special sort of enzyme inhibitor that acts as a structural analogue of a required metabolite such as a vitamin, amino acid, sugar, hormone, etc, and prevents production or use of that metabolite. Peptide antimetabolites have been useful as research tools in the past, especially those that are glutamine or glutamate analogues or that serve as prodrugs for such analogues. For instance, tabtoxin, 2-amino-4-(3-hydroxy-2-oxo-3-azetidinyl)-butanoyl-L-threonine [*40957-90-2*], $C_{11}H_{19}N_3O_6$, and bacilysin, L-alanyl-L-2,3-epoxy-

4-oxohexahydrophenylalanine [*29393-20-2*], $C_{12}H_{18}N_2O_5$, inhibit glutamine metabolism, as does bialaphos as a source of the amino acid L-phosphinothricin [*35597-44-5*], $C_5H_{12}NO_4P$.

Bialaphos, Phosphinothricin, and Glufosinate. In 1972 and 1973 two groups independently reported finding an antibacterial and antifungal tripeptide called bialaphos [*71048-99-2*], $C_{11}H_{22}N_3O_6P{\cdot}Na$, (formerly SF-1293) in *Streptomyces*

Table 7. Commercial Peptide Antibiotics[a]

Compound	CAS Registry Number	Company[b]	Use[c]
bacitracin (Zn salt) and methylenedisalicylate sodium salt	[*1405-89-6*] [*55852-84-1*]	A.L. Labs, Lundbeck, IMC, Krka, Kayaku	H (topical antibacterial), An (feed additive)
bialaphos		Meiji Seika	Ag (herbicide)
bicozamycin (bicyclomycin)		Fujisawa	An (feed additive)
bleomycin sulfate	[*9041-93-4*]	Nippon Kayaku	H (antitumor)
peplomycin sulfate	[*70384-29-1*]		
capreomycin sulfate	[*1264-72-8*]	Lilly	H (antimycobacterial)
colistin sulfate and colistimethate sodium derivative	[*1264-72-8*] [*8068-28-8*]	A.L. Labs, Kayaku, Banyu, Rhône Poulenc	H and An (topical and systemic antibacterial)
cyclosporine		Sandoz	H (antiimmune)
dactinomycin (actinomycin D)		Merck	H (antitumor)
enduracidin (enramycin)		Takeda	An (antibacterial)
enviomycin (tuberactinomycin N)	[*33103-22-9*]	Toyo Joso	H (antimyobacterial)
gramicidin (linear/ Dubos)		Biochemie, Lundbeck	H (topical antibacterial)
gramicidin S		Nikken	H (topical antibacterial)
mikamycin	[*11006-76-1*]	Kaneka	An (feed additive)
nisin		Applin and Barrett	H (food preservative)
phosphinothricin, (amino acid)		Hoechst	Ag (herbicide)
polymyxin B sulfate	[*1404-20-5*]	A.L. Labs, Pfizer	H and An (topical and systemic antibacterial)
pristinamycin	[*11006-76-1*]	Rhône Poulenc	H (systemic antibacterial)
thiopeptin	[*12609-84-6*]	Fujisawa	An (feed additive)
thiostrepton		CalBioChem	An (topical antibacterial)
tyrothricin	[*1404-88-2*]	Biochemie	H (topical antibacterial)
ubenimex (bestatin)	[*58970-76-6*]	Nippon Kayaku	H (immune stimulant)
virginiamycin		SmithKline Beecham	An (feed additive)
zinostatin (neocarzinostatin)	[*9014-02-2*]	Kayaku	H (antitumor)

[a]Japanese producers courtesy of Japan Antibiotics Research Association (306).
[b]Use in humans, H; in animals, An; plants, Ag.

viridochromogenes and in *S. hygroscopicus* (297,298). Bialaphos (L-phosphinothricyl-L-alanyl-L-alanine) yields the novel amino acid L-phosphinothricin, L-2-amino-4-methylphosphinylbutyric acid [*35597-44-5*], $C_5H_{12}NO_4P$. L-Phosphinothricin, an analogue of methionine sulfoximine, was found to be an irreversibly inactivating inhibitor, competitive with glutamate, for glutamine synthetase from a variety of sources (297–299). The inhibition of cells by L-phosphinothricin was reversed by glutamine [*56-58-9*]. Susceptible cells must be able to hydrolyze bialaphos, because the tripeptide itself has no activity on glutamine synthetase. Papers on biosynthesis and genetics may be found in the literature (299,300).

Although the antibacterial and antifungal activities of bialaphos and phosphinothricin were not found to be useful, the two agents were later used as biodegradeable, relatively nonselective, postemergent herbicides. Glutamine synthetase inhibition is toxic to plants because the enzyme is key to ammonia assimilation. There is some selectivity for individual plant species as shown by the LD_{50} for bialaphos ranging from 0.125 to 8.5 kg/ha (301–303). Bialaphos, produced by fermentation of *S. hygroscopicus,* is sold in Japan. Racemic phosphinothricin [*53369-07-6*], $C_5H_{12}NO_4P$, produced chemically (304,305), is sold in the United States and elsewhere as the monoammonium salt called glufosinate-ammonium [*77182-82-2*], $C_5H_{12}NO_4P{\cdot}H_3N$, or Total or Basta (formerly Hoe 661, Hoe 39866).

Economic Aspects

Table 7 lists the peptide antibiotics of commercial interest and their producers. Annual U.S. sales of peptide antibiotics, as wholesale price to pharmacies, can be approximated from the following estimates: bleomycin, $25 million for 2 kg; bacitracin, polymyxin, and gramicidin D, $100 million total for 4 t, 10 kg, and 400 g respectively; and cyclosporine, $130 million (3 t). Minor peptides having less than $1 million sales each, such as capreomycin and dactinomycin, are sold at a loss. U.S. agricultural sales figures for bacitracin, $20 million (>200 t), and virginiamycin, $12–15 million (ca 70 t), are indicative of the 1991 market.

BIBLIOGRAPHY

"Polypeptide Antibiotics" in *ECT* 2nd ed., Vol. 16, pp. 306–345, by S. Wilkinson, The Wellcome Research Laboratories, Beckenham, Kent, UK; "Antibiotics (Peptides)" in *ECT* 3rd ed., Vol. 2, pp. 991–1036, by D. Perlman, University of Wisconsin.

1. H. Kleinkauf and H. von Döhren, eds., *Biochemistry of Peptide Antibiotics,* W. de Gruyter, Berlin, 1990.
2. P. Jouin and co-workers, *J. Med. Chem.* **34,** 486 (1991).
3. F. J. Schmitz and co-workers, *J. Org. Chem.* **54,** 3463 (1989).
4. C. L. Bevins and M. Zasloff, *Annu. Rev. Biochem.* **59,** 395 (1990).
5. P. Elsbach, *Trends Biotechnol.* **8,** 26 (1990); C. P. Hill, J. Yee, M. E. Selsted, and D. Eisenberg, *Science* **251,** 1481 (1991).
6. H. Hofte and H. R. Whiteley, *Microbiol. Rev.* **53,** 242 (1989).

7. J. B. Neilands, *Annu. Rev. Biochem.* **50,** 715 (1981).
8. P. Demange and co-workers, *Biochemistry* **29,** 11041 (1990).
9. E. K. Dolence and co-workers, *J. Med. Chem.* **34,** 956, 968 (1991).
10. *The Use of Essential Drugs,* 6th List, Technical Report Series No. 296, World Health Organization, Geneva, Switzerland, 1990.
11. V. Lorian, ed., *Antibiotics in Laboratory Medicine,* 3rd ed., Williams & Wilkins, Baltimore, Md., 1991.
12. F. Peypoux and co-workers, *Eur. J. Biochem.* **153,** 335 (1985).
13. J. E. Walker and E. P. Abraham, *Biochem. J.* **118,** 563 (1970).
14. H. Wojciechowska, A. Konitz, and E. Borowski, *Int. J. Pept. Protein Res.* **26,** 279 (1985).
15. F. Besson, F. Peypoux, M. J. Quentin, and G. Michel, *J. Antibiot.* **37,** 172 (1984).
16. B. W. Bycroft and M. S. Gowland, *J. Chem. Soc. Chem. Commun.* **1978,** 256 (1978).
17. S. K. Das, S. Majumdar, S. Basu, S. Mukherjee, and S. K. Bose, *J. Antibiot.* **39,** 581 (1986).
18. G. Benz and co-workers, *Justus Liebig's Ann. Chem.* **1984,** 1399, 1408, 1434 (1984).
19. D. Schipper, *J. Antibiot.* **36,** 1076 (1983).
20. J. W. Lown and K. Krowicki, *J. Org. Chem.* **50,** 3774 (1985).
21. H. Bickel and co-workers, *Tetrahedron* **1966**(Suppl. 8), 171 (1966).
22. M. Nakajima and co-workers, *J. Antibiot.* **31,** 427 (1978).
23. Y. Iitaka, H. Nakamura, K. Takada, and T. Takita, *Acta Crystallogr.* **30**(Sect. B), 2817 (1974).
24. Y. Okumura and co-workers, *J. Antibiot.* **32,** 1130 (1979).
25. F. E. Hahn, in J. W. Corcoran and F. E. Hahn, eds., *Antibiotics—Mechanism of Action of Antimicrobial and Antitumor Agents,* Vol. 3, Springer-Verlag, Berlin, 1975, p. 79.
26. D. G. Martin and co-workers, *J. Antibiot.* **28,** 332 (1975).
27. T. P. Pitner and D. W. Urry, *Biochemistry* **11,** 4132 (1972).
28. N. G. Kumar and D. W. Urry, *Biochemistry* **12,** 4392 (1973).
29. O. D. Hensens and G. Albers-Schönberg, *J. Antibiot.* **36,** 832 (1983).
30. I. Oberbaumer, E. Grell, F. Raschdorf, and W. J. Richter, *Helv. Chim. Acta* **65,** 2280 (1982).
31. J. P. Springer and co-workers, *J. Am. Chem. Soc.* **106,** 2388 (1984).
32. T. Korzybski, Z. Kowszyk-Gindifer, and W. Kurylowicz, *Antibiotics: Origin, Nature, and Properties,* American Society for Microbiology, Washington, D.C., 1978.
33. J. Bérdy, *Handbook of Antibiotic Compounds,* Vol. 4/1, Vol. 4/2, CRC Press, Boca Raton, Fla., 1980.
34. A. I. Laskin and H. A. Lechevalier, eds., *CRC Handbook of Microbiology,* 2nd ed., Vol. 9, Parts A and B, CRC Press, Boca Raton, Fla., 1988.
35. B. W. Bycroft, ed., *Dictionary of Antibiotics and Related Substances,* Chapman and Hall, London, 1988.
36. *Amino Acids and Peptides,* the Royal Society of Chemistry, Cambridge, UK, Vol. 22, 1991.
37. O. L. Sebek, in D. Gottlieb and P. D. Shaw, eds., *Antibiotics—Mechanism of Action,* Vol. 1, Springer-Verlag, Berlin, 1967, p. 142.
38. H. Paulus, in D. Gottlieb and P. D. Shaw, eds., *Antibiotics—Biosynthesis,* Vol. 2, Springer-Verlag, Berlin, 1967, p. 254.
39. E. Katz and A. L. Demain, *Bacteriol Rev.* **41,** 449 (1977).
40. D. R. Storm, K. S. Rosenthal, and P. E. Swanson, *Ann. Rev. Biochem.* **46,** 723 (1977).
41. E. F. Gale, E. Cundliffe, P. E. Reynolds, M. H. Richmond, and M. J. Waring, *The Molecular Basis of Antibiotic Action,* 2nd ed., John Wiley & Sons, Inc., London, 1981, p. 175.

42. K. Vogler and R. O. Studer, *Experientia* **22,** 345 (1966).
43. H. Kleinkauf and H. von Döhren, in L. C. Vining, ed., *Biochemistry and Genetic Regulation of Commercially Important Antibiotics,* Addison-Wesley, Reading, Mass., 1983, p. 95.
44. S. Nakamura, *Iyakuhin Kaihatsu Kiso Kosa* **14,** 678 (1972).
45. S. Komura and K. Kurahashi, *J. Biochem.* **97,** 1409 (1985).
46. J. Horton and G. A. Pankey, *Med. Clin. North Am.* **66,** 135 (1982).
47. R. Fekety, in G. L. Mandel and co-eds., *Principles and Practice of Infectious Diseases,* 3rd ed., Churchill Livingston, New York, 1990, p. 323.
48. P. J. Weiss, M. L. Andrew, and W. W. Wright, *Antibiot. Chemother.* **7,** 375 (1957).
49. S. Macura, N. G. Kumar, and L. R. Brown, *J. Magnet. Reson.* **60,** 99 (1984).
50. T. Suzuki, K. Hayashi, K. Fujikawa, and K. Tsukamoto, *Report of the Institute of Protein Research,* Osaka University, Japan, 1963.
51. M. Barnett, S. R. M. Bushby, and S. Wilkinson, *Brit. J. Pharmacol.* **23,** 552 (1964).
52. R. A. Moore, L. Chan, and R. E. W. Hancock, *Antimicrob. Agents Chemother.* **26,** 539 (1984); R. E. W. Hancock, *Ann. Rev. Microb.* **38,** 237 (1984); H. Nikaido, *Handbook of Experimental Pharmacology,* Vol. 91, Springer-Verlag, Berlin, 1989, p. 1.
53. D. S. Feingold, C-C. Hsu Chen, and I. J. Sud, *Ann. N.Y. Acad. Sci.* **235,** 480 (1974).
54. S. Chihara, T. Tobita, M. Yahata, A. Ito, and Y. Koyama, *Agric. Biol. Chem.* **37,** 2455 (1973).
55. M. Vaara and T. Vaara, *Nature* **303,** 526 (1983).
56. H. Nikaido and M. Vaara, *Microbiol. Rev.* **49,** 1 (1985).
57. R. L. Danner and co-workers, *Antimicrob. Agents Chemother.* **33,** 1428 (1989).
58. Y. Kimura and co-workers, *Proc. 9th Symp. Peptide Chem. Japan, Protein Res. Found.* **9,** 159 (1972); M. Ito-Kagawa and Y. Koyama, *J. Antibiol.* **33,** 1551 (1980); H. Kitamura-Matsunaga, Y. Kimura, and T. Araki, in N. Izumiya, ed., *Peptide Chemistry, 1984, Protein Res. Found., Osaka,* 1985, p. 333; N. Yasuda, Y. Kimura, M. Miyama, and H. Matsunaga, *Agric. Biol. Chem.* **53,** 3245 (1989).
59. K. Hanasawa and co-workers, *Prog. Clin. Biol. Res.* **264,** 337 (1988); M. A. F. A. Boons, M. L. P. M. Verhoeven, L. A. A. E. Sluyterman, and H. M. Buck, *Appl. Biochem. Biotechnol.* **22,** 95 (1989); K. W. Talmadge and C. J. Siebert, *J. Chromatog.* **476,** 175 (1989).
60. B. A. Johnson, H. S. Anker, and F. L. Meleney, *Science,* **102,** 376 (1945).
61. H. S. Anker, B. A. Johnson, J. Goldberg, and F. L. Meleney, *J. Bacteriol.* **55,** 249 (1948).
62. *Bacitracin,* S. B. Penick and Co., New York, 1952.
63. A. Arriagada and co-workers, *Brit. J. Exp. Pathol.* **30,** 425 (1949).
64. E. P. Abraham and G. G. F. Newton, *Ciba Foundation Symposium: Amino Acids and Peptides with Antimetabolic Activity,* 1958, p. 205.
65. L. C. Craig, W. Kronigsberg, and R. J. Hill, in Ref. 64, p. 226.
66. R. O. Studer, in G. P. Ellis and G. B. West, eds., *Progress in Medicinal Chemistry,* Vol. 5, Plenum Press, New York, 1967, p. 1.
67. G. A. Brewer, in K. Florey, ed., *Analytical Profiles of Drug Substances,* Vol. 9, Academic Press, New York, 1980, p. 1.
68. F. L. Meleney and B. A. Johnson, *Am. J. Med.* **7,** 794 (1949); E. Jawetz, *Polymyxin, Neomycin, and Bacitracin,* Antibiotic Monograph No. 5, Medical Encyclopedia, Inc., New York, 1965, p. 85; E. Jawetz, *Pediat. Clin. North Am.* **8,** 1057 (1961).
69. D. R. Storm and W. A. Toscano, Jr. in Ref. 37, p. 1.
70. W. A. Toscano, Jr. and D. R. Storm, *Pharmac. Ther.* **16,** 199 (1982).
71. O. Frøyshov, in E. J. Vandamme, ed., *Biotechnology of Industrial Antibiotics,* Marcel Dekker, New York, 1984, p. 665.

72. R. J. Hickey, *Prog. Ind. Microbiol.* **4,** 94 (1964).
73. C. Ressler and D. V. Kashelikar, *J. Am. Chem. Soc.* **88,** 2025 (1966); R. E. Galardy, M. P. Printz, and L. C. Craig, *Biochemistry* **10,** 2429 (1971).
74. K. Tsuji, J. H. Robertson, and J. A. Bach, *J. Chromatog.* **99,** 597 (1974); H. Oka, Y. Ikai, N. Kawamura, and M. Yamada, *J. Chromatog.* **462,** 315 (1989).
75. E. Munekata, Y. Masui, T. Shiba, and T. Kaneko, *Bull. Chem. Soc. Japan* **46,** 3187 (1973).
76. P. J. Weiss, M. L. Andrew, and W. W. Wright, *Antibiot. Chemother.* **7,** 374 (1957).
77. M. L. Andrew and P. J. Weiss, *Antibiot. Chemother.* **9,** 277 (1959).
78. B. L. Damron and H. R. Wilson, *Poultry Sci.* **64,** 1451 (1985); J. H. Choi and K. S. Ryu, *Brit. Poult. Sci.* **28,** 113 (1987).
79. P. W. Waldroup and co-workers, *Poult. Sci.* **67,** 103 (1988).
80. J. V. Scudi and W. Antopol, *Proc. Soc. Exptl. Biol. Med.* **64,** 503 (1947); H. A. Zintel, R. A. Ma, A. C. Nichols, and H. Ellis, *Am. J. Med. Sci.* **218,** 439 (1949).
81. S. F. Sundlof, J. E. Riviere, and A. L. Craigmill, *Food Animal Residue Avoidance Databank (FARAD) Trade Name File, A Comprehensive Compendium of Food Animal Drugs,* University of Florida, Gainsville, 1988.
82. S. Muirhead, ed., *1990 Feed Additive Compendium,* Miller Publishing Co., Minnetonka, Minn., 1989.
83. D. R. Storm, *Ann. N.Y. Acad. Sci.* **235,** 387 (1974).
84. R. J. Dubos, *J. Exptl. Med.* **70,** 1 (1939); R. D. Hotchkiss, *Adv. Enzymol.* **4,** 153 (1944); R. P. Crease, *Science* **246,** 883 (1989).
85. G. F. Gause and M. G. Brayhnikova, *Nature* **154,** 703 (1944).
86. G. A. Brewer, in K. Florey, ed., *Analytical Profiles of Drug Substances,* Vol. 8, Academic Press, New York, p. 179.
87. E. Gross, in M. J. Weinstein and G. H. Wagman, eds., *Antibiotics: Isolation, Separation, and Purification,* Elsevier, Amsterdam, The Netherlands, 1978, p. 415.
88. B. Cornell, *J. Bioenerg. Biomemb.* **19,** 655 (1987).
89. F. R. Salemme, *Science* **241,** 145 (1988); B. A. Wallace, *Curr. Topics Membr. Transp.* **33,** 35 (1988).
90. S.-S. Sung and P. C. Jordan, *J. Theor. Biol.* **40,** 369 (1989).
91. B. A. Wallace, *Annu. Rev. Biophys. Chem.* **19,** 127 (1990).
92. F. E. Hunter, Jr. and L. S. Schwartz, in Ref. 37, p. 643.
93. K. Kurahashi, in W. Corcoran, ed., *Antibiotics—Biosynthesis,* Vol. 4, Springer-Verlag, New York, 1981, p. 325.
94. R. Sarges and B. Witkop, *J. Am. Chem. Soc.* **87,** 2020 (1965).
95. P. Daumas, F. Heitz, L. Ranjalahy-Rasoloanijaq, and R. Lazaro, *Biochemie* **71,** 77 (1989).
96. D. Lelievre, Y. Trudelle, F. Heitz, and G. Spach, *Int. J. Peptide Protein Res.* **33,** 379 (1989).
97. B. A. Wallace and K. Ravikumar, *Science* **241,** 182 (1988); D. A. Langs, *Science* **241,** 188 (1988).
98. F. E. Hunter, Jr. and L. S. Schwartz, in Ref. 37, p. 636.
99. J. Vater, in *CRC Crit. Rev. Biotechnol.* **8,** 205 (1988).
100. G. Mittenhuber, R. Weckermann, and M. A. Marahiel, *J. Bacteriol.* **171,** 4881 (1989).
101. M. Waki and N. Izumiya, in Ref. 99, p. 206.
102. R. D. Hotchkiss and R. J. Dubos, *J. Biol. Chem.* **141,** 155 (1941).
103. A. Bohg and H. Ristow, *Eur. J. Biochem.* **170,** 253 (1987); F. J. Aranda and B. de Kruijff, *Biochim. Biophys. Acta* **937,** 195 (1988).
104. W. Danders and co-workers, *Antimicrob. Agents Chemother.* **22,** 785 (1982); K. Sato and U. Nagai, in Y. Kiso, ed., *Peptide Chemistry 1985,* Protein Research Foundation, 1986,

p. 217; H. Aoyagi, S. Ando, S. Lee, and N. Izumiya, *Tetrahedron* **44,** 877 (1988); T. Katsu and co-workers, *Biochim. Biophys. Acta* **983,** 135 (1989).
105. M. Kanda and co-workers, *J. Biochem.* **105,** 653 (1989); J. Vater and co-workers, *Biol. Chem. Hoppe-Seyler* **370,** 1013 (1989); H.-J. Skarpeid, T.-L. Zimmer, B. Shen, and H. von Döhren, *Eur. J. Biochem.* **187,** 627 (1990).
106. *Chem. Eng. News* **39,** 57 (1961).
107. W. B. Sutton, R. S. Gordee, W. E. Wick, and LaV. Stanfield, *Ann. N.Y. Acad. Sci.* **135,** 947 (1966).
108. M. Lucchesi, *Antibiot. Chemother.* **16,** 27 (1970).
109. P. G. Caltrider in Ref. 37, p. 677.
110. L. Trnka, in K. Bartmann, H. Iwainsky, and co-eds., *Antituberculosis Drugs,* Springer-Verlag, Berlin, 1988, p. 150.
111. A. Nagata and co-workers, *J. Antibiot. (Tokyo)* **24,** 681 (1968).
112. T. Ando and co-workers, *J. Antibiot.* **24,** 680 (1971).
113. A. Nagata and co-workers, in M. Hejzlar and co-workers, *Advances in Antimicrobial Antineoplastic Chemotherapy,* Vol. 1/2, University Park Press, Baltimore, Md., 1972, p. 1039.
114. U.S. Pat. 3,892,732 (July 1, 1975), J. Abe and co-workers (to Toyo Joso K.K.).
115. J. M. Bond, in R. J. Schnitzer and F. Hawking, eds., *Experimental Chemotherapy,* Vol. 4, Academic Press, New York, 1966, p. 560.
116. H. Otten, in Ref. 110, p. 191.
117. U.S. Pat. 3,143,468 (Aug. 4, 1964), E. B. Herr, Jr., R. L. Hamill, and J. M. McGuire (to Eli Lilly & Co.).
118. J. F. Carter II and co-workers, *Biochemistry (U.S.)* **13,** 1221, 1227 (1974).
119. E. B. Herr, Jr. and M. O. Redstone, *Ann. N.Y. Acad. Sci.* **135,** 940 (1966).
120. S. Nomoto, T. Teshima, T. Wakamiya, and T. Shiba, *Tetrahedron* **34,** 921 (1978).
121. J. K. McClatchy, W. Kanes, P. T. Davidson, and T. S. Moulding, *Tubercle* **58,** 29 (1977).
122. M. M. Tsukamura, *Microbiol. Immunol.* **24,** 11 (1980).
123. Y. Mizuguchi, T. Udou, and T. Yamada, *Microbiol. Immunol.* **27,** 425 (1983).
124. L. Heifets and P. Lindholm-Levy, *Antimicrob. Agents Chemother.* **33,** 1298 (1989).
125. P. Marrero, M. J. Cabañas, and J. Modelell, *Biochem. Biophys. Res. Commun.* **97,** 1047 (1980); T. Yamada, T. Teshima, T. Shiba, and K. H. Nierhaus, *Nucleic Acids Res.* **8,** 5767 (1980); I. Medow, S. Rüsch-Gerdes, and K.-H. Schröder, *Zentralbl. Bakteriol. Mikrobiol. Hyg.* **A266,** 359 (1987).
126. P. DeSomer and P. Van Dijck, *Antibiot. Chemother.* **5,** 632 (1955).
127. J. Charney, W. P. Fisher, C. Curran, R. A. Machlowitz, and A. A. Tytell, *Antibiot. Chemother.* **3,** 1283 (1953).
128. J. Preud'homme, P. Tarridec, and A. Belloc, *Bull. Soc. Chim. France* **2,** 585 (1968).
129. M. Arai and co-workers, *J. Antibiot.* **11**(Ser. A), 14 (1958).
130. D. Vazquez, in Ref. 37, p. 387.
131. C. Cocito, *Microbiol. Rev.* **43,** 145 (1979).
132. E. Cundliffe, in Ref. 41, p. 402.
133. A. M. Biot, in Ref. 71, p. 695.
134. C. Cocito, in F. E. Hahn, ed., *Antibiotics—Modes and Mechanisms of Microbial Growth Inhibitors,* Springer-Verlag, Berlin, 1983, p. 296.
135. E. F. Gale, in Ref. 41, p. 176.
136. P. Crooy and R. De Neyes, *J. Antibiot.* **25,** 371 (1972).
137. J. W. Reed and D. G. I. Kingston, *J. Natl. Prod.* **49,** 626 (1986).
138. H. S. Kim and co-workers, *J. Antibiot.* **42,** 769 (1989).
139. H. Vanderhaeghe, P. Van Dijck, G. Parmentier, and P. DeSomer, *Antibiot. Chemother.*

7, 606 (1957); H. Vanderhaeghe and G. Parmentier, *J. Am. Chem. Soc.* **82,** 4414 (1960); G. R. Delpierre and co-workers, *J. Chem. Soc. (C)* **1966,** 1653; D. G. I. Kingston, Lord Todd, and D. H. Williams, *J. Chem. Soc. (C)* **1966,** 1669; J. Vanderhaeghe, G. Janssen, and F. Compernolle, *Tetrahedron Lett.* **28,** 2687 (1971).
140. F. LeGoffic, *J. Antimicrob. Chemother.* **16**(Suppl. A), 13 (1985); J. M. Paris and co-workers, *Pathol. Biol.* **33,** 493 (1985); J. C. Barriere and co-workers, *30th Interscience Conference on Antimicrobial Agenys and Chemotherapy,* American Society for Microbiology, Atlanta, Ga., 1990, p. 214.
141. J. Modai, *J. Antimicrob. Chemother.* **16**(Suppl. A), 195 (1985).
142. J. C. Lefevre and R. Bauriaud, *Antimicrob. Agents Chemother.* **33,** 2152 (1989).
143. H. LaForest, M. Fourgeaud, H. Richet, and P. H. LaGrange, *Antimicrob. Agents Chemother.* **32,** 1094 (1988).
144. R. E. Buresh, R. D. Miles, and R. H. Harms, *Poultry Sci.* **64,** 1041 (1985); R. D. Miles, D. M. Janky, and R. H. Harms, *Poultry Sci.* **64,** 139 (1985); S. A. Woodward and co-workers, *Poultry Sci.* **67,** 1222 (1988).
145. M. Di Giambattista, G. Chinali, and C. Cocito, *J. Antimicrob. Chemother.* **24,** 485 (1989).
146. D. Dubnau, *CRC Crit. Rev. Biochem.* **16,** 103 (1984).
147. J. Duval, *J. Antimicrob. Chemother.* **16**(Suppl. A), 137 (1985).
148. B. Weisblum, *J. Antimicrob. Chemother.* **16,** 63 (1985).
149. E. Cundliffe, in L. E. Bryan, ed., *Handbook of Experimental Pharmacology,* Vol. 91, Springer-Verlag, Berlin, 1989, p. 227.
150. J. F. Pagano, M. J. Weinstein, H. A. Stout, and R. Donovick, *Antibiot. Ann.* **1955/1956,** 554 (1956).
151. J. Vandeputte and J. D. Dutcher, *Antibiot. Ann.* **1955/1956,** 560 (1956).
152. S. Pestka and J. W. Bodley, in Ref. 25, p. 551; E. Cundliffe, in Ref. 41, p. 329; 492; E. Cundliffe, in Ref. 149, p. 228.
153. T.-P. Hausner, V. Geigenmüller, and K. H. Nierhaus, *J. Biol. Chem.* **263,** 13,103 (1988).
154. F. Benazet and co-workers, *Experientia* **36,** 414 (1980).
155. M. Bodanszky and co-workers, *J. Am. Chem. Soc.* **86,** 2478 (1964); M. Bodansky, J. A. Scozzie, and I. Muramatsu, *J. Am. Chem. Soc.* **91,** 4934 (1969); B. Anderson, D. C. Hodgkin, and M. A. Viswamitra, *Nature* **225,** 233 (1970).
156. K. Tori and co-workers, *Tetrahedron Lett.* **3,** 185 (1976).
157. O. D. Hensens, G. Albers-Schönberg, and B. F. Anderson, *J. Antibiot.* **36,** 799 (1983).
158. K. Florey, in K. Florey, ed., *Analytical Profiles of Drug Substances,* Vol. 7, Academic Press, New York, 1978, p. 423.
159. D. R. Houck and co-workers, *J. Am. Chem. Soc.* **109,** 1250 (1987).
160. T. Frenzel, P. Zhou, and H. G. Floss, *Arch. Biochem. Biophys.* **278,** 35 (1990).
161. S. Goto, S. Kuwahara, N. Okubo, and H. Zenyoji, *J. Antibiot.* **21,** 119 (1968); E. Higashide, K. Hatano, M. Shibata, and K. Nakazawa, *J. Antibiot.* **21,** 126 (1968); K. Tsuchiya, M. Kondo, T. Oishi, and T. Yamazaki, *J. Antibiot.* **21,** 147 (1968).
162. M. Hori, N. Sugita, and M. Miyazaki, *Chem. Pharmaceut. Bull.* **21,** 1171 (1973); M. Hori and co-workers, *Chem. Pharmaceut. Bull.* **21,** 1175 (1973); H. Iwasaki and co-workers, *Chem. Pharmaceut. Bull.* **21,** 1184 (1973).
163. M. Kawakami, Y. Nagai, T. Fujii, and S. Mitsuhashi, *J. Antibiot.* **24,** 583 (1971).
164. T. Tamura, H. Suzuki, Y. Nishimura, J. Mizoguchi, and Y. Hirota, *Proc. Nat. Acad. Sci. USA* **77,** 4499 (1980); F. Ishino and co-workers, *J. Biol. Chem.* **261,** 7024 (1986).
165. H. Yokotani and co-workers, *Takeda Kenkyusho Nempo* **28,** 76 (1969).
166. T. Miyoshi and co-workers, *J. Antibiot.* **25,** 569 (1972); S. Miyamura and co-workers, *J. Antibiot.* **25,** 610 (1972).
167. C. D. Ericsson and co-workers, *Ann. Intern. Med.* **98,** 20 (1983); C. D. Ericsson and co-

workers, *Gastroenterology* **88,** 473 (1985); C. D. Ericsson and H. L. DuPont, in P. K. Peterson and J. Verhoef, eds., *Antimicrobial Agents Annual,* Vol. 3, Elsevier, New York, 1988, p. 541.
168. N. Tanaka, in F. E. Hahn, ed., *Antibiotics—Mechanism of Action of Antibacterial Agents,* Springer-Verlag, New York, 1979, p. 18; R. M. Williams and C. A. Durham, *Chem. Rev.* **88,** 511 (1988).
169. B. Heinemann, M. A. Kaplan, R. D. Muir, and I. R. Hooper, *Antibiot. Chemother.* **111,** 1239 (1953).
170. D. E. Tisch, J. B. Hoekstra, D. S. Fisher, and H. L. Dickison, *Antibiot. Ann.* **1954/1955,** 1011 (1955).
171. U.S. Pat. 3,126,317 (Mar. 24, 1964), B. Heinemann, I. R. Hooper, and M. A. Kaplan (to Bristol-Myers Co.).
172. M. Bodanszky, G. F. Sigler, and A. Bodanszky, *J. Am. Chem. Soc.* **95,** 2352 (1973).
173. H. Tanaka, R. Oiwa, S. Matsukara, and S. Omura, *Biochem. Biophys. Res. Commun.* **86,** 902 (1979).
174. H. Tanaka and co-workers, *J. Antibiot.* **35,** 1216 (1982).
175. M. S. Kang, J. P. Spencer, and A. D. Elbein, *J. Biol. Chem.* **253,** 8860 (1978).
176. D. K. Banerjee, *J. Biol. Chem.* **264,** 2024 (1989).
177. J. H. Lakey, R. Maget-Dana, and M. Ptak, *Biochem. Biophys. Res. Commun.* **150,** 384 (1988).
178. U.S. Pat. 4,414,206 (Nov. 8, 1983), M. Gordon and G. J. Christie (to Bristol-Myers Co.).
179. H. Brockmann and G. Schmidt-Kastner, *Chem. Ber.* **88,** 57 (1955); H. Brockmann and H. Geeren, *Anal. Chemie* **603,** 216 (1957).
180. E. Gäumann and co-workers, *Experientia* **3,** 202 (1947).
181. P. A. Plattner and U. Nager, *Experientia* **3,** 325 (1947).
182. R. L. Hamill, C. E. Higgens, H. E. Boaz, and M. Gorman, *Tetrahedron Lett.* **1969,** 4255 (1969).
183. T. M. Balasubramanian and co-workers, *J. Am. Chem. Soc.* **103,** 6127 (1981); B. F. Gisin and co-workers, *J. Am. Chem. Soc.* **103,** 6373 (1981); M. K. Mathew and P. Balaram, *Molec. Cell. Biochem.* **50,** 47 (1983); E. T. Kaiser and F. J. Kézdy, *Ann. Rev. Biophys. Biophys. Chem.* **16,** 561 (1987).
184. B. C. Pressman and M. Fahim, *Ann. Rev. Pharmacol. Toxicol.* **22,** 465 (1982).
185. E. P. Bakker in Ref. 37, p. 67.
186. R. Hilgenfeld and W. Saenger, *Top. Curr. Chem.* **101,** 1 (1982).
187. G. R. Painter and B. C. Pressman, *Top. Curr. Chem.* **101,** 83 (1982).
188. M. M. Shemyakin, N. A. Aldanova, E. I. Vinogradova, and M. Y. Feigina, *Tetrahedron Lett.* **28,** 1921 (1963); M. Pinkerton, L. K. Steinrauf, and P. Dawkins, *Biochem. Biophys. Res. Commun.* **35,** 512 (1969); V. T. Ivanov and co-workers, *Biochem. Biophys. Res. Commun.* **34,** 803 (1969).
189. Yu. A. Ovchinnikov, V. T. Ivanov, and A. M. Shkrob, *Membrane-Active Complexones, BBA Library,* Vol. 12, Elsevier, Amsterdam, The Netherlands, 1974.
190. T. Anke and F. Lipmann, *FEBS Lett.* **82,** 337 (1977); H. Ristow, J. Salnikow, and H. Kleinkauf, *FEBS Lett.* **42,** 127 (1974); J. B. Perkins and co-workers, *J. Bacteriol.* **172,** 3108 (1990).
191. N. Schnell and co-workers, *Nature* **333,** 276 (1988).
192. C. Kaletta and K.-D. Entian, *J. Bacteriol.* **171,** 1597 (1989).
193. T. R. Klaenhammer, *6th International Symposium on the Genetics of Industrial Microorganisms,* Vol. 1, Strasbourg, France, 1990, p. 433.
194. *Fed. Reg.* **53**(66), 11,247 (Apr. 1988).
195. E. Lipinska, in V. M. Woodbine, ed., *Antibiotics and Antibiosis in Agriculture with Special Reference to Synergism,* Butterworths, London, 1977, p. 103; A. Hurst, *Adv. Appl. Microbiol.* **27,** 85 (1981); K. Rayman and A. Hurst in Ref. 71, p. 807.
196. L. A. Rogers, *J. Bacteriol.* **16,** 321 (1928).

197. A. T. R. Mattick and A. Hirsh, *Nature* **554,** 551 (1944).
198. E. Gross and J. L. Morell, *J. Am. Chem. Soc.* **93,** 4634 (1971).
199. M. Barber, and co-workers, *Experientia* **44,** 266 (1988).
200. K. Fukase and co-workers, *Tetrahedron Lett.* **29,** 795 (1988).
201. U.S. Pat. 2,935,503 (May 3, 1960), H. B. Hawley and R. H. Hall (to Aplin and Barrett, Ltd.).
202. E. Ruhr and H.-G. Sahl, *Antimicrob. Agents Chemother.* **27,** 841 (1985).
203. M. Debono and co-workers, *J. Antibiot.* **40,** 761 (1987).
204. G. M. Eliopoulos, C. Thauvin, B. Gerson, and R. C. Moellering, Jr., *Antimicrob. Agents Chemother.* **27,** 357 (1985).
205. L. D. Boeck, D. S. Fukuda, B. J. Abbott, and M. Debono, *J. Antibiot.* **41,** 1085 (1988).
206. M. Debono and co-workers, *J. Antibiot.* **41,** 1093 (1988).
207. G. M. Eliopoulos and co-workers, *Antimicrob. Agents Chemother.* **30,** 532 (1986); R. N. Jones and A. L. Barry, *Antimicrob. Agents Chemother.* **31,** 625 (1987).
208. P. Canepari, M. Boaretti, M. del Mar Lleó, and G. Satta, *Antimicrob. Agents Chemother.* **34,** 1220 (1990).
209. J. T. Mader and K. Adams, *Antimicrob. Agents Chemother.* **33,** 689 (1989); S. Kennedy and H. F. Chambers, *Antimicrob. Agents Chemother.* **33,** 1522 (1989); C. S. Haworth, M. W. Sobieski, W. M. Scheld, and T. S. Park, *Antimicrob. Agents Chemother.* **34,** 245 (1990).
210. M. Debono and co-workers, *Ann. N.Y. Acad. Sci.* **544,** 152 (1988); L. D. Broeck, D. S. Fukuda, B. J. Abbott, and M. Debono, *J. Antibiot.* **42,** 382 (1989).
211. R. S. Gordee and co-workers, *Ann. N.Y. Acad. Sci.* **544,** 294 (1988).
212. M. Hobbs, J. Perfect, and D. Durack, *Eur. J. Clin. Microbiol. Infect. Dis.* **7,** 77 (1988); L. H. Hanson and D. A. Stevens, *Antimicrob. Agents Chemother.* **33,** 1391 (1989); R. P. Rennie and L. Hellman, *Mycoses* **32,** 145 (1989).
213. W. L. Current and C. J. Boylan, in Ref. 140, p. 229.
214. C. S. Taft and C. P. Selitrennikoff, *J. Antibiot.* **41,** 697 (1988).
215. A. Padula and H. F. Chambers, *Antimicrob. Agents Chemother.* **33,** 1822 (1989); J. R. Perfect, M. M. Hobbs, K. A. Wright, and D. T. Durack, *Antimicrob. Agents Chemother.* **33,** 1811 (1989); C. J. Morrison and D. A. Stevens, *Antimicrob. Agents Chemother.* **34,** 746 (1990).
216. B. N. Doebbeling and co-workers, in *30th Interscience Conference on Antimicrobial Agents and Chemotherapy,* American Society for Microbiology, 1990, p. 183.
217. S. A. Waksman, ed., *Ann. N.Y. Acad. Sci.* **89,** 283 (1960).
218. S. A. Waksman, ed., *Actinomycin: Nature, Formation, and Activities,* Wiley-Interscience, New York, 1968.
219. U. Hollstein, *Chem. Rev.* **74,** 625 (1974); "Selman A. Waksman Conference on Actinomycins: Their Potential for Cancer Chemotherapy," *Cancer Chemother. Rept.* **58** (1974).
220. J. Meinhofer and E. Atherton, in D. Perlman, ed., *Structure-Activity Relationships Among the Semisynthetic Antibiotics,* Academic Press, New York, 1977, p. 427.
221. A. B. Mauger, in P. G. Sammes, ed., *Topics in Antibiotic Chemistry,* Vol. 4, Ellis Horwood, Chicester, UK, 1980, p. 224.
222. M. J. Waring, in Ref. 41, p. 314.
223. *IARC Monographs* **12,** 29 (1976).
224. R. A. Manaker, F. Gregory, L. C. Vining, and S. A. Waksman, *Antibiot. Annu.* **1954–1955,** 853 (1955).
225. A. B. Mauger and E. Katz, *J. Chromatog. Lib.* **15,** 1 (1978).
226. U.S. Pat. 2,378,876 (June 19, 1945), S. A. Waksman and H. B. Woodruff (to Merck Co.).
227. E. Katz, in Ref. 38, p. 276; F. Fawaz and G. H. Jones, *J. Biol. Chem.* **263,** 4602 (1988); C. E. Barry, III, P. G. Nayar, and T. P. Begley, *Biochemistry* **28,** 6323 (1989).
228. S. K. Sengupta and co-workers, *J. Med. Chem.* **24,** 1052 (1981); M. Yamada, K.

Nakajima, and K. Okawa, in N. Izumiya, ed., *Peptide Chemistry 1984,* Protein Research Foundation, Osaka, Japan, 1985, p. 153.
229. G. Bradley, P. F. Juranka, and V. Ling, *Biochim. Biophys. Acta* **948,** 87 (1988); *Cancer Bull.* **41**(1), 1 (1989).
230. H. M. Sobell, *Prog. Nucleic Acid Res. Mol. Biol.* **13,** 153 (1973).
231. E. V. Scott, G. Zon, L. G. Marzilli, and W. D. Wilson, *Biochemistry* **27,** 7940 (1988).
232. M. Delepierre and co-workers, *J. Biomol. Struct. Dyn.* **7,** 557 (1989).
233. H. Umezawa, K. Maeda, T. Takeuchi, and Y. Okami, *J. Antibiot. Ser. A* **19,** 200 (1966); H. Umezawa, Y. Suhara, T. Takita, and K. Maeda, *J. Antibiot. Ser. A* **19,** 210 (1966).
234. H. Umezawa, in J. M. Cassady and J. D. Douros, eds., *Anticancer Agents Based on Natural Product Models,* Academic Press, New York, 1980, p. 147.
235. N. J. Oppenheimer and co-workers, *J. Biol. Chem.* **257,** 1606 (1982); B. J. Carter and co-workers, *J. Biol. Chem.* **265,** 4193 (1990).
236. S. Oka, *Rec. Results Cancer Res.* **74,** 163 (1980).
237. U.S. Pat. 4,195,018 (Mar. 25, 1980), T. Takita and co-workers (to Nippon Kayaku KK).
238. K. Takahashi and co-workers, *Cancer Treat. Rev.* **14,** 169 (1987).
239. R. A. Newman and co-workers, *Investig. New Drugs* **8,** 33 (1990).
240. S. T. Crooke and co-workers, *Rec. Results Cancer Res.* **76,** 83 (1981).
241. K. Maeda, H. Kosaka, K. Yagishita, and H. Umezawa, *J. Antibiot. Ser. A* **9,** 82 (1956).
242. S. K. Carter, T. Ichikawa, G. Mathé, and H. Umezawa, eds., *Gann Monogr. Cancer Res.* (19) (1976).
243. C. W. Haidle and R. S. Lloyd, in Ref. 38, p. 124; H. Umezawa, in J. M. Cassady and J. D. Douros, eds., *Anticancer Agents Based on Natural Product Models,* Academic Press, New York, 1980, p. 147.
244. T. Takita, in Ref. 71, p. 595.
245. B. I. Sikic, M. Rozencweig, and S. K. Carter, *Bleomycin Chemotherapy,* Academic Press, Orlando, Fla., 1985.
246. S. M. Hecht, *Fed. Proc.* **45,** 2784 (1986); S. M. Hecht, *Acc. Chem. Res.* **19,** 383 (1986).
247. I. H. Goldberg, in C. Chagas and B. Pullman, eds., *Molecular Mechanisms of Carcinogenic and Antitumor Activity,* The Vatican Press distributed by Adenine Press, Guilderland, N.Y., 1987, p. 425.
248. J. Stubbe and J. W. Kosarich, *Chem. Rev.* **87,** 1107 (1987).
249. D. H. Petering, R. W. Byrnes, and W. E. Antholine, *Chem.-Biol. Interac.* **73,** 133 (1990).
250. T. Takita and co-workers, *J. Antibiot.* **25,** 755 (1972).
251. T. Takita and co-workers, *Tetrahedron Lett.* **23,** 521 (1982).
252. Y. Aoyagi and co-workers, *J. Am. Chem. Soc.* **104,** 5237 (1982).
253. S. Saito and co-workers, *J. Antibiot.* **36,** 92 (1983).
254. T. Takita and Y. Muraoka, in Ref. 1, p. 289.
255. J. S. Lazo, C. J. Boland, and P. E. Schwartz, *Cancer Res.* **42,** 4026 (1982); P. R. Twentyman, *Pharmacol. Ther.* **23,** 417 (1984); B. I. Sikic, *Cancer Surv.* **5,** 81 (1986); J. E. Zuckerman and co-workers, *Cancer Res.* **46,** 1748 (1986).
256. S. M. Sebti and co-workers, *Biochemistry* **28,** 6544 (1989).
257. M. Takeshita, A. P. Grollman, E. Ohtsubo, and H. Ohtsubo, *Proc. Natl. Acad. Sci. USA* **75,** 5983 (1987).
258. N. Ishida, K. Miyazaki, K. Kumagai, and M. Rikimaru, *J. Antibiot. Ser. A* **18,** 68 (1965); K. Edo, K. Saito, Y. Akiyama-Murai, and M. Mizugaki, *J. Antibiot.* **41,** 554 (1988); P. C. Dedon and I. H. Goldberg, *J. Biol. Chem.* **265,** 14,713 (1990); B. Pullman, *Adv. Drug Res.* **18,** 1 (1989); D. H. Chin, C. H. Zeng, C. E. Costello, and I. H. Goldberg, *Biochemistry* **27,** 8106 (1988); L. F. Povirk and I. H. Goldberg, *Biochemie* **69,** 815 (1987); T. Sasaki, Y. Koide, and K. Yoshigaga, *Meth. Enzymol.* **178,** 422 (1989); F. Suzuki, R. B. Pollard, S. Uchimura, T. Munakata, and H. Maeda, *Cancer Res.* **50,** 3897 (1990); E. Otsuji, T. Takahashi, T. Yamaguchi, N. Yamaguchi, and J. Imanishi, *Gastroenterol. Japan* **25,** 244 (1990); L. C. Sieker and G. G. Gnanarajah, *Proteins* **3,** 252 (1988); P. Van Roey and T. A. Beerman, *Proc. Natl. Acad. Sci. USA* **86,** 6587 (1989).

259. A. Rüegger and co-workers, *Helv. Chim. Acta* **59,** 1075 (1976).
260. J. F. Borel, C. Feurer, H. V. Gubler, and H. Stähelin, *Agents Actions* **6,** 468 (1976).
261. B. Ryffel and co-workers, *Arch. Toxicol.* **53,** 107 (1983).
262. J. P. Scott and T. W. Higenbottam, *Med. Toxicol.* **3,** 107 (1988).
263. R. Maiorca and co-workers, *Adv. Exp. Med. Biol.* **252,** 273 (1989).
264. H. Dieperink, *Dan. Med. Bull.* **36,** 235 (1989).
265. B. D. Kahan, ed., *Cyclosporine,* Vol. 1, Grune and Stratton, Inc., Orlando, Fla., 1984.
266. R. Schindler, ed., *Cyclosporin in Autoimmune Diseases,* Springer-Verlag, Berlin, 1985.
267. J. F. Borel, *Prog. Allergy* **38,** 9 (1986).
268. M. M. A. Hassan and M. A. Al-Yahya, in K. Florey, ed., *Analy. Profiles Drug Subst.* **16,** 145 (1987).
269. A. von Wartburg and R. Traber, in G. P. Ellis and G. B. West, eds., *Prog. Med. Chem.* **25,** 1 (1988).
270. G. Feutren, *Curr. Op. Immunol.* **2,** 239 (1989).
271. M. J. Mihatsch, G. Thiel, and B. Ryffel, *Toxicol. Lett.* **46,** 125 (1989).
272. A. Aszalos, *J. Med.* **19,** 297 (1988).
273. H. G. Fliri and R. M. Wenger, in Ref. 1, p. 245.
274. S. L. Schreiber, *Science* **251,** 283 (1991).
275. T. J. Petcher, H.-P. Weber, and A. Rüegger, *Helv. Chim. Acta* **59,** 1480 (1976).
276. J. D. Aebi and co-workers, *J. Med. Chem.* **33,** 999 (1990).
277. U.S. Pat. 4,117,118 (Sept. 26, 1978), E. Härri and A. Rüegger (to Sandoz, Ltd.).
278. A. Lawen and R. Zocher, *J. Biol. Chem.* **265,** 11,355 (1990).
279. G. Fischer, H. Bang, and C. Mech, *Biomed. Biochim. Acta* **43,** 1101 (1984).
280. G. Fischer and H. Bang, *Biochim. Biophys. Acta* **828,** 39 (1985).
281. G. Fischer and F. X. Schmid, *Biochemistry* **29,** 2205 (1990).
282. J. Liu, M. W. Albers, C.-M. Chen, S. L. Schreiber, and C. T. Walsh, *Proc. Natl. Acad. Sci. USA* **87,** 2304 (1990).
283. N. Takahashi, T. Hayano, and M. Suzuki, *Nature* **337,** 473 (1989); G. Fischer and co-workers, *Nature,* **337,** 476 (1989).
284. A. W. Thomson, *Immunol. Today* **10,** 6 (1989).
285. T. E. Starzl and co-workers, *Lancet* **1989ii,** 1000 (1989).
286. J. J. Siekierka and co-workers, *Nature* **341,** 755 (1989).
287. M. W. Harding, A. Galat, D. E. Uehling, and S. L. Schreiber, *Nature* **341,** 758 (1989).
288. R. K. Harrison and R. L. Stein, *Biochemistry* **29,** 3813 (1990).
289. M. K. Rosen, R. F. Standaert, A. Galat, M. Nakatsuka, and S. L. Schreiber, *Science* **248,** 863 (1990).
290. H. Umezawa and co-workers, *J. Antibiot.* **29,** 97 (1976); H. Umezawa, *Drug Exptl. Clin. Res.* **10,** 519 (1984).
291. H.-U. Schorlemmer and co-workers, *Behring Inst. Mitt.* **74,** 157 (1984).
292. K. Ota and N. Ogawa, *Biomed. Pharmacother.* **44,** 93 (1990).
293. J. S. Ricci, Jr., A. Bousvaros, and A. Taylor, *J. Org. Chem.* **47,** 3063 (1982).
294. R. Nishizawa and co-workers, *J. Med. Chem.* **20,** 510 (1977).
295. Y. Mine and co-workers, *J. Antibiot.* **36,** 1045 (1983).
296. H. Umezawa, *Rev. Infect. Dis.* **6,** 412 (1984).
297. E. Bayer and co-workers, *Helv. Chim. Acta* **55,** 224 (1972).
298. Y. Kondo and co-workers, *Meiji Seika Kenkyusho Nempo* **13,** 34 (1973).
299. R. E. Hoagland, in H. G. Cutler, ed., *Biologically Active Natural Products-Potential Use in Agriculture,* American Chemical Society, Washington, D.C., 1988, p. 182.
300. T. Hidaka and co-workers, *J. Antibiot.* **42,** 491 (1989).
301. K. L. Carlson and O. C. Burnside, *Weed Sci.* **32,** 841 (1984).
302. R. R. Bellinder, K. K. Hatzios, and H. P. Wilson, *Weed Sci.* **33,** 779 (1985).
303. S. M. Ridley and S. F. McNally, *Plant Sci.* **39,** 31 (1985).
304. U.S. Pat. 4,168,963 (Sept. 25, 1979), W. Rupp and co-workers (to Hoechst).
305. L. Maier and P. J. Lea, *Phosphorus Sulfur* **17,** 1 (1983).

306. *KYM International Directory of Pharmaceutical Companies 1988/1989,* KYM Consultancy Ltd., Kent, UK; *Corporate Technology Directory, U.S. Edition,* Corporate Technology Information Services, Wellesley Hills, Mass., 1991.

EDMUND M. WISE, JR.
The Wellcome Research Laboratories

ANTIBIOTICS, POLYENES. See ANTIPARASITIC AGENTS–ANTIMYCOTICS.

POLYETHERS

The polyether antibiotics were first recognized as a separate class with the publication of the structure of monensin in 1967 (1). Several members of the group have since found commercial application as anticoccidials in poultry farming and in improvement of feed efficiency for ruminants (see FEEDS AND FEED ADDITIVES).

These antibiotics are characterized by multiple tetrahydrofuran and tetrahydropyran rings connected by aliphatic bridges, direct C–C linkage, or spiro linkage. Other features include a free carboxyl function, many lower alkyl groups, and a variety of functional oxygen groups. These structural features enable the molecule to form a cyclic conformation with the oxygen functions at the center and the alkyl groups on the outer surface. This conformation results in lipid solubility, even for the salt forms, enabling transport of cations across lipid membranes. The potential for using these antibiotics as biochemical tools was recognized long before the structural features had been elucidated. Studies of antibiotics as inhibitors of mitochondrial enzymes (2) revealed the unique characteristics of nigericin and dianemycin, both later shown to be polyethers. The term ionophore was first used to describe compounds that can transport ions across artificial or natural membranes (3). The polyethers are sometimes referred to as carboxylic acid ionophores to distinguish these antibiotics from other compounds showing ionophoretic activity.

Individual polyethers exhibit varying specificities for cations. Some polyethers have found application as components in ion-selective electrodes for use in clinical medicine or in laboratory studies involving transport studies or measurement of transmembrane electrical potential (4). The methyl ester of monensin [*28636-21-7*] has been incorporated into a membrane slide assembly used for the assay of serum sodium (see BIOSENSORS) (5). Studies directed toward the design of a lithium selective electrode resulted in the synthesis of a derivative of monensin lactone that is highly specific for lithium (6).

Properties

A classification based first on ion specificity, then on structural features has been suggested for the polyethers (7). Another method uses the presence of unsaturation or of aromatic groups in the molecular skeleton (8). In this review the compounds are classified based on the number of carbons in the backbone according to the numbering system proposed in reference 9. The carbon backbone or skeleton refers to the longest chain of contiguous carbons between the carboxyl group and the terminal carbon.

Table 1 lists the polyether antibiotics arranged by the number of carbons in the skeleton. Many of these compounds were isolated independently in separate laboratories and thus have more than one designation. The groups are subdivided depending on the number of spiroketals. Two classes fall outside this scheme: the pyrrole ether type containing a heterocyclic ring, and the acyltetronic acid type, that has an acylidene tetronic acid instead of a carboxylic acid. These compounds are ionophores and because of their common features are included as polyethers.

Lysocellin (**1**, R = COOH, R′ = CH_2CH_3), X14889 A (**1**, R = R′ = CH_3), and X14889 C (**1**, R = COOH, R′ = CH_3) are examples of the shortest polyethers, those having 22 carbons in the backbone. X14889 A, a minor factor, is atypical in that it does not contain a carboxylic acid. The 24 carbon backbone group includes the commercially important lasalocid A (**2**), one of the first polyethers to be isolated. Monensin A (**3**, R = CH_3, R′ = CH_2CH_3, R″ = H), another widely used anticoccidial and feed efficiency enhancer, has one spiroketal and a 26 C backbone. Other closely related members of this subgroup are monensin B (**3**, R = R′ = CH_3, R″ = H) and laidlomycin (**3**, R = $COCH_2CH_3$, R′ = CH_3, R″ = H). The propionyl ester of laidlomycin [*84799-02-0*] (**3**, R = R″ = $COCH_2CH_3$, R′ = CH_3), $C_{40}H_{66}O_{13}K$, is currently at the preregistration stage of development for use as a feed additive.

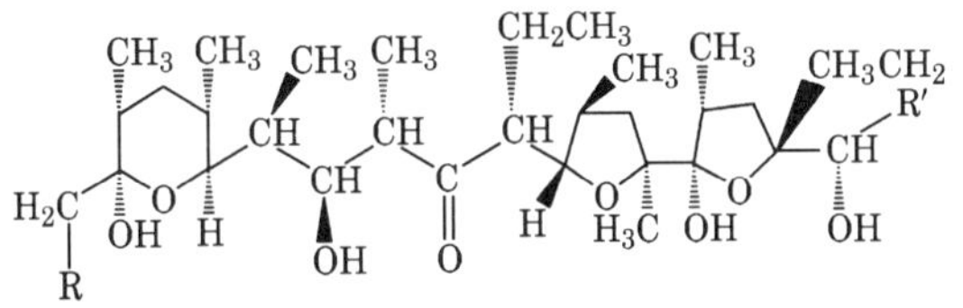

(**1**)

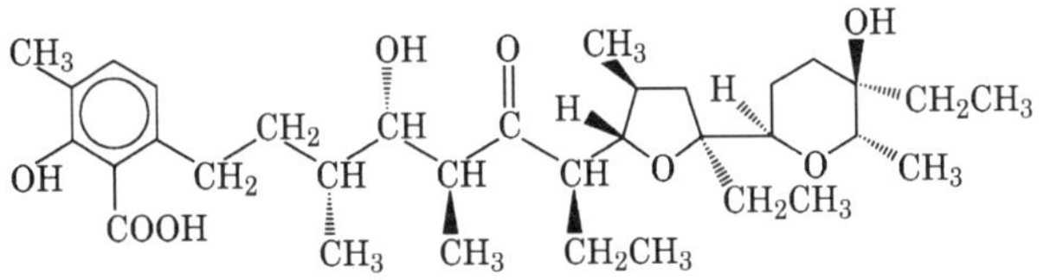

(**2**)

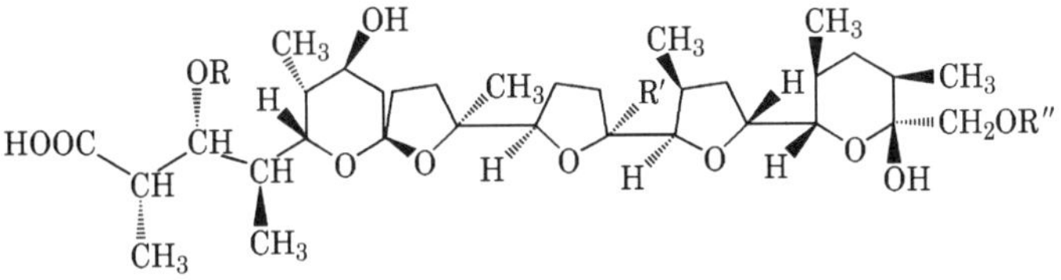

(**3**)

The C-30 skeleton group accounts for about 60% of the polyethers for which structures have been determined. Most of these contain a sugar moiety, usually 2,3,6-trideoxy-4-*O*-methyl-D-erythro pyranose [*65104-53-2*]. The single spiroketal

Table 1. Polyether Antibiotics

Name[a]	CAS Registry Number	Molecular formula	Producing organism	Mp, °C	Reference I[b]	Reference S[c]
22 to 23 Carbon backbone (22,23 C), 0-Spiroketal						
ferensimycin A (5057 A)	[*83852-59-9*]	$C_{34}H_{60}O_{10}$	*Streptomyces* sp. 5057 FERM BP-62	133–135[d]	10	10
ferensimycin B (5057 B)	[*83852-60-2*]	$C_{35}H_{62}O_{10}$	*Streptomyces* sp. 5057 FERM BP-62	143–145[d]	10	10
lysocellin (X14537 A)	[*55898-33-4*]	$C_{34}H_{60}O_{10}$	*Streptomyces cacaoi* var. *asoensis, Streptomyces longwoodenis* ATCC 29251)	158–160[d]	11,12	13
X14873 A	[*88263-37-0*]	$C_{35}H_{62}O_{11}$	*Streptomyces* sp. X 14873 ATCC 31679	154[d]	14	14
X14873 G	[*88263-35-8*]	$C_{34}H_{60}O_{8}$	*Streptomyces* sp. X 14873 ATCC 31679	152–153	14	14
X14873 H	[*88263-36-9*]	$C_{34}H_{62}O_{9}$	*Streptomyces* sp. X 14873 ATCC 31679	145–146	14	14
X14889 A	[*97671-96-0*]	$C_{33}H_{60}O_{8}$	*Streptomyces* sp. X 14889 NRRL 15517	149–150	15	15
X14889 C	[*97671-95-9*]	$C_{33}H_{58}O_{10}$	*Streptomyces* sp. X 14889 NRRL 15517	139–141[d]	15	15
X14889 D	[*97671-94-8*]	$C_{33}H_{58}O_{7}$	*Streptomyces* sp. X 14889 NRRL 15571	117	15	15
24 C, 0-Spiroketal						
inostamycin	[*129905-10-8*]	$C_{38}H_{68}O_{11}$	*Streptomyces* sp. MH816-AF15	181–181[d]	16	16
iso-lasalocid A	[*54156-67-1*]	$C_{34}H_{54}O_{8}$	*Streptomyces lasaliensis* NRRL 3382	183–185[d]	17	17
lasalocid A (X537 A)	[*25999-31-9*]	$C_{34}H_{54}O_{8}$	*Streptomyces lasaliensis* NRRL 3382	100–109	18	19

lasalocid B	[*55051-86-0*]	$C_{35}H_{56}O_8$	*Streptomyces lasaliensis* NRRL 3382	86–87	20	20
lasalocid C	[*55051-84-8*]	$C_{35}H_{56}O_8$	*Streptomyces lasaliensis* NRRL 3382	97–100	20	20
lasalocid D	[*55051-82-6*]	$C_{35}H_{56}O_8$	*Streptomyces lasaliensis* NRRL 3382	102–104	20	20
lasalocid E	[*55051-80-4*]	$C_{35}H_{56}O_8$	*Streptomyces lasaliensis* NRRL 3382	90	20	20
24 C, 1-Spiroketal						
CP54883	[*112396-63-1*]	$C_{41}H_{61}Cl_2O_{12}$	*Actinomadura routienii* Huang N365-41	330–340[d]	21	22
25 C, 0-Spiroketal						
CP78545	[*124150-86-3*]	$C_{33}H_{58}O_7$	*Streptomyces* sp. N731-45	94–98	23	23
zincophorin (Griseochelin, M144255)	[*91920-88-6*]	$C_{33}H_{30}O_7$	*Streptomyces griseus* NCIB 11504, ZIMET 43681	66–70	24,25	24,26
26 C, 1-Spiroketal						
CP47433	[*74758-62-6*]	$C_{47}H_{32}O_{14}$	*Actinomadura macra* ATCC 31286	89–99	27	28
CP47434	[*74758-61-5*]	$C_{46}H_{80}O_{14}$	*Actinomadura macra* ATCC 31286	230–238[d]	27	28
CP80585	[*128345-23-3*]	$C_{37}H_{64}O_{11}$	*Streptomyces* sp. ATCC 53862		29	29
CP82996	[*127288-12-4*]	$C_{50}H_{86}O_{16}$	*Actinomadura* sp. ATCC 53764	115–117	30	30
deoxylaidlomycin (A712)	[*102674-88-4*]	$C_{37}H_{62}O_{11}$	*Streptoverticillium eurocidicum* SANK 61484	194–196[d]	31	31
			Streptoverticillium olivoreticuli IMET 43861		32	

Table 1. (*Continued*)

Name[a]	CAS Registry Number	Molecular formula	Producing organism	Mp, °C	Reference	
					I[b]	S[c]
laidlomycin (AB78, TS822)	[*56283-74-0*]	$C_{37}H_{62}O_{12}$	*Streptoverticillium eurocidium* ATCC 31175, *Streptomyces olivoreticuli* A14939	151–153	33,34	34,35
monensin A	[*17090-79-8*]	$C_{36}H_{62}O_{11}$	*Streptomyces cinnamonensis* ATCC 15413	103–105	36	1
monensin B	[*30485-16-6*]	$C_{35}H_{60}O_{11}$	*Streptomyces cinnamonensis* ATCC 15413	227–228[d]	36	37
monensin C	[*31980-87-7*]	$C_{37}H_{64}O_{11}$	*Streptomyces cinnamonensis* ATCC 15413	212–214[d]	36	37
27 C, 1-Spiroketal						
6270 B	[*122548-05-4*]	$C_{44}H_{74}O_{13}$	*Nocardiopsis* sp. FERM BP-717		38	38
A82810 (CP84657)	[*127759-24-4*]	$C_{45}H_{78}O_{14}$	*Actinomadura fibrosa* NRRL 18348, *Actinomadura* sp. ATCC 53708	255–257[d]	39,40	39,40
cationomycin	[*82987-42-6*]	$C_{45}H_{70}O_{15}$	*Actinomadura azurea* FERM BP-83	108–112	41	41
kijimicin	[*129297-22-9*]	$C_{37}H_{64}O_{11}$	*Actinomadura* sp. MI215-NF3	217–218[d]	42	42
portmicin (A80190)	[103521-25-1]	$C_{44}H_{76}O_{14}$	*Nocardiopsis* sp. 6270 FERM BP-717, *Actinomadura oligospora* NRRL 15878	115–118	43,44	44,45
30 C, 0-Spiroketal						
UK41637	[*83532-94-9*][d]	$C_{46}H_{78}O_{16}$	*Actinomadura cremea* ATCC 31676	196–198[d]	46	46

30 C, 1-Spiroketal

6016	[*69421-39-2*]	$C_{46}H_{78}O_{16}$	*Streptomyces albus* 6016	192–195[d]	47	48
A204 A	[*43110-10-7*]	$C_{49}H_{84}O_{17}$	*Streptomyces albus* NRRL 3384	96–98	49	50
A204 B	[*65208-37-9*][d]		*Streptomyces albus* NRRL 3384		49	[e]
A28695 B	[*42617-35-6*]	$C_{48}H_{82}O_{17}$	*Streptomyces albus* NRRL 3883	122–124	51	52
A80438 (demethoxylonomycin A, SF2437)	[*108044-86-6*]	$C_{43}H_{74}O_{13}$	*Streptomyces pactum* NRRL 15970, *Streptomyces bobili* NRRL 15971, *Streptomyces* sp. FERM P-8469	160–162[d]	53,54	53,55
abierixin	[*100634-16-0*]	$C_{40}H_{68}O_{11}$	*Streptomyces albus* NRRL B-1865	83–85	56	56
BL580 beta	[*53414-72-5*]	$C_{47}H_{80}O_{16}$	*Streptomyces hygroscopicus* NRRL 5647		57	58
BL580 delta	[*66389-75-1*]	$C_{47}H_{80}O_{16}$	*Streptomyces hygroscopicus* NRRL 8180	157–161[d]	59	60
carriomycin (T42082)	[*65978-43-0*]	$C_{47}H_{80}O_{15}$	*Streptomyces hygroscopicus* ATCC 31080	120–122	61	62
CP70228	[*101621-31-2*]	$C_{53}H_{90}O_{18}$	*Actinomadura roseorufa* ATCC 39697	113–123	63	63
CP70828	[*101621-30-1*]	$C_{54}H_{92}O_{18}$	*Actinomadura roseorufa* ATCC 39697	133–136	63	63
2-epi-mutalomycin	[*124918-41-8*]	$C_{41}H_{70}O_{12}$	*Streptomyces mutabilis* NRRL 8086	166–177[d]	64	64
28-epi-mutalomycin	[*124986-36-3*]	$C_{41}H_{70}O_{12}$	*Streptomyces mutabilis* NRRL 8086	187–191[d]	64	64
epi-nigericin	[*108266-17-7*]	$C_{40}H_{68}O_{11}$	*Streptomyces hygroscopicus* NRRL B-1865	188.5[d]	65	65
etheromycin (CP38295, C20-12, T50417)	[*59149-05-2*]	$C_{48}H_{82}O_{16}$	*Streptomyces* C20-12 FERM P-2736, *Streptomyces hygroscopicus* ATCC 31050	135–138	66,67	67,68

Table 1. (*Continued*)

Name[a]	CAS Registry Number	Molecular formula	Producing organism	Mp, °C	Reference	
					I[b]	S[c]
grisorixin (K358)	[*31357-58-1*]	$C_{40}H_{68}O_{11}$	*Streptomyces griseus*	75–80	69	70
K41 A (A32887)	[*53026-37-2*]	$C_{48}H_{82}O_{18}$	*Streptomyces hygroscopicus* FERM P-1342, *Streptomyces albus* NRRL 11109	196–198[d]	90,91	92,93
K41 B	[*72017-85-7*]	$C_{54}H_{92}O_{20}$	*Streptomyces hygroscopicus* FERM P-1342	185–186[d]	94	95
LL-C23201 delta	[*107020-14-4*]	$C_{47}H_{80}O_{15}$	*Streptomyces olivaceogriseus* NRRL 15357	164–168[d]	71	71
lonomycin A (DE3936, emericid, A218)	[*58785-63-0*]	$C_{44}H_{76}O_{14}$	*Streptomyces ribosidificus* ATCC 31051, *Streptomyces hygroscopicus* FERM P-3159, *Streptomyces hygroscopicus* DS 24367	188–189[d]	72–74	75–77
lonomycin B	[*68567-60-2*]	$C_{44}H_{26}O_{14}$	*Streptomyces ribosidificus* ATCC 31051	181–182	78	79
lonomycin C	[*68537-50-8*]	$C_{44}H_{74}O_{14}$	*Streptomyces ribosidificus* ATCC 31051	186–187[d]	78	79
maduramicin alpha (LL-C23024, prinicin, X14868 A)	[*79356-08-4*]	$C_{47}H_{80}O_{17}$	*Actinomadura yumaense* NRRL 12515, *Nocardia* sp. X 14868 ATCC 31585	193–194[d]	80,81	59,81
maduramicin beta (X14868 C)	[*79331-53-6*]	$C_{46}H_{78}O_{17}$	*Actinomadura yumaense* NRRL 12515, *Nocardia* sp. X 14868 ATCC 31585	172–175[d]	80,81	59,81
27-methoxyseptamycin	[*125131-53-5*]	$C_{49}H_{84}O_{17}$	*Actinomadura* sp. ATCC 53676		82	82
mutalomycin (S11743 A)	[*62618-08-0*]	$C_{41}H_{70}O_{12}$	*Streptomyces mutabilis* NRRL 8088	163–171[d]	83	84

nigericin (K178, X464, polyetherin A, duamycin)	[*28380-24-7*]	$C_{40}H_{68}O_{11}$	*Streptomyces violaceniger* NRRL B1356, *Streptomyces hygroscopicus* E-749, 325-15	170–172	85–87	88,89
semduramicin (UK61689)	[*113378-31-7*]	$C_{46}H_{76}O_{16}$	*Actinomadura roseorufa* ATCC 53666	167[d]	96	96
septamycin (A28695 A, BL580 alpha)	[*54927-63-8*]	$C_{48}H_{82}O_{16}$	*Streptomyces hygroscopicus* NRRL 5678; NRRL 5647, *Streptomyces albus* NRRL 3883	164–166[d]	51,57,97	52,98,99
TM582	[*98791-39-0*]	$C_{40}H_{68}O_{11}$	*Streptomyces hygroscopicus* TM 582	80–82	100	100
UK58852	[*101621-29-8*]	$C_{52}H_{88}O_{18}$	*Actinomadura roseorufa* ATCC 39697	123–126	63	63
W341 C	[*110368-36-0*]	$C_{47}H_{80}O_{16}$	*Streptomyces* W341 C		101	101
X14868 D	[*79331-54-7*]	$C_{47}H_{80}O_{17}$	*Nocardia* sp. X 14868 ATCC 31585	194–195	81	81
30 C, 2-Spiroketals						
A130 B	[*73492-07-6*][d]	$C_{54}H_{90}O_{16}$	*Streptomyces hygroscopicus* ATCC 21840		102	102
A130 C	[*73522-76-6*]	$C_{47}H_{78}O_{13}$	*Streptomyces hygroscopicus* ATCC 21840		102	102
CP53607 (X14931 A)	[*84680-56-8*]	$C_{40}H_{36}O_{11}$	*Streptomyces halstedii* ATCC 31812, *Streptomyces* sp. X 14931	199–204[d]	103,104	103,104
CP80219	[*123286-64-6*]	$C_{47}H_{78}O_{14}$	*Streptomyces hygroscopicus* ATCC 53626	176–180[d]	105	106
CP82483	[*121962-58-1*]	$C_{47}H_{78}O_{14}$			107	107
dianemycin	[*38565-33-9*]	$C_{47}H_{78}O_{14}$	*Streptomyces hygroscopicus* NRRL 3444	72–74	108	109
eudusamycin (CP63517)	[*100242-41-9*]	$C_{47}H_{78}O_{14}$	*Streptomyces endus* subsp. *aureus* ATCC 39574	95–105	110	110
19-epidianemycin (CP60993)	[*93218-58-7*]	$C_{47}H_{78}O_{14}$	*Streptomyces hygroscopicus* ATCC 39205	193–205[d]	111	111

Table 1. (*Continued*)

Name[a]	CAS Registry Number	Molecular formula	Producing organism	Mp, °C	Reference I[b]	Reference S[c]
lenoremycin (A130 A, Ro 21-6150)	[*51257-84-2*]	$C_{47}H_{78}O_{13}$	*Streptomyces hygroscopicus* X14563, ATCC 21840	87–92	112,113	114,115
leuseramycin (TM531 A)	[*73537-10-7*]	$C_{47}H_{78}O_{13}$	*Streptomyces hygroscopicus* ATCC 31590	88–91	116	116
moyukamycin	[*96827-80-4*]	$C_{47}H_{76}O_{13}$	*Streptomyces hygroscopicus* FERM BP-274		43,85, 86,89, 117,118[d]	117
TM531 B	[*80118-77-0*]	$C_{46}H_{76}O_{14}$	*Streptomyces hygroscopicus* ATCC 31590	252–253[d]	119	120
TM531 C	[*80118-78-1*]	$C_{47}H_{78}O_{15}$	*Streptomyces hygroscopicus* ATCC 31590	190–191	119	120
TM531 D	[*87579-00-8*]	$C_{46}H_{76}O_{14}$	*Streptomyces hygroscopicus* ATCC 31590		121	121
TM531 E	[*87579-01-9*]	$C_{46}H_{76}O_{14}$	*Streptomyces hygroscopicus* ATCC 31590		121	121
X14934A (CP47224)	[*96998-05-9*]	$C_{48}H_{80}O_{15}$	*Streptomyces* sp. X14934 NRRL 15518, *Streptomyces hygroscopicus* ATCC 31337	156–158[d]	122,123	122
30 C, Trispiroketal						
deoxy-epi-narasin	[*70051-99-9*]	$C_{43}H_{72}O_{10}$	*Streptomyces aureofaciens* NRRL 11181		124	124
deoxy-epi-salinomycin	[*64129-77-1*]	$C_{42}H_{70}O_{10}$	*Streptomyces albus* ATCC 21838	180	125	125
deoxynarasin	[*70022-35-4*]	$C_{43}H_{72}O_{10}$	*Streptomyces aureofaciens* NRRL 11181		124	124
deoxysalinomycin	[*64003-50-5*]	$C_{42}H_{70}O_{10}$	*Streptomyces albus* ATCC 21838		125	125

narasin A (A28086)	[*55134-13-9*]	$C_{43}H_{72}O_{11}$	*Streptomyces aureofaciens* NRRL 5758	98–100	126	127
narasin B	[*58439-94-4*]	$C_{43}H_{70}O_{11}$	*Streptomyces aureofaciens* NRRL 5758	150–153	128	128
narasin D	[*58334-78-4*]	$C_{44}H_{74}O_{11}$	*Streptomyces aureofaciens* NRRL 5758		128	128
salinomycin	[*53003-10-4*]	$C_{42}H_{70}O_{11}$	*Streptomyces albus* ATCC 21838	112–113	129	130
SY4 (5-hydroxysalinomycin)	[*110947-74-5*]	$C_{42}H_{70}O_{12}$	*Streptomyces albus* FERM P-419		131	131
SY8 (18,19-dihydrosalinomycin)	[*111057-09-1*]	$C_{42}H_{72}O_{11}$	*Streptomyces albus* FERM P-419		131	131
SY9 (20-oxosalinomycin)	[*83564-04-9*]	$C_{42}H_{68}O_{11}$	*Streptomyces albus* FERM P-419		132	132
31 C, 1-Spiroketal						
hidamycin (CP51532)	[*84331-31-7*]	$C_{42}H_{72}O_{13}$	*Actinomadura* sp. MI429-38F1; *Actinomadura verrucosospora* ATCC 31466	196–199[d]	133,134	133,135
31 C, Trispiroketal						
chloronoboritomycin A (X14766 A)	[*75217-55-9*]	$C_{43}H_{63}ClO_{14}$	*Streptomyces malachitofuscus downeyi* ATCC 31547	160	136	137
noboritomycin A	[*68508-45-2*]	$C_{43}H_{64}O_{14}$	*Streptomyces noboritoensis* NRRL 8123	235–237[d]	118	84
noboritomycin B	[*68508-46-3*]	$C_{44}H_{66}O_{14}$	*Streptomyces noboritoensis* NRRL 8123	220–222[d]	118	84
32 C, 0-Spiroketal						
ionomycin	[*56092-81-0*]	$C_{41}H_{72}O_9$	*Streptomyces conglobatus* ATCC 31005	205–206[f]	138	139

Table 1. (*Continued*)

Name[a]	CAS Registry Number	Molecular formula	Producing organism	Mp, °C	Reference	
					I[b]	S[c]
32 C, Trispiroketal						
CP44161	[*66771-51-5*]	$C_{43}H_{66}O_{10}$	*Dactylosporangium salmoneum* ATCC 31222	155–157	140	141
34 C, 0-Spiroketal						
CP73064	[*118674-05-8*]	$C_{44}H_{76}O_{13}$	*Streptomyces* sp. ATCC 53523	214–216[d]	142	142
37 C, 0-Spiroketal						
alborixin (CP38986, S14750)	[*57760-36-8*]	$C_{48}H_{84}O_{14}$	*Streptomyces albus* 3840, *Streptomyces flaveolus* ATCC 31100	100–105	143,144	145
X206	[*36505-48-3*]	$C_{47}H_{82}O_{14}$	*Streptomyces* sp. X 206	133–145	18	146
Acyl tetronic acids						
A80577 (SF2487)	[*123484-11-7*]	$C_{42}H_{64}O_{12}$	*Actinomadura verrucospora* NRRL 18236, *Actinomadura* sp. FERM P-9063	250–252[d]	147,148	147,148
tetronasin (ICI139603, M139603)	[*75139-06-9*]	$C_{35}H_{54}O_8$	*Streptomyces longisporoflavus* NCIB 11426	176–178[d]	149	150
tetronomycin	[*82206-10-8*]	$C_{34}H_{50}O_8$	*Streptomyces* sp. NRRL 11266	107–110[d]	151	151

Pyrrole ethers						
A23187 (calcimycin)	[*52665-69-7*]	$C_{29}H_{37}N_3O_6$	*Streptomyces chartreusis* NRRL 3882	181–182	152	153
A83094 A (deethylindanomycin)	[*117615-33-5*]	$C_{29}H_{39}NO_4$	*Streptomyces setonii*		154	154
AC7230	[*108266-12-2*]	$C_{28}H_{34}N_2O_7$	*Dactylosporangium* sp. AC7230	>300[d]	155	155
cafamycin	[*112303-17-0*]	$C_{30}H_{41}NO_4$	*Streptomyces* sp.	152–153	156	156
CP61405	[*99623-84-4*]	$C_{26}H_{30}N_2O_7$	*Streptomyces routienii* ATCC 39446	334–335	157	157
homoindanomycin	[*129258-44-2*]	$C_{32}H_{45}NO_4$	*Streptomyces galbus* NCAIM (P) 001036		158	158
indanomycin (X14547 A)	[*66513-28-8*]	$C_{31}H_{43}NO_4$	*Streptomyces antibioticus* NRRL 8167	138–141	159	160
X14885 A	[*83917-57-1*]	$C_{27}H_{32}N_2O_4$	*Streptomyces* sp. X 14885 NRRL 12350	264–266[d]	161	161
Others						
BL580 zeta	[*69522-23-2*][d]		*Streptomyces hygroscopicus* NRRL 1108		60	*e*
UK44579	[*83589-30-4*][d]		*Actinomadura cremea* ATCC 31676		46	*e*

[a]Alternative names are given in parenthesis.
[b]References corresponding to isolations.
[c]References corresponding to structure determinations.
[d]Value given is for the sodium salt.
[e]Structure not yet determined.
[f]Value given is for the calcium salt.

subset is illustrated by two closely related compounds, semduramicin (**4**, R = H, R′ = sug 1) and maduramicin alpha (**4**, R = CH_3, R′ = sug 2). Maduramicin is marketed as an anticoccidial and semduramicin is under development for the same application.

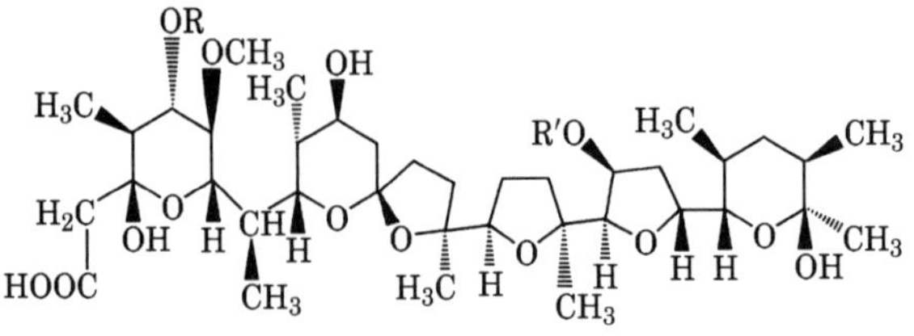

(**4**)

sug 1 = 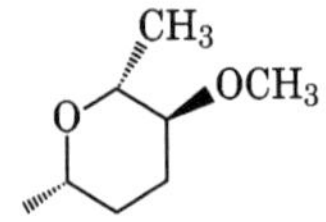 sug 2 = 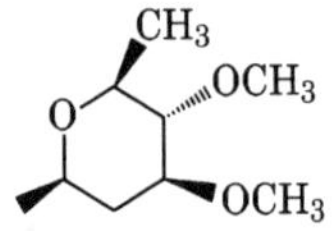

(**5**) (**6**)

The group containing two spiroketals is also characterized by an alpha–beta unsaturated ketone. Representative structures are illustrated by CP53607 (**7**, R = CH_3, R′ = OH, R″ = R‴ = R⁗ = H), which has no sugar, dianemycin (**7**, R = CH_3, R′ = OH, R″ = R⁗ = H, R‴ = sug), which has one sugar, and A130 B (**7**, R = R‴ = H, R′ = R⁗ = sug, R″ = CH_3), which has two sugars.

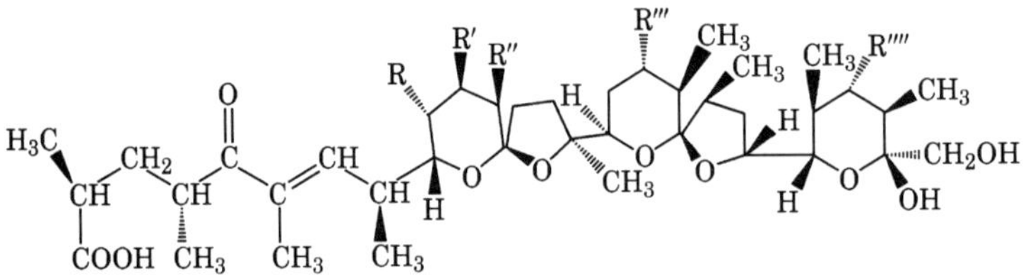

(**7**)

sug = 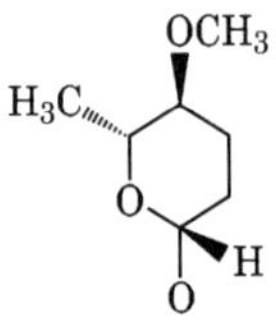

(**8**)

Two commercially important polyethers, narasin (**9**, R = CH_3) and salinomycin (**9**, R = H) have an unusual trispiroketal linkage.

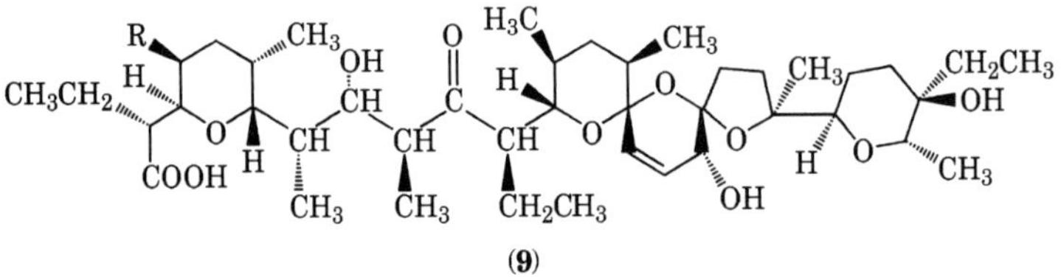

(**9**)

Ionomycin (**10**), a 32 C, 0-spiroketal, has a profound specificity for Ca^{2+} and is an important biochemical tool. The beta-carbonyl moiety provides a second charged ligation point, enabling this compound to form 1:1 complexes with divalent cations.

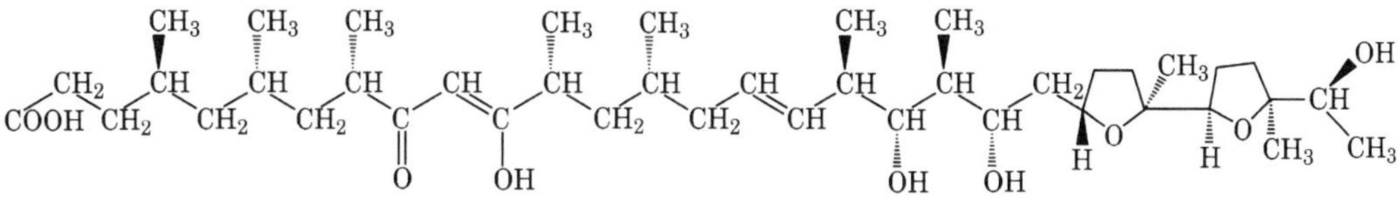

(**10**)

X206 (**11**, R = H) and alborixin (**11**, R = CH_3) are the only known representatives of the 37 C backbone subset.

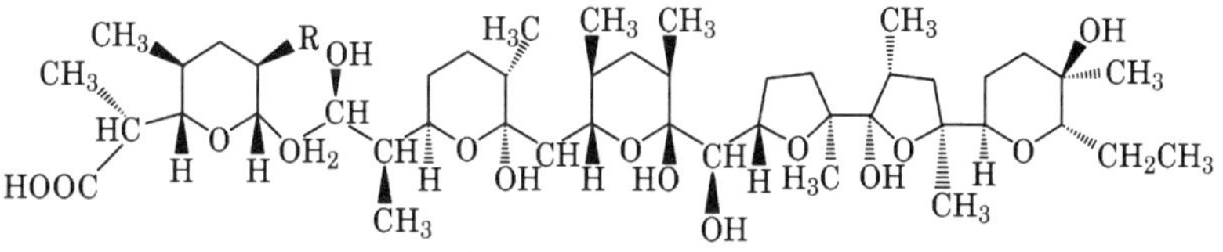

(**11**)

Tetronasin (**12**) is an example of the acyl tetronic acid class and A23187 (**13**, R = $NHCH_3$, R′ = R″ = CH_3) and related compounds, CP61405 (**13**, R = OH, R′ = R″ = H), X14885 A (**13**, R = OH, R′ = CH_3, R″ = H), and AC7230 (**13**, R = OH, R′ = R″ = CH_3) are examples of the pyrrole ether group.

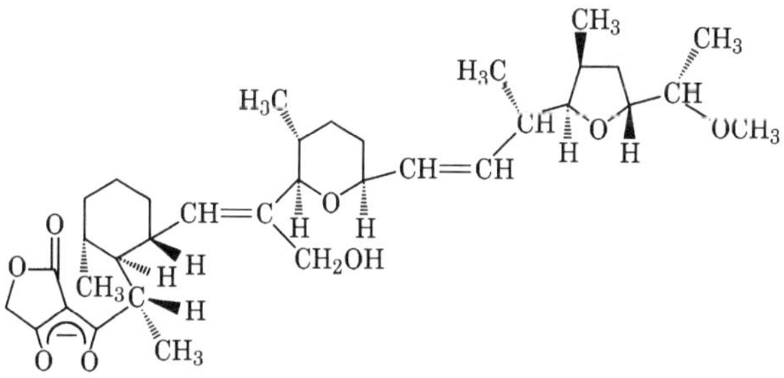

(**12**)

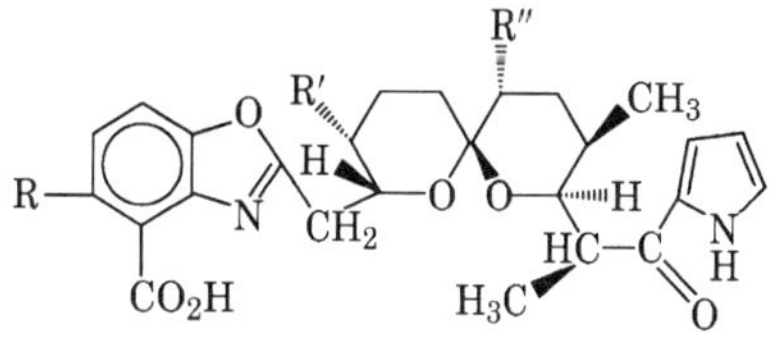

(**13**)

Purification and Production

Polyethers are usually found in both the filtrate and the mycelial fraction, but in high yielding fermentations they are mostly in the mycelium because of their low water-solubility (162). The high lipophilicity of both the free acid and the salt forms of the polyether antibiotics lends these compounds to efficient organic solvent extraction and chromatography (qv) on adsorbents such as silica gel and alumina. Many of the production procedures utilize the separation of the mycelium followed by extraction using solvents such as methanol or acetone. A number of the polyethers can be readily crystallized, either as the free acid or as the sodium or potassium salt, after only minimal purification.

Polyethers such as monensin, lasalocid, salinomycin, and narasin are sold in many countries in crystalline or highly purified forms for incorporation into feeds or sustained-release bolus devices (see CONTROLLED-RELEASE TECHNOLOGY). There are also mycelial or biomass products, especially in the United States. The mycelial products are generally prepared by separation of the mycelium and then drying by azeotropic evaporation, fluid-bed driers, continuous tray driers, flash driers, and other types of commercial driers (163). In countries allowing biomass products, crystalline polyethers may be added to increase the potency of the product.

Biological Activities

The polyether antibiotics exhibit a broad range of biological, antibacterial, antifungal, antiviral, anticoccidial, antiparasitic, and insecticidal activities. They improve feed efficiency and growth performance in ruminant and monogastric animals. Only the anticoccidial activity in poultry and cattle, and the effect on feed efficiency in ruminants such as cattle and sheep are of commercial interest.

Table 2 gives antimicrobial data for representative polyethers. In general, the polyethers are effective against gram-positive and anaerobic bacteria, but ineffective against gram-negatives (164). The antibacterial properties play an important, though poorly understood, role in the improvement of feed efficiency in ruminants (165). High toxicity has limited the use of these compounds as antibacterial and antifungal agents. The toxicity varies somewhat from species to species, and the commercial polyethers are well tolerated in poultry and cattle at the use level. Horses are subject to lethal toxicity at relatively low (1–2 mg/kg) dosages (166).

Treponema hyodysenteriae, a causative agent of swine dysentery, is sensitive to polyether antibiotics at low concentrations *in vitro*. In pigs, lasalocid was effective in controlling dysentery at levels of 0.005–0.05% in feeds (167). Several species of *Mycoplasma* are inhibited *in vitro* at a MIC range of 2.0–25 μ/mL of polyethers including narasin, carriomycin, and K41 (164).

Antiviral Activity. Tissue culture studies have demonstrated activity against a wide range of viruses including the veterinary pathogens, transmissible gastroenteritis virus, infectious canine heptatitis, Newcastle disease virus, infectious bovine rhinotracheitis, and pseudorabies (168). Monensin, narasin, septamycin, and nigericin were effective in treating an infection of transmissible gastroenteritis virus in piglets at doses of 0.1–100 mg/kg (168). The mechanism of inhibition has not been characterized, but it is probably related to the ionophoretic properties of these antibiotics. Monensin has been shown to inhibit the intracellular transport of viral membrane proteins of cells infected with Semliki Forest virus (169). The formation of syncytia, normally observed when T-lymphoblastoid cell line (CEM) cells are cocultivated with human immunodeficiency virus (HIV-1)-infected T-cell leukemia cell line (MOLT-3) cells, was significantly inhibited in the presence of monensin (170). This observation suggests that

Table 2. ***In Vitro*** **Antimicrobial Activity of the Polyether Antibiotics**

Polyether antibiotic	Minimum inhibitory concentration, μg/mL[a]								
	Staph. aureus ATCC No. 6538P	*Sarcina lutea* 9341	*Bacillus sp. E* 27859	*Bacillus subtilis* 558[b]	*Bacillus megaterium* 8011	*Bacillus sp. TA* 27860	*Mycobacterium phlei* 355	*Streptomyces cellulosae* 3313	*Paecilomyces varioti* 26820
monensin	3.1	12.5	0.4	1.6	12.5	1.6	12.5	6.3	[c]
nigericin	0.2	0.1	0.004	0.1	0.1	0.02	0.4	0.4	0.8
grisorixin	0.4	0.2	0.1	0.2	0.1	0.4	0.2	0.2	3.1
salinomycin	3.1	3.1	0.2	0.8	0.2	0.8	6.3	3.1	3.1
narasin	0.8	1.6	0.2	0.9	0.4	0.4	3.1	3.1	
lonomycin	1.6	12.5	1.6	3.1	0.8	1.6	12.5	6.3	50
X206	0.2	0.8	0.02	0.4	0.2	0.2	0.2	0.8	0.8
dianemycin	1.6	3.1	0.2	3.1	3.1	1.6	6.3	6.3	6.3
lenoremycin	0.2	0.2	0.02	0.4	0.1	0.4	0.2	1.6	0.8
septamycin	0.8	1.6	0.006	1.6	1.6	0.2	1.6	3.1	6.3
A204A	3.1	12.5	0.8	6.3	12.5	1.6	12.5	12.5	25
CP38952	12.5	6.3	1.6	6.3	3.1	3.1	3.1	12.5	12.5
lasalocid	1.6	3.1	0.2	1.6	3.1	1.6	12.5	6.3	
iso-lasalocid	12.6	6.3	1.6	1.6	1.6	3.1	6.3	25	12.5
lysocellin	0.8	0.4	0.03	0.2	0.4	0.4	3.1	1.6	1.6
A23187	0.2	0.04	0.2	0.1	0.8	0.04	1.6	6.3	
ionomycin		12.5	12.5	25	6.3				

[a]Lowest twofold dilution giving zone of inhibition in agar-well diffusion assay.
[b]NRRL collection number.
[c]Indicates no activity up to 50 μg/mL.

the viral glycoproteins in the treated cells were not transported to the cell surface from the Golgi membrane.

In Bulgaria, a preparation containing nigericin is used topically as an antiherpes drug (171).

Anticoccidial Activity. The 1968 report that claimed monensin has activity against *Eimeria* sp., particularly *E. tenella, E. maxima,* and *E. acervulina,* greatly altered the prevention and control of coccidiosis in poultry (172). It is estimated that the polyether ionophores presently constitute more than 80% of the total worldwide usage of anticoccidials (173). Lasalocid and monensin have been approved for use in control of coccidiosis in cattle.

To combat more tolerant, ie, resistant coccidia, particularly *E. acerulina* and *E. maxima,* a 1:1 combination of narasin and nicarbazin, eg, 0.45 kg premix containing 36 g of each compound, has been marketed as Maxiban by Elanco. These two compounds work synergistically against less susceptible field strains.

The ionophoretic properties of polyether antibiotics are important in their mode of action against coccidia (174). The primary coccidicidal effect of monensin involves an influx of sodium ions which results in swelling and vacuolization of the parasite because of osmotic pressure effects. A secondary effect is the stimulation of glycolysis on the part of the parasite resulting in a depletion of carbohydrate stores and eventual death. *Toxoplasma* sp., coccidia related to *Eimeria* sp., are infective in humans by two routes: fecal–oral ingestion of oocysts from the definitive host, the cat; and ingestion of tissue cysts from the intermediate host through eating undercooked meat such as lamb or pork. Monensin has been shown to be effective against toxoplasmosis in cats (175) and sheep (176). No human studies have been reported.

Enhancement of Feed Efficiency in Ruminants. Polyether antibiotics have been shown to improve feed efficiency in ruminant animals such as cattle, sheep, and goats and also to enhance weight gain. This enhancement in growth performance has been shown to correlate with increased production of propionic acid in the rumen. A decrease in the production of acetic acid, butyric acid, and methane often accompanies the increased propionic acid production. These volatile fatty acids (VFA) are produced in the rumen by the degradation of carbohydrates by microorganisms. The polyethers apparently selectively inhibit certain of the microorganisms to achieve greater propionate production. Propionate is thought to be more effectively utilized in energy metabolism in the host animal than acetate or butyrate. The effect on production efficiency and mode of action has been reviewed (177,178). (See also GROWTH REGULATORS).

Because feed comprises over 80% of the cost of producing and fattening cattle, the maximum utilization of ever increasingly expensive rations is of upmost importance (179). Monensin under the trade name of Rumensin (Elanco Products) was introduced in 1976 at a recommended level of 30 ppm in cattle feed. Lasalocid having the trade name Bovatec (Hoffmann-LaRoche, Inc.) was marketed some years later.

Salinomycin (180) and narasin (128) have been reported to be effective in improving feed efficiency but neither has been marketed for this use. Laidlomycin propionate (Syntex, Inc.) and tetronasin (Coopers Animal Health, Inc.) have been under investigation in cattle and sheep (181,182).

The polyethers have been reported to increase growth performance in monogastric animals such as horses, swine, and rabbits. No commercial use has been approved for swine diets. The sensitivity of horses to the toxic effects of these antibiotics precludes consideration of their use.

Biosynthesis

Bacteria belonging to the order Actinomycetales are the organisms reported to produce all of the polyethers. Most are secondary metabolites of *Streptomyces* sp. with the species *hygroscopicus* and *albus* accounting for about one-third of the antibiotics. Other genera represented are *Streptoverticillium, Dactylosporangium, Actinomadura, Nocardia,* and *Nocardiopsis.* The taxonomy of these producing organisms has been reviewed (183).

As is the case in other antibiotic families, the polyethers are often produced as complexes of closely related compounds. For example, lasalocid A is the principal component in a six-membered complex from *Streptomyces lasaliensis;* NRRL 3382 (20) and leuseramycin are co-produced also with four minor factors from *Streptomyces hygroscopicus* ATCC 31590 (119,121). The use of isotopically labeled substrate and subsequent analysis of the labeling patterns have established the polyketide route as the principal pathway for biosynthesis (184,185). The shortest polyether, lysocellin, is constructed from 11 short-chain fatty acid units, whereas the longest, X-206 and alborixin, are assembled from 18 subunits. Most, however, are assembled from 15 subunits giving the C-30 backbone. After cyclization of the ether rings the molecules can be modified by the addition of methyl groups or sugar moieties.

The Cane-Celmer-Westley unified stereochemical model is an elegant proposal relating the stereochemistry of thirty different polyethers (186). According to this hypothesis, the cyclic ether groups arise by a cascade of cyclization steps on a polyepoxide intermediate, which arises from a polyene precursor having double bonds in the (*E*)-configuration. However, there has been little experimental evidence to support this scheme. In studies on the biosynthesis of monensin, putative triene intermediates were synthesized but no incorporation of these compounds into monensin A or B were found (187). Other incorporation experiments using (*E*)- and (*Z*)-nerolidol suggest that the enzymes from *S. cinnamonensis* preferentially oxidize bonds of (*Z*)-configuration (188). As an extension of the work on oxidative cyclization in the biosynthesis of clavulanic acid (see ANTIBIOTICS, β-LACTAMS, β-LACTAMASE INHIBITORS), the possibility of an all (*Z*)-triene precursor for monensin has been proposed (189).

Structure Determination and Synthesis

Because of the complexity of the polyether antibiotics little progress has been made in structure determination by the chemical degradation route. X-ray methods were the techniques most successfully applied for the early structure elucidations. Monensin, X206, lasalocid, lysocellin, and salinomycin were included in nineteen distinct polyether x-ray analyses reported in 1983 (190). Use of mass spectrometry (191), and ^{1}H (192) and ^{13}C nmr (141) are also reviewed. More recently, innovative developments in these latter techniques have resulted in increased applications for structure determinations. For example, heteronuclear multiple bond connectivity (hmbc) and homonuclear Hartmann-Hahn spectroscopy were used to solve the structure of portimicin (**14**) (193). Fast atom bombardment mass spectrometry was used in solving the structures of maduramicin alpha and co-factors (58).

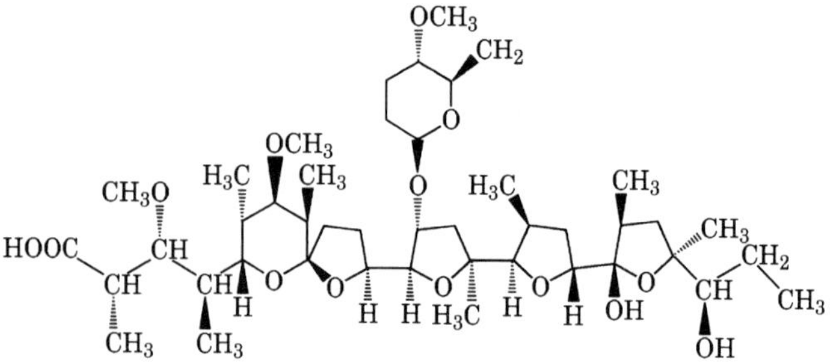

(**14**)

Total synthesis of polyether antibiotics presents a challenging task, requiring a high degree of stereo-, regio-, and chemoselectivity for each step. This subject was reviewed in 1983 (194), and since then many more compounds have been conquered. The asymmetric synthesis of a ferensimycin synthon (195) and total convergent syntheses of ionomycin (196) and X206 (197) have been reported. A method for the assembly of polypropionates with excellent diastereoselectivity has also been reported (198). Another total synthesis of ionomycin utilized a novel sulfur-assisted organocuprate displacement of a secondary tosylate with complete inversion of configuration (199). Other significant advances include the syntheses of monensin and lasalocid (200). A synthesis of zincophorin utilizes the hetero-Diels-Alder reaction in synthesis of both the polyol and pyran portions of the molecule (201,202). The stereospecific cyclization of a beta-hydroxy ketone into the spiroketal of A23187 under basic conditions has been carried out (203) and a total convergent synthesis of indanomycin uses the readily available levoglucosan as a starting material for the tetrahydropyranylacetic acid moiety (204).

Economic Aspects

The worldwide usage of polyether antibiotics for controlling coccidiosis is approximately \$190 million compared to a total market of \$210–220 million. Monensin and salinomycin represent about 65–70% of this market; lasalocid, narasin, and maduramicin make up the remainder. Other compounds for coccidiosis control include nicarbazine, halofunginone, amprolium, and robenidine. Worldwide usage is in excess of 3 million kg of product.

The total world market for the use of ionophores for feed efficiency improvement in ruminants is approximately \$80–\$90 million. The United States is the largest market. Lasalocid and monensin are the only members of this class cleared for use. Outside the United States, salinomycin is used in limited quantities. Worldwide usage is about 1.5 million kg.

The primary manufacturers are Hoffmann-LaRoche, Inc., Pfizer, Inc., Agri-Bio Corp. (a division of A. H. Robins), Elanco Products Co. (a division of Eli Lilly and Co.), Hoechst-Roussel Agri Vet Co., American Cyanamid Co., and Kaken Pharmaceutical Co.

Table 3 lists the polyether antiobitics used as poultry anticoccidial drugs in the United States. Recently, lasalocid and monensin have been approved for use in bovine coccidiosis at levels in feed of 11–33 g/t.

Table 3. Poultry Anticoccidial Agents used in the United States

Generic name	Trade name	Company	Year of first commercial use	Registered use level, g/t
maduramicin	Cygro	American Cyanamid Co.	1985	5.0–6.0
monensin	Coban	Elanco Products (division of Eli Lilly and Co.)	1971	99–120
lasalocid	Avatec	Hoffmann-LaRoche, Inc.	1977	75–125
narasin	Monteban	Elanco Products (division of Eli Lilly and Co.)	1981	60–79
salinomycin	Bio-Cox	Agri-Bio (division of A. H. Robins)	1982	44–66

BIBLIOGRAPHY

"Antibiotics, Polyethers" in *ECT* 3rd ed., Vol. 3, pp. 47–64, by J. W. Westley, Hoffmann-La Roche, Inc.

1. A. Agtarap, J. W. Chamberlin, M. Pinkerton, and L. Steinrauf, *J. Am. Chem. Soc.* **89,** 5737–5739 (1967).
2. H. A. Lardy, D. Johnson, and W. C. McMurray, *Arch. Biochem. Biophys.* **78,** 587–597 (1958).
3. B. C. Pressman, E. J. Harris, W. S. Jagger, and J. M. Johnson, *Proc. Natl. Acad. Sci. U.S.A.* **58,** 1949–1956 (1967).
4. D. Ammann, *Ion-Selective Microelectrodes: Principles, Design and Application,* Springer-Verlag, Berlin, 1986.
5. A. H. J. Maas, F. B. T. J. Boink, N. E. L. Saris, R. Sprokholt, and P. D. Wimberley, eds., *Methodology and Clinical Applications of Ion Selective Electrodes,* Vol. 7, IFCC Technical Secretariat, Copenhagen, 1986.
6. K. Tohda and co-workers, *Anal. Chem.* **62,** 936–942 (1990).
7. J. W. Westley, ed., *Polyether Antibiotics,* Vol. 1, Marcel Dekker, Inc., New York, 1982, pp. 1–20.
8. J. Berdy, *CRC Handbook of Antibiotic Compounds,* Vol. 5, CRC Press, Inc., Boca Raton, Fla., 1981, pp. 461–462.
9. J. W. Westley, *J. Antibiot.* **29,** 584–586 (1976).
10. Y. Kusakabe and co-workers, *J. Antibiot.* **35,** 1119–1129 (1982).
11. E. Ebata, H. Kasahara, K. Sekine, and Y. Inoue, *J. Antibiot.* **28,** 118–121 (1975).
12. B. L. T. Prosser and N. J. Palleroni, *Int. J. Syst. Bacteriol.* **26,** 319–322 (1976).

13. N. Otake, M. Koenuma, H. Kinashi, S. Sato, and Y. Saito, *J. Chem. Soc., Chem. Commun.*, 92–93 (1975).
14. J. W. Westley and co-workers, *J. Antibiot.* **39,** 1704–1711 (1986).
15. U.S. Pat. 4,537,956 (Aug. 27, 1985), C.-M. Liu and J. W. Westley (to Hoffmann-La Roche, Inc.).
16. M. Imoto and co-workers, *J. Nat. Prod.* **53,** 825–829 (1990).
17. U.S. Pat. 3,944,573 (Mar. 16, 1976), J. W. Westley (to Hoffmann-La Roche, Inc.).
18. J. Berger, A. I. Rachlin, W. E. Scott, L. H. Sternbach, and M. W. Goldberg, *J. Am. Chem. Soc.* **73,** 5295–5298 (1951).
19. S. M. Johnson, J. Herrin, S.-J. Liu, and I. C. Paul, *J. Am. Chem. Soc.* **92,** 4428–4434 (1970).
20. J. W. Westley and co-workers, *J. Antibiot.* **27,** 744–753 (1974).
21. W. P. Cullen and co-workers, *J. Antibiot.* **40,** 1490–1495 (1987).
22. J. Bordner, P. C. Watts, and E. B. Whipple, in Ref. 21, pp. 1496–1505.
23. J. P. Dirlam and co-workers, *J. Antibiot.* **42,** 1213–1220 (1989).
24. H. A. Brooks, D. Gardner, and J. P. Poyser, *J. Antibiot.* **37,** 1501–1504 (1984).
25. U. Grafe and co-workers in Ref. 24, pp. 836–846.
26. L. Radics, *J. Chem. Soc., Chem. Commun.*, 599–601 (1984).
27. U.S. Pat. 4,148,882 (Apr. 10, 1979), W. D. Celmer and co-workers (to Pfizer, Inc.).
28. J. Tone and co-workers, *Curr. Chemother. Infect. Dis., Proc. Int. Congr. Chemother., 11th* **1,** 469–470 (1980).
29. U.S. Pat. 4,920,050 (Apr. 24, 1990), W. P. Cullen, J. R. Hauske, G. J. Kostek, H. Maeda, and J. Tone (to Pfizer, Inc.).
30. J. P. Dirlam and co-workers, *J. Ind. Microb.* **6,** 135–142 (1990).
31. Y. Fukuzawa and co-workers, *Sankyo Kenyusho Nempo* **37,** 57–63 (1985).
32. U. Grafe, R. Schlegel, C. Stengel, W. Ihn, and L. Radics, *J. Basic Microbiol.* **29,** 149–155 (1989).
33. F. Kitame and co-workers in Ref. 20, pp. 884–888.
34. Jpn. Pat. Appl. Kokai 76-9790 (Jan. 26, 1976), A. Tamura, R. Furuta, and H. Kotani (to Dainippon Pharmaceutical Co., Ltd.).
35. F. Kitame and N. Ishida in Ref. 9, pp. 759–761.
36. M. E. Haney, Jr. and M. M. Hoehn, *Antimicrob. Agents Chemother.* **1967,** 349–352 (1968).
37. M. Gorman, J. W. Chamberlin, and R. L. Hamill, *Antimicrob. Agents Chemother.* **1967,** 363–368 (1968).
38. Eur. Pat. Appl. 293,787 (Dec. 7, 1988), Y. Kusakabe, H. Suzuki, and H. Kudo (to Kaken Pharmaceutical Co., Ltd.).
39. R. L. Hamill and co-workers, *29th Interscience Conference on Antimicrobial Agents and Chemotherapy* Houston, Tex., 1989, Abstr. 423.
40. J. P. Dirlam and co-workers, *J. Antibiot.* **43,** 668–679 (1990).
41. G. Nakamura, K. Kobayashi, T. Sakurai, and K. Isono, *J. Antibiot.* **34,** 1513–1514 (1981).
42. Y. Takahashi and co-workers in Ref. 40, pp. 441–443.
43. Y. Kusakabe, N. Takahashi, Y. Iwagawa, and A. Seino, in Ref. 21, pp. 237–238.
44. U.S. Pat. 4,582,822 (Apr. 15, 1986), R. L. Hamill and R. C. Yao (to Eli Lilly and Co.).
45. H. Seto and co-workers, *Tetrahedron Lett.* **28,** 3357–3360 (1987).
46. Brit. Pat. Appl. 2,090,243 (July 7, 1982), A. B. Penrose, J. C. Ruddock, R. Shibakawa, and J. Tone (to Pfizer, Inc.).
47. Y. Kusakabe, S. Mitsuoka, Y. Omuro, and A. Seino, *J. Antibiot.* **33,** 1437–1442 (1980).
48. H. Seto and co-workers, *J. Antibiot.* **32,** 244–246 (1979).
49. U.S. Pat. 3,705,238 (Dec. 5, 1972), R. L. Hamill and M. M. Hoehn (to Eli Lilly and Co.).

50. N. D. Jones, M. O. Chaney, J. W. Chamberlin, R. L. Hamill, and S. Chen, *J. Am. Chem. Soc.* **95,** 3399–3400 (1973).
51. U.S. Pat. 3,839,558 (Dec. 23, 1971), R. L. Hamill and M. M. Hoehn (to Eli Lilly and Co.).
52. D. E. Dorman and co-workers in Ref. 47, pp. 252–255.
53. U.S. Pat. 4,824,863 (Apr. 4, 1989), R. L. Hamill, R. Yao, and W. N. Nakatsukasa (to Eli Lilly and Co.).
54. Jpn. Pat. Appl. Kokai 60-130,394 (July 11, 1985), T. Sasaki (to Meiji Seika Kaisha, Ltd.).
55. Meiji Seika Kaisha, Mar. 31, 1988, personal communication.
56. L. David, H. L. Ayala, and J.-C. Tabet, *J. Antibiot.* **38,** 1655–1663 (1985).
57. U.S. Pat. 3,812,249 (May 21, 1974), J. H. E. J. Martin and S. Kantor (to American Cyanamid Co.).
58. M. M. Siegel, W. J. McGahren, K. B. Tomer, and T. T. Chang, *Biomed. Environ. Mass Spectrom.* **14,** 29–38 (1987).
59. U.S. Pat. 4,138,481 (Feb. 6, 1979), J. H. E. J. Martin and M. R. Hertz (to American Cyanamid Co.).
60. U.S. Pat. 4,132,779 (Jan. 2, 1979), M. R. Hertz and J. H. Martin (to American Cyanamid Co.).
61. A. Imada and co-workers, *J. Antibiot.* **31,** 7–14 (1978).
62. H. Nakayama, N. Otake, H. Miyamae, S. Sato, and Y. Saito, *J. Chem. Soc., Perkin Trans. 2,* 293–295 (1979).
63. U.S. Pat. 4,746,650 (May 24, 1988), W. P. Cullen, H. Maeda, J. C. Ruddock, and J. Tone (to Pfizer, Ltd.).
64. T. Fehr and co-workers in Ref. 23, pp. 897–902.
65. R. Barrada, G. Dauphin, and L. David, *J. Org. Chem.* **52,** 2388–2391 (1987).
66. Jpn. Pat. Appl. 76-91396 (Aug. 10, 1976), A. Seino, C. Nagatsu, H. Akutsu, and Y. Kusabe (to Kaken Chemical Co., Ltd.).
67. U.S. Pat. 4,129,578 (Dec. 12, 1978), W. D. Celmer and co-workers (to Pfizer, Inc.).
68. M. J. O. Anteunis and N. A. Rodius, *Bull. Soc. Chim. Belg.* **87,** 753–764 (1978).
69. P. Gachon, A. Kergomard, T. Staron, and C. Esteve in Ref. 11, pp. 345–350.
70. P. Gachon, A. Kergomard, H. Veschambre, C. Esteve, and T. Staron, *J. Chem. Soc., Chem. Commun.*, 1421–1423 (1970).
71. U.S. Pat. 4,628,046 (Dec. 9, 1986), D. P. Labeda, J. J. Goodman, and J. H. Martin (to American Cyanamid Co.).
72. S. Omura, S. Michinori, S. Machida, and J. Sawada in Ref. 9, pp. 15–20.
73. M. Oshoma and co-workers in Ref. 9, pp. 354–365.
74. L. Ninet and co-workers, *Experiencia* **32,** 319–321 (1976).
75. N. Otake and M. Koenuma, *Tetrahedron Lett.,* 4147–4150 (1975).
76. K. Yamazaki, K. Abe, and M. Sano in Ref. 9, pp. 9–12.
77. C. Riche and C. Pascard-Billy, *J. Chem. Soc., Chem. Commun.,* 951–952 (1975).
78. T. Mizutani and co-workers in Ref. 47, pp. 1224–1230.
79. H. Seto and co-workers in Ref. 61, pp. 929–932.
80. Belg. Pat. 894,762 (Apr. 21, 1983), D. P. Labeda and co-workers (to American Cyanamid Co.).
81. C.-M. Liu and co-workers, *J. Antibiot.* **36,** 343–350 (1983).
82. Eur. Pat. Appl. 317,231 (May 24, 1989), J. P. Dirlam, J. Tone, W. P. Cullen, and H. Maeda (to Pfizer, Inc.).
83. T. Fehr, H. D. King, and M. Kuhn, *J. Antibiot.* **30,** 903–907 (1977).
84. T. Fehr and co-workers in Ref. 48, pp. 535–536.
85. R. L. Harned, P. H. Hidy, C. J. Corum, and K. L. Jones, *Antibiot. Chemother.* **1,** 591–596 (1951).

86. J. Shoji and co-workers, *J. Antibiot.* **21,** 402–409 (1968).
87. Jpn. Pat. 70-26719 (Sept. 2, 1970), T. Misato and co-workers (to Kaken Chemical Co., Ltd.).
88. L. K. Steinrauf, M. Pinkerton, and J. W. Chamberlin, *Biochem. Biophys. Res. Commun.* **33,** 29–31 (1968).
89. T. Kubota, S. Matsutani, M. Shiro, and H. Koyama, *Chem. Commun.,* 1541–1543 (1968).
90. N. Tsuji and co-workers in Ref. 9, pp. 10–14.
91. U.S. Pat. 4,133,876 (Jan. 9, 1979), R. L. Hamill and M. M. Hoehn (to Eli Lilly and Co.).
92. M. Shiro, H. Nakai, K. Nagashima, and N. Tsuji, *J. Chem. Soc., Chem. Commun.,* 682–683 (1978).
93. J. L. Occolowitz, D. E. Dorman, and R. L. Hamill, *J. Chem. Soc., Chem. Commun.,* 683–684 (1978).
94. Jpn. Pat. Appl. Kokai 80-38352 (Mar. 17, 1980), N. Tsuji (to Shionogi and Co., Ltd.).
95. N. Tsuji, N. Nagashima, Y. Terui, and K. Tori in Ref. 48, pp. 169–172.
96. U.S. Pat. 4,912,130 (Mar. 27, 1990), A. C. Goudie and N. D. A. Walshe (to Pfizer, Inc.).
97. C. Keller-Juslen, H. D. King, Z. L. Kis, and A. von Wartburg, *J. Antibiot.* **27,** 854–859 (1975).
98. M. P. Kunstmann, 1976, personal communication.
99. T. J. Petcher and H.-P. Weber, *J. Chem. Soc., Chem. Commun.,* 697–698 (1974).
100. Jpn. Pat. Appl. Kokai 85-105679 (June 11, 1985), K. Mizogami, M. Yamagishi, S. Omura, N. Otake, and H. Seto (to Taisho Pharmaceutical Co., Ltd.).
101. R. S. Wehbie, C. Runsheng, and H. A. Lardy, *J. Antibiot.* **50,** 887–893 (1987).
102. N. Tsuji, Y. Terui, K. Nagashima, K. Tori, and L. F. Johnson in Ref. 47, pp. 94–97.
103. U.S. Pat. 4,361,649 (Nov. 30, 1982), W. D. Celmer, W. P. Cullen, R. Shibakawa, and J. Tone (to Pfizer, Inc.).
104. J. W. Westley and co-workers in Ref. 24, pp. 813–815.
105. PCT Intl. Appl. WO 88-10309 (Dec. 29, 1988), J. P. Dirlam, W. P. Cullen, H. Maeda, and J. Tone (to Pfizer, Inc.).
106. J. P. Dirlam, L. A. Presseau-Linabury, and D. A. Koss in Ref. 43, pp. 727–730.
107. J. R. Hauske and G. Kostek, *J. Org. Chem.* **54,** 3500–3504 (1989).
108. R. L. Hamill, M. M. Hoehn, G. E. Pittenger, J. Chamberlin, and M. Gorman, *J. Antibiot.* **22,** 161–164 (1969).
109. E. W. Czerwinski and L. K. Steinrauf, *Biochem. Biophys. Res. Commun.* **45,** 1284–1287 (1971).
110. J. R. Oscarson and co-workers in Ref. 23, pp. 37–48.
111. U.S. Pat. 4,707,493 (Nov. 17, 1987), W. D. Celmer, W. P. Cullen, H. Maeda, J. C. Ruddock, and J. Tone (to Pfizer, Inc.).
112. C.-M. Liu and co-workers in Ref. 9, pp. 21–28.
113. T. Kubota, H. Hinoh, M. Mayama, K. Motokawa, and Y. Yasuda in Ref. 11, pp. 931–934.
114. J. F. Blount, R. H. Evans, Jr., C.-M. Liu, T. Hermann, and J. W. Westley, *J. Chem. Soc., Chem. Commun.,* 853–855 (1975).
115. H. Koyama and K. Utsumi-Oda, *J. Chem. Soc, Perkin Trans. 2,* 1531–1536 (1977).
116. T. Mizutani and co-workers in Ref. 47, pp. 137–143.
117. H. Nakayama and co-workers in Ref. 56, pp. 1433–1436.
118. C. Keller-Juslen, H. D. King, M. Kuhn, H.-R. Loosli, and A. von Wartburg in Ref. 61, pp. 820–823.
119. U.S. Pat. 4,359,583 (Nov. 16, 1982), T. Mizutani and co-workers (to Taisho Pharmaceutical Co., Ltd.).
120. T. Mizutani and co-workers in Ref. 41, pp. 1369–1373.
121. Jpn. Pat. Appl. Kokai 83-121290 (July 19, 1983), T. Mizutani and co-workers (to Taisho Pharmaceutical Co., Ltd.).

122. U.S. Pat. 4,510,317 (Apr. 9, 1985), C.-M. Liu and J. W. Westley (to Hoffmann-LaRoche, Inc.).
123. U.S. Pat. 4,150,152 (Apr. 17, 1979), W. D. Celmer and co-workers (to Pfizer, Inc.).
124. U.S. Pat. 4,141,907 (Feb. 27, 1979), W. M. Nakatsukasa, G. G. Marconi, N. Neuss, and R. L. Hamill (to Eli Lilly Co.).
125. J. W. Westley, J. F. Blount, R. H. Evans, Jr., and C.-M. Liu in Ref. 83, pp. 610–612.
126. L. D. Boeck and co-workers, *Devel. Indus. Microbiol.* **18,** 471–485 (1977).
127. H. Seto, T. Yahagi, Y. Miyazaki, and N. Otake in Ref. 83, pp. 530–532.
128. U.S. Pat. 4,035,481 (July 12, 1977), D. H. Berg, R. L. Hamill, and M. M. Hoehn (to Eli Lilly and Co.).
129. Y. Miyazaki and co-workers in Ref. 20, pp. 814–821.
130. H. Kinashi, N. Otake, H. Yonehara, S. Sato, and Y. Saito, *Tetrahedron Lett.,* 4955–4958 (1973).
131. Jpn. Pat. Appl. Kokai 86-247398 (Nov. 4, 1986), Y. Miyazaki and co-workers (to Kaken Pharmaceutical Co., Ltd.).
132. Jpn. Pat. Appl. Kokai 82-85390 (Nov. 17, 1980), T. Nakamura and A. Shibata (to Kaken Chemical Co., Ltd.).
133. K. Dobashi and co-workers in Ref. 23, pp. 629–632.
134. U.S. Pat. 4,195,079 (Mar. 25, 1980), W. D. Celmer and co-workers (to Pfizer, Inc.).
135. J. Tone and co-workers, *21st Interscience Conference on Antimicrobial Agents and Chemotherapy,* Chicago, Ill., 1981, Abstr. 186.
136. C.-M. Liu and co-workers in Ref. 41, pp. 133–138.
137. J. W. Westley and co-workers in Ref. 41, pp. 139–147.
138. W. C. Liu and co-workers in Ref. 61, pp. 815–819.
139. B. K. Toeplitz, A. I. Cohen, P. T. Funke, W. L. Parker, and J. Z. Gougoutas, *J. Am. Chem. Soc.* **101,** 3344–3353 (1979).
140. U.S. Pat. 4,081,532 (Mar. 28, 1978), W. D. Celmer and co-workers (to Pfizer, Inc.).
141. H. Seto and N. Otake in Ref. 7, Vol. 2, 1983, p. 356.
142. U.S. Pat. 4,751,317 (June 14, 1988), W. P. Cullen, J. R. Oscarson, J. Tone, and H. Maeda (to Pfizer, Inc.).
143. C. Delhomme, A. Kergomard, G. Kergomard, and T. Staron, *J. Antibiot.* **28,** 682–695 (1976).
144. U.S. Pat. 4,022,885 (May 10, 1977), W. D. Celmer, W. P. Cullen, C. E. Moppett, J. R. Oscarson, and J. B. Routien (to Pfizer, Inc.).
145. H. Seto and co-workers in Ref. 48, pp. 970–971.
146. J. F. Blount and J. W. Westley, *J. Chem. Soc., Chem. Commun.,* 533 (1975).
147. Eur. Pat. Appl. 305,086 (Mar. 1, 1989), R. L. Hamill and R. C. F. Yao (to Eli Lilly and Co.).
148. M. Hatsu and co-workers in Ref. 40, pp. 259–266.
149. U.S. Pat. 4,279,894 (July 21, 1981), D. H. Davies and G. L. F. Norris (to Imperial Chemical Industries, Ltd.).
150. D. H. Davies, E. W. Snape, P. J. Suter, T. J. King, and C. P. Falshaw, *J. Chem. Soc., Chem. Commun.,* 1073–1074 (1981).
151. C. Keller-Juslen and co-workers in Ref. 10, pp. 142–150.
152. U.S. Pat. 3,923,823 (Dec. 2, 1975), R. M. Gale, C. E. Higgens, and M. M. Hoehn (to Eli Lilly and Co.).
153. M. O. Chaney, P. V. Demarco, N. D. Jones, and J. L. Occolowitz, *J. Amer. Chem. Soc.* **96,** 1932–1933 (1974).
154. S. H. Larsen, L. D. Boeck, F. P. Mertz, J. W. Paschal, and J. L. Occolowitz, *J. Antibiot.* **41,** 1170–1177 (1988).
155. S. Yaginuma, M. Awata, N. Muto, K. Kinoshita, and K. Mizuno in Ref. 21, pp. 239–241.
156. N. V. Murenets and co-workers, *Antibiot. Med. Biotekhnol.* **32,** 811–814 (1987).

157. U.S. Pat. 4,547,523 (Oct. 15, 1985), W. D. Celmer, W. P. Cullen, H. Maeda, and J. Tone (to Pfizer, Inc.).
158. Hung. Pat. HU T049-909 (Nov. 28, 1989), P. Toth and co-workers (to Gyogyszerkutato Interzet.).
159. C.-M. Liu and co-workers in Ref. 48, pp. 95–99.
160. J. W. Westley and co-workers in Ref. 48, pp. 100–107.
161. J. W. Westley and co-workers in Ref. 81, pp. 1275–1278.
162. R. L. Hamill and L. W. Crandall in M. J. Weinstein and G. H. Wagman, ed., *Antibiotics, Isolation, Separation and Purification,* Elsevier Scientific Publishing Co., New York, 1978, pp. 479–520.
163. Brit. Pat. 1,296,584 (Nov. 15, 1972), R. A. Kennedy and S. L. McCormick, Jr. (to Eli Lilly and Co.).
164. C.-M. Liu, in Ref. 7, pp. 43–102.
165. J. A. Russell and H. J. Strobel, *Appl. Environ. Microbiol.* **55,** 1–6 (1989).
166. T. Matsuoka, *J. Am. Vet. Med. Assoc.* **169,** 1098–1100 (1976).
167. U.S. Pat. 3,947,586 (Mar. 30, 1976), R. E. Messersmith (to Hoffmann-LaRoche, Inc.).
168. U.S. Pat. 3,995,027 (Nov. 30, 1976), C. Gale and L. R. McDougald (to Eli Lilly and Co.).
169. G. Griffiths, P. Quinn, and G. Warren, *J. Cell Biol.* **96,** 835–850 (1983).
170. R. Pal, R. C. Gallo, and M. G. Sarngadharan, *Proc. Natl. Acad. Sci., U.S.A.* **85,** 9283–9286 (1988).
171. P. Andonov and co-workers, *MBI, Med. Biol. Inf.* 6–9 (1989).
172. R. F. Shumard and M. E. Callender in Ref. 36, pp. 369–377.
173. T. K. Jeffers in P. Yvore, ed., *Coccidia and Intestinal Coccidiomorphs,* INRA Publishers, Versailles, France 1989, pp. 295–308.
174. C. K. Smith, II and R. B. Galloway, *J. Parasitol.* **69,** 666–670 (1983).
175. J. K. Frenkel and D. D. Smith, *J. Parasitol.* **68,** 851–855 (1982).
176. D. Buxton, D. A. Blewett, A. J. Trees, C. McColgan, and J. Finlayson, *J. Comp. Pathol.* **98,** 225–236 (1988).
177. W. G. Bergen and D. B. Bates, *J. Anim. Sci.* **58,** 1465–1483 (1984).
178. J. B. Russell, *J. Anim. Sci.* **64,** 1519–1525 (1987).
179. M. D. Ruff in Ref. 7, pp. 303–332.
180. N. Otake, *Drugs Pharm. Sci.* **22,** 721–742 (1984).
181. U.S. Pat. 4,431,665 (Feb. 14, 1984), A. F. Kluge and R. D. Clark (to Syntex, Inc.).
182. S. J. Bartle, R. L. Preston, and J. H. Bailie, *J. Anim. Sci.* **66,** 1502–1507 (1988).
183. B. L. T. Prosser and N. J. Palleroni, in Ref. 7, pp. 21–43.
184. J. W. Westley, *J. Nat. Prod.* **49,** 35–74 (1986).
185. D. O'Hagan, *Nat. Prod. Reports* **6,** 205–219 (1989).
186. D. E. Cane, W. D. Celmer, and J. W. Westley, *J. Am. Chem. Soc.* **105,** 3594–3600 (1983).
187. D. S. Holmes, J. A. Sherringham, U. C. Dyer, S. T. Rusell, and J. A. Robinson, *Helv. Chim. Acta* **73,** 239–259 (1990).
188. D. S. Holmes, D. M. Ashworth, and J. A. Robinson in Ref. 187, pp. 260–271.
189. C. A. Townsend and A. Basak, *Tetrahedron,* in press.
190. E. N. Duesler and I. C. Paul in Ref. 7, Vol. 2, 1983, pp. 87–195.
191. J. L. Occolowitz and R. L. Hamill in Ref. 7, Vol. 2, 1983, pp. 197–244.
192. M. J. O. Anteunis in Ref. 7, Vol. 2, 1983, pp. 245–334.
193. H. Seto and co-workers, *Tetrahedron Lett.* **28,** 3357–3360 (1987).
194. Y. Kishi, in Ref. 7, Vol. 2, 1983, pp. 1–50.
195. D. A. Evans and R. P. Polniaszek, *Tetrahedron Lett.* **27,** 5683–5686 (1986).
196. D. A. Evans, R. L. Dow, T. L. Shih, J. M. Takacs, and R. Zahler, *J. Am. Chem. Soc.* **112,** 5290–5313 (1990).
197. D. A. Evans, S. L. Bender, and J. Morris, *J. Am. Chem. Soc.* **110,** 2506–2536 (1988).

198. D. A. Evans, J. S. Clark, R. Metternich, V. J. Novack, and G. S. Sheppard, *J. Am. Chem. Soc.* **112,** 866–868 (1990).
199. S. Hanessian, N. G. Cooke, B. DeHoff, and Y. Sakito in Ref. 198, pp. 5276–5290.
200. W. C. Still and co-workers, *Lect. Heterocycl. Chem.* **9,** 33–42 (1987).
201. S. J. Danishefsky, H. G. Selnick, M. P. DeNinno, and R. P. Zelle, *J. Am. Chem. Soc.* **109,** 1572–1574 (1987).
202. S. J. Danishefsky, H. G. Selnick, R. E. Zelle, and M. P. DeNinno in Ref. 197, pp. 4368–4378.
203. D. P. Negri and Y. Kishi, *Tetrahedron Lett.* **28,** 1063–1066 (1987).
204. M. P. Edwards, S. V. Ley, S. G. Lister, B. D. Palmer, and D. J. Williams, *J. Org. Chem.* **49** (1984).

General Reference

J. W. Westley, ed., *Polyether Antibiotics,* Vols. 1 and 2, Marcel Dekker, Inc., New York, 1982–1983.

LOUISE W. CRANDALL
ROBERT L. HAMILL
Lilly Research Laboratories

TETRACYCLINES

The tetracyclines are a group of antibiotics having an identical 4-ring carbocyclic structure as a basic skeleton and differing from each other chemically only by substituent variation. Figure 1 shows the principal tetracycline derivatives now used commercially.

The first tetracycline discovered was produced by a soil organism, *Streptomyces aureofaciens,* and is now known as chlortetracycline [*57-62-5*] (**2**), $C_{22}H_{23}ClN_2O_8$ (1). This compound ushered in a new era in antibacterial chemotherapy because it was effective orally and against a broad range of gram-positive and gram-negative bacteria. Chlortetracyline was joined by the second member of the family, oxytetracycline [*79-57-2*] (**3**), $C_{22}H_{24}N_2O_9$, produced by another actinomycete, *Streptomyces rimosus,* in 1950 (2). Tetracycline [*60-54-8*] (**1**), $C_{22}H_{24}N_2O_8$, discovered in 1953, lacks the 7-chloro of chlortetracycline (**2**) and the 5-hydroxyl group of oxytetracycline (**3**). Tetracycline was produced either by reductive dechlorination of (**2**) (3,4) or by a direct fermentation (5). 6-Demethylchlortetracycline [*127-33-3*], $C_{21}H_{21}ClN_2O_8$, (**4**) was discovered (6) as a metabolite of a mutant strain of the original *Streptomyces aureofaciens* and was introduced in 1958.

The three tetracyclines most recently marketed were made by a semisynthetic pathway. The first of these were methacycline (6-methylene oxytetracycline) [*914-00-1*] (**5**), $C_{22}H_{22}N_2O_8$, (7), and its reduction product doxycycline [*564-25-2*] (**6**), $C_{22}H_{24}ClN_2O_8$ (7,8). The latter compound is a potent antibiotic which is well-absorbed and slowly excreted thus allowing small and infrequent (once or twice a day) dosage schedules. Finally, the most recent addition to the commercial tetracyclines is minocycline [*10110-90-8*] (**7**), $C_{23}H_{27}N_3O_7$ (9), which is also well-

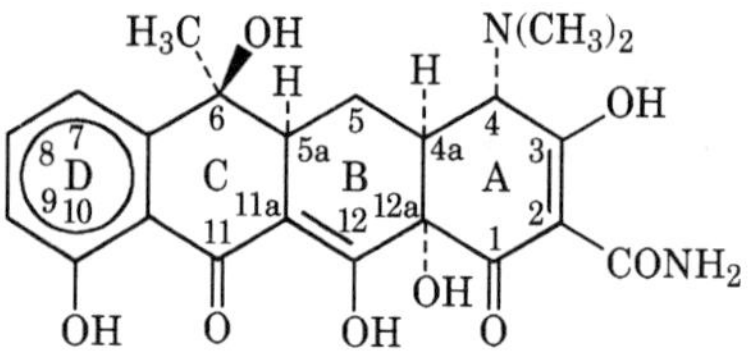

(1)

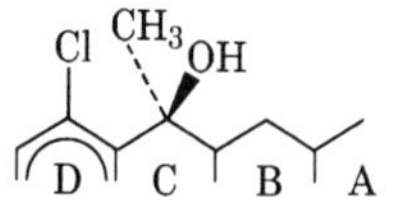

(2)

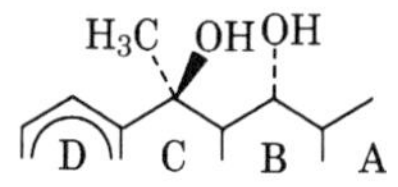

(3)

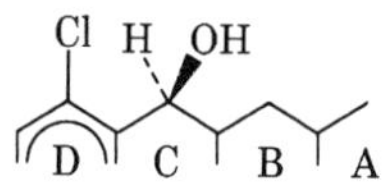

(4)

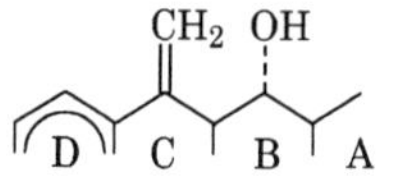

(5)

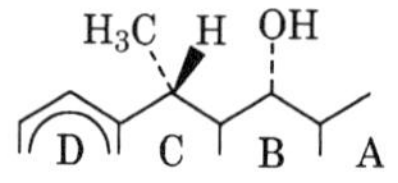

(6)

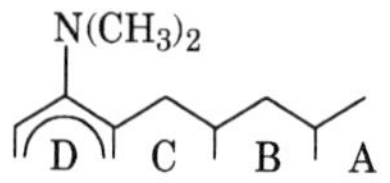

(7)

Fig. 1. Tetracycline (**1**) and its derivatives: chlortetracycline (7-chlorotetracycline) (**2**); oxytetracycline (5-hydroxytetracycline) (**3**); demeclocycline (6-demethyl-7-chlorotetracycline) (**4**); methacycline (6-demethyl-6-deoxy-5-hydroxy-6-methylenetetracycline) (**5**); doxycycline (6α-deoxy-5-hydroxytetracycline) (**6**); and minocycline (6-demethyl-6-deoxy-7-dimethylamino tetracycline) (**7**). Substituents at positions not specifically shown or mentioned remain as in (**1**). Courtesy of Blackwell Scientific Publications, Ltd.

absorbed and slowly excreted. Minocycline protects mice against staphylococcal infections which are resistant to most tetracyclines, penicillins, and many other antibiotics. In addition, it shows a marked superiority over other tetracyclines when tested against large numbers of randomly occurring, hospital-isolated gram-positive bacteria (10).

During the years from 1948 to 1952 considerable research was devoted to determining the structures of the tetracyclines. The gross structure (**3**) was determined by a combination of degradation sequences and spectral studies (11). Structure (**2**) was also published (12). Structure (**2**) was also determined by a different degradative pathway (13). The more subtle points of structure, such as the stereochemical configurations, were later determined by x-ray methods (14) and the absolute configuration by degradation. Structures of the tetracyclines developed since 1952 are easily arrived at by comparison to the structures of earlier compounds.

Physical Properties

In general, the tetracyclines are yellow crystalline compounds that have amphoteric properties (Fig. 2) (15). They are soluble in both aqueous acid and aqueous base. The acid salts tend to be soluble in organic solvents such as 1-butanol, dioxane, and 2-ethoxyethanol. In fact, 1-butanol is used to extract the salts from aqueous solution.

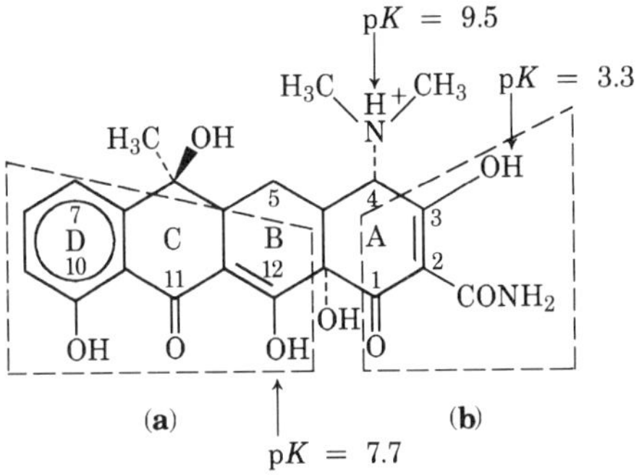

Fig. 2. Tetracycline (**1**) indicating the titratable hydrogens and showing (**a**) the BCD-chromophore and (**b**) the A-chromophore (15).

The tetracyclines are strong chelating agents. Both the A-ring and 11,12 β-diketone systems are active sites for chelation (16). This ability to chelate to metals, such as calcium, results in tooth discoloration when tetracycline is administered to children (17).

Each tetracycline possesses a characteristic uv-absorption spectrum and this property is used extensively in structure elucidation (12,13). This spectrum results from the contribution of two chromophores: the BCD ring system gives a λ_{max} at approximately 350 nm and the A-ring a λ_{max} at approximately 265 nm.

Traditionally, paper chromatography was used to monitor chemical modification reactions to determine the composition of the reaction mixture (13,18–20). The tetracyclines were usually detected by uv fluorescence on the paper strip or by bioautography where the areas on the paper strip that inhibited the growth of microorganisms were determined. The transference of paper chromatography solvent systems to column chromatography has been accomplished and since 1978, high pressure liquid chromatography (hplc) has been used with excellent success (21). Mass spectral analysis (22) and nmr (9) have become useful tools for tetracycline structure determinations.

Semisynthetic Modifications

The tetracycline molecule (**1**) presents a special challenge with regard to the study of structure–activity relationships. The difficulty has been to devise chemical pathways that preserve the BCD ring chromophore and its antibacterial properties. The lability of the 6-hydroxy group to acid and base degradation (12,13), plus the ease of epimerization (23) at position 4, contribute to chemical instability under many reaction conditions.

Under acidic conditions, dehydration to an anhydrotetracycline [*20154-34-1*] (**8**), $C_{22}H_{22}N_2O_7$, occurs; under basic ones, ring C opens to an isotetracycline [*3811-31-2*] (**9**), $C_{22}H_{24}N_2O_8$. The anhydrotetracyclines, such as (**8**), appear to exhibit a mode of antibacterial action, but it is unlike that of tetracycline (24). Epimerization (23,25,26) at C-4 occurs in a variety of solvents within the pH range 2–6, particularly in acetic acid (25). A number of anions (27) facilitate this reaction.

The reverse process, from 4-epitetracycline [*79-85-6*], $C_{22}H_{24}N_2O_8$, to tetracycline, is promoted by chelation with ions such as calcium and magnesium (28).

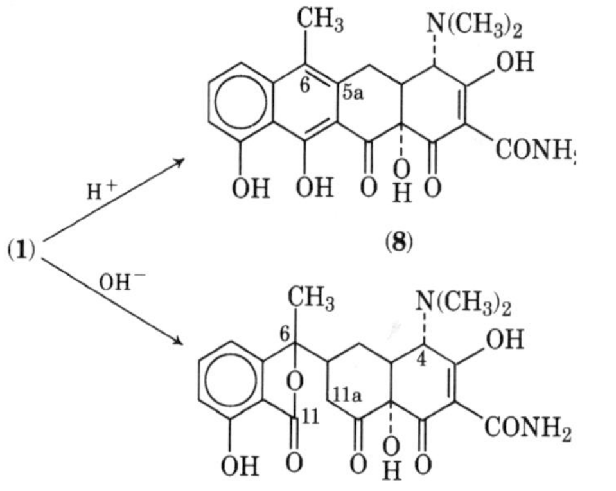

(8)

(9)

Conversion of the C-2 amide to a biologically inactive nitrile, which can be further taken via a Ritter reaction (29) to the corresponding alkylated amide, has been accomplished. When the 6-hydroxyl derivatives are used, dehydration occurs at this step to give the anhydro amide. Substituting an *N*-hydroxymethylimide for isobutylene in the Ritter reaction yields the acylaminomethyl derivative (30). Hydrolysis affords an aminomethyl compound. Numerous examples (31–35) have been reported of the conversion of a C-2 amide to active Mannich adducts which are extremely labile and easily undergo hydrolysis to the parent tetracycline. This reverse reaction probably accounts for the antibacterial activity of these tetracyclines.

Reactions at the C-4 carbon atom have been studied. With the exception of (**3**), reaction with methyl iodide (36,37) converts the 4-amino group to a quaternary amine with a concomitant loss of antibacterial activity. Treatment of this quaternary derivative with zinc in acetic acid results in a selective removal of the 4-dimethylamino grouping (36). This deamination can also be accomplished by photolysis (38). A transformation (18) involving the C-4 position has resulted in the synthesis of 4-hydroxy-6-methylpretetramid [*2011-31-6*], $C_{20}H_{15}NO_7$, an important precursor (39) in the biosynthesis of tetracycline. Another unusual reaction (40,41) yielded the 4,6-hemiketals: reactions of (**1**) or 6-demethyltetracycline [*987-02-0*] (**10**), $C_{21}H_{22}N_2O_8$, with positive halogens, or cupric or mercuric acetates yield the corresponding 4,6-hemiketals (**11**) and (**12**).

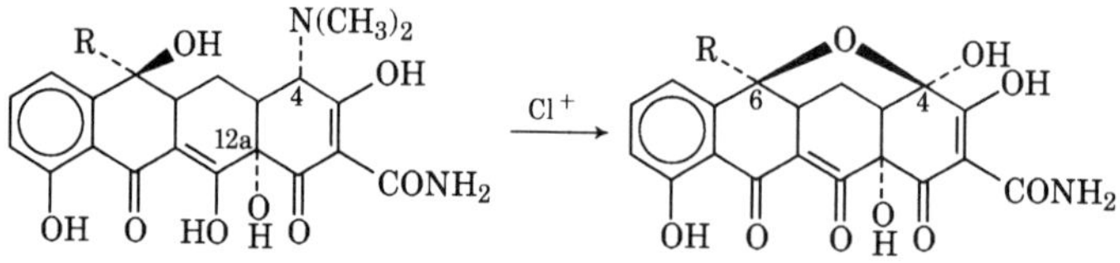

(**1**) R = CH_3
(**10**) R = H

(**11**) R = CH_3
(**12**) R = H

The hemiketal products (**11**) and (**12**) have been converted to the corresponding oximes, hydrazones, and substituted amines (40,41). Although many of these derivatives exhibit substantial antibacterial activity, they are generally less active than the parent tetracyclines.

Reactions at the C-5 position of the tetracycline molecule have been limited to the introduction of an alkoxy group (42) and the acetylation of the hydroxyl group (43) in 5-hydroxytetracyline. Neither of these modifications improved the biological activity of the molecule.

The isolation of the 6-deoxytetracyclines (44) led to other chemical modifications of (**1**). 6β-Deoxytetracycline [*5614-03-9*] (**13**), $C_{22}H_{24}N_2O_7$, was prepared by catalytic hydrogenolysis of tetracycline (**1**), resulting in an inversion (45) of the configuration at the C-6 position, but retention of antibacterial activity. Catalytic reduction (7,8) of the 6-methylene derivative (**14**) yields both the 6α-methyl (**15**) and 6β-methyl compound (**13**). The 6α-isomer (**15**) is reported (7,45) to be more active than the 6β-isomer (**13**). The α-isomer, doxycycline (**6**), is an example of a semisynthetic tetracycline that has become commercially useful.

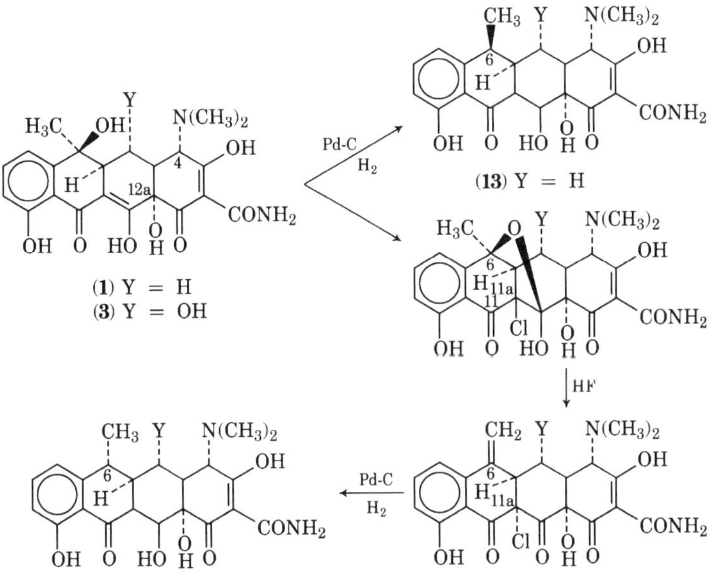

(**1**) Y = H
(**3**) Y = OH

(**13**) Y = H

(**15**) Y = H
(**6**) Y = OH

(**14**) Y = H
(**5**) Y = OH

The 6-fluoro isomers, 6-deoxy-6-demethyl-6α-fluorotetracycline [*24333-20-8*], $C_{21}H_{21}FN_2O_7$, and 6-deoxy-6-demethyl-6β-fluorotetracycline [*24333-21-9*], $C_{21}H_{21}FN_2O_7$, have been prepared and showed relatively high *in vitro* and *in vivo* biological activities compared to the parent tetracyclines.

The increased chemical stability of the 6-deoxytetracylines allows chemical modification with retention of biological activity: electrophilic substitutions have been carried out at C-7 and C-9 under strongly acidic conditions (46–53).

Reactions of 6-deoxy-6-demethyltetracycline [*808-26-4*] (**16**), $C_{21}H_{22}N_2O_7$, with electrophiles, such as nitrate ion (49), bromomium ion (46,47) (from *N*-bromosuccinimide), or *N*-hydroxymethylphthalimide (53), yielded 7-substituted tetracyclines. In the case of the nitration reaction, both the 7- and 9-nitro isomers (**17**, X = NO_2, Y = H) and (**17**, X = H, Y = NO_2) were obtained.

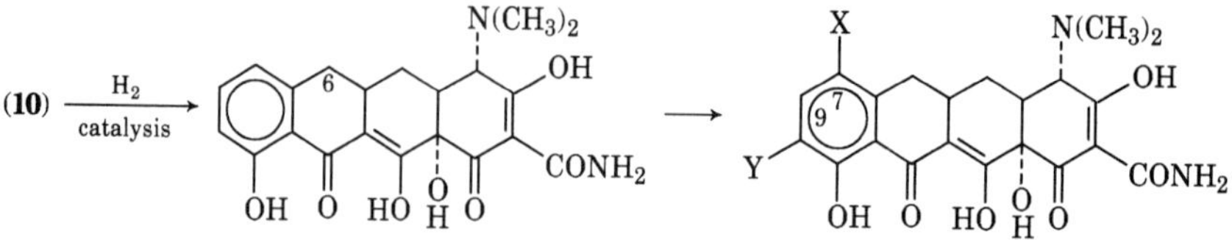

(**16**) (**17**)

A similar series of reactions can be carried out for the 6-methyl series, (**13**) and (**15**), where catalytic reduction affords the amino compounds, ie (**17**, X = NH_2, Y = H). Oxidation (54) of tetracyclines using the Udenfriend reagent has yielded 9-hydroxytetracyclines and disubstituted (C-7 and C-9) products (48) can also be obtained. Substituent assignments are made from nmr spectral interpretations. The 7- and 9-methyl tetracyclines have been prepared and reported to retain biological activity (55).

Although many of the tetracycline derivatives showed useful *in vivo* activity against a wide spectrum of pathogenic organisms, few showed any significant improvement in overall activity when compared to tetracycline. The exception is minocycline (**7**) (9) which exhibits superior activity against tetracycline-sensitive organisms and many tetracycline-resistant strains of gram-positive bacteria.

Structure–Activity Correlations

There are a number of tetracycline structural features that are prerequisites for biological activity. The linear arrangement of the rings, coupled with the phenolic β-diketone system, is essential (56,57). Any structure variation at the 11a position results in loss of activity. The C-11 to C-12 β-diketone system has exceptional chelating qualities, and probably is involved in the binding of the tetracyclines to ribosomes (58), in the interactions with bacterial repressor proteins (58), and in transport of tetracyclines into the bacterial cell. The amide hydrogen can be replaced by a methyl group, but larger residues, if not rapidly cleaved in water, bring about a reduction in activity.

The configuration at the chiral centers C-4a, C-5a, and C-12a determine the conformation of the molecule. In order to retain optimum *in vitro* and *in vivo* activity, these centers must retain the natural configuration. The hydrophobic part of the molecule from C-5 to C-9 is open to modification in many ways without losing antibacterial activity. However, modification at C-9 may be critical because steric interactions or hydrogen bonding with the oxygen atom at C-10 may be detrimental to the activity.

X-ray crystallographic studies (59) have defined the conformations and hydrogen bonding of the tetracyclines under nonpolar and polar conditions. These are shown in Figure 3. It is believed that the equilibrium between the zwitterionic and nonionized forms is of importance for the broad-spectrum antibacterial activity, membrane permeation, and pharmacokinetic properties.

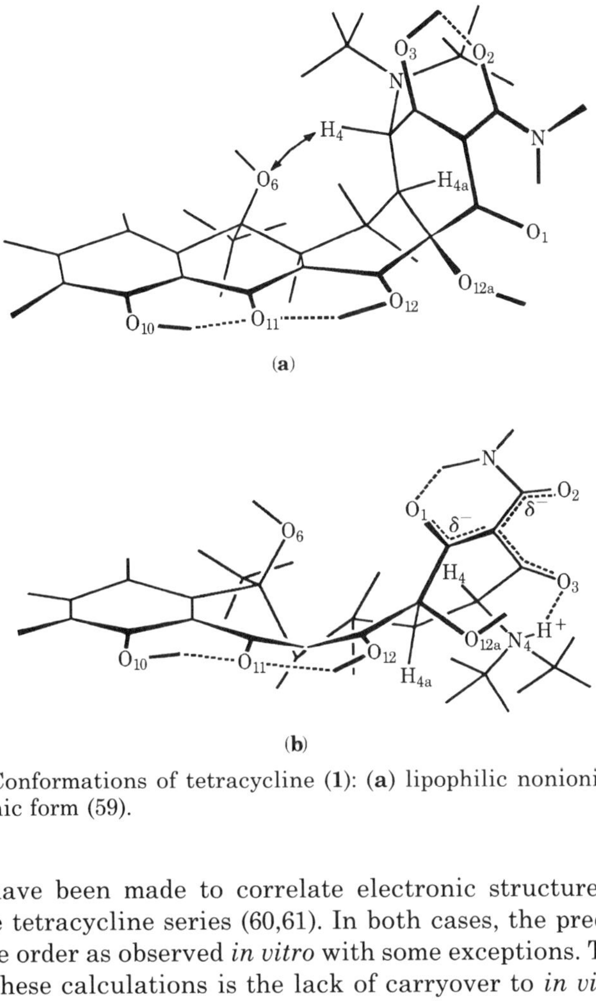

Fig. 3. Conformations of tetracycline (**1**): (**a**) lipophilic nonionized form, and (**b**) polar zwitterionic form (59).

Efforts have been made to correlate electronic structure and biological activity in the tetracycline series (60,61). In both cases, the predicted activities are of the same order as observed *in vitro* with some exceptions. The most serious drawback to these calculations is the lack of carryover to *in vivo* antibacterial activity. Attempts have also been made (62) to correlate partition coefficients and antibacterial activity. The stereochemical requirements are somewhat better defined. Thus 4-epitetracycline and 5a-epitetracycline [*65517-29-5*], $C_{22}H_{24}N_2O_8$, are inactive (63). The 6-epi compound [*19369-52-9*] is about one-half as active as the 6α (or natural) configuration.

The unexpected biological activities of tetracyclines, such as 5a-epi-6-epitetracycline [*19543-88-5*], $C_{22}H_{24}N_2O_8$, and 7-chloro-5a,11a-dehydro-6-epitetracycline [*22688-60-4*], $C_{22}H_{21}ClN_2O_8$, make predicting structure–activity relationships difficult (64). Aside from the C-2 amide Mannich-base derivatives, variation at other centers in the molecule, ie, C-4, 4a, 5a, 12a, decreases the biological activity.

Total Synthesis

The first synthesis of a tetracycline having full biological activity, 6-demethyl-6-deoxytetracycline (**16**) was reported in the early 1960s (76–79). This compound contained the important and difficult to obtain 4-dimethylamino group, but lacked the 6-hydroxyl. An alternative total synthesis of (**16**) was devised in 1965 (69) using methods that could lead to the synthesis of tetracyclines in large quantities and in good yields. The only total synthesis of a fermentation-produced tetracycline was carried out for oxytetracycline (**3**) which has six asymmetric centers (70). More recently the following nonnaturally tetracyclines have been synthesized: 5a-methyltetracyclines (59), 6-oxatetracyclines (59), and 6-thiatetracyclines (59). None of these compounds proved to be commercially important.

Manufacture

Most of the fermentation and isolation processes for manufacture of the tetracyclines are described in patents (71,72). Manufacture begins with the cultivated growth of selected strains of *Streptomyces* in a medium chosen to produce optimum growth and maximum antibiotic production. Some clinically useful tetracyclines (**2–4**) are produced directly in these fermentations; others (**5–7**) are produced by subjecting the fermentation products to one or more chemical alterations. The purified antibiotic produced by fermentation is used as the starting material for a series of chemical transformations (59).

The choice of the strain of microorganism is one of the important variables in the process. The strains to be used in manufacture are mutants of the original producer, which are chosen as the result of a planned program of mutant selection. Sometimes a spontaneous mutation occurs; usually, it is induced by mutagenic agents or irradiation of various sorts. The choice of the best strain depends on its ability to produce large amounts of the proper antibiotic in a reasonable time from ingredients that are economically feasible (73).

All steps of manufacture are monitored for antibiotic potency, purity, stability, and bioavailability by quality control methods. In addition, each batch of bulk powder and each lot of finished pharmaceutical dosage form must be certified by the FDA. During all stages of manufacture the FDA requires compliance with specified standards.

Economic Aspects

The total U.S. antibiotic market for 1990 was about \$4.73 billion, \$233 million of that was tetracyclines. The development of the semisynthetic β-lactam antibiotics (see ANTIBIOTICS, β-LACTAMS) and emergence of resistance to the tetracyclines has steadily diminished the clinical usefulness of tetracyclines.

In the United States, the manufacturers of fermentation-derived tetracyclines (**1**), (**2**), and (**3**) are the Lederle Laboratories, a division of American Cyanamid Co., Charles Pfizer Inc., Bristol Laboratories, and Rachelle Laboratories. There are also several manufacturers abroad. Tetracycline is now sold generically by many companies. Pfizer's doxycycline (**6**) and Lederle's minocycline (**7**), both semisynthetic tetracyclines, are the only members of the group that have increasing sales. Table 1 lists the commercial tetracyclines and the corresponding trade names.

Table 1. Tetracyclines Used for the Therapy of Infectious Diseases

Generic name	Trade name	Year of discovery
chlortetracyline	Aureomycin	1948
oxytetracycline	Terramycin	1948
tetracycline	Achromycin	1953
demeclocycline	Declomycin	1957
methacycline	Rondomycin	1965
doxycycline	Vibramycin	1967
minocycline	Minocin	1972

Biosynthesis

The overall biosynthetic pathway to the tetracyclines has been reviewed (74). Studies (75–78) utilizing ^{13}C labeled acetate and malonate and nmr analysis of the isolated oxytetracycline (**2**), have demonstrated the exclusive malonate origin of the tetracycline carbon skeleton, the carboxamide substituent, and the folding mode of the polyketide chain. Feeding experiments using [1-^{13}C, $^{18}O_2$] acetate and analysis of the nmr isotope shift effects, led to the location of the [$^{18}O_2$] derived oxygen substituents in oxytetracycline: carbons 1, 3, 10, 11, and 12 were labeled. The remaining oxygen substituents, those at carbons 5, 6, and 12 in oxytetracycline, originate from subsequent oxidation of the 4-hydroxy-6-methylpretetramid [*2011-31-6*].

Feeding studies using deuterated [1-^{13}C] acetate and subsequent location of the deuterated sites by the nmr isotope shift effects showed labeling at carbons 7 and 9. The absence of a detectable β-^{2}H isotope shift at carbon 4a indicates that only one of the carbon-5 hydrogens is acetate-derived and that this is stereospecifically eliminated by carbon-5α hydroxylation. 8-Methoxychlortetracyclines, with and without hydroxyl substitution at carbon-4a, have been isolated. These tetracyclines retain the original oxygen at C-8 from the polyketide chain (79–81).

Because of the continued commercial importance of the tetracyclines, a study of the genetics of the oxytetracycline (OTC) producing organism has been undertaken to improve the efficiency of antibiotic production. The biosynthetic genes of *Streptomyces rimosus* have been cloned by taking advantage of the fact that the OTC genes are clustered around at least one OTC resistance gene. Using this resistance gene as a selectable marker to detect hydridizing clones, a 30–40 kilobase piece of DNA surrounding the resistance gene, and containing some of the biosynthetic genes, was identified (82).

Anhydrotetracycline oxygenase from *Streptomyces aureofaciens,* which catalyzes the conversion of anhydrotetracycline to dehydrotetracycline, has been isolated and characterized as a flavin-dependent oxygenase (83). It consists of two subunits of mol wt $\approx$57,500 based on SDS/polyacrylamide–gel electrophoresis.

The cosynthetic factor 1 of *Streptomyces aureofaciens,* involved in the reduction of 5a,11a-dehydrochlortetracycline to chlortetracycline, has been identified as 7,8-didemethyl-8-hydroxy-5-deazariboflavin. This work was aided by comparison of spectral data with that of an authentic sample obtained from the hydrolysis of coenzyme F-420 (84).

Biological Aspects

It has been known for some time that tetracyclines are accumulated by bacteria and prevent bacterial protein synthesis (Fig. 4). Furthermore, inhibition of protein synthesis is responsible for the bacteriostatic effect (85). Inhibition of protein synthesis results primarily from disruption of codon–anticodon interaction between tRNA and mRNA so that binding of aminoacyl–tRNA to the ribosomal acceptor (A) site is prevented (85). The precise mechanism is not understood. However, inhibition is likely to result from interaction of the tetracyclines with the 30S ribosomal subunit because these antibiotics are known to bind strongly to a single site on the 30S subunit (85).

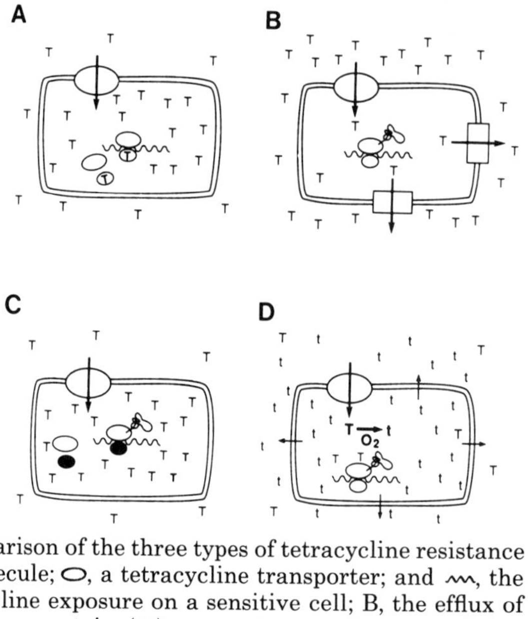

Fig. 4. Comparison of the three types of tetracycline resistance where T represents the tetracycline molecule; ⬭, a tetracycline transporter; and ∿∿, the ribosome; A shows the effect of tetracycline exposure on a sensitive cell; B, the efflux of resistance where a cytoplasmic membrane protein (□) pumps tetracycline out of the cell as fast as the tetracycline transporter takes it up; C, the ribosomal protection type of resistance where the ribosome is modified by (⬬) to block productive binding; and D, the tetracycline modification type of resistance where t is an inactive form of tetracycline. Reproduced with permission from reference 121. (See also refs. 85, 86, 88, 105–107, 109, and 111).

Studies designed to characterize the nature of the tetracycline binding domain have revealed that when bound to ribosomes, tetracycline protects base A892 in 16sRNA from reactivity toward dimethyl sulfate and enhances reactivity towards bases U1052 and C1054 (86). These results suggest that the 892–1054 region of 16sRNA contributes together with 30S ribosomal proteins (87) to the antibiotic binding domain.

Uses

The commercially available tetracyclines, listed in Table 1, have fairly similar antibacterial activities, apart from doxycycline (**6**) and minocycline (**7**), which show enhanced antibacterial activity compared to the others (59). Tetracyclines were the first prominent group of antimicrobial agents for which the term "broad spectrum" was used. That is, they exhibit activity against a wide range of gram-positive and gram-negative bacteria, including obligate anaerobes (88,89). *Mycoplasmas, rickettsiae, chlamydia,* and protozoan parasites are also, in general, susceptible to tetacyclines (88–90) (see also ANTIPARASITIC AGENTS).

Clinical Uses. The emergence of bacterial resistance to tetracyclines has limited the use of these agents as the drugs of first choice in the treatment of many infections for which they were previously effective. Nevertheless, they are still the treatment of first choice in the following cases: (*1*) for bacterial infections causing brucellosis, cholera, chancroid, granuloma inguinale, and Lyme disease (88,89,91); (*2*) for rickettsial infections, eg, typhus, scrubtyphus, and spotted fever (88,89); (*3*) for chlamydial infections, eg, psittacosis, lymphogranuloma venereum, trachoma, and inclusion conjunctivitis (88,89); (*4*) in the treatment of nonspecific urethritis because of *Chlamydia* or *Ureaplasmas* (88); and (*5*) in the treatment of acne vulgaris and rosacea (88,92).

Many sexually transmitted infections are polymicrobial in nature, ie, more than one type of microorganism is responsible for the infection (89,93,94). For example, pelvic inflammatory disease (PID) is usually caused by a concurrent infection with both *N. gonorrhoeae* and *C. trachomatis* (94). For these reasons many sexually transmitted diseases, including PID, are treated with mixtures of antibiotics to provide broad coverage for mixed, aerobic, and anaerobic infections (89). Tetracyclines play a role in such therapeutic regimens, eg, β-lactam antibiotics plus doxycycline are used in the therapy of PID (94).

It has been suggested that tetracyclines alone can be used for certain sexually transmitted syndromes, such as urethritis in which *N. gonorrhoeae, Chlamydia,* and *Ureaplasma* species are causative agents (89). Although β-lactam antibiotics remain the drugs of choice against *N. gonorrhoeae,* the tetracyclines are active against the associated pathogens. However, the increasing prevalence of tetracycline resistance in *N. gonorrhoeae* (95) is likely to seriously challenge this therapeutic strategy.

Tetracyclines are used as alternative drugs in a variety of circumstances when the patient is unable to take the drug of choice, eg, in patients allergic to penicillin (88,89). Tetracyclines are widely known to cause staining of teeth (and are therefore contra-indicated in children developing permanent teeth), photosensitivity, and, in the case of minocycline, vestibular toxicity. Details of these adverse effects and others associated with administration of tetracyclines have been comprehensively reviewed (96–101).

Veterinary Uses. Tetracyclines are widely used for veterinary therapy. The types of pathogens encountered are frequently different from those for which tetracyclines are used in humans (102). Tetracyclines are also used in animal husbandry as growth promoters (88,102) (see GROWTH REGULATORS, ANIMAL). Tetracyclines, eg, chlortetracycline, used in this way are usually administered to the animals orally, at subtherapeutic doses, either in the feed or drinking water (see FEEDS AND FEED ADDITIVES). The procedure frequently improves the efficiency of feed conversion, or the rate of weight gain in poultry and other animals reared under commercial conditions (88,102). The mechanisms responsible for growth promotion are still obscure but, in the case of chlortetracycline (**2**), probably result from suppression of deleterious organisms in the animal's intestine (102). Various antibiotics, including tetracyclines, have demonstrable biochemical effects on microorganisms at drug concentrations which are below those required to achieve complete inhibition of microbial growth (103). These effects include suppression of bacterial adhesion and interference with the secretion of extracellular toxins, both of which may relate to the growth-promoting activity of tetracyclines. Some countries, eg, the United Kingdom, have introduced legislation forbidding the addition of tetracyclines to animal feed for growth promotion purposes. See reference 102 for a discussion.

Resistance to Tetracyclines. The tetracyclines still provide inexpensive and effective treatment for several microbial infections, but the emergence of acquired resistance to this class of antibiotic has limited their clinical usefulness. Studies to define the molecular basis of resistance are underway so that derivatives having improved antibacterial spectra and less susceptibility to bacterial resistance may be developed. Tetracyclines are antibiotics of choice for relatively few human infections encountered in daily clinical practice (104), largely as a result of the emergence of acquired tetracycline-resistance among clinically important bacteria (88,105,106). Acquired resistance occurs when resistant strains emerge from previously sensitive bacterial populations by acquisition of resistance genes which usually reside in plasmids and/or transposons (88,106,107). Furthermore, resistance determinants contained in transposons spread to, and become established in, diverse bacterial species (106).

Mechanisms of Resistance. Three distinct biochemical mechanisms of resistance to tetracyclines have been identified and these are illustrated in Figure 4. The energy-dependent efflux of antibiotic mediated by resistance proteins located in the bacterial cytoplasmic membrane is illustrated in Figure 4B (85,88,105–109). The intracellular tetracycline concentration remains too low for effective binding to ribosomes. The complete sequence of the genes for several efflux proteins has been established and models for the organization of the proteins in the membrane have been proposed (109,110). Ribosomal protection, whereby tetracyclines no longer bind productively to the bacterial ribosome, is shown in Figure 4C (108,111). In a tetracycline-resistant cell, shown in Figure 4A, tetracycline accumulation within the cell is similar to that in the sensitive cell, but the ribosome is modified so that tetracycline no longer binds productively to the ribosome. Although the molecular basis of this resistance mechanism is not fully understood, it may involve modification of ribosomal RNA or protein which affects binding of antibiotic to the 30S ribosomal subunit (111). Figure 3D shows chemical alteration of the tetracycline molecule by a reaction in the cytoplasm that requires oxygen. This renders the drug inactive as an inhibitor of protein synthesis (108). The altered tetracycline then diffuses out of the cell. This mechanism of resistance is poorly characterized and may not be expressed in the natural habitat of most pathogens especially because foci of infection in the human body are usually poorly aerated (108).

Distribution of Resistance Determinants in Microorganisms. The majority of tetracycline resistance determinants are located on plasmids or transposons. These determinants have been grouped into classes defined by lack of cross-hybridization under stringent conditions (107,108). These classes, some of which are related, are given in Table 2. Classes A–D show DNA homologies of from 44–63% (108) and encode products having similar functions. Although tetracycline resistance determinants are widely distributed in microorganisms, certain classes appear to be restricted to particular groups of organisms. For instance, tetracycline efflux is the dominant mechanism of resistance in the *Enterobacteriaceae,* but this mechanism has not been reported in other organisms eg, *Neisseria* species.

Table 2. Classification and Distribution of Tetracycline Resistance Determinants in Microorganisms[a]

Representative family		
Class	Mechanism	Genus or species
A	efflux	*Enterobacteriaceae, Aeromonas, Vibrio, Pseudomonas*
B	efflux	*Enterobacteriaceae, Haemophilus, Vibrio, Yersinia*
C	efflux	*Enterobacteriaceae, Vibrio, Pseudomonas*
D	efflux	*Enterobacteriaceae, Aeromonas, Vibrio, Pasteurella*
E	efflux	*Escherichia aeromonas*
F	efflux	*Bacteroides fragilis*
G	unknown	*Vibrio anguillarum*
K	efflux	*Staphylococcus, Enterococus, Streptococcus,*[b] *Peptostreptococcus*[b]
L	efflux	*Bacillus, Staphylococcus, Streptococcus,*[b] *Enterococcus, Peptostreptococcus*[b]
M	ribosomal protection	*Neisseria, Mycoplasma, Ureaplasma, Haemophilus, Campylobacter, Clostridium, Enterococcus, Staphylococcus, Streptococcus,*[b] *Gardnerella, Kingella, Eikenella, Veillonella, Fusobacterium, Peptostreptococcus,*[b] *Clostridium difficile, Streptococcus pneumoniae*
N	unknown	*Strep agalactiae*
O	ribosomal protection	*Campylobacter, Lactobacillus, Streptococcus,*[b] *Enterococcus, Peptostreptococcus*[b]
P	unknown	*C. perfringens*
X	antibiotic modification	*Bacteroides* (cryptic)

[a]Refs. 85, 105, 107, 108, and 112.
[b]Certain isolates are reported to contain simultaneously K, L, M, and O determinants (112).

BIBLIOGRAPHY

"Tetracyclines" in *ECT* 1st ed., Vol. 13, pp. 771–810, by P. Regna, Chas. Pfizer & Co., Inc., C. H. Demos and L. M. Preuss, Lederle Laboratories, American Cyanamid Company; in *ECT* 2nd ed., Vol. 20, pp. 1–33, by R. K. Blackwood, Chas. Pfizer & Co., Inc.; in *ECT* 3rd ed., Vol. 3, pp. 64–78, by J. H. Boothe and J. J. Hlavka (Lederle Laboratories, a division of the American Cyanamid Company).

1. R. W. Broschard and co-workers, *Science* **109,** 199 (1949); B. M. Duggar, *Ann. N.Y. Acad. Sci.* **51,** 177 (1948).
2. A. C. Finlay and co-workers, *Science* **111,** 85 (1950); P. P. Regna and co-workers, *J. Am. Chem. Soc.* **73,** 4211 (1951).
3. J. H. Boothe and co-workers, *J. Am. Chem. Soc.* **75,** 4621 (1953).
4. L. H. Conover and co-workers, *J. Am. Chem. Soc.* **75,** 4622 (1953).
5. P. P. Minieri and co-workers, *Antibiot. Ann.* **1953–1954,** 81 (1953–1954).
6. J. R. D. McCormick and co-workers, *J. Am. Chem. Soc.* **79,** 4561 (1957).
7. R. K. Blackwood and co-workers, *J. Am. Chem. Soc.* **85,** 3943 (1963).
8. C. R. Stephens and co-workers, *J. Am. Chem. Soc.* **85,** 2643 (1963).
9. M. J. Martell, Jr. and J. H. Boothe, *J. Med. Chem.* **10,** 44 (1967); G. S. Redin in G. L. Hobby, ed., *Antimicrobial Agents and Chemotherapy,* Williams & Wilkins, Baltimore, Md., 1966, p. 371.
10. N. H. Steigbigel, C. W. Reid, and M. Finland, *Am. J. Med. Sci.* **255,** 179 (1968).
11. F. A. Hochstein and co-workers, *J. Am. Chem. Soc.* **75,** 5455 (1953).
12. C. R. Stephens and co-workers, *J. Am. Chem. Soc.* **76,** 3658 (1954); **74,** 4976 (1952).
13. C. W. Waller and co-workers, *J. Am. Chem. Soc.* **74,** 4981 (1952).
14. S. Hirokawa and co-workers, *Z. Kristallogr. Kristallgeom. Kristallphys. Kristallchem.* **112,** 439 (1959); J. Donohue and co-workers, *J. Am. Chem. Soc.* **85,** 851 (1963).
15. M. Schnarr, M. Matthies, and W. Lohmann, *Z. Naturforsch.* **34c,** 1156 (1979).
16. J. J. Hlavka and J. H. Boothe, eds., *Handbook of Experimental Pharmacology,* Vol. 78, Springer-Verlag, New York, 1985, p. 86.
17. H. Shwachman and A. Schuster, *Pediat. Clin. Am.* **3,** 295 (1956).
18. J. J. Hlavka and P. Bitha, *J. Am. Chem. Soc.* **87,** 1795 (1965).
19. J. J. Hlavka, P. Bitha, and J. H. Boothe, *J. Am. Chem. Soc.* **87,** 1795 (1965); **90,** 1034 (1968).
20. P. Bitha, J. J. Hlavka, and J. H. Boothe, *J. Med. Chem.* **13,** 89 (1970).
21. G. D. Mack and R. B. Ashworth, *J. Chromatog. Sci.* **16,** 93 (1978).
22. D. Hoffman, *J. Org. Chem.* **31,** 792 (1966).
23. J. R. D. McCormick and co-workers, *J. Am. Chem. Soc.* **79,** 2849 (1957).
24. K. Koschel and co-workers, *Biochem. Z.* **344,** 76 (1966); see Ref. 58.
25. C. R. Stephens and co-workers, *J. Am. Chem. Soc.* **78,** 1515 (1956).
26. A. P. Doerschuk, B. A. Bitler, and J. R. D. McCormick, *J. Am. Chem. Soc.* **77,** 4687 (1955).
27. E. G. Remers, G. Steiger, and A. Doerschuk, *J. Pharm. Sci.* **52,** 753 (1963).
28. U.S. Pat. 3,009,956 (Nov. 21, 1961), M. Noseworthy (to Chas. Pfizer & Co.).
29. C. R. Stephens and co-workers, *J. Am. Chem. Soc.* **85,** 2643 (1963); U.S. Pat. 3,028,409 (Apr. 3, 1962), C. R. Stephens (to Chas. Pfizer & Co.).
30. M. J. Martell, Jr., A. S. Ross, and J. H. Boothe, *J. Med. Chem.* **10,** 485 (1967).
31. G. C. Barrett, *J. Pharm. Sci.* **52,** 309 (1963).
32. A. Brunzell, *Acta. Chem. Scand.* **16,** 245 (1962).
33. R. Huttenrauch and J. Keiner, *Naturwissenschaften* **53,** 552 (1966).
34. W. J. Gottstein, W. F. Minor, and L. C. Cheney, *J. Am. Chem. Soc.* **81,** 1198 (1959).
35. W. Seidel, A. Soder, and F. Linder, *Munch. Med. Wschr.* **17,** 661 (1958); South Afr. Pat. Spec. 57 3169 (1957).

36. J. H. Boothe and co-workers, *J. Am. Chem. Soc.* **80,** 1654 (1958).
37. L. H. Conover, *Symposium Antibiotics of Mold Metabolites,* Spec. Publ. No. 5, The Chemical Society, London, UK, 1956.
38. J. J. Hlavka and P. Bitha, *Tetrahedron Lett.* **32,** 3843 (1966).
39. J. R. McCormick and co-workers, *J. Am. Chem. Soc.* **87,** 1793 (1965).
40. R. C. Esse and co-workers, *J. Am. Chem. Soc.* **86,** 3874, 3875 (1964); South Afr. Pat. Appl. 63 4791 (1963) (to American Cyanamid Co.); J. H. Van Den Hende, *J. Am. Chem. Soc.* **87,** 929 (1965).
41. R. K. Blackwood and C. R. Stephens, *J. Am. Chem. Soc.* **86,** 2736 (1964); *Can. J. Chem.* **43,** 1382 (1965).
42. M. Schach von Wittenau, F. Hochstein, and C. R. Stephens, *J. Org. Chem.* **28,** 2454 (1963).
43. M. Kolosov and co-workers, *Chemistry of Antibiotics,* Vol. 1, Academy of Science, Moscow, USSR, p. 180.
44. J. R. D. McCormick and co-workers, *J. Am. Chem. Soc.* **82,** 3331 (1960); South Afr. Pat. Appl. 58 513 (May 13, 1958) (to American Cyanamid Co.).
45. M. Schach von Wittenau and co-workers, *J. Am. Chem. Soc.* **84,** 2645 (1962).
46. J. J. Hlavka and co-workers, *J. Am. Chem. Soc.* **84,** 1426 (1962).
47. J. J. Hlavka and H. M. Krazinski, *J. Org. Chem.* **28,** 1422 (1963).
48. J. J. Hlavka, H. M. Krazinski, and J. H. Boothe, *J. Org. Chem.* **27,** 3674 (1962).
49. J. H. Boothe and co-workers, *J. Am. Chem. Soc.* **82,** 1253 (1960).
50. J. Petisi and co-workers, *J. Med. Pharm. Chem.* **5,** 538 (1962).
51. J. L. Spencer and co-workers, *J. Med. Chem.* **6,** 405 (1963).
52. J. J. Beereboom and co-workers, *J. Am. Chem. Soc.* **82,** 1003 (1960).
53. M. J. Martell, Jr., A. S. Ross, and J. H. Boothe, *J. Med. Chem.* **10,** 359 (1967).
54. P. Shu, *J. Am. Chem. Soc.,* 4529 (1966).
55. L. Bernardi and co-workers, *Il Farmaco Ed. Sci.* **30,** 1025 (1975).
56. M. Burstall, *Manuf. Chem.* **31,** 474 (1960).
57. J. Doluisio and A. Martin, *J. Med. Chem.* **6,** 16 (1963).
58. B. Rasmussen and co-workers, *Antimicrob. Agents Chemother.* **35,** 2306 (1991).
59. W. Rogalski in Ref. 16, p. 179.
60. A. Cammarata and co-workers, *Mol. Pharmacol.* **6,** 61 (1970).
61. F. Peradejordi, A. Martin, and A. Cammarata, *J. Pharm. Sci.* **60,** 577 (1971).
62. J. Colaizzi and P. Klink, *J. Pharm. Sci.* **58,** 1185 (1969); M. Schach von Wittenau, *Chemotherapy (Basel)* **13,** 41 (1968).
63. J. R. D. McCormick and co-workers, *J. Am. Chem. Soc.* **80,** 5572 (1958).
64. M. Martell, A. Ross, and J. Boothe, *J. Am. Chem. Soc.* **89,** 6780 (1967).
65. L. H. Conover and co-workers, *J. Am. Chem. Soc.* **84,** 3222 (1962).
66. R. B. Woodward, *Pure Appl. Chem.* **6,** 561 (1963).
67. W. Martin, H. Hartung, H. Urboch, and W. Durckheimer, *Tetrahedron Lett.,* 3513 (1973).
68. J. J. Korst and co-workers, *J. Am. Chem. Soc.* **90,** 439 (1968).
69. H. Muxfeldt and W. Rogalsky, *J. Am. Chem. Soc.* **87,** 933 (1965).
70. H. Muxfeldt and co-workers, *J. Am. Chem. Soc.* **90,** 6534 (1968).
71. R. C. Evans, *The Technology of the Tetracyclines,* Quadrangle Press, New York, 1968.
72. *Tetracycline Manufacturing Processes,* Noyes Development Corp., Park Ridge, N.J., 1969.
73. J. J. Goodman, in Ref. 16, p. 5.
74. J. A. Robinson, *Chem. Soc. Rev.* **17,** 383 (1988).
75. R. Thomas and D. J. Williams, *J. Chem. Soc., Chem. Commun.* 128, (1983).
76. *Ibid.,* 677, (1983).
77. *Ibid.,* 443, (1984).

78. *Ibid.*, 802, (1985).
79. M. Patel and co-workers, *J. Antibiot.* **40,** 1408 (1987).
80. *Ibid.*, 1414 (1987).
81. E. B. Smith and co-workers, *J. Antibiot.* **40,** 1419 (1987).
82. M. J. Butler and co-workers, *Mol. Gen. Genet.* **215,** 231 (1989).
83. I. Vancurova and co-workers, *Biochem. J.* **253,** 263 (1988).
84. J. R. D. McCormick and G. O. Morton, *J. Am. Chem. Soc.* **104,** 4014 (1982).
85. I. Chopra in Ref. 16, p. 317.
86. D. Moazed and H. F. Noller, *Nature* **327,** 389 (1987).
87. M. A. Buck and B. S. Cooperman, *Biochemistry* **29,** 5374 (1990).
88. I. Chopra and co-workers, *J. Antimicrob. Chemother.* **8,** 5 (1981).
89. B. A. Cunha in Ref. 16, p. 393.
90. T. D. Edlind, *Antimicrob. Agents Chemother.* **33,** 2144 (1989).
91. B. Cryan and D. J. M. Wright, *J. Antimicrob. Chemother.* **25,** 187 (1990).
92. J. J. Leyden, K. J. McGinley, and A. M. Kligman, *Arch. Dermatol.* **118,** 19 (1982).
93. K. E. Toomeg and R. C. Barnes, *Rev. Infect. Dis.* **12** (Suppl. 6), 645 (1990).
94. H. B. Peterson, E. I. Galaid, and J. M. Zenilman, *Rev. Infect. Dis.* **12** (Suppl. 6), 656 (1990).
95. *Morbidity Mortality Weekly Rep.* **39,** 284 (1990).
96. J. F. Noble, K. Kanegis, and D. W. Hallesy, *Toxicol. Appl. Pharmacol.* **11,** 128 (1967).
97. W. R. Wilson, *Mayo Clin. Proc.* **52,** 635 (1977).
98. E. M. Ory, in B. M. Kagan, ed., *Antimicrobial Therapy,* 3rd ed., W. B. Saunders, Philadelphia, Pa., 1980, p. 117.
99. J. Jeljasewicz in G. G. Grassi and L. D. Sabath, eds., *New Trends in Antibiotics: Research and Therapy,* Elsevier/North-Holland Biomedical Press, Amsterdam, The Netherlands, 1981, p. 45.
100. M. Jones and B. A. Cunha, *Ther. Drug Monit.* **4,** 137 (1982).
101. S. Savin and G. Houin, *Clin. Pharmacokinet.* **15,** 355 (1988).
102. R. H. Gustafson and J. S. Kiser in Ref. 16, p. 405.
103. I. Chopra and A. H. Linton, *Adv. Microb. Physiol.* **28,** 211 (1986).
104. M. Robbins, R. Marais, D. Felmingham, and G. L. Ridgway, *J. Antimicrob. Chemother.* **20,** 379 (1987).
105. S. B. Levy in L. E. Bryan, ed., *Antimicrobial Drug Resistance,* Academic Press, New York, 1984, p. 191.
106. A. D. Russell and I. Chopra, *Understanding Antibacterial Action and Resistance,* Ellis Horwood, New York, 1990.
107. S. B. Levy, *J. Antimicrob. Chemother.* **24,** 1 (1989).
108. A. A. Salyers, B. S. Speer, and N. B. Shoemaker, *Mol. Microbiol.* **4,** 151 (1990).
109. I. Chopra, *J. Antimicrob. Chemother.* **18,** 51 (1986).
110. B. Eckert and C. F. Beck, *J. Biol. Chem.* **264,** 11663 (1989).
111. E. K. Manavathu and co-workers, *Antimicrob. Agents Chemother.* **34,** 71 (1990).
112. M. C. Roberts and S. L. Hillier, *Antimicrob. Agents Chemother.* **34,** 261 (1990).

JOSEPH J. HLAVKA
GEORGE A. ELLESTAD
IAN CHOPRA
American Cyanamid Company

ANTIBLOCKING AGENTS. See RELEASE AGENTS.

ANTIBODIES. See IMMUNOASSAY; MEDICAL DIAGNOSTIC REAGENTS; VACCINE TECHNOLOGY.

ANTIDIABETIC DRUGS. See INSULIN AND OTHER ANTIDIABETIC DRUGS.

ANTIDIARRHEALS. See GASTROINTESTINAL AGENTS.

ANTIEMETICS. See GASTROINTESTINAL AGENTS.

ANTIEPILEPTICS. See HYPNOTICS, SEDATIVES, AND ANTICONVULSANTS.

ANTIFOAMING AGENTS. See DEFOAMERS.

ANTIFREEZES AND DEICING FLUIDS

An antifreeze is defined as a chemical which, when added to a water-based fluid, reduces the freezing point of the mixture. Antifreezes are used in a wide variety of mechanical equipment during the winter months to prevent freezing of aqueous heat-transfer fluids. Most commonly, antifreeze refers to the freeze-protected fluid which cools automotive engines, although antifreeze liquids are also used in ice skating rinks, refrigeration systems (as a secondary coolant), heating and air conditioning systems, solar energy units, and many other applications. Chemical antifreezes include brines, such as calcium chloride; alcohols, such as methanol, ethanol, and 2-propanol; and glycerol and glycols. Since 1960 ethylene glycol has held the majority of the antifreeze market share because of its availability and superior performance characteristics.

Colligative Properties

Many chemicals when added to water cause a freezing point depression, as shown in Table 1, and thus are termed antifreezes. The antifreeze properties of these chemicals vary widely as a function of their colligative, or concentrative, properties. The reduction in freeze point depends both on the chemical itself and the concentration of the chemical in water. The freeze point depression increases as the antifreeze chemical is added to the water, until a characteristic concentration is achieved. Further addition of the antifreeze chemical to water will either result in insolubility or serve to increase the freezing point of the mixture, as illustrated in Figure 1.

Table 1. Freeze Point Depression of Antifreeze Chemicals[a]

Component	CAS Registry Number	Molecular formula	Concentration in water, wt %	Freeze point depression, °C
calcium chloride	[*10043-52-4*]	$CaCl_2$	32[b]	−50
calcium magnesium acetate	[*76123-46-1*]	$C_2H_4O_2{\cdot}xCa{\cdot}xMg$	32.5[b]	−28
ethanol	[*64-17-5*]	C_2H_6O	50	−38
ethylene glycol	[*107-21-1*]	$C_2H_6O_2$	50	−22
glycerol	[*56-81-5*]	$C_3H_8O_3$	50	−22
methanol	[*67-56-1*]	CH_4O	50	−50
potassium acetate	[*127-08-2*]	$C_2H_4O_2{\cdot}K$	50	−50
potassium chloride	[*7447-40-7*]	KCl	13[b]	− 6.5
propylene glycol	[*57-55-6*]	$C_3H_8O_2$	50	−32
seawater			100 (6% salt)	− 3
sodium chloride	[*7647-14-5*]	NaCl	23[b]	−21
sucrose	[*57-50-1*]	$C_{12}H_{22}O_{11}$	42[b]	− 5
urea	[*57-13-6*]	CH_4N_2O	44[b]	−18

[a]Refs. 1–3.
[b]Saturated solution at this concentration. Becomes insoluble at higher concentrations.

The colligative properties of antifreeze chemicals may also result in boiling point elevation. As the chemical is added to water, the boiling point of the mixture increases. Unlike the freeze depression, the boiling elevation does not experience a maximum; the boiling point versus concentration curve is a smooth curve that achieves its maximum at the 100% antifreeze level. The boiling point elevation can be another important characteristic for antifreeze fluids in certain heat-transfer applications.

History of Antifreeze

Antifreeze solutions are found in many heat-transfer applications. For heat transfer, plain water provides superior performance, low cost, and universal availability. However, water alone has the very limited operating temperature range of 0 to 100°C. The antifreeze fluid increases this range by depressing the freezing point and elevating the boiling point, while still providing good heat transfer. This enables a water-based heat-transfer system to operate outdoors during the winter months, as does an automobile; and to operate at higher temperatures and thus higher efficiencies, as do large internal combustion engines.

Early antifreeze solutions were predominantly salt brines and alcohols.

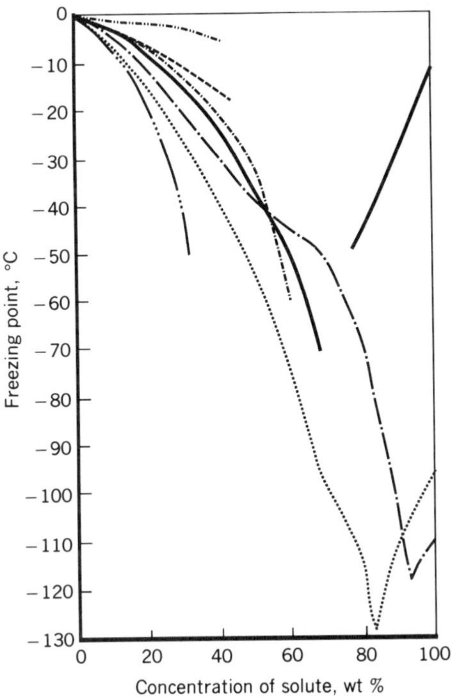

Fig. 1. Freeze point depression as a function of solute concentration (1,2). Calcium chloride (··–··–··–), sucrose (··——··—), and urea (–––––) become insoluble above maximum concentration given. Ethylene glycol (———) forms an eutectic at 65–80% solute concentration. Freezing points cannot be measured in this region. Propylene glycol (·–·–·–·–) supercools at higher concentrations. Freezing points cannot be measured. Ethanol (–·——) and methanol (·········) are also included.

Glycerol and sugar solutions were used to a limited extent. Salt brines were relatively inexpensive and widely available, and gave efficient freeze point depression. However, salt solutions proved to be corrosive to the metals of construction and caused scaling on heat-transfer surfaces. Scaling and precipitation of the salt drastically reduces heat transfer and system efficiency. Today, because of its low cost, brine solutions still find use in some antifreeze applications including refrigeration systems, off-shore heat-transfer systems (seawater), and roadway deicing; but are not used for cooling of automotive engines and other internal combustion engines.

Like brines, alcohols were readily available and widely used as antifreeze liquids in the early 1900s. Both methanol and ethanol offer excellent heat transfer and efficient freeze point depression. However, the alcohols have the distinct disadvantage of their low boiling points. During the summer months when the engines operate hot, significant amounts of the alcohols are lost because of evaporation. These evaporative losses result in costly make-up requirements. Additionally, the alcohols have very low flash points and potentially flammable vapors. These safety concerns have, particularly in recent years, caused the use of alcohols to be completely discontinued for most heat-transfer systems.

The use of ethylene glycol as an antifreeze was first initiated in 1925.

Although not as efficient a freeze point depressant as methanol, ethylene glycol has a very low vapor pressure so that evaporative losses from a cooling system are water vapor rather than glycol vapor. This reduces make-up costs and ensures that the freeze protection is maintained in the system. Thus, glycol-based antifreezes are termed permanent antifreezes. Ethylene glycol (EG) also eliminated the flammability concerns associated with methanol-based antifreeze. Ethylene glycol-based antifreeze solutions (50%) have no flash point.

For automotive engine cooling, ethylene glycol or methanol were the predominant products from the late 1920s to early 1950s. In 1927, the first formulated EG-based fluid was introduced by Union Carbide (4). *Prestone* brand antifreeze contained ethylene glycol with additives to prevent corrosion and foaming in the engine. By 1960, 90% of the automotive antifreeze market was using formulated ethylene glycol-based products. The consumption of EG-based antifreeze is shown in Table 2, as defined by the Chemical Specialty Manufacturers Association (5). In 1990 automotive antifreeze represented approximately 33% of the total ethylene glycol consumption in the United States (6). The primary antifreeze manufacturers include BASF Corp., First Brands Corp., Old World Trading Co., Shell Chemical Co., and Texaco Chemical Co.

Table 2. United States Consumption of Ethylene Glycol-Based Automotive Antifreeze[a]

Year	Consumption, 10^6 L[b]
1965	534
1970	672
1975	824
1980	738
1985	881
1986	727
1987	731
1988	818
1989	606

[a]Ref. 5.
[b]To convert from liters to U.S. gallons, divide by 3.79.

Ethylene glycol continues to be the antifreeze of choice for cooling of internal combustion engines. EG-based antifreezes may contain a few percent of other glycols, specifically diethylene and triethylene glycols. However, these other ethylene-based glycols, or the propylene-based glycols (propylene glycol (PG) and dipropylene glycol), are not widely used as the primary component of engine coolants. Ethylene glycol provides both the most effective freeze point depression, as well as the best heat-transfer characteristics (see GLYCOLS). Propylene glycol-based antifreezes do have some limited usage in Europe (specifically Switzerland and Austria), where human toxicity concerns override the performance and environmental benefits of ethylene glycol.

Engine Coolants

Besides freeze protection, antifreezes provide many other performance properties that enhance the operation of a heat-transfer system. Because the internal combustion engine is by far the largest antifreeze application, and ethylene glycol is the predominant antifreeze in use, the following focus on the performance properties of an ethylene glycol-based antifreeze and their relationship to engine cooling.

Heat Transfer. The primary role of the antifreeze liquid in an internal combustion engine is to remove heat and thus cool the engine. If heat is not removed, the engine will be destroyed. The cooling system must remove about one-third of the heat generated by the engine; one-third is removed by the exhaust system; and the balance of the heat is converted into mechanical energy. The antifreeze fluid removes heat by passing through the engine where it picks up heat, then through the radiator where it is cooled by air. The fluid operates in a closed loop system and thus is continuously reused. To provide efficient cooling, the antifreeze must have a high specific heat and thermal conductivity and low viscosity at operating temperatures. Water alone provides the most efficient heat transfer, as defined by the heat-transfer coefficient (HTC). The higher the heat-transfer coefficient, the greater the efficiency of the system. The HTC is directly proportional to the thermal conductivity (k), the specific heat (C_p), and the density of the fluid (ρ), and is inversely proportional to the viscosity (μ). The addition of antifreeze to the cooling system results in a slight penalty in heat-transfer efficiency by decreasing the HTC because of the lower specific heat and higher viscosity of the aqueous glycol solution.

Freeze Point Depression. The slight heat-transfer penalty incurred when an antifreeze is added to the aqueous heat-transfer fluid is necessitated by the need for increased operating temperature range in most internal combustion engines. Because most parts of the world achieve temperatures below freezing during some time of the year, an antifreeze fluid is required to keep equipment operational in these subfreezing temperatures.

The colligative properties of antifreeze fluids dictate the freeze point depression of the antifreeze solution. The freeze point depression offered by a chemical is dependent on the concentration of that substance in the solution; ie, as the concentration of the solute is increased, the freezing point of the solvent is depressed. Most solutes reach a maximum freezing point depression at a characteristic concentration. When this concentration is exceeded, either the solute becomes insoluble or the freezing point begins to increase. For ethylene glycol solutions, the maximum freezing point is achieved by using a 68% aqueous solution; this concentration is known as the eutectic concentration. Antifreeze concentrations in excess of 68% have higher freezing points; lower concentrations offer less freeze point depression, as indicated in Figure 1. Thus it seems ideal to use this 68% solution to give the maximum freeze protection. However, because water has the best heat transfer and heat transfer is sacrificed as the antifreeze concentration is increased, it is recommended that lower concentrations be used if local weather conditions permit. For automotive applications, a 50% solution is usually standard: the −36.7°C freezing point provides adequate freeze protection in most areas of the continental United States; heat-transfer characteristics do not compromise engine efficiency; this is typically the concentration used as the basis for design of the cooling system. In larger internal combustion engines, particularly stationary compression engines, where tremendous fuel savings can be achieved by maximizing cooling efficiency, the concentration of coolant chosen usually gives a freezing point 3–6°C (5–10°F) below the lowest historical ambient temperature for the region.

When defining the desired concentration and thus the use temperatures for an antifreeze fluid, several factors should be taken into consideration. First, the freezing point of an antifreeze is defined as the temperature at which the first ice crystal is formed in the fluid. Below this temperature, the fluid will not function efficiently as a heat-transfer fluid; however, at this temperature the fluid will not freeze to a completely solid state, thus protecting the system from the expansion that occurs upon complete freezing. In contrast, water freezes solid when the temperature is maintained at its freezing point of 0°C (32°F), and undergoes an expansion of 9% (7). The addition of an antifreeze to the water significantly reduces the amount of expansion that the fluid undergoes, thus protecting the system from bursting. For glycol solutions, the amount of expansion upon freezing decreases as the amount of glycol in solution increases (Fig. 2). For example, a 20% glycol solution expands 5% as the temperature is lowered 19.4°C below its freezing point; a 30% solution expands 3.3% at 22°C below its freezing point (8). After the first ice crystal formation at the fluid's freezing point, a continued drop in temperature is accompanied by additional crystal formation to produce a thick, slushy state. For higher glycol concentrations, a completely solid state is never achieved; rather the fluid becomes thick and taffylike. The temperature at which the fluid ceases to flow, defined as the pour point, is significantly lower than the freezing point. In systems that are not operational during the winter months, a fluid concentration is chosen to protect the system against bursting, not freezing, since some crystal formation is not harmful. However, it is important to remember that the use of a fluid down to its pour point may require significant pumping power to restart the system during these cold temperatures because of the drastic increase in viscosity as the fluid becomes slushy. Additionally, because of the poor heat transfer of these slushy fluids, systems that anticipate cold weather operation should choose a fluid having a freezing point below the coldest expected winter temperature.

Boiling Point Elevation. Boiling point elevation may also be beneficial in heat transfer applications. As engine power increases, the heat given off by the engine also increases. This increase in engine operating temperature increases engine efficiency by making more heat available to power the engine. However, if too much heat remains in the engine, overheating of both the engine and the oil may occur and engine efficiency decreases. The extra cooling required can be supplied either by increasing the cooling system capacity, or by enabling the coolant to accept more heat. An increase in the capacity of the cooling system is a costly alternative that can require physical changes to the system. Enabling the coolant to remove more heat most often requires no capital changes, and can be achieved by either increasing the flow rate of the coolant, or increasing the system pressure, and thus the boiling point of the coolant, which would allow the coolant to circulate at a higher maximum temperature. Today, most engines

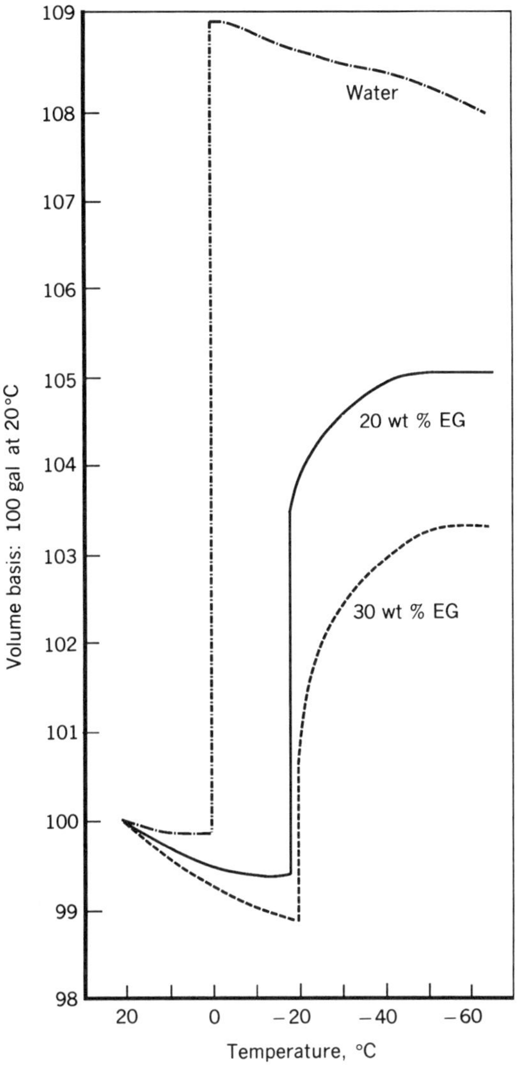

Fig. 2. Expansion of water and ethylene glycol (EG) on freezing (7,8). 100 gal = 377 L.

operate slightly pressurized to increase the maximum use temperature of the coolant and thus eliminate boiling of the coolant under extreme operating conditions. Under these operating conditions, the very low vapor pressure of glycol-based coolants significantly reduces the evaporative losses of the coolant, when compared to high vapor pressure liquids such as methanol and ethanol.

Corrosion Inhibition. Another important property of antifreeze solutions is the corrosion protection they provide. Most cooling systems contain varied materials of construction including multiple metals, elastomeric materials, and rigid polymeric materials. The antifreeze chosen must contain corrosion inhibitors that are compatible with all the materials in a system. Additionally, the fluid and its corrosion inhibitor package must be suitable for the operating temperatures and conditions of the system.

Of the commonly used chemical antifreezes, salt brines are inherently the most corrosive. The alcohols and glycols, in their pure forms, are generally noncorrosive to common metals of construction. However, as water is added to these compounds to reduce the freezing point, the aqueous antifreeze solution assumes the corrosivity of the water of dilution. Because water sources contain varying amounts of corrosive ions such as chloride and sulfate ions, antifreeze solutions contain inhibitors to counteract these ions. Additionally, in the case of glycol solutions, some corrosion inhibitors also act as buffering agents to counteract the slow oxidation of the glycol. At high temperatures, in the presence of oxygen, all glycols are slowly oxidized to form acids of degradation. These acids, if not adequately neutralized, are corrosive to the metals in the system.

The corrosive effects of water and heat-transfer fluids, if untreated, can be detrimental to the heat-transfer system. Corrosion in a system is first evidenced by the presence of corrosion products in solution, either in the form of solid particles in the fluid, or as metal ions in solution with the fluid. The corrosion products in the fluid alter the heat-transfer characteristics of the fluid and may reduce its heat-transfer efficiency. Heat transfer is further reduced because of the irregular metal surface left behind as corrosion products are entrained by the fluid. The higher surface area of these irregular surfaces require a higher concentration of corrosion inhibitors to cover and protect the metal surface. Inadequate concentrations of corrosion inhibitors result in the presence of bare metal surfaces, an open invitation to corrosion. Corrosion is further accelerated as the metal ions in solution react with the active corrosion inhibitor ions, and render them incapable of protecting the system against further corrosion. The solid corrosion products flowing with the fluid may also cause erosion corrosion, in which the solid particles impacting the metal surfaces break down the protective inhibitor layers and allow corrosion to occur. These erosive effects are most pronounced in high velocity points within the system. Another corrosion phenomena is known as transport deposition, where the corroded metal is dissolved in the fluid (usually at the engine), then deposited in another point of the system (usually the radiator), thus reducing cooling efficiency. As these various forms of corrosion continue unthwarted, the amount of solid corrosion products in the system increases, and eventually builds to the point where the solids may clog lines and restrict flow through the system. Ultimately, the lack of cooling in the engine as a result of all of these factors results in engine failure.

Corrosion reduction can be achieved either by altering the fluid or the system operation. During operation of the heat-transfer system, many factors may accelerate corrosion. Because corrosion is an oxidative process, the most critical factor is the amount of oxygen in the system. The presence of oxygen allows metal surfaces to oxidize or corrode; additionally, in glycol systems, this oxygen accelerates the oxidative degradation of the glycol to form corrosive acids. Both of these processes are accelerated by the elevated temperatures in the engine. Another factor affecting the tendency for a system to corrode is the presence of contaminants in the system. Contaminants can range from dirt and corrosion particles to oils to the ions in poor quality water. These contaminants affect the system in varied ways, such as air entrainment, erosion corrosion, reaction with the corrosion inhibitors, and glycol or inhibitor degradation. These factors make it important to operate the system as cleanly, and as oxygen-free as possible.

The first glycol-based antifreeze, preformulated with corrosion inhibitors, was introduced in the United States in 1927. Since that time, many and varied antifreeze formulations and brands have been used in automotive cooling systems. Today, all antifreezes contain inhibitors designed to minimize corrosive attack, including both organic and inorganic compounds. The most common corrosion inhibitors for glycol-based antifreezes include borates, molybdates, nitrates, nitrites, phosphates, silicates, amines, triazoles, and thiazoles (9,10). Since no single compound provides complete corrosion protection to all metals under all conditions, antifreeze formulations contain a combination of these inhibitors. Additionally, since cooling system design varies from automobile to automobile, and from year to year, the antifreeze formulation must have enough latitude to account for these changes and yet maintain adequate corrosion protection for most systems.

The choice of corrosion inhibitors for an antifreeze is also dependent on the type of service of the engine. Engine cooling systems can be categorized into three types: light-duty automotive, heavy-duty diesel, and continuous stationary engine operations. For automotive applications, the engine operates intermittently. The corrosion inhibitors must protect the system during operation and while idle. For this service, film-forming inhibitors such as silicates are predominantly used. Silicates are particularly useful for corrosion protection of heat-emitting aluminum surfaces such as that found in an aluminum head engine. Silicates have been more widely used as automobile manufacturers have tried to reduce car weight to improve fuel efficiency by increasing the use of aluminum in engines. The film-forming silicates effectively protect the system while idle (most automobiles remain idle greater than 50% of the time), but have the disadvantage of reducing the heat-transfer efficiency of the coolant. Additionally, with time, silicates react adversely with the glycol and any salts or hardness ions in the fluid (phosphates, calcium) to form a gel. The gel severely reduces heat transfer and may clog lines. If untreated, silicate gel formation may cause engine failure. Many antifreeze manufacturers now include a silicate stabilizer to retard this reaction, and recommend frequent (annual or biannual) antifreeze replacement. Nonetheless, for light-duty automotive service, silicates work effectively to inhibit corrosion.

For heavy-duty diesel service, on the other hand, high concentrations of silicates are usually not recommended because of the silicate–glycol reaction. Heavy-duty diesel service implies more continuous operation of the engine, and therefore, a heavy film-forming compound is not required to protect during long idle periods and is undesirable because of its adverse effects on heat-transfer efficiency. Also, aluminum is typically not used in heavy-duty engines. For diesel service, a low silicate formulation is used in conjunction with a supplemental coolant additive (SCA). SCAs are important in systems where cavitation of water pumps and cylinder liners is frequently evidenced. SCAs supplement the concentration levels of inhibitors in the base coolant, and are frequently replenished throughout the life of the coolant.

Inhibitor replenishment is also important in stationary engine operation, where the large volume and continuous operation of the engine prohibit frequent fluid replacement, because of the high cost and down time constraints. For stationary engines, such as those used in natural gas transmission applications, even minute concentrations of silicates must be avoided, because of the rapid gel formation and loss of heat transfer during the continuous operation of these engines. Coolant formulations for stationary engines are different from automotive and diesel formulations; they predominantly consist of phosphate for ferrous metal protection and buffering and a triazole for protection of the yellow metals. Periodic analysis of these fluids is used to determine the type and amount of inhibitor replenishment required.

Foaming. Another factor in maintaining heat-transfer efficiency and good system operation, is minimizing the foaming tendency of the fluid. Foaming or air entrainment in the fluid minimizes the amount of fluid that is in contact with the heat-transfer surface, and thus reduces the amount of heat that the fluid can remove from the system. Air entrainment may also cause a form of corrosion in the system: as the air bubbles break against the walls of the system, the pressure of them breaking may cause a phenomena called cavitation erosion corrosion. Although air entrainment alone can probably not cause system failure from cavitation, it may work in conjunction with mechanical forces such as vibration. The combination of these two forces is detrimental and frequently causes system failure, particularly of the cylinder liner.

Although glycol–water solutions are not inherently prone to foaming, mechanical and chemical conditions may cause foam to form in the system. To minimize foaming and thus the possibility of cavitation failure and the loss of heat transfer, the amount of air entrained in the coolant must be minimized. Air is often entrained in the coolant expansion tank; increasing the level in this tank to recommended levels minimizes the air entrained. Other causes of foaming include the use of organic corrosion inhibitors and the presence of various contaminants in the fluid. To counteract these effects, most antifreeze formulations contain antifoaming agents, such as silicones, polyglycols, or oils. Often the antifoamer functions by being insoluble in the glycol solution and appears as a separate phase when not circulating. In the system, the antifoamer is well-dispersed and not evident as a second phase (see DEFOAMERS).

Effect on Elastomers. The antifoamers and corrosion inhibitors used in antifreeze formulations are designed not only to function effectively in their assigned role, but also to be compatible with the other materials of construction in the system, such as the elastomers and plastics. Various components in the coolant can swell and soften elastomers or cause them to become brittle and crack. In general, ethylene glycol causes the elastomer to shrink when compared to the effect of pure water. Therefore, once a system is charged with a specific type of coolant, replacement with another chemical may require replacement of the gaskets and hoses in the system. This is most often seen when plain water is used during the summer months to achieve the maximum heat-transfer efficiency, then glycol is added for winter operations. The shrinkage of the elastomeric gaskets with the addition of the glycol may cause leakage in the system. Table 3 shows the compatibility of various elastomers with ethylene glycol.

Service Life. The service life offered by a coolant is dependent on many factors, including the initial condition of the coolant and the cooling system, the type of water used for dilution, the metals of construction in the system, the type of corrosion inhibitors and SCAs used, the system operating temperature, and the

Table 3. Compatibility of Ethylene Glycol with Elastomeric Materials[a]

Trade name or common name	Material	Temperature 25°C	80°C	160°C
Adriprene L-100[b]	urethane	good	poor	poor
Black Rubber 3773	polyisoprene, synthetic	good	poor	poor
Buna N	acrylonitrile–butadiene	good	good	
Buna S	styrene–butadiene	good	fair	poor
Butyl Rubber	isobutylene–isoprene	good	good	
EPDM	ethylene–propylene–diene	good	good	good
EPR Rubber	ethylene-propylene rubber	good	good	good
Hycar, D-24[b]	butadiene–styrene, butadiene–acrylonitrile, acrylate emulsion	good	fair	
Hypalon[b]	chlorosulfonated polyethylene	good	poor	poor
Kalrez[b]	tetrafluoroethylene, perfluoromethyl vinyl ether cure site monomer	good	good	good
Natural Rubber Gum	polyisoprene, natural	good	poor	poor
Neoprene 7797	chloroprene	good	fair	
Red Rubber #107	polyisoprene, synthetic	good	poor	poor
Silicone #65	polydimethylsiloxane	good	good	
Viton A[b]	vinylidene fluoride and hexafluoropropylene	good	good	poor

[a]Ref. 2 (see ELASTOMERS, SYNTHETIC).
[b]Registered trademarks.

type of operation the coolant is in: light-duty/intermittent, heavy-duty/diesel, or continuous stationary. Before initial charging of a cooling system with fresh coolant, the system should be free of dirt, oil, and other contaminants. New cooling systems probably only require a water flush to remove solids introduced during assembly of the system. Older systems may require chemical cleaning to remove corrosion products, residual spent coolant, and other contaminants such as scale and oils. The addition of fresh coolant to an improperly prepared system results in rapid deterioration of the condition of the fresh fluid.

Similarly, the antifreeze itself must be prepared properly to extend the life of the fluid. Because many antifreeze solutions contain inorganic corrosion inhibitors such as phosphates and molybdates, the fluids are sensitive to dilution by hard water. Hard water ions, calcium and magnesium, complex with the corrosion inhibitors and render them ineffective in inhibiting corrosion. Additionally, the calcium and magnesium salts of the corrosion inhibitors are typically insoluble in aqueous solutions, and frequently cause scale formation on the walls of the system, thus reducing heat transfer. Consequently, the use of very hard water should be avoided, particularly in the larger engines that require greater heat-transfer efficiency. In addition to the hardness ions, poor quality water of dilution can result in the introduction of corrosive ions, chloride and sulfate, to the system. The use of corrosive water must be avoided to ensure adequate corrosion protection and to extend the life of the fluid and the system.

The service life of an antifreeze can be affected both by the absolute time a coolant is in a system, and by the actual service hours of the engine. Thus, the type of engine operation and the type of corrosion inhibitors in the fluid play a significant role in determining the life of the fluid. Because of the presence of the silicate inhibitors, antifreeze in light-duty service is designed to be replaced frequently (annually or biannually). However, the size of the automotive engine cooling system ensures that the coolant replacement is not cost prohibitive. For larger engines, yearly fluid replacement would be costly in terms of both the fluid cost and the engine downtime cost. Consequently, antifreeze for heavy-duty diesel and continuous stationary operations are designed to be longer lasting and the corrosion inhibitors easily replenished. Additionally, because stationary engines have the capability of operating hotter to increase efficiency, coolants for these engines typically have greater buffering capacity to counteract the effects of the accelerated glycol degradation caused by the higher temperatures.

Despite all these safeguards to extend the service life of the antifreeze, fluid replacement is required periodically. Typically, fluids are replaced because of irreversible damage caused by one of four conditions: contamination, gel formation because of glycol/silicate reaction, extensive glycol degradation caused by overheating or excessive oxygen exposure, or inhibitor depletion.

Fluid Maintenance. The frequency of fluid replacement can be minimized and the heat-transfer efficiency and overall condition of the system maximized by proper maintenance of the engine cooling system. The most critical requirement for an antifreeze is that it provides adequate freeze protection. Because of evaporative losses of water and system makeup, it is necessary to monitor the freezing point of the antifreeze to ensure adequate protection. The freezing point of a glycol-based fluid can be determined in the field by specific gravity or refractive index methods. Both methods measure the amount of glycol in the solution, and from this the freezing point can be determined from the literature or from precalibrated measuring devices. An automotive hydrometer, measuring the fluid's specific gravity, is calibrated to read the freezing point of the ethylene glycol-based fluid directly. Because of differences in the various brands of antifreeze, the automotive hydrometer will provide the freezing point within 4.4°C (8°F) of the actual freezing point, but should not be relied on for greater accuracy. A laboratory hydrometer or pycnometer, which reads in specific gravity, provides a high degree of accuracy; however, literature values correlating specific gravity to freezing point for the specific brand of coolant are required. A handheld refractometer is another convenient method for field determinations of the coolant's freezing point. The refractometer is more accurate and more expensive than the automotive hydrometer, reading to within 1.67°C (+3°F) of the freezing point, but less accurate than a laboratory hydrometer or pycnometer. Because the handheld refractometers are also calibrated directly in freezing point, the refractive index to freezing point correlation must account for the differences in the available coolants, and thus accuracy is sacrificed.

The freezing point of the coolant should be monitored for coolants in all types of service. Additionally, maintenance of the corrosion inhibitor levels is required of the heavy-duty service coolants and the stationary engine coolants. Because corrosion inhibitors and combinations of corrosion inhibitors work most effectively at given concentrations and specific ratios to the other inhibitors, appropriate concentrations must be maintained to maximize corrosion protection. Many manufacturers of coolants for stationary engines, and manufacturers of SCAs, provide an analytical service to monitor the effective inhibitor concentrations in the system periodically. Recommendations can then be made for proper maintenance and inhibitor replenishment.

Fluid Specifications. The performance characteristics of all antifreeze solutions are governed by fluid specifications, that have been developed over the years by industry standards committees, such as the American Society for Testing and Materials (ASTM) and the Society of Automotive Engineers (SAE). Additionally, most engine and/or cooling system manufacturers have their own compositional specifications to which the fluids must conform.

ASTM D-3306 is a long-standing performance specification for engine coolants, under the jurisdiction of ASTM Committee D-15 on Engine Coolants (11). Most recently, as the varying performance requirements for light-duty versus heavy-duty engines were defined, D-3306 has been renamed and reclassified to apply specifically to coolants for light-duty and automotive applications. A separate specification, D-4985, has been developed to account for the needs of heavy-duty service. In addition to this reclassification of the performance requirements for engine coolants, several additional specifications are being developed to respond to the changing needs of the industry including a specification for prediluted coolants (diluted by the coolant manufacturer) that has been adopted by the ASTM committee. The need for specifications for propylene glycol-based coolants is also being reviewed. PG-based fluids currently find limited application in Switzerland and Austria where human toxicity concerns prohibit the use of EG-based coolants. PG is also being proposed in several developmental applications, such as anhydrous cooling systems that permit engine operation at higher temperatures and cavitation control in heavy-duty service, where the higher viscosity of PG is reported to be beneficial. Lastly, a specification for the use of recycled coolants is being explored. The many and varied modes of recycling provide vastly different performance characteristics of the recycled product, and a standard is needed to govern these fluids.

Deicing Fluids

Antifreeze chemicals, because of their characteristic freeze point depression, find additional applications in the deicing of aircraft and runways. Because of the stringent requirements of corrosion protection of aircraft surfaces and the necessity of a low flash point for use around aircraft engines, aircraft deicing fluids use glycols exclusively (over brines and alcohols) to achieve freeze point depression; runway deicing is commonly achieved with either urea or glycol-based fluids that contain urea, although two acetate salts are being studied for runway usage. Calcium magnesium acetate and potassium acetate are being reviewed by the Federal Aviation Administration (FAA) and the Air Transport Association (ATA).

Runway Deicing. Deicing compounds are applied to the runway under certain conditions to assist in the removal of frozen accumulations of snow and ice that cannot be readily removed by mechanical means such as plowing. The primary purpose of chemical deicing is not to melt the surface ice and snow, but rather to diffuse through the snow and ice to break the bond between the ice and the runway. Once the ice–surface bond is broken, the remaining frozen accumulation can then be removed by mechanical means. The use of chemical deicing often reduces the amount of time and equipment used in snow removal. Runway antiicing can also be achieved by applying certain deicing compounds in anticipation of freezing rain, snow, or icing conditions. It is critical that both deicing and antiicing be conducted in a closely controlled manner where the application rate is monitored and the runway is monitored for slipperiness. Improperly applied runway deicing fluids may create a slippery condition.

Aircraft Deicing. Aircraft deicing fluids typically contain at least 90% glycol, with the balance being water, corrosion inhibitors, wetting agents, and an orange dye. In North America, ethylene glycol is the most commonly used freeze point depressant, although other glycols, such as diethylene, triethylene, and propylene, as well as mixtures of these, are also used. The corrosion inhibitors are necessary to protect the materials of constuction of the aircraft; wetting agents ensure complete coverage of the surfaces being deiced. Deicing fluids are typically diluted prior to use, to achieve maximum freeze point depression; aqueous glycol solutions achieve their minimum freezing point at about 68 wt % deicing fluid and 32% water.

Deicing is defined as a method for removing ice, snow, and frost accumulations from the surfaces of aircraft during ground operations. These frozen deposits, if untreated, significantly affect the aerodynamic performance of the aircraft by reducing lift and altering thrust and drag characteristics. Additionally, severe damage to the engines may occur if ice buildup on the wings dislodges during the takeoff run and strikes the turbine fan blades. Removal of frozen accumulations from aircraft surfaces prior to takeoff is required by the FAA's clean aircraft concept which states that "flight following ground operations conducive to aircraft icing should not be attempted unless it is ascertained that the aircraft is free of any ice formations that may degrade the aircraft's performance or change its flight characteristics" (12).

To remove frozen accumulations, deicing fluids are applied at high temperatures ($\sim$71–82°C) and pressures to melt the ice, snow, and frost, and to break the adhesion to the surface of the aircraft. Deicing leaves the aircraft surfaces wet and therefore, to guard against refreezing, the deicing fluid must contain sufficient freeze point depressant to lower the freezing point of the residual fluid on the aircraft below the ambient or surface temperatures. The FAA Advisory Circular 20-117 recommends that "the freeze point of the residual fluids should not be greater than 11°C (20°F) below ambient or surface temperature, whichever is less" (12). In the absence of further precipitation, the low freezing residual fluid will guard against refreezing. However, during precipitation or frost-forming conditions, deicing fluids offer time-limited protection against refreezing and ice or snow buildup. Precipitation will dilute the residual fluid, raising its freezing point, and thus freezing on the surface of the aircraft may occur. Additionally, dilution of the residual fluid by precipitation lowers the viscosity of the fluid, reduces the film thickness of the fluid remaining on the aircraft, and eventually flushes the fluid from the aircraft surfaces. For these reasons, the FAA recommends that a close inspection of the aircraft be performed immediately prior to takeoff to ensure that it is aerodynamically clean and free of ice, snow, or frost accumulations.

During precipitation or frost-forming conditions, several techniques may be used to maintain an aerodynamically clean aircraft following deicing and until rotation and takeoff. Most desirably, the deicing operation should be performed as close to the takeoff point as possible. End of the runway deicing is practiced at several airports in the United States. This procedure eliminates the possibility of refreezing prior to takeoff, since the takeoff roll commences immediately following deicing. End of the runway deicing is remote deicing since the deicing operation is removed from the gate and terminal area. Remote deicing can be performed either using a mobile facility which includes several deicing trucks working in tandem, or using a fixed facility, usually a "drive thru" station. One type of fixed facility, referred to as a "car wash" system, is a large structure which surrounds the aircraft and deices using multiple spray nozzles fixed to the structure at adjustable angles. Both systems, mobile and fixed, have distinct advantages. Mobile facilities lend flexibility for reconfiguring depending on storm conditions; and fixed facilities enable collection of spent deicing fluid and thus ease the management of storm water runoff.

Although end of the runway deicing is the only reliable method for ensuring that an aircraft is aerodynamically clean at takeoff during precipitation conditions, many airports are configured such that they do not permit end of the runway deicing. At these facilities, when long taxi times and delays are expected, antiicing techniques must be employed if takeoffs during precipitation are to be permitted. Antiicing is most often performed using a two-step procedure, in which the aircraft is first deiced using a hot, high pressure fluid, and then antiiced using a cold fluid. Antiicing must only be performed following complete deicing. Antiicing fluids must only be applied to aerodynamically clean surfaces, and serve to protect the clean surfaces from further accumulations of frozen deposits for a finite period of time.

Traditionally, in North America, antiicing has been performed using cold, undiluted deicing fluids containing high glycol concentrations, so that when diluted with precipitation, the freezing point decreases and the residual fluid provides protection against refreezing. However, during certain weather conditions, because of the low film thickness and low viscosity of the fluid, the fluid can run off the surface of the aircraft before its freeze protection is depleted. Nonetheless, the use of neat deicing fluids for antiicing does provide some limited protection against refreezing during light precipitation conditions. At low volume airports where taxi times are low, neat deicing fluids work effectively to retard refreezing during taxiing and the takeoff roll under light precipitation and frost-forming conditions, and thus allow airport operations to proceed uninterrupted. Deicing fluids should never be used for antiicing purposes during heavy precipitation or freezing rain conditions.

Because of the time-limited protection that deicing fluids provide, and because of the rapidly increasing air traffic and subsequent longer taxi times, thickened, non-Newtonian antiicing fluids are being investigated for use at several of the larger airports in North America. The current antiicing fluids (AAFs) contain glycols for freeze point depression, with added water, corrosion inhibitors, wetting agents, and stabilizers. AAFs also contain a polymeric thickening agent that gives the fluid its antiicing properties. The thickening agent increases the fluid's viscosity and allows the fluid to adhere to the aircraft surfaces. Current AAFs contain a minimum of 50% glycol (usually 50–60%), and adequate water as required to solubilize the thickening agent. Because of this added water for solubilizing the thickener, the neat antiicing fluid has already achieved its minimum freezing point. Further dilution of the fluid increases the freezing point and reduces its protection against refreezing. Antiicing fluids are typically undyed; if dyed, they cannot be orange or blue, thus distinguishing AAFs from aircraft deicing fluids and runway deicing fluids.

The antiicing fluid's thickening agent gives the fluid its non-Newtonian behavior, and thus its antiicing protection. Non-Newtonian fluids are defined as fluids with a viscosity dependent on shear as shown in Figure 3. The fluid has a higher viscosity at low shear rates and a lower viscosity at higher shear rates. The non-Newtonian design is critical to its performance. When the aircraft is stationary or taxiing, the fluid is at low shear, the fluid's viscosity is high, and its high viscosity prevents flow and enables it to adhere to the aircraft surface. During the takeoff run, the fluid experiences high shear, its viscosity is reduced, and it flows from the aircraft (13).

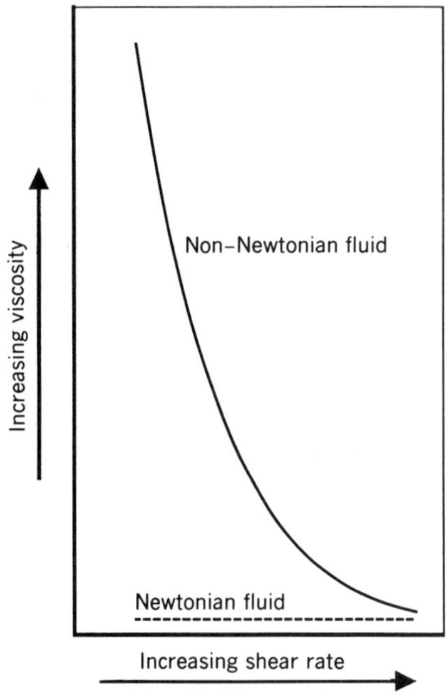

Fig. 3. Shear dependence of Newtonian versus non-Newtonian fluids (13).

Non-Newtonian antiicing fluids have been used in Europe for many years. Early non-Newtonian fluids had high viscosities of up to 50 Pa·s (500 P) at −25°C; these fluids have been found to have adverse effects on the aerodynamic performance of the aircraft, reducing the lift coefficient at takeoff by greater than 12%. Current non-Newtonian fluids have substantially lower viscosities (<10 Pa·s = 100 P) and lower lift losses of 2–7%, as compared to ethylene glycol-based deicing fluids, which have lift losses of 1–3% (14). The antiicing fluids, when applied to a clean aircraft, form a high viscosity, protective layer on the surface of the aircraft and block the adhesion of frozen deposits. However, the protection provided by these fluids is also time-limited. The amount of antiicing protection they provide, defined as the holdover time, varies tremendously with weather conditions (temperature and type and rate of precipitation), type of fluid used, and type of aircraft antiiced. Because of these varied and ever-changing conditions, the holdover times for antiicing fluids cannot be predicted. Inspection of the aircraft immediately prior to takeoff is necessary at all times to ascertain that all aerodynamic surfaces are free of adhering ice, snow, or frost.

Non-Newtonian antiicing fluids are currently being used in the United States on a limited, experimental basis. Antiicing fluids require special handling and application equipment and techniques, and therefore deicing equipment modification and personnel training are required before these fluids are used by an airline at an airport.

Materials Compatibility. In addition to the runway deicing and aircraft deicing/antiicing fluids' primary roles of removing frozen accumulations and protecting against the refreezing of aircraft and runway surfaces, these fluids are subject to many other performance criteria, as defined by the airlines, the aircraft manufacturers, and the aircraft engine manufacturers. The fluids must contain corrosion inhibitors to minimize the corrosive effects on aircraft metals of construction. Additionally, the fluids must not adversely affect the acrylics and polycarbonates used for aircraft windows by crazing or staining, and should not stain, discolor, or blister painted and unpainted aircraft surfaces.

Deicing and Antiicing Specifications. The performance criteria for deicing fluids have traditionally been governed by specifications of the Aerospace Division of Society of Automotive Engineers (SAE) (15). These specifications are widely used by most airlines in North America. The SAE specifications incorporate the materials' compatibility requirements set forth by the airframe manufacturers (Douglas CSD#1 and Boeing D6-17487), as well as the performance criteria recommended by the airlines, including minimum flash points, maximum freezing points, and hard water compatibility. A new aerodynamic performance criteria, as defined by the airframe manufacturers, is being added to all deicing and antiicing fluid specifications.

In 1987, a proposal was made to the International Standards Organization (ISO) to prepare international specifications for deicing and antiicing fluids, equipment, and procedures, in an effort to standardize winter ground operations and set minimum requirements for deicing and antiicing at airports throughout the world, and thus aid the pilot in understanding and defining the protection against refreezing as a function of the fluid used at a given airport. These specifications for worldwide standardization are currently being developed by the ISO.

Toxicity and Environmental Issues

The toxicity of antifreeze and deicing fluids is predominantly a function of the main component, the freezing point depressant. For ethylene glycol-based fluids, the toxicity is well-defined, as the toxicity of ethylene glycol has been studied extensively because of its wide usage in varied applications (16).

Ethylene glycol is acutely toxic to humans and animals by oral exposure. For humans, the lethal dose is estimated to be 1.4 mL/kg of body weight, or about 3.2 fl oz (95 mL) for a 150 lb (68 kg) person. For antifreeze applications, oral toxicity is a concern primarily because of the possibility of accidental ingestion by children and pets. Legislation is currently being proposed to study the effectiveness of adding bittering agents or emetic agents in consumer antifreeze, to reduce the likelihood of ingestion of lethal doses (17). For deicing fluid applications, oral ingestion is unlikely. Ethylene glycol is defined as an animal teratogen meaning that at high concentrations in laboratory animals, ethylene glycol has been shown to cause birth defects.

For deicing fluid applications, exposure to vapors and mists is the more likely means of exposure. Ethylene glycol has a threshold limit value (TLV) of 50 ppm for vapors. When proper deicing procedures are followed and proper protective equipment worn, the exposure of deicing personnel to vapor and mist is expected to fall well below this TLV value. This mode of exposure is unlikely for engine cooling applications.

Because of these toxicity concerns, propylene glycol is sometimes used instead of ethylene glycol in these applications. Studies on propylene glycol have shown no evidence of teratogenic effects and it has a lower oral toxicity, with an estimated lethal value of 7.0 mL/kg or 16.2 fl oz (479 mL) for a 150 lb person. No TLV has been defined for propylene glycol, but an AIHA WEEL Guide of 50 ppm has been determined. Propylene glycol has, however, been shown to be slightly more irritating to the skin than ethylene glycol (16).

Aquatic Toxicity. Aquatic toxicity is different from human toxicity and must be taken into consideration when choosing a chemical that has the potential to enter lakes, streams, or rivers, such as deicing fluids. Unlike human toxicity issues, aquatic toxicity is not strictly a function of the primary component of the fluid, ie, the freeze point depressant. In formulated deicing and antiicing fluids, the additives play a significant role in defining the aquatic toxicity. In particular, many surfactants or wetting agents are highly toxic to aquatic life; whereas both ethylene and propylene glycol are essentially nontoxic: concentrations exceeding 10,000 mg/L show no effect on aquatic life (16). Thus it is important when choosing a deicing or antiicing fluid to obtain aquatic toxicity data on the formulated product and not just the glycol itself.

Biodegradation and Oxygen Demand. Another important consideration in the use of all chemicals, including antifreezes and deicing fluids, is the rate and extent of biodegradation, or the length of time the chemical persists in the environment. Organic chemicals, such as the glycols, biooxidize to give carbon dioxide and water; however, the rate at which they degrade and the amount of oxygen used in the process determine their effect on the environment. In the case of ethylene glycol-based fluids, laboratory tests indicate complete biooxidation occurs within 20 days. The rate of biooxidation is steady throughout this 20-day period and thus exerts no undue oxygen demand on waterways (causing fish kill) or waste treatment facilities. Propylene glycol, on the other hand, biodegrades rapidly during the first five days of the test (62%), then the biooxidation rate slows down substantially such that only 79% biooxidation is indicated at the end of the 20-day test (16). The rapid initial biooxidation imposes a great oxygen demand on the receiving stream or waste treatment facility. Airports using propylene glycol-based deicing fluids must reduce the rate at which they feed the runoff to the waste treatment facility by about half when compared to the rate at which a solely ethylene glycol deicing fluid runoff can be fed to the waste treatment facility.

Human toxicity, aquatic toxicity, and the environmental impact of engine coolants and deicing fluids are typically measured on the fresh fluid only. Spent fluids contain varied contaminants that can drastically affect the toxicity and environmental impact of the fluid. Most pronounced is the impact of heavy-metal contaminants in spent antifreeze. Data on spent and recycled antifreeze, compiled by the ASTM Committee on Engine Coolants, show an average lead level 11 ppm, as well as various other metal contaminants (iron, copper, zinc) (18). The presence of these contaminants in a used fluid may require special disposal techniques for the fluids.

Fluid Recycling

When antifreeze becomes unsuitable for use, either because of depletion of inhibitors, presence of corrosion products or corrosive ions, or degradation of the fluid, recycling and reuse of the antifreeze, rather than disposal, may be considered. Although ethylene glycol is readily biodegraded in typical municipal waste treatment facilities, antifreeze disposal becomes problematic because the coolant may contain hazardous quantities of heavy metals picked up from the cooling system. Recycling may be economically preferred over coolant disposal and reduces the concern for environmental impact.

Ideally, a system for recycling spent antifreeze consists first of the removal of the deleterious contaminants such as the corrosion products, corrosive ions, degradation products, and remaining inhibitors. Then the clean fluid could be reinhibited to a known concentration of both inhibitors and glycol.

Two methods of recycling spent antifreeze, filtration or distillation, have been reported (19,20). The filtration method consists of a filter and a system for reinhibition of the filtered glycol. It may be a portable or skid-mounted recycling machine located in a shop environment. The system filters out particles and other visible contaminants, but does not remove the corrosive ions, dissolved metals, degradation products, or corrosion inhibitors. Typically, supplemental coolant additives consisting of premixed quantities of corrosion inhibitors are added to the filtered glycol. However, because all residual inhibitors are not removed in the filtration process, the addition of premixed inhibitor packages to the filtered fluid may not result in the appropriate balance of inhibitors in the recycled coolant. This is especially critical in silicated fluids, where the excess silicate may result in rapid gelation.

The other recycling method is fractional distillation. This is more complex and capital intensive than a filtering system but produces pure refined glycol which is free of all the deleterious contaminants. The refined material is then reinhibited to exactly the desired coolant formulation. Often this is accomplished through centralized locations where the spent coolant from various sources is collected, combined, then refined and reinhibited. A distillation method is also currently in limited use for the recycling of aircraft deicing fluids. The *Gantry* deicing system, a central deicing car wash-type system, collects the used deicing fluid through a porous pad, then distills the used fluid to remove many of the contaminants. The distilled product is then mixed with virgin deicing fluid before use on the aircraft for deicing.

Because of the problems with some recycling options, the ASTM committee on engine coolants is exploring a specification for recycled coolants. Some automobile manufacturers have found the in-shop recycling equipment to be ill-advised; one service bulletin warns of the deterioration of engine-coolant ethylene glycol to the extent that no additive can restore it to an acceptable state (21).

With the increasing emphasis on recycling in general and environmental awareness in the United States, further growth in the recycling of antifreezes and deicers is anticipated. Emerging technologies may allow portable shop systems to properly clean spent antifreeze to provide a good base to build an acceptable antifreeze product.

BIBLIOGRAPHY

"Antifreezes and Deicing Fluids" in *ECT* 3rd ed., Vol. 3, pp. 79–95, by R. H. Fay, The Dow Chemical Company.

1. R. C. Weast and D. R. Lide, eds., *CRC Handbook of Chemistry and Physics,* 40th ed., IRC Press, Boca Raton, Fla., 1989.
2. *Ethylene Glycol Product Information Bulletin,* Union Carbide Chemicals and Plastics Co., Danbury, Conn., 1990.
3. *Ice-B-Gon Product Literature,* Chevron Chemical Company, San Ramon, Calif., Mar. 1990.
4. *Prestone II Physical Properties Data Bulletin,* 7th ed., Union Carbide Corporation, Tarrytown, N.Y., 1974.
5. *Ethylene Glycol Base Antifreeze Survey,* Chemical Specialties Manufacturers Association, Washington, D.C., 1980–1989.
6. B. F. Greek, *Chem. Eng. News* **68**(31), 11–12 (July 30, 1990).
7. R. H. Perry and D. W. Green, *Chemical Engineers' Handbook,* 6th ed., McGraw-Hill Book Company, New York, 1984.
8. *UCARTHERM Heat Transfer Fluid Product Information Bulletin,* Union Carbide Chemicals and Plastics Co., Inc., Danbury, Conn., 1989.
9. *General Motors Engineering Standards:* Automotive Engine Coolant, Antifreeze Concentrate-Ethylene Glycol Type, GM 6038-M, Jan. 1971; Automotive Engine Coolant, Concentrate-Ethylene Glycol Type, GM 6043-M, Sept. 1983.
10. *Worldwide Trends in Engine Coolants, Cooling System Materials, and Testing,* SP-811, Society of Automotive Engineers, Warrendale, Pa., 1990.
11. *1990 Annual Book of ASTM Standards,* Vol. 15.05, ASTM, Philadelphia, Pa., 1990.
12. "Hazards Following Ground Deicing and Ground Operations in Conditions Conductive to Aircraft Icing," FAA Advisory Circular, AC 20-117, Federal Aviation Administration, Washington, D.C., Dec. 17, 1982.
13. B. A. Stefl, *Airport Technol. Int.,* 213–217 (Aug. 1989).
14. L. J. Runyan, T. A. James, A. Zierten, E. G. Hill, and H. E. Addy, *Lewis IRT Test of the Aerodynamic Effects of Aircraft Ground Deicing/Anti-Icing Fluids,* NASA technical paper, National Aeronautics and Space Administration, Fla., to be published.
15. *Aerospace Material Specifications* 1425B and 1426B (Apr. 1990), 1427A (Oct. 1981), Society of Automotive Engineers, Warrendale, Pa.
16. *A Comparison of Ethylene Glycol with Propylene Glycol,* Union Carbide Chemicals and Plastics Co., Inc., Danbury, Conn., 1989.

17. *Chem. Mark. Rep.* **238**(6), 7 (Aug. 6, 1990).
18. *ASTM Committee D-15 on Engine Coolants (Minutes),* Memphis, Tenn., American Society of Testing and Materials, Philadelphia, Pa., 1990.
19. *Radiator Rep.* **18**(2), 90-17 (Feb. 1990).
20. *Radiator Rep.* **18**(5), 90-49 (May 1990).
21. *Radiator Rep.* **18**(3), 90-33 (Mar. 1990).

BARBARA A. STEFL
KATHLEEN F. GEORGE
Union Carbide Chemicals and Plastics Company, Inc.

ANTIHISTAMINES. See HISTAMINE AND HISTAMINE ANTAGONISTS.

ANTIHYPERCHOLESTERMICS. See CARDIOVASCULAR AGENTS.

ANTIHYPERTENSIVES. See CARDIOVASCULAR AGENTS.

ANTIHYPOTENSIVES. See CARDIOVASCULAR AGENTS.

ANTIINFLAMMATORY DRUGS. See ANALGESICS, ANTIPYRETICS, AND ANTIINFLAMMATORY AGENTS.

ANTIMALARIALS. See ANTIPARASITIC AGENTS, ANTIPROTOZOALS.

ANTIMITOTICS. See CHEMOTHERAPEUTICS, ANTICANCER.

ANTIMONY AND ANTIMONY ALLOYS

Antimony [*7440-36-0*], Sb, belongs to Group 15 (VA) of the periodic table which also includes the elements arsenic and bismuth. It is in the second long period of the table between tin and tellurium. Antimony, which may exhibit a valence of +5, +3, 0, or −3 (see ANTIMONY COMPOUNDS), is classified as a nonmetal or metalloid, although it has metallic characteristics in the trivalent state. There are two stable antimony isotopes that are both abundant and have masses of 121 (57.25%) and 123 (42.75%).

History and Occurrence

Antimony and the natural sulfide of antimony were known at least as early as 4000 BC. The sulfide was used as an eyebrow paint in early biblical times, and a vase found at Tello, Chaldea, was reported to be cast antimony. Copper articles covered with a thin coating of metallic antimony, found in Egypt and dating from the period 2500–2200 BC, indicate that the early Egyptians were using antimony. Pliny (50 AD) gave it the name *stibium;* Geber used the term *antimonium;* and as late as the time of Lavoisier both terms continued to be used for the sulfide. Early alchemists referred to antimony sulfide as the "wolf of metals" because it devours all metals except gold. *Triumph Wagen des Antimonii,* written by Basil Valentine in the 15th century, is recognized as one of the first significant accounts of antimony and its chemistry. Both Agricola (1559) and Biringuccio (ca 1550) mention the liquation of antimony ores. The latter enumerates the following uses for antimony: as an alloy to increase the tone of bell metal; in the production of pewter and glass and metal mirrors; as medication for ulcers; and as yellow pigment for painting earthenware and tinting enamels and glass. A scientific treatise on the element was written by Nicolas Lemery (1645–1715).

The crustal abundance of antimony (1) is approximately 0.2 g/t. Antimony ore bodies are small and scattered throughout the world. The word antimony (from the Greek *anti* plus *monos*) means "a metal not found alone" and, in fact, native antimony is seldom found in nature because of its high affinity for sulfur and metallic elements such as copper, lead, and silver. Over a hundred naturally occurring minerals of antimony have been identified (2–4). Occasionally native metallic antimony is found; however, the most important source of the metal is the mineral stibnite [*1317-86-8*], (antimony trisulfide), Sb_2S_3. In areas where stibnite has been exposed to oxidation, it is converted to oxides of antimony. The important oxide minerals are stibiconite [*12340-12-4*], $Sb_2O_4 \cdot H_2O$, cervantite, Sb_2O_4 or $Sb_2O_3 \cdot Sb_2O_5$?, valentinite [*1317-98-2*], orthorhombic Sb_2O_3, and senarmontite [*12412-52-1*], cubic Sb_2O_3; kermesite [*12196-98-0*], 2 $Sb_2S_3 \cdot Sb_2O_3$, an oxysulfide ore, is also of commercial importance.

World reserves of antimony (5) are estimated to be from 4–5 million metric tons. Approximately 80% of the world's reserves are located in China, Bolivia, Russia, Thailand, the Republic of South Africa, and Mexico; China has the world's largest reserves. The United States possesses only 2% of the world's reserves. Most of the antimony produced in the United States is from the complex antimony deposits found in Idaho, Nevada, Alaska, and Montana. These deposits consist of stibnite and other sulfide minerals containing base metals and silver or gold. Ores of the complex deposits are mined primarily for lead, copper, zinc, or precious metals; antimony is a by-product of the treatment of these ores.

Properties

Physical properties of antimony are given in Table 1. Antimony, a silvery white, brittle, crystalline solid, is a poor conductor of electricity and heat. Two unstable allotropes, a black and a yellow modification, have been observed (6). The black modification is amorphous and forms on rapid quenching of antimony vapor. The yellow modification, covalent and similar to yellow arsenic (7,8), is formed by the low (−90°C) temperature oxidation of stibine using oxygen or chlorine. Explosive antimony, sometimes referred to as a third modification, can be produced from the electrolysis of antimony chloride, iodide, or bromide under special conditions, and is believed to be in a strained amorphous state. When scratched or bent, it explodes mildly, giving the crystalline form. Sudden heating of the electrodeposit to about 125°C produces an explosion; however, gradual heating leads safely to the crystalline form. Some electrodeposits drying at room temperature have been observed to ignite and burn, though incompletely, to form the oxide. On solidification, pure antimony contracts 0.79 ± 0.14 vol % (9).

Table 1. Physical Properties of Antimony

Property	Value
CAS Registry Number	
Sb	[*7440-36-0*]
^{121}Sb	[*14265-72-6*]
^{123}Sb	[*14119-16-5*]
at wt	121.75
mp, °C	630.8
bp, °C	1753
density at 25°C, kg/m^3	6684
crystal system	hexagonal (rhombohedral)
lattice constant, nm	
a	0.4307
c	1.1273
hardness, Mohs' scale	3.0–3.5
latent heat of fusion, J/mol[a]	19,874
latent heat of vaporization, J/mol[a]	195,250
coefficient of linear expansion at 20°C, $\mu m/(m \cdot °C)$	8–11
electrical resistivity at 0°C, $\mu\Omega \cdot cm$	39
magnetic susceptibility at 20°C, cgs	-99.0×10^{-6}
specific heat at 25°C, J/(mol·K)[a]	25.2
thermal conductivity at 0°C, W/(m·K)	25.5
capture cross-section for thermal neutrons, at 2200 m/s, 10^{-28} m^2/atom	
121.75 (at wt)	5.40 ± 0.60
121 (isotope)	6.25 ± 0.20
123 (isotope)	4.28 ± 0.16

[a]To convert J to cal, divide by 4.184.

Antimony is ordinarily quite stable and not readily attacked by air or moisture. It burns emitting a bluish light when heated to redness in air. Under controlled conditions antimony reacts with oxygen to form the oxides Sb_2O_3, Sb_2O_4, and Sb_2O_5. Antimony tetroxide [*1332-81-6*] may be considered a stoichiometric compound of composition $Sb_2O_3 \cdot Sb_2O_5$, or antimony(III) antimonate(V).

Certain conditions of pH, oxidation potential, and temperature promote the corrosion or dissolution of antimony in aqueous systems (10). In nonoxidizing, ie, unaerated solutions, antimony is stable and does not dissolve over a wide pH range. As more air, or oxidizing agent, is allowed into the solution, however, antimony begins to oxidize to the tri- and pentavalent states. In the trivalent state, antimony exists in the form SbO^+, oxoantimony [*22877-95-8*], or Sb_2O_3, antimony(III) oxide [*1309-64-4*] which dissolves in solution as antimonious acid, $HSbO_2$, and antimonite [*27264-01-3*], SbO_2^-. The pH of the oxidized solution determines which complex is predominant. Stronger oxidants such as nitric acid, mercury oxide, sodium peroxide, or hydrogen peroxide, oxidize antimony to its pentavalent state. Under these conditions, the species SbO_2^+, antimony(V) oxide [*1314-60-9*], Sb_2O_5, and SbO_3^-, antimonate [*15600-59-6*] exist.

Nitric acid oxidizes antimony forming a gelantinous precipitate of a hydrated antimony pentoxide (8). With sulfuric acid an indefinite compound of low solubility, probably an oxysulfate, is formed. Hydrofluoric acid forms fluorides or fluocomplexes with many insoluble antimony compounds. Hydrochloric acid in the absence of air does not readily react with antimony. Antimony also forms complex ions with organic acids.

Antimony reacts vigorously with chlorine to form tri- and pentachlorides. It combines with sulfur in all proportions and forms the compounds antimony red [*1345-04-6*], Sb_2S_3, and golden antimony [*1315-04-4*], Sb_2S_5. Antimony itself does not react directly with hydrogen. However, antimony hydride [*7803-52-3*] (stibine), SbH_3, which is extremely poisonous, may be formed by the reaction of metal antimonides, for example, zinc antimonide [*12039-40-6*], Zn_3Sb_2, and acid, the reduction of antimony compounds in hydrochloric acid with zinc, aluminum, or other reducing metals, and the electrolysis of acid or alkaline solutions using an antimony cathode (see ANTIMONY COMPOUNDS).

Process Metallurgy

The antimony content of commercial ores ranges from 5 to 60%, and determines the method of treatment, either pyrometallurgical or hydrometallurgical. In general, the lowest grades of sulfide ores, 5–25% antimony, are volatilized as oxides; 25–40% antimony ores are smelted in a blast furnace; and 45–60% antimony ores are liquated or treated by iron precipitation. The blast furnace is generally used for mixed sulfide and oxide ores, and for oxidized ores containing up to about 40% antimony; direct reduction is used for rich oxide ores. Some antimony ores are treated by leaching and electrowinning (4) to recover the antimony. The concentrates may be leached directly or converted into a complex matte first. The most successful processes use an alkali hydroxide or sulfide as the solvent for antimony (see METALLURGY; METALLURGY, EXTRACTIVE).

Oxide Volatilization. Removal of antimony as the volatilized trioxide is the only pyrometallurgical method suitable for low grade ores. Combustion of the sulfide components of the ore supplies some of the heat; hence, fuel requirements are minor. There are many variations of the volatilization process, the principles employed being the same but the equipment differing. In all cases, the sulfur is burned away and removed from the waste gases, whereas the volatile antimony trioxide is recovered in flues, condensing pipes, a baghouse, Cottrell precipitator, or a combination of the above. Roasting and volatilization are effected almost simultaneously by heating the ore, mixed with coke or charcoal, under controlled conditions in equipment such as a shaft furnace, rotary kiln, converter, or roaster.

If the volatilization conditions are too oxidizing, the nonvolatile antimony tetroxide may form and the recovery of antimony, as antimony trioxide, is diminished. Usually the oxide produced in this manner is impure and can be reduced to metal. However, special attention to choice of charge, volatilization conditions, and selection of product results in a high grade oxide that is suitable for use in ceramics and other applications.

A process has been developed to recover antimony and arsenic from speiss and other materials (11). The speiss is roasted along with a source of solid sulfur and coal or coke at a temperature of 482–704°C for a sufficient time to volatilize arsenic and antimony oxides. The arsenic can then be separated from the antimony through careful control of the off-gas temperature and oxygen potential (12).

Liquation. Antimony sulfide is readily but inefficiently separated from the gangue of comparatively rich sulfide ore by heating the gangue to 550–600°C in perforated pots placed in a brick furnace. The molten sulfide is collected in lower containers. A more efficient method uses a reverberatory furnace and continuous liquation; however, a reducing atmosphere must be provided to prevent oxidation and loss by volatilization. The residue, containing 12–30% antimony, is usually treated by the volatilization process to recover additional antimony. The liquated product, called crude or needle antimony, is sold as such for applications requiring antimony sulfide, or is converted to metallic antimony by iron precipitation or careful roasting to the oxide followed by reduction in a reverberatory furnace.

Oxide Reduction. The oxides of antimony are reduced with charcoal in reverberatory furnaces. An alkaline flux consisting of soda, potash, and sodium sulfate, is commonly used to minimize volatilization and dissolve residual sulfides and gangue. Part of the slag is frequently reused. Loss of antimony from the charge by volatilization is high (12–20% or more), even with use of ample slag and careful control. This necessitates the use of effective Cottrell precipitators or baghouses, and considerable recycling of oxide.

Iron Precipitation. Rich sulfide ore or liquated antimony sulfide (crude antimony) is reduced to metal by iron precipitation. This process, consisting essentially of heating molten antimony sulfide in crucibles with slightly more than the theoretical amount of fine iron scrap, depends on the ability of iron to displace antimony from molten antimony sulfide. Sodium sulfate and carbon are added to produce sodium sulfide, or salt is added to form a light fusible matte with iron sulfide and to facilitate separation of the metal. Because the metal so formed contains considerable iron and some sulfur, a second fusion with some liquated antimony sulfide and salt follows for purification.

Blast Furnace Smelting. Intermediate (25–40%) grades of oxide or sulfide or mixed ores, liquation residues, mattes, rich slags, and briquetted fines or flue dusts are processed in water-jacketed blast furnaces. In general, the blast-furnace practice used for lead is followed, employing a high smelting column, comparatively low air pressure, and separation of slag and metal in a forehearth. Slag, usually running under 1% antimony, is desired because it tends to reduce volatilization losses.

Leaching Followed by Electrolysis. At the Sunshine Mining Co. (13,14), a silver–copper–antimony concentrate containing 15–20% antimony is batch-leached in hot concentrated sodium sulfide, Na_2S, solution. Leaching is carried out in mild-steel tanks heated with steam coils and equipped with agitators. Four metric tons of sodium sulfide solution (~300 g/L) are charged with each metric ton of tetrahedrite [*12054-35-2*], $Cu_{12}S_{13}Sb_4$, concentrate. The material is leached for 14 hours at 100°C. At completion of the leach step, the slurry is pumped to a thickener. The clear solution containing sodium thioantimonate [*13776-84-6*], Na_3SbS_4, is decanted and sent to the electrowinning department. The silver and copper present in the tetrahedrite are not affected by the Na_2S leach and report to the thickener underflow. This residue is washed, filtered, and shipped to a smelter.

Electrolysis, also a batch operation, is conducted in cells each consisting of nine cathodes and eight anodes. The electrodes are fabricated from mild-steel. The busbars carry a current of 1500 amperes resulting in a current density of 3 A/m^2 (28 A/ft^2). The voltage drop across each cell is typically three volts. During electrolysis, the electrolyte can become fouled with sodium compounds including the polysulfides, thiosulfates, and sulfates which hinder the deposition of antimony on the cathode. To reduce the effect of these oxidation products, two separate electrolyte solutions, an anolyte and a catholyte, are used. The two solutions are kept separate by placing the anodes in steel baskets that have canvas sides. The canvas acts as a diaphragm that permits the flow of electric current but reduces the migration of harmful oxidation products to the catholyte. Fresh anolyte is made up of NaOH and barren electrolyte. Pregnant catholyte solution (~60 g/L Sb) is added to the electrolytic cells while withdrawing an equivalent volume of barren catholyte (~10 g/L Sb). The cathode metal obtained is 95% pure antimony.

The Bunker Hill Co. (15) and ASARCO, Inc. (16) have developed processes for the leaching and electrowinning of antimony from tetrahedrite ores. As of 1991, only Sunshine Mining Co. was electrowinning antimony metal.

The filtered sodium thioantimonate solution obtained from the leaching of stibnite with sodium sulfide may also be reduced directly to metal by elemental sodium (17). Yields in excess of 95% of 99.5 pure antimony are claimed (18).

By-Product and Secondary Antimony. Antimony is often found associated with lead ores. The smelting and refining of these ores yield antimony-bearing flue, baghouse, and Cottrell dusts, drosses, and slags. These materials may be treated to recover elemental antimony or antimonial lead from which antimony oxide or sodium antimonate may be produced.

Recycling of antimony provides a large proportion of the domestic supply of antimony. Secondary antimony is obtained from the treatment of antimony-bearing lead and tin scrap such as battery plates, type metal, bearing metal, antimonial lead, etc. The scrap are charged into blast furnaces, reverberatory furnaces, or rotary furnaces, and an impure lead bullion or lead alloy is produced. Pure lead or antimony is then added to meet the specifications of the desired lead–antimony alloy.

Refining. The metal produced by a simple pyrometallurgical reduction is normally not pure enough for a commercial product and must be refined. Impurities present are usually lead, arsenic, sulfur, iron, and copper. The iron and copper concentrations may be lowered by treating the metal with stibnite or a mixture of sodium sulfate and charcoal to form an iron-bearing matte which is skimmed from the surface of the molten metal. The metal is then treated with an oxidizing flux consisting of caustic soda or sodium carbonate and niter (sodium nitrate) to remove the arsenic and sulfur. Lead cannot be readily removed from antimony, but material high in lead may be used in the production of antimony-bearing, lead-based alloys. The yield of refined antimony from the matting and fluxing technique is 85–90%.

Impure metal may be refined by electrolysis (19), although this procedure is not as economical as the pyrometallurgical treatment. An electrolyte containing antimony fluoride and sulfuric acid gives the best result. Most of the anode impurities are lowered by electrolysis. However, if the concentrations of copper and arsenic are high in the anode they codeposit with the antimony and are present as significant impurities in the cathode metal. The arsenic and sulfur content of the cathode metal is further lowered by melting in an oxidizing flux. The purity of electrolytic antimony under favorable conditions exceeds 99.9%.

The purity of refined antimony, normally referred to as regulus, is usually judged by the appearance of a dendritic fern or starlike pattern on the surface of the metal. This is produced by casting the antimony in molds into which a small amount of a starring slag has been poured; the antimony freezes while surrounded by the still molten slag. The starring flux has a melting point below that of antimony. An effective flux contains a 60–40 mixture of sodium sulfide and potassium carbonate, although other formulations can be used satisfactorily (2,20). Though this practice is generally accepted, it is quite unreliable in judging purity as highly contaminated antimony can also be starred by controlling the casting conditions and the flux composition.

Alternative Methods of Production. In general, metal chlorides are more easily volatilized than metal oxides because of the relatively high vapor pressures. The advantages of chlorine-based pyrometallurgical processes over conventional oxidizing processes are the efficient removal of impurities and cost effectiveness. Antimony has been successfully removed from concentrates containing copper, silver, and gold using a chloridizing roast (21).

Microbiological leaching of copper and uranium has been commercially developed and research has indicated that microorganisms may be used to oxidize complex antimony sulfide minerals (22,23). If this technology is developed commercially, it may allow for the exploitation of many low grade antimony deposits.

Economic Aspects

The United States is not self-sufficient in its requirements for antimony and is heavily dependent on imports of both ore and metal (Table 2).

Production. In the 1980s, mine production in the United States accounted for only a small (<5%) percentage of the annual domestic supply of antimony. An important (>50%) component of the domestic supply is the recovery of antimony from old scrap; such as that recovered from the recycling of scrapped batteries (24). However, since the 1970s, this percentage has decreased because of the increased use of low maintenance batteries, which use lead alloys containing less or no antimony; and the downsizing of automotive batteries which require less antimony per battery. Other factors that influence the supply of secondary antimony are the percentage of available batteries being recycled, the demand for

Table 2. Antimony Statistics for the United States, t[a]

Statistic	1975	1980	1985	1986	1987	1988	1989	1990
primary production								
mine	804	311						
smelter	11,058	14,571	14,922	15,691	17,930	17,616	18,954	19,717
secondary production	16,297	18,047	13,635	14,081	15,833	16,172	19,501	20,380
exports								
metals, alloys, wastes/scrap	308	411	328	540	795	624	293	588
oxides	262	833	803	526	705	1,227	1,850	7,139
imports for consumption	16,970	16,326	18,773	23,043	24,248	30,027	25,165	29,403
industrial consumption, primary antimony	11,782	10,196	10,611	9,935	10,373	12,067	13,424	12,739
ave price, \$/kg[b]	3.89	3.33	2.89	2.69	2.44	2.29	2.08	1.81

[a]Ref. 24.
[b]New York dealer price for 99.5 to 99.6% metal, cif U.S. ports.

batteries, and the prices of lead. The remainder of the U.S. supply of antimony is imported.

Total smelter output in the United States has been growing steadily since 1982 because of the growth in antimony oxide production. Primary antimony metal output has decreased since the 1970s because of the falling demand for antimony metal, and the availability of low cost metal from China.

Producers of primary metal and oxide in the United States are Amspec Chemical Corp., Anzon America, ASARCO Inc., Laurel Industries Inc., Sunshine Mining Co., and U.S. Antimony Corp.

Imports and Exports. The availability of economical foreign sources of antimony, mainly from China, the Republic of South Africa, Mexico, Bolivia, and Russia, has resulted in an increase in the quantity of antimony imported for consumption. Since 1986, over 50% of the antimony imported by the United States originated in China (25). The majority of the antimony imported into the United States is now in the form of antimony oxide. Exports of antimony, as antimony oxide, increased during the 1980s.

Industrial Consumption. The total consumption of primary antimony fell during the period from 1970 to 1986 (Table 3) because of the declining demand for antimony in most types of metallic uses. Since 1986, the demand for primary antimony in antimonial lead has increased, probably because of an increase in demand for starting–lighting–ignition (SLI) batteries. Total consumption in nonmetallic uses has remained stable. However, an increasing proportion of this is made up of flame retardant uses. Currently, batteries and flame retardants are the two largest markets for antimony.

Prices. The price of antimony metal has declined steadily since 1984. The principal reason for the fall in antimony prices is the oversupply of material, especially from Chinese sources, in relation to demand, which has been stable or decreasing during the 1980s. Antimony prices have fallen to such an extent that some producers of antimony products now find it more economical to use antimony metal as feed material in place of antimony ore and concentrates.

Table 3. Consumption of Primary Antimony, t[a]

Product	1975	1980	1985	1986	1987	1988	1989	1990
Metal products								
ammunition	217	328	372				521	602
antimonial lead	4,144	679	515	549	1,102	1,538	1,871	1,995
bearing metal and bearings	365	202	161	139	187	178	129	90
cable covering	21	28		62				
castings	16	9	10	11	8	13	8	8
collapsible tubes and foil	8	16						
sheet and pipe	54	26		36	76	181	157	123
solder	121	122	305	253	347	256	245	208
type metal	68	19	28	8	8	6	4	3
other	109	67	95	379	584	609	80	104
Total metal products	*5,123*	*1,497*	*1,486*	*1,437*	*2,312*	*2,781*	*3,015*	*3,133*
Nonmetal products								
ammunition primers	13	18	24	21	32	34	20	23
fireworks	9	4	4	4	2	4	5	3
ceramics and glass	897	1,182	1,077	932	1,122	1,221	1,050	991
pigments	291	453	133	226	279	179	196	246
plastic	990	1,484	905	885	750	916	1,141	1,148
rubber products	415	295	23	37		29	97	27
other	597	97	128	147	199	147	159	151
Total nonmetal products	*3,212*	*3,533*	*2,294*	*2,252*	*2,384*	*2,530*	*2,668*	*2,589*
Flame retardants								
plastics	2,269	3,514	5,016	4,517	4,139	5,469	5,842	5,668
pigments	83	51	7	13	30	104	926	502
rubber	156	171	286	398	387	282	174	181
adhesives	114	418	281	154	315	251	219	189
textiles	679	855	1,140	1,163	800	643	558	460
paper	145	157	101	1				
other					6	7	22	17
Total flame retardants	*3,446*	*5,166*	*6,831*	*6,426*	*5,676*	*6,756*	*7,741*	*7,017*
Total, all uses	*11,782*	*10,196*	*10,611*	*9,935*	*10,374*	*12,067*	*13,424*	*12,739*

[a]Ref. 24.

Uses and Specifications

Antimony in the unalloyed state is extremely brittle and is not easily fabricated. For this reason the use of the pure metal is restricted to ornamental applications.

Antimony Alloys. Approximately one-half of the total antimony demand is for metal used in antimony alloys. Antimonial lead is a term used to describe lead alloys containing antimony in proportions of up to 25%. Most commercial lead–antimony alloys have antimony contents less than 11%. The compositions of several important antimony alloys are given in Table 4.

Table 4. Compositional Ranges of Antimony Alloys

Material	Composition, wt %			
	Sb	Sn	Pb	Others
type metal	2–25	2–15	balance	Cu(0–2)
battery grids				
conventional	2.5–6	0.25–1	balance	
low maintenance	1.5–2.5	0.25–1	balance	
maintenance-free	0	0.25–1	balance	Ca(0.04–0.075)
babbitt metal				
tin base	4–8.5	83–90	balance	Cu(3–8.5); As(0.1)
lead base	9–17.5	0.5–11	balance	Cu(0.5); As(0.25–1.5)
cable covering	1–6	0.25–1	balance	
sheet and pipe	4–15	0.25–1	balance	
collapsible tubes	2–3	0.25–1	balance	
solder	0–6	1–100	balance	Ag(0–6); Cu(0.1–5)
pewter	1–8	balance	0–0.05	Cu(0–3)
britannia metal	2–10	balance	0–9	Cu(0.2–5); Zn(0–5)
bullets, shrapnel	0.5–12	0.25–1.0	balance	

The largest application for antimonial lead is its use as a grid metal alloy in the lead acid storage battery (see BATTERIES, SECONDARY, LEAD–ACID). In the manufacture of grid metal, antimony imparts fluidity, increased creep resistance, and fatigue strength, as well as electrochemical stability to the lead which is particularly advantageous for battery plates required in heavy-duty cycling. A disadvantage of using antimony in grid alloy is that high antimony levels increase the self-discharge characteristics, cause high gassing and poisoning of the negative electrode resulting from ion migration. The release of gases means that the battery must be vented which then promotes the loss of electrolyte through evaporation.

Demand for high performance SLI batteries has led to the development of smaller, lighter batteries that require less maintenance. The level of antimony is being decreased from the conventional 3–5% to 1.75–2.75% to minimize the detrimental effects. Lead alloys that contain no antimony have also been introduced. Hybrid batteries use a low antimony–lead alloy in the positive plate and a calcium–lead alloy in the negative plate.

Tin–antimony–copper and lead–antimony–tin white bearing alloys, commonly referred to as babbitt, are used to reduce friction and wear in machinery and help prevent failure by seizure or fatigue. These alloys exhibit good rubbing characteristics even under extreme operating conditions such as high loads, fatigue, or high temperatures. Tin babbits have greater corrosion resistance than lead babbitts. However, lead babbitts are generally cheaper than tin babbitts. Addition of antimony to babbitts increases strength and hardness. The choice of babbitt depends on the application and resultant desired properties. The use of antimony in babbits has been declining because technological advances have reduced the thickness of babbitt on backing material, and there is competition from aluminum–tin alloys and nonmetallic substitutes.

Soldering is a method of joining two metallic surfaces by flowing between them a low melting point alloy. Many different alloys are used for this purpose; however, tin–lead alloys are the most widely used. Other metals are added in small amounts, depending on the desired properties. Antimony is added to increase the hardness of tin–lead alloys. Rising concern over the contamination of drinking water by lead has resulted in the use of lead-free alloys for soldering copper pipes. A similar trend is occurring in the canning industry. The demand for antimony in solders used in the electronic, semiconductor, and automobile industries is still strong (see SOLDERS AND BRAZING ALLOYS).

Type metal, another tin–antimony–lead alloy, is used primarily in relief or letterpress printing. Antimony is added to increase hardness, minimize shrinkage, permit sharp definition, and reduce the melting point of the alloy. There has been a substantial decrease in the use of type metals as a result of the emergence of less expensive typesetting techniques.

Antimony hardens the lead used in the manufacture of small arms ammunition. Antimony alloyed with lead is also used in cable covering, sheet and pipe, and collapsible tubes. In these applications, antimony is utilized to increase strength and inhibit corrosion.

Precision duplication, durability, and metallic beauty have made antimonial alloys, such as pewter and britannia metal, desirable for decorative castings. Several different tin-base and lead-base antimony alloys are used in the jewelry industry. These alloys are typically cast in rubber or silicone molds.

Antimony may be added to copper-base alloys such as naval brass, Admiralty Metal, and leaded Muntz metal in amounts of 0.02–0.10% to prevent dezincification. Additions of antimony to ductile iron in an amount of 50 ppm, preferably with some cerium, can make the graphite fully nodular to the center of thick castings and when added to gray cast iron in the amount of 0.05%, antimony acts as a powerful carbide stabilizer with an improvement in both the wear resistance and thermal cycling properties (26) (see CARBIDES).

Carbon (qv) impregnated with antimony gives a dense nonporous material with a low tendency to seizure or galling which may be useful in bearings and seals under high loads and velocities at temperatures up to 500°C (27).

Semiconductor and Solar Cells. High purity (up to 99.9%) antimony has a limited but important application in the manufacture of semiconductor devices (see SEMICONDUCTORS). It may be obtained by reduction of a chemically purified antimony compound with a high purity gaseous or solid reductant, or by thermal decomposition of stibine. The reduced metal may be further purified by pyrometallurgical and zone melting techniques. When alloyed with Group 13 (IIIA) elements, the Group 15 (V) semiconductors, aluminum antimonide [*25152-52-7*], AlSb, gallium antimonide [*12064-03-8*], GaSb, and indium antimonide [*1312-41-0*], InSb, are formed. These intermetallic semiconductor materials exhibit optoelectronic behavior, ie, they emit or absorb electromagnetic radiation, and are utilized in such applications as infrared devices, diodes, and Hall-effect components. More recently, high efficiency solar cells have been produced that are comprised of two layers: one of gallium arsenide and the other of gallium antimonide (see SOLAR ENERGY).

Antimony is also used as a dopant in *n*-type semiconductors. It is a common additive in dopants for silicon crystals with impurities, to alter the electrical conductivity. Interesting semiconductor properties have been reported for cadmium antimonide [*12050-27-0*], CdSb, and zinc antimonide [*12039-35-9*], ZnSb. The latter has good thermoelectric properties. Antimony with a purity as low as 99.9+% is an important alloying ingredient in the bismuth telluride [*1304-82-1*], Bi_2Te_3, class of alloys which are used for thermoelectric cooling.

In the computer industry, new read/write optical discs, capable of storing over 250 megabytes, utilize a thin coating consisting of a germanium, tellurium, and antimony compound (see IMAGING TECHNOLOGY).

Antimony Compounds. The greatest use of antimony compounds is in flame retardants (qv) for plastics, paints, textiles, and rubber. Antimony compounds used in flame retardants are antimony pentoxide, sodium antimonate [*15593-75-6*], $Na[Sb(OH)_6]$, and, most importantly, antimony trioxide. These compounds, when used alone, are poor flame retardants; however, when combined with halogen compounds, they produce mixtures that are effective.

Antimony trioxide and sodium antimonate are added to specialty glasses as decolorizing and fining agents, and are used as opacifiers in porcelain enamels. Antimony oxides are used as white pigments in paints, whereas antimony trisulfide and pentasulfide yield black, vermillion, yellow, and orange pigments. Camouflage paints contain antimony trisulfide which reflects infrared radiation. In the production of red rubber, antimony pentasulfide is used as a vulcanizing agent. Antimony compounds are also used in catalysts, pesticides, ammunition, and medicines (see ANTIMONY COMPOUNDS).

Specifications. Antimony is available as cast cakes, ingots, broken pieces, granules, shot, and single crystals. ASTM has published standards for two grades of antimony ingots (28). Grade A has a minimum antimony content of 99.8% and the following impurity maximums: arsenic 0.05%; sulfur 0.10%; lead 0.15%; and others 0.05% each. Grade B material is composed of 99.5% antimony as a minimum with maximum impurity levels of arsenic 0.1%, sulfur 0.1%, lead 0.2%, and others 0.1% each. ASTM standards are also available for pewter, babbits, and solders (29).

Health, Safety, and the Environment

Although metallic antimony may be handled freely without danger, it is recommended that direct skin contact with antimony and its alloys be avoided. Properly designed exhaust ventilation systems and/or approved respirators are required for operations that create dusts or fumes. As with other heavy metals, orderly housekeeping practice and good personal hygiene are necessary to prevent ingestion of (or exposure to) antimony.

Toxicology. Toxicity data for antimony is as follows: the lowest published lethal dosage (LDLo) is 15 mg/kg in humans, 100 mg/kg, put into the peritoneal cavity, in rats, and 150 mg/kg (peritoneal cavity) in guinea pigs (30). Standards and recommendations on a time weight average are: Occupational Safety and Health Administration (OSHA), PEL = 500 μg Sb/m^3 air, American Conference of Governmental Industrial Hygienists (ACGIH), TLV = 500 μg Sb/m^3 air, and National Institute for Occupational Safety and Health (NIOSH), recommended exposure level of 500 μg Sb/m^3 air. The Department of Transportation (DOT) and NIOSH use the identifiers 2871 and CC4025000, respectively. Antimony is reported in the EPA Toxic Substances Control Act (TSCA) Inventory; antimony and its compounds are on the EPA Community Right to Know List; and DOT classifies antimony as Poison B.

Antimony trioxide is currently designated as a possible human carcinogen by the International Agency for Research on Cancer (IARC) and ACGIH (31). However, a chronic inhalation study, conducted by the Antimony Oxide Industry Association (AOIA), found no evidence of carcinogenicity (31,32).

Antimony is not known to cause cancer, birth defects, or affect reproduction in humans. However, antimony has been shown to cause lung cancer in laboratory animals that inhaled antimony-containing dusts and prolonged exposure to antimony can cause irritation of the eyes, skin, lungs, and stomach, in the form of vomiting and diarrhea. Heart problems can also result from overexposure to antimony (33).

Tests conducted on humans have shown that people who inhale antimony in concentrations of 3 mg/m^3 for a period of eight months to two years developed heart problems and stomach ulcers. People inhaling antimony in concentrations of 9 mg/m^3 for more than nine days had lung, eye, and skin irritation (33). Antimony generally enters the body through the lungs where it is transported to the blood and then to other organs. The body eliminates antimony over several weeks through urine and feces.

Stibine (SbH_3), a highly toxic gas, can form when nascent hydrogen is present with antimony metal. OSHA and the ACGIH, have recommended a permissable exposure limit for employees exposed to stibine of 0.1 ppm as a time weighted average. Adequate safeguards against overexposure to this gas are advised when handling antimony and its alloys.

Antimony in the Environment. Antimony enters the environment through mining and processing of antimony-containing ores; in the production and working of antimony metals, alloys, compounds, and to a lesser extent, by incinerators and coal burning plants. Most of the released antimony, after being suspended in the air, ends up in the soil where it attaches strongly to iron, manganese, and aluminum (34).

Generally, the concentration of antimony in the air varies from 1×10^{-6} mg/m^3 to 170×10^{-6} mg/m^3 and can be as high as 1000×10^{-6} mg/m^3 near antimony processing facilities (33). Because of the stability of antimony in aqueous systems, the concentration of dissolved antimony in rivers is very small (about 5 ppb antimony). Also, despite the fact that antimony is in the solder used in water pipes, it does not appear to dissolve in the drinking water.

Antimony may enter the human body through the consumption of meats, vegetables, and seafood which all contain about 0.2–1.1 ppb antimony.

Disposal of Antimony. Antimony and its compounds have been designated as priority pollutants by the EPA (35). As a result users, transporters, generators, and processors of antimony-containing material must comply with regulations of the Federal Resource Conservative and Recovery Act (RCRA).

Most of the antimony used in alloys is recycled. Wastes from mining and smelting are generally disposed of in landfills and in Publicly Owned Treatment Works (POTWs). However, environmental regulations are forcing generators of these materials to evaluate techniques to recover or remove antimony from these materials prior to disposal. Plastics and fibers containing antimony oxide are generally placed in landfills or are incinerated along with normal industrial or municipal waste.

BIBLIOGRAPHY

"Antimony and Antimony Alloys" in *ECT* 1st ed., Vol. 2, pp. 50–59, by B. W. Gosner and E. M. Smith, Battelle Memorial Institute; in *ECT* 2nd ed., Vol. 2, pp. 562–570, by S. C. Carapella, Jr., ASARCO Incorporated; in *ECT* 3rd ed., Vol. 3, pp. 96–105, by S. C. Carapella, Jr., ASARCO Incorporated.

1. R. Fairbridge, ed., *Encyclopedia of Geochemistry and Environmental Sciences,* Vol. IV, Van Nostrand Reinhold Co., New York, 1972, p. 252.
2. C. Y. Wang, *Antimony,* Charles Griffin Co., London, 1952.
3. H. Quiring, *Die Metallischen Rohstoffe,* Vol. 7, Antimon Ferdinand Enke, Stuttgart, Germany, 1945.
4. *Antimony Materials Survey,* U.S. Department of the Interior, U.S. Bureau of Mines, Washington, D.C., Mar. 1960.
5. P. A. Plunkert, *Mineral Facts and Problems, Preprint of Antimony Chapter in Bulletin 675,* U.S. Department of Interior, U.S. Bureau of Mines, Washington, D.C., 1985.
6. J. W. Mellor, *Comprehensive Treatise on Inorganic and Theoretical Chemistry,* Vol. 9, Longmans, Green, & Co., Inc., New York, 1929, pp. 339–586.
7. W. J. Maeck, *U.S. Atomic Energy Commission,* NAS-NSSOSS, Clearing House Scientific & Technical Information, NBS, U.S. Dept. Commerce, Springfield, Va., Feb. 1961, p. 10.
8. M. C. Sneed and R. C. Brasted, *Comprehensive Inorganic Chemistry,* Vol. 5, D. Van Nostrand Co., Inc., Princeton, N.J., 1956, pp. 111–152.
9. A. D. Kirschenbaum and J. A. Cahill, *Trans. Am. Soc. Met.* **55,** 849 (1962).
10. M. Pourbaix, *Atlas of Electrochemical Equilibria in Aqueous Solutions,* (translated by J. A. Franklin), Pergamon Press, London, 1966.
11. U.S. Pat. 4,891,061 (Jan. 2, 1990), T. P. Clement, II, J. R. Wettlaufer, and J. A. Scott (to ASARCO Inc.).
12. U.S. Pat. 4,808,221 (Feb. 28, 1989), T. P. Clement, II, T. Li, and J. P. Hager (to ASARCO Inc.).
13. W. C. Holmes, *Eng. Min. J.* **145,** 54 (1944).
14. W. D. Gould, *Eng. Min. J.* **156,** 91 (1955).
15. *Min. World* **4**(6), 3 (1942).
16. U.S. Pat. 3,969,202 (July 13, 1976), A. E. Albrethsen, M. L. Hollander, and W. H. Wetherill (to ASARCO Inc.).
17. W. Wendt, *Met. Ind.* **77,** 276 (Dec. 15, 1950).
18. *Metal Industry Handbook and Directory 1961,* Iliffe Books Ltd., London, 1961, p. 3.
19. D. Schlain, J. D. Prater, and S. Revitz, *J. Electrochem. Soc.* **95,** 145 (1949).
20. C. Y. Wang and G. C. Riddell, *Trans. Am. Inst. Min. Metall. Pet. Eng.* **159,** 446 (1944).
21. Å. Holmström, *Scand. J. Metallurgy* **17,** 248–258 (1988).
22. G. I. Karavaiko and co-workers, in R. W. Lawrence, R. M. R. Branion, and H. G. Ebner, eds., *Fundamental and Applied Biohydrometallurgy,* Elsevier, New York, 1986, pp. 115–126.
23. I. D. Fridman and E. E. Savari, *Sov. J. Nonferrous Met.* **26**(1), 102–105 (1985).
24. T. O. Llewellyn and co-workers, *Minerals Yearbook, Antimony,* U.S. Department of the Interior, U.S. Bureau of Mines, Washington, D.C., 1974, 1976, 1981, 1989, and 1990.
25. T. O. Llewellyn, *Mineral Commodity Summaries 1991, Antimony,* U.S. Department of the Interior, U.S. Bureau of Mines, Washington, D.C., 1991.

26. R. H. Aborn, *American Foundry Society* **84,** 503 (1976).
27. V. Belogorskii and co-workers, *Russ. J. Nonferrous Met.* **16,** 59 (1975).
28. *Annual Book of ASTM Standards, Part 8—Nonferrous Metals,* American Society for Testing and Materials, Philadelphia, Pa., 1981, pp. 164–165.
29. *Annual Book of ASTM Standards, Section 2—Nonferrous Metals,* American Society for Testing and Materials, Philadelphia, Pa., 1990, pp. 9, 15, and 419.
30. W. I. Sax and R. J. Lewis, Sr., *Dangerous Properties of Industrial Materials,* 7th ed., Vol. 2, Van Nostrand Reinhold Co., New York, 1989, p. 277.
31. *A One-Year Inhalation Toxicity Study of Antimony Trioxide in the Rat (with a One-year Recovery Period),* Bio/dynamics Incorporated, East Millstone, N.J.; submitted to Antimony Oxide Industry Association (AOIA), Feb. 9, 1990.
32. V. Potkonjak and M. Pavlovich, *Int. Arch. Occup. Environmental Health,* **51,** 199–207 (1983).
33. *Toxicological Profile for Antimony and Compounds,* Syracuse Research Corporation, Agency for Toxic Substances & Disease Registry, U.S. Public Health Services, under contract #205-88-0608, Oct. 1990.
34. D. Rai and co-workers, *Electric Power Research Institute Publication,* Vol. 2, EPRI EA-3356, EPRI, Palo Alto, Calif., 1984.
35. *Analysis of Clean Water Act Effluent Guidelines Pollutants. Summary of the Chemicals Regulated by Industrial Point Source Category,* U.S. EPA, Washington, D.C., 40 CFR Parts 400–475, 1991.

General References

C. H. Mathewson, ed., *Modern Use of Non-Ferrous Metals,* 2nd ed., American Institute of Mining, Metallurgical, and Petroleum Engineers (AIME), New York, 1953.

C. R. Hayward, *An Outline of Metallurgical Practices,* 3rd ed., D. Van Nostrand Co., Inc., Princeton, N.J., 1952.

G. A. Roush, *Strategic Mineral Supplies,* 1st ed., McGraw-Hill Book Co., Inc., New York, 1939, pp. 238–273.

R. L. Kulpaca and J. C. Archibald, Jr., *J. Metals* **5**(6), 786 (1953).

Trends in Usage of Antimony, Publication NMAB274, National Academy of Sciences—National Academy of Engineering, Washington, D.C., Dec. 1970.

The Economics of Antimony, 7th ed., Roskill Information Services Ltd., London, 1990.

N. Lemery, *Cours de Chemie,* W. Kettilby, London, 1977.

T. Li
G. F. Archer
S. C. Carapella, Jr.
ASARCO Incorporated

ANTIMONY COMPOUNDS

Antimony [*7440-36-0*] is the fourth member of the nitrogen family and has a valence shell configuration of $5s^2 5p^3$. The utilization of these orbitals and, in some cases, of one or two $5d$ orbitals permits the existence of compounds in which the antimony atom forms three, four, five, or six covalent bonds.

The valence bond theory in its most elementary form predicts that trivalent compounds of antimony should have pyramidal structures derived from the $5p$ orbitals and that the $5s$ electrons should act as an inert pair. Many trivalent derivatives of antimony, however, have intervalency angles significantly larger than the 90° angle predicted by this model. The size of these angles as well as the general chemical behavior of the trivalent compounds suggest that sp^3 hybridized antimony orbitals are being employed and that the lone pair occupies one of the tetrahedral positions. The fact that the bond angles are often considerably less than the regular tetrahedral value of 109.5° may be ascribed to repulsion by the lone pair.

Pentacoordinate compounds of antimony usually exhibit trigonal bipyramidal geometry corresponding to the sp^3d hybridized antimony orbitals of valence bond theory. The antimony atom in the octahedral sp^3d^2 valence state is present in numerous complex anions of the SbX_6^- type and in neutral complexes of pentavalent halides with electron-donating molecules such as alcohols, ethers, and nitriles. Ions of the type SbX_5^{2-} are also known and possess a square pyramidal configuration in which the antimony atom is located slightly below the basal plane of the pyramid. A lone pair of electrons presumably occupies one of the octahedral positions and repels the Sb–X bonding pairs.

The annual world production capacity of antimony and its compounds as of December 31, 1988, is given in Table 1 (1). The consumption of these substances in the United States in 1988 was 12,069 metric tons (calculated as elemental antimony); about 78% of this amount was in the form of the oxide. Exports of antimony trioxide [*1309-64-4*], Sb_2O_3, from the United States increased significantly in 1988 and reached the highest levels since the Bureau of Mines began reporting such data in 1970. Thus the United States exported 849 metric tons of

Table 1. 1988 World Production Capacity for Antimony and its Compounds, t[a]

Area	Mine	Smelter
North and Central America	16,400	30,600
South America	18,900	11,300
Europe[b]	18,600	27,900
Africa	20,400	6,300
Asia[c]	38,500	34,700
Oceania (including Australia)	3,600	

[a]As elemental antimony.
[b]Includes estimates for countries having centrally planned economies in 1988.
[c]Includes estimates for China.

the oxide in 1987 and 1478 metric tons in 1988. In addition, the United States exported 882 metric tons of other antimony compounds having a value of \$2.8 million. Asarco's published price for high tint antimony trioxide in lots of 18 metric tons was \$2.98/kg at the beginning of 1988, and by the end of the year it ranged between \$3.15 and \$3.76 per kg.

A recommended exposure limit for antimony compounds of 0.5 mg/m^3 (as Sb) has been given (2). Disposal may be effected by washing residues down the drain at very high dilution unless prohibited by local regulations.

Analysis

A wide variety of titrimetric methods for the determination of antimony in the macro and semimicro range are available. Potassium bromate in strongly acid solution is probably the most widely used oxidimetric titrant. The end point can be determined either potentiometrically or by the use of an indicator. Organic dyes such as methyl red are used as indicators, but amaranth [*915-67-3*], $C_{20}H_{11}N_2Na_3O_{10}S_3$, has been reported as the indicator of choice (3). Potassium dichromate in hydrochloric acid–acetic acid solution with ferroin as the indicator has also been used (4). Other oxidimetric reagents used for the titration of antimony(III) include potassium iodate in acid solution, iodine in the presence of sodium tartrate and bicarbonate, and potassium permanganate in acid or alkaline solution. A number of organic compounds have been recommended as oxidimetric titrants. Such compounds are usually *N*-haloimides, and *N*-chlorophthalimide (5) and *N*-bromosuccinimide [*128-08-5*], $C_4H_4BrNO_2$, (6) have been used for this purpose. Sodium diethyldithiocarbamate has also been used for titrating antimony(III); the end point is determined potentiometrically (7). Because arsenic interferes with most methods used for determining antimony, the separation of the two elements is of great importance. It is possible to remove the arsenic as the trichloride by boiling a hydrochloric acid solution containing arsenic and antimony in the trivalent state. However, a method for determining both arsenic and antimony employs cerium(IV) sulfate as the titrant and ferroin as the indicator (8). A spectrofluorimetric method for determining antimony, based on the reduction of cerium(IV) to fluorescent cerium(III), has also been described (9). This method can also be used for determining antimony(III) in the presence of antimony(V).

A widely used colorimetric method for the estimation of microgram quantities of antimony is based on the reaction of antimony(V) with rhodamine B [*81-88-9*], $C_{28}H_{31}ClN_2O_3$, in hydrochloric acid solution to form a colored complex that is extracted with organic solvents and measured spectrophotometrically (10). For the determination of antimony in trace amounts, methods employing neutron activation or atomic absorption have been widely used. A comparison of the two methods has been reported (11). Both methods gave satisfactory results when used to determine specified values of antimony in several different biological materials. An excellent description of the determination of antimony involving the generation of stibine by sodium borohydride, followed by atomic absorption analysis, has also been reported (12).

Inorganic Compounds of Antimony

Stibine. Stibine [*7803-52-3*], SbH_3; mp, −88°C; bp, −18°C; density of the liquid at its bp, 2.204 g/mL; sp gr at 18°C with respect to air as 1.000, 4.344, is a colorless, poisonous gas having a disagreeable odor (13). It is the only well-characterized binary compound of antimony and hydrogen, although distibine [*14939-42-5*], Sb_2H_4, has been reported. The vapor pressure, heat of vaporization, heat capacity, density, viscosity, heat of formation, free energy of formation, surface tension, and thermal conductivity of stibine as a function of temperature have been reported (14). The formation is endothermic, 145.1 kJ/mol (34.7 kcal/mol); the compound decomposes slowly at room temperature and readily at 200°C, to give metallic antimony and hydrogen. The decomposition is autocatalytic and under certain conditions can be explosive. The molecule is trigonal pyramidal (15). Stibine is readily soluble in organic solvents such as carbon disulfide or ethanol, and is slightly soluble in water. In aqueous solution there is no measurable tendency to form a stibonium ion analogous to NH_4^+ and PH_4^+.

Stibine may be prepared by the treatment of metal antimonides with acid, chemical reduction of antimony compounds, and the electrolysis of acid or alkaline solutions using a metallic antimony cathode:

$$Zn_3Sb_2 + 6\ H_3O^+ \longrightarrow 3\ Zn^{2+} + 2\ SbH_3 + 6\ H_2O$$

$$SbO_3^{3-} + 9\ H_3O^+ + 3\ Zn \longrightarrow SbH_3 + 3\ Zn^{2+} + 12\ H_2O$$

The classical synthesis involves the dissolution of a 33% Sb–67% Zn alloy by hydrochloric acid; the evolved gases contain up to 14% stibine. A detailed procedure using a Sb–Mg alloy has also been described (16). Aluminum hydride or alkali metal borohydrides have been used to reduce antimony(III) in acidic aqueous solution to produce stibine. A 23.6% yield of stibine, based on the borohydride used, has been reported (17). A 78% yield based on Sb has been obtained by gradually adding a solution that is 0.4 M in $SbCl_3$ and saturated in NaCl, to aqueous $NaBH_4$ at mol ratios of $NaBH_4:SbCl_3 > 10$ (18).

Stibine is readily oxidized and may be ignited in the presence of air or oxygen to form water and antimony trioxide; at lower temperatures metallic antimony and water are slowly formed. Sulfur and selenium react with stibine at 100°C in the presence of light to form antimony trisulfide [*1345-04-6*], Sb_2S_3, and antimony selenide [*1315-05-5*], Sb_2Se_3, respectively. At elevated temperatures stibine reacts with most metals to give antimonides. Heavy metal salts react with stibine to produce dark, metallic-appearing precipitates. In the case of silver nitrate, silver antimonide is first formed, and this reacts in turn with additional silver nitrate to produce metallic silver and antimony trioxide:

$$2\ SbH_3 + 12\ Ag^+ + 15\ H_2O \longrightarrow 12\ Ag + Sb_2O_3 + 12\ H_3O^+$$

High purity stibine is used as an n-type, gas-phase dopant for Si in semiconductors (14). Low temperature distillation of stibine at <53.3 kPa (400 torr) yields a product that on decomposition gives metallic antimony having less than 8 × 10^{-4} % impurity (18). A method for determining quantities of stibine in the neighborhood of 0.1 mg/m^3 in air has been reported (19).

Stibine may be inadvertently formed by acidified reducing agents reacting with antimony-containing materials. It is an extremely poisonous gas which causes blood destruction and damage to the liver and kidneys (2).

Metallic Antimonides. Numerous binary compounds of antimony with metallic elements are known. The most important of these are indium antimonide [*1312-41-0*], InSb, gallium antimonide [*12064-03-8*], GaSb, and aluminum antimonide [*25152-52-7*], AlSb, which find extensive use as semiconductors. The alkali metal antimonides, such as lithium antimonide [*12057-30-6*] and sodium antimonide [*12058-86-5*], do not consist of simple ions. Rather, there is appreciable covalent bonding between the alkali metal and the Sb as well as between pairs of Na atoms. These compounds are useful for the preparation of organoantimony compounds, such as trimethylstibine [*594-10-5*], $(CH_3)_3Sb$, by reaction with an organohalogen compound.

Antimony Trioxide. Antimony(III) oxide (antimony sesquioxide) [*1309-64-4*], Sb_2O_3, is dimorphic, existing in an orthorhombic modification; valentinite [*1317-98-2*] is colorless (sp gr 5.67) and exists in a cubic form; and senarmontite [*12412-52-1*], Sb_4O_6, is also colorless (sp gr 5.2). The cubic modification is stable at temperatures below 570°C and consists of discrete Sb_4O_6 molecules. The molecule is similar to that of P_4O_6 and As_4O_6 and consists of a bowed tetrahedron having antimony atoms at each corner united by oxygen atoms lying in front of the edges. This solid crystallizes in a diamond lattice with an Sb_4O_6 molecule at each carbon position.

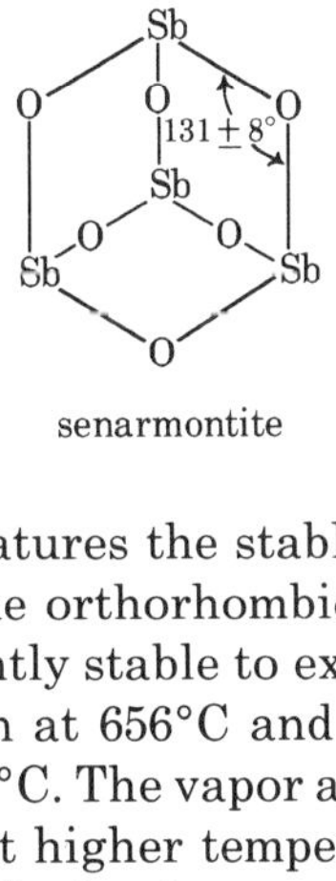

senarmontite

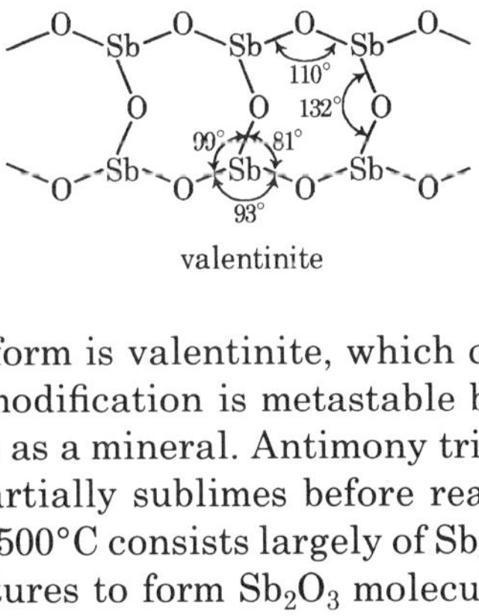

valentinite

At higher temperatures the stable form is valentinite, which consists of infinite double chains. The orthorhombic modification is metastable below 570°C; however, it is sufficiently stable to exist as a mineral. Antimony trioxide melts in the absence of oxygen at 656°C and partially sublimes before reaching the boiling temperature, 1425°C. The vapor at 1500°C consists largely of Sb_4O_6 molecules, but these dissociate at higher temperatures to form Sb_2O_3 molecules.

Common methods of preparation include direct combination of metallic antimony with air or oxygen, roasting of antimony trisulfide, and alkaline hydrolysis of an antimony trihalide and subsequent dehydration of the resulting hydrous oxide; when heated too vigorously in air, some of the Sb(III) is converted to Sb(V).

Antimony trioxide is insoluble in organic solvents and only very slightly soluble in water. The compound does form a number of hydrates of indefinite composition which are related to the hypothetical antimonic(III) acid (antimonous acid). In acidic solution antimony trioxide dissolves to form a complex series of polyantimonic(III) acids; freshly precipitated antimony trioxide dissolves in strongly basic solutions with the formation of the antimonate ion [*29872-00-2*], $Sb(OH)_6^{3-}$, as well as more complex species. Addition of suitable metal ions to these solutions permits formation of salts. Other derivatives are made by heating antimony trioxide with appropriate metal oxides or carbonates.

Antimony trioxide has numerous practical applications (1). Its principal use is as a flame retardant in textiles and plastics (see FLAME RETARDANTS; FLAME RETARDANTS IN TEXTILES). It is also used as a stabilizer for plastics, as a catalyst, and as an opacifier in glass (qv), ceramics (qv), and vitreous enamels (qv).

Antimony Tetroxide. Antimony(III,V) oxide, antimony dioxide [*1332-81-6*], SbO_2 and Sb_2O_4, occurs in two modifications. Orthorhombic antimony tetroxide has long been known as the mineral cervantite [*1332-81-6*], α-Sb_2O_4, (colorless, sp gr 4.07). More recently a monoclinic modification, β-Sb_2O_4, has been recognized. In both dimorphs half of the antimony is in the +3 oxidation state, half in the +5 state (20,21). The antimony environments are quite similar in both modifications. The Sb(V) atoms are surrounded by a slightly distorted octahedron of oxygens, and the Sb(III) atoms are coordinated to four oxygens, all on the same side of the Sb(III). The α-modification may be formed by heating Sb_2O_3, valentinite, in air between 460 and 540°C; β-Sb_2O_4 is obtained either by heating α-Sb_2O_4 at 1130°C in dry air or in oxygen (22) or by heating antimonic(V) acid above 900°C (21). At higher temperatures the solid vaporizes without first undergoing any transformation; the recondensed vapors consist of a mixture of β-Sb_2O_4 and antimony trioxide (21).

Antimony tetroxide finds use as an oxidation catalyst, particularly for the dehydrogenation of olefins.

Antimony Pentoxide Hydrates. Antimonic acid (antimony(V) acid) [*12712-36-6*], and antimony(V) oxide [*1314-60-9*], $Sb_2O_5{\cdot}nH_2O$, are both hydrates of Sb_2O_5. There is no satisfactory evidence for the existence of an anhydrous compound (21). Commercial antimony pentoxide is either hydrated Sb_2O_5 or at times β-Sb_2O_4. Material having the approximate composition $Sb_2O_5{\cdot}3.5H_2O$ may be prepared by hydrolysis of antimony pentachloride or by acidification of potassium hexahydroxoantimonate(V) [*12208-13-8*], $KSb(OH)_6$, followed by filtration and drying to constant weight in air at room temperature. This substance is a white solid which loses water upon heating and becomes yellow in color. This loss of water fails to correspond to definitive ratios of $H_2O{:}Sb_2O_5$, nor is the composition Sb_2O_5 attained. At about 700°C the material is anhydrous and white in color; this is an antimony oxide [*12165-47-8*], Sb_6O_{13}, containing both Sb(III) and Sb(V) and having a cubic pyrochlore-type structure.

Hydrated antimony pentoxide (antimonic acid) is essentially insoluble in nitric acid solutions, only very slightly soluble in water, but dissolves in aqueous KOH. Numerous hydrated antimonate(V) salts have been reported in which the Sb(V) atom is octahedrally surrounded by six OH groups. Among these are derivatives of magnesium, cobalt, and nickel that have formulas $M(SbO_3)_2{\cdot}12H_2O$, and a compound referred to as sodium pyroantimonate [*10049-22-6*], $Na_2H_2Sb_2O_7{\cdot}5H_2O$. X-ray studies show that these are actually $M(H_2O)_6[Sb(OH)_6]_2$ and sodium hexahydroxantimonate(V) [*12339-41-2*], $Na[Sb(OH)_6]$, respectively. The latter compound is one of the least soluble sodium salts known and is useful in sodium analysis. Numerous polyantimonate(V) derivatives are prepared by heat treatment of mixtures of antimony trioxide and other metal oxides or carbonates. Of these, $K_3Sb_5O_{11}$ [*12056-59-6*] and $K_2Sb_4O_{11}$ [*52015-49-3*] have been characterized by x-ray. These consist of three-dimensional networks of SbO_6 in which corners and edges are shared with K^+ ions located in tunnels through the network (23). Simple species such as SbO_4^{3-} and $Sb_2O_7^{2-}$, analogous to orthophosphate and pyrophosphate, apparently do not exist.

Antimonic acid has been used as an ion-exchange material for a number of cations in acidic solution. Most interesting is the selective retention of Na^+ in 12 *M* HCl, the retention being 99.9% (24). At lower acidities other cations are retained, even K^+. Many oxidation and polymerization catalysts are listed as containing Sb_2O_5.

Antimony Trifluoride. Antimony(III) fluoride [*7783-56-4*], SbF_3, is a white, crystalline, orthorhombic solid; vapor pressure at the mp, 26.34 kPa (0.26 atm); Sb–F bond energy, 437.4 kJ (104.5 kcal) (25). The molecule shows a very distorted octahedral arrangement. Antimony trifluoride is extremely soluble in water, the solubility being increased by the presence of hydrofluoric acid. It is also very soluble in polar solvents such as methanol, 154 g/100 mL, and acetone. Table 2 lists physical constants for the antimony halides.

Antimony(III) fluoride may be prepared by treating antimony trioxide or trichloride with hydrofluoric acid. Pure SbF_3 is then obtained by carefully evaporating all of the water from the crude product, which is subsequently sublimed. SbF_3 does not hydrolyze as readily as do the other antimony trihalides. When heated in open air at 100°C, a crystalline solid quickly forms of composition $Sb_3O_2(OH)_2F_3$, which, upon further heating, is transformed into antimony oxide fluoride [*11083-22-0*], SbOF. This compound may also be prepared by heating 1:1 mixtures of Sb_2O_3 and SbF_3. There are three known crystalline modifications.

In the presence of excess fluoride, antimony trifluoride forms numerous types of complex ions, eg, SbF_4^-, SbF_5^{2-}, $Sb_4F_{13}^-$, and $Sb_2F_7^-$; but SbF_6^{3-} is unknown (26).

Antimony trifluoride is used as a fluorinating agent to replace nonmetal chloride with fluorine. Tri- and difluoromethyl groups are readily formed from the corresponding chlorine groups, but CH_2Cl and CHCl groups are usually unaffected. Antimony trifluoride can also be used to effect the replacement of chlorine bonded to other elements; thus trichloromethylphosphonous dichloride [*3582-11-4*], CCl_3PCl_2, can be converted to trichloromethylphosphonous difluoride [*1112-03-4*], CCl_3PF_2.

Antimony Trichloride. Antimony(III) chloride [*10025-91-9*], $SbCl_3$, is a colorless, crystalline solid, readily soluble in hydrochloric acid; water, ca 9% at 25°C, increasing with temperature; $CHCl_3$, 22%; CCl_4, 13%; benzene; CS_2; and dioxane.

Antimony trichloride may be prepared by chlorination of antimony metal, Sb_2O_3, or Sb_2S_3, or by reaction of Sb_2O_3 with concentrated HCl. $SbCl_3$ hydrolyzes readily, giving hydrous Sb_2O_3 with excess water, but when limited quantities of water are used, a large number of partially hydrolyzed products has been claimed, eg, SbOCl, Sb_2OCl_4, $Sb_4O_5Cl_2$, $Sb_4O_3(OH)_3Cl_2$, $Sb_8O_{11}Cl_2$, and Sb_8OCl_{22}, some of which are listed in Table 3. The hydrolysis product most frequently obtained and best characterized is tetraantimony dichloride pentoxide [*12182-69-3*], $Sb_4O_5Cl_2$, which is initially precipitated as a thick white solid, changing to well-defined

Table 2. Physical Constants of the Antimony Halides

Parameter	Antimony trifluoride	Antimony trichloride	Antimony tribromide	Antimony triiodide[a]	Antimony pentafluoride	Antimony pentachloride
formula	SbF_3	$SbCl_3$	$SbBr_3$	SbI_3	SbF_5	$SbCl_5$
CAS Registry Number	[*7783-56-4*]	[*10025-91-9*]	[*7789-61-9*]	[*7790-44-5*]	[*7783-70-2*]	[*7647-18-9*]
mp, °C	291 ± 1	73.2	96.0 ± 0.5	170.5	6	3.2 ± 0.1
bp, °C	346 ± 10	222.6	287	401	150	68[b], 140[c]
ΔH_f° at 298°C, kJ/mol[d]	−915.5	−382.2	−259.4	−100.4		−450.8 ± 6.2
S° at 298°C, J/(mol·K)[d]	127	184	207	216 ± 1		263 ± 12
ΔH_{fusion}, kJ/mol[d,e]	21.4			22.7_{444} ± 0.2		
ΔS_{fusion}, J/(mol·K)[d,e]	38.2			51.5_{444} ± 0.4		
ΔH_{vap}, kJ/mol[d,e]	102.8_{298} ± 1.3	46.72_{496}	53.2_{560}			43.45_{449}
ΔS_{vap}, J/(mol·K)[d,e]	175.8_{298} ± 2.5	93.3_{496}	94.9_{560}			95.44_{449}
C_p, J/mol·K[d]		108[f]		96[f], 144[g]		

[a]The ΔH_{subl} at 298°C is 101.6 ± 0.4 kJ/mol (24.3 ± 0.1 kcal/mol).
[b]At a pressure of 1.82 kPa.
[c]Decomposes at atmospheric pressure, 101.3 kPa.
[d]To convert from J to cal, divide by 4.184.
[e]At the temperature in °C indicated by the subscript.
[f]Value given is for solid.
[g]Value given is for liquid.

Table 3. Inorganic Antimony Compounds and Adducts

Compound	CAS Registry Number	Formula
trifluorotetraoxotriantimonic(III) acid	*[65229-25-6]*	$Sb_3O_2(OH)_2F_3$
antimony chloride oxide	*[7791-08-4]*	$SbOCl$
diantimony tetrachloride oxide	*[65229-26-7]*	Sb_2OCl_4
tetraantimony dichloride pentoxide	*[12182-69-3]*	$Sb_4O_5Cl_2$
tetraantimony trihydroxydichlorotrioxide		$Sb_4O_3(OH)_3Cl_2$
antimony trichloride·diethyl ether	*[10025-91-9]*	$SbCl_3{\cdot}(C_2H_5)_2O$
antimony trichloride·aniline	*[21645-17-0]*	$SbCl_3{\cdot}H_2NC_6H_5$
antimony trichloride·trimethylamine	*[65186-11-0]*	$SbCl_3{\cdot}(CH_3)_3N$
bis(antimony trichloride)·trimethylamine	*[65186-12-1]*	$2SbCl_3{\cdot}(CH_3)_3N$
antimony trichloride·bisacetone	*[65186-13-2]*	$SbCl_3{\cdot}2CH_3COCH_3$
bis(antimony trichloride)·benzene	*[1123-15-5]*	$2SbCl_3{\cdot}C_6H_6$
(antimony trichloride)tricarbonylnickel	*[65208-44-8]*	$Ni(CO)_3SbCl_3$
bis(antimony trichloride)tricarbonyliron	*[65208-43-7]*	$Fe(CO)_3(SbCl_3)_2$
antimony pentachloride·bis(iodine chloride)	*[65186-14-3]*	$SbCl_5{\cdot}2ICl$
antimony pentachloride·tris(iodine chloride)	*[65186-15-4]*	$SbCl_5{\cdot}3ICl$
antimony pentachloride·sulfur tetrachloride	*[15597-82-7]*	$SbCl_5{\cdot}SCl_4$
bis(hexachloroantimonic(III) acid)·nonahydrate	*[65208-45-9]*	$HSbCl_6{\cdot}4.5H_2O$
antimony pentabromide·diethyl ether	*[29702-86-1]*	$SbBr_5{\cdot}O(C_2H_5)_2$

colorless crystals. By carefully controlled hydrolysis SbOCl is obtained, which, upon further dilution with water, changes to $Sb_4O_5Cl_2$.

In many situations $SbCl_3$ behaves as a Lewis acid. In the presence of excess Cl^- and suitable cations, numerous chloroantimonate(III) ions are formed, eg, $SbCl_6^{3-}$, $SbCl_5^{2-}$, $SbCl_4^-$, $Sb_2Cl_7^{2-}$, $Sb_2Cl_7^{3-}$, $Sb_2Cl_9^{3-}$, and $Sb_2Cl_{11}^{5-}$. The first two of these are simple discrete ions. A large number of adducts have been formed by reaction of $SbCl_3$ with organic bases, eg, $SbCl_3{\cdot}(C_2H_5)_2O$, $SbCl_3{\cdot}H_2NC_6H_4NO_2$, $SbCl_3{\cdot}(CH_3)_3N$, $2SbCl_3{\cdot}(CH_3)_3N$, and $SbCl_3{\cdot}2CH_3COCH_3$. Isolable adducts are also formed with aromatic hydrocarbons, eg, $2SbCl_3{\cdot}C_6H_6$ and $SbCl_3{\cdot}C_6H_6$. In a few situations $SbCl_3$ apparently acts as an electron donor; thus the carbonyl complexes $Ni(CO)_3SbCl_3$ and $Fe(CO)_3(SbCl_3)_2$ have been isolated (27). Adducts and compounds are listed in Table 3.

Antimony trichloride is used as a catalyst or as a component of catalysts to effect polymerization of hydrocarbons and to chlorinate olefins. It is also used in hydrocracking of coal (qv) and heavy hydrocarbons (qv), as an analytic reagent for chloral, aromatic hydrocarbons, and vitamin A, and in the microscopic identification of drugs. Liquid $SbCl_3$ is used as a nonaqueous solvent.

Antimony Tribromide and Triiodide. Antimony(III) bromide *[7789-61-9]*, $SbBr_3$, is a colorless, crystalline solid having a pyramidal dimorphic molecular structure and an acicular (α-$SbBr_3$) and a bipyramidal (β-$SbBr_3$) habit.

Antimony(III) iodide *[7790-44-5]*, SbI_3, forms red rhombohedral crystals, intermediate in structure between a molecular and an ionic crystal. In SbI_3 vapor there is no indication of association.

Both antimony tribromide and antimony triiodide are prepared by reaction of the elements. Their chemistry is similar to that of $SbCl_3$ in that they readily hydrolyze, form complex halide ions, and form a wide variety of adducts with ethers, aldehydes, mercaptans, etc. They are soluble in carbon disulfide, acetone, and chloroform. There has been considerable interest in the compounds antimony bromide sulfide [*14794-85-5*], antimony iodide sulfide [*13868-38-1*], ISSb, and antimony iodide selenide [*15513-79-8*] with respect to their solid-state properties, ferroelectricity, pyroelectricity, photoconduction, and dielectric polarization.

Antimony Pentafluoride. Antimony(V) fluoride [*7783-70-2*], SbF_5, is a colorless, hygroscopic, viscous liquid that has SbF_6 units with *cis*-fluorines bridging to form polymeric units. ^{19}F nmr shows that at low temperatures there are three different types of F atoms (28). Contamination with a small amount of HF markedly decreases the extent of polymerization. The vapor density at 150°C corresponds to the trimer. The solid is a *cis*-fluorine-bridged tetramer (29).

Antimony pentafluoride may be prepared by fluorination of SbF_3 or by treatment of $SbCl_5$ with HF. In the latter method the fifth chlorine is removed with difficulty; failure to remove the chlorine completely results in contamination of the distilled SbF_5 with Sb(III) (27).

Antimony pentafluoride is a strong Lewis acid and a good oxidizing and fluorinating agent. Its behavior as a Lewis acid leads to the formation of numerous simple and complex adducts. It reacts vigorously with water to form a clear solution from which antimony pentafluoride dihydrate [*65277-49-8*], $SbF_5{\cdot}2H_2O$, may be isolated. This is probably not a true hydrate, but may well be better formulated as $[H_3O][SbF_5OH]$.

Antimony pentafluoride reacts with iodine to form bis(antimony pentafluoride) iodide [*12324-61-7*], $Sb_2F_{10}I$, and antimony pentafluoride iodide [*12324-57-1*], SbF_5I; with nitrosyl fluoride to give a very stable compound nitrosyl hexafluoroantimonate(V) [*16941-06-3*], $NOSbF_6$; with sulfur to give a dark blue solution from which antimony pentafluoride sulfur can be isolated; and with NO_2 to form nitrosyl pentafluoronitratoantimonate(V) [*26117-73-7*], $NO[SbF_5(NO_3)]$. Combinations of Sb(V) and Sb(III) fluorides give fluorides of the general formula $SbF_5(SbF_3)_n$ where n may be 2, 3, 4, or 5. In combination with hydrofluoric acid or other fluorides the hexafluoroantimonate(V) ion [*17111-95-4*], SbF_6^-, is formed. This is frequently used as a negative counterion for compounds containing rather unstable, highly oxidizing, or highly fluorinated cations. The SbF_6^- anion has been shown to have an octahedral structure and may be hydrolyzed to $Sb(OH)_6^-$. Combinations of SbF_5, with either HSO_3F alone or with HSO_3F and SO_3, have extremely high acidities (30,31). A 1:1 mixture of HSO_3F and SbF_5 is frequently used for stabilization of carbocations and has been referred to as magic acid.

Treatment of graphite with SbF_5 liquid or vapor results in intercalation of SbF_5 between the graphite layers, and at 70°C a blue-black solid, antimony graphite fluoride [*56126-99-9*], $C_{13}Sb_2F_{10}$, is formed (32). This modifies the fluorinating properties of SbF_5, and such materials are used as specific fluorinating agents. Several mixed pentahalides are known. Thus fluorination of $SbCl_3$ yields antimony trichloride difluoride [*31244-70-9*], $SbCl_3F_2$, and chlorination of SbF_3 gives antimony dichloride trifluoride [*7791-16-4*], $SbCl_2F_3$. The latter compound has been shown to consist of $SbCl_4^+$ and $Sb_2Cl_2F_9^-$ (33).

The principal use of antimony pentafluoride is as a fluorinating agent. It readily replaces all chlorines with fluorine in organic compounds, and it fluorinates double bonds and aromatic rings.

Antimony Pentachloride. Antimony(V) chloride [*7647-18-9*], $SbCl_5$, is a colorless, hygroscopic, oily liquid that is frequently yellow because of the presence of dissolved chlorine; it cannot be distilled at atmospheric pressure without decomposition, but the extrapolated normal boiling point is 176°C. In the solid, liquid, and gaseous states it consists of trigonal bipyramidal molecules with the apical chlorines being somewhat further away than the equatorial chlorines (27).

Antimony pentachloride is usually prepared by chlorination of molten $SbCl_3$. It undergoes partial dissociation to Cl_2 and $SbCl_3$; ΔH_{496} for this equilibrium is −76.69 kJ/mol (−18.33 kcal/mol) and ΔS_{496} is −145 J/mol (−34.7 cal/mol) (34).

Antimony pentachloride is a strong Lewis acid and a useful chlorine carrier. Chlorine is lost in a number of chemical reactions resulting in the formation of adducts. Thus iodine is chlorinated to form ICl, which in turn combines with additional $SbCl_5$ to give $SbCl_5{\cdot}2ICl$ and $SbCl_5{\cdot}3ICl$. A similar reaction occurs with sulfur, giving $SbCl_5{\cdot}SCl_4$. These adducts are listed in Table 3. In the presence of excess hydrochloric acid or metal chlorides the hexachloroantimonate(V) ion [*17949-89-2*], $SbCl_6^-$, is formed. The strong acid, $HSbCl_6{\cdot}4.5H_2O$, as well as many hexachloroantimonate salts can be isolated. With a 1:1 mol ratio of water to $SbCl_5$, antimony pentachloride monohydrate [*14215-03-3*], $SbCl_5{\cdot}H_2O$, which is insoluble in $CHCl_3$, is formed; with a 4:1 mol ratio, chloroform soluble antimony pentachloride tetrahydrate [*52940-44-0*], $SbCl_5{\cdot}4H_2O$, can be isolated. Numerous examples can be cited where the Lewis acid $SbCl_5$ forms 1:1 adducts with oxygen- and nitrogen-containing bases, eg, sulfur nitride, ethers, nitriles, alcohols, or esters. Some of these are given in Table 3.

Derivatives of Antimony Pentabromide and Pentaiodide. The existence of $SbBr_5$ and SbI_5 is in doubt, although from time to time they are reported in the literature (35). The existence of a 1:1 adduct, $SbBr_5{\cdot}O(C_2H_5)_2$, however, is generally accepted. In addition, the $SbBr_6^-$ ion is known, and from x-ray studies has been found to have a slightly distorted octahedral structure (36). Indeed, there are quite a number of complex bromoantimony compounds with alkali metals and organic bases, some of which contain Sb(V). Thus the quinuclidinium salt $(C_7H_{13}NH)_4Sb_2Br_{16}$ is actually made up of four quinuclidinium cations, a hexabromoantimonate(III) [*31204-78-1*], $SbBr^{3-}$, a hexabromoantimonate(V) ion [*22994-42-9*], $SbBr_6^-$, and two bromine molecules (37).

Antimony Trisulfide. Antimony(III) sulfide (antimony sesquisulfide) [*1345-04-6*], SbS_3, exists as a black crystalline solid, stibnite [*1317-86-8*], and as an amorphous red to yellow-orange powder. Stibnite melts at 550°C and has $\Delta H^\circ_{f,298}$, −175 kJ/mol (−41.8 kcal/mol); S°_{298}, 182 J/(mol·K) [43.5 cal/(mol·K)]; for the amorphous solid $\Delta H^\circ_{f,298}$ is −147 kJ/mol (−35.1 kcal/mol) (38). The crystal structure of stibnite contains two distinctly different antimony sites and consists of two parallel Sb_4S_6 chains that are linked together to form crumpled sheets (two per unit cell).

Amorphous Sb_2S_3 can be prepared by treating an $SbCl_3$ solution with H_2S or with sodium thiosulfate, or by heating metallic antimony or antimony trioxide with sulfur. Antimony trisulfide is almost insoluble in water but dissolves in concentrated hydrochloric acid or in excess caustic. In the absence of air, Sb_2S_3 dissolves in alkaline sulfide solutions to form the thioantimonate(III) ion [*43049-98-5*]; SbS_2^-, in the presence of air the tetrathioantimonate(V) ion [*17638-29-8*], SbS_4^{3-}, is formed. The lemon-yellow crystalline salt, $Na_3SbS_4{\cdot}9H_2O$, known as Schlippe's salt [*1317-86-8*], contains the tetrahedral tetrathioantimonate(V) ion.

Antimony trisulfide is used in fireworks, in certain types of matches, as a pigment, and in the manufacture of ruby glass.

Antimony Pentasulfide. Antimony pentasulfide [*1315-04-4*], Sb_2S_5, is a yellow to orange to red amorphous solid of indefinite composition. It is frequently given the formula Sb_2S_5, but actually consists of Sb(III) with a variable quantity of sulfur (39). The product is prepared commercially by the conversion of Sb_2S_3 to tetrathioantimonate(V) by boiling with sulfur in alkaline solution. The antimony pentasulfide is liberated as a yellowish orange precipitate when the resulting mixture is acidified with hydrochloric acid. The product is commercially known as golden sulfide of antimony, and is used in vulcanization to produce a red variety of rubber. The material is also used as a pigment and in fireworks.

Antimony(III) Salts. Concentrated acids dissolve trivalent antimony compounds. From the resulting solutions it is possible to crystallize normal and basic salts, eg, antimony(III) sulfate [*7446-32-4*], $Sb_2(SO_4)_3$; antimonyl sulfate [*14459-74-6*], $(SbO)_2SO_4$; antimony(III) phosphate [*12036-46-3*], $SbPO_4$; antimony(III) acetate [*6923-52-0*], $Sb(C_2H_3O_2)_3$; antimony(III) nitrate [*20328-96-5*], $Sb(NO_3)_3$; and antimony(III) perchlorate trihydrate [*65277-48-7*], $Sb(ClO_4){\cdot}3H_2O$. The normal salts all hydrolyze readily.

Compounds Containing Sb–O–C or Sb–S–C Linkages. A large number of compounds have been prepared in which the antimony atom is linked to carbon through an oxygen or sulfur atom. The simplest of these compounds are esters of the hypothetical antimonic acid [*13453-11-7*], H_3SbO_3, or thioantimonic acid [*65277-44-3*], H_3SbS_3. The esters of H_3SbO_3 can be prepared by refluxing antimony trioxide with the appropriate alcohol and removing the water formed by means of anhydrous copper sulfate. They may also be obtained by ester exchange. For example, tributyl antimonate [*2155-74-0*], $C_{12}H_{27}O_3Sb$, is formed by the interaction of the triethyl ester, triethyl antimonate [*10433-06-4*], $C_6H_{15}O_3Sb$, and butyl alcohol. Esters of thioantimonic(III) acid are easily synthesized by the reaction of antimony trichloride and a mercaptan. A series of these thioantimonic esters have been prepared in the search for compounds of medicinal value (40).

By far the largest group of compounds containing the Sb–O–C linkage are those obtained by reaction of an antimony oxide with an α-hydroxy acid, *o*-dihydric phenol, sugar alcohol, or some other polyhydroxy compound containing at least two adjacent hydroxyl groups. The best known compound of this type is antimony potassium tartrate (tartar emetic) [*28300-74-5*] prepared by refluxing potassium hydrogen tartrate with freshly precipitated antimony trioxide. Tartar emetic has been used as an antiparasitic agent in medicine, as an insecticide, and as a mordant in the textile and leather industries.

Tartar emetic was the subject of controversy for many years, and a variety of incorrect structures were proposed. In 1966, x-ray crystallography showed that tartar emetic contains two antimony(III) atoms bridged by two tetranegative D-tartrate residues acting as double bidentate ligands to form dipotassium bis[D-μ-(2,3-dihydroxybutanedioato)]diantimonate [*28300-74-5*] (41).

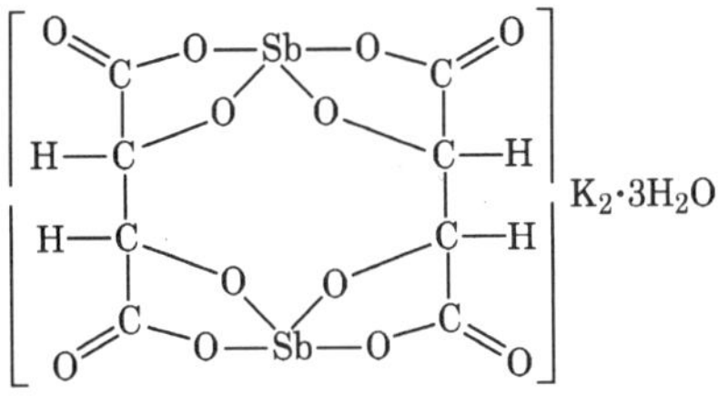

The four 5-membered chelate rings are nearly planar, and the oxygen atoms about the antimony atoms occupy four corners of a distorted square pyramid, the apex of which is presumably occupied by an unshared electron pair. Later work (42) has also shown that the three water molecules in the structure are hydrogen bonded to each other and to the carboxyl oxygen atoms and they connect the anions in infinite sheets. Essentially the same tartrato-bridged binuclear anion has been found in the racemic salts $(NH_4)_2Sb_2$(D,L-$C_4H_2O_6)_2$·$4H_2O$ (43) and K_2Sb_2(D,L-$C_4H_2O_6)_2$·$3H_2O$ (44), and in the ammonium (45) and the tris(*o*-phenanthroline)iron(II) (46) analogues of tartar emetic. It has been suggested that the formation of a bridged *meso*-tartrato dimer of antimony(III) requires an unfavorable eclipsed conformation for the bridging ligands (47), as *meso*-tartrate complexes similar to tartar emetic have never been prepared.

Two five-membered chelate rings per antimony atom are present in antimony hydrogen bis(thioglycolate) [*65277-45-4*], $C_4H_7O_4S_2Sb$, a compound prepared by the interaction of antimony trioxide and thioglycolic acid in aqueous solution (48). The coordination around the antimony has been described as a distorted trigonal bipyramid in which the two axial apices are occupied by oxygen atoms, and two of the equatorial apices are occupied by sulfur atoms. The third equatorial position is presumably occupied by an unshared electron pair which is responsible for the deformation of the bipyramid. Each unit of antimony hydrogen bis(thioglycolate) is joined to two other units by hydrogen bonds, forming endless zig-zag chains.

Lactic, malic, mandelic, and oxalic acids also give antimony(III) derivatives in which two molecules of acid are associated with one atom of antimony. Studies of the reaction between antimony trioxide and lactic or oxalic acid as a function of pH have suggested the following structures for the oxalate and lactate complexes (49,50).

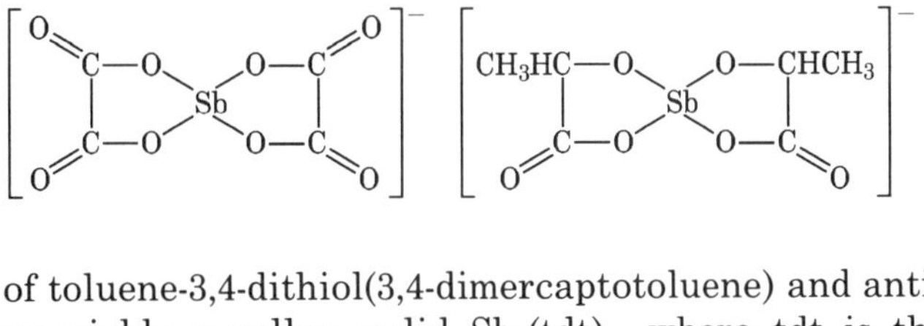

The reaction of toluene-3,4-dithiol(3,4-dimercaptotoluene) and antimony trichloride in acetone yields a yellow solid $Sb_2(tdt)_3$, where tdt is the toluene-3,4-dithiolate anionic ligand (51). With the disodium salt of maleonitriledithiol ((*Z*)-dimercapto-2-butenedinitrile), antimony trichloride gives the complex ion $[Sb(mnt)_2]^-$, where mnt is the maleonitriledithiolate anionic ligand. This complex has been isolated as a yellow, crystalline, tetraethylammonium salt. The structures of these antimony dithiolate complexes have apparently not been unambiguously determined.

Antimony pentoxide also reacts with a variety of dihydroxy compounds. Thus pyrocatechol yields a crystalline substance in which three molecules of the diol are associated with one atom of antimony (52). The configuration of this substance has not been established, but the following structure seems reasonable:

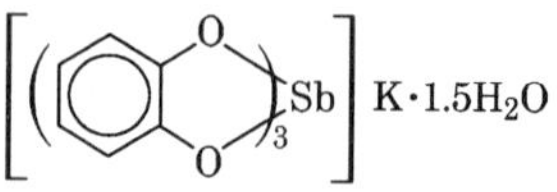

A number of complex derivatives of antimony pentoxide with polyhydroxy compounds have been investigated as drugs. The most important of these substances is known as antimony sodium gluconate [*16037-91-5*], $C_{12}H_{20}O_{17}Sb_2{\cdot}9H_2O{\cdot}3Na$, which is prepared by the reaction of antimony pentoxide, gluconic acid, and sodium hydroxide (53).

Organoantimony Compounds

A wide variety of compounds containing the Sb–C bond is known. Organoantimony compounds can be broadly divided into Sb(III) and Sb(V) compounds. The former may contain from one to four organic groups, and the Sb(V) compounds from one to six organic groups. With a few exceptions, the nomenclature used here is that proposed by the International Union of Pure and Applied Chemistry (54). There are a number of heterocyclic compounds in which one or more antimony atoms are members of the heterocycle. Such compounds have been thoroughly discussed (55). A more recent but less comprehensive report on heterocyclic antimony compounds has also been published (56). The synthesis of organoantimony compounds has been described (57). Organoantimony(III) compounds (58,59) and organoantimony(V) compounds containing four, five, or six C–Sb bonds (60) and three C–Sb bonds (61), respectively, have been summarized, and a summary of organoantimony(V) compounds containing one, two, or three C–Sb bonds has been recently published (62). Two lists of all organoantimony compounds prepared or studied between 1937 and 1968 have been published (63,64). Another monograph, published in 1970, critically discusses organoarsenic, -antimony, and -bismuth chemistry (65). Work on organoantimony compounds is reviewed annually, and the use of organoantimony and organobismuth compounds in organic synthesis has been reviewed (66).

Primary and Secondary Stibines. Relatively few primary ($RSbH_2$) and secondary (R_2SbH) stibines are known. Methylstibine [*23362-09-6*], CH_5Sb, ethylstibine [*68781-03-3*], C_2H_7Sb, isopropylstibine, C_3H_9Sb, and butylstibine [*68781-04-4*], $C_4H_{11}Sb$, have been prepared by the reduction of the corresponding alkyldichlorostibines using lithium aluminum hydride or sodium borohydride (67). All of the alkylstibines are thermally unstable, easily oxidizable, colorless liquids with strong alliaceous odors. Decomposition products include hydrogen and nonvolatile black solids analyzing for $(RSb)_x$. Reaction of the alkylstibines with hydrogen chloride produces lustrous, pale green polymeric solids also analyzing for $(RSb)_x$ (68). Diethylstibine, $C_4H_{11}Sb$, (69), di-*tert*-butylstibine, $C_8H_{19}Sb$,

(70), and dicyclohexylstibine [*1011-94-5*], $C_{12}H_{23}Sb$, (71) have been obtained by reduction of the corresponding dialkylhalostibines with lithium aluminum hydride. Dimethylstibine [*23362-10-9*], C_2H_7Sb, was first prepared by the interaction of dimethylbromostibine [*53234-94-9*], C_2H_6BrSb, and $LiHB(OCH_3)_2$ at temperatures below $-40°C$ (72). Treatment of dimethylstibine with hydrogen chloride yields hydrogen:

$$(CH_3)_2SbH + HCl \longrightarrow H_2 + (CH_3)_2SbCl$$

Phenylstibine [*58266-50-5*], C_6H_7Sb, has been obtained by the reduction of phenyldiiodostibine [*68972-61-2*], $C_6H_5I_2Sb$, (73) or phenyldichlorostibine [*5035-52-9*], $C_6H_5Cl_2Sb$, (74) with lithium borohydride. It has also been prepared by the hydrolysis or methanolysis of phenylbis(trimethylsilyl)stibine [*82363-95-9*], $C_{12}H_{23}Si_2Sb$ (75). Diphenylstibine [*5865-81-6*], $C_{12}H_{11}Sb$, can be prepared by the interaction of diphenylchlorostibine [*2629-47-2*], $C_{12}H_{10}ClSb$, with either lithium borohydride (76) or lithium aluminum hydride (77). It is also formed by hydrolysis or methanolysis of diphenyl(trimethylsilyl)stibine [*69561-88-2*], $C_{15}H_{19}SbSi$ (75). Dimesitylstibine [*121810-02-4*] has been obtained by the protonation of lithium dimesitylstibide with trimethylammonium chloride (78). The x-ray crystal structure of this secondary stibine has also been reported.

The aromatic primary and secondary stibines are readily oxidized by air, but they are considerably more stable than their aliphatic counterparts. Diphenylstibine is a powerful reducing agent, reacting with many acids to liberate hydrogen (79). It has also been used for the selective reduction of aldehydes and ketones to the corresponding alcohols (80). At low temperatures, diphenylstibine undergoes an addition reaction with ketene (81):

$$(C_6H_5)_2SbH + CH_2{=}C{=}O \longrightarrow (C_6H_5)_2Sb\underset{\underset{O}{\|}}{C}CH_3$$

Tertiary Stibines. A large number of trialkyl- and triarylstibines are known (58). They are usually prepared by the interaction of a reactive organometallic compound and an antimony trihalide, a halostibine, or a dihalostibine. The type of organometallic compound most widely employed in these syntheses is the Grignard reagent (82,83). Organolithium (84,85), organocadmium (86,87), organoaluminum (88), and organomercury (89) compounds have also been used (see ORGANOMETALLICS, σ-BONDED ALKYLS AND ARYLS). Triarylstibines can be readily prepared from an aryl halide, an antimony trihalide, and sodium (90).

Another excellent method for preparing tertiary stibines involves the interaction of an organostibide and an alkyl or aryl halide (91,92). This method is of particular value in preparing unsymmetrical tertiary stibines. For example, an interesting hybrid ligand has been obtained by the following reaction carried out in liquid ammonia (93):

$$CH_3SeC_6H_4Br + NaSb(CH_3)_2 \longrightarrow CH_3SeC_6H_4Sb(CH_3)_2 + NaBr$$

Trialkylstibines are sensitive to oxygen, and in some cases they ignite spontaneously in air. Trimethylstibine [*594-10-5*], C_3H_9Sb, may explode on contact with atmospheric oxygen (94). Triarylstibines usually do not react with air, and they are quite stable thermally (95). Trialkylstibines are powerful reducing agents and can convert halides of phosphorus or antimony to the corresponding elements (96). Triarylstibines are much less reactive as reducing agents, but they are readily oxidized by halogens, interhalogens, pseudohalogens, sulfur, and fuming nitric acid (97,98).

Tertiary stibines have been widely employed as ligands in a variety of transition metal complexes (99), and they appear to have numerous uses in synthetic organic chemistry (66), eg, for the olefination of carbonyl compounds (100). They have also been used for the formation of semiconductors by the metal–organic chemical vapor deposition process (101), as catalysts or cocatalysts for a number of polymerization reactions (102), as ingredients of light-sensitive substances (103), and for many other industrial purposes.

Halostibines, Dihalostibines, and Related Compounds. Alkyldichloro- and alkyldibromostibines are readily prepared by the alkylation of the corresponding antimony trihalide with an organolead reagent (67,104):

$$R_4Pb + 3\ SbX_3 \longrightarrow 3\ RSbX_2 + PbX_2 + RX$$

The alkylation can also be accomplished using tetraalkyltin compounds. Alkyldiiodostibines are formed in about 20% yield via the interaction of alkylmagnesium iodides and antimony trichloride (67). Dialkylchlorostibines are obtained in good yields by the cleavage of tetraalkyldistibines using sulfuryl chloride (91):

$$R_2SbSbR_2 + SO_2Cl_2 \longrightarrow 2\ R_2SbCl + SO_2$$

Dialkylbromo- and dialkyliodostibines can similarly be prepared by the cleavage of tetraalkyldistibines using equimolar amounts of bromine or iodine (X_2) (105):

$$R_2SbSbR_2 + X_2 \longrightarrow 2\ R_2SbX$$

The thermal decomposition of trialkylantimony dihalides has been used for the preparation of chloro-, bromo-, and iodostibines (106):

$$R_3SbX_2 \xrightarrow{\Delta} R_2SbX + RX$$

The interaction of triarylstibines and antimony trichloride or tribromide is a convenient and efficient method for preparing aryldihalo- and diarylhalostibines (104,107,108):

$$Ar_3Sb + 2\ SbX_3 \longrightarrow 3\ ArSbX_2$$

$$2\ Ar_3Sb + SbX_3 \longrightarrow 3\ Ar_2SbX$$

Compounds $ArSbX_2$ and Ar_2SbX, in which X is Cl or Br and Ar is an aryl group, have also been obtained by the reduction of the corresponding stibonic or stibinic acids in hydrochloric or hydrobromic acid. The usual reducing agent is sulfur dioxide catalyzed by iodide ion, although stannous chloride has also been employed (109). The reduction of unsymmetrical diarylantimony trihalides is probably the best method for the synthesis of unsymmetrical chloro- and bromostibines (110). Aryldiiodostibines can be prepared by the reaction of the corresponding oxides with hydriodic acid (111):

$$\frac{1}{x}\,(ArSbO)_x + 2\ HI \longrightarrow ArSbI_2 + H_2O$$

Diaryliodostibines are usually obtained by the metathetical reaction of the chlorostibines with sodium iodide (111,112). Diphenylfluorostibine [*6651-55-4*], $C_{12}H_{10}FSb$, can be prepared from an organosilicon species (113):

$$2\ (NH_4)_2(C_6H_5SiF_5) + SbF_3 \longrightarrow (C_6H_5)_2SbF + 2\ (NH_4)_2SiF_6$$

Alkyldihalo- and dialkylhalostibines are highly reactive substances which are rapidly oxidized in air. Some are spontaneously inflammable (114). The aromatic counterparts are less susceptible to air oxidation but are readily oxidized by halogens. Alkaline hydrolysis (109,115,116) of the dihalo- and halostibines yields compounds of the types $(RSbO)_x$ and $R_2SbOSbR_2$, respectively, whereas reaction of the stibine with sodium sulfide (115) gives the analogous sulfur compounds. An interesting method for obtaining the bis(diarylantimony) oxides is by thermal disproportionation of the corresponding polymeric oxides (109,117):

$$\frac{4}{x}\,(ArSbO)_x \longrightarrow (Ar_2Sb)_2O + Sb_2O_3$$

A number of compounds of the types $RSbY_2$ and R_2SbY, where Y is an anionic group other than halogen, have been prepared by the reaction of dihalo- or halostibines with lithium, sodium, or ammonium alkoxides (118,119), amides (120), azides (121), carboxylates (122), dithiocarbamates (123), mercaptides (124,125), or phenoxides (118). Dihalo- and halostibines can also be converted to compounds in which an antimony is linked to a main group (126) or transition metal (127).

Distibines and Distibenes. A considerable number of tetraalkyl- and tetraaryldistibines have been investigated. These are usually obtained by the reduction of a dialkyl- or diarylhalostibine with sodium hypophosphite (111,112) or magnesium (108,116). Distibines can also be prepared by the treatment of a metal dialkyl- or diarylstibide with a 1,2-dihaloethane (70,71,77,85,91,128)

$$2\ R_2SbM + XCH_2CH_2X \longrightarrow R_2SbSbR_2 + CH_2{=}CH_2 + 2\ MX$$

where M = Li, Na, or K and X = Cl or Br. Distibines undergo a variety of interesting reactions (69–72,85,105,108,111) and have also attracted attention because a number of these substances are thermochromic (85,128,129).

Although distibenes, the antimony analogues of azo compounds, have never been isolated as free, monomeric molecules (130), a tungsten complex, tritungsten pentadecacarbonyl[μ_3-η^2-diphenyldistibene] [*82579-41-7*], $C_{27}H_{10}O_{15}Sb_2W_3$, has been prepared by the reductive dehalogenation of phenyldichlorostibine (131):

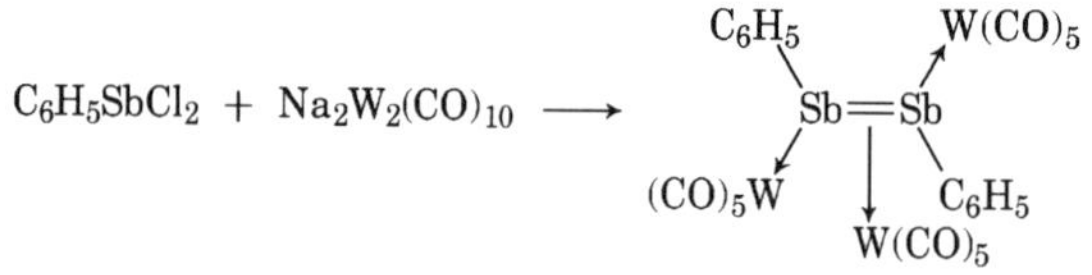

As expected, the Sb–Sb bond distance in this complex is significantly shorter than the Sb–Sb single bond distance. Chromium and tungsten complexes of dialkyldistibines have also been isolated (132).

Cyclic and Polymeric Substances Containing Antimony–Antimony Bonds. A number of organoantimony compounds containing rings of four, five, or six antimony atoms have been prepared. The first such compound to be adequately characterized, tetrakis-1,2,3,4-*tert*-butyltetrastibetane [*47191-73-5*], $C_{16}H_{36}Sb_4$, was obtained by the interaction of a dialkylstibide and iodine (70):

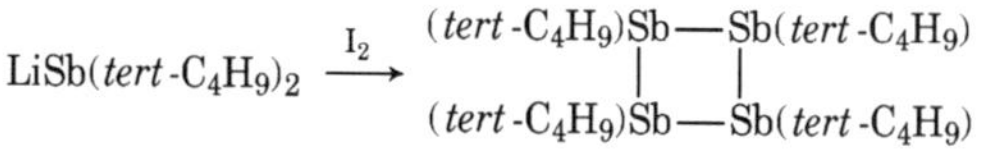

More recently, it has been prepared by the dehalogenation of *tert*-butyldichlorostibine [*67877-43-4*], $C_4H_9Cl_2Sb$, with magnesium (133). The corresponding five-membered ring compound, (*tert*-$C_4H_9Sb)_5$, is also formed in this reaction, but it has not yet been isolated in pure form (134). A tristibirane can also be formed by the dehalogenation of a dichlorostibine, but this substance has likewise been obtained only in admixture with a tetrastibetane (135). Compounds containing rings of six antimony atoms have been prepared by the slow air oxidation of bis(trimethylsilyl)phenylstibine [*82363-95-9*], $C_{12}H_{23}SbSi_2$, dissolved in 1,4-dioxane, benzene, or toluene (136):

$$6\ C_6H_5Sb[Si(CH_3)_3]_2 + 3\ O_2 + \text{solvent} \longrightarrow (C_6H_5Sb)_6\cdot\text{solvent} + 6\ [(CH_3)_3Si]_2O$$

Polymeric substances, $(RSb)_x$ or $(ArSb)_x$, have been obtained by the decomposition of primary stibines (67,74), the reaction of primary stibines with hydrogen chloride (68), the treatment of primary stibines with dibenzylmercury (137), or the reduction of dihalostibines (138–140). Most of these polymers have not been well characterized.

Antimonin and Its Derivatives. Antimonin(stibabenzene) [*289-75-8*], C_5H_5Sb, the antimony analogue of pyridine, can be prepared by the dehydrohalogenation of a cyclic chlorostibine using 1,5-diazabicyclo[4.3.0]non-5-ene (141):

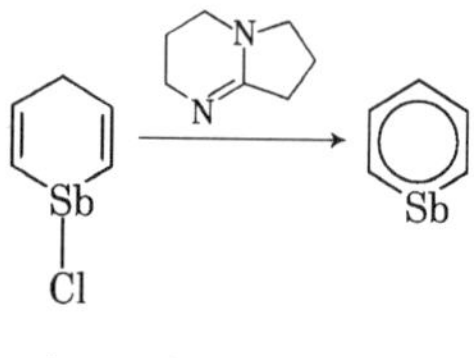

A number of derivatives of antimonin are also known (141,142). The potential aromaticity of this ring system has aroused considerable interest and has been investigated with the aid of spectroscopy as well as *ab initio* molecular orbital calculations (143). There seems to be no doubt that antimonin does possess considerable aromatic character.

Stibonic and Stibinic Acids. The stibonic acids, $RSbO(OH)_2$, and stibinic acids, $R_2SbO(OH)$, are quite different in structure from their phosphorus and arsenic analogues. The stibonic and stibinic acids are polymeric compounds of unknown structure and are very weak acids. IUPAC classifies them as oxide hydroxides rather than as acids. Thus $C_6H_5SbO(OH)_2$ is named phenylantimony dihydroxide oxide [*535-46-6*]; the *Chemical Abstracts* name is dihydroxyphenylstibine oxide [*535-46-6*], $C_6H_7O_3Sb$.

Methylstibonic acid, the only alkylstibonic acid known with certainty, was not reported until 1990 (144). Previous attempts to obtain alkylstibonic acids were unsuccessful (144). The methyl compound was prepared by two methods, the hydrolysis of a tetraalkoxymethylantimony compound or the oxidation of dimethoxymethylstibine [*54553-25-2*], $C_3H_9O_2Sb$, with hydrogen peroxide:

$$CH_3Sb(OR)_4 + 3\ H_2O \xrightarrow[0^\circ C]{CH_2Cl_2} CH_3SbO_3H_2 + 4\ ROH$$

$$CH_3Sb(OCH_3)_2 + H_2O_2 + H_2O \longrightarrow CH_3SbO_3H_2 + 2\ CH_3OH$$

Dialkylstibinic acids can be readily prepared. The preferred method is the aqueous hydrolysis of trialkoxydialkylantimony compounds (145):

$$R_2Sb(OR')_3 + 2\ H_2O \longrightarrow R_2SbO(OH) + 3\ R'OH$$

Dimethylstibinic acid [*35952-95-5*], $C_2H_7O_2Sb$, diethylstibinic acid [*35952-96-6*], $C_4H_{11}O_2Sb$, dipropylstibinic acid [*35952-97-7*], $C_6H_{15}O_2Sb$, and dibutylstibinic acid [*35952-98-8*], $C_8H_{19}O_2Sb$, have been prepared in this manner. Except for the dimethyl compound, they can be readily recrystallized from organic solvents.

Aromatic stibonic acids are readily prepared by the diazo reaction:

$$ArN_2Cl + SbCl_3 \longrightarrow ArSbCl_4 + N_2$$

$$ArSbCl_4 + 3\ H_2O \longrightarrow ArSbO(OH)_2 + 4\ HCl$$

where Ar represents aryl.

The arylstibonic acids prepared by this procedure are invariably contaminated with inorganic antimony compounds. Purification is usually effected by dissolving the crude product in hydrochloric acid and adding ammonium chloride or an amine hydrochloride, whereupon crystalline salts of the type $[R_4N][ArSbCl_5]$ separate from solution. These may be recrystallized and then hydrolyzed to the pure stibonic acids. The diarylstibinic acids are also usually prepared by the diazo reaction by substituting an aryldihalostibine for antimony trichloride:

$$ArN_2Cl + Ar'SbCl_2 \longrightarrow ArAr'SbCl_3 + N_2$$

$$ArAr'SbCl_3 + 2\ H_2O \longrightarrow ArAr'SbO(OH) + 3\ HCl$$

Diphenylstibinic acid [*22811-63-8*], $C_{12}H_{11}O_2Sb$, can be readily prepared in good yield from triphenylstibine by an oxidative cleavage reaction in which the stibine is heated with alkali and hydrogen peroxide (146). Stibonic and stibinic acids have had few industrial uses. A patent covering the use of stibonic or stibinic acids as catalysts for the condensation-polymerization of ethylene glycol and terephthalic acid has been issued (147). Another patent describes the use of diphenylstibinic acid as a cocatalyst with triisobutylaluminum for the polymerization of epichlorohydrin (148). The use of stibonic acids as catalysts for the epoxidation of alkenes by hydrogen peroxide is the subject of another patent (149). Anhydrides of arylstibonic acids, $(ArSbO_2)_n$, obtained by heating stibonic acids *in vacuo* at 100°C, have proved to be effective catalysts for the polymerization of oxiranes (150).

Stibine Oxides and Related Compounds. Both aliphatic and aromatic stibine oxides, R_3SbO, or their hydrates, $R_3Sb(OH)_2$, are known. Thus both dihydroxotrimethylantimony [*19727-41-4*], $C_3H_{11}O_2Sb$, and trimethylstibine oxide [*19727-40-3*], C_3H_9OSb, have been prepared. The former may be readily obtained by passing an aqueous solution of dichlorotrimethylantimony [*13059-67-1*], $C_3H_9Cl_2Sb$, through an anionic-exchange resin (151). When heated to 110°C, the dihydroxy compound loses one mole of water to form the oxide. Aliphatic stibine oxides containing larger alkyl groups have been obtained by oxidation of trialkylstibines using mercury(II) oxide (152). The trialkylstibine oxides are hygroscopic crystalline solids. Molecular weights in solution are usually two to three times the values calculated for monomeric structures. The Mössbauer spectrum of trimethylstibine oxide was consistent with a trigonal-bipyramidal structure having three methyl groups in equatorial positions (153). The use of dihydroxotrimethylantimony as a catalyst for the epoxidation of alkenes by aqueous hydrogen peroxide has been the subject of a patent (154).

The structure of triphenylstibine oxide [*4756-75-6*], $C_{18}H_{15}OSb$, has been the subject of considerable controversy. Apparently it can exist in two different forms; as prismatic crystals, mp 221–222°C, and as an amorphous powder. The structure of the crystalline form was shown by x-ray diffraction to be a dimer containing a planar four-membered ring with Sb–O–Sb bonds (155,156). It can be prepared by a number of synthetic methods including the thermal decomposition of hydroxotetraphenylantimony [*19638-16-5*], $C_{24}H_{21}OSb$, (157) or methoxotetraphenylantimony [*14090-94-9*], $C_{25}H_{23}OSb$ (155). The amorphous form of triphenylstibine oxide, termed poly(triphenylstibine oxide) [*36562-85-3*], $(C_{18}H_{15}OSb)_x$, is insoluble in water and in organic solvents; its structure is unknown. A number of papers have been published on the use of triphenylstibine oxide as a catalyst for various reactions which may be of considerable commercial value. It has been used as a catalyst for the condensation of diamines and carbon dioxide to form cyclic urea compounds (158), for the ring-opening polymerization of ethylene oxide or propylene oxide (159), and for the formation of 2-oxazolidinones from carbon dioxide and 2-aminoalcohols (160). It has been claimed that triphenylstibine oxide and triphenylstibine sulfide are of value as catalyst modifiers in the polymerization of propene by Ziegler catalysts (161,162), and a number of patents have been issued on the use of the oxide as a catalyst in various industrial processes. In addition to triphenylstibine oxide, other triarylstibine oxides have been synthesized by the oxidation of the corresponding triarylstibines with hydrogen peroxide (163). Dihydroxotriphenylantimony [*896-29-7*], $C_{18}H_{17}O_2Sb$, has been reported a number of times in the chemical literature. There has been, however, no investigation utilizing analytical instrumental techniques on this compound. Dihydroxotris(2,4,6-trimethylphenyl)antimony [*113002-54-3*], $C_{27}H_{35}O_2Sb$, is the only compound of this class, the structure of which has been confirmed by x-ray diffraction (163).

Both trialkyl- and triarylstibine sulfides and selenides are known. Trimethylstibine sulfide [*15082-93-6*], C_3H_9SSb, has been prepared from trimethylstibine oxide and hydrogen sulfide (164). It is monomeric in benzene and chloroform. Trialkylstibine sulfides and selenides have been prepared from trialkylstibines and sulfur or selenium, respectively (165). Unlike triphenylstibine oxide, the structure of triphenylstibine sulfide is tetrahedral, as shown by both Mössbauer and x-ray diffraction studies (153). A patent covering the synergistic use of triphenylstibine sulfide with aromatic amines as antioxidants in lubricating oils has been issued (166).

Pentacovalent Antimony Halides and Related Compounds. Antimony halides of the types $RSbX_4$, R_2SbX_3, R_3SbX_2, and R_4SbX, where X is a halogen, are known, but compounds of the first type have only been isolated and characterized where R is aryl. Tetrachloromethylantimony has been prepared at −70°C, but not isolated (167):

$$CH_3SbCl_2 + SO_2Cl_2 \xrightarrow{-70°C} CH_3SbCl_4 + SO_2$$

$$CH_3SbCl_4 + (CH_3)_4NN_3 \xrightarrow{-70°C} [CH_3SbCl_4N_3][(CH_3)_4N]$$

The aryl compounds are unstable substances which decompose on standing and are hydrolyzed in moist air. The chlorides are readily prepared by the action of hydrochloric acid on the corresponding arylstibonic acids. Tetraacetatophenylantimony [*116122-86-27*], $C_{14}H_{17}O_8Sb$, has been prepared (168):

$$C_6H_5SbCl_2 + SO_2Cl_2 + 4\ Ag(O_2CCH_3) \xrightarrow{CH_2Cl_2} C_6H_5Sb(O_2CCH_3)_4 + 4\ AgCl + SO_2$$

The Sb atom in this tetraacetato compound is hexacoordinate with three monodentate acetate groups and one symmetrically bonded bidentate acetate group.

Both dialkyl- and diaryltrihaloantimony compounds are known, although only a few dialkyl compounds have been described. The trichlorides have been obtained by the chlorination of either dialkylchlorostibines (169) or tetraalkyldistibines (170) with sulfuryl chloride. Dimethyltrichloroantimony [*7289-79-4*], $C_2H_6Cl_3Sb$, is dimeric in the solid state but is monomeric in solution (171). The dimer exists in two different forms, covalent and ionic (172). The covalent form contains bridging chlorine atoms; the ionic form possesses the structure $[(CH_3)_4Sb]\ [SbCl_6]$. The diaryl compounds, both symmetrical and unsymmetrical, are best prepared from a diazonium halide and an aryldihalostibine:

$$ArN_2Cl + Ar'SbCl_2 \longrightarrow ArAr'SbCl_3 + N_2$$

When a diazonium salt is allowed to react with antimony pentachloride or an aryltetrachloroantimony compound, the onium salts $[ArN_2][SbCl_6]$ or $[ArN_2][Ar'SbCl_5]$, respectively, are formed. These can be decomposed in an organic solvent by the addition of a powdered metal such as iron or zinc, with the formation of a diaryltrichloroantimony compound:

$$2\ [ArN_2][SbCl_6] + 3\ Fe \longrightarrow Ar_2SbCl_3 + 2\ N_2 + SbCl_3 + 3\ FeCl_2$$

This is known as the Nesmeyanov reaction. The trichloro compounds are somewhat more stable than the tetrachloro compounds and can usually be readily recrystallized. These are also formed from diarylstibinic acids and hydrochloric acid, and, because they usually possess sharp melting points, they can be used for the characterization of the corresponding stibinic acids. Trichlorodiphenylantimony [*21907-22-2*], $C_{12}H_{10}Cl_3Sb$, crystallizes from hydrochloric acid as trichlorophenylantimony monohydrate [*18762-79-9*], $(C_6H_5)_2SbCl_3{\cdot}H_2O$, but this readily loses water on heating to form the anhydrous trichloro compound which exists in the solid state as a dimer (173). The hydrate can also be prepared from $SbCl_5$ and $(C_6H_5)_4Sn$ (174). In addition to the trichlorides and tribromides, the mixed compounds, for example, dibromochlorodiphenylantimony [*71191-19-0*], $(C_6H_5)_2SbBr_2Cl$, and bromodichlorodiphenylantimony [*71191-18-9*], $(C_6H_5)_2SbBrCl_2$, have been prepared (175):

$$(C_6H_5)_2SbCl + Br_2 \xrightarrow[CH_2Cl_2]{-196°C} (C_6H_5)_2SbBr_2Cl$$

$$(C_6H_5)_2SbBr + Cl_2 \xrightarrow[CH_2Cl_2]{-90°C} (C_6H_5)_2SbBrCl_2$$

In contrast to the trichloro compound, these mixed halo compounds, as well as tribromodiphenylantimony [*62170-61-0*], $(C_6H_5)_2SbBr_3$, are monomeric in the solid state.

In addition to the trihalo compounds, triacetatodiphenylantimony [*93833-20-6*], $C_{18}H_{19}O_2Sb$, has been prepared (176):

$$(C_6H_5)_2SbCl_3 + 3\ Ag(O_2CCH_3) \longrightarrow (C_6H_5)_2Sb(O_2CCH_3)_3 + 3\ AgCl$$

The best known of the halides are the trialkyldihalo- and triaryldihaloantimony compounds. The dichloro, dibromo, and diiodo compounds are generally prepared by direct halogenation of the corresponding tertiary stibines. The difluoro compounds are obtained by metathasis from the dichloro or dibromo compounds and silver fluoride. The diiodo compounds are the least stable and are difficult to obtain in a pure state. The trialkyl- and triaryldichloro- and dibromoantimony compounds are all crystalline solids which are stable at room temperature but decompose on heating:

$$R_3SbX_2 \xrightarrow{\Delta} R_2SbX + RX$$

The difluoro compounds, however, do not undergo this thermal decomposition.

Dichlorotriphenylantimony has been suggested as a flame retardant

(177,178) and as a catalyst for the polymerization of ethylene carbonate (179). Dibromotriphenylantimony has been used as a catalyst for the reaction between carbon dioxide and epoxides to form cyclic carbonates (180) and for the oxidation of α-keto alcohols to diketones (181).

In addition to the trialkyldihalo- and triaryldihaloantimony compounds, mixed dihalo compounds such as chloroiodotriphenylantimony [*7289-82-9*], $(C_6H_5)_3SbClI$, have been reported (182). It has been shown, however, that such compounds disproportionate in solution to give a mixture of starting material plus products (183):

$$2\ R_3SbXY \rightleftharpoons R_3SbX_2 + R_3SbY_2$$

Other trialkyl and triaryl compounds of the type R_3SbY_2, where Y is a pseudohalide or a group such as NO_3, ClO_4, or ½ SO_4, have been prepared. They are usually obtained from the dihalides by metathesis with a silver salt, eg:

$$R_3SbCl_2 + 2\ AgNO_3 \longrightarrow R_3Sb(NO_3)_2 + 2\ AgCl$$

Compounds of the type $R_3Sb(OSO_2R')_2$, and $R_3Sb(O_2CR')_2$, where R is an alkyl or an aryl group, have been prepared from a dihydroxide or oxide and the appropriate acid:

$$R_3Sb(OH)_2 + 2\ R'SO_3H \longrightarrow R_3Sb(OSO_2R')_2 + 2\ H_2O$$

$$(Ar_3SbO)_2 + 4\ R'CO_2H \longrightarrow 2\ Ar_3Sb(O_2CR')_2 + 2\ H_2O$$

Hydrolysis of trialkyl- and triaryldihaloantimony compounds generally leads to the isolation of compounds of the type $(R_3SbX)_2O$ rather than compounds of the type $R_3Sb(OH)X$. However, hydroxoiodobis(2,6-dimethylphenyl)antimony [*112515-20-5*], $(2,6(CH_3)_2C_6H_3)_2Sb(OH)I$, (184) and four cyclohexyl compounds have been prepared (185): chlorohydroxotricyclohexylantimony [*85362-32-9*], $C_{18}H_{34}ClOSb$, bromohydroxotricyclohexylantimony [*85362-33-0*], $C_{18}H_{34}BrOSb$, acetatohydroxotricyclohexylantimony [*85362-34-1*], $C_{20}H_{37}O_3Sb$, and hydroxonitratotricyclohexylantimony [*85362-35-2*], $C_{18}H_{34}NO_4Sb$.

Tetraalkyl and tetraaryl compounds, R_4SbX, are well-known and are often referred to as stibonium salts. There is evidence, however, that most of the tetraaryl compounds contain pentacovalent antimony. The perchlorate $[(C_6H_5)_4Sb](ClO_4)$, however, is ionic (186). The tetraalkyl halides are readily prepared by quaternization of the corresponding tertiary stibines:

$$R_3Sb + RI \longrightarrow R_4SbI$$

The tetraaryl compounds can be prepared by employing anhydrous aluminum chloride (187):

$$(C_6H_5)_3Sb + C_6H_5Cl \xrightarrow[230°C]{AlCl_3} (C_6H_5)_4SbCl$$

Both the tetraalkyl and tetraaryl compounds can be prepared by cleavage of the corresponding pentaalkyl- or pentaarylantimony by a halogen or a hydrogen halide:

$$R_5Sb + HX \longrightarrow R_4SbX + RH$$

The cleavage of pentamethylantimony [*15120-50-0*], $C_5H_{15}Sb$, with HN_3, HCN, or HSCN gives the corresponding tetramethylantimony azides, cyanides, or thiocyanates (188); cleavage with one molecular equivalent of a carboxylic acid gives compounds of the type $(CH_3)_4SbO_2CR$ (189). Tetraphenylantimony iodide [*13903-91-8*], $C_{24}H_{20}ISb$, has been found to catalyze the condensation of oxetanes with carbon dioxide to give dioxanones (190) and the cycloaddition of oxiranes or oxetanes with diphenylketene to give γ- or δ-lactones (191).

Stibonium Ylids and Related Compounds. In contrast to phosphorus and arsenic, only a few antimony ylids have been prepared. Until quite recently triphenylstibonium tetraphenylcyclopentadienylide [*15081-36-4*], $C_{47}H_{35}Sb$, was the only antimony ylid that had been isolated and adequately characterized (192). A new method, utilizing an organic copper compound as a catalyst, has resulted in the synthesis of a number of new antimony ylids (193):

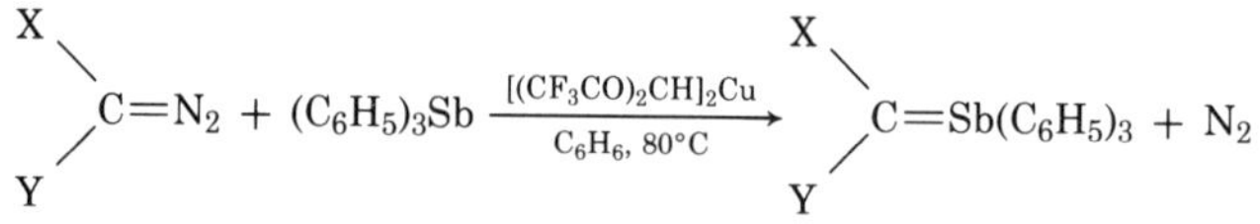

Among the ylids prepared by this method are those in which X and Y are $C_6H_5SO_2$, 4–$CH_3C_6H_4SO_2$, or CH_3CO or where X is CH_3CO and Y is C_6H_5CO. These ylids are solids, stable in a dry atmosphere, but readily hydrolyzed in protic solids by traces of moisture. Attempts to carry out the Wittig reaction with the stibonium ylids containing sulfonyl or carbonyl substituents, using highly reactive 2,4-dinitrobenzaldehyde as the substrate, were unsuccessful (194). Closely related to the ylids are imines of the type $R_3Sb{=}NR'$, where R is either an alkyl or an aryl group. The alkyl compounds, where R is ethyl or propyl and Ar is phenyl or 4-tolyl, have been prepared from trialkylstibines and arenesulfonyl azides (195):

$$R_3Sb + ArSO_2N_3 \longrightarrow R_3Sb{=}NSO_2Ar + N_2$$

The reaction of triphenylstibine with chloramine-T leads to the formation of a tosylimine (196):

$$(C_6H_5)_3Sb + CH_3C_6H_4SO_2N(Na)Cl \longrightarrow (C_6H_5)_3Sb{=}NSO_2C_6H_4CH_3 + NaCl$$

Organoantimony Compounds with Five Sb–C Bonds. A number of pentaalkyl- and pentaalkenylantimony compounds have been prepared from tetraalkyl- or tetraalkenylstibonium halides and alkyl or alkenyllithium or Grignard reagents, for example:

$$(CH_3)_4SbBr + CH_3Li \longrightarrow (CH_3)_5Sb + LiBr$$

Rather than using the stibonium halide, a trialkyl- or trialkenyldihaloantimony compound can be used, as in the preparation of pentavinylantimony [*65277-46-5*], $C_{10}H_{15}Sb$:

$$(CH_2{=}CH)_3SbBr_2 + 2\ CH_2{=}CHMgBr \longrightarrow (CH_2{=}CH)_5Sb + 2\ MgBr_2$$

Pentaarylantimony compounds can be readily prepared in a similar fashion:

$$(C_6H_5)_3SbF_2 + 2\ C_6H_5MgBr \longrightarrow (C_6H_5)_5Sb + MgBr_2 + MgF_2$$

Pentaphenylantimony [*2170-05-0*], $C_{30}H_{25}Sb$, has attracted considerable attention because it possesses square-pyramidal rather than the expected trigonal-bipyramidal geometry, both in the solid state and in solution. The cyclohexane solvate $(C_6H_5)_5Sb{\cdot}\frac{1}{2}C_6H_{12}$ and penta-4-tolylantimony [*51017-91-5*], $C_{35}H_{35}Sb$, however, both possess trigonal-bipyramidal geometry. In addition to compounds of the type R_5Sb, mixed compounds of the type $R_4R'Sb$ or $R_3R'_2Sb$, where R and R′ may be alkyl, alkenyl, alkynyl, or aryl groups, are known. Thus triethyldimethylantimony [*67576-92-5*], $C_8H_{21}Sb$, has been prepared (197):

$$(C_2H_5)_3SbCl_2 + 2\ CH_3Li \xrightarrow[-10°C]{(C_2H_5)_2O} (C_2H_5)_3Sb(CH_3)_2 + 2\ LiCl$$

Methyltetraphenylantimony [*33756-93-3*], $C_{25}H_{23}Sb$, is readily prepared (198):

$$(C_6H_5)_4SbF + CH_3MgBr \longrightarrow (C_6H_5)_4SbCH_3 + MgBrF$$

Compounds containing aryl and alkynyl groups have also been prepared (199):

$$Ar_3SbBr_2 + 2\ C_6H_5C{\equiv}CLi \longrightarrow Ar_3Sb(C{\equiv}CC_6H_5)_2 + 2\ LiBr$$

Antimony Compounds Used in Medicine

Compounds of antimony have been used as therapeutic agents for thousands of years (200). There is evidence that the ancient Egyptians used a preparation of naturally occurring stibnite as a cosmetic and as a prophylactic agent against diseases of the eyes. References to the use of antimonials as drugs occur in the Bible and in ancient writings from India, China, and Mexico. During the Middle Ages antimony compounds were widely employed in European medicine. The toxicity of these substances was noted at an early date, and in 1566 their use as drugs was condemned by the physicians of Paris and banned by an act of the Paris Parliament. This prohibition was withdrawn in 1667 when Louis XIV became convinced that his recovery from a serious illness (possibly typhoid) resulted from the administration of an antimony compound. The therapeutic use of antimonials reached a peak during the eighteenth or early nineteenth century. By the beginning of the twentieth century, antimony compounds were virtually obsolete in medical practice and were used only rarely for veterinary purposes.

In 1912, however, (201) it was discovered that espundia (American mucocutaneous leishmaniasis) can be cured by tartar emetic. It was soon learned that kala-azar (visceral leishmaniasis) and oriental sore (a cutaneous form of the disease occurring in the Middle East) also respond to antimonial therapy, especially when compounds of pentavalent antimony are employed. Treatment of leishmaniasis with the latter type of antimonials is safe and effective in over 90% of the cases (202). In 1918, it was demonstrated that tartar emetic is of value in the treatment of schistosomiasis (203). Pentavalent antimonials proved to be less effective. The introduction of antimony compounds for the treatment of parasitic diseases is undoubtedly one of the important milestones in the history of therapeutics (see ANTIPARASITIC AGENTS).

Antimony potassium tartrate (tartar emetic) has the advantage of being low in cost. It has been called the drug of choice for *Schistosoma japonicum* infection (204) even though it fails to cure the disease in many patients. However, trivalent antimonials are no longer recommended for the treatment of helminthic infections because these compounds have an unacceptable toxicity and are too difficult to administer (205).

Antimony(V) sodium gluconate [*16037-91-5*] (sodium stibogluconate, Pentostam) is a well-tolerated and potent antileishmanial agent . It is one of the best drugs in clinical use for the treatment of all types of leishmaniasis (206). In the United States this substance is available only from the Centers for Disease Control in Atlanta, Georgia. Its main disadvantages are the long course of therapy required, the necessity for intravenous or intramuscular administration, and the relatively high cost. Although less hazardous than the trivalent antimonials, it is not an innocous substance and may give rise to cardiotoxic effects.

Meglumine [*6284-40-8*] antimonate [*133-51-7*], $C_7H_{17}NO_5{\cdot}HO_3Sb$, (Glucantime) is a pentavalent antimonial used for the treatment of leishmaniasis in Latin America (207) and in the French-speaking areas of Europe and Africa (206). It is administered by deep intramuscular injection. Although relatively safe and usually well tolerated, numerous side effects have been observed. It is not available in the United States.

BIBLIOGRAPHY

"Antimony Compounds," in *ECT* 1st ed., Vol. 2, pp. 59–64, by I. E. Campbell; in *ECT* 2nd ed., Vol. 2, pp. 570–588, by G. O. Doak, Leon Freedman, and G. Gilbert Long, North Carolina State of the University of North Carolina at Raleigh; in *ECT* 3rd ed., Vol. 3, pp. 105–128, by Leon Freedman, G. O. Doak, and G. Gilbert Long, North Carolina State University.

1. T. O. Llewellyn, *Mineral Yearbook for 1988,* Vol. 1, *Metals and Minerals,* U.S. Department of the Interior, Bureau of Mines, U.S. Government Printing Office, Washington, D.C., 1990, pp. 125–132.
2. L. Bretherick, *Hazards in the Chemical Laboratory,* 4th ed., The Royal Society of Chemistry, London, UK, 1986, p. 181.
3. H. S. Gowda and S. Gurumurthy, *Indian J. Chem.* **21A,** 550 (1982).
4. S. G. Viswanath and M. K. N. Yenkie, *Chem. Anal. (Warsaw)* **28,** 43 (1983).
5. N. Jayasree and P. Indrasenan, *Talanta* **32,** 1067 (1985).

6. H. S. Gowda, R. Shakunthala, and U. Subrahmanya, *Indian J. Chem.* **20A,** 823 (1981).
7. W. S. Selig and G. L. Roberts, *Microchem J.* **34,** 140 (1986).
8. K. Sriramam, B. S. R. Sarma, N. R. Sastry, and A. R. K. V. Prasad, *Talanta* **28,** 963 (1981).
9. M. A. Al-Hajjaji and co-workers, *Anal. Lett.* **19,** 283 (1986).
10. S. Williams, ed., *Official Methods of Analysis of the Association of Official Analytical Chemists,* 14th ed., Association of Official Analytical Chemists, Arlington, Va., 1984, p. 449.
11. P. P. Coetzee and H. Pieterse, *S. Afr. J. Chem.* **39,** 85 (1986).
12. N. E. Parisis and A. Heyndrickx, *Analyst (London)* **111,** 281 (1986).
13. D. T. Hurd, *An Introduction to the Chemistry of the Hydrides,* John Wiley & Sons, Inc., New York, 1952, pp. 132–134.
14. C. L. Yaws and co-workers, *Solid State Technol.* **17,** 47 (1974).
15. A. W. Jache, G. S. Blevins, and W. Gordy, *Phys. Rev.* **97,** 680 (1955).
16. P. W. Schenk, *G. Brauer's Handbook of Preparative Inorganic Chemistry,* Vol. I, 2nd ed., Academic Press, New York, 1963, p. 591.
17. L. H. Berka, *The Chemistry of Stibine,* M.S. dissertation UCRL-8781, University of California, Berkeley, 1959, pp. 3–16.
18. A. D. Zorin and co-workers, *J. Appl. Chem. USSR (Engl. Transl.)* **47,** 1233 (1974).
19. G. Kh. Sorokin and S. A. Lomonosov, *Ind. Lab. USSR (Engl. Transl.)* **40,** 28 (1974).
20. G. G. Long, J. G. Stevens, and L. H. Bowen, *Inorg. Nucl. Chem. Lett.* **5,** 799 (1969).
21. D. J. Stewart and co-workers, *Can. J. Chem.* **50,** 690 (1972).
22. D. Rogers and A. C. Skapski, *Proc. Chem. Soc. (London),* 400 (1964).
23. H. Y.-P. Hong, *Acta Crystallogr.* **B30,** 945 (1974).
24. S. S. Krishnan and D. R. Crapper, *Radiochem. Radioanal. Lett.* **20,** 279 (1975).
25. D. Cubicciotti, *High Temp. Sci.* **1,** 268 (1969).
26. C. J. Adams and A. J. Downs, *J. Chem. Soc., A,* 1534 (1971).
27. L. Kolditz, *Halogen Chem.* **2,** 115 (1967).
28. T. K. Davies and K. C. Moss, *J. Chem. Soc., A,* 1054 (1970).
29. A. J. Edwards and P. Taylor, *Chem. Commun.,* 1376 (1971).
30. R. C. Thompson and co-workers, *Inorg. Chem.* **4,** 1641 (1965).
31. R. J. Gillespie and T. E. Peel, *J. Am. Chem. Soc.* **95,** 5173 (1973).
32. J. Melin and A. Herold, *C. R. Acad. Sci., Ser. C* **280,** 641 (1975).
33. H. Preiss and L. Kolditz, *Proceedings of the 2nd Seminar on Crystallochemistry of Coordination and Metallorganic Compounds,* 1973, p. 90.
34. H. Oppermann, *Z. Anorg. Allg. Chem.* **356,** 1 (1967).
35. I. Lefkowitz, B. Brodman, and M. A. Shields, *J. Cryst. Growth* **6,** 13 (1969).
36. M. L. Hackert, R. A. Jacobson, and T. A. Keiderline, *Inorg. Chem.* **10,** 1075 (1971).
37. S. L. Lawton and R. A. Jacobson, *Inorg. Chem.* **10,** 709 (1971).
38. D. D. Wagman and co-workers, *N.B.S. Technical Note 270-3,* U.S. Dept. of Commerce, Government Printing Office, Washington, D.C., 1968, pp. 99–102.
39. G. G. Long, J. G. Stevens, and L. H. Bowen, *Inorg. Nucl. Chem. Lett.* **5,** 21 (1969).
40. L. W. Clemence and M. T. Leffler, *J. Am. Chem. Soc.* **70,** 2439 (1948).
41. H.-C. Mu, *K'o Hsueh T'ung* **17,** 502 (1966).
42. M. E. Gress and R. A. Jacobson, *Inorg. Chim. Acta* **8,** 209 (1974).
43. G. A. Kiosse, N. I. Golovastikov, and N. V. Belov, *Dokl. Akad. Nauk SSSR* **155,** 545 (1964).
44. B. Kamenar, D. Grdenić, and C. K. Prout, *Acta Crystallogr.* **B26,** 181 (1970).
45. G. A. Kiosse and co-workers, *Dokl. Akad. Nauk SSSR* **177,** 329 (1967).
46. A. Zalkin, D. H. Templeton, and T. Ueki, *Inorg. Chem.* **12,** 1641 (1973).
47. R. E. Tapscott, R. L. Belford, and I. C. Paul, *Coord. Chem. Rev.* **4,** 323 (1969).
48. I. Hansson, *Acta Chem. Scand.* **22,** 509 (1968).

49. A. C. Nanda and S. Pani, *J. Indian Chem. Soc.* **31,** 588 (1954).
50. B. C. Mohanty and S. Pani, *J. Indian Chem. Soc.* **31,** 593 (1954).
51. G. Hunter, *J. Chem. Soc., Dalton Trans.,* 1496 (1972).
52. R. Weinland and R. Scholder, *Z. Anorg. Allg. Chem.* **127,** 343 (1923).
53. S. Datta and T. N. Ghosh, *Sci. Cult.* **11,** 699 (1946).
54. J. Rigaudy and S. P. Klesney, *IUPAC Nomenclature of Organic Chemistry, Sections A–F, and H,* Pergamon Press, Oxford, UK, 1979, pp. 382–408.
55. F. G. Mann, *The Heterocyclic Derivatives of Phosphorus, Arsenic, Antimony and Bismuth,* 2nd ed., Wiley-Interscience, New York, 1970.
56. R. E. Atkinson in A. R. Katritzky and C. W. Rees, eds., *Comprehensive Heterocyclic Chemistry,* Vol. 1, Pergamon Press, Oxford, UK, 1984, pp. 539–561.
57. S. Samaan, *Houben–Weyl Methoden der Organischen Chemie: Metallorganischen Verbindungen, As, Sb, Bi,* Georg Thieme Verlag, Stuttgart, Germany, 1978, Band XIII, Teil 8.
58. M. Wieber, *Gmelin Handbook of Inorganic Chemistry, Sb Organoantimony Compounds, Part 1,* 8th ed., Springer-Verlag, Berlin, Germany, 1981.
59. Ref. 58, Part 2, 1981.
60. Ref. 58, Part 3, 1982.
61. Ref. 58, Part 4, 1986.
62. M. Mirbach and M. Wieber in Ref. 58, Part 5, 1990.
63. M. Dub, *Organometallic Compounds,* Vol. III, 2nd ed., Springer-Verlag, New York, 1968.
64. Ref. 63, 1st Supplement, 1972.
65. G. O. Doak and L. D. Freedman, *Organometallic Compounds of Arsenic, Antimony, and Bismuth,* John Wiley & Sons, Inc., New York, 1970.
66. L. D. Freedman and G. O. Doak, *Chem. Met.-Carbon Bond,* **5,** 397 (1989).
67. A. L. Rheingold, P. Choudhury, and M. F. El-Shazly, *Synth. React. Inorg. Met.-Org. Chem.* **8,** 453 (1978).
68. P. Choudhury, M. F. El-Shazly, C. Spring, and A. L. Rheingold, *Inorg. Chem.* **18,** 543 (1979).
69. K. Issleib and B. Hamann, *Z. Anorg. Allg. Chem.* **339,** 289 (1965).
70. K. Issleib, B. Hamann, and L. Schmidt, *Z. Anorg. Allg. Chem.* **339,** 298 (1965).
71. K. Issleib and B. Hamann, *Z. Anorg. Allg. Chem.* **332,** 179 (1964).
72. A. B. Burg and L. R. Grant, *J. Am. Chem. Soc.* **81,** 1 (1959).
73. E. Wiberg and K. Mödritzer, *Z. Naturforsch Teil B* **B12,** 128 (1957).
74. K. Issleib and A. Balszuweit, *Z. Anorg. Allg. Chem.* **418,** 158 (1975).
75. M. Ates, H. J. Breunig, and S. Gülec, *Phosphorus, Sulfur Silicon Relat. Elem.* **44,** 129 (1989).
76. E. Wiberg and K. Mödritzer, *Z. Naturforsch. Teil B* **B12,** 131 (1957).
77. K. Issleib and B. Hamann, *Z. Anorg. Allg. Chem.* **343,** 196 (1966).
78. A. H. Cowley, R. A. Jones, C. M. Nunn, and D. L. Westmoreland, *Angew Chem., Int. Ed. Engl.* **28,** 1018 (1989).
79. A. N. Nesmeyanov, A. E. Borisov, and N. V. Novikova, *Izv. Akad. Nauk SSSR, Ser. Khim.,* 815 (1967).
80. Y.-Z. Huang, Y. Shen, and C. Chen, *Tetrahedron Lett.* **26,** 5171 (1985).
81. B. N. Laskorin and V. V. Yakshin, *Dokl. Akad. Nauk SSSR* **206,** 653 (1972).
82. W. J. C. Dyke, W. C. Davies, and W. J. Jones, *J. Chem. Soc.,* 463 (1930).
83. G. S. Hiers, *Organic Synthesis, Collective Volumes,* Vol. 1, 2nd ed., John Wiley & Sons, Inc., New York, 1941, p. 550.
84. T. V. Talalaeva and K. A. Kocheshkov, *Zh. Obshch. Khim.* **16,** 777 (1946).
85. A. J. Ashe III, C. M. Kausch, and O. Eisenstein, *Organometallics* **6,** 1185 (1987).
86. A. N. Nesmeyanov and L. G. Makarova, *Zh. Obshch. Khim.* **7,** 2649 (1937).

87. D. Naumann, W. Tyrra, and F. Leifeld, *J. Organomet. Chem.* **333,** 193 (1987).
88. L. I. Zakharkin and O. Yu. Okhlobystin, *Dokl. Akad. Nauk SSSR* **116,** 236 (1957).
89. E. A. Ganja, C. D. Ontiveros, and J. A. Morrison, *Inorg. Chem.* **27,** 4535 (1988).
90. S. P. Olifirenko, *Visn. L'viv. Derzh. Univ. Ser. Khim.* **6,** 100 (1963).
91. H. A. Meinema, H. F. Martens, and J. G. Noltes, *J. Organomet. Chem.* **51,** 223 (1973).
92. G. O. Doak and L. D. Freedman, *Synthesis,* 328 (1974).
93. E. G. Hope, T. Kemmitt, and W. Levason, *J. Chem. Soc., Perkin Trans. 2,* 487 (1987).
94. J. Seifter, *J. Pharmacol. Exp. Ther.* **66,** 366 (1939).
95. J. I. Harris, S. T. Bowden, and W. J. Jones, *J. Chem. Soc.,* 1568 (1947).
96. R. R. Holmes and E. F. Bertaut, *J. Am. Chem. Soc.* **80,** 2983 (1958).
97. G. O. Doak, G. G. Long, and L. D. Freedman, *J. Organomet. Chem.* **4,** 82 (1965).
98. P. Raj, R. Rastogi, and Firojee, *Indian J. Chem.* **26A,** 682 (1987).
99. C. A. McAuliffe and W. Levason, *Studies in Inorganic Chemistry,* Vol. 1, *Phosphine, Arsine and Stibine Complexes of the Transition Elements,* Elsevier, Amsterdam, The Netherlands, 1979.
100. C. Chen, Y.-Z. Huang, Y. Shen, and Y. Liao, *Heteroatom Chem.* **1,** 49 (1990).
101. R. D. Dupuis, *Science* **226,** 623 (1984).
102. T. Yoshimura, T. Masuda, and T. Higashimura, *Macromolecules* **21,** 1899 (1988).
103. A. S. Kholmanskii, E. A. Kuz'mina, and V. F. Tarasov, *Zh. Fiz. Khim.* **58,** 2095 (1984).
104. M. Wieber, D. Wirth, and I. Fetzer, *Z. Anorg. Allg. Chem.* **505,** 134 (1983).
105. H. J. Breunig and H. Jawad, *J. Organomet. Chem.* **243,** 417 (1983).
106. H. J. Breunig and W. Kanig, *Phosphorus Sulfur* **12,** 149 (1982).
107. M. Nunn, D. B. Sowerby, and D. M. Wesolek, *J. Organomet. Chem.* **251,** C45 (1983).
108. M. Ates, H. J. Breunig, A. Soltani-Neshan, and M. Tegeler, *Z. Naturforsch., B: Anorg. Chem., Org. Chem.* **41B,** 321 (1986).
109. G. O. Doak and H. H. Jaffé, *J. Am. Chem. Soc.* **72,** 3025 (1950).
110. I. G. M. Campbell and A. W. White, *J. Chem. Soc.,* 1184 (1958).
111. F. F. Blicke and U. O. Oakdale, *J. Am. Chem. Soc.* **55,** 1198 (1933).
112. F. F. Blicke, U. O. Oakdale, and F. D. Smith, *J. Am. Chem. Soc.* **53,** 1025 (1931).
113. S. P. Bone and D. B. Sowerby, *J. Chem. Soc., Dalton Trans.,* 1430 (1979).
114. G. T. Morgan and G. R. Davies, *Proc. Roy. Soc., Ser. A* **110,** 523 (1926).
115. H. J. Breunig and H. Kischkel, *Z. Naturforsch, B: Anorg. Chem., Org. Chem.* **36B,** 1105 (1981).
116. H. J. Breunig and H. Jawad, *Z. Naturforsch, B: Anorg. Chem., Org. Chem.* **37B,** 1104 (1982).
117. H. H. Jaffé and G. O. Doak, *J. Am. Chem. Soc.* **71,** 602 (1949).
118. N. Baumann and M. Wieber, *Z. Anorg. Allg. Chem.* **408,** 261 (1974).
119. V. L. Foss, N. M. Semenenko, N. M. Sorokin, and I. F. Lutsenko, *J. Organomet. Chem.* **78,** 107 (1974).
120. H. A. Meinema and J. G. Noltes, *Inorg. Nucl. Chem. Lett.* **6,** 241 (1970).
121. J. Müller and co-workers, *Z. Naturforsch. Teil B* **40B,** 1320 (1985).
122. W. T. Reichle, *J. Organomet. Chem.* **18,** 105 (1969).
123. E. J. Kupchik and C. T. Theisen, *J. Organomet. Chem.* **11,** 627 (1968).
124. H. Preut, F. Huber, and K.-H. Hengstmann, *Acta Crystallogr., Sect. C: Cryst. Struct. Commun.* **C44,** 468 (1988).
125. D. N. Kravtsov, B. A. Kvasov, S. I. Pombrik, and É. I. Fedin, *Izv. Akad. Nauk SSSR, Ser. Khim.,* 927 (1974).
126. H. Schumann, T. Östermann, and M. Schmidt, *J. Organomet. Chem.* **8,** 105 (1967).
127. A. M. Arif, A. H. Cowley, N. C. Norman, and M. Pakulski, *Inorg. Chem.* **25,** 4836 (1986).
128. A. J. Ashe III, E. G. Ludwig Jr., and H. Pommerening, *Organometallics* **2,** 1573 (1983).
129. A. J. Ashe III, *Adv. Organomet. Chem.* **30,** 77 (1990).

130. F. Klages and W. Rapp, *Chem. Ber.* **88,** 384 (1955).
131. G. Huttner, U. Weber, B. Sigwart, and O. Scheidsteger, *Angew. Chem. Suppl.,* 411 (1982).
132. U. Weber, G. Huttner, O. Scheidsteger, and L. Zsolnai, *J. Organomet. Chem.* **289,** 357 (1985).
133. H. J. Breunig and W. Kanig, *Chem.-Ztg.* **102,** 263 (1978).
134. H. J. Breunig and H. Kischkel, *Z. Anorg. Allg. Chem.* **502,** 175 (1983).
135. H. J. Breunig and A. Soltani-Neshan, *J. Organomet. Chem.* **262,** C27 (1984).
136. H. J. Breunig, A. Soltani-Neshan, K. Häberle, and M. Dräger, *Z. Naturforsch. Teil B* **41B,** 327 (1986).
137. A. L. Rheingold and P. Choudhury, *J. Organomet. Chem.* **128,** 155 (1977).
138. G. Chobert and M. Devaud, *Electrochim. Acta* **25,** 637 (1980).
139. H. J. Breunig and W. Kanig, *J. Organomet. Chem.* **186,** C5 (1980).
140. M. Ates and co-workers, *Chem. Ber.* **122,** 473 (1989).
141. A. J. Ashe III, *Top. Curr. Chem.* **105,** 125 (1982).
142. A. J. Ashe III, T. R. Diephouse, and M. Y. El-Sheikh, *J. Am. Chem. Soc.* **104,** 5693 (1982).
143. K. K. Baldridge and M. S. Gordon, *J. Am. Chem. Soc.* **110,** 4204 (1988).
144. M. Wieber and J. Walz, *Z. Naturforsch., B: Chem. Sci.* **45B,** 1615 (1990); Ref. 65, p. 284.
145. H. A. Meinema and J. G. Noltes, *J. Organomet. Chem.* **36,** 313 (1972).
146. G. O. Doak and J. M. Summy, *J. Organomet. Chem.* **55,** 143 (1973).
147. U.S. Pat. 3,642,702 (Feb. 15, 1972), J. J. Ventura and J. G. Natoli (to M & T Chemicals, Inc.).
148. Ger. Offen. 2,102,102 (Sept. 7, 1972), H. G. J. Overmars and A. Van Elven (to Schering A.-G).
149. Eur. Pat. Appl. EP 74,259 (Mar. 16, 1983), Y. Kuriyama, M. Kakuda, and S. Nitoh (to Mitsubishi Gas Chemical Co., Inc.).
150. R. Nomura, Y. Wada, and H. Matsuda, *J. Polym. Sci., Part A: Polym. Chem.* **26,** 627 (1988).
151. G. G. Long, G. O. Doak, and L. D. Freedman, *J. Am. Chem. Soc.* **86,** 209 (1964).
152. G. N. Chremos and R. A. Zingaro, *J. Organomet. Chem.* **22,** 637 (1970).
153. J. Pebler, F. Weller, and K. Dehnicke, *Z. Anorg. Allg. Chem.* **492,** 139 (1982).
154. Ger. Offen. 2,605,041 (Aug. 26, 1976), M. Pralus, J. P. Schirmann, and S. Y. Delavarenne (to Ugine Kuhlmann).
155. J. Bordner, G. O. Doak, and T. S. Everett, *J. Am. Chem. Soc.* **108,** 4206 (1986).
156. G. Ferguson and co-workers, *Acta Crystallogr., Sect. C: Cryst. Struct. Commun.* **C43,** 824 (1987).
157. W. E. McEwen, G. H. Briles, and D. N. Schulz, *Phosphorus* **2,** 147 (1972).
158. R. Nomura, M. Yamamato, and H. Matsuda, *Ind. Eng. Chem. Res.* **26,** 1056 (1987).
159. R. Nomura, Y. Shiomura, A. Ninagawa, and H. Matsuda, *Makromol. Chem.* **184,** 1163 (1983).
160. H. Matsuda and co-workers, *Ind. Eng. Chem. Prod. Res. Dev.* **24,** 239 (1985).
161. N. M. Karayannis and S. S. Lee, *Makromol. Chem.* **186,** 1871 (1985).
162. N. M. Karayannis, H. M. Khelghatian, and S. S. Lee, *Makromol. Chem.* **187,** 863 (1986).
163. F. Huber, T. Westhoff, and H. Preut, *J. Organomet. Chem.* **323,** 173 (1987).
164. M. Shindo, Y. Matsumura, and R. Okawara, *J. Organomet. Chem.* **11,** 299 (1968).
165. R. A. Zingaro and A. Merijanian, *J. Organomet. Chem.* **1,** 369 (1964).
166. U.S. Pat. 4,032,462 (June 28, 1977), B. W. Hutten and J. M. King (to Chevron Research Co.).
167. K. Dehnicke and H.-G. Nadler, *Z. Anorg. Allg. Chem.* **426,** 253 (1976).
169. H.-G. Nadler and K. Dehnicke, *J. Organomet. Chem.* **90,** 291 (1975).

170. H. A. Meinema, H. F. Martens, and J. G. Noltes, *J. Organomet. Chem.* **51,** 223 (1973).
171. K. Dehnicke and H.-G. Nadler, *Chem. Ber.* **109,** 3034 (1976).
172. W. Schwarz and H.-J. Guder, *Z. Naturforsch., B: Anorg. Chem. Org. Chem.* **33B,** 485 (1978).
173. J. Bordner, G. O. Doak, and J. R. Peters Jr., *J. Am. Chem. Soc.* **96,** 6763 (1974).
174. I. Haiduc and C. Silvestru, *Inorg. Synth.* **23,** 194 (1985).
175. S. P. Bone and D. B. Sowerby, *J. Chem. Soc., Dalton Trans.,* 715 (1979).
176. D. B. Sowerby, M. J. Begley, and P. L. Millington, *J. Chem. Soc., Chem. Commun.,* 896 (1984).
177. O. Horak, J. Havranek, and J. Vladyka, *Kunststoffe* **72,** 493 (1982).
178. O. Horak, *Plasty Kauc.* **24,** 271 (1987).
179. T. Otsu, K. Endo, K. Hozawa, and M. Komatsu, *Mem. Fac. Eng., Osaka City Univ.* **26,** 101 (1985).
180. A. Ninagawa, H. Matsuda, and R. Nomura, *Kenkyu Hokoku-Asahi Garasu Kogyo Gijutsu Shoreikai* **39,** 117 (1981).
181. K. Ohkata, H. Ohnari, and K. Akiba, *Nippon Kagaku Kaishi,* 1267 (1987).
182. A. D. Beveridge, G. S. Harris, and F. Inglis, *J. Chem. Soc. A,* 520 (1966).
183. C. G. Moreland, M. H. O'Brien, C. E. Douthit, and G. G. Long, *Inorg. Chem.* **7,** 834 (1968).
184. G. Ferguson, G. S. Harris, and A. Khan, *Acta Crystallogr., Sect. C: Cryst. Struct. Commun.* **C43,** 2078 (1987).
185. Y. Kawasaki, Y. Yamamoto, and M. Wada, *Bull. Chem. Soc. Jpn.* **56,** 145 (1983).
186. G. Ferguson, C. Glidewell, D. Lloyd, and S. Metcalfe, *J. Chem. Soc., Perkin Trans. 2,* 731 (1988).
187. G. O. Doak, G. G. Long, and L. D. Freedman, *J. Organomet. Chem.* **12,** 443 (1968).
188. H. Schmidbaur, K.-H. Mitschke, J. Weidlein, and St. Cradock, *Z. Anorg. Allg. Chem.* **386,** 139 (1971).
189. H. Schmidbaur, K.-H. Mitschke, and J. Weidlein, *Z. Anorg. Allg. Chem.* **386,** 147 (1971).
190. H. Matsuda and A. Baba, *Kenkyu Hokoku-Asahi Garasu Kogyo Gijutsu Shoreikai* **50,** 195 (1987).
191. M. Fujiwara, M. Imada, A. Baba, and H. Matsuda, *J. Org. Chem.* **53,** 5974 (1988).
192. D. Lloyd and M. I. C. Singer, *Chem. Ind. (London),* 787 (1967).
193. C. Glidewell, D. Lloyd, and S. Metcalfe, *Tetrahedron* **42,** 3887 (1986).
194. G. Ferguson and co-workers, *J. Chem. Soc., Perkin Trans. 2,* 1829 (1988).
195. Z. I. Kuplennik, Zh. N. Belaya, and A. M. Pinchuk, *Zh. Obshch. Khim.* **51,** 2711 (1981).
196. G. Wittig and D. Hellwinkel, *Chem. Ber.* **97,** 789 (1964).
197. N. Tempel, W. Schwarz, and J. Weidlein, *J. Organomet. Chem.* **154,** 21 (1978).
198. G. Doleshall, N. A. Nesmeyanov, and O. A. Reutov, *J. Organomet. Chem.* **30,** 369 (1971).
199. K. Akiba, T. Okinaka, M. Nakatani, and Y. Yamamoto, *Tetrahedron Lett.* **28,** 3367 (1987).
200. H. Schmidt and F. M. Peter, *Advances in the Therapeutics of Antimony,* Thieme, Leipzig, Germany, 1938.
201. G. Vianna, *Arch. Bras. Med.* **2,** 426 (1912); L. T. Morton in F. H. Garrison and L. T. Morton, eds., *Medical Bibliography,* Grafton, London, UK, 1954, p. 464.
202. L. T. Webster, Jr. in A. G. Gilman, T. W. Rall, A. S. Niews, and P. Taylor, eds., *Goodman and Gilman's The Pharmacological Basis of Therapeutics,* 8th ed., Pergamon Press, New York, 1990, p. 1009.
203. J. B. Christopherson, *Lancet* **195,** 325 (1918).
204. AMA Division of Drugs, *AMA Drug Evaluations,* 5th ed., W. B. Saunders Company, Philadelphia, Pa., 1983, p. 1833.

205. Ref. 202, pp. 959–960.
206. Ref. 202, pp. 1013–1014.
207. J. D. Chulay in G. T. Strickland, ed., *Hunter's Tropical Medicine,* 6th ed., W. B. Saunders Company, Philadelphia, Pa., 1984, p. 592.

LEON D. FREEDMAN
G. O. DOAK
G. GILBERT LONG
North Carolina State University

ANTINAUSEA AGENTS. See HYPNOTICS, SEDATIVES, AND ANTICONVULSANTS.

ANTIOBESITY DRUGS

Obesity is a complex and widespread disease with multiple etiology. As of 1985, an estimated 34 million adults in the United States were 20% or more above their ideal body weight and could benefit from weight loss therapy (1). Stress and environment are important contributing factors to the development of obesity, but there is also a significant genetic component as indicated in a number of adoption and twin studies (2–5). For example, it has recently been suggested that an abnormality in skeletal muscle metabolism among obese individuals contributes to their excessive weight gain (6). In addition to being a cosmetic problem, obesity is associated with an enhanced likelihood of developing chronic conditions including hypertension, coronary heart disease (7–9), diabetes (10), cancer, gall bladder disease, and arthritis (1).

In order to define the obese state in a clinical setting, it is necessary to have a means of estimating the amount of adipose (fat) tissue relative to lean body mass. Whereas highly accurate determinations of body composition require complex laboratory procedures, large clinical studies typically employ measures of skin-fold thickness (11) or more commonly, body mass index (BMI) as a quantitative measure of obesity.

$$\text{BMI} = \frac{\text{weight in kilograms}}{(\text{height in meters})^2}$$

Normal individuals have a BMI of 20–25 and those with a BMI in the range of 25–30 are generally not at sufficiently increased risk to justify medical intervention. People having a higher BMI and particularly those having a BMI of over 40 may be considered morbidly obese and are candidates for increasingly aggressive treatment (12). An important consideration is the location of the excess fat. Although obesity is a serious health risk factor for both sexes (13,14), the abdominal or android pattern of fat distribution typical of overweight males carries a higher risk of life threatening complications than the gluteal or gynoid fat distribution typical of overweight females (15,16).

Obesity is a difficult condition to treat. Dietary restriction of caloric intake is the first line therapy and is optimally combined with an exercise program to promote loss of fat relative to lean body mass (17). For the grossly obese (BMI > 40), invasive mechanical measures such as jaw wiring, gastric banding, and gastric by-pass have been attempted with at least limited success (18).

Appetite suppressants have been widely used as an adjunct to dietary restriction and sympathomimetic amines have traditionally been used for this purpose. These agents have not proven particularly useful and frequently cause unacceptable side effects, hence their popularity has been waning for several years. The most promising newer drugs work through a serotoninergic mechanism and hold considerable promise at least for certain obese patients.

The majority of formerly obese patients eventually regain excess weight lost and thus a truly successful program must include behavior modification. Whereas no drugs are available for long-term use in a maintenance program, some agents are being studied clinically, such as the thermogenic ones. Among individuals who have been overweight for long periods, metabolic set points, such as lipoprotein lipase activity, may be altered favoring reestablishment of the obese state (19). Lipoprotein lipase is responsible for the hydrolysis of circulating triglycerides into free fatty acids, facilitating fatty acid uptake into adipocytes prior to reesterification and storage as lipid droplets. An elevation of messenger RNA (mRNA) levels of this enzyme occurs in obese patients following weight reduction (19). This finding may provide a partial explanation at the molecular level for this set point alteration.

Sympathomimetic Agents

The sympathetic or adrenergic nervous system operates in juxtaposition to the parasympathetic nervous system to maintain homeostasis in response to physical activity and physical or psychological stress. Sympathomimetic neurotransmission is generally mediated by norepinephrine [*51-41-2*] (**1**), $C_8H_{11}NO_3$, released from presynaptic storage granules upon stimulation. A second endogenous sympathomimetic agent, epinephrine [*51-43-4*] (**2**), $C_9H_{13}NO_3$, is released systemically from the adrenal glands during emergencies as part of the "fight or flight" response. Based on the discovery of selective ligands, two broad subclasses of receptors termed α- and β-adrenergic receptors have been defined. Each receptor subclass has two or more subtypes. A large variety of peripheral organ functions are affected by the sympathetic nervous systems including heart rate, cardiac contractile force, blood pressure, bronchopulmonary tone, and metabolism. In the central nervous system, sympathetic nervous stimulation results in increases in wakefulness, psychomotor activity, and reduction of appetite (20).

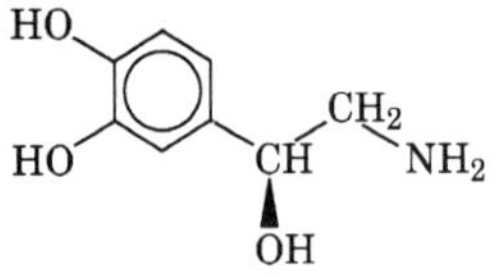

(1)

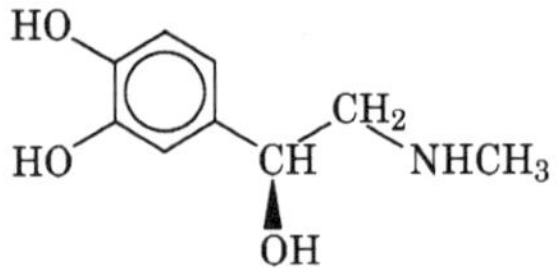

(2)

Compounds structurally related to the endogenous sympathomimetic amines epinephrine and norepinephrine have classically been employed as appetite suppressants. These agents, of which amphetamine [*300-62-9*], $C_9H_{13}N$, is the prototypical example, generally retain the phenethylamine, but lack the catechol moiety present in the natural substances. As a consequence, they are well absorbed after oral administration and readily distribute into the central nervous system where they exert their anorectic effects at hypothalamic appetite control centers. A component of their efficacy in promoting weight loss may be a result of a drug induced increase in metabolic rate through stimulation of thermogenesis or physical activity. These compounds act primarily by indirect mechanisms involving displacement of norepinephrine from presynaptic nerve storage vesicles or by prevention of its reuptake rather than by a direct effect at the receptor level (20). To a lesser extent, certain agents of this class affect dopaminergic or serotoninergic neurons. The overall pharmacological profile of members of the noncatechol sympathomimetic amine class depends on individual tissue distribution and mechanism of action.

Compounds available in the United States are listed in Table 1. Whereas they vary in degree, all of them share similar liabilities of cardiovascular side effects, the potential for central nervous system (CNS) stimulation, the development of tolerance, and abuse potential. All, with the exception of mazindol, are derivatives of phenethylamine. The introduction of an oxygen atom on the β-carbon of the side chain tends to reduce CNS stimulant properties without decreasing the anorectic activity. Following the Federal Controlled Drug Act of 1970, drugs were classified into one of five schedules according to medical utility and abuse potential.

Amphetamine and the other drugs in Schedule II have small sales and no legitimate role in antiobesity therapy, because of the pronounced potential for misuse. The compounds in Schedules III and IV, considered progressively safer, are prescribed more widely. They do not differ greatly in efficacy. However, the drug of choice for an individual depends on the patient's general health status with respect to subtle differences in side effect profiles. These agents are relatively effective over the course of at least several weeks and although tolerance is considered to be a problem, it does not always occur (12). Despite the prevalence of obesity in the United States, the use of prescription anorectic drugs has not increased in the past decade primarily as a result of concerns over abuse and general dissatisfaction with the long-term efficacy of sympathomimetics. As there are currently few alternatives for the drug therapy of obesity, the overall U.S. prescription market for antiobesity drugs has remained stagnant for several years and in 1989 amounted to only $75 million (21).

As indicated in Table 1, phenylpropanolamine [*14838-15-4*], $C_9H_{13}NO$, is the one agent listed which is available over the counter (OTC). It is present in a number of common diet aids and nasal decongestants having such names as Dexatrim, Contac, and Dimetapp (see also HISTAMINE AND HISTAMINE ANTAGONISTS). Because it has two asymmetric carbon atoms, phenylpropanolamine has the four possible stereoisomers (**3–6**) shown as Fischer projections. The compound sold as phenylpropanolamine in the United States is (±)-norephedrine and consists of equal amounts of D-(−) and L-(+)-norephedrine, (**3**) and (**4**), respectively, which have the erythro relationship between the amino and hydroxyl groups. The diastereomers with the opposite (threo) relationship between these groups are referred to as D-(−) and L-(+)-norpseudoephedrine, (**5**) and (**6**), respectively. In Europe and Australia the material sold as cathine and often referred to as phenylpropanolamine in the literature is (**6**). Although also used as an anorectic agent, (**6**) has greater CNS stimulant effects and a different toxicological profile from (±)-norephedrine (23). This situation has led to considerable confusion in the literature over side effects and efficacy of individual isomers (24).

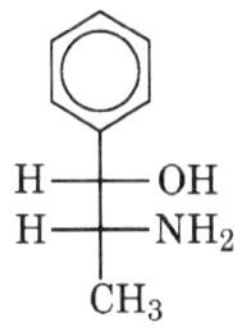

(**3**)

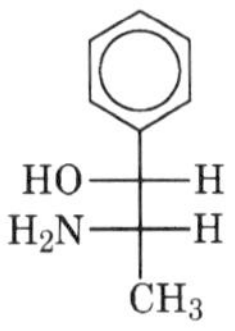

(**4**)

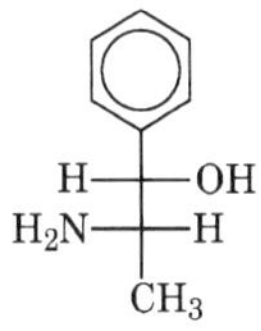

(**5**)

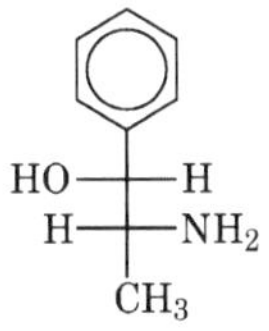

(**6**)

(±)-Norephedrine is pharmacologically similar to other indirect acting sympathomimetic amines, but has little CNS stimulant activity. After high doses, it does affect peripheral α- and β-sympathetic receptors resulting in cardiovascular symptoms, but these are generally not troublesome at normal clinical doses. (Its efficacy as a nasal decongestant is related to vasoconstrictor effects on blood vessels in the nasopharynx.) (±)-Norephedrine is well absorbed after oral administration and is excreted largely unchanged in the urine over the course of 24 h (24). A number of clinical trials have focused on its efficacy in appetite suppression (25,26) indicating effectiveness in promoting weight loss by suppression of appetite. Because most of these trials were of short duration and rarely were run in comparison with other anorectic drugs, it is difficult to assess the relative utility of (±)-norephedrine with respect to the pharmacologically related prescription drugs in the longer term treatment of morbid obesity. Informal surveys indicate the product is largely used by nonmorbidly obese people to assist in short term weight loss programs following a period of unusual weight gain, as might occur for example during a vacation (25). U.S. drugstore and supermarket sales of (±)-norephedrine for the weight loss indication were estimated to be in the range of $100 million in 1989.

Serotoninergic Agents

Serotoninergic neurons utilize serotonin [*50-67-9*], $C_{10}H_{12}N_2O$, which is 5-hydroxytyrptamine (5-HT), as a neurotransmitter. Central serotoninergic neurons play an

Table 1. Sympathomimetic Agents in Use as Appetite Suppressants

Generic name[a]	CAS Registry Number[b]	Molecular formula	Chemical name	Trade name	Manufacturer[c]	Structure	Anorexic dose
			Schedule II				
amphetamine sulfate	[*60-13-9*]	$C_9H_{14}SO_4N$	(±)-α-methylbenzene-ethanamine sulfate	Benzedrine	Lannett	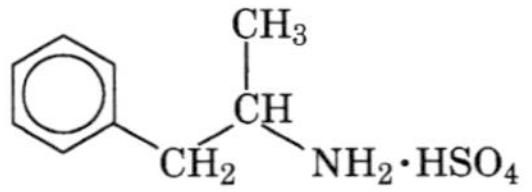	5–10 mg 3 × daily
dextroamphetamine sulfate	[*51-63-8*]	$C_9H_{14}NO_4S$	(*S*)-α-methylbenzene-ethanamine sulfate	Dexedrine	SKB[d]	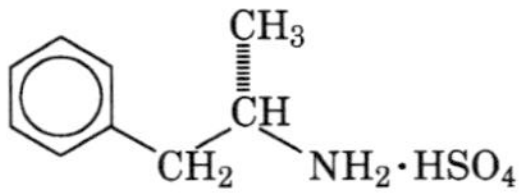	5–10 mg 3 × daily
methamphetamine hydrochloride	[*51-57-0*]	$C_{10}H_{16}ClN$	(*S*)-*N*,α-dimethyl-benzeneethanamine hydrochloride	Desoxyn	Abbott	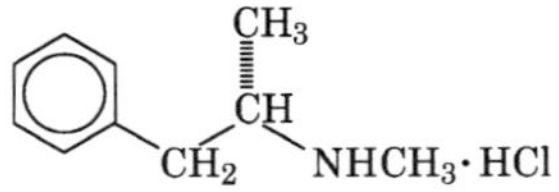	5–10 mg 1 × daily
phenmetrazine hydrochloride	[*1707-14-8*]	$C_{11}H_{16}ClNO$	2-phenyl-3-methyl-morpholine hydrochloride	Preludin	Boehringer Ingelheim	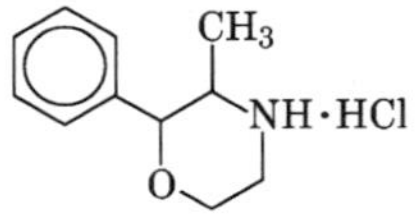	75 mg 1 × daily
			Schedule III				
phendimetrazine tartrate	[*50-58-8*]	$C_{16}H_{23}NO_7$	(2*S*, *trans*)-3,4-dimethyl-2-phenyl-morpholine tartrate	Plegine	Wyeth-Ayerst	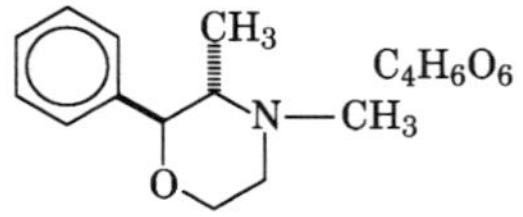	35 mg 2–3 × daily
benzphetamine hydrochloride	[*5411-22-3*]	$C_{17}H_{22}ClN$	(*S*)-*N*,α-dimethyl-*N*-(phenylmethyl)ben-zeneethanamine hydrochloride	Didrex	UpJohn	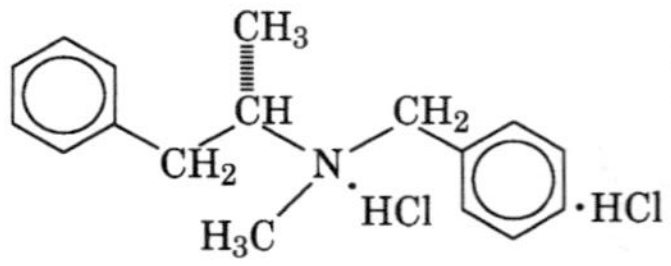	25–50 mg 1–3 × daily

				Schedule IV			
phenteramine hydrochloride	[*1197-21-3*]	$C_{10}H_{16}NCl$	α,α-dimethylbenzene-ethanamine hydrochloride	Fastin Adipex-P Ionamin	SKB[d] Lemmon Fisons	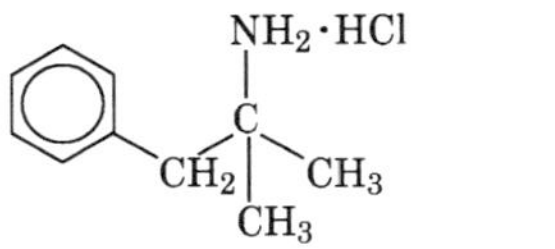	30 mg 1 × daily
diethylpropion hydrochloride	[*134-80-5*]	$C_{13}H_{20}ClNO$	2-(diethylamino)-1-phenyl-1-propanone hydrochloride	Tenuate Tepanil	Lakeside Rikker	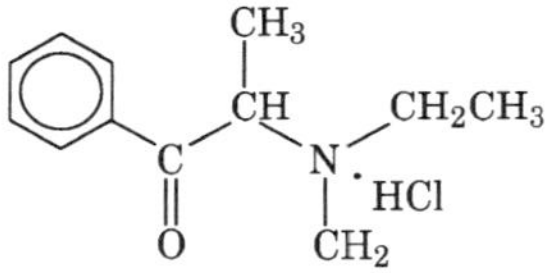	25 mg
mazindol	[*22232-71-9*]	$C_{16}H_{13}ClN_2O$	5-(4-chlorophenyl)-2,5-dihydro-3*H*-imidazo[2,1-a]iso-indol-5-ol	Sanorex Mazanor	Sandoz Wyeth-Ayerst	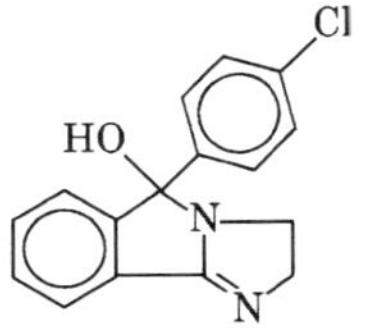	1 mg 1–3 × daily
				OTC[e]			
phenylpropanolamine hydrochloride[f]	[*154-41-6*]	$C_9H_{14}ClNO$	(±)-(*R*,*S*)-α-(1-amino-ethyl)benzene-methanol hydrochloride	Dexatrim Acutrim	Thompson Medical Company CIBA Consumer Pharmaceuticals	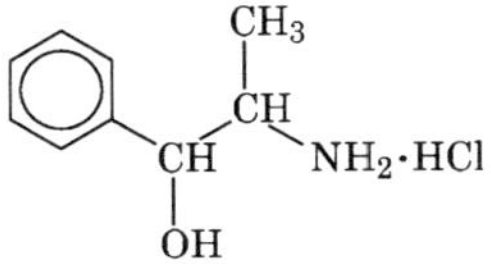	75 mg 1 × daily

[a]For physical properties and preparation, see Ref. 22.
[b]Most of these compounds have multiple registry numbers assigned to different salt forms and nomenclature variations. The registry number listed is that of the chemical name given.
[c]This listing is intended to be informative, not exhaustive, many of these compounds are distributed by several manufacturers.
[d]SmithKline Beecham, formerly SmithKline Beckman.
[e]OTC is over the counter.
[f]Also known as (±)-norephedrine hydrochloride.

important role in determining mood and food intake. Work with selective agonists and antagonists has led to the identification of a variety of receptor subtypes classified as 5-HT_{1a}, 5-HT_{1b}, 5-HT_{1c}, 5-HT_2, and 5-HT_3. Selective serotoninergic agents are receiving increased attention as antidepressants, antiemetics, and treatments for migraine. Stimulation of postsynaptic 5-HT_{1b} and 5-HT_{1c} receptors in the ventromedial hypothalamus has been implicated in the hypophagic response to serotonin agonists (27,28).

Fenfluramine, the structure of which is shown in Table 2, is a close analogue of amphetamine. However, fenfluramine acts as a serotoninergic agonist primarily by inhibiting serotonin reuptake and to a lesser extent by stimulation of serotonin release. This topic has been reviewed (29–31). Fenfluramine is a racemate. The (+)-enantiomer is twice as effective as the racemate as an appetite suppressant and has been shown in clinical trials to be relatively free of side effects (32,33). Dexfenfluramine [*3239-44-9*], $C_{12}H_{16}F_3N$, is currently in phase II clinical trials in the United States and is marketed in Europe by Servier. The U.S. license is Interneuran Pharmaceuticals. The (−)-enantiomer of fenfluramine, unlike the (+)-enantiomer, stimulates norepinephrine release and increases norepinephrine and dopamine metabolism (31). The catecholaminergic effects of the (−)-enantiomer may be largely responsible for the side effects encountered with fenfluramine. Rat studies indicate that whereas amphetamine tends to delay the onset of meals and decrease protein intake, fenfluramine decreases the rate of eating and promotes early meal termination with no effect on meal pattern or nutrient selection (34,35). Like many *N*-alkylphenethylamines, *N*-dealkylation is the primary metabolic pathway for both enantiomers of fenfluramine leading to norfenfluramine [*1886-26-6*] (**7**), $C_{10}H_{12}F_3N$. The (+)-metabolite is more potent as a stimulator of 5-HT release and less potent as an uptake inhibitor than its parent. It is also more persistent in the CNS and is thus in part responsible for the overall pharmacological profile of fenfluramine.

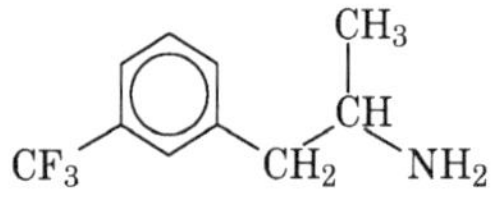

(**7**)

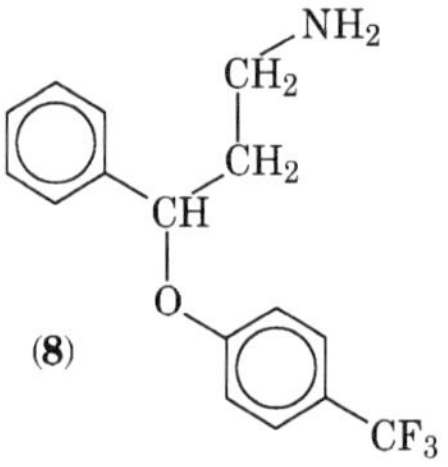

(**8**)

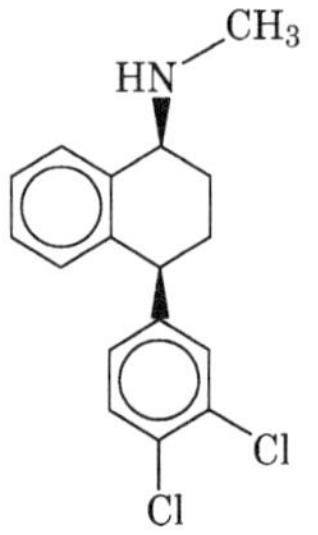

(**9**)

Fluoxetine, also shown in Table 2, is a serotonin reuptake inhibitor. It is more selective than fenfluramine and is being marketed by Eli Lilly as an antidepressant (36–38) (see PSYCHOPHARMACOLOGICAL AGENTS). Introduced in 1987, the 1990 sales were estimated at $700 million (39). Since it was observed in animal studies that fluoxetine decreased meal size and promoted weight loss, these findings have been extended to humans in a number of clinical trials. Fluoxetine has been consistently found to be effective in nondepressed humans and exhibits minimal side effects (40–43). The half-life of fluoxetine is 1–3 days and that of its

Table 2. Serotoninergic Agents as Appetite Suppressants

Generic name[a]	CAS Registry Number[b]	Molecular formula	Chemical name	Trade name	Manufacturer	Structure[c]	Anorexic dose
fenfluramine hydrochloride	[*404-82-2*]	$C_{12}H_{17}ClF_3N$	(±)-*N*-ethyl-α-methyl-3-trifluoromethylbenzene-ethanamine hydrochloride	Pondinim	Robbins	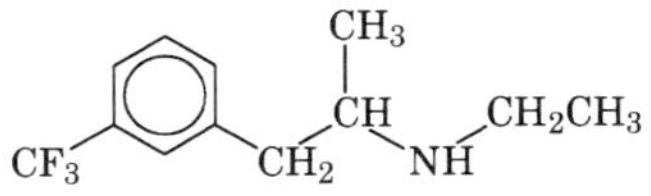	20 mg 3 × daily
fluoxetine hydrochloride	[*56296-78-7*]	$C_{17}H_{19}ClF_3NO$	(±)-*N*-methyl-γ-[4-(trifluoro-methyl)phenoxy]benzene-propanamine hydrochloride	Prozac	Lilly	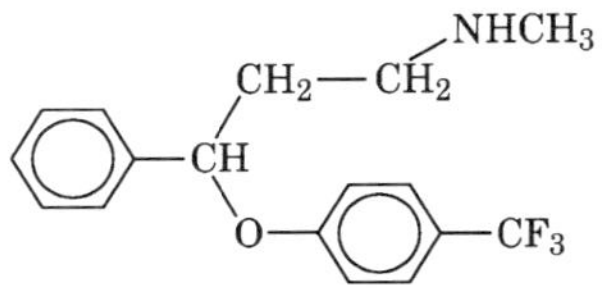	20 mg 3 × daily[d]

[a]For physical properties and preparation, see Ref. 22.
[b]These compounds have multiple registry numbers assigned to different salt forms and nomenclature variations.
[c]Structure does not reflect hydrochloride.
[d]Recommended antidepressant dose is 20 mg, 1–2 × daily.

principal metabolite, norfluoxetine [*56161-73-0*] (**8**), $C_{16}H_{16}F_3NO$, is 7–15 days (36). Fluoxetine is racemic and evidence has been presented indicating that both enantiomers are approximately equally active (44). An excellent synthesis relying on a catalytic chiral hydroboration reaction has been reported (45). It is likely that fluoxetine is being prescribed as an appetite suppressant although it is not yet approved for this indication. Other serotonin uptake inhibitors, such as sertraline [*79617-96-2*] (**9**), $C_{17}H_{17}Cl_2N$, which is being developed by Pfizer as an antidepressant, also has antiobesity activity and may eventually capture a part of this market (46,47).

Gut Hormones

Cholecystokinin (CCK) [*9011-97-6*] represents a family of gut and neurohormones of which the 33 amino acid peptide CCK-33 and the equipotent 8 amino acid peptide CCK-8 (**10**), are the most abundant circulating bioactive forms. Peripheral activities of CCK include stimulation of enzyme secretion from pancreatic acinar cells, inhibition of gastric emptying, and stimulation of vagal afferents terminating in the nucleus of the solitary tract and the area postrema. Exogenous administration of CCK to animals (48) or to humans (49,50) decreases meal size and in rats, elicits a sequence of behavior mimicking normal satiety. These effects have prompted the proposal that CCK released endogenously from the small intestine is a signaling hormone mediating postprandial satiety (51–53). Work with selective CCK antagonists (54) indicates that there are at least two distinct CCK receptors, a peripheral (CCK-A) type responsive to CCK-8 and a central (CCK-B) type responsive to the C-terminal tetrapeptide CCK-4. The precise role of CCK in terminating food intake remains to be defined, although animal studies suggest that both peripheral and central receptors are involved. Research is focused on developing selective CCK agonists which may be useful as adjuncts to dietary therapy by limiting meal size. Other peptides which have been suggested to play a role in regulating food intake include bombesin [*31362-50-2*], calcitonin, glucagon [*9007-92-5*], and corticotropin-releasing factor (55–57). However, the data suggesting a physiological role in regulating feeding behavior is not as strong for these compounds.

$$H_2N—Asp—Tyr(SO_3H)—Met—Gly—Trp—Met—Asp—Phe—NH_2$$

(10)

Lipid Adsorption Inhibitors

Dietary fat occurs largely in the form of triglycerides which must be hydrolyzed through the action of lipases, primarily pancreatic lipase, in the digestive tract prior to absorption. In a situation where maintenance or induction of weight loss is desired with a minimal impact on meal composition, an attractive approach to limiting caloric intake is to minimize the absorption of fat by inhibiting pancreatic lipase. Lipstatin is a pancreatic lipase inhibitor isolated from *Streptomyces toxytricini* (58–59). The corresponding tetrahydro derivative [*96829-58-2*] (**11**) has been shown to decrease fat absorption in mice, rats, and monkeys and to lower the rate of body weight gain in rats on a high calorie diet (60–62). This compound is undergoing clinical trials and potentially represents a new approach to antiobesity therapy.

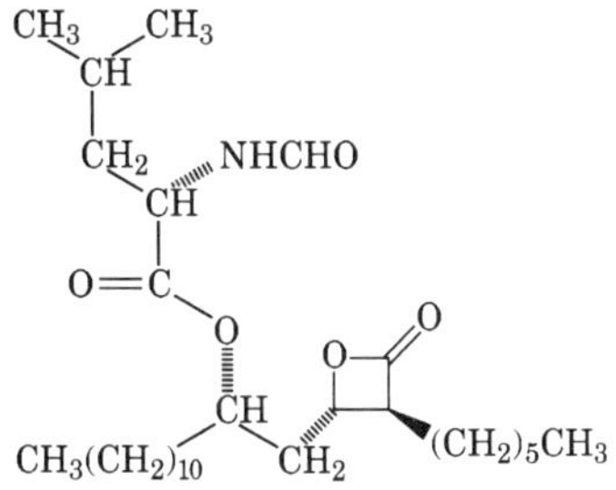

(11)

Thermogenic Agents

Drugs which stimulate or block peripheral adrenergic receptors have been in widespread use as, for example, antihypertensives, antianginal agents, and bronchodilators (see ANTIASTHMATIC AGENTS; CARDIOVASCULAR AGENTS). Detailed investigations with some of these has led to the proposal of a novel type of β-adrenergic receptor, termed the β_3-adrenergic receptor, which mediates catecholamine induced lipolysis in brown adipose tissue leading to increases in thermogenesis. The human gene coding this receptor has been identified and expressed in eukaryotic cells (63). Compounds such as ICI 201,651 [*107332-57-0*], $C_{19}H_{23}NO_6$, **(12)** (64) and BLR 37,344 [*90730-96-4*], $C_{19}H_{22}ClNO_4$, **(13)** (65) have been identified which selectively stimulate this receptor and lead to increases in metabolic rate and thermogenesis in rodents with little effect on food intake. In humans such compounds could be expected to decrease the efficiency of food utilization and could be particularly helpful to those patients whose metabolic set-points favor an inappropriate body mass.

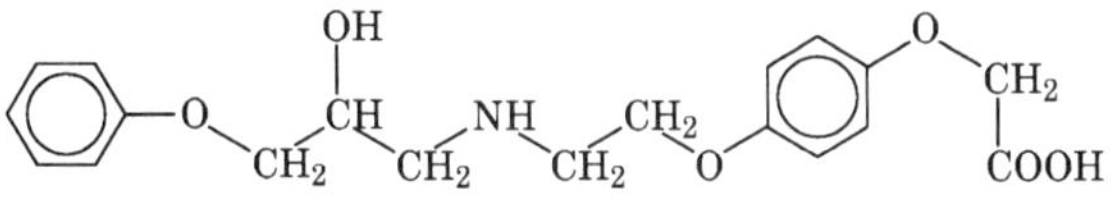

(12)

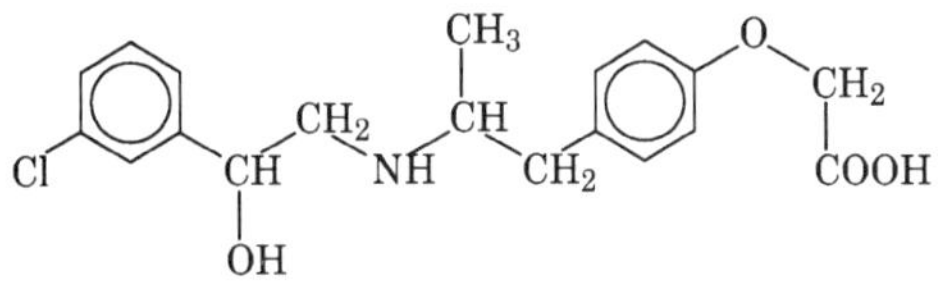

(13)

Economic Aspects

The most striking aspect of the antiobesity market is the relatively small fraction of dollars which are spent on drugs in comparison with diet foods, meal replacements, and other treatments for obesity such as diet centers. The overall market for the prescription adrenergic agents listed in Table 1 was about $75 million in 1989 and is not growing. The somewhat larger sales of phenylpropanolamine-based products reflect use by a wider audience than the morbidly obese as a consequence of their OTC availability. Several promising new agents are currently being developed. Foremost among these is the serotoninergic agent fluoxetine. Whereas it is not approved in the U.S. for use in obesity, some percentage of its approximately $700 million in 1990 sales were derived from this off-label use.

BIBLIOGRAPHY

"Appetite-Suppressing Agents" in *ECT* 3rd ed., Vol. 3, pp. 184–193, by S. T. Ross (Smith, Kline & French Laboratories).

1. *Ann. Int. Med.* **103,** 1073–1077 (1985).
2. C. Bouchard and co-workers, *N. Engl. J. Med.* **322,** 1477–1482 (1990).
3. A. J. Stunkard, J. R. Harris, N. L. Pedersen, and G. E. McClearn, *N. Engl. J. Med.* **322,** 1483–1487 (1990).
4. R. A. Price, R. J. Cadoret, A. J. Stunkard, and E. Troughton, *Am. J. Psychiatry* **144,** 1003–1008 (1987).
5. A. J. Stunkard and co-workers, *N. Engl. J. Med.* **314,** 193–198 (1986).
6. A. J. Wade, M. M. Marbut, and J. M. Round, *Lancet* **335,** 805–808 (1990).
7. R. J. Garrison, M. Feinleib, W. P. Castelli, and P. M. McNamara, *J. Am. Med. Assoc.* **249,** 2199–2203 (1983).
8. H. B. Hubert, P. M. Feinleib, P. M. McNamara, and W. P. Castelli, *Circulation* **67,** 968–977 (1983).
9. F. W. Ashley and W. B. Kannel, *J. Chronic Dis.* **27,** 103–114 (1974).
10. E. Barrett-Connor, *Epidemiol. Rev.* **11,** 172–181 (1989).
11. A. S. Jackson and M. L. Pollock, *Phys. Sportmed.* **13,** 76–90 (1985).
12. G. A. Bray, *Dis.-a-Mon.* **35,** 449–540 (1989).
13. J. E. Manson and co-workers, *N. Engl. J. Med.* **322,** 882–889 (1990).
14. A. Rissanen and co-workers, *Brit. Med. J.* **301,** 835–837 (1990).
15. A. H. Kissebah and co-workers, *J. Clin. Endocrinol. Metab.* **54,** 254–260 (1982).
16. M. Krotkiewski, P. Björntorp, L. Sjöström, and U. Smith, *J. Clin. Invest.* **72,** 1150–1162 (1983).
17. I. D. Caterson, *Drugs* **39**(Suppl. 3), 20–32 (1990).
18. J. F. Munro and co-workers, *Ann. N.Y. Acad. Sci.* **499,** 305–312 (1987).
19. P. A. Kern, J. M. Ong, B. Saffari, and J. Carty, *N. Engl. J. Med.* **322,** 1053–1059 (1990).
20. N. Weiner, in A. G. Gilman, L. S. Goodman, T. W. Rall, and F. Murad, eds., *The Pharmacological Basis of Therapeutics,* 7th ed., Macmillan, New York, 1985, pp. 145–169.
21. IMS Drug Store and Hospital Audit, IMS America, Plymouth Meeting, Pa., 1989.
22. *Merck Index,* 11th ed., Merck Corporation, Rahway, N.J., 1989.
23. J. P. Morgan, in J. P. Morgan, D. V. Kagan, and J. S. Brody, eds., *Phenylpropanolamine: Risks, Benefits, and Controversies,* Greenwood Press, Westport, Conn., 1985, pp. 180–194.
24. M. Weiner, in Ref. 23, pp. 25–36.
25. M. Weintraub, in Ref. 23, pp. 53–79.

26. H. Blumberg and J. P. Morgan, in Ref. 23, pp. 80–93.
27. G. A. Kennett, C. T. Dourish, and G. Curzon, *Eur. J. Pharmacol.* **141,** 429–435 (1987).
28. G. A. Kennett and G. Curzon, *Psychopharmacology* **96,** 93–100 (1988).
29. N. E. Rowland and J. Carlton, *Prog. Neurobiol.* **27,** 13–62 (1986).
30. C. Nathan and Y. Rolland, *Ann. N.Y. Acad. Sci.* **499,** 277–296 (1987).
31. S. Garattini and co-workers, *Appetite* **7,** 15–38 (1986).
32. N. Finer, D. Craddock, R. Lavielle, and H. Keen, *Curr. Ther. Res. Clin. Exp.* **38,** 847–854 (1985).
33. G. Enzi and co-workers, *Clin. Neuropharmacol.* **11**(Suppl. 1), S173–S178 (1988).
34. J. J. Wurtman and R. J. Wurtman, *Curr. Med. Res. Opin.* **6**(Suppl. 1), 28–33 (1979).
35. J. E. Blundell and co-workers, *Curr. Med. Res. Opin.* **6**(Suppl. 1), 34–52 (1979).
36. L. Lemberger, N. A. Farid, R. F. Bergstrom, and R. L. Wolen, *Int. J. Obesity* **11**(Suppl. 3), 157–161 (1987).
37. R. W. Fuller and D. T. Wong, *Drug Dev. Res.* **17,** 1–15 (1989).
38. R. W. Fuller, D. T. Wong, and D. W. Robertson, *Med. Res. Rev.* **11,** 17–34 (1991).
39. M. Waldholz, *Wall St. J.* B1 (July 18, 1990).
40. J. M. Ferguson and J. P. Feighner, *Int. J. Obesity* **11**(Suppl. 3), 163 (1987).
41. L. R. Levine, S. Rosenblatt, and J. Bosomworth, *Int. J. Obesity* **11**(Suppl. 3), 185 (1987).
42. R. L. Zerbe, *Int. J. Obesity* **11**(Suppl. 3), 191 (1987).
43. L. R. Levine and co-workers, *Int. J. Obesity* **13,** 635–645 (1989).
44. D. W. Robinson, J. H. Krushinski, R. W. Fuller, and J. D. Leander, *J. Med. Chem.* **31,** 1412–1417 (1988).
45. E. J. Corey and G. A. Reichard, *Tetrahedron Lett.* **30,** 5207–5210 (1989).
46. B. K. Koe, *J. Clin. Psychiatry* **51**(Suppl. B), 13–17 (1990).
47. J. Heym and B. K. Koe, *J. Clin. Psychiatry* **49**(Suppl. 40–45) (1988).
48. J. J. Antin and co-workers, *J. Comp. Physiol. Psychol.* **89,** 784–790 (1975).
49. H. R. Kissileff, F. X. Pi-Sunyer, J. Thornton, and G. P. Smith, *Am. J. Clin. Nutr.* **34,** 154–160 (1981).
50. N. Muurahainen, H. R. Kissileff, A. J. Derogatis, and F. X. Pi-Sunyer, *Physiol. Behav.* **44,** 645–649 (1988).
51. A. J. Silver and J. E. Morley, *Prog. Neurobiol.* **36,** 23–34 (1991).
52. G. P. Smith, in A. J. Stunkard and E. Stellar, eds., *Eating and Its Disorders,* Raven, New York, 1984, pp. 67–75.
53. C. A. Baile, C. L. McLaughlin, and M. A. Della-Fera, *Physiol. Rev.* **66,** 172–234 (1986).
54. R. F. Freidinger, *Med. Res. Rev.* **9,** 271 (1989).
55. S. F. Leibowitz, *Fed. Proc.* **45,** 1396–1403 (1986).
56. J. E. Morley and co-workers, *Peptides* **6**(Suppl. 2), 181–192 (1985).
57. J. E. Morley, *Endocr. Rev.* **8,** 256–287 (1987).
58. E. K. Weibel and co-workers, *J. Antibiot.* **40,** 1081–1085 (1987).
59. E. Hochuli and co-workers, *J. Antibiot.* **40,** 1085–1089 (1987).
60. S. Hogan and co-workers, *Int. J. Obesity* **11**(Suppl. 3), 35–42 (1987).
61. E. Fernandez and B. Borgström, *Biochim. Biophys. Acta* **1001,** 249–255 (1989).
62. P. Hadvary, H. Lengsfeld, and H. Wolfer, *Biochem. J.* **256,** 357–361 (1988).
63. L. J. Emorine and co-workers, *Science* **245,** 1118–1121 (1989).
64. B. R. Halloway, in H. Lardy and F. Stratman, eds., *Hormones, Thermgensis, and Obesity,* Elsevier, New York, 1989, pp. 477–484.
65. J. R. S. Arch and co-workers, *Nature (London)* **309,** 163–165 (1984).

JEFFERSON W. TILLEY
Hoffmann-LaRoche Corporation

ANTIOXIDANTS

Antioxidants are used to retard the reaction of organic materials with atmospheric oxygen. Such reaction can cause degradation of the mechanical, aesthetic, and electrical properties of polymers; loss of flavor and development of rancidity in foods; and an increase in the viscosity, acidity, and formation of insolubles in lubricants. The need for antioxidants depends upon the chemical composition of the substrate and the conditions of exposure. Relatively high concentrations of antioxidants are used to stabilize polymers such as natural rubber and polyunsaturated oils. Saturated polymers have greater oxidative stability and require relatively low concentrations of stabilizers. Specialized antioxidants which have been commercialized meet the needs of the industry by extending the useful lives of the many substrates produced under anticipated conditions of exposure. The sales of antioxidants in the United States were approximately $730 million in 1990 (1,2).

Mechanism of Uninhibited Autoxidation

The mechanism by which an organic material (RH) undergoes autoxidation involves a free-radical chain reaction (3–5):

Initiation

$$\text{RH} \longrightarrow \text{Free radicals, eg, R}\cdot\text{, ROO}\cdot\text{, RO}\cdot\text{, HO}\cdot \qquad (1)$$

$$\text{ROOH} \longrightarrow \text{RO}\cdot + \text{OH}\cdot \qquad (2)$$

$$2\ \text{ROOH} \longrightarrow \text{RO}\cdot + \text{ROO}\cdot + \text{H}_2\text{O} \qquad (3)$$

$$\text{ROOR} \longrightarrow 2\ \text{RO}\cdot \qquad (4)$$

Propagation

$$\text{R}\cdot + \text{O}_2 \longrightarrow \text{ROO}\cdot \qquad (5)$$

$$\text{ROO}\cdot + \text{RH} \longrightarrow \text{ROOH} + \text{R}\cdot \qquad (6)$$

Termination

$$2\ \text{R}\cdot \longrightarrow \text{R—R} \qquad (7)$$

$$\text{ROO}\cdot + \text{R}\cdot \longrightarrow \text{ROOR} \qquad (8)$$

$$2\ \text{ROO}\cdot \longrightarrow \text{nonradical products} \qquad (9)$$

Initiation. Free-radical initiators are produced by several processes. The high temperatures and shearing stresses required for compounding, extrusion, and molding of polymeric materials can produce alkyl radicals by homolytic chain cleavage. Oxidatively sensitive substrates can react directly with oxygen, particularly at elevated temperatures, to yield radicals.

It is virtually impossible to manufacture commercial polymers that do not contain traces of hydroperoxides. The peroxide bond is relatively weak and cleaves homolytically to yield radicals (eqs. 2 and 3). Once oxidation has started, the concentration of hydroperoxides becomes appreciable. The decomposition of hydroperoxides becomes the main source of radical initiators.

The absorption of uv light produces radicals by cleavage of hydroperoxides and carbonyl compounds (eqs. 10–12)

$$ROOH \xrightarrow{hv} RO\cdot + OH\cdot \tag{10}$$

$$R{-}\overset{\overset{O}{\|}}{C}{-}R \xrightarrow{hv} R{-}\overset{\overset{O}{\|}}{C}\cdot + R\cdot \tag{11}$$

$$R{-}\overset{\overset{O}{\|}}{C}\cdot \longrightarrow R\cdot + CO \tag{12}$$

Most polymer degradation caused by the absorption of uv light results from radical-initiated autoxidation.

Direct reaction of oxygen with most organic materials to produce radicals (eq. 13) is very slow at moderate temperatures. Hydrogen-donating antioxidants (AH), particularly those with low oxidation–reduction potentials, can react with oxygen (eq. 14), especially at elevated temperatures (6).

$$RH + O_2 \longrightarrow R\cdot + HO_2\cdot \tag{13}$$

$$AH + O_2 \longrightarrow A\cdot + HO_2\cdot \tag{14}$$

Propagation. Propagation reactions (eqs. 5 and 6) can be repeated many times before termination by conversion of an alkyl or peroxy radical to a nonradical species (7). Homolytic decomposition of hydroperoxides produced by propagation reactions increases the rate of initiation by the production of radicals.

The reaction rate of molecular oxygen with alkyl radicals to form peroxy radicals (eq. 5) is much higher than the reaction rate of peroxy radicals with a hydrogen atom of the substrate (eq. 6). The rate of the latter depends on the dissociation energies (Table 1) and the steric accessibility of the various carbon–hydrogen bonds; it is an important factor in determining oxidative stability.

Table 1. Dissociation Energies of Carbon–Hydrogen Bonds[a]

R–H	D_{R-H} kJ/mol[b]	Bond type
$CH_2{=}CHCH_2{-}H$	356	allylic
$(CH_3)_3C{-}H$	381	tertiary
$(CH_3)_2CH{-}H$	395	secondary

[a]Ref. 8.
[b]To convert kJ to kcal, divide by 4.184.

Polybutadiene and polyunsaturated fats, which contain allylic hydrogen atoms, oxidize more readily than polypropylene, which contains tertiary hydrogen atoms. A linear hydrocarbon such as polyethylene, which has secondary hydrogens, is the most stable of these substrates.

Autocatalysis. The oxidation rate at the start of aging is usually low and increases with time. Radicals, produced by the homolytic decomposition of hydroperoxides and peroxides (eqs. 2–4) accumulated during the propagation and termination steps, initiate new oxidative chain reactions, thereby increasing the oxidation rate.

Metal-Catalyzed Oxidation. Trace quantities of transition metal ions catalyze the decomposition of hydroperoxides to radical species and greatly accelerate the rate of oxidation. Most effective are those metal ions that undergo one-electron transfer reactions, eg, copper, iron, cobalt, and manganese ions (9). The metal catalyst is an active hydroperoxide decomposer in both its higher and its lower oxidation states. In the overall reaction, two molecules of hydroperoxide decompose to peroxy and alkoxy radicals (eq. 5).

$$\text{ROOH} + \text{M}^{n} \longrightarrow \text{ROO}\cdot + \text{M}^{(n-1)+} + \text{H}^{+} \qquad (15)$$

$$\text{ROOH} + \text{M}^{(n-1)+} \longrightarrow \text{RO}\cdot + \text{M}^{n+} + \text{OH}^{-} \qquad (16)$$

$$2\ \text{ROOH} \longrightarrow \text{ROO}\cdot + \text{RO}\cdot + \text{H}_2\text{O} \qquad (3)$$

Termination. The conversion of peroxy and alkyl radicals to nonradical species terminates the propagation reactions, thus decreasing the kinetic chain length. Termination reactions (eqs. 7 and 8) are significant when the oxygen concentration is very low, as in polymers with thick cross-sections where the oxidation rate is controlled by the diffusion of oxygen, or in a closed extruder. The combination of alkyl radicals (eq. 7) leads to cross-linking, which causes an undesirable increase in melt viscosity.

Radical Scavengers

Hydrogen-donating antioxidants (AH), such as hindered phenols and secondary aromatic amines, inhibit oxidation by competing with the organic substrate (RH) for peroxy radicals. This shortens the kinetic chain length of the propagation reactions.

$$\text{ROO}\cdot + \text{AH} \xrightarrow{k_{17}} \text{ROOH} + \text{A}\cdot \qquad (17)$$

$$\text{ROO}\cdot + \text{RH} \xrightarrow{k_{6}} \text{ROOH} + \text{R}\cdot \qquad (6)$$

Because k_{17} is much larger than k_6, hydrogen-donating antioxidants generally can be used at low concentrations. The usual concentrations in saturated thermoplastic polymers range from 0.01 to 0.05%, based on the weight of the polymer. Higher concentrations, ie, ca 0.5–2%, are required in substrates that are highly sensitive to oxidation, such as unsaturated elastomers and ABS.

Hindered Phenols. Even a simple monophenolic antioxidant, such as 2,6-di-*tert*-butyl-*p*-cresol [*128-37-0*] (**1**), has a complex chemistry in an autooxidizing substrate (Fig. 1).

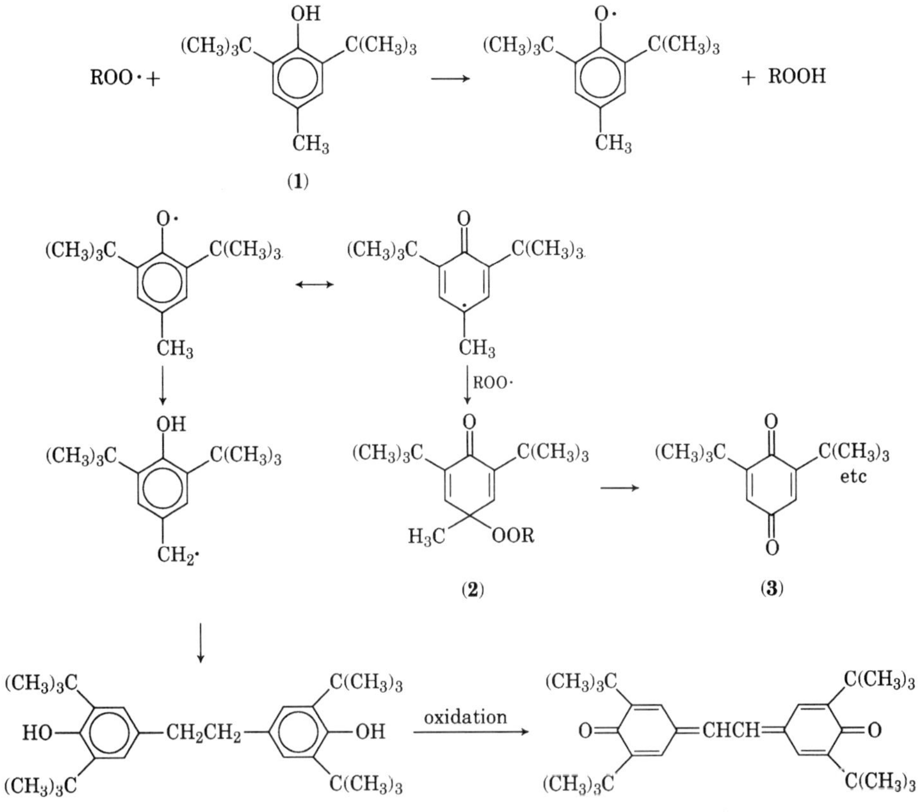

Fig. 1. Chemical transformations of 2,6-di-*tert*-butyl-*p*-cresol in an oxidizing medium (10).

Stilbenequinones such as (**5**) absorb visible light and cause some discoloration. However, upon oxidation phenolic antioxidants impart much less color than aromatic amine antioxidants and are considered to be nondiscoloring and nonstaining.

The effect substitution on the phenolic ring has on activity has been the subject of several studies (11–13). Hindering the phenolic hydroxyl group with at least one bulky alkyl group in the ortho position appears necessary for high antioxidant activity. Nearly all commercial antioxidants are hindered in this manner. Steric hindrance decreases the ability of a phenoxyl radical to abstract a hydrogen atom from the substrate and thus produces an alkyl radical (14) capable of initiating oxidation (eq. 18).

$$A\cdot + RH \longrightarrow AH + R\cdot \qquad (18)$$

Replacing a methyl with a tertiary alkyl group in the para position usually decreases antioxidant effectiveness. The formation of antioxidants such as (**4**) by dimerization is precluded because all benzylic hydrogen atoms are replaced by methyl groups. A strong electron-withdrawing group on the aromatic ring, like cyano or carboxy, decreases the ability of the phenol to donate its hydrogen atom to a peroxy radical of the substrate and reduces antioxidant effectiveness.

The usefulness of a hindered phenol for a specific application depends on its radical-trapping ability, its solubility in the substrate, and its volatility under test conditions. Table 2 shows the importance of volatility to stabilizer performance. Equimolar quantities of alkyl esters (**6**) of 3,5,di-*tert*-butyl-4-hydroxyhydrocinnamic acid were evaluated in polypropylene at 140°C using two different procedures (15). When tested in an air stream, only the octadecyl ester, (**6**) where $n = 18$, was effective in stabilizing the polymer. Under these conditions, the lower homologues were lost by volatilization. The oxygen-uptake test, carried out in a closed system that minimizes evaporative loss, showed that homologues were effective to varying degrees. The differences in effectiveness can probably be attributed to differences in the solubility of various homologues in the amorphous phase of the polypropylene. When dodecane, a liquid in which all the compounds are soluble, was used as a substrate instead of polypropylene, the antioxidant activities were relatively close.

Table 2. Influence of Antioxidant (6) Volatility on Effectiveness at 140°C

		Time to failure in PP, h		Time to failure, h
n[a]	$t_{1/2}$, h[b]	O_2 Uptake	Air stream	In dodecane
1	0.28	95	2	25
6	3.60	312	2	23
12	83.0	420	2	20
18	660.0	200	165	20

[a] n in

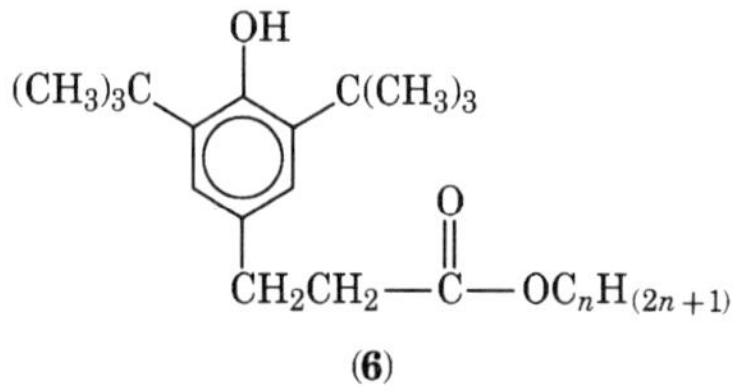

(**6**)

[b] Antioxidant half-life in polypropylene exposed to a nitrogen stream at 140°C.

Introducing long aliphatic chains into a stabilizer molecule decreases volatility and increases solubility in hydrocarbon polymers. This improves performance. However, it also increases the equivalent weight of the active moiety. Di- and polyphenolic antioxidants combine relatively low equivalent weights with low volatility.

Commercially important di- and polyphenolic stabilizers include 2,2′-methylenebis(6-*tert*-butyl-*p*-cresol) [*85-60-9*] (**7**), 1,3,5-trimethyl-2,4,6-tris(3′5′-

di-*tert*-butyl-4-hydroxybenzyl)benzene [*1709-70-2*] (**8**), and tetrakis[methylene(3,5-di-*tert*-butyl-4-hydroxyhydrocinnamate)]methane [*6683-19-8*] (**9**).

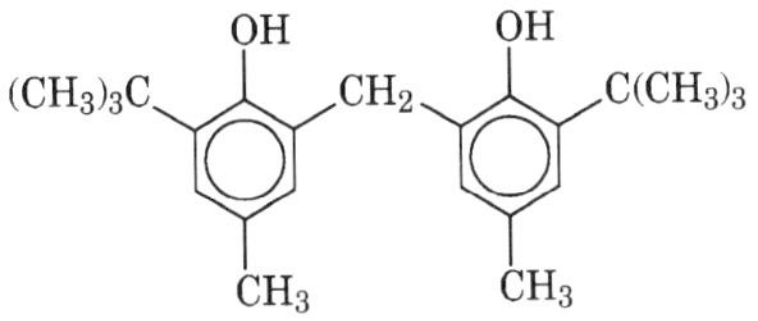

(**7**)

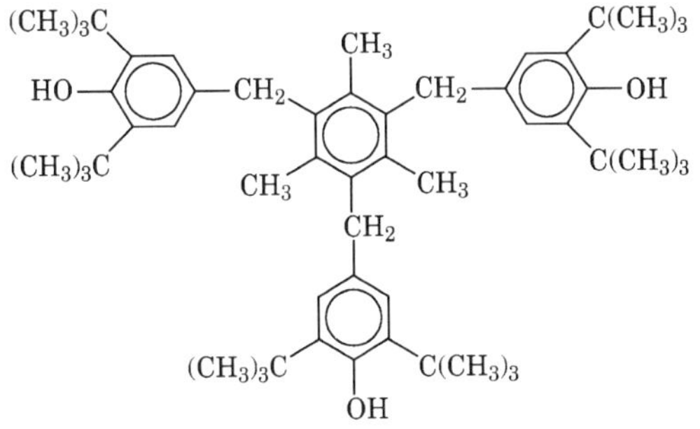

(**8**)

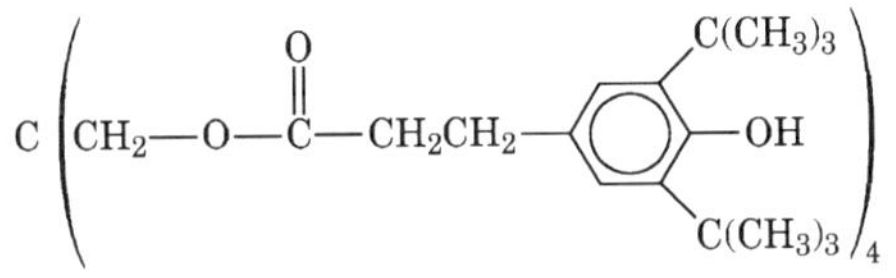

(**9**)

Aromatic Amines. Antioxidants derived from *p*-phenylenediamine and diphenylamine are highly effective peroxy radical scavengers. They are more effective than phenolic antioxidants for the stabilization of easily oxidized organic materials, such as unsaturated elastomers. Because of their intense staining effect, derivatives of *p*-phenylenediamine are used primarily for elastomers containing carbon black (qv).

N,N'-Disubstituted-*p*-phenylenediamines, such as *N*-phenyl-*N'*-(1,3-dimethylbutyl)-*p*-phenylenediamine [*793-24-8*] (**10**), are used in greater quantities than other classes of antioxidants. These products protect unsaturated elastomers against oxidation as well as ozone degradation (see ANTIOZONANTS).

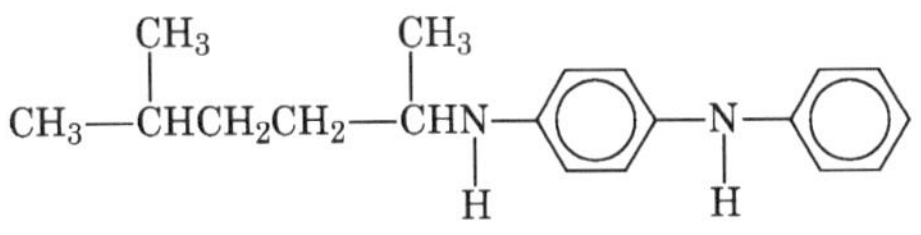

(**10**)

Low concentrations of alkylated paraphenylenediamines such as *N,N'*-di-*sec*-butyl-*p*-phenylenediamine [*69796-47-0*] are added to gasoline to inhibit oxidation (see AMINES, AROMATIC-PHENYLENEDIAMINES).

Radical Trapping. Figure 2 shows some of the reactions of aromatic amines that contribute to their activity as antioxidants and to their tendency to form highly colored polyconjugated systems.

Alkylated diphenylamines (**11**) and derivatives of both dihydroquinoline (**12**) and polymerized 2,2,4-trimethyl-1,2-dihydroquinoline [*26780-96-1*] (**13**) develop less color than the *p*-phenylenediamines and are classified as semistaining antioxi-

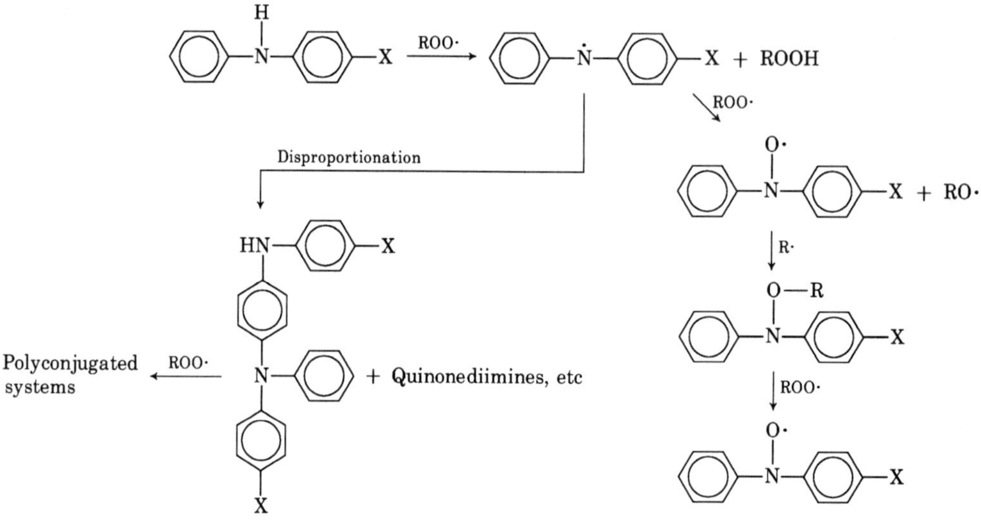

where X = H, NHR′

Fig. 2. Oxidation of aromatic amine antioxidants (10).

dants. Derivatives of dihydroquinoline are used for the stabilization of animal feed and spices.

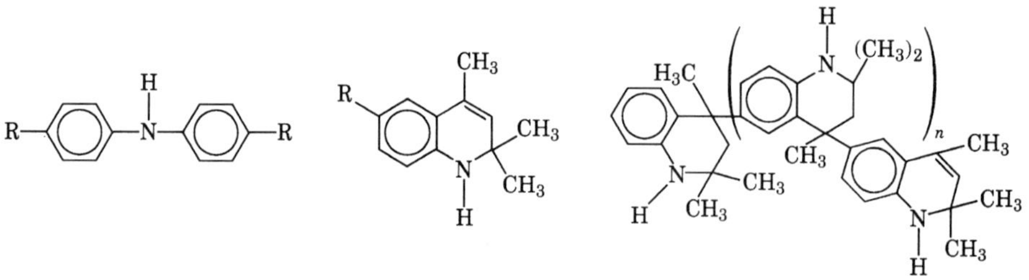

(11) (12) (13)

where n = 0–6

4,4′-Bis(α,α-dimethylbenzyl)diphenylamine [*1008-67-1*] (**14**) has only a slight tendency to stain and has been recommended for use in plastics as well as elastomers.

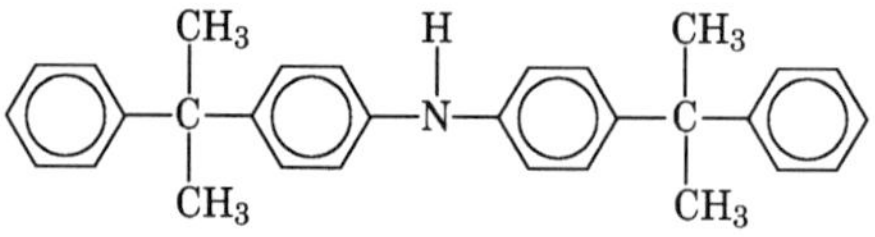

(**14**)

Hindered Amines. Hindered amines are extremely effective in protecting polyolefins and other polymeric materials against photodegradation. They usually are classified as light stabilizers rather than antioxidants.

Most of the commercial hindered-amine light stabilizers (HALS) are derivatives of 2,2,6,6-tetramethylpiperidine [*768-66-1*] (**15**).

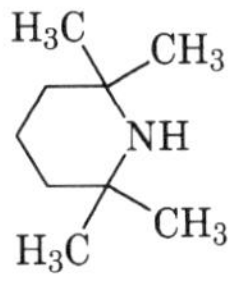

(**15**)

These stabilizers function as light-stable antioxidants to protect polymers. Their antioxidant activity is explained by the following sequence (16):

$$>NH \xrightarrow{\text{radicals } O_2} >NO\cdot \qquad (19)$$

$$>NO\cdot + R\cdot \longrightarrow >NOR \qquad (20)$$

$$>NOR + R'OO\cdot \longrightarrow >NO\cdot + R{=}O + R'OH \qquad (21)$$

According to this mechanism, hindered-amine derivatives terminate propagating reactions (eqs. 5 and 6) by trapping both the alkyl and peroxy radicals. In effect, NO competes with O_2, and NOR competes with RH. Since the nitroxyl radicals are not consumed in the overall reactions, they are effective at low concentrations.

Peroxide Decomposers

Thermally induced homolytic decomposition of peroxides and hydroperoxides to free radicals (eqs. 2–4) increases the rate of oxidation. Decomposition to nonradical species removes hydroperoxides as potential sources of oxidation initiators. Most peroxide decomposers are derived from divalent sulfur and trivalent phosphorus.

Divalent Sulfur Deivatives. A dialkyl ester of thiodipropionic acid (**16**) is capable of decomposing at least 20 moles of hydroperoxide (17). Some of the reactions contributing to the antioxidant activity of these compounds are shown in Figure 3.

According to Figure 3, hydroperoxides are reduced to alcohols, and the sulfide group is oxidized to protonic and Lewis acids by a series of stoichiometric reactions. The sulfinic acid (**21**), sulfonic acid (**23**), sulfur trioxide, and sulfuric acid are capable of catalyzing the decomposition of hydroperoxides to nonradical species.

When used alone at low temperatures, dialkyl thiodipropionates are rather weak antioxidants. However, synergistic mixtures with hindered phenols are highly effective at elevated temperatures and are used extensively to stabilize polyolefins, ABS, impact polystyrene (IPS), and other plastics.

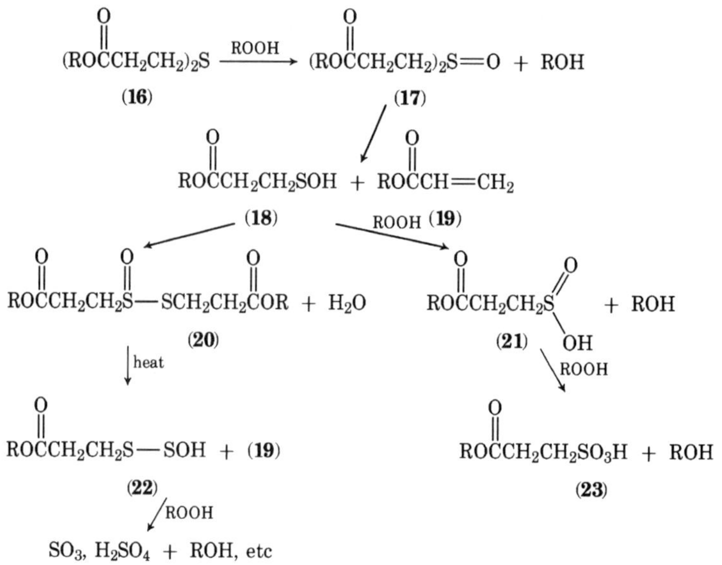

Fig. 3. Decomposition of hydroperoxides by esters of thiodipropionic acid (18).

Esters of thiopropionic acid tend to decompose at high processing temperatures, and their odor makes them unsuitable for some food-packaging applications.

Trivalent Phosphorus Compounds. Trivalent phosphorus compounds reduce hydroperoxide to alcohols:

$$\mathrm{ROOH} + \mathrm{PR'_3} \longrightarrow \mathrm{ROH} + \mathrm{O{=}PR'_3} \qquad (22)$$

These compounds are used most frequently in combination with hindered phenols for a broad range of applications in rubber and plastics. They are also able to suppress color development caused by oxidation of the substrate and the phenolic antioxidant. Unlike phenols and secondary aromatic amines, phosphorus-based stabilizers generally do not develop colored oxidation products.

Esters of phosphorous acid derived from aliphatic alcohols and unhindered phenols, eg, tris(nonylphenyl)phosphate (**24**), hydrolyze readily and special care must be taken to minimize decomposition by exposure to water or high humidity. The phosphorous acid formed by hydrolysis is corrosive to processing equipment, particularly at high temperatures.

The hydrolysis of phosphites is retarded by the addition of a small amount of a base such as triethanolamine. A more effective approach is the use of hindered phenols for esterification. Relatively good resistance to hydrolysis is shown by two esters derived from hindered phenols: tris(2,4-di-*tert*-butylphenyl)phosphite [*31570-04-4*] (**25**) and tetrakis(2,4-di-*tert*-butylphenyl)4,4′-biphenylenediphosphonite [*38613-77-3*] (**26**). The hindered fluorophosphite [*118337-09-0*] (**27**) has excellent resistance to hydrolysis.

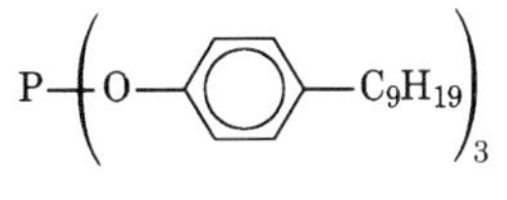

(24)

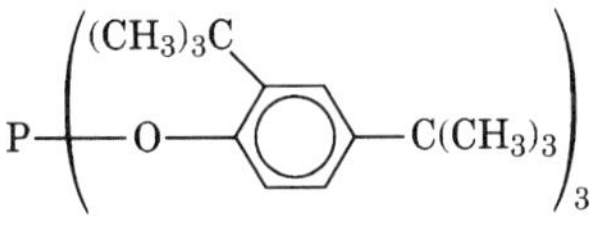

(25)

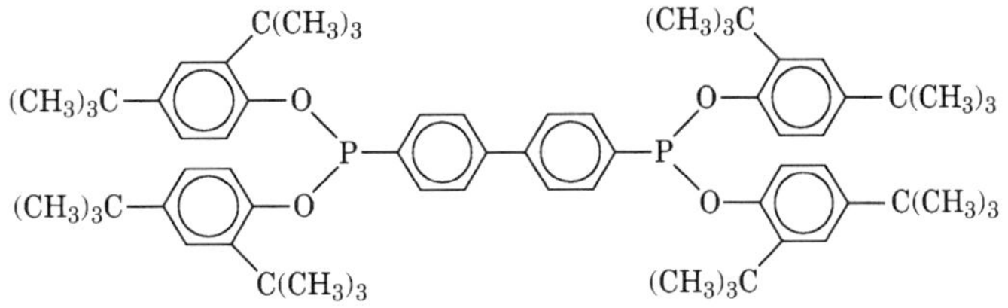

(26)

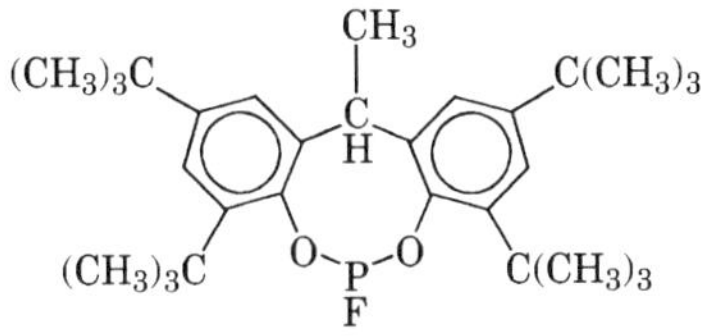

(27)

Synergism. A mixture of antioxidants that function by different mechanisms might be synergistic and provide a higher degree of protection than the sum of the stabilizing activities of each component. The most frequently used synergistic mixtures are combinations of radical scavengers and hydroperoxide decomposers.

Metal Deactivators. The ability of metal ions to catalyze oxidation can be inhibited by metal deactivators (19). These additives chelate metal ions and increase the potential difference between the oxidized and reduced states of the metal ions. This decreases the ability of the metal to produce radicals from hydroperoxides by oxidation and reduction (eqs. 15 and 16). Complexation of the metal by the metal deactivator also blocks its ability to associate with a hydroperoxide, a requirement for catalysis (20).

Examples of commercial metal deactivators used in polymers, gasoline, and foods are oxalyl bis(benzylidene)hydrazide [*6629-10-3*] (**28**), *N,N′*-bis(3,5-di-*tert*-butyl-4-hydroxyhydrocinnamoylhydrazine [*32687-78-8*] (**29**), 2,2′-oxamidobis-ethyl(3,5-di-*tert*-butyl-4-hydroxyhydrocinnamate) [*70331-94-1*] (**30**), *N,N′*-(disalicylidene)-1,2-propanediamine [*94-91-7*] (**31**), and ethylenediaminetetra-acetic acid [*60-00-4*] (**32**) and its salts and citric acid (**33**) (Fig. 4).

Application of Antioxidants in Polymers

Nearly all polymeric materials require the addition of antioxidants to retain physical properties and to ensure an adequate service life. The selection of an antioxidant or system of antioxidants is dependent upon the polymer and the anticipated end use. A product that will not be exposed to the elements for a long period of time such as polyethylene grocery bags does not need a long term stabilizer; polyethylenes used to insulate communication cable must be stabilized for many years of service.

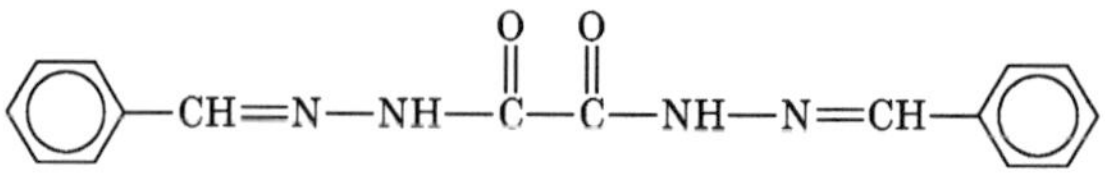

(28)

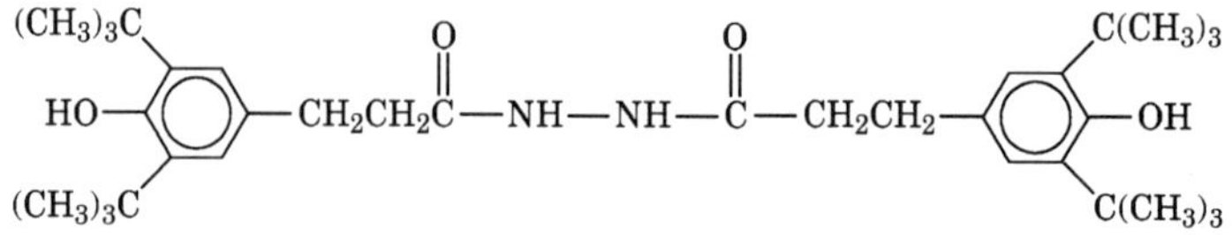

(29)

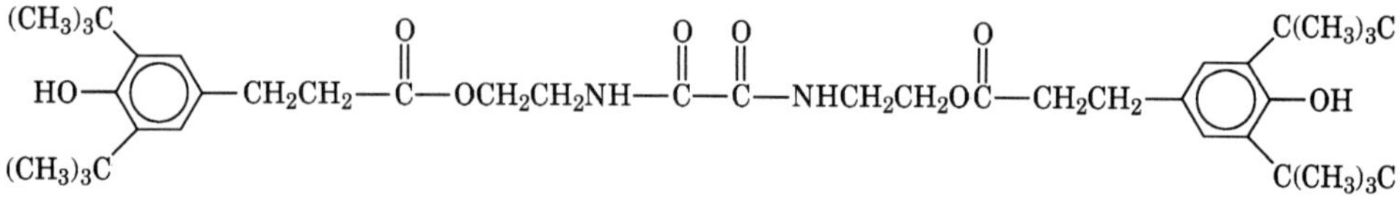

(30)

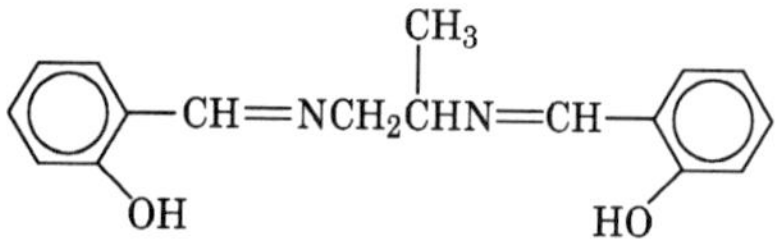

(31)

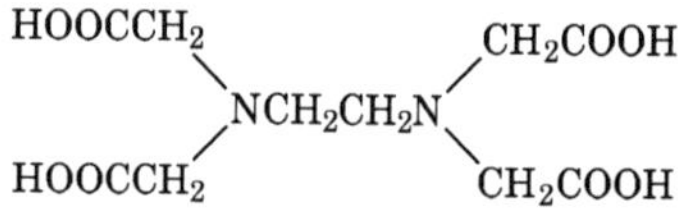

(32)

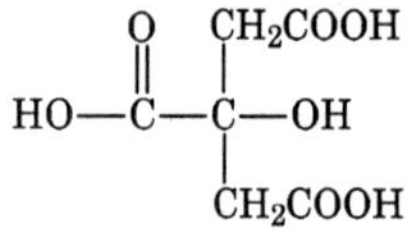

(33)

Fig. 4. Commercial metal deactivators.

Polyolefins. Low concentrations of stabilizers (<0.01%) are added to polyethylene and polypropylene after synthesis and prior to isolation to retard oxidation of the polymers exposed to air.

These polymers are subjected to high temperatures, ca 300°C, during extrusion and injection molding. Processing stabilizers are used to decrease both the change in viscosity of the polymer melt and the development of color. A phosphite, such as tris(2,4-di-*tert*-butylphenyl)phosphite (**25**) or bis(2,4-di-*tert*-butylphenyl)pentaerythritol diphosphite [*26741-53-7*], in combination with a phenolic antioxidant such as octadecyl 3,5-di-*tert*-butyl-4-hydroxyhydrocinnamate (**6**), may be used. Concentrations usually range from 0.01 to 0.5% depending on the polymer and the severity of the processing conditions. For long-term exposure, a persistent antioxidant like tetrakis[methylene(3,5-di-*tert*-butyl-4-hydroxyhydrocinnamate)]methane (**9**), at a concentration of 0.1–0.5%, may be added to the base stabilization package.

A sulfur-containing synergistic mixture can be used to obtain an extended service life at a decreased cost.

The synergistic effect of a hydroperoxide decomposer, eg, dilauryl thiodipropionate [*123-28-4*] (**34**), and a radical scavenger, eg, tetrakis[methylene(3,5-di-*tert*-butyl-4-hydroxyhydrocinnamate)]methane (**9**), in protecting polypropylene during an oxygen-uptake test at 140°C is shown in Table 3.

$$S(CH_2CH_2{-}\overset{\overset{\displaystyle O}{\|}}{C}{-}OC_{12}H_{25})_2$$

(34)

The sum of the individual activities of these antioxidants was 20 days, whereas a mixture of the two stabilizers protected the polymer for 45 days.

Table 3. Synergism Between a Hindered Phenol and a Thiosynergist[a]

Additive, %		
Radical scavenger[b]	Hydroperoxide decomposer[c]	Induction period, d
0.0	0.3	4
0.1	0.0	16
0.1[d]	0.3[d]	45

[a]Ref. 21.
[b]Tetrakis[methylene(3,5-di-*tert*-butyl-4-hydroxyhydrocinnamate)]methane.
[c]Dilauryl thiodipropionate.
[d]Mixture.

Oligomeric hindered amine light stabilizers are effective thermal antioxidants for polypropylene. Thus 0.1% of *N,N'*-bis(2,2,6,6-tetramethyl-4-piperadinyl)-1,6-hexanediamine polymer, with 2,4,6-trichloro-1,3,5-triazine and 2,4,4-trimethyl-2-pentaneamine [*70624-18-9*] (**35**) (Fig. 5), protects polypropylene multifilaments against oxidation when exposed at 120°C in a forced-air oven (22) for 47 days. 3,5-Di-*tert*-butyl-4-hydroxytoluene [*128-37-0*] (0.1%) affords protection for only 14 days.

The stabilization of polyolefins used to insulate copper conductors requires the use of a long-term antioxidant plus a copper deactivator. Both *N,N'*-bis(3,5-di-*tert*-butyl-4-hydroxycinnamoyl)hydrazine (**29**) and 2,2'-oxamidobisethyl(3,5-di-*tert*-butyl-4-hydroxycinnamate) (**30**) are bifunctional. They are persistent antioxidants that have built-in metal deactivators. Oxalyl bis(benzylidenehydrazide) (**28**) is an effective copper deactivator when part of an additive package that includes an antioxidant.

Stabilization of Elastomers. Polyunsaturated elastomers are sensitive to oxidation. Stabilizers are added to the elastomers prior to vulcanization to protect the rubber during drying and storage. Nonstaining antioxidants such as butylated hydroxytoluene (**1**), 2,4-bis(octylthiomethyl)-6-methylphenol [*110553-27-0*], 4,4'-bis(α,α-dimethylbenzyl)diphenylamine (**14**), or a phosphite such as tris(nonylphenyl)phosphite (**24**) may be used in concentrations ranging from 0.01% to 0.5%.

Staining antioxidants such as *N*-isopropyl-*N'*-phenyl-*p*-phenylenediamine [*101-72-4*] (**36**) are preferred for the manufacture of tires (see also AMINES, AROMATIC, PHENYLENEDIAMINES). These potent antioxidants also have antiozonant

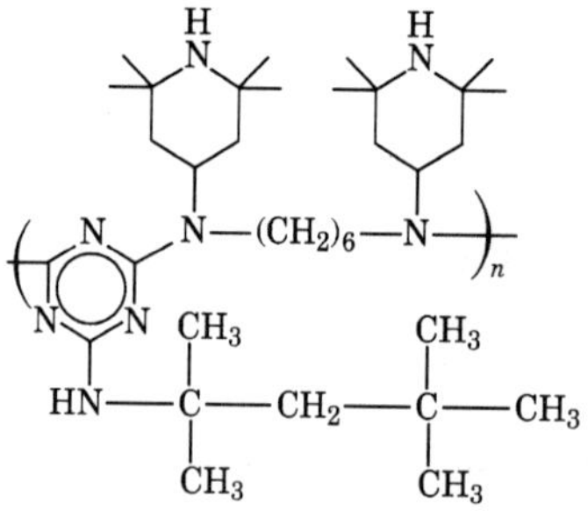

(35)

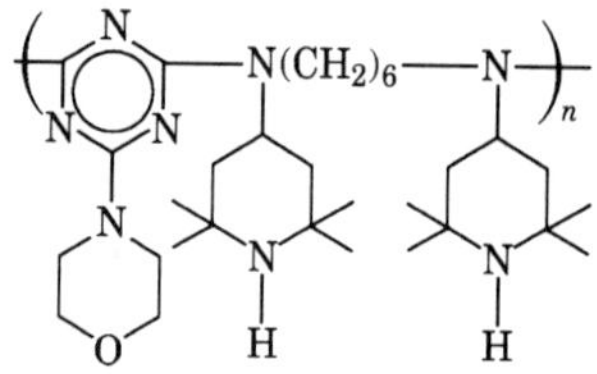

(44)

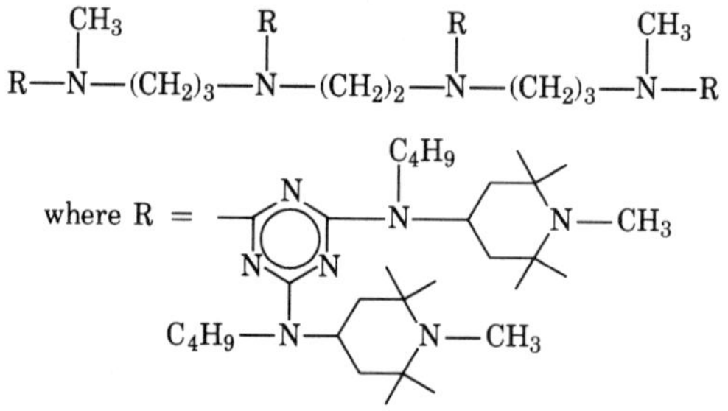

(45)

Fig. 5. Hindered 1,6-hexanediamine antioxidants (Table 4).

activity and retard stress cracking of the vulcanized rubber. Carbon black (qv), used in tires for reinforcement, hides the color developed by the antioxidant. According to use requirements, up to 3% of an amine antioxidant having antiozonant activity is added prior to vulcanization.

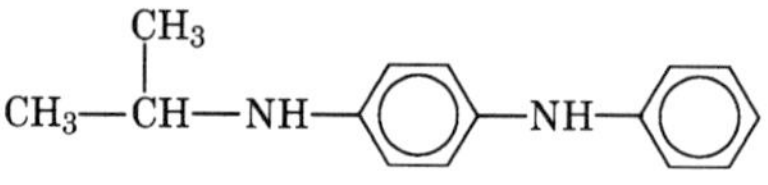

(36)

When staining tendencies of the substituted paraphenylene diamines cannot be tolerated, semistaining amine antioxidants are used to provide some protection to the elastomers. The semistaining antioxidants include polymerized 2,2,4-trimethyl-1,2-dihydroquinolines (**13**) and substituted diphenylamines such as 4,4′-bis(α,α-dimethylbenzyl)diphenylamine (**14**). These compounds protect unsaturated elastomers against oxidation but provide no protection against ozone.

Antioxidants resistant to extraction by lubricants and gasoline are preferred for the stabilization of elastomers used in automotive applications such as gaskets and tubing. Aromatic amine antioxidants, such as *N*-phenyl-*N*′-(*p*-toluenesulfonyl)-*p*-phenylenediamine [*100-93-6*] (**37**), with low solubility in hydrocarbons, are extracted slowly from elastomers and are used for these applications.

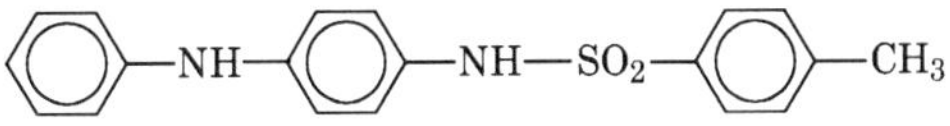

(37)

However, binding the antioxidant chemically to the elastomer chain by copolymerization or grafting is a better solution to this problem. The addition of *N*-(4-anilino-phenyl)methacrylamide [*22325-96-8*] (**38**) to a polymerization recipe for NBR rubber produces a polymer with a built-in antioxidant resistant to extraction (23).

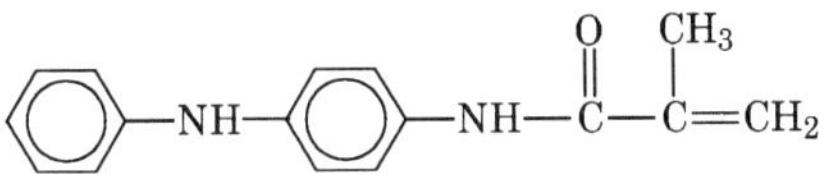

(38)

Raw NBR containing 1.5% of the built-in antioxidant retained 92% of its original resistance to oxidation after exhaustive extraction with methanol. NBR containing a conventional aromatic amine antioxidant (octylated diphenylamine) retained only 4% of its original oxidative stability after similar extraction.

It is also possible to graft an aromatic amine antioxidant bearing a sulfhydryl group on to the backbone of an elastomer.

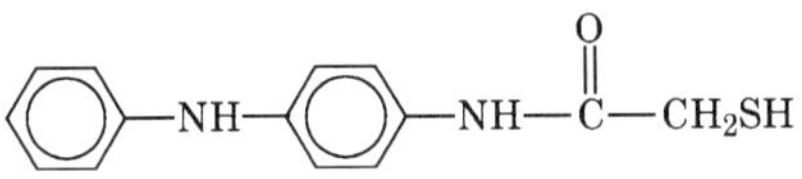

(39)

When 4-(mercaptoacetamido)diphenylamine [*60766-26-9*] (**39**) is added to EPDM rubber and mixed in a torque rheometer for 15 minutes at 150°C, 87% of it chemically binds to the elastomer (24). The mechanical and thermal stress placed on the polymer during mixing ruptures the polymer chain, producing radicals that initiate the grafting process.

Stabilization of Fuels and Lubricants. Gasoline and jet engine fuels contain unsaturated compounds that oxidize on storage, darken, and form gums and deposits. Radical scavengers such as 2,4-dimethyl-6-*tert*-butylphenol [*1879-09-0*], 2,6-di-*tert*-butyl-*p*-cresol (**1**), 2,6-di-*tert*-butylphenol [*128-39-2*], and alkylated paraphenylene diamines are used in concentrations of about 5–10 ppm as stabilizers.

The catalytic activity of copper as an oxidant can be inhibited by the use of a metal deactivator such as *N,N'*-disalicylidene-1,2-diaminopropane (**31**) at a concentration of 5–10 ppm.

Lubricants for gasoline engines are required to withstand harsh conditions. The thin films of lubricants coating piston walls are exposed to heat, oxygen, oxides of nitrogen, and shearing stress. Relatively high concentrations of primary antioxidants and synergists are used to stabilize lubricating oils. Up to 1% of a mixture of hindered phenols, of the type used for gasoline, and secondary aromatic amines, such as alkylated diphenylamine and alkylated phenyl-α-napththylamine, are used as the primary antioxidants. About 1% of a synergist, zinc dialkyldithiophosphonate, is added as a peroxide decomposer. Zinc dialkyldithiophosphates (**40**) are cost effective multifunctional additives. They interrupt oxidative chains by trapping radicals by electron donation, decompose peroxides, and serve as a corrosion and wear inhibitos.

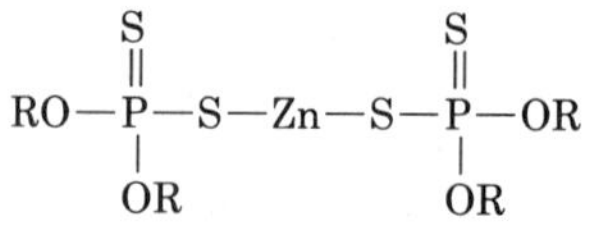

(**40**)

$R = C_3H_7, C_8H_{17}$

Both zinc and phosphorus deactivate the catalysts used to control emissions and governmental regulations limit their concentrations to below 0.1%.

Stabilization of Foods. The oxidation of food containing fats and oils results in a loss of sensory appeal and eventually rancidness. FDA regulations covering the use of direct food additives are stringent and few new materials have been regulated. A number of products in use today, such as citric acid (**33**), and α-tocopherol [*59-02-9*] (**41**), are in the GRAS list. The most commonly used materials are butylated hydroxytoluene [*128-37-0*] (**1**), *n*-propyl gallate [*129-79-9* (**42**), α-tocopherol, and butylated hydroxyanisole [*25013-16-5*] (**43**). The concentrations allowed in food are less than 0.02%. Metal deactivators such as citric acid (**33**), ethylenediaminetetraacetic acid (**32**) and its salts, and calcium chelate are used to deactivate transition metal oxidation catalysts.

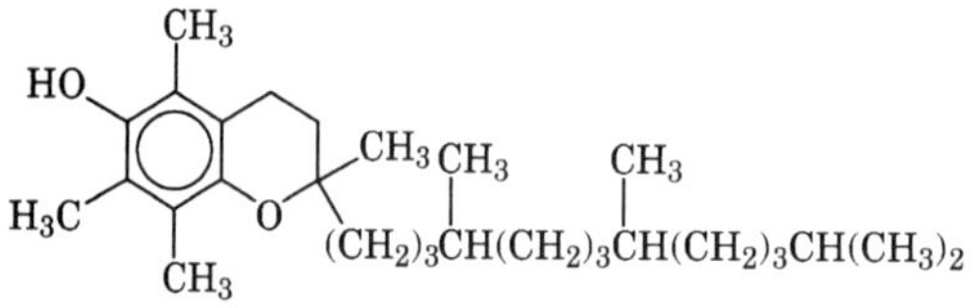

(**41**)

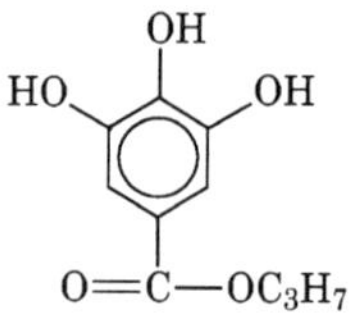

(**42**)

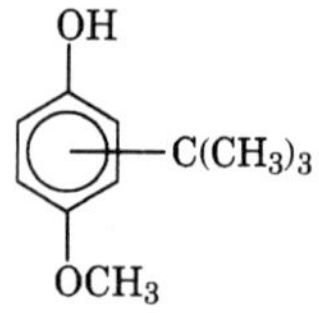

(**43**)

Test Methods. The oxidative stability of a polymer is evaluated best under the conditions to which it will be exposed during processing and use.

The stability of a plastic subjected to the thermal and mechanical stress of processing is determined by extruding it one or more times in laboratory equipment and measuring changes in melt viscosity and color after each extrusion. An effective stabilizer protects the polymer from changes in melt viscosity and prevents color development. A decrease in melt viscosity indicates that the polymers have undergone chain scission, whereas an increase indicates crosslinking. The conditions of these elevated-temperature, short-term tests are similar enough to actual manufacturing conditions to be relevant.

Estimation of oxidative stability under conditions of use is usually much more difficult. A polymer stable to oxidation might last for many years at ambient temperatures before it fails because of loss of essential properties such as tensile strength, elongation, impact resistance, etc. To decrease the time required for oxidative failure, specimens are exposed at temperatures considerably higher than those anticipated in use. If the apparent activation energy remains constant, a plot of the logarithm of failure time against the reciprocal of the absolute temperature results in a straight line. Extrapolation to temperature of use provides an estimate of failure time.

Because the inhibition of oxidation depends on a number of different reactions as well as the physical states of the polymer and the additives, the apparent activation energy of the overall process can deviate considerably from linearity and an extrapolation can lead to serious errors in estimating failure time. Based on an extrapolation of an Arrhenius plot of test data obtained at temperatures ranging from 140 to 180°C, it was estimated that stabilized polyethylene used as insulation for copper wire would not fail at 40°C for at least 100 years. However, aging tests made at lower temperatures indicated that the time to failure would probably be less than 10 years (25).

Increasing surface-to-volume ratio increases susceptibility to oxidation. Thin film and fiber are much more sensitive to oxidation than thick specimens (26). The effectiveness of an antioxidant for products with high surface-to-volume ratios is determined not only by its inherent activity in a particular polymer, but also by the rate of loss by volatilization.

Oxygen-uptake measurements at elevated temperatures are also used to determine oxidation resistance. Generally, the time required to reach the induction period is an indication of oxidative stability. The time required to absorb a specified volume of oxygen is an indication of failure when the change in the rate of oxygen absorption is not sufficient to permit an accurate measurement of the induction period. To be of practical significance, the amount of oxygen absorbed and the loss of desired properties must be correlated. In the case of most elastomers, useful properties have deteriorated when 1–2% oxygen is absorbed.

Since oxygen uptake is tested in a closed system, loss by volatilization is minimal. This type of test is not reliable for estimating service life under conditions in which the polymer is exposed to air movement.

A sequence of tests has been devised to evaluate antioxidants for use in automotive crankcase lubricants. The Indiana Stirring Oxidation Test (ISOT) JISK2514 is an example of a laboratory screening test. The oil is stirred at 165.5°C in the presence of air. Copper and iron strips are used as metal catalysts. The development of sludge, viscosity, and acidity are determined periodically. Failure time is determined when the development of acidity requires 0.4 mg KOH/g for neutralization. Formulated lubricants are then evaluated for performance in engines that are mounted on test platforms and run under a variety of stressed conditions (Oldsmobile Sequence III D) (27).

Finally, candidate lubricants containing the antioxidants are tested in fleets of automobiles for thousands of miles. The engines are dismantled and examined for wear and coatings of varnish and other deposits. The lubricant is evaluated for sludge, viscosity, acidity, etc. Since evaluation in automobiles is expensive and time-consuming it is reserved for only the most promising candidates. Oxygen uptake tests are also used to measure oxidative stability (28).

The effectiveness of antioxidants as preservatives for fats and oils is evaluated by determining the rate of peroxide development using the Active Oxygen Method (AOM) (29). The development of a rancid odor is used to evaluate the stability of food items (Schaal Oven Stability test) (30).

Table 4. Commercial Antioxidants

Compound	CAS Registry Number	Molecular formula	Suggested[a] substrates	Suppliers
Monophenols				
2,6-di-*tert*-butylphenol	*[128-39-2]*	$C_{14}H_{22}O$	F, GAS, PO	Ethyl Corp.; Schnectady Chemicals, Inc.
2,6-di-*tert*-butyl-4-methylphenol	*[128-37-0]*	$C_{15}H_{24}O$ (**1**)	F, GAS, LUB, PA, PES, PO, POM, PUR, PVC, RU, PS	Koppers Co., Inc.; Mobay Chemical Co.; R. T. Vanderbilt Co., Inc.; Shell Chemical; Uniroyal Chemical; UOP, Inc., PMC Specialties Group, Inc.
2,6-di-*tert*-butyl-4-ethylphenol	*[4130-42-1]*	$C_{16}H_{26}O$	PO	PMC Specialties Group, Inc.
2,6-di-*tert*-butyl-4-nonylphenol	*[4306-88-1]*	$C_{23}H_{40}O$		Rhone-Poulenc, Inc.
2-*tert*-butyl-4,6-dimethylphenol	*[1879-09-0]*	$C_{12}H_{18}O$	GAS	DuPont
2,4-bis[(octylthio)methyl]-6-methylphenol	*[110553-27-0]*	$C_{25}H_{44}OS_2$	RU	CIBA-GEIGY Corp.
styrenated phenol	*[61788-44-1]*	variable[b]	PO, PVC, RU	Ferro Corp.; Goodyear Tire and Rubber Co.; Harwick Chemical Corp.; PMC Specialties Group, Inc.; Reichold Chemicals, Inc.; Uniroyal Chemical
2,6-di-*tert*-butyl-4-*sec*-butylphenol	*[17540-75-9]*	$C_{18}H_{30}O$	PES, PO, PVC, PS	Schenectady Chemicals, Inc.; R. T. Vanderbilt Co., Inc.
octadecyl 3,5-di-*tert*-butyl-4-hyroxyhydrocinnamate	*[2082-79-3]*	$C_{35}H_{62}O_3$ (**6**)	CE, PA, PO, PUR, PVC	CIBA-GEIGY Corp.; Enichem Synthesis; GE Specialty Chemicals
N-stearoyl-*p*-aminophenol	*[103-99-1]*	$C_{24}H_{41}NO_2$	PO, RU	Hexel Corp.
2,6-di-*tert*-butyl-4-(dimethylamino-methyl)phenol	*[88-27-7]*	$C_{17}H_{29}NO$	PO, RU	Ethyl Corp.
2,4,bis[*n*-octylthio-6-(4-hydroxy-3,5-di-*tert*-butylanilino)]1,3,5-triazine	*[991-84-4]*	$C_{33}H_{56}N_4OS_2$	RU, PS	CIBA-GEIGY Corp.
Diphenols				
2,2′-methylenebis(4-methyl-6-*tert*-butylphenol)	*[119-47-1]*	$C_{23}H_{32}O_2$	CE, PA, POM, PUR, PVC, RU, PS	Aceto Chemical Co., Inc.; American Cyanamid Co.; Ferro Corp.; Chemical Mobay Co.; Monsanto Co.; GE Specialty Chemical Co.; R. T. Vanderbilt Co., Inc.; PMC Specialties Group, Inc.

2,2′-methylenebis(4-ethyl-6-*tert*-butylphenol)	[*88-24-4*]	$C_{25}H_{36}O$	PA, PO, PVC	American Cyanamid Co.
4,4′-methylenebis(2,6-di-*tert*-butylphenol)	[*118-82-1*]	$C_{29}H_{44}O_2$	CE, PES, PO, RU	Aceto Chemical Co., Inc.; Ethyl Corp.; Ferro Corp.; PMC Specialties Group, Inc.
2,2′-ethylidenebis(4,6-di-*tert*-butylphenol)	[*35958-30-6*]	$C_{30}H_{46}O_2$	PES, PO, PUR, PVC, PS	R. T. Vanderbilt Co., Inc.; Schenectady Chemicals, Inc.
triethylene glycol bis(3-*tert*-butyl-4-hydroxy-5-methyl-hydrocinnamate)	[*36443-68-2*]	$C_{34}H_{50}O_8$	PA, POM, PVC, PS	CIBA-GEIGY Corp.
4,4′-butylidenebis(6-*tert*-butyl-3-methylphenol)	[*85-60-9*]	$C_{26}H_{38}O_2$	CE, EVA, PA, PES, PO, PVC, RU, PS	Monsanto Co.; PMC Specialties Group, Inc.; R. T. Vanderbilt Co., Inc.
4,4′-thiobis(6-*tert*-butyl-3-3-methylphenol)	[*96-69-5*]	$C_{22}H_{30}O_2S$	PA, PES, PO, PVC, RU, PS	Aceto Chemical Co. Inc.; Monsanto Co.
4,4′-thiobis(2-methyl-6-*tert*-butylphenol)	[*96-66-2*]	$C_{22}H_{30}O_{25}$	PO, RU	Ethyl Corp.; GE Specialty Chemicals, Inc.
1,6-hexamethylene bis(3,5-di-*tert*-butyl-4-hydroxyhydrocinnamate)	[*35074-77-2*]	$C_{40}H_{62}O_6$	CE, PES, PO, POM, PUR, PVC, PS	CIBA-GEIGY Corp.
thiodiethylene bis(3,5-di-*tert*-butyl-4-hydroxyhydrocinnamate)	[*41484-35-9*]	$C_{38}H_{58}O_{65}$	CE, PA, PO, PUR, PVC, RU, PS	CIBA-GEIGY Corp.
N,N′-hexamethylene bis(3,5-di-*tert*-butyl-4-hydroxyhydrocinnamide)	[*23128-74-7*]	$C_{40}H_{64}N_2O_4$	PA, PES, POM, RU	CIBA-GEIGY Corp.
calcium bis[*O*-ethyl(3,5-di-*tert*-butyl-4-hydroxybenzyl)phosphonate]	[*65140-91-2*]	$C_{17}H_{29}O_4P \cdot \frac{1}{2}Ca$	PO, RU	CIBA-GEIGY Corp.
nickel bis[*O*-ethyl(3,5-di-*tert*-butyl-4-hydroxybenzyl)phosphonate]	[*30947-30-9*]	$C_{17}H_{29}O_4P \cdot \frac{1}{2}Ni$	PO	CIBA-GEIGY Corp.
2,2′-oxamidobisethyl(3,5-di-*tert*-butyl-4-hydroxyhydrocinnamate)	[*70331-94-1*]	$C_{40}H_{60}N_2O_8$ (**30**)	PO, PS, PA, PES, EVA	Uniroyal Chemical
N,N′-bis(3,5-di-*tert*-butyl-4-hydroxyhydrocinnamoyl)hydrazine	[*32687-78-8*]	$C_{34}H_{52}N_2O_4$ (**29**)	CE, PA, PES, PO, PVC, RU, PS	CIBA-GEIGY Corp.; Fairmount Chemical
bis[3,3-bis(4-hydroxy-3-*tert*-butylphenyl)butanoic acid],glycol ester	[*32509-66-3*]	$C_{50}H_{66}O_8$	PA, PES, PO, PUR, PS	Hoechst Celanese Corp.
2,2′-bis[4-2-3(3,5-di-*tert*-butyl-4-hydroxyhydrocinnamoyloxy)ethoxyphenyl]propane	[*105350-68-3*]	$C_{53}H_{72}O_8$	PO, PS	ICI Americas, Inc.

Table 4. (*Continued*)

Compound	CAS Registry Number	Molecular formula	Suggested[a] substrates	Suppliers
Polyphenols				
butylated reaction product of *p*-cresol and dicyclopentadiene	*[68610-51-5]*	variable[b]	CE, PA, PES, POM, PVC, RU, PS	Goodyear Tire and Rubber Co.; R. T. Vanderbilt Co., Inc.
tetrakis[methylene(3,5-di-*tert*-butyl-4-hydroxyhydrocinnamate)]methane	*[6683-19-8]*	$C_{73}H_{108}O_{12}$ (**9**)	CE, PA, PES, PO, POM, PUR, RU, PS	CIBA-GEIGY Corp.; Enichem Synthesis
1,3,5-trimethyl-2,4,6-tris-(3′,5′di-*tert*-butyl-4′-hydroxybenzyl)benzene	*[1709-70-2]*	$C_{54}H_{78}O_3$ (**8**)	CE, PA, PES, PO, POM, PVC, RU	CIBA-GEIGY Corp.; Ethyl Corp.
1,3,5-tris(3,5-di-*tert*-butyl-4-hydroxybenzyl)isocyanurate	*[27676-62-6]*	$C_{48}H_{69}N_3O_6$	CE, PA, PES, PO, POM, PUR, PVC, RU, PS	CIBA-GEIGY Corp.; R. T. Vanderbilt Co., Inc.; PMC Specialties Group, Inc.
3:1 condensate of 3-methyl-6-*tert*-butylphenol with crotonaldehyde	*[1843-03-4]*	$C_{37}H_{52}O_3$	CE, PA, PES, PO, POM, PUR, PVC, RU, PS	ICI Americas Inc.; Fairmount Chemicals, Inc.
3,5-di-*tert*-butyl-4-hydroxy-hydrocinnamic triester with 1,3,5-tris(2-hydroxyethyl)s-triazine-2,4,6-(1*H*,3*H*,5*H*)-trione	*[34137-09-2]*	$C_{60}H_{87}N_3O_{12}$	CE, PA, PO, PUR, PVC, RU, PS	BF Goodrich Co.; R. T. Vanderbilt Co., Inc.; CIBA-GEIGY Corp.
1,3,5-tris(4-*tert*-butyl-3-hydroxy-2,6-dimethylbenzyl)s-triazine-2,4,6-(1*H*,3*H*,5*H*)trione	*[40601-76-1]*	$C_{42}H_{57}N_3O_6$	CE, PA, PES, PO, POM, PUR, PVC, RU, PS	American Cyanamid Co.
Hydroquinones				
2,5-di-*tert*-amylhydroquinone	*[79-74-3]*	$C_{16}H_{26}O_2$	PES, RU	Monsanto Co.
tert-butylhydroquinone	*[1948-33-0]*	$C_{10}H_{14}O_2$	F, PES	Eastman Chemical Products
butylated hydroxyanisole	*[121-00-6]*	$C_{11}H_{16}O_2$	F	Eastman Chemical Products
Diarylamines				
N-phenyl-α-naphthylamine	*[90-30-2]*	$C_{14}H_{13}N$	PS, PES, PO, PVC, PS	Akron Chemical Co.; Mobay Chemical Co.; Uniroyal Chemical
p-oriented styrenated diphenylamine	*[68442-68-2]*	variable[b]	PES, PA, PO, POM, PUR, RU, LUB	Goodyear Tire and Rubber Co.; Mobay Chemical Co.

octylated *N*-phenyl-α-naphthylamine	[*68259-36-9*]	$C_{24}H_{29}N$	LUB	CIBA-GEIGY Corp.
octylated diphenylamines	[*101-67-7*]	$C_{28}H_{43}N$	RU, PUR, LUB	American Cyanamid Co.; Akron Chemical; CIBA-GEIGY Corp.; Mobay Chemical Co.; Monsanto Chemical Co.; Pennwalt Corp.; R. T. Vanderbilt Co., Inc.; Uniroyal Chemical
4,4′-bis(α,α-dimethylbenzyl)diphenylamine	[*10081-67-1*]	$C_{30}H_{31}N$ (**14**)	CE, PES, PO, PVC, PS, PA, PUR, RU	Uniroyal Chemical
Alkylated p-phenylenediamines				
N-phenyl-*N*′-(1,3-dimethylbutyl)-*p*-phenylenediamine	[*793-24-8*]	$C_{13}H_{24}N_2$ (**10**)	RU	Akron Chemical Co.; Goodyear Tire and Rubber Co.; Mobay Chemical Co.; Monsanto Co.; Pennwalt Corp.; R. T. Vanderbilt Co., Inc.; Uniroyal Chemical; UOP, Inc.
N-phenyl-*N*′-isopropyl-*p*-phenylenediamine	[*101-72-4*]	$C_{15}H_{18}N_2$ (**36**)	RU	Akron Chemical Co.; Mobay Chemical Co.; Monsanto Co.; Pennwalt Corp.; Uniroyal Chemical. R. T. Vanderbilt Co., Inc.
N,N′-di-*sec*-butyl-*p*-phenylenediamine	[*69796-47-0*]	$C_{14}H_{24}N_2$	GAS	UOP, Inc.
N-phenyl-*N*′-(1-methylheptyl)-*p*-phenylenediamine	[*15233-47-3*]	$C_{20}H_{28}N_2$	RU	UOP, Inc.
N-phenyl-*N*′-cyclohexyl-*p*-phenylenediamine	[*101-87-1*]	$C_{18}H_{22}N_2$	RU	Uniroyal Chemical; R. T. Vanderbilt Co., Inc.
N,N′-diphenyl-*p*-phenylenediamine	[*74-31-7*]	$C_{18}H_{16}N_2$	PO, RU	R. T. Vanderbilt Co., Inc.; Uniroyal Chemical
N,N′-di-β-naphthyl-*p*-phenylenediamine	[*93-46-9*]	$C_{26}H_{20}N_2$	PO, RU	R. T. Vanderbilt Co., Inc.
N,N′-bis(1,4-dimethylpentyl)phenylenediamine	[*3081-14-9*]	$C_{20}H_{36}N_2$	RU	Monsanto Co.; Uniroyal Chemical; UOP, Inc.
N,N′-bis(1-ethyl-3-methylpentyl)-*p*-phenylenediamine	[*139-60-6*]	$C_{20}H_{36}N_2$	RU	R. T. Vanderbilt Co., Inc.; Uniroyal Chemical; UOP, Inc.
N-1,3-dimethylbutyl-*N*′-phenyl-*p*-phenylenediamine	[*793-24-8*]	$C_{22}H_{40}N_2$		Monsanto Co.; Uniroyal Chemical; R. T. Vanderbilt Co., Inc.
N-N′-bis(1-methylheptyl)-*p*-phenylenediamine	[*103-96-8*]	$C_{22}H_{40}N_2$	RU	R. T. Vanderbilt Co., Inc.; Uniroyal Chemical; UOP, Inc.
N-phenyl-*N*′-(*p*-toluenesulfonyl)-*p*-phenylenediamine	[*100-93-6*]	$C_{19}H_{18}N_2O_2S$ (**37**)	RU	Uniroyal Chemical

Table 4. ***(Continued)***

Compound	CAS Registry Number	Molecular formula	Suggested[a] substrates	Suppliers
Dihydroquinolines				
polymerized 2,2,4-trimethyl-1,2-dihydroquinoline	*[26780-96-1]*	$(C_{12}H_{15}N)_x$ (**13**)	RU	American Cyanamid Co.; Akron Chemical Co.; Monsanto Co.; Pennwalt Corp.; R. T. Vanderbilt Co., Inc.; Uniroyal Chemical
6-dodecyl-2,2,4-trimethyl-1,2-dihydroquinoline	*[89-28-1]*	$C_{24}H_{39}N$	RU	Monsanto Co.
6-ethoxy-2,2,4-trimethyl-1,2-dihydroquinoline	*[91-53-2]*	$C_{14}H_{19}NO$	RU	Mobay Chemical Co.; Monsanto Co.
Thioethers				
dioctadecyldisulfide	*[2500-88-1]*	$C_{36}H_{74}S_2$	PO, CE, EVA, PO, PVC, RU	Hoechst Celanese Corp.; American Cyanamid Co.
didodecyl 3,3′-thiodipropionate	*[123-28-4]*	$C_{30}H_{58}O_4S$ (**34**)	PS	Argus Chemical Corp.; Carstab Corp.; Evans Chemetics; ICI Americas Inc.
distearyl thiodipropionate	*[693-36-7]*	$C_{42}H_{82}O_4S$	CE, EVA, PO, PVC, RU, PS	American Cyanamid Co.; Argus Chemical Corp.; Evans Chemetics; Hoechst Celanese Inc.; Morton Thiokol Inc.; ICI Americas Inc.
dimyristyl thiodipropionate	*[16545-54-3]*	$C_{34}H_{66}O_4S$	CE, EVA, PO, PVC, RU, PS	American Cyanamid Co.; Argus Chemical Corp.; Evans Chemetics; Morton Thiokol Inc.
ditridecyl thiodipropionate	*[10595-72-9]*	$C_{32}H_{62}O_4S$	EVA, PO, PVC, RU, PS	American Cyanamid Co.; Argus Chemical Corp.; Evans Chemetics
pentaerythritol tetrakis(3-dodecylthio)propionate	*[29598-76-3]*	$C_{65}H_{124}O_8S_4$	CE, EVA, PO, RU, PS	Argus Chemical Corp.
2-mercaptoluimidazole	*[53988-10-6]*	$C_8H_8N_2S$	RU	Mobay Chemical Co.; R. T. Vanderbilt Co., Inc.
zinc 2-mercaptotoluimidazole	*[61617-00-3]*	$C_8H_8N_2S{\cdot}\frac{1}{2}N$	RU	Mobay Chemical Co.; R. T. Vanderbilt Co., Inc.

Trivalent Phosphorus Compounds				
tris(nonylphenyl)phosphite[c]	[*26523-78-4*]	$C_{45}H_{69}O_3P$ (**24**)	CE, PES, PO, PUR, PVC, RU, PS	Argus Chemical Corp.; GE Specialty Chemicals Inc.; Olin Interstab Chemical Co.; Uniroyal Chemical
bis(2,4-di-*tert*-butylphenyl)pentaerythritoldiphosphite[c]	[*26741-53-7*]	$C_{33}H_{50}O_6P_2$	PA, PES, PO, PVC, RU, PS	GE Specialty Chemicals Inc.; CIBA-GEIGY Corp.
distearyl pentaerythritoldiphosphite[c]	[*3806-34-6*]	$C_{41}H_{82}O_6P_2$	PES, PO, PVC, PS	GE Specialty Chemicals Inc.
tris(2,4,di-*tert*-butylphenyl)phosphite[c]	[*31570-04-4*]	$C_{42}H_{63}O_3P$ (**25**)	CE, PES, PO, PUR, PS	Argus Chemical Corp.; CIBA-GEIGY Corp.; Enichem Synthesis; Hoechst Celanese Corp.; Uniroyal Chemical
tetrakis(2,4-di-*tert*-butylphenyl)-4,4′-biphenylene disphosphite[c]	[*38613-77-3*]	$C_{68}H_{92}O_4P_2$ (**26**)	PA, PES, PO, PVC, PS	Sandoz Color and Chemicals Inc.
2,2′-ethylidenebis(4,6-di-*tert*-butylphenyl)fluorophosphite	[*118337-09-0*]	$C_{30}H_{44}FO_2P$ (**27**)	PO	Ethyl Corp.
Hindered Amines[d]				
polymer[e] formed from *N,N′*-bis(2,2,6,6-tetramethyl-4-piperidinyl)-1,6-hexanediamine, with 2,4,6-trichloro-1,3,5-triazine and 2,4,4-trimethyl-2-pentanamine (**35**)	[*70624-18-9*]	$(C_{24}H_{50}N_4 \cdot C_8H_{19}N \cdot C_3Cl_3N_3)_x$	PO	CIBA-GEIGY Corp.
polymer[e] formed from *N,N′*-bis(2,2,6,6-tetramethyl-4-piperidinyl)-1,6-hexanediamine, with 2,4-dichloro-6-(4-morpholinyl)-1,3,5-triazine (**44**)	[*82451-48-7*]	$(C_{24}H_{50}N_4 \cdot C_7H_8Cl_2N_4O)_x$	PO	American Cyanamid Co.
(**45**)[e]	[*106990-43-6*]	$C_{132}H_{250}N_{32}$	PO	CIBA-GEIGY Corp.

[a]CE = cellulosics, EVA = ethylene–vinyl acetate copolymers, F = foods, GAS = gasoline, LUB = lubricants, PA = polyamides, PES = polyesters, PO = polyolefins, POM = poly(oxymethylene)s, PUR = polyurethanes, RU = rubbers, and PS = polystyrenes.
[b]Mixture.
[c]Accepted by the FDA as an indirect food additive in April 1990. Title 21 of the Code of Federal Regulations provides details.
[d]Hindered amines are generally classified as light stabilizers. The three selected products are derivatives of 1,6-hexanediamine and are effective antioxidants for polyolefins.
[e]Structure shown in Figure 5.

Health and Safety Factors

Safety is assessed by subjecting the antioxidant to a series of animal toxicity tests, eg, oral, inhalation, eye, and skin tests. Mutagenicity tests are also carried out to determine possible or potential carcinogenicity. Stabilizers are being granulated and liquid products are receiving greater acceptance to minimize the inhalation of dust and to improve flow characteristics.

A number of antioxidants have been accepted by the FDA as indirect additives for polymers used in food applications. Acceptance is determined by subchronic or chronic toxicity in more than one animal species and by the concentration expected in the diet, based on the amount of the additive extracted from the polymer by typical foods or solvents that simulate food in their extractive effects. Only materials of insignificant risk to the consumer are regulated by the FDA for use in plastics contacted by food stuffs.

Commercial Antioxidants

Table 4 includes the main classes of antioxidants sold in the United States and the supplier's suggested applications. Some of these are mixtures rather than single substrates. This is especially true of alkylated amines and alkylated phenols. The extent of alkylation and the olefins used for alkylation can vary among manufacturers. Table 4 is not a complete listing of available antioxidants in the United States.

BIBLIOGRAPHY

"Antioxidants" in *ECT* 1st ed., Vol. 2, pp. 69–75, by A. E. Bailey; in *ECT* 2nd ed., Vol. 2, pp. 588–604, by L. R. Dugan, Jr., Michigan State University; "Antioxidants and Antiozonants" in *ECT* 3rd ed., Vol. 2, pp. 128–148, by P. P. Nicholas and A. M. Luxeder, The B.F. Goodrich Co., L. A. Brooks, R. T. Vanderbilt Co. and P. A. Hammes, Merck and Co.

1. "Sales of Antioxidants in the United States in 1990," *Chem. Mark. Rep.,* 26 (Nov. 18, 1991).
2. "Synthetic Organic Chemicals," *United States Production and Sales 1988,* U.S. Government Printing Office, Washington D.C., 1989.
3. S. L. Bolland, *Quart. Rev. London* **3,** 1 (1949).
4. L. Bateman, *Quart. Rev. London* **8,** 147 (1954).
5. K. U. Ingold, *Chem. Rev.* **61,** 563 (1961).
6. J. R. Shelton and W. L. Cox, *Rubber Chem. Technol.* **27,** 671 (1954).
7. D. Barnard, L. Bateman, J. I. Cunneen, and J. F. Smith in L. Bateman, ed., *Chemistry and Physics of Rubber-like Substances,* Maclaren, London, 1963, p. 593.
8. J. A. Kerr, *Chem. Rev.* **66,** 465 (1966).
9. A. J. Chalk and J. F. Smith, *Trans. Faraday Soc.* **53,** 1214 (1957).
10. J. P. Pospisil in E. Scott, ed., *Developments in Polymer Stabilization,* Vol. 1, Applied Science Publishers, Ltd., London, 1979, pp. 6–11.
11. A. C. Nixon, H. B. Minor, and G. M. Calhoun, *Ind. Eng. Chem.* **48,** 1874 (1956).
12. R. H. Rosenwald, J. R. Hoatson, and J. A. Chenicek, *Ind. Eng. Chem.* **41,** 162 (1950).
13. J. I. Wasson and W. M. Smith, *Ind. Eng. Chem.* **45,** 197 (1953).

14. G. Scott, *Atmospheric Oxidation and Antioxidants,* Elsevier, Amsterdam, The Netherlands, 1965, p. 109.
15. G. Scott, *J. Appl. Polym. Sci.* **35,** 131 (1979).
16. P. P. Klemchuk and M. E. Gande, *Makromol. Chem. Macromols. Symp.* **28,** 117–144 (1989).
17. J. R. Shelton in W. L. Hawkins, ed., *Polymer Stabilization,* Wiley-Interscience, New York, 1972, pp. 97–98.
18. G. Scott, ed., *Development in Polymer Stabilization,* Vol. 4, Applied Science Publishers, Ltd., London, 1981, pp. 16–17; C. Armstrong, M. J. Husbands, and G. Scott, *Eur. Polym. J.* **15,** 241 (1979).
19. Ref. 9, p. 1235.
20. Ref. 17, p. 83.
21. D. Rysavy, *Kunststoffe* **60,** 118 (1970).
22. A. Tozzi, G. Cantatore, and F. Masina, *Text. Res. J.* **48,** 434 (1978).
23. G. E. Meyer, R. W. Kavchik, and F. J. Naples, *Rubber Chem. Technol.* **46,** 106 (1973).
24. G. Scott and Setoudeh, *Polym. Degrad. Stability* **5,** 81 (1983).
25. R. H. Hansen, D. R. Falcone, W. M. Martin, and W. I. Vroom, *Org. Coat. Plast. Prepr.* **34,** 97 (1974).
26. H. Gysling, *Adv. Chem. Ser.* **85,** 239 (1968).
27. *ASTM Bull.,* 19103 (1990).
28. D. Chasan in J. Pospisil and P. P. Klemchuk, eds., *Oxidation Inhibition in Organic Materials,* Vol. I, CRC Press, Inc., Boca Raton, Fla., **33** and **31,** pp. 291–326, 1990.
29. "Active Oxygen Method," *Technical Data Bulletin ZG-159c,* Eastman Chemical Products, Inc., Kingsport, Tenn., Mar. 1985.
30. "Storage Stability Tests for Evaluating Antioxidants in Fats and Oils," *Technical Data Bulletin ZG-194c,* Eastman Chemical Products, Inc., Kingsport, Tenn., Oct. 1987.

General References

Refs. 14, 17, and 27 are good general references.

M. Dexter in J. I. Kroschwitz, ed., *Encyclopedia of Polymer Science and Engineering,* Vol. 2, Wiley-Interscience, New York, 1985, pp. 73–91.

J. Voigt, *Die Stabilisierung der Kunstoffe Gegen Licht and Warme,* Springer Verlag, Heidelberg, 1966.

K. Schwarzenbach in R. Gachter and H. Muller, eds., *Plastics Additives Handbook,* Carl Hanser Verlag, Munich, 1990, pp. 1–104.

J. Pospisil and P. P. Klemchuk, eds., *Oxidation Inhibition in Organic Materials,* Vols. I and II, CRC Press Inc., Boca Raton, Fla., 1990.

MARTIN DEXTER
Consultant

ANTIOZONANTS

The term antiozonant, in its broadest sense, denotes any additive that protects an elastomer against ozone deterioration. Most frequently, the protective effect results from a reaction with ozone, in which case the term used is chemical antiozonant. Ozone is generated naturally by electrical discharge and also by solar radiation in the stratosphere (1). These sources produce ground-level ozone concentrations of 1–5 parts per hundred million (pphm). Ozone is also produced in urban centers by ultraviolet photolysis, which can yield concentrations up to ≈25 pphm. Only a few pphm ozone in air can cause rubber cracking, which may destroy the usefulness of elastomer products. Ozone attacks any elastomer with backbone unsaturation. Degradation results from the reaction of ozone with rubber double bonds. Unstretched rubber reacts with ozone but is not cracked. Ozone cracking in stretched rubber is always perpendicular to the direction of the applied stress. Commercial antiozonants have been available since the early 1950s. Since that time, the ozone degradation problem has worsened as atmospheric ozone concentrations have gradually increased, especially in urban industrial areas.

Chemistry of Ozone Attack

Ozone cracking is a physico-chemical phenomenon, and many factors are involved in explaining the effect of ozone attack on elastomers (2–7). From a chemical point of view, ozone attack on olefinic double bonds causes chain scission and the formation of the decomposition products shown in Figure 1 (8,9). These chemical reactions are believed to be similar for both small olefins and unsaturated rubbers. The first step is the formation of a relatively unstable primary ozonide (or molozonide) (**1**), which cleaves to an aldehyde or ketone (**2**) and a carbonyl oxide (or zwitterion) (**3**). Subsequent recombination of (**2**) and (**3**) produces a secondary ozonide (or just ozonide) (**4**). In small olefins, ozonide

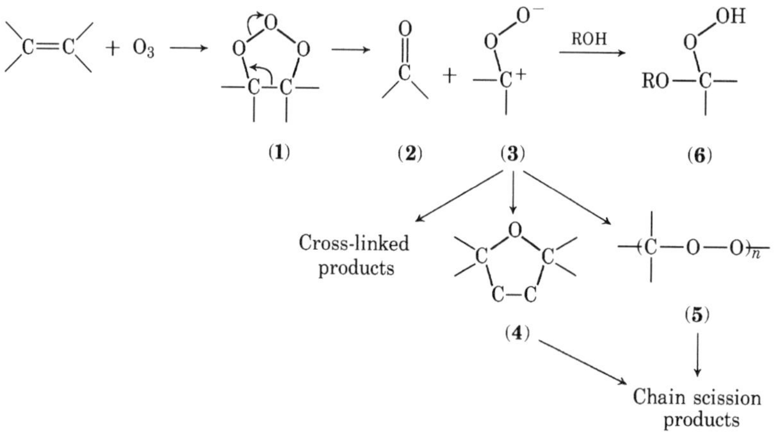

Fig. 1. Alkene–ozone reaction scheme.

formation is generally a facile process. In stretched rubber, however, ozonide formation is more difficult, since the cleaved intermediates (**2**) and (**3**) may be forcefully separated to relieve the stress. Ozonides are reasonably stable in neutral environments, but they decompose readily under the influence of heat or various reducing agents to yield such chain scission products as aldehydes, ketones, acids, and alcohols. Polymeric peroxides (**5**) may be formed initially from the carbonyl oxide, but these are unstable and eventually decompose to yield chain scission products. The rate of chain scission is increased in the presence of active hydrogen (eg, water), probably because of the reaction with carbonyl oxides to form reactive hydroperoxides (**6**).

Physical Factors. Unsaturated elastomers must be stretched for ozone cracking to occur. Elongations of 3–5% are generally sufficient. Crack growth studies (10–18) have shown that some minimum force, called the critical stress, rather than a minimum elongation is required for cracking to occur. Critical stress values are nearly the same for most unsaturated rubbers. However, polychloroprene has a higher critical stress value than other diene rubbers, consistent with its better ozone resistance. It has been found that temperature, plasticization, and ozone concentration have little effect on critical stress values.

Typically, ozone cracking initiates at sites of high stress (flaws) on the rubber surface. Thus, in general, rubber articles should be designed to minimize potential sites of high elongation such as raised lettering. Similarly, the use of clean molds helps to reduce the incidence of surface flaws.

Desirable Properties of Antiozonants

A physical antiozonant must provide an effective barrier against the penetration of ozone at the rubber surface. A chemical antiozonant, on the other hand, must first of all be extremely reactive with ozone.

The antiozonant should possess adequate solubility and diffusivity characteristics (19). Since ozone attack is a surface phenomenon, the antiozonant must migrate to the surface of the rubber to provide protection. The antiozonant should have no adverse effects on the rubber processing characteristics, eg, mixing, fabrication, vulcanization, or physical properties.

The antiozonant should be effective under both static and dynamic conditions.

It should persist in the rubber over its entire life cycle. For noncarbon black filled rubbers, the antiozonant must be nondiscoloring and nonstaining. The antiozonant should have a low toxicity and should be nonmutagenic, and, the antiozonant should be acceptable economically, eg, have low manufacturing and end use costs.

Hydrocarbon Waxes

Waxes are one of the two general classes of commercial antiozonants. Waxes are derived from petroleum and are of two common types, paraffin and microcrystalline (20–23). Typical carbon numbers are n = 20–50 for paraffin waxes and n =

30–70 for microcrystalline materials. If a wax is present in a vulcanizate at a concentration exceeding its solubility, some of it will migrate to the rubber surface where it can form a physical barrier to prevent the penetration of ozone. Waxes, of course, are essentially unreactive towards ozone so that there is no appreciable chemical protection. Commercial waxes are usually blends of paraffin and microcrystalline waxes. Such blends provide protection over a wide temperature range (≈10–60°C). Advantages of waxes are their relatively low cost; they are nonstaining; and they generally have no adverse effects on rubber processing and vulcanization. Unfortunately, waxes have a number of shortcomings: they are ineffective under dynamic stress conditions; their protection capability is highly dependent on exposure temperature; and waxes may easily be lost during storage or use by embrittlement, scuffing, or delamination (6).

Chemical Antiozonants

Chemical antiozonants comprise the second general class of commercial antiozonants. Of the many compounds reported to be chemical antiozonants, nearly all contain nitrogen. Compound classes include derivatives of 1,2-dihydro-2,2,4-trimethylquinoline, *N*-substituted ureas or thioureas, substituted pyrroles, and nickel or zinc dithiocarbamate salts (see also ANTIOXIDANTS). The most effective antiozonants, however, are derivatives of *p*-phenylenediamine [*106-50-3*] (*p*-PDA):

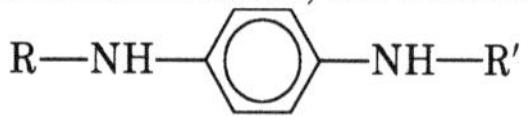

(see AMINES-AROMATIC, PHENYLENEDIAMINES). The commercial materials fall into three classes (Table 1): *N,N′*-dialkyl-*p*-PDAs, *N*-alkyl-*N′*-aryl-*p*-PDAs, and *N,N′*-diaryl-*p*-PDAs.

The *N,N′*-dialkyl-*p*-PDAs (where the alkyl group may be 1-methylheptyl, 1-ethyl-3-methylpentyl, 1,4-dimethylpentyl or cyclohexyl) are the most effective in terms of their reactivity towards ozone (24–26). These derivatives increase the critical stress required for the initiation of crack growth, and they also reduce the rate of crack growth significantly (15). The *sec*-alkyl group is most active, for reasons that are as yet not completely clear. The drawbacks of these derivatives are: their rapid destruction by oxygen, ie, shorter useful lifetimes; their activity as vulcanization accelerators, and hence increased scorchiness; their tendencies to cause dark red or purple discoloration; and the difficulty in handling because they are liquids. The dialkyl-*p*-PDAs are seldom used alone in rubber compounds, although they can be used effectively when blended with *N*-alkyl-*N′*-aryl-*p*-PDAs.

The *N*-alkyl-*N′*-aryl-*p*-PDAs (where the aryl is phenyl and the alkyl may be cyclohexyl, 1,3-dimethylbutyl or 1-methylethyl) are the most widely used *p*-PDAs. These derivatives reduce the rate of crack growth and also the number of cracks. The alkyl–aryl-*p*-PDAs are in general excellent antiozonants, particularly in dynamic environments. These derivatives are destroyed only slowly by oxygen and increase the scorchiness of the stock only slightly. These are intermediate in staining among the three classes of *p*-PDAs.

The *N,N′*-diaryl-*p*-PDAs (where the aryl group may be phenyl, methylphenyl, or napthalenyl) are only moderately active antiozonants, that can only be used at low concentrations (generally less than 2 phr) because of their poor solubility. However, they have minimal scorching effects, are the most stable towards

Table 1. Commercial Antiozonants

p-Phenylenediamine[a]	CAS Registry Number	Abbreviation	Melting point, °C	Trade names[b]
N,N'-bis(1,4-dimethylpentyl)	[*3081-14-9*]	DMPPD	liquid	Eastozone 33, Santoflex 77, Flexzone 4L, UOP 788
N,N'-bis(1-ethyl-3-methylpentyl)	[*139-60-6*]	DEMPD	liquid	Flexzone 8L, UOP 88, Antozite 2
N,N'-bis(1-methylheptyl)	[*103-96-8*]	DMHPD or DOPPD	low melting	UOP 288, Antozite 1
N-(1,3-dimethylbutyl)-*N'*-phenyl	[*61931-82-6*]	HPPD or 6PPD	48–50	Antiozonant PD-2, $Anto_3$ "E", Antozite 67P, Flexzone 7L/7F, Permanax 6PPD, Santoflex 13/13F, UOP 588, Vulkanox 4020
N-(1-methylethyl)-*N'*-phenyl	[*101-72-4*]	MEPD or IPPD	70–78	Antiozonant PD-1, $Anto_3$ "H", Flexzone 3C, Permanax IPPD, Santoflex IP, UOP 388, Vanox 3C, Vulkanox 4010 NA
N,N'-dicyclohexyl	[*4175-38-6*]	DCHPD	106	UOP 26
N-cyclohexyl-*N'*-phenyl	[*101-87-1*]	CHPD	110	Antioxidant 4010, Flexzone 6H, UOP 36, Vanox 6H
mixed-*N,N'*-diaryl	[*68953-84-4*]	DTPD	90–115	Wingstay 100 [*50864-70-5*], Wingstay 200 [*88026-77-1*], Vanox 2
N,N'-diphenyl	[*74-31-7*]	DPPD	144–153	AgeRite DPPD, J-Z-F, Permanax DPPD
N,N'-bis(2-naphthalenyl)	[*93-46-9*]	DNPD	224–230	AgeRite White, Vulkanox DNP

[a]1,4-Diaminobenzene [*106-50-3*].

[b]Manufacturers' trade names: Akron Chemical (Antiozonant), Eastman (Eastozone), Goodyear (Wingstay), Mobay (Vulkanox, Antioxidant), Monsanto (Santoflex), Pennwalt ($Anto_3$), Uniroyal (Flexzone, J-Z-F), Universal Oil Products (UOP), R. T. Vanderbilt (Antozite, AgeRite, Vanox), and Vulnax (Permanax).

oxygen, and stain the least. Their main advantage is high resistance to loss by consumption and vaporization (2), and in combination with more reactive antiozonants, they can offer a degree of increased protection in such long term applications as radial passenger tires (27).

The principal objection to *p*-PDA antiozonants is their staining characteristics. The lack of suitable alternative antiozonants for light-colored diene rubber articles is one of the outstanding problems in rubber technology. Few chemical antiozonants outside the class of *p*-PDAs have much commercial importance. One of the few exceptions to this rule is 6-ethoxy-1,2-dihydro-2,2,4-trimethylquinoline [*91-53-2*], one of the first commercial antiozonants.

Mechanism of Action of Chemical Antiozonants

Several theories have appeared in the literature regarding the mechanism of protection by *p*-PDA antiozonants. The scavenger theory states that the antiozonant diffuses to the surface and preferentially reacts with ozone, with the result that the rubber is not attacked until the antiozonant is exhausted (25,28,29). The protective film theory is similar, except that the ozone–antiozonant reaction products form a film on the surface that prevents attack (28). The relinking theory states that the antiozonant prevents scission of the ozonized rubber or recombines severed double bonds (14). A fourth theory states that the antiozonant reacts with the ozonized rubber or carbonyl oxide ((**3**) in Fig. 1) to give a low molecular weight, inert self-healing film on the surface (3).

The literature suggests that more than one mechanism may be operative for a given antiozonant, and that different mechanisms may be applicable to different types of antiozonants. All of the evidence, however, indicates that the scavenger mechanism is the most important. All antiozonants react with ozone at a much higher rate than does the rubber which they protect. The extremely high reactivity with ozone of *p*-phenylenediamines, compared to other amines, is best explained by their unique ability to react free-radically. The chemistry of ozone–*p*-PDA reactions is known in some detail (30,31). The first step is believed to be the formation of an ozone–*p*-PDA adduct (32), or in some cases a radical ion. Four competing fates for dissociation of the initial adduct have been described: amine oxide formation, side-chain oxidation, nitroxide radical formation, and amino radical formation.

Although all antiozonants must react rapidly with ozone, not all highly reactive materials are antiozonants. Something else in addition to the scavenging effect is required. The protective film theory contends that ozonized products, to a considerable extent, prevent ozone from reaching the rubber. There is visual and microscopic evidence for formation of a protective film on the rubber surface (33). Spectroscopic characterization has shown that this film consists of unreacted antiozonant and many of the same components observed in ozonized liquid antiozonant (30,33,34). The components of the film are polar and tend not to diffuse back into the rubber bulk.

The relinking (14) and self-healing film (3) theories require chemical interaction between the antiozonant and ozonized rubber. The evidence for these interactions is meager (35,36). Overall, there seems to be no clear evidence in the literature for *p*-PDA derivatives becoming attached to rubber chains as a result of ozone attack. Much fundamental work in this area remains to be done, however. It seems clear at this point that any antiozonant–rubber interaction must be much less important than the scavenging effect of the antiozonant. In summary, the scavenger model is believed to be the principal mechanism of antiozonant action. Ozone–antiozonant reaction products form a surface film that provides additional protection (37).

Manufacture and Production

The *N,N′*-dialkyl-*p*-PDAs are manufactured by reductively alkylating *p*-PDA with ketones. Alternatively, these compounds can be prepared from the ketone and *p*-nitroaniline with catalytic hydrogenation. The *N*-alkyl-*N′*-aryl-*p*-PDAs are made by reductively alkylating *p*-nitro-, *p*-nitroso-, or *p*-aminodiphenylamine with ketones. The *N,N′*-dialkyl-*p*-PDAs are made by condensing various anilines with hydroquinone in the presence of an acid catalyst (see AMINES-AROMATIC, PHENYLENEDIAMINES).

Alternatives to Antiozonants

Any diene rubber article subjected to flexing, bending, or folding requires protection against ozone. Chemical antiozonants, primarily *p*-PDAs, are used in general purpose commercial applications (mainly tires, hoses, flat belts, and transmission belts) where staining and discoloration are not serious problems. Waxes act as antiozonants for diene rubbers under conditions requiring little or no flexing, such as tie-down straps. Under these conditions, waxes are more effective than the *p*-PDA antiozonants (21). However, in applications involving dynamic stress where discoloration cannot be tolerated, alternatives to the traditional antiozonants must be used. Typical applications include white tire sidewalls, gaskets, weather stripping, gloves, and sporting goods. The two principal alternatives to antiozonants are ozone-resistant elastomers and blends of ozone-resistant elastomers and diene rubbers.

Ozone-resistant elastomers which have no unsaturation are an excellent choice when their physical properties suit the application, for example, polyacrylates, polysulfides, silicones, polyesters, and chlorosulfonated polyethylene (38). Such polymers are also used where high ozone concentrations are encountered. Elastomers with pendant, but not backbone, unsaturation are likewise ozone-resistant. Elastomers of this type are the ethylene–propylene–diene (EPDM) rubbers, which possess a weathering resistance that is not dependent on environmentally sensitive stabilizers. Other elastomers, such as butyl rubber (IIR) with low double-bond content, are fairly resistant to ozone. As unsaturation increases, ozone resistance decreases. Chloroprene rubber (CR) is also quite ozone-resistant.

Blends of diene and backbone-saturated rubbers are frequently used in applications where discoloration by chemical antiozonants cannot be tolerated, yet where cost is still a primary consideration (eg, white sidewalls of tires). Disadvantages are that physical properties have to be compromised and the two rubbers usually differ greatly in their rates of vulcanization. Usually, at least a 25% replacement by the ozone-resistant rubber is needed for an appreciable enhancement in ozone protection (6).

Testing

Since antiozonants are affected by most compounding ingredients, each new rubber compound requires the development of a cost-effective antiozonant system. Outdoor as well as accelerated ozone-chamber tests are available (39–45). Laboratory tests involve exposing a statically or dynamically elongated test sample to ozone and measuring the time to crack formation, the severity of cracking, or the decay of 100% modulus with time. Cracking is affected by ozone concentration and flow rate, temperature, humidity, sample shape, and type of strain (static, dynamic, or both). Cracking is accelerated by increasing ozone concentration. Normal test concentrations are 10–25 pphm, but they can be as high as 50 pphm in accelerated tests. Above 70°C ozone decomposes, and therefore testing is usually carried out at 30–50°C. Humidity can accelerate ozone cracking, and the maximum recommended value is ≈65%.

Strain affects both the number and size of cracks. Low strain (10–20%) causes fewer but larger cracks; higher strain gives numerous small cracks. Testing is usually done at low strain. Most antiozonants are used in applications where flexing is involved, and dynamic tests are needed. However, since actual field conditions involve a combination of static and dynamic stress, an intermittent static–dynamic ozone exposure often provides a more realistic test regimen. Laboratory tests are best used as screening procedures for new additives and compounds. Only field testing of rubber articles can provide a true judgment of the protective capability of a particular formulation.

Health and Safety Factors

The first *p*-PDA antiozonants were low molecular weight *N,N′*-dialkyl-*p*-PDAs which caused skin irritations. Current higher molecular weight *N,N′*-dialkyl or *N*-alkyl-*N′*-aryl derivatives are not primary skin irritants. A notable exception is *N*-(1-methylethyl)-*N′*-phenyl-*p*-PDA, which causes dermatitis. However, since some individuals are more sensitive than others, antiozonants should always be handled with care (46). When skin contact does occur, the affected area should be washed with mild soap and water. In case of eye contact, flush well with water. Inhalation of rubber chemicals should be avoided, and respiratory equipment should be used in dusty areas.

Uses and Formulations

Chemical antiozonants are routinely used to protect diene rubbers (NR, IR, BR, SBR, NBR) against atmospheric ozone for extended periods of time. Large volumes are used in tire, belt, and hose applications. The *N*-alkyl-*N′*-aryl-*p*-PDAs have largely displaced the *N,N′*-dialkyl-*p*-PDAs as the materials of choice since they are less scorchy, more persistent, and more easily handled. The *N,N′*-diaryl-*p*-PDAs are more effective antiozonants for chloroprene rubber (CR). Antiozonants (*p*-PDAs) are added to raw rubber stocks during mixing at 1–5 phr concentrations and, most frequently, at 2–4 phr. Paraffin waxes (1–2 phr) may also be added to rubber formulations containing *p*-PDAs for increased static ozone protection (21).

Certain fillers may absorb or oxidize *p*-PDA antiozonants, slightly increasing the antiozonant requirement. The *p*-PDA antiozonants are generally compatible with sulfur-curing systems. Extender oils and plasticizers increase the ozone cracking of unprotected rubbers. This is because of increased chain mobility and a more facile separation of the ozonized rubber fragments (14,47). General purpose antioxidants extend the service life of antiozonants and reduce the overall cost of the rubber protective system. Water can leach antiozonants and shorten the life of the product (48,49). Leaching of *N,N'*-dialkyl-*p*-PDAs decreases with increasing size of the alkyl group and becomes negligible for groups containing more than seven carbon atoms.

BIBLIOGRAPHY

"Antioxidants" in *ECT* 1st ed., Vol. 2, pp. 69–75, by A. E. Bailey; in *ECT* 2nd ed., Vol. 2, pp. 588–604, by L. R. Dugan, Jr., Michigan State University; "Antioxidants and Antiozonants" in *ECT* 3rd ed., Vol. 3, pp. 128–148, by P. Nicholas and A. M. Luxeder, BFGoodrich Co., L. A. Brooks, R. T. Vanderbilt Co., and P. A. Hammes, Merck and Co.

1. *Air Quality Criteria for Photochemical Oxidants,* National Air Pollution Administration Pub. No. AP-63, U.S. Government Printing Office, Washington, D.C., 1970.
2. J. C. Ambelang and co-workers, *Rubber Chem. Technol.* **36,** 1497 (1963).
3. L. D. Loan, R. W. Murray, and P. R. Story, *J. Inst. Rubber Ind.* **2,** 73 (1968).
4. P. M. Lewis, *NR Technol.* **1,** 1 (1972).
5. S. D. Razumovskii and G. E. Zaikov, *Dev. Polym. Stab.* **6,** 239 (1983).
6. P. M. Lewis, *Polym. Degrad. Stab.* **15,** 33 (1986).
7. J. Pospisil, *Dev. Polym. Stab.* **7,** 1 (1984).
8. P. S. Bailey, *Chem. Rev.* **58,** 925 (1958).
9. R. Criegee, *Angew. Chem. Int. Ed. Eng.* **14,** 745 (1975).
10. M. Braden and A. N. Gent, *J. Appl. Polym. Sci.* **3,** 90 (1960).
11. M. Braden and A. N. Gent, *J. Appl. Polym. Sci.* **3,** 100 (1960).
12. M. Braden and A. N. Gent, *Trans. Inst. Rubber Ind.* **37,** 88 (1961).
13. E. H. Andrews and M. Braden, *J. Polym. Sci.* **55,** 787 (1961).
14. M. Braden and A. N. Gent, *Rubber Chem. Technol.* **35,** 200 (1962).
15. M. Braden and A. N. Gent, *J. Appl. Polym. Sci.* **6,** 449 (1962).
16. E. H. Andrews and M. Braden, *J. Appl. Polym. Sci.* **7,** 1003 (1963).
17. A. N. Gent and J. E. McGrath, *J. Polym. Sci. Part A* **3,** 1473 (1965).
18. A. N. Gent and H. Hirakawa, *J. Polym. Sci. Part A* **5,** 157 (1967).
19. G. Emin, *Proc. Int. Rubber Conf.,* 1055 (1979).
20. F. Jowett, *Elastomerics* **111**(9), 48 (1979).
21. D. A. Lederer and M. A. Fath, *Rubber Chem. Technol.* **54,** 415 (1981).
22. F. Jowett, *Rubber World* **188**(2), 24 (1983).
23. R. M. Mavrina, L. G. Angert, I. G. Anisimov, and A. V. Melikhova, *Kauch. Rezina* **31**(12), 29 (1972).
24. R. F. Shaw, Z. T. Ossefort, and W. J. Touhey, *Rubber World* **130,** 636 (1954).
25. W. L. Cox, *Rubber Chem. Technol.* **32,** 364 (1959).
26. R. W. Layer, *Rubber Chem. Technol.* **39,** 1584 (1966).
27. D. E. Miller, R. W. Dessent, and J. L. Kuczkowski, *Rubber World* **193**(1), 31 (1985).
28. E. R. Erickson, R. A. Berntsen, E. L. Hill, and P. Kusy, *Rubber Chem. Technol.* **32,** 1062 (1959).
29. S. D. Razumovskii and L. S. Batashova, *Rubber Chem. Technol.* **43,** 1340 (1970).
30. R. P. Lattimer and co-workers, *Rubber Chem. Technol.* **53,** 1170 (1980).

31. R. P. Lattimer, E. R. Hooser, R. W. Layer, and C. K. Rhee, *Rubber Chem. Technol.* **56,** 431 (1983).
32. P. S. Bailey, J. E. Keller, D. A. Mitchard, and H. M. White, *Adv. Chem. Ser.* **77,** 58 (1968).
33. J. C. Andries and co-workers, *Rubber Chem. Technol.* **52,** 823 (1979).
34. J. C. Andries, D. B. Ross, and H. E. Diem, *Rubber Chem. Technol.* **48,** 41 (1975).
35. O. Lorenz and C. R. Parks, *Rubber Chem. Technol.* **36,** 194 (1963).
36. O. Lorenz and C. R. Parks, *Rubber Chem. Technol.* **36,** 201 (1963).
37. R. W. Layer and R. P. Lattimer, *Rubber Chem. Technol.* **63,** 426 (1990).
38. R. O. Babbit, ed., *The Vanderbilt Rubber Handbook,* R. T. Vanderbilt Co., Norwalk, Conn., 1978.
39. A. Hartmann and F. Glander, *Rubber Chem. Technol.* **29,** 166 (1956).
40. C. S. Amsden, *Trans. Inst. Rubber Ind.* **13T91** (1966).
41. J. Crabtree and A. R. Kemp, *Ind. Eng. Chem. Anal. Ed.* **18,** 769 (1946).
42. J. R. Beatty and A. E. Juve, *Rubber World* **131**(2), 232 (1954).
43. A. G. Veith, *ASTM Spec. Tech. Pub. No. 229,* American Society for Testing and Materials, Philadelphia, Pa., 1958, pp. 113–130.
44. K. M. Davies and D. G. Lloyd, *Dev. Polym. Stab.* **4,** 111 (1981).
45. *Annual Book of ASTM Standards,* American Society for Testing and Materials, Philadelphia, Pa., 1980, parts 37 and 38.
46. W. E. McCormick, *Rubber Chem. Technol.* **45,** 627 (1972).
47. D. J. Buckley, *Rubber Chem. Technol.* **32,** 1475 (1959).
48. G. R. Browning and R. R. Barnhart, *Rubber Chem. Technol.* **44,** 1441 (1971).
49. A. G. John and co-workers, *Kautsch. Gummi Kunstst.,* **37**(2), 115 (1984).

ROBERT P. LATTIMER
C. K. RHEE
ROBERT W. LAYER
The BFGoodrich Company

ANTIPARASITIC AGENTS

ANTHELMINTICS

Human infections caused by worms (helminths) represent one of the most important public health problems in the world. Helminths form three main categories or phyla: Platyhelminths, flatworms; Aschelminthes, roundworms; and Nemathelminthes, thorny-headed worms. Platyhelminths consist of Trematoda and Cestoda. Members of the class Trematoda, or flukes (schistosomes), are slender leaf-shaped organisms that possess attachment organs in the form of cup-shaped hooks called suckers. Members of the class Cestoidea, or tapeworms, are flat and ribbonlike. These worms have serially repeated sections behind the neck and an attachment organ called the scolex. The Aschelminthes (class Nematoda) are roundworms, which have a basic cylindrical shape with major variations in proportions, size, and structure. The Nemathelminthes (Acanthocephala), or thorny-headed worms, are widely distributed among animals and generally do not involve a human host.

Over the last several years, substantial progress in the discovery and development of anthelmintic drugs has been made. Effective agents are available for most human gastrointestinal infections (Table 1); however, drugs that are effective in treating the extraintestinal complications of many helminthic infections are still needed.

Frequently, the treatment of helminthic diseases requires adjunct medication. Allergic reactions are commonly seen as a result of tissue invasion by worms or as a consequence of anthelmintic therapy. Antihistamines and corticosteroids may be necessary adjuncts to therapy. Anemia, indigestion, and secondary bacterial infections can also occur and may require concomitant therapy with hematopoietic drugs and appropriate antibiotics.

Treatment of Trematode Infections

Blood Flukes (Schistosomiasis). Three main species of blood flukes cause schistosomiasis in humans: *Schistosoma haematobium, S. mansoni,* and *S. japonicum.* The effectiveness of antischistosomal drugs depends on the reduction or arrest of egg production. Infection only occurs as a result of the penetration of the intact skin by free-swimming cercaria. After they develop to pre-adult forms in the skin and lung, the parasites migrate through host blood vessels to various tissues. Adult worms that are approximately 2 cm in length remain paired within the portal or intestinal vasculature of the host. The female, however, travels as far as possible toward the capillary beds to lay eggs, a portion of which embolize to the liver and induce granuloma formation. During the early stages of the disease, patients may experience fever, gastrointestinal distress, headache, and fatigue. In later stages of the disease, signs of hepatic fibrosis and ascites are seen. The number of worm pairs found in a patient may vary from a few to over 100. In untreated patients, adult worms are now generally assumed to live less than 5–10 years.

Praziquantel. This drug (**1**), $C_{19}H_{24}N_2O_2$, can be used to treat schistosomiasis caused by any one of the three major species. Praziquantel is an acylated pyrazino–isoquinoline derivative that has replaced the traditional (and more toxic) trivalent antimonial compounds.

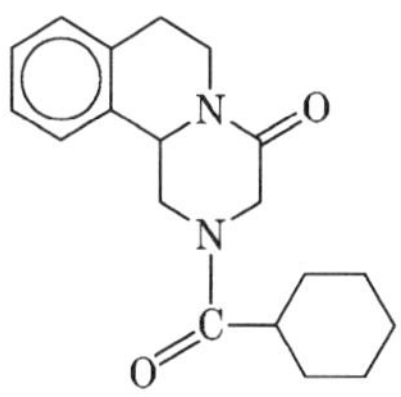

(**1**)

Table 1. Anthelmintics: Uses and Properties

Drug	Structure number	CAS Registry Number	Trade name	Physical properties	Solubility	Disease (organism)	Reference
praziquantel	**(1)**	*[55268-74-1]*	Biltricide solid, Cesol, Livera, Pontel, Prazi, Pyquiton	mp 136–138°C	practically insoluble in water; sparingly in alcohol, chloroform, and ether	flukes (*C. sinensis, F. buski, P. westermani, S. haematobium, S. mansoni, S. japonicum*), tapeworms (*H. nana, T. solium, T. saginata, D. latum*), and larval stage (cysticercosis)	1,2
oxamniquine	**(2)**	*[21738-42-1]*	Mansil, Vansil	pale yellow crystals	sparingly soluble in water; soluble in acetone, chloroform, and methyl alcohol	flukes (*S. mansoni*)	3
metrifonate	**(3)**	*[52-68-6]*	Bilarcil	white crystals, mp 83–84°C	1 g in 6.5 mL water, 33 mL chloroform, 5.9 mL ether	blood flukes (*Schistosoma haematobium*)	4
bithionol[a]	**(4)**	*[97-18-7]*	Bitin	crystals or gray-white powder, mp 188°C	insoluble in water, acetone, and 4% NaOH	lung flukes (*Paragonimu westermani*), sheep liver flukes (*Fasciola hepatica*)	5,6
tetrachloroethylene	**(5)**	*[127-18-4]*	Nema worm capsules, Perklone	colorless, nonflammable fluid, ethereal odor, deteriorates in warm climates, bp 121°C	soluble 1:10,000 in water; miscible with ethanol	intestinal fluke (*Fasicolopsis buski*)	7

niclosamide	(**6**)	[*50-65-7*]	Cestocida, Mansonil, Nasemo, Niclocide, Sulqui, Yomesan	pale yellow crystals, or yellow-white powder, mp 225–230°C	insoluble in water; sparingly soluble in ethanol	tapeworms (*T. saginata, T. solium,* and *D. latum*)	8
quinacrine	(**7**)	[*83-89-6*]	Atabrine hydrochloride[b]	yellow crystals,[b] mp ca 250°C	1 g in 36 mL water; slightly soluble in ethanol	pork tapeworm (*Taenia solium*)	9
piperazine citrate[c]	(**8**)	[*144-29-6*]	Antepar, Helmezine, Pinozan, Pinrou, Uvilon	white crystalline powder, mp ca 185°C (dec)	freely soluble in water; insoluble in ethanol	roundworms (*Ascaris lumbricoides*)	10
diethylcarbamazine citrate	(**9**)	[*1642-54-2*]	Banocide, Filarex, Hetrazan, Notezine	white crystalline powder, mp 141–143°C	soluble in water and hot ethanol; insoluble in acetone	lymphatic filariases (*Wuchereria bancrofti, Loa loa,* and other filaria)	11
pyrantel pamoate	(**10**)	[*22204-24-6*]	Antiminth, Cobantil, Combantin, Helmex, Pyrantin	white or yellow or tan crystalline powder, melts and decomposes, mp ca 250°C	insoluble in water and nearly so in ethanol	pinworms (*Enterobius vermiculus*), roundworms (*Ascaris lumbricoides*), hookworms (*Ancylostoma duodenale, Necator americanus*)	12
mebendazole	(**11**)	[*31431-39-7*]	Meben, Totamin, Vermox, Vorme	crystals or yellow amorphous powder, mp 289°C	very slightly soluble in water and most organic solvents	whipworms (*Trichuris trichiura*), hookworms (*Ancylostoma duodenale, Nector americanus*), filariasis (*Mansonella perstans*), roundworm (*Ascaris lumbricoides*), Trichinosis (*Trichinella spiralis*)	13

Table 1. ***(Continued)***

Drug	Structure number	CAS Registry Number	Trade name	Physical properties	Solubility	Disease (organism)	Reference
thiabendazole	**(12)**	[*148-79-8*]	Foldan, Lombrstop, Mintezol, Minzolum, Nomoxiur, Thibenzole, Tiabenda, Triasox, TBZ	white crystals, mp ca 300°C	almost insoluble in water; slightly soluble in ethanol	hookworms (*Trichostrongylus sp.*, cutaneous larva migrans, *Angiostrongylus costaricensis*, and *Strongyloides stercoralis*)	14
ivermectin[a]	**(13)**	[d]	Eqvalan, Ivomec, Mectizan	off-white powder, mp 155–157°C	4 μg/mL water, insoluble in saturated hydrocarbons, very soluble in propylene glycol	filariasis (*Onchocerca volvulus*)	15

[a]Available from the CDC Drug Service.
[b]The dihydrochloride dihydrate [*69-05-6*].
[c]Formed from 3 mol piperazine (**8**) and 2 mol citric acid.
[d]See Figure 1.

Praziquantel is thought to alter the permeability of the worm's cell membrane to calcium, resulting in tetanic paralysis and the dislodging of schistosomes from their attachment sites. The drug also causes the disintegration of the schistosome tegument, which is followed by host phagocytosis (16). Adverse reactions include dizziness, headache, and gastrointestinal distress. Cure rates of approximately 85% have been achieved with a single treatment regimen. The principal producer of praziquantel in the United States is Miles Pharmaceutical of West Haven, Connecticut. Other manufactures include Bayer A.G. of Leverkusen, Germany and Chinoin of Budapest, Hungary.

Oxamniquine. This tetrahydroquinoline (**2**), $C_{14}H_{21}N_3O_3$, has been proven effective for the treatment of *S. mansoni* infections. It is clinically ineffective against *S. haematobium* and *S. japonicum* infections.

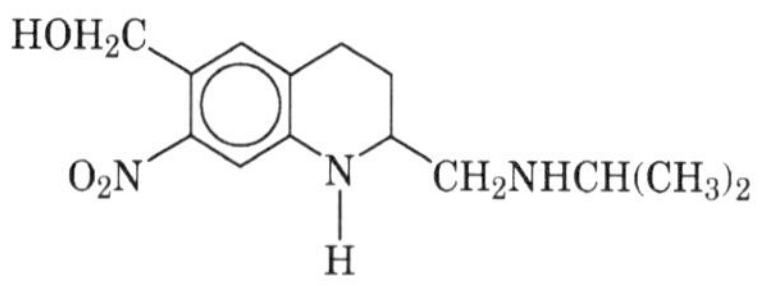

(**2**)

Its precise mode of action is not known, although it does exhibit anticholinergic properties (17). Oxamniquine is generally well tolerated, although dizziness, with or without drowsiness, does occur in at least one-third of the patients taking the drug. Allergic-type reactions, including fever, increased circulating immune complexes, pulmonary infiltrates, and pruritic skin rashes, may also occur. These allergic symptoms may be due to the death of the infecting parasites (18). A single dose is usually sufficient for strains of *S. mansoni* acquired in the Western Hemisphere; those acquired in East Africa may require longer periods of treatment. Oxamniquine is solely manufactured by Pfizer, Inc. in the United States, United Kingdom, and France.

Metrifonate. This organosphosphorus compound (**3**), $C_4H_8Cl_3PO_4$, was first used as an insecticide.

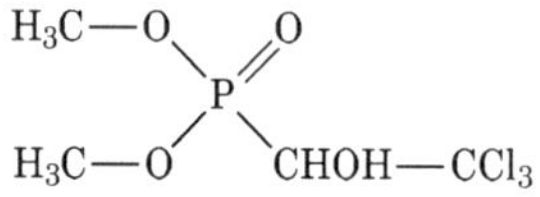

(**3**)

In the body, metrifonate converts to the active metabolite dichlorvos, (2,2-dichlorovinyl dimethyl phosphate), which is responsible for the inhibition of the enzyme acetylcholinesterase in the susceptible worm. This effect alone is unlikely to explain the antischistosomal properties of metrifonate (19). Clinically, metrifonate is effective only against infection caused by *S. haematobium*. Metrifonate is administered in three doses at 2-wk intervals (17). The drug is well tolerated. Side effects such as mild vertigo, nausea, and cramps are dose-related. This product is not available in the United States. The only manufacturer of metrifonate is Bayer A.G. of Leverkusen, Germany.

Lung and Liver Flukes. Paragonimus organisms (lung flukes) are about 1 cm long and normally live encapsulated within lung cysts. They primarily infect lung tissue, but may be found in other viscera or in the brain, producing tumorlike symptoms.

Fasciola organisms (liver flukes) are about 3 cm long. After extensive migration, young flukes invade and mature in the bile ducts of the liver, where heavy infections may result in fibrosis of the bile duct wall. Flukes migrating through the peritoneal cavity may become lodged in other locations, causing intense tissue reactions.

Bithionol. This compound (**4**), $C_{12}H_6Cl_4O_2S$, can be used to treat persons infected with either *Paragonimus* or *Fasciola* organisms. It is a phenolic compound similar in structure to hexachlorophene.

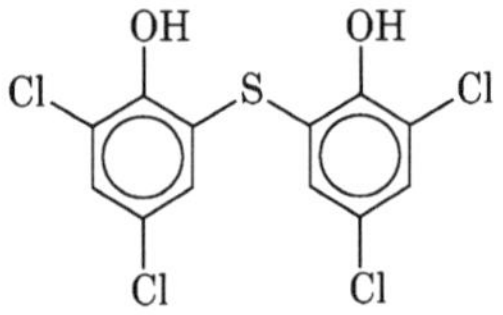

(**4**)

Bithionol and hexachlorophene have been used as anthelmintics in veterinary medicine. In the past, because of its antibacterial activity, bithionol was incorporated into at least 20 medicated skin cleansers manufactured in the United States; however, this agent proved to cause skin irritation, and its use as an antibacterial cleanser was discontinued.

Bithionol interferes with the neuromuscular physiology of helminths, impairs egg formation, and may cause defects in the protective cuticle covering the worm. At the biochemical level, the oxidative phosphorylation of the worm is inhibited.

Bithionol is an iron chelator and therefore may inactivate iron-containing enzyme systems (20). A 10-d oral regimen of bithionol may resolve the pathology ie, abnormalities of lung disease in about three months, but improvement in cerebral paragonimiasis and *Fasciola hepatica* (sheep liver fluke) infections are variable. Adverse effects include abdominal pain, diarrhea, vomiting, rashes, and photosensitivity. The drug is available in the United States through the CDC Drug Service of Atlanta, Georgia. The only manufacturer of bithionol is Tanabe Seiyaku of Osaka, Japan.

Praziquantel (**1**) (previously mentioned for the treatment of schistosomiasis) is also effective against all *paragonimus* species.

Chinese Liver Fluke. The adult worm of the Chinese liver fluke (*Clonorchis sinensis*) can grow to be 2 cm long. Worms infect the biliary tree where they cause local inflammation, diarrhea, and hepatomegaly in the acute infection. Progressive biliary obstruction and cirrhosis can occur in the more advanced disease state. The presence of 20–200 worms is common, but they may number over 20,000. Infection is the consequence of eating raw fish that contain viable parasites. Untreated worms can live for up to 30 years. Treatment is with praziquantil (**1**).

Intestinal Fluke. The intestinal fluke, *Fasciolopsis buski,* which can reach a length of 8 cm, lives attached to the wall of the small intestine. Rarely, untreated heavy infections may cause ulceration, toxemia, and death in children. Praziquantel (**1**) or tetrachloroethylene (**5**) are both effective in the treatment of *Fasciolopsis* infection.

Tetrachloroethylene. This chlorocarbon (**5**), $Cl_2C{=}CCl_2$, has been used as a solvent in dry cleaning and a metal degreaser as well as an anthelmintic. Tetrachloroethylene, C_2Cl_4, has replaced carbon tetrachloride, CCl_4, which is very hepatotoxic, in the treatment of intestinal flukes. Investigators do not agree on the mechanism of action of tetrachloroethylene or on its basic pharmacology. Some investigators believe that tetrachloroethylene dissolves the cell lipids in the helminthic neuromuscular system and interferes with neuromuscular function. In the treatment of flukes, a single oral dose is given on an empty stomach preceded by a day of low fat diet. A 24-h period of low fat intake is necessary because fat in the gastrointestinal tract increases drug absorption, thereby increasing its toxicity. This drug is primarily used in veterinary medicine (21).

Treatment of Cestode Infections

Fish tapeworm (*Diphyllobothrium latum*) grows to a length of 10 m with 4000 proglottides, or segments, each containing reproductive organs. The proglottides tend to remain attached to the intestinal wall and discharge eggs, which are passed into the feces. Besides ingesting host nutrients, this worm avidly sequesters vitamin B_{12} and folic acid. The closer to the stomach that the intestinal fluke is attached, the more vitamin the worm ingests. Depletion of host B_{12} supplies leads to the development of pernicious anemia, with associated neurological symptoms. There is usually only one worm in a patient and it can survive for 10 years.

Beef tapeworm (*Taenia saginata*) is found worldwide in people who eat rare beef and is most common in developing countries. Worms are often 5–10 m long with about 1000 proglottides. Usually a patient carries only one worm. Infected persons are frequently asymptomatic, but intestinal disturbances can occur. Detached sections of worm 0.5 cm × 2 cm may creep out of the end of the digestive tract.

Both the adult and the larval cysticerci (bladderworm) of *Taenia solium* (pork tapeworm) are able to live in humans; the parasite is found sporadically in uncooked pork. In the stomach, the larva is digested out of the pork flesh; it then grows and attaches to the wall of the small intestine. Maturity is reached in 5–12 weeks. The adult is 5 m long, and untreated adult worms may survive for 25 years.

Dwarf tapeworm (*Hymenolepis nana*) is the only human tapeworm that does not utilize an intermediate host. Infection is transmitted directly from person to person. *H. nana* is only 2–4 cm long, it is of universal distribution in mice and humans in temperate zones, where children, especially those in institutions, are most frequently infected. Although an infected person is often symptomless, this tapeworm can cause abdominal discomfort and diarrhea if infection is heavy.

Niclosamide. This drug (**6**) is a halogenated salicylanilide derivative, $C_{13}H_8Cl_2N_2O_4$.

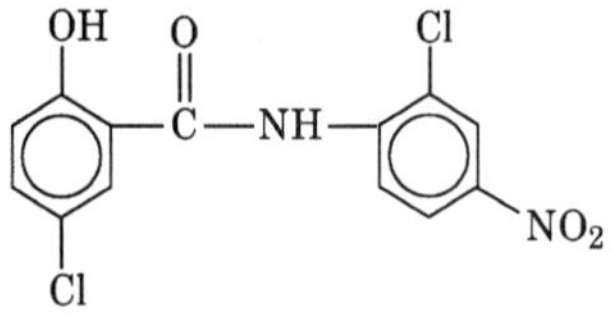

(**6**)

The modes of action for niclosamide are interference with respiration and blockade of glucose uptake. It uncouples oxidative phosphorylation in both mammalian and taenioid mitochondria (22,23), inhibiting the anaerobic incorporation of inorganic phosphate into adenosine triphosphate (ATP). Tapeworms are very sensitive to niclosamide because they depend on the anaerobic metabolism of carbohydrates as their major source of energy. Niclosamide has selective toxicity for the parasites as compared with the host because little niclosamide is absorbed from the gastrointestinal tract. Adverse effects are uncommon, except for occasional gastrointestinal upset.

Niclosamide causes the tapeworm head to disengage from the intestinal wall of the host and the body wall of the parasite to disintegrate. The first dose of niclosamide disposes of the worms in the intestinal cavity; however, any worms in developing stages in the tissues of the intestinal wall are unharmed. Therefore, dosage must be repeated in 5–7 days to destroy the worms that will finish their development and reenter the intestinal lumen.

Niclosamide given in a single oral dose is effective for the treatment of tapeworm infections. A large percentage of patients are cured, and adjunct therapy is not required when used against beef tapeworm (*T. saginata*), the dwarf tapeworm (*H. nana*), or the fish tapeworm of humans (*D. latum*) and most other human intestinal cestode infections. It is also effective in the treatment of pork tapeworm (*T. solium*), but not against infection with larval *T. solium* (cysticercosis) for which praziquantel (**1**) is preferred. Niclosamide is available in the United States and is manufactured by Miles Pharmaceutical of West Haven, Connecticut. Other manufacturers include Bayer A.G. of Leverkusen, Germany and Bellon Laboratories of Neuilly-sur-Seine, Cedex, France.

Quinacrine. Quinacrine (**7**), $C_{23}H_{30}ClN_3O$, is an acridine derivative. It is used in the form of the dihydrochloride dihydrate, $C_{23}H_{32}Cl_3N_3O{\cdot}2H_2O$, [*69-05-6*].

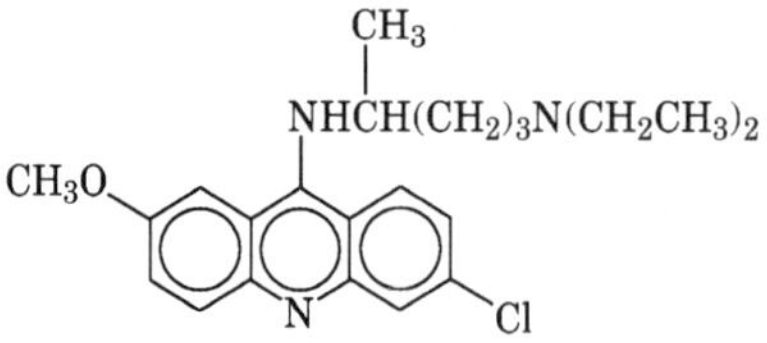

(**7**)

This drug was used prior to the availability of niclosamide and is considered less satisfactory for the treatment of tapeworms then niclosamide (24). It causes more side effects and produces severe nausea. Quinacrine, however, is preferred by some clinicians for the treatment of *Taenia solium* infection because, unlike niclosamide, it expels the worms intact, thus reducing the theoretical risk of cysticercosis (25).

Quinacrine concentrates in the scolex of the parasite and causes the muscles needed for holding onto the intestinal wall to relax. The worms are stained yellow and pass from the body, still alive. Quinacrine can intercalate with DNA and inhibit nucleic acid synthesis. It creates fluorescent bands in deoxyadenylate–deoxythmidylate-rich regions of DNA and has been used as a stain in the study of human genetics.

The drug is readily absorbed from the gastrointestinal tract and may persist in human tissues for two months after administration. Quinacrine is given as a single dose for the treatment of tapeworms; it is manufactured by Winthrop-Breon of New York, New York.

Paromomycin. Paromomycin [*7542-37-2*] is a broad-spectrum antibiotic that has been used in the treatment of *H. nana* and *T. saginata,* and amebiasis, and also as an antibacterial agent in cases of diarrhea and dysentery (see ANTIBIOTICS, OLIGOSACCHARIDES).

Treatment of Nematode Infections

Female roundworms of *Ascaris lumbricoides* are about 30 cm × 0.5 cm. The adults live and feed in the small intestine where they can cause abdominal discomfort or pain. Frequently, light infections are symptomless. The worm develops through four larval stages, with a molt between each successive stage. The eggs of the adult are deposited in the feces. Eggs do not hatch in the soil or enter the body through the cutaneous route. Infective eggs are swallowed and the larvae hatch in the stomach. Hatched larvae actively migrate into the lungs. Larvae continue to develop in the lungs, then migrate through the respiratory passages to the throat and are swallowed. In the stomach they molt and grow to the mature adult form. Untreated adult worms may live for a year and a half.

The pinworm (*Enterobius vermicularis*) is distributed worldwide in temperate zones. In cities, it is frequently a disease of households or institutions. Female worms are about 1 cm long. While growing to maturity, they live in the gut lumen, attaching by their mouths to the intestinal mucosa. Itching in the perianal region occurs when the mature female migrates out the anus to discharge her eggs. Usually an intestinal population is less than 100 worms; however, more than 5000 have been recovered from a single patient. The eggs are resistant to drying and may get into household dust and be inhaled or swallowed by household members, causing light infections with few worms in the adult relatives of infected children.

Whipworm (*Trichuris trichiura*) adult females are 5 cm long. These worms thread their entire body into the epithelium of the colon, where they feed on tissue juice and small amounts of blood. Infections of several hundred worms may cause irritation and inflammation of the mucosa, with abdominal pain, diarrhea, and gas. Eggs are discharged and passed into the feces. Infections result from the swallowing of eggs that are obtained directly from contaminated soil. Untreated adult worms live for years.

The two most common hookworms in humans are *Ancylostoma duodenale* and *Necator americanus*. *A. duodenale* adult worms are 8–13 mm in length. The filariform larva found in moist soils may be either ingested or penetrate the skin of its host. It is then carried through the circulatory system to the lungs and migrates up the respiratory tree into the digestive tract. The worms feed on intestinal tissue and blood. Some worms may persist in humans as long as nine years. Infestations cause cutaneous reactions, pulmonary lesions, intestinal ulcerations, and anemia.

American hookworm (*N. americanus*) adult worms, 7–11 mm in length, may live up to 15 years. The life cycle of the *N. americanus* is similar to *A. duodenale* except that *N. americanus* requires a pulmonary migration with development in the lungs before it can develop to the adult worm in the intestine.

Piperazine Salts. Piperazine citrate is highly effective against both *A. lumbricoides* (large roundworm) and *E. vermicularis* (pinworm). A large number of substituted piperazine derivatives exhibit anthelmintic activity; however, only diethylcarbamazine (**9**) is used clinically (26).

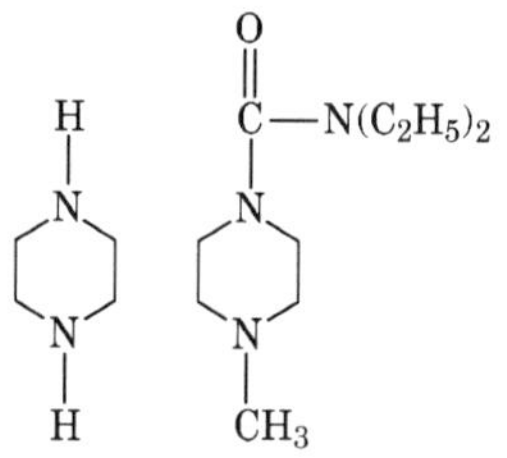

(**8**) (**9**)

Piperazine is taken orally by nonfasting subjects. For large roundworms, one dose a day on two consecutive days reportedly cures approximately 90% of cases treated (27). This drug can be used for pinworms with a 7-d course of treatment.

Piperazine causes flaccid paralysis of *Ascaris* by blocking the worms' ability to respond to acetylcholine, thus dislodging the worms from their position in the digestive tract. The worms are passed through the digestive tract still alive. The drug is well absorbed from the human intestine, and the reason for its selective toxicity is still uncertain.

Occasional side effects from piperazine include abdominal cramps, nausea, vomiting, and diarrhea. Rarely, patients may experience headache, vertigo, and tremors, or they may feel weak and have difficulty concentrating. Red patches can appear on the skin, sometimes with the flat, elevated, itching welts of urticaria. Piperazine is manufactured by Burroughs Wellcome Company in Research Triangle Park, North Carolina and Wallace Laboratories of Cranbury, New Jersey. Other manufacturers include Laboratories Clin Midy of Paris, France, Regent Laboratories Ltd. of London England, and Lennon Ltd. of Saxonwold, South Africa.

Pyrantel Pamoate. This drug (**10**), $C_{34}H_{30}N_2O_6S$, cures pinworm infections and is close to 100% effective against *Ascaris* when given in a single dose (28,29). Pyrantel pamoate is also a principal drug for the treatment of hookworm infections; thus it is useful in patients with mixed worm infections.

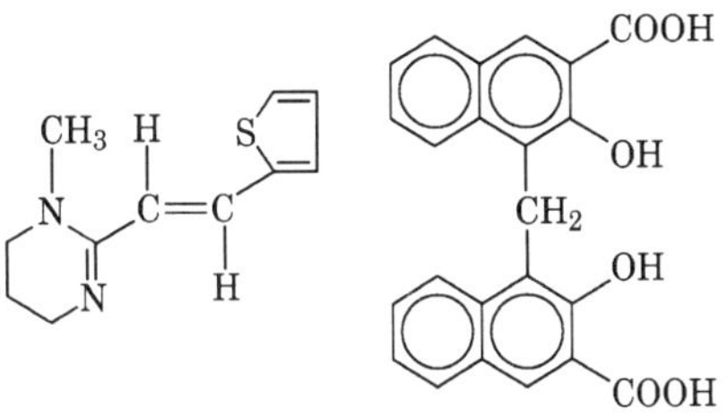

(10)

For hookworms, treatment is continued for three days. In roundworms, muscle tissue is stimulated continuously, resulting in spastic paralysis. The drug was introduced in veterinary medicine in 1966 and then applied to clinical medicine three years later.

Since about 85% of the administered dose is passed unchanged in the feces of the patient, selective toxicity of the drug can be attributed primarily to poor absorption. Side effects include abdominal pain, nausea, vomiting, diarrhea, loss of appetite, headaches, and vertigo or drowsiness. Skin rashes can also develop. Pyrantel pamoate is produced by Pfizer, Inc., New York, New York.

Mebendazole. In therapeutic dosages administered orally for three or four consecutive days, mebendazole **(11)** is highly effective against whipworm (*Trichuris trichiura*). In addition, a single dose usually cures pinworm infections. Mebendazole is also effective against trichinosis (*Trichinella spiralis*), hookworms (*Ancylostoma duodenale* and *Necator americanus*), and *Ascaris*, partially effective against threadworms (*Strongyloides*), and taeniid tapeworms. This benzimidazole derivative, $C_{16}H_{13}N_3O_2$, is a broad-spectrum anthelmintic (30) and is used in veterinary medicine as well as in clinical practice. It belongs to the same series of drugs as thiabendazole, $C_{10}H_7N_3S$ **(12)**.

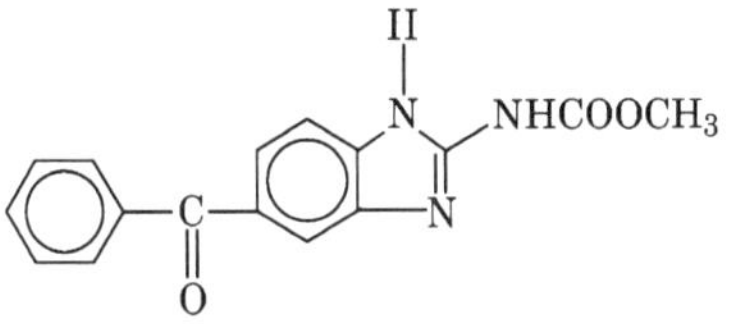

(11)

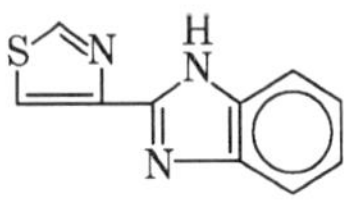

(12)

These drugs are closely related to fungicides such as benomyl [*17804-35-2*] **(13)**, $C_{14}H_{18}N_4O_3$.

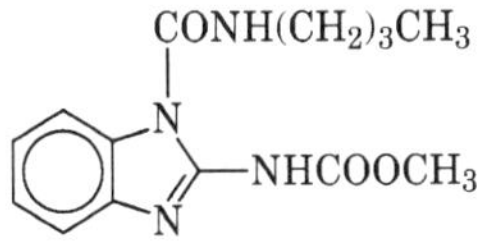

(13)

Mebendazole interferes with the glucose metabolism of helminths, irreversibly inhibiting glucose uptake. When the glycogen stores in the worms are depleted, ATP supplies fail. The worms are slowly immobilized, die, and are excreted over a period of three days. In developing *Trichuris* eggs, larvae fail to develop normally. Selective toxicity of mebendazole may be due to its antimicrotubule action on nematode intestinal cells with no effect on mammalian cells. Furthermore, relatively little drug is absorbed from the human gastrointestinal tract. Adverse effects in patients include abdominal pain and diarrhea. Mebendazole is manufactured by Janssen Pharmaceutica, Inc., Piscataway, New Jersey. Other manufacturers include Esteve Drug of Barcelona, Spain and Cadila Laboratories of Maninagar, Ahmedabad, India.

Treatment of Tissue Roundworm (Nematode) Infections

Many parasitic worms cause systemic infections outside the gastrointestinal tract. These include: *Strongyloides stercoralis* (threadworm), *Trichinella spiralis, Dracunculus medinensis,* and the several species of nematodes that cause filariasis (*Mansonella perstans* and *Onchocerca volvulus*).

S. stercoralis mature adult worms are between 2 and 3 mm long. Threadworms have two separate developmental stages. In the first stage, the worm has a complete free-living nonparasitic life cycle, and in the second stage, a filariform larvae penetrates the skin of its host and moves into cutaneous blood vessels. The larvae are then carried to the lungs. In the lungs, the migrating larvae break out of the pulmonary capillaries and migrate through the respiratory passages to the throat where they are swallowed. Once they are inside the intestine, they become imbedded in intestinal wall.

Trichinosis is the condition caused by the adult worm of *T. spiralis* which is between 1 and 2 mm long. The cysts are found in contaminated meat, primarily pork. When ingested, cysts release immature larvae, which invade the intestinal lining and develop into adult worms. Mature worms mate in the submucosa and deposit mobile larvae, which migrate into intestinal lymphatics. They are then disseminated throughout the body and become encapsulated within striated muscle fibers. Adult worms live about one month. Mebendazole (**10**) is highly effective against *T. spiralis* infections. Treatment with mebendazole for trichinosis is for 13 days.

Thiabendazole. This benzimidazole derivative (**12**) is an effective oral drug used in the treatment of intestinal roundworms and selected tissue parasites. Infections with *S. stercoralis* are commonly treated with thiabendazole for two days. Disseminated strongyloidiasis is treated for at least five days. Satisfactory results have been reported when used for *T. spiralis* infections; however, its effectiveness on encapsulated muscle larvae has not been clinically demonstrated. Extensively used in veterinary medicine, thiabendazole was the first broad-spectrum benzimidazole anthelmintic. It is widely used in the United States. The mechanism of action is not clearly understood, but it has been shown that thiabendazole inhibits the enzyme fumarate reductase, which is found specifically in the mitochondria of helminths (31).

One ppm of thiabendazole prevents *Ascaris* eggs from maturing. In therapeutic topical concentrations, it is used to prevent the development of dog and cat hookworm larvae that have invaded human skin, causing cutaneous larva migrans. In laboratory animals, thiabendazole has antipyretic, analgesic, and antiinflammatory properties, as well as anthelmintic properties. These properties are probably also exhibited in humans and may account for the symptomatic improvement in persons with early systemic trichinosis.

The adverse effects include digestive disturbances, neurological symptoms, and manifestations of allergic responses. As many as half of the patients taking it are incapacitated by some of these adverse reactions for several hours. Whether these symptoms are caused by hypersensitivity to the drug, the parasite, or by a manifestation of the disease is not known. Overall, effects are dose-related and transient.

Guinea Worm. Thiabendazole (**12**) and metronidazole (**14**) [*443-48-1*], $C_6H_9N_3O_3$, the broad-spectrum antiprotozoal agent

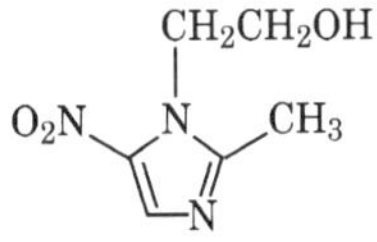

(**14**)

are used in the treatment of the guinea worm (*Dracunculus medinensis*) (see also ANTIPARASITIC AGENTS, ANTIPROTOZOALS). These stringlike worms, about 1 m long but less than 2 mm in diameter, can be seen and felt below the skin surface. The head of the worm is usually exposed in an ulcer on the foot. The far end of the worm is hooked, so that although it is possible to grasp the front end and pull, the worm does not slide out easily and is apt to break, causing infectious, allergic, and toxic consequences as the remainder of the worm under the skin dies and deteriorates. After drug therapy, the worm either exits or it can be pulled easily without danger of breakage. Although metronidazole and thiabendazole have no direct effect on guinea worms, their effectiveness against the guinea worm may be due to their antiinflammatory effects. If left untreated, within a month the worm may come out naturally, withdraw from the opening, and be resorbed or become calcified in the body.

Filarias. Diethylcarbamazine (**9**) (orally administered) has been successful for decades in the treatment of filariasis, eg, elephantiasis (32). It kills microfilariae in the blood of the filarial worms *Wuchereria bancrofti, Brugia malayi,* and *Loa loa*.

Diethylcarbamazine has limited antimicrofilarial activity against *Onchocerca volvulus*. Adults of *W. bancrofti,* the filarial worm causing elephantiasis, coil in the lymph system. Here females can attain a length of 10 cm. Over the years, tissue reactions result in obstruction to lymph return. Lymph nodes, lymph vessels, and the spleen become enlarged. The condition of elephantiasis is a late and unusual complication of filariasis, where the lower extremities of the body become edematous, enlarge, and over a period of time harden with a rough nodular skin.

Untreated adult worms live for 5–10 years. *Loa loa* females are about 6 cm long and migrate constantly through connective tissues. Calabar swelling occurs as a result of local responses to worms. The swelling lasts two or three days and then subsides. If a worm migrates into the anterior chamber of the eye, the patient may be able to see it. Sometimes one eye is swollen shut when a worm is in the vicinity. Untreated worms will live as long as 10 years.

Microfilariae in the tissues of the lung (ie, eosinophilic lung or tropical eosinophilia) produce symptoms of chest pain, cough, and asthma, which are often relieved by the administration of diethylcarbamazine.

Diethylcarbamazine. This derivative of piperazine (**9**), causes the microfilariae to become susceptible to phagocytosis by the macrophages in the reticuloendothelial system. Its pharmacological effects enhance the body's immunologic response. The mechanism of the filaricidal action on adult *Wuchereria* and *Loa* is unknown. The selective toxicity of diethylcarbamazine is as yet unexplained. It is well absorbed and widely distributed throughout the body and does not accumulate, even after repeated doses. The adverse effects of nausea, fever, headache, and dizziness may occur during diethylcarbamazine therapy. The simultaneous administration of corticosteroids and antihistamines is intended to reduce the occurrence of rash, itching, edema, and other severe allergic reactions to the disintegration of the worms. Diethylcarbamazine is used in veterinary and clinical medicine. It is manufactured by Lederle Laboratories of Pearl River, New York. Other manufacturers include the Wellcome Foundation Ltd. in Cheshire, England, Cidan of Castellon, Spain, and Specia S.A. in Paris, France.

Mansonella perstans adult worms live encysted in various peritoneal tissues (eg, pericardium, pleura, and peritoneum). *M. perstans* microfilariae are not affected by diethylcarbamazine but mebendazole has been successful in treating *M. perstans* filariasis (33).

Ivermectin. This drug (**15**) is a semisynthetic derivative of one of the avermectins, which are macrocyclic lactones produced by the actinomycete *Streptomyces avermitilis* (34) (Fig. 1). This antiparasitic agent has a broad spectrum of activity against nematode worms and insects in animals and is widely used in veterinary medicine (see ANTIPARASITIC AGENTS, AVERMECTINS). The drug acts on susceptible organisms by potentiating the release of γ-aminobutyric acid [*56-12-2*] (GABA) and binding the GABA at postsynaptic sites on the neu-

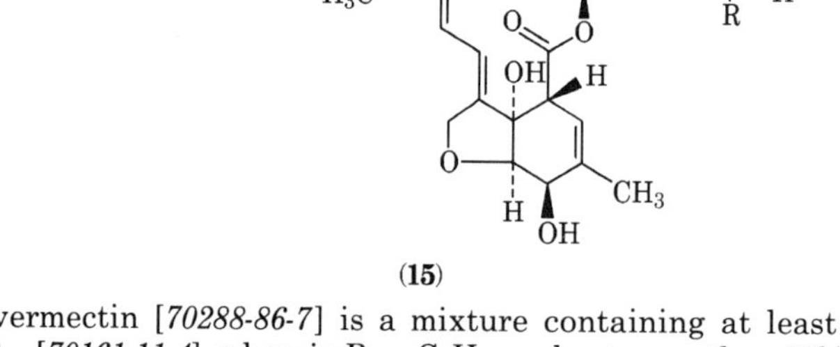

(**15**)

Fig. 1. Ivermectin [*70288-86-7*] is a mixture containing at least 80% 22,23-dihydroavermectin B_{1a} [*70161-11-4*], wherein R = C_2H_5, and not more than 20% B_{1b} [*70209-81-3*], wherein R = CH_3.

romuscular junction, thus paralyzing the nematode. Ivermectin acts slowly and has a long duration of action. Ivermectin has no effect on adult filariid nematodes, but it has proved to be superior to diethylcarbamazine as a microfilaricidal agent in the treatment of onchocerciasis. Neither drug completely cures the infection. More recent medical studies have also shown that ivermectin has a high activity against various intestinal nematodes (eg, *A. lumbricoides, T. trichiura,* and *E. vermicularis*) (35).

Systemic reactions are less severe than with diethylcarbamazine. The most commonly seen reactions are fever, rash, and lymph-node pain or swelling. Suppressive ivermectin therapy consists of a single oral dose every 6–18 months. The required duration of suppressive therapy is unknown, probably at least three years (36). Ivermectin is available from the CDC Drug Service on request. It is manufactured by Merck Sharp and Dohme in the United States and England.

BIBLIOGRAPHY

"Chemotherapeutics, Anthelmintic" in *ECT* 3rd ed., Vol. 5, pp. 451–468, by J. W. Ingalls, John Wiley & Sons, Inc.

1. U.S. Pat. 4,001,411 (Jan. 4, 1977), P. Andrews, J. Seubert, and H. Thomas (to Merck & Co.).
2. Ger. Pat. 2,362,539 (Dec. 17, 1973), P. Andrews, J. Seubert, and H. Thomas (to Merck & Co.).
3. U.S. Pat. 3,903,283 (Sept. 2, 1975), H. Richards (to Pfizer, Inc.).
4. U.S. Pat. 2,701,225 (Feb. 1, 1955), A. Lorenz (to Bayer AG).
5. Ger. Pat. 583,055 (Aug. 28, 1933), F. Muth (to I.G. Farbenind AG).
6. U.S. Pat. 2,849,494 (Aug. 26, 1958), R. H. Cooper and K. Goldberg (to Monsanto).
7. U.S. Pat. 3,040,109 (June 9, 1959), R. E. Feathers and R. H. Rogerson (to Pittsburgh Plate Glass).
8. U.S. Pat. 3,079,297 (Feb. 26, 1963), E. Schranfstraller and R. Gonnert (to Bayer AG).
9. U.S. Pat. 2,113,357 (Apr. 15, 1939), F. Mietzsch and H. Mauss (to Winthrop Chemical Co.).
10. U.S. Pat. 2,901,482 (Aug. 25, 1959), G. F. Mackenzie and K. L. Turbin (to The Dow Chemical Company).
11. U.S. Pat. 2,467,893; 2,467,894; 2,467,895 (Apr. 19, 1949), S. Kushner and L. Brancone (to American Cyanamid Co.).
12. South Afr. Pat. 6,800,516 (June 27, 1968), R. V. Kasubrick and J. W. McFarland (to Chas. Pfizer, Inc.).
13. U.S. Pat. 3,657,267 (Feb. 18, 1971), J. C. H. VanGelder (to Janssen).
14. U.S. Pat. 3,017,415 (June 16, 1962), L. Sarett (to Merck & Co.).
15. U.S. Pat. 4,199,569 (Aug. 4, 1979), J. Chabala and M. Fisher (to Merck & Co.).
16. P. Andrews and co-workers, *Med. Res. Rev.* **3,** 147 (1983).
17. *Control of Schistosomiasis,* Technical report Series, WHO No. 728, World Health Organization, Washington, D.C., 1985.
18. N. Katz, F. Zicker, and J. P. Pereira, *Rev. Inst. Med. Trop. Sao Paulo* **18,** 371–377 (1976).
19. A. Bloom, *Acta. Pharmacol. Toxicol.* **49,** 109–113 (1981).
20. *Informational Material for Physicians, Bithionol,* Centers for Disease Control, Atlanta, Ga., 1986, pp. 1–18.
21. M. Abramovicz, ed., *Med. Lett.* **30,** 18 (1988).
22. J. Putter and Z. Parasitenk, **34,** 23 (1970).

23. J. Putter and P. Andrews, *Conference of the German Society for Parasitology,* Wuppertal, Apr. 9–11, 1970.
24. F. Richards and P. M. Schantz, *Lancet* **1,** 1264 (1985).
25. Ref. 21, p. 15.
26. O. D. Standen in R. J. Schnitzer and F. Harding, eds., *Experimental Chemotherapy,* Vol. 1, Academic Press, Inc., New York, 1963, pp. 702–892.
27. R. H. R. White and J. W. Scopes, *Lancet* **1,** 256–258 (1960).
28. N. E. Pitts and J. R. Migliardi, *Clin. Pediatr.* **13,** 87–94 (1974).
29. T. S. Bumbalo, D. J. Fugazzotto, and J. V. Wyczalek, *Am. J. Trop. Med. Hyg.* **18,** 50–52 (1969).
30. H. VandenBossche, *Biochem. Pharmacol.* **29,** 1981–1990 (1980).
31. H. M. Robinson and C. S. Samorodin, *Arch. Dermatol.* **112,** 1757–1760 (1976).
32. F. Hawking, *Adv. Pharmacol. Chemother.* **16,** 129–194 (1979).
33. M. Wahlgren and J. Frolov, *Trans. R. Soc. Trop. Med. Hyg.* **77,** 422 (1983).
34. W. C. Campbell and co-workers, *Science* **212,** 823–828 (1983).
35. C. Naquira and co-workers, *Am. J. Trop. Med. Hyg.* **40,** 304–309 (1989).
36. *Informational Material for Physicians, Ivermectin,* Centers for Disease Control, Atlanta, Ga., 1986, pp. 1–23.

General References

P. C. Beaver, R. C. Jung, and E. W. Cupp, *Clinical Parasitology,* 9th ed., Lea and Febiger, Philadelphia, Pa., 1984, 825 pp.

P. E. C. Manson-Bahr and D. R. Bell, *Mansons's Tropical Diseases,* 19th ed., Bailliere Tindall, Philadelphia, Pa., 1987, 1557 pp.

A. S. Benson, ed., *Control of Communicable Diseases in Man,* 14th ed., American Public Health Association, Washington, D.C., 1985, 485 pp.

E. Braunwald and co-workers, eds., *Harrison's Principles of Internal Medicine,* 11th ed., McGraw-Hill, New York, 1987, pp. 805–829.

W. C. Campbell and R. S. Rew, eds., *Chemotherapy of Parasitic Diseases,* Plenum Press, New York, 1986, 655 pp.

T. Cheng, *General Parasitology,* Academic Press, New York, 1973, 355 pp.

L. L. Gustafsson, B. Beermann, and Y. A. Abdi, *Handbook of Drugs for Tropical Parasitic Infections,* Taylor and Francis, Philadelphia, Pa., 1987, 151 pp.

Health Information for International Travel 1989, U.S. Department of Health and Human Services, Publication No. (CDC) 89-8280, 1989.

K. F. Lampe and co-workers, *AMA Drug Evaluations,* W. B. Saunders, Philadelphia, Pa., pp. 1565–1613.

E. K. Markell and M. Voge, *Medical Parasitology,* 5th ed., W. B. Saunders, Philadelphia, Pa., 1984, 374 pp.

J. F. Reynolds and co-workers, *Martindale, The Extra Pharmacopeia,* 29th ed., The Pharmaceutical Press, London, 1989, pp. 47–69, 505–521.

G. D. Schmidt and L. Roberts, *Foundations of Parasitology,* 3rd. ed., C. V. Mosby, St. Louis, Mo., 1985.

E. A. Swinyard in A. R. Gennaro, ed., *Remington's Pharmaceutical Sciences,* 17th ed., Mack Publishing Company, Easton, Pa., 1985, pp. 1234–1240.

M. Verderame and J. Mackiewicz in M. Verderame, ed., *CRC Handbook of Chemotherapeutic Agents,* Vol. 2, CRC Press, Boca Raton, Fla., 1986, pp. 95–178.

L. T. Webster in L. S. Goodman, A. Gilman, T. W. Rall, and F. Murad, eds., *The Pharmacological Basis of Therapeutics,* 7th ed., Macmillan, New York, 1985, pp. 1004–1028.

JOHN BECHER
CDC Drug Service

ANTIMYCOTICS

Since the 1960s, there has been such a rapid evolution in the pharmacotherapy of mycoses that three phases of the development of antimycotic agents can be distinguished: (*1*) compounds that existed before griseofulvin (1939); (*2*) compounds presently available on the market, including itraconazole (1987); (*3*) new antimycotics that are being studied now and will be prescribed tomorrow. Both milestones (griseofulvin and itraconazole) are intended for oral administration. Topical treatment of superficial dermatomycoses (trichophytosis, favus, and microsporosis) has been complemented by systemic oral treatment.

The world market for antimycotics in 1980 was valued at a total of $350 million. About 10% of this came from the sale of systemic (intravenous or oral) antifungals. In 1990, the antifungal world market was estimated at $1560 million of which almost $280 million was accounted for by systemic antifungals. This important increase in market value has induced more interest in research activities in the field of antifungals leading to more and more costly new treatment modalities. The five most prescribed antifungals today are miconazole, clotrimazole, ketoconazole, nystatin, and econazole. As this list indicates, among antifungal drugs the azoles are most significant; roughly 70% of all antifungals used today belong to this chemical category.

Though highly pathogenic, fungi represent only a minority; the importance of opportunistic infections is increasing from year to year. Human resistance against fungal infections is undermined by a number of factors and both exogenous and endogenous factors are capable of increasing the pathogenicity of certain fungi. Fungal growth may be stimulated when the host is treated with antibiotics, oral contraceptives, or cytostatics. Exogenous factors such as protheses, catheters, and valves may give rise to an increased incidence of mycoses. In immunocompromised patients and diabetics, saprophytic fungi may become pathogenic due to disturbances in the internal environment. These patients may present chronic mucocutaneous candidosis (CMC) due to *Candida albicans*. *Aspergillus fumigatus* and *A. niger* may become moderately to highly pathogenic in the presence of previously mentioned predisposing factors.

To mycologists, each fungal infection has something specific, either in its symptomatology or its etiology. However, this is less obvious to practitioners. The incidence and the severity of the pathology are sometimes underestimated. Mycoses may be classified as follows:

Superficial	Deep	
dermatophytosis	aspergillosis	histoplasmosis
trichophytosis	blastomycosis	maduromycosis
microsporosis	candidosis	paracoccidioidomycosis
epidermophytosis	chromomycosis	sporotrichosis
pityriasis	coccidioidomycosis	
candidosis	cryptococcosis	

The physician must decide whether to opt for systemic or topical treatment and which arguments will convince the patient to use one or both forms for the minimum period recommended. It is typical for mycoses that the period of treatment generally exceeds one week and may even last a few months. There is still a strong tendency to treat the organ or the part of the organ affected by the pathogenic microorganism. This is justifiable in the case of localized unifocal mycoses of the skin caused by dermatophytes.

Oral treatment offers the advantage of bringing all the lesions at all sites under control, in addition to the absence of unpleasant cosmetic effects. In certain cases, it may be preferable to use oral treatment for *C. albicans* vaginitis and for extensive and persistent pityriasis versicolor, a skin disorder caused by *Pityrosporum orbiculare*. In the case of onychomycosis, a combination treatment, topical plus systemic, is required. It is preferable to use oral treatment for deep and systemic mycoses, though intravenous or intrathecal treatment is sometimes required.

The final choice of a suitable antimycotic and the route of administration are determined by many factors: safety of the antimycotic, easy administration, broad-spectrum activity, and rapid clinical improvement associated with mycological cure.

Topical Antimycotics

Older Compounds. Some antimycotics have been used for a long time; Whitfield's ointment is a typical example (1907). The ointment usually contains 6% benzoic acid [*65-85-0*], $C_7H_6O_2$, and 3% salicylic acid [*69-72-7*], $C_7H_6O_3$. The action is attributed to the keratolytic effect of the salicylic acid and the direct effect of benzoic acid on the fungus.

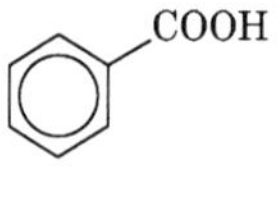

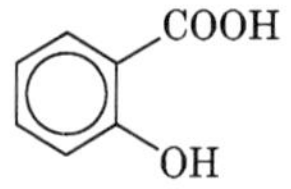

The main advantage of this ointment is its low price. This combination has now been replaced by more active modern antimycotics. Long ago, the antimycotic effect of aliphatic carboxylic acids with an increasing number of C-atoms was discovered. The optimum is 11 C-atoms and $CH_2{=}CH(CH_2)_8COOH$, undecylenic acid [*112-38-9*], $C_{11}H_{20}O_2$, is used in several ointments.

Other drugs are a combination of a keratolytic agent (mesulfen [*135-58-0*], $C_{14}H_{12}S_2$) and the well-known antimycotic zinc undecylenate [*557-08-4*], $C_{22}H_{38}O_4Zn$.

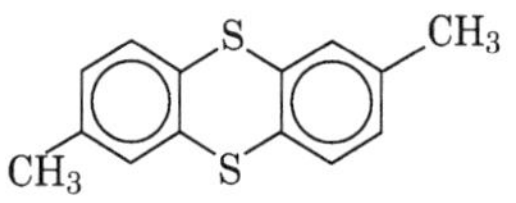

mesulfen

This cream also contains zinc naphthenate, terpineol, chlorcresol, and methyl salicylate [*119-36-8*]. The latter has a deeply penetrative effect. It combines antimycotic and antibacterial properties and is used to treat swimmer's eczema. The cream must not be used on open skin areas infected by bacteria. Some foot powders are a combination of zinc undecylenate, boric acid, hexachlorophene, and salicylic acid.

Tolnaftate [*2398-96-1*], $C_{19}H_{17}NOS$, which is active against dermatophytes, is the active component in another antifungal powder; it also contains cetylpyridinium chloride and talcum venetum.

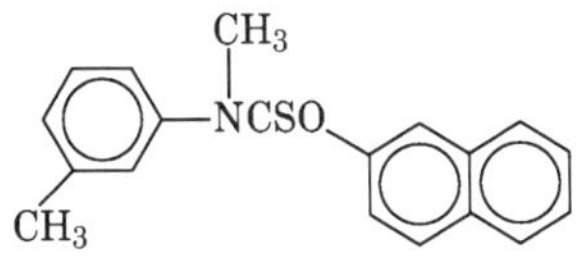

tolnaftate

In addition to tolnaftate and cetylpyridinium chloride, the ointment also contains poly(ethylene glycol). This preparation is not active against *C. albicans.*

ANTIMYCOTIC ANTIBIOTICS

Nystatin. This compound (**1**) belongs to the group of polyene antimycotic antibiotics (Table 1). Since the early 1960s, approximately 200 polyenes have been isolated from different *Streptomyces* strains.

Table 1. Polyene Antimycotic Antibiotics[a]

Structure number	Name	CAS Registry Number	Molecular formula	Trade name
(**1**)[b]	nystatin	[*1400-61-9*]	[b]	Fungicidin, Biotanal, Diastatin, Candex Mycostatin[c], Mioronal, Multilind, Nystan, Nystarescent
(**2**)	natamycin	[*7681-93-8*]	$C_{33}H_{47}NO_{13}$	Pimaricin, Mycophyt, Myprozine, Natacyn, Pimatucin, Synogil
(**3**)	amphotericin B	[*1397-89-3*]	$C_{47}H_{73}NO_{17}$	Amphozone, Fungizone[c], Fungilin, Ampho-Mioronal
(**4**)[c]	candicidin	[*1403-17-4*]	[d]	Levorin, Candeptin, Candimon, Vanobid
	filipin	[*11078-21-0*]	[e]	Filimarisin
	homycin	[*1403-71-0*]	[e]	Primamycin
(**5**)	etruscomycin[f]	[*13058-67-8*]	$C_{36}H_{53}NO_{13}$	Etruscomycin
	trichomycin[g]	[*1394-02-1*]	[e]	Cabimicina, Trichonat

[a]See Figure 1.
[b]Nystatin has three biologically active components: A_1, A_2, and A_3. Figure 1 depicts A_1.
[c]Squibb is the U.S. producer.
[d]A 4-component mixture; candicidin D shown in Figure 1 is the primary component.
[e]Multicomponent mixture.
[f]Also known as lucensomycin.
[g]Also known as hachimycin.

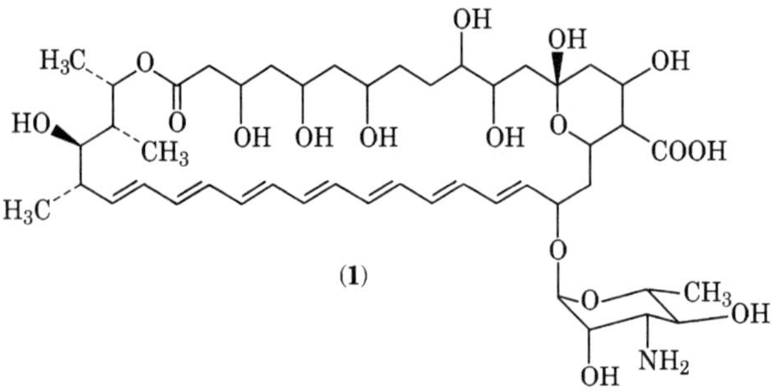

(**1**)

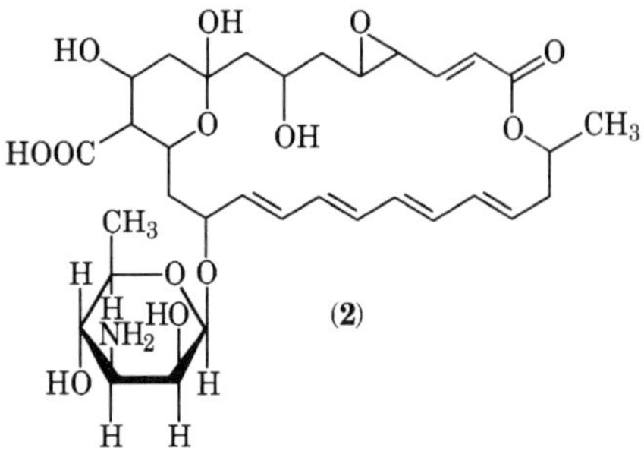

(**2**)

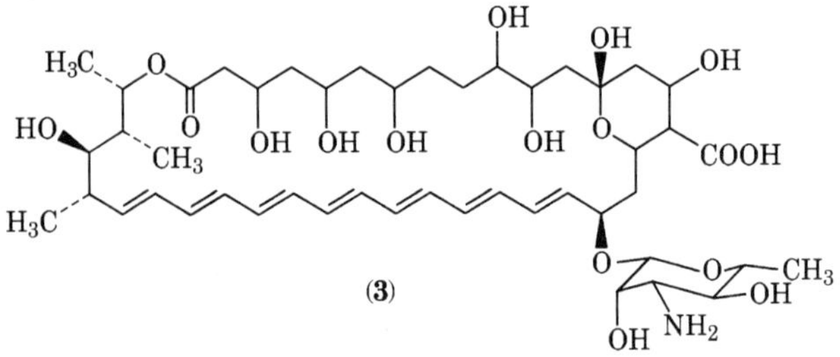

(**3**)

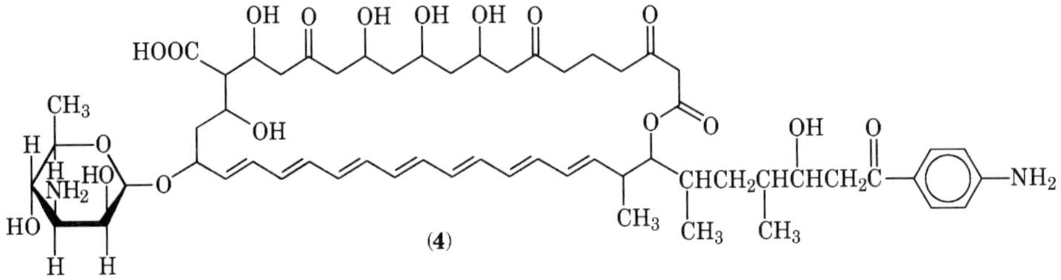

(**4**)

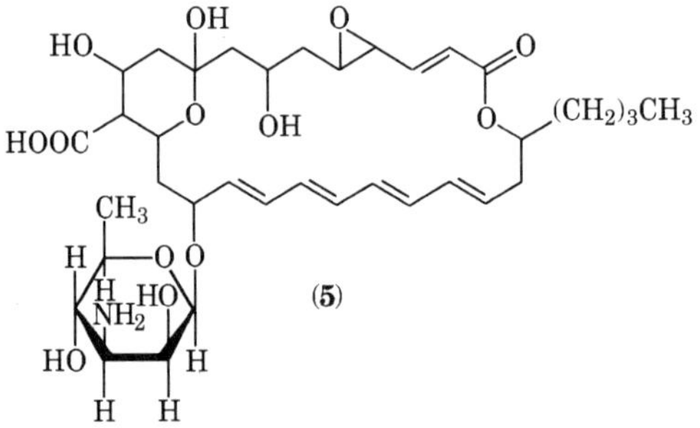

(**5**)

Fig. 1. Polyene antibiotics (see Table 1).

Nystatin is obtained from *Streptomyces noursei*. The polyenes are characterized by a macrolide ring and differ from one another by the number (12–37) of carbon atoms in the ring structure, the number (6–14) of hydroxyl groups, and the presence or absence of a carbohydrate (1). Polyenes alter the membrane permeability of sensitive cells by forming a complex with the sterol present in the membranes. Due to this binding to sterols, potassium ions are lost. Both nystatin and amphotericin B bind more strongly to ergosterol than to cholesterol.

Ergosterol is the principal sterol in the membrane of yeasts and fungi, whereas the cell membrane in mammals contains mainly cholesterol. This difference in binding capacity is probably responsible for the selectivity (1,2). Since bacteria, with the exception of *Mycoplasma* and *Acholeplasma*, contain no sterols; polyenes have no antibacterial activity.

Nystatin has a local fungicidal and fungistatic action against *C. albicans*. This polyene is not water-soluble. It is absorbed to a limited extent from the digestive tract, which limits the field of indications after oral administration to candidoses of the oral cavity, the esophagus, and the intestines. Oral administration of large doses may cause gastrointestinal disorders (nausea, vomiting, diarrhea). Nystatin is too toxic for parenteral administration. More information concerning chemical, mycological, and clinical properties of the polyenes is available (1,3,4).

Nystatin is mainly used to treat vaginal and oral infections and localized skin lesions, including *Candida* intertrigo and *Candida* nappy dermatitis. It may also be used as prophylaxis during treatment with antibiotics.

Nystatin (100,000 IU) is also available in combination with neomycin sulfate [*1405-10-3*] (35,000 IU), polymyxin B sulfate [*1405-20-5*] (35,000 IU), acetarsol [*97-44-9*] (150 mg), and dimethicone [*8050-81-5*] (2,500 mg). One or two ovules per day are inserted vaginally for at least 6–12 days. This combination has an antibacterial, antimycotic, and antitrichomonas action (see also ANTIBIOTICS, OLIGOSACCHARIDES; ANTIPARASITIC AGENTS, ANTIPROTOZOALS).

Natamycin. Natamycin or pimaricin (**2**) is a polyene with only four conjugated double bonds. This tetraene is obtained from *Streptomyces natalensis*. Like the other polyenes, pimaricin induces K^+ release from cells with membranes containing 20–40 mol % ergosterol. This is also the ergosterol concentration in the membrane of *Saccharomyces cerevisiae*.

Natamycin is not water-soluble or soluble in most organic solvents. It is not absorbed in the digestive tract. It is indicated for skin and nail infections with *C. albicans*, intertrigo and fissures at the corners of the mouth (perleche) caused by *C. albicans, Candida vulvitis*, and vaginitis. Natamycin plays an important role in the treatment of mycotic keratitis. Natamycin also appears to be active against the protozoan *Trichomonas vaginalis*. Side effects are nausea or diarrhea during oral treatment. Dosage is application of cream several times a day.

Combination creams or ointments contain 3.5 mg neomycin base and 10 mg hydrocortisone [*50-23-7*] per g, in addition to 10 mg natamycin. The combination as a lotion contains 1.75 mg neomycin and 5 mg hydrocortisone/g, in addition to 10 mg natamycin. This combination has an antiinflammatory, antibacterial, and antimycotic action. It is applied 2–4 times per day.

Amphotericin B. Amphotericin B (**3**), an important polyene antibiotic, is administered almost exclusively via the intravenous route and is therefore discussed in more detail under the systemic antimycotics. The vaginal tablets contain 50 mg amphotericin B, and 100 mg tetracycline base per tablet (see also ANTIBIOTICS, TETRACYCLINES). The tablets for oral use contain 50 mg amphotericin B, 250 mg tetracycline base, and 125 mg sodium hexametaphosphate. A combination ointment contains 1 mg fludrocortisone acetate, 2.5 mg neomycin, 0.25 mg gramicidin, and 1 g plastibase in addition to 30 mg amphotericin B (see also ANTIBIOTICS, PEPTIDES).

SYNTHETIC ANTIMYCOTICS

Because of their limited activity, small spectrum, and side effects, the older topical antimycotics have generally been surpassed by newer antimycotic chemotherapeutic agents. These newer antimycotics for topical use include the imidazole derivatives clotrimazole, miconazole, econazole, isoconazole, sulconazole, fenticonazole, oxiconazole, bifonazole, butoconazole, zinoconazole, tioconazole, and the triazole derivative, terconazole (Table 2) (5–7). The introduction of the azole derivatives represents a milestone in the treatment of mycoses.

Clotrimazole. The imidazole derivative clotrimazole (**6**) was introduced in 1969. Clotrimazole [*23593-75-1*] or 1-(*o*-chloro-α,α-diphenylbenzyl)imidazole is a water-insoluble antimycotic for topical application, with a broad-spectrum activity against mycoses of the skin and the vagina.

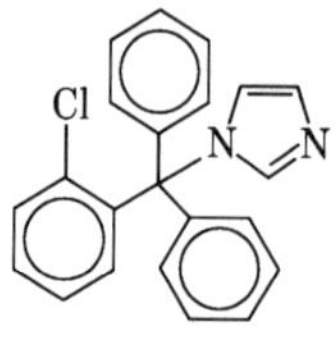

(**6**)

This antimycotic, which was introduced for topical use, shows *in vitro* activity against dermatophytes, species of *Candida, Aspergillus, Coccidioides immitis, Histoplasma capsulatum, Cryptococcus neoformans,* and *Madurella* species. *Petriellidium (Allescheria) boydii* and *Phialophora* species are less sensitive.

Clotrimazole and other azole derivatives have a different mode of action than the polyenes, eg, amphotericin B. The latter bind to the ergosterol present in the membranes of yeasts and fungi, but azole derivatives inhibit the cytochrome P-450 dependent biosynthesis of ergosterol (8–11). This inhibition not only results in a reduction of ergosterol, but also in an accumulation of C-14 methyl sterols. They disturb membrane permeability, inhibit cell replication, and are basically responsible, in combination with the reduction of ergosterol levels, for the antifungal action.

Clotrimazole is only partly absorbed in the digestive tract. It is a microsomal enzyme (cytochrome P-450) inductor and induces its own metabolism. This explains why it is not active *in vivo* against the etiologic agents of deep mycoses, including *H. capsulatum, C. immitis,* and *C. neoformans*. Enzyme induction also produces increased levels of other liver enzymes, such as transaminases. Because of this and possible side effects such as nausea, vomiting, and hallucinations (cerebral toxicity), the use of clotrimazole is limited to topical application. Topical application of clotrimazole is well tolerated; local allergic reactions and generalized skin irritations may occur occasionally.

Miconazole. Miconazole nitrate [*22832-87-7*] (Fig. 2), the 1-phenethylimidazole derivative first described in 1969, interferes at low doses with the cytochrome P-450 dependent ergosterol biosynthesis in yeasts and fungi. The result is accumulation of C-14 methylated sterols on the one hand and reduction of the ergosterol levels in the membranes on the other hand (12). In analogy to clotrimazole, this leads to a disturbance in the membranes; it results in inhibition

Table 2. Azole Antimycotics[a]

Name	CAS Registry Number	Structure number	Molecular formula	Trade name
clotrimazole	[*23593-75-1*]	**(6)**	$C_{22}H_{17}ClN_2$	Canesten, Canifug, Expecid, Lotrimin, Mycofug, Pedisafe, Rimazole, Tibatin, Trimysten
miconazole	[*22916-47-8*]	**(7a)**	$C_{18}H_{14}Cl_4N_2O$	Fungisdin, Aflorix[b], Albistat[b], Andergin[b], Brentan[b], Conoderm[b], Constite[b], Daktor[b], Deralbine[b], Dermonistat[b], Florid[b], Micatin[b], Monistat[b], Prilagin[b], Vodol[b]
econazole	[*27220-47-8*]	**(7b)**	$C_{18}H_{15}Cl_3N_2O$	Ecostatin[b], Ifenec[b], Micofugal[b], Micogin[b], Palavale[b], Pargin[b], Pevaryl[b], Spectazole[b]
isoconazole	[*27523-40-6*]	**(7c)**	$C_{18}H_{14}Cl_4N_2O$	Fazol[b], Travogen[b]
ketoconazole	[*65277-42-1*]	**(10)**	$C_{26}H_{28}Cl_2N_4O_4$	Fungarest, Fungoral, Ketoderm, Ketoisdin, Nizoral, Orifungal M, Panfungol
butoconazole	[*64872-76-0*]	**(11)**	$C_{19}H_{17}Cl_3N_2S$	Exelgyn[b], Femstat[b], Gynomyk[b]
oxiconazole	[*64211-46-7*][b]	**(12)**	$C_{18}H_{13}Cl_4N_3O$	Myfungar[b], Oceral[b], Oxistat[b]
omoconazole	[*74512-12-2*]	**(13)**	$C_{20}H_{17}Cl_3N_2O_2$	Fangorex[b], Fongarex[b]
sulconazole	[*61318-90-9*]	**(14)**	$C_{18}H_{15}Cl_3N_2S$	Exelderm[b], Myk[b], Sulcosyn[b]
itraconazole	[*84625-61-6*]	**(18)**	$C_{35}H_{38}Cl_2N_8O_4$	Sporanox
fluconazole	[*86386-73-4*]	**(19)**	$C_{13}H_{12}F_2N_6O$	Diflucan, Triflucan
saperconazole	[*110558-57-3*]			
fenticonazole	[*72479-26-6*]		$C_{24}H_{20}Cl_2N_2OS$	Falvin[b], Lomexin[b]
bifonazole	[*60628-96-8*]		$C_{22}H_{18}N_2$	Amycor, Amolmen, Bedriol, Mycospor, Mycosporan
terconazole	[*67915-31-5*]		$C_{26}H_{31}Cl_2N_5O_3$	Fungistat, Panlomyc, Terazol, Tercospor
tioconazole	[*65899-73-2*]		$C_{16}H_{13}Cl_3N_2OS$	Fungibacid, Trosyd, Trosyl, Zoniden
zinoconazole	[*80168-44-1*]			

[a] U.S. producers include Ortho Pharmaceutical (eg, Monistat), Schering-Plough (eg, Lotrimin), and Syntex (Femstat).
[b] The nitrate.

of cell replication, mycelium development (in *C. albicans*), and finally, cell death. High concentrations of miconazole, which may be achieved with topical use, disturb the orientation of phospholipids in the membranes, which produces leaks (13).

Miconazole [*22916-47-8*] has a potent antifungal action against dermatophytes and *Candida* species. It is also active against gram-positive bacilli and

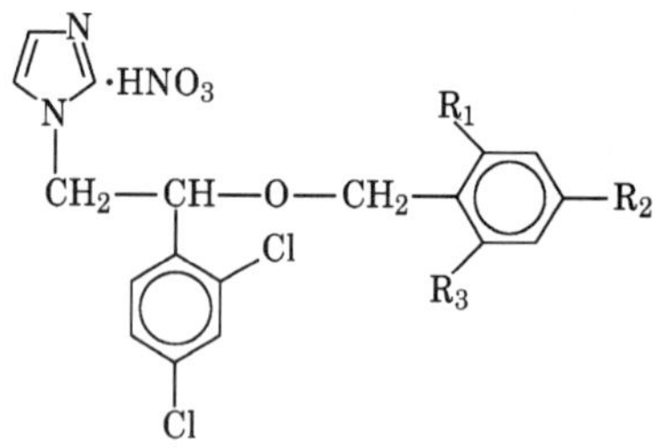

(**7**)

Fig. 2. Miconazole nitrate (**7a**) $R_1 = R_2 = Cl$, $R_3 = H$; econazole nitrate (**7b**) $R_1 = R_3 = H$, $R_2 = Cl$; isoconazole nitrate (**7c**) $R_1 = R_3 = Cl$, $R_2 = H$.

cocci. *In vitro,* it is also effective against *Trichomonas, Leishmania,* and *Plasmodium.* Miconazole is only moderately absorbed in the digestive tract. After intravenous administration, it is rapidly metabolized.

This drug is indicated for infections of skin and nails due to dermatophytes or *Candida* species and vulvovaginal infections due to *Candida* species. Miconazole is used orally for prophylactic purposes in patients who have been treated with cytostatics or immunosuppressants, in particular, to prevent candidosis of the mouth and digestive tract. Side effects include possible nausea during oral treatment. Miconazole tablets may increase the anticoagulant effect of coumarin derivatives. Topical treatment with cream or powder is well tolerated. Local irritation and allergic reactions of the skin and mucosa occur only rarely.

Daktacort is a combination product. It combines a broad-spectrum antimycotic with antibacterial action against gram-positive bacilli and cocci, miconazole and hydrocortisone. This corticosteroid was added for its anti-inflammatory and antipruriginous properties. Daktacort is indicated for skin infections due to dermatophytes or *Candida* species associated with inflammatory symptoms. For this reason, daktacort is used mainly during the initial phase of treatment. Afterwards, it is possible to switch to a cream containing only miconazole nitrate.

Econazole. Econazole and isoconazole are structurally very similar to miconazole; they differ only in the pattern of chlorine substitution on the terminal phenyl ring. Chemically, econazole (Fig. 2, **7b**) is closely related to miconazole. The clinical indications are almost identical. Econazole [*27220-47-9*] is available as a vaginal cream and ovule containing respectively 1% and 150 mg econazole nitrate. For mycoses of the skin, several forms are available including cream, spray-powder, spray-solution, and lotion. All these formulations contain 1% econazole nitrate [*68797-31-9*]. The dosages are practically identical to those of miconazole.

Isoconazole. Like econazole, both the chemical structure and the clinical indications of this imidazole derivative (Fig. 2, **7c**) closely resemble those of miconazole. It is available as a vaginal tablet (300 mg isoconazole nitrate per tablet). In some countries isoconazole [*27523-40-6*] is also available as a cream for dermatological use and as a cream and ovules for treatment of vaginal candidosis.

Naftifine. Naftifine (8) belongs to the allylamines, a new class of antimycotics (14). It is used to treat superficial mycoses and is particularly active against dermatophytes.

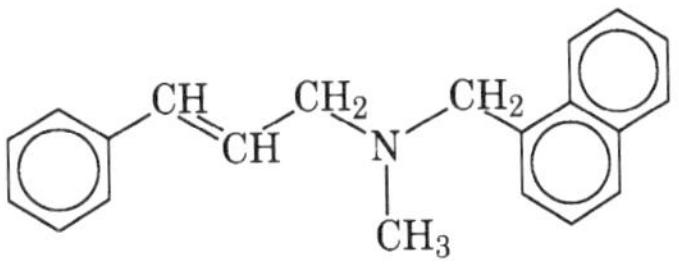

(8)

Like the azole derivatives, it inhibits the biosynthesis of ergosterol. However, naftifine [*65472-88-0*] does not inhibit the cytochrome P-450 dependent C-14-demethylase, but the epoxidation of squalene. Squalene epoxidase catalyzes the first step in the conversion of squalene via lanosterol to ergosterol in yeasts and fungi or to cholesterol in mammalian cells. The squalene epoxidase in *C. albicans* is 150 times more sensitive to naftifine, $C_{21}H_{21}N$, than the enzyme in rat liver (15). Naftifine is available as a 1% cream.

Ciclopiroxolamine. This ethanolamine salt, $C_{14}H_{24}N_2O_3$, of ciclopirox (**9**) is a topically active antimycotic, available as a cream and powder (16).

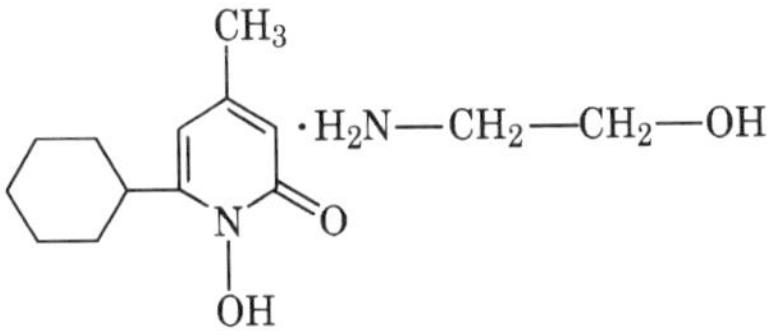

(9)

The biochemical basis for its action in fungi has not been totally elucidated. Ciclopiroxolamine [*41621-49-2*] is used mainly in the treatment of mycoses of the skin and nails.

Ketoconazole. Initial observations indicating that oral administration of ketoconazole (**10**) produced good results in seborrheic eczema and dandruff, led to the development of a 2% cream and a 2% shampoo (scalp gel) of this antimycotic (17,18). Naturally, these two topical forms of ketoconazole [*65277-42-1*] are highly active against superficial mycoses.

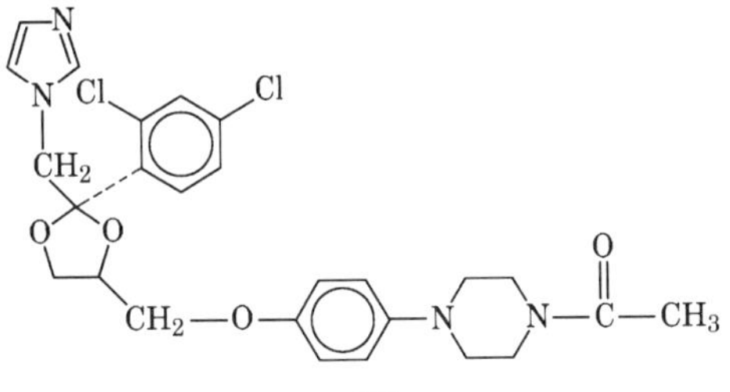

(10)

The main application of these two topical forms is the treatment of seborrheic eczema and dandruff. Ketoconazole's potent pityrosporicidal effect at concentrations of 100 ng/mL to 1 μg/mL forms the rationale of this activity.

Haloprogin. 3-Iodo-2-propynyl 2,4,5-trichlorophenyl ether, $C_9H_4Cl_3IO$, is an almost insoluble antifungal for topical use, available in the United States. Haloprogin [*777-11-7*] was developed in 1963 and it is active against dermatophytes. The drug's effect against *Candida* sp. is rather limited.

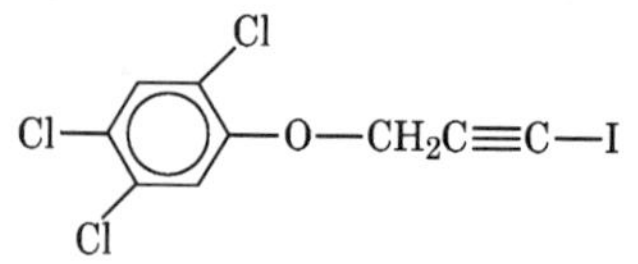

Developmental Topical Antimycotics. Most of the topical antimycotics still being developed are azole derivatives (mostly imidazole), including butoconazole, oxiconazole, omoconazole, zinoconazole, and sulconazole (Fig. 3). Generally speaking, these newer azoles offer no additional therapeutic advantages compared to existing substances. These drugs are already available in certain countries, but are not yet commercialized on a worldwide scale.

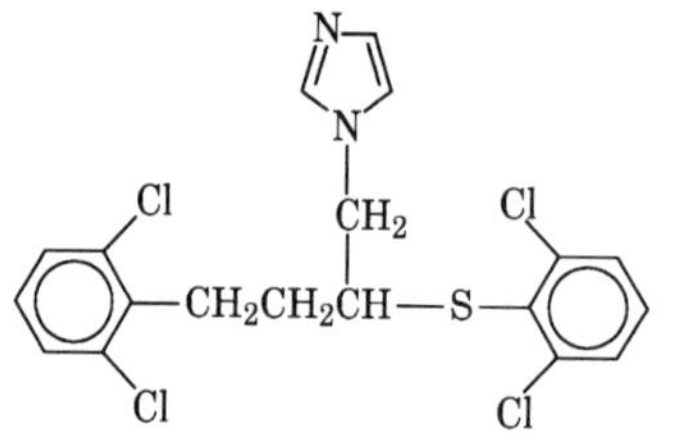

(**11**)

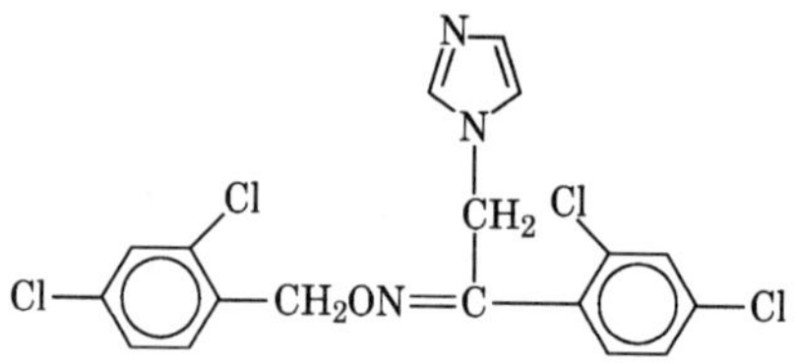

(**12**)

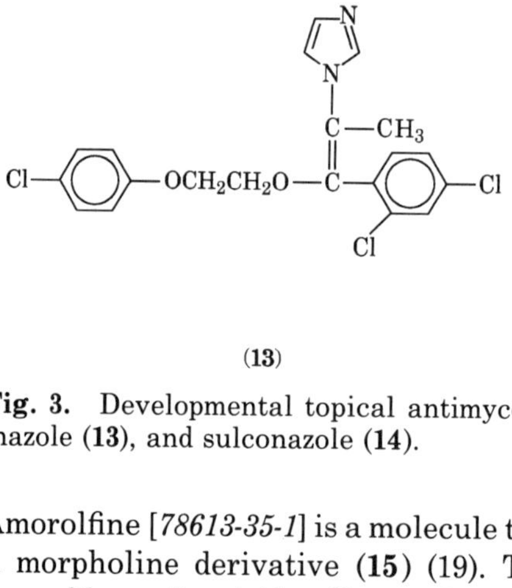

(**13**)

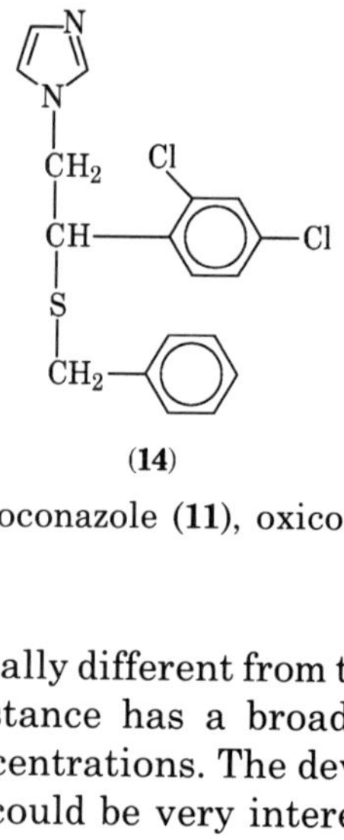

(**14**)

Fig. 3. Developmental topical antimycotics. Butoconazole (**11**), oxiconazole (**12**), omoconazole (**13**), and sulconazole (**14**).

Amorolfine [*78613-35-1*] is a molecule that is totally different from this group; it is a morpholine derivative (**15**) (19). This substance has a broad-spectrum activity and has a fungicide effect at fairly low concentrations. The development of a nail varnish for treatment of onychomycoses could be very interesting.

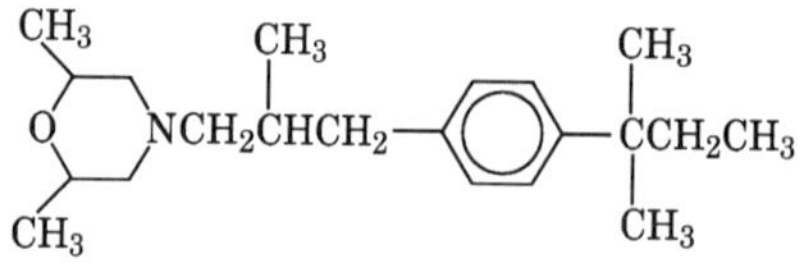

(**15**)

Systemic Antimycotics

Potassium Iodide. When potassium iodide [*7681-11-0*] is administered orally for several (6–8) weeks, a therapeutic effect may be obtained in the subcutaneous form of sporotrichosis. Amphotericin B is used intravenously to treat systemic sporotrichosis. The KI dosage is usually a saturated solution in water (1 g/mL). The usual oral dose is 30 mg/kg/d. Children should receive five droplets, three times a day (after meals); the dose may be increased to 15–20 droplets. Side effects include digestive disorders, swelling of the salivary glands, and lacrimation. Thyroid function tests may be disturbed.

Griseofulvin. Although this antibiotic was first isolated in 1939 and was described as a metabolite of *Penicillium griseofulvum* (20), its action against *Microsporum* and *Trichophyton* was only discovered in 1958. Several modes of action were suggested for (+) 7 chloro-4,6-dimethoxycoumaran-3-one-2-spiro-1′-(2′-methoxy-6′-methylcyclohex-2′-en-4′-one) ($C_{17}H_{17}ClO_6$) (**16**), this antibiotic derived from *P. griseofulvum*.

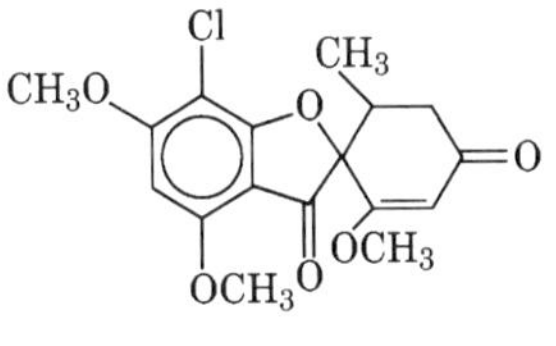

(**16**)

It interferes with the synthesis of the hyphal walls, the biosynthesis of nucleic acids, and the synthesis of chitin. The interaction with microtubules has also been described. The sensitivity of a cell seems to depend particularly on the ability to form griseofulvin–nucleic acid complexes. Further information concerning griseofulvin is available (21).

After oral administration, griseofulvin [*126-07-8*] is absorbed moderately. The absorption is increased when it is administered together with oil or fat. Micronization also increases the absorption greatly in the digestive tract; griseofulvin with a particle diameter of 2.7 μm reaches twice the plasma levels of griseofulvin with a particle diameter of 10 μm. Griseofulvin is found in the outer layers of the stratum corneum. Perspiration probably plays an important role in bringing it there.

Indications are mycoses of the skin, hair, and nails due to species of *Trichophyton, Epidermophyton floccosum,* and *Microsporum.* Yeasts and bacteria are not sensitive. Griseofulvin has a very weak effect against *Aspergillus.* Side effects may include headaches and gastrointestinal disorders, but they are usually only temporary. Occasionally, urticaria, erythema, or photosensitivity are observed. Griseofulvin is a strong inducer of the mixed function oxidase in the liver. It therefore has a stimulating effect on the conversion of various other drugs, including oral anticoagulants. It also affects the biosynthesis of porphyrin; this may result in accumulation of protoporphyrin in hepatocytes.

In the case of tinea corporis, treatment may last longer than six weeks. The treatment of onychomycosis may last up to twelve months, certainly if the toenails are involved. Caution is required in patients with impaired kidney or liver function or hematological disorders. Also, use is contraindicated during pregnancy.

Amphotericin B. This heptaene macrolide antibiotic (**3**) is produced by *Streptomyces nodosus*. It has fungistatic and fungicide properties and is effective against *Candida* species and the etiologic agents of systemic mycoses, including *Histoplasma capsulatum, Blastomyces dermatitidis, Coccidioides immitis,* and *Cryptococcus neoformans*. It is also active against *Sporothrix schenckii*. Like nystatin, this polyene has a greater affinity for ergosterol than cholesterol. Consequently, much lower levels of amphotericin B are required to induce potassium release in *C. albicans* than in mammalian cells. Potassium release is secondary to the increased flux of protons (1). In yeasts, the proton gradient plays an important role in the function of the plasma membrane. The transport of amino acids and other nutrients depends on this proton gradient (22).

Absorption is minimal after oral and topical administration. In order to obtain a systemic effect, the compound must be administered intravenously. Plasma levels are fairly stable during the first 24 hours; only a small fraction is eliminated via urine during this period. The polyene can be detected in urine up to three weeks after treatment. Diffusion of amphotericin B into cerebrospinal fluid is only moderate. More information concerning polyene antibiotics is available (1–4,23). Several side effects should be taken into account during intravenous administration. The main side effects are noted in the glomeruli of the renal parenchyma and include thickening of the basal membrane with hyalinisation, and in the tubuli, where focal or generalized degeneration and nephrocalcinosis may occur. Hypokalemia is an important consequence of nephrotoxicity. Headache, nausea, and vomiting are usually the first signs of toxicity. The dosage should be reduced until these side effects disappear. Vomiting, fever, and headache may be prevented by the administration of antihistamines and corticosteroids. Amphotericin B is administered over a 6-h period by a slow intravenous infusion. The recommended concentration is 0.1 mg/mL. The initial dose is 0.1–0.2 mg/kg/d. The dose is then increased to 1 mg/kg/d. The daily dose must never exceed 1.5 mg/kg. The duration of the treatment depends on the nature and the extent of the infection.

A marked improvement is generally noted after 4–8 weeks of treatment. Treatment is often continued until a total dose of 3 g is reached. In the case of coccidioidomycosis, for example, treatment with 0.4–0.8 mg/kg/d may last months. The polyene is administered intrathecally to treat *Coccidioides* meningitis. However, the results are only moderate. It is very important to check renal and hepatic function during treatment with amphotericin B.

Flucytosine. Flucytosine (**17**) or 5-fluorocytosine (4-amino-5-fluoro-2-pyrimidone, 5-FC), $C_4H_4FN_3O$, is a pyrimidine derivative, that is efficient against *Candida albicans, Cryptococcus neoformans,* and *Torulopsis glabrata*.

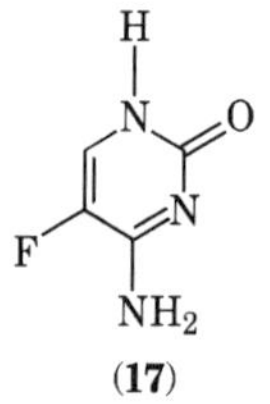

(**17**)

It is also active against *Aspergillus, Phialophora,* and *Cladosporium.* The protozoa *Acanthamoeba culbertsoni* and *Leishmania* are sensitive to flucytosine (1,24).

Flucytosine [*2022-85-7*] is well absorbed in the digestive tract, which is why oral administration is preferable. Plasma levels of 30–40 mg/L are obtained after a dose of 30 mg/kg body weight. Approximately 90% of the pyrimidine derivative is found unaltered in urine, indicating that it is highly suitable for the treatment of renal candidosis. High concentrations were also noted in cerebrospinal fluid; the average concentration is approximately 75% of the plasma concentration.

With the aid of cytosine permease, flucytosine reaches the fungal cell where it is converted by cytosine deaminase into 5-fluorouracil [*51-21-8*]. Cytosine deaminase is not present in the host, which explains the low toxicity of 5-FC. 5-Fluorouracil is then phosphorylated and incorporated into RNA and may also be converted into 5-fluorodeoxyuridine monophosphate, which is a potent and specific inhibitor of thymidylate synthetase. As a result, no more thymidine nucleotides are formed, which in turn leads to a disturbance of the DNA-synthesis. These effects produce an inhibition of the protein synthesis and cell replication (1,23,24). 5-Fluorouracil cannot be used as an antimycotic. It is poorly absorbed by the fungus to begin with and is also toxic for mammalian cells.

Flucytosine-resistant strains can develop very rapidly. These mutants may have a disturbed 5-FC-metabolism, or a compensatory mechanism for the disturbed nucleic acid functions. No cytosine permease was found in a resistant *Cryptococcus neoformans* strain, whereas cytosine deaminase was absent in resistant *C. albicans* strains. A deficiency of uridine monophosphate pyrophosphorylase occurs frequently in resistant *C. albicans* strains (1).

The dosage of flucytosine is 150–200 mg/kg orally in four portions every six hours. A 1% flucytosine solution has been developed for intravenous administration. In some countries, a 10% ointment is also available. In patients with normal renal function, flucytosine is seldom toxic, but occasionally severe toxicity may be observed (leukopenia and thrombocytopenia). Plasma levels should be determined and the dose in patients with impaired renal function should be checked. Liver function tests (transaminases and alkaline phosphatase) should be performed regularly. In some patients with high flucytosine plasma levels, hepatic disorders have been observed (24).

Miconazole. Miconazole (Fig. 2, **7a**) is also available as a sterile solution for intravenous infusion. Miconazole has a therapeutic effect on systemic mycoses due to *C. albicans, Aspergillus fumigatus, Cryptococcus neoformans, Blastomyces dermatitidis, Histoplasma capsulatum, Coccidioides immitis, Paracoccidioides brasiliensis,* and *Petriellidum boydii.*

The recommended dosage produces plasma levels that exceed the *in vitro* MIC-values for the previously mentioned fungi. Doses >9 mg/kg generally produce plasma levels of ≥1 mg/L. Large doses may cause loss of appetite, nausea, and vomiting. Very rapid infusion may lead to transitory cardiac arrhythmia and tachycardia. During prolonged and repeated intravenous administration, liver function tests are recommended. Cremophor, the solvent, may induce hypertriglyceridemia.

Ketoconazole. For treatment of systemic mycoses with amphotericin B or miconazole, the patient must be admitted to a hospital. This is not always possible, particularly in areas where systemic mycoses occur frequently, nor is it always desirable, because of the expense. For these reasons, it was desirable to find an antimycotic that combined safety and broad-spectrum activity with oral administration. Ketoconazole (**10**), which is orally active, met most of these requirements. This inhibitor of the ergosterol biosynthesis is an *N*-substituted imidazole, that differs from its precursors by the presence of a dioxolane ring (6,7). Ketoconazole is rapidly absorbed in the digestive system after oral administration. Sufficient gastric acid is required to dissolve the compound and for absorption. Therefore, medication that affects gastric acidity (for example, cimetidine and antacids) should not be combined with ketoconazole.

Plasma levels of 3–5 μg/mL are obtained two hours after administration of 200 mg ketoconazole. No accumulation in the bloodstream was noted after a 30-wk treatment with this dose. The half-life is approximately eight hours. When ketoconazole is taken with meals, higher plasma levels are obtained. Distribution studies using radioactive ketoconazole in rats show radioactivity mainly in the liver and the connective tissue. Radioactivity is also present in the subcutaneous tissue and the sebaceous glands. After one dose of 200 mg in humans, ketoconazole is found in urine, saliva, sebum, and cerumen. Like miconazole, the mode of action is based on inhibition of the cytochrome P-450 dependent biosynthesis of ergosterol. This results in disturbed membrane permeability and membrane-bound enzymes (8,10,23,25).

Ketoconazole is active against dermatophytes (*Microsporum, Trichophyton, Epidermophyton*), yeasts (*Candida* species, *Cryptococcus*), and the dimorphous fungi (*Coccidioides, Histoplasma,* and *Paracoccidioides*). Activity has also been demonstrated in patients with aspergillosis and in a limited number of patients with chromomycosis and sporotrichosis; promising results were obtained. Some *Leishmania* species are sensitive.

It is indicated for infections of the skin and nails due to dermatophytes and/or yeasts, yeast infection of the gastrointestinal tract, chronic recurring vaginal candidosis, and systemic fungal infections. Ketoconazole is well tolerated by most patients. Gastrointestinal symptoms or pruritus have been noted in 1–3% of the patients. Occasionally, liver enzymes may be increased temporarily. Normalization usually occurs during treatment. Of the estimated 4,000,000 patients treated, 300 developed hepatitis, ie, one per 15,000 patients (26). Treatment with ketoconazole should be discontinued when clinical and/or laboratory indications of hepatitis are present (7,27). In the case of prolonged treatment, it is advisable to check liver function twice a month during the first month of treatment, followed by monthly checks. With large doses (600 mg–1.2 g/d) a reversible inhibition of testosterone synthesis is observed (28). The period of treatment, with the exception of vaginal candidosis, depends on the clinical and mycological results. In the case of recurrent vaginal candidosis, 400 mg/d is prescribed for five days, but more recent studies with *Pityriasis versicolor* have demonstrated that shorter treatments also produce good results. Ketoconazole is contraindicated during pregnancy.

Itraconazole. Itraconazole (**18**) is a highly lipophilic compound with a triazole structure. Compared to ketoconazole, itraconazole has a broader spectrum (including *Aspergillus* spp.) (29,30) and an *in vitro* activity that is 10 times higher than ketoconazole for most species.

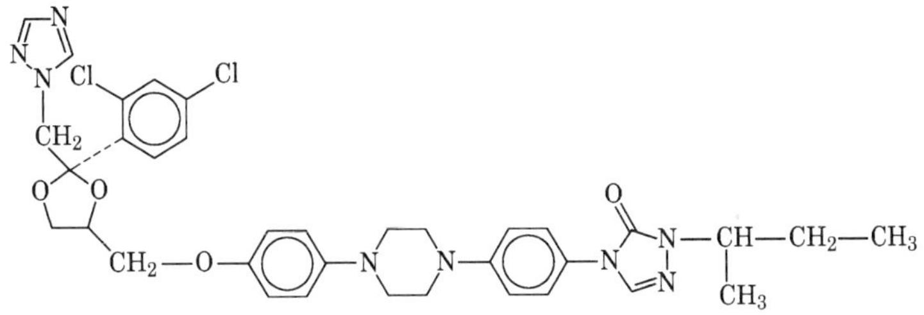

(**18**)

This *in vitro* superiority is also confirmed by animal experiments. Compared to ketoconazole, itraconazole [*84625-61-6*] is a much more selective inhibitor of the cytochrome P-450 dependent ergosterol biosynthesis in yeasts and fungi (29,31). Pharmacokinetically, itraconazole has a much longer bioavailability ($\pm$24-h) and a greater tissue affinity. In general, itraconazole tissue levels are significantly higher than the corresponding plasma levels. It has been established toxicologically that therapeutic doses of itraconazole do not interfere with steroid metabolism in humans. Itraconazole does not affect the liver as much as ketoconazole.

There are few side effects including headache, nausea, vertigo, intestinal disorders, etc. Itraconazole is used to treat vaginal candidosis, pityriasis versicolor, tinea corporis–tinea cruris, tinea pedis–tinea manus, and most of the systemic mycoses including aspergillosis cryptococcal meningitis. Itraconazole is contraindicated during pregnancy (32).

Fluconazole. This substance (**19**) is a water-soluble bis-triazole tertiary alcohol (33). Fluconazole [*86386-73-4*] has a broad spectrum, but has little activity *in vitro*. However, animal experiments reveal a broad spectrum and a potent effect.

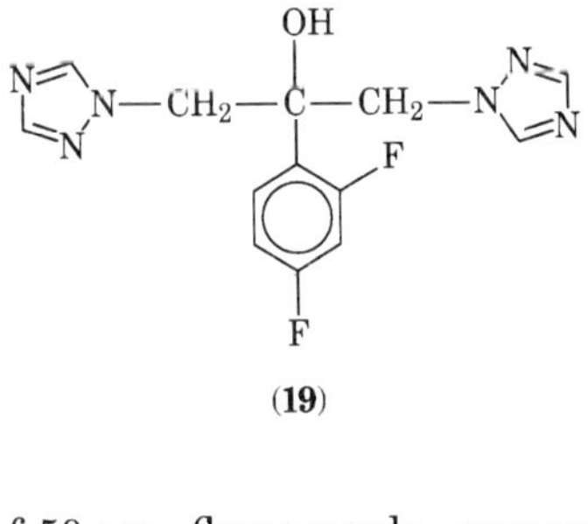

(**19**)

At the tested daily dose of 50 mg, fluconazole appears to be slightly less active against dermatophytes. The substance is used mainly to treat vaginal candidosis (a single capsule of 150 mg) and oral and esophageal candidosis (50 mg od for 14 d). In a number of countries, ie, England, the maximal period of treatment is 14 d.

Toxicological studies have demonstrated that there are no important problems with fluconazole. Therapeutic doses of fluconazole may cause enzyme induction in the liver. This suggests that interactions with other drugs cannot be excluded. The side effects are similar to those of itraconazole and include nausea, headache, and vertigo. Occasionally, increased liver enzymes may be noted. Like itraconazole, fluconazole is contraindicated during pregnancy.

Future Antimycotics for Systemic Treatment. Two new antimycotics for systemic use have now reached the stage of clinical development. The first is a triazole and fluoride analogue of itraconazole. This compound (saperconazole) is extremely active against *Aspergillus* spp. and slightly more soluble. Consequently, intravenous administration might be possible (34). The second molecule is terbinafine [*91161-71-6*], an allylamine, $C_{21}H_{25}N$, that appears to be particularly active against dermatophytes, just like topical naftifine (35).

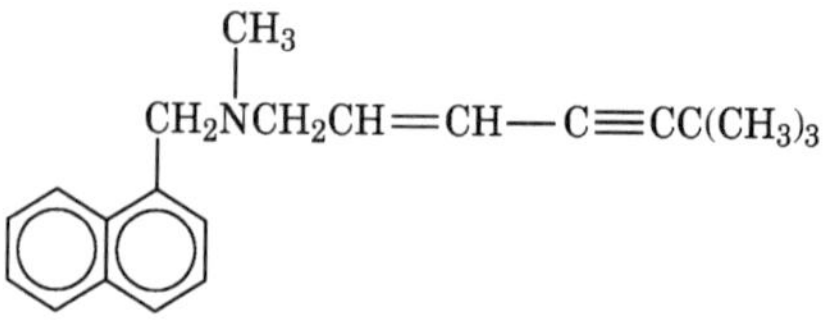

Both substances will probably become available for more widespread clinical use in the following three to four years.

BIBLIOGRAPHY

1. D. Kerridge and L. E. Whelan in A. P. J. Trinci and J. F. Ryley, eds., *Mode of Action of Antifungal Agents,* British Mycological Society Symposium 9, Cambridge University Press, Cambridge, UK, 1984, pp. 343–375.
2. E. F. Gale, E. Cundliffe, P. E. Reynolds, M. H. Richmond, and M. J. Waring, *The Molecular Basis of Antibiotic Action,* 2nd ed., John Wiley & Sons, Inc., 1981, pp. 201–219.
3. G. Medoff and G. A. Kobayashi in D. C. E. Speller, ed., *Antifungal Chemotherapy,* John Wiley & Sons, Inc., New York, 1980, pp. 3–33.
4. G. Medoff, J. Brajtburg, and G. S. Kobayashi, *Ann. Rev. Pharmacol. Toxicol.* **23,** 303–330 (1983).
5. R. J. Holt in Ref. 3, pp. 107–147.
6. H. B. Levine, *Ketoconazole in the Management of Fungal Disease,* Adis Press, Sydney, Australia, 1982.
7. H. E. Jones in R. Fleischmajer, ed., *Progress in Diseases of the Skin,* Vol. 2, Grune & Stratton, New York, 1984, pp. 217–249.
8. H. Vanden Bossche and co-workers, *Pestic. Sci.* **15,** 188–198 (1984).
9. H. Vanden Bossche, G. Willemsens, P. Marichal, W. Cools, and W. Lauwers in Ref. 1, pp. 21–41.
10. H. Vanden Bossche in M. McGinnes, ed., *Current Topics in Medical Mycology,* Vol. 1, Springer-Verlag, New York, 1985, pp. 313–351.
11. H. Vanden Bossche and co-workers, *Pestic. Sci.* **21,** 289–306 (1987).
12. H. Vanden Bossche, G. Willemsens, W. Cools, W. F. J. Lauwers, and L. Le Jeune, *Chem. Biol. Interact.* **21,** 59–78 (1978).
13. H. Vanden Bossche and co-workers, *Biochem. Pharmacol.* 2609–2617 (1982).
14. G. Petranyi, N. S. Ryder, and A. Stütz, *Science* **224,** 1239–1241 (1984).
15. N. S. Ryder in C. Nombela, ed., *Microbial Cell Wall Synthesis and Autolysis,* Elsevier Science Publishers, Amsterdam, The Netherlands, 1984, pp. 313–312.
16. B. Farkas, L. Kem;aaeny, and M. Csato in E. Tümbay, ed., *The Efficacy of Cy-*

clopyroxolamine in the Treatment of Mycosis and Its Influence on the Keratinocyte Killing Activity, The Turkish Microbiological Society No. 5, 1986, p. 35.
17. J. D. Stratigos and co-workers, *J. Am. Acad. Dermatol.* **19**(5) (part 1), 850–853 (1987).
18. G. Cauwenbergh, P. De Doncker, P. Schrooten, and H. Degreef, *Int. J. Dermatol.* **25**(8), 541 (1986).
19. A. Polak and D. M. Dixon in R. A. Fromtling, ed., *Recent Trends in the Discovery, Development and Evaluation of Antifungal Agents,* J. R. Prous Science Publishers, SA, 1987, pp. 555–573.
20. A. E. Oxford, H. Raistrick, and P. Simonart, *Biochem. J.* **33,** 240–248 (1939).
21. R. R. Davies in Ref. 3, pp. 49–82.
22. F. Foury and A. Goffeau, *J. Biol. Chem.* **250,** 2354–2362 (1974).
23. H. Vanden Bossche, G. Willemsens, and P. Marichal, *CRC Crit. Rev. Microbiol.* **15,** 57–72 (1987).
24. H. J. Scholer in Ref. 3, pp. 35–106.
25. H. Vanden Bossche and co-workers, *Antimicrob. Agents Chemother.* **17,** 922–928 (1980).
26. J. H. Lewis, H. J. Zimmerman, G. D. Benson, and K. G. Ishak, *Gastroenterology* **86**(3), 503–513 (1986).
27. P. A. J. Janssen and J. E. Symoens, *Am. J. Med.* 80–85 (1983).
28. H. Vanden Bossche, W. Lauwers, G. Willemsens, and W. Cools in F. Schroeder, ed., *European Organisation for Research on Treatment of Cancer,* Monograph 2, Alan E. Liss Inc., New York, 1985, pp. 187–196.
29. J. Van Cutsem, F. Van Gerven, M-A. Van De Ven, M. Borgers, and P. A. J. Janssen, *Antimicrob. Agents Chemother.* **26,** 527–534 (1984).
30. H. Vanden Bossche, D. W. R. Mackenzie, and G. Cauwenbergh, eds., *Aspergillus and Aspergillosis,* Plenum Press, New York, 1988.
31. H. Vanden Bossche in Ref. 19, pp. 207–221.
32. G. Cauwenbergh and P. De Doncker, *Drug Dev. Res.* **8,** 317–323 (1986).
33. P. F. Troke in Ref. 19, pp. 103–112.
34. J. Van Cutsem and P. A. J. Janssen, *28th Interscience Conference on Antimicrobial Agents and Chemotherapy,* Los Angeles, Calif., 1988, p. 141.
35. N. S. Ryder in Ref. 19, pp. 451–459.

GEERT CAUWENBERGH
Janssen Research Foundation

ANTIPROTOZOALS

Diseases caused by protozoa affect more people than those brought on by any other biological cause (1). There are over 60,000 species of protozoa, of which some 10,000 are parasitic. In humans, protozoa chiefly infect the gastrointestinal tract, vagina, urethra, blood, and blood-forming organs. Malaria is the most widespread of the protozoan diseases, and is responsible for the greatest number of deaths due to infection. Although protozoan diseases occur throughout the world, they impact most severely on people of tropical areas where there is widespread malnutrition, minimal health education, and poor sanitation. Fatal protozoal infections are occurring to an increased extent in the more developed countries among immunosuppressed individuals, especially those with AIDS. Domesticated and wild animals are also extensively affected and can harbor and propagate infectious protozoa. Protozoan diseases of poultry and ruminants deprive large numbers of people, especially in Africa, of much needed food.

The word protozoa is derived from the Greek *protos* and *zoon,* meaning first and animal, respectively. Protozoa are a highly varied group of one-celled animals that reproduce by both sexual and asexual means. Given suitable conditions, ie, host susceptibility and lack of immunity, their numbers can increase by the tens of thousands in the course of one day. They are eukaryotic (having a well-defined nucleus), are surrounded by a membrane through which nutrients can pass, and their cytoplasm contains organelles that enable them to generate energy, digest nutrients, respire, grow, and reproduce. Protozoa are capable of motility and propel themselves by means of cilia, flagella, or pseudopods. Accurate differentiation and identification of infective protozoa facilitates the choice of treatment, but is frequently a difficult task.

Progress in finding new chemotherapeutic agents to combat protozoal disease has been slow. The urgency of the problem has been dramatized by the realization that AIDS patients usually die from opportunistic diseases, many of which are protozoan in origin. There are no satisfactory treatments for many of the protozoal diseases; where there are proven drugs available, parasites in many parts of the world are developing resistance to them. This resistance has necessitated the continual design, synthesis, and testing of new classes of prophylactic and therapeutic agents. Despite the alarming view by a World Health Organization spokesman in 1990 that the war against tropical diseases is being lost, only a modest proportion of the world's research resources is being devoted to devising new treatments. Unfortunately, most pharmaceutical firms do not anticipate adequate financial reward from solving the health problems of developing countries.

This article discusses the main diseases caused by protozoa and the important chemotherapeutic antiprotozoal agents currently in use or in an advanced state of development. Not all alternative proprietary names are given for each drug. There are excellent sourcebooks on protozoology and parasitology (2–8).

Amebiasis

Amebiasis, a widespread disease of humans, causes an estimated 500 million cases (without the inclusion of China) annually. It is believed to be the third leading cause of death due to parasites. The disease is caused by pathogenic strains of the protozoan *Entamoeba histolytica,* which exists both in a stable infective cyst form and in a more fragile, potentially invasive, trophozoite form. Amebiasis is transmitted by drinking water contaminated with fecal matter from infected individuals or by eating food contaminated by infected humans or flies. Symptoms of the disease are generally centered around the gastrointestinal tract. They include amebic dysentery and sometimes appendicitis or fulminant colitis. The disease of the intestine may range from acute dysentery with chills, fever, and blood or mucoid diarrhea (amoebic dysentery), to mild abdominal discomfort with diarrhea, containing blood or mucus. Its infection is often asymptomatic. Blood-borne propagation can cause liver abscesses, and less commonly, infection of the brain or lungs. Skin ulceration may occur by extension of intestinal lesions. Recently, it was shown that *E. histolytica* has both pathogenic and nonpathogenic strains, with characteristic isoenzyme electrophoresis patterns, that occur in symptomatic and asymptomatic persons, respectively.

Acute intestinal amebic dysentery is most commonly treated with metronidazole (**1**, Flagyl) (Table 1). An alternative drug is iodoquinol (**2**, diiodohydroxyquin, diiodohydroxyquinoline [*83-73-8*]), which has also been used advantageously in combination with metronidazole or oxytetracycline (**3**, Terramycin [*79-57-2*]). For patients too sick to take iodoquinol orally, emetine (**4**) or dehydroemetine (**5**, 2,3-dehydroemetine) hydrochloride is administered either subcutaneously or intramuscularly. The latter sometimes is followed by chloroquine (**6**, Aralen) phosphate or iodoquinol. Because emetine and dehydroemetine cause cardiac arrhythmias, muscle weakness, and inflammation at the injection site, these toxic compounds are used primarily for patients whose lives are threatened by the disease. Antibiotics, such as tetracycline (**7**, Achromycin [*60-54-8*]) and paromomycin (**8**, Humatin), are also effective against moderate intestinal amebiasis.

Extraintestinal (eg, hepatic) amebiasis is treated with metronidazole and can be followed by iodoquinol or a combination of dehydroemetine or emetine hydrochloride with chloroquine phosphate. Iodoquinol is the drug of choice for asymptomatic amebiasis, whereas diloxanide furoate (**9**, Furamide) has been used successfully to treat symptomatic and asymptomatic intestinal amebic cyst carriers.

Other compounds that have shown promising properties *in vitro* against *E. histolytica* are niridazole (**10**), tinidazole (**11**, Fasigyn, Simplotan), gossypol (**12**), bithionol [(**13**), 2,2′-thiobis(4,6-dichlorophenol)], and pimaricin (**14**, Natamycin [*7681-93-8*]). The chemotherapy of amebiasis has been reviewed (9).

Related amoeba that cause infections of lower incidence but greater severity are the invasive organisms *Naegleria fowleri* and species of *Acanthamoeba* such as *A. castellanii* and *A. polyphaga*. These amoeba are free-living in the soil or warm fresh water, occur widely, and affect many animals but have no known animal reservoir. Entry of *Naegleria* into the body usually is by the nasal passages, whereas *Acanthamoeba* can enter via the eye, skin, or lung. They cause primary amebic meningoencephalitis (PAM), an infection of the central nervous system that is usually fatal. Symptoms in humans include severe headache, confusion, nausea, seizures, and coma. Frequently the olfactory tract is affected. These infections are considered opportunistic and are seen to a greater extent in immunosuppressed humans (due to AIDS, steroids, radiotherapy, or chemotherapy) than in those who are healthy.

Naegleria is treatable with intravenous amphotericin B (**15**, Fungizone), a toxic drug that must be used with caution.

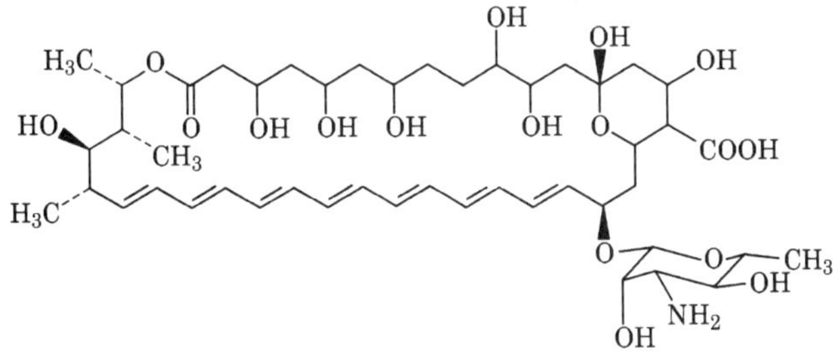

(**15**)

Table 1. Amebiasis Antiprotozoal Agents[a]

Structure number	Compound name	CAS Registry Number	Molecular formula	Structure
(1)	metronidazole[b,c,d,e,f]	*[443-48-1]*	$C_6H_9N_3O_3$	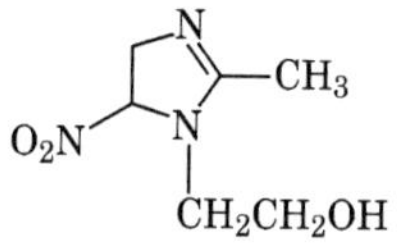
(2)	iodoquinol[b,d]	*[83-73-8]*	$C_9H_5I_2NO$	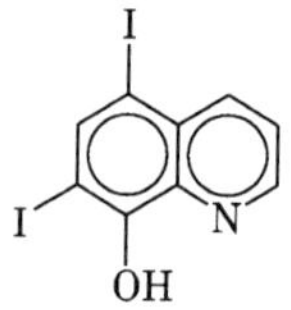
(3)	oxytetracycline[b,g,h]	*[79-57-2]*	$C_{22}H_{24}N_2O_9$	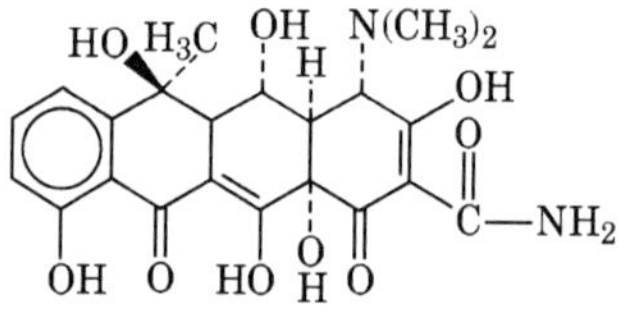
(4)	emetine (HCl)	*[316-42-7]*	$C_{29}H_{40}N_2O_4 \cdot 2ClH$	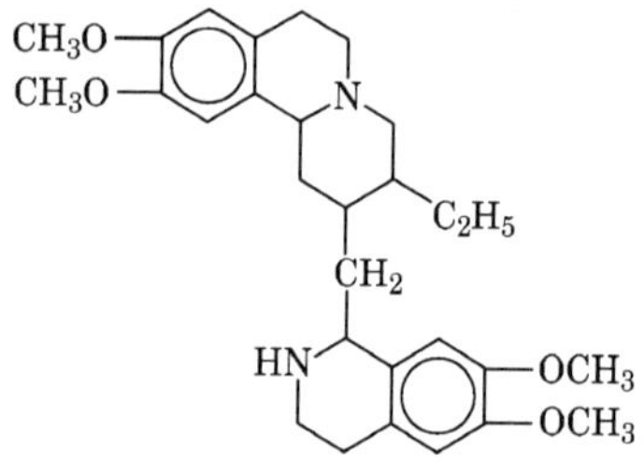
(5)	dehydroemetine	*[4914-30-1]*	$C_{29}H_{38}N_2O_4$	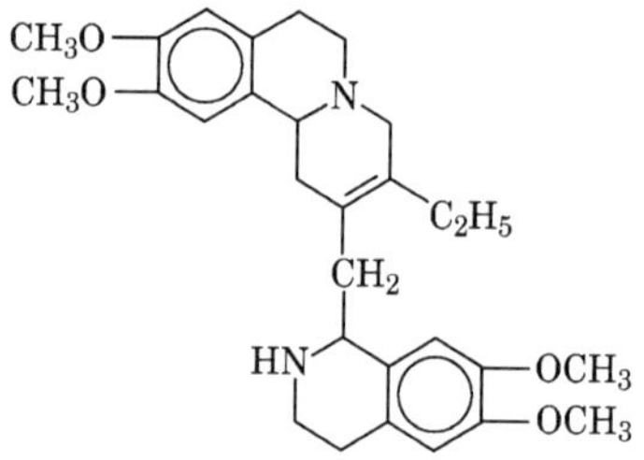
(6)	chloroquine [g,i,j] (diphosphate)	*[50-63-5]*	$C_8H_{26}ClN_3 \cdot 2H_3O_4P$	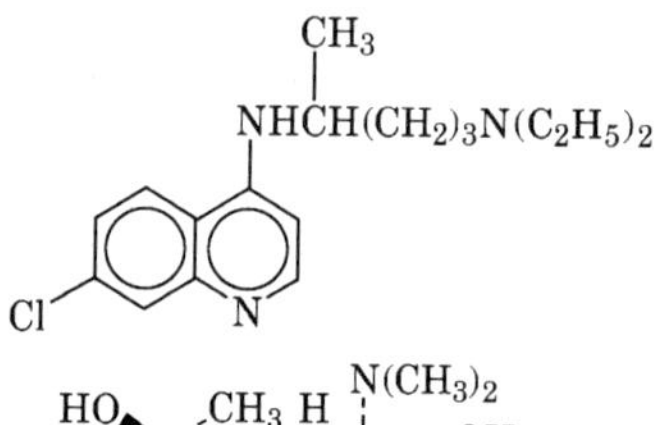
(7)	tetracycline[b,j]	*[60-54-8]*	$C_{22}H_{24}N_2O_8$	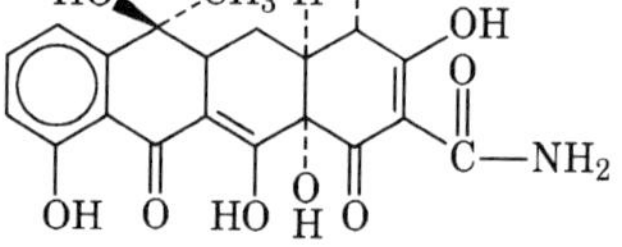

Table 1. ***(Continued)***

Structure number	Compound name	CAS Registry Number	Molecular formula	Structure
(8)	paromomycin[b,c,k]	*[59-04-1]*	$C_{23}H_{45}N_5O_{14}$	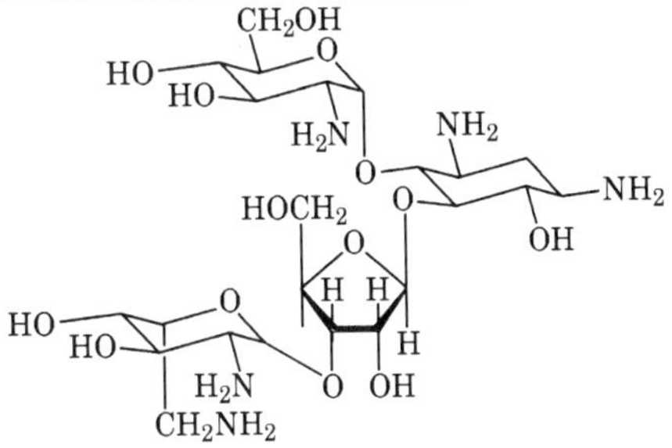
(9)	diloxanide furoate	*[3736-81-0]*	$C_{14}H_{11}Cl_2NO_4$	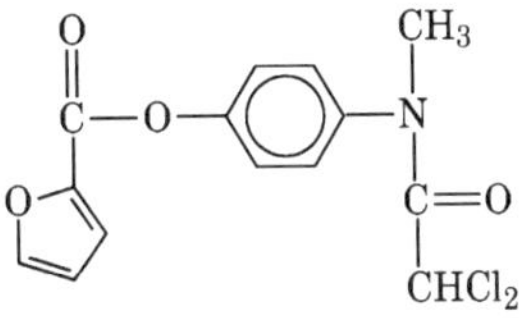
(10)	niridazole[f]	*[61-57-4]*	$C_6H_6N_4O_3S$	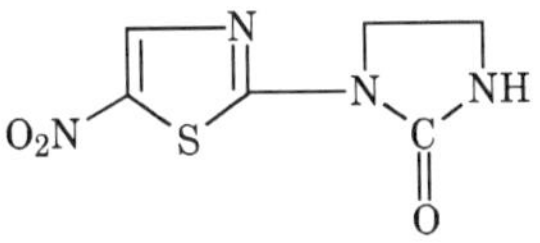
(11)	tinidazole[b,c,d,e,f]	*[19387-91-8]*	$C_8H_{13}N_3O_4S$	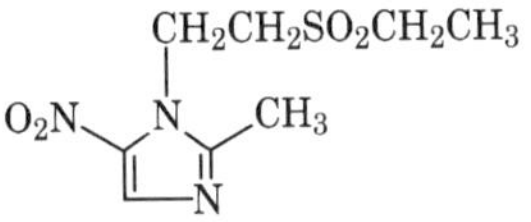
(12)	gossypol	*[303-45-7]*	$C_{30}H_{30}O_8$	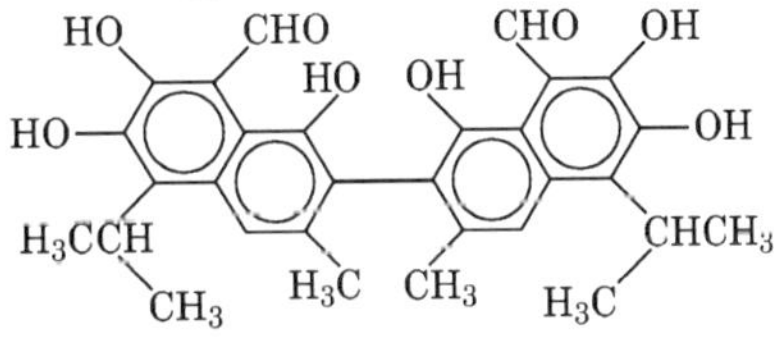
(13)	bithionol	*[97-18-7]*	$C_{12}H_6Cl_4O_2S$	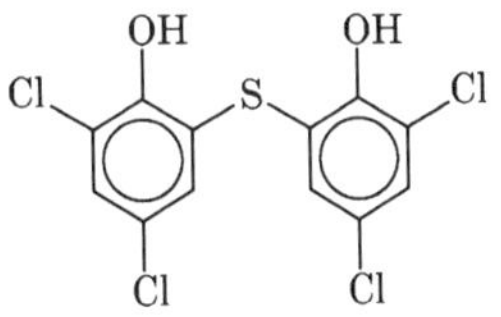
(14)	pimaricin	*[7681-93-8]*	$C_{33}H_{47}NO_{13}$	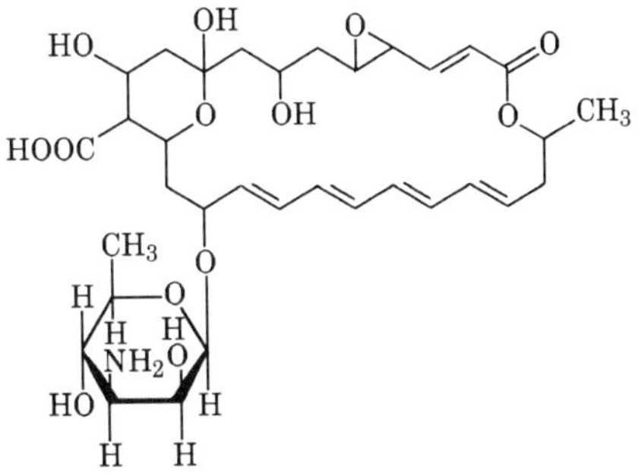

Table 1. (*Continued*)

Structure number	Compound name	CAS Registry Number	Molecular formula	Structure
(**15**)	amphotericin B[k]	[*1397-89-3*]	$C_{47}H_{73}NO_{17}$	in text
(**16**)	miconazole	[*22916-47-8*]	$C_{18}H_{14}Cl_4N_2O$	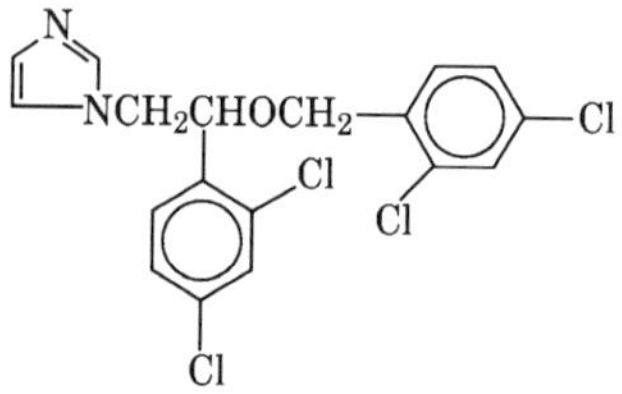
(**17**)	rifampin[j]	[*13292-46-1*]	$C_{43}H_{58}N_4O_{12}$	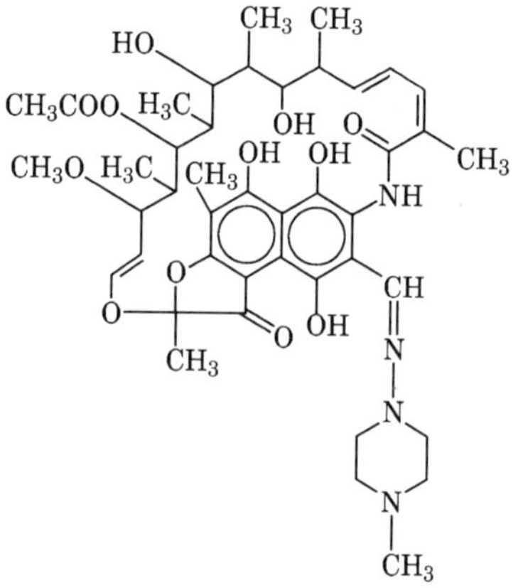
(**18**)	minocycline (HCl)	[*13614-98-7*]	$C_{28}H_{27}N_3O_7 \cdot ClH$	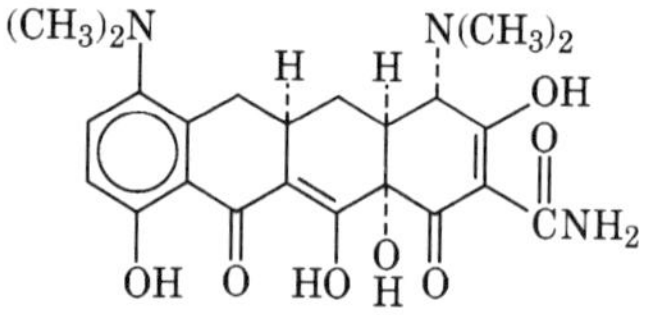

[a]Other applications of these agents are indicated in footnotes to the table.
[b]Balantidiasis.
[c]Giardiasis.
[d]Hexamitosis.
[e]Histomoniasis.
[f]Trichomoniasis.
[g]Anaplasmosis.
[h]Theileriasis.
[i]Babesiasis.
[j]Malaria.
[k]Leishmaniasis.

A combination of amphotericin B, miconazole (**16**), and rifampin (**17**) was used to successfully cure one patient. In addition, tetracycline (**7**) and minocycline (**18**) have been recommended although their clinical efficacy have not been established. No proven therapeutic agents exist for treating *Acathamoeba* infections; however, the phenothiazines, trifluoperazine [*117-89-5*] and chlorpromazine [*50-53-3*], show promise *in vitro*.

Anaplasmosis

Anaplasmosis is a tick-borne disease of cattle and other ruminants, such as deer in the western United States and elk in the former USSR. It is caused by the protozoan *Anaplasma marginale,* so named because the organism appears to be devoid of cytoplasm and resides at the margin within an erythrocyte. At least 30 species of ticks have been shown to be capable of transmitting anaplasmosis under laboratory conditions. Horseflies are also vectors of the disease. Anaplasmosis is mild in young calves and increases in severity, being frequently fatal to cattle above the age of three. Although the disease may be asymptomatic, it is sometimes manifest by anorexia, weakness, fever, anemia, jaundice, reduced milk production, depression, dehydration, difficult respiration, and abortion. Infection can also cause infertility in bulls. The disease in its acute form occurs frequently in purebred animals and high producing milk cows, which may die in a few hours after the start of infection.

Imidocarb (**19**) dihydrochloride [*5318-76-3*], the most effective drug against anaplasmosis, is reported to be curative in a single dosage if given early in the infection during the rise in parasitemia. Multiple doses are required if the animals are treated at a later stage or if a relapse occurs (Table 2). Some antibiotics active against the disease include the tetracyclines chlortetracycline (**20**, Aureomycin [*57-62-5*]) and oxytetracycline (**3**), which are now given in long-acting pharmaceutical dosage forms because of the need for extended treatment. Many

Table 2. Anaplasmosis Antiprotozoal Agents[a]

Structure number	Compound name	CAS Registry Number	Molecular formula	Structure
(**19**)	imidocarb[b]	[*27885-92-3*]	$C_{19}H_{20}N_6O$	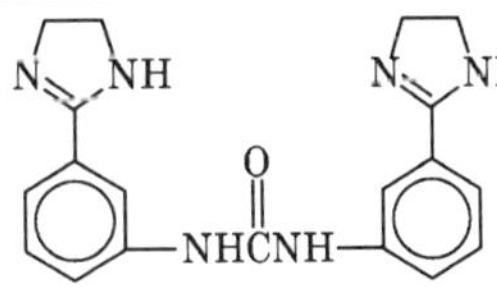
(**20**)	chlortetracycline[b,c,d]	[*57-62-5*]	$C_{22}H_{23}ClN_2O_8$	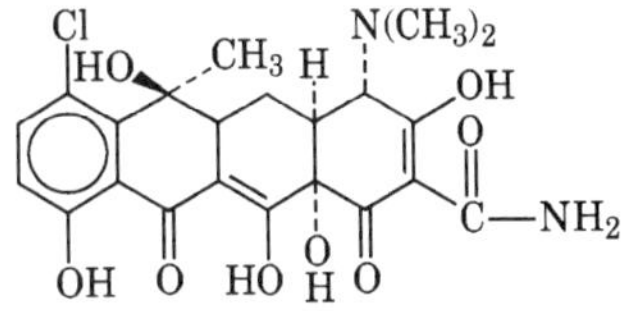
(**21**)	amodiaquine[e]	[*86-42-0*]	$C_{20}H_{22}ClN_3O$	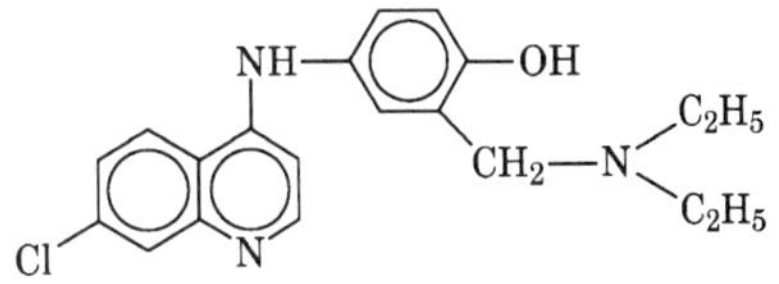

[a]Other applications of these agents are indicated in footnotes.
[b]Theileriasis.
[c]Balantidiasis.
[d]Hexamitosis.
[e]Malaria.

other drugs, including arsenicals, antimonials, antimalarials, eg, chloroquine (**6**) and amodiaquine (**21**, the dihydrochloride dihydrate Camoquin [*6398-98-7*]), and dyes have been employed in the past in an effort to cure or prevent anaplasmosis. α-Ethoxyethylglyoxal dithiosemicarbazone (Gloxazone) has a marked inhibitory effect on *A. marginale* but was found to be too toxic to lactating cattle.

Babesiasis

Babesiasis (babesiosis, piroplasmosis), primarily a disease of cattle, is caused by a protozoan related to the malaria parasite. The protozoan, *Babesia microti,* after ingestion by its vector, the ixodid hard-bodied tick, grows inside the gut of the insect and spreads throughout its body. Infection of the vertebrate host, which can be either a wild or domesticated animal, occurs when the tick obtains a blood meal from it. Humans do not appear to be a reservoir for the infection and are not commonly affected by the disease (10). However, they can acquire babesiasis by blood transfusion. The consequences of infection by *B. microti* in humans can range from being asymptomatic to manifestation of a severe and prolonged illness. Typically, there is a gradual onset of fever, chills, sweating, generalized myalgia, fatigue, anemia, and renal insufficiency and failure. Infections by *Babesia divergens* are associated with cattle and are characterized in humans by chills, high fever, nausea, vomiting, and severe hemolytic anemia. Individuals infected with *B. divergens* generally lack functioning spleens and the disease has been fatal in a large percentage of cases. Related infections include *Babesia bigemina,* the cause of Texas cattle fever, and *Babesia bovis,* responsible for hemoglobinuric fever in European cattle.

Babesiasis in humans is frequently self-limiting. Quinine [*130-95-0*] (**22**) plus clindamycin (**23**) is considered the treatment of choice and has also been used effectively to treat transfusion babesiasis (Table 3). Quinine, either alone or in combination with pyrimethamine (**24**), has failed to eliminate experimental *B. microti* in hamsters. Babesiasis contracted in the United States has been successfully treated with chloroquine (**6**). However, when tested in infected hamsters, chloroquine failed to clear the infection, as did the other antimalarials sulfadiazine (**25**) and pyrimethamine (**24**). There was only a limited response to pentamidine (**26**) and diminazene aceturate [**27**, 4,4′-(diazoamino)dibenzamidine aceturate, Berenil, also diminazene dilactate Babesin]. Pentamidine has been used successfully to treat the symptoms of the disease in humans and in cats infected with *Babesia felis.* Diminazene aceturate is effective against experimental *B. microti* in jirds, but not *in vitro* against *Babesia equi.* Imidocarb (**19**) dipropionate is a therapeutic and prophylactic treatment for babesiasis in cattle, gerbils, and other domestic animals. The urea, amicarbalide (**28**) diisethionate [*3671-72-5*] (3,3′-diamidinocarbanilide diisethionate, Diampron) is also used for babesiasis in cattle. *Babesia gibsoni* in dogs has been treated effectively with phenamidine isethionate (**29**). Another urea, Babesan, (**30**), 1,3-di-6-quinolinylurea as the bismethosulfate salt ie, quinuronium sulfate, $C_{23}H_{26}N_4O_9S_2$, is used for veterinary purposes.

Table 3. Babesiasis Antiprotozoal Agents[a]

Structure number	Compound name	CAS Registry Number	Molecular formula	Structure
(22)	quinine (sulfate)[b]	*[804-63-7]*	$C_{20}H_{24}N_2O_2{\cdot}\frac{1}{2}H_2O_4S$	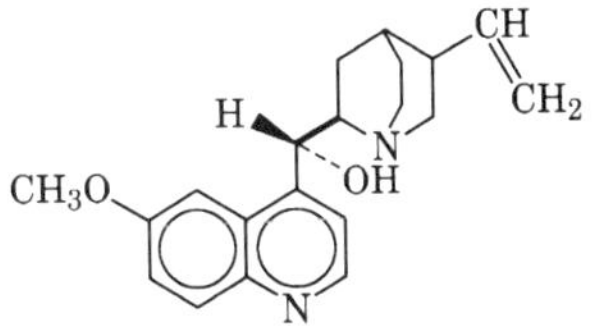
(23)	clinda-mycin[b,c,d]	*[18323-44-9]*	$C_{18}H_{33}ClN_2O_5S$	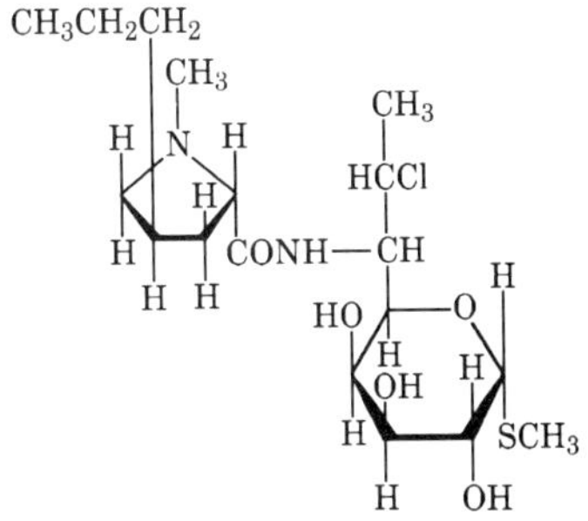
(24)	pyrimeth-amine[b,c,d,e]	*[58-14-0]*	$C_{12}H_{13}ClN_4$	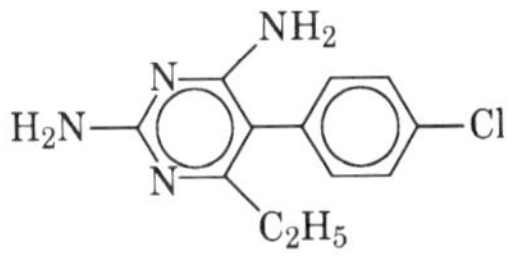
(25)	sulfadi-azine[c,d,f]	*[68-35-9]*	$C_{10}H_{10}N_4O_2S$	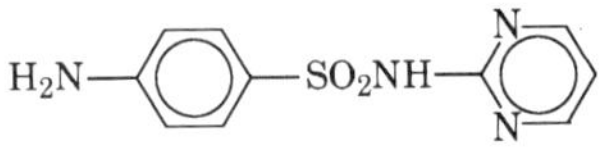
(26)	pentamidine[g]	*[100-33-4]*	$C_{19}H_{24}N_4O_2$	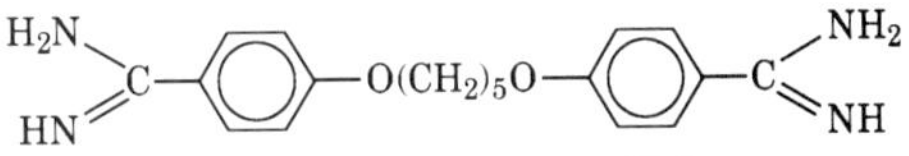
(27)	diminazene aceturate[c,h]	*[908-54-3]*	$C_{22}H_{29}N_9O_6$	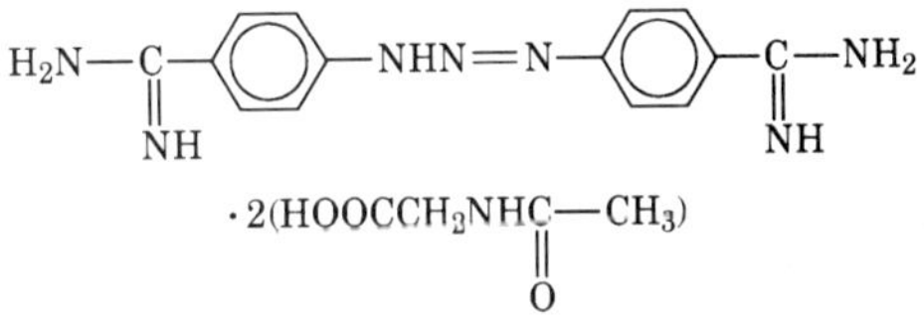
(28)	amicarbalide	*[3459-96-9]*	$C_{15}H_{16}N_6O$	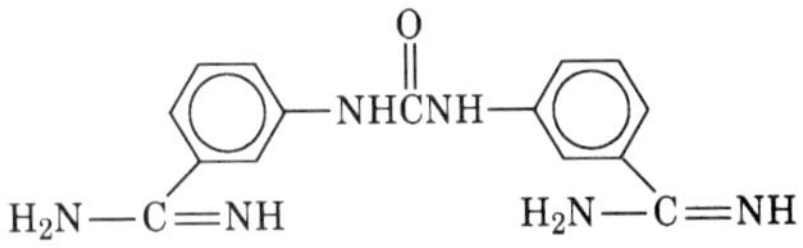
(29)	phenamidine isethionate	*[620-90-6]*	$C_{18}H_{26}N_4O_9S_2$	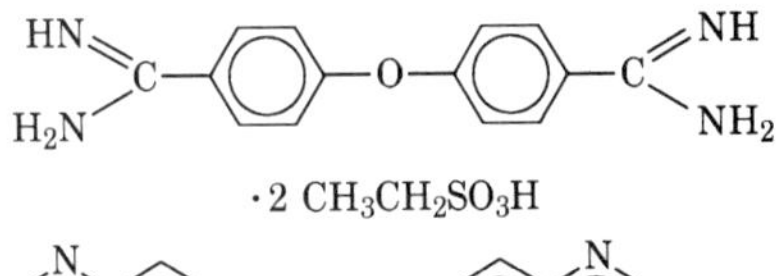
(30)	babesan	*[532-05-8]*	$C_{19}H_{14}N_4O$	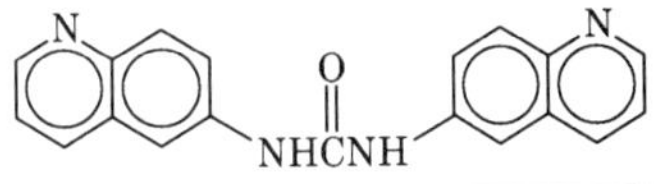

[a]Other applications are indicated by footnotes.
[b]Malaria.
[c]Pneumocystosis.
[d]Toxoplasmosis.
[e]Coccidiosis.
[f]Balantidiasis.
[g]Leishmaniasis.
[h]Trypanosomiasis (African trypanosomiasis).

Balantidiasis

Balantidiasis (balantidiosis, balantidial dysentery), an intestinal disease seen almost worldwide, is caused by the large ciliated protozoan, *Balantidium coli*. The organism is usually found in the lumen of the large intestine of humans and animals. Cysts formed in the lumen of the colon or in freshly evacuated feces of humans or domesticated and wild animals, can colonize the colon and terminal ileum of new hosts by the latter's ingestion of contaminated food or water. The hog has been found to be the most heavily parasitized host. Its association with the rat may be a means for maintaining a reservoir infection in the two animals.

Balantidiasis in humans is manifest by chronic episodes of intermittent diarrhea and constipation, symptoms similar to those of amebiasis. The patient may also have abdominal pain, tenderness over the colon, anorexia, nausea, severe weight loss, and weakness. The disease may be fatal and, before the availability of a treatment, was the cause of death in approximately 30% of infected individuals.

The infection can be cured most readily with tetracycline (**7**), oxytetracycline (**3**), or chlorotetracycline (**20**). Metronidazole (**1**) and iodoquinol (**2**) are also effective. Additional effective drugs include paromomycin (**8**), tinidazole (**11**), sulfadiazine (**25**), and carbarsone [*121-59-4*] (**31**, *p*-ureidobenzenearsonic acid, $C_7H_9AsN_2O_4$, Ameban).

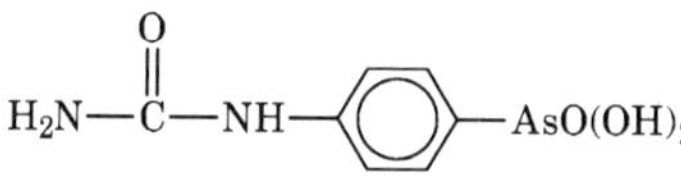

(**31**)

Coccidiosis

Coccidiosis is a widespread disease that occurs most often in fowl, such as chickens and turkeys, and other farm animals (cows, sheep, swine, horses, and rabbits) (11). In chickens the disease has caused severe economic losses. Coccidiosis also occurs in ox, water buffalo, zebu, bighorn sheep, wild goat, alpaca, lion, puma, fox, mink, parakeet, Canada goose, snow goose, and camel, among others. It is seen only rarely in humans, and dogs and cats are only occasionally infected.

Coccidiosis in chickens, as with cattle, is frequently caused by a combination of infective *Eimeria* protozoa. *Eimeria tenella* is the most important, but *Eimeria nectrix* is more common. *Eimeria maxima, Eimeria acervulina,* and *Eimeria mivati* are moderately pathogenic; *Eimeria brunetti* is highly pathogenic, but relatively uncommon. In chickens, young birds become infected by ingesting with their food oocysts shed in the feces of older birds. Diseased birds exhibit diarrhea, emaciation, and anorexia. The disease results in a marked decrease in the production of eggs, as well as a ca 10% mortality caused by the ingestion of a large number of oocysts.

Anticoccidial drugs are administered in the drinking water or feed of poultry and are primarily prophylactic, rather than curative, in nature. The most common ones in use are salinomycin (**32**, Coxistac [*53003-10-4*]) and monensin (**33**), a carboxylic acid which is used as its water-soluble sodium salt (Coban, Rumensin). These polyether ionophoric antibiotics are capable of complexing and interfacing with specific alkali metal cations to render them lipid-soluble and diffusible across parasite mitochondrial membranes. Related useful polyether antibiotics are narasin [*55134-13-9*] and lasalocid (**34**, Lasalocid A) (see ANTIBIOTICS, POLYETHERS). The potent ionophoric antibiotic maduramicin (**35**, Cygro [*84878-61-5*]) is currently being evaluated as an alternative for use against resistant strains of *Eimeria* inasmuch as resistance to the entire class of ionophoric compounds is presenting serious problems. Strains resistant to all coccidiostats arise rapidly because of the widespread use of these drugs. Sulfa drugs, such as sulfadimethoxine (**36**) and sulfaquinoxaline (**37**) are members of the first nontoxic class of anticoccidials (see ANTIBACTERIAL AGENTS, SYNTHETIC, SULFONAMIDES). Other types of compounds that have been considered treatments for the control of coccidiosis are the arsenicals, such as roxarsone (**38**, 4-hydroxy-3-nitrophenylarsonic acid [*121-19-7*]); the alkylidene diphenols; 3,3′-dinitrodiphenyl disulfide; the nitrofurans exemplified by nitrofurazone (**39**, 5-nitro-2-furaldehyde semicarbazone [*59-87-0*], Amifur, Furacin); the triazinones, diclazuril (**40**) and toltrazuril (**41**, Hycox, Maxicox); the carbanilide complexes such as nicarbazin [**42**, bis(4-nitrophenyl) urea 1:1 complex with 4,6-dimethyl-2(1*H*)-pyrimidone, Nicarb]; pyrimethamine (**24**); clopidol (**43**, Coyden); robenidine {**44**, 1,3-bis[(*p*-chlorobenzylidene)amino]guanidine}; and halofuginone (**45**) and its hydrobromide [*64924-67-0*], Stenorol (Table 4).

In cattle, infections by a single species of protozoan are unusual; combinations of the protozoa such as *Eimeria zuernii, Eimeria bovis,* and *Eimeria auburnensis* are more typical. Bovine coccidiosis occurs mainly in calves from three weeks to six months old. Adult animals are usually the carriers but tend to be asymptomatic. Infection in calves occurs by ingestion of the oocysts with their feed and/or water. As with poultry, the severity of the disease is a function of the number of oocysts ingested. Small numbers may cause no disease, whereas a large profusion ultimately may cause death. Severe disease is marked by anemia, weakness, emaciation, and anorexia. Monensin (**33**) is effective against coccidia in both cattle and sheep. Clopidol (**43**), halofuginone (**45**), and amprolium (**46**) are administered as feed additives. Other coccidiostats that have been employed are nitrofurazone (**39**), lincomycin (**47**), lasalocid (**34**), and salinomycin (**32**). Sulfonamides are only partially effective.

Coccidiosis is seen frequently in puppies and kittens and is responsible for diarrhea and even death if the animals are maintained in unsanitary conditions. For dogs and cats there exists no satisfactory treatment; however, the disease is self-limiting. Human infections can frequently be traced to these animals.

Cryptosporidiosis, an intestinal infection caused by protozoa of *Cryptosporidium* species is a taxonomically related disease (12). The disease affects animals, such as calves, lambs, and chickens, and infects humans worldwide, especially infants and children in developing countries. Symptoms range from mild self-limiting diarrhea and abdominal pain to a potentially fatal extreme diarrhea that results in weight loss and poor nutritional absorption. Increasingly, this opportunistic disease occurs in those with a suppressed immune system, especially those with AIDS. The only effective treatment for cryptosporidiosis in AIDS patients who do not respond readily to therapy is spiramycin (**48**).

Table 4. Coccidiosis Antiprotozoal Agents

Structure number	Compound name	CAS Registry Number	Molecular formula	Structure	Applications
(32)	salinomycin	*[53003-10-4]*	$C_{42}H_{70}O_{11}$	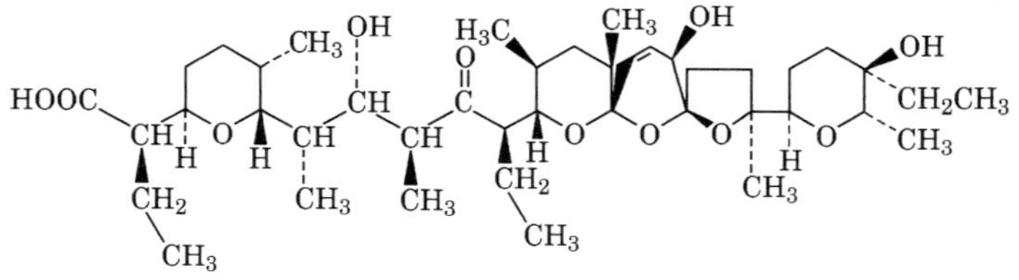	fowl, cattle, sheep
(33)	monensin[a]	*[17090-79-8]*	$C_{36}H_{62}O_{11}$	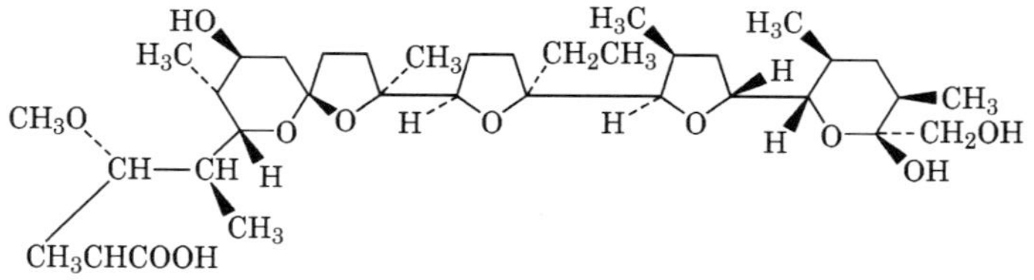	fowl, cattle, sheep
(34)	lasalocid A	*[25999-31-9]*	$C_{34}H_{54}O_8$	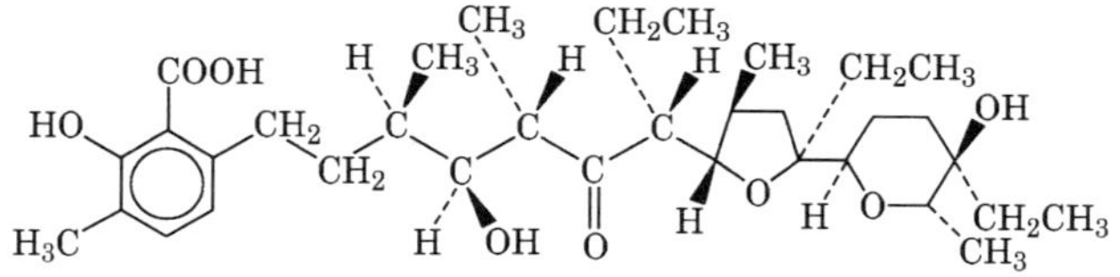	fowl, cattle, sheep
(35)	maduramicin	*[84878-61-5]*	$C_{47}H_{83}NO_{17}$	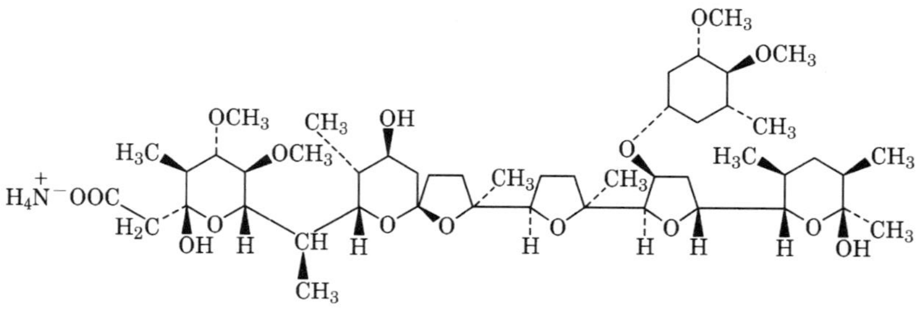	fowl

(**36**)	sulfadimethoxine	[*122-11-2*]	$C_{12}H_{14}N_4O_4S$	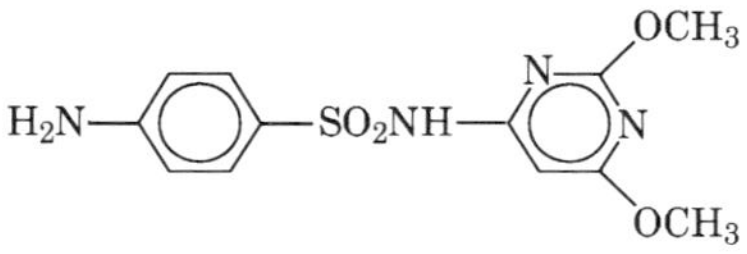	fowl
(**37**)	sulfaquinoxaline	[*59-40-5*]	$C_{14}H_{12}N_4O_2S$	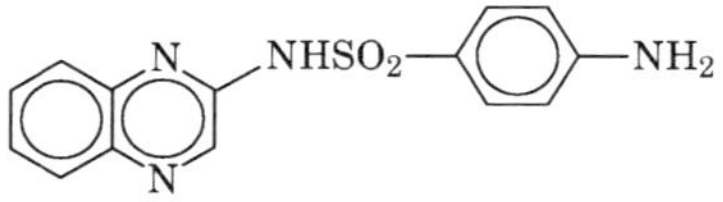	fowl
(**38**)	roxarsone	[*121-19-7*]	$C_6H_6AsNO_6$	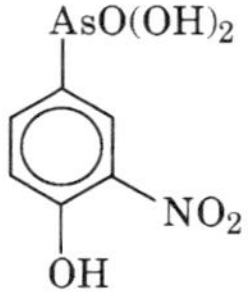	fowl
(**39**)	nitrofurazone	[*59-87-0*]	$C_6H_6N_4O_4$	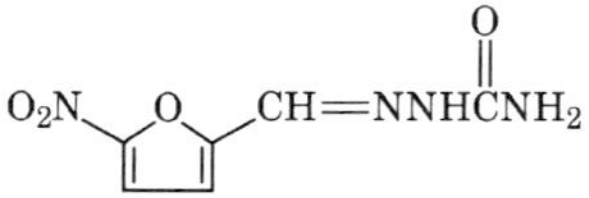	fowl, cattle, sheep
(**40**)	diclazuril	[*101831-37-2*]	$C_{17}H_9Cl_3N_4O_2$	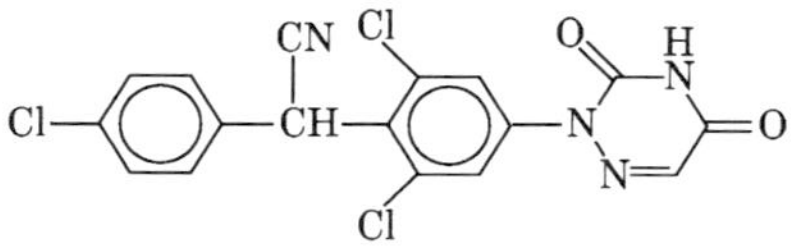	fowl
(**41**)	toltrazuril	[*69004-03-1*]	$C_{18}H_{14}F_3N_3O_4S$	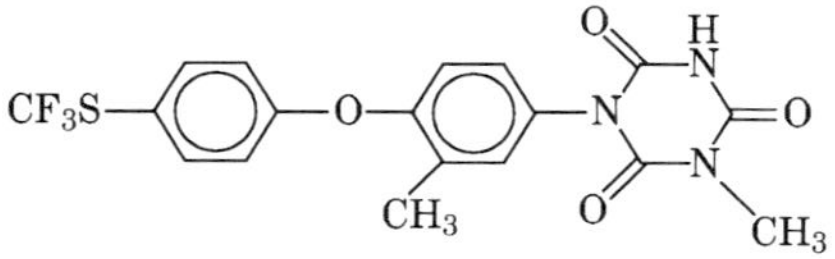	fowl
(**42**)	nicarbazin	[*330-95-0*]	$C_{13}H_{10}N_4O_5 \cdot C_6H_8N_2O$	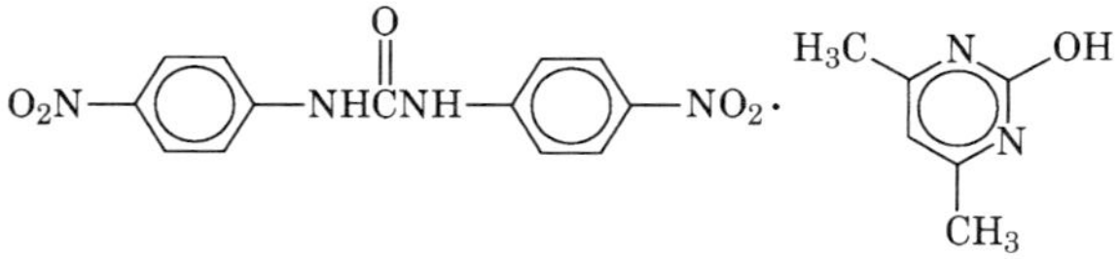	fowl

Table 4. (*Continued*)

Structure number	Compound name	CAS Registry Number	Molecular formula	Structure	Applications
(43)	clopidol	[*2971-90-6*]	$C_7H_7Cl_2NO$	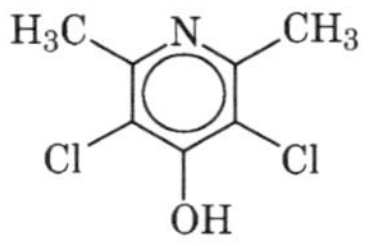	fowl, cattle, sheep
(44)	robenidine	[*25875-51-8*]	$C_{15}H_{13}Cl_2N_5$	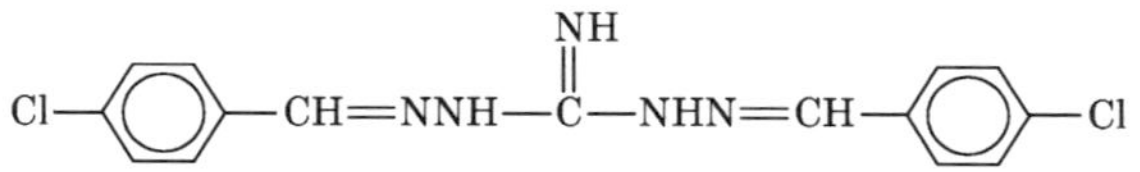	fowl
(45)	halofuginone[b]	[*55837-20-2*]	$C_{16}H_{17}BrClN_3O_3$	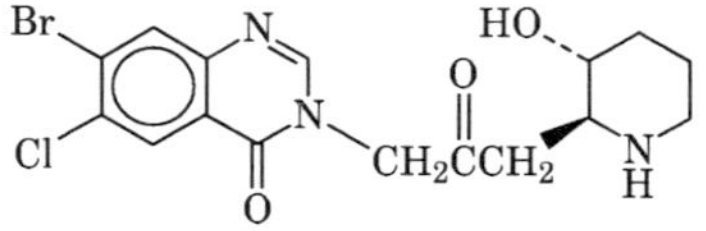	fowl, cattle, sheep
(46)	amprolium	[*121-25-5*]	$C_{14}H_{19}N_4Cl$	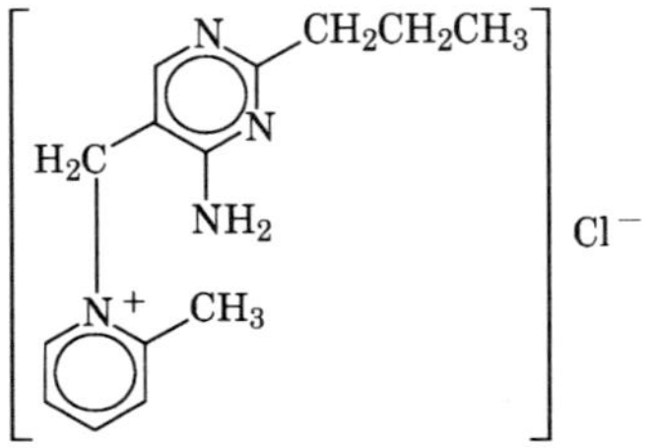	cattle, sheep

(**47**)	lincomycin	[*154-21-2*]	$C_{18}H_{34}N_2O_6S$	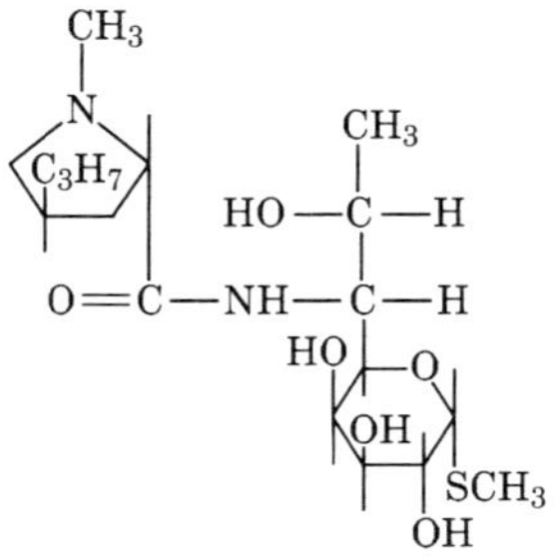	cattle, sheep
(**48**)	spiramycin[c]	[*8025-81-8*]	mixture	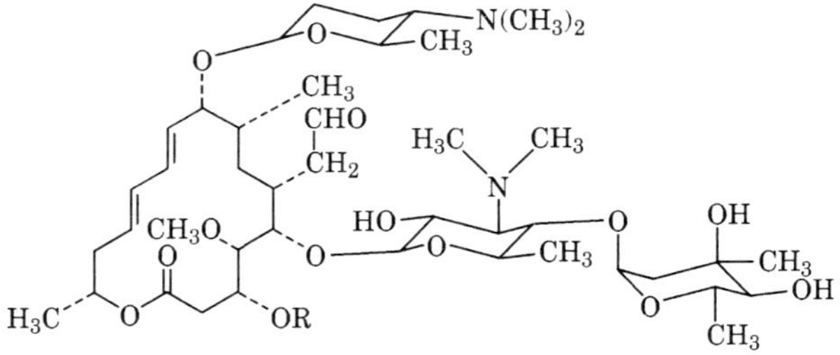	

[a]Also used to treat histomoniasis.
[b]Also used to treat theileriasis.
[c]Also used to treat toxoplasmosis.

Giardiasis

Giardiasis is a water-borne enteric disease of protozoan origin that occurs throughout the world. It is the most prevalent protozoal disease found in humans in the United States; the Rocky Mountain region is a particularly highly endemic area. It is also common in Central America, the former USSR, and India. The infection of the small intestine is caused by the flagellated protozoan, *Giardia lamblia*. The disease is transmitted by drinking water that has been fecally contaminated and poorly purified, or by person-to-person transmission of fecally contaminated food. Besides humans, the host for the protozoan includes the domesticated dog and certain wild animals, especially the beaver. The disease is more prevalent in children than in adults, and is especially common in those attending daycare centers. It is increasing in prevalence among male homosexuals.

Untreated giardiasis may be self-limiting and may produce no apparent symptoms, regardless of the duration of the infection. Alternatively, it may continue for numerous weeks or months and be extremely debilitating with symptoms that include severe and chronic diarrhea, abdominal cramps, weakness, anorexia, vomiting, fatigue, and weight loss. Because fats and fat-soluble nutrients are malabsorbed, the disease may be responsible for some of the mortality seen in underdeveloped countries. In severe giardiasis, there is inflammation and damage to the duodenal and jejunal mucosa.

Treatment of giardiasis is most successful with quinacrine (**49**, mepacrine) or its hydrochloride eg, Atabrine dihydrochloride. The drug may cause side effects such as dizziness, headache, nausea, vomiting, and reversible yellow staining of the skin. Alternatively, metronidazole (**1**) is recommended by the Centers for Disease Control, but because of its potential for mutagenesis or carcinogenesis should not be administered to children or pregnant women. Failures that occur with both drugs necessitate repetition of the therapy. Another drug reported to be effective is furazolidone (**50**), perhaps even more so in children than quinacrine (**49**) (Table 5). The aminoglycoside paromomycin (**8**) is recommended for pregnant women because the usual drugs are contraindicated. Tinidazole (**11**) has also been used successfully, but its activity could not be confirmed in a murine model. The disease has also been treated with acranil {**51**, 1-[(6-chloro-2-methoxy-9-acridyl)amino]-3-(diethylamino)-2-propanol dihydrochloride}. Further details pertaining to giardiasis are available (13).

Hexamitosis

Hexamitosis is a disease of chickens, turkeys, quail, and pheasants in which there is an infectious catarrhal enteritis in the duodenum and small intestine. The disease, caused by the protozoan *Hexamita meleagridis,* occurs in the United States, the United Kingdom, South America, and parts of Europe. It is primarily a disease of young birds under 10 weeks old in which mortality may be as high as 80%. The infected animals develop diarrhea, lose weight rapidly, appear listless and weak, and may eventually die. Hexamitosis is transmitted through contaminated food and water. Carrier birds are adults that have survived earlier in-

Table 5. Giardiasis Antiprotozoal Agents[a]

Structure number	Compound name	CAS Registry Number	Molecular formula	Structure
(49)	quinacrine[b]	*[83-89-6]*	$C_{23}H_{30}ClN_3O$	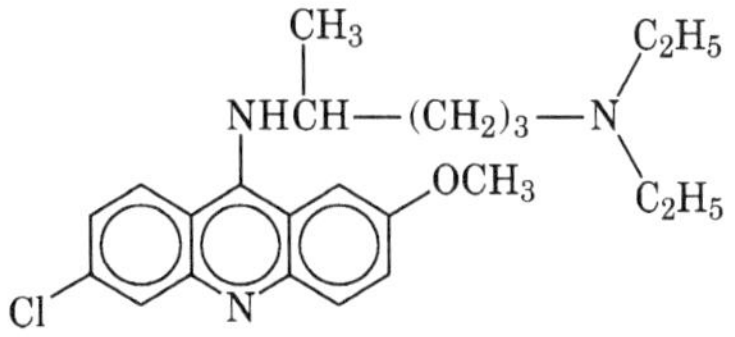
(50)	furazolidone[c]	*[67-45-8]*	$C_8H_7N_3O_5$	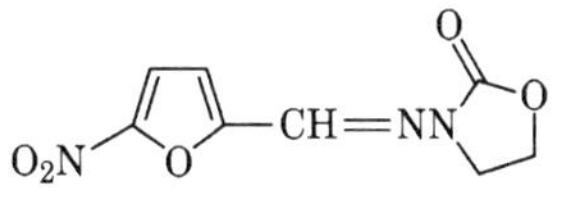
(51)	acranil[d]	*[1684-42-0]*	$C_{21}H_{28}Cl_3N_3O_2$	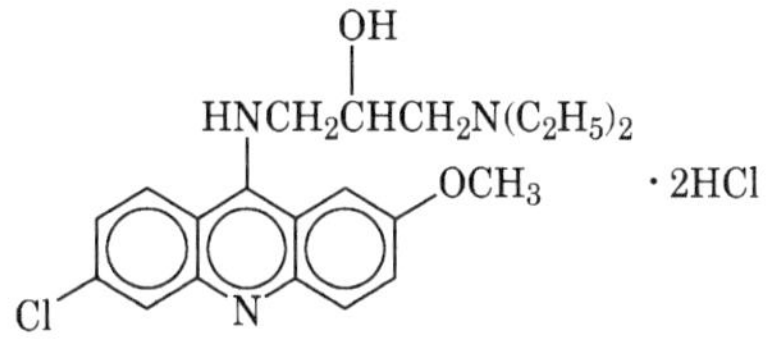

[a]Other applications are indicated in footnotes.
[b]Malaria.
[c]Histomoniasis.
[d]Hexamitosis.

fections. Hot weather and overcrowding may exacerbate the severity of the outbreak.

There is no effective treatment for hexamitosis. Penicillin (**52**), oxytetracycline (**3**), chlorotetracycline (**20**), Enheptin (**53**, 2-amino-5-nitrothiazole *[121-66-4]*), and streptomycin (**54**) sulfate *[3810-74-0]* were found to have limited value. Hexamitosis in carrier pigeons caused by *Hexamita columbae* was treated successfully with ronidazole (**55**, Dugro) (Table 6). An experimental infection of nude mice with *Hexamita muris* was treated with dimetridazole (**56**, 1,2-dimethyl-5-nitro-1*H*-imidazole *[557-92-8]*), metronidazole (**1**), tinidazole (**11**), and acranil (**51**). All compounds lowered or suppressed the fecal discharge of cysts, but the latter reappeared when the 1–3 week treatment terminated. In a related study, dimetridazole (**56**) controlled the clinical disease in mice but did not eliminate the infection.

Histomoniasis

Histomoniasis (histomonas, enterohepatitis, blackhead disease) is primarily an affliction of chickens and turkeys, but it also affects wild populations of peafowl, guinea fowl, pheasant, grouse, quail, and partridge. The disease occurs throughout the world and extensive outbreaks have been responsible for enormous losses of domesticated fowl; however, good sanitation reduces the number of affected animals. Because chickens are carriers of the disease, they must be separated from turkeys, which are more readily infected. Young turkeys from 3 to 12 weeks

Table 6. Hexamitosis and Histomoniasis Antiprotozoal Agents

Structure number	Compound name	CAS Registry Number	Molecular formula	Structure
(52)	penicillin	[*1406-05-9*]		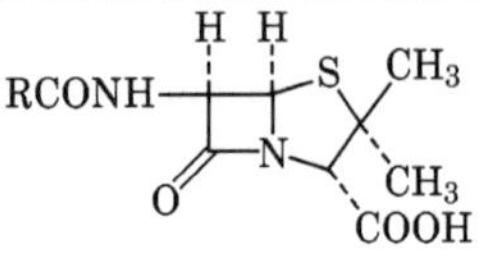
(53)	enheptin	[*121-66-4*]	$C_3H_3N_3O_2S$	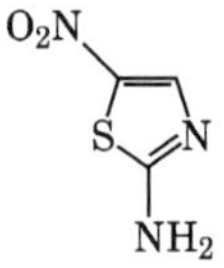
(54)	streptomycin	[*57-92-1*]	$C_{21}H_{39}N_7O_{12}$	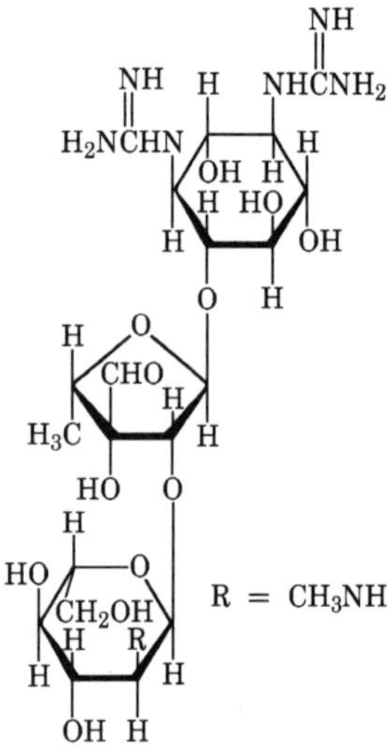
(55)	ronidazole	[*7681-76-7*]	$C_6H_8N_4O_4$	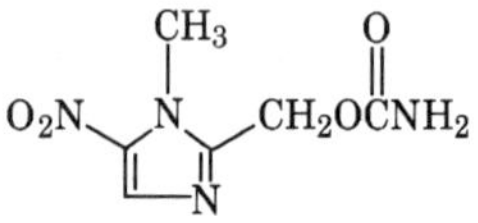
(56)	dimetridazole	[*551-92-8*]	$C_5H_7N_3O_2$	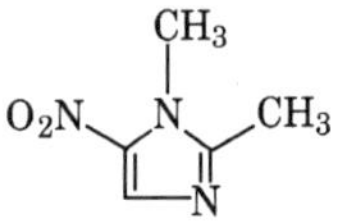
(57)	acetylenheptin	[*140-40-9*]	$C_5H_5N_3O_3S$	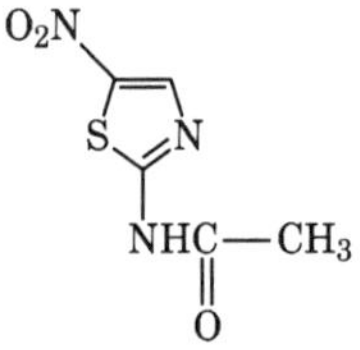
(58)	nithiazide	[*139-94-6*]	$C_6H_8N_4O_3S$	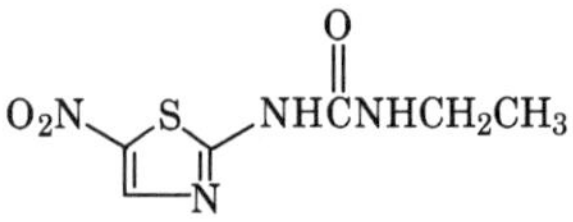
(59)	ipronidazole	[*14885-29-1*]	$C_7H_{11}N_3O_2$	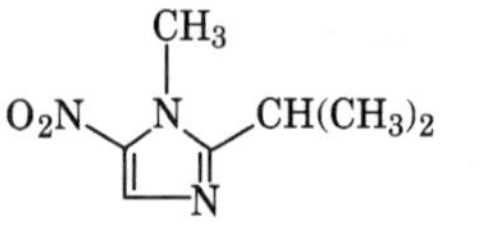

of age are particularly susceptible to the disease and may die within three days of the appearance of the first symptoms. Mortality decreases with age. Diseased birds look weak and droopy, may have diarrhea due to enterohepatitis, the head may become darkened, and death may ensue.

Histomoniasis is caused by the protozoan, *Histomonas meleagridis,* found in the liver and cecum of birds. The protozoans from the cecum are more infective than those from the liver. Transmission primarily occurs by ingestion of the eggs of the cecal worm, *Heterakis gallinarium,* which carries the protozoan and can survive several years in the soil.

A favored treatment for histomoniasis is dimetridazole (**56**) (Table 6). Other useful antiprotozoal agents include the thiazoles, Enheptin (**53**), acetylenheptin (**57**, 2-acetamido-5-nitrothiazole [*140-40-9*], aminitrozole, Enheptin-A), and nithiazide [**58**, 1-ethyl-3-(5-nitro-2-thiazolyl)urea [*139-94-6*], Hepzide], which are administered continuously into the feed for prevention and suppression. Furazolidone (**50**) and ipronidazole (**59**, Ipropran [*14885-29-1*]) have also been used. The latter has been evaluated, in addition, as a growth promotant. In an *in vitro* study, metronidazole (**1**), monensin (**33**), and tinidazole (**11**) showed good antihistomonal activity. Arsenicals that have shown activity are 4-nitrophenylarsonic acid and carbarsone (**31**). The drugs mentioned above, in order to prevent relapses, typically require continuous administration in the drinking water of the poultry until a few days prior to their slaughter, and are responsible for reduced egg production.

Leishmaniasis

Leishmaniasis affects some 12 million humans annually in an area where 350 million are at risk. It is a complex of at least two protozoan diseases, consisting primarily of cutaneous and visceral forms. A mucocutaneous form is considered by some to be another distinct variety. Clinical manifestations of the disease range from an asymptomatic infection to an infection in which there is considerable destruction of cutaneous tissue and mucous membranes. Leishmaniasis can often be fatal, especially in the visceral form. The seriousness of the disease depends on the state of the immunological system of the host and the species of parasite that inflicts the damage (14). The different species of *Leishmania* that are responsible for the disease can be difficult to distinguish from one another and methods to identify them have occupied the efforts of many researchers. The various species may be localized geographically or may overlap, and each is responsible for somewhat different symptoms. Although the human clinical manifestations may be categorized into different types, infections by more than a single species may be present at one time.

The disease is initiated by a bite from an infected female sand fly of the genus *Phlebotomus*. The injected parasites usually multiply near the site of the original infection, causing a lesion. The sand fly vector becomes infected when it acquires a blood meal from the skin or peripheral blood of a parasitized host and ingests the protozoans in the amastigote stage. The amastigotes develop into promastigotes in the intestine of the sand fly and eventually migrate to its proboscis. The saliva of the sand fly appears to enhance the survival of the injected promastigotes which, after phagocytosis, are transformed into the amastigote phase. The numerous reservoirs for the protozoans include humans, dogs, foxes, cats, rodents, and horses.

Cutaneous leishmaniasis is characterized by one or more slowly healing superficial ulcers that may be painful. These lesions are liable to further infection and may remain as open sores or become hard, wartlike nodules. The form of cutaneous leishmaniasis referred to as New World disease is caused by species of the *Leishmania (Viannia) braziliensis* complex (*L. braziliensis, L. guyanensis, L. panamanesis,* and *L. peruviana*) and species of the *Leishmania (Leishmania) mexicana* complex (*L. mexicana, L. amazonensis,* and *L. venezuelensis*) in the western hemisphere. Old World disease (oriental sore, Delhi or Baghdad boil) is caused by *Leishmania tropica* and *Leishmania major* in Asia, Africa, and southern Europe, and *Leishmania aethiopica,* restricted to eastern Africa. The incubation period for cutaneous leishmaniasis ranges from a few weeks to several months. Spontaneous cures can take place in a period ranging from one month to several years.

Visceral leishmaniasis is also known as kala azar, meaning black fever or black sickness in Hindi, because of the characteristic pigmentary changes that occur in the skin. The disease is endemic in much of the tropical and subtropical regions of southern Europe and the Mediterranean, the Middle East, India, Africa, and South and Central America. It is caused by species of the *Leishmania (Leishmania) donovani* complex [*L. donovani* (Old World), *L. infantum* (Old World), and *L. chagasi* (New World)]. In this systemic form of leishmaniasis, the parasites invade internal organs. It is characterized in patients by an enlarged liver and spleen, fever, weight loss, hemorrhage, leukopenia, and anemia. There is abdominal discomfort due to the spleen and liver involvement. Death due to severe diarrhea, pneumonia, and gastrointestinal bleeding may ensue if the disease is not treated. Following treatment, patients may relapse with a leishmanial form that solely affects the skin (post-kala azar dermal leishmaniasis). Rarer forms of leishmaniasis include chronic relapsing or recidivans forms of the cutaneous disease and diffuse cutaneous leishmaniasis.

Certain species of the *L. braziliensis* complex may cause mucocutaneous leishmaniasis, a form seen most frequently in northern and central South America. The patient first develops skin infections which later metastasize to produce highly disfiguring mucocutaneous lesions of the oronasopharynx.

Antimony compounds have been used to treat leishmaniasis ever since tartar emetic (antimony potassium tartrate) was discovered early in the 20th century to have efficacy against the mucocutaneous form of the disease. The cutaneous form has been treated with tartar emetic formulated in an ointment. Many side effects have been seen with this trivalent antimonial, some of which can be ascribed to the difficulty of obtaining pure antimony for its manufacture. These side effects include toxicity to the heart, liver, and kidneys. Other promising trivalent antimonials have been abandoned in favor of pentavalent antimonials with lower toxicity.

Pentavalent antimony preparations constitute the primary treatments for all forms of leishmaniasis, the most important of which are sodium stibogluconate (**60**, Pentostam [*16037-91-5*]) and glucantime (*N*-methylglucamine antimonate, $C_7H_{18}NO_8Sb$, meglumine antimonate) (Table 7). The actual chemical structure of

Table 7. Leishmaniasis Antiprotozoal Agents

Structure number	Compound name	CAS Registry Number	Molecular formula	Structure
(60)	sodium stibogluconate	[*16037-91-5*]	$C_{12}H_{20}O_{17}Sb_2{\cdot}9H_2O{\cdot}3Na$	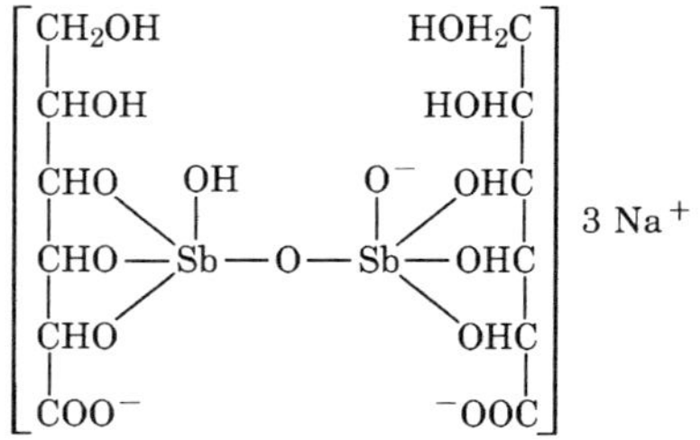
(61)	stilbamidine	[*122-06-5*]	$C_{16}H_{16}N_4$	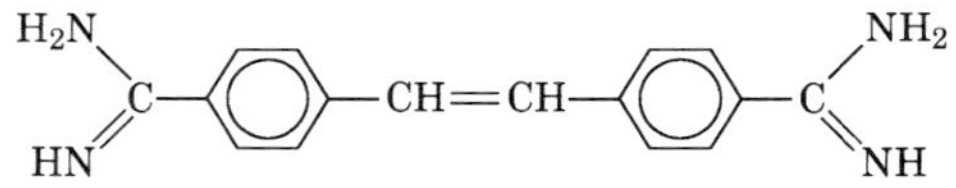
(62)	allopurinol	[*315-30-0*]	$C_5H_4N_4O$	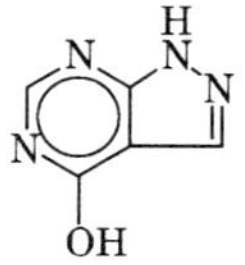
(63)	ketoconazole	[*65277-42-1*]	$C_{26}H_{28}Cl_2N_4O_4$	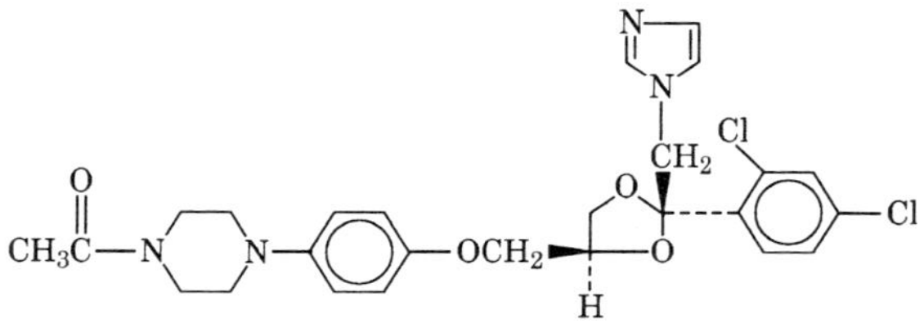

Sodium stibogluconate [*16037-91-5*
Stilbamidine [*122-06-5*], *3*:000
Allopurinol [*315-30-0*], *3*:000
Ketoconazole [*65277-42-1*], *3*:000

the antimonials is unknown. Studies indicate that they are mixtures of compounds with widely differing molecular weights. Use of glucantime is favored in Latin America and French-speaking countries. Cardio- and hepatotoxicity have been observed when the drug is administered over the typical 3–4 weeks of treatment. Liposome preparations of Pentostam and glucantime substantially increased the activity of these antimonials under experimental conditions. In 1990, an estimated 10,000 out of 200,000 cases of kala azar reported in India were unresponsive to treatment with antimonials; in Kenya, the strains were even more resistant to antimony. Although antimony-resistant strains of leishmania can be treated with diamidines, the latter compounds are more toxic. Pentamidine (**26**) as the isethionate, is the most widely used diamidine and has a high cure rate. Another diamidine, stilbamidine (**61**, 4,4′-diamidinostilbene [*122-06-5*]), is also effective.

Amphotericin B (**15**) is an antifungal macrolide antibiotic produced by *Streptomyces nodosus* that has been used as an alternative, albeit more toxic, drug to the antimonials. It acts as a leishmanicide against the visceral and mucocutaneous forms of the disease. To overcome its potentially severe nephrotoxicity, the drug must be administered over an extended period of time.

Because of the outbreak of antimony-resistant leishmaniasis and the need to develop an orally-administered therapy, the use of many other compounds has been considered. Those that appear to have clinical utility are allopurinol (**62**), ketoconazole (**63**), and both systemically and topically applied paromomycin (**8**) (see ANTIPARASITIC AGENTS, ANTIMYCOTICS).

Malaria

Malaria affects an estimated 270 million people and causes 2–3 million deaths annually, approximately one million of which occur in children under the age of five. While primarily an affliction of the tropics and subtropics, it has occurred as far north as the Arctic Circle. The disease essentially has been eradicated in most temperate-zone countries, but some 1100 cases of malaria in U.S. citizens returning from abroad were reported to the Centers for Disease Control during 1990. Malaria is seen today in Southeast Asia, Africa, and Central and South America. It is on the increase in Afghanistan, Brazil, China, India, Mexico, the Philippines, Sri Lanka, Thailand, and Vietnam. Escalation of the disease is because of the discontinued use of the insecticide DDT which effectively kills mosquito larvae, but has been found to be toxic to livestock and wildlife. Also, chloroquine (**6**), a reliable drug for the prophylaxis and treatment of falciparum malaria, is ineffective in many parts of the world because of the spread of drug-resistant strains.

Malaria is transmitted by the bite of an infected female *Anopheles* mosquito, one of the few species of the insect capable of carrying the human malaria parasite. The responsible protozoa are from the genus *Plasmodium,* of which only four of some 100 species can cause the disease in humans. The remaining species affect rodents, reptiles, monkeys, birds, and livestock. The species that infect humans are *Plasmodium falciparum, Plasmodium vivax, Plasmodium malariae,* and *Plasmodium ovale.* Note that concomitant multiple malaria infections are commonly seen in endemic areas, a phenomenon that further complicates choice of treatment.

An infection by *P. falciparum* has an incubation period (time from mosquito bite to clinical symptoms) of 6–25 days (average 12). Infection from this parasite can result in severe anemia, renal failure, pulmonary edema, shock, jaundice, cerebral malaria, and death. The life cycle of *P. falciparum* begins with a bite by an infected female mosquito in need of a blood meal for the reproductive stage of her life. Some parasites in the sporozoite stage are injected with her saliva, which contains material to prevent rapid clotting of the wound. The sporozoites enter the circulatory system and reach the liver in about one hour. The parasites grow and multiply by asexual division inside liver cells for 5–7 days. Then, as merozoites, having multiplied some 40,000-fold, they leave the liver to enter the bloodstream and invade red blood cells where they grow and further multiply over a period of 1–4 days. The merozoites emerge from the ruptured erythrocytes in a synchronous manner in about 48 hours. This results in the patient experiencing the clinical symptoms of the disease, namely, chills with rising temperatures, followed by fever and intense sweating. In addition, there can be severe headache, fatigue, dizziness, nausea, vomiting, anorexia, and diarrhea. Repetition of these symptoms occur on alternate days in this so-called tertian malaria. The falciparum parasites also cause the red cells to adhere to blood vessels, resulting in clumps that diminish the flow of blood to organs such as the brain, liver, and kidneys. Some released blood forms reinfect other red cells, where they again multiply leading to a further concerted rupturing and a return of the malaria symptoms. Some other merozoites develop into the sexual forms, ie, male or female gametocytes. Repetition of the malaria life cycle begins when gametocytes are acquired by an uninfected mosquito upon biting an infected individual. The gametocytes enter the mosquito stomach, undergo fertilization, and the zygotes thus formed migrate to the salivary glands of the insect where they are capable of being transmitted to another human by a bite. Chloroquine-resistant strains of *P. falciparum* have been found in Southeast Asia, Africa, and South America.

Plasmodium vivax, responsible for the most prevalent form of malaria (benign tertian), has an incubation period of 8–27 days (14 average). A variety seen in northern and northeastern Europe has an incubation period as long as 8–10 months. The disease can cause splenic rupture and anemia. Relapses (renewed manifestations of erythrocytic infection) can occur with this type of malaria. Overall, *P. vivax* is still susceptible to chloroquine; however, resistant strains have been reported from Papua New Guinea and parts of Indonesia. *Plasmodium malariae,* the cause of quartan malaria, has an incubation period of 15–30 days and its asexual cycle is 72 hours. This mildest form of malaria can cause nephritis in addition to the usual symptoms. It is a nonrelapsing type of malaria but the red blood cell infection can last for many years. No resistance to chloroquine by this plasmodium has been reported. *Plasmodium ovale,* responsible for ovale tertian malaria, has an incubation period of 9–17 days (15 average). Relapses can occur in people infected with this plasmodium. No chloroquine resistance has been reported for this parasite.

Antimalarials. Antimalarials can be categorized according to their mode of action (15).

Tissue Schizonticides. These eradicate the liver stages of the parasite and thereby prevent their entry into the blood. As a class, therefore, they are useful for prophylaxis. Some tissue schizonticides can act on the long-lived tissue forms (hypnozoites) of *P. vivax* and *P. ovale* and thus can cure the latter infections by preventing relapses.

Blood Schizonticides. These destroy the erythrocytic stages of the parasites and are useful for the clinical cure of falciparum malaria or suppression of relapsing infections.

Gametocytocides. These annihilate the sexual forms of the plasmodia (gametocytes) and also destroy the stages of the parasites in the *Anopheles* mosquito.

Sporontocides. These act against the sporozoites and oocysts in the mosquito and thereby prevent the transmission of the disease.

It should be noted that drugs may operate by more than one mechanism, and may possess a specific mode of action against one species of plasmodium but lack efficacy against others. In addition, antimalarial drugs may be classified according to their structural types.

Quinine (**22**), the first known antimalarial, is a 4-quinolinemethanol that has served as a model for the design of numerous antiplasmodial drugs. Its history began in mid-seventeenth century Peru when the Incas told the Jesuits of the medicinal properties of the bark of an evergreen mountain tree they called quinaquina (later called cinchona). The bark, when made into an aqueous infusion, was capable of curing malaria. The use of cinchona spread to Europe and the alkaloid from it, quinine, was isolated in 1820. Quinine, an extremely bitter substance, has been used by millions of malaria sufferers. In recent times, it has been employed successfully to treat chloroquine-resistant strains of *P. falciparum* but it frequently fails to provide a complete cure of the infection. Quinine acts on the asexual blood forms of the plasmodia in a manner slower than many synthetic drugs. Overdoses cause tinnitus and visual disturbances, side effects that disappear on withdrawal of the drug. It can also cause premature contractions in women in late stages of pregnancy. Although quinine is a favored antimalarial for parenteral administration, it is nevertheless hazardous by this route. Quinidine (**64**), has been shown to be even more effective in combatting the disease (Table 8). However, it has undesirable cardiac side effects that reduce its suitability as an antimalarial. Mixtures of cinchona alkaloids, known as totaquine, are easier to produce and have been employed in treatment. Totaquine has been standardized to contain a minimum of 15% quinine.

The success of quinine inspired the search for other antimalarials. The greatest impetus for the development of synthetic drugs came this century when the two World Wars interrupted the supply of cinchona bark to the combatants. A structurally related 4-quinolinemethanol is mefloquine (**65**, Lariam [*51773-92-3*]), which now serves as an effective alternative agent for chloroquine-resistant *P. falciparum.* This is a potent substance that requires less than one-tenth the dose of quinine to effect cures. There are some untoward side effects associated with this drug such as gastrointestinal upset and dizziness, but they tend to be transient. Mefloquine is not recommended for use by those using beta-blockers, those whose job requires fine coordination and spatial discrimination, or those with a history of epilepsy or psychiatric disorders. A combination of mefloquine with Fansidar (a

Table 8. Malaria Antiprotozoal Agents[a]

Structure number	Compound name	CAS Registry Number	Molecular formula	Structure
		4-Quinolinemethanols		
(64)	quinidine (sulfate)	[*50-54-4*]	$C_{20}H_{24}N_2O_2{\cdot}\frac{1}{2}H_2O_4S$	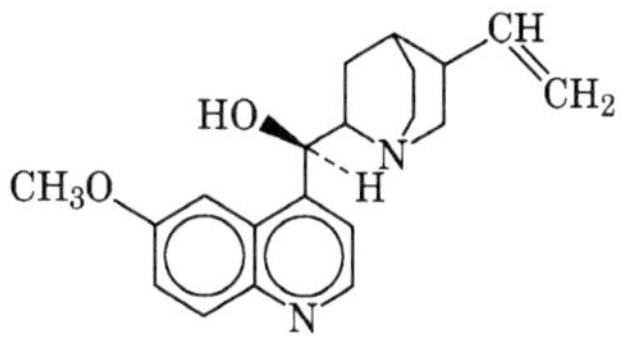
(65)	mefloquine	[*53230-10-7*]	$C_{17}H_{16}F_6N_2O$	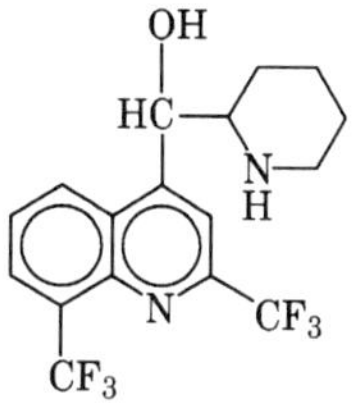
		Phenanthrene-methanol		
(66)	halofantrine	[*69756-53-2*]	$C_{26}H_{30}Cl_2F_3NO$	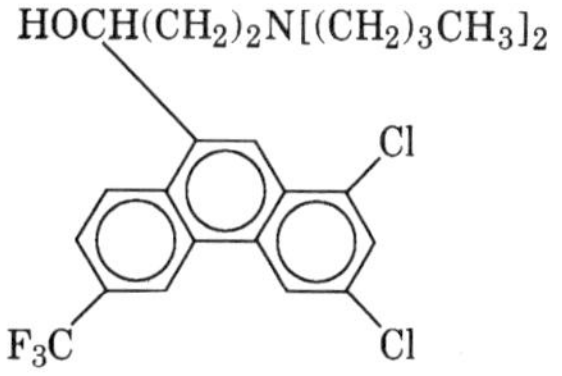
		4-Aminoquinoline		
(67)	hydroxy-chloroquine	[*118-42-3*]	$C_{18}H_{26}ClN_3O$	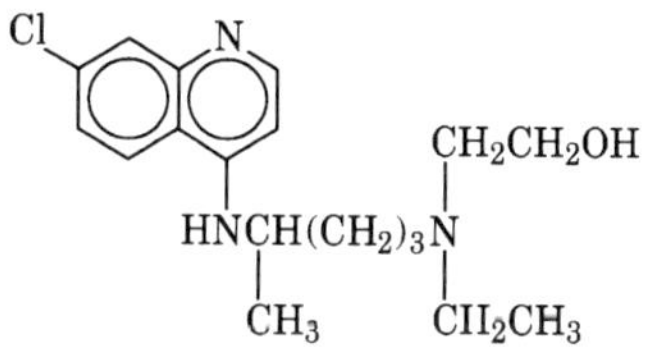
		8-Aminoquinolines		
(68)	pamaquine[b]	[*491-92-9*]	$C_{19}H_{29}N_3O$	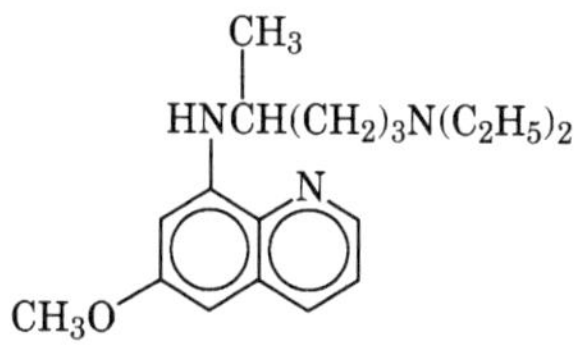
(69)	primaquine (phosphate)[c,d]	[*63-45-6*]	$C_{15}H_{21}N_3O{\cdot}2H_3O_4P$	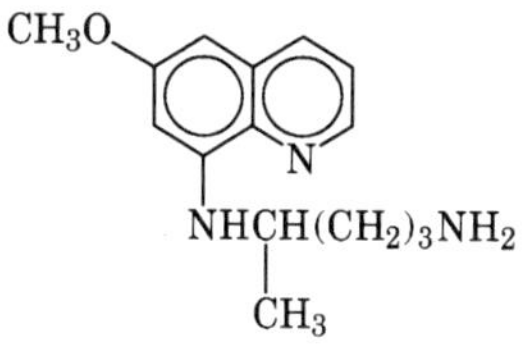
		Antifolates		
(70)	dapsone	[*80-08-0*]	$C_{12}H_{12}N_2O_2S$	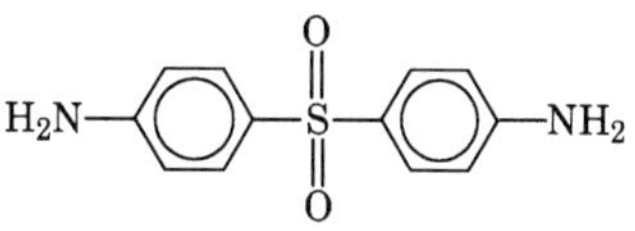

Table 8. ***(Continued)***

Structure number	Compound name	CAS Registry Number	Molecular formula	Structure
(71)	sulfadoxine[c]	*[2447-57-6]*	$C_{12}H_{14}N_4O_4S$	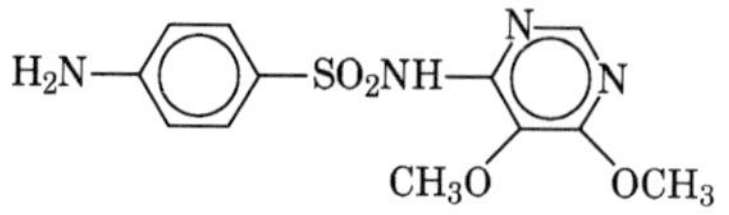
(72)	sulfalene[e]	*[152-47-6]*	$C_{11}H_{12}N_4O_3S$	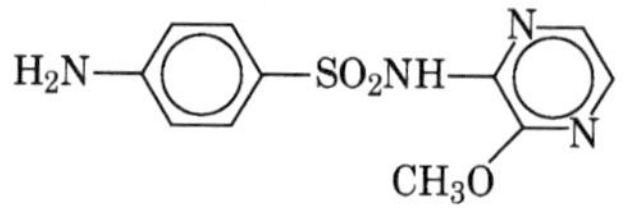
(73)	trimethoprim[c,e]	*[738-70-5]*	$C_{14}H_{18}N_4O_3$	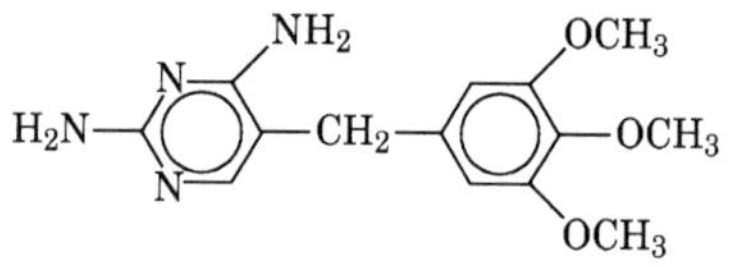
(74)	chlorguanide	*[500-92-5]*	$C_{11}H_{16}ClN_5$	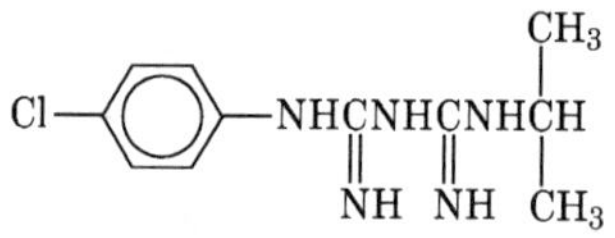
(75)	cycloguanil	*[516-21-2]*	$C_{11}H_{14}ClN_5$	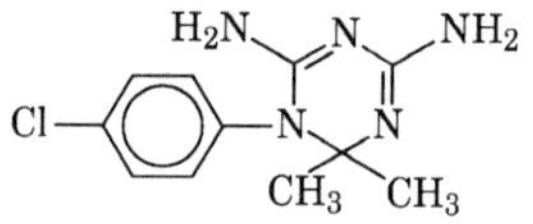
		Others		
(76)	menoctone[b]	*[14561-42-3]*	$C_{24}H_{32}O_3$	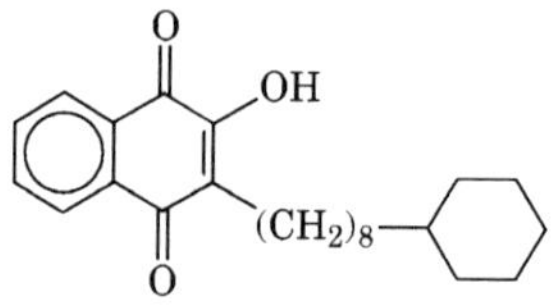
(77)	pyronaridine	*[74847-35-1]*	$C_{29}H_{32}ClN_5O_2$	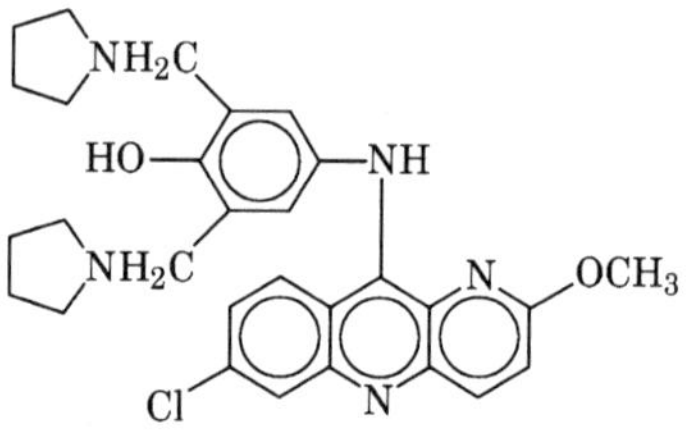
(78)	doxycycline	*[564-25-0]*	$C_{22}H_{24}N_2O_8$	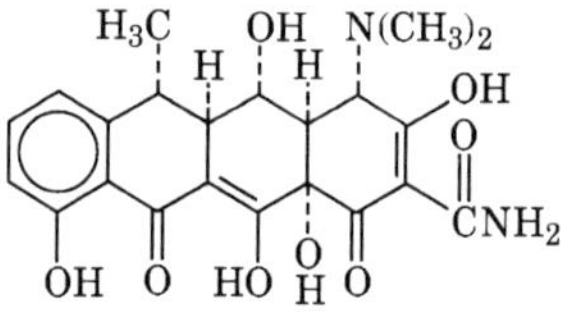
(79)	chloram-phenicol	*[56-75-7]*	$C_{11}H_{12}Cl_2N_2O_5$	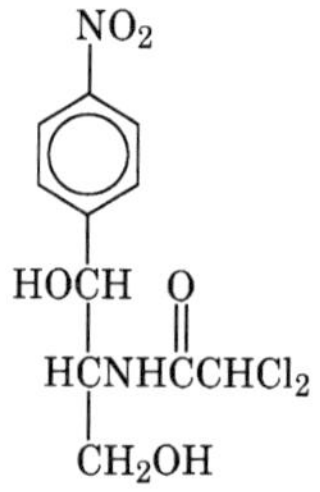

Table 8. (*Continued*)

Structure number	Compound name	CAS Registry Number	Molecular formula	Structure
(80)	febrifugine	[*24159-07-7*]	$C_{16}H_{19}N_3O_3$	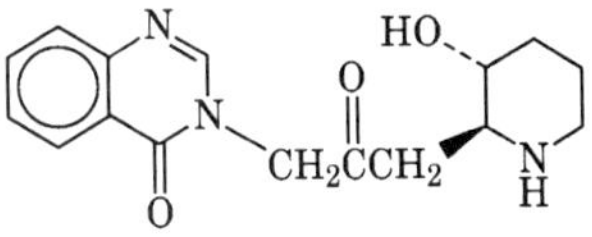
(81)	artemisinin	[*63968-64-9*]	$C_{15}H_{22}O_5$	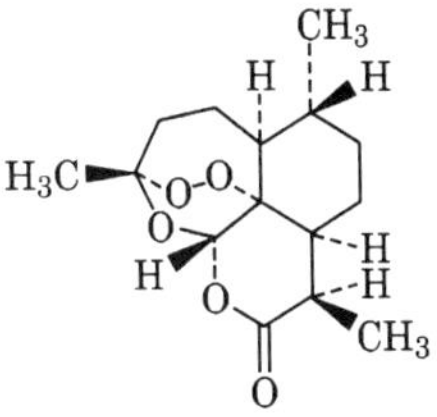
(82)	artemether	[*71963-77-4*]	$C_{16}H_{26}O_5$	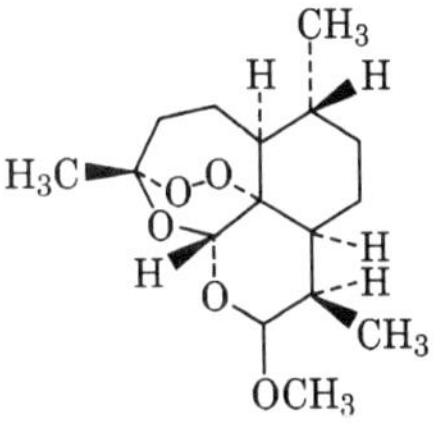
(83)	arteether	[*75887-54-6*]	$C_{17}H_{28}O_5$	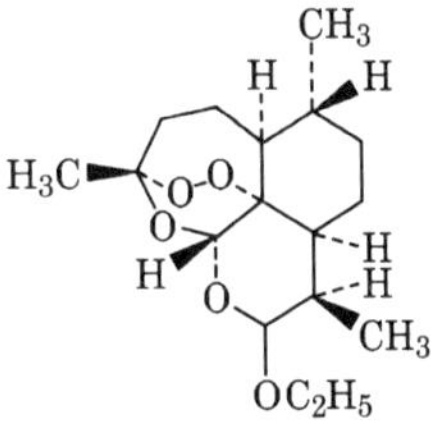
(84)	artesunic acid	[*88495-63-0*]	$C_{19}H_{28}O_8$	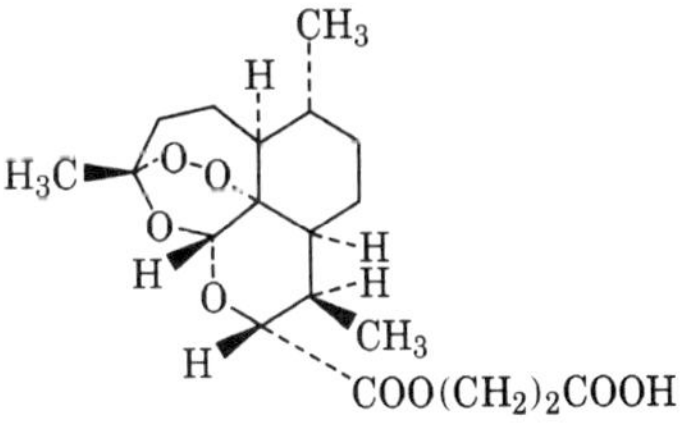
(85)	artelinic acid	[*109637-83-4*]	$C_{23}H_{30}O_7$	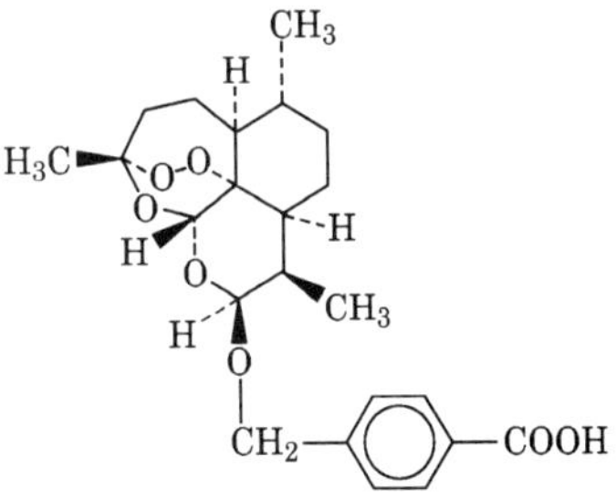

[a]Other applications are indicated in footnotes.
[b]Theileriasis.
[c]Pneumocystosis.
[d]American trypanosomiasis.
[e]Toxoplasmosis.

mixture of pyrimethamine and sulfadoxine) is known as Fansimef but its use is not recommended. Resistance to mefloquine has been reported even though the compound has not been in wide use.

The best example of the class of phenanthrene-methanols is halofantrine (**66**, Halfan [*36167-63-2*]), a drug that is effective against chloroquine-resistant malaria and is now being evaluated in Africa. It produces temporary gastrointestinal disturbances.

Chloroquine (**6**) is the most effective of the hundreds of synthesized 4-aminoquinolines and was the antimalarial of choice until resistance developed to it in Colombia and Southeast Asia in 1962. It is still one of the most important antimalarial agents in use for suppression (prophylaxis) in endemic areas where there remains sensitivity to the drug. It acts as a blood schizonticide and rarely produces serious side effects in individuals taking it prophylactically. Ironically, it was the widespread prophylactic use that most likely led to drug resistance. For treatment of the disease, the phosphate salt is administered orally, or the chloride, parenterally. The minor side effects include gastrointestinal upset, headache, dizziness, blurred vision, and pruritus (itch), but discontinuance of the drug is seldom necessary. Those who have used chloroquine continually for more than 6 years, either for prophylaxis or in the treatment of arthritis, are recommended to have ophthalmologic exams to check for possible retinal damage. It was demonstrated that chloroquine resistance can be reversed *in vitro* by treating *P. falciparum* parasitized erythrocytes with the calcium channel blocking drug verapamil (16). Subsequently, other calcium channel blockers and antihistamines have been found to be effective in overcoming chloroquine resistance *in vitro* and in rodents. Desipramine (Norpramin) and other antidepressants have been shown to reverse chloroquine resistance in monkeys. Alternatives to chloroquine for treatment are amodiaquine (**21**) and hydroxychloroquine (**67**), a chloroquine modification in which one of the ethyl groups on the tertiary nitrogen atom of the side chain is replaced with hydroxyethyl. These compounds are less effective, and amodiaquine is more toxic than chloroquine. The 4-aminoquinolines also act as gametocytocidal agents against *P. vivax, P. ovale, P. malariae* and immature gametocytes of *P. falciparum.*

The 8-aminoquinolines are effective against both the primary and secondary tissue forms and sexual blood forms of the parasites. At toxic levels they are also active against asexual blood forms in humans. Pamaquine (**68**, plasmochin, plasmoquine), the oldest useful member of the class of 8-aminoquinolines, was synthesized in Germany in the mid-1920s. Primaquine [*90-34-6*] (**69**), which is less toxic and more effective, is the most widely used of the 8-aminoquinolines. It is gametocytocidal against all species of human malaria parasites and is an antirelapsing drug, but has the undesirable ability to cause severe hemolysis in glucose-6-phosphate dehydrogenase (G6PD) deficient individuals. This accounts for its recommended administration in limited doses over 1–2 weeks and its low utilization. During pregnancy, it is considered inadvisable for primaquine to be taken as it could be passed on to a G6PD-deficient fetus, thereby causing hemolytic anemia. Primaquine is generally used in conjunction with a blood schizonticide such as chloroquine (**6**), amodiaquine (**21**), or pyrimethamine (**24**) and is used for the prevention of relapses only against *P. vivax* and *P. ovale.*

The drugs known as antifolates act as effective blood schizonticides. Unfortunately, the parasites readily develop resistance to them. Most antifolates show poor oral tolerance, absorption, and host toxicity. They fall into two types depending on the mechanisms by which they operate.

The first are competitors of PABA (*p*-aminobenzoic acid) and thus interrupt host *de novo* formation of the tetrahydrofolic acid required for nucleic acid synthesis. Examples of drugs that fall into this group are the sulfones and sulfonamides. The most well-known of the sulfones is dapsone (**70**, 4,4′-diaminodiphenyl sulfone, DDS), whose toxicity has discouraged its use. Production of folic acid, which consists of PABA, a pteridine unit, and glutamate, is disturbed by the substitution of a sulfonamide (structurally similar to PABA). The antimalarial sulfonamides include sulfadoxine (**71**, Fanasil [*2447-57-6*]), sulfadiazine (**25**), and sulfalene (**72**, sulfamethoxypyrazine [*152-47-6*], Kelfizina). Compounds of this group are rapidly absorbed but are cleared slowly.

The second type of antifolates bind preferentially with, and thus selectively inhibit, the enzyme dihydrofolate reductase contained in the plasmodia. This interferes with the ability of the malaria parasites to convert dihydrofolate to tetrahydrofolic acid. In the erythrocyte host, however, dihydrofolate reductase is considerably less sensitive to these drugs and the blood cells are capable of utilizing exogenous folate. Dihydrofolate reductase inhibitors are potent blood schizonticides that act on the asexual blood stages. Members of this class include the pyrimidines, pyrimethamine (**24**), and trimethoprim (**73**). Pyrimethamine is a slow-acting drug that is not recommended for use in acute attacks and is potentiated by combination with sulfadoxine (**71**) (mixture is called Fansidar), dapsone (**70**) (mixture is called Maloprim), sulfalene (**72**) (mixture is called Kelfimeta or Metakelfin), or chloroquine (**6**) (mixture is called Darachlor). It is an effective suppressant against *P. falciparum*. The mode of action of trimethoprim (**73**) is similar to that of pyrimethamine (**24**). Because it is also slow-acting, it is generally used in combination with the fast-acting sulfonamide, sulfalene (**72**), to give an effective treatment against *P. falciparum*. Related structurally and functionally are the so-called open and cyclic triazines, exemplified by chlorguanide (**74**, proguanil [*500-92-5*], or the hydrochloride, Paludrine), a biguanide, and cycloguanil (**75**, Camolar [*609-78-9*]), respectively. The compounds have low toxicity and are also effective as causal prophylactics, ie, they destroy parasites before they enter the red blood cells.

Quinones and naphthoquinones were explored during the World War II Antimalarial Drug Program. Now that chloroquine resistance is a serious problem, compounds of this group such as menoctone (**76**) are being reinvestigated.

Quinacrine (**49**) is an acridine that was used extensively from the mid-1920s to the end of World War II. It acts much like chloroquine and is reasonably effective. Because it causes the skin to turn yellow and, in high doses, causes yellow vision, the drug is no longer in use as an antimalarial. Pyronaridine (**77**), a 1-azaacridine developed in China, appears to be effective against mefloquine-resistant, but not entirely against chloroquine-resistant, strains of *P. falciparum*.

Some antibiotics, such as the tetracyclines, tetracycline (**7**), doxycycline (**78**), and minocycline (**17**), chloramphenicol (**79**), and clindamycin (**23**) have modest antimalarial properties, but are slow-acting.

A Chinese traditional herbal treatment for malaria obtained from the roots of *Dichroa febrifuga* is called Ch'ang Shan and was investigated in the 1940s.

Febrifugine (**80**), the alkaloid responsible for its activity, was isolated and found to be considerably more active than quinine in experimental infections. Unfortunately, the drug caused nausea and vomiting in humans. Synthesized analogues were generally less effective than the parent.

The most promising group of new antimalarials is based on the compound artemisinin (**81**), an unusual sesquiterpene lactone endoperoxide first isolated in China in 1972 and then in the United States from the weed *Artemisia annua* (17). Artemisinin, also called qinghaosu, meaning in Chinese "extract of green herb," is a rapid-acting and relatively nontoxic therapeutic agent. Its low oil and water solubility has been increased by making certain structural modifications to the molecule. Oil-soluble artemether (**82**), the methyl ether of the lactol dihydroartemisinin, has been administered clinically in China by the intramuscular route. The closely related ethyl ether, arteether (**83**), is being developed by the World Health Organization as an alternative oil-soluble analogue. The sodium salts of artesunic acid (**84**) and artelinic acid (**85**) are the most interesting water-soluble modifications of artemisinin. If administered intravenously, these sodium salts could be advantageously applied in the treatment of cerebral malaria where rapid return to consciousness of the comatose patient and reduction of the parasitemia are the primary clinical goals. Although the water-soluble compound sodium artelinate is more stable in aqueous solution than sodium artesunate, only the latter has been tested clinically. Artelinic acid and related compounds effectively eliminated or prevented the establishment of parasitemia in *P. berghei*-infected mice when they were administered by the transdermal route. Artemisinin and its derivatives have been used successfully in China with several thousand *P. falciparum* and *P. vivax* patients; because this class of drugs is new, parasite resistance has not yet been encountered.

Pneumocystosis

The organism responsible for pneumocystosis in humans, *Pneumocystis carinii,* having the characteristics of both protozoan and fungi, has defied simple taxonomic classification. Recent evidence suggests that it falls into the fungi category. Acute pneumocystosis rarely strikes healthy individuals, although the organism is harbored by a wide variety of animals and most people without any apparent adverse effect. *P. carinii* becomes active only in those individuals who have a serious impairment of their immunologic systems (18). Typically, this opportunistic disease occurs in AIDS patients, 80% of whom ultimately contract *P. carinii* pneumonia, and is one of the main causes of mortality. The disease also occurs in those being administered immunosuppressive drugs for, eg, organ transplantation or treatment of malignant disease. In addition, pneumocystosis is seen in malnourished infants whose immunological systems are impaired. The disease is characterized by a severe pneumonia caused by a rapid multiplication of the organisms almost exclusively in lung tissue. If left to run its course, the acute disease is generally fatal.

The drug of choice is the antifolate mixture, trimethoprim (**73**)-sulfamethoxazole (**86**) (mixture is called co-trimoxazole, Bactrim, Septra) given orally or intravenously (Table 9). It is well tolerated if given in low doses. Also used is pyrimethamine (**24**) combined with sulfadiazine (**25**) or sulfadoxine (**71**). Pentamidine (**26**) isethionate is administered parenterally but is generally not prescribed for prophylactic use because there is a high incidence of hypotension, renal failure, and pain and tissue injury at the injection site. However, there have been successful prophylactic trials with the drug in the aerosolized form for inhalation. The polyamine-inhibitor eflornithine, (**87**, α-difluoromethylornithine, DFMO; [*67037-37-0*] the hydrochloride monohydrate is Ornidyl) is undergoing evaluation as a treatment for *P. carinii*. A combination of clindamycin (**23**) and primaquine (**69**) shows great promise against this protozoan in clinical trials when administered for prophylaxis or for therapy in mild to moderately severe pneumocystis pneumonia. Either drug alone is not effective.

Theileriasis

Theileriasis (theileriosis) is a tick-transmitted protozoan disease of cattle seen primarily in central and eastern Africa. The disease not only affects domesticated cattle, but also ox, zebu, water buffalo, and African buffalo. The most significant form of the disease is caused by the protozoan *Theileria parva (Piroplasma kochi, Piroplasma parvum, Theileria kochi, Theileria lawrencei)* and is known as African Coast fever, East Coast fever, bovine theileriasis, or Corridor disease. The most important vector of the disease is the tick, *Rhipicephalus appendiculatus,* which harbors the protozoan through one molt. About 90–100% adult cattle infected with *T. parva* die, but calves seem to be less susceptible. After receiving a bite from an infected tick, there is an incubation period of 8–25 days before the appearance of the disease. In the acute form, the animals have a fever, cease to eat, may have a nasal discharge and lachrymate, have general weakness, and diarrhea. Rapid and labored breathing precede death.

There are other milder forms of theileriasis that resemble the disease caused by *T. parva* infection. They also primarily affect cattle and are transmitted by ticks. Causative organisms include *Theileria annulata,* responsible for tropical thieleriasis, tropical piroplasmosis, Egyptian fever, and Mediterranean Coast fever. These affect ox, zebu, and water buffalo. Endemic areas include northern Africa, southeastern Europe, the southern part of the former USSR, and Asia. *Theileria mutans* is the cause of benign bovine thieleriasis, Marico calf disease, and mild gall sickness. It affects ox and zebu in Africa, Asia, southern Europe, England, the former USSR, Australia, and North America.

There are no fully effective therapeutic agents for the treatment of theileriasis. Chlorotetracycline (**20**) and oxytetracycline (**3**) have therapeutic activity during the incubation period. Pamaquine (**68**) was reported to have a specific effect on the erythrocyctic forms. Other drugs with limited efficacy are imidocarb (**19**) on *T. annulata,* halofuginone (**45**) on both *T. annulata* and *T. parva,* and the naphthoquinones menoctone (**76**), parvaquone (**88**), and buparvaquone (**89**) on *T. parva*. Methotrexate (**90**) has been found to be active *in vitro* (Table 9).

Table 9. Various Antiprotozoal Agents

Structure number	Compound name	CAS Registry Number	Molecular formula	Structure
			Pneumocystosis	
(86)	sulfamethoxazole[a]	*[723-46-6]*	$C_{10}H_{11}N_3O_3S$	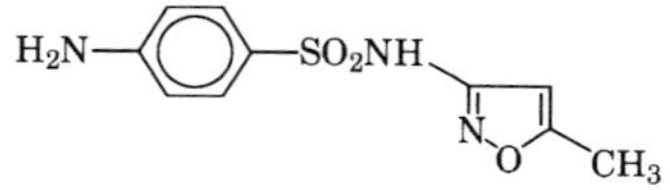
(87)	eflornithine[b]	*[67037-37-0]*	$C_6H_{12}F_2N_2O_2$	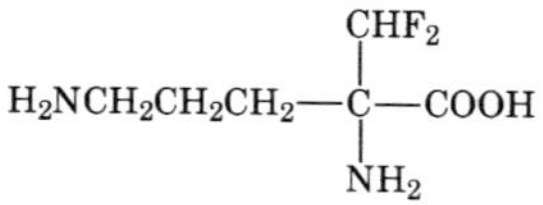
			Theileriasis	
(88)	parvaquone	*[4042-30-2]*	$C_{16}H_{16}O_3$	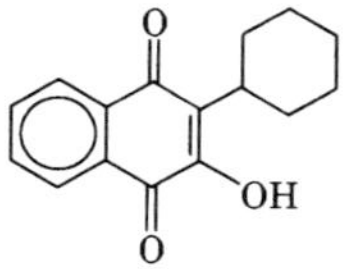
(89)	buparvaquone	*[88426-33-9]*	$C_{21}H_{26}O_3$	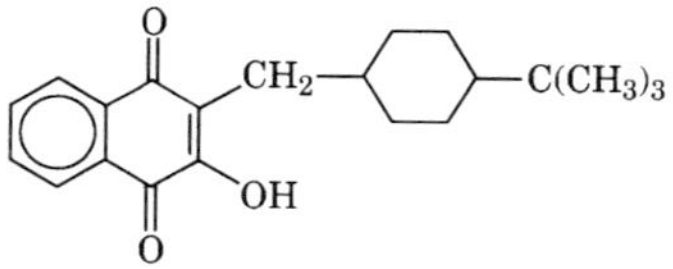
(90)	methotrexate	*[59-05-2]*	$C_{20}H_{22}N_8O_5$	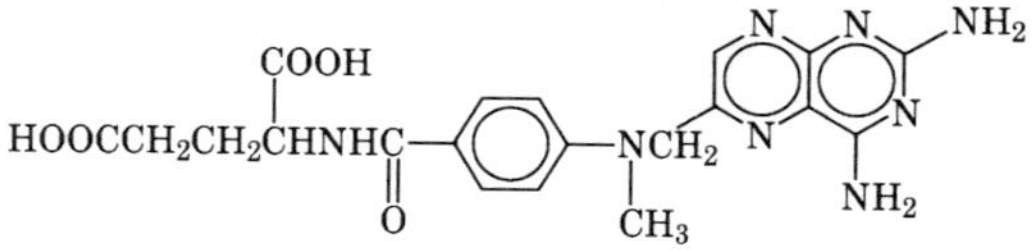
			Trichomoniasis	
(91)	ornidazole	*[16773-42-5]*	$C_7H_{10}ClN_3O_3$	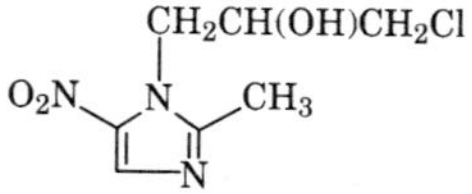

[a]Also used for toxoplasmosis.
[b]Also used for African trypanosomiasis.

Toxoplasmosis

Toxoplasmosis, a coccidial disease, affects both humans and animals throughout the world. Approximately one-third of the population has antibodies to it but is asymptomatic. Although rare in the past, toxoplasmosis is an increasingly important opportunistic disease that afflicts immunocompromised hosts such as post-transplant patients taking immunosuppressive drugs and AIDS patients. The disease is responsible for a central nervous system infection with symptoms in humans that include fever, headache, swollen lymph glands, cough, sore throat, nasal congestion, anorexia, and skin rash. Complications include neurological abnormalities, seizures, and fatal toxoplasmic encephalitis. In pregnant women, toxoplasmosis can cause miscarriage, premature birth, or blindness in the fetus and is a leading cause of birth defects. In the newborn child, the disease may manifest itself primarily in the eye as chorioretinitis (fetal inflammation of the retina), but it may also cause hydrocephalus, microcephalus, carditis, and hepatitis.

Toxoplasmosis is caused by the protozoan *Toxoplasma gondii*. In humans, the disease is generally acquired by eating inadequately cooked meat or by association with infected felines. Meat becomes infected when cows or sheep graze in pastures contaminated by fecal matter from infected felines. Cats, the most important reservoir of the disease, become carriers by eating infected rodents or by the ingestion of infected feline feces. Pregnant women are advised to avoid eating insufficiently cooked meat, avoid having close contact with cats, and, especially, avoid cleaning feline litter boxes.

The cyst of *T. gondii* is persistant, making long-term therapy a requirement in patients whose immune systems are compromised. Treatment of cerebral or ocular toxoplasmosis is generally accomplished with antifolate compounds consisting of a combination of pyrimethamine (**24**) and a long-acting sulfonamide such as sulfadiazine (**25**). Sometimes the latter is replaced by triple sulfonamides or sulfalene (**72**). There is evidence that the related mixture of trimethoprim (**73**) plus sulfamethoxazole (**86**) (mixture is called Bactrim) is nearly as effective but less toxic than the pyrimethamine combinations. Additional drugs for toxoplasmosis treatment are the antibiotics spiramycin (**48**), used alone or in combination with pyrimethamine or a sulfonamide, and clindamycin (23). Clindamycin has been used by itself (reportedly not with great success) and in combination with pyrimethamine to treat chorioretinitis and toxoplasmic encephalitis in AIDS patients. It is especially useful for patients who cannot tolerate extended regimens of sulfonamides.

Trichomoniasis

Trichomoniasis is a widespread disease of the urogenital tract caused by the protozoan *Trichomonas vaginalis*. This sexually transmitted disease affects humans only and has no known animal reservoirs (19). While it primarily affects women, *T. vaginalis* can infect men as well. The disease is marked in women by the development of purulent, frothy, vaginal discharge with a highly disagreeable odor. The role of *T. vaginalis* in producing inflammation of the male genital tract is not fully understood. The protozoan primarily invades the urethra, but may also enter the seminal vesicles or prostate. In a vast majority of infected males, the disease is asymptomatic, perhaps because they are protected by the high zinc content of prostatic secretions. Successful invasion of the prostate by the protozoan can lead to painful urination.

Treatment by the 5-nitroimidazole class of antiprotozoal agents can destroy most strains of *T. vaginalis;* however, some strains have been reported to be resistant to these drugs. The agent of choice is orally administered metronidazole (**1**), the first systemic antitrichomonal agent, introduced in 1960. Other members of this therapeutic group include niridazole (**10**), tinidazole (**11**), and ornidazole (**91**) but, in general, they offer no distinct advantage over metronidazole (Table 9). It is recommended that both female and male sexual partners be treated simultaneously to avoid reinfecting one another. Nitroimidazoles have been demonstrated to be mutagenic in bacteria and carcinogenic in high doses in laboratory animals. It is advised that pregnant women in the first trimester of pregnancy refrain from taking these drugs even though there is no clinical evidence of fetal abnormalities being caused by them. Other nitro-heterocycles have been synthesized in large number and many show promising *in vitro* activity.

Trypanosomiasis

The disease trypanosomiasis has two geographically delineated forms, the so-called African and American varieties. Because of their distinct differences they will be discussed separately (20).

African Trypanosomiasis. According to the World Health Organization, African trypanosomiasis affects some 25,000 people annually in 36 subSaharan countries where 50 million inhabitants are at risk. It is the cause of sleeping sickness, a disease transmitted by the bite of either the female or male tsetse fly. The tsetse fly is infected by ingesting blood from an animal or human containing trypanosomes. The elimination of these vectors is difficult because they thrive on both domestic cattle and wild animals. Trypanosomiasis in cattle, known as nagana, has seriously limited the production of this primary food source and has thwarted economic growth in endemic areas. African trypanosomiasis is caused by *Trypanosoma brucei* whose subspecies, *Trypanosoma brucei gambiense* and *Trypanosoma brucei rhodesiense,* have identical ultrastructures in humans. *T. b. gambiense* is responsible for an unevenly advancing sickness, whereas *T. b. rhodesiense* leads to a more acute form with rapid progression.

A human bitten by an infected tsetse fly develops a small nodule (primary or trypanosomal chancre) at the site of skin penetration from which the organisms may be isolated. These protozoa, whose life cycle is complex, change form, multiply, and invade the bloodstream via the lymphatic system in a matter of weeks or months. The trypanosomes are responsible for causing the clinical form of the disease, which includes severe headache, anemia, local edema, skin rashes, fever, and liver and/or spleen damage. Lymph nodes become enlarged and, in the advanced form of the disease, damage occurs to the central nervous system that causes the sleeping sickness phase. Lesions develop in the brain and, in some instances, in the myocardium. The disease has a high rate of mortality if left untreated. In the case of *T. b. rhodesiense,* death may ensue in several months.

The early stages of African trypanosomiasis, ie, before central nervous system involvement and the sleeping sickness take effect, can be best treated with intravenous suramin sodium (**92**), a compound with a dyelike structure (Table 10). Suramin is most effective against *T. b. rhodesiense,* but has also been used against *T. b. gambiense* infection. The compound causes side effects such as nausea, photophobia, and peripheral neuropathy which disappear shortly after conclusion of administration. Because the drug is unable to pass the blood–brain barrier, prompt treatment of patients is essential. Suramin in combination with tryparsamide (**93**, Tryparsone [*554-72-3*]) is an alternative that has been investigated. Pentamidine (**26**), as the isethionate or methanesulfonate salt, has been used successfully against *T. b. gambiense,* but the compound also does not enter the central nervous system and is not effective in advanced disease. It can cause severe side effects such as syncope, hypotension, hypoglycemia, vomiting, abdominal pain, and a peripheral neuropathy. Pentamidine (**26**) has been administered to large populations prophylactically against *T. b. gambiense.* This probably accounts for the appearance of resistant strains.

For late-stage disease, in which the central nervous system is implicated, the compound of choice until recently was melarsoprol (**94**, Mel B, Arsobal [*494-79-1*]) for *T. b. gambiense* or *T. b. rhodesiense.* The drug, administered intravenously, is a solution containing a combination of BAL (2,3-dimercaptopropanol) and the trivalent arsenic compound, melarsen oxide. Not only can the drug cause serious side effects such as intense dermal irritation, myocarditis, and renal and hepatic damage, but it is also responsible for death in 5% of patients. Resistance has developed to melarsoprol.

The first new treatment in 40 years that is showing great promise is eflornithine (**87**). It is an irreversible inhibitor of ornithine decarboxylase that blocks putrescine biosynthesis in *T. brucei in vitro* and inhibits the growth and multiplication of the parasite *in vivo.* The drug, administered orally or parenterally, is relatively nontoxic and is effective in eliminating the parasite *T. b. rhodesiense* and, to a greater extent, *T. b. gambiense.* Late-stage disease is rapidly reversed, causing comatose sleeping sickness patients to regain consciousness. Eflornithine was approved for human use by the FDA in late 1990. Side effects associated with oral administration are diarrhea, nausea, and vomiting. There is a significant relapse rate. A mixture of eflornithine (**87**) and nifurtimox (**95**, Lampit, Bayer 2502) shows good activity against *T. b. gambiense.* The former compound is also reported to potentiate the effect of melarsoprol in infected mice.

A combination of salicylhydroxamic acid (SHAM) and glycerol, although capable of destroying bloodstream trypanosomiasis, is less effective against infections involving the central nervous system.

American Trypanosomiasis. American trypanosomiasis, known as Chagas' Disease, is limited to South and Central America, where it affects 16–18 million people annually in an area where 90 million are at risk. Although only an estimated 1% of infected individuals contract the disease, poverty and poor housing exacerbate it. There is a particularly high incidence of the disease in children.

The protozoan causative agent, *Trypanosoma cruzi,* is harbored in domesticated animals, such as dogs, cats, and pigs, as well as wild animals, eg, rats, bats, foxes, opossums, and monkeys. These reservoirs can infect the vector, the night-feeding, blood-sucking triatomid bug. This insect, after biting a person usually on

Table 10. Trypanosomiasis Antiprotozoal Agents

Structure number	Compound name	CAS Registry Number	Molecular formula	Structure
			African trypanosomiasis	
(92)	suramin sodium	[*129-46-4*]	$C_{51}H_{34}N_6Na_6O_{23}S_6$	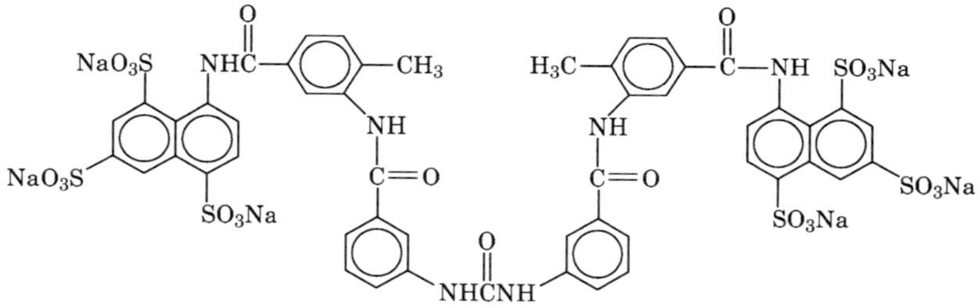
(93)	tryparsamide	[*554-72-3*]	$C_8H_{10}AsN_2NaO_4$	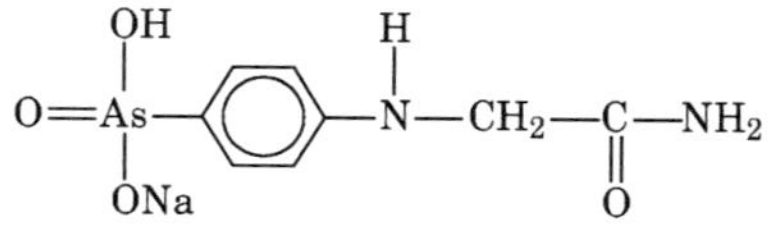
(94)	melarsoprol	[*494-79-1*]	$C_{12}H_{15}AsN_6OS_2$	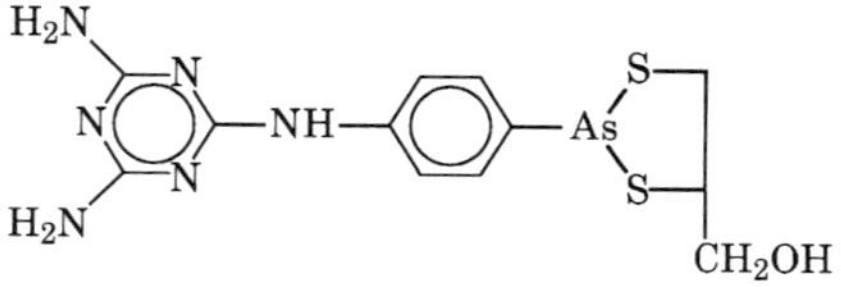
(95)	nifurtimox[a]	[*23256-30-6*]	$C_{10}H_{13}N_3O_5S$	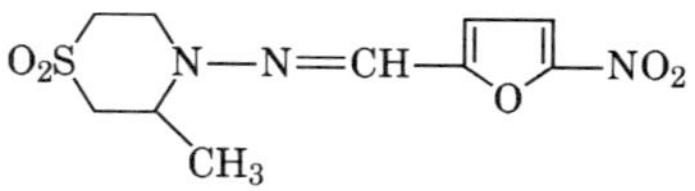
			American trypanosomiasis	
(96)	benznidazole	[*22994-85-0*]	$C_{12}H_{12}N_4O_3$	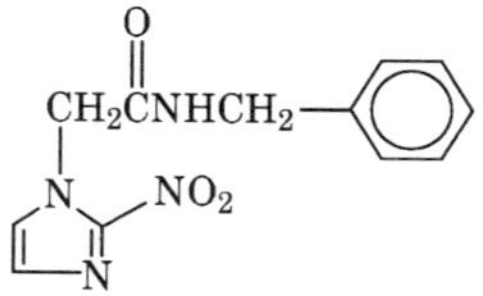

[a]Also used to treat American trypanosomiasis.

the face, defecates near the location of the wound. Parasites from the insect feces may either enter the human host at the point of the bite and/or by fecal contamination of mucus membranes, usually those of the mouth or eyes. This causes a skin lesion (called a primary chagoma) that consists of a small, reddish, slightly painful nodule. One of the eyelids may become swollen. The injected trypanosomes enter many kinds of cells of the host where they multiply and become amastigotes. They then differentiate into trypomastigotes that circulate in the bloodstream and are distributed to all parts of the body. They do not replicate until they enter another cell or are ingested by another blood-sucking insect vector.

The disease has two phases. The acute phase is of approximately two months' duration in which the parasite spreads. Although the disease may be asymptomatic, more typically it is marked by mild fever, colonic and esophageal dilation, liver and spleen enlargement, erythromatous rash, and acute myocarditis. In children, this phase of the disease can be serious and even fatal. During the chronic phase there is slow tissue destruction that can persist over many years. There may be cardiac and/or digestive involvement, with the former being the primary cause of death in Chagas' Disease. Despite its potential seriousness, the chronic form can also be asymptomatic.

The nitrofuran nifurtimox (**95**) is the most effective drug against *T. cruzi*, being cidal against the trypomastigote and amastigote forms. It is active against both the extra- and intracellular form of the parasite. In combination with corticosteroids it has prevented myocardial inflammation and destroyed the parasites within the heart. Side effects of the drug tend to be mild and include nausea, vomiting, insomnia, nervous excitation, vertigo, and skin rashes. Cardiac symptoms are treated symptomatically. Benznidazole (**96**, *N*-benzyl-2-nitro-1-imidazoleacetamide, RO 7-1051), an alternative drug, can prevent the spread of the parasites from one tissue to another although relapses are common. Primaquine (**69**) can destroy the extracellular trypanosomes in the blood, but is not effective against the intracellular forms of the parasite.

BIBLIOGRAPHY

"Chemotherapeutics, Antiprotozoal" in *ECT* 3rd ed., Vol. 5, pp. 513–542, by E. J. Martin and H. C. Zell, Food and Drug Administration, and B. T. Poon, Walter Reed Army Institute of Research.

1. G. T. Strickland, ed., *Hunter's Tropical Medicine,* 7th ed., W. B. Saunders, Philadelphia, Pa., 1991, p. 547.
2. K. S. Warren and A. A. F. Mahmoud, eds., *Tropical and Geographical Medicine,* 2nd ed., McGraw-Hill, New York, 1989.
3. M. Katz, D. D. Despommier, and R. W. Gwadz, *Parasitic Diseases,* 2nd ed., Springer-Verlag, New York, 1988.
4. J. H. Leech, M. A. Sande, and R. K. Root, eds., *Parasitic Infections,* Churchill Livingston, New York, 1988.
5. G. T. Strickland, ed., *Hunters's Tropical Medicine,* 7th ed., W. B. Saunders, Philadelphia, Pa., 1991, pp. 546–679.
6. W. C. Campbell and R. S. Rew, *Chemotherapy of Parasitic Diseases,* Plenum Press, New York, 1986.

7. G. C. Cook, *Parasitic Disease in Clinical Practice,* Springer-Verlag, London, 1990.
8. N. D. Levine, *Veterinary Protozoology,* Iowa State Press, Ames, 1985.
9. R. Knight, *J. Antimicrob. Chemother.* **6,** 577 (1980).
10. T. K. Ruebush, *Trans. Roy. Soc. Trop. Med. Hyg.* **74,** 149 (1980).
11. P. L. Long, ed., *Coccidiosis in Man and Domestic Animals,* CRC Press, Boca Raton, Fla., 1990.
12. S. Tzipori, *Adv. Parasitol.* **27,** 63 (1988).
13. S. L. Earlandsen and E. A. Meyer, eds., *Giardia and Giardiasis,* Plenum Press, New York, 1984.
14. J. D. Berman, *Rev. Infect. Dis.* **10,** 560 (1988); W. Peters and R. Killick-Kendrick, eds., *The Leishmaniases in Biology and Medicine,* Academic Press, New York, 1987.
15. W. Peters and W. H. G. Richards, eds., *Antimalarial Drugs,* Vols. 1 and 2, Springer-Verlag, New York, 1984; L. J. Bruce-Chwatt, ed., *Chemotherapy of Malaria,* WHO, Geneva, Switzerland, 1981; *Practical Chemotherapy of Malaria,* WHO Technical Report 805, 1990.
16. S. K. Martin, A. M. Oduola, and W. K. Milhous, *Science (Washington, D.C.),* **235,** 899 (1987).
17. D. L. Klayman, *Science (Washington, D.C.),* **228,** 1049 (1985).
18. W. T. Hughes, *Pneumocystis carinii,* Vols. 1 and 2, CRC Press, Boca Raton, Fla., 1987.
19. B. M. Honigberg, ed., *Trichomonads Parasitic in Humans,* Springer-Verlag, New York, 1989.
20. T. M. Leach and C. J. Roberts, *Pharmac. Therap.* **13,** 91 (1981); D. H. Molyneux and R. W. Ashford, *The Biology of Trypanosoma and Leishmania, Parasites of Man and Domestic Animals,* Taylor and Francis, London, 1983.

DANIEL L. KLAYMAN
Walter Reed Army Institute of Research

AVERMECTINS

In 1976 scientists at the Merck Corporation discovered a complex of eight closely related natural products, subsequently named avermectins A_{1a} through B_{2b}, in a culture of *Streptomyces avermitilis* MA-4680 (NRRL8165) originating from a soil sample collected at Kawana, Ito City, Shizuoka Prefecture, Japan and isolated by the Kitasato Institute. Their structures are shown in Figure 1 (1–6). They are among the most potent anthelmintic, insecticidal, and acaricidal compounds known.

The avermectins are closely related to another group of pesticidal natural products, the milbemycins. First described by Japanese workers, milbemycins were later found to be more abundant in nature than the avermectins (7–12). Both the avermectins and milbemycins are sixteen-membered lactones, with a spiroketal system containing two six-membered rings. The principal difference between them is that the avermectins have an α-L-oleandrosyl-α-L-oleandrosyl disaccharide attached at the 13-position whereas the milbemycins have no 13-substituent. Milbemycin structures are shown in Figure 2.

Two avermectins, abamectin (the avermectin B_1) [*71751-41-2*] and ivermectin [*70288-86-7*], which is saturated at C22–C23, have been commercialized to date. These two marketed avermectins have been described in considerable detail (Table 1) (13).

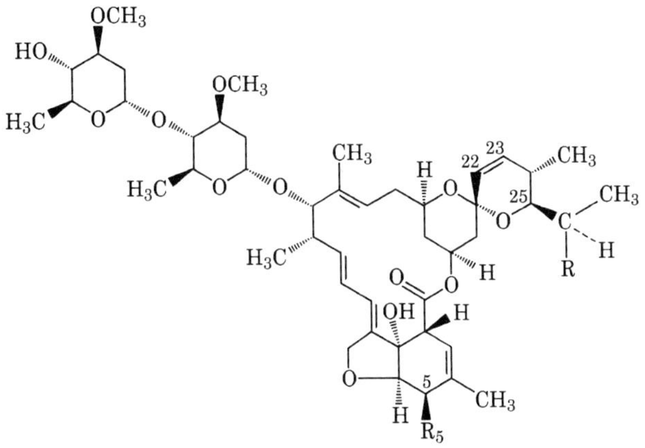

Fig. 1. The avermectins show variations at positions 5, 22–23, and 25. In the A series $R_5 = OCH_3$; for the B series $R_5 = OH$. The designation 1 corresponds to $\underset{22}{—CH}{=}\underset{23}{CH—}$ as shown. The designation 2 indicates $\underset{22}{—CH_2}—\underset{23}{CHOH}$. In ivermectin, C-22–C-23 is saturated. The avermectins are isolated as mixtures wherein the major component a has a *sec*-butyl group at position 25 ($R = C_2H_5$), and component b has an isopropyl group ($R = CH_3$) at C-25.

Avermectin	CAS Registry Number	Molecular formula
A_{1a}	*[65195-51-9]*	$C_{49}H_{74}O_{14}$
A_{1b}	*[65195-52-0]*	$C_{48}H_{72}O_{14}$
A_{2a}	*[65195-53-1]*	$C_{49}H_{76}O_{15}$
A_{2b}	*[65195-54-2]*	$C_{48}H_{74}O_{15}$
B_{1a}	*[65195-55-3]*	$C_{48}H_{72}O_{14}$
B_{1b}	*[65195-56-4]*	$C_{47}H_{70}O_{14}$
B_{2a}	*[65195-57-5]*	$C_{48}H_{74}O_{15}$
B_{2b}	*[65195-58-6]*	$C_{47}H_{72}O_{15}$

Table 1. Trade Name Registrations by the Merck Corporation[a]

Trade name	Delivery form	Recipient
Abamectin		
Avid	foliar spray	
Agri-Mek	foliar spray	
Vertimek	foliar spray	
Affirm	bait	fire ants
Avomec	injection	cattle
Duotin	injection	cattle
Ivermectin		
Ivomec	injection	cattle, sheep, goats, pigs, camels
	oral liquid	cattle, sheep
	paste, pour on	cattle
Ivomec F	injection	cattle
Oramec	oral liquid	sheep, goats
Eqvalan	paste	horses
	oral liquid	horses
Heartguard 30	tablet or chewable	dogs
Cardomec	tablet	dogs
Cardotek	tablet	dogs
Mectizan	tablet	humans

[a]Ref. 12.

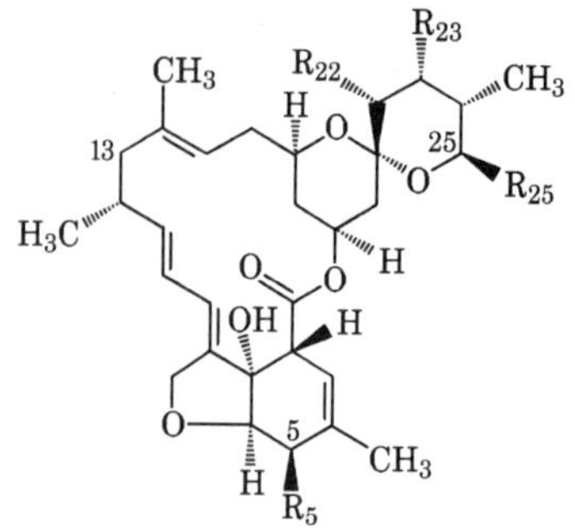

Fig. 2. Milbemycin structures in the milbemycin Alfa series. R_5 = OH, OCH_3; R_{22} = H, OH; R_{23} = H, $OCOCH(CH_3)(CH_2)_3CH_3$; R_{25} = CH_3, C_2H_5. In the anthelmintic F-28249, antibiotics S541 series R_5 = OH, OCH_3; R_{22} = H; R_{23} = OH; R_{25} =

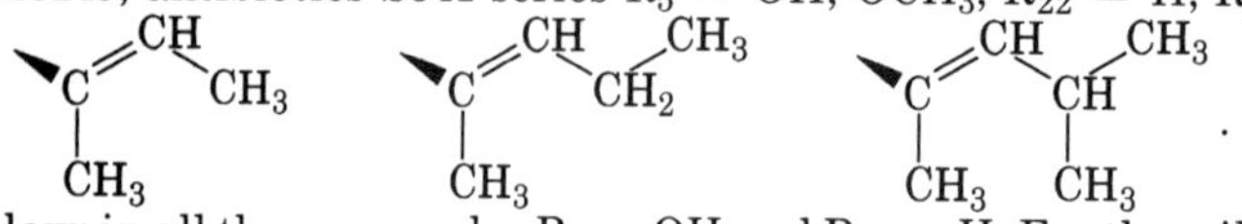

. Some specific examples follow; in all these examples R_5 = OH and R_{22} = H. For the milbemycins, R_{23} = H; R_{23} = OH for nemadectin.

Name	CAS Registry Number	R_{25}	Molecular formula
milbemycin Alfa$_1$ (A_3)	*[51596-10-2]*	CH_3	$C_{31}H_{44}O_7$
milbemycin Alfa$_3$ (A_4)	*[51596-11-3]*	CH_2CH_3	$C_{32}H_{46}O_7$
milbemycin D	*[77855-81-3]*	$CH(CH_3)_2$	$C_{33}H_{48}O_7$
anthelmintic F-28249-Alpha (nemadectin)		$C(CH_3)CHCH(CH_3)_2$	$C_{36}H_{52}O_8$

Abamectin

Avermectin B_1 is the most effective of the avermectin family of natural products against agriculturally important insects and mites (14). It has been commercialized for agricultural use under the nonproprietary name abamectin. This mixture of avermectins contains at least 80% of avermectin B_{1a} ($C_{48}H_{72}O_{14}$) and not more than 20% of avermectin B_{1b} ($C_{47}H_{70}O_{14}$). Abamectin for use in foliar spray applications is formulated as an emulsifiable concentrate. It is being developed for use in a number of horticultural and agronomic crops to control mites and insects. A summary of its biological activity is shown in Table 2 and a summary of its agricultural applications is shown in Table 3.

Abamectin is also used to control the imported red fire ant *Solenopsis invicta*. For this use abamectin is formulated as a bait together with soybean oil and corn grits. Worker ants transport the bait to the colony, the queen becomes sterile, and the colony is eliminated after 12 to 21 weeks. Similar effects on the fecundity of other female insects at nonlethal doses have been reported (16,17).

Ivermectin

Selective reduction of the 22,23-olefin of avermectin B_1 yields the 22,23-dihydro derivative assigned the nonproprietary name ivermectin (18). The structure

Table 2. Biological Activity of Abamectin (Avermectin B_1) Against Mites and Insects[a]

Common name	Species name	LC_{90}(ppm)
Mite species[b]		
citrus rust mite	*Phyllocoptruta oleivora*	0.02
two-spotted spider mite	*Tetranychus urticae*	0.03
strawberry mite	*Tetranychus turkestani*	0.08
European red mite	*Panonychus ulmi*	0.04
citrus red mite	*Panonychus citri*	0.24
broad mite	*Polyphagotarsonemus*	0.03
Insect species[c]		
Colorado potato beetle	*Leptinotarsa decemlineata*	0.03
tomato hornworm	*Manduca secta*	0.02
Mexican bean beetle	*Epilachna varivestes*	0.20
pea aphid	*Acyrthosiphon pisum*	0.40
cabbage looper	*Trichoplusia ni*	1.0
corn earworm	*Heliothis zea*	1.5
southern armyworm	*Spodoptera eridania*	6.0

[a]Ref. 15.
[b]Contact effect against adult mites.
[c]Foliar residue bioassay.

Table 3. Principal Agricultural Applications of Abamectin Foliar Spray

Common name	Species name	Application
armyworm	*Spodoptera exigua*	vegetables[a]
	Spodoptera eridania	
broad mite	*Polyphagotarsonemus latus*	citrus
citrus red mite	*Panonychus citri*	citrus
citrus rust mite	*Phyllocoptruta oleivora*	citrus
citrus thrip	*Scirtothrips citri*	citrus
cotton aphid	*Aphis gossypii*	cotton
deciduous tree nut		miscellaneous crops
European red mite	*Panonychus ulmi*	pears
leafminer	*Liriomyza trifolii*	ornamentals, vegetables[a]
	Liriomyza sativae	vegetables[a]
Lepidoptera		cotton
pear psylla	*Psylla pyricola*	pears
pear rust mite	*Epitrimerus pyri*	pears
strawberry		miscellaneous crops
spider mite	*Tetranychus urticae*	cotton
	Tetranychus turkestani	cotton
	Tetranychus pacificus	cotton
	Tetranychus cinnabarinus	cotton
thrips species		ornamentals
tomato pinworm	*Keiferia lycopersicella*	vegetables[a]
tomato russet mite	*Aculops lycopersici*	vegetables[a]
two-spotted spider mite	*Tetranychus urticae*	vegetables[a], pears, ornamentals

[a]Celery, tomato, pepper.

shown depicts the 25-*sec*-butyl derivative [*70161-11-4*], but it should be noted that both commercial products contain up to 20% of the 25-isopropyl analogue [*70209-81-3*].

Avermectin B_{1a} $\xrightarrow[\text{Wilkinson's catalyst}]{H_2}$

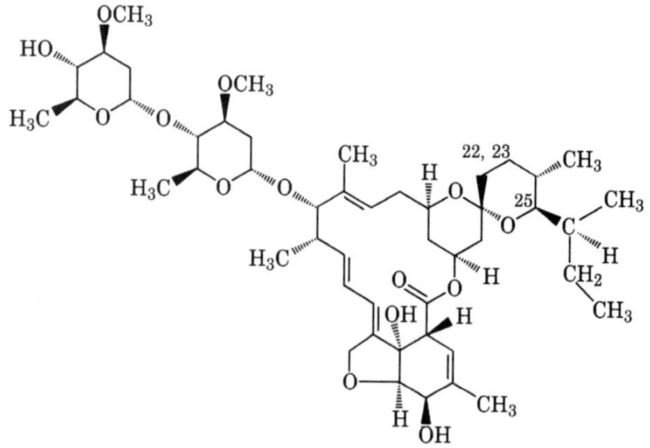

Ivermectin is active against two significant phyla of animal parasite: the Nemathelminthes or nematodes (roundworms) and the Arthropoda (insects, ticks, and mites). Ivermectin is inactive against platyhelminthes (flukes and tapeworms).

Ivermectin has an extremely broad spectrum of antinematodal activity in a variety of domestic animals. Indeed, among the many nematodes against which it has been tested, none has been found that is not affected by ivermectin during at least one stage of the life cycle. In all but a few instances the drug is highly active against both immature and mature worms (19).

Perhaps the most striking aspect of the antinematodal action of ivermectin is its potency. This varies widely from species to species and stage to stage, but in all cases the minimum effective dosage is much less than that of other anthelmintics. Among the most sensitive parasites are immature *Dirofilaria immitis* in dogs (0.001 mg/kg) and *Oesophagostomum radiatum* and *Dictyocaulus viviparus* in cattle (0.05 mg/kg). Among the least sensitive nematodes are *Nematodirus helvetianus* in cattle, and *Toxascaris leonina,* and perhaps immature *Trichuris vulpis,* in dogs. Two parasite life cycle forms against which activity has not been demonstrated even at high dosage (more than 1.0 mg/kg) are adult *Dirofilaria immitis* (heartworm) in dogs and immature *Trichinella spiralis* in the muscle of mice and pigs even though other life cycle stages are affected. The vast majority of nematodes, regardless of host species or stage of worm development, are highly susceptible to ivermectin when the drug is given as a single oral or parenteral dose at 0.1–0.2 mg/kg.

The activity of ivermectin against the filarial parasite *Dirofilaria immitis* in dogs suggested a possible role for the control of filarial parasites of humans (20). It has been extensively tested in human onchocerciasis and is now considered to be the drug of choice. In a single yearly oral dose, it suppresses microfilariae in the skin and eyes and, in most cases, prevents the progression of the disease to blindness. Table 4 shows the results of a 30-patient double-blind study recorded over one year.

Table 4. Ivermectin and Diethylcarbamazine Against *Onchocerca volvulus* Infections[a]

Study day	Placebo[b]	Ivermectin[c]	DEC[d]
−1	99.4	130.4	100.3
2	108.2	38.8	27.0
4	99.7	14.1	14.4
8	105.1	6.6	4.1
14	125.9	2.2	6.8
28	102.6	0.6	9.2
90	84.5	1.0	18.0
180	65.3	2.9	21.8
270	80.8	5.0	27.5
360	93.0	11.8	45.1

[a]Numbers represent the skin density of microfilariae (21).
[b]Ten patients received placebo.
[c]Ten patients received a single oral dose of ivermectin, 12 mg.
[d]Ten patients received diethylcarbamazine [*90-89-1*] (DEC) daily orally for eight days: total dose 1.3 g.

Two forms of lymphatic filariasis are found in India. The Bancroftian form is the most common and accounts for more than 90% of the disease whereas Brugian filariasis accounts for the rest. In a study carried out in India (6) in 40 patients with *Wuchereria Bancrofti* filariasis treated with single oral doses, all of the dose levels chosen (25, 50, 100, and 200 mg/kg) were efficacious in clearing microfilariae from the blood of all patients treated. However, after three months some microfilaria recurred in the blood of most patients (Table 5). Further studies are planned and some are underway using different doses and regimens. Ivermectin still appears to hold promise as a new treatment for lymphatic filariasis.

Table 5. Efficacy[a] of Ivermectin in the Treatment of *Wuchereria Bancrofti* Filariasis[b]

Single oral dose, mg/kg	Day						
	0	1.5	5	12	30	90	180
25	761	2.9	<1	<1	5.2	42.9	98
50	1154	3.3	<1	<1	3.5	103.6	92.3
100	610	3.0	<1	<1	<1	19.9	95.9
200	478	<1	<1	<1	1.5	43.7	70.8

[a]Geometric mean microfilariae/mL.
[b]Ref. 22.

Ivermectin is widely used as an endectocide for cattle as an injectable, oral, topical, or slow release bolus; for sheep as an injectable or oral formulation; for swine as an injectable; for horses as a paste or drench; and for goats as an injectable or oral formulation. Ivermectin has recently been introduced for heartworm prophylaxis in dogs and it is being studied for use with cats, many other mammals, birds, fish, and reptiles.

Ivermectin is used in cattle, sheep and horses at 0.2 mg/kg; swine at 0.3 mg/kg; dogs at 0.006 mg/kg; and man at 0.05–0.2 mg/kg. It is effective against parasitic nematodes, grubs, lice, mites, ticks, and bots. Ivermectin is not active against tapeworms, flatworms, bacteria, or fungi.

Biosynthesis

The proposed pathway for the biosynthesis of the avermectins (Fig. 3) has been described in a review (23). Some of the details are yet to be elucidated, although the steps, in general, are based on firm evidence from four types of studies: incorporation of labeled precursors, conversion of putative intermediates by producing strains and blocked mutants, *in vitro* measurement of biosynthetic enzymes, and studies with enzyme inhibitors. The biosynthesis of the oleandrose units was elucidated from studies using $^{3}H/^{14}C$ and ^{13}C labeled glucose, which indicated a direct conversion of glucose to oleandrose.

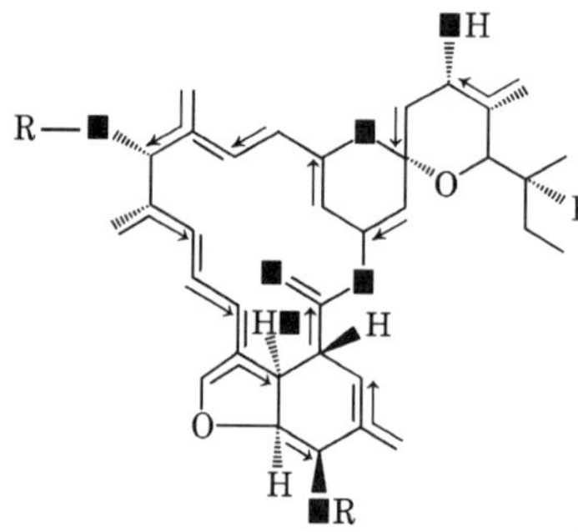

Fig. 3. Biosynthetic scheme of avermectins, ■ = ^{18}O, $\overrightarrow{CH_3COOH}$, CH_3CH_2COOH.

Chemistry

Stability. Avermectins are highly lipophilic substances and dissolve in most organic solvents such as chloroform, methylene chloride, acetone, alcohols, toluene, cyclohexane, dimethylformamide, dimethyl sulfoxide, and tetrahydrofuran. Their solubility in water is correspondingly low, only 0.006–0.009 ppm (= mg/L).

Chemical stability studies are monitored by silica gel thin-layer chromatography (tlc) or by high performance liquid chromatography (hplc) using a reverse-phase C_{18} coated column (24). Hplc peaks or tlc spots are visualized by their uv absorption at 245 nm; the tlc spots can also be detected by ceric sulfate or phosphomolybdic acid staining.

Avermectins are acid-sensitive. Although stable in anhydrous glacial acetic acid solution at room temperature, aqueous CH_3COOH and stronger acids, such as dilute HCl, H_2SO_4, and *p*-toluenesulfonic acid in methanol solution, more rapidly cleave off the first sugar. Stronger concentrations of methanolic or aqueous acids are necessary to hydrolyze off the second sugar to give the aglycone.

Strong bases such as methanolic potassium hydroxide, sodium methoxide, or 1,8-diazabicyclo [5.4.0] undec-7-ene (DBU), cause epimerization at the C-2 carbon or shift the beta–gamma double bond into conjugation with the lactone carbonyl (Fig. 4) (25,26).

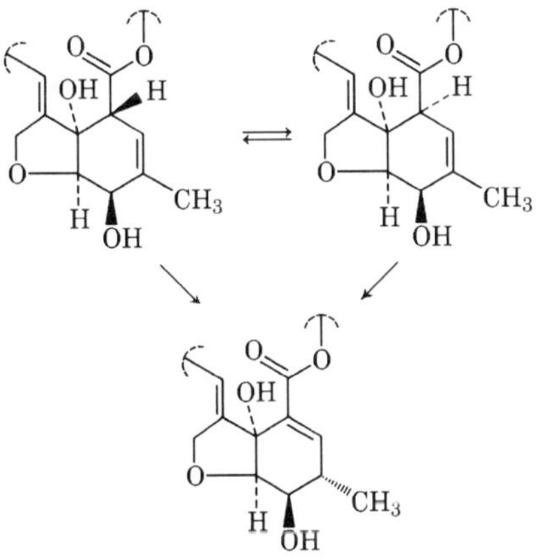

Fig. 4. Treatment of ivermectin with methanolic potassium hydroxide solution (25).

Uv light below 280 nm rapidly isomerizes the (*E*) (*trans*) 8,9- and 10,11-double bonds to the 8,9- and 10,11-(*Z*)-isomers (Fig. 5) (27).

Reactivity. Avermectin B_1 and ivermectin both contain two secondary and one tertiary hydroxy group. The two secondary alcohols are readily acetylated with acetic anhydride in pyridine, although at 0°C, a fair amount of the 4″-monoacetate, but not the 5-monoacetate, is obtained. With more reactive acid chlorides, however, it is not possible to obtain the 4″-*O*-monoacyl derivative, and it is necessary to protect the 5-OH as a 5-*O*-*tert*-butyldimethysilyl (TBDMS) derivative. This can be prepared selectively because of the hindered C-4″ position (28). Removal of the 5-protecting group is accomplished with *p*-toluenesulfonic acid monohydrate (p-TsOH·H_2O) in CH_3OH (1% at RT for 20–30 min; longer reaction times generate substantial amounts of monosaccharide) or with an

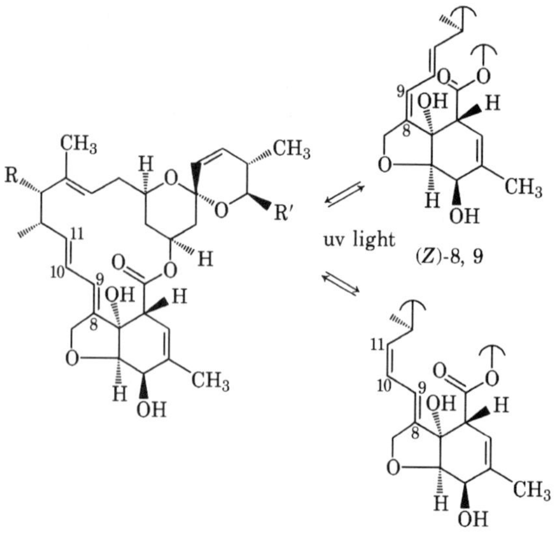

Fig. 5. Photoisomerization of avermectins. R = α-L-oleandrosyl-α-L-oleandrosyl; R′ = *sec*-C_4H_9 or *i*-C_3H_7 (see Fig. 1) (27).

HF–pyridine–THF mixture (29). It is sometimes possible to prepare 5-monoacyl derivatives directly from unprotected ivermectin by careful use of one equivalent of acylation reagent, particularly when the latter is bulky (eg, bis(trichloroethoxy)phosphoryl chloride). Many 4″-*O*-substituted derivatives have been prepared as they retain considerable bioactivities; however, 5-*O*-substitution products lose most anthelmintic activities.

Analytical Procedures

Since the avermectins exhibit unprecedented potency, they are used at unusually low doses of 6–300 μg/kg, which makes the detection and isolation of residues and metabolites from animal tissue a new challenge. For this reason a sensitive analytical assay requires a derivative suitable for detection at concentrations down to 1/10 or 1/100 of one ppm. Ivermectin and avermectin B_1 are therefore converted into an aromatic derivative which allows detection by fluorescence absorbance. To achieve this derivatization, avermectin B_1, ivermectin, or their derivatives are heated with acetic anhydride in pyridine at 100°C for 24 h (30). The reaction time can be reduced to 1 h by using *N*-methylimidazole as a catalyst (31). The resultant 4″-*O*-acetyl-2,5,6,7-bisdehydro analogue is detected after hplc separation and characterized by a distinct retention time. Using a C_2-extraction cartridge, with 22,23-dihydro avermectin B_{1a} monosaccharide as an internal standard, it is possible to detect as little as 1 ng/mL of ivermectin in plasma or milk (32). This procedure was adapted as a confirmatory assay by hydrolysis of the avermectin derivatives to aglycone, monosaccharide, and parent disaccharide, subsequent aromatization, and finally detection of three distinct hplc peaks corresponding to the acetylated bisdehydro aglycone, monosaccharide, and parent compound peaks via fluorescence detection. A procedure using chemical ionization mass spectrometry–mass spectrometry (ms/ms) was developed for a confirmatory assay of ivermectin tissue residues in cattle (33). It is also possible to determine avermectins directly after extraction, using a synthetically isomerized avermectin derivative as an internal standard for hplc with uv detection (25). A hplc-reverse isotope dilution assay is available for the determination of ivermectin residues in animal tissue (34). A similar fluorescent derivative containing a 5-phenol group obtained by dehydration of the 5-keto derivative of ivermectin with ammonium acetate was also reported (35). An extraction procedure allowing direct hplc analysis with uv detection of ivermectin in bovine serum at concentrations as low as 2 ppb is available (36). Monoclonal antibodies for immunoassay of avermectins have been prepared (37).

Monosaccharides and Aglycones

The macrocyclic lactone of all avermectins has an α-L-oleandrosyl-α-L-oleandrosyloxy substituent at carbon 13, which is a 2-deoxy sugar glycoside, relatively sensitive to acid hydrolysis or alcoholysis. A solution of ivermectin in methanol containing a strong acid such as 1% sulfuric acid readily gives a good yield of the aglycone after 16–24 h at RT. When 2-propanol is substituted for methanol in this reaction, a good yield of the monosaccharide is obtained, as the bulkier 2-propanol preferentially attacks the sterically less hindered 1″-position over the 1′-position, which is attached directly to the C-13 of the macrocycle, which in turn is flanked on either side by the C-12 and C-14 methyl groups, respectively (Fig. 6) (38). These procedures readily yield the monosaccharides of ivermectin and avermectin B_1 and the aglycone of ivermectin. The preparation of the aglycone of avermectin B_1, however, is complicated: under the reaction conditions, addition of methanol to the 22,23-double bond occurs, yielding mainly the two epimeric 22,23-dihydro-23-methoxy monosaccharides and/or aglycones. The aglycone of the avermectin B_1 therefore must be prepared with aqueous acid; it can be obtained as a 1:1 mixture of aglycone and monosaccharide using 10% sulfuric acid in aqueous THF solution (38).

Removal of the 13-hydroxy group from avermectin aglycones gives 13-deoxyavermectin aglycones which are closely related to certain milbemycins. For example, ivermectin was converted into its aglycones (22,23-dihydroavermectin B_{1a} and B_{1b} aglycones); the 5-hydroxy group was then protected as an *O-tert*-butyldimethylsilyl group. The 13-hydroxy group was substituted by chloro; the 13-chloro group was removed by reduction; C-5-OH was deprotected; and, finally, the C-25 B_{1a} and B_{1b} homologues were separated into 13-deoxy-22,23-dihydroavermectin B_{1a} and B_{1b} aglycones (39). Interestingly, the latter was subsequently also obtained from a milbemycin fermentation and given the name milbemycin B41-D (see Fig. 2 for structure B_{1b}).

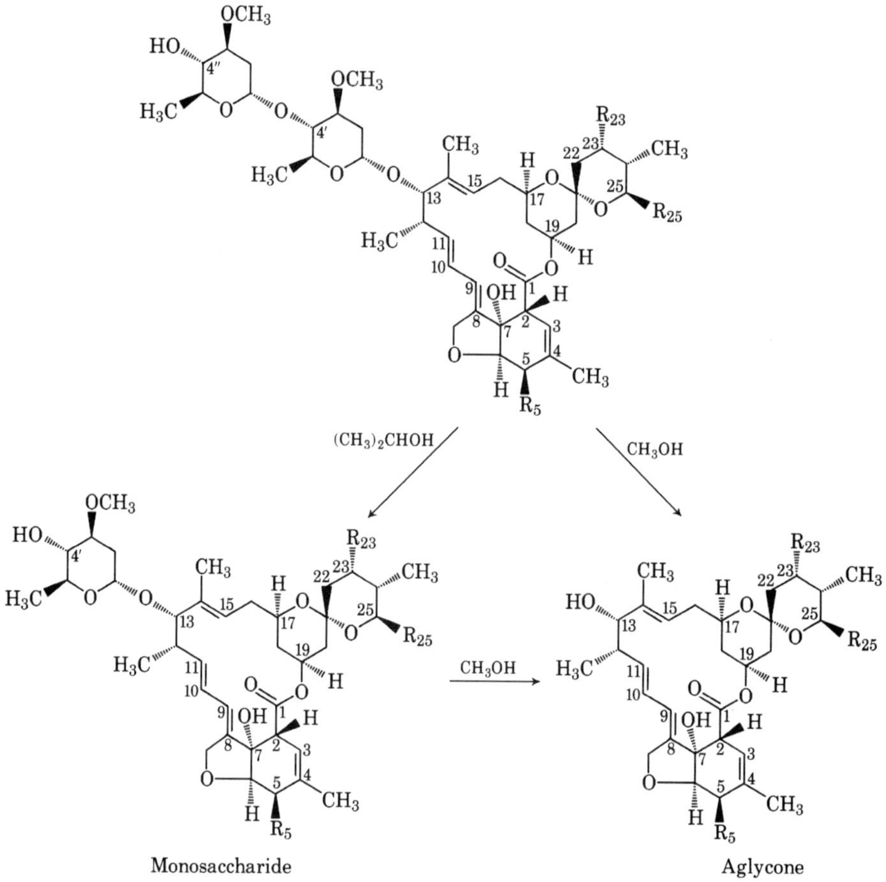

Fig. 6. Avermectin monosaccharides and aglycones. R_{25} = *sec*-C_4H_9; R_5 = OCH_3 or OH; R_{23} = OH or H. The reaction conditions are 18°C, 16 h, 1% H_2SO_4 in the solvent indicated (38).

Chemical Reactions

Oxidations. Avermectin B_1 and ivermectin have two secondary hydroxy groups susceptible to oxidation to the corresponding ketones. Because of the allylic nature of the C-5 hydroxy group its selective oxidation to 5-oxoavermectin is readily accomplished with manganese dioxide (40). Although tautomerization to the enol and elimination of the remaining tertiary C-7 hydroxy group leads to a phenol (35), the 5-oxo derivatives are nonetheless stable enough to allow further chemical reactions. Particularly important is the stereospecific reduction with $NaBH_4$ to the 5-beta alcohol with natural configuration, which is used to prepare tritiated avermectin derivatives. Reaction with trifluoroacetic anhydride, however, gives the 2,5,6,7-tetradehydro phenol analogue. Use of mild oxidation procedures, such as oxalyl chloride–DMF (Swern oxidation), gives 4″,5-dioxo avermectin derivatives or, after protection of the C-5 hydroxy group as a *tert*-butyldimethylsilyl derivative, 5-*O*-*tert*-butyldimethylsilyl-4″-oxoavermectins, which are valuable intermediates for further 4″-modified analogues. Sodium borohydride reduction of the 4″-keto function gives a mixture of 4″-epimeric hydroxy compounds, with the unnatural axial 4″-hydroxy group the major product and the natural epimer the minor one. Reductive amination with $NaCNBH_3$ and NH_4OOCCH_3 likewise yields an epimeric mixture of 4″-aminoderivatives with interesting biological properties (41).

The avermectins also possess a number of allylic positions that are susceptible to oxidative modification. In particular the 8a-methylene group, which is both allylic and alpha to an ether oxygen, is susceptible to radical oxidation. The primary product is the 8a-hydroperoxide, which has been isolated occasionally as an impurity of an avermectin B_1 reaction (such as the catalytic hydrogenation of avermectin B_1 with Wilkinson's rhodium chloride-triphenylphosphine catalyst to obtain ivermectin). An 8a-hydroxy derivative can also be detected occasionally as a metabolite (42) or as an impurity arising presumably by air oxidation. An 8a-oxo-derivative can be obtained by oxidizing 5-*O*-protected avermectins with pyridinium dichromate (43). This also can arise by treating the 8a-hydroperoxide with base.

The allylic 4a-methyl group can be oxidized with SeO_2-catalyzed *tert*-butylhydroperoxide to a 4a-hydroxy analogue (44).

The double bonds of avermectins react with *m*-chloroperbenzoic acid to give 3,4-, 8,9-, and 14,15-epoxides. The 8,9-epoxide is the primary product and can be isolated in good yield (45). The 8,9-epoxide was opened by aqueous acids to the 8,9-diol (46). The 3,4-diol can be obtained readily and regiospecifically by osmium tetroxide oxidation. Neither peracids nor OsO_4 will attack the 22,23-double bond.

Reductions Including Ivermectin Preparation. Reductions of avermectin B_1 containing five double bonds with Pd or Pt catalysts proceed at almost comparable rates at the two disubstituted 10,11- and 22,23-double bonds to give a mixture of 10,11- and 22,23-dihydro derivatives. Further hydrogenation leads to 10,11,22,23-tetrahydro- and 3,4,10,11,22,23-hexahydro-derivatives; exhaustive hydrogenation leads to 3,4,8,9,10,11,22,23-octahydro analogues. Under no circumstances is the 14,15-double bond hydrogenated. For the preparation of ivermectin (22,23-dihydro avermectin B_1), Wilkinson's homogeneous $[(C_6H_5)_3P]_3RhCl$ catalyst has been employed for the regiospecific hydrogenation of the avermectin B_1 22,23-double bond, which is the only disubstituted cis double bond (18). To prepare 10,11-dihydro avermectin B_1, the regioselective reaction with *N*-bromacetamide was used to make a 11,10-bromohydrin, which can then be dehalogenated and deoxygenated to the desired product (47).

Radiolabeled Derivatives. Since ivermectin (= 22,23-dihydroavermectin B_1) is obtained by catalytic reduction of avermectin B_1, the same procedure using tritium gas conveniently affords tritiated ivermectin (22,23-$^3[H]$-22,23-dihydroavermectin B_1). The preparation of a tritiated derivative containing a 22,23-double bond starts with the readily available 5-ketone, which is reduced with $^3[H]$-sodium borohydride stereospecifically to a 5-$^3[H]$-derivative (40). Carbon-14 labeled avermectins can be obtained by a biosynthetic process using sodium (1-^{14}C)propionate as labeled precursor (48).

Interconversions of Avermectins. Methylation of avermectins B_1 and B_2 leads to the corresponding derivatives of the A series (49). A procedure involving the oxidation of the 5-methoxy group with mercuric acetate and $NaBH_4$ reduction of the 5-keto-intermediate allows the conversion of the A to the B components (50). The 23-hydroxy group of the "2" components, after selective protection of the other secondary hydroxy groups, is converted to a thionocarbonate, which can be eliminated to give the 22,23-double bond of the "1" components; alternatively it can be reduced with tributyltin hydride to the 22,23-dihydro derivatives (= ivermectins) (51).

Syntheses

Avermectin aglycones, monosaccharides, and the naturally occurring disaccharides themselves have been further modified by attaching various sugars to the different hydroxy groups. Most of these methods have used 1-bromo sugars via the Konigs-Knorr procedure (52,53). More recently, the use of 1-phenylthio and 1-fluoro sugars has resulted in better yields of those glycosides (54). A mixture of the two anomeric methyl glycosides of oleandrose is obtained by acid methanolysis of avermectins, and these have been converted to L-oleandrose (55) and the 1-phenylthio glycoside (54). Total synthesis of avermectins including the glycosylation of avermectin aglycones by interesting, new methods have been reported recently by several research groups (56–59). Two excellent reviews of partial and total synthesis of avermectins and milbemycins have been published (60,61).

Economic Aspects

The worldwide acceptance of ivermectin in livestock production and health care of companion animals has made it the largest selling animal health drug. Abamectin is in commercial use as an agricultural pesticide and its applications are continuing to expand. Ivermectin under the trademark Mectizan is now considered the drug of choice for treatment of human onchocerciasis. There is no commercial return since it is provided free of charge to the World Health Organization. Ivermectin also shows considerable promise for the treatment of human lymphatic filariasis and strongyloidiasis.

BIBLIOGRAPHY

1. R. W. Burg and co-workers, *Antimicrob. Agents Chemother.* **15,** 361 (1979).
2. G. Albers-Schönberg and co-workers, *J. Am. Chem. Soc.* **103,** 4216 (1981).
3. J. P. Springer, B. H. Arison, J. M. Hirschfield, and K. Hoogsteen, *J. Am. Chem. Soc.* **103,** 4221 (1981).
4. M. H. Fisher and H. Mrozik, in S. Omura, ed., *Macrolide Antibiotics,* Academic Press, New York, 1984, pp. 553–606.
5. H. G. Davies and R. H. Green, *Natl. Prod. Rep.* **3,** 87 (1986).
6. M. H. Fisher, in N. F. Janes, ed., *Recent Advances in the Chemistry of Insect Control,* Royal Society of Chemistry, London, 1985, pp. 53–72; H. Mrozik, *BCPC Mono* No. 32, *Biotechnology and its Application to Agriculture,* 1985, pp. 133–143; H. Mrozik, *Topics in Medicinal Chemistry,* 1987, pp. 245–254; T. Blizzard, M. H. Fisher, H. Mrozik, and T. L. Shih, in G. Lukacs and M. Ohno, eds., *Recent Progress in the Chemical Synthesis of Antibiotics,* Springer-Verlag, Berlin, Heidelberg, Germany, 1990, pp. 65–102.

7. Y. Takiguchi and co-workers, *J. Antibiot.* **33,** 1120 (1980).
8. H. Mishima, J. Ide, S. Muramatsu, and M. Ono, *J. Antibiot.* **36,** 980 (1983).
9. Eur. Pat. Appl. 170,006 (1986), J. B. Wood and co-workers (to American Cyanamid Co.).
10. Brit. Pat. Appl. 2,166,436A (1986), J. B. Wood and co-workers (to Glaxo Corp.).
11. G. T. Carter, J. A. Nietsche, and D. B. Borders, *J. Chem. Soc. Chem. Commun.* 402 (1987).
12. M. J. V. Ramsay and co-workers, *Tetrahedron Lett.* **28,** 5353 (1987).
13. W. C. Campbell, ed., *Ivermectin and Abamectin,* Springer, New York, 1989, pp. 1–363.
14. J. Putter and co-workers, *Experientia* **37,** 963 (1981).
15. R. A. Dybas and A. St. J. Green, *Proc. 1984 Brit. Crop Protect. Conf.-Pests Dis.,* **31,** 947–954 (1984).
16. C. S. Lofgren and D. F. Williams, *J. Econ. Entolmol.* **75,** 798–803 (1982).
17. L. Strong and T. A. Brown, *Bull. Entomol. Res.* **77,** 357–389 (1987).
18. J. C. Chabala and co-workers, *J. Med. Chem.* **23,** 1134 (1980).
19. Ref. 13, pp. 215–259.
20. Ref. 13, pp. 311–323.
21. M. Lariviere and co-workers, *Lancet* **2,** 174 (1985).
22. V. Kumaraswami and co-workers, *J. Am. Med. Assoc.* **259,** 3150 (1988).
23. T. S. Chen, O. D. Hensens, and M. D. Schulman, in Ref. 13, pp. 55–72.
24. T. W. Miller and co-workers, *Antimicrob. Agents Chemother.* **15,** 368–371 (1979).
25. J. V. Pivnichny, J. S. K. Shim, and L. A. J. Zimmerman, *Pharm. Sci.* **72,** 1447 (1983).
26. J. V. Pivnichny and co-workers, *J. Agric. Food Chem.* **36,** 826 (1988).
27. H. Mrozik and co-workers, *J. Org. Chem.* **53,** 1820 (1988).
28. H. Mrozik and co-workers, *J. Med. Chem.* **25,** 658–663 (1982).
29. C. Bliard and co-workers, *J. Chem. Soc., Chem. Commun.,* 368 (1987).
30. J. W. Tolan and co-workers, *J. Chromatogr.* **190,** 367–376 (1980).
31. P. C. Tway, J. S. Wood, and G. V. Downing, *J. Agric. Food Chem.* **29,** 1049–1063 (1981).
32. R. Chiou, R. J. Stubbs, and W. F. Bayne, *J. Chromatogr.* **416,** 196–202 (1987).
33. P. C. Tway and co-workers, *Biomed. Mass Spectrom.* **11,** 172–176 (1984).
34. S. H. L. Chiu and co-workers, *J. Agric. Food Chem.* **33,** 99–102 (1985).
35. J. D. Stong, *Anal. Chem.* **59,** 266–270 (1987).
36. D. D. Oehler and J. A. Miller, *J. Assoc. Off. Anal. Chem.* **72,** 59 (1989).
37. D. J. Schmidt and co-workers, *J. Agric. Food Chem.* **38,** 1763 (1990).
38. H. Mrozik and co-workers, *J. Org. Chem.* **47,** 489–492 (1982).
39. H. Mrozik and co-workers, *Tetrahedron Lett.* **24,** 5333 (1983).
40. J. C. Chabala, A. Rosegay, and M. A. R. Walsh, *J. Agric. Food Chem.* **29,** 881 (1981).
41. H. Mrozik and co-workers, *Experientia* **45,** 315 (1989).
42. D. L. Bull and co-workers, *J. Agric. Food Chem.* **32,** 94–102 (1984).
43. U.S. Pat. 4,547,491 (1985), H. Mrozik and F. S. Waksmunski (Merck & Co., Inc.).
44. U.S. Pat. 4,457,920 (1984), H. Mrozik (Merck & Co., Inc.).
45. U.S. Pat. 4,530,921 (1985), H. Mrozik (Merck & Co., Inc.).
46. A. B. Smith, III and A. S. Thompson, *Tetrahedron Lett.* **26,** 4279 (1985).
47. T. L. Shih, H. Mrozik, J. Ruiz-Sanchez, and M. H. Fisher, *J. Org. Chem.* **54,** 1459 (1989).
48. C. C. Ku and S. C. Hwang, *J. Labelled Compd. Radiopharm.* **22,** 451–459 (1985).
49. U.S. Pat. 4,200,581 (1980), M. H. Fisher, A. Luisi, and R. L. Tolman (Merck & Co., Inc.).
50. H. Mrozik, P. Eskola, and M. H. Fisher, *J. Org. Chem.* **51,** 3058 (1986).
51. H. Mrozik, P. Eskola, and M. H. Fisher, *Tetrahedron Lett.* **23,** 2377–2378 (1982).
52. U.S. Pat. 4,134,973 (1979), M. H. Fisher and R. L. Tolman (Merck & Co., Inc.).
53. U.S. Pat. 4,203,976 (1980), M. H. Fisher and R. L. Tolman (Merck & Co., Inc.).
54. K. C. Nicolaou, R. E. Dolle, D. P. Papahatjis, and J. L. Randall, *J. Am. Chem. Soc.* **106,** 4189 (1984).
55. H. Els, W. D. Celmer, and K. Murai, *J. Am. Chem. Soc.* **80,** 3777–3782 (1958).
56. S. Hanessian, D. Dube, and P. J. Hodges, *J. Am. Chem. Soc.* **109,** 7063 (1987).

57. S. J. Danishefsky and co-workers, in Ref. 56, p. 8117.
58. S. J. Danishefsky and co-workers, in Ref. 56, p. 8119.
59. M. J. Ford, J. G. Knight, S. V. Ley, and S. Vile, *SYNLETT.* 331–332 (1990).
60. M. T. Crimmins, W. G. Hollis, Jr., and R. O'Mahony, in Atta-ur-Rahman, ed., *Studies Directed Toward the Total Synthesis of the Milbemycins and Avermectins in Studies in Natural Products Chemistry,* Vol. 1, Part A, Elsevier, Amsterdam, The Netherlands, 1988, p. 435.
61. T. Blizzard, M. H. Fisher, H. Mrozik, and T. L. Shih, in G. Lukacs and M. Ohno, eds., *Recent Progress in the Chemical Synthesis of Antibiotics,* Springer-Verlag, Berlin, Heidelberg, Germany, 1990.

MICHAEL H. FISHER
HELMUT MROZIK
Merck, Sharp and Dohme Research Laboratories

ANTIPERSPIRANTS. See COSMETICS.

ANTIPYRETICS. See ANALGESICS, ANTIPYRETICS, AND ANTIINFLAMMATORY AGENTS.

ANTIRHEUMATIC AGENTS. See ANALGESICS, ANTIPYRETICS, AND ANTIINFLAMMATORY AGENTS; IMMUNOTHERAPEUTIC AGENTS; VETERINARY DRUGS.

ANTISEPTICS. See DISINFECTANTS AND ANTISEPTICS.

ANTISTATIC AGENTS

Electrification is the process of producing an electric charge on an object. If the charge is confined to the object it is said to be electrostatic. The term static electricity refers to accumulated, immobile, electrical charges in contrast to charges in rapid flow, which is the subject of electrodynamics.

Static electricity was described as early as 600 BC by the Greek scientist Thales of Miletus, who observed that amber attracts small particles of dust when rubbed with animal fur. William Gilbert (1540–1603), an English scientist, discovered that many other materials behave like amber when rubbed. There have been many attempts made at investigating the origin and nature of electrostatic charges in various materials. As far back as 1757 Wilcke (1) noted that various substances such as amber, silk, wool, etc, could be arranged in a triboelectric series. It was assumed earlier that electric charges are generated by the mechanical work dissipated in friction, hence the name triboelectricity. A triboelectric series can be defined as a grouping of materials arranged according to their electrostatic susceptibility, the first in the series being positively charged with respect to all the others, and the last in the series being negatively charged with respect to all the others. Wilcke showed that rubbing any two of the substances together produces opposite charges, the one that precedes in the series becoming positively charged. Since that time many such lists have been published (2–7), some of which are shown in Table 1. A widely accepted triboelectric series is that listed in the Electronics Industry Association (EIA) Standard 541, which is also listed in Table 1 (8). There are certain similarities among the tables, eg, wool and polyamides are located on the positive end of the series whereas polyesters appear on the negative end; however, several inconsistencies also exist. Overall, it is possible to predict the signs of the charges for a given pair of materials when rubbed together, especially when these materials are sufficiently far apart in the series; however, the magnitude of charge generated is difficult to predict. Kinetic effects play an important role in static generation and render it unpredictable. Many other factors such as differences in the molecular surface structure of materials, presence of surface impurities, intensity of contact, electrochemical and thermochemical effects, etc, may also be responsible for the inconsistency of the triboelectric series (9).

Static is a surface phenomenon. When two surfaces are brought into contact, electrons pass across the interface in both directions. This exchange is not symmetrical, and even with two identical bodies one of the two acquires an excess of electrons at the expense of the other. As long as contact is maintained, there is no further effect. However, when the surfaces are separated part of the charge is discharged into air, and the remainder is left on the material. One of the two bodies has an excess of electrons and the other has a shortage of electrons, ie, the objects are electrically charged with the same magnitude of charge but opposite signs. The magnitude of the charge depends on the rate of separation of the two bodies. The electrons in a conductor move freely, thus the excess of electrons is eliminated by backflow. However, the electrons in an insulator are not mobile, and the phenomenon of static electricity becomes noticeable.

Coulomb (1736–1806) stated the law of repellency between similarly charged bodies and attraction between oppositely charged bodies, and Faraday (1791–1867) described the laws of electrostatic induction. The inductive principle known as Faraday's ice-pail method is still in use in modern measuring equipment.

The density of static charge accumulated by an insulator is limited due to leakage through air, and therefore the attraction and repulsion of charged bodies are also limited. The breakdown potential of air at atmospheric pressure is 30,000 V/cm; thus the maximum charge that can exist on a plane surface is about 3.3×10^{-9} C/cm^2 (10 esu/cm^2) (10). Such charges are sufficient to hold only small

Table 1. Triboelectric Series

Reference 2	Reference 3	Reference 4	Reference 7	Reference 8
wool	glass	wool	platinum	quartz
nylon	human hair	nylon	formvar	glass
silk	nylon yarn	viscose	filter paper	mica
viscose	nylon polymer	cotton	cellulose	human hair
polyethylene	wool	silk	acetate	celluloid
Cordura[a]	silk	acetate	cellulose	nylon
human skin	viscose	Lucite[a]	triacetate	wool
fiber glass	cotton	poly(vinyl alcohol)	polyethylene	fur
glass	paper	Dacron[a]	aluminum	silk
acetate	ramie	Dynel[b]	polystyrene	aluminum
Dacron[a]	steel	Velon[c]	natural rubber	paper
chromium	hard rubber	polyethylene		cotton
Orlon[a]	acetate	Teflon[a]		steel
	Orlon[a], Saran[d]			wood
	polyethylene			amber
				sealing wax
				hard rubber
				nickel, copper, brass, silver, old platinum
				sulfur
				acetate rayon
				polyester
				Orlon[a], Saran[d]
				polyurethane
				polyethylene
				polypropylene
				PVC
				silicon
				Teflon[a]

[a]DuPont trademark.
[b]Union Carbide trademark.
[c]Firestone trademark.
[d]Dow trademark.

particles. Increasing the density of the charge beyond the breakdown potential of air causes a discharge through an electric spark.

During the first half of the nineteenth century one of the few significant researchers of static was Helmholtz (1821–1894), who introduced a comprehensive theory of electricity and also developed the contact potential theory of the electrification of insulators by contact. Also, P. E. Shaw showed the influence of chemical and physical conditions of surfaces on static generation. Many problems have arisen in recent decades in the textile, paper, and plastic industries because of the introduction of new insulating materials, particularly synthetic fibers and plastics, as well as increased processing speeds. This has led to an accelerated pace of research into the generation and discharge of static electricity in insulators.

Nature of Charge Generation

Objects are thought of as developing static charges either by tribocharging or by an induction mechanism. Tribocharging involves the rubbing together of two objects to form the charge. Induction is a process by which an outside electrical field charges an object.

Tribocharging. The root word "tribo" is from the Greek word meaning "to rub." When two objects contact they have the potential for the transfer of electrons from the surface of one object to the surface of the other object. Electron transfer can occur simply by touching the objects together and then quickly separating the objects. The more the objects are rubbed against each other the more the electrons are excited, and the greater the potential for electron transfer. Electrons are removed from the surface atoms of one material more readily than from the other material; the surface with the electron deficit thus develops a positive charge. The surface accepting these electrons has an excess of electrons and therefore acquires a negative charge. Whether a material is an acceptor or a donor is determined by their relative positions along the triboelectric series (11).

The magnitude of the static electric charge developed by tribocharging is dependent on the intimacy of contact between the two surfaces, the amount of rubbing, the coefficient of friction of the two surfaces, the speed at which the two surfaces are separated, and the conductivity of the materials. The first three factors are somewhat interrelated. The closer the contact between the two surfaces, the more opportunity exists for the transfer of electrons. When the two surfaces are in contact the overall charge is neutral; a static charge does not establish until the two surfaces are separated. If the surfaces are separated slowly, the electrons have time to return to their normal state. If they are separated quickly, the electrons do not have time to return to their normal state and remain on the surface that is closer to the negative charge end of the triboelectric series.

The rubbing of the two surfaces need not be a continuous, repetitive motion. Actions as simple as walking across a room or sitting in a chair can generate a static charge.

Induction. An object can acquire a static charge from another charged object by induction. If a charged object A is brought into the vicinity of a neutral object B and the neutral object is grounded, the neutral object acquires a charge of opposite polarity by induction (Fig. 1). In this example, the electrons on object B are repelled by the negative charge on object A. The electrons flow to the ground, leaving a positive charge on the object. If the ground is removed, C, the object retains the positive charge even after the inducing object A is removed. If the object remains grounded after the inducing object is removed, the electrons flow from the ground and return the neutrality of the object. In the example, if object A had been positively charged, the object B would have attracted electrons from the ground and would have been negatively charged.

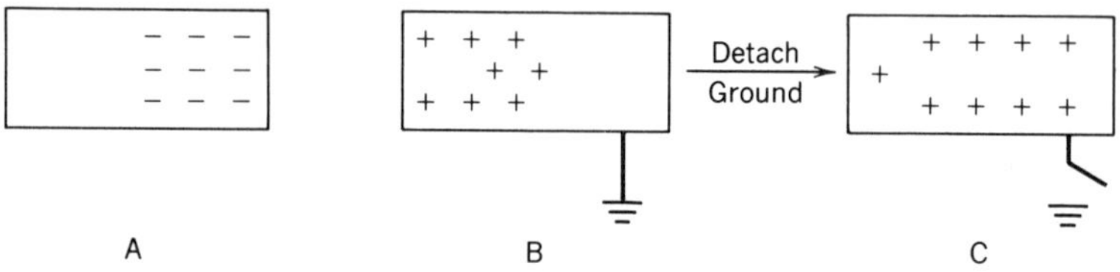

Fig. 1. Acquisition of static charge by induction.

Effects of Static Electricity

Repellency of Similarly Charged Objects. Mutual repulsion of fibers having similar charges causes difficulties in many textile processing procedures. Repulsion causes filaments in a charged warp to bend away from one another, ballooning of a bundle of slivers (12), excessive fly (very short fibers) during processing, and increases in hairiness of the yarn produced. Difficulty in folding fabrics into layers is especially disturbing in dyeing and finishing plants where long lengths of dry fabrics must be handled.

Attraction of Oppositely Charged Objects. Mutual attraction of oppositely charged fibers leads to difficulties in carding, drawing, weaving, knitting, roving, spinning, etc, due to tangled fibers and yarns. Charged fibers are attracted to adjacent parts of the spinning machinery and thus interfere with processing (13–15). Static attraction of dirt and fly from the air produces dark localized stains on fabrics left on looms during weaving (fog marking) and soiling of packages of nylon yarn (16). For the consumer, many annoyances occur frequently as a result of mutual attraction of oppositely charged fabrics. Examples are clinging of apparel to the body and to undergarments (17–20). Static can contribute to the soiling of garments, draperies, and other textile products by attracting oppositely charged particles of dust and dirt from the atmosphere to a charged fabric (21–23). One of the most severe outcomes of mutual attraction of oppositely charged fabrics is failure of parachutes to open.

Static phenomena due to the attraction of oppositely charged objects are evident in many other industries. Electrostatic charges tend to accumulate on the surface of plastic articles during fabrication, and these charges attract dust and dirt particles to the surface, causing fabrication problems. Dust attracted to finished consumer products makes them unattractive for sale. The generation of electrostatic charges is particularly troublesome in the fabrication of sheets, films, or filaments because the static charges tend to cause the articles to cling together or to the processing equipment (24). Electrostatic charges are also accumulated during the subsequent use of the fabricated article and cause excessive pickup of dust and dirt particles by bottles, film wrap, and resin powders. Static buildup on phonograph records attracts dirt and dust, consequently impairing the quality of sound reproduction. It also causes accelerated wear of the stylus and the record (25). Electrically charged resin powders are troublesome since the charge causes them to adhere to processing equipment, packaging materials, and other surfaces, thus causing various difficulties in their preparation and utilization. Accumulated electrostatic charge on photographic films can cause adhesions of surrounding dust onto the surface of the film, which impairs the quality of the film. Discharge of such films during processing causes fogging. In the paper industry, electrostatic charges developed on paper during printing prevent proper layering of the sheets.

Static Electric Discharge. When a charged object comes into contact with an object of lower charge or into contact with a ground, a static electric discharge occurs. If the charged object has come into contact with an ungrounded object, the discharge continues until the charge on each object is equivalent. If the object comes into contact with a grounded conductor, then it completely discharges.

The contact between the two objects is not required to be a physical touching of the two objects. If the charge on the object is greater than the dielectric strength of air, the charge will pass over the air gap to the other object.

The static electric discharge can be a small annoyance, such as shock received when a person touches a doorknob after walking on a charged carpet, or the discharge can be catastrophic. Static electric discharge can cause explosions in areas where flammable gases are in use and in grain elevators when the grain dust reaches combustible levels. Perhaps the most famous example of the dangers of static electric discharge is that of the German airship, the Hindenberg, which ignited its hydrogen, causing the airship to burn.

Static electric discharge is a serious problem in the electronics industry. Electronic devices are extremely sensitive to static electric discharges. Examples of the sensitivity to electrostatic discharge (ESD) are given in Table 2 (26).

Table 2. ESD Susceptibility of Various Electronic Devices

Device type	Range of ESD susceptibility: Potential, V	Energy, μJ
MOSFET	100–200	0.5–2.0
EPROM	100	0.5
JFET	140–7000	0.98–2.45
OP-AMP	190–2500	1.62–312.5
CMOS	250–3000	3.13–450
Schottky diodes	300–2500	4.5–312.5
film resistors (thick, thin)	300–3000	4.5–450
bipolar transistors	300–7000	4.5–2.45
Schottky TTL	1000–2500	50–312.5

The energy from an electrostatic discharge on an integrated circuit chip causes localized heating in the chip, which melts areas of silicon in the chip. This melting destroys the circuits engraved on the silicon. Temperatures have been known to reach 760°C at certain points on the chip. The temperature is highest at positions of restricted current flow on the chip (27).

Electrostatic charges are also generated when liquids move in contact with other materials, liquid or solid, eg, during pumping of gasoline. Serious industrial hazards caused by static in chemical and related fields have been described (28), and a study of accidents in the chemical industry revealed that 115 out of 1600 accidents, or 7%, were ascribed to static electricity (29) (see PLANT SAFETY).

Accumulation of electrical charges on textiles, plastics, and other materials of insulating character causes problems which occur with greater frequency as the relative humidity decreases, particularly during winter months. Heated rooms in winter provide an environment in which static phenomena are frequently apparent.

Useful Applications. Suggestions have been made for uses of static electricity in the textile (9) and other industries, but the number of commercially successful applications is few. One well-known use of static electricity in the textile industry is the process of flocking, in which short fibers (flocks) are embedded vertically and cured in an adhesive-backed substrate, usually fabric or paper, to make a large range of velvetlike materials. Static charges are generated by a high voltage source and are used to control the deposition of the short fibers onto the backing (30). Other successful applications are in the lay-down of fibers in the process of spunbonded nonwoven fabrics (31,32) and in electrostatic polishing of pilo fabrics (see NONWOVEN TEXTILES).

Electrostatic fields generated by high voltage electrodes are employed in electrostatic spray painting where the charge first attracts the spray to the object and then repels the excess spray from it (see COATING PROCESSES). Avoiding soiling due to the attraction of dust and dirt particles by static is applied in air filters (33) (see AIR POLLUTION CONTROL METHODS). The Shirley Institute has developed perhaps the most novel application of static phenomena, the detection of footprints on carpets and other floor coverings (31).

Other suggested applications include the use of electrostatic fields in electrostatic printing (see PRINTING PROCESSES) (34), shearing machines (35), and measurements of yarn hairiness and moisture content of fabrics (31). Other examples are utilization of static electricity in fiber processing (35), cleaning of cotton and wool fibers from trash (35–38), separation of long and short fibers (35,59), separation of wool from synthetic fibers (40), open-end spinning of yarns (41–42), and straightening of filling thread in shuttleless looms (41). It has been claimed that negative static charges generated on undergarments made of synthetic fibers have considerable therapeutic effects on rheumatic illnesses such as arthritis, rheumatism, etc (43–44).

Methods of Controlling Static Charges

The amount of electrostatic charge developed on a material represents a balance between the rate of generation and the rate of dissipation. Good conductors disperse a static charge quickly; however, textiles and plastics have a high surface resistivity and charge decay can occur only at a low rate. For example, it has been calculated that a given charge on a polystyrene surface would take three years to decay to a negligible value (45). It is feasible to reduce accumulation of static charges by either reducing the rate of generation or by increasing the rate of dissipation.

Preventing Static Generation. The most logical remedy to static problems is prevention of static generation. Reduction of friction generally results in reduction of static, but friction is essential for many processes and cannot be excessively reduced. Moreover, it cannot be regarded as a true remedy since charges are actually generated whenever two materials in contact are separated, regardless of rubbing. Also, charges can be generated by induction from another charged object, although static-shielding containers that provide protection from static charging by induction can be purchased.

Another method reported to reduce charge generation is changing the electrical potential of the contacting elements (31,46,47). Perhaps the most popular suggestion to eliminate static generation utilizes the rank of materials in the triboelectric series. This includes either impressing countercharges on the charged material by contact with materials on the other side of the triboelectric series (48–50), or neutralizing the charge generation by blending two kinds of fibers with a tendency to attain opposite charge (2,7). The blending technique, however, is generally not effective since it is practically impossible to predict the exact composition of the optimum blend of two fibers that will generate equal but opposite static charges.

Increasing Rate of Static Dissipation. Since it is not feasible to prevent the generation of static electricity, all practical remedial measures are based on increasing the rate of charge dissipation or charge leakage. Static charges on polymers dissipate or decay through two processes, namely surface and volume conductivity of charges through the substrate, and loss of charge by radiation to the air (51,52). The radiation process has a significant effect on charge decay and in some polymers, ie, nylon-6,6, the evidence strongly suggests that the loss of charge by radiation is greater than by the conductivity process. Even for polymers that are conducting, ie, wool and cellulose, it is found that the rate constant for radiation is approximately equal to that for conduction (51). The following comprise the main practical methods for increasing the rate of static dissipation.

Grounding. Use of a grounded bar, rod, or roll to discharge static charge buildup is effective for conductive materials which contain a charge. Grounding is ineffective for insulators since the electrons on an insulator are not sufficiently mobile to travel to ground. For partially conductive materials, grounding can be somewhat effective for the area of the object in contact with the ground. Grounding of a partially conductive material is best attempted by contacting the entire surface of the object, eg, using a string of copper tinsel stretched across a roll of paper on a computer printer.

Ionization. If a charge is located on an insulator, there is in principle no way by which the charge may be removed. If, however, the charged insulator is completely surrounded by a conducting fluid in contact with all points of the surface, the charge or the field from the charge may be neutralized by oppositely charged ions being attracted to the insulator. Although this scenario in principle could be established by using a conducting liquid, the only practical solution is to surround the charged body with ionized air (53).

Static eliminating devices that ionize the atmosphere around the device are available. Static eliminators are divided into two main groups: silent (corona) discharge eliminators, such as inductive eliminators, or high voltage eliminators which initiate an impact ionization of the air by applying strong electrostatic fields, and radioactive eliminators that provide a multitude of ions from independent ion sources (9).

Increasing Conductivity. Increasing the electrical conductivity of the material by increasing its electrolytic (ionic) or electronic conductivity is another means of controlling static charges. The electrolytic conductivity can be improved by increasing the moisture content of the surrounding atmosphere (humidification), by application of internal or external antistatic agents, or by chemically modifying the material. Chemical modification is especially used for textile fibers and involves surface modification to increase hygroscopic properties (54,55) or grafting of functional groups leading to ionic configurations or greater moisture sensitivity (56). The electrical conductivity of the material can also be improved by increasing its electronic conductivity. This can be achieved by coating fibers with a thin layer of a conductor such as silver or carbon black, by blending fine metal fibers with static-prone fibers in the textile material, or by incorporating carbon black in plastics.

Humidification. Adding moisture to air has long been used to control static and to help dissipate static electricity in textile mills (see AIR CONDITIONING). Moisture does not improve the electrical conductivity of the air, but it increases the electrical conductivity of materials absorbing moisture. A definite limit to humidification is imposed by discomfort and corrosion problems, and in some cases high humidity has to be avoided due to its deleterious effects on the materials or on the processing performance. Also, humidification alone has effective limits; for example, lightning, which is a static electric discharge, occurs during rainstorms. Humidification within a tolerable range is the most widely used antistatic process in the processing of materials of limited static propensity such as cotton and rayon, but this does not suffice with other materials such as wool, synthetic fibers, plastics, etc; the use of suitable antistatic agents in the processing of such materials is essential. Antistatic agents are also used to eliminate static problems that arise in consumer use. Antistatic agents prevent the bulk of static problems, but almost all of them rely on atmospheric moisture for their effectiveness; humidity control therefore remains a requirement for the smooth processing of hydrophobic materials.

Measuring Static Charges

Static charges can be characterized by several techniques (57). A simple test is to suspend an object 2 cm above a shallow dish containing cigarette ashes. A charged object attracts the ashes to its surface; the amount of ashes attracted is proportional to the charge on the object. More scientifically vigorous tests are available by measuring surface resistance, field voltage, coulombic charge, and charge decay rate.

Field Voltage. A field voltmeter measures a static charge by putting a conductive plate in the field of charge around a charged object. The plate is always positioned the same distance from the charged object. By estimating the dielectric constant of the object and knowing the area and the distance, the capacitance of the object can be estimated

$$C = \frac{KE_0A}{x}$$

where C = capacitance, K = dielectric constant, $E_o = 8.85 \times 10\text{–}14$ F/cm, A = area, and x = distance. After capacitance and charge are determined, voltage can be calculated by the formula: voltage = charge/capacitance.

For an accurate measurement using a field voltmeter, the meter must be calibrated (or zeroed) before a test. The charge on a grounded conductor is measured. A grounded conductor should have a charge equal to zero, which means the voltage would be zero. Therefore, the field meter should read zero (Fig. 2**a**).

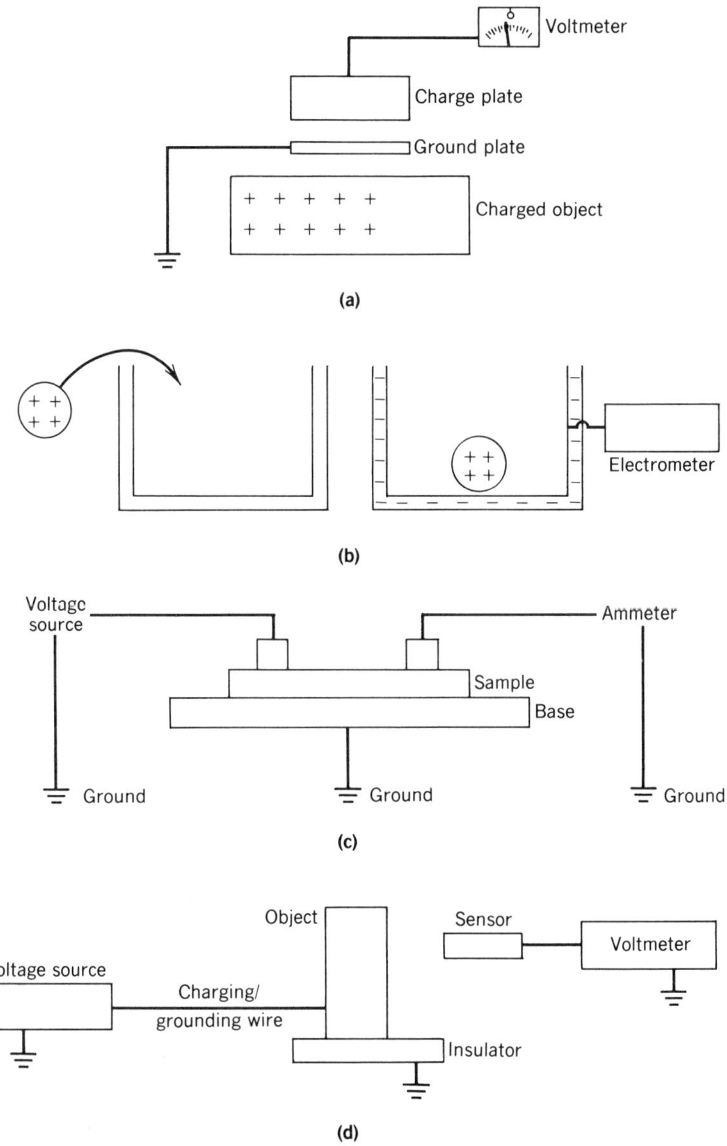

Fig. 2. Measuring static charge. (**a**) Field voltage measurement; (**b**) Faraday cage; (**c**) surface resistivity measurement; and (**d**) static decay test.

In practice, field voltage is a quick and easy test. Several companies manufacture small handheld voltmeters. However, the test is error-prone. The capacitance of the object is needed for an exact measurement, and the meter must be held an exact distance from the object being measured. If the distance is different from the one used by the manufacturer to calculate the voltage, then the voltage reading on the meter will be different from the actual voltage. The voltmeter does not measure the entire charge, but only the charge directly in front of the plate. Voltmeters have the advantage of being easy to use and can measure voltages on large or immovable objects that would not fit into other test devices.

Faraday Cage. One of the best means of measuring a static charge, typically in coulombs, is by means of a Faraday Cage. When a charged object is put inside a conductive box, the Faraday Cage, a charge is induced on the box. This charge is of opposite polarity but equal magnitude to the charge on the object. The Faraday Cage can simply be connected to an electrometer, and the charge on the cage measured. The cage should be grounded before the test to bring its charge to zero, and the ground should be removed before the object is put into the cage. In practice the Faraday Cage should be inside of but insulated from a second Faraday Cage. The second Faraday Cage shields any outside charges from interfering with the test. If the Faraday Cage is deeper than it is wide, it does not require a lid to give an accurate reading. An object put into the Cage falls to the bottom of the cage so that very little of its charge is lost through the air out of the top of the cage (Fig. 2**b**).

Faraday Cage measurements have the advantage of measuring the total charge on an object and also are equally effective for testing conductive as well as insulating objects. One disadvantage of the test is that the object to be tested must fit into the cage.

Surface Resistivity. Resistance is the measure of the conductivity of a material. Conductive materials have the ability to dissipate static charges. Therefore, resistance is a good measure of the ability of an object to dissipate a static charge. Since static electricity is primarily a surface phenomenon, the resistance along the surface of the object is of most interest.

Surface resistance is measured by placing two electrodes on the surface of the object to be tested. A d-c voltage is applied to one electrode and an ammeter is attached to the other electrode. The current travels from one electrode to the other by traveling across the surface of the object. When the current is measured and the applied voltage is known, resistance can be calculated by the formula: resistance = voltage/current.

Resistance measurements change with an alteration in voltage, and the voltage used to run a test should be compared when evaluating resistivity results. The proper voltage to use is often listed in the test organizational procedures that specify surface resistance test methods. The most common test method is provided by the American Society of Testing and Materials (ASTM) in standard D257. The Electrical Overstress/Electronics Industry Association has found that ASTM D257 is a good procedure for conductive materials and insulators but is not accurate when testing static dissipative materials. They have proposed a variation on ASTM D257 in their standard DS11.11 (58). ASTM D257 is run with a 500 V d-c charge. EOS/ESD DS11.11 is run with a 100 V d-c charge for materials with a resistance above $1.0 \times 10^6\ \Omega$ and a 10 V d-c charge when the resistance is below $1.0 \times 10^6\ \Omega$.

The terms resistance and resistivity are both used when referring to the resistance of an object to current flow. Surface resistance is the measure of electrical resistance along the surface of an object. However, the current flow is not limited to the surface of the object. Some of the current passes through the bulk of the object from one electrode to the other electrode. Surface resistivity includes the dimensions of the object in its measurement (Fig. 2**c**).

$$\text{Surface resistivity} = \frac{\text{Area}}{\text{Thickness}} \times \text{Resistance}$$

Static Decay. A static decay test measures the length of time required for the charge on an object to decay once the object has been grounded. The test suspends an object from conductors or ground. A charge is induced on the object using a d-c voltage source. The charge on the object is determined by a voltmeter. The charged object is grounded and the drop in voltage is measured by the voltmeter (Fig. 2**d**). The test is usually run in a Faraday Cage to prevent outside charge sources from interfering with the test. A typical charging voltage is 5000 V d-c. The test method is described in Federal Test Method Standard 101B Method 4046 (59).

The static decay test is a good measure of the static dissipative nature of an object. However, the test is controlled by the object's resistance to ground. Objects that have a high resistance, such as insulators, will not readily accept an induced charge or decay an induced charge. The static decay test is most suited for objects with resistances of 10^{12} or less (52).

Tests Simulating Use Conditions. Several simulation tests involving wear and use testing have been developed. The first and only one to be adopted as a standard in the United States is AATCC Test Method 134-1975 for measuring the electrostatic propensity of carpets (60). In this test a person wearing standard neoprene and leather sole shoes and carrying a probe connected to an electrometer walks on a carpet for 30 to 60 s until the measured body voltage builds up to a maximum.

A test that does not involve a human test subject has been adopted as AATCC Standard Test Method 115-1973 called Electrostatic Clinging of Fabrics: Fabric-to-Metal Test (61). This test is used for evaluating lightweight fabrics, and is based on the principle of frictional charge generation by rubbing with a fabric of different composition. The charged fabric is allowed to cling to a grounded metal plate and the time for declinging is measured.

Subjective tests using humans have also been described, such as the sail test (62) and others (63–65). As might be expected, the results of these tests are not very reproducible.

Tribocharging Tests. Since a common source of a static charge is by tribocharging, a lot of effort has been expended in developing tribocharging tests. In general, the results of such tests are not reproducible, and the tests are effective only for the narrow range of materials used in developing the tests. The tests usually involve rubbing a cloth (usually wool) against an object or sliding an object down a quartz or Teflon rod; if the object is not suitable for sliding down a rod, then a quartz or Teflon disk is slid down the object. The charge on the object or the disk is then measured. The EOS/ESD Association is attempting to develop a standard reproducible tribocharging test under standard DS11.2 (53).

Significance of Static Charge Measurements. Attempts have been made to correlate static charge measurements to a set of standards that would ensure that articles meeting the standard would not pose static electric problems. Most of the tests have only provided general guidelines for static protection. For problems such as static-induced dust attraction no standards exist, and performance measures are strictly subjective. The electronics industry and the National Fire Protection Association have developed standards based on the surface resistivity and static decay tests. These standards have evolved over time and will no doubt be modified in the future. The current standards are as follows: EIA-541 for static dissipative materials requires a surface resistivity of 10^5–10^{12} Ω and a static decay rate of 99% charge decay in less than 2.0 s (66). The National Fire Protection Association Standard 99 for materials used in areas containing flammable gases requires a surface resistivity of 1.0×10^{13} Ω or less and a static decay rate of a 90% charge decay in less than 0.5 s (67). Since it is recognized that most antistatic agents function by attracting atmospheric moisture, the standards specify the test to be run in a 15% rh atmosphere.

Static Dissipative Materials

The term antistatic traditionally referred to materials having a surface resistance between the resistance of conductors and insulators. The resistance values of an antistatic material were considered to be in the range of 10^9 to 10^{14} Ω. The antistatic property of a material is not a dependent function of material resistivity, which is an intrinsic property used to define the material's degree of conductivity without regard to other materials. A material's propensity to form a static charge depends on the nature of the material itself, the material with which it is in contact, and the means of surface separation. In the Electronics Industry Association Standard 541, the term antistatic no longer refers to a material's ability to resist triboelectric charge generation (68).

Materials that have the ability to dissipate a charged formed, by any means including tribocharging and induction, on the material are referred to as static dissipative. There is a correlation between static dissipation and surface resistance. EIA-541 currently defines the static dissipative range as 10^5 to 10^{12} Ω.

The term antistatic agent is generally used to describe a substance that is added to a material to make that material static dissipative. The two biggest markets for antistatic agents are textiles and plastics.

ANTISTATIC AGENTS FOR TEXTILE MATERIALS

Antistatic finishes are defined as antistatic agents in combination with water, mineral oil, composite finishes applied by producers during fiber manufacture, oils used to facilitate throwing or coning, textile softeners or lubricants, and hand building compounds. The terms antistatic agents and antistatic finishes have often been used interchangeably in the literature.

Effective antistatic agents must act at a relative humidity below 40%, preferably below 15%. The agent must form a film on various surfaces and be applied from a solution or dispersion in water or other inexpensive solvents. The antistatic agent must not interfere with subsequent processing of the product, impair the hand, or affect color, odor, appearance, and performance properties of the substrate. It should be nontoxic and nonflammable.

Function. Antistatic agents can function either by reducing the generation of charge, by increasing the rate of charge dissipation, or by both mechanisms. The way in which rate of generation is diminished is not completely understood; however, it seems likely that the presence of the antistatic agent at the interface reduces the intimacy of contact between surfaces and therefore the net charge transfer (9). Evidently most lubricants function this way, and it is often desirable to combine antistatic action with lubrication (69) (see LUBRICATION AND LUBRICANTS).

Most antistatic agents operate by increasing the rate of charge dissipation. Good correlation has been shown between the surface resistivity of a material and its charge decay rate (70). Moreover, there is a direct correlation between the effectiveness of an antistatic agent and its ability to lower the surface resistivity of a polymeric material (70,71). It is believed, however, that static charges on polymers are lost through two processes, namely surface and volume conductivity of the charge through the substrate, and loss of charge through air by radiation. The radiation process is at least as important as the conduction process in effecting charge dissipation (51,72).

Chemical Classification. Substances of high electrical conductivity are effective antistatic agents. Numerous chemicals have been proposed and reviewed (73–75). In general, most antistatic agents belong to one of the following classes (76): nitrogen compounds such as long-chain amines, amides, and quaternary ammonium salts; esters of fatty acids and their derivatives; sulfonic acids and alkyl aryl sulfonates; polyoxyethylene derivatives; polyglycols and their derivatives; polyhydric alcohols and their derivatives; phosphoric acid derivatives; solutions of electrolytes in liquids with high dielectric constants; molten salts; metals; carbon black; semiconductors; and liquids with high dielectric constant, such as water, which are usually volatile and have a temporary effect only.

Role of Atmospheric Moisture in Static Protection. Water is outstanding among liquids because of its low cost, absence of toxicity, and nonflammability. Furthermore, although volatile, water can be continuously replenished from the atmosphere. Most nonmetallic antistatic agents utilize the conductivity of water for charge dissipation and ensure its presence by their hygroscopic nature. The amount of moisture absorbed in equilibrium with atmospheres of various relative humidities at a given temperature represents a moisture absorption isotherm for the antistatic material. A reduction of atmospheric humidity reduces the effectiveness of antistatic agents. A small quantity of adsorbed water is essential for ionization in the process of antistatic protection (77,78). In contrast, substances such as carbon black and metallic filaments do not rely on the presence of moisture to dissipate the static charge.

Most organic polymers absorb significant amounts of moisture from the atmosphere; the amount absorbed is a function of the relative humidity. The absorption isotherms of polymers depend on the affinity for, and accessibility to, water molecules. Pure water has a specific resistance of 10^8 Ω/cm; in general, therefore, it has a resistance 10^8 times lower than an organic polymer. Moreover, water usually contains dissolved electrolytes that increase its conductivity a thousandfold. The conductivity of a polymer increases significantly by absorption of even small quantities of moisture. The efficiency of an antistatic agent depends on the degree of its hygroscopicity, the distribution of moisture, and its ability to supply mobile ions to the aqueous layer it forms with absorbed water (79). Hygroscopicity alone is sufficient for antistatic action of a finish, as demonstrated by the effectiveness of nonionic antistatic agents, since stray electrolytes are usually present in the form of impurities. However, the degree of efficiency of an antistatic agent can be increased by the addition of electrolytes.

Experimental observations (80) relating moisture content to specific resistance R_s for polymeric fibrous materials are shown in Figure 3. Most natural fibers such as cotton (qv), silk (qv), and wool (qv), are hydrophilic in nature, and synthetic polymers such as polyolefins (see OLEFIN POLYMERS), polyamides, polyesters, and polyacrylonitrile are comparatively hydrophobic (see FIBERS, ACRYLIC; FIBERS, POLYAMIDES; FIBERS, POLYESTERS). Therefore, under identical

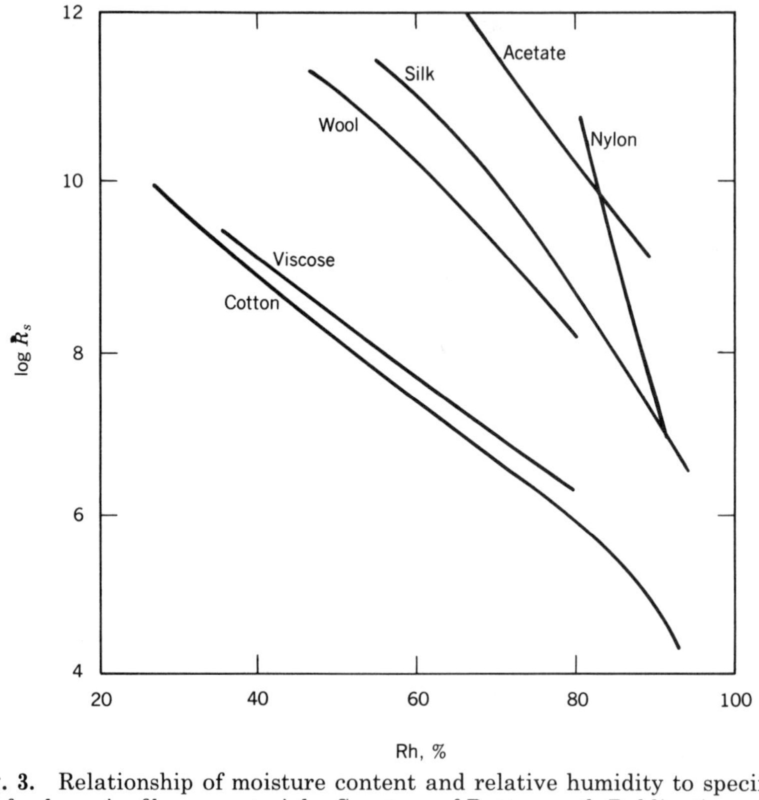

Fig. 3. Relationship of moisture content and relative humidity to specific resistance R_s of polymeric, fibrous materials. Courtesy of Butterworth Publications, Ltd.

conditions, synthetic fibers as a group are better insulators and are more static-prone than natural fibers. In a certain range, however, nylon shows lower resistance than the more hydrophilic viscose (rayon) for the same moisture content (see FIBERS, REGENERATED CELLULOSICS). The range of resistivity, even in the incomplete range of moisture content, embraces eight orders of magnitude. Under the same moisture conditions, hydrophobic fibers tend to show lower conductivity compared to hydrophilic fibers. This lower conducting efficiency is perhaps related to a lower dielectric constant and, consequently, a lower degree of ionization of the hydrophobic polymer. However, both synthetic and natural fibers become increasingly static-prone as the moisture content of the fiber is reduced. The exceptional position of wool indicates that moisture is preferably absorbed by the interior of the wool (cortex) rather than the periphery (cuticle or epicuticle) of the fiber. Static propensity of fiber assemblies, such as of yarns and fabrics, is related to the properties of the component polymers; however, this is complicated by a form factor, mainly by the varying density of the structure (81).

Static electrification may not be a property of the basic structure, but of a new surface formed by a monomolecular layer of water (82). All textile fibers at a relative humidity, at which a continuous monomolecular layer is formed, actually do have the same charge density. This is attributed to the absence of ionic transport which cannot occur in a monomolecular layer. At higher moisture levels than required to form a monomolecular layer, ionic conductivity can occur because of excess water molecules and by hydration of the ions. At very low moisture-regain levels, all materials acquire the same charge (83).

The antistatic efficiency of absorbed water in a hydrophilic fiber is due to the submicroscopic, continuous channels it forms in the fiber. In contrast, a hydrophobic fiber that has been coated with an antistatic agent has a high surface conductivity. The increased conductivity of the treated hydrophobic fiber is due to a surface layer, even though a larger portion of the absorbed moisture may still be inside the fiber. The ratio of water to material is higher in a good antistatic layer than in a hygroscopic fiber so that the conductivity effect of moisture on an antistatic surface can be higher than in a hygroscopic fiber. An effective antistatic agent should therefore form a surface layer of aqueous solution, preferably of high ionic concentration; however, there is no clear distinction between the surface and volume conductivity of textile fibers.

Charge decay rates of polymers have been correlated to their chemical structures (51). The presence of groups such as $-Cl$, $-CN$, and $-COOCH_3$ in a polymer contributes less to antistatic protection than hydrophilic groups such as $-COONa$, $-SO_3Na$, and $-OCH_2CH_2-$ contribute. However, it is difficult to explain the superiority of $-CON(CH_3)_2$ groups over $-SO_2NH_2$ groups. A study of such factors as the mode of distribution of absorbed water, mobility of ions inside the swollen polymer, dissociation of ionizable groups, skin and core effects in polymers, etc, might provide a more general correlation between chemical structure and charge decay.

Quantitative Relationship of Conductivity and Antistatic Action. Assuming that an antistatic finish forms a continuous layer, the conductance it contributes to the fiber is proportional to the volume or weight and specific conductance of the finish. As long as the assumption of continuity is fulfilled it does not matter whether the finish surrounds fine or coarse fibers. Assuming a cylindrical filament of length 1 cm and radius r, denoting the thickness of the finish layer as Δr and the specific conductance of the finish k, the conductance K of the finish layer is given by the equation (84):

$$K = k\,\frac{2\pi r \Delta r}{1}$$

The specific conductance of the finish on the filament k is not necessarily the specific conductance it exhibits in its bulk condition. For instance, absorption of ions from the finish by the fiber can reduce the conductivity. The specific conductance greatly depends on the amount of moisture present. Figure 4**a** shows the experimentally observed resistance of yarn as a function of the amount of antistatic agent applied in comparison to the calculated resistance. Below 0.05% of antistatic agent the experimental values show a lower conductivity than calculated; this may be due to a lack of continuity of the antistatic agent.

Unless some measures are taken to ensure uniformity, about 0.1% is needed for antistatic protection of textile fabrics. The exact amount depends on the efficiency of the agent, relative humidity, temperature, diameter of fiber, fabric structure, and degree of antistatic protection desired. Figure 4**b** shows that the observed surface conductivity of a nonionic agent is higher than the calculated value. This might be attributed to impurities or to the absorption of excess moisture. Even the experimentally observed conductivities are approximately one order of magnitude lower than the values for the nonionic antistatic agents considered in Figure 4**a** at the same concentration.

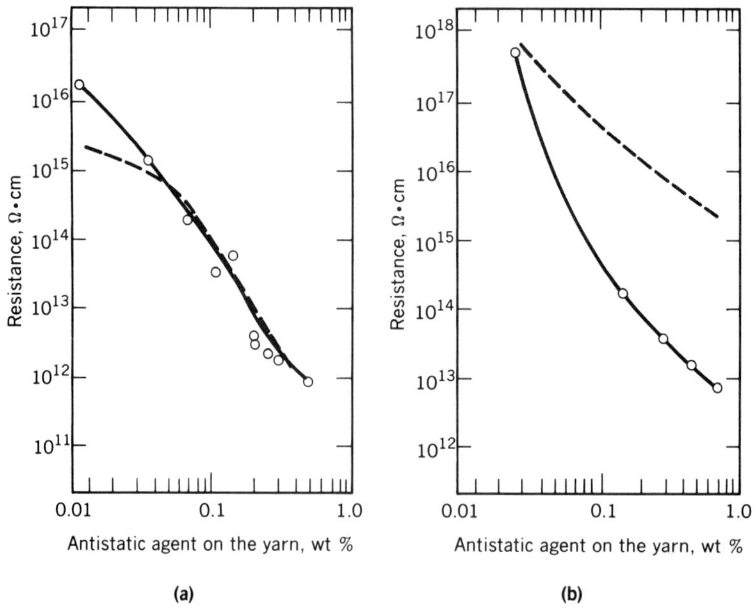

Fig. 4. (**a**) Yarn resistance in Ω·cm vs amount of antistatic agent on the yarn. The agent is the ethyl sulfate salt of an amine. (**b**) Resistance vs amount of nonionic, hygroscopic agent on the yarn. Dotted lines are calculated from the specific resistance of the dry bulk solution; solid lines are experimental yarn resistance. Courtesy of *Textile Research Journal*.

In another investigation (71), nine different antistatic agents were applied at various concentrations to a nylon fabric, and the magnitude of charges generated by rubbing against aluminum plates in a reproducible manner was compared with the fabric resistivity. A linear relation holds between the logarithms of charge and resistivity. In view of the broad confidence limit of the charge values, these results confirm the relationship between antistatic protection and resistivity.

From a quantitative study of the effect of lubricating oils in static generation by yarns, the role of dynamic coefficient of friction was compared with that of electrical conductivity (85). In order to avoid static difficulties in yarn processing, the yarns must have a low coefficient of friction and, more importantly, have the highest possible conductivity. Whatever the coefficient of friction may be, the voltage approaches zero if the conductivity becomes high enough (85).

Although it has been generally demonstrated (15,77,86–89) that antistatically treated fabrics exhibit increased surface conductivities, many examples have been found where static behavior is not always related to the conductivity of the fabric (90) or the material. One of the main reasons for this is the fact that static charges decay not only by conduction but also by radiation.

Durability of Finishes. The required extent of durability depends on the use of the treated article. For example, an antistatic finish that is removed after several launderings would be termed nondurable for blankets to be used in hospitals and motels where static protection through over fifty launderings is desirable. However, the same finish applied to blankets for household use which would be laundered once or twice per year could be termed durable. Antistatic finishes have been widely reviewed (73–75,77,86,90–95).

Nondurable Antistatic Finishes. Most surface-active agents reduce static properties of materials to which they are applied and are classified as nondurable antistatic agents. They appear to work on the basis of moisture absorption, and their effectiveness decreases as the humidity (and temperature) in the atmosphere is reduced, reaching a cutoff in effectiveness when ionization is inhibited. Antistatic agents are often combined with lubricating oils to allow emulsification of the oils and to facilitate their removal during scouring and dyeing of textile materials (69). Surface-active agents of the nonionic or cationic type are commonly used as the antistat, and compatibility with lubricants becomes an added requirement.

The structures that render a substance surface-active are its hydrophobic hydrocarbon residue and a terminal hydrophilic group. Synthetic fibers treated with the separate components or a mixture fail to achieve the antistatic efficiency provided by the surface-active molecule derived from the two components. The hydrophobic portion of the surface-active agent is attached to the hydrophobic surface, and the hydrophilic groups are repelled. As a result of this orientation, the fiber surface is covered with a radiating hedgehog structure. Orientation of these molecules at the fiber surface with the formation of an electrostatic shield probably has a specific effect on the static behavior (20,90,96). This is difficult to prove because the mechanism of charge generation is not completely known. Regardless, surface-active agents are generally hygroscopic, nonvolatile compounds with good lubricating power that can provide useful antistatic protection to various substrates.

Surface-active agents increase the conductivity of oils quite significantly (97), and addition of water, probably dissolved at the interface with the surfactant, further increases the conductivity. Nonionic and cationic surface-active agents are preferred to anionic surface-active agents probably because of their higher solubility in oils and higher hygroscopicity. Many anionic surfactants have adequate antistatic efficiency, but they are used less frequently.

The relationship between chemical constitution and antistatic action of an array of surface-active agents has been evaluated (92). Residual amounts of surface-active detergents in hydrophobic garments after laundering can also be effective in aiding charge dissipation. Some examples of commercial nondurable antistatic agents are given in Table 3. Nondurable antistatic agents are suitable

Table 3. Antistatic Agents for Textiles

Trade name	Manufacturer	Chemical type
	Durable antistats	
Aerotex Antistatic D	American Cyanamid Corp.	
Aston 123	Refinex Onyx	polyamine resin
Cirrasol Z	ICI Americas	
Nonax 1166	Henkel	polyamine resin
Permalose T	ICI Americas	nonionic
Stanax 1166	Standard Chemical Products	polyamine resin
	Cationic antistatic softeners	
Ampitol VAC	Dexter Chemical Co.	quaternary ammonium
Avitex DN	DuPont	cationic
Kemamine Q9702C	Humko Chemical	quaternary ammonium compound
	Unclassified	
Antistatin D	BASF	
Arkostat AC	Hoechst	
Aston MS	Refinex Onyx	quaternized (cationic) fatty amine condensate
Aston AP	Refinex Onyx	amine condensate
Cassastat	Cassella	cationic
Catanac SN	American Cyanamid Corp.	cationic
Elfugin UW	Sandoz	
Igepal CO 430	GAF Corp.	nonylphenol ethylene oxide
Neutro-stat[a]	Simco Co.	
Siligen APE[b]	BASF	quaternary ammonium compound
Statexan HA	Bayer	
Tinorex TC	CIBA-GEIGY	
Zelec DP	DuPont	cationic
Zero C	Lutex Chemical Corp.	nitrogeneous polymer

[a]Spray.
[b]For polyester.

for textile processing aids in such operations as spinning, winding, weaving, and fiber manufacture, but cannot provide the material with antistatic protection in consumer use.

Durable Antistatic Finishes. The difficulty with nondurable finishes, as far as the consumer is concerned, is that they are water-soluble and thus easily removed by washing. An effective antistatic finish must be durable and capable of withstanding repeated laundering and dry-cleaning cycles. Only a small number of durable antistatic agents are available for textiles. Semidurable antistatic finishes for textile materials based on compounds of limited solubility and moderate resistance to wet treatments were known in the early 1950s.

Some commercial durable antistatic finishes have been listed in Table 3 (98). Early patents suggest that amino resins (qv) can impart both antislip and antistatic properties to nylon, acrylic, and polyester fabrics. Cyclic polyurethanes, water-soluble amine salts cross-linked with styrene, and water-soluble amine salts of sulfonated polystyrene have been claimed to confer durable antistatic protection. Later patents included dihydroxyethyl sulfone [*2580-77-0*], hydroxyalkylated cellulose or starch, poly(vinyl alcohol) [*9002-86-2*] cross-linked with dimethylolethylene urea, chlorotriazine derivatives, and epoxy-based products. Other patents claim the use of various acrylic polymers and copolymers. Essentially, durable antistats are polyelectrolytes, and the majority of useful products involve variations of cross-linked polyamines containing polyethoxy segments (92,99–101).

Polyelectrolytes (qv) carry ionizable repeating groups and combine the characteristics of polymers and electrolytes in one molecule. These polymers are linear and soluble, but upon cross-linking they form an insoluble three-dimensional network. The giant polyelectrolyte molecule is insoluble in water but still hygroscopic and capable of ionization. A three-dimensional network cannot be deposited on textiles from solution but the monomers can be deposited together with a suitable catalyst and then polymerized and insolubilized by drying and curing *in situ*. Cross-linked polyelectrolytes are also known as ion-exchange resins (qv). The resin produced must be hygroscopic but, when swollen with water, it must have sufficient mechanical strength to withstand the abrasive action of laundering and drying. In addition, the resin must display high ion-exchange capacity in the neutral pH range.

N-Alkylation of amines provides one route to polyelectrolytes (100). Alkylation of the amine with a hygroscopic polymer not only provides hygroscopicity, but also ionizable groups required for efficient antistatic performance. The ratio of hygroscopic to ionizable groups can be controlled by the proper choice of raw materials. For instance, alkylating a primary amine with poly(ethylene glycol dihalide) as the hygroscopic component produces a polyamine that contains oxyethylene interspacing groups. The polyamine contains preponderately tertiary amine groups along the polymer chain, and the end groups may be tertiary amines and/or halide groups. This polymer is capable of undergoing further reaction, resulting in cross-linking of the individual polymer chains to a giant three-dimensional thermosetting resin. Cross-linking occurs mainly at the tertiary nitrogen-forming polyquaternary groups, with some tertiary and quaternary groups at the chain ends. The resulting structures are, in effect, ion-exchange resins. Amines that have been used in this manner include diethylenetriamine

[*111-40-0*] and hexamethylenediamine [*124-09-4*]. A very suitable dihalide is the diiodide of poly(ethylene glycol). Other polyfunctional alkylating agents can be used in place of halides, eg, compounds containing oxirane groups (polyepoxides) and polyfunctional compounds containing both oxirane and halide groups, eg, epichlorohydrin.

Whether a polyelectrolyte is dissolved in water or is immobilized in it by cross-linking, its conductivity is not appreciably affected. In a solution, transport of electrical charges is accomplished by migration of the ions and their compensating ions (gegenions) to the electrodes. In the cross-linked ion-exchange resin, however, the polymeric ion is fixed but its gegenions and water can move. In this case the transport of charges occurs about as fast as if the polymeric ion were dissolved. The conductivity of a water-swollen polymer is therefore determined not only by the amount of dissolved electrolytes present, but also by the amount of ionizable groups linked to the polymer by covalent bonds or van der Waals forces. Atmospheric moisture also plays an important role in the performance of a durable antistatic finish. The effectiveness of a durable antistatic treatment decreases as the temperature and humidity decrease, and cutoff occurs where the moisture content of the atmosphere is low, usually at some temperature in the range of −30 to 0°C.

Cross-linked finishes are not permanent in the true sense of the word; however, under optimum conditions the finish can last for the useful life of the material. Wet abrasion during laundering is probably the principal cause of gradual removal of the finish. In order to retain antistatic protection for extended use, an excess of finish is often applied. The extent of chemical interaction between the durable antistatic agents and the fiber substrates to which they are applied is not perfectly understood. Certain oxidizing agents such as hypochlorite bleaches tend to depolymerize and remove some durable antistatic finishes. Some of the durable finishes have also produced undesirable side effects on textile materials, ie, harsh hand, discoloration, and loss of tensile properties.

Application of Antistatic Finishes. Antistatic finishes are commonly applied to textile materials by padding (dipping and squeezing off excess finish), exhausting (absorption from solution due to affinity of the antistat for the textile material), spraying, and coating (see TEXTILES, FINISHING). For instance, exhaust application is possible with cationic finishes which have an affinity for the anionic groups in polymeric materials. After application, the textile is dried. Durable antistatic finishes require cross-linking of the resin. Cross-linking is usually achieved by subjecting the treated, dried material to heat curing. A catalyst is often incorporated to accelerate insolubilization.

Effect of Aging on Antistatic Protection. Aging is known to improve the laundering resistance of durable antistatic finishes that have been incompletely cross-linked by curing, but a gradual reduction of static protection on aging has been observed with some nondurable antistats (77). Although volatilization of the antistatic agent was thought to be the cause, migration of the antistatic agent from the surface to the interior of the polymer fiber may be responsible for this effect. Migration into the interior dilutes the antistatic agent and destroys its continuity. Migration in the opposite direction, from the interior of the polymer to its surface, is the action mechanism of durable antistats for plastics, and will be discussed later. Exposure to high temperatures can accelerate this migrating-aging effect. Extraction and reapplication with a solvent can restore the static protection. This verifies that loss of effectiveness can be due to migration rather than volatilization or decomposition (90).

Soiling of Antistatic Finishes. Soiling of fabrics having a tendency for accumulation of charges has been assumed to be an electrostatic phenomenon, and therefore it follows that if static is eliminated, soiling will be reduced. However, most antistatic agents have been developed and used for reasons other than the reduction of soiling.

The relationship between static and soiling as understood prior to 1952 has been reviewed (102). The conclusions drawn from this survey were that most soil particles are uncharged or only lightly charged; the charged particles may be attached to fabrics and held mechanically by a soft tacky layer; uncharged particles may be strongly attracted to charged fabrics. Even though two-thirds of all airborne soil particles may be negatively charged, airborne soil pickup is only slightly influenced by static (103).

The effect of nontacky antistats such as lithium chloride [*7447-41-8*] on the soiling of clothing has also been studied (22). No difference was noted on the cuffs and collars in soiling between treated and untreated polyester and polyacrylonitrile shirts; however, on polyester and nylon slips significant reduction of soiling was noted near the hemline of the treated garment. The effect of nonionic, anionic, and cationic antistats, including durable and nondurable antistats on the soiling and soil removal properties of textile materials, has been examined (21,103–105). One study showed that a significant percentage of soiling of women's slips was due to static, especially at the hemline. Antistatic agents had no effect on areas where contact soiling occurred. Some antistatic agents, because of tackiness, increased soiling, and this varied from agent to agent (105). The evaluated nondurable agents helped soil removal in contrast to durable antistatic treatment where soil removal was more difficult. Several other workers have attempted to establish the relationship between static and soiling (106–108).

Various hydrophilic finishes have been used in connection with soil release and soil redeposition during laundering. These finishes can impart a degree of antistatic protection to the treated materials; however, the main functions of these finishes are to release soil and to prevent the redeposition during laundering.

Inherently Static-Free Textile Fibers and Fabrics. The antistatic agents discussed previously are applied to the surface of the material to provide a conductant surface layer, and their application increases the electrolyte (ionic) conductivity of the surface, thus reducing the propensity of the material to accumulate electrostatic charges. They are therefore called external antistatic agents, and even the so-called durable antistatic agents do not provide permanent protection. Thus there have been many efforts to develop fibers that are inherently static-free. One method involves incorporation of internal antistatic additives in the bulk of the polymer before spinning, and a second method consists of coating the fiber with conducting metals or carbon black and incorporating fine metal fibers in the textile material. Both methods increase the electrical conductivity of the whole material rather than the surface conductivity alone. However, the first method increases ionic conductivity, whereas the second increases the electronic conductivity.

Internal Antistatic Agents. Most of the commercially available antistatic fibers which contain internal antistatic agents are polyamides. Permanently antistatic nylon carpet yarns were introduced in the mid-1960s by the DuPont Company under the trade name of Antron. The first static-resistant nylon fibers containing internal additives and intended for apparel use became commercially available in late 1967 when the Monsanto Textiles Company introduced Ultron (109). The antistatic protection in Ultron fiber products is permanent and withstands textile processing as well as multiple launderings and dry-cleanings for the product's lifetime. Similar modifications were built into Monsanto's nylon carpet yarns (110) where the antistatic protection is also permanent to mill processing and subsequent cleaning treatments. Similar nylon fibers have been developed by other fiber producers. Examples include Antistat Celon (Courtaulds), Perlon Antistatic (Bayer), and Ecstatic (Enkalon), which are all nylon-6 type, and Counterstat (ICI), which is a nylon-6,6 (see FIBERS, POLYAMIDES). Other examples of carpet antistatic fibers are Anso X (Allied Chemical), Enkalon IT and Enkastat (Akzo), and Nylon L (Toray) which are all nylon-6 types (111).

The type of internal additive and the exact composition of these inherently antistatic fibers have not been published. However, the patent literature (112,113) suggests that these fibers contain 2–10% of a polyethoxylated compound, such as poly(ethylene glycol), mol wt 20,000. The additive is uniformly dispersed in the bulk of the polymer before spinning and may partially form chemical linkages with reactive groups in the polymer. During spinning, the additive is trapped inside the spun fiber, thus only the surface antistat is removed during subsequent processing and washings. Consequently, the fibers maintain their antistatic properties after laundering, even though the internal additives are water-soluble and can be partially extracted. These internal antistatic agents appear to work on the basis of moisture absorption from the atmosphere to increase the electrolyte conductivity of the fiber. For melt-spun fibers, the internal additive has to be thermally stable at the spinning temperature. Also, it must not adversely affect the dyeing characteristics of the fibers; obviously these requirements limit the list of potential additives.

The incorporation of metal salts of amphoteric surface active agents (Mostat Series) as internal antistatic agents in polypropylene fibers has been reported (95). Metal salts of alanine, amidoamine, and imidazoline-type amphoteric surface-active agents show excellent performance as internal antistatic agents and also improve the dyeing ability of the fibers with acid dyes.

Incorporation of Metals and Carbon Black. The incorporation of metals and carbon black increases the electrical conductivity of the whole material, and their effectiveness does not depend on the moisture content of the fiber. They can easily dissipate electrostatic charges. Early methods involved the blending of small percentages of metallic fibers with organic fibers. The metallic fibers dissipate electrostatic charges even when these fibers fail to provide a continuous, electrically conducting path because of low concentration in the blend. They apparently act as dipoles that interact with the electrostatic field of the static charges, thus increasing delocalization of the charges and accelerating their dissipation (56). Typically these fibers are included in amounts varying form 0.1–2.0 wt % of the fiber content, and they reduce the static propensity of carpets made from such blends by 50–75% (114). Stainless chrome–nickel steel fibers were developed in the United States in the mid-1960s; since then others have been produced in Europe and Japan. The three main commercially available types are (*1*) Staple stainless steel fibers, such as Brunsmet (Brunswick Corporation) and Bekinox (Bekaert), that can be blended with static-prone fibers prior to spinning to eliminate static problems in carpets (111,115) and in garments used in explosive atmospheres (29,116); (*2*) spun yarns consisting of a blend of about 87% polyamide and 13% stainless steel fibers, such as Brunslon (Brunswick Corporation) and Bekitex (Bekaert), that can be combined with continuous filament carpet yarns and tufted to produce static-resistant carpets (111,117). Also available is Tenavelle (Teijin) a spun yarn of polyester and stainless steel fiber, which is used in industrial fabrics (118); and (*3*) monofilament yarns that have an aluminum core coated with either a plastic material (Zefstat, Dow Badische) (111,119) or with polyamide (MFD, Allied Chemical) (112) and are primarily used in carpets. For the best results, the steel fibers should be distributed evenly in both yarn and pile, and care must be taken that mechanical or chemical damage during processing does not result in loss of conductivity.

The principle of blending a conduction fiber with a static-prone fiber has been known for years. A mixture of a substantial quantity (30–40%) of a hydrophilic fiber such as cotton or rayon with a hydrophobic static-prone fiber such as a polyester can produce a static-free blend under ordinary conditions. However, blocking the hydrophilic groups by cross-linking of the cotton with bifunctional reagents such as dimethylolethylene urea or addition of a water-repellent finish such as a silicone resin increases the static propensity of such a blend.

Durable antistatic nylon fibers consisting of a filament or a staple fiber coated with a thin layer (on the order of 1 cm) of a conductor such as silver or nickel have been developed (31), eg, Metalian (Teijin) and X-Static (Rohm & Haas) (111). Ultron advanced-generation BCF nylon carpet yarns were introduced to offer static protection and soil-hiding characteristics (119). These properties result from a combination of fiber properties and a fine carbon black-containing conductive filament. The carbon black appears as a hairline stripe on the side of the conductive filament. One conductive filament in the yarn bundle offers complete antistatic protection and is invisible in the finished carpet. Other fiber producers have developed similar nylon carpet fibers containing conductive carbon black, such as Antron III (DuPont) and F901 (The Dow Chemical Company). These fibers are primarily used in carpets by incorporating one conducting filament into a regular nylon yarn; carpets containing as little as 0.1 wt % of conductive filaments show satisfactory antistatic properties (31). Also available are polyester fibers containing conductive carbon black, such as Epitropic (ICI) and Dacron III (DuPont) (112). The principal advantages of these fibers are their high effectiveness in dry conditions and their resistance to cleaning and wear (111) (see CARBON, CARBON BLACK).

Special types of carbon black may be incorporated in latexes used for backing of carpets to enhance their conductivity. Nylon carpets backed with such conductive latex offer excellent antistatic properties even at 10% rh (110).

Consumption of Antistatic Agents in Textile Applications. In textile finishing of synthetic fibers and their blends, durable antistatic agents represent a potential market of a million kg/yr or more; however, volumes have not been disclosed. Many thousands of metric tons of nondurable antistatic agents are also used in textile processing, and the volume increases as use of synthetic fibers and production speeds increase. Antistatic action in textile finishing is often combined with lubrication and softening. Almost all nondurable antistatic lubricating oils are formulated with surface-active agents, usually of the cationic or nonionic types. Moreover, surface-active agents are frequently used by themselves as nondurable antistatic agents. Thus the complexity of antistatic formulations and their multipurpose uses in textile finishing agents make it difficult to determine the actual volume of this market. One estimate is that approximately 250 t/yr of antistats are used in textile mill applications. This does not include antistats formulated for use on synthetic fibers during manufacturing and for home laundry application.

Antistatic compositions are being sold for application during household laundering to be added in the rinse cycle or during tumble drying. Most of these products, listed as antistatic agents for textiles, are nondurable and based on cationic and nonionic surface-active agents (see SURFACTANTS). Durable antistatic agents for textiles are few and most of them are variations of cross-linked polyamines containing polyethoxy segments.

ANTISTATIC AGENTS FOR PLASTICS

Plastics are excellent insulators and have a great tendency to generate and retain static charges. Static charges on bottles and other packaging products charge dust particles in the air by induction. The dust particles acquire a charge of the opposite polarity than the plastic package, and so are attracted to the package. The dust attracted to the packages of consumer products makes them less attractive for selling. Dust attracted to phonograph records wears down the record and the phonograph needle when the record is played. Plastic resin powders are subject to an explosion hazard when they generate a static charge. Static charged sheets of plastic are difficult to separate. Static charged plastic products are an explosion hazard in areas where flammable gases are used. Electronic circuit chips are susceptible to damage from static charged plastic packaging.

Antistatic agents may be applied to the surface of the finished article or incorporated in the bulk of the polymer during processing (see PLASTICS TECHNOLOGY). They function by decreasing the rate of charge generation, by increasing the rate of charge dissipation, or by both mechanisms (73,76).

The way in which the rate of charge generation is reduced is not entirely understood, but it was found (95) that metal salts of amphoteric surface-active agents (Mostat Series) form a molecular layer at the interface, and the presence of this layer reduces the intimacy of contact and therefore the net charge transfer. There is also an indication that mold lubricants can function in this manner (45), and poly(ethylene glycol) oleates were found to be more effective than poly(ethylene glycol) laurates in reducing the rate of charge generation on poly(vinyl chloride) surfaces (120). Fatty amides reportedly reduce charge buildup on plastics. Fatty amides are lubricants and are thought to reduce tribocharging (see AMIDES, FATTY ACID). They are often used in conjunction with a static dissipative-type antistatic agent to both reduce the tribocharging and dissipate any charge that might form (121). Combining the lubricant and the static dissipative additive are necessary because propensity to tribocharge and surface resistivity do not correlate. Studies show that ethylene vinyl acetate (EVA) film with an amide lubricant has a very low propensity to tribocharge, but has a surface resistivity higher than the accepted static dissipative range. EVA film containing a static dissipating amount of carbon black exhibits large triboelectric voltages. By combining an amide lubricant and the carbon black, the EVA film becomes both static dissipating and produces smaller triboelectric charges (Table 4) (122).

Table 4. Lubricant–Antistat Combinations in EVA Films

Film composition	Surface resistivity, Ω/sq	Triboelectric voltage
EVA 0.4% fatty amide wax	2.1×10^{13}	823
EVA 59% carbon black	3.2×10^{11}	3199[a]
EVA 59% carbon black 0.4% fatty amide wax	5.0×10^{10}	1850

[a]Detector saturated.

Types of Antistats. Plastics are rendered static dissipative by either adding a conductive material to the plastic in sufficient quantity that there is a three-dimensional conductive pathway through the plastics, or by adding to the plastic a surfactant-type chemical that will attract moisture to the surface of the plastic. Since water is a conductor, the surface layer of moisture then dissipates a static charge. The source of conductive material can be either an inherently conductive polymer, a metal or metal-containing material, or carbon black. The conductive additive can, if desired, be added to the plastic at a concentration sufficient to make it static dissipative, but not completely conductive. Surfactant-type additives are of several different chemical types. They can be applied to the surface of the finished plastic article or incorporated in the bulk of the polymer during processing (see PLASTICS TECHNOLOGY).

Inherently Conducting Polymers. Conducting polymers are polymers with a pi-electron backbone capable of passing an electrical current. These polymers generally are not sufficiently conductive as neat polymers but require the inclusion of an oxidizing or reducing agent (dopant) to render them conductive.

Common conductive polymers are polyacetylene, polyphenylene, poly(phenylene sulfide), polypyrrole, and polyvinylcarbazole (123) (see ELECTRICALLY CONDUCTIVE POLYMERS). A static-dissipative polymer based on a polyether copolymer has been announced (124). In general, electroconductive polymers have proven to be expensive and difficult to process. In most cases they are blended with another polymer to improve the processibility. Conductive polymers have met with limited commercial success.

Metal-Containing Polymers. Metal-containing polymers function simply by adding sufficient quantities of a metal to form a three-dimensional conduction pathway through the plastic. The metal is in the form of a powder, micrometer-sized needles, or as a thin coating on glass spheres, carbon fibers, or mica. The metals normally employed are nickel, zinc, stainless steel, copper, and aluminum. At higher levels these materials make the polymer capable of shielding electromagnetic impulses (EMI) (125).

The metal fillers act as a reinforcing material that results in added strength and stiffness (126). They color the plastic gray for nickel, zinc, stainless steel, and aluminum, and brown for copper. Metal additives are more expensive than carbon black or surface-active agents, but they get extensive use in EMI shielding applications.

Carbon Blacks. The high electrical conductivity of carbon black is utilized where its color is not objectionable and its reinforcing action is used (see FILLERS; COMPOSITES). Carbon black increases the electrical conductance of the polymer to which it is added, and therefore its effectiveness does not depend on moisture absorption (see CARBON, CARBON BLACK).

Vulcanized rubber is an insulator (volume resistivity is 10^{15} Ω-cm), and the static generated by rubber tires created serious problems in vehicles until the introduction of electrically conductive carbon black as a reinforcing pigment. An excellent correlation was found between the potential generated and the resistivity of the tires (127,128) (see RUBBER).

Airplanes in flight acquire the same potential as the surrounding atmosphere, and, upon landing, the plane might be at high potential with respect to the ground. To bring the airplane to ground potential before the passengers step out, highly conductive tail-wheel tires have been developed. Some of these have rim-to-tread resistivities of only 2×10^3 Ω as compared to about 10^{10} Ω for ordinary tires.

A minimum of 10 to 35 parts carbon black to 100 parts of rubber is required to obtain a resistivity in the order of 10^4 Ω-cm. At that loading the carbon black particles, which have an average radius of 10 nm, form grapelike aggregates that provide continuous paths for the electrical current. Special purpose rubbers containing even more carbon black have resistances as low as 1 Ω-cm (129). The electrical resistivity of rubber with carbon black is sensitive to strain history, probably because of temporary disruptions of the continuity of the carbon black aggregates.

Other rubber products that require high conductivity are belts, hoses, footwear, and rubber products used in hospital operating rooms. Fabrics have been woven with viscose yarns containing carbon black for operating room use (130). Conductive linoleum or asphalt tiles are also based on high loads of carbon black. Too high a conductivity of flooring, however, increases the danger of electrical shock and fire hazards caused by circuit insulation defects (131).

The carbon blacks used in plastics are usually different from the carbon blacks used in rubber. The effect of carbon black is detrimental to the physical properties of plastics such as impact strength and melt flow. Electroconductive grades of carbon black have much higher surface areas than conventional carbon blacks. The higher surface areas result in a three-dimensional conductive pathway through the polymer at much lower additive levels of the carbon black.

The additive concentrations of electroconductive carbon blacks is usually ¼ to ⅓ that of a regular carbon black (132).

Carbon black finds use in lower cost products. Vinyl audio records are a commonly recognized use of carbon black, but it is also useful in vinyl video albums (133).

Surfactant-Type Antistats. Inherently conductive antistats have the advantage of not being dependent on atmospheric moisture to function. Their drawbacks include expense, coloration of the plastic, and alteration of the mechanical properties of the plastic. The added stiffness caused by conductive fillers may not be a problem with a rigid container, but it can be a problem for a flexible bag.

Surfactant-type antistats find the widest use because of their low cost and minimal effect on the mechanical properties of the plastic. Ease of use is another favorable aspect to surfactants. They can be mixed with the bulk of the plastic prior to processing or can be applied to the surface of the finished plastic article as the need dictates.

Internal surfactant antistats are physically mixed with the plastic resin prior to processing. When the resin is melted, the antistat distributes evenly in the polymer matrix. The antistat usually has some degree of solubility in the molten polymer. However, when the polymer is processed (extruded, molded, etc) into its final form and allowed to cool, the antistat migrates to the surface of the finished article due to its limited solubility in the solidified resin. The molecule of a surface-active agent is composed of a polar hydrophilic portion and a nonpolar hydrophobic portion. The hydrophilic portion of the surfactant at the surface attracts moisture from the atmosphere; it is the moisture that has the static dissipative effect.

Because the antistatic effect only occurs after the surfactant has migrated to the surface, the solubility of the surfactant in the polymer is an important consideration. Surfactants can generally be classified as one of four chemical types: cationic, where the hydrophilic portion has a positive charge; anionic, where the hydrophilic portion has a negative charge; nonionic, where the hydrophile does not have a charge; and amphoteric, where the molecule contains both positive and negative charges. Figure 5 shows the effect of changing surfactant type on the surface resistivity of two different polymers (134).

Cationic, anionic, and amphoteric surfactants derive their water solubility from their ionic charge, whereas the nonionic hydrophile derives its water solubility from highly polar terminal hydroxyl groups. Cationic surfactants perform well in polar substrates like styrenics and polyurethane. Examples of cationic surfactants are quaternary ammonium chlorides, quaternary ammonium methosulfates, and quaternary ammonium nitrates (see QUARTERNARY AMMONIUM COMPOUNDS). Anionic surfactants work well in PVC and styrenics. Examples of anionic surfactants are fatty phosphate esters and alkyl sulfonates.

Nonionic surfactants perform well in nonpolar polymers such as polyethylene and polypropylene. Examples of nonionic surfactants are ethoxylated fatty amines, fatty diethanolamides, and mono- and diglycerides (see AMINES, FATTY AMINES; ALKANOLAMINES). Amphoteric surfactants find little use in plastics (134).

Varying the fatty acid portion of the surfactant can have an effect on antistatic performance. In the example given in Table 5, ethoxylated coconut

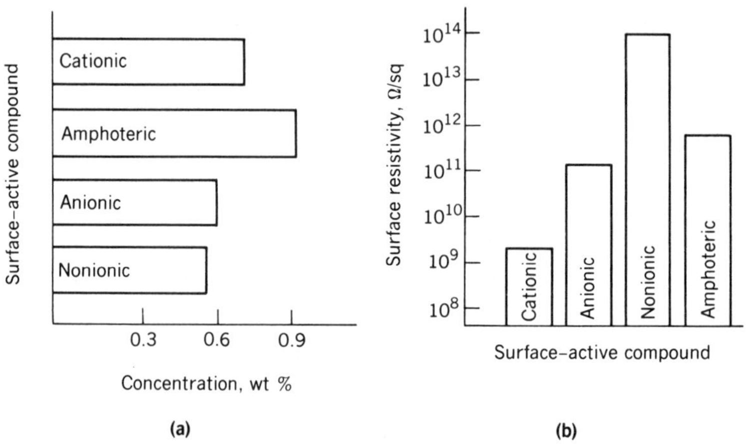

Fig. 5. Effect of surfactant type on surface resistivity. (**a**) Concentration of surface-active compound in low density polyethylene (LDPE) required to achieve 10^{10} Ω/sq surface resistivity; and (**b**) effect on surface resistivity of an acrylic polymer. Concentration of surface-active compound is 0.3%.

Table 5. Static Decay Times of Ethoxylated Fatty Amines in High Density Polyethylene

Ethoxylated amine[a]	Time, s
coconut	0.93
tallow	2.01
oleyl	1.26
stearyl	>100

[a]At 1500 ppm.

amine results in the shortest static decay time, whereas ethoxylated stearyl amine results in the longest static decay time (135).

In the same class of polymers, an antistat can exhibit different degrees of effectiveness. As seen in Table 6 the performance of ethoxylated oleyl amine varies among polyolefins. The data for polypropylene (PP) also shows the concentration dependence of antistats.

Table 6. Effect of Ethoxylated Oleyl Amines on Static Decay in Polyolefins

	Static decay time, s			
Antistat, ppm	LDPE	LLDPE	HDPE	PP
1000	0.37	1.07		>100
1500			1.26	
3000	0.23		1.01	7.93
4000				0.65

Since surfactant-type antistats function by attracting atmospheric moisture to the plastic, the relative humidity (rh) has a significant effect on antistat performance (Fig. 6). Relative humidity also has an effect on charge generation (Table 7).

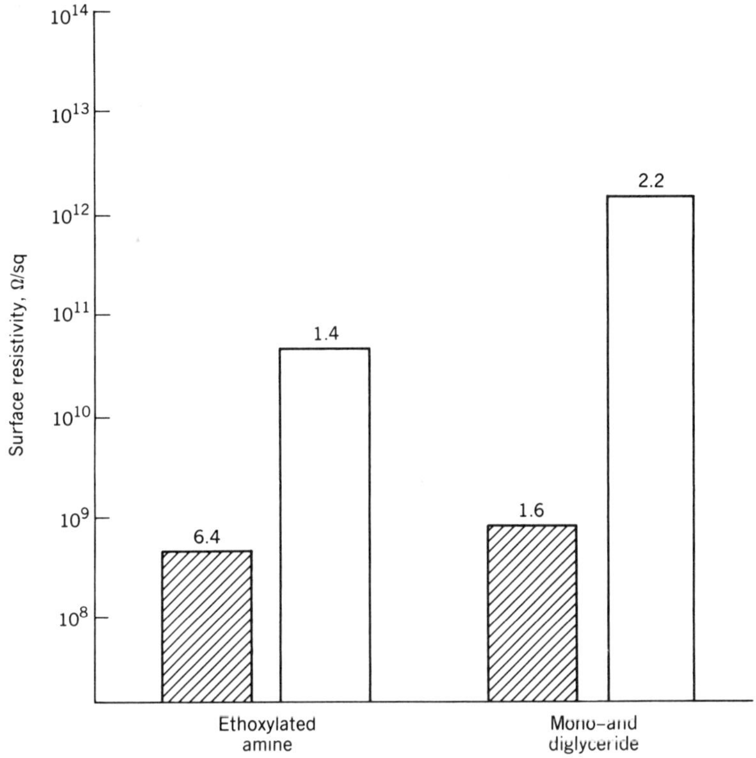

Fig. 6. Effect of humidity on antistat performance. ▨, 50% rh; □, 15% rh.

Table 7. Effect of Relative Humidity on Charge Generation

	Electrostatic voltage at	
Charge generation source	10–20% rh	65–90% rh
walking across carpet	35,000	1,500
walking over vinyl floor	12,000	250
working at bench	6,000	100
pick up polymeric bag	20,000	1,200
chair padded with polyurethane foam	18,000	1,500

There are many commercial surface and internal antistatic additives (136). Table 8 lists several of the internal antistats along with the polymers for which they are recommended. Surface-applied antistats are not included because they function independent of polymer type.

Table 8. Commercially Available Internal Antistatic Agents and the Polymers for Which They Are Recommended[a]

Trade name[b]	Manufacturer	ABS	LDPE	HDPE	PP	PS	Concentration, %
		Amines					
Armostat 310	Akzo (Noury) USA		x	x	x		0.05–0.5
Armostat 410	Akzo (Noury) USA	x	x	x	x	x	0.05–2.0
Atmer 163	ICI Americas	x	x	x	x	x	0.5–5.0
Hexcel 273C	Hexcel		x	x	x		0.05–0.2
Hexcel 273E	Hexcel		x	x	x		0.05–0.3
Hostastat FA14	Hoechst Celanese		x	x	x		0.1–0.5
Hosastat FA18	Hoechst Celanese		x	x	x		0.1–0.5
Kemamine AS-650	Humko Chemical	x	x	x	x	x	0.05–2.0
Kemamine AS-975	Humko Chemical		x	x	x		0.05–0.5
Kemamine AS-989	Humko Chemical		x	x			0.1–0.3
Kemamine AS-990	Humko Chemical		x				0.1–0.5
Varstat T22[c]	Sherex		x	x	x	x	0.05–0.5
		ABS	Acrylic	PS	RPVC	PU	
		Quaternary ammonium compounds[d]					
Cyastat 609[c]	American Cyanamid	x	x		x		0.5–2.0
Cyastat LS[c]	American Cyanamid	x		x	x		0.5–2.0
Cyastat SN	American Cyanamid	x	x		x		1.0–2.0
Cyastat SP	American Cyanamid				x		1.0–2.0
Larostat 377 DPG	Lonza Chemical					x	1.0–2.0
Larostat SC	Lonza Chemical	x	x		x		
Markastat AI-12[e]	Argus					x	3.0–5.0
Markastat AL-33	Argus	x	x			x	0.5–1.5
		LDPE	HDPE	PP	PS	FPVC	
		Esters					
Atmos 150	Humko Chemical	x	x	x			0.2–1.5
Dimul SK	Humko Chemical	x	x	x	x		0.2–2.0
Glycostat	Lonza Chemical	x	x	x	x	x	0.5–2.0
Hostastat FE-20[f]	Hoechst Celanese	x	x	x	x	x	0.5–2.0
Hostastat HS1[f,g]	Hoechst Celanese				x	x	0.1–3.0
Markastat AL-14	Argus					x	3.0–7.0
Myrj 45[d]	ICI Americas	x	x				0.5–2.0

[a]Ref. 136. LDPE = low density polyethylene; HDPE = high density polyethylene; PP = polypropylene; PS = polystyrene; FPVC = flexible PVC; RPVC = rigid PVC; and PU = polyurethane.
[b]Has some FDA approvals unless otherwise noted.
[c]Also useful for FPVC.
[d]Not approved by FDA.
[e]Also useful for polyacetate.
[f]Also useful for ABS and RPVC.
[g]FDA approved for PVC.

Internal antistats are considered permanent antistats. This permanence is based on the concept that most plastic products are disposable, so that the antistat is not required to last long. The antistatic effectiveness of an internal antistat can decrease over time. One study showed large increases in surface resistivity on antistatic bags stored at 71°C for six months. Antistatic bags stored at room temperature showed only a small increase in surface resistivity (137). Loss of antistatic effectiveness is attributed to the volatility of the antistatic agent. The antistat does not easily wear off the plastic, but it can be removed with solvents and/or repeated wear.

Surface Applied Surfactants. Antistat agents can be applied directly to the surface of a plastic part. Usually the antistat is diluted in water or in a solvent. The antistat solution is applied by spraying, dipping, or wiping on the surface. The water or solvent dries leaving a thin film that attracts moisture. Since it is applied to the surface, migration through the resin is not a factor. In practice, the quaternary ammonium compounds find the most use. They are soluble in water and effective at low concentrations.

The antistatic protection provided by surface treatment is excellent while it lasts. However, surface treatments provide only temporary protection. The antistat layer is extremely vulnerable to rubbing and wiping, especially when wet (45). Loss of antistatic agent can also occur by diffusion from the surface into the bulk of the material on storage (77,78). Several attempts have been made to obtain more durable antistatic coatings with varying degrees of success using reactive resins and surface grafting (122,138–144).

Consumption of Antistatic Agents for Use in Plastics Applications. Antistatic agents are widely used in the plastic industry, and their economic significance is greater than in the textile industry. Food and drug packaging accounts for well over half the market, and FDA acceptable products such as polyethylene, poly(vinyl chloride), and polypropylene film and bottles are predominant. The use of antistatic agents in medical and surgical applications of flexible poly(vinyl chloride) and polyethylene film and sheet has increased, providing higher safety in places where a buildup of static charges can trigger explosions (145). The use of antistatic agents in plastics used for the electronics industry has grown to the point that virtually all of these plastic products contain some type of antistatic agent (see ELECTRONIC MATERIALS).

Determining the exact consumption of antistatic products is difficult due to commercial secrecy. The world consumption for hydrophilic-type antistatic agents has been estimated for 1983 at 11,400 t, with an average growth rate of

Table 9. Estimated USA Volume of Antistatic Agents Used in Plastics[a]

Year	t
1960	182
1963	318
1966	681
1969	1135
1972	1300
1973	1790
1974	1830
1975	1600
1976	1800
1983	3300

[a]See refs. 129, 145, and 146.

about 6.0% per y (146). The estimated average consumption of hydrophilic-type antistatic agents in the United States over time is given in Table 9.

BIBLIOGRAPHY

"Antistatic Agents" in *ECT* 3rd ed., Vol. 3, pp. 149–183, by Zev Gur-Arieh and B. J. Reuben, Monsanto Textiles Company.

1. J. C. Wilcke, *Disputatio Physica Experimentalis de Electricitatibus Contraris,* Rostock, 1757.
2. J. W. Ballou, *Text. Res. J.* **24,** 146 (1954).
3. D. J. Lehmicke, *Am. Dyest. Rep.* **38**(24), 853 (1949).
4. S. P. Hersh and D. J. Montgomery, *Text Res. J.* **25,** 279 (1955).
5. H. Frotscher, *Z. Gesamte Textilind.* **62**(18), 746 (1960).
6. H. Gruner, *Faserforsch. Textiltech.* **4,** 249 (1953).
7. P. S. H. Henry, *Brit. J. Appl. Phys.* **4**(Suppl. 2), S31 (1953).
8. N. Sclater, *Electrostatic Discharge Protection for Electronics,* Tab Books, Blue Ridge, Pa., 1990.
9. J. Gayler, R. E. Wiggins, and J. B. Arthur, *Static Electricity, Generation, Measurement, and Its Effects on Textiles,* University of North Carolina, Raleigh, 1965.
10. W. E. Morton and J. W. S. Hearle, *Physical Properties of Textile Fibers,* Butterworth & Co., Ltd. and The Textile Institute, UK, 1962.
11. R. Walker, *Electrical Overstress/Electrostatic Discharge Symposium,* EOS/ESD Association, Rome, N.Y., 1983.
12. A. Webster, *Text. Manuf.* **78**(932), 425 (1952).
13. A. G. Beaumont, *J. Text. Inst. Proc.* **48,** P100 (1957).
14. E. Kirschner, *Melliand Textilber.* **37,** 892 (1956).
15. J. A. Medley, *J. Text. Inst. Trans.* **45,** T123 (1954).
16. D. Finlayson, *J. Text. Inst. Proc.* **48,** P55 (1957).
17. K. H. Barnard, *Am. Dyest. Rep.* **44,** 111 (1955).
18. H. Weigand, *Text. Prax.* **10,** 810 (1955).
19. J. W. S. Hearle, *Skinner's Silk Rayon Rec.* **30**(2), 148, 150; (3), 267 (1956).
20. M. Buhler, *Text. Prax.* **12,** 1142, 1234 (1957).
21. D. Wilson, *J. Text. Inst. Trans.* **53,** T1 (192).
22. J. Lord and W. H. Rees, *J. Text. Inst. Trans.* **51,** T419 (1962).
23. A. Sippel, *Text. Prax.* **8,** 1030 (1953).
24. W. W. Levy, *Soc. Plast. Eng. J.* **18,** 1288 (1962).
25. J. C. Martin, *Soc. Plast. Eng. J.* **18,** 392 (1962).
26. M. H. Downing, *EDS/ESD Symposium Proceedings,* EOS/ESD Association., Rome, N.Y., 1983.
27. B. Unger, *ESD Basics,* EOS/ESD Tutorial, EOS/ESD Association, Rome, N.Y., 1991.
28. R. Beach, *Fact. Manage. Maint.* **115**(7), 106 (1957).
29. R. R. Varin, *Mod. Text.* **52**(2), 28 (1971).
30. S. M. Suchecki, *Text. Ind.* **128**(5), 59 (1964).
31. S. P. Hersh, *Polym. Plast. Technol. Eng.* **3**(1), 29 (1974).
32. U.S. Pat. 3,117,055 (Jan. 7, 1964), E. Parrish and M. Katz (to E. I. du Pont de Nemours & Co., Inc.).
33. L. Silverman, E. W. Conners, Jr., and D. M. Anderson, *Ind. Eng. Chem.* **47,** 952 (1955).
34. *Daily News Record* (26) 22 (Feb. 8, 1965).

35. A. P. Little, *Text. J. Austr.* **24**(7), 634, 668 (Sept. 1949).
36. H. M. Brown, *The Cotton Research Clinic,* National Cotton Council of America, Memphis, Tenn., 1953, p. 66.
37. C. E. Kirkwood, Jr., N. S. Kendrick, and H. M. Brown, *Text. Res. J.* **24,** 841 (1954).
38. C. E. Kirkwood, Jr. and co-workers, *Text. Res. J.* **24,** 853 (1954).
39. J. J. LaFranca, Jr., M. Mayer, Jr., and H. W. Weller, Jr., *Text. World* **116**(6), 58 (1966).
40. *L'Industrie Textile* (921), 126 (Feb. 1964).
41. P. R. Lord, *Text. Manuf.* **88**(1055), 490 (1962).
42. J. Shaw, in *New Ways to Produce Textiles,* The Textile Institute, Manchester, 1972, pp. 231–250.
43. L. Bak, E. Bator, and G. Gatai, *Magy. Textiltech.* **17,** 83 (1965).
44. C. Froger, *Text. Res. J.* **32,** 168 (1962).
45. P. C. Woodland and E. E. Ziegler, *Mod. Plast.* **28**(9), 95 (1951).
46. N. Murasaki, *J. Text. Mach. Soc. Japan* **7**(1), 37 (1961).
47. J. A. Medley, *Brit. J. of Appl. Phys.* **4**(Suppl. 2), S28 (1953).
48. G. Neuhof, *Text. Prax.* **5,** 224 (1950).
49. H. F. Schwenkhagen, *Melliand Textilber.* **34,** 1182 (1953).
50. A. Sippel, *Melliand Textilber.* **35,** 831 (1954).
51. V. E. Shashoua, *J. Polym. Sci.* **33,** 65 (1958); 1A, 169 (1963).
52. J. L. Rogers, *Soc. Plast. Eng. J.* **29,** 28 (1973).
53. N. Jonassen, *EOS/ESD Symposium Proceedings,* EOS/ESD Association, 1991.
54. G. Quintalier, M. Warzee, and R. Sioncke, *J. Text. Inst. Proc.* **48,** P26 (1957).
55. R. E. Lacy and G. C. Ward, *Can. Text. J.* **75**(2), 49 (Jan. 24, 1958).
56. F. J. Rizzo, *Electrostatic Phenomena in Textiles and Clothing Systems,* U.S. Army Natick Laboratories, Natick, Mass., Oct. 1973.
57. J. Crowley, *Basic Materials Measurements,* EOS/ESD Tutorial, EOS/ESD Association, 1991.
58. *Surface Resistance Measurement of Static Dissipative Planar Materials* DS11.11, EOS/ESD Association, 1991.
59. E. I. Valko and G. C. Tesoro, *Text. Res. J.* **29,** 21 (1959).
60. *AATCC Test Method 134-1975 Technical Manual,* American Association of Textile Chemists and Colorists, Research Triangle Park, N.C., 1975.
61. *AATCC Test Method 115-1973 Technical Manual,* American Association of Textile Chemists and Colorists, Research Triangle Park, N.C., 1973.
62. P. R. Wilkinson, *Mod. Text.* **51**(12), 35 (1970).
63. P. S. H. Henry, *Proc. 3rd Conf. on Static Electrification;* Inst. Phys. Conf. Ser. No. 11, 1971, p. 212.
64. R. E. Thompson, *Saf. Eng.* **3,** 17 (1969).
65. A. M. Crugnola and H. M. Robinson, *Textile Series Report 110,* Quartermaster Research and Engineering Command, U.S. Govt. Printing Office, Washington, D.C., 1959.
66. *Electronics Industry Association Standard 541.*
67. *Standard for the Use of Inhalation Anesthetics,* NFPA 99, National Fire Protection Association.
68. C. R. Hynes, *EOS/ESD Technology Part II: from MIL-P-81705 (AS) to MIL-B-81705C,* Miller Freeman Publications, Aug./Sept. 1990.
69. M. vom Hove, *Text. Rundsch.* **15,** 109 (1960).
70. D. Wilson, *J. Text. Inst. Trans.* **54,** T97 (1965).
71. F. H. Steiger, *Text. Res. J.* **28,** 721 (1958).
72. J. L. Rogers, *Soc. Plast. Eng. J.* **29,** 28 (1973).
73. *Antistatic Additives and Surface Treatments for Fibers, Films, and Moldings, A Literature Survey 1907–1967,* Science Surveys, Mountainside, N.J., 1967.

74. K. Johnson, *Antistatic Agents, Technology and Applications,* Noyes Data Corp., Park Ridge, N.J., 1972.
75. K. Johnson, *Antistatic Compositions for Textiles and Plastics,* Noyes Data Corp., Park Ridge, N.J., 1976.
76. N. R. Cansdale and J. M. Heaps, *Plastics,* 68 (June 1964).
77. G. R. Ward, *Am. Dyestuff Rep.* **44,** 220 (1955).
78. J. Diemunsh and J. Chabert, *Bull. Inst. Text. France* **102,** 887 (1962).
79. E. I. Valko, G. C. Tesoro, and W. Ginilewicz, *Am. Dyest. Rep.* **47,** 403 (1958).
80. J. W. S. Hearle, in J. W. S. Hearle and R. H. Peters, eds., *Moisture in Textiles,* Butterworth Publications Ltd., London, UK, 1960, p. 123.
81. S. P. Hersh and D. J. Montgomery, *Text. Res. J.* **22,** 805 (1952).
82. P. J. Sereda and R. F. Feldman, *J. Text. Inst. Trans.* **55,** T288 (1964).
83. J. F. Keggin, G. Morris, and A. M. Yuill, *J. Text. Inst. Trans.* **40,** T702 (1949).
84. G. J. Sprokel, *Text. Res. J.* **27,** 501 (1957).
85. G. Quintalier, M. Warzee, and R. Sioncke, *J. Text. Inst. Proc.* **48,** P26 (1957).
86. M. Hayek, *Am. Dyest. Rep.* **43,** 368 (1954).
87. N. A. Teixeira and S. M. Edelstein, *Am. Dyest. Rep.* **43,** 195 (1954).
88. M. Hayek and F. C. Chromey, *Am. Dyest. Rep.* **40,** 225 (1951).
89. J. Turner and J. K. Smith, *Am. Dyest. Rep.* **51,** 41 (1962).
90. A. E. Henshall, *J. Soc. Dyers Color.* **76,** 525 (1960).
91. S. M. Edelstein, *Am. Dyest. Rep.* **41,** 518 (1952).
92. E. I. Valko and G. C. Tesoro, *Text. Res. J.* **29,** 21 (1959).
93. S. A. Heap and F. H. Holmes, *J. Soc. Dyers Color.* **83,** 12 (1967).
94. D. Gardiner and F. H. Holmes, *J. Soc. Dyers Color.* **83,** 43 (1967).
95. H. Marumo, *The Internal Antistatic Treatment of Thermoplastic Polymers,* 1st International Congress of Static Electricity, Vienna, Austria, May 3–7, 1970.
96. G. W. Graham, 3rd Canadian Textile Seminar, Queens University, Kingston, Ontario, Canada, Sept. 11–13, 1952, p. 32.
97. J. A. Medley, *J. Text. Inst. Proc.* **48,** P112 (1957).
98. H. Mark, N. S. Wooding, and S. M. Atlas, eds., *Chemical Aftertreatment of Textiles,* Wiley-Interscience, New York, 1971.
99. E. I. Valko and G. C. Tesoro, *Mod. Text. Mag.* **38**(7), 62 (1957).
100. E. I. Valko and G. C. Tesoro, *Text. Res. J.* **29,** 21 (1959).
101. Brit. Pat. 880,897 (Oct. 25, 1961), (to Bohme Fettchemie); Brit. Pat. 911,495 (Nov. 28, 1962), (to Imperial Chemical Ind.); U.S. Pat. 2,965,517 (Dec. 20, 1960), O. Albrecht and M. Maeder (to CIBA A.G.); U.S. Pat. 3,070,552 (Dec. 25, 1962), E. I. Valko, G. C. Tesoro, and A. Freiman (to Onyx Oil & Chemical Co.); U.S. Pat. 3,021,232 (Feb. 13, 1962), J. E. Pretka (to E. I. du Pont de Nemours & Co., Inc.); U.S. Pat. 3,063,870 (Nov. 13, 1962), R. L. Wakeman and W. L. Ginilewicz (to Onyx Oil & Chemical Co.); and Ger. Pat. 1,131,404 (June 14, 1962), (to Farbwerke Hoechst).
102. E. A. Leonard and E. K. Schwartz, New York Section AATCC, *Am. Dyest. Rep.* **41,** 322 (1952).
103. J. M. Salsbury, T. F. Cooke, E. S. Pierce, and P. B. Roth, *Am. Dyest. Rep.* **45,** 190 (1956).
104. W. P. Utermohlen, Jr., M. E. Ryan, and D. O. Young, *Text. Res. J.* **21,** 510 (1951).
105. J. C. Harris, *Text. Res. J.* **29,** 99 (1959).
106. J. W. Schappel, *Text. Res. J.* **26,** 211 (1956).
107. A. S. Weatherburn and C. H. Bayley, *Text. Res. J.* **27,** 199 (1957).
108. W. H. Rees, *J. Text. Inst. Proc.* **45,** P612 (1954).
109. G. T. Barnes, *Mod. Text.* **51**(7), 45 (1970).
110. *Cadon Contract Carpets,* Technical Publication N-11, Monsanto Textiles Division, 1971.

111. N. Wilson, *Text. J. Austr.* **51**(7), 10 (1976).
112. U.S. Pat. 3,475,898 (Nov. 4, 1969), E. E. Magat and W. H. Sharkey (to E. I. du Pont de Nemours & Co., Inc.).
113. U.S. Pat. 3,329,557 (July 4, 1967), E. E. Magat and D. Tanner (to E. I. du Pont de Nemours & Co., Inc.).
114. *Int. Dyer, Text. Printer, Bleacher Finish.* **149,** 560 (1973).
115. G. F. Barry, *Mod. Text.* **50**(5), 64 (1969).
116. R. E. Thompson, *Saf. Eng.* **3,** 17 (1969).
117. H. F. Lewis, *Text. Inst. Ind.* **10,** 199 (1972).
118. *Japan Text. News* (264), 72 (Nov. 1976).
119. *Technical Publication TT57,* Monsanto Textiles Company, 1976.
120. B. D. Solley, *Brit. Plast.* **29,** 103 (1956).
121. A. McKenna, *Fatty Amides-Synthesis, Properties, Reactions, and Applications,* Humko Chemical Division of Witco Corp., Memphis, Tenn.,1982.
122. M. Havens, EOS/ESD Symposium Proceedings, EOS/ESD Association, 1991.
123. J. Frommer, *The Potential for Applications of Conducting Polymers,* 43rd Annual Technical Conference Proceedings, Society of Plastics Engineers, Brookfield, Conn., 1985.
124. B. L. Lee, 48th Annual Technical Conference Proceedings, Society of Plastics Engineers, Brookfield, Conn., 1990.
125. S. R. Gerteisen, *Modern Plastics Encyclopedia,* McGraw-Hill, New York, 1988.
126. D. Janeczek, 43rd Annual Technical Conference Proceedings, Society of Plastics Engineers, Brookfield, Conn., 1985.
127. J. W. Liska and E. E. Hanson, *Ind. Eng. Chem.* **34,** 618 (1942).
128. D. Bulgin, *Brit. J. of Appl. Phys.* **4**(Suppl. 2), S83 (1953).
129. *Plast. World,* 40 (July 1975).
130. J. Boyd and D. Bulgin, *J. Text. Inst. Proc.* **48,** P66 (1957).
131. P. G. Guest, V. W. Sihora, and B. Lewis, *U.S. Bur. Mines Bull.,* 520 (1953).
132. J. D. Van Drumpt, 45th Annual Technical Conference, Society of Plastics Engineers, 1987.
133. K. Palit, 43rd Annual Technical Conference, Society of Plastics Engineers, 1985.
134. B. Anderson, EOS/ESD Symposium Proceedings, EOS/ESD Assoc., 1991.
135. L. Walp and co-workers, 44th Annual Technical Conference Proceedings, Society of Plastics Engineers, 1986.
136. *Modern Plastics Encyclopedia,* McGraw-Hill, New York, 1990.
137. J. Kolyer and W. Anderson, EOS/ESD Symposium Proceedings, EOS/ESD Assoc., 1983.
138. U.S. Pat. 2,690,407 (Sept. 28, 1954), L. Pessel (to Radio Corporation of America).
139. Brit. Pat. 787,940 (Dec. 18, 1957), Y. Noguchi and co-workers (to Mitsubishi Rayon Co., Ltd.).
140. U.S. Pats. 2,832,696 and 2,832,699 (Apr. 29, 1958), W. Walles (to Dow Chemical Co.).
141. U.S. Pat. 2,937,066 (May 17, 1960), W. Walles (to Dow Chemical Co.).
142. U.S. Pat. 2,891,878 (June 23, 1959), R. Chamberlain (to American Cyanamid Co.).
143. U.S. Pat. 2,955,954 (Oct. 11, 1960), R. Collins (to E. I. du Pont de Nemours & Co., Inc.).
144. Brit. Pat. 883,793 (Dec. 6, 1961), (to Polyplastic).
145. *Mod. Plast.* **53**(9), 52 (Sept. 1976).
146. D. Klein, in *Chemical Economics Handbook,* SRI International, Menlo Park, Calif., 1985.

LEONARD E. WALP
Humko Chemical
Division of the Witco Corporation

ANTISYPHILITICS. See ANTIBACTERIAL AGENTS, SYNTHETIC.

ANTITUMOR AGENTS. See CHEMOTHERAPEUTICS, ANTICANCER.

ANTITUSSIVES. See EXPECTORANTS, ANTITUSSIVES, AND RELATED AGENTS.

ANTIVIRAL AGENTS

Viral infections are among the greatest causes of human morbidity, and it is estimated that in developed countries more than 60% of all the episodes of human illness result from viral infections (1). High virus infection rates also occur among pets, livestock, and plants. The high morbidity and the resulting economic loss caused by these infections have generated tremendous efforts in recent years to develop means to combat viral infections using antiviral agents.

Since viruses propagate only within living cells, the development of antiviral drugs which would disrupt the viral replication without affecting the metabolism of the host cell was initially believed to be difficult, if not impossible. However, dramatic progress in viral molecular biology has now made it possible to identify enzymatic processes which are unique to virus-infected cells. As a consequence, it is becoming feasible to design chemical compounds which identify infected cells, block a specific step in viral replication, and leave uninfected cells unharmed.

Viruses are one of the smallest biological entities (except viroids and prions) that carry all the information necessary for their own reproduction. They are unique, differing from procaryotes and eucaryotes in that they carry only one type of nucleic acid as genetic material, which can be transported by the virus from one cell to another. Viruses are composed of a shell of protein enclosing a core of nucleic acid, either ribonucleic acid (RNA) or deoxyribonucleic acid (DNA), that codes for viral reproduction. The outer shell serves as a protective coat to keep the nucleic acid intact and safe from enzymatic destruction. In addition to their protein coat, some viruses contain an outer covering known as an outer envelope. This outer envelope consists of a lipid or polysaccharide material.

The principal obstacle encountered by a virus is gaining entry into the cell. The cells are protected by a cell wall of a thickness comparable to the size of the virus. The virus must first become attached to the cell surface. Much of the specificity of a virus for a certain type of cell lies in its ability to attach to the surface of that specific cell. Durable contact is thus important if the virus is to infect the host cell. The ability of the virus and the cell surface to interact is a property of both the virus and the host cell. The fusion of viral and host-cell membranes, based on unique chemical recognition factors at the cellular surface, allows the intact viral particle or, in certain cases, only its infectious nucleic acid to enter the cell. This process also may involve a type of enzymatic digestion or breakdown of the host plasma membrane at the site of attachment. It is not only during initial infection of a cell that viruses have to cross cell membranes, however. Although the phenomenon of syncitia formation (for HIV infection) and lysing (for picornavirus infection) are observed, cell-to-cell transfer must also occur if the virus is to effectively multiply and cause a substantial infection.

Once inside the host cell, the virus must replicate its own nucleic acid. To do this, it often uses part of the normal synthesizing machinery of the host cell. If the virus is to continue its growth cycle, viral nucleic acid and viral protein must be properly transported within the cell, assembled into the infective virus particle, and ultimately released from the cell. All of these fundamental processes involve an intimate utilization of both cellular and viral enzymes. Certain enzymes that are involved in this process are specifically supplied by the invading virus. It is this type of specificity that can provide the best basis for antiviral chemotherapy. Thus an effective antiviral agent should specifically inhibit the viral-encoded or virus-induced enzymes without inhibition of the normal enzymes involved in the biochemical process of the host cell. Virus-associated enzymes have been reviewed (2,3) (Table 1).

Table 1. Virus-Associated Enzymes

Active enzymes	Viruses
neuraminidase	influenza virus, paramyxovirus
protein kinase	orthopoxvirus, herpes virus, oncovirus
nucleases	mastadenovirus, orthopoxvirus, oncovirus
nucleotide phosphohydrolases	orthopoxvirus
DNA-dependent RNA polymerase	orthopoxvirus
RNA-dependent RNA polymerase	influenza virus, paramyxovirus, vesiculovirus
RNA-dependent DNA polymerase (reverse transcriptase)	oncornavirus, human immunodeficiency virus (HIV)

Because viruses multiply only within the host cells and are totally dependent on these cells for nutrition and reproduction, effective antiviral agents must penetrate the cell to prevent viral multiplication. The ideal antiviral agent should not interfere with the cell's normal defense mechanisms against viral infection but should complement the normal cellular immunity and humoral antibody response to viral reproduction. During the early stages of infection, the activity of several enzymes already present in the host cell increases because of the need of the virus for increased synthesis of the nucleic acid and proteins required by its progeny. This increased activity may stimulate certain other host-controlled enzymic activity.

Viruses contain either RNA or DNA, and this nucleic acid composition forms the basis for their classification. Although viruses are known to infect bacteria, insects, plants, animals, and humans, this discussion is restricted to the important viruses of vertebrates. The relevant viruses are summarized in Table 2,

Table 2. Important[a] Viruses of Vertebrates

Family	Genus	Virus	Type of infection induced
		DNA group	
Adenoviridae	*Mastadenovirus*	numerous adeno types	respiratory and ophthalmic
Herpetoviridae	*Herpesvirus*	herpes types 1 and 2	ophthalmic, central nervous system, genital cutaneous, oral, upper respiratory
		cytomegalo	generalized of infant respiratory, glandular
		varicella (herpes zoster)	nervous, cutaneous
		pseudorabies	central nervous system of livestock
		infectious rhinotracheitis	respiratory of livestock, fowl
Poxviridae	*Orthopoxvirus*	variola (smallpox)	generalized
		vaccinia	cutaneous
Papovaviridae	*Polyomavirus*	polyoma, SV 40	tumors of rodents
		RNA group	
Orthomyxoviridae	*Influenzavirus*	influenza	respiratory
Paramyxoviridae	*Paramyxovirus*	parainfluenza	respiratory
		mumps	glandular
		Newcastle disease	respiratory and nervous of fowl
	Morbillivirus	measles	generalized
		canine distemper	generalized in dogs
	Pneumovirus	respiratory syncytial	respiratory
Picotnaviridae	*Enterovirus*	polio	nervous
		coxsackie	respiratory, nervous, cardiovascular
		echo	nervous, intestinal
	Rhinovirus	numerous rhino types	respiratory
		foot and mouth disease	generalized of livestock
	Calicivirus	feline calici	respiratory of cats
		vesicular exanthema	generalized of swine
Rhabdoviridae	*Vesiculovirus*	vesicular stomatitis	generalized of livestock
	Lyssavirus	rabies	nervous
Coronaviridae	*Coronavirus*	corona	respiratory
		infectious bronchitis	respiratory of fowl
Togaviridae	*Alphavirus*	eastern, western, Venezuelan equine encephalomyelitis	nervous, respiratory
		Semliki Forest	nervous
	Flavivirus	yellow fever	nervous
		numerous encephalitides, including Japanese B, St. Louis, Russian spring-summer	nervous
		dengue	joints, cutaneous
	Rubivirus	rubella	skin, generalized nervous

Table 2. *(Continued)*

Family	Genus	Virus	Type of infection induced
Bunyaviridae	*Bunyavirus*	numerous encephalitides, including bunyamwera and California	nervous
Arenaviridae	*Arenavirus*	lymphocytic choriomeningitis	nervous
		lassa	generalized
Retroviridae	Types B and C *Oncovirus*	numerous leukemias, including friend, Gross Moloney, Rauscher	oncogenic of lower animals
		rous sarcoma, HIV	oncogenic of fowl, aids

[a]Based on disease incidence.

using the nomenclature and taxonomy recommended by the International Committee on Taxonomy of Viruses (4,5).

Extensive efforts have been made over the years to develop drugs for the safe and effective treatment of viral infections. Numerous such compounds have been synthesized and evaluated against several viruses. A large number of these compounds have been found active in cell culture and/or experimental animals. Only a few of these compounds have shown efficacy in clinical trials. The majority of antiviral drugs which are under clinical development today generally interrupt viral nucleic acid synthesis. These compounds often do not affect host cell metabolism and possess considerable selectivity against virus-induced enzymes. This article discusses agents exhibiting significant antiviral activity against viral infections in animal model systems.

Agents Active Against DNA Viruses

One of the simplest molecules found to inhibit the replication of DNA viruses in animals is phosphonoformic acid [*4428-95-9*] (PFA, **1**) CH_3O_5P. Both PFA (as the trisodium salt CNa_3O_5P, foscarnet [*63585-09-1*]) and its homologue phosphonoacetic acid [*4408-78-0*] (PAA, **2**) $C_2H_5O_5P$, were developed by Astra Pharmaceuticals (6) and show selective inhibition of DNA polymerase in various herpes viruses.

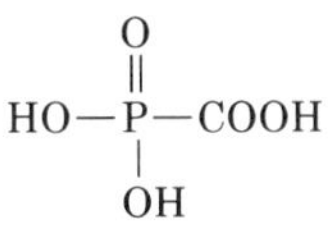
(1)

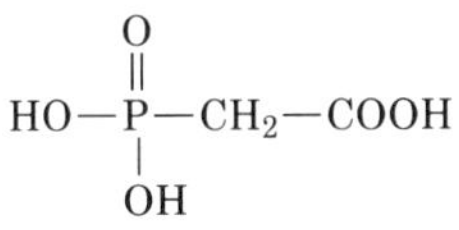
(2)

PFA has recently undergone clinical evaluation in humans for the treatment of recurrent genital herpes, hepatitis B viral infection, and acquired immunodeficiency syndrome (AIDS), as well as cytomegalovirus (CMV) infection of bone marrow and renal transplant patients (7). PFA was found to be effective against genital herpes when applied topically as a 3% cream, herpes labiales (cold sore of the lip), and CMV. However, the results have not been generally encouraging when the compound was tried against AIDS. The thio-analogues of PFA [*102805-86-7*], CH_3O_4PS and PAA [*102805-84-5*], $C_2H_5O_4PS$, (**3** and **4**, respectively) are also effective inhibitors of HSV-1 DNA polymerase (8). Carbonyl bisphosphonic acid [*14255-62-0*], $CH_4O_7P_2$, (**5**) has been shown (9) to inhibit the reverse transcriptase of human immunodeficiency virus (HIV).

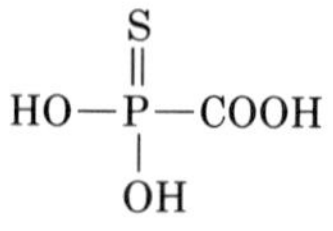

(**3**)

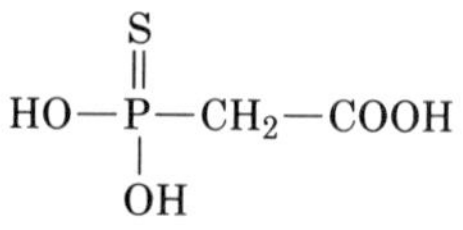

(**4**)

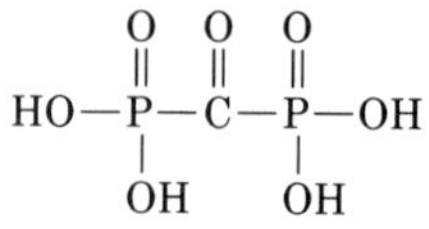

(**5**)

Thiosemicarbazones have long been used as antiviral agents, principally against pox viruses of the vaccinia family. One compound of this series, the isatin derivative (**6**) $C_9H_8N_4OS$, has been used prophylactically to prevent outbreaks of smallpox in humans (10) and to inhibit the protein synthesis in poxvirus-infected cells. The molecular mechanics relating to this property are still not known (11), though the binding of a metal ion may be a key factor (12). Methisazone [*1910-68-5*] $C_{10}H_{10}N_4OS$ (1-methyl-3-thiosemicarbazone of 2-oxoindole, **7**), one of the more active in the isatin-3-thiosemicarbazone [*487-16-1*] series, has been used in the treatment and prevention of smallpox and vaccinia infections that develop as complications of smallpox vaccination (13). Replacement of the sulfur atom in the thiosemicarbazone moiety by an oxygen atom leads to a loss of antiviral activity. Methisazone has no significant effect on vaccinia virus DNA synthesis (14) but seems to inhibit late protein synthesis by a mechanism that remains to be elucidated.

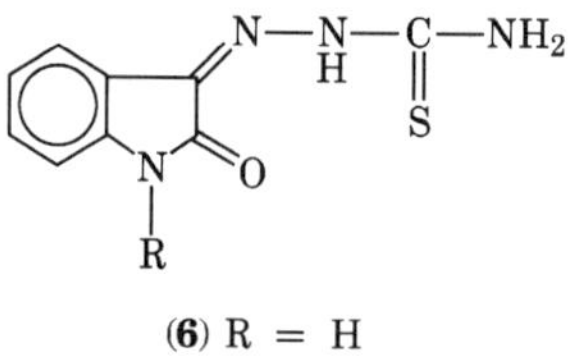

(**6**) R = H
(**7**) R = CH_3

Levamisole [*14769-73-9*] (6-phenyl-2,3,5,6-tetrahydroimidazo[2,1-*b*]thiazole, **8**) $C_{11}H_{12}N_2S$, was found to be effective against herpes virus infections in humans (15,16). It acts as a modulator of host resistance mechanisms, especially as an enhancer of cellular immunity.

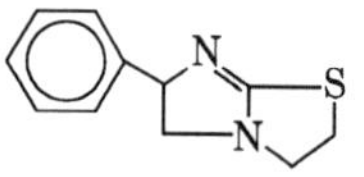

(**8**)

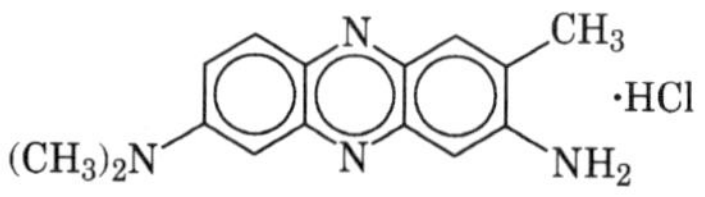

(**9**)

Certain heterocyclic dyes such as neutral red [*553-24-2*], $C_{15}H_{17}ClN_4$, (**9**), acridine orange [*65-61-2*], and proflavine [*92-62-6*], which act by binding to viral nucleic acid, absorbing visible light, and inactivating the virus by oxidation, have shown antiviral activity of clinical potential against cutaneous herpes virus infections (17). However, concern has been raised that photoinactivation of herpes viruses may be clinically dangerous because the inactivated virus may be capable of transforming normal cells to a malignant state (18). Moreover, these dyes have a strong affinity for the nuclei of normal cells and may damage their genetic makeup (19,20).

Several pyrimidine bases have been found to inhibit herpes virus-induced keratitis in rabbits (21,22), eg, 2-amino-4,6-dichloropyrimidine [*56-05-3*] $C_4H_3Cl_2N_3$, (**10**) and 1-allyl-6-chloro-3,5-diethyluracil [*20938-38-9*], $C_{11}H_{15}ClN_2O_2$, (**11**). Compound (**11**) exhibits efficacy in treating herpetic skin and mucus diseases in humans (22). An important aspect of these aglycones is their continued antiviral activity late in the viral replicative cycle. Mechanistic studies with these compounds suggest that they inhibit the assembly of viral proteins into procapsides, subsequently preventing the coating of available infectious viral nucleic acids which form mature virus particles (21).

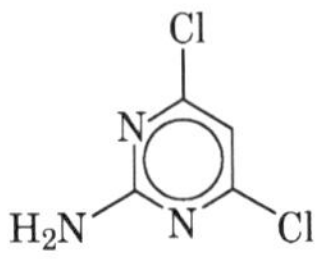

(**10**)

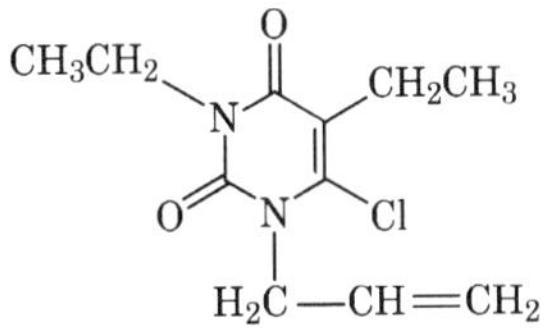

(**11**)

The most successful clinical antiviral agents belong to the nucleoside category. Nucleoside analogues with potent antiviral activity have been known since idoxuridine [*54-42-2*] and trifluridine [*70-00-8*] were shown to be efficacious against herpes keratitis more than 30 years ago. However, despite their antiviral efficacy the therapeutic usefulness of these early nucleosides was limited because of mutagenic, teratogenic, carcinogenic, cytostatic, or cytotoxic side effects. Recently the specificity of antiviral action of this class of compounds has been significantly improved; potent and highly selective nucleoside antiviral agents have now been developed. These are either antiviral nucleosides which are only activated by virus-infected cells, or antiviral agents which specifically inhibit virus-induced enzymes required for viral replication.

Idoxuridine. The first clinically effective nucleoside antiviral agent was 5-iodo-2′-deoxyuridine [*54-42-2*] (idoxuridine, IdU, **12**) $C_9H_{11}IN_2O_5$, synthesized (23) in 1959 (Fig. 1). It was used for the treatment of herpes keratitis in humans (24) and was the first successful use in humans of a synthetic antiviral drug against a herpes virus. It unequivocally demonstrated that IdU prevents the development of severe ocular infections and blindness. FDA approval of IdU for this use quickly followed; the drug was sold by SmithKline Beecham for 20 years for use by ophthalmologists. Recently it has been replaced by more effective and less toxic nucleoside analogues. Moreover, IdU's utility for treatment of systemic herpes infections has been limited by the drug's toxicity. In any event, the human use of IdU became a prototype for the search for other synthetic nucleoside antiviral agents. The other halogenated 2′-deoxyuridines have had mixed reports of activity in experimental systems and have not found clinical usefulness as antiviral agents (25,26).

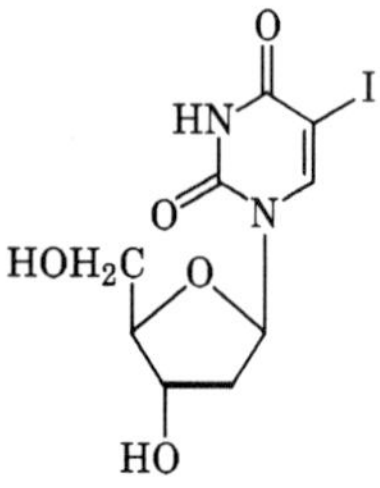

(12)

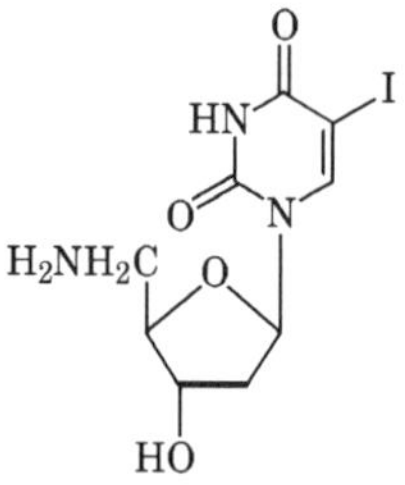

(13)

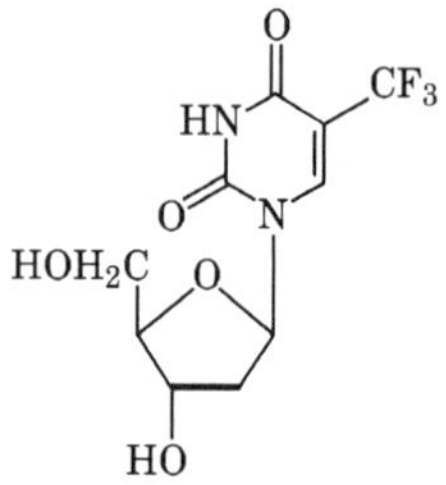

(14)

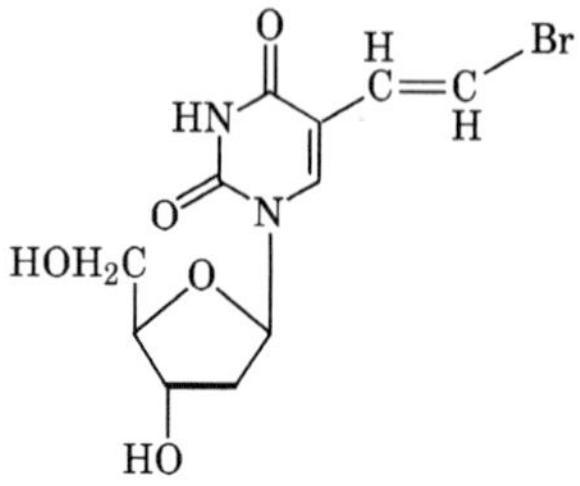

(15)

Fig. 1. Halogenated 2′-deoxyuridine derivatives.

5′-Amino-5-iodo-2′,5′-dideoxyuridine [*56045-73-9*] (**13**) $C_9H_{12}IN_3O_4$, was synthesized in 1975 (27) and was found effective against herpes keratitis in rabbits (28). This compound is markedly less cytotoxic than IdU, indicating that it may have a safer and more specific mode of antiviral activity. A potential limitation of this group of nucleosides is their specificity, for they fail to inhibit all strains of herpes viruses. The specific antiviral activity of (**13**) is considered to be a result of the incorporation of the 5′-*N*-phosphate into both viral and host DNA in infected cells, but not into the DNA of normal cells. Phosphorylation of (**13**) occurs only in herpes virus-infected cells, brought about by a virus-induced thymidine kinase (29).

Trifluridine, $C_{10}H_{11}F_3N_2O_5$, (5-trifluoromethyl-2′-deoxyuridine [*70-00-8*], F_3TdU, **14**) was first prepared (30) in 1962. It is used for topical therapy of herpes virus-infected eyes. It is especially useful for treating infections that are resistant to IdU therapy. Like IdU, trifluridine is incorporated into DNA in place of thymidine in both infected and uninfected cells. But it is 10 times more potent than IdU against herpes keratitis in rabbits and 10 times more soluble in water. Trifluridine is also useful in treating human cytomegalovirus (HCMV), but its toxicity to bone marrow may limit its clinical use.

Another 2′-deoxyuridine derivative, (*E*)-5-(2-bromovinyl)-2′-deoxyuridine [*69304-47-8*], BVdU, (**15**) $C_{11}H_{13}BrN_2O_5$, was reported in 1979 (31). BVdU differs from IdU and F_3TdU by being specifically phosphorylated in the 5′-position by herpes simplex virus type-1 (HSV-1) induced thymidine kinase. This restricts its action to cells infected by HSV-1. It is less active against genital herpes (HSV-2). HSV-1-induced thymidine kinase converts BVdU to the corresponding 5′-mono- and diphosphate, but HSV-2-induced thymidine kinase stops at the stage of the 5′-phosphate of BVdU. Apparently, cellular kinases phosphorylate BVdU-5′-diphosphate to the corresponding 5′-triphosphate, which inhibits HSV-1 DNA polymerase to a greater extent than similar cellular DNA polymerases.

BVdU is incorporated into DNA, but since it is phosphorylated specifically in the virally infected cell, incorporation is mainly confined to such cells. Evidence suggests that BVdU also inhibits the biosynthesis of HSV-1 glycoproteins, which may contribute to the inhibition of the HSV-1 virus (32). BVdU is degraded by thymidine phosphorylase more rapidly than the natural substrate, thymidine. This rapid enzymic degradation may present a problem in its clinical use. Moreover, herpes viruses develop resistance to BVdU, apparently because of mutant viruses that have lower thymidine kinase activity. G. D. Searle has dropped further development of BVdU because of increased animal tumor incidence induced by prolonged dosing (1).

Cytosine Derivatives. 1-β-D-Arabinofuranosylcytosine [*147-94-4*] (ara-C, **16**), $C_9H_{13}N_3O_5$, reportedly has had significant therapeutic effects in patients with localized herpes zoster, herpes eye infections, and herpes encephalitis (33), although several negative results have also been reported (34) (Fig. 2). Ara-C, also known as cytarabine, is quite toxic and is only recommended for very severe viral infections. It is rapidly deaminated in humans to the relatively inactive ara-U. Ara-C is converted in the cell to the 5′-monophosphate by deoxycytidine kinase, followed by formation of the corresponding di- and triphosphate. The triphosphate has been shown to inhibit DNA polymerase.

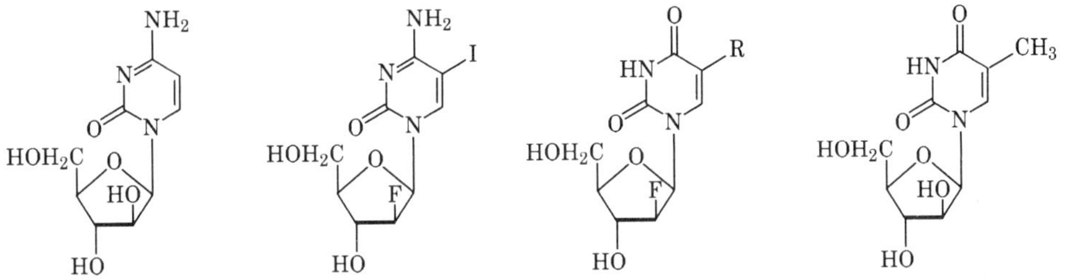

(**16**) (**17**) (**18**) R = CH_3 (**19**) R = CH_2CH_3 (**20**)

Fig. 2. 1-β-D-Arabinofuranosylcytosine (ara-C) and derivatives.

The 3′,5′-cyclic phosphate of ara-C has shown significant antiviral activity *in vivo* with an efficacy greater than that exhibited by ara-C itself (35). It was speculated that the 3′,5′-cyclic phosphate moiety may inhibit the deamination of ara-C, thus causing the increased *in vivo* potency. A number of derivatives of ara-C have been prepared in an effort to improve on antiviral activity and to reduce the toxicity. One such derivative is 2′-fluoro-5-iodo-1-β-D-arabinofuranosylcytosine [*69123-90-6*] (FIAC, **17**), synthesized (36) in 1979. It is active against certain DNA viruses. FIAC, $C_9H_{11}FIN_3O_4$, strongly inhibits HSV-1 and HSV-2 *in vivo*. This selective activity is due in large part to the HSV-encoded thymidine kinase. Clinical studies with FIAC against herpes infections in immunosuppressed cancer patients have shown evidence that this nucleoside can abort cutaneous varicella zoster. The remarkable metabolic stability of the *N*-glycosyl linkage of FIAC may account for the potent activity *in vivo*. When given to mice, FIAC is substantially deaminated to 2′-fluoro-5-iodo-1-β-D-arabinofuranosyluracil [*69123-98-4*], $C_9H_{10}FIN_2O_5$, which is also a significant antiviral agent active against herpes viruses.

FIAC also strongly inhibits HCMV and Epstein-Barr virus (EBV) *in vitro*, the two viruses known not to induce a specific viral thymidine kinase for their replication. However, HCMV may stimulate cellular kinases that can anabolize FIAC to its 5′-triphosphate, which specifically inhibits the HCMV-encoded DNA polymerase. This selective activity suggests that FIAC should be evaluated against HCMV infections. FIAC-triphosphate incorporated into DNA has shown strong *in vitro* activity against the DNA polymerases of human hepatitis B virus (HBV) and of woodchuck hepatitis virus (WHV) (37).

Several uracil nucleoside analogues of FIAC also have been prepared. Among these are 2′-fluoro-5-methylarabinosyluracil [*69256-17-3*] (FMAU, **18**), $C_{10}H_{13}FN_2O_5$, and the corresponding 5-ethyl derivative FEAU (**19**) [*83546-42-3*], $C_{11}H_{15}FN_2O_5$. Both are effective *in vivo* against HSV-1 and HSV-2. FMAU has shown dramatic results in the treatment of WHV in woodchucks and thus may prove effective against HBV infections in humans. However, FMAU is associated with severe toxicity. By contrast, FEAU induced a sustained, although less dramatic, decrease of viral replication without apparent toxic effect (38).

Yet another pyrimidine nucleoside is ara-T (1-β-D-arabinofuranosylthymine [*605-23-2*], $C_{10}H_{14}N_2O_6$, **20**) isolated from Caribbean sponges in 1950 and first synthesized in 1957 (39). Ara-T inhibits herpes DNA replication and is selectively toxic to cells infected by herpes virus but does not inhibit DNA synthesis in uninfected cells. *In vivo* studies with ara-T against experimental HSV-1-induced encephalitis in mice showed remarkable efficacy; ara-T is as potent as the monophosphate of ara-A (**23**) which will be discussed. Ara-T is being developed as an antiherpes agent by Yamasa Shoyu Co. in Japan, although some neurotoxicity has been noted in animal studies with this agent.

An interesting azapyrimidine nucleoside reported to have significant antiviral activity *in vivo* is 6-azauridine [*54-25-1*] (**21**). 6-Azauridine, $C_8H_{11}N_3O_6$, exhibits activity *in vivo* against vaccinia and herpes virus infections (40), and has been reported to be effective against herpes simplex and smallpox virus infections in humans, although side effects have curtailed the use of this drug (25). The basis for the antiviral activity of 6-azauridine has appeared to be conversion of this nucleoside to the 5′-monophosphate, which inhibits orotidylic acid decarboxylase (41). Another active azapyrimidine nucleoside with antiherpes virus activity *in vivo* is 5,6-dihydro-5-azathymidine [*57350-36-4*] (**22**), which was isolated from the fermentation broth of *Streptomyces platensis* var clarensis (42). The antiviral activity of (**22**), $C_9H_{15}N_3O_5$, may be due to inhibition of thymidine phosphorylation in the infected cell (42).

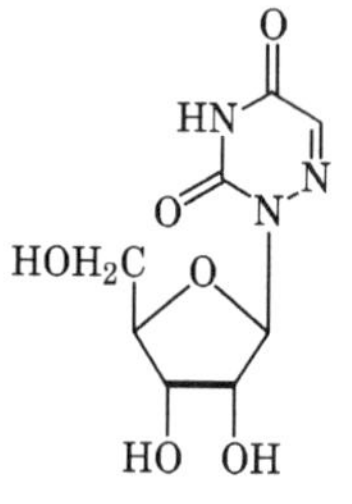

(**21**)

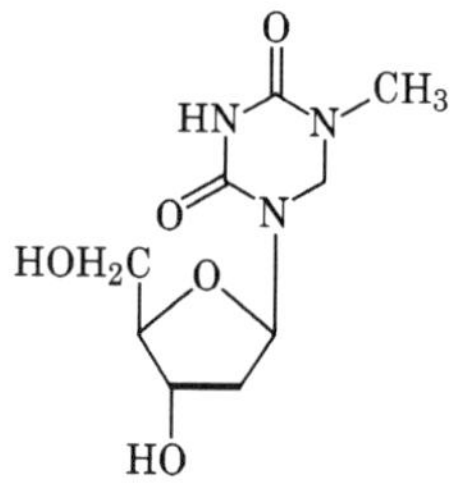

(**22**)

Purine Nucleoside Derivatives. A number of purine nucleoside analogues are also found to be active against several DNA viruses (Fig. 3). The clinically active antiviral drug ara-A (9-β-D-arabinofuranosyladenine [*5536-17-4*], vidarabine, **23**) is active against a number of DNA viruses *in vivo* and also inhibits certain RNA tumor viruses which replicate through a DNA intermediate (43). Ara-A, $C_{10}H_{13}N_5O_4$, was first synthesized in 1960 (44) and later discovered as a

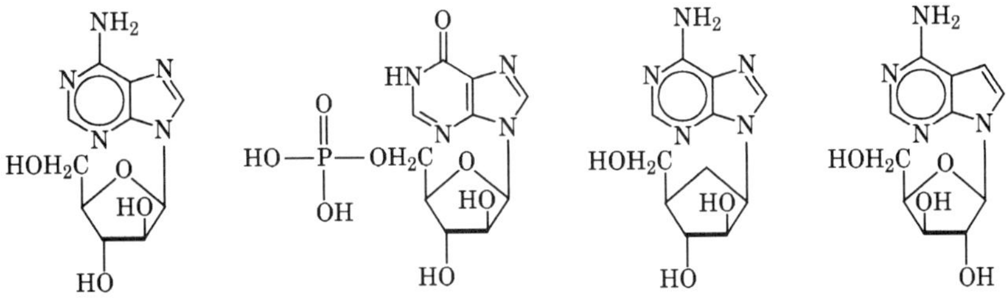

(23) (24) (25) (26)

Fig. 3. Purine nucleoside analogues found to be active against DNA viruses.

nucleoside antibiotic by Warner-Lambert researchers. The *in vivo* antiviral activity of ara-A against intracerebral vaccinia infection in mice and herpes simplex viral keratitis in hamsters, as well as intracerebral HSV-1 infections in mice, was reported by the Southern Research Institute. Ara-A is now marketed in the United States by Warner-Lambert for the systemic treatment of the life-threatening viral illness herpes encephalitis. It also is effective in treating neonatal herpes and varicella zoster (chicken pox), in addition to replacing idoxuridine in treating herpes keratitis (1).

Because it is relatively insoluble, ara-A must be administered by slow infusion, requiring large volumes of fluids. It is also rather toxic, causing nausea, vomiting, and diarrhea. In addition, chromosomal damage caused by ara-A has been observed in blood, and the compound is teratogenic, carcinogenic, and mutagenic in certain animals.

Ara-A is rapidly deaminated by adenosine deaminase in animals to give 9-β-D-arabinofuranosylhypoxanthine [*7013-16-3*] (ara-Hx), $C_{10}H_{12}N_4O_5$, which is approximately 10 times less active than ara-A in cell culture experiments. It has been shown that various adenosine deaminase inhibitors greatly increase the effectiveness of ara-A in culture. However, ara-A-2′,3′-di-*O*-acetate [*65174-96-1*], $C_{14}H_{17}N_5O_6$, is resistant to deamination and is hydrolyzed by tissue esterases to the active ara-A (45). The minimum inhibitory concentration for HSV-1 or HSV-2 plaque reduction by ara-A-2′,3′-di-*O*-acetate was 6 to 10 times higher than that of ara-A or ara-AMP (46). In addition, inhibition of virus-induced lesions was observed in the HSV-2 genital infection of the guinea pig after topical application of 10% ara-A-2′,3′-di-*O*-acetate. On the basis of these results ara-A-2′,3′-di-*O*-acetate is considered an antiviral agent suitable for topical therapy of primary genital infection with herpes virus (47).

Ara-A-5′-monophosphate [*29984-33-6*] (ara-AMP), $C_{10}H_{14}N_5O_7P$, is more water-soluble than ara-A, and therefore can be used in higher dosage during the first hours of treatment of viral infections. Ara-AMP has been shown to decrease virion-associated DNA polymerase concentrations in ground squirrels carrying ground squirrel hepatitis virus. The hypoxanthine derivative, ara-HxMP [*54656-49-4*] (**24**) is more water-soluble, appears to have a similar antiviral spectrum to ara-A, and is considerably less toxic (48).

Topical application of ara-HxMP, $C_{10}H_{13}N_4O_8P$, significantly inhibited the development of keratitis-induced HSV-1, HSV-2, or vaccinia virus in the eyes of rabbits. Ara-HxMP also significantly controlled the development of HSV-1 or vaccinia viral-induced encephalitis in mice and was also active in preventing equine abortion viral deaths in hamsters. Clinical trials with ara-HxMP have not yet been reported.

Ara-A is phosphorylated in mammalian cells to ara-AMP by adenosine kinase and deoxycytidine kinase. Further phosphorylation to the di- and triphosphates, ara-ADP and ara-ATP, also occurs. In HSV-1 infected cells, ara-A also is converted to ara-ATP. Levels of ara-ATP correlate directly with HSV replication. It has recently been suggested that ara-A also may exhibit an antiviral effect against adenovirus by inhibiting polyadenylation of viral messenger RNA (mRNA), which may then inhibit the proper transport of the viral mRNA from the cell nucleus.

It is likely that ara-HxMP similarly exerts its antiviral activity in the form of the triphosphate, ara-HxTP, since ara-HxTP inhibits HSV-1 DNA polymerase (49). Another possible explanation of the antiviral activity of ara-HxTP is that it is metabolically converted to ara-AMP. In fact, it has been shown at Wellcome Research Laboratories that ara-HxMP is a substrate for adenylosuccinate synthetase, and that the resulting arabinofuranosyladenylosuccinate is cleaved to ara-AMP by adenylosuccinate lyase (1). The selective action of ara-A against HSV appears to be a consequence of the preferential inhibition of ara-ATP against HSV-1 and HSV-2 polymerases. Ara-ATP also inhibits normal cellular DNA polymerases, which may be the reason for its cellular toxicity. Also, it has been observed that ara-A is incorporated uniformly throughout the HSV-1 genome, which may result in defective viral DNA (50).

Another interesting adenosine analogue active against DNA viruses is cyclaradine *[69979-46-0]* (**25**), $C_{11}H_{15}N_5O_3$, the carbocyclic analogue of ara-A (51). In this nucleoside the furanose ring oxygen of ara-A is replaced by a methylene group. Cyclaradine is resistant to adenosine deaminase and, in contrast to ara-A, is highly active against HSV-2. Topical (intravaginal) treatment of genital herpes infections in guinea pigs with the 5′-methoxyacetyl derivative of cyclaradine beginning three days after infection largely prevented the development of viral lesions (52).

Of considerable interest is the xylofuranosyl analogue (**26**) *[64526-29-0]*, $C_{11}H_{14}N_4O_4$, of tubercidin *[69-33-0]* (53). When (**26**) was administered intraperitoneally at 50 mg/kg per day, a significant reduction of the mortality rate of mice infected with HSV-2 was achieved (54). Compound (**26**) applied topically in 1% DMSO to intracutaneous HSV-2 in hairless mice gave 100% survivors without observable toxicity. This nucleoside would appear to have considerable potential for topical treatment of HSV-2 infections.

Adenine Derivatives. A number of aliphatic adenosine analogues (Fig. 4) which exhibit broad-spectrum antiviral activity have been prepared in recent years. One of the first such nucleoside analogues shown to have antiviral activity is (*S*)-9-(3-hydroxy-2-phosphonylmethoxypropyl)adenine *[92999-29-6]*, $C_9H_{14}N_5O_5P$ [(*S*)-HPMPA, (**27**)]. This nucleotide analogue has been prepared (55) and reported (56) to be active against a number of DNA viruses *in vitro*. When given orally, (*S*)-HPMPA was active against a cutaneous HSV-1 model infection in hairless mice and a lethal systemic HSV-1 infection in mice (56). It is also active in the herpetic eye infection model in rabbits (57). (*S*)-HPMPA is not enzymatically dephosphorylated and penetrates cells directly as the phosphonate. It appears to

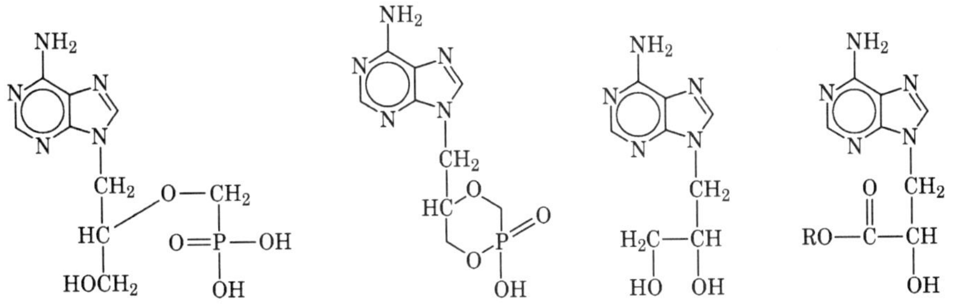

(27) (28) (29) (30)

Fig. 4. Aliphatic adenosine analogues with broad-spectrum antiviral activity.

be incorporated as a chain terminator. It is active against herpes thymidine kinase which lacks mutants resistant to acyclovir and BVdU.

The cyclic phosphonate of (*S*)-HPMPA, [(*S*)-cHPMPA [*106941-26-8*], $C_9H_{12}N_5O_4P$, **28**] also is active against HSV-1 and HSV-2 infection in mice. Both (**27**) and (**28**) are reported to be efficacious against adenovirus, vaccinia virus, as well as varicella-zoster virus infections (58).

The first synthesis of racemic [9-(2,3-dihydroxypropyl)adenine] DHPA, $C_8H_{11}N_5O_2$, was reported in 1965 (59), and the synthesis of the (*R*)- and (*S*)-enantiomers of DHPA was reported later (60). Subsequently (*S*)-DHPA [*54262-83-8*] (**29**) was shown to be active against a broad range of DNA and RNA viruses including vaccinia, HSV-1, HSV-2, measles, and vesicular stomatitis viruses. The racemic mixture, (*R,S*)-DHPA [*55904-02-4*] was almost as active as the (*S*)-enantiomer. However, the (*R*)-isomer of DHPA was inactive as an antiviral agent.

The antiviral activity of (*S*)-DHPA *in vivo* was assessed in mice inoculated intranasally with vesicular stomatitis virus; (*S*)-DHPA significantly increased survival from the infection. (*S*)-DHPA did not significantly reduce DNA, RNA, or protein synthesis and is not a substrate for adenosine deaminase of either bacterial or mammalian origin. However, (*S*)-DHPA strongly inhibits deamination of adenosine and ara-A by adenosine deaminase. Its mode of action may be inhibition of *S*-adenosyl-L-homocysteine hydrolase (61). Inhibition of SAH hydrolase results in the accumulation of SAH, which is a product inhibitor of *S*-adenosylmethionine-dependent methylation reactions. Such methylations are required for the maturation of vital mRNA, and hence inhibitors of SAH hydrolase may be expected to block virus replication by interference with viral mRNA methylation.

Adeninylhydroxypropanoic acid alkyl esters [(*R,S*)-AHPA esters, (**30**)] represent a new class of broad-spectrum antiviral agents, which are, like (*S*)-DHPA, targeted at SAH hydrolase (62). The free acid, (*R,S*)-AHPA, is only weakly active as an antiviral agent. Thus the alkyl moiety merely serves as a protecting group to facilitate uptake of AHPA by the cells. Within the cells, the AHPA alkyl esters would be hydrolyzed to release the free acid, which should then interact with SAH hydrolase.

Acyclic Purine Nucleosides. As a consequence of earlier studies (63–65) involving adenosine deaminase as a model enzyme for examining structural features required for antiviral activity, a number of acyclic purine nucleosides have emerged as antiviral agents (Fig. 5). The forerunner in this group of compounds is the guanosine analogue, acyclovir [*59277-89-3*] [9-(2-hydroxyethoxymethyl)guanine, **31**] (66). Acyclovir, $C_8H_{11}N_5O_3$, is proven to be active against herpes viruses, especially HSV-1 and HSV-2, and remains the drug of choice for control of HSV infections. It has been approved by FDA for general topical as well as oral and intravenous use in treating HSV-1 and HSV-2 infections. It has been marketed by Burroughs Wellcome Co. since 1982.

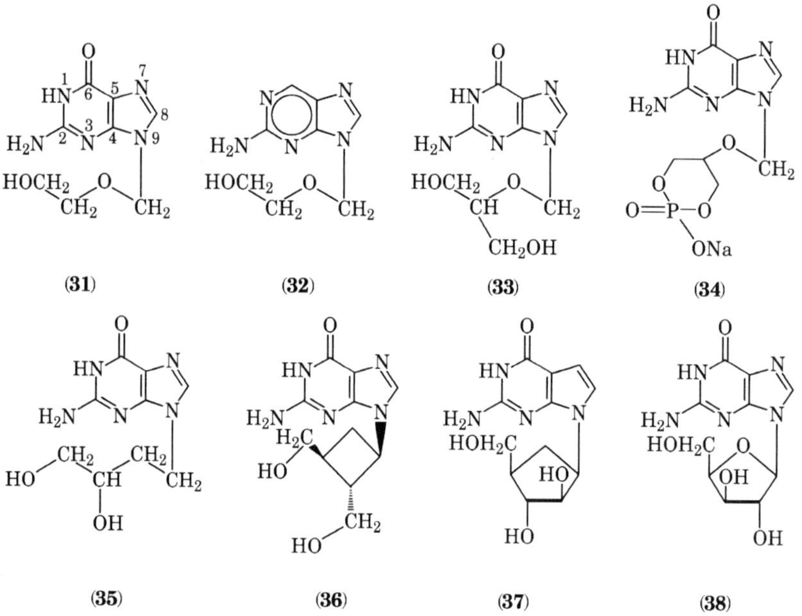

Fig. 5. Acyclovir and its analogues.

Acyclovir is more effective the more serious the disease and the earlier it is given. It has been shown to be efficacious when used systemically in the prophylaxis of HSV infections in immunosuppressed patients, ie, bone marrow transplant recipients (67). Acyclovir therapy appears to be superior to ara-A in the treatment of herpes simplex encephalitis in humans (68).

Acyclovir has a low toxicity and is excreted unchanged in the urine (69). In patients with normal renal function, 10% of the acyclovir dose is recovered in the urine as 9-carboxymethoxymethylguanine, the only significant metabolite (70). It has a half-life of 2.5–3.0 h after intravenous administration (71). Renal toxicity occurred in early studies when the drug was given intravenously as a bolus, especially in poorly hydrated patients or in those with renal disease. Although its toxicity is low, acyclovir is quite insoluble in body fluids. A number of prodrugs of acyclovir have been studied, eg, 2′-*O*-glycylacyclovir and 6-deoxyacyclovir. These have greater solubility in aqueous medium and are better absorbed orally. 6-Deoxyacyclovir [*75128-58-4*] (**32**) $C_8H_{11}N_5O_2$, is metabolized rapidly by xanthine oxidase to acyclovir. When given orally, compound (**32**) gives levels of acyclovir similar to those reached with intravenous acyclovir.

The antiviral mechanism of action of acyclovir has been reviewed (72). Acyclovir is converted to the monophosphate in herpes virus-infected cells (but only to a limited extent in uninfected cells) by viral-induced thymidine kinase. It is then further phosphorylated by host cell guanosine monophosphate (GMP) kinase to acyclovir diphosphate [*66341-17-1*], which in turn is phosphorylated to the triphosphate by unidentified cellular enzymes. Acyclovir triphosphate [*66341-18-2*] inhibits HSV-1 viral DNA polymerase but not cellular DNA polymerase. As a result, acyclovir is 300 to 3000 times more toxic to herpes viruses in an HSV-infected cell than to the cell itself. Studies have shown that a once-daily dose of acyclovir is effective in prevention of recurrent HSV-2 genital herpes (1). HCMV, on the other hand, is relatively uninhibited by acyclovir.

The incorporation of acyclovir triphosphate into calf thymus DNA primer template has been shown to be much more rapid and extensive with HSV-1 DNA polymerase than with vero cell DNA polymerase α. This incorporation of acyclovir ceased after 15 min since the template is chain terminated by the acyclovir incorporation, as there is no 3′-hydroxyl group on which to continue elongation. The viral DNA polymerase is also inactivated by tight binding to the terminated template.

Several promising analogues of acyclovir have been discovered. Two clinically approved guanosine analogues are DHPG [9-(1,3-dihydroxy-2-propoxymethyl)guanine [*82410-32-0*], Ganciclovir, **33**] and DHPG cyclic phosphate (**34**). The synthesis and antiviral activity of DHPG has been reported simultaneously by several groups (73–77). DHPG, $C_9H_{13}N_5O_4$, is more soluble than acyclovir and has greater bioavailability. Oral administration of DHPG is fiftyfold more efficacious than acyclovir in the treatment of systemic or local HSV-1 or HSV-2 intravaginal infections in mice. Its mode of antiviral action seems to be generally similar to that of acyclovir, involving phosphorylation to the monophosphate by herpes viral thymidine kinase followed by phosphorylation by cellular kinases to the triphosphate, which is a potent inhibitor of herpes viral DNA polymerase and thus inhibits DNA viral replication. However, DHPG is reported to be a thirtyfold more efficient substrate for HSV-1 thymidine kinase than acyclovir and cellular GMP kinase phosphorylates DHPG monophosphate more readily than acyclovir monophosphate. These differences result in seven times more DHPG triphosphate in HSV-1 infected cells than acyclovir triphosphate 6 h after infection. DHPG is being developed by Syntex.

DHPG also is clinically effective against HCMV infections in immunosuppressed patients. It has been shown (78) that DHPG triphosphate inhibits cellular DNA polymerase at levels about 10 times higher than that required to inhibit HCMV-induced DNA polymerase. Treatment with DHPG of patients with AIDS and serious CMV infections resulted in suppression of CMV replication and improvement in CMV retinitis in seven out of eight patients (79). However, treatment of CMV viral pneumonia with DHPG has been disappointing (80).

The sodium salt of the cyclic phosphate of DHPG [*91516-85-7*], $C_9H_{11}N_5O_6P{\cdot}Na$ (**34**), (81), is much more soluble than DHPG and has the broadest spectrum of activity against DNA viruses of the acyclovir analogues (82,83). This cyclic phosphate inhibits HSV-1, HSV-2, HCMV, vaccinia virus, and adenovirus, but is inactive against RNA viruses. Its *in vitro* antiviral activity is about 10 times greater than that of DHPG. This cyclic nucleotide, which is not dependent on virus-specified thymidine kinase for antiviral activity, appears to inhibit DNA viruses by a mechanism different from those of acyclovir or DHPG. The synthesis of the derivatives of (**34**) has been reviewed (84).

3,4-Dihydroxybutylguanine (DHBG, **35**), [*86304-28-1*] $C_9H_{13}N_5O_3$, is another analogue of acyclovir, reported by the Astra Pharmaceutical group (85). DHBG is similar to acyclovir in its activity against HSV-1 keratitis in rabbits. With regard to HSV-2 infections in mice, DHBG appears somewhat superior to acyclovir; the *R* enantiomer (Buciclovir) is more active than the *S*. DHBG, however, is less effective than acyclovir against HSV-1-induced encephalitis in mice. Currently, DHBG is under clinical study as an antiherpetic agent by Astra Pharmaceuticals in Sweden.

(±)-(1α,2β,3α)-9-[2,3-Bis(hydroxymethyl)cyclobutyl]guanine [(±)-BHCG, **36**] is a recently synthesized (86) nucleoside analogue with potent and selective antiviral activity against members of the herpesvirus group, including human cytomegalovirus. (±)-BHCG [*126062-18-8*], $C_{11}H_{15}N_5O_3$, is readily phosphorylated by purified HSV-1 thymidine kinase, and BHCG triphosphate synthesized enzymatically is a selective inhibitor of HSV-1 DNA polymerase (86). (±)-BHCG did not inhibit host cell growth at concentrations at least one thousandfold higher than HSV-2 inhibitory concentrations. The activity against a thymidine kinase deficient HSV-2 mutant was twenty-fivefold poorer than against the parent virus, suggesting that phosphorylation is an important prerequisite for antiviral activity against HSV-2 (86). Subcutaneous administration of (±)-BHCG was protective against HSV-1 systemic infections in mice.

The ara-carbocyclic analogue of 7-deazaguanosine {(±)-2-amino-3,4-dihydro-7-[(1α,2α,3β,4α)-2,3-dihydroxy-4-(hydroxymethyl)-1-cyclopentyl]pyrrolo[2,3-*d*]-pyrimidin-4-one, **37**} [*97570-35-9*], $C_{12}H_{16}N_4O_4$, has been synthesized (87) and studies done against an HSV-1 encephalitis model at a total dose of 100 mg/kg given intraperitoneally over a 48 h period. This resulted in 4 of 15 survivors as compared to zero of 30 for the controls (88). In this model, in comparison, BVdU was only marginally active with 1 of 15 survivors. Compound (**37**) at 25 μg/mL inhibited viral DNA synthesis fourfold over cellular DNA synthesis (88).

Moderate *in vivo* antiherpes virus activity was demonstrated by 9-β-D-xylofuranosylguanine [*27462-39-1*] (xylo-G, **38**), $C_{10}H_{13}N_5O_5$, and the 5′-mono- and 3′,5′-cyclic phosphates of (**38**), although none was as active as ara-A (89). Generally, guanine base-modified analogues of acyclovir are less active than acyclovir because they are not readily phosphorylated by herpes thymidine kinase.

Viruses become resistant to these guanosine analogues by several different mechanisms. The most common resistant variant is deficient in viral thymidine kinase. Virus strains that are resistant to idoxuridine also have been found to be resistant to acyclovir because both drugs depend on viral thymidine kinase for activation. Acyclovir-resistant strains of HSV-1 have been isolated from certain patients receiving acyclovir therapy. With necessary repeated dosing against recurrent HSV-1, resistance could prove to be a primary clinical problem. Certain HSV-2 mutants with an altered viral DNA polymerase also are markedly resistant to DHPG. The problem of drug resistance and antiviral agents has been reviewed (90).

Agents Active Against RNA Viruses

A large number of α-hydroxybenzylbenzimidazole [*50-97-5*] (HBB, **39**), $C_{14}H_{12}N_2O$, derivatives has been prepared and extensively studied as selective inhibitors of the RNA-containing enteroviruses (91). Although none of these derivatives have shown any antiviral activity in animals, 1,2-bis(5-methoxy-2-benzimidazol-2-yl)-1,2-ethanediol [*16656-27-2*] (**40**), $C_{18}H_{18}N_4O_4$, was found to be active against an experimentally induced rhino virus infection in chimpanzees (92). However, the *in vivo* antiviral efficacy was accompanied by significant toxicity.

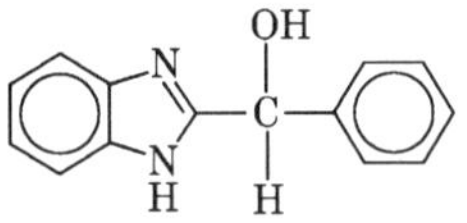

(**39**)

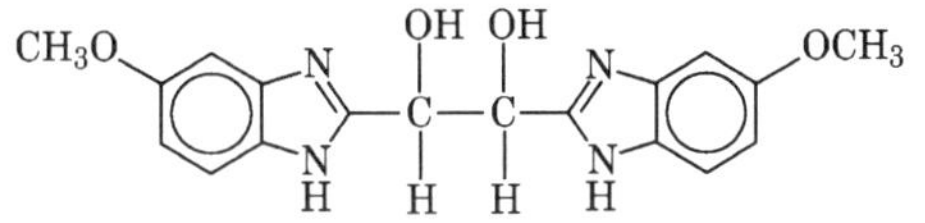

(**40**)

Amantadine hydrochloride [*665-66-7*] (1-adamantanamine hydrochloride, **41**), $C_{10}H_{17}N{\cdot}HCl$, (93) is a good example of a narrow-spectrum agent active only against influenza A virus. It became the first antiviral drug available for systemic use in the United States when it was approved by the FDA in 1966 for use against Asian influenza. In 1976, FDA approval was extended to the use of amantadine for the relief of symptoms of all influenza A strains. Amantadine is marketed by Du Pont de Nemours & Co., Inc. A structurally related drug, rimantadine hydrochloride [*1501-84-4*] $C_{12}H_{21}N{\cdot}HCl$, (α-methyl-1-adamantanemethylamine hydrochloride, **42**), is widely used in Russia to treat influenza A virus (94).

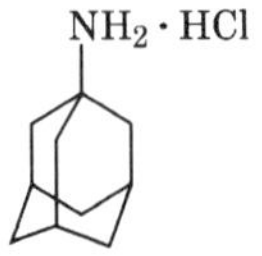

(**41**)

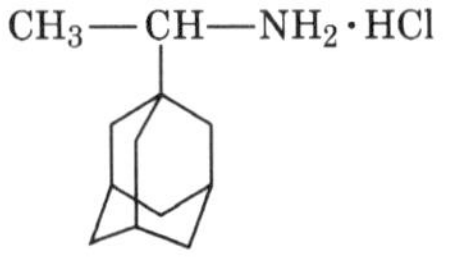

(**42**)

Both amantadine and rimantadine have been found to reduce the duration of influenza A-induced fever and malaise, and to lessen viral shedding. Prophylactic treatment has been recommended for high risk patients (95). It has been suggested that, in the presence of amantadine, the influenza virus attaches normally to cells, but once inside the cell the virus fails to initiate replication. Thus amantadine appears to inhibit the initiation of transcription at an early stage between uncoating and viral-specific RNA synthesis (96).

In most clinical trials, both (**41**) and (**42**) have provided protection against influenza A viruses in approximately 70% of the cases. Neither agent is effective against influenza B viral infections, and influenza A virus becomes resistant to both agents. This may account for the fact that both compounds have had difficulty being accepted in the United States as a valid treatment for influenza infections. In addition, amantadine causes mild transient side effects involving the central nervous system, such as insomnia, lightheadedness, and difficulty in concentration. The incidence of side effects is less with rimantadine. A number of derivatives of amantadine and rimantadine have been prepared and tested, some in clinical trials, but none has given overall results better than the original drugs (97).

Lipophilic β-Diketones. Arildone [*56219-57-9*] (**43**), $C_{20}H_{29}ClO_4$, and several other lipophilic β-diketones (Fig. 6) have exhibited significant *in vivo* activity against a number of RNA viruses. The most potent action is against poliovirus where arildone prevents the uncoating of the viral capsid (98). Arildone is moderately effective against the rhinoviruses. Extensive structure–activity studies with variation of the substituents on the phenyl ring and the β-diketone side chain of (**43**) led to WIN 51711 [*87495-31-6*] (**44**) with potent *in vitro* antirhinovirus and antipoliovirus activity (99,100). WIN 51711 (Disoxaril) is effective in a prophylactic regime against polio-induced death in mice at 100 mg/kg. However, at higher doses formation of crystalline urea was observed. Recently, a number of 2,6-disubstituted derivatives of WIN 51711, $C_{20}H_{26}N_2O_3$, have been prepared and evaluated against several rhinovirus serotypes (101). The 2,6-dichloro analogue was found to be highly effective *in vitro* against rhinoviruses with an MIC_{80} of 0.3 μM, and was also effective in preventing paralysis in mice infected with coxsackievirus A-9 (101).

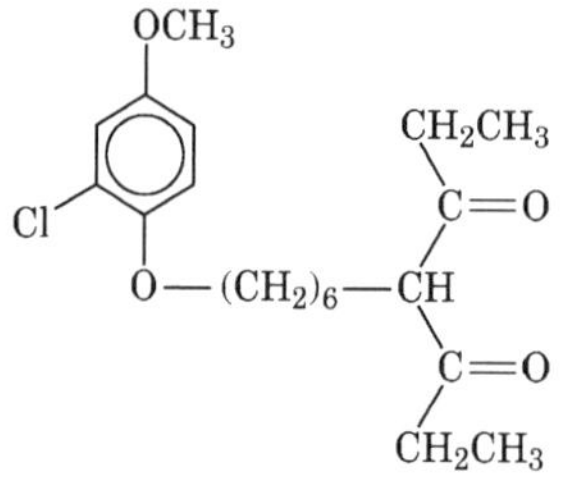

(**43**)

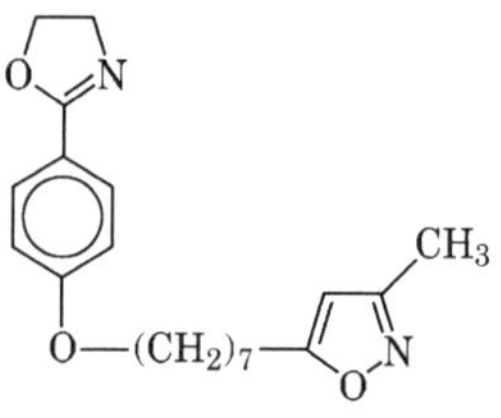

(**44**)

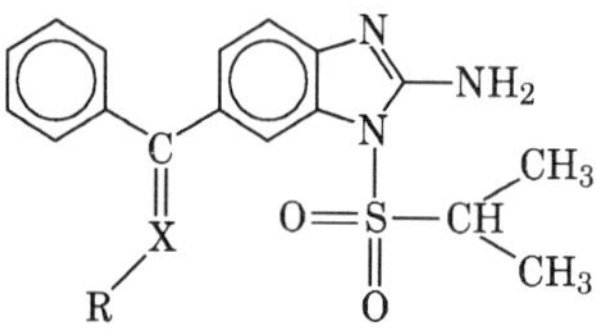

(**45**) X = N, R = OH
(**46**) X = CH, R = CH_3

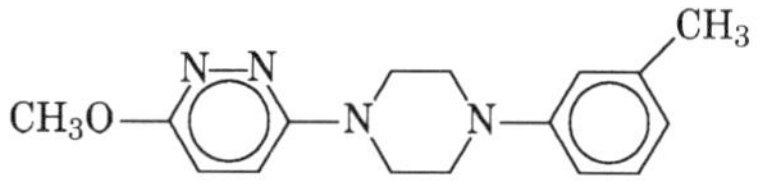

(**47**)

Fig. 6. Lipophilic β-diketones and derivatives active against RNA viruses, especially rhinovirus.

Enviroxime [*72301-79-2*] (**45**), $C_{17}H_{18}N_4O_3S$, a benzimidazole derivative, is found to be highly effective against several different strains of rhinovirus. When given intranasally to volunteers challenged with rhinovirus type-9, symptoms of rhinovirus infection and the virus titre of nasal wash specimens were reduced, and also there was a significant decrease in rhinorrhoea (102). Although enviroxime is active clinically, further studies are needed. The enviroxime surrogate, namely enviradene [*66334-86-9*] (**46**) $C_{19}H_{21}N_3O_2S$, has been shown to possess excellent *in vitro* antirhinovirus activity (103). A new antirhinoviral compound 3-methoxy-6-[4-(3-methylphenyl)-1-piperazinyl]pyridazine [*100241-46-1*] (R61837, **47**) $C_{16}H_{20}N_4O$, has been reported (104). R61837 is active against rhinoviruses and coxsackievirus A21, with MICs as low as 0.018 μM for some serotypes. R61838 has shown prophylaxis against HRV-9 using intranasal delivery (104).

Adenosine and Guanosine Analogues. Several adenosine analogues have been found to be active against both RNA and DNA viruses (Fig. 7). The prototype nucleoside in this group is 3-deazaadenosine [*6736-58-9*] (**48**), $C_{11}H_{14}N_4O_4$, first synthesized in 1966 (105). 3-Deazaadenosine inhibits the replication of RNA type C virus, HL-23, and Rous sarcoma virus (106). Avian sarcoma virus B77, grown in chicken embryo fibroblasts, is inhibited by 90% at 100 μM of (**48**). 3-Deazaadenosine is also active in cell culture against vesicular stomatitis (VSV), parainfluenza 3, and reovirus at less than 10 μg/mL, which is five to twentyfold lower than the host cell cytotoxic concentration. The broad-spectrum antiviral activity of this nucleoside is attributed (107) to its inhibition of *S*-adenosylhomocysteine hydrolase. This inhibition prevents the regeneration of the *S*-adenosylmethionine [*29908-03-0*] required to transfer a methyl group to viral mRNA. Methylation reactions, such as the methylation of the 5′-cap of viral mRNA, are essential to viral replication. 3-Deazaadenosine is not enzymatically deaminated or phosphorylated and can act both as an inhibitor of *S*-adenosylhomocysteine hydrolase and as a substrate of the *S*-adenosylhomocysteine hydrolase to yield 3-deazaadenosylhomocysteine, which has been shown to inhibit methylations *in vivo,* and which acts as a feedback inhibitor of the viral methyltransferases.

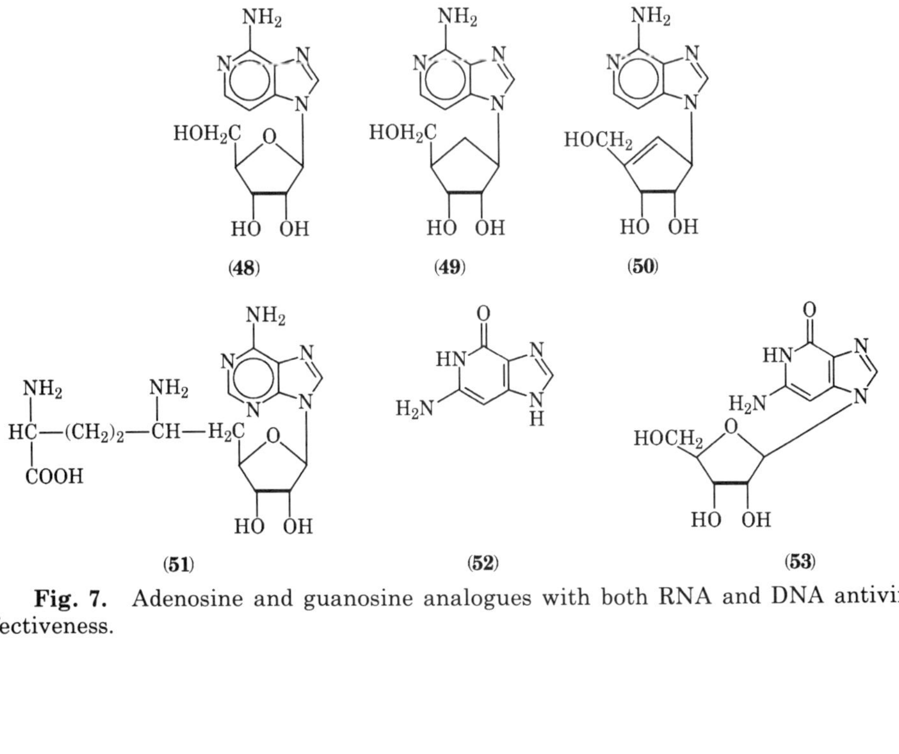

Fig. 7. Adenosine and guanosine analogues with both RNA and DNA antiviral effectiveness.

Other adenosine analogues that inhibit both RNA and DNA viruses are (±)-3-deazaaristeromycin, [*81601-33-4*], neplanocin A [*72877-50-0*], and sinefungin [*58944-73-3*]. Their mode of antiviral action is believed to be similar to that of 3-deazaadenosine. The carbocyclic analogue of 3-deazaadenosine, [(±)-3-deazaaristeromycin, **49**] $C_{12}H_{16}N_4O_3$, was first synthesized (108) in 1982 and reported to be active against HSV-1, vaccinia virus, and HL-23 *in vitro*. Subsequently, it has been shown (109) that (**49**) protects newborn mice against a lethal infection of VSV. When given 1 h after infection, mortality was reduced from 90% in the controls to 48% in the treated mice. Nucleoside (**49**) is also active *in vitro* against the RNA viruses, Sindbis, measles, and parainfluenza 3 and reo-1 viruses; and the activity in general is superior to that of 3-deazaadenosine.

Neplanocin A (**50**), $C_{11}H_{13}N_5O_3$, is a naturally occurring nucleoside antibiotic (110) which was synthesized (111) in 1983. Neplanocin A inhibits vaccinia virus multiplication of mouse L-cells. The inhibition of vaccinia virus by (**50**) coincides with the inhibition of *S*-adenosylhomocysteine hydrolase in infected cells. Thus the resultant inhibition of macromolecular methylation reactions inhibits the production of new viral particles by preventing methylation of new viral mRNA. Although neplanocin A shows excellent antiviral activity against vaccinia, influenza, measles, and vesicular stomatitis viruses, it gives only marginal protection against a lethal infection of VSV in mice (112).

Sinefungin (**51**), $C_{15}H_{23}N_7O_5$, an antifungal antibiotic (113), is an analogue of *S*-adenosylhomocysteine. It is a potent inhibitor of both Newcastle disease virus (NDV) and vaccinia virus in mouse L-cells. Sinefungin is also active against cutaneous HSV-1 and vaginal HSV-2 in guinea pigs, and systemic herpes viral infections in mice (114). It inhibits methylation of NDV and vaccinia viral mRNA. Sinefungin is effective in preventing Rous sarcoma virus-induced transformation of chick embryo fibroblasts. Unfortunately, sinefungin is also an inhibitor of tRNA and protein methyltransferases in CEF cells, suggesting it may be too toxic for human use.

Both 3-deazaguanine [*41729-52-6*] (**52**) and 3-deazaguanosine [*56039-11-3*] (**53**), prepared by ICN Pharmaceuticals (115), have shown significant *in vitro* antiviral activity against the RNA viruses, parainfluenza 1 and 3, five strains of rhino viruses, vesicular stomatitis virus, and a number of DNA viruses including adenovirus, HSV-1, HSV-2, CMV, vaccinia virus, and pseudorabies virus (116). Although (**52**) and (**53**) show significant antiviral activity against parainfluenza and influenza A_2 infections in mice, the cumulative toxicity may limit their use for treatment of human viral infections. 3-Deazaguanine, $C_6H_6N_4O$, and 3-deazaguanosine, $C_{11}H_{14}N_4O_5$, are metabolized to the 5′-nucleotides (117) and are incorporated into DNA, which is believed to be responsible for the observed toxicity.

The naturally occurring nucleoside antibiotic pyrazofurin [*30868-30-5*] (**54**), $C_9H_{13}N_3O_6$, shows broad-spectrum antiviral activity and broad safety margin against both RNA and DNA viruses *in vitro* (118). However, in mice inoculated with Rift Valley Fever virus, 0.25 mg/kg of pyrazofurin administered daily for five days with treatment initiated a day before infection resulted in a survival rate of only 20%. Similar results were obtained for the treatment of Pichende virus infections in guinea pigs. Thus neither the potent efficacy nor the broad margin of safety noted *in vitro* are reflected in animal models (118). Attempts to improve on this antiviral activity by the synthesis of a number of pyrazofurin derivatives were unsuccessful (119).

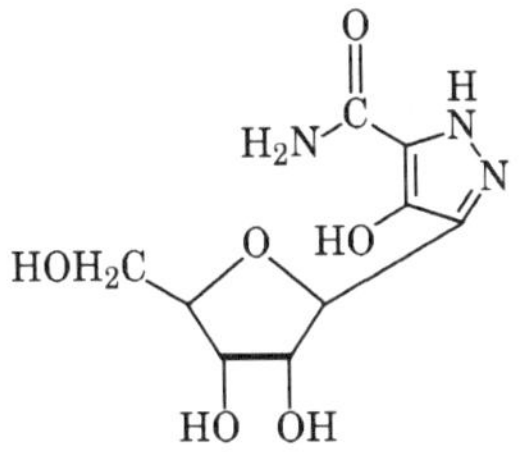

(54)

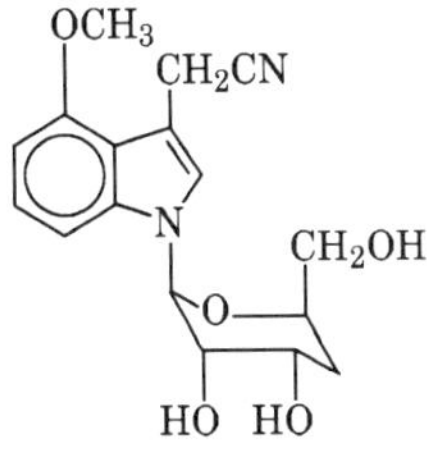

(55)

Another naturally occurring nucleoside antibiotic SF-2140 [*93207-27-3*], $C_{17}H_{20}N_2O_5$, a 3-cyanomethyl-4-methoxyindole nucleoside (**55**) is found to be as active as amantadine in protecting mice against an APR-8 strain of influenza A (120). This suggests that it may be possible to alter both the heterocyclic base and the carbohydrate moiety for greater specificity since no acute toxicity was observed by intraperitoneal injection of SF-2140 at 2 g/kg. These studies have not been confirmed by other laboratories.

Ribavirin and Structural Analogues. One of the broad-spectrum antiviral agents that emerged from ICN Pharmaceuticals is an azole ribonucleoside 1-β-D-ribofuranosyl-1,2,4-triazole-3-carboxamide, designated as ribavirin [*36791-04-5*] (**56**) (121) (Fig. 8). Ribavirin, $C_8H_{12}N_4O_5$, has been studied in more animals and against more viruses than any other antiviral agent known today (122). It is active in cell culture against approximately 85% of all viruses studied, although it is inactive against polio virus, certain coxsackie viruses, most corona viruses, pseudorabies virus, and HBV.

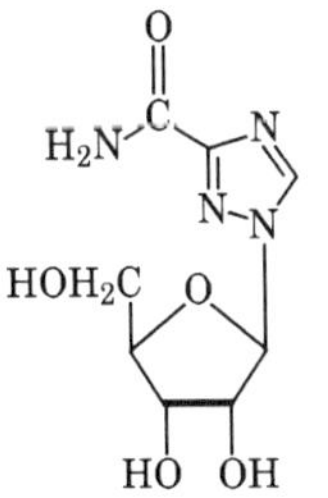

(56)

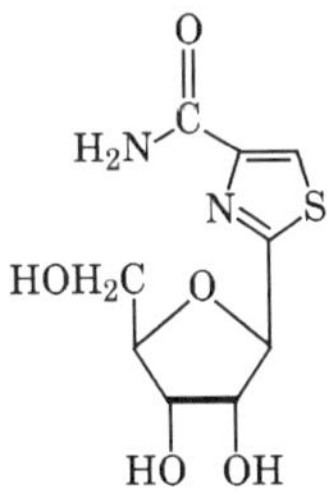

(57)

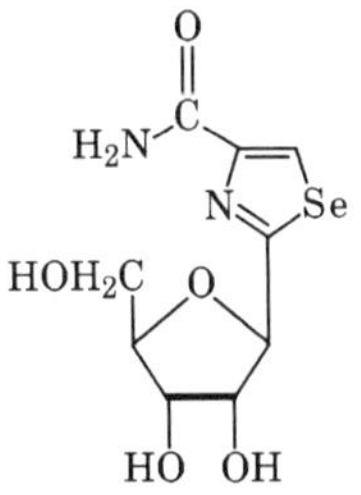

(58)

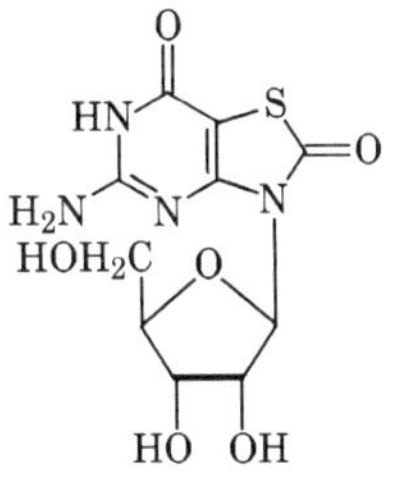

(59)

Fig. 8. Ribavirin (**56**) and active structural analogues.

Ribavirin is not incorporated into the DNA or RNA of either mammalian or viral systems. It has been shown (123), however, that a high dosage of ribavirin given over a prolonged period to Rhesus monkeys results in anemia of red blood cells. This effect is dose related and reversible upon cessation of treatment. Guanosine partially reverses the antiviral effect of ribavirin against certain viruses.

The antiviral mode of action of ribavirin was originally thought to be based on the inhibition of inosine monophosphate (IMP) dehydrogenase by ribavirin

5′-monophosphate, which results in a depletion of cellular guanosine triphosphate (GTP) pools. This lowering of cellular GTP is a self-potentiating effect since GTP competes with ribavirin 5′-triphosphate (RTP), the most abundant form of ribavirin in the cell, in the inhibition of viral-specific RNA polymerase by acting as a GTP or ATP substrate analogue. With certain RNA and DNA viruses, RTP also inhibits the viral-specific mRNA capping enzymes, guanylyl transferase, and N^7-methyl transferase, so that either no viral protein is transcribed or, if it is synthesized, it is nonfunctional (Fig. 9).

This multisite inhibition of viral mRNA processing by RTP appears to be a result of biochemical differences between the viral and the natural cellular enzymes involved in capping mRNA. The viral enzymes are much more strongly inhibited, consequently, preventing viral protein synthesis or giving rise to short segments of viral protein that are nonfunctional. All capped viral mRNAs contain a purine exclusively in the 5′ penultimate position (B_1 in Fig. 9) and initiation of viral mRNA synthesis appears to be purine specific. In mammalian cell mRNA, B_1 may be a pyrimidine as well as a purine, and most often B_1 is cytosine and B_2 uracil.

The nucleotide form of ribavirin does not manifest its antiviral activity simply by lowering the GTP levels, but may indeed participate directly in binding to specific G proteins (124). Ribavirin has recently been studied as an inhibitor of vesicular stomatitis virus and La Crosse virus (125). Of the phosphorylated forms of the drug, ribavirin-5′-diphosphate was by far the most potent inhibitor of viral replication for these two viruses.

Clinical trials in Brazil in 1977 of the oral administration of ribavirin prompted the Brazilian government to approve its use for treating infectious hepatitis A (126). Because ribavirin is water-soluble, it can be dispersed readily in a fine-particle aerosol. Clinical trials (127) indicate that such aerosols reduce the fever, the systemic illness, and the presence of virus in respiratory secretions in patients infected with influenza A or B viruses. It has been found that aerosolized ribavirin used to treat both young adults and infants suffering from RSV infections significantly reduces the severity of the disease (128). The same type of treatment has been used successfully for infants with severe combined immunodeficiency disease, life-threatening RSV, and parainfluenza-3 infections. Ribavirin aerosol has been used successfully to treat adenoviral pneumonia in young children. In 1985, the FDA approved the use of ribavirin aerosol for treating severe infections of RSV, a disease often fatal to infants and children. Ribavirin is being marketed by ICN Pharmaceuticals.

In monkeys, ribavirin has potent antiviral activity against human toga, bunya, and arena viruses. When used intravenously to treat comatose patients infected with Lassa Fever virus in Sierra Leone, ribavirin reduced mortality from an expected 80% to only 10% (129). Ribavirin also is active against various viruses that cause hemorrhagic fevers in mice and monkeys. In model infections with arenaviruses in guinea pigs and monkeys, ribavirin has shown both prophylactic and therapeutic efficacy (130). In the People's Republic of China, ribavirin significantly reduced mortality in hemorrhagic fever patients with renal syndrome (130). Ribavirin is highly effective against sandfly fever virus in humans (131). Its only adverse effect in humans is manageable, reversible anemia. Intra-

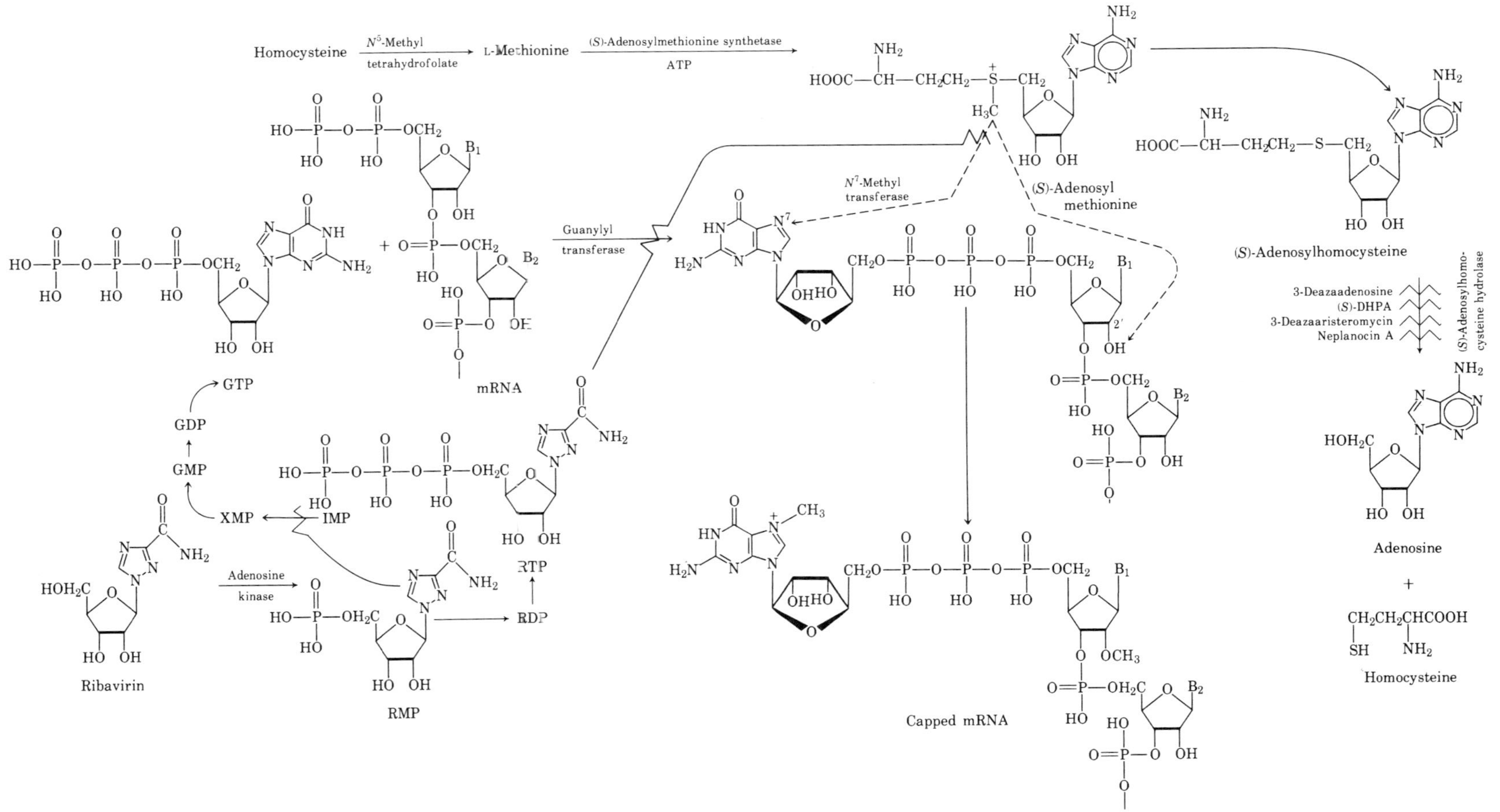

Fig. 9. Ribavirin 5′-monophosphate (RMP) strongly inhibits inosine monophosphate (IMP) dehydrogenase, preventing the conversion of IMP to xanthosine monophosphate (XMP). This, in turn, results in the lowering of guanosine triphosphate (GTP) levels within the cell. Ribavirin 5′-triphosphate (RTP) inhibits viral guanylyl transferase and viral N^7-methyl transferase, probably binding as a substrate analogue of GTP. The low levels of GTP accentuate the viral enzyme inhibition of RTP by competing poorly with RTP for the natural guanylate binding sites of viral-specific guanylyl transferase and N^7-methyl transferase. B_1 and B_2 are purines or pyrimidines as described in the text.

peritoneal treatment with ribavirin 2′,3′,5′-triacetate effected significant increases in both mean survival time and survival rate of dengue virus type-2 infected mice.

Unlike idoxuridine, BVdU, and acyclovir, viral strains susceptible to ribavirin have not been found to develop a resistance to the drug. The resistance against ribavirin is less likely because the drug exhibits multiple sites of antiviral action.

An azole nucleoside structurally related to ribavirin is tiazofurin [*60084-10-8*] (2-β-D-ribofuranosylthiazole-4-carboxamide, **57**) (Fig. 8). It was synthesized in 1977 (132). Tiazofurin, $C_9H_{12}N_2O_5S$, has shown good activity against tomato-spotted wilt virus and good *in vitro* activity against several RNA viruses (132,133), but the *in vivo* experiments against RNA viral-infected animals have been disappointing. However, tiazofurin is found to be efficacious in the treatment of yellow fever virus-infected mice. A combination of ribavirin and tiazofurin showed significant synergistic activity against a number of togaviruses, bunyaviruses, and arenaviruses *in vitro* (133).

Another azole nucleoside that was prepared in 1983 is selenazofurin [*83705-13-9*] (2-β-D-ribofuranosylselenazole-4-carboxamide, **58**) (134), which is structurally similar to ribavirin. *In vitro,* selenazofurin, $C_9H_{12}N_2O_5Se$, has a broad antiviral spectrum similar to ribavirin. It shows definite but marginal therapeutic effects against influenza A and B in mice (134). A combination of ribavirin and selenazofurin, used to treat guinea pigs infected with Pichinde virus, is superior to ribavirin alone and results in the survival of 80% of infected animals. Similarly, ribavirin in combination with selenazofurin has been shown to exhibit synergistic activity *in vitro* against mumps, measles, parainfluenza-3, and vaccinia viruses, as well as HSV-1 (135). It has been observed that selenazofurin has a positive effect against murine hepatitis infections in mice (136). Similarly, selenazofurin had significant effect when injected into rabies-infected mice by the intramuscular route (137).

Although the structures of ribavirin and selenazofurin are similar, they appear to exert their antiviral action at different enzyme sites along the same biochemical pathway. Selenazofurin forms the nicotinamide adenosine dinucleotide (NAD) analogue, which inhibits IMP dehydrogenase by binding in place of the NAD cofactor, and hence this potent reduction of guanylate pools is responsible for the antiviral effect of selenazofurin.

7-Thia-8-oxoguanosine {5-amino-3-β-D-ribofuranosylthiazolo[4,5-*d*]pyrimidine-2,7(3*H*,6*H*)-dione, **59**} [*122970-40-5*] synthesized by ICN Pharmaceuticals, (138) has shown broad-spectrum antiviral activity. 7-Thia-8-oxoguanosine, $C_{10}H_{12}N_4O_6S$, is highly active in mice and rats against Semliki Forest, San Angelo, benzi, rat corona, and encephalomyocarditis viruses when administered intraperitoneally before exposure to the virus (139). The compound was moderately effective in mice infected intraperitoneally with HSV-2 or intranasally with vesicular stomatitis virus. The mode of antiviral action of (**59**) *in vivo* may be due in part to the induction of interferon α (138).

The treatment of Semliki Forest, San Angelo, and benzi viruses in mice suggests 7-thia-8-oxoguanosine (**59**) may be useful in treating arthropod-borne diseases that are endemic to Asia, the Pacific, and Africa.

Agents Active Against Persistent Viral Infections

Persistent viral infection is a difficult challenge for antiviral chemotherapy. Retroviruses as a class are often found to be responsible for persistent viral infections. Retroviruses are unique RNA viruses characterized by the transcription of their single-stranded RNA into the double-stranded DNA of the host cell using the viral enzyme reverse transcriptase. AIDS is an example of such a persistent and latent human viral infection. Such viral infections may indeed be responsible for multiple sclerosis, Alzheimer's disease, and many other dementia of unknown etiology (140). The high incidence of AIDS dementia is due to direct brain infection by human immunodeficiency virus (HIV) (141). Controlling such persistent viral infections by antiviral chemotherapy would require penetration of the blood–brain barrier and prolonged treatment with a relatively nontoxic agent.

Following the identification of a retrovirus, HIV, as the etiological agent of AIDS (142–144), an intense effort has been made to identify drugs for the treatment or prevention of this debilitating, lethal disease (145–147). Several 2′,3′-dideoxyribonucleosides of purine and pyrimidine were discovered to be potent inhibitors of HIV replication *in vitro* (148–150). Considerable data has also accumulated on *in vitro* antiHIV testing of acyclic and carbocyclic nucleoside analogues (151). Although more than 100 nucleosides have been shown to have an *in vitro* antiHIV activity commensurate with development to clinical trials (150,152), only 3′-azido-3′-deoxythymidine [*30516-87-1*] (retrovir, zidovudine, AZT, **60**) $C_{10}H_{13}N_5O_4$, has become widely available for the treatment of AIDS (153,154), and approved by the FDA for treatment of advanced AIDS cases. Although clinically useful in many settings, AZT is associated with substantial toxicity, especially in myelosuppression, leading to serious hematological side effects (155). AZT was first prepared in 1964 (156), and the *in vitro* activity against HIV was initially described in 1985 (157). AZT-resistant HIV strains have been isolated from AIDS patients (158).

Basically, AZT is anabolically phosphorylated to AZT mono-, di-, and triphosphates by various enzymes (kinases) of a target cell (159). AZT-triphosphate competes with other phosphorylated pyrimidine nucleosides for incorporation into HIV DNA by the viral reverse transcriptase. Incorporation of the AZT-triphosphate into reverse transcriptase results in viral DNA chain termination. Reverse transcriptase is essential in the replicative cycle of HIV.

Optimism certainly exists that similar 2′,3′-dideoxynucleosides (Fig. 10) also will prove clinically effective in the treatment of HIV infection but complete clinical data are not yet available. It has been shown that at 10 μM 2′,3′-dideoxyadenosine [*4097-22-7*] (DDA, **61**), first prepared in 1964 (160), completely protected ATH8 cells exposed to HIV and enabled the cells to survive and grow. Although DDA, $C_{10}H_{13}N_5O_2$, is a potent DNA chain terminator of *E. coli,* resistance develops rapidly. Similar inhibition is also shown by 2′,3′-dideoxyguanosine [*85326-06-3*], $C_{10}H_{13}N_5O_3$, (161). 2′,3′-Dideoxycytidine [*7481-89-2*] (DDC, **63**), first synthesized in 1967 (162), reportedly (161) inhibits HIV at 0.5 μM and is presently undergoing clinical trials at NIH in AIDS patients. The principal problem with dideoxynucleosides is that they may be converted within certain cells to the

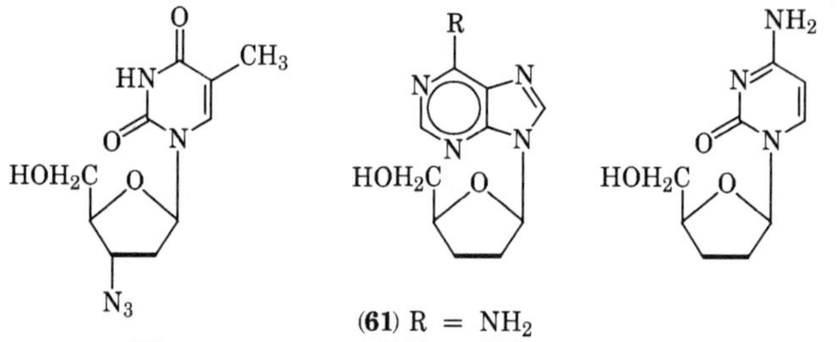

(60) **(61)** R = NH_2 **(62)** R = OH **(63)**

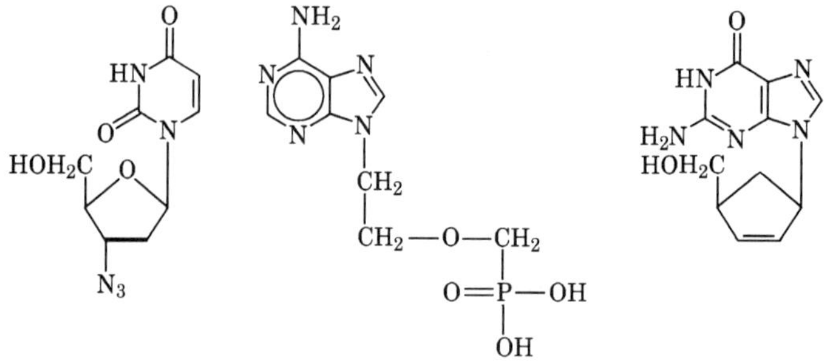

(64) **(65)** **(66)**

Fig. 10. 2′,3′-Dideoxynucleosides active against human immunodeficiency virus (HIV).

corresponding 5′-triphosphates, which could terminate host DNA synthesis or be incorporated into host DNA (163). The conversion of DDC, $C_9H_{13}N_3O_3$, to the 5′-triphosphate takes place in several mammalian cells and is not solely a property of HIV infected T-lymphocytes (164). Unlike AZT, DDC does not have significant hematologic toxicity and has not shown cross-resistance with AZT *in vitro*. However, DDC has been associated with a severe dose-related peripheral neuropathy. A combination therapy of AZT and DDC in patients with AIDS and advanced AIDS-related complex has been reported (165).

In a long-term clinical study (166) with 58 patients, the experimental drug 2′,3′-dideoxyinosine [*69655-05-6*] (DDI, **62**), was found to improve signs of immune functions in most patients with AIDS and AIDS-related complex without causing severe side effects, and it is at least as efficacious as AZT. In this study, 80% of AIDS patients were alive after 21 months of DDI treatment, compared with about 50% similarly treated with AZT. The common dosage level of DDI, $C_{10}H_{12}N_4O_3$, is 500–700 mg/d. Although none of the patients died from drug toxicity, DDI shows mild side effects such as insomnia, anxiety, irritability, and headaches. Other side effects associated with DDI include painful nerve damage to the feet, and occasionally pancreatitis, an inflammation of the pancreas. DDI, however, causes little or no bone marrow suppression. A combination of DDI and AZT to fight AIDS may be useful in achieving desired effects with lower doses of each drug and lessening overall side effects. DDI is being developed by Bristol-Myers Squibb Co. After an expedited review, the U.S. Food and Drug Administration (FDA) recently granted approval to Bristol-Myers Squibb Co. to market DDI (Videx) for the treatment of AIDS. The drug, approved initially for narrow use, can be used to treat both adult and pediatric AIDS patients who are intolerant to AZT.

Another dideoxypyrimidine nucleoside active against human immunodeficiency virus is 3′-azido-2′,3′-dideoxyuridine [*84472-85-5*] (AZDU or CS-87, **64**) $C_9H_{11}N_5O_4$. Since its synthesis, (167) CS-87 has been identified as a promising antiHIV agent (168) and is currently undergoing phase I clinical trials in patients with AIDS and AIDS-related complex. It appears to be less potent than AZT against HIV in a peripheral blood mononuclear (PBM) cell screening system and in MT-4 cell lines. This lower activity in PBM cells appears to be related to a lower affinity of CS-87 for the enzyme responsible for its initial phosphorylation (169). However, CS-87 has significantly lower toxicity on bone marrow cells than AZT (170) and penetration of the CNS as a 5′-dihydropyridine derivative.

All 2′,3′-dideoxynucleoside analogues are assumed to be intracellularly phosphorylated to their active form (5′-triphosphate), and then targeted at the virus-associated reverse transcriptase. The rate and extent of the 2′,3′-dideoxynucleosides phosphorylate to the 5′-triphosphates may be of equal or greater importance than the differences in the relative abilities of these 5′-triphosphates to inhibit the viral reverse transcriptase (171). At the level of viral reverse transcriptase, the 5′-triphosphate of AZT and other dideoxynucleosides may either serve as a competitive inhibitor with respect to the natural substrates or may act as an alternate substrate, thus leading to chain termination (172).

9-(2-Phosphonylmethoxyethyl)adenine [*106941-25-7*] (PMEA, **65**) (173), synthesized in 1987 (174), is foremost among the acyclic nucleoside analogues proven to be effective inhibitors of HIV-1 replication. The *in vitro* potency and selectivity of PMEA is comparable to the antiHIV-1 potency and selectivity of 2′,3′-dideoxyadenosine (175). Although less potent than AZT *in vitro,* PMEA, $C_8H_{12}N_5O_4P$, is markedly more potent than AZT as an *in vivo* inhibitor of retrovirus replication (176). In fact, PMEA has proven efficacious in the treatment of murine, feline, and simian retrovirus infections in mice, cats, and monkeys, respectively.

The mode of action of PMEA may be quite similar to the mechanism by which (*S*)-HPMPA accomplishes its selective inhibitory activity against herpes viruses. For PMEA to reach its active triphosphate form, it needs only two phosphorylation steps. The triphosphate derivative of PMEA has a much stronger affinity for HIV-1 reverse transcriptase than for cellular DNA polymerases (175). Whether it is actually incorporated into DNA and terminates the growing DNA chain is currently under investigation.

Carbocyclic 2′,3′-didehydro-2′,3′-dideoxyguanosine [*118353-05-2*] (carbovir, CBV, **66**), $C_{11}H_{13}N_5O_2$, synthesized in 1988 (177), is a promising candidate for the chemotherapy of AIDS. CBV inhibits HIV replication and HIV-induced cytopathic effects in a variety of human T-lymphoblastoid cell lines at concentrations approximately two hundred- to four hundredfold below its cytotoxic concentrations (177). CBV is as effective as AZT and DDC in reducing the expression of viral antigen in HIV-infected CEM cells (177). The antiviral potency and selectivity of carbovir is comparable to the anti-HIV-1 potency and selectivity of 2′,3′-dideoxyadenosine (178). The exact mode of antiviral action of carbovir has not yet been elucidated, but may be the modulating effect of intracellular nucleotides on 5′-nucleotidase activity (179).

The initial antiretroviral activity observed with ribavirin (180,181) prompted a study (182) of ribavirin against HIV *in vitro*. Viral suppression at 50 μg/mL and considerable lowering of reverse transcriptase activity in the presence of ribavirin was reported. Phase I oral dose finding clinical studies of ribavirin in 23 patients with lymphadenopathy syndrome (LAS) showed decreased viral replication and enhanced immune function (183) at several dose levels. Phase II clinical trials of ribavirin in the treatment of patients with LAS have shown that at 800 mg/d of ribavirin in a randomized, double-blind, placebo-controlled trial, 53 patients did not develop AIDS during a 24 week study, whereas 9 of 56 patients developed AIDS in the placebo group during the same period (184). Open clinical studies in five AIDS patients treated with 1200 mg, twice daily for two weeks, showed that in four of five AIDS patients, T4 cells doubled and then returned to the baseline level once the drug was discontinued (185). Also, six of nine patients initially positive for HIV in blood became negative during ribavirin treatment (186). A prolonged regimen of a three-day loading dose followed by 600 mg of ribavirin was given to patients for one year. Ribavirin may suppress the HIV virus and produce clinical benefits without the significant bone marrow suppressive effects caused by other drugs such as zidovudine (185).

The search for new antiviral agents is ongoing and extensive; not all viruses have been included here, and new viruses pathogenic to humans will continue to be identified. Novel nucleosides as well as nonnucleosidic compounds which possess greater potency and enzymic specificity for the future treatment of both acute and persistent human viral infection will continue to be discovered.

NOMENCLATURE

acyclovir	9-(2-hydroxyethoxymethyl)guanine
AHPA	adeninylhydroxypropanoic acid
AIDS	acquired immunodeficiency syndrome
ara-A	9-β-D-arabinofuranosyladenine
ara-AMP	5′-monophosphate of ara-A
ara-C	1-β-D-arabinofuranosylcytosine
ara-Hx	9-β-D-arabinofuranosylhypoxanthine
ara-HxMP	5′-monophosphate of ara-Hx
ara-T	1-β-D-arabinofuranosylthymine
ara-U	1-β-D-arabinofuranosyluracil
AZT	3′-azido-3′-deoxythymidine
BHCG	(1α,2β,3α)-9-[2,3-bis(hydroxymethyl)cyclobutyl]guanine
BVdU	(*E*)-5-(2-bromovinyl)-2′-deoxyuridine
carbovir	carbocyclic 2′,3′-didehydro-2′,3′-dideoxyguanosine
CS-87	3′-azido-2′,3′-dideoxyuridine
DDA	2′,3′-dideoxyadenosine
DDC	2′,3′-dideoxycytidine
DDI	2′,3′-dideoxyinosine
DHBG	9-(3,4-dihydroxybutyl)guanine
DHPA	9-(2,3-dihydroxypropyl)adenine
DHPG	9-(1,3-dihydroxy-2-propoxymethyl)guanine
DMSO	dimethyl sulfoxide
DNA	2′-deoxyribonucleic acid
EBV	Epstein-Barr virus

FEAU	2′-deoxy-2′-fluoro-5-ethylarabinosyluracil
FIAC	2′-deoxy-2′-fluoro-5-iodoarabinosylcytosine
FMAU	2′-deoxy-2′-fluoro-5-methylarabinosyluracil
FMDV	foot and mouth disease virus
F_3TdU	5-trifluoromethyl-2′-deoxyuridine
GMP	guanosine 5′-monophosphate
GTP	guanosine 5′-triphosphate
HBB	α-hydroxybenzylbenzimidazole
HBV	hepatitis B virus
HCMV	human cytomegalovirus
HIV	human immunodeficiency virus
HPMPA	9-(3-hydroxy-2-phosphonylmethoxypropyl)adenine
HSV-1	herpes simplex type-1 virus
HSV-2	herpes simplex type-2 virus
IdU	5-iodo-2′-deoxyuridine
IMP	inosine 5′-monophosphate
LAS	lymphadenopathy syndrome
NAD	nicotinamide adenosine dinucleotide
NDV	Newcastle disease virus
PAA	phosphonoacetic acid
PFA	phosphonoformic acid
PMEA	9-(2-phosphonylmethoxyethyl)adenine
ribavirin	1-β-D-ribofuranosyl-1,2,4-triazole-3-carboxamide
RNA	ribonucleic acid
mRNA	messenger RNA
RSV	respiratory syncytial virus
RTP	ribavirin 5′-triphosphate
selenazofurin	2-β-D-ribofuranosylselenazole-4-carboxamide
tiazofurin	2-β-D-ribofuranosylthiazole-4-carboxamide
VSV	vesicular stomatitis virus
WHV	woodchuck hepatitis virus
xylo-G	9-β-D-xylofuranosylguanine

BIBLIOGRAPHY

1. R. K. Robins, *Chem. Eng. News* **64,** 28 (1986).
2. W. M. Mitchell, in W. A. Carter, ed., *Selective Inhibitors of Viral Functions,* CRC Press, Cleveland, Ohio, 1973, p. 51.
3. B. R. McAuslan, *Life Sci.* **14,** 2085 (1974).
4. F. Fenner, *Intervirology* **7,** 4 (1976).
5. R. W. Sidwell and J. T. Witkowski in M. E. Wolff, ed., *Burger's Medicinal Chemistry,* 4th ed., John Wiley & Sons, Inc., New York, 1979, pp. 543–593.
6. E. Helgstrand and co-workers, *Science* **201,** 819 (1978).
7. B. Öberg, *Pharmac. Ther.* **40,** 213 (1989).
8. D. W. Hutchinson and S. Masson, *IRCS Med. Sci.* **14,** 176 (1986).
9. L. Vrang and B. Öberg, *Antimicrob. Agents Chemother.* **29,** 867 (1986).
10. C. J. Pfau, *Handbook Exp. Pharmacol.* **61,** 147 (1982).

11. J. A. Cooper, B. Moss, and E. Katz, *Virology* **96,** 381 (1979).
12. D. W. Hutchinson, *Antiviral Res.* **5,** 193 (1985).
13. D. M. McLean, *Ann. N.Y. Acad. Sci.* **284,** 118 (1977).
14. K. B. Easterbrook, *Virology* **17,** 245 (1962).
15. A. Kint and L. Verlinden, *New Engl. J. Med.* **291,** 308 (1974).
16. R. J. O'Reilly, A. Chibbaro, R. Wilmot, and C. Lopez, *Ann. N.Y. Acad. Sci.* **284,** 161 (1977).
17. J. L. Melnick and C. Wallis, *Ann. N.Y. Acad. Sci.* **284,** 171 (1977).
18. F. Rapp, *J. Am. Med. Assoc.* **225,** 456 (1973).
19. L. Weinstein and T. Chang, *New Engl. J. Med.* **289,** 725 (1973).
20. E. G. Friedrick, *J. Obstetr. Gynecol.* **43,** 304 (1974).
21. P. La Colla and co-workers, *Ann. N.Y. Acad. Sci.* **284,** 294 (1977).
22. K. K. Gauri and B. Rohde, *Klin.-Ther. Wochenschr.* **47,** 375 (1969).
23. W. H. Prusoff, *Biochim. Biophys. Acta.* **32,** 295 (1959).
24. H. E. Kaufman, *Proc. Soc. Exp. Biol. Med.* **109,** 251 (1962).
25. W. H. Prusoff and D. C. Ward, *Biochem. Pharmacol.* **25,** 1233 (1976).
26. J. Sugar and H. E. Kaufman in Ref. 2, p. 295.
27. Y-C. Cheng and co-workers, *J. Virol.* **15,** 1284 (1975).
28. D. M. Albert and co-workers, *J. Invest. Ophthalmol.* **15,** 470 (1976).
29. M. S. Chen, D. F. Ward, and W. H. Prusoff, *J. Biol. Chem.* **251,** 4833 (1976).
30. C. Heidelberger, D. G. Parsons, and D. C. Remy, *J. Am. Chem. Soc.* **84,** 3597 (1962).
31. A. S. Jones, G. Verhelst, and R. T. Walker, *Tetrahedron Lett.,* 4415 (1979).
32. S. Olofsson, M. Lundström, and R. Datema, *Virology* **147,** 201 (1985).
33. B. E. Juel-Jensen, *Brit. Med. J.* **1,** 406 (1973).
34. C. B. Lauter, C. J. Bailey, and A. M. Lerner, *Antimicrob. Agents Chemother.* **6,** 598 (1974).
35. R. W. Sidwell and co-workers, *Nature New Biol.* **242,** 204 (1973).
36. K. A. Watanabe and co-workers, *J. Med. Chem.* **22,** 21 (1979).
37. O. Hantz and co-workers, *Antiviral Res.* **4,** 187 (1984).
38. I. Fourel and co-workers, *Antimicrob. Agents Chemother.* **34,** 473 (1990).
39. J. J. Fox, N. Yung, and A. Bendich, *J. Am. Chem. Soc.* **79,** 2775 (1957).
40. G. A. Galegov, R. M. Bikbulatov, K. A. Vanag, and R. M. Shen, *Vop. Virusol.* **13,** 18 (1968).
41. R. E. Handschumaker and C. A. Pasternak, *Biochim. Biophys. Acta* **30,** 451 (1958).
42. H. E. Renis, *Antimicrob. Agents Chemother.* **13,** 613 (1978).
43. B. J. Sloan in D. Pavan-Langston, R. A. Buchanan, and C. A. Alford Jr., eds., *Adenine Arabinoside: An Antiviral Agent,* Raven Press, New York, 1975, p. 45.
44. W. W. Lee, A. Benetz, L. Goodman, and B. R. Baker, *J. Am. Chem. Soc.* **82,** 2648 (1960).
45. D. C. Baker, T. H. Haskell, S. R. Putt, and B. J. Sloan, *J. Med. Chem.* **22,** 273 (1979).
46. J. T. Richards, E. R. Kern, C. J. Overall Jr., and L. A. Glasgow, *Antiviral Res.* **2,** 27 (1982).
47. W. M. Shannon and co-workers, *Antimicrob. Agents Chemother.* **24,** 706 (1983).
48. G. R. Revankar and co-workers, *J. Med. Chem.* **18,** 721 (1975).
49. D. Derse, K. F. Bastow, and Y. C. Cheng, *J. Biol. Chem.* **257,** 10251 (1982).
50. J. C. Drach, in E. De Clercq and R. T. Walker, eds., *Targets for Design of Antiviral Agents,* Plenum Press, New York, 1984, p. 234.
51. R. Vince and S. Daluge, *J. Med. Chem.* **20,** 612 (1977).
52. R. Vince and co-workers, *Science* **221,** 1405 (1983).
53. E. De Clercq and co-workers, *J. Med. Chem.* **30,** 481 (1987).
54. E. De Clercq and M. J. Robins, *Antimicrob. Agents Chemother.* **30,** 719 (1986).
55. A. Holy, *Chemica Scripta* **26,** 83 (1986).
56. E. De Clercq and co-workers, *Nature* **323,** 464 (1986).

57. P. C. Maudgal, E. De Clercq, and P. Huyghe, *Invest. Ophthalmol. Vis. Sci.* **28,** 243 (1987).
58. E. De Clercq, in E. De Clercq and R. T. Walker, eds., *Antiviral Drug Development—A Multidisciplinary Approach,* Plenum Press, New York, 1987, pp. 97–122.
59. H. J. Schaeffer, D. Vogel, and R. Vince, *J. Med. Chem.* **8,** 502 (1965).
60. A. Holy, *Collect. Czech. Chem. Commun.* **40,** 187 (1975).
61. I. Votruba and A. Holy, *Collect. Czech. Chem. Commun.* **45,** 3039 (1980).
62. E. De Clercq and A. Holy, *J. Med. Chem.* **28,** 282 (1985).
63. L. M. Lerner and R. R. Rossi, *Biochemistry* **11,** 2272 (1972).
64. H. J. Schaeffer, S. Gurwara, R. Vince, and S. Bittner, *J. Med. Chem.* **14,** 367 (1971).
65. H. J. Schaeffer and D. F. Schwender, *J. Med. Chem.* **17,** 6 (1974).
66. H. J. Schaeffer and co-workers, *Nature* **272,** 583 (1978).
67. D. Gold and L. Corey, *Antimicrob. Agents Chemother.* **31,** 361 (1987).
68. R. J. Whitley and co-workers, *New Eng. J. Med.* **314,** 144 (1986).
69. O. L. Laskin and co-workers, *Antimicrob. Agents Chemother.* **21,** 393 (1982).
70. P. de Miranda and co-workers, *Clin. Pharmacol. Ther.* **26,** 718 (1979).
71. O. L. Laskin, *Clin. Pharmacokinet.* **8,** 187 (1983).
72. G. B. Elion, *Am. J. Med.* **73,** 7 (1982).
73. K. K. Ogilvie and co-workers, *Can. J. Chem.* **60,** 3005 (1982).
74. K. O. Smith, K. S. Galloway, K. K. Ogilvie, and U. O. Cherujan, *Antimicrob. Agents Chemother.* **22,** 1026 (1982).
75. J. C. Martin and co-workers, *J. Med. Chem.* **26,** 759 (1983).
76. D. F. Smee, J. C. Martin, J. P. H. Verheyden, and T. R. Matthews, *Antimicrob. Agents Chemother.* **23,** 676 (1983).
77. A. K. Field and co-workers, *Proc. Natl. Acad. Sci. USA* **80,** 4139 (1983).
78. K. B. Frank, J. F. Chiou, and Y. C. Cheng, *J. Biol. Chem.* **259,** 1556 (1984).
79. O. L. Laskin, C. M. Stahl-Bayliss, C. M. Kalman, and L. R. Rosecan, *J. Infect. Dis.* **155,** 323 (1987).
80. E. C. Reed, P. S. Dandliker, and J. D. Meyers, *Ann. Intern. Med.* **105,** 214 (1986).
81. R. L. Tolman and co-workers, *Biochem. Biophys. Res. Commun.* **128,** 1329 (1985).
82. J. Germershausen and co-workers, *Antimicrob. Agents Chemother.* **29,** 1025 (1986).
83. J. D. Karkas and co-workers, *Biochim. Biophys. Acta* **911,** 127 (1987).
84. M. MacCoss and co-workers, *Chemica Scripta* **26,** 113 (1986).
85. A. Larsson and co-workers, *Antimicrob. Agents Chemother.* **23,** 664 (1983).
86. A. K. Field and co-workers, *Antiviral Res.* **13,** 41 (1990).
87. M. Legraverend, R-M. N. Ngongo-Tekam, E. Bisagni, and A. Zerial, *J. Med. Chem.* **28,** 1477 (1985).
88. A. Zerial, M. Zerial, M. Legraverend, and E. Bisagni, *Ann. Inst. Pasteur/Virol.* **137E,** 317 (1986).
89. G. R. Revankar and co-workers, *J. Med. Chem.* **19,** 1026 (1976).
90. B. Loddo, *Pharmac. Ther.* **10,** 431 (1983).
91. I. Tamm and L. A. Caliguiri, in D. J. Bauer, ed., *International Encyclopedia of Pharmacology and Therapeutics,* Vol. 1, Pergamon Press, Oxford, UK, 1972, p. 115.
92. N. L. Shipkowitz and co-workers, *Appl. Microbiol.* **23,** 117 (1972).
93. K. Gerzon and co-workers, *J. Med. Chem.* **6,** 760 (1963).
94. A. W. Galbraith, *Brit. Med. Bull.* **41,** 381 (1985).
95. S. L. Barriere, *Pharm. Int.,* 139 (1984).
96. A. J. Hay, A. J. Wolstenholme, J. J. Skehel, and M. H. Smith, *EMBO J.* **4,** 3021 (1985).
97. D. L. Swallow and G. L. Kampfner, *Brit. Med. Bull.* **41,** 322 (1985).
98. J. J. McSharry, L. A. Caliguiri, and H. J. Eggers, *Virology* **97,** 307 (1979).
99. G. D. Diana and co-workers, *J. Med. Chem.* **28,** 748 (1985).
100. G. D. Diana, M. J. Otto, and M. A. McKinlay, *Pharmac. Ther.* **29,** 287 (1985).

101. G. D. Diana and co-workers, *J. Med. Chem.* **32,** 450 (1989).
102. R. J. Phillpotts and co-workers, *Lancet,* 1342 (1981).
103. S. M. Roberts, in Ref. 58, pp. 37–54.
104. K. Andries and co-workers, *Arch. Virol.* **101,** 155 (1988).
105. R. J. Rousseau, L. B. Townsend, and R. K. Robins, *Biochemistry* **5,** 756 (1966).
106. A. J. Bodner, G. L. Cantoni, and P. K. Chiang, *Biochem. Biophys. Res. Commun.* **98,** 476 (1981).
107. J. P. Bader, N. R. Brown, P. K. Chiang, and G. L. Cantoni, *Virology* **89,** 494 (1978).
108. J. A. Montgomery and co-workers, *J. Med. Chem.* **25,** 626 (1982).
109. E. De Clercq and J. A. Montgomery, *Antiviral Res.* **3,** 17 (1983).
110. M. Hayashi, S. Yaginuma, H. Yoshioka, and K. Nakatsu, *J. Antibiot.* **34,** 675 (1981).
111. M. Arita and co-workers, *J. Am. Chem. Soc.* **105,** 4049 (1983).
112. E. De Clercq, *Antimicrob. Agents Chemother.* **28,** 84 (1985).
113. R. L. Hamill and M. M. Holhn, *J. Antibiot.* **26,** 463 (1973).
114. U.S. Pat. 4,158,056 (July 13, 1978), R. Nagarajan.
115. P. D. Cook and co-workers, *J. Am. Chem. Soc.* **98,** 1492 (1976).
116. L. B. Allen and co-workers, *Antimicrob. Agents Chemother.* **12,** 114 (1977).
117. T. M. Page and co-workers, *Int. J. Biochem.* **18,** 957 (1986).
118. P. G. Canonico, P. B. Jahrling, and W. L. Pannier, *Antiviral Res.* **2,** 331 (1983).
119. C. R. Petrie and co-workers, *J. Med. Chem.* **29,** 268 (1986).
120. T. Ito and co-workers, *J. Antibiot.* **37,** 931 (1984).
121. J. T. Witkowski, R. K. Robins, R. W. Sidwell, and L. N. Simon, *J. Med. Chem.* **15,** 1150 (1972).
122. R. W. Sidwell, G. R. Revankar, and R. K. Robins, in D. Shugar, ed., *Intl. Enc. Pharmacol. Ther.* Sec. 116, Vol. 2, Pergamon Press, New York, 1985, pp. 49–108.
123. P. G. Canonico and co-workers, *Toxicol. Applied Pharmacol.* **74,** 155, 163 (1984).
124. R. K. Robins and co-workers, *Adv. Enzyme Reg.* **29,** 97 (1989).
125. P. Toltzis, K. O'Connell, and J. L. Patterson, *Antimicrob. Agents Chemother.* **32,** 492 (1988).
126. P. A. A. Galvao and I. O. Castro, *Ann. N.Y. Acad. Sci.* **284,** 278 (1977).
127. V. Knight and co-workers, *Lancet,* 945–949 (Oct. 31, 1981).
128. C. B. Hall and co-workers, *New Eng. J. Med.* **308,** 1443 (1983).
129. J. B. McCormick and co-workers, *New Engl. J. Med.* **314,** 20 (1986).
130. J. W. Huggins, *Rev. Inf. Dise.* **11**(Suppl. 4), S750 (1989).
131. J. W. Huggins, G. R. Kim, and K. T. McKee Jr., *VI International Congress of Virology,* Sendai, Japan, Sept. 1984.
132. P. C. Srivastava and co-workers, *J. Med. Chem.* **20,** 256 (1977).
133. J. W. Huggins, R. K. Robins, and P. G. Canonico, *Antimicrob. Agents Chemother.* **26,** 476 (1984).
134. P. C. Srivastava and R. K. Robins, *J. Med. Chem.* **26,** 445 (1983).
135. J. J. Kirsi and co-workers, *Antimicrob. Agents Chemother.* **26,** 466 (1984).
136. R. W. Sidwell and co-workers, *Antiviral Res.* **6,** 343 (1986).
137. F. Bussereau, M. Picard, J. Blancou, and P. Sureau, *Acta Virol.* **32,** 33 (1988).
138. K. Nagahara and co-workers, *J. Med. Chem.* **33,** 407 (1990).
139. D. F. Smee and co-workers, *Antimicrob. Agents Chemother.* **33,** 1487 (1989).
140. A. T. Haase, *J. Infect. Dis.* **153,** 441 (1986).
141. B. A. Navia, B. D. Jordan, and R. W. Price, *Ann. Neurol.* **17,** 517 (1986).
142. F. Barr'e-Sinoussi and co-workers, *Science* **220,** 868 (1983).
143. R. C. Gallo and co-workers, *Science* **224,** 500 (1984).
144. J. A. Levy and co-workers, *Science* **225,** 840 (1984).
145. S. Broder, H. Mitsuya, R. Yarchoan, and G. N. Pavlakis, *Annals Int. Med.* **113,** 604 (1990).

146. E. De Clercq, ed., in *Design of Anti-Aids Drugs,* Elsevier, New York, 1990, pp. 1–24.
147. H. Mitsuya, T. Shirasaka, and S. Broder, in Ref. 146, pp. 25–61.
148. P. Pauwels and co-workers, *Biochem. Pharmacol.* **37,** 1317 (1988).
149. P. Herdewijn and E. De Clercq, in Ref. 146, pp. 141–174.
150. M. Nasr, C. Litterst, and J. McGowan, *Antiviral Res.* **14,** 125 (1990).
151. J. Balzarini and E. De Clercq, in Ref. 146, pp. 175–194.
152. M. S. Hirsch and J. C. Kaplan, *Antimicrob. Agents Chemother.* **31,** 839 (1987).
153. H. D. Langtry and D. M. Campoli-Richards, *Drugs* **37,** 408 (1989).
154. S. Broder, *Med. Res. Reviews* **10,** 419 (1990).
155. D. D. Richman and co-workers, *N. Engl. J. Med.* **317,** 192 (1987).
156. J. P. Horwitz, J. Chua, and M. Noel, *J. Org. Chem.* **29,** 2076 (1964).
157. H. Mitsuya and co-workers, *Proc. Natl. Acad. Sci. USA* **82,** 7096 (1985).
158. B. A. Larder, G. Darby, and D. D. Richman, *Science* **243,** 1731 (1989).
159. R. Yarchoan and S. Broder, *N. Engl. J. Med.* **316,** 557 (1987).
160. M. J. Robins and R. K. Robins, *J. Am. Chem. Soc.* **86,** 3585 (1964).
161. H. Mitsuya and S. Broder, *Proc. Natl. Acad. Sci. USA* **83,** 1911 (1986).
162. J. P. Horwitz, J. Chua, M. Noel, and J. T. Donatti, *J. Org. Chem.* **32,** 817 (1967).
163. M. C. Starnes and J. C. Cheng, *J. Biol. Chem.* **262,** 988 (1987).
164. D. A. Cooney and co-workers, *Biochem. Pharmacol.* **35,** 2065 (1986).
165. T-C. Meng, M. A. Fischl, and D. D. Richman, *Am. J. Med.* **88**(5B), 27S (1990).
166. R. Yarchoan and co-workers, *Lancet* **336,** 526 (1990).
167. T-S. Lin and W. R. Mancini, *J. Med. Chem.* **26,** 544 (1983).
168. C. K. Chu and co-workers, *J. Med. Chem.* **32,** 612 (1989).
169. B. F. H. Eriksson, C. K. Chu, and R. F. Schinazi, *Antimicrob. Agents Chemother.* **32,** 1729 (1989).
170. C. K. Chu and co-workers, *J. Med. Chem.* **33,** 2188 (1990).
171. Z. Hao and co-workers, *Mol. Pharmacol.* **34,** 432 (1988).
172. M. H. St. Clair and co-workers, *Antimicrob. Agents Chemother.* **31,** 1972 (1987).
173. R. Pauwels and co-workers, *Antimicrob. Agents Chemother.* **32,** 1025 (1988).
174. A. Holy and I. Rosenberg, *Collect. Czech. Chem. Commun.* **52,** 2801 (1987).
175. E. De Clercq, in V. Claassen, ed., *Trends in Drug Research,* Elsevier Science Publishers, Amsterdam, The Netherlands, 1990, p. 133.
176. J. Balzarini and co-workers, *Proc. Natl. Acad. Sci. USA* **86,** 332 (1989).
177. R. Vince and co-workers, *Biochem. Biophys. Res. Commun.* **156,** 1046 (1988).
178. R. Vince and M. Hua, *J. Med. Chem.* **33,** 17 (1990).
179. L. L. Bondoc, Jr. and co-workers, *Biochemistry* **29,** 9839 (1990).
180. W. M. Shannon, *Ann. NY Acad. Sci.* **284,** 472 (1977).
181. R. W. Sidwell and D. F. Smee, *Antiviral Res.* **1,** 47 (1981).
182. J. B. McCormick, S. W. Mitchell, J. P. Getchell, and D. R. Hicks, *Lancet,* 1367 (Dec. 15, 1984).
183. R. B. Roberts and co-workers, *Clin. Res.* **34,** 678 (1986).
184. R. B. Roberts and co-workers, *J. AIDS* **3,** 884 (1990).
185. C. Crumpacker and co-workers, *Int. Conf. on AIDS,* Paris, France, June 1986, Abstr. No. 84, s14b.
186. C. Crumpacker and co-workers, *Ann. Int. Med.* **107,** 664 (1987).

Ganapathi R. Revankar
Triplex Pharmaceutical Corporation

Roland K. Robins
ICN Nucleic Acid Research Institute

AQUACULTURE CHEMICALS

Intensive or extensive culture of aquatic animals requires chemicals that control disease, enhance the growth of cultured species, reduce handling trauma to organisms, improve water quality, disinfect water, and control aquatic vegetation, predaceous insects, or other nuisance organisms. The aquacultural chemical need for various species have been described for rainbow trout, *Oncorhynchus mykiss* (1); Atlantic and Pacific salmon, *Salmo salar* and *Oncorhynchus* sp. (2); channel catfish, *Ictalurus punctatus* (3); striped bass, *Morone saxatilis* (4); milkfish, *Chanos chanos* (5); mollusks (6); penaeid (*Penaeus* sp.) shrimp (7); and a variety of other marine species (8).

Laws and regulations on the use of chemicals in aquaculture vary by country and serve to ensure safe and effective use and protection of humans and the environment. Regulations and therapeutants or other chemicals that are approved or allowed for use in the United States, Canada, Europe, and Japan are presented below.

Regulation of Aquacultural Chemicals in the United States

In the United States, the application of chemicals to organisms or to their environments is regulated by the U.S. Food and Drug Administration (FDA) and the U.S. Environmental Protection Agency (EPA). FDA controls the use of drugs and anesthetics and EPA controls the application of chemicals and pesticides to the environment. In cases that involve water treatments to control pathogens, the jurisdiction becomes unclear and has been changed over time. Each agency develops appropriate guidelines and policies to implement the laws for its field of responsibility (9,10).

Therapeutant approval requires information on human safety, efficacy against target organisms, toxicity to nontarget organisms, residues in food-producing animals, and effects on the environment. Drugs for food-producing animals require information on metabolites and residue dynamics to establish tolerance levels and withdrawal times. Withdrawal time is the period of time that must pass after the last treatment or exposure to a certain drug, chemical, or pesticide before an animal can be consumed. Residue studies may not be necessary if the actual holding time of an aquatic animal is significantly longer than the required withdrawal period for a major species; examples include the treatment of eggs, fry, or small fingerlings long before they are harvested or slaughtered for food as adults. Studies to generate the required data must be conducted according to good laboratory practices (GLP) requirements and each drug must be produced and formulated with good manufacturing practices (GMP). Requirements of GLP and GMP significantly increase the cost of developing new drugs (9,10).

Most pharmaceutical and chemical companies do not register aquacultural chemicals because of increased costs for registration requirements, limited sales and profit potential, diversity of cultured aquatic animals and diseases, and lack of uniform testing guidelines and data requirements among countries (9). To be profitable, an aquacultural drug must be marketed worldwide, but few nations accept data on aquacultural drugs that were tested in a foreign country and the required tests are too expensive to repeat in every country. Data on mammalian safety, environmental fate, residues, and metabolism are the most expensive to obtain and are not available for most of the drugs known to be effective on aquatic organisms. In fact about $3 million of the estimated $3.5 million spent to register a new drug in the United States is for these types of studies.

Registered Aquacultural Chemicals in the United States

Antibacterials. Few therapeutants are registered in the United States for use on any cultured aquatic species (Table 1). In the most critical area of antibacterials, only two (Terramycin for Fish and Romet-30) are fully registered and available. Only the diseases listed on the label may be treated. Terramycin (oxytetracycline) is legal to use against *Aeromonas, Hemophilus,* and *Pseudomonas* on salmonids and catfishes (10,11). Romet-30 (sulfadimethoxine, ormetoprim) is labeled for the control of furunculosis in salmon and enteric septicemia in catfishes (ESC) (10,11). Data are available for the control of enteric redmouth disease with Romet but have not been submitted to FDA for label inclusion. Unfortunately, a range of bacteria have developed resistance to both compounds, and broad spectrum antibacterial agents are desperately needed as replacements.

Fungicides. Formalin as Paracide-F is the only registered fungicide and is labeled for use on eggs of trout, salmon, and esocids at 1,000–2,000 mg/L for 15 min (9–11). Delivery apparatus has been developed to reduce human exposure to formalin. Fish culturists in the Pacific Northwest have successfully used formalin to control fungal infections on salmon broodstock, but controlled research that indicates formalin is a good fungicide on adult fish of many species is lacking.

When it approved the New Animal Drug Application (NADA) of formalin, FDA ruled that use of formalin for fisheries was safe for humans and the environment. They ruled that effluents from fish treatments at 250 mg/L should be diluted 10 times and from egg treatments 75 times if 1,000–2,000 mg/L were used (10,11). Before registering the compound, FDA also addressed carcinogenicity by stating it was not concerned about human exposure from either water or fish treated with formalin. The U.S. Fish and Wildlife Service (USFWS) has procedural guidelines that should protect workers from harmful levels of formalin. Calculations based on treatment levels demonstrated that a fishery worker is exposed to not more than 0.117 mg/L formalin in the air, well below the levels established by the U.S. Occupational Safety and Health Administration to protect workers.

Salt applied as equal parts of unionized sodium chloride and calcium chloride at 20 g total per L for 1 h, three times a week, has also been used to control fungal infections on eggs. The salt combination is first applied one day after fertilization to the first pick of eggs. These compounds are categorized as generally recognized as safe (GRAS).

Parasiticides. Formalin (Paracide-F) is the only parasiticide currently registered and available. It is only registered for use on trout, salmon, catfish, largemouth bass (*Micropterus salmoides*), and bluegill (*Lepomis macrochirus*) (10,11). However, when FDA accepted the toxicity data on formalin in 1981, data

Table 1. Chemicals Registered[a] or Allowed[b] for Use in Aquaculture in the United States by the FDA or the EPA[c]

Product (trade or alternative name)	CAS Registry Number	Molecular formula	Use pattern	Tolerance[d]	Withdrawal time[e]	Comments
			Therapeutants			
acetic acid, glacial (vinegar)	*[64-19-7]*	$C_2H_4O_2$	parasiticide for fish, etc.—1000–2000 mg/L for 1–10 min	exempted		generally recognized as safe
formalin (Paracide-F)	*[50-00-0]*	CH_2O	parasiticide for trout, salmon, catfish, largemouth bass, and bluegill—25 mg/L in ponds; up to 250 mg/L for 1 h in tanks and raceways; fungicide for trout, salmon, and escoid eggs—1000–2000 mg/L for 15 min in egg-treatment tanks	none required	none required	registration on shrimp pending
nifurpirinol (Furanace Capsules)	*[13411-16-0]*	$C_{12}H_{10}N_2O_4$	antibacterial for aquarium fish—3.8 mg capsule for ~38 L (10 gal) of water for 1 h	none established		nonfood fish use; no longer available
oxytetracycline (Terramycin for Fish)	*[79-57-2]*	$C_{22}H_{24}N_2O_9$	antibacterial for fish and lobster—2.75–4.125 g/50 kg[f] per day for 10 d in feed	0.1 ppm in salmonids, catfish, and lobsters	21 d	some data generated for penaeid shrimp
sodium chloride calcium chloride (salt)	*[7647-14-5]* *[10043-52-4]*	NaCl $CaCl_2$	osmoregulatory enhancer for fish—0.5–1% for indefinite period; 3% for 10–30 min; fungicide for fish eggs—20 g/L for 1 h, 3 times a week	exempted		generally recognized as safe

sulfadimethoxine and ormetoprim (Romet-30, Romet-B)	[*122-11-2*] [*6981-18-6*]	$C_{12}H_{14}N_4O_4S$ $C_{14}H_{18}N_4O_2$	antibacterial for salmonids and catfish—50 mg/kg per day for 5 d	0.1 ppm in salmonids and catfish	6 wks for salmonids, 3 d for catfish	has potential for use in other aquatic species
sulfamerazine (Sulfamerazine in Fish Grade)	[*127-79-7*]	$C_{11}H_{12}N_4O_2S$	antibacterial for salmonids—11 g/50 kg[g] per day for 14 d in feed; discontinued after 14 d	zero in edible tissues of trout	3 wks	no longer available
trichlorfon (Masoten)	[*52-68-6*]	$C_4H_8C_{13}O_4P$	parasiticide for goldfish and baitfish—0.25 mg/L active ingredient for indefinite period	none established		nonfood fish use; currently not available
			Disinfecting agents			
calcium hypochlorite (Olin HTH Dry Chlorinator Granular)	[*7778-54-3*]	$Ca(ClO)_2$	disinfectant and sanitizer in fish tanks, raceways, and on utensils—200 mg/L for 1 h; control of algae and bacteria in ponds—5–10 mg/L residual chlorine for 12–24 h	exempted		
didecyldimethyl-ammonium chloride (Sanaqua)	[*7173-51-5*]	$C_{22}H_{48}ClN$	disinfection of aquarium and fish-holding equipment—59 mL in 15 L[h] water for 10 min; disinfection in fish disease control institutions—104 mL in 15 L for 10 min	none required	none required	do not use directly on fish or other aquatic life
povidone–iodine compounds	[*25655-41-8*]		disinfection of fish eggs—10–200 mg/L for 10 min	none established		exempted from registration by FDA
quaternary ammonium compounds (benzalkonium chloride) (benzethonium chloride)	[*68424-85-1*] [*121-54-0*]	$C_{27}H_{42}ClNO_2$	disinfection of water, gear, and tanks—2 mg/L for 1 h	none established		exempted from registration by FDA

Table 1. (*Continued*)

Product (trade or alternative name)	CAS Registry Number	Molecular formula	Use pattern	Tolerance[d]	Withdrawal time[e]	Comments
			Water treatment compounds and dyes			
fluorescein sodium	*[518-47-8]*	$C_{20}H_{12}O_5$	dye to check water flows or dilution—0.1 mg/L	none established		exempted from registration by EPA
lime, slaked lime (calcium hydroxide) (calcium oxide) (calcium carbonate)	*[1305-62-0]* *[1305-78-8]* *[471-34-1]*	$Ca(OH)_2$ CaO $CaCO_3$	pond sterilant—0.33 L/m^2 (1.338 L/acre) of quick lime; 0.2 kg/m^2 (1.784 lb/acre) of slaked lime	exempted	none required	generally recognized as safe
oxytetracycline (Terramycin)	*[79-57-2]*	$C_{22}H_{24}N_2O_9$	dye to mark fish	0.1 ppm in salmonids and catfish	7 d for oral, 15 d for injectable compounds	FDA ruled that there is no public health concern when used as directed
potassium permanganate	*[7722-64-7]*	$KMnO_4$	oxidizer and detoxifier—2 mg/L	none established		exempted from registration by EPA
rhodamine B and WT	*[81-88-9]*	$C_{28}H_{31}ClN_2O_3$	dye to check water flows or dilution—20 μg/L	exempted		exempted from registration by EPA
sodium methanesulfonate (Amquel)	*[2386-57-4]*	$CH_4O_3S{\cdot}Na$	detoxifier of chlorine, ammonia, and chloramines	exempted		FDA ruled that the compound is not a drug and therefore is not under FDA jurisdiction
tetracycline	*[60-54-8]*	$C_{22}H_{24}N_2O_8$	dye to mark fish	none established	15 d for injectable compounds	FDA ruled that there is no public health concern when used as directed

			Anesthetics			
carbon dioxide (carbonic acid)	[*124-38-9*]	CO_2	anesthetic on fish—200–400 mg/L for 4 min	exempted		generally recognized as safe
sodium bicarbonate (baking soda)	[*144-55-8*]	$NaHCO_3$	anesthetic for fish—142–642 mg/L for 5 min	exempted		generally recognized as safe
tricaine (Finquel; MS-222)	[*886-86-2*]	$C_{10}H_{15}NO_5S$	anesthetic for fish and amphibians—15–66 mg/L for 6–48 h for sedations; 50–330 mg/L for 1–40 min for anesthesia	none established	21 d	food fish use
			Herbicides and algicides			
acid blue and acid yellow (Aquashade)			algicide and herbicide	exempted		also used to control off-flavors in fish
aluminum sulfate	[*10043-01-3*]	$Al_2(SO_4)_3$	algicide and herbicide	exempted		
calcium sulfate	[*7778-18-9*]	$CaSO_4$				
boric acid (Clean-Flow Lake Cleanser)	[*10043-35-3*]	H_3BO_3				
copper, elemental (Aquatrine, etc)	[*7440-50-8*]	Cu	algicide and antibacterial	exempted	fish and shrimp may be harvested immediately	do not use in water containing trout; used as antibacterial on penaeid shrimp
copper sulfate (Calco Copper Sulfate, etc)	[*7758-98-7*]	$CuSO_4$	algicide and herbicide	exempted		low alkalinity could cause the chemical to be hazardous to fish
2,4-D (Aquacide, Aqua-Kleen, etc)	[*94-75-7*]	$C_8H_6Cl_2O_3$	herbicide	1 ppm in fish and shellfish		

Table 1. (*Continued*)

Product (trade or alternative name)	CAS Registry Number	Molecular formula	Use pattern	Tolerance[d]	Withdrawal time[e]	Comments
dichlobenil (Acme Norosac G-10, Casoron-10G)	*[1194-65-6]*	$C_7H_3Cl_2N$	herbicide	none established		nonfood fish use
diquat dibromide (Aqua-Clear, etc)	*[2764-72-9]*	$C_{12}H_{12}Br_2N_2$	algicide and herbicide	0.1 ppm in fish and shellfish		EPA proposed a maximum contaminant level (MCL) of 0.02 ppm for drinking water
endothall (Aquathol Granular, Aquathol K, etc)	*[45-73-2]*	$C_8H_{10}O_5$	herbicide	0.2 ppm in potable water	3 d	EPA proposed a MCL of 0.1 ppm for drinking water
fluridone (Sonar)	*[59156-60-4]*	$C_{19}H_{14}F_3NO$	herbicide	0.5 ppm in fish and crayfish		not for use in tidewater or brackish water
glyphosate (Rodeo)	*[1071-83-6]*	$C_3H_8NO_5P$	herbicide	0.25 ppm in fish		EPA proposed a MCL of 0.1 ppm for drinking water
potassium ricinoleate (Solricin 135)	*[7492-30-0]*		algicide	exempted	4 wks	also used to control off-flavors in fish
simazine (Aquazine)	*[122-34-9]*	$C_7H_{12}ClN_5$	algicide and herbicide	12 ppm in fish		EPA proposed a MCL of 0.001 ppm for drinking water
xylene (water weed killer)	*[1330-20-7]*	C_8H_{10}	herbicide	exempted		not for use in potable water

			Piscicides		
antimycin (Fintrol Concentrate)	[*27220-56-0*]	$C_{26}H_{36}N_2O_9$	piscicide	none established	general fish toxicant; can also be used to selectively remove scaled fish from catfish ponds
rotenone (Chem-fish Synergized, Noxfish, Prentox, etc)	[*83-79-4*]	$C_{23}H_{22}O_6$	piscicide	pending	general fish toxicant
niclosamide (Bayluscide)	[*1420-04-8*]	$C_{13}H_8Cl_2N_2O_4$	piscicide; molluscicide	none established	piscicide only for use on sea lamprey in the Great Lakes area; molluscicide to control snails (vectors of swimmers itch)

[a] A registered compound is an available commercial product bearing an EPA or FDA label specifying its allowed uses.
[b] An allowed product does not necessarily have an EPA or FDA label because some other classification or designation may allow its use in aquatic situations.
[c] Ref. 9.
[d] Tolerance refers to residue levels of a drug or chemical that are permitted by regulatory agencies in food eaten by humans.
[e] Withdrawal time is the period of time that must pass after the last treatment or exposure to a certain drug, chemical, or pesticide before an animal can be consumed. None has been established if there is no notation.
[f] 2.75–4.125 g/50 kg = 2.5–3.75 g/100 lb.
[g] 11 g/50 kg = 10 g/100 lb.
[h] 59 mL in 15 L $\approx$ 2 fl oz in 4 gal.

on other fish species were either not available or not acceptable. A second chemical, trichlorfon (Masoten) was registered for use on nonfood fishes by EPA but is not currently available. Vinegar (glacial acetic acid) and salt (sodium chloride) are also used to control external parasites on fishes and these compounds are classified as GRAS (9,11).

Disinfectants. Several disinfecting agents can be used in hatcheries and two are of particular interest. Because they are not considered drug or food additive uses by FDA, povidone–iodine compounds can be used to disinfect the surface of eggs (9). Benzalkonium chloride [*68424-85-1*] and benzethonium chloride (quaternary ammonium compounds), are allowed at 2 mg/L by FDA to disinfect water containing fish. These compounds are also known to have therapeutic properties, especially against external bacteria (9).

Water Treatment Compounds. Like the disinfecting agents, several water treatment compounds are used in aquaculture. Of particular interest is potassium permanganate which is exempted from registration by EPA when used as an oxidizer or detoxifier and can control certain parasites, external bacteria, and possibly fungi (9).

Anesthetics. Tricaine is the only currently registered anesthetic and requires a 21-day withdrawal time (9,11). The withdrawal time for MS-222 is of special concern to FDA when the broodfishes of salmon or other species are taken immediately after spawning for pet or human food. Both carbon dioxide and sodium bicarbonate [*144-55-8*], have also been used as anesthetics; however, both chemicals are difficult to use with consistent results and involve long induction and recovery periods (9).

Herbicides. An array of herbicides are registered for use in aquatic sites, but copper sulfate and diquat dibromide are of additional interest because they also have therapeutic properties (9,10). Copper sulfate has been used to control bacteria, fungi, and certain parasites, including *Ichthyophthirius* (ich). Diquat dibromide can control columnaris disease, but it also exhibits fungicidal properties (9,10). EPA recently proposed to limit the amount of diquat dibromide, endothall, glyphosate, and simazine that can be present in drinking water; therefore, the use of these compounds may be reduced if they cannot be removed from the effluent.

Piscicides. The two piscicides, antimycin and rotenone, are both used in ponds to control nuisance fish. Antimycin is used selectively to remove scaled fishes from catfish ponds, and rotenone is used as a general fish toxicant (9,10). Recent observations by catfish farmers indicate that antimycin at low concentrations also acts as a therapeutant against external parasites.

Regulation and Registration of Aquacultural Chemicals Outside the United States

The control of aquacultural drugs varies among countries from no regulation to restrictive regulations. Generally, few requirements are needed for a therapeutant to be licensed or registered in South America, Africa, and most of Asia. Seafood-exporting countries are increasingly concerned because importing countries may no longer accept products without a guarantee that the products contain no chemical residues of concern.

Canada. Except for environmental studies, requirements for registration data in Canada are similar to requirements in the United States. However, Canada has significantly different regulations and approval processes. Canadian aquaculturalists use drugs (Table 2) that are either licensed for other food animals and prescribed by veterinarians or used in an emergency under the direction of the Canadian Bureau of Veterinary Drugs (BVD) (12). The BVD is concerned about the lack of data on the pharmacokinetics of fishes, especially the difference in uptake of drugs at a range of temperatures (13). Chloramphenicol and tributyltin compounds are the only two classes of compounds that cannot be used in Canadian aquaculture. Canadian regulations also differ from the United States in that they have no minor-use policy or classifications such as GRAS compounds.

Table 2. Chemicals Authorized or Allowed for Use in Aquaculture in Canada

Product (trade or alternative name)	CAS Registry Number	Molecular formula
	Chemicals specially authorized	
oxytetracycline[a]	*[79-57-2]*	$C_{22}H_{24}N_2O_9$
sulfadimethoxine and ormetoprim[b]	*[122-11-2]*	$C_{12}H_{14}N_4O_4S$
(Romet-30 and Romet-B)	*[6981-18-6]*	$C_{14}H_{18}N_4O_2$
	Chemicals authorized with an emergency-release permit	
erythromycin	*[114-07-8]*	$C_{37}H_{67}NO_{13}$
dichlorvos	*[62-73-7]*	$C_4H_7Cl_2O_4P$
(Nuvan)		
oxolinic acid	*[14698-29-4]*	$C_{13}H_{11}NO_5$
oxytetracycline	*[79-57-2]*	$C_{22}H_{24}N_2O_9$
penicillin G	*[61-33-6]*	$C_{16}H_{18}N_2O_4S$
sulfamerazine	*[127-79-7]*	$C_{11}H_{12}N_4O_2S$
	Chemicals sold freely	
chloramine-T	*[127-65-1]*	$C_7H_7ClNNaO_2S$
formalin	*[50-00-0]*	CH_2O
sodium chloride	*[7647-14-5]*	NaCl
	Chemical not usually released, but not banned	
malachite green	*[569-64-2]*	$C_{23}H_{25}ClN_2$

[a]Licensed only for use in the treatment of fish and lobster. Withdrawal time is at least 40 d at > 10°C and at least 80 d at < 10°C.
[b]Withdrawal time is 42 d at > 10°C and 81 d at < 10°C.

Europe. The 12 countries of the European Economic Community (EEC) will implement uniform regulations in 1992. Licensing of drugs will be directed by the EEC Committee on Veterinary Medicinal Products with guidelines on safety to the target species, efficacy to control target pathogens, risks to the consumer for exposure to residues in food, hazards to the workers handling the chemical,

Table 3. Chemicals Authorized or Allowed for Use in Aquaculture in Certain European Countries

Product (trade or alternative name)	CAS Registry Number	Molecular formula	Withdrawal time
		Therapeutants	
ampicillin	[*69-53-4*]	$C_{16}H_{19}N_3O_4S$	[a]
chloramphenicol	[*56-75-7*]	$C_{11}H_{12}Cl_2N_2O_5$	31 d only in Germany
chlortetracycline	[*57-62-5*]	$C_{22}H_{23}ClN_2O_8$	30 d only in Germany
dibutyltin oxide	[*818-08-6*]	$C_8H_{18}SnO$	[a]
dimetridozole	[*551-92-8*]	$C_5H_7N_3O_2$	90 d only in Germany
estomycine			[a]
flumequine	[*42835-25-6*]	$C_{14}H_{12}FNO_3$	2 d in France
furaltadone	[*139-91-3*]	$C_{13}H_{16}N_4O_6$	
furazolidone	[*67-45-8*]	$C_8H_7N_3O_5$	90 d
nifurpracine hydrochloride			90 d only in Germany
3-nitro-2-thiazolylamine			3 d only in Germany
nitrofurazone (Tricofuron)	[*59-87-0*]	$C_6H_6N_4O_4$	
oxolinic acid (Aqualinic Powder, etc)	[*14698-29-4*]	$C_{13}H_{11}NO_5$	6 d in France 200 degree d in UK[b]
oxytetracycline (Terramycin, etc)	[*79-57-2*]	$C_{22}H_{24}N_2O_9$	30–40 d at >9 or >10°C; 60 d at <9 or <10°C 400 degree d in UK[b]
paramomycinsulfate			[a]
sulfadiazine and trimethoprim (Tribrissen, etc)	[*68-35-9*] [*738-70-5*]	$C_{10}H_{10}N_4O_2S$ $C_{14}H_{18}N_4O_3$	40 d at >10–12°C; 60–90 d at <8–10°C 200 degree d in the UK[b]
sulfamerazine	[*127-79-7*]	$C_{11}H_{12}N_4O_2S$	30 d at >9°C; 60 d at <9°C 200 degree d in the UK[b]
sulfadimethoxine and trimethoprim	[*122-11-2*] [*738-70-5*]	$C_{12}H_{14}N_4O_4S$ $C_{14}H_{18}N_4O_3$	49 d only in Germany
sulfamethazine	[*57-68-1*]	$C_{12}H_{14}N_4O_2S$	[a]
sulfadoxine	[*2447-57-6*]	$C_{12}H_{14}N_4O_4S$	49 d only in Germany
trimethoprim	[*738-70-5*]	$C_{14}H_{18}N_4O_3$	[a]
		Topicals	
chloramine-T	[*127-65-1*]	$C_7H_7ClNNaO_2S$	
copper sulfate	[*7758-98-7*]	$CuSO_4$	
dichlorvos (Nuvan EC, Aquaguard)	[*62-73-7*]	$C_4H_7Cl_2O_4P$	
formalin	[*50-00-0*]	CH_2O	
potassium permanganate	[*7722-64-7*]	$KMnO_4$	

Table 3. (*Continued*)

Product (trade or alternative name)	CAS Registry Number	Molecular formula	Withdrawal time
povidone–iodine compounds	[*25655-41-8*]		
sodium chloride	[*7647-14-5*]	NaCl	
trichlorfon	[*52-68-6*]	$C_4H_8C_{13}O_9P$	14–30 d
		Anesthetics	
chlorobutanol	[*57-15-8*]	$C_4H_7Cl_3O$	
carbon dioxide	[*124-38-9*]	CO_2	
tricaine	[*886-86-2*]	$C_{10}H_{15}NO_5S$	

[a]Compounds used only in conjunction with other drugs.
[b]Degree days are number of days past treatment multiplied by the water temperature (°C).

and potential for polluting the environment. Residue and pharmacokinetic studies will be required to determine the pattern of residue depletion at different temperatures and to develop withdrawal times, expressed in degree days, that are based on temperature-related metabolism. Mammalian safety studies will also be required to determine chronic toxicity, mutagenicity, carcinogenicity, reproduction, and teratogenicity of a product. If a product is registered in one country, it should be easily registered in other EEC countries, although different environmental conditions may require additional field trials. Most other European countries are interacting with the EEC to develop similar regulations and guidelines (14–16).

Under current regulations, chemicals allowed for use in various European countries (Table 3) are either fully licensed for aquacultural use (oxytetracycline, oxolinic acid) or can be prescribed by veterinarians if they are licensed for use on other food animals (14–16). In addition, previously unlicensed chemicals that are applied to the water (topicals) may now be used under a grandfather clause if no one questions their safety. The question of whether a chemical is a medicine or a pesticide has also been addressed. For example, dichlorvos (Nuvan 500 EC) was initially designated as a pesticide in the United Kingdom, but was later categorized as a medicine. A similar product, trichlorfon (Masoten), was treated the same way in the United States.

Japan. In Japan, registration of drugs for aquatic species requires the same data as those required for drugs on other animals. The Ministry of Agriculture, Forests, and Fisheries and the Ministry of Welfare control the use of chemicals in aquaculture in Japan (17). The preclinical data requirements include product chemistry, toxicity (acute, subacute, special) using rats and mice, safety to target animals, and metabolism. The requirements for clinical data include availability and residues. As of July 1990, more chemicals were registered for aquacultural use in Japan than in any other country (Table 4).

Table 4. Chemicals Registered or Approved for Use in Aquatic or Fishery Situations in Japan, July 1990[a]

Product	CAS Registry Number	Molecular formula	Use pattern, mg/kg per day[b]	Withdrawal time, d
		Therapeutants		
amoxicillin	*[26787-78-0]*	$C_{16}H_{19}N_3O_5S$	20–40	
ampicillin	*[69-53-4]*	$C_{16}H_{19}N_3O_4S$	5–20	5
colistin	*[1066-17-7]*	$C_{53}H_{100}N_{16}O_{13}$	1.3332[c]	3
doxycycline	*[17086-28-1]*	$C_{22}H_{24}N_2O_8H_2O$	20–50	20
erythromycin	*[114-07-8]*	$C_{37}H_{67}NO_{13}$	25–50	30
florfenicol	*[73231-34-2]*	$C_{12}H_{14}Cl_2FNO_4S$	10	
flumequine	*[42835-25-6]*	$C_{14}H_{12}FNO_3$	20	
josamycin	*[16846-24-5]*	$C_{42}H_{69}NO_{15}$	30–50	
lincomycin	*[1392-21-8]*	$C_{18}H_{34}N_2O_6S$	20–40	20
miloxacin	*[37065-29-5]*	$C_{12}H_9NO_6$	5–30	20
nalidixic acid	*[389-08-2]*	$C_{12}H_{12}N_2O_3$	20	7
nifurstylenic acid			50	2
novobiocin	*[303-81-1]*	$C_{31}H_{36}N_2O_{11}$	50	
oleandomycin	*[3922-90-5]*	$C_{35}H_{61}NO_{12}$	25	30
oxolinic acid	*[14698-29-4]*	$C_{13}H_{11}NO_5$	25–30[d]	14–30
oxytetracycline	*[79-57-2]*	$C_{22}H_{24}N_2O_9$	50	25–30
piromidic acid	*[19562-30-2]*	$C_{14}H_{16}N_4O_3$	10–20	20
spiramycin	*[8025-81-8]*	$C_{43}H_{74}N_2O_{14}$	25–40	30
sulfadimethoxine	*[122-11-2]*	$C_{12}H_{14}N_4O_4S$	50–100	30
sulfamonomethoxine	*[1220-83-3]*	$C_{11}H_{12}N_4O_3S$	100–200	15–30
sulfamonomethoxine	*[1220-83-3]*	$C_{11}H_{12}N_4O_3S$	10–20	15
and ormetoprim	*[6981-18-6]*	$C_{14}H_{18}N_4O_2$		
sulfisozole	*[73247-57-1]*		100–200	10–15
tetracycline	*[60-54-8]*	$C_{22}H_{24}N_2O_8$	55–110	10
thiamphenicol	*[15318-45-3]*	$C_{12}H_{15}Cl_2NO_5S$	20–50	15
		Anesthetics		
eugenol	*[97-53-0]*	$C_{10}H_{12}O_2$	20–50[c]	
phenthiazamine	*[34161-31-4]*	$C_9H_8N_2S{\cdot}BrH$	20–50[c]	
tricaine	*[886-86-2]*	$C_{10}H_{15}NO_5S$	20–100[c]	
		Insecticides		
hydrogen peroxide	*[7722-84-1]*	H_2O_2		
trichlorfon	*[52-68-6]*	$C_4H_8Cl_3O_4P$	0.2–0.3[c]	5

[a]Ref. 18.
[b]Unless otherwise noted.
[c]mg/L solution.
[d]25–30 mg/kg per day or 5–10 mg/L.

Promising Chemicals for Registration for Aquaculture

More therapeutants and vaccines may soon be added to the medicine chest of fish farmers. A variety of chemicals have potential for registration and use in aquaculture (19,20).

Antibacterials. Research has been conducted on three important antibacterial compounds in the United States. Various registration data are or have been generated by USFWS on chloramine-T with funds from both USFWS and the U.S. Department of Agriculture's Interregional Research Project No. 4 (IR-4). Chloramine-T is used mainly to control bacterial gill disease on fry and fingerlings of salmonids but may also be effective for other external bacterial and parasitic diseases of a variety of fishes (9,20). After it is registered, the use of chloramine-T may be expanded. The proposed registration date may be postponed if more toxicological and metabolism research is required than was originally anticipated (see CHLORAMINES AND BROMAMINES).

Sarafloxacin [*98105-99-8*], $C_{20}H_{17}F_2N_3O_3$, a broad-spectrum antibacterial in a class of compounds called quinolones, is under development by Abbott Laboratories (Chicago, Ill.) for worldwide use by the aquaculture industry. It is effective against *Edwardsiella, Aeromonas, Yersinia, Vibrio,* and other gram-negative bacteria (9,20). Resistance tests indicate that sarafloxacin will probably not cause development of resistant bacterial strains. Registration of sarafloxacin is especially important because economic success might attract other chemical companies to aquaculture. The anticipated registration dates for sarafloxacin are 1992 in Norway, Canada, and other countries, and 1993 in the United States (see ANTIBACTERIAL AGENTS, SYNTHETIC, QUINOLONES).

The University of Idaho and USFWS, with funds from the Bonneville Power Administration, are also gathering data for registration of erythromycin. Erythromycin is intended for control of bacterial kidney disease in salmonid fingerlings that can also be transmitted by broodstock to the eggs (9). If research is completed on schedule and data indicate that the compound is safe and effective, registration is scheduled for 1994 (see ANTIBIOTICS, MACROLIDES).

Several promising antibacterial compounds are available and may be considered for use in U.S. aquaculture. In addition to sarafloxacin, other quinolones, flumequine and oxolinic acid, are already registered in Europe. However, resistance to both of these compounds developed in bacteria in just a few years (20). Enrofloxacin, [*93106-60-6*], $C_{19}H_{22}FN_3O_3$, a quinolone product of Bayer A.G. (Germany), is also a candidate for aquaculture registration in Europe and may be considered for registration in the United States.

Fungicides and Parasiticides. USFWS is working on several fronts to improve the availability of fungicides and parasiticides. Promising candidates for replacement of malachite green are being screened and tested for use as both fungicides and parasiticides (9).

Fungicide testing at the National Fisheries Research Center (La Crosse, Wis.) is being funded by Bonneville Power Administration. Although several compounds show promise for controlling fungi, the best fungicide candidate has not been identified (20,21). Any potential fungicide must be both effective and inexpensive if it is to be accepted and used by aquaculturists. Fungal infections can be controlled by other methods, and aquaculturists might choose to change cultural methods rather than use a product they consider to be too expensive. Such an option is usually not available for controlling bacterial diseases.

The Fish Farming Experimental Laboratory (Stuttgart, Ark.) has screened several compounds to replace malachite green for the control of external protozoans and has identified dichlorophen [*97-23-4*], $C_{13}H_{10}Cl_2O_2$, as the most promising compound (9). No funding has yet been allocated to continue investigating and gathering data for its registration (see ANTIPARASITIC AGENTS, ANTHELMINTICS).

Mobay Corporation (Kansas City, Mo.) has agreed to support the FDA registration of trichlorfon (Masoten) as a parasiticide if public funding is available. Masoten was formerly registered with EPA for use in aquaculture as a pesticide to control diseases. Uniroyal Chemical Company is interested in registering Dimilin [*35367-38-5*] (diflubenzuron), $C_{14}H_9ClF_2N_2O_2$, for control of external parasites on finfishes, but the company is currently more involved with reregistering it as a pesticide with EPA (see INSECT CONTROL TECHNOLOGY). Researchers in Norway are pursuing the registration of praziquantel [*55268-74-1*] (Droncit), $C_{19}H_{24}N_2O_2$, for use as a parasiticide. Ivermectin [*70288-86-7*] and an experimental, pyrethrum-based compound (Py-Sal 25) are both involved in clinical trials in Europe to determine if they can control sea lice on salmon (see ANTIPARASITIC AGENTS, AVERMECTINS).

Disinfectants. Promising disinfectants include ultraviolet (uv) light and ozone [*10028-15-6*], O_3. High doses of uv light have prevented the transmission of the epizootic epitheliotropic virus (EEV) disease among lake trout (*Salvelinus namaycush*) (20). Uv sterilization can only be successful when water volumes and suspended solids are low (22). Ozone is also considered to be a good candidate to both control EEV and other diseases, and remove chemicals and wastes from aquacultural effluents (20). However, ozone is considered to be a food additive by FDA and, therefore, must meet registration requirements that are almost as stringent as the registration requirements for drugs.

Anesthetics. Ethyl aminobenzoate [*94-09-7*] (benzocaine), $C_9H_{11}NO_2$, is the only anesthetic candidate that might allow spawned-out broodstock carcasses to be used for pet or human food. Studies are still required to determine which residues remain in the carcasses (9). Electronarcosis is an alternative to chemical anesthesia that uses varying electrical frequencies to rapidly anesthetize fishes and allow gentle recovery. Electronarcosis has been used effectively on tilapia (*Oreochromis* sp.) and the common carp (*Cyprinus carpio*) and the technique is being tested with other fishes (23,24).

Herbicides and Piscicides. Sponsors are currently reregistering aquatic herbicides under the 1988 Pesticide Law. Reregistration of piscicides is funded mainly by public agencies. Data collection for the reregistration of rotenone has been completed by USFWS. Antimycin may soon require reregistration and funding must be solicited. Control of Bayluscide (niclosamide) was recently transferred from the Mobay Corp. to USFWS after Mobay determined that the estimated \$2.5–3.5 million for reregistration was not economically feasible. Bayluscide is used by USFWS to control the sea lamprey (*Petromyzon marinus*) in the Laurentian Great Lakes and to manage and control nuisance mollusks in both aquaculture and natural waters. Bayluscide also has the potential to control digenetic trematodes and cestodes of finfishes (20).

BIBLIOGRAPHY

"Aquaculture," in *ECT* 3rd ed., Vol. 3, pp. 194–213, H. P. Clemens and M. Conway, University of Oklahoma.

1. R. A. Busch, *Vet. Hum. Toxicol.* **29**(Suppl. 1), 45 (1987).
2. L. W. Harrell, *Vet. Hum. Toxicol.* **29**(Suppl. 1), 49 (1987).
3. M. H. Beleau and J. A. Plumb, *Vet. Hum. Toxicol.* **29**(Suppl. 1), 52 (1987).
4. J. P. McCraren, *The Aquaculture of Striped Bass,* Sea Grant Publication No. UM-SG-MAP-84-01, University of Maryland, College Park, Md., 1984.
5. G. Lio-Po in J. V. Juario, R. P. Ferraris, and L. V. Benitz, eds., *Proceedings of the Second International Milkfish Aquaculture Conference, October 4–6, 1983,* Iloilo City, Philippines, 1984.
6. C. Brown, *Vet. Hum. Toxicol.* **29**(Suppl. 1), 35 (1987).
7. R. R. Williams and D. V. Lightner, *J. World Aquacult. Soc.* **19,** 188 (1988).
8. C. J. Sindermann and D. V. Lightner, *Disease Diagnosis and Control in North American Marine Aquaculture,* Elsevier, New York, 1988.
9. R. A. Schnick, *Prog. Fish-Cult.* **50,** 190 (1988).
10. R. A. Schnick, F. P. Meyer, and L. D. Gray, *A Guide to Approved Chemicals in Fish Production and Fishery Resource Management,* MP241-5M-3-89RV, U.S. Fish and Wildlife Service and University of Arkansas Cooperative Extension Service, Little Rock, Ark., 1989.
11. U.S. Food and Drug Administration (FDA), *Food and Drugs, Code Fed. Reg.* **21,** Parts 500–599 (1990).
12. G. E. Brooks, *Aquacult. Assoc. Canada Bull.* **89-4,** 39 (1989).
13. G. R. Johnson and D. J. Rainnie, *Aquacult. Assoc. Canada Bull.* **89-4,** 43 (1987).
14. J. Shepherd, paper presented to the *Aquaculture Feed and Veterinary Products: Worldwide Business Opportunities for Feed, Pharmaceutical, and Chemical Companies Conference,* Stamford, Conn., June 6–8, 1990.
15. L. A. Brown, *Vet. Hum. Toxicol.* **29**(Suppl. 1), 54 (1987).
16. E.-M. Bernoth, *J. Vet. Med. Ser. B.* **37,** 401 (1990).
17. T. Sano and H. Fukuda, *Aquaculture* **67,** 59 (1987).
18. T. Arai, personal communication, July 1990.
19. D. J. Alderman in J. F. Muir and R. J. Roberts, eds., *Recent Advances in Aquaculture,* Vol. 3, Timber Press, Portland, Oreg., 1988, p. 1.
20. F. P. Meyer and R. A. Schnick, *Rev. Aquat. Sci.* **1,** 693 (1989).
21. T. A. Bailey and S. M. Jeffrey, *U.S. Fish Wild. Serv. Invest. Fish Control* **99,** 1 (1989).
22. G. Barnabe', *Aquaculture,* Vol. 1, Ellis Horwood, London, UK, 1990.
23. W. T. Barham, A. J. Schoonbee, and J. G. J. Visser, *Onderstepoort J. Vet. Res.* **54,** 617 (1987).
24. *Ibid.,* **56,** 215 (1989).

ROSALIE A. SCHNICK
U.S. Fish and Wildlife Service

ARAMID FIBERS. See HIGH PERFORMANCE FIBERS.

ARGON. See HELIUM-GROUP GASES.

ARSENIC AND ARSENIC ALLOYS

Arsenic [*7440-38-2*], although often referred to as a metal, is classified chemically as a nonmetal or metalloid and belongs to Group 15 (VA) of the periodic table (as does antimony). The principal valences of arsenic are +3, +5, and −3. Only one stable isotope of arsenic having mass 75 (100% natural abundance) has been observed.

Elemental arsenic normally exists in the α-crystalline metallic form which is steel-gray in appearance and brittle in nature, and in the β-form, a dark-gray amorphous solid. Other allotropic forms, ie, yellow, pale reddish-brown to dark brown, have been reported (1), but the evidence supporting some of these allotropes is meager. Metallic arsenic, heated under ordinary conditions, does not exhibit a discrete melting point but sublimes. Molten arsenic can be obtained by heating under pressure.

Properties

Physical properties of α-crystalline metallic arsenic are given in Table 1. The properties of β-arsenic are not completely defined. The density of β-arsenic is 4700 kg/m^3; it transforms from the amorphous to the crystalline form at 280°C; and the electrical resistivity is reported to be 107 Ω·cm.

Metallic arsenic is stable in dry air, but when exposed to humid air the surface oxidizes, giving a superficial golden bronze tarnish that turns black upon

Table 1. Physical Properties of Arsenic

Property	Value
atomic weight	74.9216
mp at 39.1 MPa[a], °C	816
bp, °C	615[b]
density at 26°C, kg/m^3	5,778
latent heat of fusion, J/(mol·K)[c]	27,740
latent heat of sublimation, J/(mol·K)[c]	31,974
specific heat at 25°C, J/(mol·K)[c]	24.6
linear coefficient of thermal expansion at 20°C, μm/(m·°C)	5.6
electrical resistivity at 0°C, μΩ·cm	26
magnetic susceptibility at 20°C	-5.5×10^{-6} [d]
nuclear absorption cross section[e]	4.3 ± 0.2
crystal system	hexagonal (rhombohedral)
lattice constants at 26°C, nm	a = 0.376, c = 1.0548
hardness, Mohs' scale	3.5

[a]To convert MPa to psi, multiply by 145.
[b]Sublimes.
[c]To convert to cal/(mol·K), divide by 4.184.
[d]From cgs system units.
[e]Thermal neutrons 2200 m/s of arsenic mass 75.

further exposure. The amorphous form is more stable to atmospheric oxidation. Upon heating in air, both forms sublime and the vapor oxidizes to arsenic trioxide [*1327-53-3*], As_2O_3. Although As_4O_6 represents its crystalline makeup, the oxide is more commonly referred to as arsenic trioxide. A persistent garliclike odor is noted during oxidation.

Elemental arsenic combines with many metals to form arsenides. When heated in the presence of halogens it forms trihalides; however, pentahalides with the exception of AsF_5 (2) and the unstable $AsCl_5$ are not readily formed. It reacts with sulfur to form the compounds As_2S_3, AsS, As_2S_5, and complex mixtures in various proportions (see ARSENIC COMPOUNDS).

Arsenic vapor [*12187-88-5*], As_4, does not combine directly with hydrogen to form hydrides. However, arsine (arsenic hydride) [*7784-42-1*], AsH_3, a highly poisonous gas, forms if an intermetallic compound such as AlAs is hydrolyzed or treated with HCl. Arsine may also be formed when arsenic compounds are reduced using zinc in hydrochloric acid. Heating to 250°C decomposes arsine into its elements.

Metallic arsenic is not readily attacked by water, alkaline solutions, or nonoxidizing acids. It reacts with concentrated nitric acid to form orthoarsenic acid [*7778-39-4*], H_3AsO_4. Hydrochloric acid attacks arsenic only in the presence of an oxidant.

Arsenic may be detected qualitatively as a yellow sulfide, As_2S_3, by precipitation from a strongly acidic HCl solution. Other members of this group that are normally precipitated with hydrogen sulfide do not interfere if the solution contains 25% or more hydrochloric acid. Trace quantities of arsenic may be detected by first converting to arsine (3). The arsine is decomposed by heating the gas in a small tube and an arsenic mirror is formed (Marsh test). Alternatively, the arsine may be allowed to react with test paper impregnated with mercuric chloride (Gutzeit test).

Arsenic is widely distributed about the earth and has a terrestrial abundance of approximately 5 g/t (4). Over 150 arsenic-bearing minerals are known (1). Table 2 lists the most common minerals. The most important commercial source of arsenic, however, is as a by-product from the treatment of copper, lead, cobalt, and gold ores. The quantity of arsenic usually associated with lead and copper

Table 2. Naturally Occurring Arsenic-Bearing Minerals

Name	CAS Registry Number	Formula	Name	CAS Registry Number	Formula
loellingite	[*12255-65-1*]	$FeAs_2$	sperrylite	[*12255-87-7*]	$PtAs_2$
safforlite	[*12044-43-8*]	$CoAs_2$	arsenopyrite (mispickel)	[*1303-18-0*]	FeAsS
niccolite	[*1303-13-5*]	NiAs	cobaltite	[*1303-15-7*]	CoAsS
rammelsbergite	[*1303-22-6*]	$NiAs_2$	enargite	[*14933-50-7*]	Cu_3AsS_4
realgar	[*12044-30-3*]	AsS	gersdorffite	[*12255-11-7*]	NiAsS
orpiment	[*12255-89-9*]	As_2S_3	glaucodot	[*12198-14-0*]	(Co,Fe)AsS

ores may range from a trace to 2–3%, whereas the gold ores found in Sweden contain 7–11% arsenic. Small quantities of elemental arsenic have been found in a number of localities.

Metallurgy

Metallic arsenic can be obtained by the direct smelting of the minerals arsenopyrite or loellingite. The arsenic vapor is sublimed when these minerals are heated to about 650–700°C in the absence of air. The metal can also be prepared commercially by the reduction of arsenic trioxide with charcoal. The oxide and charcoal are mixed and placed into a horizontal steel retort jacketed with firebrick which is then gas-fired. The reduced arsenic vapor is collected in a water-cooled condenser (5). In a process used by Boliden Aktiebolag (6), the steel retort, heated to 700–800°C in an electric furnace, is equipped with a demountable air-cooled condenser. The off-gases are cleaned in a scrubber system. The yield of metallic arsenic from the reduction of arsenic trioxide with carbon and carbon monoxide has been studied (7) and a process has been patented describing the gaseous reduction of arsenic trioxide to metal (8).

The demand for metallic arsenic is limited and thus arsenic is usually marketed in the form of the trioxide, referred to as white arsenic, arsenious oxide, arsenious acid anhydride, and also by the generally accepted misnomer arsenic.

Arsenic trioxide is readily volatilized during the smelting of copper and lead concentrates, and is therefore concentrated with the flue dust. Crude flue dust may contain up to 30% arsenic trioxide, the balance being oxides of copper or lead, and perhaps of other metals such as antimony and zinc. This crude flue dust is further upgraded by mixing with a small amount of pyrite or galena concentrate and roasting (9). The pyrite or galena is added to prevent arsenites from forming during roasting and to obtain a clinkered residue which can be returned for additional processing. The gases and vapors are passed through a cooling flue which consists of a series of brick chambers or rooms called kitchens. The temperature of the gas and vapor is controlled so that they enter the first kitchen at 220°C and by the time the gas and vapor reach the last kitchen they are cooled to 100°C or less. The arsenic trioxide vapor which condenses in these chambers is of varying purity analyzing from 90–95%. A higher purity product is obtained by resubliming the crude trioxide, an operation normally carried out in a reverberatory furnace. The vapors pass first through a settling chamber and then through approximately 39 kitchens that cover a length of about 68.6 m. The temperature of the settling chamber is kept at approximately 295°C, which is above the condensation temperature of the trioxide. A black, amorphous mass containing about 95% As_2O_3 condenses in the kitchens nearest the furnace and is reprocessed. The bulk of the trioxide is condensed in the kitchens having temperature ranges of 180–120°C giving a product of 99–99.9% purity. The dust which exits from the kitchens at a temperature of 90–100°C is caught in the baghouse. It assays about 90% As_2O_3 and may be sold as a crude grade or reprocessed. (A typical flow sheet for the production of arsenic trioxide at a copper smelter is depicted in Figure 1.)

In addition to pyro-refining, crude oxide analyzing 80–95% can be refined by

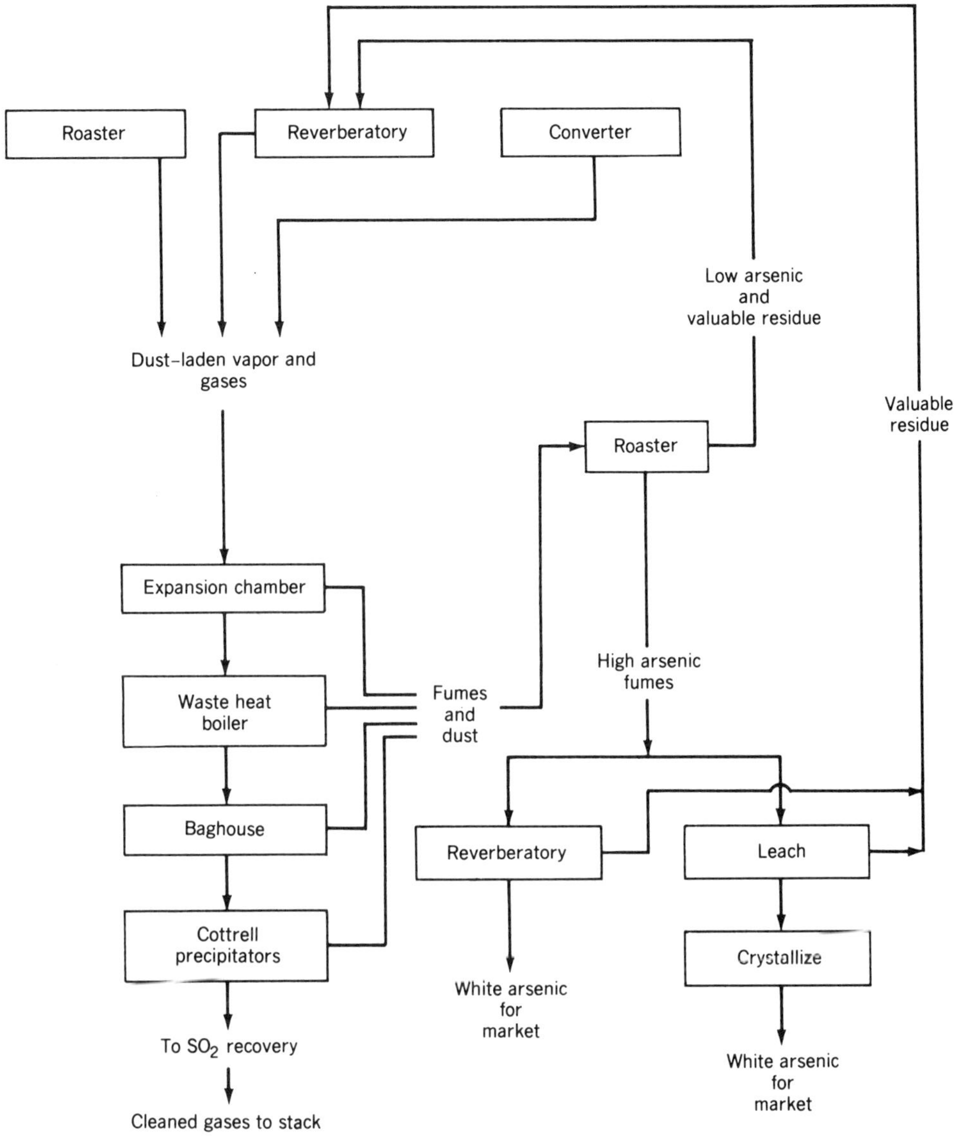

Fig. 1. Arsenic trioxide copper smelter-flow sheet.

a wet process (6). In this process advantage is taken of the fact that the solubility of arsenic trioxide in water increases with temperature as follows:

Temperature, °C	Solubility, g/L
0	12.1
20	18.1
98.5	81.8

Most of the common impurities present in crude arsenic have a low solubility.

The crude oxide is pressure-leached in a steam-heated autoclave using water or circulating mother liquor. The arsenic trioxide dissolves, leaving behind a residue containing a high concentration of heavy metal impurities and silica. The solution is vacuum-cooled and the crystallization is controlled so that a coarse oxide is obtained which is removed by centrifuging. The mother liquor is recycled. The oxide (at least 99% purity) is dried and packaged in a closed system.

The refined arsenic trioxide is analyzed for purity and also tested for solubility, a term referring to its rate of reactivity with nitric acid; this test is important if the arsenic is used in the manufacture of insecticides and herbicides. The product is graded for marketing as white soluble having 99% min As_2O_3, white insoluble, or crude having 95% min As_2O_3.

Minor quantities of arsenic trioxide have been obtained from the roasting of arsenopyrite, but the presence of copious amounts of SO_2 in the gas and vapor stream requires the use of lead-lined kitchens (9).

Economic Aspects

The demand for arsenic metal is limited. The 1990 U.S. requirement for metallic arsenic was supplied by the People's Republic of China. Arsenic was formerly supplied domestically (1974–1986) by ASARCO Inc., which dismantled its operation because of economic and environmental pressures, and by imports from Boliden Metall, Sweden, which suspended production in late 1987 because of the low price offered by China.

Commercial arsenic metal is sold at a typical purity of 99% in fragment or lump (5–7.5 cm) form. Shipments of arsenic metal from China were sporadic, and also temporarily suspended, in 1989. The metal is not sold on a published price basis and can vary with each shipment. Shipments in early 1990 had leveled to an estimated price of about $2.20/kg.

Arsenic metal is also offered in high (ranging from 99.99% to in excess of 99.999%+) purity form for semiconductor applications. It is normally packaged in fragmented form in evacuated and sealed glass ampoules to prevent oxidation. The United States is not self-reliant in its requirements for high purity arsenic and depends on imports primarily from Japan, Canada, and the United Kingdom. A typical price for semiconductor grade arsenic in kilogram quantities is around $.10/g, but price varies depending on requirements.

Estimated imports of both commercial grade and high purity arsenic metal (10,11) are listed in Table 3.

Table 3. Imports of Commercial and High Purity Grades of Arsenic Metal[a]

	Commercial grade			High purity grade		
Year	Quantity, t	Total value, $	Av value, $/kg	Quantity, t	Total value, $	Av value, $/g
1986	378	1,260,000	$3.33	17	1,385,000	0.081
1987	604	1,281,000	$2.12	27	2,179,000	0.081
1988	574	951,000	$1.66	25	1,674,000	0.067
1989	925	2,014,000	$2.18	3	136,000	0.045

[a]Estimated from Refs. 10 and 11.

Uses

The predominant use of arsenic in the United States is in the manufacture of chemicals. During the 1980s, the market for arsenic chemicals had shifted from cotton farming, where its use is now restricted because of environmental considerations, to wood (qv) preservatives for the protection of lumber and other wood products. Arsenic trioxide is the basic commodity of commerce from which a number of important chemicals are manufactured.

Monosodium methanearsonate (MSMA) [*2163-80-6*], $CH_3AsONaOOH$, disodium methanearsonate (DSMA) [*144-21-8*], $CH_3AsO(NaO)_2$, cacodylic acid [*75-60-5*], $(CH_3)_2AsOOH$, and arsenic acid [*7778-39-4*], H_3AsO_4, are used in agriculture applications (11,12). MSMA, DMSA, and cacodylic acid are used as herbicides (qv) especially in cotton (qv) fields for the control of Johnson and nutsedge grass and other weeds. Arsenic acid (13) and cacodylic acid may be used as a desiccant for the defoliation of the cotton boll prior to harvesting. Calcium arsenate [*7778-44-1*], $Ca_3(AsO_4)_2$, once an important chemical for the control of the boll weevil and cotton worm, has disappeared from application and the use of lead arsenate [*7784-40-9*], Pb_3AsO_4, for fruit crops is currently restricted.

Refined (99% min) arsenic trioxide, As_2O_3, formerly used as a decolorizer and fining agent for the production of bottleglass and other types of glassware (14), is being replaced by arsenic acid for environmental reasons (15). Arsenic acid is used in the formulation of wood preservative salts (16,17), mainly chrome copper arsenate [*11125-95-4*], $CrO_3{\cdot}CuO{\cdot}As_2O_5$. Arsenilic acid (*p*-aminophenylarsonic acid) [*98-50-0*], $C_6H_4NH_2AsO(OH)_2$, is useful as a feed additive for poultry and swine (see FEEDS AND FEED ADDITIVES). Sodium arsenite [*7784-46-5*], $NaAsO_2$, is useful for cattle and sheep dips.

The application of many arsenical chemicals are subject to registration and must comply with federal and local government environmental regulations.

Until the late 1980s the United States was a principal supplier of arsenic trioxide for domestic use. However, it must now rely entirely on imports. Table 4 gives the annual world production and United States imports of arsenic trioxide for the latter part of the 1980s.

Table 4. World Production and U.S. Imports of Arsenic Trioxide[a]

Year	World production, t	U.S. imports, t	Value, $	Price, $/t
1986	53,173	25,728	16,347,000	635.38
1987	52,351	26,843	16,800,000	625.86
1988	52,047	28,056	16,461,000	586.72
1989	52,390	28,348	13,526,000	477.14

[a]Refs. 10, 11.

Alloys. Arsenic metal is used primarily in alloys in combination with lead and, to a lesser extent, copper.

Trace quantities of arsenic are added to lead–antimony grid alloys used in lead–acid batteries (18) (see BATTERIES, LEAD–ACID). The addition of arsenic permits the use of a lower antimony content, thus minimizing the self-discharging characteristics of the batteries that result from higher antimony concentrations.

No significant loss in hardness and casting characteristics of the grid alloy is observed (19,20).

Arsenic added in amounts of 0.1–3% improves the properties of lead-base babbitt alloys used for bearings (see BEARING MATERIALS). Arsenic (up to 0.75%), has been added to type metal to increase hardness and castability (21). Addition of arsenic (0.1%) produces a desirable fine-grain effect in electrotype metal without appreciably affecting the hardness or ductility. Arsenic (0.5–2%) improves the sphericity of lead ammunition. Automotive body solder of the composition 92% Pb, 5.0% Sb, and 2.5% Sn, contains 0.50% arsenic (see SOLDERS AND BRAZING ALLOYS).

Minor additions of arsenic (0.02–0.5%) to copper (qv) and copper alloys (qv) raise the recrystallization temperature and improve corrosion resistance. In some brass alloys, small amounts of arsenic inhibit dezincification (22), and minimize season cracking.

Phosphorized deoxidized arsenical copper (alloy 142 (23)) is used for heat exchangers and condenser tubes. Copper–arsenical leaded Muntz metal (alloy 366), Admiralty brass (alloy 443), naval brass (alloy 465), and aluminum brass (alloy 687), all find use in condensers, evaporators, ferrules, and heat exchanger and distillation tubes. The composition of these alloys is listed in Table 5.

Table 5. Arsenical Copper Alloys

Alloy number	Composition, wt %						
	Cu	As	Pb	Fe	Sn	P	Al
142	99.4	0.015–0.50				0.015–0.040	
366[a]	58–61	0.02–0.10	0.40–0.9	0.15	0.25		
443[a]	70–73	0.02–0.10	0.07	0.06	0.9–1.2		
465[a]	59–62	0.02–0.10	0.20	0.10	0.5–1.0		
687[a]	76–79	0.02–0.10	0.07	0.06			1.8–2.5

[a]Zinc constitutes the remainder of composition.

Semiconductor Applications. A limited but important demand for metallic arsenic of 99.99% and greater (exceeding 99.999+%) purities exists in semiconductor applications (see SEMICONDUCTORS). This high purity arsenic may be prepared by the reduction of a highly purified arsenic compound using a high purity gaseous or solid reductant.

High purity (HP) arsenic, when alloyed with aluminum, gallium, and indium, form the III–V semiconductor compounds, aluminum arsenide [*22831-42-1*], AlAs, gallium arsenide [*1303-00-0*], GaAs, and indium arsenide [*1303-11-3*], InAs, respectively. These compounds or variations such as GaAlAs, GaAsP, and InGaAs, are used in device manufacture. Gallium aluminum arsenide, GaAlAs, is used in the manufacture of solar cells having efficiencies exceeding 20% (see SOLAR ENERGY). $GaAs_xP_y$ is used in the manufacture of light emitting diodes (LEDs) of red light; yellow LEDs are produced by increasing the phosphorus content (see LIGHT GENERATION, LIGHT EMITTING DIODES). GaAs infrared emitters and infrared detectors have use in fiber optic applications (see FIBER OPTICS).

GaAs is also used in the manufacture of microwave devices (see MICROWAVE TECHNOLOGY), integrated circuits (qv), lasers (qv), laser windows, and optoelectronic devices. Indium arsenide has been used to produce Hall effect and infrared devices (see INFRARED AND RAMAN SPECTROSCOPY; MAGNETOHYDRODYNAMICS). InGaAs is used as lasers and photodetectors (qv).

HP arsenic is used in the manufacture of photoreceptor arsenic–selenium alloys for xerographic plain paper copiers (see ELECTROPHOTOGRAPHY; REPROGRAPHY). The level of arsenic may be 0.5%, 5.0%, or 35% present as arsenic triselenide [*1303-36-2*], As_2Se_3.

Arsenic from the decomposition of high purity arsine gas may be used to produce epitaxial layers of III–V compounds, such as InAs, GaAs, AlAs, etc, and as an *n*-type dopant in the production of germanium and silicon semiconductor devices. A group of low melting glasses based on the use of high purity arsenic (24–27) were developed for semiconductor and infrared applications.

Health and Safety Factors

The toxicity of arsenic ranges from very low to extremely high depending on the chemical state. Metallic arsenic and arsenious sulfide [*1303-33-9*], As_2S_3, have low toxicity. Arsine is extremely toxic. The toxicity of other organic and inorganic arsenic compounds varies (28).

Arsenic is classified as a carcinogen by the International Agency for Research on Cancer (IARC) (29). An association between high and lengthy exposures to inorganic arsenic compounds and cancer has been reported (30), but evidence supporting this relationship is equivocal (31). Ulceration and perforation of the nasal septum is caused by airborne As_2O_3 if proper precautions are not observed. However, these injuries have not been associated with malignancy (32).

The handling of arsenic in the workplace should be in compliance with the Occupational Safety and Health Administration (OSHA) regulations: the maximum permissible exposure limit for arsenic in the workplace is 10 $\mu g/m^3$ of air as determined as an average over an 8-h period (33).

Precaution should be taken to avoid accidental generation of arsine gas; the maximum permitted exposure is 0.05 ppm in air per 8-h period five days per week (34). Disposal of arsenical products should be in compliance with Federal and local government environmental regulations.

BIBLIOGRAPHY

"Arsenic" in *ECT* 1st ed., Vol. 2, pp. 113–118, by G. A. Roush, "Mineral Industry, Arsenic;" in *ECT* 2nd ed., Vol. 2, pp. 711–717, by S. C. Carapella, Jr., American Smelting and Refining Company; in *ECT* 3rd ed., Vol. 3, pp. 243–250, by S. C. Carapella, Jr., American Smelting and Refining Company.

1. J. W. Mellor, *Comprehensive Treatise on Inorganic and Theoretical Chemistry,* Vol. 9, Longmans, Green & Co., Inc., New York, 1930, pp. 3–9, 16–19.
2. M. C. Sneed and R. C. Brasted, *Comprehensive Inorganic Chemistry,* Vol. 5, D. Van Nostrand Co., Inc., Princeton, N.J., 1956, p. 135.

3. N. H. Furman, ed., *Scott's Standard Methods of Chemical Analysis,* Vol. 1, 6th ed., D. Van Nostrand Co., Inc., Princeton, N.J., 1962, pp. 106–137.
4. *American Institute of Physics Handbook,* McGraw-Hill Book Co., Inc., New York, 1957, Sect. 7, Chapt. 9.
5. C. H. Jones, *Chem. Met. Eng.* **23,** 957–960 (1920).
6. S. Wallden and H. Hilmer, *Ulmanns Encyclopadie der Technischen Chemie,* Verlag Chemie, GmbH, Weinheim, Germany, 1974, pp. 53–55.
7. R. C. Vickery and R. W. Edwards, *Metallurgia* **36,** 3 (1947).
8. U.S. Pat. 3,567,370 (March 2, 1971), S. T. Henriksson (to Boliden Akliebolag).
9. W. C. Smith in D. M. Liddell, ed., *Handbook of Non Ferrous Metallurgy,* Vol. 2, McGraw-Hill Book Co., Inc., New York, 1945, pp. 94–103.
10. *Minerals Yearbook, Minor Metals Bulletin (Arsenic Section),* Bureau of Mines, U.S. Dept. of the Interior, Washington D.C., 1988.
11. J. R. Loebenstein, *Mineral Industry Surveys, Arsenic in 1989 (Annual Review),* Bureau of Mines, U.S. Dept. of Interior, Washington D.C., 1989.
12. *Information Bulletin,* No. 5 and 6, Arsenic Development Committee, Rue LaFayette, Paris, France, 1965.
13. U.S. Pat. 3,130,035 (Apr. 21, 1964), W. H. Culver (to Pennsalt Chemicals Corp.).
14. S. deLajarte, *Arsenic in Glass,* Arsenic Development Committee, Rue LaFayette, Paris, France, Mar. 1969.
15. R. J. Bauer, in W. H. Lederer and R. J. Fensterheim, eds., *Arsenic: Industrial, Biomedical, Environmental Perspectives,* Van Nostrand Reinhold Co., New York, 1983, pp. 45–55.
16. D. D. Nicholas, ed., *Preservative and Preservative Systems,* Vol. II, Syracuse University Press, Syracuse, N.Y., 1973, pp. 66–84.
17. W. J. Baldwin in Ref. 15, pp. 99–110.
18. W. Hofmann, *Lead and Lead Alloys,* English translation by Lead Development Association, Springer-Verlag, Berlin, Germany, 1970, pp. 349–356.
19. U.S. Pat. 3,801,310 (Apr. 2, 1974), S. Nijhawan (to Varta Aktiegelsellschaft).
20. Ger. Offens. 2,312,322; 2,412,320; 2,412,321 (Sept. 19, 1974), K. Peters (to Electric Power Storage Ltd.).
21. *Metals Handbook,* 8th ed., American Society for Metals, Ohio, 1961, pp. 1061–1062.
22. E. E. Langenegger and F. Robinson, *Corrosion* **25,** 137 (1969).
23. *Copper and Copper Alloy Data Permanent File No. 1,* Copper Development Association, London, England, 1968.
24. S. Flaschen, D. Pearson, and W. Northover, *J. Am. Ceram. Soc.* **43,** 274 (1960); *J. Appl. Phys.* **31,** 219 (1960).
25. U.S. Pat. 2,883,292; 2,883,295 (Apr. 25, 1959) and U.S. Pat. 3,241,986 (Mar. 22, 1966), J. Jerger, Jr. (to Servo Corp.).
26. U.S. Pat. 3,154,424 (Oct. 27, 1964), L. Bailey and co-workers (to Texas Instruments).
27. R. Hilton, *Appl. Optics* **5,** 1877 (1966).
28. *Occupational Exposure to Inorganic Arsenic,* U.S. Dept. of HEW National Institute of Safety and Health, Washington, D.C., 1973.
29. "IARC Monographs on the Evaluation of the Carcinogenic Risks of Chemicals to Humans" **23,** *Some Metals and Metallic Compounds,* IARC, Lyon, France, 1980.
30. *Fed. Regist.* **40**(14), OSHA, Department of Labor, Jan. 21, 1975.
31. Ref. 15, pp. 151–154, 166–169, 203–209, 245–254.
32. P. Drinker and T. Hatch, *Industrial Dust,* 2nd ed., McGraw-Hill Book Co., Inc., New York, 1954.
33. *Code of Federal Regulations,* Title **29,** Part 1910.1018, U.S. Food and Drug Administration, Washington, D.C., revised July 11, 1988.
34. *Code of Federal Regulations,* Title **29,** Part 1910, OSHA, Washington, D.C., May 21, 1971.

General References

W. H. Lederer and R. J. Fensterheim, eds., *Arsenic: Industrial, Biomedical, Environmental Perspectives,* Van Nostrand Reinhold Co., Inc., N.Y., 1983.

R. Reddy, ed., *Arsenic Metallurgy-Fundamental and Applications,* The Metallurgical Society (AIME), Warrendale, Pa., 1988.

H. Quiring, *Die Metallischen Rohstoffe-Arsen,* Vol. 8, Ferdinand Enke, Stuttgart, Germany, 1946.

H. C. Beard, *The Radiochemistry of Arsenic,* NAS-NS 3002, U.S. Atomic Energy Commission, Washington D.C., Jan. 1960.

O. Herneryd, O. Sundstrom, and A. Norro, *J. Metals* **6**(3), 330 (Mar. 1954).

S. C. Carapella, Jr.
Consultant

ARSENIC COMPOUNDS

Arsenic is the third member of the nitrogen family of elements and hence possesses an outermost shell having the electron configuration of $4s^24p^3$. The most common oxidation states of arsenic are -3, $+3$, and $+5$, although compounds containing the simple As^{3-}, As^{3+}, and As^{5+} ions are unknown. In the majority of arsenic compounds the arsenic atom is in the tetrahedral valence state. Compounds in which the arsenic atom is three coordinate are assumed to contain the tetrahedrally hybridized arsenic atom with a lone pair of electrons in one of the hybrid orbitals. Arsenic also forms a number of hypervalent compounds with five or six ligands attached to the central atom. The geometry of such species is generally trigonal–bipyramidal or octahedral. The existence of these compounds has been ascribed to the presence of *d*-orbitals in the valence shell of the arsenic atom. This view has, however, been challenged (1). It has been suggested that the three equatorial ligands in the trigonal–bipyramidal molecules are attached by conventional two-center-two-electron bonds and that the axial ligands are attached by three-center bonds. Participation of *d*-orbitals may slightly stabilize the molecule. Similar reasoning may apply to hypervalent compounds with octahedral geometry.

Arsenic compounds have numerous practical applications. Although a variety of inorganic and organic arsenicals are used in commerce, arsenic trioxide [*1327-53-3*], As_2O_3, accounted for 98% of the arsenic consumed in 1988 (2).

The commercial uses of arsenic compounds in 1988, measured in terms of elemental arsenic, are wood (qv) preservatives, 69%; agricultural products (herbicides (qv) and desiccants (qv)), 23%; glass (qv), 4%; nonferrous alloys and electronics, 2%; and animal feed additives and pharmaceuticals (qv), 2% (see Feeds and feed additives). Chromated copper arsenate (CCA) [*11125-95-4*] is the most widely used arsenic-based wood preservative. The Environmental Protection Agency has, however, restricted the use of arsenical wood preservatives to certified applicators.

The principal sources of imported arsenicals by the United States for the years 1986, 1987, and 1988 are shown in Table 1.

Table 1. Annual Imports of Arsenicals, Metric Tons

	Year		
Exporting country	1986	1987	1988
Canada	1,924	2,012	2,086
Chile	1,659	4,800	6,709
France	6,274	5,341	6,909
Mexico	4,408	4,457	4,187
Sweden	7,069	4,824	3,664
others[a]	4,394	5,409	4,501
Total	*25,728*	*26,843*	*28,056*

[a]Others includes Belgium-Luxembourg, China, Hong Kong, Japan, Philippines, South Africa, and Germany.

The primary world producers of arsenic trioxide for 1988 were Chile, 7,000 t; France, 10,000 t; Sweden, 10,000 t; and Russia, 8,100 t. The December 31, 1988 total world production capacity was 70,000 t. The price of arsenic trioxide (95% pure) in 1986 was \$0.726/kg, fob, Tacoma, Washington; the price of Mexican, 99.13% arsenic trioxide in 1986 and 1987 was \$0.968/kg, fob, Laredo, Texas.

Arsenic compounds must be considered extremely poisonous. Dust or fumes irritate mucous membranes and lead to arsenical poisoning. When swallowed they irritate the stomach and affect the heart, liver, and kidneys. Nervousness, thirst, vomiting, diarrhea, cyanosis, and collapse are among the symptoms of arsenical poisoning (3). In spite of the toxicity of arsenic compounds, there is evidence that arsenic is an essential nutrient for several animal species (4).

Analysis

Although arsenic in quantities in excess of 1 mg can be determined either gravimetrically or titrimetrically, the former method is now seldom used. Arsenic in the +3 oxidation state can be titrated using iodine or potassium bromate, iodate, or permanganate. A number of organic oxidimetric titrating agents have been used. Such reagents are usually *N*-haloamides or *N*-haloimides (see CHLORAMINES AND BROMAMINES). Thus bromamine-B (5), chlorobromamine-B (6), bromohydantoin (1-bromo-5,5-dimethylhydantoin) (7), and *N*-chlorosuccinimide (8) are among the compounds suggested for this purpose. Although these titrimetric methods can be adapted to the determination of micro amounts of arsenic, it is more common to determine arsenic colorimetrically, especially in food, by the formation of a diethyldithiocarbamate or molybdenum blue complex (9). Trace amounts of arsenic are usually determined by neutron activation or atomic absorption methods. A detailed description of the determination of arsenic in foods by atomic absorption is available (10). A comparison of the following methods for the determination of traces of arsenic has been made: inductively coupled plasma atomic emission spectrometry, flow injection hydride generation atomic emission spectrometry, graphite furnace atomic absorption, combined furnace–flame atomic absorption, and nondestructive neutron activation (11). For determining arsenic in biological materials, graphite furnace atomic absorption is the most suitable. Another comparison of graphite furnace atomic absorption and neutron activation for the determination of arsenic in biological materials has been reported (12).

Inorganic Compounds

Arsenic Hydrides. Although there are occasionally reports of other arsenic hydrides, eg, As_2H_4, As_2H_2 (or AsH), and As_4H_2, the only well-characterized binary compound of arsenic and hydrogen is arsine.

Arsine [*7784-42-1*], AsH_3, is a colorless, exceedingly poisonous gas with an unpleasant garliclike odor; mp, −116.3°C; bp, −62.4°C; density of liquid at −64.3°C, 1.640 g/mL; $\Delta H^\circ_{f,298}$ 66.44 kJ/mol (16 kcal/mol); ΔS°_{298} 222.7 J/(mol·K) (53 cal/(mol·K)) (13). It is trigonal pyramidal in shape with As—H bond distances of 151.9 pm and H—As—H bond angles of 91.8°. Arsine is soluble to the extent of 20 mL at 101 kPa (1 atm) per 100 g of water RT. It shows no tendency to accept a proton from water to form an onium ion as does ammonia. At temperatures below −10°C or under pressure, arsine hexahydrate [*65423-90-7*], $AsH_3{\cdot}6H_2O$, is formed.

Arsine is formed when any inorganic arsenic-bearing material is brought in contact with zinc and sulfuric acid. The arsenides of the electropositive metals are decomposed with the formation of arsine by water or acid. Calcium arsenide [*12255-53-7*], Ca_3As_2, treated with water gives a 14% yield of arsine. Better yields (60–90%) are obtained by decomposing a solution of sodium arsenide [*12044-25-6*], Na_3As, in liquid ammonia with ammonium bromide (14,15). Arsine may be accidentally formed by the reaction of arsenic impurities in commercial acids stored in metal tanks, so that a test should be made for arsine before entry is made into such vessels. It is injurious in 1:20,000 dilution, and a few inhalations may cause death from anoxia or pulmonary edema. Although arsine is toxic to higher living forms, some strains of bacteria and fungi are capable of producing arsine or arsine derivatives.

Arsine is not particularly stable and starts to decompose into its elements well below 300°C. If moisture is present, light effects the decomposition. Arsine is a good reducing agent, capable of reducing many substances. It is not oxidized by air at room temperatures but may be ignited with the formation of arsenic, arsenic trioxide, or arsenic pentoxide, depending upon the supply of air. It reacts with dilute silver nitrate solution with the formation of metallic silver, and with mercuric chloride to give mercuric arsenide [*65496-41-5*], As_2Hg_3. Chlorine reacts with arsine to give hydrogen chloride and arsenic [*7440-38-2*]. However, at low temperatures the action of chlorine upon arsine produce chloroarsines, AsH_2Cl and $AsHCl_2$. These are relatively unstable yellow solids (16).

Arsine is used for the preparation of gallium arsenide [*1303-00-0*], GaAs, (17), and there are numerous patents covering this subject (see ARSENIC AND ARSENIC ALLOYS). The conversion of a monomeric arsinogallane to gallium arsenide has also been described (18). Gallium arsenide has important applications in the field of optoelectronic and microwave devices (see LASERS; MICROWAVE TECHNOLOGY; PHOTODETECTORS).

Other Arsenic Hydrides. Diarsine [*15942-63-9*], As_2H_4, occurs as a by-product in the preparation of arsine by treatment of a magnesium aluminum arsenide alloy with dilute sulfuric acid and also may be prepared by passing arsine at low pressure through an ozonizer-type discharge tube (19). Diarsine is fairly stable as a gas but quite unstable (above −100°C) in condensed phases. The ΔH_f° for diarsine is +117 ± 4 kJ/mol (28 ± 1 kcal/mol) and the As—As bond strength is 167 kJ/mol (40 kcal/mol) (19). In addition, two other hydrides of arsenic have been reported but their chemical individuality awaits clarification. Reduction of arsenic(III) compounds by stannous chloride in hydrochloric acid yields a brown amorphous powder corresponding to the formula of arsenic monohydride [*14452-84-7*], As_2H_2 or AsH (20). The other solid arsenic hydride is reported to be hydrogen diarsenide [*65496-40-4*], As_4H_2 or As_2H, and is formed by oxidation of arsine with stannic chloride.

Arsenic Halides. Arsenic forms a complete series of trihalides, but arsenic pentafluoride is the only well-known simple pentahalide. All of the arsenic halides, the physical properties of which are given in Table 2, are covalent compounds that hydrolyze in the presence of water. The trihalides form pyramidal molecules similar to the trivalent phosphorus analogues and may be prepared by direct combination of the elements.

Table 2. Physical Properties of Arsenic Halides

Arsenic halide	CAS Registry Number	Color and physical state at 25°C	Mp, °C	Bp, °C	Specific gravity[a]	Heat of formation, ΔH°_{298}, kJ/mol[b,c]	Entropy, S°_{298}, J/(mol·K)[b,c]
arsenic trifluoride (AsF_3)	[*7784-35-2*]	colorless liquid	−6.0	62.8	2.666^0	−956.25	
arsenic pentafluoride (AsF_5)	[*7784-36-3*]	colorless gas	−79.8	2.8	2.33^{-53}		
arsenic trichloride ($AsCl_3$)	[*7784-34-1*]	colorless liquid	−16.2	130.2	2.205^0	−305	208
arsenic tribromide ($AsBr_3$)	[*7784-33-0*]	yellow solid	31.2	221	3.66^{15}	−197	363.8[d]
arsenic triiodide (AsI_3)	[*7784-45-4*]	red solid	140.4	ca 400	4.39^{15}	−58.2	213.0

[a]Temperature, °C, of measurement given as a superscript.
[b]Ref. 13.
[c]To convert J to cal, divide by 4.184.
[d]Gaseous phase.

Arsenic trifluoride (arsenic(III) fluoride), AsF_3, can be prepared by reaction of arsenic trioxide with a mixture of sulfuric acid and calcium fluoride or even better with fluorosulfonic acid. Chlorine reacts with ice-cold arsenic trifluoride to produce a hygroscopic solid compound, arsenic dichloride trifluoride [*14933-43-8*],

$AsCl_2F_3$, consisting of $AsCl_4^+$ and AsF_6^- ions (21). Arsenic trifluoride forms a stable adduct, $2AsF_3{\cdot}3SO_3$, with sulfur trioxide and reacts with nitrosyl fluoride to give nitrosonium hexafluoroarsenate(V) [*18535-07-4*], $[NO][AsF_6]$.

Arsenic pentafluoride (arsenic(V) fluoride), AsF_5, is a colorless gas that condenses to a yellow liquid; its dielectric constant is 12.8 at 20°C. It is formed by reaction of a mixture of bromine and antimony pentafluoride with arsenic trifluoride. The molecule is a trigonal bipyramid and is somewhat dissociated as indicated by vapor density measurements.

Arsenic trichloride (arsenic(III) chloride), $AsCl_3$, is the most common and important halide of arsenic. It may be formed by spontaneous combination of the elements and, in addition, by the following reactions: (*1*) chlorine with arsenic trioxide; (*2*) sulfur monochloride, S_2Cl_2, or a mixture of S_2Cl_2 and chlorine, with arsenic trioxide; and (*3*) arsenic trioxide with concentrated hydrochloric acid or with a mixture of sulfuric acid and a chloride.

Arsenic pentachloride [*22441-45-8*] (arsenic(V) chloride), $AsCl_5$, is produced by irradiation of a solution of $AsCl_3$ in chlorine at −105°C using ultraviolet light (22). Arsenic pentachloride is a pale yellow solid (presumably because of an entrained chlorine impurity) which melts with partial decomposition at −50°C. Raman spectra show that it is a trigonal–bipyramidal molecule both in solution and in the solid state.

Arsenic tribromide (arsenic(III) bromide), $AsBr_3$, is similar to the trichloride. The dielectric constant at 35°C is 8.33. The compound is usually made by treating arsenic with a solution of bromine in carbon disulfide.

Arsenic triiodide (arsenic(III) iodide), AsI_3, can be precipitated from a hot solution of trivalent arsenic in hydrochloric acid by the addition of potassium iodide, or it can be formed by treating elemental arsenic with a solution of iodine in carbon disulfide. It is not as easily hydrolyzed as the other arsenic halides, but it decomposes slowly in air at 100°C (rapidly at 200°C) to give a mixture of iodine, arsenic trioxide, and elemental arsenic. Solutions of AsI_3 are unstable, particularly in the presence of moisture.

Also formed by the direct combination of the elements is a red solid compound, arsenic diiodide [*13453-17-3*], As_2I_4 or AsI_2, which melts at 130°C and dissolves in organic solvents. Treatment of this compound with water causes disproportionation.

$$3\ As_2I_4 \longrightarrow 4\ AsI_3 + 2\ As$$

Arsenic Oxides and Acids. The only arsenic oxides of commercial importance are the trioxide and the pentoxide. These are readily soluble in alkaline solution, forming arsenites and arsenates, respectively.

Arsenic trioxide [*1327-53-3*] (arsenic(III) oxide, arsenic sesquioxide, arsenous oxide, white arsenic, arsenic), As_2O_3, is the most important arsenic compound of commerce. The octahedral or cubic modification, arsenolite [*1303-24-8*], $\Delta H^\circ_{f,298}$, −1313.9 kJ/mol (−314 kcal/mol); S°_{298}, 214 J/(mol·K) (51 cal/(mol·K)), is the most common form and has been known from early times. The monoclinic form, claudetite [*13473-03-5*], $\Delta H^\circ_{f,298}$, −1310 kJ/mol (−313 kcal/mol); S°_{298}, 230 J/(mol·K) (55 cal/(mol·K)) (13), is the thermodynamically stable modification consisting of sheets of AsO_3 pyramids sharing oxygen. The octahedral variety is a white solid that sublimes freely above 135°C and melts at 275°C under its own vapor pressure.

The octahedral crystal consists of As_4O_6 molecules arranged in a diamond-type lattice. At temperatures above 800°C dissociation to As_2O_3 can be detected, and at 1800°C the molecular weight corresponds to As_2O_3. In nitrobenzene solution the molecular weight corresponds to the formula As_4O_6. Condensation of the vapor above 250°C results usually in the formation of a glassy phase. This is a supercooled liquid which readily devitrifies to form octahedral crystals.

Arsenic trioxide may be made by burning arsenic in air or by the hydrolysis of an arsenic trihalide. Commercially, it is obtained by roasting arsenopyrite [*1303-18-0*], FeAsS. It dissolves in water to a slight extent (1.7 g/100 g water at 25°C) to form a weakly acidic solution which probably contains the species H_3AsO_3, orthoarsenous acid [*36465-76-6*]. The oxide is amphoteric and hence soluble in acids and bases. It is frequently used as a primary analytical standard in oxidimetry because it is readily attainable in a high state of purity and is quantitatively oxidized by many reagents commonly used in volumetric analysis, eg, dichromate, nitric acid, hypochlorite, and iron(III).

Arsenous Acids and the Arsenites. Arsenous acid [*13464-58-9*], AsH_3O_3, is known to exist only in solution. It is a weak acid with a dissociation constant of 8×10^{-16} at 25°C. The free acid apparently has three OH groups attached directly to the arsenic atom and hence is not analogous to phosphorous acid. A number of complex arsenites are known, among which are copper acetate arsenite [*12002-03-8*], $Cu_2(C_2H_3O_2)(AsO_3)$, Paris green, and cupric hydrogen arsenite [*10290-12-7*], $CuHAsO_3$, Scheele's green.

Arsenic pentoxide [*1303-28-2*] (arsenic oxide, arsenic(V) oxide), As_2O_5, is made up of equal numbers of AsO_6 octahedra and AsO_4 tetrahedra sharing corner oxygens to give cross-linked strands (23). The compound is thermally unstable and begins to decompose near the melting point, ca 300°C. The vapor obtained is wholly dissociated into oxygen and arsenic trioxide. The pentoxide can be prepared by reaction of arsenic or of arsenic trioxide with oxygen under pressure. The best method of preparation seems to be the dehydration of crystalline arsenic acid at temperatures above 200°C. Arsenic pentoxide is an oxidizing agent capable of liberating chlorine from hydrogen chloride. It is deliquescent and dissolves only slowly in water, although it is soluble, 230 g/100 g of water at 20°C.

Arsenic Acids and the Arsenates. Commercial arsenic acid, corresponds to the composition, one mole of arsenic pentoxide to four moles of water, and probably is the arsenic acid hemihydrate [*7774-41-6*], $H_3AsO_4{\cdot}0.5H_2O$. It is obtained by treatment of arsenic trioxide with concentrated nitric acid. Solutions of this substance or of arsenic pentoxide in water behave as triprotic acids with successive dissociation constants: $K_1 = 5.6 \times 10^{-3}$, $K_2 = 1.7 \times 10^{-7}$, and $K_3 = 3.0 \times 10^{-12}$. Hydrated arsenic acid loses water when heated at about 100°C with the formation of pyroarsenic acid [*13453-15-1*], $H_4As_2O_7$. At higher temperatures additional water is lost to yield metaarsenic acid [*10102-53-1*], $HAsO_3$. Arsenates derived from each of these acids are known. Treatment of a meta- or pyroarsenate with cold water gives the orthoarsenate.

Arsenates are oxidizing agents and are reduced by concentrated hydrochloric acid or sulfur dioxide. Treatment of a solution of orthoarsenate with silver nitrate in neutral solution results in the formation of a chocolate-brown precipitate of silver orthoarsenate, Ag_3AsO_4, which may be used as a test to distinguish arsenates from phosphates. With hydrofluoric acid, orthoarsenate solutions yield hexafluoroarsenates, eg, potassium hexafluoroarsenate [17029-22-0], $(KAsF_6)_2{\cdot}H_2O$. Arsenates of calcium or lead are used as insecticides; sodium arsenate is used in printing inks and as a mordant.

Arsenous arsenate [12505-65-6] (arsenic dioxide, arsenic tetroxide), As_2O_4, is known and probably corresponds to $As(AsO_4)$.

Arsenic Sulfides. The physical properties of the common arsenic sulfides are given in Table 3. Numerous arsenic sulfides have been reported as well as compounds containing the $As_3S_4^+$ cation [77825-63-9] (24).

Table 3. Physical Properties of Arsenic Sulfides

Arsenic sulfides	CAS Registry Number	Molecular formula	Color and physical state at 25°C	Heat of formation, ΔH_{298}, kJ/mol[a,b]
arsenous sulfide (orpiment)[c]	[12255-89-9]	As_2S_3	yellow solid	−338
arsenic sulfide (realgar)	[12279-90-2]	As_4S_4	red or orange solid	−285
arsenic pentasulfide	[1303-34-0]	As_4S_{10}	yellow solid	
tetraarsenic trisulfide	[1303-41-9]	As_4S_3	orange-yellow	
tetraarsenic pentasulfide	[25114-28-7]	As_4S_5		

[a]Ref. 13.
[b]To convert J to cal, divide by 4.184.
[c]The entropy, S°_{298}, is 327 J/(mol·K)[b] (13).

Arsenous sulfide (arsenic(III) sulfide, arsenic sesquisulfide, arsenic red), As_2S_3, starts to sublime before melting at 320°C. The bp is 707°C. The solid as well as the vapor consists of discrete molecules similar to those of arsenic trioxide. A red modification is known, but it has not been definitely determined whether this represents another crystal configuration or is because of the presence of impurities. Arsenous sulfide is almost insoluble in water but dissolves in liquid ammonia. It may be prepared by heating arsenic trioxide and elemental sulfur or by passing hydrogen sulfide through a solution of trivalent arsenic in dilute hydrochloric acid. The compound is insoluble in acid, but it dissolves readily in many bases or in alkali metal sulfides to give dithioarsenites, eg, in sodium sulfide solution, sodium dithioarsenite [19102-63-7], $NaAsS_2$ is formed. In polysulfide solutions, arsenous sulfide dissolves forming the thioarsenate ion, AsS_4^{3-}. Chlorine converts arsenous sulfide to arsenic trichloride and sulfur chloride, S_2Cl_2. Arsenous sulfide burns in air yielding arsenic trioxide and sulfur dioxide; it is oxidized by nitric acid. Arsenic sulfide [65113-28-2], As_4S_6, is also known.

Tetraarsenic trisulfide (dimorphite), As_4S_3, exists in α- and β- modifications. The compound may be synthesized by direct combination of appropriate quantities of the elements. The β- form is stable at room temperature, but changes to the α- form upon heating to 130°C. The As_4S_3 molecule consists of three As atoms in a triangle. Above each arsenic is a sulfur, and then these three sulfur atoms are capped by an arsenic (25–27). As_4S_3 is sparingly soluble in CS_2. If the CS_2 contains some dissolved sulfur, As_4S_3 reacts to form As_4S_4 if the solution is illuminated for a long period of time. The compound forms a broad metastable series of solid solutions with P_4S_3 that contain six different cage molecules (28).

Arsenic sulfide (arsenic monosulfide), As_4S_4, occurs naturally as the mineral realgar. At 267°C it changes into a black allotropic modification; the compound melts at 307°C. The vapor consists of As_4S_4 molecules at 550°C, boils at 565°C, but it begins to dissociate at 800°C, and at 1000°C the molecular weight corresponds to As_2S_2. The compound is made commercially by heating a mixture of iron pyrites and arsenopyrite or by heating arsenic trioxide with sulfur. It is also made by prolonged treatment of arsenous sulfide with boiling aqueous sodium carbonate or by heating a sodium bicarbonate–arsenous sulfide mixture in a sealed tube. It is unaffected by water, but inflames in chlorine, and is oxidized by nitric acid. It is insoluble in hot concentrated hydrochloric acid, but dissolves when warmed with solutions of alkali hydroxides or sulfides with the formation of thioarsenites and some elemental arsenic. The compound is used in pyrotechnics and at times as a depilatory in the leather (qv) industry.

Arsenic pentasulfide (arsenic(V) sulfide), As_4S_{10}, is stable in air up to 95°C, but at higher temperatures begins to dissociate into arsenous sulfide and sulfur. It is prepared by the fusion of arsenic with sulfur followed by extraction with ammonia and reprecipitation at low temperatures by addition of hydrochloric acid. Arsenic pentasulfide is precipitated at low temperatures from strongly acidic arsenate solutions by a rapid stream of hydrogen sulfide. It is hydrolyzed by boiling with water, yielding arsenous acid and sulfur. Salts derived from a number of thioarsenic acids are formed from arsenic pentasulfide and alkali metal sulfides.

Other Arsenic Compounds. Arsenic trisulfate [*65496-42-6*], $As_2(SO_4)_3$, may be formed by the treatment of arsenic trioxide with SO_3 at 100°C. Arsenyl sulfate [*65423-91-8*], $(AsO)_2SO_4$, an easily hydrolyzed, hygroscopic crystalline compound, is formed when arsenic trioxide is dissolved in concentrated sulfuric acid. Arsenic triacetate [*5128-94-9*], $As(C_2H_3O_2)_3$, mp 82°C, bp 165–170°C, is easily hydrolyzed to a mixture of arsenous and acetic acids.

Organic Compounds

Arsenic combines readily with carbon to form a wide variety of compounds containing one or more As—C bonds. A number of examples are given in Table 4. These may be broadly divided into As(III) and As(V) compounds. The As(III) compounds contain from one to four organic groups; the As(V) compounds from one to six organic groups. The nomenclature used here, with a few minor exceptions, is that proposed by the International Union of Pure and Applied Chemistry (29). Treatises on organoarsenic compounds are available (30,31). Two noncritical lists of all organoarsenic compounds prepared or studied between 1937 and 1964 (32) and between 1965 and 1968 (33) have been published. Heterocyclic compounds containing arsenic in the ring are described in two books (34,35) and several older monographs on organoarsenic compounds exist that are largely of historic interest (36–38).

Primary and Secondary Arsines. Compared to the vast number of known organoarsenic compounds, relatively few primary ($RAsH_2$) and secondary (R_2AsH) arsines have been described. Primary arsines are commonly prepared by the zinc–hydrochloric acid reduction of substances containing one organic group

Table 4. Organic Arsenic Compounds

Compound	CAS Registry Number	Molecular formula	Structural formula
ethylarsine	[*593-59-9*]	C_2H_7As	$C_2H_5AsH_2$
diethylarsine	[*692-42-2*]	$C_4H_{11}As$	$(C_2H_5)_2AsH$
triphenylarsine	[*603-32-7*]	$C_{18}H_{15}As$	$(C_6H_5)_3As$
dimethylbromoarsine	[*676-71-1*]	C_2H_6AsBr	$(CH_3)_2AsBr$
methyldifluoroarsine	[*420-24-6*]	CH_3AsF_2	CH_3AsF_2
oxophenylarsine	[*637-03-6*]	C_6H_5AsO	C_6H_5AsO
phenylarsonous acid	[*25400-22-0*]	$C_6H_7AsO_2$	$C_6H_5As(OH)_2$
dimethylarsinous cyanide	[*683-45-4*]	C_3H_6AsN	$(CH_3)_2AsCN$
methyl diphenylarsinite	[*24582-54-5*]	$C_{13}H_{13}AsO$	$(C_6H_5)_2AsOCH_3$
tetrakis(trifluoromethyl)diarsine	[*360-56-5*]	$C_4As_2F_{12}$	$(CF_3)_2AsAs(CF_3)_2$
pentamethylpentaarsolane	[*20550-47-4*]	$C_5H_{15}As_5$	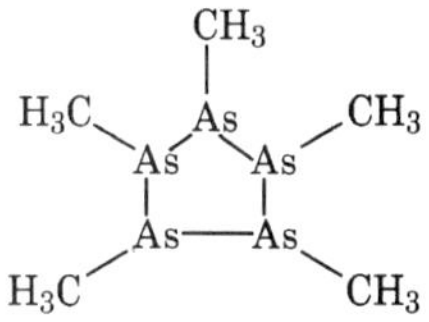
4-ethylarsenin	[*76782-94-0*]	C_7H_9As	
1-chloroarsolane	[*30077-24-8*]	C_4H_8AsCl	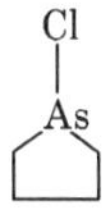
1*H*-arsole	[*4542-21-6*]	C_4H_5As	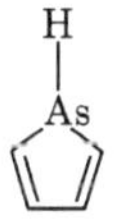
phenylarsonic acid	[*98-05-5*]	$C_6H_7AsO_3$	$C_6H_5AsO(OH)_2$
diphenylarsinic acid	[*4656-80-8*]	$C_{12}H_{11}AsO_2$	$(C_6H_5)_2AsO(OH)$
arsonoacetic acid	[*107-38-0*]	$C_2H_5AsO_5$	$H_2O_3AsCH_2CO_2H$
diethyl methylarsonate	[*14806-25-8*]	$C_5H_{13}AsO_3$	$CH_3AsO(OC_2H_5)_2$
triphenylarsine oxide	[*1153-05-5*]	$C_{18}H_{15}AsO$	$(C_6H_5)_3AsO$
tetrachlorophenylarsorane	[*29181-03-1*]	$C_6H_5AsCl_4$	$C_6H_5AsCl_4$
tetramethylarsonium perchlorate	[*84742-76-7*]	$C_4H_{12}AsClO_4$	$(CH_3)_4AsClO_4$
triphenylarsonium 2-propenylide	[*88329-28-6*]	$C_{21}H_{19}As$	$(C_6H_5)_3As{=}CH{-}CH{=}CH_2$

attached to arsenic (such as arsonic acids, dihaloarsines, and compounds with arsenic—arsenic bonds); the zinc is often amalgamated (39) or coated with copper (40). For example, trifluoromethylarsine [*420-42-8*], CH_2AsF_3, has been obtained in 98% yield from trifluoromethylarsonous diiodide [*353-91-3*], CF_3AsI_2, by using copper-coated zinc and hydrochloric acid:

$$CF_3AsI_2 + 2\ H_3O^+ + 2\ Zn \longrightarrow CF_3AsH_2 + 2\ Zn^{2+} + 2\ I^- + 2\ H_2O$$

Good yields of phenylarsine [*822-65-1*], C_6H_7As, have been obtained by the reaction of phenylarsenic tetrachloride [*29181-03-1*], $C_6H_5AsCl_4$, or phenyldichloroarsine [*696-28-6*], $C_6H_5AsCl_2$, with lithium aluminum hydride or lithium borohydride (41). Electrolytic reduction has also been used to convert arsonic acids to primary arsines (42). Another method for preparing primary arsines involves the reaction of arsine with calcium and subsequent addition of an alkyl halide. Thus methylarsine [*593-52-2*], CH_5As, is obtained in 80% yield (43):

$$Ca + 2\ AsH_3 \xrightarrow[NH_3]{\text{liquid}} Ca^{2+} + 2\ AsH_2^- + H_2$$

$$CH_3Cl + AsH_2^- \longrightarrow CH_3AsH_2 + Cl^-$$

Secondary arsines, which can be synthesized by methods analogous to those used for primary arsines, are obtained in good yields by the reduction of arsinic acids (44) or haloarsines (45) with amalgamated zinc and hydrochloric acid. They can also be prepared by the alkylation of primary arsenides (46):

$$RAsHLi + RX \longrightarrow R_2AsH + LiX$$

Another good method for preparing secondary arsines involves cleavage of an alkyl or aryl group from a tertiary arsine and subsequent hydrolysis of the resulting secondary arsenide (47,48):

$$R_3As + 2\ Na \xrightarrow[NH_3]{\text{liquid}} R_2AsNa + NaR$$

$$R_2AsNa + H_2O \longrightarrow R_2AsH + Na^+ + OH^-$$

Finally, secondary arsines can be obtained by the reductive cleavage of diarsines with mercury and hydrogen iodide (49) or with lithium aluminum hydride (50).

Methylarsine, trifluoromethylarsine, and bis(trifluoromethyl)arsine [*371-74-4*], C_2HAsF_6, are gases at room temperature; all other primary and secondary arsines are liquids or solids. These compounds are extremely sensitive to oxygen, and in some cases are spontaneously inflammable in air (45). They readily undergo addition reactions with alkenes (51), alkynes (52), aldehydes (qv) (53), ketones (qv) (54), isocyanates (55), and azo compounds (56). They also react with diborane (43) and a variety of other Lewis acids. Alkyl halides react with primary and secondary arsines to yield quaternary arsenic compounds (57).

Tertiary Arsines. An enormous number of trialkyl- and triarylarsines are known (32,33). They are usually prepared by the reaction of an organometallic compound with an arsenic trihalide, a haloarsine, or a dihaloarsine. The Grignard reagent is the organometallic compound most widely employed in these syntheses (58,59). Organolithium (60), organoaluminum (61), organotin (62), organocadmium (63), and organomercury (64) compounds have also been employed (see ORGANOMETALLICS, σ-BONDED ALKYLS AND ARYLS). Some triarylarsines (Ar_3As) can be prepared by the sodium-promoted interaction of an aryl halide and an arsenic trihalide (AsX_3) (65):

$$3\ ArX + AsX_3 + 6\ Na \longrightarrow Ar_3As + 6\ NaX$$

This type of reaction is sometimes considered a variant of the Wurtz-Fittig reaction.

Another excellent method for preparing tertiary arsines involves the interaction of a secondary arsenide and an alkyl or aryl halide (66). This method is of particular value in preparing unsymmetrical tertiary arsines (67).

Other methods for preparing tertiary arsines have been employed, but they have limited usefulness. These methods include the cleavage of quaternary arsonium compounds (68), the cyclodehydrohalogenation of haloarsines (69), the reaction of diazonium salts with arsenic trichloride in the presence of a metal such as zinc or iron (70), and the disproportionation of halo- or dihaloarsines (71).

Trimethylarsine [*593-88-4*], C_3H_9As, has been identified as the toxic volatile arsenical, once known as "Gosio gas," produced by the reaction of certain molds that grow on wallpaper paste and react with inorganic arsenic compounds present in the paper. A number of microorganisms can methylate arsenic trioxide and other arsenic-containing compounds to yield trimethylarsine. These microorganisms include *Scopulariopsis brevicaulis, Candida humicola,* and *Gliocladium roseum* (72).

Most trialkylarsines are volatile liquids with intensely disagreeable odors. They react readily with oxygen, and in some cases they ignite spontaneously when exposed to air. Triarylarsines are solids that can usually be handled in air without danger of oxidation. They are, however, easily converted to triarylarsine oxides with suitable oxidizing agents (73).

Tertiary arsines have been widely employed as ligands in a variety of transition metal complexes (74), and they appear to be useful in synthetic organic chemistry, eg, for the olefination of aldehydes (75). They have also been used for the formation of semiconductors (qv) by vapor-phase epitaxy (76), as catalysts or cocatalysts for a number of polymerization reactions (77), and for many other industrial purposes.

Haloarsines, Dihaloarsines, and Related Compounds. Halo- and dihaloarsines (R_2AsX and $RAsX_2$, where X = Cl or Br) are easily obtained by the reduction of the corresponding arsinic or arsonic acids in hydrochloric or hydrobromic acid. The usual reducing agent is sulfur dioxide catalyzed by iodide ion (78). Iodo- and diiodoarsines are similarly prepared by the reduction of arsinic or arsonic acids with hydriodic acid (79). Dimethylchloroarsine [*557-89-1*], C_2H_6AsCl, has been obtained by the reduction of dimethylarsinic acid [*75-60-5*], $C_2H_7AsO_2$, with sodium hypophosphate in hydrochloric acid (80). Fluoro- and difluoroarsines are best prepared from the corresponding chloro- or dichloroarsines by metathesis with an inorganic fluoride. Thus dimethylfluoroarsine [*420-23-5*], $C_2H_6AsF_2$, has been obtained by the following reaction (81):

$$(CH_3)_2AsCl + NH_4F \xrightarrow[85°C]{SbF_3} (CH_3)_2AsF + NH_4Cl$$

Other halo- and dihaloarsines can also be prepared by suitable metathetical reactions (82).

The alkylation of arsenic trihalides with organometallic reagents is another method of preparing halo- and dihaloarsines. The metals used in these syntheses include tin (62), cadmium (83), lithium (83), zinc (84), aluminum (85), and magnesium (86). These reactions are not commonly employed because they often give mixtures of haloarsines, dihaloarsines, and tertiary arsines. Good yields of chloro- or bromoarsines can be obtained, however, by the following procedure (86):

$$2\ RMgX + R'_2NAsCl_2 \longrightarrow R_2AsNR'_2 \xrightarrow{HX} R_2AsX$$

A Friedel-Crafts type of reaction between arsenic trichloride and diarylamines or diaryl ethers is a good method of preparing certain heterocyclic chloroarsines (34,35,87). For example, a good yield of 10-chloro-5,10-dihydrophenarsazine [*578-94-9*], $C_{12}H_9AsClN$, often known as Adamsite, is obtained by refluxing a mixture of diphenylamine and arsenic trichloride for several hours. Although this chloroarsine is a solid, it can be dispersed in air in the form of minute particles and employed as a chemical warfare agent (see CHEMICALS IN WAR). 2-Chlorovinyldichloroarsine [*541-25-3*], $C_2H_2AsCl_3$, is obtained in admixture with the corresponding chloroarsine and tertiary arsine by the interaction of acetylene and arsenic trichloride in the presence of a catalyst such as aluminum chloride or mercuric chloride (88). The common name for this dichloroarsine is Lewisite. It is an extremely toxic vesicant and has been proposed as a chemical warfare agent.

Dihalo- and haloarsines are very reactive substances. They are readily oxidized by halogens to compounds of the types $RAsX_4$ or R_2AsX_3 (89). Hydrolysis of dihaloarsines yields either arsonous acids, $RAs(OH)_2$, or their anhydrides, $(RAsO)_x$ (90). Methyloxoarsine, $(CH_3AsO)_x$, has been found to act as an oxygen transfer agent (91). Thus it reacts with compounds of the type $[C_5(CH_3)_5M(CO)_2]_2$, where M is Mo or W and $C_5(CH_3)_5$ is pentamethylcyclopentadienyl, to form the neutral metal-oxo clusters $[C_5(CH_3)_5]_2W_6O_{17}$ and $[C_5(CH_3)_5]_6Mo_8O_{16}$. Haloarsines on hydrolysis give arsinous acids, R_2AsOH, or their anhydrides, $(R_2As)_2O$ (92). In addition, dihalo- and haloarsines react with Grignard or lithium reagents to give tertiary arsines (93), with certain reducing agents to give primary or secondary arsines (41,42), with alcohols or alkoxides to give esters (94), with amines to give aminoarsines (95), with thiols to give sulfides, often known as thioarsenites (96), and with selenols to give selenides (97). They also undergo metathetical reactions with ammonium or metallic halides (81,82), pseudohalides (98), and carboxylates (99).

Diarsines and Diarsenes. Under certain conditions, the reduction of compounds with two organic groups attached to arsenic may give rise to tetraalkyl- or tetraaryldiarsines. Thus a number of diarsines have been obtained by the reduction of arsinic acids with phosphorous or hypophosphorous acid (100). Diarsines can also be prepared by the treatment of a metal dialkyl- or diarylarsenide with iodine (101) or a 1,2-dihaloethane (102).

Diarsines are extremely reactive compounds. Tetramethyldiarsine (cacodyl) [*471-35-2*], $C_4H_{12}As_2$, and tetraethyldiarsine [*612-08-8*], $C_8H_{20}As_2$, are spontaneously flammable in air. Tetraallyldiarsine is said to explode violently on being dropped into oxygen (103). Tetraphenyldiarsine [*2215-36-31*], $C_{24}H_{20}As_2$, is not spontaneously flammable in air. It does, however, react essentially quantitatively with moist air to give diphenylarsinic acid [*4656-80-8*], $C_{12}H_{11}AsO_2$, and bis(diphenylarsenic) oxide [*34262-48-1*], $C_{24}H_{20}As_2O$ (104):

$$2\ (C_6H_5)_2AsAs(C_6H_5)_2 + 2\ O_2 + H_2O \longrightarrow 2\ (C_6H_5)_2AsO_2H + (C_6H_5)_2AsOAs(C_6H_5)_2$$

Unlike many distibines and dibismuthines, diarsines do not appear to exhibit thermochromic effects (105).

Although the preparation of diarsenes, the arsenic analogues of azo compounds, was first claimed during the nineteenth century, it was not until 1983 that such a compound was unequivocally characterized (106). It was obtained in good yield by the base-promoted interaction of a primary arsine and a dichloroarsine, both of which were sterically hindered:

$$2,4,6\text{-}(t\text{-}C_4H_9)_3C_6H_2AsH_2 + [(CH_3)_3Si]_2CHAsCl_2 + 2\ DBU \xrightarrow{THF} 2,4,6\text{-}(t\text{-}C_4H_9)_3C_6H_2As{=}AsCH[Si(CH_3)_3]_2 + 2\ DBU{\cdot}HCl$$

where DBU = 1,5-diazabicyclo[5.4.0]undec-5-ene. The arsenic—arsenic distance in the diarsene is 0.22 nm, the shortest such distance ever reported.

Cyclic and Polymeric Substances Containing Arsenic—Arsenic Bonds. A number of organoarsenic compounds containing rings of four, five, or six arsenic atoms have been prepared (cyclic polyarsines). The first such four-membered ring compound to be adequately characterized, tetrakis (trifluoromethyl) tetrarsetane [*7547-15-1*], was obtained by the interaction of a diiodoarsine and mercury (107,108):

$$4\ CF_3AsI_2 + 4\ Hg \longrightarrow \begin{matrix} CF_3As\text{-}AsCF_3 \\ |\quad\ | \\ CF_3As\text{-}AsCF_3 \end{matrix} + 4\ HgI_2$$

Other compounds containing rings of four or five arsenic atoms can be prepared by the treatment of dichloroarsines with sodium (109). The compound $(C_6H_5As)_6$, once known as arsenobenzene [*20738-31-2*], is readily synthesized by the reduction of phenylarsonic acid with hypophosphorous acid (110). This substance contains a six-membered ring of arsenic atoms arranged in the chair form. The As—As distances are 0.24 nm, and the As As—As angles average 91°. The phenyl groups are arranged equatorially. The preparation of pentamethylpentaarsolane [*20550-47-4*], $(CH_3As)_5$, by two different methods has been described (111).

Cyclic polyarsines undergo a number of reactions with transition metal compounds to form complexes containing both As—As and As—metal bonds. The structural chemistry of these complexes has been the subject of a recent review (112).

A large number of polymeric substances, $(RAs)_x$ or $(ArAs)_x$, are also known (113). They are usually prepared by the reduction of arsonic acids with hypophosphorous acid (100,114) or sodium dithionite (115). Most of these polymers have not been well characterized. An insoluble, purple material, poly(methylarsinidene) [*26403-94-1*], $(CH_3As)_x$, prepared by the interaction of methylarsine and a dihalomethylarsine, however, has been shown by an x-ray investigation to have a ladderlike polymeric structure in which the inter-rung distances correspond to one-electron bonds (116):

$$\begin{matrix} \vdots & & \vdots \\ CH_3As & \text{—} & AsCH_3 \\ \vdots & & \vdots \\ CH_3As & \text{—} & AsCH_3 \\ \vdots & & \vdots \\ CH_3As & \text{—} & AsCH_3 \\ \vdots & & \vdots \end{matrix}$$

The substance appears to be a semiconductor with pronounced photoconductive properties (117).

Arsenin and Its Derivatives. Arsenin (arsabenzene) [*289-31-6*], C_5H_5As, the arsenic analogue of pyridine, can be prepared by the treatment of 1,4-dihydro-1,1-dibutylstannabenzene with arsenic trichloride (118):

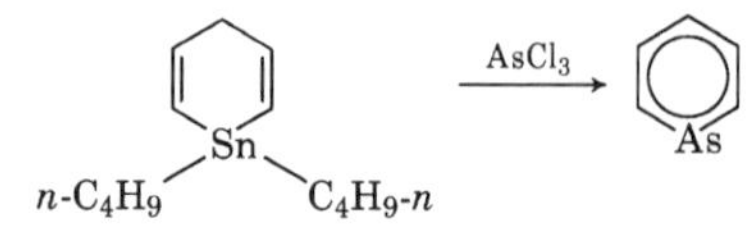

A large number of arsenin derivatives have also been studied. The potential aromaticity of this ring system has aroused considerable interest and has been investigated with the aid of nmr, uv, photoelectron, and microwave spectroscopy as well as by *ab initio* molecular orbital calculations (119). Arsenin does possess aromatic character.

Arsonic and Arsinic Acids. The arsonic acids, compounds of the type $RAsO(OH)_2$, are among the most important organic arsenicals. The aliphatic arsonic acids are generally prepared by the Meyer reaction; ie, heating an alkyl halide with As_4O_6 in alkaline solution:

$$As(ONa)_3 + RX \longrightarrow RAsO(ONa)_2 + NaX$$

Where X is Br or Cl, the free acids may be obtained by acidification of the alkaline solution, but where X is I, the acids must be isolated as salts to avoid reduction of the arsonic acids by HI. Rather than using alkyl halides, alkyl or dialkyl sulfates or alkyl arenesulfonates can be used. Primary alkyl halides react rapidly and smoothly, secondary halides react only slowly, whereas tertiary halides do not give arsonic acids. Allyl halides undergo the Meyer reaction, but vinyl halides do not. Substituted alkyl halides can be used; eg, ethylene chlorohydrin gives 2-hydroxyethylarsonic acid [*65423-87-2*], $C_2H_7AsO_4$. Arsinic acids, $R_2AsO(OH)$, are also readily prepared by substituting an alkali metal arsonite, $RAs(OM)_2$, for sodium arsenite:

$$RAs(ONa)_2 + R'X \longrightarrow RR'As(O)ONa + NaX$$

A number of long-chain dialkylarsinic acids have been prepared by the following series of reactions (120):

$$R_2NAsCl_2 + 2\ R'MgX \longrightarrow R'_2AsNR_2 + MgCl_2 + MgX_2$$

$$R'_2AsNR_2 + H_3O^+ \longrightarrow R'_2AsOH + R_2NH_2^+$$

$$R'_2AsOH + H_2O_2 \longrightarrow R'_2AsO_2H + H_2O$$

Aromatic arsonic acids are generally prepared by the Bart reaction from an aromatic diazonium salt and sodium arsenite:

$$ArN_2X + As(ONa)_3 \longrightarrow ArAsO(ONa)_2 + N_2 + NaX$$

A number of substituted phenylarsonic acids have been prepared by means of the Bart reaction (121). For the preparation of arsinic acids a sodium arylarsonite is used, and mixed diaryl or alkylarylarsinic acids can be prepared:

$$ArN_2X + RAs(ONa)_2 \longrightarrow ArRAs(O)ONa + N_2 + NaX$$

The Bart reaction is successful with a wide variety of aromatic and heterocyclic amines. A variation in which an aromatic amine, in the presence of arsenic trichloride, is diazotized in an organic solvent (the Scheller reaction) has also found wide application. Both arsonic and arsinic acids can be prepared by the Scheller reaction which often gives better yields than the Bart reaction with electron-attracting substituents on the aromatic ring. For the commercial preparation of 4-aminophenylarsonic acid [*98-50-0*] (arsanilic acid), $C_6H_8AsNO_3$, and 4-hydroxyphenylarsonic acid [*98-14-6*], $C_6H_7AsO_4$, the Béchamp reaction is used:

$$C_6H_5NH_2 + AsO(OH)_3 \longrightarrow 4\text{-}H_2NC_6H_4AsO(OH)_2 + H_2O$$

$$C_6H_5OH + AsO(OH)_3 \longrightarrow 4\text{-}HOC_6H_4AsO(OH)_2 + H_2O$$

Although the yields in the Béchamp reaction are not large, the starting materials can be recovered and recycled. In addition to aniline, substituted anilines such as 2-, 3-, and 4-methylanilines, 2- and 4-chloroanilines, and 2- and 4-nitroanilines have been arsonated under the conditions of the Béchamp reaction. Similarly, 2- and 3-methylphenol give the 3-methyl-4-hydroxyphenylarsonic acid and 2-methyl-4-hydroxyphenylarsonic acid [*26738-23-8*], $C_7H_9AsO_4$. By-products of the Béchamp reaction are the corresponding arsinic acids. Thus from aniline, bis(4-aminophenyl)arsinic acid [*6309-25-7*], $C_{12}H_{13}AsN_2O_2$, and from phenol, bis(4-hydroxyphenyl)arsinic acid are formed. Under high temperature conditions the arsinic acids become the principal products of the reaction.

Arsonic and arsinic acids have found a number of industrial uses. They have been used as corrosion inhibitors for iron and steel, additives to motor fuel, agricultural bactericides, herbicides, and fungicides. 3-Nitro-4-hydroxyphenylarsonic acid (roxarsone) [*121-19-7*], $C_6H_6AsNO_6$, has found widespread use as an additive to poultry feed for the control of coccidiosis and other poultry diseases (122,123). Arsanilic acid [*98-50-0*], $C_6H_8AsNO_3$, has also been used for this purpose (124) and for growth promotion in swine. It is manufactured and supplied by Duphar Nutrition Co., Inc., Fleming Laboratories, Inc., Dr. Mayfield Laboratories, Inc., and Whitmoyer Laboratories, Inc. 4-Nitrophenylarsonic acid [*98-72-6*], $C_6H_6AsNO_5$ and 4-[(aminocarbonyl)amino]phenylarsonic acid (carbarsone) [*121-59-5*], $C_7H_9AsN_2O_4$, have been used as additives to turkey feed to control histomoniasis and to improve feed efficiency. Both monosodium methylarsonate (MSMA) [*2163-80-6*], and disodium methylarsonate (DSMA) [*144-21-8*] have been widely used as selective herbicides, particularly for the control of crabgrass in established turf and for grassy weeds in cotton fields. In 1986 the U.S. consumption of MSMA was 2,700 metric tons and of DSMA was 900 metric tons. A mixture of dimethylarsinic acid (cacodylic acid) and its sodium salt [*59164-68-0*] have been used as herbicides and for the defoliation of cotton. Over one thousand metric tons of cacodylic acid were consumed in the United States in 1986. In 1987 the United States exported 490 metric tons of cacodylic acid and its salts.

Both arsonic and arsinic acids give precipitates with many metal ions, a property which has found considerable use in analytical chemistry. Of particular importance are certain azo dyes (qv) containing both arsonic and sulfonic acid groups which give specific color reactions with a wide variety of transition, lanthanide, and actinide metal ions. One of the best known of these dyes is 3,6-bis[(2-arsonophenyl)azo]-4,5-dihydroxy-2,7-naphthalenedisulfonic acid [*1668-00-4*], $C_{22}H_{18}As_2N_4O_{14}S_2$, generally called Arsenazo III. This compound forms stable inner complexes with certain metal ions enabling spectrophotometric determination in strong acid solution. A number of other azo-substituted arsonic acids have been introduced. A review article has been written on the use of these compounds (125).

A number of esters of arsonic and arsinic acids have been prepared. One method involves the oxidation of dialkyl alkylarsonites with selenium dioxide:

$$RAs(OR')_2 \xrightarrow{SeO_2} RAs(O)(OR')_2$$

A better method involves the use of silver salts of arsonic acids:

$$RAs(O)(OAg)_2 + 2\ R'X \longrightarrow RAs(O)(OR')_2 + 2\ AgX$$

Finally, both arsonic and arsinic acids can be esterified directly using the appropriate alcohol by removal of the water formed in the esterification. Azeotropic distillation using benzene or toluene is generally used for this purpose. On heating slowly, arsonic acids form anhydrides, $(RAsO_2)_n$. Thioarsonic acids, $RAsS(SH)_2$, and thioarsinic acids, $R_2AsS(SH)$, are unknown, but sodium salts, $RAsS(SNa)_2$ and $R_2AsS(SNa)$ have been prepared. Esters of the type $R(C_6H_5)AsS(OR')$ have been prepared by heating alkyl arsinites with sulfur (126):

$$R(C_6H_5)AsOR' + S \longrightarrow R(C_6H_5)AsS(OR')$$

Arsine Oxides. Both aliphatic and aromatic arsine oxides, compounds of the type R_3AsO, are well known. These compounds have been prepared by the oxidation of trialkylarsines using mercury(II) oxide or 30% hydrogen peroxide, with the rigid exclusion of oxygen (127). The products, which are purified by sublimation, are extremely hygroscopic and can only be handled under dry box conditions. The aromatic arsine oxides are generally prepared by the alkaline hydrolysis of dihaloarsoranes, R_3AsX_2, or by oxidation of triarylarsines using potassium permanganate, hydrogen peroxide, or iodine. Mixed aliphatic–aromatic arsine oxides, eg, $(C_6H_5)_2CH_3AsO$ (127), diethylphenylarsine oxide [*17767-98-5*], $C_6H_5(C_2H_5)_2AsO$, (63) and $C_6H_5(C_2H_5)As(O)CH_2CO_2H$ [*32295-95-7*] (128), have also been prepared. When dihaloarsoranes are hydrolyzed under alkaline conditions, hydrated compounds are obtained. These were at one time considered to be dihydroxides, $Ar_3As(OH)_2$. X-ray studies on the triphenyl compound, however, have shown that this exists as a dimer $[(C_6H_5)_3AsO{\cdot}H_2O]_2$, with each AsO group hydrogen bonded to two water molecules and each water molecule bonded to two AsO groups (129). The hydrolysis of dihaloarsoranes under milder conditions (aqueous ethanol or aqueous acetone) gives compounds which are usually listed as hydroxyhalides, $R_3As(OH)X$. These can also be prepared by the action of hydrohalic acids on arsine oxides. In addition to the halides, compounds of the type $R_3As(OH)Y$, where Y is NO_3, HSO_4, or ClO_4, are known. Chlorohydroxytriphenylarsenic [*15736-85-3*], $(C_6H_5)_3As(OH)Cl$, and bromohydroxytriphenylarsenic [*15736-84-2*], $(C_6H_5)_3As(OH)Br$, have been shown by x-ray diffraction to possess the structure $(C_6H_5)_3AsO$---H---X, with very short hydrogen bonds (130). Most of these compounds are weak electrolytes in acetonitrile, but the perchlorate is a strong electrolyte in this solvent (131). With strong acids, $HClO_4$, HBF_4, CF_3CO_2H, and $HSbCl_6$, compounds of the type $[(C_6H_5)_3AsO]_2HY$ can be prepared (132,133). The arsine oxides are Lewis bases and form coordination compounds with transition and actinide metal compounds. Solutions of arsine oxides in nitromethane are useful for extracting salts of U(VI), Pu(IV), Pu(VI), Np(IV), and Np(VI) from nitric acid solutions of these elements (134,135). The arsine oxides form 1:1 adducts with a variety of phenols (136).

Trialkyl- and triarylarsine sulfides have been prepared by several different methods. The reaction of sulfur with a tertiary arsine, with or without a solvent, gives the sulfides in almost quantitative yields. Another method involves the reaction of hydrogen sulfide with a tertiary arsine oxide, hydroxyhalide, or dihaloarsorane. X-ray diffraction studies of triphenylarsine sulfide [*3937-40-4*], $C_{18}H_{15}AsS$, show the arsenic to be tetrahedral; the arsenic–sulfur bond is a true double bond (137). Triphenylarsine sulfide and trimethylarsine sulfide [*38859-90-4*], C_3H_9AsS, form a number of coordination compounds with salts of transition elements (138,139). Both trialkyl- and triarylarsine selenides have been reported. The trialkyl compounds have been prepared by refluxing trialkylarsines with selenium powder (140). The preparation of triphenylarsine selenide [*65374-39-2*], $C_{18}H_{15}AsSe$, from dichlorotriphenylarsorane and hydrogen selenide has been reported (141), but other workers could not duplicate this work (140).

Haloarsoranes. Halides of the types $RAsX_4$, R_2AsX_3, R_3AsX_2, and R_4AsX are known. The R_4AsX compounds are ionic in nature and are discussed under arsonium salts. The tetrahalides are unstable compounds which frequently decompose on standing and are readily hydrolyzed in moist air to form arsonic acids. Aryltetrachloroarsoranes are readily prepared from aryldichloroarsines in an atmosphere of dry chlorine (142). Tetrafluorophenylarsorane [*650-44-2*], $C_6H_5AsF_4$, has been prepared from phenylarsonic acid and sulfur tetrafluoride (143):

$$C_6H_5AsO_3H_2 + 3\ SF_4 \longrightarrow C_6H_5AsF_4 + 3\ SOF_2 + 2\ HF$$

The ^{19}F nmr spectrum of this compound gives only one signal over a wide range of temperatures, a result attributed to Berry pseudorotation (144). No alkyl- or aryltetrabromoarsorane has been reported. There is, however, an early report on the preparation of tetraiodomethylarsorane from methylarsonic acid and hydriodic acid (145).

Dialkyl- and diaryltrihaloarsoranes are more stable than the tetrahalo compounds, but can be readily decomposed at comparatively low temperatures. The trichloro compounds are usually prepared by the action of chlorine on dialkyl- or diarylchloroarsines. Tribromodiphenylarsorane [*38338-43-1*], $C_{12}H_{10}AsBr_3$, has been prepared by means of a bromodiphenylarsine [*3095-87-2*], $C_{12}H_{10}AsBr$, intermediate (146). The starting material is diphenyl(phenylthio)arsine [*15367-57-4*], $C_{18}H_{15}AsS$:

$$(C_6H_5)_2AsSC_6H_5 + Br_2 \longrightarrow (C_6H_5)_2AsBr + C_6H_5SBr$$

$$(C_6H_5)_2AsBr + Br_2 \longrightarrow (C_6H_5)_2AsBr_3$$

No triiodoarsorane has been reported. Trifluorodiphenylarsorane [*2357-18-8*], $C_{12}H_{10}AsF_3$, has been prepared by several methods. One method involves the reaction of benzene with arsenic pentafluoride and subsequent treatment of the resulting salt with cesium fluoride (147):

$$2\ C_6H_6 + 2\ AsF_5 \longrightarrow [(C_6H_5)_2AsF_2][AsF_6] + 2\ HF$$

$$[(C_6H_5)_2AsF_2][AsF_6] + CsF \longrightarrow (C_6H_5)_2AsF_3 + CsAsF_6$$

It has also been prepared from diphenylarsinic acid and sulfur tetrafluoride (147), and by the direct fluorination of diphenylarsine [*829-83-4*] using fluorine diluted with argon.

Trialkyl- and triaryldihaloarsoranes have been studied to a much greater extent than the tri- and tetrahaloarsoranes. The dihalo compounds are stable crystalline species, although they decompose on heating:

$$R_3AsX_2 \xrightarrow{\Delta} R_2AsX + RX$$

The usual method of preparing these compounds is by direct halogenation of tertiary arsines. Other halogenating agents, eg, lead tetrachloride or sulfuryl chloride, have also been used. Although the difluorides have been prepared by direct fluorination using fluorine diluted with argon (148), they are usually prepared from the dibromides or dichlorides by reaction with silver fluoride. Mixed diarylalkyl- and aryldialkyldihaloarsoranes are also well known. When iodine or bromine is used in excess, tetrahalides of the type R_3AsX_4 are obtained. These compounds are ionic in the solid state and are strong electrolytes in acetonitrile solution. Dibromo- and dichlorotriphenylarsoranes react with interhalogens to form such compounds as $[(C_6H_5)_3AsCl][ICl_2]$ [*76002-28-31*], $[(C_6H_5)_3AsCl][ICl_4]$ [*76002-97-1*], and $[(C_6H_5)_3AsCl][IBr_2]$ [*76002-28-3*]. The structure of the dihaloarsoranes R_3AsX_2 varies with the nature of both R and X. Thus, trimethyldifluoroarsorane [*64811-88-7*], $C_3H_9AsF_2$, and trimethyldichloroarsorane [*17756-06-8*], $C_3H_9AsCl_2$, contain pentacovalent arsenic as shown by conductivity measurements and infrared spectra, and they appear to be trigonal bipyramids. Dibromotrimethylarsorane [*17756-09-1*], $C_3H_9AsBr_2$, is a stronger electrolyte than the dichloro compound in acetonitrile. Spectroscopic evidence is in accord with an ionic structure for this compound (149). Diiodotrimethylarsorane [*17756-08-0*], $C_3H_9AsI_2$, and diiodotriphenylarsorane [*3964-45-2*], $C_{18}H_{15}AsI_2$, are both strong electrolytes in acetonitrile solution. It has been concluded that the high conductance of the triphenyl compound results from disproportionation of the following type (150):

$$2\ (C_6H_5)_3AsI_2 \rightleftharpoons (C_6H_5)_3As + (C_6H_5)_3AsI^+ + I_3^-$$

The strong conductance of the trimethyl compound may be because of a similar disproportionation.

In addition to the trialkyl- and triaryldihaloarsoranes there are a number of compounds of the type R_3AsY_2, where Y is a pseudohalide, nitrate, carboxy, or other negative group. They are usually prepared from trialkyl- or triaryldibromo- or dichloroarsoranes by metathesis. Thus diazidotrimethylarsorane [*21121-84-6*], $C_3H_9AsN_6$, has been prepared in the following manner (151):

$$(CH_3)_3AsCl_2 + 2\ NaN_3 \longrightarrow (CH_3)_3As(N_3)_2 + 2\ NaCl$$

Dinitratotriphenylarsorane [*18513-99-0*], $C_{18}H_{15}AsN_2O_6$, has similarly been obtained from dibromotriphenylarsorane and silver nitrate (152):

$$(C_6H_5)_3AsBr_2 + 2\ AgNO_3 \longrightarrow (C_6H_5)_3As(NO_3)_2 + 2\ AgBr$$

The nitrato compound is a weak electrolyte in acetonitrile solution.

Arsonium Salts. Arsonium salts are compounds of the type R_4AsY, where R may be either an alkyl or aryl group and Y is a wide variety of negative groups, such as halogen, nitrate, sulfate, and perchlorate. Both triaryl- and trialkylarsines are readily quaternized by treatment with an alkyl halide. Thus triphenylarsine and methyl bromide give methyltriphenylarsonium bromide [*14002-59-6*], $C_{19}H_{18}AsBr$ (153). The order of increasing reactivity of alkyl halides is $Cl < Br < I$ and primary > secondary > tertiary. Tetraphenylarsonium bromide [*507-27-7*], $C_{24}H_{20}AsBr$, is prepared from triphenylarsine oxide and phenylmagnesium bromide (154). Another method involves the reaction of triphenylarsine, bromobenzene, and anhydrous aluminum chloride. Potassium iodide is added to the reaction mixture so that the product is isolated as tetraphenylarsonium iodide [*7422-32-4*], $C_{24}H_{20}AsI$ (155). Arsonium salts other than halides can be readily obtained from the halides by reaction of the appropriate reagent. For example, reaction of arsonium halides with acids such as nitric or perchloric acids yields arsonium nitrates or perchlorates, respectively. Another method involves the reaction of an arsonium hydroxide with the appropriate acid. Arsonium salts can also be obtained by the cleavage of pentaalkyl or pentaarylarsenic compounds. Thus reaction of pentaphenylarsorane [*19376-61-5*], $C_{30}H_{25}As$, with excess chlorine, bromine, or iodine gives the corresponding tetraphenylarsonium trihalides and halobenzenes. Pentaarylarsenic compounds react with hydrohalic acids to yield tetraarylarsonium halides and aryl halides (156).

Arsonium salts have found considerable use in analytical chemistry. One such use involves the extraction of a metal complex in aqueous solution with tetraphenylarsonium chloride in an organic solvent. Titanium(IV) thiocyanate [*35787-79-2*] (157) and copper(II) thiocyanate [*15192-76-4*] (158) in hydrochloric acid solution have been extracted using tetraphenylarsonium chloride in chloroform solution in this manner, and the Ti(IV) and Cu(II) thiocyanates determined spectrophotometrically. Cobalt, palladium, tungsten, niobium, and molybdenum have been determined in a similar manner. In addition to their use for the determination of metals, anions such as perchlorate and perrhenate have been determined as arsonium salts. Tetraphenylarsonium permanganate is the only known insoluble salt of this anion.

Arsonium halides containing one or more alkyl groups lose an alkyl halide on pyrolysis to give a tertiary arsine. This reaction has been used for the synthesis of compounds containing arsenic as a member of the heterocyclic ring. For example, pyrolysis of trimethyl-4-(1,7-dibromoheptyl)arsonium bromide, $[(CH_3)_3AsCH(CH_2CH_2CH_2Br)_2]Br$, resulted in the loss of three moles of methyl bromide and the formation of 1-arsabicyclo[3.3.0]octane [*4432-50-2*], $C_7H_{13}As$ (159)

Arsonium Ylides. Arsonium ylides were first prepared by reaction between an arsonium halide and phenyllithium. Thus methyltriphenylarsonium iodide [*1499-33-8*], $C_{19}H_{18}AsI$, and phenyllithium give triphenylarsonium methylide [*19365-61-8*], $C_{19}H_{17}As$:

$$[(C_6H_5)_3AsCH_3]I + C_6H_5Li \longrightarrow (C_6H_5)_3As{=}CH_2 + LiI + C_6H_6$$

In this case the ylide was not isolated but allowed to react with benzophenone to give, after hydrolysis with hydrochloric acid, 1,1-diphenylethylene, diphenylacetaldehyde, and triphenylarsine (160). An excellent method for preparing arsonium ylides involves the reaction between a stable diazo compound and triphenylarsine in the presence of a copper catalyst such as bis(acetylacetonato)copper(II) (161). Rather than a diazo compound, an iodonium ylide can be used; again a copper catalyst is necessary for an optimum yield of product. An example of the use of a diazo compound is shown in the formulation of triphenylarsonium 2,3,4-triphenylcyclopentadienylide [*29629-32-1*], $C_{41}H_{31}As$:

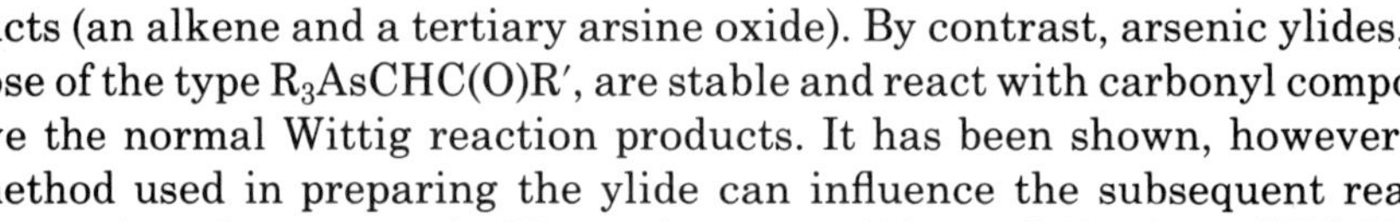

Arsenic ylides of the types $R_3As{=}CH_2$ and $R_3As{=}CHR'$ are unstable and generally react with carbonyl compounds to give either a tertiary arsine and an epoxide, or a mixture of these two compounds and the normal Wittig reaction products (an alkene and a tertiary arsine oxide). By contrast, arsenic ylides, such as those of the type $R_3AsCHC(O)R'$, are stable and react with carbonyl compounds to give the normal Wittig reaction products. It has been shown, however, that the method used in preparing the ylide can influence the subsequent reaction with a carbonyl compound. Thus, the two ylides, $(C_6H_5)_3As{=}CHC_6H_5$ and $(C_6H_5)_3As{=}CHC_6H_4Cl\text{-}4$, when prepared by two different methods, gave only the normal Wittig products with carbonyl compounds when prepared by one method, and only an epoxide and triphenylarsine when prepared by the other method (162). Arsenic ylides react several hundred times more rapidly than phosphorus ylides in the Wittig reaction. Furthermore, arsenic ylides often react with carbonyl compounds that do not react with phosphorus ylides. For this reason arsenic ylides are frequently preferred for the Wittig reaction (163).

Pentaalkyl- and Pentaarylarsoranes. Compounds of the type R_5As, where R may be aliphatic or aromatic, have assumed considerable importance in arsenic chemistry. Pentamethylarsorane [*51043-92-6*], $C_5H_{15}As$, has been prepared from dichlorotrimethylarsorane [*17756-06-8*], $C_3H_9AsCl_2$, and two equivalents of methyllithium in dimethyl ether solution at $-60°C$ (164). It is stable in air and only slowly hydrolyzes to tetramethylarsonium hydroxide [*34618-96-7*], $C_4H_{13}AsO$, and methane. Pentaphenylarsorane [*19376-61-5*], $C_{30}H_{25}As$, has been prepared by the reaction of phenyllithium with tetraphenylarsonium bromide [*507-27-7*], $C_{24}H_{20}AsBr$, triphenylarsine oxide [*1153-05-5*], $C_{18}H_{15}AsO$ or 4-methylphenyliminotriphenylarsorane [*65423-89-4*], $4\text{-}CH_3C_6H_4N{=}As(C_6H_5)_3$. Although only a few pentaalkylarsoranes have been described in the chemical literature, a large number of pentaarylarsoranes have been prepared, including a number of spirocyclic compounds. These compounds are of particular interest in studies on the stereochemistry of five-covalent compounds (165).

Arsenic Compounds Used in Medicine

Arsenic compounds have been employed as therapeutic agents for at least 2400 years (166,167). During the fifth century BC, Hippocrates recommended the use of an ointment containing arsenic sulfide (probably realgar) for the treatment of ulcers. Galen (138–201 AD) and various Roman authors prescribed similar arsenic preparations for skin diseases, tuberculosis, asthma, and leprosy. The great Persian physician, Avicenna, believed that arsenic compounds were valuable for the treatment of a number of infectious diseases. During the Middle Ages, arsenic trioxide (white arsenic) was extensively used both by physicians and by professional poisoners. However, at the close of the twentieth century, the therapeutic use of inorganic arsenicals must be considered obsolete.

Modern work on arsenical drugs can be said to have started in 1905; it was demonstrated that sodium hydrogen 4-aminophenylarsonate (Atoxyl) [*127-85-5*] could cure experimental trypanosomiasis (168). This finding prompted a systematic study of the chemical and biological properties of aromatic arsenicals (169). Using Atoxyl as a starting material, about a thousand organic derivatives of arsenic were prepared during a period of ten years. This research led to several drugs that can cure syphilis.

By 1932 a total of 12,500 compounds of arsenic had been synthesized and some had been clinically tested. Since 1943, when penicillin was shown to be effective for the therapy of syphilis (170) (see ANTIBIOTICS, β-LACTAMS, PENICILLIN AND OTHERS), there has been much less work on the use of organoarsenic compounds in medicine. No important new arsenical drug has been introduced. However, arsenicals are still important for the treatment of African trypanosomiasis (166,171–173); they are probably indispensable for the late neurological stage of the disease (see ANTIPARASITIC AGENTS-ANTHELMINTICS). Toxic reactions caused by arsenicals can often be successfully treated using 2,3-dimercapto-1-propanol (dimercaprol, BAL) [*59-52-9*], $C_3H_8OS_2$.

Melarsoprol (Mel B, arsobal) [*494-79-1*], $C_{12}H_{15}AsN_6OS_2$, is a thioarsenite that is active against human trypanosomiasis caused by *T. gambiense* or *T. rhodesiense,* and it has been successfully employed for the treatment of infections of the nervous system (171–173). It often causes toxic effects, however, and it should be employed only under strict medical supervision (173). The drug is much too toxic for prophylactic use. Another disadvantage is that trypanosomes are becoming resistant to melarsopol (172). In the United States this drug is available only from the Centers for Disease Control, Atlanta, Georgia.

Melarsonyl potassium (Mel W, Trimelarsen) [*13355-00-5*] is a thioarsenite closely related to melarsoprol, and it also has been used for the treatment of trypanosomiasis (172). However, it appears to be more toxic and less effective than melarsoprol. The only advantage of melarsonyl potassium is that it is water-soluble and can be administered intramuscularly or subcutaneously. This property is useful when the intravenous route cannot be employed.

Carbarsone (Amebarsone) was once widely used for the treatment of intestinal amebiasis. Like other arsonic acids, however, carbarsone may cause skin rashes and even damage to the vision. Although it is still available for medicinal use, it is really obsolete as an amebicide because less toxic and more effective nonarsenicals are now available (174).

Arsenamide (thiacetarsamide) [*531-72-6*], $C_{11}H_{12}AsNO_5S_2$, is a thioarsenite that is employed in veterinary medicine. Although relatively toxic, it has proved useful for the treatment of *Dirofilaria immitis* (heartworm) infestation in dogs (175). It is in fact the only drug recommended for killing the adult worms associated with this disease. It should not be used, however, in critically ill dogs.

BIBLIOGRAPHY

"Arsenic Compounds" in *ECT* 1st ed., Vol. 2, pp. 119–123, by I. E. Campbell, Battelle Memorial Institute; "Arsenic Preparations" in Vol. 2, pp. 123–133, by H. G. Steinman, U.S. Public Health Service, Laboratory of Experimental Therapeutics, Johns Hopkins School of Hygiene; "Arsonic Acids and Other Arsenic Compounds" in Vol. 2, pp. 133–134, by C. K. Banks, Parke, Davis & Co.; "Arsenic Compounds" in *ECT* 2nd ed., Vol. 2, pp. 718–733, by G. O. Doak, L. D. Freedman, and G. G. Long, North Carolina State of The University of North Carolina at Raleigh; "Arsenic Compounds" in *ECT* 3rd ed., Vol. 3, pp. 251–266, G. O. Doak, G. G. Long, and L. D. Freedman, North Carolina State University.

1. T. A. Albright, J. K. Burdett, and M. H. Whangbo, *Orbital Interactions in Chemistry,* John Wiley & Sons, Inc., New York, 1985, Chapt. 14.
2. *Minerals Yearbook for 1988, Metals and Minerals,* Vol. 1, U.S. Department of the Interior, Washington, D.C., 1990, pp. 1059–1062.
3. L. Bretherick, *Hazards in the Chemical Laboratory,* 4th ed., The Royal Society of Chemistry, London, UK, 1986, p. 182.
4. E. O. Uthus, W. E. Cornatzer, and F. H. Nielsen, in W. H. Lederer and R. J. Fensterheim, eds., *Arsenic,* Van Nostrand Reinhold Company, New York, 1983, pp. 173–189.
5. H. S. Gowda and H. N. A. Khan, *Talanta* **29,** 787 (1982).
6. B. N. Usha, Rangaswamy, and H. S. Yathirajan, *Talanta* **29,** 803 (1982).
7. K. Radhakrishnan and P. Indrasenan, *J. Indian Chem. Soc.* **62,** 84 (1985).
8. N. Jayasree and P. Indrasenan, *Talanta* **32,** 1067 (1985).
9. K. W. Boyer, in S. Williams, ed., *Official Methods of Analysis of the Association of Official Analytical Chemists,* 14th ed., Association of Official Analytical Chemists, Inc., Arlington, Va., 1984, p. 450.

10. Ref. 9, p. 445.
11. A. Brzezinska-Paudyn, J. Van Loon, and R. Hancock, *At. Spectrosc.* **7,** 72 (1986).
12. P. P. Coetzee and H. Pieterse, *S. Afr. J. Chem.* **39,** 85 (1986).
13. D. D. Wagman and co-workers, *National Bureau of Standards Technical Note 270–3,* U.S. Dept. of Commerce, Washington, D.C., 1968, pp. 95–98.
14. W. C. Johnson and A. Pechukas, *J. Am. Chem. Soc.* **59,** 2065 (1937).
15. P. W. Schenk, *G. Brauer's Handbook of Preparative Inorganic Chemistry,* Vol. 1, 2nd ed., Academic Press, Inc., New York, 1963, p. 591.
16. A. Stock, *Ber. Dtsch. Chem. Ges.* **53,** 837 (1920).
17. C. Blaauw and co-workers, *Can. J. Phys.* **63,** 664 (1985).
18. E. K. Byrne, L. Parkanyi, and K. H. Theopold, *Science* **241,** 332 (1988).
19. S. R. Gunn, *Inorg. Chem.* **11,** 796 (1972).
20. E. J. Weeks and J. G. F. Druce, *Chem. News* **129,** 31 (1924).
21. H. Preiss, *Z. Anorg. Allg. Chem.* **380,** 45 (1971).
22. K. Seppelt, *Angew. Chem., Int. Ed. Engl.* **15,** 377 (1976).
23. M. Jansen, *Angew. Chem., Int. Ed. Engl.* **16,** 314 (1977).
24. B. H. Christian, R. J. Gillespie, and J. F. Sawyer, *Inorg. Chem.* **20,** 3410 (1981).
25. H. J. Whitfield, *J. Chem. Soc., Dalton Trans.,* 1737 (1973).
26. H. J. Whitfield, *J. Chem. Soc. A,* 1800, (1970).
27. H. J. Whitfield, *J. Chem. Soc., Dalton Trans.,* 1740 (1973).
28. R. Blachnik, A. Hoppe, U. Rabe, and U. Wickel, *Z. Naturforsch. B: Anorg. Chem., Org. Chem.* **36B,** 1493 (1981).
29. J. Rigaudy and S. P. Klesney, *IUPAC Nomenclature of Organic Chemistry,* Section D, Pergamon Press, Oxford, UK, 1979, pp. 382–408.
30. S. Samaan, *Houben-Weyl Methoden der Organischen Chemie. Metallorganische Verbindungen As, Sb, Bi,* Georg Thieme Verlag, Stuttgart, Germany, 1978.
31. G. O. Doak and L. D. Freedman, *Organometallic Compounds of Arsenic, Antimony, and Bismuth,* John Wiley & Sons, Inc., New York, 1970.
32. M. Dub, *Organometallic Compounds,* Vol. 3, 2nd ed., Springer-Verlag, Inc., New York, 1968.
33. M. Dub, *Organometallic Compounds,* Vol. 3, 2nd ed., Suppl. 1, Springer-Verlag, Inc., New York, 1972.
34. F. G. Mann, *The Heterocyclic Derivatives of Phosphorus, Arsenic, Antimony and Bismuth,* 2nd ed., Wiley-Interscience, New York, 1970.
35. A. Tzschach and J. Heinicke, *Arsenheterocyclen,* VEB Deutscher Verlag für Grundstoffindustrie, Leipzig, Germany, 1978.
36. G. T. Morgan, *Organic Compounds of Arsenic and Antimony,* Longmans, Green and Co., London, UK, 1918.
37. G. W. Raiziss and J. L. Gavron, *Organic Arsenical Compounds,* The Chemical Catalog Co., New York, 1923.
38. A. E. Goddard, in J. N. Friend, ed., *A Text-Book of Inorganic Chemistry,* Vol. 11, Part 2, *Derivatives of Arsenic,* Charles Griffin & Co., London, UK, 1930.
39. F. G. Mann and A. J. Wilkinson, *J. Chem. Soc.,* 3336 (1957).
40. H. J. Emeléus, R. N. Haszeldine, and E. G. Walaschewski, *J. Chem. Soc.,* 1552 (1953).
41. E. Wiberg and K. Mödritzer, *Z. Naturforsch., B: Anorg. Chem., Org. Chem., Biochem., Biophys., Biol.* **B11,** 750 (1956).
42. A. Tzschach, H. Matschiner, and H. Biering, *Z. Anorg. Allg. Chem.* **436,** 60 (1977).
43. F. G. A. Stone and A. B. Burg, *J. Am. Chem. Soc.* **76,** 386 (1954).
44. W. R. Cullen and W. R. Leeder, *Can. J. Chem.* **47,** 2137 (1969).
45. E. J. Cragoe, Jr. and co-workers, *J. Am. Chem. Soc.* **69,** 925 (1947).
46. T. R. Carlton and C. D. Cook, *Inorg. Chem.* **10,** 2628 (1971).
47. F. G. Mann and M. J. Pragnell, *J. Chem. Soc.,* 4120 (1965).

48. S. C. Grocott and S. B. Wild, *Inorg. Chem.* **21,** 3526 (1982).
49. R. G. Cavell and R. C. Dobbie, *J. Chem. Soc. A,* 1308 (1967).
50. K. Issleib, A. Tzschach, and R. Schwarzer, *Z. Anorg. Allg. Chem.* **338,** 141 (1965).
51. P. Aslanidis and J. Grobe, *Z. Naturforsch. B: Anorg. Chem., Org. Chem.* **38B,** 289 (1983).
52. G. Märkl, W. Weber, and W. Weiss, *Chem. Ber.* **118,** 2365 (1985).
53. J. Grobe, N. Krummen, and Le Van Duc, *Z. Naturforsch., B: Anorg. Chem., Org. Chem.* **39B,** 1711 (1984).
54. A. Tzschach, D. Drohne, and J. Heinicke, *J. Organomet. Chem.* **60,** 95 (1973).
55. A. Tzschach and R. Schwarzer, *Justus Liebigs Ann. Chem.* **709,** 248 (1967).
56. K.-H. Linke and W. Brandt, *Z. Anorg. Allg. Chem.* **433,** 119 (1977).
57. A. Tzschach and G. Pacholke, *Chem. Ber.* **97,** 419 (1964).
58. W. J. C. Dyke and W. J. Jones, *J. Chem. Soc.,* 2426 (1930).
59. Yu. F. Gatilov, F. D. Yambushev, and N. Kh. Tenisheva, *Zh. Obshch. Khim.* **43,** 2273 (1973).
60. F. D. Yambushev, Yu. F. Gatilov, N. Kh. Tenisheva, and V. I. Savin, *Zh. Obshch. Khim.* **44,** 1730 (1974).
61. R. S. Dickson and B. O. West, *Aust. J. Chem.* **15,** 710 (1962).
62. P. Jutzi and K. Deuchert, *Angew. Chem., Int. Ed. Engl.* **8,** 991 (1969).
63. B. D. Chernokal'skii, A. S. Gel'fond, and G. Kh. Kamai, *Zh. Obshch. Khim.* **37,** 1396 (1967).
64. J. F. Morgan, E. J. Cragoe, Jr., B. Elpern, and C. S. Hamilton, *J. Am. Chem. Soc.* **69,** 932 (1947).
65. W. R. Cullen and P. E. Yates, *Can. J. Chem.* **41,** 1625 (1963).
66. G. O. Doak and L. D. Freedman, *Synthesis,* 328 (1974).
67. E. G. Hope, T. Kemmitt, and W. Levason, *J. Chem. Soc., Perkins Trans. 2,* 487 (1987).
68. L. Horner, J. Röder, and D. Gammel, *Phosphorus* **3,** 175 (1973).
69. G. H. Cookson and F. G. Mann, *J. Chem. Soc.,* 2888 (1949).
70. W. E. Hanby and W. A. Waters, *J. Chem. Soc.,* 1029 (1946).
71. H. Bauer, *J. Am. Chem. Soc.* **67,** 591 (1945).
72. W. R. Cullen and co-workers, *J. Organomet. Chem.* **139,** 61 (1977).
73. F. F. Blicke and S. R. Safir, *J. Am. Chem. Soc.* **63,** 575 (1941).
74. C. A. McAuliffe and W. Levason, *Studies in Inorganic Chemistry,* Vol. 1, Elsevier, Amsterdam, The Netherlands, 1979.
75. L. Shi, W. Wang, Y. Wang, and Y.-Z. Huang, *J. Org. Chem.* **54,** 2027 (1989).
76. G. B. Stringfellow, *Organometallic Vapor-Phase Epitaxy,* Academic Press, San Diego, Calif., 1989.
77. N. Stransky, R. Herzschuh, J.-P. Gehrke, and R. Taube, *J. Organomet. Chem.* **270,** 357 (1984).
78. S. T. Chow and C. A. McAuliffe, *J. Organomet. Chem.* **77,** 401 (1974).
79. R. D. Feltham, A. Kasenally, and R. S. Nyholm, *J. Organomet. Chem.* **7,** 285 (1967).
80. F. Kober and W. J. Rühl, *Z. Anorg. Allg. Chem.* **406,** 52 (1974).
81. E. G. Claeys, *J. Organomet. Chem.* **5,** 446 (1966).
82. W. R. Cullen and N. K. Hota, *Can. J. Chem.* **42,** 1123 (1964).
83. D. Hellwinkel and G. Kilthau, *Chem. Ber.* **101,** 121 (1968).
84. M. A. Weiner and G. Pasternack, *J. Org. Chem.* **32,** 3707 (1967).
85. L. I. Zakharkin and O. Yu. Okhlobystin, *Dokl. Akad. Nauk SSSR* **116,** 236 (1957).
86. M. Green and D. Kirkpatrick, *J. Chem. Soc. A,* 483 (1968).
87. L. D. Freedman and V. L. Styles, *J. Org. Chem.* **40,** 2684 (1975).
88. G. N. Jarman, *Adv. Chem. Ser.* **23,** 328 (1959).
89. H. H. Sisler and C. Stratton, *Inorg. Chem.* **5,** 2003 (1966).

90. K. Sommer, *Z. Anorg. Allg. Chem.* **377,** 120 (1970).
91. J. R. Harper and A. L. Rheingold, *J. Am. Chem. Soc.* **112,** 4037 (1990).
92. K. Sommer, *Z. Anorg. Allg. Chem.* **383,** 136 (1971).
93. L. Horner and S. Samaan, *Phosphorus* **4,** 1 (1974).
94. J. P. Casey and K. Mislow, *J. Chem. Soc. D,* 999 (1970).
95. F. Kober, *Z. Anorg. Allg. Chem.* **401,** 243 (1973).
96. G. Kh. Kamai, N. A. Chadaeva, and K. A. Mamakov, *Izv. Akad. Nauk SSSR, Ser. Khim.,* 1092 (1970).
97. B. L. Chamberland and A. G. MacDiarmid, *J. Am. Chem. Soc.* **83,** 549 (1961).
98. Yu. F. Gatilov, L. B. Ionov, A. P. Korovyakov, and L. G. Kokorina, *Zh. Obshch. Khim.* **43,** 2677 (1973).
99. W. R. Cullen and L. G. Walker, *Can. J. Chem.* **38,** 472 (1960).
100. F. Knoll, H. C. Marsmann, and J. R. Van Wazer, *J. Am. Chem. Soc.* **91,** 4986 (1969).
101. K. Issleib and A. Tzschach, *Angew. Chem.* **73,** 26 (1961).
102. A. Tzschach and W. Lange, *Chem. Ber.* **95,** 1360 (1962).
103. K. V. Vijayaraghavan, *J. Indian Chem. Soc.* **22,** 141 (1945).
104. P. Borgstrom and M. M. Dewar, *J. Am. Chem. Soc.* **44,** 2915 (1922).
105. A. J. Ashe, III and E. G. Ludwig, Jr., *J. Organomet. Chem.* **303,** 197 (1986).
106. A. H. Cowley, J. G. Lasch, N. C. Norman, and M. Pakulski, *J. Am. Chem. Soc.* **105,** 5506 (1983).
107. A. H. Cowley, A. B. Burg, and W. R. Cullen, *J. Am. Chem. Soc.* **88,** 3178 (1966).
108. N. Mandel and J. Donohue, *Acta Crystallogr. B* **27,** 476 (1971).
109. A. Tzschach and V. Kiesel, *J. Prakt. Chem.* **331,** 259 (1971).
110. S. E. Rasmussen and J. Danielsen, *Acta Chem. Scand.* **14,** 1862 (1960).
111. A. L. Rheingold, *Organomet. Synth.* **3,** 618 (1986).
112. A.-J. Dimaio and A. L. Rheingold, *Chem. Rev.* **90,** 169 (1990).
113. A. S. Levinson, *J. Chem. Ed.* **54,** 98 (1977).
114. C. S. Palmer, *J. Am. Chem. Soc.* **45,** 3023 (1923).
115. W. G. Christiansen, A. J. Norton, and J. B. Shohan, *J. Am. Chem. Soc.* **47,** 2712 (1925).
116. J. J. Daly and F. Sanz, *Helv. Chim. Acta* **53,** 1879 (1970).
117. A. L. Rheingold, J. E. Lewis, and J. M. Bellama, *Inorg. Chem.* **12,** 2845 (1973).
118. A. J. Ashe, III, *Top. Curr. Chem.* **105,** 125 (1982).
119. K. K. Baldridge and M. S. Gordon, *J. Am. Chem. Soc.* **110,** 4204 (1988).
120. K. Irgolic, R. A. Zingaro, and M. R. Smith, *J. Organomet. Chem.* **6,** 17 (1966).
121. F. D. Yambushev and co-workers, *Zh. Obshch. Khim.* **51,** 2232 (1981).
122. E. G. Schildknecht and co-workers, *Poult. Sci.* **59,** 268 (1980).
123. P. W. Waldroup and co-workers, *Nutr. Rep. Int.* **30,** 1079 (1984).
124. M. A. Villadelgado and P. F. Reyes, *Philipp J. Vet. Med.* **14,** 64 (1975).
125. S. B. Savvin, *Usp. Khim.* **32,** 195 (1963).
126. L. B. Ionov and co-workers, *Zh. Obshch. Khim.* **47,** 1569 (1977).
127. A. Merijanian and R. A. Zingaro, *Inorg. Chem.* **5,** 187 (1966).
128. Yu. F. Gatilov, L. B. Ionov, and F. D. Yambushev, *Zh. Obshch. Khim.* **40,** 2250 (1970).
129. V. K. Belsky, *J. Organomet. Chem.* **213,** 435 (1981).
130. G. Ferguson and E. W. Macaulay, *J. Chem. Soc. Chem. Commun.,* 1288 (1968).
131. G. S. Harris and F. Inglis, *J. Chem. Soc. A,* 497 (1967).
132. M. Grundwald-Wyspianska and M. Szafran, *Pol. J. Chem.* **55,** 1705 (1981).
133. F. Yu. Akhmadullina and co-workers, *Zh. Obshch. Khim.* **54,** 2249 (1984).
134. B. N. Laskorin, V. V. Yakshin, E. A. Filippov, and N. A. Lyubosvetova, *Dokl. Akad. Nauk SSSR* **249,** 651 (1979).
135. B. N. Laskorin and co-workers, *Radiokhimiya* **20,** 702 (1978); *Chem. Abstr.* **90,** 41812t (1979).

136. H. Schindlbauer and H. Stenzenberger, *Monatsh. Chem.* **99,** 2468 (1968).
137. P. M. Boorman, P. W. Codding, and K. A. Kerr, *J. Chem. Soc. Dalton Trans.* 1482 (1979).
138. P. E. Nicpon, *Diss. Abstracts B.* **27,** 3836 (1967); see ref. 33, pp. 306–307.
139. A. M. Brodie, S. H. Hunter, G. A. Rodley, and C. J. Wilkins, *J. Chem. Soc. A* 987 (1968).
140. R. A. Zingaro and A. Merijanian, *Inorg. Chem.* **3,** 580 (1964).
141. K. A. Jensen and P. H. Nielsen, *Acta Chem. Scand.* **17,** 1875 (1963).
142. G. I. Kokorev, F. D. Yambushev, and S. V. Kut'in, *Zh. Obshch. Khim.* **56,** 482 (1986).
143. W. C. Smith, *J. Am. Chem. Soc.* **82,** 6176 (1960).
144. E. L. Muetterties, W. Mahler, K. J. Packer, and R. Schmutzler, *Inorg. Chem.* **3,** 1298 (1964).
145. H. Klinger and A. Kreutz, *Justus Liebigs Ann. Chem.* **249,** 147 (1888).
146. N. H. Chadaeva, G. Kh. Kamai, and K. A. Mamakov, *Izv. Akad. Nauk, Ser. Khim.,* 1612 (1972).
147. F. L. Tanzella and N. Bartlett, *Z. Naturforsch. B: Anorg. Chem., Org. Chem.* **36B,** 1461 (1981).
148. I. Ruppert and V. Bastian, *Angew. Chem., Int. Ed. Engl.* **17,** 214 (1978).
149. L. Verdonck and G. P. Van der Kelen, *Spectrochim. Acta A* **33A,** 601 (1977).
150. A. D. Beveridge and G. S. Harris, *J. Chem. Soc.,* 6076 (1964).
151. A. Schmidt, *Chem. Ber.* **101,** 3976 (1968).
152. G. C. Tranter, C. C. Addison, and D. B. Sowerby, *J. Organomet. Chem.* **12,** 369 (1968).
153. W. Richter, Y. Yamamoto, and H. Schmidbaur, *Chem. Ber.* **110,** 1312 (1977).
154. F. F. Blicke, H. H. Willard, and J. T. Taras, *J. Am. Chem. Soc.,* **61,** 88 (1939).
155. J. Chatt and F. G. Mann, *J. Chem. Soc.,* 1192 (1940).
156. G. Wittig and K. Clauss, *Justus Liebigs Ann. Chem.* **577,** 26 (1952).
157. B. Tamhina, V. Maul, and M. J. Herak, *Microchem. J.* **22,** 275 (1977).
158. B. Tamhina, *Microchem. J.* **25,** 235 (1980).
159. D. M. Heinekey, I. T. Millar, and F. G. Mann, *J. Chem. Soc.,* 725 (1963).
160. M. C. Henry and G. Wittig, *J. Am. Chem. Soc.* **82,** 563 (1960).
161. J. N. C. Hood, D. Lloyd, W. A. MacDonald, and T. M. Shepherd, *Tetrahedron* **38,** 3355 (1982).
162. R. S. Tewari and S. C. Chaturvedi, *Tetrahedron Lett.* **43,** 3843 (1977).
163. N. A. Nesmeyanov, E. V. Binshtok, and O. A. Reutov, *Dokl. Akad. Nauk SSSR* **198,** 1102 (1971).
164. K.-H. Mitschke and H. Schmidbaur, *Chem. Ber.* **106,** 3645 (1973).
165. R. Luckenbach, *Dynamic Stereochemistry of Pentaco-ordinated Phosphorus and Related Elements,* Georg Thieme, Stuttgart, Germany, 1973.
166. C. D. Klaassen, in A. G. Gilman, T. W. Rall, A. S. Niews, and P. Taylor, eds., *Goodman and Gilman's The Pharmacological Basis of Therapeutics,* 8th ed., Pergamon Press, New York, 1990, p. 1602.
167. W. D. Buchanan, *Toxicity of Arsenic Compounds,* Elsevier Publishing Company, Amsterdam, The Netherlands, 1962.
168. H. W. Thomas, *Br. Med. J.* **I,** 1140 (1905).
169. M. Marquardt, *Paul Ehrlich,* Schuman, New York, 1951.
170. J. F. Mahoney, R. C. Arnold, and A. Harris, *Vener. Dis. Inf.* **24,** 355 (1943).
171. AMA Department of Drugs, *Drug Evaluations,* 6th ed., W. B. Saunders Company, Philadelphia, Pa., 1986, pp. 1575, 1579.
172. H. C. Spencer, in G. T. Strickland, ed., *Hunter's Tropical Medicine,* 6th ed., W. B. Saunders Company, Philadelphia, Pa., 1984, pp. 562–563.
173. L. T. Webster, in ref. 166, pp. 1009–1010.

174. J. E. F. Reynolds, ed., *Martindale, The Extra Pharmacopoeia,* 29th ed., The Pharmaceutical Press, London, UK, 1989, pp. 660–661.
175. E. L. Roberson, in N. H. Booth and L. E. McDonald, eds., *Veterinary Pharmacology and Therapeutics,* 5th ed., Iowa State University Press, Ames, Iowa, 1982, pp. 845, 846.

G. O. Doak
G. Gilbert Long
Leon D. Freedman
North Carolina State University

ARTIFICIAL INTELLIGENCE. See Expert systems.

ARTIFICIAL ORGANS. See Prosthetic and biomedical devices.

ASBESTOS

The term asbestos is a generic designation referring usually to six types of naturally occurring mineral fibers which are or have been commercially exploited. These fibers are extracted from certain varieties of hydrated alkaline silicate minerals comprising two families: serpentines and amphiboles. The serpentine group contains a single fibrous variety: chrysotile; five fibrous forms of amphiboles are known: anthophyllite, amosite, crocidolite, tremolite, and actinolite.

These fibrous minerals share several properties which qualify them as asbestiform fibers: (*1*) they are found in large clusters which can be easily separated from the host matrix or cleaved into thinner fibers (1); (*2*) the fibers exhibit high tensile strengths (1); (*3*) they show high length:diameter ratios, from a minimum of 20 up to >1000 (1); (*4*) they are sufficiently flexible to be spun; and (*5*) macroscopically, they resemble organic fibers such as cellulose (2).

Since asbestos fibers are all silicates, they exhibit several other common properties, such as incombustibility, thermal stability, resistance to biodegradation, chemical inertia toward most chemicals, and low electrical conductivity.

The usual definition of asbestos fiber excludes numerous other fibrous minerals which could be qualified as asbestiform following the criteria listed above. However, it appears the term asbestos has traditionally been attributed only to those varieties which are commercially exploited (1,2).

The mineralogical designations of the various asbestos fibers, their most common alternative designations, and the main producing countries of these fibers are reported in Table 1. The fractional breakdown of the recent world production of the various fiber types shows that the industrial applications of asbestos fibers have now shifted almost exclusively to chrysotile. Two types of amphiboles, commonly designated as amosite and crocidolite are still being used, but their combined production is currently less than 2% of the total world production. The other three amphibole varieties, anthophyllite, actinolite, and tremolite, have no significant industrial applications presently. This statement excludes asbestiform amphiboles which may occur in other industrial minerals.

Table 1. Asbestos Fiber Production

Mineral species[a]	Other designations	Main producers[b]	World production, 1988, %
chrysotile	white asbestos	Russia, Canada, Brazil, Zimbabwe, China, South Africa	>98%
cummingtonite-grunerite	amosite (brown asbestos)	South Africa	<1%
riebeckite	crocidolite (blue asbestos)	South Africa	<1%
anthophyllite		deposits in Finland, South Africa, Bulgaria, United States	insignificant
actinolite		deposits in Taiwan, South Africa	insignificant
tremolite		deposits in South Africa, Italy, Pakistan, Korea	insignificant

[a]Chrysotile is in the serpentine mineral group; all others are amphiboles.
[b]In order of production.

History

Early utilizations of asbestos have been reported in which the matrix reinforcement and thermal properties of asbestos fibers were exploited. The first recorded application can be traced to Finland (approximately 2500 BC), where anthophyllite from a local deposit was used to reinforce clay utensils and pottery (3). Numerous early references can also be found describing the use of asbestos fibers for the fabrication of lamp wicks and crematory clothing. Other applications of asbestos fibers in heat- or flame-resistant materials have been sporadically reported, for example, by Marco Polo. At the end of the seventeenth century, Peter the Great of Russia initiated the fabrication of asbestos paper, using chrysotile fibers extracted from deposits in the Ural Mountains. The use of asbestos fibers on a true industrial scale began in Italy early in the nineteenth century with the development of asbestos textiles. By the end of the nineteenth century, significant asbestos deposits had been identified throughout the world and their exploitation had begun in Canada (1878), South Africa (1893, 1908–1916), and the USSR.

From the beginning of this century, the demand for asbestos fibers grew in a spectacular fashion for numerous applications, in particular for thermal insulation in steam engines and technologies (4). Moreover, the development of the Hatschek machine in 1900 for the continuous fabrication of sheets from an asbestos–cement composite opened an important field of industrial application for asbestos fibers.

World War II supported the growth of asbestos fiber production for military applications, typically in thermal insulation and fire protection. Such applications were later extended into residential or industrial constructions for several decades following the war.

During the late 1960s and 1970s, the finding of health problems associated with heavy exposure to airborne asbestos fibers led to a strong reduction (or ban) in the use of asbestos fibers for thermal insulation. In most of the current applications, asbestos fibers are contained within a matrix, typically cement or organic resins.

The world production of asbestos fibers reached a maximum in 1979 of 5×10^6 t; it then decreased until 1986 to 4×10^6 t, after which it increased slightly to 4.3×10^6 t in 1989 (5). The main extraction sites of chrysotile asbestos are located in Russia (60%), Canada (16%), Brazil, Zimbabwe, China, and South Africa (5); the main producing sites of crocidolite and amosite are in South Africa (6). Currently, active mining operations of asbestos fibers are found in 21 countries (5).

Geology and Fiber Morphology

The genesis of asbestos fibers as mineral deposits required certain conditions with regard to chemical composition, nucleation, and fiber growth; such conditions must have prevailed over a period sufficiently long and perturbation-free to allow a continuous growth of the silicate chains into fibrous structures (7). Some of the important geological or mineralogical features of the industrially significant asbestos fibers are summarized in Table 2. Only three varieties of amphibole fibers are discussed for the following reasons: (*1*) crocidolite and amosite are the only amphiboles with significant industrial uses; (*2*) tremolite, although having no industrial application, may be found as a contaminant in other fibers or in other industrial minerals (eg, talc). Also, in view of the relative industrial importance of amphiboles and chrysotile, more emphasis is given to the latter.

Chrysotile. Chrysotile belongs to the serpentine group of minerals, varieties of which are found in most of the important mountain ranges and precambrian shields (8). Only a small part of these serpentine occurrences are in the asbestiform chrysotile variety. Chrysotile fibers are found as veins in serpentines or related minerals in serpentinized ultramafic rocks and in serpentinized dolomitic marbles (9). It has been suggested that the ultrabasic rocks (forsterite, Mg-rich pyroxenes, and amphiboles) are first attacked in an hydrothermal process and transformed in serpentines; in a later hydrothermal

Table 2. Geological Occurence of Asbestos Fibers

	Chrysotile [*12001-29-5*]	Amosite [*12172-73-5*]	Crocidolite [*12001-28-4*]	Tremolite [*14567-73-8*]
mineral species	chrysotile	cummingtonite-grunerite	riebeckite	tremolite
structure	as veins in serpentine	lamellar, coarse to fine, fibrous and asbestiform	fibrous in ironstones	long, prismatic, and fibrous aggregates
origin	alteration and metamorphism of basic igneous rocks rich in magnesium silicates	metamorphic	regional metamorphism	metamorphic
essential composition	hydrous silicates of magnesia	hydroxy silicate of Fe and Mg	hydroxy silicate of Na, Mg, and Fe	hydroxy silicate of Ca and Mg

event, the serpentines are partially redissolved and crystallized as chrysotile fibers (9). Clearly, the genesis of each chrysotile deposit must have involved specific features related to the composition of the precursor minerals, the stress and deformations in the host matrix, the water content, the temperature cycles, etc. Nonetheless, it is generally observed that the chemical composition of the fibrous phase is closely related to that of the surrounding rock matrix (9).

The growth of chrysotile fibers at right angles to the walls of cracks in massive serpentine formations led to the most common type of chrysotile deposit called cross-vein. Most of the industrial chrysotile fibers are extracted from such deposits where fiber lengths can reach several centimeters, but most often do not exceed 1 cm (9). Figure 1 illustrates the typical aspect of cross-vein chrysotile fibers separated from the host serpentine rock; a nonfibrous variety of serpentine, antigorite [*12135-86-3*], having a chemical composition nearly identical to that of chrysotile, is also pictured in Figure 1. In some occurrences, the relative motions (slipping) of blocks in the host rock, during or after fiber growth, lead to veins in which the fibers are inclined or parallel to the vein axes (slip fibers). In still other local conditions, dispersed aggregates of short fibers are found with no preferential orientation; in such cases the fiber content of the rock can be very high, up to 50%.

Chrysotile is a hydrated magnesium silicate and its stoichiometric chemical composition may be given as $Mg_3Si_2O_5(OH)_4$ [*12001-29-5*]. However, the geothermal processes which yield the chrysotile fiber formations usually involve the co-deposition of various other minerals. These mineral contaminants comprise: brucite [*1317-43-7*] ($Mg(OH)_2$), magnetite [*1309-38-2*] (Fe_3O_4), calcite [*13397-26-7*] ($CaCO_3$), dolomite [*16389-88-1*] ($Mg,CaCO_3$), chlorites ($(Mg,Al,Fe)_{12}Si_8O_{20}(OH)_{16}$), and talc [*14807-96-6*] ($Mg_6Si_8O_{20}(OH)_4$). Other iron-containing minerals may also be found in chrysotile, for example, pyroaurite, brugnatellite, and pyroxenes. In commercial fibers, dust particles from the host rock generated during the mining

Fig. 1. Asbestos fibers (chrysotile, crocidolite, and amosite) as separated from host rock and their massive varieties (antigorite, riebeckite, cummingtonite–grunerite) (8). Courtesy of R. T. Vanderbilt Company, Inc.

and milling processes are most inevitably present (10). Elemental analysis data for several chrysotiles and amphiboles are collected in Table 3.

Chrysotile fibers can be extremely thin, the unit fiber having a diameter of approximately 25 nm (0.025 μm). Industrial chrysotile fibers are aggregates of these unit fibers which usually exhibit diameters from 0.1 to 100 μm; their lengths range between a fraction of a mm to several cm, though most of the chrysotile fibers used are shorter than 1 cm.

The Amphiboles. The amphibole group of minerals is widely spread in the earth's mantle and their chemical composition can vary widely. Of the amphiboles, only a few varieties are asbestiform and the latter occur in relatively low quantities. The geological origin of amphibole asbestos fibers appears to be quite varied (10). In the case of the crocidolite deposit of South Africa (Transvaal), the amphibole fibers originated from a gel of iron hydroxide and colloidal silica, later consolidated by secondary reactions to yield formations of banded ironstone. Crocidolite fiber veins form in such rocks, presumably with magnetite particles acting as nucleating agents. The presence of mechanical stress appears to be necessary for the formation of crocidolite as well as of riebeckite. The amosite deposit found in similar rock formations is the result of a high temperature metamorphic process (10).

Table 3. Elemental Analysis of Asbestos Fibers[a]

Variety and source	SiO_2 (silica)	FeO (ferrous oxide)	Fe_2O_3 (ferric oxide)	Al_2O_3 (alumina)	MgO (magnesia)	CaO (lime)	MnO (manganese oxide)	Na_2O (sodium oxide)	K_2O (potassium oxide)	H_2O, adsorbed	H_2O^+, combined
Chrysotile											
Quebec	40.2	1.0	0.5	2.9	39.9	1.1	0.1	0.1	0.1	0.8	13.4
So. Rhodesia	39.7	0.7	0.3	3.2	40.3	1.1	0.3	0.1	0.1	0.6	12.2
Ural Mts.	38.1	1.3	1.4	5.0	37.7	2.2	0.1	0.1	0.1	0.8	11.1
Crocidolite											
Cape Province	50.9	20.5	16.9		1.1	1.5	0.1	6.2	0.2	0.2	2.2
Australia	52.8	14.9	18.6	0.2	4.6	1.1	trace	6.0	0.1	0.2	2.8
Bolivia	55.7	3.8	13.0	4.0	13.1	1.5	trace	6.9	0.4	trace	1.8
Amosite											
Transvaal	49.4	40.6	0.1		6.7	0.7	0.7	0.1	0.2	0.1	1.9
Tremolite											
Pakistan	55.1	2.0	0.3	1.1	25.7	11.5	0.1	0.3	0.2	3.5	0.2

[a]Ref. 11.

Amphibole minerals, in general, are characterized by prismatic cleavage planes which intersect at an angle of about 55°. Hence, in the crushing of massive, nonfibrous amphiboles, microscopic fragments are frequently found having the appearance of asbestos fibers. However, the statistical average of their aspect ratio is considerably lower than that of the asbestiform amphiboles.

The complexity of amphibole geology and structure is readily understood from their chemical composition. The average chemical composition of amphibole minerals may be represented as (2):

$$W_{0-1},\ X_2Y_5Z_8O_{22}\ (OH,\ O,\ F,\ Cl)_2$$

$$\begin{aligned} \text{where } W &= Na \\ X &= Na,\ Ca,\ Mg,\ Fe^{2+},\ Mn \\ Y &= Al,\ Fe^{2+},\ Fe^{3+},\ Ti \\ Z &= Si,\ Al \end{aligned}$$

W, X, Y each represent cationic sites of different geometries (12). From this generic representation, the chemical composition of the amphibole asbestos can be given as follows:

Name	*Composition*	*CAS Registry Number*
grunerite (amosite)	$(Fe^{2+})_2(Fe^{2+}, Mg)_5Si_8O_{22}(OH)_2$	[*12172-73-5*]
riebeckite (crocidolite)	$Na_2(Fe^{2+}, Mg)_3Fe^{3+}{}_2Si_8O_{22}(OH)_2$	[*12001-28-4*]
anthophyllite	$Mg_7Si_8O_{22}(OH)_2$	[*17068-78-9*]
tremolite	$Ca_2Mg_5Si_8O_{22}(OH)_2$	[*14567-73-8*]
actinolite	$Ca_2(Mg, Fe^{2+})_5Si_8O_{22}(OH)_2$	[*12172-67-7*]

From their respective compositions, the amphibole fibers can be viewed as a series of minerals in which one cation is progressively replaced by another at a given site. For example, tremolite and actinolite may be seen related: the magnesium in tremolite is partly replaced by divalent iron in the X position to yield actinolite. Similarly, in the cummingtonite–grunerite amphiboles, the substitution involves the replacement of magnesium in positions X and Y by divalent iron (2,12). The change of designation from cummingtonite to grunerite is, by convention, at a given Mg:Fe ratio (7).

The two main amphibole asbestos fibers are amosite and crocidolite, and both are hydrated silicates of iron, magnesium, and sodium. The appearance of these fibers and of the corresponding nonfibrous amphiboles is shown in Figure 1. Although the macroscopic visual aspect of clusters of various types of asbestos fibers is similar, significant differences between chrysotile and amphiboles appear at the microscopic level. Under the electron microscope, chrysotile fibers are seen as clusters of fibrils, often entangled, suggesting loosely bonded, flexible fibrils (Fig. 2**a**). Amphibole fibers, on the other hand, usually appear as individual needles with a crystalline aspect (Fig. 2**b**).

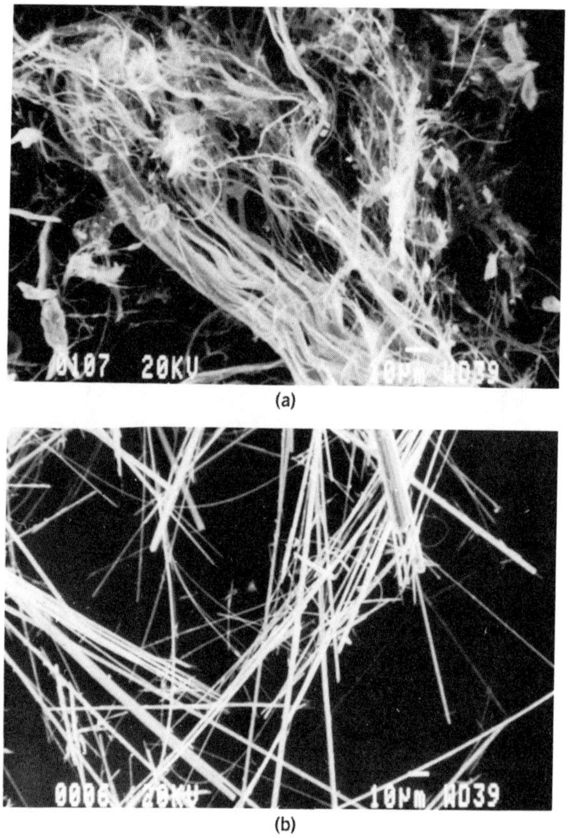

Fig. 2. Electron micrographs of asbestos fibers: (**a**) chrysotile; (**b**) crocidolite.

Crystal Structure of Asbestos Fibers

The microscopic and macroscopic properties of asbestos fibers stem from their intrinsic, and sometimes unique, crystalline features. As with all silicate minerals, the basic building blocks of asbestos fibers are the silicate tetrahedra which may occur as double chains $(Si_4O_{11})^{6-}$, as in the amphiboles, or in sheets $(Si_4O_{10})^{4-}$, as in chrysotile (12) (Fig. 3).

Chrysotile. In the case of chrysotile, an octahedral brucite layer having the formula $(Mg_6O_4(OH)_8)^{4-}$ is intercalated between each silicate tetrahedra sheet, as illustrated in Figure 4. The silicate and brucite layers share oxygen atoms which would normally be separated by distances of 0.305 nm in the silicate layer and 0.342 nm in the brucite layer (12). This mismatch of O–O distances induces a curvature of the sheets, the ideal radius of which has been calculated as 8.8 nm (13). The curvature of the sheets propagates along a preferred axis leading to the formation of the tubular structure found in chrysotile. The concentric sheets forming the fibers have a curvature radius from 2.5 to 3.0 nm for the internal layers up to approximately 25 nm for the external layers, yielding unit

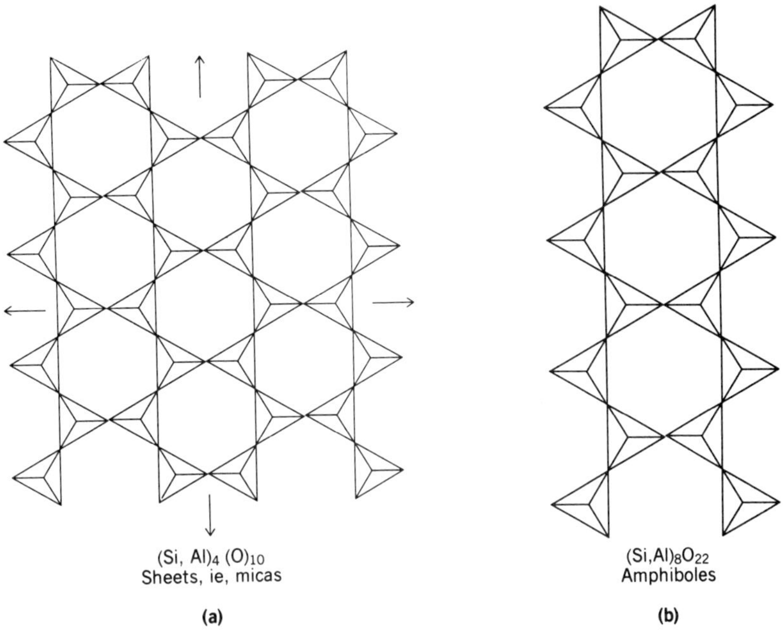

Fig. 3. Silicate backbones of chrysotile and amphiboles: (**a**) the sheet silicate structure of chrysotile, analogous to that of micas; (**b**) the double-chain silicate structure found in amphiboles (12). Courtesy of Oxford University Press.

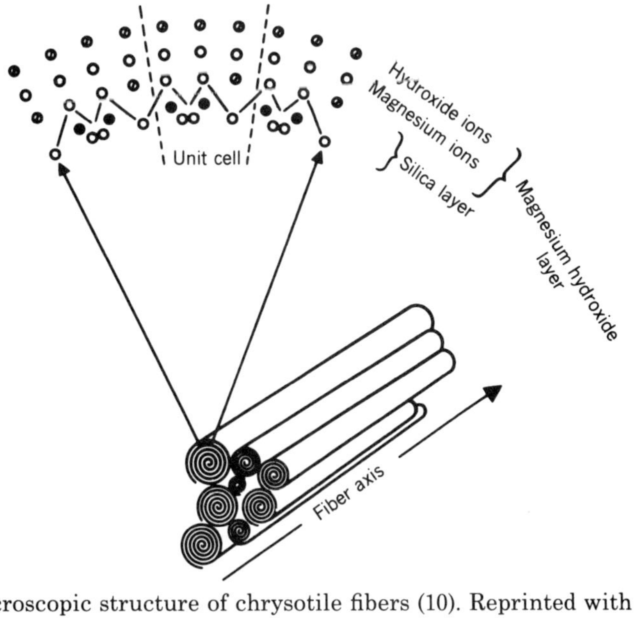

Fig. 4. Microscopic structure of chrysotile fibers (10). Reprinted with permission.

fibers (fibrils) with external diameters ranging between 20 and 50 nm (14). Electron microscopy studies (14) have also shown that, in the unit chrysotile fiber cross section, the layers may appear in a concentric or spiral arrangement.

The stacking of the tetrahedral and octahedral sheets in the chrysotile structure has been shown to yield three types of chrysotile fibers (10):

clino-chrysotile: monoclinic stacking of the layers, most abundant form
ortho-chrysotile: orthorhombic stacking of the sheets
para-chrysotile: a combination of the clino- and ortho-chrysotile stacking modes

The extent of substitution of magnesium and silicon by other cations in the chrysotile structure is limited by the structural strain that would result from replacement with ions having inappropriate radii. In the octahedral layer (brucite), magnesium can be substituted by several divalent ions, Fe^{2+}, Mn^{2+}, or Ni^{2+}. In the tetrahedral layer, silicon may be replaced by Fe^{3+} or Al^{3+}, leaving an anionic vacancy. Most of the other elements which are found in vein fiber samples, or in industrial asbestos fibers, are associated with interstitial mineral phases. Typical compositions of bulk chrysotile fibers from different locations are given in Table 3.

Amphiboles. The crystalline structure common to amphibole minerals consists of two ribbons of silicate tetrahedra placed back to back as shown in Figure 5. The plane of anionic valency sites created by this double ribbon arrangement is neutralized by the metal cations. The unit cell has seven cationic sites of three different types; these sites can host a large variety of metal cations without substantial disruption of the lattice.

In contrast to chrysotile fibers, the atomic crystal structure of amphiboles does not inherently lead to fiber formation. The formation of asbestiform amphiboles must result from multiple nucleation and specific growth conditions. Also, whereas the difference between asbestiform and massive amphibole minerals is obvious on the macroscopic scale, the crystalline structures of the two varieties do not exhibit substantial differences. Nonfibrous amphiboles also exhibit preferential cleavage directions, yielding fiber-shaped fragments.

Properties of Asbestos Fibers

Asbestos fibers used in most industrial applications consist of aggregates of smaller units (fibrils). This is most evident with chrysotile which exhibits an inherent, well-defined unit fiber. Typical diameters of fibers in bulk industrial samples may reach several tens of micrometers; fiber lengths are on the order of one to ten millimeters.

The mechanical processes employed to extract the fibers from the host matrix, or to further separate (defiberize, open) the aggregates, can impart significant morphological alterations to the resulting fibers. Typically, microscopic observations on mechanically opened fibers reveal fiber bends and kinks, partial separation of aggregates, fiber end-splitting, etc. The resulting product thus

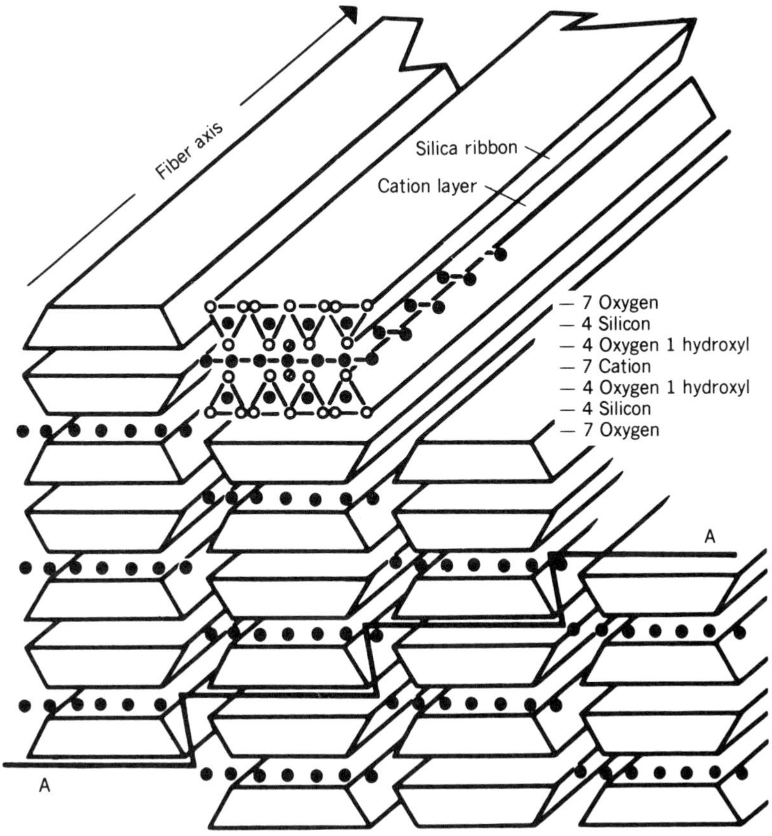

Fig. 5. Microscopic structure of amphibole fibers (10). Reprinted with permission.

exhibits a wide variety of morphological features, some of which can be seen in Figure 2. The consequences of the peculiar morphology of fiber shapes are difficult to assess, but it is quite obvious that a proper dimensional characterization of these fibers requires a shape factor in addition to diameter and length.

The morphological variance appears more important with chrysotile than with amphiboles. The intrinsic structure of chrysotile, its higher flexibility, and interfibril adhesion (10) allow a variety of intermediate shapes when fiber aggregates are subjected to mechanical shear. Amphibole fibers are generally more brittle and accommodate less morphological deformation during mechanical treatment.

Fiber Length Distribution. For industrial applications, the fiber length and length distribution are of primary importance because they are closely related to the performance of the fibers in matrix reinforcement. Various fiber classification methods have thus been devised. Representative distributions of fiber lengths and diameters can be obtained through measurement and statistical analysis of microphotographs (14); fiber length distributions have also been obtained recently from automated optical analyzers (15). Typical fiber length distributions obtained from these approaches are illustrated in Figure 6 for

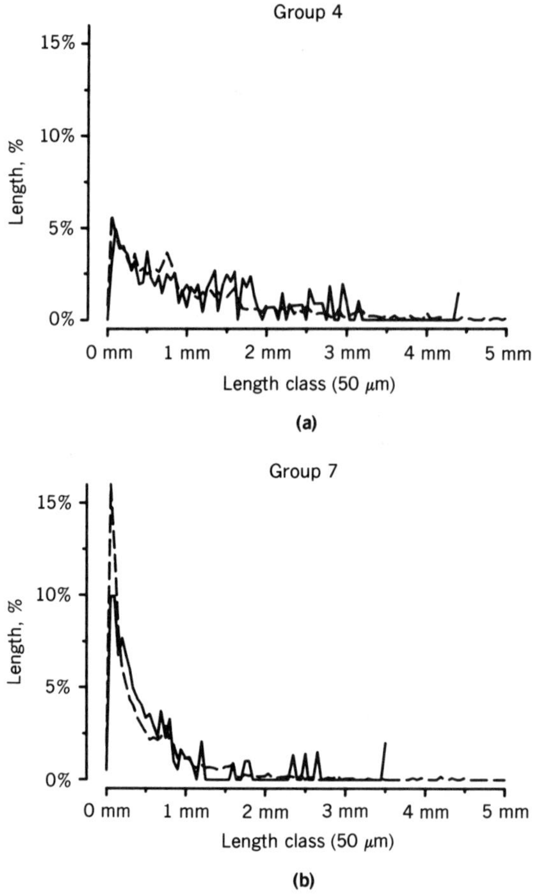

Fig. 6. Fiber length distribution for (**a**) a long sample (group 4) and (**b**) a short sample (group 7) of chrysotile; successive length classes separated by 50 μm. ---, Automated measurement; —, microphotographic measurement.

chrysotile fibers. As in the cases shown there, industrial asbestos fiber samples usually contain a rather broad distribution of fiber lengths.

Physico-Chemical Properties. The industrial applications of chrysotile fibers were developed taking advantage of their particular combination of properties: fibrous morphology, high tensile strength, resistance to heat and corrosion, low electrical conductivity, and high friction coefficient. In many applications, the surface properties of the fibers also play an important role; in such cases, distinction between chrysotile and amphiboles can be observed because of their differences in chemical composition and surface microstructure. Technologically relevant physical and chemical properties of asbestos fibers are given in Table 4.

Thermal Behavior. Asbestos fiber minerals are hydrated alkaline silicates, and thus their behavior as a function of temperature is related first to dehydration (or dehydroxylation) reactions. In the case of chrysotile, the crystalline structure is stable up to approximately 550°C (depending on the heating period (16)), where the dehydroxylation of the brucite layer begins. This process is

Table 4. Physical and Chemical Properties of Asbestos Fibers

	Chrysotile	Amosite	Crocidolite	Tremolite
color	usually white to grayish green; may have tan coloration	yellowish gray to dark brown	cobalt blue to lavender blue	gray-white, green, yellow, blue
luster	silky	vitreous to pearly	silky to dull	silky
hardness, Mohs	2.5–4.0	5.5–6.0	4.0	5.5
specific gravity	2.4–2.6	3.1–3.25	3.2–3.3	2.9–3.2
optical properties	biaxial positive parallel extinction	biaxial positive parallel extinction	biaxial oblique extinction	biaxial negative oblique extinction
refractive index	1.53–1.56	1.63–1.73	1.65–1.72	1.60–1.64
flexibility	high	fair	fair to good	poor, generally brittle
texture	silky, soft to harsh	coarse but somewhat pliable	soft to harsh	generally harsh
spinnability	very good	fair	fair	poor
tensile strength, MPa[a]	1100–4400	1500–2600	1400–4600	<500
resistance to:				
acids	weak, undergoes fairly rapid attack	fair, slowly attacked	good	good
alkalies	very good	good	good	good
surface charge, mV (zeta potential)	+13.6 to +54[b]	−20 to −40	−32	
decomposition temperature, °C	600–850	600–900	400–900	950–1040
residual products	forsterite, silica, eventually enstatite	Fe and Mg pyroxenes, magnetite, haematite, silica	Na and Fe pyroxenes, haematite, silica	Ca, Mg, and Fe pyroxenes, silica

[a]To convert MPa to psi, multiply by 145.
[b]Chrysotile fibers tend to become negative after weathering and/or leaching.

completed near 750°C and is characterized by a total weight loss of 13%. The resulting magnesium silicate recrystallizes to form forsterite and silica in the temperature range 800–850°C, as an exothermic process. The weight loss as a function of heating temperature (thermogravimetry) and the reaction heats (differential thermal analysis) are illustrated for chrysotile in Figure 7**a**. The strongly endothermic dehydration process contributes to enhance the high temperature thermal insulation properties of chrysotile asbestos.

The behavior of amphibole fibers under continuous heating is similar to that observed with chrysotile, although the temperatures of dehydroxylation and recrystallization processes are different. The amphiboles have a lower water (hydroxyl) content and their dehydroxylation reaction begins between 400–600°C depending on the amphibole type; this reaction leads to a weight loss of approximately 2%. The latter is illustrated in Figure 7**b**, together with a dta recording of the associated thermal events. The products resulting from the thermal decomposition of the amphiboles are pyroxenes, magnetite, hematite, and silica.

In the presence of oxygen (air), the thermal decomposition of amphiboles is associated with an oxidation of divalent iron to trivalent iron, which may lead to an increase in the sample weight; the oxidation process also induces an obvious color alteration, the fibers acquire the characteristic ferric oxide color.

Asbestos fibers, in particular chrysotile fibers, can undergo substantial thermal degradation during mechanical grinding. In high energy attrition equipment, such as ball mills, the high localized impact energies can lead to amorphization of the crystalline structure. This is particularly true for dry milling or milling in organic solvents. However, milling in aqueous slurries is less detrimental, water offering some protection toward high impact degradation (17).

Tensile Strength. The inherent tensile strength of a single asbestos fiber, based on the strength of Si—O—Si bonds in the silicate chain, should be near 10 GPa (1.45×10^6 psi) (18). However, industrial fibers exhibit substantially lower values, because of the presence of various types of structural or chemical defects.

The measured tensile strength of chrysotile fibers has been reported in the range 1.1–4.4 GPa (160,000–640,000 psi). The accurate determination of this parameter is difficult since the measurement performed on a fiber aggregate is influenced by interfibril adhesion, discontinuities in some of the unit fibers, mineral inclusions, etc. Consequently, higher tensile strength results are obtained for measurements done on short and thin fibers (19,20). The tensile strengths of amphiboles (crocidolite, amosite) are comparable to that of chrysotile. With amphiboles, the tensile strength is highly influenced by the iron content since iron–oxygen bonds located in the fiber axes, particularly those involving trivalent Fe, are particularly strong. The trend observed in the tensile strength of amphiboles from tremolite, to amosite, to crocidolite is directly related to the iron content of these fibers (21).

The variation in tensile strength of asbestos fibers as a function of temperature also sharply distinguishes chrysotile and amphiboles. Chrysotile retains (and even slightly increases) its tensile strength up to 500°C, until the dehydroxylation reaction begins; it drops sharply at higher temperatures. Amphiboles, on the other hand, exhibit a decreasing tensile strength beginning at approximately 200°C. For example, at 350°C, crocidolite has lost 50% of its initial tensile strength.

Asbestos Fibers in Aqueous Media. Although asbestos fibers cannot be viewed as water-soluble silicates, prolonged exposure of chrysotile or amphiboles to water (especially at high temperature) leads to slow progressive leaching of both their metal and silicate components (22). In the case of chrysotile fibers (in a given amount of water), the brucite layer will, fairly rapidly, dissolve in part, with concomitant increase in the pH of the solution. The equilibrium pH value for an aqueous chrysotile slurry reaches 10.0–10.5 Free brucite, present as contaminant

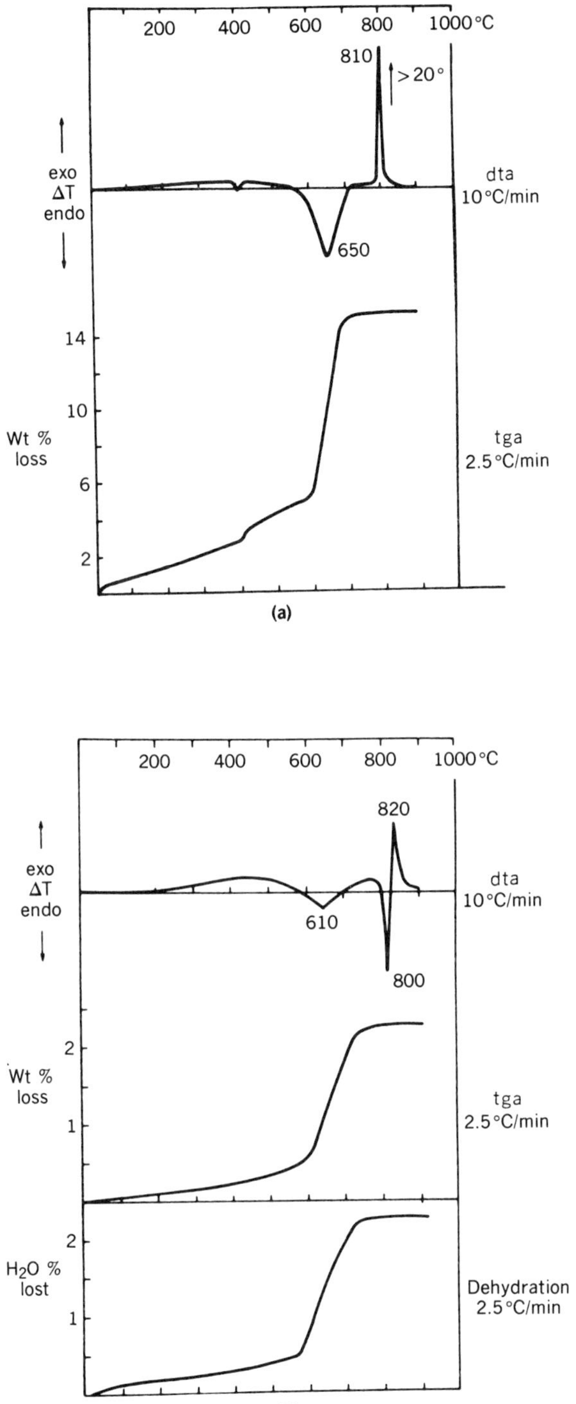

Fig. 7. Thermal analysis curves for asbestos: (**a**) chrysotile; (**b**) crocidolite (in inert atmosphere) (10). Reprinted with permission.

in the fibers, also contributes to the pH increment. Concurrently, magnesium ions are released into the aqueous phase.

In contact with solutions of mineral acids, organic acids, or magnesium complexing agents, the rate of dissolution of the brucite layers is increased. When carried out to the extreme, the leaching of magnesium leads to a weight loss of 58%. The residual silica is largely amorphous and, although the fibers retain an apparent fibrous morphology, their mechanical resistance is drastically reduced.

The dispersion of amphiboles in concentrated HCl solutions also leads to partial leaching, the rate of which depends on the metal cations present. With crocidolite, only small amounts of magnesium and sodium are extracted in these conditions, whereas amosite liberates substantial quantities of iron and magnesium. Overall, tremolite appears to exhibit the highest resistance to acid leaching.

On the other hand, both chrysotile and the amphiboles exhibit a high degree of chemical inertia towards strong alkalies over extended periods. At high temperatures, reactions with alkalies (NaOH, KOH, $Ca(OH)_2$) become significant over relatively short periods; for example, crocidolite was reported to be attacked by potassium hydroxide above 100°C (22).

Other Bulk Physical Properties. The hardness of asbestos fibers is comparable to that of other crystalline or glassy silicates. Compared to glass fibers, amphiboles have similar hardness values, while chrysotile shows lower hardness values.

The friction coefficients of asbestos fibers are also different for chrysotile and amphiboles (when measured against the same material). Compared to glass fibers, the friction coefficients decrease in the order: chrysotile, amphiboles, glass fibers.

The high electrical resistivity of asbestos fibers is well-known, and has been widely exploited in electrical insulation applications. In general, the resistivity of chrysotile is lower than that of the amphiboles, particularly in high humidity environments (because of the availability of soluble ions). For example, the electrical resistivity of chrysotile decreases from 1 to 2100 MΩ/cm in a dry environment to values of 0.01 to 0.4 MΩ/cm at 91% relative humidity. Amphiboles, on the other hand, exhibit resistivity between 8,000 and 900,000 MΩ/cm.

With respect to magnetic properties, the intrinsic magnetic susceptibility of pure chrysotile is very weak. However, the presence of associated minerals such as magnetite, as well as substitution ions (Fe, Mn), increases the magnetic susceptibility to values around 6×10^{-8} m^3/kg. With amphiboles, the magnetic susceptibility is much higher, mainly because of the high iron content; typically, amosite and crocidolite exhibit susceptibility values of 100 and 75×10^{-8} m^3/kg, respectively (23).

Surface Properties. *Surface Area.* The specific surface area of industrial asbestos fibers obviously depends on the extent of their defiberization (opening), and is usually between 1 and 30 m^2/g. As measured by BET nitrogen adsorption, chrysotile fibers exhibit surface areas between 15 and 30 m^2/g. With the amphibole fibers, surface areas of 1.8–9 m^2/g have been reported for crocidolite and 1.3–5.5 m^2/g for amosite (24). Such a difference originates, in part, in the relative sizes of the unit fibers of chrysotile and amphiboles; however, the results also reflect the ability of nitrogen molecules to diffuse between the unit fibers of a larger aggregate. In relation to magnesium acid leaching of chrysotile discussed earlier, the surface area of an amorphous silica resulting from extensive acid leaching may reach 450 m^2/g (25).

Also, the adsorption of anionic or neutral surfactants on chrysotile fibers in aqueous dispersions enhances fiber separation, with a concomitant increase of surface area (26). Such effects have not been reported for amphibole fibers.

Surface Charge in Aqueous Media. Because of dissolution–ionization effects, the surface of asbestos fibers in aqueous dispersions adopts an electrostatic charge. In the case of chrysotile, partial dissolution of the brucite layer leads to a positive surface charge (or potential), which is strongly influenced by the solution pH. The isoelectric point of chrysotile (pH of zero surface charge) has been reported as 11.8. In the case of amphibole fibers, the surface charge seems dominated by the silica component, and is generally observed to be negative, increasing towards 0 as the pH is decreased. Since the progressive leaching of magnesium from the external brucite layer of chrysotile gradually exposes silica, the surface potential rapidly decreases early in the leaching reaction.

Adsorption and Surface Chemical Grafting. As with silica and many other silicate minerals, the surface of asbestos fibers exhibit a significant chemical reactivity. In particular, the highly polar surface of chrysotile fibers promotes adsorption (physi- or chemisorption) of various types of organic or inorganic substances (22). Moreover, specific chemical reactions can be performed with the surface functional groups (OH groups from brucite or exposed silica).

The chemical reaction of coupling agents, such as organosilane compounds, on chrysotile yields fibers with a highly hydrophobic surface. Through an adequate choice of coupling agents (organosilanes or others), the fiber matrix interactions in composite materials can thus be optimized (27). Surface chemical modifications can be carried out in fiber slurries or through gas–solid reactions; for example the gas-phase reaction of $POCl_3$ with chrysotile, leads to a surface coverage with insoluble magnesium phosphate (28). The surface properties and reactivity of chrysotile fibers can also be modified by cationic substitution, for example, replacing magnesium by aluminum (29) or other metal cations (30).

Analytical Methods and Identification of Asbestos Fibers

In a general way, the identification of asbestos fibers can be performed through morphological examination, together with specific analytical methods to obtain the mineral composition and/or structure. Morphological characterization in itself usually does not constitute a reliable identification criteria (1). Hence, microscopic examination methods and other analytical approaches are usually combined.

Microscopic Methods. The use of microscopic methods is preferred in cases where limited quantities of sample are available, typically in analyzing fibers recovered from sampling of airborne dust. With fibers having lengths in excess of 5 μm, optical microscopy with light polarization (Optical Polarization Light Microscopy) has proven a very powerful technique. The optical properties of the different types of asbestos fibers (refractive index, dispersion of refractive indexes, birefringence, extinction angles), combined with information on fiber shapes, enables positive identification of all varieties of asbestos fibers. This can be carried out even when the fibers are mixed with their nonfibrous analogue or with various other materials (31).

Asbestos fiber identification can also be achieved through transmission or scanning electron microscopy (tem, sem) techniques which are especially useful with very short fibers, or with extremely small samples (see MICROSCOPY). With appropriate peripheral instrumentation, these techniques can yield the elemental composition of the fibers using energy dispersive x-ray fluorescence, or the crystal structure from electron diffraction, selected area electron diffraction (saed).

Instrumental Methods for Bulk Samples. With bulk fiber samples, or samples of materials containing significant amounts of asbestos fibers, a number of other instrumental analytical methods can be used for the identification of asbestos fibers. In principle, any instrumental method that enables the elemental characterization of minerals can be used to identify a particular type of asbestos fiber. Among such methods, x-ray fluorescence (xrf) and x-ray photo-electron spectroscopy (xps) offer convenient identification methods, usually from the ratio of the various metal cations to the silicon content. The x-ray diffraction technique (xrd) also offers a powerful means of identifying the various types of asbestos fibers, as well as the nature of other minerals associated with the fibers (9).

Thermoanalytical methods (tga, dta) often enable definite identification of the type of asbestos fibers (Fig. 7). For example, the strong exotherm observed with chrysotile at 830°C can be used as a routine indicator for determining the chrysotile content of talc (4,10). Thermal methods are also useful for determining certain mineral contaminants of asbestos fibers, for example brucite and calcite in chrysotile.

Infrared spectroscopy can also be used incisively to identify the six main varieties of asbestos fibers. Specific absorption bands in the infrared spectrum can be associated with the asbestos fibers, first in the 3600–3700 cm^{-1} range (specific hydroxyl bands) and, second, in the ranges 600–800 and 900–1200 cm^{-1} (specific absorption bands for various silicate minerals (10)).

Production

The evolution in the world production of asbestos fibers since 1950 is illustrated in Table 5 (5); after a peak near 1980, production leveled off after 1985 at 4.2–4.3 ×

Table 5. World Production[a] of Asbestos by Principal Producing Countries, 1950–1989, 10^3 t

Year	Canada[b]	USSR[b]	World total[c]
1950	794		1300
1960	1015	599	2214
1970	1507	1066	3490
1980	1323	2070	4699
1985	750	2500	4243
1989 (est)	691	2600	4314

[a]Ref. 5.
[b]Chrysotile.
[c]Chrysotile and amphiboles.

10^6 t. Changes in the production of the two main producers, Canada and the former USSR, over the same period are also illustrated. These figures show a substantial decrease in the Canadian production with a concomitant increase in the former USSR production. During recent years, several other countries, namely Brazil, Zimbabwe, and China, have substantially increased their production of chrysotile. Most of China's production, as well as the limited production of many other countries, is used in local industrial applications. South Africa is the only country where the three main types of asbestos are produced (chrysotile, crocidolite, and amosite), and the only significant producer of amphibole fibers.

Mining and Milling Technologies

The finding and mapping of chrysotile asbestos ore deposits usually relies on magnetometric surveys. Indeed, magnetite is generally associated with asbestos fiber deposits, except in the case of ore bodies located in sedimentary formations. As in other mining operations, core drilling is used for a precise evaluation of the grade and volume of the ore body (32).

The choice of a particular mining method depends on a number of parameters, typically the physical properties of the host matrix, the fiber content of the ore, the amount of sterile materials, the presence of contaminants, and the extent of potential fiber degradation during the various mining operations (33). However, most of the asbestos mining operations are of the open pit type, using bench drilling techniques.

The fiber extraction (milling) process must be chosen so as to optimize recovery of the fibers in the ore, while minimizing reduction of fiber length. Since the asbestos fibers have a chemical composition similar to that of the host rock, the separation processes must rely on differences in the physical properties between the fibers and the host rock rather than on differences in their chemical properties (33).

In dry milling operations, which are currently the most widely used, the ore is first crushed to a nominal size and then dried (Fig. 8). Fiber extraction then begins through a series of crushing operations, each followed by a vacuum aspiration of the ore running on a vibrating screen. On the latter, the fibers released from the ore have a tendency to move to the surface and, because of their large hydrodynamic volumes (low density), they can be readily collected into a vacuum system. The fibers recovered from consecutive vibrating screens are brought to cyclone separators, and the air is filtered to remove the finer suspended fibers.

Generally, the consecutive crushing–aspiration steps liberate fibers which are increasingly shorter. The wider veins which contain the longer fibers are easier to break and thus come out in the early phases of the milling operation. In this way, a primary classification, or grading of the fibers can be achieved. Various secondary processing systems may be used to further separate the fiber aggregates and to remove nonfibrous mineral dust. All fiber extraction and classification operations are usually carried out under negative pressure to minimize airborne dust in the working environment.

In recent years, several wet milling operations have been initiated with obvious advantages in dust control and potential advantages in the separation of

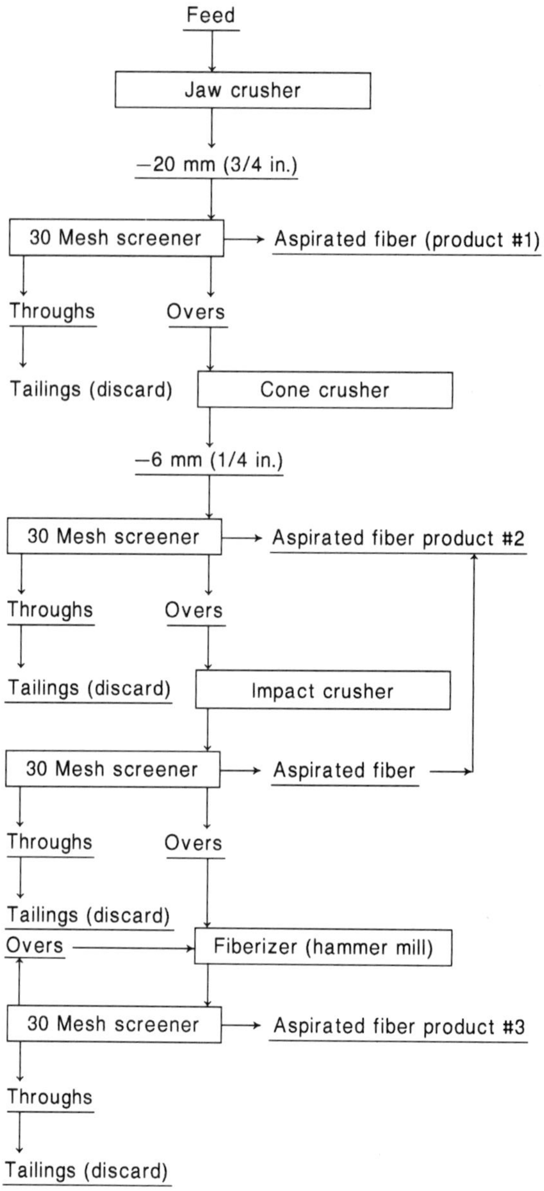

Fig. 8. Schematic of a typical asbestos milling flowline (30 mesh = 590 μm) (32).

mineral contaminants from the fiber product. On the other hand, large-scale industrial wet classification methods are not in use at present.

Fiber Classification and Standard Testing Methods

In the production, or industrial applications, of asbestos fibers, several parameters are considered critically important and are used as standard evaluation criteria (34): length (or length distribution), degree of opening and surface area, performance in cement reinforcement, and dust and granule content. The measurement of fiber length is important since the length determines the product category in which the fibers will be used and, to a large extent, their commercial value.

Dry Classification Method. The most widely accepted method for chrysotile fiber length characterization in the industry is the Quebec Standard test (QS). This is a dry sieving method (on vibrating screens) which enables the fractionation of a particular asbestos fiber sample into four fractions of decreasing sizes (+2, +4, +10, −10 M). This standard test has been used as a basis for developing a grade system comprising nine main fiber grades. The grading system identifies the longest fibers as grade 1 and the shortest as grade 9. Grades 1 and 2 designate crude asbestos fibers which are hand-picked and essentially contain large clusters or aggregates of fibers; the respective length of grades 1 and 2 are defined as >¾ in. (1.9 cm) and >⅜ in. (0.95 cm).

Fiber grades between 3 and 9 (decreasing lengths) are known as milled asbestos fibers and represent most of the bulk production. The fiber grades 3–7 are determined using the QS test, whereas the grades 8 and 9, containing very short fibers, are usually determined according to their bulk densities.

Other fiber classification schemes have been devised for chrysotile fibers, but historically the QS grade system has been used as a reference; other classification schemes usually have correspondence scales for conversion to the QS values. Amosite can be classified according to the QS grade system, but crocidolite requires a different scheme (mainly due to the harshness of these fibers).

Wet Classification Method. A second industrially important fiber length evaluation technique is the Bauer-McNett (BMN) classification. In this method, a fiber slurry is circulated through a series of four grids with decreasing opening size (positioned vertically), thus yielding five fractions (+4, +14, +35, +200, −200 M). A similar method, using smaller samples and horizontal grids has also been developed and referred to as the Turner-Newall classification.

Currently, the Bauer-McNett classification and the QS test are the most widely used fiber classification techniques. Whereas there are qualitative relationships between QS and BMN, there is no quantitative correspondence. It is readily understood that these standard tests do not provide accurate definition of the fiber lengths; the classification also reflects the hydrodynamic behavior (volumes) of the fibers, which, because of their complex shapes, is not readily predictable.

Other classification techniques have been developed which provide some insight on fiber lengths, typically the Ro-Tap test, the Suter-Webb Comb, and the Wash test.

Other Fiber Evaluation Methods. The extent of fiber separation (fiber openness) is an important evaluation criteria that is commonly measured by several techniques, namely air permeability, adsorbed gas volume, bulk density, and resilience (compression and recovery). The adsorption and retention of kerosene is also used as a measure of fiber openness and fiber adsorption capacity (34).

The filtration behavior of asbestos fibers is evaluated through a drainage test using, generally, a saturated calcium sulfate solution. This test is designated as "freeness" and reflects the water drainage rate as relevant to asbestos–cement fabrication processes (34).

The reinforcing capacity of asbestos fibers in a cement matrix constitutes another key criteria for the evaluation of asbestos fibers. This property is assessed by preparing samples of asbestos–cement composites which, after a standard curing period, are tested for flexural resistance. The measured rupture moduli are converted into a parameter referred to as the fiber strength unit (FSU) (34).

Finally, other properties of asbestos fibers may be evaluated depending on the envisaged application. Typically, the grits and spicule content, the magnetic susceptibility (magnetic rating), the content in soluble chlorides, and the humidity level may be of a particular interest in specific applications.

Industrial Applications

Asbestos fibers have been used in a broad variety of industrial applications. In the peak period of asbestos consumption in industrialized countries, some 3000 applications, or types of products, have been listed. Because of recent restrictions, many of these applications have now been abandoned and others are pursued under strictly regulated conditions.

The main characteristic properties of asbestos fibers that can be exploited in industrial applications (8) are their thermal, electrical, and sound insulation; nonflammability; matrix reinforcement (cement, plastic, and resins); adsorption capacity (filtration, liquid sterilization); wear and friction properties (friction materials); and chemical inertia (except in acids). These properties have led to several main classes of industrial products or applications (4).

Loose asbestos fibers, or formulations containing asbestos fibers for spray coatings, have been widely used in the building industry for fire protection and heat or sound insulation. Such applications used mainly chrysotile or amosite but, because of health concerns, this practice has been discontinued.

Asbestos fibers have also been widely used for the fabrication of papers and felts for flooring and roofing products, pipeline wrapping, electrical insulation, etc. Asbestos textiles, comprising yarn, thread, cloth, tape, or rope, also found wide application in thermal and electrical insulation, friction products in brake or clutch pads, etc. In recent years, some of these applications have decreased to various extents, although others remain fairly active, typically in friction materials.

The reinforcing properties of asbestos fibers have been widely exploited in asbestos–cement products mostly for the construction industry and sanitation (sheets, pipes). Into the 1990s, asbestos–cement products represent by far ($\approx 70\%$) the largest industrial consumption of asbestos fibers.

Asbestos fibers have likewise been used in reinforcement of plastics such as poly(vinyl chloride), phenolics, polypropylene, nylon, etc. Reinforcement of both thermoset and thermoplastic resins by asbestos fibers has been practiced to develop products for the automotive, electronic, and printing industries.

The combination of asbestos fibers with various types of natural or synthetic resins has led to the development of a variety of products and applications. Among those, the incorporation of asbestos fibers (mainly chrysotile) into rubber matrices yields materials that are widely used for fabrication of packings and gaskets. Complex formulations, comprising short asbestos fibers (usually chrysotile), resins, and other fillers and modifiers, have been developed as friction materials for brake linings and pads. Asbestos fibers have also found broad application as reinforcing agents in fillers (qv), coatings, sealings, and adhesive formulations.

Finally, the combined reinforcing effect and high absorption capacity of asbestos fibers have been exploited in a variety of applications to increase dimensional stability, typically in vinyl or asphalt tiles and asphalt road surfacing. Figure 9 summarizes, as of 1984, the various classes of application for

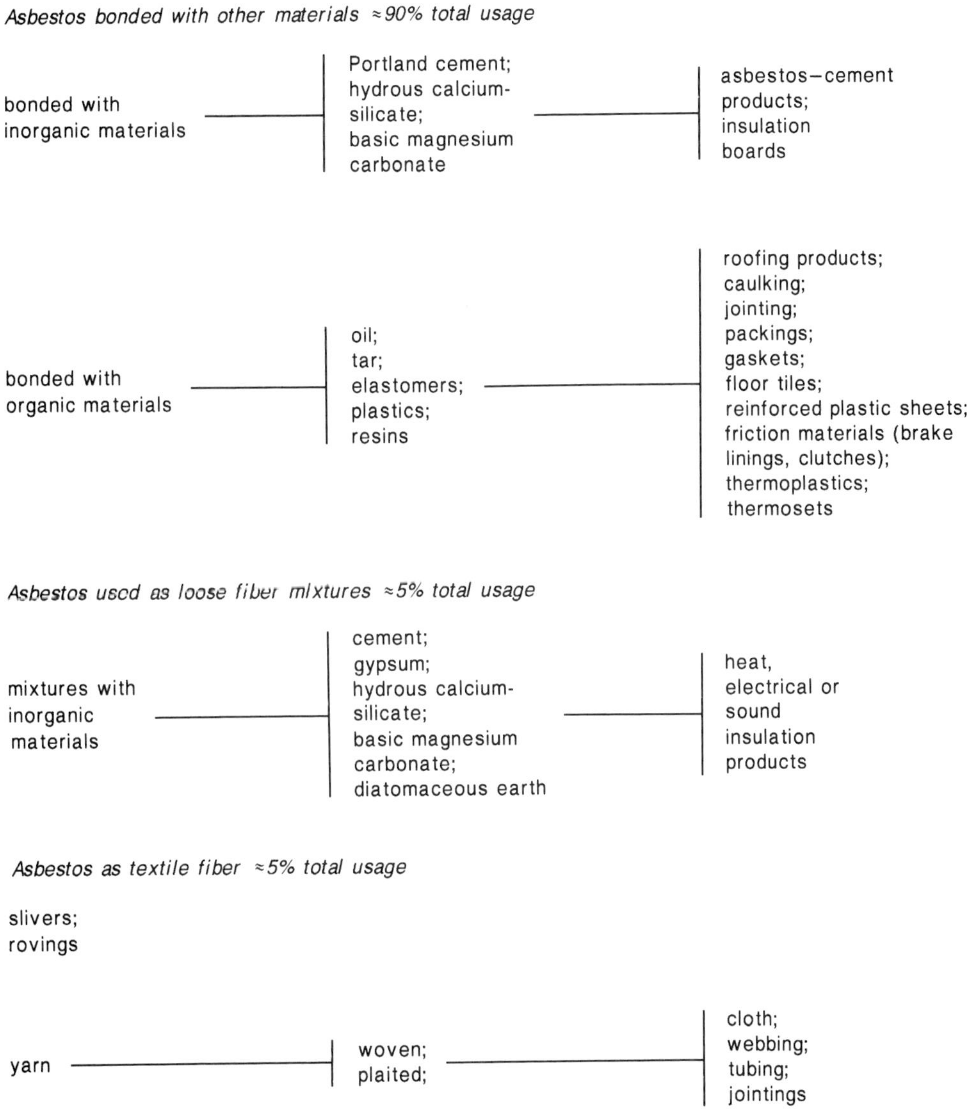

Fig. 9. Utilization of asbestos fibers by process, leading to various end products (35).

asbestos fibers in combination with other materials. The diagram shows that in recent years, most industrial applications have evolved towards composite materials where the fibers are bonded within an organic or inorganic matrix.

In the various applications, each type or group of products usually requires a selected asbestos fiber grade (or range of grades). Table 6 lists various types of chrysotile applications in relation to fiber grades (Quebec Standard grade system).

Table 6. Uses of Asbestos[a]

Grade[b]	Length specifications, mm	Uses
long fiber		
no. 1 crude	19	textiles
no. 2 crude	9.5–19	textiles, insulation
no. 3	6–9.5	textiles, packings, brake linings, clutch facings, electrical, high pressure and marine insulation
medium fiber		
no. 4	3–6	asbestos–cement pipe
no. 5	3–6	asbestos–cement pipe and sheets, molded products, paper products, brake linings and gaskets
no. 6	3	asbestos–cement products, brake linings and gaskets, plaster, backing for vinyl sheets
short fiber		
no. 7	3	molded brake linings and clutch facings, plastics, filler in vinyl and asphalt floor tiles, asphalt compounds, caulking compounds, paints and drilling mud additive
no. 8	[c]	similar to no. 7

[a]Ref. 33.
[b]Quebec Standard.
[c]Loose density specifications.

Alternative Industrial Fibers and Materials

Considerable effort has been devoted to finding alternative fibers or minerals to replace asbestos fibers in their applications. Such efforts have been motivated by various reasons, typically, availability and cost, and more recently, health concerns. During World War I, some countries lost access to asbestos fiber supplies and had to develop substitute materials. Also, in the production of fiber reinforced cement products, many developing countries focused on alternatives to asbestos fibers, in particular on cellulose fibers readily available locally at minimal cost. Since the 1980s however, systematic research has been pursued in several industrialized countries to replace asbestos fibers in all of their current applications because of perceived health risks.

The substitution of asbestos fibers by other types of fibers or minerals must, in principle, comply with three types of criteria (36): the technical feasibility of the substitution; the gain in the safety of the asbestos-free product relative to the asbestos-containing product; and the availability of the substitute and its comparative cost.

In some applications, particularly those which rely simultaneously on several characteristic features of asbestos fibers, the substitution has presented a significant challenge. For example, in fiber–cement composites, the fibers must exhibit high tensile strength, good dispersibility in Portland cement pastes, and high resistance to alkaline environments. Likewise, in friction materials, several properties of the asbestos fibers are important: high affinity toward resins, good heat resistance, high friction coefficient, and low abrasion of the opposing surface. In such applications, the replacement of asbestos fibers usually requires a combination of several materials.

Table 7 lists some of the materials and fibers that have been suggested or used in the development of asbestos-free products. These are listed by categories, and no attempt is made to provide a detailed coverage of the respective properties of these materials. Table 7 also provides an estimate of the cost ranges of asbestos fibers and several types of substitution materials.

Table 7. Asbestos Substitutes and Relative Costs[a]

Minerals	Synthetic mineral fibers	Synthetic organic fibers
	<2 $/kg[b]	
attapulgite	mineral wool	
diatomite	glass wool	
mica		
perlite		
sepiolite		
talc		
vermiculite		
wollastonite		
asbestos, grades 3–7		
	2–10 $/kg	
	steel fibers	polypropylene (PP)
	continuous filament glass	poly(vinyl alcohol) (PVA)
	alkali-resistant glass	polyacrylonitrile (PAN)
	aluminosilicates	
	10–20 $/kg	
	continuous filament glass	polytetrafluoroethylene (PTFE)
	>20 $/kg	
	alumina fibers	polybenzimidazole (PBI)
	silica fibers	aramid fibers
		pitch and PAN carbon fibers

[a]In U.S. $, 1989. Ref. 36.
[b]The natural organic fiber, cellulose (pulp), also falls in the <$2/kg range.

Various asbestos substitution strategies have been followed and a number of asbestos-free products have been proposed. Whereas little data is available in most cases on the performance of the products and their long-term behavior, the asbestos substitution is said to be effective from the point of view of the technology, as well as from the economics.

For example, in bulk applications as thermal insulation, synthetic mineral fibers (glass or slag fibers) have adequately replaced natural asbestos fibers. In sprayed insulation coatings, asbestos fibers have been replaced, for example, by vermiculite. As replacement for asbestos textiles, clothing made from aramid fibers or aluminized glass fibers is being offered (see HIGH PERFORMANCE FIBERS).

In fiber–cement construction materials, several alternatives are being practiced, either using cellulosic fibrous products or synthetic organic fibers such as polypropylene or polyacrylonitrile.

For friction material applications, composites (qv) comprising glass or metallic fibers with other minerals have been developed. In such applications also, aramid and graphite fibers are effective, although the cost of these materials restricts their use to heavy duty or high technology applications (see CARBON FIBERS).

The search for asbestos replacement materials is obviously an ongoing process and further developments in this area are to be expected. The extent of substitution of asbestos fibers by other fibers or other materials has been limited by several factors, typically: the availability of adequate replacement materials, the cost:performance ratio of such materials, and the uncertainty of long-term health risks of these replacement materials. From the data currently available, it may be estimated that between 10–20% of the industrial consumption of asbestos fibers was diverted to other materials during the 1980s.

Health and Safety

The relationship between workplace exposure to airborne asbestos fibers and respiratory diseases is one of the most widely studied subjects of modern epidemiology (37–39). Asbestos-related health concerns were first raised at the beginning of the century in the UK and the latter appears to have been the first country to regulate the asbestos-user industry (40). However, at that time, infectious respiratory diseases were a much greater concern than those arising from poor industrial hygiene practices.

After antibiotics succeeded in virtually eliminating mortality from bacterial infections, the mortality rate from lung cancer became more apparent (eg, during the mid-1950s) and causes for the latter were examined. Cigarette smoking was readily identified as the main cause of lung cancer in the general population, but in the early 1960s, researchers established a correlation between worker exposure to asbestos fibers and respiratory diseases (41). This finding triggered a significant research effort to unravel many important issues such as: the influence of fiber size, shape, and chemical composition; the relationship between exposure levels and diseases; the consequence of exposure to asbestos fibers in different types of industries, or from different types of products, and the development of technologies to avoid worker exposure.

The research efforts resulted in significant consensus in some areas, although strong controversies remain in other areas. Typically, it is widely recognized that the inhalation of long (considered usually as $>5\ \mu m$), thin, and durable fibers can induce or promote lung cancer. It is also widely accepted that asbestos fibers can be associated with three types of diseases (42): asbestosis: a lung fibrosis resulting from long-term, high level exposures to airborne fibers; lung cancer: usually resulting from high level exposures and often correlated with asbestosis; mesothelioma: a rare form of cancer of the lining of the thoracic and abdominal cavities (mesothelium).

A further consensus developed within the scientific community regarding the relative carcinogenicity of the different types of asbestos fibers. There is strong evidence that the genotoxic and carcinogenic potentials of asbestos fibers are not identical; in particular mesothelial cancer is mostly, if not exclusively, associated with amphibole fibers (43).

The replacement of asbestos fibers by other fibrous materials has raised similar health issues in relation to substitute materials. However, since lung cancer has a latency period of approximately 25 years, and since the fiber exposure levels in contemporary industries is far lower than those which prevailed half a century ago, the epidemiological data on most substitutes is insufficient. A possible exception is slag fibers for which several studies on worker populations are available over extended periods (44); some results show a substantial increase in lung cancer occurrence. Consequently, the toxicity of asbestos substitute fibers remains a subject of active investigation.

Regulation. The identification of health risks associated with asbestos fibers, together with the fact that huge quantities of these minerals were used (≈ 5 million tons yearly) in a variety of applications, has prompted strict regulations to limit the maximum exposure of airborne fibers in workplace environments. These exposure limits may be defined as averaged or peak values, measured either as a weight or as a number of-fibers-per-unit-volume (cm^3, m^3), for fibers having lengths $>5\ \mu m$ and diameters $<3\ \mu m$. The International Labor Organization has adopted the following definition (Convention 162, article 2d): "the term respirable asbestos fibers means asbestos fibers having a diameter of less than 3 μm and a length–to–diameter ratio greater than 3:1. Only fibers of a length greater than 5 μm shall be taken into account for purposes of measurement."

In accordance with demonstrated differences between the various asbestos fiber types, the workplace regulation in many countries specifies different exposure limits for chrysotile and the amphiboles (45). Moreover, to alleviate established, or apprehended, risk from substitute fibers, the regulation often specifies maximum exposure limits for synthetic fibers (46); values of exposure limits adopted in leading industrial countries are collected in Table 8.

Some countries have opted for a broader approach and have adopted regulation to minimize exposure of the general public to environmental asbestos fibers, ie, by banning or restricting asbestos imports and types of applications. Because the environmental levels of airborne asbestos fibers are generally found to be three to four orders of magnitude lower than the maximum allowed workplace exposure levels, health risks from environmental asbestos are not immediately obvious. The outcome of ongoing reviews on these matters (in particular by the U.S. Health Effect Institute) will likely have a strong impact on future uses of

Table 8. Exposure Limits for Asbestos Fibers, Synthetic Mineral Fibers (SMF) and Nuisance Dust in the Workplace[a]

Country	Asbestos exposure limits: f/cm³ [b]	Asbestos exposure limits: Fiber type	Exposure limits for SMF and nuisance dust
EEC	0.5	crocidolite	
	1.0	other fibers	
Canada	2.0	chrysotile	
	0.5	amosite	
	0.2	crocidolite	
Denmark	0.3	asbestos excluding crocidolite	2 f/cm³ for SMF (rock, slag, and glasswool)
	0.3[c]	crocidolite	5 mg/m³ [d] total dust for nonstationary workplaces
Germany	0.25	chrysotile	6 mg/m³ general dust
Finland	0.50	all types	5 mg/m³ for wollastonite (recommended)
Japan	2.0	except crocidolite	2.9 mg/m³ for mineral dusts (without free silica)
	0.2	crocidolite	
New Zealand	1.0	chrysotile	5 mg/m³ SMF total dust less than 1 f/cm³ for SMF fiber diameter less than 3 μm
	0.1	crocidolite and amosite	
Sweden	0.5	chrysotile only	1 f/cm³ for SMF (during a full working day)
United Kingdom			
	Averaged over 4 h:		
	0.2	crocidolite and amosite	5 mg/m³ for SMF, total dust
	0.5	other types	
	For any continuous period of 10 minutes:		
	0.6	crocidolite and amosite	
	1.5	other types	
United States	0.2	all types	5 mg/m³ for SMF, total dust 3 f/cm³ for SMF (both recommended)
Russia	2 mg/m³ for dust containing chrysotile and anthophyllite 6 mg/m³ for asbestos–cement dust 8 mg/m³ for asbestos–reinforced plastic and rubber		2 mg/m³ for SMF, total dust

[a]Refs. 45, 46.
[b]Fibers per cubic centimeter, unless otherwise noted.
[c]Ceiling value only.
[d]For glass fiber, 1.0 mg/m³ corresponds approximately to 0.1 f/cm³; this value is higher for smaller fibers.

asbestos fibers. However, it seems likely that since safe technologies have been developed for industrial processing of fibers, asbestos fibers will remain an attractive option based on cost:performance ratio, wherever mineral fibers are required in composite materials applications.

BIBLIOGRAPHY

"Asbestos" in *ECT* 1st ed., Vol. 2, pp. 134–142, by M. S. Badollet, Johns-Manville Research Center; in *ECT* 2nd ed., Vol. 2, pp. 734–747 by M. S. Badollet, Consultant; in *ECT* 3rd ed., Vol. 3, pp. 267–283, by W. C. Streib, Johns-Manville Corporation.

1. M. Ross, R. A. Kuntze, and R. A. Clifton, in B. Levadie, ed., *Definition for Asbestos and Other Health Related Silicates,* ASTM STP 834, American Society for Testing and Materials, Philadelphia, Pa., 1984, pp. 139–147.
2. W. J. Campbell and co-workers, *Selected Silicates Minerals and Their Asbestiform Varieties,* IC 8751, U.S. Bureau of Mines, Washington, D.C., 1977, pp. 5–17, 33.
3. A. Europaeus-Äyräpää, *Acta Archaeol.* **1,** 169 (1930).
4. M. A. Bernarde, ed., *Asbestos: The Hazardous Fiber,* CRC Press, Boca Raton, Fla., 1990, pp. 4, 30–38, 41.
5. *The Economics of Asbestos,* 6th ed., Roskill Information Services, Ltd., London, 1990, p. 11.
6. C. J. van der Merwe, in *South Africa's Mineral Industry,* Republic of South Africa Dept. of Mineral and Energy Affairs, Braamfontein, S.A., 1989, pp. 141–145.
7. E. J. W. Whittaker, in R. L. Ledoux, ed., *Short Course in Mineralogical Techniques of Asbestos Determination,* Vol. 4, Mineralogical Association of Canada, Quebec, Canada, May 1979, pp. 16, 30.
8. *Report of the Royal Commission on Matters of Health and Safety Arising from the Use of Asbestos in Ontario,* J. S. Dupré, chairman, Vol. 1, Ontario Government Book-Store, Toronto, Ontario, Canada, 1984, pp. 87, 89–93.
9. F. J. Wicks, in Ref. 7, pp. 58, 65–72.
10. A. A. Hodgson, *Scientific Advances in Asbestos, 1967 to 1985,* Anjalena Publication, Crowthorne, UK, 1986, pp. 10, 23, 42, 76, 95–99, 107–117.
11. *Asbestos Factbook,* Asbestos, Willow Grove, Pa., 1970.
12. H. C. W. Skinner, M. Ross, and C. Frondel, *Asbestos and Other Fibrous Materials,* Oxford Press, New York, 1988, pp. 21–23, 25, 31, 34, 35.
13. E. J. W. Whittaker, *Acta Crystallogr.* **10,** 149 (1957).
14. K. Yada, *Acta Crystallogr.* **27,** 659 (1971).
15. J.-F. Alary, Y. Côté, and C. Jolicoeur, unpublished data, 1991.
16. C. Jolicoeur and D. Duchesne, *Can. J. Chem.* **59**(10), 1521 (1981).
17. E. Papirer and P. Roland, *Clays Clay Miner.* **29,** 161 (1981).
18. E. Orowan, *Rep. Prog. Phys.* **12,** 186 (1948–1949).
19. R. G. Bryans and B. Lincoln, *Paper 6.24, 3rd International Conference on the Physics and Chemistry of Asbestos Minerals,* Québec, Canada, 1975.
20. R. G. Bryans and B. Lincoln, *Paper 3.6, 2nd International Conference on the Physics and Chemistry of Asbestos Minerals,* Université de Louvain, Louvain, Belgium, 1971.
21. A. A. Hodgson, in L. Michaels and S. S. Chissick, eds., *Asbestos: Properties, Applications and Hazards,* Vol. 1, John Wiley & Sons, Inc., New York, 1979, p. 90.
22. S. Speil and J. P. Leinerveber, *Environ. Res.* **2**(3), 166 (1969).
23. G. Stroink, R. A. Dunlop, and D. Hutt, *Can. Mineral.* **19,** 519 (1981).
24. A. A. Hodgson and C. A. White, *Papers 2–10, The Physics and Chemistry of Asbestos Minerals, Oxford Conference on Asbestos Minerals,* Oxford, UK, 1967.
25. Z. Johan and co-workers, *Adv. Org. Geochem.,* 883–903 (1973).

26. T. Otouma and S. Take, *Paper 5.21,* in Ref. 19.
27. L. Zapata and co-workers, *Paper 3.4,* in Ref. 20.
28. U.S. Pat. 4,356,057 (Oct. 26, 1982), J. M. Lalancette and J. Dunnigan (to Société Nationale de L'amiante).
29. E. Papirer, *Clays Clay Miner.* **29,** 69 (1981).
30. U.S. Pat. 3,689,430 (Sept. 5, 1972), P. C. Yates (to E. I. du Pont de Nemours & Co., Inc.).
31. W. C. McCrone, *The Asbestos Particle Atlas,* Ann Arbor Science Publishers, Ann Arbor, Mich., 1980, p. 33.
32. M. Cossette and P. Delvaux, in Ref. 7, pp. 79–80.
33. *Appraisal of Minerals Availability for 34 Commodities,* U.S. Bureau of Mines, Washington, D.C., 1987, p. 35.
34. *Chrysotile Asbestos Test Manual,* 3rd ed., Asbestos Textile Institute and the Quebec Asbestos Mining Association, Quebec, Canada, 1974 (revised 1978).
35. J. Harrod and V. Thorpe, *Asbestos, Politics and Economics of a Lethal Product,* International Federation of Chemical, Energy and General Workers' Unions, Geneva, Switzerland, 1984.
36. A. A. Hodgson, ed., *Alternatives to Asbestos, The Pros and Cons,* John Wiley & Sons, Inc., New York, 1989, p. xi.
37. *Asbestos and Other Natural Mineral Fibers,* WHO Environmental Health Criteria 53 World Health Organization, Geneva, Switzerland, 1986.
38. J. Bignon, J. Peto, and R. Saracci, eds., *Nonoccupational Exposure to Mineral Fibres,* IARCC Scientific Publication No. 90, International Agency for Research on Cancer, Lyon, France, 1989.
39. *Occupational Exposure Limit for Asbestos,* Report of a WHO Meeting, Oxford, UK, Apr. 10–11, 1989.
40. R. Murray, *Br. J. Ind. Med.* **47,** 361 (1990).
41. I. J. Selikoff, J. Churg, and E. C. Hammond, *J. Am. Med. Assoc.* **188,** 142 (1964).
42. J. Peto and co-workers, *Environ. Health Perspect.* **70,** 51 (1986).
43. M. J. Gardner and C. A. Powell, *J. Soc. Occup. Med.* **36**(4), 124 (1986).
44. "International Symposium on Man-made Mineral Fibers in the Working Environment," Copenhagen, Denmark, Oct. 28–29, 1986, *Ann. Occup. Hyg.* **31,** 4B (1987).
45. *Summary of Main Features of Asbestos/Health Regulations at the Workplace,* AIA Information Memorandum (AIM) No. 3/80, Asbestos International Association, Epson, Surrey, UK, Nov. 1990.
46. *Safety in the Use of Mineral and Synthetic Fibers,* Working Documents (Part 1) of ILO Meeting of Experts, Geneva, Switzerland, Apr. 17–25, 1989, International Labor Organization, 1989, p. 47.

CARMEL R. JOLICOEUR
JEAN-FRANÇOIS ALARY
ASTA SOKOV
Université de Sherbrooke, Canada

ASBESTOS–CEMENT PRODUCTS. See CEMENT.

ASBESTOSIS. See INDUSTRIAL HYGIENE AND PLANT SAFETY; TOXICOLOGY.

ASCORBIC ACID. See VITAMINS.

ASPARAGINE. See AMINO ACIDS.

ASPARTIC ACID. See AMINO ACIDS.

ASPHALT

Asphalt [*8052-42-4*] is defined by the American Society for Testing and Materials (ASTM) (1) as a dark brown to black cementitious material in which the predominating constituents are bitumens that occur in nature or are obtained in petroleum processing. Bitumen is a generic term defined by ASTM as a class of black or dark-colored (solid, semisolid, or viscous) cementitious substances, natural or manufactured, composed principally of high molecular weight hydrocarbons, of which asphalts, tars, pitches, and asphaltites are typical.

Canadian and European practice (2–4) and geologists and archaeologists in the United States use bitumen or asphaltic bitumen as a synonym for asphalt, and apply asphalt to the mixture of bitumen and inorganic matter that is used for paving purposes. On the other hand, pitches and tars are derived from the destructive distillation of coal, crude oils, and other organic materials.

ASTM (1) further classifies asphalts or bituminous materials as solids, semisolids, or liquids. Semisolid and liquid asphalts predominate in commercial practice today. Prior to 1907, the principal portion of asphalt used occurred naturally; since that time most asphalts have been produced from the refining of petroleum. Air-blown petroleum asphalts (5) of diverse hardness became available in the early 1900s, and as their use increased, the demand for native products diminished.

More recently, asphalts have been defined as the manufactured materials that are produced during petroleum processing whereas bitumens have been defined as the naturally occurring materials (5,6). That is, petroleum and related materials can be divided into various class subgroups:

Natural Materials	*Manufactured Materials*	*Derived Materials*
petroleum	wax	oils
heavy oil	residuum	resins
mineral wax	asphalt	asphaltenes
bitumen (native asphalt)	tar	carbenes
asphaltite	pitch	carboids
asphaltoid	coke	
migrabitumen	synthetic crude oil	
bituminous rock		
bituminous sand		
kerogen		
natural gas		

The term tar sands is a misnomer; tar is a product of coal processing. Oil sands is also a misnomer but equivalent to usage of "oil shale." Bituminous sands is more correct; bitumen is a naturally occurring asphalt. Asphalt is a product of a refinery operation, usually made from a residuum. Residuum is the nonvolatile portion of petroleum and often further defined as atmospheric (bp $>350°C$) or vacuum (bp $>565°C$). For convenience, the terms "asphalt" and "bitumen" will be used interchangeably in this article.

Asphalts characteristically contain very high molecular weight molecular polar species, called asphaltenes, which are soluble in carbon disulfide, pyridine, aromatic hydrocarbons, chlorinated hydrocarbons, and tetrahydrofuran.

Naturally Occurring Materials

Native Asphalt (Bitumen). Construction activities in ancient times depended on native asphalts (2,3,7–13). The product found on the surface of the Dead Sea may have been used by the Egyptians for mummification. In fact, the word mummy is thought to have derived from mumiia or Bitumen of Judea. Standard grades of bitumens apparently were known to the Sumerians; olive oil was used as the flux and fibers were added, if necessary, to prevent flow of the mixture. Clay-stabilized mastics were used as flooring and for waterproofing aprons. The presence of asphalt seepages signals a probable crude-oil reservoir. Several centuries later, mention is made of Trinidad asphalt by Raleigh during his visit to the island in 1595.

Lake Asphalt. Trinidad Lake asphalt was first reported used for paving in the United States about 1874 (14), and the use of the asphalt from Bermudez Lake in Venezuela soon followed. Trinidad asphalt (3,7) exists in a number of deposits on the island of Trinidad, near the northeast coast of Venezuela. The largest deposit, and the one of chief commercial importance, is an asphalt lake occupying about 0.4 km^2 and of uniform composition to a depth of 87 m. As the bitumen is mined, the openings fill in, the level appears constant and an overturning movement of the lake from center to edge keeps the composition constant. This asphalt has been used for pavement constructions and other applications throughout the world.

Trinidad asphalt has a relatively uniform composition of 29% water and gas, 39% bitumen soluble in carbon disulfide, 27% mineral matter on ignition, and 5% bitumen that remains adsorbed on the mineral matter. Refining is essentially a process of dehydration by heating the crude asphalt to ca 165°C. The refined product averages 36% mineral ash with a penetration at 25°C of about 2 (0.2 mm), a softening point (ring and ball method) of 99°C, a flash point (Cleveland open cup) of 254°C, a sulfur content of 3.3%, and a saponification value of 45 mg KOH/g. The mineral matter typically contains particle sizes from 40% finer than 10 μm to 10% from the 30–75 μm range. Because of the influence of the fine mineral particles, many of them colloidal in size, the viscosity of Trinidad Lake asphalt is higher than that of normal bitumens.

Asphalt from Bermudez Lake (7) previously was exported in bulk in cargo vessels to the United States. In the early 1900s, it was used extensively for paving and waterproofing, but has not been available since the 1940s.

Rock Asphalt. Rock asphalt (15) deposits contain from 5–25% asphalt (Table 1). When extracted the asphalt varies from a soft-flux material to hard penetration. The mineral content, usually of a sandstone or limestone nature, has many gradations. Rock asphalt mixtures, ground and sized, are placed and rolled to form pavement surfaces. Some have superior antiskid properties in wearing surfaces, particularly the sandstone type. Rock asphalts are found in Texas, Alabama, Oklahoma, Colorado, California, and Kentucky. The composition of Kentucky asphalt is suitable for paving in its natural condition if selected at bitumen contents of 7–9% (16). Some of the well-known and long-used asphalt deposits of Europe (3,7) are located at Seyssel, France; Ragusa in Sicily; Val-de-Travers, Switzerland; and Vorwohle, Germany.

Table 1. Typical Composition and Properties of Rock Asphalts

Property	Kentucky[a]	Oklahoma[b]	Alabama[c]	Texas[d]	Sicily[e]	Germany[f]
bitumen content, wt %	7–9	12.5	4.5	10–20	8–9	4.7–9.5
softening point (ring and ball), °C		28			38	30–36
penetration of 100 g at 25°C for 5 s, mm/10	too soft	over 360	15–25	4	too soft	soft
type of mineral	sand to sandstone	sand	oolitic limestone	limestone	marly limestone	limestone

[a]Crushed to fine size, less than 6.4 mm. Spread and rolled at temperatures above 15.6°C for pavements.
[b]Product blended from various deposits or sand added.
[c]Quarried, crushed, sized from 0.95 cm to dust and fluxed with softer asphalt for use as cold-applied pavement surfacing.
[d]Mixed with trap rock and asphalt flux to obtain mix for cold rolling as pavement surfacing.
[e]Selected grades ground and used for paving purposes.
[f]Enriched by adding asphalt and marketed for use as pavements or mastic.

Asphalt (bitumen) also occurs in various oil sand (also called tar sand) deposits which occur widely scattered through the world (17) and the bitumen is available by means of various extraction technologies. A review of the properties and character of the bitumen (18) suggests that, when used as an asphaltic binder, the bitumen compares favorably with specification-grade petroleum asphalts and may have superior aging characteristics and produce more water-resistant paving mixtures than the typical petroleum asphalts.

Manufactured Materials

Since the early 1900s most of the asphalts produced from the refining of petroleum have been used primarily in paving and roofing applications. The advent of motorized transportation led to increased asphalt manufacture from petroleum in order to provide binders for hard-surfaced pavements.

Petroleum asphalts, compared to native asphalts, are organic with only trace amounts of inorganic materials. They derive their characteristics from the nature of their crude origins with some variation possible by choice of manufacturing process. Although there are a number of refineries or refinery units whose prime function is to produce asphalt, petroleum asphalt is primarily a product of integrated refineries (Fig. 1). Crudes may be selected for these refineries for a variety of other product requirements and the asphalt (or residuum) produced may vary somewhat in characteristics from one refinery-crude system to another and even by cut-point (Table 2) and asphalt content (Fig. 2) (5,6). The approximate asphalt yields (%) from various crude oils are as follows:

Nigerian light	7	L. A. Basin	44
Arabian light	14	California Valley	55
North Slope	21	California Coastal	65
Arabian heavy	31	Boscan	79

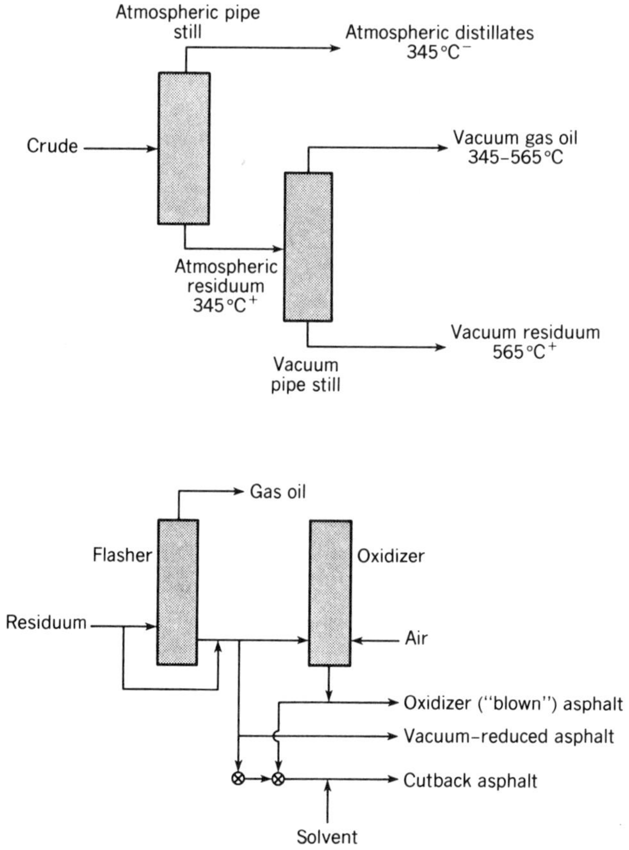

Fig. 1. Residua are obtained by removal of the volatile constituents of the feedstock at atmospheric pressure or at reduced pressure and can be converted to asphalt by various methods.

Straight Run Asphalt. In crude-oil refining, the crude oil at 340–400°C is injected into a fractionating column (5,6,19,20). The lighter fractions are separated as overhead products and the residuum is straight-reduced asphalt.

Crude oil containing about 30% asphalt can be refined completely in an

Table 2. Properties of Various Residua Produced from Tia Juana, Venezuela Light Crude Oil

Property	Values						
boiling range, °C	whole crude	>220	>295	>345	>400	>455	>505
yield on crude, vol %	100.0	70.2	57.4	48.9	39.7	31.2	17.9
gravity, API	31.6	22.5	19.4	17.3	15.1	12.6	7.1
specific gravity	0.8676	0.9188	0.9377	0.9509	0.9652	0.9820	1.0209
sulfur, wt %	1.08	1.42	1.64	1.78	1.93	2.12	2.59
carbon residue (Conradson), wt %		6.8	8.1	9.3	11.2	13.8	21.6
nitrogen, wt %				0.33	0.39	0.45	0.60
pour point, °C	−20	−9	−1	7	16	24	49
viscosity							
kinematic, mm^2/s (= cSt)							
38°C	10.2	83.0	315	890	3100		
99°C		9.6	19.6	35.0	77.0	220	7959
Furol (SFS), s							
50°C			70.6	172	528		
99°C					37.6	106	3760
Universal (SUS), s at 99°C		57.8	96.8	165	359	1025	
metals							
vanadium, ppm				185			450
nickel, ppm				25			64
iron, ppm				28			48

atmospheric unit to an asphalt product. However, most crude oil cannot be distilled satisfactorily to an asphalt product at atmospheric pressure because of the presence of substantial proportions of high boiling gas oil fractions. Thus, as a supplement to the atmospheric process, a second fractionating tower (a vacuum tower) is added (Fig. 1).

This two-stage process is particularly applicable to crude oils containing 15–30% asphalt. Vacuum distillation units are continuous flow-through units and include a heating unit, the vapor–liquid flash separation zone, the fractionating zone, and the vacuum producing equipment. The heated feedstock is pumped into a flash zone where the more volatile constituents vaporize. The asphalt is the nonvolatile residual fraction which is removed continuously. Manufacturing asphalt by straight reduction does not change the chemical nature of asphalt, other than its viscosity.

Propane Asphalt. As noted above, crude oils contain different quantities of residuum (Fig. 2) and, hence, asphalt. Asphalt is also a product of the propane deasphalting and fractionation process (5,6,21,22) which involves the precipitation of asphalt from a residuum stock by treatment with propane under controlled conditions. The petroleum charge stock is usually atmospheric-reduced residue from a primary distillation tower.

Propane is usually used in this process although propane–butane mixtures and pentane have been used with some variation in process conditions and hardness of the product. Propane deasphalting is used primarily for crude oils of relatively low asphalt content, generally <15%. Asphalt produced from this

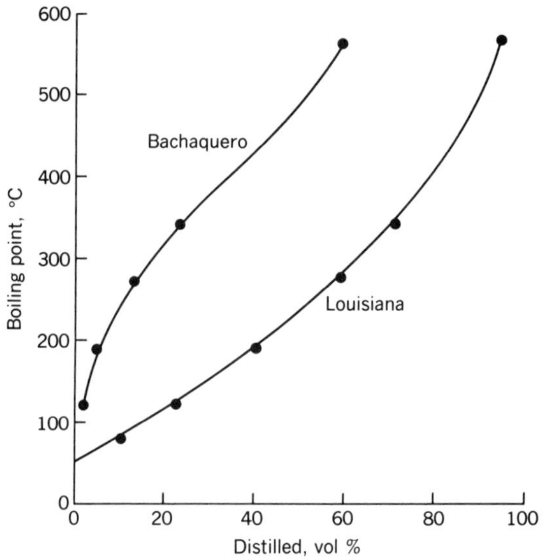

Fig. 2. Different crude oils contain different amounts of residua and the properties of specific "cut point" residua may be different, eg, for the 565°C residua:

	Louisiana	Bachaquero
Gravity, API	13.1	2.8
Sulfur, wt %	0.9	3.7
Nitrogen, wt %	0.4	0.6
Carbon, wt %	15.8	27.5
Nickel, ppm	20	100
Vanadium, ppm	8	900
Pour point, °C		55

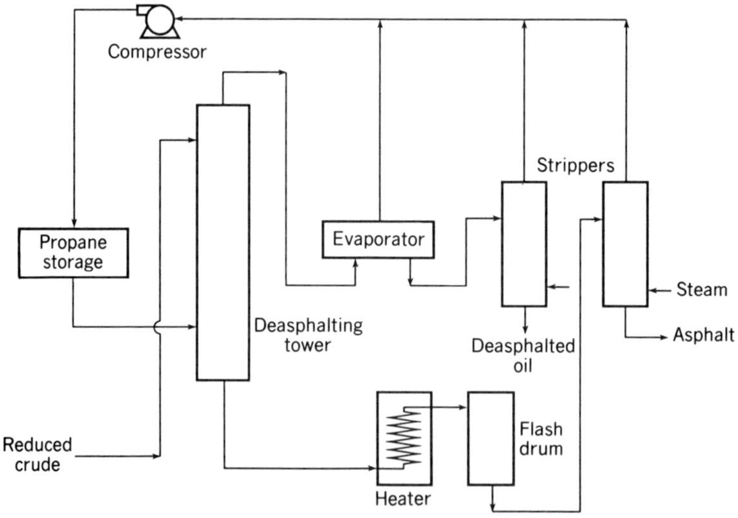

Fig. 3. Schematic representation of propane deasphalting and fractionation process.

process is normally blended with other asphaltic residua for making paving asphalt.

The process (Fig. 3) is a countercurrent liquid–liquid extraction. The feedstock is introduced near the top of an extraction tower and the liquid propane near the bottom, using solvent-to-oil ratios from 4:1 to 10:1. The deasphalted oil–propane solution is withdrawn overhead and the asphalt from the bottom, and each is subsequently stripped of propane.

Temperature, solvent ratio, and pressure each have an effect upon the split point or yield of the oil and asphalt components (Table 3). Contrary to straight reduction which is a high temperature and low pressure process, propane deasphalting is a low temperature and high pressure process.

There are small differences in the properties of asphalts prepared by propane deasphalting and those prepared by vacuum distillation from the same stock

Table 3. Typical Yields and Characteristics of Products Obtained from the Propane Deasphalting Process

	Crude source				
Process	Pa.	Okla.	Ill.	Le Duc, Canada	Iraq
reduced crude					
gravity, °API[a]	25.1	19.3	13.7	12.1	7.1
kinematic viscosity at 98.9°C, mm^2/s (= cSt)	47.2	82.8	378	441	1050
Saybolt viscosity, s	220	385	1760	2050	4880
crude, vol %	24.0	26.1	11.3	15.2	16.4
deasphalted oil					
yield, vol %	83.7	75.6	50.7	47.1	34.1
gravity, °API[a]	26.6	23.6	23.4	22.7	20.2
kinematic viscosity at 98.9°C, (mm^2/s (= cSt))	30.4	29.8	40.7	42.6	38.2
Saybolt viscosity, s	144	141	190	199	179
precipitated asphalt					
yield, vol %	16.3	24.4	49.3	52.9	65.9
specific gravity at 15.6°C	0.953	1.023	1.030	1.054	1.064
softening point (ring and ball), °C		62.2	62.8	80	62.8
penetration of 100 g at 25°C for 5 s, mm/10		23	21	0	16
operating conditions					
throughput, kL/d[b]			800		
charge stock			reduced crude		
gravity charge, °API[a]			20.0		
temperature of deasphalting tower, °C			65		
solvent-to-oil ratio, by volume			6:1		
pressure, kPa[c]			2170 (300)		

[a] $°\text{API gravity at } 15.6°\text{C} = \dfrac{141.5}{\text{sp gr at } 15.6/15.6°\text{C}} - 131.5.$

[b] To convert kL/d to barrels/d, multiply by 6.25.

[c] To convert kPa to psi, multiply by 0.145.

(Fig. 4). Propane deasphalting also has the ability to reduce a residuum even further and produce an asphalt product having a lower viscosity, higher ductility, and higher temperature susceptibility than other asphalts (23). However, such properties might be anticipated to be very much crude oil dependent. Propane deasphalting is conventionally applied to low asphalt-content crudes, which are generally different in type and source from those processed by distillation of higher yield crudes.

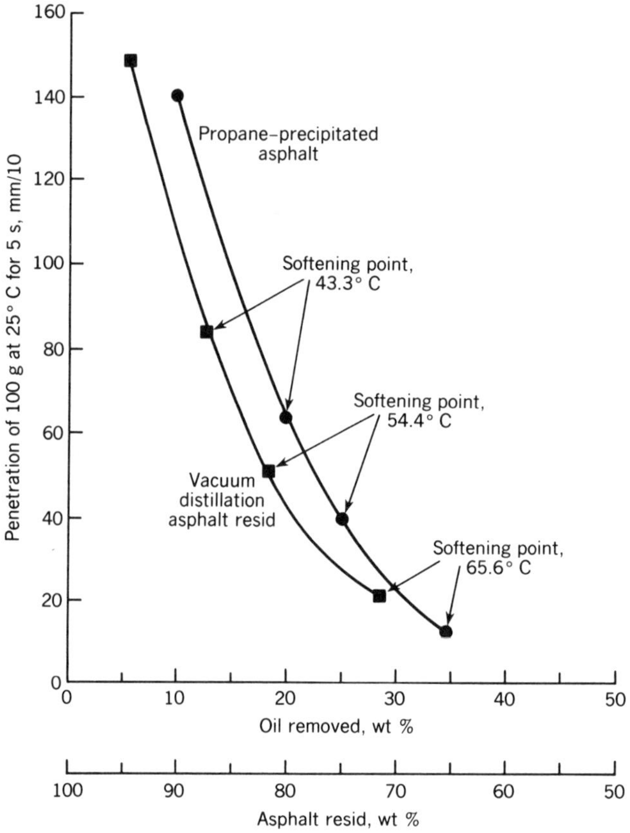

Fig. 4. Comparison of propane deasphalting with vacuum distillation of asphalt from Lagunillas, Venezuela crude.

Air-Blown Asphalt. Air-blowing is an exothermic process that is dependent upon several variable process parameters (24). In the process, an asphalt (flux) is converted to a harder product by air contact at 200–275°C. Dehydrogenation and polymerization are involved, and oxygen is not retained in the asphalt product except in a very minor amount (25). In theory, the oxygen in the air combines with the hydrogen in the asphalt to evolve water vapor. This leaves unsaturation for cross-linking. However, the chemistry may be much more complex than the simple theory predicts (24). For example, asphaltenes are produced during the air-blowing process (Table 4) and the evidence suggests that those constituent species of the asphalt which have higher proportions of heteroatoms (ie, nitrogen, oxygen, and sulfur) react preferentially with oxygen leaving less polar substances unreacted (26,27). This also raises the specter of incompatibility of the resulting products. For example, if the more polar species, ie, the resins, are preferentially oxidized to produce more asphaltenes, then the means by which the asphaltenes are dispersed in the oily medium is removed. Incompatibility should be anticipated.

Table 4. Fractional Changes in Hardness of Asphalt[a] upon Air-Blowing

	Initial	Air-blown	Air-blown	Air-blown
Properties				
softening point (ring and ball), °C	54.4	85	96.1	173.3
penetration of 100 g at 25°C for 5 s, mm/10	36	13	9	1
Fractions, wt %				
asphaltenes (hexane insolubles)	14.8	26.9	31.4	51.3
hard resins	45.5	36.6	36.1	19.6
soft resins	25.0	22.3	20.9	16.9
oils	12.3	11.9	10.0	11.1
waxes	2.5	2.0	1.8	1.6
Total	*100.1*	*99.7*	*100.2*	*100.5*

[a]Straight-reduced Arkansas asphalt.

A variety of other substances can provide the same reaction: sulfur yields hydrogen sulfide; chlorine yields hydrogen chloride. In some cases, some of the bonds created are quite weak resulting in, after an induction period, a phenomenon termed "fallback." When fallback occurs, usually at a time when the hardened or oxidized asphalt is stored at or near the original processing or reaction temperature, softening of the asphalt is the result (28–31).

Air-blown asphalts, more resistant to weather and changes in temperature than the types mentioned previously are produced by batch and continuous methods. Air-blown asphalts, of diverse viscosities and flow properties with added fillers, polymers, solvents, and in water emulsions, provide products for many applications in the roofing industry.

Batch Process. In the batch process (Fig. 5), the feedstock is preheated in a tube furnace or heater placed between the feedstock storage and the blowing vessel. The air supply is provided by a variety of blowers or compressors and a vertical-tower vessel is preferable for air-blowing. Knockout drums, water scrubbers, incinerators, furnaces, and catalytic burning units have been used for fume disposal (32). Steam is used for safety and to ensure positive fume flow to the incinerator.

Temperature is the most important variable and preheating is generally necessary to 200–230°C. After air has been introduced, there is a gradual temperature rise because of the exothermic reaction, until some means is applied to hold the temperature such as a water or steam spray on the asphalt surface to maintain

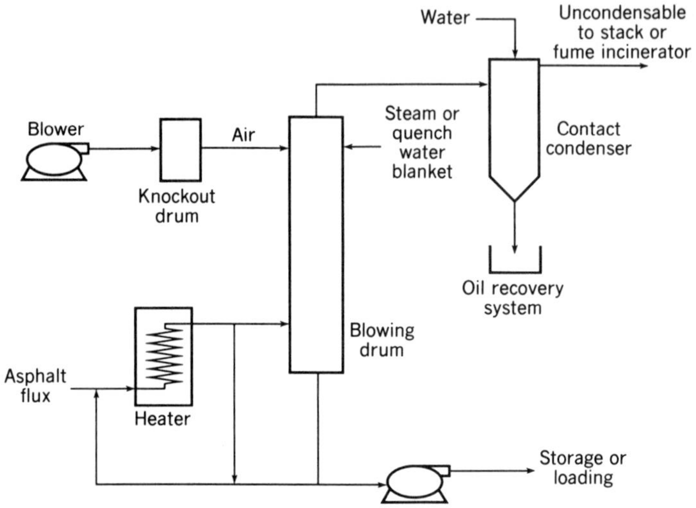

Fig. 5. Schematic representation of the batch process for air-blowing.

a temperature of approximately 260°C. The end point can be predicted by periodic testing of the softening point.

Continuous Process. Continuous air-blowing (Fig. 6) offers the following potential advantages: lower equipment and maintenance costs for the same production capacity; shorter blowing times with air-efficient equipment; less preheat capacity requirements; better control and operation; lower blowing losses; and lower operating labor requirement. It is particularly preferable when several principal products, such as coatings, saturants, and built-up roofing asphalts, can be processed from a single feedstock. When a comparison is made with a batch

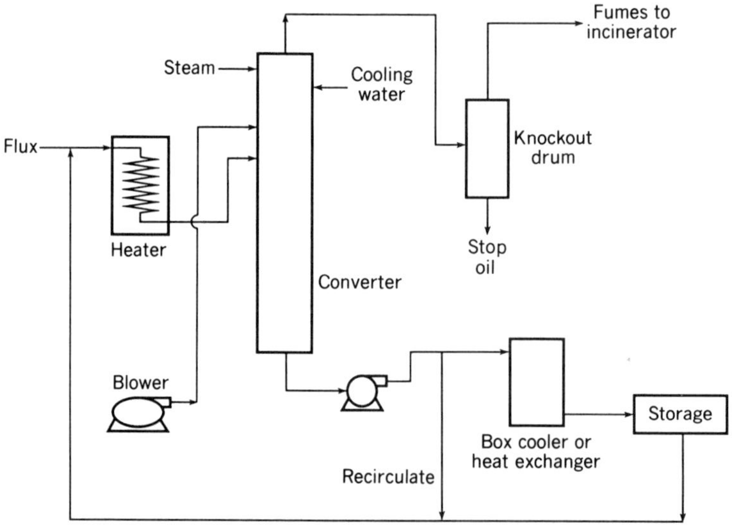

Fig. 6. Schematic representation of a continuous air-blowing process.

operation on the basis of the same daily capacity, the saving in blowing time is most apparent. Practical use of continuous air-blowing has been made in the manufacture of paving binders from soft vacuum residua and in the manufacture of roofing asphalts. Blends of intermediate grades from a harder air-blown asphalt product such as unfilled coating asphalt (100–115°C softening point) and a softer air-blown product such as roll saturant (40–45°C softening point) are often used.

Continuous oxidizers are usually operated at a constant temperature (260°C) and a constant liquid level with the production rate and product characteristics controlled by air rate and charging rate.

Catalytic Asphalt. The term catalytic asphalt has been accepted although this is not truly a catalytic process. The agents used are consumed in the reaction. The general effect is the reduction of blowing time as well as a change of the softening point–penetration relationship (33), which would normally be obtained from the same feedstock. Reduction of blowing time is an economic incentive, whereas the change in product properties permits manufacture to specifications not possible otherwise and the use of a wider variety of crude sources for flux.

Many agents have been proposed and patented including copper sulfate (34), zinc chloride (35), ferric chloride (36), aluminum chloride (36), and phosphorus pentoxide (37); ferric chloride, zinc chloride, and phosphorus pentoxide have been most widely used. The addition of these agents may vary from 0.1 to 3%, depending upon the feedstock and the desired characteristics of the product (Table 5) and all asphalt feedstocks do not respond to catalysts in the same way. Differences in feedstock composition are important qualifiers in determining the properties of the asphalt product. The important softening point–penetration relationship, which describes the temperature susceptibility of an asphalt, also varies with the source of the feedstock. Straight-reduced, air-blown, and air-blown catalytic asphalts from the same crude feedstock also vary considerably.

Thermal Asphalt. Thermal or cracked asphalts differ from other asphalts in that they are products of a cracking process. They have relatively high specific gravity, low viscosity, and high temperature susceptibility, and they contain

Table 5. Air-Blowing of Fluxes With and Without Ferric Chloride and Phosphorus Pentoxide Catalysts

	California stock			Arkansas stock		
Property	No catalyst	0.8% $FeCl_3{\cdot}6H_2O$	0.5% P_2O_5	No catalyst	2.2% $FeCl_3{\cdot}6H_2O$	1.0% P_2O_5
blowing temperature, °C	221.1	221.1	221.1	254.4	237.8	237.8
blowing time, min	135	44	103	105	46	128
softening point[a], °C	87.8	86.1	86.1	86.7	83.3	87.8
penetration of 100 g at 25°C for 5 s, mm/10	27	51	53	27	62	59
penetration of 200 g at 0°C for 60 s, mm/10	13	30	28	15	38	38
penetration of 50 g at 46°C for 5 s, mm/10		74		48	109	110
ductility at 25°C, cm	3.5	3.4	4.8	3.3	3.9	5.3

[a]Ring and ball method.

cokelike bodies (carbenes) as indicated by the Oliensis test (38). Thermal asphalts have significant application as saturants for cellulosic building products such as insulation boards, brick-finish siding, and fiber soil pipes.

The thermal cracking process (Fig. 7) involves preheating of the charge to 480–600°C and then discharging into a reaction vessel or chamber under pressures up to 1480 kPa (200 psig). Cracking decomposition takes place to form both lighter and heavier products, including a certain amount of fixed carbon. Distillation is used to separate the overall products into gas, gasoline, middle distillate, and an asphaltic residuum.

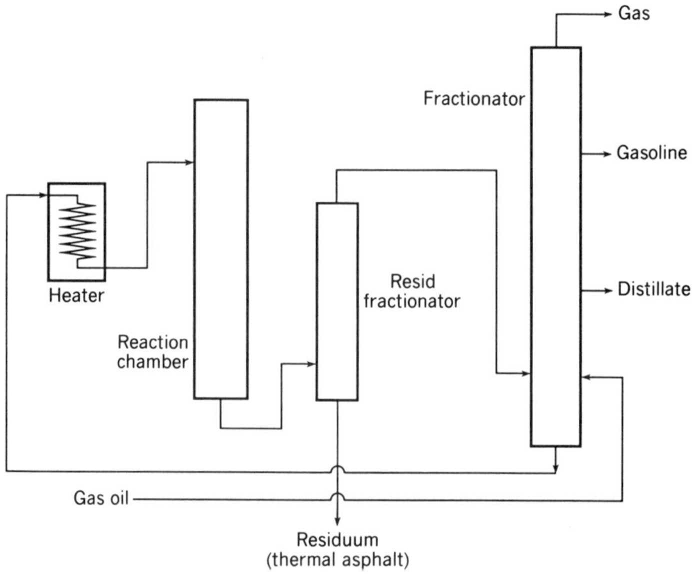

Fig. 7. Schematic representation of thermal cracking process.

Blended Asphalt. Any particular refinery may stock two grades of asphalt, one at each end of the viscosity spectrum of the entire product grade requirements. Intermediate grades are prepared by blending (proportioning) the extremes (39). The preparation of asphalts in liquid form by blending (cutting back) an asphalt with a petroleum distillate fraction is customary. There are three general types of cutback asphalt which differ mainly in the diluent used (Table 6). The slow-curing type is often called road oil, which is made by direct

Table 6. Three Types of Liquid Asphalt Made by Cutting Back with Petroleum Diluents

Type asphalt	Diluent type	Diluent, %	Viscosity range[a], mm^2/s (= cSt)
slow curing (SC)	gas oil	0–50	70–6000
medium curing (MC)	kerosene	15–45	30–6000
rapid curing (RC)	naphtha	15–45	70–6000

[a]The liquid cutback asphalts are prepared in a number of viscosity grades, ranging generally from 70 to 6000. The grade number indicates the viscosity at 60°C.

reduction, but most are commonly blended. The medium curing and rapid curing products usually have four grades within a given type and differ mainly in the amount of diluent used and in the kinematic viscosity of the blended product.

Preparation is accomplished by simple blending of the diluent into the hot base asphalt. This is generally accomplished in tanks equipped with coils for air agitation or with a mechanical stirrer or a vortex mixer. Line blending in a batch circulation system or in a continuous fashion (40) is used where the volume produced justifies the extra facilities. A continuous, line-blending system is applicable to the manufacture of cutback asphalts and asphalt cements (Fig. 8).

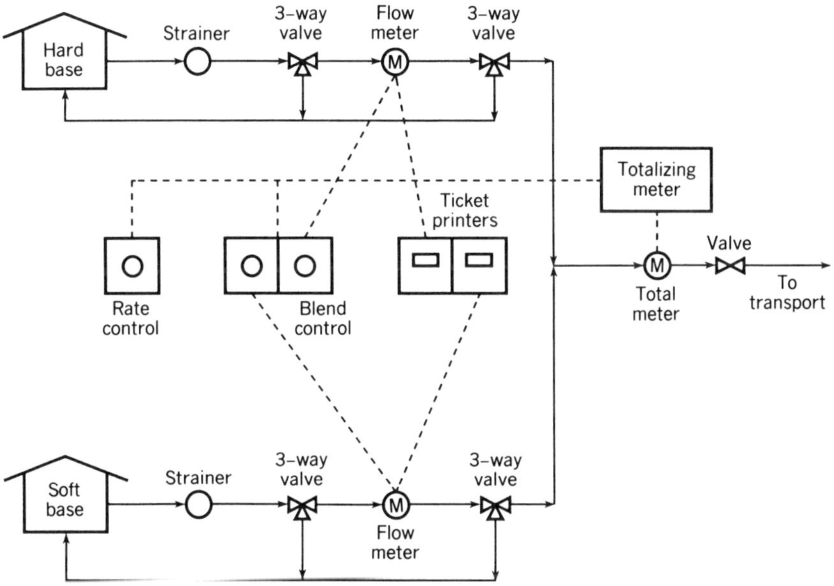

Fig. 8. Schematic representation of a continuous asphalt blender.

Asphalt Emulsions. An emulsion is one immiscible liquid dispersed in the other in the form of very fine droplets from about 1–25 μm and an average of 5 μm diameter. In the most common asphalt emulsion, the oil-in-water type, the asphalt is the dispersed (internal) phase, and water is the continuous (external) phase. Most of the important properties of an emulsion are dependent upon the amount and type of emulsifying agent used. The more popular cationic emulsions (41,42) are made with cation-active agents which provide emulsion droplets with positive charges. The anionic agent is usually the sodium or potassium salt of a fatty acid (43). Nonionic cellulose derivatives are also used to increase the viscosity of the emulsion if needed. The acid number of asphalt is an indicator of its emulsifiability and reflects the presence of high molecular weight asphaltogenic or naphthenic acids. Diamines, frequently used as cationic agents, are made from the reaction of tallow acid amines with acrylonitrile, followed by hydrogenation.

Emulsion processes (Fig. 9) vary and the emulsions are made in rapid-, medium-, and slow-setting types for diverse application techniques in the road-building industry. Emulsions are most commonly formulated with rapid-setting properties and are designated as a spraying type, ie, for road seal-coating opera-

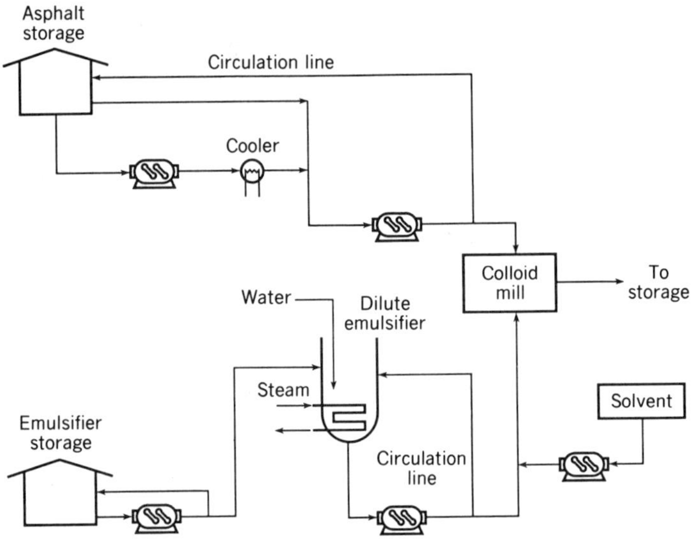

Fig. 9. Flow diagram of an asphalt emulsion plant.

tions. Slow-setting emulsions are usually used for soil stabilization or in mulching applications.

Industrial emulsions have applications outside the road-building industry. They are made with harder grades of asphalt and contain clays, casein, gelatin, or blood albumin as peptizing agents. Certain clays, such as bentonite, are good emulsion dispersants and impart a buttery consistency to the emulsion. These emulsions have found a wide variety of applications, such as in surface coating of asphalt pavements, for built-up roofs, and for other weatherproof coverings.

Composition and Properties

Determination of the components of asphalts has always presented a challenge because of the complexity and high molecular weights of the molecular constituents (6,44,45). The principle behind composition studies is to evaluate asphalts in terms of composition and performance. Such studies are always suspect when this is not the overriding goal of the work.

The influence of the composition of asphalt has been recognized, for many years, as being an important factor in controlling the performance of such materials. Furthermore, rheological properties have always been associated with composition but, in order to utilize compositional data effectively, more definitive correlations between composition and properties are needed (46–48).

The methods employed can be conveniently arranged into a number of categories: (*1*) fractionation by precipitation; (*2*) fractionation by distillation; (*3*) separation by chromatographic techniques; (*4*) chemical analysis using spectrophotometric techniques (infrared, ultraviolet, nuclear magnetic resonance, x-ray fluorescence, emission, neutron activation), titrimetric and gravimetric techniques, elemental analysis; (*5*) molecular weight analysis by mass spectrometry, vapor pressure osmometry, and size-exclusion chromatography.

However, for the past 30 years fractional separation has been the basis for most asphalt composition analysis (Fig. 10). The separation methods that have been used divide asphalt into operationally defined fractions. Four types of asphalt separation procedures are now in use: (*1*) chemical precipitation in which *n*-pentane separation of asphaltenes is followed by chemical precipitation of other fractions with sulfuric acid of increasing concentration (ASTM D2006); (*2*) solvent fractionation: separation of an "asphaltene" fraction by the use of 1-butanol followed by dissolution of the 1-butanol solubles in acetone. The acetone solution is chilled, forcing the precipitation of paraffinics from the cold acetone-soluble cyclics; (*3*) adsorption chromatography using a clay–gel procedure where, after removal of the asphaltenes, the remaining constituents are separated by selective adsorption–desorption on an adsorbent (ASTM D2007 and D4124); (*4*) size-exclusion chromatography in which gel-permeation chromatographic (gpc) separation of asphalt constituents occurs based on their associated sizes in dilute solutions (ASTM D3593).

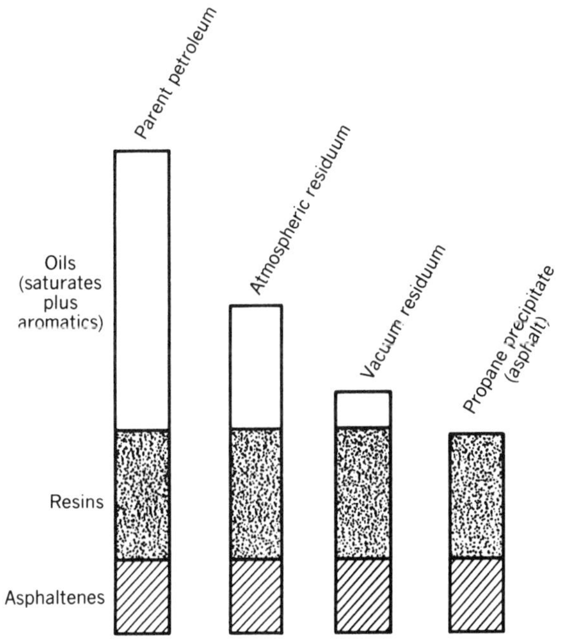

Fig. 10. The relationship between a crude oil, two residua from the crude oil, and the propane asphalt.

The fractions obtained in these schemes are defined operationally or procedurally. The amount and type of asphaltenes in an asphalt are, for instance, defined by the solvent used for precipitating them. Fractional separation of asphalt does not provide well-defined chemical components. The materials separated should only be defined in terms of the particular test procedure.

The component of highest carbon content is the fraction termed carboids

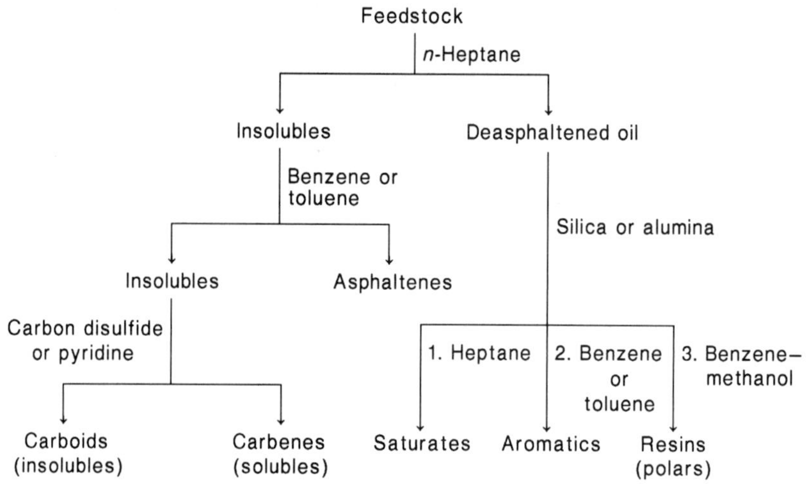

Fig. 11. Simplified representation of the separation of petroleum into six principal fractions.

and consists of species that are insoluble in carbon disulfide or in pyridine (5,6,49). The fraction that has been called carbenes contains molecular species that are soluble in carbon disulfide and soluble in pyridine but which are insoluble in carbon tetrachloride and in benzene (Fig. 11) (5,6,49).

Asphaltenes seem to be relatively constant in composition in residual asphalts, despite the source, as determined by elemental analysis (6). Determination of asphaltenes is relatively standard, and the fractions are termed *n*-pentane, *n*-hexane, *n*-heptane, or naphtha-insoluble, depending upon the precipitant used (5,6,49). After the asphaltenes are removed, resinous fractions are removed from the maltenes–petrolenes usually by adsorption on activated gels or clays. Recovery of the resin fraction by desorbtion is usually nearly quantitative.

The forerunner of the modern methods of asphalt fractionation was first described in 1916 (50) and the procedure was later modified by use of fuller's earth (attapulgite [*1337-76-4*]) to remove the resinous components (51). Further modifications and preferences led to the development of a variety of fractionation methods (52–58). Thus, because of the nature and varieties of fractions possible and the large number of precipitants or adsorbents, a great number of methods can be devised to determine the composition of asphalts (5,6,44,45). Fractions have also been separated by thermal diffusion (59), by dialysis (60), by electrolytic methods (61), and by repeated solvent fractionations (62,63).

The composition data should always be applied to in-service performance (64,65) in order to properly evaluate the behavior of the asphaltic binder under true working conditions.

Colloidal State. The principal outcome of many of the composition studies has been the delineation of the asphalt system as a colloidal system at ambient or normal service conditions. This particular concept was proposed in 1924 and described the system as an oil medium in which the asphaltene fraction was dispersed. The transition from a colloid to a Newtonian liquid is dependent on temperature, hardness, shear rate, chemical nature, etc. At normal service temperatures asphalt is viscoelastic, and viscous at higher temperatures. The disperse phase is a micelle composed of the molecular species that make up the asphaltenes and the higher molecular weight aromatic components of the petrolenes or the maltenes (ie, the nonasphaltene components). Complete peptization of the micelle seems probable if the system contains sufficient aromatic constituents, in relation to the concentration of asphaltenes, to allow the asphaltenes to remain in the dispersed phase.

Many attempts have been made to characterize the stability of the colloidal state of asphalt at ordinary temperature on the basis of chemical analysis in generic groups. For example, a colloidal instability index has been defined as the ratio of the sum of the amounts in asphaltenes and flocculants (saturated oils) to the sum of the amounts in peptizers (resins) and solvents (aromatic oils) (66):

$$I_c = \frac{(\text{asphaltenes} + \text{saturated oils})}{(\text{resins} + \text{aromatic oils})}$$

As this ratio increases so does the gel character of the asphalt cement, and its colloidal stability increases. The properties of the asphalt (softening point, ductility, embrittlement temperature, thermal susceptibility, elastic recovery, shearing susceptibility, etc) vary significantly with the colloidal instability index and hence with composition.

Many investigations of relationships between composition and properties take into account only the concentration of the asphaltenes, independently of any quality criterion. However, a distinction should be made between the asphaltenes which occur in straight run asphalts and those which occur in blown asphalts. Remembering that asphaltenes are a solubility class rather than a distinct chemical class means that vast differences occur in the make-up of this fraction when it is produced by different processes.

Elemental Analysis. Asphalt is not composed of a single chemical species, but is rather a complex mixture of organic molecules that vary widely in composition from nonpolar, saturated hydrocarbons to highly polar, highly condensed aromatic ring systems. Although asphalt molecules are composed predominantly of carbon and hydrogen, most molecules contain one or more of the heteroatoms nitrogen, sulfur, and oxygen, together with trace amounts of metals, principally vanadium and nickel. The heteroatoms, although a minor component compared to the hydrocarbon moiety, can vary in concentration over a wide range depending on the source of the asphalt. Because the heteroatoms often impart functionality and polarity to the molecules, their presence may make a disproportionately large contribution to the differences in physical properties among asphalts from different sources.

Generally, most asphalts are 79–88 wt % C, 7–13 wt % H, trace–8 wt % S, 2–8 wt % O, and trace–3 wt % N (Table 7). Trace metals such as iron, nickel, vanadium, calcium, titanium, magnesium, sodium, cobalt, copper, tin, and zinc, occur in crude oils. Vanadium and nickel are bound in organic complexes and, by virtue of the concentration (distillation) process by which asphalt is manufactured, are also found in asphalt.

Table 7. Elemental Analyses of Representative Petroleum Asphalts

Code source	B-2959 Mexican Blend	B-3036 Arkansas–Louisiana	B-3051 Boscan	B-3602 California
carbon, %	83.77	85.78	82.90	86.77
hydrogen, %	9.91	10.19	10.45	10.94
nitrogen, %	0.28	0.26	0.78	1.10
sulfur, %	5.25	3.41	5.43	0.99
oxygen, %	0.77	0.36	0.29	0.20
vandium, ppm	180	7	1380	4
nickel, ppm	22	0.4	109	6

Asphalt typically contains oxygen (<2%), nitrogen (<2%), and sulfur (<7%), as can be measured by many methods of elemental analysis. The heteroatom content tends to increase with increasing boiling point within an asphalt residue. The heteroatom content of a typical asphalt is significant, especially when considered on the molecular level. For example, if the mean molecular mass of the asphaltene fraction is 1000, then one heteroatom per molecule in the asphaltene would require the following proportions of heteroatoms: sulfur 3.2%, nitrogen 1.4%, and oxygen 1.6%. Such concentrations of heteroatoms are commonly found in whole asphalts and are typically exceeded in the asphaltene fractions. The heteroatom content seems to vary more than the hydrogen:carbon (H:C) ratio of the asphaltene fraction from different crudes (5,6).

Techniques are available for quantitative identification of the six principal types of heteroatom compounds in asphalt: carboxylic acids, 2-quinolones, phenols, pyrroles, amides, and pyridines (67).

Many investigators have also measured the trace metal content of asphalts (68). The catalytic behavior of vanadium has prompted studies of the relation between vanadium content and an asphalt's sensitivity to oxidation (viscosity ratio). The significance of metals in the behavior of asphalts is not yet well understood or defined.

Molecular Weight. The molecular weights of the individual fractions of asphalt have received more attention, and have been considered to be of greater importance, than the molecular weight of the asphalt itself. Asphaltenes display a wide range of molecular weights, from 500 to at least 2500, depending upon the method of measurement. Asphaltenes associate in dilute solution in nonpolar solvents (69) giving higher molecular weights than is actually the case on an individual molecule basis. The molecular weights of the resins are somewhat lower than those of the asphaltenes and usually fall within the range 500 to 1000 (70). This is not only because of the absence of association but also because of a lower absolute molecular size. The molecular weights of the oil fractions (ie, the asphalt minus the asphaltenes and minus the resins) is usually less than 500, often 300 to 400.

Acid Number. Asphalt contains a small amount of organic acids and saponifiable material and the acid values of asphalt, 0.1–2.8 mg potassium hydroxide per gram of asphalt, are related to emulsification conditions and ease of dispersion. The acid content is largely determined by the percentage of naphthenic (cycloparaffinic) acids of higher molecular weight that are originally present in the crude oil. With increased hardness, asphalt from a particular crude oil normally decreases in acid number as more of the naphthenic acids are removed during the distillation process.

Rheology. Asphalt is a viscoelastic material whose rheological properties reflect crude type and, to a lesser extent, processing. The ability of the asphalt to perform under many conditions depends on flow behavior (71,72). Asphalt films or coatings showing no appreciable change from original conditions are usually desired, ie, they should allow some structural movement without permanent deformation. Some procedures for measuring viscoelastic properties of asphalt have been described (2,73–75). Liquid asphalts, in effect, substitute diluent (or water emulsification) for heat to allow pumping, spraying, or application of the material at lower temperature.

The viscosity of hydrocarbons and temperature are related by the Walther equation (76):

$$\log \log (100\ \eta_0) = a - m \log T$$

where η_0 is the limiting viscosity at low shear rates, T is the absolute temperature, a and m are constants reflecting the intercept level and slope (or measure of susceptibility of viscosity to temperature, respectively). On the other hand, the general relationship for viscosity, temperature, and shear rate is

$$\log \log 100\ \eta_0\ (1 \times CD^N) = a - m \log T$$

where D is shear rate; C is a function of limiting viscosity, limiting slope, and other constants; and N is the limiting slope.

Typical profiles for the different general families of asphalts (cutbacks or liquid materials, paving asphalt cements and the harder roofing and industrial materials which are usually graded by softening point) can be generated by the use of the equation (Fig. 12). At lower temperatures (60°C and lower) and/or higher shear rates, which are typical of asphalt service conditions after incorporation in a roof or pavement, semisolid and solid asphalts display an increasing elastic component which relates viscosity with shear rate (73,77,78). The constant high viscosity at lower shear rates is the limiting viscosity. Viscosities in the area where viscosity changes with shear rate are generally termed apparent viscosities.

Curves for the viscosity data, when displayed as a function of shear rate with temperature, show the same general shape with limiting viscosities at low shear rates and limiting slopes at high shear rates. These curves can be combined in a single master curve (for each asphalt) employing vertical and horizontal shift factors (77–79). Such data relate reduced viscosity (from the vertical shift) and reduced shear rate (from the horizontal shift).

A number of viscometers have been developed for securing viscosity data at temperatures as low as 0°C (58,59). The most popular instruments in current use are the cone plate (ASTM D3205), parallel plate, and capillary instruments (ASTM D2171 and ASTM D2170). The cone plate can be used for the determination of viscosities in the range of 10 to over 10^9 Pa·s (10^{10} P) at temperatures of

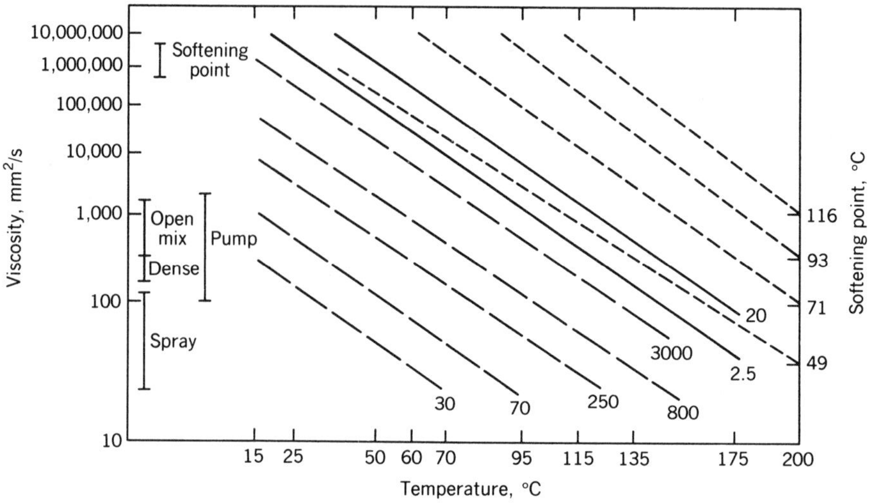

Fig. 12. A viscosity–temperature chart, mm^2/s = cSt; – – –, industrial asphalts; —— cutback asphalts; —— paving asphalts.

0–70°C and at shear rates from 10^{-3} to 10^{2} s^{-1}. Capillary viscometers are commonly used for the determination of viscosities at 60–135°C.

Tests developed for measurement of viscoelastic properties are directly usable in engineering relations. Properties can be related to the inherent structure of bituminous materials. The fraction of highest molecular weight, the asphaltenes, is dispersed within the asphalt and is dependent upon the content and nature of the resin and oil fractions. Higher aromaticity of the oil fractions or higher temperatures leads to viscous (sol) conditions. A more elastic (gel) condition results from a more paraffinic nature and is indicated by large elastic moduli or, empirically, by a relatively high penetration at a given softening point. Empirically, the penetration index *PI*(2) and penetration temperature susceptibility *PTS* have been used to measure the degree of dispersion.

$$PI = \frac{30}{1 + 90 \times PTS} - 10$$

$$PTS = \frac{\log 800 - \log \text{penetration at } 25°\text{C, 100 g, 5 s}}{(\text{softening point, ring and ball, °C}) - 25°\text{C}}$$

This equation is based on the approximation that the penetration is 800 at the softening point, but the approximation fails appreciably when a complex flow is present (80,81). However, the penetration index has been, and continues to be, used for the general characteristics of asphalt; for example asphalts with a *PI* less than −2 are considered to be the pitch type, from −2 to +2, the sol type, and above +2, the gel or blown type (2). Other empirical relations that have been used to express the rheological–temperature relation are fluidity factor; a Furol viscosity *V*, at 135°C and penetration *P*, at 25°C, relation of $(V\text{–}P)P/100$; and penetration viscosity number PVN again relating the penetration at 25°C and kinematic viscosity at 135°C (82,83).

Asphalts develop an internal structure with age, steric hardening (3), in which viscosity can increase upon aging without any loss of volatile material (73,83). Those with a particularly high degree of gel structure exhibit thixotropy.

Durability. The term "durable" has several meanings, but in the present context it is used to describe an asphalt that possesses the necessary chemical and physical properties required for the specified pavement performance, being resistant to change during the in-service conditions that are prevalent during the life of the pavement.

Asphalts are used as protective films, adhesives, and binders because of their waterproof and weather-resistant properties. One particularly valuable property, the ability of the asphalt to undergo movement without fracture, can occur because of the viscous (sol) nature. In addition, asphalts have long and continuous satisfactory service because of their slow rate of hardening from heat, oxidation, fatigue, and weathering (84–87). Exposed asphalt films harden partially from a loss of volatile oils and to a greater extent from the formation of additional asphaltene material, with the concurrent loss of petrolene–maltene constituents through oxidation. Such chemical change undoubtedly is catalyzed by uv irradiation (86,88–90). The oxidation may form water-soluble degradation products which are removed from the asphalt film (89,90), but the conversion of the petrolene–maltene constituents to lower molecular weight asphaltene material (27) cannot be ignored.

Failure of an asphalt film is reflected in the appearance of typical crack patterns because of a decrease in the plasticity of the asphalt through insufficient lower molecular weight constituents remaining in the continuous phase. Other manifestations of failure are loosening of the bond, such as loosening of granules in roofing shingles to actual loss of the film, or the binder in pavement mixtures might become friable and crack. A stiffer asphalt, under uniform loading conditions, could reduce pavement deflection, extend fatigue life, and allow less flow deformation. A softer material would normally allow a longer weathering life before the maltene–asphaltene composition becomes critical in service. Usually the softest material allowed by initial service needs is selected.

The water resistance of asphalt films is also a manifestation of durability. Asphalts that have a low content of soluble salts show a low water absorption. The pickup of water is primarily a surface phenomenon that causes the film to soften leading to blistering. Even with a high rate of absorption, asphalt films show little loss of bond to surfaces on continued immersion in water, and continue to protect metals from corrosion for long periods of time. Bacteria and fungi can attack the very low molecular weight portion of bituminous materials. However, deterioration of the harder bituminous films as a result of such agents seems insignificant (91,92).

Mineral fillers are often added to asphalts to influence their flow properties, reduce costs, and are commonly used as stabilizers in roofing coatings at concentrations up to 60 wt % (93). Mineral-filled films show improved resistance to flow at elevated temperatures, improved impact resistance, and better flame-spread resistance. Fillers may increase the water absorption of asphalts. Fillers commonly used are ground limestone, slate flours, finely divided silicas, trap rocks, and mica; they are often produced as by-products in rock-crushing operations. Opaque fillers offer protection from weathering. Asbestos (qv) filler has special properties because of its fibrous structure, high resistance to flow, and toughness. It has been used in asphalt paving mixes to increase the resistance to movement under traffic (94) and in roofing materials for fire-retardant purposes.

Specifications and Tests

In 1903 an American Society for Testing and Materials (ASTM) Committee on Road and Paving Materials was formed to develop test methods and specifications for highway materials. Test methods for volatilization, penetration, and bitumen were developed by the Office of Public Roads and were adopted by ASTM in 1911.

In the volatilization test, a 20-g sample was heated for 5 h at 163°C in a tin box 6 cm in diameter and 2 cm deep. The loss in weight was determined and the consistency of the residue was an optional requirement of that time. To provide a greater depth of sample the method was revised to use a 50-g sample and a tin box 5.5 cm in diameter and 3.5 cm deep.

For the asphalt cements produced at that time the adoption of the volatilization and penetration tests provided some degree of control of excessive changes during plant mixing that might be reflected in more durable asphalts. The adoption of the method for bitumen was intended to provide a means for identifying Trinidad asphalt by observing the amount and color of the insoluble ash.

During the 1920s the following three national specifications for asphalt cements were published: (*1*) Federal specifications, adopted in 1925; (*2*) the American Association of State Highway Transportation Officials (AASHTO) specifications, adopted in 1924, revised in 1926; and (*3*) ASTM specifications, adopted in 1922 to 1926, withdrawn in 1939, and re-issued in 1947.

The Federal specifications stipulated that only those asphalts that had been demonstrated by service tests as satisfactory for the intended use would be accepted. The specifications also indicated the type and location of construction and the relative amount of traffic for each of the grades. The AASHTO specifications indicated that the use of each grade depended on the type of road, climate, and traffic. The ASTM suggested the type of construction for which each grade would be used.

With minor exceptions the requirements for the physical and chemical properties of asphalt were essentially the same for the three national specifications and included: penetration and ductility at 25°C; flash point; % loss at 163°C; penetration of residue as a % of original; solubility in carbon disulfide; solubility in carbon tetrachloride; specific gravity at 25°C; and softening point.

Remembering that the properties of residua vary with cut-point (Table 2), ie, the vol % of the crude oil (Fig. 13) helps the refiner produce asphalt of a specific type or property. There are several properties that are usually controlled in asphalt specifications:

Property	*Test*
safety	flash point
purity	solubility, ash, water content
composition	naphtha insolubles, various separation techniques, distillation tests, homogeneity, wax content

rheology	penetration, viscosity, softening point, float, ductility, flow, impact or shock, break point, temperature susceptibility ratios
durability	thin-film oven, aging index, weathering
density or specific gravity	specific gravity
special properties	bond or adhesion, compatibility, stain, storage stability, chemical resistance, etc

Asphalts are usually specificed in several grades for the same industry, differing in hardness or viscosity (95). However, with the changing nature and composition of crude oil feedstocks over the past two decades, performance and supply factors are also an important consideration (95).

Specifications for paving asphalt cements usually include five grades differing in either viscosity or penetration level at 60°C (Table 8) (ASTM D496). Susceptibility of viscosity to temperature is usually controlled in asphalt cement by viscosity limits at a higher temperature such as 135°C and a penetration or viscosity limit at a lower temperature such as 25°C.

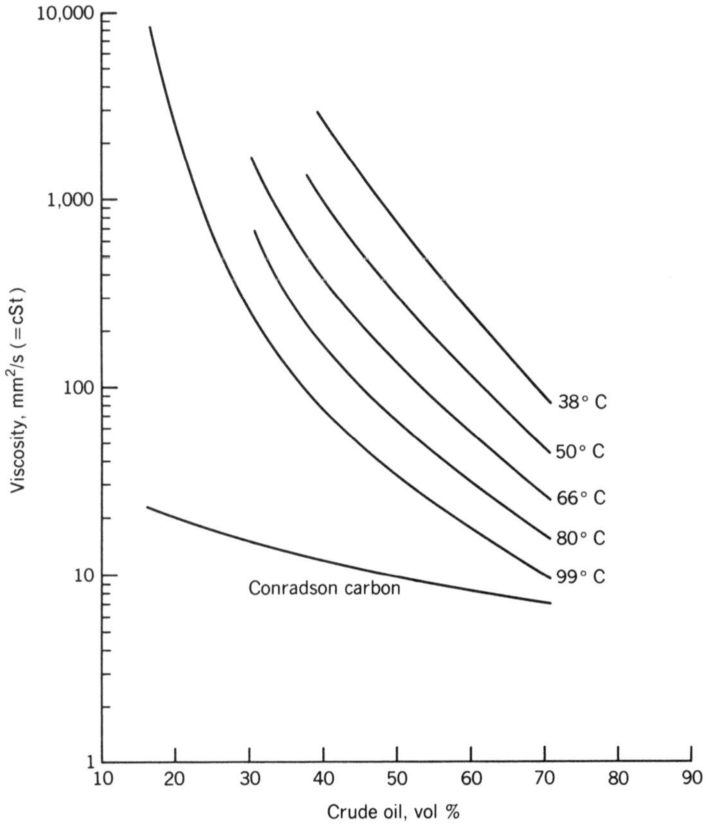

Fig. 13. Relation of physical properties to the yield of residuum from Tia Juana, Venezuela light crude oil.

Table 8. ASTM Requirements for Asphalt Cement Viscosity

Tests on residue from rolling thin-film oven test[a]	Viscosity grade[b]				
	AR-1000	AR-2000	AR-4000	AR-8000	AR-16000
viscosity, 60°C, Pa·s[c]	100 ± 25	200 ± 50	400 ± 100	800 ± 200	1600 ± 400
viscosity, 135°C, min, mm^2/s (= cSt)	140	200	275	400	550
penetration of 100 g at 25°C for 5 s, min, mm/10	65	40	25	20	20
original penetration at 25°C, min %		40	45	50	52
ductility, 25°C, 5 cm/min, min, cm	100[d]	100[d]	75	75	75
tests on original asphalt:					
flash point, Pensky-Martens closed tester, min, °C	205	219	227	232	238
solubility in trichloroethylene, min %	99.0	99.0	99.0	99.0	99.0

[a]Thin-film oven test may be used but the rolling thin-film oven test shall be the referee method.
[b]Grading based on residue from rolling thin-film oven test.
[c]To convert Pa·s to P, multiply by 10.
[d]If ductility is less than 100, material will be accepted if ductility at 15.5°C is 100 minimum at a rate of 5 cm/min.

Paving cutbacks are also graded at 60°C by viscosity, with usually four to five grades of each type. For asphalt cements, the newer viscosity grade designation is the midpoint of the viscosity range. The cutback's grade designation is the minimum kinematic viscosity for the grade, with a maximum grade viscosity of twice the minimum.

Roofing and industrial asphalts are also generally specified in various hardnesses usually with a combination of softening point and penetration to distinguish grades. Asphalts for roof construction are specified in ASTM D312 grades and asphalts for damp-proofing and waterproofing in ASTM D449. Temperature susceptibility is usually controlled in these requirements by specifying penetration limits or ranges at 25°C and other temperatures as well as softening point ranges at higher temperatures. The ASTM D312 asphalts for built-up roof constructions are differentiated according to application depending primarily on pitch of the roof and to some extent on whether or not mineral surfacing aggregates are specified. The damp-proofing grades reflect above or below grade construction, primarily, and whether or not a self-healing property is incorporated.

Test Procedures. Most tests applied to petroleum asphalts are empirical in nature (45). The ASTM tests are not the only ones applied to asphalt testing. Private tests of a propriety nature have been used within companies. Some of these tests are now being made public and occur from time to time in literature reports. Such tests may become a part of the ASTM standards.

The significance of a particular test is not always apparent by reading the procedure, and sometimes can only be gained through working familiarity with the test (96). The following tests are commonly used to characterize asphalts.

Bitumen Insoluble in Paraffin Naphtha (AASHTO T46). This test designated by the American Association of State Highway and Transportation Officials (AASHTO) is used to indicate the content of naphtha-insoluble asphaltenes in an asphalt. Other solvents such as *n*-heptane (ASTM D3279), *n*-hexane, and *n*-pentane have been substituted for the naphtha solvent.

Bitumen Soluble in Carbon Disulfide (ASTM D4). Asphalt is defined as a mixture of hydrocarbons that are completely soluble in carbon disulfide. Trichloroethylene or 1,1,1-trichloroethane have been used in recent years as safer solvents for this purpose. The procedure for these and other solvents for asphalt with little or no mineral matter are described in ASTM D2042.

Breaking Point (FRAAS, IP 80/53). This test of the Institute of Petroleum is an approximate indication of the temperature at which an asphalt possesses no ductility and would reflect brittle fracture conditions.

Bond and Adhesion (ASTM D1191). This test, designed for use on crack and joint sealers, is used primarily to determine whether a jointing material possesses an arbitrary amount of bonding strength at low temperatures where portland cement concrete is being used.

Compatibility (ASTM D1370). When a coating asphalt and a saturating-grade asphalt are used together, as in prepared roofing, this test indicates whether they are likely to bleed, show strike-through, or disbond under stress at the coating felt interface.

Distillation (ASTM D402). Approximate amounts of volatile constituents are determined by this test which is particularly applicable to cutback asphalt and road oils.

Ductility (ASTM D113). The ductility of an asphalt is expressed as the distance in cm which a standard briquet can be elongated before breaking. Ductility is a combination of flow properties and reflects both cohesion and shear susceptibility.

Emulsified Asphalts (ASTM D244). This standard covers a variety of tests for the composition, handling, nature and classification, storage, use, and specifying of asphalt emulsions used primarily for paving purposes.

Flash Point (ASTM D92). The Cleveland open cup method is most commonly used although the Tag open cup (ASTM D3143) is applicable to cutbacks. Flash point is an indication of fire hazard and the test is frequently used to indicate whether a given product has been contaminated with materials of lower flash point.

Float Test (ASTM D139). The consistency properties of an asphalt at a very low applied force are indicated by this test. The test is normally used for those asphalts that are too soft for the penetration test.

Homogeneity. Visual tests for homogeneity are generally used to judge uniformity.

Penetration (ASTM D5). This is a commonly used consistency test. It involves the determination of the extent to which a standard needle penetrates a properly prepared sample of asphalt under definitely specified conditions of temperature, load, and time. The distance that the needle penetrates in units of mm/10 measured from 0 to 300, is the penetration value. Soft asphalts have high penetration values.

Sampling (ASTM D140). This standard provides guidance for the sampling of asphalts, liquid and semisolid, at point of manufacture, storage, or delivery.

Softening Point (Ring and Ball Method, ASTM D36). The softening point of an asphalt may be defined as that temperature at which an asphalt attains a particular degree of softness under specified conditions of test. It is commonly used to classify industrial and roofing asphalt grades. A method (ASTM D2398) has been developed to provide a single bath medium for the full range covered by the two media in ASTM D36.

Specific Gravity (ASTM D70). For solid and semisolid asphalts a pycnometer is generally used and a hydrometer is applicable to liquid asphalts (ASTM D3142).

Spot Test (AASHTO T102). The test distinguishes asphalts that contain bodies poorly tolerated by the asphalt system.

Stain (ASTM D1328). This test measures the amount of stain on paper or other cellulosic materials by asphalt. Variations of the cigarette paper stain procedure include the Barber stain, usually conducted at 54.4°C and 3.9 MN (400 g-force). Talc stain tests are also used.

Temperature Susceptibility. This is the term used to designate the change in consistency of an asphalt with changes in temperature. The Walther equation slope constant m can be used as a fundamental measure.

Temperature–Volume Correction (ASTM D1250). Tables are provided to allow the conversion of volumes of asphaltic materials from one temperature to another or, as generally used, to adjust volumes to a temperature of 15.6°C, the standard basis of measurement in the United States. The value commonly taken for mean coefficient of expansion is 0.00036 in the range 15.6–121.1°C.

Viscosity. This is a measure of resistance to flow. A number of instruments are in common use with asphalt for this purpose. The vacuum capillary (ASTM D2171) is usually used to classify paving asphalt cements at 60°C, although it is applicable to materials in the range 4.2–20,000 Pa·s (42–200,000 P). Kinematic capillary instruments (ASTM D2170) are used in the 60–135°C temperature range for both liquid and semisolid asphalts in the range of 30–100,000 mm^2/s (= cSt). Saybolt tests (D88) are also used in this temperature range and at higher temperature (ASTM E102). At lower temperatures the cone and plate instrument (ASTM D3205) has been used extensively in the viscosity range 10^2–10^9 Pa·s (10^3–10^9 P). Other techniques include use of the sliding plate microviscometer and the rheogoniometer (see RHEOLOGY).

Thin-Film Oven Test (ASTM D1754). This test has the purpose of determining the hardening effect of heat and air on a static film of asphalt when exposed in a thin film. An analogous procedure is the Rolling Thin-Film Test (ASTM D2872) which has the same purpose but utilizes a moving film exposed for 75 min at 163°C.

Water Content (ASTM D95). This test covers the water content of asphalt by distillation using a Dean-Stark receiver.

Wax Content. The Deutsche Industrie Normen (DIN) method utilizes destructive distillation of the asphalt, followed by freezing out of the wax in the distillate fractions.

Weathering (ASTM D529). This test evaluates the relative weather resistance of asphalts used for protective-coating applications, especially for roofing. No direct measure of outdoor life or service can be obtained from this test. Methods for preparing test panels (ASTM D1669) and failure end-point testing (D1670) are available.

Finally, these are not the only tests used for determining the property and behavior of an asphaltic binder. As in the petroleum industry (5,6), a variety of tests are employed that evolved through local, or company, usage (97–99) without having passed through the ASTM system of approval. Tests are also necessary to determine the most practical design of asphalt–aggregate mixtures (100).

Uses

Properties, and therefore uses, depend upon the method of manufacture (Table 9). The Asphalt Institute lists a multitude of uses for asphalt (101), including hydraulics (dam facings, canal linings, pond linings) (102), recreation (substrate for artificial surfaces, tennis courts, running tracks) (103), agriculture (mulches, underground water barriers, stockyard paving) (104–107), transportation (railroad ballast treatment and roadbeds) (108), and metals (ore leaching pads, ore and coal briquetting, etc).

Straight-Run Asphalt. The largest use of straight-run asphalts is in the paving industry (101) where they serve primarily as binders in paving mixes and as bases in liquid asphalts used as seal coatings, surface treatments, road mixes, and soil stabilizers. The most important technical innovation in asphalt paving has been to utilize asphalt throughout the entire pavement structure (termed total asphalt) to provide more efficient distribution of traffic stresses to the subgrade and provide better protection of the base from intrusion of outside materials such as water and soil. It has been established that 2.5 cm of asphaltic concrete can replace at least 5 cm of aggregate base for this purpose. In these mixes the binder contents usually vary from 4.0–10% depending on the aggregate gradation and nature and the construction.

The hot mixes are designed by using a standard laboratory compaction procedure to develop a composition reflecting established criteria for volume percent air voids, total volume percent voids between aggregate particles, flow and stability, or compressive strength. Tests such as the Marshall, Unconfined Compression, Hubbard-Field, Triaxial Procedure, or the Hveem stabilometer method are used (109).

If straight-run asphalts are reduced to a hardness below 300 penetration (30 mm), they are termed asphalt cements. For hot-mix paving, either the AC-10 viscosity grade or the next harder AC-20 grade is commonly specified.

Propane-precipitated asphalts have properties similar to the straight-run asphalts, assuming the same crude source, but are also obtainable in harder grades with higher softening points. The possible degree of reduction is much greater than with the distillation process. A practical limitation is the ability to remove and handle the precipitated asphalt fraction. These materials are often employed in fesco board manufacture and as a component in self-sealing roofing shingle adhesive.

Table 9. Properties of Asphalts

Property	Straight-run, residual	Thermal	Air-blown residual
softening point (ring and ball), °C	46	113	93
penetration of 100 g at 25°C for 5 s, mm/10	90	0	20
ductility at 25°C, 5 cm/min, cm[a]	150+	too hard	3.2
specific gravity, 15.6/15.6°C	1.03	1.12	1.05
mean coefficient of cubical expansion/°C			
15.6–65.6°C	0.00063	0.00058	0.00063
15.6–232°C	0.00068	0.00063	0.00068
specific heat, J/(kg·K)[b]			
4.4°C	1675	1549	1633
93.3°C	1968	1842	1926
204.4°C	2345	2177	2303
thermal conductivity at 26.7°C, W/(m·K)	0.16	0.16	0.16
permeability constant at 25°C, $kg{\cdot}m/(m^2{\cdot}s{\cdot}N/m^2)$			
water vapor	$0.62–1.93 \times 10^{-15}$	1.1×10^{-15}	$1.25–2.4 \times 10^{-15}$
oxygen			0.08×10^{-15}
water absorption of 10-mil[c] films on aluminum panels, wt %			
50 wks			1.5–10
100 wks			2.5–16.5
surface tension, mN/m (= dyn/cm)			
25°C	34		32
100°C	27		28
dielectric strength, spherical electrodes, V/m	$11–45 \times 10^6$	36×10^6	$30–35 \times 10^6$
dielectric constant, 50 Hz at 20°C	2.7	3.0	2.7

[a]A + sign after a number indicates a minimum value.
[b]To convert J to cal, divide by 4.184.
[c]10 mil = 6.254 mm.

Air-Blown Asphalts. Air-blown asphalts can be used in prepared roofings (110,111) as well as for saturating the felts used in asphalt shingles and mineral-surfaced roll roofing. The shingle saturant is commonly a flux air-blown to the 50–60°C softening point level with a 50–90 penetration (5–9 mm) at 25°C, and the coating is the same flux air-blown to the 100–115°C softening point level with a penetration of 16–25 (1.6–2.5 mm). The coating asphalt is commonly filled with 50–60% mineral stabilizer, most of which passes a 74 μm (200 mesh) screen, and then is applied at approximately 35% of the finished weight of roofing shingles. This coating is also lightly applied to the back of the felt for protection. In addition to providing the waterproof surfacing, the coating also bonds the colored granules to the saturated felt. The opacity of the granule protects the coating from the hardening influence of light catalyzed reactions. The combination provides economical but highly durable mineral-surface shingles or roll roofing.

Built-up roofing constitutes several plies of a saturated roofing felt (low melt, flexible asphalt saturant) with each ply mopped in place and the structure covered by air-blown asphalts of from 60° to 105°C softening point, with the hardness selected depending primarily on roof slope. These roofs are usually surfaced with mineral aggregates.

A wide variety of industrial products are made from air-blown asphalts, such as laminants, pipeline coatings, waterproofings, potting compounds, and joint fillers. Additionally, the oxidized asphalts can be cut back with various solvents to produce a variety of protective coatings and paints. These coatings are often reinforced with mineral fillers to produce cold-applied roof coatings, adhesives, sound deadeners, railroad protective coatings, and undercoatings for automobiles.

Thermal Asphalt. Thermal asphalt products are in low supply because the thermal process has been virtually replaced by catalytic cracking processes. Thermal pitches, because of their high viscosity temperature susceptibility, are very hard at ordinary temperatures (Table 9), but become quite fluid when heated. They are especially useful as binders and as saturants for fiberboards, and as sizings for fiberboard when ground or dispersed and introduced directly into the wet furnish. Their weatherability on exposure in thin films is such that they are usually not satisfactory for use as protective coatings.

Liquid Asphalt. Liquid asphalt products comprise cutback asphalts and emulsions. A number of grades of different viscosities are available, which permit application from ambient temperatures to 150°C. The lower viscosity products are used for dust-laying purposes and as tack coats, prior to laying asphalt surface courses. The heavier grades are used for mix-in-place road mixes.

The properties of asphalt emulsions (ASTM D977 and D2397) allow a variety of uses. The rapid-setting grades are used for surface treatment and seal coat or surface treatments requiring cover aggregate. The medium-setting grade has higher stability to break on contact with surfaces and is used for mix-in-place macadam aggregates. It may also be used in plant mixes to produce a cold-laid macadam aggregate. Both of these types are less stable than the slow-setting type which has sufficient stability so that a finely divided material such as cement may be mixed without breaking the emulsion. The slow-setting grade is suitable for mixing with very finely divided dense-graded aggregates, or for the so-called slurry seal. The slow-setting grade is also used as a tack coat, and as a prime for dust laying and for soil treatment mixtures (Table 10).

In addition to the paving grades of emulsions, a variety of products are available that are suitable for industrial uses, for paper sizings, and for use in the roofing application field. Asphalts emulsified with colloidal clays are especially suitable for outdoor applications because of their excellent weathering properties. Some of these products have been adapted for use with roofing systems, either with brush or spray application.

Modified Asphalts

The deterioration of asphalt pavements under heavy traffic loads and/or under hostile weather conditions, coupled with the inability of many modern crude oil

Table 10. General Uses of Emulsified Asphalt[a]

Type of construction	D977[b] RS-1, CRS-1[c]	RS-2, CRS-2[c]	MS-1, HFMS-1	MS-2h, HFMS-2h	SS-1, CSS-1[c]	SS-1h, CSS-1h[c]
Bituminous-aggregate mixtures						
for pavement bases and surfaces						
plant mix (hot) D2629				X[d]		
plant mix (cold)						
open-graded aggregate				X		
dense-graded aggregate					X	X
travel plant (mixed-in-place)						
open-graded aggregate			X	X		
dense-graded aggregate					X	X
sand					X	X
sandy soil					X	X
slurry seal					X	X
Bituminous-aggregate applications						
treatments and seals						
single-surface treatment (chip seal)	X	X				
multiple-surface treatment	X	X				
sand seal	X	X	X			
penetration macadam						
large voids		X				
small voids	X					
Bituminous applications						
fog seal			X[e]		X[f]	X[f]
prime coat-penetrable surface					X[f]	X[f]
tack coat			X[e]		X[f]	X[f]
dust binder					X[f]	X[f]
mulch treatment					X[f]	X[f]
crack filler				X	X	X
Maintenance mix						
immediate use			X	X		

[a]Only those grades of emulsified asphalt in general use have been indicated herein. It is possible that under certain variation of aggregates and/or climatic conditions, additional selections might be appropriate. Where the use of emulsified asphalt for applications other than those listed in the table are contemplated, the emulsion supplier should be consulted.

[b]Unless otherwise noted.

[c]Cationic D2397.

[d]ASTM D2629 permits use of other emulsion grades by note; grades of emulsion other than MS-2h may be used where experience has shown that they give satisfactory performance.

[e]Diluted with water by manufacturer.

[f]Diluted with water.

feedstocks to meet the challenge of producing specification-grade asphalts, has resulted in the search for asphalt modifiers. These are materials that can be added to the asphaltic binder to improve service life through enhancement of a particular property or providing a stronger bond between the binder and the aggregate thereby reducing the susceptibility of the mix to, for example, water damage (112,113).

A variety of materials has been proposed to modify the properties of asphaltic binders to enhance the properties of the mix (112), including fillers and fibers to reinforce the asphalt–aggregate mixture (114), sulfur to strengthen or harden the binder (115,116), polymers (98,117–121), rubber (122), epoxy–resin composites (123), antistripping agents (124), metal complexes (125,126), and lime (127,128). All of these additives serve to improve the properties of the binder and, ultimately, the properties of the asphalt–aggregate mix.

The addition of fossil fuel derived materials has also been investigated. Of particular interest are the materials which contain functional groups that are capable of enhancing the asphalt–aggregate bond. Modifiers such as those isolated from the products of coal liquefaction (129) and the components of shale oil and even shale oil residua (130) have been used. The latter is especially interesting since the concept focuses on the use of basic nitrogen functions in the shale oil which enhance the adhesion of the asphalt to the aggregate. Test strips of shale oil modified asphalt have been laid on highways in various parts of the United States.

Aggregates

Although much of the focus tends to be on the composition and behavioral characteristics of the asphalt (ie, the binder or cement), the properties of the aggregate also play a significant role in determining the ultimate properties of the asphalt–aggregate mix. There are many factors to consider and they include: the functionalities in the asphalt (131), absorption of the asphalt into the pores of the aggregate (132), the aging properties of the asphalt when it is on the aggregate (133), as well as methods for coating the asphalt on to the aggregate (134).

Attempts have also been made to carry out surface modifications of the aggregate to enhance interactions with the asphalt (135) and other workers have made attempts to measure or predict the strength and type of asphalt–aggregate bonds (136,137). However, it must also be remembered that mix design parameters play an important role in determining the performance of asphalt–aggregate mixes (138–142).

Health and Safety

The word asphalt has been carelessly used in that it is not adequately differentiated from thermally degraded materials, especially coal tar and derivatives. It is essential to differentiate asphalt from these materials which contain known carcinogens and health hazards. For this reason, the use of cracked asphalts must be treated with caution.

There are a variety of tests for asphalt emissions (Table 11). No significant air pollution problems are associated with emissions from hot paving operations using several asphalt cements (143). In fact, the concentrations of gaseous substances and emissions from paving asphalt cement have been found to be in very low concentration and within existing EPA and OSHA standards, even when the ambient air sampling was done under confined conditions (144–146). In addition, the high molecular weight polynuclear aromatic constituents that occur in asphalt have been studied as health hazards (147–151). The general conclusion from these studies is that surveillance should be continued although asphalt has not been shown to be a material of significant hazard.

Table 11. Analysis Techniques for Asphalt Emissions

Sample	Analytical technique
air	mass spectrometry
volatile hydrocarbons (C1–C6)	gas chromatography
carbon monoxide	infrared spectroscopy
volatile sulfur compounds	gas chromatography
nitrogen dioxide	uv-vis spectroscopy at 550 nm
aldehydes	uv spectroscopy
phenols	uv spectroscopy
particulates	optical microscopy
polynuclear aromatics (PNA)	gas chromatography

Steps to minimize potential safety hazards in the handling of asphalt are set forth by the American Petroleum Institute (152). These include: (*1*) sudden pressure increases from hot asphalt in contact with moisture in enclosed tanks or transports; (*2*) exposure to air at 150°C or above; (*3*) local overheating above heating coils; flashing of asphalt volatiles in the presence of an ignition source or possible autoignition; and (*4*) hydrogen sulfide from high temperature operations.

BIBLIOGRAPHY

"Asphalt" in *ECT* 1st ed., Vol. 2, pp. 164–199, by A. P. Anderson, Shell Oil Company, and V. A. Endersby, Shell Development Company; in *ECT* 2nd ed., Vol. 2, pp. 762–806, by A. J. Hoiberg, The Flintkote Company, L. W. Corbett, Esso Research & Engineering Company, and R. B. Lewis, Shell Oil Company; in *ECT* 3rd ed., Vol. 3, pp. 284–327, by J. V. Evans, Amoco Oil Company.

1. *Terms Relating to Materials for Roads and Pavements,* ASTM Designation D8, American Society for Testing and Materials, Philadelphia, Pa., 1976.
2. P. H. Pfeiffer, *The Properties of Asphaltic Bitumen,* Elsevier Publishing Company, Amsterdam, The Netherlands, 1950.
3. H. Abraham, *Asphalts and Allied Substances,* 6th ed., Vol. 1, Van Nostrand Co. Inc., Princeton, N.J., 1960, Chapt. 2.
4. *A Glossary of Petroleum Terms,* The Institute of Petroleum, London, UK, 1961.

5. J. G. Speight, *Fuel Science and Technology Handbook,* Marcel Dekker, Inc., New York, 1990.
6. J. G. Speight, *The Chemistry and Technology of Petroleum,* 2nd ed., Marcel Dekker, Inc., New York, 1991.
7. J. S. Miller, Jr., in E. A. Dunstan, ed., *The Science of Petroleum,* Vol. 4, Oxford University Press, New York, 1938.
8. B. Redwood, *Treatise on Petroleum,* 5th ed., Vol. 1, Griffin, London, 1976, plate 7, pp. 53, 208–212.
9. Herodotus, *Persian Wars,* Vol. 6, Modern Library, New York, p. 119.
10. A. Sequin, *Proc. 2nd World Pet. Congr.* **4,** 56 (1937).
11. R. J. Forbes, *Bitumen* 5, 9, 41, 63 (1935).
12. R. J. Forbes, *Bitumen and Petroleum in Antiquity,* E. J. Brill, Leiden, The Netherlands, 1936.
13. W. A. Attwooll and D. C. Broome, *Trinidad Lake Asphalt,* 2nd ed., Baynard Press, London, UK, 1954.
14. *Proc. Assoc. Asphalt Paving Technol.* **43A,** (1974).
15. *Statistical Abstracts of the United States,* U.S. Department of Commerce, Washington, D.C., 1961, p. 716.
16. W. R. Jillson, *Pan Am. Geol.* **41,** 251 (1924).
17. N. Berkowitz and J. G. Speight, *Fuel* **54,** 138 (1975).
18. J. C. Petersen, *Fuel Sci. Technol. Int.* **6,** 255 (1988).
19. W. L. Nelson, *Petroleum Refining Engineering,* 4th ed., McGraw-Hill Book Co., Inc., New York, 1958.
20. C. H. Nielson, *Distillation in Practice,* Chapman and Hill Ltd., London, UK, 1956.
21. H. M. Chelton and R. N. Traxler, *Proc. 5th World Pet. Congr.,* New York, 1959, p. 247.
22. A. E. Leybourne, III and H. E. Schweyer, *Ind. Eng. Chem. Process Des. Dev.* **1,** 127 (1962).
23. I. Ishai and Y. A. Tuffour, *Proc. Assoc. Paving Technol.* **56,** 599 (1987).
24. M. A. Quddus and F. Khan, *Fuel* **69,** 509 (1990).
25. L. W. Corbett and R. E. Swarbrick, *Proc. Assoc. Asphalt Paving Technol.* **29,** 104 (1960).
26. A. J. Hoiberg and W. E. Garris, Jr., *Ind. Eng. Chem. Anal. Ed.* **16,** 294 (1944).
27. S. E. Moschopedis and J. G. Speight, *Fuel* **57,** 235 (1978).
28. W. L. Nelson, *Oil Gas J.* **53**(16), 158 (Aug. 23, 1954).
29. D. B. Pruess, *Pet. Eng.* **30,** C-6 (Oct. 1958).
30. R. L. Purvin, *Pet. Eng.* **22,** C-28 (Apr. 1950).
31. W. H. Shearon and A. J. Hoiberg, *Ind. Eng. Chem.* **41,** 2672 (1949).
32. W. L. Nelson, *Oil Gas J.* **46**(25), 99 (Oct. 25, 1947).
33. A. J. Hoiberg, *Proc. Assoc. Asphalt Paving Technol.* **19,** 225 (1950).
34. Aust. Pat. 153,506 (June 10, 1938), R. Lichtenstein.
35. U.S. Pat. 1,782,186 (Nov. 18, 1930), G. Abson (to Chicago Paving Laboratory Inc.).
36. U.S. Pat. 2,200,914 (May 14, 1940), R. E. Burk and C. H. Whitacre (to Standard Oil Company of Ohio).
37. U.S. Pat. 2,450,756 (Oct. 5, 1948), A. J. Hoiberg (to Lion Oil Company).
38. G. L. Oliensis, *Proc. Assoc. Asphalt Paving Technol.* **6,** 88 (1935).
39. D. W. Gagle, *Refining Eng.* **30,** C-15 (Mar. 1958).
40. L. Lowry, *Pet. Eng.* **29,** C-50 (1957).
41. E. W. Mertens, L. D. Coyne, and E. D. Rogers, *Am. Soc. Test. Mater. Spec. Tech. Publ.* **294,** 68 (1960).
42. H. Al-Abdulwahhab, F. Bayomy, and A. Al-Halhouli, *Trans. Res. Rec.* **1106**(1), 71 (1987).
43. W. A. McIntosh, *Ind. Eng. Chem.* **44,** 1656 (1952).

44. K. H. Altgelt, *Die Makromolekulare Chemie* **88,** 75 (1965).
45. J. B. Green, J. W. Reynolds, and S. K-T Yu, *Fuel Sci. Technol. Int.* **7,** 1327 (1989).
46. D. A. Anderson, J. C. Petersen, and D. W. Christensen, *Proc. Assoc. Asphalt Paving Technol.* **55,** 250 (1986).
47. D. S. Decker and J. L. Goodrich, *Proc. Assoc. Asphalt Paving Technol.* **58,** 503 (1989).
48. U.S. Pat. 4,904,305 (1990), L. Zanzotto and co-workers (to Nova Husky Research Corp., Canada).
49. W. A. Gruse and D. R. Stevens, *Chemical Technology of Petroleum,* 3rd ed., McGraw-Hill Book Co., Inc., New York, 1960, p. 585.
50. J. Marcusson, *Z. Agnew Chem.* **29,** 1, 21 (1916).
51. L. W. Corbett, *Anal. Chem.* **36,** 1967 (1964).
52. L. W. Corbett, *Anal. Chem.* **41,** 576 (1969).
53. W. R. Middleton, *Anal. Chem.* **39,** 1839 (1967).
54. F. S. Rostler and H. W. Sternberg, *Ind. Eng. Chem.* **41,** 598 (1949).
55. F. S. Rostler and R. M. White, *Proc. Assoc. Asphalt Paving Technol.* **31,** 35 (1962).
56. R. N. Traxler and H. E. Schweyer, *Oil Gas J.* **52,** 158 (1953).
57. A. C. Pitchford and W. M. Axe, *Pet. Rep.* **6**(3), B-43 (1961).
58. E. C. Knowles and H. Levin, *Ind. Eng. Chem. Anal. Ed.* **13,** 314 (1941).
59. R. A. Gardner, *J. Chem. Eng. Data* **4,** 155 (1959).
60. R. N. Traxler, *Chem. Rev.* **19,** 119 (1936).
61. L. H. Csanyi and B. S. Bassi, *Proc. Assoc. Asphalt Paving Technol.* **27,** 52 (1958).
62. E. S. Hillman and B. Barnett, *Proc. Am. Soc. Test. Mater.* **37,** 559 (1937).
63. Grader, *Oe. u Kohle* **38,** 867 (1942).
64. J. V. Evans and L. E. Ott, in Ref. 46, p. 353.
65. P. W. Jennings and J. A. S. Pribanic, in Ref. 45, p. 1269.
66. C. Gaestel, R. Smadja, and K. Lamminan, *Rev. Gen. des Routes et Aerodromes* **466,** 85 (1971).
67. J. C. Petersen and H. Plancher, *Anal. Chem.* **47,** 112 (1975).
68. R. Fish, *Anal. Chem.* **56,** 2452 (1984) and references cited therein.
69. S. E. Moschopedis, J. F. Fryer, and J. G. Speight, *Fuel* **55,** 227 (1976).
70. J. A. Koots and J. G. Speight, *Fuel* **54,** 179 (1975).
71. W. O. Yandell, in O. E. Briscoe, ed., *Asphalt Rheology: Relationship to Mixture,* ASTM Special Publication No. 941, ASTM, Philadelphia, Pa., 1985.
72. M. Tia and B. E. Ruth, in Ref. 71.
73. A. W. Sisko, *Highw. Res. Rec.* **67,** 27 (1965).
74. K. A. Lammiman and J. E. Roberts, *Lab. Pract.* **10,** 816 (1961).
75. R. Roque, M. Tia, and B. E. Rutto. in O. E. Briscoe, ed., *Asphalt Rheology: Relationship to Mixture,* ASTM Special Publication No. 941, ASTM, Philadelphia, Pa., 1985.
76. C. Walther, *Oil Kohle* **1,** 71 (1933).
77. J. F. Brodnyan, *Highw. Res. Board Bull.* **192,** 1 (1958).
78. F. H. Gaskins, *Trans. Soc. Rheol.* **4,** 265 (1960).
79. J. L. Williams, R. F. Landel, and J. D. Ferry, *J. Am. Chem. Soc.* **77,** 3701 (1955).
80. R. N. Traxler, *Ind. Eng. Chem.* **30,** 322 (1938).
81. R. H. Lewis and J. Y. Welborn, *Public Roads* **21,** 1 (1940).
82. R. N. Traxler, H. E. Schweyer, and J. W. Romberg, *Ind. Eng. Chem.* **36,** 823 (1944).
83. C. Van der Poel, in M. Reiner, ed., *Building Materials, Their Elasticity and Inelasticity,* Interscience Publishers, New York, 1954, p. 373.
84. R. N. Traxler, *Asphalt, Its Composition, Properties and Uses,* Reinhold Publishing Co., New York, 1961, Chapt. 6.
85. A. J. Hoiberg, *Bituminous Materials: Asphalts, Tars, and Pitches,* Vols. 1, 2, and 3, Wiley-Interscience, New York, 1964–1966.
86. K. G. Martin, *Am. Chem. Soc., Div. Pet. Chem. Prepr.* **16,** D111 (1971).

87. J. C. Petersen, in Ref. 86, p. D27.
88. L. R. Kleinschmidt and H. R. Snoke, *J. Res. Nat. Bur. Stand.* **63C,** 31 (1959).
89. S. H. Greenfield, *J. Res. Nat. Bur. Stand.* **63C,** 287 (1960).
90. J. W. H. Oliver and H. Gibson, in Ref. 86, p. D60.
91. J. O. Harris, *Microbiological Studies Reveal Significant Factors in Oil and Gas Pipeline Back-Filled Ditches,* Technical Bulletin 102, Kansas State University, Manhattan, Kans., 1959.
92. K. G. Martin, *Deterioration of Bituminous Roofing Fabrics, Commonwealth Scientific and Industrial Research Organization,* Technical Paper 11, Division of Building Research, Melbourne, Aust., 1961.
93. S. H. Greenfeld, *The Effects of Mineral Additives on the Durability of Coating Grade Roofing Asphalts,* Building Materials and Structures Report 147, National Bureau of Standards, Washington, D.C., 1956.
94. D. A. Tamburro, A. T. Blekicki, and J. H. Kietzman, in Ref. 55, p. 151.
95. J. L. Goodrich and H. L. Dimpfl, *Proc. Assoc. Asphalt Paving Technol.* **55,** 57 (1986).
96. R. L. Dunning, in Ref. 46, p. 33.
97. A. S. Noureldin and L. E. Wood, in Ref. 47, p. 256.
98. J. Zielinski and A. Bukowski, *J. Thermal Anal.* **32,** 1797 (1987).
99. K. W. Lee and M. I. Al-Jarallah, *Trans. Res. Rec.* **1096,** 81 (1986).
100. J. Bonnot, *Trans. Res. Rec.* **1096,** 42 (1986).
101. *A Brief Introduction to Asphalt and Some of Its Uses,* Manual Series No. 5, The Asphalt Institute, College Park, Md., 1975.
102. *Asphalt in Hydraulics,* Manual Series No. 12, The Asphalt Institute, College Park, Md., 1976.
103. *Athletics and Recreation on Asphalt,* Information Series No. 147, The Asphalt Insitute, College Park, Md.
104. C. M. Hansen and A. E. Erickson, *Am. Chem. Soc. Div. Pet. Chem. Prepr.* **12,** C164–C169 (1968).
105. L. C. Brunstrum, L. E. Ott, and T. L. Speer, *Proc. 7th World Pet. Congr.* **7** (1968).
106. L. C. Brunstrum, L. E. Ott, and T. L. Speer, *Proc. Am. Pet. Inst.* Sect. 3, (1968).
107. *Asphalt Mulch Treatment,* Information Series IS-161, The Asphalt Institute, College Park, Md., 1960.
108. *Railway Roadbeds for Tomorrow,* Information Series IS-137, The Asphalt Institute, College Park, Md.
109. *Mix Design Methods for Asphaltic Concrete,* Manual Series No. 2, The Asphalt Institute, College Park, Md., 1974.
110. R. N. Traxler, *Asphalt,* Reinhold Publishing Co., New York, 1961.
111. *Roofing and Shingles; Asphalt-Prepared, Mineral Surfaced,* Federal Specification SS-R-521, General Services Administration, Washington, D.C.
112. W. A. Higgins, *Trans. Res. Rec.* **1115,** 89 (1987).
113. J. A. Scherocman, K. A. Mesch, and J. J. Proctor, in Ref. 46, p. 213.
114. U.S. Pat. 4,634,622 (1987), K. L. Jenkins and co-workers (to Manville Corp.).
115. U.S. Pat. 4,769,288 (1988), D. Saylak (to Texas A & M University).
116. W. Akili and G. J. Courval, in Ref. 71.
117. U.S. Pat. 4,639,273 (1987), D. W. Gilmore and T. G. Kugele (to Morton Thiokol Inc.).
118. L. D. Coyne, *Proc. Assoc. Asphalt Paving Technol.* **57,** 545 (1988).
119. C. Lenoble, *Am. Chem. Soc. Div. Pet. Prepr.* **35**(3), 541 (1990).
120. T. S. Shuler, J. H. Collins, and J. P. Kirkpatrick, in Ref. 71.
121. H. W. Murray, G. N. King, and J. B. Prudhomme, in Ref. 71.
122. H. B. Takallon, R. G. Hicks, and D. C. Esch, in Ref. 99, p. 68.
123. C. H. Lhymn, *J. Mater. Sci.* **8,** 1019 (1989).
124. H. W. Busching and co-workers, in Ref. 46, p. 120.

125. U.S. Pat. 4,721,529 (1988), T. E. Mullins (to Unichem International Inc.).
126. J. Epps and co-workers, in Ref. 99, p. 106.
127. T. W. Kennedy, *Bulletin No. 325,* National Lime Association, Arlington, Va.
128. M. Stroup-Gardiner and J. Epps, in Ref. 112, p. 12.
129. H. H. Yoon and A. R. Tarrer, *Am. Chem. Soc. Div. Fuel Chem. Prepr.* **33**(4), 975 (1988).
130. U.S. Pat. 4,325,738 (1982), H. Plancher and J. C. Petersen (to U.S. Department of Energy).
131. C. W. Curtis, Y. W. Jeon, and D. J. Clapp, in Ref. 45, p. 1225.
132. J. A. Guin, H. S. Chang, and S. C. Yen, *Am. Chem. Soc. Div. Fuel Chem. Prepr.* **35**(3), 370 (1990).
133. J. F. McKay, in Ref. 132, p. 496.
134. U.S. Pat. 4,692,350 (1987), A. R. Clarke, C. L. Martin, and R. H. Bowering (to Mobil Oil Corp.).
135. R-F. Ju, Y-L. Lee, S. H. Lin, and W. J. Burke, *Colloids and Surfaces* **49,** 395 (1990).
136. R. T. Podoll and K. C. Irwin, in Ref. 132, p. 346.
137. A. R. Tarrer and V. P. Wagh, in Ref. 132, p. 361.
138. F. M. Bayomy and J. A. Naji, in Ref. 47, p. 362.
139. R. L. Davis, in Ref. 71.
140. R. Roque, M. Tia, and B. E. Ruth, in Ref. 71.
141. N. W. McLeod, in Ref. 71.
142. P. S. Kandhal and W. C. Koehler, in Ref. 71.
143. *Asphalt Hot-Mix Emission Study,* Research Report RR-75-1, The Asphalt Institute, College Park, Md., 1975.
144. R. P. Hangebranck, D. J. von Lehmden, and J. E. Meaker, *Sources of Polynuclear Hydrocarbons in the Atmosphere,* Publication No. 999-AP-33, U.S. Public Health Service, Washington, D.C., 1967.
145. *Particulate Polycyclic Organic Matter,* National Academy of Sciences, Washington, D.C., 1972.
146. J. M. Colucci and C. R. Begeman, *Environ. Sci. Technol.* **5,** 145 (Feb. 1971).
147. *Petroleum Asphalt and Health,* API Medical Research Rep. EA 7103, American Petroleum Institute, Washington, D.C., Dec. 1, 1971.
148. *Health Lab. Sci.* **11,** 218 (July 1974).
149. C. H. Baylor and N. K. Weaver, *Arch. Environ. Health* **17,** (Aug. 1968).
150. W. Luinsky, *Anal. Chem.* **32,** 684 (1960).
151. W. C. Hueper, *Occupational Tumors and Allied Diseases,* Charles C Thomas, Springfield, Ill., 1942, p. 69.
152. *Guide for the Safe Storage and Handling of Heated Petroleum-Derived Asphalt Products and Crude Oil Residue,* Publication 2023, American Petroleum Institute, Washington, D.C., Mar. 1977.

JAMES G. SPEIGHT
Western Research Institute

ASPHALT TILE. See BUILDING MATERIALS, PLASTIC.

ASPIRIN. See ANALGESICS, ANTIPYRETICS, AND ANTIINFLAMMATORY AGENTS; SALICYCLIC ACID AND RELATED COMPOUNDS.

ATMOSPHERIC MODELING

Mathematical air quality models provide a powerful framework for understanding the dynamics of pollutants in the atmosphere and for assessing the impact emission sources have on pollutant concentrations. Two classes of models are commonly used. Empirical models provide an understanding of source impacts by statistically analyzing historical air quality data. Diagnostic models provide a comprehensive description of the detailed physics and chemistry of compounds in the atmosphere, following the evolution of pollutant emissions to their ultimate fate. At the heart of the model is a system of mathematical routines that integrate the effects of individual processes. The complexity and computational intensity of modern models have necessitated the development of algorithms for fast and accurate mathematical solution techniques. Air quality models are being applied to solving such problems as urban smog, acid deposition, regional ozone, haze in scenic regions, and the destruction of the protective stratospheric ozone layer.

Mathematical models have grown increasingly detailed in descriptions of air pollutant dynamics and are thus key tools for gaining scientific understanding of atmospheric processes. These models also are the most practical and scientifically defensible means of relating pollutant emissions to air quality. They are widely used in regulatory planning and analysis as indicated in Figure 1. A wide variety of models are used to address problems ranging from indoor air pollution to regional acid deposition and global climate change. The models used are remarkably similar. This article focuses on the models used to understand air pollution dynamics.

Development of a mathematical air quality model proceeds through four stages. In the conceptual stage, a working set of relationships approximating the

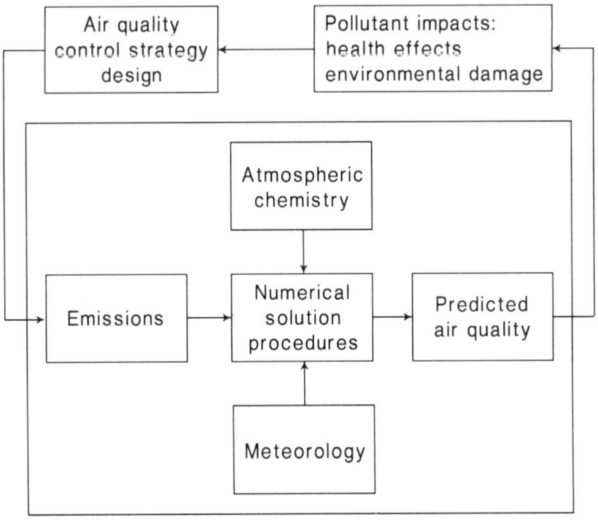

Fig. 1. Schematic of the role of an air quality model, in the air quality control planning process. Some studies, for example, the prediction of indoor air pollutant concentrations, require more than one model.

physical system is derived. Next, these relationships are expressed as mathematical equations, giving a formal description of the idealized system. The third step is the computational implementation of the model, including development of the algorithms and computer code needed to solve the equations given various inputs. The final step is the application of the model, including acquisition and processing of the necessary input data, and evaluation of model results.

Air Pollutants

Air pollution (qv) problems are characterized by their scale and the types of pollutants involved. Pollutants are classified as being either primary, that is emitted directly, or secondary, ie, formed in the atmosphere through chemical or physical processes. Examples of primary pollutants are carbon monoxide [*630-08-0*] (qv), CO, lead [*7439-92-1*] (qv), Pb, chlorofluorocarbons, and many toxic compounds. Notable secondary pollutants include ozone [*10028-15-6*] (qv), O_3, which is formed in the troposphere by reactions of nitrogen oxides (NO_x) and reactive organic gases (ROG), and sulfuric and nitric acids.

Models are used extensively to understand a wide range of problems, including indoor air pollution, such as elevated levels of radon [*10043-92-2*], Rn, and formaldehyde [*50-00-0*], CH_2O, high concentrations of carbon monoxide and particulate matter in the vicinity of congested roadway intersections, spills of volatile toxic chemicals, urban smog, acid deposition, global climate change, and stratospheric ozone depletion. Each of these problems involves pollutant transport via advection and turbulent diffusion. Several of them also involve atmospheric chemistry and processes of wet or dry deposition may also be important. Thus the models used to describe the dynamics in each case primarily vary in temporal and spatial resolution, the chemical transformations being tracked, and the deposition processes included. If a pollutant is strictly primary in nature and is long-lived in the atmosphere, the models used to describe its dynamics need only follow transport. Transport-only models are used when dealing with carbon monoxide, many toxic chemicals, and most components of particulate matter. Following the evolution of secondary or highly reactive pollutants is generally much more complicated because chemical reactions between many compounds must be considered in addition to transport and removal. Chemically active systems include urban and regional smog, acid deposition, and stratospheric ozone depletion.

Air Quality Models

Models used in air pollution analysis fall into two classes: empirical–statistical and deterministic as shown in Figure 2. In the former, the model statistically relates observed air quality data to the accompanying emission patterns, and chemistry and meteorology are included only implicitly. In the latter, analytical or numerical expressions describe the complex transport and chemical processes that affect air pollutants. Pollutant concentrations are determined as explicit functions of meteorology, topography, chemical transformation, surface deposi-

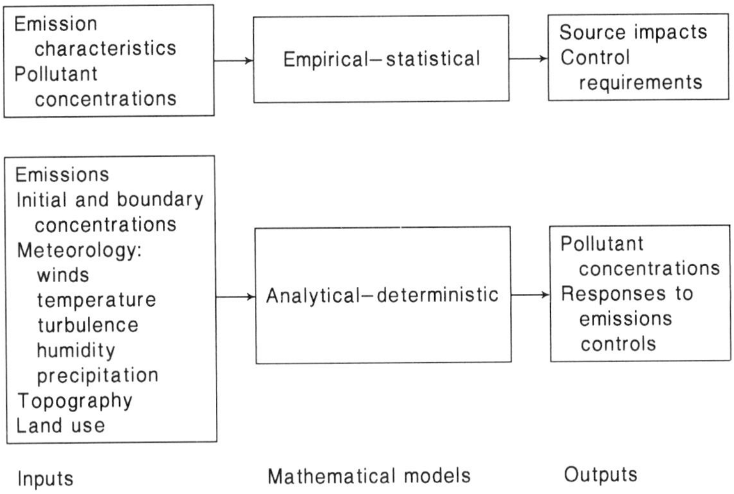

Fig. 2. Inputs, types of models, and outputs used in air quality modeling studies.

tion, and source characteristics. A typical schematic is shown in Figure 3. A listing of many of the air quality models, including status, applications, and the model formulations can be found in ref. 2. Each formulation involves approximations and has certain strengths and limitations.

Empirical–Statistical Models. Empirical–statistical models are based on establishing a relationship between historically observed air quality and the corresponding emissions. The linear rollback model is simple to use and requires few data, and for these reasons has been widely applied (3,4). Linear rollback models assume that the highest measured pollutant concentration is proportional to the basinwide emission rate, plus the background value; that is,

$$c_{\max} = aE + c_b \tag{1}$$

where $c_{\max}$ is the maximum measured pollutant concentration, E is the emission rate, c_b is the background concentration resulting from sources outside the modeling region, and a is a constant of proportionality. The constant implicitly accounts for the dispersion, transport, deposition, and chemical reactions of the pollutant. The allowable emission rate E_a necessary to reach a desired ambient air quality goal c_d can be calculated from

$$\frac{E_a}{E_o} = \frac{c_d - c_b}{c_{\max} - c_b} \tag{2}$$

where E_o is the emission rate that prevailed at the time that $c_{\max}$ was observed. Presumably, pollutant concentrations at other times decrease toward background levels as emissions are reduced. The linear rollback model is a very simplified approach and its application is limited. Nonlinear processes, such as chemical reactions and spatial or temporal changes in emission patterns, are not accounted for in the rollback model.

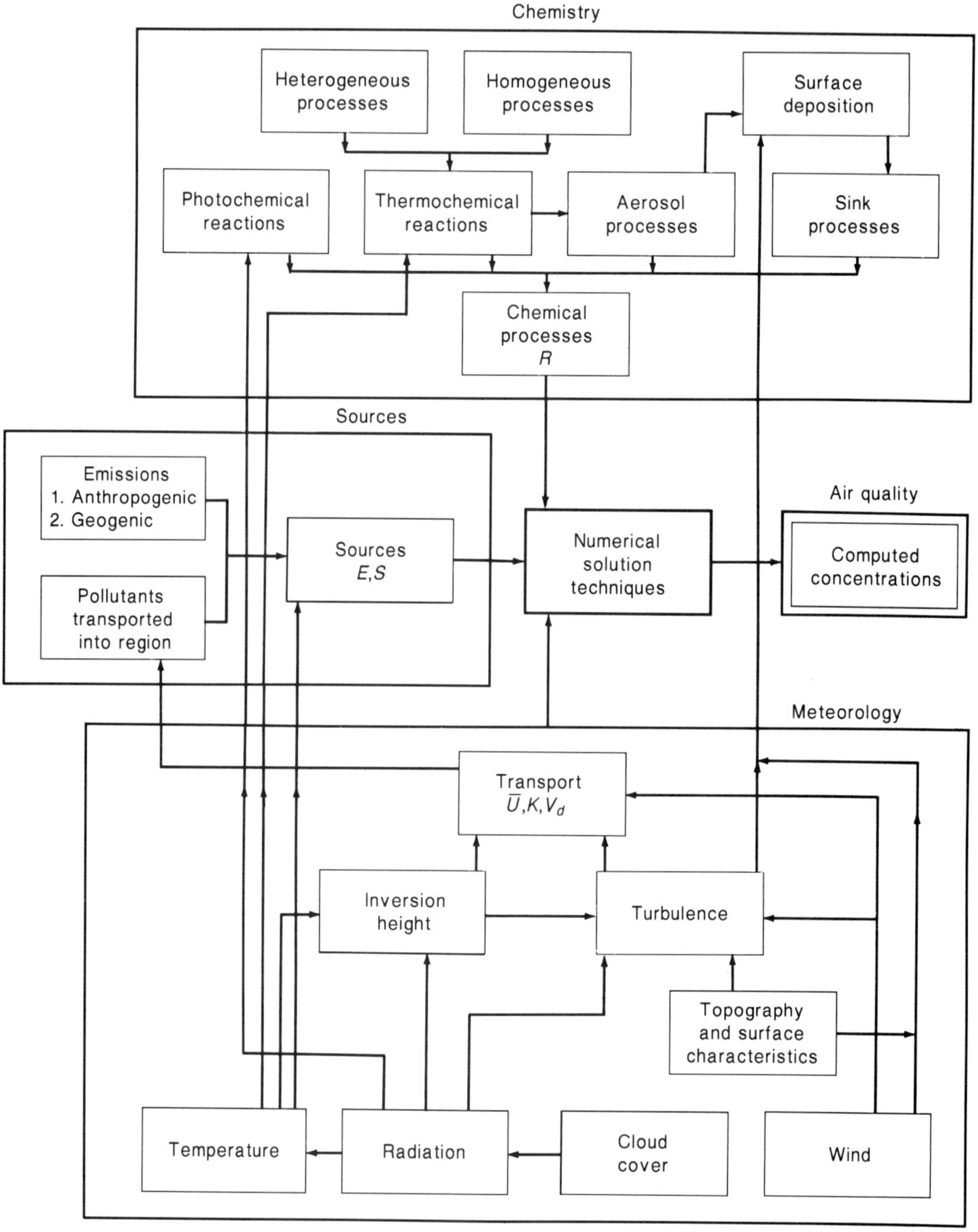

Fig. 3. Schematic diagram of a deterministic air quality model, showing the model components and interactions (1) where each of the boxes involves a large number of individual processes. Terms are defined in text.

A second class of empirical–statistical models is the receptor-oriented model, which has been used extensively for estimating the contributions that distinguishable sources such as automobiles or municipal incinerators make to particulate matter concentrations (5–14). Attempts have also been made to track nonreacting gases (15), and under special conditions, reactive organic compounds (16–18), to their sources using receptor modeling methods. Receptor models compare the measured chemical composition of particulate matter at a receptor site to the chemical composition of emissions from the primary sources to identify the source contributions at the monitoring location. There are three principal categories of receptor models: chemical mass balance, multivariate, and microscopic. Hybrid analytical and receptor (or combined source–receptor) models have been used (19), but further investigation into their capabilities is required.

Receptor models are powerful tools for source apportionment of particulates because a vast amount of particulate species characterization data have been collected at many sampling sites worldwide, and because many aerosol species are primary pollutants. Most of the information available is for elemental concentrations, eg, lead, nickel, and aluminum, although more recent measurements have provided data on concentrations of ionic species and carbonaceous compounds. At a sampling (or receptor) site, the aerosol mass concentration of each species i is

$$c_i = \sum_{j=1}^{n} a_{ij} S_j \qquad i = 1, 2, \ldots m \tag{3}$$

where c_i is the mass concentration of species i at the receptor site; S_j is the total mass concentration of all species at the receptor site that is attributable to source j; a_{ij} is the fraction that species i constitutes of the total mass concentration arriving at the sampling site from source j; m is the total number of species measured; and n is the total number of sources. The mass concentration at the receptor site and the coefficients a_{ij} that describe the chemical composition for the sources are the inputs from which S_j, the mass apportioned to source j, is determined. Because a_{ij} characterizes the source, it is referred to as the source fingerprint and should be unique to the source. When the chemical compositions of the emissions from two source categories are similar, it is extremely difficult for receptor models to distinguish between the sources. The categories of receptor models are differentiated by the techniques used to determine S_j.

If the source fingerprints, a_{ij}, for each of n sources are known and the number of sources is less than or equal to the number of measured species ($n \leq m$), an estimate for the solution to the system of equations (3) can be obtained. If $m > n$, then the set of equations is overdetermined, and least-squares or linear programming techniques are used to solve for S_j. This is the basis of the chemical mass balance (CMB) method (20,21). If each source emits a particular species unique to it, then a very simple tracer technique can be used (5). Examples of commonly used tracers are lead and bromine from mobile sources, nickel from fuel oil, and sodium from sea salt. The condition that each source have a unique tracer species is not often met in practice.

Microscopic identification models are similar to the CMB methods except that additional information is used to distinguish the source of the aerosol. Such chemical or morphological data include particle size and individual particle composition and are often obtained by electron or optical microscopy.

Multivariate models, including factor analysis models (14,22–24), rely on finding the underlying structure of large sets of air quality data in order to determine sources. Models based on factor analysis are the most widely used. Multivariate models operate by identifying groups of elements or species, the concentrations of which fluctuate together from sampling period to sampling period, implying that these groups come from a single "source". When the composition of the hypothetical source is compared to the known composition of specific sources, it often becomes obvious what the group of cofluctuating chemical elements represents. For example, lead and bromine concentrations are usually highly correlated because they are emitted primarily by the same sources, ie, automobiles burning leaded gasoline. Multivariate techniques do not rely on a detailed knowledge of the source fingerprint a_{ij} and can be used to refine estimates of the fingerprint.

Deterministic Models. Deterministic air quality models describe in a fundamental manner the individual processes that affect the evolution of pollutant concentrations. These models are based on solving the atmospheric diffusion–reaction equation, which is in essence the conservation-of-mass principle for each pollutant species (25):

$$\frac{\delta c_i}{\delta t} + \overline{U}\cdot\nabla c_i = \nabla\cdot D_i\nabla c_i + R_i(c_1, c_2, c_3, \ldots c_n) + S_i(\overline{x}, t) \quad i = 1, 2, 3, \ldots n \tag{4}$$

where c_i is the concentration of species i; $\overline{U}$ is the wind velocity vector; D_i is the molecular diffusivity of species i; R_i is the net production (depletion if negative) of species i by chemical reaction; S_i is the emission rate of i; and n is the number of species. R can also be a function of the meteorological variables. Equation 4 states that the time rate of change of a pollutant depends on convective transport (term 2), diffusion (term 3), chemical reactions (term 4), and emissions (term 5), together with initial and boundary conditions. Deposition and surface level emissions enter as boundary conditions at the ground:

$$E - v_d c = -K_{zz}\frac{\delta c}{\delta z}$$

where E is the ground level emissions, v_d is the deposition velocity, and K_{zz} is the vertical diffusivity. The turbulence closure problem makes it necessary to approximate the atmospheric diffusion equation, usually by K-theory (26,27):

$$\frac{\delta c_i}{\delta t} + \langle\overline{U}\rangle\cdot\nabla\langle c_i\rangle = \nabla\cdot K\nabla\langle c_i\rangle + R_i(\langle c_1\rangle, \langle c_2\rangle, \ldots \langle c_n\rangle) + \langle S_i(x, t)\rangle \quad i = 1, 2, 3, \ldots n \tag{5}$$

where the braces $\langle\ \rangle$ indicate an ensemble average, and K is the turbulent (eddy) diffusivity tensor. Known analytical solutions exist only for the simplest source distributions and chemical reaction mechanisms, $\langle S_i \rangle$ and R, in equation 5.

Examination of equation 5 shows that if there are no chemical reactions, ($R = 0$), or if R is linear in $\langle c_i \rangle$ and uncoupled, then a set of linear, uncoupled differential equations are formed for determining pollutant concentrations. This is the basis of transport models which may be transport only or transport with linear chemistry. Transport models are suitable for studying the effects of sources of CO and primary particulates on air quality, but not for studying reactive pollutants such as O_3, NO_2, HNO_3, and secondary organic species.

Lagrangian Models. There are two distinct reference frames from which to view pollutant dynamics. The most natural is the Eulerian coordinate system which is fixed at the earth's surface and in which a succession of different air parcels are viewed as being carried by the wind past a stationary observer. The second is the Lagrangian reference frame which moves with the flow of air, in effect maintaining the observer in contact with the same air parcel over extended periods of time. Because pollutants are carried by the wind, it is often convenient to follow pollutant evolution in a Lagrangian reference frame, and this perspective forms the basis of Lagrangian trajectory and Lagrangian marked-particle or particle-in-cell models. In a Lagrangian marked-particle model, the center of mass of parcels of emissions are followed, traveling at the local wind velocity, while diffusion about that center of mass is simulated by an additional random translation corresponding to the atmospheric diffusion rate (28,29).

Lagrangian trajectory models can be viewed as following a column of air as it is advected in the air basin at the local wind velocity. Simultaneously, the model describes the vertical diffusion of pollutants, deposition, and emissions into the air parcel as shown in Figure 4. The underlying equation being solved is a simplification of equation 5:

$$\frac{\delta c_i}{\delta t} = \frac{\delta}{\delta z} K_{zz} \frac{\delta c_i}{\delta z} + S_i(t) + R(\bar{c}, t) \qquad (6)$$

Trajectory models require spatially and temporally resolved wind fields, mixing-height fields, deposition parameters, and data on the spatial distribution of emissions. Lagrangian trajectory models assume that vertical wind shear and horizontal diffusion are negligible. Other limitations of trajectory and Eulerian models have been discussed (30).

Gaussian Plume Model. One of the most basic and widely used transport models based on equation 5 is the Gaussian plume model. Gaussian plume models for continuous sources can be obtained from statistical arguments or can be derived by solving:

$$\overline{U} \frac{\delta c}{\delta x} = K_{yy} \frac{\delta^2 c}{\delta y^2} + K_{zz} \frac{\delta^2 c}{\delta z^2} \qquad (7)$$

where $\overline{U}$ is the temporally and vertically averaged wind velocity; x, y, and z are the distances in the downwind, crosswind, and vertical directions, respectively;

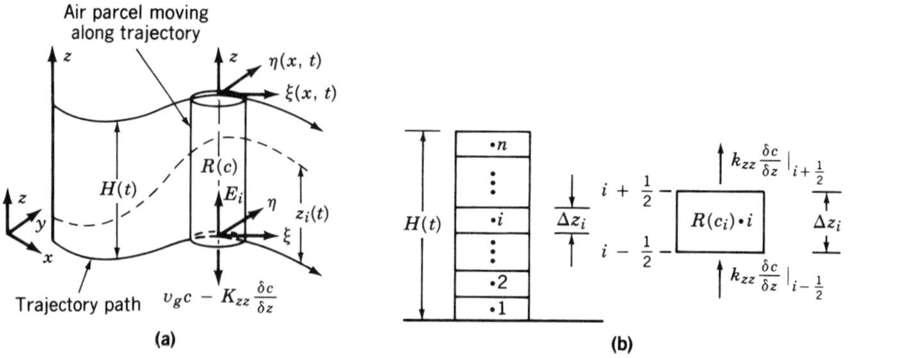

Fig. 4. Schematic diagram of a Lagrangian trajectory model where $H(t)$ represents the air column height in both Eulerian (x,y,z) and Lagrangian (ξ,η,z) coordinates; c, the pollutant concentration; R, chemical reactions; v_g, deposition of pollutants; and K_{zz}, vertical diffusion. **(a)** The column of air being modeled is advected at the local wind velocity along a trajectory path across the modeling region; $z_i(t)$ represents the mixing height variation along the trajectory path; E_i, the emissions of the ith pollutant. Within the moving air parcel, the model describes the important processes affecting the pollutant evolution and concentration. **(b)** Vertical resolution is gained by dividing the column into a number of cells in the vertical direction; $R(c_i)$ represents the concentration produced by chemical reactions within the ith cell of which Δz_i is the height.

and K_{yy} and K_{zz} are the horizontal and vertical turbulent diffusivities, respectively. For a source with an effective height H, emission rate Q, and a reflecting (nonabsorbing) boundary at the ground, the solution is

$$c(x,\,y,\,z) = \frac{Q}{2\pi\overline{U}\sigma_y(x)\sigma_z(x)} \exp\left[\frac{-y^2}{2\sigma_y^2(x)}\right]\left[\exp\frac{-(z-H)^2}{2\sigma_z^2(x)} + \exp\frac{-(z+H)^2}{2\sigma_z^2(x)}\right]$$

This solution describes a plume with a Gaussian distribution of pollutant concentrations, such as that in Figure 5, where $\sigma_y(x)$ and $\sigma_z(x)$ are the standard deviations of the mean concentration in the y and z directions. The standard deviations are the directional diffusion parameters, and are assumed to be related simply to the turbulent diffusivities, K_{yy} and K_{zz}. In practice, $\sigma_y(x)$ and $\sigma_z(x)$ are functions of x, $\overline{U}$, and atmospheric stability (2,31–33).

Gaussian plume models are easy to use and require relatively few input data. Multiple sources are treated by superimposing the calculated contributions of individual sources. It is possible to include the first-order chemical decay of pollutant species within the Gaussian plume framework. For chemically, meteorologically, or geographically complex situations, however, the Gaussian plume model fails to provide an acceptable solution.

Eulerian Models. Of the Eulerian models, the box model is the easiest to conceptualize. The atmosphere over the modeling region is envisioned as a well-mixed box, and the evolution of pollutants in the box is calculated following conservation-of-mass principles including emissions, deposition, chemical reactions, and atmospheric mixing.

Eulerian "grid" air quality models are the most complex, but potentially the most powerful, involving the least-restrictive assumptions. They are also the

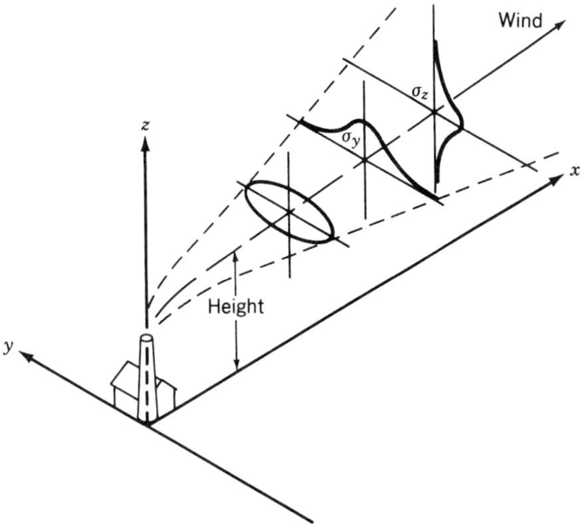

Fig. 5. Diffusion of pollutants from a point source. Pollutant concentrations have separate Gaussian distributions in both the horizontal (y) and vertical (z) directions. The spread is parameterized by the standard deviations (σ) which are related to the diffusivity (K).

most computationally intensive. Grid models solve a finite approximation to equation 5, including temporal and spatial variation of the meteorological parameters, emission sources, and surface characteristics. Grid models divide the modeling region into a large number of cells, horizontally and vertically, which interact with each other by simulating diffusion, advection, and sedimentation (for particles) of pollutant species. Input data requirements for grid models are similar to those for Lagrangian trajectory models, but, in addition, data on background concentrations (boundary conditions) at the edges of the grid system are required. Eulerian grid models predict pollutant concentrations throughout the entire airshed. Over successive time periods the evolution of pollutant concentrations and how they are affected by transport and chemical reaction can be tracked.

Modeling Chemically Reactive Compounds. A number of compounds are formed or destroyed in the atmosphere by a series of complex, nonlinear chemical reactions, eg, stratospheric ozone. Models that not only describe pollutant transport, but also account for complex chemical transformations, $R(\bar{c},t)$ in equation 5, are necessary for these systems. Such models are also required to study the dynamics of chemically reactive primary pollutants such as nitric oxide [*10102-43-9*], NO, and pollutants that are primary as well as secondary in origin, for example, nitrogen dioxide [*10102-44-0*], NO_2, and formaldehyde. Addition of the capability to describe a series of interconnected chemical reactions greatly increases requirements for computer storage as well as computing time and input data requirements. Increased computational demands arise because the evolution of the interacting species must be followed simultaneously, leading to a system of coupled, nonlinear differential equations.

Box, Lagrangian trajectory, and Eulerian grid models have all been developed to include nonlinear chemical reactions. Box models assume that the pollutants are mixed homogeneously within the modeling region, an assumption that is often inappropriate. Trajectory and grid models resolve pollutant dynamics spatially and have been used widely and with success particularly for studying photochemical smog and acid deposition problems (26,34–45).

Temporal and Spatial Resolution. The temporal and spatial resolutions of models can vary from minutes to a year and from meters to hundreds of kilometers. The minimum meaningful resolution of a model is determined by the input data resolution and the structure of the model. Statistical models generally rely on several years' worth of measurements of hourly or daily pollutant concentrations. The resolution of the input data represents the minimum resolution of a statistical model. Resolution of analytical models is limited by the spatial and temporal resolution of the emissions inventory, the meteorological fields, and the grid size chosen for model implementation. For modeling urban air basins, the size of individual grid cells is on the order of a few kilometers per side, whereas for modeling street canyons, the cell size must be reduced to a few meters on each edge. At the other extreme, regional models have horizontal resolutions varying from 20 to 100 km, and global models may have a resolution as coarse as a few thousand km. The temporal resolution of models ranges from about 15 minutes to a few hours or days.

The information desired from modeling studies often depends on processes that occur on spatial scales much smaller than the resolution of most air quality models. The modeling of NO_x air quality in street canyons involves small-scale processes of this sort. Introduction of point-source emissions into grid-based air quality models likewise involves a mismatch between the high concentrations that exist near the source versus the lower concentrations computed by a model that immediately mixes those emissions throughout a grid cell of several kilometers on each side. Because of computational time constraints, it has often been considered impractical to describe fully the processes that take place on a scale smaller than the main model grid, ie, subgrid scale. Nevertheless, values obtained from large-scale calculations should be accurate over the spatial averaging scale adopted by the model.

Model Components

A model's ability to correctly predict pollutant dynamics and to apportion source contributions depends on the accuracy of the individual process descriptions and input data, and the fidelity with which the framework reflects the interactions of the processes.

Turbulent Transport and Diffusion. There are two pollutant transport terms in equation 5: an advection term, in which pollutants are carried along with the time-averaged mean wind flow; and a dispersion term representing transport resulting from local turbulence. The averaging time that determines the mean winds is related to the spatial scale of the system being modeled. Minutes may be appropriate for urban-scale simulations, multihour averages for the regional scale, and daily to weekly averages for determining long-term concentrations of nonreactive pollutants.

Turbulent transport is determined by complex interactions between meteorological conditions and topography. In addition to gross topographical features, surface "roughness" scales have been devised to parameterize surface characteristics according to land-use categories. Very small values are assigned to smooth water or ice, and increasingly higher values to grasslands, croplands, residential areas, and urban–industrial centers. In general, the rougher the surface, the greater the local turbulence. Often, particularly during sunny days, atmospheric turbulence is generated by the heating of the earth's surface.

Turbulent fluxes are difficult to measure, and hence there are uncertainties in the various methods employed by models to describe them. As in equation 5, these fluxes are typically parameterized as being proportional to the gradient of mean pollutant concentrations. Further simplifying the problem, the components of the eddy diffusivity tensor, K, are assumed to be constant along their respective axes. An obvious need when applying K-theory is some algorithm for establishing the value of K. As a result of the large variety of processes involved, there are also a number of methods to parameterize the horizontal and vertical diffusion coefficients (46). The usual limitation to the accuracy of diffusion calculations in a practical application is determined by the extent of measurements of the atmospheric structure taken during the period to be simulated. For most model applications, the number of observed factors relating to atmospheric turbulence are few and include only ground-level winds and temperatures, surface roughness, and cloud cover. At a few locations and times, the inversion base, or mixing height, wind speeds aloft, and vertical temperature gradient may also be known. As the amount and accuracy of information characterizing atmospheric structure increases, confidence in model predictions of dispersion increases.

Removal Processes. Pollutant removal processes, particularly dry deposition and scavenging by rain and clouds, are a primary factor in determining the dynamics and ultimate fate of pollutants in the atmosphere.

Dry Deposition. Dry deposition occurs in two steps: the transport of pollutants to the earth's surface, and the physical and chemical interaction between the surface and the pollutant. The first is a fluid mechanical process (see FLUID MECHANICS), the second is primarily a chemical process, and neither is completely characterized at the present time. The problem is confounded by the interaction between the pollutants and biogenic surfaces where pollutant uptake is enhanced or retarded by plant activity that varies with time (47,48). It is very difficult to measure the depositional flux of pollutants from the atmosphere, though significant advances were made during the 1980s and early 1990s (49,50).

Many factors affect dry deposition, but for computational convenience air quality models resort to using a single quantity called the deposition velocity, designated v_d or v_g, to prescribe the deposition rate. The deposition velocity is defined such that the flux F_i of species i to the ground is

$$F_i = v_d c_i(z_r) \tag{9}$$

where $c_i(z_r)$ is the concentration of species i at some reference height z_r, typically from one to several meters. For a number of pollutants, v_d has been measured under various meteorological conditions and for a number of surface types (50).

Early models used a value for v_d that remained constant throughout the day. However, measurements show that the deposition velocity increases during the day as surface heating increases atmospheric turbulence and hence diffusion, and plant stomatal activity increases (50–52). More recent models take this variation of v_d into account. In one approach, the first step is to estimate the upper limit for v_d in terms of the transport processes alone. This value is then modified to account for surface interaction, because the earth's surface is not a perfect sink for all pollutants. This method has led to what is referred to as the resistance model (52,53) that represents v_d as the analogue of an electrical conductance

$$v_d = (r_a + r_b + r_s)^{-1} \tag{10}$$

where r_a is the aerodynamic resistance controlled primarily by atmospheric turbulence, r_b is the resistance to transport in the fluid sublayer very near the plant surface, and r_s is the surface (or canopy) resistance. Of the three resistances, r_a is essentially the same for all species, r_b is the same for gaseous species with the same diffusivities, though it can be considerably greater for aerosols, and r_s depends greatly on the surface affinity for the diffusing species. For example, nitric acid [*7697-37-2*], HNO_3, which reacts rapidly with most surfaces, has a very low surface resistance, usually taken as zero (49,54,55), whereas CO is not very reactive and has a high r_s value. Significant differences between the deposition of gases and aerosols are that aerosols have a much lower diffusivity, the rate of gravitational settling can be significant for larger particles, and the surface resistance for aerosols is not determined by species reactivity, alone. Recent models account for the variation of surface resistance and diurnal change in fluid mechanical transport. These parameterizations have been used to quantify the deposition flux of various compounds (49,56).

Scavenging by Rain, Fog, and Clouds. Wet removal, or precipitation scavenging, can be effective in cleansing the atmosphere of pollutants, and depends on the intensity and size of the raindrops (57). Fog and cloud droplets can also absorb gases, capture particles, and promote chemical reactions. Precipitation scavenging is not as important on an urban scale as on a regional scale and is not included in most urban-scale models. Fog chemistry can be important to human health on an urban scale, as evidenced in London in 1952 when thousands of persons died during an episode of excess industrial air pollution and fog (27). Cloud, rain, and fog processes are often important in regional and global scale modeling.

Representation of Atmospheric Chemistry Through Chemical Mechanisms. A complete description of atmospheric chemistry within an air quality model would require tracking the kinetics of many hundreds of compounds through thousands of chemical reactions. Fortunately, in modeling the dynamics of reactive compounds such as peroxyacetyl nitrate [*2278-22-0*] (PAN), $C_2H_3NO_5$, O_3, and NO_2, it is not necessary to follow every compound. Instead, a compact representation of the atmospheric chemistry is used. Chemical mechanisms represent a compromise between an exhaustive description of the chemistry and computational tractability. The level of chemical detail is balanced against computational time, which increases as the number of species and reactions increases. Instead of the hundreds of species present in the atmosphere, chemical mechanisms include on the order of 50 species and 100 reactions.

Three different types of chemical mechanisms have evolved as attempts to simplify organic atmospheric chemistry: surrogate (58,59), lumped (60–63), and carbon bond (64–66). These mechanisms were developed primarily to study the formation of O_3 and NO_2 in photochemical smog, but can be extended to compute the concentrations of other pollutants, such as those leading to acid deposition (40,42).

Surrogate mechanisms use the chemistry of one or two compounds in each class of organics to represent the chemistry of all the species in that class. For example, the explicit chemistry of butane [*106-97-8*], C_4H_{10}, might be used to describe the chemistry of the alkanes.

Lumped mechanisms are based on the grouping of chemical compounds into classes of similar structure and reactivity. For example, all alkanes might be lumped into a single class, the reaction rates and products of which are based on a weighted average of the properties of all the alkanes present. For example, as shown in Table 1, the various alkanes, C_nH_{2n+2}, react with OH in a similar manner to form alkyl radicals, $C_nH_{2n+1}\cdot$. When expressed explicitly, over 30 species and 20 reactions are involved. By lumping, the series of reactions can be reduced to one, and the number of required organic compounds is reduced to two. Thus lumping yields a tremendous savings in computational time, yet maintains the necessary chemical detail.

Table 1. Alkane–Hydroxide Radical Reactions

Explicit reactions

Reaction	
$C_2H_6 + OH \rightarrow C_2H_5^{\bullet} + H_2O$	initial reaction with OH
$C_3H_8 + OH \rightarrow C_3H_7^{\bullet} + H_2O$	
$\vdots$	
$C_nH_{2n+2} + OH \rightarrow C_nH_{2n+1}^{\bullet} + H_2O$	
$C_2H_5^{\bullet} + O_2 \rightarrow C_2H_5O_2^{\bullet}$	alkyl radical oxidation reaction
$C_3H_7^{\bullet} + O_2 \rightarrow C_3H_7O_2^{\bullet}$	
$\vdots$	
$C_nH_{2n+1}^{\bullet} + O_2 \rightarrow C_nH_{2n+1}O_2^{\bullet}$	

Lumped representation

$$\text{alkane} + OH \rightarrow RO_2^{\bullet}$$

The carbon bond mechanism (64–66), a variation of a lumped mechanism, splits each organic molecule into functional groups using the assumption that the reactivity of the molecule is dominated by the chemistry of each functional group.

Aerosol Dynamics. Inclusion of a description of aerosol dynamics within air quality models is of primary importance because of the health effects associated with fine particles in the atmosphere, visibility deterioration, and the acid deposition problem. Aerosol dynamics differ markedly from gaseous pollutant dynamics in that particles come in a continuous distribution of sizes and can coagulate, evaporate, grow in size by condensation, be formed by nucleation, or be deposited by sedimentation. Furthermore, the species mass concentration alone does not fully characterize the aerosol. The particle size distribution, which changes as a function of time, and size-dependent composition determine the fate of particulate air pollutants and their environmental and health effects. Particles of about 1 μm in diameter or smaller penetrate the lung most deeply and represent a substantial fraction of the total aerosol mass as shown in Figure 6. The origin of these fine particles is difficult to identify because much of the fine particle mass is formed by gas-phase reaction and condensation in the atmosphere.

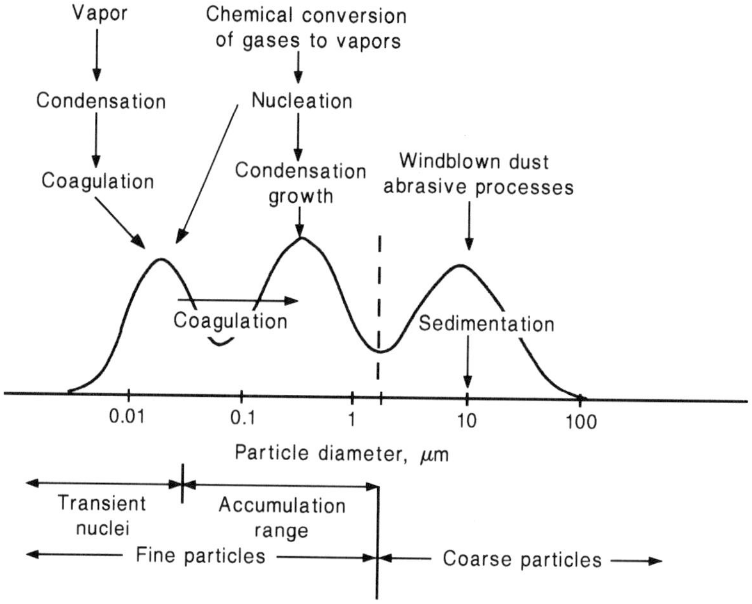

Fig. 6. Size distribution of an urban aerosol showing the three modes containing much of the aerosol mass. The fine mode contains particles produced by condensation of low volatility gases. The mid-range, or accumulation mode, results from coagulation of smaller aerosols and condensation of gases on preexisting particles. Coarse particulates, the largest aerosols, are usually generated mechanically.

Simulation of aerosol processes within an air quality model begins with the fundamental equation of aerosol dynamics which describes aerosol transport (term 2), growth (term 3), coagulation (terms 4 and 5), and sedimentation (term 6):

$$\frac{\delta n}{\delta t} + \nabla \cdot \overline{U} n + \frac{\delta I}{\delta v} = \frac{1}{2} \int_0^v \beta(\overline{v}, v - \overline{v}) n(\overline{v}) n(v - \overline{v}) d\overline{v} - \int_0^\infty \beta(\overline{v}, v) n(\overline{v}) n(v) d\overline{v} - \nabla \cdot C n \quad (11)$$

where n is the particle size distribution function; $\overline{U}$ is the fluid velocity; I is the droplet current that describes particle growth and nucleation resulting from gas-to-particle conversion; v is the particle volume; β is the rate of particle coagulation; and C is the sedimentation velocity. Modeling the formation and growth of aerosols is done by sectioning the size distribution n into discrete ranges. Then the size and chemical composition of an aerosol is followed as it evolves by condensation, coagulation, sedimentation, and nucleation.

Air Quality Model Inputs. Inputs to analytical air quality models can be broadly grouped as those dealing with meteorology, emissions, topography, and atmospheric concentrations. Meteorological inputs generally control the transport rate of pollutants and are used to determine reaction rates and the depositional flux of compounds. Topography influences transport and deposition. Observed compound concentrations are used to specify both initial and boundary conditions for model simulations. Especially for pollution problems involving organic compounds, emissions are a key input subject to considerable uncertainty. Although emissions from primary industrial facilities or utilities may be reasonably well known, emissions from residential or commercial facilities, mobile sources, and natural sources are often roughly estimated and difficult to verify.

The data requirements for applying models differ greatly among model types. For a Gaussian plume model, the required data could include as little as the mean wind velocity, source emission rate, atmospheric stability (and hence diffusivity), effective source height and air quality data for comparison with model predictions (67). At the other extreme, a large grid model that incorporates chemical kinetics requires considerably more information. Ideally a comprehensive model evaluation study would incorporate spatially and temporally resolved meteorological data, eg, winds, temperature, and solar insolation, temporally varying emissions for every species in each cell of the modeling region, topographical data, eg, land use and elevation, initial species concentrations, boundary conditions for each species, and concentration data for comparison against model predictions (44).

Meteorological Inputs. Meteorological conditions determine how fast and to where pollutants are dispersed, rates of chemical reactions, and losses resulting from deposition. Inputs vary depending on the type of model being applied. Simpler models, eg, Gaussian plume models with no chemical reactions, may require only a single wind velocity and cloud cover. On the other hand, a complex, three-dimensional, chemically active model requires hourly vertically and horizontally resolved wind fields, as well as hourly temperature, humidity, mixing depth, and solar insolation fields. The spatial and temporal resolution of the meteorological inputs must match the resolution of the air quality model. Much of the time the necessary inputs are determined directly from observations and relatively little mathematical modeling is used. In other cases, it is necessary to interpolate relatively sparse observations over a modeling domain using objective analysis techniques. Because of the sparseness of the data, newer air quality model applications have used dynamic, or prognostic, meteorological models. Use of dynamic models based on fundamental equations is intended to improve the accuracy of the meteorological inputs over the modeling domain.

Objective analysis techniques use a prescribed method with observations collected at a few discrete locations and times to calculate meteorological variables over the modeling domain. Usually this involves two- and three-dimensional spatial interpolation, and does not involve solving for any time dependency of the flow. Particular care must be taken in developing wind fields from sparse data because the wind field should be mass consistent. Objective analysis, including diagnostic procedures, is used to reduce the divergence of interpolated wind fields and account for some topographical features (68–71). A field of input values generated by interpolation over a large geographic area from sparse data is intrinsically uncertain and leads to uncertainty in model predictions. Upper-level variables such as temperature structure (mixing depths) and wind fields are particularly susceptible to this uncertainty, and have led to the wider use of dynamic models.

Dynamic meteorological models, much like air pollution models, strive to describe the physics and thermodynamics of atmospheric motions as accurately as is feasible. Besides being used in conjunction with air quality models, they are also used for weather forecasting. Like air quality models, dynamic meteorological models solve a set of partial differential equations (also called primitive equations). This set of equations, which are fundamental to the fluid mechanics of the atmosphere, are referred to as the Navier-Stokes equations, and describe the conservation of mass and momentum. They are combined with equations describing energy conservation and thermodynamics in a moving fluid (72):

mass
$$\frac{\partial \rho}{\partial t} + \nabla \cdot \rho \overline{U} = 0 \tag{12}$$

momentum
$$\frac{\partial \rho \overline{U}}{\partial t} + \overline{U} \cdot \nabla \rho \overline{U} = D \nabla^2 \overline{U} + \overline{F} - \nabla P \tag{13}$$

energy
$$\frac{\partial \rho e}{\partial t} + \nabla \cdot (\rho e \overline{U}) = \dot{Q} - \dot{W} \tag{14}$$

thermodynamics
$$\frac{p}{\rho} = RT \tag{15}$$

where ρ is the local density of the atmosphere, $\overline{U}$ is the wind velocity vector, D is the molecular diffusivity, F represents external forces such as gravity, p is pressure, e is the local internal energy of the atmosphere, and $\dot{Q}$ is the heat flux in and $\dot{W}$ is the work done by, the fluid, friction. The last equation is the Ideal Gas Law. Often it is also necessary to follow the transport of water, eg, for predicting clouds,

$$\frac{\partial q}{\partial t} + \overline{U} \cdot \nabla q = S \tag{16}$$

where q is the water concentration and S is the source or sinks of water, including rainout, etc. These equations form a set of coupled, nonlinear partial differential equations that are as formidable as those describing the chemical and physical dynamics of trace pollutants.

Experience in the solution of these governing equations is extensive, primarily because they are used in weather forecasting. This experience has led to useful simplifications and to the realization that the system is very sensitive to initial conditions. The sensitivity to initial conditions leads to the temporal amplification of any errors in the initial or boundary conditions, or of computational errors. The growth of these errors seriously limits the period of time that can be simulated for use in air pollution studies unless a separate mechanism is used to dampen error growth. This same limitation makes longer range weather forecasts increasingly uncertain. A method that has been developed to reduce the growth of errors arising from initial and boundary conditions when reconstructing historical conditions is to use observations to adjust the solution back towards the actual. This is referred to as "Newtonian nudging" or four-dimensional data assimilation (FDDA) (73). In essence, the predicted solution from the dynamic model is averaged, using various weightings, and the meteorological fields developed by objective analysis.

The equations governing air motion are generally assumed to be independent of those describing the chemical pollutant dynamics. This is because for problems such as urban smog and acid deposition, pollutant concentrations are so low that they do not significantly impact radiative transfer, and hence future weather. This may not be the case for problems such as stratospheric ozone depletion, and is not true for global climate change. In the latter case, the chemical evolution of the greenhouse gases, eg, carbon dioxide [*124-38-9*], CO_2, and chlorofluorocarbons (CFC), is currently assumed to be slow, and the pollutant distribution is found using little or no description of the chemistry. The radiative forcing is then found using expected concentrations.

Mathematical and Computational Implementation. Solution of the complex systems of partial differential equations governing both the evolution of pollutant concentrations and meteorological variables, eg, winds, requires specialized mathematical techniques. Comparing the two sets of equations governing pollutant dynamics (eq. 5) and meteorology (eqs. 12–14) shows that in both cases they can be put in the form:

$$\underbrace{\frac{\partial A}{\partial t}}_{\text{response}} + \underbrace{\overline{U}\cdot\nabla A = \nabla\cdot K\nabla A}_{\text{convective terms}} + \underbrace{S}_{\text{source}} \tag{17}$$

where A is the variable being modeled, eg, pollutant concentrations, wind velocities, temperature, and water vapor, $\overline{U}$ is the wind velocity vector, K is the turbulent diffusivity, and S is a generalized source or sink, eg, chemical production–destruction and emissions for modeling pollutant concentrations; pressure gradients, gravity, and coriolis forces for momentum conservation; thermal heating for energy conservation; and evaporation for water vapor modeling. Equation 17 is the general transport equation, which can be complex and nonlinear for most atmospheric systems. The nonlinearity and coupling arise as a result of the second and fourth terms. Because the systems of equations used in pollutant dynamics and meteorological modeling have the same general form, similar mathematical techniques are used to solve them.

The generalized transport equation, equation 17, can be dissected into terms describing bulk flow (term 2), turbulent diffusion (term 3) and other processes, eg, sources or chemical reactions (term 4), each having an impact on the time evolution of the transported property. In many systems, such as urban smog, the processes have very different time scales and can be viewed as being relatively independent over a short time period, allowing the equation to be "split" into separate operators. This greatly shortens solution times (74). The solution sequence is

$$\left.\frac{\partial A}{\partial t}\right|_H = -U{\cdot}\nabla A + \nabla{\cdot}K\nabla A = L_H(A) \tag{18}$$

$$\left.\frac{\partial A}{\partial t}\right|_V = -W\frac{\partial A}{\partial z} + \frac{\partial}{\partial z}K_z\frac{\partial A}{\partial z} = L_V(A) \tag{19}$$

$$\left.\frac{\partial A}{\partial t}\right|_S = S = L_S(A) \tag{20}$$

$$\left.\frac{\partial A}{\partial t}\right|_{Total} = (L_H + L_V + L_S)(A) \tag{21}$$

where L_H is the horizontal transport operator, L_V is the vertical transport operator, and L_S is the operator describing other processes, such as chemistry. Modifications of the splitting process can be used to improve accuracy and computational speed. Besides leading to smaller systems of coupled equations, splitting also allows use of solution techniques that are designed to effectively describe specific processes. For example, in a photochemical air quality model, one routine is used for horizontal transport resulting from bulk winds, another for vertical motions, which are generally diffusive processes, and a third for the chemistry.

Historically, numerical schemes used to calculate the rate of transport have been based on finite difference and, more recently, finite element techniques. Spectral methods have also shown promise, but nonlinear chemical effects tend to degrade solution accuracy in some tests (75,76). A problem with solving the set of equations is that the spatial discretization of the modeling region leads to artificial numerical dispersion, which is manifested by the formation of spurious waves and by pollutant peaks being spread out. For example, Figure 7 shows how various proposed schemes perform when following the rotation of a cone-shaped pollutant puff. In one instance, the original cone is virtually lost as a result of numerical errors. This remains a classic problem in computational fluid mechanics.

Tracking the chemical dynamics, eg, R in the atmospheric diffusion equation (eq. 5) and S (in eq. 20), is particularly difficult and time-consuming. This is because of the wide range of compound lifetimes, or more specifically, the characteristic reaction times. Some species, such as the O atom, have atmospheric lifetimes on the order of microseconds whereas others last for weeks. It is impractical to use a standard technique, eg, Runga Kutta, having a time-step corresponding to the shortest-lived species. This would be both excessively time-consuming and lead to the accumulation of numerical errors, making the computed solution invalid. Gear methods can be used, but are computationally slow. Instead, the chemical nature of the process is used to motivate a solution proce-

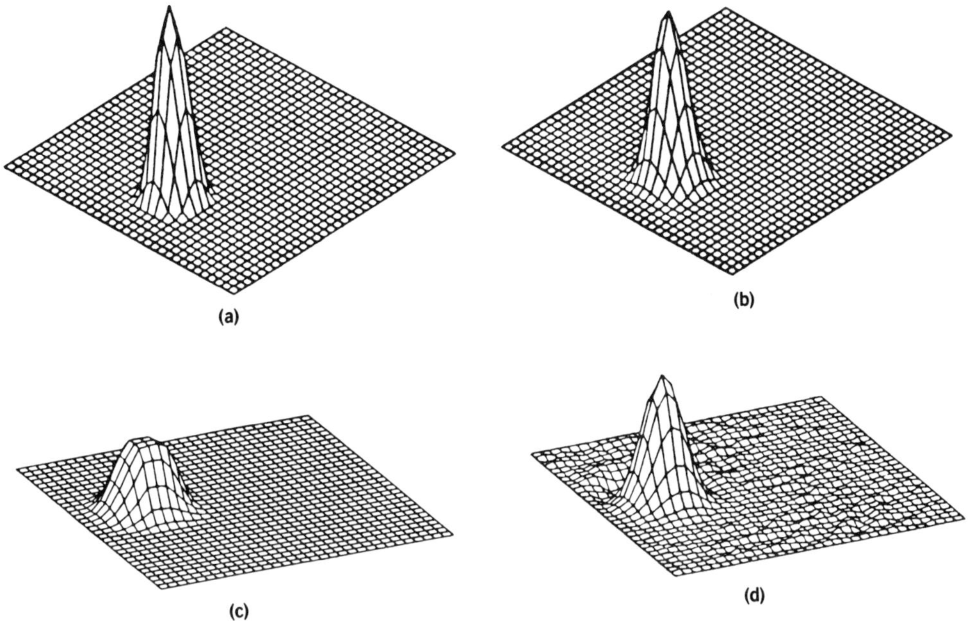

Fig. 7. Comparison of various transport schemes for advecting a cone-shaped puff in a rotating windfield after one complete rotation: (**a**), the exact solution; (**b**), obtained by an accurate numerical technique; (**c**), the effect of numerical diffusion where the peak height of the cone has been severely truncated; and (**d**), where the predicted concentration field is very bumpy, showing the effects of artificial dispersion. In the case of (**d**), spurious waves are formed that impair the predictions, though the peak height is better maintained.

dure. For a compound experiencing chemical production and destruction, the species concentration can be found using:

$$\frac{\partial c_i}{\partial t} = S_i = A_i - B_i c_i$$

where A_i is the production of species i from chemical reactivity and direct emissions, and $B_i c_i$ is the loss of i from chemical reactivity and deposition. In general, A_i and B_i are time varying and depend on concentrations of other compounds. As it turns out, even though species i might react very rapidly, A_i and B_i are rather constant over periods of a minute or so. Thus an approximate analytical solution to the above equation is used. Such techniques have been shown to be very accurate and an order of magnitude faster than others (74,77). However, solution of the chemical kinetics is still the most computationally intensive part of a chemically active air quality model, consuming about 85% of the computer time.

Historically, the computational intensity of the more complex chemically active models have limited their application. For example, modeling only the gas-phase dynamics over an area like Los Angeles requires solving about 500,000 simultaneous, nonlinear equations. Until the late 1980s, computational power severely limited the use of chemically active models, and particularly inhibited the development and use of regional oxidant and acid deposition models. The rapid increase in computational power is making it possible to address much larger problems, for example, regional and global scale pollution. One aspect of air quality models is that they are highly parallelizable. The development of parallel computers should allow for significantly more detailed modeling of large systems.

Application of Air Quality Models

Both receptor and analytical air quality models have proven to be powerful tools for understanding atmospheric pollutant dynamics and for determining the impact of sources on air quality.

Receptor Models. Receptor models, by their formulation, are effective in determining the contributions of various sources to particulate matter concentrations. In classic studies, sources contributing to airborne particle loadings have been identified in Washington, D.C. (78), St. Louis (9,24), Los Angeles (7,12), Portland, Oregon (78), and Boston (79–81), as well as other areas including the desert (82).

A receptor model (78) and a source-oriented deterministic model were combined as part of a particulate air quality control strategy analysis in Portland, Oregon. Using CMB techniques, source contributions to the ambient aerosol were identified and then dispersion modeling was used to confirm the source contributions. The results obtained with the two models were compared, and a revised particulate emissions inventory was input into the source–dispersion model. Finally, the revised emissions inventory was used in dispersion modeling of emission control strategy alternatives. This approach utilized the strengths of both types of models. Receptor models are suitable for predicting the outcome of perturbations in some sources but not others. They are, however, good for determining the sources of particulate matter when an accurate emissions inventory is not available. Disperson models, on the other hand, are well-suited for modeling the impact of a wide variety of emissions changes that would result from changed emission control regulations, but rely totally on an input emissions inventory, which may be uncertain or difficult to obtain.

Analytical Modeling Studies. Analytical air quality models have been used the most in modeling the dynamics of pollutants at local and urban scales. Nonreactive, mass conservation models based on solving equation 3, including Gaussian plume models, have been used extensively for source apportionment, control strategy analysis, and source impact modeling of nonreactive pollutants, such as CO, and of carbonaceous aerosol. Models of this type have also been used to study the sources that contribute to secondary sulfate aerosol formation and they have been used to follow releases of toxic gases in local spills. Variants of the Gaussian plume model are often employed for regulatory purposes. The modest input requirements of these models are the key advantage. Frequently the goal is to estimate the maximum likely impact, so "worst case" meteorological conditions, ie, conditions that cause local pollutant accumulation, are modeled along with the maximum emissions rate that is allowed at the facility.

Of great interest is the use of chemically active air quality models for studying and controlling urban air pollution. This is because of the high costs of controls, the complexity of the system, and the historic lack of success in reducing ozone levels in urban areas. Interest in extending the application of chemically active models to the regional scale has heightened because the ozone problem has been recognized as extending well beyond urban areas.

The lack of success in reducing ozone levels can be partially ascribed to the nonlinearity of the system. An incremental change in the emissions of precursors to a secondary pollutant such as ozone need not lead to a proportional change in the pollutant concentration, or any change at all. For example, both NO_x and organics are precursors to the formation of O_3, but increasing the emissions of one can have a very different result from increasing those of the other. In fact, decreasing NO_x emissions may increase O_3 concentrations nearby while at the same time decreasing O_3 concentrations downwind. On the other hand, reducing emissions of reactive organic gases (ROGs) can lead to significant decreases in ozone in some cities, such as New York, but have little effect in other cities and in rural areas. A common representation of the relationship between maximum O_3 concentrations and initial concentrations of NO_x and ROGs is an O_3 isopleth diagram as shown in Figure 8. Regions of both high and low sensitivities to NO_x and ROG emissions are shown. Analysis of the effect of emission controls on O_3 air quality is further complicated by the fact that the effect of controlling two emission sources together is not necessarily equal to the sum of the incremental improvements from controlling each source alone. Thus definitively assessing the impact that a single source has on air quality is clouded in that its impact

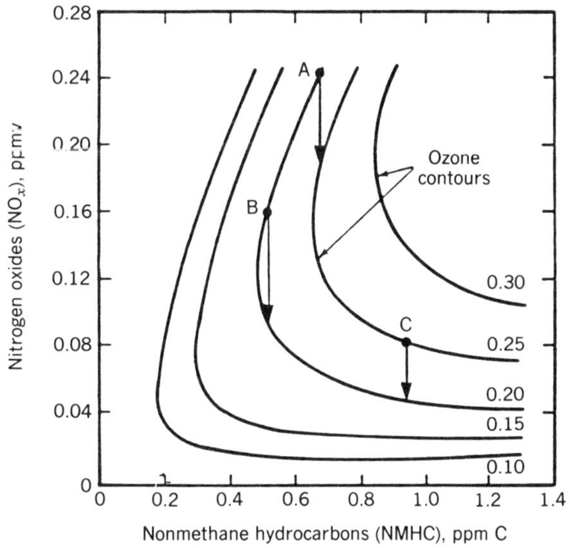

Fig. 8. O_3 isopleth diagram showing the response of O_3 concentrations to changes in initial NO_x and nonmethane hydrocarbon concentrations expressed as parts per million of carbon atoms (ppm C). The response to NO_x reductions is dependent on the particular initial concentrations. At point A, the graph indicates that decreasing NO_x would increase O_3 formation, and at point C, a decrease in the NO_x concentration results in a much larger decrease in O_3 concentrations than a similar decrease at point B on the same isopleth.

responds dynamically to changes in other sources and to varying meteorological conditions.

Because of the severity of the smog problem in Los Angeles, this location is the most studied area with regard to the role of NO_x and ROG emissions in the formation of ozone and other photochemical pollutants. The effect of NO_x and ROG emissions on ozone formation was found to vary throughout the region (83). Near downtown Los Angeles, ozone formation is inhibited by increasing NO_x emissions, and is effectively reduced by lowering ROG emissions. Downwind, however, where the highest ozone levels are found, ozone is relatively insensitive to ROG emissions, and is effectively reduced by lowering NO_x emissions. Modeling studies have also shown that concentrations of associated pollutants, such as nitric acid, aerosol nitrate, NO_2, and peroxyacetyl nitrate (PAN), are lowered when NO_x emissions are controlled (45). Another application of chemically active air quality models has been to study the effectiveness of using alternative automotive fuels, such as methanol (see ALCOHOL FUELS), compressed natural gas (see GAS, NATURAL), and electric power (see BATTERIES), for reducing smog (84).

Regional oxidant and acid deposition models came into use later than urban photochemical models because of the increased computational intensity, the need to describe more physical and chemical processes, and the later regulatory mandate for development and use. Regional models are very similar to urban-scale photochemical models, differing primarily in their horizontal resolution, 20–100 km vs 4–5 km, and in their treatment of cloud processes and liquid-phase chemistry. Applications of regional models also cover longer periods because of the increased residence time of polluted air masses within their domain. For example, applications have followed the evolution of regional pollution episodes of over a week in the northeastern United States, compared to local episodes of two or three days in Los Angeles.

Regional ozone modeling of the Northeast has shown that air pollution problems across cities within this region are linked (85). Air masses starting in Washington, D.C. can travel up the coast, impacting downwind cities such as Baltimore and New York. Emissions in New York further impact Connecticut and Massachusetts. These studies have shown that the type of controls that are most effective in one city may be counterproductive in another. For example, modeling studies have found NO_x control to be effective for controlling ozone in Boston, Washington D.C., and Philadelphia, but lead to increased concentrations in New York (85). The horizontal resolution of the model used for this study was about 20 km, giving some detail in urban areas but not as much as urban-scale models provide.

A variety of models have been developed to study acid deposition. Sulfuric acid is formed relatively slowly in the atmosphere, so its concentrations are believed to be more uniform than ozone, especially in and around cities. Also, the impacts are viewed as more regional in nature. This allows an even coarser horizontal resolution, on the order of 80 to 100 km, to be used in acid deposition models. Atmospheric models of acid deposition have been used to determine where reductions in sulfur dioxide emissions would be most effective. Many of the ecosystems that are most sensitive to damage from acid deposition are located in the northeastern United States and southeastern Canada. Early acid deposition models helped to establish that sulfuric acid and its precursors are transported over long distances, eg, from the Ohio River Valley to New England (86–88). Models have also been used to show that sulfuric acid deposition is nearly linear in response to changing levels of emissions of sulfur dioxide (89).

In the mid-1980s, the destruction of the ozone layer above the Antarctic was recognized to be a potential environmental disaster, and chemically active transport models were used to identify the most important processes leading to depleted stratospheric ozone levels. As with ground-level chemistry, the gas-phase chemistry that occurs in the stratosphere is relatively well understood, but the incorporation of heterogeneous chemistry is an ongoing challenge. In addition, the models that have incorporated stratospheric chemistry in the greatest detail have oversimplified the transport and dynamics of stratospheric circulations. Conversely, three-dimensional models that focus on the dynamics of the atmosphere have used limited treatments of stratospheric chemistry.

Zero- through three-dimensional models have been used to simulate the stratospheric ozone problem. Zero-dimensional models focus on radiative transfer and chemistry at a single point. One-dimensional models incorporate vertical diffusive transport as well. Two-dimensional models currently represent the best compromise between computational tractability and chemical detail (90), and are resolved by latitude, as well as in the vertical dimension. Stronger flows and greater homogeneity over changes in longitude make two-dimensional treatments reasonable, except in the polar regions. Three-dimensional models are needed to study the dynamics of polar regions, including exchanges with air at lower latitudes.

Meridional circulation in two-dimensional stratospheric models has been specified based on observations or general circulation model calculations; recently efforts have been undertaken to calculate circulations from first principles, within the stratospheric models themselves. An important limitation of using models in which circulations are specified is that these cannot be used to study the feedbacks of changing atmospheric composition and temperature on transport, factors which may be important as atmospheric composition is increasingly perturbed.

The key gas-phase reactions occurring in the stratosphere are generally known. Comprehensive reviews of kinetic data have led to general consensus on the rate parameters that should be used in stratospheric models (91). Nevertheless, discrepancies are still apparent when the chemical components of different stratospheric models are compared. Disagreements arise from differing estimates of photolysis conditions in the stratosphere, depending in part on whether or not light scattering in the troposphere is included in the model (91).

Heterogeneous chemistry occuring on polar stratospheric cloud particles of ice and nitric acid trihydrate has been established as a dominant factor in the aggravated seasonal depletion of ozone observed to occur over Antarctica. Preliminary attempts have been made to parameterize this chemistry and incorporate it in models to study ozone depletion over the poles (91) as well as the potential role of sulfate particles throughout the stratosphere (92).

Models can be used to study human exposure to air pollutants and to identify cost-effective control strategies. In many instances, the primary limitation on the accuracy of model results is not the model formulation, but the accuracy of the available input data (93). Another limitation is the inability of models to account for the alterations in the spatial distribution of emissions that occurs when controls are applied. The more detailed models are currently able to describe the dynamics of unreactive pollutants in urban areas.

Because of the expanded scale and need to describe additional physical and chemical processes, the development of acid deposition and regional oxidant models has lagged behind that of urban-scale photochemical models. An additional step up in scale and complexity, the development of analytical models of pollutant dynamics in the stratosphere is also behind that of ground-level oxidant models, in part because of the central role of heterogeneous chemistry in the stratospheric ozone depletion problem. In general, atmospheric liquid-phase chemistry and especially heterogeneous chemistry are less well understood than gas-phase reactions such as those that dominate the formation of ozone in urban areas. Development of three-dimensional models that treat both the dynamics and chemistry of the stratosphere in detail is an ongoing research problem.

BIBLIOGRAPHY

1. K. L. Demerjian, "Photochemical Diffusion Models for Air Quality Simulation: Current Status," in *Proceedings of the Conference on the State of the Art of Assessing Transportation-Related Air Quality Impacts,* Oct. 22–24, 1975, Transportation Research Board, Washington, D.C., 1975.
2. P. Zanetti, *Air Pollution Modeling,* Van Nostrand Reinhold, New York, 1990.
3. D. S. Barth, *J. Air Pollut. Control Assoc.* **20,** 519–523 (1970).
4. *Final Air Quality Management Plan,* 1982 rev., South Coast Air Quality Management District and Southern California Association of Governments, El Monte, Calif., 1982.
5. S. K. Friedlander, *Environ. Sci. Technol.* **7,** 235–240 (1973).
6. S. L. Heisler, S. K. Friedlander, and R. B. Husar, *Atmos. Environ.* **7,** 633–649 (1973).
7. G. Gartrell and S. K. Friedlander, *Atmos. Environ.* **9,** 279–299 (1975).
8. D. F. Gatz, *Atmos. Environ.* **9,** 1–18 (1975).
9. D. F. Gatz, *J. Appl. Meteorol.* **17,** 600–608 (1978).
10. G. E. Gordon, *Environ. Sci. Technol.* **14,** 792–800 (1980).
11. J. G. Watson, R. C. Henry, J. A. Cooper, and E. S. Macias, in E. S. Macias and P. K. Hopke, eds., *Atmospheric Aerosol: Source/Air Quality Relationships,* Symp. Ser. No. 167, American Chemical Society, Washington, D.C., 1981, pp. 89–106.
12. G. R. Cass and G. J. McRae, *Environ. Sci. Technol.* **17,** 129–139 (1983).
13. J. G. Watson, *J. Air Pollut. Control Assoc.* **34,** 619–623 (1984).
14. P. K. Hopke, *Receptor Modeling in Environmental Chemistry,* John Wiley & Sons, Inc., New York, 1985.
15. R. J. Yamartino, in S. C. Dattner and P. K. Hopke, eds., *Receptor Models Applied to Contemporary Pollution Problems,* Air Pollution Control Association, Pittsburgh, Pa., 1983, pp. 285–295.
16. P. F. Nelson, S. M. Quigley, and M. Y. Smith, *Atmos. Environ.* **17,** 439–449 (1983).
17. W. J. O'Shea and P. A. Scheff, *JAPCA* **38,** 1020–1026 (1988).
18. P. F. Aronian, P. A. Scheff, and R. A. Wadden, *Atmos. Environ.* **23,** 911–920 (1989).
19. J. E. Core, P. L. Hanrahan, and J. A. Cooper, in Ref. 11, pp. 107–123.
20. M. S. Miller, S. K. Friedlander, and G. M. Hidy, *J. Colloid Interface Sci.* **39,** 165–176 (1972).
21. J. A. Cooper and J. G. Watson, *J. Air Pollut. Control Assoc.* **30,** 1116–1125 (1980).
22. S. K. Friedlander, *Smoke, Dust and Haze,* Wiley-Interscience, New York, 1977.

23. R. C. Henry and G. M. Hidy, *Atmos. Environ.* **16,** 929–943 (1982).
24. P. K. Hopke, in Ref. 11, pp. 21–49.
25. R. G. Lamb and J. H. Seinfeld, *Environ. Sci. Technol.* **7,** 253–261 (1973).
26. R. G. Lamb, *Atmos. Environ.* **7,** 257–263 (1973).
27. J. H. Seinfeld, *Atmospheric Chemistry and Physics of Air Pollution,* Wiley-Interscience, New York, 1986.
28. R. G. Lamb and M. Neiburger, *Atmos. Environ.* **5,** 239–264 (1971).
29. G. R. Cass, *Atmos. Environ.* **15,** 1227–1249 (1981).
30. M.-K. Liu and J. H. Seinfeld, *Atmos. Environ.* **9,** 555–574 (1975).
31. F. A. Gifford, *Nucl. Saf.* **2,** 47–57 (1961).
32. D. B. Turner, *J. Appl. Meteorol.* **3,** 83–91 (1964).
33. D. B. Turner, *A Workbook of Atmospheric Dispersion Estimates,* Public Health Service Publication No. 999-AP-26, U.S. Government Printing Office, Washington, D.C., 1967.
34. S. D. Reynolds, P. W. Roth, and J. H. Seinfeld, *Atmos. Environ.* **7,** 1033–1061 (1973).
35. A. C. Lloyd and co-workers, *Development of the ELSTAR Photochemical Air Quality Simulation Model and Its Evaluation Relative to the LARPP Data Base,* Environmental Research and Technology Report, No. P-5287–500, West Lake Village, Calif., 1979.
36. S. D. Reynolds, T. W. Tesche, and L. E. Reid, *An Introduction to the SAI Airshed Model and Its Usage,* Report No. SAI-EF79–31, Systems Applications, Inc., San Rafael, Calif., 1979.
37. J. H. Seinfeld, *J. Air Waste Management* **38,** 616–645 (1988).
38. D. P. Chock, A. M. Dunker, S. Kumar, and C. S. Sloane, *Environ. Sci. Technol.* **15,** 933–939 (1981).
39. G. R. Carmichael, T. Kitada, and L. K. Peters, *Atmos. Environ.* **20,** 173–188 (1986).
40. G. R. Carmichael, L. K. Peters, and R. D. Saylor, *Atmos. Environ.* **25A,** 2077–2090 (1991).
41. A. G. Russell and G. R. Cass, *Atmos. Environ.* **20,** 2011–2025 (1986).
42. J. G. Chang and co-workers, *J. Geophys. Res.* **92,** 14,681–14,700 (1987).
43. S. A. McKeen and co-workers, *J. Geophys. Res.* **96,** 10,809–10,845 (1991).
44. G. J. McRae and J. H. Seinfeld, *Atmos. Environ.* **17,** 501–522 (1983).
45. A. G. Russell, K. McCue, and G. R. Cass, *Environ. Sci. Technol.* **22,** 1336–1347 (1988).
46. T. W. Yu, *Mon. Weather Rev.* **105,** 57–66 (1977).
47. B. B. Hicks and M. L. Wesely, *Boundary-Layer Meteorol.* **20,** 175–185 (1981).
48. B. B. Hicks and co-workers, in H. R. Pruppacher, R. G. Semonin, and W. G. N. Slinn, eds., *Precipitation Scavenging, Dry Deposition and Resuspension,* Elsevier, New York, 1983, pp. 933–942.
49. C. J. Walcek and co-workers, *Atmos. Environ.* **20,** 949–964 (1986).
50. M. L. Wesely, D. R. Cook, R. L. Hart, and R. E. Speer, *J. Geophys. Res.* **90,** 2131–2143 (1985).
51. D. M. Whelpdale and R. W. Shaw, *Tellus* **26,** 196–205 (1974).
52. M. L. Wesely and B. B. Hicks, *J. Air Pollut. Control Assoc.* **27,** 1110–1116 (1977).
53. D. Fowler, *Atmos. Environ.* **12,** 369–373 (1978).
54. B. J. Huebert, in Ref. 48, pp. 785–794.
55. B. J. Huebert and C. H. Robert, *J. Geophys. Res.* **90,** 2085–2090 (1985).
56. G. J. McRae and A. G. Russell, in B. B. Hicks, ed., *Deposition Both Wet and Dry,* Acid Precipitation Series, Vol. 4, Butterworth, Boston, 1983, pp. 153–193.
57. A. Martin, *Atmos. Environ.* **18,** 1955–1961 (1984).
58. T. E. Graedel, L. A. Farrow, and T. A. Weber, *Atmos. Environ.* **10,** 1095–1116 (1976).
59. M. C. Dodge, *Combined Use of Modeling Techniques and Smog Chamber Data to Derive Ozone-Precursor Relationships,* Report No. EPA-600/3-77-001a, U.S. Environmental Protection Agency, Research Triangle Park, N.C., 1977.
60. A. H. Falls and J. H. Seinfeld, *Environ. Sci. Technol.* **12,** 1398–1406 (1978).

61. R. Atkinson, A. C. Lloyd, and L. Winges, *Atmos. Environ.* **16,** 1341–1355 (1982).
62. W. R. Stockwell, *Atmos. Environ.* **20,** 1615–1632 (1986).
63. F. W. Lurmann, W. P. L. Carter, and L. A. Coyner, *A Surrogate Species Chemical Reaction Mechanism for Urban-Scale Air Quality Simulation Models,* Report No. EPA/600/3-87-014, U.S. Environmental Protection Agency, Research Triangle Park, N.C., 1987.
64. J. P. Killus and G. Z. Whitten, *A New Carbon Bond Mechanism for Air Quality Modeling,* Report No. EPA 60013-82-041, U.S. Environmental Protection Agency, Research Triangle Park, N.C., 1982.
65. G. Z. Whitten and co-workers, *Modeling of Simulated Photochemical Smog with Kinetic Mechanism,* Vols. I and II, Report No. EPA-600/3-79-001a, U.S. Environmental Protection Agency, Research Triangle Park, N.C., 1979.
66. M. W. Gery, G. Z. Whitten, J. P. Killus, and M. C. Dodge, *J. Geophys. Res.* **94,** 12,925–12,956 (1989).
67. E. Weber, *Air Pollution, Assessment Methodology and Modeling,* Plenum Press, New York, 1982.
68. M. H. Dickerson, *J. Appl. Meteorol.* **17,** 241–253 (1978).
69. W. R. Goodin, G. J. McRae, and J. H. Seinfeld, *J. Appl. Meteorol.* **19,** 98–108 (1979).
70. G. Freuhauf, P. Halpern, and P. Lester, *Environ. Software* **3,** 72–80 (1988).
71. S. G. Douglas and R. C. Kessler, *User's Guide to the Diagnostic Wind Model (Version 1.0),* Systems Applications, Inc., San Rafael, Calif., 1988.
72. R. A. Pielke, *Mesoscale Meteorological Modeling,* Academic Press, Orlando, Fla., 1984.
73. D. R. Stauffer and N. L. Seaman, *Mon. Weather Rev.* **118,** 1250–1277 (1990).
74. G. J. McRae, *Mathematical Modeling of Photochemical Air Pollution,* Ph.D. dissertation, California Institute of Technology, Pasadena, Calif., 1981.
75. O. Hov and co-workers, *Atmos. Environ.* **23,** 967–983 (1990).
76. D. P. Chock, *Atmos. Environ.* **19,** 571–586 (1985).
77. M. T. Odman, N. Kumar, and A. G. Russell, *Atmos. Environ.,* in press.
78. G. E. Gordon, W. H. Zoller, G. S. Kowalczyk, and S. W. Rheingrover, in Ref. 11, pp. 51–74.
79. P. K. Hopke and co-workers, *Atmos. Environ.* **10,** 1015–1025 (1976).
80. D. J. Alpert and P. K. Hopke, *Atmos. Environ.* **14,** 1137–1146 (1980).
81. *Ibid.,* **15,** 675–687 (1981).
82. P. D. Gaarenstroom, S. P. Perone, and J. L. Moyers, *Environ. Sci. Technol.* **11,** 795–800 (1977).
83. J. B. Milford, A. G. Russell, and G. J. McRae, *Environ. Sci. Technol.* **23,** 1290–1301 (1989).
84. T. Y. Chang, R. H. Hammerle, S. M. Japan, and I. T. Salmeen, *Environ. Sci. Technol.* **25,** 1190–1197 (1991).
85. N. C. Possiel, L. B. Milich, and B. R. Goodrich, eds., *Regional Ozone Modeling for Northeast Transport (ROMNET),* Report No. EPA-450/4-91-002a, U.S. Environmental Protection Agency, Research Triangle Park, N.C., 1991.
86. H. Rodhe, P. Crutzen, and A. Vanderpol, *Tellus* **33,** 132–141 (1981).
87. A. Eliassen, *J. Appl. Meteor.* **19,** 231–240 (1980).
88. *U.S.–Canada Memorandum of Intent on Transboundary Air Pollution, Atmospheric Sciences and Analysis Work Group 2 Phase III Final Report,* U.S. Environmental Protection Agency, Washington, D.C., 1983.
89. J. S. Chang and co-workers, *The Regional Acid Deposition Model and Engineering Model,* State-of-Science–Technology Report 4, National Acid Precipitation Assessment Program, Washington, D.C., 1989.
90. *Scientific Assessment of Stratospheric Ozone: 1989,* United Nations Environment Program and World Meteorological Organization, New York, 1989.

91. W. B. DeMore and co-workers, *Chemical Kinetics and Photochemical Data for Use in Stratospheric Ozone Modeling,* Evaluation No. 8, JPL Publ. 87–41, Jet Propulsion Laboratory, Pasadena, Calif., 1987.
92. D. J. Hoffmann and S. Solomon, *J. Geophys. Res.* **94,** 5029 (1989).
93. W. R. Pierson, A. W. Gertler, and R. L. Bradow, *J. Air Waste Manage. Assoc.* **40,** 1495–1504 (1990).

ARMISTEAD G. RUSSELL
Carnegie Mellon University

JANA B. MILFORD
University of Connecticut

ATOMIC ABSORPTION SPECTROSCOPY. See SPECTROSCOPY.

ATTRACTANTS. See INSECT CONTROL TECHNOLOGY.

AUTOMATED INSTRUMENTATION

CLINICAL CHEMISTRY

Clinical chemistry, initiated in the 1940s, involves the biochemical testing of body fluids to provide objective information on which to base clinical diagnosis. Between 1946 and 1975 clinical testing had a period of rapid growth in the United States, when the volume of tests increased by about 15% a year. The ever-increasing demand for high quality, routine clinical testing stimulated the development of automated techniques, and as early as the 1960s automation in the clinical laboratory was the rule rather than the exception.

Automation made the tests easier and less costly to perform, resulting in still greater demand. The growth rate, about 10% per year throughout the 1980s and into the 1990s, resulted not from a proliferation of laboratories as much as from increased productivity. Whereas the volume of tests more than quadrupled, the total number of United States hospital and private laboratories was reduced by half from 1975 to 1985 (1). Although growth was initially driven by automation, in the 1990s the growth in U.S. laboratory testing may be attributed to several factors. Among them are the discovery of new diseases and the introduction of new therapies as well as better understanding of body chemistry, and the aging of the U.S. population, which increases the risk of contracting age-related illnesses requiring clinical chemistry analyses (2).

Assay Automation

Clinical chemistry analyzers are automated instruments used for measuring concentrations of the various chemical constituents of blood or other body fluids. For a discussion of the related category of instruments used for the measurement of blood cell parameters, see AUTOMATED INSTRUMENTS, HEMATOLOGY.

Before the advent of automation, the steps a clinical technologist had to follow in performing a single clinical chemistry assay included: sample preparation, identification, centrifugation, and filtering; reagent preparation; manual metering and addition of sample and reagents into a reaction vessel; mixing and incubation in the reaction vessel; optical measurement of the mixture in a separate cuvette; and calculation and recording of the results. With the exception of the sample and reagent preparation procedure, which is sometimes done separately, these steps are all performed by an automated analyzer such as that shown in Figure 1.

Tubes or cups containing the liquid samples to be analyzed are placed by the operator into an area of the analyzer known as the sampler. The sampler, which in this case is a circular tray, moves each specimen to a position where it can be accessed by the sample aspirating probes. A sampling probe aspirates an amount of sample and dispenses it into a reaction cuvette, which is most often made of clear glass or plastic. Precise metering and transfer of the sample is controlled by the sample aspirating–dispensing system. The chemical reagents necessary for performing an assay on that sample are aspirated from reagent containers and dispensed into the same cuvette by the reagent handling system. Immediately after dispensing the reagents into the cuvette, sample and reagents are thoroughly mixed, eg, by a motor-driven mixing paddle, in order to obtain a uniform reaction rate within each cuvette. The reaction cuvettes, most often arranged at the periphery of a circular reaction tray, are held for several minutes at a constant temperature of 37°C or 30°C in the incubator, or reaction chamber. During this time chemical reactions in the cuvettes are monitored by an optical detection system. At the end of the incubation period each cuvette is evacuated, washed, and dried at a washing station, to be ready for the next assay. The optical measurement results are converted into metabolite concentration units by the analytical processor computer. Collating the results of different assays into patient reports and storing the test results are usually done by a separate computer, known as results processor or data manager.

In some systems, known as continuous-flow analyzers, the reaction develops as the sample–reagent mixture flows through a conduit held at constant temperature. In such systems, the reaction cuvettes are replaced by optical reading stations called flow cells. In most analyzers, whether of discrete- or continuous-flow type, determination of electrolyte tests, eg, sodium and potassium levels, is

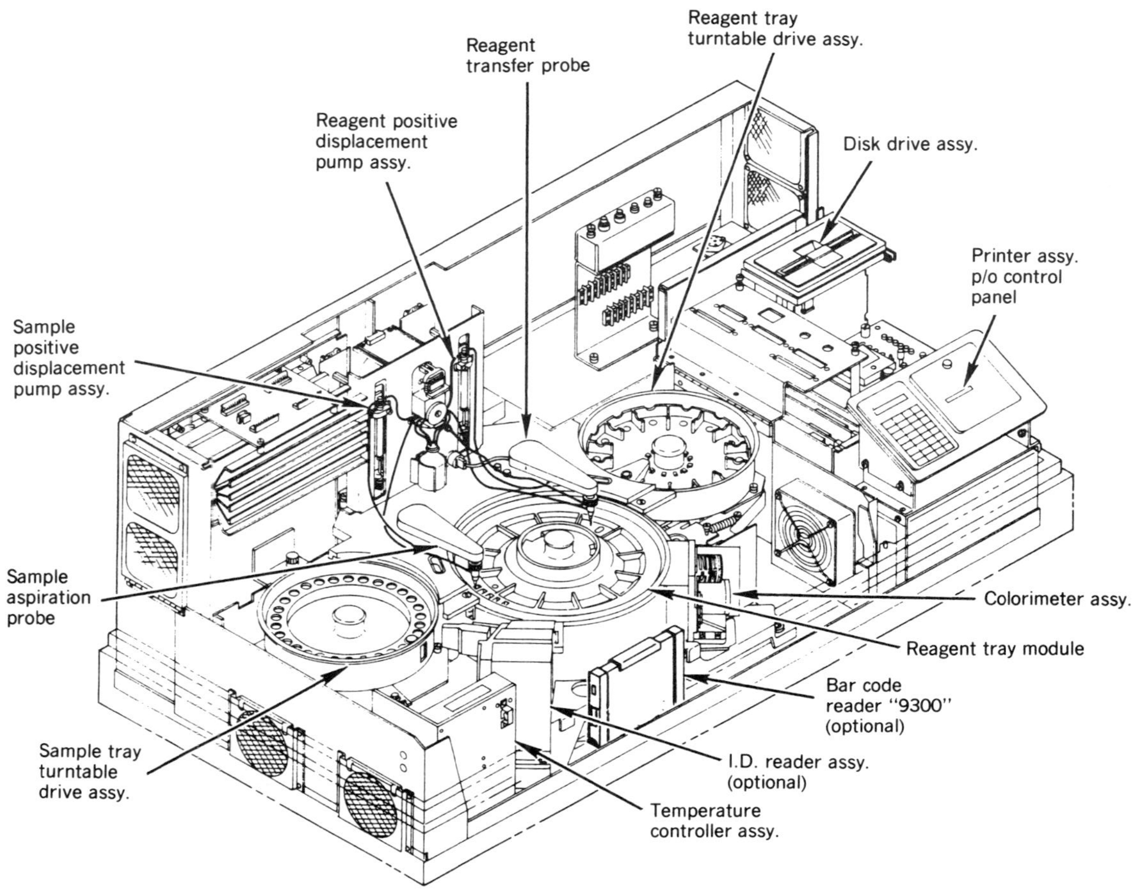

Fig. 1. Internal view of a clinical chemistry analyzer (Technicon RA-XT). Courtesy of Miles Inc.

done by a separate unit using the technique of ion-selective electrodes (ISE) rather than optical detection.

Automated methods are more reliable and much more precise than the average manual method: dependence on the technique of the individual technologist is eliminated. The relative precision, or repeatability, measured by the consistency of the results of repeated analyses performed on the same sample, ranges between 1% and 5% on automated analyzers. The accuracy of an assay, defined as the closeness of the result or of the mean of replicate measurements to the true or expected value (4), is also of importance in clinical medicine.

Technology

An automated system for clinical analysis consists of the instrument (hardware), the reagents, and the experimental conditions (time, temperature, etc) required for each determination. The reagents plus the experimental conditions are sometimes referred to as the chemistry of the system. The chemistry employed is generally similar to that used in manual assays because most automated assay methods have been adapted from the manual ones. However, automated analyzers rarely afford the flexibility of experimental procedure that is possible in manual analysis.

Table 1 lists several of the chemical determinations and the corresponding reactions utilized, which are available on automated clinical analyzers. With the exception of assays for various electrolytes, eg, Na^+, K^+, Cl^-, and CO_2, determination is normally done by photometric means at wavelengths in the ultraviolet and visible regions. Other means of assay include fluorescence, radioisotopic assay, electrochemistry, etc. However, such detection methods are normally required only for the more difficult assays, particularly those of serum or urine constituents at concentrations below $\mu g/L$. These latter assays are discussed more fully in the literature (3,4).

Essential features of an automated method are the specificity, ie, the assay should be free from interference by other serum or urine constituents, and the sensitivity, ie, the detector response for typical sample concentration of the species measured should be large enough compared to the noise level to ensure assay precision. Also important are the speed, ie, the reaction should occur within a convenient time interval (for fast analysis rates), and adequate range, the result for most samples should fall within the allowable range of the assay.

The assay methods listed in Table 1 for the various biochemical species can be classified according to reaction rate behavior, eg, end point vs kinetic methods, blanking schemes, or reaction principle and type of reagents employed.

End Point vs Kinetic Methods. Samples may be assayed for enzymes, ie, biocatalysts, and for other substances, all of which are referred to as substrates. The assay reactions for substrates and enzymes differ in that substrates themselves are converted into some detectable product, whereas enzymes are detected indirectly through their conversion of a starting reagent A into a product B. The corresponding reaction curves, or plots of detector response vs time, differ for these two reaction systems, as shown in Figure 2. Figure 2**a** illustrates a typical

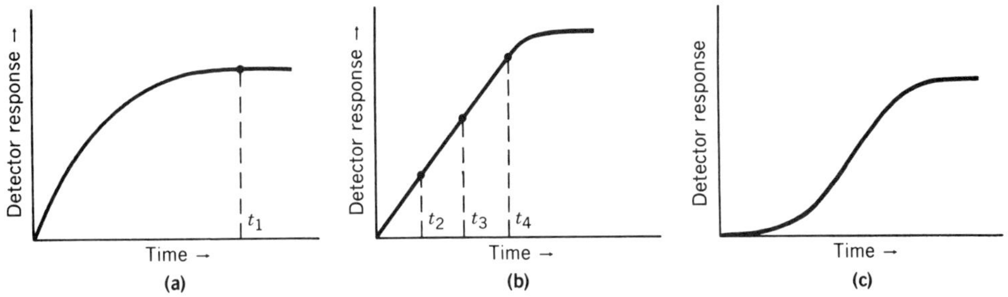

Fig. 2. Assay reaction curves for (**a**) substrates, (**b**) enzymes, and (**c**) enzymes exhibiting a lag phase or reduced reaction rate; where t_x is measurement time.

substrate reaction curve; Figure 2**b** shows a typical enzyme reaction curve (see ENZYME APPLICATIONS).

In substrate assays, the reaction product is measured at time t_1 (Fig. 2**a**) when the reaction is complete. The concentration of the species to be assayed is usually proportional to the net detector response at t_1, and generally there is no problem in relating this measurement to substrate concentration. These so-called end point assays correspond to a constant value and are less subject to errors occurring as a result of variations in experimental conditions (temperature, reagent concentrations, etc) than measurements made under rapidly changing conditions. In some cases, t_1 can be as long as 5–10 min, which would severely limit maximum analysis rates if such assays were performed individually. Automated analyzers address this problem in various ways. For example, most analyzers allow a series of samples to pass through the system one after the other. In this way each sample may spend up to 5–10 min within the system, but if n samples are in the system at any given time, the net analysis rate, expressed in samples per minute, is n divided by 5–10. Alternatively, all samples might be loaded into the system for parallel processing in the same 5–10 min. Continuous-flow analyzers are an example of the first scheme, whereas cuvette-based analyzers may work in the latter manner. A more important constraint in end point assays by automated analyzers is the need in some analytical reactions for multiple addition of different reagents at different times. Cuvette-based analyzers, for example, do not normally allow for more than two reagent additions.

For enzyme determinations, kinetic assay procedures are generally favored. These refer to the determination of the slope of the reaction curve. However, not all enzyme reaction curves are as simple as that shown in Figure 2**b**. Some reactions show a so-called lag phase, or reduced reaction rate, at the beginning of the reaction, as shown in Figure 2**c**. Other reactions level off quickly, much before the average time t_4, shown in Figure 2**b**. As a result, many clinical chemists prefer enzyme assays that are based on monitoring the reaction curve over a wide range in time so that the linear zero-order region can be readily identified and the slope calculated in this region. However, this method of monitoring represents a serious constraint on many analyzers; ie, if there is one detector, then the maximum analysis rate is $1/t_4$. Because t_4 is commonly 5–10 min, this would limit

Table 1. Clinical Tests Offered on Automated Analyzers[a]

Assay	Reaction
	Chemical
albumin	albumin + bromcresol green (BCG) $\longrightarrow$ albumin BCG[b]
calcium	calcium + cresolphthalein complexone $\longrightarrow$ complex[b]
chloride	$2Cl^- + Hg(SCN)_2 \longrightarrow HgCl_2 + 2\,SCN^-$
	$3\,SCN^- + Fe^{3+} \longrightarrow Fe(SCN)_3$[b]
creatinine	creatinine + sodium picrate $\longrightarrow$ colored product[b]
iron	iron + Ferene-S $\longrightarrow$ colored complex
total bilirubin	bilirubin + sulfanilic acid + NO_2^- $\longrightarrow$ azobilirubin[b]
total protein	protein + $CuSO_4 \xrightarrow{OH^-}$ cuproproteinate complex
	Biochemical
alanine aminotransferase (ALT)	1-alanine + oxoglutarate $\xrightarrow{\text{ALT}}$ pyruvate + 1-glutamate
	pyruvate + NADH $\xrightarrow[\text{dehydrogenase}]{\text{lactate}}$ lactate + NAD
alkaline phosphatase	*p*-nitrophenyl monophosphate $\xrightarrow[\text{phosphatase}]{\text{alkaline}}$ *p*-nitrophenol[b] + PO_4^{3-}
aspartate transaminase (AST)	glutamate + *a*-ketoglutarate $\xrightarrow{\text{aspartate transaminase}}$ oxaloacetate
	oxaloacetate + NADH $\xrightarrow[\text{dehydrogenase}]{\text{malate}}$ malate + NAD
carbon dioxide	phosphoenolpyruvate + HCO_3^- $\xrightarrow[\text{phosphoenolpyruvate carboxylase}]{}$ oxalacetate + PO_3^{3-}
	oxalacetate + NADH $\xrightarrow[\text{dehydrogenase}]{\text{malate}}$ malate + NAD
cholesterol	cholesterol esters $\xrightarrow[\text{esterase}]{\text{cholesterol}}$ free cholesterol + fatty acids

	$\text{free cholesterol} + O_2 \xrightarrow[\text{oxidase}]{\text{cholesterol}} \text{cholest-4-en-3-one} + H_2O_2$
	$H_2O_2 + \text{phenol} + \text{4-aminoantipyrine} \xrightarrow{\text{peroxidase}} \text{quinoneimine dye}$
creatine kinase (CK)	$\text{creatine phosphate} + \text{ADP} \xrightarrow[\text{kinase}]{\text{creatine}} \text{creatine} + \text{ATP}$
	$\text{ATP} + \text{glucose} \xrightarrow{\text{hexokinase}} \text{ADP} + \text{glucose-6-phosphate}$
	$\text{NAD} + \text{glucose-6-phosphate} \xrightarrow[\text{dehydrogenase}]{\text{glucose-6-phosphate}} \text{NADH} + \text{6-phosphogluconate}$
gamma glutamyl transferase (GGT)	$\text{glutamyl-}p\text{-nitroanalide} + \text{glycylglycine} \xrightarrow{\text{GGT}} \text{glutamylglycylglycine} + p\text{-nitroaniline}$
glucose	$\text{ATP} + \text{glucose} \xrightarrow{\text{hexokinase}} \text{ADP} + \text{glucose-6-phosphate}$
	$\text{glucose-6-phosphate} + \text{NAD} \xrightarrow[\text{dehydrogenase}]{\text{glucose-6-phosphate}} \text{6-phosphogluconate} + \text{NADH}$
lactate dehydrogenase (LD)	$\text{lactate} + \text{NAD} \xrightarrow[\text{dehydrogenase}]{\text{lactate}} \text{pyruvate} + \text{NADH}^b$
triglycerides	$\text{triglycerides} \xrightarrow{\text{lipase}} \text{glycerol} + \text{fatty acids}$
	$\text{glycerol} + \text{ATP} \xrightarrow[\text{kinase}]{\text{glycerol}} \text{glycerol-1-phosphate} + \text{ADP}$
	$\text{ADP} + \text{phosphoenolpyruvate} \xrightarrow[\text{kinase}]{\text{pyruvate}} \text{ATP} + \text{pyruvate}$
	$\text{pyruvate} + \text{NADH}^b \xrightarrow[\text{dehydrogenase}]{\text{lactate}} \text{lactate} + \text{NAD}$
urea nitrogen	$\text{urea} + H_2O \xrightarrow{\text{urease}} 2\ NH_3 + CO_2$
	$NH_3 + \text{NADH} + \text{oxoglutarate} \xrightarrow[\text{dehydrogenase}]{\text{glutamate}} \text{NAD} + \text{glutamate}$

[a]Coenzymes such as adenosine diphosphate (ADP), adenosine 5′-triphosphate (ATP), nicotinamide adenine dinucleotide (NAD), and nicotinamide adenine dinucleotide, reduced (NADH), are involved in some reactions (4).
[b]Chromophore, optical detection used.

maximum analysis rates to only 6–12 samples per hour, which would be much too slow for large clinical laboratories.

Early instruments ignored the assay time problem, made a single measurement at some time, as in end point analysis, and calculated an assumed slope dividing the net detector response by the time after initiation of the reaction. For many assays, this was satisfactory. However, frequent interferences were encountered with these single-point procedures, as use of so-called 340 nm chemistries increased for various enzymes. The 340 nm chemistry is based on the production of the coenzyme reduced nicotinamide adenine dinucleotide (NADH), which absorbs light at a wavelength of 340 nm. One solution is the use of parallel blank reactions, where one of the reaction reagents is missing. Subtraction of the resulting blank reaction slope from the total reaction slope gives the desired enzyme reaction curve. However, this approach greatly increases the cost of both equipment and reagents. More importantly, it does not allow for detection of reagent depletion, where high enzyme concentrations cause a bending of the reaction curve in a very short time, as shown in Figure 2**b** after t_4. Newer cuvette-based analyzers, such as that shown in Figure 1, use multiple detector readings or multiple detectors. In these analyzers, a batch of samples for the same assay is rotated through a common detector station so that multiple points along the reaction curve are collected for each sample. This multiplexing of the detector results in an overall analysis rate of n/t_4, where n is the number of samples in the batch. Thus these analyzers achieve high throughput rates, including a fairly complete definition of the total reaction curve, plus reliable determination of the zero-order slopes for all samples.

Blanking Schemes. Whereas most clinical assays rely on the specificity of the assay reaction to avoid interferences that would require a blank correction, certain ubiquitous interferences, eg, lipemic turbidity, hemoglobin in hemolyzed sera, bilirubin [*635-65-4*], $C_{33}H_{36}N_4O_6$, in icteric specimens, present special problems in many otherwise specific assays. Sample interferences can be classified as follows: (*1*) fixed absorbance in the absence of reagents, eg, turbidity; (*2*) fixed absorbances as a result of reaction with reagents, eg, free-glycerol interference in assays for triglycerides; and (*3*) changing absorbance as the result of slow reaction with reagents.

Each interference type may require a different elimination strategy, the simplest of which is to remove the potential interference before beginning the assay. Removal is commonly done in manual analysis, usually by deproteinization, eg, by addition of some denaturing agent such as trichloroacetic acid. This technique also tends to remove other large molecules such as bilirubin and hemoglobin, as well as eliminating sample turbidity. However, such procedures are not easily adapted to automated analysis, and in fact are generally not used. Modern assay methods generally employ mild reaction conditions which not only do not cause sample protein precipitation; these methods, because they frequently use highly specific enzymes as reagents, also have significantly reduced interference.

In cases where sample blank correction is required, a variety of techniques are employed. For interferences of type 1, a common technique is to use a separate determination in which an essential reagent for the primary reaction is eliminated; eg, the sodium nitrite, $NaNO_2$, in the usual reaction for bilirubin. The blank, or interference, response in the detector mimics that in the assay reaction, but the assay reaction itself does not occur. Subtraction of the blank from the assay result on a given sample then provides a correction for interferences of type 1. For interferences of type 2, a blank is run using the reagent that reacts with the interfering species. A modification of this technique uses two reagents. The sample is added to the first reagent, and the blank is measured. No assay reaction occurs. The second reagent, added later, contains the active component which produces the measurable assay response. A subtraction of the two measurements cancels out the unwanted contribution of the blank reaction to the desired response.

Separate sample blanking requires an additional analytical channel, and is therefore wasteful of both reagents and hardware. An alternative approach that is used on several automated systems, eg, Du Pont ACA, BM-Hitachi 704, Technicon RA-1000, is that of bichromatic analysis (5) where absorbance measurements are taken at two, rather than one, wavelength. When the spectral curves for the interference material and the chromogen of the species measured differ sufficiently, this can be an effective technique for reducing blank contributions to assay error. Bichromatic analysis is effective for blanks of both the first and second type.

Blanks of the third type are the most difficult to handle. Examples are found in the usual Jaffe reaction for creatinine, and in certain enzyme reactions. In the creatinine [*57-00-1*], $C_4H_9N_3O_2$, assay, if potential interferences are not removed by dialysis or some similar procedure, they exhibit both slower and faster reactions compared to those of creatinine. A two-point analysis can be utilized here with the measurement times selected to be later than the time required for fast-reacting interferences and before the time for significant reaction of slow-reacting interferences. This procedure is utilized in the creatinine assay on many automated chemistry analyzers.

Reactions and Reagents. The three general categories of reactions are chemical, biochemical, and immunological. Automated analyzers mainly utilize the assays listed in Table 1, which fall into one of the first two categories; trace-level assays are carried out by immunological methods (see IMMUNOASSAY). Chemical methods are based on reactions that do not normally take place in living systems, eg, the assay for calcium by the cresolphthalein complexone reaction. These assays are often relatively nonspecific, and may require removal of sample interferences or the use of blanking techniques. Chemical methods can be used by most types of automated clinical analyzers. Biochemical methods rely on reactions that play a role in the functioning of living systems; eg, the assay of glucose by the coupled enzyme system hexokinase plus glucose-6-phosphate dehydrogenase. Biochemical assays are often highly specific because the corresponding biochemical reactions in the body are specific. Automatic analyzers utilizing biochemical methods predominate in the marketplace because of greater specificity and reduction in cost through increased use.

Immunological methods suffer from many disadvantages, and automation has been attained only in the 1980s and 1990s. Immunological reactions utilize antibodies as reagents, and are often competitive binding procedures. Radioimmunoassay was the first technique employed for immunological methods. The use of nephelometric, turbidimetric, and fluorometric means of detection have gained ground; however, use of radioimmunoassay is decreasing in part because of the problems of radioactive waste disposal.

Measurement Methods

The majority of the various analyte measurements made in automated clinical chemistry analyzers involve optical techniques such as absorbance, reflectance, luminescence, and turbidimetric and nephelometric detection means. Some of these are illustrated in Figure 3. The measurement of electrolytes such as sodium and potassium have generally been accomplished by flame photometry or ion-selective electrode sensors (qv). However, the development of chromogenic ionophores permits these measurements to be done by absorbance photometry also.

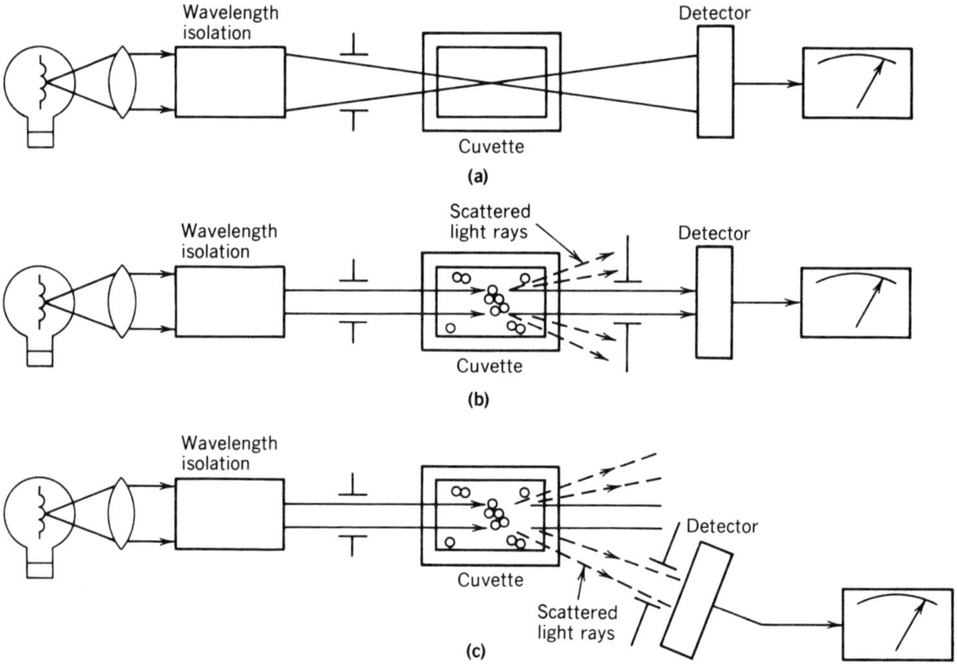

Fig. 3. Schematic of optical detection methods, (**a**) absorbance or colorimetric; (**b**) turbidimetric; and (**c**) nephelometric.

Absorbance. Analyte measurements in clinical analyzers using liquid reagents are most commonly performed by transmission of light, ie, by absorbance photometry or colorimetry (Fig. 3**a**). The liquid to be analyzed is either held in a cuvette or passed through a flowcell having transparent walls. A light beam traverses the liquid compartment and then falls on an optical detector (see PHOTODETECTORS). Prior to striking the detector, a narrow wavelength band of radiant power centered at the wavelength of peak absorbance, the analytical wavelength, is isolated. Wavelength isolation, shown in Figure 3 as occurring before the light impinges upon the sample, may take place after the sample. In dilute solutions, analyte concentrations are proportional to the measured absorbance at the analytical wavelength in accordance with the Beer-Lambert law: $T = P/P_o = 10^{-\epsilon cl}$ where T is the relative transmittance, P_o is the beam power for a nul concentration sample, and P is the power for the sample being measured, c is concentration in moles per liter (M), l is the path length in centimeters of the absorber, and ϵ is the molar extinction coefficient having units $M^{-1}\text{cm}^{-1}$. The absorbance A is defined as $-\log_{10} T$, thus $A = \epsilon cl$ where l the path length, representes the length of the light path through the liquid chromophore in the flowcell or cuvette.

Reflectance. In dry chemistry analyzers, the diffuse reflectance of the chromophore is measured in the supporting solid matrix. The output response is approximated by the Kubelka-Munk function in which the diffuse reflectance of the sample depends only on the ratio of the absorption and scattering coefficients and not on their absolute values. The scattering coefficient is generally considered to be constant for a given matrix. Typically the liquid sample is deposited on one side of the reagent-impregnated strip or slide, and the optical measurement is made on the opposite side. In making diffuse reflectance measurements, it is important to minimize the specular component of reflection from erroneously affecting the reading.

Turbidimetry and Nephelometry. In contrast to classical absorbance methods, immunoassay reactions frequently involve agglutination in which the optical scatter signal of the agglutinated particles is measured by turbidimetric or nephelometric means. The principles of light scattering as it relates to analytical methods is discussed in reference 6.

Turbidimetry is considered a transmission type of measurement in which the detector is optically in line with the light source (Fig. 3**b**). Particles in the sample scatter the incoming beam away from the in-line detector, and consequently cause a loss in light power falling on the detector. This decrease in light intensity is measured and converted to optical density units relative to a solution without particles. In the absence of any absorbing species in the sample, the turbidity is proportional to the concentration of the particles, the scattering cross section, and the optical pathlength of the cuvette. The scattering cross section is a function of the relative particle size to wavelength ratio and to the refractive indexes of the particle and surrounding medium.

In nephelometry the detector is located at an angle to the light path and to the cuvette centerline. Essentially, only scattered light is collected by the detector (Fig. 3**c**). Although the detector signal level is considerably lower than in turbidimetric measurements, the signal is measured against an essentially zero background rather than a high signal level as in the turbidimetric technique. This results in an increase in the ratio of the maximum to minimum signals and generally is advantageous, providing the signal level is adequate, eg, where the electronic or shot noise is not significant, for the precision required.

In both turbidimetric and nephelometric measurements used for agglutination reactions, the relationship of the concentration of the analyte to the measured optical density or scatter signal is a complex one. Various optical and physical factors, such as the rate of formation of two- or three-particle agglutinates generated from the monomers, ie, unagglutinated particles, their orientation, and degree of multiple scattering, affect the signal to concentration relationship. Obviously the chemistry of the immunologic reaction is paramount to determining the kinetics and sensitivity of the reaction.

Fluorescence. The fluorescence detection technique is often used in clinical chemistry analyzers for analyte concentrations that are too low for the simpler absorbance method to be applied. Fluorescence measurements can be categorized into steady-state and dynamic techniques. Included in the former are the conventional simultaneous excitation–emission method and fluorescence polarization.

In time-resolved fluorescence, rare earths are frequently used as fluorescent labels. The fluorophores have large Stokes shifts, ie, shifts of the emitted light to a higher wavelength relative to the absorption wavelength, and comparatively long decay times, approximately 0.5 ms. This simplifies the optical detection and minimizes serum background effects, as the background fluorescence has essentially decayed to zero when the longer lifetime signal is being read out. The more research-oriented instruments, based on fluorescence lifetime measurements, utilize the more sophisticated impulse–time decay and the frequency and phase domain techniques.

The vast majority of detectors, used in absorbance and fluorescence measurements, operate in the analogue mode, in which the average of a large number of time-overlapped electron pulses constitute the measured current. In this technique the amplified electron pulses, generated initially by the photons striking the photocathode, are counted individually as one single count per photon, irrespective of their amplitudes. For very low radiant power measurements, such as may be found in some fluorescence applications, a photon counting method may be used advantageously.

Ion Selective Electrodes Technique. Ion selective (ISE) methods, based on a direct potentiometric technique (7) (see ELECTROANALYTICAL TECHNIQUES), are routinely used in clinical chemistry to measure pH, sodium, potassium, carbon dioxide, calcium, lithium, and chloride levels in biological fluids.

In potentiometry, the potential difference in volts is measured between two electrodes, one a reference electrode having a stable potential under controlled conditions, and the other a working or measuring electrode. In the working ISE, an ion-selective membrane is used to separate two solutions of different ionic strength. As the sample containing the analyte, eg, sodium ions, to be measured flows past the ISE, changes in electrical potential take place between the ionically constant inner surface, saturated KCl solution, and the ionically variable outer surface, the sample. These changes in electrical potential are automatically measured against those of the reference electrode. The potential difference (voltage) between the two electrodes is related to the activity of the ions in solutions, and the activity is proportional to the logarithm of the respective analyte concentration of the sample, according to the Nernst (8) equation. The analogue signal generated is digitized and then converted to reportable concentration units by the instrument's computer processing unit (CPU).

Classification of Clinical Chemistry Analyzers

Automated clinical analyzers can be divided into discrete and continuous-flow systems. In discrete systems, the sample–reagent mixtures from different specimens are kept in individual reaction cuvettes during incubation and optical reading. A notable exception in this category are the dry chemistry analyzers, where the cuvettes were replaced by individual slides coated with dry reagent films. In continuous-flow analyzers, the sample–reagent mixtures form liquid segments flowing through a tube. Adjacent segments are separated by air bubbles. Optical readings are taken at different locations along the tube, which is kept at the incubation temperature.

Clinical analyzers can also be classified according to their degree of flexibility. Most of the modern systems are random access analyzers, for which the tests on various specimens are performed in any order programmed by the operator. Some analyzers operate in batch or profile mode, ie, they perform the same test or group of tests on every sample until the system is reset for another test or group of tests.

The throughput range, number of assays performed per hour, of clinical analyzers reflects the diversity of the market. Systems that perform less than 400 tests per hour are usually referred to as small analyzers; medium analyzers cover the range from about 500 tests per hour to 1,500 tests per hour, while high throughput analyzers can process up to 10,000 tests per hour. Table 2 lists a number of representative automated systems and their characteristics. Whereas small analyzers, eg, the Ames CLINISTAT, can easily fit on a desktop, high throughput analyzers, eg, the Technicon DAX-96, can occupy a good portion of a laboratory.

Functional Elements of Clinical Chemistry Analyzers

Sample Handling System. Venous or capillary blood, urine, and cerebrospinal fluid are specimens routinely used in medical diagnostic testing. Of these biological fluids, the use of venous blood is by far the most prevalent. Collection devices such as syringes and partial vacuum test tubes, eg, Vacutainer, are used to draw ten milliliters or less of venous blood. At collection time, the test tubes are carefully labeled for later identification.

The specimen, as drawn, contains cells, platelets, fibrin, and particulates. For most chemistry tests, eg, determination of glucose, cholesterol, etc, it is necessary to first separate the cellular blood fraction which, if present during the assay, would interfere with the determination and adversely affect the accuracy of the measurement. The cellular components are separated by centrifugation. The centrifuge is generally not part of the automated analyzer. If anticoagulants are added to the sample, the resultant supernatant layer is called plasma; if not, the sample must be allowed to stand for approximately ten minutes before centrifugation to allow completion of the clotting process. In the latter case, the supernatant is referred to as serum (see FRACTIONATION, BLOOD).

After centrifugation, the specimens are placed in a holder specifically de-

Table 2. Features of Automated Analyzers

Manufacturer and system	Throughput per hour		Number of resident methods	Amount of sample needed per test, μL	Incubator temperature, °C	Optical system[a]	Distinctive features
	Samples	Tests					
Roche COBAS MIRA	variable	125	30	2–95	37	AMW,P	disposable cuvettes
Technicon RA-1000	variable	240	12	2–30	30, 37	AMW,P	no probe wash
Ames CLINISTAT	up to 80	up to 80		10	37	R	dry chemistry
Abbott SPECTRUM	variable	400–600	23 3	1.25–25	25, 30, 37	AMW,P,T	polychromatic sample blanking
Beckman SYNCHRON CX3	up to 75	600	8	122/8 tests	37	AMW,P	STAT analyzer[b] Beckman liquid calibrators and controls
Kodak EKTACHEM 700	300 max	600	26	10, 11	37	R	dry chemistry slides, routine/STAT[b]
Baxter PARAMAX	240	720	32	2–50	37	AMW,P	closed container sampling option; unit dose dry tablet reagent
BM Hitachi 737	300	1,200	23	3–20	30, 37	AMW,P	
Technicon CHEM 1	variable	1,800	35	1	30, 37	AMW,P	continuous-flow capsule chemistry
BM Hitachi 736-50	300	8,100	27	10	30, 37	AMW,P	
Technicon DAX-96	300	10,200	34	3–67	37	AMW,P	

[a] AMW = absorption, multiple wavelength per test; N = nephelometry; T = turbidimetry; P = potential; and R = reflectance.
[b] A specimen that has to be processed with priority vs other specimens is known as STAT sample.

signed to work with a particular instrument. Some clinical analyzers utilize sample carrier blocks; others use a rotary tray or a continuous flexible chain link. In some cases, an aliquot of the sample has to be poured into a secondary container, small cup, or tube which is then labeled and placed on the sample holder. Direct or primary tube sampling is preferable because it involves less sample handling and less risk of loss of patient identification. A trend in primary tube sampling is the introduction of closed tube sampling, which allows the proper amount of serum or plasma to be aspirated directly from the primary collection tube without having to first remove the stopper. One such technique is described schematically in Figure 4. The obvious advantage of such techniques is the significant reduction of risks associated with sample handling such as operator exposure to hepatitis B and HIV viruses.

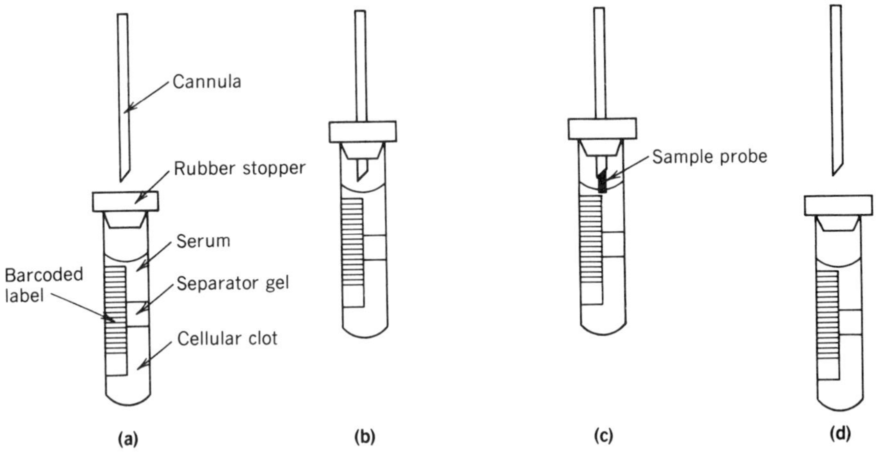

Fig. 4. Schematic of the Closed Container Sampling technique used in the Baxter PARAMAX analyzer showing (**a**) the collection tube with bar-coded label being brought into sampling position under the cannula; (**b**) the tube raised so that the cannula has penetrated the stopper; (**c**) the sample sensing probe coming through the cannula to aspirate the exact volume required for each assay; and (**d**) after sampling, where the tube is lowered away from the cannula. The stopper removes any sample from the cannula and reseals itself.

Once loaded, the sample holder is placed onto the carrier transport mechanism of the system, which is referred to as the sampler. The function of the sampler is to bring each specimen into a position where a sample of it can be taken by the aspiration probe, then to transport it further to the unloading area. Most analyzers allow continuous feeding, or loading, and unloading of samples. Priority sampling, which allows the sampler to temporarily interrupt sampling of routine specimens in order to aspirate from emergency, or STAT, specimens, is also a common feature. The analyzer has to keep track of sample identity. Some systems use optical detection devices to read a bar-coded label on each test tube or sample cup before aspirating from it. This approach is known as positive sample identification. In other systems, the sample holders, eg, carrier blocks, are assigned unique numbers that can be read by an optical device, then the specimens are identified by their position on the carrier block.

When a test result for a particular specimen is found to have an elevated, out of method range value, some analyzers, eg, Beckman CX3, can automatically repeat the sampling from the same specimen. For elevated concentrations, the precision of the optical system is reduced so an automatic dilution of the sample, eg, by aspiration of a reduced amount of sample, is provided during the second sampling.

Sample Aspiration and Dispensing. Clinical chemistry analyzers perform tests mostly on blood serum or plasma, and less often on other body fluids. Precisely metered amounts of the sample have to be aspirated rapidly and without allowing intersample contamination, known as sample carry-over. Typical sample volume aspirated per test ranges between 1 μL and 25 μL. Generally, pipetting is accomplished by motor-driven syringes connected through a fluid line to a thin aspiration probe (Fig. 1).

In most analyzers, sample carry-over is minimized by one or more of the following techniques: sensing the liquid level in the sample container, and limiting the probe immersion in the liquid to a few millimeters; or rinsing the aspiration probe inside and out after each immersion. Probe rinsing has been eliminated in a few analyzers by the use of an aspiration probe employing disposable tips. After aspiration of all the aliquots needed from a particular specimen, the tip is discarded, and a new one is loaded for the next sample. Yet another approach to carry-over prevention is where the aspirating probe is coated inside and out with a thin film of a chemically inert fluorocarbon oil. During immersion this coating prevents the sample from wetting the probe wall while allowing aspiration to take place.

Reagent Handling System. Many analyzers can be programmed to perform a wide variety of assays. However, reagents for only a limited number of tests, usually referred to as resident tests, are available on the instrument at any one time. The main reason for this limitation is that, for infrequently requested tests, the time period until reagent depletion may exceed the chemical stability time limit. Refrigerated compartments for the reagent containers, usually kept between about 4°C to 15°C, are generally provided on the medium to high throughput systems in order to extend the stability of the resident reagents. For most analyzers using liquid reagents, the reagent stability on the system ranges between 2 days to 30 days. In some systems, such as the Baxter Paramax, reagents are stored in tablet form and reconstituted as needed, thus extending on-board reagent stability to 60 days.

Precise metering of the amount of liquid reagent needed for a test is generally done using a motor-driven syringe. Other variants, such as peristaltic pumps or pneumatic syringes, are also employed. The amount of reagent used per test ranges between 0.1 mL and 0.7 mL. In some analyzers, a dedicated conduit links each reagent container to a dispensing nozzle above the cuvette. Other instruments use a single reagent probe to aspirate the reagents from the reagent containers and dispense them into the reaction cuvette. If the latter approach is used, reagent cross-contamination by the probe has to be avoided. This is usually achieved by a thorough rinsing of the probe after each reagent dispense. Reagent-to-reagent carry-over typically has to be kept to less than 1 part per 20,000.

Coating the reagent probe with an inert liquid eliminates the need for the rinsing operation.

Two notable technological developments related to the reagent handling system occurred in the last 1980s. One of these was the introduction of the dry slide reagent technology where the reagents needed for a particular assay are deposited in thin layers on a slide (Fig. 5). The reaction starts when a drop of sample, deposited on the upper slide surface, dissolves the reagents. A top coating layer, covering the dry reagents, helps the uniform spread of the sample drop and filters out interfering substances. The dye formed in the reaction is measured by reflectance spectrophotometry from the lower side of the slide, the light being reflected by the top coating layer. The dry slide technology eliminates the preparation and handling of liquid reagents, as well as the need to meter the amount of reagents for each test. A slide cartridge is provided for each resident test. After opening the cartridge, the reagents are stable for up to seven days.

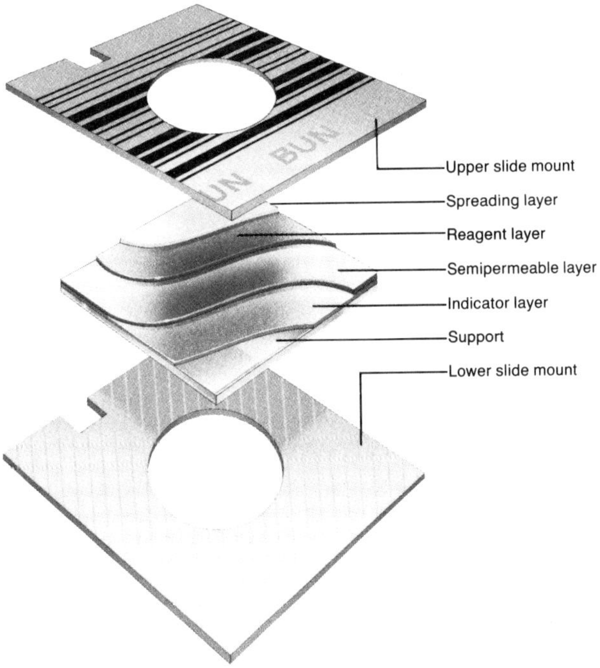

Fig. 5. Schematic of a dry reagent slide used in Kodak EKTACHEM analyzers. Courtesy of Kodak Instruments.

Another unique development, which significantly reduced the amount of reagent necessary for an assay, is an extension of the continuous-flow analyzer technology called capsule chemistry. The core of the system is a capillary Teflon tube coated on the inside with a thin, flowing film of fluorocarbon oil. Sample and liquid reagents, alternating with air bubbles, are aspirated into the tube forming a segmented flow. Each liquid segment surrounded by the film of immiscible fluorocarbon oil, forms a discrete capsule which does not contaminate the neighboring test capsules. Because of the small size of the capillary tube, the amount of liquid reagents per test is reduced to 14 μL, and the amount of sample to 1 μL per test. The small reagent volumes used allow on-board storage of reagent sufficient for up to 1200 tests for any of the resident methods. The reaction takes place while the liquid capsules are flowing through the capillary tube, and the photometric reading is taken through the tube wall, eliminating the need for reaction cuvettes (Fig. 6).

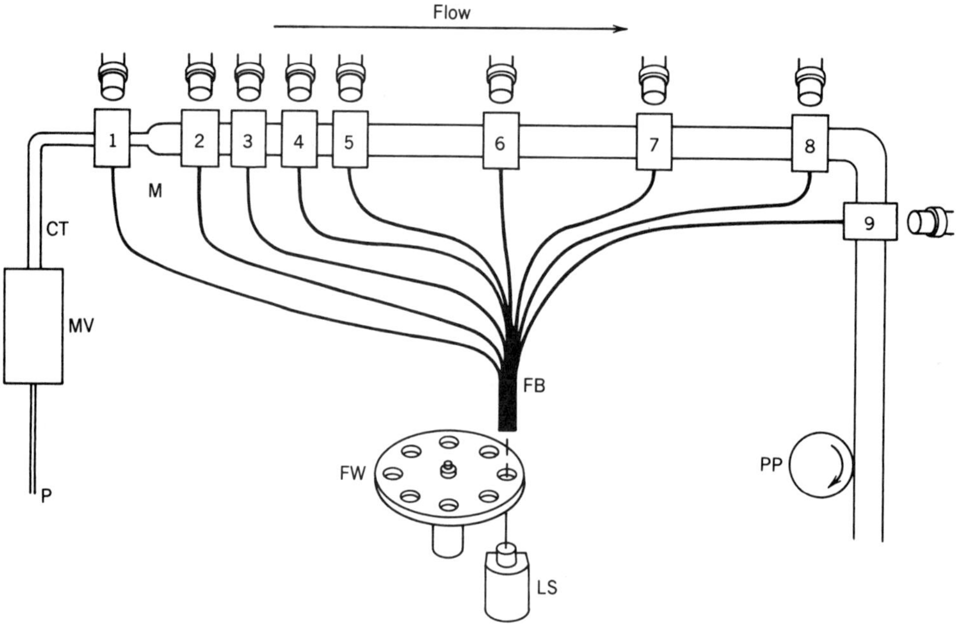

Fig. 6. Schematic of Technicon CHEM I Analyzer where P = aspiration; MV = metering valves; CT = capillary Teflon tube; M = reagent mixing zone; FW = filter wheel; FB = fiber optic bundle; LS = light source; PP = peristaltic pump. The colorimetric reading stations are marked 1 through 9.

Optical Detection System. The optical detection system includes dedicated or time-multiplexed arrangements of the various elements such as the light source, cuvette or flowcell, wavelength isolation elements, and detector. A common layout may have a single light source with light distribution via fiber optics to many cuvettes or flowcell read stations. On the output side of the reaction cuvette or flowcell, fiber optics pipe the light signals to an optical chopper, through dedicated interference filters, and to a common multiplexed photomultiplier. A variation of this arrangement uses a rotating mirror device to multiplex the light output of many cuvettes into the input of a single grating–array photodiode.

Light Sources. The majority of clinical chemistry methodologies result in analyte reactant products which are analyzed using either ultraviolet or visible light. Light sources include the tungsten halogen lamp, lasers, light-emitting diodes, and arc lamps such as deuterium, xenon, and xenon–mercury of the continuous or the flash lamp type. The tungsten lamp generally has greater short- and long-term stability than the arc lamps, and consequently is the preferred source for wavelengths above 340 nm. Lasers (qv) and light-emitting diodes, which operate at a single wavelength, with the exception of tunable dye lasers, are special-purpose sources used in single wavelength applications (see LIGHT GENERATION).

Reaction Cuvette/Flowcell. This critical element is often unique to the system, and differs to the greatest extent among various instruments. Wet chemical analyzers typically use a cuvette or flowcell to contain the reactant to be measured. The volume of this fluid may range from as little as 15 μL in flowcell analyzers to several hundred microliters in cuvette-based instruments. In absorbance measuring instruments, the light path lengths generally range from 1 to 10 mm. Because the path length of the cell affects the measured absorbance directly, the choice of this design parameter is carefully considered with regard to the system volume constraints.

Wavelength Isolation. There are several means of isolating the analytical wavelength needed for the measurement of the analyte. One of the most common methods is the use of a multilayer interference filter illustrated in Figure 7. When coupled with suitable blocking filters, this optical element has an extremely good rejection ratio of the out-of-band to in-band power, minimizing the stray light effects which can contribute to nonlinearity. The half-bandwidth of the filter, spectral bandwidth at half the peak transmission, is the fundamental filter specification, and this value should be small relative to the chromophore bandwidth. Bandwidths range from 2 to 20 nm. Where a large number of wavelengths are required for the system, a multiwavelength dispersive device such as a prism or grating is more suitable.

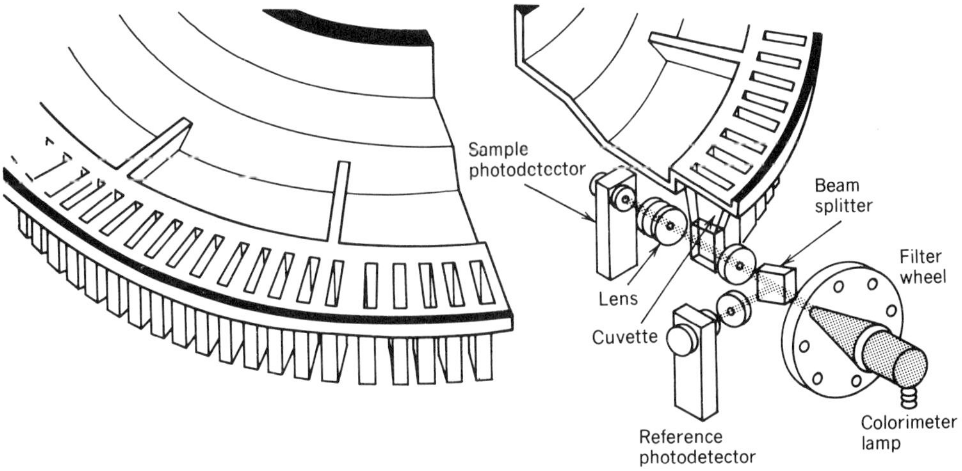

Fig. 7. Schematic of a reaction tray and colorimetric system.

Prisms and gratings have historically been used for spectral band isolation in spectrophotometers. These optical elements provide a wide range of wavelengths and have very narrow bandwidth characteristics when coupled with suitable entrance and exit slits. The mechanically scanned prisms or gratings provide sequential wavelength information, whereas the more recent grating/multielement array detector combinations produce almost simultaneous (in the millisecond range) information. Acousto-optical tunable filters are newer devices which are used to isolate narrow spectral bands. They can operate at scanning speeds at least 100 times faster than the mechanically scanned devices.

Detectors. The solid-state silicon photodiode has become ubiquitous in clinical instruments and is the detector of choice for the near ultraviolet–visible spectral range. Among the detector's attributes are small size, high quantum efficiency (especially in the visible–near infrared region), low noise equivalent power, wide linear range, excellent response times, and simplicity of interfacing to the data acquisition hardware. Silicon photodiodes are available in single-element packages and in multielement arrays.

Another class of detectors consists of the vacuum phototube devices. Included in this group is the photomultiplier detector, which has high gain but requires a high voltage power supply. It has an advantage over the silicon detector in that it has extremely low noise, and thus is useful in applications requiring extremely low level light detection and/or extremely fast response times. Additionally, the ultraviolet quantum efficiency of multialkali photocathodes is generally superior to that of the silicon photodiode.

The photoconductive detector is primarily used in the visible–infrared region rather than the ultraviolet–visible range.

Silicon charge coupled devices (CCDs), commonly used in solid-state video cameras and in research applications, are being applied to low light level spectroscopy applications. The main advantage of area array CCDs over linear photodiode detectors is the two-dimensional format, which provides simultaneous measurements of spatial and spectral data.

Computer System. The brain of the modern clinical chemistry analyzer is its computer system. The part of the computer system that controls the functional aspects of the analyzer is known as the process control computer or analytical processor (AP); the test results are handled by the data management computer, also known as the results processor (RP).

The analytical processor stores the sample identification information, and, checking against the worklist input by the operator, determines which tests should be performed on the current sample. The AP controls the aspiration of the right amount and type of reagents, dispenses the reagents together with the sample in the reaction cuvette, and collects the raw data from the detection system. Some tasks, eg, temperature control and motion profiles, are actually delegated to dedicated microprocessors reporting to the AP. These microprocessors are often referred to as embedded control processors. The analytical processor also monitors the status of various sensors in the analyzer and warns the operator of abnormal conditions by displaying messages on the computer monitor, or by triggering audible alarms. Other functions controlled by the AP include start-up and shut-down procedures, calibration procedures, and, in some analyzers, self-diagnosis of failure modes.

The results processor computes the test results from the raw data furnished by the AP and collates these results together with the demographic patient data into test reports. Test results falling outside normal limits are flagged on the report to speed up the diagnosis process. These data managers can also store thousands of patient reports in their current memory. Some of the more sophisticated systems also store the actual reaction curves used to determine the test results.

Whereas the analytical unit's form and function is somewhat determined by a particular technology, the results processor's approach may be influenced by a larger system strategy. The RP might be part of a single analyzer, a stand alone entity, or it could act as a single results processor for two or more analytical units. In another configuration, the RP could act as an interface between one or more analytical units or between one or more analytical units and a larger, more centralized computer system.

Economic Aspects

It is estimated that the worldwide clinical chemistry diagnostics market is about \$3 billion. This amount includes an estimated \$700 million in instrument sales, and \$2.3 billion in sales of reagents and consumables. Some of the principal instrument manufacturers are Hitachi (Japan), Miles Laboratories/Technicon Instruments (United States), E. I. du Pont de Nemours (United States), Beckman Instruments (United States), Eastman Kodak (United States), Abbott Laboratories (United States), Olympus (Japan), Toshiba (Japan), Hoffmann-La Roche (Switzerland), and Ciba Corning Diagnostics (United States).

The proliferation of automated analyzers in the 1980s enticed an increased number of manufacturers to enter the market, increasing competition and lowering selling prices. In 1991 list prices for automated clinical chemistry analyzers ranged from \$30,000 to \$80,000 for small (120–400 tests per hour) analyzers, \$100,000 to \$175,000 for medium (500–1500 tests per hour) throughput, and \$200,000 to \$350,000 for very high throughput systems capable of processing up to 10,000 tests per hour.

Automated clinical analyzers became affordable, not only to small hospital laboratories, but to doctors' group practices and to individual doctor's offices as well.

BIBLIOGRAPHY

1. *Report #7076,* Biomedical Business International, 1987.
2. R. Gambino, *14th International Congress of Clinical Chemistry,* San Francisco, Calif., July 22–26, 1990.
3. N. W. Tietz, ed., *Textbook of Clinical Chemistry,* W. B. Saunders Co., Philadelphia, Pa., 1986.
4. L. J. Kaplan and A. J. Pesce, *Clinical Chemistry, Theory, Analysis, and Correlation,* The C. V. Mosby Co., 1989.
5. G. Kessler and L. R. Snyder, *Technicon International Congress,* Vol. 1, Mediad Press, Tarrytown, N.Y., 1977, pp. 28–35.
6. I. M. Kolthoff, P. J. Elving, and E. J. Meehan, *Optical Methods of Analysis,* Vol. 8, *Treatise on Analytical Chemistry,* Part I, John Wiley & Sons, Inc., New York, 1986.
7. G. Eisenman and co-workers, *Science* **126,** 831 (1957).
8. G. Eisenman, *Glass Electrodes for Hydrogen and Other Cations, Principles and Practices,* Marcel Dekker, Inc., New York, 1967, p. 2.

PAUL GHERSON
HOWARD LANZA
MILTON PELAVIN
DAN VLASTELICA
Miles Inc.

HEMATOLOGY

Hematology analyzers provide information about blood cells and their constituents. The three basic blood cell types are erythrocytes or red blood cells, first described in 1674 by Van Leeuwenhoek using a microscope of his own design; leukocytes or white blood cells, noted by Hewson in 1773; and thrombocytes or platelets, described in 1841 by Addison, who also noted the granules present in certain of the leukocytes. Hemoglobin, the principal nonaqueous component of red blood cells, was identified by Hoppe-Seyler in 1864 when it was shown that hemoglobin is a complex of four Fe^{2+} protoporphyrins, called hemes, and a protein, or globin, consisting of two alpha and two beta chains. The absorption spectrum of the oxidized form of hemoglobin showing that hemoglobin can absorb and release oxygen was also reported (1). Hemoglobin's physiological importance gives it the status of a primary hematologic constituent.

The first hematologic cell-counting chamber, the hemocytometer, was developed in 1873. A drop of a diluted suspension of blood cells was inserted by capillary action between a glass cover slip and the finely cross-hatched surface of a glass chamber. The blood cells were allowed to settle to a common focal plane. Then, using a microscope, the red blood cells (RBC) within the cross-hatched areas were counted. Eventually, technologists used the hemocytometer to count white blood cells (WBC) and platelets (PLT) as well, first destroying the red blood cells which predominated. Hemocytometers are still used on occasion.

In 1877 Ehrlich developed a triacid stain that distinguished the nucleus and cytoplasm of cells in a stained blood film and originated the white blood cell differential count (Diff). By applying the triacid stain to a blood film slide and examining the slide under a microscope, five white cell types were distinguishable by the shapes of the nuclei and by the varying affinity of cell granules for the stain (2). The five white blood cell types are lymphocytes (LYMPHS) which have round nuclei and no granules; monocytes (MONOS) which have indented nuclei and sparse granules; neutrophils (NEUTS) which have polymorphonuclear (PMN) or segmented nuclei and granules that have affinity for neutral dyes such as picric acid-rosaniline; eosinophils (EOS) which have PMNs and granules having affinity for acidophilic dyes, such as eosin; and basophils (BASOS) which have PMNs and granules having affinity for basophilic dyes, such as fuchsin. Most clinical laboratories still routinely use Ehrlich's method for white blood cell differential counts, determining the number of each cell type in a 100 cell count, albeit using other stains.

Technologists have measured a sample's hemoglobin content in various ways. In 1878 a method for measuring hemoglobin concentration in whole blood (HGB) using visual photometry was introduced. Hemoglobin concentration may be defined as the number of grams of hemoglobin per deciliter of whole blood. Samples of blood were diluted and the color compared to that of a solution of picrocarmine. Later samples were converted to carboxyhemoglobin and the color compared to that of a carboxyhemoglobin standard or converted to acid hematin through the addition of dilute hydrochloric acid and the brown color compared to that of a standard. In 1920 a method to convert hemoglobin to cyanmethemoglobin by lysing the red cells in distilled water and adding potassium ferricyanide and potassium cyanide was introduced. The ferricyanide oxidized the heme to the

Fe^{3+} form, yielding methemoglobin, and the cyanide converted methemoglobin to cyanmethemoglobin. This method was later modified by adding sodium bicarbonate, in dilute form, to reduce the turbidity of the lysed product. Hemoglobin concentration was determined by comparing cyanmethemoglobin's color to that of a reference material or by measuring its photometric absorption at 540 nm and converting to hemoglobin concentration. Drabkin's solution, the ferricyanide, cyanide, bicarbonate mixture, was used for many years for photometric hemoglobin measurements because it effectively converted all forms of cellular hemoglobin to a single, stable form for photometric measurement. In 1966 the International Committee for Standardization of Hematology (ICSH) adopted the cyanmethemoglobin photometric method, which is related to the use of Drabkin's solution. Technologists also use physical properties of hemoglobin, such as specific gravity and refractive index, to determine concentration.

A method for determining a sample's total hemoglobin content was developed in 1929. A tube of whole blood was centrifuged, packing the red cells at the bottom of the tube, and the percentage of total sample volume occupied by the RBC was determined by dividing the height of the packed cells by the total height of the sample. This percentage was called packed cell volume (PCV) or hematocrit (HCT). In combination with RBC measurements, the hematocrit provided a sample's mean red blood cell volume (MCV) (3).

$$MCV = (HCT/RBC) \times 10$$

Two additional parameters, the mean cellular hemoglobin concentration (MCHC) and the mean cellular hemoglobin (MCH), can be obtained.

$$MCHC = (HGB/HCT) \times 100$$

$$MCH = (HGB/RBC) \times 10$$

By 1935 a technologist could perform a complete blood count (CBC) consisting of: WBC in $10^3/\mu L$, RBC in $10^6/\mu L$, HGB in g/dL, HCT in %, MCV in fL (10^{-15} L), MCH in pg, MCHC in g/dL, and PLT in $10^3/\mu L$; and a complete white blood cell differential count, consisting of all white blood cell types, ie, the five normal types plus any other cells not normally found in peripheral blood.

The first commercially successful automated blood cell counter, the Model A Coulter Counter, was introduced in 1956. The Model A counted cells by using electrical impedance properties, and the Coulter Counter quickly became the instrument of choice for counting red and white blood cells. When it was later demonstrated that cell volume was roughly proportional to the electrical impedance signal amplitude, the Coulter Counter was modified to provide MCV as the mean of the individually measured red cell volumes. Impedance counters calculated HCT as (RBC $\times$ MCV)/10 and, by the early 1970s, these instruments also counted platelets. By the mid-1970s, impedance counters combined with photometric hemoglobinometers to produce CBCs.

In 1965 Technicon introduced the SMA 4/A Autocounter, which counted red and white cells by using light-scattering properties and determined a sample's HCT by measuring electrical conductivity. The instrument included a photometer for automated HGB measurements. The Autocounter was the basis for the first fully automated CBC analyzer including PLT, the Hemalog 8, introduced in 1970. The first commercial automated five-part Diff analyzer, the Hemalog D, introduced in 1971 by Technicon Instruments Corporation, produced five-part white blood cell differential counts by combining measurements of the light-scattering and cytochemical properties of white cells. By 1975 Geometric Data, Corning Medical and Scientific, Abbott Laboratories, and Perkin-Elmer each had instruments on the market. The latter instruments were automated slide readers equipped with computerized image analyzers (4). Commercial success was limited, however, because these instruments were only slightly faster than manual differential counting methods. Additionally, they were expensive to operate and suffered from questionable quality of the prepared blood film slides.

The first automated CBC/five-part Diff analyzer was introduced in 1981. This instrument used modified Hemalog D 90 technology as well as other light-scattering measurements and photometric hemoglobinometry. It was followed by an improved version, the H*1 (Technicon), that incorporated a novel method for measuring certain red blood cell parameters. Then in 1989 TOA introduced the NE-8000, which produces five-part differentials by combining impedance technology with cytochemistry; Unipath introduced the Cell-Dyn 3000, which uses light scattering only to produce five-part differentials; and Coulter introduced the Coulter STKS, which performs five-part differential counts by using impedance and light scattering.

Aperture Impedance Instruments

The aperture impedance principle of blood cell counting and sizing, also called the Coulter principle (5), exploits the high electrical resistivity of blood cell membranes. Red blood cells, white blood cells, and blood platelets can all be counted. In the aperture impedance method, blood cells are first diluted and suspended in an electrolytic medium, then drawn through a narrow orifice (aperture) separating two electrodes (Fig. 1). In the simplest form of the method, a d-c current flows between the electrodes, which are held at different electrical potentials. The resistive cells reduce the current as the cells pass through the aperture, and the current drop is sensed as a change in the aperture resistance.

Aperture impedance counters are flow cytometers, as are all current automated cell counters. Flow cytometers measure cells as the cells flow hydraulically, one at a time, through sensing zones. All flow cytometers, including aperture impedance counters, count fewer cells than actually pass through their sensing zones (6–11). When two or more cells are in a sensing zone simultaneously, by coincidence, the counter senses only one cell. A number of approaches may be taken to correct for count loss resulting from coincidence. The simplest is to dilute the sample to the point where the probability of coincidence is acceptably small (12,13). However, because this approach typically requires dilution factors of a million or more, it introduces dilution errors. Also, because dilution significantly reduces the number of cells counted, this approach also increases sampling error. An alternative approach is to model the probability of coinci-

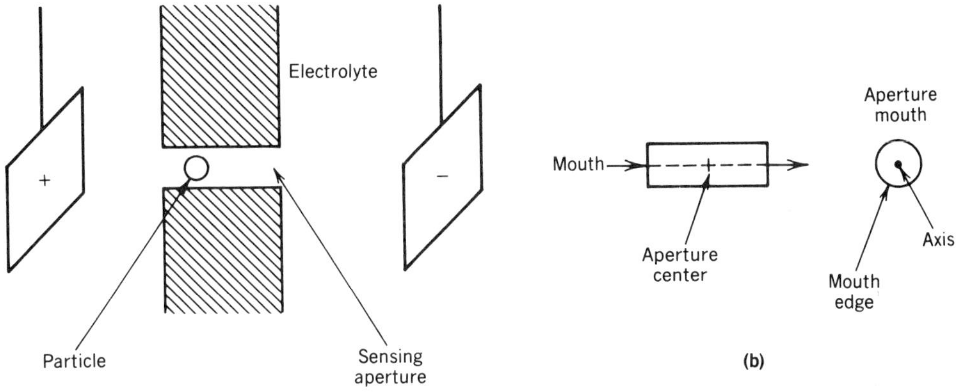

(a)

Fig. 1. (**a**) Schematic diagram of the electrical impedance method for counting and sizing blood cells. The electrodes are represented as charged rectangles. As particles flow through an aperture, the impedance and, hence, the flow of current are modified; (**b**) schematic diagram of the sensing aperture of a flow cytometer.

dence as a function of either sample concentration or the fraction of time the sensing zone is occupied by passing cells (dead time) (6–11). The model produces a coincidence-correction factor which is applied to the count. Still another approach is to reduce the size of the sensing zone, which reduces the probability of coincidence (7). In practice, commercial counters combine the latter two approaches.

Flow cytometer cell counts are much more precise and more accurate than hemocytometer counts. Hemocytometer cell counts are subject both to distributional (13) and sampling (14–16) errors. The distribution of cells across the surface of a hemocytometer is sensitive to the technique used to charge the hemocytometer, and nonuniform cell distribution causes counting errors. In contrast, flow cytometer counts are free of distributional errors. Statistically, count precision improves as the square root of the number of cells counted increases. Flow cytometer counts usually involve 100 times as many cells per sample as hemocytometer counts. Therefore, flow cytometry sampling imprecision is one-tenth that of hemocytometry.

Aperture impedance counters provide cell volume information as well as cell counts. The relationship between aperture resistance change and cell volume may be expressed as (17–20):

$$\frac{DR}{R} = f\frac{v}{V}$$

where DR = change in resistance resulting from the presence of a cell in the aperture, R = aperture resistance of the electrolyte alone, f = function of particle shape and orientation, v = cell volume, and V = effective aperture volume. Aperture impedance counters provide mean red cell volume (MCV) and mean platelet volume (MPV), both in femtoliters, as:

$$MCV = \frac{\Sigma \text{ red cell volumes}}{RBC}$$

$$MPV = \frac{\Sigma \text{ platelet volume}}{PLT}$$

Aperture impedance and most other automated counters measure MCV and RBC independently, in contrast to the manual methods where MCV and MCH accuracies depend on hemocytometer red cell count accuracy.

Automated counters can also provide cell volume distribution information because they measure volume on a cell-by-cell basis. Manual methods measure MCV indirectly. Automated counters provide red cell volume distribution width (RDW) and platelet volume distribution width (PDW) where RDW is the standard deviation of red cell volumes/MCV and PDW is the standard deviation of platelet volumes. RDW measured as a percentage and PDW in units of femtoliters have descriptive value. For example, a high RDW value suggests the presence of microcytes (small red cells) and/or macrocytes (large red cells). Microcytes may be associated with red blood cell iron deficiency and macrocytes with vitamin B-12 or folate deficiency. Automated counters also group cell volumes by frequency to form cell volume histograms. The histograms provide integrated views of cell volume distribution (Fig. 2).

Several factors affect aperture impedance cell volume measurements. DR/R depends on f, the shape and orientation of the cells passing through the aperture. For spherical objects $f = 1.5$, whereas for needlelike objects oriented along the cylindrical axis of the aperture, $f = 1$ (20–22). Additionally, as needle-shaped objects tumble, the f values change. Red cells are especially subject to shape and orientation effects in apertures because these cells are highly deformable and their shape in the aperture depends on the shear forces therein. Photographs show that as red cells pass through apertures at typical flow rates, they deform from their native biconcave disk shape to one approximating an oblate ellipsoid (cigar-shape) (22). If all red cells deformed to the same extent, deformability could be factored into f. However, red cell deformability depends on cellular hemoglobin concentration. High cellular hemoglobin concentration reduces red cell deformability thereby reducing elongation from shear forces. Less deformed red cells have higher f values than normally deformed cells. Therefore, aperture impedance counters often assign erroneously high cell volume values to red cells having high cellular hemoglobin concentration, and erroneously low cell volume values to cells where the cellular hemoglobin concentration is low. A fraction of the red cells may tumble about their long (deformation) axis as they pass through the aperture along the aperture wall (21–23), because of discontinuous changes in hydrodynamic forces near the wall. As the cells tumble, f increases and skews MCV upward.

Aperture impedance measurements of cell volume are affected by variations in the electric field strength that occur in the region of the aperture. The field is strongest at the edge of the aperture mouth and weakens as the aperture axis is

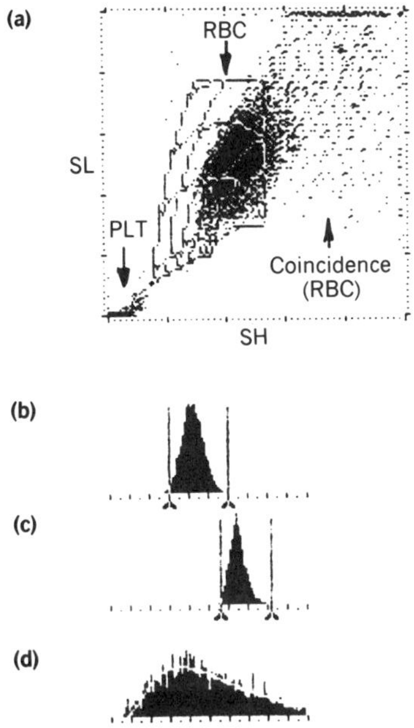

Fig. 2. Histograms representing (**a**), the frequency of occurrence of platelets (PLT) and red blood cells (RBC) and illustrating coincidence where SL is low angle scatter, 2°–3°, and SH is high angle scatter, 5°–15°; (**b**), the volume of RBC, from 0 to 200 fL; (**c**), the hemoglobin (HGB) concentration from 0 to 50 g/dL; and (**d**), the platelet volume from 0 to 20 fL.

approached (21,22) (Fig. 1**b**). The field also weakens from the aperture entrance to the center of the aperture. Therefore, DR/R and computed cell volumes are greater for cells passing along the aperture wall than for cells passing along the aperture axis.

A number of investigators (24–26) demonstrated that aperture impedance cell volume measurements could be improved by focusing or restricting cell flow to the aperture axis, thus reducing variations resulting from inhomogeneous electric and hydrodynamic fields. Counters that caused the cell suspensions and electrolyte solutions to flow concentrically through the apertures were designed. The electrolytes sheathed the cell suspensions, narrowing them and centering the cells on the aperture axis.

Aperture impedance measurements of cell volume must take into account the osmolality and pH of the medium. A hypotonic medium causes cells to swell; a hypertonic medium causes them to shrink. Some manufacturers of aperture impedance counters deliberately provide hypertonic electrolytic media for red blood cell measurements. The shrunken red cells not only become more nearly spherical and thus less affected by orientation, but also less deformable than cells in isotonic media and thus less affected by differences in hemoglobin content.

The basic aperture impedance method can produce three-part white cell differential counts. Impedance counters can distinguish three white cell types by size: the LYMPHS, mid-range cells including MONOS and BASOS, and granulocytes including NEUTS and EOS.

A refined version of the aperture impedance method which combines radio frequency (r-f) aperture currents and d-c currents, can produce four-part differentials. R-f aperture impedance is sensitive to the internal structure of white cells (27). R-f aperture impedance measurements distinguish BASOS from LYMPHS by virtue of BASOS lower nucleus-to-cytoplasm ratio. Therefore, the combined d-c/r-f aperture impedance counter distinguishes LYMPHS, MONOS, BASOS, and NEUTS + EOS. To distinguish EOS from NEUTS, and thereby produce five-part differentials, requires the use of light-scattering measurements.

In summation, aperture impedance counters provide information on WBC, RBC, HCT, MCV, PLT, MPV, RDW, PDW, and three-part Diff, for d-c current only, or four-part Diff using a d-c/r-f combination.

Light-Scattering Instruments

The light-scattering principle of cell counting is based on the observation that microscopic particles, such as blood cells, scatter into small (0–15°) angles, most of the visible light incident upon them (28). This principle is used to count red blood cells, white blood cells, and platelets. In the basic form of the light-scattering method, a dilute suspension of cells and a sheathing fluid having a matching refractive index flow concentrically through an optical flow cell onto which light is focused. The cells scatter the incident light as they pass through the flow cell, and the light scattered into a small-angle interval, typically 1–3 degrees, is sensed by an optical detector. The purpose of the sheath is to narrow the cell stream and reduce the cell passage to single file while centering the stream in the flow cell. The unscattered component of focused light is much larger than the scattered component and must be blocked from the detector so that it does not swamp the desired small-angle scatter signal (Fig. 3).

The basic single-angle interval light-scattering method cannot accurately measure individual red blood cell or platelet volumes, but it can provide MCV and MPV. Red cells are bi-concave disks, and platelets are rod to disk shaped. Scattering intensities depend on the orientation in the flow cell. Because the cells can interrupt the optical path in random orientations, individual scattering intensities are not proportional to cell volume. However, because thousands of cells of each type pass through the flow cell, the effects of orientation can be averaged. To a first approximation HCT and platelet crit (PCT), the percentage of blood sample volume occupied by platelets, is proportional to the sums of the scattering intensities of the red cells and platelets, respectively. MCV can be computed from HCT and RBC, whereas MPV can be computed from PCT and PLT. The accuracy of MCV determined by this method is tied to the RBC accuracy, as is the case for the manual MCV method. Ortho Instruments Corporation's ELT-8 uses these counting and sizing methods.

Light-scattering measurements of sphered cells are not subject to orientation effects, and a method for the rapid sphering and fixing of red cells for the purpose of measuring them in a light-scattering flow cytometer has been developed (29). The reagents used to lyse red blood cells, typically surfactants (qv), can also be used at lower concentrations to sphere them. In fact, the sphering process is part of the overall process of hemolysis (30). The red cell's sphered state is the

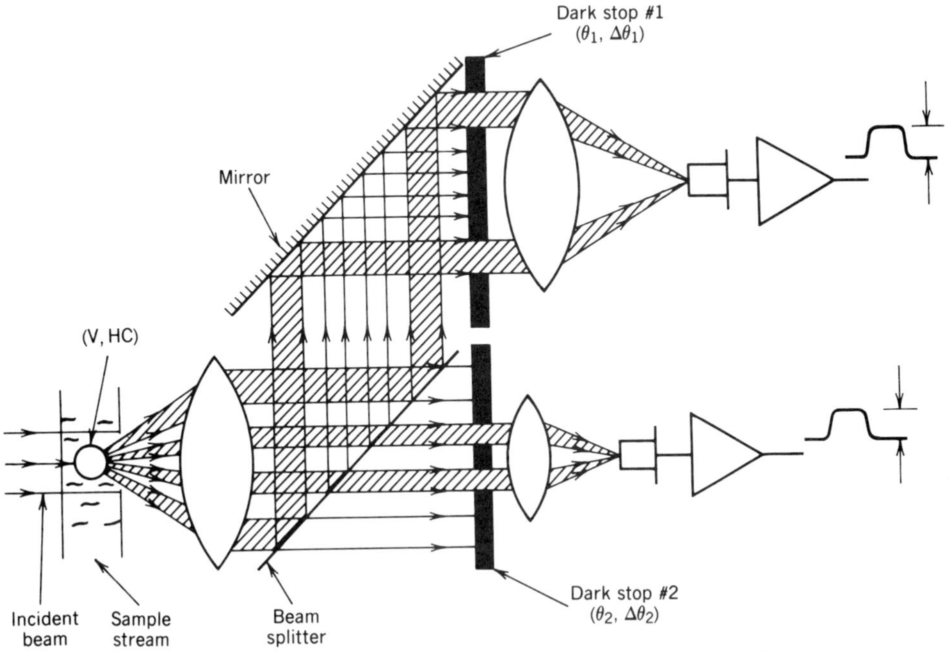

Fig. 3. Schematic diagram of a dark field system for measuring the light scattering by a spherical red blood cell where V = volume of sample and HC = hemoglobin concentration. Optical flow cell system having double-angular interval detection at angles θ_1 and θ_2.

last intact state before it is hemolyzed. Hemolysis is, however, undesirable because, depending on the extent of hemolysis, red cells are reduced in size or totally destroyed. The sphering agent concentration should not be too high, because it can hemolyze the red cells in samples having low red blood cell concentrations. However, the concentration of sphering agent must be high enough to sphere all the red cells in samples having high red blood cell concentrations. Fixing the cells after they are sphered prevents hemolysis in samples having low RBC values, yet permits concentrations of sphering agents that are high enough to sphere all red cells in concentrated red cell suspensions.

A single reagent capable of rapidly sphering and fixing all the red blood cells in a sample over a wide range of red blood cell concentrations has been developed. The reagent composition ensures that red cells are not fixed before they are sphered, and that the cells do not lyse before they are fixed. The sphering and fixing reagent contains sodium dodecyl sulfate (SDS) [*151-21-3*], a buffer, and gluteraldehyde [*111-30-8*], $C_5H_8O_2$. SDS is the sphering agent, the buffer provides an isotonic medium in which to isovolumetrically sphere red cells, and gluteraldehyde is the fixing agent. The Technicon H*6000 Hematology Analyzer incorporated this method in order to measure red cell volume on a cell-by-cell basis. In contrast to the light-scattering method for sphered and fixed red cells, the aperture impedance method for determining the MCV of unsphered red cells is subject to orientational effects.

Light-scattering measurements of red cell volume are accurate only if the measurements account for the effects of cell refractive index on scattering intensity. Basic single-angle interval light-scattering measurements of red cell volume, even for sphered and fixed red cells, are only accurate to within approximately 10%. Mie Scattering Theory predicts that the angular scattering intensity pattern for sphered red cells depends on cell refractive index as well as cell size (31,32). Red cell refractive index is linearly related to cellular hemoglobin concentration (33). Therefore, two normal red cells of the same volume but of different cellular hemoglobin concentration scatter light differently. A sphered cell's volume and the hemoglobin concentration can both be accurately determined from Mie Scattering Theory calculations by making simultaneous measurements of light-scatter intensity over two suitably chosen angle intervals (34). Two suitable angle intervals for red cells are 2–3 degrees, known as low angle scatter (SL), and 5–15 degrees, high angle scatter (SH) (Fig. 2). In contrast to the aperture impedance method, this method for determining red cell volume does not suffer from inaccuracies as a result of cellular hemoglobin concentration variability.

In principle, the two-angle interval method can produce all CBC parameters within a single measurement channel, uniquely providing cell-by-cell hemoglobin concentration. The mean of the concentrations provides an alternative (and direct) measurement of MCHC. The method also provides an alternative HGB measurement, because HGB may be set equal to (RBC × MCV × MCHC)/1000. This method, like the basic light-scattering method, uses the same flow cell to measure platelets and red cells with the result that the method is capable of providing the CBC parameters RBC, HGB, HCT, MCV, MCHC, MCH, and PLT. The method can also count a sample's white blood cells if the sample's red blood cells have been lysed.

The distribution of cellular hemoglobin concentrations within a blood sample, also known as hemoglobin distribution width (HDW), can be provided. HDW is the standard deviation of cellular hemoglobin concentrations measured in units of g/dL. HDW has descriptive value: a high HDW value suggests the presence of abnormally high numbers of hyperchromic and/or hypochromic red cells, ie, either high or low cellular hemoglobin concentration cells. Hypochromic cells are usually associated with anemia, whereas hyperchromic cells may be associated with hereditary spherocytosis or spherical blood cells, a condition in which the survival rate of the spherocytes is decreased. Instruments using the two-angle interval scatter method can display cellular hemoglobin concentration histograms which provide integrated views of cellular hemoglobin concentration distributions (Fig. 2).

In summation, the two-angle interval scattering method can provide CBC, RDW, PDW, PCT, and HDW. This two-angle interval method has been incorporated into the Technicon H*1 system which provides RBC, HCT, MCV, PLT, MPV, RDW, PDW, PCT, and HDW.

The single-angle interval light-scattering method counts white cells but cannot produce five-part Diffs. White cell light-scattering intensity depends on cell size, cell refractive index, and internal cell structure. White cell size ranges overlap and the cells differ in refractive index. Scattering measurements can distinguish among cell types based on differences in internal structure alone, but only if the cell types have the same overall size and refractive index.

Light-scattering measurements made over two suitably chosen angle intervals can combine with depolarized light-scattering measurements to provide five-part Diffs (35). Unipath's Cell-Dyn 3000 instrument measures white cell light scatter at 1–3 degrees and at 3–11 degrees in order to resolve LYMPHS, BASOS, MONOS, and NEUTS + EOS. The two-angle interval light-scattering method distinguishes the cell types based on size and refractive index and is similar to the Technicon H*1 RBC/PLT method. The Cell-Dyn 3000 instrument uses the scattering behavior of EOS granules to distinguish EOS from NEUTS. EOS granules are refractile, numerous, large, and closely packed. Light incident on the granules is scattered a number of times, from granule to granule, before reaching the detectors. If the incident light is polarized, multiple scattering depolarizes the light scattered onto the detectors. The other white cell types scatter polarized incident light once, but do not depolarize it. Depolarization as a result of multiple scattering is enhanced at high scattering angles. Thus the Cell-Dyn 3000 measures the depolarized and polarized components of light scattered at 90 degrees to incidence by white cells and plots depolarized scattering intensity vs polarized scattering intensity for each cell. A distinct EOS cluster forms on the plot.

Basic light-scattering measurements can be combined with other measurements to provide five-part Diffs. Coulter's STKS combines d-c/r-f aperture impedance and light-scattering measurements to provide five-part Diffs. The STKS lyses a sample's red cells, then shrinks the white cells by suspending them in a hypertonic medium. The system distinguishes among the shrunken cells based on cell size differences and internal structure differences related to r-f impedance and light-scattering characteristics. D-c/r-f measurements distinguish among LYMPHS, BASOS, MONOS, and NEUTS + EOS based on cell size and r-f impedance. Small-angle scattering measurements resolve EOS and NEUTS based on EOS granules scattering behavior.

Cytochemistry

Cytochemical techniques can be combined with light-scattering and absorption measurements to provide five-part Diffs. Cytochemistry concerns the chemical reactions of cell components. The reactions for automated white blood cell differential analysis include those that bind chromophores to the granules of white cell types, based on the presence of various substrates and enzymes in the granules. These reactions yield products suitable for light-scattering and optical absorption measurements. Other reactions exploit the differential resistance of white cell types to cytoplasmic stripping by the lysing action of surfactants. The reaction products are suitable for light-scattering and aperture impedance measurements. Optical absorption measurements made on intact cells in suspension are different from those made on dissolved hemoglobin. Measurements made on intact cells involve reduced overall scattering intensity and do not obey Beer's Law; measurements made on solutions involve dissolved chromophores and do obey Beer's law (see SPECTROSCOPY).

Although both the manual and light-scatter–cytochemistry methods of counting white cell differentials involve the staining of white cells, the staining and measurement techniques are different (36). Classic stains used in manual differential analyses, such as Wright's Stain and Giemsa Stain, consist mainly of azure dyes, which are basic, and eosin dyes, which are acidic. The DNA of white cell nuclei are azurophilic (basophilic) and are stained blue to magenta by the azure dyes, as are the granules of NEUTS and the DNA contained in the cytoplasm of LYMPHS and MONOS. The heparin in BASOS granules is also basophilic. The acidophilic (eosinophilic) proteins in EOS granules and nonazurophilic NEUT granules, however, are stained red by eosin. The white blood cell types are not distinguishable by azure–eosin staining specificity and intensity alone. In order to distinguish among the white blood cell types, a visual examination of differences in their size, shape, granular content, and nuclear structure is also necessary.

The Technicon Hemalog D was the first hematology analyzer to combine cytochemistry with light-scattering–optical absorption measurements to produce five-part Diffs. This instrument used cell-specific staining properties of the leukocytes as well as their light-scattering behavior to distinguish among the five normal leukocyte types (37,38). BASOS are unique in that the heparin in their granules reacts with alcian blue to precipitate a blue-green dye in the granules. MONOS alone contain lipase, an enzyme catalyzing the reactions of esters, and the substrates α-naphthyl butyrate and basic fuchsin combine, in the presence of lipase and a diazonium salt, to precipitate a red-brown azo dye. Only EOS are peroxidase-active below pH 3. Between pH 3 and pH 5, NEUTS are also peroxidase-active, though not as strongly as EOS, as are MONOS weakly peroxidase-active in this pH range. Therefore, EOS, NEUTS, and LYMPHS + BASOS + MONOS are distinguishable by peroxidase staining alone. By combining the three staining techniques, the Hemalog D could have distinguished among the five normal white cell types based on stain absorption alone. However, the Hemalog D achieved improved resolution of LYMPHS and MONOS by adjusting peroxidase staining conditions and measuring white cell light scattering as well as absorption in the peroxidase channel.

In order to use the leukocytyes' enzymatic properties, the Hemalog D carried out reactions under specific conditions. Because enzymes are easily damaged or released from cells, the cells were fixed to stop metabolism and to immobilize the enzymes without significantly changing their catalytic activity. Fixation was controlled so that plasma proteins did not precipitate and generate spurious particle counts, clogging the system. Red cells were lysed at elevated temperatures to destroy red cell and serum catalase and hemoglobin pseudoperoxidase activity, which would otherwise have completely consumed the peroxidase reaction substrates. Then plasma esterase inhibitor was inactivated by adding concentrated formaldehyde and ethylene glycol, in which the inhibitor is labile. The esterase inhibitor would otherwise have stopped the MONO reaction. In order to specifically bind alcian blue to heparin and not all nuclei, the BASO reaction was carried out at pH 2.2. The three staining reactions are

1. Peroxidase reaction for NEUTS, EOS, (MONOS)

$$\text{4-chloro-1-naphthol} + H_2O_2 \xrightarrow[\text{peroxidase}]{} \text{dark precipitate}$$

2. BASO reaction

$$\text{alcian blue} + \text{heparin (in BASO granules)} \longrightarrow \text{blue-green dye}$$

3. MONO reaction

$$\alpha\text{-naphthyl butyrate} + \text{basic fuchsin} \xrightarrow[\text{lipase}]{\text{MONOS}} \text{butyric acid} + \alpha\text{-naphthol}$$

$$\alpha\text{-naphthol} + \text{diazonium salt} \longrightarrow \text{red-brown azo dye}$$

The system measured small-angle light-scattering intensity and/or cell stain absorption in each of three reaction channels. In the MONOS channel, an absorption detector registered as MONOS those white cells that absorbed light at the MONOS stain-absorption wavelength as the cells passed through a flow cell. In the BASOS channel, white cells interrupted a broad wavelength band of focused light as they passed through a flow cell. Two scattering detectors, one with a green filter, the other with a red filter, registered scattering intensity. The analyzer counted as BASOS those cells that scattered less green light than red light. In the peroxidase channel, white cells scattered and absorbed light differentially, depending on their peroxidase activity, as they passed through a flow cell. The analyzer plotted small-angle light-scattering intensity versus absorption, for each cell. The plotted points formed clusters by cell type. Cluster-analysis algorithms distinguished among LYMPHS + BASOS, NEUTS, EOS, and, to a limited extent, MONOS. The Hemalog D also used the sum of the small-angle scatter signals in the peroxidase channel to provide WBC.

The Technicon H*6000 system used a different combination of cytochemistry and light-scattering–absorption measurements to produce five-part Diffs. The system retained the Hemalog D peroxidase channel but eliminated the MONO channel, thus reducing the number of system flow cells by one. The system relied on modified staining conditions and peroxidase channel cluster analysis to distinguish MONOS as well as LYMPHS + BASOS, NEUTS, and EOS (39). A series of reagents were developed that better fix monocytes, as well as the other white cell types, than did the Hemalog D reagents. The well-fixed cells are better able to retain peroxidase stain, stain more uniformly, and have narrower size distributions than the cells fixed by Hemalog D reagents. The resultant peroxidase channel clusters are tighter and thus easier to distinguish.

The Technicon H*1 and H*2 systems use a different method to distinguish BASOS, exploiting the BASOS resistance to cytoplasmic stripping. The BASOS reagent contains surfactant and phthalic acid plus hydrochloric acid. The surfactant lyses the red cells. Additionally, surfactant and phthalic acid, at low pH, strip the cytoplasm from all white cell types except BASOS. The reaction mixture then contains intact BASOS and the nuclei of the other white cell types. BASOS and nuclei pass through the same flow cell as the one used for RBC/PLT measurements. Intact BASOS and nuclei scatter light into the SL and SH detectors. BASOS scatter much more light into the SL detector than do nuclei because the intact BASOS are much larger. Thus the analyzer distinguishes BASOS by the high SL scattering intensity.

The H*1 and H*2 method of performing five-part Diffs is superior to the H*6000 method because a single stain is used instead of two; the same flow cell is used for BASOS measurements as for the RBC/PLT ones, thus reducing the number of flow cells from three to two; and the BASO channel provides more information about white cells because the BASO channel produces the lobularity index (LI), a measure of granulocyte maturity, in addition to producing a percentage of BASO and absolute BASO count.

The H*1 and H*2 systems plot SL intensity versus SH intensity for the nuclei. Two clusters form on the BASOS cytogram as shown in Figure 4. The left-hand cluster is comprised of LYMPHS and MONOS, which are both mononuclear (MN), or nonsegmented. The right-hand cluster is comprised of NEUTS and EOS, which are polymorphonuclear (PMN), or segmented. SH intensity is sensitive to refractive index and number of nuclear segments. The higher the refractive index or the greater the number of nuclear segments, the higher the SH intensity. Mature granulocytes have more nuclear segments than younger granulocytes. In addition, the more mature nuclei have higher refractive indexes than the less mature ones, because the more mature ones are denser. Therefore, the more mature segmented nuclei appear further to the right on the BASO cytogram than the less mature segmented nuclei. Technicon H*1 and H*2 systems define the lobularity index (LI) as the ratio of the PMN histogram mode channel to the MN histogram mode channel (Fig. 4).

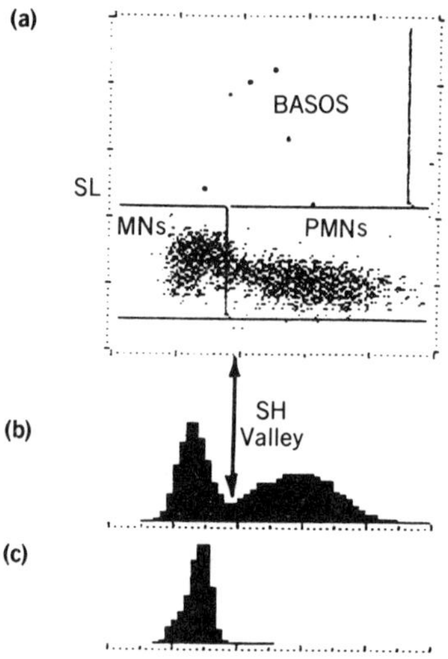

Fig. 4. (**a**), BASOS cytogram showing intensities for BASOS, mononuclear (MNs), and polymorphonuclear (PMNs) cells; (**b**), intensity at SH, high angle scatter; (**c**), intensity at SL, low angle scatter.

The TOA NE-8000 system combines d-c/r-f aperture impedance measurements with cytochemistry to produce five-part Diffs. The d-c/r-f impedance measurements are used to distinguish LYMPHS, mid-range cells, and granulocytes. The NE-8000 system lyses all white cell membranes except those of BASOS and identifies the BASOS by their higher d-c aperture impedance. Then, in a separate reaction, the NE-8000 lyses all white cell membranes except those of EOS and identifies EOS by their higher d-c aperture impedance.

Various method combinations can be used to produce five-part Diffs including d-c/r-f aperture impedance plus light scattering; two-angle interval light-scattering plus depolarized light scattering; light scattering plus cytochemistry;

and d-c/r-f aperture impedance plus cytochemistry. Flow cytometric methods produce five-part Diffs that are generally more precise and accurate than manual Diffs. However, manual methods can provide more information. White cell differential counts based on microscopic examination of blood smears are subject to sampling errors (14–16). Manual white cell differential counts are usually based on 100 cells. Therefore, for basophils which comprise 0–2% of the differential count, and for eosinophils (1–4%), and even for monocytes (2–8%), statistical sampling error is large. Flow cytometer differential counts, which are typically based on 5,000–10,000 cells, have much smaller sampling errors for these white cell types. Also, flow cytometers provide both direct absolute counts of each white cell type and differential counts, whereas manual methods provide only indirect absolute counts as the product of the differential count and WBC. Manual white cell differential counts are subject to distributional errors (40–43). Drawing a film of blood across the surface of a glass slide tends to force the larger, less dense blood cells to the edges of the slide. Uneven cell distribution across the surface of the slide can cause counting errors. Flow cytometric differential counts are not subject to distributional error. On the other hand, a technologist performing manual white blood cell differential counts can identify abnormal cell types that the automated analyzers cannot identify.

Photometric Hemoglobinometry

Most automated hemoglobinometry methods are derivatives of the manual ICSH reference method (44). The manual method follows three steps: (*1*) The blood sample is diluted in a reagent containing Triton X-100, a nonionic surfactant. The surfactant lyses the RBC membrane, releasing hemoglobin into solution. (*2*) Ferricyanide present in the diluent diffuses into the hemoglobin molecule and oxidizes heme from the Fe^{2+} to the Fe^{3+} form, yielding methemoglobin. (*3*) Cyanide diffuses into methemoglobin's interior and reacts with heme to yield cyanmethemoglobin (HiCN), which has a characteristic absorption spectrum that is measured spectrophotometrically.

The Technicon H*1 HGB automated method follows four steps: (*1*) The blood sample is diluted in a reagent containing lauryldimethylamine *N*-oxide (LO), an ionic surfactant and more powerful lytic agent than Triton X-100. The RBC membrane is lysed as is Step 1, above. (*2*) The combined action of surfactant and a pH 11.3 releases hemes from globin as a mixture of ferrous (Fe^{2+}) and ferric (Fe^{3+}) hemes. (*3*) Hemes are air-oxidized to ferric (Fe^{3+}) form. (*4*) The ferric hemes are ligated by cyanide and micellized by LO, forming reaction products with characteristic absorption spectra that are measured spectrophotometrically.

The automated method differs from the ICSH method chiefly in that oxidation and ligation of heme iron occur after the hemes have been released from globin. Therefore, ferricyanide and cyanide need not diffuse into the hemoglobin and methemoglobin, respectively. Because diffusion is rate-limiting in this reaction sequence, the overall reaction time is reduced from approximately three minutes for the manual method to 3–15 seconds for the automated method. Reaction sequences in the Coulter S+II and the Technicon H*1 and H*2 are similar. Moreover, similar reactions are used in the other Coulter systems and in the TOA and Unipath instruments.

Summary

All automated hematology analyzers are comprised of flow cytometers, which count and size red cells, white cells, and platelets, and photometric hemoglobinometers, which measure hemoglobin concentration. All analyzers produce CBCs and most also produce MPV, RDW, and PDW. Automated CBC analyzers use either the basic aperture impedance method or one of the light-scattering methods to count and size cells. Automated analyzers that also produce five-part Diffs use some combination of d-c/r-f aperture impedance, light scattering, and cytochemistry to do so. One disadvantage of automated systems is the inability to identify abnormal cell types.

BIBLIOGRAPHY

1. M. M. Wintrobe, *Blood Pure and Eloquent,* Vol. 11, McGraw-Hill Book Co., Inc., New York, 1980.
2. J. L. Binet, *Blood Cells* **6,** 507–511 (1980).
3. M. M. Wintrobe, *J. Lab. Clin. Med.,* (15), 287–289 (1929).
4. *Geometric Data: Hematrack Automated Differential System: Model-590, Operator's Manual,* Geometric Data, Wayne, Pa., 1982.
5. U.S. Pat. 2,656,508 (1953), W. H. Coulter.
6. J. L. Grant and co-workers, *Amer. J. Clin. Path.* **33,** 138–143 (1960).
7. M. Wales and co-workers, *Rev. Sci. Inst.* **32,** 1132–1136 (1961).
8. L. H. Princen and co-workers, *Rev. Sci. Inst.* **36,** 646–653 (1965).
9. W. B. Mercer, *Rev. Sci. Inst.* **37,** 1515–1520 (1966).
10. P. W. Helleman, *The Coulter Electronic Particle Counter: Aspects and Views in Counting and Sizing of Erythrocytes,* De Bilt Holland, the Netherlands, 1972, Chapt. 7–11.
11. U.S. Pat. 4,447,883 (1984), G. A. Farrell and co-workers.
12. A. R. Jones, in O. W. Van Assendelft and J. M. England, eds., *Advances in Hematological Methods: The Blood Count,* CRC Press, Boca Raton, Fla., 1982, Chapt. 5.
13. C. F. T. Mattern and co-workers, *J. Appl. Physiol.* **10,** 56–70 (1957).
14. C. L. Rumke, *Blood Cells* **11,** 137–140 (1985).
15. *Ibid.,* pp. 141–144.
16. D. W. Ross, *Blood Cells,* **11,** 145–157 (1985).
17. D. C. Golibersuch, *Biophys. J.* **13,** 265–280 (1973).
18. N. B. Grover and co-workers, *Biophys. J.* **9,** 1398–1414 (1969).
19. *Ibid.,* pp. 1415–1425.
20. N. B. Grover and co-workers, *Biophys. J.* **12,** 1099–1211 (1975).
21. C. S. Waterman and co-workers, *Clin. Chem.* **21**(9), 1201–1211 (1975).
22. B. B. Shank and co-workers, *J. Lab. Clin. Med.* **74**(4), 630–641 (1969).
23. J. Gutman, PhD dissertation, Max Planck Institute, Munich, Germany, 1965, p. 17.
24. V. Kachel, *J. Histochem. Cytochem.* **24**(1), 211–230 (1976).
25. J. T. Merrill and co-workers, *Rev. Sci. Inst.* **42**(8), 1157–1163 (1971).
26. R. Thom, in G. Izak, ed., *Modern Concepts in Hematology,* Academic Press, New York, 1972, pp. 191–200.
27. U.S. Pat. 3,502,974 (1970), W. H. Coulter.
28. H. C. van de Hulst, *Light Scattering by Small Particles,* John Wiley & Sons, Inc., New York, 1957, Chapt. 8.
29. U.S. Pat. 4,574,490 (1986), L. Ornstein.

30. E. Ponder, *Hemolysis and Related Phenomena,* Grune and Stratton, New York and London, 1971, pp. 10–49.
31. Ref. 28, Chapts. 9 and 12.
32. M. Kerker, *The Scattering of Light and Other Electromagnetic Radiation,* Academic Press, Inc., Orlando, Fla., 1969, Chapts. 3 and 4.
33. R. Barer, in A. W. Pollister, ed., *Physical Techniques in Biological Research* IIIa, Academic Press, New York, 1966, Chapt. 1.
34. U.S. Pat. 4,735,504 (1988), D. H. Tycko.
35. U.S. Pat. 5,017,497 (1991), G. de Grooth and co-workers.
36. H. Ansley and L. Ornstein, *Advances in Automated Analysis,* Vol. 1, Thurman Associates, Miami, Fla., 1971, pp. 437–446.
37. A. M. Saunders, *Clin. Chem.* **18**(8), 783–788 (1972).
38. H. P. Mansberg and co-workers, *J. Histochem and Cytochem* **22**(7), 711–724 (1974).
39. U.S. Pat. 4,801,549 (1989), J. Cremins and co-workers.
40. J. V. Dacie and co-workers, *Practical Hematology,* 5th ed., Churchill Livingstone, 1975, pp. 48–51.
41. A. Simmons, *Technical Hematology,* 2nd ed., J. B. Lippincott Co., 1976, pp. 100–103.
42. C. E. Seiverd, *Hematology for Medical Technologists,* 5th ed., Lea and Febiger, 1983, p. 255.
43. R. E. Miller, *Clinics in Laboratory Medicine,* Vol. 1, 1st ed., 1985, pp. 143–147.
44. M. J. Malin and co-workers, *J. Clin. Path.* **92**(3), 286–294 (1989).

DAVID ZELMANOVIC
ROQUE A. GARCIA
JOHN TURRELL
Technicon Instruments Corporation

AUTOMATION. See PROCESS CONTROL; ROBOTS IN THE SCIENTIFIC LABORATORY; ROBOTS, PROCESSING.

AUTOMOBILE EXHAUST CONTROL. See EXHAUST CONTROL, AUTOMOTIVE.

AUTOXIDATION. See ANTIOXIDANTS; ANTIOZONANTS.

AUXINS. See GROWTH REGULATORS, PLANT.

AVIATION AND OTHER GAS TURBINE FUELS

The gas turbine power plant which has revolutionized aviation derives basically from the steam turbine adapted to a different working fluid. The difference is crucial with respect to fuel because steam can be generated by any heat source, whereas the gas turbine requires a fuel that efficiently produces a very hot gas stream and is also compatible with the turbine itself. The hot gas stream results from converting chemical energy in fuel directly and continuously by combustion in compressed air. It is expanded in a turbine to produce useful work in the form of jet thrust or shaft power.

The jet engine combines three basic steps (Fig. 1). Air is compressed in a series of stages, fuel is burned continuously and intensively in the compressed air, and the hot gas is then expanded through a turbine which extracts energy to run the compressor and the fan, and also to provide shaft power. The residual exhaust mixes with fan air and exits through a nozzle as a high velocity jet which provides forward thrust to the system. Whittle is credited with the vision in his 1930 patent to adapt the gas turbine concept to aircraft. He realized that at high speed and altitude this type of engine improved in efficiency and made jet propulsion competitive with propeller drive (1). His W-1 engine of 1937 has since proliferated into much more efficient aircraft power plants ranging in size from small auxil-

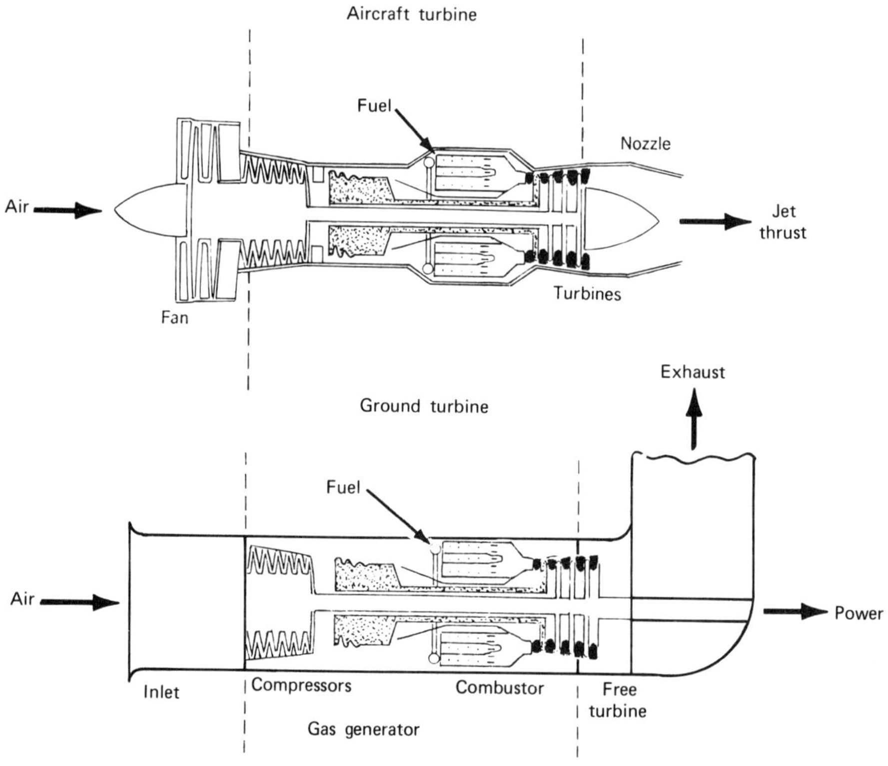

Fig. 1. Schematic of aircraft and ground gas turbines.

iary power units to the giant turbofans which propel wide-bodied airliners. In its most advanced form, the aviation gas turbine for supersonic aircraft, a second combustion step is added to the three basic steps of compression, combustion, and turbine work extraction. Additional fuel is burned in the turbine exhaust to increase the velocity of the jet from the nozzle and to gain more thrust.

The first application of the gas turbine principle to a ground system was in 1906 when waste heat from furnace gases was used to operate a turboblower which compressed air for a blast furnace. Developments in metallurgy and compressor design after World War I led, in 1937, to the first industrial gas turbine for generating electrical power (2). These developments also produced successful turbocompressors for increasing the air supply and power output of aircraft piston engines, a vital advantage for miliary aircraft during World War II.

Ground turbines carry out the same three basic steps of the jet engine, but residual energy of the turbine exhaust is extracted by a waste heat boiler or regenerator. Industrial turbines have heavy compressors, external combustion systems, and massive turbine sections. Lightweight design for aircraft required new high temperature metals and advanced concepts for cooling, compression, and combustion. Today these lightweight types have been adapted for many ground applications in gas compression, pumping, electrical power generation, and general-purpose utility units. Because the gas turbine is fundamentally a heat-conversion device, the combustor or gas generator is the heart of the engine. Combustion of fuel with a portion of compressed air is continuous. Early developers of gas turbines depended on the experience with oil-fired furnaces to design fuel nozzles and combustion chambers (3).

The first fuels burned in a turbine combustor were petroleum liquids, kerosene in Whittle's prototype jet engine and diesel fuel in the Brown-Boveri industrial turbine, selected because of their availability and convenience. After 50 years of experience the gas turbine has been adapted to a wide variety of combustible gases and liquids, including crude oil (4). Although such a wide-ranging appetite for fuels capable of releasing heat energy would appear to classify the gas turbine as fuel-insensitive, clearly defined limits on fuel properties have developed during a half century to optimize its performance. When gas turbine applications are classified according to fuels, as in Table 1, the aircraft propulsion unit is the most demanding, accepting only certain liquid distillates; the industrial turbines (low inlet temperatures) are the least demanding. For ground turbines a wide range of gases and liquids are suitable, although advanced units with turbine inlet temperature above 650°C are limited in accepting residual fuel components (2).

The special requirements of gas turbine fuels that have developed relate to the individual processes that take place in the engine itself and in the support systems for handling fuels. It is evident that aviation gas turbines place more severe demands on liquid fuels than ground turbines do, and that gaseous fuels are much easier to utilize since they only need to supply heat energy. Whittle chose a kerosene for turbine research in order to conserve gasoline for the impending war effort and because it had better low temperature properties than a diesel fuel. This decision tended to establish a dividing line between air and ground turbines; subsequent developments led to lighter fuels for jet engines and heavier fuels for industrial turbines. Aircraft have continued to fly on kerosene

Table 1. Fuels Used in Gas Turbine Applications

Fuel	Aircraft propulsion	Ground power: Aircraft type	Ground power: Advanced industrial	Ground power: Heavy duty
hydrogen	[a]	X	X	
methane (natural gas)	[a]	X	X	
synthetic gas		X	X	
LPG		X	X	
light naphtha		X	X	
methanol (ie, nonhydrocarbons)		X	X	
gasoline	[b]			
wide-cut fuel	X	X	X	
kerosene	X	X	X	
No. 2 fuel oil (diesel)		X	X	X
other gas oils		X	X	X
residual fuel			[c]	X
crude oil				X

[a]Future application as cryogenic liquid.
[b]Emergency fuel only.
[c]Limited by metal content of fuels.

fuel or on a wide-cut blend of heavy naphtha and kerosene. Supersonic transports and advanced military vehicles are also operated on kerosene fuel. Most ground turbines utilize liquid fuel systems based on standard No. 2 diesel or on home burner fuel.

Gas Turbine Products

Composition. Because the jet aircraft is a weight-limited vehicle, a high premium is assigned to hydrocarbon fuels with a maximum gravimetric heat content or hydrogen-to-carbon ratio. Paraffinic fuels have high mass-heat contents (about 45 MJ/kg or 10,800 kcal/kg) which decrease very slowly as chain length (mol wt) increases. However, because the density of these fuels is low the heat content on a volumetric basis is less than that of nonparaffinic fuels. Since fuels are measured and sold on a volumetric basis, the aircraft user obtains less energy for each cubic meter of the preferred paraffinic fuel but more energy per unit of mass. Ground turbine users are not concerned with weight limitations and instead seek maximum energy per unit of cost, which means that heavier fuels are more economical for ground application. Fuel properties are largely determined by the nature of the crude oil from which they are derived. This is illustrated by the typical density-paraffin content curve for the kerosene fractions (150–288°C) of different crudes (Fig. 2). Aromatics in this cut are of special concern because of their effects on combustion and elastomers. The mass spectrometric data in Table 2 illustrate that a wide range of single and multiring aromatics is possible even within a total aromatic concentration of 25%, the maximum permitted by most specifications.

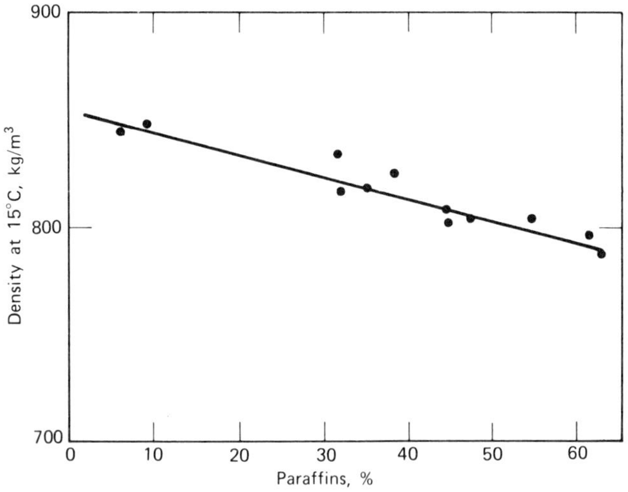

Fig. 2. Density vs paraffin content of gas turbine fuels (150–288°C fractions).

Table 2. Composition[a] of 150–288°C Kerosenes

Crude source	Middle East			North Africa	United States			South America				
saturates, wt %	78.8	82.4	80.3	83.1	85.5	81.0	76.8	81.7	78.8	76.8	76.1	74.5
paraffins	63.0	61.7	54.0	47.4	44.4	37.9	35.3	44.0	34.9	31.3	9.0	6.0
cycloparaffins, single ring	10.6	12.0	14.5	23.8	24.1	22.9	27.3	26.9	27.3	29.0	33.3	31.7
cycloparaffins, two-rings	4.7	7.6	8.7	9.8	12.0	17.9	11.3	9.6	12.8	12.6	24.9	28.1
cycloparaffins, 2 + rings	0.9	1.1	3.2	3.1	5.0	2.3	2.9	1.2	3.8	3.9	8.9	8.7
aromatics, wt %	18.4	16.3	18.2	14.7	14.0	17.7	21.9	17.0	19.9	22.0	23.7	25.5
single-ring	16.9	14.1	14.8	11.3	11.8	13.9	19.0	15.1	14.8	16.6	16.4	16.8
two-rings	1.4	2.0	3.0	3.2	2.1		2.6	1.9	4.7	5.1	3.6	7.7
2 + rings	0.1	0.1	0.2	0.2	0.1	3.8	0.3		0.5	0.4	0.7	1.0

[a]By mass spectrometric analysis.

Specifications. The global nature of jet aircraft operations has mandated that aviation fuel quality be closely controlled in every part of the world. Specifications tend to be industry standards issued by a consensus organization, like ASTM, or by a government body rather than by manufacturer's requirements. Table 3 lists some of the requirements for the principal grades of civil and military aviation turbine fuels in use throughout the world. Domestic and international airlines use fuel modeled on ASTM specification grades. Specifications issued by the British Ministry of Defense, the recommendations of the International Air Transport Association, and ASTM specifications are coordinated as a joint system checklist to control quality at many worldwide airports where jointly-operated fueling systems prevail. Specifications issued by Eastern European govern-

Table 3. Selected Specification Properties of Civil and Military Aviation Gas Turbine Fuels

Characteristic	ASTM D1655[a] Jet A kerosene A/L and	ASTM D1655[a] Jet B wide-cut Gen Avn	Mil-T-5624-N[a] JP-4 wide-cut USAF	Mil-T-5624-N[a] JP-5 kerosene USN	Mil-T-83133C[a] JP-8 kerosene USAF
composition					
aromatics, vol % max	25[b]	25[b]	25	25	25
sulfur, wt % max	0.3	0.3	0.4	0.4	0.3
volatility					
dist. temp max °C, 10% rec'd	205			205	205
dist. temp max °C, 50% rec'd		190	190		
dist. temp max °C, end pt	300		270	300	300
flash pt, °C min	38			60	38
vapor pressure at 38°C, kPa max (psi)		21 (3)	14–21 (2–3)		
density at 15°C, kg/m^3	775–840	751–802	751–802	788–845	775–840
fluidity					
freezing pt, °C max	−40[c]	−50	−58	−46	−47
viscosity at −20°C, mm^2/s (= cSt) max	8.0			8.5	8.0
combustion					
heat content, MJ/kg, min[d]	42.8	42.8	42.8	42.6	42.8
smoke pt, mm, min	18[e]	18[e]	20	19	19
H_2 content, wt % min			13.5	13.4	13.4
stability					
test temp[f], °C min	245	245	260	260	260

[a]Full specification requires other tests.
[b]For aromatics above 20% (22% for Jet A1) users must be notified.
[c]International airlines use Jet A1 with −47°C freeze point.
[d]To convert MJ to kcal, multiply by 239.
[e]Plus naphthalenes 3 vol % max unless smoke point exceeds 25.
[f]Thermal stability test by D3241 to meet 3.3 kPa (25 mm Hg) pressure drop and Code 3 Deposit Rating.

ments are similar to other world aircraft fuels but are not identical in all respects. International airlines that fly to Eastern European countries can utilize both types of fuels without difficulty. Except for severe arctic areas where Jet B fuel is used, essentially all civil aviation uses kerosene fuel. About half of the demand is Jet A fuel marketed in the United States, and half is Jet A1, the international fuel marketed outside the United States, similar to Jet A except for freezing point. A lower temperature limit of −47°C for the (wax) freezing point of Jet A1 is required for long-duration international flights. The same fuel is labeled JP-8 by the U.S. Air Force and is used at most bases outside the United States to replace JP-4, the wide-cut fuel originally developed after World War II to ensure maximum availability. JP-4 is still the primary U.S. Air Force fuel used at bases and at certain NATO installations. The other military fuel of importance is JP-5, the high flash point kerosene used by the U.S. Navy to ensure handling safety aboard aircraft carriers. Although all three military fuels are used in supersonic aircraft, a special kerosene fuel called JP-7 developed for the Mach 3 SR-71 reconnaissance aircraft has been retired from service, like the SR-71 itself.

Liquid fuels for ground-based gas turbines are best defined today by ASTM Specification D2880. Table 4 lists the detailed requirements for five grades which cover the volatility range from naphtha to residual fuel. The grades differ primarily in basic properties related to volatility; eg, distillation, flash point, and density of No. 1 GT and No. 2 GT fuels correspond to similar properties of kerosene and diesel fuel respectively. These properties are not limited for No. 0 GT fuel, which allows naphthas and wide-cut distillates. For heavier fuels, No. 3 GT and No. 4 GT, the properties that must be limited are viscosity and trace metals.

Table 4. Specifications for Ground Gas Turbine Fuels

Property	0-GT[a] Naphtha	1-GT[a] Light distillate	2-GT[a] Medium distillate	3-GT[a] Heavy distillate	4-GT Heavy residual
90% distillation, °C		288 max	282–338		
density, kg/m^3, min		850	876		
flash pt, °C max		38	38	55	66
pour pt, °C max		−18	−6		
carbon residue, % max	0.15	0.15	0.35		
ash, wt % max	0.01	0.01	0.01	0.03	
viscosity at 40°C, mm^2/s (= cSt)		1.3–2.4	1.9–4.1		
viscosity at 50°C, mm^2/s (= cSt) max				638	638
water and sediment, vol % max	0.05	0.05	0.05	1.0	1.0

[a]Trace metal limits at point of delivery are established at 0.5 ppm by wt maximum for vanadium, sodium and potassium, calcium, and lead.

Manufacture. Aviation turbine fuels are primarily blended from straight-run distillates rather than cracked stocks because of specification limitations on olefins and aromatics. However, in refineries where heavy gas oil is hydrocracked, ie, a process that involves catalytic cracking in a hydrogen atmosphere, aviation fuels can include hydrocracked components since they are free of olefins, are low in sulfur, and are stable. Hydrocracking provides a means to extend availability of gasolines and distillates by converting most of the heavy ends in crudes to lighter products. Ground turbine fuels are equivalent to their fuel oil counterparts and are manufactured as dual-purpose products. No. 0 GT fuel is a light naphtha, the closest aviation counterpart of which is JP-4. No. 1 GT fuel is kerosene blended to Jet A or A1 quality and sold as dual-purpose kerosene (DPK). No. 2 GT corresponds to No. 2 diesel fuel or No. 2 heating oil and is usually blended from straight-run and cracked components since olefins and aromatics are not limiting. No. 3 GT and No. 4 GT fuels bracket the viscosity range of No. 4, 5, and 6 fuels and No. 4D diesel and are blended from gas, oils, and residua from pipestills and vacuum units with lighter distillates. A schematic of a typical refinery processing sequence appears in Figure 3. In this example the basic processes after distillation are catalytic reforming for gasoline and hydrogen processing of naphtha, kerosene, and gas oil using the hydrogen from reforming.

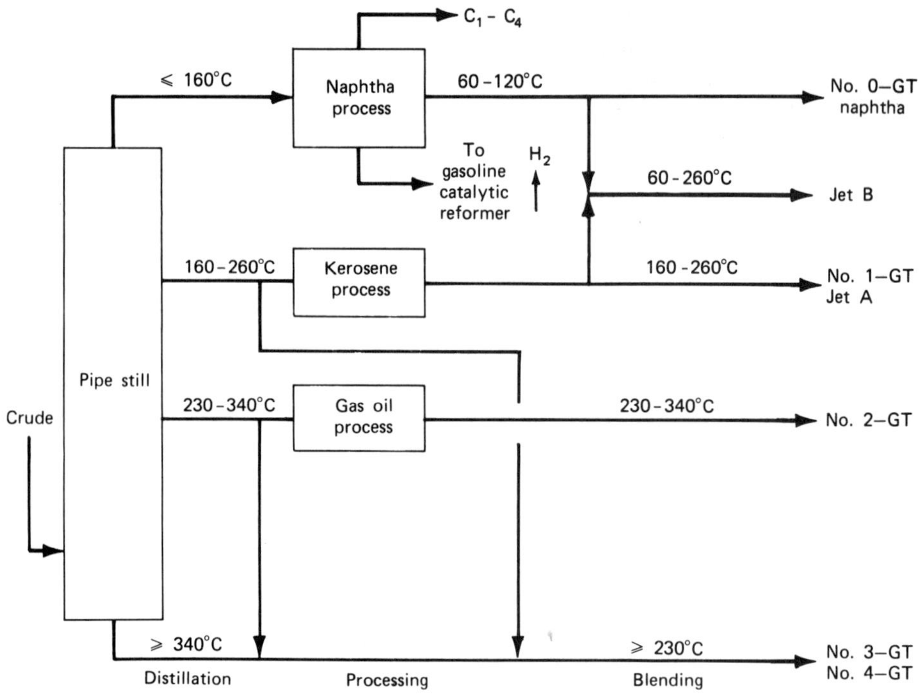

Fig. 3. Refinery process for gas turbine fuels. Processing may include hydrogen treating.

A hydrocracking unit usually requires a separate source of hydrogen. The approximate boiling ranges of the sidestream cuts and of the finished gas turbine fuels are shown.

The distillation cut points must be closely controlled to yield a product that meets the requirements of flash point on one hand and freezing point on the other. In practical terms, a jet fuel requires a fraction of about 60°C initial boiling point and a final boiling point not exceeding 300°C. The lighter portion of this fraction contains gasoline components and meets the specification for JP-4. The heavier portion above 160°C must be tailored to either the Jet A or Jet A1 freezing point, but the final boiling point is dependent on crude composition. With higher aromatic crudes, undercutting is required to meet compositional limits or a test such as smoke point. Lower boiling components compensate for aromatic or freezing point limitations, but sometimes it is impossible to take advantage of them because a dual-purpose kerosene usually requires a higher flash point than jet fuel. The distilled fractions from the crude are apt to contain mercaptans or organic acids in excess of specification limits. A caustic wash is normally used to control acidity and to remove traces of hydrogen sulfide. Removal or conversion of odorous mercaptans is carried out in a sweetening process. The most widely used chemical treatment today, Merox sweetening, utilizes dissolved air to oxidize mercaptans to disulfides over a fixed-bed cobalt chelate catalyst (5). It has the advantage of simplicity and minimum waste disposal problems but does not lower the sulfur content. A modern version of the old doctor process, Bender sweetening, uses lead plumbite deposited on a fixed bed to carry out the mercaptan oxidation and the spent doctor regeneration processes simultaneously, but product quality is more difficult to control, and waste disposal problems are greater.

Most of the crudes available in greatest quantity today are high in sulfur, and yield products that must be desulfurized to meet specifications or environmental standards. The process most widely used is mild catalytic hydrogenation. Hydrofining uses a fixed-bed catalyst, usually cobalt molybdate on alumina, to convert all but ring sulfur compounds to H_2S, which is then removed (and recovered) from recycled gas. At reactor temperatures below 300°C only mercaptans react. This technique is called hydrosweetening. At reactor temperatures of 400–500°C extents of desulfurization are 80% or better and other reactive species, including some multiring sulfur compounds, are hydrogenated. The product of the hydrofiner is caustic washed to remove traces of dissolved H_2S and is then passed through sand or clay to blend tanks.

Hydrogen treating removes oxygen-containing species, such as phenols and naphthenic acids, that are found in some crudes. The former tend to perform as natural inhibitors of hydrocarbon oxidation by trapping peroxy radicals, while the latter act as natural lubricating agents. Recognition of this side effect of hydrogen treatment led refiners to add antioxidants to hydrotreated components; this procedure is now a specification requirement. Blending of two or more components is carried out to match as closely as practical the various specification constraints. At this point, additives are usually introduced, eg, an antioxidant to inhibit gum formation in storage or a metal deactivator to deactivate any dissolved metal ions (Table 5). Military fuels have made the addition of antiicing agents and corrosion inhibitors mandatory, the former because military aircraft

Table 5. Additives Used in Aviation Gas Turbine Fuels

Class of additive[b]	Chemical types	Purpose	Specification status[a]	
			Civil	Military
antioxidant	alkyl phenylene diamines, hindered alkylphenols	improve oxidation stability–storage and high temperature	O	O
metal deactivator	disalicylidene alkyl diamine	remove metal ions that catalyze oxidation	O	O
antiicing	glycol ether	prevent low temperature filter icing		R
corrosion inhibitor	dimer acid, phosphate ester	reduce development of corrosion products, improve lubrication		R
antistatic	alkyl chromium salt and calcium sulfosuccinate, olefin–sulfur dioxides and polyamines	increase conductivity	OR	R
biocide	boron ester	inhibit organic growth in fuel tanks	A	

[a]O = optional, R = required, A = airline use in maintenance.

[b]See also ANTIOXIDANTS; ANTIFREEZES AND DEICING FLUIDS; ANTISTATIC AGENTS; CORROSION AND CORROSION INHIBITORS; INDUSTRIAL ANTIMICROBIAL AGENTS.

fuel systems do not include fuel filter heaters to prevent ice formation, and the latter to reduce the tendency of fuels to pick up rust particles from pipelines and tanks. The Air Force also requires that antistatic additive be included in JP-4 to reduce the hazard of static generation and discharge in handling fuel and in operating aircraft with foam-filled tanks. It was discovered that the antiicing additives inhibit growth of microorganisms in aircraft and storage tanks and that certain corrosion inhibitors are excellent lubricity agents for prolonging the life of engine pumps.

With civil fuels, antiicing agents are not required since air transports are equipped with fuel-filter heaters to prevent ice formation. Corrosion inhibitors are not normally used in civil fuels since they tend to degrade the efficiency of filter-coalescers and would be removed in the clay filters that are commonly installed at receiving terminals and airports. However, antistatic additives are widely used in civil fuels, particularly Jet A1 at international airports. Antistatic additive is introduced into some Jet A to safeguard against static discharge in truck filling, but the widespread use of clay filters in the United States to clean up fuel before airport delivery makes refinery addition impractical. Table 5 also lists a boron-containing biocide used as a shock treatment additive to inhibit the tendency of organisms to form fungal mats in aircraft fuel tanks.

Distribution and Handling. Preserving the quality of gas turbine fuels between the refinery and the point of use is an important but difficult task. The difficulty arises because of the complicated distribution systems of multiproduct pipelines which move fuel and sometimes introduce contaminants. The importance is reflected by the sensitivity of gas turbine engines and fuel systems to water, corrosion products, metal salts, microorganisms, and other extraneous materials that can be introduced by the distribution system. Most products moved through pipelines contain additives (eg, detergents, deicers, antifouling dispersants, and rust inhibitors). Gasoline contains corrosion inhibitors, heating oil contains dispersants, and diesel fuel contains ignition promotors. Traces of these additives left on pipeline walls are picked up by nonadditive jet fuel, which makes it difficult for pipeline operators to deliver gas turbine fuel of the same quality as that introduced to the line.

The principal means of removing contaminants are tank settling and filtration. With aviation fuels it is common practice to install several stages of cartridge-type filter-coalescers between the storage tank and the aircraft delivery point. Figure 4 is a schematic of a typical airport fueling system which transfers fuel by underground hydrant lines. Filter elements of fiberglass covered with synthetic fabric are designed to coalesce water and to remove particulates at high flow rates. A hydrophobic barrier filter prevents coalesced water from passing with the fuel.

Tank settling as a means of contaminant removal is not very efficient with fuels having the viscosity of kerosene. It is common practice to design tanks with cone-down drains and floating suctions to facilitate water and solids removal.

Contamination control by filtration would be relatively easy if it were not for surface-active materials produced during fuel manufacture or picked up in distribution systems from pipe walls and tank bottoms. These surfactants stabilize water-in-oil emulsions, disarm filter elements, coat solid particles, and promote growth of organisms at fuel–water interfaces. When they reach the aircraft

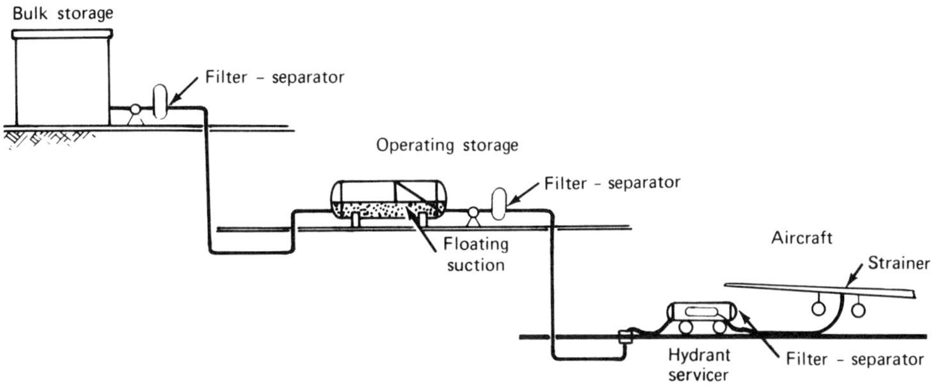

Fig. 4. Typical airport hydrant system.

tanks, the surfactants tend to trap water containing dissolved salts and particulates, causing fungal growths, coating attack, and metal corrosion.

Gas turbine fuels can contain natural surfactants if the crude fraction is high in organic acids, eg, naphthenic (cycloparaffinic) acids of 200–400 mol wt. These acids readily form salts that are water-soluble and surface-active. Older treating processes for sulfur removal can leave sulfonate residues which are even more powerful surfactants. Refineries have installed processes for surfactant removal. Clay beds to adsorb these trace materials are widely used, and salt towers to reduce water levels also remove water-soluble surfactants. In the field, clay filters designed as cartridges mounted in vertical vessels are also used extensively to remove surfactants picked up in fuel pipelines, in contaminated tankers, or in barges.

Specifications for gas turbine fuels prescribe test limits that must be met by the refiner who manufactures fuel; however, it is customary for fuel users to define quality control limits for fuel at the point of delivery or of custody transfer. These limits must be met by third parties who distribute and handle fuels on or near the airport. Tests on receipt at airport depots include appearance, distillation, flash point (or vapor pressure), density, freezing point, smoke point, corrosion, existing gum, water reaction, and water separation. Tests on delivery to the aircraft include appearance, particulates, membrane color, free water, and electrical conductivity.

The extensive processing and cleanup steps carried out on gas turbine fuels produce a purified liquid dielectric of high resistivity which is capable of retaining electrical charges long enough for buildup of large surface voltages. Because fuel is filtered at high flow rates through filter media of extensive surface area, the ionic species that adsorb on the filter surface generate a static charge. The current carried by fuel increases sharply as fuel flows through a filter and then decreases at a rate determined by the fuel conductivity. The charge remaining in the fuel at any given point is determined by the residence time available for charges to recombine. In a hose delivering fuel to an aircraft, only a few seconds are available for charge relaxation. The result is a possible hazard—a tank filled with charged fuel that under some circumstances can discharge its energy to ground in a spark capable of igniting fuel mists or vapors (6).

This aspect of fuel handling has received much attention because of a number of accidents that have resulted in tank explosions, most often in filling tank trucks but also in fueling aircraft. Several approaches have been taken to reduce the risk of static discharge. The most common method requires introduction of an additive to increase the electrical conductivity of the fuel, ie, to speed up charge relaxation to a fraction of a second. Charge generation can be decreased by using low charging screens instead of filters or by reducing flow rates. The conductivity additives used in fuel are readily removed in clay filters or are lost in pipeline movement and must be monitored by conductivity tests. The additives can also degrade water separation properties and lower filter efficiency. Nevertheless, the safety record since antistatic additives were introduced far outweighs the negative effects on other fuel properties.

Performance

Combustion. The primary reaction carried out in the gas turbine combustion chamber is oxidation of a fuel to release its heat content at constant pressure. Atomized fuel mixed with enough air to form a close-to-stoichiometric mixture is continuously fed into a primary zone. There its heat of formation is released at flame temperatures determined by the pressure. The heat content of the fuel is therefore a primary measure of the attainable efficiency of the overall system in terms of fuel consumed per unit of work output. Table 6 lists the net heat content of a number of typical gas turbine fuels. Net rather than gross heat content is a more significant measure because heat of vaporization of the water formed in combustion cannot be recovered in aircraft exhaust. The most desirable gas turbine fuels for use in aircraft, after hydrogen, are hydrocarbons. Fuels that are liquid at normal atmospheric pressure and temperature are the most practical and widely used aircraft fuels; kerosene, with a distillation range from 150 to 300°C, is the best compromise to combine maximum mass–heat content with other desirable properties. For ground turbines, a wide variety of gaseous and heavy fuels are acceptable.

Liquid fuel is injected through a pressure-atomizing or an air-blast nozzle.

Table 6. Net Heat Content of Gas Turbine Fuels

Fuel	Heat content, MJ/kg[a]
hydrogen	120
natural gas (methane)	50
light naphtha (0-GT)	44
wide-cut fuel (JP-4)	43
Jet A kerosene (1-GT)	43
No. 2-GT fuel	42
No. 3-GT fuel	41
No. 4-GT fuel	40
methanol	20

[a]To convert MJ to kcal, multiply by 239.

This spray is sheared by air streams into laminae and droplets that vaporize and burn. Because the atomization process is so important for subsequent mixing and burning, fuel-injector design is as critical as fuel properties. Figure 5 is a schematic of the processes occurring in a typical combustor.

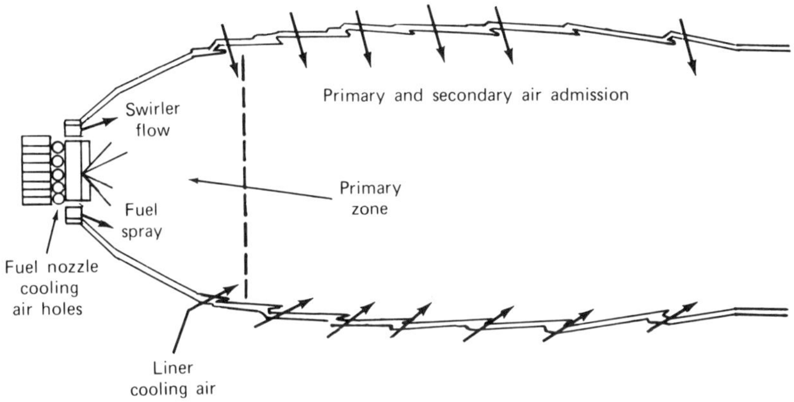

Fig. 5. Gas turbine combustion chamber.

Droplet size, particularly at high velocities, is controlled primarily by the relative velocity between liquid and air and in part by fuel viscosity and density (7). Surface tension has a minor effect. Minimum droplet size is achieved when the nozzle is designed to provide maximum physical contact between air and fuel. Hence primary air is introduced within the nozzle to provide both swirl and shearing forces. Vaporization time is characteristically related to the square of droplet diameter and is inversely proportional to pressure drop across the atomizer (7).

The vapor cloud of evaporated droplets burns like a diffusion flame in the turbulent state rather than as individual droplets. In the core of the spray, where droplets are evaporating, a rich mixture exists and soot formation occurs. Surrounding this core is a rich mixture zone where CO production is high and a flame front exists. Air entrainment completes the combustion, oxidizing CO to CO_2 and burning the soot. Soot burnup releases radiant energy and controls flame emissivity. The relatively slow rate of soot burning compared with the rate of oxidation of CO and unburned hydrocarbons leads to smoke formation. This model of a diffusion-controlled primary flame zone makes it possible to relate fuel chemistry to the behavior of fuels in combustors (7).

Aromatics readily form soot in the fuel-rich spray core as their hydrogens are stripped, leaving a carbon-rich benzenoid structure to condense into large molecular aggregates. Multi-ring aromatics form soot more readily than single-ring aromatics, and they exhibit greater smoking tendency and greater flame luminosity. At the other extreme, *n*-paraffins undergo little cyclization in the fuel-rich spray core, holding soot formation and flame radiation to a minimum. Other hydrocarbon structures exhibit smoke and radiation characteristics between the extremes of multiring aromatics and *n*-paraffins as illustrated by Figure 6, which

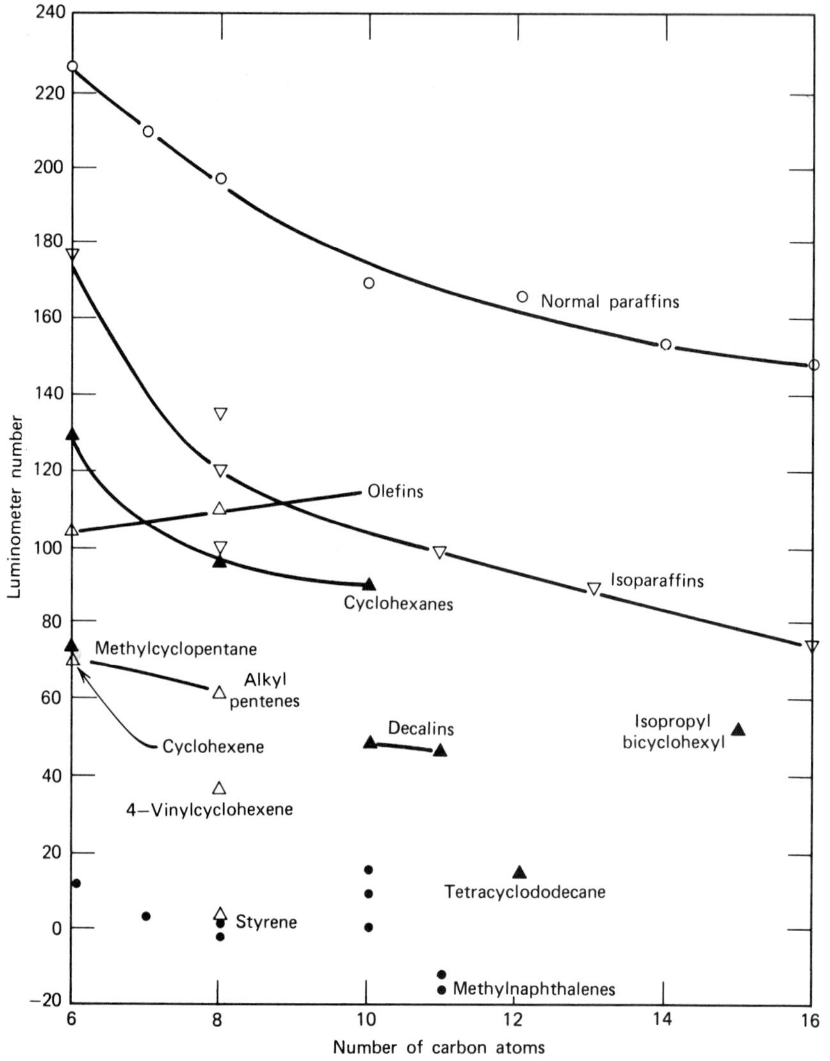

Fig. 6. Luminometer ratings of hydrocarbons (7), ○, *n*-paraffins; ▽, *i*-paraffins; ▲, cycloparaffins; △, olefins; ●, aromatics.

shows the Luminometer number (LN) ratings of different types of hydrocarbons (8). High LN numbers are synonymous with minimum flame radiation and clean (soot-free) combustion. In mixed fuels, two-ring aromatics have been shown to have a tenfold greater effect in lowering LN rating than single-ring aromatics (9). Luminous flames in a gas turbine raise temperatures of the metal in the combustor liner and turbine inlet by the direct transmission by radiation of the heat energy in the flame core.

Air in the gas turbine combustor is carefully divided between primary and secondary uses. Primary air promotes the atomization process by adding swirl to the fuel exiting the nozzle; it must be limited to ensure that the flame core remains fuel-rich to avoid flame blow-out and to facilitate reignition. Most of the air in the gas turbine combustor is directed either toward secondary reactions beyond the primary flame core or toward dilution of the hot exhaust products so that a uniform temperature gas is presented to the turbine section. The secondary reactions involve intensely turbulent mixing under stoichiometric or fuel-lean mixtures to complete the combustion to CO_2 and burn up the soot particles formed in the fuel-rich core. A combustor design that causes proper mixing is critical for ensuring smoke-free exhaust products. Nevertheless, in any given combustor there is a relationship between smoke output and the hydrogen content of fuel, eg, as shown in Figure 7 (10).

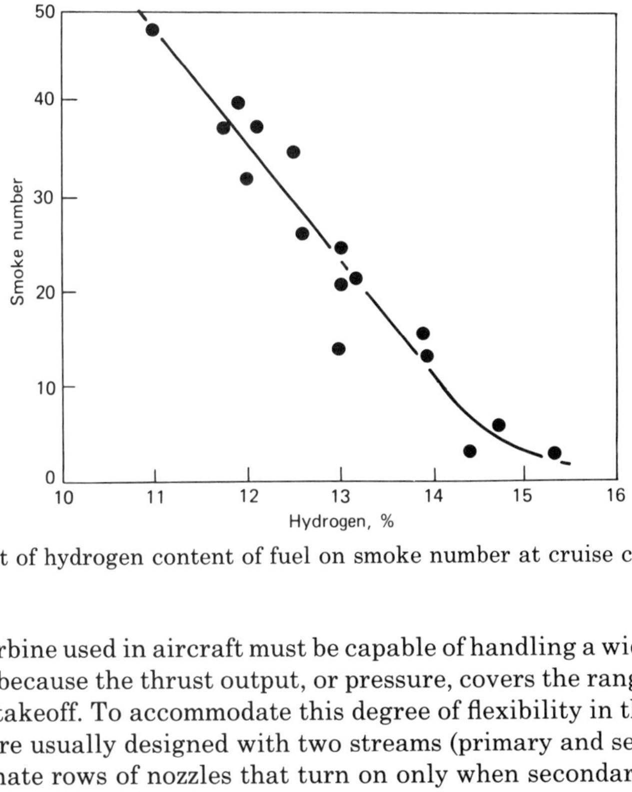

Fig. 7. Effect of hydrogen content of fuel on smoke number at cruise conditions (9).

A gas turbine used in aircraft must be capable of handling a wide span of fuel and air flows because the thrust output, or pressure, covers the range from idle to full-powered takeoff. To accommodate this degree of flexibility in the combustor, fuel nozzles are usually designed with two streams (primary and secondary flow) or with alternate rows of nozzles that turn on only when secondary flow (or full thrust power) is needed. It is more difficult to vary the air streams to match the different fuel flows and, as a consequence, a combustor optimized for cruise conditions (most of the aircraft's operation) operates less efficiently at idle and full thrust.

Unlike the aircraft turbine, the ground-based gas turbine operates continuously at the same power setting, usually 60–80% of maximum power output except when starting. Air and fuel flow patterns are constant and the combustor can be tuned to minimize smoke formation and high metal temperatures.

Exhaust emissions of CO, unburned hydrocarbons, and nitrogen oxides reflect combustion conditions rather than fuel properties. The only fuel component that degrades exhaust is sulfur; the SO_2 concentrations in emissions are directly proportional to the content of bound sulfur in the fuel. Sulfur concentrations in fuel are determined by crude type and desulfurization processes. Specifications for aircraft fuels impose limits of 3000–4000 ppm total sulfur but the average is half of these values. Sulfur content in heavier fuels is determined by legal limits on stack emissions.

Control of nitrogen oxides in aircraft exhaust is of increasing concern because nitrogen oxides react with ozone in the protective layer of atmosphere which exists in the altitude region where supersonic aircraft operate. Research is under way to produce a new type of combustor which minimizes NO_x formation. It is an essential component of the advanced propulsion unit needed for a successful supersonic transport fleet.

Another class of pollutant in exhaust does serious damage to the turbine itself. In this case the damage is hot corrosion of the metal blades and vanes by alkali metal (primarily sodium) salts which remove blade coatings and promote intergranular corrosion. The source of the salts can be fuel (as metal salts dissolved or entrained in water) or the air that enters the gas turbine in huge quantities. Sulfur oxides in the exhaust react with alkali to form sulfates which produce a low melting flux on the oxide coatings. The problem is especially serious with gas turbines used in ground (or marine) locations because heavier fuels are apt to contain metal contaminants. For example, some crudes contain soluble vanadium which forms V_2O_5, leading to hot corrosion. Metal specification limits on gas turbine fuels are noted in Table 4 and are discussed fully in appendixes to ASTM D2880.

Stability. Aviation fuel is exposed to a wide range of thermal environments, and these greatly influence required fuel properties. In the tank of a long-range subsonic aircraft, fuel temperatures can drop so low that ice crystals, wax formation, or viscosity increase may affect the fuel system performance. On its way to the engine the fuel then absorbs heat from airframe or engine components and in fact is used as a coolant for engine lubricant. Current high thrust engines expose fuel in the main engine control, the pump, and the manifold to an intense thermal environment, and fuel temperatures steadily rise as flow reaches the nozzle. Fuel stability assumes primary importance since freedom from deposits within the fuel system is essential for both performance and life.

The fuel systems of ground-based turbines are far less critical, since coolants other than fuel can be used and fuel lines can be well insulated. The tendency for deposit formation in fuel is not a concern in ground systems.

Deposits sometimes block fuel nozzles and distort fuel spray patterns, leading to skewed temperature distribution with the possibility of burnout of turbine parts by a "hot streak" exhaust. These deposits are sometimes associated with metal-containing particulates, but in general are another manifestation of fuel instability.

Oxidation of hydrocarbons by the air dissolved in fuel is catalyzed by metals and leads to polymer formation, ie, varnish and sludge deposits, by a chain reaction mechanism involving free radicals. Since it is impossible to exclude air dissolved in fuel, oxidation stability is controlled by eliminating species prone to form free radicals and by introducing antioxidants (qv). An accelerated test, the Jet Fuel Thermal Oxidation Test (JFTOT), simulates the critical temperature regimes of the gas turbine engine fuel manifold and nozzles, ASTM D3241, and has been developed to measure the oxidation stability of aircraft gas turbine fuels. In the JFTOT test, deposits, formed in the heated test section in 2.5 h at a specified temperature, are assessed. This relative rating ranks fuels that would normally be expected to operate for thousands of hours in an engine without deposit formation. In those few instances when deposits have been observed in service, the fuel has been shown to fail JFTOT tests at the specification temperature.

The reactive species that initiate free-radical oxidation are present in trace amounts. Extensive studies (11) of the autoxidation mechanism have clearly established that the most reactive materials are thiols and disulfides, heterocyclic nitrogen compounds, diolefins, furans, and certain aromatic-olefin compounds. Because free-radical formation is accelerated by metal ions of copper, cobalt, and even iron (12), the presence of metals further complicates the control of oxidation. It is difficult to avoid some metals, particularly iron, in fuel systems.

It is possible to deactivate a metal ion by adding a compound such as disalicylidene alkyl diamine, which readily forms a chelate with most metal atoms to render them ineffective. Metal deactivator has been shown to reduce oxidation deposits dramatically in the JFTOT test and in single tube heat exchanger rigs. The role of metal deactivator in improving fuel stability is complex, since quantities beyond those needed to chelate metal atoms act as passivators of metal surfaces and as antioxidants (13).

Oxidation deposits in aircraft engines are related to the thermal stresses imposed by heat soakback, which results from the off/on cycles associated with many landings and takeoffs. In contrast, ground turbines tend to operate longer at constant flow and do not subject fuel to the same degree of thermal stress. Fuel stability testing has not been established as a quality requirement, although a carbon residue test (ASTM D524) provides a rough screening tool to protect against the tendency for carbonization in a fuel injection system.

Volatility. The volatility of aircraft gas turbine fuel is controlled primarily by the aircraft itself and by its operating environment. For example, limits of the vapor pressure of aviation gasoline were dictated by the vapor and liquid entrainment losses that could occur in a piston aircraft capable of climbing to an altitude of about 6000 m; the vapor pressure of the warm fuel exceeds atmospheric pressure at that altitude (48 kPa or 7 psi). Since early military jet aircraft could climb even higher (12,000 m) and faster, it was necessary to further limit the vapor pressure of military gas turbine fuels to 21 kPa (3 psi). Originally, the United States followed the British lead and selected a kerosene of very low vapor pressure for military jet engines. Because projected production of this fuel was limited, fuel availability was the eventual determining factor for the selection of the wide-cut gasoline type fuel JP-4 as the fuel for military jet aircraft. The volatility of kerosene fuel is measured not by its vapor pressure but by its temperature at the point where its vapors just prove to be flammable, ie, the flash point.

Commercial aviation utilizes low volatility kerosene defined by a flash point minimum of 38°C. The flammability temperature has been invoked as a safety factor for handling fuels aboard aircraft carriers; Navy JP-5 is a low volatility kerosene of minimum flash point of 60°C, similar to other Navy fuels.

The dependence of vapor pressure on temperature for the fuels most commonly used in gas turbines appears in Figure 8 (14). The points on the abscissa reflect the flash point temperatures used to define the volatility of higher molecu-

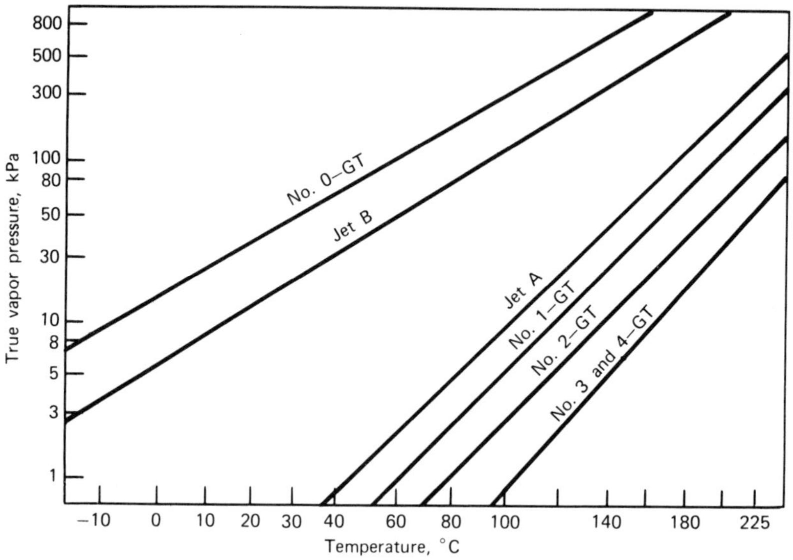

Fig. 8. Variation of vapor pressure with temperature for gas turbine fuels (14). To convert kPa to psi, divide by 6.9.

lar weight fuels. When vapor pressure itself is limited, as with JP-4 or Jet B, a test temperature of 38°C is specified.

Ground turbine fuels are not subject to the constraints of an aircraft operating at reduced pressures of altitude. The temperature of fuel in ground tanks varies over a limited range, eg, 10–30°C, and the vapor pressure is defined by a safety-handling factor such as flash point temperature. Volatile fuels such as naphtha (No. 0-GT) are normally stored in a ground tank equipped with a vapor recovery system to minimize losses and meet local air quality codes on hydrocarbons.

A minimum volatility is frequently specified to assure adequate vaporization under low temperature conditions. It can be defined either by a vapor pressure measurement or by initial distillation temperature limits. Vaporization promotes engine start-up. Fuel vapor pressure assumes an important role particularly at low temperature. For example, if fuel has cooled to −40°C, as at arctic bases, the amount of vapor produced is well below the lean flammability limit. In this case a spark igniter must vaporize enough fuel droplets to initiate combustion. Start-up under the extreme temperature conditions of the arctic is a major constraint in converting the Air Force from volatile JP-4 to kerosene-type JP-8, the military counterpart of commercial Jet A1.

The lower volatility of JP-8 is a significant factor in the U.S. Air Force conversion from JP-4, since fires and explosions under both combat and ordinary handling conditions have been attributed to the use of JP-4. In examining the safety aspects of fuel usage in aircraft, a definitive study (15) of the accident record of commercial and military jet transports concluded that kerosene-type fuel is safer than wide-cut fuel with respect to survival in crashes, in-flight fires, and ground fueling accidents. However, the difference in the overall accident record is small because most accidents are not fuel-related.

Low Temperature Fuel Flow. The decrease in the temperature of fuel in the tank of an aircraft during a long duration flight produces a number of effects which can influence flight performance. Fuel viscosity increases, necessitating more pumping energy in the tank boost pump and in the engine pump. Fuel becomes saturated with water, and droplets of free water form and settle. Those carried with fuel may form ice on the cold filter which protects the fuel control. For this reason filter heaters are used in civil aircraft to avoid ice blockage and fuel starvation. Military aircraft avoid the complication of a filter heater and depend instead on an antiicing additive in the fuel to depress the freezing point of water. At still lower temperatures, crystals of wax form in fuel. The temperature at which these wax crystals disappear is called the freezing point and distinguishes Jet A1, used by long-range international airlines, from Jet A, used by domestic airlines for relatively short duration flights.

The increase in fuel viscosity with temperature decrease is shown for several fuels in Figure 9. The departure from linearity as temperatures approach the pour point illustrates the non-Newtonian behavior created by wax matrices. The freezing point appears before the curves depart from linearity. It is apparent that the low temperature properties of fuel are closely related to its distillation range as

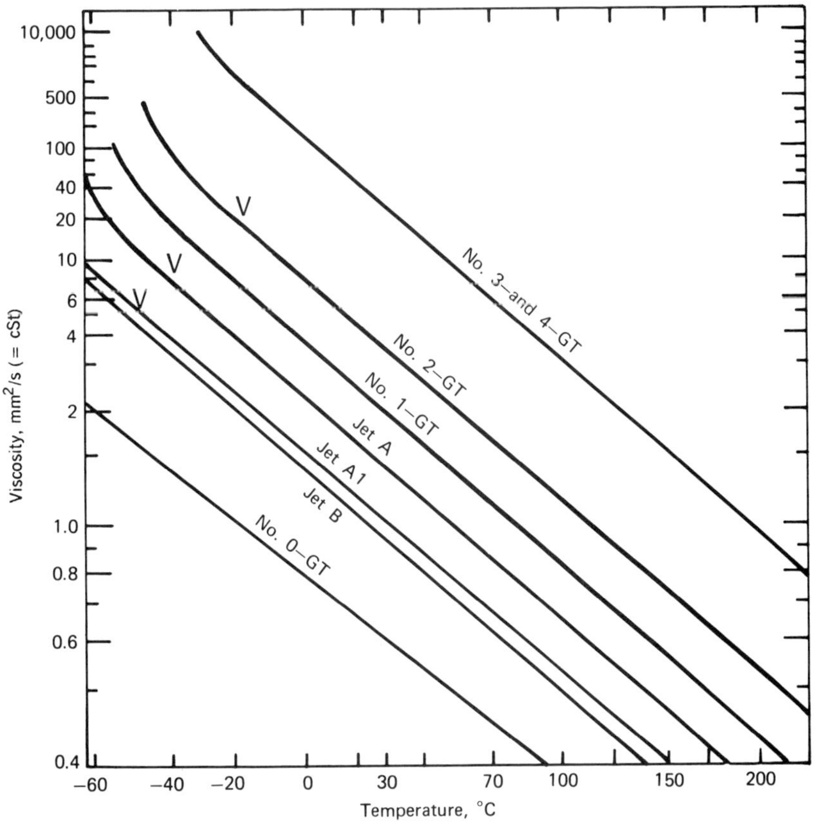

Fig. 9. Variation of viscosity with temperature of gas turbine fuels (14). V = specified limit for freezing (wax) point.

well as to hydrocarbon composition. Wide-cut fuels have lower viscosities and freezing points than kerosenes, whereas heavier fuels used in ground turbines exhibit much higher viscosities and freezing points.

Low temperature viscosities have an important influence on fuel atomization, and they affect engine starting. Cycloparaffinic and aromatic fuels reach unacceptably high viscosities at low temperatures. A kinematic viscosity of 35 mm^2/s (= cSt) represents the practical upper limit for pumps on aircraft, whereas much higher limits are acceptable for ground installations.

Water in Fuel. The solubility of water in a hydrocarbon is related to the molecular structure of the hydrocarbon and to its temperature. Aromatics dissolve about five times more water than corresponding paraffins and, in turn, lower molecular weight paraffins dissolve more water than longer chain compounds. Figure 10 is a plot of water solubility for certain pure hydrocarbons and typical jet fuels. The water saturation values in fuel vary considerably because of composition effects. For example, wide-cut fuel normally dissolves more water than does kerosene, but because the latter may contain twice as many aromatics, the higher molecular weight fuel may in fact exhibit an equivalent water-saturation curve. At a temperature of 23°C, a Coordinating Research Council program on Jet A fuel showed a range of 56–120 ppm of soluble water (16).

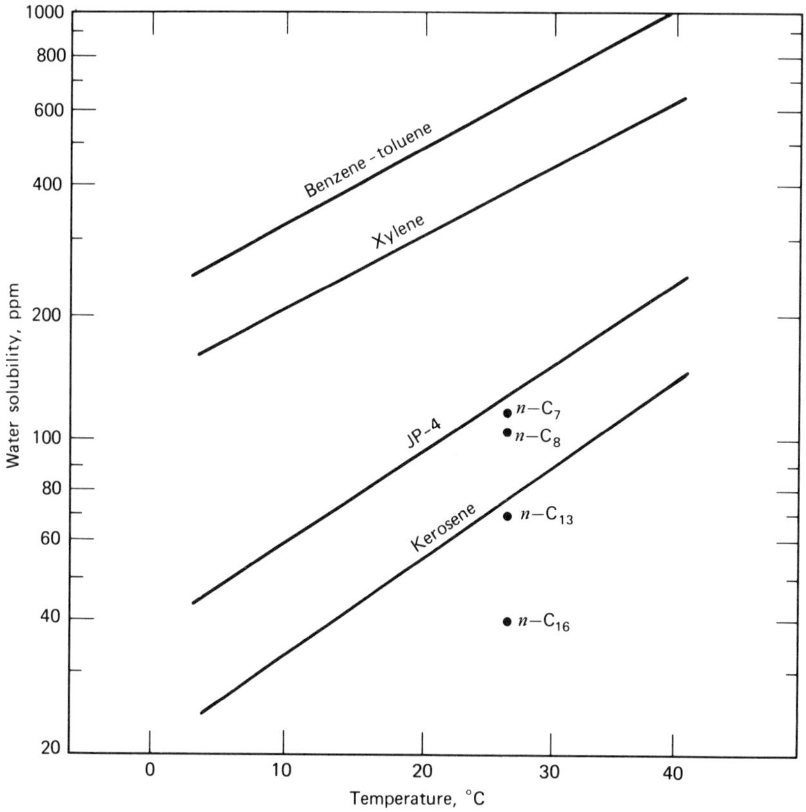

Fig. 10. Variation of water solubility of hydrocarbons with temperature.

The slope of the water solubility curves for fuels is about the same, and is constant over the 20–40°C temperature range. Each decrease of 1°C decreases water solubility about 3 ppm. The sensitivity of dissolved water to fuel temperature change is important. For example, the temperature of fuel generally drops as it is pumped into an airport underground hydrant system because subsurface temperatures are about 10°C lower than typical storage temperatures. This difference produces free water droplets, but these are removed by pumping fuel through a filter-coalescer and hydrophobic barrier before delivery into aircraft.

The amount of water actually dissolved in fuel at any given temperature is determined by equilibration with the water in the atmosphere above the fuel. If a tank is vented to the atmosphere, fuel may enter saturated with water but lose half its dissolved water when the relative humidity is 50%. Conversely, a tank of fuel in a humid environment rapidly picks up water. Many storage tanks contain floating roofs which effectively eliminate any opportunity for dissolved water to be removed as vapor. The tank in the aircraft behaves like a storage tank on the ground. In the upper atmosphere, the partial pressure of water is very low and fuel tends to dry out; dissolved water leaves fuel and exits from the vent. An aircraft descending through clouds picks up water and saturates its fuel. At the end of a flight, fuel in a tank is apt to be both cold and cloudy with free water.

A stable cloud of water in fuel usually means that a surfactant is present to form a stable water-in-oil emulsion. Smaller droplets resist natural coalescence processes. A surfactant that is potent as an emulsifying agent is apt to disarm the coalescing filters, allowing excess water to be delivered with fuel.

The effects of water in fuel inside a tank, particularly an aircraft tank, are important because of the demonstrated proclivity of free water to form undrainable pools where microorganisms can flourish. In ground storage tanks an attempt is made to exclude water bottoms by proper design and regular draining practices. This is more difficult in aircraft because pools of water freeze in flight and ice may not melt when fueling turnarounds are rapid. In a large storage tank, organisms (usually fungi and yeasts) develop at fuel–water interfaces, forming a cuff of growth which can plug filters. In the aircraft the growth usually takes place on tank surfaces, forming a fungal mat under which metabolic products such as organic acids penetrate polymeric coatings to attack the aluminum skin itself. The growth may also affect the capacitance gauge used to read liquid level.

It is common to curb growth of organisms by biocidal treatment. In storage tanks a water-soluble agent is used. Aircraft tanks are opened periodically for hand cleaning and subsequent treatment with a fuel-soluble boron-containing biocide. The antiicing additive used by the military to lower the freezing point of water, 2-methoxyethanol, happens also to inhibit growth of organisms effectively. Therefore, aircraft tank treatment to remove fungal mats is needed only for commercial transports.

Water plays a primary role in corrosion of the metal walls of tanks and pipes (17), and increases the tendency for high speed pumps to produce wear particles and to exhibit shortened life. Formation of corrosion products can be controlled by addition of corrosion inhibitors, a mandatory additive in military fuels. However, corrosion inhibitors may also degrade other fuel properties and adversely affect ground filtration equipment. Thus they are not generally acceptable in commercial fuels where rigorous attention is given to clean and dry fuels upon aircraft fueling.

Compatibility and Corrosion. Gas turbine fuels must be compatible with the elastomeric materials and metals used in fuel systems. Elastomers are used for O-rings, seals, and hoses as well as pump parts and tank coatings. Polymers tend to swell and to improve their sealing ability when in contact with aromatics, but degree of swell is a function of both elastomer-type and aromatic molecular weight. Rubbers can also be attacked by peroxides that form in fuels that are not properly inhibited (see ELASTOMERS, SYNTHETIC; RUBBER, NATURAL).

Attack on metals can be a function of fuel components as well as of water and oxygen. Organic acids react with cadmium plating and zinc coatings. Traces of H_2S and free sulfur react with silver used in older piston pumps and with copper used in bearings and brass fittings. Specification limits by copper and silver strip corrosion tests are required for fuels to forestall these reactions.

Boundary lubrication of rubbing surfaces such as those found in high speed pumps and fuel controls has been found to be related to the presence in fuel of species capable of forming a chemisorbed film that reduces friction and wear (18). The lubricity of fuel is attributed to polar materials (19), multiring aromatics (19), and a thiohydroindane species (20). Oxygen-containing compounds, particularly long-chain acids, tend to adsorb on metal surfaces and act as lubricity agents. Corrosion inhibitors, which tend to form tenacious films on metal surfaces, are generally excellent lubricity agents. Refining processes to reduce sulfur, remove olefins, and control acidity tend to degrade a fuel's lubrication propensity. Corrosion inhibitors are mandatory in military fuels since they also improve the lubricity of fuels by extending the life of pumps and controls.

Both friction and wear measurements have been used to study boundary lubrication of fuel because sticking fuel controls and pump failures are primary field problems in gas turbine operation. An extensive research program of the Coordinating Research Council has produced a ball-on-cylinder lubricity test (BOCLE), standardized as ASTM D5001, which is used to qualify additives, to investigate fuels, and to assist pump manufacturers (21).

Economic Aspects

The exacting list of specification requirements for aviation gas turbine fuels and the constraints imposed by delivering clean fuel safely from refinery to aircraft are the factors that affect the economics. Compared with other distillates such as diesel and burner fuels, kerosene jet fuels are narrow-cut specialized products, and usually command a premium price over other distillates. The prices charged for jet fuels tend to escalate with the basic price of crude, a factor which seriously undermined airline profits during the Persian Gulf war as crude prices increased sharply.

Availability and cost of jet fuels are also affected by the interrelationships of the market for other petroleum products. An excellent example of these interrelationships was developed for the National Aeronautic and Space Administration (NASA) in a study on the affects of jet fuel property changes on producibility and cost in the United States, Canada, and Europe (22). For example, the western

United States is a unique market compared with the rest of the country because demand for gasoline and jet fuel is high, but demand for burner and boiler fuels is low. This product distribution encourages high conversion refineries with particular emphasis on hydrocracking to make lighter products. At the same time, environmental concerns over urban air quality are pressuring refineries to market reformulated gasolines and low sulfur diesel fuels. Refining these products and also producing specification jet fuel has tended to limit availability and to increase the cost of aviation fuels in the western United States.

The majority of refineries operated by petroleum companies in different parts of the world to make local products, such as gasoline and burner fuels, also produce jet fuels. Even a small refinery with simple equipment can make suitable jet fuel if it has access to the right crude. However, the principal supply of both civil and military jet fuels is produced in large refineries. Many are located near major cities and airports and are frequently connected by pipeline directly to the airport. Modern airports have extensive storage and handling facilities operated by local authorities, petroleum companies or consortia, or the airlines themselves.

Worldwide demand for the jet fuels specified in Table 3 amounted to about 477,000 m^3/d (3 million barrels per day) in 1990. About one-half of this demand was kerosene Jet A sold in the United States. One-third represented kerosene Jet A1 for delivery to international airlines outside the United States; the balance comprised various military fuels used by air forces around the world.

Future Trends

Aircraft Fuels. Demand for aviation gas turbine fuels has been growing more rapidly than demand for other petroleum products since 1960, about 3–5% per year compared with 1% for all oil products. This strong demand reflects a current and predicted growth in worldwide air traffic of 4–7% annually until the end of the century. Total world oil demand will be up by 15% by the year 2000, but aviation fuel demand will increase by 50–125%. However, the fraction of the oil barrel devoted to aviation, now about 8%, will increase only slightly.

Changes in the crude supply outlook are, in fact, decreasing the potential yield of the straight-run kerosene fraction, needed not only for aviation, but also for blending into diesel and heating fuels. For example, Arabian Gulf crudes, the most abundant worldwide supply, yield less kerosene than the primary U.S. crude. This situation requires more conversion processes at refineries such as hydrocracking of gas oil, a process that yields high quality but more expensive aviation kerosene.

The vulnerability of the aviation market was exposed in the winter of 1973–1974 when the Arab oil embargo caused actual shortages of jet fuel (23). Coupled with the dramatic rise in crude and product prices, the supply crisis precipitated a new generation of fuel-efficient aircraft and a relaxation of specification limits, such as higher freezing points to −47°C and higher aromatics to 25%. It has been estimated that 18% of the world's crudes produce a 150–250°C fraction containing more than 20% aromatics. Two of these, Heavy Arabian and North Slope, represent the important new sources of marginal jet fuel.

Operation of aircraft gas turbines on a wider cut than the 160–260°C fraction of crude to expand availability has been considered (24). Boiling range is limited at the low boiling end by flammability, ie, flash point, and at the high boiling end by low temperature needs, ie, freezing point. In the case of Jet A, a reduction of flash point from 38°C to 32°C would increase yield by 17%; an increase in Jet A1 freezing point to −40°C would add about 25% to the kerosene pool.

Synthetic fuels derived from shale or coal will have to supplement domestic supplies from petroleum someday, and aircraft gas turbine fuels producible from these sources have been assessed. Shale-derived fuels can meet current specifications if steps are taken to reduce the nitrogen levels. However, extracting kerogen from shale rock and denitrogenating the jet fuel are energy-intensive steps compared with petroleum refining; it has been estimated that shale jet fuel could be produced at about 70% thermal efficiency compared with 95% efficiency for petroleum (25). Such a difference represents much higher cost for a shale product.

Synthetic jet fuel derived from coal is even more difficult and expensive, since the best of the conversion processes produces a fuel very high in aromatics. With hydrogenation, overall thermal efficiency is only 50%. Without additional hydrogenation, the gas turbine fuels would contain 60–70% aromatics.

Production of liquid jet fuel from processing of abundant natural gas is a more promising and cheaper source of high quality product than shale or coal.

Fuels for Advanced Aircraft. The area of greatest traffic growth is predicted to be the Pacific basin where distances are great and the greatest incentive exists for an advanced aircraft that could greatly reduce travel time. Research at NASA is proceeding on a high speed commercial transport (HSCT) that would satisfy this market by the year 2010. The aircraft would carry about 300 passengers nonstop over 8000 miles at three times the speed of sound, and would reduce travel time from 14 to 6 hours. To be economically and environmentally viable, the HSCT would have to produce less nitrogen oxides in exhaust, cause lower sonic boom, and operate on the same kerosene as subsonic aircraft.

The first commercial supersonic transport, the Concorde, operates on Jet A1 kerosene but produces unacceptable noise and exhaust emissions. Moreover, it is limited in capacity to 100 passengers and to about 3000 miles in range. At supersonic speed of Mach 2, the surfaces of the aircraft are heated by ram air. These surfaces can raise the temperature of fuel held in the tanks to 80°C. Since fuel is the coolant for airframe and engine subsystems, fuel to the engine can reach 150°C (26). An HSCT operated at Mach 3 would place much greater thermal stress on fuel. To minimize the formation of thermal oxidation deposits, it is likely that fuel delivered to the HSCT would have to be deoxygenated.

NASA is also considering a more advanced aircraft such as Mach 5 to cut Pacific travel time to about three hours, but in this case kerosene fuel is no longer acceptable, and liquefied natural gas or liquefied hydrogen would be needed to provide the necessary cooling and stability. However, a completely new fueling system would be required at every international airport to handle these cryogenic fluids.

In the speed regime of Mach 6 and beyond (hypersonic vehicles), hydrogen is the fuel of choice because of its high energy content and its combustion kinetics. For example, a first stage to orbit reusable booster for space vehicles would employ a supersonic combustor ramjet engine burning hydrogen, the only fuel capable of maintaining combustion in the shock wave. A booster that replaced 80% of the present weight of oxidizer would revolutionize space transportation by replacing the current space shuttle as a delivery vehicle.

Use of kerosene fuel rather than air as a coolant focuses attention on another fuel property, specific heat, which is a measure of its efficiency to absorb thermal energy. Heat capacity increases with temperature but decreases with density, which favors a low density paraffinic fuel from petroleum or natural gas sources.

Because of tank heating, fuel volatility is also more critical in supersonic aircraft. For example, the Concorde tank is pressurized to prevent vapor losses which could be significant at high altitude where fuel vapor pressure may equal atmospheric pressure. The tank can reach 6.9 kPa (1 psi) at the end of a flight. The need to deoxygenate fuel for thermal stability in the HSCT will doubtless require a similar pressurized system.

Ground Turbine Fuels. Unlike the outlook for aviation fuels, the demand for ground turbine fuels has stabilized because the ground turbine can be replaced by a more fuel-efficient engine for passenger cars and trucks. On the other hand, the ground turbine is unique in its ability to provide peak electrical power on a few seconds notice and to accept a variety of fuels as a prime mover in pipeline pumping and gas compression. The fuels described in Table 1 from gas to crude will continue to provide the energy source for ground turbines depending on location. Cogeneration plants which provide both electricity and heat simultaneously will undoubtedly continue to be a prime application of ground turbines to many complexes serving suburban and rural areas.

BIBLIOGRAPHY

"Aviation and Other Gas Turbine Fuels" in *ECT* 3rd ed., Vol. 3, pp. 328–351, by W. G. Dukek, Exxon Research and Engineering Company.

1. W. G. Dukek, A. R. Ogston, and D. R. Winans, *Milestones in Aviation Fuels,* AIAA No. 69–779, American Institute of Aeronautics and Astronautics, New York, July 1969.
2. W. A. Herbst and N. V. Hakala, *7th World Petroleum Congress,* Mexico City, Mexico, Apr. 1967.
3. R. S. Schlaefer and S. D. Heron, *Development of Aircraft Engines and Fuels,* Harvard Business School, Boston, Mass., 1950.
4. S. M. DeCorso, R. A. Symonds, and G. Vermes, *Crude Oil Firing in the Utility Gas Turbine,* ASME 71–WA/GT–11, American Society of Mechanical Engineers, New York, 1971.
5. K. M. Brown, *Commercial Results with UOP MEROX Process for Mercaptan Extraction in U.S. and Canada,* UOP Booklet 267, UOP, Inc., Des Plaines, Ill., 1960.
6. W. M. Bustin and W. G. Dukek, *Electrostatic Hazards in the Petroleum Industry,* John Wiley & Sons, Inc., Chichester, Sussex, PO19 IUD, UK, 1983.
7. N. A. Chigler, *Progress in Energy and Combustion Science,* Vol. 21, No. 4, Pergamon Press, Inc., Elmsford, N.Y., 1976.
8. *Microburner Studies of Flame Radiation as Related to Hydrocarbon Structure,* Report

3752-64R, Navy Buweps Contract NOw 63-0406d, Phillips Petroleum Co., Bartlesville, Okla., May 1964.
9. M. Smith, *Aviation Fuels,* G. T. Foulis & Co., Ltd., Henley-on-Thames, Oxfordshire, UK, 1970.
10. H. F. Butze and R. C. Ehlers, *Effect of Fuel Properties on Performance of a Single Aircraft Turbojet Combustor,* NASA TM X-71789, National Aeronautics and Space Administration, Lewis Research Center, Cleveland, Ohio, Oct. 1975.
11. W. F. Taylor, *Development of High Stability Fuels,* Final report, Contract NO0140-74-C-0618, Naval Air Propulsion Center, Trenton, N.J., Dec. 1975.
12. W. F. Taylor, *J. Appl. Chem.* **18,** 25 (1968).
13. R. H. Clark, *3rd International Conference on Stability and Handling of Liquid Fuels,* London, UK, Sept. 1988.
14. H. C. Barnett and R. R. Hibbard, *Properties of Aircraft Fuels,* NASA TN 3276, NASA, Lewis Research Center, Cleveland, Ohio, Aug. 1956.
15. *Aviation Fuel Safety—1975,* CRC Report No. 482, Coordinating Research Council, Inc., Atlanta, Ga., Nov. 1975.
16. *Survey of Electrical Conductivity and Charging Tendency Characteristics of Aircraft Turbine Fuels,* CRC Report No. 478, Coordinating Research Council, Inc., Atlanta, Ga., Apr. 1975.
17. H. R. Porter and co-workers, *Salt Driers Assure Really Dry Fuel to Aircraft,* SAE 710440, Society of Automotive Engineers, Warrendale, Pa., 1971.
18. J. J. Appeldoorn and W. G. Dukek, *Lubricity of Jet Fuels,* SAE 660712, Society of Automotive Engineers, Warrendale, Pa., 1966.
19. J. K. Appeldoorn and F. F. Tao, *Wear* **12** (1968).
20. R. A. Vere, *SAE Trans.* **78,** 2237 (1969).
21. W. G. Dukek, *Ball-on-Cylinder Testing for Aviation Fuel Lubricity,* SAE 881537, Society of Automotive Engineers, Warrendale, Pa., 1988.
22. G. M. Varga and A. J. Avella, *Jet Fuel Property Changes and Their Effect on Producibility and Cost in U.S., Canada, and Europe,* NASA Contract Report 174840, Lewis Research Center, Cleveland, Ohio, Feb. 1985.
23. A. G. Robertson and R. E. Williams, *Shell Aviation News,* Shell Oil Co., Houston, Tex., 1976, p. 435.
24. *Factors in Using Kerosene Jet Fuel of Reduced Flash Point,* ASTM STP 688, American Society for Testing Materials, Philadelphia, Pa., 1979.
25. W. Dukek and J. P. Longwell, *Alternative Hydrocarbon Fuels for Aviation,* Esso Air World(b)29 No. 4, Exxon International Co., Florham Park, N.J., 1977.
26. H. Strawson and A. Lewis, *Predicting Fuel Requirements for the Concorde,* SAE 689734, Society of Automotive Engineers, Warrendale, Pa., Oct. 1968.

General References

M. Smith, *Aviation Fuels,* G. T. Foulis & Co., Ltd., Henley-on-Thames, Oxfordshire, UK, 1970.

W. M. Bustin and W. G. Dukek, *Electrostatic Hazards in the Petroleum Industry,* John Wiley & Sons, Inc., Chichester, Sussex, PO19 IUD UK, 1983.

W. F. Taylor, *Development of High Stability Fuels,* Final report, Contract NO0140-74-C-0618, Naval Air Propulsion Center, Trenton, N.J., Dec. 1975.

W. G. DUKEK
Consultant

AZELAIC ACID. See DICARBOXYLIC ACIDS.

AZINE DYES

Azine, oxazine, and thiazine dyes were among the earliest of synthetic dyes. The names are derived from the 6-member heterocyclic ring system present in all dyes of these classes: 1,4-diazine [*290-37-9*] (**1**), 1,4-oxazine [*290-47-1*] (**2**), and 1,4-thiazine [*290-57-3*] (**3**).

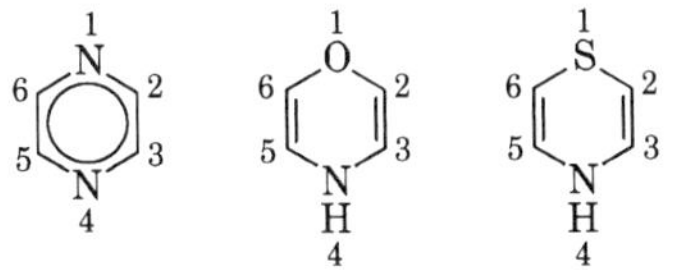

(**1**) (**2**) (**3**)

The parent rings are always included in phenazine [*92-82-0*] (**4**), phenoxazine [*135-67-1*] (**5**), or phenothiazine [*92-84-2*] (**6**) ring systems.

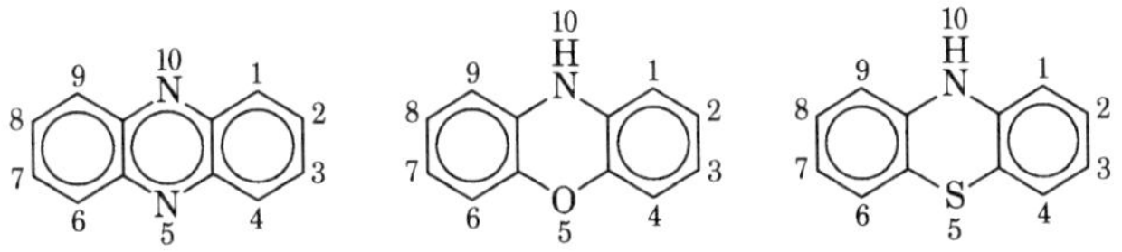

(**4**) (**5**) (**6**)

The first of these dyes to be made synthetically was Perkin's mauveine [*6373-22-4*], which is a substituted phenazine of structure (**7**).

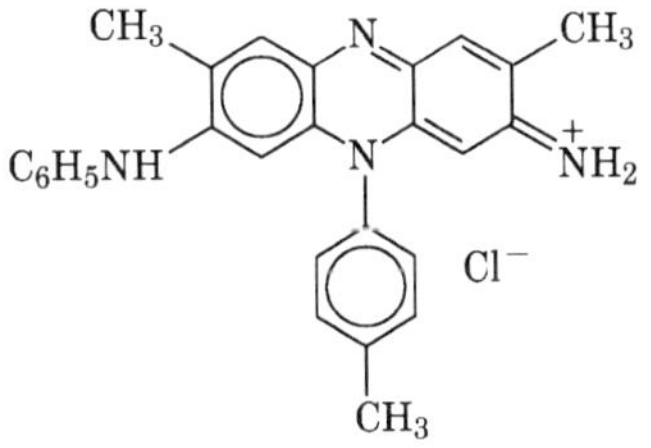

(**7**)

Mauveine is in a group of azines termed safranine dyes; ie, it is a *N*-phenylphenazonium chloride. Although the structures of these dyes are often written to show a positive charge on a particular hetero atom, the charge is in fact distributed through resonance throughout the molecule, thus accounting for their deep color.

Perkin's discovery of mauveine in 1856 led to a flurry of activity in synthetic organic chemistry in England, which spread to Germany, and flourished through the last half of the nineteenth century, one of the richest periods for synthetic organic chemistry. During this time many synthetic organic methods were developed which are still used today.

Azine, oxazine, and thiazine dyes were historically more important than they are at present. However, at least one example of each, introduced more than 100 years ago, is still offered commercially today (1,2). Azo and anthraquinone dyes have largely displaced them in commercial application. Azo dyes (qv) offer better fastness and broader shade ranges at more economical prices.

All classes of azine dyes are vattable, ie, they are reduced to "colorless" forms, then oxidized back to the dye (2). They therefore offer good fastness to oxidation (3). Because of this property, many find uses as redox indicators in titrations.

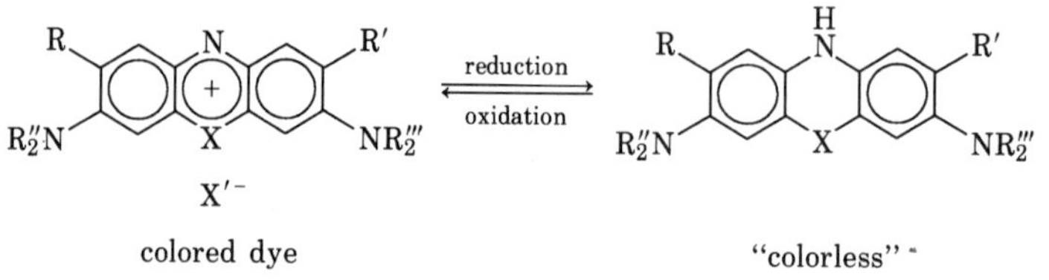

Dyes in these classes are generally basic dyes; ie, the chromophore is cationic. Some structures have been sulfonated to acid dyes, eg, the Nigrosine, (CI Solvent Black 5; CI 50415), (**8**) to CI Acid Black 2 [*8005-03-6*] (CI 50420) (**9**).

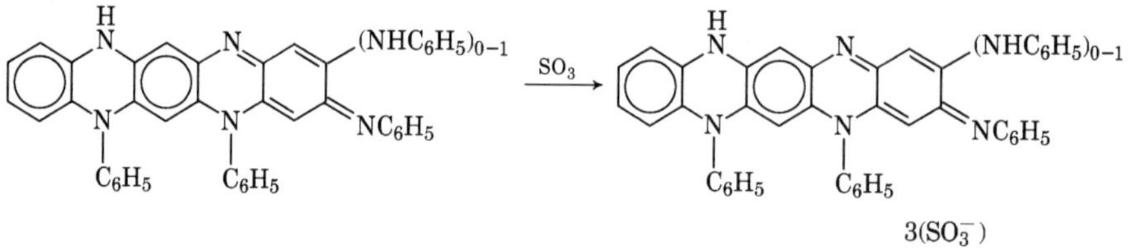

(**8**) (**9**)

Both Nigrosine and Acid Black 2 are mixtures, and the structures shown are only typical components. The purer azine dyes of this structure are prepared by controlled stepwise synthesis by reaction of an azobenzene with aniline and aniline hydrochloride. Nigrosine, prepared by oxidizing aniline and aniline hydrochloride with nitrobenzene in the presence of ferric chloride, is similar in structure to nongreening Aniline Black [*13007-86-8*] (CI Pigment Black 1; CI 50440) (**10**), prepared by oxidizing aniline and aniline hydrochloride with dichromate or sodium chlorate in the presence of an oxygen carrier such as a copper, vanadium, or iron salt. Aniline blacks are prepared in the dyebath with the material to be dyed.

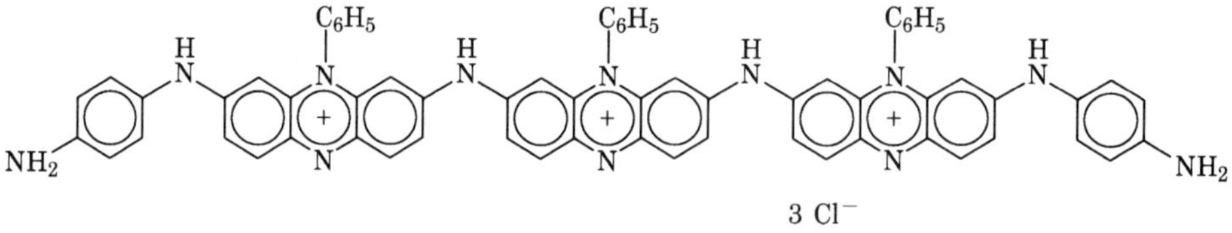

(**10**)

Phenylrosinduline (**11**) has been sulfonated to CI Acid Red 101, CI 50085, (**12**); such dyes are no longer commercial items.

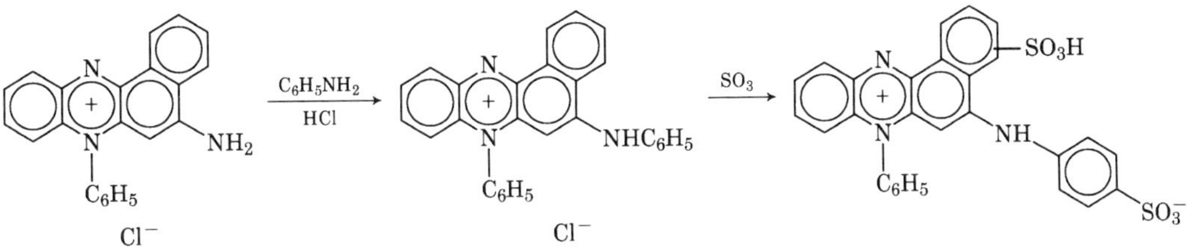

(11) (12)

There are oxazine disperse dyes reported (4) and several reactive dyes containing the oxazine ring and a few containing thiazine (5) (see DYES, REACTIVE).

Azines

Azines were the first of the synthetic dyes. Perkins oxidized a crude mixture of aniline containing toluidines and obtained mauveine (**7**).

Azine dyes are relatively unimportant as a class of dyes but are used extensively as biological stains. Colors are mostly yellow to red. Durability of some of these dyes is supported by the 1990 *AATCC Buyers Guide,* published by the American Association of Textile Chemists and Colorists, which lists Basic Red 2, Safranine T [*477-73-6*], (**13**) and Basic Red 5, Neutral Red [*553-24-2*], (**14**), discovered in 1859 and 1879, respectively (1). Basic Red 2 is a safranine similar to mauveine (**7**).

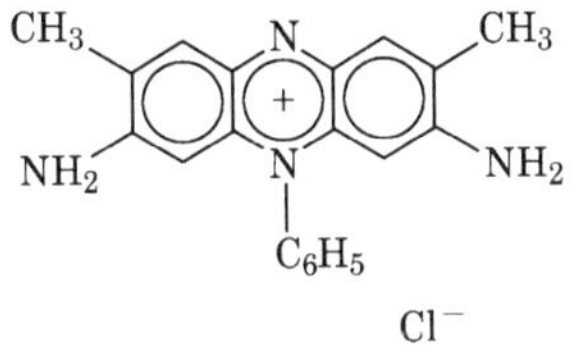

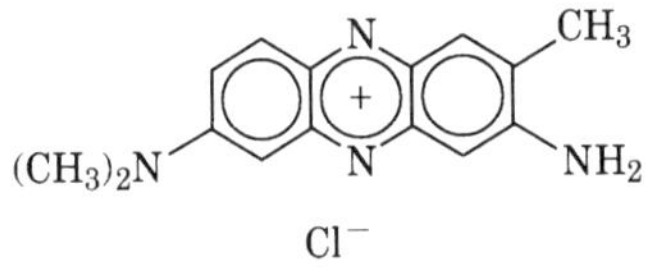

(13) (14)

There are approximately one dozen references to azine dyes reported in *Chemical Abstracts* traceable literature since 1976. Of these references, all but one refers to titration indicators. One basic dye is reported, but it is not a commercial product (6):

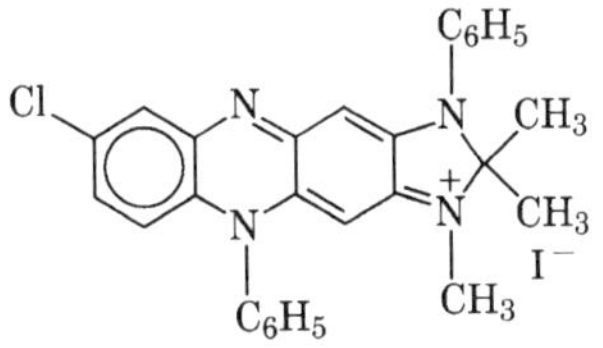

Synthesis. Methods of synthesis have changed little since the synthesis of Basic Red 2. One method starts with *o*-toluidine [*95-53-4*], C_7H_9N, which is diazotized and coupled to form amino azotoluene [*97-56-3*], $C_{14}H_{15}N_3$. Amino azotoluene is reduced to one mole each of *o*-toluidine and 2,5-diaminotoluene [*95-70-5*],

$C_7H_{10}N_2$. Condensation and oxidation with aniline [*62-53-3*], C_6H_7N, gives the desired dye (**13**).

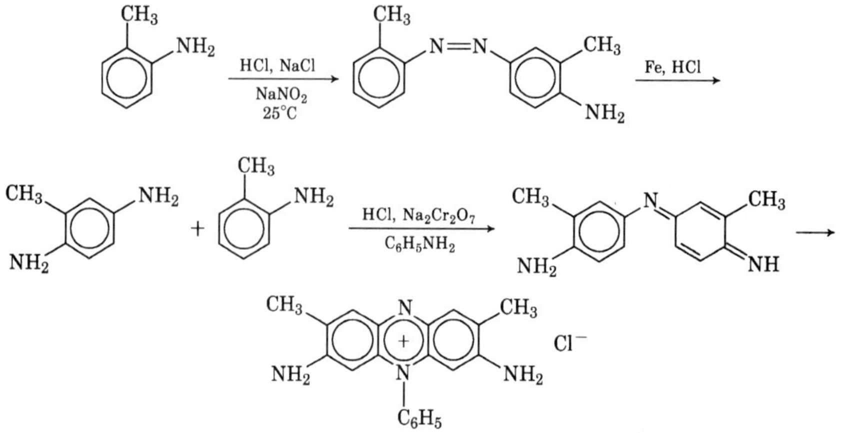

(**13**)

The method described in reference 6 differs only in that quaternization is an additional step.

Oxazine Dyes

There have been approximately 80 references to oxazine dyes in the past 15 years. Several are references to their use as laser dyes and titration indicators. Thirty of these references are to oxazine dyes for either acrylic fibers or leather (qv) (7) (see FIBERS, ACRYLIC).

There are also examples of reactive dyes having only oxazine heterocyclic rings in the chromophore such as (**15**) [*58104-86-2*], as well as those having both oxazine and thiazine rings such as (**16**) [*97140-65-3*].

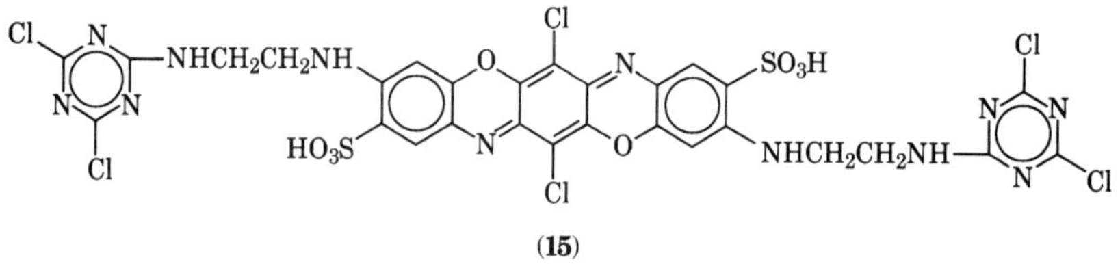

(**15**)

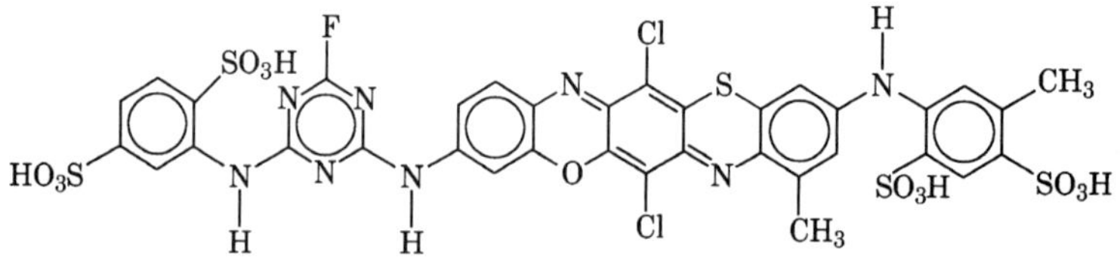

(**16**)

CI Direct Blues 106 (CI 51300) (**17**, R = H) and 190 (CI 51305) (**17**, R = Cl) are oxazine dyes.

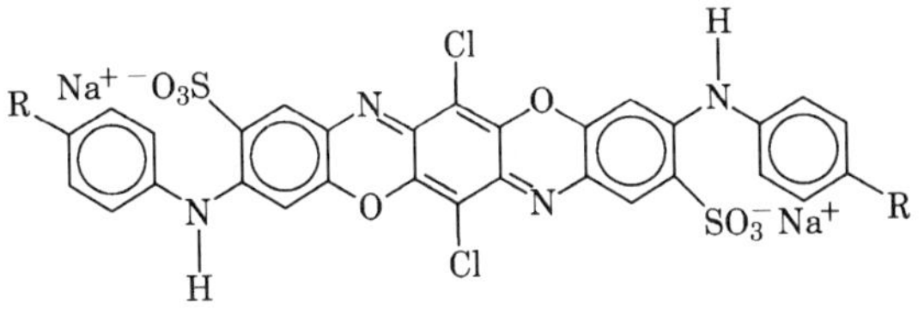

(**17**)

Additional examples have been reported (5). Disperse dyes which contain an oxazine ring include the yellow 1,3-dinitro-10*H*-phenoxazine [*26103-32-2*] (**18**) and the bright orange red (**19**) [*66115-93-3*] (4).

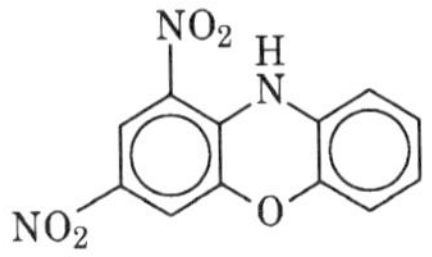

(**18**)

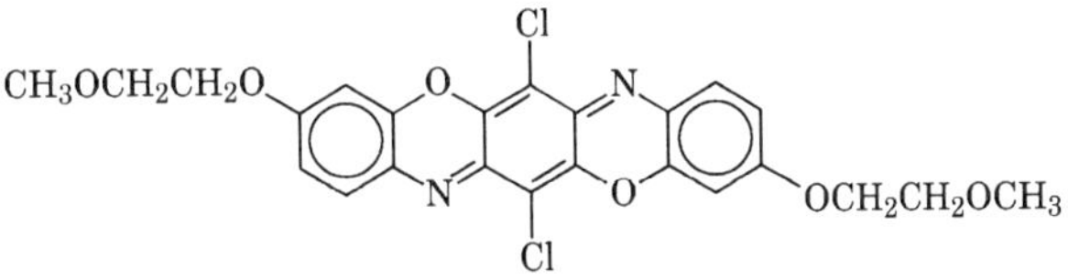

(**19**)

Neither of these dyes has been reported as a commercial item.

Oxazine dyes containing sulfonic acid groups and claimed to be suitable for dyeing leather in brown shades include [*89477-76-9*] (**20**) and [*87617-03-6*] (**21**) (8).

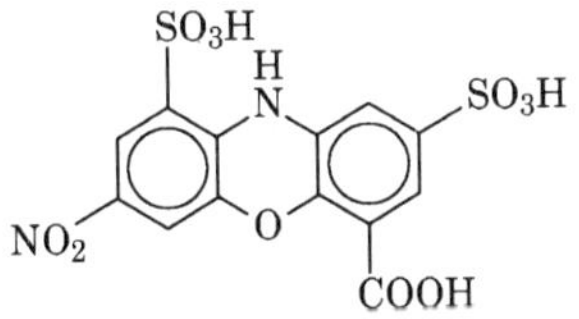

(**20**)

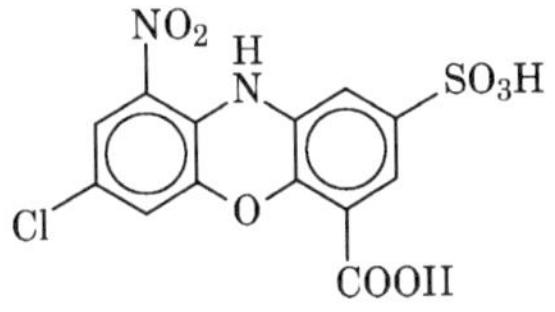

(**21**)

Cationic oxazine dyes suitable for dyeing acrylic fibers have been reported (7).

Synthesis. Several synthetic methods have been reported for oxazine dyes. The following are representative:

Condensation of phenol [*108-95-2*], C_6H_6O, with *p*-nitroso-*N*,*N*-dimethyl-aniline [*138-89-6*], $C_8H_{10}N_2O$, in acetic acid (9).

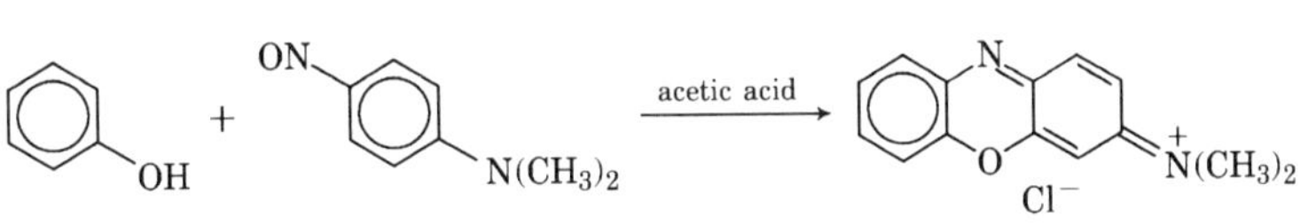

The following condensation produces the dye [*58378-00-00*], a trichlorozincate.

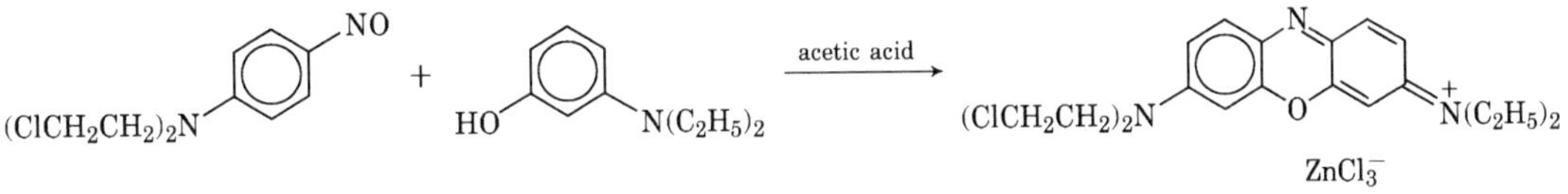

Condensation of an *o*-nitrosophenol with an amine hydrochloride in acetic acid (10): Nile Blue [*2381-85-3*] is formed from the hydrochloride of 1-naphthylamine.

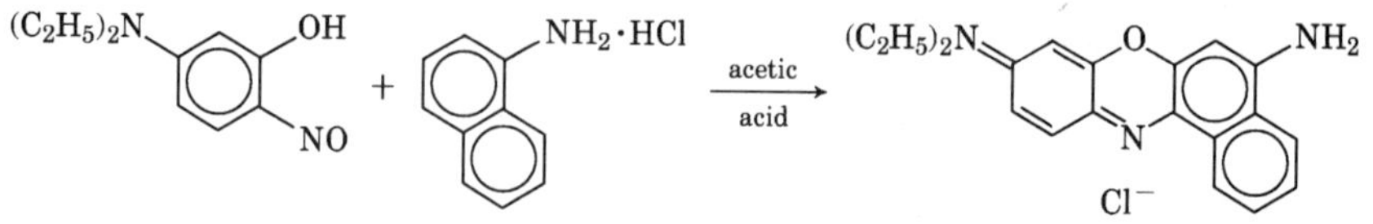

The dye [*57713-93-6*] is formed from 2,5-diamino-diphenylamine:

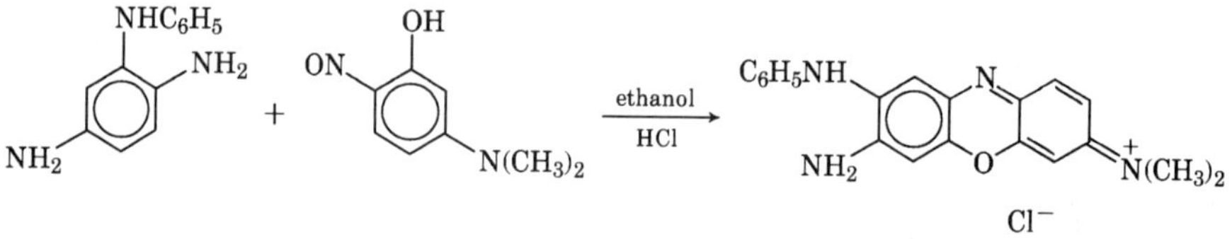

Condensation of 2,3,5,6-tetrachloro-2,5-cyclohexadiene [*118-75-2*], $C_6Cl_4O_2$, with 9-ethyl-9*H*-carbazol-3-amine [*132-32-1*], $C_{14}H_{14}N_2$, (11) leads first to (**22**) [*6358-30-1*]. Treatment with *N*-methylol-α-chloroacetamide [*2832-19-1*] and sulfuric acid produces (**23**) [*87564-36-1*], which forms the pyridinium salt (**24**) [*87564-37-2*].

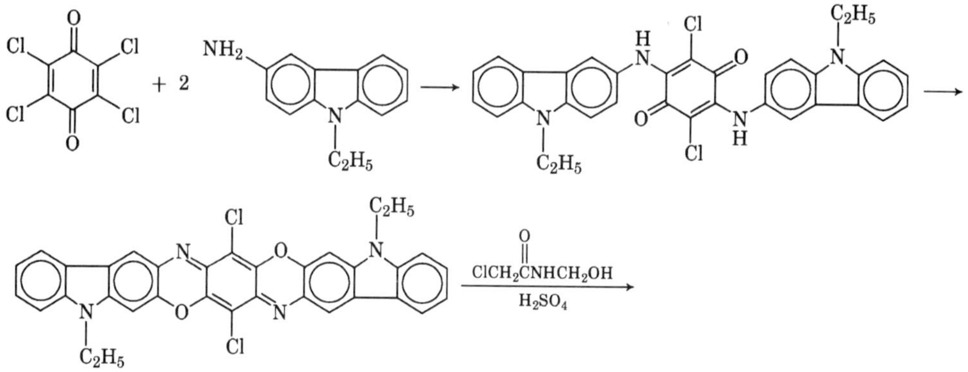

(**22**)

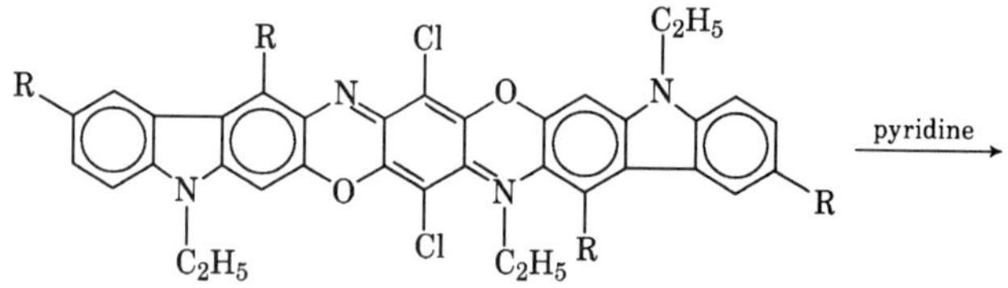

(**23**) R = $CH_2NHCOCH_2Cl$

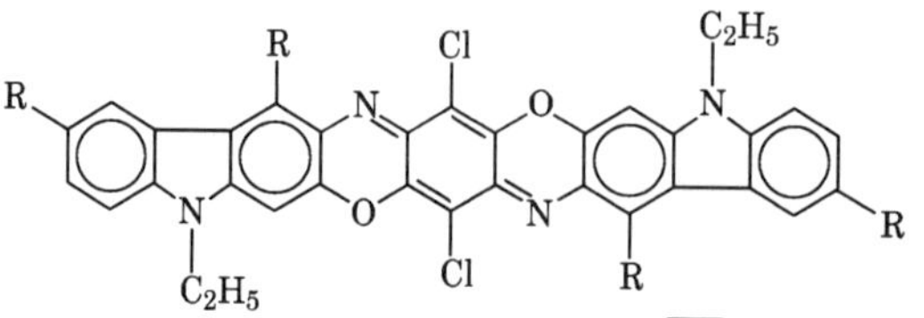

(**24**) R = $CH_2NHCOCH_2\overset{+}{N}C_5H_5$

Concerns have arisen recently over potential presence of dioxins in dyes produced from 2,3,5,6-tetrachloro-2,5-cyclohexadiene [*118-75-2*].

It has often been necessary to isolate oxazine dyes as insoluble zinc salts. Addition of urea [*57-13-6*], CH_4N_2O, or thiourea [*62-56-6*], CH_4N_2S, to the reaction mixture is reported to give good yields without using zinc (12), eg, the salt (**25**) [*33203-82-6*] is formed.

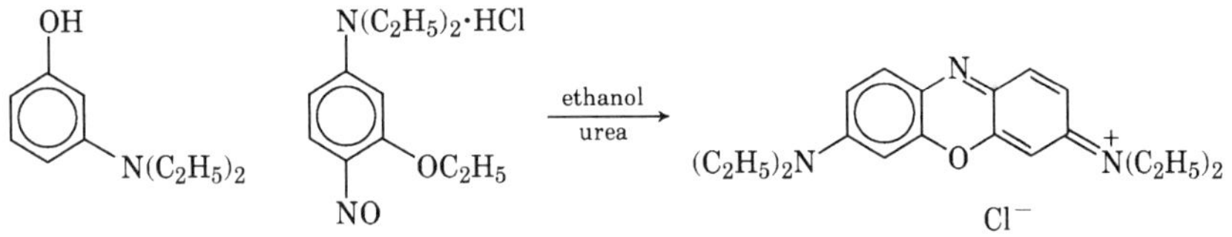

(**25**)

Thiazine Dyes

Methylene Blue [*61-73-4*] CI Basic Blue 9 (CI 52015), (**26**), is the classic thiazine dye still in use today. This dye was first reported in 1876 (1).

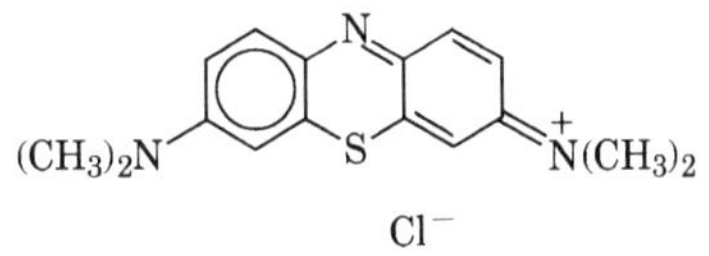

(**26**)

There have been approximately 60 references to thiazine dyes in the past 15 years of *Chemical Abstracts*. Although most of these references are to titration indicators, photophysical evaluation, and spectral properties, a few refer to structures for use as dyes (13).

Thiazine is not important in dyes as such, but it is a part of some reactive dyes (5). Thiazine is important in sulfur dyes (qv).

Synthesis. The method of synthesis for Methylene Blue described in reference 14 is still the stepwise method of choice for thiazine dyes. *N,N*-Dimethyl-*p*-phenylene diamine [*99-98-9*], $C_8H_{12}N_2$, reacts with sodium thiosulfate [*7772-98-7*] to form the thiosulfonic acid which condenses with *N,N*-dimethylaniline [*121-69-7*], $C_8H_{11}N$, in the presence of sodium dichromate [*10588-01-9*] to the indamine, then with copper sulfate [*18939-61-2*] and sodium dichromate to Methylene Blue (**26**).

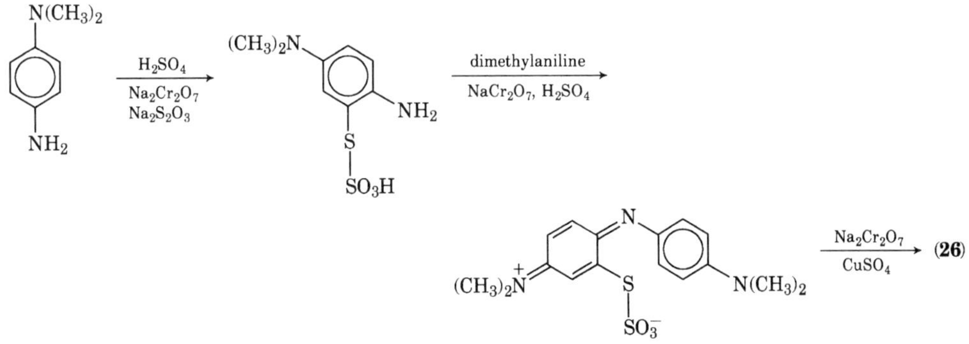

BIBLIOGRAPHY

"Azine Dyes" in *ECT* 1st ed., Vol. 2, pp. 214–223, by A. W. Dawes and J. C. Moessinger, General Aniline Works Division, General Aniline & Film Corp.; in *ECT* 2nd ed., Vol. 2, pp. 859–868, by H. E. Nursten, Massachusetts Institute of Technology; in *ECT* 3rd ed., Vol. 3, pp. 378–386, by J. C. Conger, Sr., American Cyanamid Company.

1. *The Colour Index,* 3rd ed., Vol. 4, Society of Dyers and Colorists, Bradford, Yorkshire, UK, 1971, pp. 4445–4473.
2. *Toxic Substances Control Act Chemical Substances Inventory,* U.S. Environmental Protection Agency, Office of Toxic Substances, United States Printing Office, Washington D.C., 1985.
3. K. Venkataraman, *The Chemistry of Synthetic Dyes and Pigments,* Vol. 2, Academic Press, Inc., New York, 1952, pp. 761–795.
4. Ger. Pat. 224,856 (July 17, 1985), H. Modrow and co-workers; Ger. Pat. 236,105 (May 28, 1986), G. Wallaschek and co-workers (to VEB Bitterfeld); Jpn. Pat. 78 26,826 (Mar. 13, 1978), S. Imahori, Y. Murata, J. Yoshiharu, and S. Suzuki (to Mitsubishi).
5. Eur. Pat. Appl. 84,718 (Aug. 3, 1983), A. H. M. Renfrew (to ICI); Ger. Pat. 3,330,547 (Mar. 21, 1985), W. Harms, G. Franke, and K. Wunderlich (to Bayer A.G.); Eur. Pat. Appl. 135,381 (Mar. 27, 1985), R. D. McClelland and A. Renfrew (to ICI); Ger. Pat. 3,336,362 (Apr. 18, 1985), H. Jaeger (to Bayer A.G.).
6. Ger. Pat. 3,531,774 (Mar. 12, 1987), G. Franke (to Bayer A.G.).
7. Ger. Pat. 2,518,587 (Nov. 13, 1975), P. Moser and A. N. Nicopoulos (to CIBA-GEIGY); Ger. Pat. 2,458,347 (June 16, 1976), R. Mohr and E. Fleckenstein (to Hoechst A.G.); Ger. Pat. 2,065,887 (Oct. 21, 1976), E. Mundlos, R. Mohr, and L. Herz (to Hoechst A.G.); Brit. Pat. 1,488,609 (Oct. 12, 1977) (to Hoechst A.G.); Ger. Pat. 2,631,040 (Jan. 27, 1977), E. Brunn (to CIBA-GEIGY); Ger. Pat. 2,631,207 (Jan. 19, 1978), K. Lehment, F. Klaus, and U. Trense (to Bayer A.G.); U.S. Pat. 4,116,622 (Sept. 26, 1978), S. Koller and J. Koller (to CIBA-GEIGY); U.S. Pat. 4,196,286 (Apr. 1, 1980), E. Brunn (to CIBA-GEIGY); Ger. Pat. 3,033,439 (Sept. 5, 1980), U. Mayer (to BASF); U.S. Pat. 4,288,227 (Sept. 8, 1981), B. Gertiser and B. Henz (to Sandoz); Eur. Pat. Appl. 55,223 (June 30, 1982), P. Loew (to CIBA-GEIGY); Switz. Pat. 642,987 (May 15, 1984), P. Moser (to CIBA-GEIGY).
8. Ger. Pat. 200,890 (June 22, 1983), E. Jauer and co-workers (to VEB Bitterfeld); Ger. Pat. 203,065 (Oct. 12, 1983), E. Jauer and co-workers (to VEB Bitterfeld).
9. Brit. Pat. 1,410,815 (Oct. 22, 1975), J. L. Leng (to ICI).
10. Ger. Pat. 2,411,761 (Oct. 16, 1975), E. Fleckenstein (to Hoechst A.G.).
11. Eur. Pat. Appl. 83,309 (July 6, 1983), J. M. Adams (to CIBA-GEIGY).
12. Ger. Pat. 2,353,987 (Apr. 30, 1975), H. Psaar (to Bayer A.G.).
13. Ger. Pat. 2,518,587 (Nov. 13, 1975), P. Moser and A. Nicopoulos (to CIBA-GEIGY); Ger. Pat. 2,516,920 (Oct. 21, 1976), W. Loehr and co-workers; Ger. Pat. 139,268 (Dec. 19, 1979), G. Mann and co-workers.
14. *FIAT,* Final Rep. No. 1313, Vol. 2, Feb. 1948, pp. 371, 376.

Roy E. Smith
CIBA-GEIGY Corporation

AZINES. See Azine dyes.

AZIRIDINE. See Imines, cyclic.

AZO DYES

The term azo dyes is applied to those synthetic organic colorants that are characterized by the presence of the chromophoric azo group (–N=N–). This divalent group is attached to sp^2 hybridized carbon atoms: on one side, to an aromatic or heterocyclic nucleus; on the other, it may be linked to an unsaturated molecule of the carbocyclic, heterocyclic, or aliphatic type. No natural dyes contain this chromophore. Commercially, the azo dyes are the largest and most versatile class of organic dyestuffs. There are more than 10,000 Colour Index (CI) generic names assigned to commercial colorants, approximately 4500 are in use and over 50% of these belong to the azo class (1). U.S. dye manufacturers produced 127,000 t of dyes in 1988, valued at $766 million, which is still 4% less in terms of value than in 1980. In terms of quantity, sales of dyes in 1988 were about 14% higher than in 1980. Synthetic dyes are derived in whole or in part from cyclic intermediates. Approximately two-thirds of the dyes consumed in the United States are used by the textile industry to dye natural and synthetic fiber or fabrics, about one-sixth is used for coloring paper, and the rest is used chiefly in the production of organic pigments and in the the dyeing of leather and plastic. Dyes are sold as pastes, powders, and liquids; concentrations vary from 6 to 100%. The concentration, form, and purity of a dye is determined largely by the use for which it is intended.

Classification and Designations. The most authoritative compilation covering the constitution, properties, preparations, manufacturers, and other coloring data is the publication *Colour Index,* which is edited jointly by the Society of Dyers and Colourists and the American Association of Textile Chemists and Colorists (AATCC). In the *Colour Index* a dual classification system is employed to group dyes according to area of usage and chemical constitution. Because of the ease of synthesis of azo, disazo, and polyazo dyes, and their wide range of applications, azo dyes comprise the largest chemical class in numbers, monetary value, and tonnage produced. There are more than 2200 chemical structures of azo dyes disclosed in the *Colour Index* (Vols. 4–8).

Nearly all dye manufacturers use letters and numerals in the names of their products to define the hue. Thus B is blue; G, yellow (gelb in German) or green; R, red; and Y, yellow. Numerals, ie, 2G (or GG), 3G, 4G, etc indicate, in this case, a successively yellower or greener shade. Occasionally, suffixed letters are used to feature other properties such as solubility, lightfastness, brightness, and use on synthetic fibers.

Chemically, the azo class is subdivided according to the number of azo groups present into mono-, dis-, tris-, tetrakis-, etc. Mono- and disazo dyes are essentially equal in importance, trisazo dyes are less important, and tetrakisazo dyes, except for a few, are much less important. For this reason, substances with more than three azo linkages are generally included under the heading of polyazo dyes. The *Colour Index* lists the azo dyes as follows:

Chemical Class	*Colour Index Number Range*
monoazo	11,000–19,999
disazo	20,000–29,999
trisazo	30,000–34,999
polyazo	35,000–36,999

The Chemistry of Synthesis

Peter Griess synthesized the first azo dye soon after his discovery of the diazotization reaction in 1858. The two reactions which form the basis for azo dye chemistry are diazotization (eq. 1) and coupling (eq. 2) (2).

$$ArNH_2 + 2\ HX + NaNO_2 \longrightarrow ArN_2^+X^- + NaX + 2\ H_2O \qquad (1)$$

$$\text{where } X = Cl^-, Br^-, NO_3^-, HSO_4^-, BF_4^-, \text{etc.}$$

$$ArN_2^+X^- + RH \longrightarrow ArN{=}N{-}R + HX \qquad (2)$$

where R represents an alkyl or aryl radical whose conjugate acid RH is capable of coupling.

Diazotization. Practically every aromatic primary amine is a potential diazo component. The value of an amine as a diazo component is determined chiefly by the properties of the dye prepared from it. Cost of amine, ease of diazotization, stability of the diazonium salt, and final cost of dye are factors that influence the selection of an amine. There are many methods (3) described for the synthesis of diazo compounds. A technically important one involves treating a primary arylamine with sodium nitrite in the presence of mineral acid at 0–5°C to form a diazonium salt as shown in equation 1. Since the mechanism of diazotization involves attack by the nitrosating species on the free amine and not the amine salt, it is often necessary to choose a higher acid strength in order to complete diazotization of weakly basic amines such as nitroanilines and trihalogenoanilines. The reaction medium should also provide stabilization of the diazonium salt by solvation effects. Negatively substituted and, therefore, very weakly basic amines, for example, 2,4-dinitro-6-halogenoaniline, 2,4-dinitroaniline [*97-02-9*] or 2,6-dihalogeno-4-nitroaniline, are dissolved in a concentrated acid (sulfuric, phosphoric, or glacial acetic acid) and diazotized with nitrosylsulfuric acid [*7782-78-7*] $HSO_4{\cdot}NO$. These diazos are relatively more stable in solution. Suspensions of diazonium salts in higher concentrations are dangerous and can form explosive mixtures (4). Amines containing sulfonic acid groups (zwitterions) are difficult to dissolve in acid medium. Such amines, for example sulfanilic acid [*121-57-3*], metanilic acid [*121-47-1*], naphthionic acid [*84-86-6*], or diaminostilbenedisulfonic acid [*81-11-8*], can preferably be diazotized by the so-called indirect method. In this case, the sodium salt of the acid may be treated with the required amount of nitrite in aqueous solution, and this solution may be run into ice-cooled hydrochloric acid solution. Sometimes diazotization of aminophenols (qv) and aminonaphthols is carried out in the presence of catalytic amounts of copper or zinc salts (5). This suppresses the oxidation of the hydroxy groups to the corresponding quinones by nitrous acid. The diazotization of *o*-aminophenols and naphthols is important in relation to the synthesis of *o,o′*-dihydroxyazo metal complex dyes.

Dyes with heterocyclic diazo components have received much attention because of their high tinctorial power and excellent brightness. A few technically important heterocyclic amines used in their preparation are shown in Figure 1. Heterocyclic azo dyes from some of these amines offer excellent fastness properties and are commercially competitive with more expensive anthraquinone dyes.

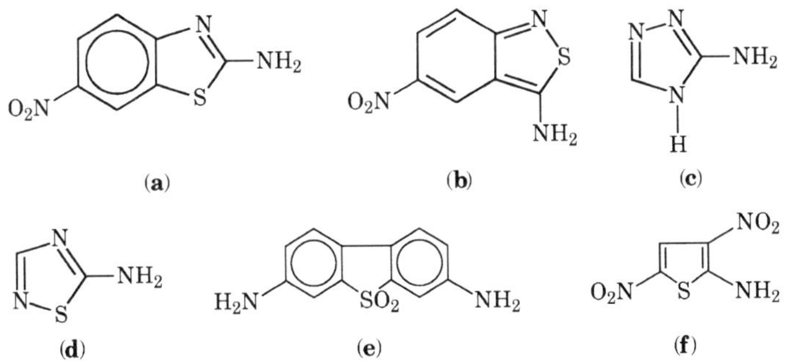

Fig. 1. Heterocyclic amines used in azo dyes. (**a**) 2-Amino-6-nitrobenzothiazole [*6285-57-0*]; (**b**) 3-amino-5-nitro-2,1-benzisothiazole [*14346-19-1*]; (**c**) 3-amino-4*H*-1,2,4-triazole [*65312-61-0*]; (**d**) 5-amino-1,2,4-thiadiazole [*7552-07-0*]; (**e**) 4,4′-diamino-2,2′-biphenylsulfone [*6259-19-4*]; (**f**) 2-amino-3,5-dinitrothiophene [*2045-70-7*].

During the 1980s, a large number of patents were issued for dyes based on 2-aminothiophenes (6). The diazotization of heterocyclic amines has been reviewed (7), but no systematic methods are available. Protonation at the hetero ring can reduce the nucleophilicity of the primary amino group leading to an equilibrium between the amine and the diazonium ion. This equilibrium favors the protonated amine when the concentration of acid is increased (8). 2-Aminothiazoles and 2-aminobenzothiazoles are readily diazotized to diazonium salts with sodium nitrite or nitrosylsulfuric acid in strong acid such as sulfuric acid, phosphoric acid, or nitric acid. In concentrated hydrochloric acid the diazonium salts react rapidly and 2-chlorothiazoles are formed. When aminothiadiazoles (5-amino-1,2,4-thiadiazoles and 2-amino-1,3,4-thiadiazoles) are treated with sodium nitrite in phosphoric acid solution, diazonium salts are the products. These exhibit a higher coupling reactivity compared to diazotized 2,4-dinitroaniline (9). Among the various types of heterocyclic amines, it is evident that nitrosoamines are favored under dilute acid conditions and diazonium formation is favored under strong acid conditions. Diazotization is improved by the addition of a mixture of concentrated acetic acid and propionic acid (10). Further methods of diazotization of aromatic amines have been described (11,12).

Diazotizations in organic solvents followed by coupling or simultaneous diazotization and coupling in one step have gained importance (13). The process can be conducted batchwise or continuously. When the dye is soluble in the reaction mixture, the product can be packed out in its entirety. This offers an advantage in reducing the loss of dye during processing. The liquid dye also offers ease of handling, accuracy of measurement, and uniformity. Besides sodium nitrite [*7632-00-0*] and nitrosylsulfuric acid, nitrites of glycol and glycol derivatives (14) can be employed as diazotizing agents. These compounds in comparison to nitrites of low molecular weight alcohols, such as amyl nitrite, are much less dangerous in handling, physiological properties, and explosiveness. These types of processes have been employed for a limited number of commercial azo dyes. The mechanism of the diazotization reaction has appeared (15) and other results have been reviewed (16,17). Nitrosation of the amino group is the rate determining step; transformation to the diazonium ion follows:

$$\mathrm{ArNH_2 + HNO_2 \longrightarrow Ar\overset{+}{N}H_2{-}NO \longrightarrow ArNH{\cdot}NO \longrightarrow ArN{=}N{-}OH \longrightarrow Ar\overset{+}{N}{\equiv}N \longrightarrow ArN{=}N^+} \quad (3)$$

Before coupling, excess nitrous acid must be destroyed. Nitrite can react with coupling components to form nitroso compounds causing deliterious effects on the final dyestuff. The presence of nitrite can be detected by 4,4′diamino-diphenylmethane-2,2′-sulfone [*10215-25-5*] (Green reagent) or starch–iodide. Removal of nitrite is achieved by addition of sulfamic acid or urea [*57-13-6*], however, sulfamic acid [*5329-14-6*] has been more effective in kinetic studies of nine nitrous acid scavangers (18).

$$NH_2SO_3H + HNO_2 \longrightarrow N_2 + H_2SO_4 + H_2O \qquad (4)$$

$$NH_2CONH_2 + HNO_2 \longrightarrow N_2 + NH_2COOH + H_2O \qquad (5)$$

Azo Coupling. The coupling reaction between an aromatic diazo compound and a coupling component is the single most important synthetic route to azo dyes. Of the total dyes manufactured, about 60% are produced by this reaction. Other methods include oxidative coupling, reaction of arylhydrazine with quinones, and oxidation of aromatic amines. These methods, however, have limited industrial applications.

All coupling components used to prepare azo dyes have the common feature of an active hydrogen atom bound to a carbon atom. Compounds of the following types can be used as azo coupling components: (*1*) aromatic hydroxy compounds such as phenols and naphthols; (*2*) aromatic amines; (*3*) compounds that possess enolizable ketone groups of aliphatic character, ie, compounds that have active methylene groups, where X is an electron attracting group such as –COR, –COOH, –CN, R is alkyl or aryl, and Y is usually a substituted or unsubstituted amino group;

$$X{-}CH_2{-}\overset{\overset{\displaystyle O}{\|}}{C}{-}Y \rightleftharpoons X{-}CH{=}\overset{\overset{\displaystyle OH}{|}}{C}{-}Y$$

and (*4*) heterocyclic compounds such as those containing pyrrole [*109-97-7*], indole [*120-72-9*], pyridine [*110-86-1*], pyrimidine [*289-95-2*], and similar ring systems, eg, 5-pyrazolones.

Analogous to aromatic halogenation, nitration, and sulfonation, the azo coupling reaction is an electrophilic aromatic substitution. The effect of the reaction rate of substituents on both the diazo and the coupler components is in agreement with this mechanism. Thus the reaction is facilitated by electron-attracting groups in the diazo components, and by electron-donating groups in phenol and aromatic amine-type coupler components. The reactivity of coupling components (nucleophilic substrate) increases with increasing basicity. The phenoxide ion (ArO^-) and free amine ($ArNH_2$) are more basic than corresponding free phenol and the ammonium ion ($C_6H_5NH_3^+$) and, therefore, react more easily. A mechanistic explanation of this acid–base equilibria has been discussed (19). The effect of pH on the coupling reaction changes from one set of conditions to another. In the case of the coupling of a simple benzenediazonium salt to a naphtholsulfonic acid whose hydroxyl group has a pK_a of 9.0 at pH below 8, an increase in alkalinity by one pH unit increases the coupling rate tenfold because of the increase in the concentration of naphtholate anion. Above pH 10, the naphthol–naphtholate equilibrium is shifted completely to the right and the coupling rate is independent of hydroxyl ion concentration. However, at high pH the diazonium–diazotate equilibrium is involved and the rate of coupling declines because the diazotate does not couple. Therefore, an optimum pH exists for each azo coupling reaction that is limited by the acidities, expressed in pH units, that correspond to the pK_a value of the coupling component or the constants of the diazonium–diazotate equilibria. This optimum pH range lies approximately between pH 4 and pH 8 for aromatic amines as couplers; for enols, eg, acetoacetanilide [*102-01-2*], 3-methyl-1-phenyl-5-pyrazolone [*89-25-8*], and phenols, the range is pH 7–9 and pH 9–10, respectively.

The role played by coupling components is somewhat more complex. Electron-attracting α-substituents, eg, acetoacetanilide, 5-pyrazolones, speed the coupling reaction of methylene compounds in alkaline media. On the other hand, the presence of groups, such as nitro and sulfonate, depress the coupling capability of phenols and aromatic amines. Electron-attracting groups promote the shifts of the phenol–phenolate anion equilibrium to the side of the reactive phenolate species; however, the lowering of electron density on the aromatic ring at the reaction site is the overriding factor. Thus nitronaphthols, in which both substituents are in the same nucleus, do not, or only very sluggishly, couple. This is especially true of nitrophenols. A technically important example is salicyclic acid, in which the effect of the carboxyl group on the π-electrons of the ring and the lower acidity of the phenolic proton because of hydrogen bonding, are responsible for depressing the ability to couple. On the other hand, since electron-donating groups such as alkyl, alkoxy, and amino residues, facilitate reaction, the second hydroxyl group of resorcinol enables this component to be used for syntheses that could not be carried out with phenol. Normally, coupling occurs at the position para to the hydroxyl group. If this is occupied by a substituent not readily eliminated, as in *p*-cresol [*1319-77-3*] or 2-naphthol [*135-19-3*], the diazo component attacks the ortho position. When both ortho and para positions are free, two arylazo residues can be introduced, but with phenols the second coupling occurs with great difficulty. A few well-known coupling components are shown in Figure 2 together with the favored positions of coupling under the conditions normally used. The complete ionization of sodium sulfonate groups in aqueous solution, as in H-acid [*90-20-0*] (**6**), decisively influences solution properties, and particularly the dyeing behavior of such compounds.

In coupling there are always two competing reactions:

$$Ar{-}N_2^+X^- + RH \longrightarrow Ar{-}N{=}N{-}R + HX \tag{2}$$

$$Ar{-}N_2^+X^- + HOH \longrightarrow Ar{-}OH + N_2 + HX \tag{6}$$

where Ar is an aryl radical, RH a coupling component, and X^- a halide ion. Factors that increase the rate of the desired reaction (eq. 2) also increase the rate of diazo decomposition (eq. 6). Since the rates of the two reactions do not parallel each other, the yield of equation 2 can be maximized by a judicious choice of conditions.

Increasing the temperature of the coupling reaction usually has an unfavorable effect because the diazo decomposition reactions have greater energies of activation and hence larger temperature gradients than azo coupling. For every

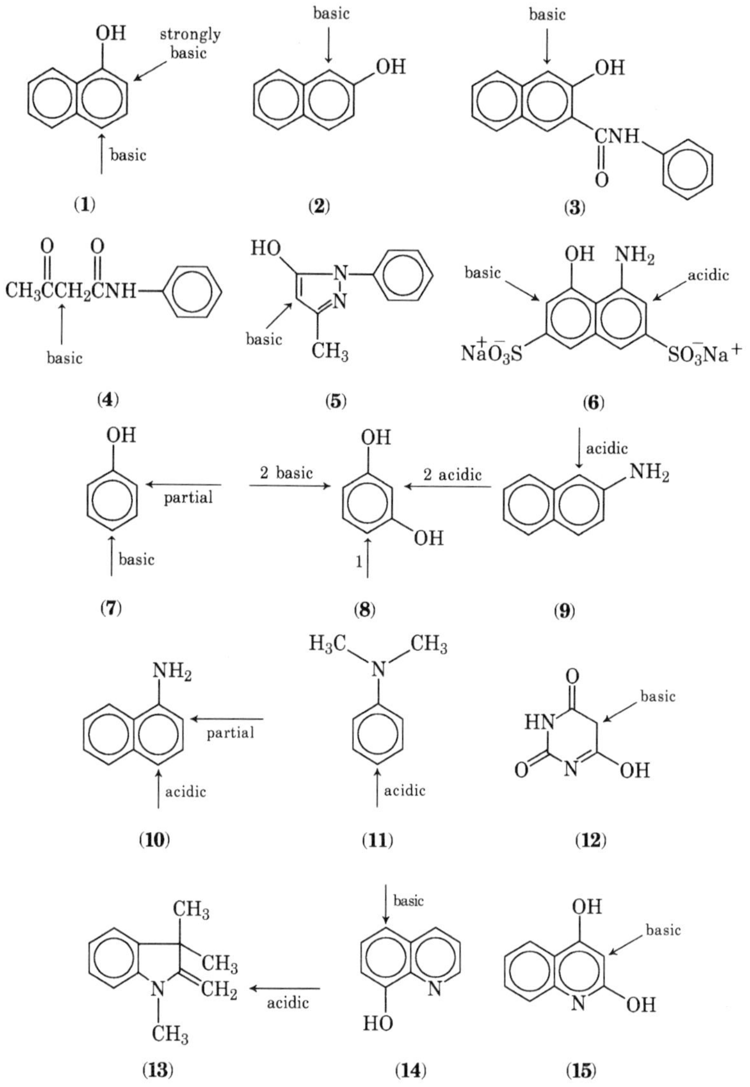

Fig. 2. Coupling components in azo dyes. The preferred positions of coupling are indicated by arrows and the usual pH conditions are noted. (**1**) 1-Naphthol [*90-15-3*]; (**2**) 2-naphthol [*135-19-3*]; (**3**) 3-hydroxy-2-naphthanilide [*92-77-3*]; (**4**) acetoacetanilide [*102-01-2*]; (**5**) 3-methyl-1-phenyl-5-pyrazoline [*89-25-8*] (enol form); (**6**) 8-amino-1-naphthol-3,6-disulfonic acid [*90-20-0*] (H-acid); (**7**) phenol [*108-95-2*]; (**8**) resorcinol [*108-46-3*] (1 and 2 indicate the first and second substitution sites); (**9**) 2-naphthylamine [*91-59-8*]; (**10**) 1-naphthylamine [*134-32-7*]; (**11**) *N,N*-dimethylaniline [*121-69-7*]; (**12**) barbituric acid [*67-52-7*]; (**13**) Fischer's base [*118-12-7*]; (**14**) 8-hydroxyquinoline [*148-24-3*]; (**15**) 2,4-dihydroxyquinoline [*86-95-3*].

10°C rise in temperature, the rate of coupling increases 2.0–2.4 times, whereas that of the decomposition rises 3.1–5.3 times.

The acid–base equilibria are fundamental to the kinetics of azo coupling and of practical significance for azo technology. Thus it is important that coupling reactions be carried out in a medium such that the acid–base equilibria of the diazo and coupling components favor as much as possible the diazonium ions and the phenolate ions or the free amine, respectively.

Broad principles governing the activity of coupling components may be summarized as follows:

1. Diazo coupling follows the rules of orientation of substituents in aromatic systems in accordance with the mechanism of electrophilic aromatic substitution and the concept of resonance.
2. Generally, phenols (as the phenolate anion) couple more readily than amines, and members of the naphthalene series more readily than the members of the benzene series.
3. Electron-attracting substituents in the coupling components such as halogen, nitro, sulfo, carboxyl, and carbonyl, are deactivating and tend to retard coupling.
4. A lower alkyl or alkoxy group substituted in the ortho or meta position to an amino group may promote coupling. Good couplers are obtained from dimethylaniline when lower alkyl, lower alkoxy, or both groups are present in the 2- and 5-position.
5. It is possible for diazo compounds to attack both the ortho and para position of hydroxyl and amino coupling components when these positions are not already occupied.

Technologically, the most important examples of such couplers are 1-naphthylamine, 1-naphthol, and sulfonic acid derivatives of 1-naphthol (Fig. 2). Of great importance in the dyestuff industry are derivatives of 1-naphthol-3-sulfonic acid, such as H-acid (8-amino-1-naphthol-3,6-disulfonic acid [*90-20-0*]) (**6**), J-acid (6-amino-1-naphthol-3-sulfonic acid [*87-02-5*]) (**16**), and gamma acid (7-amino-1-naphthol-3-sulfonic acid [*90-51-7*]) (**17**).

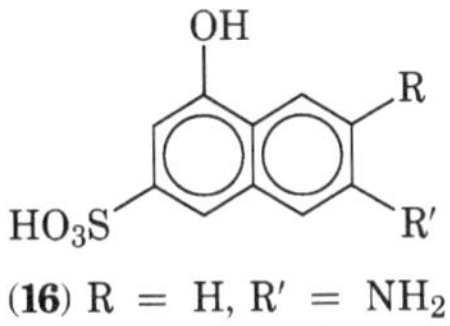

(**16**) R = H, R′ = NH_2
(**17**) R = NH_2, R′ = H

The ortho:para isomer ratio depends on several factors, including the nature of the diazo component, the nature of the solvent, the pH of the medium, the temperature of coupling, the presence of catalysts, and the position of substituents. Simple benzenediazonium compounds couple to 1-naphthol almost exclusively in the para position under acidic and weakly alkaline conditions; however, in strongly basic media considerable quantities of the 2,4-disazo dye are formed (20).

The ratio between the isomers obtained in coupling with 1,3- and 1,5-naphtholsulfonic acids depends on the reactivity of the diazo component. Energetic ones, such as the 2,4-dinitrobenzenediazonium compound, essentially couple only with 1-naphthol-3-sulfonic acid [*3771-14-0*] in the para position, but 4-chlorobenzenediazonium salt (a weaker diazo) attacks the ortho position. Both isomers result when mononitrobenzenediazonium compounds are used. The tendency to couple para is greater in 1-naphthol-5-sulfonic acid [*117-59-9*], $C_{10}H_8O_4S$ (21). For the combination of 2-nitrobenzenediazonium salt and 1-naphthol-3-sulfonic acid an increase in temperature from 10°, to 20° and 30°C increases the ortho:para ratio from 3.2 to 4.35 and 7.55 (22). In one kinetic study, the azo coupling of gamma acid with strongly electrophilic diazo components such as 3-trifluoromethyl- and 4-nitrobenzenediazonium ion in concentrated aqueous alkaline solutions gave ratios of aminoazo to hydroxyazo compounds, which are much higher than expected on the basis of pre-equilibria (23). Relative increase of the aminoazo dyes in an alkaline system is attributed to a micromixing effect. The same effect is observed when a benzenediazonium ion is coupled with 5-hydroxy-2-naphthalenesulfonic acid [*6500-22-4*] (a stronger nucleophile than gamma acid) and a considerable amount of 5-hydroxy-6,8-bisphenylazo-2-naphthalene sulfonic acid is formed (24). Different solvents can affect azo coupling in different ways (25). The coupling of benzenediazonium tetrafluoroborate [*446-46-8*] with 2-naphthol is very rapid when the solvent system is changed from water to acetonitrile (26).

The azo coupling reaction proceeds by the electrophilic aromatic substitution mechanism. In the case of 4-chlorobenzenediazonium compound with 1-naphthol-4-sulfonic acid [*84-87-7*], the reaction is not base-catalyzed, but that with 1-naphthol-3-sulfonic acid and 2-naphthol-8-sulfonic acid [*92-40-0*] is moderately and strongly base-catalyzed, respectively. The different rates of reaction agree with kinetic studies of hydrogen isotope effects in coupling components. The magnitude of the isotope effect increases with increased steric hindrance at the coupler reaction site. The addition of bases, even if pH is not changed, can affect the reaction rate. In polar aprotic media, reaction rate is different with alkyl–ammonium ions. Cationic, anionic, and nonionic surfactants can also influence the reaction rate (27).

Diazophenols, ie, *o*-hydroxyaryldiazonium salts, couple to 1-naphthol in weakly basic solution primarily in the para position, but as the hydroxyl ion concentration is increased, formation of the ortho isomer is favored and is frequently the sole product. Pyridine and pyridine derivatives, urea, and acetate, etc, used as buffers can also catalyze azo coupling reactions (28). 1-amino-2-naphthol-4-sulfonic acid [*116-63-2*] (1,2,4-acid) and 1-naphthol yield the important Eriochrome Black A [*3564-14-5*] (**18**a, R = H) (CI Mordant Black 3; CI 14640) which is reportedly (20) a mixture of ortho and para isomers.

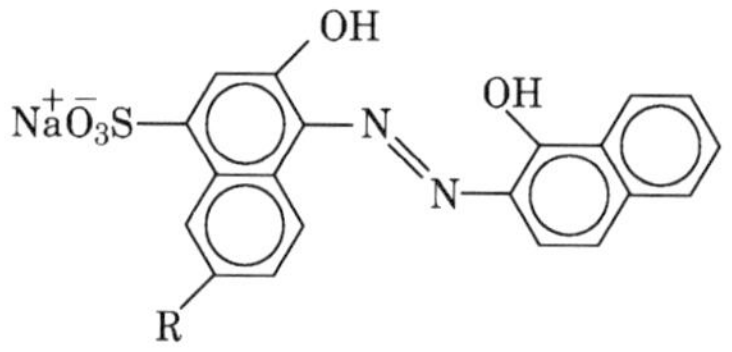

(**18**)

The corresponding diazo component with a 6-nitro substituent gives Eriochrome Black T [*1787-61-7*] (**18**b, R = NO_2) (CI Mordant Black 11; CI 14645) which is supposed to consist almost exclusively of the ortho compound.

1,5-Dihydroxynaphthalene [*83-56-7*] behaves similarly to 1-naphthol; coupling takes place mainly in the 4-position by simple diazonium compounds, and in the 2-position with diazophenols. Diazotized 2-aminophenol-4-sulfonic acid [*98-37-3*] couples with 1,5-dihydroxynaphthalene to produce the important mordant dye Diamond Black PV [*2052-25-7*] (see structure **53**) (CI Mordant Black 9; CI 16500).

6-Nitro-1-diazo-2-naphthol-4-sulfonic acid prefers the 2-position in spite of the nitro group, and increasing alkalinity favors ortho coupling with diazophenols. 1-Naphthalenesulfamic acid [*24344-19-2*] ($ArNHSO_3H$) and *N*-nitro-1-naphthylamine [*4323-69-7*] ($ArNHNO_2$) couple exclusively in the para position. The substitution of resorcinol [*108-46-3*] and *m*-phenylenediamine [*108-45-2*] is complicated and has been discussed (29,30). The first azo dyes from aniline, eg, Aniline Yellow [*60-09-3*] (**19**) (CI Solvent Yellow 1; CI 11000) were manufactured in 1861 and Bismark Brown [*10114-58-6*] (**20**) (CI Basic Brown 1; CI 21000) appeared in 1863. The reaction is as follows:

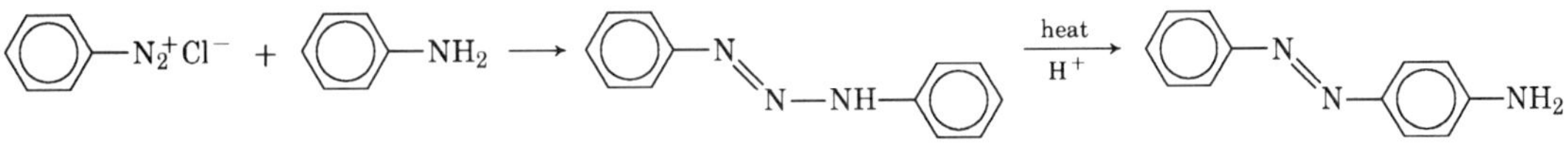

(**19**)

The dye (**19**) has little value for dyeing any textile fiber, but it is still manufactured for use as a diazo component for disazo dyes. In this type of reaction it is more usual to block the amino group, eg, by substituting anilinomethanesulfonic acid [*103-06-0*], $C_6H_5NHCH_2SO_3H$ for the aniline; the coupling then occurs exclusively in the position para to the amino group, and the sulfomethyl group is hydrolyzed giving the free amine. CI Basic Brown I is a disazo dye manufactured by adding two moles of sodium nitrite to three moles of *m*-phenylenediamine in a solution containing an excess of hydrochloric acid. The product is a mixture of aminoazo compounds, but the main component is (**20**).

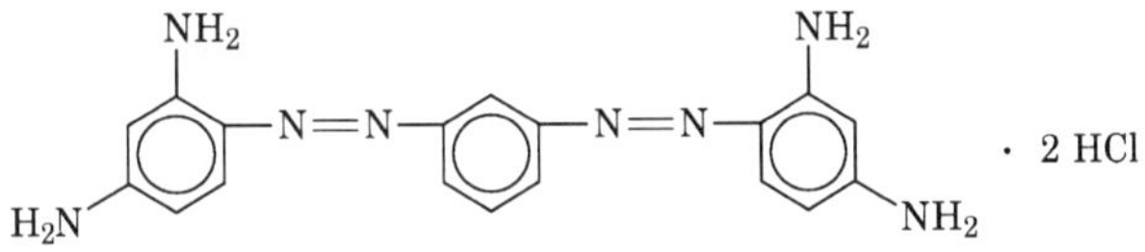

(**20**)

Alkaline Coupling Process. Orange II [*633-96-5*] (**21**) (CI Acid Orange 7; CI 15510), a monoazo dye discovered in 1876, serves as an example of the production of an azo dye by alkaline coupling. A suspension of diazotized sulfanilic acid (0.1 mol) is added to a solution (cooled to about 3°C) of 14.4 g 2-naphthol dissolved in 15 g 30% sodium hydroxide, 25 g sodium carbonate, and 200 mL of water. The temperature should not be allowed to rise above 5°C. The reaction is heated until solution occurs and the dye is precipitated with 100 g sodium chloride. The mixture is cooled and filtered, and the product is dried.

Acid Coupling Process. On the other hand, Chrysoidine Y [*532-82-1*] (**22**) (CI Basic Orange 2; CI 11270) is an example of a dye produced by acid coupling. Aniline, 9.3 g, is diazotized and the diazo solution is added to a solution of 11 g *m*-phenylenediamine in 200 mL water; then the solution is made slightly acidic with hydrochloric acid. The reaction mixture is maintained with stirring until coupling is complete as indicated by spot test where, upon addition of an alkaline solution of 2-naphthol, there is no color change if the diazo is exhausted. The solution is warmed to about 60°C. A concentrated solution of salt is added (about 120 g). The reaction mixture is then allowed to cool and the dye is filtered and dried. Reddish brown water-soluble crystals are obtained.

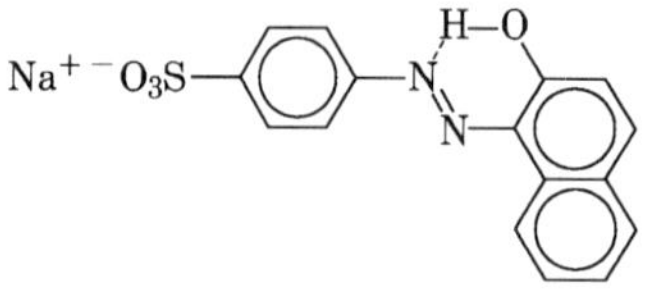

(21)

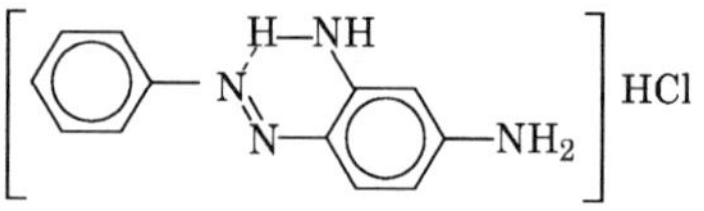

(22)

The dyes (**21**) and (**22**) were the first dyes manufactured by two separate reaction steps, ie, diazotization in one pot and the coupling reaction in another vessel, at BASF in 1875. Anthosin Orange 35 (BASF) is an Acid Orange 7 (**21**) listed along with 18 other commercial names in *AATCC Buyer's Guide* (31). The product is available in powder and concentrated liquid form. Its main application is for dyeing paper and to a lesser extent, leather and wool. The U.S. production for the paper industry is estimated at 270 tons of liquid dye per year. Chrysoidine (**22**) is no longer an important cotton dye; U.S. sales in 1988 were nearly 135 tons at an average cost of $1.27 per kg. For detailed information on azo coupling, mechanism, and properties of azo compounds, see references 32–36.

Other Routes to Azo Dyes. *Oxidation of Aromatic Amines.* The technically important dye Direct Yellow 28 (**23**) [*10114-47-3*] (CI 19555) for cotton usage is manufactured by oxidation of dehydrothio-*para*-toluidinesulfonic acid sodium salt with sodium hypochlorite in aqueous alkaline solution.

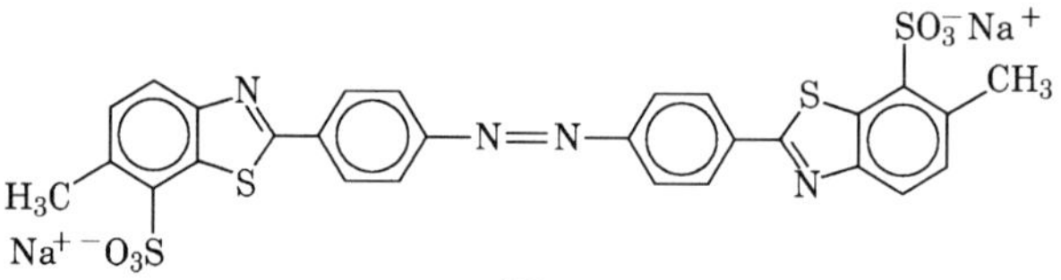

(23)

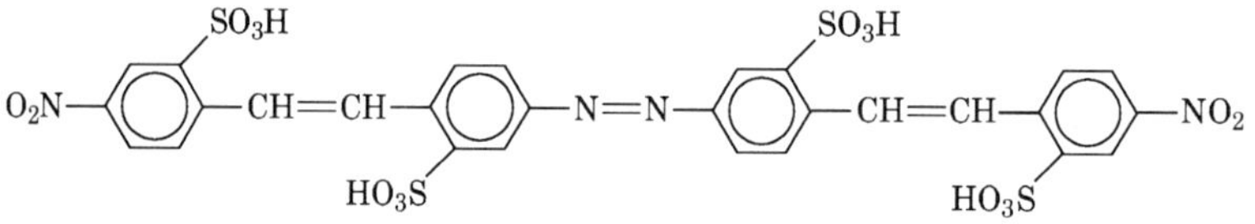

(24)

Condensation of Nitro Compounds. The stilbene dyes derived from condensation of 4-nitrotoluene-2-sulfonic acid [*121-03-9*] (para acid) in aqueous caustic solution either alone or with arylamines, are mixtures of dyes containing azo and azoxy groups with indefinite structures. The primary products are 4,4′-dinitro-2,2′-stilbenedisulfonic acid [*128-42-7*] and 4,4′dinitrodibenzyl-2,2′-

disulfonic acid [*6404-60-0*] which condense further to form the dye. The properties of the stilbene dyes vary with the amounts and concentrations of the reactants and with the time and temperature of heating. The hue range is mostly yellow, orange, and brown, classified in the CI under stilbene dyes. The CI numbers in these cases refer not to any specific chemical structure, but to groups of commercial products which possess closely similar dyeing and fastness properties. A process was patented for making stable liquid directly using lithium hydroxide (37). Direct Yellow 11 [*13253-77-3*] (**24**) (CI 40000) is a stilbene dye sold under trade names such as Pontamine Yellow GXG (Mobay) or Fastusol Yellow 76 (BASF) and has an estimated U.S. market of nearly 2000 tons liquid for paper dyeing alone. The most probable structure is (**24**).

Oxidative Coupling of Heterocyclic Hydrazones. This method has opened the way to the preparation of azo derivatives of diazo compounds unobtainable by other means, ie, heterocyclic compounds in which the diazotizable amino group is conjugated with the heterocyclic nitrogen atom as in 2- and 4-aminopyridine, compounds which do not normally yield stable diazonium salts (38). The reaction occurs as illustrated by equation 7 for the interaction of (*N*-methylcarbostyryl)hydrazone [*28219-37-6*] and dimethylaniline; the overall process is oxidation.

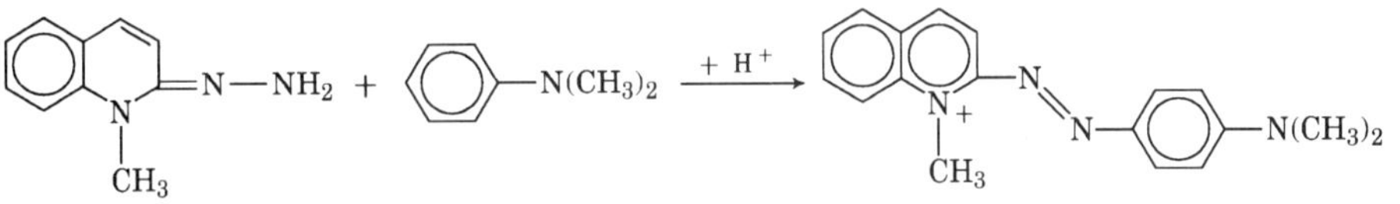

(7)

The most suitable oxidizing agent is potassium ferricyanide, but ferric chloride, hydrogen peroxide in the presence of ferrous salts, ammonium persulfate, lead dioxide, lead tetraacetate or chromate, or silver and cupric salts may be useful. Water mixed, eg, with methanol, dimethylformamide, or glycol ethers, is employed as reaction medium.

Classification of Azo Dyes

In addition to classification according to the number of azo groups, further subdivision is achieved, first according to whether the compound is water-soluble and, secondly, according to the types of component used. Another system of classification is based on dyeing classes. All colorants are divided to indicate the chief method of application or to indicate principal use. Azo dyes are found in the acid, basic (cationic), direct, disperse, mordant, and reactive dyeing classes. Although the two systems of classification overlap with each other, the chemical structure method is important from a chemistry viewpoint, whereas dyeing class method is important from an industrial viewpoint. In industry, a product can be offered competitively based on similar properties and function regardless of structure.

In the disazo group of azo dyes, primary and secondary types are distinguished. The former covers compounds made from two molecules of a diazo derivative and one molecule of a bifunctional coupling component. In both cases, the monofunctional reagent may consist of two molecules of one compound or one molecule of each of two substances used stepwise, the first alternative yielding symmetrical products. In stepwise reactions, it is the rule to carry out the coupling occurring with greater difficulty first (usually acid coupling), since generally when components contain one or more azo groups they become progressively more inert.

Structural Representation. To indicate the constitution of azo coloring matter in the conventional manner becomes much too cumbersome, particularly with polyazo dyes. Apart from using trade names, the color chemist has developed a kind of shorthand, which simultaneously provides some information about the synthesis of the azo compound concerned. A system of arrows is employed linking the intermediates used, the arrow pointing from the amino compound to be diazotized to the substance acting as the coupling component. With polyazo compounds, numbers show the order in which the processes are carried out and it is usually indicated whether the reaction is performed in acidic or basic solution when aminohydroxyl compounds are used as coupling components. For example, Amidine Benzo Dark Green N [*3626-28-6*] (CI Direct Green 1; CI 30280) is simply written as follows:

$$\text{phenol} \xleftarrow{3} \text{benzidine} \xrightarrow[\text{acid}]{1} \text{H-acid} \xleftarrow[\text{basic}]{2} \text{aniline}$$

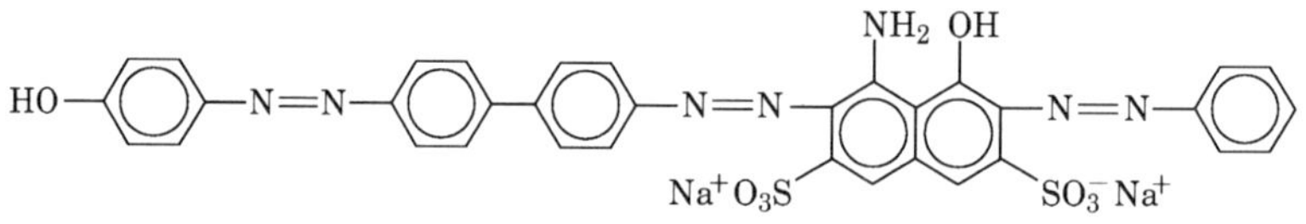

The *Chemical Abstract Index* name is 2,7-naphthalene-disulfonic acid, 4-amino-5-hydroxy-3-[[4′-[(4-hydroxyphenyl)azo][1,1′-biphenyl]-4-yl]azo]-6-(phenylazo)-, disodium salt.

Primary Disazo Dyes. The following dyes are examples of the four types of primary disazo compounds. Their structures appear in Figure 3.

Fastusol Yellow 66 [*91-34-9*] (**25**) (CI Direct Yellow 4; CI 24890) is a symmetrical primary disazo dye from a tetrazo component (4,4′diamino-2,2′stilbene-disulfonic acid [*81-11-8*].

Amidine Fast Red F [*2429-84-7*] (**26**) (CI Direct Red 1; CI 22310) is an unsymmetrical primary disazo dye from a tetrazo component (benzidine [*92-87-5*]).

Resorcine Brown R [*5850-16-8*] (**27**) (CI Acid Brown 14; CI 20195) is a symmetrical primary disazo dye with bifunctional coupling component (resorcinol).

Calcocid Blue Black Ex [*1064-48-8*] (**28**) (CI Acid Black 1; CI 20470) is an unsymmetrical primary disazo dye with bifunctional coupling component (H-acid).

Secondary Disazo Dyes. There are about 250 dyes of known constitutions in this group. They are made by diazotizing an aminoazo compound, the amino

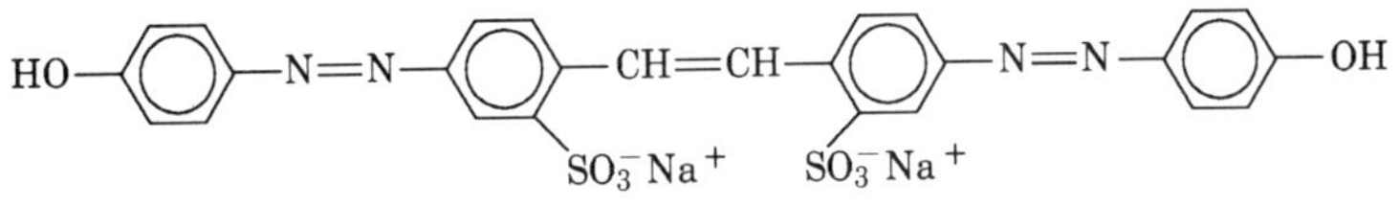

(25)

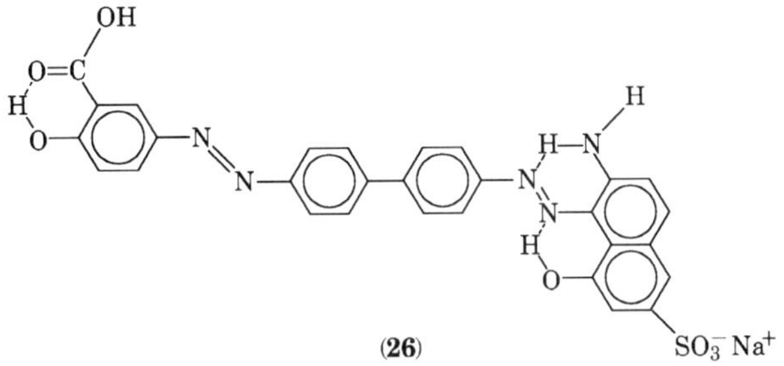

(26)

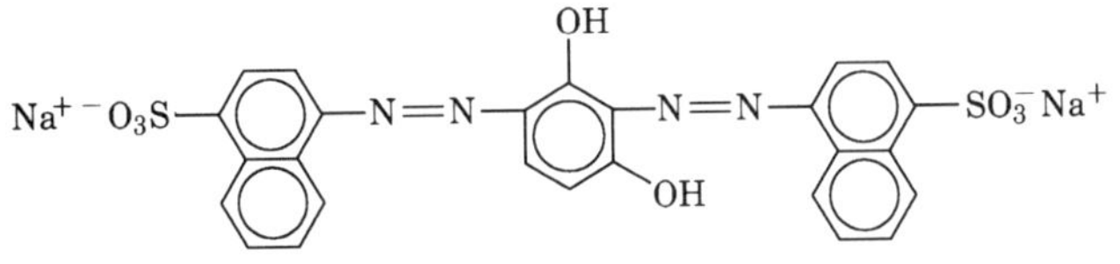

(27)

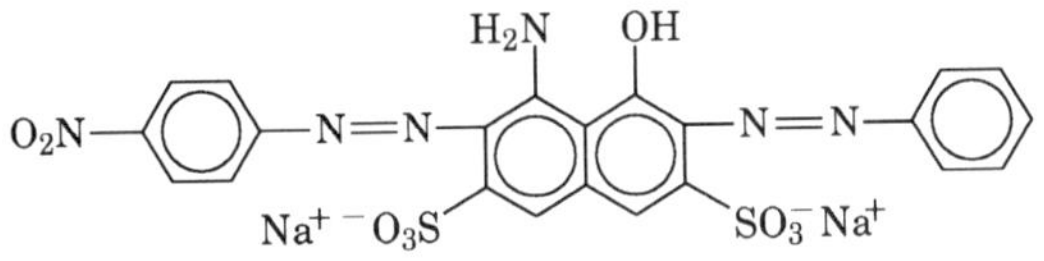

(28)

Fig. 3. Primary disazo dyes.

group of which derives from the original coupling component and coupling it to a suitable intermediate. Intrasil Orange YBLH [*19800-42-1*] (CI Disperse Orange 29; CI 26077) is an example as follows:

p-nitroaniline ⟶ *o*-anisidine ⟶ phenol

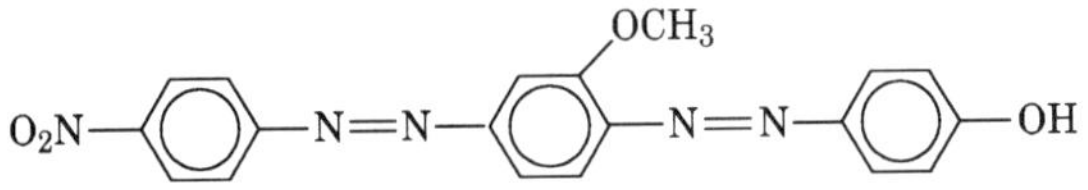

Sulphocyanone 5R [*3351-05-1*] (CI Acid Blue 113; CI 26360) (see structure **41**) is another example.

Miscellaneous Disazo Dyes. Another group of disazo dyes is prepared by condensation of two identical or different aminoazo compounds commonly with phosgene, cyanuric chloride, or fumaryl dichloride, the fragments of which act as blocking groups between chromophores.

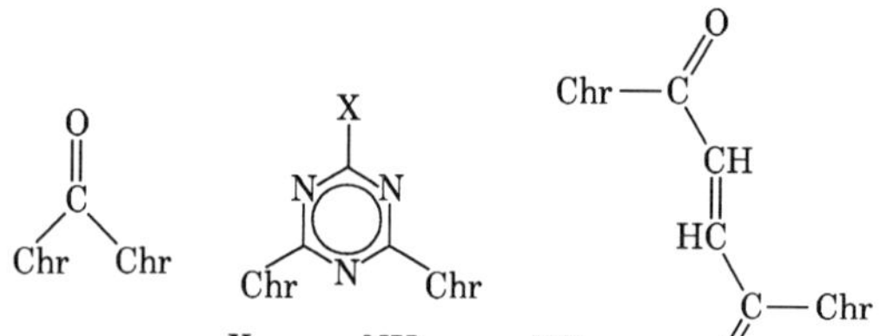

X = $-NH_2$ or $-NR_2$
Chr = chromophore

Other common blocking groups are

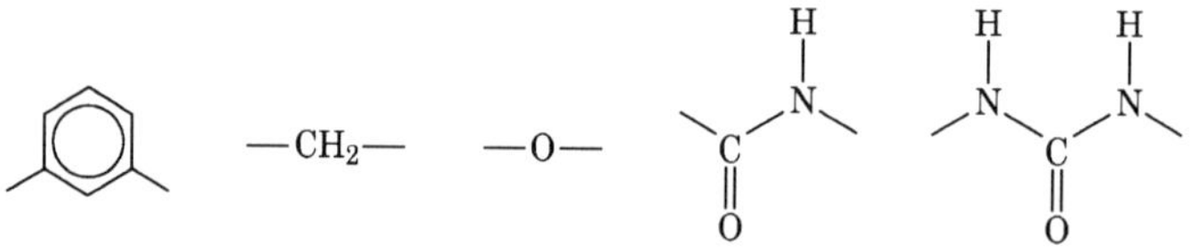

In this way, many green azo dyes have been made by combining separate conjugated systems in the same molecule, eg, one yellow and the other blue. The blocking or the insulating group prevents the electronic interaction of one chromophore system with the second. Chloramine Fast Scarlet 4BS [*3441-14-3*] (**29**) (CI Direct Red 23; CI 29160) and Fastusol Orange 85 [*6420-41-3*] (**30**) (CI Direct Orange 102; CI 29156) are examples.

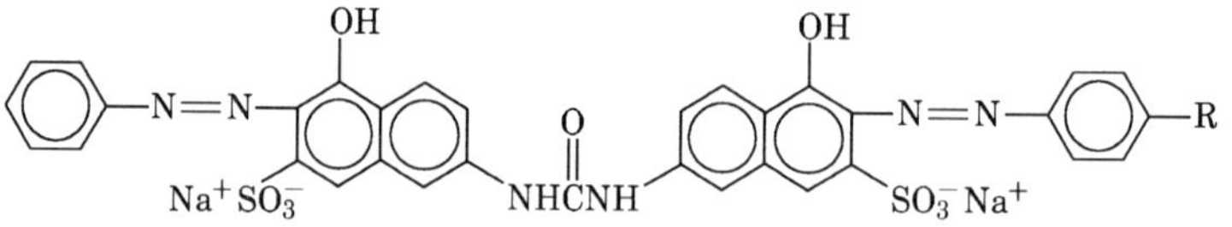

(**29**) R = $NHCOCH_3$
(**30**) R = COONa

Trisazo and Polyazo Dyes. These are mostly direct dyes, the hues are predominantly brown, black, or dark blue or green. Some are leather dyes. Benzidine, which used to be an important bisdiazo component, has been replaced by 4,4′-diaminobenzanilide [*785-30-8*], 4,4′-diaminodiphenylamine-2-sulfonic acid [*119-70-0*], etc. Benzidine dyes are almost never produced any longer because of their carcinogenicity. A triszazo dye, CI Direct Black 166 [*57131-19-8*], CI 30026 can be represented as follows:

$$m\text{-phenylenediamine} \xleftarrow{3} \text{4,4}'\text{-diaminobenzanilide} \xrightarrow[\text{acid}]{1} \text{H-acid} \xleftarrow[\text{alk}]{2} \text{aniline}$$

CI Direct Black 22 [*6473-13-8*] (CI 35435) is an example of a tetrakisazo dye, (4,4′-diaminodiphenylamine-2-sulfonic acid bisdiazotized and coupled to 2 mol gamma acid; which is then bisdiazotized and coupled to 2 mol *m*-phenylenediamine) which had production of 552 t in 1985.

Dyeing Classes: Structure, Application, Uses

Of all classes of dyestuffs, azo dyes have attained the widest range of usage because variations in chemical structure are readily synthesized and methods of application are generally not complex. There are azo dyes for dyeing all natural substrates such as cotton, paper, silk, leather, and wool; and there are azo dyes for synthetics such as polyamides, polyesters, acrylics, polyolefins, viscose rayon, and cellulose acetate; for the coloring of paints, varnishes, plastics, printing inks, rubber, foods, drugs, and cosmetics; for staining polished and absorbed surfaces; and for use in diazo printing and color photography. The shades of azo dyes cover the whole spectrum.

ACID DYES

Commercial acid dyes contain one or more sulfonate groups, thereby providing solubility in aqueous media. These dyes are applied in the presence of organic or mineral acids (pH 2–6). Such acids protonate any available cationic sites on the fiber, thereby making possible bonding between the fiber and the anionic dye molecule. Wool, an animal fiber, is an amphoteric colloid, possessing both basic and acidic properties because of the amino and carboxylic groups of the protein structure. In order to dye such a system, coulombic interactions between the dye molecule and the fiber must take place; ie, H_3N^+-wool-COO^- + H^+ $\rightarrow$ H_3N^+-wool-COOH. The term acid dye is applied to those that are capable of such interactions. Acid dyes themselves are not generally acids (ie, dye-SO_3H) but are sulfonate salts (dye-SO_3Na) of strong sulfonic acid groups.

There are three general classifications of acid dyes depending on their method of application: acid dyes that dye directly from the dyebath, mordant dyes that are capable of forming metallic lakes on the fiber when aftertreated with metallic salts, and premetallized dyes.

The data in Table 1, as reported by the U.S. International Trade Commission (39), show the extent of acid dye production and sales in the United States in 1988. In addition, the Trade Commission Import (40) data shows that 2753 t of acid dyes valued at $24.9 million was imported in 1983. Table 2 shows the important acid dyes for which production and sales data are revealed by the manufacturers.

Acid Yellow 23 (**31**), commonly known as Tartrazine, still maintains sales of

Table 1. Acid Dyes: U.S. Production and Sales, 1988

Color	Production, t	Sales, t	Sales, 10^3 $
yellow	1,344	1,086	9,380
orange	2,127	1,459	8,243
red	1,128	1,282	17,883
violet	40	34	704
blue	2,219	1,968	26,986
green	83	76	1,486
brown	344	314	3,269
black	779	754	7,387
Total	*8,064*	*6,973*	*75,338*

Table 2. Commercial Acid Dyes,[a] U.S. Production and Sales, 1988

CI name	CAS Registry Number	CI number	Production, t	Sales, t	Sales, 10^3 $
Acid Yellow 17	[*6359-98-4*]	18965	42	38	634
Acid Yellow 23	[*1934-21-0*]	19140	47	46	490
Acid Orange 8	[*5850-86-2*]	15575	49	49	319
Acid Red 1	[*3734-67-7*]	18050	97	98	707
Acid Red 137	[*6222-63-5*]	17755	17	19	361
Acid Red 182[b,c]	[*58302-43-5*]		275	186	2,471
Acid Blue 324[b,d]	[*88264-80-6*]		860	758	10,058

[a]Chemical type is monoazo, unless otherwise noted.
[b]Structure is not disclosed.
[c]Metallized monoazo.
[d]Anthraquinone.

nearly $0.5 million/yr in the United States. It was first discovered in 1884 and is made by coupling equimolar quantities of diazotized sulfanilic acid to 3-carboxy-1-(*p*-sulfophenyl)-5-pyrazolone. Other monoazopyrazolone dyes of commercial importance include Acid Yellow 17 (**32**) (sulfanilic acid → 1-(2,5-dichloro-4-sulfophenyl)-3-methyl-5-pyrazolone and Acid Yellow 40 [*6372-96-9*] (**33**) (CI 18950) (*p*-aminophenol → 1-(4-chloro-2-sulfophenyl)-3-methyl-5-pyrazolone) followed by esterification of the phenolic hydroxy group with *p*-toluenesulfonyl chloride.

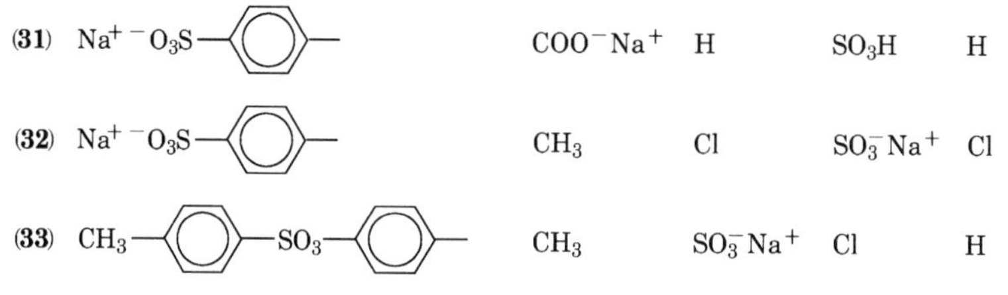

With Orange I [*574-69-6*] (**34**) (CI Acid Orange 20; CI 14600) the naphthalene moiety was introduced to azo chemistry. Basacid Red 340 [*1658-56-6*] (**35**) (CI Acid Red 88; CI 15620) the first red azo dye of technical value was discovered by BASF in 1876. Its previous name was Fast Red AV and it is still produced in large amounts in the United States because of its low cost and good dyeing and fastness properties. This dye became the prototype of a large number of red azo dyes that were developed simultaneously with the introduction of new derivatives of naphthalene.

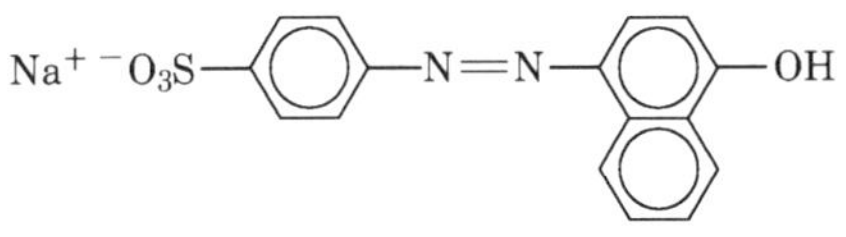

(34)

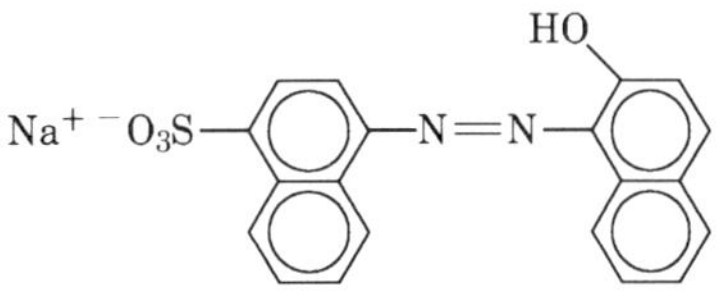

(35)

In textile dyes, it is important that hydroxyl and amino groups be ortho to azo, carboxyl, or sterically bulky groups. In dyes such as Anthosin Orange 35 (**21**), Chrysoidine (**22**), and Basacid Red 340 (**35**) hydrogen bonding shifts the acid base equilibrium to pH regions (pH <2 or >11) outside those ordinarily employed in textile processing. In Acid Orange 20 (**34**) hydroxyl group dissociation leads to tint changes in alkaline washing tests. If coupling in the ortho position to a hydroxyl or amino group cannot be attained (eg, in the case of unsubstituted phenol), these groups must be alkylated, esterified, or acylated afterward. Phenolic hydroxyl groups can be converted into esters by *p*-toluenesulfonyl chloride, as in Acid Yellow 40 (**33**). Coupling components with a second amino or hydroxyl group must also be acetylated or tosylated to get dyes of good alkaline and acid fastness. An important example is Amido Naphthol Red 6B [*4321-69-1*] (**36**) (CI Acid Violet 7; CI 18055) from acetyl H-acid. The corresponding compound with a free amino group is of no technical value.

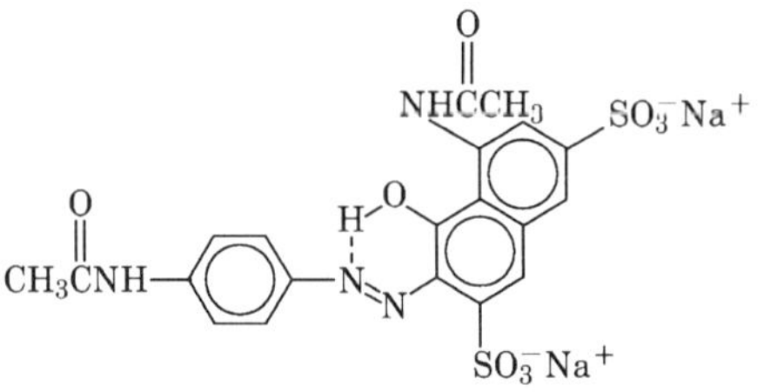

(**36**)

Instead of acetyl groups, other acyl and aroyl groups can be utilized which, conveniently and characteristically, change the dyeing properties (Fig. 4). To get good leveling properties, these dyes require a dyeing pH of 4–6.5. Longer alkyl chains in the diazo component also increase the wash fastness. This effect is caused by hydrophobic interactions (41). Other commercially important azo acid dyes, for which production and sales data have not been published separately but structures are available, are listed in Tables 3 and 4.

Of these dyes, Acid Yellow 151 (**37**) still has the greatest market among the yellows. As reported by USITC, production had increased to 1989 tons in 1985 from 706 tons in 1975. It is produced by coupling diazotized 2-amino-1-phenol-4-

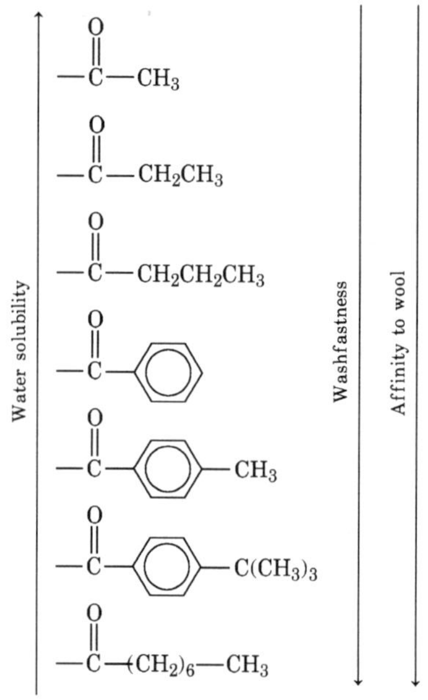

Fig. 4. Dyeing characteristics as affected by various blocking groups. Water solubility decreases; washfastness and affinity increase.

Table 3. Yellow and Orange Commercial Azo Acid Dyes, U.S. 1988

CI name	Structure number	CAS Registry Number	CI number	Chemical type
Acid Yellow 34		*[6359-90-6]*	18890	monoazo
Acid Yellow 36		*[587-98-4]*	13065	monoazo
Acid Yellow 49		*[12239-15-5]*	18640	monoazo
Acid Yellow 59		*[5601-29-6]*	18690	monoazo, metallized
Acid Yellow 65		*[6408-90-8]*	14170	monoazo
Acid Yellow 99		*[10343-58-5]*	13900	monoazo, metallized
Acid Yellow 135		*[74082-21-6]*	14255	monoazo
Acid Yellow 151	(**37**)	*[12715-61-6]*	13906	monoazo, metallized
Acid Yellow 200		*[6359-95-1]*	18930	monoazo
Acid Orange 7	(**21**)	*[633-96-5]*	15510	monoazo
Acid Orange 10		*[1936-15-8]*	16230	monoazo
Acid Orange 24	(**38**)	*[1320-07-6]*	20170	disazo
Acid Orange 60		*[12731-53-2]*	18732	monoazo, metallized
Acid Orange 156		*[68555-86-2]*	26501	disazo

sulfonamide to acetoacetanilide followed by metallizing with cobalt to obtain a 1:2 cobalt complex. Acid Orange 24 (**38**), which is sulfanilic acid coupled to resorcinol to which diazotized mixed xylidines have been coupled, is an unsymmetrical primary diasazo dye with a bifunctional coupling component.

Table 4. Other Commercial Azo Acid Dyes, U.S. 1988

CI name	Structure number	CAS Registry Number	CI number	Chemical type
Acid Red 4		[*5858-39-9*]	14710	monoazo
Acid Red 14	(**52**)	[*3567-69-9*]	14720	monoazo
Acid Red 18		[*2611-82-7*]	16255	monoazo
Acid Red 73		[*5413-75-2*]	27290	disazo
Acid Red 85		[*3567-65-5*]	22245	disazo
Acid Red 88	(**35**)	[*1658-56-6*]	15620	monoazo
Acid Red 114	(**40**)	[*6459-94-5*]	23635	disazo
Acid Red 137		[*6222-63-5*]	17755	monoazo
Acid Red 151	(**39**)	[*6406-56-0*]	26900	disazo
Acid Red 186		[*52677-44-8*]	18810	monoazo, metallized
Acid Red 266		[*12217-37-7*]	17101	monoazo
Acid Violet 3		[*1681-60-3*]	16580	monoazo
Acid Violet 7	(**36**)	[*4321-69-1*]	18055	monoazo
Acid Violet 12		[*6625-46-3*]	18075	monoazo
Acid Blue 92		[*3861-73-2*]	13390	monoazo
Acid Blue 113	(**41**)	[*3351-05-1*]	26360	disazo
Acid Blue 118		[*6406-32-2*]	26410	disazo
Acid Green 20	(**42**)	[*5850-39-5*]	20495	disazo
Acid Brown 14	(**27**)	[*5850-16-8*]	20195	disazo
Acid Black 1	(**28**)	[*1064-48-8*]	20470	disazo
Acid Black 52	(**43**)	[*5610-64-0*]	15711	monoazo, metallized
Acid Black 60		[*12218-95-0*]	18165	monoazo, metallized
Acid Black 63	(**44**)	[*32517-36-5*]	12195	monoazo, metallized

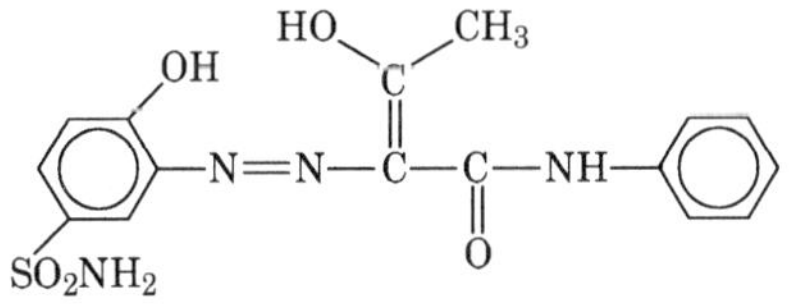

(**37**)

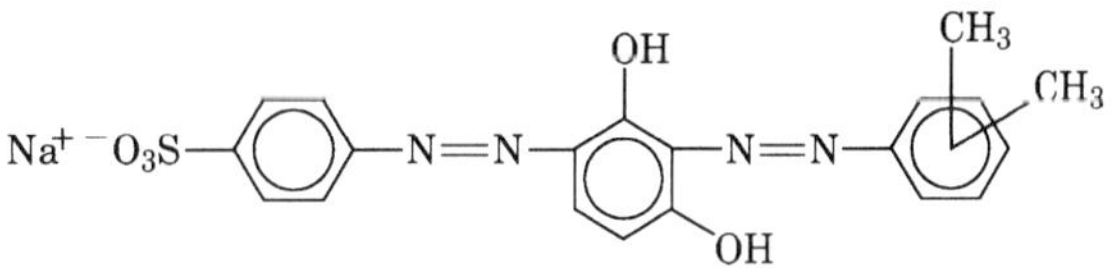

(**38**)

Economically, the most important acid red is Acid Red 151. This disazo dye (**39**) is obtained by coupling *p*(*p*-aminophenylazo)benzenesulfonic acid to 2-naphthol.

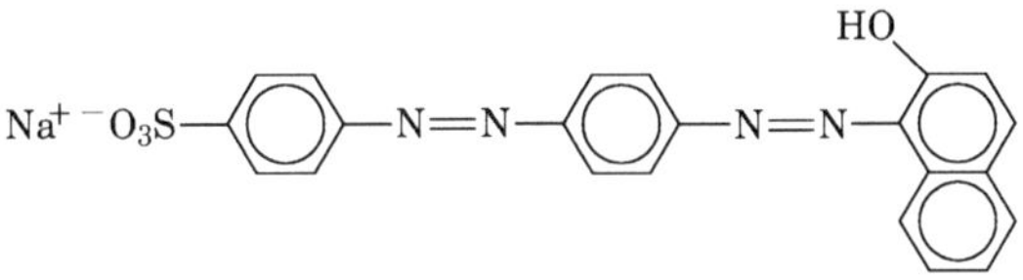

(**39**)

Other disazo dyes with good substantivity and high wet-fastness properties on polyamides are Acid Red 114 (**40**), made by coupling *o*-tolidine to phenol which is then coupled to G-acid, followed by reaction of the phenolic hydroxyl group with *p*-toluenesulfonyl chloride, and Acid Blue 113 (**41**) (metanilic acid → 1-naphthylamine → 8-anilino-1-naphthalenesulfonic acid).

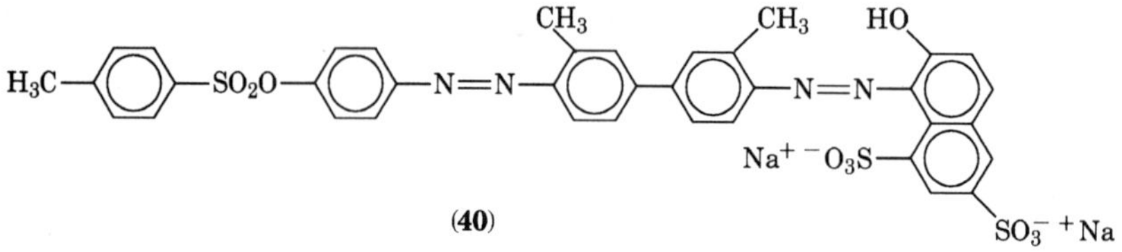

(**40**)

The largest volume acid blues are either of triphenylmethane (TPM) or anthraquinone chemical constituency (see DYES, ANTHRAQUINONE).

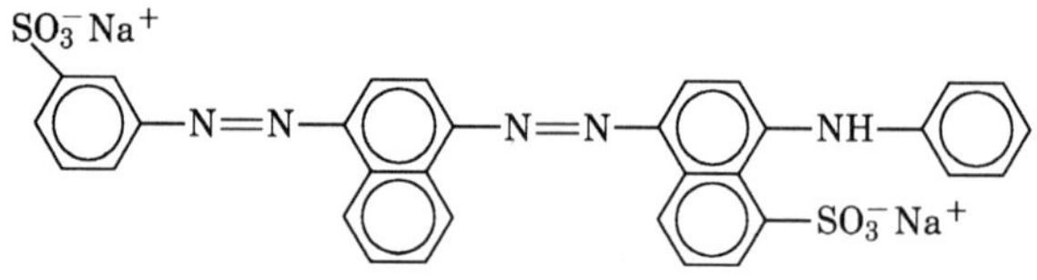

(**41**)

Among the acid colors of green, brown, and black shades, three disazo dyes are important: Acid Green 20 (**42**) (*p*-nitroaniline $\xrightarrow{\text{1-acid}}$ H-acid $\xleftarrow{\text{basic}}$ aniline, followed by reduction of the nitro group with sodium sulfide);

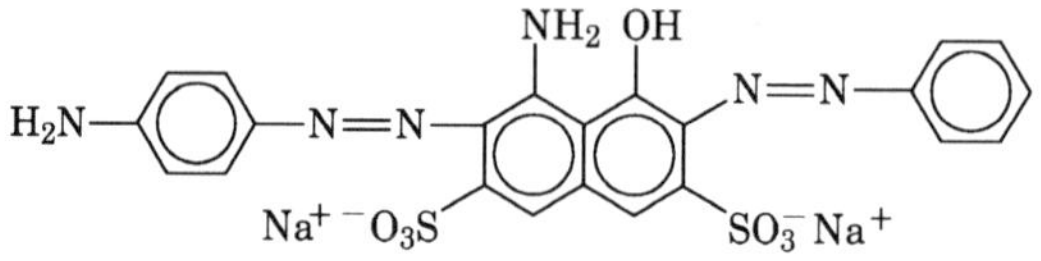

(**42**)

Acid Brown 14 (**27**) (2 mol of naphthionic acid, ie, 4-amino-1-naphthalenesulfonic acid ⟹ 1 mol resorcinol); and Acid Black 1 (**28**). Two other azo acid blacks, Acid Black 52 (**43**) and Acid Black 63 (**44**) are both metallized with chromium.

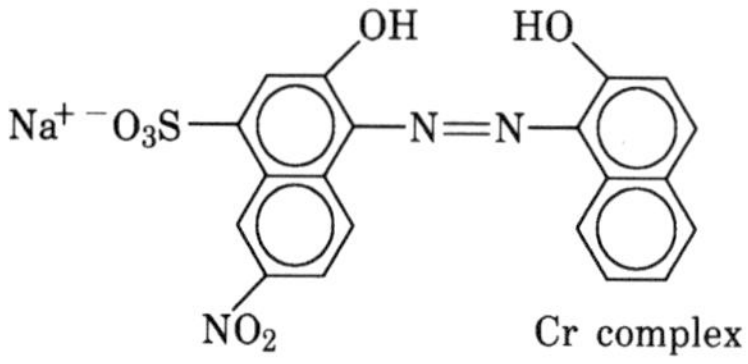

(**43**)

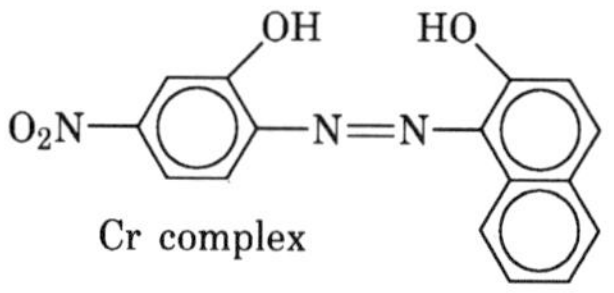

(**44**)

Other modifications of acid wool dyes have groups which react by nucleophilic substitution of basic groups in protein fibers (NH_2 groups, etc). The introduction of Procion dyes by ICI in 1956 was the most important development in the field of technological azo dye chemistry. Chemically, many of the reactive dyes are prepared from the group of the anionic monoazo dyes (see DYES, REACTIVE).

METAL COMPLEXES OF AZO DYES

Metal complexes of certain *o,o′*-dihydroxyazo, *o*-carboxy-*o′*-hydroxyazo, *o*-amino-*o′*hydroxyazo, arylazosalicyclic acid, and formazan compounds are used as dyes for wool, nylon, and cotton with generally much improved washfastness and lightfastness properties when compared to their respective unmetallized precursors. Dyes that are chelated with the metal on the substrate during the dyeing process are termed metallizable or mordant dyes. Conversely, those dyes that have been metallized by the dye manufacturer prior to use by the dyer, are classified as premetallized dyes. The two important types of premetallized dyes are the 1:1 and 2:1 complexes, eg, complexes with 1:1 and 2:1 ligand-to-metal ratios, respectively.

Chromium is the principal metal used with mordant dyes for wool, whereas both chromium and cobalt are used extensively in premetallized types for wool and nylon. Copper(II) is employed almost exclusively as the chelating metal ion in both metallizable and premetallized direct dyes for cotton.

An excellent review of metal complex azo dye chemistry has been provided (42). More recent developments in the field of metal-containing dyes have also been presented (43,44).

Mordant Dyes. Metallizable azo dyes are applied to wool by the method used for acid dyes and then treated with metal salts such as sodium chromate [*7775-11-3*], sodium dichromate [*10588-01-9*], and chromium fluoride [*1488-42-5*] to form the metal complex *in situ*. This treatment usually produces a bathochromic shift in shade, decreases the solubility of the coloring matter, and yields dyeings with improved fastness properties. The chromium salts can be applied to the substrate before dyeing (chrome-mordant or chrome-bottom method), together with the dye in a single bath procedure (metachrome process), or as a treatment after dyeing (afterchrome process).

Most mordant dyes are monoazo structures. The most important feature of this class of dyes is excellent fastness to light and washing. Mordant dyes are available in all shades of the spectrum with the exception of bright violets, blues, and greens. To be useful, the metal complexes must be stable, ie, must not demetallize when subjected to dyebath conditions and all aftertreatment processes, especially repeated washings. Chromium forms stable chelate rings with mordant dyes which are not affected by treatment with either weak acid or alkali (see COORDINATION COMPOUNDS).

Certain chemical configurations required to produce suitable mordant dyes are shown in the following examples of monoazo dyes. Substitution in the two positions ortho to the azo group may be illustrated by structure (**45**), in which X represents –OH, $-OCH_3$, $-OCH_2COOH$, –COOH, $-COOC_2H_5$, $-NH_2$, and similar groups. With *o,o*-dihydroxyazo dyes, hydrogen bonding involving both hydroxyl groups is possible as indicated in (**46**).

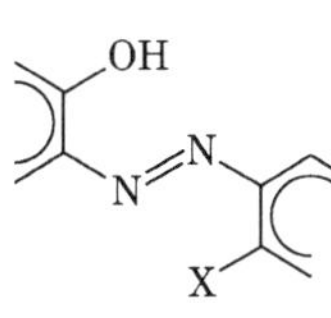

(**45**)

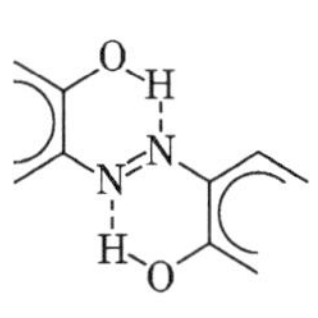

(**46**)

From acidity measurements it is known that one of the hydroxyl hydrogen atoms is much more strongly bound, ie, much less acidic than the other. For Eriochrome Black T [*1787-61-7*] (**18b**), the following equilibria apply (45), where HO-dye-OH represents the sulfonic acid salt (**18b**) as shown earlier with neither of the two OH groups neutralized.

$$\underset{\text{wine red}}{\text{HO-dye-OH}} \rightleftharpoons \underset{\text{blue}}{\text{HO-dye-O}^-} \rightleftharpoons \underset{\text{orange}}{^-\text{O-dye-O}^-}$$

$$pK_{a2} = 6.3;\ pK_{a3} = 11.5$$

It has not been determined which hydroxyl group contains the more acidic hydrogen atom corresponding to pK_{a2}. Dyes of this type would behave as indicators and exhibit large color shifts with the pH range normally encountered in textile processing. Hence they are always stabilized by coordination with metal ions.

By far the most important metal containing dyes are derived from *o,o*-dihydroxyazo structures in which one of the two azo nitrogen atoms and the two hydroxyl oxygen atoms are involved in bonding with the metal ion. Thus these dyes serve as terdentate ligands. In the case of metal ions with a coordination number of four, eg, Cu(II), the fourth position is usually occupied by a solvent molecule (**47**).

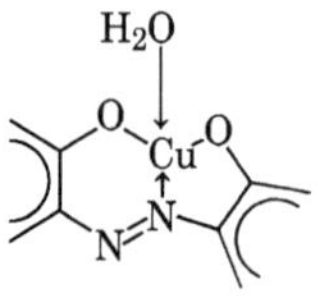

(**47**)

With Cr(III) and Co(III) ions where the coordination number is six, the predominant configurations are 1:1 complexes in which solvent molecules (H_2O) or ions (OH^-) as monodentate ligands occupy three of the positions, or a monodentate ligand and bidentate ligand such as ethylenediamine (1,2-diamino ethane) occupy three of the positions, the remaining three being occupied by the terdentate dye, and 2:1 complexes in which two terdentate dye molecules occupy all six valences. Thus, especially with metallized dyes containing chromium, it is necessary to distinguish between 1:1 and 2:1 complexes. An example of a premetallized 2:1 chromium complex azo dye is Irgalan Brown Violet DL [*16073-16-8*] (**48**) (CI Acid Violet 78; CI 12205).

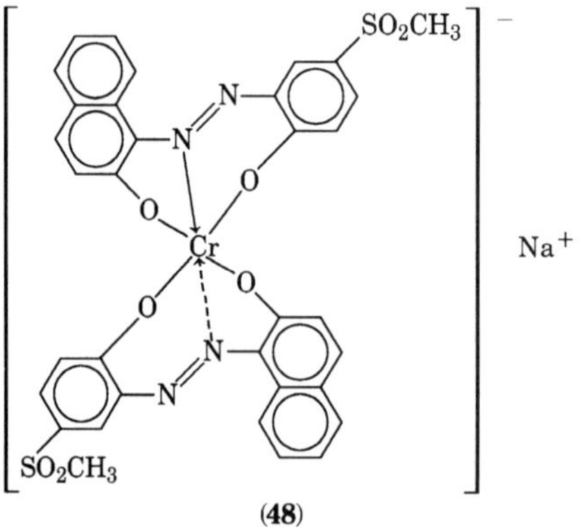

(**48**)

Using x-ray structural analyses both the meridional configuration, Drew-Pfitzner (46), and the facial configuration, Schetty-Pfeiffer (47), ligands in 2:1 chromium complexes have been confirmed. Definite coordination by one of the two azo nitrogen atoms is supported by x-ray studies, but the question of which azo nitrogen atom is involved in specific instances has not been resolved. The greater the number of annelated 5- and 6-membered chelate rings, the more stable the metal complex. Stability is also enhanced with more strongly basic ligand anions.

In mordant dyes, phenols, naphthols, and enolizable carbonyl compounds, such as pyrazolones, are generally the couplers. As a rule, 2:1 metal complexes are formed in the afterchroming process. A typical example of a mordant dye is Eriochrome Black T (**18**b) which is made from the important dyestuff intermediate nitro-1,2,4-acid, 4-amino-3-hydroxy-7-nitro-1-naphthalenesulfonic acid [*6259-63-8*]. Eriochrome Red B [*3618-63-1*] (**49**) (CI Mordant Red 7; CI 18760) (1,2,4-acid → 1-phenyl-3-methyl-5-pyrazolone) is another example. The equilibrium of the two tautomeric forms depends on the nature of the solvent.

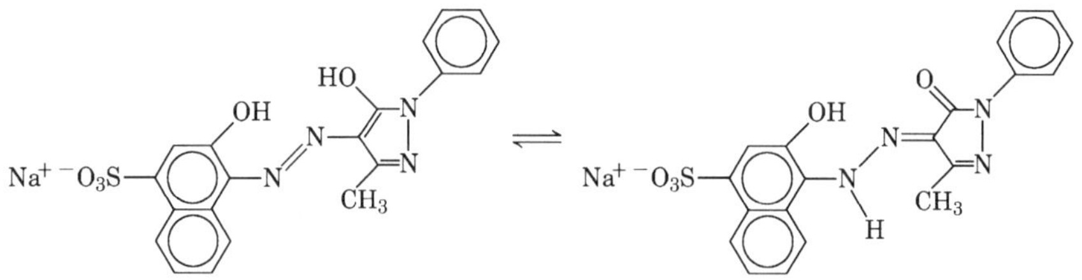

(**49**)

A hydroxyl group is situated ortho to a carboxyl group which as a bidentate ligand is terminally metallized on the fiber when aftertreated with dichromate. An example is Alizarine Yellow GG [*584-42-9*] (**50**) (CI Mordant Yellow 1; CI 14025). Cr(III) has a coordination number of six, and therefore normally two dye molecules of the salicylic type are chelated to the metal ion.

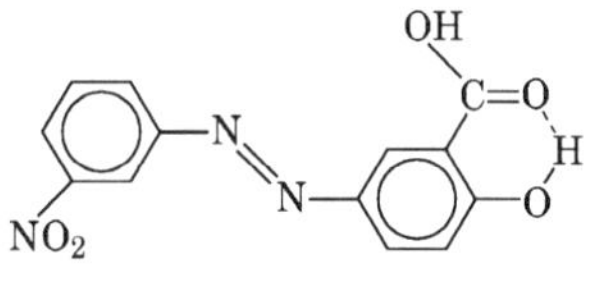

(**50**)

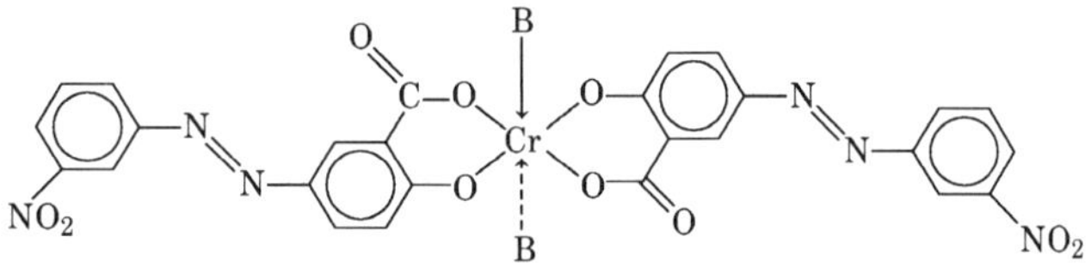

(**51**)

The other two coordination positions are available for groups of the fiber having free electron pairs, as in (**51**) where B is $-NH_2$, $-NH-$, or $-OH$.

This is apparently the reason for the high washfastness of these complexes as compared to their unmetallized precursors (48). The dichromate-acid aftertreatment that is used technically oxidizes primarily cystine configurations in the wool, which converts Cr(VI) to Cr(III). Alternatively, the fiber may be pretreated with dichromate whereby essentially the same final shade is obtained, although it is not known whether the same complexes are formed.

In contrast to metal complexes of *o,o*′hydroxyazo types, these terminally metallized dyes derived from the *o*-hydroxy carboxyl structures undergo little color change on metallization and exhibit little improvement in lightfastness (49).

Certain dyes when applied by the afterchrome method are oxidized prior to metal complex formation. Examples include, Chromotrope FB [*3567-69-9*] (**52**) (CI Acid Red 14; CI 14720) (naphthionic acid → 1-naphthol-4-sulfonic acid) in which a hydroxyl group is introduced by oxidation at the position marked with an asterisk, after which the metal complex is formed in the usual way, and Diamond Black PV [*2052-25-7*] (**53**) (CI Mordant Black 9; CI 16500) in which a naphthoquinone is formed by oxidation without alteration of the original tridentate ligand.

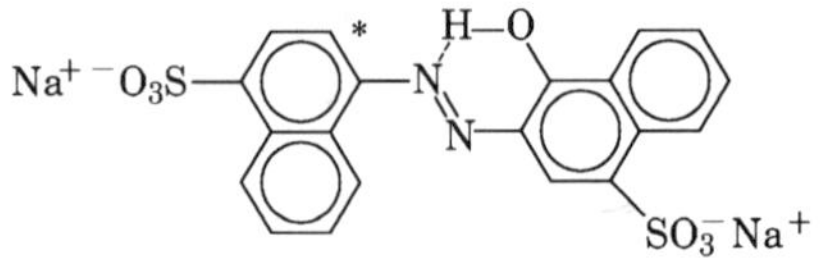

(**52**)

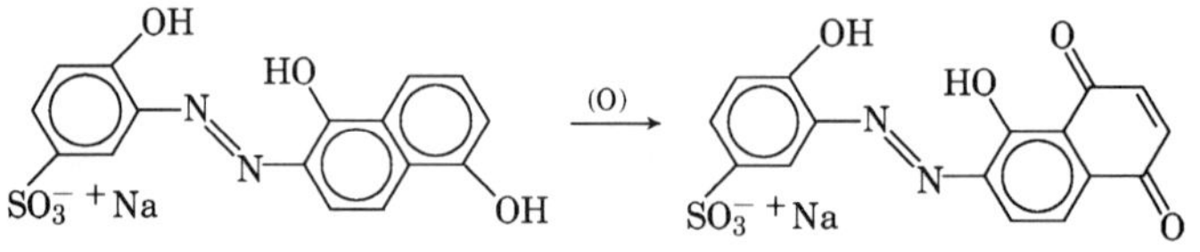

(**53**)

With azo dyes derived from 4,5-dihydroxy-2,7-naphthalenedisulfonic acid [*148-25-4*] (chromotropic acid) as the coupling component, metal complex formation occurs with the perihydroxy groups without oxidation.

Dyes with *o,o′*-hydroxycarboxyazo and *o,o′*-hydroxyaminoazo ligands are important for yellow shades. Anthranilic acid [*118-92-3*] is used advantageously with various couplers. *o,o′*-Hydroxyaminoazo dyes are also used to obtain green and brown shades. An example is Monochrome Brown EB [*3564-15-6*] (**54**) (CI Mordant Brown 1; CI 20110), an unsymmetrical primary disazo dye.

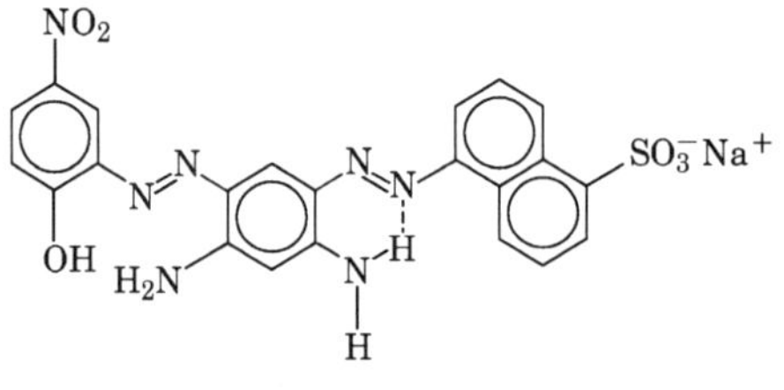

(**54**)

Disazo Mordant Dyes. Examples of disazo dyes are Diamond Alizarine Black SN [*3258-74-0*] (**55**) (CI Mordant Black 25; CI 21725) and a secondary disazo dye, Diamond Black F [*8027-29-0*] (**56**) (CI Mordant Black 5; CI 26695).

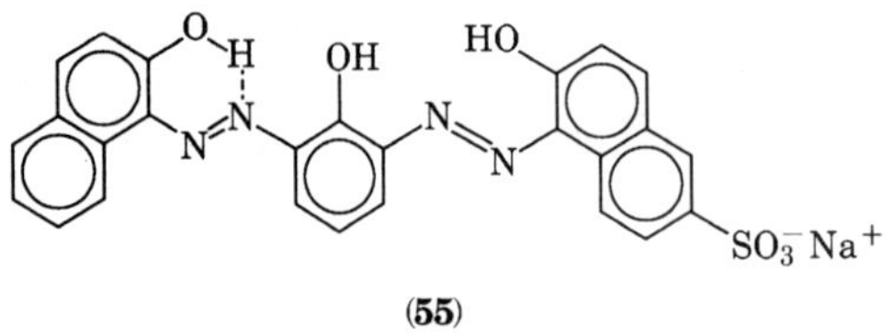

(**55**)

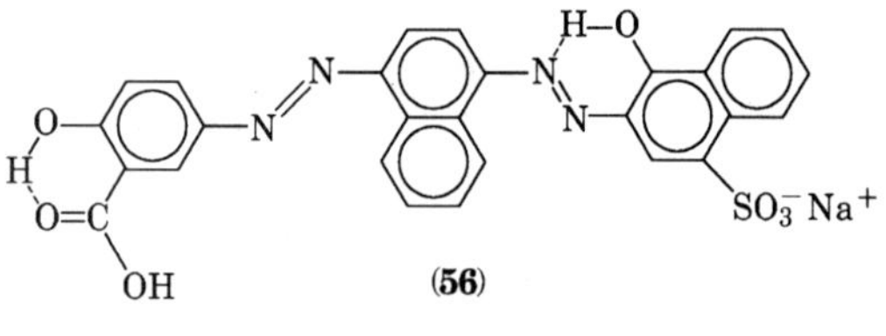

(**56**)

Dyeing Procedures Employing Mordant Dyes. In the chrome-mordant or chrome-bottom method, the wool is mordanted by immersion in a bath containing about 3% sodium dichromate and a small amount of, preferably, cream of tartar, or oxalic or formic acid, and by bringing to the boil; the chromate is reduced to the chromic state which combines with the wool. The mordanted goods are rinsed and immersed in a dyebath containing 1–8% acetic acid and are slowly brought to the boil. In the afterchrome or topchrome method, wool, after being dyed with a mordant dye, is aftertreated in acid solution with a chromium compound, generally dichromate. The metachrome or monochrome method, being a one-bath process, is simple and less expensive to use. Sodium chromate is added to the dyebath together with a salt of a strong acid and a weak base, such as ammonium sulfate, just before introduction of the fiber; mordanting and dyeing take place simultaneously. It has been found that the suitability for dyes of the one-bath (metachrome) method diminishes as the number of solubilizing groups increases (50).

Premetallized Dyes. Although discovered in 1912, the 1:1 chromium complexes known as Palatine Fast (BASF) and Neolan (Ciba) dyes had little practical use as wool dyes until 1920 when it was found that a strongly acidic dyebath (pH ca 2.0) (51) was required to obtain satisfactory dyeing and acceptable fastness properties. Dyes of this type exemplified by Neolan Blue 2G [*6370-12-3*] (**57**) (CI Acid Blue 158A; CI 15050) are still in use despite the damage to the wool caused by the strong acid in the dyebath.

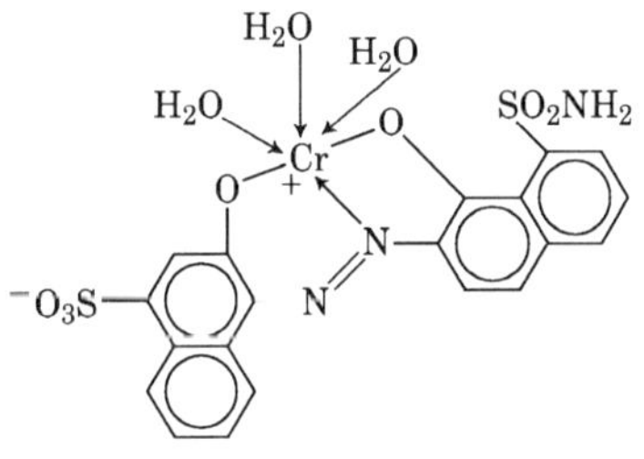

(**57**)

The use of acetoacetanilides and 2,4-dihydroxyquinolines as coupling components is demonstrated in the 1:1 chromium complexes Neolan Yellow GR [*10343-58-5*] (**58**) (CI Acid Yellow 99; CI 13900) (4-nitro-2-aminophenol → acetoacetanilide) and Palatine Fast Red BEN [*6656-02-6*] (**59**) (CI Acid Red 214; CI 19355) (2-amino-6-nitrophenol-4-sulfonic acid → 2,4-dihydroxyquinoline).

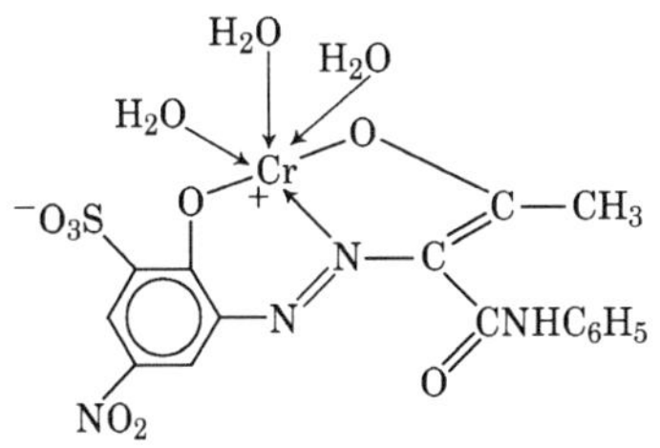

(**58**)

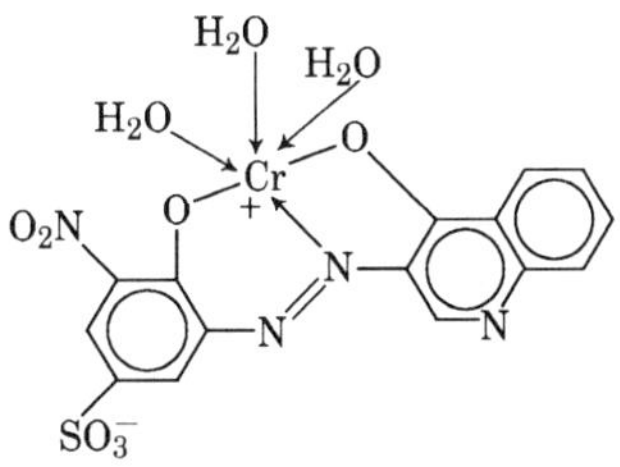

(**59**)

Under the low pH dyebath conditions, 2:1 complexes of the sulfo-containing types, disproportionate into a 1:1 complex and metal-free dye. At higher pH, 2:1 complexes are stable, but they do not yield level dyeings of dye to their ionized substituents (SO_3^-).

A notable advancement was the introduction in 1949 of Irgalan dyes by Geigy (52); a detailed historical review of this development was published (53). These dyes have excellent leveling properties and fastness to washing. Since they are dyed from neutral or slightly acidic (pH 5–7) dyebaths there is no adverse effect on wool. Chemically, these dyes are 2:1 chromium and cobalt complexes derived from *o,o'*-dihydroxyazo precursors having no substituents that are ionized in the 5–7 pH range (sulfo and carboxyl groups). To promote water solubility, this class of dyes generally contains nonionic water-solubilizing groups, such as alkylsulfonyl, sulfamyl, and acylamino. An example showing the use of methylsulfonyl groups to enhance water solubility is Irgalan Brown Violet DL (**48**). So-called mixed complexes (**60**) (1 mol dye A + 1 mol dye B + 1 mol metal ion) are valuable for obtaining homogeneous browns (54). The color of a cobalt complex is shifted hypsochromically relative to that of the analogous chromium complex. The Neopalatine Fast dyes (BASF) are mixed complexes from one colored terdentate ligand and one colorless bidentate ligand such as salicyclic acid (55).

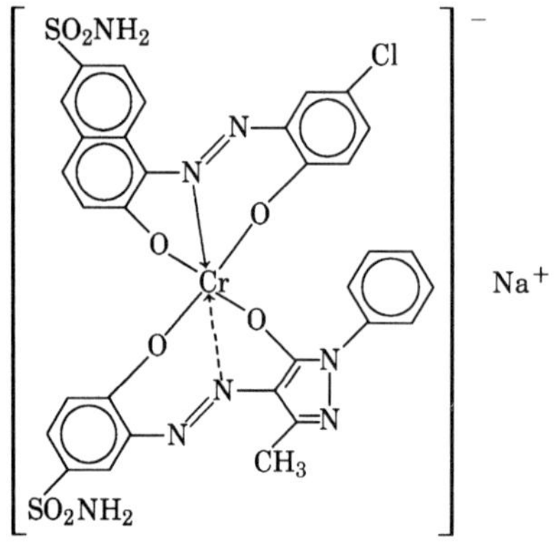

(**60**)

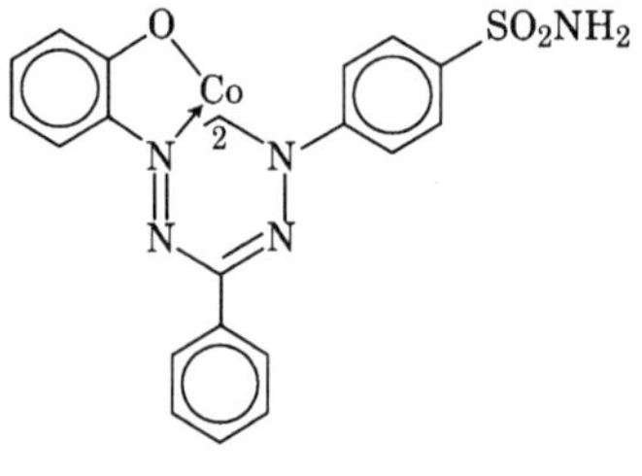

(**61**)

Another class of metal complex dyes is derived from the formazan structure. These dyes are applied to wool and nylon from a neutral or weakly acidic dyebath analogous to the 2:1 premetallized *o,o'*-dihydroxyazo complexes. The bluish-gray dye CI Acid Black 180 [*11103-91-6*] (**61**) (CI 13710) is a 2:1 cobalt complex of the formazan type.

The process and theory of dyeing is very complex (see DYES, APPLICATION AND EVALUATION). Dyes are held to the fiber by chemical interactions, by mechanical forces, as solid solutions, as insoluble particles dispersed in various media, or by a combination of these factors. The two most commonly occurring natural fibers, wool and cotton, exemplify the differences in dyeing theory and technique and the differences in dyestuff molecules required to meet these needs. Acid dyes, which are good wool dyes, as compared to the direct dyes for cotton show the following characteristics: acid dyes are quite soluble in dilute solutions and are much more soluble than direct dyes; solutions of direct dyes contain a large proportion of more complex micelles; and acid dyes have little or no affinity for pure cotton. The molecular weight of direct dyes is on the average much higher than that of acid dyes.

DIRECT DYES

The extent of direct dye production and sales in the United States can be judged by the following data supplied by the USITC (39) for domestic usage in 1988:

Color	*Production, t*	*Sales, t*	*Sales, 10^3 $*
yellow	7,537	7,334	34,034
orange	859	609	3,201
red	3,510	2,699	20,622
blue	3,508	4,113	22,766
violet and green	99	67	740
brown	112	115	584
black	3,252	3,231	16,187
Total	*18,877*	*18,168*	*98,134*

In 1983, 1,728 t of direct dyes was imported (40).

Direct dyes are defined as anionic dyes substantive to cellulosic fibers (cotton, viscose, etc), when applied from an aqueous bath containing an electrolyte. Before the discovery of Congo Red in 1884, only mordanted cotton could be dyed. Congo Red [*573-58-0*] (**62**) (CI Direct Red 28; CI 22120), a primary symmetrical disazo dye, which is made readily from bisdiazotized benzidine and naphthionic acid [*84-86-6*] (4-amino-1-naphthalenesulfonic acid), was the precursor of a most important line of dyes, including all shades, derived from benzidine and its homologues. Today, no benzidine dye is produced because benzidine is carcinogenic.

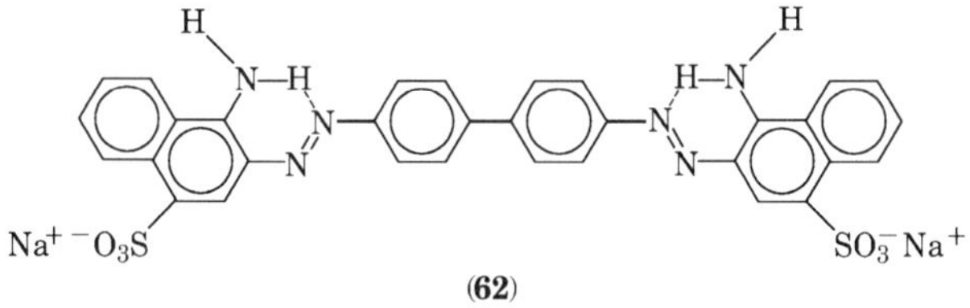

(**62**)

Direct dyes are one of the most versatile classes of dyestuff. U.S. production in 1988 was nearly 18,900 t valued at $100 million. In worldwide usage for cellulosic textiles, direct dyes are the second largest class of dyestuff. The AATCC *Buyers Guide* (July 1991) lists over 180 different CI categories for direct dyes representing nearly 850 commercially available products. U.S. production figures are not released for most of these dyes; the important direct yellows and oranges of revealed chemical composition are listed in Table 5.

Three important direct yellows of revealed chemical composition belong to the stilbene class. Direct Yellow 11 (**24**) was mentioned earlier in the discussion of

Table 5. Yellow and Orange Shade Commercial Direct Dyes, U.S. 1988

CI name	Structure number	CAS Registry Number	CI number	Chemical type
Direct Yellow 4	(**63**)	[*91-34-9*]	24890	disazo
Direct Yellow 6		[*1325-38-8*]	40001–40006	stilbene
Direct Yellow 11	(**24**)	[*1325-37-7*]	40000	stilbene
Direct Yellow 28	(**23**)	[*8005-72-9*]	19555	thiazole
Direct Yellow 34		[*15999-06-1*]	29060	disazo
Direct Yellow 44	(**64**)	[*8005-52-5*]	29000	disazo
Direct Yellow 106		[*12222-60-5*]	40300	stilbene
Direct Yellow 118		[*12217-75-3*]	29042	disazo
Direct Orange 6		[*61814-85-5*]	23375	disazo
Direct Orange 8		[*2429-79-0*]	22130	disazo
Direct Orange 15		[*1375-35-5*]	40002–40003	stilbene
Direct Orange 26	(**67**)	[*3626-36-6*]	29150	disazo
Direct Orange 34		[*12222-37-6*]	40215–40220	stilbene
Direct Orange 39		[*1325-54-8*]	40215	stilbene
Direct Orange 72	(**66**)	[*12217-64-0*]	29058	disazo
Direct Orange 102	(**65**)	[*6598-63-6*]	29156	disazo

the condensation of nitro compounds. Another stilbene dye, Direct Yellow 4 (**63**), an old dyestuff discovered in 1886, is produced by coupling bisdiazotized 4,4′-diamino-2,2′-stilbenedisulfonic acid to two moles of phenol.

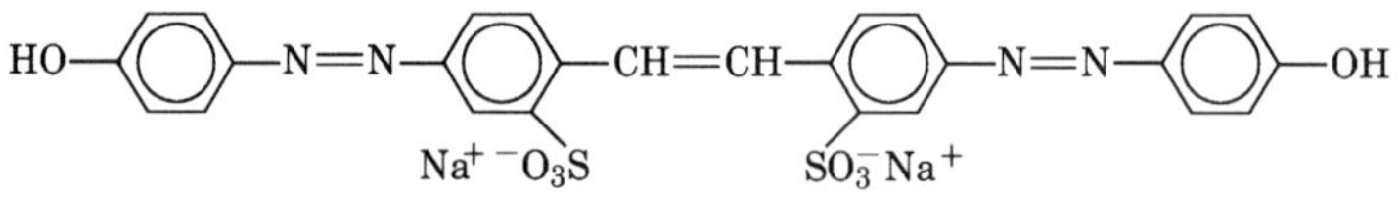

(**63**)

Its production was 621 t and the average price $0.75/kg in 1987. Direct Yellow 44 (**64**) is prepared by phosgenation of an equimolar mixture of metanilic acid coupled to *o*-anisidinomethanesulfonic acid (with subsequent hydrolysis of the methanesulfonic acid group) and *p*-nitroaniline coupled to salicyclic acid (with subsequent reduction of the nitro group).

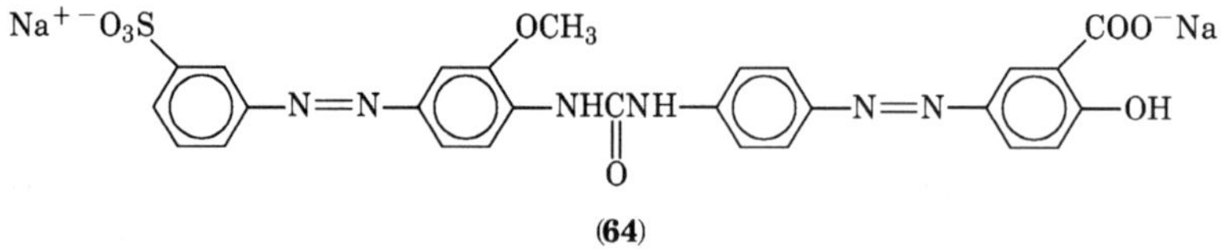

(**64**)

Direct Oranges. All principal commercially produced direct oranges are of disazo or stilbene chemical composition (Table 5). Direct Orange 102 (**65**) (R = COO^-Na^+), is manufactured by first coupling aniline to 6,6′-ureylenebis-1-naphthol-3-sulfonic acid followed by a second coupling with *p*-aminobenzoic acid [*150-13-0*]. 6,6′-Ureylenebis-1-naphthol-3-sulfonic acid [*134-47-4*] is the coupling component in many of the most important direct colors. It is produced by phosgenation of two moles of J-acid (6-amino-1-naphthol-3-sulfonic acid).

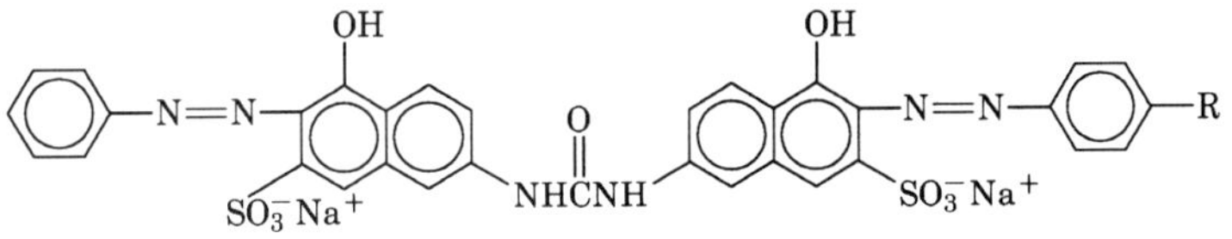

(65), **(66)**

Other commercially important direct oranges include Direct Orange 26 (**66**) (R = H) (2 mol aniline coupled to 6,6′-ureylenebis-1-naphthol-3-sulfonic acid), and Direct Orange 72 (**67**) which is obtained by phosgenating two moles of 7-amino-1,3-naphthalenedisulfonic acid [*86-65-7*] coupled to cresidine [*120-71-8*].

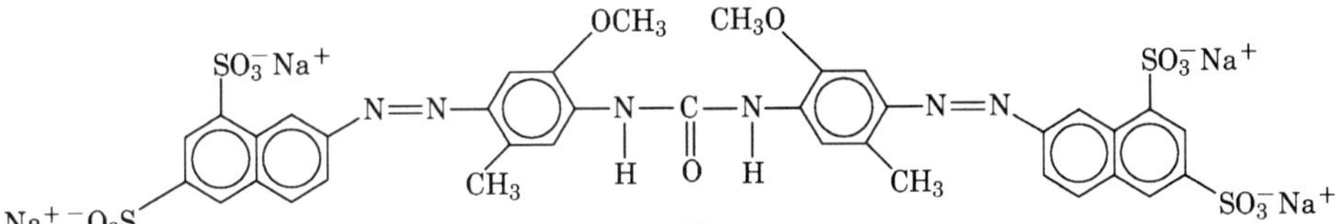

(67)

Direct Reds. The principal commercially produced direct reds, with revealed chemistry, are of disazo or polyazo composition (Table 6). Direct Red 81 (**68**) is manufactured by coupling *p*-(*p*-aminophenylazo)benzenesulfonic acid to *N*-benzoyl J-acid. Other important direct red dyes include those shown in Figure 5.

Direct Blues. Direct Blue 86, a phthalocyanine direct dye, represents a small but important segment of the direct dye structure groups. The dyes are brilliant greenish blue or turquoise shades. Sales of nearly $2 million were reported in the United States in 1988. Among the disazo blues, Direct Blue 80 (**74**) and Direct Blue 98 (**75**) were sold in 152 t and 107 t quantities at an average cost of $2.22 and $1.81/kg, respectively, in 1988. Table 7 and Figure 6 show some direct blues.

Table 6. Red Shade Commercial Direct Dyes[a]

CI name	Structure number	CAS Registry Number	CI number
Direct Red 2	(**69**)	[*992-59-6*]	23500
Direct Red 4		[*6598-63-6*]	29165
Direct Red 16		[*6227-02-7*]	27680
Direct Red 23	(**70**)	[*3441-14-3*]	29160
Direct Red 24	(**71**)	[*6420-44-6*]	29185
Direct Red 26		[*3687-80-7*]	29190
Direct Red 72	(**72**)	[*8005-64-9*]	29200
Direct Red 73		[*6460-01-1*]	29180
Direct Red 80[b]	(**73**)	[*2610-10-8*]	35780
Direct Red 81	(**68**)	[*2610-11-9*]	28160
Direct Red 83		[*15418-16-3*]	29225

[a]Disazo, unless otherwise noted.
[b]Polyazo.

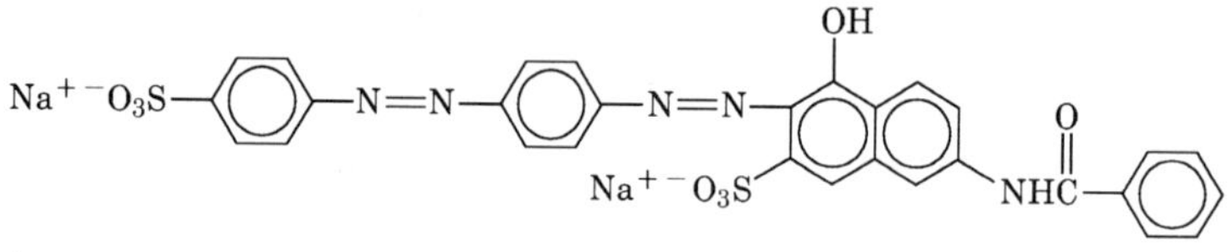

(68)
(a)

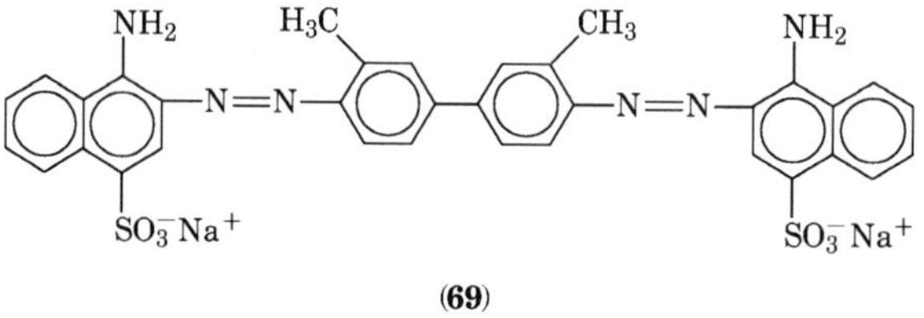

(69)
(b)

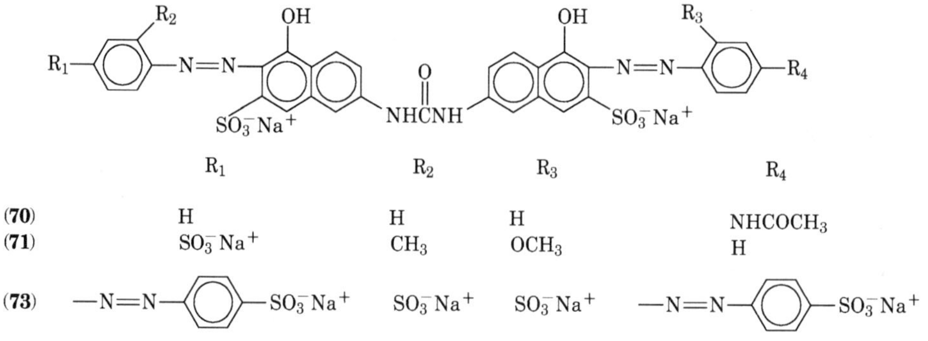

(c)

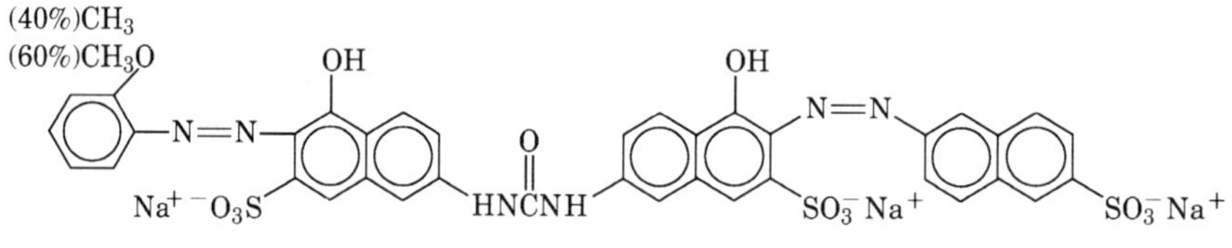

(72)
(d)

Fig. 5. Direct red dyes. (**a**) Direct Red 81 described in text (**68**); (**b**) Direct Red 2 (*o*-tolidine coupled to two moles of naphthionic acid) (**69**); (**c**) Direct Red 23 (aniline coupled to 6,6′-ureylenebis-1-naphthol-3-sulfonic acid with a second coupling with *p*-aminoacetanilide) (**70**); and Direct Red 80 (2 mol 6-amino-3,4′-azobenzenedisulfonic acid coupled twice to 6,6′-ureylenebis-1-naphthol-3-sulfonic acid) (**73**). Direct Red 24 (4-amino-*m*-toluenesulfonic acid coupled under acidic conditions to 6,6′-ureylenebis-1-naphthol-3-sulfonic acid followed by an alkaline coupling of *o*-anisidine) (**71**); (**d**) Direct Red 72 (Broenner's acid, ie, 6-amino-2-naphthalenesulfonic acid coupled under acidic conditions to 6,6′-ureylenebis-1-naphthol-3-sulfonic acid followed by an alkaline coupling of a 3:2 mole ratio of *o*-anisidine and *o*-toluidine) (**72**).

Table 7. Blue Shade Commercial Direct Dyes

CI name	Structure number	CAS Registry Number	CI number	Chemical type
Direct Blue 15	(**76**)	[*2429-74-5*]	24400	disazo
Direct Blue 22	(**77**)	[*2586-57-4*]	24280	disazo
Direct Blue 25		[*2150-54-1*]	23790	disazo
Direct Blue 75	(**78**)	[*72245-54-6*]	34220	trisazo
Direct Blue 76		[*16143-79-6*]	24410	disazo
Direct Blue 80	(**74**)	[*12222-00-3*]	24315	disazo
Direct Blue 86		[*1330-38-7*]	74180	phthalocyanine
Direct Blue 98	(**75**)	[*6656-03-7*]	23155	disazo
Direct Blue 108		[*1324-58-9*]	51320	oxazine
Direct Blue 218		[*10401-50-0*]	24401	disazo metallized

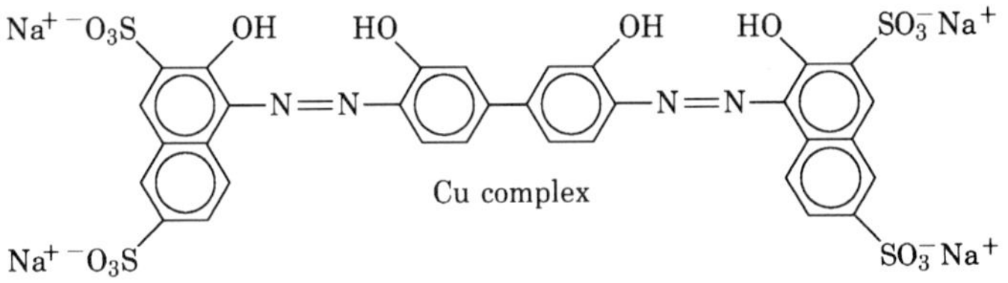

(**74**)

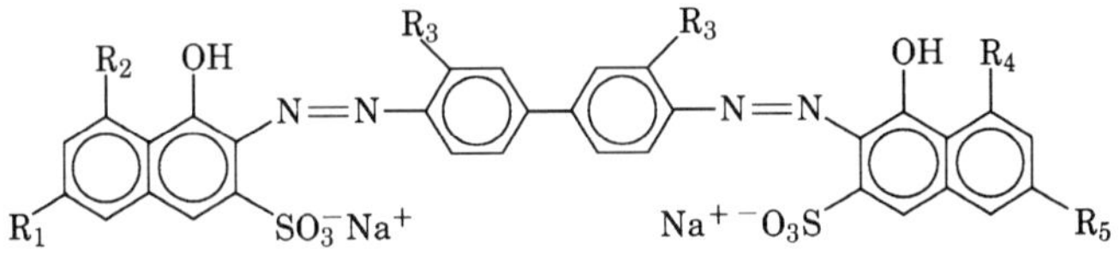

	R_1	R_2	R_3	R_4	R_5
(**75**)	H	$SO_3^-Na^+$	OH	H	NHC_6H_5
(**76**)	$SO_3^-Na^+$	NH_2	OCH_3	NH_2	$SO_3^-Na^+$

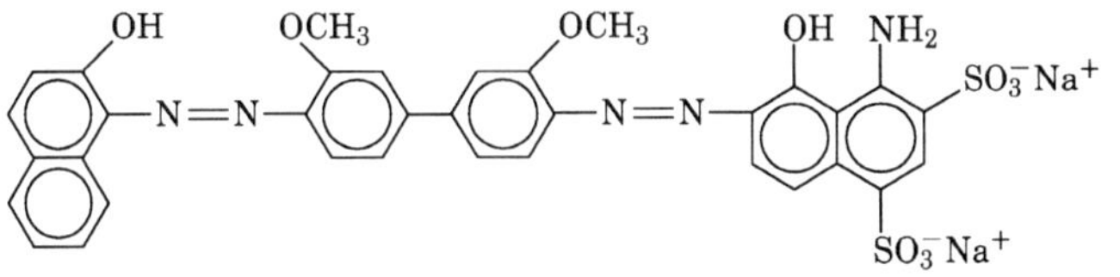

(**77**)

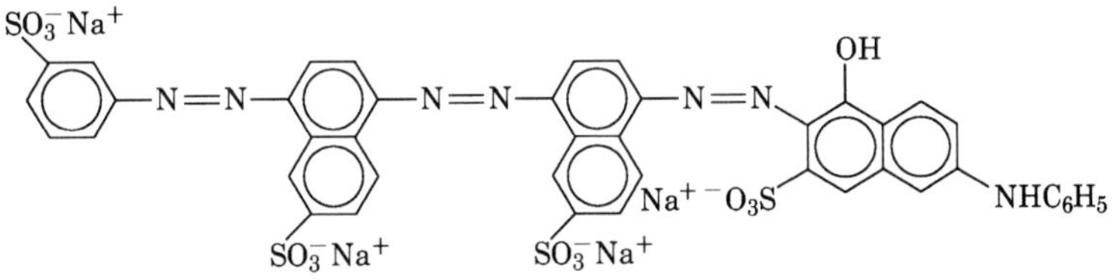

(**78**)

Fig. 6. Direct blue dyes.

Direct Blue 218 had reported sales of 623 t valued at \$4.4 million in 1987. It is produced from Direct Blue 15 (**76**) by metallizing and elimination of methyl groups from the methoxide to form the copper complex. Direct Blue 15 (**76**) is prepared by coupling *o*-dianisidine [*119-90-4*] to two moles of H-acid (4-amino-5-hydroxy-2,7-naphthalenedisulfonic acid) under alkaline pH conditions. Other important direct blues include Direct Blue 80 (**74**), *o*-dianisidine coupled to two moles of R-acid (3-hydroxy-2,7-naphthalenedisulfonic acid [*148-75-4*]) followed by metallizing to form a bis copper complex, and Direct Blue 22 (**77**), an asymmetrical disazo dye, prepared by coupling *o*-dianisidine to Chicago acid [*82-47-3*] and 2-naphthol. Direct Blue 75 (**78**) is an example of a trisazo dye represented as metanilic acid → 1,6-Cleve's acid → 1,6-Cleve's acid → (alk) *N*-phenyl J-acid.

Direct Violets, Greens, Browns, and Blacks. Direct violets and greens are small volume products, with combined production of nearly 100 t in 1988.

The *Colour Index* (up to June 1991) lists 21 direct violets with disclosed chemical constitutions. Commercially important are CI Direct Violet 9 [*6227-14-1*] (**79**) (CI 27885) (sulfanilic acid coupled to cresidine followed by alkaline coupling to *N*-phenyl J-acid) and CI Direct Violet 66 [*6798-03-4*] (**80**) (CI 29120) (a copper complex of 2-amino-1-phenol-4-sulfonamide (2 mol) coupled to 6,6′-iminobis-1-naphthol-3-sulfonic acid).

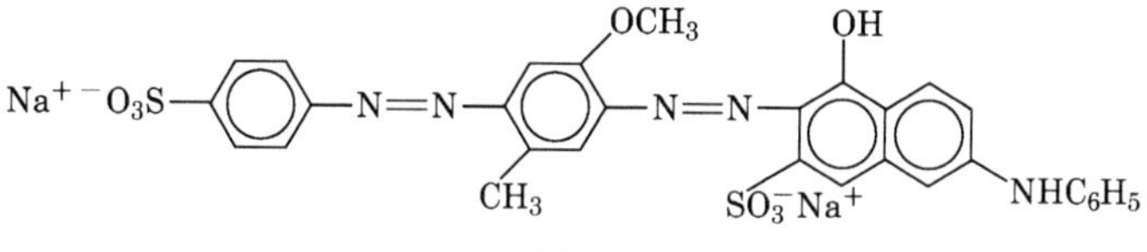

(**79**)

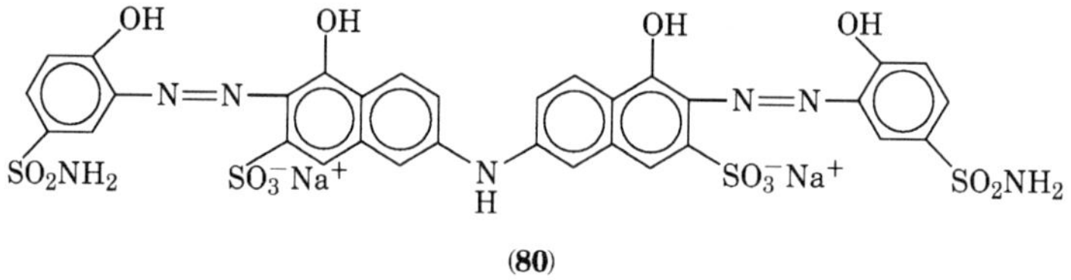

(**80**)

CI Direct Green 1 [*3626-28-6*] is of trisazo chemical composition shown previously. CI Direct Green 26 [*6388-26-7*] (**81**) (CI 34045) is produced through a combination of a blue and yellow dye components (H-acid is coupled to cresidine which is coupled to a coupling component; this coupling component is prepared by condensation of cyanuric chloride successively with H-acid, the azo dye obtained by reducing the nitro group in *p*-nitroaniline coupled to salicyclic acid, and aniline). Here the triazinyl group, similar to the –NH–CO–NH– group acts as a bridge and increases the substantivity and lightfastness. Brightness also results from the intramolecular combination as the triazinyl group protects the two different chromophoric systems from resonance interactions (56).

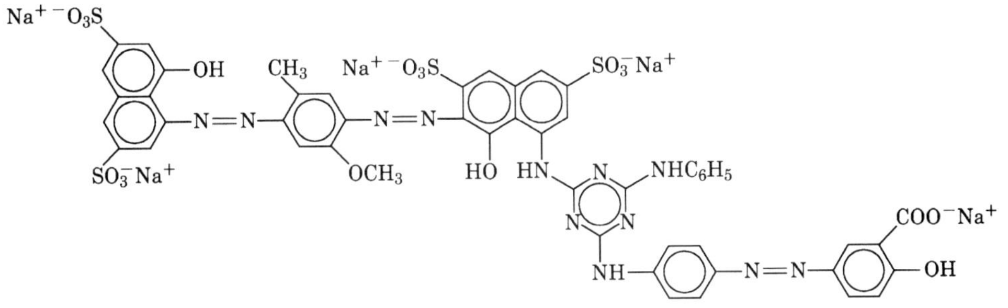

(81)

Two important browns, other than benzidine derivatives, are of azo chemical composition. Direct Brown 30 [*6222-60-2*] (**82**) (CI 17630) is produced by coupling Primuline [*30133-37-2*] (CI 49000) to *N*-phenylgamma acid under alkaline conditions. Direct Brown 44 [*6252-62-6*] (CI 35005–35010) is produced by coupling two moles of sulfanilic acid to Bismark Brown [*10114-58-6*] (Basic Brown 1) (CI 21000) forming a heterogeneous product with a significant component (**83**).

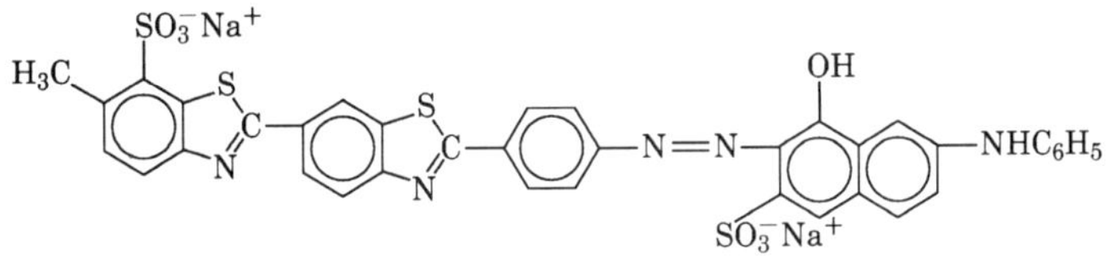

(82)

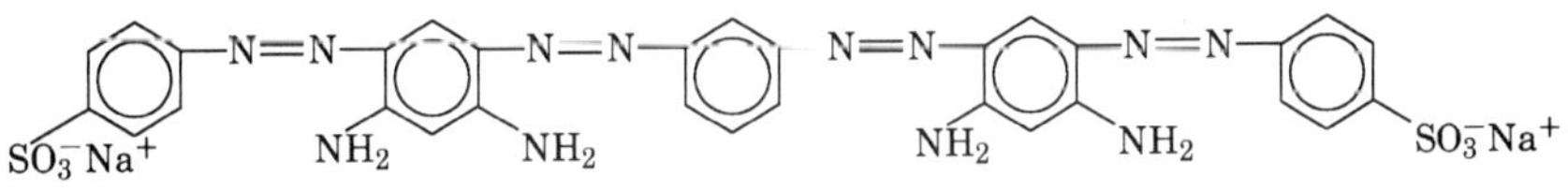

(83)

Direct Black 22 [*6473-13-8*] (**84**) (CI 35435) is produced first by coupling 4,4′diaminodiphenylamine-2-sulfonic acid to two moles of gamma acid to form a primary disazo dye, which is then bisdiazotized and coupled to two moles of *m*-phenylenediamine (MPD).

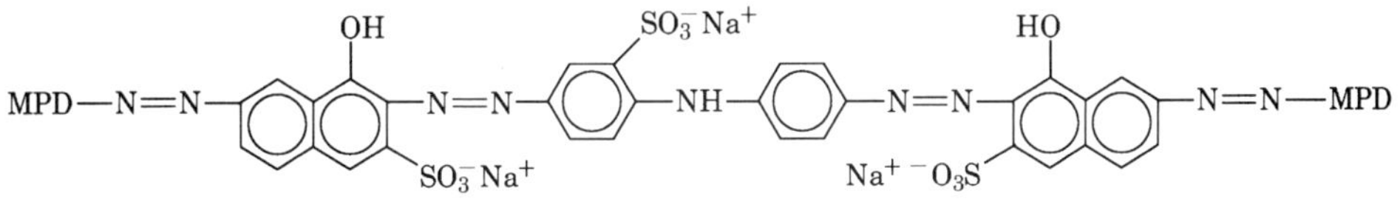

(84)

Table 8 lists some other direct blacks. Progress in the field of direct dyes is the

Table 8. Black Shade Commercial Direct Dyes

CI name	CAS Registry Number	CI number	Chemical type
Direct Black 19	[*6428-31-5*]	35255	polyazo
Direct Black 22	[*6473-13-8*]	35435	polyazo
Direct Black 150	[*6897-38-7*]	32010	trisazo
Direct Black 166	[*57131-19-8*]	30026	trisazo

development of reactant fixable dyes (indosol dyes). These dyes and the economics of dyeing processes have been reviewed (57,58).

Azoic or Naphthol Dyes. Azoic dyes (known also as ice colors and ingrain colors) are water-insoluble azo pigments, free from solubilizing groups, formed on the fiber by reaction of a diazo component with a coupling component, a so-called Naphthol AS compound, such as an arylide of 3-hydroxy-2-naphthoic acid. The idea of producing insoluble azo dyes on the fiber was first patented in 1880, by chemists at Read Holliday & Sons, Ltd., in England, who impregnated cotton cloth with an alkaline solution of 2-naphthol and then coupled it with diazotized 2-naphthylamine to produce a bluish-red dyed fiber. The introduction in 1912 of Naphthol AS, so designated from Anilid Säure, German for acid anilide, was an event of great importance in the dyeing and printing industry. The discovery that 3-hydroxy-2-naphthoic acid arylides have greater substantivity for cotton than 2-naphthol, tremendously increased the range of bright and fast shades, and led to the introduction of an extensive line of Naphthol AS derivatives, fast color salts, rapid fast dyes, and the rapidogen dyes from diazoamino compounds.

Naphthol AS Coupling Components. Naphthol AS components are the arylides of either *o*-hydroxyarylcarboxylic acids or acylacetic acids. They are free of sulfo and carboxyl groups, but form salts with bases; these salts dissolve in water to give colloidal solutions, which couple with diazo components to form colored pigments. The whole class derives from the anilide of 3-hydroxy-2-naphthoic acid [*92-70-6*], Naphthol AS (**85**) (CI Azoic Coupling Component 2).

Naphthols of the *o*-hydroxyarylcarboxylic acid type are manufactured according to the equation:

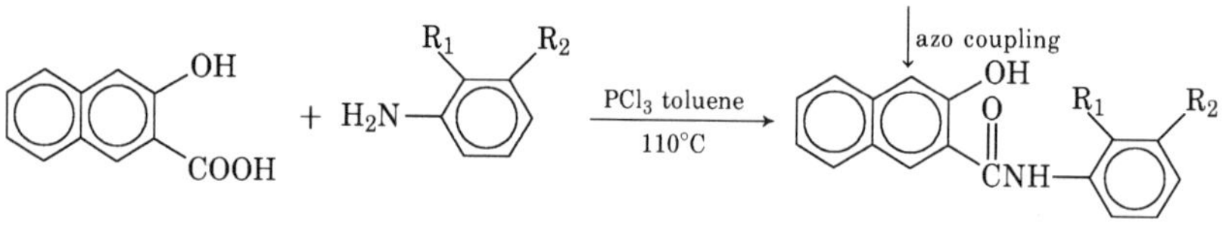

	(**85**)	(**86**)	(**87**)	(**88**)
R_1	H	H	CH_3	OCH_3
R_2	H	NO_2	H	H

In the production of arylides of acetoacetic acid, two methods are employed:

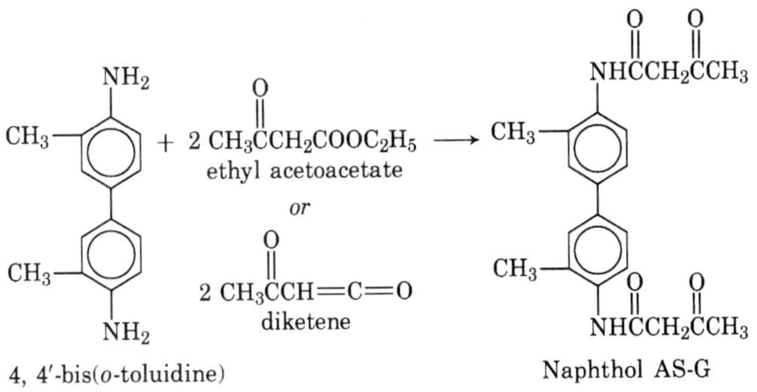

Based on the U.S. International Trade Commission Import (39) data, nearly \$3.7 million worth of Naphthol AS and derivatives (551 t) were imported in 1983. Commercially important coupling components include: Azoic Coupling Component 2 (3-hydroxy-2-naphthanilide [*92-77-3*]) (**85**); Azoic Coupling Component 17, Naphthol AS-BS, (3-hydroxy-3′-nitro-2-naphthanilide) (**86**); Azoic Coupling Component 18, Naphthol AS-D, (3-hydroxy-2′-methyl-2-naphthanilide) (**87**); and Azoic Coupling Component 20, Naphthol AS-OL, (3-hydroxy-2′-methoxy-2-naphthanilide) (**88**) (Table 9).

Table 9. Naphthol AS and Its Derivatives

CI azoic coupling component	CAS Registry Number	CI number	Commercial name
2	[*92-77-3*]	37505	Naphthol AS
3	[*91-92-9*]	37575	Naphthol AS-BR
4	[*132-68-3*]	37560	Naphthol AS-BO
7	[*135-64-8*]	37565	Naphthol AS-SW
9	[*4273-88-5*]	37625	naphthanilide L4G
12	[*92-72-8*]	37550	Naphthol AS-ITR
14	[*92-74-0*]	37558	Naphthol AS-PH
16	[*132-62-7*]	37605	Naphthol AS-DB
17	[*135-65-9*]	37515	Naphthol AS-BS
18	[*135-61-5*]	37520	Naphthol AS-D
20	[*135-62-6*]	37530	Naphthol AS-OL
29	[*92-75-1*]	37527	Naphthol AS-MX
34	[*137-52-0*]	37531	Naphthol AS-CA (naphthanilide EL)

Fast Color Salts. In order to simplify the work of the dyer, diazonium salts, in the form of stable dry powders, were introduced under the name of fast color salts. When dissolved in water they react like ordinary diazo compounds. These diazonium salts, derived from amines, free from solubilizing groups, are prepared by the usual method and are salted out from the solutions as the sulfates, the metallic double salts, or the aromatic sulfonates. The isolated diazonium salt is sold in admixture with anhydrous salts such as sodium sulfate or magnesium sulfate.

Commercially, the most important fast color bases (Table 10) and salts include: Azoic Diazo Component 3, produced by the reduction of 1,4-dichloro-2-nitrobenzene; Azoic Diazo Component 5, produced by nitration of *o*-acetanisidide with subsequent hydrolysis; Azoic Diazo Component 8, produced by nitration of *p*-acetotoluidide in H_2SO_4 and subsequent saponification; Azoic Diazo Component 9, produced by selective ammonolysis of 1,4-dichloro-2-nitrobenzene; Azoic Diazo Component 11, produced by chlorination of *o*-acetotoluidide and subsequent saponification; Azoic Diazo Component 13, produced either by partial reduction of 2,4-dinitroanisole or nitration of *o*-anisidine; Azoic Diazo Component 20 (4′-amino-2′,5′-diethoxybenzanilide) (**89**), produced by condensation of 2,5-diethoxyaniline with benzoyl chloride, followed by nitration and reduction; Azoic Diazo Component 32, produced by nitration of *p*-chlorotoluene followed by reduction of the resulting 4-chloro-2-nitrotoluene; and Azoic Diazo Component 34, produced by nitration of *o*-benzenesulfonotoluidide followed by hydrolysis. Variamine Blue B the half sulfate [*6254-98-4*] of (**90**), ie, Azoic Diazo Component 35 used for dark blue or green shades is marketed as the free base or half sulfate. It is produced by condensation of *p*-anisidine with 2-chloro-5-nitrobenzenesulfonic acid, reduction, and desulfonation of the resulting 5-amino-2-*p*-anisidinobenzenesulfonic acid.

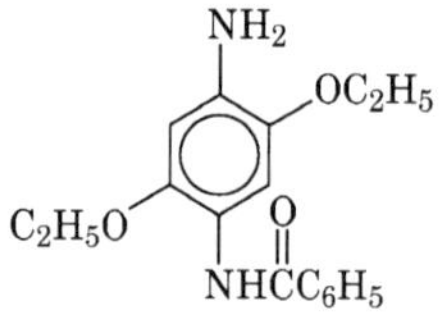

(**89**)

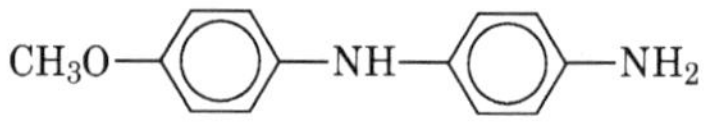

(**90**)

Applications of Naphthol AS Compounds and Fast Color Salts. For dyeing, the Naphthol AS is dissolved in water with base and diluted to the proper concentration. The cloth is passed through the solution and squeezed. The degree of squeezing together with the concentration of the Naphthol AS determines the depth of the final color. After impregnation, the material, which may or may not be dried, is passed into a cold solution of a fast color salt or diazotized amine, where practically instantaneous coupling takes place and the insoluble dye is formed on the material. After coupling, the material is passed through squeeze rolls, washed with cold water to remove all unreacted diazo compound, and washed in boiling soap solution to develop the true shade and improve fastness.

For printing, the naphthol and the fast color salts are coupled as for dyeing, but by different mechanical means. In one important method the cloth is naphtholated as for dyeing. The material is then dried and passed through the printing machine. To obtain multicolor prints, each roller of the printing machine prints a paste containing a different fast color salt. The cloth is dried again, aged in either neutral or acid steam, washed, soaped to remove the uncombined Naphthol AS from the unprinted part of the material, and dried, leaving a colored print on a white background.

In order to eliminate the expensive naphtholating and drying operations

Table 10. Fast Color Bases

CI azoic diazo component	Amine structure	CAS Registry Number	CI number	Fast base or salt
1		[*96-96-8*]	37135	Bordeaux GP
3	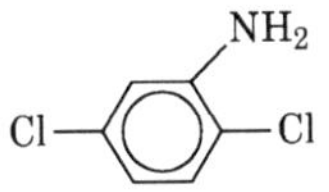	[*95-82-9*]	37010	Scarlet GG
5	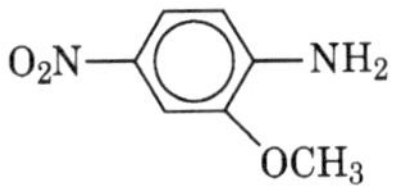	[*97-52-9*]	37125	Red B
8	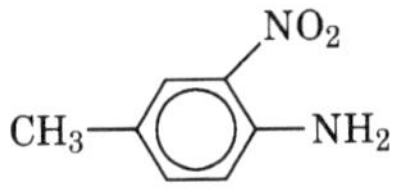	[*89-62-3*]	37110	Red GL
9	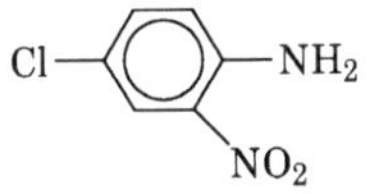	[*89-63-4*]	37040	Red 3GL
10		[*4274-03-7*]	37120	Red RC
11	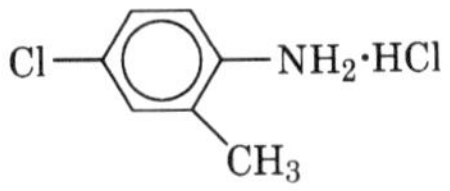	[*3165-93-3*]	37085	Red TR
12		[*99-55-8*]	37105	Scarlet G
13	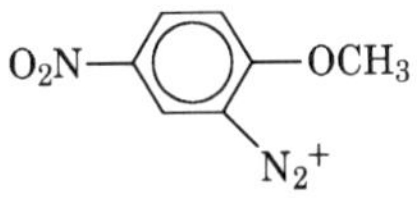	[*27165-17-9*]	37130	Scarlet R
14		[*80-22-8*]	37175	Red PDC
20	**(89)**	[*120-00-3*]	37175	Blue BB
32	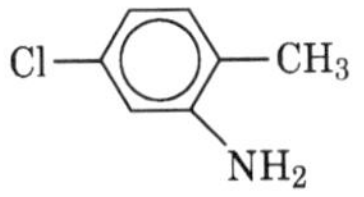	[*6259-42-3*]	37090	Red KB
34	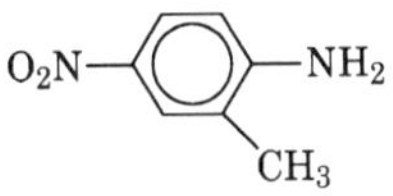	[*99-52-5*]	37100	Red RL
35	**(90)**	[*6254-98-4*]	37255	Blue VB[a]
41		[*99-21-8*]	37165	Violet B
42		[*97-35-8*]	37150	Red ITR
44		[*137-04-2*]	37000	Yellow GC
48		[*20282-70-6*]	37235	Blue B
49		[*121-50-6*]	37050	Orange RD
109		[*6369-04-6*]	37245	Black BN

[a]Variaminblue B.

some further important developments were introduced by Chemische Fabrik Griesheim-Elektron under the names of Rapid Fast, Rapidazol, and Rapidogen combinations (mixtures of substantive coupling components with antidiazotates, diazosulfonates, and diazoamino compounds, respectively).

DISPERSE AZO DYES

Disperse dyes are coloring substances having very low aqueous solubility which are applied to hydrophobic fibers from an aqueous system in which the dye is present in a highly dispersed state. The sharp increase in the importance of disperse dyes in the 1970s and 1980s, can be attributed directly to the emergence of polyester and nylon as the principal synthetic fibers. As shown in Table 11, U.S. disperse dye production and sales increased sevenfold from 1960 to 1980 and the dollar value rose 1150%. Of all types of dyes produced in the United States, the market share for disperse dyes grew from 4.2% in 1960 to 19% in 1980. The high volume of U.S. textile imports, increasing governmental controls, environmental constraints, and increase in the use of polyester–cotton blends have reduced the market share to about 11%. From 1971 when the third edition of the CI was issued until the first supplement to the third edition, Vol. 6, appeared in 1975, there was an increase from 554 to 840 (52%) in the number of active CI generic names. During the next decade, rationalization by many dye producers and fewer introductions of new products reduced the number of current commercial products of disperse dyes. Volume 5 of the *Colour Index,* (third revision, 1987) lists 591 disperse CI generic names.

Table 11. U.S. Production and Sales of Disperse Dyes[a]

Year	Production, t	Sales, t	Value, 10^6 $
1960	3,000	3,200	14.2
1965	7,000	6,100	32.9
1970	13,100	11,500	65.7
1975	15,800	16,200	122.7
1980	21,200	18,500	178.1
1985	11,400	11,700	94.3
1988	13,640	9,920	99.7

[a]Data from U.S. International Trade Commission (formerly, U.S. Tariff Commission).

Table 12 compares disperse dye production and sales in the United States in 1980 and 1988 by color.

Azo and anthraquinone compounds comprise the two principal structural types which are used as disperse dyes. Other compounds used to a much lesser extent include methines, cyanostyryls, hydroxyquinophthalones, and nitrodiarylamines.

Disperse dyes are used mainly in the coloring of polyester, cellulose acetate, and triacetate fibers in textile applications. Another important use is the dyeing

Table 12. U.S. Production of Disperse Dyes[a]

Color	1980		1988	
	Production, t	Sales, 10^3 $	Production, t	Sales, 10^3 $
yellow	2,457	19,925	1,021	8,133
orange	2,342	13,257	2,541	9,938
red	4,823	47,694	2,484	25,877
violet	421	3,210	161	2,206
blue	9,933	84,412	6,186	43,960
black, brown, green	1,216	9,599	1,243	9,575
Total	*21,192*	*178,097*	*13,636*	*99,689*

[a]Ref. 39.

of nylon, especially carpets and hosiery in which disperse dyes effectively cover barré defects in the fiber caused by processing inconsistencies. Acrylics and modacrylics are also dyed with disperse dyes generally in light to medium depths. A summary of methods for dyeing and printing disperse dyes on polyester, acetate, triacetate, nylon, and acrylic fibers can be found in the *Colour Index* (59) (see also FIBERS, ACRYLIC; FIBERS, CELLULOSE ESTERS; FIBERS, POLYAMIDE; and FIBERS, POLYESTER).

Commercial Disperse Azo Dyes. The first proposal to use insoluble dyes in suspension in an aqueous foam bath, ie, disperse dyes, to dye cellulose acetate was in 1921 (60). Commercialization of disperse dyes began in 1924 with the introduction of the Duranol dyes by British Dyestuffs Corporation (61) and the SRA dyes by British Celanese Company (62). In contrast to the acid monoazo dyes, derivatives of benzene rather than of naphthalene are of the greatest importance as coupling components. Among these components mono- and dialkylanilines (especially *N*-β-hydroxyethyl- and *N*-β-acetoxyethylaniline derivatives) are widely used couplers. Nitrodiazobenzenes are widely used as diazo components. A typical example is Celliton Scarlet B [*2872-52-8*] (**91**) (CI Disperse Red 1; CI 11110).

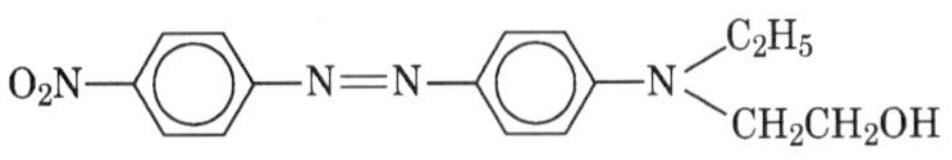

(**91**)

Generally speaking, disperse dyes for acetate possess only moderate fastness to gas, light, sublimation (heat), and washing. There were no suitable blue azo disperse dyes available in early years and, as a consequence, aminoanthraquinone types gained prominence in this shade range. The latter, however, were subject to gas fading, ie, on exposure to oxides of nitrogen and ozone acetate dyed blue became pink. This deficiency led to the development of many antigas fading additives. Commercially important disperse yellow and orange dyes for which chemical structures are revealed by the producers are listed in Table 13.

The early yellow disperse dyes were based on phenolic coupling components, eg, CI Disperse Yellow 3 (**92**) (diazotized 4-aminoacetanilide coupled to *p*-cresol) which is still used today for the coloration of cellulose acetate and nylon fibers.

Table 13. Yellow and Orange Shade Commercial Disperse Dyes

CI name	Structure number	CAS Registry Number	CI number
Disperse Yellow 3	(**92**)	[*2832-42-8*]	11855
Disperse Yellow 4		[*6407-80-3*]	12770
Disperse Yellow 5		[*6439-53-8*]	12790
Disperse Yellow 7	(**93**)	[*6300-37-4*]	26090
Disperse Yellow 8		[*6358-49-2*]	12690
Disperse Yellow 10		[*8005-71-8*]	12795
Disperse Yellow 23	(**94**)	[*6250-23-3*]	26070
Disperse Yellow 60		[*15790-15-5*]	12712
Disperse Orange 1		[*2581-69-3*]	11080
Disperse Orange 3		[*730-40-5*]	11005
Disperse Orange 5		[*6232-56-0*]	11100
Disperse Orange 13	(**95**)	[*6253-10-7*]	26080
Disperse Orange 25	(**97**)	[*31482-56-1*]	11227
Disperse Orange 29	(**96**)	[*19800-42-1*]	26077
Disperse Orange 30		[*5261-31-4*]	11119
Disperse Orange 56		[*67162-11-2*]	12650
Disperse Orange 62		[*37672-70-1*]	11239
Disperse Orange 138		[*43047-20-7*]	11145

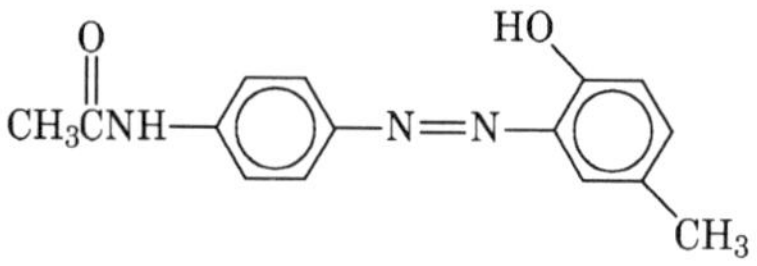

(**92**)

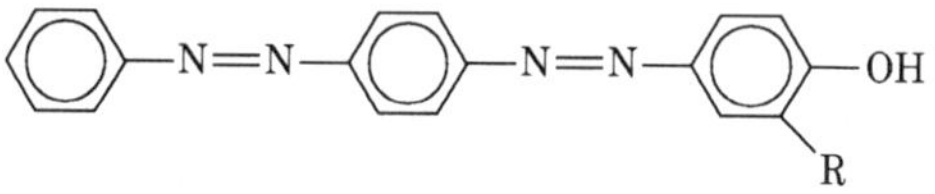

(**93**), (**94**)

Disperse Yellow 7 (**93**, R = CH_3) and Disperse Yellow 23 (**94**, R = H) are two disazo dyes which are manufactured by diazotizing *p*-phenylazoaniline [*60-09-3*] (CI 11000) followed by coupling to *o*-cresol and phenol, respectively. Among the azo colors of orange shades, Disperse Orange 13 (**95**) and Disperse Orange 29 (**96**) are two disazo dyes containing phenol as coupling components.

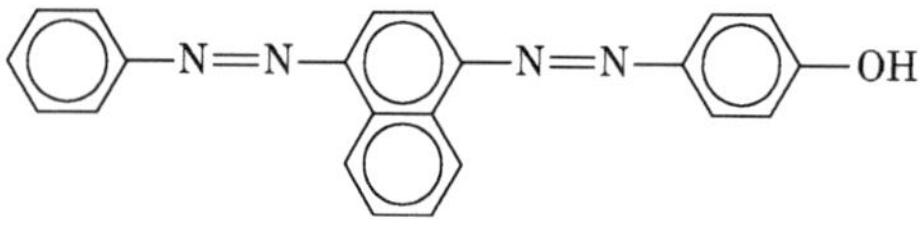

(**95**)

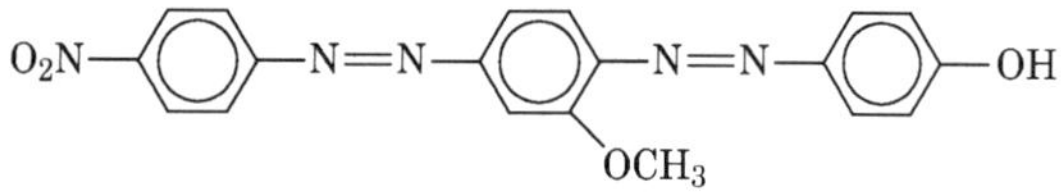

(**96**)

Disperse Orange 25 (**97**, R = H) had sales in excess of $700,000 (114 tons) in 1988. Other disperse azo colors produced in the United States of revealed chemical constitution are listed in Table 14.

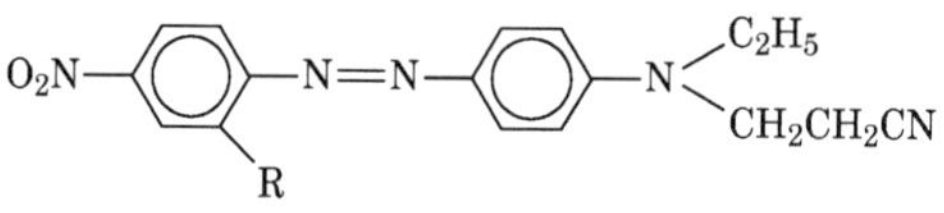

(**97**), (**98**)

Table 14. Commercial Disperse Dyes

CI name	Structure number	CAS Registry Number	CI number
Disperse Red 1	**(91)**	[*2872-52-8*]	11100
Disperse Red 5		[*3769-57-1*]	11215
Disperse Red 7		[*4540-00-5*]	11150
Disperse Red 13		[*3180-81-2*]	11115
Disperse Red 17		[*3179-89-3*]	11210
Disperse Red 19		[*3003-33-6*]	11130
Disperse Red 31		[*2475-43-6*]	11250
Disperse Red 32		[*3084-21-7*]	11190
Disperse Red 58		[*6373-93-9*]	11135
Disperse Red 65		[*16586-43-9*]	11228
Disperse Red 72		[*12223-39-1*]	11114
Disperse Red 73	**(98)**	[*16889-10-4*]	11116
Disperse Red 90		[*27767-98-2*]	11117
Disperse Violet 24	**(99)**	[*6374-03-4*]	11200
Disperse Violet 33		[*66882-16-4*]	11218
Disperse Blue 11	**(102)**	[*6358-51-6*]	11260
Disperse Blue 79	**(100)**	[*12239-34-8*]	11345
Disperse Blue 165		[*41642-51-7*]	11077
Disperse Blue 183		[*2309-94-6*]	11078
Disperse Brown 1	**(101)**	[*23355-64-8*]	11152
Disperse Black 1	**(103)**	[*6054-48-4*]	11365

Disperse reds are second only to blues as the most important disperse color manufactured. All commercial disperse reds are monoazo dyes. In 1988, Disperse Red 73 (**98**, R = CN) had production of 270 tons valued at nearly $1.6 million. Disperse Violet 24 (**99**) is produced from diazotized 2-bromo-4,6-dinitroaniline by coupling with 2-(*N*-butyl-*m*-toluidine)ethanol.

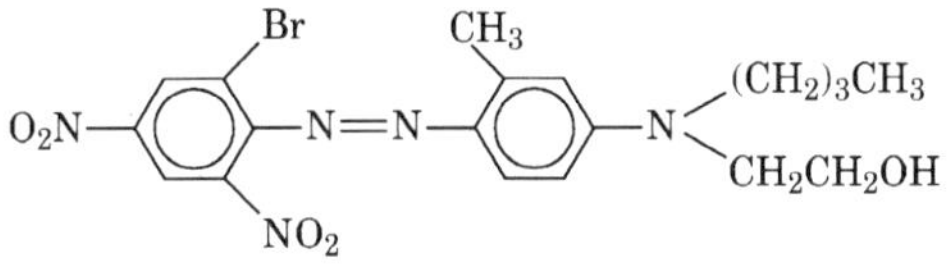

(99)

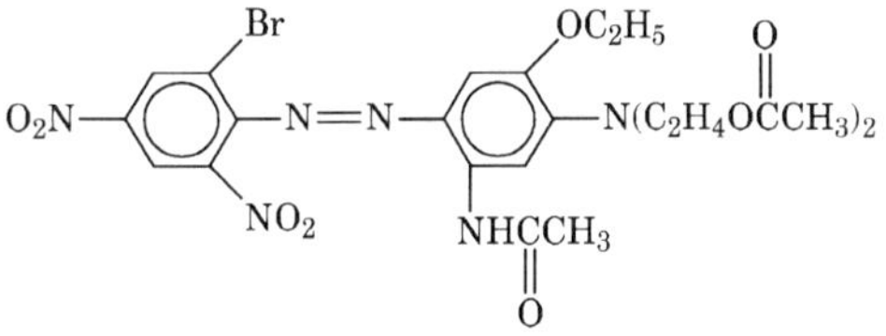

(100)

Blues are the single most important color in the disperse class, both in terms of amounts produced and dollar sales volume. Among all dyes, Disperse Blue 79 (**100**) (6-bromo-2,4-dinitroaniline [*1817-73-8*] coupled to 3-bis-(2-acetoxyethyl)-amino-*p*-acetophenetidine [*20249-05-2*]) was the highest volume dye in the late 1980s and continues. As reported in 1985, the United States produced 2872 tons valued at $7.7 million.

The only disperse brown of commercial importance is Disperse Brown 1 (**101**), which in 1988 had sales in excess of $2.3 million (production 317 tons). A method still in use to produce navy and black shades economically on acetate involves formation of the dye directly on the substrate (azoic dyeing).

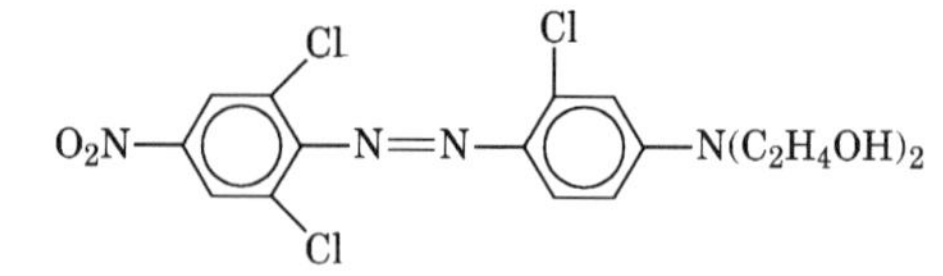

(101)

Monoazo dyes such as Disperse Blue 11 (**102**) and Disperse Black 1 (**103**) are applied to cellulose acetate as a dispersion and dyed in the usual way, then diazotized *in situ* and coupled in this instance, to 3-hydroxy-2-naphthoic acid forming the blue and black shades, respectively.

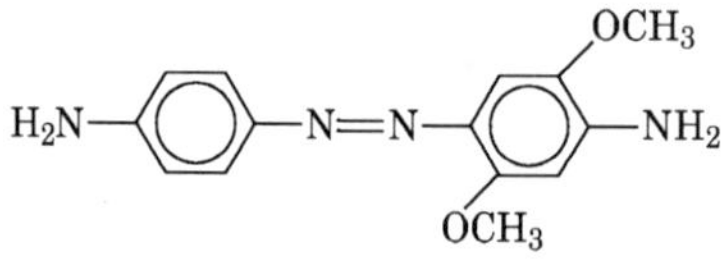

(102)

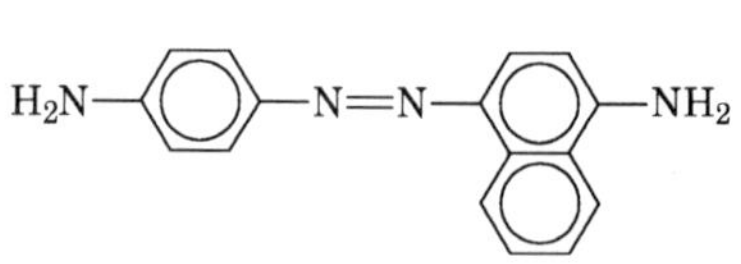

(103)

Dispersion Technology. Substantial advancements in dispersion technology have been made since the initial introduction in 1923 of disperse dyes in paste form for cellulose acetate. Dyes were dissolved in sulfonated fatty acids such as sulforicinoleic acid [*36634-48-7*] (SRA), 12-hydroxy-9-octadecenoic acid sulfate (**104**), and then formed into a dispersed paste by mixing the solution with water giving rise to an entire line of SRA dyes for acetate (63). The Duranol dyes (61) were prepared by grinding or dissolving aminoanthraquinone dyes in a solvent followed by drowning into water containing soap or Turkey Red Oil.

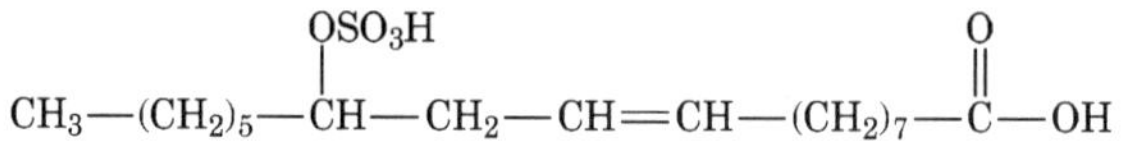

(104)

Manufacturing procedures for producing dye dispersions are generally not disclosed. The principal dispersants in use include long-chain alkyl sulfates, alkaryl sulfonates, fatty amine–ethylene oxide condensates, fatty alcohol–ethylene oxide condensates, naphthalene–formaldehyde–sulfuric acid condensates, and the lignin sulfonic acids.

All dispersions are thermodynamically unstable, since the interfacial area and hence the surface energy tend to decrease, ie, agglomeration occurs. The primary function of dispersing agents is to stabilize dispersions. Anionic dispersing agents, such as the lignosulfonates, are adsorbed on the minute dye particles creating negatively charged surfaces. On close approach the particles then tend to repel each other preventing rapid agglomeration and thereby promoting dispersion stability. A significant development has been production of dispersing agents which confer improved dispersion stability at elevated temperatures and in the case of lignin dispersants (64) products with less tendency to stain the fiber or cause reduction of the dye have been introduced. An excellent review of developments in disperse dye technology has been given (65).

Various mechanical means of reducing the dye particle size have been employed to improve the quality of dispersions. An early process in Germany, and now outmoded, involved kneading a viscous taffylike aqueous paste mixture of dyestuff and dispersing agent in a mixer containing two heavy, closely fitting sigmoid blades. The blades turning at different speeds in opposite directions created a shearing action to break down the dye crystals. This method was largely replaced by the ball mill. Generally, the wet dye filter cake, dispersing agent, and some water are added into the ball mill which is a large horizontal, steel, cylindrical tank partially filled with ceramic balls. The cylinder is rotated for a period of time sufficient for the balls to grind the dye to the desired particle size.

A highly efficient device now in widespread use is the sand mill, which consists of a vertically cylindrical tank equipped with a specially designed impeller and containing sand. The aqueous dye slurry being pumped into the cylinder is subjected to the high shearing action of the swirling sand (66). Other grinding media, such as glass beads, carborundum, and zircon, can be used in place of sand. Sand mills can be the open type with the dye slurry flowing out by gravity or the closed type in which the dye slurry throughput is constantly under pressure. Modern dispersion machinery and methods for the evaluation of dye dispersions have been described (67).

Disperse dyes are sold in liquid form as aqueous pastes, or in powder form (by spray drying). The particle size distribution of commercial disperse dyes is mostly in the range of 1–2 μm which is a decided improvement when compared with the fineness of the early disperse dyes for acetate. Disperse dyes are now formulated to withstand longer term exposure to the prevailing higher dyeing temperatures and must pass stringent filtration and dyeing tests before they are offered for sale. Disperse dyes in liquid form contain additives and stabilizers to prevent settling, freezing and drying out, and control foaming. Since practically all dyes are dispersed by wet-milling processes, spray drying is used to obtain dye powders. In addition to the additives and stabilizers in the liquid types, the powder brands are either treated or prepared in the form of grains to minimize dusting in handling (68).

Application Techniques, Structural Variations, and Fastness Properties. When applied to polyester fiber, many of the disperse dyes originally developed for cellulose acetate were found to be deficient in lightfastness, build-up properties, and especially fastness to the high temperatures employed in the newer dyeing and finishing, printing, and Thermosol (dry heat) processes.

Since 1950, there has been a steady development of new disperse dyes to meet the demands imposed by the changing application methods and to provide the much needed improvement in fastness properties. Six different methods of applying disperse dyes have been developed since the introduction of polyester fibers (69): (*1*) Dyeing at the boil in the presence of a carrier is used for delicate fabrics, polyester–wool blends, etc. (*2*) Dyeing at 120–135°C in pressurized vessels gives better exhaustion and often improved fastness to light, rubbing, and perspiration. (*3*) Thermofixation techniques at 190–220°C are used for the continuous processing of certain types of fabrics. (*4*) Transfer printing, generally at 210°C for 30 seconds, is an important development. (*5*) Solvent dyeing methods are available, but are not popular. (*6*) Printing and continuous dyeing processes have been developed for polyester–cotton blends using specialist dyes and application techniques such as the Dybln (DuPont), Cellestren (BASF), and Dispersol (ICI) ranges.

Disperse dyes are classified as high energy or low energy types (70). The use of higher dyeing temperatures for polyester fibers compared with those used for cellulose acetate has made possible the use of dyes of higher molecular weight, the so-called high energy dyes. The sublimation fastness has been improved by increasing the molecular size. For example, an existing dye, CI Disperse Yellow 9 [*6373-73-5*] (**105**) (CI 10375) has been diazotized and coupled to a substituted phenol to give CI Disperse Yellow 70 [*12223-91-5*] (**106**) (CI 11900) which has better sublimation fastness. Low energy dyes have lower fastness to sublimation, diffuse rapidly into the fiber, and readily produce level dyeings at lower temperatures.

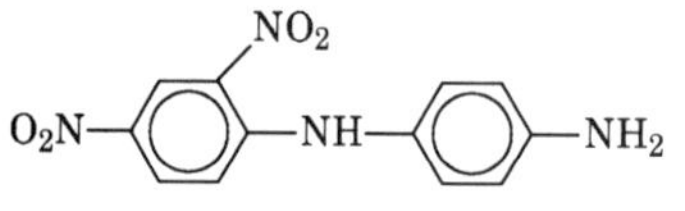

(**105**)

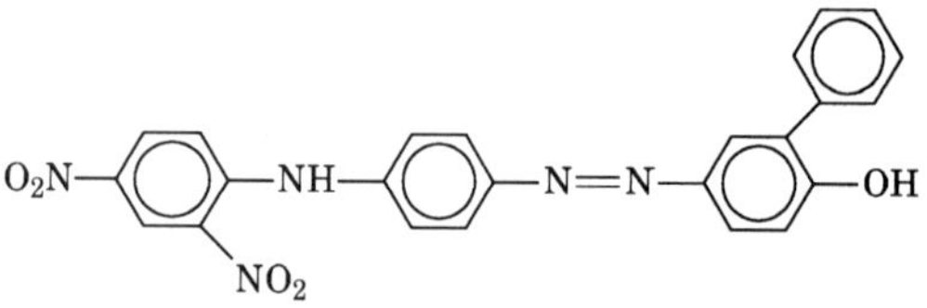

(**106**)

High energy dyes are required in the Thermosol application and in those instances in which the dyed fabric is subjected to a heat-setting treatment. Each energy type has a characteristic rate of exhaust and, as a consequence, only disperse dyes of the same energy class are used in mixes for shade matching purposes.

A use for disperse dyes which has undergone rapid growth since 1970 is in inks for the heat-transfer printing of polyester, especially double knit fabrics. This simple method consists of printing the desired design on paper and then transferring the design from the paper to the fabric with heat. In the heat-transfer process the dye volatilizes, is adsorbed onto the fiber surface, and then diffuses into the fiber. Generally low and medium energy types are used for this purpose. To make color mixes, dyes with the same transfer rate, not necessarily the same energy class, must be selected. There are three main groups of azo disperse dyes. A selection of aminoazobenzene and disazo dye structures (**107–115**) for coloring polyester appears in Figure 7 (71).

All of the disperse dyes for polyester shown in Figure 7 are reported to have high resistance to sublimation, good lightfastness, and good build-up of color. Generally the dye structures have been modified to achieve these desirable properties by either increasing the molecular weight or introducing more polar groups, or both, and by decreasing further their slight solubility in water. The hydrophilic *N*-β-hydroxyethyl group present in each of the acetate disperse dyes (**91**), (**99**), and (**101**) is not to be found in the structures (**107–115**). At the higher temperatures employed in the dyeing of polyester, dyes containing the *N*-β-hydroxyethyl moiety have increased water-solubility, do not exhaust completely from the dye liquor, and as a consequence lack good build-up. Also, aminoazobenzene disperse dyes containing *N*-β-hydroxyethyl groups have relatively poor lightfastness on polyester. These shortcomings can be overcome by

Monoazo dyes

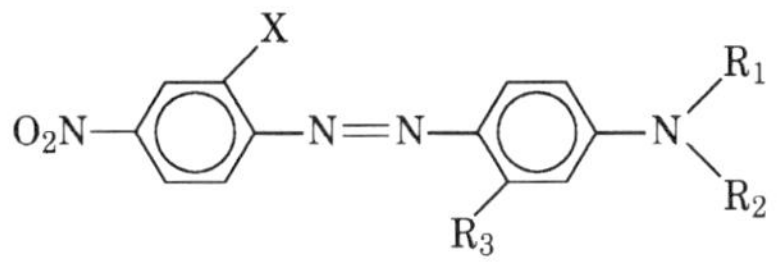

Structure Number	Color	X	R_1	R_2	R_3	CAS Registry Number	Reference
(**107**)	orange	H	CH_2CH_2CN	CH_2CH_2CN	H	[*4234-72-4*]	72
(**108**)	scarlet	Cl	CH_2CH_2CN	$CH_2CH_2OCOCH_3$	H	[*6021-61-1*]	73
(**109**)	red	SO_2CH_3	$CH_2CH_2OCOCH_3$	$CH_2CH_2OCOCH_3$	H	[*26692-47-7*]	74
(**110**)	dark red	Cl	CH_2CH_2CN	CH_2CH_3	$NHCOCH_3$	[*65605-39-2*]	75
(**111**)	rubine	CN	CH_2CH_2CN	$CH_2CH_2OCOCH_3$	H	[*65605-40-5*]	73

Disazo dyes

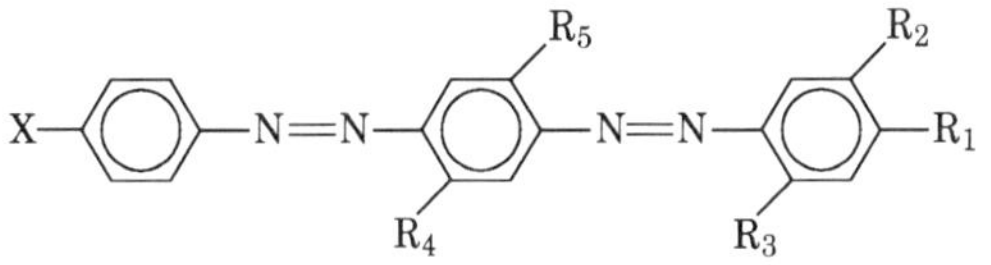

Structure Number	Color	X	R_1	R_2	R_3	R_4	R_5	Reference
(**112**)	yellow	$NHCOCH_3$	OH	H	H	CH_3	H	76
(**113**)	orange	NO_2	OH	CH_3	H	CH_3	H	77
(**114**)	yellowish red	H	N(CH_2CH_2CN)($CH_2CH_2OCOCH_3$)	H	$NHCOCH_3$	H	H	78
(**115**)	red	NO_2	N(CH_2CH_2CN)($CH_2CH_2OCOCH_3$)	H	H	H	H	79

Fig. 7. Azo disperse dyes for polyester.

acylation of the hydroxyl group to form, eg, the *N*-acetoxyethyl group present in a number of the structures in Figure 7. The introduction of the polar *N*-β-cyanoethyl group into the molecule yields dyes with good fastness to light and sublimation. The combination of *N*-β-cyanoalkyl and *N*-β-acyloxyalkyl groups in a coupling component for disperse azo dye is particularly advantageous (80).

The use of three electron-withdrawing groups in the 2-, 4-, and 6-positions of the diazo component and two electron-donating groups in the 2- and 5-positions of the aminobenzene coupler has afforded blue azo disperse dyes for polyester with excellent build-up, good fastness to sublimation, and moderate to good lightfastness. Compared with aminoanthraquinone blue dyes, these dyes are relatively dull in shade and not as lightfast. However, the azo types are used with success in heavy navy and black shades (80). Besides Disperse Blue 79 (**100**), other representative chemical structures (**116–119**) for blue azo disperse dyes are as follows:

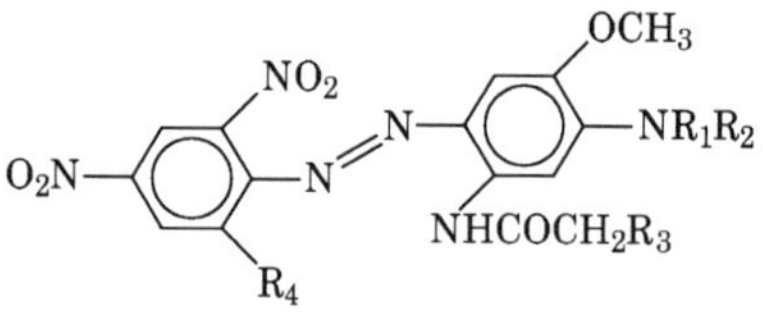

	R_1	R_2	R_3	R_4	CAS Registry Number
(**116**)	$CH_2CH_2OCH_2CH_2CN$	$CH_2CH_2OCH_2CH_2CN$	H	Cl	[*16539-95-0*]
(**117**)	CH_2— 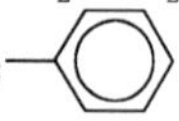	CH_2— 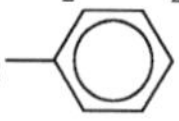	H	CN	[*52583-51-4*]
(**118**)	$CH_2CH_2OCH_2CH_2CN$	CH_2— 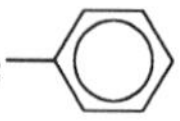	CH_3	Cl	[*65605-44-9*]
(**119**)	CH_2CH_2CN	C_2H_5	H	Br	[*22578-86-5*]

The patented method of preparation of the blue dye (**120**) [*19187-01-0*] (81) involves treating the analogous dibromo substituted azo dye with cuprous cyanide in dimethylformamide or *N*-methylpyrrolidinone at 50°C to effect replacement of the two bromo substituents by cyano groups.

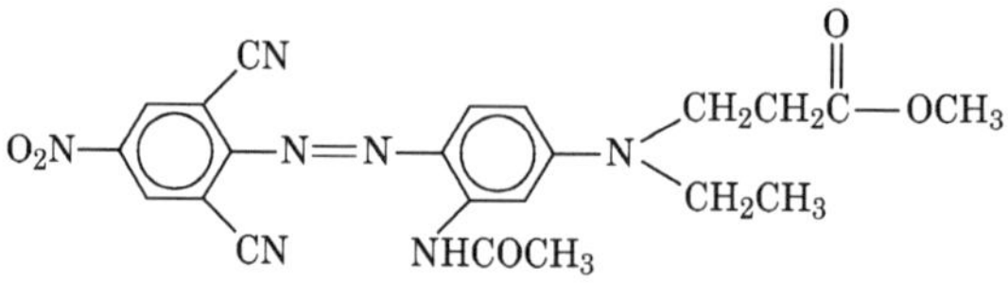

(**120**)

The greenish-blue dye (**117**) (82) is prepared in a similar fashion, replacing bromo with cyano by using cuprous cyanide, pyridine, and 2-methoxyethanol as solvent at 85°C.

Heterocyclic Disperse Dyes. Diazotizable aminoheterocyclic compounds are also used in the production of disperse dyes. Examples of the important class of 2-aminobenzothiazole dyes follow. Red shades include (**121–123**):

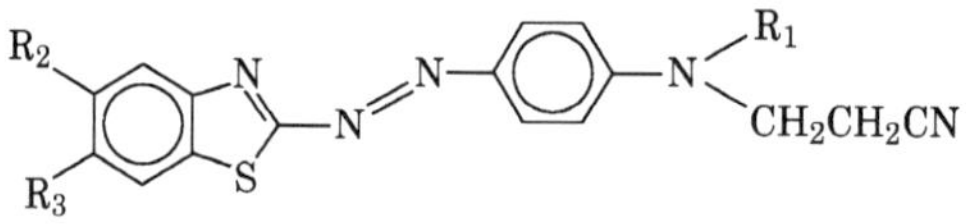

	R_1	R_2	R_3	CAS Registry Number	Reference
(**121**)	C_2H_5	Cl	Cl	[*25176-89-0*]	83
(**122**)	CH_2— 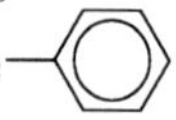	H	NO_2	[*65605-45-0*]	84
(**123**)	$CH_2CH_2OOCCH_3$	H	SO_2CH_3	[*54983-30-1*]	85

(**124**) is bluish red and (**125**) is bright blue.

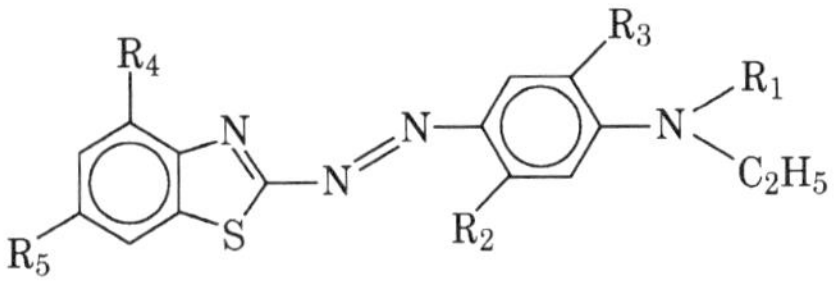

	R_1	R_2	R_3	R_4	R_5		Reference
(**124**)	$CH_2CH_2OOCC_6H_5$	CH_3	H	H	SO_2CH_3	[*43042-04-2*]	86
(**125**)	C_2H_5	$NHCOCH_3$	OCH_3	OCH_3	NO_2	[*19433-87-5*]	87

The absorption spectra, and fastness properties of a number of substituted aniline dyes have been compared with those of the similarly substituted benzothiazolylazo dyes (88). For instance, the dye prepared from 4-nitroaniline [*100-01-6*] and a given coupler was compared with the dye made from 2-amino-6-nitrobenzothiazole [*6285-57-0*] and the same coupler since 4-nitroaniline yields 2-amino-6-nitrobenzothiazole on treatment with potassium thiocyanate and bromine in acetic acid. The benzothiazolylazo dyes are deeper in shade (bathochromic) and have higher absorptivities, ie, greater tinctorial strength than their phenylazo counterparts. Further, the benzothiazole types generally have higher fastness to light fading and exhibit definitely superior sublimation fastness.

A large bathochromic shift is observed in the blue-green dye (**126**) [*65605-46-1*] (89) prepared from the diazo component 3-amino-5-nitro-2,1-benzisothiazole.

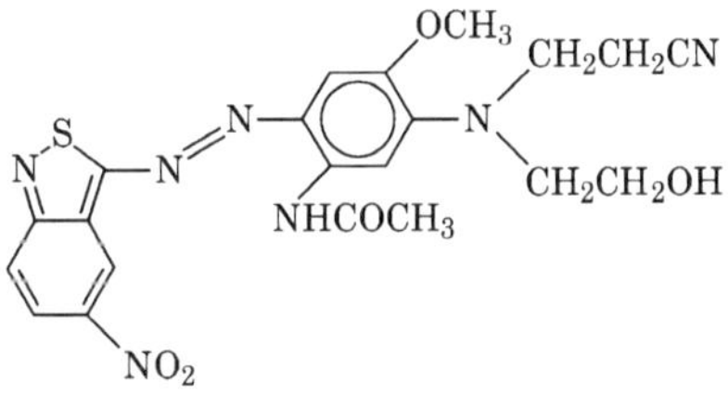

(**126**)

Couplers which form scarlet dyes with 4-nitroaniline and red dyes with 2-amino-6-nitrobenzothiazole yield blue dyes with 3-amino-5-nitro-2,1-benzisothiazole [*14346-19-1*].

A number of other heterocyclic diazo components such as thiazole, indazole, thiophenes, and thiadiazole types (see Fig. 1), as well as heterocyclic couplers, ie, 6-hydroxy-2-pyridinone [*626-06-2*], barbituric acid [*67-52-7*], and tetrahydroquinoline [*25448-04-8*] have been cited in the literature (90,91). Reviews on disperse dyes have been published (92,93).

OIL SOLUBLE AZO DYES

The oil soluble, water-insoluble, azo dyes dissolve in oils, fats, waxes, etc. Generally, yellow, orange, red, and brown oil colors are azo structures and greens, blues, and violets are primarily anthraquinones (see DYES, ANTHRAQUINONE). Blacks are usually nigrosines and indulines of the azine type (see AZINE DYES). An example is Oil Red [*85-83-6*] (**127**) (CI Solvent Red 24; CI 26105). Uses include the coloring of hydrocarbons, waxes, oils, candles, etc.

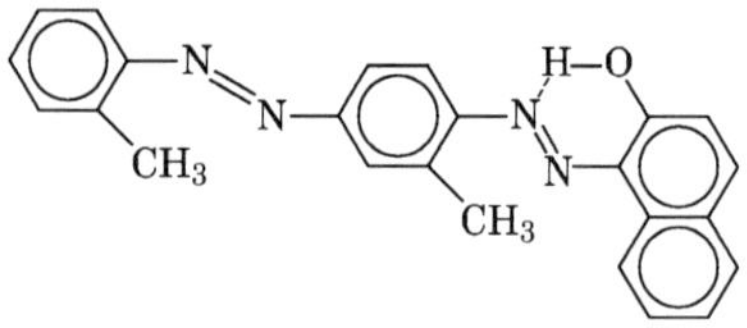

(**127**)

Substitution by chloro, nitro, and similar groups increases the molecular weight and improves sublimation fastness but lowers the oil solubility of this group of dyes.

SPIRIT SOLUBLE AZO DYES

Spirit-soluble azo dyes dissolve in polar solvents, such as alcohol and acetone, and find application in the coloring of lacquers, plastics, printing inks, and ball-point pen inks. Of the two principal types of azo structures used, the most important are the insoluble salts of azo dyes containing sulfo groups and relatively complex organic amines. Mono- and dicyclohexylamine, isoamylamine, and the arylguanidines (94) often serve as the amine, and the anionic component is chosen from the class of acid dyes for wool. An example is Zapon Fast Scarlet CG [*5413-75-2*] (**128**) (CI Acid Red 73; CI 27290).

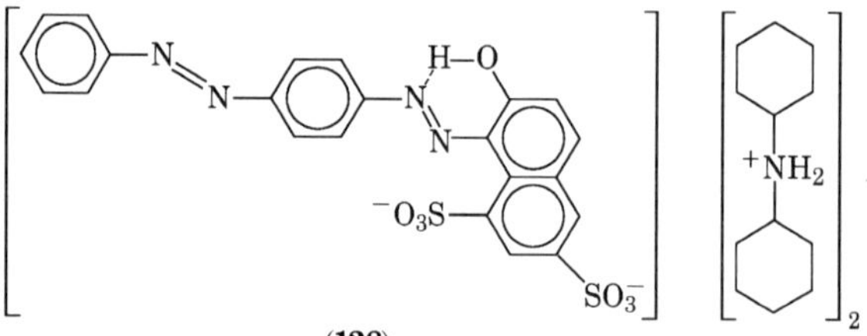

(**128**)

The second type is comprised of 2:1 metal complexes of *o,o′*-dihydroxy azo dyes which generally do not contain sulfo or other strongly hydrated groups as found in the premetallized 2:1 complexes for wool. Thus their solubility in esters, ketones, and alcohols is relatively increased. CI Solvent Violet 1 [*6421-64-3*] (**129**) (CI 12196), a 2:1 cobalt complex, illustrates this second type.

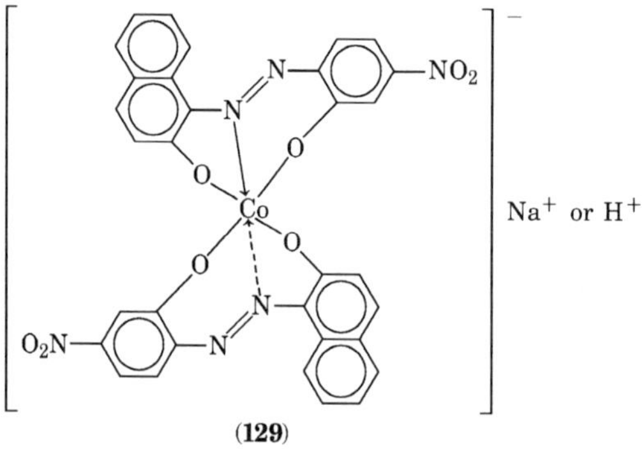

(**129**)

BASIC (CATIONIC) AZO DYES

Basic dyes of the azo class are the simplest and oldest known synthetic dyes. Earlier dyes, which were used for paper, leather, and in dyeing and printing of tanning-mordanted cotton, had poor to moderate lightfastness and washfastness. When polyacrylonitrile fibers were developed, these dyes gained importance because they showed superior lightfastness. Current cationic dyes are used for modified acrylics, modified nylons, modified polyesters, leather, unbleached papers, and inks. An important application is for conversion into pigments. Acrylics, nylon, and polyester fibers are modified by incorporation of acidic groups as dye sites in the fiber to increase its ability to be dyed with basic dyes. Principal chemical classes include azo, anthraquinone, triarylmethane, methine, thiazine, oxazine, etc (95). The dyes are applied in acidic dyebaths to fibers made of negatively charged polymer molecules.

Cationic azo dyes carry a positive charge in the chromophore portion of the molecule. The salt-forming counterion is usually a chloride or acetate. CI basic dyes are ammonium, sulfonium, or oxonium salts. Commercial basic azo dyes for which chemical structures are revealed by U.S. producers are listed in Table 15.

Basic Orange 1 (**130**) (aniline coupled to 2,4-diaminotoluene) and Basic Orange 2 (**22**) (aniline coupled to *m*-phenylenediamine) are examples of amine salt

Table 15. Commercial Basic Azo Dyes, U.S. 1988

CI name	Structure number	CAS Registry Number	CI number
Basic Yellow 15		*[72208-25-4]*	11087
Basic Yellow 24		*[52435-14-0]*	11480
Basic Yellow 25	(**136**)	*[41025-67-6]*	11450
Basic Yellow 57		*[68391-31-1]*	12719
Basic Orange 1	(**130**)	*[4438-16-8]*	11320
Basic Orange 2	(**22**)	*[532-82-1]*	11270
Basic Red 18	(**131**)	*[14097-03-1]*	11085
Basic Red 22	(**134**)	*[12221-52-2]*	11055
Basic Red 24		*[37216-10-7]*	11088
Basic Red 29	(**135**)	*[42373-04-6]*	11460
Basic Red 39		*[12221-63-5]*	11465
Basic Red 76		*[68391-30-0]*	12245
Basic Blue 41	(**133**)	*[12270-13-2]*	11105
Basic Blue 54	(**132**)	*[15000-59-6]*	11052
Basic Blue 65		*[12221-37-3]*	11076
Basic Blue 66		*[12221-38-4]*	11075
Basic Blue 67		*[12221-39-5]*	11185
Basic Brown 1		*[10114-58-6]*	21000
Basic Brown 2		*[6358-83-4]*	21030
Basic Brown 4		*[5421-66-9]*	21010
Basic Brown 16		*[26381-41-9]*	12250
Basic Brown 17		*[71134-97-9]*	12251
Basic Black 2		*[4443-99-6]*	11825

type cationic azo dyes. The cation is formed by protonation under acidic conditions. Under neutral or alkaline conditions, these dyes behave more like disperse dyes. In 1988 the U.S. production of CI Basic Orange 2 amounted to 132 tons.

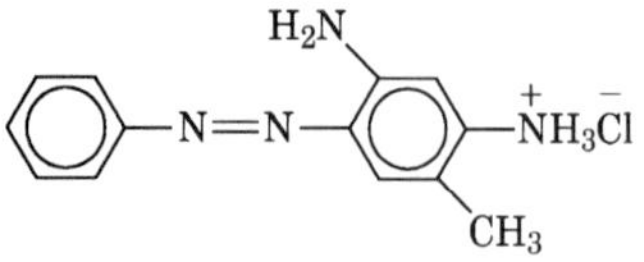

(130)

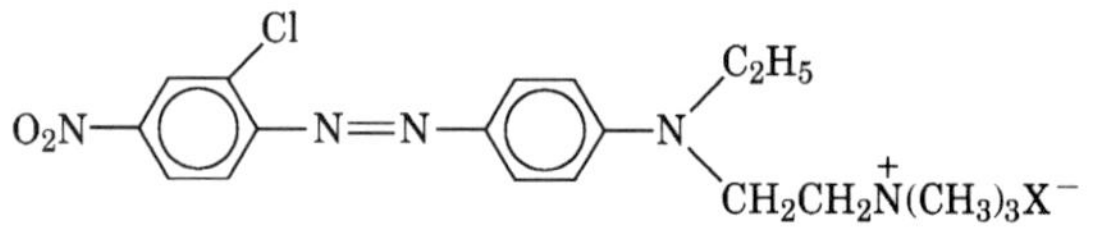

(131)

Basic Red 18 (**131**), Basacryl Red X-NL [*14097-03-1*] is an example of a pendant cationic azo dye, in which a localized positive charge is not conjugated with the chromophoric system. *N*-Ethyl-*N*-(2-chloroethyl)aniline [*92-49-9*] reacts with trimethylamine to form the ammonium salt coupler. The diazo component in Basic Red 18 is 2-chloro-4-nitroaniline [*121-87-9*].

So-called delocalized cationic azo dyes are another type in which the positive charge is delocalized (or distributed) across the dye cation. Basic Blue 54 (**132**) (2-amino-6-methoxybenzothiazole coupled to *N,N*-dimethylaniline and then quarternized with dimethyl sulfate) is an example of this class which can also be considered as a diazamethine dye.

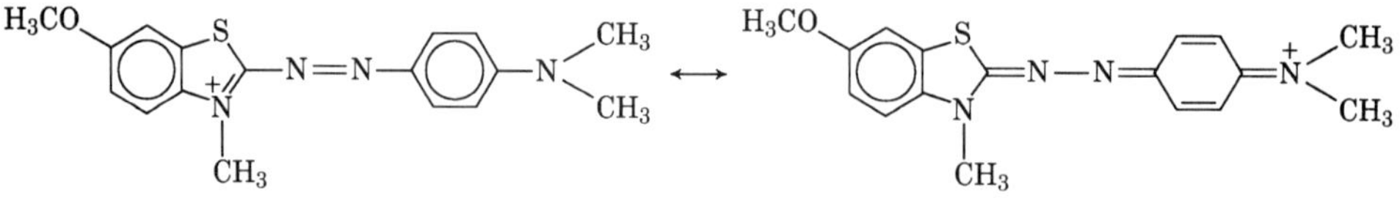

(132)

Another bright blue dye from diazotized 2-amino-6-methoxybenzothiazole [*1747-60-0*] by azo coupling, eg, with 2(*N*-ethylanilino)ethanol is Basacryl Blue X-3GL [*12270-13-2*] (**133**) (CI Basic Blue 41; CI 11105). After coupling, the water-insoluble dye is methylated at the thiazole nitrogen.

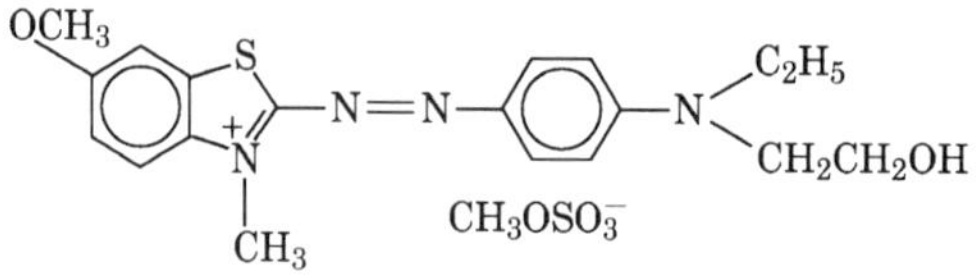

(133)

Basic Red 22 (**134**), which contains 1 part in 7 of the yellowish red 1,4-dimethyl isomer, Basic Red 29 (**135**), and Basic Yellow 25 (**136**) are all examples of delocalized cationic azo dyes. Dyes of this type can also be synthesized by Hünig's oxidative coupling reaction of heteroaromatic hydrazones with tertiary aromatic amines.

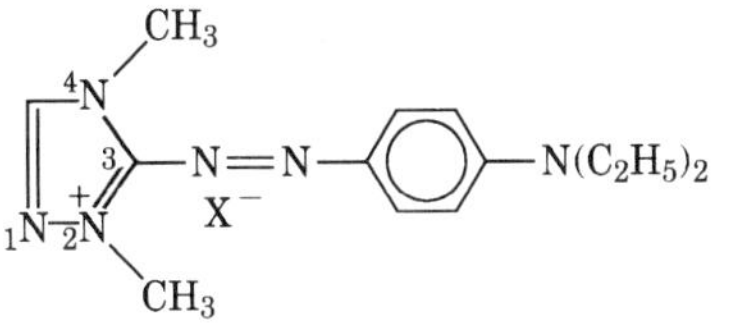

(134)

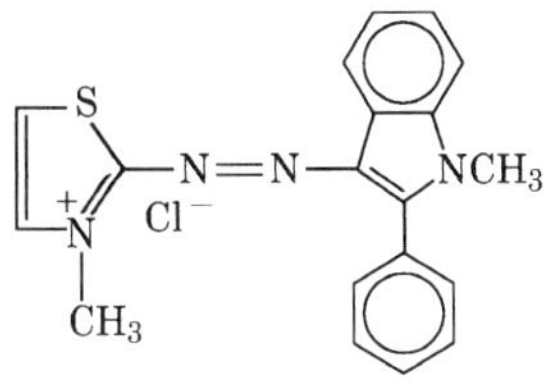

(135)

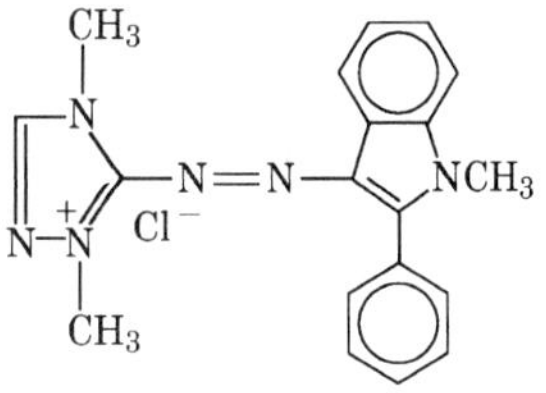

(136)

AZO PIGMENTS

Organic pigments are an important class of organic colorants. Expanding areas of usage include the mass coloration of synthetic fibers and textile printing in the textile field, and in the nontextile area, plastics. A pigment is insoluble in the medium in which it is used. The physical properties of pigments are of great significance since the coloring process does not involve solution of the colorant. Azo pigments can be grouped as metal toners, metal chelates, and metal-free azo pigments.

The first synthetic organic pigments were used to shade or tone the weaker colorants and became known as toners. Metal toners usually contain one sulfonic acid group and often a carboxylic acid group. The pigment is rendered insoluble, ie, *laked* with a heavy metal cation. An example of a calcium salt is Lithol Rubine BK [*5858-81-1*] (**137**) (CI Pigment Red 57; CI 15850).

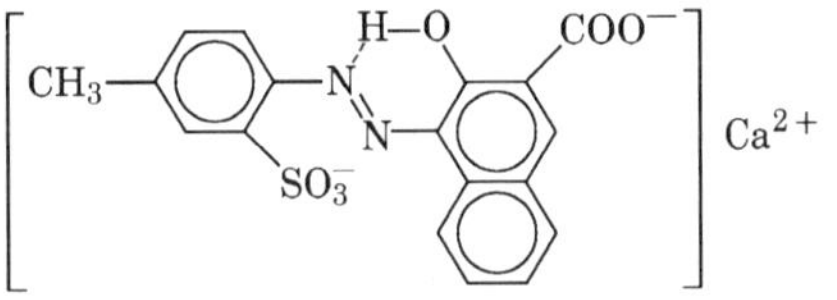

(137)

Toners derived from 6-chlorometanilic acid [*88-43-7*], 6-amino-4-chloro-*m*-toluenesulfonic acid [*88-51-7*], and 6-amino-*m*-toluenesulfonic acid [*88-44-8*] have improved fastness properties and find use in paints, inks, and plastics.

Metal chelation is also a means of insolubilizing organic molecules. For example, CI Pigment Green 10 [*51931-46-5*], (**138**) (CI 12775) is a 2:1 nickel complex of a bidentate *o*-hydroxyazo ligand.

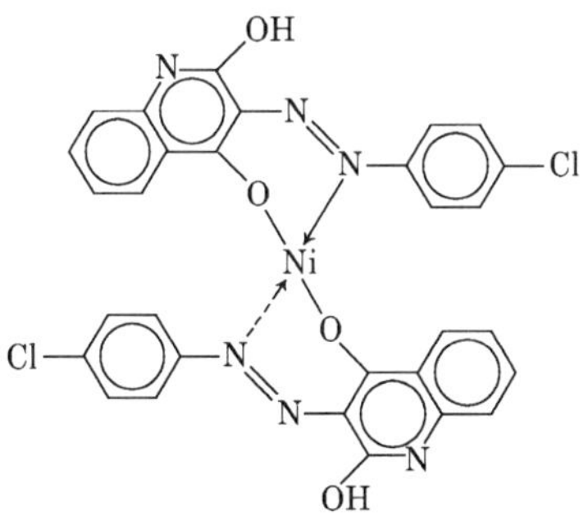

(138)

In the metal-free class of azo pigments it is remarkable that the simplest derivatives of 2-naphthol such as Hansa Red B [*2425-85-6*] (**139**) (CI Pigment Red 3; CI 12120, Toluidine Red) and Para Red [*6410-10-2*], both known since 1905, are still of importance.

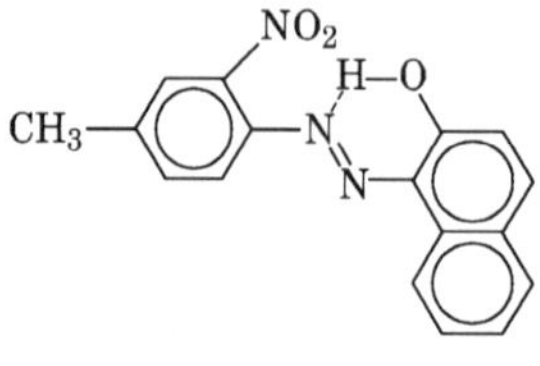

(**139**)

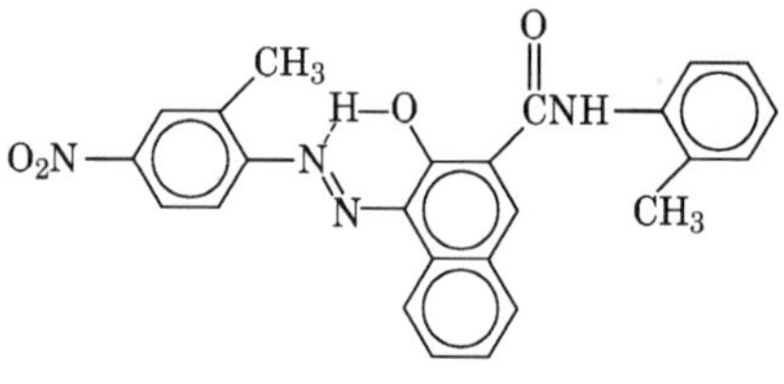

(**140**)

Because these pigments are organic in nature they tend to bleed in resins and solvents. Increasing the molecular weight often reduces this tendency. For this reason, I.G. Farbenindustrie introduced a line of pigments based on Naphthol AS as the coupling component, eg, Permanent Bordeaux FRR Extra [*6410-32-8*] (**140**) (CI Pigment Red 12; CI 12385).

In the same way acetoacetanilides and pyrazolones can be used as coupling components. Larger molecules can be formed in the pigment series by using a benzidine derivative as the middle component as, eg, in Vulcan Fast Red B [*6358-87-8*] (**141**) (CI Pigment Red 38; CI 21120).

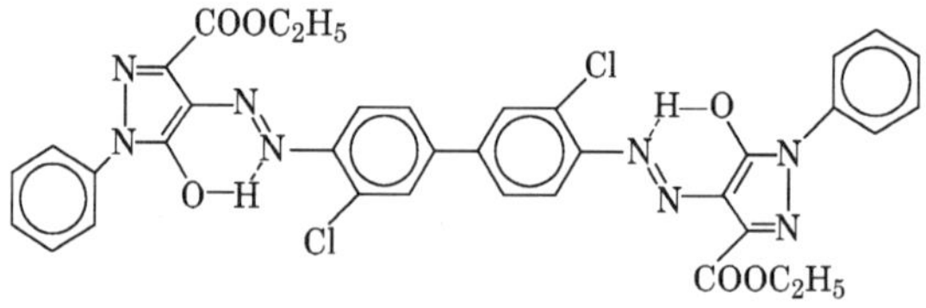

(**141**)

A survey of organic pigments has been done (96). A more recent publication on organic pigments (97), describes their syntheses, chemical and physical properties, and the technology of the various applications.

BIBLIOGRAPHY

"Azo Dyes" in *ECT* 1st ed., Vol. 2, pp. 226–268, by J. C. Moessinger and Shirley K. Morse; in *ECT* 2nd ed., Vol. 2, pp. 868–910, by R. F. Johnson, A. Zenhaüsern, and H. Zollinger, Swiss Federal Institute of Technology; in *ECT* 3rd ed., Vol. 3, pp. 387–433, by Sigmund C. Catino and Russell E. Farris (GAF Corporation).

1. *Colour Index,* 3rd ed., Vol. 8, The Society of Dyers and Colourists, Bradford, Yorkshire, UK, 1987.
2. K. Venkataraman, *The Chemistry of Synthetic Dyes,* Vols. 1 and 2, Academic Press, New York, 1952.
3. H. E. Fierz-David and L. Blangey, *Grundlegende Operationen der Farbenchemie,* 8th ed., Vienna, Austria, 1952.

4. P. Bersier, L. Valpiana, and H. Zubler, *Chem. Ing. Tech.* **43,** 1311 (1971).
5. R. Pütter, in *Houben-Weyl, Methoden der Organischen Chemie,* 4th ed., Vol. 10, Part 3, Georg Thieme, Stuttgart, Germany, 1965, p. 51.
6. U.S. Pat. 4,255,326 (May 12, 1981), R. Giles and M. Weaver (to Eastman Kodak Co.).
7. R. N. Butler, *Chem. Rev.* **75,** 241 (1975).
8. J. Kavalek, K. Janak, and V. Sterba, *Collect. Czech. Chem. Commun.* **44,** 3102 (1979).
9. J. Goerdeler and H. Haubrich, *Chem. Ber.* **93,** 397 (1960).
10. O. Dann, *Chem. Ber.* **82,** 72 (1949).
11. K. H. Saunders and R. L. M. Allen, *Aromatic Diazo Compounds,* Edward Arnold, London, UK, 1985.
12. H. Zollinger, *Diazo and Azo Chemistry,* Interscience Publishers Inc., New York, 1961.
13. U.S. Pat. 3,793,305 (Apr. 20, 1972), W. J. Balon (to E. I. du Pont de Nemours & Co., Inc.).
14. U.S. Pat. 3,792,077 (Feb. 12, 1974), R. Bermes and K. Schmeidl (to BASF A.G.).
15. E. D. Hughes, C. K. Ingold, and J. H. Ridd, *J. Chem. Soc.* **58–86** (1958).
16. J. H. Ridd, *Quart. Rev. (London)* **15,** 418 (1961); *J. Soc. Dyers Colour.* **81,** 355 (1965); *Adv. Phys. Org. Chem.* **16,** 1 (1978).
17. D. L. H. Williams, *Adv. Phys. Org. Chem.* **19,** 381 (1983).
18. J. Fitzpatrick and co-workers, *J. Chem. Soc. Perkin Trans. 2,* 927 (1984).
19. H. Zollinger and C. Wittwer, *Helv. Chim. Acta* **35,** 1209 (1952).
20. H. E. Fierz-David and H. Brütsch, *Helv. Chim. Acta* **4,** 380 (1921).
21. L. Gattermann and H. Liebermann, *Ann. Chem.* **393,** 198 (1912).
22. O. A. Stamm and H. Zollinger, *Helv. Chim. Acta* **40,** 1105, 1955 (1957).
23. R. Kaminski, U. Lauk, P. Skrabal, and H. Zollinger, *Helv. Chim. Acta* **66,** 2002 (1983).
24. H. Belevi, J. R. Bourne, and P. Rys, *Helv. Chim. Acta* **64,** 1599, 1618 (1981).
25. J. R. Penton and H. Zollinger, *Helv. Chim. Acta* **54,** 573 (1971).
26. Y. Hasida, H. Ishida, S. Sekiguchi, and K. Matsui, *Bull. Chem. Soc. Jpn.* **47,** 1224 (1974).
27. B. Gloor and H. Zollinger, *Helv. Chim. Acta* **54,** 553, 563 (1971); F. Snyckers and H. Zollinger, *Tetrahedron Lett.* **32,** 2759 (1970).
28. Z. J. Allan and F. Muzik, *Collect. Czech. Chem. Commun.* **23,** 1927 (1958).
29. T. S. Gore and K. Venkataraman, *Proc. Indian Acad. Sci.* **34**A, 368 (1951).
30. H. F. Hodson, O. A. Stamm, and H. Zollinger, *Helv. Chim. Acta* **41,** 1816 (1958).
31. *Textile Chemist and Colorist Buyer's Guide,* Vol. 21, No. 7, American Association of Textile Chemists and Colorists, Research Triangle Park, N.C., July, 1989.
32. C. V. Stead, in K. Venkataraman, ed., *The Chemistry of Synthetic Dyes,* Vol. 3, Academic Press, New York, 1970, pp. 249–301.
33. I. Szele and H. Zollinger, *Curr. Chem.* **112,** 1 (1983).
34. H. Zollinger, *Acc. Chem. Res.* **6,** 335 (1973).
35. J. M. Tedder, in Ref. 32, pp. 223–248.
36. T. Asano and T. Okada, *J. Org. Chem.* **49,** 4387 (1984).
37. U.S. Pat. 3,905,949 (Sept. 16, 1975), M. A. Perkins and H. K. Urion (to E. I. du Pont de Nemours & Co., Inc.).
38. S. Hünig, *Angew. Chem.* **70,** 215 (1958).
39. U.S. International Trade Commission, *Synthetic Organic Chemicals, Dyes: Production and Sales* (Sect. 4), Government Printing Press, Washington D.C., 1980–1988.
40. U.S. International Trade Commission, *Imports of Benzenoid Chemicals & Products, Section: Benzenoid Dyes,* Government Printing Press, Washington D.C., 1983.
41. J. Meybeck and P. Galafassi, *Appl. Polym. Symp.* **18,** 463 (1971).
42. R. Price, in Ref. 32, pp. 303–383.
43. F. Beffa and E. Steiner, *Rev. Progr. Color Relat. Top.* **4,** 60 (1973).
44. F. Beffa and G. Back, *Rev. Progr. Color Relat. Top.* **14,** 33 (1984).
45. G. Schwarzenbach and W. Biedermann, *Helv. Chim. Acta* **35,** 716 (1952).

46. H. Pfitzner, *Angew. Chem.* **62,** 242 (1950).
47. H. Jaggi, *Helv. Chim. Acta* **51,** 580 (1968).
48. E. Race, F. M. Rowe, and J. B. Speakman, *J. Soc. Dyers Colour.* **62,** 372 (1946).
49. H. E. Fierz-David and E. Mannhard, *Helv. Chim. Acta* **20,** 1024 (1937).
50. G. Schetty, *Helv. Chim. Acta* **50,** 1039 (1967).
51. R. Casty, *Melliand Textilber.* **33,** 950 (1952).
52. U.S. Pat. 2,551,056 (May 1, 1957), G. Schetty (to J.R. Geigy A.G.).
53. G. Schetty, *J. Soc. Dyers Colour.* **71,** 705 (1955).
54. U.S. Pat. 2,885,392 (May 5, 1959), E. Csendes and I. W. Dobratz (to E. I. du Pont de Nemours & Co., Inc.).
55. H. Pfitzner, *Melliand Textilber.* **35,** 650 (1954).
56. J. Boulton, *J. Soc. Dyers Colour.* **67,** 522 (1951).
57. T. Robinson and W. B. Egger, *Text. Chem. Color.* **15**(10), 189 (1983).
58. J. Shore, *Rev. Progr. Color. Relat. Top.* **11,** 58 (1981).
59. Ref. 1, Vol. 2, 1971, p. 2480.
60. Brit. Pat. 182,830 (Jan. 12, 1921), R. Clavel.
61. Brit. Pat. 211,720 (Mar. 8, 1923), J. Baddiley and A. Shepherdson.
62. Brit. Pat. 219,349 (Jan. 27, 1923), G. Ellis.
63. G. H. Ellis, *J. Soc. Dyers Colour.* **40,** 255 (1924).
64. G. Prazak, *Am. Dyest. Rep.* **59,** 44 (1970).
65. S. Heimann, *Rev. Progr. Color. Relat. Top.* **11,** 1 (1981).
66. G. D. Parfitt, *Dispersions of Powders in Liquids,* Elsevier Publishing Co., Inc., New York, 1969, p. 259.
67. H. Leube and H. Uhrig, *Can. Text. J.* **91,** 57 (1974).
68. A. N. Derbyshire and A. T. Leaver, *J. Soc. Dyers Colour.* **91,** 253 (1975).
69. J. F. Dawson, *J. Soc. Dyers Colour.* **99,** 183 (1984).
70. P. L. Munier and S. N. Boyd, Jr., *Text. Chem. Color.* **3,** 278 (1971).
71. J. M. Straley, in Ref. 32, pp. 385–462.
72. U.S. Pat. 2,782,187 (Feb. 19, 1957), M. F. Sartori (to E. I. du Pont de Nemours & Co., Inc.).
73. U.S. Pat. 2,891,942 (June 23, 1959), E. Merian (to Saul and Co.).
74. U.S. Pat. 3,497,495 (Feb. 24, 1970), D. G. Carmichael (to Martin Marietta Corp.).
75. U.S. Pat. 2,971,953 (Feb. 14, 1961), P. Rhyner (to Ciba, Ltd.).
76. Brit. Pat. 128,804 (1969) (to Yorkshire Chemicals, Ltd.).
77. Brit. Pat. 1,533,121 (1975) (to Imperial Chemicals Industries).
78. Brit. Pat. 1,395,791 (1971) (to Sandoz, Ltd.).
79. U.S. Pat. 3,661,888 (May 9, 1972), J. Sagal (to Martin Marietta Corp.).
80. S. Al-kassim and A. T. Peters, *J. Soc. Dyers Colour.* **89,** 358 (1973).
81. Brit. Pat. 1,125,683 (Aug. 28, 1968), A. Gottschlich and K. Leverenz (to Farbenfabriken Bayer A.G.).
82. U.S. Pat. 3,821,195 (June 28, 1974), D. E. Putzig (to E. I. du Pont de Nemours & Co., Inc.).
83. U.S. Pat. 3,502,645 (Mar. 24, 1970), D. G. Carmichael (to Martin Marietta Corp.).
84. U.S. Pat. 3,483,180 (Dec. 9, 1969), V. Ramanathan (to Ciba, Ltd.).
85. U.S. Pat. 2,785,157 (Mar. 12, 1957), J. M. Straley and D. J. Wallace (to Eastman Kodak Co.).
86. Brit. Pat. 1,324,235 (July 25, 1973), J. F. Dawson (to Yorkshire Chemicals, Ltd.).
87. U.S. Pat. 3,405,118 (Oct. 8, 1968), M. F. Sartori (to E. I. du Pont de Nemours & Co., Inc.).
88. M. F. Sartori, *J. Soc. Dyers Colour.* **83,** 114 (1967).
89. Brit. Pat. 1,112,146 (May 1, 1968); Neth. Appl. 6,608,032 (Dec. 12, 1968), M. Seefelder, H. Wippel, J. Dehnert, and H. Armbrust (to Badische Anilin. & Soda Fabrik A.G.).

90. H. R. Schwander, *Dyes Pigm.* **3,** 133–160 (1982).
91. M. A. Weaver and L. Shutteworth, *Dyes Pigm.* **3,** 81–121 (1982).
92. J. F. Dawson, *Rev. Prog. Colour. Relat. Top.* **14,** 90 (1984).
93. G. Hallas, in J. Griffiths, ed., *Developments in the Chemistry and Technology of Organic Dyes,* Blackwell, Scientific Publishing, Oxford, UK, 1984.
94. U.S. Pat. 1,674,125 (June 19, 1928), R. E. Rose (to E. I. du Pont de Nemours & Co., Inc.).
95. D. R. Baer, in Ref. 32, Vol. 4, 1971.
96. F. M. Smith, *J. Soc. Dyers Colour.* **78,** 222 (1962).
97. W. Herbst and K. Hunger, *Industrielle Organische Pigmente,* VCH Verlagsgesselschaft, Weinheim, Germany 1986.

RASIK J. CHUDGAR
BASF Corporation

B

BABBIT METALS. See BEARING MATERIALS.

BACITRACINS. See ANTIBIOTICS, PEPTIDES.

BAGASSE. See CHEMURGY; SUGAR.

BAKERY PROCESSES AND LEAVENING AGENTS

YEAST-RAISED PRODUCTS

The U.S. baking industry comprises an important portion of the food industry, the nation's largest business. Sales of various bakery foods by the baking industry currently exceed $41 billion per year (1). The industry is diverse in both products made and size of production units. Large manufacturing complexes are increasing, but much production takes place in relatively small plants since distribution of highly perishable bakery foods to the consumer is feasible only within a limited area around the manufacturing facility.

Bakery foods are produced in over 48,000 bakeries throughout the country. Over 45,000 of these facilities are made up of retail and in-store bakeries where products are typically baked and sold directly to the consumer at the same location. Only about 33% of the dollar volume of the baking industry is attributed to these small producing facilities. Of the bakery foods produced and consumed in the United States, about 67% are made in approximately 3000 wholesale bakeries. Although many companies involved in wholesale bakery production operate single plants, many larger regional and national companies operate from two to 53 bakeries, usually located in large population centers. Increasingly, the larger bakeries are becoming more automated and computerized.

Another source of bakery foods is the food service industry. Over 200,000 establishments, including restaurants, hospitals, and prisons, produce bakery foods with a combined value exceeding $8 billion.

Differentiated from the quickly perishable bakery foods are the dry bakery products such as cookies, crackers, pretzels, and ice cream cones. These latter items possess a much longer shelf life and may be distributed over a wider area from typically very large manufacturing facilities. According to the 1987 Census of Manufacturers (2), there are 380 establishments producing these dry-type bakery foods, and the value added by such manufacturing facilities amounts to over $4 billion.

Per capita consumption of all bakery products in the United States was about 50 kg in 1990 (3). Bread accounted for about 45% of this consumption, while all yeast-leavened items comprised approximately 74% of the total. The U.S. Dept. of Commerce's *1991 U.S. Industrial Outlook* projects that per capita consumption of bakery foods will increase 2.2% annually from 1991 to 1996.

Yeast. Bread and other yeast-raised bakery foods are widely consumed and desired, in part, because of their appealing flavor and light, porous texture, properties which depend largely on yeast fermentation. Leavening by the action of yeast was known to the Egyptians as early as 2000 BC; mummified yeast-raised loaves have been found in early Egyptian tombs as well as in the Roman ruins of Pompeii (4,5). The Hebrews also understood the art of leavening, presumably by yeast, because laws concerning the use of both leavened and unleavened bread are given in the book of Exodus in the Bible.

The role of baker's yeast (*Saccharomyces cerevisiae*) in producing leavened bread depends on two factors: the ability of yeast to generate carbon dioxide and alcohol through the breakdown of simple sugars, and the unique ability of wheat flour proteins to form films in dough that trap evolved gases (see FERMENTATION; YEAST). Basic bread is made with flour, water, salt, and yeast. Product variety is achieved by incorporating varying amounts of additional ingredients; by altering the breadmaking process; by shaping or cutting or putting toppings on the dough prior to baking; or by the method of baking. Each of these practices may be utilized alone or in combination to produce a virtually limitless number of yeast-raised products. Thus raisin bread is made from a white-type bread dough with the addition of raisins and possibly spice. French bread is made from a lean white dough, shaped into a cylinder and baked on the hearth of a steam-filled oven. Pita bread is also made from a lean white dough, but it is shaped into thin, round pieces before baking in a very hot (400°C) oven, to achieve the typical pocket formation. Sweet roll doughs are made with relatively high levels of sugar, fat, eggs, and yeast, compared to ordinary bread dough, and are shaped in various ways, with additional fillings and toppings to obtain the desired product.

Many other products belong to the category of yeast-raised bakery foods (6–9). Some that may be cited include various kinds of specialty breads, coffee cakes and danish pastries, bagels, croissants, yeast-raised doughnuts, some types of crackers, English muffins, and rolls. Of the total annual sales of the baking industry, yeast-raised goods constitute about 61%.

Chemical Leavening. Chemical leaveners (primarily baking powders) are utilized in cakes, cookies, some crackers, refrigerated dough, and quick bread manufacturing (9–12). These products make up roughly 39% of the value of the baking industry's total production (see BAKERY PROCESSES AND LEAVENING AGENTS, CHEMICAL LEAVENING AGENTS). Both yeast and chemical leaveners are occasionally used together in the same formulation, for example in the production of certain frozen doughs.

Ingredients

A great many ingredients may be utilized in the production of innumerable bakery foods. The following discussion is limited to a relatively few key ingredients.

Flour. The primary ingredient of most bakery foods is wheat flour (6,13–15). This is especially true in breadmaking, where flour may comprise up to 95% of the ingredients, excluding water, in a lean bread dough. When the flour comes in contact with water in the dough, and mixing energy is applied, some of the proteins form elastic, gas-retaining films known as gluten (16) (see WHEAT). Mixing the dough is a critical processing step since it not only serves to distribute ingredients homogeneously, but also develops the gluten protein strands to a proper balance of viscoelastic properties that have the unique ability to retain fermentation gases in numerous small gas cells. During fermentation, the dough, which is tough and relatively elastic after mixing, mellows and becomes more extensible so that it is readily manipulated and shaped by appropriate machines and then baked into foods of good volume and quality.

The quality of flour used in the production of yeast-raised products depends on several factors, including the quality and characteristics of the wheat variety, the environmental conditions under which the wheat was grown, the skill of the miller in separating the endosperm of the wheat kernel from the germ and bran, and the proper treatment of the flour in the mill.

Flour quality analyses performed by the producer and user of flour include moisture, protein, ash, enzymatic activity, physical dough properties, and baking tests, all of which evaluate flour for specific end uses. The ash content of flour serves as an index of refinement. Enzymatic activity tests, which indirectly measure the effects of alpha-amylase on the hot-paste viscosity of flour suspensions, are estimated chiefly with the amylograph or Falling Number methods (17); optimum alpha-amylase activity in bread flour is important for overall quality and staling properties of bread. Physical properties of dough are generally evaluated by means of either the farinograph or mixograph. These instruments are small recording mixers which measure work input during mixing, and provide information such as mixing time, amount of water required (absorption) to yield a dough of appropriate consistency, mixing stability, and overall gluten strength.

Bread flours are primarily milled from hard red winter wheats grown in the Great Plains and hard red spring wheats grown in the north central region of the United States. Most white pan breads are made with flour having a protein content of 11.0–12.5% (14% moisture basis). Certain products, such as mixed-grain or high fiber breads, or crusty hard rolls, may require stronger bread flours having about 1–2% higher protein. Clear flours, which are darker in color than typical bread flours but have a protein content of 13% or higher, are often used to bolster flour strength in nonwhite rye or wheat bread doughs.

Cake flours are produced from soft red or white wheats. Good quality cake flours are more highly refined than bread flours, and have lower ash and protein contents. Cake flours are usually treated with low concentrations of chlorine, which bleaches the flour and improves its baking performance such that higher amounts of sugar, shortening, and liquid may be used in formulations. The resulting "high ratio" cakes, which contain more sugar than flour, have good volumes, a fine grain, and tender, moist-eating properties. Soft wheat flours, less refined than cake flours, are used in the manufacture of biscuits, crackers, pies, and certain pastries. These flours are occasionally treated with chlorine gas in order to reduce spreading of certain cookies during production.

Yeast. Most of the yeast (6,18,19) used by wholesale bakers is available in the form of fresh compressed yeast, granular, or as cream yeast (a slurry). Fresh or creamed yeasts are delivered at frequent intervals and refrigerated until used. Yeast fermentation leads to gas production which, in turn, leads to leavened dough and bread as well as the development of fermentation flavors. The quantity of yeast added to dough is directly related to the time required for fermentation. Currently, most bread doughs are made with 2–3% fresh compressed yeast, based on flour. This amount varies depending on the form of yeast utilized. High sugar levels in sweet yeast-raised doughs tend to inhibit yeast activity through increased osmotic pressure and higher yeast levels are necessary to compensate.

Active dry yeasts of improved quality have been available for many years, and more recently instant active dry yeast has been introduced (15,20). This instant yeast exhibits more activity than regular active dry yeast due to improved drying techniques, and can replace compressed yeast at a rate of 33–40%. Dried yeasts, which are stable for long periods of time at room temperature, are of interest to bakers because of the high distribution cost of fresh compressed yeast. This is especially true for those away from distribution centers and for smaller bakers whose usage rate of yeast is low.

Yeast Foods. The so-called yeast foods are universally employed by wholesale bakers in quantities ranging from about 0.25–0.50% based on flour. Yeast foods are multifunctional proprietary materials, containing principally an ammonium salt, a calcium salt, and an oxidant. The only food in yeast foods is the ammonium ion, which serves as a source of nitrogen for the yeast. The bivalent calcium ion has a beneficial strengthening effect on the colloidal structure of the gluten, and the oxidizing agent has an improving action in dough through some largely unknown mechanism. A widely used yeast food is composed of 30% calcium sulfate, 9.4% ammonium chloride, 0.3% potassium bromate, 35% sodium chloride, and 25.3% starch or flour added as a filler.

Sugar. Sugar usage in white pan bread has increased over the years to a present concentration of about 8%, based on flour. Sugar is utilized to modify flavor, to help support yeast fermentation, and as a contributor to crust color and toasting properties through browning and caramelization reactions (6,15,21). High fructose corn syrups are predominantly used in bread production, although sucrose (cane or beet sugar) and corn syrups see some usage in breadmaking. In products requiring greater sweetness, sucrose is preferred. For frostings and fillings sucrose in the form of powdered sugars or fondant is the sweetener of choice because of its contribution to structure and its solubility properties. For cakemaking, sucrose is also preferred; replacement of sucrose in cake systems with high fructose corn syrup is limited because the latter causes flour starch to swell differently in a baking cake (22) and it also leads to excessive browning.

Shortening. Animal and vegetable fats and oils are used in many bakery foods to produce tenderness and to perform many other specific functions in the finished product (6,15) (see FATS AND FATTY OILS; VEGETABLE OILS). In breadmaking, fat addition yields a 20% volume increase besides improving keeping properties and tenderness. Fat usage has undergone considerable changes over the years; lard gave way to hydrogenated vegetable shortenings some years ago, and now liquid oils are mostly employed by wholesale bakers. Liquid soybean oil, used in conjunction with dough-strengthening surfactant systems (23), is economical, easily handled in bulk systems, and is desirable from a nutritional perspective. Fats in bread doughs have decreased from about 3–4% to 2–2.5%. Many cakes, icings, and fillings are made with specialty plasticized shortenings; however, wholesale snack cake bakers now widely use liquid soybean oil, which can make good quality cakes when employed with emulsifiers tending to form alpha crystals (24). Fluid shortenings, made with a base oil stock and small quantities of crystalline fat and surfactants, are also readily handled in bulk systems and are reported to yield superior quality bakery foods (23).

Butter is used in some, usually more expensive, bakery foods, and is prized for its flavor contribution. Fats are used in some products such as pie crust, croissants, or puff pastry, up to 60%, based on flour. Stability of fats and oils in perishable items such as breads, cakes, or pastries is of no consequence because shelf life is so limited that rancidity does not occur. In cookies and crackers, however, stable fats must be used in the formula since prolonged shelf life could lead to product deterioration with fats that develop rancidity.

A recent trend in the baking industry is to produce bakery foods with either no or reduced fat, to achieve perceived nutritional benefits (25). The functions of fat are achieved, to a degree, with materials such as maltodextrins and modified starches in combination with gums and emulsifiers (see FAT SUBSTITUTES).

Surfactants. Surfactants are widely used in bread and other bakery foods (6,15,26). In breadmaking, surfactants perform as crumb softeners or dough strengtheners, and they apparently do so by forming complexes. Monoglycerides, for example, complex with starch during baking to minimize subsequent retrogradation and staling, while ethoxylated monoglycerides complex with gluten proteins to provide strength. Other surfactants utilized in baking include sodium or calcium salts of fatty acids and lactic acid, polysorbates, succinylated monoglycerides, and diacetyl tartaric acid esters of monoglycerides. Surfactants are incorporated in cakes, icings, and fillings to promote emulsion stability, and to aid air incorporation and fat dispersion (see EMULSIONS; SURFACTANTS).

Milk and Milk Replacers. White pan bread was long made with about 3–4% nonfat dry milk (NFDM) in the United States, for reasons of enhanced nutrition, increased dough absorption, improved crust color, fermentation buffering, and better flavor. For some years, however, sharply increased milk prices have led to a decline in its use in breadmaking. Many bakers have turned to the use of milk replacers to control the costs of their products, and these ingredients are now commonly utilized. Milk replacers were designed to duplicate some of the functions and nutrition of milk. These blends may contain soy flour or cereals, with whey, buttermilk solids, sodium or calcium caseinate, or NFDM. Milk replacers or NFDM used in bread dough amount to about 1–2%, based on flour.

NFDM and whey constitute the most important dairy ingredients, but whole dry milk, fluid milk, condensed or evaporated milks, buttermilk, and cream cheese are employed to lesser extents in the production of specialty bakery items. For use in yeast-leavened products, milk must first be specially heat treated to inactivate some unknown factor or factors that would otherwise adversely affect dough handling properties and cause the production of poor bread quality (see MILK PRODUCTS). Milk is used in many cakes, where it is quite functional in terms of stabilizing the batter and improving structure of the baked cake; it is generally more difficult to replace the functions of milk in cakemaking than it is in breadmaking.

Eggs. Eggs are not used much in breadmaking, except for specialty egg breads. Egg whites are occasionally used on the surface of hard rolls to impart a crispy crust. Yeast-leavened sweet doughs or danish doughs often contain egg, up to 20%, based on flour, to achieve richness and to influence color and flavor. Some bakery foods, eg, sweet goods, croissants, and puff pastry, are often washed with egg wash (a mixture of egg and water or milk) prior to baking to obtain a rich, golden brown color in the finished product.

Eggs are employed in most cakes, at levels of 35–100%, based on flour, where they are functionally important (6,15). Unique properties of whippability, heat coagulation, and emulsification are critical factors in many bakery foods. Films of proteins surrounding gas bubbles in batter, provided by egg, are important in the leavening process and ultimately to the cellular structure of the finished cake. Egg whites are used to make white and angel food cakes; for other cakes, and for sweet yeast-raised goods, whole eggs, whole eggs fortified with yolks, or occasionally yolks alone, are used.

Bakers may use egg products in liquid, dried, or frozen forms. Liquid eggs, which have been churned, filtered, and pasteurized by the egg producer, are available in refrigerated tank cars to the large wholesale bakeries. The quality of dried egg products have been immensely improved in recent years, and many types of dried eggs are widely used by U.S. bakers in their production of bakery foods. Frozen egg products have lost ground to dried eggs because the former require careful thawing, a tedious, labor-intensive, and sometimes uncertain process.

Salt. Virtually all bakery foods contain salt (sodium chloride), except for a small number of products made for individuals on low sodium diets. Salt is quite functional in yeast-leavened bakery products. In breadmaking, salt, used at roughly 2% based on flour, provides flavor, moderates yeast fermentation by increasing the osmotic pressure in dough, and toughens gluten proteins. Crystal size and solubility are important quality factors in salt. Many bakery foods are best made with fine-flaked salt that dissolves readily; on the other hand, pretzels, and some hard rolls and bread sticks, require salt crystals of a larger size as a topping for purposes of appearance and flavor. For long shelf life foods such as crackers, it is important to use salt that is refined to minimize trace levels of iron and copper, since these tend to promote fat oxidation.

Water. The components of yeast-raised doughs and chemically leavened batters are dispersed in water. Water for dough mixing is generally not softened because the minerals in hard water may be beneficial, tending to strengthen gluten proteins. Sulfide waters have a deleterious action because the gluten protein is weakened by the sulfhydryl groups. Alkaline water, especially well-buffered alkaline water, also adversely affects bread quality since it interferes with the normal acidity development in dough, a necessary factor for flavor and overall bread quality (27).

Enzymes. Adequate amylolytic enzymes must be present in dough for several reasons. They break down available starch into fermentable sugars that constitute the energy source for yeast and degrade some of the starch to simpler structures during the early stages of gelatinization in baking to improve volume, texture, and keeping properties (26). Most wheats harvested in the United States are deficient in alpha-amylase, and flours milled from these wheats may be supplemented with enzymatic preparations. This is most typically done with the addition of malted barley flour. Fungal amylase preparations derived from *Aspergillus oryzae* are also permitted by the U.S. standards of identity for use in flour or bread, but fungal amylases are more often added at the bakery rather than the mill level. The amylase from *B. subtilis,* which is relatively heat stable and partly survives the baking process for further starch modifications, is utilized by some wholesale bakers to slow down the firming of bread.

Proteolytic enzymes, also derived from *A. oryzae,* may be used in yeast-raised dough production to reduce mixing time of dough or to make doughs more pliable. These functional effects are achieved through selective hydrolysis of gluten protein bonds, thus modifying dough and causing it to be more extensible.

Mold Inhibitors (Antimycotics). Antimycotics play an important role in extending the shelf life of many bakery foods (28). Calcium propionate [*4075-81-4*] is the most widely used antimycotic in breadmaking. It is often utilized at about 0.2%, flour basis; higher concentrations lead to flavor problems and begin to inhibit yeast fermentation. Sodium propionate [*137-40-6*] and potassium sorbate [*590-00-1*] are used to inhibit mold growth in chemically leavened products. Potassium sorbate suspensions may be used to spray the surface of yeast-leavened bakery foods for antimycotic protection; the sorbates cannot be used in doughs because of a strong yeast-inhibiting effect. Other agents sometimes used to control microbiological growth in bakery foods include vinegar, sodium diacetate [*126-96-5*], monocalcium phosphate monohydrate [*10031-30-8*], and lactic acid. Proprietary materials based on the fermentation of whey are also now available for antimycotic purposes; such materials are apparently effective on the basis of propionic acid that is formed during fermentation of the whey.

Flavorings. Various spices are employed to provide distinctive flavors in many bakery foods. Similarly, flavors and colors, both natural and artificial, are used to enhance bakery products in terms of both eating properties and appearance (6,15). Cocoa, chocolate, and many varieties of fruit, as well as some vegetables, (fresh, frozen, canned, and dried) are used in the food product or in fillings or icings.

Enriching Ingredients. Most commercially produced white breads are enriched with added thiamin, riboflavin, and iron. When breads or rolls are labeled enriched, these nutrients must be present in amounts prescribed by federal standards of identity (29). In g/kg product these are thiamin 4.0, riboflavin 2.4, niacin 33, and iron 27.5. Calcium may optionally be added at 1.3 g/kg. Although federal regulations relate to interstate commerce, most states also require enrichment of white bread and rolls. Enrichment nutrients may be added by the baker to the dough in the form of tablets, as a mixture of powdered nutrients in water-soluble pouches, or as a proprietary salt-enrichment blend. Alternatively, many bakers prefer to purchase flour that has been enriched to prescribed levels during the flour milling operation. Products labeled as enriched must also carry a nutritional label. Typically, a serving of one slice, approximately 1 oz (28 g), of enriched bread provides the following percentages of the U.S. Recommended Daily Allowance: thiamin 8.0%, riboflavin 4.0%, niacin 6.0%, iron 4.0%, and calcium 4.0%.

Procedures and Equipment

Bulk Ingredient Handling. Mills ship flour in bulk directly to large commercial bakeries via specially designed railroad cars or trucks. Upon receipt, this flour is transferred pneumatically from the cars or trucks to bins in the plant, from which it is conveyed (also pneumatically) to the mixing process as required. Smaller plants may instead use a tote bin system of bulk handling. In this system, several bins, each holding about 1360 kg of flour, are carried on a flatbed truck. When unloaded at the bakery these serve as storage bins and are eventually returned to the mill for refilling. Flour used in small quantities, or flour used by bakers with no bulk handling facilities, is delivered in multiwalled paper bags and is handled on skids with forklift trucks.

Most of the sugar used by commercial bread bakers is in syrup form; high fructose corn syrup is primarily used. Following shipment in tank cars, it is piped into bakery storage tanks from where it is metered into mixers upon demand.

Shortening and oil can also be handled in bulk. If the shortening is plastic, it is melted so that it can be pumped through pipes. These pipes must be heated throughout the plant to prevent the shortening from congealing.

Yeast (compressed and cream), and other perishable ingredients, are stored under refrigeration; freezers are required for frozen eggs and fruits. Yeast may be suspended into a slurry and stored under refrigeration for short-time intervals to expedite its transport and metering into mixers. Water is metered into mixers rather than weighed. Minor ingredients may be weighed directly from their containers or they may be suspended in water slurries which are subsequently metered into mixers (1,3–5).

DOUGH PROCESSES FOR BREAD PRODUCTION

Principal process categories used in the manufacture of yeast-raised products include the sponge and dough method, the straight dough method, and highly accelerated short-time methods that include frozen dough processing, continuous mix, and liquid ferment processes. Considerable variation exists among commercial bakeries within each of these categories. Rolls, buns, and sweet yeast-raised products are produced in ways analogous to those described for bread production. Principal differences are that special makeup machinery is required to produce the various shapes, and finishing equipment is required to fill and ice sweet roll and coffee cake units.

Sponge and Dough System: Conventional Breadmaking. Most domestically produced bread is manufactured by the two-mixing stage sponge and dough process (Fig. 1) (6,30–33).

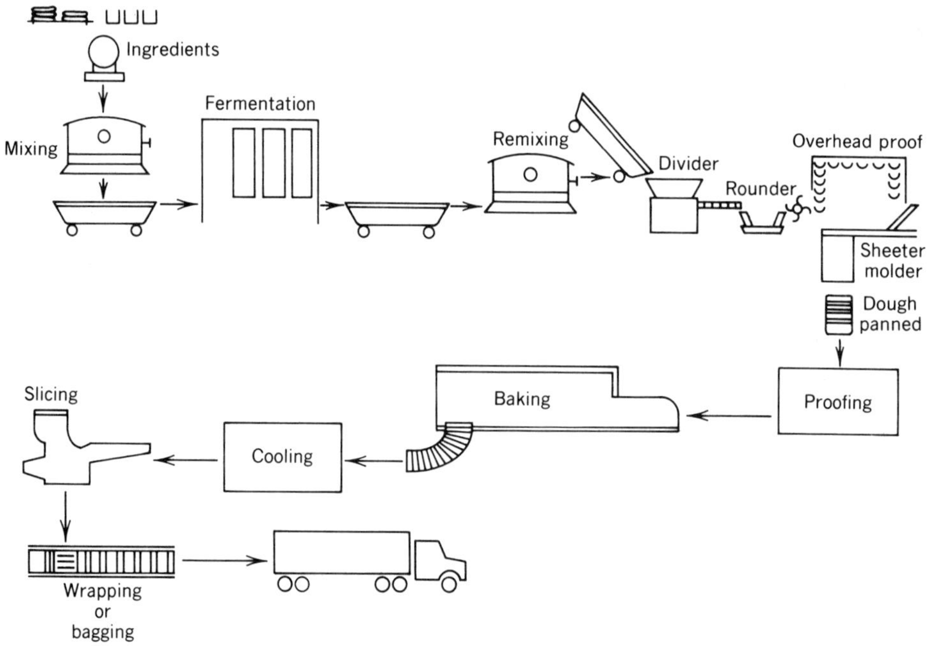

Fig. 1. Conventional dough process. Courtesy of Union Machinery Division. American Machine & Foundry Co.

Mixing and Fermentation of the Sponge. Sifted flour is pneumatically conveyed to a weighing hopper (scale) above a horizontal mixer. About 60–80% of the total flour required for a batch is used in mixing the sponge. Yeast, the yeast food, and the enrichment ingredients, usually presuspended in water slurries, are metered on top of the flour. The proper amount of water is metered into the bowl, and these ingredients are mixed into a sponge. The sponge may also include enzymes used as flour supplements and a fraction of the formula salt. Mixing of the sponge is intended to produce a homogeneous mass, but is not intended to produce full mechanical development of the wheat gluten. The sponge is dumped from the mixer into greased troughs which are transferred to a fermentation room. The fermentation room provides an optimum environment for sponge bulk fermentation of about 27°C and 75–80% rh. In practice, sponge fermentation time ranges from 2.5 to 6.0 hours, with best results usually found using 3.5–4.5 hours. As fermentation proceeds, the sponge volume increases dramatically until it reaches such extension that it drops spontaneously. This drop point is used as a reference indicating completion of 66–70% of total sponge fermentation. This bulk fermentation not only generates gas, but also generates fermentation by-products that contribute to final product flavor, develops the viscoelastic properties of dough that improve its gas retention capabilities, and reduces pH to a range that is optimum for final breadbaking results.

After the sponge has fermented sufficiently, it is returned to the mixer. The remaining flour and water, plus sugar, salt, shortening, nonfat dry milk (or milk replacer), and any other minor ingredients are added to the sponge and the final dough is mixed. Dough stage mixing is intended both to produce a homogeneous mass and also to mechanically develop the hydrated wheat gluten by repeated stretching and folding. This mixing to full development improves loaf volume, grain, and texture of the resultant loaf. There is a definite optimum mixing time range; deviation too far off the optimum in either direction harms end product quality. Mixers are jacketed so that temperatures of sponges and doughs may be partially controlled either by circulating refrigerant or brine in the jacket, or using direct-expansion coils. The properly developed dough, with a temperature of about 27°C, is ejected from the mixer into a greased trough where it is permitted a rest period (floor time) of 15–45 min before makeup.

Makeup. Following the floor time, the trough is hoisted and dumped to transfer the dough to the hopper of a divider. Several types of dividers are used, depending on dough consistency at this processing stage. A common type consists of pistons of adjustable size and/or "throw" which operate in cylinders. Piston movement draws dough into a cylinder, the cylinder door closes to cut off the piece of dough, and the cut dough piece is forced out of the chamber onto a conveyor belt. Ejected pieces are checkweighed frequently, and the size of the piston throw is adjusted accordingly.

The conveyor belt moves the rough, ragged dough pieces to a rounder, where they are converted to smooth, spherical shapes with consequent further mechanical development of the gluten network from the stretching and folding encountered. The rounder consists of a large rotating cone and a pressure board. Dough pieces are rolled and compressed as they are forced to travel up the pressure board by the rotating cone. Rounded dough pieces are transferred to an overhead or intermediate proofer, which is a large enclosed chamber through which the dough pieces are carried on trays suspended on a chain drive. The trip through the overhead proofer takes 7–15 min. The gluten network, which was tightened by the previous mechanical manipulation, relaxes during this period so that the dough pieces become pliable enough to successfully undergo sheeting and molding.

Dough pieces are conveyed out of the overhead proofer to a molder, usually of the cross-grain variety. In the molder, a series of reduction rollers sheet the relaxed ball-shaped dough pieces to thin flattened ovals. The resulting sheets pass through a chain curler to be formed into loosely formed cylinders. Each cylinder is tightened by passage under a compression plate. Cross-grain molders sheet each dough unit in two axes at right angles to each other, creating mechanically elongated and strengthened air cells that culminate in improved interior grain of the final baked bread loaves. At the end of the molder, these dough cylinders are automatically deposited into baking pans that are joined together in straps of four to six by means of cross pieces. Strapping facilitates handling and alignment through the next stages in the process.

Proofing. The panned dough cylinders are allowed to rise a final time in a controlled environment proof room or proofer at 35–43°C, 85–95% rh. Proofing is usually designed to take about one hour. During this period, the dough piece expands to approximately six times its original volume. Proofing may be accomplished by continuously conveying the pans through a proofer or alternatively, pans may be manually loaded onto racks which are transferred into a proofer.

Baking. If a continuous proofer was used, pans of proofed dough are mechanically conveyed into the oven. If proofing was done on racks, strapped pans are transferred to the oven by hand. Most large-scale bakeries use traveling tunnel or lap ovens in which bread pans are loaded at one end and the baked loaves exit the other. Such ovens generally consist of several chambers at different temperatures. One pound (0.45 kg) loaves generally bake at 215–225°C for 17–23 min, depending on formulation and desired end product. Smaller bakeries may use reel ovens, in which trays are rotated about a horizontal axis reel, or one of a number of ovens using heated air blown at high speed including convection ovens, rotating rack ovens, and air-jet impingement ovens (6,34–36).

Cooling, Slicing, and Wrapping. Baked bread is depanned automatically and conveyed through a cooler to lower loaf interior temperature to 40.5°C. Loaves are conveyed to a slicer where they are sliced by bandsaw blades. Sliced bread loaves are then bagged in plastic bags, or less frequently today, in waxed paper or transparent film. Packages are twist-tied or plastic clipped, or alternately heat sealed.

The sponge and dough process is used to manufacture a wide variety of yeast-leavened products that extend beyond the scope of the pan bread manufacturing process discussed here. To illustrate, sweet roll doughs often utilize the sponge and dough process. Such sweet doughs are characterized by their richer formulas, the equipment used to create their specific shapes, and equipment used to fill, ice, and package these products.

Straight Dough Method. The straight dough method differs from the sponge and dough method in that all ingredients are mixed in a single mixing stage to full development. Following mixing, the dough undergoes a bulk fermentation period of about two hours at 27°C. Fermentation is followed by a "punch" in which the risen dough is folded over on itself in its trough. The punch deflates the dough, expelling excess carbon dioxide gas and contributing to the strength (elasticity) of the maturing dough via the mechanical manipulation of the gluten network. A floor time of about 30 min follows, after which the dough is divided into desired unit pieces, rounded and rested briefly, then molded into loaves or other desired shapes. Subsequent processing is identical to that described in the sponge and dough method. Domestically, the straight dough method is used primarily in smaller retail-scale bakeries. It is sometimes used in larger commercial bakeries for short production runs of specialty products. In general, straight dough processes are more sensitive (less tolerant) to deviations in the processing steps than sponge and dough versions utilizing the same formula.

A variation on the straight dough method is the category of remixed straight doughs in which most or all ingredients are mixed to homogeneity, bulk fermented, then remixed to full development, given a short floor time, and subsequently processed as normal. In a sense, a remixed straight dough can also be considered a 100% flour sponge and dough system. Such methods are useful in smaller bakeries desiring maximum flexibility with respect to production scheduling.

No-Time Doughs. No-time or short-time dough processes are a special subset of the straight dough method. Increased amounts of yeast and such fast-acting oxidants as ascorbic acid and azodicarbonamide [*123-77-3*] (ADA) enable the elimination of most of the normal straight dough bulk fermentation period. Though attractive from an economic and production planning viewpoint, products of these processes tend to have open grain, less flavor, and shorter shelf lives. No-time processes are used advantageously in the production of short-lived hard-crusted hearth breads in smaller bakeries, and in the larger-scale production of frozen dough products.

Frozen Dough Products. Frozen dough products have found increasing acceptance as the numbers of skilled baker/artisans available to small bakeries have declined. End users of frozen dough products are typified by the growing number of grocery store in-store bakeries.

Doughs intended for frozen dough production usually include reduced water percentages and increased amounts of water-binding ingredients to minimize the amount of free water available for ice crystallization. No-time dough processes are generally used for frozen dough in order to minimize yeast activation prior to freezing. With minimized activation, yeast cells appear to be less damaged by the ice crystallization that occurs in frozen storage. The no-time dough process is thought to yield dough products that are least damaged by freezing and by frozen storage relative to gas production and gas retention capability. Dough strengthener additives, such as SSL and DATEM, are used to partially counter the structure weakening effect of glutathione leached from yeast cells that have been damaged during freezing. Frozen doughs that are produced from no-time doughs necessarily include high concentrations of oxidants including ascorbic acid, azodicarbonamide, and to a lesser degree, potassium bromate (37–41).

After mixing, doughs are immediately divided and rounded, rested briefly, then molded. Molded units are rapidly frozen in blast (or other) freezers until reaching a core temperature of about −7 to −5°C. Units are boxed and stored at −18°C, then shipped to end users. End users slowly thaw these products in 2°C retarders over 16–18 h, proof, and bake the products as needed (37–41).

Continuous-Mix Breadmaking Process. Continuous breadmaking uses equipment which continuously mixes and deposits dough in pans. This manufacturing method differs markedly from the conventional sponge and dough method discussed above, and produces loaves with characteristics substantially different from the bread produced by the conventional methods (6,30,31,33). Early versions of the continuous-mix process accomplished fermentation in a liquid water brew containing water, yeast, sugar, yeast food, and enrichment nutrients. Subsequent continuous-mix methods featured the use of some percentages of the total flour (up to 70%) in the brew. Proponents of flour usage in brews claimed improvements in bread flour and overall bread quality. The fermented brew is metered into a premixer along with flour, shortening, additional sugar, a dairy ingredient, other optional ingredients, and an oxidizing agent solution of potassium bromate and azodicarbonamide (ADA, 1,1′1-azobisformamide) to form a more or less homogeneous dough. This is pumped into a continuous developer where rapidly rotating agitators mechanically develop the dough. The developed dough is extruded under pressure as a ribbon which is automatically cut into properly sized pieces. These drop into pans and are subsequently proofed and baked.

Continuous-mix bread processes are economically attractive alternatives to conventional batch breadmaking processes. Because of this it is estimated that continuous-mix plants accounted for nearly 40% of domestic white pan bread production in the late 1960s and early 1970s. However, the end product lacks the grain and bite that the market expects of its bread, as well as the subtle yet complex flavor of breads produced from doughs that have undergone bulk fermentation. Some continuous-mix bakeries, as noted above, found that improved bread flavor could be obtained by including a percentage of the formula flour in the fermenting brew, but grain remained unchanged. Consumer preference has pulled most manufacturers back to more conventional breadmaking processes, and continuous-mix bread today accounts for perhaps only 5% of domestic bread production.

Liquid Ferment Processes. The second most prevalent commercial dough making process today is the liquid ferment category, accounting for about 20% of domestic bread production and nearly all bun production. Liquid ferment processes use either a water brew, or more commonly today, a fermented flour-containing brew, or liquid sponge, in conjunction with a batch-mixing process using conventional horizontal mixers. After mixing, dough pieces are divided, rounded, rested, and molded as in the conventional breadmaking process.

The higher the percentage of flour included in the fermenting brew, the more improved is final product flavor, and the lower the oxidant addition requirements. It is possible to use as high as 70% of formula flour in a brew, but this requires all of the formula water to still have a pumpable slurry. In practice, most bakeries use flour brews that include about 40% of formula flour.

Flour brews are usually mixed and fermented in batches. Alternatively, these brews may be mixed and fermented on a continuous basis in which the flour brew, or liquid sponge, is moved progressively through a fermentator tank system. Depending on the equipment design and the consistency of the brew system used, movement through the fermentator tank is accomplished either by means of brew density changes through time or by rotors (6,30,33).

Other Yeast-Leavened Product Categories

Buns and Rolls. Hamburger and hot dog buns and rolls may be processed in a manner directly analogous to those used for bread production. Almost all domestically produced bun products are produced by a batch liquid ferment process. After mixing, doughs are immediately divided and rounded to small unit-sized balls. Following a brief rest, these are given their final shape, proofed, baked, cooled, and packaged.

Sweet Yeast Goods. Sweet roll and coffee cake production is likewise directly analogous to bread-production methods. Chief differences are in makeup equipment used to produce the specific shapes and sizes of units. Also, following baking and cooling, sweet goods are "finished," ie, iced and/or filled prior to packaging.

English Muffins. English muffins are fairly tough, chewy, and honeycombed with holes and have a somewhat sour flavor. Typically made by the sponge and dough method, the final dough of English muffins has a comparatively high water content which makes the dough soft enough to flow. These soft dough units flow to fill griddle cups into which they have been deposited. Units are baked in these covered griddle cups on a highly automated line (26).

Bagels and Pretzels. Though their production processes are not necessarily typical of other yeast-leavened products, the principal distinguishing feature of these products is that they are both cooked briefly in a boiling water bath just before transfer to the hearth of an oven. The boiling water bath gelatinizes starch on the surface of the units, yielding a glossy crisp surface on the finished product (14).

Croissants and Danish. Both product categories are produced from laminated doughs. Croissant dough consists of a white bread straight dough into which butter is folded to create 27–54 layers. Similarly, Danish doughs are lean sweet roll doughs laminated with butter and/or other solid fats. In the Scandinavian countries, the weight of roll-in butter is generally about half the total weight of the dough being laminated. Following the lamination process, these doughs are retarded to facilitate subsequent sheeting and makeup. After makeup and proofing, these laminated dough units additionally puff gently upon oven heating, creating a crisp, friable, flaky, and buttery texture to the final product.

Raised Doughnuts. Mixing and fermentation of raised doughnut doughs, like that of sweet roll doughs, is virtually identical to processes used in bread manufacture and sweet roll dough production. The only significant difference is that the proofed units are fried instead of baked for approximately 1 min per side in 191°C fat. Fried units are cooled and drained briefly, then glazed or finished as desired.

Crackers and Cookies. These products have long shelf lives and so may be distributed over wide geographic areas. For this reason, cracker and biscuit plants are generally larger than those producing bread and cake products. This large volume from a single plant warrants more expensive and intricate mechanical equipment, and these plants are the most highly automated in the baking industry. Their products are generally produced from soft wheat flours, usually straight grade; saltine crackers may be made with some hard wheat flour to provide strength during dough processing (10–12,32,42,43).

Saltine cracker manufacture begins with mixing of a sponge consisting of 65% of the formula flour (10% protein), a small amount of yeast, and 27% water, considerably less than is used in a sponge for bread production, plus a lactobacillus inoculum provided by a portion of an old dough "starter." The sponge is fermented for 18 h. By the end of sponge fermentation, sponge pH has reached 4.0, activating the proteolytic enzymes naturally found in flour to yield greatly increased extensibility. The remaining 35% flour (8.5–9% protein), fat, salt, plus sufficient sodium bicarbonate to raise final dough pH to 7.5 are added at the dough mixing stage. This dough is allowed to ferment for 6 h. The dough is sheeted and laminated, then sheeted and scored into the shapes and sizes desired. Crackers are baked on a moving steel band in a tunnel oven (12,44).

Snack crackers are made from nonfermented straight doughs that obtain required extensibility via chemical or enzymatic means. They have harder, denser, and less friable texture than saltines, and obtain their flavor mostly from artificial flavors incorporated in the dough or sprayed onto the surface of baked crackers. Cookies are also made from a nonfermented dough. Cookie doughs contain eggs, sugar, flour, shortening, milk, salt, leavening agent, flavoring, and possibly fruits and nuts. Shapes are obtained by extrusion or by passing the sheeted dough under rollers with appropriate dies, and these are likewise baked in a band oven.

Nutritionally Modified Yeast-Leavened Products. Current growth segments in bakery foods markets include products with formula modifications that provide high fiber, reduced food calories, reduced total and saturated fats, and/or reduced cholesterol (45–47). The potential for percentage calorie reduction in bread products is somewhat limited by their relatively low fat and sugar content and their relatively low total caloric value. The need for high fiber foods in diets is today widely recognized (47). High fiber breads are often reduced in calories; this reduction is largely achieved through the use of various nondigestible fibers and increased water in the formula (45,48). In richer dough products, ingredients may be replaced with materials that attempt to simulate the function of fats. This substitution may be either in the dough itself, as with reduced fat sweet dough items and low fat crackers, or in the roll-in fat of laminated doughs, as with reduced fat Danish products. Some typical fat substitutes (qv) include modified food starches, gums, emulsifiers, polydextrose, microparticulate protein, and β-glucan fiber (45,46). High fiber ingredients with low calorie content may be used as bulking agents to replace higher calorie ingredients (sugar, flour), but mostly these are nonfunctional "dead weight." High intensity sweeteners such as sugar substitutes will be increasingly useful as workable bulking agents are developed (45).

Standards of Identity for Bakery Foods

Federal standards of identity exist for white bread, enriched bread, milk bread, raisin bread, and whole wheat bread that move in interstate commerce (29). Many states have adopted the federal standards or have promulgated nearly identical laws. Most bakery foods sold in the United States are produced by large commercial bakeries engaged in interstate commerce. As a result, nearly all bread and rolls necessarily conform to the federal standards. The FDA has taken the position that if an ingredient or packaging material has crossed state boundaries, then products containing that ingredient or wrapped in that packaging material are covered under regulations pertaining to interstate commerce. Most specialty breads, such as rye, multigrain, pita, and French, are not covered by standards of identity. All ingredients permitted in standardized bakery foods are considered optional ingredients and therefore must be declared in ingredient legends on bread wrappers. Since 1975, the FDA has required that all enriched bread-type products bear nutrition labeling.

BIBLIOGRAPHY

"Bakery Processes and Products" in *ECT* 1st ed., Vol. 2, pp. 273–287, by O. Skovholt, Quality Bakers of America; "Bakery Processes and Leavening Agents (Yeast-Raised Products)" in *ECT* 2nd ed., Vol. 3, pp. 41–55, by W. B. Bradley, American Institute of Baking; in *ECT* 3rd ed., Vol. 3, pp. 438–449, by K. Kulp, American Institute of Baking, and F. N. Hepburn, U.S. Department of Agriculture.

1. D. Malovany, *Bakery Prod. and Marketing* **25,** 44 (June 1990).
2. Census of Manufacturers, *Bakery Products,* U.S. Dept. Commerce, Bureau of the Census, Washington, D.C., May 1990.
3. R. Lahvic and D. Malovany, *Bakery Prod. and Marketing* **26** (Feb. 1990).
4. H. E. Jacob, *Six Thousand Years of Bread,* Greenwood Press Publishers, Westport, Conn., 1970.
5. H. G. Muller, *Baking and Bakeries,* Shire Publications Ltd., Aylesbury, Bucks, UK, 1986.
6. E. J. Pyler, *Baking Science and Technology,* 3rd ed., Vol. 1 and 2, Sosland Publishing Co., Merriam, Kans., 1988.
7. W. J. Sultan, *Practical Baking,* 5th ed., Van Nostrand Reinhold, New York, 1990.
8. B. S. Miller, *Variety Breads in the United States,* American Association of Cereal Chemists, St. Paul, Minn., 1981.
9. S. A. Matz, *Formulas and Processes for Bakers,* Pan-Tech International, Inc., McAllen, Tex., 1987.
10. P. Wade, *Biscuits, Cookies, and Crackers,* Vol. 1, Elsevier Applied Science, New York, 1988.
11. N. Almond, *Cookies and Crackers,* Vol. 2, Elsevier Applied Science, New York, 1989.
12. N. Almond, M. H. Gordon, P. Reardon, and P. Wade, *Biscuits, Cookies, and Crackers,* Vol. 3, Elsevier Applied Science, New York, 1991.
13. Y. Pomeranz, *Wheat Chemistry and Technology,* 3rd ed., American Association of Cereal Chemists, St. Paul, Minn., 1988.
14. N. L. Kent, *Technology of Cereals,* 3rd ed., Pergamon Press, Oxford, UK, 1983.
15. S. A. Matz, *Ingredients for Bakers,* Pan-Tech International, Inc., McAllen, Tex., 1987.
16. W. Bushuk and R. Tkachuk, *Gluten Proteins,* American Association of Cereal Chemists, St. Paul, Minn., 1990.
17. *Approved Methods of the AACC,* Methods 22-10A and 56-81B, American Association of Cereal Chemists, St. Paul., Minn., 1983.
18. J. G. Ponte, Jr. and G. Reed, in *Prescott and Dunn's Industrial Microbiology,* 4th ed., AVI Publishing Co., Westport, Conn., 1982.
19. N. B. Trivedi and G. Jacobson, in *Progress in Industrial Microbiology,* Vol. 23, Elsevier Science Publishers, Amsterdam, The Netherlands, 1986.
20. N. Trivedi, J. Hauser, T. Nagodawithana, and G. Reed, *AIB Technical Bulletin XI* Issue 2, American Institute of Baking, Manhattan, Kans., Feb. 1989.
21. J. G. Ponte, Jr., in *Sugar—A User's Guide to Sucrose,* Van Nostrand Reinhold, New York, 1990.
22. M. M. Bean, W. T. Yamazaki, and D. H. Donelson, *Cereal Chem.* **55,** 945 (1978).
23. T. Petricca, *Baker's Dig.* **50**(5), 39 (Oct. 1976).
24. H. Birnbaum, *Baker's Dig.* **50**(1), 28 (Feb. 1978).
25. J. D. Dziezak, *Food Technol.* **42,** 66 (July, 1989).
26. C. E. Stauffer, *Functional Additives for Bakery Foods,* Van Nostrand Reinhold, New York, 1990.
27. R. McFaul, *Am. Soc. Bakery Eng. Proc.,* 70 (1981).
28. J. G. Ponte, Jr. and C. C. Tsen, in *Food and Beverage Mycology,* 2nd ed., Van Nostrand Reinhold, New York, 1987.

29. *Code of Federal Regulations, Title 21 Food and Drugs,* Part 136, U.S. Government Printing Office, Washington, D.C., Apr. 1991.
30. S. A. Matz, *Bakery Technology and Engineering,* 2nd ed., AVI Publishing Co., Inc., Westpoint, Conn., 1972.
31. D. K. Tressler and W. J. Sultan, *Cereal, Bakery Goods, Dairy and Egg Products,* Vol. 2, AVI Publishing Co., Inc., Westport, Conn., 1975.
32. S. A. Matz, *Cookie and Cracker Technology,* AVI Publishing Co., Inc., Westport, Conn., 1968.
33. S. A. Matz, *Equipment for Bakers,* Pan-Tech International, Inc., McAllen, Tex., 1988.
34. C. E. Walker, *AIB Technical Bulletin IX,* Issue 11, American Institute of Baking, 1987.
35. C. E. Walker, *AIB Technical Bulletin XI,* Issue 11, American Institute of Baking, 1989.
36. C. A. Stear, *Handbook of Breadmaking Technology,* Elsevier Applied Science, New York, 1990.
37. B. L. Bruinsma and J. Geisenschlag, *Bakers' Dig.* **58**(6), 6 (1984).
38. D. E. Fuhrmann, *Am. Soc. Bakery Engineers Proc.,* 149 (1985).
39. D. K. Dubois and D. Blockcolsky, *AIB Technical Bulletin VIII,* Issue 4, American Institute of Baking, 1986.
40. *Ibid.,* Issue 6.
41. M. Evenson, *Am. Soc. Bakery Engineers Proc.,* 85 (1987).
42. W. H. Smith, *Biscuits, Crackers and Cookies,* Vol. 2, Magazines for Industry, New York, 1972.
43. D. J. R. Manley, *Technology of Biscuits, Crackers and Cookies,* Ellis Horwood Limited, Chichester, UK, 1983.
44. R. C. Hoseney, *Principles of Cereal Science and Technology,* American Association of Cereal Chemists, Inc., St. Paul, Minn., 1986.
45. J. L. Vetter, Caloric and Fat-Modified Bakery Products, *AIB Technical Bulletin XIII,* Issue 5, American Institute of Baking, 1991.
46. J. P. Zallie, *Fat-Free Products, Proceedings,* American Society of Bakery Engineers, Chicago, Ill., 1991.
47. B. L. Bruinsma, *Fiber—What's Hot, What's Not, Proceedings,* American Society of Bakery Engineers, Chicago, Ill., 1990.
48. C. E. Stauffer, "Fiber and Baking," *1991 Reference Source,* Sosland Publishing Co., Kansas City, Mo., 1991.

JOSEPH G. PONTE, JR.
JOEL D. PAYNE
Kansas State University

CHEMICAL LEAVENING AGENTS

Chemical leavening involves the action of an acid on bicarbonate to release carbon dioxide gas for aeration of a dough or batter during mixing and baking. The aeration provides a light, porous cell structure, fine grain, and a texture with desirable appearance along with palatability to baked goods.

Leavening of a dough with carbon dioxide from yeast fermentation was known to ancient civilization before the advent of recorded history. In contrast, the first use of chemical leavening dates back only to the early nineteenth century. Sodium bicarbonate [*144-55-8*] was used either alone or with sour milk in home baking around 1830 (1). At about the same time, alum was used as baking acid. Monocalcium phosphate was introduced in the form of an impure baking acid in 1856 (2,3). Almost 80 years later, in 1937, the first free-flowing powdered monocalcium phosphate monohydrate was introduced, followed by a milestone discovery of slow-acting coated anhydrous monocalcium phosphate in 1939 (4). Meanwhile, anhydrous sodium aluminum sulfate (SAS) was developed around 1892 (2). Used in combination with the fast-acting monocalcium phosphate, it forms a double-acting baking powder. Sodium acid pyrophosphate was introduced in the early 1900s, and sodium aluminum phosphates also joined the list of dry food-grade leavening acids in 1951 (5) (Table 1).

Composition of Chemical Leavening Systems

There are essentially two components in a chemical leavening system: bicarbonate that supplies carbon dioxide gas, and an acid which triggers the liberation of carbon dioxide from bicarbonate upon contact with moisture (see CARBON DIOXIDE).

Sodium bicarbonate (baking soda) is the primary source of carbon dioxide gas in practically all chemical leavening systems. This compound is stable and is obtainable as a highly purified dry powder at relatively low production cost. Two types of baking soda differing in granulation are available to cover practically all applications. Powdered soda is preferred for self-rising flour and prepared biscuit mixes. A fine granular soda is used for baking powder, prepared cake mixes, and frozen or refrigerated dough products or any product subject to storage for six months or longer.

Other bicarbonates of considerable commercial importance are ammonium bicarbonate [*1066-33-7*] and potassium bicarbonate [*298-14-6*]. These compounds are decomposed by the oven heat, liberating ammonia, carbon dioxide, and water to facilitate leavening action. Their uses are limited to low moisture products such as cookies and crackers.

The prevalent baking acids in modern chemical leavening systems are sodium or calcium salts of ortho, pyro, and complex phosphoric acids in which at least two active hydrogen ions are attached to the molecule (Table 1). Alum, although not a protonic acid, is used in retail baking powder formulations. Some organic acids are also used in refrigerated dough products. The discussion here focuses mainly on systems involving acidic phosphates and baking soda, since practically all chemical leavening systems today contain these components.

Characteristics of Leavening Agents

The evolution of carbon dioxide essentially follows the stoichiometry of acid–base reactions. Baking soda determines the amount of carbon dioxide evolved, whereas the type of acid controls the speed of liberation. The reaction equations for some acids with baking soda are as follows:

$$3\ Ca(H_2PO_4)_2 + 4\ NaHCO_3 \longrightarrow 4\ CO_2 + Ca_3(PO_4)_2 + 4\ NaH_2PO_4 + 4\ H_2O \tag{1}$$

$$3\ Ca(H_2PO_4)_2 + 8\ NaHCO_3 \longrightarrow 8\ CO_2 + Ca_3(PO_4)_2 + 4\ NaHPO_4 + 8\ H_2O \tag{2}$$

$$Na_2H_2P_2O_7 + NaHCO_3 \longrightarrow CO_2 + Na_3HP_2O_7 + H_2O \tag{3}$$

Table 1. Baking Acids Commercially Produced for Leavening Systems

Common name [Chemical Abstracts name]	CAS Registry Number	Formula	Neutralizing value	Uses[a]
monocalcium phosphate monohydrate, MCP [phosphoric acid, calcium salt (2:1), monohydrate]	*[10031-30-8]*	$CaH_4(PO_4)_2{\cdot}H_2O$	80	H, L, C
monocalcium phosphate anhydrous, AMCP coated [phosphoric acid, calcium salt (2:1)]	*[7758-23-8]*	$CaH_4(PO_4)_2$	83	(H)[b], L
sodium acid pyrophosphate, SAPP [diphosphoric acid, disodium salt]	*[7758-16-9]*	$Na_2H_2P_2O_7$	72	C, L
sodium aluminum phosphate, SALP				
[phosphoric acid, aluminum sodium salt (8:3:1)]	*[10305-76-7]*	$NaH_{14}Al_3(PO_4)_8{\cdot}4H_2O$	100	C, L
[phosphoric acid, aluminum sodium salt (8:2:3)]	*[10279-59-1]*	$Na_3H_{15}Al_2(PO_4)_8$		
[phosphoric acid, aluminum sodium salt]	*[7785-88-8]*	$Na_xAl_x(PO_4)_x$		
dicalcium phosphate, dihydrate, DPD [phosphoric acid, calcium salt (1:1)]	*[7789-77-7]*	$CaHPO_4{\cdot}2H_2O$	33	L
sodium aluminum sulfate, SAS (soda alum) [sulfuric acid, aluminum sodium salt (2:1:1)]	*[10102-71-3]*	$Al_2(SO_4)_3{\cdot}Na_2SO_4$	100	H
glucono-delta-lactone, GDL [D-gluconic acid-δ-lactone]	*[90-80-2]*	$C_6H_{10}O_6$	50	L
potassium hydrogen tartrate (cream of tartar) [butanedioic acid, 2,3-dihydroxy-, [*R*-(*R**,*R**)]-monopotassium salt]	*[868-14-4]*	$KHC_4H_4O_6$	45	(H)[b]

[a]C = commercial baking powders; H = household baking powders; L = leavening agents for preleavened products.
[b]( ) = some use, but limited.

$$Na_2H_2P_2O_7 + 2\ NaHCO_3 \longrightarrow 2\ CO_2 + Na_4P_2O_7 + 2\ H_2O \quad (4)$$

$$2\ NaAl_3H_{14}(PO_4)_8{\cdot}4\ H_2O + 23\ NaHCO_3 \longrightarrow 23\ CO_2 + Na_5Al_6(PO_4)_6(OH)_5\ 12\ H_2O + 10\ Na_2HPO_4 + 14\ H_2O \quad (5)$$

$$Na_2SO_4{\cdot}Al_2(SO_4)_3 + 6\ NaHCO_3 \longrightarrow 6\ CO_2 + Al_2O_3{\cdot}3H_2O + 4\ Na_2SO_4 \quad (6)$$

The neutralizing value (NV) of leavening acid determines the number of parts by weight of baking soda that will be neutralized by 100 parts of the acid to impart neutral pH in baked goods.

$$NV = a/b \times 100$$

where a = grams of baking soda and b = grams of leavening acid required to complete the reaction.

These equations are based on the reactions that occur when the materials are placed in boiling water. In a dough or batter there appear to be complicated reactions of phosphate, involving ion exchange with ingredients such as milk and calcium salts, and formation of complexes with proteins and starch. In the case of a blend containing different types of leavening acid, the reactions are complex and cannot be expressed in NV equations. Note for instance that more than one chemical equation can be written for a given phosphate acid, yielding different by-products. For this reason the neutralizing values given are often empirically determined.

Most of the acids react with baking soda, instantly liberating all the available carbon dioxide within a minute. Leavening of baked goods, however, requires both fast and slow liberation of carbon dioxide at a controlled speed; and the action must continue until just before the crumb structure of the baked goods is set by the heat of the oven. The required volume of carbon dioxide to be liberated at a particular stage during baking varies depending on the type of bakery product and processing conditions involved.

A number of acidic phosphates which vary in their rate of reaction are available for use in a wide variety of bakery applications. These acids, which include monocalcium phosphate, sodium aluminum phosphate, and sodium acid pyrophosphate, release carbon dioxide at a controlled rate to give a certain fraction prior to baking; the remaining fraction is released at a specific time during baking. Controlled releasing of carbon dioxide at the time it is needed can also be achieved by a mixture of different types of leavening acids.

The dough rate of reaction (DRR) is a term that defines the speed of carbon dioxide evolved during mixing and holding of a dough at bench prior to baking. It is determined by measuring the volume of carbon dioxide evolved from a standard dough formula containing known quantities of leavening acid and baking soda under a constant temperature of 27°C (Fig. 1). The DRR is often used as a guide for selecting the type of leavening acid that is best suited for a particular product application.

Leavening Agents for Preleavened Mixes and Doughs

By far, the greatest use of leavening in the home is in preleavened mixes. Pancake mixes, biscuit mixes, and cake mixes have long been established as commercial

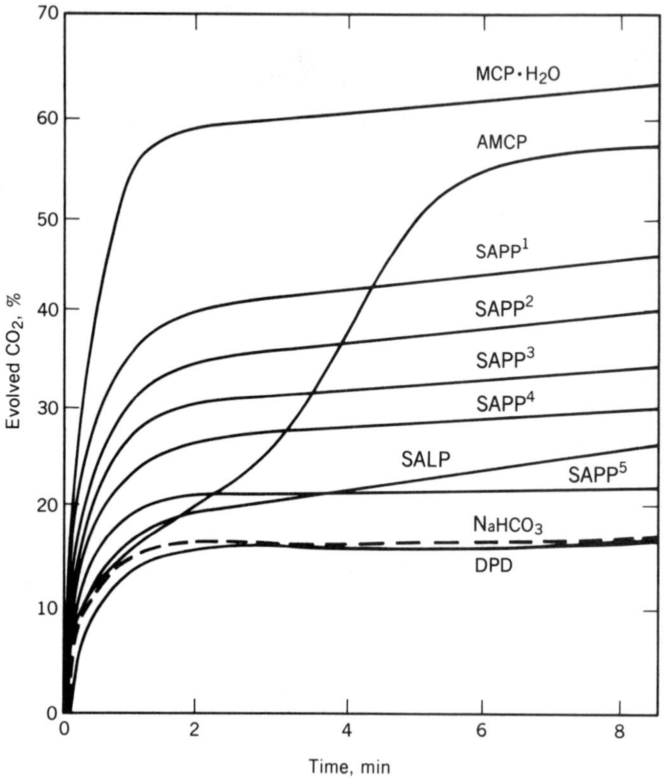

Fig. 1. Typical DRR of chemical leavening acids. The SAPP superscripts designate various speed grades: $SAPP^1$ = fastest; $SAPP^2$ = fast; $SAPP^3$ = medium fast; $SAPP^4$ = slow; $SAPP^5$ = slowest at 27°C. See Table 1 for definitions of abbreviations.

products. The advent of powerful emulsifiers coupled with components produced in rapidly dispersible forms produce bakery mixes that require very little mixing and have a wide tolerance for variability in handling. Some of these mixes may be prepared in the pan in which they are to be baked. Many mixes are also produced for commercial use. Because of the complexity of the doughnut system, many commercial bakeries buy preformulated doughnut mixes. This approach has spread to commercial use of cake mixes and mix concentrate and has been adopted by institutional baking operations. Most of the mixes on the market require the consumer to add only milk or water and possibly an egg. Certain types of cake mixes may be given a final treatment through a high speed shearing to disperse the fat properly.

Sales of frozen or refrigerated doughs and batters have increased rapidly in recent years. This is attributed to increased refrigeration and freezer capacity in stores and to the development of new and improved ingredients. The latter includes leavening acids, emulsifiers, gums, and starches which provide greater dough and batter stability during storage. For many of these products, leavening is the most critical component. The leavening acids are incorporated into the mix separately or with the baking soda. The type, quantity, and combination are all tailored to each product to give maximum stability, tolerance, performance, and quality. Practically all of the leavening acids used in the mix industry today are of the phosphate type.

Monocalcium Phosphate. The monohydrate of monocalcium phosphate (MCP) was the first of the phosphates to be used as a leavening acid, and has since been in commercial use for more than a century. Used alone, it reacts with baking soda very rapidly, releasing 60–70% of the available carbon dioxide in a two-minute mixing of dough. The remaining gas is not evolved at room temperature since the reaction generates some dicalcium phosphate [*7789-77-7*], and the heat of baking is required to complete the reaction. Because of its fast reaction, MCP is used at low concentration in combination with a slow-acting acid such as sodium acid pyrophosphate in double-acting baking powder.

The primary advantage of MCP is to create a large number of gas cells rapidly during mixing. These gas cells serve as nuclei for a greater expansion later in the oven. MCP also finds many uses in the products in which fast release of carbon dioxide and a low bench action are required. Examples are pancake mixes, cookie mixes, and angel food cakes. The milling industry uses MCP in the manufacture of phosphated flour.

A coated anhydrous monocalcium phosphate (AMCP) was developed in 1939 (3). The reaction with baking soda at room temperature is delayed by coating each AMCP crystal with a relatively insoluble condensed phosphate element. Retardations of the reaction (3–5 min) at onset enables the dough or batter to develop a greater resilience for a sustained oven spring. This is especially advantageous in developing desirable tenderness in biscuits and pancakes prepared from self-rising flour and in corn bread prepared from self-rising corn meal.

Sodium Acid Pyrophosphate (SAPP). Sodium acid pyrophosphate is prepared by thermal dehydration of monosodium phosphate. By varying processing conditions, a product with a wide range of reactivity can be prepared. Various manufacturing techniques are used to adjust the reactivity of SAPP for use in widely differing applications (6–11). SAPP is classified as a slow-reacting acid, but its function is dependent on its secondary reaction with calcium ion and protein in flour and milk. SAPP is also the only leavening acid that can be affected by granulation. In a batter, the finer the size of SAPP particles, the slower the reaction with baking soda. When treated with water and baking soda in the absence of a flour ingredient, all of the SAPP grades are fast and similar, releasing 60–70% of carbon dioxide from baking soda in a few minutes. This interference in reactivity is presumably caused by *in situ* coating of SAPP particles by relatively insoluble calcium pyrophosphate from the reaction of SAPP with calcium ions from batter ingredients. A number of grades of SAPP, ranging from fast to very slow reactivity, are available for a wide range of baking applications (see Fig. 1).

The fast-acting SAPP releases 40–43% of carbon dioxide in a two-minute mixing. It is often blended with a slower acting SAPP to provide intermediate reaction. The SAPP for cake doughnut production requires a certain quantity of gas to be released during mixing of the batter, followed by a rapid and uniform response to heating for proper expansion of the doughnut during frying. A special grade of SAPP is available to meet rigid specification for cake doughnut production.

Refrigerated dough or batter for canned biscuits, muffins, and cakes requires a SAPP with very slow reactivity at room temperature. A slow reactivity is needed before and during mixing, make-up, and packing processes; but a rapid and controlled gassing during proofing is essential to fill and seal the can with dough to prevent the entrance of air into the can. A specially designed, very slow-acting grade of SAPP that conforms to the most rigid specifications is available for this particular application.

A SAPP with intermediate reactivity is used in combination with fast-acting MCP for the manufacturer of industrial baking powder and for retail and wholesale prepared cake mixes. SAPP imparts a bitter aftertaste which is often characterized as a mild burning sensation, especially when used in a product of low sweetness. SAPP is normally used at an NV of 72. However, it may be used at slightly higher or lower NV to obtain specific effects in certain types of baked goods.

Sodium Aluminum Phosphate (SALP). This leavening acid was introduced in 1951, somewhat later than SAPP and MCP (5). It quickly gained a dominant position in self-rising flour, prepared cake mixes, pancakes, waffles, and refrigerated and frozen dough or batter products because of a number of outstanding features (12–15). SALP has a relatively low initial release of carbon dioxide. It exhibits retarded reaction in cold, even with extended holding of dough or batter. It has a bland flavor in baked goods and an excellent buffering action that provides greater tolerance of dough or batter against the influence of other ingredients and flour variety. SALP has a high neutralizing value of 100, which is an advantage for bakers. The baked products made from SALP or its blend are notable for firm yet tender crumb texture due to the great number of thin-walled cells which are formed.

One modified form is especially efficient in improving the tolerance of retail cake mixes containing highly emulsified shortening (16). A mixture of SALP and a small amount of coated AMCP has been adopted by the self-rising flour industry for the production of biscuit mixes ever since the introduction of SALP in the early 1960s (17). This system offers increased shelf life for the flour, and it enables one to hold biscuit doughs and pancake batters for extended periods without loss in baking response or change in viscosity. This imparts considerable versatility and convenience to the use of self-rising flour. This development has been further extended to frozen and refrigerated pancake batters with an extended stability in the refrigerated state. These batters may be conveniently marketed in dairy or frozen food cases. The relatively inactive state of this acid in batter and dough and its efficient response to oven heat make it ideally suited for this application.

Dicalcium Phosphate Dihydrate (DPD). Dicalcium phosphate dihydrate is completely nonreactive at room temperature. At 65–71°C and in the presence of water, it dehydrates and decomposes into hydroxyapatite and acidic monocalcium phosphate, or a free phosphoric acid (18). It is used to some extent in cake mixes in combination with faster acting acid. Its primary function is to provide acidity late in the baking cycle and thus produce a neutral and palatable product. DPD has an NV of 33. It provides sufficient acidity only in products requiring long baking times.

Sodium Aluminum Sulfate (SAS). Sodium aluminum sulfate is a dehydrated double salt of aluminum and sodium sulfate. It does not react with baking soda in cold, but in the heat of oven 1 mol of SAS produces 6 mol of carbon dioxide from reacting with baking soda. Historically, SAS was one of the first materials used to liberate carbon dioxide from baking soda. Today its primary use is in household baking powder production. It is used either alone or in combination with MCP. SAS is not recommended for use in prepared mixes due to its lack of compatibility with other ingredients in a mix.

Glucono-delta-lactone (GDL). Glucono-delta-lactone is an inner ester of gluconic acid. In a water solution, GDL slowly hydrates to become acidic and thus acts as a leavening acid. Hydrolysis is slow in cold but is accelerated by heat, making GDL ideal for refrigerated or frozen dough products (19). A drawback is the high cost, which, combined with a low NV of 47, make GDL prohibitive for use in most baked goods except in some refrigerated or frozen dough formulations.

Baking Powder. Modern baking powder consists of a mixture of baking soda, one or more acid components, and an inert ingredient which serves to keep the reactive components physically separated and thus minimizes premature reaction in dry mixtures. The inert ingredient is usually starch dried to 5–7% moisture. Calcium sulfate and calcium carbonate have sometimes been substituted for part of the starch. In addition to stabilizing the active ingredients, the inert substance provides a means for adjusting and standardizing the composition to a sodium bicarbonate concentration of approximately 30%. Table 2 lists typical formulations for both household and commercial baking powders.

Of the household baking powders in general use, the type containing sodium aluminum sulfate (SAS) is most prevalent. A small amount of monocalcium phosphate monohydrate (MCP) is used in combination with the SAS to make a double-acting baking powder. The MCP serves to generate gas cells during the mixing of dough or batter so that uniform expansion can be attained in the oven. This is necessary since SAS is almost completely nonreactive until heat is applied. Commercial baking powders contain sodium acid pyrophosphate (SAPP), which is superior to SAS in stability and performance. SAPP has an inherent aftertaste which is effectively masked by sweetness of sugar in cakes, doughnuts, cookies, and other sweet goods. Household baking powders are often used in biscuits where the pyro taste precludes the use of SAPP. Considerable variation exists among baking powders produced by various manufacturers. The concentration of baking soda may vary 28–30%; the acids vary proportionately to provide neutralized products. The use of cream of tartar as an acid component in baking powder is quite minor today because acids of equal performance are available at a lower cost.

Manufacture of Baking Powders

The blending of baking powder is essentially a physical mixing of the various components in a large-scale batch mixer and is often carried out in automated plants. The order in which mixing occurs may influence the stability of the product. Mixing of starch or other inert components with individual reactive components tends to maintain the separation of reactive components so as to

Table 2. Baking Powder Compositions, %

Constituents	Pure phosphate[a]			Phosphate-SAS double-acting types[a]			Cream of tartar	Commercial baking powders			
	1	2	3	1	2	3		1	2	3	4
baking soda, fine granular	28.0	28.0	28.0	30.0	30.0	30.0	27.0	30.0	30.0	30.0	30.0
monocalcium phosphate monohydrate, MCP	35.0			8.7	12.0	5.0		5.0		5.0	
monocalcium phosphate anhydrous, AMCP		34.0									
sodium aluminum phosphate, SALP			30.5								32.7
corn starch, redried	37.0	38.0	36.5	26.6	37.0	19.0	20.0	24.5	26.0	27.0	32.3
sodium aluminum sulfate, SAS				21.0	21.0	26.0					
sodium acid pyrophosphate, SAPP								38.0	44.0	38.0	
calcium sulfate [*7778-18-9*]				13.7							
calcium carbonate [*471-34-1*]						20.0					
cream of tartar							47.0				
tartaric acid [*133-37-9*]							6.0				
calcium lactate [*814-80-2*]								2.5			
tricalcium phosphate [*7758-87-4*]			5.0								5.0

[a]Household baking powders.

prevent premature reaction during storage. Rigid specifications for purity, granulation, and moisture content of the ingredients must be adhered to if a uniform, stable, and reliable product is to be obtained. The proper kind and speed of blending is essential to attain uniform distribution of ingredient particles. The baking powder is usually packaged in airtight metal or fiber cans. Hermetically sealed packages are not used because of the possibility of buildup of high internal pressures.

Baking powder has declined in home use since the 1940s. However, leavening acids are used in an increasing array of modern food products. These products include self-rising flour, self-rising corn meal, prepared cake mixes, pancake mixes, cookie mixes, doughnut mixes, and refrigerated or frozen doughs and batters.

BIBLIOGRAPHY

"Baking Powder" in *ECT* 1st ed., Vol. 2, pp. 289–293, by C. S. Bryan, Rumford Chemical Works; "Chemical Leavening" under "Bakery Processes and Leavening Agents" in *ECT* 2nd ed., Vol. 3, pp. 51–59, by James W. Tucker, Victor Chemical Works, a division of Stauffer Chemical Company; in *ECT* 3rd ed., Vol. 3, pp. 450–457, by W. B. Chess, Stauffer Chemical Company.

1. E. B. Bennion and co-workers, *Cake Making,* Leonard Hill Books, London, UK, 1966, p. 76.
2. L. H. Bailey, *Development and Use of Baking Powder and Baking Chemicals,* Circular No. 138, United States Department of Agriculture, May 1940.
3. U.S. Pat. 14,772 (Apr. 22, 1856), E. N. Horsford.
4. U.S. Pat. 2,160,232 (May 30, 1939), W. H. Knox and J. R. Schlaeger (to Victor Chemical Works/Stauffer Chemical Co.) (now Rhone-Poulenc Basic Chemicals Co.).
5. U.S. Pat. 2,957,750 (Oct. 25, 1960), W. H. Knox and J. Blanch (to Stauffer Chemical Co.) (now Rhone-Poulenc Basic Chemicals Co.).
6. U.S. Pat. 2,021,012 (Nov. 12, 1935), C. R. McCullough (to Swann Research).
7. U.S. Pat. 2,408,258 (Sept. 24, 1946), E. N. Hetzel and G. E. Taylor (to Monsanto Chemical Co.).
8. U.S. Pat. 2,630,372 (Mar. 3, 1953), F. H. Wright Jr. (to Monsanto Chemical Co.).
9. U.S. Pat. 2,636,808 (Apr. 28, 1953), F. E. Hubbard and F. H. Wright, Jr. (to Monsanto Chemical Co.).
10. U.S. Pat. 2,844,437 (July 22, 1958), L. A. Kramer and L. E. Netherton (to Victor Chemical Works/Stauffer Chemical Co.) (now Rhone-Poulenc Basic Chemicals Co.).
11. U.S. Pat. 3,034,899 (May 15, 1962), J. W. Tucker (to Stauffer Chemical Co.) (now Rhone-Poulenc Basic Chemicals Co.).
12. U.S. Pat. 3,205,073 (Sept. 7, 1965), J. E. Blanch and F. McCollough (to Stauffer Chemical Co.) (now Rhone-Poulenc Basic Chemicals Co.).
13. U.S. Pat. 3,311,448 (Mar. 28, 1967), J. E. Blanch, L. B. Post, and G. I. Klein (to Stauffer Chemical Co.) (now Rhone-Poulenc Basic Chemicals Co.).
14. U.S. Pat. 3,501,314 (Mar. 17, 1970), T. P. Kichline and N. E. Stuhlhebec (to Monsanto Chemical Co.).
15. U.S. Pat. 3,736,151 (May 29, 1973), L. B. Post, H. J. Rosen, and J. H. Zeh (to Stauffer Chemical Co.) (now Rhone-Poulenc Basic Chemicals Co.).

16. U.S. Pat. 3,041,177 (June 26, 1962), R. Lauck and J. W. Tucker (to Stauffer Chemical Co.) (now Rhone-Poulenc Basic Chemicals Co.).
17. U.S. Pat. 3,109,738 (Dec. 5, 1963), J. W. Tucker (to Stauffer Chemical Co.) (now Rhone-Poulenc Basic Chemicals Co.).
18. A. D. F. Toy, *Phosphorus Chemistry in Everyday Living,* American Chemical Society, Washington, D.C., 1976, p. 36.
19. C. Feldberg, *Cereal Sci. Today* 4(4), 96 (1959).

FRANK H. Y. CHUNG
Rhône-Poulenc Inc.

BARBITURIC ACID AND BARBITURATES. See HYPNOTICS, SEDATIVES, AND ANTICONVULSANTS.

BARIUM

Barium [*7440-39-3*], Ba, is a member of Group 2 (IIA) of the periodic table where it lies between strontium and radium. Along with calcium and strontium, barium is classed as an alkaline earth metal, and is the densest of the three. Barium metal does not occur free in nature; however, its compounds occur in small but widely distributed amounts in the earth's crust, especially in igneous rocks, sandstone, and shale. The principal barium minerals are barytes [*13462-86-7*] (barium sulfate) and witherite [*14941-39-0*] (barium carbonate) which is also known as heavy spar. The latter mineral can be readily decomposed via calcination to form barium oxide [*1304-28-5*], BaO, which is the ore used commercially for the preparation of barium metal.

In 1774, Scheele determined that barium oxide was a distinct oxide or "earth," and named it terra ponderosa because of its high density (1). Later, this name was changed to barote from the Greek word meaning heavy. Later still, the name of the oxide was modified to baryta to conform to the nomenclature recommended by Lavoisier, and from this the name barium was derived.

After many unsuccessful attempts, Davy produced barium as a mercury amalgam in 1808 by electrolyzing barium chloride in the presence of a mercury cathode. Attempts were made to isolate the pure metal by distilling the mercury, but it is doubtful that metal of high purity was ever obtained. Early in the twentieth century, high purity barium was prepared by heating barium amalgam in a stream of hydrogen, thereby converting the barium to a hydride and simultaneously volatilizing the last traces of mercury. The pure metal was then obtained by thermal decomposition of the hydride followed by condensation of the volatile barium vapor (1).

Barium is prepared commercially by the thermal reduction of barium oxide with aluminum. Barium metal is highly reactive, a property which accounts for its principal uses as a getter for removing residual gases from vacuum systems and as a deoxidizer for steel and other metals.

Physical Properties

Pure barium is a silvery-white metal, although contamination with nitrogen produces a yellowish color. The metal is relatively soft and ductile and may be worked readily. It is fairly volatile (though less so than magnesium), and this property is used to advantage in commercial production. Barium has a bcc crystal structure at atmospheric pressure, but undergoes solid-state phase transformations at high pressures (2,3). Because of such transformations, barium exhibits pressure-induced superconductivity at sufficiently low temperatures (4,5).

Additional physical properties of barium are given in Table 1. Owing to the high chemical reactivity of barium, it is not easy to obtain and store highly pure

Table 1. Physical Properties of Barium[a]

Property								Value
atomic weight (^{12}C = 12.000)								137.33
stable isotopes								
weight	130	132	134	135	136	137	138	
natural abundance, %	0.101	0.097	2.42	6.59	7.81	11.32	71.66	
density or specific gravity at 20°C, kg/m^3								3.51×10^3
melting point, °C								729
boiling point, °C								1640
heat of fusion, ΔH_f, kJ/mol[b]								7.66
heat of vaporization, ΔH_v, kJ/mol[b]								149.20
vapor pressure data								
pressure, kPa[c]	0.00133	0.0133	0.133	1.33	13.3	53.3	101.3	
temperature, °C	629	730	860	1050	1300	1520	1640	
specific heat at 20°C, J/(g·K)[b]								0.192
coefficient of thermal expansion, m/(m·°C)								1.8×10^{-5}
Gruneisen parameter								−0.2
electrical resistivity, μΩ-cm								
commercial purity at 0°C								34.8
ultra-high purity at 0°C								29.4
liquid barium at mp								314
electron work function, eV								2.11
crystal structure								bcc
lattice constant, nm								0.5025
superconducting temperature at 8.8×10^6 kPa[d], K								3.05

[a]Refs. 4–11.
[b]To convert J to cal, divide by 4.184.
[c]To convert kPa to mm Hg, multiply by 7.5.
[d]Increases with increasing pressure at the rate of 1.3×10^{-7} K/kPa. To convert kPa to atm, divide by 101.3.

samples of the metal. Therefore, accurate measurement of many physical properties of barium is difficult.

Chemical Properties

Barium has a valence electron configuration of $6s^2$ and characteristically forms divalent compounds. It is an extremely reactive metal, and its compounds possess large free energies of formation. At room temperature, it combines readily and exothermically with oxygen and the halogens. It reacts vigorously with water, liberating hydrogen and forming barium hydroxide [*17194-00-2*], $Ba(OH)_2$. At elevated temperatures, barium combines with hydrogen to form barium hydride [*13477-09-3*], BaH_2, and with nitrogen to form barium nitride [*12047-79-9*], Ba_3N_2. With nitrogen and carbon, barium forms a cyanide that is thermally stable.

Finely divided barium is susceptible to rapid, violent combination with atmospheric oxygen. Therefore, in powdered form it must be considered pyrophoric and very dangerous to handle in the presence of air or other oxidizing gases. Barium powder must be stored under dry argon or helium to avoid the possibility of violent explosions. Massive pieces of barium, however, oxidize relatively slowly and present no explosion hazard if kept dry.

Most barium compounds are not as thermodynamically stable as the corresponding compounds of magnesium and calcium and therefore can be reduced by these metals. However, rather than producing pure barium, barium alloys are formed. Barium combines with most metals, forming a wide range of alloys and intermetallic compounds. Among the phase systems that have been better characterized are those with Ag, Al, Bi, Hg, Pb, Sn, Zn, and the other Group 2 metals (12).

Barium reduces the oxides, halides, and sulfides of most of the less reactive metals, thereby producing the corresponding metal. It has reportedly been used to prepare metallic americium via reduction of americium trifluoride (13). However, calcium metal can, in most cases, be used for similar purposes and is usually preferred over barium because of lower cost per equivalent weight and nontoxicity (see ACTINIDES AND TRANSACTINIDES).

Commercially produced barium is analyzed for metallic impurities by means of emission spectroscopy. Carbon content can be determined by combustion, and nitrogen by the Kjeldahl method. Total barium can best be determined gravimetrically by precipitation as the sulfate.

Manufacture and Processing

Because of its high reactivity, production of barium by such processes as electrolysis of barium compound solution or high temperature carbon reduction is impossible. Electrolysis of an aqueous barium solution yields $Ba(OH)_2$, whereas carbon reduction of an ore such as BaO produces barium carbide [*50813-65-5*], BaC_2, which is analogous to calcium carbide (see CARBIDES). Attempts to produce barium by electrolysis of molten barium salts, usually $BaCl_2$, met with only limited success (14), perhaps because of the solubility of Ba in $BaCl_2$ (15).

Barium metal is produced commercially by the reduction of barium oxide with a less reactive, nonvolatile element, usually aluminum (16–22). Depending on initial stoichiometry, two overall reactions occur in the BaO reduction:

$$6\ BaO + 2\ Al \longrightarrow 3BaO{\cdot}Al_2O_3 + 3\ Ba\ (gas)$$

$$4\ BaO + 2\ Al \longrightarrow BaO{\cdot}Al_2O_3 + 3\ Ba\ (gas)$$

These reactions are thermodynamically unfavorable at temperatures below ca 1500°C. However, at temperatures in the range from 1000 to 1200°C a small but finite equilibrium pressure of barium vapor is formed at the reaction site. By means of a vacuum pump, the barium vapor can be transported to a cooled region of the reactor where condensation takes place. This destroys the equilibrium at the reaction site and allows more barium vapor to be formed. The process is completely analogous to that used in the thermal reduction of CaO with aluminum to produce metallic calcium (see CALCIUM AND CALCIUM ALLOYS).

In the commercial process, ground BaO, which is usually produced by calcination of witherite, $BaCO_3$, and finely divided aluminum are dry blended and briquetted to ensure good contact. The briquettes are placed in a horizontal metal tube known as a retort. The retort, made of heat-resistant steel, is heated in a furnace to 1100–1200°C. The open end of the retort protrudes from the furnace and is cooled by means of a water jacket for condensation of the barium vapor. After being sealed, the retort is evacuated to a pressure of less than 13 Pa (0.1 mm Hg), and the reaction is allowed to proceed for about 24 h. The vacuum is then broken with argon, and the condensed block of metal (known as a crown) and the barium aluminate residue are removed from the retort. Barium production requires large amounts of energy for two reasons: the high temperatures required in the process itself and the energy-intensive raw materials employed, the calcined BaO and the electrolytically produced aluminum.

Shipment and Storage

The barium crowns are usually broken into smaller pieces and can be sold in this form or cast or extruded into bars or wire. Usually the metal is packaged in argon-filled plastic bags inside argon-filled steel containers. Barium is classed as a flammable solid and cannot be mailed. It should be stored in a well-ventilated area so as to remove any hydrogen formed through reaction with water vapor. It should not be stored where contact with water is possible.

Specification and Standards

The purity of commercial barium depends to a large extent on the purity of the raw material used. Impurities such as calcium, magnesium, and alkali metal compounds are reduced in the manufacturing process and can contaminate the barium. The typical composition of commercial barium is shown in Table 2.

Table 2. Composition of Commercial Barium Metal, wt %

Constituent	%
barium	99.5
calcium	0.10
magnesium	0.08
aluminum	0.01
iron	0.03
nitrogen	0.10

Production of ultra-pure barium metal has been investigated on a laboratory scale. Redistillation (23,24), zone recrystallization (25,26), and combinations of these techniques (27) have been studied. Impurity levels of less than 100 ppm have been attained.

Economic Aspects

Barium is no longer produced in the United States or Canada. In Europe it is produced by Degussa AG, Hanau, Germany, in the form of rods. Saes Getters SpA, Italy, produces barium but its entire output of the metal is dedicated to alloy production for getters. Production facilities doubtlessly exist in Russia as Russian barium is offered in the United States by United Mineral Chemical Corp., Lyndhurst, New Jersey, in the form of ingots, but no details are known. A Polish firm, Zaklau Aparatory Prozniowej of Boloslawiec is also known to supply barium in the form of rods. Production figures are not available. Yearly worldwide production of barium, however, probably does not exceed 20,000–30,000 kg. Prices in 1991 ranged around $80–140/kg depending on the metal form, quantity, and quality.

Health and Safety Factors

Barium metal and most barium compounds are highly poisonous. A notable exception is barium sulfate which is nontoxic because of its extreme insolubility in water. Barium ion acts as a muscle stimulant and can cause death through ventricular fibrillation of the heart. Therefore, care must be taken to avoid contact with open areas of the skin. Workers must wear respirators (of type approved for toxic airborne particles), goggles, gloves, and protective clothing at all times. The toxic barium aluminate residue obtained from barium production is detoxified by reaction with a solution of ferrous sulfate and converted into nontoxic barium sulfate. According to OSHA standards, the TWA value for Ba and Ba compounds in air is 0.5 mg/m^3.

Barium also presents a hydrogen explosion hazard if allowed to come into contact with water or atmospheric moisture, and must always be kept dry and preferably sealed in the shipping containers.

Uses

The largest use for barium is as a getter to remove the last traces of gases from vacuum and television picture tubes. It is ideal for this use because of its combination of high chemical reactivity and low vapor pressure (28–32). In some cases it is used as powder obtained by vaporization in an electric arc (33). It can also be used as an aluminum alloy (see VACUUM TECHNOLOGY).

Barium improves the performance of lead alloy grids of acid batteries (see BATTERIES) (34). In the form of thin films, barium has been found to be a good high temperature lubricant on the rotors of anodes operating at 3500 rpm in vacuum x-ray tubes (35).

Barium is also used as a component in various proprietary nodularizing and deoxidizing alloys. Nodularizing alloys are added to molten gray and ductile irons to control the shape of precipitated graphite inclusions (36,37), whereas the deoxidizing alloys take advantage of barium's chemical reactivity to lower the oxygen content of molten steel (38), copper (39), and other metals. Numerous other uses are patented or mentioned in the literature (40).

BIBLIOGRAPHY

"Barium" treated in *ECT* 1st ed. under "Alkaline Earth Metals and Alkaline Earth Metal Alloys," Vol. 1, pp. 458–463, by C. L. Mantell, Consulting Chemical Engineer; in *ECT* 2nd ed., Vol. 3, pp. 77–80, by L. M. Pidgeon, University of Toronto; "Barium" in *ECT* 3rd ed., Vol. 3, pp. 457–463, by C. J. Kunesh, Pfizer, Inc.

1. J. W. Mellor, *Comprehensive Treatise on Inorganic and Theoretical Chemistry,* Vol. 3, Longmans, Green & Co., Inc., New York, 1923, pp. 619–631.
2. J. P. Bastide, C. Susse, and R. Epain, *C. R. Acad. Sci.* **267,** 857 (1968).
3. S. Akimoto and co-workers, *High Temp. High Pressures* **7,** 287 (1975).
4. M. A. Il'ina and E. S. Itskevich, *JETP Lett.* **11,** 32 (1970).
5. M. A. Il'ina, E. S. Itskevich, and E. M. Dizhur, *Zh. Eksp. Teor. Fiz.* **61,** 2357 (1971).
6. R. C. Weast, ed., *Handbook of Chemistry and Physics,* 57th ed., CRC Press, Cleveland, Ohio, 1976, pp. B1, B3, B10, B307–308, D62, D165, D182, D211, and E81.
7. D. T. Peterson and M. Indig, *J. Am. Chem. Soc.* **82,** 5645 (1960).
8. F. Emley, in D. M. Considine, ed., *Chemical and Process Technology Encyclopedia,* McGraw-Hill, New York, 1974, p. 151.
9. G. K. White, *J. Phys. F.* **2,** 865 (1972).
10. M. S. Rashid and F. X. Kayser, *J. Less Common Met.* **24,** 253 (1971).
11. H. J. Güntherodt, E. Hauser, and H. U. Künzi, *J. Phys. F* **5,** 889 (1975).
12. W. G. Moffatt, *Handbook of Binary Phase Diagrams,* Genium Publishing Corporation, New York, update July 1990.
13. C. A. Hampel, in C. A. Hampel, ed., *The Encyclopedia of the Chemical Elements,* Reinhold Book Corp., New York, 1968, pp. 45–46.
14. C. L. Mantell, in C. A. Hampel, ed., *Rare Metals Handbook,* 2nd ed., Reinhold Publishing Corp., London, 1961, p. 26.
15. K. Grjotheim and H. G. Nebell, *Acta Chem. Scand.* **22,** 1159 (1968).
16. W. J. Kroll, *U.S. Bur. Mines, Inf. Circ.,* 7327 (1945).
17. E. Fukuda and H. Yokomizo, *J. Electrochem. Soc. Jpn.* **20,** 430 (1952).
18. M. Orman and E. Zembala, *Prac. Inst. Met.* **4,** 437 (1952).

19. G. G. Gvelisiani and V. A. Pazukhin, *Ref. Zh. Metall.* (1018), (1956).
20. H. Sawamoto, T. Oki, and T. Umemura, *Mem. Fac. Eng. Nagoya Univ.* **12,** 123 (1960).
21. G. N. Kozhevnikov, *Tsvetn. Metall.* **36,** 53 (1963).
22. G. G. Gvelisiani and N. P. Mgaloblishvili, *Tr. Gruz. Inst. Met.* **14,** 205 (1965).
23. T. E. Brown and K. A. McEwen, *J. Physics D* **3,** 980 (1970).
24. J. Evers and co-workers, *J. Less Common Met.* **30,** 83 (1974).
25. V. N. Vigdorovich and co-workers, *Izv. Vyssh. Uchebn. Zaved. Tsvetn. Metall.* (1), 86 (1973).
26. A. V. Vokhobov, V. G. Khudaiberdiev, and M. K. Nasyrova, *Russ. Metall.* (4), 116 (1974).
27. A. V. Vokhobov, V. N. Vigdorovich, and V. G. Khudaiberdiev, *Izv. Vyssh. Uchebn. Zaved. Tsvetn. Metall.* (4), 115 (1973).
28. L. F. Ehrke and C. M. Slack, *J. Appl. Phys.* **11,** 129 (1940).
29. J. Bailleul Langlais, *Vacuum* **10,** 143 (1960).
30. P. della Porta and F. Ricca, *Vacuum* **10,** 215 (1960).
31. P. della Porta and E. Argano, *Vacuum* **10,** 223 (1960).
32. J. Verhoeven and H. van Doveren, *J. Vac. Sci. Technol.* **20,** 64 (1982).
33. Brit. Pat. 1,068,529 (May 10, 1967), D. E. Bobo (to Union Carbide Corp.).
34. N. L. Parr, A. Muscott, and A. J. Cocker, *J. Inst. Met.* **87,** 321 (1959).
35. Z. Jeffries, *Min. Metall.* **22,** 560 (1970).
36. D. M. Stefanescu and G. Bolintineanu, *Metallurgia* **22,** 560 (1970).
37. *Ibid.,* p. 702.
38. E. Plöckinger, W. Holzgruber, and G. Kühnelt, *Radex Rundsch.* (2), 508 (1969).
39. U.S. Pat. 2,503,377 (Apr. 11, 1950), H. S. Cooper (to Acme Aluminum Alloys, Inc.).
40. *Barium Bibliography,* Mineral Products Division, Food Machinery and Chemical Corporation, New York, 1961, pp. 199–238.

General References

Barium Bibliography, Mineral Products Division, Food Machinery and Chemical Corporation, New York, 1961 (esp. pp. 95–155 and 199–238).

W. J. Kroll, *U.S. Bur. Mines Inf. Circ.,* 7327 (1945).

C. L. Mantell, in C. A. Hampel, ed., *Rare Metals Handbook,* 2nd ed., Reinhold Publishing Corp., London, 1961, pp. 15–31.

J. W. Mellor, *Comprehensive Treatise on Inorganic and Theoretical Chemistry,* Vol. 3, Longmans, Green & Co., Inc., New York, 1923, pp. 619–652.

CLAUDIO BOFFITO
Saes Getters SpA

BARIUM COMPOUNDS

The first report concerning barium compounds occurred in the early part of the seventeenth century when it was noted that the ignition of heavy spar gave a peculiar green light. A century later, Scheele reported that a precipitate formed when sulfuric acid was added to a solution of barium salts. The presence of natural barium carbonate, witherite [*14941-39-0*], $BaCO_3$, was noted in Scotland by Withering.

In its natural form, barium [*7440-39-3*], Ba, never occurs as the metal but is almost always found as the ore barite [*13462-86-7*], $BaSO_4$. More than 90% of all barium is actually used as the ore, albeit after preliminary beneficiation. The petroleum industry consumed 90% of the ore for oil- and gas-well drilling fluids (muds). The other 10% is used as filler and/or for extender uses and the manufacture of all other barium chemicals. Witherite, the only other significant natural barium ore, is not mined commercially (1). The quantity of U.S. barium chemicals produced in 1986 and 1987 is given in Table 1.

Table 1. Barium Chemicals Produced in the United States[a]

		Sold or used by processors	
Year	Production, t	Quantity, t	Value, \$ $\times 10^3$
1986	23,705	23,133	16,871
1987	25,861	25,462	16,466

[a]Ref. 2.

Barium is a member of the alkaline-earth group of elements in Group 2 (IIA) of the periodic table. Calcium [*7440-70-2*], Ca, strontium [*7440-24-6*], Sr, and barium form a closely allied series in which the chemical and physical properties of the elements and their compounds vary systematically with increasing size, the ionic and electropositive nature being greatest for barium (see CALCIUM AND CALCIUM ALLOYS; CALCIUM COMPOUNDS; STRONTIUM AND STRONTIUM COMPOUNDS). As size increases, hydration tendencies of the crystalline salts increase; solubilities of sulfates, nitrates, chlorides, etc, decrease (except fluorides); solubilities of halides in ethanol decrease; thermal stabilities of carbonates, nitrates, and peroxides increase; and the rates of reaction of the metals with hydrogen increase.

In metallic form, barium is very reactive, reacting readily with water to release hydrogen. In aqueous solution it is present as an ion with a +2 charge. Barium acetate, chloride, hydroxide, and nitrate are water-soluble, whereas barium arsenate, chromate, fluoride, oxalate, and sulfate are not. Most water-insoluble barium salts dissolve in dilute acids; barium sulfate, however, requires strong sulfuric acid.

Compared to the hydroxides of calcium and strontiuim, barium hydroxide is the most water-soluble and also the strongest base. Additionally, barium hydroxide is more difficult to convert to the oxide by heating than are the corresponding hydroxides of calcium and strontium. Barium oxide is more readily converted to the peroxide than are the oxides of the other alkaline earths.

The large size, ionic radius = 0.143 nm, and electronic configuration [Xe] $6s^2$, of the barium(II) ion [*22541-12-4*], Ba^{2+}, makes isomorphous substitution possible only with strontium, Sr^{2+}, 0.127 nm, and generally not with other members of Group IIA such as Ca^{2+}, 0.106 nm, and Mg^{2+}, 0.078 nm. Among the other elements that occur with barium in nature, substitution is common only for potassium, K^+, 0.144 nm, but not for the smaller ions of Na, Fe, Mn, Al, and Si (3). For a discussion of barium bromate [*13967-90-3*], $Ba(BrO_3)_2$, see BROMINE COMPOUNDS; for barium chlorate [*13477-000-4*], $Ba(ClO_3)_2{\cdot}H_2O$, see CHLORINE OXYGEN ACIDS AND SALTS, CHLORIC ACID AND CHLORATES; for barium chromate [*10294-40-3*], $BaCrO_4$, see CHROMIUM COMPOUNDS; and for barium cyanide [*542-62-1*], $Ba(CN)_2$, see CYANIDES. For a discussion of barium ferrite [*11138-11-7*] and [*12409-27-7*] ($BaFe_{12}O_{13}$), see FERRITES; for barium fluoride [*7787-32-8*], BaF_2, and barium fluorsilicate [*17125-80-3*], $BaSiF_6$, see FLUORINE COMPOUNDS, INORGANIC; and for barium hydride [*13477-09-3*], BaH_2, see HYDRIDES.

Barite

Barite [*13462-86-7*], natural barium sulfate, $BaSO_4$, commonly known as barytes, and sometimes as heavy spar, till, or cawk, occurs in many geological environments in sedimentary, igneous, and metamorphic rocks. Commercial deposits are of three types: vein and cavity filling deposits; residual deposits; and bedded deposits. Most commercial sources are replacement deposits in limestone, dolomitic sandstone, and shales, or residual deposits caused by differential weathering that result in lumps of barite enclosed in clay. Barite is widely distributed and has minable deposits in many countries.

Mineralogically, barite crystallizes in the dipyramidal class of the orthorhombic system. Although the barite of many deposits fractures unevenly or has an apparent cleavage along planes because of separation between successively deposited layers, well-formed crystals are mostly tabular and cleave along three different planes. The barite in most deposits occurs in irregular masses, nodules, rosettelike aggregates, and in laminated to massive beds of fine crystallinity. Barite is most commonly associated with quartz [*14808-60-7*], chert, jasperoid, calcite [*13397-26-7*], dolomite [*17069-72-6*], siderite [*14476-16-5*], rhodochrosite [*14476-12-1*], celestite [*14291-02-2*], gluorite, various sulfide minerals, and their oxidation products. In most mines, barite is the primary material being mined, yet barite is also a common gangue mineral in many types of ore deposits including those for lead, zinc, gold, silver, fluorite, and rare-earth minerals (4).

Barite is a moderately soft crystalline mineral, Mohs' hardness 3–3.5; sp gr 4.3–4.6; n_D 1.64. The ore is white opaque to transparent, but impurities can produce pale shades of yellow, green, blue, brown, red, or gray-black. The most important impurities are Fe_2O_3, Al_2O_3, SiO_2, and $SrSO_4$, all of which are undesirable in chemical-grade barite. When the barite is used for drilling mud, the iron content can be permitted to be much higher than for other uses.

Residual barite is usually mined by open-pit methods. Bedded and vein deposits may be mined by either open-pit or underground methods, depending on the characteristics of each deposit. Some barite is sufficiently high, up to 96% $BaSO_4$, grade that it can be shipped without beneficiation. Many deposits require concentration and beneficiation can involve any one or a combination of gravity separations. "Primary barite," the first marketable product, includes crude run-of-mine barite, flotation concentrates, and material concentrated by other beneficiated processing such as washing, jigging, or magnetic separation. Chemical and glass manufacturers prefer coarser material; for chemicals, ca 4760–840 μm (4–20 mesh) is preferred, and for glass, ca 590–105 μm (30–140 mesh). Barite to be used in well drilling is ground dry to 44 μm (−325 mesh).

Production and Consumption. About 80% of the world's barite production is used as a weighting agent for the muds circulated in rotary drilling of oil and gas wells (see PETROLEUM, DRILLING FLUIDS AND OTHER OIL RECOVERY CHEMICALS). Table 2 shows the U.S. production–consumption balance. The 1988 demand for barite increased nearly 40% over that recorded in 1987. However, by the end of 1988, oil prices had declined and renewed economic uncertainties depressed exploration and development activity. Barite demand fell accordingly and imports of lower cost foreign product exceeded domestic production.

World mine production, reserves, and reserve base is listed in Table 3.

Table 2. U.S. Barite Production for 1986–1988[a]

	1986	1987	1988
production, mine, 10^3 t	270	407	369
imports for consumption, 10^3 t	677	750	909
exports, 10^3 t	6	8	1
consumption, apparent, 10^3 t	941	1149	1277
consumption, ground and crushed, 10^3 t	1105	1304	1798
price, average value per mine, $/t	37.73	32.11	30.63
employment, mine and mill	400	400	400
net import reliance of apparent consumption, %	71	65	71

[a]Ref. 5.

Uses. Drilling muds are aqueous suspensions of clay and barite used in the petroleum industry. During drilling operations, the muds are pumped into a well through the hollow drill stem, passing through the tip of the bit, and back up the space between the stem and the walls of the hole. The mud is effective in lubricating and cooling the drill bit, in sealing the walls to prevent the caving of the hole, in suspending the drill cuttings and carrying them to the surface, and, by establishing a hydrostatic column head of weighted fluid, helps to restrain high gas and oil pressures, reducing the tendency for blowouts. For this last reason, muds having a specific gravity as great as 2.5 are used and barite is the material of choice because of its high density, chemical inertness, relative nonabrasiveness, and widespread availability at reasonable cost.

Table 3. World Barite Production, Reserves, and Reserve Base, 10^3 t[a]

	Mine production			
	1987	1988	Reserve	Reserve base
United States	407	369	27,273	50,000
Canada	41	41	2,727	6,364
France	141	136	1,818	2,273
Germany, Federal Republic	182	182	909	909
India	273	295	27,273	29,091
Ireland	150	136	909	1,364
Italy	114	109	1,818	1,818
Mexico	405	318	6,364	7,727
Morocco	144	136	9,090	10,000
Peru	30	27	1,818	1,818
Thailand	34	45	6,364	7,727
Yugoslavia	36	36	1,818	8,182
other market economy countries	955	727	17,273	80,000
China	1,000	909	36,364	136,364
Russia	541	455	9,090	68,182
other centrally planned economies	227	205		18,182
World Total	*4,636*	*4,182*	*150,909*	*430,000*

[a]Ref. 5.

Finely ground barite which may be bleached, usually by sulfuric acid, or unbleached, is used as a filler or extender in paints (qv), especially in automotive undercoats, where its low oil absorption, easy wettability in oils, and good sanding properties are advantageous (see FILLERS). It is also used as a filler in plastics and rubber products. In nonasbestos brake linings, the barite filler acts as a heat sink (see BRAKE LININGS AND CLUTCH FACINGS). In floor mats and carpet backings made of polyurethane foam, barite imparts sound-deadening characteristics and improves processing qualities (see NOISE POLLUTION AND ABATEMENT METHODS; URETHANE POLYMERS). In furniture manufacture where polyurethane foam is used for recoil and density properties, the unique chemical inertness of barite in combination with its high density plays a primary role.

In the glass (qv) and ceramic industry (see CERAMICS), barite can be used both as a flux, to promote melting at a lower temperature or to increase the production rate, and as an additive to increase the refractive index of glass. The viscosity of barite-containing glass often needs to be raised. Alumina in the form of feldspar is sometimes used. To offset any color produced by iron from the barite addition, more decolorizer may be needed. When properly used, barytes help reduce seed, increase toughness and brilliancy, and reduce annealing time. Barite is also a raw material for the manufacture of other barium chemicals.

Barium Acetate

Barium acetate [*543-80-6*], $Ba(C_2H_3O_2)_2$, crystallizes from an aqueous solution of acetic acid and barium carbonate or barium hydroxide. The level of hydration depends on crystallization temperature. At <24.7°C the trihydrate, density 2.02 g/mL is formed; from 24.7 to 41°C barium acetate monohydrate [*5908-64-5*], density 2.19 g/mL precipitates; and above 41°C the anhydrous salt, density 2.47 g/mL results. The monohydrate becomes anhydrous at 110°C. At 20°C, 76 g of the monohydrate dissolves in 100 g of water. Barium acetate is used in printing fabrics, lubricating grease, and as a catalyst for organic reactions.

Barium Bromide

Barium bromide [*10553-31-8*], $BaBr_2$, mp 854°C, density 4.781 g/mL, also exists as barium bromide dihydrate [*7791-28-8*], $BaBr_2 \cdot 2H_2O$, dehydration temperature 120°C, density 3.58 g/mL. The solubility, wt %, of $BaBr_2$ in water is

Temperature, °C	−20	0	20	40	60	80
Solubility, wt % $BaBr_2$	45.6	49.5	51.0	53.2	55.1	57.4

Barium bromide is very soluble in methanol, yet almost insoluble in ethanol. Reported uses of barium bromide include: fabrication of phosphors, for example from BaF_2, $BaBr_2 \cdot 2H_2O$ and $EuBr_3$ (6); as a crystallization nucleating agent to control supercooling of $CaBr_2$ soluions (7); and in the production of halide glasses having ir transmission properties (8). Glass-transition temperature is sometimes influenced by $BaBr_2$ content.

Barium Carbonate

Most barium compounds are prepared from reactions of barium carbonate [*513-77-9*], $BaCO_3$, which is commercially manufactured by the "black ash" process from barite and coke in a process identical to that for strontium carbonate production. Depending on the co-product, soda ash and/or carbon dioxide are also consumed.

Precipitated or synthetic barium carbonate is the most commercially important of all the barium chemicals except for barite. Barium carbonate is an unusually dense material, that is almost insoluble in water and only slightly soluble in carbonated water. It does dissolve in dilute hydrochloric, nitric, and acetic acids and is also soluble in ammonium nitrate and ammonium chloride solutions.

Manufacture. An outline of the black ash process for $BaCO_3$ manufacture is shown in Figure 1. It is from the appearance of the product exiting the thermal reduction step that the process derives its name.

Barite ore is reduced in size to between 20 and 200 mesh (840 to 74 μm). Sizing represents a compromise between increases in reaction rate and dust loss. Coke is fed to the kiln at a coke to ore ratio of about 0.20:0.25.

Thermal Reduction. Thermal reduction is usually accomplished in a high temperature countercurrent rotary kiln. "Hot zone," a region near the kiln spill, temperature is usually controlled at 1100–1200°C. The reaction rate has been shown to be only slightly lower at 1050°C than at 1130°C (9). About 6% of the feed $BaSO_4$ remains unreacted after 30 min at 1050°C. Reaction completion is approached in less than 10 min at 1100°C (10).

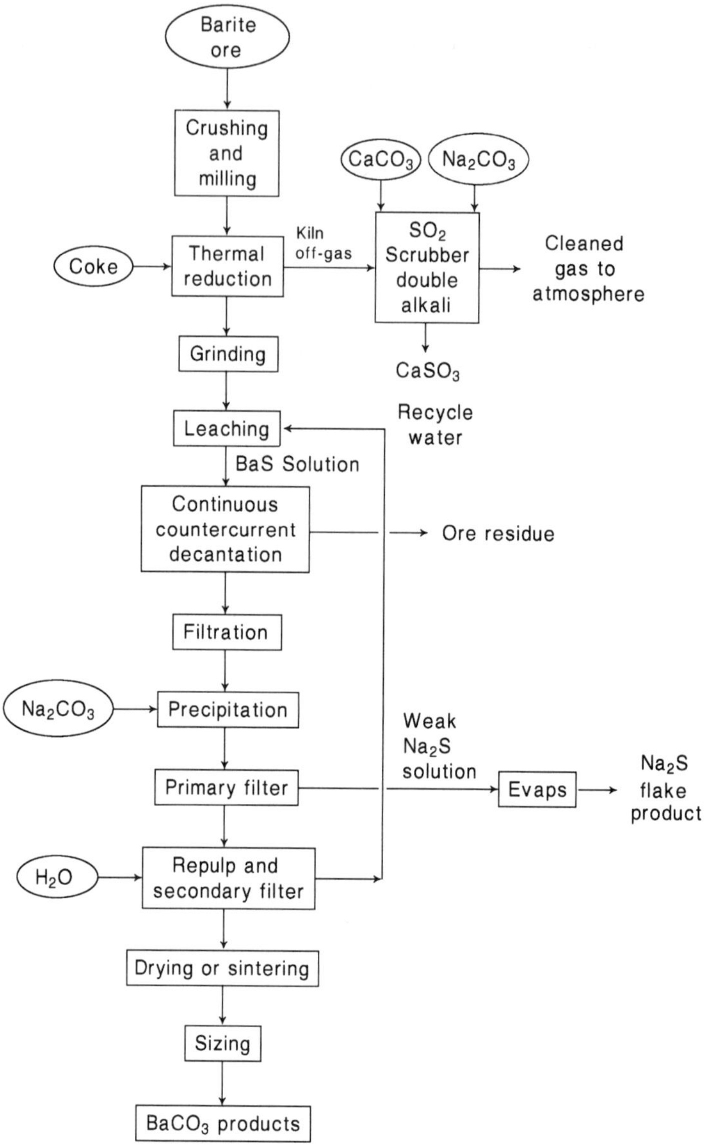

Fig. 1. Flow diagram for the black ash process.

The chief reduction reaction in the kiln is

$$BaSO_4 + 4\ CO \rightleftharpoons BaS + 4\ CO_2 \tag{1a}$$

BaS is leached from the black ash by hot water. The resulting solution is filtered and then treated with soda ash or carbon dioxide or a combination of the two to precipitate fine $BaCO_3$ crystals, which in turn are filtered and dried. Sulfide values can be recovered as H_2S, NaHS, Na_2S, or elemental sulfur.

Carbon dioxide, detrimental to high BaS yields, is repressed according to the

Boudouard equilibrium

$$C + CO_2 \rightleftharpoons 2\ CO \tag{1b}$$

Side reactions of barium with silicate impurities in the ore have been noted (1,11–13). These reactions can cause appreciable loss of barium values by forming water-insoluble barium compounds (14). For example, barium sulfate can form the orthosilicate

$$2\ BaSO_4 + SiO_2 \longrightarrow Ba_2SiO_4 + 2\ SO_3 \tag{2}$$

whereas if carbon is not present, this reaction does not occur below 1100°C. In the presence of carbon the reaction initiates at 1005°C.

Any $BaCO_3$ formed in the kiln can undergo the following reactions at temperatures below 1100°C to form the metasilicate (13)

$$BaCO_3 + SiO_2 \longrightarrow BaSiO_3 + CO_2 \tag{3}$$

and subsequently the orthosilicate

$$BaSiO_3 + BaCO_3 \longrightarrow Ba_2SiO_4 + CO_2 \tag{4}$$

In the hot leaching step, barium metasilicate is insoluble, and the orthosilicate yields only one-half of its barium value to give barium hydroxide

$$Ba_2SiO_4 + H_2O \longrightarrow BaSiO_3 + Ba(OH)_2 \tag{5}$$

If coke is not present in sufficient quantity, more silicates are formed, presumably because of increased CO_2 concentration which leads to greater amounts of $BaCO_3$ being formed (1).

Barium is reported in the kiln spill in three different forms: water-soluble, eg, BaS, BaO; acid-soluble, eg, $BaCO_3$, aluminate, silicate, ferrate; and insoluble, eg, unreduced barite.

Reduction of $BaSO_4$ appears to begin about 900°C (15). The presence of iron or iron oxide can catalyze the barium (9) and also strontium reduction reaction rates. However, iron impurity can also increase the acid-soluble content of the black ash (9).

Calcareous minerals such as gypsum [*13397-24-5*], when added in stoichiometric amounts relative to the barite impurities, reduce acid-soluble barium losses (16).

Off-gases from the kiln are scrubbed for SO_2, SO_3, dust, and organic volatiles. In addition to equation 2, reactions which can generate SO_x are

$$BaS + 3\ CO_2 \longrightarrow BaO + 3\ CO + SO_2 \tag{6}$$

and

$$BaSO_4 + CO \longrightarrow BaO + CO_2 + SO_2 \tag{7}$$

Black ash wt % composition is typically BaS, 65–72, SiO_2, 10, $BaSO_4$, 3, metal oxides, 0.5, $BaSiO_3$, 16, and carbon, 5.5. The amount of black ash produced is about 65% by weight of the total ore and coke fed.

Significant waste heat may be recovered from the high (about 600°C) kiln off-gas. Pre-heating combustion air or feed ore improves the energy efficiency of the process. Reduction of barite in a fluid bed with CO and/or hydrogen has been performed on an experimental scale.

Leaching. Details of a typical countercurrent decantation system used in the leaching of barium black ash are shown in Figure 2. Black ash from the ball milling step is fed to a series of staged make-up tanks in which the dissolution of BaS begins. The dissolving media is an aqueous solution of 6–7% BaS from the overflow of the second of three clarifiers. Barium sulfide on contact with water hydrolyzes to the hydroxide and hydrosulfide according to

$$2\ BaS + 2\ H_2O \longrightarrow Ba(OH)_2 + Ba(SH)_2 \quad (8)$$

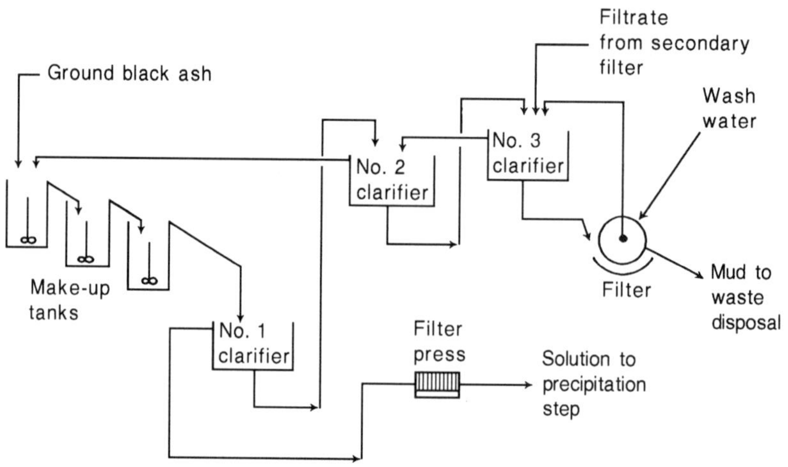

Fig. 2. Schematic of leaching circuit.

Leaching is 90–95% complete as the solids exit in the underflow of the first clarifier. Final washing of the solids (mud) using fresh water takes place in a rotary filter from whence the mud, consisting of 65% H_2O, 10–15% $BaSO_4$ and silicates, and 10–15% coke, is delivered to landfill at an approved site.

Overflow from the first clarifier, typically a 20% BaS solution, is filtered and sent on to the precipitation department. Settling of solids in the clarifiers can be enhanced by various flocculating agents (qv), preferably weakly anionic polyacrylamides (17).

The net percentage of solids in the clarifier underflows increases in going from the first to the second to the third clarifier. Typical values are 11, 15, and 21% solids, respectively. Clarifier operations have been investigated by computer simulation studies (17).

Precipitation. Filtered overflow from the first clarifier, 20% BaS solution, is fed to an agitated tank where, on tight control, carbonate values are added in slight excess of stoichiometric requirements. The excess carbonate suppresses soluble barium which would otherwise later precipitate in equipment.

Carbonate values are usually supplied from soda ash or carbon dioxide or an equimolar mix. Possible reactions are

$$BaS + Na_2CO_3 \longrightarrow BaCO_3 + Na_2S \tag{9}$$

or

$$BaS + 0.5\ Na_2CO_3 + 0.5\ CO_2 + 0.5\ H_2O \longrightarrow BaCO_3 + NaHS \tag{10}$$

or

$$BaS + CO_2 + H_2O \longrightarrow BaCO_3 + H_2S \tag{11}$$

An additional possibility is

$$BaS + NaHCO_3 \longrightarrow BaCO_3 + NaHS \tag{12}$$

The H_2S in equation 11 can, if desired, be converted into elemental sulfur in a Claus furnace.

A photomicrograph of barium carbonate formed by precipitation using pure soda ash (eq. 9), is shown in Figure 3. The average particle size is 1.2 μm. The exclusive use of soda ash results in a barium carbonate having included sodium that cannot be reduced below a certain level by repeated washings. The sodium can be detrimental if the $BaCO_3$ is to be used for barium titanate production.

Filtration. The slurry resulting from the precipitation step is filtered in a primary stage on rotary filters separating the product, $BaCO_3$, from the co-

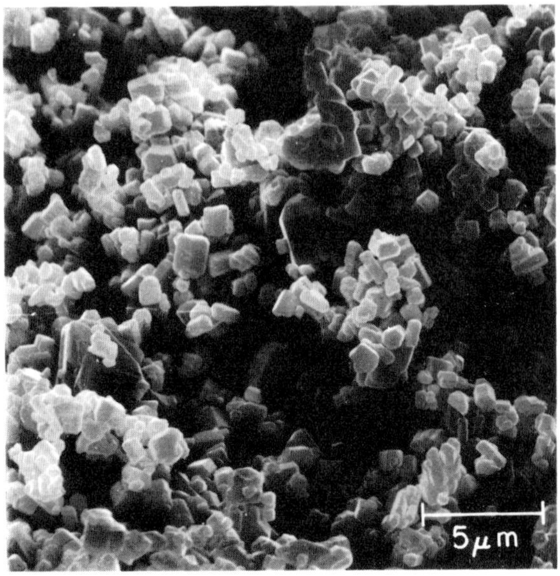

Fig. 3. Photomicrograph of precipitated $BaCO_3$.

product, eg, Na_2S or NaHS in solution. The $BaCO_3$ wet cake is repulped in fresh water and filtered on a second-stage rotary filter. The aqueous stream from this separation, carrying some sulfide values, is then used in the third clarifier of the leaching step (see Figs. 1 and 2).

Drying and Calcining. A rotary dryer operating at a 600–800°C spill temperature may be used to form a hard, sintered, granular barium carbonate product. This form finds favor in glass manufacture. More powdery products can be produced at rotary dryer spill temperatures of less than 400°C (14). Spray-dried barium carbonate, including added dispersants (qv), provides a finer pigment-grade material which is used in ferrite production (see FERRITES).

Sulfide Solution Concentration. The by-product of the black ash process has considerable bearing on the ultimate economic viability of barium production. It is desirable to be able to vary by-products to maximize earnings. Sodium sulfide [*1313-82-2*], Na_2S, sold as 60% flake, and sodium bisulfide [*16721-80-5*], NaHS, sold as 45% solution or as 70% flake, are typically co-produced.

When solid soda ash is used to supply all the carbonate values in the precipitation step (eq. 9), a ca 10% Na_2S solution results from the primary filtration step which can be concentrated to 40% Na_2S in a three-effect evaporator train. Final concentration to 60% Na_2S occurs in a high vacuum single-effect evaporator. This concentrated solution can then be fed to a flaker to produce a 60% sodium sulfide flake which is sold as co-product.

Both sodium sulfide and the bisulfide are used in the flotation process for copper minerals and as a depilatory for animal hides (see COPPER; COPPER ALLOYS; LEATHER). Also, sodium polysulfide can be produced from Na_2S, and elemental sulfur can be produced if H_2S is generated as an intermediate.

Supply and Demand. The barium carbonate supply and demand in the United States for the years 1985 through 1989 is given in Table 4.

Table 4. United States Barium Carbonate Supply and Demand, t[a]

Year	Estimated production	Imports	Exports	Apparent demand
1985	18,000	11,304		28,300[b]
1986	18,000	10,313	1,032	27,281
1987	18,000	11,404	1,524	27,880
1988	18,000	9,653	1,791	25,862
1989	18,000	14,401	527	31,874

[a]Trade data is from U.S. Department of Commerce Reports EM 545 and IM 145.
[b]Value is estimated.

Uses. There are several different grades of barium carbonate manufactured to fit the specific needs of a wide variety of applications: very fine, highly reactive grades are made for the chemical industry; coarser and more readily handleable grades are mainly supplied to the glass industry.

In 1989, 30% of the barium carbonate produced was used in glass manufacturing, 30% in brick and clay products, 20% in the manufacture of barium chemicals, 5% in the manufacture of barium ferrites, 5% in the production of photographic papers, and 10% in miscellaneous uses.

Barium carbonate acts as a flux for silica. It is added to textile fiber glass, crown and flint optical glass, laboratory glassware, decorative glass, and frits for ceramic material. Barium carbonate has the lowest melting point of all the alkaline earths, and, in the presence of sodium carbonate, glass melting reactions may begin at less than 400°C. Barium oxide from the carbonate becomes part of the silicate structure and imparts many useful properties. It increases the glass durability, adds weight and density, increases the refractive index imparting a brilliance to the glass, and perhaps most significantly, absorbs x rays. Hence, barium carbonate is used in television glass manufacturing as an x-ray screening agent. Barium carbonate is used especially in black and white television face plates, and in small quantities in color face plates, where strontium carbonate is the principal screening agent.

Glass-grade barium carbonate has a high bulk density so that it does not become airborne when charged to the furnace. It also has a particle size distribution which has less tendency to segregate from other materials in the glass mixture.

Barium carbonate prevents formation of scum and efflorescence in brick, tile, masonry cement, terra cotta, and sewer pipe by insolubilizing the soluble sulfates contained in many of the otherwise unsuitable clays. At the same time, it aids other deflocculants by precipitating calcium and magnesium as the carbonates. This reaction is relatively slow and normally requires several days to mature even when very fine powder is used. Consequently, often a barium carbonate emulsion in water is prepared with carbonic acid to further increase the solubility and speed the reaction.

In the oil-well drilling industry, the barite suspension used as drilling mud can be destabilized by the presence of soluble materials such as gypsum. Addition of barium carbonate precipitates the gypsum, inhibits coagulation, and thus permits the mud to retain the desired consistency and dispersion.

Barium carbonate of finely controlled particle size reacts in the solid state when heated with iron oxide to form barium ferrites. Magnetically aligned barium ferrite [*11138-11-7*] powder can be pressed and sintered into a hard-core permanent magnet which is used in many types of small motors. Alternatively, ground up magnetic powder can be compounded into plastic strips which are used in a variety of appliances as part of the closure mechanism.

Barium carbonate also reacts with titania to form barium titanate [*12047-27-7*], $BaTiO_3$, a ferroelectric material with a very high dielectric constant (see FERROELECTRICS). Barium titanate is best manufactured as a single-phase composition by a solid-state sintering technique. The asymmetrical perovskite structure of the titanate develops a potential difference when compressed in specific crystallographic directions, and vice versa. This material is most widely used for its strong piezoelectric characteristics in transducers for ultrasonic technical applications such as the emulsification of liquids, mixing of powders and paints, and homogenization of milk, or in sonar devices (see PIEZOELECTRICS; ULTRASONICS).

Barium carbonate is also used in the manufacture of photographic paper to generate barium sulfate which imparts a flat white appearance (see PHOTOGRAPHY).

Barium Chloride

Both anhydrous barium chloride [*10361-37-2*], $BaCl_2$, mol wt 208.25, density 3.856 g/mL, and barium chloride dihydrate [*10326-27-9*], $BaCl_2{\cdot}2H_2O$, mol wt 244.28, density 3.097 g/mL, are produced from a filtered aqueous solution formed by the reaction of hydrochloric acid and $BaCO_3$ or BaS. If BaS is used, the H_2S generated must be appropriately handled.

The solubility of $BaCl_2$ in water is

Temperature, °C	0	20	40	60	80	100
Solubility, wt % $BaCl_2$	24.0	26.3	28.9	31.7	34.4	26.2

Anhydrous $BaCl_2$ exists as monoclinic or cubic crystals. The transition to cubic occurs at 925°C. Barium chloride melts at 962°C; the dihydrate, which has monoclinic crystals, loses water at 113°C. Barium chloride, which is very hygroscopic, is sold in moisture-proof bags and steel or fiber drums.

Barium chloride finds use in the production of barium colors, such as the diazo dyes barium lithol red [*50867-36-2*] and barium salt of Red Lake C [*5160-02-1*], a mordant for acid dyes and dying of textiles. Other uses include aluminum refining and boiler water treatment.

$BaCl_2$ is used in heat treating baths because of the eutectic mixtures it readily forms with other chlorides. The melting points of some eutectic mixtures are: $BaCl_2{\cdot}2KCl$, 672–680°C; $BaCl_2$–NaCl, 39 mol % $BaCl_2$, 654°C; $BaCl_2$–$CaCl_2$, 631°C. $BaCl_2$ is also used to set up porcelain enamels for sheet steel (see ENAMELS, PORCELAIN OR VITREOUS; STEEL), and it is used to produce blanc fixe [*7727-43-7*].

Barium 2-Ethylhexanoate

Barium 2-ethylhexanoate [*4696-54-2*], $Ba(C_8H_{16}O_2)_2$, also known as barium octanoate or barium octoate is usually used in synergistic combination with cadmium or zinc organic salts as a thermal stabilizer for PVC (18,19). It is often available preformulated with other additives such as the cadmium or zinc salts. Owing to cadmium's toxicity, the development of noncadmium stabilizers is growing, and use of barium octanoate is declining. Barium ethylhexanoate is a liquid; barium stearate [*6865-35-6*], $Ba(C_{18}H_{36}O_2)_2$, is a powder that also serves as a thermostabilizer for PVC. Organophosphites are often added to barium–zinc stabilizers to improve stabilizing effectiveness.

Barium Hydrosulfide

Barium hydrosulfide [*25417-81-6*], $Ba(HS)_2$, is formed by absorption of hydrogen sulfide into barium sulfide solution. On addition of alcohol, barium hydrosulfide tetrahydrate [*12230-74-9*], $Ba(HS)_2{\cdot}4H_2O$, crystallizes as yellow rhombic crystals that decompose at 50°C. Solid barium hydrosulfide is very unstable. Its solubility in water is

Temperature, °C	0	20	40	60	80	100
Solubility, wt % $Ba(HS)_2$	32.6	32.8	34.5	36.2	39.0	43.7

Barium Hydroxide

Barium hydroxide is the strongest base and has the greatest water-solubility of the alkaline-earth elements. Barium hydroxide (barium hydrate, caustic baryta) exists as the octahydrate [*12230-71-6*], $Ba(OH)_2{\cdot}8H_2O$, the monohydrate [*22326-55-2*], $Ba(OH)_2{\cdot}H_2O$, or as the anhydrous [*17194-00-2*] material, $Ba(OH)_2$. The octahydrate and monohydrate have sp gr 2.18 and 3.74, respectively. The mp of the octahydrate and anhydrous are 77.9°C and 407°C, respectively.

Solubility of $Ba(OH)_2$ in water is strongly temperature dependent in the range >40°C.

Temperature, °C	0	20	40	60	80
Solubility, wt % $Ba(OH)_2$	1.65	3.74	7.60	17.32	50.35

These solutions are highly alkaline and can effectively remove CO_2 or other acidic gases from ambient atmosphere. The octahydrate is also soluble in methanol, but only slightly soluble in ethanol.

The octahydrate is prepared by dissolving BaO in hot water for several hours, filtering off undissolved impurities, then cooling the solution to effect crystallization. Vapor pressure data for the octahydrate (20) is

Temperature, °C	17.6	32.1	40.7	50.7	57.8	64.8
Pressure, kPa	0.48	1.32	2.36	5.08	8.01	12.56

The monohydrate can be produced by vacuum drying of the octahydrate (21).

Barium hydroxide is used in the manufacture of barium greases and plastic stabilizers such as barium 2-ethylhexanoate, in papermaking, in sealing compositions (see SEALANTS), vulcanization accelerators, water purification, pigment dispersion, in a formula for self-extinguishing polyurethane foams (see FIRE-EXTINGUISHING AGENTS), and in the protection of objects made of limestone from deterioration (see FINE ART EXAMINATION AND CONSERVATION; LIME AND LIMESTONE). Uses of the octahydrate include: use as a low temperature latent heat storage material in combination with Na or KNO_3 or NaOAc (22); use as a nucleating agent to reduce supercooling of $CaBr_2$ solution (7); and removal of CO_2 and $^{14}CO_2$ by passing through a bed of solid mono- or octahydrate.

Barium Iodide

Barium iodide dihydrate [*7787-33-9*], $BaI_2{\cdot}2H_2O$, crystallizes from hot aqueous solution. Below 30°C, barium iodide hexahydrate [*13477-15-1*], $BaI_2{\cdot}6H_2O$, crystallizes. The dihydrate, sp gr 5.15, loses water to form anhydrous barium iodide [*13718-50-8*], BaI_2, at 150°C, mp 740°C. Barium iodide discolors on exposure to air containing carbon dioxide, and decomposes to barium carbonate and iodine. BaI_2 is soluble in water.

Temperature, °C	−20	0	25	60	120
Solubility, wt % BaI_2	58.60	62.5	68.8	70.70	74.30

BaI_2 is also soluble in alcohol and acetone. BaI_2 is useful in making other iodides. BaI_2 has been cited for producing ir transparent glasses that are useful in power transmission from CO and CO_2 lasers (qv) from the ZnI_2–CsI–BaI_2 system (23); as a catalyst promoter, for such catalysts as rhodium(III) chloride, in carbonylation reactions (24,25); as being useful in chemical vapor deposition (see FILM DEPOSITION TECHNIQUES) as a precursor in forming the superconducting composition $YBa_2Cu_3O_{7-x}$; as a sintering aid for aluminum nitride (26), and in phosphor formulations for cathode-ray tubes.

Barium Metaborate

Barium metaborate monohydrate [*23436-05-7*], $Ba(BO_2)_2{\cdot}H_2O$, has a sp gr of 3.25–3.35, a fusion pt of 900–1050°C, n_D of 1.55–1.60, solubility of 0.3% in water at 21°C, and pH 9.8–10.3 for the saturated solution. It can be prepared from the reaction of a solution of BaS and sodium tetraborate.

$Ba(BO_2)_2{\cdot}H_2O$, used in flame retardant plastic formulations as a synergist for phosphorus or halogen compounds and as a partial or complete replacement for antimony oxide (see FLAME RETARDANTS), is excellent as an afterglow suppressant. The low refractive index of $Ba(BO_2)_2$ results in greater transparency and brighter colors in formulated plastics (27). Barium metaborate has been reported in paint formulations to convey insecticidal properties (28) (see INSECT CONTROL TECHNOLOGY). Barium metaborate crystals (the low temperature beta form) have been grown by Czocharski techniques (29,30) and are used in nonlinear optics for high power uv lasers and can be applied to uv photolithography (see LITHOGRAPHY; NONLINEAR OPTICAL MATERIALS). $Ba(BO_2)_2$ has been reported to be used in antibacterial coatings for aluminum heat exchanger surfaces of air conditioners (31). Use of the metaborate as a sintering aid for $BaTiO_3$ has been studied (32).

The barium oxide borate [*12007-55-5*], $BaO{\cdot}2B_2O_3$, and dibarium oxide triborate [*13840-10-3*], $2BaO{\cdot}3B_2O_3$, have also been reported.

Barium Nitrate

Barium nitrate [*10022-31-8*], $Ba(NO_3)_2$, occurs as colorless crystals; mp 592°C; sp gr 3.24. Its solubility in water is

Temperature, °C	0	25	40	60	80	100	135
Solubility, wt % $Ba(NO_3)_2$	4.72	9.27	12.35	16.9	21.4	25.6	32.0

Barium nitrate is prepared by reaction of $BaCO_3$ and nitric acid, filtration and evaporative crystallization, or by dissolving sodium nitrate in a saturated solution of barium chloride, with subsequent precipitation of barium nitrate. The precipitate is centrifuged, washed, and dried. Barium nitrate is used in pyrotechnic green flares, tracer bullets, primers, and in detonators. These make use of its property of easy decomposition as well as its characteristic green flame. A small amount is used as a source of barium oxide in enamels.

Barium Nitrite

Barium nitrite [*13465-94-6*], $Ba(NO_2)_2$, crystallizes from aqueous solution as barium nitrite monohydrate [*7787-38-4*], $Ba(NO_2)_2 \cdot H_2O$, which has yellowish hexagonal crystals, sp gr 3.173, solubility 54.8 g $Ba(NO_2)_2$/100 g H_2O at 0°C, 319 g at 100°C. The monohydrate loses its water of crystallization at 116°C. Anhydrous barium nitrite, sp gr 3.234, melts at 267°C and decomposes at 270°C into BaO, NO, and N_2. Barium nitrite may be prepared by crystallization from a solution of equivalent quantities of barium chloride and sodium nitrite, by thermal decomposition of barium nitrate in an atmosphere of NO, or by treating barium hydroxide or barium carbonate with the gaseous oxidiation products of ammonia. It has been used in diazotization reactions.

Barium Oxide

Barium oxide [*1304-28-5*], BaO, occurs as colorless cubic or hexagonal crystals; mp 1923°C; sublimation ca 2000°C; bp ca 3088°C; sp gr (cubic) 5.72, (hexagonal) 5.32. Barium oxide is highly reactive toward carbon dioxide in the presence of water vapor, and is converted to the hydroxide and carbonate when exposed to air. Upon additions of small amounts of water, or in a carbon dioxide atmosphere, absorption is so rapid that large amounts of heat are liberated, raising the temperature to a red heat. The heat of formation of barium hydroxide from barium oxide and water is 102 kJ/mol (24.4 kcal/mol) and that of barium carbonate from barium oxide and carbon dioxide is 264 kJ/mol (63.1 kcal/mol). Consequently, accumulation of barium oxide and the peroxide, dust, and dirt presents a fire hazard. When a fire occurs, large concentrations of carbon dioxide from burning organic matter cause further incandescence because the carbon dioxide is absorbed by the barium oxide. Such fires spread rapidly and are difficult to extinguish.

Of the alkaline-earth carbonates, $BaCO_3$ requires the greatest amount of heat to undergo decomposition to the oxide. Thus carbon in the form of coke, tar, or carbon black, is added to the carbonate to lower reaction temperature from about 1300°C in the absence of carbon to about 1050°C. The potential for the reverse reaction is decreased by removing the CO_2 as shown in equation 1b.

Barium oxide, which can react directly with oxygen to give the peroxide (33), is soluble in methanol and ethanol forming the alkoxides (see ALKOXIDES, METAL).

BaO is used to impart improved strength to porcelain (34), as a solid base catalyst, in specialty cements, and for drying gases.

Barium Peroxide

When heated in air or oxygen to 500°C, barium oxide is converted readily to barium peroxide [*1304-29-6*], BaO_2. Upon further heating to 700°C, the peroxide decomposes to BaO and oxygen. This reaction was used for many years to make pure oxygen via the Brin process. Other preparations for the peroxide such as the reaction between $BaCl_2$ and H_2O_2 in alkaline solution, or between BaO and H_2O_2 have been reported.

Reported uses of BaO_2 include in the cathodes of fluorescent lamps, formation of $YBa_2Cu_3O_{7-x}$ superconducting phase from CuN_3, BaO_2, and Y_2O_3 (24), and as a drying agent for lithographic inks (qv).

Barium Sodium Niobium Oxide

Barium sodium niobium oxide [*12323-03-4*], $Ba_2NaNb_5O_{15}$, finds application for its dielectric, piezoelectric, nonlinear crystal and electro-optic properties (35,36). It has been used in conjunction with lasers for second harmonic generation and frequency doubling. The crystalline material can be grown at high temperature, mp ca 1450°C (37).

Barium Sulfate

Barium sulfate [*7727-43-7*], $BaSO_4$, occurs as colorless rhombic crystals, mp 1580°C (dec); sp gr 4.50; solubility 0.000285 g/100 g H_2O at 30°C and 0.00118 at 100°C. It is soluble in concentrated sulfuric acid, forming an acid sulfate; dilution with water reprecipitates barium sulfate. Precipitated $BaSO_4$ is known as blanc fixe, prepared from the reaction of aqueous solutions of barium sulfide and sodium sulfate.

Because of its extreme insolubility, barium sulfate is not toxic; the usual antidote for poisonous barium compounds is to convert them to barium sulfate by administering sodium or magnesium sulfate. In medicine, barium sulfate is widely used as an x-ray contrast medium (see IMAGING TECHNOLOGY; X-RAY TECHNOLOGY). It is also used in photographic papers, filler for plastics, and in concrete as a radiation shield. Commercially, barium sulfate is sold both as natural barite ore and as a precipitated product. Blanc fixe is also used in making white sidewall rubber tires or in other rubber applications.

Barium Sulfide

Impure barium sulfide with 20–35% contaminants is produced in large volume by the black ash kiln. Pure barium sulfide [*21109-95-5*], BaS, occurs as colorless cubic crystals, sp gr 4.25 and as hexagonal plates of barium sulfide hexahydrate [*66104-39-0*], $BaS{\cdot}6H_2O$. BaS melts at 2227°C. Solubility in water is reported (21) as

Temperature, °C	0	20	40	60	80	90	100
Solubility, wt % BaS	2.8	7.3	13.0	21.7	33.3	40.2	37.6

BaS hydrolyzes to $Ba(OH)_2$ and barium hydrosulfide. Cooling of an aqueous BaS solution can precipitate the double salt barium hydroxide sulfide hydrate [*42821-46-5*], $Ba(OH)_2{\cdot}Ba(SH)_2{\cdot}xH_2O$.

Barium sulfide solutions undergo slow oxidation in air, forming elemental sulfur and a family of oxidized sulfur species including the sulfite, thiosulfate, polythionates, and sulfate. The elemental sulfur is retained in the dissolved liquor in the form of polysulfide ions, which are responsible for the yellow color of most BaS solutions. Some of the more highly oxidized sulfur species also enter the solution. Sulfur compound formation should be minimized to prevent the compounds made from BaS, such as barium carbonate, from becoming contaminated with sulfur.

BaS is used in the manufacture of lithophone [*8006-32-4*], useful as a white pigment in paints, according to

$$BaS + ZnSO_4 \longrightarrow BaSO_4\ (ppt) + ZnS\ (ppt)$$

BaS has been used in the production of thin-film electroluminescent phosphors, often activated with Eu^{2+} (38,39). Similarly, infrared-triggered phosphors may be fabricated from Ba or Sr sulfide, a dopant such as EuO and a fusible salt such as LiF (40). BaS has also been mentioned as a nucleating agent in combination with $BaCl_2{\cdot}2H_2O$ for a $CaCl_2{\cdot}6H_2O$ heat storage system (41), and in vulcanization of carbon black-filled neoprene rubbers (42).

Barium Titanate

The basic crystal structure of barium titanate [*12047-27-7*], $BaTiO_3$, the so-called perovskite structure, after the mineral, $CaTiO_3$, leads to unique, outstanding dielectric properties. The barium and oxygen ions together form a face-centered cubic lattice, the titanium ions fitting into octahedral interstices. The minimum energy positions for the titanium ion are off-center and give rise to an ordered electrical dipole, which can be made permanent during production by the proper heating and cooling regime (see CERAMICS; FERROELECTRICS; TITANIUM COMPOUNDS).

Barium titanate has widespread use in the electronics industry. Its high dielectric constant and the ease with which its electrical properties can be modified by combination with other materials make it exceptionally suitable for a variety of items, ie, miniature capacitors (see CERAMICS AS ELECTRICAL MATERIALS).

Barium titanate is usually produced by the solid-state reaction of barium carbonate and titanium dioxide. Dielectric and piezoelectric properties of $BaTiO_3$ can be affected by stoichiometry, microstructure, and additive ions that can enter into solid solution. In the perovskite lattice, substitutions of Pb^{2+}, Sr^{2+}, Ca^{2+}, and Cd^{2+} can be made for part of the barium ions, maintaining the ferroelectric characteristics. Similarly, the Ti^{4+} ion can partially be replaced with Sn^{4+}, Hf^{4+}, Zr^{4+}, Ce^{4+}, and Th^{4+}. The possibilities for forming solution alloys in all these structures offer a range of compositions, which present a wide range of dielectric, temperature dependence, and other characteristics. At the same time, impurities such as SiO_2, Al_2O_3, Na_2O, etc, or anything affecting the crystal lattice, can also alter the dielectric properties and must be carefully controlled, both in raw materials and within the production process itself. Nonstoichiometry originating from the BaO:TiO_2 ratio or additional impurities may result in semiconductive ceramics upon firing.

Barium titanate ceramic products are used for application in numerous sonic and ultrasonic devices, such as underwater sonar, guided missiles, acoustic mines, ultrasonic cleaning, measuring instruments, eg, for flaw detection, liquid-level sensing, and thickness gauges, sound reproduction, filters, and ultrasonic machining (see LIQUID-LEVEL MEASUREMENTS; NONDESTRUCTIVE TESTING; SURFACE AND INTERFACE ANALYSIS). Its most important application is for ceramic capacitors of disk, tube, and multilayer designs (43,44).

Yttrium–Barium–Copper Oxide

Yttrium–barium–copper oxide, $YBa_2Cu_3O_{7-x}$, is a newly developed high T_c material which has been found to be fully superconductive at temperatures above 90 K, a temperature that can be maintained during practical operation. The foremost challenge is to be able to fabricate these materials into a flexible form to prepare wires, fibers, and bulk shapes. Ultrapure powders of yttrium–barium–copper oxide that are sinterable into single-phase superconducting material at low temperatures are required (see ULTRAPURE MATERIAL), creating a worldwide interest in high purity barium chemicals.

A number of promising new routes to chemically synthesize ultrapure, ultrahomogeneous particles of controlled particle size and particle size distribution are currently under development. A modified sol–gel method has been developed to produce thermoplastic gels that are compatible with fiber spinning technology (see SOL–GEL TECHNOLOGY), as has a process which avoids the formation of barium carbonate, a troublesome impurity at the grain boundaries which seriously detracts from the critical current. In this latter process, hyponitrites and hydrated oxides derived from the hydrolysis of organometallic solutions are subjected to a series of thermal treatments involving decomposition, oxidation, and finally annealing. Materials which can be decomposed and converted to the orthorhombic superconducting form are produced (45,46).

Health and Safety

The average adult human body contains 22 mg Ba, of which 93% is present in bone (47). The remainder is widely distributed throughout the soft tissues of the body in very low concentrations. Accumulation of barium also takes place in the pigmented parts of the eyes.

Environmental Levels and Exposures. Barium constitutes about 0.04% of the earth's crust (47). Agricultural soils contain Ba^{2+} in the range of several micrograms per gram. The Environmental Protection Agency, under the Safe Drinking Water Act, has set a limit for barium of 1 mg/L for municipal waters in the United States.

Generally, barium content of food parallels calcium content in a ratio from $1/10^2$ to $1/10^5$ (47). Milk contains about 45–136 micrograms Ba per gram Ca; wheat and oatmeal contain 1300 and 2320–8290 micrograms Ba per gram Ca, respectively. The average daily intake of barium may be as high as 1.33 mg in the diet of the general population. Dietary barium intake has been estimated to originate 25% from milk, 25% from flour, 25% from potatoes, and 25% from miscellaneous high barium foods, such as nuts, consumed in small quantities.

Toxicity. The toxicity of barium compounds depends on solubility (47–49). The free ion is readily absorbed from the lung and gastrointestinal tract. The mammalian intestinal mucosa is highly permeable to Ba^{2+} ions and is involved in the rapid flow of soluble barium salts into the blood. Barium is also deposited in the muscles where it remains for the first 30 h and then is slowly removed from the site (50). Very little is retained by the liver, kidneys, or spleen and practically none by the brain, heart, and hair.

Soluble Compounds. The mechanism of barium toxicity is related to its ability to substitute for calcium in muscle contraction. Toxicity results from stimulation of smooth muscles of the gastrointestinal tract, the cardiac muscle, and the voluntary muscles, resulting in paralysis (47). Skeletal, arterial, intestinal, and bronchial muscle all seem to be affected by barium.

Oral Exposure. Following oral exposure to high doses of barium compounds, symptoms of nausea, vomiting, colic, and diarrhea result. Other symptoms include excessive salivation, hypertension, tremors, tingling of the extremities, twitching of facial muscles, and paralysis of the tongue and pharynx with loss or impairment of speech (49,51). Barium stimulation of arterial muscles causes an elevation in blood pressure resulting from vasoconstriction. Severe cases of poisoning cause loss of tendon reflexes, heart fibrillation, and general muscular paralysis, including the respiratory muscles, leading to death (47). Acute barium toxicity results in low serum potassium levels and leukocytosis in both experimental animals and humans (49,52). Symptoms of chronic barium poisoning are similar, but of lesser severity.

The more soluble forms of barium such as the carbonate, chloride, acetate, sulfide, oxide, and nitrate, tend to be more acutely toxic (50). Mean lethal doses for ingested barium chloride were 300–500 mg/kg in rats and 7–29 mg/kg in mice (47).

Administration of 5 ppm barium, the acetate, to mice in the drinking water in a life-time study had no observable effects on longevity, mortality, and body weights, or on the incidence of tumors (53). Long-term studies in rats exposed to Ba^{2+} in drinking water containing 5 mg/L, as acetate, or 10–250 mg/L, as chloride, resulted in no measurable toxic effects (47).

Water-insoluble barium salts are poorly absorbed. In fact, barium sulfate is used as a contrast material for x-ray examination of the gastrointestinal tract based on its limited solubility and low toxicity (52). Barium sulfate fed to mice at various levels up to 8 ppm dietary Ba ($\sim$1.14 mg/kg·d as Ba^{2+}) for three generations had no significant effects on growth, mortality, morbidity, or reproductive performance (53).

Inhalation Exposure. Workers exposed to barium carbonate dust for 7–27 years did not reveal any specific chronic poisoning (21). The American Conference of Governmental Industrial Hygienists (ACGIH) and the Occupational Safety and Health Administration (OSHA) have recommended a threshold limit value (TLV) of 0.5 mg/m^3 as Ba, as a time weighted average, for soluble compounds of barium. This time weighted average is an average airborne exposure in any 8-h work shift of a 40-h work week which should not be exceeded.

In humans, inhaled insoluble barium salts are retained in the lung (47,49). Inhalation of high concentrations of the fine dusts of barium sulfate can result in the formation of harmless nodular granules in the lungs, a condition called baritosis (49). Baritosis produces no specific symptoms and no changes in pulmonary function. The nodulates disappear upon cessation of exposure to the barium salt. However, it is possible that barium sulfate may produce benign pneumoconiosis because, unlike barium carbonate, barium sulfate is poorly absorbed (21).

The threshold of a toxic dose in adult humans is about 0.2–0.5 g Ba; the lethal dose in untreated cases is 3–4 g Ba, LD_{50} about 66 mg/kg (47). The fatal dose of barium chloride for humans is reported to be between 0.8 and 0.9 g (0.55–0.60 g of Ba) (50). However, for most of the acid-soluble salts of barium, doses greater than 1 g have been tolerated (51). Lethal doses are summarized in Table 5. Dusts of barium oxide are considered potential dermal and nasal irritants (52).

Table 5. Acute Lethal Doses of Soluble Barium Compounds[a]

Compound	Route of exposure	Species	Toxicity, LD_{50}, mg/kg
$BaCO_3$	oral	rat	418
	oral	mouse	200
	oral	human	11[b]
			29[b]
	oral	human	57[c]
$BaNO_3$	oral	rat	355
	intravenous	mouse	8.5
$BaCl_2$	oral	rat	118
	oral	human	11.4[c]
BaO	subcutaneous	mouse	50
BaF_2	oral	rat	250

[a]Ref. 54.
[b]Value is toxic dose low, TD_{LO}, the lowest dose of a substance introduced by any route other than inhalation, over any given period of time to which humans or animals have been exposed and reported to produce any nonsignificant toxic effect in humans or to produce nonsignificant tumorigenic or reproductive effects in animals or humans.
[c]Value is lethal dose low, LD_{LO}, the lowest dose of a substance introduced by any other route other than inhalation, over any given period of time and reported to have caused death in humans or animals.

Treatment. Treatment of poisoning from soluble barium salts may be preventive or curative (47,51). Preventive treatment involves inhibition of intestinal absorption by administering such soluble sulfates as magnesium or sodium, causing precipitation of barium sulfate in the alimentary tract. Curative treatment involves counteracting the paralytic effect of the Ba^{2+} ion on the muscle by intravenous infusion of a potassium salt.

Analysis

The classical analytical method of determination of barium ion is gravimetric, by precipitating and weighing insoluble barium sulfate. Barium chromate, which is more insoluble than strontium chromate in a slightly acidic solution, gives a fairly good separation of the two elements.

Alkaline-earth metals are often determined volumetrically by complexometric titration at pH 10, using Eriochrome Black T as indicator. The most suitable complexing titrant for barium ion is a solution of diethylenetriaminepentaacetic acid (DTPA). Other alkaline earths, if present, are simultaneously titrated, and in the favored analytical procedure calcium and strontium are determined separately by atomic absorption spectrophotometry, and their values subtracted from the total to obtain the barium value.

Barium can also be determined by x-ray fluorescence (XRF) spectroscopy, atomic absorption spectroscopy, and flame emission spectroscopy. Prior separation is not necessary. XRF can be applied directly to samples of ore or products to yield analysis for barium and contaminants. All crystalline barium compounds can be analyzed by x-ray diffraction.

BIBLIOGRAPHY

"Barium Compounds" in *ECT* 1st ed., Vol. 2, pp. 307–323, by L. Preisman, Pittsburgh Plate Glass Co., and "Barium Hydride," by L. W. Davis, Metal Hydrides, Inc.; in *ECT* 2nd ed., Vol. 3, pp. 80–98, by L. Preisman, Pittsburgh Plate Glass Co.; in *ECT*, 3rd ed., Vol. 3, pp. 463–479, by T. Kirkpatrick, Sherwin-Williams Co.

1. W. Lobunez, personal communication, Apr. 10, 1973.
2. *Minerals Year Book.*
3. F. A. Cotton and G. Wilkinson, *Advanced Inorganic Chemistry,* John Wiley & Sons, Inc., New York, 1980, pp. 271–273.
4. *Industrial Minerals and Rocks,* 4th ed., American Institute of Mining, Metallurgical, and Petroleum Engineers, Inc., New York, 1975, pp. 427–442.
5. *U.S. Bureau of Mines Mineral Commodity Summaries,* 1989.
6. U.S. Pat. 4,534,884 (Aug. 13, 1985), A. Satoshi, N. Takashi, and T. Kenji (to Fuji Photo Film Co.).
7. U.S. Pat. 4,690,769 (Sept. 1, 1987), G. A. Lane and H. E. Rossow (to The Dow Chemical Company).
8. Jpn. Pat. 60,246,242 (Dec. 5, 1985) (to Nippon Sheet Glass KK).
9. W. Lobunez, *FMC Report No. 5729-R,* Princeton, N.J., July 13, 1972.
10. W. Lobunez, *FMC Report No. 5686-R,* Princeton, N.J., Mar. 13, 1973.
11. W. E. Brownell, *J. Am. Ceram. Soc.* **46,** 1225–1228 (1963).
12. T. G. Akhmetov, *Chem. Ind.* **48,** 288 (1973).
13. T. Yamaguchi and co-workers, *J. Inorg. Nucl. Chem.* **34,** 2739 (1972).
14. D. J. Muyskens, D. W. Tunison, and E. Rau, *Engineering Encyclopedia.*
15. D. W. Tunison, *FMC Report No. 5527-R,* Princeton, N.J., 1971.
16. K. Kadic and M. Klofec, *Chem. Prum.* **31**(10), 519–521 (1981).
17. J. L. Manganaro, R. C. Schmidt, and W. K. Lau, *FMC Report ICG/T-79-075,* Princeton, N.J., June 12, 1979.
18. H. Andreas, in R. Gachter and H. Muller, eds., *Plastics Additives Handbook,* Hanser Publications, 1985, Chapt. 4.

19. O. S. Kauder, in J. T. Lutz, ed., *Thermoplastics Polymer Additives,* Marcel Dekker Inc., New York, 1989, Chapt. 12.
20. E. W. Washburn, ed., *International Critical Tables of Numerical Data, Physics, Chemistry, and Technology (ICT),* Vol. 7, McGraw-Hill, New York, 1930.
21. W. Gerhatz, ed., *Ullmann's Encyclopedia of Industrial Chemistry,* Vol. A3, VCH, New York, 1985.
22. R. Kniep, S. Mann, and A. Zachos, *Sol. Energy* **36,** 291 (1986).
23. K. Kadonao and co-workers, *J. Non-Cryst. Solids* **116**(1), 33–38 (1990).
24. K. Shinoda, K. Yasuda, and D. Miyatani, *Nippon Kayaku (Kaishi)* **5,** 724–729 (1988).
25. U.S. Pat. 4,705,890 (Nov. 10, 1987), G. Steinmetz and M. Rule (to Eastman Kodak Co.).
26. Jpn. Pat. 61,021,977 (Jan. 30, 1986), H. Taniguchi and N. Kuramoto (to Tokuyama Soda KK).
27. J. Green and co-workers, in H. S. Katz and J. V. Milewski, eds., *Handbook of Fillers for Plastics,* Van Nostrand Reinhold Co., New York, 1987, Chapt. 18.
28. A. Z. Gomaz, *Pigm. Resin Technol.* **18**(4), 4–8 (1989).
29. Jpn. Pat. 63,215,598 (Sept. 8, 1988), N. Oonishi (to Agency of Ind. Sci. Tech.).
30. U.S. Pat. 4,793,894 (Dec. 27, 1988), J. C. Jacco and G. M. Loiacono (to North American Philips Corp.).
31. Jpn. Pat. 61,168,675 (July 30, 1986), T. Watanabe (to Mitsubishi Heavy Ind KK, Nakabishi Eng KK).
32. D. Kolar, M. Trontely, and L. Marsel, *J. Phylics. Colloq.* **C11, C1**(447), **Ci**(450) (1986).
33. T. Moeller, *Inorganic Chemistry,* Wiley-Interscience, New York, 1952.
34. H. Schubert, *Sprechsael* **120**(6), 529–531 (1987).
35. Jpn. Pat. 61,106,463 (May 24, 1986), K. Nagata and K. Hikita (to Mitsubishi Mining Cement).
36. S. V. Kruzhalov and co-workers, *Pis'ma Zh. Tech. Fiz.* **8**(12), 756 (1982).
37. L. G. Van Uitert, *Platinum Metals Rev.* **14,647,** 118–121 (1970).
38. W.O. 8,908,921 A1 (Sept. 21, 1989), H. Hirano, N. Iwase, and N. Koshino (to Fujitsu Ltd.).
39. Jpn. Pat. 63,000,995 A2 (Jan. 5, 1988), K. Takahaski, Y. Onuki, and A. Kondo (to Toyo Soda M. Lq. KK).
40. U.S. Pat. 4,705,952 (Nov. 10, 1987), J. Lindmayer (to Quantex Corp.).
41. EP 240,583 (Oct. 14, 1987), J. Lindmayer (to Kubota Tekko KK).
42. A. A. Nosuikov, C. A. Blokh, and V. V. Bondovenko, *Izv. Vyssh. Uchebn. Zaved., Khim. Khim. Tekhnol.* **29**(12), 121–125 (1986).
43. W. D. Kingery, H. K. Bowen, and D. R. Uhlmann, *Introduction to Ceramics,* 2nd ed., John Wiley & Sons, Inc., New York, 1960.
44. *Ceramic Industry,* 33 (Jan. 1989).
45. J. R. Gaines, Jr., *Am. Ceram. Soc. Bull.* **68**(4), 857–869.
46. D. R. Ulrich, "Chemical Processing of Ceramics," *C & EN* (Jan. 1, 1990).
47. A. L. Reeves, in L. Friberg, G. F. Nordberg, and V. B. Vouk, eds., *Handbook on the Toxicology of Metals,* Vol. 2, 2nd ed., Elsevier, New York, 1986, pp. 84–94.
48. C. D. Klaaseen, M. O. Amdur, and J. Doull, eds., *Casarett and Doull's Toxicology: The Basic Science of Poisons,* 3rd ed., Macmillan Publishing Co., New York, 1986, pp. 623–624.
49. B. Venugopal and T. D. Luckey, *Metal Toxicity in Mammals,* Vol. 2, Plenum Press, New York, 1978, pp. 63–67.
50. G. D. Clayton and F. E. Clayton, eds., *Patty's Industrial Hygiene and Toxicology,* Vol. 2A, 3rd ed., John Wiley & Sons, Inc., New York, 1981, pp. 1531–1537.
51. R. E. Gosselin, R. P. Smith, and H. C. Hodge, *Clinical Toxicology of Commercial Products,* Williams and Wilkins, Baltimore, Md., 1984, Section III-61-3.
52. *Documentation of the Threshold Limit Values and Biological Exposure Indices,* Ameri-

can Conference of Governmental Industrial Hygienists (ACGIH), 5th ed., 1986, pp. 47–48.
53. E. J. Underwood, *Trace Elements in Human and Animal Nutrition,* 4th ed., Academic Press, New York, 1977, pp. 434–436.
54. *Registry of Toxic Effects of Chemical Substances,* 1985–1986.

Patrick M. DiBello
James L. Manganaro
Elizabeth R. Aguinaldo
FMC Corporation

BARRIER POLYMERS

Barrier polymers are used for many packaging and protective applications. As barriers they separate a system, such as an article of food or an electronic component, from an environment. That is, they limit the introduction of matter from the environment into the system or limit the loss of matter from the system or both. In many cases the environment is simply room air, but the environment can be very different, such as in the case of protecting a submerged system from water.

All polymers are barriers to some degree; however, no polymer is a perfect barrier. Polymers only limit the movement of substances. When a polymer limits the movement enough to satisfy the requirements of a particular application, it is a barrier polymer for that application. Hence, barrier polymer finds definition in the application. A given polymer may be a good barrier for protecting one system from a component of an environment, but a poor barrier for protecting another system from a different component of the environment. This description of a barrier polymer is much wider than the traditional operational definitions that focus narrowly on oxygen permeabilities and water-vapor transmission rates. This article discusses barrier polymers and the permeation process within the context of barrier applications. It also covers units of measure, physical factors affecting permeation, typical barrier structures, prediction and measurement of barrier properties, uses, and health and safety factors.

The Permeation Process

Barrier polymers limit movement of substances, hereafter called permeants. The movement can be through the polymer or, in some cases, merely into the polymer. The overall movement of permeants through a polymer is called permeation, which is a multistep process. First, the permeant molecule collides with the polymer. Then, it must adsorb to the polymer surface and dissolve into the polymer bulk. In the polymer, the permeant "hops" or diffuses randomly as its own thermal kinetic energy keeps it moving from vacancy to vacancy while the polymer chains move. The random diffusion yields a net movement from the side of the barrier polymer that is in contact with a high concentration or partial pressure of the permeant to the side that is in contact with a low concentration of permeant. After crossing the barrier polymer, the permeant moves to the polymer surface, desorbs, and moves away.

Permeant movement is a physical process that has both a thermodynamic and a kinetic component. For polymers without special surface treatments, the thermodynamic contribution is in the solution step. The permeant partitions between the environment and the polymer according to thermodynamic rules of solution. The kinetic contribution is in the diffusion. The net rate of movement is dependent on the speed of permeant movement and the availability of new vacancies in the polymer.

A few simple equations describe most applications of barrier polymers. Equation 1 is an adaptation for films of Fick's First Law.

$$\frac{\Delta M_x}{\Delta t} = \frac{PA\Delta p_x}{L} \tag{1}$$

where $\Delta M_x/\Delta t$ is the steady-state rate of permeation of permeant x through a polymer film, P is the permeability coefficient (commonly called the permeability), A is the area of the film, Δp_x is the difference in pressure of the permeant on the two sides of the film, and L is the thickness of the film.

The permeability is the product of the diffusion coefficient D and the solubility coefficient S.

$$P = DS \tag{2}$$

The diffusion coefficient, sometimes called the diffusivity, is the kinetic term that describes the speed of movement. The solubility coefficient, which should not be called the solubility, is the thermodynamic term that describes the amount of permeant that will dissolve in the polymer. The solubility coefficient is a reciprocal Henry's Law coefficient as shown in equation 3.

$$C_{\text{eq}} = Sp_x \tag{3}$$

where C_{eq} is the equilibrium concentration of permeant that will dissolve in a polymer when the permeant has a partial pressure of p_x.

A polymer can have a low permeability because it has a low value of D or a low value of S or both. A low value of D can result from either static or dynamic effects. Static effects include molecular packing in the amorphous phase, orientation, and the amount of crystallinity. Molecular packing affects the way permeant molecules move through the free volume or vacanies in the polymer host. When there is a small amount of free volume, movement is limited. Symmetric monomers lead to good packing and hence lower diffusion rates. For example, vinylidene chloride, which is symmetric 1,1-dichloroethene [*75-35-4*], leads to a good barrier polymer because the adjacent molecular chains can pack together well. The analogue vinyl chloride [*75-01-4*], which is asymmetric monochloroethene, leads to a polymer that does not pack as well. The simple chain of high density polyethylene (HDPE) packs much better than the branched chain of low density polyethylene (LDPE). Hence, HDPE has better barrier properties than LDPE.

Orientation sometimes leads to lower permeability values (better barrier properties). Orientation can increase packing density, which lowers the diffusion coefficient *D*; it can also increase the difficulty of hopping or diffusing in a direction perpendicular to the film. In the latter case, movement in general may be fast, but movement through the film is limited. However, mere stretching does not always lead to orientation of the molecular chains. In fact, stretching can lead to void formation, which increases permeability.

Increased crystallinity can reduce permeability values because the crystal regions of a polymer are impenetrable in most semicrystalline polymers. Hence, the average value of the solubility coefficient *S* is reduced. It also means that movement must occur around the crystallites, which means that a longer distance must be traveled. This lowers the effective value of *D*.

The dynamic effects on the diffusion coefficient are related to polymer chain mobility. Permeant movement is enhanced when there are fluctuations in the free volume of the polymer. As chains move with thermal motion, opportunities for diffusion or hopping increase. Spaces between polymer molecules that are too small for passage at one point in time can be adequate later as the molecules move. Ultimately, this effect is related to temperature in general and often to the glass-transition temperature T_g in particular. As the temperature increases, thermal motion increases, and the probability of larger openings occurring increases. Below T_g, thermal motion is limited to short-range; small openings occur at low rates. Above T_g, thermal motion is enhanced, longer-range motion is common, and larger openings occur at usable rates. All the polymer–polymer interactions that affect T_g affect the permeability, including dipole–dipole interactions and efficient packing.

The common model of a polymer as a collection of noodles in a bowl is not adequate for diffusion. A better model is a collection of long, active worms. Movement through the collection is enhanced if the worms are widely spaced (static effect) and if they have agitated motion (dynamic effect).

A low solubility coefficient can result when the permeant does not condense readily or does not interact strongly with the polymer. Generally, those permanent gases that have low boiling temperatures have low solubility coefficients in polymers. They are inherently adverse to existing in a condensed state either as a liquid or in a polymer solution. Hence, molecules like oxygen have solubility coefficients that are many orders of magnitude lower than the solubility coefficients of heavy flavor molecules. When no specific interactions, such as dipole–dipole interactions or hydrogen bonding, occur between the polymer and the permeant, solution is minimal. All those effects that control solutions in general apply to the polymer solution. These effects are not fully understood, but progress is being made.

Units of Measure

To understand permeability or barrier property values, it is necessary to define the units of measure. These units are complicated and many different sets of units are in common use. Furthermore, from time to time the units of permeability are presented in confused or incorrect fashion in the literature.

The dimensions of permeability become clear after rearranging equation 1 to solve for P. The permeability must have dimensions of quantity of permeant (either mass or molar) times thickness in the numerator with area times a time interval times pressure in the denominator. Table 1 contains conversion factors for several common unit sets with the permeant quantity in molar units. The unit nmol/(m·s·GPa) is used herein for the permeability of small molecules because this unit is SI, which is preferred in current technical encyclopedias, and it is only a factor of 2, different from the commercial permeability unit, (cc(STP)·mil)/(100 in.^{2}·d·atm). The molar character is useful for oxygen permeation, which could ultimately involve a chemical reaction, or carbon dioxide permeation, which is often related to the pressure in a beverage bottle.

For permeation of flavor, aroma, and solvent molecules another metric combination of units is more useful, namely, (kg·m)/(m^2·s·Pa). In this unit the permeant quantity has mass units. This is consistent with the common practice of describing these materials. Permeability values in these units often carry a cumbersome exponent; hence, a modified unit, an MZU (10^{-20} kg·m)/(m^2·s·Pa), is used herein. The conversion from this permeability unit to the preferred unit for small molecules depends on the molecular weight of the permeant. Equation 4 expresses the relationship where MW is the molecular weight of the permeant in daltons (g/mol).

$$\text{permeability in MZU} \times (10/MW) = \text{permeability in nmol/(m·s·GPa)} \qquad (4)$$

The solubility coefficient must have units that are consistent with equation 3. In the literature S has units cc(STP)/(cm^3·atm), where cc(STP) is a molar unit for absorbed permeant (nominally cubic centimeters of gas at standard temperature and pressure) and cm^3 is a volume of polymer. When these units are multiplied by an equilibrium pressure of permeant, concentration units result. In preferred SI units, S has units of nmol/(m^3·GPa).

$$\begin{aligned}&\text{solubility coefficient in cc(STP)/(cm}^3\text{·atm)} \times (4.04 \times 10^{14})\\ &\quad = \text{solubility coefficient in nmol/(m}^3\text{·GPa)}\end{aligned} \qquad (5)$$

In the mass units for flavor, aroma, and solvent molecules, the solubility coefficient has units kg/(m^3·Pa). Equation 6 shows how to convert from the mass units to the molar units.

$$\begin{aligned}&\text{solubility coefficient in kg/(m}^3\text{·Pa)} \times (10^{21}/MW)\\ &\quad = \text{solubility coefficient in nmol/(m}^3\text{·GPa)}\end{aligned} \qquad (6)$$

The diffusion coefficient has been commonly reported in cm^2/s in many applications. Hereafter, m^2/s will be used since it uses the basic SI units.

The water-vapor transmission rate (WVTR) is another descriptor of barrier polymers. Strictly, it is not a permeability coefficient. The dimensions are quantity times thickness in the numerator and area times a time interval in the denominator. These dimensions do not have a pressure dimension in the denomi-

Table 1. Permeability Units[a] with Conversion Factors

Multiply→ to obtain↓	$\frac{\text{nmol}}{\text{m·s·GPa}}$	$\frac{\text{cc·mil}}{\text{100 in.}^2\text{·d·atm}}$	$\frac{\text{cc·mil}}{\text{m}^2\text{·d·atm}}$	$\frac{\text{cc·cm}}{\text{cm}^2\text{·s·atm}}$	$\frac{\text{cc·cm}}{\text{cm}^2\text{·s·cm Hg}}$	$\frac{\text{cc·20 }\mu\text{m}}{\text{m}^2\text{·d·atm}}$
$\frac{\text{nmol}}{\text{m·s·GPa}}$	1	2	0.129	4.390×10^{10}	3.336×10^{12}	0.1016
$\frac{\text{cc·mil}}{\text{100 in.}^2\text{·d·atm}}$	0.50	1	6.452×10^{-2}	2.195×10^{10}	1.668×10^{12}	5.08×10^{-2}
$\frac{\text{cc·mil}}{\text{m}^2\text{·d·atm}}$	7.75	15.50	1	3.402×10^{11}	2.585×10^{13}	0.787
$\frac{\text{cc·cm}}{\text{cm}^2\text{·s·atm}}$	2.278×10^{-11}	4.557×10^{-11}	2.939×10^{-12}	1	76.00	2.315×10^{-12}
$\frac{\text{cc·cm}}{\text{cm}^2\text{·s·cm Hg}}$	2.998×10^{-13}	5.996×10^{-13}	3.860×10^{-14}	1.316×10^{-2}	1	3.046×10^{-14}
$\frac{\text{cc·20 }\mu\text{m}}{\text{m}^2\text{·d·atm}}$	9.84	19.68	1.27	4.32×10^{11}	3.283×10^{13}	1

[a]Throughout the *Encyclopedia* cm^3 (or mL) is used in preference to cc. However, the advantage of using cc here is an obvious visual aid in the complex units and there are further comments regarding cc vs cm^3 in the text.

nator as does the permeability. Common commercial units for WVTR are (g·mil)/(100 in.^{2}·d). Table 2 contains conversion factors for several common units for WVTR. This text uses the preferred nmol/(m·s). The WVTR describes the rate that water molecules move through a film when one side has a humid environment and the other side is dry. The WVTR is a strong function of temperature because both the water content of the air and the permeability are directly related to temperature. For the WVTR to be useful, the water-vapor pressure difference for the value must be reported. Both these facts are recognized by specifying the relative humidity and temperature for the WVTR value. This enables the user to calculate the water-vapor pressure difference. For example, the common conditions are 90% relative humidity (rh) at 37.8°C, which means the pressure difference is 5.89 kPa (44 mm Hg). The WVTR may be converted to a permeability by dividing by the specified pressure difference of the water vapor.

Table 2. Water Vapor Transmission Rate Units with Conversion Factors

Multiply→ to obtain↓	$\frac{\text{nmol}}{\text{m·s}}$	$\frac{\text{g·mil}}{100\ \text{in.}^2\text{·d}}$	$\frac{\text{g·cm}}{\text{m}^2\text{·d}}$
$\frac{\text{nmol}}{\text{m·s}}$	1	0.253	6.43
$\frac{\text{g·mil}}{100\ \text{in.}^2\text{·d}}$	3.95	1	25.40
$\frac{\text{g·cm}}{\text{m}^2\text{·d}}$	0.155	3.94×10^{-2}	1

Small Molecule Permeation

PERMANENT GASES

Table 3 lists the permeabilities of oxygen [*7782-44-7*], nitrogen [*7727-37-9*], and carbon dioxide [*124-38-9*] for selected barrier and nonbarrier polymers at 20°C and 75% rh. The effect of temperature and humidity are discussed later. For many polymers the permeabilities of nitrogen, oxygen, and carbon dioxide are in the ratio 1:4:14.

The traditional definition of a barrier polymer required an oxygen permeability less than 2 nmol/(m·s·GPa) (originally, less than (1 cc·mil)/(100 in.^{2}·d·atm)) at room temperature. This definition was based partly on function and partly on conforming to the old commercial unit of permeability. The old commercial unit of permeability was created so that the oxygen permeability of SaranWrap brand plastic film, a trademark of The Dow Chemical Company, would have a numerical value of 1.

However, the traditional definition of a barrier polymer is a good starting point for food packaging. Table 4 contains the approximate oxygen tolerances of several food groups. This table and equation 1 can be used to analyze a hypothetical packaging situation. For a package with an area of 0.045 m^2 (70 in.^{2}) and a wall

Table 3. Permeabilities of Selected Polymers[a]

Polymer	Gas permeability nmol/m·s·GPa		
	Oxygen	Nitrogen	Carbon dioxide
vinylidene chloride copolymers	0.02–0.30	0.005–0.07	0.1–1.5
ethylene–vinyl alcohol copolymers, dry	0.014–0.095		
at 100% rh	2.2–1.1		
nylon-MXD6[b]	0.30		
nitrile barrier polymers	1.8–2.0		6–8
nylon-6	4–6		20–24
amorphous nylon (Selar[c] PA 3426)	5–6		
poly(ethylene terephthalate)	6–8	1.4–1.9	30–50
poly(vinyl chloride)	10–40		40–100
high density polyethylene	200–400	80–120	1200–1400
polypropylene	300–500	60–100	1000–1600
low density polyethylene	500–700	200–400	2000–4000
polystyrene	500–800	80–120	1400–3000

[a]Ref. 1; see Table 1 for unit conversion.
[b]Trademark of Mitsubishi Gas Chemical Co.
[c]Trademark of E. I. du Pont de Nemours & Co., Inc.

thickness of 50.8 μm (2 mil), the oxygen permeability needs to be less than 1.9 nmol/(m·s·GPa) to give 100 days of protection in air (2.1×10^{-5} GPa O_2) to 500 g of contents that could withstand up to 20 ppm (wt/wt) oxygen. A lower oxygen permeability is needed if the application requires greater oxygen protection, a longer storage time, or a thinner wall.

Table 4. Estimated Maximum Oxygen Tolerance of Various Foods[a]

Food or beverage	Maximum oxygen tolerance, ppm
beer (pasteurized)	1–2
typical autoclaved low acid foods	1–3
canned milk	
canned meats and vegetables	
canned soups	
baby foods	
fine wine	2–5
coffee (fresh ground)	2–5
tomato-based products	3–8
high acid fruit juices	8–20
carbonated soft drinks	10–40
oils and shortenings	20–50
salad dressings, peanut butter	30–100
liquor, jams, jellies	50–100+

[a]Ref. 2.

Poly(ethylene terephthalate) [*25038-59-9*] (PET), with an oxygen permeability of 8 nmol/(m·s·GPa), is not considered a barrier polymer by the old definition; however, it is an adequate barrier polymer for holding carbon dioxide in a 2-L bottle for carbonated soft drinks. The solubility coefficients for carbon dioxide are much larger than for oxygen. For the case of the PET soft drink bottle, the principal mechanism for loss of carbon dioxide is by sorption in the bottle walls as 500 kPa (5 atm) of carbon dioxide equilibrates with the polymer (3). For an average wall thickness of 370 μm (14.5 mil) and a permeability of 40 nmol/(m·s·GPa), many months are required to lose enough carbon dioxide (15% of initial) to be objectionable.

The diffusion and solubility coefficients for oxygen and carbon dioxide in selected polymers have been collected in Table 5. Determination of these coefficients is neither common, nor difficult. Methods are discussed later. The values of S for a permeant gas do not vary much from polymer to polymer. The large differrences that are found for permeability are due almost entirely to differences in D.

Table 5. Diffusion and Solubility Coefficients for Oxygen and Carbon Dioxide in Selected Polymers at 23°C, Dry[a]

	Oxygen		Carbon dioxide	
Polymer	D, m^2/s	S, $nmol/(m^3 \cdot GPa)$[b]	D, m^2/s	S, $nmol/(m^3 \cdot GPa)$[b]
vinylidene chloride copolymer	1.2×10^{-14}	1.01×10^{13}	1.3×10^{-14}	3.2×10^{13}
ethylene–vinyl alcohol copolymer[c]	7.2×10^{-14}	2.4×10^{12}		
acrylonitrile barrier polymer	1.0×10^{-13}	1.0×10^{13}	9.0×10^{-14}	4.4×10^{13}
poly(ethylene terephthalate)	2.7×10^{-13}	2.8×10^{13}	6.2×10^{-14}	8.1×10^{14}
poly(vinyl chloride)	1.2×10^{-12}	1.2×10^{13}	8.0×10^{-13}	9.7×10^{13}
polypropylene	2.9×10^{-12}	1.1×10^{14}	3.2×10^{-12}	3.4×10^{14}
high density polyethylene	1.6×10^{-11}	7.2×10^{12}	1.1×10^{-11}	4.3×10^{13}
low density polyethylene	4.5×10^{-11}	2.0×10^{13}	3.2×10^{-11}	1.2×10^{14}

[a]Refs. 4–9.
[b]For unit conversion, see equation 5.
[c]42 mol % ethylene.

Polymers With Good Barrier to Permanent Gases. Those polymers that are good barriers to permanent gases, especially oxygen, have important commercial significance. They represent a diverse set of chemical structures shown in Figure 1.

Vinylidene chloride copolymers are available as resins for extrusion, latices for coating, and resins for solvent coating. Comonomer levels range from 5 to 20 wt %. Common comonomers are vinyl chloride, acrylonitrile, and alkyl acrylates.

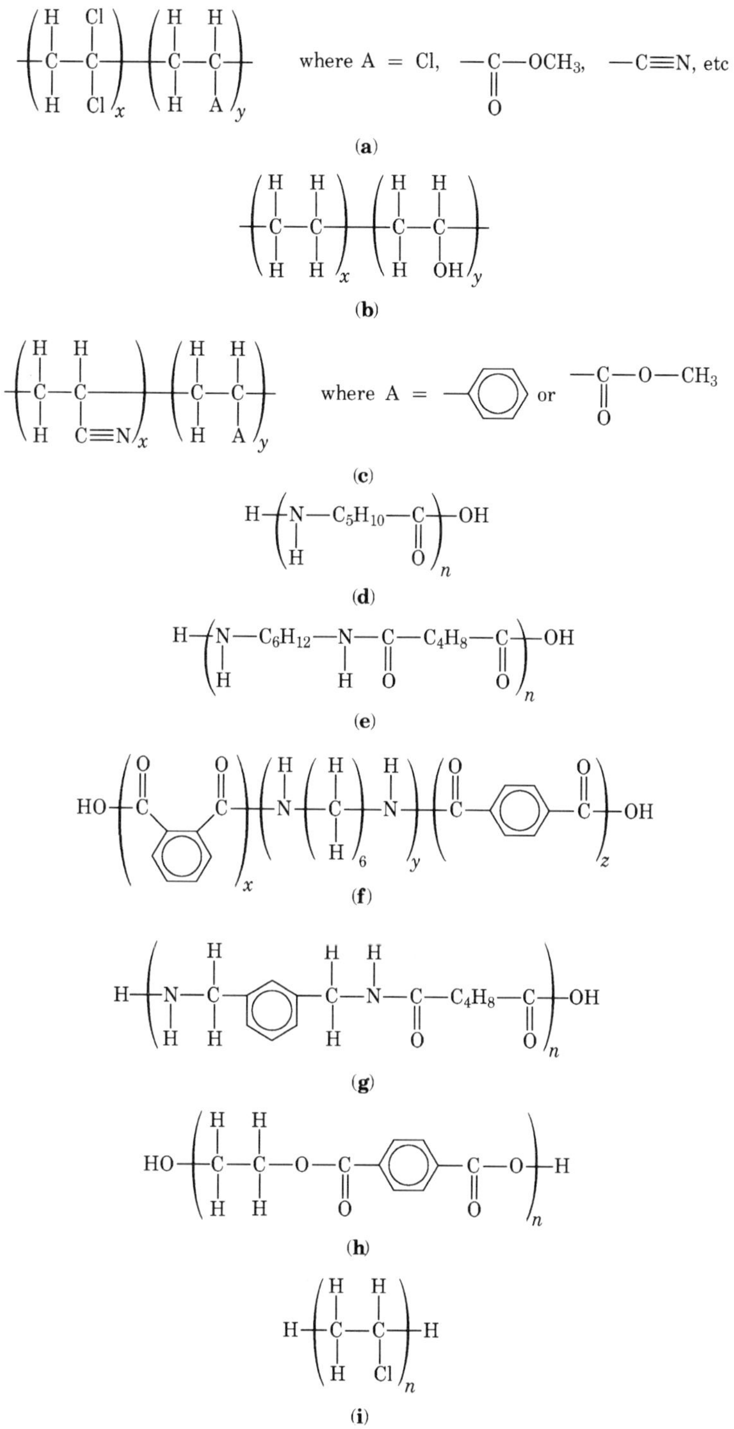

Fig. 1. Chemical structures of barrier polymers. (**a**) Vinylidene chloride copolymers; (**b**) hydrolyzed ethylene–vinyl acetate (EVOH); (**c**) acrylonitrile barrier polymers; (**d**) nylon-6; (**e**) nylon-6,6; (**f**) amorphous nylon (Selar PA 3426), $y = x + z$; (**g**) nylon-MXD6; (**h**) poly(ethylene terephthalate); and (**i**) poly(vinyl chloride).

The permeability of the polymer is a function of type and amount of comonomer. As the comonomer fraction of these semicrystalline copolymers is increased, the melting temperature decreases and the permeability increases. The permeability of vinylidene chloride homopolymer has not been measured.

Vinylidene chloride copolymers are marketed under a variety of trade names. Saran is a trademark of The Dow Chemical Company for vinylidene chloride copolymers. Other trade names include Daran (W.R. Grace), Amsco Res (Union Oil), and Serfene (Morton Chemical) in the United States; Haloflex (Imperial Chemical Industries, Ltd.), Diofan (BASF), Ixan (Solvay and Cie SA), and Polyidene (Scott-Bader) in Europe.

Hydrolyzed ethylene–vinyl acetate copolymers [*24937-78-8*], commonly known as ethylene–vinyl alcohol (EVOH) copolymers [*25067-34-9*], are usually used as extrusion resins, although some may be used in solvent-coating applications. As the ethylene fraction of these semicrystalline copolymers increases, the melting temperature decreases, the permeabilities increase, and the sensitivity to humidity decreases. The permeabilities as a function of polymer composition and humidity are shown in Figure 2. Vinyl alcohol homopolymer [*9002-89-5*] has a very low oxygen permeability in dry conditions; however, the polymer is water-soluble. Trade names for these barrier polymers include Eval, Soarnol, Selar OH, and Clarene. Table 6 lists the compositions of some commonly used EVOH copolymers.

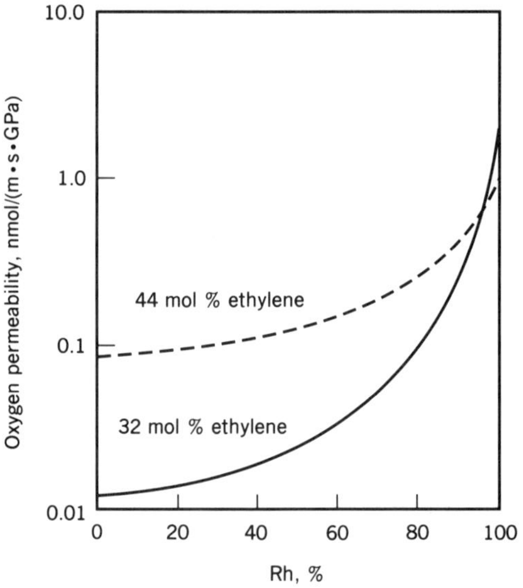

Fig. 2. Oxygen permeabilities of ethylene–vinyl alcohol copolymers at 20°C (10). See Table 1 for unit conversions.

Copolymers of acrylonitrile [*107-13-1*] are used in extrusion and molding applications. Commercially important comonomers for barrier applications include styrene and methyl acrylate. As the comonomer content is increased, the permeabilities increase as shown in Figure 3. These copolymers are not moisture-

Table 6. Composition of Ethylene Vinyl Alcohol (EVOH) Copolymers

Manufacturer	Name[a]	Ethylene, mol %
E. I. du Pont de Nemours & Co., Inc.	Selar OH 3003, 3007	30
	Selar OH 3803	38
	Selar OH 4416	44
Nippon Goshei	Soarnol D, DT	29
	Soarnol E, ET	38
Eval Company of America	Eval L	27
	Eval F	32
	Eval H	38
	Eval E	44
	Eval K[b]	38
Solvay	Clarene L	30
	Clarene P	36

[a]Names are trademarks of the individual manufacturer.
[b]Blend of 32 mol % and 44 mol % copolymers.

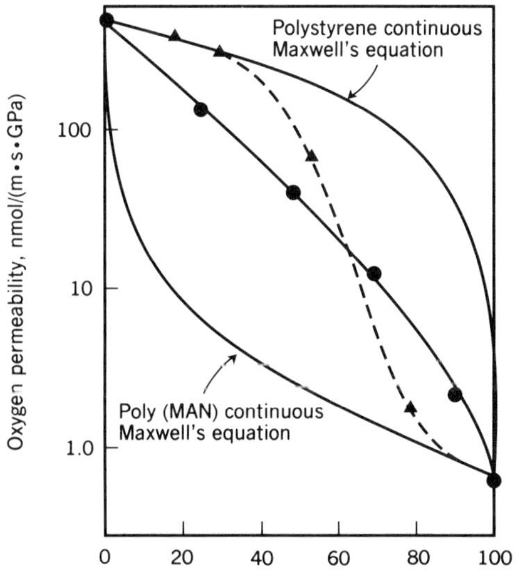

Fig. 3. Oxygen permeabilities of block (▲) and random (●) styrene–methacrylonitrile (MAN) copolymers [*33961-16-9*] (11). See Table 1 for unit conversions.

sensitive. Table 7 contains descriptions of three high nitrile barrier polymers. Barex and Cycopac resins are rubber-modified to improve the mechanical properties.

Polyamide polymers can provide good-to-moderate barrier to permeation by permanent gases. Nylon-6 [*25038-54-4*] and nylon-6,6 [*32131-17-2*] have been available for many years. Nylon-6 and nylon-6,6 are moderate oxygen barriers that are slightly humidity sensitive. They are, however, excellent barriers against migration of flavor and aroma compounds. New polyamides are being introduced. One

Table 7. Compositions of High Nitrile Barrier Polymers

Polymer[a]	Manufacturer	Chemical composition[b]
Lopac	Monsanto Co.	100% copolymer of 70% acrylonitrile and 30% styrene
Barex	Sohio	90% copolymer of 74% acrylonitrile and 26% methyl acrylate + 10% butadiene rubber graft
Cycopac	Borg-Warner Chemicals	90% copolymer of 74% acrylonitrile and 26% styrene + 10% butadiene rubber graft

[a]Names are trademarks of the individual manufacturer.
[b]Data from FDA regulations for corresponding materials.

of these is an amorphous nylon, Selar PA (E. I. du Pont de Nemours & Co., Inc.), which is a terpolymer of hexamethylenediamine [*124-09-4*] and a mixture of phthalic acids or acid chlorides (12). Nylon-MXD6 (Mitsubishi Gas Chemical Company) is a copolymer of *m*-xylylenediamine [*1477-55-0*] and adipic acid [*124-04-9*]. The amorphous nylons are unusual because the oxygen permeability actually decreases slightly as the humidity increases. Nylon-MXD6 [*25805-74-7*] is more typical; its oxygen permeability increases as the humidity increases. Research is active in the area of polyamide chemistry for new barrier polymers.

Two often used polymers have adequate properties for some applications. Poly(ethylene terephthalate) (PET) is used to make films and bottles. This polymer is commonly made from ethylene glycol and dimethyl terephthalate [*120-61-6*] or from ethylene glycol [*107-21-1*] and terephthalic acid [*100-21-0*]. PET is a moderate barrier to permanent gases; however, it is an excellent barrier to flavors and aromas. The oxygen permeability decreases slightly with increasing humidity. Poly(vinyl chloride) (PVC) is a moderate barrier to permanent gases. Plasticized poly(vinyl chloride) is used as a household wrapping film. The plasticizers greatly increase the permeabilities. Table 8 lists the permeabilities of selected permeants in the three main household wrapping films.

Table 8. Permeability of Household Films at 25°C to Selected Permeants

	Film type		
Permeant[a]	Plasticized vinylidene chloride copolymer[b]	Plasticized poly(vinyl chloride)[c]	Polyethylene[d]
oxygen, nmol/(m·s·GPa)	1.9	220	640
water vapor[e], nmol/(m·s)	0.055	0.30	0.19
d-limonene, MZU[f]	130	1.1×10^5	3.3×10^5
dipropyl disulfide, MZU[f]	1.1×10^4	3.3×10^6	6.8×10^6

[a]See Tables 1 and 2 for unit conversions.
[b]SaranWrap brand plastic film (trademark of The Dow Chemical Company).
[c]Reynolds plastic wrap (trademark of Reynolds Metals Co.).
[d]Handi-Wrap II brand plastic film (trademark of DowBrands, Inc.).
[e]Measured at 37.8°C, 90% rh.
[f]MZU = $(10^{-20}$ kg·m)/(m^2·s·Pa).

WATER VAPOR TRANSMISSION

Table 9 lists WVTR values for selected polymers. Comparison of Tables 3 and 9 shows that often there is a reversal of roles. Those polymers that are good oxygen barriers are often poor water-vapor barriers and vice versa. This can be rationalized as follows. Barrier polymers often rely on dipole–dipole interactions to reduce chain mobility and, hence, diffusional movement of permeants. These dipoles can be good sites for hydrogen bonding. Water molecules are attracted to these sites, leading to high values of S. Furthermore, the water molecules enhance D by interrupting the attractions and chain packing. Polymer molecules without dipole–dipole interactions, such as polyolefins, dissolve very little water and have low WVTR and permeability values. The low values of S more than compensate for the naturally higher values of D.

Table 9. Water-vapor Transmission Rates of Selected Polymers[a]

Polymer	WVTR, nmol/(m·s)[b]
vinylidene chloride copolymers	0.005–0.05
high density polyethylene (HDPE)	0.095
polypropylene	0.16
low density polyethylene (LDPE)	0.35
ethylene–vinyl alcohol, 44 mol % ethylene[c]	0.35
poly(ethylene terephthalate) (PET)	0.45
poly(vinyl chloride) (PVC)	0.55
ethylene–vinyl alcohol, 32 mol % ethylene[c]	0.95
nylon-6,6, nylon-11	0.95
nitrile barrier resins	1.5
polystyrene	1.8
nylon-6	2.7
polycarbonate	2.8
nylon-12	15.9

[a]At 38°C and 90% rh unless otherwise noted (13).
[b]See Table 2 for unit conversions.
[c]Measured at 40°C.

Large Molecule Permeation

The permeation of flavor, aroma, and solvent molecules in polymers follows the same physics as the permeation of small molecules. However, there are two significant differences. For these larger molecules, the diffusion coefficients are much lower and the solubility coefficients are much higher. This means that steady-state permeation may not be reached during the storage time of some packaging situations. Hence, large molecules from the environment might not enter the contents, or loss of flavor molecules would be limited to sorption into the polymer. However, since the solubility coefficient is large, the loss of flavor could be important solely from sorption in the polymer. Furthermore, the large solubility coefficient can lead to enough sorption of the large molecule that plasticization occurs in the polymer, which can increase the diffusion coefficient.

Table 10 contains some selected permeability data including diffusion and solubility coefficients for flavors in polymers used in food packaging. Generally, vinylidene chloride copolymers and glassy polymers such as polyamides and EVOH are good barriers to flavor and aroma permeation whereas the polyolefins are poor barriers. Comparison to Table 5 shows that the large molecule diffusion coefficients are 1000 or more times lower than the small molecule coefficients. The solubility coefficients are as much as one million times higher. Equation 7 shows how to estimate the time to reach steady-state permeation t_{ss} if the diffusion coefficient and thickness of a film are known.

$$t_{ss} = \frac{L^2}{4D} \tag{7}$$

For *d*-limonene diffusion in a 50-μm thick vinylidene chloride copolymer film, steady-state permeation is expected after 2000 days. For a 50-μm thick LDPE film, steady-state permeation is expected in less than one hour. If steady-state permeation is not achieved, the effective penetration depth L^* for simple diffusion, after time t has elapsed, can be estimated with equation 8.

$$L^* = 2(Dt)^{1/2} \tag{8}$$

For a food container, the amount of sorption could be estimated in the following way. For simple diffusion the concentration in the polymer at the food surface could be estimated with equation 3. This would require a knowledge of the partial pressure of the flavor in the food. This is not always available, but methods exist for estimating this when the food matrix is water-dominated. The concentration in the polymer at the depth of penetration is zero. Hence the average concentration C_{avg} is as from equation 9.

$$C_{avg} = 0.5Sp_x \tag{9}$$

The quantity sorbed ΔM_{sorb}, is simply the volume of polymer affected times the average concentration. The volume affected is the package area times L^*. The result is equation 10.

$$\Delta M_{sorb} = A(Dt)^{1/2}Sp_x \tag{10}$$

In this equation A and t are observable. The pressure of the flavor must be obtained from an external reference, and D and S must be measured or estimated. If steady state is achieved, the quantity sorbed is given by equation 11. In either case the quantity sorbed can be important.

$$\Delta M_{sorb} = 0.5ALSp_x \tag{11}$$

The scalping of flavor and aroma by a package can be minimized by placing a barrier material as near as possible to the food. The ingress of undesirable

Table 10. Examples of Permeation of Flavor and Aroma Compounds in Selected Polymers at 25°C[a], Dry[b]

Flavor/aroma compound	Permeant formula	P, MZU[c]	D, m^2/s	S, $kg/(m^3 \cdot Pa)$[d]
		Vinylidene chloride copolymer		
ethyl hexanoate [*123-66-0*]	$C_8H_{16}O_2$	570	8.0×10^{-18}	0.71
ethyl 2-methylbutyrate [*7452-79-1*]	$C_7H_{14}O_2$	3.2	1.9×10^{-17}	1.7×10^{-3}
hexanol [*111-27-3*]	$C_6H_{14}O$	40	5.2×10^{-17}	7.7×10^{-3}
trans-2-hexenal [*6728-26-3*]	$C_6H_{10}O$	240	1.8×10^{-17}	0.14
d-limonene [*5989-27-5*]	$C_{16}H_{16}$	32	3.3×10^{-17}	9.7×10^{-3}
3-octanone [*106-68-3*]	$C_8H_{16}O$	52	1.3×10^{-18}	0.40
propyl butyrate [*105-66-8*]	$C_7H_{14}O_2$	42	4.4×10^{-18}	9.4×10^{-2}
dipropyl disulfide [*629-19-6*]	$C_6H_{14}S_2$	270	2.6×10^{-18}	1.0
		Ethylene–vinyl alcohol copolymer		
ethyl hexanoate		0.41	3.2×10^{-18}	1.3×10^{-3}
ethyl 2-methylbutyrate		0.30	6.7×10^{-18}	4.7×10^{-4}
hexanol		1.2	2.6×10^{-17}	4.6×10^{-4}
trans-2-hexenal		110	6.4×10^{-17}	1.8×10^{-2}
d-limonene		0.5	1.1×10^{-17}	4.5×10^{-4}
3-octanone		0.2	1.0×10^{-18}	2.0×10^{-3}
propyl butyrate		1.2	2.7×10^{-17}	4.5×10^{-4}
		Low density polyethylene		
ethyl hexanoate		4.1×10^{6}	5.2×10^{-13}	7.8×10^{-2}
ethyl 2-methylbutyrate		4.9×10^{5}	2.4×10^{-13}	2.3×10^{-2}
hexanol		9.7×10^{5}	4.6×10^{-13}	2.3×10^{-2}
trans-2-hexenal		8.1×10^{5}		
d-limonene		4.3×10^{6}		
3-octanone		6.8×10^{6}	5.6×10^{-13}	1.2×10^{-1}
propyl butyrate		1.5×10^{6}	5.0×10^{-13}	3.0×10^{-2}
dipropyl disulfide		6.8×10^{6}	7.3×10^{-14}	9.3×10^{-1}
		High density polyethylene		
d-limonene		3.5×10^{6}	1.7×10^{-13}	2.5×10^{-1}
menthone [*1074-95-9*]	$C_{10}H_{18}O$	5.2×10^{6}	9.1×10^{-13}	4.7×10^{-1}
methyl salicylate [*119-36-8*]	$C_8H_8O_3$	1.1×10^{7}	8.7×10^{-14}	1.6
		Polypropylene		
2-butanone [*78-93-3*]	C_4H_8O	8.5×10^{3}	2.1×10^{-15}	4.0×10^{-2}
ethyl butyrate [*105-54-4*]	$C_6H_{12}O_2$	9.5×10^{3}	1.8×10^{-15}	5.3×10^{-2}
ethyl hexanoate		8.7×10^{4}	3.1×10^{-15}	2.8×10^{-1}
d-limonene		1.6×10^{4}	7.4×10^{-16}	2.1×10^{-1}

[a] Values for vinylidene chloride copolymer and ethylene–vinyl alcohol are extrapolated from higher temperatures.
[b] Permeation in the vinylidene chloride copolymer and the polyolefins is not affected by humidity; the permeability and diffusion coefficient in the ethylene–vinyl alcohol copolymer can be as much as 1000 times greater with high humidity (14–17).
[c] MZU = $(10^{-20}\ kg \cdot m)/(m^2 \cdot s \cdot Pa)$; see equation 4 for unit conversions.
[d] See equation 6 for unit conversions.

permeants from the environment can be minimized by placing a barrier polymer between the food and the environment, not necessarily near the food.

Figures 4 and 5 show how the diffusion coefficient and solubility coefficient vary for a series of linear esters in low density polyethylene film. The trends are generally true for other permeants in other films. As the size of the permeant increases, the diffusion coefficient decreases and the solubility coefficient increases. Since the increase in solubility coefficient is larger than the decrease in the diffusion coefficient, the permeability actually increases as the permeant size increases.

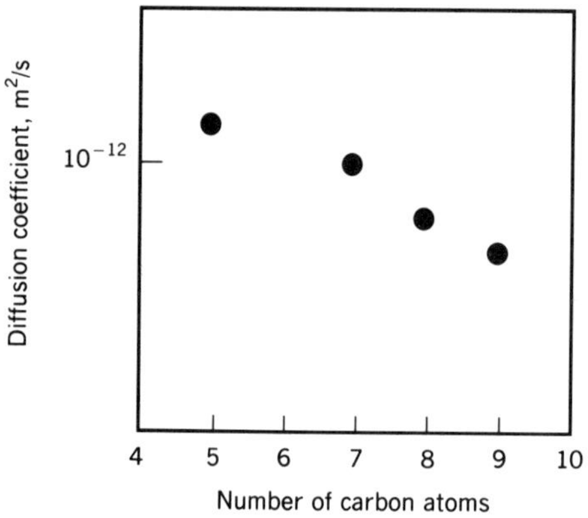

Fig. 4. Diffusion coefficient at 30°C of esters in a low density polyethylene film (18).

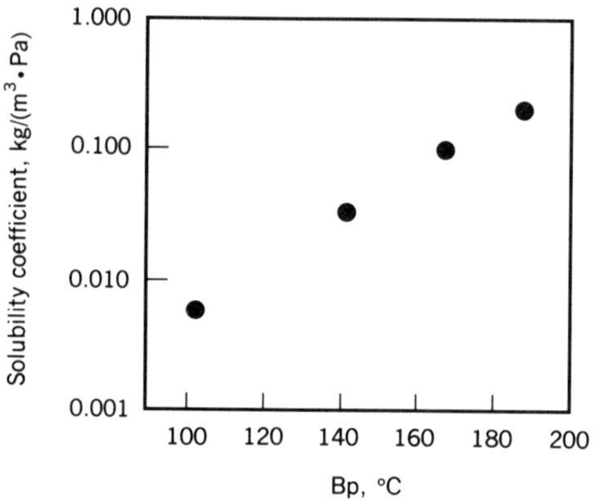

Fig. 5. Solubility coefficient at 30°C versus boiling point of ester in a low density polyethylene film (18). For unit conversion see equation 6.

Physical Factors Affecting Permeability

Several physical factors can affect the barrier properties of a polymer. These include temperature, humidity, orientation, and cross-linking.

Temperature. The permeability varies with temperature according to equation 12 where P_o is a constant, E_p is the activation energy for permeation, R is the gas constant, and T is the absolute temperature.

$$P = P_o \exp(-E_p/RT) \tag{12}$$

The temperature dependence of the permeability arises from the temperature dependencies of the diffusion coefficient and the solubility coefficient. Equations 13 and 14 express these dependencies where D_o and S_o are constants, E_d is the activation energy for diffusion, and ΔH_{sol} is the heat of solution for the permeant in the polymer.

$$D = D_o \exp(-E_d/RT) \tag{13}$$

$$S = S_o \exp(-\Delta H_{sol}/RT) \tag{14}$$

The activation energy for diffusion is always positive and increases with increasing permeant size. Figure 6 shows this for diffusion in poly(vinyl chloride). The heat of solution is near zero for small permanent gases such as oxygen; however, it becomes increasingly negative with increasing permeant size because the main contribution to the heat of solution is the heat of condensation. Equation 15 shows the relationship among the three exponential factors.

$$E_p = E_d + \Delta H_{sol} \tag{15}$$

Although, in principle, this sum could be negative and give a negative activation energy for the permeability, no examples are known. The permeability increases with increasing temperature.

Figure 7 shows generally how the permeability varies with temperature. Equation 12 is useful above and below the glass-transition temperature, but not at the glass-transition temperature. The main contribution to this effect is the

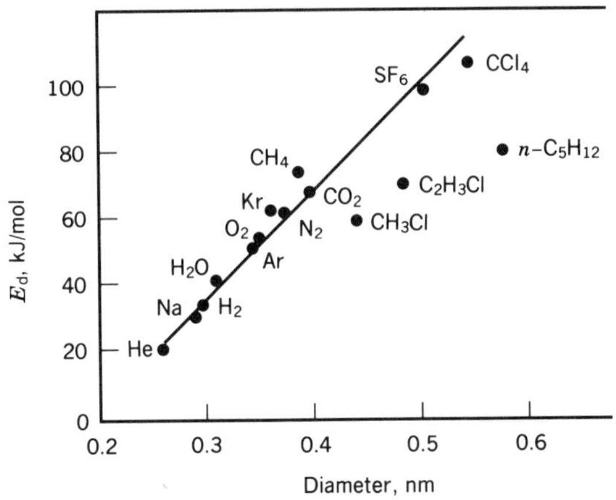

Fig. 6. Activation energy for diffusion in poly(vinyl chloride) as a function of penetrant mean diameter (19). To convert J to cal, divide by 4.184.

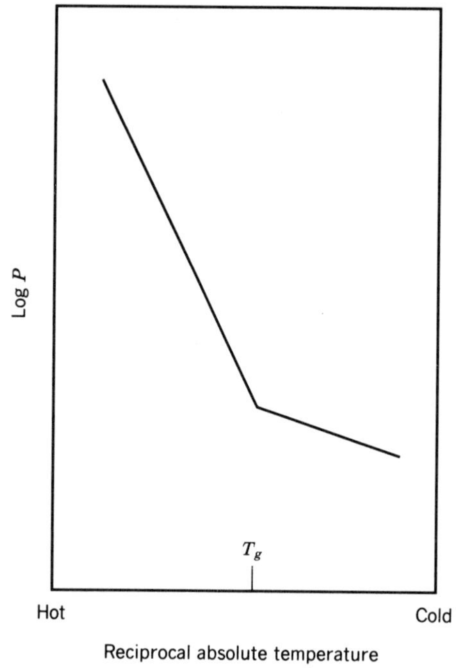

Fig. 7. Relationship of permeability and temperature.

diffusion coefficient. Above T_g, the diffusion process is controlled by motions of the polymer chains; hence, the activation energy is large. Below T_g, the diffusion process is controlled by the movement of the permeant; therefore, the activation energy is smaller. There is growing evidence with theoretical support that, above T_g, E_d decreases slightly with increasing temperature. Hence, the curve in Figure 7 would be concave downward above T_g.

For oxygen, the permeabilities increase about 10% per degree in polymers that are above their T_g such as vinylidene chloride copolymers and polyolefins. The permeabilities increase about 5% per degree in polymers that are below their T_g such as acrylonitrile copolymers, EVOH, and PET.

Humidity. When a polymer equilibrates with a humid environment, it absorbs water. The water concentration in the polymer might be very low as in polyolefins or it might be several weight percent as in ethylene–vinyl alcohol copolymers. Absorbed water does not affect the permeabilities of some polymers including vinylidene chloride copolymers, acrylonitrile copolymers, and polyolefins. Absorbed water increases the permeabilities in some polymers including ethylene–vinyl alcohol copolymers and most polyamides. Figure 2 shows how the oxygen permeability in an ethylene–vinyl alcohol copolymer increases with increasing humidity. A few polymers show a slight decrease in the oxygen permeability with increasing humidity. These include PET and the new amorphous nylon.

Orientation. The effect of orientation on the permeability of polymers is difficult to assess because the words orientation and elongation or strain have been used interchangeably in the literature. Diffusion in some polymers is unaffected by orientation; in others, increases or decreases are observed. The circumstances of the orientation (elongation) are important. Table 11 shows the effect of orientation on the oxygen permeability of four polymers. In each case a slight reduction in permeability occurs with orientation. In another study, however, no difference in oxygen or nitrogen permeability was observed between biaxially oriented polystyrene films and films that were cast from toluene solutions (20). Vinylidene chloride copolymer films show no difference in oxygen permeabilities between biaxially oriented (700%) and unoriented films. When increases in permeability have been observed, the creation of microcracks has been suspected.

Table 11. Effect of Orientation on Oxygen Permeability for Certain Polymers

Polymer	Degree of orientation, %	Oxygen permeability,[a] nmol/(m·s·GPa)[b]
polypropylene	0	300
	300	160
polystyrene	0	840
	300	600
polyester	0	20
	500	10
copolymer of 70% acrylonitrile	0	2.0
and 30% styrene	300	1.8

[a]At 23°C.
[b]See Table 1 for unit conversions.

Cross-Linking. Cross-linking has been shown to decrease the diffusion coefficient. The effect is greater for large permeants. To be effective, the cross-links need to be closer than ~30–40 nm. Such a high extent of cross-linking is a severe handicap for commercial applications; hence, little research is being done in this area for barrier materials.

Barrier Structures

Barrier polymers are often used in combination with other polymers or substances. The combinations may result in a layered structure either by coextrusion, lamination, or coating. The combinations may be blends that are either miscible or immiscible. In each case, the blend seeks to combine the best properties of two or more materials to enhance the value of a final structure.

Layered Structures. Whenever a barrier polymer lacks the necessary mechanical properties for an application or the barrier would be adequate with only a small amount of the more expensive barrier polymer, a multilayer structure via coextrusion or lamination is appropriate. Whenever the barrier polymer is difficult to melt process or a particular traditional substrate such as paper or cellophane [*9005-81-6*] is necessary, a coating either from latex or a solvent is appropriate. A layered structure uses the barrier polymer most efficiently since permeation must occur through the barrier polymer and not around the barrier polymer. No short cuts are allowed for a permeant. The barrier properties of these structures are described by the permeance P^* which is described in equation 16 where P_i and L_i are the permeabilities and thicknesses of the layers.

$$\frac{1}{P^*} = \frac{L_1}{P_1} + \frac{L_2}{P_2} + \cdots \frac{L_i}{P_i} \tag{16}$$

The permeance can be used in equation 17 (which is a modification of eq. 1) to estimate package performance.

$$\frac{\Delta M_x}{\Delta t} = P^* A \Delta p_x \tag{17}$$

Figure 8 shows how the permeance varies for a hypothetical two-layer sheet as the relative thicknesses shift for a barrier layer with a permeability of 0.1 units and a nonbarrier layer with a permeability of 100 units. Figures 9 and 10 show profiles of coextruded sheet. Figure 9 shows a typical symmetric five-layer sheet. This is the simplest case and three extruders are necessary. One extruder supplies the outer layers that give structural integrity and toughness, eg, polypropylene. Another extruder supplies the barrier layer, and a third extruder supplies the adhesive tie layers that hold the structure together. A fourth extruder would be necessary if the outermost layers are different. Figure 10 shows an asymmetric sheet with a recycle layer. An additional extruder is needed to supply this layer to the coextrusion die. Both Figures 9 and 10 are drawn to scale. Figure 10 represents a recycle layer that is 30–50% of the total structure. Recycle at 30–50% has been demonstrated and reported for containers with a barrier layer of a vinylidene chloride copolymer (21–23). Development with other barrier polymers is continuing.

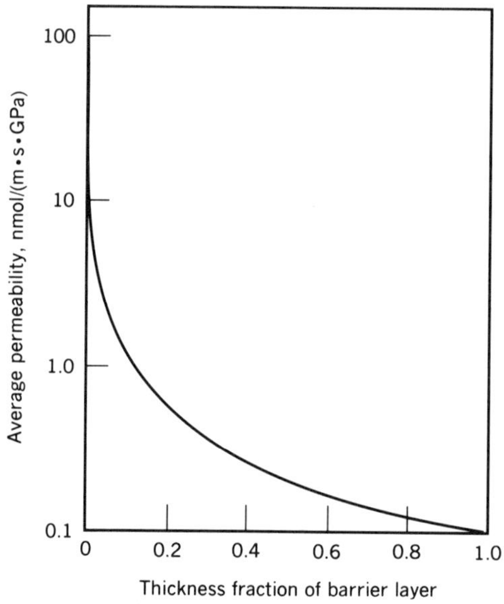

Fig. 8. Average permeance of a two-layer structure. See Table 1 for unit conversions.

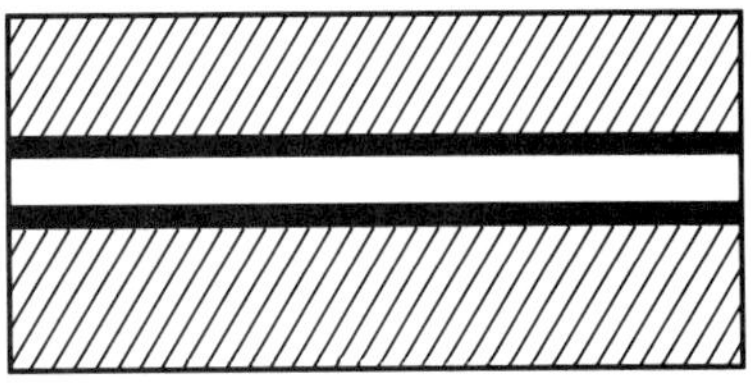

Fig. 9. Symmetric five-layer barrier coextruded sheet.

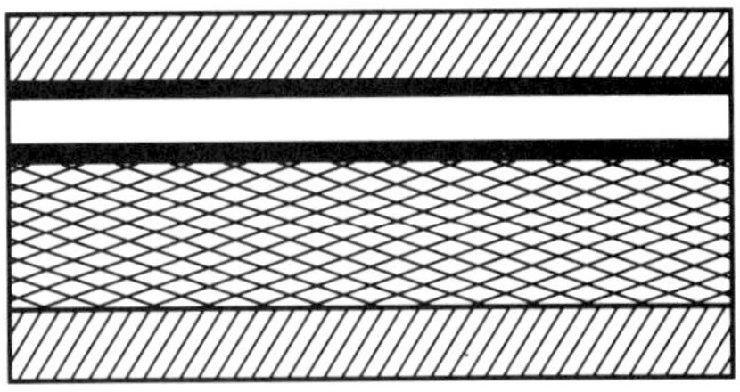

Fig. 10. Barrier sheet with recycle layer.

A typical coextruded sheet has a total thickness of 1270 μm (50 mil). After the container is formed by one of several methods, the sheet might then yield a wall thickness of 510 μm. The barrier layer in the container is typically in the range of 25 to 75 μm.

A thin layer of a barrier polymer can be coated onto a substrate. For example, a water-borne latex can be used to coat paper or a polymer such as polypropylene or PET. Commonly, two coats are applied to give a total barrier thickness of 5 μm. Two coats are useful so that minor holes in the first layer can be covered with the second layer. Adhesion of the coating is important for all substrates and hold out (minimizing wasteful soaking-in) is important for paper. Hence, sometimes the first layer is different from the second layer or the substrate receives a surface treatment before coating. A typical application is a vinylidene chloride copolymer latex on oriented polypropylene film for packaging snack foods. The most significant commercial application of solvent coating with barrier polymers is a special vinylidene chloride resin dissolved in a polar solvent coated onto cellophane or PET. Ethylene–vinyl alcohol polymers are potentially applicable for solvent coating. Inorganic coatings on polymers have the potential to enhance the barrier. However, commercial development is not significant.

Immiscible Blends. When two polymers are blended, the most common result is a two-phase composite. The most interesting blends have good adhesion between the phases, either naturally or with the help of an additive. The barrier properties of an immiscible blend depend on the permeabilities of the polymers, the volume fraction of each, phase continuity, and the aspect ratio of the discontinuous phase. Phase continuity refers to which phase is continuous in the composite. Continuous for barrier applications means that a phase connects the two surfaces of the composite. Typically, only one of the two polymer phases is continuous, with the other polymer phase existing as islands. It is possible to have both polymers be continuous. The aspect ratio L/W refers to the shape of the particles in the discontinuous phase. It is the average dimension of this phase parallel to the plane of the film L divided by the average dimension perpendicular to the film *W*. Plates in the plane of the film would have a high aspect ratio. Spheres or cubes would have an aspect ratio equal to 1.

Figure 11 shows the theoretical permeabilities that are expected for a two-phase blend of polymers. The two solid curves represent calculations based upon Maxwell's equation (24) for an aspect ratio of 1 for the discontinuous phase. The dotted line is a prediction of the permeability using Nielsen's model (25) when a barrier polymer with an aspect ratio of 8 is discontinuous in a nonbarrier matrix. Figure 12 shows the expected result of a phase inversion for a two-polymer blend. The discontinuous phase is assumed to have an aspect ratio of 1. At some critical composition, the composite switches from being continuous in one polymer to being continuous in the other. Figure 12 is really a special case of Figure 11. Selar RB is a blend of polyethylene and nylon-6. Polyethylene is the majority constituent and forms the continuous phase. The product has its best barrier when it can be used in processes that impart orientation to the product. This gives a high aspect ratio to the nylon-6 and enhanced barrier to the article. Blends of polyethylene and EVOH are being developed.

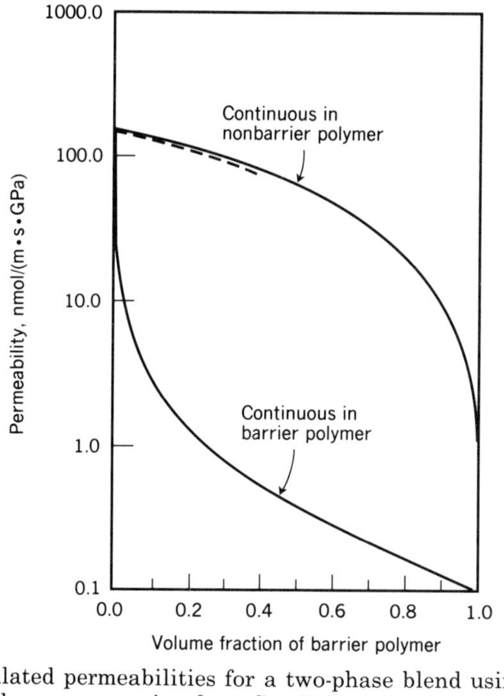

Fig. 11. Calculated permeabilities for a two-phase blend using Maxwell's result. Discontinuous phase has aspect ratio of 1.0. See Table 1 for unit conversion.

Block copolymers of styrene and methacrylonitrile appear to behave like immiscible blends (11). Another case of immiscible blends is an inert filler in a polymer matrix. Although the inorganic fillers are assumed to have permeabilities nearly equal to zero, the filler must be wetted by the polymer to be effective at lowering the permeability of the host polymer. Table 12 shows the effect of a surface treatment on calcium carbonate to enhance wetting by polyeth-

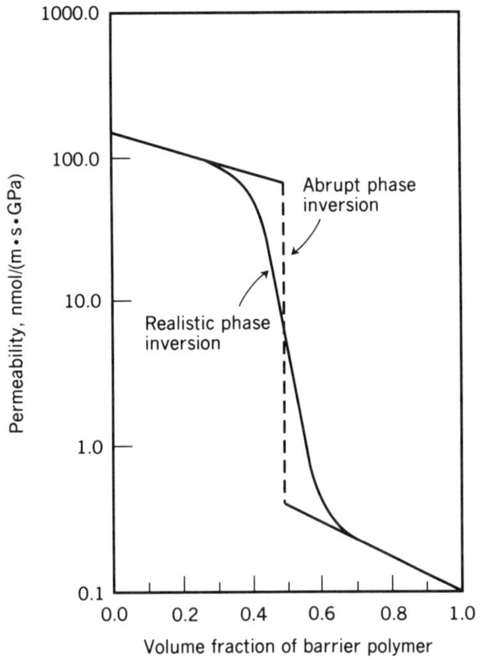

Fig. 12. Permeabilities for a two-phase blend with a phase inversion. Discontinuous phase has aspect ratio of 1.0. See Table 1 for unit conversion.

ylene. With good wetting the permeability is reduced; with poor wetting the permeability is increased. Selar OH Plus resins are an example of filled polymers. By using 23.1 wt % mica flakes in an ethylene–vinyl alcohol copolymer, the permeability can be reduced with optimum processing to as low as 20% of the permeability of the unfilled polymer. Crystallinity in polymers acts like inert filler and lowers the permeability. However, since the amount of crystallinity in a given polymer is rarely under the control of the user, it is not considered here. Some orientation efforts can lower the permeability if the aspect ratio of the crystallites is increased.

Table 12. Effect of Calcium Carbonate Fillers on Oxygen Permeability of Low Density Polyethylene

$CaCO_3$, volume %	Type $CaCO_3$	Oxygen permeability,[a] nmol/(m·s·GPa)[b]
0		960
15	untreated	~2000
25	untreated	~4000
15	treated	500
25	treated	300

[a]At 23°C.
[b]See Table 1 for unit conversion.

Miscible Blends. Sometimes a miscible blend results when two polymers are combined. A miscible blend has only one amorphous phase because the polymers are soluble in each other. There may also be one or more crystal phases. Simple theory (26) has supported the empirical relation for the permeability of a miscible blend. Equation 18 expresses this relation where P_{mb} is the permeability of the miscible blend and ϕ_1 and ϕ_2 are the volume fractions of polymer 1 and 2.

$$\ln P_{mb} = \phi_1 \ln P_1 + \phi_2 \ln P_2 \tag{18}$$

This relationship is shown in Figure 13 where Polymer 1 has a permeability 1000 times higher than that of Polymer 2. Published data have small negative deviations from this theoretical relationship. Part of the deviation can be explained by densification of the blend relative to the starting components. Random copolymers, which are forced (by covalent bonds) to imitate combinations of two materials, have permeabilities that are similar to miscible blends. However, the deviations from equation 18 tend to be positive. A series of styrene–methacrylonitrile copolymers were studied (11) and slight positive deviations were found. Figure 14 shows the oxygen permeabilities of a series of vinylidene chloride–*n*-butyl acrylate copolymers [*9011-09-0*].

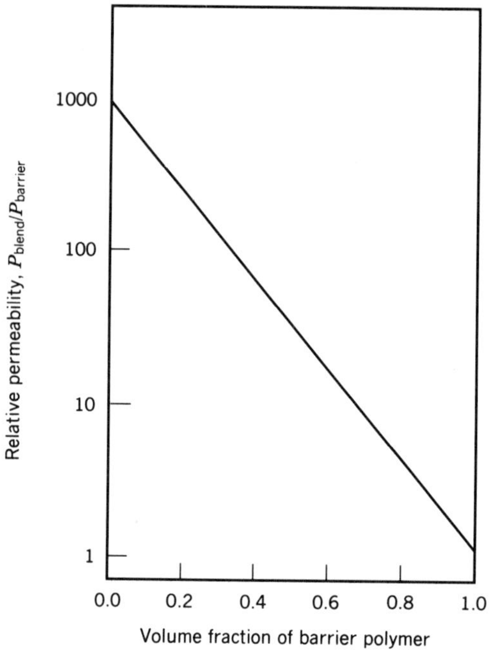

Fig. 13. Theoretical permeabilities for miscible blends of two polymers.

Plasticized polymers have been observed to behave like miscible blends. The permeabilities of oxygen, carbon dioxide, and water vapor in a vinylidene chloride copolymer increase exponentially with increasing plasticizer (4,5,28). About 1.6 parts plasticizer per hundred parts polymer is enough to double the permeability.

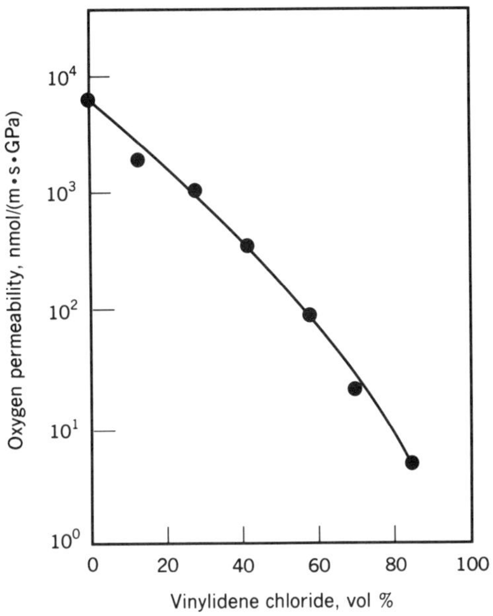

Fig. 14. Oxygen permeabilities of vinylidene chloride–*n*-butyl acrylate copolymers (27). See Table 1 for unit conversion.

Predicting Permeabilities

Reasonable prediction can be made of the permeabilities of low molecular weight gases such as oxygen, nitrogen, and carbon dioxide in many polymers. The diffusion coefficients are not complicated by the shape of the permeant, and the solubility coefficients of each of these molecules do not vary much from polymer to polymer. Hence, all that is required is some correlation of the permeant size and the size of holes in the polymer matrix. Reasonable predictions of the permeabilities of larger molecules such as flavors, aromas, and solvents are not easily made. The diffusion coefficients are complicated by the shape of the permeant, and the solubility coefficients for a specific permeant can vary widely from polymer to polymer.

The permachor method is an empirical method for predicting the permeabilities of oxygen, nitrogen, and carbon dioxide in polymers (29). In this method a numerical value is assigned to each constituent part of the polymer. An average number is derived for the polymer, and a simple equation converts the value into a permeability. This method has been shown to be related to the cohesive energy density and the free volume of the polymer (2). The model has been modified to liquid permeation with some success.

A less empirical model based solely on the specific free volume of the polymer has been proposed (30). Once the density and the intrinsic volumes of the polymer components are known, the specific free volume can be calculated. A correlation between the specific free volume and the oxygen permeabilities of several polymers is shown in Figure 15. The model seems to give better predictions for polymers with higher permeabilities. This model is for low molecular weight gases such as oxygen, nitrogen, and carbon dioxide.

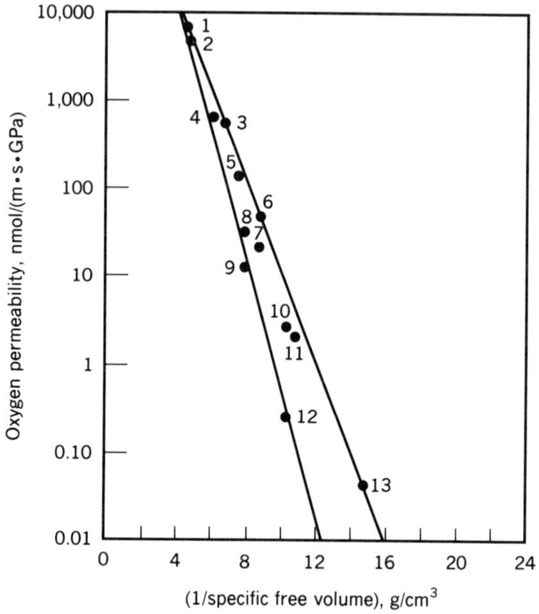

Fig. 15. Oxygen permeability versus 1/specific free volume at 25°C (30). 1. Polybutadiene; 2. polyethylene (density 0.922); 3. polycarbonate; 4. polystyrene; 5. styrene-*co*-acrylonitrile; 6. poly(ethylene terephthalate); 7. acrylonitrile barrier polymer; 8. poly(methyl methacrylate); 9. poly(vinyl chloride); 10. acrylonitrile barrier polymer; 11. vinylidene chloride copolymer; 12. polymethacrylonitrile; and 13. polyacrylonitrile. See Table 1 for unit conversions.

For larger molecules, independent predictions of the diffusion coefficients and the solubility coefficients are required. Figure 16 shows how the diffusion coefficient varies as a function of permeant size in poly(vinyl chloride) (PVC). The two sets of data represent glassy PVC, which is below its glass-transition temperature and plasticized PVC, which is above its glass-transition temperature. Other glassy polymers show the steep slope, and other rubbery polymers show the shallow slope. The points near the lines represent spherical permeants whereas the points above the lines represent linear permeants. Predicting the diffusion coefficient for a permeant in a polymer requires knowing one other diffusion coefficient in the polymer. Figure 4 shows how the diffusion coefficient varies for permeant molecules slightly larger than shown in Figure 16.

The solubility coefficients are more difficult to predict. Although advances are being made, the best method is probably to use a few known solubility coefficients in the polymer to predict others with a simple plot of S vs $(\delta_{\text{poly}} - \delta_{\text{perm}})^2$ where δ_{poly} and δ_{perm} are the solubility parameters of the polymer and permeant respectively. When insufficient data are available, S at 25°C can be estimated with equation 19 where $\kappa = 1$ and the resulting units of cal/cm^3 are converted to kJ/mol by dividing by the polymer density and multiplying by the molecular mass of the permeant and by 4.184 (16).

$$S = \kappa(\delta_{\text{poly}} - \delta_{\text{perm}})^2 \tag{19}$$

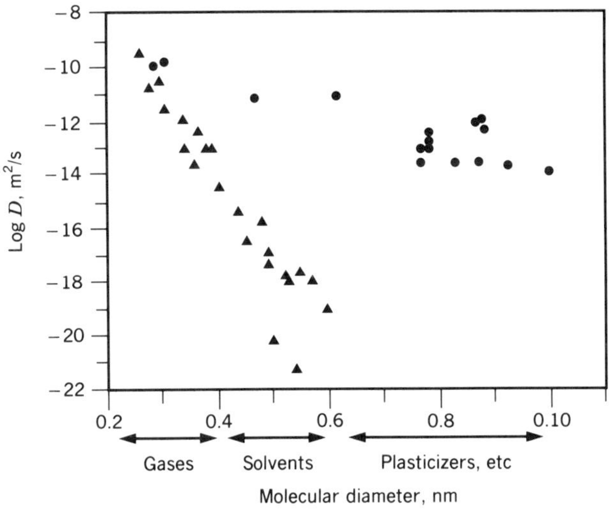

Fig. 16. Diffusivities of penetrants in rigid (▲) and plasticized (●) poly(vinyl chloride) versus molecular diameter at 30°C (31).

The boiling temperature of a permeant can be used to predict the solubility coefficient only when the solubility coefficients of other permeants of the same chemical family are known.

Measuring Barrier Properties

Measuring the barrier properties of polymers is important for several reasons. The effects of formulation or process changes need to be known, new polymers need to be evaluated, data are needed for a new application before a large investment has been made, and fabricated products need to have performance verified. For some applications a full range of data is necessary, including P, D, and S plus the effects of temperature and humidity.

Oxygen Transport. The most widely used methods for measuring oxygen transport are based upon the Ox-Tran instrument (Modern Controls, Inc.). Several models exist, but they all work on the same principle. The most common application is to measure the permeability of a film sample. Typically, oxygen is introduced on one side of the film, and nitrogen gas sweeps the other side of the film to a coulometric detector. The detector measures the rate that oxygen comes through the film. The detector response at steady state can easily be converted to $\Delta M_{\text{oxygen}}/\Delta t$ (eq. 1). Simple algorithms come with the instruction manual for calculating the permeability. The detector response can be monitored as a function of time to determine D. This is demonstrated in Figure 17. During the transient time before steady state is achieved, the transmission rate increases. First $t_{1/2}$, the time for the transmission rate to rise to 50% of the steady-state rate, must be found. Then equation 20 (32) is used where L is the film thickness. After P and D are known, S can be calculated.

$$D = \frac{L^2}{7.2 t_{1/2}} \tag{20}$$

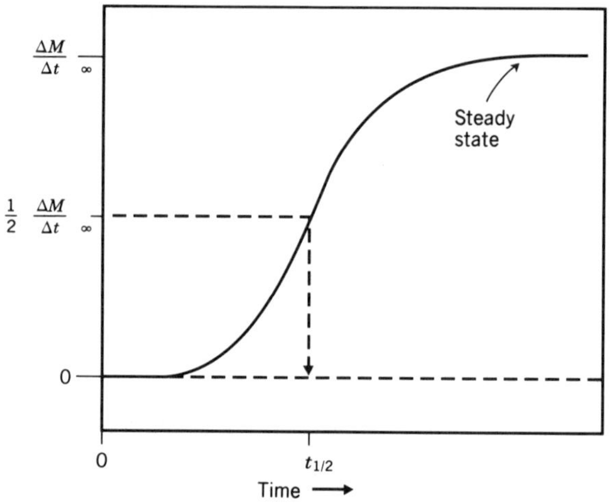

Fig. 17. Determining the diffusion coefficient from instantaneous transmission rates.

The instrument has a heating element that allows for measurements at selected temperatures. The humidity can be controlled by passing the test gases through bubblers, monitoring the humidity, and adjusting the flow rates. This method requires some sophistication and is preferred when such testing is routine. The sandwich method is preferred for infrequent testing (33). With this method, pads that have been treated with saturated salt solutions are placed beside the film and encapsulated with thin polyethylene layers. The polyethylene offers almost no barrier to the oxygen and does a good job of holding the humidity near the test film.

Containers can be tested with the Ox-Tran instrument. However, care must be taken to obtain a good seal around the base of the container. A generous amount of epoxy glue is recommended.

Water Transport. Two methods of measuring water-vapor transmission rates (WVTR) are commonly used. The newer method uses a Permatran-W (Modern Controls, Inc.). In this method a film sample is clamped over a saturated salt solution, which generates the desired humidity. Dry air sweeps past the other side of the film and past an infrared detector, which measures the water concentration in the gas. For a calibrated flow rate of air, the rate of water addition can be calculated from the observed concentration in the sweep gas. From the steady-state rate, the WVTR can be calculated. In principle, the diffusion coefficient could be determined by the method outlined in the previous section. However, only the steady-state region of the response is serviceable. Many different salt solutions can be used to make measurements at selected humidity differences; however, in practice, $CaSO_4{\cdot}7H_2O$ is used nearly always because it gives a humidity of 90% as required by the traditional commercial applications. Typical experiments take between several hours and a day to complete.

The other method is the ASTM cup method (34). In this method a desiccant is placed in a waterproof dish. The dish is covered with the experimental film and placed in an environmental chamber. The temperature and humidity are set for the conditions of interest, typically 37.8°C and 90% rh. At regular intervals, the dish is removed and weighed. After a few days enough data have been gathered to describe a steady-state rate of weight gain, and the WVTR can be calculated. Typical experiments take about a week to complete.

Carbon Dioxide Transport. Measuring the permeation of carbon dioxide occurs far less often than measuring the permeation of oxygen or water. A variety of methods are used; however, the simplest method uses the Permatran-C instrument (Modern Controls, Inc.). In this method, air is circulated past a test film in a loop that includes an infrared detector. Carbon dioxide is applied to the other side of the film. All the carbon dioxide that permeates through the film is captured in the loop. As the experiment progresses, the carbon dioxide concentration increases. First, there is a transient period before the steady-state rate is achieved. The steady-state rate is achieved when the concentration of carbon dioxide increases at a constant rate. This rate is used to calculate the permeability. Figure 18 shows how the diffusion coefficient can be determined in this type of experiment. The time lag t_L is substituted into equation 21. The solubility coefficient can be calculated with equation 2.

$$D = \frac{L_2}{6t_L} \tag{21}$$

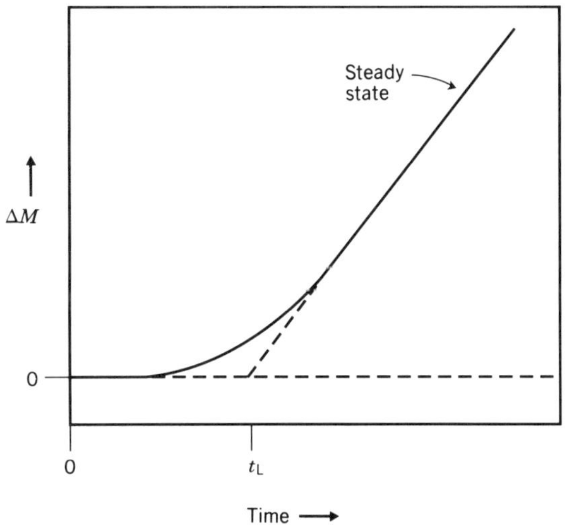

Fig. 18. Determining the diffusion coefficient from cumulative transmission.

Flavor and Aroma Transport. Many methods are used to characterize the transport of flavor, aroma, and solvent molecules in polymers. Each has some value, and no one method is suitable for all situations. Any experiment should obtain the permeability, the diffusion coefficient, and the solubility coefficient. Furthermore, experimental variables might include the temperature, the humidity, the flavor concentration, and the effect of competing flavors.

Any of the mathematical methods discussed previously to calculate P, D, and S in a single experiment are applicable to flavor permeation. Two cautions are necessary. First, since the diffusion coefficients are typically low, the time to reach steady state can be long. The instrument must give stable responses for long times or something must be done to speed the experiment. The latter can be accomplished by using a thinner film or by raising the temperature. Second, the solubility coefficients are usually high. This means that the supply of permeant molecules could become depleted or the film could become altered. Recognizing the potential for depleting the flavor reservoir or for altering the experimental film is important. Using a thinner film or a constant supply of fresh permeant can help. Also working at high flavor concentrations should be done only when modeling performance that involves high concentrations.

Experimental data that show the effect of environmental variables have been given earlier. The most commonly misconstrued facet of the data is the effect of concentration on the permeability. The permeabilities of flavors in polymers have been correctly reported to be both independent and strongly dependent on the flavor concentration. However, to assume that there is disagreement would be incorrect. At very low concentrations, as commonly observed in foods, the permeabilities are independent of concentration. However, at high concentrations, as observed in some foods and in neat flavor oils, the permeabilities are quite concentration-dependent. The entire trend has been observed (35,36). Figure 19 shows that the permeability of ethyl propionate [*105-37-3*] in dry poly(vinyl alcohol) is not affected by permeant concentration until the partial pressure of the permeant exceeds about 30% of the equilibrium vapor pressure. Above that concentration, the permeability is strongly dependent on the permeant concentration. This topic is important to include here because many techniques lack the sensitivity to work at low permeant concentrations. Then, at high concentrations, great variation is observed from experiment to experiment and from experimenter to experimenter. The variation does not necessarily diminish the quality of the measurement; it merely reflects the strong concentration dependence.

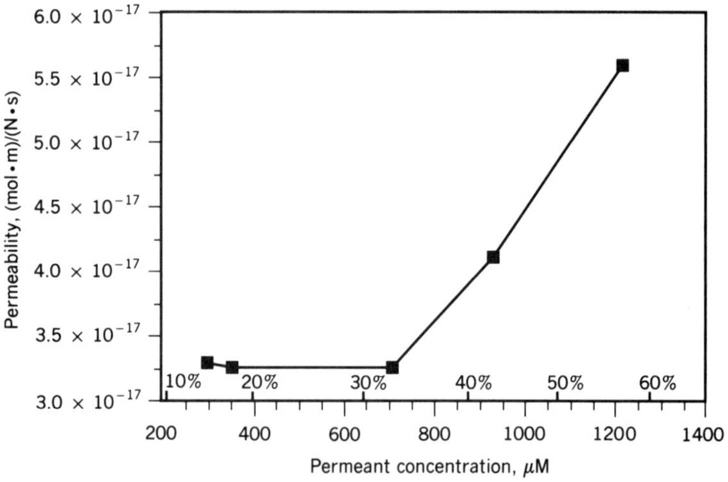

Fig. 19. Permeation of ethyl propionate at various concentrations through poly(vinyl alcohol) at 25°C, 0% rh. Percentages on x-axis represent partial pressure of the permeant as a fraction of the equilibrium vapor pressure.

Applications

The primary application for barrier polymers is food and beverage packaging. Barrier polymers protect food from environmental factors that could compromise both taste and shelf life. They also help retain desirable flavors and aroma. Barrier polymers are also used for packaging medical products, agricultural products, cosmetics, and electronic components and in moldings, pipe, and tubing.

Safety and Health Factors

The use of safe materials is vital for barrier applications, particularly for food, medical, and cosmetics packaging. Suppliers of specific barrier polymers can provide the necessary details, such as material safety data sheets, to ensure safe processing and use of barrier polymers.

BIBLIOGRAPHY

"Barrier Polymers" in *ECT* 3rd ed., Vol. 3, pp. 480–502, by S. Steingiser, S. P. Nemphos, and M. Salame, Monsanto Company.

1. *Permeability of Polymers to Gases and Vapors,* Brochure P302-335-79, The Dow Chemical Company, Midland, Mich., 1979.
2. M. Salame, *J. Plast. Film Sheeting* **2,** 321–325 (1986).
3. P. T. DeLassus, D. L. Clarke, and T. Cosse, *Mod. Plast.* 86–88 (Jan. 1983).
4. P. T. DeLassus, *J. Vinyl Technol.* **1,** 14 (1979).
5. P. T. DeLassus and D. J. Grieser, *J. Vinyl Technol.* **2,** 195 (1980).
6. R. Sezi and J. Springer, *Colloid Polym. Sci.* **259,** 1170–1176 (1981).
7. M. Salame, *J. Polym. Sci., Symposium No. 41,* 1–15 (1973).
8. D. W. Van Krevelen, *Properties of Polymers,* 3rd ed., Elsevier, Amsterdam, The Netherlands, 1990, Chapt. 18, pp. 535–583.
9. H. Iwasaki and K. Hoashi, *Kobunshi Ronbunshu,* **34**(11), 785–791 (1977).
10. A. L. Blackwell, *J. Plast. Film Sheeting* **1,** 205–214 (1985).
11. A. E. Barnabeo, W. S. Creasy, and L. M. Robeson, *J. Polym. Sci. Polym. Chem. Ed.* **13,** 1979–1986 (1975).
12. T. D. Krizan, J. C. Coburn, and P. S. Blatz, in W. J. Koros, ed., *Barrier Polymers and Structures, ACS Symposium Series No. 423,* American Chemical Society, Washington, D.C., 1990, pp. 111–125.
13. L. E. Gerlowski, in W. J. Koros, ed., *Barrier Polymers and Structures, ACS Symposium Series No. 423,* American Chemical Society, Washington, D.C., 1990, Chapt. 8, pp. 177–191.
14. P. T. DeLassus and co-workers, in J. H. Hotchkiss, ed., *Food and Packaging Interactions, ACS Symposium Series No. 365,* American Chemical Society, Washington, D.C., 1988, Chapt. 2, pp. 11–27.
15. P. T. DeLassus, G. Strandburg, and B. A. Howell, *TAPPI Journal* **71**(11), 177–181 (1988).
16. G. Strandburg, P. T. DeLassus, and B. A. Howell, in W. J. Karos, ed., *Barrier Polymers and Structures, ACS Symposium Series No. 423,* American Chemical Society, Washington, D.C., 1990, pp. 333–350.

17. G. Strandburg, P. T. DeLassus, and B. A. Howell, in S. A. Risch and J. H. Hotchkiss, eds., *Food and Packaging Interactions II, ACS Symposium Series,* No. 473, American Chemical Society, Washington, D.C., 1991, pp. 133–148.
18. Ref. 16, Chapt. 18.
19. A. R. Behrens and H. B. Hopfenberg, *J. Memb. Sci.* **10,** 283–303 (1982).
20. A. F. Burmester, T. A. Manial, J. S. McHattie, and R. A. Wessling, *Polym. Prepr. Am. Chem. Soc. Div. Polym. Chem.* **27**(2), 414–415 (1986).
21. K. Krumm, *Proc. TAPPI Polym. Laminations Coatings Conf.,* 313 (1987).
22. M. Yamada, *Proc. Europak '87,* 121 (1987).
23. A. J. Westlie and G. D. Oliver, *Proc. Future Pak '87,* 213 (1987).
24. J. C. Maxwell, *Electricity and Magnetism,* Vol. 1, 3rd ed., Dover, New York, 1981, p. 440.
25. L. E. Nielsen, *J. Macromol. Sci. Chem.* **A1**(5), 929–942 (1967).
26. D. R. Paul, *J. Membrane Soc.* **18,** 75–86 (1984).
27. P. T. DeLassus, K. L. Wallace, and H. J. Townsend, *Polym. Prepr. Am. Chem. Soc. Div. Polym. Chem.* **26**(1), 116–117 (1985).
28. P. T. DeLassus, *J. Vinyl Technol.* **3,** 240 (1981).
29. M. Salame, *Polym. Prepr. Am. Chem. Soc. Div. Polym. Chem.* **8,** 137 (1967).
30. W. M. Lee, *ACS Organic Coatings and Plastics Chemistry* **39,** 341 (1978).
31. A. R. Behrens, *Polym. Prepr. Am. Chem. Soc. Div. Polym. Chem.* **30**(1), 5–6 (1989).
32. K. D. Ziegel, H. K. Frensdorff and D. E. Blair, *J. Polym. Sci.* Part A-2 **7,** 809 (1969).
33. R. A. Wood, *J. Test. Eval.* **12**(3), 149–151 (1984).
34. *1990 Annual Book of ASTM Standards,* ASTM Designation: E96-80, Vol. 15.09, ASTM, Philadelphia, Pa., pp. 811–818.
35. M. G. R. Zobel, *Polymer Testing* **5,** 153–165 (1985).
36. J. Landois-Garza and J. H. Hotchkiss, in J. H. Hotchkiss, ed., *Food and Packaging Interactions, ACS Symposium Series No. 365,* American Chemical Society, Washington, D.C., 1988, Chapt. 4, pp. 42–58.

General References

W. A. Combellick, in J. I. Kroschwitz, ed., *Encyclopedia of Polymer Science and Engineering,* 2nd ed., Vol. 2, Wiley-Interscience, New York, 1985, pp. 176–192.

J. Comyn, ed., *Polymer Permeability,* Elsevier Applied Science Publishers, Ltd., Barking, UK, 1985.

J. Crank and G. S. Park, eds., *Diffusion in Polymers,* Academic Press, London, 1968.

R. M. Felder and G. S. Huvard, in R. A. Fava, ed., *Methods of Experimental Physics,* Vol. 16, Part C, Academic Press, New York, 1980, Chapt. 17, pp. 315–377.

PHILLIP DELASSUS
The Dow Chemical Company

BATTERIES

INTRODUCTION

Batteries, storehouses for electrical energy "on demand," range in size from large house-sized batteries for utility storage, cubic meter-sized batteries for automotive starting, lighting, and ignition, down to tablet-sized batteries for hearing aids and paper-thin batteries for memory protection in electronic devices. The historical development of batteries starting with Galvani and Volta has been summarized (1–5) and useful texts are available for a more detailed discussion of the topics covered herein (5–9).

In bulk chemical reactions, an oxidizer (electron acceptor) and fuel (electron donor) react to form products resulting in direct electron transfer and the release or absorption of energy as heat. By special arrangements of reactants in devices called batteries, it is possible to control the rate of reaction and to accomplish the direct release of chemical energy in the form of electricity on demand without intermediate processes.

Figure 1 schematically depicts an electrochemical reactor in which the chemical energy stored in the electrodes is manifested directly as a voltage and current flow. The electrons involved in the chemical reactions are transferred from the active materials undergoing oxidation to the oxidizing agent by means of an external circuit. The passage of electrons through this external circuit generates an electric current providing a direct means for energy utilization without going through heat as an intermediate step. As a result, electrochemical reactors can be significantly more efficient than Carnot Cycle heat engines.

The relationship between current flow and chemical reactions was established by Faraday who demonstrated that the amount of chemical change was directly proportional to the quantity of charge passed (It) and to the equivalent weight of the reacting material.

$$g = \frac{MIt}{nF} \tag{1}$$

where g is the mass of material (grams) that reacts during electrolysis, M is the molecular weight (g/mol), n is the number of charges transferred in the electrode reaction (equiv/mol), t is the time of electrolysis, F is Faraday's constant (26.8 A·h/g-equiv), and I is the current (amperes) passed. Equation 1 can be expressed in current per unit area i,

$$i = I/A = \frac{nFg}{MtA} \tag{2}$$

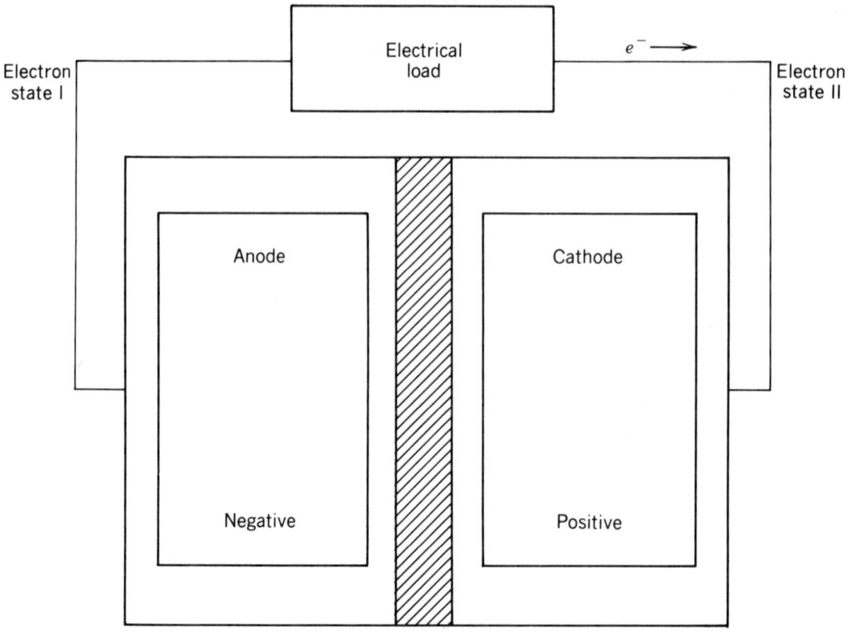

Fig. 1. Schematic representation of a battery system also known as an electrochemical transducer where the anode, also known as electron state I, may be comprised of lithium, magnesium, zinc, cadmium, lead, or hydrogen, and the cathode, or electron state II, depending on the composition of the anode, may be lead dioxide, manganese dioxide, nickel oxide, iron disulfide, oxygen, silver oxide, or iodine.

The name electrode is taken from the Greek *hodes* (pathway) for the path of electron flow. The anode, from the Greek *anodos* (up and out), or negative electrode of the cell is defined as the electrode from which an electron leaves the cell and oxidation occurs. The cathode, from the Greek *kathodos* (down and in or descent), or positive electrode of the cell is the electrode from which electrons enter the cell and reduction occurs. These terms, anode and cathode, work well for primary batteries but are not adequate for secondary or rechargeable ones. Because electron flow changes direction for recharging whereas the electrodes retain the polarity, the terms negative and positive electrode are best used to describe the terminals of rechargeable systems.

The three main types of batteries are primary, secondary, and reserve. A primary battery is used or discharged once and discarded. Secondary or rechargeable batteries can be discharged, recharged, and used again. Reserve batteries are normally special constructions of primary battery systems that store the electrolyte apart from the electrodes, until put into use. They are designed for long-term storage before use. Fuel cells (qv) are not discussed herein.

Useful definitions include:

Battery is one or more electrically connected electrochemical cells having terminals/contacts to produce electrical energy.

Primary battery is an electrolytic cell or group of cells for the generation of electric energy intended to be used until exhausted and then discarded.

Secondary battery is an electrolytic cell or group of cells for the generation of electric energy in which the cell, after being discharged, may be restored to its original charged condition by an electric current flowing in the direction opposite to the flow of current when the cell was discharged.

Reserve battery is a primary battery that is stored dry to improve the shelf life of the cell or battery and activated just before use by addition of electrolyte. The electrolyte is usually stored in the battery.

Anode is the negative electrode of a primary cell associated with chemical reactions that release electrons into the external circuit.

Cathode is the positive electrode of a primary cell associated with chemical reactions that gain electrons from the external circuit.

Electrolyte is a material that provides ionic conductivity between the positive and negative electrodes of a cell.

Separator is a physical barrier between the positive and negative electrodes incorporated into most cell designs to prevent electrical shorting. The separator can be a gelled electrolyte or a microporous plastic film or other porous inert material filled with electrolyte. Separators must be permeable to ions and inert in the battery environment.

Active mass is the material which generates electrical current by means of a chemical reaction within the battery.

Open circuit voltage is the voltage across the terminals of a cell or battery when no external current flows. It is usually close to the thermodynamic voltage for the system.

Closed circuit voltage is the voltage of a cell or battery when the battery is producing current into the external circuit.

Discharge is an operation in which a battery delivers electric energy to an external load.

Charge is an operation in which the battery is restored to its original charged condition by reversal of the current flow.

Internal resistance or *impedance* is the resistance or impedance that a battery or cell offers to current flow.

Faraday constant F is the amount of charge that transfers when an equivalent weight of active mass reacts, 96,485.3 C/g-equiv = 26.8015 A·h/g-equiv).

Economic Aspects

The U.S. primary battery market is divided according to the chemical system used in the batteries, whereas the secondary battery market is usually divided according to usage. The 1989 estimate of the total battery market is given in Table 1. The lead–acid battery accounts for over 85% of the secondary battery market.

Thermodynamics

Batteries are miniature chemical reactors that convert chemical energy into electrical energy on demand. The thermodynamics of battery systems follow

Table 1. Estimated 1989 U.S. Battery Market, $ \times 10^6$

Primary batteries		Secondary batteries	
Segment	Sales	Segment	Sales
alkaline	1700	automotive	3000
carbon–zinc	350	industrial	450
lithium	125	sealed	450
specialty	125	stationary	350
Total	*2300*	*Total*	*4250*

directly from that for bulk chemical reactions (10). For the general reaction

$$aA + bB \rightleftharpoons cC + dD \tag{3}$$

the basic thermodynamic equations for a reversible electrochemical transformation are given as

$$\Delta G = \Delta H - T\Delta S \tag{4}$$

$$\Delta G^\circ = \Delta H^\circ - T\Delta S^\circ \tag{5}$$

where ΔG is the Gibbs free energy, or the energy of a reaction available for useful work, ΔH is the enthalpy, or the energy released by the reaction, ΔS is the entropy, or the heat associated with the organization of material, and T is the absolute temperature. The superscript ° is used to indicate that the value of the function is for the material in the standard state at 25°C and unit activity. Although the Helmholtz free energy ΔA is used to describe constant volume situations found in battery systems, the use of the Gibbs free energy ΔG is adequate to describe practical battery systems.

The terms ΔG, ΔH, and ΔS are state functions and depend only on the identity of the materials and the initial and final state of the reaction. Tables of thermodynamic quantities are available for most known materials (see also THERMODYNAMIC PROPERTIES) (11,12).

Because ΔG is the net useful energy available from a given reaction, in electrical terms, the net available electrical energy from a reaction is given by

$$-\Delta G = nFE \tag{6}$$

and

$$-\Delta G^\circ = nFE^\circ \tag{7}$$

where n is the number of electrons transferred in the reaction, F is Faraday's constant, E is the voltage or electromotive force (emf) of the cell, and E° is that voltage at 25°C and unit activity. The voltage is unique for each group of reactants comprising the battery system. The amount of electricity produced is determined by the total amount of materials involved in the reaction. The voltage may be thought of as an intensity factor, and the term nF may be considered a capacity factor.

The more negative the value of ΔG, the more energy or useful work can be obtained from the reaction. Reversible processes yield the maximum output. In irreversible processes, a portion of the useful work or energy is used to help carry out the reaction. The cell voltage or emf also has a sign and direction. Spontaneous processes have a positive emf; the reaction, written in a reversible fashion, goes in the forward direction.

From equations 4 and 6 it can be shown that:

$$\Delta G = \Delta H - nFT(\partial E/\partial T)_P \tag{8}$$

$$\Delta S = nF(\partial E/\partial T)_P \tag{9}$$

$$-\Delta H = nF[E - T(\partial E/\partial T)_P] \tag{10}$$

where $(\partial E/\partial T)_P$ is the temperature coefficient of the electromotive force (emf) of a reversible cell at constant pressure. Equations 8, 9, and 10 permit the calculation of heats of reaction from simple measurements of the voltage of an electrochemical cell as a function of temperature.

Once the values of thermodynamic functions, ΔH_{T_1}, ΔS_{T_1}, are known at a given temperature T_1 the value for the function can be calculated at any other temperature T_2 by:

$$\Delta H_{T_2} = \Delta H_{T_1} + \int_{T_1}^{T_2} \Delta C_P \, dT \quad \text{and} \quad S_{T_2} = S_{T_1} + \int_{T_1}^{T_2} (\Delta C_P/T) \, dT \tag{11}$$

where ΔC_P is the heat capacity at constant pressure.

The relationship between the chemical equilibrium constant K and the Gibbs free energy is

$$-RT \ln K = \Delta G^\circ \tag{12}$$

The Van't Hoff isotherm identifies the free energy relationship for bulk chemical reactions.

$$\Delta G = \Delta G^\circ + RT \ln \frac{\Pi A^s \text{ (products)}}{\Pi A^s \text{ (reactants)}} \tag{13}$$

Combining equation 6 and 7 with the Van't Hoff isotherm the Nernst equation for electrochemical reactions is obtained:

$$E = E^\circ - \frac{RT}{nF} \ln \frac{\Pi A^s \text{ (products)}}{\Pi A^s \text{ (reactants)}} \tag{14}$$

where ΠA^s (products) is the product of the activities of the products, each raised to its stoichiometric power, and ΠA^s (reactants) is the corresponding reactants' product, R is the gas constant, T is the absolute temperature, and E° is the reversible cell voltage under standard conditions. It is important to realize that only the surface composition of the electrode in contact with the electrolyte should be considered in calculating $E°$. Material buried deep within an electrode, completely covered with active material, and without direct electronic contact to the current collector, does not affect $E°$ and the measured values of cell voltage.

The activity or effective concentration of the reactants and products often differs from the actual concentration of the material. The activity is related to the chemical potential μ_i of the species i by

$$\mu_i = (\partial G/\delta n_i)_{T,P,n_{j\neq i}} \tag{15}$$

where n_i is the mole fraction of species i. Then for species A

$$\mu_A = u_A^0 + RT \ln A_A \tag{16}$$

where $A_A = f_A x_A$, the activity of species A; f_A is the activity coefficient; and x_A is the concentration. In dilute solutions, $f_A \rightarrow 1$ as $x_A \rightarrow 0$. In concentrated solutions the activity coefficient often decreases through a minimum as the concentration increases.

The modern definition of electricity involves the flow of current from a higher potential to a lower one. This places materials having a high electron energy level as the negative material (anodes) in an electrochemical cell and those having a lower electron energy level as the positive material (cathodes). Whereas the potentials of electrodes can be calculated using a Born-Haber cycle, the absolute electrode potentials of single electrodes cannot be measured experimentally. Only differences in potential between two electrodes can be measured. For convenience, the potential of the hydrogen electrode at 25°C, 101.3 kPa (1 atm) H_2 pressure, and unit activity of hydrogen ions is chosen as the reference point for potential measurements. The standard hydrogen electrode (SHE) is defined as the zero point on the potential scale.

$$H_2 - 2\,e^- \rightleftharpoons 2\,H^+ \qquad E° = 0.000 \text{ V} \tag{17}$$

Figure 2 illustrates the potential scale for selected battery reactions. The sign for the electrode potential indicates whether the electrode is negative or positive to the potential of the hydrogen electrode. Reactions in both acid and basic media are included to indicate the influence of pH on the electrode potential. In a battery, the electrode with the more negative potential becomes the battery's negative electrode. The voltage of a battery is the algebraic difference between the individual electrode potentials on the scale. For instance, the voltage of a zinc [*7440-66-6*]–silver(I) oxide [*20667-12-3*] cell in alkaline media is 1.597 V = 0.342 V − (−1.255) V. Several excellent compilations of electrode potentials are available (13–15). Similar scales of electrode potentials can be constructed for each nonaqueous electrolyte.

When the reactants and products of an electrode reaction share a single-phase or crystal habitat, or form an intercalate, the voltage of that electrode depends on the ratio of concentration of product to reacting species. This results in a sloping discharge curve that is typical of cells having manganese dioxide

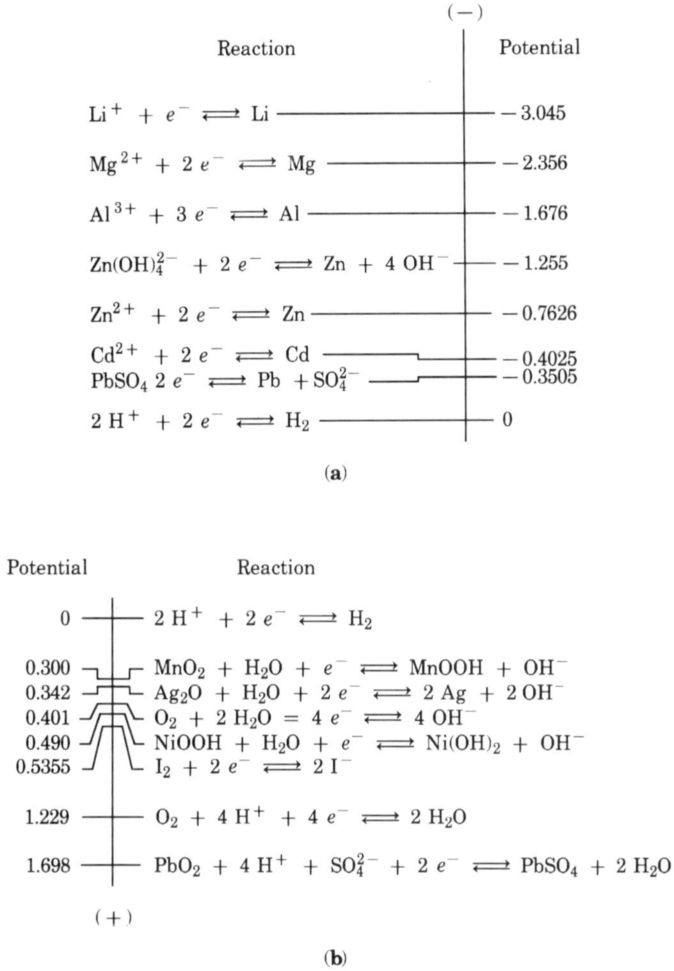

Fig. 2. Standard potentials of battery (**a**) negative electrode and (**b**) positive electrode reactions (13). Potentials are given in volts.

[*1313-13-9*], MnO_2, titanium disulfide [*12039-13-3*], TiS_2, molybdenum disulfide [*1317-33-5*], MoS_2, etc, electrodes. When the reactants and products of an electrode reaction form separate phases, the voltage of the electrode is constant during discharge. This is illustrated in electrodes of mercury [*7439-97-6*]–mercuric oxide [*21908-53-2*], ie, Hg/HgO, cadmium [*7440-43-9*]–cadmium hydroxide [*21041-95-2*], ie, $Cd/Cd(OH)_2$, and lead dioxide [*1309-60-0*]–lead sulfate [*7446-14-2*], $PbO_2/PbSO_4$. These electrodes are often used as reference electrodes in experimental studies because of their invariant voltages.

Because batteries directly convert chemical energy to electrical energy in an isothermal process, they are not limited by the Carnot efficiency. The thermodynamic efficiency ϵ for electrochemical processes is given by:

$$\epsilon = \frac{\Delta G^\circ}{\Delta H^\circ} = 1 - \frac{T\Delta S^\circ}{\Delta H^\circ} \tag{18}$$

In electrochemical units, equation 18 becomes

$$\epsilon = 1 - \frac{nFT(\partial E^\circ/\partial T)_P}{\Delta H^\circ} \tag{19}$$

Whenever energy is transformed from one form to another, an inefficiency of conversion occurs. Electrochemical reactions having efficiencies of 90% or greater are common. In contrast, Carnot heat engine conversions operate at about 40% efficiency. The operation of practical cells always results in less than theoretical thermodynamic prediction for release of useful energy because of irreversible (polarization) losses of the electrode reactions. The overall electrochemical efficiency is, therefore, defined by:

$$Eff_{\text{electrochem}} = \int_{t=o}^{t=t} \frac{E'I\ dt}{\Delta G} \tag{20}$$

where E' is the closed circuit terminal voltage when the net current i is flowing at time t.

The performance of a battery is often designed to be limited by one electrode in order to achieve special performance characteristics, such as overcharge protection and safety. The coulombic efficiency of the active mass is of particular interest in battery design and performance.

$$Eff_{\text{coulombic}} = It/Q \tag{21}$$

where Q is the quantity of coulombs expected from a given amount of active mass based on the equivalent weight and the amount of material. Utilization of the active mass may vary during the course of battery operation and depends on the cutoff voltage, rate of discharge–charge, and the nature of the electrode structure.

In addition to reversible heat absorbed or released as a result of the entropy of reaction, heat is released in batteries during operation because of the irreversibility of the reaction processes involved in converting chemical into electrical energy. The amount of heat q generated during battery operation is given by:

$$q = \frac{T\Delta S}{nF} + \int I(E - E')dt \tag{22}$$

The total heat released is the sum of the entropy contribution plus the irreversible contribution. This heat is released inside the battery at the reaction site. Heat release is not a problem for low rate applications; however, high rate batteries must make provisions for heat dissipation. Failure to accommodate heat can lead to thermal runaway and other catastrophic situations.

Electrolytes

Electrolytes are a key component of electrochemical cells and batteries. Electrolytes are formed by dissolving an ionogen into a solvent. When salts are dissolved in a solvent such as water the salt dissociates into ions through the action of the dielectric, water. Strong electrolytes, ie, salts of strong acids and bases, are completely dissociated in solution into positive and negative ions. The ions are solvated but positive ions tend to interact more strongly with the solvent than do the anions. The ions of the electrolyte provide the path for the conduction of electricity by movement of charged particles through the solution. The electrolyte also provides the physical separation of the positive and negative electrodes needed for electrochemical cell operation.

Electrical conduction in electrolytic solutions follows Ohm's Law:

$$I = \kappa V \tag{23}$$

where I is the current in amperes, V is the voltage drop in volts across the electrolytic resistor of conductivity κ in $(\text{ohm}\cdot\text{cm})^{-1}$. The conductivity κ is the reciprocal of the resistivity ρ in ohm·cm. The resistance of the electrolyte depends on the length l and cross-sectional area A of the conductor.

$$R = \rho l/A \tag{24}$$

The Debye-Hückel theory of electrolytes based on the electric field surrounding each ion forms the basis for modern concepts of electrolyte behavior (16,17). The two components of the theory are the relaxation and the electrophoretic effect. Each ion has an ion atmosphere of equal opposite charge surrounding it. During movement the ion may not be exactly in the center of its ion atmosphere, thereby producing a retarding electrical force on the ion. A finite time is required to reestablish the ion atmosphere at any new location. Thus the ion atmosphere produces a drag on the ions in motion and restricts their freedom of movement. This is termed a relaxation effect. When a negative ion moves under the influence of an electric field, it travels against the flow of positive ions and solvent moving in the opposite direction. This is termed an electrophoretic effect. The Debye-Hückel theory combines both effects to calculate the behavior of electrolytes. The theory predicts the behavior of dilute (≤ 0.05 molal) solutions but does not portray accurately the behavior of concentrated solutions found in practical batteries.

Ions of an electrolyte are free to move about in solution by Brownian motion and, depending on the charge, have specific direction of motion under the influence of an external electric field. The movement of the ions under the influence of an electric field is responsible for the current flow through the electrolyte. The velocity of migration of an ion v_i is given by:

$$v_i = -z_i u_i F d\phi/dx \tag{25}$$

where z_i is the charge number of the ion, $d\phi/dx$ is the electric field gradient, and u_i is the mobility of the ion. Solutions are electrically neutral so:

$$\Sigma_i \; z_i c_i = 0 \tag{26}$$

When current I is passed through an electrolyte, the total current is given by:

$$I = \Sigma_i \; z_i c_i F v_i \tag{27}$$

Each ion has its own characteristic mobility. The total conductivity of the electrolyte is the sum of the conductivities of the positive and negative ions. This is known as Kohlrausch's Law of Independent Migration of Ions.

Because both positive and negative ions move under the influence of an electric field, albeit in opposite directions, the fraction of the current carried by an ion is given by the ratio of the mobility of the ion and the total mobility of all the ions in solution. This ratio is called the transference number t of the ion. Thus:

$$t_+ = u_+/(u_+ + u_-) \qquad t_- = u_-/(u_+ + u_-) \tag{28}$$

where

$$t_+ + t_- = 1 \tag{29}$$

The term equivalent conductance Λ is often used to describe the conductivity of electrolytes. It is defined as the conductivity of a cube of solution having a cross-section of one square centimeter and containing one equivalent of dissolved electrolyte.

$$\Lambda = k/C^* \qquad \text{or} \qquad \Lambda = 1000\, k/C \tag{30}$$

where C^* and C are the concentration in equiv/mL and equiv/L, respectively. It can be easily shown that:

$$\Lambda = F(u_+ + u_-) \tag{31}$$

Transport properties of the electrolye, as well as electrode reactions, have a significant impact on battery operation. The electrode reactions and ionic transference that occur during discharge result in considerable modifications to the solution composition at each electrode compartment. The negative and positive electrode compartments can lose or gain electrolyte and solvent depending on the transference numbers of the ions and the electrode reactions. The composition of the electrolyte in the separator between the two compartments generally remains unchanged.

Battery electrolytes are concentrated solutions of strong electrolytes and the Debye-Hückel theory of dilute solutions is only an approximation. Typical values for the resistivity of battery electrolytes range from about 1 ohm·cm for sulfuric acid [*7664-93-9*], H_2SO_4, in lead–acid batteries and for potassium hydroxide [*1310-58-3*], KOH, in alkaline cells to about 100 ohm·cm for organic electrolytes in lithium [*7439-93-2*], Li, batteries.

The physical picture in concentrated electrolytes is more aptly described by the theory of ionic association (18,19). It was pointed out that as the solutions become more concentrated, the opportunity to form ion pairs held by electrostatic attraction increases (18). This tendency increases for ions with smaller ionic radius and in the lower dielectric constant solvents used for lithium batteries. A significant amount of ion-pairing and triple-ion formation exists in the high concentration electrolytes used in batteries. The ions are solvated, causing solvent molecules to be highly oriented and polarized. In concentrated solutions the ions are close together and the attraction between them increases ion-pairing of the electrolyte. Solvation can tie up a considerable amount of solvent and increase the viscosity of concentrated solutions.

Lithium batteries must use nonaqueous electrolytes, usually combinations of solvents, for stability because lithium reacts readily with water. Many of these electrolytes dissolve a high concentration of solute but are relatively poor conductors. The cause appears to be related to the increase of ion-pair and triple-ion formation in the lower dielectric constant, relative to water, solvents and to the increased viscosity of concentrated solutions. For instance, a 1 *M*, lithium perchlorate [*7791-03-9*], $LiClO_4$, solution in propylene carbonate (PC) [*108-32-7*], $C_4H_6O_3$, has only about 20% of the conductivity expected because approximately 80% of the ions are involved in ion-pairing and triple-ion formation. Tables of electrolyte properties are available for most aqueous electrolytes (20,21) and selected nonaqueous electrolytes (22,23).

Each electrolyte is stable only within certain voltage ranges. Exceeding these limits results in decomposition. The stable range depends on the solvent, electrolyte composition, and purity level. In aqueous systems, hydrogen and oxygen form if the voltage limit is exceeded. In the nonaqueous organic solvent-based systems used for lithium batteries, exceeding the voltage limit can result in polymerization or decomposition of the solvent system. It is especially important to remove traces of water from the nonaqueous electrolytes as water can catalyze the electrolytic decomposition of the organic solvent.

In addition to the liquid conductors described above, two types of solid-state ionic conductors have been developed; one involves inorganic compounds and the other is based on polymeric materials. Several inorganic solids have been found to have excellent conductivity resulting wholly from ionic motion in the crystal lattice. Conductivity is related to specific crystal structures in which one ion, usually the cation, can move freely through the lattice. In the case of the solid, silver rubidium iodide, materials, $AgRb_4I_5$, the silver ion moves freely through the lattice and $t_{Ag^+} = 1$. The voltage of batteries constructed with these materials is limited only by the decomposition voltage of the electrolyte. One solid electrolyte, lithium iodide [*10377-51-2*], LiI, has found application in heart-pacer batteries even though it has a fairly low conductivity.

A second type of solid ionic conductors based around polyether compounds such as poly(ethylene oxide) [*25322-68-3*] (PEO) has been discovered (24) and characterized. These materials follow equations 23–31 as opposed to the electronically conducting polyacetylene [*26571-64-2*] and polyaniline type materials. The polyethers can complex and stabilize lithium ions in organic media. They also dissolve salts such as $LiClO_4$ to produce conducting solid solutions. The use of these materials in rechargeable lithium batteries has been proposed (25).

Electrical Double Layer

When two conducting phases come into contact with each other, a redistribution of charge occurs as a result of any electron energy level difference between the phases. If the two phases are metals, electrons flow from one metal to the other until the electron levels equilibrate. When an electrode, ie, electronic conductor, is immersed in an electrolyte, ie, ionic conductor, an electrical double layer forms at the electrode–solution interface resulting from the unequal tendency for distribution of electrical charges in the two phases. Because overall electrical neutrality must be maintained, this separation of charge between the electrode and solution gives rise to a potential difference between the two phases, equal to that needed to ensure equilibrium.

On the electrode side of the double layer the excess charges are concentrated in the plane of the surface of the electronic conductor. On the electrolyte side of the double layer the charge distribution is quite complex. The potential drop occurs over several atomic dimensions and depends on the specific reactivity and atomic structure of the electrode surface and the electrolyte composition. The electrical double layer strongly influences the rate and pathway of electrode reactions. The reader is referred to several excellent discussions of the electrical double layer at the electrode–solution interface (26–28).

Figure 3 schematically depicts the structure of the electrode–solution interface. The inner Helmholtz plane (IHP) refers to the distance of closest approach of specifically adsorbed ions, generally anions to the electrode surface. In aqueous systems, water molecules adsorb onto the electrode surface. The outer Helmholtz plane (OHP) refers to the distance of closest approach of nonspecifically adsorbed ions, generally cations. The interactions of the ions of the OHP with the surface are not specific and have the character of longer range coulombic interactions. Cations that populate the outer Helmholtz plane are usually solvated and are generally larger in size than the anions.

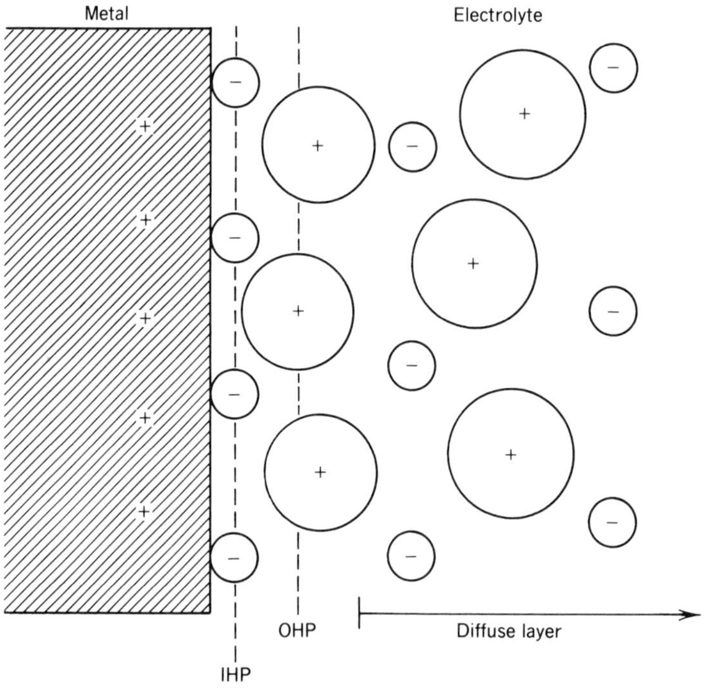

Fig. 3. Representation of the electrical double layer at a metal electrode–solution interface for the case where anions occupy the inner Helmholtz plane (IHP) and cations occupy the outer Helmholtz plane (OHP).

To a first approximation, the ions in both Helmholtz layers can be considered point charges. They induce an equal and opposite image charge inside the conductive electrode. When the electrode is negative to the point of zero charge, cations populate the inner Helmholtz layer.

When the electrode is positive to the point of zero charge, anions occupy the IHP. Anions are smaller in size and are solvated to a lesser degree than are the cations. Anions are more likely to adsorb specifically or bond to the electrode surface. If this is the case, the electrode can reach a considerable negative potential before strongly bonded anions are repelled from the inner layer. When an ion bonds to the surface it partially transfers charge to the surface. This charge transfer reduces the effective charge of the ion and causes the potential difference produced by the ion at the surface to be less than would be expected by a point charge an atomic distance from the surface.

The region of the gradual potential drop from the Helmholtz layer into the bulk of the solution is called the Gouy or diffuse layer (29,30). The Gouy layer has similar characteristics to the ion atmosphere from electrolyte theory. This layer has an almost exponential decay of potential with increasing distance. The thickness of the diffuse layer may be approximated by the Debye length of the electrolyte.

Electrical double layers are not confined to the interface between conducting phases. Solid particles of active mass, or of conductive additives of colloidal size, can acquire an electric charge by specific adsorption of cations or anions from the electrolyte or by reaction of surface moieties with components of the solution. The resulting excess charge on the particle is neutralized by a diffuse or Gouy layer in the solution. The electrokinetic properties of the interface and zeta potential concepts are based on the characteristics of the Gouy layer. Migration of the colloidal-sized solid particles can occur under the influence of an applied electric field during battery operations.

Electrically, the electrical double layer may be viewed as a capacitor with the charges separated by a distance of the order of molecular dimensions. The measured capacitance ranges from about two to several hundred microfarads per square centimeter depending on the structure of the double layer, the potential, and the composition of the electrode materials. Figure 4 illustrates the behavior of the capacitance and potential for a mercury electrode where the double layer capacitance is about 16 $\mu F/cm^2$ when cations occupy the OHP and about 38 $\mu F/cm^2$ when anions occupy the IHP. The behavior of other electrode materials is judged to be similar.

Kinetics and Transport

Activation Processes. To be useful in battery applications reactions must occur at a reasonable rate. The rate or ability of battery electrodes to produce current is determined by the kinetic processes of electrode operations, not by thermodynamics, which describes the characteristics of reactions at equilibrium when the forward and reverse reaction rates are equal. Electrochemical reaction kinetics (31–35) follow the same general considerations as those of bulk chemical reactions. Two differences are a potential drop that exists between the electrode and the solution because of the electrical double layer at the electrode interface;

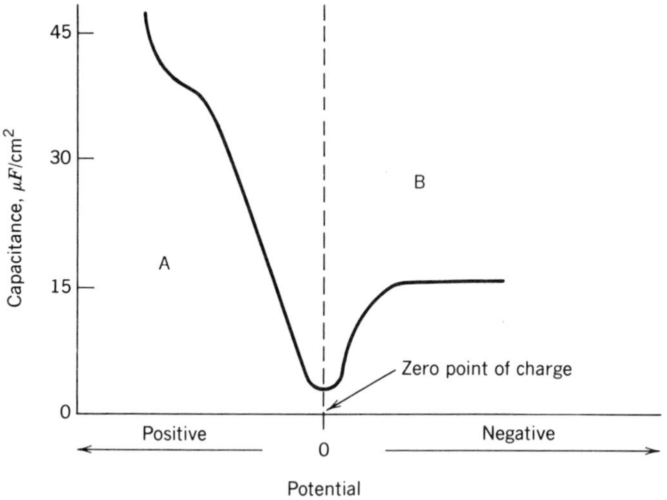

Fig. 4. Capacitance–potential relationship at a mercury electrode for a nonspecific absorbing electrolyte where regions A and B represent inner layer anions and cations, respectively.

and the reaction that occurs at interfaces that are two-dimensional rather than in the three-dimensional bulk.

Electrode kinetics lend themselves to treatment using the absolute reaction rate theory or the transition state theory (36,37). In these treatments, the path followed by the reaction proceeds by a route involving an activated complex where the element determining the reaction rate, ie, the rate limiting step, is the dissociation of the activated complex. The general electrode reaction may be described as:

$$A = C^{n+} + n\,e^- \tag{32}$$

where n is the number of electrons in the reaction. Using reaction rate theory, it can be shown that the current flow results from a change in the activation energy barrier introduced by a departure from the equilibrium potential drop across the electrode–solution interface. The free energies of activation for the forward ΔG_f and reverse ΔG_r reactions are given by:

$$\Delta G_f = \Delta G^* - \alpha nFE' \tag{33}$$

$$\Delta G_r = \Delta G^{**} + (1 - \alpha)nFE' \tag{34}$$

where E' is the potential of the electrode in its operating environment, α is the transfer coefficient or the fraction of the potential drop through the electrical double layer that operates on the activated complex, and $\Delta G^{**} = \Delta G^* + \Delta G^\circ$. The rate of electrochemical reactions are expressed as a current flow I in coulombs per second (amperes), rather than as a change in concentration per unit time. The direction of the potential change from the equilibrium potential E determines the direction of current flow. In terms of net current flow where $i =$

I/cm^2, it can be shown that:

$$i = i_f - i_r = nFk_1 A_A e^{(-\Delta G^* - \alpha nFE'/RT)} - nF_{k-1} A_c e^{(-\Delta G^{**} + (1-\alpha)nFE'RT)} \quad (35)$$

where A_A and A_C are the activities of reactants and products, k_1 and k_{-1} are the specific rate constants, and ΔG^* and ΔG^{**} are the free energy of activation for the forward and reverse reactions, respectively. The free energy of the activation is strongly influenced by the fraction of the potential drop across the electrode–solution interface that acts on the activated complex. Thus the structure of the electrical double layer plays a key role in the kinetics and rate of electrode reactions.

At equilibrium there is no net current flow, the rate of the forward and reverse reactions are equal and $E' = E$. The rate of the forward and reverse reactions at equilibrium

$$i_f = i_r = i_o \quad (36)$$

expressed as a current flow i_o are given the special term exchange current. Because electrode reactions are in dynamic equilibrium, the term i_o is the rate of the reaction at the equilibrium potential E.

The exchange current is directly related to the reaction rate constant, to the activities of reactants and products, and to the potential drop across the double layer. The larger i_o, the more reversible the reaction and, hence, the lower the polarization for a given net current flow. Electrode reactions having high exchange currents are favored for use in battery applications.

Taking the polarization η as the departure from equilibrium potential ($\eta = E' - E$), it follows from equations 35 and 36 that

$$i = i_o \left\{ e^{-(\alpha nF\eta/RT)} - e^{((1-\alpha)nF\eta/RT)} \right\} \quad (37)$$

The expected behavior of a current–potential diagram is given in Figure 5. These plots constitute the classic method for investigating electrode kinetics and for characterizing battery reactions. The value of α, and the relationship of i_o with concentration, provide valuable insight into the kinetics of the electrode reactions. At low current drains, the polarizations in battery reactions are almost always activation energy (charge-transfer) controlled. When the electrode reaction occurs and the polarization is larger than about 0.05 V, the term for the reverse reaction in equation 37 may be neglected and equation 37 can be simplified.

$$i = i_o \left\{ e^{-(\alpha nF\eta/RT)} \right\} \quad (38)$$

or

$$\eta = a - b \log i \quad (39)$$

where $a = (2.303\ RT/\alpha nF) \log i_o$ and $b = 2.303\ RT/\alpha nF$. Equation 39 is the well-known Tafel equation that describes the voltage–current behavior for processes

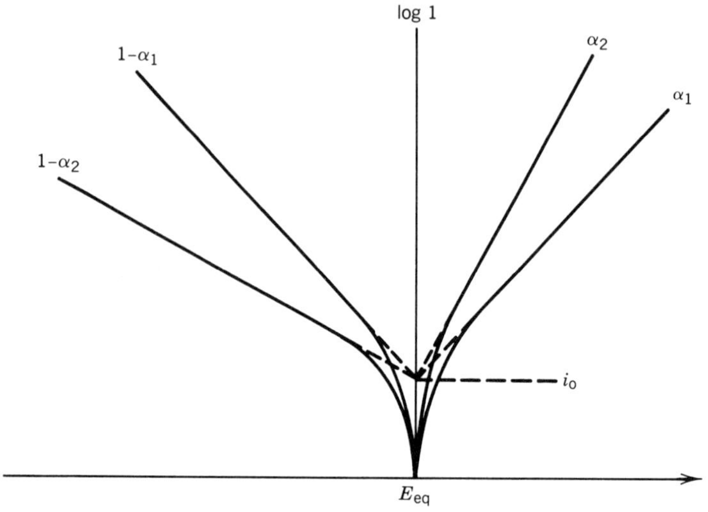

Fig. 5. Current–potential behavior of an electrode reaction based on equation 37. Tafel behavior is noted at high currents for two different values of α.

under activation control, and is illustrated in Figure 5 by the linear portion of the plot. The kinetic parameters for many electrode reactions have been summarized (38).

Tranport Processes. The velocity of electrode reactions is controlled by the charge-transfer rate of the electrode process, or by the velocity of the approach of the reactants, to the reaction site. The movement or transport of reactants to and from the reaction site at the electrode interface is a common feature of all electrode reactions. Transport of reactants and products occurs by diffusion, by migration under a potential field, and by convection. The complete description of transport requires a solution to the transport equations. A full account is given in texts and discussions on hydrodynamic flow (39,40) (see FLUID MECHANICS). Molecular diffusion in electrolytes is relatively slow. Although the process can be accelerated by stirring, enhanced mass transfer (qv) by stirring or convection is not possible in most battery designs. Natural convection from density changes does occur but does not greatly enhance transport in battery operation. Lead–acid batteries, used for motive power and stationary applications, are given a gassing overcharge on a regular basis. The gas evolution stirs up the electrolyte and equalizes the sulfuric acid concentration in the electrolyte.

Whenever the local concentration of a reacting component in a battery departs significantly from its equilibrium value, the rate of reaction becomes controlled by the transport of that component to the reaction site. The polarization resulting from these concentration changes η_c is given by:

$$\eta_c = \frac{RT}{nF} \ln (C_e/C) \tag{40}$$

where C_e is the concentration at the electrode surface and C is the concentration in the bulk of the electrolyte.

The limiting current density i_l for the transport of species i to the reacting site is given from Fick's law by:

$$i_l = \frac{DnFC_B}{\delta} \qquad (41)$$

where δ is the thickness of the stationary diffusion layer, D is the diffusion coefficient, C_B is the concentration on the bulk of the solution, and n and F are as previously defined. Allowing for electromigration, equation 41 becomes:

$$i_l = \frac{DnFC_B}{\delta(1 - t_i)} \qquad (42)$$

where t_i is the transference number of the reacting species. The limiting current calculation corresponds to the case where the concentration at the electrode surface is zero and the thickness of the diffusion layer is very thin. For many aqueous electrolytes the limiting current can be estimated by:

$$i_l = 0.025\ C \qquad (43)$$

where C is in g-equiv/L. The relationship between concentration polarization and limiting current density is given by:

$$\eta_c = \frac{RT}{nF} \ln\ (1 - i/i_l) \qquad (44)$$

Whereas the above discussion on concentration polarization was developed for electrolyte-side supply of reactants, concentration polarization can also arise from surface diffusion, diffusion into the solid structure (intercalation) of the active mass, and diffusion of products away from the reaction site. The deposition and dissolution reactions can involve movement of surface atoms to and from the deposition site to the equilibrium position in the lattice.

The detailed mechanism of battery electrode reactions often involves a series of chemical and electrochemical or charge-transfer steps. Electrode reaction sequences can also include diffusion steps on the electrode surface. Because of the high activation energy required to transfer two electrons at one time, the charge-transfer reactions are believed to occur by a series of one electron-transfer steps illustrated by the reactions of the zinc electrode in strongly alkaline medium (41).

$$Zn + OH^- \rightleftharpoons Zn(OH) + e^- \qquad (45)$$

$$Zn(OH) + OH^- \rightleftharpoons Zn(OH)_2^- \qquad (46)$$

$$Zn(OH)_2^- + OH^- \rightleftharpoons Zn(OH)_3^- + e^- \qquad (47)$$

$$Zn(OH)_3^- + OH^- \rightleftharpoons Zn(OH)_4^{2-} \qquad (48)$$

In this reaction sequence, equation 47, the formation of the complex ion, $Zn(OH)_3^-$, is the rate determining step. Once the solubility of zincate, $Zn(OH)_4^{2-}$, is exceeded, zinc hydroxide [*20427-58-1*], $Zn(OH)_2$, precipitates. The crystal form that falls out of solution depends on the concentration of the alkali.

Experimental Techniques

In the thermodynamic treatment of electrode potentials, the assumption was made that the reactions were reversible, which implies that the reactions occur infinitely slowly. This is never the case in practice. When a battery delivers current, the electrode reactions depart from reversible behavior and the battery voltage decreases from its open circuit or equilibrium voltage E. Thus the voltage during battery use or discharge E' is lower than the voltage measured under open circuit or reversible conditions E by a quantity called the polarization η.

$$\eta = E' - E \tag{49}$$

Likewise, the battery voltage is increased from its open circuit voltage on charge.

Three sources for the departure of battery operation from equilibrium voltage are: activation, concentration, and ohmic polarizations. The regions controlled by each of these change as the battery is discharged. *Activation polarization* arises from a kinetic hinderance in one or more of the detailed reaction processes. Slow or rate limiting steps of the reaction mechanism may occur in the charge-transfer process or in the chemical steps preceding or following the charge-transfer process. An activation barrier and an activation energy characterize these processes. On initiation or cessation of current flow, activation polarization generally builds up or decays in an exponential manner. *Concentration polarization* is associated with the decrease in availability of reacting species at the reaction site in the electrode–solution interface. Changes in concentration occur when the transport rate of reactants and/or products is slower than the rate of reaction. On initiation or cessation of current flow, the concentration polarization builds up or decays in a complex manner and is slower to recover than either activation and ohmic polarization. *Ohmic polarization* arises from the electrical resistances of the electrolyte, current collectors, the active mass, the conductive additives, and the contact of the collector to the active mass. It may also be associated with a resistive film formation on the electrode surface. Ohmic polarization follows Ohm's law. On cessation of current flow, the ohmic polarization disappears instantaneously.

A variety of experimental techniques are used to study electrochemical and battery reactions (34,42–44). The most common are the direct measurement of the instantaneous current-voltage characteristics or power curve and the discharge curve at various discharge rates. Indeed, the discharge performance of a typical battery is characterized by these curves as depicted in Figure 6. The characteristics of the power curve varies depending on the state of charge of the battery.

The impedance behavior of a battery can reveal a significant amount of information about battery operation characteristics (34,44). The impedance of an electrode or battery is given by:

$$Z = R + jX \tag{50}$$

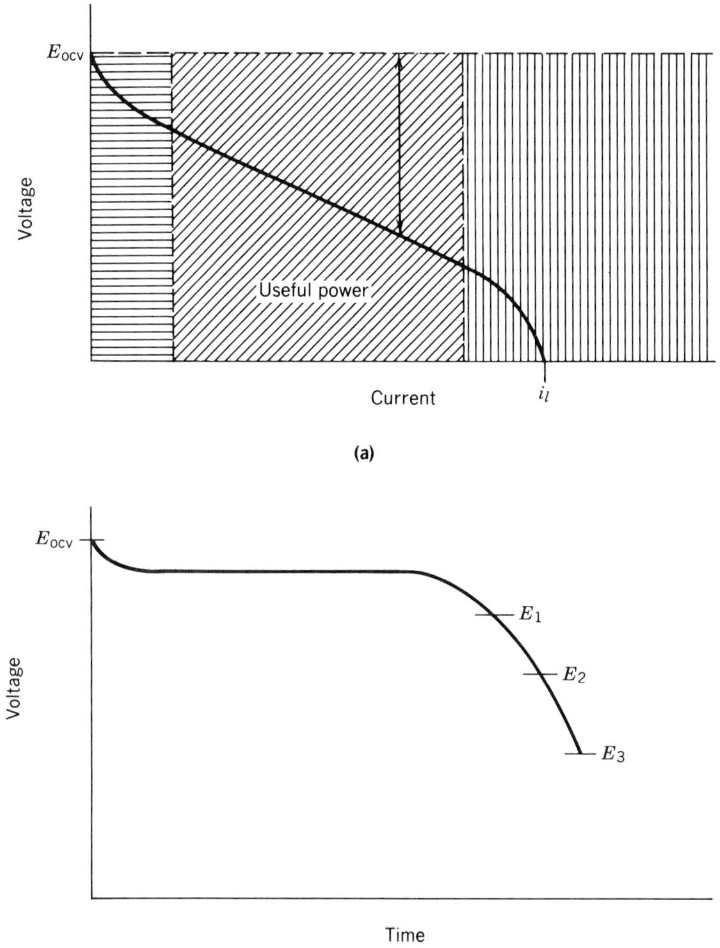

Fig. 6. Discharge behavior of a battery where E_{ocv} is the open circuit voltage; (**a**) current–potential or power curve showing ⊟ activation, ▨ ohmic, and ▥ concentration polarization regions where the double headed arrow represents polarization loss and (**b**) voltage–time profile.

where $X = \omega L - 1/(\omega C)$, $j = \sqrt{-1}$, ω is the angular frequency $(2\pi f)$, L the inductance, and C the capacitance. Each of the characteristic polarizations has a distinctive impedance behavior. A schematic of a battery circuit and the corresponding Argand diagram, illustrating the behavior of the simple processes, are shown in Figure 7. In ideal behavior, activation processes exhibit a semicircular behavior with frequency that is characteristic of relaxation processes; concentration processes exhibit a 45° behavior characteristic of diffusion processes; and ohmic polarizations are independent of frequency. Battery electrodes have large surface areas and, therefore, exhibit large capacitances. It is common for larger cells to have a capacitance of farads and a resistance of milliohms in measurements of battery impedances.

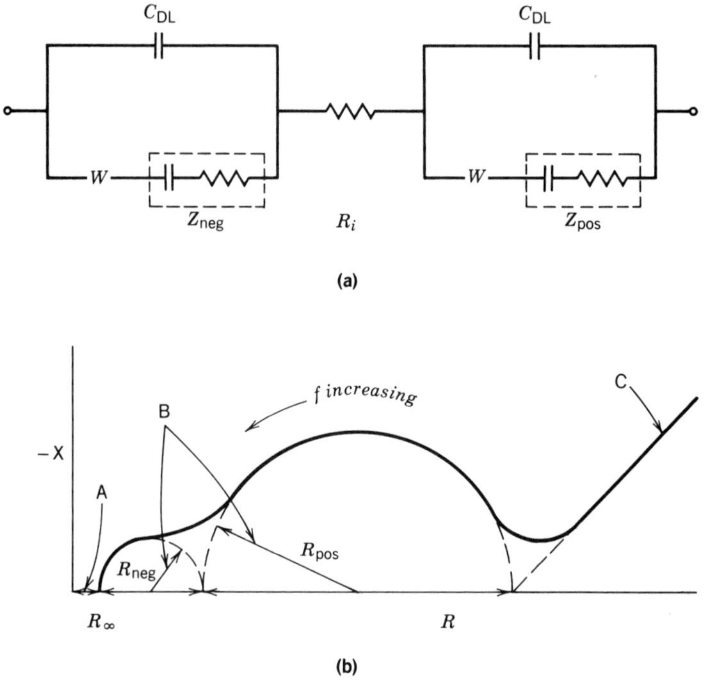

Fig. 7. (**a**) Simple battery circuit diagram where C_{DL} represents the capacitance of the electrical double layer at the electrode–solution interface, W depicts the Warburg impedance for diffusion processes, and R_i is internal resistance; and (**b**) the corresponding Argand diagram of the behavior of impedance with frequency, f, for an idealized battery system, where the characteristic behavior of A, ohmic; B, activation; and C, diffusion or concentration (Warburg behavior) processes are shown. X is as defined in equation 50.

Practical Battery Systems

Most battery electrodes are porous structures in which an interconnected matrix of solid particles, consisting of both nonconductive and electronically conductive materials, is filled with electrolyte. When the active mass is nonconducting, conductive materials, usually carbon or metallic powders, are added to provide electronic contact to the active mass. The solids occupy 50% to 70% of the volume of a typical porous battery electrode. Most battery electrode structures do not have a well defined planar surface but have a complex surface extending throughout the volume of the porous electrode. Macroscopically, the porous electrode behaves as a homogeneous unit.

When a battery produces current, the sites of current production are not uniformly distributed on the electrodes (45). The nonuniform current distribution lowers the expected performance from a battery system, and causes excessive heat evolution and low utilization of active materials. Two types of current distribution, primary and secondary, can be distinguished. The primary distribution is related to the current production based on the geometric surface area of the battery construction. Secondary current distribution is related to current production sites inside the porous electrode itself. Most practical battery constructions have nonuniform current distribution across the surface of the electrodes. This primary current distribution is governed by geometric factors such as height (or length) of the electrodes, the distance between the electrodes, the resistance of the anode and cathode structures; by the resistance of the electrolyte; and by the polarization resistance or hinderance of the electrode reaction processes.

Cell geometry, such as tab/terminal positioning and battery configuration, strongly influence primary current distribution. The monopolar construction is most common. Several electrodes of the same polarity may be connected in parallel to increase capacity. The current production concentrates near the tab connections unless special care is exercised in designing the current collector. Bipolar construction, wherein the terminal or collector of one cell serves as the anode and cathode of the next cell in pile formation, leads to greatly improved uniformity of current distribution. Several representations are available to calculate the current distribution across the geometric electrode surface (46–50).

Whereas current producing reactions occur at the electrode surface, they also occur at considerable depth below the surface in porous electrodes. Porous electrodes offer enhanced performance through increased surface area for the electrode reaction and through increased mass transfer rates from shorter diffusion path lengths. The key parameters in determining the reaction distribution include the ratio of the volume conductivity of the electrolyte to the volume conductivity of the electrode matrix, the exchange current, the diffusion characteristics of reactants and products, and the total current flow. The porosity, pore size, and tortuosity of the electrode all play a role. Figure 8 illustrates the reaction distribution in porous electrodes. The location of the reaction sites is seen to be strongly dependent on the characteristics of the electrode structure and reactions.

The effectiveness of a porous electrode over a plane surface electrode is

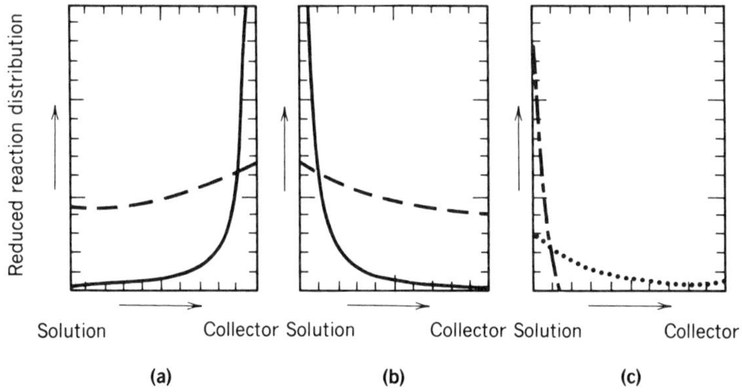

Fig. 8. Representation of the current distribution in porous electrodes showing the effect of conductivities of the electrolyte and electrodes where for (**a**) $K_{electrolyte} \gg K_{matrix}$ and (**b**) $K_{matrix} \gg K_{electrolyte}$ (48) (— — —) represents low current and (———) high current; and for (**c**) showing the effect of exchange current (49) where (• • •) represents low i_0 and (— - —) high i_0.

given by the product of the active surface area S in cm^2/mL and the penetration depth L_P of the reaction process into the porous electrode.

$$\text{Effectiveness} = SL_P \tag{51}$$

An effectiveness value greater than one indicates that the porous electrode is more effective than an electrode of the same geometric surface area, and that the reaction extends into the porous electrode structure.

Mathematical formulations of the models of primary and secondary current distribution permit rapid optimization in the design of new battery configurations. Models to describe and predict porous electrode performance in the lead–acid battery system have been developed (51–53). The high rate performance of the present starting, lighting and ignition (SLI) automotive batteries have evolved directly from coupling collector designs with the porous electrode compositions identified from modeling studies.

In practice, the Peukert equation, equation 52, finds wide use in reporting and comparing battery performance (54).

$$It^n = \text{constant} \tag{52}$$

where I is the current, t the time to a given discharge voltage, and n is a constant. If the system is reversible, $n = 1$. However, because battery systems have limitations in reaction rate, diffusion, and internal resistance, $n > 1$ for most practical systems. Figure 9 shows the Peukert curve constructed from the performance of a given battery to different cutoff voltages. Other equations for describing and predicting battery performance have also been developed (55–56).

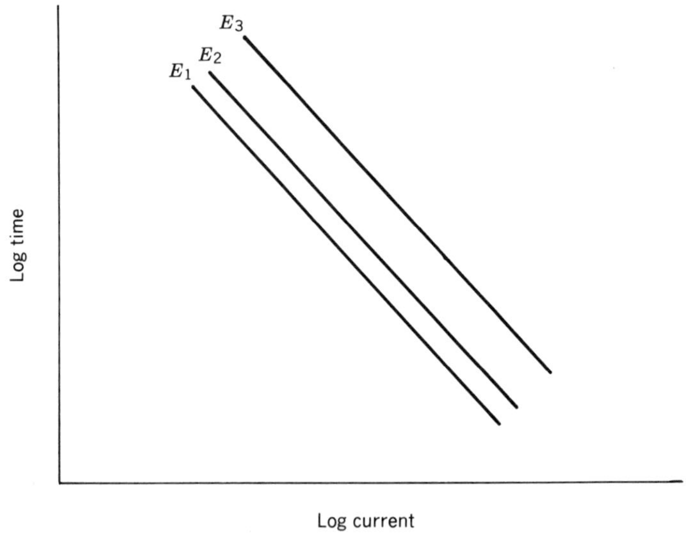

Fig. 9. Peukert diagram of the performance of a battery based on the discharge curves from Figure 6(**b**) where $E_1 > E_2 > E_3$.

The positive electrode in a battery system is most often a metal oxide, but it may also be a metal sulfide or halide. Generally, these materials are relatively poor electrical conductors and exhibit extremely high ohmic polarizations (impedances) if not combined with supporting electronic conductors such as graphite [*7782-42-5*], lead [*7439-92-1*], silver [*7440-22-4*], copper [*7440-50-8*], or nickel [*7440-02-0*] in the form of powder, rod, mesh, wire, grid, or other configurations. In almost all cases the negative electrode is a metallic element of sufficient conductivity to require only minimal supporting conductive structures. Exceptions are the oxygen (air) positive and hydrogen gas negative electrodes which require a substantial conductive, catalytically active, surface support which also serves as current collector.

Although there are a multitude of chemical reactions that release energy, only a few reactions have the characteristics requisite for use in commercial batteries. A set of criteria can be established to characterize reactions suitable for battery development (57,58). The principal features necessary for battery reactions are: (*1*) *Mechanical and chemical stability*. The reactants or active masses and cell components must be stable over time (5 years or more) in the operating environment. The reactants must not change crystal structure or surface properties during storage. For the case of rechargeable batteries, an additional condition is that on charging the battery, the reaction products must return the reactants, exactly, to their original charged state. The materials problems in batteries arise from irreversibility in phase changes and chemical interconversion; isolation, electrolytically or electronically, of active materials; and local, poor conductivity of materials in the discharged state (59). (*2*) *Energy content*. The reactants must have sufficient energy content to provide a useful voltage and current level, measured in W·h/L or W·h/kg. This is usually taken to be a unit cell voltage of at least one volt. The carbon–zinc and lead–acid systems set the lower limit of storage capability for primary and rechargeable batteries, respectively. (*3*) *Power density*. The reactants must be capable of reacting at rates sufficient to deliver useful rates of electricity, measured in terms of W/L or W/kg. This implies that the kinetics of the reaction are fast and without significant kinetic hindrances. (*4*) *Temperature range*. The reactants must be able to maintain energy, power, and stability over a normal operating environment. The military often specifies $-50°C$ to $75°C$. The average consumer has a less severe range of operating requirements usually $-10°C$ to $50°C$. (*5*) *Safety*. The battery must be safe in the normal operating environment as well as under mild abusive conditions. It should not leak, vent hazardous materials, or explode (*6*) *Cost*. The reactants and the materials of construction should be inexpensive and in good supply. The marketplace rules out unreasonable cost for a unit of energy. This may be somewhat subjective in that expensive materials such as silver, may be used in small cells or for special purposes where performance is more important than cost. In addition other criteria for commercial rechargeable battery systems include: cycle life, overchargeability, cell reversal, range of operation, and shape of discharge curve.

Tables 2 and 3 contain characteristics of various primary and secondary battery systems. The alkaline Zn–MnO_2 battery sets the standard for primary battery systems. This battery was first introduced commercially in 1958 and

Table 2. Commercial Primary Battery Systems

Common name	Cell reactions	Nominal voltage	Energy content[a] W·h/mL	Energy content[a] W·h/kg	Comments	Manufacturers
Leclanché	$Zn + 2\ MnO_2 + 2\ NH_4Cl \rightarrow Zn(NH_3)_2Cl_2 + H_2O + Mn_2O_3$	1.5	0.20	80	low cost; general purpose, wide range of sizes	Eveready Rayovac Bright Star
zinc chloride	$4\ Zn + 8\ MnO_2 + ZnCl_2 + 9\ H_2O \rightarrow 8\ MnOOH + ZnCl_2{\cdot}4ZnO{\cdot}5H_2O$	1.5	0.20	125	intermediate cost and performance	Eveready Rayovac
alkaline	$2\ Zn + 3\ MnO_2 \rightarrow 2\ ZnO + Mn_3O_4$	1.5	0.25	130	sets standard for cylindrical cells	Duracell Eveready Rayovac
silver	$Zn + Ag_2O \rightarrow 2\ Ag + ZnO$	1.6	0.5	200	good pulse, higher voltage than mercury or Zn–air cells	Duracell Eveready Rayovac
mercury	$Zn + HgO \rightarrow Hg + ZnO$	1.35	0.5	165	sets standard for button cells; environmental problems	Duracell Rayovac Alexander
air	$2\ Zn + O_2\ (air) \rightarrow 2\ ZnO$	1.4	1.0	300	twice capacity of mercury and silver cells, limited active stand	Duracell Rayovac
Li–CuO	$2\ Li + CuO \rightarrow Li_2O + Cu$	1.5	0.6	300	potential replacement for Leclanché and zinc chloride	Panasonic
Li–FeS	$2\ Li + FeS \rightarrow Li_2S + Fe$	1.6	0.4	160	replacement for mercury and silver cells, no mercury	Eveready
Li–CF_x	$x\ Li + CF_x \rightarrow x\ LiF + C$	2.7	0.5	300	high voltage, long shelf life, wide operating temperature	Rayovac Eagle Picher
Li–MnO_2	$Li + MnO_2 \rightarrow LiMnO_2$	2.8	0.5	330	high voltage, long shelf life, wide operating temperature	Duracell Eveready Power Conversion
Li–I_2	$2\ Li + I_2 \rightarrow 2\ LiI$	2.8	0.9	290	solid electrolyte, 10+ year life welded construction, used in most of heart pacers, iodine charge transfer complex	Medtronic Catalyst Wilson Greatbatch
Li–SO_2	$2\ Li + 2\ SO_2 \rightarrow Li_2S_2O_4$	2.8	0.4	330	military battery, low temperature, excellent storage	SAFT Power Conversion
Li–$SOCl_2$	$4\ Li + 2\ SOCl_2 \rightarrow 4\ LiCl + SO_2 + S$	3.6	1.0	530	high voltage, high energy density	Electrochem Industries Eagle Picher Power Conversion

[a]Approximate values.

Table 3. Commercial Rechargeable Battery System

Common name	Cell reaction	Nominal voltage	Energy content[a] W·h/L	Energy content[a] W·h/kg	Comments	Manufacturers
lead–acid	$Pb + PbO_2 + 2\ H_2SO_4 \rightleftarrows 2\ PbSO_4 + 2\ H_2O$	2.10	80	35	lowest cost, largest sales, available sealed	Johnson Controls Delco Exide GNB
nickel–cadmium	$Cd + 2\ NiOOH + 2\ H_2O \rightleftarrows Cd(OH)_2 + 2\ Ni(OH)_2$	1.35	80	38	high rate, available sealed	Gates SAFT
nickel–metal hydride	$H_2(M) + 2\ NiOOH \rightleftarrows 2\ Ni(OH)_2 + M$	1.35	160	55	hydrogen absorbing alloy, good cycle life, high self-discharge	Ovonics Gates
nickel–iron	$Fe + 2\ NiOOH + 2\ H_2O \rightleftarrows Fe(OH)_2 + 2\ Ni(OH)_2$	1.25	90	30	limited production, very long cycle life, almost indestructible, old technology	SAB Eagle Picher
nickel–hydrogen	$H_2(g) + 2\ NiOOH \rightleftarrows 2\ Ni(OH)_2$	1.35	90	45	special space battery, very long cycle life, high self discharge	Hughes Eagle Picher
silver–zinc	$Zn + 2\ AgO \rightleftarrows ZnO + Ag_2O$ $Zn + Ag_2O \rightleftarrows ZnO + 2\ Ag$	1.86 1.60	200	100	two step discharge, limited cycle life, high energy density	Yardney
lithium–MoS_2	$Li + MoS_2 \rightleftarrows LiMoS_2$	2.3	150	80	under development, small sealed cell	Moli Energy
lithium–MnO_2	$Li + MnO_2 \rightleftarrows LiMnO_2$	3.2	140	55	sealed coin cell	Sanyo
lithium–V_2O_5	$Li(C) + V_2O_5 \rightleftarrows LiV_2O_5 + C$	3.0	75	30	sealed coin cell	Toshiba
lithium–carbon	$Li + C \rightleftarrows Li(C)$ (intercalate)	3.0	6	2.1	sealed coin cell, "supercapacitor"	Panasonic

[a]Approximate values.

Table 4. Rechargeable Battery Systems in Various Stages of Research and Development

Common name	Cell reaction	Nominal voltage	Energy content[a]		Comments
			W·h/L	W·h/kg	
nickel–zinc	$Zn + 2\ NiOOH + H_2O \rightleftarrows ZnO + 2\ Ni(OH)_2$	1.65	95	60	limited cycle life, high rate capability
zinc–bromine	$Zn + Br_2 \rightleftarrows ZnBr_2$	1.85	75	65	bromine complex with quaternary ammonium salt, circulating electrolyte
lithium–FeS	$2\ LiAl + FeS \rightleftarrows Li_2S + 2\ Al + Fe$	1.33	90	95	low cost, high temperature fused salt, second step possible
sodium–sulfur	$2\ Na + 3\ S \rightleftarrows Na_2S_3$	2.1	120	160	solid β-Al_2O_3 separator, high temperature operation
aluminum–air	$4\ Al + 3\ O_2 + 2\ H_2O \rightleftarrows 2\ Al_2O_3{\cdot}H_2O$	1.6	360	250	circulating electrolyte, low cost, low energy efficiency, replaceable negative electrode
lithium–V_6O_{13}	$8\ Li + V_6O_{13} \rightleftarrows Li_8V_6O_{13}$ (intercalate)	2.5	200	200	solid polymer electrolyte

[a]Approximate values.

replaced the carbon–zinc as the principal primary battery system in about 1985. Lithium batteries were first developed in the early 1970s for military applications and the commercial market began to develop in the late 1970s, largely in Japan. The lead–acid battery dominates the rechargeable battery market: sales of lead–acid batteries account for over 85% of the rechargeable battery market. Table 4 contains performance parameters for promising rechargeable battery systems in various stages of research and commercial development. Table 5 gives the theoretical energy content of selected high energy systems that are under consideration for use in electric vehicle propulsion and other applications where these high energy battery systems are required. Several of the systems are more energetic than gasoline. The primary lithium–thionyl chloride battery system and the rechargeable sodium–sulfur battery system are included for comparison.

Table 5. Theoretical High Energy Systems

System/Reaction	Cell voltage, V	Energy content, W·h/kg
$2\ Li + H_2O_2 \rightleftarrows 2\ LiOH$	3.98	4400
$4\ Li + O_2 + 2\ H_2O \rightleftarrows 4\ LiOH$	4.47	3240
$2\ Mg + O_2 + 2\ H_2O \rightleftarrows 2\ Mg(OH)_2$	3.09	2850
$4\ Al + 3\ O_2 + 2\ H_2O \rightleftarrows 2\ Al_2O_3{\cdot}H_2O$	2.7	2791
$2\ Li + Cl_2 \rightleftarrows 2\ LiCl$	3.95	2510
gasoline		2300
$2\ H_2 + O_2 \rightleftarrows 2\ H_2O$	1.23	1940
$4\ Li + 2\ SOCl_2 \rightleftarrows 4\ LiCl + SO_2 + S$	3.6	1450
$2\ Na + 3\ S \rightleftarrows Na_2S_3$	2.01	760

BIBLIOGRAPHY

"Batteries, Primary (Introduction)" in *ECT* 2nd ed., Vol. 3, pp. 99–110, by E. B. Yeager, Western Reserve University, and J. F. Yeager, Union Carbide Corp., "Batteries and Electric Cells, Secondary (Introduction)" in *ECT* 2nd ed., Vol. 3, pp. 161–171, by W. W. Jakobi, Gould-National Batteries, Inc.; "Batteries and Electric Cells, Primary (Introduction)" in *ECT* 3rd ed., Vol. 3, pp. 569–591, by J. B. Doe, ESB Technology Co. "Batteries and Electric Cells, Secondary (Introduction)" in *ECT* 3rd ed., Vol. 3, pp. 569–591, by J. B. Doe, ESB Technology Co.

1. G. W. Heise and N. C. Cahoon, eds., *The Primary Battery,* John Wiley & Sons, Inc., New York, 1965, Vol. 1; 1976, Vol. 2.
2. S. U. Falk and A. J. Salkind, *Alkaline Storage Batteries,* John Wiley & Sons, Inc., New York, 1969.
3. H. Bode, *Lead Acid Batteries,* John Wiley & Sons, Inc., New York, 1976.
4. D. Linden, ed., *Handbook of Batteries and Fuel Cells,* McGraw-Hill Book Co., Inc., New York, 1984.
5. C. A. Vincent, B. Scrosati, M. Lazzari, and F. Bonino, *Modern Batteries,* Edward Arnold, Ltd., London, 1984.
6. J. O'M. Bockris and A. K. N. Reddy, *Modern Electrochemistry,* Plenum Press, New York, 1972, Vols. 1 and 2.
7. J. Koryta and J. Dvorak, *Principles of Electrochemistry,* John Wiley & Sons, Inc., New York, 1987.
8. A. J. Salkind and R. J. Brodd, "Primary and Secondary Batteries" in W. Smyrl and F.

McLarnon, eds., *Proc. Symposium Teaching Electrochem.,* The Electrochemical Society, Pennington, N.J., 1987, Vol. 87-3, pp. 127–145.
9. J. O'M. Bockris, B. E. Conway, E. B. Yeager, R. E. White, eds., *Comprehensive Treatise of Electrochemistry,* Plenum Press, New York, 1980, Vols. 1–9.
10. I. M. Klotz, *Chemical Thermodynamics,* Benjamin Press, New York, 1967.
11. M. W. Chase, Jr., C. A. Davies, J. R. Downey, Jr., D. J. Frurip, R. A. McDonald, and A. N. Syverud, *JANAF Thermochemical Tables,* 3rd ed., *J. Phys. Chem. Ref. Data,* **14,** Suppl. No. 1 (1985).
12. D. D. Wagman, W. H. Evans, V. B. Parker, R. H. Schuum, I. Halow, S. M. Bailey, K. L. Churney, and R. L. Nuttall, "The NBS Tables of Chemical Thermodynamic Properties," *J. Phys. Chem. Ref. Data* **11,** Suppl. 2 (1982).
13. A. J. Bard, R. Parsons, and J. Jordan, eds., *Standard Potentials in Aqueous Solution,* Marcel Dekker, Inc., New York, 1985.
14. W. M. Latimer, *Oxidation Potentials,* 2nd ed., Prentice-Hall, Inc., Englewood Cliffs, N.J., 1952.
15. M. Pourbaix, *Atlas of Electrochemical Equilibria in Aqueous Solutions,* Pergamon Press, Inc., Elmsford, N.Y., 1969.
16. P. Debye and E. Hückel, *Z. Phys. (Lepizig)* **24,** 185,305 (1923).
17. R. A. Robinson and R. H. Stokes, *Electrolytic Solutions,* 2nd ed., Butterworths, London, 1959.
18. N. Bjerrum, *Kgl. Danske Vidensk. Selskab* **7**(9) (1926).
19. R. M. Fuoss and F. Accascina, *Electrolytic Conductance,* Interscience Publishers, New York, 1959.
20. H. S. Harned and B. B. Owen, *The Physical Chemistry of Electrolyte Solutions,* 3rd ed., Reinhold Publishing Corp., New York, 1958.
21. B. E. Conway, *Electrochemical Data,* Elsevier, Amsterdam, The Netherlands, 1952.
22. H. V. Venkatasetty, ed., *Lithium Battery Technology,* Wiley-Interscience, New York, 1984.
23. G. J. Janz and R. P. T. Thomkins, *Nonaqueous Electrolyte Handbook,* Academic Press, Inc., 1972, Vol. 1; 1974, Vol. 2.
24. D. E. Furton, J. M. Parker, and P. V. Wright, *Polymer* **14,** 589 (1973).
25. M. Armand, J. M. Chabagno, and M. Duclot, in P. Vashishta, ed., *Fast Ion Transport in Solids,* North Holland, New York, 1979, p. 131.
26. J. A. V. Butler, *Electrical Phenomena at Interfaces,* Butterworths, London, 1951.
27. D. C. Grahame, *Chem. Rev.* **41,** 441 (1947).
28. D. M. Mohilner, in A. Bard, ed., *Electroanalytical Chemistry,* Marcel Dekker, Inc., New York, 1966, Vol. 1, p. 241.
29. G. Gouy, *J. Phys.* **9,** 457 (1910).
30. D. L. Chapman, *Philos. Mag.* **25,** 475 (1913).
31. K. Vetter, *Electrochemical Kinetics,* Academic Press, New York, 1967.
32. R. A. Marcus, *Ann. Rev. Phys. Chem.* **15,** 155 (1964).
33. T. Erdey-Gruz, *Kinetics of Electrode Processes,* Wiley-Interscience, New York, 1972.
34. A. J. Bard and L. R. Faulkner, *Electrochemical Methods, Fundamentals and Applications,* John Wiley & Sons, Inc., New York, 1980.
35. W. J. Albery, *Electrode Kinetics,* Clarendon, Oxford, 1975.
36. S. Glasstone, K. J. Laidler, and H. Eyring, *The Theory of Rate Processes,* McGraw-Hill Book Co., Inc., New York, 1941.
37. W. C. Gardner, Jr., *Rates and Mechanisms of Chemical Reactions,* Benjamin, New York, 1969.
38. N. Tanaka and R. Tamamushi, *Electrochim. Acta* **9,** 963 (1964).
39. V. C. Levich, *Physicochemical Hydrodynamics,* Prentice-Hall, Inc., Englewood Cliffs, N.J., 1962.
40. J. Newman, *Electrochemical Systems,* Prentice-Hall, Inc., Englewood Cliffs, N.J., 1973.

41. J. O'M. Bockris, Z. Nagy, and A. Damjanovic, *J. Electrochem. Soc.* **119,** 285 (1972).
42. R. J. Brodd and A. Kozawa, "Primary Batteries," in E. B. Yeager and A. J. Salkind, eds., *Techniques of Electrochemistry,* John Wiley & Sons, Inc., New York, 1978, Vol. 3.
43. G. Halpert, "A Review of Secondary Batteries and Evaluation Techniques," in E. B. Yeager and A. J. Salkind, eds., *Techniques of Electrochemistry,* John Wiley & Sons, Inc., New York, 1978, Vol. 3.
44. R. J. Brodd, *Proc. 5th Ann. Battery Conf.,* California State University, Long Beach, Calif., 1990.
45. R. J. Brodd, *Electrochem. Tech.* **6,** 289 (1968).
46. C. W. Shephard, *J. Electrochem. Soc.* **112,** 252 (1965).
47. C. Wagner, *J. Electrochem. Soc.* **98,** 116 (1951).
48. J. Newman and C. W. Tobias, *J. Electrochem. Soc.* **109,** 1183 (1962).
49. E. A. Grens II and C. W. Tobias, *Ber. Bunsenges. Phys. Chem.* **68,** 236 (1964).
50. J. A. Klingert, S. Lynn, and C. W. Tobias, *Electrochim. Acta* **9,** 297 (1961).
51. J. Newman and W. Tiedemann, *AIChE Journal* **21,** 25 (1975).
52. W. H. Tiedemann and J. S. Newman, in S. Gross, ed., *Proc. Symp. Batt. Design Optim.* The Electrochemical Society, Pennington, N.J., Vol. 79-1, 1979, pp. 23–49.
53. W. H. Tiedemann and J. S. Newman, in K. Bullock and D. Pavlov, eds., *Proc. Symp. Adv. Lead–Acid Batt.* The Electrochemical Society, Pennington, N.J., Vol. 84-14, 1984, pp. 336–349.
54. W. Peukert, *Z. Electrotech.* **18,** 287 (1897).
55. R. Selim and P. Bro, *J. Electrochem. Soc.* **118,** 829 (1971).
56. C. M. Shephard, *J. Electrochem. Soc.* **112,** 657 (1965).
57. R. J. Brodd and R. M. Wilson, in G. W. Heise and N. C. Cahoon, eds., *The Primary Battery,* John Wiley & Sons, Inc., New York, 1976, Vol. 2, pp. 369–428.
58. R. J. Brodd, in U. Landau, E. B. Yeager, and D. Kortan, eds., *Electrochemistry in Industry,* Plenum Press, New York, 1982, pp. 153–180.
59. B. E. Conway "Lecture Notes" *Rechargeable Battery Seminar,* Florida Atlantic University, March 3, 1986.

RALPH BRODD
Gould Inc.

PRIMARY CELLS

Primary cells are galvanic cells designed to be discharged only once and attempts to recharge them can present possible safety hazards. The cells are designed to have the maximum possible energy in each cell size because of the single discharge. Thus, comparison between battery types is usually made on the basis of the energy density in W·h/cm^3. The specific energy, W·h/kg, is often used as a secondary criterion for primary cells especially when the application is weight sensitive as in space applications. The main categories of primary cells are carbon–zinc, known as heavy-duty and general purpose; alkaline, cylindrical and miniature; lithium; and reserve or specialty cells.

Carbon–Zinc Cells

Carbon–zinc batteries are the most commonly found primary cells worldwide and are produced in almost every country. Traditionally there is a carbon [*7440-44-0*] rod, for cylindrical cells, or a carbon-coated plate, for flat cells, to collect the current at the cathode and a zinc [*7440-66-6*] anode. There are two basic versions of carbon–zinc cells: the Leclanché cell and the zinc chloride [*7646-85-7*], $ZnCl_2$, or heavy-duty cell. Both have zinc anodes, manganese dioxide [*1313-13-9*], MnO_2, cathodes, and include zinc chloride in the electrolyte. The Leclanché cell also has an electrolyte saturated with ammonium chloride [*12125-02-9*], NH_4Cl. Additional undissolved ammonium chloride is usually added to the cathode, whereas the zinc chloride cell has at most a small amount of ammonium chloride added to the electrolyte. Both types are dry cells in the sense that there is no excess liquid electrolyte in the system. The zinc chloride cell is often made using synthetic manganese dioxide and gives higher capacity than the Leclanché cell, which uses inexpensive natural manganese dioxide for the active cathode material. The MnO_2 is only a modest conductor. Thus the cathodes in both types of cell contain 10–30% carbon black in order to distribute the current. Because of the ease of manufacture and the long history of the cell, this battery system can be found in many sizes and shapes.

Leclanché cells are usually made using paste separators in which electrolyte solution and various types of cereal are cooked until thick. Some designs use a cold-set paste however. The paste is metered into the cell and the prepressed cathode body is inserted, forcing the paste into a separating layer. Figure 1**a** shows a cutaway view of a typical pasted cylindrical cell. Paper-lined cells have a starch or modified starch-coated paper separator, which is much thinner and more conductive than the starch paste separator. This starch-coated paper separator is typically used in premium Leclanché cells and zinc chloride cells. In these cells, the separator is inserted into the zinc can and the cathode mix is extruded into the can, followed by insertion of the carbon rod into the cathode mix. Figure 1**b** shows a cutaway of a typical paper-lined cell. Flat cells are used to stack up in multicell batteries, most commonly for 6- or 9-volt batteries. Figure 2 shows a cutaway of a typical flat cell, as used in a 9-volt battery. Note that in this cell the zinc plate is carbon-coated to serve as the collector for the cathode of the cell beneath as the cells are stacked.

The Leclanché cell was patented in 1866 (2). It consisted of an amalgamated zinc rod for the anode, a clay pot filled with a mix of manganese dioxide and carbon, and a carbon rod or plate driven into the mix as the current collector. The electrolyte was an aqueous solution of ammonium chloride that filled the glass container for the whole assembly. The cathode was open to the air allowing workers to replace the zinc rod whenever necessary yet continue to operate the cell. The MnO_2 was regarded as simply a depolarizer, ie, as acting to reduce the polarization of the cathode reaction that was thought to be the production of hydrogen. The term depolarizer is sometimes used in the battery field in this incorrect sense. In a later development (3) a zinc can was used as a container for the entire contents of the cell and thickeners were placed in the electrolyte, making the first version of the dry cell.

Cell Chemistry. Work on the mechanism of the carbon–zinc cell has been summarized (4), but the dynamics of this system are not entirely understood. The electrochemical behavior of electrolytic (EMD), chemical (CMD), and natural (NMD) manganese dioxide is slightly different. Battery-grade NMD is most commonly in the form of the mineral nsutite [*12032-72-3*], $H_2O{\cdot}xMnO_2$, which is a

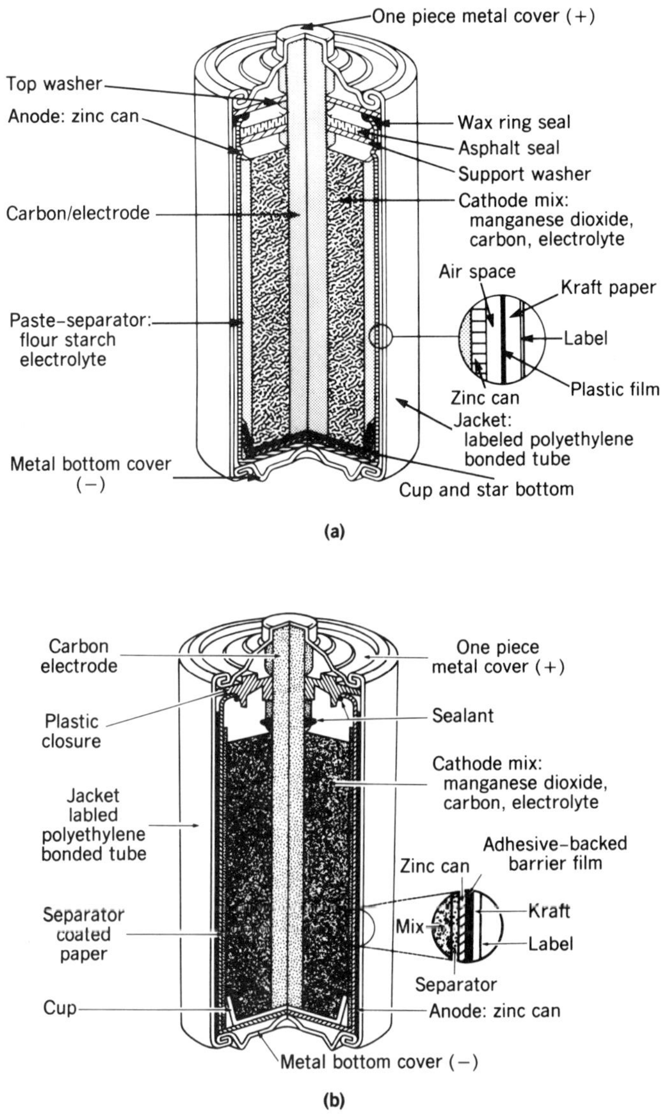

Fig. 1. Cutaway view of (**a**) a "D"-size pasted Leclanché cell, and (**b**) a "D"-size paper-lined carbon–zinc cell (1). Courtesy of Eveready Battery Co.

structural intergrowth of the minerals ramsdellite [*12032-73-4*], MnO_2, and pyrolusite [*14854-26-3*], MnO_2 (5). Occasionally, the pyrolusite or cryptomelane [*12325-71-2*], KMn_8O_{16}, form is used in batteries. The pure ramsdellite form is quite rare and expensive and is not used. The mineral as well as the synthetic types of MnO_2 have manganese oxidation states less than four and contain bound protons, generally considered to be present as hydroxyl ions, as well as internal and adsorbed water. NMD normally contains higher levels of impurities than the other types, so that the overall activity or theoretical capacity is considerably lower. However, the cost of NMD is considerably less; the mineral is relatively

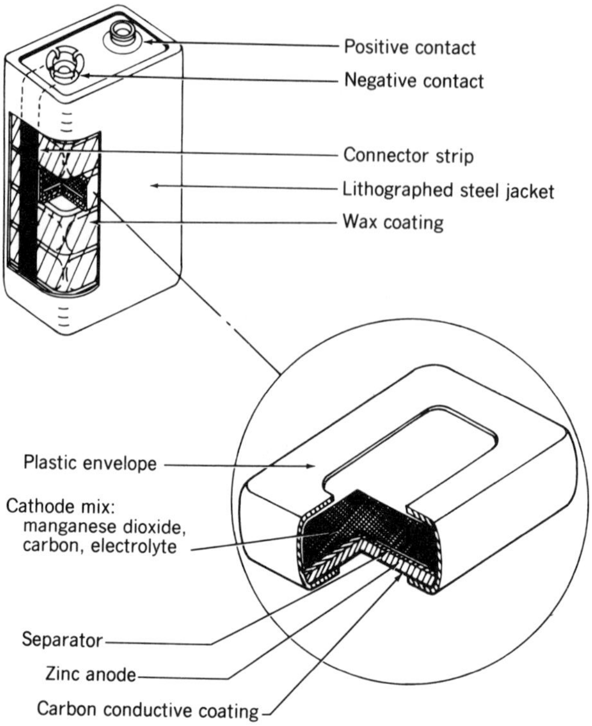

Fig. 2. Cutaway view of a Leclanché flat cell used in multicell batteries (1). Courtesy of Eveready Battery Co.

abundant and no chemical or electrochemical processing is required. Thus NMD is the material of choice for low priced carbon–zinc cells. Premium cells are made with some proportion of the higher priced, higher activity CMD or EMD in the cathode. CMD, which is made by a rather complex chemical process, often has comparable activity to EMD on a weight basis and has a lower cost, but its high surface area and poor packing capability makes it less favorable in some applications. The most active form is EMD, which is also the most expensive. It is made by electrolytic deposition of MnO_2 from a bath of $MnSO_4$ in sulfuric acid (see MANGANESE COMPOUNDS).

The mechanism of the cathode reaction for all three types of MnO_2 can best be described by two approximately one-electron steps.

$$MnO_2 + H_3O^+ \text{ (or } NH_4^+) + e^- \longrightarrow MnOOH + H_2O \text{ (or } NH_3) \quad (1)$$

and

$$MnOOH + H_3O^+ \text{ (or } NH_4^+) + e^- \longrightarrow Mn(OH)_2 + H_2O \text{ (or } NH_3) \quad (2)$$

where NH_4^+ and NH_3 are appropriate to Leclanché cells and H_3O^+ and H_2O to zinc chloride cells. Equation 1 is believed to occur in a single overall phase with the gradual incorporation of protons and electrons. It appears that the end product of equation 1 is an intergrowth material in which the manganese ions occupy the same octahedral sites as in MnO_2 and the remaining octahedral sites are filled with protons, although the hexagonal close packed oxide ion sublattice is somewhat distorted. This kind of homogeneous reaction is similar to a reaction occurring in solution and the potential is related to the proton activity, expressed in mole fraction, within the MnO_2.

As the reduction of the cathode proceeds through equation 1 and the potential decreases, the reduction equation 2 becomes more favorable. First the divalent $Mn(OH)_2$ is solubilized to some extent in the mildly acidic electrolyte.

$$Mn(OH)_2 + 2\ H_3O^+ \longrightarrow Mn^{2+} + 4\ H_2O \tag{3}$$

The manganous ion [*16397-91-4*], Mn^{2+}, in solution then reacts with higher valent manganese oxide and zinc ions in solution to form a new phase called hetaerolite [*12163-55-2*], $ZnMn_2O_4$, or hydrohetaerolite, a poorly defined hydrated form.

$$MnO_2 + Mn^{2+} + Zn^{2+} + 2\ H_2O \longrightarrow ZnMn_2O_4 + 4\ H^+ \tag{4}$$

Hetaerolite is the most stable form of trivalent manganese at low cathode potentials and this sequence of reactions contributes to the familiar recovery of potential on open circuit that gives rise to higher cell capacity on intermittent discharge compared to continuous discharge. A second mode of potential recovery is the equilibration of the pH, resulting from the slow kinetics of zinc complex formation in solution and the dependence of the potential of the MnO_2 electrode on the pH.

In most cylindrical carbon–zinc cells, the zinc anode also serves as the container for the cell. The zinc can is made by drawing or extrusion. Mercury [*7439-97-6*] has traditionally been incorporated in the cell to improve the corrosion resistance of the anode, but the industry is in the process of removing this material because of environmental concerns. Corrosion prevention is especially important in cylindrical cells because of the tendency toward pitting of the zinc can which leads to perforation and electrolyte leakage. Other cell types, such as flat cells, do not suffer as much from this problem.

The anode reaction depends on the electrolyte used, but the charge-transfer step is

$$Zn \longrightarrow Zn^{2+} + 2\ e^- \tag{5}$$

In Leclanché cells, the high concentration of ammonium chloride leads to the formation of insoluble diammine zinc chloride through the reaction in the electrolyte.

$$Zn^{2+} + 2\ NH_3 + 2\ Cl^- \longrightarrow Zn(NH_3)_2Cl_2 \tag{6}$$

where the sources of the ammonia are the cathode reactions of equations 1 and 2. The initial form of the zinc ions is predominantly $ZnCl_4^{2-}$ because of the high concentration of ammonium chloride. As the reaction proceeds and NH_4Cl is consumed, however, lower chlorocomplexes, $ZnCl_3^-$, $ZnCl_2$, and $ZnCl^+$, are formed. The final stage in the anode reaction is the precipitation of basic zinc chloride.

$$5\ Zn^{2+} + 2\ Cl^{-} + 17\ H_2O \longrightarrow ZnCl_2{\cdot}4Zn(OH)_2{\cdot}H_2O + 8\ H_3O^{+} \quad (7)$$

In the zinc chloride cell, precipitated basic zinc chloride is the primary anode product because of the low concentration of ammonium chloride in the cell. Water and zinc chloride are consumed in equations 1 and 7 and must be provided in adequate amounts for the cell to discharge efficiently. Usually more carbon is used in zinc chloride cells than in Leclanché cells in order to increase the electrolyte absorptivity of the cathode and thus allow the use of a larger volume of electrolyte. Also, the use of a thin paper separator, which decreases internal resistance, allows less space for water storage than the thick, pasted separator construction traditionally used in Leclanché cells.

Corrosion and other unwanted reactions cause difficulties in the design of carbon–zinc cells. Fortunately, the overpotential for hydrogen evolution on zinc is quite high in both mildly acidic and alkaline solutions. In some cases, basic ZnO is added to zinc chloride electrolyte to raise the pH and lower the corrosion rate. Good control of heavy metal impurities in the cathode is also important, to avoid these metals being solubilized by the electrolyte and subsequently deposited on the zinc to become low overpotential sites for hydrogen evolution. Another type of corrosion is the direct reaction of oxygen from air entering the cell. To protect against this loss, the cell is sealed so that access of air is restricted. The cell must not be hermetically sealed, however, as hydrogen and other gases, mainly carbon dioxide, must be able to escape from the cell as they are formed. The source of CO_2 is the cathode and results from the excellent oxidative properties of MnO_2. Thus, if a starch paste separator is used, some glucose becomes available from starch hydrolysis and readily reacts with MnO_2. The carbon used as conductor in the cathode also reacts with MnO_2, but more slowly. The higher the voltage the more easily the MnO_2 reacts with these materials. Thus premium cells have special separator materials such as methyl cellulose-coated paper to minimize this reaction.

The porous carbon rod is often the main pathway of escape for the gases formed in the cell. This pathway also allows ingress of oxygen to the cell limiting the shelf life of the system. The use of shrink tube outer wrapping and other devices have, however, improved the leakage property dramatically over prior generation cells.

Performance. Carbon–zinc cells perform best under intermittent use and many standardized tests have been devised that are appropriate to such applications as light and heavy flashlight usage, radios, cassettes, and motors (toys). The most frequently used tests are American National Standards Institute (ANSI) tests (6). The tests are carried out at constant resistance and the results reported in minutes or hours of service. Figure 3 shows typical results under a light load for different size cells, whereas Figure 4 shows results for different types of R20 "D"-size cells under a heavy intermittent load.

To compare one battery with another, it is useful to compute the energy density from these data. Because the voltage declines with capacity, the average voltage during the discharge is used to compute an average current which is then

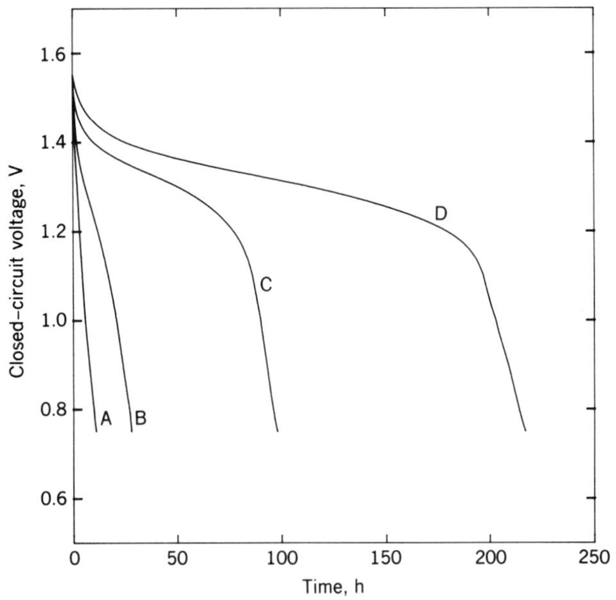

Fig. 3. Hours of service on 40-Ω discharge for 4 h/d radio test at 21°C for A, RO3 "AAA"; B, R6 "AA"; C, R14 "C"; and D, R20 "D" paper-lined, heavy-duty zinc chloride cells.

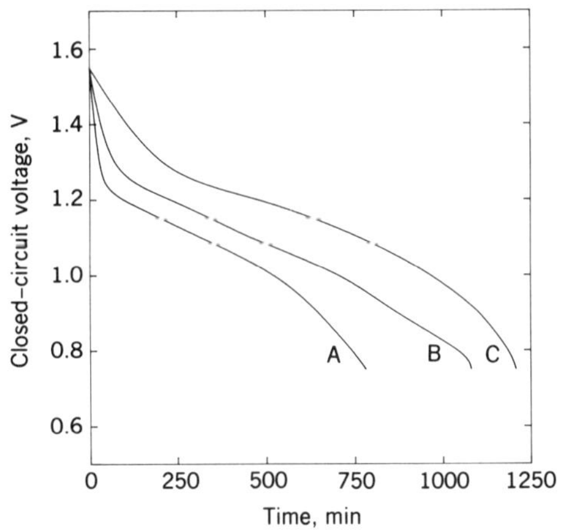

Fig. 4. Hours of service on 4-Ω discharge, 15 min/h, 8 h/d, LIF test (see Table 1) at 21°C for A, general purpose (pasted Leclanché) "D"-size; B, premium Leclanché (paper-lined Leclanché); and C, heavy-duty (paper-lined $ZnCl_2$) cells.

multiplied by the service in hours to give the ampere-hours of capacity. Watt-hours of energy can be obtained by multiplying again by the average voltage.

The voltage at the end of discharge is also important in defining these quantities because of the sloping discharge curve. Typical computed values for R20 "D" and R6 "AA" cells of the Leclanché and zinc chloride battery types are given in Table 1. For heavy duty cells, the specific energies (W·h/kg) are higher for the larger R20 cells than for the R6 cells, whereas the energy densities (W·h/mL) are higher for the R6 cells. This is because the can weight makes a much greater contribution to the total weight of the system for small cells. Thus the smaller cathode of the R6 cell allows a more efficient design of the cell. These relationships are not exactly preserved in the data for the general purpose cells in Table 1, because the R6 cell used in this test is made from the more efficient zinc chloride cell design.

Table 1. Specific Energies and Energy Densities of Carbon–Zinc Cells[a]

Battery designation and type	Test, Ω[b,c]	W·h/mL	W·h/kg
R20 "D"-size cells			
general purpose Leclanché	40 Ω radio	0.12	73
	LIF	0.07	41
heavy-duty zinc chloride	40 Ω radio	0.16	100
	LIF	0.15	90
R6 "AA"-size zinc chloride cells			
general purpose	75 Ω radio	0.13	72
	LIF	0.10	51
heavy-duty	75 Ω radio	0.17	81
	LIF	0.15	68

[a]Calculations are based on data from reference 7.
[b]The radio test is one 4-h continuous discharge daily.
[c]Each discharge performed at 2.25 Ω on the light industrial flashlight (LIF) test: one 4-min discharge at 1-h intervals for eight consecutive hours each day, with 16-h rest periods; ie, 32 min of discharge per day.

The effect of the discharge rate is especially pronounced for the general purpose cells. On intermittent tests, the heavy-duty cell operates at high efficiency even at high rate. On continuous test at high rate, heavy-duty cells provide 60–70% of the intermittent service, whereas general purpose cells give only 30–50% of the intermittent service values.

Cylindrical Alkaline Cells

Primary alkaline cells use sodium hydroxide [*1310-73-2*] or potassium hydroxide [*1310-58-3*] as the electrolyte. They can be made using a variety of chemistries and physical constructions. Early alkaline cells were of the wet cell type, but the alkaline cells of the 1990s are mostly of the limited electrolyte, dry cell type. Most primary alkaline cells are made using zinc as the anode material; a variety of cathode materials can be used. Primary alkaline cells are commonly divided into two classes, based on type of construction: the larger, cylindrically shaped batteries, and the miniature, button-type cells. Cylindrical alkaline batteries are mainly produced using zinc–manganese dioxide chemistry, although some cylindrical zinc–mercury oxide cells are made.

Cylindrical alkaline cells are zinc–manganese dioxide cells having an alkaline electrolyte, which are constructed in the standard cylindrical sizes, R20 "D", R14 "C", R6 "AA", RO3 "AAA", as well as a few other less common sizes. They can be used in the same types of devices as ordinary Leclanché and zinc chloride cells. Moreover, the high level of performance makes them ideally suited for applications such as toys, audio devices, and cameras.

Figure 5 shows a cross-sectional diagram of a typical R20 size cylindrical alkaline cell. The battery is contained in a steel can, which also serves as the current collector for the cathode. Inside the can is a dense compacted cathode mass containing manganese dioxide, carbon, and sometimes a binder. The cathode has a hollow center lined with a separator to isolate the cathode from the anode. Within the separator-lined cavity is placed the anode mix consisting of zinc powder and alkaline electrolyte, together with a small amount of gelling material to immobilize the electrolyte and suspend the zinc powder. Contact to the anode is made by a metal pin or leaf inserted into the anode mix. The cell has a plastic seal assembly, to keep electrolyte from leaking out of the cell and to keep air from getting in. Contact to the anode collector is made through this seal. The seal contains a fail-safe vent that is activated when the battery internal pressure exceeds a certain level.

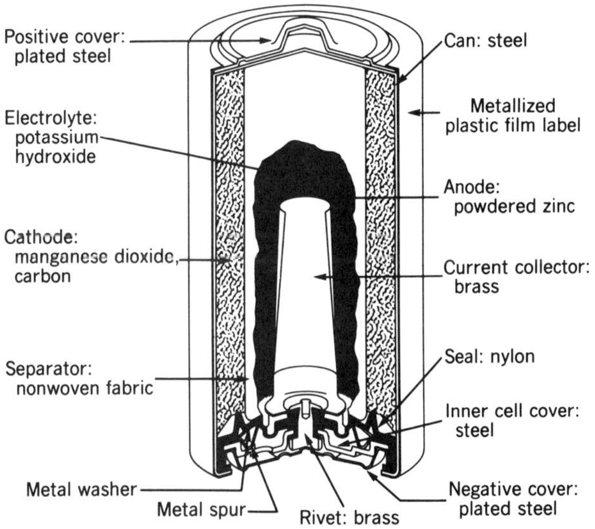

Fig. 5. Cutaway view of typical R20 or "D"-size alkaline manganese battery showing components and the corresponding materials of construction (8). Courtesy of Eveready Battery Co.

Cells using alkaline electrolytes have been known since the early days of batteries. Early alkaline cells included the zinc–copper oxide cell in the 1880s (9), the rechargeable iron–nickel oxide cells (10,11), and an early version of the zinc–manganese dioxide alkaline cell (12). Commercial zinc–mercuric oxide batteries were developed in the 1940s (13). Work on the use of the zinc anode in alkaline cells was very valuable to the subsequent development of the cylindrical alkaline general purpose cell in the late 1950s as was research into the use of manganese dioxide as a cathode material in small alkaline cells (14). Commercial "B" batteries for portable radios came out of those investigations. "B" batteries contained a large number of individual cells connected in series to produce a high voltage but only a relatively low current. In the late 1950s the first multipurpose, cylindrical, zinc–manganese dioxide batteries were commercialized (15). Such batteries having high output capacity and high current carrying ability are now made by many manufacturers throughout the world. There is ongoing competition among manufacturers to improve the performance of cylindrical alkaline batteries.

Chemistry. The alkaline cell derives its power from the reduction of the manganese dioxide cathode and the oxidation of the zinc anode. The reactions

$$Zn + 2\ OH^- \longrightarrow ZnO + H_2O + 2\ e^- \quad \text{anode} \tag{8}$$

$$2\ e^- + 2\ MnO_2 + 2\ H_2O \longrightarrow 2\ MnOOH + 2\ OH^- \quad \text{cathode} \tag{9}$$

give, as the overall reaction,

$$Zn + 2\ MnO_2 + H_2O \longrightarrow ZnO + 2\ MnOOH \tag{10}$$

Both the anode and cathode equations are more complicated than is indicated by the equations shown.

Anode. *Anode Corrosion Reaction.* Zinc might at first appear to be an unusual choice for battery anode material, because the metal is thermodynamically unstable in contact with water

$$Zn + H_2O \longrightarrow ZnO + H_2 \quad \Delta G = -81.142\ \text{kJ/mol}\ (-19.39\ \text{kcal/mol}) \tag{11}$$

However, the reaction with water can be made to be extremely slow. Because the alkaline electrolyte is corrosive toward human tissue as well as toward the materials in devices, it is more important to have a good seal toward preventing electrolyte leakage in an alkaline battery than in a carbon–zinc cell. The formation of a good seal is, however, incompatible with the formation of a noncondensable gas like hydrogen.

The reaction of zinc and water is not a simple homogeneous one. Rather it is a heterogeneous electrochemical reaction, involving a mechanism similar to that of a battery. There are two steps to the reaction: zinc dissolves at some locations as shown in equation 8 while hydrogen gas is generated at other sites.

$$2\ e^- + 2\ H_2O \longrightarrow H_2 + 2\ OH^- \quad \text{cathode} \tag{12}$$

These two reactions then add up to the overall zinc corrosion reaction (eq. 11).

In a battery, the anode and cathode reactions occur in different compartments, kept apart by a separator that allows only ionic, not electronic conduction. The only way for the cell reactions to occur is to run the electrons through an external circuit so that electrons travel from the anode to the cathode. But in the corrosion reaction the anode and cathode reactions, equations 8 and 12 respectively, occur at different locations within the anode. Because the anode is a single, electrically conductive mass, the electrons produced in the anode reaction travel easily to the site of the cathode reaction and the zinc acts like a battery where the positive and negative terminals are shorted together.

The rate at which the corrosion of the zinc proceeds depends on the rates of the two half reactions (eqs. 8 and 12). Equation 8, a necessary part of the desired battery reaction, fortunately represents a reaction that proceeds rather rapidly, whereas the reaction represented by equation 12 is slow. Ie, the generation of hydrogen on pure zinc is a sluggish reaction and thus limits the overall corrosion reaction rate.

Certain impurities on zinc can act as catalysts for the generation of hydrogen, thereby greatly increasing the corrosion rates. For this reason, zinc in alkaline cells must be of high purity, and careful control exercised over the level of the harmful impurities. Moreover, other components of the cell must not contain harmful levels of these impurities that might dissolve and migrate to the zinc anode.

Alternatively, there are also inhibitors that decrease the rate of hydrogen generation and thus decrease corrosion. Mercury, effective at inhibiting zinc corrosion, has long been used as an additive to zinc anodes. More recently, however, because of increased interest in environmental issues, the amount of mercury in alkaline cells has been reduced.

Zinc Oxidation Mechanism. The oxidation reaction for the zinc anode (eq. 8) takes place in several steps (16,17), ultimately resulting in the zincate ions [*16408-25-6*], $Zn(OH)_4^{2-}$, that dissolves in the electrolyte.

$$Zn + 4\ OH^- \longrightarrow Zn(OH)_4^{2-} + 2\ e^- \tag{13}$$

As a battery discharges the concentration of zincate in the electrolyte increases. Eventually, the concentration exceeds the solubility of zinc oxide [*1314-13-2*], ZnO, which can precipitate from the solution

$$Zn(OH)_4^{2-} \longrightarrow ZnO + H_2O + 2\ OH^- \tag{14}$$

This precipitation can be sluggish and in some cases the zincate concentration can increase to three or four times the equilibrium solubility value, after which precipitation of zinc hydroxide [*20427-58-1*] can occur.

$$Zn(OH)_4^{2-} \longrightarrow Zn(OH)_2 + 2\ OH^- \tag{15}$$

Dehydration of zinc hydroxide can then lead to formation of the thermodynamically more stable zinc oxide.

$$Zn(OH)_2 \longrightarrow ZnO + H_2O \tag{16}$$

In a cylindrical alkaline cell, the amount of electrolyte is limited and the electrolyte becomes saturated with zincate rather early in the discharge. Thus the cell produces the zinc oxide reaction product through most of its discharge.

Cathode Reaction. There are many different types of manganese dioxide (18), having varying activity in batteries. The only type suitable for alkaline batteries is γ-MnO_2, the mineral form of which is nsutite. The chemical composition of γ-MnO_2 has been described (19) by the general formula $(MnO_2)_{2n-3}\cdot(MnOOH)_{4-2n}\cdot mH_2O$, where the value of n ranges from 1.5 to 2. γ-MnO_2 is useful in batteries because of the unusual nature of its discharge reaction. When γ-MnO_2 is discharged the reaction product α-MnOOH (groutite structure) is formed and this material, when formed during discharge, has a structure so similar to that of the starting MnO_2 that the two substances can form a solid solution in which equilibrium is rapidly achieved. Thus when MnOOH is formed at the surface of MnO_2, there is a rapid diffusion of protons from the surface MnOOH into the interior, coupled with electron transfer; ie, the result is that the lower valent material initially formed on the surface effectively diffuses into the solid particle. In this way the entire particle is homogeneously reduced and the surface tends to have the same composition as the overall particle (20). The cathode material starts as nearly all manganese(IV) and ends as manganese(III).

The homogeneous phase reduction of γ-MnO_2 results in a discharge curve that differs from that of many other battery systems. In any battery system there can be a voltage drop from the open circuit value resulting from polarization effects. But for a typical heterogeneous (two-phase) reaction as exists in lead–acid batteries, Zn–HgO batteries, Zn–Ag_2O batteries, nickel cadmium batteries, etc, the open circuit recovery voltage is substantially constant throughout the discharge. In the homogeneous phase reduction of γ-MnO_2, however, the open circuit voltage itself changes as the cathode is reduced (Fig. 6). Thus the operating voltage of the alkaline zinc–manganese dioxide battery starts around 1.5 volts, and drops to below 1 volt by the time a 1-electron reduction of the MnO_2 is complete. At this point, when the cathode material approaches 100% manganese(III), the solubility of the manganese increases and this soluble manganese

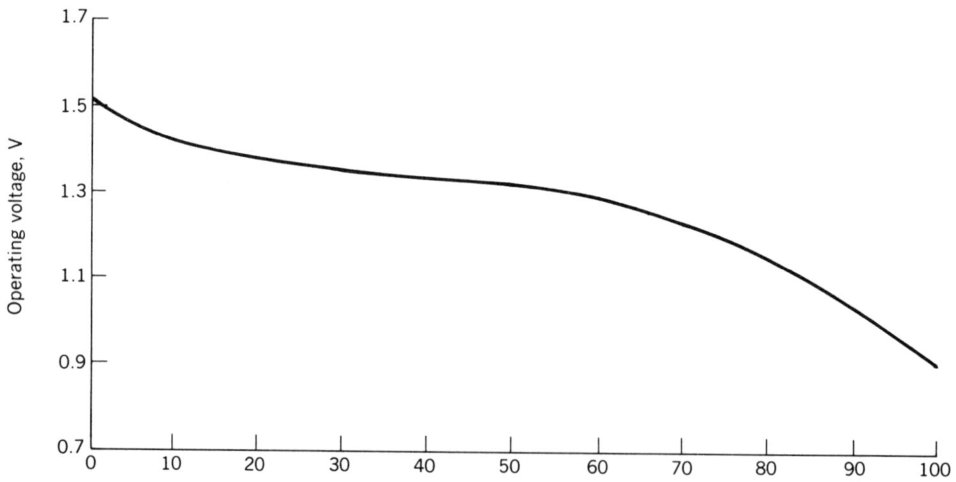

Fig. 6. Alkaline–manganese battery operating voltage as a function of remaining capacity (21). Courtesy of Eveready Battery Co.

then combines with the zincate present in the electrolyte to form the insoluble hetaerolite ($ZnMn_2O_4$). In the process, water is released.

$$2\ MnOOH + Zn(OH)_4^{2-} \longrightarrow ZnMn_2O_4 + 2\ H_2O + 2\ OH^- \quad (17)$$

For heavy drain discharges of alkaline cells, there is no useful capacity after this point because the rate of discharge of the $ZnMn_2O_4$ is quite slow. But for lighter drain discharges, further reduction of the cathode is possible. The reaction mechanisms are not entirely clear, but there is some evidence for the formation of a final reaction product resembling hausmannite [*1309-55-3*], Mn_3O_4.

Performance. Alkaline manganese-dioxide batteries have relatively high energy density, as can be seen from Table 2. This results in part from the use of highly pure materials, formed into electrodes of near optimum density. Moreover, the cells are able to function well with a rather small amount of electrolyte. The result is a cell having relatively high capacity at a fairly reasonable cost.

Table 2. Characteristics of Aqueous Primary Batteries

Parameter	Carbon–zinc (Zn/MnO_2)	Alkaline manganese dioxide ($Zn–MnO_2$)	Mercuric oxide ($Zn–HgO$)	Silver oxide ($Zn–Ag_2O$)	Zinc–air ($Zn–O_2$)
nominal voltage, V	1.5	1.5	1.35	1.5	1.25
working voltage, V	1.2	1.2	1.3	1.55	1.25
specific energy, W·h/kg	40–100	80–95	100	130	230–400
energy density, W·h/mL	0.07–0.17	0.15–0.25	0.40–0.60	0.49–0.52	0.70–0.80
temperature range, °C					
storage	−40–50	−40–50	−40–60	−40–60	−40–50
operating	−5–55	−20–55	−10–55	−10–55	−10–55

The performance of the cells is influenced not only by the relatively high theoretical capacity, but also by the fact that the cells can provide good efficiency at various currents over a wide variety of conditions. The high conductivity and low polarization of the alkaline electrolyte, combined with the good electronic and ionic conductivity in the electrodes, lead to high performance even under heavy drain conditions. Alkaline zinc–manganese dioxide batteries show less variation in output capacity with variation in discharge rate than do Leclanché or zinc chloride batteries. Thus alkaline cells have a moderate advantage over Leclanché cells in capacity on light drains (Fig. 7**a**) and they have a much greater advantage on heavy drains (Fig. 7**b**).

Whereas Leclanché and zinc chloride batteries perform much better on intermittent high rate discharge than on continuous high rate discharge, alkaline

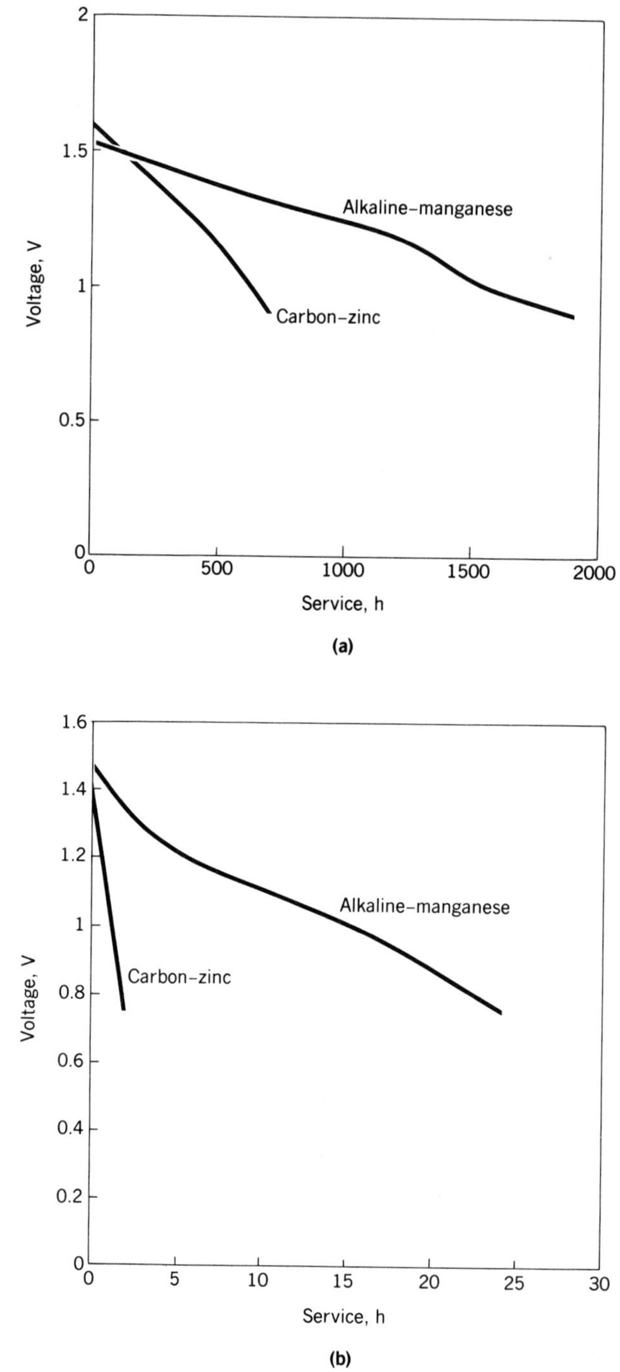

Fig. 7. Performance comparison of "D"-size alkaline–manganese vs carbon–zinc batteries at 21°C on (**a**) a light drain 150-Ω continuous test at 21°C, and (**b**) a heavy drain 2.2-Ω continuous test (8). Courtesy of Eveready Battery Co.

zinc–manganese dioxide cells do not show this effect to any substantial degree. The output of these alkaline cells at a given drain rate is similar regardless of whether the discharge is continuous or intermittent. Thus alkaline cells have an advantage over Leclanché cells on high rate intermittent tests, but have an even greater advantage on high rate continuous tests.

Batteries tend to perform more poorly as the operating temperature is decreased because of decreased conductivity of the electrolyte and slower electrode kinetics. Ultimately, freezing of the electrolyte occurs and the battery fails. Batteries tend to perform better at higher temperatures only up to the point that loss of performance occurs because of cell venting and drying out or parasitic reactions in the cell. Overall, alkaline cells have less performance loss at low and high temperatures than do Leclanché cells as can be seen in Figure 8.

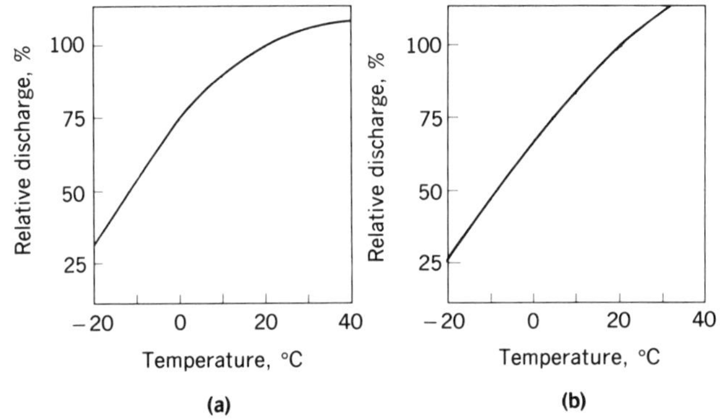

Fig. 8. Effect of temperature on relative discharge performance of a fresh "D"-size battery for service on simulated radio use, 25-Ω 4-h/d test for **(a)** an alkaline–manganese battery undergoing 260 h of service, and **(b)** a carbon–zinc battery undergoing 70 h of service (22). Courtesy of Eveready Battery Co.

The ability of a cell to retain capacity after long-term storage is another important aspect of performance. Cells can lose capacity during storage because of various parasitic reactions, such as zinc corrosion. Also, if the cell is not tightly sealed, there can be transfer of water vapor into or out of the cell, transfer of carbon dioxide into the cell, etc. Alkaline zinc–manganese dioxide cells have good capacity retention on long-term storage. The use of high purity materials ensures very low rates of parasitic reactions in the cells. Moreover, the alkaline cells are well sealed so that there is minimal vapor transmission into or out of the cell. Figure 9 shows the capacity retention of alkaline cells and that of Leclanché cells after prolonged storage at different temperatures.

Miniature Alkaline Cells

Miniature alkaline cells are small, button-shaped cells, which use alkaline NaOH or KOH electrolyte, generally have zinc anodes but may have a variety of cathode materials. They are used in watches, calculators, cameras, hearing aids, and other

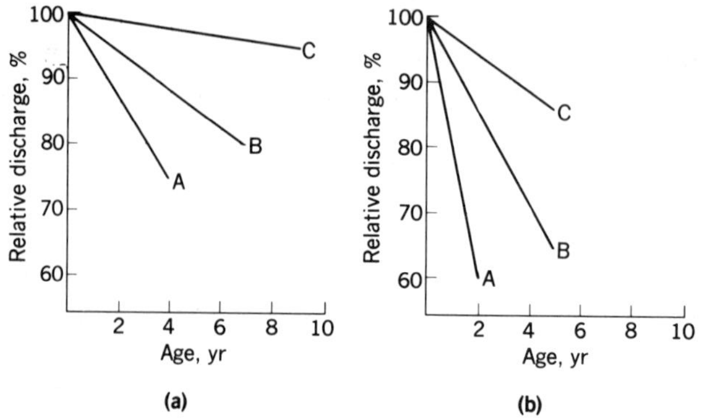

Fig. 9. Effect of time and storage temperatures, A, 40°C; B, 20°C; and C, 0°C, on relative discharge performance of fresh and aged "D"-size cells on simulated radio use, 25-Ω 4-h/d test for (**a**) alkaline–manganese, and (**b**) carbon–zinc batteries (22). Courtesy of Eveready Battery Co.

miniature devices. Figure 10 shows the construction of a typical miniature alkaline cell. The cathode mix is placed into the can, followed by the insertion of a separator layer. The type of separator materials used and the number of layers of material vary according to the type of cathode used in the cell. The anode mix is then placed in contact with the anode cup. An insulating, sealing gasket electrically separates the anode cup from the can and prevents leakage of electrolyte from the cell. The miniature cells have a similar basic construction, but differ in cathode materials, separator materials, electrolyte, etc.

Cylindrical alkaline cells are made in only a few standard sizes and have only one important chemistry. In contrast, miniature alkaline cells are made in a

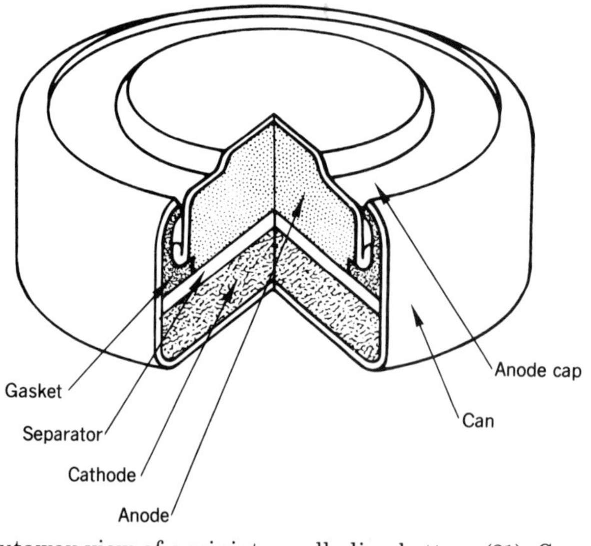

Fig. 10. Cutaway view of a miniature alkaline battery (21). Courtesy of Eveready Battery Co.

large number of different sizes, using many different chemical systems. Whereas the cylindrical alkaline batteries are multipurpose batteries, used for a wide variety of devices under a variety of discharge conditions, miniature alkaline batteries are highly specialized, with the cathode material, separator type, and electrolyte all chosen to match the particular application.

Zinc–Mercuric Oxide Batteries. Miniature zinc–mercuric oxide batteries have a zinc anode and a cathode containing mercuric oxide [*21908-53-2*], HgO. The cathode reaction

$$HgO + H_2O + 2\,e^- \longrightarrow Hg + 2\,OH^- \tag{18}$$

gives, when combined with the anode reaction (eq. 8), the overall reaction

$$Zn + HgO \longrightarrow ZnO + Hg \tag{19}$$

Water is not used in the reaction. Therefore, these cells have a very high capacity, exceeding that of zinc–manganese dioxide batteries (Table 2).

The cathode reaction is rather simple: mercuric oxide is reduced to mercury metal which is a liquid and does not block off the surface of the cathode. Therefore, the reaction is a normal heterogeneous two-phase reaction, producing a substantially constant voltage during discharge, in contrast to the sloping voltage of a cell containing a manganese dioxide cathode. Thus the zinc–mercuric oxide cell has high capacity and a flat discharge curve as shown in Figure 11. Thick separators are used to keep the liquid mercury from bridging to the anode and short circuiting the cell.

Some cells are made using small amounts of manganese dioxide added to the cathode. These cells have somewhat more sloping discharge curves (Fig. 11) and

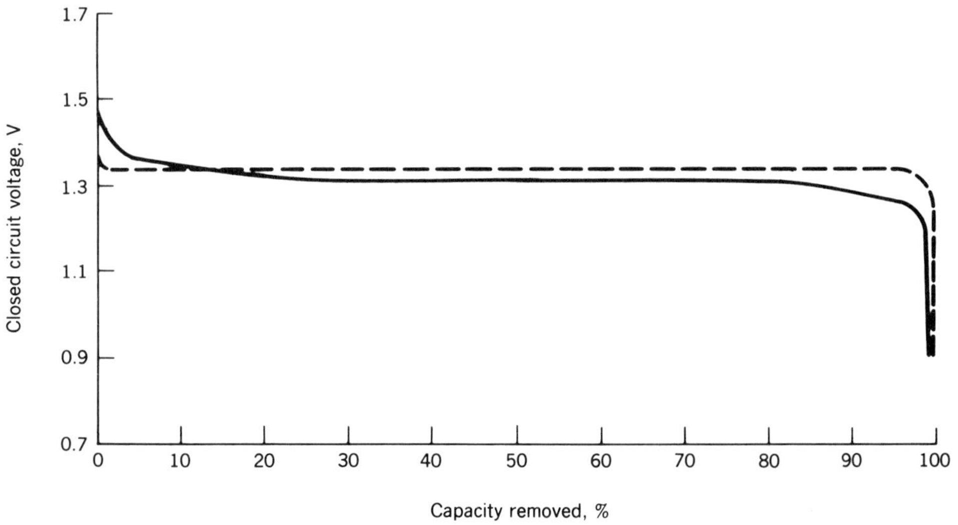

Fig. 11. Typical discharge curve comparison for zinc–mercuric oxide batteries: (— — —) model 325, HgO, and (———) model E312E, HgO–MnO_2, cathodes (21). Courtesy of Eveready Battery Co.

are used in some devices which do not require as constant a voltage. The added manganese dioxide provides for improved high rate capability for the cells. In addition, such cells have a slightly more gradual voltage drop at the end of battery life, giving the consumer some advance warning that the battery should be replaced.

Miniature zinc–mercuric oxide batteries may be made with either KOH or NaOH as the electrolyte. Cells having KOH operate more efficiently than those having NaOH at high current drains (Fig. 12) because of the higher conductivity of KOH. On the other hand, batteries with KOH are more difficult to seal, cells with NaOH are more resistant to leakage.

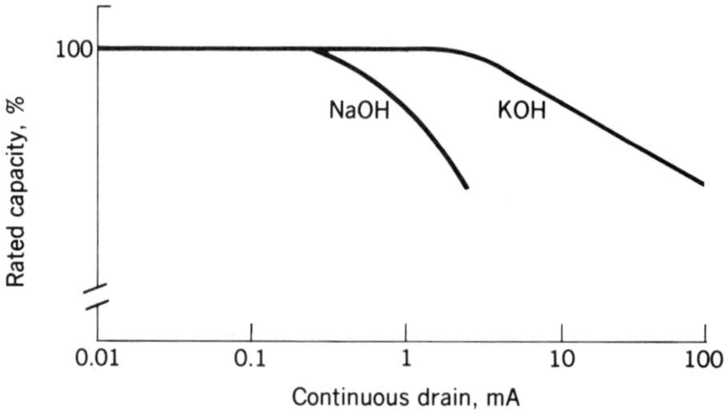

Fig. 12. Comparison of battery efficiency for miniature zinc–mercuric oxide cells containing KOH or NaOH electrolyte (21). Courtesy of Eveready Battery Co.

Miniature zinc–mercuric oxide batteries function efficiently over a wide range of temperatures. The maximum useful temperature is about 55°C, and the minimum is −28°C for batteries having KOH electrolyte, or −10°C for batteries having NaOH electrolyte. Figure 13 shows the effect of temperature on efficiency of discharge for a typical zinc–mercuric oxide cell. Miniature zinc–mercuric oxide batteries also have good storage life. As shown in Figure 14, they maintain 90% of capacity even after storage at 20°C for five years.

Although the zinc–mercuric oxide battery has many excellent qualities, increasing environmental concerns has led to a de-emphasis in the use of this system. The main environmental difficulty is in the disposal of the cell. Both the mercuric oxide in the fresh cell and the mercury reduction product in the used cell have long-term toxic effects.

Zinc–Silver Oxide Batteries. Miniature zinc–silver oxide batteries have a zinc anode, and a cathode containing silver oxide [*20667-12-3*], Ag_2O. The cathode reaction

$$Ag_2O + H_2O + 2\,e^- \longrightarrow 2\,Ag + 2\,OH^- \qquad (20)$$

gives, when combined with the anode reaction (eq. 8), the overall reaction

$$Zn + Ag_2O \longrightarrow ZnO + 2\,Ag \qquad (21)$$

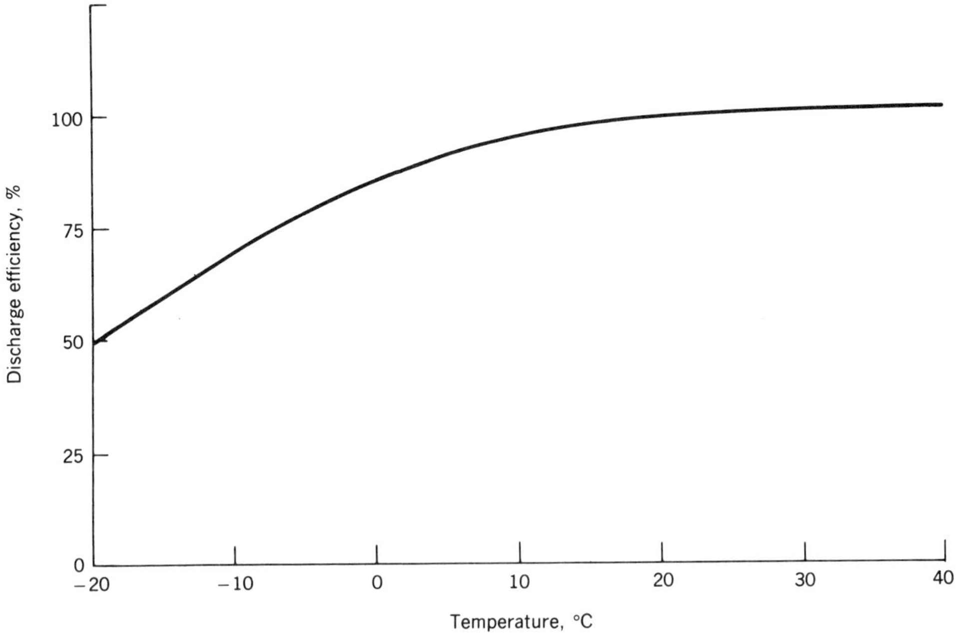

(a)

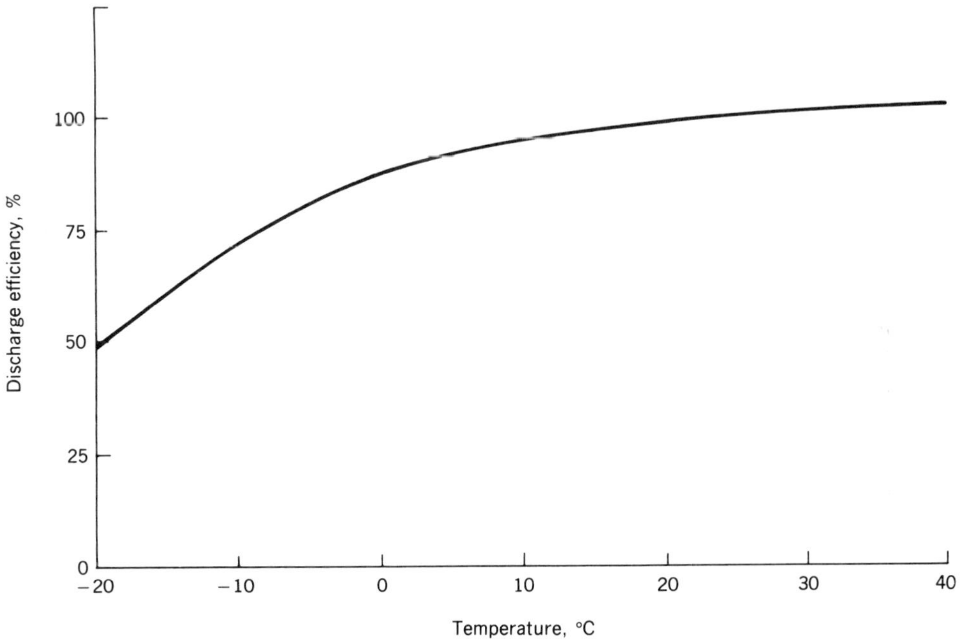

(b)

Fig. 13. Effect of temperature on discharge efficiency (**a**) at 270 mA·h of miniature zinc–mercuric oxide batteries type EP675E, and (**b**) at 175 mA·h of miniature zinc–silver oxide batteries type 357 (21). Courtesy of Eveready Battery Co.

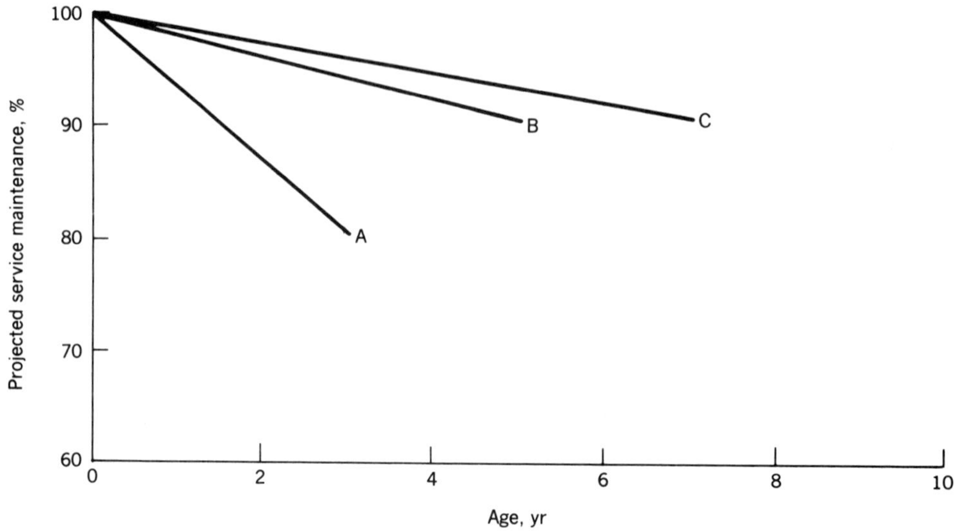

Fig. 14. Retention of discharge capacity of miniature zinc–mercuric oxide batteries after storage at temperatures of A, 40°C; B, 20°C; and C, 0°C (21). Courtesy of Eveready Battery Co.

The overall reaction neither consumes nor produces water. As in the case of mercury cells these cells have very high capacity.

The cathode reaction involves reduction of silver oxide to metallic silver [*7440-22-4*]. The reaction is a two-phase, heterogeneous reaction producing a substantially constant voltage during discharge. Some manganese dioxide may be added to the cathode, as in the case of mercury oxide cells.

Unlike some other cathode materials, such as manganese dioxide, which are quite insoluble, silver oxide has a fair degree of solubility in alkaline electrolyte. If the soluble silver species were allowed to be transported to the zinc anode it would react directly with the zinc, and as a result the cell would self-discharge. In order to prevent this from happening, zinc–silver oxide cells use special separator materials such as cellophane [*9005-81-6*], that are designed to inhibit migration of soluble silver to the anode.

Zinc–silver oxide cells have high energy density, nearly as high as for mercury cells. They operate at higher voltages than mercury cells, 1.5 vs 1.3 V, but for somewhat less time as shown in Figure 15. Miniature zinc–silver oxide batteries may be made using either KOH or NaOH electrolyte. Again, KOH-containing cells operate more efficiently than NaOH ones at high current drains (Fig. 16), whereas batteries having NaOH are easier to seal and are more resistant to leakage.

Miniature zinc–silver oxide batteries are commonly used in electronic watches and in other applications where high energy density, a flat discharge profile, and a higher operating voltage than that of a mercury cell are needed. These batteries function efficiently over a wide range of temperatures and are comparable to mercury batteries in this respect (Fig. 13**b**). Miniature zinc–silver oxide batteries have good storage life as shown in Figure 17. The use of barrier

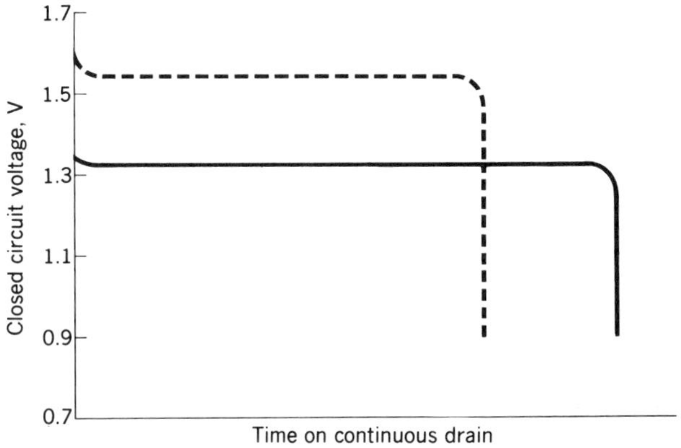

Fig. 15. Relative discharge curves for (---) zinc–silver oxide, and (—) zinc–mercuric oxide batteries. Cells are of equal volume (21). Courtesy of Eveready Battery Co.

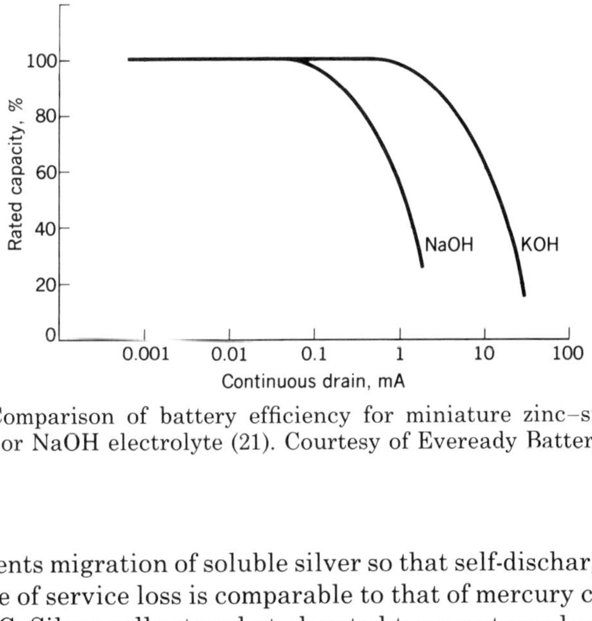

Fig. 16. Comparison of battery efficiency for miniature zinc–silver oxide cells containing KOH or NaOH electrolyte (21). Courtesy of Eveready Battery Co.

separators prevents migration of soluble silver so that self-discharge is kept quite low, and the rate of service loss is comparable to that of mercury cells for storage at 20°C or at 0°C. Silver cells stored at elevated temperatures lose capacity at a slightly higher rate than do mercury cells.

Divalent Silver Oxide Batteries. It is possible to produce a silver oxide in which the silver has a higher oxidation state, approaching a composition of AgO. This material can provide both higher capacity and higher energy density than Ag_2O. Alternatively, a battery can be made with the same capacity as a monovalent silver cell, but with cost savings. However, some difficulties with regard to material stability and voltage regulation must be addressed.

The cathode reaction for divalent silver oxide [*1301-96-8*], AgO

$$AgO + H_2O + 2\,e^- \longrightarrow Ag + 2\,OH^- \tag{22}$$

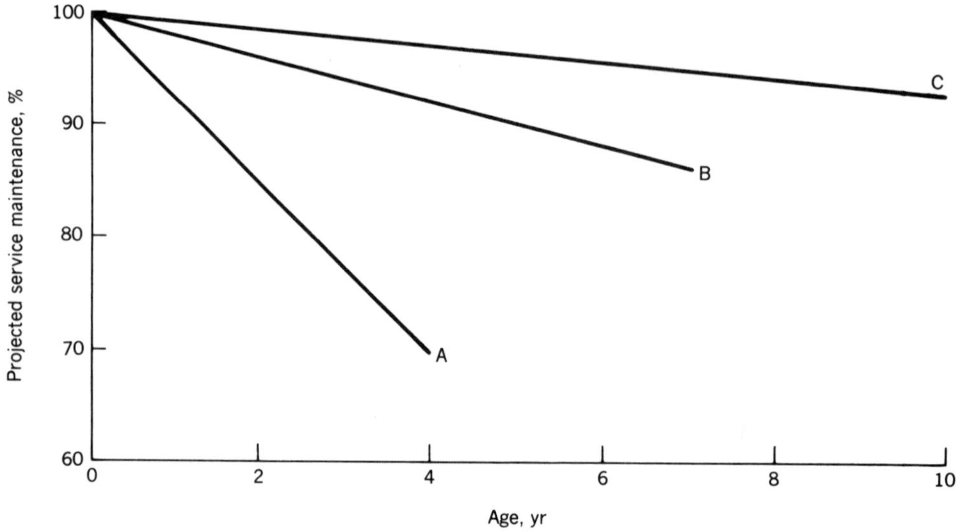

Fig. 17. Retention of discharge capacity of miniature zinc–silver oxide batteries after storage at temperatures of A, 40°C; B, 20°C; and C, 0°C (21). Courtesy of Eveready Battery Co.

gives the overall reaction when combined with equation 8

$$Zn + AgO \longrightarrow ZnO + Ag \tag{23}$$

The reaction of divalent silver oxide would normally tend to occur in two steps. In the first step divalent silver oxide would be reduced to monovalent silver oxide. Then, in the second step, the monovalent silver oxide would be further reduced to silver metal. These two reactions occur at different voltages so the battery would normally operate at a higher voltage during the first part of the discharge than during the latter part. Such discharge behavior, however, is unacceptable for many devices using the battery and it is desirable to make the cell discharge at a constant voltage. The material at the cathode reaction surface is caused to be all Ag_2O, and the interior AgO is protected from contact with the electrolyte. The voltage level of the monovalent silver oxide is thus obtained, but with the enhanced coulombic capacity of the divalent material. Several schemes have been developed to accomplish this behavior. Construction features of divalent silver oxide batteries are similar to those of monovalent silver oxide batteries.

Zinc–Manganese Dioxide Batteries. The combination of a zinc anode and manganese dioxide cathode, which is the dominant chemistry in large cylindrical alkaline cells, is used in some miniature alkaline cells as well. Overall, this type of cell does not account for a large share of the miniature cell market. It is used in cases where an economical power source is wanted and where the devices can tolerate the sloping discharge curve shown in Figure 18.

The chemistry is the same as for alkaline manganese–dioxide batteries. The construction features are typical of the other miniature alkaline batteries.

Zinc–Air Batteries. Zinc–air batteries offer the possibility of obtaining extremely high energy densities. Instead of having a cathode material placed in

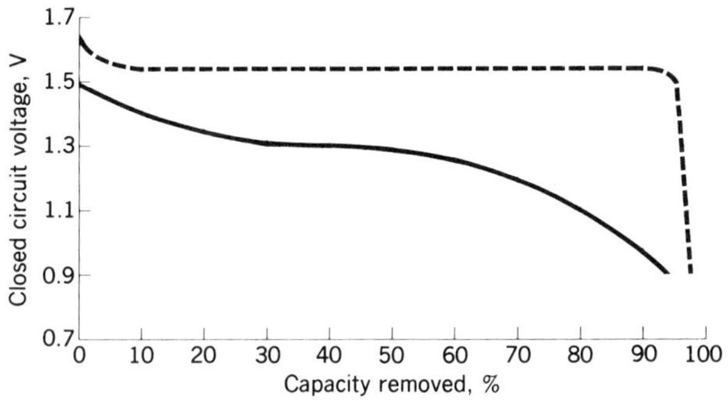

Fig. 18. Discharge curves for miniature zinc–silver oxide batteries (---), and zinc–manganese dioxide batteries (—) (21). Courtesy of Eveready Battery Co.

the battery when manufactured, oxygen from the atmosphere is used as cathode material, allowing for a much more efficient design. The construction of a miniature air cell is shown in Figure 19. From the outside, the cell looks like any other miniature cell, except for the air access holes in the can. On the inside, however, the anode occupies much more of the internal volume of the cell. Rather than the thick cathode pellet, there is a thin layer containing the cathode catalyst and air distribution passages. Air enters the cell through the holes in the can and the oxygen reacts at the surface of the cathode catalyst. The air access holes are often covered with a protective tape, which is removed when the cell is placed in service.

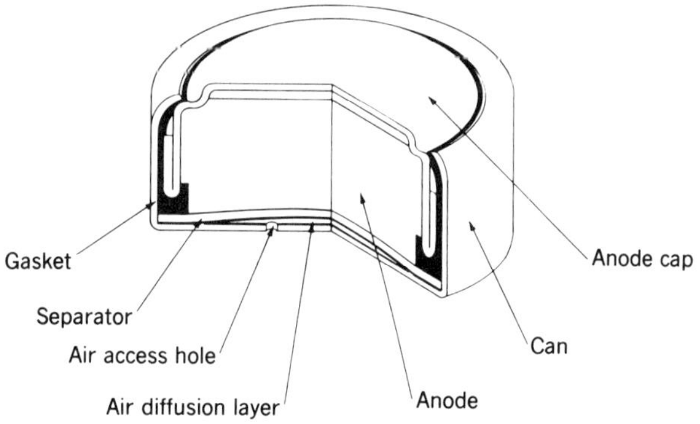

Fig. 19. Cutaway view of a miniature air cell battery (21). Courtesy of Eveready Battery Co.

The cathode reaction is

$$O_2 + 2\ H_2O + 4\ e^- \longrightarrow 4\ OH^- \tag{24}$$

and the overall cell reaction, after combining with equation 8, is

$$2\ Zn + O_2 \longrightarrow 2\ ZnO \tag{25}$$

The oxygen reaction is quite complex. Complete reduction from oxygen gas to hydroxide ion involves four electrons and requires several steps. Initially, oxygen is reduced to peroxyl ion [*14691-59-9*]

$$O_2 + H_2O + 2\,e^- \longrightarrow HO_2^- + OH^- \tag{26}$$

The peroxyl may be further reduced:

$$HO_2^- + H_2O + 2\,e^- \longrightarrow 3\,OH^- \tag{27}$$

or it may be catalytically decomposed, producing hydroxide and oxygen.

$$2\,HO_2^- \longrightarrow 2\,OH^- + O_2 \tag{28}$$

In a typical miniature air cell, the reaction proceeds according to equations 26 and 28. The performance of the air cathode cell depends largely on the ability to catalyze the reaction of equation 28. If that reaction is too slow, then during discharge large amounts of peroxide builds up in the cathode, resulting in large polarization, and low operating voltage. If the reaction is well catalyzed, the battery can operate at a higher, more useful voltage. The miniature air cell cathode typically contains carbon to provide a surface for the initial reduction of oxygen (eq. 26). Some types of carbon are also particularly effective at catalyzing the peroxyl decomposition and are used in air cells. Alternatively, small amounts of metal oxides can be used as catalysts.

Figure 20 shows a typical discharge curve for a miniature air cell which discharges at 1.2–1.3 V, with a flat discharge profile. Its discharge voltage is slightly lower than that of a mercury cell, and considerably lower than that of a silver cell, but its capacity is far greater than for any of the other miniature cells, including lithium cells (23).

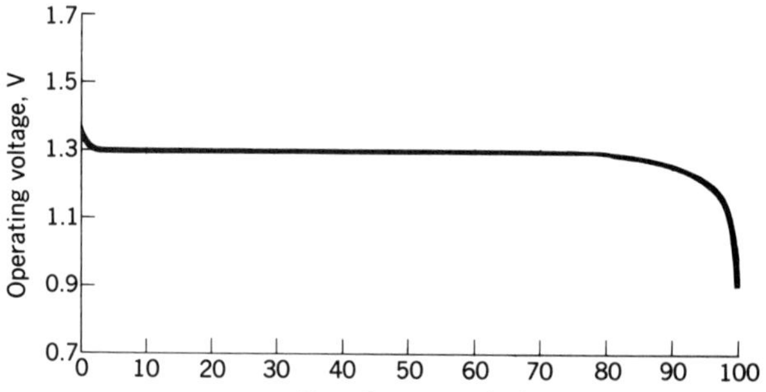

Fig. 20. Discharge curve for miniature zinc–air battery (21). Courtesy of Eveready Battery Co.

The performance level of air cells is exceptional, but these are not general purpose cells. They must be used in applications where the usage is largely continuous, and where the discharge level is relatively constant and well-defined. The reason for these limitations lies in the fact that the cell must be open to the atmosphere and the holes that allow oxygen into the cell also allow other gases to enter or leave the cell. Carbon dioxide from the air can enter an unsealed cell and react with the electrolyte to form potassium carbonate which results in decreased cell performance. Water vapor can enter the cell if the outside humidity is high or leave the cell if the outside humidity is low. In the former case, the cell fills up with water and eventually bulges or leaks. In the latter case, the cell dries out and can no longer function.

The relationship between the rate of water-vapor diffusion into or out of the cell and the rate of oxygen diffusion into the cell has been studied (24). If the cell openings are large then larger amounts of oxygen diffuse into the cell, and the cell maintains relatively high discharge rates. But the same large openings also allow more rapid diffusion of water vapor and carbon dioxide, so that the useful life of the cell is shortened. Alternatively, if the cell openings are small, the diffusion of water into or out of the cell, and the diffusion of carbon dioxide into the cell, is slower, and the cell operates for a longer period of time. However, the smaller holes limit oxygen diffusion so that the cell is limited to lower currents. Therefore, an air cell must be carefully designed for a particular use such that the air access passages are just big enough to handle the required oxygen flux. The cell must be used essentially continuously after activation, and disposed of promptly after discharge. Miniature air cells are mainly used in hearing aids, where they are required to produce a relatively high current for a relatively short time period such as a few weeks. In this application they provide exceptional performance compared to other batteries.

Air cells are packaged with sealing tape over the air holes or in sealed containers. In the sealed state, they maintain their capacity well during storage as shown in Figure 21. Once the sealing tape is removed, or the cell is taken from its sealed package, it must be put into use.

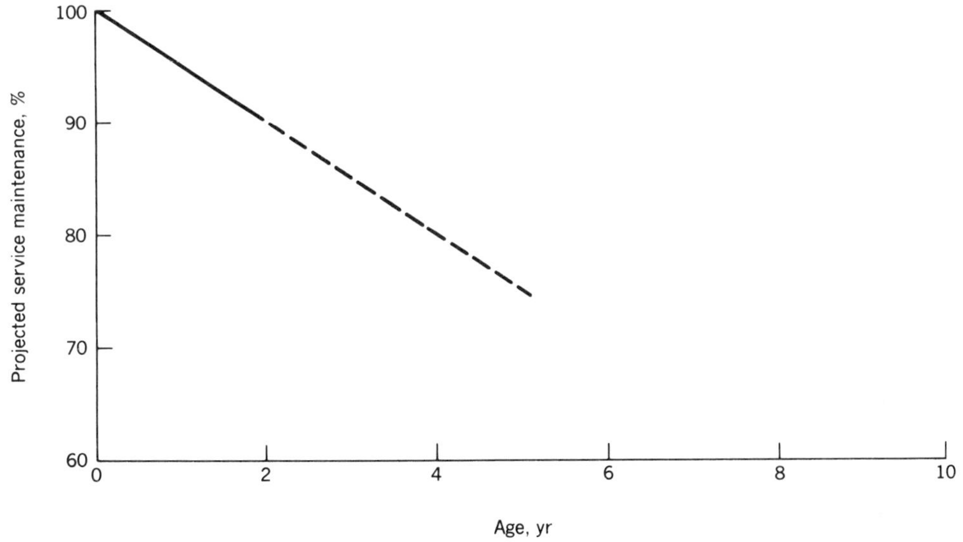

Fig. 21. Retention of discharge capacity of miniature zinc–air battery having an unopened sealed cell after storage at 20°C (---) projected data (21). Courtesy of Eveready Battery Co.

Miniature air cells can be used for continuous service over a temperature range of −10°C to +55°C. However, performance at high temperatures suffers as a result of drying out of the cells. Figure 22 shows typical air cell service as a function of temperature.

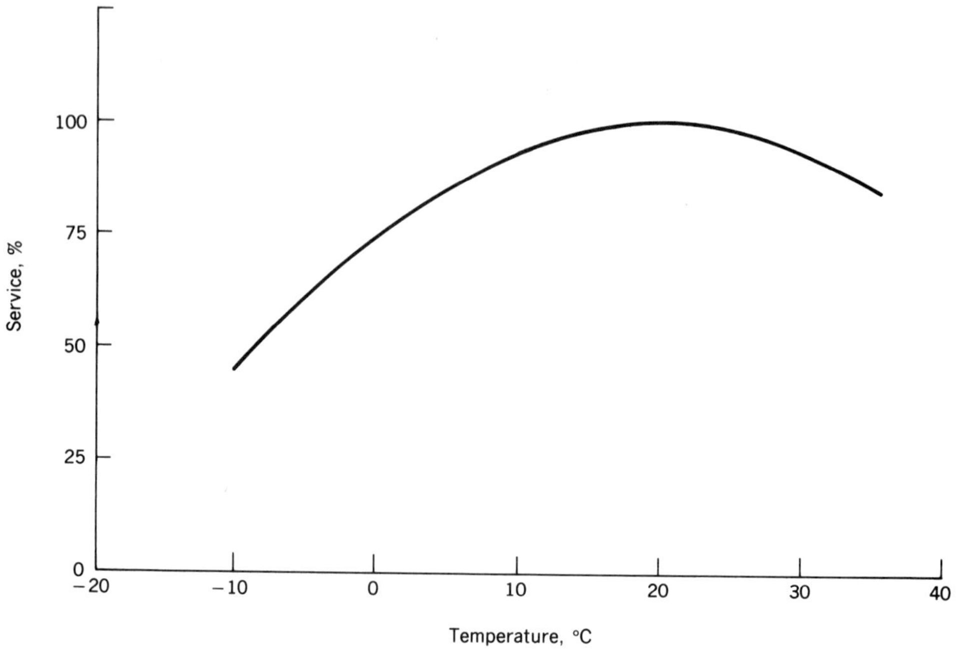

Fig. 22. Effect of temperature on discharge efficiency of miniature zinc–air batteries (21). Courtesy of Eveready Battery Co.

Lithium Cells

Cells having lithium [*7439-93-2*] anodes are generally called lithium cells regardless of the cathode. They can be conveniently separated into two types: cells having solid cathodes and cells having liquid cathodes. Cells having liquid cathodes also have liquid electrolytes and in fact, at least one component of the electrolyte solvent and the active cathode material are one and the same. Cells having solid cathodes may have liquid or solid electrolytes but, except for the lithium–iodine system, those having solid electrolytes are not yet commercial.

All of the cells take advantage of the inherently high energy of lithium metal and its unusual film-forming property. This latter property, discussed in detail in reference 25, allows for lithium compatibility with solvents and other materials with which it is thermodynamically unstable, yet permits high electrochemical activity when the external circuit is closed. This is possible because the film formed with certain organic and inorganic materials is conductive to lithium ions, but not to electrons. Thus corrosion reactions occur only to the extent of forming thin, electronically insulating films on lithium which are both coherent and adherent to the base metal. The thinness of the film is important in order to allow reasonable rates of transport of lithium ions through the film when the circuit is closed. Many materials, such as water and alcohols, which are thermodynamically unstable to lithium, do not form this kind of passivating film (see THIN FILMS).

Much analytical study has been required to establish the materials for use as solvents and solutes in lithium batteries. References 26 and 27 may be consulted for discussions of electrolytes. Among the best organic solvents are cyclic esters, such as propylene carbonate [*108-32-7*] (PC), $C_4H_6O_3$, ethylene carbonate [*96-49-1*] (EC), $C_3H_4O_3$, and butyrolactone [*96-48-0*], $C_4H_6O_2$, and ethers, such as dimethoxyethane [*110-71-4*] (DME), $C_4H_{10}O_2$, the glymes, tetrahydrofuran (THF), and 1,3-dioxolane [*646-06-0*], $C_3H_6O_2$. Among the most useful electrolyte salts are lithium perchlorate [*7791-03-9*], $LiClO_4$, lithium trifluoromethanesulfonate [*33454-82-9*], $LiCF_3SO_3$, lithium tetrafluoroborate [*14283-07-9*], $LiBF_4$, and lithium hexafluoroarsenate [*29935-35-1*], $LiAsF_6$. A limitation of these organic electrolytes is the relatively low conductivity, compared to aqueous electrolytes. This limitation combined with the generally slow kinetics of the cathode reactions has forced the use of certain designs such as thin electrodes and very thin separators, in all lithium batteries. This usage led to the development of coin cells rather than button cells for miniature batteries and jelly or Swiss roll designs rather than bobbin designs for cylindrical cells. Figure 23 shows cutaway views of typical coin cells (Fig. 23**a**) and cylindrical jelly roll (Fig. 23**b**) lithium batteries.

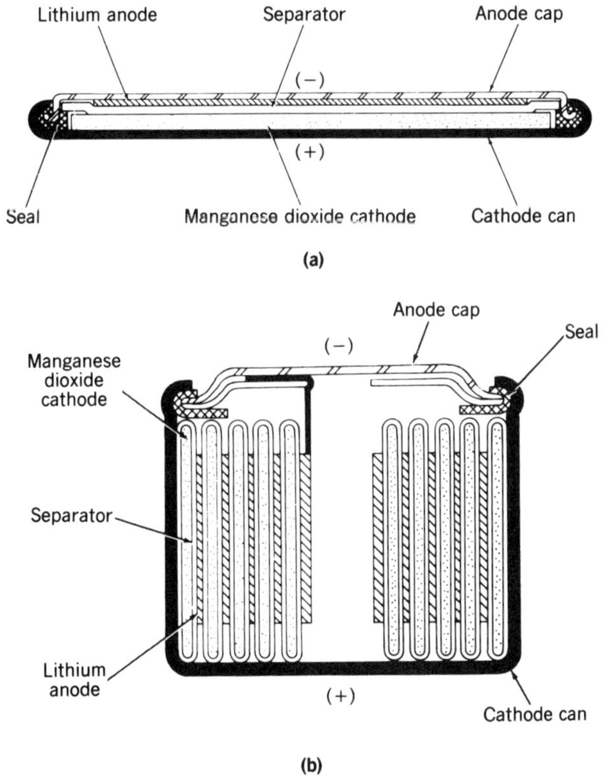

Fig. 23. Cutaway view of (**a**) 2016-size Eveready lithium manganese dioxide coin cell, and (**b**) jelly roll cylindrical lithium manganese dioxide cell (28). Courtesy of Eveready Battery Co.

Many of the cylindrical cells have glass-to-metal hermetic seals, although this is becoming less common because of the high cost associated with this type of seal. Alternatively, cylindrical cells have compression seals carefully designed to minimize the ingress of water and oxygen and the egress of volatile solvent. These construction designs are costly and the high price of the lithium cell has limited its use. However, the energy densities are superior.

Solid Cathode Cells. Solid cathode cells were investigated first and the very hydrophobic material, carbon monofluoride [*3889-75-6*], CF, was among the first cathode materials to be produced in commercial cells (29). The hydrophobicity minimized the need for drying the cathode material and the layerlike structure of the material facilitated the cathode reaction. Shortly thereafter, the solid electrolyte system, $Li–I_2$, which allowed *in situ* formation of a lithium iodide [*10377-51-2*], LiI, interface between the two materials, was developed (30) and eventually became the principal system to be used in heart pacers and other low drain medical applications. This system also was intrinsically hydrophobic. The $Li–MnO_2$ system, unsuccessfully investigated in the 1960s and early 1970s, was shown in 1975 to be viable (31) if certain types of MnO_2 were heat-treated to at least 300°C. The large amounts of water within and on the surface of the MnO_2 had caused cathode inefficiencies as well as problems with the lithium, because of migration through the electrolyte. The other solid cathode material in use in lithium cells is iron pyrite [*1309-36-0*], FeS_2. Whereas this material had been widely studied in high temperature systems, its use in room temperature systems relied heavily on the introduction of new electrolytes (32).

Lithium–Manganese Dioxide Cells. The $Li–MnO_2$ cell is becoming the most widely used 3-V lithium battery. The critical step in obtaining good performance is the heat treatment of the MnO_2 prior to incorporation in the cathode. There is still some controversy about the exact nature of the reduction of the cathode, although the reduction appears to occur in two steps. The first step, in close analogy with the aqueous reduction, is a homogeneous process that concludes with the formation of a partially lithiated material.

$$x\,Li^+ + x\,e^- + MnO_2 \longrightarrow Li_xMnO_2 \tag{29}$$

where $x \approx 0.4$. This step is followed by a heterogeneous process to a new phase, the structure of which has not been determined.

$$(1 - x)\,Li^+ + (1 - x)\,e^- + Li_xMnO_2 \longrightarrow LiMnO_2 \tag{30}$$

The anodic reaction is conventionally written as

$$Li \longrightarrow Li^+ + e^- \tag{31}$$

where it is understood that the Li^+ ion is solvated predominantly by the strongest Lewis base among the solvents and that, prior to the dissolution/solvation step, the lithium ion must diffuse through the passivating film. The most common electrolytes used with the system are $LiClO_4$ or $LiCF_3SO_3$ in a mixture of PC and DME (usually 1:1).

The kinetics are not very sensitive to the electrolyte so the choice is largely dependent on safety, toxicity, and cost. The relatively slow kinetics of the system has necessitated the use of thin electrodes in order to obtain sufficient current carrying capability and these cells are designed as coin cells (Fig. 23**a**) or as jelly rolls (Fig. 23**b**) with alternating anode, separator, cathode, and another separator layer. These 3-V batteries are made in sizes not used for aqueous 1.5-V cells to help prevent their insertion in circuits designed for 1.5 V.

The energy density of the system depends on the type of cell as well as the current drain. Table 3 gives the specification for the various lithium systems. These coin cells have already been widely used in electronic devices such as calculators and watches, whereas the cylindrical cells have found applications in cameras.

Table 3. Characteristics of Lithium Primary Batteries

Cathode (cell system)	Voltage, V		Specific energy, W·h/kg	Energy density, W·h/mL	Operating temperature range, °C
	Nominal	Working			
Solid cathodes					
carbon monofluoride (Li–CF)	3	2.5–2.7	200	0.40	−20–60
manganese dioxide (Li–MnO_2)	3	2.7–2.9	200	0.40	−20–55
iron disulfide (Li–FeS_2)	1.5	1.3–1.7	125	0.40	−20–50
iodine (Li–I_2)	3	2.4–2.8	250	0.92	37
Liquid cathodes					
sulfur dioxide (Li–SO_2)	3	2.7–2.9	250	0.44	−55–70
thionyl chloride (Li–$SOCl_2$)	3.6	3.3–3.5	300	0.95	−50–70

Lithium–Carbon Monofluoride Cells. Although the Li–CF cell was the first lithium cell to be manufactured in quantity for consumer applications, it is becoming less popular because of the relatively high cost of the cathode material and the ability of the less expensive Li–MnO_2 cells to perform most of the same functions. The cathodes have traditionally used carbon as a conductor because carbon monofluoride is itself a poor conductor. Some interesting hybrid cathode cells using mixtures of MnO_2 and CF have also been studied (33). Carbon monofluoride has a laminar structure, which is similar to graphite, and when graphite reacts with fluorine to make CF, the flat planes of graphite become puckered as layers of fluorine atoms become interspersed among the carbon layers. Another pure compound, C_2F, is also known and this has been shown to have battery activity as well (34).

The CF cathode reaction is believed to be a heterogeneous process, initiated by the insertion of lithium ions between the CF planes. It is completed by the extrusion of LiF and the collapse of the structure to carbon.

$$Li^+ + CF + e^- \longrightarrow LiF + C \qquad (32)$$

Various lithium salts and butyrolactone or PC–DME mixtures are usually used as electrolytes. The close competitive performance of CF_x and MnO_2 cathodes is evidenced in Table 3. The construction of cells is also similar for the two systems. In addition to uses mentioned for the lithium manganese dioxide system, some unique applications such as lighted fishing bobbers have been developed for the Japanese market.

Lithium–Iron Disulfide Cells. These cells were first manufactured in button cell sizes and because of the voltage similarity, they are used as direct replacements for Zn–Ag_2O cells (35). The lower conductivity of the organic electrolyte in Li–FeS_2 systems as compared to the KOH electrolyte in the Zn–Ag_2O cell, precludes use in the highest drain applications, but such uses are relatively few. Recently, "AA"-size Li–FeS_2 cylindrical batteries with a high surface area jelly roll construction and a plastic gasket compression seal have been introduced. The high area construction permits the use of these cells in almost all applications for which alkaline Zn–MnO_2 cells can be used and gives higher energy than that system. The compression seal helps to keep the cost at reasonable levels.

The chemistry of the Li–FeS_2 system is quite complex. There are at least two steps to the reaction at low discharge rates. The first reaction is an approximately two-electron reduction to a new phase which is a lithiated FeS_2 compound.

$$2\,Li^+ + FeS_2 + 2\,e^- \longrightarrow Li_2FeS_2 \qquad (33)$$

The second step is also heterogeneous and involves the breakdown of the intermediate compound with further lithiation into lithium sulfide [*12136-58-2*] and finely divided iron [*7439-89-6*] particles.

$$2\,Li^+ + Li_2FeS_2 + 2\,e^- \longrightarrow 2\,Li_2S + Fe \qquad (34)$$

The crystal structure of the intermediate is not well understood. The final iron phase is termed superparamagnetic because the particle size is too small to support ferromagnetic domains. At low rates, the discharge occurs in two steps separated by a small voltage difference. At high rates, however, the two steps become one, indicating that the first step is rate limiting, ie, the second step (eq. 34) occurs immediately after formation of the intermediate (eq. 33).

The performance of the cylindrical cell is superior to alkaline cells at all rates of discharge, but, because of the high area electrode construction, the performance advantage is magnified during high rate discharge. Table 3 shows the energy density and a comparison to Tables 1 and 2 shows the superiority of the cells to aqueous alkaline and Leclanché cells and the parity for Zn–Ag_2O button cells.

Lithium–Iodine Cells. Iodine [*7553-56-2*] forms charge-transfer complexes with unsaturated compounds, many of which have reasonably high electronic conductivity. The iodine cathode takes advantage of this property by combining iodine and poly(2-vinylpyridine) (PVP) or other copolymers having unsaturated repeating units. In one process for making a Li–I_2 cell, the iodine and polymer, using excess iodine, are heated together to form the molten charge-transfer complex. Then the liquid is poured into the waiting cell which already contains a lithium anode and a cathode collector screen. After cooling and solidification of the cathode, the cell is hermetically sealed. In another process, the iodine–PVP mixture is reacted and then pelletized. The pellets are then pressed into the lithium to form a sandwich structure which is inserted into the waiting cell and hermetically sealed. The direct contact of the excess iodine and the lithium results in an immediate reaction to form an *in situ* LiI film which then protects the lithium from further chemical degradation.

The cathodic reaction is the reduction of iodine to form lithium iodide at the carbon collector sites as lithium ions diffuse to the reaction site. The anode reaction is lithium ion formation and diffusion through the thin lithium iodide electrolyte layer. If the anode is corrugated and coated with PVP prior to adding the cathode fluid, the impedance of the cell is lower and remains at a low level until late in the discharge. The cell eventually fails because of high resistance, even though the drain rate is low.

The Li–I_2 battery system has allowed simple heart pacers to operate for as long as 10 years compared to the 1.5–2 years of operation for alkaline batteries. This is mostly because of the very high stability of the lithium batteries allowing for trouble-free operation. Many other lithium batteries have been tried in this application, but the lithium iodine system is now almost universally used and the surgical gain is high. Table 3 gives the energy density of the system and Figure 24 shows a cutaway view of the letter "D" shaped cell which is used to fit the electronics pacer package. These cells are extremely expensive because of the great care taken in manufacturing and the careful testing, labeling, and monitoring of each cell. The system is very sensitive to moisture because corrosion of both the lithium anode and the stainless steel container occur in the presence of moist iodine vapor. However, a completely dry cell does not form any gas and has no fluids to leak and it is this stability which has made the Li–I_2 cell so useful for medical batteries.

Less expensive coin-type cells have also been developed for consumer elec-

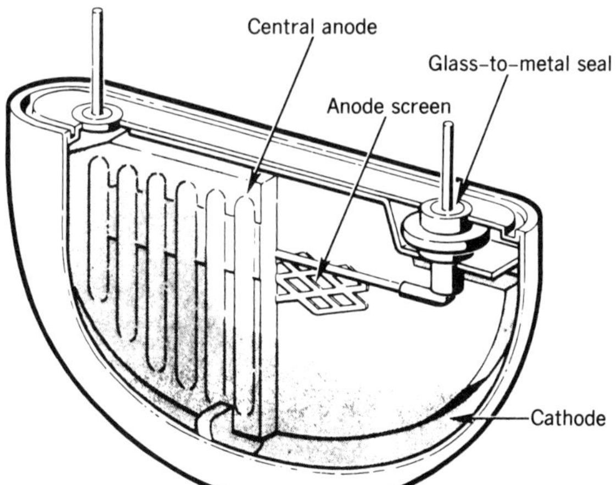

Fig. 24. Cutaway view of lithium–iodine pacemaker cell in case-grounded, central anode configuration (36). Courtesy of Plenum Press.

tronics applications, but the severe current limitations have restricted the use of the cells.

Liquid Cathode Cells. Liquid cathode cells were discovered at almost the same time as the successful solid cathode cells. A strongly oxidizing liquid such as SO_2, was determined to be suitable for direct contact with the strongly reducing lithium, because an excellent passivating film forms spontaneously on the lithium (37). Subsequently, a number of other liquid cathode cells were discovered, the most successful of which is the Li–$SOCl_2$ cell (38). Although liquid cathode cells have been used primarily in military applications, most recently a number of memory backup applications have been served by low rate lithium–thionyl chloride cells. These cells have high energy, a long shelf life, and a constant voltage profile.

Lithium–Sulfur Dioxide Cells. Lithium–sulfur dioxide cells (39) generally use either acetonitrile [*75-05-8*] AN, C_2H_3N, or PC or a mixture of the two as cosolvents with the SO_2, usually in amounts as high as 50 vol %. This has the advantage of lowering the vapor pressure of the sulfur dioxide [*7446-09-5*] which at 25°C is about 300 kPa (3 atm). The liquids are usually chilled, however, to make management of the toxic SO_2 gas easier during cell filling and sealing, which are the last steps in the construction of cells. AN can be used as a cosolvent only because of the excellent film-forming properties of SO_2. Lithium is known to react extensively with AN in the absence of sulfur dioxide. The other properties of AN, relatively high dielectric constant and very high fluidity, enhance the electrolyte conductivity and permit construction of high rate cells. The cell reaction proceeds according to

$$2\ Li + 2\ SO_2 \longrightarrow Li_2S_2O_4 \tag{35}$$

where the cathode reaction, reduction of SO_2, takes place on a highly porous, >80%, carbon electrode. The lithium dithionite [*59744-77-3*] product is insoluble in the electrolyte and precipitates in place in the cathode as it is formed. The reaction involves only one electron per SO_2 molecule which, along with the high volume of cosolvent in the cell, limits the capacity and energy density. Lithium bromide [*7550-35-8*], LiBr, is the most commonly used salt, although sometimes $LiAsF_6$ is used, especially in reserve cell configurations. Because of the high rate capability, the cell has been widely used for military applications.

Safety has been a primary focus for the Li–SO_2 system, partly because of the high toxicities of the materials. Also, the potentially high rate of reaction during an accidental short circuit or other incident has led to some runaway reactions of high energy. However, tight control has allowed usage and the Li–SO_2 battery has become an important battery for military communications. In addition, the careful engineering of electrodes and vent mechanisms has controlled incidents. The cells have been subjected to extensive abuse testing.

All sulfur dioxide batteries have hermetic glass-to-metal seals to prevent loss of the highly volatile sulfur dioxide. They also have pressure operated vents to accomplish emergency evacuation of the cell under abuse to prevent explosions. Safety has also been improved by using a balanced cell design, ie, an equal capacity of lithium and sulfur dioxide, so that low rate discharge uses up all of the lithium. The performance of these cells is generally optimized for high current density operation and the battery is usually limited by blockage of the porous cathode structure with the insoluble lithium dithionite reaction product. Figure 25 gives a comparison of the energy output for lithium sulfur dioxide cells with that for lithium thionyl chloride liquid cathode cells. Both types of batteries have high energy density and specific energy (Table 3), because of the efficient use of space characteristic of a liquid cathode cell. Many cell types and sizes have been manufactured and many of these cells have been combined in series-parallel networks in batteries. They have been mostly made under contract for government use or, in some cases, for original equipment use for electronic device manufacturers. It is difficult for a consumer to purchase these cells separately.

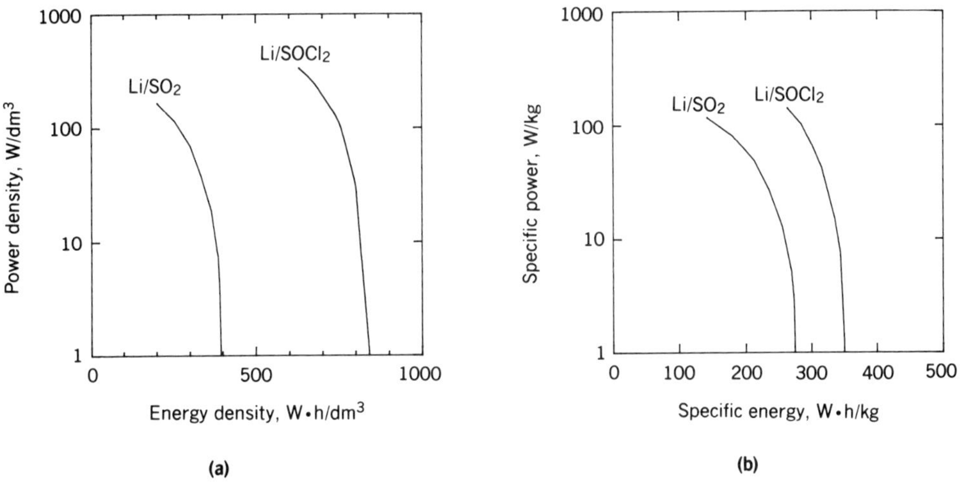

Fig. 25. Lithium–sulfur dioxide and lithium–thionyl chloride high rate batteries profile with (**a**) power density vs energy density, and (**b**) specific power vs specific energy.

Lithium–Thionyl Chloride Cells. Lithium–thionyl chloride cells have very high energy density. One of the main reasons is the nature of the cell reaction.

$$4\ Li + 2\ SOCl_2 \longrightarrow 4\ LiCl + S + SO_2 \tag{36}$$

The reaction involves two electrons per thionyl chloride [*7719-09-7*] molecule (40). Also, one of the products, SO_2, is a liquid under the internal pressure of the cell, facilitating a more complete use of the reactant. Finally, no cosolvent is required for the solution, because thionyl chloride is a liquid having only a modest vapor pressure at room temperature. The electrolyte salt most commonly used is lithium aluminum chloride [*14024-11-4*], $LiAlCl_4$. Initially, the sulfur product is also soluble in the electrolyte, but as the composition changes to a higher SO_2 concentration and sulfur [*7704-34-9*] builds up, a saturation point is reached and the sulfur precipitates.

The end of discharge occurs as the anode is used up because the cells are usually anode limited. However, if very high discharge rates are used, the collector, which is usually made of a porous carbon such as that in the SO_2 cell, becomes blocked with the insoluble lithium chloride [*7447-41-8*], LiCl. In either case, the discharge curve is flat over time, then declines abruptly at the end of life. Because of the high activity of the cathode and the good mass transfer of the liquid catholyte, the system is capable of very high rate discharges and operates at 3.6 V in open circuit. The anode is also very active except for a tendency to form thick films of LiCl during open circuit stand, much like the lithium iodine system. The film can cause delay effects during start-up of the cell.

Delay effects can be largely eliminated by treating the anode with various types of polymeric films, which decrease the rate of corrosion and LiCl formation markedly. One of the best films for this purpose is poly(vinyl chloride) [*9002-86-2*]. The wide liquid range of the catholyte solutions gives the system a good operating range, from very low (−50°C) to unusually high (>100°C) temperatures. The cells should not be stored at high temperatures, however, because corrosion reactions can occur which reduce the cell life or cause an enhanced delay effect.

Several applications are developing for the Li–$SOCl_2$ battery system. Because of the excellent voltage control, high energy density, and high voltage, the battery is finding increasing use on electronic circuit boards to supply a fixed voltage for memory protection and other standby functions. These cells are designed in low rate configuration which maximizes the energy density and cell stability. Military and space applications are also developing because of the wide range of temperature/performance capability as well as the high specific energy.

Reserve Batteries

Reserve batteries have been developed for applications that require a long inactive shelf period followed by intense discharge during which high energy and power, and sometimes operation at low ambient temperature, are required. These batteries are usually classified by the mechanism of activation which is employed. There are water-activated batteries that utilize fresh or seawater; electrolyte-activated batteries, some using the complete electrolyte, some only the solvent; gas-activated batteries where the gas is used as either an active cathode material or part of the electrolyte; and heat-activated or thermal batteries which use a solid salt electrolyte activated by melting on application of heat.

Activation of these batteries involves adding the missing component which can be done in a simple way, such as pouring water into an opening in the cell, for water-activated cells, or in a more complicated way by using pistons, valves, or heat pellets activated by gravitational or electric signals for the case of the electrolyte- or thermal-activation types. Such batteries may be stored for 10–20 years while awaiting use. Reserve batteries are usually manufactured under contract for various government agencies such as the Department of Defense, although occasional industrial or safety uses have been found. Many of the electrochemical systems involved in these batteries are beyond the scope of this article. Reference 41 contains further details on these systems.

The lithium–thionyl chloride, or the lithium–sulfur dioxide, system is often used in a reserve battery configuration in which the electrolyte is stored in a sealed compartment which upon activation may be forced by a piston or inertial forces into the interelectrode space. These high energy density systems have gained some of the applications of the older liquid ammonia [*7664-41-7*] reserve batteries, which usually had a magnesium [*7439-95-4*] anode and an *m*-dinitrobenzene [*99-65-0*], $C_6H_4N_2O_4$, cathode. Much of the engineering relating to flowing liquid ammonia into the battery was applied to the manufacture of lithium cells. An even closer analogy to the ammonia battery is the lithium–vanadium pentoxide battery which performs as a very high rate, high energy cell. Most applications for such batteries are in mine and fuse applications in military ordnance.

One variant of the liquid cathode reserve battery is the lithium–water cell in which water serves as both the liquid cathode and the electrolyte. A certain amount of corrosion occurs, but sufficient lithium is provided to compensate. The reaction product is soluble lithium hydroxide [*1310-65-2*], Li(OH). Sometimes a solid cathode material, such as silver oxide, or another liquid reactant, such as hydrogen peroxide [*7722-84-1*], H_2O_2, is used in combination with the lithium anode and aqueous electrolyte to improve the rate or decrease the amount of gas given off by the system. These cells are mostly used in the marine environment where water is available or compatible with the cell reaction product. Common applications are for torpedo propulsion and to power sonobuoys and submersibles. An older system still used for these applications is the magnesium–silver chloride–seawater-activated battery. This cell is much heavier than the corresponding lithium cell, but the buoyancy of the sea makes this less of a detriment than might first be thought. The magnesium–silver chloride cell is also useful for powering emergency communication devices for airplane crews whose planes have come down in the sea.

The last type of reserve cell is the thermally activated cell. The older types use calcium [*7440-70-2*] or magnesium anodes; newer types use lithium alloys as anodes. Lithium forms many high melting alloys such as those with aluminum, silicon, and boron. Furthermore, lithium can diffuse rapidly within the alloy phase permitting high currents to flow. The electrolyte for both the older and newer chemistries is usually the eutectic composition of lithium chloride and potassium chloride which melts at 352°C. This electrolyte has temperature dependent conductivities which are an order of magnitude higher than the best aqueous electrolytes. The high conductivity and the enhanced kinetics and mass transport allow the battery to be discharged at a very high rate of several A/cm^2 with complete discharge in 0.5 s. The cathodes for the older calcium anode cells are typically metal chromates such as calcium chromate [*13765-19-0*], $CaCrO_4$. The anode reaction product is calcium chloride [*10043-52-4*], whereas the cathode product is a mixed calcium chromium oxide of uncertain composition. One of the best cathodes for lithium alloy cells is FeS_2 which forms a system similar in reaction mechanism to that used in miniature Li–FeS_2 cells.

The heat pellet used for activation in these batteries is usually a mixture of a reactive metal such as iron or zirconium [*7440-67-7*], and an oxidant such as potassium perchlorate [*7778-74-7*]. An electrical or mechanical signal ignites a primer which then ignites the heat pellet which melts the electrolyte. Sufficient heat is given off by the high current to sustain the necessary temperature during the lifetime of the application. Many millions of these batteries have been manufactured for military ordnance as they have been employed in rockets, bombs, missiles, etc.

Economic Aspects

In 1989 the breakdown in sales of primary cells in the United States by category was alkaline, 65%; carbon–zinc, 30%, 21% of which was heavy-duty, 9% general purpose; specialty, 4%; and lithium, 1% (7). The total primary cell market in that year was $2813 million; with a unit volume of 2514 million cells. Over five years the average annual growth rate was 9.2% in terms of sales and 4.5% in terms of units. The growth of sales has been tied mostly to growth in the electronics industry, although the sales in battery-operated toys has also shown substantial growth.

The market outside of the United States reflects the historical dominance of carbon–zinc cells. For example, in Japan nearly half of all sales are carbon–zinc cells, about 33% are alkaline, and about 17% are lithium. The high proportion of lithium cells relative to U.S. sales reflects the important photographic market in Japan. Western European sales are similar to those in Japan, and Third World sales are almost totally dominated by carbon–zinc batteries.

There are hundreds of primary battery manufacturers, most of which are limited to a specialty product or a limited national market. However, a number of multinational battery suppliers have manufacturing facilities in many countries and a broad line of products. Table 4 lists the companies having sales of primary batteries of greater than $100 million per annum together with the regions of their headquarters and their product lines.

Table 4. Battery Companies Manufacturing Primary Cells

	Cell type		
Company	Carbon–zinc	Alkaline	Lithium
North America			
Duracell International		x	x
Eastman Kodak		x	
Eveready Battery Co.	x	x	x
Ray-O-Vac International	x	x	x
European Economic Community			
Ever Ready (BEREC)	x	x	
SAFT			x
Varta Batterie AG	x	x	x
Far East			
Matsushita	x	x	x
Sanyo	x	x	x
Toshiba	x	x	x
Yuasa	x	x	x

BIBLIOGRAPHY

"Batteries, Electric (Batteries, Primary)," in *ECT* 1st ed., Vol. 2, pp. 324–340, by J. J. Coleman and O. W. Storey, Burgess Battery Company; "Cells, Electric," Vol. 3, pp.

292–342, by W. J. Hamer, National Bureau of Standards; and "Cells, Electric," Suppl. 2, pp. 126–161, by C. K. Morehouse, R. Glicksman, and G. S. Lozier, Radio Corporation of America; "Batteries and Electric Cells, Primary (Primary Cells)," in *ECT* 2nd ed., Vol. 3, pp. 111–139, by J. F. Yeager, Union Carbide Corporation, E. B. Yeager, Western Reserve University, A. F. Daniel (Military Types), U.S. Army Signal Research and Development Laboratory; "Batteries, Primary (Primary Cells)," in *ECT* 3rd ed., Vol. 3, pp. 515–545, by H. Gu and D. N. Bennion, University of California, Los Angeles.

1. *Eveready Battery Engineering Data,* Vol. 3, Eveready Battery Co., St. Louis, Mo., 1984.
2. Fr. Pat. 71,865 (1866), G. Leclanché.
3. U.S. Pat. 373,064 (1887), C. Gassner.
4. F. L. Tye, in T. Trans and M. Skyllas-Kazeos, eds., *Proceedings of the 7th Australian Electrochemistry Conference,* The Electrochemistry Division, The Royal Australian Chemical Institute, 1988, pp. 37–48.
5. R. G. Burns and V. M. Burns, in A. Kozawa and R. J. Brodd, eds., *Proceedings of the Manganese Dioxide Symposium,* Vol. 1, Cleveland, I. C. Sample Office, Cleveland, Ohio, 1975, p. 306.
6. *American National Standard Specification for Dry Cells and Batteries,* ANSI C18.1-1979, American National Standards Institute, Inc., New York, May 1979.
7. *The Dry Cell Battery Market, Packaged Facts,* The International/Research Co., New York, 1990.
8. *Eveready Battery Engineering Data,* Vol. 2A, Eveready Battery Co., St. Louis, Mo., 1990.
9. Fr. Pat. 143,644 (1881), F. de Lalande and G. Chaperon; U.S. Pat. 274,110 (1883).
10. Ger. Pat. 157,290 (1901), T. A. Edison; U.S. Pat. 678,722 (1901).
11. Sw. Pat. 15,567 (1901), W. Jungner; Ger. Pat. 163,170 (1901).
12. Ger. Pat. 24,552 (1882), G. Leuchs.
13. U.S. Pat. 2,422,045 (1947) and 2,576,266 (1951), S. Ruben.
14. W. S. Herbert, *J. Electrochem. Soc.* **99,** 190C (1952).
15. C. G. Saxe and R. J. Brodd, in A. J. Salkind, ed., *Proceedings of the Symposium on History of Battery Technology,* Vol. 87-14, The Electrochemical Society, Pennington, N.J., 1987, p. 47.
16. J. O'M. Bockris, Z. Nagy, and A. Damjanovic, *J. Electrochem. Soc.* **119,** 285 (1972).
17. J. Hendrikx, A. van der Putten, W. Visscher, and E. Barendrecht, *Electrochim. Acta* **29,** 81 (1984).
18. R. G. Burns and V. M. Burns, in B. Schumm, Jr., H. M. Joseph, and A. Kozawa, eds., *Proceedings of the Manganese Dioxide Symposium,* Vol. 2, Tokyo, I. C. Sample Office, Cleveland, Ohio, 1980, p. 97.
19. P. Ruetschi, R. Giovanoli, and P. Burki, in Ref. 5, p. 12.
20. A. Kozawa and J. F. Yeager, *J. Electrochem. Soc.* **112,** 959 (1965); A. Kozawa and R. A. Powers, *J. Electrochem. Soc.* **113,** 870 (1966).
21. *Eveready Battery Engineering Data,* Vol. 1A, Eveready Battery Co., St. Louis, Mo., 1988.
22. *Eveready Battery Engineering Data, Temperature Effects,* BE-282, 1988.
23. R. A. Putt and A. I. Attia, *Proceedings of the 31st Power Sources Symposium,* The Electrochemical Society, Pennington, N.J., 1984, p. 339.
24. J. W. Cretzmeyer, H. R. Espig, and R. S. Melrose, in D. H. Collins, ed., *Power Sources 6,* Academic Press, New York, 1966, p. 269.
25. E. Peled in J. P. Gabano, ed., *Lithium Batteries,* Academic Press, New York, 1983, p. 43.
26. G. E. Blomgren, Ref. 24, p. 13.
27. H. V. Venkatasetty, ed., *Lithium Battery Technology,* John Wiley & Sons, Inc., New York, 1984, pp. 1 and 13.

28. *Eveready Lithium Battery Engineering Data,* Eveready Battery Co., St. Louis, Mo., 1987.
29. U.S. Pat. 3,536,532 (1970) and 3,700,502 (1972), N. Watanabe and M. Fukuda (to Matsushita Electric Co.).
30. A. A. Schneider, J. R. Moser, T. H. E. Webb, and J. E. Desmond, *Proc. Ann. Power Sources Conf.* **24,** 27 (1970).
31. H. Ikeda, T. Saito, and H. Tamura, in Ref. 5, p. 384.
32. U.S. Pat. 3,996,069 (1976), M. L. Kronenberg (to Union Carbide Corp.).
33. U.S. Pat. 4,327,166 (1982), V. Z. Leger (to Union Carbide Corp.).
34. N. Watanabe, R. Hagiwara, and T. Nakajima, *J. Electrochem. Soc.* **131,** 1980 (1984).
35. M. B. Clark, in Ref. 25, p. 115.
36. C. F. Holmes, in B. B. Owens, ed., *Batteries for Implantable Biomedical Devices,* Plenum Press, New York, 1986, p. 156.
37. U.S. Pat. 3,567,515 (1971), D. L. Maricle and J. P. Mohns (to American Cyanamid).
38. Ger. Pat. 2,262,256 (1972), G. E. Blomgren and M. L. Kronenberg (to Union Carbide Corp.).
39. C. R. Walk, in Ref. 25, p. 281.
40. C. R. Schlaikjer, in Ref. 25, p. 303.
41. D. Linden, ed., *Handbook of Batteries and Fuel Cells,* McGraw-Hill, New York, 1984.

General References

G. W. Heise and N. C. Cahoon, eds., *The Primary Battery,* Vol. 1 and 2, John Wiley & Sons, Inc., New York, 1971, 1976.

D. Linden, ed., *Handbook of Batteries and Fuel Cells,* McGraw-Hill, New York, 1984.

K. V. Kordesch, ed., *Batteries, Manganese Dioxide,* Vol. 1, Marcel Dekker, New York, 1974.

T. R. Crompton, *Battery Reference Book,* Butterworths, London, 1990.

J. P. Gabano, ed., *Lithium Batteries,* Academic Press, New York, 1983.

H. V. Venkatasetty, ed., *Lithium Battery Technology,* John Wiley & Sons, Inc., New York, 1984.

B. Boone Owens, ed., *Batteries for Implantable Biomedical Devices,* Plenum Press, New York, 1986.

R. J. Brodd, *Batteries for Cordless Appliances,* John Wiley & Sons, Inc., New York, 1987.

GEORGE BLOMGREN
JAMES HUNTER
Eveready Battery Company, Inc.

SECONDARY CELLS

ALKALINE

Alkaline electrolyte storage battery systems are more suitable than others in applications where high currents are required, because of the high conductivity of the electrolyte. Additionally, in almost all of these battery systems, the electrolyte which is usually an aqueous solution containing 25–40% potassium hydroxide [*1310-58-3*], KOH, does not enter into the chemical reaction. Thus concentration and cell resistance are invariant with state of discharge and these battery systems give high performance and have long cycle life. The annual production value of alkaline storage batteries is growing: it was over $2 billion (worldwide) in 1990, representing approximately 20% of the value for all secondary batteries.

Positive electrode active materials have been made from the oxides or hydroxides of nickel, silver, manganese, copper, mercury, and from oxygen. Negative electrode active materials have been fabricated from various geometric forms of cadmium [*7440-43-9*], Cd, iron [*7439-89-6*], Fe, and zinc [*7440-66-6*], Zn, and from hydrogen [*1333-74-0*]. Two different types of hydrogen electrode designs are common: those used in space, which employ hydrogen as a gas, and those used in consumer batteries, where the hydrogen is used as a metallic hydride. As indicated in Table 1 and Figure 1, nine electrode combinations exist in some scale of commercial production. Five system combinations are in the research/development stage as of this writing, and two have been abandoned before or after commercial production for reasons such as short life, high cost, low voltage, low energy density, and excessive maintenance.

The annual production value of small, sealed nickel–cadmium cells is over

Table 1. Rechargeable Alkaline Storage Battery Systems

System[a]	Historical name	Voltage, V	Production[b]
nickel–cadmium	Jungner	1.30	vl
nickel–iron	Edison	1.37	s
nickel–zinc	Drumm	1.70	vs
nickel–hydrogen		1.30	l
silver–cadmium		1.38 and 1.16[c]	vs
silver–iron	Jirsa	1.45 and 1.23[c]	s
silver–zinc	Andre	1.86 and 1.60[c]	s
silver–hydrogen		1.38 and 1.16[c]	vs
manganese–zinc[d]		1.52	vs
mercury–cadmium		0.92	r
air(oxygen)–zinc		1.60	r
air(oxygen)–iron		1.40	r
air(oxygen)–aluminum			r
copper–lead		1.20	r
copper–cadmium	Darrieus	0.45	n
copper–zinc	Waddell-Entz, Edison-LeLande, Lelande-Chaperon	0.85	n

[a]The substance named first represents the positive electrode; the substance named second is the negative electrode. In all cases except for air(oxygen) systems, the active electrode material is the oxide or the hydroxide of the named species.
[b]vl = >100 × 10^6 A·h/yr product; l = >25 × 10^6 A·h/yr; s = >5 × 10^6 A·h/yr; vs = <5 × 10^6 A·h/yr; r = research and development phase; and n = no longer in production.
[c]Silver electrodes have two voltage plateaus.
[d]Secondary system.

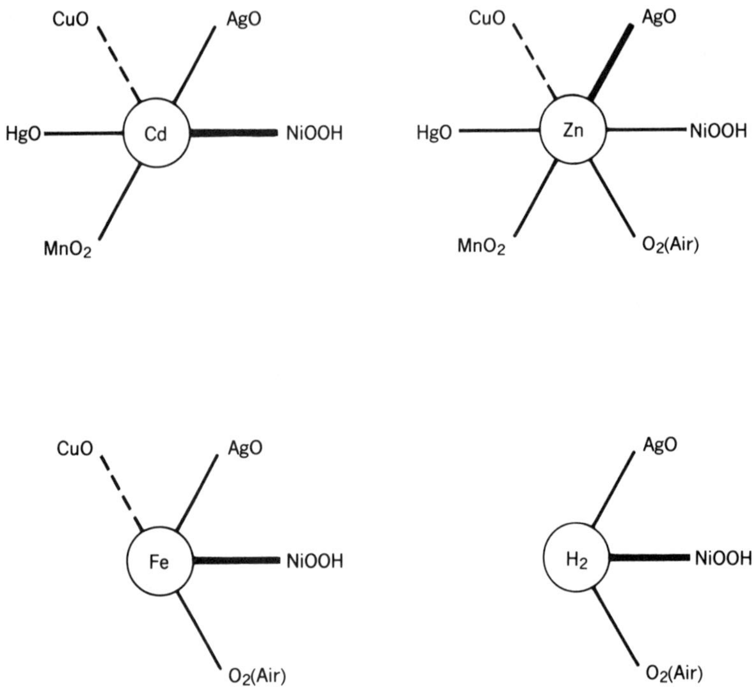

Fig. 1. Electrode combinations for alkaline storage batteries where the substance within the circle comprises the negative electrode and the combinations include (—) commercial products, (—) systems in research and development or limited production, and (---) historic systems. The air (O_2)–H_2 system is commonly called a fuel cell.

$1.2 billion. However, environmental considerations relating to cadmium are necessitating changes in the fabrication techniques, as well as recovery of failed cells. Battery system designers are switching to nickel–metal hydride (MH) cells for some applications, typically in "AA"-size cells, to increase capacity in the same volume and avoid the use of cadmium.

There are many methods of fabricating the electrodes for these cell systems. The earliest commercially successful developments used nickel hydroxide [*12054-48-7*], $Ni(OH)_2$, positive electrodes. These electrodes are commonly called nickel electrodes, disregarding the actual chemical composition. Alkaline cells using the copper oxide–zinc couple preceeded nickel batteries but the CuO system never functioned well as a secondary battery. It was, however, commercially available for many years as a primary battery (see BATTERIES-PRIMARY CELLS).

The original alkaline battery, designed by Edison around 1900, contained nickel hydroxide mixed with graphite [*7782-42-5*] in perforated steel pockets as the positive active material and a high surface area iron powder as the negative. Both positive and negative electrodes were "pocket" plates, where the active material was contained in small, rectangular, boxlike pockets formed from finely perforated sheet steel. These plates very closely resembled the pocket plates in use in some 1990 designs of nickel–cadmium batteries.

Because the nickel–iron cell system has a low cell voltage and high cost compared to those of the lead–acid battery, lead–acid became the dominant automotive and industrial battery system except for heavy-duty applications. Renewed interest in the nickel–iron and nickel–cadmium systems, for electric vehicles started in the mid-1980s using other cell geometries.

In the early 1930s, production of nickel–cadmium batteries having thinner pocket-type plates and lower internal resistence became available. This grew to be an important design for truck, locomotive, and marine engine starting applications. Very high rate cell designs of nickel–cadmium cells became possible with the development of the sintered nickel substrate in 1928 (1) and then in the late 1940s and early 1950s, the shortcomings of the sealed nickel–cadmium designs were overcome (2,3). The use of sealed nickel–cadmium cells grew rapidly and beginning in the early 1990s, designs of nickel–hydrogen, the hydrogen as hydride, cells captured a strong commercial and technical interest. Table 2 details the advances in alkaline battery technology from 1948 to 1990.

Table 2. Technology and Processing Contributions to Alkaline Battery Development

Year	Procedure	Reference
1948	impregnation procedures for nickel and cadmium electrodes	6
1956	x-ray structure of nickel hydroxide electrode	7
1958	*in situ* x-rays of nickel and cadmium electrodes	8
	slurry-processed nickel-sintered electrode (SAFT)	9
1960	crystal structures of silver oxide	10,11
	structure of nickel hydroxide electrode	12
	in situ study of the nickel electrode	13
1961	microporous plastic-reinforced zinc electrode	14
1962	microporous plastic-bound cadmium electrode	15
1965	solid-state chemistry of nickel hydroxide	16
	crystal structures of nickel hydroxide	17
1966–1970	impregnation procedures for nickel electrodes	18–20
1972	electrochemical impregnation method for nickel electrodes	21
1974	nickel foam substrates	22
1975	plastic-bonded nickel electrodes	23
1978	controlled microgeometry electrodeposited electrodes	24
1980	nickel composite electrode	25
1981	nickel fiber mat substrates	26
1984	nickel felt substrates	27
	production of nickel fiber electrodes	28
1987	rechargeable MnO_2–zinc cell	29
1989	modified manganese dioxide for deep-cycling	30
1990	metal–hydride-type hydrogen electrodes used in commercial nickel–MH cells	31–35

The nickel–zinc combination has a high cell voltage (about 1.75 V), which results in a very favorable energy density compared to that of nickel–cadmium or lead–acid. Additionally, zinc is relatively inexpensive and, in the absence of mercury additive, is environmentally benign. The nickel–zinc system was discussed as early as 1899 (4). There has been a resurgence of interest in the system for electric vehicles, but the problems of limited cycle life have not been completely overcome.

Silver [*7440-22-4*], Ag, as an active material in electrodes was first used by Volta, but the first intensive study using silver as a storage battery electrode was reported in 1889 (5) using silver oxide–iron and silver oxide–copper combinations. Work on silver oxide–cadmium followed. In the 1940s, the use of a semipermeable membrane combined with limited electrolyte was introduced by André in the silver oxide–zinc storage battery.

Many of the most recent applications for alkaline storage batteries require higher energy density and lower cost designs than previously available. Materials such as foam and/or fiber nickel [*7440-02-0*], Ni, mats as substrates, and new processing techniques including plastic bounded, pasted, or electroplated electrodes, have enabled the alkaline storage battery to meet these new requirements, while reducing environmental problems in the manufacturing plants. In addition, substantial technical efforts have been devoted to the recovery of used batteries. The most recent innovations in materials relate to the development of metal–hydride alloys for the storage and electrochemical utilization of hydrogen. Modifications to the chemical structure and/or the cell design of manganese dioxide [*1313-13-9*], MnO_2, electrodes have resulted in sufficient improvement to allow the reintroduction of the rechargeable MnO_2–zinc cell to the market as a lower cost, albeit lower performance, alternative to nickel–cadmium consumer size cells. Improvements in materials science and electrical circuits have lead to better separators, seals, welding techniques, feedthroughs, and charging equipment.

Nickel–Cadmium Cells

Electrodes. A number of different types of nickel oxide electrodes have been used. The term nickel oxide is common usage for the active materials that are actually hydrated hydroxides at nickel oxidation state 2+, in the discharged condition, and nickel oxide hydroxide [*12026-04-9*], NiO·OH, nickel oxidation state 3+, in the charged condition. Nickelous hydroxide [*12054-48-7*], $Ni(OH)_2$, can be precipitated from acidic solutions of bivalent nickel either by the addition of sodium hydroxide or by cathodic processes to cause an increase in the interfacial pH at the solution–electrode surface (see NICKEL AND NICKEL ALLOYS; NICKEL COMPOUNDS).

Several investigators have used combined approaches, particularly in the *in situ* precipitation of active material in the pores of sintered substrates, using cathodic polarization and caustic precipitation in simultaneous or nearly simultaneous steps. A considerable amount of the reported information on the chemistry, electrochemistry, and crystal structure of the nickel electrode has been obtained on thin films (qv) made by the anodic corrosion of nickel surfaces. However, such films do not necessarily duplicate the chemical and/or crystallographic condition of active material in practical electrodes. In particular, the high surface area, space charge region, and lattice defect structure are different. Some of the higher (3.5+) valence state electrochemical behavior seen in thin films has rarely been reproduced in practical electrodes.

The many varieties of practical nickel electrodes can be divided into two main categories. In the first, the active nickelous hydroxide is prepared in a separate chemical reactor and is subsequently blended, admixed, or layered with an electronically conductive material. This active material mixture is afterwards contained in a confining porous metallic structure or pasted onto a metallic mat or grid. Electrodes for pocket, tubular, pasted, and most button cells are made this way. The porous metallic structures such as the pocket or tube, ensure good particulate contact, prevent shedding, and confine expansion. Less expensive alternatives, those resulting from pasting on expanded metal, foam metal, punched corrugated sheet, or fiber metal nickel mat, do not confine expansion as well or provide the same amount of mechanical strength. Plastic reinforcement to minimize shedding and promote adherence to the substrate is often employed. The plastic mix can be achieved either by milling or extrusion of mixes at plastic flow temperatures or by using plastic–solvent combinations. The latter includes simple gels of water, KOH, and methylcellulose [*9004-67-5*].

The other type of nickel electrode involves constructions in which the active material is deposited *in situ*. This includes the sintered-type electrode in which nickel hydroxide is chemically or electrochemically deposited in the pores of a 80–90% porous sintered nickel substrate that may also contain a reinforcing grid.

Almost all the methods described for the nickel electrode have been used to fabricate cadmium electrodes. However, because cadmium, cadmium oxide [*1306-19-0*], CdO, and cadmium hydroxide [*21041-95-2*], $Cd(OH)_2$, are more electrically conductive than the nickel hydroxides, it is possible to make simple pressed cadmium electrodes using less substrate (see CADMIUM AND CADIUM ALLOYS; CADMIUM COMPOUNDS). These are commonly used in button cells.

Electrochemistry and Crystal Structure. The solid-state chemistry of the nickel electrode is complex. Nickel hydroxide in the discharged state has a hexagonal layered lattice, where planes of Ni^{2+} ions are sandwiched between planes of OH^- as shown in Figure 2. This structure, similar to that of cadmium iodide [*7790-80-9*], CdI_2, is common to seven metal hydroxides including those of cadmium and cobalt. There are various hydrated and nonhydrated nickel hydroxides that have slightly different crystal habitats and electrochemical potentials. The most common form of charged material observed in batteries is NiOOH, density = 4.6 g/mL. In comparison, $Ni(OH)_2$ has a density of 4.15 g/mL. Thus the theoretical change in density on charge–discharge is only 9%, and the kinetics involve only a proton transfer. This reaction can be written in simplified form as:

$$2\ \beta\text{-NiOOH}^* + Cd + 2\ H_2O \rightleftharpoons 2\ Ni(OH)_2^* + Cd(OH)_2$$

where the asterisks represent differing amounts of hydrated or absorbed water and/or electrolyte. It has also been established that species such as Li^+ and K^+ enter the nickel hydroxide structure to form a space–charge region and the actual reaction is believed to be more complicated than that shown. The interlayer separation of planes of nickel ions is increased by insertion of water or ions from the electrolyte. Different preparative and cycling conditions result in variations in defect crystal structures that affect electrochemical activity and the ability to retain charge (36). Additionally, electrode material oxidation states do not range precisely between +2 and +3, and because of hysteresis in the charge–discharge curves, direct measurement of an equilibrium potential is not possible.

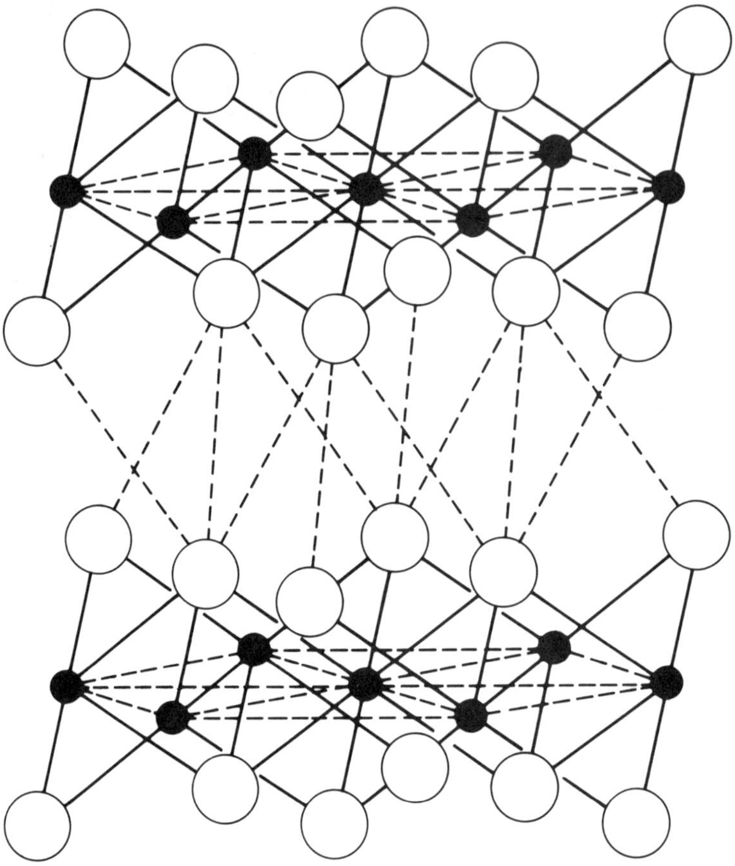

Fig. 2. Nickel hydroxide structure where (●) represents nickel and (○) represents oxygen or hydroxyl.

The existence of $Ni(OH)_2$ and β-NiOOH as the usual discharged and charged materials was confirmed through x-ray diffraction powder patterns (7,8,37). The presence of other structural forms, in different electrolytes and under unusual cycling conditions, has also been observed. The crystalline-structural variability of $Ni(OH)_2$ has been described as being dependent on preparative procedure (16). Other phases form after prolonged overcharging of β-NiOOH in concentrated sodium hydroxide [*1310-73-2*], NaOH, or KOH. At elevated temperatures, in lithium hydroxide [*1310-65-2*], LiOH, electrolytes, $Ni(OH)_2$ is oxidized to a trivalent phase having the lithium nickelate [*12031-65-1*], $LiNiO_2$, crystalline structure and lithium additions to KOH electrolytes eliminate the formation of α-phase material. The sequence of these electrochemical processes has been summarized as: (*1*) an electrochemical exchange at the solid electrolyte interface involving a set of ionic-electronic lattice imperfections; (*2*) electron transport through the oxidized phase region to the metallic contact; (*3*) mass transport through the oxidized phase to the lower valence phase at the second interface; and (*4*) electrochemical exchange at the second solid-phase interface functioning as a source or sink for the ionic-electronic imperfections necessary to maintain the mass and charge balance.

A sequence of events taking place in the charge process has been described (16) demonstrating that during constant current charge, the surface double-layer region first produces a local interior space–charge field. As the semiconductor charge phase is extended into the interior, a distributed space–charge field is formed that is characterized by a combination of the semiconductor band structure and the mobility of dopant imperfections. When the state of charge increases, the magnitude of the space charge decreases. At the end of charge, there is only an ir (voltage) loss in the solid, and the applied field is mainly at the solid electrolyte interface. If charging is continued until a steady-state oxygen evolution is attained, the distributed space charge disappears. The semiconductor is characterized by a flat band potential and, if the nickel electrode is left on open circuit, the space charge-field gradually decays through atom movements in the solid phase, approximating a flat band potential condition.

The charge–discharge process cannot be satisfactorily represented by one equation (17). At least two reactions, based on different starting materials as well as different products, can be formulated. Both reactions are heterogeneous. In each of the reaction chains two distinct states exist in the oxidized phase. One of these can only be charged whereas the cathodic current are blocked. The other state, existing at a lower potential, can only be discharged but was blocked in the anodic direction. When the observed potential differences are extrapolated into the region of very low current densities, the difference between the two states is ca 60 mV in one of the reaction chains and ca 100 mV in the other. Thermodynamically this means that neither reaction series can be described by a single, reversible potential as in the manner of the Cd–$Cd(OH)_2$ or PbO_2–$PbSO_4$ electrodes. Rather, the semiconductor properties of oxidized nickel hydroxides play some part.

The description of the process is best illustrated as (17):

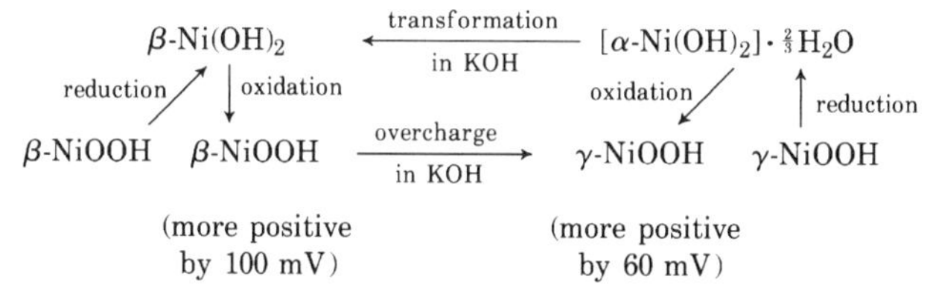

The nickel oxide modification obtained electrochemically in KOH electrolyte contained potassium ion and its nickel oxidation level are higher than that of $NiO_{1.5}$. Conclusions regarding the transitions between the reduced and oxidized products within the two series are that the redox process was not reversible and although the oxidized phases of the β- and the γ-nickel hydroxides differ in energy contents, differences in analyses and x-ray patterns are not significant.

Some γ-NiOOH has been shown to be formed in sintered nickel electrodes (38), and changes in water and KOH concentration during the cycling of nickel electrodes has been studied (12,39–41). Although there is some disagreement on the movement of water, KOH is adsorbed on the nickel electrode when the cell is charged and desorbed from the electrode when the cell is discharged.

The chemistry, electrochemistry, and crystal structure of the cadmium electrode is much simpler than that of the nickel electrode. The overall reaction is generally recognized as:

$$Cd + 2\ OH^- \underset{\text{charge}}{\overset{\text{discharge}}{\rightleftharpoons}} Cd(OH)_2 + 2\ e^-$$

However, there is a strong likelihood of a soluble intermediate in the formation of $Cd(OH)_2$. Cadmium has an appreciable solubility in alkaline solutions: $\sim 2 \times 10^{-4}$ mol/L in 8 *M* potassium hydroxide at room temperature. In general it is believed that the solution process consists of anodic dissolution of cadmium ions in the form of complex hydroxides (see CADMIUM COMPOUNDS).

In more recent studies involving cyclic line scan voltammetry of the nickel electrode, it was suggested that nickel can exist in the positive 2, 3, and 4 oxidation states (42). The structural parameters (43) using transmission extended x-ray absorption fine structure (exafs) and *in situ* electrodes confirmed earlier x-ray data (7,8), showing that the presence of cobalt [*7440-48-4*], Co, does not change crystal lattice parameters. The density and compressibility of nickel electrodes have been found to be highly variable (44). Cobalt additions appear to reduce the compressibility of nickel hydroxide, resulting in a firmer attachment of the active material to the substrate. However, a felt metal grid has been shown (45) to reduce shear failures of the electrode structure and minimize the need for stabilizing additives, such as cobalt.

In addition to the normal charge–discharge reaction, properly fabricated sealed nickel–cadmium cells have a mechanism for absorbing infinite amounts of overcharge. The cell must be fabricated with an excess amount of uncharged active material [$Cd(OH)_2$] in the cadmium electrode. When the nickel electrode nears full charge oxygen is evolved. This oxygen diffuses through open areas of the separator and reacts with the charged cadmium species forming cadmium oxide that hydrates to cadmium hydroxide.

$$Cd(OH)_2 + 2\ e^- \longrightarrow Cd + 2\ OH^-$$

Therefore, the cadmium electrode is being electrically charged and chemically discharged at the same rate.

$$2\ Cd(OH)_2 \longleftarrow 2\ Cd + O_2 + 2\ H_2O$$

Sealed Cells. Most sealed cells are based on the principles appearing in patents of the early 1950s (2,3) where the virtues of limiting electrolyte, a separator that would absorb and retain electrolyte, and leaving free passage for the oxygen from the positive to the negative plate were described. First, the negative electrode has a surplus of uncharged active material so that the positive plate starts to produce oxygen before the negative plate is fully charged. The oxygen reacts with the negative active material, so that the negative electrode never becomes fully charged and consequently never evolves hydrogen. Second, the amount of electrolyte used is generally lower than can normally be absorbed in the electrodes and separators, facilitating the transfer of oxygen from the positive to the negative plate. Oxygen transport, at least to a certain extent is carried out in the gaseous stage. Third, the separators generally used can pass oxygen to the gaseous state for rapid transfer to the negative plate. Although both pocket and sintered electrodes of the nickel–cadmium type have been used in sealed-cell construction, the preponderant majority of cells in commercial production use sintered positive (nickel) electrodes, and either sintered or pasted negative (cadmium) electrodes.

Although the charged form of the anode is metallic cadmium, some traces of the discharged form always remain in the electrode even at prolonged overcharge. Indeed, as the anode is oxidized (discharged), x-ray diffraction peaks show lines from both Cd and $Cd(OH)_2$. These lines have sharper peaks than those obtained using the cathodic nickel hydroxides, confirming the more crystalline nature of cadmium.

Cadmium electrodes maintain a flat $E°$ potential throughout discharge, as is true of most electrodes in which the materials of the charged and discharged state form distinct independent crystalline forms. Because the conductivities of cadmium electrodes are high, and no ir drops or films are generated during discharge, they also display a flat working voltage curve.

The nickel–cadmium cell, unlike the lead–acid system, has a negative temperature voltage coefficient that is in the range of $-(0.2–0.4)$ mV/°C, depending on design factors and the nature and doping of the nickel active materials. Thus at higher temperatures the cell has a slightly lower open circuit voltage and this, combined with a reduction of internal resistance at higher temperatures, can result in a reduced back emf in charging. Therefore, special precautions must be taken in charging under constant voltage conditions to avoid the so-called runaway condition; ie, an increase in charge current that comes from a reduction in back emf resulting in an increase in temperature and then a new increase in charge current. Cells can be destroyed in this runaway condition unless current limiters are provided in the charge circuit. The runaway condition problem is not limited to the nickel–cadmium battery. However, it is more likely in those systems where the voltage change with temperature is negative and in those designs where there is very low internal resistance.

The capacity utilization of active material depends on cell design, discharge rate, temperature, and charging conditions. High rate ability depends on the degree of conducting support provided for the active material surface that is in contact with electrolyte, separator resistance, and state-of-charge. The maximum capacity that can normally be achieved from nickel active materials is 0.30 A·h/g calculated on the basis of $Ni(OH)_2$. The capacity of negative active material is ca 0.37 A·h/g based on $Cd(OH)_2$. In actual cell use, working capacities range from 60 to 80% of these values, depending on discharge rate and temperature.

Cell Fabrication Methods. *Pocket Cells.* A view of a pocket electrode nickel–cadmium cell is shown in Figure 3. The essential steps of positive (nickel) electrode construction are (*1*) cold-rolled steel ribbon is cut to proper width and is perforated using either needles or rolls; (*2*) the perforated steel ribbon is nickel-plated and usually annealed in hydrogen. The ribbon is formed into a trough shape, is filled with active material by either a briquetting or a powder-filling technique; (*3*) a second strip is formed into a lid that covers and locks with the filled trough; (*4*) the filled strips are cut to length and are arranged to form an

Fig. 3. View of pocket electrode nickel–cadmium cell.

electrode sheet by interleaving. This operation, carried out by means of rollers in a forming roll, is often combined with the pressing of a pattern into the electrode sheet in order to ensure good contact between ribbon and active material and to add mechanical strength to the construction; and (*5*) the electrode sheet is then cut to pieces of appropriate size and side bedding and lugs attached to form a metallic frame. The frame material is usually also cold-rolled steel ribbon.

The pockets are usually arranged horizontally in the electrodes as shown in Figure 3, but in a few cases vertical pockets are used. No significant difference has been observed between the two arrangements.

Pocket-type cadmium electrodes are made by a procedure similar to that described for the positive electrode. Because cadmium active material is more dense than nickel active material, and because cadmium has a 2+ valence, cadmium electrodes, when fabricated to equal thicknesses, have almost twice the working capacity of the nickel electrode. A cell having considerably greater negative capacity provides for loss of negative capacity during life and avoids generation of hydrogen during charging. Thus in actual practice plates of equal thickness are used. Some manufacturers prefer to make the electrolyte transfer area larger for the negative electrode by increasing the area of perforation, usually by increasing the hole size. Pocket electrode plates usually have thicknesses of 0.7–6.3 mm.

After the individual pocket electrodes are fabricated, they are assembled into electrode groups. Electrodes of the same polarity are electrically and mechanically connected to each other and to a pole bolt as illustrated in Figure 3. Plates of opposite polarity are interleaved with separators. Ordinarily single-plates are used for each leaf in a plate group, but some manufacturers prefer to use two thin plates back to back to form one leaf in the positive group. Both bolting and welding methods are used in the assembly of electrodes into groups.

To complete the assembly of a cell, the interleaved electrode groups are bolted to a cover and the cover is sealed to a container. Originally, nickel-plated steel was the predominant material for cell containers but, more recently plastic containers have been used for a considerable proportion of pocket nickel–cadmium cells. Polyethylene, high impact polystyrene, and a copolymer of propylene and ethylene have been the most widely used plastics.

Steel containers are mechanically stronger than plastic and easier to fabricate in large sizes. Also they dissipate heat better and tend to keep the electrodes cooler during high temperature or high rate operations. However, cells assembled in steel containers must not have contact with each other during assembly to prevent intercell shorts. Plastic containers are the better option for most small and medium-sized cells because they require no protection against corrosion, they permit visual observation of the electrolyte level, they are lighter than steel containers, they can be closely packed into a battery, and small cells can be cemented or taped into batteries, eliminating a tray.

The cells are usually filled with an electrolyte solution of potassium hydroxide of density 1.18–1.23 g/mL, which may also contain lithium hydroxide. A potassium hydroxide solution of 1.20 g/mL freezes at $\sim -27°C$. Thus cells intended for climates colder than $-27°C$ have an electrolyte density of 1.25 g/mL or greater. In some designs there is a large volume of excess electrolyte above the electrodes in order to reduce the need for rewatering to once every few years. In these cells, the initial electrolyte density is lower than normal because the solution concentrates during operation. Concentrations might be as high as 1.26 g/mL at the time of rewatering (topping up). Lithium hydroxide has been shown to increase the life of positive pocket electrodes in cycling operation. However, the addition also increases the electrolyte resistivity, and is not ordinarily used in high rate starter batteries.

Individual cells are usually precycled before assembling into batteries. These early charge–discharge cycles, often called formation cycles, improve the capacity of the cell by increasing the surface area of the active material and effecting crystal structure changes.

The individual cells are assembled into batteries after a leakage test to check for faulty welding joints in steel containers or cracks and improper seals in the plastic encased cells. Cells that are to be delivered without electrolyte are emptied after the formation cycles. The steel-cased cells have to be separated by mechanical means to prevent intercell shorts and are often assembled into wooden crates. The cells in plastic cases can be cemented together or strapped with tape.

Tubular Cells. Although the tubular nickel electrode invented by Edison is almost always combined with an iron negative electrode, a small quantity of cells is produced in which nickel in the tubular form is used with a pocket cadmium electrode. This type of cell construction is used for low operating temperature environments, where iron electrodes do not perform well or where charging current must be limited.

Sintered Cells. The fabrication of sintered electrode batteries can be divided into five principal operations: preparation of sintering-grade nickel powder; preparation of the sintered nickel plaque; impregnation of the plaque with active material; assembly of the impregnated plaques (often called plates) into electrode groups and into cells; and assembly of cells into batteries.

A good powder for sintering purpose should be very pure (Fe, Cu, and S should be especially avoided) and should have a very low apparent density, in the range of 0.5–0.89 g/mL. An excellent powder for this purpose is made by the decomposition of nickel carbonyl [*13463-39-3*], $Ni(CO)_4$, (see CARBONYLS). Nickel carbonyl is a poisonous vapor obtained by passing carbon monoxide [*630-08-0*] over finely divided metallic nickel at about 200°C in rotary kilns using special gas-tight seals. The gas is condensed and the liquid carbonyl distilled to eliminate impurities. The purified liquid nickel carbonyl is injected into large decomposers that have heated walls. The nickel carbonyl decomposes into carbon monoxide, which can then be recirculated, and a finely divided nickel powder. The structure of such powders is shown in Figure 4.

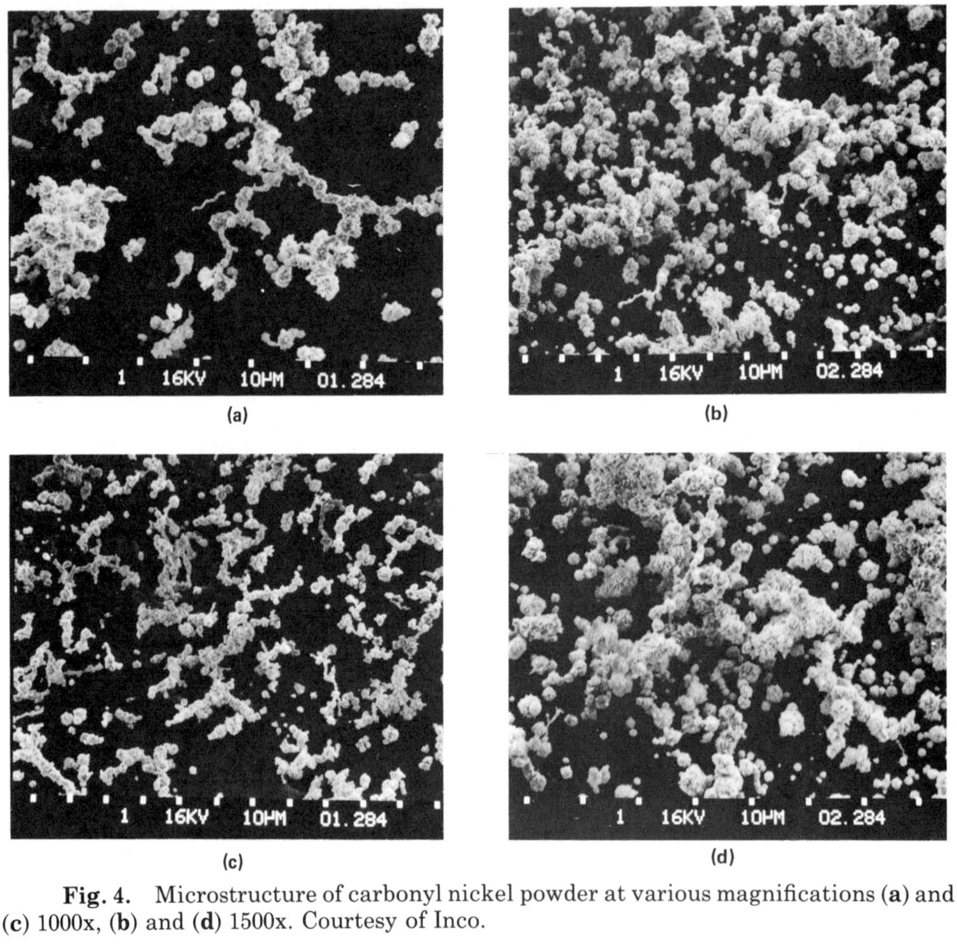

Fig. 4. Microstructure of carbonyl nickel powder at various magnifications (**a**) and (**c**) 1000x, (**b**) and (**d**) 1500x. Courtesy of Inco.

Sintering is a thermal process through which a loose mass of particles is transformed to a coherent body. It usually takes place at a temperature equal to two-thirds the melting point, or ca 800–1000°C for nickel. The sintered nickel structure without active material is called a plaque and it can be prepared by either dry or wet processes (see METALLURGY, POWDER).

In the dry process, nickel powders are sifted into ceramic or carbon molds with a reinforcing layer of wire mesh cloth or electroformed nickel. In most production processes part of the powder is sifted into the mold, a grid is inserted, and a second layer of sifted powder is added. The reinforcing grid is provided to reduce shrinkage during the sintering, reduce electrical resistivity, and provide extra mechanical strength. The molds are transferred to a sintering furnace that contains a reducing or inert atmosphere. Usually a reducing atmosphere is maintained by cracking ammonia ($2\ NH_3 \rightarrow N_2 + 3\ H_2$) or partially dissociating natural gas. After 5–10 min in a hot zone, the molds are transferred to a cold zone in the furnace. Most furnaces for the dry processes are horizontal and of the moving-belt type.

To reduce labor and other expenses, most sintered nickel plaques are produced by a wet-slurry method. A nickel slurry is prepared by mixing a low density nickel powder with a viscous aqueous solution such as carboxymethylcellulose [*9004-42-6*] (CMC). Pure nickel gauze, a nickel-plated gauze, or a nickel-plated perforated steel strip is continuously carried through a container filled with the nickel paste and sintering is done in a horizontal furnace. The time of the sinter in the furnace is ca 10–20 min.

Usually the plaques produced by either method are coined (compressed) in those areas where subsequent welded tabs are connected or where no active material is desired, eg, at the edges. The uncoined areas usually have a Brunauer-Emmet-Teller (BET) area in the range of 0.25–0.5 m^2/g and a pore volume $>80\%$. The pores of the sintered plaque must be of suitable size and interconnected. The mean pore diameter for good electrochemical efficiency is 6–12 μm, determined by the mercury-intrusion method.

The process by which porous sintered plaques are filled with active material is called impregnation. The plaques are submerged in an aqueous solution, which is sometimes a hot melt in a compound's own water of hydration, consisting of a suitable nickel or cadmium salt and subjected to a chemical, electrochemical, or thermal process to precipitate nickel hydroxide or cadmium hydroxide. The electrochemical (46) and general (47) methods of impregnating nickel plaques have been reviewed.

In the original process for the positive electrode, the plaques were placed in a metal vessel, which was evacuated to <5.3 kPa (40 mm Hg), and a nearly saturated solution of nickel nitrate (density 1.6 g/mL) admitted. After a 5–15 min soaking period, the plaques were transferred at 101 kPa (1 atm) to a polarizing unit where they were cathodically polarized in hot caustic solution. After polarization the plates were washed and dried. These four steps were repeated four or five times until the desired weight gain of active material was achieved.

For the negative electrolyte, cadmium nitrate solution (density 1.8 g/mL) is used in the procedure described above. Because a small (3–4 g/L) amount of free nitric acid is desirable in the impregnation solution, the addition of a corrosion inhibitor prevents excessive contamination of the solution with nickel from the sintered mass (see CORROSION AND CORROSION INHIBITORS). In most applications for sintered nickel electrodes the optimum positive electrode performance is achieved when one-third to one-half of the pore volume is filled with active material. The negative electrode optimum has one-half of its pore volume filled with active material.

Other processes have been developed in which the impregnation is accomplished in one or two steps; the most promising is electrodeposition directly from nitrate solutions having pH controlled at 4–5. After electrodeposition, the plaques are either cathodically polarized in sodium hydroxide solution or electrochemically formed in sodium hydroxide to eliminate all traces of nitrate. The latter steps must proceed at low current densities to avoid blistering and shedding of the loaded plaques.

Some manufacturers add a small (10–20% of the positive loading) amount of cadmium to positive plates as an antipolar mass to prevent some of the problems of reversal in sealed cells. This practice may, however, create as many problems as it solves in that positive capacity is reduced proportionally to the quantity of antipolar mass added.

In most cases, the impregnation process is followed by an electrochemical formation where the plaques are assembled into large temporary cells filled with 20–30% sodium hydroxide solution, subjected to 1–3 charge–discharge cycles, and subsequently washed and dried. This eliminates nitrates and poorly adherent particles. It also increases the effective surface area of the active materials.

Formation also offers a convenient means of regulating the state of charge of the plates prior to cell assembly. This is important in sealed cell manufacture, where cell performance is optimized when 10–15% of the negative active mass is in the charged condition prior to initial cell charging. Some manufacturers have found that elimination of formation is feasible by extensive conversion of the nitrates in the impregnation process, followed by meticulous washing. However, charge retention can suffer to some degree because of traces of nitrate in the finished cells.

Cell Assembly. The methods for cell assembly, starting with the processed plaques depend on whether the cells are to be vented or sealed. For vented cells, processed plaques are usually compressed to 85–90% of their processed thickness allowing sufficient porosity for electrolyte retention and strengthening the plate structure. For sealed cells, sizing of the negative plaques is usually avoided because maximum surface area is important to oxygen recombination.

The next operation is the cutting of plaques into individual plates, usually through the coined areas (Fig. 5). Plate edges are sometimes coated with an adherent plastic film to smooth any rough edges that could otherwise penetrate the thin separators used in sintered-plate cells. A tab of nickel or nickel-plated steel is welded to each plate. In some cases, this is accomplished by spot or projection welding at the coined area. For plates provided with perforated sheet grids the tabs are spot welded directly to the sheet at the proper point at the edge of the sheet. For both types of cells most of the requirements of the separators are high electronic resistance and low electrolytic resistance in the usual electrolyte (30% KOH solution). Separators should be as thin as possible and have small and evenly distributed pores. Additionally, they should be resistant to heat, KOH, gases formed on charge, and other components of the system.

The most important separator material, property difference for vented and sealed cells is permeability to oxygen. In the charging of the vented cells, it is important to minimize oxygen transport from the positive to the negative plates. In float-charge applications, oxygen recombination at the negative plates results in generation of heat and ultimately to thermal runaway if excessive charge current is drawn with rising temperature. The opposite is required in sealed cells,

Fig. 5. Coined master plaque.

where oxygen recombination is necessary to prevent an increase of internal pressure. Glycerol-free cellophane is used as part of the separator in the vented cells. Other components in typical vented cell separators are woven nylon, felted nylon, felted nylon–cellulose, felted (vinyl chloride–acrylonitrile) copolymer, microporous polypropylene, and irradiated polyethylene. Most sealed-cell separators are single-layer felted materials, usually nylon or polypropylene, chosen for the ability to retain electrolyte and prepared in such a manner as to allow oxygen permeability, eg, calendering of such materials should be avoided.

Another difference between vented and sealed cell manufacture is found in the ratio of negative to positive active material. Although both types of cells require an excess of negative material, in the vented type a ratio of 1.5 is sufficient for the high rates of discharge and low temperature conditions encountered. In the sealed type a typical ratio is 2, to provide for sufficient sites for the oxygen recombination, to provide for the residual negative capacity mentioned earlier, and to preclude the possibility of the negative plate reaching the fully charged state exemplified by H_2 evolution.

Both vented and sealed cells use the same basic electrolyte (30% KOH), but different amounts are required. The vented cell contains a considerable amount of free electrolyte to allow for decomposition and loss of water on charge and to allow for maximum performance on discharge. The sealed cell contains only enough electrolyte to fill the plate pores and to completely saturate the separator; excess electrolyte inhibits oxygen recombination by reducing the number of sites, ie, oxygen–cadmium–electrolyte interfaces, for oxygen recombination. In certain applications, 15–30 g/L lithium hydroxide is added to the electrolyte. Where elevated temperature operation is encountered, this addition improves charge acceptance, especially in sealed cells. Larger (50 g/L) amounts of lithium hydroxide are used in repeated cycling with constant voltage charging and in float-charge applications (vented cells). This larger amount of lithium hydroxide maintains capacity better than KOH alone but increases electrolyte resistance and adversely affects low temperature performance.

The presence of certain forms of cellulose (qv) in the electrolyte is beneficial for negative electrodes, probably by preventing the formation of large grain sizes of cadmium on charge. Cellophane in vented cell separators usually provides sufficient cellulose for this effect. Early sealed cells using a single layer of cellulosic filter paper as separator gave excellent performance but had reduced cycle life as a result of the degradation of the cellulose. In later sealed cells, where felted nylon or felted polypropylene was used as separator, the addition of small amounts of cellulose to the electrolyte, improved negative plate performance, but led to early separator degradation.

Assembly of vented cells begins by interleaving the electrodes and separators. The most widely used separator is a sandwich of woven nylon (0.08–0.10 mm), cellophane (0.05 mm), and woven nylon (0.08–0.10 mm) of the proper width, usually 0.6 cm wider than the plate height, and of a length to form a continuous barrier around the edges of all the plates. The electrode tabs of each polarity are brought together and connected to the cell terminals by spot welding or, in a few cases, by bolting. The stack is attached to the cell cover by inserting terminals through holes that are marked for polarity. When threaded terminal posts are used, sealing is accomplished by compressing a neoprene gasket or O-ring between a flat or groove on the horizontal surface of the terminal and the inside surface of the cover, with the proper torque being applied by means of a nut on the outside of the cover. In the case of a smooth terminal post, compression is effected by means of a snap ring and belleville washer arrangement.

The most common cell case and cover materials are nylon and styrenic (ABS). After inserting the cell element, with cover attached, into the cell case, cover and case are cemented. For nylon, a most satisfactory adhesive is phenol. For styrene copolymers, either a solvent seal or an ultrasonic seal is effective (see ADHESIVES).

The cell filler caps in the covers are designed to prevent spillage and usually contain a venting mechanism to allow for escape of gas during charge and to prevent the carbon dioxide of the air from contacting the electrolyte. Most filler caps are a type of rubber-ring screw vent assembly provided with holes or slits that are sealed by the rubber ring. These are usually made of nylon or polystyrene, but in certain applications they are made of nickel-plated steel.

In filling cells with electrolyte, vacuum techniques are employed for uniformity and for hastening the complete wetting of plates and separator. A properly manufactured cell is ready for use immediately after the first charge following electrolyte fill. The first charge is ideally at the 10-h rate for 20 h. Higher rates can be used but should not exceed the 5-h rate.

Plastic-cased cells are assembled into batteries by placing them a small distance apart or in contact with heat-conducting fins in a specially treated steel container. The cell spacing depends on the battery application. For example, in aircraft starting batteries provision is required for circulating cooling air between cells. The steel boxes (battery cases) are previously prepared by coating all surfaces with a tough, insulating plastic (usually epoxy). Excellent results are obtained using a fluidized bed technique of application, which results in a uniformly thick coating. Intercell connectors are normally of nickel-plated copper, but in small sizes connectors are often of nickel or nickel-plated steel.

The prismatic sealed cells are made in a manner similar to the above. Most prismatic sealed cells use a metal case and cover. The most desirable case material is stainless steel, although nickel-plated steel can be used. Terminal feed-through is effected by ceramic-to-metal or glass-to-metal sealing techniques; and case-to-cover seal, by inert gas welding. Most prismatic sealed cells incorporate a high release, resealable vent in the cover that usually consists of a metallic spring working on an elastomer sealing disk or ring.

The prismatic sealed cells are not self-supporting and are normally used in battery operation where the battery case is used to constrain cell cases because internal cell pressures in the range of 690 kPa (100 psi) are common.

By far the majority of sealed cells are of the small cylindrical self-supporting type in the familiar "AA," "C," and "D" commercial sizes such as that shown in Figure 6. The element for these cells contains only one plate of each polarity; thus the electrodes are relatively long. The plates and the single-layer separator are rolled into a tight spiral, the plates arranged so that the positive tabs protrude from the element in one direction and the negative tabs in the opposite direction. The top and bottom of the element are protected and insulated by circular plastic disks having slits for the protruding tabs.

Elements are then inserted in nickel-plated steel cans, normally with the negative tab bent 90° to make contact with the can bottom. The tab is spot welded to the can bottom employing a long slender electrode that fits through the center hole of the element. The cell cover is welded to the positive tab. The cover is nickel-plated steel with a nylon gasket around the edges. After adding electrolyte, the cover is pressed into the can so that the nylon gasket rests on a shelf formed into the can well by scoring (after insertion of element). The rim of the can and sometimes the upper part of the gasket is folded over either in a press or by a rolling, rotating tool.

The cover assembly usually contains a safety vent. The most common type consists of a steel or nickel diaphragm built into the cover and a bent point cut from the cover. At a certain internal cell pressure, the diaphragm moves to the bent point and is pierced. Some cylindrical sealed cells use one or more terminal feed-throughs employing glass-to-metal or ceramic-to-metal techniques.

Battery assembly using cylindrical cells varies, and cell-to-cell connections are spot welded after using either flat tabs or cup tabs. Cell-to-cell insulation is effected either by using plastic cell jackets (shrink-on) or by inserting cells in plastic modules with each cell occupying its own cavity.

Other Cells. Other methods to fabricate nickel–cadmium cell electrodes include those for the button cell, used for calculators and other electronic devices. This cell, the construction of which is illustrated in Figure 7, is commonly made using a pressed powder nickel electrode mixed with graphite that is similar to a pocket electrode. The cadmium electrode is made in a similar manner. The active material, graphite blends for the nickel electrode, are almost the same as that used for pocket electrodes, ie, 18% graphite.

Lower cost and lower weight cylindrical cells have been made using plastic

Fig. 6. Partial cutaway of a coiled, sintered-plate, nickel–cadmium cell ("D"-size).

bound or pasted active material pressed into a metal screen. These cells suffer slightly in utilization at high rates compared to a sintered-plate cylindrical cell, but they may be adequate for most applications. The effect of temperature and discharge rate on the capacity of sealed nickel-cadmium cells is illustrated in Figure 8 and Table 3.

Applications. Nickel–cadmium cells represent almost 20% of the market for all storage batteries, including lead–acid, manufactured in the world. Uses are divided into three categories: pocket cells are used in emergency lighting, diesel starting, and stationary and traction applications where the reliability, long life, medium-high rate capability, and low temperature performance characteristics warrant the extra cost over lead–acid storage batteries; sintered, vented cells are used in extremely high rate applications, such as jet engine and large diesel engine starting; and sealed cells, both the sintered and button types, are used in computors, phones, cameras, portable tools, electronic devices, calculators, cordless razors, toothbrushes, carving knives, flashguns, and in space applications, where nickel–cadmium is optimum because it can be recharged a great

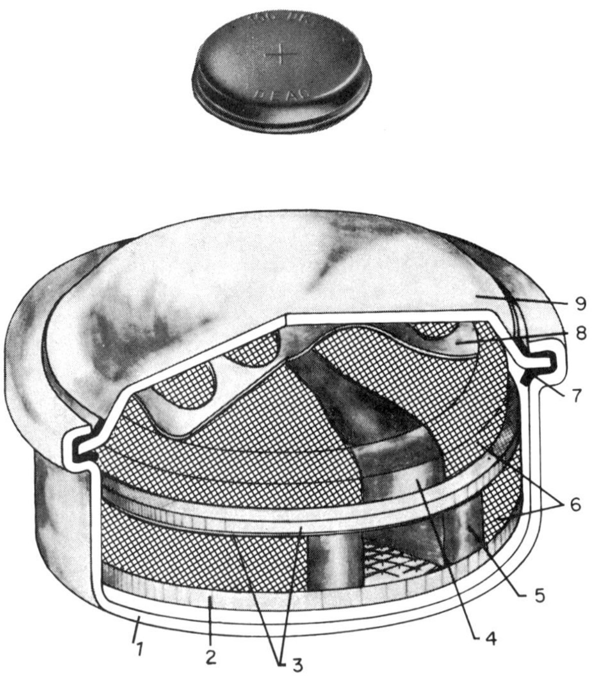

Fig. 7. Section of disk-type cell where: 1, is the cell cup; 2, is the bottom insert; 3, is the separator; 4, is the negative electrode; 5, is the positive electrode; 6, is the nickel wire gauze; 7, is the sealing washer; 8, is the contact spring; and 9, is the cell cover.

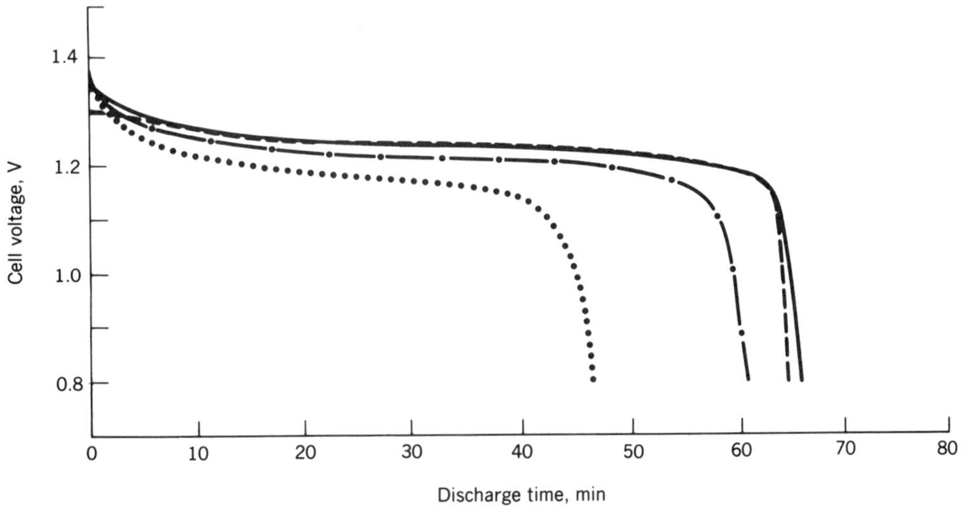

Fig. 8. Discharge capacity of small sealed nickel–cadmium cells where the initial charge is 0.1 C × 16 h at 20°C and the discharge is 1 C at temperatures of (• • •) −20°C, (— • —) 0°C, (———) 20°C, and (– – –) 60°C. C is the current required to discharge the cell in one hour.

Table 3. Nominal Capacities of Consumer Nickel–Cadmium Cells

	Capacity, A·h	
Size	Normal cells	High capacity cells
"AA"	0.5	0.7–0.8
"Sub-C"	1.2	1.7
"C"	2.0	2.3
"D"	4.0	5.0
"F"	7.0	

number of cycles and given prolonged trickle overcharge. Cells of this category are generally made in sizes comparable to conventional dry cells, such as "D", "C", "AA", etc. In sizes larger than the "D" cell, sealed lead–acid cells offer a useful economic alternative for applications where the extra weight and space of the lead–acid is not critical.

In order to reduce costs, achieve higher energy density, and minimize environmental problems, a growing percentage of prismatic and cylindrical cells are now made by pasting on substrate. The newer substrates include nickel fiber mat, nickel foam, and nickel-plated graphite fiber. One manufacturer uses a nickel-plated plastic fiber substrate for the fabrication of large nickel–cadmium cells in prismatic configurations (28). These cells are reported to require less maintenance than the older design pocket cells (48). In small industrial sizes, pasted cells are offered as sealed batteries.

Charger Technology. Alkaline storage batteries are commonly charged from rectified d-c equipment, solar panels, or other d-c sources and have fairly good tolerance to ripple and transient pulses. Because the voltage of the nickel electrode is variable, the cutoff voltage is not a good control parameter for nickel–cadmium cells. It is, however, often used in vented cell chargers. For sealed nickel–cadmium cells, and other systems where a combination mechanism exists, a negative voltage slope detector is often incorporated into the charger control circuit. As the charge of a cell or battery progresses there is a slow rise in the unit voltage. When the nickel electrode approaches full charge, the oxygen evolved combines with the cadmium electrode reducing the overpotential on the cadmium electrode. This slight change in cell voltage can be detected by electronic voltage slope detectors and used to reduce the charge current or shut off the charge.

Nickel–Iron Cells. The original tubular design nickel–iron battery developed by Edison has little commercial application. However, there has been renewed interest in the system for electric vehicles (EV). The EV design is based on a high rate, usually sintered, iron electrode as well as high rate nickel electrodes. Design and performance characteristics have been reported (49–54) as has data on pulse characteristics (55). The battery for electric vehicles is usually used in a 100–220 V array with approximately a 200 A·h capacity. The system is capable of high discharge pulses even at low states of charge. Testing data on EV regimes has been given (56–58).

Electrochemistry and Kinetics. The electrochemistry of the nickel–iron battery and the crystal structures of the active materials depends on the method of preparation of the material, degree of discharge, the age (life cycle), concentration of electrolyte, and type and degree of additives, particularly the presence of lithium and cobalt. A simplified equation representing the charge–discharge cycle can be given as:

$$2\ NiOOH^* + \alpha\text{-}Fe + 2\ H_2O \rightleftharpoons 2\ Ni(OH)_2^* + Fe(OH)_2$$

where the asterisks indicate adsorbed water and KOH. However, the discharge can be carried to a lower plateau for the iron electrode, although this is usually undesirable for life cycle, represented by:

$$8\ NiOOH^* + 3\ Fe + 4\ H_2O \rightleftharpoons 8\ Ni(OH)_2^* + Fe_3O_4$$

When discharges are carried out beyond the voltage range of the ferrous hydroxide [*18624-44-7*], $Fe(OH)_2$, plateau which is ca −0.90 to −0.85 V vs HgO, a second reaction in the voltage range of −0.65 to −0.5 V takes place (59,60).

$$3\ Fe(OH)_2 + 2\ OH^- \rightleftharpoons Fe_3O_4 + 4\ H_2O$$

Discharging to this lower cell voltage usually results in shorter cycle life. Enough excess iron should be provided in the cell design to avoid this problem.

Active iron in the metallic state is slowly attacked by the alkaline electrolyte according to

$$Fe + 2\ H_2O \longrightarrow Fe(OH)_2 + H_2$$

This reaction is accelerated by increased temperature, increased electrolyte concentration, and by the use of sodium hydroxide rather than potassium hydroxide in the electrolyte. It is believed that the presence of lithium and sulfur in the electrode suppress this problem. Generally, if the cell temperature is held below 50°C, the oxidation and/or solubility of iron is not a problem under normal cell operating conditions.

Electrode Structures. The classical iron active material for pocket and pasted iron electrodes was formed by roasting recrystallized ferrous sulfate [*7720-78-7*], $FeSO_4$, in an oxidizing atmosphere to ferric oxide [*1309-37-1*], Fe_2O_3, and then reducing the latter in hydrogen. The α-iron formed was then heated to a mixture of Fe_3O_4 and Fe. As such it was pure enough to be used for pharmaceutical purposes. For battery use, a small amount of sulfur, as FeS, was added as were other additives which were believed to increase the cycle life by acting as depassivating agents, ie, helping to reduce the tendency of iron to evolve hydrogen upon standing in alkaline electrolyte. Extensive studies on the stability of iron active material have been reported (61–63). Addition of cadmium and antimony salts have been claimed to decrease hydrogen evolution on stand by 50% (62). However, some blends also reduce electrode capacity. A blend of materials such as mucic acid [*526-99-8*], $C_6H_{10}O_8$, with indium [*7440-74-6*], In, increases the iron stability on stand without changing electrode capacity (63). The effects of electrolyte makeup and concentration on iron corrosion have also been studied (64).

A study of sintered iron electrodes claimed advantages of high rate capability, long life, and low hydrogen evolution (52). Fine carbonyl generated iron powder was coated on a nickel screen and sintered in a hydrogen–nitrogen atmosphere. Sintered raw plaques were believed suitable if the porosity was between 50 and 80%. The sinter had to be strong enough to leave a conductive skeleton after part of the iron was corroded to active material. This process is unlike sintered-type nickel electrodes in which active material is deposited directly into the pores of a very porous plaque. Sintered iron electrodes (49) are now in pilot production in Sweden and the United States.

Sintered nickel electrodes used in nickel iron cells are usually thicker than those used in Ni/Cd cells. These result in high energy density cells, because very high discharge rates are usually not required.

Performance Characteristics. The sintered nickel-sintered iron design battery has outstanding power characteristics at all states of discharge making them attractive to the design of electric vehicles (EV) which must accelerate with traffic even when almost completely discharged. Although the evolution of hydrogen is a problem preventing sealed cell design, introduction of automatic watering systems have ameliorated the maintenance time requirements. Typical performance curves are given in Figures 9 and 10 and the design characteristics of an advanced nickel–iron battery are given in Table 4 (49).

Silver–Zinc Cells

The silver–zinc battery has the highest attainable energy density of any rechargeable system in use as of this writing. In addition, it has an extremely high rate capability coupled with a very flat voltage discharge characteristic. Its use, in the early 1990s, is limited almost exclusively to the military for various aerospace applications such as satellites and missiles, submarine and torpedo propulsion applications, and some limited portable communications applications. The main drawback of these cells is the rather limited lifetime of the silver–zinc system. Life is normally limited to less than 200 cycles with a total wet-life of no more than about two years. The silver–zinc system also carries a very high cost and applications are justified only where cost is a minor factor. The high cost of silver battery systems is attributable to the cost of the active silver material used in the positive electrodes. In 1991 silver prices were in the range of $128/kg ($4/troy oz).

Cellophane or its derivatives have been used as the basic separator for the silver–zinc cell since the 1940s (65,66). Cellophane is hydrated by the caustic electrolyte and expands to approximately three times its dry thickness inside the cell exerting a small internal pressure in the cell. This pressure restrains the zinc anode active material within the plate itself and renders the zinc less available for dissolution during discharge. The cellophane, however, is also the principal limitation to cell life. Oxidation of the cellophane in the cell environment degrades the separator and within a relatively short time short circuits may occur in the cell. In addition, chemical combination of dissolved silver species in the electrolyte may form a conductive path through the cellophane.

A second lifetime limitation is the zinc anode. In spite of the separator and cell designs, some zinc material is solubilized during the charge–discharge reaction. Over a period of cycling there is a shift of active material, originally

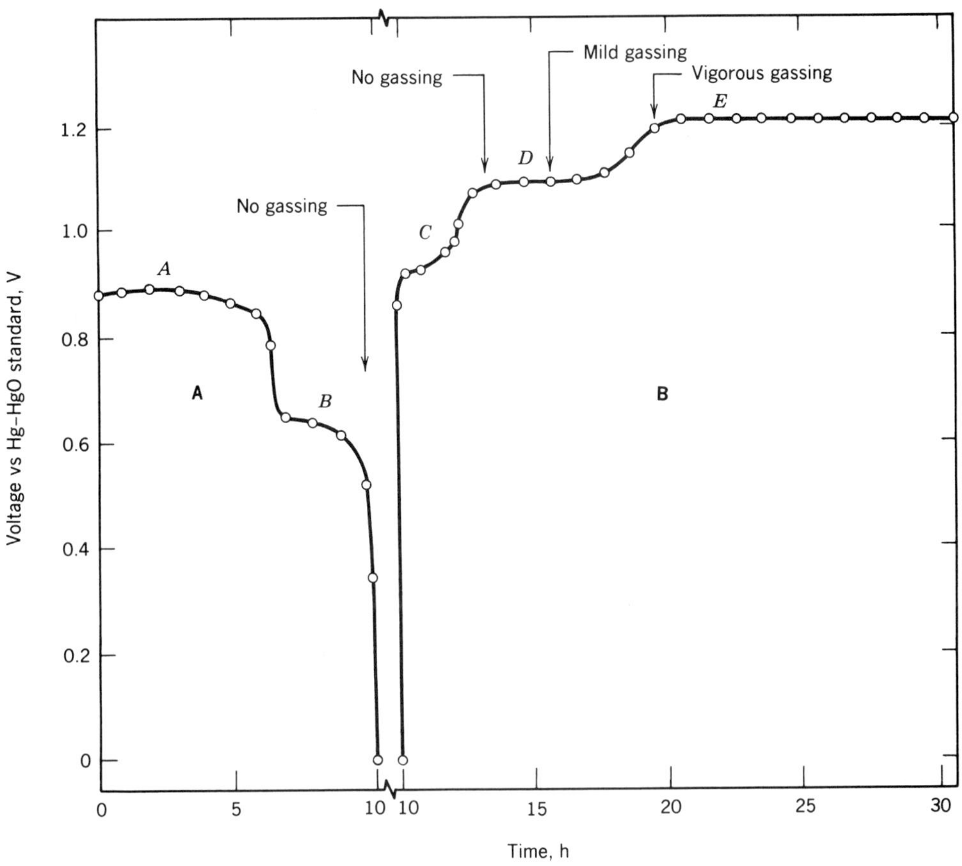

Fig. 9. Discharge and charging curves for a sintered iron electrode at a constant current of 0.2 A where the apparent geometrical surface area is 36 cm^2 and porosity is 65%. **A** and **B** represent the discharging and charging regions, respectively. Overall electrode reactions, midpoint potentials, and, in parentheses, theoretical potentials at pH 15 are *A*, α-Fe + 2 OH^- → $Fe(OH)_2$ + 2 e^-, 0.88 V (1.03 V); *B*, $Fe(OH)_2$ → FeOOH + H^+ + e^-, 0.63 V (0.72 V); *C*, FeOOH + H^+ + e^- → $Fe(OH)_2$, 0.93 V (0.72 V); *D*, $Fe(OH)_2$ + 2 e^- → α-Fe, 1.08 V (1.03 V); and *E*, 2 H_2O + 2 e^- → H_2 + 2 OH^-, 1.20 V (0.99 V).

distributed evenly over the face of the electrode, to the center and bottom areas of the electrode (50). This shape change limits the life of the cell as exemplified by a fading of the capacity and a build-up of internal pressure that may eventually lead to a short circuit.

Reaction Mechanisms. There is considerable difference of opinion concerning the specific cell reactions that occur in the silver–zinc battery. Equations that are readily acceptable are

$$2\,AgO + Zn + H_2O \rightleftharpoons Ag_2O + \begin{matrix} Zn(OH)_2 \\ (ZnO + H_2O) \end{matrix} \qquad E^\circ = 1.85\ V$$

and

$$Ag_2O + Zn + H_2O \rightleftharpoons 2\,Ag + \begin{matrix} Zn(OH)_2 \\ (ZnO + H_2O) \end{matrix} \qquad E^\circ = 1.59\ V$$

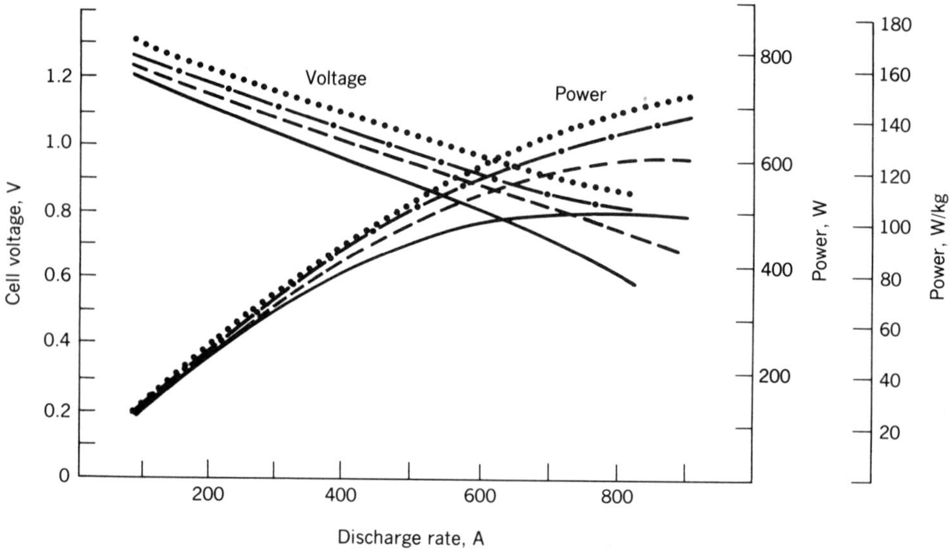

Fig. 10. Power and voltage characteristics of the nickel–iron cell where the internal resistance of the cell, R_i, is 0.70 mΩ, at various states of discharge: (• • •) 8%; (— • —) 32%; (– – –) 52%; and (———) 72%. Courtesy of Westinghouse Electric Corp.

Table 4. Component Data and Energy Density for 300 A·h Nickel Oxide–Iron Cells

	Cell types		
Parameter	Prototype	Near-term	Advanced
nickel–oxide electrodes			
thickness, mm	3.2	3.1	4.8 3.2
capacity, A·h	20	25	36 24
number $cell^{-1}$	15	12	7 2
iron electrodes			
thickness, mm	1.2	1.4	1.9
capacity, A·h	22	28	38
number $cell^{-1}$	14	11	8
separation			
distance, mm	0.8	0.8	0.5
weight, g			
nickel oxide electrodes	3000	2500	2500
iron electrodes	1330	1180	1090
electrolyte	1280	1100	920
separators	150	120	60
connections	320	320	320
cell case	300	280	250
Total	*6380*	*5500*	*5140*
volume, L	2.85	2.48	2.14
cell voltage, V[a]	1.18	1.21	1.21
energy density			
per weight, W·h/kg	55.5	66.0	70.6
per volume, W·h/L	124	146	170

[a]Mean value (C/4); where C is the current required to discharge the cell in one hour.

The charge and discharge of silver–zinc cells occurs on two voltage levels, generating silver peroxide and silver monoxide [*20667-12-3*]. Silver peroxide [*25455-73-6*] is often represented as Ag_2O_2, and the zinc hydroxide [*20427-58-1*] or zinc oxide [*1314-13-2*], ZnO, is represented as potassium zincate [*59271-74-8*], $K_2Zn(OH)_4$, or various other forms of hydrated zinc oxide. The two-level voltage of the silver–zinc system plays an important part in both charging and discharging of cells and batteries.

Electrochemistry. Silver–zinc cells have some unusual thermodynamic properties. The equations indicate that the higher valence silver oxide is AgO, silver(II) oxide [*1301-96-8*]. However, in the crystallographic unit cell, which is monolithic, there are four silver atoms and four oxygen atoms, and none of the Ag–O bonds conforms to a silver(II) bond length. Instead there are two Ag–O bonds of 0.218 nm corresponding to silver(I) and two Ag–O bonds of 0.203 nm corresponding to silver(III) (67). This structure has also been proposed on the basis of magnetic and semiconductor properties (67) and confirmed using neutron diffraction (68,69).

For the Ag–O material a reversible voltage of 1.856 V is obtained and the $(\partial E/\partial T)_P$ is positive, $+5.7 \times 10^{-5}$ V/°C. For the Ag_2O material a reversible voltage of 1.602 V is obtained in 11.6 *N* KOH but the $\partial E/\partial T$ is negative, -16.9×10^{-5} V/°C. This is about one-third the value of nickel–cadmium cells. However, because the negative coefficient compound is not present in the fully charged state, the high risk of thermal runaway described for nickel–cadmium cells does not exist. The high conductivity of the silver active material and the low internal resistance of the remaining components do, however, make thermal runaway a possibility to be considered in charger and system design.

Electrodes. All of the finished silver electrodes have certain common characteristics: the grids or substrates used in the electrodes are exclusively made of silver, although in some particular cases silver-plated copper is used. Material can be in the form of expanded silver sheet, silver wire mesh, or perforated silver sheet. In any case, the intent is to provide electronic contact of the external circuit of the battery or cell and the active material of the positive plate. Silver is necessary to avoid any possible oxidation at this junction and the increased resistance that would result.

Finished electrodes need fairly good physical strength so that they can be handled easily during separator wrapping and cell assembly. This is usually accomplished by a sintering process. Finished electrodes should also have a relatively high surface area per unit weight of active material coupled with an apparent porosity of about 50–60% based on the active material. The high surface area of the active material is attained by high surface area starting materials. These can be finely divided powders of metallic silver or either the monovalent or divalent oxides of silver. Silver electrodes can attain coulombic efficiencies of up to 85% of theoretical when manufactured from high surface area active material.

There are three methods of silver electrode fabrication: (*1*) the slurry pasting of monovalent or divalent silver oxide to the grid, drying, reducing by exposure to heat, and then sintering to agglomerate the fine particles into an integral, strong structure; (*2*) the dry processing of fine silver powders by pressing in a mold or by a continuous rolling operation onto a silver grid followed by sintering; and, (*3*) the use of plastic-bonded active material formed by imbedding the active material (fine silver powder) in a plastic vehicle such as polyethylene, which can then be milled into flexible sheets. These sheets are cut to size, pressed in a mold on both sides of a conductive grid, and the pressed electrode subjected to sintering where the plastic material is fired off, leaving the metallic silver. Silver electrodes produced by these processes range from 0.18 to 1.52 mm in thickness. Electrodes prepared by method (*1*) usually cannot be below 0.76 mm in thickness whereas the thinner electrodes can be prepared by methods (*2*) and (*3*).

Silver electrodes prepared by any of the three methods are almost always subjected to a sintering operation prior to cell or battery assembly. Sintering is basically a heating operation at temperatures well below the melting point of pure silver, 960.5°C, during which an agglomeration of the particles occurs, greatly strengthening the electrodes produced. Sintering is normally carried out in electric muffle furnaces, either on an individual batch process or a continuous conveyor belt-type operation. Reducing atmospheres are not necessary for silver electrode sintering because the process is carried out well above the decomposition temperatures of the various silver oxides. Sintering process parameters vary, however, examples are: 537°C for 30 min and 732°C for 3 min. The longer exposure at a somewhat lower temperature is said to produce electrodes that are less susceptible to shedding. Such a procedure, however, requires a furnace with a long heating zone, is more expensive, and is not practical to most manufacturers.

Zinc electrodes for secondary silver–zinc batteries are made by one of three general methods: the dry-powder process, the slurry-pasted process, or the electroformed process. Current-carrying grids for zinc electrodes can be the same regardless of the process of plate manufacture chosen. Expanded metal, screen, or perforated metal is the generally accepted form for these grids. Silver is the material of preference. However, cost considerations often dictate that copper be used. In these cases silver-plated copper forms are usually employed to avoid the possibility of copper dissolving in the caustic solution during over-discharge of a cell.

The active material used in any of the processes for the manufacture of electrodes is a finely divided zinc oxide powder, USP grade 12. The zinc oxide active material is usually blended with from 1–4% mercuric oxide in the dry state for any of the dry processing procedures. Mercuric oxide is converted to mercury during charge which then amalgamates the zinc formed at the same time. This tends to suppress the evolution of hydrogen on the zinc electrode during charged stand, and is required in most military applications to avoid a hydrogen hazard.

In the dry powder process the active material mix is spread evenly in a mold, the grid and lead assembly inserted, and the entire electrode then pressed in a hydraulic press. Often binder additives such as poly(vinyl alcohol) (PVA) or fine Teflon powder are added to the active material mix to increase the cohesiveness of the finished electrodes. Cellulosic paper liners are also used, and they are inserted into the mold prior to introduction of the active mix and grid. After the active mix and grid are introduced, the paper liner is then folded over the top of the electrode and the electrode pressed as before. Porosities obtained by pressing electrodes in this manner are usually in the range of 50%, although in actual production these figures can range from about 35–60% depending on the rate of discharge required. In the slurry or paste method a paste is prepared by mixing water with the active material mix (zinc oxide plus mercuric oxide). Sometimes a binder such as CMC, or a flock such as short rayon or Dynel fibers is added to the paste to increase the cohesiveness of the pasted electrode. Pasting is usually done on large strips of grid material.

After pasting, the strips are air dried at relatively low heats, individual electrodes are cut from the strip, and the electrodes are pressed to desired thicknesses. The porosities and densities of the active material made by the pasting processes are approximately the same as those indicated for the dry pressed powder processes. One danger in pasting electrodes is the occurrence of sharp grid edges on the electrodes after cutting to size. Often a secondary operation of smoothing or trimming is required to avoid this problem.

The electroforming or deposition of zinc from solution uses a slurry of zinc oxide in strong caustic solution or actual metallic zinc anodes in caustic solution. In the slurry method, the zinc oxide is deposited on silver or copper grids that comprise the grids of the finished electrodes. In the metallic zinc anode method, the solutions are not depleted but rather the anodes are depleted. Both methods utilize large sheets onto which the active zinc material is deposited. Often individual plate leads are welded in position prior to deposition. In other cases active zinc must be scraped from the grid in a secondary operation and the plate leads then welded into position. Zinc plates prepared by these deposition methods usually have the active material in a voluminous, mossy state. After plating, the sheets of deposited material must be washed to completely free them of any traces of caustic solution to avoid fire hazards during production and the subsequent drying operation.

After the plates have been washed and dried thoroughly, they are pressed in a preliminary operation to the desired thickness. Individual electrodes are then cut from the sheets and a secondary pressing operation to final thickness is done. Often a secondary operation is required to remove sharp edges of electrodeposited zinc electrodes.

Silver–Zinc Separators. The basic separator material is a regenerated cellulose (unplasticized cellophane) which acts as a semipermeable membrane allowing ionic conduction through the separator and preventing the migration of active materials from one electrode to the other.

A stronger separator is one made of sausage casing material (FSC), a regenerated cellulose similar to cellophane but including some fibrous material. FSC is usually extruded in tubes and electrodes are inserted into each end of the tube. The tube is folded to form the so-called U wrap.

Another method of extending the life of the cellulosic separators has been to incorporate a silver organic compound, eg, silver xanthate [*6333-67-1*], into the cellulosic separator. This is done by passing the separator material through a hot caustic solution containing a silver salt. The result is a deposition of a silver organic compound within the structure of the cellophane or FSC separator. This type of material resists the degradative effects of oxidation more than the untreated material. In general the positive electrodes are wrapped in the layers of separator. Normally, an absorber is used around each positive electrode to maintain an adequate supply of electrolyte. Absorbers are usually of nonwoven materials, such as polyamide or Dynel felts. In some cases, nonwoven cellulosic felts are also used, but these tend to degrade more rapidly. The absorber wrapped positive electrodes are usually wrapped in several layers of cellophane or FSC and then folded to form Us. Negative electrodes are alternately stacked forming the cell assembly. Normally ca three or four layers of PUDO (battery-grade) cellophane 0.025 mm thick are used in so-called high rate designs. For lower rate cells where longer life is required the number of layers of cellophane may be as many as ten. Occasionally cells are constructed with the negative electrodes wrapped in the separator material. This is said to reduce the shape change effect on the negative electrodes and increase life of the cell. These so-called reverse-wrap cells are usually reserved for low rate applications only.

Electrolyte. The electrolyte in silver–zinc cells is 30–45% KOH. The lower concentrations in this range have higher conductivities and are preferred for high rate cells. Higher concentrations have a less deleterious effect on cellulosic separators and are preferable for extended life characteristics. The higher concentrations also have a greater capacity for dissolving zinc oxide and accelerate change in shape of the zinc electrode. In most cases, a concentration of about 40% KOH is considered as the optimum. Occasionally, some manufacturers use as electrolyte a saturated zincate solution at the particular concentration desired. This supposedly slows the effect of zinc dissolution. However, in actual use such additives have not been shown to have much beneficial effect. Other additives, such as LiOH, which are used in other alkaline systems, have no beneficial effect on silver–zinc batteries. In practice the electrolyte is added to the cell in the discharged condition, and a period of soaking is allowed for the separators to absorb electrolyte and the electrodes to become thoroughly impregnated with electrolyte.

Cell Hardware. Cell jars are constructed almost exclusively of injection-molded plastics, which are resistant to the strong alkali electrolyte. The most generally used materials are modified styrenes or copolymers of styrene and acrylonitrile (SAN). Another material that has been found to increase shock resistance of cells is ABS plastic (acrylonitrile–butadiene–styrene). All of these plastics can be injection-molded, are solvent-sealable and, in general, meet operating temperature ranges up to about 70°C. For applications that require greater resistance to temperature, some of the more recent plastics such as polysulfone and poly(phenylene oxide) (PPO) injection-moldable materials able to withstand operating temperatures up to 150°C are used.

Cell terminal connections are usually brought out by two-threaded terminals that protrude through the cell jar cover. They are usually steel, brass, or copper with a hollow construction. The plate leads are soldered in place in the center hollow portion of the terminal to effect an electrical contact and cell seal. The terminal itself is potted into the jar cover using epoxy-type potting compounds. Normally, terminal hardware is silver-plated. However, for corrosion resistance nickel-plating has been used.

Performance. *Charging.* Charging of silver–zinc cells can be done by one of several methods. The constant-current method which is most common consists of a single rate of current usually equivalent to a full input within the 12–16-h period. A typical two-level constant-current charge curve results as shown in Figure 11. The first 20–30% of the A·h input is attained at slightly above the monovalent silver level of ca 1.65 V. After this input a sharp rise to the divalent level takes place, and the remainder of the charge is completed at a voltage somewhat above the divalent silver level of 1.90–1.95 V. When the silver electrodes have reached full charge, there is a sharp rise to about 2.0–2.05 V; at

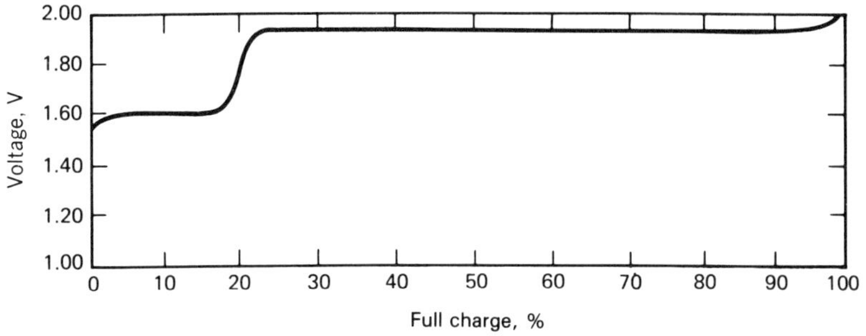

Fig. 11. Constant-current charge curve for a high rate Ag–Zn cell at room temperature. Charging carried out at the 10-h rate.

this point oxygen is liberated at the positive electrode. Normally the charge is terminated to avoid excessive overcharge of the zinc electrodes and further oxygen evolution. If the charge is continued beyond this point, charging voltage remains at approximately a 2.05–2.1 V level until all of the zinc material is fully charged. Then another rise in voltage occurs to ca 2.2–2.3 V, and both hydrogen and oxygen are liberated. Further charging results in electrolysis of the water in the electrolyte.

Other acceptable charging methods that have been used are the two-level charging, modified constant potential, and constant potential methods. The two-level method is one in which a relatively high rate of charge is used until a specified input is obtained, at which time the charging rate is lowered to a second level until a voltage cut-off is reached. In the modified constant potential method an initial charging rate is set, somewhat higher than would be used for a single-level constant, current charge; and then the charging current is allowed to drift downward as the battery voltage rises during charge. The constant potential systems are usually current-limited to avoid excessive inrush currents, and are essentially similar to modified constant potential methods in that the initial current is a very high value which rapidly decreases and approaches zero during final stages of charge.

Charge acceptance of the silver–zinc system is normally on the order of 95–100% efficient based on coulombic (ampere-hour output over input) values. This is true of any of the charging methods when carried out in the proper manner. Thus overcharge is rarely necessary in charging silver–zinc cells and batteries.

Discharge. Silver–zinc cells have one of the flattest voltage curves of any practical battery system known. However, there are two voltage plateaus. Even at rates as high as 10 minutes a fairly flat characteristic is obtained. The actual level of voltage is, of course, rate dependent. However, because of the high conductivity of the silver electrode, derating of voltage with higher rates is less than other systems. Figure 12 gives typical discharge curves for a high rate silver–zinc cell. At the low rates the initial part of the voltage discharge curve exhibits the higher level or peroxide voltage. After about 20–25% of the A·h capacity has been discharged, this drops to the monovalent level. As the rate increases, the proportion of the discharge capacity at the higher voltage level becomes less, and at very

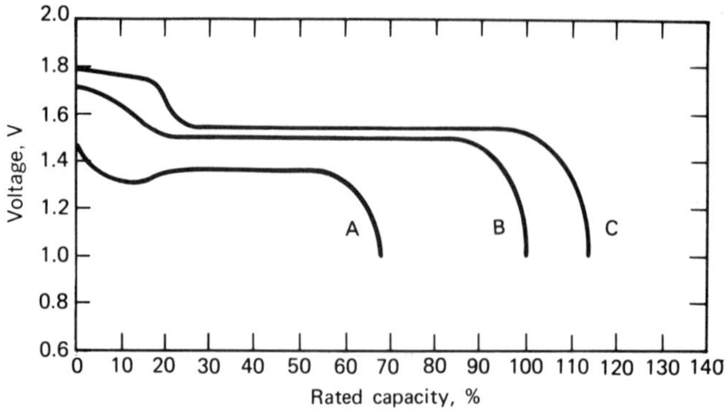

Fig. 12. Silver–zinc cell discharge curves at rates of A, 10 min; B, 1 h; and C, 10 h.

high rates a dip often occurs at the beginning of discharge as in the 10-min curve. The capacity of the silver–zinc cells is also rate dependent, but less derating occurs here than in most other batteries. For example, from Figure 12 it can be seen that even at the 10-min rate a high rate silver–zinc cell provides over 60% of its rated capacity.

Performance of silver–zinc cells is normally considered to be adequate in the temperature range of 10–38°C. If a wider temperature range is desired silver–zinc cells and batteries may be used in the range 0–71°C without any appreciable derating. Lower temperatures result in some reduction of cell voltage capacities at medium to high rates, and higher temperatures curtail life because of deterioration of the separator materials. Operation at temperatures below 0°C results in more serious derating and normally external heat is provided for operation in this range. Long use at temperatures above 71°C seriously curtails life.

Cell Life. Silver–zinc cells are usually manufactured as either low or high rate cells. Low rate cells contain fewer and thicker electrodes and have many layers of separator (up to the equivalent of 10 layers of cellophane). High rate cells, on the other hand, contain many thinner electrodes and have separator systems of the equivalent of three to four layers of cellophane. Approximately 10–30 cycles can be expected for high rate cells depending on the temperature of use, the rate of discharge, and methods of charging. Low rate cells have been satisfactorily used for 100–300 cycles under the proper conditions. In general, the overall life of the silver–zinc cell with the separator systems normally in use is approximately 1–2 yr.

Other Silver Positive Electrode Systems

Silver–Cadmium Cells. The first silver–cadmium batteries were manufactured in 1900 for motor cars. Use of this electrochemical system was quite limited. Then in the late 1950s, there was interest for applications such as appliances, power tools, and scientific satellites when it was hoped that silver–cadmium batteries could offer an energy density close to silver–zinc batteries and a life characteristic approaching that of the nickel–cadmium system. In satellite applications the nonmagnetic property of the silver–cadmium battery was of utmost importance because magnetometers were used on satellites to measure radiation and the effects of magnetic fields of energetic particles. Satellites had to be constructed of nonmagnetic components in sealed batteries.

The overall reactions are

$$2\,AgO + Cd + H_2O \rightleftharpoons Ag_2O + Cd(OH)_2 \qquad E^\circ = 1.38\text{ V}$$

$$Ag_2O + Cd + H_2O \rightleftharpoons 2\,Ag + Cd(OH)_2 \qquad E^\circ = 1.16\text{ V}$$

The silver–cadmium cell exhibits a two-plateau voltage characteristic on charge and discharge. At moderate discharge rates at 25°C, approximately 20% of the A·h capacity is delivered at a nominal 1.2 V; and the remaining capacity is delivered at 1.08–0.9 V. The discharge voltage characteristics are very sensitive to current rates, amount of cycling, charged stand time, float-charging, and temperature. For example, cells that are continuously float-charged exhibit a flat discharge voltage and no loss of capacity. At moderate charge rates, the voltage characteristic consists of two levels, 1.3 V and 1.5 V, nominal. Either constant current or constant potential charging is used. During cycling, A·h efficiencies are >95% and W·h efficiencies are <75%.

The positive plates are sintered silver on a silver grid and the negative plates are fabricated from a mixture of cadmium oxide powder, silver powder, and a binder pressed onto a silver grid. The main separator is four or five layers of cellophane with one or two layers of woven nylon on the positive plate. The electrolyte is aqeous KOH, 50 wt %. In the aerospace applications, the plastic cases were encapsulated in epoxy resins. Most useful cell sizes have ranged from 3 to 15 A·h, but small (0.1 A·h) and large (300 A·h) sizes have been evaluated. Energy densities of sealed batteries are 26–31 W·h/kg.

Silver–cadmium satellite batteries have been used in cyclic periods of five hours or more with discharge times of 30–60 min. Based on nominal capacities, depths of discharge have been 8–30%. The electrical performance of silver–cadmium batteries degrades below 0°C and above 40°C. At low temperatures the main problems are capacity maintenance during cycling and a dip in voltage on initiation of discharge. Operational and test programs have shown cycle life periods of 3 yr at low temperatures. At temperatures of 40°C and 50°C, the cycle life is 1 yr and 0.2 yr, respectively. The cycle life at intermediate temperatures is 1.4–2.0 yr.

Another application for silver–cadmium batteries is propulsion power for submarine simulator-target drones. High current drains are required (average C, pulses up to 6C), and greater recyclability than the silver–zinc counterparts used in torpedo propulsion. Batteries designed for this use utilize vented cells, high temperature plastic cell jars, and cell designs having a large number of thin electrodes to maximize electrode surface area.

Silver–Iron Cells. The silver–iron battery system combines the advantages of the high rate capability of the silver electrode and the cycling characteristics of the iron electrode. Commercial development has been undertaken (70) to solve problems associated with deep cycling of high power batteries for ocean systems operations.

Cells consist of porous sintered silver electrodes and high rate iron electrodes. The latter are enclosed with a seven-layered, controlled-porosity polypropylene bag which serves as the separator. The electrolyte contains 30% KOH and 1.5% LiOH.

Initial experiments were conducted with 350-A·h cells that maintained capacity over 200 cycles. Conventional silver–zinc cells lose capacity in similar deep-cycle operations. Cell tests conducted under a variety of temperature and pressure conditions revealed no discernable effects up to pressures of 69 MPa (10,000 psi) using a flooded electrolyte system. Tests at 0–25°C showed less than 5% capacity loss resulting from temperature when tested at the 8-h discharge rate. The cells as produced have a discharge voltage of ca 1.1 V. Voltage capacity characteristics of a nominal 140-A·h silver–iron cell are shown in Figure 13. The discharge capacity of a typical cell is 145–155 A·h at discharge rates of 80–10 A. A battery of 24 series-connected cells (containing excess electrolyte) weighs 40 kg and provides an energy density of 3.5 kW·h. At the 3-h rate the energy density is 100 W·h/kg, or about three and a half times the energy density of comparable lead–acid batteries.

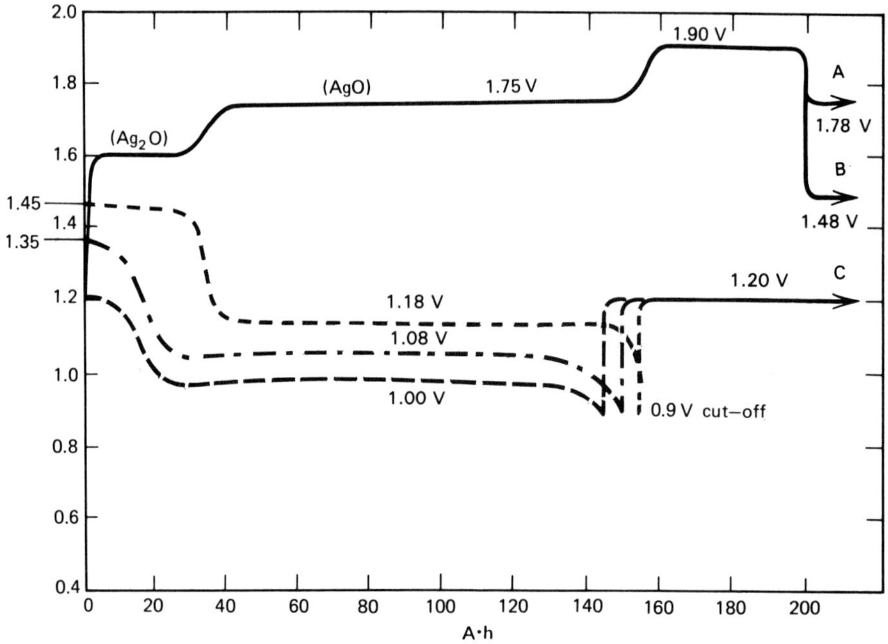

Fig. 13. Charge–discharge characteristics of a nominal 140-A·h silver–iron cell where the charge (———) is at 25 A for 8 h, A represents a 0.25 A float charge; B an open circuit; and the discharge at open circuit after 1 h is shown for (- - -) 10 A, (— - —) 45 A, and (———) 80 A; and C represents the open circuit discharge.

Applications have been found for these batteries in emergency power applications for telecommunications systems in tethered balloons. Unfortunately, the system is expensive because of the high cost of the silver electrode. Applications are, therefore, generally sought where recovery and reclamation of the raw materials can be made.

Small silver–iron sealed button cells have been produced in Sweden (70) for long-life operation of hearing aids, calculators, and electric razors. These units have energy densities similar to silver–zinc button cells but are reported to be capable of being fully charged and discharged up to five hundred times without leakage or change in performance. Comparable performance silver–zinc cells have limited cycle life of the order of 75–100 cycles.

Nickel–Zinc Cells

Nickel–zinc cells offer potential advantages over other rechargeable alkaline systems. The single-level discharge voltage, 1.60–1.65 V/cell is approximately 0.35–0.45 V/cell higher than nickel–cadmium or nickel–iron and approximately equal to that of silver–zinc. In addition, the use of zinc as the negative electrode should result in a higher energy density battery than either nickel–cadmium or nickel–iron and a lower cost than silver–zinc. In fact, nickel–zinc cells having energy densities in the range of 40–60 W·h/kg have been successfully demonstrated.

Work in the 1930s resulted in a rechargeable nickel–zinc railroad battery utilizing standard Edison tubular or Jungner pocket nickel positives and electrodeposited zinc negatives. Cells were constructed utilizing physical separations between the plates and containing a large excess of KOH electrolyte. This Drumm battery exhibited a limited life caused by negative shape change and premature short circuits. Shape change is a phenomenon caused by zinc replating in a nonuniform manner during charge and resulting in a loss of negative capacity. Short circuits resulted from zinc dendrites formed during discharge that grew perpendicular to the face of the negative electrode until electrical contact was made with the positive. In the late 1950s and 1960s interest in the nickel–zinc system was again aroused as a possible substitute for silver–zinc batteries (71–75). By then the technology of nickel and zinc electrodes was well developed and led to a higher energy density nickel–zinc battery having tightly packed electrodes, barrier separators, and limited electrolyte. Most of this work related to vented batteries for low to medium rate applications, such as portable military communication applications.

Some efforts toward sealed battery development (76) were made. However, a third electrode, an oxygen recombination electrode was required to reduce the cost of the system. High rate applications such as torpedo propulsion were investigated (77) and moderate success achieved using experimental nickel–zinc cells yielding energy densities of 35 W·h/kg at discharge rates of 8 C. A commercial nickel–zinc battery is considered to be the most likely candidate for electric vehicle development (78,79). If the problems of limited life and high installation cost ($100–150/kW·h) are solved, a nickel–zinc EV battery could provide twice the driving range for an equal weight lead–acid battery. Work is developmental; there is no commercial production of nickel–zinc batteries.

Reaction Mechanism. The overall reactions in the nickel–zinc cell can be represented by

$$2\ \text{NiOOH} + \text{Zn} + 2\text{H}_2\text{O} \rightleftharpoons 2\ \text{Ni(OH)}_2 + \text{Zn(OH)}_2 \qquad E^\circ = +1.73\ \text{V}$$

Alternatively the discharged state of the zinc electrode is represented as ZnO.

Cell Construction. Nickel–zinc batteries are housed in molded plastic cell jars of styrene, SAN, or ABS material for maximum weight savings. Nickel electrodes can be of the sintered or pocket type, however, these types are not cost effective and several different types of plastic-bonded nickel electrodes (78–80) have been developed.

Nickel hydrate, usually 5–10% cobalt added, serves as the active material and is mixed with a conductive carbon, eg, graphite. The active mass is mixed with an inert organic binder such as polyethylene or poly(tetrafluoroethylene) (TFE). The resultant mass is rolled into sheets on a compounding mill or pressed into electrodes as a dry powder on a nickel grid. The resultant electrodes offer a high energy density and low cost of fabrication. In performance, the plastic-bonded electrodes can support rates up to C/2, discharge capacity in 2 h, at voltages equivalent to those of sintered electrodes, but at higher rates some derating occurs. Life of plastic-bonded electrodes has not been fully evaluated, but a life of 500 cycles appears attainable.

Negative electrodes are fabricated of zinc oxide by any of the methods (pasting, pressing, etc) described. Binders, usually TFE, are used to reduce the solubility of the electrode in KOH. In addition, other techniques such as extended edges, inert extenders, contouring, and variable density have been tried in an effort to reduce shape change of the negative electrode upon cycling. The electrolyte is KOH, usually a 30–35% aqueous solution with the addition of LiOH at a level of 10–25 g/L to enhance nickel capacity throughout life.

Separators are both of the organic and inorganic type. During the 1960s most cells were built using cellophane or FSC (fibrous sausage casing) separators. Developments by NASA (81) and others in the area of inorganic films, consisting of a layered or film structure of a heavy metallic oxide such as zirconia, ZrO_2, or magnesia, MgO, bonded by an inert organic film and often layered with an organic resin-impregnated asbestos mat led to use of these materials as separators. In Ni–Zn electric vehicle batteries (82), over 800 cycles at 50% depth are claimed for these inorganic separators. Drawbacks are the relatively high cost of manufacture and the increased resistance per layer, which in effect limits discharge rate.

Performance. The limited life of nickel–zinc batteries is the principal drawback to widespread use. Normally the nickel cathode is not a factor. Even plastic-bonded nickel electrodes perform for a greater number of cycles than either the zinc counter electrodes or the separator used and to a great extent it is the zinc electrode that limits the life of a nickel–zinc cell. In order to charge nickel electrodes sufficiently, a discreet amount of overcharge is necessary. Unfortunately, overcharge is not desirable for the zinc electrodes, as it promotes dendritic growth which will eventually penetrate the separator and cause short circuits. This can be alleviated somewhat by an overdesign in zinc capacity, but that reduces energy density. Typical cell and electrode voltage curves are shown in Figure 14 and cycle life data as a function of depth of discharge in Figure 15.

Nickel–zinc batteries containing a vibrating zinc anode has been reported (83). In this system zinc oxide active material is added to the electrolyte as a slurry. During charge the anode substrates are vibrated and the zinc is electroplated onto the surface in a uniform manner. The stationary positive electrodes (nickel) are encased in a thin, open plastic netting which constitutes the entire separator system.

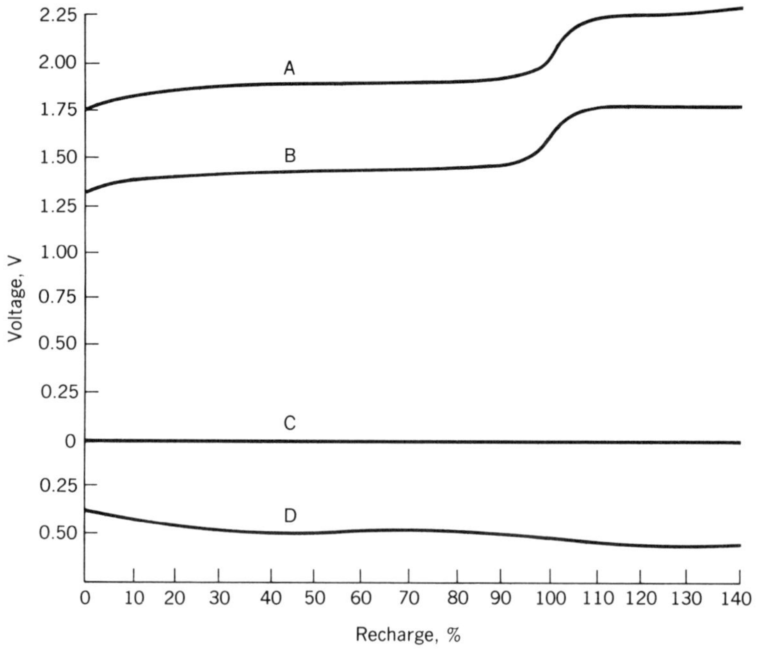

Fig. 14. Nickel–zinc electrode potential on charge A, the Ni–Zn cell; B, the Zn vs Hg ref.; C, the Hg–HgO ref.; and D, the Ni vs Hg ref.

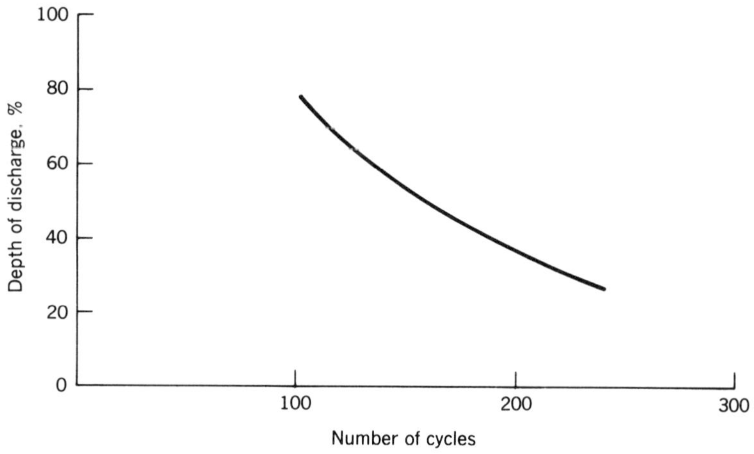

Fig. 15. Cell performance as a function of the number of cell recharge cycles.

The vibration serves a dual purpose. First, a macroturbulence is created which keeps the ZnO uniformly dispersed in the electrolyte. Second, the microturbulence created at the surface of the negative electrode minimizes the zincate concentration gradients, discouraging the formation of zinc dendrites and causing a uniform deposit of zinc. Any zinc dendrites that may form are brushed off against the plastic netting by the vibratory motion. Thus the problems of shape change, dendritic shorting, and separator failure are in theory solved by this system; the zinc electrode dissolves during discharge, replates during charge, and no separator is required. A disadvantage is reduced energy density, especially on a volume basis, as a result of the increased electrode spacing and the quantity of electrolyte required. The vibration hardware imposes an additional 5% weight penalty as well as increased cost. Alternatively, similar benefits have been reported from experiments in which the electrolyte is pulsed.

Nickel–Hydrogen Cells

There are two types of nickel–hydrogen cells; those that employ a gaseous H_2 electrode and those that utilize a metal hydride, MH.

Gaseous Hydrogen Systems. The nickel–hydrogen cell incorporating a gaseous hydrogen electrode is a hybrid consisting of one gaseous and one solid electrode. The nickel electrode is of the type used in a nickel–cadmium battery and the hydrogen electrode is a gas diffusion electrode of the type used in alkaline fuel cells (qv). These two electrodes are capable of extremely long, stable life. This system was developed to serve as a long life, lightweight battery for satellite applications that would be superior to the aerospace nickel–cadmium batteries (84–86).

The couple has a theoretical energy density of 172 W·h/kg and complete cells are capable of delivering 55–66 W·h/kg. The cell reaction is

$$NiOOH + \tfrac{1}{2}\,H_2 \rightleftharpoons Ni(OH)_2$$

However, the generation and migration of water in the half-cell reactions must be considered in the cell design. At the nickel electrode:

$$NiOOH + H_2O + e^- \rightleftharpoons Ni(OH)_2 + OH^-$$

and at the hydrogen electrode:

$$\tfrac{1}{2}\,H_2 + OH^- \rightleftharpoons H_2O + e^-$$

During charge the nickel hydroxide is converted to NiOOH, the charged state of nickel, and on the surface of the hydrogen electrode, hydrogen gas is evolved. By placing the electrode stack in a sealed container, the hydrogen is captured for subsequent reuse. During discharge the same hydrogen is reconsumed on the same electrode surface and the nickel electrode is reduced to provide electric energy.

Another desirable feature of this battery system is its capability of high rate of overcharge. During overcharge oxygen gas is generated on the surface of the nickel electrode. Simultaneously hydrogen continues to be evolved on the surface of the hydrogen electrode. Because there is a large area of catalyzed hydrogen electrode and ready access of the oxygen to diffuse to that surface, the oxygen readily recombines within the cell to form water and the cell pressure remains constant.

The primary packaging arrangements for this chemistry has been single cylindrical cells as shown in Figure 16, where a stack of disk electrodes are placed within a cylindrical outer housing having domed end caps. The housing also

Fig. 16. Cutaway view of a typical construction of a nickel–hydrogen cell.

serves as a lightweight pressure vessel for hydrogen containment with all the free volume inside the housing used for hydrogen storage. Two insulated feed throughs are provided in the cell housing for electric contact to the positive and negative electrodes. The electrode stack is a repeating sequence of sintered nickel, absorber separator, typically asbestos, or other inorganic stable materials, teflon bonded platinum black fuel cell-type electrodes, and gas spacers. The electrolyte is typically 30–35% KOH with a LiOH additive for the nickel electrode.

Figure 17 shows a typical charge–discharge voltage and pressure profile for a 50 A·h cell. In this design the cell is precharged at 517 kPa (75 psi) hydrogen and the operating pressure range is from 517 to 4100 kPa (75–600 psi). Monitoring cell pressure, which is typically done with pressure transducers, enables the user to follow the cells state of charge. This is an additional desirable feature of this system. Figure 18 shows cell discharge characteristics as a function of rate, demonstrating the high rate capability of this battery. Figure 19 gives open circuit stand characteristics. This cell exhibits a greater self-discharge than the nickel–cadmium chemistry because of the reaction of pressurized hydrogen with the active nickel material.

Because this battery is only produced for special satellite applications, production quantities are limited. Rigorous quality inspections, expensive lightweight housing and seals, and the use of high loading platinum hydrogen electrodes all make this type of battery very expensive. For typical satellite applica-

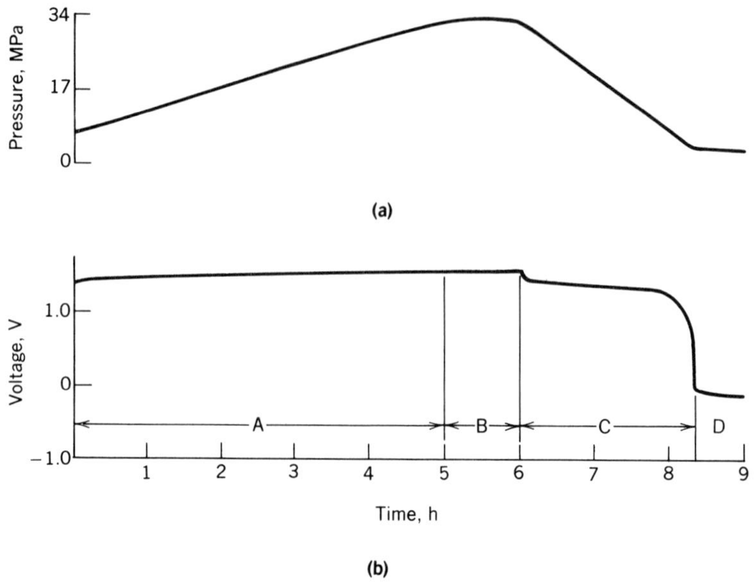

Fig. 17. 50 A·h nickel–hydrogen performance showing (**a**) pressure and (**b**) voltage curves where region A represents charging at 10 A, region B represents overcharge at 10 A, region C represents discharge at 25 A, and region D represents reversal at 25 A. To convert MPa to psi, multiply by 145.

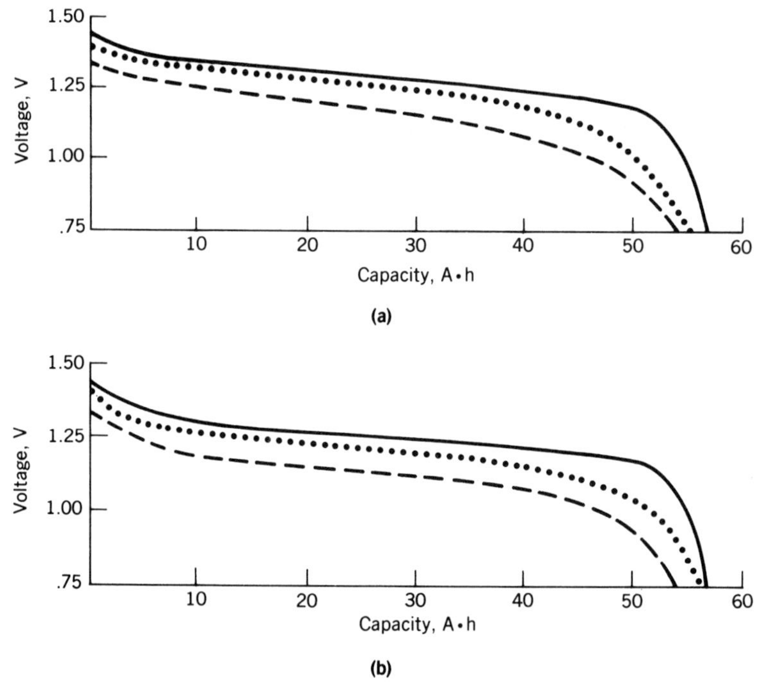

Fig. 18. Discharge characteristics for 50 A·h nickel–hydrogen batteries at (——) 10 A, (• • •) 25 A, and (– – –) 50 A at (**a**) 20°C and (**b**) 0°C.

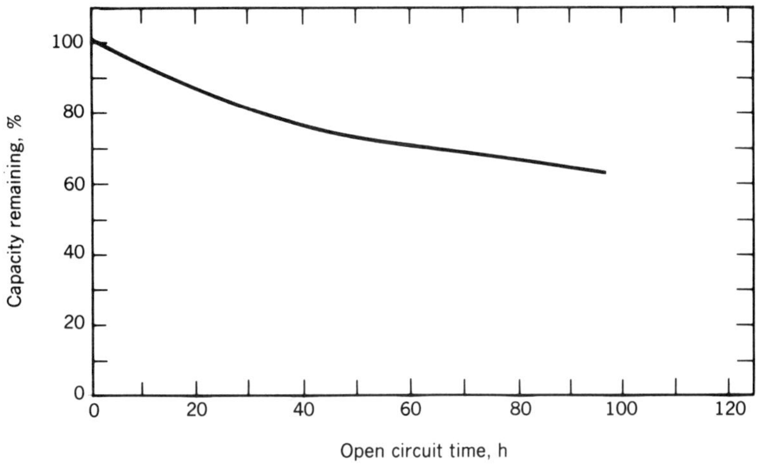

Fig. 19. Self-discharge at ambient temperatures for a 35 A·h cell, NTS-2 prototype Sanyo Electric Co. cell.

tions multiple single-cells are connected in series and are packaged in an egg crate mounting. In this arrangement the waste heat generated during discharge and overcharge is conducted from the stack out the cylinder wall through conduction rings to a radiator plate of the satellite.

Limited development efforts have been undertaken to develop a lower cost battery for terrestrial use utilizing reduced catalyst quantities and multicell packaging. Multicell common pressure vessel arrangements have also been considered for aerospace applications, but these configurations have not found market acceptance.

Metal Hydride Systems. The success of the gaseous nickel–hydrogen system led to the investigation of replacing the gaseous hydrogen with metal hydrides in order to reduce the cell pressure and the volume required for hydrogen storage. A number of metal hydrides were developed for reversible hydrogen storage. Of particular interest were $LaNi_5$ and $MnNi_5$ (Misch metal [*8049-20-5*]) the isotherms of which are shown in Figures 20 and 21, and FeTi. In the initial efforts these materials were packed into the free volume of standard nickel hydrogen cells using standard catalyzed gas diffusion hydrogen electrodes. The metal hydrides absorb up to one hydrogen atom per metal atom and cells incorporating hydrides could be fabricated that were more compact than gas pressure cells. However these hydrides deactivated upon repeated deep-cycling and gradually lost their hydrogen absorption ability presumably because of attack on the hydrides by water vapor or oxygen gas within the cell environment.

An alternative approach utilizes the hydride material as the hydrogen electrode. In this case, as the hydrogen is generated on the hydrogen electrode, it enters the hydride lattice for storage. On discharge the hydrogen leaves the hydride for reaction. The cell reactions are essentially the same as gaseous nickel hydrogen cells.

$$Ni(OH)_2 + OH^- \rightleftharpoons NiOOH + H_2O + e^-$$

$$M + H_2O + e^- \rightleftharpoons MH + OH^-$$

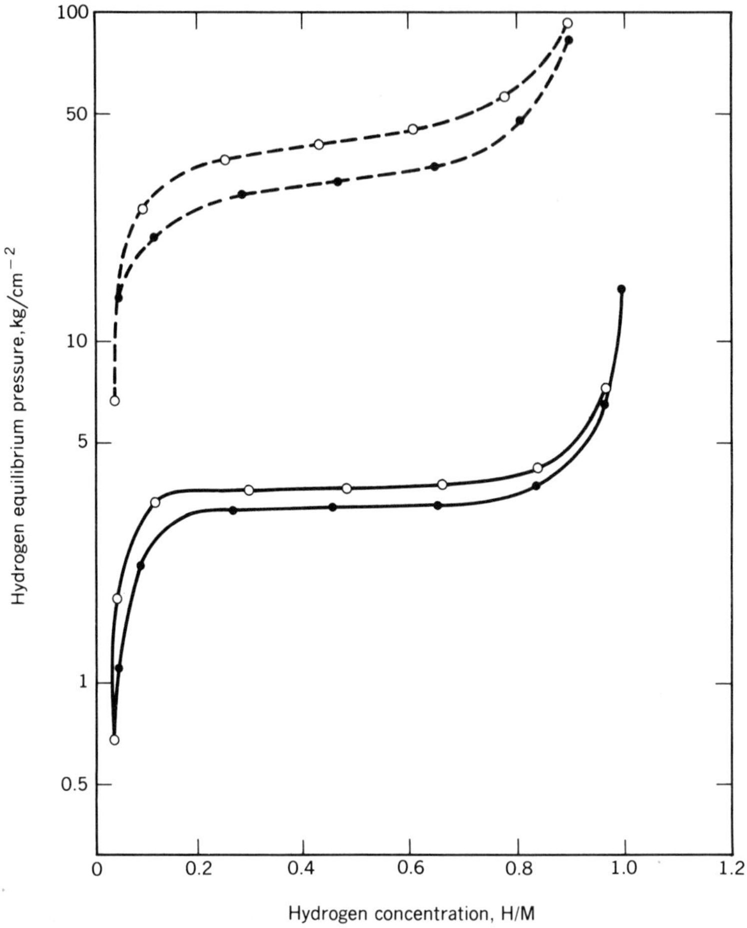

Fig. 20. Pressure concentration curves of $MmNi_5$ (---) and $LaNi_5$ (—) at 45°C where open circles denote absorption and closed circles desorption of hydrogen. H/M represents the ratio in the hydride of the mole fraction of hydrogen to the mole fraction of the metal.

and overall

$$Ni(OH)_2 + M \rightleftharpoons NiOOH + MH$$

The overcharge reactions for the cell are the same as for nickel–cadmium and nickel–hydrogen cells. The oxygen generated on the nickel electrode at the end of charge and overcharge finds its way to the anode and reacts to form water in the $Ni-H_2$ case and $Cd(OH)_2$ in the Ni–Cd case.

A critical issue is the stability of the hydride electrode in the cell environment. A number of hydride formulations have been developed. Table 5 shows hydride materials that are now the focus of attention. Most of these are Misch metal hydrides containing additions of cobalt, aluminum, or manganese. The hydrides are prepared by making melts of the formulations and then grinding to fine powers. The electrodes are prepared by pasting and or pressing the powders

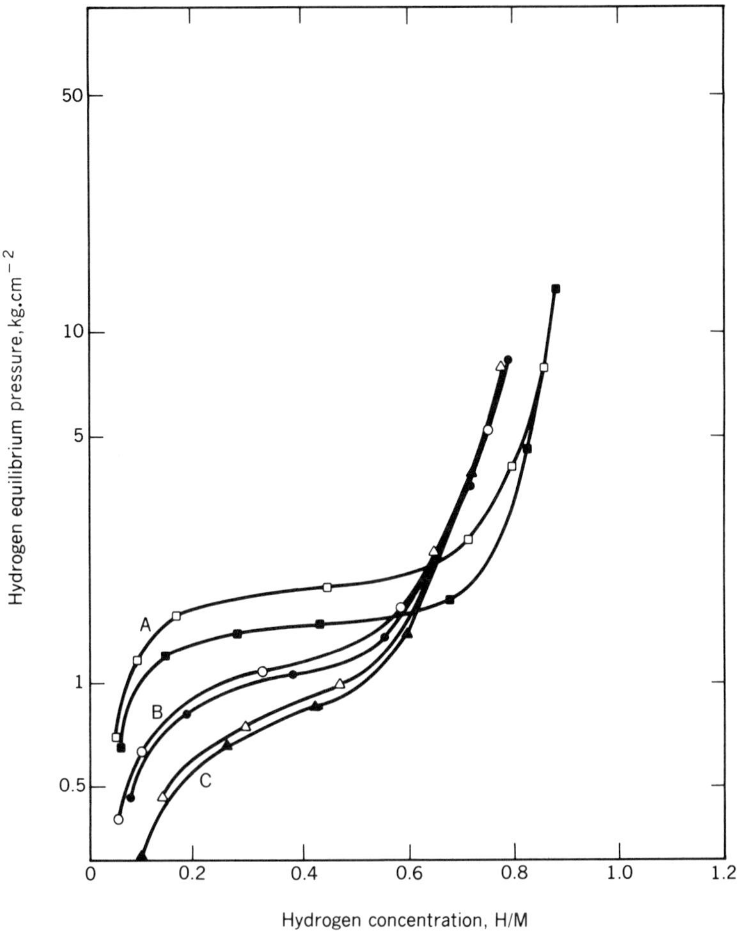

Fig. 21. Pressure constant temperature (PCT) curves of $MmNi_5$ alloy system where open symbols represent absorption and closed symbols represent desorption for A, $MmNi_{4.3}Mn_{0.4}Al_{0.3}$; B, $MmNi_{3.8}Mn_{0.4}Al_{0.3}Co_{0.5}$; and C, $MmNi_{3.5}Mn_{0.4}Al_{0.3}Co_{0.75}$. H/M represents the ratio in the hydride of the mole fraction of hydrogen to the mole fraction of the metal.

Table 5. International MH Alloys

Designation[a]	Composition[b]
IBA MH no. 1	$MmNi_{3.55}Co_{0.75}Mn_{0.4}Al_{0.3}$
IBA MH no. 2	$MmNi_{3.5}Co_{0.7}Al_{0.8}$
IBA MH no. 3	LmNi alloy
IBA MH no. 4	$Ti_{1.6}V_{2.2}Zr_{1.6}Ni_{4.2}Cr_{0.7}$
IBA MH no. 5	$MmNi_{3.5}Co_{0.7}Al_{0.8}$
IBA MH no. 6	$MmNi_{3.5}Co_{0.8}Mn_{0.4}Al_{0.3}$

[a]Standard battery materials are maintained by The International Battery Materials Association (IBA) (87).
[b]Mm is Misch metal; Lm is lanthanum.

into metal screens or felt. The additives are reported to retard the formation of passive oxide films on the hydrides.

A number of manufacturers started commercial production of nickel–MH cells in 1991 (31–35). The initial products are "AA"-size, "Sub-C", and "C"-size cells constructed in a fashion similar to small sealed nickel–cadmium cells. Table 6 compares the Ovonics experimental cell and a similar sized nickel–cadmium cell. Ovonics also delivered experimental electric vehicle cells, 22 A·h size, for testing. The charge–discharge of "AA" cells produced in Japan (Matsushita) are compared in Figure 22.

Table 6. Specification of "C" Cells

Parameter	Ni–MH cell[a]	Ni–Cd cell
capacity A·h (nominal)	3.5	2.0
voltage, V	1.2	1.2
cycles[b]	>450	>450
memory effect limitation	none	severe
toxicity of material	No, Cd, Pb, Hg, Li	Cd
operating temperature, °C	−20 to +50	−40 to +60[c]
discharge rate, A	up to 12–15	up to 15
internal resistance, Ω (ANSI method)	25	13–20
charge retention, %[d]	50	60–70
overcharge capability[e]	continuous up to C/5 rate (700 mA)	continuous up to C/5 rate (400 mA)
over–discharge capability	20%	3%

[a]From Ovonics.
[b]One hour charge/one hour discharge at 100% depth of discharge.
[c]Special design required at +60.
[d]Capacity retained in 30 d at 25°C.
[e]Ni–MH cell 1-h charge (3.5 A) to thermal cutout followed by trickle charge of 0.350 A is acceptable. The Ni–Cd charger can be used with these cells. Ni–Cd: 1-h charge (2.0 A) to thermal cutout followed by trickle charge of 0.200 A is acceptable.

From these data, the hydride cells contain approximately 30–50% more capacity than the Ni–Cd cells. The hydride cells exhibit somewhat lower high rate capability and higher rates of self-discharge than nickel–cadmium cells. Life is reported to be 200–500 cycles. Though not yet in full production it has been estimated that these cells should be at a cost parity to nickel–cadmium cells on an energy basis.

Other Cell Systems

Silver–Hydrogen Cells. With the development of the nickel–hydrogen system limited attention was directed to the development of a silver–hydrogen cell (89,90). The main characteristics of interest were the potential for a higher gravametric energy density based on the ligher weight of the silver electrode vs

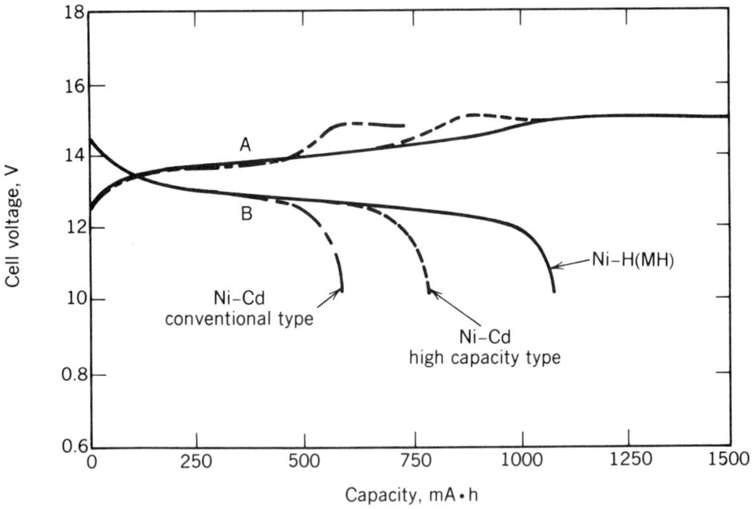

Fig. 22. Charge A and discharge B curves of (——) a Ni–H(MH) cell employing $MmNi_{3.55}Mn_{0.4}Al_{0.3}Co_{0.75}$ alloy compared with those of (— — —) conventional and (- - -) high capacity types of Ni–Cd cells (88).

that of the nickel. The cell reactions for this couple are

Hydrogen electrode

$$2\,H_2O + 2\,e^- \rightleftharpoons H_2 + 2\,OH^-$$

Silver electrode

$$Ag + OH^- \rightleftharpoons \tfrac{1}{2}\,Ag_2O + \tfrac{1}{2}\,H_2O + e^-$$

$$\tfrac{1}{2}\,Ag_2O + OH^- \rightleftharpoons AgO + \tfrac{1}{2}\,H_2O + e^-$$

Overall reaction

$$H_2 + AgO \rightleftharpoons Ag + H_2O$$

The packaging approach utilized for this battery is similar to that for nickel–hydrogen single cylindrical cells as shown in Figure 23. The silver electrode is typically the sintered type used in rechargeable silver–zinc cells. The hydrogen electrode is a Teflon-bonded platinum black gas diffusion electrode.

Because the silver oxide electrode is slightly soluble in the potassium hydroxide electrolyte the separator is of a barrier type to minimize silver diffusion to the opposite electrode. Figure 24 shows a charge–discharge profile of a 25 A·h cell that exhibits the two voltage plateaus seen for silver electrodes in the silver–zinc battery system. The silver–hydrogen cell exhibits a lower self-discharge than nickel–hydrogen cells as a result of the slower rate of reaction of hydrogen with silver oxide. The actual cells do not exhibit substantial differences in energy density when compared to those of nickel–hydrogen. Therefore this system is not being actively pursued.

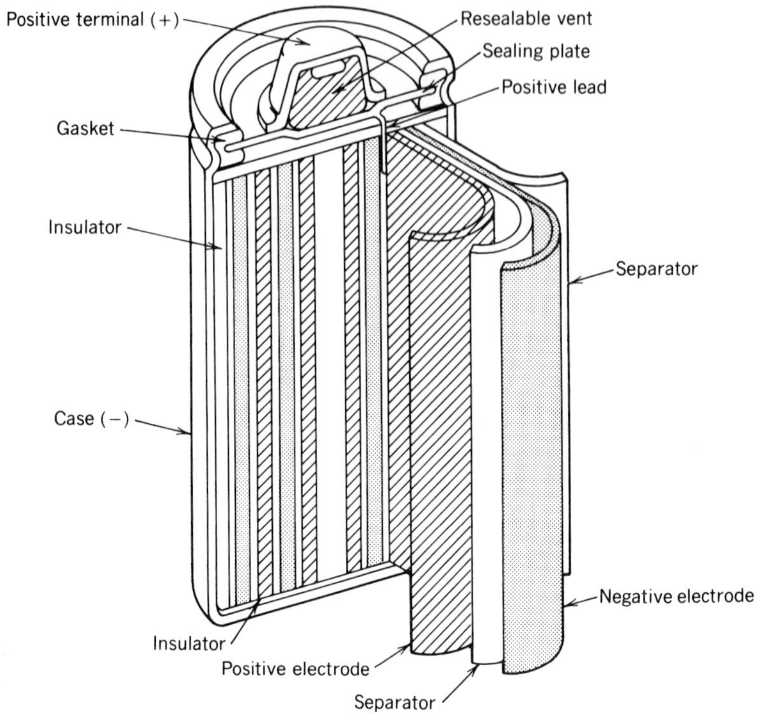

Fig. 23. Schematic diagram of a Ni–H(MH) cell (88).

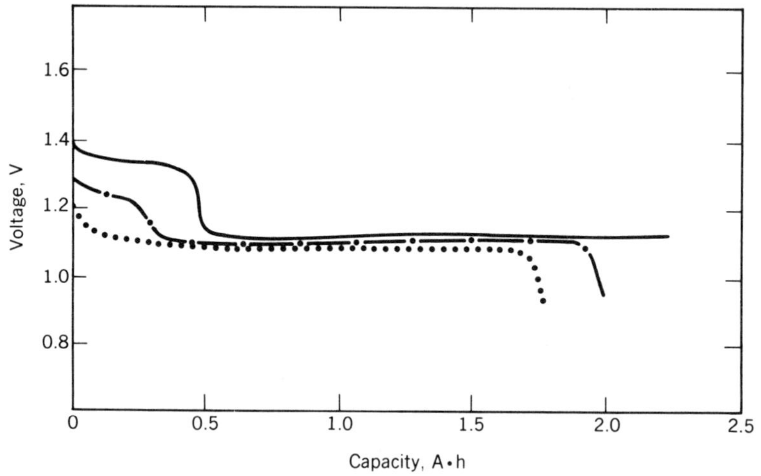

Fig. 24. Silver–hydrogen cell discharge characteristics where (• • •) represents a 0.5 h rate at 4 A or 80 mA/cm^2; (— • —) represents a 1 h rate at 2 A; and (———) represents a 4 h rate at 0.5 A or 10 mA/cm^2.

Zinc–Oxygen Cells. On the basis of reactants the zinc–oxygen or air system is the highest energy density system of all the alkaline rechargeable systems with the exception of the $H_2 \cdot O_2$ one. The reactants are cheap and abundant and therefore a number of attempts have been made to develop a practical rechargeable system. The reactions of this system are as follows:

Zinc electrode

$$Zn + 2\ OH^- \rightleftharpoons Zn(OH)_2 + 2\ e^-$$

Oxygen electrode

$$\frac{1}{2}\ O_2 + H_2O + 2\ e^- \rightleftharpoons 2\ OH^-$$

Overall reaction

$$Zn + H_2O + \frac{1}{2}\ O_2 \rightleftharpoons Zn(OH)_2$$

In open cycle systems the oxygen reactant is released into the surrounding air during charge, and during the subsequent discharge oxygen from the surrounding air is consumed on a fuel cell-type electrode. The oxygen is delivered by forced or free confection depending on the system design. Whereas the zinc–oxygen cell has significant potential, it also has a number of inherent problems. First is the poor rechargeability of the normal zinc electrode; then there is the poor stability of the oxygen electrode when used in a bifunctional, ie, charge and discharge mode, contamination of the electrolyte by carbon dioxide from the air when used as an open cycle system, and poor retention of energy efficiency because of the irreversibility of the oxygen electrode reaction.

The system shown in Figure 25, in which the electrolyte was circulated through a multicell stack to improve the rechargeability of the zinc electrode has been studied by Sanyo Electric. However, such circulating systems are prone to electrolyte leakage and have common manifolds that cause internal self-discharge. Systems in which zinc particles or zinc-coated beads are circulated through the electrode stack and are utilized as a fluidized electrode have also been investigated by many organizations. In these systems the zinc particles contact the anode current collector to undergo reaction. In order to dissolve all

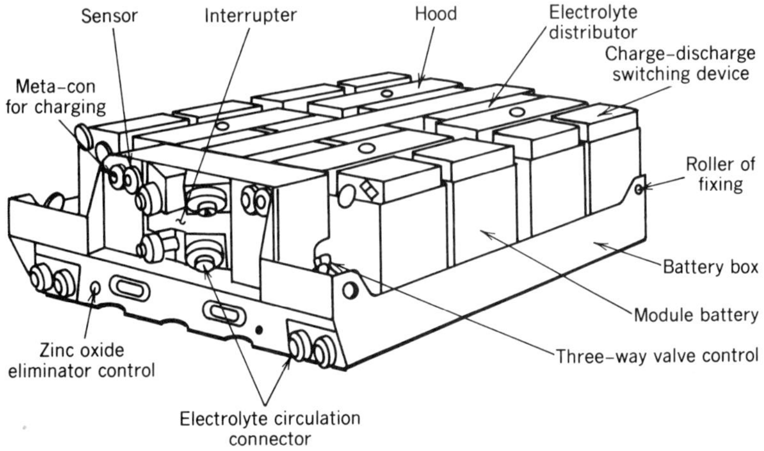

Fig. 25. Zinc–air 124-V battery system. Meta-con is a connector. Courtesy of Sanyo Electric Co.

the reaction product during discharge, it is necessary to use relatively large quantities of electrolyte so that a zincate solubility of approximately 200 g/L is not exceeded. The CGE (91) (Fig. 26) system utilized a tubular cell and a separate regeneration stack to deposit new zinc particles from the discharged electrolyte, which is saturated with dissolved zincate. By utilizing a separate regeneration stack the stability problem of the bifunctional oxygen electrode is avoided and regeneration can be carried out at any location so that the battery can be refueled rapidly. A sealed system in a configuration similar to single cylindrical gaseous nickel–hydrogen cells has also been studied (92).

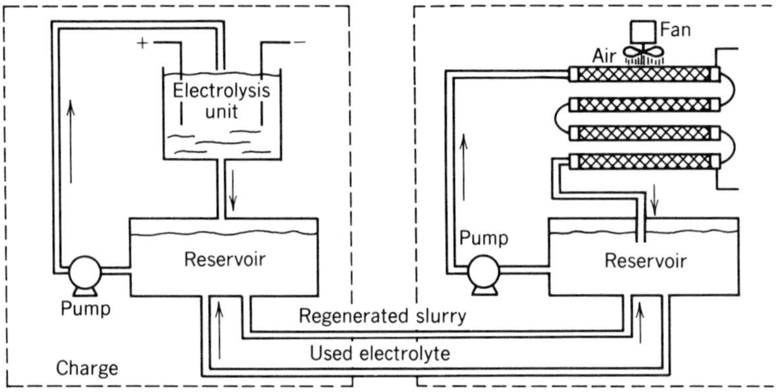

Fig. 26. Schematic diagram of the separate charge and discharge modules of the Génerale d'Electricité circulating zinc–air battery (91).

Iron–Air Cells. The iron–air system is a potentially low cost, high energy system being considered mainly for mobile applications. The iron electrode, similar to that employed in the nickel–iron cell, exhibits long life and therefore this system could be more cost effective than the zinc–air cell. Reactions include:

Iron electrode

$$Fe + 2\,OH^- \rightleftharpoons Fe(OH)_2 + 2\,e^-$$

Oxygen electrode

$$\tfrac{1}{2}\,O_2 + H_2O + 2\,e^- \rightleftharpoons 2\,OH^-$$

Overall reaction

$$Fe + \tfrac{1}{2}\,O_2 + H_2O \rightleftharpoons Fe(OH)_2$$

In the experimental systems studied the iron electrode has been of the sintered type and the oxygen–air electrodes have been of the bifunctional type.

A system as shown in Figure 27, which incorporates circulating electrolyte for thermal management and removal of gases generated during charge, has been developed (93). The iron electrode has a low hydrogen overvoltage and therefore

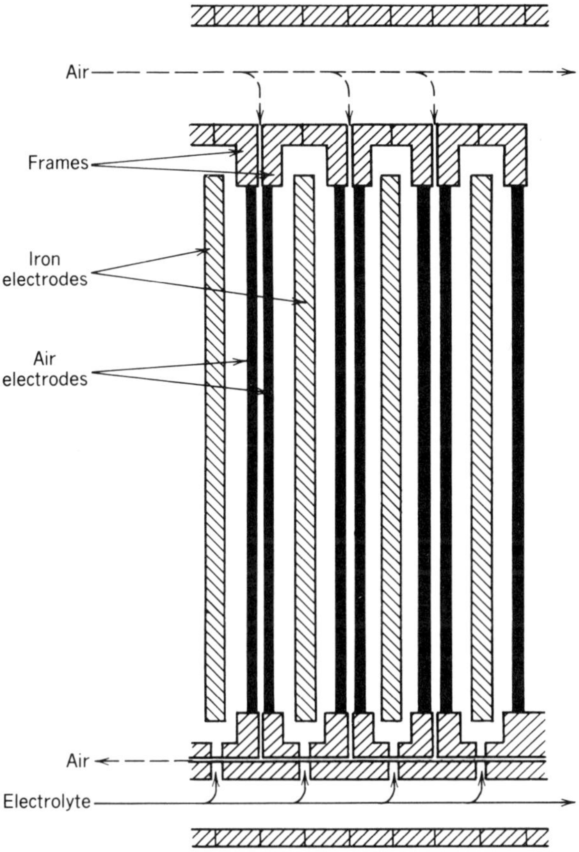

Fig. 27. Cross-section of SNDC iron–air battery pile (93).

the electrode evolves hydrogen during charge and to some extent during open circuit stand. A similar system, where the primary emphasis is on the development of a stable bifunctional air electrode is under investigation (94). A Teflon-bonded formulation consisting of a carbon-base, catalyzed with silver and other additives, is reported to be stable for up to 500 cycles. Because of the inefficiency of the iron electrode, and the irreversibility of the oxygen electrode, this system exhibits recharge energy efficiencies of less than 50%.

Hydrogen–Oxygen Cells. The hydrogen–oxygen cell can be adapted to function as a rechargeable battery, although this system is best known as a primary one (see FUEL CELLS). The electrochemical reactions involve:

Electrodes

$$H_2 + 2\,OH^- \rightleftharpoons 2\,H_2O + 2\,e^-$$

$$\tfrac{1}{2}\,O_2 + H_2O + 2\,e^- \rightleftharpoons 2\,OH^-$$

Overall

$$H_2 + \tfrac{1}{2}\,O_2 \rightleftharpoons H_2O$$

During charge, water is electrolyzed to produce hydrogen and oxygen which are stored as pressurized gas. During discharge those gases electrochemically react to produce water. The reactants have a theoretical energy content of 770 W·h/kg and the interest that has been directed at this system has been primarily for light batteries for military and/or aerospace applications. In the 1960s a multicell bipolar stack contained inside a cylindrical pressure vessel was studied (95). The stack was internally manifolded to feed hydrogen and oxygen to separate compartments within the pressure vessel at operating pressures that ranged between 0.7 and 2.4 MPa (100–350 psi). Figure 28 shows a charge–discharge profile of the system.

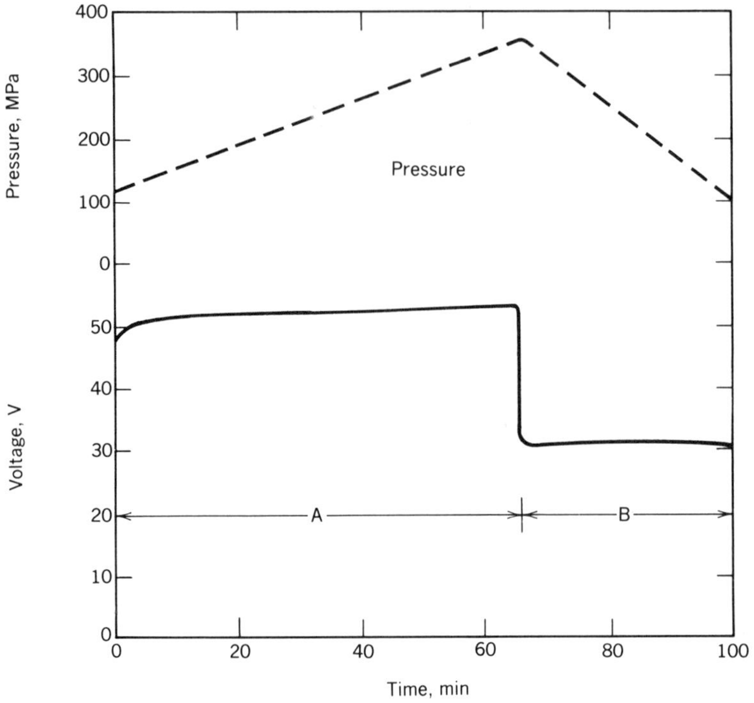

Fig. 28. Cycle of 34-cell regenerative hydrogen–oxygen fuel cell where A represents the charging region at 10 A, B represents discharging at 18.2 A. Both (—) voltage and (---) pressure changes are shown. To convert MPa to psig, multiply by 145.

Because of cross-gas leakage and other complexities, a single-cell was developed. In this configuration the electrodes were constructed in the form of a cylinder; the hydrogen gas was stored in the central compartment, and the oxygen gas in the space between the electrode core and the outer pressure vessel. These two compartments were sized in a two to one volume ratio to maintain the two gases at the same pressure as they are generated during charge. A flexible bellows was placed between the two compartments to compensate for slight differences in the compartments' volume and temperature effects. This system was not pursued however; it was dropped in favor of the simpler single-gas nickel–hydrogen system.

As primary alkaline fuel cells were developed for space applications, consideration was given to separate stack rechargeable designs. In these approaches, the product water formed during the discharge of a primary fuel cell is stored, then fed to a separate electrolizer stack during charge. The hydrogen and oxygen gas generated during charge is stored in separate pressure vessels. This approach overcomes the stability problem of the bifunctional oxygen electrode and the respective stacks can be optimized for their function. This system is still rather complex and bulky and has not yet been applied.

Mechanically Rechargeable Batteries. To avoid the time required for electric recharge, the problems of *in situ* electric recharge, or to utilize anodes that are not electrically rechargeable in aqueous electrolytes, mechanically rechargeable batteries have been studied. These systems are metal–air couples. The anodes that have received attention are zinc, lithium, and aluminum (97–99). Mechanically rechargeable zinc–air batteries were developed for military portable electronic equipment in the 1970s. These batteries were never deployed because of difficulties of leakage, problems of repeatedly replacing the anodes, and the development of lithium primary batteries having superior performance.

Lithium as an anode in alkaline electrolyte has been considered in the battery system shown in Figure 29. Even though lithium reacts directly with water, it was possible to operate the battery because of a protective lithium hydroxide film that forms on the anode. However, the film was not totally protective and units exhibited poor efficiency and were very complex.

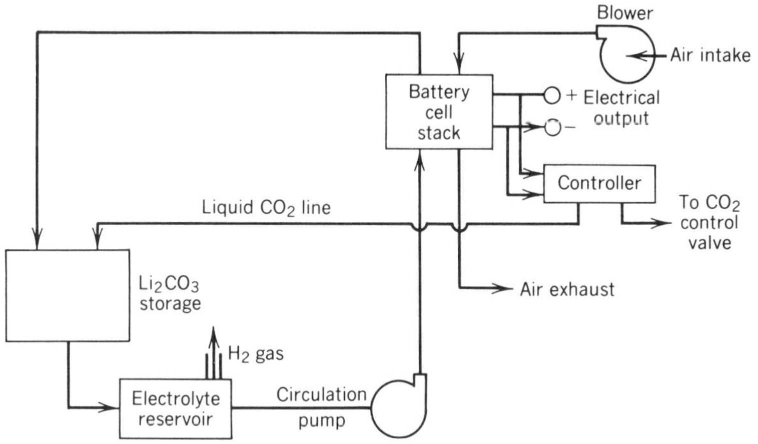

Fig. 29. Schematic of lithium–air automotive propulsion system.

The most significant results with these battery types focus on aluminum as the anode. Figure 30 shows an aluminum–air cell being developed for electric vehicle applications. The aluminum hydroxide reaction product would be returned to the factory to be reprocessed into fresh aluminum anodes. One set of anodes could yield up to 500 miles range before replacement. However, the corrosion reaction of the aluminum with the electrolyte is still a problem. Additionally,

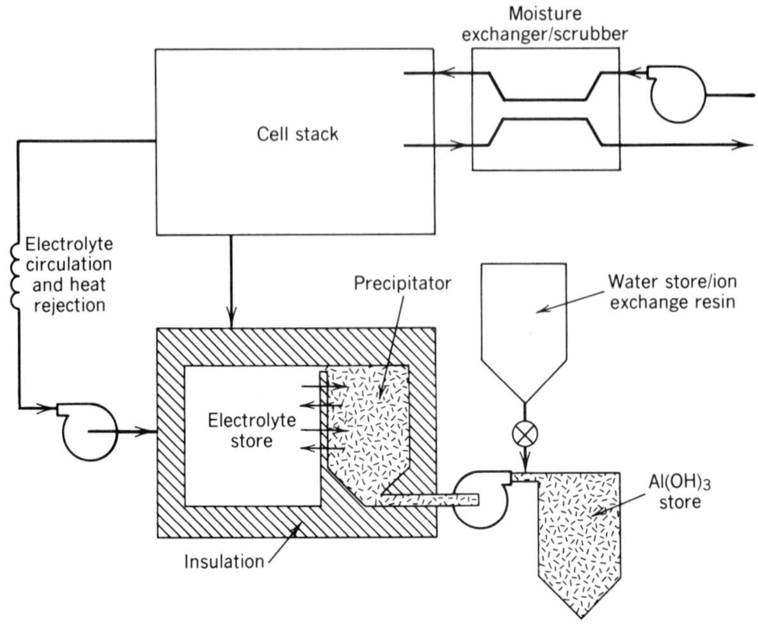

Fig. 30. Aluminum–air power cell system. The design provides for forced convection of air and electrolyte, heat rejection, electrolyte concentration control via $Al(OH)_3$ precipitation, and storage for reactants and products.

the system is complex and it is anticipated that replacing the anodes repeatedly will be as problematic as the zinc systems. The system has poor energy efficiency when consideration is given to the full cycle of electric generation, aluminum production, battery efficiency, and reprocessing of the battery reaction product back to aluminum.

Electrolyte

Potassium hydroxide is the principal electrolyte of choice for the above batteries because of its compatibility with the various electrodes, good conductivity, and low freezing point temperature. Potassium hydroxide is a white crystalline substance having a mol wt = 56.10; density = 2.044 g/mL, and mp = 360°C (see POTASSIUM COMPOUNDS). It is hygroscopic and very soluble in water. The most conductive aqueous solution at 25°C is at 27% KOH, but the conductivity characteristics are relatively flat over a broad range of concentrations.

The characteristics for aqueous KOH (97–99) solutions vary somewhat for battery electrolytes when additives are used. Furthermore, potassium hydroxide reacts with many organics and with the carbon dioxide in air to form carbonates. The build-up of carbonates in the electrolyte is to be avoided because carbonates reduce electrolyte conductivity and electrode activity in some cases.

Safety and Disposal

The potassium hydroxide electrolyte used in alkaline batteries is a corrosive hazardous chemical. It is a poison and if ingested attacks the throat and stomach linings. Immediate medical attention is required. It slowly attacks skin if not rapidly washed away. Extreme care should be taken to avoid eye contact that can result in severe burns and blindness. Protective clothing and face shields or goggles should be worn when filling cells with water or electrolyte and performing other maintenance on vented batteries.

Alkaline batteries generate hydrogen and oxygen gases under various operating conditions. This can occur during charge, overcharge, open circuit stand, and reversal. In vented batteries free ventilation should be provided to avoid hydrogen accumulations surrounding the battery. A vented battery must never be placed in a sealed container for which it was not designed. As a result of operation, hydrogen–oxygen mixtures that are flammable or explosive can exist in the cells' head space. Ignition of this mixture, which is rare, can result in blowing off cell lids. This can occur from internal short circuits or external sparks that propagate back into the cell housing.

Alkaline batteries are capable of high current discharges and accidental short circuits should be avoided. Short circuiting can result in significant heat generation, electrolyte boiling, and cell rupture. High voltage cell strings also present an electric shock hazard; therefore tools should be insulated and operators should not wear rings.

Spontaneous low resistance internal short circuits can develop in silver–zinc and nickel–cadmium batteries. In high capacity cells heat generated by such short circuits can result in electrolyte boiling, cell case melting, and cell fires. Therefore cells that exhibit high resistance internal short circuits should not continue to be used. Excessive overcharge that can lead to dry out and short circuits should be avoided.

Because of increasing environmental concerns, the disposal of all batteries is being reviewed (70–76). Traditionally silver batteries were reclaimed for the silver metal and all other alkaline batteries were disposed of in landfills or incinerators. Some aircraft and industrial nickel–cadmium batteries are rebuilt to utilize the valuable components.

To reduce or eliminate the scattering of cadmium in the environment, the disposal of nickel–cadmium batteries is under study. Already a large share of industrial batteries are being reclaimed for the value of their materials. Voluntary battery collection and reclaiming efforts are under way in both Europe and Japan. However the collection of small batteries is not without difficulties. Consideration is being given to deposit approaches to motivate battery returns for collection and reclamation.

BIBLIOGRAPHY

The Alkaline Secondary Cells are treated in *ECT* 1st ed., under "Cells, Electric," Vol. 3, pp. 292–342, by W. J. Hamer, National Bureau of Standards, and "Cells, Electric," Suppl. 2, pp.

126–161, by C. K. Morehouse, R. Glicksman, and G. S. Lozier, Radio Corporation of America; "Batteries and Electric Cells, Secondary, Secondary Cells, Alkaline," in *ECT* 2nd ed., pp. 172–249, by W. W. Jakobi, Gould-National Batteries, Inc.; "Batteries and Electric Cells, Secondary, Alkaline Cells," in *ECT* 3rd ed., Vol. 3, pp. 592–639, by A. J. Salkind and E. Pearlman, ESB Technology Company.

1. Ger. Pat. 491,498 (July 6, 1928), K. Ackermann, P. Gomelin, G. Pfleiderer, and F. Spouna (to I. G. Farbenindustrie, A. G.); Brit. Pat. 331,540 (Jan. 4, 1929), 332,052 (July 8, 1929), 380,242 (1930), 339,645 (1930), to (I. G. Farbenindustrie, A. G.).
2. U.S. Pat. 2,571,927 (Oct. 16, 1951), U. Gottesman and G. Newmann (to Bureau Technique Gautrat Sarl).
3. Brit. Pat. 784,851 (1952), (to Bureau Technique Gaurat).
4. Ger. Pat. 112,351 (1899); Brit. Pat. 15,370 (1899), T. De Michalowski.
5. Swed. Pat. 11,132 (1889), W. Jungner.
6. A. Fleischer, *J. Electrochem. Soc.* **94,** 289 (1948). A. Fleischer, *Proc. 10th Ann. Battery Res. Conf.* **73** (1956). U.S. Pat. 2,771,499 (1951), A. Fleischer.
7. G. W. Briggs, E. Jones, and W. F. K. Wynne-Jones, *Trans. Faraday Soc.* **51,** 1433 (1955); G. W. Briggs and W. F. K. Wynne-Jones, *Trans. Faraday Soc.* **52,** 1972 (1956).
8. A. J. Salkind and P. F. Bruins, *Nickel–Cadmium Sandia Project Reports,* Polytechnic University of Brooklyn, N.Y., 1956–1958; *J. Electrochem. Soc.* **108,** 356–360 (1962).
9. U.S. Pat. 2,819,962 (1958), (to SAFT).
10. J. A. McMillan, *Acta Cryst.* **7,** 640 (1954); *J. Inorg. Nucl. Chem.* **13,** 28 (1960).
11. V. Scatturin, P. Bellon, and A. J. Salkind, *J. Electrochem. Soc.* **108,** 819 (1961).
12. P. L. Bourgault and B. E. Conway, *Can. J. Chem.* **38,** 1557 (1960).
13. S. U. Falk, *J. Electrochem. Soc.* **107,** 661 (1960).
14. U.S. Pat. 3,003,015 and 3,121,029 (1961), J. C. Duddy.
15. A. J. Salkind and J. C. Duddy, *J. Electrochem. Soc.* **109,** 360 (1962).
16. D. Tuomi, *J. Electrochem. Soc.* **112,** 1 (1965).
17. H. Bode, K. Dehmelt, and J. Witte, *Crystal Structures of Nickel Hydroxides,* CITCE Meeting, Strasbourg, France, 1965.
18. Brit. Pat. 917,291 (1963), L. Kandler; U.S. Pat. 3,282,808 (1966), L. Kandler; E. Hausler, *5th Int. Power Sources Symp.* **287** (1966).
19. E. J. McHenry, *Electrochem. Tech.* **5,** 275–279 (1965).
20. U.S. Pat. 3,653,967 (1972), R. L. Beauchamp.
21. U.S. Pat. 3,873,368 (1972), D. F. Pickett.
22. M. A. Gutjahr, H. Buckner, K. D. Beccu, and H. Saufferer, in D. H. Collins, ed., *Power Sources,* Vol. 4, Oriel Press, UK, 1973, p. 79.
23. U.S. Pat. 3,898,099 (1975), B. Baker and M. Klein.
24. J. Edwards, T. Turner, and J. Whittle, *Proc. Int. Power Sources* **7,** (1970). U.S. Pat. 3,785,867 (1974), J. Edwards and J. Whittle.
25. U.S. Pat. 4,215,190 (1980), R. A. Ferrando and R. Satula.
26. U.S. Pat. 4,298,383 (1981); 4,312,670 (1981), J. F. Joyce and S. L. Colucci.
27. B. Bugnet and D. Doniat, *Proceedings of the 31st Power Sources Conference* 1984.
28. *DAUG Co. literature,* Germany, 1985.
29. K. Kordesch, J. Gsellmann, W. Harer, W. Taucher, and K. Tomantschger, *Prog. Batt. Solar Cells* **7,** 194–204 (1988); K. Kordesch and co-workers, *Electrochim. Acta* **26,** 1495 (1981).
30. U.S. Pat. 4,451,543 (1984), M. A. Dzieciuch, N. Gupta, H. Wroblowa, and J. T. Kummer; U.S. Pat. 4,520,005 (1985), Y. Y. Yao.
31. *Commercial literature,* Ovonic Battery Co., Troy, Mich, 1991.
32. *Commercial literature,* Matsushita Battery Co., Seattle, Wash., Oct. 1990.

33. N. Furukawa, *JEC Battery Newsletter #2,* Sanyo Battery Co., JEC Press, Brunswick, Ohio, 1990.
34. *Commercial literature,* Toshiba Co., Seattle, Wash., Oct. 1990.
35. *Commercial literature,* Gates Battery Co., Gainsville, Fla., 1991.
36. P. C. Milner and U. B. Thomas, in P. Delahay and C. W. Tobias, eds., *Advances in Electrochemistry and Electrochemical Engineering,* Vol. 5, Interscience Publishers, New York, 1967.
37. Swed. Pat. 15,567 (1901), W. Jungner; Ger. Pat. 163,170 (1901), W. Jungner.
38. J. P. Harivel and co-workers, *Proceedings of the 5th International Power Sources Symposium,* Brighton, England, Sept. 1966.
39. B. C. Bradshaw, *Proceedings of the Annual Power Sources Conference* **12,** (1958).
40. F. Kornfeil, in Ref. 39.
41. B. V. Ershler, G. S. Tyvrikov, and A. S. Smirnova, *J. Phys. Chem. (USSR)* **14,** 985 (1940).
42. G. Halpert and L. May, *Extended Abstracts 169th Meeting of the Electrochemical Society,* Washington, D.C., Oct. 1983.
43. J. McBreen, W. E. O'Grady, K. I. Panda, R. W. Hoffman, and D. E. Sayers, *Langmuir* **3,** 428–433 (1987).
44. A. H. Zimmerman and P. K. Effa, *Extended Abstracts, 168th Meeting of the Electrochemical Society,* Oct. 1985.
45. D. T. Fritts, *Extended Abstracts, 164th Electrochemical Society,* Oct. 1983.
46. A. Estelle, *Tekn. Tidskizift* **39,** 105 (1909).
47. G. Halpert, *Proc. 176 Electrochem. Soc.* Oct. 1989.
48. *Commercial literature,* Hoeppecke and Acme Electric Battery Co., 1991.
49. L. Ojefors, *Proceedings of the 9th International Society for Electrochemistry,* Marcoussis, France, 1975; B. Andersson and L. Ojefors, *11th International Power Sources Symposium,* Brighton, England, 1978; *J. Electrochem. Soc.* **123,** 824 (1976).
50. G. Kramer and V. A. Olinpunom, in Ref. 49.
51. A. Nilsson, private communication, 1991.
52. E. R. Bowerman, "Sintered Iron Electrode," *Proc. 22 Power Sources Conf.* (1968).
53. W. A. Bryant, *J. Electrochem. Soc.* **126,** 1899–1901 (1979).
54. Project reports, Eagle Picher Co., Joplin, Miss., 1988–1991.
55. A. J. Salkind, J. J. Kelley, and J. B. Ockerman, *Proc. 176 Electrochem. Soc. Meet.* (Oct. 1989).
56. Reports on EPRI Nickel–Iron Electric Vehicle Project, Argonne National Labs, Argonne, Ill., 1988–1991.
57. R. Swaroop, *Reports on Electric Vehicle Batteries,* Electric Power Research Institute (EPRI), Palo Alto, Calif., 1989–1991.
58. T. Iwaki, T. Mitsumata, and H. Ogawa, *Prog. in Batt. Solar Cells* **4,** 255–259 (1982).
59. S. Hills and A. J. Salkind, "Alkaline Iron Electrode," *Proc. 22 Power Sources Conf.* (1968).
60. A. J. Salkind, C. J. Venuto, and S. U. Falk, *J. Electrochem. Soc.* **111,** 493–495 (1964).
61. P. Hersch, *Trans. Faraday Soc.* **51,** 1442–1448 (1955).
62. U.S. Pat. 2,644,022 (1953), J. Moulton and E. F. Schweitzer.
63. U.S. Pat. 3,483,291 (Dec. 16, 1969), M. J. Mackenzie and A. J. Salkind.
64. D. S. Poa, J. F. Miller, and N. P. Yao, *Argonne National Labs Report #ANL/OEPM-85-2,* Apr. 1985.
65. H. G. André, *Bull. Soc. Franc. Electriciens* **6,** 1, 132 (1941).
66. U.S. Pat. 2,317,711 (Apr. 27, 1943), H. G. André.
67. J. A. McMillan, *J. Inorg. Nucl. Chem.* **13,** 28 (1960).
68. V. Scatturin, P. Bellon, and A. J. Salkind, *J. Electrochem. Soc.* **108,** 819 (1961).

69. A. J. Salkind "Crystal Structures of the Silver Oxides," in A. Fleischer and J. J. Lander, eds., *Zinc–Silver Oxide Batteries,* John Wiley & Sons, Inc., New York, 1971.
70. *The Silver Institute Letter,* Vol. VII, No. 9, Oct. 1977.
71. N. A. Zhulidov and F. I. Yefrennov, *Vestn. Elektropromsti.* **2,** 74 (1963); M. C. Tisgankov, *Electrical Accumulators,* Moscow (1959).
72. V. N. Flerov, *Zh. Prikl. Khim.* **32,** 1306 (1959).
73. P. Goldberg, "Nickel–Zinc Cells Part 1," *Proceedings of the 21st Annual Power Sources Conference,* 1967.
74. P. V. Popat, E. J. Rubin, and R. B. Flanders, "Nickel–Zinc Cells Part 3," *Proceedings of the 21st Annual Power Sources Conference,* 1967.
75. M. J. Sulkes, "Nickel–Zinc Secondary Batteries," *Proceedings of the 23rd Annual Power Sources Conference,* 1969.
76. A. Charkey, "Sealed Nickel–Zinc Cells," *Proceedings of the 25th Annual Power Sources Conference,* Atlantic City, N.J., 1972.
77. A. Charkey, "Nickel–Zinc Cells for Sustained High-Rate Discharge," *Proceedings of the 7th Intersociety Energy Confersion Engineering Conference,* 1972.
78. *Design and Cost Study Zinc–Nickel Oxide Battery for Electric Vehicle Propulsion in Final Report, ANL Contract No. 109-38-3543,* Yardney Elect. Corp., Oct. 1976.
79. Final Report, *Design and Cost Study of Nickel–Zinc Batteries for Electric Vehicles, ANL Contract No. 31-109-38-3541,* Energy Research Corp., Oct. 1976.
80. D. P. Boden and E. Pearlman, in D. H. Collins, ed., *Power Sources,* Vol. 4, Academic Press, New York, 1972.
81. J. M. Parry, C. S. Leung, and R. Wells, *Structure and Operating Mechanisms of Inorganic Separators, Report NASA CR-134692,* Mar. 1974.
82. Brit. Pat. 365,125 (1930), J. J. Drumm.
83. Ger. Pat. 2,150,005 (May 25, 1972), O. von Krusenstierna (to AGA Aktiebolaget); U.S. Pat. 3,923,550 (Dec. 2, 1975), O. von Krusenstierna (to AGA Aktiebolaget); O. von Krusenstierna, *SAE Meeting,* Mar. 4, 1977, paper 21.
84. J. Giner and J. D. Dunlop, *Electrochem. Soc.* **122,** 4 (1975).
85. M. Klein and M. George, *Proceedings of the 26th Power Sources Symposium,* Atlantic City, N.J., 1974, p. 18.
86. J. E. Clifford and E. W. Brooman, *Assessment of Nickel–Hydrogen Batteries for Terrestrial Solar Applications,* SAND80-7191, Sandia National Laboratories, 1981.
87. A. J. Salkind, *Samples of Metal Hydrides,* International Battery Materials Association.
88. H. Ogawa, M. Ikoma, H. Kawano, and I. Matsumoto, "Metal Hydride Electrode for High Energy Density Sealed Nickel–Metal Hydride Battery," *Proceedings of the 16th International Power Sources Conference,* UK, 1988.
89. M. Klein, in D. H. Collins, ed., *Power Sources,* Vol. 5, Academic Press, New York, 1974.
90. P. Antoine and P. Fougtere, in J. Thompson, ed., *Power Sources,* Vol. 7, Academic Press, New York, 1979.
91. A. J. Appleby and J. M. Jacquier, *J. Power Sources* **1** (1976).
92. M. Klein and A. Charkey, *Zinc–Oxygen Battery Development,* Electrochemical Society, Atlanta, Ga., Oct. 1977.
93. L. Ojefors and L. Carlson, *J. Power Sources* **2** (1977–1978).
94. B. G. Demezyk, W. A. Bryant, C. T. Liu, and E. S. Buzzelli, "Performance and Structural Characteristics of the Iron–Air Battery System," *15th IECEC Conference,* 1980.
95. E. Findl and M. Klein, "Electrolytic Hydrogen–Oxygen Fuel, Cell Battery," *Proceedings of the 20th Power Sources Conference,* Red Bank, N.J., 1966.
96. M. Klein and R. Costa, "Electrolytic Regenerative H_2–O_2 Secondary Fuel Cell," *Proceedings of Space Technology Conference, ASME,* June 1970.
97. A. L. Almerini and S. J. Bartosh, "Simulated Field Tests on Zinc–Air Batteries," *Proceedings of the 26th Power Sources Symposium,* Atlantic City, N.J., 1974.

98. W. R. Momyer and E. L. Litauer, "Development of a Lithium Water–Air Primary Battery," *Proceedings of the 15th Intersocietal Energy Conversion Engineering Conference,* Seattle, Wash., Aug. 1980.
99. J. F. Cooper, R. V. Homsy, and J. H. Landrum, "The Aluminum–Air Battery for Electric Vehicle Propulsion," in Ref. 98.

ALVIN J. SALKIND
MARTIN KLEIN
Rutgers University

LEAD–ACID

The lead–acid battery is one of the most successful electrochemical systems and the most successful storage battery developed. In 1988 total battery sales, excluding Eastern European central economy countries, were more than $17 billion (1). Lead–acid battery sales accounted for about 57% of that figure. About 80% of the lead [*7439-92-1*] (qv), Pb, consumption in the United States was for batteries in that year.

The lead–acid battery consists of a number of cells in a container. These cells contain positive (PbO_2) and negative (Pb) electrodes or plates, separators to keep the plates apart, and sulfuric acid [*7664-93-9*], H_2SO_4, electrolyte. The battery reactions are highly reversible, so that the battery can be discharged and charged repeatedly. The number of charge–discharge cycles that can be obtained depends strongly on the use mode and can vary from several hundred to thousands of cycles.

Each cell has a nominal voltage of 2 V and capacities typically vary from 1 to 2000 ampere-hours. Lead–acid cells can be operated with coulombic efficiencies as high as 95% and with energy efficiencies greater than 80%. The many cell designs available for a wide variety of uses can be divided into three main categories: automotive, industrial, and consumer. Automotive batteries, starting, lighting, and ignition (SLI) for cranking of internal combustion engines accounted for nearly 73% of the lead–acid battery sales in 1988 (1). Industrial batteries are used for heavy-duty application such as motive and standby power. More recently, the use of batteries for utility peak shaving has been increasing. Consumer batteries for emergency lighting, security alarm systems, cordless convenience devices and power tools, and small engine starting. This is one of the fastest growing markets for the lead–acid battery (1).

In Figure 1, the cutaway view of the automotive battery shows the components used in its construction. An industrial motive power battery, shown in Figure 2 (2), is the type used for lift trucks, trains, and mine haulage. Both types of batteries have the standard free electrolyte systems and operate only in the vertical position. Although a tubular positive lead–acid battery is shown for industrial applications, the flat plate battery construction (Fig. 1) is also used in a comparable size.

Two types of batteries having immobilized electrolyte systems are also made. They are most common in consumer applications, but their use in industrial and

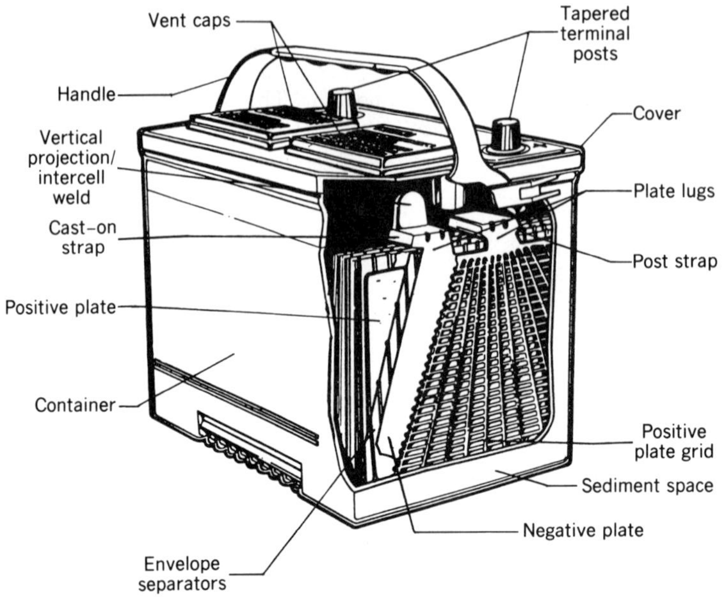

Fig. 1. Cutaway view of an automotive SLI lead–acid battery container and cell element. Courtesy of Johnson Controls, Inc.

SLI applications is increasing. Both types have low maintenance requirements and usually can be operated in any position. They are sometimes called valve regulated or recombinant batteries because they are equipped with a one-way pressure relief vent and normally operate in a sealed condition with an oxygen recombination cycle to reduce water loss.

In the gelled electrolyte battery, the sulfuric acid electrolyte has been immobilized by a thixotropic gel. This is made by mixing an inorganic powder such as silicon dioxide [*7631-86-9*], SiO_2, with the acid (3). Other cells, such as the one shown in Figure 3, use a highly absorbent separator to immobilize the electrolyte (4).

History

Gaston Planté developed the first working model of the lead–acid battery (5) in 1859, but electrochemical energy storage was not practical then because there was no efficient way of recharging a battery. The lead–acid battery generated interest because it could provide higher currents than primary batteries. The use of the system as a capacitor was also of great interest. Between 1860 and 1880, the principal commercial application of the Planté battery was in the telegraphic industry (6).

The first commercial application for a rechargeable energy storage device was electric lighting. Edison had developed a long-lasting incandescent bulb and improved the dynamo in 1879 (7). The invention of pasted plates (8) and the perforated lead current collector (9) in 1881 improved the energy storage capacity

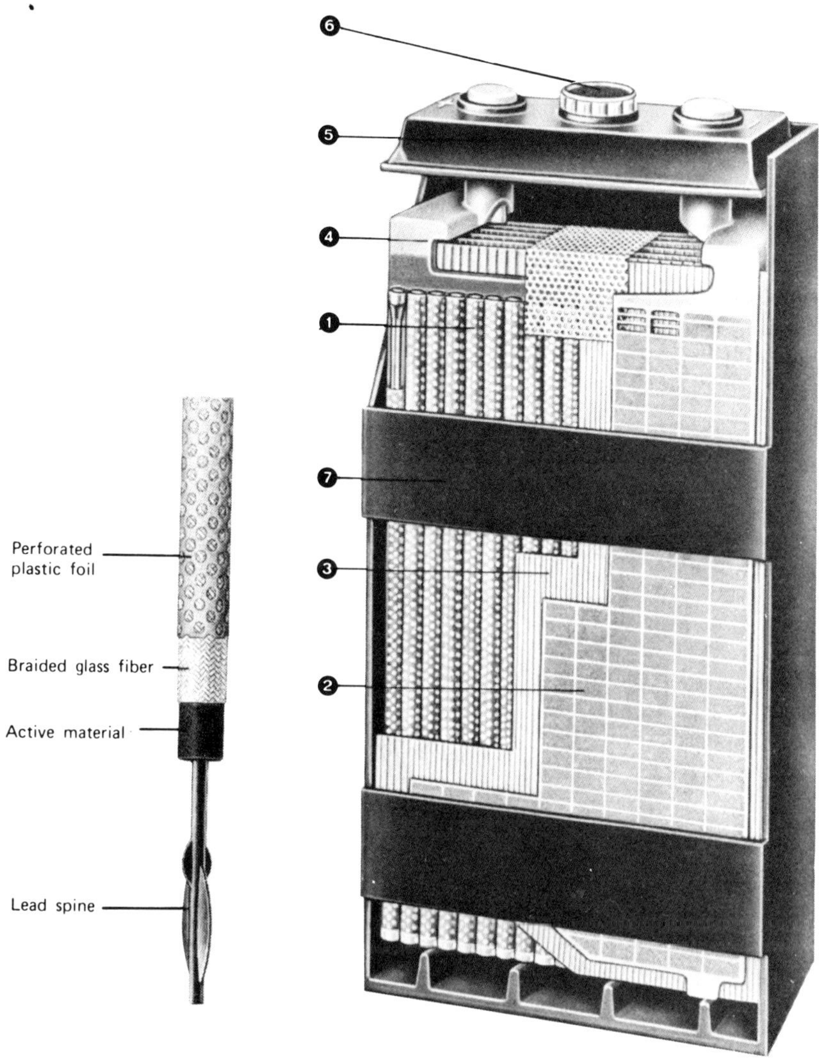

Fig. 2. Cutaway view of a tubular positive lead–acid battery. (1) Positive tubular plate; (2) negative plate; (3) separator; (4) connecting strap; (5) cell cover; (6) cell plug; and (7) cell container. Courtesy of A. B. Tudor, Sweden.

of the lead–acid battery and in the 1880s European and American companies were organized to commercialize electric lighting. Lighting provided a strong incentive for improvements in battery technology. Better grids, flat plates, and lead–antimony alloys soon followed (6). Electric vehicles, including boats, submarines, cars, and trams were developed and batteries were an important factor in the development of the power industry. Batteries were first used as transformers in electric power stations and later for load leveling. They provided the flexibility that was necessary for expansion of the industry without excessive capital investment. As power production improved and expanded, however, batteries were phased out because the maintenance costs were too high.

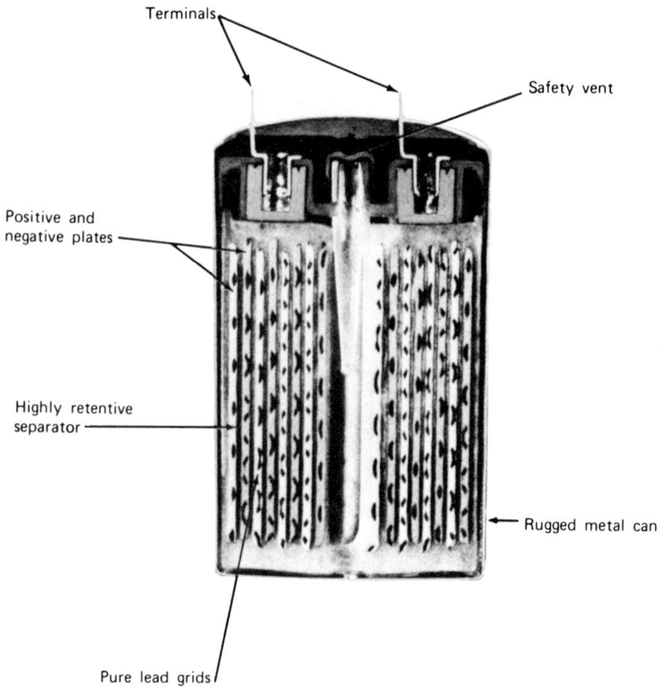

Fig. 3. Sealed cylindrical lead–acid cell. Courtesy of Gates Energy Products, Inc.

The invention of the self-starter in 1911 impacted heavily on the battery market (10). The battery-started, gasoline-driven vehicle was so successful that electric vehicles could not compete. Except for submarines and a few delivery trucks, electric vehicles disappeared and automotive starting became the largest market for the lead–acid battery. Indeed, the SLI application has not only grown steadily but it has also financed the majority of lead–acid battery research and development. Over the years, substantial advances have been made by using new materials such as rubber and plastics for battery containers, separators, and other components, to increase the power and energy density of batteries. Maintenance requirements have been reduced substantially, and machine automation has accelerated production rates and lowered costs. Better electronics for charge and discharge control have improved battery performance and life.

In the 1990s, the use of batteries in electric vehicles and for load leveling is being revived partly for environmental reasons and partly because of scarce energy resources. Improvements in battery performance and life, fewer maintenance requirements, and automatic control systems are making these applications feasible. Research and development is ongoing all over the world to develop improved lead–acid batteries as well as other systems to meet these needs.

Cell Thermodynamics

The chemical reaction of the lead-acid battery was explained as early as 1882 (11). The double sulfate theory has been confirmed by a number of methods (12–14) as the only reaction consistent with the thermodynamics of the system. The thermodynamics of the lead–acid battery has been reviewed in great detail (15).

At the cathode, or positive electrode, lead dioxide [*1309-60-0*], PbO_2, reacts with sulfuric acid to form lead sulfate [*7446-14-2*], $PbSO_4$, and water in the discharging reaction

$$PbO_2 + 3\ H^+ + HSO_4^- + 2\ e^- \rightleftharpoons PbSO_4 + 2\ H_2O \tag{1}$$

and the reverse occurs as the battery charges. At the anode, or negative electrode, metallic lead reacts with bisulfate ion to form lead sulfate in the discharging reaction

$$PbSO_4 + H^+ + 2\ e^- \rightleftharpoons Pb + HSO_4^- \tag{2}$$

and the reverse occurs as the battery charges. The sum of these two half-cell reactions is called the double sulfate reaction.

$$Pb + PbO_2 + 2\ H_2SO_4 \rightleftharpoons 2\ PbSO_4 + 2\ H_2O \tag{3}$$

Lead sulfate is formed as the battery discharges, sulfuric acid is regenerated as the battery is charged. The open circuit voltage of the lead–acid battery is a function of the acid concentration and temperature. A review of this subject is available (16). The Nernst equation may be used to calculate the open circuit cell voltage. The battery voltage is then obtained by multiplying the cell voltage by the number of cells.

The Nernst equation for the cell voltage V is

$$V = V^\circ + 2.303(RT/F)\ (\log a_{H_2SO_4} - \log a_{H_2O}) \tag{4}$$

Because Pb, PbO_2, and $PbSO_4$ are all solids having low solubilities, the activities of these substances are unity. At 25°C, the absolute temperature T is 298.15 K. The value of R, the gas constant, used is 8.3144 J/(mol·K). F, the Faraday constant, is 96,485 C/mol. The standard cell voltage V° for the double sulfate reaction must be known as well as the activities of sulfuric acid and water at any given concentration or temperature.

The dependence of cell voltage on pressure is small and can be neglected in normal applications. At 25°C and an acid concentration of 3.74 m, $dV/dP = -3.32 \times 10^{-5}\ \mu V/Pa\ (-4.81 \times 10_{\mu}^{-3} V/psi)$. Tables and equations for the pressure dependence are available (15).

Sulfuric acid dissociates only partially according to the following equilibria.

$$H_2SO_4 \rightleftharpoons H^+ + HSO_4^- \qquad K_1 = a_{H^+}a_{HSO_4^-}/a_{H_2SO_4} \tag{5}$$

$$HSO_4^- \rightleftharpoons H^+ + SO_4^{2-} \qquad K_2 = a_{H^+}a_{SO_4^{2-}}/a_{HSO_4^-} \tag{6}$$

Here the values of a are the activities of the designated ions in solution, and K_1 and K_2 are the equilibrium constants for the dissociation reactions. K_1 is infinity because dissociation to hydrogen and bisulfate ions is essentially complete. The best value for K_2 is probably 0.0102 (17). Thus sulfuric acid contains a mixture of hydrogen, bisulfate, and sulfate ions where the ratios of these ions vary with concentration and temperature.

The activity of any ion, $a = \gamma m$, where γ is the activity coefficient and m is the molality (mol solute/kg solvent). Because it is not possible to measure individual ionic activities, a mean ionic activity coefficient, $\gamma_{\pm}$, is used to define the activities of all ions in a solution. The convention used in most of the literature to report the mean ionic activity coefficients for sulfuric acid is based on the assumption that the acid dissociates completely into hydrogen and sulfate ions. This assumption leads to the following formula for the activity of sulfuric acid.

$$a_{H_2SO_4} = (a_{H^+})^2 a_{SO_4^{2-}} = (\gamma_{\pm}\ 2m)^2(\gamma_{\pm}\ m) = 4\gamma_{\pm}^3 m^3 \tag{7}$$

where m is the molality of sulfuric acid (mol acid/kg water).

To adjust the activities of sulfuric acid to the convention which assumes that the acid dissociates only partially into hydrogen and bisulfate ions, the following formula is used:

$$a'_{H_2SO_4} = a_{H_2SO_4}/K_2 \tag{8}$$

The activity coefficient of water is related to the osmotic coefficient by the formula:

$$2.303 \log a_{H_2O} = -3m\phi/55.5 \tag{9}$$

where ϕ is the osmotic coefficient of sulfuric acid of molality m.

The activity coefficients of sulfuric acid have been determined independently by measuring three types of physical phenomena: cell potentials, vapor pressure, and freezing point. A consistent set of activity coefficients has been reported from 0.1 to 8 m at 25°C (14), from 0.1 to 4 m and 5 to 55°C (18), and from 0.001 to 0.02 m at 25°C (19). These values are all based on cell potential measurements. The activity coefficients based on vapor pressure measurements (20) agree with those from potential measurements when they are corrected to the same reference activity coefficient.

To calculate the open circuit voltage of the lead–acid battery, an accurate value for the standard cell potential, which is consistent with the activity coefficients of sulfuric acid, must also be known. The standard cell potential for the double sulfate reaction is 2.048 V at 25°C. This value is calculated from the standard electrode potentials: for the $(Pt)H_2|H_2SO_4(m)|PbSO_4|PbO_2(Pt)$ electrode 1.690 V (14), for the $Pb(Hg)|PbSO_4|H_2SO_4(m)|H_2(Pt)$ electrode 0.3526 V (19), and for the $Pb|Pb^{2+}|Pb(Hg)$ 0.0057 V (21).

Table 1 gives the calculated open circuit voltages of the lead–acid cell at 25°C at the sulfuric acid molalities shown. The corrected activities of sulfuric acid from vapor pressure data (20) are also given.

The temperature dependence of the open circuit voltage has been accurately determined (22) from heat capacity measurements (23). The temperature coefficients E_T are given in Table 2. The accuracy of these temperature coefficients does not depend on the accuracy of the open circuit voltages at 25°C shown in Table 1. Using the data in Tables 1 and 2, the open circuit voltage can

Table 1. Thermodynamic Values for the Lead-acid Cell at 25°C

Molality, m	$\gamma_{\pm}$	α_{H_2O}[a]	$\alpha_{H_2SO_4}$[a]	V
0.1	0.245	9.963×10^{-1}	5.882×10^{-5}	1.798
0.2	0.193	9.928×10^{-1}	2.296×10^{-4}	1.833
0.3	0.169	9.892×10^{-1}	5.167×10^{-4}	1.854
0.5	0.144	9.819×10^{-1}	1.483×10^{-3}	1.881
0.7	0.131	9.743×10^{-1}	3.067×10^{-3}	1.900
1.0	0.121	9.618×10^{-1}	7.164×10^{-3}	1.922
1.5	0.117	9.387×10^{-1}	2.137×10^{-2}	1.951
2.0	0.118	9.126×10^{-1}	5.224×10^{-2}	1.975
2.5	0.123	8.836×10^{-1}	1.158×10^{-1}	1.996
3.0	0.131	8.516×10^{-1}	2.440×10^{-1}	2.016
3.5	0.143	8.166×10^{-1}	4.989×10^{-1}	2.035
4.0	0.157	7.799×10^{-1}	9.883×10^{-1}	2.054
4.5	0.173	7.422×10^{-1}	1.888×10	2.072
5.0	0.192	7.032×10^{-1}	3.541×10	2.090
5.5	0.213	6.643×10^{-1}	6.463×10	2.106
6.0	0.237	6.259×10^{-1}	1.148×10^{1}	2.123
6.5	0.263	5.879×10^{-1}	2.002×10^{1}	2.139
7.0	0.292	5.509×10^{-1}	3.421×10^{1}	2.154
7.5	0.323	5.152×10^{-1}	5.685×10^{1}	2.169
8.0	0.356	4.814×10^{-1}	9.255×10^{1}	2.183
8.5	0.393	4.488×10^{-1}	1.492×10^{2}	2.197
9.0	0.431	4.180×10^{-1}	2.334×10^{2}	2.211
9.5	0.472	3.886×10^{-1}	3.617×10^{2}	2.224
10.0	0.516	3.612×10^{-1}	5.490×10^{2}	2.236
11.0	0.610	3.111×10^{-1}	1.208×10^{3}	2.260
12.0	0.711	2.681×10^{-1}	2.480×10^{3}	2.283
13.0	0.819	2.306×10^{-1}	4.835×10^{3}	2.304
14.0	0.938	1.980×10^{-1}	9.072×10^{3}	2.324
15.0	1.065	1.698×10^{-1}	1.630×10^{4}	2.343
16.0	1.200	1.456×10^{-1}	2.828×10^{4}	2.361
17.0	1.338	1.252×10^{-1}	4.708×10^{4}	2.378
18.0	1.484	1.076×10^{-1}	7.621×10^{4}	2.394
19.0	1.634	9.250×10^{-2}	1.198×10^{5}	2.410
20.0	1.790	7.960×10^{-2}	1.836×10^{5}	2.424
21.0	1.951	6.860×10^{-2}	2.750×10^{5}	2.439
22.0	2.122	5.890×10^{-2}	4.072×10^{5}	2.453
23.0	2.302	5.060×10^{-2}	5.940×10^{5}	2.466
24.0	2.460	4.410×10^{-2}	8.233×10^{5}	2.478
26.0	2.805	3.310×10^{-2}	1.552×10^{6}	2.502
28.0	3.159	2.500×10^{-2}	2.767×10^{6}	2.524
30.0	3.499	1.910×10^{-2}	4.627×10^{6}	2.544

[a]From Ref. 20.

be calculated from 0 to 60°C at concentrations of sulfuric acid from 0.1 to 13.877 m.

The mercurous sulfate [*7783-36-0*], Hg_2SO_4, mercury reference electrode, $(Pt)H_2|H_2SO_4(m)|Hg_2SO_4(Hg)$, is used to accurately measure the half-cell potentials of the lead–acid battery. The standard potential of the mercury reference

Table 2. Temperature Coefficients of the Lead–Acid Cell, E_T[a], mV

Concentration,	Temperature, °C											
m	0	5	10	15	20	30	35	40	45	50	55	60
0.1	−6.102	−4.713	−3.410	−2.191	−1.055	0.975	1.871	2.689	3.431	4.098	4.692	5.213
0.2	−3.872	−2.943	−2.092	−1.319	−0.622	0.548	1.024	1.428	1.762	2.026	2.223	2.353
0.3	−2.414	−1.785	−1.231	−0.749	−0.339	0.270	0.472	0.607	0.676	0.681	0.621	0.498
0.4	−1.317	−0.914	−0.581	−0.319	−0.126	0.059	0.053	−0.017	−0.151	−0.347	−0.605	−0.923
0.5	−0.400	−0.193	−0.050	0.029	0.045	−0.106	−0.271	−0.495	−0.776	−1.115	−1.509	−1.958
0.6	0.351	0.399	0.387	0.316	0.187	−0.243	−0.542	−0.894	−1.301	−1.760	−2.272	−2.834
0.7	1.005	0.913	0.766	0.564	0.308	−0.360	−0.772	−1.233	−1.744	−2.304	−2.911	−3.566
0.8	1.606	1.385	1.113	0.791	0.420	−0.468	−0.982	−1.543	−2.150	−2.801	−3.496	−4.235
0.9	2.135	1.801	1.421	0.993	0.519	−0.563	−1.171	−1.821	−2.513	−3.248	−4.023	−4.839
1.0	2.620	2.180	1.698	1.173	0.607	−0.647	−1.334	−2.059	−2.823	−3.624	−4.462	−5.337
1.110	3.063	2.528	1.953	1.339	0.688	−0.725	−1.485	−2.281	−3.112	−3.977	−4.875	−5.807
1.388	4.020	3.280	2.508	1.704	0.867	−0.898	−1.826	−2.784	−3.771	−4.787	−5.831	−6.903
1.850	5.175	4.194	3.185	2.150	1.088	−1.113	−2.251	−3.414	−4.601	−5.812	−7.046	−8.303
2.220	5.702	4.612	3.496	2.355	1.189	−1.213	−2.449	−3.709	−4.990	−6.294	−7.620	−8.967
2.775	6.081	4.912	3.720	2.503	1.263	−1.286	−2.593	−3.923	−5.274	−6.647	−8.040	−9.454
3.172	6.168	4.982	3.771	2.537	1.280	−1.302	−2.627	−3.973	−5.341	−6.729	−8.138	−9.568
3.700	6.002	4.854	3.679	2.478	1.252	−1.276	−2.577	−3.901	−5.249	−6.620	−8.014	−9.431
4.626	5.256	4.270	3.250	2.199	1.115	−1.146	−2.322	−3.527	−4.762	−6.026	−7.318	−8.638
5.551	4.134	3.437	2.674	1.846	0.954	−1.016	−2.092	−3.228	−4.423	−5.675	−6.984	−8.349
6.167	3.651	3.056	2.393	1.661	0.863	−0.928	−1.919	−2.972	−4.087	−5.262	−6.497	−7.790
6.938	3.171	2.674	2.107	1.472	0.769	−0.834	−1.732	−2.693	−3.717	−4.801	−5.945	−7.149
7.929	2.760	2.340	1.854	1.301	0.682	−0.745	−1.553	−2.422	−3.350	−4.338	−5.384	−6.487
8.539	2.641	2.238	1.772	1.242	0.651	−0.711	−1.481	−2.308	−3.192	−4.131	−5.126	−6.175
9.291	2.602	2.192	1.725	1.201	0.628	−0.681	−1.413	−2.195	−3.028	−3.909	−4.838	−5.815
10.092	2.647	2.207	1.721	1.191	0.617	−0.659	−1.360	−2.102	−2.884	−3.706	−4.567	−5.466
11.101	2.725	2.249	1.737	1.192	0.612	−0.645	−1.322	−2.030	−2.769	−3.539	−4.339	−5.168
12.335	2.797	2.290	1.756	1.197	0.611	−0.636	−1.296	−1.981	−2.689	−3.421	−4.176	−4.953
13.877	2.843	2.313	1.764	1.195	0.607	−0.626	−1.269	−1.931	−2.611	−3.308	−4.022	−4.753

[a] $E = E_{25°C} - E_T$ (mV) of the lead storage cell from the third law of thermodynamics.

electrode is 0.6125 V (14). The potentials of the lead dioxide, lead sulfate, and mercurous sulfate, mercury electrodes versus a hydrogen electrode have been measured (24,25). These data may be used to calculate accurate half-cell potentials for the lead dioxide, lead sulfate positive electrode from temperatures of 0 to 55°C and acid concentrations of from 0.1 to 8 *m*.

Lead Grid Corrosion

The corrosion of the lead grid at the lead dioxide electrode is one of the primary causes of lead–acid battery failure. Grid corrosion is complex for several reasons. First, the acid concentration and voltage at the lead dioxide electrode vary widely during battery operation. Second, the voltage, pH, and other ionic concentrations vary across the corrosion film. Different conditions across the film favor different reactions. Changing conditions mean multiple corrosion products form including lead sulfate, basic lead sulfates, and lead oxides (26). Third, the lead oxides form a nonstoichiometric series which can be represented by the general formula, PbO_n, where $1 < n < 2$. Two polymorphs have been identified in lead corrosion films for both lead(II) oxide [*1317-36-8*], PbO, orthorhombic and tetragonal, and PbO_2, rhombic and tetragonal. PbO_2 may also be amorphous. A variety of intermediates, such as the intermediate lead oxide [*1314-41-6*], Pb_3O_4, and lead(III) oxide [*1314-27-8*], Pb_2O_3, have also been found.

The mechanisms of lead corrosion in sulfuric acid have been studied and good reviews of the literature are available (27–30). The main techniques used in lead corrosion studies have been electrochemical measurements, x-ray diffraction, and electron microscopy. More recently, laser Raman spectroscopy and photoelectrochemistry have been used to gain new insight into the corrosion process (30,31).

Charge–Discharge Processes

An excellent review covers the charge and discharge processes in detail (30) and ongoing research on lead–acid batteries may be found in two symposia proceedings (32,33). Detailed studies of the kinetics and mechanisms of lead–acid battery reactions are published continually (34). Although many questions concerning the exact nature of the reactions remain unanswered, the experimental data on the lead–acid cell are more complete than for most other electrochemical systems.

Significant progress is also being made in simulating battery charge and discharge processes using electrochemical models. Models of the lead–acid cell based on porous electrode theory have been developed (35–39). Discharge behavior can be predicted within 10% accuracy over a wide range of discharge rates. Recharge behavior has also been simulated (40); a detailed model of the changes in acid concentration resulting from convection in free electrolyte cells has been presented (41); a model for distribution of acid immobilized in a porous separator has been developed (42); and a model of the effect of lead sulfate supersaturation on battery behavior has been published (43). Such models can be used to predict the performance of large batteries from data on small cells, develop improved battery designs, and simulate the performance of any design in a new application. Battery development is therefore accelerated by reducing the time required to build and test battery prototypes.

At high discharge rates, such as those required for starting an engine, the voltage drops sharply primarily because of the resistance of the lead current collectors. This voltage drop increases with the cell height and becomes significant even at moderate discharge rates in large industrial cells. Researchers have measured this effect in industrial cells and have developed a model which has been used to improve grid designs for automotive batteries (44,45).

Self-Discharge Processes. The shelf life of the lead–acid battery is limited by self-discharge reactions, first reported in 1882 (46), which proceed slowly at room temperature. High temperatures reduce shelf life significantly. The reactions which can occur are well defined (47) and self-discharge rates in lead–acid batteries having immobilized electrolyte (48) and limited acid volumes (49) have been measured.

The lead current collector in the positive lead–dioxide plate corrodes and the compounds which form are a function of the acid concentration and positive electrode voltage. Other reactions which take place at the positive electrode are shown.

Oxygen evolution

$$PbO_2 + H_2SO_4 \rightleftharpoons PbSO_4 + H_2O + \tfrac{1}{2}\,O_2 \tag{10}$$

Oxidation of organics, R,

$$2\,PbO_2 + R + 2\,H_2SO_4 \rightleftharpoons 2\,PbSO_4 + 2\,H_2O + CO_2 + R' \tag{11}$$

Sulfation of PbO (in new cells)

$$PbO + H_2SO_4 \rightleftharpoons PbSO_4 + H_2O \tag{12}$$

Oxidation of additives such as antimony, in the grid alloy:

$$5\,PbO_2 + 2\,Sb + 6\,H_2SO_4 \rightleftharpoons (SbO_2)_2SO_4 + 5\,PbSO_4 + 6\,H_2O \tag{13}$$

Similar reactions can be written for other metallic additives. At the negative electrode two more reactions can occur.

Hydrogen evolution

$$Pb + H_2SO_4 \rightleftharpoons PbSO_4 + H_2 \tag{14}$$

Oxygen recombination

$$Pb + H_2SO_4 + \tfrac{1}{2}\,O_2 \rightleftharpoons PbSO_4 + H_2O \tag{15}$$

The rate of the oxygen recombination reactions is very fast depending only on the rate at which the oxygen is transported to the lead in the negative electrode. If the cell electrode stack is fully saturated with acid, the oxygen recombination reaction proceeds very slowly. The rate depends on the solubility of oxygen in sulfuric acid. However, in batteries which are stored with limited acid volume, oxygen can recombine very rapidly. A small leak in the cell container results in rapid self-discharge of the negative electrode by atmospheric oxygen. Although hydrogen recombination at the positive electrode has been proposed, its rate appears to be insignificant (50).

The rates of the other self-discharge reactions are primarily determined by the acid concentration (48,49). High acid concentrations accelerate the gas evolution reactions at both electrodes. In contrast, grid corrosion appears to be faster at low acid concentrations and voltages. Sulfation of unconverted lead oxide is so rapid that it generally occurs before the battery is shipped from the factory. Batteries are sometimes recharged just before shipment to complete the conversion of lead oxide to lead dioxide in the positive plate. Organics are necessary additives to the separator and negative plate but are kept to a minimum to avoid excessive self-discharge and grid corrosion.

Overcharge Reactions. Water electrolysis during overcharge is an irreversible process. Oxygen forms at the positive electrode:

$$H_2O \rightleftharpoons 2\,H^+ + \tfrac{1}{2}\,O_2(g) + 2\,e^- \tag{16}$$

At the negative electrode, hydrogen ions react to form hydrogen gas:

$$2\,H^+ + 2\,e^- \rightleftharpoons H_2(g) \tag{17}$$

The net reaction being

$$H_2O \rightleftharpoons H_2(g) + \tfrac{1}{2}\,O_2(g) \tag{18}$$

Theoretically, water should decompose at a voltage below the voltage required to recharge a lead–acid battery. However, the rate of water electrolysis is much slower than the rate of the recharge reaction. Thus the lead–acid battery can operate with as little as 5% excess charge to compensate for water electrolysis. Use of lead–antimony alloys for the current collectors in lead–acid batteries increases water loss. Some of these batteries need regular maintenance by addition of water to replace the water lost on overcharge. Many newer designs, however, use either lower concentrations of antimony in the alloy or lead–calcium alloys to reduce water loss (see LEAD ALLOYS). This is the basis for the maintenance-free batteries.

Some battery designs have a one-way valve for pressure relief and operate on an oxygen cycle. In these systems the oxygen gas formed at the positive electrode is transported to the negative electrode where it reacts to reform water. Hydrogen evolution at the negative electrode is normally suppressed by this reaction. The extent to which this process occurs in these valve regulated lead–acid batteries is called the recombination-efficiency. These processes are reviewed in the literature (50–52).

The oxygen cycle is often described by equation 16 at the positive electrode, and the reverse of equation 16 at the negative electrode,

$$\tfrac{1}{2}\,O_2 + 2\,H^+ + 2\,e^- \rightleftharpoons H_2O \tag{19}$$

When these half-reactions are summed, there is no net reaction. Thus the material balance of the cell is not altered by overcharge. At open circuit, equation 19 at the negative electrode is the sum of a two-step process, represented by equation 15 and

$$PbSO_4 + 2\,H^+ + 2\,e^- \rightleftarrows Pb + H_2SO_4 \tag{20}$$

These equations are based on the thermodynamically stable species. Further research is needed to clarify the actual intermediate formed during overcharge. In reality, the oxygen cycle can not be fully balanced because of other side reactions, that include grid corrosion, formation of residual lead oxides in the positive electrode, and oxidation of organic materials in the cell. As a result, some gases, primarily hydrogen and carbon dioxide (53), are vented.

Material Fabrication and Manufacturing Processes

The lead–acid battery is comprised of three primary components: the element, the container, and the electrolyte. The element consists of positive and negative plates connected in parallel and electrically insulating separators between them. The container is the package which holds the electrochemically active ingredients and houses the external connections or terminals of the battery. The electrolyte, which is the liquid active material and ionic conductor, is an aqueous solution of sulfuric acid.

Element. The process of fabricating lead–acid battery elements as depicted in Figure 4 involves numerous chemical and electrochemical reactions and several mechanical assembly operations. All of the processes involved must be carefully controlled to ensure the quality and reliability of the product.

Plates. Plates are the part of the cell that ultimately become the battery electrodes. The plates consist of an electrically conductive grid pasted with a lead oxide–lead sulfate paste which is the precursor to the electrode active materials which participate in the electrochemical charge–discharge reactions.

Lead is the basic raw material for the two components, both the active material and grid, of both the positive and negative plates or electrodes. A finely divided mixture of lead monoxide powder and metallic lead particles is the key ingredient in preparing positive and negative active material. The physical and chemical characteristics of this leady litharge, battery oxide, or grey oxide as it is called, have a profound impact on the performance and life expectancy of the battery. Battery oxide is manufactured from primary or secondary (recycled) lead of extremely high purity using either the Barton or ball mill process.

In the Barton (54) process, molten lead is metered into a reactor or a Barton pot where it impinges on a rotating paddle. The resulting droplets of lead are oxidized and conveyed out in an airstream. The ball mill process converts solid lead pieces directly into oxide through attrition by causing them to rub against each other within a rotating reactor or ball mill. The lead in the ball mill is the grinding media as well as a reactant. The particles or agglomerates produced by the molten lead Barton process are droplike or spherical whereas those produced by the solid state ball mill grinding process are flat or flakelike. The oxidation

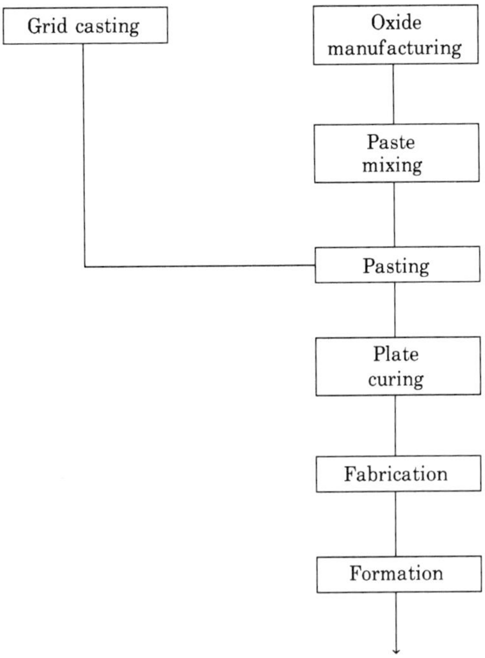

Fig. 4. Flow sheet of plate processing and battery manufacturing.

reaction in either process is controlled such that approximately 75% of the product is oxidized to lead monoxide, leaving about 25% as finely divided metallic lead to be oxidized in subsequent processes. In addition to chemical composition, particle size or reactivity of the material is measured and controlled. Process control variables include reactor temperature, air flow, and lead feed rate.

The oxide exiting either the Barton or ball mill reactor is conveyed by an air stream to separating equipment, ie, settling tank, cyclone, and baghouse, after which it is stored in large hoppers or drummed for use in paste mixing. Purity of the lead feed stock is extremely critical because minute quantities of some impurities can either accelerate or slow the oxidation reaction markedly. Detailed discussions of the oxide-making process and product are contained in references 55–57.

Paste Mixing. The active materials for both positive and negative plates are made from the identical base materials. Lead oxide, fibers, water, and a dilute solution of sulfuric acid are combined in an agitated batch mixer or reactor to form a pastelike mixture of lead sulfates, the normal, tribasic, and tetrabasic sulfates, plus PbO, water, and free lead. The positive and negative pastes differ only in additives to the base mixture. Organic expanders, barium sulfate [*7727-43-7*], $BaSO_4$, carbon, and occasionally mineral oil are added to the negative paste. Red lead [*1314-41-6*] or minium, Pb_3O_4, is sometimes added to the positive mix. The paste for both electrodes is characterized by cube weight or density, penetration, and raw plate density.

Additives, ie, carbon or Pb_3O_4, are used to improve pasted plate conductivity and formed or charged plate characteristics. The expander used in the negative paste keeps it porous or spongelike throughout its life, rather than becoming a dense, tightly packed sheet (58,59). The barium sulfate in the negative paste is believed to act to stimulate the formation of fine seed crystals of $PbSO_4$ during discharge of the electrode (60–62). The physical characteristics of the paste are controlled by the mixing process (63,64). The amounts and order of constituent addition also play a role in the makeup of the finished paste.

During the paste mixing process the lead oxides are partially converted to basic lead sulfate (65). Some of these compounds are monobasic lead sulfate [*12036-76-9*], $PbO{\cdot}PbSO_4$, dibasic lead sulfate [*65589-92-6*], $2PbO{\cdot}PbSO_4$, predominantly tribasic lead sulfate [*12202-17-4*], $3PbO{\cdot}PbSO_4$, and, if high temperatures are encountered during the mixing step, tetrabasic lead sulfate [*12065-90-6*], $4PbO{\cdot}PbSO_4$. These compounds, because of their crystal morphology, bind tightly together on the grid, and facilitate plate processing and battery cycling. It is the degradation of this binding property resulting from an increased crystal growth (66) that causes failure of the positive plate by shedding of the active material. This active material drops through the separator ribs and collects in the sediment space (Fig. 1). This material is lost during electrochemical conversion and, subsequently, performance of the positive plate deteriorates over repeated cycling. The use of antimony [*7440-36-0*], Sb, in the positive grid increases the paste adhesion and cohesion. Some designs use tubular positive plates or separator-glass mat combinations to retard shedding.

The water added during the mixing operation functions as a lubricant and bulking agent to produce a paste of the proper consistency. During the plate drying operation, the evaporation of water gives the desired plate porosity. In addition to forming the binding material, the sulfuric acid expands the paste and thus gives it greater porosity.

During the acid addition step of the paste mixing process, considerable heat is generated because of the exothermic reaction of lead oxide and sulfuric acid. Care must be taken to prevent excessive loss of water before the paste can be applied to the grids. The mixer is usually cooled by water or air. After the paste has reached the desired uniformity and consistency for applying to the grid, the mixing is terminated. The paste density range is generally 3.7–4.0 g/mL for high initial capacity battery designs and for long cycle life product it may be as high as 4.3–4.6 g/mL. Muller or plough type mixers are used and a typical batch size is approximately 1100 kg. An excellent discussion of paste preparation and curing processes is available (67) and comprehensive review (68) describes the morphological features of the compounds used or evolved in the initial paste material.

Grids. The grid in a battery plate performs two vital functions. It acts as the mechanical support framework of the plate during manufacturing, and provides uniform, efficient current flow to and from the plate during formation and use. Grids are designed to have a lug or current collection point, current carrying arms, structural frame, and usually feet (Fig. 5).

The grid must possess sufficient stiffness to prevent damage or distortion during the casting, plate pasting, and battery assembly operations. During the life of the battery the grid must bear significant loads such as active material weight, its own weight, and stresses resulting from corrosion and volumetric changes caused by the sulfation of lead and lead dioxide. The most severe environment for the grid is in the positive plate. The positive active material operates in a potential range where lead is thermodynamically unstable. Thus the lead grid is

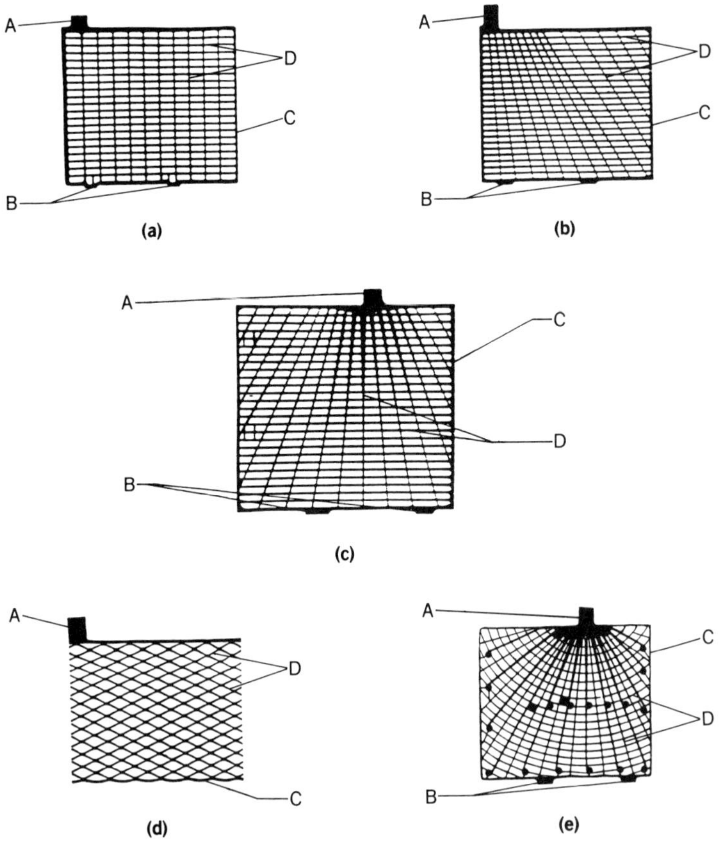

Fig. 5. Lead–acid battery grid design variations showing A lugs, B feet, C frames, and D current carrying wire for (**a**) rectilinear design, (**b**) corner lug radial, (**c**) center lug radial, (**d**) corner lug expanded metal, and (**e**) plastic/lead composite.

continuously oxidizing to PbO and PbO_2 which affects the grid by reducing its strength and conductivity, and sometimes causing elongation of its members, commonly called grid growth.

Most grids are cast or mechanically expanded from cast or wrought strip. Historically, negative and positive grids have been the same; however, with the advent of specialized batteries and extreme performance demands, this is less likely. However if these plates are different, it is generally in configuration, alloy, and/or thickness.

The conductivity of the grid plays a substantial role in a battery's ability to meet high current demands. The importance of grid conductivity for lead–acid batteries has been discussed (1,69). Composition and configuration are important design factors impacting grid conductivity.

Potential distribution in grids (69–72) has been evaluated and the results put into practice (73). Innovative grid structures which incorporate element members positioned to optimize current flow patterns have been proposed (69,74,75) as well as economical lightweight grids for the negative plate made from plastics coated or woven with lead (76,77), or with plastic strengthening members and metallic conductors (78). All incorporate lead terminal connectors in their design for easily soldering to one another, or to current collectors.

Grid Material. Pure lead has low tensile and creep strength, and thus is easily deformed at room temperature. Because the grid supplies plate strength during manufacturing and life, it must be durable. This is accomplished by alloying lead with other elements. The primary alloying ingredients are antimony, calcium [*7440-70-2*], and tin [*7440-31-5*]. Numerous other elements, eg, arsenic [*7440-38-2*], selenium [*7782-49-2*], etc can and are being used in lesser amounts.

High antimony (2–12%) content lead alloys have been used widely in the manufacture of grids where antimony increases the structural integrity of the resulting grid. Antimony not only improves its manufacturability, but also retards positive grid growth and stress deformation caused by corrosion and active material sulfation. Antimony migrates into, and alters the morphology of, the positive active mass which improves the PbO_2 adhesion and cyclability (79).

Antimony from the grid also diffuses into the electrolyte and migrates to the negative electrode and plates out. This plating or poisoning increases the self-discharge rate of the negative plate. It also reduces the negative plate's hydrogen gassing overpotential, which increases charge inefficiency and water loss resulting from gas generation. These adverse effects have been minimized by using low antimony (1–2%) content alloys (80,81). Such alloys use low percentages of other additives, eg, selenium, tin, sulfur [*7704-34-9*], and arsenic (82) to improve their strength. The low antimony content alloy reduces the detrimental gassing effects of its high antimony counterpart. It also has improved castability and enhanced corrosion resistance.

Lead alloy containing calcium (0.075–0.1%) as its stiffening agent is used in some maintenance-free batteries. This alloy is more difficult to cast than antimonial–lead alloys. The grids formed from this alloy have somewhat lower self-discharge and water loss than lead–antimony alloy grids, but batteries containing lead–calcium alloy positive grids do not give as much cycle service as those made with lead–antimony alloy positive grids (81). However, this difference is nullified if the battery is in float (constant voltage) or limited cycling applications (83).

Other alloying ingredients in lead, eg, arsenic (0.5–0.7%) and silver [*7440-22-4*] (0.1–0.15%), inhibit grid growth on overcharge and reduce positive grid corrosion. Tin added to a lead alloy produces well-defined castings that are readily adapted to mass production techniques (84).

The proper selection of the lead alloy depends on the intended use and the economics of the lead–acid battery application. The metallurgical and electrochemical aspects of the lead are discussed in the literature in a comprehensive manner (81,85–87) as are trends of lead alloy use for manufacture of battery grids (88).

Grid Casting. The grid casting machine consists of a center parting grid, or book mold, trimming mechanism, and melting pots. The grid mold consists of two cast iron parts, each having a grid design for a face. The mold design can also be a single grid or multiple grid configuration. The grid mold is heated to 135–180°C to prevent premature solidification of the lead alloy and holes in the mold release trapped air while the mold is being filled with the molten metal.

The melting pot is heated either electrically or by gas to 427–524°C. The pot capacity is typically over 100 kg of lead alloy, and periodically the top of the molten metal must be skimmed to remove the dross. The pot fumes must be removed by adequate ventilation (forced suction). When the molten metal has reached the proper temperature and flow characteristics, it is transported by pump to the grid mold.

Prior to casting, the grid mold halves are sprayed using a mold release agent. The releasing agent also serves to control grid thickness and weight during casting. After filling with a slight excess of molten lead alloy, the mold is cooled for several seconds before opening. The cast grid is trimmed to remove rough edges, gates, and minor imperfections before stacking for use in the plate processing operation.

Another common manufacturing technique is expanding cast or wrought lead strip. The strip is typically a calcium alloy of lead, although antimonial alloys have been used. The process consists of casting or rolling a continuous strip and coiling it into large rolls which are subsequently expanded. The expanding process is a multistage process in which the strip is perforated with a cutting tool and progressively stretched apart to form a diamond pattern grid, punched to form lugs, and trimmed to form the singular grids (Fig. 5), before or after pasting.

The advent of higher compression, more powerful engines in compact engine compartments and more electrical loads has precipitated the need for smaller, more powerful batteries. This has driven battery manufacturers to look at thinner plate technology. Grid designs and materials are being reevaluated to establish components to meet future demands. Technologies such as dual material (69), bipolar (89), and folded plate designs (56) are broadening the role of the grid and what it is expected to do.

Plate Pasting. Two types of machines are commonly used to apply paste to grids. They are the belt type and the fixed orifice type. Each of these types of pasters has a paste hopper that supplies paste to the machine and is fed by the paste mixer. The belt paster presses paste into the grid which is being transported by a belt into the paste stream. A fixed orifice paster squeezes the paste into the grid and then pushes the grid and paste through an orifice to smooth both plate surfaces. In both instances the pasted plate is flash dried and/or papered to remove surface moisture, allowing the plates to be stacked and transported to a curing area. In flash drying the plate passes quickly through a high temperature oven which effectively dries only the surface of the plate. When paper is used, a thin film is applied to both surfaces of the plate directly after pasting. Moisture content of the paste in plates after flash drying is typically 10%.

The tubular positive plate uses rigid, porous fiber glass tubes covered with a perforated plastic foil as the active material retainer (Fig. 2). Dry lead oxide, PbO, and red lead, Pb_3O_4, are typically shaken into the tubes which are threaded over the grid spines. The open end is then sealed by a polyethylene bar. Patents describe a procedure for making a type of tube for the tubular positive plate (90) and a method for filling tubular plates of lead–acid batteries (91). Tubular positive plates are pickled by soaking in a sulfate solution and are then cured. Some proceed directly to formation and do not require the curing procedure.

Plate Curing. Curing strengthens or hardens and dries the paste and establishes a cohesive paste–grid bond. It is a critical process in battery fabrication because it increases the plate's structural integrity and creates the electrode's active material morphology. A typical curing process consists of placing pasted plates in an elevated temperature, high humidity environment for some extended period of time such as from 16 to 48 hours. There, the plate is subjected to a controlled drying scenario. Unreacted lead particles oxidize in an exothermic reaction, creating heat that slowly drives off plate moisture. The rate of reaction of lead to lead oxide is dependent on the moisture content of the paste and the manner in which this happens affects the formation of the lead sulfate–lead oxide crystal structure (92). A normal curing environment results in small crystals of tribasic lead sulfate, $3PbO{\cdot}PbSO_4{\cdot}H_2O$. If the temperature is elevated to >57°C, the result is coarse tetrabasic lead sulfate, $4PbO{\cdot}PbSO_4$, crystals. This is especially critical for the positive electrode where tribasic sulfate converts readily to PbO_2 during battery formation but tetrabasic sulfate does not (93).

After curing, the plates are allowed to finish the drying process in ambient or elevated temperature air. The moisture and metallic lead content of the cured plates should be substantially reduced to less than 2%.

Separators. The separator's purpose is to isolate the positive and negative electrodes electronically while allowing ionic exchange between the two. This is accomplished using a microporous, electronically nonconductive material. Separators are generally designed with a backweb that has ribs on one side. The microporous backweb is kept thin to reduce resistance to ionic diffusion and migration. The ribs, which are normally placed on the positive electrode side, act as standoffs, allowing a reservoir of electrolyte to exist between the plates.

The separator must be structurally sound to withstand the rigors of battery manufacturing, and chemically inert to the lead–acid cell environment. Numerous materials have been used for separators ranging from wood, paper, and rubber to glass and plastic. The majority of separators used are either nonwoven–bound glass or microporous plastic such as PVC or polyethylene.

Separator manufacturing procedures and materials are discussed in the literature (94,95) and separator test procedures are also available (96).

Strap. Plates of similar polarity within a cell, whether positive or negative, are connected in parallel by a strap. Most batteries use either a burned-on or a cast-on strap (COS). In the case of COS, a stacked element is aligned and the lugs are placed in a mold where molten lead is cast around them to form a strap and vertical projection. When burning is used, the lugs are placed into a precast comblike strap and fused by melting the lugs and comb together. The strap has a large vertical projection on one end (Fig. 1) that is used to connect the elements in series once inside the container.

Container. The battery container is made up of a cover, vent caps, lead bushings, and case. Cost and application are the two primary factors used to select the materials of construction for container components. The container must be fabricated from materials that can withstand the abusive environment the battery is subjected to in its application. It must also be inert to the corrosive environment of the electrolyte and solid active materials, and weather, vibration, shock, and thermal gradients while maintaining its liquid seal.

The case is the largest portion of the container. The case is divided into compartments which hold the cell elements. The cores normally have a mud-rest area used to collect shed solids from the battery plates and supply support to the element. Typical materials of construction for the battery container are polypropylene, polycarbonate, SAN, ABS, and to a much lesser extent, hard rubber. The material used in fabrication depends on the battery's application. Typical material selections include a polypropylene–ethylene copolymer for SLI batteries; polystyrene for stationary batteries; polycarbonate for large, single cell standby power batteries; and ABS for certain sealed lead–acid batteries.

Covers for the battery designs in Figures 1 and 2 are typically molded from materials identical to that of the respective case, and vent plugs are frequently made of molded polypropylene. Other combinations are possible, eg, containers molded of polyethylene or polypropylene may be mated with covers of high impact rubber for use in industrial batteries. After the cover is fitted over the terminal post, it is sealed onto the case. The cover is heat bonded to the case, if it is plastic; it is sealed with an epoxy resin or other adhesive, if it is vulcanized rubber. Vent caps are usually inserted into the cover's acid fill holes to facilitate water addition and safely vent gasses, except for nonaccessible maintenance-free or recombinant batteries. In nonaccessible batteries, the vent is fabricated as part of the cover.

Electrolyte. *Sulfuric Acid.* Sulfuric acid is a primary active material of the battery. It must be present to provide sufficient sulfate ions during discharge and to retain suitable conductivity. Lead–acid batteries generally use an aqueous solution of acid in either a free-flowing or in an immobilized state.

Sulfuric acid solutions are quantified by density or specific gravity. Concentrated sulfuric acid has a density ≈ 1.840 g/mL, sp gr $= 1.840$ at room temperature; water is 1.000 g/mL, at room temperature; mixtures vary between these values. The temperature of the electrolyte affects its specific gravity, thus temperature should be noted and correction factors applied when measurements are made. At room temperature a common range of specific gravities for lead–acid batteries is 1.210 to 1.300. The electrolyte's specific gravity changes as the battery is charged and discharged. This specific gravity change occurs as sulfate ions transfer from liquid H_2SO_4 to solid $PbSO_4$. Thus the specific gravity of the electrolyte is often used as a means of approximating battery condition or state of charge.

When acid is used in the immobilized state, it is either absorbed into a high void-volume separator, or it is gelled. Electrolyte is gelled using one of several procedures (97–99). Materials such as silicon dioxide and sodium silicate [*1344-09-8*], Na_2SiO_4, are used as gelling agents. The components are mixed and poured into the battery cell where the gel is allowed to set for some preset time. In this case the battery has normally been formed and its liquid electrolyte removed, after which the gelled electrolyte is added.

When a battery is designed with absorbed electrolyte, it is built with dry separators, filled with acid and formed (Fig. 3). Excess electrolyte may then be removed.

Battery Assembly

The cell element (Fig. 1) is normally constructed from groupings of positive and negative plates. The number and size of plates of each type is determined by the desired performance level for the battery. Positive and negative plates are alternately stacked using separators in between to form the proper plate count. When the element is assembled, this piece moves to the final assembly.

The elements are inserted into containers such as that shown in Figure 1. The voltage of each charged cell is approximately 2 V, thus three elements are connected in series in 6 V batteries, six in 12 V batteries, and so forth. Once in the container, the strap vertical projections are fused together in a series arrangement to produce an unformed battery. Normal fusing techniques include resistance welding, burning, or lead extrusion. When the fusing operation is complete, the cover is sealed to the container. Most batteries use heat sealing, although adhesives, friction welding, and other methods have application.

The assembled unit is then subjected to a pressure test to ensure leak-free seals around the periphery and between the cells. Finally, positive and negative external terminations (located in the two end cells) are fused to the cover bushings. The battery is filled with electrolyte by immersing it in a bath of diluted sulfuric acid or by pouring electrolyte into the cells. After filling, the battery is placed on charge as quickly as possible to retard the conversion of tribasic lead sulfate to normal lead sulfate. These neutralization reactions cause undesired heat buildup and create products that are difficult to form. The reactions also dilute the acid, increasing the solubility of expanders, allowing them to leach out. This is detrimental to finished battery performance.

Formation. During formation the cured plate material is electrochemically converted into electrode active material, ie, spongy lead in the negative, PbO_2 in the positive, utilizing current supplied by power sources. Formation reactions create heat which, when coupled with ohmic heat and heat of neutralization, can harm electrode active materials. Thus formation is often done using evaporative coolers, water baths, or air tables. The efficiency of formation is highly dependent on the precursor cured plate morphology (94), temperature, formation current, and acid concentration (57). These processes are explained in great detail (60,93,100–102) for the positive and negative plate active materials. Normally between 110% and 130% of the battery's capacity (A·h), dependent on the formation current, is required to form it and formation times range from eight hours to beyond 48 hours.

The majority of batteries manufactured are supplied as wet, ie, acid in the formed battery, battery products, and thus utilize in-container formation. Some dry charged product where plates are washed and dried after formation, and moist, dumped and/or centrifuged, batteries are produced for some applications and markets (103).

Testing. The finished battery may be tested for voltage, capacity, charge rate acceptance, cycle life, accelerated life, storage stability, overcharge, normal temperature discharge operation, low temperature cranking, shock and vibration, and gas generation (104). The detailed test procedures for automotive battery testing are contained in the specifications of the Battery Council International (BCI) (96) and the Society of Automotive Engineers (SAE) (105). Other battery test procedures are indicative of a specialized application or general all-around use. After completion of testing, the test specimens are frequently disassembled and critically examined by chemical, physical, and metallurgical means to manufacturing methods. The SAE (106) has produced a different test procedure for battery powered vehicles.

Economic Aspects

World production of lead–acid batteries in 1988, excluding the Eastern European central economy countries, has been estimated at $9.45 billion. The automotive market was $6743 million or 211.6 million units. Industrial battery sales were $2082 million and consumer battery sales were $454 million. Motorcycle batteries accounted for an additional $170 million or 25 million units. Most batteries are produced in the United States, Western Europe, and Japan, but smaller numbers are produced worldwide. The breakdown in sales for the three production areas follows. Automotive battery sales were $2304 million in the United States, $1805 in Western Europe, and $945 million in Japan. Industrial battery sales were $525 million in the United States, $993 million in Western Europe, and $266 million in Japan. Consumer battery sales were $104 million in the United States, $226 million in Japan, and $82 million in Western Europe. More than half of all motorcycle batteries are produced in Japan and Taiwan (1).

Whereas automotive batteries have the majority of the market, other types of lead–acid batteries, such as sealed and small maintenance-free varieties, are making inroads into various applications. The automotive battery's operating environment has changed substantially in the last 10 years. Underhood temperature has risen and electrical loads have increased. This trend is expected to continue as car manufacturers reevaluate their design strategies and objectives. Battery design is changing to meet these needs.

BIBLIOGRAPHY

"Batteries, Electric (Lead-Acid Storage)" in *ECT* 1st ed., Vol. 2, pp. 340–360, by Joseph A. Orsino and Thomas C. Lynes, National Lead Company; "Secondary Cells, Lead–Acid" under "Batteries and Electric Cells, Secondary" in *ECT* 2nd ed., Vol. 3, pp. 249–271, by Joseph A. Orsino, National Lead Company; "Batteries and Electric Cells, Secondary (Lead–Acid)" in *ECT* 3rd ed., Vol. 3, pp. 640–663, by James B. Doe, ESB Technology Company.

1. J. H. B. George, *The World Battery Industry—A Multiclient Study,* George Consulting International, Inc., Concord, Mass., Nov. 1989.
2. *Storage Battery Technical Service Manual,* 10th ed., Battery Council International, Chicago, 1987.
3. K. Eberts and O. Jache, in *Lead 65, Proceedings of the 2nd International Conference on Lead-Arnheim, Belgium,* Pergamon Press, Inc., Elmsford, N.Y., 1967, p. 199.
4. D. H. McClelland, *Prog. B&S* **3,** 194 (1980).
5. G. Planté, *Compt. Rend.* **49,** 402 (1859); *Recherches sur l'electricite',* Gauthier-Villars, Paris, 1883, pp. 35, 36, 40.
6. R. H. Schallenberg, *Bottled Energy: Electrical Engineering and the Evolution of Chemical Energy Storage,* American Philosophical Society, Philadelphia, Pa., 1982.
7. T. A. Edison, *The Beginning of the Incandescent Lamp and Lighting System,* The Edison Institute, Dearborn, Mich., 1976.
8. Brit. Pat. 129 (Jan. 11, 1881), C. Fauré.
9. Brit. Pat. 3926 (Sept. 10, 1881), J. S. Sellon.
10. K. R. Bullock, "The Development and Applications of Storage Batteries—Historical Perspectives and Future Prospects," in *Proceedings, 7th Australian Electrochemistry Conference,* 1988.

11. J. H. Gladstone and A. Tribe, *Nature (London)* **27,** 583 (1883).
12. G. W. Vinal, *Storage Batteries,* 4th ed., John Wiley & Sons, Inc., New York, 1955.
13. W. H. Beck and W. F. K. Wynne-Jones, *Trans. Faraday Soc.* **50,** 136 (1954).
14. A. K. Covington, J. W. Dobson, and W. F. K. Wynne-Jones, *Trans. Faraday Soc.* **61,** 2050 (1965).
15. H. Bode, *Lead–Acid Batteries* (translated by R. J. Brodd and K. V. Kordesch), John Wiley & Sons, Inc., New York, 1977.
16. K. R. Bullock, *J. Power Sources* **35,** 197 (1991).
17. H. E. Wirth, *Electrochim. Acta* **50,** 1345 (1971).
18. W. L. Gardner, R. E. Mitchell, and J. W. Cobble, *J. Phys. Chem.* **73,** 2021 (1969).
19. T. H. Lilley and C. C. Briggs, *Electrochim. Acta* **20,** 257 (1975).
20. R. A. Robinson and R. H. Stokes, *Electrolyte Solutions,* 2nd ed., Butterworths, London, 1959, p. 477.
21. R. H. Gerke, *J. Am. Chem. Soc.* **44,** 1684 (1922); *Chem. Rev.* **1,** 377 (1925).
22. J. A. Duisman and W. F. Giauque, *J. Phys. Chem.* **72,** 562 (1968).
23. W. F. Giauque, E. W. Hornung, J. E. Kunzler, and T. R. Rubin, *J. Am. Chem. Soc.* **82,** 62 (1960).
24. W. H. Beck, K. P. Singh, and W. F. K. Wynne-Jones, *Trans. Faraday Soc.* **55,** 331 (1959).
25. W. H. Beck, J. V. Dobson, and W. F. K. Wynne-Jones, *Trans. Faraday Soc.* **56,** 1172 (1960).
26. K. R. Bullock, *J. Electrochem. Soc.* **127,** 662 (1980).
27. J. Burbank, A. C. Simon, and E. Willihnganz, in C. W. Tobias, ed., *Advances in Electrochemistry and Electrochemical Engineering,* Vol. 8, Wiley-Interscience, New York, 1971, p. 157.
28. J. L. Dawson, in A. T. Kuhn, ed., *The Electrochemistry of Lead,* Academic Press, London, 1979, p. 309.
29. A. T. Kuhn, in Ref. 28, p. 365.
30. D. Pavlov, in B. D. McNicol and D. A. J. Rand, eds., *Power Sources for Electric Vehicles,* Elsevier, Amsterdam, 1984, p. 111.
31. K. R. Bullock, *J. Electroanal. Chem.* **222,** 347 (1987).
32. K. R. Bullock and D. Pavlov, eds., *Proceedings of the Symposium on Advances in Lead Acid Batteries,* 84-14, The Electrochemical Society, Pennington, N.J., 1984.
33. D. A. J. Rand and D. Pavlov, eds., *Proceedings of LABAT-89,* Varna, 1989; *J. Power Sources,* Vols. 30 and 31, 1990.
34. D. A. J. Rand, *J. Power Sources* **15,** B1 (1985); **18,** B31 (1986); **21,** B1 (1987); **27,** B1 (1989).
35. D. Simonsson, *J. Appl. Electrochem.* **3,** 261 (1973); **4,** 109 (1974); *J. Electrochem. Soc.* **120,** 151 (1973).
36. K. Micka and I. Rousar, *Electrochim. Acta.* **18,** 629 (1973); **19,** 499 (1974); **21,** 599 (1976); *Colln. Czech. Chem. Commun.* **40,** 921 (1975).
37. J. Newman and W. Tiedemann, *J.A.I.Ch.E.* **21,** 25 (1975).
38. W. Tiedemann and J. Newman, *J. Electrochem. Soc.* **122,** 70 (1975); S. Gross, ed., *Proceedings of the Symposium on Battery Design and Optimization,* 79-1, The Electrochemical Society, Inc., Princeton, N.J., 1979, pp. 23, 39.
39. W. Tiedemann and J. Newman, in K. R. Bullock and D. Pavlov, eds., *Proceedings of the Symposium on Advances in Lead–Acid Batteries,* 84-14, The Electrochemical Society, Inc., Princeton, N.J., 1984, p. 336.
40. B. L. McKinney, W. Tiedemann, and J. Newman, in Ref. 39, p. 360.
41. A. Eklund, D. Simonsson, R. Karlsson, and F. N. Bark, "Theoretical and Experimental Studies of Free Convection and Stratification of Electrolyte in a Lead–Acid Cell," LABAT-89 Conference, Varna, Bulgaria, 1989.

42. S. Atlung and B. Fastrup, *J. Power Sources* **13,** 39 (1984).
43. D. M. Bernardi, *Extended Abstracts* 88-2, The Electrochemical Society, Inc., Princeton, N.J., 1988, p. 1.
44. N. E. Bagshaw, K. P. Bromelow, and J. Eaton, in D. H. Collins, ed., *Power Sources 6, Proceedings 10th International Power Sources Symposium,* Academic Press, Inc., New York, 1977, p. 1.
45. W. Tiedemann, J. Newman, and E. DeSua, in D. H. Collins, ed., *Power Sources 6,* 1977, p. 15.
46. J. H. Gladstone and A. Tribe, *Nature* **25,** 221 (1882).
47. P. Ruetschi and R. T. Angstadt, *J. Electrochem. Soc.* **105,** 555 (1958).
48. K. R. Bullock and D. H. McClelland, *J. Electrochem. Soc.* **123,** 327 (1976).
49. K. R. Bullock and E. C. Laird, *J. Electrochem. Soc.* **129,** 1393 (1982).
50. J. S. Symanski, B. K. Mahato, and K. R. Bullock, *J. Electrochem. Soc.* **138,** 548 (1988).
51. C. S. C. Bose and N. A. Hampson, *J. Power Sources* **19,** 261 (1987).
52. J. Mrha, K. Micka, J. Jindra, and M. Musilova, *J. Power Sources* **27,** 91 (1989).
53. R. F. Nelson and D. H. McClelland, in S. Gross, ed., *Battery Design and Optimization,* The Electrochemical Society, Princeton, N.J., 1979, p. 13.
54. U.S. Pat. 3,244,562 (Apr. 5, 1966), F. M. Coppersmith and G. J. Vahrenkamp (to National Lead Co.).
55. Ref. 15, pp. 162–176, 196–199, 207–216.
56. U.S. Pat. 4,900,643 (Feb. 13, 1990), M. D. Eskra, W. C. Delaney, G. K. Bowen.
57. V. M. Halsall and J. R. Pierson, "Plate Processing—The Heart of the Lead–Acid Battery," *BCI Proceedings,* June 1972.
58. B. K. Mahato, *J. Electrochem. Soc.* **127**(8), 1679–1687 (1980).
59. B. K. Mahato, *J. Electrochem. Soc.* **128**(7), 1416–1422 (1981).
60. J. R. Pierson, P. Gurlusky, A. C. Simon, and S. M. Caulder, *J. Electrochem. Soc.* **117**(12), 1463–1469 (1970).
61. N. A. Hampson, J. B. Lakeman, and K. S. Sodhi, *Surface Technol.* **5,** 377–384 (1980).
62. N. A. Hampson and J. B. Lakeman, *J. Electroanal. Chem.* **119,** 3–15 (1981).
63. J. A. Orsino, "Mixing & Pasting—Part I," *The Battery Man,* Apr. 1982.
64. J. F. Dittmann and H. R. Harner, *Storage Battery Research,* Form No. A-391, Chemical Division, Eagle-Picher Co., Cincinnati, Ohio, 1956.
65. G. W. Vinal, *Storage Batteries,* 4th ed., John Wiley & Sons, Inc., New York, 1955, pp. 33 and 34.
66. S. Hattori and co-workers, *Final Report (ILZRO Project LE-197),* International Lead–Zinc Research Organization, Inc., New York, 1976.
67. Ref. 15, pp. 216–244.
68. J. Perkins, J. L. Pokorny, and M. T. Coyle, *Naval Postgraduate School (NPS) Report,* NPS-69PS76101, Monterrey, Calif., Oct. 1976, pp. 17–19, 23, 24.
69. J. R. Pierson and R. T. Johnson, *SAE Technical Paper Series 830277,* SAE International Congress & Exposition, Detroit, Mich., 1983.
70. Ref. 45, p. 15.
71. J. E. Puzey and W. M. Orriel, in D. H. Collins, ed., *Power Sources 2, Proceedings 6th International Power Sources Symposium,* Pergamon Press, Inc., Oxford, UK, 1968, p. 121.
72. H. Silverman, *Environmental Protection Agency Final Report,* Contract No. 68-04-0028, EPA, Washington, D.C., 1972.
73. *Prod. Eng.* **43** (1977).
74. U.S. Pat. 3,989,539 (Nov. 2, 1976), N. G. Grabb (to Varta Batteries, Ltd.).
75. U.S. Pat. 3,959,015 (May 25, 1976), J. Brinkmann, G. Trippe, and W. Heissman (to Varta Batterie AG).

76. Fr. Pat. 2,239,020 (Feb. 21, 1975), (to Société Fulmen et Compagnie Europeénne d Accumulateurs).
77. U.S. Pat. 3,956,012 (May 11, 1976), W. R. Scholle (to Scholle Corp.).
78. J. R. Pierson and C. E. Weinlein, "Development of Unique Lightweight, High Performance Lead–Acid Batteries," in J. Thompson, ed., *Power Sources 9,* Academic Press, London, 1983.
79. D. Marshall and W. Tiedemann, *J. Electrochem. Soc.* **123**(12), 1849–1855 (1976).
80. D. L. Douglas and G. W. Mao, in D. H. Collins, ed., *Power Sources 4, Proceedings 8th International Power Sources Symposium,* Academic Press, Inc., New York, 1973, p. 379.
81. R. T. Johnson and J. R. Pierson, "The Impact of Grid Composition on the Performance Attributes of Lead–Acid Batteries," in L. J. Pearce, ed., *Power Sources 11,* International Power Sources Symposium Committee, 1987.
82. D. Berndt and S. C. Nijhawan, *J. Power Sources* **1,** 3 (1976–1977).
83. D. E. Koontz and co-workers, *Bell Syst. Tech. J.* **49,** 1253 (1970).
84. Ref. 15, pp. 251–255.
85. T. J. Hughel and R. H. Hammar, in D. H. Collins, ed., *Power Sources 3, Proceedings 7th International Power Sources Symposium,* Pergamon Press, Inc., Elmsford, N.Y., 1970, p. 209.
86. N. E. Bagshaw, *Lead 68, Proceedings 3rd International Conference on Lead,* Venice, Italy, Pergamon Press, Inc., Elmsford, N.Y., 1969, p. 209.
87. D. L. Douglas and G. W. Mao, in D. H. Collins, ed., *Power Sources 4, Proceedings 8th International Power Sources Symposium,* Academic Press, Inc., New York, 1973, p. 379.
88. M. Torralba, *J. Power Sources* **1,** 301 (1976–1977).
89. U.S. Pat. 4,275,130 (June 23, 1981), W. E. Rippel and B. Edwards.
90. Ger. Pat. 2,346,517 (Apr. 3, 1975), O. Metzler.
91. Ger. Pat. 2,315,984 (Oct. 17, 1974), (to Varta Batterie AG).
92. J. R. Pierson, *Power Sources,* Pergamon Press, New York, 1968, pp. 103–118.
93. J. R. Pierson, *Electrochem. Tech.* **5,** 323–327 (1967).
94. K. Murata and S. Hattori, "Advantageous Features of Microporous Thin Embossed Waffle Shape Separator," *89th Battery Council International Meeting,* Washington, D.C., Apr. 1977.
95. N. I. Palmer, *Proceedings of the 87th Battery Council International Meeting,* Hollywood-by-the-sea, Fla., April 1975, p. 105.
96. *Battery Technical Manual,* 2nd ed., Battery Council International.
97. U.S. Pat. 3,716,412 (Feb. 13, 1973), K. Peters (to Electric Power Storage Ltd.).
98. U.S. Pat. 3,765,942 (Oct. 16, 1973), O. Jache.
99. U.S. Pat. 3,930,881 (Jan. 6, 1976), J. P. Cestaro and L. J. Crosby (to National Lead Industries, Inc.).
100. D. Pavlov, G. Papazov, and V. Iliev, *J. Electrochem. Soc.* **119**(1), 8–19 (1972).
101. D. Pavlov, V. Iliev, G. Papazov, and E. Bashtavelova, *J. Electrochem. Soc.* **121**(7), 854–860 (1974).
102. D. Pavlov and Papazov, *J. Electrochem. Soc.* **127**(10), (1980).
103. V. M. Halsall and R. R. Wiethaup, "A New Manufacturing Method for Lead–Acid Storage Batteries," SAE 720041, 1972.
104. J. R. Pierson, C. E. Weinlein, and C. E. Wright, "Determination of Acceptable Contaminant Ion Concentration Levels in a Truly Maintenance-Free Lead–Acid Battery," in D. H. Collins, ed., *Power Sources 5-1974,* Pergamon, London and New York, 1975.
105. "SAE Standard Test Procedure for Storage J537—Jun '86," *SAE Recommended Practices,* SAE, New York, June 1986.
106. "SAE J227, Electric Vehicle Test Procedure," *SAE Recommended Practices,* SAE, New York, Mar. 1971.

General References

B. L. McKinney, G. L. Wierschem, and E. N. Mrotek, *Society of Automotive Engineers Technical Paper Series 830229,* International Congress and Exposition, Detroit, Mich., 1983.

W. C. Delaney, B. L. McKinney, E. N. Mrotek, and C. E. Weinlein, *Proceedings of the 19th Intersociety Energy Conversion Conference,* Vol. 2, San Francisco, Calif., 1984, p. 172.

U.S. Pat. 4,963,444 (June 1990), W. C. Delaney (to JCI Corp.).

D. T. Ferrell, Jr. and co-workers, *Electric Power Research Institute Final Report, EPRI EM-375,* Research Project 219-1, EPRI, Palo Alto, Calif., Feb. 1977.

G. H. Brilmyer and R. Hammann, "A Turn-Key System for Battery Based Energy Management," *Proceedings of the 2nd International Conference of EPRI, NEDO, BEWAG—Batteries for Utility Energy Storage,* Newport Beach, Calif., July 1989.

K. K. Humphreys and D. R. Brown, *Life Cycle Cost Comparison of Advanced Storage Batteries and Duel Cells for Utility Stand-Alone and Electric Vehicle Application,* Pacific Northwest Laboratory, Battelle Memorial Institute, 1990.

KATHRYN R. BULLOCK
JOHN R. PIERSON
Johnson Controls, Inc.

OTHER

The proliferation of portable electronic devices has fueled rapid market growth for the rechargeable battery industry. Miniaturization of electronics coupled with consumer demand for lightweight batteries providing ever longer run times continues to spur interest in advanced battery systems. Interest also continues to run strong in electric vehicles (EVs) and the large auto manufacturers continue to develop prototype EVs. It is clear that advances in battery technology are required for a widely acceptable EV. Advanced batteries continue to play a strong role in other applications such as load leveling for the electric utility industry and satellite power systems for aerospace.

The goal of advanced battery research is to develop batteries that supply a high number of watt·hours in a small volume (volumetric energy density) and at low weight (gravimetric energy density). Obviously this increase in energy density must be achieved in a manner that is manufacturable, safe, and of acceptable cost. There has been a significant growth in the number of advanced battery systems in development and several systems are either nearing commercial viability or have been introduced as commercial products in specific applications.

Ambient Temperature Lithium Systems

Traditionally, secondary battery systems have been based on aqueous electrolytes. Whereas these systems have excellent performance, the use of water imposes a fundamental limitation on battery voltage because of the electrolysis of water, either to hydrogen at cathodic potentials or to oxygen at anodic potentials. The application of nonaqueous electrolytes affords a significant advantage in terms of achievable battery voltages. By far the most actively researched field in nonaqueous battery systems has been the development of practical rechargeable lithium batteries (1). These are systems that are based on the use of lithium [*7439-93-2*] metal, Li, or a lithium alloy, as the negative electrode (see LITHIUM AND LITHIUM COMPOUNDS).

The use of lithium as a negative electrode for secondary batteries offers a number of advantages (2). Lithium has the lowest equivalent weight of any metal and affords very negative electrode potentials when in equilibrium with solvated lithium ions resulting in very high theoretical energy densities for battery couples. These high theoretical energy densities have prompted a wealth of research activity in a wide variety of experimental battery systems. However, realization of the technology to commercialize these systems has been slow.

A key technical problem in developing practical lithium batteries has been poor cycle life attributable to the lithium electrode (3). The highly reactive nature of freshly plated lithium leads to reactions with electrolyte and impurities to form passivating films that electrically isolate the lithium metal (4). This isolation results in less than 100% current efficiency for cycling of the lithium electrode, hence limiting cycle life (5). A goal of research for practical rechargeable lithium systems has been to improve the cycling efficiency of the lithium electrode in aprotic electrolytes.

The choice of battery electrolyte is of paramount importance to achieving acceptable cycle life because of the high reducing power of the metallic lithium. The formation of surface films on the lithium electrode (6) imparts the apparent stability of the electrolyte to the electrode. It is critical to determining lithium cycling efficiency. In addition to providing a stable film in the presence of lithium, the electrolyte must satisfy additional requirements including good conductivity, being in the liquid range over the battery operating temperature, and electrochemical stability over a wide voltage range. Solubility of the electrolyte salt in the solvent system is important in achieving good conductivity (7). In order to satisfy the various electrolyte system requirements, the use of mixed solvent electrolytes has become common in practical cells. Examples are tetrahydrofuran [*109-99-9*], C_4H_8O, -based electrolytes (8) or ethylene carbonate [*96-49-1*], $C_2H_4O_3$, –propylene carbonate [*108-32-7*], $C_4H_6O_3$, mixed solvent systems (9). Typical electrolyte salts include lithium perchlorate [*7791-03-9*], $LiClO_4$, lithium hexafluoroarsenate [*29935-35-1*], $LiAsF_6$, or lithium tetrafluoroborate [*14283-07-9*], $LiBF_4$ (10). The progress in the development of liquid electrolyte systems that provide near 100% current efficiencies for cycling of lithium has played a critical role in the emergence of rechargeable lithium as a commercially viable technology.

A second class of important electrolytes for rechargeable lithium batteries are solid electrolytes. Of particular importance is the class known as solid polymer electrolytes (SPEs). SPEs are polymers capable of forming complexes with lithium salts to yield ionic conductivity. The best known of the SPEs are the lithium salt complexes of poly(ethylene oxide) [*25322-68-3*] (PEO), $-(CH_2CH_2O)_n-$, and poly(propylene oxide) [*25322-69-4*] (PPO) (11–13). Whereas a number of experimental battery systems have been constructed using PEO and PPO electrolytes, these systems have not exhibited suitable conductivities at or near room temperature. Advances in the 1980s included a new class of SPE based on polyphosphazene complexes suggesting that room temperature SPE batteries may be achievable (14,15).

The lithium or lithium alloy negative electrode systems employing a liquid electrolyte can be categorized as having either a solid positive electrode or a liquid positive electrode. Systems employing a solid electrolyte employ solid positive electrodes to provide a solid-state cell. Another class of lithium batteries are those based on conducting polymer electrodes. Several of these systems have reached advanced stages of development or initial commercialization such as the Seiko Bridgestone lithium polymer coin cell.

The most important rechargeable lithium batteries are those using a solid positive electrode within which the lithium ion is capable of intercalating. These intercalation, or insertion, electrodes function by allowing the interstitial introduction of the Li^+ ion into a host lattice (16,17). The general reaction can be represented by the equation:

$$x\,Li^+ + x\,e^- + MY_n \rightleftharpoons Li_xMY_n$$

where MY_n represents a layered compound. A large number of inorganic compounds have been investigated for their ability to function as a reversible positive electrode in a lithium battery. Intercalation electrodes have found wide application in systems employing both liquid or solid electrolytes. The theoretical electrochemical characteristics of the most advanced of these electrode systems are described in Table 1.

Table 1. Theoretical Energy Densities for Rechargeable Lithium Systems

Battery couple	Operating voltage range, V	Energy density W·h/kg	Energy density W·h/L
Li–CoO_2	4.4–2.5	766	2700
Li–MnO_2	3.5–2.0	415	553
Li–V_2O_5	3.5–2.0	541	1304
Li–TiS_2	2.6–1.6	473	1187
Li–$NbSe_3$	2.1–1.5	436	1600
Li–MoS_2	2.3–1.3	233	882

Solid Electrolyte Systems. Whereas there has been considerable research into the development of solid electrolyte batteries (18–21), development of practical batteries has been slow because of problems relating to the low conductivity of the solid electrolyte. The development of an all solid-state battery would offer significant advantages. Such a battery would overcome problems of electrolyte leakage, dendrite formation, and corrosion that can be encountered with liquid electrolytes.

The general configuration of one system that has reached an advanced stage of development (22) is shown in Figure 1. The negative electrode consists of thin lithium foil. The composite cathode is composed of vanadium oxide [*12037-42-2*], V_6O_{13}, mixed with polymer electrolyte. Demonstration batteries have been con-

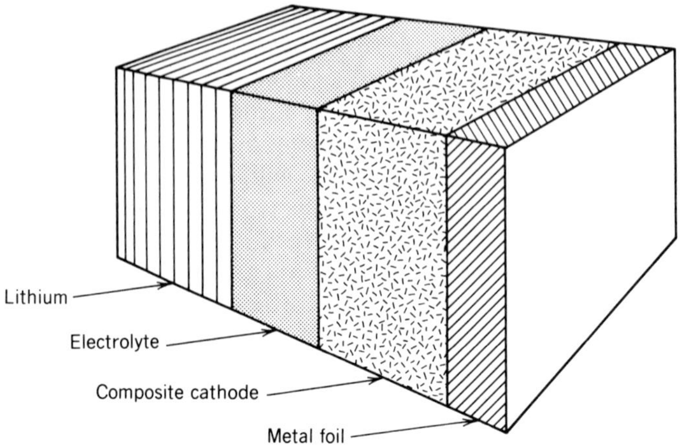

Fig. 1. Configuration for a solid polymer electrolyte rechargeable lithium cell where the total thickness is 100 μm. Courtesy of Mead Corp.

structed having energy densities of 320 W·h/L. A key to the technology is a unique radiation cross-linked polymer electrolyte which shows good conductivity at ambient temperatures (23).

A new all solid-state lithium battery employing a positive electrode comprised of organosulfur polymers, $-(SRS)_n-$, has been reported (24). During discharge of the battery, current is produced by cleavage of the sulfur–sulfur bonds in the polymer, depolymerizing the polymer. On charge, the process is reversed and the disulfides are polymerized back to their original form. This use of a polymerization–depolymerization reaction for a battery electrode is unique and this electrode is expected to offer significantly improved rate capability over intercalation electrodes. The organosulfur electrodes provide excellent stability and reversibility, which should result in long cycle life. Cells constructed to date have employed a PEO electrolyte and hence require operation at elevated temperature (25).

Coin and Button Cell Commercial Systems. Initial commercialization of rechargeable lithium technology has been through the introduction of coin or button cells. The earliest of these systems was the Li–C system commercialized by Matsushita Electric Industries (MEI) in 1985 (26,27). The negative electrode consists of a lithium alloy and the positive electrode consists of activated carbon [*7440-44-0*], carbon black, and binder. The discharge curve is not flat, but rather slopes from about 3 V to 1.5 V in a manner similar to a capacitor. Use of lithium alloy circumvents problems with cycle life, dendrite formation, and safety. However, the system suffers from generally low energy density.

More recently, commercial introduction of a rechargeable Li–V_2O_5 coin cell having significantly higher energy density than the previous Li–C cell has been announced (28). This system employs a Li–Al alloy negative electrode coupled with a vanadium pentoxide [*1314-62-1*], V_2O_5, positive electrode. The cell voltage on discharge is 3 V. This cell is claimed to have twice the energy density of a conventional Ni–Cd cell (29).

A rechargeable coin cell that employs a lithium salt electrolyte but avoids the use of a metallic lithium negative electrode has been commercially introduced by Toshiba. This cell employs a material called linear-graphite hybrid (LGH) in lieu of a lithium electrode to avoid formation of lithium dendrites. LGH is synthesized by carbonizing organic aromatic polymers. The LGH electrode is coupled with an amorphous V_2O_5 positive electrode to give a battery with a working voltage of 3 V to 1.5 V (30). A similar system employing pyrolytic carbon as a negative electrode has also been reported (31).

A significant development in commercial acceptance of rechargeable lithium was the introduction by Sanyo of rechargeable Li–MnO_2 coin and button batteries (32,33). Manganese dioxide [*1313-13-9*] is an attractive positive electrode system because of its relatively high and flat discharge voltage and the low cost of MnO_2. The widespread use of Li–MnO_2 primary batteries demonstrates the utility of this system. However, the rechargeability of the MnO_2 conventionally employed in primary cells was not sufficient to yield practical secondary cells. The breakthrough came in the modification of the crystal structure of the MnO_2 to improve rechargeability by a lithiation reaction (34). This modified MnO_2, referred to as composite dimensional manganese oxide (CMDO), is coupled with a Li–Al alloy having special additives to give excellent cycle life and stress resistance to prevent cracking and blistering of the alloy during cycling (35).

Advanced Systems. Applications for the coin and button secondary lithium cells is limited. However, researchers are working to develop practical "AA"-sized and larger cells. Several systems have reached advanced stages of development.

The first commercially available lithium "AA" cell was introduced in the early 1980s by Moli Energy Ltd. This cell, known as the Molicel, employs a molybdenum disulfide [*1317-33-5*], MoS_2, positive electrode coupled with a pure Li negative one (36). The cell is constructed in a spirally wound configuration using a microporous separator (Fig. 2). The cells have an open circuit voltage of about 2.3 V when fully charged. The discharge curve slopes, with a midpoint voltage of 1.2 V when discharged. The sloping discharge curve allows indication of state of charge (37). Cell capacity is about 600 mA for "AA" cells. Cycle life was found to be strongly dependent on operating conditions including depth of discharge, recharge and discharge currents, and voltage limits on charge and discharge. However, cycle lives of 300 cycles or more were achieved under normal cycling conditions (38).

One of the most widely studied intercalation electrode materials is titanium disulfide [*12039-13-3*], TiS_2. A number of factors make TiS_2 attractive for secondary lithium cells including good rate capability, high theoretical energy density, and a highly reversible intercalation reaction (39). Very high cycle lives have been demonstrated for TiS_2 electrodes (40). One of the earliest efforts to develop commercial lithium cells employed TiS_2 (41).

The development of "AA" Li–TiS_2 cells combines the excellent properties of TiS_2 and a proprietary method for preparing polymer bonded electrodes (42) to give a flexible, high capacity bonded electrode. Using this technology, "AA" cells have been constructed giving 1.05 A·h capacity at a voltage of 2.3 V to 1.6 V on discharge. Using excess Li for the negative electrode, cycle lives of 200 cycles are claimed for this cell. Other advantages are a good high rate capability (up to C rate), good gravimetric energy density, and very low self-discharge (43).

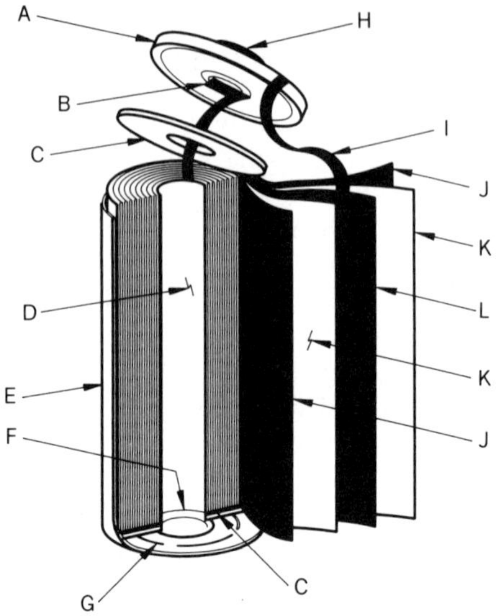

Fig. 2. Configuration for spirally wound rechargeable lithium cell. A, Cap; B, cathode tab; C, insulating disk (2); D, mandrel; E, can; F, ball; G, safety vent; H, glass-to-metal seal (with center pin); I, anode tab; J, cathode; K, separator; L, anode. Courtesy of Moli Energy Ltd.

A rechargeable lithium "AA" cell employing niobium triselenide [*12034-78-5*], $NbSe_3$, as the positive electrode material has been developed (44,45). The key to this system was a method for thermally growing $NbSe_3$ fibers that can then be pressed onto a current collector to provide a very high energy density electrode. No binder or additional conductive material is required. The excellent properties of this electrode also allow for higher discharge currents than those typically available for rechargeable lithium systems. Discharge rates in excess of C rate have been demonstrated. "AA" cells employing this technology give capacities of 1.1 A·h when cycled between 2.4 and 1.4 V. Cycle lives of over 250 cycles have been demonstrated for this cell (46).

An "AA" Li–MnO_2 cell was announced in the late 1980s, however, this system was not commercialized (47,48). This cell is claimed to offer considerable advantages over the conventional Ni–Cd system, offering higher energy density, high operating voltage, high rate capability, and excellent cycle life. A comparison of the parameters of these cells is shown in Table 2 (47).

An advanced Li–MnO_2 battery under the trade name Molicel[2] has been developed by Moli Energy Ltd. (49). The cell has a nominal voltage of 3 V, allowing replacement of two NiCd cells with one Molicel[2], hence significantly reducing battery pack size and volume. Production "AA" cells demonstrate a nominal capacity of 600 mA·h for an energy density of 100 W·h/kg. Cycle life for this cell is reported to be typically 200 cycles (50).

Other solid cathode systems that have been widely investigated include those containing lithium cobalt oxide [*12190-79-3*], $LiCoO_2$ (51), vanadium pentoxide [*1314-62-1*], V_2O_5, and higher vanadium oxides, eg, V_6O_{13} (52,53).

In addition to cells employing solid positive electrodes, a rechargeable lith-

Table 2. Comparison of "AA"-Size Li–MnO_2 and NiCd Cells

Parameter	Cell system: Li–MnO_2	Ni–Cd	Parameter ratio Li–MnO_2/Ni–Cd
cell weight, g	16.0	25.0	0.65
operating voltage, V	2.8[a]	1.2[b]	2.3
cutoff voltage, V	2.0	1.0	
energy at 0.8 W discharge, W·h	2.03	0.85	2.4
self-discharge after 1 mo	1%	25%	0.04
cycles (% DOD[c])			
at 0.85 W·h/cycle	1000 (40%)	500 (100%)	2
at 1.3 W·h/cycle	400 (64%)		
at 1.7 W·h/cycle	200 (84%)		

[a]Average voltage.
[b]Nominal voltage.
[c]DOD = depth of discharge.

ium cell employing sulfur dioxide [*7446-09-5*], SO_2, as a liquid electrode has been developed (54). The widespread use of primary Li–SO_2 batteries, particularly by the U.S. military, led to a strong interest in developing a rechargeable version. The cell electrolyte for this system consists of liquid sulfur dioxide and lithium tetrachloroaluminate [*14024-11-4*], $LiAlCl_4$, salt. The overall cell chemistry of the rechargeable system involves the reduction and oxidation of a complex which forms between $LiAlCl_4$, SO_2, and the carbon that is employed as the current collector for the positive electrode. Specifications for a prototype Li–SO_2 "C"-sized cell are shown in Table 3 (54). Advantages for this system include very high

Table 3. Specifications for a Prototype Li–SO_2 "C" Cell[a]

Parameter	Value
open circuit voltage, V	3.2
operating voltage, at C/6,[b] V	3.0
energy, W·h	5.4
volumetric energy density, W·h/L	220
gravimetric energy density, W·h/kg	134
operating temperature, °C	−30 to 40
cycle life, cycles	50
discharge rate	C/1.8[b,c]
charge rate	C/18[b,d]
charge retention after 9 mo	100%

[a]Where the nominal capacity is 1.8 A·h to 2 V.
[b]Where C is the current required to discharge the cell in 1 hour.
[c]C/1.8 = 1 A.
[d]C/18 = 0.1 A.

energy densities, a flat running voltage, and the ability to sustain limited overcharge. The principal drawback has been safety concerns over potential cell ruptures on cycling resulting from internal short circuits.

A related system employs a copper(II) chloride [*7447-39-4*], $CuCl_2$, electrode in a sulfur dioxide electrolyte (55). "AA" cells of nominal capacity of 500 mA·h at a discharge voltage of 3.4–3.0 V were constructed. Cycle life for this cell is claimed to be 200 cycles. The high voltage of this system offers a significant energy density advantage over conventional Ni–Cd systems.

The discharge performance characteristics of advanced rechargeable lithium "AA" cells and the Ni–Cd cell are illustrated in Figure 3. Whereas it is clear that a number of lithium systems offer energy capacities of significant improvement over conventional Ni–Cd cells, several critical issues remain for cell development before lithium technology gains widespread application. Cycle life continues to be a limitation to most technologies. Unlike Ni–Cd, there is no overcharge or overdischarge protection for the rechargeable lithium cells which generally must be recharged only within closely defined voltage limits. This presents a problem particularly in the design of multicell battery packs. Experiments using electrolyte additives to provide overcharge protection have been carried out (56), but this concept has not been demonstrated in advanced cell designs. Additionally, the high reactivity of lithium metal continues to pose serious safety concerns among battery manufacturers and users. Incidences of lithium cells exploding and/or igniting or venting with flame during use have highlighted the need for extensive safety testing of lithium-containing batteries prior to the widescale commercialization of this technology.

A lithium ion rechargeable cell has apparently been developed in response to safety concerns over the Li–MnO_2 cell. This technology is expected to be used

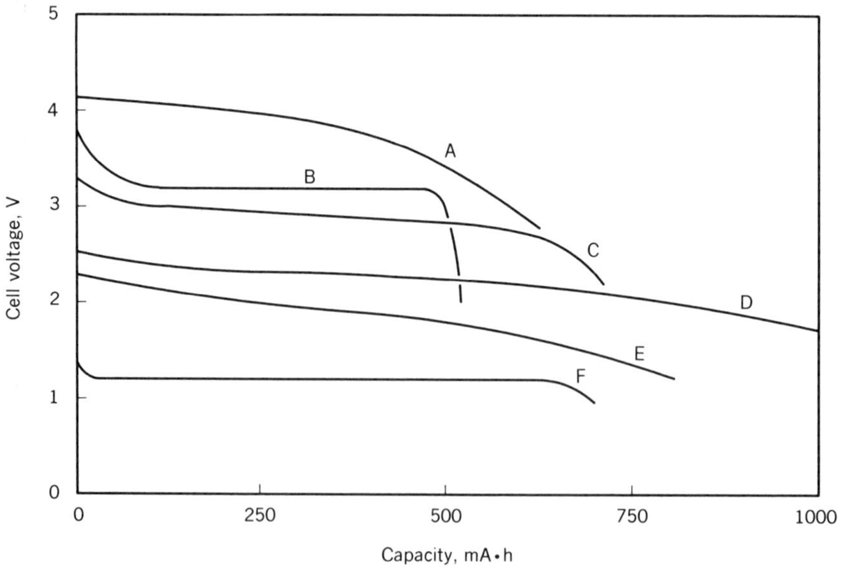

Fig. 3. Cell voltage profiles as a function of discharge capacity for rechargeable "AA" cells: A, Li ion; B, Li–SO_2; C, Li–MnO_2; D, Li–TiS_2; E, Li–MoS_2; F, Ni–Cd.

for camcorder battery packs. The technical details of this battery have not been reported, but it is understood that the negative electrode consists of a carbon material capable of undergoing an insertion reaction with Li^+ ion (57). This electrode is coupled with an insertion positive electrode such as lithium cobalt oxide (58). The resulting cell has a nominal operating voltage of 3.6 V, three times that of NiCd, providing a significant advantage in energy density (see Fig. 3) for multicell battery packs. The cell is claimed to be capable of rapid charge, discharge rates up to 2C, and 1200 cycles (59).

High Temperature Systems

Lithium–Aluminum/Metal Sulfide Batteries. The use of high temperature lithium cells for electric vehicle applications has been under development since the 1970s. Advances in the development of lithium alloy–metal sulfide batteries have led to the Li–Al/FeS system, where the following cell reaction occurs.

$$2\ LiAl_x + FeS \rightleftharpoons Fe + Li_2S + 2x\ Al$$

The cell voltage is 1.33 V to give a theoretical energy density of 458 W·h/kg (60). The cell employs a molten salt electrolyte, most commonly a lithium chloride [*7447-41-8*]/potassium chloride [*7447-40-7*], LiCl–KCl eutectic mixture. The cell is generally operated at 400–500°C. The negative electrode is composed of lithium–aluminum alloy, which operates at about 300 mV positive of pure lithium. The positive electrode is composed of iron sulfide [*1317-37-9*] mixed with a conductive agent such as carbon or graphite. Electrodes are constructed by cold pressing powder onto current collectors (61).

Development of practical and low cost separators has been an active area of cell development. Cell separators must be compatible with molten lithium, restricting the choice to ceramic materials. Early work employed boron nitride [*10043-11-5*], BN, but a more desirable separator has been developed using magnesium oxide [*1309-48-4*], MgO, or a composite of MgO powder–BN fibers. Corrosion studies have shown that low carbon steel or stainless steel are suitable for the cell housing as well as for internal parts such as current collectors (62).

Li–Al/FeS cells have demonstrated good performance under EV driving profiles and have delivered a specific energy of 115 W·h/kg for advanced cell designs. Cycle life expectancy for these cells is projected to be about 400 deep discharge cycles (63). This system shows considerable promise for use as a practical EV battery.

A similar system under development employs iron disulfide [*12068-85-8*], FeS_2, as the positive electrode. Whereas this system offers a higher theoretical energy density than does Li–Al/FeS, the FeS_2 cell is at a lower stage of development (64,65).

Sodium–Sulfur. The best known of the high temperature batteries is the sodium [*7440-23-5*]–sulfur [*7704-34-9*], Na–S, battery (66). The cell reaction is best represented by the equation:

$$2\ Na + 3\ S \rightleftharpoons Na_2S_3$$

occurring at a cell voltage of 1.74 V, to give a specific energy of 760 W·h/kg. The cell is constructed using a solid electrolyte typically consisting of β-alumina [*1344-28-1*], β-Al_2O_3, ceramic, although borate glass fibers have also been used. These materials have high conductivities for the sodium ion. The negative electrode consists of molten sodium metal and the positive electrode of molten sulfur. Because sulfur is not conductive, a current collection network of graphite is required. The cell is operated at about 350°C. A typical cell design is shown in Figure 4. The outer cell case is constructed from mild steel. The β-alumina separator tube fits inside the cell case with a small gap in between. The positive electrode of molten sulfur is placed inside the alumina tube with a carbon current collector. The molten sodium is stored below the cell and is wicked into the space between the separator and outer cell casing. The mild steel case acts as the negative electrode current collector (67).

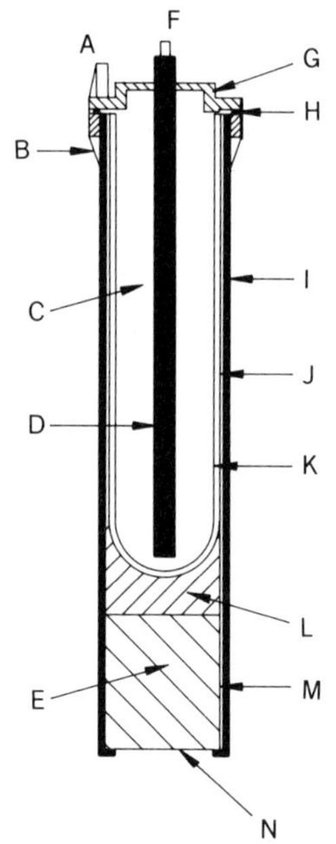

Fig. 4. Construction of a sodium–sulfur battery. A, Negative terminal; B, springs plus graphite felt; C, sulfur; D, carbon; E, sodium reservoir; F, positive terminal; G, insulator; H, aluminum sealing gaskets; I, steel case; J, film of sodium; K, β-alumina tube; L, carbon felt; M, wick; N, aluminum can (67).

The Na–S battery couple is a strong candidate for applications in both EVs and aerospace. Projected performance for a sodium–sulfur-powered EV van is shown in Table 4 for batteries having three different energies (68). The advantages gained from using a Na–S system rather than the conventional sealed lead–acid batteries are evident.

The Na–S system is expected to provide significant increases in energy density for satellite battery systems (69). In-house testing of Na–S cells designed

Table 4. Electric Vehicle Battery Performance

	Battery			
Parameter	Lead–acid	Sodium–sulfur		
battery energy, kW·h	40.0	40.0	60.0	85.0
range, km	84.0	113.0	169.0	242.0
max payload, t	0.9	1.7	1.6	1.6
battery weight, kg	1250.0	330.0	424.0	580.0

to simulate midaltitude (MAO) and geosynchronous orbits (GEO) demonstrated over 6450 and over 1400 cycles, respectively.

Difficulties with the Na–S system arise in part from the ceramic nature of the alumina separator: the specific β-alumina is expensive to prepare; and the material is brittle and quite fragile. Separator failure is the leading cause of early cell failure. Cell failure may also be related to performance problems caused by polarization at the sodium/solid electrolyte interface. Lastly, seal leakage can be a determinant of cycle life. In spite of these problems, however, the safety and reliability of the Na–S system has progressed to the point where pilot plant production of these batteries is anticipated for EV and aerospace applications.

A battery system closely related to Na–S is the Na–metal chloride cell (70). The cell design is similar to Na–S; however, in addition to the β-alumina electrolyte, the cell also employs a sodium chloroaluminate [*7784-16-9*], $NaAlCl_4$, molten salt electrolyte. The positive electrode active material consists of a transition metal chloride such as iron(II) chloride [*7758-94-3*], $FeCl_2$, or nickel chloride [*7791-20-0*], $NiCl_2$, (71,72) in lieu of molten sulfur. This technology is in a younger state of development than the Na–S.

Miscellaneous Systems

Rechargeable cells employing aluminum [*7429-90-5*], Al, as a negative electrode in room temperature molten salts have been investigated. Aluminum has a very high electrochemical equivalent weight producing a high energy density, yet aluminum is much less reactive than lithium, alleviating many of the safety issues of lithium electrode systems. The aluminum electrode has been shown to be very reversible in room temperature molten salts based on aluminum chloride [*7446-70-0*], $AlCl_3$–organic quaternary ammonium salt mixtures (73). Positive electrodes include intercalation (74) and metal sulfide ones (75,76).

Redox flow batteries, under development since the early 1970s, are still of interest primarily for utility load leveling applications (77). Such a battery is shown schematically in Figure 5. Unlike other batteries, the active materials are not contained within the battery itself but are stored in separate tanks. The reactants each flow into a half-cell separated one from the other by a selective membrane. An oxidation and reduction electrochemical reaction occurs in each

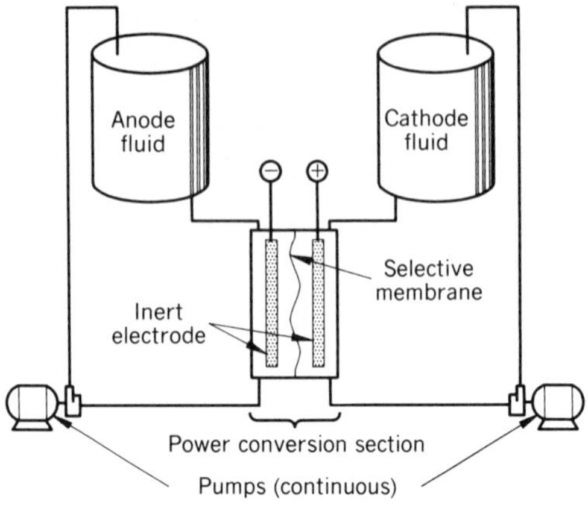

Fig. 5. Schematic for a redox flow battery (78). Courtesy of Plenum Press.

half-cell to generate current. Examples of this technology include the iron–chromium, Fe–Cr, battery (79) and the vanadium redox cell (80).

Other flow batteries investigated for both electric vehicle application and utility load leveling include zinc [*7440-66-6*]–chlorine [*7782-50-5*], Zn–Cl_2, and zinc–bromine [*7726-95-6*], Zn–Br_2, batteries (78,81,82).

Economic Aspects

As of this writing, there is little commercialization of advanced battery systems. Small rechargeable lithium button cells have been commercialized, however, by Sanyo, Matsushita (Panasonic), and Toshiba. These cells are intended for original equipment manufacturer (OEM) use in applications such as memory backup and are not available to the general consumer.

Efforts to commercialize larger versions of rechargeable lithium cells have been frustrated by concerns over product safety. Moli Energy Ltd. briefly introduced "AA" Li–MoS_2 cells for OEM use in laptop computers and cellular phones. However, safety issues resulted in a product recall and a halt to commercialization of this product.

Efforts to develop commercially viable EV versions of advanced battery systems continue. The ultimate goal is to develop battery technology suitable for practical, consumer-acceptable electric vehicles. The United States Advanced Battery Consortium (USABC) has been formed with the express purpose of accelerating development of practical EV batteries (83).

BIBLIOGRAPHY

"Batteries and Electric Cells, Secondary (Other Cells)" in *ECT* 3rd ed., Vol. 3, pp. 663–670, by J. B. Doe, ESB Technology, Co.

1. M. Hughes, N. A. Hampson, and S. A. G. R. Karunathilaka, *J. Power Sources* **12,** 83–144 (1984).

2. E. Yeager, in E. Yeager and co-eds., *Proceedings of the Workshop on Lithium Nonaqueous Battery Electrochemistry,* Vol. 80-7, The Electrochemical Society, Inc., Cleveland, Ohio, 1980, pp. 1–12.
3. V. R. Koch, *J. Power Sources* **6,** 357–370 (1981).
4. M. Garreau, *J. Power Sources* **20,** 9–17 (1987).
5. A. Dey, *Thin Solid Films* **43,** 131–171 (1977).
6. S. B. Brummer, in Ref. 2, pp. 130–142.
7. Y. Matsuda, *J. Power Sources* **20,** 19–26 (1987).
8. K. M. Abraham, *J. Power Sources* **14,** 179–191 (1985).
9. J. R. Stiles, *New Mater. New Processes* **3,** 89–91 (1985).
10. J. Gabano, *Prog. Batt. Solar Cells* **8,** 149–162 (1989).
11. M. Armand, *Solid State Ionics* **9, 10,** 745–754 (1983).
12. A. Hooper, in K. M. Abraham and B. B. Owens, eds., *Materials and Processes for Lithium Batteries,* Vol. 89-4, The Electrochemical Society, Inc., Cleveland, Ohio, 1980, pp. 15–32.
13. P. E. Harvey, *J. Power Sources* **26,** 23–32 (1989).
14. *Chem. Eng. News,* 23, 24 (Nov. 5, 1984).
15. K. M. Abraham, M. Alamgir, and S. J. Perrotti, *J. Electrochem. Soc.* **135,** 535, 536 (1988).
16. K. M. Abraham, *J. Power Sources* **7,** 1–43 (1981 and 1982).
17. A. D. Yoffe, *Solid State Ionics* **9, 10,** 59–70 (1983).
18. B. Scosati, A. Selvaggi, and B. Owens, *Prog. Batt. Solar Cells* **8,** 135–137 (1989).
19. A. Hammou and A. Hammouche, *Electrochimica Acta* **33,** 1719, 1720 (1988).
20. F. Bonino, M. Ottaviani, and B. Scrosati, *J. Electrochem. Soc.* **135,** 12–15 (1988).
21. A. Hooper and J. North, *Solid State Ionics* **9, 10,** 1161–1166 (1983).
22. D. Shackle, *4th International Seminar on Lithium Battery Technology and Applications,* Deerfield Beach, Fla., Mar. 1989.
23. U.S. Pat. 4,830,939 (May 1989), M. Lee, D. Shackle, and G. Schwab.
24. *Science News* **136,** 342 (1989).
25. M. Liu, S. Visco, and L. DeJonghe, in S. Subbarao and co-workers, eds., *Rechargeable Lithium Batteries,* Vol. 90-5, The Electrochemical Society, Inc., Cleveland, Ohio, 1980, pp. 233–244.
26. N. Koshiba and K. Momose, in Y. Matsuda and C. Schlaikjer, eds., *Practical Lithium Batteries,* JEC Press, Inc., 1988, pp. 79–81.
27. K. Momose, H. Hayakawa, N. Koshiba, and T. Ikehata, *Prog. Batt. Solar Cells* **6,** 56, 57 (1987).
28. *JEC Battery Newsletter,* (1), 7 (Jan./Feb. 1989).
29. *Japan Component News* **7**(2), 4–6 (Feb. 1990).
30. K. Inada and co-workers, in Ref. 26, pp. 96–99.
31. M. Mohiri and co-workers, *J. Power Sources* **26,** 545–551 (1989).
32. U.S. Pat. 4,748,484 (July 19, 1988), N. Furukawa, T. Saito, and T. Nohma (to Sanyo Electric Co., Ltd.).
33. U.S. Pat. 4,904,552 (Feb. 27, 1990), N. Furukawa, T. Saito, and T. Nohma (to Sanyo Electric Co., Ltd.).
34. T. Nohma, T. Saito, and N. Furukawa, *J. Power Sources* **26,** 389–396 (1989).
35. J. Carcone, *3rd International Rechargeable Battery Seminar,* Deerfield Beach, Fla., Mar. 1990.
36. *Technical Brochure,* Moli Energy Ltd., Burnaby, British Columbia, Canada.
37. J. Stiles, in Ref. 26, pp. 74–78.
38. F. C. Laman and K. Brandt, *J. Power Sources* **24,** 195–206 (1988).
39. E. J. Frazer and S. Phang, *J. Power Sources* **6,** 307–317 (1981).
40. D. H. Shen and co-workers, *Extended Abstract 49,* Vol. 87-2, *172nd Meeting of the Electrochemical Society,* Honolulu, Hawaii, Oct. 1987.

41. U.S. Pat. 4,084,046 (Apr. 1978), M. S. Whittingham (to Exxon Corp.).
42. U.S. Pat. 4,731,310 (Mar. 1988), M. Anderman and J. Lundquist (to W. R. Grace & Co.).
43. M. Anderman and J. Lundquist, *Extended Abstract 49,* Vol. 87-2, 172nd Meeting of the Electrochemical Society, Honolulu, Hawaii, Oct. 1987.
44. U.S. Pat. 3,864,167 (Feb. 4, 1975), J. Broadhead, F. J. DiSalvo, and F. A. Trumbore.
45. B. Vyas, *33rd International Power Sources Symposium,* Cherry Hill, N.J., June 1988.
46. F. Trumbore, *4th International Meeting on Lithium Batteries,* Vancouver, Canada, May 1988.
47. *JEC Battery Newsletter* (2), 26, 27 (1988).
48. U.S. Pat. 4,828,834 (May 1989), T. Nagaura and M. Yokokawa (to Sony Corp.).
49. U.S. Pat. 4,959,282 (Sept. 1990), J. Dahn and B. Way (to Moli Energy, Ltd.).
50. *Product Literature,* Moli Energy Ltd., Burnaby, British Columbia, Canada.
51. E. Plichta and co-workers, *J. Power Sources* **21,** 25–31 (1987).
52. K. Wiesener and co-workers, *J. Power Sources* **20,** 157–164 (1987).
53. K. West, B. Zachua-Christiansen, M. Ostergard, and T. Jacobsen, *J. Power Sources* **20,** 165–172 (1987).
54. A. N. Dey, K. C. Kuo, P. Piliero, and M. Kallianidis, *J. Electrochem. Soc.* **135,** 2115–2120 (1988).
55. C. Cecilio and co-workers, *4th International Seminar on Lithium Battery Technology and Applications,* Deerfield Beach, Fla., Mar. 1989.
56. W. Behl and D. Chin, *J. Electrochem. Soc.* **135,** 16–25 (1988).
57. U.S. Pat. 4,668,595 (May 1987), A. Yoshino, K. Sanechika, and T. Nakajima (to Sony Corp.).
58. Jpn. Pat. 01 122 562, (to Sony Corp.).
59. *JEC Battery Newsletter,* (2), 9–12 (1990).
60. E. J. Cairns, in J. Bockris and co-eds., *Comprehensive Treatise of Electrochemistry,* Vol. 3, Plenum Press, New York, 1981, Chapt. 11.
61. R. W. Glazebrood and M. J. Willars, *J. Power Sources* **8,** 327–339 (1982).
62. P. A. Nelson, *4th International Seminar on Lithium Battery Technology and Applications,* Deerfield Beach, Fla., Mar. 1989.
63. P. A. Nelson and H. Shimotake, *Prog. Batt. Solar Cells* **6,** 150–154 (1987).
64. T. D. Kaun, in Ref. 25, pp. 294–306.
65. T. D. Kaun and co-workers, *Proc. of the 25th IECEC* **3,** 335–340 (1990).
66. J. L. Sudworth and A. R. Tilley, *The Sodium Sulfur Battery,* Chapman and Hall Ltd., 1985, and references therein.
67. D. Pletcher, *Industrial Electrochemistry,* Chapman and Hall Ltd., 1984, pp. 272–274.
68. P. Bindin, *Proc. 20th IECEC Conf.* **2,** 1111–1114 (Aug. 1985).
69. S. Wolanczyk and S. Vukson, *Proc. 25th IECEC Conf.* **3,** 122–124 (Aug. 1990).
70. R. J. Bones, J. Coetzer, R. C. Galloway, and D. A. Teagle, *J. Electrochem. Soc.* **134,** 2379–2382 (1987).
71. R. C. Galloway, *J. Electrochem. Soc.* **134,** 256, 257 (1987).
72. R. J. Bones, S. D. Brooker, and F. L. Cullen, *J. Electrochem. Soc.* **136,** 1274–1277 (1989).
73. J. S. Wilkes and co-workers, *Frank J. Seiler Laboratory Technical Report,* FJSRL-TR-81-011, ADA 107,989, Oct. 1981.
74. P. R. Gifford and J. B. Palmisano, *J. Electrochem. Soc.* **135,** 650–654 (1988).
75. V. R. Koch, *3rd International Rechargeable Battery Seminar,* Deerfield Beach, Fla., Mar. 1990.
76. N. Talami and N. Koura, *J. Electrochem. Soc.* **136,** 730, 731 (1989).
77. M. Bartolozzi, *J. Power Sources* **27,** 219–234 (1989) and references therein.
78. J. McBreen, in Ref. 60, Chapt. 10.
79. R. F. Gahn, N. H. Hagedorn, and J. A. Johnson, *Proc. 20th IECEC* **2,** 91–97 (Aug. 1985).

80. M. Skyllas-Kazacos and F. Grossmith, *J. Electrochem. Soc.* **134,** 2950–2953 (1987).
81. F. Will and H. S. Spacil, *J. Power Sources* **5,** 173–188 (1980).
82. A. Leo and A. Charkey, *Proc. 20th IECEC* **2,** 63–69 (Aug. 1985).
83. F. Jamerson, *DOE Automotive Technology Development Contractors' Coordination Meeting,* Vol. 2, Dearborn, Mich., Oct. 28, 1991.

PAUL R. GIFFORD
Ovonic Battery Company

BATTERIES AND ELECTRIC CELLS, PRIMARY-FUEL CELLS. See FUEL CELLS.